PHYSICAL METALLURGY HANDBOOK

Anil Kumar Sinha

McGRAW-HILL

New York   Chicago   San Francisco   Lisbon   London   Madrid
Mexico City   Milan   New Delhi   San Juan   Seoul
Singapore   Sydney   Toronto

The McGraw·Hill Companies

Cataloging-in-Publication Data is on file with the Library of Congress

Copyright © 2003 by The McGraw-Hill Companies, Inc. All rights reserved.
Printed in the United States of America. Except as permitted under the
United States Copyright Act of 1976, no part of this publication may
be reproduced or distributed in any form or by any means, or stored in a
data base or retrieval system, without the prior written permission of the
publisher.

1 2 3 4 5 6 7 8 9 0   DOC/DOC   0 9 8 7 6 5 4 3 2

ISBN 0-07-057986-5

*The sponsoring editor for this book was Kenneth P. McCombs,
the editing supervisor was David E. Fogarty, and the
production supervisor was Pamela A. Pelton. It was set in
Times Roman by SNP Best-set Typesetter Ltd., Hong Kong.*

Printed and bound by RR Donnelley.

This book is printed on acid-free paper.

McGraw-Hill books are available at special quantity discounts to use as pre-
miums and sales promotions, or for use in corporate training programs. For
more information, please write to the Director of Special Sales, Professional
Publishing, McGraw-Hill, Two Penn Plaza, New York, NY 10121-2298. Or
contact your local bookstore.

CONTENTS

Chapter 7. Pearlite and Proeutectoid Phases   7.1

Chapter 8. Martensite   8.1

Chapter 9. Bainite   9.1

Chapter 10. Austenite   10.1

Chapter 15. Thermomechanical Treatment   15.1

Chapter 16. Surface Hardening Treatments   16.1

Chapter 17. Defects and Distortion in Heat-Treated Parts   17.1

Chapter 18. Surface Modification and Thin-Film Deposition   18.1

Chapter 19. Thermal Spray Coatings   19.1

Appendix A. Conversion Table for Units, Constants, and Factors in Common Use   A.1

Appendix B. Temperature Conversions   B.1

Index follows Appendix B

PREFACE

Physical Metallurgy Handbook is an enlarged edition of my earlier *Ferrous Physical Metallurgy* (Butterworths, 1989). Four new chapters of increased significance, namely, diffusion in metals and alloys, solidification, surface modification and thin film deposition, and thermal spray coatings have been added in addition to the complete revision of all 15 chapters. As before this Handbook focuses on both the theoretical elements, such as those dealing with diffusion, solidification, deformation, annealing phenomena, nucleation in solids, phase transformation in solids, kinetics of phase transformations, structure-property relationships, and the processing elements such as dealing with heat treating operations, surface modification and thin film deposition, and thermal spray coatings. This Handbook covers mostly ferrous and nonferrous alloys.

Chapter 1 is devoted to iron-carbon alloys and their relevant phase diagrams, effects of alloying elements, and classification of steels and cast irons. Chapter 2 is on diffusion in metals and alloys, and it deals with the diffusion equation, diffusion mechanisms, the effect of key variables on diffusivity, self-diffusion, diffusion in substitutional and concentrated alloys, electro- and thermomigration; diffusion along short circuits, diffusion in ionic solids and semiconductors, and radiation effects and diffusion. Chapter 3 describes heat transfer in solidification; nucleation and growth; plane front solidification of alloys; cellular and dendritic growth; eutectics, monotectics, and peritectics; segregation; solidification processes and cast structures; single crystal growth; and grain refinement and eutectic modification. Chapter 4 treats tensile properties, yielding phenomena, flow stress, cold working, slip and twinning, strain aging, and deformation texture in engineering materials. Chapter 5 focuses on release of stored energy, recovery, recrystallization, laws of recrystallization, recrystallization texture, and grain growth. Chapter 6 deals with classical nucleation theories, nonclassical nucleation (i.e., spinodal decomposition), precipitation hardening in a variety of ferrous and nonferrous alloys, and strength of precipitation hardened alloys. Chapter 7 explains both the mechanisms and kinetics of transformation of austenite into pearlite, interphase precipitation, fibrous carbide, proeutectoid phases, ferrite morphology, proeutectoid cementite, and ferrite-pearlite and pearlitic steels. Chapter 8 introduces general and typical characteristics of martensitic transformation, ferrous and nonferrous martensites, nucleation and growth in martensitic transformation, thermoelastic martensitic transformation, strengthening mechanisms, toughness of martensite, and omega transformation. Chapter 9 emphasizes three definitions of bainite based on microstructure, surface relief, and kinetics; overall mechanisms; acicular ferrite; and bainitic steels. Chapter 10 elaborates on the formation of austenite and retained austenite, austenitic grain size, austenitic and superaustenitic stainless steels, duplex and superduplex stainless steels, and physical properties and machinability of stainless steels. Chapter 11 details isothermal and continuous cooling transformation diagrams. Chapter 12 discusses the definition, importance, selection, and classification of heat treatment; annealing, normalizing, decarburization, and graphitization of steels; annealing and normalizing of cast irons; engineering properties and applications of cast irons; and structure-property relations in gray iron. Chapter 13 expounds quenching, quench-hardening, and inverse-quench hardening of steels; direct quenching; intense quenching and martempering of steels; austempering of steels and ductile iron; quench cracking; hardenability and hardenabilty steels; and alloy steel selection based on hardeanbility. Chapter 14 offers a detailed treatment on tempering, the structural and mechanical property changes associated with tempering of hardened steels, secondary hardening of steels, the tempering parameter, tempering methods, strengthening mechanisms of tempered martensite and bainite, various types of embrittlement phenomena occurring in low alloy quenched and tempered steels, and maraging steels. Chapter 15 gives an account of ferrous and nonferrous ther-momechanical treatments, superplasticity largely found in ferrous and nonferrous alloys, and potential applications in aerospace industries. Chapter 16 thoroughly covers various surface hardening heat treatments and their advantages, disadvan-tages, and applications as well as newer processes such as supercarburizing, borid-ing, and the thermoreactive deposition/diffusion (TRD) processes. Chapter 17 concentrates on overheating and burning of low alloy steels, residual stresses, dis-tortion in heat treatment, and the importance of correct design to lessen distortion and the danger of cracking. Chapters 18 dwells on, in detail, ion beam processes, physical vapor deposition, molecular beam epitaxy, and chemical vapor deposition. The final chapter describes the advantages, disadvantages, important processes, recent developments, coating characteristics, and applications of thermal spray techniques.

The purpose of this Handbook is to present to the readers the latest infor-mation on fundamental principles, alloy design, and technologically useful microstructures, properties, forms, and applications of ferrous and nonferrous ma-terials. This Handbook, which describes physical metallurgy with a novel approach and a comprehensive treatment, will serve as a valuable tool in understanding the interplay between microstructure, properties, and performance of a variety of engineering materials; in the selection of materials, treatments, and processes for specific applications; in solving heat treatment, surface modification, and other processing problems; in the tradeoff decisions that are often made in the automo-tive, aerospace, and other metalworking industries; and in the design, product devel-opment, and materials engineering of components that must operate reliably under service conditions.

The intended audience of this Handbook includes practicing materials scientists; practicing manufacturing, mechanical, metallurgical, and product engineers; design engineers; researchers; heat treaters; sophisticated coaters; senior undergraduate stu-dents; and beginning graduate students. Academic courses for which the book might be useful as a text or for collateral readings are physical metallurgy, ferrous physi-cal metallurgy, phase transformations, heat treatment of ferrous and nonferrous alloys, surface modification and thin film deposition, and thermal spray coatings.

The exhaustive lists of references provided at the end of each chapter will enable readers to pursue the subject in still greater detail. The abundance of figures and tables provided in the text will be useful for better comprehension of the concepts of physical metallurgy.

Anil Kumar Sinha

ACKNOWLEDGMENTS

I acknowledge with gratitude the helpful comments and valuable advice on various sections of the Handbook provided by Professor W. C. Leslie and C. R. Brooks (Chapter 1); Professors R. W. Balluffi, P. G. Shewmon, U. Gösele, and T. Y. Tan and Drs. M. C. Petri and E. P. Simonen (Chapter 2); Professors M. C. Flemings, R. K. Trivedi, J. D. Verhoeven, W. Kurz, and A. Hellawell and Dr. K. P. Young (Chapter 3); Professor F. B. Pickering and Drs. M. A. Imam, C. S. Pande, H. Jones, Z. Zimerman (Chapter 4); Professors R. W. Cahn, R. D. Doherty, T. Gladman, and C. L. Briant and Drs. B. B. Rath, and B. P. Bewlay (Chapter 5); Professors T. H. Sanders, W. A. Soffa, and A. J. Ardell and Drs. J. F. Grubb and Terry Tebold (Chapter 6); Professors G. J. Shiflet, F. B. Pickering, Paul Clayton, M. R. Notis, and T. Gladman and Drs. Bruce L. Bramfitt and Roger K. Steele (Chapter 7); Professor G. B. Olson and F. B. Pickering and Dr. L. McD. Schetky (Chapter 8); Professors H. I. Aaronson, H. K. D. H. Bhadeshia, and F. B. Pickering and Drs. Bruce L. Bramfitt and Roger K. Steele (Chapter 9); Drs. Riad Asfahani, and J. F. Grubb (Chapter 10); Dr. R. Vishwanathan (Chapter 12); Professor J. S. Kirkaldy, Drs. B. M. Kapadia and R. W. Foreman, and Messrs. R. R. Blackwood, R. Keogh, and Rick Houghton (Chapter 13); Professors R. A. Oriani and C. J. McMahon, Jr. and Drs. A. M. Sherman, K. A. Taylor, Michael L. Schmidt, Terry Tebold, James M. Dahl, and J. H. Bulloch (Chapter 14); Professors C. M. Sellars, T. Gladman, D. C. Dunand, F. B. Pickering, and J. C. Pilling and E. M. Taleef and Drs. Jeffrey Wadsworth, Steve Madeiro, Noshir M. Bhathena, and Rolf G. Sundberg (Chapter 15); Drs. Tohru Arai, V. S. Nemkov, V. I. Rudnev, C. A. Stickels, R. W. Foreman, David Pye, H.-J. Hunger, R. Bakish, Joarchim Bosslet, and W. K. Liliental and Messrs. R. C. Goldstein, Tom Sterner, Steven Verhoff, Mike Ives, Rick Houghton, M. M. Stirrine, Joseph Greene, and J. R. Easterday (Chapter 16); Messrs. G. Parrish and W. T. Cook (Chapter 17); Professors R. L. Boxman, Markus Pessa, Deepak G. Bhat, and William Rees, Jr., and Drs. Gary Tompa, Robert Aharonov, D. M. Mattox, Bruce Sartwell, Dennis Teer, A. J. Armini, and Angel Sanjurjo (Chapter 18); and Professor Lech Pawlowski and Drs. R. C. Tucker, Jr. and Richard Knight (Chapter 19).

I also acknowledge many societies and publishers for their generous permission to use figures, photographs, and tables in this Handbook. Thanks are due to the management and staff of McGraw Hill for their editorial and administrative contributions to the production of this book. The author would like to express his appreciation to Messrs. Mark J. Eriksen and Roy Smith of Winona State University Library for their dedicated help in getting articles, books, and a majority of reference materials. Finally, I wish to pay tribute to my wife Priti, son Manish, daughter-in-law Rashmi, and daughter Shruti for their understanding, love, moral support, and sacrifice, without which this book would not have been completed.

ABOUT THE AUTHOR

Anil Kumar Sinha, Ph.D., M. Tech, is a former professor of metallurgical engineering at Notre Dame University, Cornell University, the University of Wisconsin, and Ranchi University. He also worked at Peerless Chain Company as Staff Metallurgist, Thompson Steel Company, Inc. as Manager of Metallurgy and Quality Control, Bohn Engine & Foundry as Senior Metallurgist and Chief Metallurgist, and National Metallurgical Laboratory as Senior Research Fellow. Currently president of Computer Wire EDM Corporation and a consultant, he is the author of an earlier version of this book, *Ferrous Physical Metallurgy*, as well as another book, *Powder Metallurgy*, and sixteen research papers. Dr. Sinha earned his Ph.D. at the University of Minnesota and his Master of Technology in Physical Metallurgy from the Indian Institute of Technology.

CHAPTER 1
IRON-CARBON ALLOYS

1.1 INTRODUCTION

Steels are the most complex and widely used engineering materials because of the abundance of iron in the earth's crust, high melting temperature of iron (1534°C), and wide range of mechanical properties and associated microstructures produced by solid-state phase transformations by varying the cooling rate from the austenitic condition. The iron-cementite phase diagram is the very useful foundation on which analysis of all steel heat treating processes depends, whereas both iron-cementite and iron-graphite diagrams are useful for the heat treatment of cast iron. The phase diagram is a map showing structures or phases and phase boundaries present as the temperature and overall composition of the alloy are varied under constant pressure (usually 1 atm). This chapter deals with structures of iron and iron-carbon alloys, iron-cementite and iron-graphite phase diagrams, critical temperatures, solubilities of carbon and nitrogen in ferrite and austenite, effects of alloying elements, and classification of steels and cast irons.

1.2 CRYSTAL STRUCTURES OF IRON AND IRON-CARBON ALLOYS

1.2.1 Alpha-Iron

Pure iron, the carbon steels, and other metals such as V, Cr, Mo, and W have the body-centered cubic (bcc) structure at room temperature, which is characterized by the unit cell shown in Fig. 1.1.[1] The bcc structure of pure iron at room temperature, called either α-iron or ferrite, has one atom at the center of the cube and an atom at each corner of the unit cell and constitutes $1 + (8 \times {}^{1}\!/_{8}) = 2$ atoms per unit cell. The atomic packing factor for this structure is 0.68 and represents the volume fraction of the unit cell occupied by two atoms. The lattice parameter of α-iron at room temperature is 2.86 Å. The ferrite is a more open or less dense structure than the other structural modification of iron, called either gamma-iron or austenite. The difference in atomic packing of α- and γ-iron is responsible for volume contraction that takes place on heating low-density ferrite to higher-density austenite. Austenite is properly used for γ-iron with carbon in solid solution. Both ferrite and austenite are quite soft and ductile. Their average properties include tensile strength, 40,000 psi; elongation, 40%; hardness, 150 BHN, RC-0, or less than RB-90. Ferrite

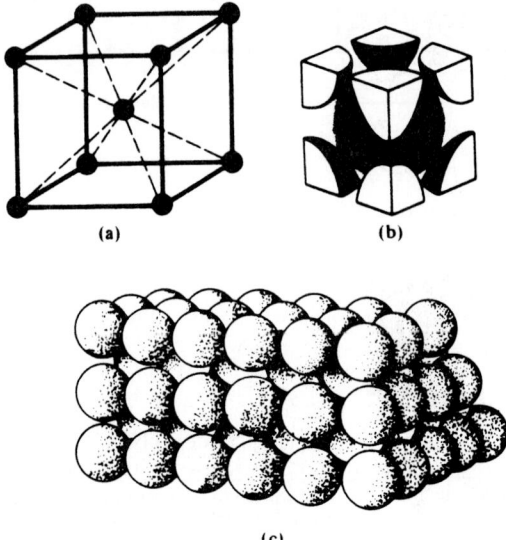

(a)   (b)

(c)

FIGURE 1.1 Body-centered cubic structure of a metal showing (a) the atomic-site unit cell, (b) the isolated unit cell,[1] (c) model made from hard balls. [(a) and (b) reprinted by permission of McGraw-Hill, New York; (c) reprinted with permission from B. A. Rogers, The Nature of Metals, 2d edition, ASM, Metals Park, Ohio.]

is ferromagnetic below 768°C (1414°F) and paramagnetic in the temperature range of 768 to 910°C (1414 to 1670°F). The temperature at which this magnetic transformation takes place is called the *Curie temperature*.

1.2.2 Gamma-Iron

Austenite as well as other metals such as Al, Ni, Cu, Ag, Pt, and Au have the close-packed face-centered cubic (fcc) structure. Its unit cell is shown in Fig. 1.2[1] and has an atom at each corner and an atom at the center of each face. Each corner atom is shared by eight unit cells that come together at the corner, while each face atom is shared by two adjacent unit cells. Thus there are $(8 \times \frac{1}{8})$ + $(6 \times \frac{1}{2}) = 1 + 3 = 4$ atoms per unit cell. The atomic packing factor for this structure is 0.74. The lattice parameter of austenite is 3.57 Å, which is larger than that of ferrite. Gamma-iron is the stable form of pure iron in the temperature range of 910 to 1393°C (1670 to 2540°F). Unlike ferrite, austenite is paramagnetic. It is also soft and ductile.

1.2.3 Delta-Iron

The third phase that occurs in pure iron is δ-iron or ferrite with a bcc structure which is crystallographically similar to alpha-iron. Delta-iron is stable at temperature between 1393 and 1534°C (2540 and 2793°F). Its lattice parameter is 2.89 Å; it is also soft and ductile, and its hardness and elongation are similar to those of ferrite and austenite in their stable forms. The δ-ferrite is of no significance in heat treating practice used for plain carbon and low-alloy steels.

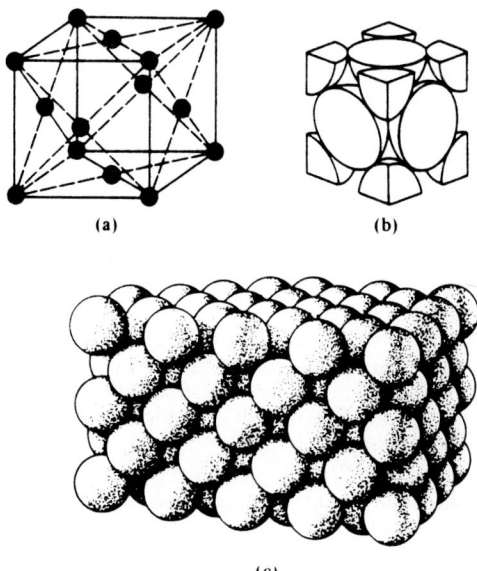

(a)   (b)

(c)

FIGURE 1.2 Face-centered cubic structure of a metal showing (a) the atomic-site unit cell, (b) the isolated unit cell,[1] (c) model made from hard balls. [(a) and (b) reprinted by permission of McGraw-Hill, New York; (c) reprinted with permission from B. A. Rogers, The Nature of Metals, 2d edition, ASM, Metals Park, Ohio.]

1.2.4 Cementite

Cementite, represented by the formula Fe_3C, is a metastable Fe-C intermetallic compound. It has a negligible solubility limit in α-iron and contains 6.67 wt % (29 at %) carbon. It is ferromagnetic with a Curie temperature of 215°C (419°F). In sharp contrast to ferrite and austenite, cementite is hard (BHN over 700, VPH 1300) and brittle (0% elongation). This is an interstitial compound of low tensile strength (~5 ksi) but high compressive strength. It is the hardest structure that appears on the Fe-Fe$_3$C phase diagram. It plays an important role in the hardening of many commercial steels. It has an orthorhombic crystal structure with lattice limit: $a = 4.52$ Å, $b = 5.09$ Å, and $c = 6.74$ Å and 12 iron and 4 carbon atoms per unit cell.

1.3 PHASE DIAGRAM

A *phase* is a portion of a system whose properties, composition, and crystal structure are uniform or homogeneous and which is separated from the remainder by distinct bounding surfaces. By the word *system* we mean an isolated and homogeneous portion of matter, and the components of a system are the metallic elements* that constitute or form the system. A one-component system is a single metallic element (e.g., pure iron); a two-component system is a mixture of two metallic elements, called *binary alloys*; and three-component systems are mixtures of three metallic elements, called *ternary alloys*.[2]

In a phase diagram (also called *equilibrium diagram* or *constitutional diagram*), temperature is plotted vertically and composition horizontally at constant (atmos-

---

* The elements need not be metallic; however, this term has been used to emphasize the metallic system that is of immediate significance to us.

pheric) pressure. Figure 1.3*a* and *b* shows the conventional and modified versions, respectively, of the phase diagram, where each part shows both the metastable Fe-Fe$_3$C and stable or equilibrium Fe-graphite diagram; the former is indicated by full lines, and the latter is indicated by dashed lines.[3,4] The phases present in Fe-Fe$_3$C and Fe-C diagrams are molten alloy, austenite, α-ferrite, δ-ferrite, cementite, and graphite. These phases are alternatively called *constituents*. However, not all constituents are single phases, but rather are a mixture of two phases—ferrite and cementite like pearlite and bainite. There is a slight difference between the two sets of diagrams in both temperatures and compositions corresponding to the critical points and the reaction curves. Readers may use either version. However, the conventional version of these diagrams will be used hereafter in the entire text because of its wider use and greater familiarity.

The iron-iron carbide diagram is not a true equilibrium diagram but rather a metastable equilibrium diagram because cementite is a metastable phase. Given a very long period, cementite will decompose into more stable, or equilibrium, phases of graphite and iron. However, once cementite is formed, it is very stable and may be treated for all practical purposes as an equilibrium phase.[2] A study of Fe-Fe$_3$C and Fe-C diagrams is valuable in understanding the heat treatment, accurate determination of phase compositions, control of their properties and solid-state reactions in general, and the basic differences among iron alloys in particular.

The liquidus curve *ABCD* or *ABC′D′* shown in Fig. 1.3*a* represents the boundary between the two-phase regions and the liquid. The liquidus curve *AB* is practically a straight line joining the melting temperature of iron and the endpoint of the first isothermal reaction (point *B*). This horizontal line, *HJB*, at 1493°C represents the *peritectic temperature*. The *peritectic reaction*, occurring during solidification of carbon steels, may be expressed in the form:

liquid (0.51 wt % C, point *B*) + solid (δ-ferrite phase, point *H*)

$\xrightarrow[\text{heating}]{\text{cooling}}$ austenite phase (0.16 wt % C, point *J*)

Point *J* is called a *peritectic point*.

The liquidus curve *BC* of the austenite phase ends at the *eutectic* horizontal: liquid (4.3 wt % C, point *C*) $\xrightarrow[\text{heating}]{\text{cooling}}$ γ (2.06 wt % C, point *E*) + Fe$_3$C (6.67 wt % C, point *F*). Thus the horizontal line *ECF* at 1147°C represents the *eutectic temperature*, and point *C* represents the *eutectic point*. The eutectic mixture of γ and Fe$_3$C is called *ledeburite*.

The solidus curve *AHJECF* represents the boundary between the two-phase region and the solid. The horizontal line *PSK* at 723°C corresponds to the eutectoid reaction: γ (0.8 wt % C, point *S*) $\xrightarrow[\text{heating}]{\text{cooling}}$ α-ferrite (0.02 wt % C, point *P*) + Fe$_3$C (6.67 wt % C, point *K*). This eutectoid mixture is a lamellar structure comprising alternate lamellae of ferrite and cementite to which the name *pearlite* has been given. The eutectoid point is represented by 0.8 wt % C content, point *S*. It is thus clear that the portion of the diagram that lies between the liquidus and solidus lines (*ABCD* and *AHJECF*, respectively) represents the solidification of the liquid solution, whereas the areas between *GSECF* and *PSK* as well as below *PSK* represent the decomposition of austenite on slow cooling.

Although the Fe-Fe$_3$C diagram extends from a temperature of 1925°C (3500°F) down to room temperature, the left-hand bottom portion of the diagram (Fig. 1.3*a*) which lies below 1035°C (1900°F) is commonly used for the heat treatment of steel because the steel heat treating practice seldom involves temperatures beyond this value.[5] The large phase field of austenite shows solubility of carbon in this structure ranging from 0 to 2.06 wt % through 0.8 wt % carbon at 723°C (Fig. 1.3*a*). This

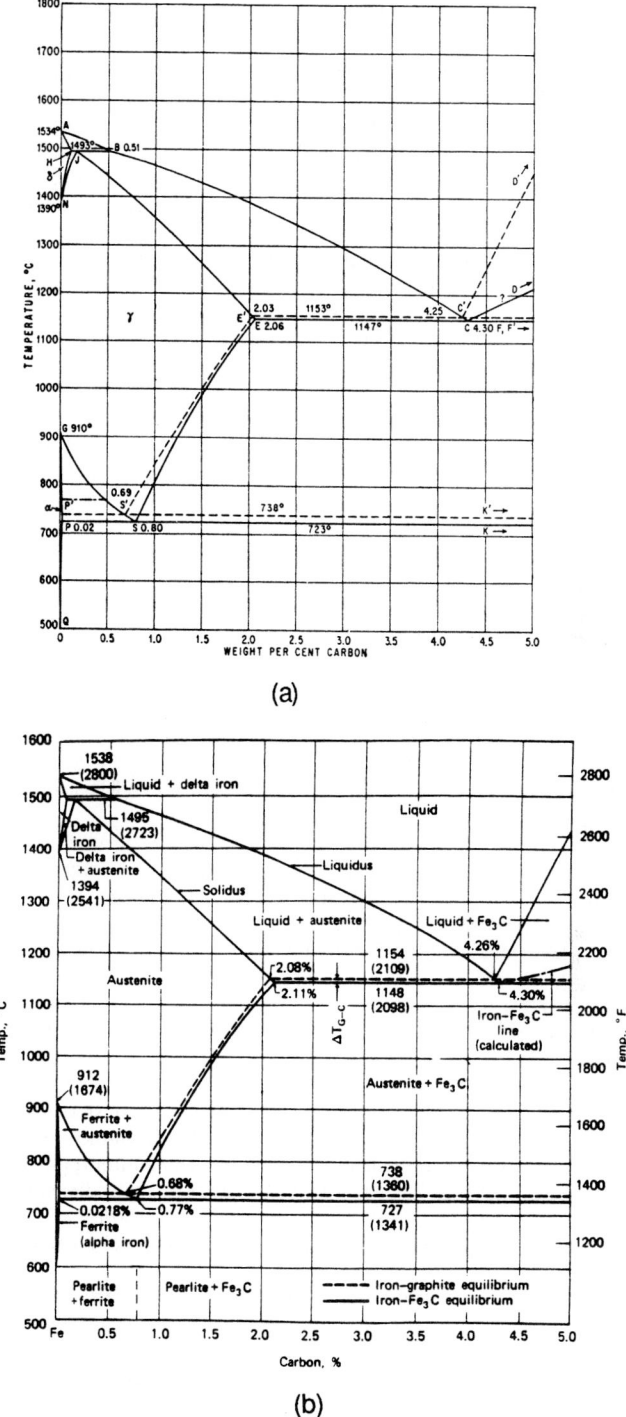

FIGURE 1.3 The Fe-Fe₃C and Fe-C phase diagrams: (*a*) conventional[3] and (*b*) modified.[4] [(*a*) *reprinted by permission of McGraw-Hill, New York; (b) reprinted by permission of ASM International, Materials Park, Ohio.*]

maximum solubility of carbon content of 2.06 wt % corresponds to the boundary between steels and cast irons. Fe-C alloys with carbon content up to 2.06 wt % are arbitrarily classed as *steels*, and those beyond this amount are called *cast irons*. Actually, only in rare instances is steel used with more than 1.1 wt % carbon.[2] The solubility of carbon in the α-iron phase field is very low, with a maximum of about 0.02 wt % carbon at the eutectoid temperature, 723°C, and it decreases with decreasing temperature until it is about 0.008 wt % at 0°C. This large difference in solubility of carbon in the two phases of iron is of great significance in the heat treatment of steel.[6]

It is customary to subdivide the steel range into *hypoeutectoid* and *hypereutectoid* depending on whether composition lies below or above the eutectoid composition. Likewise, the cast iron range may be subdivided into *hypoeutectic* and *hypereutectic*, if the carbon content in cast iron is below or above the eutectic composition, respectively. Steels and white cast irons obey the metastable Fe-Fe$_3$C phase diagram, whereas other cast irons (consisting of graphite precipitates in a solid metal matrix, similar to steel) obey both the equilibrium Fe-graphite and metastable Fe-Fe$_3$C phase diagrams. Usually the carbon content varies from 2.2 to 4.5 wt % for the cast irons. The microstructures of the cast irons, which are dependent on the carbon content and cooling rate, may be deduced from both of the phase diagrams. However, it was realized later that the analysis of structures of cast irons is much more complex than that of steel and is much more sensitive to the processing conditions employed in their manufacture.

It is noted here that commercial cast irons are not simple alloys of iron and carbon but, instead, contain appreciable amounts of other elements which have important and powerful effects on the structure of cast iron. The most important additional element is silicon ranging from 1.0 to 3.0%. Thus we may treat cast irons as the ternary alloy of Fe, C, and Si. However, cast irons usually contain minor additions of S, Mn, P, and trace elements such as Al, Sn, Sb, and Bi as well as gaseous elements H, N, and O.

Cast irons may also differ in many respects from steels. For example, cast irons have low melting temperature, low ductility, and poor impact properties, which may restrict their applications. The molten cast iron is more fluid, is less reactive with air and molding materials, and is of relatively low cost when compared to steels or other common alloys. In addition, it can be readily machined. Thus it is an excellent engineering material.[7,8]

Another difference between cast irons and steels is the fact that cast iron properties are determined by four factors, i.e., chemical composition, inoculation, solidification rate, and cooling rate, while steel properties are controlled primarily by the chemical composition.

The addition of Si promotes graphitization in cast iron. In other words, Si is the catalytic agent that permits free carbon (flakes, nodules, etc.) to appear in the microstructure. High temperature and the presence of silicon greater than 1% speed up the dissociation reaction of Fe$_3$C, which may be written as follows:[9]

$$\text{Fe}_3\text{C} \longrightarrow 3\text{Fe (austenite)} + \text{C (graphite)} \qquad (1.1)$$

As a consequence, cast irons may contain carbon in free form as graphite and in combined form as cementite. It is clear that cast iron can solidify according to either the stable Fe-graphite system (gray iron) or to metastable Fe-Fe$_3$C system (white iron). As a result, the eutectic may be γ-graphite or γ-Fe$_3$C (*ledeburite*). This also differentiates cast irons from steels because the latter possess only combined carbon as cementite. The formation of lower-density graphite during the solidification of iron castings causes the reduced or negligible volume change of the metal from liquid to solid. This permits the formation of very complex castings such as one-piece water-jacketed internal combustion engine blocks without any shrinkage voids in the metal.[8] The shape and distribution of free graphite, rather than variations in composition, are commonly used to classify cast irons.

1.4 CRITICAL TEMPERATURES

There are three transformation temperatures, often referred to as *critical temperatures,* which are of interest in heat treatment of steels. The temperature A_1 is the eutectoid temperature of 723°C in the binary phase diagram which is the boundary between ferrite-cementite field and the austenite-ferrite or austenite-cementite field. Temperature A_3 is the temperature at which α-iron transforms to γ-iron, which, for pure iron, occurs at 910°C. The A_3 line represents the boundary between the ferrite-austenite and austenite fields. Similarly, the A_{cm} line is the boundary between the cementite-austenite and the austenite fields. The temperature difference between A_1 and A_3 is called the *critical* (temperature) *range.* Sometimes $A_1, A_3,$ and A_{cm} are written as $Ae_1, Ae_3,$ and Ae_{cm}, respectively, denoting equilibrium conditions. These critical temperatures are detected by thermal analysis or dilatometry during heating or cooling cycles, and some thermal hysteresis (lag) is observed. The thermal hysteresis that occurs on heating is indicated by the letter c, representing the French word *chauffage,* meaning heating. Similarly, thermal hysteresis on cooling is indicated by the letter r, representing the French word *refroidissement,* meaning cooling. Thus there are two sets of critical temperatures: $Ac_1, Ac_3,$ and Ac_{cm} for heating and $Ar_1, Ar_3,$ and Ar_{cm} for cooling. These sets of critical temperatures are shown in Fig. 1.4.[9] The faster the rate of heating, the higher the Ac point: the faster the rate of cooling, the lower the Ar point. Thus the faster the heating and cooling rates, the larger the difference between the Ac and Ar points of the reversible equilibrium point A.

Usually the critical temperatures which are necessary for the heat treatment of carbon and alloy steels can be known experimentally.[10] However, empirical formulas that show the effects of alloying elements on the critical temperatures have been developed by regression analysis of large amounts of experimental data by Andrews,[11] Grange,[12] Kunitake et al.[13,14] and Miyoshi et al.[15] for calculating the practical Ac_1 and Ac_3 temperatures.

For 0.08 ~ 1.4% C steel:[11]

$$Ac_1(°C) = 723 - 10.7\text{Mn} - 16.9\text{Ni} + 29.1\text{Si} + 16.9\text{Cr} + 290\text{As} + 6.38\text{W} \quad (1.2a)$$

For 0.3 ~ 0.6% C low-alloy steel:[12]

$$Ac_1(°C) = 723 - 13.9\text{Mn} - 14.4\text{Ni} + 22.2\text{Si} + 23.3\text{Cr} \quad (1.2b)$$

For 0.25 ~ 0.45% C low-alloy steel:[13]

$$Ac_1(°C) = 755 - 32.3\text{C} - 17.8\text{Mn} + 23.3\text{Si} + 17.1\text{Cr} + 4.5\text{Mo} + 15.6\text{V} \quad (1.2c)$$

For 0.10 ~ 0.55% C low-alloy steel:[14]

$$Ac_1(°C) = 751 - 16.3\text{C} - 27.5\text{Mn} - 5.5\text{Cu} - 15.9\text{Ni} + 34.9\text{Si} + 12.7\text{Cr} + 3.4\text{Mo} \quad (1.2d)$$

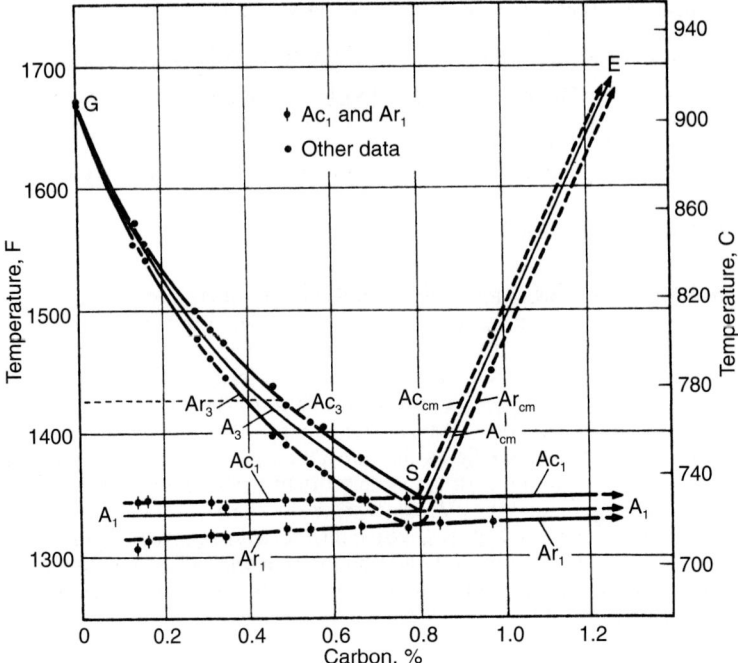

FIGURE 1.4 A portion of Fe-Fe$_3$C diagram showing two sets of critical cooling temperatures: Ac_1, Ac_3, and Ac_{cm} for heating and Ar_1, Ar_3, and Ar_{cm} for cooling. Rate of heating and cooling at 0.125°C/min.[9] (*Reprinted by permission of ASM International, Materials Park, Ohio.*)

For 0.07 ~ 0.22% C low-alloy steel:[15]

$$Ac_1(°\text{C}) = 751 - 26.6\,\text{C} - 11.1\,\text{Mn} - 22.9\,\text{Cu} - 23.0\,\text{Ni} + 17.6\,\text{Si}$$
$$+ 24.1\,\text{Cr} + 22.5\,\text{Mo} - 39.7\,\text{V} + 223\,\text{Nb} - 169\,\text{Al} - 895\,\text{B} \qquad (1.2e)$$

For 0.08 ~ 1.4% C steel:[11]

$$Ac_3(°\text{C}) = 910 - 203\sqrt{\text{C}} - 30\,\text{Mn} - 20\,\text{Cu} - 15.2\,\text{Ni} - 11\,\text{Cr} - 700\,\text{P}$$
$$+ 44.7\,\text{Si} + 31.5\,\text{Mo} + 104\,\text{V} + 460\,\text{Al} + 13.1\,\text{W} + 120\,\text{As} \qquad (1.3a)$$

For 0.3 ~ 0.6% C low-alloy steel:[12]

$$Ac_3(°\text{C}) = 854 - 179\,\text{C} - 13.9\,\text{Mn} - 17.8\,\text{Cu} - 1.7\,\text{Ni} + 44.4\,\text{Si} \qquad (1.3b)$$

For 0.25 ~ 0.45% C-Si-Cr-Mo-V low-alloy steel:[13]

$$Ac_3(°\text{C}) = 930 - 395\,\text{C} - 14.4\,\text{Mn} + 55\,\text{Si} + 5.8\,\text{Ni} + 24.5\,\text{Cr} + 83.4\,\text{Mo} \qquad (1.3c)$$

For 0.10 ~ 0.55% C low-alloy steel:[14]

$$Ac_3(°\text{C}) = 881 - 206\,\text{C} - 15\,\text{Mn} - 26.5\,\text{Cu} - 20.1\,\text{Ni} - 0.7\,\text{Cr} + 53.1\,\text{Si} + 41.1\,\text{V} \qquad (1.3d)$$

For 0.07 ~ 0.22% C low-alloy steel:[15]

$$Ac_3(°C) = 937 - 476C - 19.7\,Mn - 16.3\,Cu - 26.6\,Ni - 4.9\,Cr$$

$$+ 56\,Si + 38.1\,Mo + 12.5\,V - 19\,Nb + 198\,Al + 3315\,B \qquad (1.3e)$$

Among these empirical relations, the ones formulated by Andrews are widely adopted. There is good agreement between the calculated and observed temperatures.

1.5 SLOWLY COOLED PLAIN-CARBON STEELS

1.5.1 Eutectoid Steel

Figure 1.5 shows the enlarged section of the Fe-Fe$_3$C diagram. When a eutectoid steel is heated to the austenitizing temperature and held there for a sufficient time, its structure will become homogeneous austenite. On very slow cooling under conditions approaching equilibrium, the structure will remain that of austenite until just above the eutectoid temperature. At the eutectoid temperature or just below it, the entire structure of austenite will transform into pearlite. The ferrite and cementite that are incorporated in the pearlite are called *eutectoid ferrite* and *eutectoid cementite*, respectively. Figure 1.6 shows a light micrograph of pearlite advancing into the unstable austenite.[16]

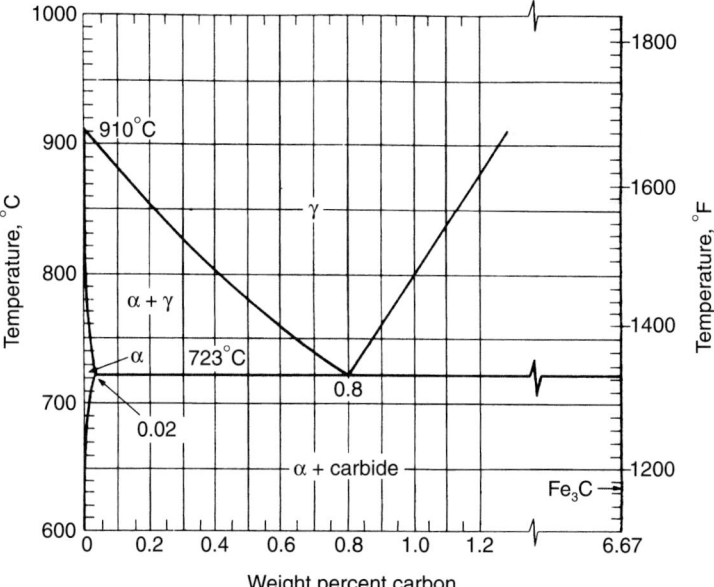

FIGURE 1.5 The eutectoid portion of the Fe-Fe$_3$C diagram. (*Reprinted by permission of Addison-Wesley Publishing Co., Reading, Massachusetts.*)

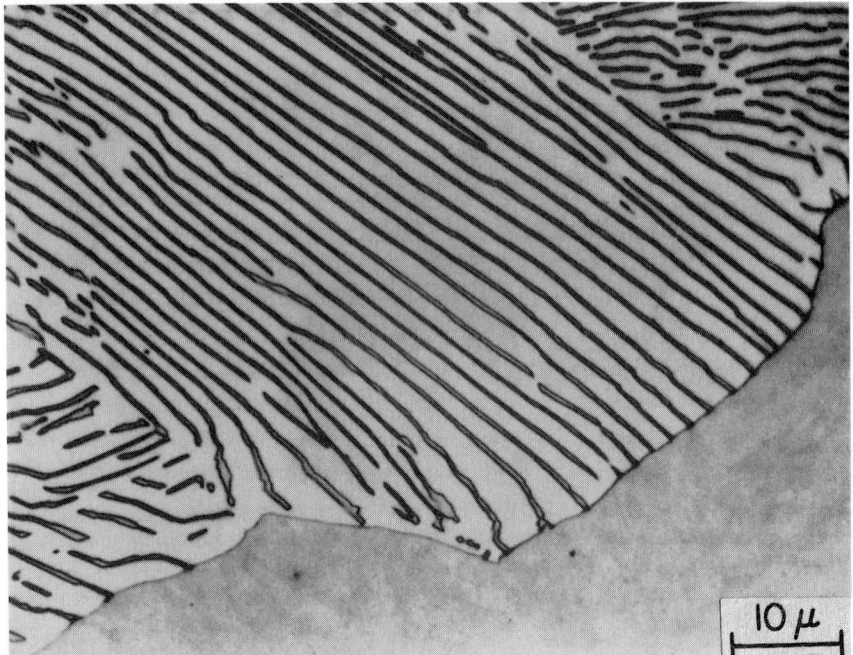

FIGURE 1.6   Light micrograph of pearlite colony advancing into an austenite grain.[16] (*Courtesy of the Metallurgical Society, Warrendale, Pennsylvania; after J. R. Vilella.*)

The amount of phases present in a two-phase field of a binary phase diagram can be determined by applying the lever rule. The alloy composition represents the fulcrum of a lever with the horizontal line, called a *tie line*, touching the two-phase field representing its length. The ends of the tie line fix the compositions of the coexisting phases, and the relative amounts of the phases are directly proportional to the length of the "opposite lever arm." Thus, the amount of phases present in the pearlite formed just below 723°C can be easily computed by using the lever rule:

$$\text{wt \% ferrite} = \frac{6.67 - 0.8}{6.67 - 0.02} \times 100 = \frac{5.87}{6.65} \times 100 = 88.27 \approx 88 \qquad (1.4a)$$

$$\text{wt \% cementite} = \frac{0.8 - 0.02}{6.67 - 0.02} \times 100 = \frac{0.78}{6.65} \times 100 = 11.73 \approx 12 \qquad (1.4b)$$

Since the densities of ferrite and cementite, being 7.87 and 7.70 g/cm³, respectively, are very close, the lamellae of ferrite and cementite have respective widths of about 7.5 to 1.

1.5.2   Hypoeutectoid Steel

When a hypoeutectoid steel is allowed to cool slowly after heating into the austenite phase field, that is, above the *GS* line (or A_3 line) (Fig. 1.3), the primary or proeu-

FIGURE 1.7 Structures of slowly cooled steels, 500×: (a) hypoeutectoid steel, 0.45% C, showing ferrite (white areas) and pearlite (resolved and unresolved); (b) hypereutectoid steel, 0.9% C. The white network is cementite. [(b) *reprinted by permission of Wadsworth, Inc., Belmont, California. Source: D. S. Clark and W. R. Varney, Physical Metallurgy for Engineers, Van Nostrand, New York, 1962.*]

tectoid ferrite will begin to precipitate at the austenite grain boundaries at a temperature indicated by a point on the $(\alpha + \gamma)/\gamma$ phase boundary for the alloy composition concerned. As the temperature decreases further (proeutectoid) ferrite is formed and the carbon content of austenite coexisting in the two-phase $(\alpha + \gamma)$ region continuously increases (due to rejection of excess carbon at the austenite/ferrite interface from the ferrite formed) until the eutectoid composition is reached at 723°C. On passing through a temperature of 723°C, the austenite will transform to pearlite. The final structure of slowly cooled hypoeutectoid steel will consist of primary or proeutectoid ferrite and pearlite (Fig. 1.7a); the proportion of the latter increases with the carbon content until, at 0.8% C, the structure will be completely pearlitic.

Figure 1.5 shows that the proeutectoid ferrite in slowly cooled 0.4% plain carbon steel begins to form at about 800°C. It forms at the austenite grain boundaries. As the alloy is continuously cooled to a temperature just above the eutectoid temperature (say, 724°C), the weight percent proeutectoid ferrite and weight percent austenite can be determined by using the lever rule:

$$\text{wt \% proeutectoid ferrite} = \frac{0.80 - 0.40}{0.80 - 0.02} \times 100 \approx 50$$

$$\text{wt \% austenite} = \frac{0.40 - 0.02}{0.80 - 0.02} \times 100 \approx 50$$

Since all the remaining austenite will transform to pearlite at the eutectoid temperature, the weight percent of pearlite just below 723°C (say, 722°C) will be 50%, if conditions approaching equilibrium exist.

Since the solubility of carbon in ferrite decreases from 723°C to room temperature, further slow cooling to room temperature results in the precipitation of carbide from ferrite; however, its amount is small, and therefore it increases the overall hardness of the steel only to a slight extent which cannot be easily measured. Also, the slightly increased amount of carbide is difficult to detect in the microstructure. Hence, the amount of ferrite and carbide analyzed and calculated at just below 723°C can be considered to be valid at room temperature.[17]

1.5.3 Hypereutectoid Steel

Consider the slow cooling of a hypereutectoid steel (say, 1.2% plain-carbon steel) from an austenitizing temperature of 920°C. The separation of proeutectoid cementite occurs at the grain boundaries of austenite beginning at a temperature of about 880°C and is completed at a temperature of 723°C. An increasing amount of primary cementite precipitates with decreasing temperature until a temperature of 723°C is reached, where the remaining austenite, depleted in carbon content, reaches its minimum level of 0.80%. The microstructure of slowly cooled hypereutectoid steel will consist of primary cementite and pearlite at any temperature below 723°C.

At a temperature slightly above the eutectoid temperature, the weight percentage of proeutectoid cementite is $[(1.20 - 0.80)/(6.67 - 0.80)] \times 100 = 6.8\%$, and that of the remainder austenite is $[(6.67 - 1.20)/(6.67 - 0.80)] \times 100 = 93.2\%$. This austenite transforms to pearlite at or below 723°C. Thus the whole structure of a hypereutectoid steel will consist of primary cementite and pearlite at any temperature below 723°C. Figure 1.7*b* shows the microstructure of a slowly cooled hypereutectoid steel containing 0.9% C.

1.6 SOLUBILITY OF CARBON AND NITROGEN IN FERRITE AND AUSTENITE

The atomic radii of carbon (0.8 Å) and nitrogen (0.7 Å) are much smaller than that of iron (1.28 Å), which allows these solute elements to enter into the interstices or "holes" of the α-iron and γ-iron crystal lattices. On a hard sphere model for fcc austenite, the largest hole at an octahedral site is 0.52 Å in radius which is surrounded by six atoms located at the corners of a regular octahedron (Fig. 1.8*a*); the next largest hole is 0.28 Å in radius at a tetrahedral site which is surrounded by a tetrahedron of four atoms (Fig. 1.8*b*). In spite of the fact that the bcc ferrite is not densely packed, the octahedral site is only 0.19 Å in radius and is surrounded by six atoms at the corners of a slightly compressed octahedron (Fig. 1.9*a*) while the tetrahedral site is 0.36 Å in radius (Fig. 1.9*b*).[18] Evidently, octahedral holes in α-iron are much smaller than those in γ-iron. Since C and N are larger than the available interstices in either lattice, it is apparent that some distortion must ensue when they occupy the interstices of iron lattices. However, these interstitial atoms reside in octahedral sites, with an expansion caused by the displacement of two nearest-neighbor iron atoms rather than four, as in tetrahedral sites.

It is thus clear that the maximum solubility of carbon (2.0 wt %) and nitrogen (2.8 wt %) in γ-iron is much greater than in α-iron as a result of much larger octahedral holes in austenite. This large difference in solubilities is of great significance in the heat treatment of steels and is fully exploited to improve strength. Nitrogen

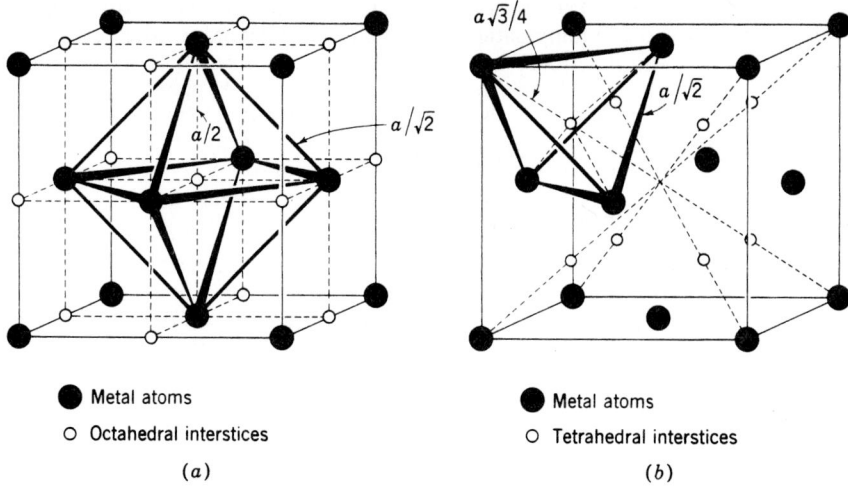

FIGURE 1.8 (*a*) Octahedral and (*b*) tetrahedral interstitial voids in fcc structure.[18] (*Courtesy C. S. Barrett and T. B. Massalski.*)

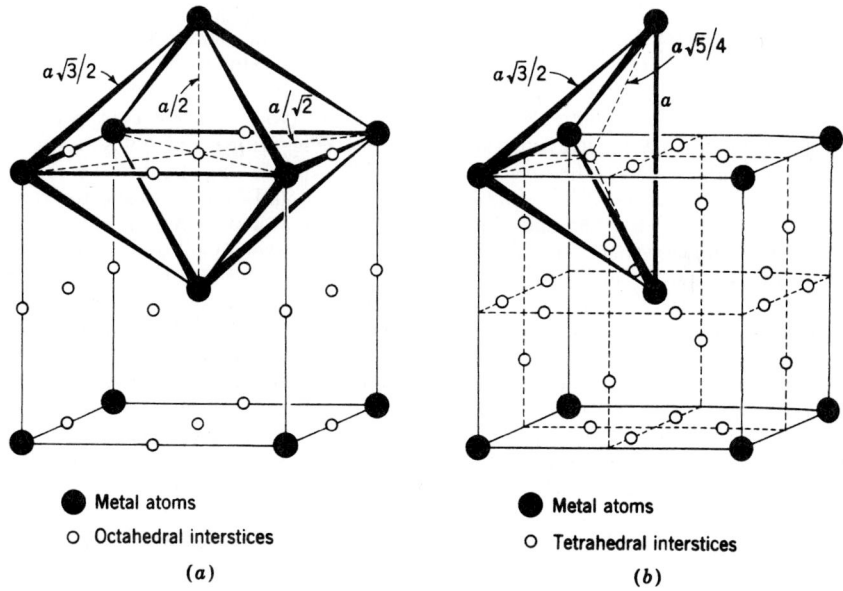

FIGURE 1.9 (*a*) Octahedral and (*b*) tetrahedral interstitial voids in bcc structure.[18] (*Courtesy C. S. Barrett and T. B. Massalski.*)

has more solubility than carbon in ferrite. That is why the nitriding heat treatment is usually performed at a temperature lower than the eutectoid temperature.

1.7 EFFECTS OF ALLOYING ELEMENTS

1.7.1 The γ- and α-Phase Fields

Steels contain alloying elements and impurities that must be associated with austenite, ferrite, and cementite. It has been pointed out that we can divide the alloying elements into two groups based on their influence on the phase diagram.[19,20]

1. By widening (or opening) the γ-phase field and promoting the formation of austenite over larger compositional range. These elements are called *austenite stabilizers* or *formers*.

2. By shrinking (or closing) the γ-phase field and promoting the formation of ferrite over a larger compositional range. These elements are called *ferrite stabilizers* or *formers*. Based on the alloying elements, the iron binary phase diagrams can be divided into four categories, as follows.[19,21]

   Type 1: *Open γ-phase field*. In this group the alloying elements, for example, Mn, Ni, Co, Ru, Rh, Pd, Os, Ir, and Pt, expand the temperature range for stable austenite by lowering the two-phase $(\alpha + \gamma)$ region toward room temperature and raising the two-phase $(\delta + \gamma)$ zone to the melting range (Fig. 1.10*a*); that is, both Ae_1 and Ae_3 are depressed. It is also easier to obtain metastable γ by quenching from the γ region to room temperature. Consequently Mn and Ni are useful elements in the formulation of stainless steels.

   Type 2: *Expanded γ-phase field*. This is the same as type 1 above except its range is shortened by iron-rich compound formation (Fig. 1.10*b*). Examples are C, N, Cu, Zn, and Au. Thus the presence of C and N expands the γ-field to the extent that the solid solubility of C and N increases to 2.0 and 2.8 wt %, respectively, in the austenite. This effect underlines the whole of the heat treatment of steels.

   Type 3: *Closed γ-phase field*. In this group the alloying elements restrict the temperature range for stable austenite, with the results that a smaller area of γ-phase (called the *gamma loop*) and the continuous and wider δ- and α-phase fields are obtained (Fig. 1.10*c*). This means that the alloying elements in this category promote the formation of ferrite, which include Si, Al, Be, and P, together with the strong carbide-forming elements Cr, Ti, V, Mo, and W.

   Type 4: *Contracted γ-phase field*. In this group the α- and γ-phase fields are bounded by a miscibility gap. That is, α- and γ-solid solutions are in equilibrium with an intermetallic compound or solid solution (Fig. 1.10*d*). Boron is the most significant element in this group together with the carbide formers Ta, Zr, and Nb.

These phenomena are associated with the crystal structure of the alloying elements since no fcc alloying element stabilizes ferrite and likewise no bcc element stabilizes austenite. Thus it appears from these phase diagrams that the crystal structure of solid solutions of iron at room temperature is the important basis for classifying steels. If austenite is predominant at room temperature because of the addition of sufficiently large amounts of Ni and Mn, it is called an *austenitic steel*. Examples are *Hadfield steel* containing 13% Mn, 1.2% Cr, and 1% C; 18% Cr-8% Ni *austenitic stainless steel*; and *precipitation-hardening stainless steels* with fine dis-

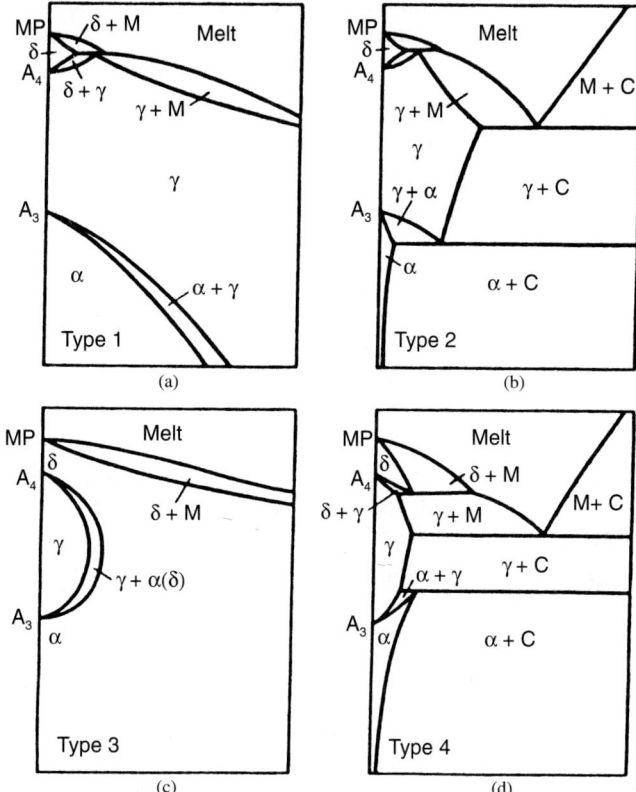

FIGURE 1.10 Classification of equilibrium diagram for iron alloys: (*a*) open γ-phase field; (*b*) expanded γ-phase field; (*c*) closed γ-phase field; (*d*) contracted γ-phase field.[6,21] (*Reprinted by permission of ASM International, Materials Park, Ohio.*)

persion of stable or metastable coherent, ordered fcc γ' (Ni_3Al, Ti) phase in the fcc iron-rich matrix. This type of mixed microstructure is also observed in *age-hardenable nickel-base superalloys.*

On the other hand, if the room-temperature structure consists mostly of α-iron solid solution that is made possible by the ferrite-forming elements (e.g., Cr, Si, Mo, W, and Al), it is called *ferritic steel.* Examples of ferritic steels are Fe-Cr alloys containing more than 13% Cr and low-carbon transformer steel containing about 3% Si.

1.7.2 Purpose of Alloying Elements

Alloy additions are made to fulfill the following functions: (1) increase the hardenability (or strength in large sections), (2) reduce distortion due to heat treatment, (3) provide improved toughness at a particular hardness level, (4) increase the abrasion resistance at a given hardness level, and (5) increase the elevated temperature

TABLE 1.1 The Effect of Alloying Elements on Some Specific Properties[22]

Property	Elements (in order of decreasing effectiveness)
Hardenability	Mn, Mo, Cr, Si, Ni, V
Minimum distortion	Mo (with Cr), Cr, Mn
Toughness	Ni (produces general toughness), V, W, Mo, Mn, Cr
Wear resistance	V, W, Mo, Cr, Mn
Hot hardness	W, Mo, Co, V, Cr, Mn

Source: After G. A. Roberts, J. C. Hamaker, and A. R. Johnson, *Tool Steels*, ASM, Materials Park, Ohio, 1971.

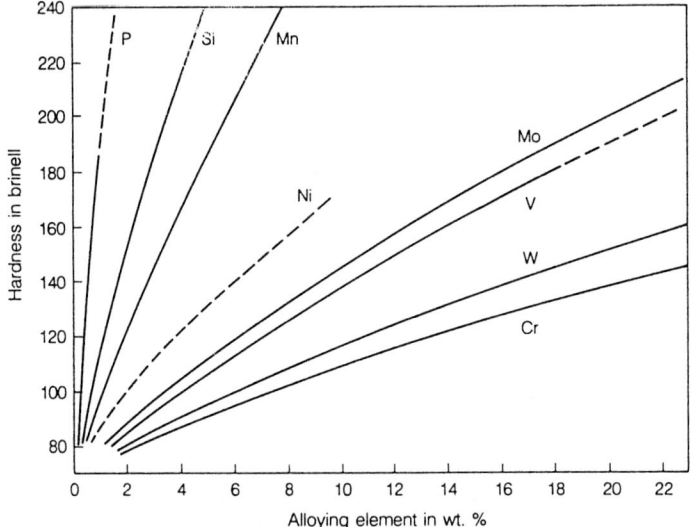

FIGURE 1.11 Solid-solution hardening effects of various alloying elements dissolved in α-iron.[20] (*Reprinted by permission of ASM International, Materials Park, Ohio.*)

strength and hardness. Table 1.1 shows the effects of various alloying elements on some specific properties.[22,23] This approach has some drawbacks in complex alloys because of the mutual interactions between two or more elements.

1.7.3 Distribution of Alloying Elements

If only steels in which γ transforms to ferrite and carbide on slow cooling are considered, the alloying elements can be divided into three groups: (1) elements entering only in the ferrite phase, (2) elements forming stable carbides and entering the ferrite phase, and (3) elements entering only the carbide phase.

In the first group belongs Cu, Ni, P, and Si, elements that are usually found in solid solution in the ferrite phase. Their solid solubility in cementite or alloy carbides is negligible. The relative strengthening effects of some substitutional solutes in solid solution in α-iron are shown in Fig. 1.11.[20] The carbide formers Cr, W, V, and Mo appear to be relatively ineffective.

The majority of alloying elements belong to the second group. They are carbide formers as well as ferrite formers with respect to iron. At higher concentrations most will form thermodynamically more stable alloy carbides than cementite. At low concentrations they go into solid solution in cementite and also form solid solutions in ferrite. Typical examples are Mn, Cr, Mo, V, Ti, W, and Nb. Manganese does not form a separate carbide in steel; rather Mn can dissolve readily in Fe_3C. The carbide-forming elements are usually present in greater amounts than required in the carbide phase, which are determined mainly by the carbon content of the steel. The remainder enter into solid solution in the ferrite with the non-carbide-forming elements Ni and Si. Some of these elements, particularly Ti, W, and Mo, produce considerable solid solution hardening of ferrite.

In the third group are those elements which enter directly into the carbide phase. The affinity of elements for carbon increases from left to right: Mn, Cr, W, Mo, V, Ti, Nb, Ta, to Zr. All carbide formers are also nitride formers. Nitrogen is the most significant element, and it forms carbonitrides with iron and many alloying elements. However, in the presence of certain very strong nitride-forming elements, such as Ti and Al, a separate alloy nitride phase can form. The affinity of elements for nitrogen decreases from left to right: Al, Ti, Mo, Cr, V, and Ni.

1.7.4 Alloy Carbides

At the carbon-rich side we find hard metastable cementite (Fe_3C) or M_3C with a complex orthorhombic crystal structure, where M stands for a metal atom or some combination of metal atoms. Now M_3C is the predominant carbide type in low-carbon low-alloy steels in the absence of strong carbide-forming elements. Thus, M_3C occurs in annealed steels containing low W, V, and Mo contents. Any substitution of Fe in the carbide is done mainly by Mn or Cr.[23] There are three other metastable iron carbides with iron-carbon ratios $\leq 3:1$ which may be produced by low-temperature carburization of iron, iron oxide, or iron nitrides. These are (1) Hägg carbide, χ-carbide or iron percarbide, Fe_5C_2 or M_5C_2 phase; (2) Fe_7C_3; and (3) ε-carbide [formed during low-temperature (175 to 250°C) carburization of iron or tempering of certain high-carbon ferrous martensites].[24]

M_5C_2 Carbides. In steels M_5C_2 has been observed to precipitate as rodlike carbides during tempering of martensitic carbon steels, in the AISI 4340 steels used for gun barrels, and in the 1 Cr–0.5 Mo steels. It is a monoclinic phase and appears to nucleate preferentially on intraferrite M_2C carbides and replace the needlelike M_2C carbides after prolonged service. This appears to be a more stable phase thermodynamically than M_2C under typical service conditions.[25]

In iron-based alloys or stainless steels, the predominant carbide is $M_{23}C_6$. In Cr-Mo-V steels, precipitation of several types of carbides such as MC, M_2C, M_7C_3, $M_{23}C_6$, and M_6C has been reported. Several researchers have observed carbide transformation types MC $\rightarrow M_2C$, $M_3C \rightarrow M_7C_3$, $M_2C \rightarrow M_6C$, and $M_{23}C_6 \rightarrow M_6C$. In low and more highly alloyed steels, transitional carbides may form during aging before the stable occurs (see Chap. 8 for more details).[24,26] Generally, the sequence of carbide precipitation in all steels during martensite aging may be written as $M_3C \rightarrow MC + M_2C + M_7C_3 \rightarrow M_{23}C_6 \rightarrow M_6C$.[27] It is further suggested that prolonged service exposure gives rise to precipitation of one or more carbides M_2C, M_7C_3, M_6C, and $M_{23}C_6$. In Si-containing steel, the sequence of formed carbides is (ε-carbide) $\rightarrow M_3C \rightarrow M_2C \rightarrow M_7C_3 \rightarrow M_{23}C_6$.[28]

$M_{23}C_6$ Carbide. $M_{23}C_6$ represents single alloy carbide as well as double and complex carbides containing iron and carbide-forming elements. Examples are

$Cr_{23}C_6$, $Mn_{23}C_6$, $(CrFe)_{23}C_6$, $(Fe_{21}Mo_2)C_6$, $(Fe_{21}W_2)C_6$, and $(FeMnVNbMoW)_{23}C_6$. $M_{23}C_6$ carbides readily form in alloys with moderate to high Cr content. Their formation, along with that of γ' (gamma prime), occurs at lower aging temperature according to Eq. (10.15).

M_6C Carbides. M_6C represents double carbides containing iron and carbide-forming elements such as Mo and W. Examples are Fe_4Mo_2C and $Fe_4(MoW)_2C$, Fe_4W_2C, and $Fe_3(MoW)_3C$. This carbide can also dissolve moderate quantities of Cr, V, and Co. They form when Mo and/or W content is >6 to 8 at %, typically in the range of 815 to 980°C (1500 to 1800°F). Thus it is the main carbide in high-speed steels and is resistant to solution during austenitizing, leaving undissolved abrasion-resistant particles which are also growth-resistant during tempering.[23] It is also observed in stainless steels containing >6% Mo or 0.8 to 2% Nb. In types 316 and 316L stainless steels, M_6C appears to form from $M_{23}C_6$ after a prolonged aging time (>1500 hr) in a limited temperature range around 650°C according to Eq. (10.16). Also, M_6C and $M_{23}C_6$ interact, forming one from the other:

$$M_6C + M' \rightarrow M_{23}C_6 + M'' \quad \text{or} \quad Mo_3(NiCo)_3C + Cr \rightleftharpoons Cr_{21}Mo_2C_6 + NiMoCo \quad (1.5)$$

Because M_6C carbides are stable at higher temperatures than are $M_{23}C_6$ carbides, M_6C is more commercially important as a grain boundary precipitate for controlling grain size during the processing of wrought alloys. This also forms in neutron-irradiated type 316 stainless steel.

M_7C_3 Carbides. Chromium-rich M_7C_3, a hexagonal structure, probably forms in Fe-Cr-C or Fe-Cr-Ni-C alloys where carbon concentrations are considerably larger than those specified for the 300 series.

M_2C, like M_6C, is W- or Mo-rich but has hexagonal crystal structure, for example, W_2C and Mo_2C. It dissolves in Cr but not Fe, and is mainly associated with secondary hardening. Tempering causes its transformation into either M_6C or $M_{23}C_6$ and is not commonly present in annealed steels.[23]

MC Carbides. MC represents iron-free VC, V_4C_3, NbC, TiC, (Ti, Nb)C, TaC, or ZrC carbide, which is coherent with the γ-iron or nickel-base matrix and has a fcc NaCl-type structure. All these elements are very strong carbide formers. Thus MC always occurs if V is present. For the formation of MC carbides, usually sufficient amounts of Nb and Ti are added to exceed the stoichiometric (i.e., atomic weight) ratios of 4:1 and 8:1, respectively.[19] The high yield strength of microalloyed steels (i.e., high-strength low-alloy steels) is attributed to the ultrafine dispersion of these carbides in α-iron solid solution. MC carbides are a major source of carbon for subsequent phase reactions during heat treatment and service. MC formation is favored more by the presence of Mo than by W, and after heat treatment undissolved MC particles are very abrasion-resistant and play a significant role in improving wear resistance.[23] In V-bearing steels, fine precipitates of MC (or VC) mostly contribute to the secondary hardening and their associated high tempering resistance. (See also Chap. 12 for more details.) The preferred order of formation (in order of decreasing stability) in superalloys for these carbides is HfC, TaC, NbC, and TiC.

1.7.5 Effect on the Eutectoid Composition and Temperature

Austenite and ferrite stabilizers widen the respective phase fields. If alloying elements are added to the iron-carbon alloy (steel), the position of A_1, A_3, and A_{cm}

boundaries as well as the eutectoid composition are changed. Classical diagrams, introduced by Bain,[5] illustrate the influence of increasing the amount of a selected number of alloying elements on eutectoid carbon content and A_1 (eutectoid temperature) (Fig. 1.12a and b). Figure 1.12a shows the influence of alloying addition on eutectoid temperature, and Fig. 1.12b shows the related influence on eutectoid carbon content.[9,20] The austenite stabilizers lower the eutectoid temperature, thereby widening the temperature range over which austenite is stable. Similarly, the ferrite formers raise the eutectoid temperature, thereby restricting the γ-phase field. The effects of alloying elements, particularly Ti and Cr, on the γ-phase field in the Fe-Ti-C and Fe-Cr-C systems, respectively, are shown in Fig. 1.13,[29] from which it is evident that just over 1% Ti is required to eliminate the γ-loop, whereas 20%

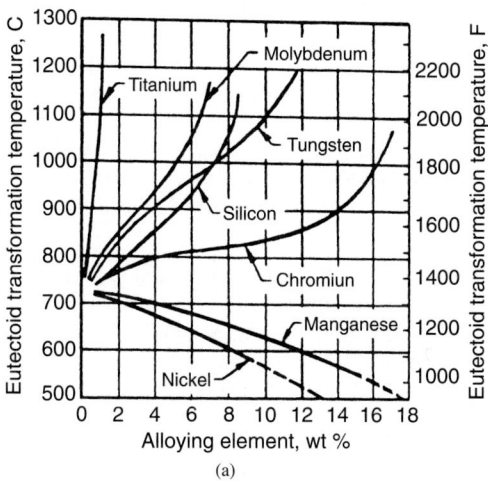

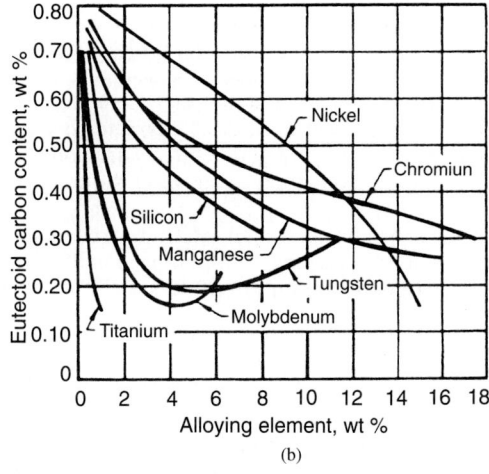

FIGURE 1.12 Effects of alloying elements on (a) the eutectoid reaction temperature and (b) the eutectoid carbon content.[9,20] (*Reprinted by permission of ASM International, Materials Park, Ohio.*)

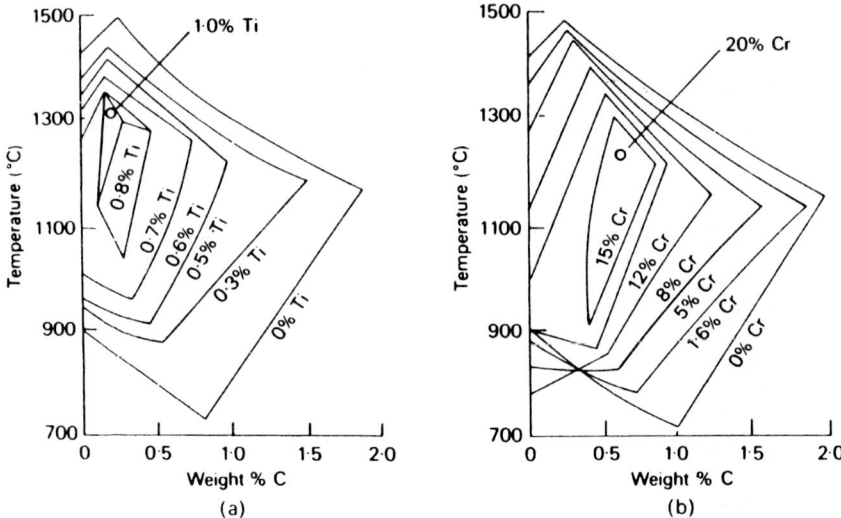

FIGURE 1.13 Effect of alloying additions on the γ-phase field: (*a*) titanium, (*b*) chromium.[20,29] (*Reprinted by permission of ASM International, Materials Park, Ohio.*)

Cr is necessary to achieve this result. Associated changes in eutectoid temperature and composition are also illustrated. The effect on the bainitic and martensitic transformation and tempering processes will be discussed in later chapters.

1.7.6 Steelmaking Practices and Characteristics

Steels contain alloying elements and impurities that must be associated with austenite, ferrite, and cementite. The combined effect of alloying elements and heat treatment produces an enormous variety of microstructures and properties.

In this section, the effects of various alloying elements, residual (or impurity) elements (such as P, S, As, Sb, Sn, N, H, and O), and deoxidants (such as Al, Si, and Ca) commonly found in steels are summarized. It should be noted that the effects of a single alloying element on either steelmaking practice or steel characteristics are modified by the presence of other elements. Such interactive effects are complex; these interreactions must be considered when evaluating a change in the composition of a steel.

Carbon. The amount of C required in the finished steel limits the type of steel that can be made. As the C content of *rimmed steels** increases, surface quality deteriorates. In contrast, *killed steels†* in the approximate range of 0.15 to 0.30% C may

---

* Rimmed steels are cast into ingots without deoxidation by Al or Si. They have less shrinkage pipes than killed steel ingots, some shrinkage porosity, and lower C, S, and P near the surface region than the average composition of the ingot.

† Killed steels are produced by adding deoxidizing elements, mostly Al and Si, to the ladle before pouring. They have a much larger shrinkage pipe in the center of the ingot, but usually have uniform chemical composition and mechanical properties throughout the ingot.

have poorer surface quality and require special processing to attain surface quality comparable to steels with higher or lower carbon contents. Carbon has a moderate tendency to segregate, and C segregation is often more significant than any other alloying elements. Carbon is the main hardening element in all steels, except the austenitic PH stainless steels and maraging steels. Tensile strength (in the as-rolled conditions), hardness, and hardenability increase as the C content increases up to about 0.85%. However, toughness, ductility, and weldability decrease with the increasing C content.[30]

Manganese. Managanese is present in virtually all steels in amounts of 0.3% or more.[31] Manganese is essentially a deoxidizer and desulfurizer.[32] It has a lesser tendency for macrosegregation than any of the common elements. Steels above 0.60% Mn cannot be readily rimmed.

Manganese is beneficial to surface quality in all carbon ranges (except extremely low-carbon, rimming steels) and is especially beneficial in resulfurized and free-cutting steels due to a reduction in the risk of *red* or *hot shortness** by forming dispersed MnS inclusions. Manganese addition contributes to the strength and hardness of steel, but to a lesser extent than carbon and, in addition, favorably affects forgeability and weldability. Manganese is a solid-solution strengthener in steel and is very effective in increasing the hardenability, but contributes to temper embrittlement (see also Chap. 14 for details). However, large quantities (>2%) result in increased tendency toward cracking and distortion during quenching.[33]

Phosphorus. Phosphorus segregates, but to a lesser extent than C and S. A small amount of P dissolves in the ferrite and slightly increases the strength and hardness of steel. A large quantity decreases the ductility and (notch) impact toughness in the as-rolled condition, imparting *cold shortness* (or *brittleness under impact*) to the steel, particularly in quenched and tempered higher-carbon steels. Higher P content is often specified in low-carbon (free-machining) steels to improve machinability. In low-alloy structural steels containing ~0.1% C, P increases strength and atmospheric corrosion resistance (rust-resistant steels). In austenitic Cr-Ni steels, P addition can cause precipitation effects and increase in yield point.[33]

Sulfur. Increased amount of S has a detrimental effect on transverse ductility, notch-impact toughness, weldability, and surface quality (particularly in low-carbon and low-manganese steels), but has only a slight effect on longitudinal tensile properties. It can cause reduction in hot working properties (i.e., increased red/hot shortness) due to the low-melting sulfide eutectics surrounding the grains in a network fashion.[33,34] Higher S grades (>0.05%) are more susceptible to quench cracking than the low-S grades. The reasons are as follows: (1) S, mainly present in the form of sulfide inclusions, has a greater segregation tendency than any other common elements. Obviously, greater frequency of sulfide inclusions can be expected in the resulfurized grades. (2) The surface of the hot-rolled high-sulfur-containing steel has a greater tendency to form *seams*, which act as stress raisers during quenching. (3) They are usually coarse-grained for better machinability, which increases brittleness and, therefore, promotes quench cracking.[35] Hence, only a low S content (<0.05%) is maintained in most carbon steels, where good weldability, fabrication properties, and minimum quench cracking tendency during hardening are desired. However, S in the range of 0.08 to 0.35% is added intentionally to steels (called *free-machining grades*) for increased machinability.[36]

---

* Any excess S present as FeS at the grain boundaries fuses at the forging temperature, thereby destroying the continuity of the grains.

Sulfur improves fatigue life of bearing steels,[37] because (i) MnS inclusions produce compressive stresses in the surrounding matrix, (ii) thermal coefficient of expansion of these inclusions is higher than that of steel, and (iii) they coat or cover other inclusions (notably silicate and alumina), thereby reducing the tensile stresses.[38,39]

Silicon. Silicon is one of the two principal deoxidizers used in steelmaking; therefore, Si content is directly related to the type of steel being produced. Killed carbon steels may contain any amount of Si up to 0.60% maximum. Semikilled steels may contain moderate amounts of Si, although there is a definite maximum amount that can be tolerated in these steels. Rimmed and capped steels contain no significant amounts of Si. For example, in rimmed steel, the Si content is generally less than 0.10%.

Silicon is somewhat less effective than Mn in increasing as-rolled strength and hardness, but generally impairs machinability and cold forming. Silicon has only slight tendency to segregate; it is usually detrimental to the surface quality in low-carbon steels, and this condition is more pronounced in low-carbon resulfurized grades.

When Si content in steel is below 0.30%, it dissolves completely in ferrite, increasing its strength without greatly decreasing ductility. Beyond 0.40%, a marked decrease in ductility is noticed in plain-carbon steels.[1] However, it increases wear resistance in Si-Mn heat-treated steels, elastic limit in spring steels, and scale resistance in air up to 260°C (or 500°F) in heat-resistant steels, but decreases the magnetic hysteresis loss. Silicon is also used as electrical-quality steel sheet.[33]

Higher Si levels in new spring steels significantly improve the *spring sag resistance.* (*Note*: Spring sag is a relaxation phenomenon which leads to a reduced load-carrying capability. It occurs from both static and dynamic loading service and is logarithmically dependent on time.)[40]

Cobalt. Cobalt inhibits grain growth at high temperatures and significantly improves the retention of temper and high-temperature strength, resulting in an increase in tool life. Cobalt is not an element commonly added to alloy steels. Cobalt does not form carbides. In low-carbon chromium steels, Co decreases hardenability. It marginally hardens ferrite and has only a small effect on the transformation temperature of iron.[41] The use of Co is generally restricted to high-speed steels, hot-forming tool steels, and creep-resistant and high-temperature materials.[31,33]

Copper. Copper addition has a moderate tendency to segregate. Copper is usually considered as a detrimental element in steel, being associated with severe hot shortness. Copper raises the ITT by about 22 K/wt %, whereas Ni lowers the ITT by an average of 26 K/wt % in structural steels. Small additions (e.g., 0.25%) of Cu are beneficial in retarding hydrogen-induced cracking in line pipe grades.[42] Copper above 0.30% can cause precipitation hardening. It increases hardenability of low-alloy steels with a potency similar to that of Ni.[42] If present in appreciable amounts, it is detrimental to hot-working operations and forge welding, but it has no adverse effect on arc or oxyacetylene welding. Copper is detrimental to surface quality and exaggerates the surface defects inherent in resulfurized steels. Copper, however, improves the atmospheric corrosion resistance of strip and structural steels (when it exists in excess of 0.20%) and the tensile properties in alloy and low-alloy steels, and reportedly helps the adhesion of paint.[30,32] In acid-resistant high-alloy steels, a Cu content above 1% results in improvements in resistance to HCl and H_2SO_4 acids.[33] Copper is a versatile alloying element in the case of austenitic stainless steels where up to 4% Cu additions are incorporated to provide significant benefit to both austenite stability and corrosion resistance in sulfuric acid.[42]

Lead. Lead is sometimes added (in the range of 0.2 to 0.5%) to carbon and alloy steels through mechanical dispersion during teeming to improve machinability. At temperatures near its melting point, it can result in liquid embrittlement.

Boron. Boron, in very small amounts (0.001 to 0.003%), has a startling effect on the hardenability of fully killed steel. Boron also improves the hardenability of other alloying elements, and in the United States at least, it is being used as a very economical substitute for some of the more expensive elements. A 0.003% soluble B producing an increase in hardenability is equivalent to about 0.5% of elements such as Mn, Cr, and Mo. However, the use of B as a hardenability agent is usually limited to steels containing less than 0.4% C, and this effect decreases significantly with increasing carbon content. The beneficial effects of B are only realized in lower- and medium-carbon steels, there being no real increase in hardenability above 0.6% C.[32] The weldability of B alloyed steels is another principal reason for their use. Boron can cause hot shortness and can reduce toughness. However, large amounts of B result in brittle, unworkable steels.

Boron additions are beneficial to both the hot workability and rupture ductility of austenitic stainless steels. Boron in excess of 0.006% produces hot-working problems due to the formation of a eutectic phase. However, B is used at about 1% to obtain special properties in austenitic steels. In the nuclear power and fuel processing industries, stainless steels containing 1% B are now used as neutron absorption materials for storage and transportation vessels.[43]

Chromium. Chromium is a strong carbide former and stabilizer when present in small amounts (~0.5%). Chromium increases hardenability, corrosion, heat, and oxidation resistance; improves high-temperature strength (in conjunction with Mo) and high-pressure hydrogenation properties; and enhances abrasion resistance in high-carbon grades. Chromium carbides are hard and wear-resistant and increase the edge-holding quality. Complex chromium-iron carbides slowly go into solution in austenite; therefore, a longer time at temperature is necessary to allow solution to take place before quenching is accomplished.[30,32] Chromium is, in fact, the most efficient of the common hardening elements and is frequently used with a toughening element such as Ni to produce superior mechanical properties.

Nickel. Nickel is a ferrite strengthener and toughener. As a result of the open austenite-phase field, Ni contents >7% produce austenitic structure to chemically resistant steels down to well below room temperature.[33] Nickel alloy steels also have superior low-temperature strength and toughness.[37] The good ductility, toughness, and flexible heat treatment of low-carbon nickel steels make them good case hardening materials.[37] In combination with Cr, Ni produces alloy steels with increased hardenability, impact strength, and fatigue resistance than are possible with carbon steels.

Molybdenum. Molybdenum is a pronounced carbide former. Molybdenum addition produces fine-grained steels, increases the hardenability, and improves the fatigue strength. It is added in constructional steels, usually in the range of 0.10 and 0.60%. Molybdenum can induce secondary hardening during the tempering of quenched steels and improves the creep strength of low-alloy steels at elevated temperatures. Alloy steels containing 0.15 to 0.30% Mo and/or V minimize the susceptibility of steel to temper embrittlement (TE) due to the slow precipitation of alloy carbides of increasing stability. It increases corrosion resistance and is thus used greatly with high-alloy Cr steels and with austenitic Cr-Ni steels. High Mo contents reduce the susceptibility to pitting.[33] It retards pearlitic transformation from γ far more than it does for bainitic transformation from γ; hence, bainite can be formed by continuous cooling of molybdenum-containing steels.

Tungsten.   Tungsten is a very important carbide former. Tungsten in steel forms very hard, abrasion-resistant carbides. It promotes hot strength and red hardness, and thus the cutting ability. It improves toughness and prevents grain growth. This combination of properties makes it very useful in high-speed cutting tools.[31] It has been suggested as a replacement for molybdenum in reduced-activation ferritic steels for nuclear applications. However, W impairs scaling resistance.

Vanadium.   Vanadium is an excellent deoxidizer, carbide former, and grain refiner, but it is very expensive and scarce.[36] It dissolves to some extent in ferrite, imparting strength and toughness. Vanadium increases the deep drawing characterics for hot-band low-carbon steel and prevents an excessive ferrite grain growth, if the coiling temperature is high.[43a] Vanadium increases the fatigue strength on one hand, but improves the notch sensitivity on the other hand; it has no appreciable effect on the corrosion resistance. Vanadium also forms nitrides and is present in most nitriding steels. Vanadium addition up to ~0.05% increases the hardenability of steel; larger additions tend to reduce the hardenability, probably due to the formation of vanadium carbides which have difficulty dissolving in austenite. However, a typical V content of 0.1% can increase the yield strength of a controlled-rolled Nb-containing microalloyed plate steel due to a combination of both grain refinement and precipitation strengthening. Vanadium additions up to 0.2% are made into microalloyed medium-carbon forging grades.

Vanadium increases abrasion wear resistance, edge-holding quality, and high-temperature strength. It is used, therefore, mainly as an additional alloying element (of 1% and above) in high-speed, hot-forming, and creep-resistant steels. Vanadium additions of up to 0.75% are incorporated into low-alloy Cr-Mo-V steels to improve secondary hardening and good creep strength up to 565°C temperature.[44] It promotes the weldability of heat-treatable steels.

Vanadium steels exhibit a much finer structure than steels of a similar composition without V. It provides other important alloying effects such as increased hardenability, secondary hardening during tempering (through precipitation hardening), and increased elevated-temperature hardness. The presence of V retards the rate of temper embrittlement in molybdenum bearing steels by a factor of 10; the mechanism has not yet been established.

Niobium and Tantalum.   These are ferrite formers and, therefore, reduce the austenite phase. Small additions of Nb increase the yield strength and, to a lesser extent, the tensile strength of carbon steel. A 0.02% Nb addition can increase the yield strength of medium-carbon steel by 70 to 100 MPa (10 to 15 ksi). This increased strength may be accompanied by considerably reduced notch toughness unless special measures are employed to refine grain size during hot rolling. Grain refinement during hot rolling involves special thermomechanical treatment techniques such as controlled-rolling practices, low finishing temperature for final reduction passes, and accelerated cooling after the completion of rolling (see also Chap. 15).

Aluminum.   Aluminum, in small amounts (0.015 to 0.060 wt %), is mostly used as the other principal deoxidizer in steelmaking; however, it also performs as a grain refiner.[10] It has the drawback of a tendency to promote graphitization, if present, in excess of 0.06% (e.g., ~0.35%), and dramatically reduces the creep strength and is therefore undesirable in steels to be used for high-temperature applications. As Al forms very hard nitrides with nitrogen, it is usually an alloying element in nitriding steels. It increases scaling resistance and is therefore often added to alloy ferritic heat-resistant steels. Aluminum combines with N in the solid state to minimize austenite and to minimize the effects of strain aging. This combination also helps control the plastic strain ratio of sheet products.

Calcium. Calcium is sometimes used to deoxidize steels to improve machinability and control shape and distribution of nonmetallic (sulfide) inclusions, thereby improving the toughness.[45]

Titanium. The effects of Ti are similar to those of V and Nb, but Ti is only beneficial in fully killed (aluminum-deoxidized) steels due to its strong deoxidizing effects. Titanium also lowers the soluble carbon and nitrogen to very low levels and is used to produce the interstitial-free (IF) steels which have improved ductility and extremely high cold formability.[46] Microalloying with Ti improves drawability in low-carbon wire rod steels.[46a] Titanium is used widely in stainless steels as a carbide former for stabilization against intergranular corrosion. Titanium increases creep rupture strength through the formation of special nitrides and tends significantly to segregation and banding.[33] Titanium addition is made in boron-treated steels because it combines instantly with any oxygen and nitrogen in steel, thereby increasing the effectiveness of B in improving the steel hardenability.[45]

Titanium, Zr, and V are effective grain growth inhibitors; however, for structural steels that require heat treatment (quenching and tempering), these three elements may have adverse effects on hardenability, because their carbides are very stable and difficult to dissolve in austenite prior to quenching.

Zirconium. Zirconium addition to killed high-strength low-alloy steels is made to obtain improvements in inclusion characteristics, particularly sulfide inclusions where modifications in inclusion shape improve ductility in transverse bending. Zirconium increases the life of heating conductor materials and produces contracted gamma-phase field.[33] Its main use is to improve hot-rolled properties in HSLA steels. Zirconium in solution improves slightly the hardenability.[47]

Arsenic and Antimony. They are ferrite stabilizers. They can render steel susceptible to temper embrittlement.

Tin. Tin in relatively small amounts is harmful to steels for deep drawing, but for most uses, the effects of tin in the amounts usually present are negligible.[48] It tends toward increased segregation, is a ferrite stabilizer, and limits the gamma-phase field.[33] It can increase susceptibility of a steel to temper embrittlement and hot shortness.[45]

Hydrogen. Hydrogen dissolved in steel during manufacturing has an embrittling effect which can result in flaking during cooling from hot-rolling temperatures. However, dissolved hydrogen rarely affects the finished mill products because reheating of the steel prior to hot forming bakes out nearly the entire hydrogen.

Nitrogen. Nitrogen increases the strength, hardness, and machinability of steel, but it lowers the ductility and toughness (i.e., raises the ITT) of ferrite-pearlite steels and can give rise to strain aging. In Al-killed steels, N forms AlN particles that control the grain size of the steel, thereby improving both strength and toughness. It can decrease the effect of B on the hardenability of steels.

Nitrogen in solid solution is deleterious to the cold formability of low-carbon strip steel, causing low $\bar{r}$ values, the mean plastic strain ratio (see also Chap. 4). Nitrogen produces a considerable solid-solution hardening and precipitation strengthening reactions which form the foundation of many high-strength steels. Nitrogen additions are also advantageous to the constitution and pitting resistance of austenitic stainless steels.[46]

Oxygen. Oxygen, which is widely observed in rimmed steels, can slightly increase the strength of steel, but adversely affects the toughness.

1.8   STEEL CLASSIFICATIONS

Steels can be classified by several different systems depending on (1) the compositions, such as carbon, low-alloy, alloy, or stainless steels; (2) the manufacturing
methods, such as basic and acid open hearth, or electric furnace methods; (3) the
finishing methods, such as hot rolling or cold rolling; (4) the product shape, such as
bar, plate, strip, tubing, or structural shape; (5) the application, such as structural,
spring, and high tensile steels; (6) the deoxidation practice, such as killed, semikilled,
capped, and rimmed steels; (7) the microstructure, such as ferritic, pearlitic, and
martensitic; (8) the required strength level, as specified in ASTM Standards; (9) heat
treatment, such as annealing, quenching and tempering, and thermomechanical processing; and (10) quality descriptors/classifications, such as forging quality and commercial quality.[30,45,49]

Among the above classification systems, quality descriptors and chemical compositions are the widely used basis for designation and will be described in this
chapter. Classification systems based on deoxidation practice will be discussed in
Chap. 3.

1.8.1   Quality Descriptors/Classifications

Quality descriptors are names applied to various steel products to indicate that the
particular products possess certain characteristics that make them especially well
suited for specific applications or fabrication processes. The quality designations/
descriptors for various carbon steel products and alloy steel plates are listed in
Table 1.2. Forging quality and cold-extrusion quality descriptors for carbon steels
are self-explanatory. However, others are not explicit; for example, merchant-quality
hot-rolled carbon steel bars are made for noncritical applications requiring modest
strength and mild bending or forming, but not requiring forging or heat treating
operations. The quality classification for one particular steel commodity is not necessarily extended to subsequent products made from the same commodity—for
example, standard-quality cold-finished bars are produced from special-quality hot-
rolled carbon steel bars. Alloy steel plate qualities are described by structural,
drawing, cold working, pressure vessel, and aircraft qualities.[49]

The various physical and mechanical characteristics indicated by a quality
descriptor result from the combined effects of several factors, such as (1) internal
soundness; (2) uniformity of chemical composition; (3) number, size, and distribution of nonmetallic inclusions; (4) relative freedom from harmful surface imperfections; (5) extensive testing during manufacture; (6) size of the discard cropped from
the ingot; and (7) hardenability requirements.

Control of these factors during manufacture is essential to achieve mill products
with the desired characteristics. The degree of control over these and other related
factors is another segment of information conveyed by the quality descriptor.

Some, but not all, of the basic quality descriptors may be modified by one or
more additional requirements as may be appropriate, namely, macroetch test,
special discard, restricted chemical composition, maximum incidental (residual)
alloying elements, austenitic grain size, and special hardenability. These limitations
could be applied to forging-quality alloy steel bars, but not to merchant-quality bars.

Understanding the various quality descriptors is difficult because most of the
prerequisites for qualifying a steel for a specific descriptor are subjective. Only

TABLE 1.2 Quality Descriptions* of Carbon and Alloy Steels[51]

Carbon steels		Alloy steels
Semifinished for forging	Tin mill products	Alloy steel plates
Forging quality	Specific quality descriptions are not	Drawing quality
Special hardenability	applicable to tin mill products	Pressure vessel quality
Special internal soundness		Structural quality
Nonmetallic inclusion requirement	Carbon steel wire	Aircraft physical quality
Special surface	Industrial quality wire	Hot-rolled alloy steel bars
Carbon steel structural sections	Cold extrusion wires	Regular quality
Structural quality	Heading, forging, and roll-threading	Aircraft quality or steel subject to
Carbon steel plates	wires	magnetic particle inspection
Regular quality	Mechanical spring wires	Axle shaft quality
Structural quality	Upholstery spring construction	Bearing quality
Cold-drawing quality	wires	Cold-heading quality
Cold-pressing quality	Welding wire	Special cold-heading quality
Cold-flanging quality	Carbon steel flat wire	Rifle barrel quality, gun quality,
Forging quality	Stitching wire	shell or A.P. shot quality
Pressure vessel quality	Stapling wire	Alloy steel wire
Hot-rolled carbon steel bars	Carbon steel pipe	Aircraft quality
Merchant quality	Structural tubing	Bearing quality
Special quality	Line pipe	Special surface quality
Special hardenability	Oil country tubular goods	Cold-finished alloy steel bars
Special internal soundness	Steel specialty tubular products	Regular quality
Nonmetallic inclusion requirement	Pressure tubing	Aircraft quality or steel subject to
Special surface	Mechanical tubing	magnetic particle inspection
Scrapless nut quality	Aircraft tubing	Axle shaft quality
Axle shaft quality	Hot-rolled carbon steel wire rods	Bearing shaft quality
Cold extrusion quality	Industrial quality	Cold-heading quality
Cold-heading and cold-forging quality	Rods for manufacture of wire	Special cold-heading quality
Cold-finished carbon steel bars	intended for electric welded	Rifle barrel quality, gun quality,
Standard quality	chain	shell or A.P. shot quality
Special hardenability	Rods for heading, forging, and roll-	Line pipe
Special internal soundness	threading wire	Oil country tubular goods
Nonmetallic inclusion requirement	Rods for lock washer wire	Steel specialty tubular goods
Special surface	Rods for scrapless nut wire	Pressure tubing
Cold-heading and cold-forging quality	Rods for upholstery spring wire	Mechanical tubing
Cold extrusion quality	Rods for welding wire	Stainless and head-resisting pipe,
Hot-rolled sheets		pressure tubing, and
Commercial quality		mechanical tubing
Drawing quality		Aircraft tubing
Drawing quality special killed		Pipe
Structural quality		
Cold-rolled sheets		
Commercial quality		
Drawing quality		
Drawing quality special killed		
Structural quality		
Porcelain enameling sheets		
Commercial quality		
Drawing quality		
Drawing quality special killed		
Long terne sheets		
Commercial quality		
Drawing quality		
Drawing quality special killed		
Structural quality		
Galvanized sheets		
Commercial quality		
Drawing quality		
Drawing quality special killed		
Lock-forming quality		
Electrolytic zinc-coated sheets		
Commercial quality		
Drawing quality		
Drawing quality special killed		
Structural quality		
Hot-rolled strip		
Commercial quality		
Drawing quality		
Drawing quality special killed		
Structural quality		
Cold-rolled strip		
Specific quality descriptions are		
not provided in cold-rolled		
strip because this product is		
largely produced for specific		
end use		

* In the case of certain qualities, P and S are usually finished to lower limits than the specified maximum.
Reprinted by permission of Society of Automotive Engineers, Warrendale, Pa.

1.27

limitations on chemical composition ranges, residual alloying elements, nonmetallic inclusion count, austenitic grain size, and special hardenability are quantifiable. The subjective evaluation of the other attributes depends on the experience and skill of the individuals who make the evaluation. Although the use of these subjective quality descriptors might appear impractical and imprecise, steel products made to meet the requirements of a specific quality descriptor can be relied upon to have those characteristics necessary for that product to be used in the suggested application or fabrication operation.[30]

1.8.2 Steel Classification Based on Chemical Composition

1.8.2.1 Carbon and Carbon-Manganese Steels. In addition to carbon, plain-carbon steels contain the following elements: Mn up to 1.65%, S up to 0.05%, P up to 0.04%, Si up to 0.60%, and Cu up to 0.60%. The effects of each of these elements in plain-carbon steels have been summarized in Sec. 1.7.5.

Carbon steels can be classified according to various deoxidation practices (see Sec. 3.10.1.1). Deoxidation practice and steelmaking process have effects on the characteristics and properties of the steel. However, variations in C have the greatest effect on mechanical properties; increased C addition leads to increased hardness and strength. As such, carbon steels are generally grouped according to their C content. In general, carbon steels contain up to 2% total alloying elements and can be subdivided into low-carbon steels, medium-carbon steels, high-carbon steels, ultrahigh-carbon steels, and boron-treated steels; each of these designations is discussed below.

As a group, carbon steels are most widely used. Table 1.3 lists various grades of standard carbon and low-alloy steels with SAE-AISI designations. Tables 1.4 through 1.7 show some representative standard plain (nonresulfurized, 1% Mn maximum) carbon steel, free-cutting (resulfurized) carbon steel, free-cutting (resulfurized and rephosphorized) carbon steel, and high-manganese (nonresulfurized) carbon steel compositions, respectively, with SAE-AISI and corresponding Unified Numbering System (UNS) designations.[30,50,51]

Low-Carbon Steels. They contain up to 0.25% C. The largest category of this class is flat-rolled products (sheet or strip), bar, rod, wire, nut, bolt, tube, and numerous machined parts that are subjected to low stresses. The carbon content for high-formability and high-drawability steels is very low (<0.10% C) with up to 0.40% Mn. These lower-carbon steels are used in automobile body panels, tin plates, appliances, and wire products.

The low-carbon steels (0.10 to 0.25% C) in this group have increased strength and hardness and reduced cold formability compared to the lowest-carbon group. They are commonly called carburizing or case hardening steel grades. Selection of these grades for carburizing applications depends on the nature of the part, the properties required, and the processing practices preferred. An increase in carbon content of the base steel results in greater core hardness for a given quench. However, an increase in Mn increases the hardenability of both the core and the case.

A typical application for carburized plain-carbon steel is for parts with hard wear-resistant surface, but without any need for increased mechanical properties in the core, e.g., small shafts, plungers, lightly loaded gears.[51] Rolled structural steels in the form of plates and sections contain ~0.25% C, 1.5% Mn maximum, and Al, if improved toughness is required. When used for stampings, forgings, seamless tubes, and boiler plate, Al addition should be avoided. An important type of this category

TABLE 1.3 SAE-AISI System for Carbon and Low-Alloy Steels

Numerals and digits*	Type of steel and nominal alloy content, %
Carbon steels	
10xx	Plain carbon (Mn 1.00 max)
11xx	Resulfurized
12xx	Resulfurized and rephosphorized
15xx	Plain carbon (max Mn range: 1.00–1.65
Manganese steels	
13xx	Mn 1.75
Nickel steels	
23xx	Ni 3.50
25xx	Ni 5.00
Nickel-chromium steels	
31xx	Ni 1.25; Cr 0.65 and 0.80
32xx	Ni 1.75; Cr 1.07
33xx	Ni 3.50; Cr 1.50 and 1.57
34xx	Ni 3.00; Cr 0.77
Molybdenum steels	
40xx	Mo 0.20 and 0.25
44xx	Mo 0.40 and 0.52
Chromium-molybdenum steels	
41xx	Cr 0.50, 0.80, and 0.95; Mo 0.12, 0.20, 0.25, and 0.30
Nickel-chromium-molybdenum steels	
43xx	Ni 1.82; Cr 0.50 and 0.80; Mo 0.25
43BVxx	Ni 1.82; Cr 0.50; Mo 0.12 and 0.25; V 0.03 min
47xx	Ni 1.05; Cr 0.45; Mo 0.20 and 0.35
81xx	Ni 0.30; Cr 0.40; Mo 0.12
86xx	Ni 0.55; Cr 0.50; Mo 0.20
87xx	Ni 0.55; Cr 0.50; Mo 0.25
88xx	Ni 0.55; Cr 0.50; Mo 0.35
93xx	Ni 3.25; Cr 1.20; Mo 0.12
94xx	Ni 0.45; Cr 0.40; Mo 0.12
97xx	Ni 0.55; Cr 0.20; Mo 0.20
98xx	Ni 1.00; Cr 0.80; Mo 0.25
Nickel-molybdenum steels	
46xx	Ni 0.85 and 1.82; Mo 0.20 and 0.25
48xx	Ni 3.50; Mo 0.25
Chromium steels	
50xx	Cr 0.27, 0.40, 0.50, and 0.65
51xx	Cr 0.80, 0.87, 0.92, 0.95, 1.00, and 1.05
Chromium (bearing) steels	
50xxx	Cr 0.50 ⎫
51xxx	Cr 1.02 ⎬ C 1.00 min
52xxx	Cr 1.45 ⎭
Chromium-vanadium steels	
61xx	Cr 0.60, 0.80, and 0.95; V 0.10 and 0.15 min
Tungsten-chromium steel	
72xx	W 1.75; Cr 0.75

TABLE 1.3 SAE-AISI System for Carbon and Low-Alloy Steels (*Continued*)

Numerals and digits*	Type of steel and nominal alloy content, %
Silicon-manganese steels	
92*xx*	Si 1.40 and 2.00; Mn 0.65, 0.82, and 0.85; Cr 0 and 0.65
High-strength low-alloy steels	
9*xx*	Various SAE grades
Boron steels	
*xx*B*xx*	B denotes boron steel
Leaded steels	
*xx*L*xx*	L denotes leaded steel
Vanadium steels	
xxVxx	V denotes vanadium steel

* The *xx* in the last two digits of these designations represents the insertion of carbon content (in hundredths of a percent).
Reprinted by permission of ASM International, Materials Park, Ohio.

TABLE 1.4 Standard Carbon Steel Compositions with SAE-AISI and Corresponding UNS Designations[45]

Applicable to semifinished products for forging, hot-rolled and cold-finished bars, wire rods, and seamless tubing

Designation UNS number	SAE-AISI number	Cast or heat chemical ranges and limits, %* C	Mn	P max	S max
G10050	1005	0.06 max	0.35 max	0.040	0.050
G10060	1006	0.08 max	0.25–0.40	0.040	0.050
G10080	1008	0.10 max	0.30–0.50	0.040	0.050
G10100	1010	0.08–0.13	0.30–0.60	0.040	0.050
G10120	1012	0.10–0.15	0.30–0.60	0.040	0.050
G10130	1013	0.11–0.16	0.50–0.80	0.040	0.050
G10150	1015	0.13–0.18	0.30–0.60	0.040	0.050
G10160	1016	0.13–0.18	0.60–0.90	0.040	0.050
G10170	1017	0.15–0.20	0.30–0.60	0.040	0.050
G10180	1018	0.15–0.20	0.60–0.90	0.040	0.050
G10190	1019	0.15–0.20	0.70–1.00	0.040	0.050
G10200	1020	0.18–0.23	0.30–0.60	0.040	0.050
G10210	1021	0.18–0.23	0.60–0.90	0.040	0.050
G10220	1022	0.18–0.23	0.70–1.00	0.040	0.050
G10230	1023	0.20–0.25	0.30–0.60	0.040	0.050
G10250	1025	0.22–0.28	0.30–0.60	0.040	0.050
G10260	1026	0.22–0.28	0.60–0.90	0.040	0.050
G10290	1029	0.25–0.31	0.60–0.90	0.040	0.050
G10300	1030	0.28–0.34	0.60–0.90	0.040	0.050
G10350	1035	0.32–0.38	0.60–0.90	0.040	0.050
G10370	1037	0.32–0.38	0.70–1.00	0.040	0.050
G10380	1038	0.35–0.42	0.60–0.90	0.040	0.050
G10390	1039	0.37–0.44	0.70–1.00	0.040	0.050
G10400	1040	0.37–0.44	0.60–0.90	0.040	0.050
G10420	1042	0.40–0.47	0.60–0.90	0.040	0.050
G10430	1043	0.40–0.47	0.70–1.00	0.040	0.050
G10440	1044	0.43–0.50	0.30–0.60	0.040	0.050
G10450	1045	0.43–0.50	0.60–0.90	0.040	0.050

TABLE 1.4 Standard Carbon Steel Compositions with SAE-AISI and Corresponding UNS Designations[45] (*Continued*)

UNS number	SAE-AISI number	C	Mn	P max	S max
		Cast or heat chemical ranges and limits, %*			
G10460	1046	0.43–0.50	0.70–1.00	0.040	0.050
G10490	1049	0.46–0.53	0.60–0.90	0.040	0.050
G10500	1050	0.48–0.55	0.60–0.90	0.040	0.050
G10530	1053	0.48–0.55	0.70–1.00	0.040	0.050
G10550	1055	0.50–0.60	0.60–0.90	0.040	0.050
G10590	1059	0.55–0.65	0.50–0.80	0.040	0.050
G10600	1060	0.55–0.65	0.60–0.90	0.040	0.050
G10640	1064	0.60–0.70	0.50–0.80	0.040	0.050
G10650	1065	0.60–0.70	0.60–0.90	0.040	0.050
G10690	1069	0.65–0.75	0.40–0.70	0.040	0.050
G10700	1070	0.65–0.75	0.60–0.90	0.040	0.050
G10740	1074	0.70–0.80	0.50–0.80	0.040	0.050
G10750	1075	0.70–0.80	0.40–0.70	0.040	0.050
G10780	1078	0.72–0.85	0.30–0.60	0.040	0.050
G10800	1080	0.75–0.88	0.60–0.90	0.040	0.050
G10840	1084	0.80–0.93	0.60–0.90	0.040	0.050
G10850	1085	0.80–0.93	0.70–1.00	0.040	0.050
G10860	1086	0.80–0.93	0.30–0.50	0.040	0.050
G10900	1090	0.85–0.98	0.60–0.90	0.040	0.050
G10950	1095	0.90–1.03	0.30–0.50	0.040	0.050

* When Si ranges or limits are required for bars and semifinished products, the following ranges are commonly used: 0.10% max; 0.10 to 0.20%, 0.15 to 0.35%; 0.20 to 0.40%; or 0.30 to 0.60%. For rods, the following ranges are commonly used: 0.10 max; 0.07 to 0.15% 0.10 to 0.20%; 0.15 to 0.35%; 0.20 to 0.40%; and 0.30 to 0.60%. Steels listed in this table can be produced with the addition of lead or boron. Leaded steels typically contain 0.15 to 0.35% Pb and are recognized by inserting the letter L in the designation (10L40); boron steels usually contain 0.005 to 0.003% and are identified by inserting the letter B in the designation (10B20).
Reprinted by permission of ASM International, Materials Park, Ohio.

TABLE 1.5 Free-Cutting (Resulfurized) Carbon Steel Compositions[45]

Applicable to semifinished products for forging, hot-rolled and cold-finished bars, wire rods, and seamless tubing

UNS number	SAE-AISI number	C	Mn	P max	S
		Cast or heat chemical ranges and limits, %*			
G11080	1108	0.08–0.13	0.50–0.80	0.040	0.08–0.13
G11100	1110	0.08–0.13	0.30–0.60	0.040	0.08–0.13
G11170	1117	0.14–0.20	1.00–1.30	0.040	0.08–0.13
G11180	1118	0.14–0.20	1.30–1.60	0.040	0.08–0.13
G11370	1137	0.32–0.39	1.35–1.65	0.040	0.08–0.13
G11390	1139	0.35–0.43	1.35–1.65	0.040	0.13–0.20
G11400	1140	0.37–0.44	0.70–1.00	0.040	0.08–0.13
G11410	1141	0.37–0.45	1.35–1.65	0.040	0.08–0.13
G11440	1144	0.40–0.48	1.35–1.65	0.040	0.24–0.33
G11460	1146	0.42–0.49	0.70–1.00	0.040	0.08–0.13
G11510	1151	0.48–0.55	0.70–1.00	0.040	0.08–0.13

* When Pb ranges or limits are required, or when Si ranges or limits are required for bars or semifinished products, the values in Table 1.4 apply. For rods, the following ranges and limits for Si are commonly used up to SAE 1110 inclusive, 0.10% max; SAE 1117 and over, 0.10% max, 0.10 to 0.20%, or 0.15 to 0.35%.
Reprinted by permission of ASM International, Materials Park, Ohio.

TABLE 1.6 Free-Cutting (Resulfurized and Rephosphorized) Carbon Steel Compositions[45]

Applicable to semifinished products for forging, hot-rolled and cold-finished bars, wire rods, and seamless tubing

| Designation | | Cast or heat chemical ranges and limits, %* | | | | |
UNS number	SAE-AISI number	C max	Mn	P	S	Pb
G12110	1211	0.13	0.60–0.90	0.07–0.12	0.10–0.15	...
G12120	1212	0.13	0.70–1.00	0.07–0.12	0.16–0.23	...
G12130	1213	0.13	0.70–1.00	0.07–0.12	0.24–0.33	...
G12150	1215	0.09	0.75–1.05	0.04–0.09	0.26–0.35	...
G12144	12L14	0.15	0.85–1.15	0.04–0.09	0.26–0.35	0.15–0.35

* When Pb ranges or limits are required, the values in Table 1.4 apply. It is the usual practice to produce 12xx series of steels to specified limits for Si because of its adverse effect on machinability. For rods, the following ranges and limits for Si are commonly used up to SAE 1110 inclusive, 0.10% max; SAE 1117 and over, 0.10% max, 0.10 to 0.20%, or 0.15 to 0.35%.
Reprinted by permission of ASM International, Materials Park, Ohio.

TABLE 1.7 High-Manganese (Nonresulfurized) Carbon Steel Compositions[45]

Applicable to semifinished products for forging, hot-rolled and cold-finished bars, wire rods, and seamless tubing

| Designation | | Cast or heat chemical ranges and limits, %* | | | |
UNS number	SAE-AISI number	C	Mn	P max	S max
G15130	1513	0.10–0.16	1.10–1.40	0.040	0.050
G15220	1522	0.18–0.24	1.10–1.40	0.040	0.050
G15240	1524	0.19–0.25	1.35–1.65	0.040	0.050
G15260	1526	0.22–0.29	1.10–1.40	0.040	0.050
G15270	1527	0.22–0.29	1.20–1.50	0.040	0.050
G15360	1536	0.30–0.37	1.20–1.50	0.040	0.050
G15410	1541	0.36–0.44	1.35–1.65	0.040	0.050
G15480	1548	0.44–0.52	1.10–1.40	0.040	0.050
G15510	1551	0.45–0.56	0.85–1.15	0.040	0.050
G15520	1552	0.47–0.55	1.20–1.50	0.040	0.050
G15610	1561	0.55–0.65	0.75–1.05	0.040	0.050
G15660	1566	0.60–0.71	0.85–1.15	0.040	0.050

* When Si, Pb, and B ranges or limits are required, the values in Table 1.4 apply.
Reprinted by permission of ASM International, Materials Park, Ohio.

is the low-carbon free-cutting steels containing 0.15% C maximum, 1.2% Mn maximum, a minimum of Si, and 0.35% S maximum with or without 0.30% Pb. These steels are suited to automotive mass manufacturing methods.[1]

Medium-Carbon Steels. Medium-carbon steels containing 0.30 to 0.55% C and 0.60 to 1.65% Mn are used where higher mechanical properties are desired. They are usually hardened and strengthened by quenching and tempering or by cold work. Lower carbon and manganese contents in this group of steels find wide applications for certain types of cold-formed parts which need annealing, normalizing, or quenching and tempering treatment prior to their use. The higher-carbon grades are often cold drawn to specified mechanical properties for use without heat treatment for some applications.

All these steels can be used for forgings, the selection being dependent on the section size and the mechanical properties needed after heat treatment.[51] These grades, generally produced as killed steels, are used for a wide range of applications that include automobile parts for body, engines, suspensions, and steering, engine torque converter, and transmission. Some Pb and/or S additions make them free-cutting grades, whereas Al addition produces grain refinement and improved toughness. In general, steels containing 0.4 to 0.60% C are used as rails, railway wheels, tires, and axles; parts for lathes and presses; machined parts requiring moderate to high strength; heavy stamped or pressed products; crankshafts, connecting rods, gears, and many other automotive parts. In addition, it is used in agricultural equipment and petroleum industries.

High-Carbon Steels. High-carbon steels containing 0.55 to 1.00% C and 0.30 to 0.90% Mn have more restricted applications than the medium-carbon steels because of higher production cost and poor formability (or ductility) and weldability. High-carbon steels find applications in the spring industry (as light and thicker flat springs, laminated springs, and heavier coiled springs), farm implement industry (as plow beams, plowshares, scraper blades, disks, mower knives, and harrow teeth), and stamping industry. They are also used for high-strength cable, rope, music wire, and cutting tools such as drills, reamers, taps and dies, and cutlery. They are usually purchased in the annealed condition; the manufactured parts are then heat-treated to achieve the desired properties.

Ultrahigh-Carbon Steels. Ultrahigh-carbon (UHC) steels are experimental plain-carbon steels with 1.0 to 2.1% C (15 to 32 vol % cementite).[52–54] Optimum superplastic elongation has been found at about 1.6% C content. These steels have the capability of emerging as important technological materials, because they exhibit superplasticity. The superplastic behavior of these materials is attributed to the structure consisting of uniform distribution of very fine, spherical, discontinuous particles (0.1 to 1.5 μm in diameter) in a very fine-grained ferrite matrix (0.5 to 2.0 μm in diameter) that can be readily achieved by any of four thermomechanical treatment routes described in Chap. 15.

Boron Steels. Boron is a strong carbide- and nitride-forming element and increases strength in quenched and tempered low-carbon steels through the formation of martensite and the precipitation strengthening of ferrite. Boron-containing killed carbon steels are available as low-cost replacements for the high-carbon and low-alloy steels used for sheet and strip. The low-carbon boron steels have better cold-forming characteristics and can be heat treated to equivalent hardness and greater toughness for a wide variety of applications, such as tools, machine components, and fasteners. (See Chaps. 9 and 13 for further details).

1.8.2.2 Ultralow- or Extralow-Carbon (ULC or ELC) Steels.
Ultralow-carbon (ULC) steels or interstitial-free (IF) steels (containing C ≈ 0.003 wt %) can be regarded as dilute microalloyed (HSLA) steels (C ≈ 0.04 to 0.1 wt %).[55]

Bake-Hardening (BH) Steels. Dent-resistant bake-hardening steel parts can be produced by batch annealing, continuous annealing processing including continuous annealing on galvanizing or aluminizing lines. BH steels are Al-killed, containing dissolved C and N levels of about 0.001 wt %. They are characterized by their ability to exhibit accelerated (carbon and nitrogen) strain aging, i.e., an increase in yield strength (up to 50 MPa or 4 to 8 ksi) during paint baking (or curing) operation of press-formed (or stamped) auto body parts at 150 to 250°C for about 15 to 20 min. The extent of bake-hardening effect increases with the decrease of the ferrite grain size.[56] In batch annealing process, P and Si usually increase the bake-hardening tendency while Mn has a negative influence. Good drawability is controlled by appropriate Al and N content, low coiling temperature, batch annealing heating rate, soak temperature, and time. Reduced carbon and increased annealing temperature also favor bake-hardening tendency. A batch-annealed, bake-hardening steel with a yield strength below 150 N/mm^2 can be achieved in stabilized interstitial-free (IF) steel, where remaining excess C provides bake-hardening effect.[57] In the continuous annealing method, bake hardenability is achieved due to the abundance of C remaining during the rapid cooling from the annealing temperature; higher coiling temperature prior to cold rolling, slower cooling prior to overaging, and presence of P and Mn content favor the strain aging, thereby bake-hardening behavior.[58,59]

Interstitial-Free (IF) Steels. IF steels are low-carbon steels from which the remaining C and N in solution are scavenged as precipitates by the addition of Ti and/or Nb for achieving high drawability and non-aging characteristics for steel sheets.[60] Available in cold-rolled and coated grades, IF steel is produced by P and Mn additions, vacuum degassing, and agitation of the molten steel to reduce carbon, nitrogen, and oxygen. There are five types of IF steels: (1) "ordinary" Ti-stabilized IF steel (typically ~0.006 to 0.010% C maximum, ~0.07 to 0.12% Ti, i.e., highly alloyed; (2) "ordinary" Nb-stabilized grade (typically ~0.006 to 0.010% C maximum, ~0.08 to 0.12% Nb, i.e., highly alloyed; (3) Ti-stabilized IF steel, ELC type (typically ~0.003% C maximum, ~0.002% N, ~0.05 to 0.07% Ti, i.e., lightly alloyed; (4) Nb-stabilized IF steel, ELC type (~0.03% Nb), i.e., lightly stabilized; and (5) Nb + Ti-stabilized IF steel, ELC type (typically ~0.003% C maximum, ~0.002% N, ~0.010% Nb, ~0.03% Ti, i.e., lightly alloyed.[61] In these grades, adequate stabilizing elements such as 0.06 to 0.08% Ti or Nb tie up residual C and N from the steel melting process and develop non-strain aging and non-quench aging properties. IF steels exhibit a tensile strength of about 450 MPa, low yield strength, excellent press formability, high dent resistance and elongation, good deep drawability, and improved normal and planar anisotropies (i.e., ~2.00 r_m value and reduced earing).[45] Advantage of stabilized IF steels includes continuous annealing that permits high annealing temperature without undesirable coil sticking.[62] (See also Chap. 13.)

1.8.2.3 Low-Alloy Steels.

Alloy steels may be defined as those steels which owe their improved properties to the presence of one or more special elements or to the presence of larger proportions of elements such as Mn and Si than are ordinarily present in carbon steels.[48] Alloy steels contain Mn, Si, or Cu in quantities greater than the maximum limits (e.g., 1.65% Mn, 0.60% Si, and 0.60% Cu) of carbon steels or they contain specified ranges or minimums for one or more other alloying elements. The alloying elements improve the mechanical and fabrication properties as well as hardenability. Broadly, alloy steels can be divided into: (a) low-alloy steels containing 2 to 8 wt % total non-carbon alloy addition and (b) alloy steels with more than 8 wt % total non-carbon alloy addition. In some cases, an overlap exists between the low-alloy and alloy steels. Table 1.8 lists the low-alloy steel compositions with SAE-AISI and corresponding UNS designations.

Low-alloy steels constitute a group of steels that exhibit superior mechanical properties compared to plain carbon steels as the result of addition of such alloying elements as Ni, Cr, and Mo. For many low-alloy steels, the main function of the alloying elements is to increase hardenability in order to optimize mechanical properties and toughness after heat treatment. In some instances, however, alloying elements are used to reduce environmental degradation under certain specified conditions.

Low-alloy steels can be classified according to: (i) chemical composition such as nickel steels, nickel-chromium steels, molybdenum steels, chromium-molybdenum steels and so forth, based on the principal alloying element(s) present and as described in Table 1.2; (ii) heat treatment such as quenched and tempered, normalized and tempered, annealed, and so on, and (iii) weldability.

Because of the large variety of chemical compositions possible and the fact that some steels are employed in more than one heat-treated condition, some overlap exists among the low-alloy steel classifications. However, these grades can be addressed into four major groups such as (1) low-carbon quenched and tempered (QT) steels (Table 1.9), (2) medium-carbon ultrahigh-strength steels (Table 1.10), (3) bearing steels (Table 1.11), and (4) heat-resistant Cr-Mo steels (Table 1.12).

Low-carbon quenched and tempered steels are characterized by high yield strength (345 to 895 MPa or 50 to 125 ksi), high tensile strength (485 to 1035 MPa, or 70 to 150 ksi), good notch toughness, ductility, atmospheric corrosion resistance, and weldability. They typically contain <0.25% C and <5% alloy content. The various steels have different combinations of these characteristics depending on their intended applications. The chemical compositions of typical QT low-carbon steels are listed in Table 1.9. The majority of these steels are Cr-Mo and Cr-Mo-V steels and are covered by ASTM designations, but some steels such as HY-80, HY-100, and HY-130 are Ni-Cr-Mo steels and are covered by Military specifications. These steels are primarily available in the form of plate, sheet, bar, structural shape, forgings, and castings. They are widely used in such applications as pressure vessels, earth-moving and mining equipment, and as major members of large steel structures. They also find applications for cold-headed and cold-forged parts such as fasteners or pins and are heat treated to the desired properties.[48]

Ultra high-strength steels. Structural steels with very high strength ($\geq$1380 MPa or 200 ksi) and plain-strain fracture toughness are often referred to as ultrahigh-strength steels. Four types of ultrahigh-strength steels are: medium-carbon low-alloy steels, medium-alloy air-hardening steels, high-fracture toghness steels (see Table 1.10), and maraging steels. (See Chap. 14 for more detail.)

Medium-carbon ultrahigh-strength steels are structural steels with very high strength and high fracture toughness. These steels exhibit a minimum yield strength of 1380 MPa (200 ksi). Table 1.10 lists typical compositions such as AISI/SAE 4130, high-strength 4140, deeper hardening higher-strength 4340, 300M (a 1.6% Si modified 4340 steel) (to prevent embrittlement when the steel is tempered at the low temperatures required for very high strength), and Ladish D–6a and Ladish D–6ac steels (another modified 4340 with V grain refiner and higher C, Cr, and Mo contents, developed for aircraft and missile structural applications). Certain proprietary steels such as Hy Tuf (a Si-modified steel similar to 300M) exhibit excellent toughness (similar to maraging steel) at strengths up to or slightly above 1380 MPa (200 ksi). Other less prominent steels included in this family are AISI/SAE 6150 steel (a tough shock-resisting, shallow-hardening Cr-V steel with high fatigue and impact resistance in the heat-treated condition) and 8640 steel (an oil-hardening steel exhibiting properties similar to those of 4340 steel).[63] Product forms included in this category are billet, bar, rod, forging, plate, sheet, tubing, and welding wire.

AF1400 and AeroMet 100 steels belonging to high-fracture-toughness steels are discussed in Chap. 14.

TABLE 1.8 Low-Alloy Steel Compositions Applicable to Billets, Blooms, Slabs, and Hot-rolled and Cold-finished Bars. (Slightly wider ranges of compositions apply to plates.)[45]

	Designation		Ladle chemical composition limits, %*								
UNS number	SAE number	Corresponding AISI number	C	Mn	P	S	Si	Ni	Cr	Mo	V
G13300	1330	1330	0.28–0.33	1.60–1.90	0.035	0.040	0.15–0.35	...	...	...	...
G13350	1335	1335	0.33–0.38	1.60–1.90	0.035	0.040	0.15–0.35	...	...	...	...
G13400	1340	1340	0.38–0.43	1.60–1.90	0.035	0.040	0.15–0.35	...	...	...	...
G13450	1345	1345	0.43–0.48	1.60–1.90	0.035	0.040	0.15–0.35	...	...	...	...
G40230	4023	4023	0.20–0.25	0.70–0.90	0.035	0.040	0.15–0.35	...	...	...	...
G40240	4024	4024	0.20–0.25	0.70–0.90	0.035	0.035–0.050	0.15–0.35	...	...	0.20–0.30	...
G40270	4027	4027	0.25–0.30	0.70–0.90	0.035	0.040	0.15–0.35	...	...	0.20–0.30	...
G40280	4028	4028	0.25–0.30	0.70–0.90	0.035	0.035–0.050	0.15–0.35	...	...	0.20–0.30	...
G40320	4032	...	0.30–0.35	0.70–0.90	0.035	0.040	0.15–0.35	...	...	0.20–0.30	...
G40370	4037	4037	0.35–0.40	0.70–0.90	0.035	0.040	0.15–0.35	...	...	0.20–0.30	...
G40420	4042	4042	0.40–0.45	0.70–0.90	0.035	0.040	0.15–0.35	...	...	0.20–0.30	...
G40470	4047	4047	0.45–0.50	0.70–0.90	0.035	0.040	0.15–0.35	...	...	0.20–0.30	...
G41180	4118	4118	0.18–0.23	0.70–0.90	0.035	0.040	0.15–0.35	...	0.40–0.60	0.08–0.15	...
G41300	4130	4130	0.28–0.33	0.40–0.60	0.035	0.040	0.15–0.35	...	0.80–1.10	0.15–0.25	...
G41350	4135	...	0.33–0.38	0.70–0.90	0.035	0.040	0.15–0.35	...	0.80–1.10	0.15–0.25	...
G41370	4137	4137	0.35–0.40	0.70–0.90	0.035	0.040	0.15–0.35	...	0.80–1.10	0.15–0.25	...
G41400	4140	4140	0.38–0.43	0.75–1.00	0.035	0.040	0.15–0.35	...	0.80–1.10	0.15–0.25	...
G41420	4142	4142	0.40–0.45	0.75–1.00	0.035	0.040	0.15–0.35	...	0.80–1.10	0.15–0.25	...
G41450	4145	4145	0.41–0.48	0.75–1.00	0.035	0.040	0.15–0.35	...	0.80–1.10	0.15–0.25	...
G41470	4147	4147	0.45–0.50	0.75–1.00	0.035	0.040	0.15–0.35	...	0.80–1.10	0.15–0.25	...
G41500	4150	4150	0.48–0.53	0.75–1.00	0.035	0.040	0.15–0.35	...	0.80–1.10	0.15–0.25	...
G41610	4161	4161	0.56–0.64	0.75–1.00	0.035	0.040	0.15–0.35	...	0.70–0.90	0.25–0.35	...
G43200	4320	4320	0.17–0.22	0.45–0.65	0.035	0.040	0.15–0.35	1.65–2.00	0.40–0.60	0.20–0.30	...
G43400	4340	4340	0.38–0.43	0.60–0.80	0.035	0.040	0.15–0.35	1.65–2.00	0.70–0.90	0.20–0.30	...
G43406	E4340†	E4340	0.38–0.43	0.65–0.85	0.025	0.025	0.15–0.35	1.65–2.00	0.70–0.90	0.20–0.30	...
G44220	4422	...	0.20–0.25	0.70–0.90	0.035	0.040	0.15–0.35	...	...	0.35–0.45	...
G44270	4427	...	0.24–0.29	0.70–0.90	0.035	0.040	0.15–0.35	...	...	0.35–0.45	...
G46150	4615	4615	0.13–0.18	0.45–0.65	0.035	0.040	0.15–0.25	1.65–2.00	...	0.20–0.30	...

G46170	4617	⋯	0.15–0.20	0.45–0.65	0.035	0.040	0.15–0.35	1.65–2.00	⋯	0.20–0.30	⋯
G46200	4620	4620	0.17–0.22	0.45–0.65	0.035	0.040	0.15–0.35	1.65–2.00	⋯	0.20–0.30	⋯
G46260	4626	4626	0.24–0.29	0.45–0.65	0.035	0.040 max	0.15–0.35	0.70–1.00	⋯	0.15–0.25	⋯
G47180	4718	4718	0.16–0.21	0.70–0.90	⋯	⋯	⋯	0.90–1.20	0.35–0.55	0.30–0.40	⋯
G47200	4720	4720	0.17–0.22	0.50–0.70	0.035	0.040	0.15–0.35	0.90–1.20	0.35–0.55	0.15–0.25	⋯
G48150	4815	4815	0.13–0.18	0.40–0.60	0.035	0.040	0.15–0.35	3.25–3.75	⋯	0.20–0.30	⋯
G48170	4817	4817	0.15–0.20	0.40–0.60	0.035	0.040	0.15–0.35	3.25–3.75	⋯	0.20–0.30	⋯
G48200	4820	4820	0.18–0.23	0.50–0.70	0.035	0.040	0.15–0.35	3.25–3.75	⋯	0.20–0.30	⋯
G50401	50B40‡	⋯	0.38–0.43	0.75–1.00	0.035	0.040	0.15–0.35	⋯	0.40–0.60	⋯	⋯
G50441	50B44‡	50B44	0.43–0.48	0.75–1.00	0.035	0.040	0.15–0.35	⋯	0.40–0.60	⋯	⋯
G50460	5046	⋯	0.43–0.48	0.75–1.00	0.035	0.040	0.15–0.35	⋯	0.20–0.35	⋯	⋯
G50461	50B46‡	50B46	0.44–0.49	0.75–1.00	0.035	0.040	0.15–0.35	⋯	0.20–0.35	⋯	⋯
G50501	50B50‡	50B50	0.48–0.53	0.75–1.00	0.035	0.040	0.15–0.35	⋯	0.40–0.60	⋯	⋯
G50600	5060	⋯	0.56–0.64	0.75–1.00	0.035	0.040	0.15–0.35	⋯	0.40–0.60	⋯	⋯
G50601	50B60‡	50B60	0.56–0.64	0.75–1.00	0.035	0.040	0.15–0.35	⋯	0.40–0.60	⋯	⋯
G51150	5115	⋯	0.13–0.18	0.70–0.90	0.035	0.040	0.15–0.35	⋯	0.70–0.90	⋯	⋯
G51170	5117	5117	0.15–0.20	0.70–0.90	0.035	0.040	0.15–0.35	⋯	0.70–0.90	⋯	⋯
G51200	5120	5120	0.17–0.22	0.70–0.90	0.040	0.040	0.15–0.35	⋯	0.70–0.90	⋯	⋯
G51300	5130	5130	0.28–0.33	0.70–0.90	0.035	0.040	0.15–0.35	⋯	0.80–1.10	⋯	⋯
G51320	5132	5132	0.30–0.35	0.60–0.80	0.035	0.040	0.15–0.35	⋯	0.75–1.00	⋯	⋯
G51350	5135	5135	0.33–0.38	0.60–0.80	0.035	0.040	0.15–0.35	⋯	0.80–1.05	⋯	⋯
G51400	5140	5140	0.38–0.43	0.70–0.90	0.035	0.040	0.15–0.35	⋯	0.70–0.90	⋯	⋯
G51470	5147	5147	0.46–0.51	0.70–0.95	0.035	0.040	0.15–0.35	⋯	0.85–1.15	⋯	⋯
G51500	5150	5150	0.48–0.53	0.70–0.90	0.035	0.040	0.15–0.35	⋯	0.70–0.90	⋯	⋯
G51550	5155	5155	0.51–0.59	0.70–0.90	0.035	0.040	0.15–0.35	⋯	0.70–0.90	⋯	⋯
G51600	5160	5160	0.56–0.64	0.75–1.00	0.035	0.040	0.15–0.35	⋯	0.70–0.90	⋯	⋯
G51601	51B60‡	51B60	0.56–0.64	0.75–1.00	0.035	0.040	0.15–0.35	⋯	0.70–0.90	⋯	⋯
G50986	50100†	⋯	0.98–1.10	0.25–0.45	0.025	0.025	0.15–0.35	⋯	0.40–0.60	⋯	⋯
G51986	51100†	E51100	0.98–1.10	0.25–0.45	0.025	0.025	0.15–0.35	⋯	0.90–1.15	⋯	⋯
G52986	52100†	E52100	0.98–1.10	0.25–0.45	0.025	0.025	0.15–0.35	⋯	1.30–1.60	⋯	⋯
G61180	6118	6118	0.16–0.21	0.50–0.70	0.035	0.040	0.15–0.35	⋯	0.50–0.70	⋯	0.10–0.15
G61500	6150	6150	0.48–0.53	0.70–0.90	0.035	0.040	0.15–0.35	⋯	0.80–1.10	⋯	0.15 min
G81150	8115	8115	0.13–0.18	0.70–0.90	0.035	0.040	0.15–0.35	0.20–0.40	0.30–0.50	0.08–0.15	⋯
G81451	81B45‡	81B45	0.43–0.48	0.75–1.00	0.035	0.040	0.15–0.35	0.20–0.40	0.35–0.55	0.08–0.15	⋯
G86150	8615	8615	0.13–0.18	0.70–0.90	0.035	0.040	0.15–0.35	0.40–0.70	0.40–0.60	0.15–0.25	⋯

TABLE 1.8 Low-Alloy Steel Compositions Applicable to Billets, Blooms, Slabs, and Hot-rolled and Cold-finished Bars. (Slightly wider ranges of compositions apply to plates.)[45] *(Continued)*

UNS number	SAE number	Corresponding AISI number	Ladle chemical composition limits, %*								
			C	Mn	P	S	Si	Ni	Cr	Mo	V
G86170	8617	8617	0.15–0.20	0.70–0.90	0.035	0.040	0.15–0.35	0.40–0.70	0.40–0.60	0.15–0.25	...
G86200	8620	8620	0.18–0.23	0.70–0.90	0.035	0.040	0.15–0.35	0.40–0.70	0.40–0.60	0.15–0.25	...
G86220	8622	8622	0.20–0.25	0.70–0.90	0.035	0.040	0.15–0.35	0.40–0.70	0.40–0.60	0.15–0.25	...
G86250	8625	8625	0.23–0.28	0.70–0.90	0.035	0.040	0.15–0.35	0.40–0.70	0.40–0.60	0.15–0.25	...
G86270	8627	8627	0.25–0.30	0.70–0.90	0.035	0.040	0.15–0.35	0.40–0.70	0.40–0.60	0.15–0.25	...
G86300	8630	8630	0.28–0.33	0.70–0.90	0.035	0.040	0.15–0.35	0.40–0.70	0.40–0.60	0.15–0.25	...
G86370	8637	8637	0.35–0.40	0.75–1.00	0.035	0.040	0.15–0.35	0.40–0.70	0.40–0.60	0.15–0.25	...
G86400	8640	8640	0.38–0.43	0.75–1.00	0.035	0.040	0.15–0.35	0.40–0.70	0.40–0.60	0.15–0.25	...
G86420	8642	8642	0.40–0.45	0.75–1.00	0.035	0.040	0.15–0.35	0.40–0.70	0.40–0.60	0.15–0.25	...
G86450	8645	8645	0.43–0.48	0.75–1.00	0.035	0.040	0.15–0.35	0.40–0.70	0.40–0.60	0.15–0.25	...
G86451	86B45‡	...	0.43–0.48	0.75–1.00	0.035	0.040	0.15–0.35	0.40–0.70	0.40–0.60	0.15–0.25	...
G86500	8650	...	0.48–0.53	0.75–1.00	0.035	0.040	0.15–0.35	0.40–0.70	0.40–0.60	0.15–0.25	...
G86550	8655	8655	0.51–0.59	0.75–1.00	0.035	0.040	0.15–0.35	0.40–0.70	0.40–0.60	0.15–0.25	...
G86600	8660	...	0.56–0.64	0.75–1.00	0.035	0.040	0.15–0.35	0.40–0.70	0.40–0.60	0.15–0.25	...
G87200	8720	8720	0.18–0.23	0.70–0.90	0.035	0.040	0.15–0.35	0.40–0.70	0.40–0.60	0.20–0.30	...
G87400	8740	8740	0.38–0.43	0.75–1.00	0.035	0.040	0.15–0.35	0.40–0.70	0.40–0.60	0.20–0.30	...
G88220	8822	8822	0.20–0.25	0.75–1.00	0.035	0.040	0.15–0.35	0.40–0.70	0.40–0.60	0.30–0.40	...
G92540	9254	...	0.51–0.59	0.60–0.80	0.035	0.040	1.20–1.60	...	0.60–0.80	...	...
G92600	9260	...	0.56–0.64	0.75–1.00	0.035	0.040	1.80–2.20	...	...	...	...
G93106	9310†	...	0.08–0.13	0.45–0.65	0.025	0.025	0.15–0.35	3.00–3.50	1.00–1.40	0.08–0.15	...
G94151	94B15‡	...	0.13–0.18	0.75–1.00	0.035	0.040	0.15–0.35	0.30–0.60	0.30–0.50	0.08–0.15	...
G94171	94B17‡	94B17	0.15–0.20	0.75–1.00	0.035	0.040	0.15–0.35	0.30–0.60	0.30–0.50	0.08–0.15	...
G94301	94B30‡	94B30	0.28–0.33	0.75–1.00	0.035	0.040	0.15–0.35	0.30–0.60	0.30–0.50	0.08–0.15	...

* Small amounts of certain elements that are not specified or required may be found in alloy steels. These elements are considered as incidental and are acceptable to the following maximum amount: Cu to 0.35%, Ni to 0.25%, Cr to 0.20%, and Mo to 0.06%.

† Electric furnace steel.

‡ Boron content is 0.0005 to 0.003%.

Reprinted by permission of ASM International, Materials Park, Ohio.

TABLE 1.9 Selected Quenched and Tempered Low-Alloy Structural Steels[30]

Steel	Composition, wt%*								
	C	Si	Mn	P	S	Ni	Cr	Mo	Other
A 514/A 517 grade A	0.15–0.21	0.40–0.80	0.80–1.10	0.035	0.04	...	0.50–0.80	0.18–0.28	0.05–0.15 Zr† 0.0025 B
A 514/A 517 grade F	0.10–0.20	0.15–0.35	0.60–1.00	0.035	0.04	0.70–1.00	0.40–0.65	0.40–0.60	0.03–0.08 V 0.15–0.50 Cu 0.0005–0.005 B
A 514/A 517 grade R	0.15–0.20	0.20–0.35	0.85–1.15	0.035	0.04	0.90–1.10	0.35–0.65	0.15–0.25	0.03–0.08 V
A 533 type A	0.25	0.15–0.40	1.15–1.50	0.035	0.04	...	...	0.45–0.60	...
A 533 type C	0.25	0.15–0.40	1.15–1.50	0.035	0.04	0.70–1.00	...	0.45–0.60	...
HY-80	0.12–0.18	0.15–0.35	0.10–0.40	0.025	0.025	2.00–3.25	1.00–1.80	0.20–0.60	0.25 Cu 0.03 V 0.02 Ti
HY-100	0.12–0.20	0.15–0.35	0.10–0.40	0.025	0.025	2.25–3.50	1.00–1.80	0.20–0.60	0.25 Cu 0.03 V 0.02 Ti

* Single values represent the maximum allowable.
† Zr may be replaced by Ce. When Ce is added, the Ce/S ratio should be approximately 1.5/1, based on heat analysis.

TABLE 1.10 Compositions of the Ultrahigh-Strength Steels[5,78]

Designation or trade name	UNS number	C	Mn	Si	Cr	Ni	Mo	V	Co
					Composition, wt%*				
Medium-carbon low-alloy steels									
4130	G41300	0.28–0.33	0.40–0.60	0.20–0.35	0.80–1.10	...	0.15–0.25	...	...
4140	G41400	0.38–0.43	0.75–1.00	0.20–0.35	0.80–1.10	...	0.15–0.25	...	...
4340	G43400	0.38–0.43	0.60–0.80	0.20–0.35	0.70–0.90	1.65–2.00	0.20–0.30	...	...
AMS 6434		0.31–0.38	0.60–0.80	0.20–0.35	0.65–0.90	1.65–2.00	0.30–0.40	0.17–0.23	...
300M	K44220	0.40–0.46	0.65–0.90	1.45–1.80	0.70–0.95	1.65–2.00	0.30–0.45	0.05 min	...
D-6a		0.42–0.48	0.60–0.90	0.15–0.30	0.90–1.20	0.40–0.70	0.90–1.10	0.05–0.10	...
6150	G61500	0.48–0.53	0.70–0.90	0.20–0.35	0.80–1.10	...	...	0.15–0.25	...
8640	G86400	0.38–0.43	0.75–1.00	0.20–0.35	0.40–0.60	0.40–0.70	0.15–0.25	...	...
Medium-alloy air-hardening steels									
H11 mod	T20811 mod	0.37–0.43	0.20–0.40	0.80–1.00	4.75–5.25	...	1.20–1.40	0.40–0.60	...
H13	T20813	0.32–0.45	0.20–0.50	0.80–1.20	4.75–5.50	...	1.10–1.75	0.80–1.20	...
High fracture toughness steels									
AF1410†	K92571	0.13–0.17	0.10 max	0.10 max	1.80–2.20	9.50–10.50	0.90–1.10	...	13.50–14.50
HP 9-4-30‡	K91283	0.29–0.34	0.10–0.35	0.20 max	0.90–1.10	7.0–8.0	0.90–1.10	0.06–0.12	4.25–4.75
Aer Met 100§	K92580	0.21–0.25	0.10 max	0.10 max	2.85–3.35	10.5–12.5	1.20	...	12.50–14.50

* P and S contents may vary with steelmaking practice. Usually, these steels contain no more than 0.035 P and 0.040 S.
† AF1410 is specified to have 0.008 P and 0.005 S composition. Ranges utilized by some producers are narrower.
‡ HP 9-4-30 is specified to have 0.10 max P and 0.10 max S. Ranges utilized by some producers are narrower.
§ Contains 0.015 max Al, 0.0015 N max; 0.015 O max; 0.008 P max; 0.005 S max; 0.015 max Ti.
Source: Refs. 5 and 78.

TABLE 1.11 Compositions of Bearing Steels[64-66]

UNS	AISI	C, %	Mn, %	Si, %	Ni, %	Cr, %	Mo, %
			Carburizing bearing steels*[†]				
...	3310	0.08–0.13	0.45–0.60	0.15–0.35	3.25–3.75	1.40–1.75	...
G41180	4118	0.18–0.23	0.70–0.90	0.15–0.30	...	0.40–0.60	0.08–0.18
G43200	4320	0.17–0.22	0.45–0.65	0.15–0.35	1.65–2.00	0.40–0.60	0.20–0.30
G46200	4620	0.17–0.23	0.45–0.65	0.15–0.35	1.65–2.00	...	0.15–0.25
G47200	4720	0.17–0.22	0.50–0.70	0.15–0.35	0.90–1.20	0.350–0.55	0.15–0.25
G48200	4820	0.18–0.23	0.50–0.70	0.15–0.35	3.25–3.75	...	0.20–0.30
G51200	5120	0.18–0.23	0.70–0.90	0.15–0.30	...	0.70–0.90	...
G86200	8620	0.18–0.23	0.70–0.90	0.15–0.35	0.40–0.70	0.40–0.60	0.15–0.25
...	9310	0.08–0.13	0.45–0.65	0.15–0.35	3.00–3.50	1.00–1.40	0.08–0.15
			Through-hardened bearing steels				
...	52100	0.98–1.10	0.25–0.45	0.15–0.35	0.25 max	1.30–1.60	0.10 max
...	51100	0.98–1.10	0.25–0.45	0.15–0.35	0.25 max	0.90–1.15	0.10 max
...	50100	0.98–1.10	0.25–0.45	0.15–0.35	0.25 max	0.40–0.60	0.10 max
...	5195	0.90–1.03	0.75–1.00	0.15–0.35	0.25 max	0.70–0.90	0.10 max
K19526	...	0.89–1.01	0.50–0.80	0.15–0.35	0.25 max	0.40–0.60	0.10 max
...	1070M	0.65–0.75	0.80–1.10	0.15–0.35	0.25 max	0.20 max	0.10 max
...	5160	0.56–0.64	0.75–1.00	0.15–0.35	0.25 max	0.70–0.90	0.10 max
			Corrosion-resistant bearing stainless steel				
S44004	440C	0.95–1.20	1.00 max	1.00 max	0.75 max	16.00–18.00	0.40–0.75
	440C mod	1.00–1.10	0.30–1.00	0.20–1.00	0.75 max	13.00–15.00	3.75–4.25

* The P and S limitations for special quality bearings shall be 0.015% maximum.
† When not specified, the following elements are considered incidental and may be present to the following maximum amounts: Cu, 0.35%; Ni, 0.25%; Cr, 0.20%; Mo, 0.10%.

These steels are used for gears, aircraft landing gear, airframe parts, pressure vessels, bolts, springs, screws, axles, studs, fasteners, machinery parts, connecting rods, crankshafts, oil well drilling bits, high-pressure tubing, flanges, wrenches, sprockets, etc.[63]

Bearing Steels. Bearing steels used for race, ball, and roller bearing applications are made from low-carbon carburizing steels, medium-carbon steels, or high-carbon through-hardening steels (Table 1.11). Standard bearings are made from steel melted in electric arc furnace or vacuum induction melting furnace, followed by refining to high levels of steel cleanliness in secondary steelmaking plants incorporating vacuum arc-remelting (VAR), electroflux refining (EFR), vacuum degassing, and other processing steps designed to eliminate nonmetallic impurities and to lower gas contents to a minimum.

Carburizing grades (such as 4118, 4320, 4620, 4720, 4820, 5120, 8617, 8620, and 9310) are used for off-highway earthmoving equipment or rolling mill backup bearings, and corrosion-resistant stainless steels such as AISI 440C and 440C modified are used in severe corrosive environments. Re-refined high-speed steels are used for the very demanding applications such as gas turbine aeroengines to provide the temperature and fatigue capability.[64,65]

TABLE 1.12 Nominal Chemical Compositions for Heat-Resistant Chromium–Molybdenum Steels[30]

Type	UNS designation	Composition, %*						
		C	Mn	S	P	Si	Cr	Mo
$\frac{1}{2}$Cr-$\frac{1}{2}$Mo	K12122	0.10–0.20	0.30–0.80	0.040	0.040	0.10–0.60	0.50–0.80	0.45–0.65
1Cr-$\frac{1}{2}$Mo	K11562	0.15	0.30–0.60	0.045	0.045	0.50	0.80–1.25	0.45–0.65
1$\frac{1}{4}$Cr-$\frac{1}{2}$Mo	K11597	0.17	0.30–0.60	0.030	0.030	0.50–1.00	1.00–1.50	0.45–0.65
1$\frac{1}{4}$Cr-$\frac{1}{2}$Mo	K11592	0.10–0.20	0.30–0.80	0.040	0.040	0.50–1.00	1.00–1.50	0.45–0.65
2$\frac{1}{4}$Cr-1Mo	K21590	0.15	0.30–0.60	0.040	0.040	0.50	2.00–2.50	0.87–1.13
3Cr-1Mo	K31545	0.15	0.30–0.60	0.030	0.030	0.50	2.65–3.35	0.80–1.06
3Cr-1MoV†	K31830	0.18	0.30–0.60	0.020	0.020	0.10	2.75–3.25	0.90–1.10
5Cr-$\frac{1}{2}$Mo	K41545	0.15	0.30–0.60	0.030	0.030	0.50	4.00–6.00	0.45–0.65
7Cr-$\frac{1}{2}$Mo	K61595	0.15	0.30–0.60	0.030	0.030	0.50–1.00	6.00–8.00	0.45–0.65
9Cr-1Mo	K90941	0.15	0.30–0.60	0.030	0.030	0.50–1.00	8.00–10.00	0.90–1.10
9Cr-1MoV‡	...	0.08–0.12	0.30–0.60	0.010	0.020	0.20–0.50	8.00–9.00	0.85–1.05

* Single values are maximums.
† Also contains 0.02 to 0.030% V, 0.001 to 0.003% B, and 0.015 to 0.035% Ti.
‡ Also contains 0.40% Ni, 0.18 to 0.25% V, 0.06 to 0.10% Nb, 0.03 to 0.07% N, and 0.04% Al.
Reprinted by permission of ASM International, Materials Park, Ohio.

The through-hardening C-Cr steels such as 52100, 51100, 50100, 5195, 1070M, 5160, and UNS K19526 are designed to produce a hard martensitic structure throughout the section so as to provide the maximum load-bearing capacity, good wear resistance, and high fatigue life. This finds applications for electric motors and gearboxes and for high-performance applications such as gas turbine aeroengine, jet engine main shaft, helicopter, and steel rolling mill support bearings.[64] Readers are advised to refer to ASTM A534, A535, A756, and A866 standards for chemical composition of bearing steel grades.[64,66]

Chromium-Molybdenum Heat-Resistant Steels. They contain 0.5 to 9% Cr, 0.5 to 1.0% Mo and usually <0.20% C and are used up to temperatures of 650°C (1250°F). They are ordinarily supplied in the normalized and tempered, quenched and tempered, or annealed condition. The Cr-Mo steels are extensively used as structural components in the oil refineries, oil and gas industries, chemical industries, electric power generation, and fossil fuel and nuclear power plants for piping, heat exchangers, superheater tubes, and pressure vessels, because of their excellent hardenability, high strength, good toughness, increased resistance to oxidation and corrosion, and resistance to hydrogen embrittlement. Nominal chemical compositions are given in Table 1.12. ASTM specifications for various product forms are also available elsewhere.[30]

Silicon Steels. Also called *electrical steels*, these are soft magnetic (iron-silicon) alloys (with a reasonable hardness of 100 to 150 HV) that are widely used for low- and intermediate-frequency applications. The silicon addition to iron (1) increases electrical resistivity (lowers eddy current loss) and permeability and lowers core loss; (2) suppresses the γ loop, causing stabilization of α- (ferrite-) phase to higher temperature; (3) allows special heat treatment necessary to develop a preferred orientation (grain structure); (4) reduces magnetocrystalline anisotropy; (5) reduces magnetostriction constants to nearly zero at ~6.5% silicon, producing maximum high permeability and low hysteresis losses; (6) lowers magnetic saturation and Curie temperature; and (7) drastically decreases the tensile ductility.[67]

Silicon bearing steels are divided into two categories—nonoriented (or isotropic) and grain-oriented (or anisotropic) grades. Nonoriented electrical flat-rolled steel sheets are mostly used in less critical applications as core laminations or punched ring cores for small and medium-size motors and transformers, radio and television transformers, alternators, amplifiers, transducers, generators, and magnetic circuits of industrial machinery, where magnetic flux is not forced along any one direction. They are available in the fully processed (AISI M15 through M47) grades and the semiprocessed (AISI M27 through M47) grades. The semiprocessed materials are low-carbon (below 0.030%) silicon-iron, or silicon-aluminum-iron alloys containing up to 2.5% silicon, 0.8 to 1% aluminum, below 0.025% sulfur, 0.10 to 0.70% Mn, and phosphorus, antimony, and tin up to 0.15%. The full magnetic properties are developed by customer stress-relief anneal on stamped/punched/sheared parts prior to assembly into machinery.[68,69] The basic factors affecting magnetic properties are strip thickness; concentration of main alloying elements, notably, silicon and aluminum; grain size; impurity content; and crystallographic texture.[70] Excellent magnetic properties after stress-relief annealing and good punchability for increased productivity are essential for electrical steel sheets to be used as high-efficiency motor cores.

The production method for semiprocessed nonoriented electrical steels involves hot rolling, pickling, cold rolling, and annealing. An additional annealing operation may precede or follow the pickling operation. Sometimes a final temper-rolling (6 to 10% reduction) process is provided (ASTM A683). They are supplied with a tightly adherent surface oxide or coating (coating types C–0, C–4–AS, and C–5–AS according to ASTM A976 specification) to provide adequate insulating ability for small cores (ASTM A683 specification). When necessary, thicker applied coatings (coatings types C–4 or C–5 according to ASTM A976 specification) are supplied for higher levels of insulating ability.

Fully processed nonoriented silicon steel grades are low-carbon (below 0.02%), silicon-iron, or silicon-aluminum-iron alloys with up to about 3.5% silicon, up to about 0.6% aluminum, less than 0.025% sulfur, and 0.10 to 0.40% manganese. Phosphorus, copper, nickel, chromium, molybdenum, antimony, and tin are usually present only in residual amounts except in the higher-numbered core loss types where phosphorus up to 0.15% and antimony and tin up to 0.10% may be present. Their processing sequence comprises hot rolling, annealing, pickling, cold rolling, and decarburization annealing to develop large, clean ferrite grains with a nonoriented texture, followed by organic (such as enamels or varnishes) coating or inorganic (phosphate-type) coating to reduce eddy current in lamination stacks.[67,72] Organic coatings can improve punchability to a marked extent, but they cannot survive a stress-relief anneal. However, inorganic coating is designed to survive stress-relief anneal at temperatures below 801°C in a neutral or slightly reducing atmosphere (ASTM A677M).

Nonoriented silicon steels are to be distinguished from motor lamination steels that are widely employed in fractional-horsepower motors and constitute the largest single tonnage electrical steel. Motor lamination steels contain below 0.60% total silicon plus aluminum and are considered low-carbon steel.[67]

The grain-oriented Si steels (AISI M2 through M6) are basically high-purity Fe-Si alloys with a constant Si content of about 3.25% and have very low core losses. They are subdivided into hot-rolled semiprocessed; cold-rolled semiprocessed; hot-rolled, fully processed; cold-rolled fully processed; and low-stress type. Special coatings give electrical insulation and induced tensile stresses in the steel substrate. In this instance, the induced stresses reduce core loss and minimize noise in transformers. They are used in fractional-horsepower motors, power and distribution transformers, and relays. The preferred orientations developed in silicon steels are the cube-on-face (100)[001] orientation and the cube-on-edge (110)[001] Goss orientation where (100) and (110) planes, respectively, lie in the plane of the sheet. Conventional grain-oriented 3.15% Si steel has grains about 3 mm (0.12 in.) in diameter. The high-permeability silicon steels appear to have grains about 8 mm (0.31 in.) or larger in diameter.

The thin gauge electrical steel strips are usually produced by thermal treatment process with insulation on both sides of the strip and are characterized by low core losses and high magnetic induction. They find applications as high-frequency cores, radio-frequency cores, and other specialized uses. Readers are required to consult various sources for more details.[48,71,72] (See also Chap. 5.)

Steels for Low-Temperature and Cryogenic Service. The alloy steels that are used at extremely cold climates and cryogenic temperatures are characterized by a combination of high yield and tensile strengths and good ductility and fracture toughness at 77 K. For specific applications, other criteria such as low coefficient of thermal expansion, low magnetic permeability, low heat conductivity, low emissivity, cleanliness, good fatigue limit, and weldability may be required to be fulfilled. These steels are divided into three categories: (1) Ferritic steels with 2.25 to 3.5% Ni grade (such as HY 80 and 100) and 5 to 9% Ni grades (such as HY-130 and ASTM A645 and ASTM A353). High-nickel grades are used for applications involving exposure to temperatures from 0 to −195°C (or 32 to −320°F); (2) 300-series austenitic stainless steels such as 304LN and 316LN with a combination of high strength and toughness at 4 K with good weldability; (3) high-strength, stable austenitic alloys for structural applications at 4 K. These are Fe-Ni-Cr alloys (e.g., JK1: 15Ni-25Cr-4Mn-0.4N-0.3Si-0.01C) and Fe-Mn-Cr-N alloys (e.g., JK1: 22Mn-13Cr-5Ni-0.2N-0.002C-0.5Si), and nickel-base superalloy, Inconel 908 (48Ni-4.5Cr-3Nb-1.5Ti-1Al-0.02C).[48] These alloys find applications in transportation and storage vessels for liquefied gases (such as propane, anhydrous NH_3, CO_2, ethane, methane, O_2, N_2, H_2, Ar, and He), high-field semiconducting magnets, vacuum-transfer lines, and structures and machinery designed for use in cold regions, steelmaking and chemical-process and other major industries including space and medical fields.[45,48,73]

1.8.2.4 High-Strength Low-Alloy Steels. High-strength low-alloy (HSLA) steels are low-carbon steels containing Mn up to 1.8% and a small amount of Al, and microalloying elements such as Nb, V, or Ti that exhibit and develop significant strength, toughness, formability, and corrosion resistance. Table 1.13 lists HSLA steels according to chemical composition, mill forms, and characteristics described in SAE J410C. (See Chap. 15 for more details.)

HSLA steels are successfully used as sheet, strip, plate, bar, structural section, and forged bar products; they find structural applications in several diverse fields such as in oil and gas pipelines; automotive, enamel, transportation (such as railroad tank cars, railway tracks, trucks, trailers), agricultural, and pressure vessel industries; as off-shore structures and platforms; and in the construction of cranes, bridges, buildings, mining equipment, earthmoving equipment, shipbuilding, power transmission and TV towers, electricity pylons, penstocks, steel pilings, and reinforcing bars.[74]

1.8.2.5 Low-Alloy Ultrahigh-Strength Steels. Low-alloy ultrahigh-strength steels (UHSS) are defined as those steels with an incoming minimum tensile strength > 100 ksi (~700 MPa). Over the past decade, their use has increased significantly for automotive applications in safety-critical parts and components such as door impact beam, bumper reinforcement beams, side sills, cross car beams, and seat structures.[75]

For most grades, accelerated cooling methods are used to achieve the requisite strength levels through development of martensitic, ferritic martensitic (dual-phase), or multiphase microstructures. Furthermore, UHSS can be produced as bare or galvanized through various coating processes. The UHSS family consists of several products with varying grades, basic characteristics, properties, and application fields.[75]

1.8.2.6 Tool Steels. A tool steel is any steel used to shape other metals by cutting, forming, machining, shearing, battering, die casting, or to shape and cut wood, paper, rock, or concrete. Hence they are designed to have high hardness and durability under severe service conditions. Tool steels comprise a wide range of plain-carbon steels up to 1.2% carbon without appreciable amounts of alloying elements to the highly alloyed steels in which alloying additions reach 50%. Although some carbon tool steels and low-alloy tool steels have a wide range of carbon content, most of the higher-alloy tool steels have a comparatively narrow carbon range. A mixed classification system is used to classify tool steels based on the use, composition, special mechanical properties, or method of heat treatment.

TABLE 1.13 Compositions, Mill Forms, and Characteristics of HSLA Steels Described in SAE J410C[30]

Grade*	C (max)	Mn (max)	P (max)	Other elements‡	Available mill forms	Special characteristics
		Heat compositional limits, %†				
942X	0.21	1.35	0.04	Nb, V	Plate, bar, and shapes ≤100 mm (4 in.) in thickness	Similar to 945X and 945C except for better weldability and formability
945A	0.15	1.00	0.04	…	Sheet, strip, plate, bar, and shapes ≤75 mm (3 in.) in thickness	Excellent weldability, formability, and notch toughness
945C	0.23	1.40	0.04	…	Sheet, strip, plate, bar, and shapes ≤75 mm (3 in.) in thickness	Similar to 950C except that lower carbon and manganese content improve weldability, formability, and notch toughness
945X	0.22	1.35	0.04	Nb, V	Sheet, strip, plate, bar, and shapes ≤75 mm (3 in.) in thickness	Similar to 945C except for better weldability and formability
950A	0.15	1.30	0.04	…	Sheet, strip, plate, bar, and shapes ≤75 mm (3 in.) in thickness	Good weldability, notch toughness, and formability
950B	0.22	1.30	0.04	…	Sheet, strip, plate, bar, and shapes ≤75 mm (3 in.) in thickness	Fairly good notch toughness and formability
950C	0.25	1.60	0.04	…	Sheet, strip, plate, bar, and shapes ≤75 mm (3 in.) in thickness	Fair formability and toughness
950D	0.15	1.00	0.15	…	Sheet, strip, plate, bar, and shapes ≤75 mm (3 in.) in thickness	Good weldability and formability; phosphorus should be considered in conjunction with other elements
950X	0.23	1.35	0.04	Nb, V	Sheet, strip, plate, bar, and shapes ≤40 mm (1.5 in.) in thickness	Similar to 950C except for better weldability
955X	0.25	1.35	0.04	Nb, V, N	Sheet, strip, plate, bar, and shapes ≤40 mm (1.5 in.) in thickness	Similar to 945X and 950X except that progressively higher strengths are obtained by increasing the carbon and manganese contents, or by adding nitrogen ≤0.015%; formability and weldability generally decrease with increased strength; toughness varies with composition and mill practice
960X	0.26	1.45	0.04	Nb, V, N	Sheet, strip, plate, bar, and shapes ≤40 mm (1.5 in.) in thickness	
965X	0.26	1.45	0.04	Nb, V, N	Sheet, strip, plate, bar, and shapes ≤20 mm (0.75 in.) in thickness	
970X	0.26	1.65	0.04	Nb, V, N	Sheet, strip, plate, bar, and shapes ≤20 mm (0.75 in.) in thickness	
980X	0.26	1.65	0.04	Nb, V, N	Sheet, strip, plate, bar, and shapes ≤10 mm (0.38 in.) in thickness	

* Fully killed steel made to fine-grain practice may be specified by adding a second suffix, K, for instance, 945XK. Steels made to K practice are normally specified only for applications requiring better toughness at low temperatures than steels made to normal semikilled practice. † 0.05% P (max) and 0.90% Si (max), all grades. ‡ Elements normally added singly or in combination to produce specified mechanical properties and other characteristics. Other alloying elements such as copper, chromium, and nickel may be added to enhance atmospheric-corrosion resistance.
Reprinted by permission of ASM International, Materials Park, Ohio.

According to AISI specifications, there are seven main groups of wrought tool steels. Table 1.14 lists the compositions of these tool steels with corresponding designated symbols,[76] which are discussed herein.

1. *High-speed steels* are used for applications requiring long life at relatively high operating temperatures such as for heavy cuts or high-speed machining. High-speed steels are the most important alloy tool steels because of their very high hardness and good wear resistance in the heat-treated condition and their ability to retain high hardness at elevated temperatures often encountered during the operation of the tool at high cutting speeds. This red- or hot-hardness property is an important feature of a high-speed steel.[66]

High-speed steels are grouped into molybdenum type M and tungsten type W. Type M tool steels contain Mo, W, Cr, V, Co, and C as major alloying elements while type T tool steels contain W, Cr, V, Co, and C as the main alloying elements. In the United States, type M steels comprise 95% of the high-speed steels produced. There is also a subgroup consisting of intermediate high-speed steels in the M group. The most popular grades among molybdenum types are M1, M2, M4, M7, M10, and M42, while those among tungsten types are T1 and T15.

The main advantage of type M steel is its lower initial cost (approximately 40% cheaper than that of similar type T steels), but it is more susceptible to decarburizing, thereby necessitating better temperature control than type T steels. By using salt baths and sometimes surface coatings, decarburization can be controlled. The mechanical properties of type M and type T steels are similar except that type M steels have slightly greater toughness than type T steels at the same hardness level.

2. *Hot-work tool steels (AISI H series)* fall into three major groups: (1) chromium base, types H1 to H19; (2) tungsten base, types H20 to H39; and (3) molybdenum base, types H40 to H59. The distinction is based on the principal alloying additions; however, all classes have medium-carbon content and Cr content varying from 3 to 12.75%. Among these steels H11, H12, and H13 are produced in large quantities. These steels possess good red-hardness and retain high hardness (~50 Rc) after prolonged exposures at 500 to 550°C. They are used extensively for hot-work applications which include parts for aluminum and magnesium die casting and extrusion, plastic injection molding, and compression and transfer molds.[66]

3. *Cold-work tool steels* include three categories: (1) air-hardening, medium-alloy, tool steels (AISI A series); (2) high-carbon, high-chromium tool steels (AISI D series); and (3) oil-hardening tool steels (AISI O series). AISI A series tool steels have high hardenability and harden readily on air cooling. In the air-hardened and tempered condition, they are suitable for applications where improved toughness and reasonably good abrasion resistance are required, such as for forming, blanking, and drawing dies. The most popular grade is A2. AISI D series tool steels possess excellent wear resistance and nondeforming properties, thereby making them very useful as cold-work die steels. They find applications in blanking and cold-forming dies, drawing and lamination dies, thread rolling dies, shear and slitter blades, forming rolls, and so forth. Among these steels, D2 is by far the most popular grade.[66] AISI O series tool steels are used for blanking, coining, drawing, and forming dies and punches, shear blades, gauges, and chuck jaws after oil quenching and tempering.[51] Among these grades, O1 is the most widely used.

4. *Shock-resisting tool steels (AISI S series)* are used where repetitive impact stresses are encountered such as with hammers, chipping and cold chisels, rivet sets, punches, driver bits, stamps, and shear blades in quenched and tempered conditions.

TABLE 1.14 Composition Limits for Principal Types of Tool Steels[76]

| Designation | | Composition, %* | | | | | | | | |
AISI	UNS	C	Mn	Si	Cr	Ni	Mo	W	V	Co
					Molybdenum high-speed steels					
M1	T11301	0.78–0.88	0.15–0.40	0.20–0.50	3.50–4.00	0.30 max	8.20–9.20	1.40–2.10	1.00–1.35	...
M2	T11302	0.78–0.88; 0.95–1.05	0.15–0.40	0.20–0.45	3.75–4.50	0.30 max	4.50–5.50	5.50–6.75	1.75–2.20	...
M3, class 1	T11313	1.00–1.10	0.15–0.40	0.20–0.45	3.75–4.50	0.30 max	4.75–6.50	5.00–6.75	2.25–2.75	...
M3, class 2	T11323	1.15–1.25	0.15–0.40	0.20–0.45	3.75–4.50	0.30 max	4.75–6.50	5.00–6.75	2.75–3.75	...
M4	T11304	1.25–1.40	0.15–0.40	0.20–0.45	3.75–4.75	0.30 max	4.25–5.50	5.25–6.50	3.75–4.50	...
M7	T11307	0.97–1.05	0.15–0.40	0.20–0.55	3.50–4.50	0.30 max	8.20–9.20	1.40–2.10	1.75–2.25	...
M10	T11310	0.84–0.94; 0.95–1.05	0.10–0.40	0.20–0.45	3.75–4.50	0.30 max	7.75–8.50	...	1.80–2.20	...
M30	T11330	0.75–0.85	0.15–0.40	0.20–0.45	3.50–4.25	0.30 max	7.75–9.00	1.30–2.30	1.00–1.40	4.50–5.50
M33	T11333	0.85–0.92	0.15–0.40	0.15–0.50	3.50–4.00	0.30 max	9.00–10.00	1.30–2.10	1.00–1.35	7.75–8.75
M34	T11334	0.85–0.92	0.15–0.40	0.20–0.45	3.50–4.00	0.30 max	7.75–9.20	1.40–2.10	1.90–2.30	7.75–8.75
M35	T11335	0.82–0.88	0.15–0.40	0.20–0.45	3.75–4.50	0.30 max	4.50–5.50	5.50–6.75	1.75–2.20	4.50–5.50
M36	T11336	0.80–0.90	0.15–0.40	0.20–0.45	3.75–4.50	0.30 max	4.50–5.50	5.50–6.50	1.75–2.25	7.75–8.75
M41	T11341	1.05–1.15	0.20–0.60	0.15–0.50	3.75–4.50	0.30 max	3.25–4.25	6.25–7.00	1.75–2.25	4.75–5.75
M42	T11342	1.05–1.15	0.15–0.40	0.15–0.65	3.50–4.25	0.30 max	9.00–10.00	1.15–1.85	0.95–1.35	7.75–8.75
M43	T11343	1.15–1.25	0.20–0.40	0.15–0.65	3.50–4.25	0.30 max	7.50–8.50	2.25–3.00	1.50–1.75	7.75–8.75
M44	T11344	1.10–1.20	0.20–0.40	0.30–0.55	4.00–4.75	0.30 max	6.00–7.00	5.00–5.75	1.85–2.20	11.00–12.25
M46	T11346	1.22–1.30	0.20–0.40	0.40–0.65	3.70–4.20	0.30 max	8.00–8.50	1.90–2.20	3.00–3.30	7.80–8.80
M47	T11347	1.05–1.15	0.15–0.40	0.20–0.45	3.50–4.00	0.30 max	9.25–10.00	1.30–1.80	1.15–1.35	4.75–5.25
M48	T11348	1.42–1.52	0.15–0.40	0.15–0.40	3.50–4.00	0.30 max	4.75–5.50	9.50–10.50	2.75–3.25	8.00–10.00
M62	T11362	1.25–1.35	0.15–0.40	0.15–0.40	3.50–4.00	0.30 max	10.00–11.00	5.75–6.50	1.80–2.10	...
					Tungsten high-speed steels					
T1	T12001	0.65–0.80	0.10–0.40	0.20–0.40	3.75–4.50	0.30 max	...	17.25–18.75	0.90–1.30	...
T2	T12002	0.80–0.90	0.20–0.40	0.20–0.40	3.75–4.50	0.30 max	1.00 max	17.50–19.00	1.80–2.40	...
T4	T12004	0.70–0.80	0.10–0.40	0.20–0.40	3.75–4.50	0.30 max	0.40–1.00	17.50–19.00	0.80–1.20	4.25–5.75
T5	T12005	0.75–0.85	0.20–0.40	0.20–0.40	3.75–5.00	0.30 max	0.50–1.25	17.50–19.00	1.80–2.40	7.00–9.50
T6	T12006	0.75–0.85	0.20–0.40	0.20–0.40	4.00–4.75	0.30 max	0.40–1.00	18.50–21.00	1.50–2.10	11.00–13.00
T8	T12008	0.75–0.85	0.20–0.40	0.20–0.40	3.75–4.50	0.30 max	0.40–1.00	13.25–14.75	1.80–2.40	4.25–5.75
T15	T12015	1.50–1.60	0.15–0.40	0.15–0.40	3.75–5.00	0.30 max	1.00 max	11.75–13.00	4.50–5.25	4.75–5.25

Intermediate high-speed steels										
M50	T11350	0.78–0.88	0.15–0.45	0.20–0.60	3.75–4.50	0.30 max	3.90–4.75	...	0.80–1.25	...
M52	T11352	0.85–0.95	0.15–0.45	0.20–0.60	3.50–4.30	0.30 max	4.00–4.90	0.75–1.50	1.65–2.25	...
Chromium hot-work steels										
H10	T20810	0.35–0.45	0.25–0.70	0.80–1.20	3.00–3.75	0.30 max	2.00–3.00	...	0.25–0.75	...
H11	T20811	0.33–0.43	0.20–0.50	0.80–1.20	4.75–5.50	0.30 max	1.10–1.60	...	0.30–0.60	...
H12	T20812	0.30–0.40	0.20–0.50	0.80–1.20	4.75–5.50	0.30 max	1.25–1.75	1.00–1.70	0.50 max	...
H13	T20813	0.32–0.45	0.20–0.50	0.80–1.20	4.75–5.50	0.30 max	1.10–1.75	...	0.80–1.20	...
H14	T20814	0.35–0.45	0.20–0.50	0.80–1.20	4.75–5.50	0.30 max	...	4.00–5.25	...	...
H19	T20819	0.32–0.45	0.20–0.50	0.20–0.50	4.00–4.75	0.30 max	0.30–0.55	3.75–4.50	1.75–2.20	4.00–4.50
Tungsten hot-work steels										
H21	T20821	0.26–0.36	0.15–0.40	0.15–0.50	3.00–3.75	0.30 max	...	8.50–10.00	0.30–0.60	...
H22	T20822	0.30–0.40	0.15–0.40	0.15–0.40	1.75–3.75	0.30 max	...	10.00–11.75	0.25–0.50	...
H23	T20823	0.25–0.35	0.15–0.40	0.15–0.60	11.00–12.75	0.30 max	...	11.00–12.75	0.75–1.25	...
H24	T20824	0.42–0.53	0.15–0.40	0.15–0.40	2.50–3.50	0.30 max	...	14.00–16.00	0.40–0.60	...
H25	T20825	0.22–0.32	0.15–0.40	0.15–0.40	3.75–4.50	0.30 max	...	14.00–16.00	0.40–0.60	...
H26	T20826	0.45–0.55	0.15–0.40	0.15–0.40	3.75–4.50	0.30 max	...	17.25–19.00	0.75–1.25	...
Molybdenum hot-work steels										
H42	T20842	0.55–0.70[†]	0.15–0.40	0.15–0.40	3.75–4.50	0.30 max	4.50–5.50	5.50–6.75	1.75–2.20	...
Air-hardening, medium-alloy, cold-work steels										
A2	T30102	0.95–1.05	1.00 max	0.50 max	4.75–5.50	0.30 max	0.90–1.40	...	0.15–0.50	...
A3	T30103	1.20–1.30	0.40–0.60	0.50 max	4.75–5.50	0.30 max	0.90–1.40	...	0.80–1.40	...
A4	T30104	0.95–1.05	1.80–2.20	0.50 max	0.90–2.20	0.30 max	0.90–1.40	...	...	...
A6	T30106	0.65–0.75	1.80–2.50	0.50 max	0.90–1.20	0.30 max	0.90–1.40	...	...	...
A7	T30107	2.00–2.85	0.80 max	0.50 max	5.00–5.75	0.30 max	0.90–1.40	0.50–1.50	3.90–5.15	...
A8	T30108	0.50–0.60	0.50 max	0.75–1.10	4.75–5.50	0.30 max	1.15–1.65	1.00–1.50	...	...
A9	T30109	0.45–0.55	0.50 max	0.95–1.15	4.75–5.50	1.25–1.75	1.30–1.80	...	0.80–1.40	...
A10	T30110	1.25–1.50[‡]	1.60–2.10	1.00–1.50	...	1.55–2.05	1.25–1.75	...	...	...
High-carbon, high-chromium, cold-work steels										
D2	T30402	1.40–1.60	0.60 max	0.60 max	11.00–13.00	0.30 max	0.70–1.20	...	1.10 max	...
D3	T30403	2.00–2.35	0.60 max	0.60 max	11.00–13.50	0.30 max	...	1.00 max	1.00 max	...
D4	T30404	2.05–2.40	0.60 max	0.60 max	11.00–13.00	0.30 max	0.70–1.20	...	1.00 max	...
D5	T30405	1.40–1.60	0.60 max	0.60 max	11.00–13.00	0.30 max	0.70–1.20	...	1.00 max	2.50–3.50
D7	T30407	2.15–2.50	0.60 max	0.60 max	11.50–13.50	0.30 max	0.70–1.20	...	3.80–4.40	...

TABLE 1.14 Composition Limits for Principal Types of Tool Steels[76] (*Continued*)

Designation		Composition, %*								
AISI	UNS	C	Mn	Si	Cr	Ni	Mo	W	V	Co
Oil-hardening cold-work steels										
O1	T31501	0.85–1.00	1.00–1.40	0.50 max	0.40–0.60	0.30 max	...	0.40–0.60	0.30 max	...
O2	T31502	0.85–0.95	1.40–1.80	0.50 max	0.50 max	0.30 max	0.30 max	...	0.30 max	...
O6	T31506	1.25–1.55‡	0.30–1.10	0.55–1.50	0.30 max	0.30 max	0.20–0.30	...	...	...
O7	T31507	1.10–1.30	1.00 max	0.60 max	0.35–0.85	0.30 max	0.30 max	1.00–2.00	0.40 max	...
Shock-resisting steels										
S1	T41901	0.40–0.55	0.10–0.40	0.15–1.20	1.00–1.80	0.30 max	0.50 max	1.50–3.00	0.15–0.30	...
S2	T41902	0.40–0.55	0.30–0.50	0.90–1.20	...	0.30 max	0.30–0.60	...	0.50 max	...
S5	T41905	0.50–0.65	0.60–1.00	1.75–2.25	0.50 max	...	0.20–1.35	...	0.35 max	...
S6	T41906	0.40–0.50	1.20–1.50	2.00–2.50	1.20–1.50	...	0.30–0.50	...	0.20–0.40	...
S7	T41907	0.45–0.55	0.20–0.90	0.20–1.00	3.00–3.50	...	1.30–1.80	...	0.20–0.30§	...
Low-alloy special-purpose tool steels										
L2	T61202	0.45–1.00†	0.10–0.90	0.50 max	0.70–1.20	...	0.25 max	...	0.10–0.30	...
L6	T61206	0.65–0.75	0.25–0.80	0.50 max	0.60–1.20	1.25–2.00	0.50 max	...	0.20–0.30§	...
Low-carbon mold steels										
P2	T51602	0.10 max	0.10–0.40	0.10–0.40	0.75–1.25	0.10–0.50	0.15–0.40	...	...	...
P3	T51603	0.10 max	0.20–0.60	0.40 max	0.40–0.75	1.00–1.50	...	...	...	...
P4	T51604	0.12 max	0.20–0.60	0.10–0.40	4.00–5.25	...	0.40–1.00	...	...	...
P5	T51605	0.10 max	0.20–0.60	0.40 max	2.00–2.50	0.35 max	...	...	...	...
P6	T51606	0.05–0.15	0.35–0.70	0.10–0.40	1.25–1.75	3.25–3.75	...	...	...	...
P20	T51620	0.28–0.40	0.60–1.00	0.20–0.80	1.40–2.00	...	0.30–0.55	...	...	...
P21	T51621	0.18–0.22	0.20–0.40	0.20–0.40	0.50 max	3.90–4.25	...	...	0.15–0.25	1.05–1.25Al
Water-hardening tool steels										
W1	T72301	0.70–1.50¶	0.10–0.40	0.10–0.40	0.15 max	0.20 max	0.10 max	0.15 max	0.10 max	...
W2	T72302	0.85–1.50¶	0.10–0.40	0.10–0.40	0.15 max	0.20 max	0.10 max	0.15 max	0.15–0.35	...
W5	T72305	1.05–1.15	0.10–0.40	0.10–0.40	0.40–0.60	0.20 max	0.10 max	0.15 max	0.10 max	...

* All steels except group W contain 0.25 max Cu, 0.03 max P, and 0.03 max S; group W steels contain 0.20 max Cu, 0.025 max P, and 0.025 max S. Where specified, sulfur may be increased to 0.06 to 0.15% to improve machinability of group A, D, H, M, and T steels. † Available in several carbon ranges. ‡ Contains free graphite in the microstructure. § Optional. ¶ Specified carbon ranges are designated by suffix numbers.
Reprinted by permission of ASM International, Materials Park, Ohio.

In these steels, high toughness is the major concern, and hardness is the secondary concern. Among these grades, S5 and S7 are perhaps the most widely used.

5. *Low-alloy special-purpose tool steels (AISI L series)* are similar in composition to the type W tool steels, except that the addition of Cr and other elements renders greater hardenability and wear resistance properties. Type L6 and a low-carbon version of L2 are commonly used for a large number of machine parts.

6. *Mold steels (AISI P series)* are mostly used in low-temperature die casting dies and in molds for the injection or compression molding of plastics.[76]

7. *Water-hardening tool steels (AISI W series).* Among the three compositions listed, W1 is the most widely used as cutting tools, punches, dies, files, reamers, taps, drills, razors, woodworking tools, and surgical instruments in the quenched and tempered condition.

1.8.2.7 Stainless Steels.
Stainless may be defined as complex alloy steels containing a minimum of 10.5% Cr with or without other elements to produce austenitic, ferritic, duplex (ferritic-austenitic), martensitic, and precipitation-hardening grades. AISI uses a three-digit code for stainless steels. Among these steels, austenitic steels and duplex stainless steels are treated in detail in Chap. 10, while precipitation-hardening stainless steels are described in Chap. 6.

1.8.2.8 Maraging Steels.
Maraging steels are a specific class of low-carbon ultrahigh-strength steels which derive their strength not from carbon but from precipitation of intermetallic compounds.[50–52] The commonly available maraging steels contain 17 to 19% Ni, 8 to 12% Co, 3 to 5% Mo, and 0.2 to 1.6% Ti. Since these steels develop very high strength by martensitic transformation and subsequent age-hardening, they are termed maraging steels. (See Chap. 14 for more details.)[77]

1.9 DESIGNATIONS FOR STEELS

A designation is the specific identification of each grade, type, or class of steel by a number, letter, symbol, name, or suitable combination thereof unique to a certain steel; it is used in a specific document as well as in a particular country. In the steel industries, these terms have very specific uses: *Grade* is used to describe chemical composition; *type* is used to denote deoxidation practice; and *class* is used to indicate some other attributes such as tensile strength level or surface quality.[51]

In ASTM specifications, however, these terms are used somewhat interchangeably. For example, in ASTM A 434, *grade* identifies chemical composition and *class* indicates tensile properties. In ASTM A 515, *grade* describes strength level; the maximum carbon content allowed by the specification is dependent upon both the plate thickness and the strength level. In ASTM A 533, *type* indicates chemical analysis, while *class* denotes strength level. In ASTM A 302, *grade* identifies requirements for both chemical composition and tensile properties. ASTM A 514 and A 517 are specifications for high-strength quenched and tempered alloy steel plate for structural and pressure vessel applications, respectively; each has a number of *grades* identifying chemical composition, capable of developing the required mechanical properties. However, all *grades* of both designations have the same composition limits.

By far the most widely used basis for classification and/or designation of steels is the chemical composition. The most commonly used system of designating carbon and alloy steels in the United States is that of the American Iron and Steel Institute and the Society of Automotive Engineers (AISI and SAE) numerical designation. The Unified Numbering System (UNS) is also being employed with increasing rate. Other designations used in the specialized fields include Aerospace Materials Specification (AMS), American Petroleum Institute (API), etc. Two designation systems are discussed below, and others can be found elsewhere.[74]

1.9.1 SAE-AISI Designations

As stated above, the SAE-AISI system is the most widely used designation for carbon and alloy steels. The SAE-AISI system is applied to semifinished forgings, hot-rolled and cold-finished bars, wire rod, seamless tubular goods, structural shapes, plates, sheet, strip, and welded tubing. Table 1.3 lists the SAE-AISI system of numerical designations for both carbon and low-alloy steels.

With few exceptions, the SAE-AISI system uses a four-digit number to designate carbon and alloy steels, specified to chemical composition ranges. There are certain types of alloy steels which are designated by five digits (numerals). Table 1.3 shows the abbreviated listing of four-digit designations of the SAE-AISI carbon and alloy steels. The first digit, 1, of this designation indicates a carbon steel; i.e., carbon steels comprise 1xxx groups in the SAE-AISI system and are subdivided into four different series due to the variance in certain fundamental properties among them. Thus, the plain-carbon steels comprise 10xx series grades (containing 1.00% Mn maximum) (Table 1.4); resulfurized (free-cutting) carbon steels comprise the 11xx series grades (SAE-AISI 1108 to 1151) (Table 1.5); resulfurized and rephosphorized(free-cutting) carbon steels comprise the 12xx series grades (SAE-AISI 1211 to 1215) (Table 1.6); and nonresulfurized high-manganese (up to 1.65%) carbon steels comprise the 15xx series grades (SAE-AISI 1513 to 1560) (Table 1.7). Both 11xx and 12xx groups of steels are produced for applications requiring improved machinability.

Carbon and alloy steel designations showing the letter B inserted between the second and third digits indicate that the steel has 0.0005 to 0.003% B. Likewise, the letter L inserted between the second and third digits indicates that the steel has 0.15 to 0.35% Pb for enhanced machinability. Sometimes, the prefix M is used for merchant-quality steels, and the suffix H is used to comply with standard hardenability requirements. In alloy steels, the prefix letter E is used to designate steels which are produced by the basic electric furnace process with special practices.

The major alloying elements in an alloy steel are indicated by the first two digits of the designation (Table 1.3). Thus, the first digit 2 denotes a nickel steel; 3 a nickel-chromium steel; 4 a molybdenum, chromium-molybdenum, nickel-molybdenum, or nickel-chromium-molybdenum steel; 5 a chromium steel; 6 a chromium-vanadium steel; 7 a tungsten-chromium steel; 8 a nickel-chromium-molybdenum steel; and 9 a silicon-manganese steel or a nickel-chromium-molybdenum steel. In the case of a simple alloy steel, the second digit represents the approximate percentage of the predominant alloying element. For example, 2520 grade indicates a nickel steel of approximately 5% Ni (and 0.2% carbon).

The last two digits of four-numeral designations and the last three digits of five-numeral designations indicate the approximate carbon content of the allowable carbon range in hundredths of a percent. For example, 1020 steel indicates a plain-carbon steel with an approximate mean 0.20% carbon varying within acceptable carbon limits of 0.18 and 0.23%. Similarly, 4340 steels are Ni-Cr-Mo steels and contain an approximate mean of 0.40% carbon varying within an allowable carbon range of 0.38 and 0.43%; and 51100 steel is a chromium steel with an approximate mean of 1.00% carbon varying within an acceptable carbon range of 0.98 and 1.10%.[5,45,50]

1.9.2 UNS Designations

The Unified Numbering System has been developed by the American Society for Testing and Materials (ASTM E 527), the Society of Automotive Engineers (SAE J1086), and several other technical societies, trade associations, and United States government agencies. A UNS number, which is a designation of chemical composition and not a specification, is assigned to each chemical composition of the standard carbon and alloy steel grades for which controlling limits have been established by the SAE-AISI.[48,50,78]

The UNS designation consists of a single-letter prefix followed by five numerals (digits). The letters denote the broad class of alloys; the numerals define specific alloys within that class. The prefix letter G signifies standard grades of carbon and alloy steels; the prefix letter H indicates standard grades that meet certain hardenability requirements (limits) (SAE-AISI H steels); the prefix T includes tool steels, wrought and cast; the prefix letter S relates to heat- and corrosion-resistant steels (including stainless steel), valve steels, and iron-base superalloys; the prefix letter J is used for cast steels (except tool steels); the prefix letter K identifies miscellaneous steels and ferrous alloys; and the prefix W denotes welding filler metals (for example, W00001 to W59999 series represent a wide variety of steel compositions).[74] The first four digits of the UNS number usually correspond to the standard SAE-AISI designations, while the last digit (except zero) of the five-numeral series denotes some additional composition requirements, such as boron, lead, or nonstandard chemical ranges. Tables 1.4 to 1.8 list the UNS numbers corresponding to SAE-AISI numbers for various standard carbon and alloy steels, respectively, with composition ranges. Tables 1.10 to 1.12 list the UNS numbers for ultrahigh-strength steel, bearing steel, and heat-resistant Cr-Mo steels, respectively.

1.10 CAST IRON CLASSIFICATIONS

There are six generic types of cast irons. In each type there are several grades, such as (1) white iron, (2) gray iron, (3) ductile iron, (4) compacted graphite iron, (5) malleable iron, and (6) high-alloy irons. Table 1.15 lists typical composition ranges for five types of unalloyed cast irons, and Fig. 1.14 illustrates the ranges in carbon and silicon contents for four types of cast irons and for steel. The lower dashed line defines the upper limit of carbon contents (or the limit of carbon solubility) in steel.[7]

Each of these unalloyed types may be moderately alloyed or heat-treated without affecting its basic classifications. The high-alloy irons, commonly with alloy-

TABLE 1.15 Chemical Composition Range for Typical Unalloyed Cast Irons[7]

Type of Iron	Percent				
	Carbon	Silicon	Manganese	Sulfur	Phosphorous
White	1.8–3.6	0.5–1.9	0.25–0.8	0.06–0.2	0.06–0.2
Malleable (cast white)	2.2–2.9	0.9–1.9	0.15–1.25	0.02–0.2	0.02–0.1
Gray	2.5–4.0	1.0–3.0	0.2–1.0	0.02–0.25	0.02–1.0
Ductile	3.0–4.0	1.8–3.0	0.1–1.0	0.005–0.035	0.015–0.1
Compacted graphite	2.5–4.0	1.0–3.0	0.2–1.0	0.01–0.03	0.01–0.1

Courtesy American Cast Metals Association, Des Plaines, Illinois.

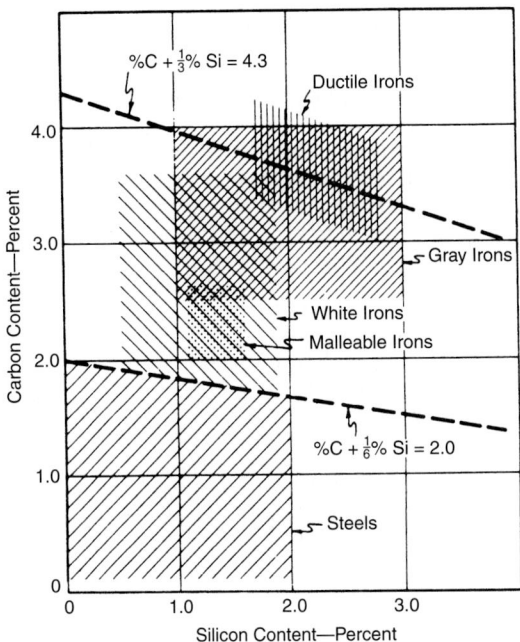

FIGURE 1.14 The approximate range in carbon and silicon content of ferrous alloys.[7] (*Courtesy of American Cast Metals Association, Des Plaines, Illinois.*)

ing additions in excess of 3%, can also be classified commercially as white, gray, or ductile irons. (See Sec. 3.11.5 for more details.)

In addition to the above six basic types, there are other specific forms of cast irons to which specific names have been assigned. *Chilled iron* is white iron that is produced in a selected or entire area of the casting by very rapid cooling through the solidification range. The principal difference in microstructure between chilled iron and white iron is that the former is fine-grained and displays directionality normal to the chilled surface, while the latter is usually coarse-grained, randomly oriented, and white throughout the cross section, even in substantially heavy sections. This difference suggests the composition difference between these two types of abrasion-resistant irons. In chilled iron, the gray iron composition is cooled so rapidly at one or more faces through the eutectic temperature that the white iron solidifies, growing inward from the chilled surface. In the case of white iron, the composition is so low in carbon equivalent (CE) or very rich in alloy content that the production of gray iron is impossible even at slow cooling that is commonly present in the heaviest section of the casting.[79]

An area of the casting that solidifies at a rate intermediate between those of chilled iron and gray iron consists of both graphite and carbide. This structure is called *mottled iron* and exhibits mixed microstructural and fracture-surface characteristics.

1.10.1 The Iron–Iron Carbide–Silicon System

To understand the nature of the solidification behavior of cast irons, it is necessary to know some features of a section of the ternary Fe-Fe$_3$C-Si phase diagram at 2% Si which approximates the silicon content in many cast irons (Fig. 1.15a). Figure 1.15b shows the estimated Fe-graphite-silicon and Fe-Fe$_3$C-silicon diagrams at 2.5% Si.[79a] A comparison of Figs. 1.3 and 1.15 shows that silicon addition significantly shifts the eutectoid composition, the eutectic composition, and the limit of carbon solubility in austenite to lower carbon content. The introduction of silicon also changes the eutectoid and eutectic reaction temperatures from single values and at lower temperatures in the Fe-Fe$_3$C system to a range of temperatures and at higher temperatures in the Fe-Fe$_3$C-Si system. In general, the temperature range over which these reactions occur in the Fe-Fe$_3$C-Si system increases as the carbon and/or silicon content increases.

The solidification of certain compositions occurs solely in the stable system, producing iron and graphite as the reaction products, and it sometimes occurs partly in the metastable system, producing iron and iron carbide as the reaction products. These compositions include the gray, ductile, and compacted graphite cast irons.

1.10.1.1 Carbon Equivalent. The upper dashed line in Fig. 1.14 shows the eutectic composition for Fe-C-Si systems. The eutectic occurs at 4.3% carbon in the absence of silicon and at decreasing carbon content as the silicon content in iron is increased. This linear relationship can be expressed by the following equation:

$$\%\,C + \frac{1}{3}\%\,Si = 4.3 \qquad (1.6)$$

To simplify the evaluation of the effect of silicon in unalloyed cast irons, the carbon equivalent (CE) term is often incorporated and is expressed as

$$CE = \%\,C + \frac{1}{3}\%\,Si \qquad (1.7)$$

The CE in cast iron denotes how near a given analysis is to that of the eutectic composition. A CE of 4.3 indicates a eutectic cast iron. A CE of 3.8 indicates a cast iron of lower carbon and silicon contents (hypoeutectic) than the eutectic composition, while a CE of 4.7 indicates a cast iron of greater carbon and silicon contents (hypereutectic) than the eutectic composition. When CE is near the eutectic, the solidification occurs within a narrow temperature range. In hypoeutectic irons, the lower the CE, the greater the tendency for the formation of white or mottled iron

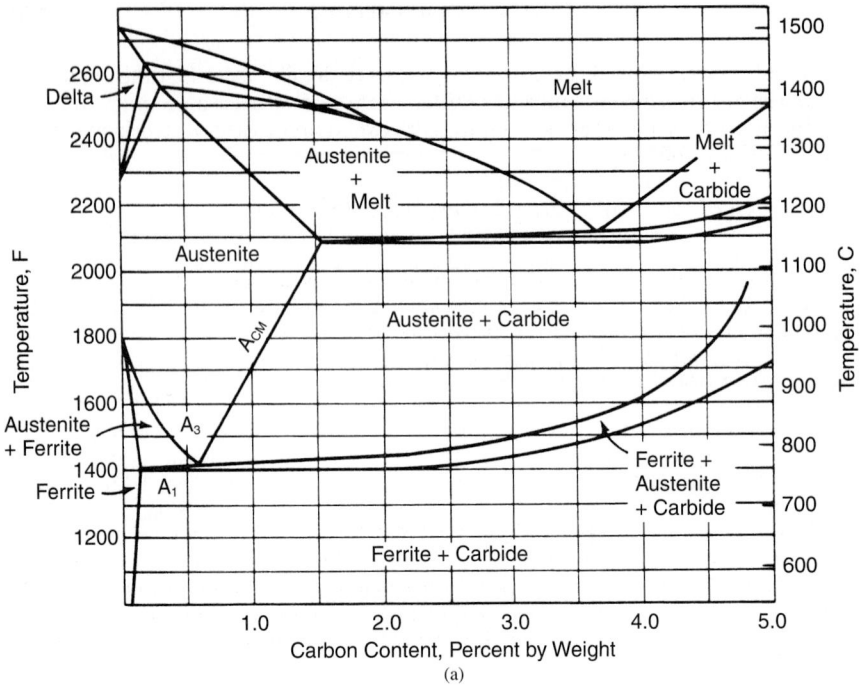

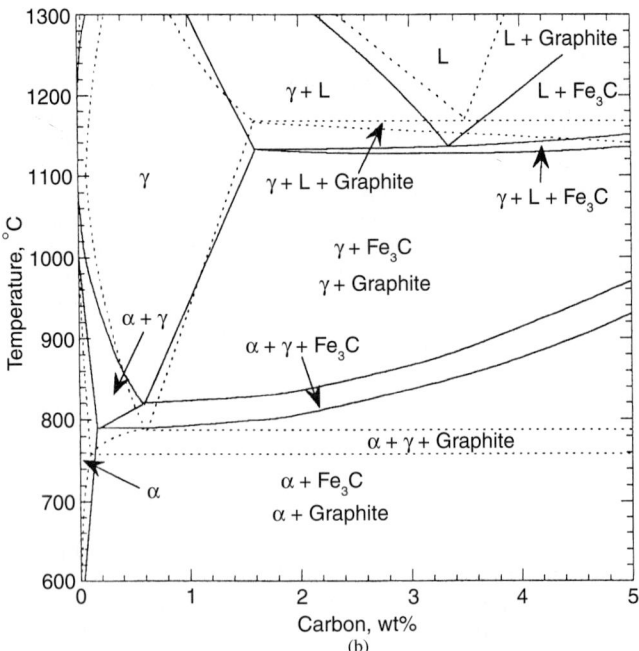

FIGURE 1.15 (*a*) The iron-iron carbide-silicon ternary phase diagram sectioned at 2% silicon.[7] (*b*) Estimated phase diagram for the Fe-Fe₃C (solid lines) and Fe-graphite (dotted lines) system containing 2.5% silicon.[79a] [(*a*) *Courtesy of American Cast Metals Association, Des Plaines, Illinois;* (*b*) *Courtesy of C. K. Syn, D. R. Lesuer, and O. D. Sherby.*]

on solidification. In hypereutectic irons, the larger the CE, the greater the tendency for the precipitation of proeutectic graphite on solidification.

The CE can also be expressed in the presence of appreciable quantities of phosphorus in cast irons as

$$CE = \%C + \frac{\%Si + \%P}{3} \tag{1.8}$$

The total carbon and silicon contents of the alloy, as related to the CE value, establish both the solidification temperature range of the alloy and the foundry characteristics of the alloy and its properties. It must be noted that irons of a constant CE, but with quite different carbon and silicon contents, will have different casting properties.[79]

1.10.2 Eutectic Cell

Cast iron solidification involves the initial formation of graphite nuclei from molten iron into eutectic solids. Subsequently, the growth of a spherical lump of eutectic occurs from individual graphite nuclei, which are usually called *eutectic cells* or *eutectic colonies*. Mostly their size is influenced by the solidification behavior of cast ingot. No direct relation between cell size and service properties has yet been established.[7,80]

1.10.3 White Cast Iron

If the chemical composition of the alloy lies in the white cast iron range (Table 1.15) and the solidification rate is quite rapid, a white cast iron will be produced. In white cast iron, the carbon in the molten iron combines with the iron and, upon solidification, forms iron carbide or cementite, which is a hard and brittle compound and dominates the microstructure of white iron (Fig. 1.16).[8] It is named from the white crystalline fractured surface that appears because of its freedom from graphite. It has an exceptionally high compressive strength and a very high abrasive wear resistance, and it retains its hardness for limited periods even up to red heat.[8] The iron-carbon diagram of Fig. 1.3 and the kinetic diagram of Fig. 3.91 show the difference in thermodynamic equilibrium and the growth kinetics of the two eutectic structures found in white and gray irons.

1.10.4 Gray Cast Iron

When the composition of the cast iron is in the gray cast iron range (Table 1.15) and the solidification rate is appropriate, the carbon in the iron separates and forms distinct graphite flake morphologies. These graphite flakes become interconnected within each eutectic cell as the gray iron solidifies, as shown in the scanning electron micrograph of a deeply etched surface (Fig. 1.17a). This characteristic flake morphology exerts a marked influence on the mechanical and physical properties of gray irons. Because of these interconnected flakes, gray iron can be considered as a composite material, and it really shows many of the properties characteristic of composites.[81]

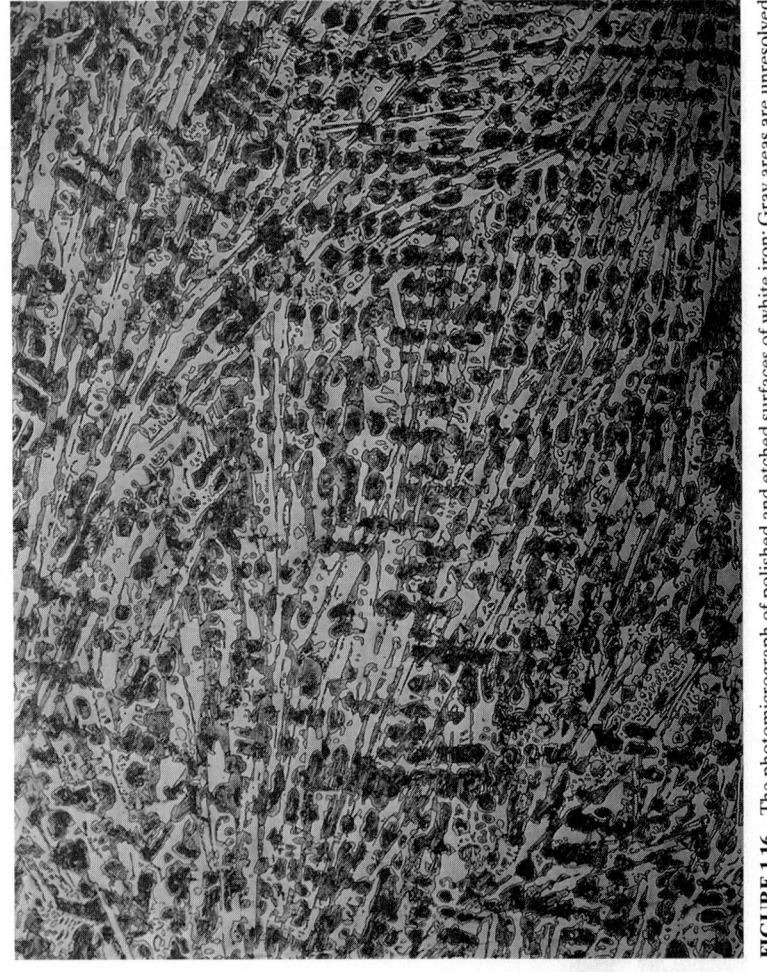

FIGURE 1.16 The photomicrograph of polished and etched surfaces of white iron: Gray areas are unresolved pearlite, and white areas are iron carbide.[7] (*Courtesy of American Cast Metals Association, Des Plaines, Illinois.*)

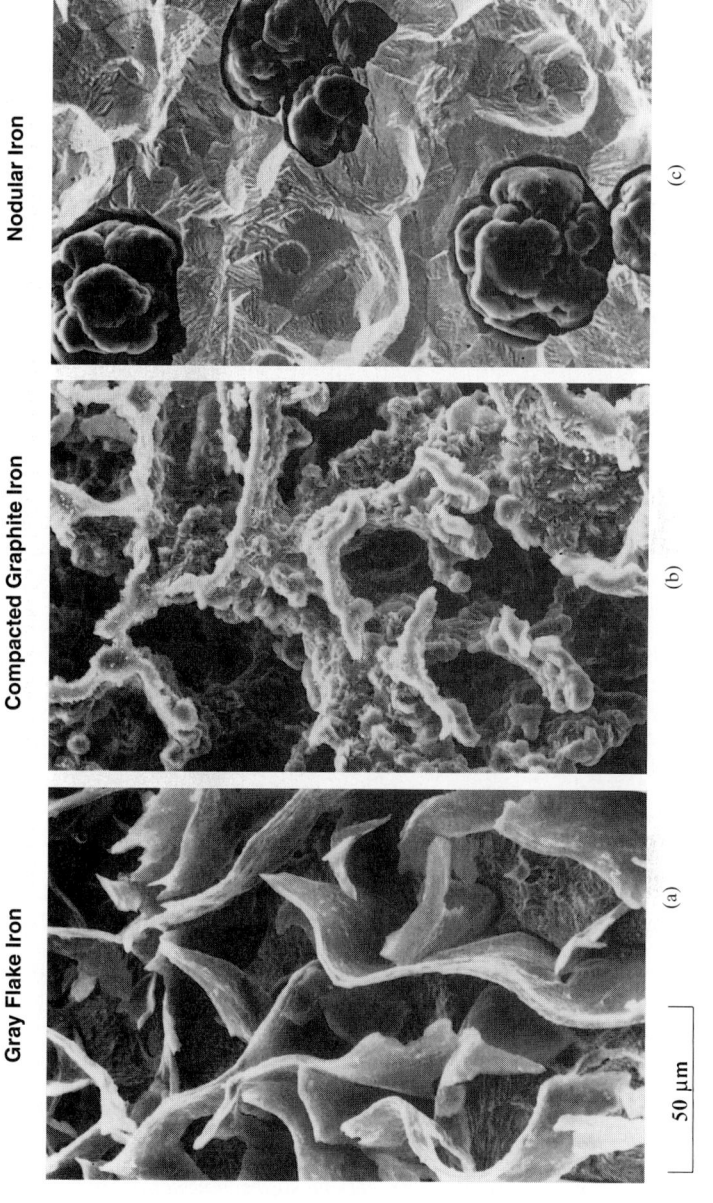

Gray Flake Iron

Compacted Graphite Iron

Nodular Iron

50 μm

(a)

(b)

(c)

FIGURE 1.17 Deep-etched scanning electron micrographs showing the true three-dimensional morphology of graphite particles in (a) gray cast iron, (b) compacted graphite cast iron, and (c) ductile (or nodular) iron. (*Courtesy of SinterCast S.A.*)

The number of positions at which graphite nucleation occurs during solidification and the number of eutectic cells are increased by inoculation of the molten iron.[8] When gray iron is subsequently broken, most of the fractured surfaces occur along the graphite flakes, thereby producing its characteristic gray appearance. Since the production of gray iron castings is the largest of all the iron castings, the generic term *cast iron* is often improperly used to mean gray iron.

Several factors influence the nucleation and growth of the graphite flakes. The foundryman employs special techniques to obtain a particular form of graphite that will improve the required properties. A standard classification system for variations in graphite structures together with the length of the flakes has been developed by the ASTM in specification A-247, as shown in Fig. 1.18. According to this system, graphite flakes are classified progressively into the Type A, coarsest, completely random distribution of *flake graphite* of uniform size and orientation and occurring at higher temperature; Type B, a coarse, radially growing Type A flake, called *rosette or cluster pattern*; Type C, random, thick, coarse flakes, primary *Kish graphite* (found in hypereutectic irons); Type D, fine *undercooled graphite*, forming at low temperature; and Type E, same as Type D except that the graphite exhibits a degree of preferred orientation. Type E forms in strongly hypoeutectic irons of low CE. The amount of graphite present, its size, and its distribution have important bearings on the properties of iron (see also Chap. 12).

It is customary to specify the desired properties and not the factors influencing them. Gray irons are usually classified in terms of minimum tensile strength expressed in kilopounds per square inch by the ASTM specification A-48, which has a range from class 20 (minimum 20-ksi or 138-MPa tensile strength) to class 60 (minimum 60-ksi or 414-MPa tensile strength).

The following properties of gray cast irons increase with increasing tensile strength from class 20 to class 60: (1) all strengths at elevated temperatures, (2) ability to be machined to a fine finish, (3) modulus of elasticity, and (4) ability to be cast in thin sections.[79]

1.10.5   Ductile Iron

Ductile iron, developed in 1940s, has grown in relative importance over the last two decades and now constitutes about 25% of cast iron production in most industrialized countries.[82] Ductile iron has its free carbon formed as graphite spheroids (or spherulites) rather than as flakes. These nodules act as "crack arresters" and make ductile iron "ductile."[83] It is often called *nodular iron* in the United States and *spherulitic* or *spheroidal graphite* (SG) iron in the United Kingdom. The spheroidal graphite in these irons is obtained by adding a very small but definite amount of magnesium (most widely used spheroidizing element) and/or rare earths (such as optimum cerium in the range of 0.006 to 0.010% for low-cerium rare earths and ~0.015 to 0.02% for high-cerium rare earths, lanthanum, yttrium, etc.) to molten iron of a proper composition [with CE of at least 4.3, low S level (<0.01%), and controlled trace and minor elements] in a processing step called *nodulizing*.

In the nodulizing treatment, magnesium reacts with sulfur and oxygen present in the melt to avoid interference of these elements in the formation of spheroidal graphite. The nodulizing step is usually accompanied by, or followed by, an inoculating (nucleating) addition (in the range of 0.25 to 1%), usually ferrosilicon alloy (in wire or powder form) containing 60 to 80% Si plus minor elements such as Ca,

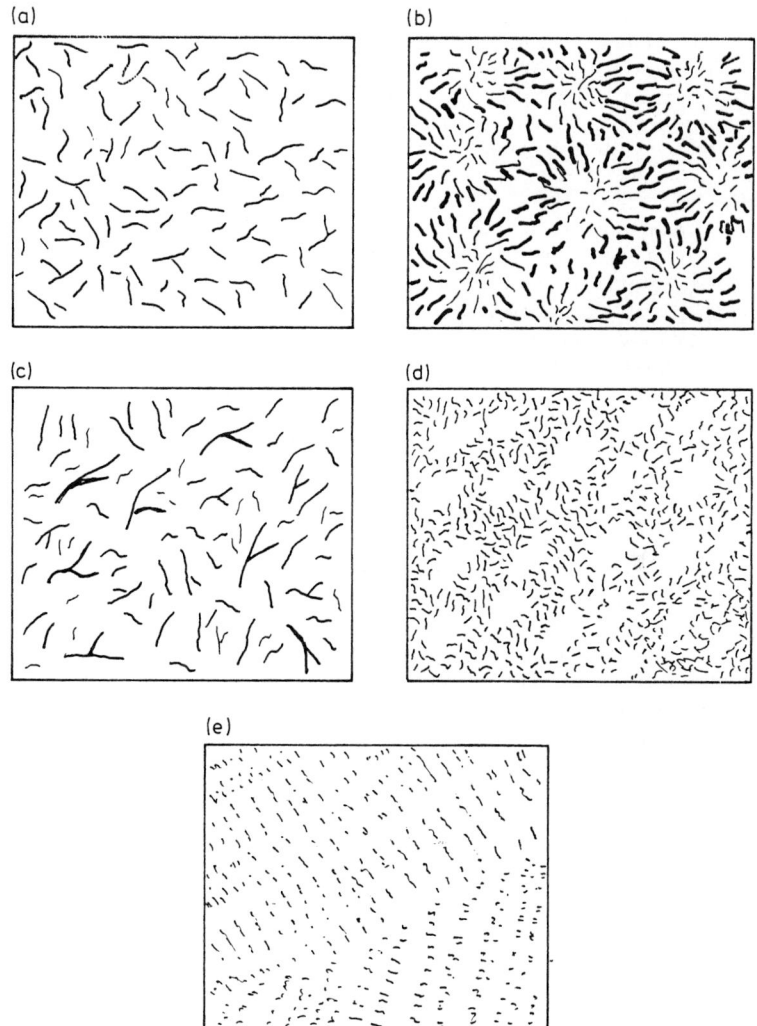

FIGURE 1.18 Types of graphite flakes in gray iron defined by ASTM Standard A-247. (*a*) Type A, uniform distribution, random orientation. (*b*) Rosette grouping, random orientation. (*c*) Type C, superimposed flake size, random orientation. (*d*) Type D, interdendritic segregation, random orientation. (*d*) Type E, interdendritic segregation, preferred orientation.

Ce, Al, etc. to realize the iron's graphitizing potential and to control the size and distribution of the graphite spherulites.

The precise level of residual Mg, $Mg_{residual}$ (usually 0.03 to 0.05%) required to produce spheroidal graphite is a function of the cooling rate; a higher cooling rate requires less Mg. The amount of Mg to be added in the iron depends on the initial sulfur content $S_{initial}$ and the recovery of Mg, η, in the specific process employed:[79]

FIGURE 1.19 Unetched graphite spherulite.[84] (*Courtesy of BCIRA, International Center for Cast Metal Technology, England.*)

$$Mg_{added} = \frac{0.75 S_{initial} + Mg_{residual}}{\eta} \qquad (1.9)$$

A very low $Mg_{residual}$ level produces inadequate nodularity, thereby resulting in a deterioration of its mechanical properties. A very high $Mg_{residual}$ content favors carbide precipitation.

Each graphite spherulite present in SG irons consists of a radial array of graphite crystallites growing from a common nucleus. Usually the basal planes of the graphite are perpendicular to the growth direction (i.e., the radii) of the spherulites. Figures 1.17c and 1.19 show the deep-etched and unetched graphite spherulite, respectively, at a high magnification.[84] The matrix microstructure consists of ferrite, ferrite and pearlite, pearlite, martensite, tempered martensite, ferrite and austenite (as austempered bainite), and austenite (Fig. 1.20).[83] Consistent nodularity, nodule count, and low porosity and carbide content determine the mechanical properties (such as tensile strength) of ductile irons.

ASTM A-576 defines seven basic morphologies of ductile iron. Types I and II are desirable nodule shapes and are usually specified to be >85% in an iron. Type III is cluster graphite observed in malleable irons. Type IV is typical of compacted graphite cast irons. Types V and VI are degenerate types and are undesirable. Type VII is the flake graphite defined earlier in Fig. 1.18. Minor elements promoting undesirable graphite morphology are listed in Table 1.17. Quasi-flake graphite structures occur with inadequate addition of spheroidizing element. Intercellular structures form from the segregation of flake graphite promoting elements to cell boundaries; Mg exceeding 0.1 wt % may act in this manner. Chunky graphite structure, being intermediate between spheroidal and flake graphite, can be eliminated by, and nodular graphite can be restored by, the controlled addition of elements from column 2.

The relatively high strength and toughness, better or similar castability, surface hardenability, stiffness, and corrosion resistance of ductile iron over gray iron or malleable iron make it versatile and superior in many structural applications. The different grades of ductile iron associated with a wide range of properties can be

MATRIX

Ferritic Grade 5	Ferritic- pearlitic Grade 3	Pearlitic Grade 1	Martensitic (With retained austenite)	Tempered Martensitic	ADI Grade 150	ADI Grade 230	Austenitic
60,000 p.s.i. (414 MPa)	80,000 p.s.i. (552 MPa)	100,000 p.s.i. (690 MPa)	N.A. *	115,000 p.s.i. (793 MPa)	150,000 p.s.i. (1050 MPa)	230,000 p.s.i. (1600 MPa)	45,000 p.s.i. (310 MPa)

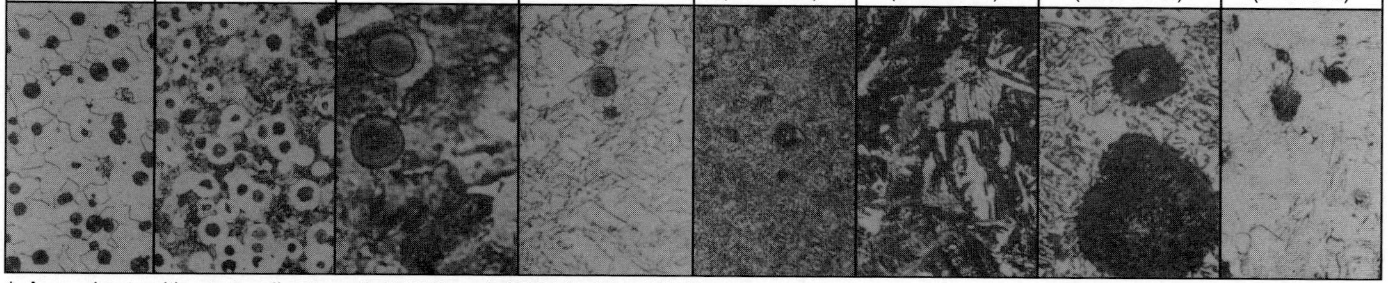

* Approximate ultimate tensile strength 87,000 p.s.i. (600 MPa) hard, brittle.

FIGURE 1.20 Microstructure and tensile strengths for various types of ductile iron. (Note that the magnifications are different.) (*Courtesy of Ductile Iron Society, QIT America.*)

1.63

produced either by subsequent heat treatment that modifies the matrix structure or by addition of special alloying elements such as nickel, chromium, copper, and molybdenum in the casting.[37] (See also Chaps. 3, 12, and 13.)

Thus, small amounts of REM (Ce, La) are often used to restore the graphite nodule count and nodularity in ductile iron containing subversive elements such as Pb, Sb, and Ti. On the other hand, rare earths in large concentrations may result in the problem with chunky graphite in large cast products and chill formation in thin section irons, with subsequent deterioration in the mechanical properties.[85]

The ASTM specifications A-536 and A-476 used to designate the grade of ductile iron by mechanical properties include the first two numbers representing tensile strength in ksi, the second two numbers representing yield strength in ksi, and the last two numbers representing elongation in percent. The common grades are (1) 60-40-18, (2) 60-42-10, (3) 65-45-12, (4) 70-50-05, (5) 80-55-06, (6) 80-60-03, (7) 100-70-03, and (8) 120-90-02. Grade 80-60-03 is extensively used in applications where high ductility is not a concern. Grades 65-45-12 and 60-40-18 are used when high ductility and impact strength are important. Grades 60-42-10 and 70-50-05 are used for special applications such as annealed pipe or as-cast pipe fittings.

1.10.6 Compacted Graphite (CG) Iron

A recently developed type of cast iron is called *compacted graphite iron* or *vermicular graphite iron*. CG iron has a graphite structure intermediate between that of gray iron and ductile iron (Fig. 1.17b).[86] The "worm-shaped" compacted graphite particles are elongated and interconnected within each eutectic cell (similar to flake graphite morphology of gray iron), thereby providing high thermal conductivity and vibration damping, and have the rounded or blunt edges (similar to spheroidal graphite structure of ductile iron) providing high strength, stiffness, and excellent fatigue property. Thus it is used to fill the property gap between flake and nodular graphite irons. They exhibit superior tensile strength, stiffness, ductility, surface finish, fatigue life, durability, impact and elevated-temperature properties, and wear resistance to gray irons. Their castability, machinability, dimensional stability, and thermal conductivity are superior to those for ductile iron. Their combined resistance to crazing, cracking, and distortion exceeds that of gray and ductile irons. No specifications have been formulated for CG irons.[82]

Like ductile iron, CG iron is usually produced by the addition of cerium or other rare earth elements, the presence of ~0.06 to 0.13% titanium, and maintaining a sulfur level ≤0.03% in the treated iron with a small excess of uncombined magnesium (in the range of 0.015 to 0.025%). Undertreatment with magnesium or very high initial sulfur level results in the production of flake graphite of gray iron in the structure, whereas the converse leads to the production of graphite spheroids of ductile iron in the structure[86] (see also Chap. 10). Their successful production requires careful melt treatment to ensure that the flake graphite is absent and spheroidal graphite is not more than 20%; i.e., there should at least 80% of the eutectic compacted graphite or ASTM Standard A247 Type IV.[82]

1.10.7 Malleable Iron

The necessary requirements for *malleable cast iron* production are low CE and graphitization potential to ensure that it solidifies as white iron with a metastable carbide structure in a pearlitic matrix. Malleable cast iron is characterized by having most of its carbon present as irregularly shaped nodules of graphite. This form of graphite is called temper carbon nodules (Type III in ASTM Standard A247) because it is formed by the decomposition of Fe_3C in the solid state after an extended heat treatment of white cast iron of suitable composition in a controlled-atmosphere furnace to a temperature above the eutectoid temperature, usually 900°C (1650°F).[7,87] It is shown in scanning electron micrograph of a deeply etched surface (Fig. 1.21). Based on the heat treatment practice, the matrix structure around temper carbon may be ferrite, pearlite, or tempered martensite or a combination of these (see Chap. 12).

The rapid solidification requirement for white cast iron restricts the thickness of the castings that can be produced into malleable iron by the malleabilizing process, and the high solidification shrinkage of white iron restricts the production of complex malleable iron castings. The allowable thickness of white iron castings appropriate for proper malleabilizing process can, however, be increased by the balanced addition of trace elements such as Bi (0.001%) or B as ferroboron (0.001%). Te addition of 0.0005 to 0.001% is made to suppress mottle formation. It is more effective if added together with Cu or Bi. Residual B should not exceed 0.0035% in order to avoid graphite cluster alignment and carbide formation. Addition of 0.005% Al to the pouring ladle remarkably improves graphitization without favoring mottle.[88] The chemical composition of malleable iron generally conforms to the ranges given in Table 1.15. Small amounts of Cr (0.01 to 0.03%), B (0.002%), Cu (~1.0%), Ni (0.5 to 0.8%), and Mo (0.35 to 0.5%) are also sometimes present.

1.10.8 High-Alloy Iron

This group of cast irons includes high-alloy white irons, high-alloy gray irons, and high-alloy ductile irons. Usually, malleable irons are not highly alloyed because

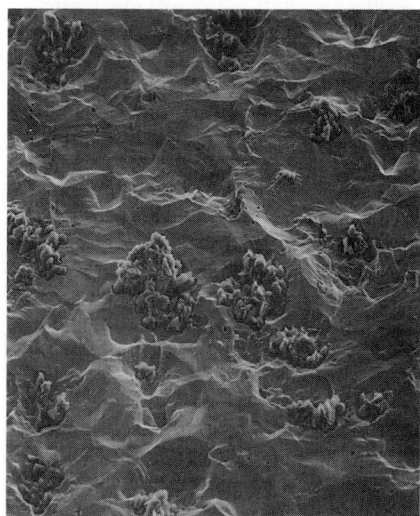

FIGURE 1.21 Scanning electron micrograph of a deeply etched surface of a malleable cast iron, showing the graphite nodules.[7] (*Courtesy of American Cast Metals Association, Des Plaines, Illinois.*)

there is interference with the malleabilizing process. CG irons have not been studied well in the highly alloyed condition.

High-alloy irons are used in applications requiring high strength, hardness, and hardenability or improved resistance to abrasive wear, heat, and chemical corrosion properties and whenever special-purpose physical properties such as low thermal expansion and nonmagnetic properties are in demand. These irons are produced by foundries specializing in their production because alloy content in the range of 3 to 30% or more requires the use of special melting equipment, casting technique, and quality control. These irons are normally specified by their chemical composition; however, mechanical property requirements may also be employed.

Table 1.16 lists ranges of alloy content for various types of alloy cast irons which can be grouped into abrasion-resistant white irons, corrosion-resistant cast irons, and heat-resistant gray and ductile irons. The listed ranges act only to distinguish the types of alloy used in specific types of applications.[79]

Corrosion-resistant irons derive their resistance to chemical attack mainly from their high-alloy content. They are called high Si, high Cr, high Ni-Cr (or Ni-resist) gray, and high Ni-Cr (or Ni-resist) ductile irons. Three alloying elements dominate the composition (Si up to 15%, Cr up to 28%, and Ni up to 35%). On the basis of the composition, inoculation practice, and cooling rate, a corrosion-resistant iron can be white, gray, or nodular, and matrix may be ferritic, pearlitic, martensitic, or austenitic.

Heat-resistant irons derive their resistance to high-temperature oxidation and scaling from their high-alloy content and initial microstructure plus the stability of the carbon-containing phase. They are usually ferritic or austenitic gray or ductile irons. Seldom are they ferritic and austenitic white iron grades.

Alloy white cast irons such as high Ni-Cr (or Ni-hard), high Cr-Mo, and high Cr white cast irons find wide applications for their excellent abrasive wear resistance. Their wear properties depend on the extent of formation of two very hard eutectic carbides $(FeCr)_7C_3$ and $(FeCr)_3C$ in a strong supporting matrix during solidification.[89] In case of high Ni-Cr white irons, C is held in the range of 3.2 to 3.6% for optimum abrasion resistance, but it is restricted to 2.7 to 3.2% for optimum impact loading applications. The most recently developed iron in this group is Ni-hard 4. High Cr-Mo white irons are available either as cast with an austenitic or austenitic/martensitic matrix, or heat-treated to give a predominantly martensitic matrix for maximum abrasion resistance and toughness. Heat-treated heavy castings are used in the mining, coal, and mineral processing industries.[82] Superior abrasive wear resistance to low-alloy and Ni-Cr white irons, combined with moderate impact and corrosion resistance, and low production costs, makes high Cr white cast irons attractive candidates in the mining and mineral industry for crushing, processing, cement production, and paper and pulp manufacturing industries for applications in the grinding, crushing, screening, milling, and pumping apparatuses used to process hard materials such as ore, coal, gravel, and cement.[89–91]

1.11 ALLOYING ELEMENTS IN GRAY CAST IRON

Addition of alloying elements in common cast iron influences graphitization potential, structure and properties of the matrix, and, to some extent, mechanical

TABLE 1.16 Ranges of Alloy Content for Various Types of Alloy Cast Irons[79]

Description	Composition, %[a]									Matrix structure, as cast[c]
	TC[b]	Mn	P	S	Si	Ni	Cr	Mo	Cu	
Abrasion-resistant white irons										
Low-carbon white iron[d]	2.2–2.8	0.2–0.6	0.15	0.15	1.0–1.6	1.5	1.0	0.5	e	CP
High-carbon, low-silicon white iron	2.8–3.6	0.3–2.0	0.30	0.15	0.3–1.0	2.5	3.0	1.0	e	CP
Martensitic nickel-chromium iron	2.5–3.7	1.3	0.30	0.15	0.8	2.7–5.0	1.1–4.0	1.0	...	M,A
Martensitic nickel, high-chromium iron	2.5–3.6	1.3	0.10	0.15	1.0–2.2	5–7	7–11	1.0	...	M,A
Martensitic chromium-molybdenum iron	2.0–3.6	0.5–1.5	0.10	0.06	1.0	1.5	11–23	0.5–3.5	1.2	M,A
High-chromium iron	2.3–3.0	0.5–1.5	0.10	0.06	1.0	1.5	23–28	1.5	1.2	M
Corrosion-resistant irons										
High-silicon iron[f]	0.4–1.1	1.5	0.15	0.15	14–17	...	5.0	1.0	0.5	F
High-chromium iron	1.2–4.0	0.3–1.5	0.15	0.15	0.5–3.0	5.0	12–35	4.0	3.0	M,A
Nickel-chromium gray iron[g]	3.0	0.5–1.5	0.08	0.12	1.0–2.8	13.5–36	1.5–6.0	1.0	7.5	A
Nickel-chromium ductile iron[h]	3.0	0.7–4.5	0.08	0.12	1.0–3.0	18–36	1.0–5.5	1.0	...	A
High-resistant gray irons										
Medium-silicon iron[i]	1.6–2.5	0.4–0.8	0.30	0.10	4.0–7.0	...	...	...	...	F
Nickel-chromium iron[g]	1.8–3.0	0.4–1.5	0.15	0.15	1.0–2.75	13.5–36	1.8–6.0	1.0	7.5	A
Nickel-chromium-silicon iron[j]	1.8–2.6	0.4–1.0	0.10	0.10	5.0–6.0	13–43	1.8–5.5	1.0	10.0	A
High-aluminum iron	1.3–2.0	0.4–1.0	0.15	0.15	1.3–6.0	...	20–25 Al	...	...	F
Heat-resistant ductile irons										
Medium-silicon ductile iron	2.8–3.8	0.2–0.6	0.08	0.12	2.5–6.0	1.5	...	2.0	...	F
Nickel-chromium ductile iron[h]	3.0	0.7–2.4	0.08	0.12	1.75–5.5	18–36	1.75–3.5	1.0	...	A
Heat-resistant white irons										
Ferritic grade	1–2.5	0.3–1.5	...	...	0.5–2.5	...	30–35	...	...	F
Austenitic grade	1–2.0	0.3–1.5	...	...	0.5–2.5	10–15	15–30	...	...	A

[a] Where a single value is given rather than a range, that value is a maximum limit. [b] Total carbon. [c] CP, coarse pearlite; M. martensite; A, austenite; F, ferrite. [d] Can be produced from a malleable iron base composition. [e] Copper can replace all or part of the nickel. [f] Such as Duriron, Durichlor 51, Superchlor. [g] Such as Ni-resist austenitic iron (ASTM A 436). [h] Such as Ni-resist austenitic ductile iron (ASTM A439). [i] Such as Silal. [j] Such as Nicrosilal.
Reprinted by permission of ASM International, Materials Park, Ohio.

TABLE 1.17 Effect of Minor Elements on the Nodular Graphite Structure[82]

Beneficial	Intercellular flake	Chunky	Other deleterious elements
Mg, Ce, La, Ca	Bi, Cu, Al	Ce, Ca	Zr, Zn, Se
Other spheroidizers	Pb, Sb, Sn, Ag, Cd	Si, Ni	Ti, N, S, O

properties. The main elements in terms of their positive graphitization potential (in descending order from left to right) are C, Sn, P, Si, Al, Cu, and Ni. Similarly the main elements in terms of negative graphitizing potential (in ascending order from left to right) are Mn, Cr, Mo, and V.

Sulfur. Sulfur present in small amounts forms sulfide, which acts as a sulfide substrate for graphite nucleation. However, sulfur content in excess of that required for the formation of substrates hinders the growth rate of graphite flakes and, in extreme cases, leads to adequate undercooling and carbide formation,[79] thereby tending to produce white iron.

Manganese. When manganese is not present in the iron, the sulfur will combine with iron to form sulfide, FeS, usually at the grain boundaries on solidification. When an approximate amount of manganese is present, precipitation of manganese sulfide MnS or manganese-iron-sulfide occurs as randomly dispersed angular particles. These particles have little effect on the solidification behavior and mechanical properties of cast irons. An excess amount of manganese favors undercooling and carbide precipitation, which, in turn, promotes pearlitic structure and increases strength and hardness.

Phosphorus. A sufficient amount of phosphorus present in cast iron forms steadite, a eutectic of iron and phosphorus, on solidification. This is a hard and brittle constituent, which segregates at the grain boundaries. An increase in the amount of phosphorus (with 0.5% or above) in a cast iron can increase the cellular network of steadite, which can increase its hardness, brittleness, and wear resistance but decrease the machinability and impact strength.

Other Minor Elements. Rare earths promote graphite nucleation when present with sulfur; however, in excess quantities, they can cause large undercooling and carbide formation. Tellurium, lead, and bismuth have similar effects.

In moderately alloyed gray iron, compositions lie within the following ranges: 0.2 to 0.6% Cr, 0.2 to 1% Mo, 0.1 to 0.2% V, 0.6 to 1% Ni, 0.5 to 1.5% Cu, and 0.04 to 0.08% Sn. These alloying elements may be used to meet the particular hardness and microstructural requirements or to provide the properties needed for the specified service conditions.

1.12 ALLOYING ELEMENTS IN DUCTILE IRON

Usually, alloying elements have the same effect on structure and properties of ductile iron as gray iron. A good graphite morphology permits more efficient use of mechanical properties of the matrix; hence, alloying is more common in ductile iron than in gray iron.[79]

Sulfur. Sulfur present in nodular iron should not exceed 0.015%; otherwise, excess magnesium addition or use of other desulfurizing agents will be needed.[84]

Subversive Elements. Typical maximum limits of deleterious (antispheroidizing) minor elements to cause graphite shape deterioration, up to complete graphite degeneration, and to be accepted in the composition of ductile cast iron are 0.05% Al, 0.02% As, 0.002% Bi, 0.01% Cd, 0.002% Pb, 0.001% Sb, 0.03% Se, 0.02% Te, 0.03% Ti, and 0.1% Zr.

Very small amounts of certain impurities present in the melt, such as Pb, Sb, Bi, Ti, and Cu (called subversive elements), can prevent the required graphite spheroidizing effect of magnesium and may result in the formation of undesirable flakes, either entirely or in part, thereby adversely affecting its mechanical properties. The harmful effect of these subversive elements can, however, be nullified by the addition of a small amount of cerium and other rare earth elements.[81]

REFERENCES

1. W. F. Smith, *Principles of Materials Science and Engineering*, McGraw-Hill, New York, 1986.

2. R. E. Reed-Hill and G. J. Abbaschian, *Physical Metallurgy Principles*, PWS-Kent, Boston, 1992.

3. M. Hansen (ed.), *Constitution of Binary Alloys*, 2d ed., McGraw-Hill, New York, 1958.

4. *Metals Handbook*, vol. 8, 8th ed., ASM, Metals Park, Ohio, 1972.

5. *Heat Treating Guide*, ASM International, Materials Park, Ohio, 1995.

6. C. A. Siebert, *Met. Treating* (Rocky Mount NC), vol. VII, no. 5, 1956, pp. 14–16.

7. C. F. Walton (ed.), *Iron Castings Handbook*, Iron Casting Society, Inc., Des Plaines, Ill., 1995.

8. C. F. Walton, in *Encyclopaedia of Materials Science and Engineering*, Pergamon, Oxford, 1986, pp. 529–537.

9. G. Krauss, *Steel—Heat Treatment and Processing Principles*, ASM International, Materials Park, Ohio, 1989.

10. H. Ohtani, in *Constitution and Properties of Steels*, vol. ed. F. B. Pickering, VCH, Weinheim, Germany, 1992, pp. 147–181.

11. K. W. Andrews, *JISI*, July 1965, pp. 721–727.

12. R. A. Grange, *Met. Progress.*, vol. 79, 1961, p. 474.

13. T. Kunitake and T. Kato, *Tetsu-to-Hagane*, vol. 50, 1964, p. 666.

14. T. Kunitake and H. Ohtani, *Tetsu-to-Hagane*, vol. 53, 1967, p. 1280.

15. E. Miyoshi, T. Kunitake, T. Okada, and T. Kato, *Tetsu-to-Hagane*, vol. 51, 1965, p. 2006.

16. L. S. Darken and R. M. Fisher, *Decomposition of Austenite by Diffusional Processes*, eds. V. F. Zackay and H. I. Aaronson, Interscience, New York, 1962, p. 249.

17. C. R. Brooks, *Heat Treatment of Ferrous Alloys*, Hemisphere Publishing, New York, 1979.

18. C. S. Barrett and T. B. Massalski, *Structure of Metals*, 3d rev. ed., McGraw-Hill, New York, 1980.

19. R. W. K. Honeycombe and H. K. D. H. Bhadeshia, *Steels—Microstructure and Properties*, Edward Arnold, London, 1995.

20. E. C. Bain and H. W. Paxton, *Alloying Elements in Steel*, ASM, Metals Park, Ohio, 1966.

21. F. Wever, *Arch. Eisenhuttenwes.*, vol. 2, 1928–1929, p. 739.

22. H. W. Rayson, in *Constitution and Properties of Steels*, vol. ed. F. B. Pickering, VCH, Weinheim, Germany, 1992, pp. 583–640.

23. G. A. Roberts, J. C. Hamaker, and A. R. Johnson, *Tool Steels*, ASM, Metals Park, Ohio, 1971.

24. H. L. Yakel, *Intl. Metall. Rev.*, vol. 30, no. 1, 1985, pp. 17–40.

25. S. D. Mann, D. G. McCulloch, and B. C. Muddle, *Met. and Mats. Trans.*, vol. 26A, 1995, pp. 509–520.

26. E. Hornbogen, in *Physical Metallurgy*, eds. R. W. Cahn and P. Haasen, North-Holland Physics Publishing, Amsterdam, 1983, pp. 1076–1138.

27. G. D. Pigrova, *Met. Trans.*, vol. 27A, 1996, pp. 498–502.

28. J. Yu, *Met. Trans.*, vol. 20A, 1989, pp. 1561–1564.

29. W. Tofaute and A. Buttinghais, *Arch. Eisenhuttenwe.*, vol. 12, 1938, p. 331.

30. *Classification and Designation of Carbon and Low-Alloy Steels*, ASM Handbook, vol. 1, 10th ed., ASM International, Materials Park, Ohio, 1990, pp. 140–194.

31. H. E. Boyer, in *Fundamentals of Ferrous Metallurgy*, Course 11, Lesson 12, Materials Engineering Institute, ASM International, Metals Park, Ohio, 1981.

32. Robert B. Ross, *Metallic Materials Specification Handbook*, 4th ed., Chapman and Hall, London, 1992.

33. C. W. Wegst, Stahlschlussel (Key to Steel), Verlag Stahlschlüssel Wegst GmbH, 1998.

34. W. J. McG. Tegart and A. Gittins, in *Sulfide Inclusions in Steel*, eds. J. T. Debarbadillo and E. Snape, American Society for Metals, Metals Park, Ohio, 1975, p. 198.

35. R. Kern, *Heat Treating*, vol. 17, no. 4, 1985, pp. 38–42.

36. C. W. Kovach, in *Sulfide Inclusions in Steel*, eds. J. T. Debarbadillo and E. Snape, American Society for Metals, Metals Park, Ohio, 1975, p. 459.

37. C. M. Lyne and A. Kazak, *Trans. ASM*, vol. 61, no. 10, 1968.

38. W. C. Leslie, *The Physical Metallurgy of Steel*, McGraw-Hill, New York, 1981.

39. D. Brovoksbank and K. W. Andrews, *JISI*, vol. 206, 1968, p. 595.

40. W. E. Heritmann, T. G. Oakwood, and G. Krauss, in *Fundamentals and Applications of Microalloying Forging Steels*, eds. C. J. Van Tyne, G. Krauss, and D. K. Matlock, The Metallurgical Society, Warrendale, Pa., 1996, pp. 431–453.

41. *Cobalt Facts*, The Cobalt Development Institute, England.

42. D. T. Llwellyn, *Ironmaking and Steelmaking*, vol. 22, no. 1, 1995, pp. 25–34.

43. D. T. Llwellyn, *Ironmaking and Steelmaking*, vol. 20, no. 5, 1993, pp. 338–343.

43a. J. Dille, J. Charlier, Ch. Ducarne, and Y. Riquier, in *International Conf.: Thermomechanical Processing of Steels and Other Materials*, eds., T. Chandra and T. Sakai, TMS, Warrendale, Pa., 1997, pp. 555–561.

44. D. T. Llwellyn, *Ironmaking and Steelmaking*, vol. 23, no. 5, 1996, pp. 397–405.

45. *Carbon and Alloy Steels*, ASM International, Materials Park, Ohio, 1996, pp. 3–40.

46. D. T. Llwellyn, *Ironmaking and Steelmaking*, vol. 20, no. 1, 1993, pp. 35–41.

46a. M. P. Staiger et al., *Mater. Forum*, 1998, pp. 575–582.

47. *Tools and Manufacturing Engineers Handbook*, vol. 3: *Materials, Finishes and Coating*, eds. C. Wick and R. F. Vielleux, 4th ed., SME, Dearborn, MI, 1985, chap. 1, pp. 1.1–1.32.

48. *The Making, Shaping and Treating of Steel*, eds. W. D. Lankford and H. E. McGannon, 10th ed., United States Steel, Pittsburgh, 1985.

49. *Steel Products Manual Plates, Rolled Floor Plates: Carbon, High Strength Low Alloy Steel*, American Iron and Steel Institute, Washington, D.C., August 1985.

50. *Numbering System, Chemical Composition*, 1993 SAE Handbook, vol. 1: *Materials*, Society of Automotive Engineers, Warrendale, Pa., pp. 1.01–1.189.

51. *Carbon and Alloy Steels*, SAE J411, Nov. 89, 1998 SAE Handbook, vol. 1: *Materials, Fuels, Emissions, and Noise*, Society of Automotive Engineers, Warrendale, Pa., 1998, pp. 2.01–2.04.

52. O. D. Sherby, B. Walser, C. M. Young, and E. M. Cady, *Scr. Metall.*, vol. 9, 1975, p. 569.

53. E. S. Kayali, H. Sunada, T. Oyama, J. Wadsworth, and O. D. Sherby, *J. Mater. Sci.*, vol. 14, 1979, pp. 2688–2692.

54. D. W. Kum, T. Oyama, O. D. Sherby, O. A. Ruano, and J. Wadsworth, *Superplastic Forming, Conference Proceedings, 1984*, American Society for Metals, 1985, pp. 32–42.

55. M. Hua, C. I. Garcia, and A. J. DeArdo, *Met. and Mats. Trans.*, vol. 28A, 1997, pp. 1769–1780.

56. H. Takechi, in *HSLA Steels '95*, eds. G. Liu, H. Stuart, W. Zhang, and C. Li, Chinese Science Tech. Press, 1995, p. 72.

57. K. Yamazaki, T. Horita, Y. Umehara, and T. Morishita, in *Proc: Micro-Alloying, '88*, ASM, Materials Park, Ohio, 1988, p. 327.

58. R. C. Hudd, in *Constitution and Properties of Steels*, vol. ed. F. B. Pickering, VCH, Weinheim, Germany, 1992, pp. 219–284; D. T. Llewellyn and R. D. Hudd, *Steels: Metallurgy and Applications*, 3d ed., Butterworth-Heinemann, London, 1998.

59. M. Abe, in *Constitution and Properties of Steels*, vol. ed. F. B. Pickering, VCH, Weinheim, Germany, 1992, pp. 219–284.

60. H. Takechi, *Bull. Jpn. Inst. Met.*, vol. 30, 1991, p. 677.

61. R. K. Ray, J. J. Jona, and R. E. Hook, *Inter. Mater. Reviews*, vol. 39, no. 4, 1994, pp. 129–172.

62. F. B. Fletcher, in *The Encyclopedia of Advanced Materials*, Pergamon Press, Oxford, 1994, pp. 2650–2653.

63. T. A. Philip, in *ASM Handbook*, vol. 1, 10th ed., ASM International, Materials Park, Ohio, 1990, pp. 430–448.

64. *Steels—Bars, Forgings, Chains, Springs*, vol. 01.05, Annual Book of ASTM Standards, ASTM, Philadelphia, Pa., 1998.

65. J. M. Hampshire, in *The Encyclopedia of Advanced Materials*, Pergamon Press, Oxford, 1992, pp. 1277–1285.

66. M. G. H. Wells, in *Encyclopedia of Materials Science and Engineering*, Pergamon Press, Oxford, 1986, pp. 5115–5120.

67. D. W. Dietrich, in *Metals Handbook*, vol. 2, 10th ed., ASM, Materials Park, Ohio, 1993, pp. 761–781.

68. Y. Kurosaki et al., *ISIJ Int.*, vol. 39, no. 6, 1999, pp. 607–613.

69. M. Takashima, N. Morito, A. Honda, and C. Maeda, *IEEE Trans. on Magnetics*, vol. 35, no. 1, 1999, pp. 557–561.

70. H. Huneus, K. Gunther, T. Kochmann, V. Plutniok, and A. Shoppa, *J. Mats. Engr. and Performance*, vol. 2, no. 2, April 1993, pp. 199–204.

71. W. C. Leslie and E. Hornbogen, *Physical Metallurgy*, 4th ed., Elsevier Science BV, Amsterdam, 1996, pp. 1555–1620.

72. J. G. Benford and S. D. Washko, *Encyclopedia of Materials Science and Engineering*, Supplement, vol. 3, Pergamon Press Plc, Oxford, 1993, pp. 2650–2666.

73. J. W. Morris, in *Encyclopedia of Advanced Materials*, Pergamon, Oxford, 1992, pp. 2654–2658.

74. A. K. Sinha, in *Steel Heat Treatment Handbook*, eds., G. E. Totten and M. A. H. Howes, Marcel Dekker, New York, 1997, pp. 1061–1118.

75. *High Strength Steel Bulletin*, edition 17, Fall 1997.

76. A. M. Bayer and L. R. Walton in *ASM Handbook*, vol. 1, ASM International, Materials Park, Ohio, 1990, pp. 757–779.

77. T. Morrison, *Metall. Mater. Technol.* vol. 8, 1976, pp. 80–85; S. Floreen, in *Encyclopedia of Materials Science & Engineering*, Pergamon, Oxford, 1986, pp. 5171–5177.

78. *Metals and Alloys in the Unified Numbering System*, 8th ed., Society of Automotive Engineers, Warrendale, Pa., 1999.

79. *Cast Irons*, ASM International, Materials Park, Ohio, 1996, pp. 3–15.

79a. C. K. Syn, D. R. Lesuer, and O. D. Sherby, *Met. and Mats. Trans.*, vol. 28A, 1997, pp. 1213–1218.

80. J. D. Verhoeven, *Fundamentals of Physical Metallurgy*, Wiley, New York, 1975.

81. J. F. Wallace, in *Encyclopedia of Materials Science and Engineering*, Pergamon, Oxford, 1986, pp. 2059–2062.

82. R. Elliott, in *Constitution and Properties of Steels*, vol. ed. F. B. Pickering, VCH, Weinheim, Germany, 1992, pp. 693–737.

83. *Ductile Iron Data for Design Engineers*, QIT America, Chicago, 1990.

84. H. Morrough, in *Encyclopedia of Materials Science and Engineering*, Pergamon, Oxford, 1986, pp. 4539–4543.

85. M. I. Onsoian, O. Gong, T. Skaland, and K. Jorgensen, *Mats. Sc. and Technol.*, vol. 15, March 1999, pp. 253–259.

86. J. F. Wallace, in *Encyclopedia of Materials Science and Engineering*, Pergamon, Oxford, 1986, pp. 743–746.

87. L. Jenkins, in *Encyclopedia of Materials Science and Engineering*, Pergamon, Oxford, 1986, pp. 2725–2729.

88. *ASM Handbook*, vol. 1, 10th ed., ASM International, Materials Park, Ohio, 1990, pp. 71–84.

89. I. R. Sare and B. K. Arnold, *Met. Trans.*, vol. 26A, 1995, pp. 357–370.

90. O. N. Dogan, J. A. Hawk, and G. Laird, II, *Met. Trans.*, vol. 28A, 1997, pp. 1315–1328.

91. C. P. Tabrett, I. R. Sare, and M. Ghomashchi, *Int. Mater. Reviews*, vol. 41, no. 2, 1996, pp. 59–82.

CHAPTER 2
DIFFUSION IN METALS AND ALLOYS

2.1 INTRODUCTION

The study of diffusion can be traced back to the formulation of Fick's law in 1855 and Einstein's equations for Brownian motion in the early 1900s. However, increased research activity in the field of the mechanism and thermodynamics of diffusion processes has been established since 1945 after the easy availability of high specific activity radioactive tracer isotopes from nuclear reactors. This led to the rapid development of our understanding of diffusion mechanisms in the last five decades.

Diffusion is an irreversible, spontaneous, and thermally activated atomic migration phenomenon. Atomic movements in solids are quite slow, compared to those in liquids and gases, due to atomic bonding; however, thermal vibrations do allow some atomic movement. This is closely connected with the study of defects in solids, notably point defects, dislocations, and grain boundaries.

Diffusion is a ubiquitous mass transport mechanism in solids that controls the rate or kinetics of many solid-state transformations in most metallurgical processes that are of great technological importance in many fields. Some examples are heat treating such as annealing, recrystallization, grain growth; surface hardening of steel, precipitation hardening of alloys; instability of solid phases at elevated temperatures; materials for high-temperature applications; homogenizing of cored solid solution; phase transformations; solid-state chemical reactions; crystal growth; formation of coatings or deposits of alloys on steel; oxidation; claddings; diffusion of hydrogen in metals; corrosion; sintering; creep; metal joining by solid-state welding; ionic conductivity; deterioration of transistors through solute segregation induced by an applied electric field; preparation of impurity transistors; migration of vacancies and voids to the center of a nuclear fuel element due to the temperature gradient; and radiation-induced defects, void swelling, diffusion, segregation and precipitation, inverse Kirkendall effects, and phase transformations. In ionic crystals and semiconductors, diffusion can be explained on the basis of relations developed for metals and alloys and the thermodynamics of defects in these materials.[1] Diffusion can cause degradation of properties and appearance of a deposit at the base metal interface. Diffusion can influence even the thermodynamic properties and structures.

In this chapter we will present a concise and relatively complete treatment of diffusion in metals and alloys. This includes the diffusion equations, atomic theory of diffusion, vacancy diffusion, random walk theory, effects of key variables on diffusion coefficients, types of diffusion coefficients, self-diffusion in pure metals, diffusion in dilute and concentrated alloys, electro- and thermomigration; diffusion along short circuits (or high-diffusivity paths), application of thin films to diffusion study, diffusion in ionic crystals and semiconductors, and radiation effects and diffusion.

2.2 DIFFUSION EQUATION

Diffusion processes may be grouped into two categories: steady state and non-steady state. A steady state is one that takes place at a constant rate; that is, once the process starts, the number of atoms crossing a given interface remains constant with time. In non-steady-state diffusion, the rate of diffusion depends on time. These types of diffusion may be described by Fick's law of diffusion, which relates the atom flux to a diffusion coefficient. Several diffusion coefficients can be considered: (1) Chemical diffusion coefficient D which can be expressed in terms of intrinsic or partial diffusion coefficient D_A and D_B (at infinite dilution) for a binary system. (2) Self-diffusion coefficient of a chemically pure solute or the same element comprising radioactive isotopes or tracers which are denoted by D_{A*} and D_{B*} in a chemically homogeneous solvent or in the solvent dilute solution or in a concentrated solution which is nonradioactive of the same element or in the homogeneous alloy AB which are denoted by D_{A*}^{AB} and D_{B*}^{AB}. (3) Tracer diffusion coefficient in which the system is homogeneous or in equilibrium.

2.2.1 Fick's First Law

Fick's first law is the basic phenomenological law of diffusion that deals with the diffusion of solute atoms under steady-state conditions. A simple example is the loss of hydrogen under constant pressure through the sidewalls of the container.

Let us consider the diffusion of solute atoms in one direction between two parallel atomic planes (1) and (2) of unit area, perpendicular to the plane of the paper and separated by a distance α, as shown in Fig. 2.1. If the number of solute atoms per unit area is n_1 on plane (1) and that on plane (2) is n_2 where $n_1 > n_2$, the atomic concentration of solute atoms in number per unit volume on plane (1) is $C_1 = n_1/\alpha$ and on plane (2) $C_2 = n_2/\alpha$. Thus a concentration gradient

$$\frac{\partial c}{\partial x} = \frac{C_2 - C_1}{\alpha} \tag{2.1}$$

exists along the x direction.

Let us assume that the atomic jump frequency from each plane is Γ and that jumps are in a random direction; i.e., jumps in all directions are equally probable. Then $n_1\Gamma/2$ atoms from plane (1) will jump in the unit time to plane (2), and $n_2\Gamma/2$ atoms will jump in the unit time from plane (2) to plane (1). The factor of $\frac{1}{2}$ implies that atoms may jump in either the $+x$ or $-x$ direction. The instantaneous net flux of atoms J from plane (1) to plane (2) is the number of diffusing solute atoms which pass per unit time through a unit area of a plane in the direction of diffusion; i.e.,

$$J = \frac{1}{2}(n_1 - n_2)\Gamma = \frac{\alpha}{2}(C_1 - C_2)\Gamma \tag{2.2}$$

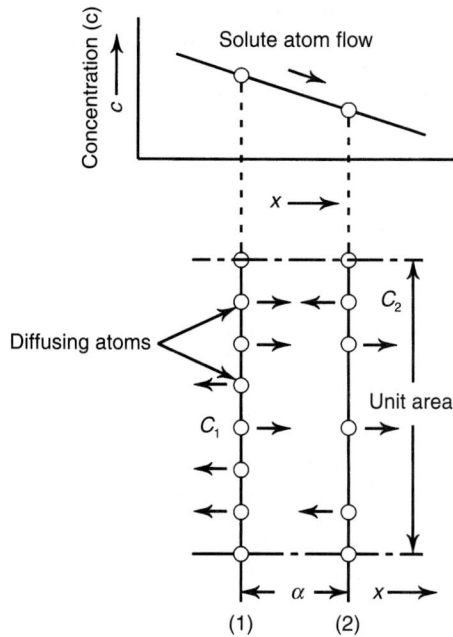

FIGURE 2.1 Relation between atomic jumps and diffusion in a concentration gradient.

The unit of J is atoms·m^{-2}·s^{-1} or atoms·cm^{-2}·s^{-1}. Since atomic jumps in a simple cubic lattice take place along x, y, and z axes, we can write Eq. (2.2), for three dimensions, as

$$J = \frac{\alpha}{6}(C_1 - C_2)\Gamma \tag{2.3}$$

Combining Eqs. (2.1) and (2.3) gives

$$J = -\Gamma\left(\frac{\alpha^2}{6}\right)\left(\frac{\partial c}{\partial x}\right) \tag{2.4}$$

The ratio of $-J$ and $\partial c/\partial x$ is defined as the intrinsic diffusion coefficient or diffusivity D which is expressed as

$$D = -\frac{J}{\partial c/\partial x} \tag{2.5a}$$

The minus sign in Eq. (2.5a) denotes the opposite direction of the flux relative to the concentration gradient. Equation (2.5a) is only valid for self-diffusion. For systems with n components, there exists a matrix of diffusion coefficients D_{ij} that relates a component's intrinsic flux J_i to the various concentration gradients

$$J_i(n) = -\sum_i^n D_{ij}\left(\frac{\partial c}{\partial x}\right)_t \qquad (i = 1, 2, \ldots, n) \tag{2.5b}$$

Thus Fick's first law of diffusion provides a formal definition of the diffusivity D. The unit of D is $m^2 \cdot s^{-1}$ or $cm^2 \cdot s^{-1}$ while the unit of concentration is atoms$\cdot m^{-3}$ or atoms$\cdot cm^{-3}$. Like flux, the diffusivity may be dependent upon position and time, or simply concentration.[2] This equation is similar to the heat flow equation which is used to define the thermal conductivity where k is the ratio of heat loss (or thermal flux) to the temperature gradient ($\partial T/\partial x$). The driving force for the heat flow is the temperature gradient. Fick's first law for three-dimensional intrinsic self-diffusion can be generalized to

$$J = -D\frac{\partial c}{\partial x} = -D \text{ grad } c = -D\nabla c \qquad (2.6)$$

where D is a second-rank tensor.[2] It is assumed that J is determined only by grad c, and ∇ is the gradient vector operator of the form[4,5] and

$$D = \frac{1}{6}\Gamma\alpha^2 \qquad (2.7)$$

For interdiffusion in a binary system, Eq. (2.6) can be written as $\tilde{J} = -\tilde{D}\,(\partial c/\partial x)_t$. Equation (2.7) holds for any randomly diffusing atoms in a simple cubic lattice, i.e., for jumps in three dimensions. In fcc structures, $\alpha = (a/2)\sqrt{2}$, where a is the lattice parameter; each site has 12 near-neighbor sites, and $D = \frac{1}{12}(\Gamma a^2)$. In bcc structures, $\alpha = (a/2)\sqrt{3}$; each site has eight near neighbors. In this case, $D = \frac{1}{8}(\Gamma a^2)$. Equation (2.6) is called *Fick's first law* of diffusion. This basic phenomenological law of diffusion states that, for steady-state diffusion conditions, the specimen becomes homogeneous as the diffusant flux ($\partial c/\partial t$) $\rightarrow 0$ for all x. Equation (2.6) gives the information concerning the direction and the rate of diffusion. For example, the diffusion occurs in the direction of decreasing diffusant concentration. The rate of diffusion is proportional to the concentration gradient (i.e., the slope of concentration-versus-distance curve). The rate of diffusion is positive if the slope is negative, i.e., if the concentration decreases with increasing x. Table 2.1 lists typical values of atomic diffusivities of selected interstitial and substitutional diffusion in metallic systems.[6]

When an external force such as an electric field also acts on the system, we can write a more general expression

$$J = -D\left(\frac{\partial c}{\partial x}\right) + \langle v\rangle c \qquad (2.8)$$

$$= -D\nabla c + \langle v\rangle c \qquad (2.9)$$

where $\langle v\rangle$ is the average velocity of the center of mass resulting from the application of external force such as an electric field or a thermal gradient on the particles. The first term in Eq. (2.8) is thus the diffusive term, and the second one is the drift term. Flynn applied such a general consideration of diffusion under stress gradients.[7]

By itself, Fick's first law, Eq. (2.6), is not particularly useful for the measurement of diffusivity in the solid state because it is extremely difficult to determine the atomic flux unless steady state is attained. Moreover, since solid-state diffusivity is normally small, a very long time is required to reach such a steady state. In some cases, where solid-state diffusivity is high (e.g., C diffusion in γ-iron), the steady-state flux and the concentration gradient can be measured and the diffusivity calculated directly from Eq. (2.6).[8]

TABLE 2.1 Diffusivity Data for Selected Interstitial and Substitutional Diffusion in Metallic Systems[6]

Diffusing metal	Matrix metal	Diffusion temperature (°C)	Coefficient (cm²/hr)
Ag	Al	466	6.84–8.1×10^{-7}
		500	7.2–3.96×10^{-8}
		573	1.26×10^{-5}
	Pb	220	5.40×10^{-5}
		250	1.08×10^{-4}
		285	3.29×10^{-4}
	Sn	500	1.73×10^{-1}
Al	Cu	500	6.12×10^{-9}
		850	7.92×10^{-6}
As	Si		$0.32e^{-82,000/RT}$
Au	Ag	456	1.76×10^{-9}
		491	0.92–2.38×10^{-13}
		585	3.6×10^{-8}
		601	3.96×10^{-8}
		624	2.5–5×10^{-11}
		717	1.04–2.25×10^{-9}
		729	1.76×10^{-9}
		767	1.15×10^{-6}
		847	2.30×10^{-6}
		858	3.63×10^{-8}
		861	3.92×10^{-8}
		874	3.92×10^{-8}
		916	5.40×10^{-6}
		1040	1.17×10^{-6}
		1120	2.29×10^{-5}
		1189	5.42×10^{-6}
	Au	800	1.17×10^{-8}
		900	9×10^{-8}
		1020	5.4×10^{-7}
	Bi	500	1.88×10^{-1}
	Cu	970	5.04×10^{-6}
	Hg	11	3×10^{-2}
	Pb	100	8.28×10^{-8}
		150	1.80×10^{-4}
		200	3.10×10^{-4}
		240	1.58×10^{-3}
		300	5.40×10^{-3}
		500	1.33×10^{-1}
			$0.001e^{-25,000/RT}$
	Sn	500	1.94×10^{-1}
B	Si		$10.5e^{-85,000/RT}$
Ba	Hg	7.8	2.17×10^{-2}
Bi	Si		$1030e^{-107,000/RT}$
	Pb	220	1.73×10^{-7}
		250	1.33×10^{-6}
		285	1.58×10^{-6}
C	W	1700	1.87×10^{-3}
	Fe	930	7.51–9.18×10^{-9}
Ca	Hg	10.2	2.25×10^{-2}

TABLE 2.1   Diffusivity Data for Selected Interstitial and Substitutional Diffusion in Metallic Systems[6] (*Continued*)

Diffusing metal	Matrix metal	Diffusion temperature (°C)	Coefficient (cm²/hr)
Cd	Ag	650	9.36×10^{-7}
		800	4.68×10^{-6}
		900	2.23×10^{-5}
	Hg	8.7	6.05×10^{-2}
		15	6.51×10^{-2}
		20	5.47×10^{-2}
		99.1	1.23×10^{-1}
	Pb	200	4.59×10^{-7}
		252	3.10×10^{-6}
Cd, 1 atom %	Pb	167	1.66×10^{-7}
Ce	W	1727	3.42×10^{-6}
Cs	Hg	7.3	1.88×10^{-2}
	W	27	4.32×10^{-3}
		227	5.40×10^{-4}
		427	2.88×10^{-2}
		540	1.44×10^{-1}
Cu	Al	440	1.8×10^{-7}
		457	2.88×10^{-7}
		540	5.04×10^{-6}
		565	$4.68 – 5.00 \times 10^{-4}$
	Ag	650	1.04×10^{-6}
		760	1.30×10^{-6}
		895	3.38×10^{-6}
	Au	301	5.40×10^{-10}
		443	8.64×10^{-9}
		560	3.38×10^{-7}
		604	5.10×10^{-7}
		616	7.92×10^{-7}
		740	3.35×10^{-6}
	Cu	650	1.15×10^{-5}
		750	2.34×10^{-8}
		830	1.44×10^{-7}
		850	9.36×10^{-7}
		950	2.30×10^{-6}
	Ge	700–900	$1.01 \pm 0.1 \times 10^{-1}$
	Pt	1041	$7.83 – 9 \times 10^{-8}$
		1213	5.04×10^{-7}
		1401	6.12×10^{-6}
Fe	Au	753	1.94×10^{-6}
		1003	2.70×10^{-5}
			$0.0062e^{-20,000/RT}$
Ga	Si		$3.6e^{-81,000/RT}$
Ge	Al	630	3.31×10^{-1}
	Au	529	1.84×10^{-1}
		563	2.80×10^{-1}
	Ge	766–928	$7.8e^{-68,509/RT}$
		1060–1200 K	$87e^{-73,000/RT}$
Hg	Cd	156	9.36×10^{-7}
		176	2.55×10^{-6}

TABLE 2.1 Diffusivity Data for Selected Interstitial and Substitutional Diffusion in Metallic Systems[6] (*Continued*)

Diffusing metal	Matrix metal	Diffusion temperature (°C)	Coefficient (cm²/hr)
		202	9×10^{-6}
	Pb	177	8.34×10^{-8}
		197	2.09×10^{-5}
In	Ag	650	1.04×10^{-6}
		800	6.84×10^{-6}
		895	4.68×10^{-5}
			$16.5e^{-90,000/RT}$
K	Hg	10.5	2.21×10^{-2}
	W	207	2.05×10^{-2}
		317	3.6×10^{-1}
		507	$1.1 \times 10^{+1}$
Li	Hg	8.2	2.75×10^{-2}
Mg	Al	365	3.96×10^{-8}
		395	$1.98\text{--}2.41 \times 10^{-7}$
		420	$2.38\text{--}2.74 \times 10^{-7}$
		440	1.19×10^{-7}
		447	9.36×10^{-7}
		450	6.84×10^{-6}
		500	$3.96\text{--}7.56 \times 10^{-6}$
		577	1.58×10^{-5}
	Pb	220	4.32×10^{-7}
Mn	Cu	400	7.2×10^{-10}
		850	4.68×10^{-7}
Mo	W	1533	9.36×10^{-10}
		1770	4.32×10^{-9}
		2010	7.92×10^{-8}
		2260	2.81×10^{-7}
Na	W	20	2.88×10^{-2}
		227	1.80
		417	9.72
		527	1.19×10^{-1}
Ni	Au	800	2.77×10^{-6}
		1003	2.48×10^{-5}
	Cu	550	2.56×10^{-9}
		950	7.56×10^{-7}
		320	1.26×10^{-6}
	Pt	1043	1.81×10^{-8}
		1241	1.73×10^{-6}
		1401	5.40×10^{-6}
Ni, 1 atom %	Pb	285	8.34×10^{-7}
Ni, 3 atom %	Pb	252	1.25×10^{-7}
Pb	Cd	252	2.88×10^{-8}
	Pb	250	5.42×10^{-8}
		285	2.92×10^{-7}
	Sn	500	1.33×10^{-1}
Pb, 2 atom %	Hg	9.4	6.46×10^{-9}
		15.6	5.71×10^{-2}
		99.2	8×10^{-2}
Pd	Ag	444	4.68×10^{-9}

TABLE 2.1 Diffusivity Data for Selected Interstitial and Substitutional Diffusion in Metallic Systems[6] (*Continued*)

Diffusing metal	Matrix metal	Diffusion temperature (°C)	Coefficient (cm²/hr)
		571	1.33×10^{-7}
		642	4.32×10^{-7}
		917	4.32×10^{-6}
	Au	727	2.09×10^{-8}
		970	1.15×10^{-6}
	Cu	490	3.24×10^{-9}
		950	$9.0\text{–}10.44 \times 10^{-7}$
Po	Au	470	4.59×10^{-11}
	Al	20	1.08×10^{-9}
		500	1.80×10^{-7}
	Bi	150	1.80×10^{-7}
	Pb	150	4.59×10^{-11}
		200	4.59×10^{-9}
		310	5.41×10^{-7}
Pt	Au	740	1.69×10^{-8}
		986	$6.12\text{–}10.08 \times 10^{-7}$
	Cu	490	2.01×10^{-9}
		960	$3.96\text{–}8.28 \times 10^{-7}$
	Pb	490	7.04×10^{-2}
Ra	Au	470	1.42×10^{-8}
	Pt	470	3.42×10^{-8}
Ra ($\beta + \gamma$)	Ag	470	1.57×10^{-8}
Rb	Hg	7.3	1.92×10^{-9}
Rh	Pb	500	1.27×10^{-1}
Sb	Ag	650	1.37×10^{-6}
		760	5.40×10^{-6}
		895	1.55×10^{-5}
			$5.6e^{-91,000/RT}$
Si	Al	465	1.22×10^{-6}
		510	7.2×10^{-6}
		600	3.35×10^{-5}
		667	1.44×10^{-1}
		697	3.13×10^{-1}
	Fe + C*	1400–1600	$3.24\text{–}5.4 \times 10^{-2}$
Sn	Ag	650	2.23×10^{-6}
		895	2.63×10^{-6}
	Cu	400	1.69×10^{-9}
		650	2.48×10^{-7}
		850	1.40×10^{-5}
	Hg	10.7	6.38×10^{-2}
	Pb	245	1.12×10^{-7}
		250	1.83×10^{-7}
		285	5.76×10^{-7}
Sr	Hg	9.4	1.96×10^{-2}
Th	Mo	1615	1.30×10^{-6}
		2000	3.60×10^{-3}
	Tl	285	8.76×10^{-7}
	W	1782	3.96×10^{-7}
		2027	4.03×10^{-6}

TABLE 2.1 Diffusivity Data for Selected Interstitial and Substitutional Diffusion in Metallic Systems[6] (*Continued*)

Diffusing metal	Matrix metal	Diffusion temperature (°C)	Coefficient (cm²/hr)
		2127	1.29×10^{-5}
		2227	2.45×10^{-5}
Th (β)	Pb	165	2.54×10^{-12}
		260	2.54×10^{-8}
		324	5.84×10^{-6}
Tl	Hg	11.5	3.63×10^{-2}
	Pb	220	1.01×10^{-7}
		250	7.92×10^{-7}
		270	3.96×10^{-7}
		285	1.12×10^{-6}
		315	2.09×10^{-6}
			$16.5e^{-85,000/RT}$
U	W	1727	4.68×10^{-8}
Y	W	1727	6.55×10^{-5}
Zn	Ag	750	1.66×10^{-5}
		850	4.37×10^{-5}
	Al	415	9×10^{-7}
		473	1.91×10^{-6}
	Hg	11.5	9.09×10^{-2}
		15	8.72×10^{-2}
		99.2	1.20×10^{-1}
	Pb	285	5.84
Zr	W	1727	1.17×10^{-5}

* Saturated FeC Alloy.
For diffusion coefficients in m²/s, multiply values in cm²/hr by 2.778×10^{-8}.
Source: Data from R. Loebel, in *Handbook of Chemistry and Physics*, 51st ed., ed. R. C. Weast, Chemical Rubber, Cleveland, 1970, F-55. Reprinted by permission of CRC Press, Boca Raton, Fla.

2.2.1.1 Steady-State Methods of Measuring Diffusion Coefficient. Steady-state methods of measuring diffusion are based directly on Fick's first law. An experimental study to check the validity of Eq. (2.6) was made by Smith[9] by passing a carburizing atmosphere ($H_2 + CH_4$) through the inside of the hollow iron tube, maintained at a constant temperature, and a decarburizing gas, of lower carbon activity, over the outside tube and measuring the resulting steady rate of flow J. He found that a steady state $\partial c/\partial t = 0$ was reached when carbon concentration passing through the tube per unit time q/t no longer changed with time (i.e., became constant). Thus,

$$J = \frac{q}{At} = \frac{q}{2\pi r l t} \qquad (2.10)$$

where r is the radius and l the cylindrical length over which carbon diffusion occurs. A combination of Eqs. (2.6) and (2.10) yields

$$q = -D(2\pi l t)\frac{dc}{d(\ln r)} \qquad (2.11)$$

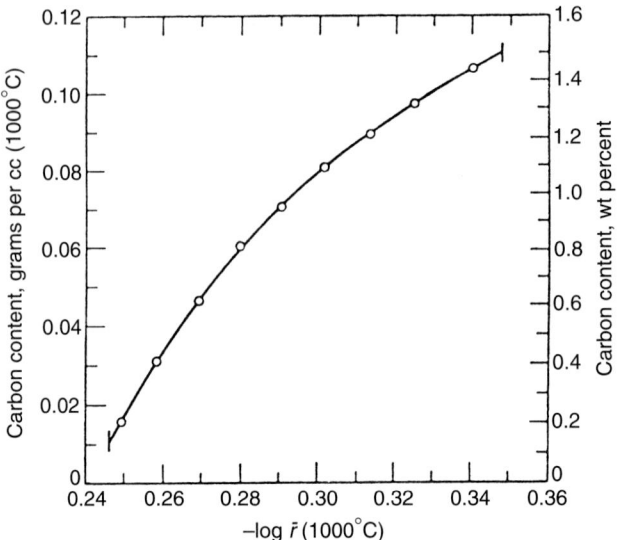

FIGURE 2.2 Steady-state carbon concentration profile through a hollow cylinder of iron at 1000°C.[9] (*Reprinted by permission of McGraw-Hill, New York.*)

Thus a plot of c versus $\ln r$ should be a straight line[9,10] if the diffusion coefficient does not vary with composition. However, for carbon diffusing in γ-iron, the slope of such a plot, as shown in Fig. 2.2, became smaller on passing from the low-carbon side to the high-carbon side. Therefore, the diffusion coefficient must be a function of composition.

Another study has been accomplished by maintaining unequal concentrations of diffusants (such as H_2, O_2, N_2 gases) on opposite sides of a specimen, usually a thin metal foil/sheet membrane or thin-walled tube and measuring a steady-state flux and thickness of the diaphragm Δx.

Permeability. Note that concentration of hydrogen in the metal at each gas-metal interface is taken as the solubility S that would exist in equilibrium with the gas because of the extreme difficulty in measuring concentration as a function of distance through the foil. Moreover, the solubility S of the dissolved gas is proportional to the square root of the applied partial pressure of the gas, according to the Sievert's law, provided S is small. Thus, the flux of, say, H_2 through the metal J_H, assuming diffusivity to be a constant, is given by[11]

$$J_H = -D\frac{C_1 - C_2}{\Delta x} = -D\frac{S_1 - S_2}{\Delta x} = -Dk\frac{\sqrt{p_1} - \sqrt{p_2}}{\Delta x} \tag{2.12}$$

where C_1 and C_2 are the concentrations, S_1 and S_2 are the solubilities, and p_1 and p_2 are the partial pressures of hydrogen maintained on opposite sides of the specimen. This method is often used to measure average diffusivity for interstitial solutes such as H_2, O_2, N_2 gases in solids. The flux in Eq. (2.12) for a given pressure drop is often termed the *permeability P*, which is defined by

$$P = DS = Dk\sqrt{p} \qquad (2.13)$$

or
$$P = Ap^{1/2}\exp\!\left(\frac{-Q_p}{RT}\right) \qquad (2.14)$$

where both S and D are temperature-dependent and A and Q_p are constants. We can rewrite Eq. (2.12) in the form

$$J_H = -\frac{P_1 - P_2}{\Delta x} \qquad (2.15)$$

Sometimes permeability is also defined as

$$J_H = -P^*\frac{\sqrt{p_1} - \sqrt{p_2}}{\Delta x} \qquad (2.16)$$

so that $P^* = P = DS$ *only at 1 atm pressure*, or

$$P^* = Dk \qquad (2.17)$$

In this situation, we have

$$P^* = P_0^*\exp\!\left(\frac{-Q_p}{RT}\right) \qquad (2.18)$$

where $P_0^* = $ cm^3(STP)$\cdot$s$^{-1}\cdot$cm$^{-2}\cdot$atm$^{-1/2}$ measured for a centimeter thickness and at 1 atm pressure and $Q_p = $ cal/mol (activation energy for permeation).[11]

The third method, called the *time-delay method*, consists of measuring just the time t_0 required to reach a steady-state flux J across a parallel-sided plate of thickness Δx. The average D (for interstitial atom such as H$_2$) is given by the relation

$$D = \frac{(\Delta x)^2}{6t_0} \qquad (2.19)$$

where t_0 is defined in such a way that, after time $t \gg t_0$, the total quantity diffused via the plate becomes $J(t - t_0)$. This method does not require the determination of concentration.[13]

2.2.2 Fick's Second Law

Mostly non-steady-state situations are set up during the solid-state transformations where concentration varies with both distance and time. For simplicity, let us consider a region of solid bar where a concentration profile exists only along one dimension (x), as shown in Fig. 2.3a. The flux due to the concentration gradient at individual values of x is illustrated in Fig. 2.3b. Let us further assume a volume element of material with thickness Δx and cross-sectional area A, as shown in Fig. 2.3c. The flux J_1 into this element across plane (1) is greater than J_2 out of plane (2). The rate of accumulation of diffusible substance in this volume element is the difference between the inward and outward flux. In other words, whatever comes in and does not go away, remains there. The net increase of concentration of diffusible atoms in a given volume element, $A\,\Delta x$, in a time increment ∂t is given by

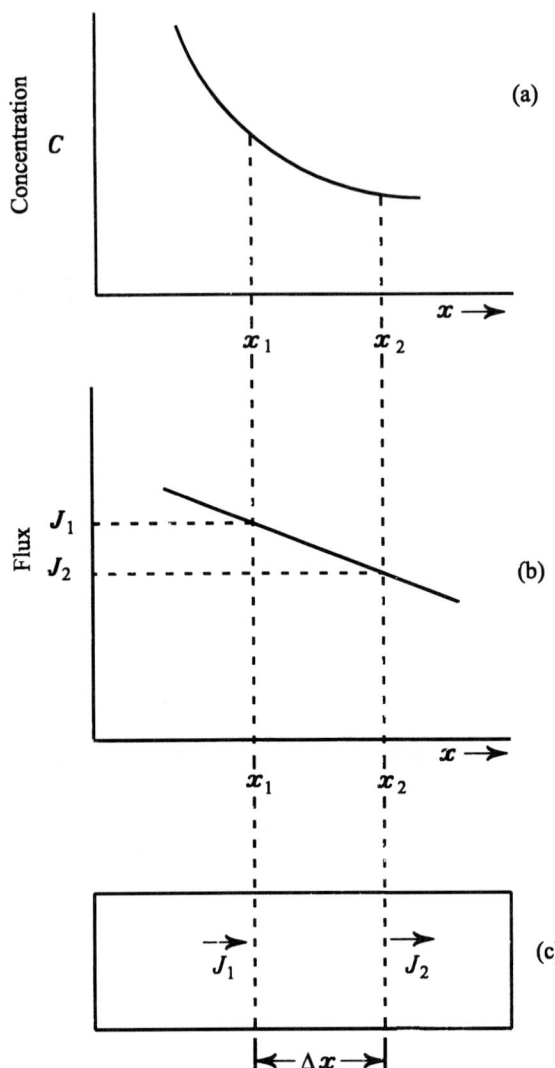

(a)

(b)

(c) **FIGURE 2.3** The derivation of Fick's second law, showing (a) an assumed $C(x)$ plot, (b) $J(x)$ for this plot, and (c) the element of volume with the flux J_1 entering and J_2 leaving.

$$J_1 - J_2 = \Delta x \left(\frac{\partial c}{\partial t} \right) = -\Delta x \left(\frac{\partial J}{\partial x} \right) \tag{2.20}$$

It is clear from Eq. (2.20) that in the limit of $\partial t \rightarrow 0$

$$\frac{\partial c}{\partial t} = -\frac{\partial J}{\partial x} = -\nabla J \tag{2.21}$$

Substituting Eq. (2.6) for the flux gives

$$\frac{\partial c}{\partial t} = \frac{\partial}{\partial x} \frac{D \partial c}{\partial x} \qquad (2.22)$$

If D is assumed to be a constant (however, this assumption is not typically true), independent of concentration and, therefore, position x in the sample, or if the sample is chemically homogeneous as in self-diffusion, Eq. (2.22) reduces to

$$\frac{\partial c}{\partial t} = D \frac{\partial^2 c}{\partial x^2} \qquad (2.23)$$

The differentials, Eqs. (2.21) through (2.23), are often called (one-dimensional) *Fick's second law* of diffusion or simply the *diffusion equation*. Equation (2.23) has a simple graphical interpretation. If $C(x)$ plot is concave upward (Fig. 2.3a), the concentration at all points on such a curve increases with time and $\partial C/\partial t$ becomes positive (that is, $\partial^2 c/\partial x^2 > 0$) at every point. In contrast, if $C(x)$ plot is convex upward, the concentration decreases with time and $\partial c/\partial t$ is negative (that is, $\partial^2 c/\partial x^2 < 0$).[10]

Equation (2.8) can also be developed in the same way to give

$$\frac{\partial c}{\partial t} = \frac{\partial}{\partial x} \frac{D \partial c}{\partial x} - \frac{\partial}{\partial x}(\langle v \rangle c) \qquad (2.24a)$$

Thus the general diffusion equation is a second-order partial differential equation. It cannot be solved analytically if D and $\langle v \rangle$ are functions of concentration and therefore of x and t, which is the situation for systems that remain chemically homogeneous (e.g., self-diffusion).[3] Then Eq. (2.24a) can be written

$$\frac{\partial c}{\partial t} = D \frac{\partial^2 c}{\partial x^2} - \langle v \rangle \frac{\partial c}{\partial x} \qquad (2.24b)$$

If there is no drift term, Eq. (2.24b) reduces to Eq. (2.23). For three-dimensional diffusion, Eq. (2.23) can be expressed by

$$\frac{\partial c}{\partial t} = D_x \frac{\partial^2 c}{\partial x^2} + D_y \frac{\partial^2 c}{\partial y^2} + D_z \frac{\partial^2 c}{\partial z^2} \qquad (2.25)$$

where the three diffusion coefficients allow for anisotropy of diffusion in noncubic metals.[14] More usually they can be written in vectorial form

$$\frac{\partial c}{\partial t} = -\nabla \boldsymbol{J} = \text{div}\,(D\,\text{grad}\,c) = \nabla(\boldsymbol{D}\,\nabla c) = \boldsymbol{D}\,\nabla^2 \boldsymbol{c} \qquad (2.26)$$

where $\nabla^2 \boldsymbol{c}$ is the Laplacian of $\boldsymbol{c}$.[10]

The above differential equation of Fick's second law can also be transformed into other coordinates. For example, in spherical coordinates, Fick's second law becomes

$$\frac{\partial c}{\partial t} = D\left(\frac{\partial^2 c}{\partial r^2} + \frac{2}{r} \frac{\partial c}{\partial r} \right) \qquad (2.27)$$

In cylindrical coordinates, this diffusion equation is given by

$$\frac{\partial c}{\partial t} = D\left(\frac{\partial^2 c}{\partial r^2} + \frac{1}{r}\frac{\partial c}{\partial r}\right) \tag{2.28}$$

For a hollow cylinder, this diffusion equation is given by

$$\frac{1}{D}\frac{\partial c}{\partial t} = \frac{\partial^2 c}{\partial r^2} + \frac{1}{r}\frac{\partial c}{\partial r} \tag{2.29}$$

where $C = C(r,t)$. The initial condition at $t = 0$ is $C(r,0) = 0$ at $a < r < b$.

Fick's Second Law for Concentration-Dependent Diffusion Coefficient. In general, the diffusion coefficient is concentration-dependent which implies that the diffusion coefficient changes with position in the sample. In this situation, according to Eq. (2.22), Fick's second law must be written in terms of chemical diffusion coefficient $D(c)$:[12]

$$\frac{\partial c}{\partial t} = \frac{\partial}{\partial x}\left(\tilde{D}\frac{\partial c}{\partial x}\right) = \frac{\partial \tilde{D}}{\partial x}\frac{\partial c}{\partial x} + \tilde{D}\frac{\partial^2 c}{\partial x^2} \tag{2.30}$$

2.2.2.1 Non-steady-state Solutions. In a specific diffusion problem, Eq. (2.23) has to be solved for the given initial and boundary conditions. The equation can be integrated only if the diffusivity is independent of position. We will outline here the solution to the one-dimensional diffusion equation (2.23) for three types of situations which are of technical importance. The first concerns the situation where a complete homogenization is reached within a finite time and the second is for a semi-infinite system where the concentration remains unchanged within a finite period in the regions at large x.[13] The third is an application of a thin-film solution or thin-layer method. The practical significance of the degree of homogeneity of alloys is well known throughout the metal industry.

1. *Homogenization.* It is often necessary to know the time required for an inhomogeneous alloy or casting to reach complete homogeneity. Here we will discuss the diffusive processes in solids of the practical problems of the elimination of periodic microsegregation arising from alloy solidification.

Let us consider the situation in which the variation of solute concentration with distance in one dimension is sinusoidal (Fig. 2.4). Here solute atoms diffuse down the concentration gradients such that the solute concentration decreases in the regimes between $x = 0$ and $x = l$ and increases in the regimes between $x = l$ and $x = 2l$. The curvature is zero at $x = 0, l$, and $2l$; hence, the concentration at these points remains unchanged with time, assuming D is independent of C, as given in Eq. (2.23). Initially (i.e., at time $t = 0$) the concentration at any point along the line $C(x,0)$ can be described by

$$C(x,0) = C_0 + C_m \sin\frac{\pi x}{l} \tag{2.31}$$

where C_0 is the overall or average alloy content or concentration, C_m is the amplitude of the initial concentration profile, and l is the average distance between adjacent maxima and minima. The solution to Eq. (2.23), using interdiffusion coefficient D (discussed later), which shows many of the salient features of homogenization in binary systems and satisfies this initial condition, is given by

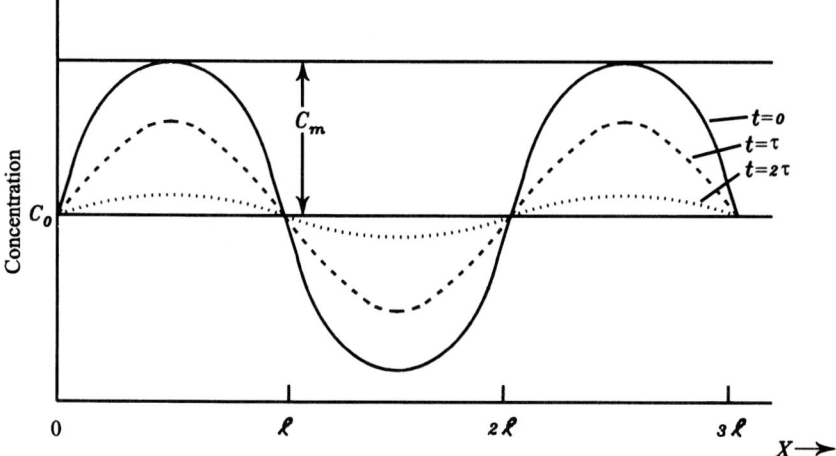

FIGURE 2.4   The effect of diffusion on a sinusoidal distribution of solute toward homogenization.

$$C(x,t) = C_0 + C_m \cos\left(\frac{\pi x}{l}\right)\exp\left(\frac{-Dt\pi^2}{l^2}\right) \tag{2.32}$$

That is, the amplitude C_m of a sinusoidal initial distribution will decay exponentially in time; the rate depends on the wavelength $2l$. The time to reach $1/e$ of the initial amplitude, called *relaxation time*, τ is equal to $l^2/(\pi^2 D)$. Thus, we can rewrite Eq. (2.32) as

$$C(x,t) = C_0 + C_m \cos\left(\frac{\pi x}{l}\right)\exp\left(\frac{-t}{\tau}\right) \tag{2.33}$$

Similarly, the concentration profile at $x = l/2$ and time t is given by

$$C\left(\frac{l}{2},t\right) = C + C_{0.5m} \cos\left(\frac{\pi x}{l}\right)\exp\left(\frac{-t}{\tau}\right) \tag{2.34}$$

where
$$C_{0.5m} = C_m \exp\left(\frac{-t}{\tau}\right) \tag{2.35}$$

Equation (2.33) also shows that the deviation of concentration C at any point from C_0 (that is, $C - C_0$) decreases exponentially with time and after a long time approaches zero, i.e., becomes completely homogenized, or $C = C_0$. The rate at which this occurs is a function of τ, which, in turn, is a function of l and D. The solute distribution at this stage should appear as a dashed curve in Fig. 2.4. After a time $t = 2\tau$, $C - C_0$ is reduced by a total of $1/e^2$.[14,15]

   In the foregoing derivations it is assumed that the initial solute concentration varies sinusoidally along the x direction. However, usually the initial concentration distribution should be considered as a superposition of a series of different sinusoidal wavelength $2l_i$, associated with a characteristic time τ, proportional to the wavelength squared. Thus, the shortest wavelength terms will decay quickly, and the homogenization time will ultimately depend on the largest wavelength present in the initial solute distribution (or spectrum).

It can be seen from Eq. (2.33) that the rate of homogenization increases quickly with the short-range variations in concentration: the longer-range variations would produce a 16-fold decrease in the time needed for equivalent homogenization $(\tau \propto l^2)$.[15] This process can also be augmented by raising the temperature and thus increasing D. Alternative methods of decreasing the magnitude of l include rapid cooling to provide fine-scale segregation and hot working of castings.

The following practical conclusions can be derived from the homogenization of alloys:[11]

- In commercial alloys, with large dendritic arm spacing (DAS) (200 to $400\,\mu$), substitutional elements do not homogenize until excessively high temperatures and long diffusion times are used.

- It is possible to significantly homogenize substitutional elements at reasonable temperatures and times only if the material has fine DAS ($<50\,\mu$) that is obtained by rapid solidification.

- Interstitial elements (e.g., carbon in steel) diffuse very rapidly at austenitizing temperature.

2. *Semi-infinite system.* When the movement of a change in concentration into a specimen dimension is considerably larger than the mean diffusion distance (i.e., when the diffusion flow does not reach the end of the specimen, as is the case in a thermochemical surface treatment), the system is said to be *semi-infinite*. An example of this type of solution is applicable to the case of a gas-solid reaction such as the carburizing of steel where a solid specimen is in contact with an infinite reservoir of gas. This results in the surface composition remaining constant. The solid is assumed to be semi-infinite; i.e., its length in $+x$ direction is infinity.

In gas carburizing, low-carbon steels are heated to the austenitizing temperature (called the carburizing temperature) in contact with a carburizing atmosphere to produce an increased carbon content at the surface. The most important carburizing reaction that occurs at the surface in the austenite range is

$$2CO \rightleftharpoons CO_2 + C \tag{2.36}$$

The carbon dissolves at the surface and diffuses into the surface to produce a carbon concentration gradient between the surface and the interior of the component. To maintain a constant carbon concentration C_s (at $x = 0$) at the surface of the steel, the CO/CO_2 ratio is properly determined and controlled. Figure 2.5 shows the concentration profiles developed at four different times $t_1(2\,hr)$, $t_2(4\,hr)$, $t_3(8\,hr)$, and $t_4(16\,hr)$ at a given temperature 927°C (1700°F) in a direction perpendicular to the surface. This figure shows the increase in the depth of penetration of the diffusant with the increasing diffusing times. A mathematical equation for these profiles can be found by solving Fick's second law and using boundary conditions: $C_{x=0} = C_s$ and $C_{x=\infty} = C_0$ (the base composition). The area under a particular curve and for a particular time represents the total amount of carbon added to the specimen. The slope $(\partial c/\partial x)_{x=0}$ times D is the rate at which carbon is supplied at the surface to maintain the surface concentration C_s.

In general, the diffusivity of carbon in austenite D_c^γ is a function of concentration; it increases to some extent with carbon content, but here a mean value of D is assumed to be a constant at a given temperature (and independent of composition), which provides an approximate concentration profile. It is assumed that no

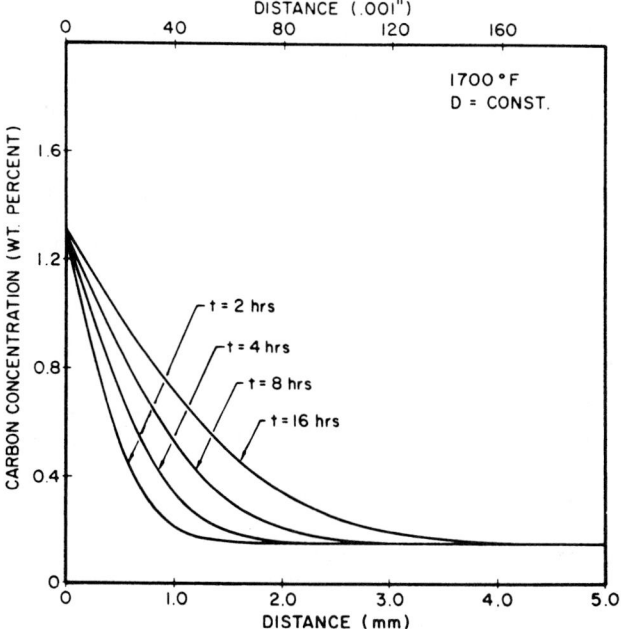

FIGURE 2.5   Carbon concentration-versus-distance curves for varying times at 1700°F, $D \neq f(C)$, binary Fe-C. (*Source: Met. Trans., vol. 9A, 1978, p. 1515.*)

volume changes occur in the lattice during diffusion.[16] A simple solution to the diffusion equation is thus obtained: [10]

$$C(x,t) = C_s - (C_s - C_0)\mathrm{erf}\left(\frac{x}{2\sqrt{Dt}}\right) \qquad (2.37)$$

or

$$\frac{C_s - C}{C_s - C_0} = \mathrm{erf}\left(\frac{x}{2\sqrt{Dt}}\right) \qquad (2.38)$$

or

$$\frac{C - C_0}{C_s - C_0} = 1 - \mathrm{erf}\left(\frac{x}{2\sqrt{Dt}}\right) \qquad (2.39)$$

where $C(x,t)$ is the carbon concentration as a function of distance x from the surface of a steel at time t (in seconds) of the diffusion treatment, $C_0[= C(x,0)]$ is the initial carbon concentration of the uncarburized steel, $C_s[= C(0,t)]$ is the surface carbon concentration, and erf is the Gaussian or complimentary error function. This is an indefinite integral which is defined by the equation

$$\mathrm{erf}(z) = \frac{2}{\sqrt{\pi}}\int_0^z e^{-y^2}dy \qquad (2.40)$$

Because the concentration terms appear in a ratio, concentration may be used in any form, e.g., mole fraction, atomic fraction, or weight percent. The error function values are listed in Table 2.2 and may be represented graphically.

If we define the case depth as the depth into the bar at which the composition is midpoint between C_s and C_0, called $C_{0.5} = C_s/2$, we can write

$$\frac{C_s - C_{0.5}}{C_s - C_0} = 0.5 = \text{erf}\left[\frac{x_{0.5}}{2(Dt)^{1/2}}\right] = \text{erf}(0.477) \tag{2.41}$$

or

$$x_{0.5} = 0.954(Dt)^{1/2} \tag{2.42}$$

where $x_{0.5}$ is the case depth for $C = 0.5C_s$. If we had $C = 0.25C_s$, we would have observed $x_{0.25} = 1.6(Dt)^{1/2}$. Thus, a general expression may be given by

$$x_c = (\text{constant})(Dt)^{1/2} \tag{2.43}$$

Equation (2.43) gives a rough estimate of the time and temperature necessary to achieve an appreciable diffusion or solute distribution over a distance x. Thus the case thickness, penetration distance, depth of the carburized layer, or depth of any isoconcentration line is directly proportional to $(Dt)^{1/2}$. In other words, to obtain a 2 times increase in penetration or depth requires a 4 times increase in time. This relation between the penetration distance, time, and temperature has many applications in materials science and is perhaps the most widely used relation in the practical applications of diffusion concepts.[16]

The total quantity of material absorbed by the specimen, or lost from it, in a time t by diffusion through the surface area A is given by[11]

$$q(t) = 2|C_0 - C_s|A \frac{(Dt)^{1/2}}{\pi} \tag{2.44}$$

where absorption and loss (or desorption) of diffusant through the component surface correspond, respectively, to the carburization and decarburization process of steel.

3. *Decarburization of steel.* The process of decarburization of high-carbon steel in an oxidizing atmosphere consists of three steps: (1) oxygen transport within the gas to the metal surface, (2) carbon exchange at the gas-metal interface, and (3) diffusion of carbon out of the metal specimen. There are several variations of this type of solution as given by Eq. (2.37), (2.38), or (2.39). For example, during decarburization of high-carbon steel sample, the surface concentration of carbon is decreased to a very low level mainly due to the diffusion of carbon out of the specimen. That is, the direction of the diffusion process is opposite to that in the carburizing process discussed above (i.e., the formation of ferrite layer at the surface). The carbon profile in this case, assuming the carbon content of the surface to drop to zero, is represented by

$$C(x,t) = C_0 \, \text{erf}\left(\frac{x}{2(Dt)^{1/2}}\right) \tag{2.45}$$

In this situation also, the depth of the decarburized layer is proportional to $(Dt)^{1/2}$. Generally, the diffusion of carbon within the metal is the most important parameter in controlling the rate of decarburization, which after a short initial period follows a parabolic time law. The model used to represent decarburization of the

TABLE 2.2 Error Function

y	$\mathrm{erf}(y)$	y	$\mathrm{erf}(y)$	y	$\mathrm{erf}(y)$
0.000	0.000	0.500	0.520	0.960	0.825
0.050	0.0564	0.510	0.529	0.970	0.830
0.060	0.0676	0.520	0.538	0.980	0.834
0.070	0.0789	0.530	0.546	0.990	0.839
0.080	0.0901	0.540	0.555	1.00	0.843
0.090	0.101	0.550	0.563	1.01	0.847
0.100	0.112	0.560	0.572	1.02	0.851
0.110	0.124	0.570	0.580	1.03	0.855
0.120	0.135	0.580	0.588	1.04	0.859
0.130	0.146	0.590	0.596	1.05	0.862
0.140	0.157	0.600	0.604	1.06	0.866
0.150	0.168	0.610	0.612	1.07	0.870
0.160	0.179	0.620	0.619	1.08	0.873
0.170	0.190	0.630	0.627	1.09	0.877
0.180	0.201	0.640	0.635	1.10	0.880
0.190	0.212	0.650	0.642	1.11	0.884
0.200	0.223	0.660	0.649	1.12	0.887
0.210	0.234	0.670	0.657	1.13	0.890
0.220	0.244	0.680	0.664	1.14	0.893
0.230	0.255	0.690	0.671	1.15	0.896
0.240	0.266	0.700	0.678	1.16	0.899
0.250	0.276	0.710	0.685	1.17	0.902
0.260	0.287	0.720	0.691	1.18	0.905
0.270	0.297	0.730	0.698	1.19	0.908
0.280	0.308	0.740	0.705	1.20	0.910
0.290	0.318	0.750	0.711	1.21	0.913
0.300	0.329	0.760	0.718	1.22	0.916
0.310	0.339	0.770	0.724	1.23	0.918
0.320	0.349	0.780	0.730	1.24	0.921
0.330	0.359	0.790	0.736	1.25	0.923
0.340	0.369	0.800	0.742	1.26	0.925
0.350	0.379	0.810	0.748	1.27	0.928
0.360	0.389	0.820	0.754	1.28	0.930
0.370	0.399	0.830	0.760	1.29	0.932
0.380	0.409	0.840	0.765	1.30	0.934
0.390	0.419	0.850	0.771	1.31	0.936
0.400	0.428	0.860	0.776	1.32	0.938
0.410	0.438	0.870	0.781	1.33	0.940
0.420	0.447	0.880	0.787	1.34	0.942
0.430	0.457	0.890	0.792	1.35	0.944
0.440	0.466	0.900	0.797	1.36	0.946
0.450	0.475	0.910	0.802	1.40	0.952
0.460	0.485	0.920	0.807	1.50	0.966
0.470	0.494	0.930	0.812	2.00	0.995
0.480	0.503	0.940	0.816	∞	1.000
0.490	0.512	0.950	0.821		

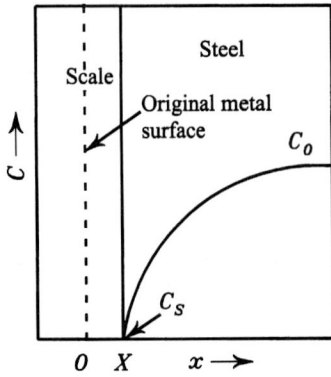

FIGURE 2.6 Model for decarburization in fully austenitic state.[17]

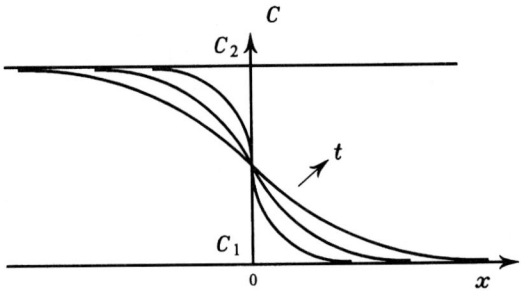

FIGURE 2.7 Concentration profiles at successive times ($t_2 > t_1 > 0$) when the two semi-infinite bars of different compositions are annealed after welding.

fully γ-iron with the formation of α-Fe on the surface in the temperature range of 723 to 910°C is shown in Fig. 2.6.[17,18]

4. *Infinite (or diffusion) couple method.* Another variation of Eq. (2.45) occurs when two homogeneous alloys of unequal solute concentrations C_1 and C_2 ($<C_1$) are joined across a plane interface (by welding) and subsequently annealed at a constant temperature for a time t to allow the interdiffusion to occur. Here one specimen serves as the source of the solute and the other as the sink. The distribution of concentrations in the "couple" $C(x)$ is then determined by removal and subsequent analysis of successive thin layers cut parallel to the initial interface. Alternatively, the sample is cut perpendicular to interface and analyzed by microprobe or similar technique. Such profiles for various times are shown in Fig. 2.7.[19] When the couple is effectively *infinite* (or sufficiently thick that the concentration remains unchanged at the outer surfaces and the diffusion zone does not expand to either end), D is assumed to be constant over the range C_1 to C_2. The appropriate relevant solution to the diffusion equation (2.22) is

$$C(x) = \frac{C_1 + C_2}{2} - \frac{C_1 - C_2}{2}\,\mathrm{erf}\left[\frac{x}{2(Dt)^{1/2}}\right] \qquad (2.46)$$

$$= C_2 + \frac{C_1 - C_2}{2} \mathrm{erf}\left[\frac{x}{2(Dt)^{1/2}}\right] \tag{2.47}$$

or

$$\frac{C - C_2}{C_1 - C_2} = \frac{1}{2}\left[1 - \mathrm{erf}\frac{x}{2(Dt)^{1/2}}\right] \tag{2.48}$$

As mentioned above, D is considered to be constant over the range C_1 to C_2, and the distance at any given composition advanced from $x = 0$ after time t is proportional to $(Dt)^{1/2}$. This analysis is applicable to both $+x$ and $-x$ values.

Seldom is the diffusion couple method employed to determine self-diffusion coefficients, one-half of the couple being normal metal, the other rich in one of its active or normal isotopes. With analytical solutions like Eq. (2.47), D can be calculated by measuring C at one position only.

5. *Thin layer (or instantaneous source) methods.*[11,20] These methods are now primarily used to measure self-diffusion and tracer diffusion coefficients. An infinitesimally small or very thin layer [$<<(Dt)^{1/2}$] (say, 100 Å) of tracer (or radioactive) diffusant, of total amount M per unit area, is deposited on a plane surface of a semi-infinite [$>>(Dt)^{1/2}$] solid specimen, usually, by electrodeposition, evaporation, sputtering, or another appropriate method. After isothermal diffusion annealing for time t in a protective atmosphere, the concentration of the tracer $C(x,t)$ at a depth x from the surface after a diffusion annealing of time t (i.e., concentration profile) is given by the *Gaussian distribution*

$$C(x) = C(x,t) = \frac{M}{(\pi Dt)^{1/2}} \exp\left(\frac{-x^2}{4Dt}\right) \tag{2.49}$$

This equation shows that for homogeneous volume diffusion in the x direction, the isoconcentration contours are represented by straight lines parallel to the $x = 0$ surface. This condition is easy to satisfy because a very small amount is adequate to study the diffusion due to the very high sensitivity of methods of detecting and measuring radioactive substances. Similarly there is a negligible variation in the chemical composition of the sample, so D is constant and Eq. (2.23), of which Eq. (2.49) is the solution for this case, is applicable.[5] After diffusion, the activity of each of a series of slices cut from the sample may be measured, and D can be determined from the slope of $\ln C(x,t)$ versus x^2 or by the expression

$$\frac{d\ln C(x,t)}{d(x^2)} = -\frac{1}{4Dt} \tag{2.50}$$

obtained from differentiation of Eq. (2.49). If a very thin layer of the diffusing radioactive isotope of total amount M per unit area is deposited as a sandwich layer at the boundary ($x = 0$) between two identical samples of "infinite" thickness, after diffusion for a time t, the concentration $C(x,t)$ becomes one-half and is described by the following equation (and shown in Fig. 2.8a):

$$C(x,t) = \frac{M}{2(\pi Dt)^{1/2}} \exp\left(\frac{-x^2}{4Dt}\right) \tag{2.51}$$

provided the thickness of the deposited layer is $<<2\sqrt{Dt}$. Equation (2.51) is usually called either the *instantaneous source solution* or *Gaussian concentration profile*. Figure 2.8b is a plot of Eq. (2.51) in linear scales for four different values of

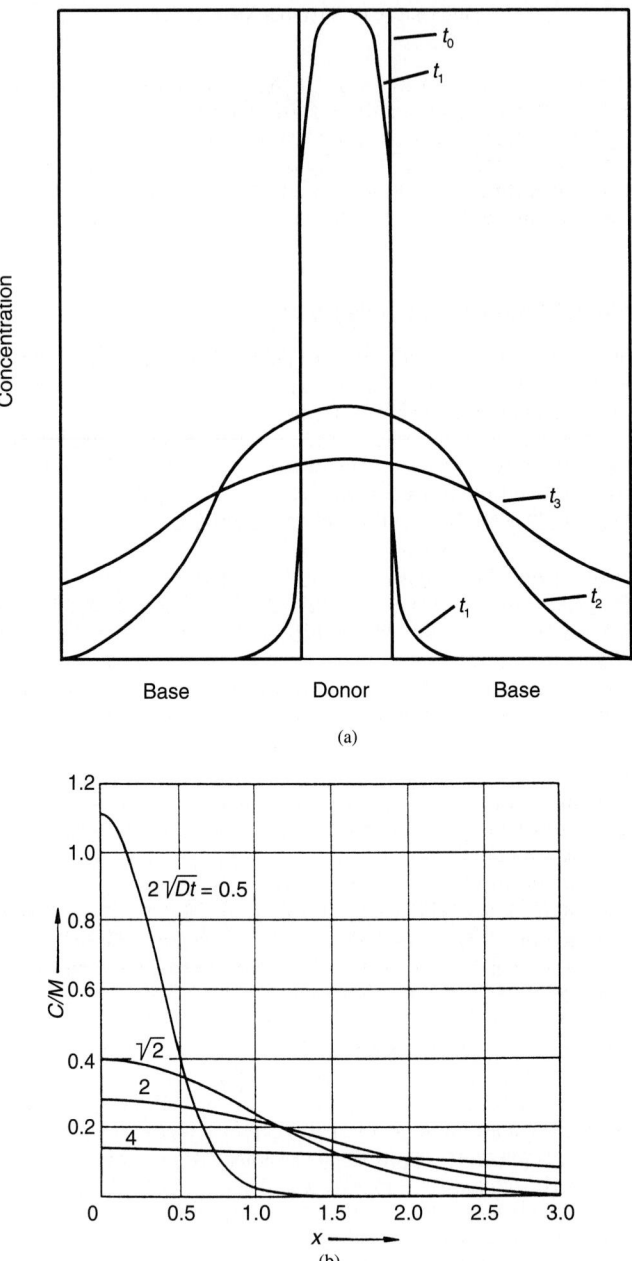

FIGURE 2.8 (*a*) Time evolution of Eq. (2.51). (*b*) Instantaneous source (Gaussian) diffusion profiles. The concentration normalized to the total amount M is plotted versus penetration distance x for four different values of the diffusion length $2\sqrt{Dt}$.[12] [*Part (b) reprinted by permission of Springer-Verlag, Berlin.*]

$2\sqrt{Dt}$. The quantity $2\sqrt{Dt}$ is a measure of the penetration and occurs in most diffusion problems. It is often called the *diffusion length*.[12]

Recently the electron microprobe analysis (EMPA) and other sensitive methods have been used to measure even impurity diffusion coefficients. In fact instruments are now available with a sensitivity sufficient to monitor diffusion from deposited layers of inactive diffusant thin enough to meet the requirements for the use of Eq. (2.49).

2.2.3 Indirect Methods, Not Based on Fick's Laws

Indirect (or microscopic) methods use measurements of some phenomena in solids that depend for their occurrence on the thermally activated atomic mobility, rather than macroscopic phenomena entailing concentration changes or fluxes. In the last two decades the studies of atomic transport have benefited greatly by these new nontraditional or novel measurement methods, especially when one considers the new experimental and theoretical techniques to investigate the diffusion phenomena. We will briefly state here these methods and the degree to which they have been used.

1. *Intensity and shape of quasi-elastic neutron scattering (QENS) spectra* are routinely used in the study of diffusion in liquids and solids. Most solid-state applications have been best suited for fast diffusion, like hydrogen diffusion in metals, superionic / fast ion conductors, and diffusion of alkali atoms in graphite intercalation compounds. Requirements include diffusion coefficients greater than about $10^{-11} \, m^2 \, s^{-1}$ and elements exhibiting reasonably high scattering cross section. For a detailed discussion the reader is referred to reviews made by Zabel[21] and Petry et al.[22]

2. *Nuclear magnetic resonance (NMR) methods* are based on a phenomenon which is very sensitive to interactions of the nuclear magnetic moments with fields produced by their local environments. These methods entail the determination of the effect of temperature on NMR line widths and on the closely associated spin-lattice relaxation (T_1) and spin-spin relaxation (T_2) times. These are influenced by diffusion through the effect of the relative movements of atoms on the magnetic dipolar interaction of their nuclei. It is applied to measure the self-diffusion coefficient in a few metals (e.g., Li, Al, and Cu) and solute diffusion in a few alloys (e.g., of Al) but is less accurate than the tracer techniques.[13,23–25]

3. *Mössbauer effect.* We can assume the Mössbauer effect as a resonant filter. The gamma rays can be emitted or absorbed by excited nuclei. If one of the emitting or absorbing nuclei is moving, by either thermal vibration or diffusion jumps, it will result in a complete loss of coherence, and consequently a broadening of the Mössbauer line occurs due to self-interference effects (see Fig. 2.9). The characteristic time that determines the line width is the residence time τ_0 which becomes smaller than the nuclear lifetime τ_N. More precisely, Singwi and Sjölander have given the theory of diffusion broadening in the case of liquid and solid-state diffusion situation which can be expressed as

$$\Gamma = 2\hbar DK^2 \tag{2.52}$$

where $\hbar$ is Planck's constant divided by 2π and K is the scattering vector.[25] When the net broadening arising from diffusion becomes equal to that due to nuclear lifetime τ_N, this results in a lower limit in diffusion coefficient of $10^{-13} \, m^2 \, s^{-1}$ (in solids) for ^{57}Fe.[25] Similar studies have now been performed with synchrotron X-rays.

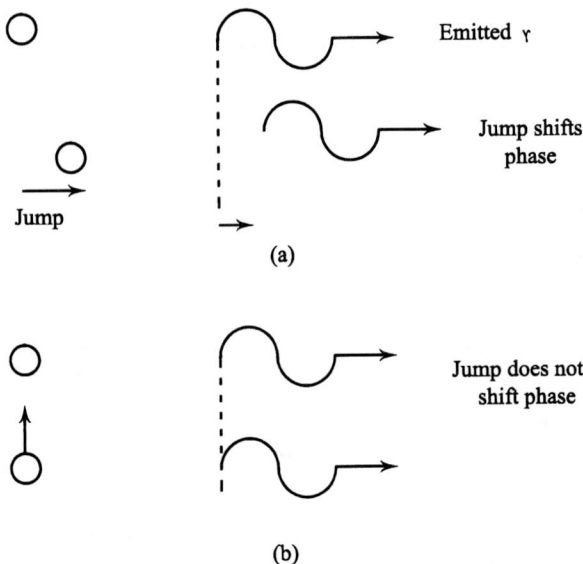

FIGURE 2.9   Schematic representation of (*a*) a translational jump in the direction of emission, causing a loss of coherence and a resulting broadening of the Mössbauer line, and (*b*) a jump normal to the direction of emission which preserves phase coherence and results in no broadening.[27]

General prerequisites are that the diffusion coefficients of the Mössbauer active isotope should be greater than about $10^{-13}\,m^2\,s^{-1}$. Mostly this has been applied to solids with the rather ideal isotope ^{57}Fe. We refer to reviews made by Wertheim and Mullen for a detailed discussion.[25–27]

4. *Relaxation methods* comprising anelasticity using torsion pendulum, in either *after-effect mode* or *internal friction mode*, and the resonant bar for high frequency; *Snoek effect*; *Zener effect*; and *Gorsky effect*. The Snoek effect has been extensively and reliably used to measure D for dilute interstitial solutes such as C, N, O in bcc metals, especially Fe, V, Ta, and Nb. It is an anelastic relaxation which exhibits itself most significantly as a peak in internal friction (loss of mechanical energy in vibration) as a function of frequency or of temperature.[28,29] When a tensile stress is applied along the x axis, the atomic separation in that direction is slightly increased. Accordingly, the occupation of x-type interstitial sites is favored slightly over the other two types. This redistribution of population provides a time-dependent or anelastic strain involving a relaxation time $\tau_r^{-1} = 3\Gamma_i/2$. Under oscillatory stress, the internal friction peak that results, allows τ_r to be known, since $\omega\tau_r = 1$ at the peak (where ω is the angular frequency of oscillation). But the diffusion coefficient D_i is also related to Γ_i and, in fact, is given by the equation

$$D_i = \frac{1}{24}a^2\Gamma_i = \frac{a^2}{36\tau_r} \tag{2.53}$$

Thus, the determination of τ_r directly provides D_i value. The main advantage of the Snoek effect is that it allows D_i to be measured at temperatures much lower than

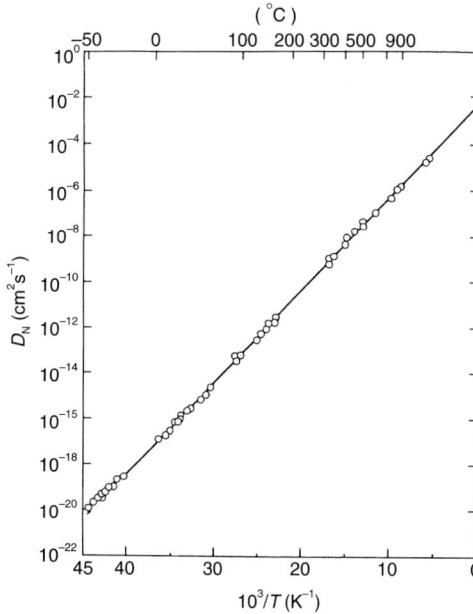

FIGURE 2.10 Arrhenius plot for diffusion of N in α-iron, obtained by combining diffusion and Snoek effect measurements by various authors. (*After D. N. Beshers, in Diffusion, ASM, Metals Park, Ohio, 1973, pp. 209–240.*)

those at which direct diffusion measurements are possible. In reality, combination of direct D_i and Snoek measurements leads to the most extensive Arrhenius plots ever observed. Figure 2.10 illustrates such a plot for diffusion of N in Fe where the range of D_i is about 14 decades and the temperature range is from −50 to 1500°C. This figure suggests a simple mechanism where the activation parameters are not temperature-dependent.[28]

The Snoek effect also provides a means to study interstitials generated by irradiation and gives information especially on the configuration of dissociated interstitials, their mobility, and their reaction with lattice defects.[3]

Contrary to Snoek relaxation, the Zener effect (or directional ordering) is used to study diffusion in substitutional AB solid solution (fcc metals) and concentrated alloys due, presumably, to stress-induced changes in short-range order because of its dependence on the square of the substitutional solute atom pairs.[30,31] Although the precise relation of the diffusion coefficients of solute and solvent to the Zener effects measured is still not distinct, this effect is of great interest in the study of the behavior of point defects in alloys in, or out, of equilibrium. Their reorientation under stress produces an anelastic relaxation which can be noticed in all lattices of high symmetry.[30]

While the above two phenomena involve diffusion over atomic distance only, the Gorsky effect requires a long-range diffusion over the sample dimension. It is chiefly employed to study the hydrogen diffusion in metals; its main advantage is the absence of surface effects.[3,32] This involves migration of solute B atoms which has a strong size effect in the A matrix, under the influence of a gradient of deformation produced, for example, by bending. The transport produces a relaxation of elastic stresses, by the migration of centers of dilation from the regions in compression to those regions in tension. A relaxation time τ_r is measured as in the elastic after-effect, and D is measured from it according to $D_B = (1/\phi\tau_r)(d/\pi)^2$, where d is the zone dimension over which diffusion occurs (e.g., thickness of a bar in bending) and ϕ is the thermodynamic factor. Wire or thin-foil samples are employed in a pendulum in order to produce bending.[3]

5. *Magnetic relaxation* in ferromagnetic alloys. In ferromagnetic alloys the local interactions between a magnetic momentum and local order generate relaxation phenomena similar to those noticed under stress, but attributed to the induced anisotropy energy.[33,34] Like the Zener effect, the relation between relaxation time and diffusion is difficult to establish. However, the sensitivity is very high and allows the study of diffusion at a very low defect concentration (10^{-8} to 10^{-10} vacancy fraction).[30]

6. *Nuclear reaction analysis* that occurs only within a narrow energy range has been used to measure the concentration of all elements lighter than chlorine without suitable radioisotopes. Various types of nuclear reactions are available to measure the concentration profiles. For detailed discussion see Lanford et al.[35]

7. *Rutherford back scattering (RBS) analysis* makes use of nuclear interactions with an ion beam (usually ^{4}He ions, typical energies, several megaelectron volts, from an accelerator), but the incident beam particles reemerge unchanged except for a reduction in energy. It works best for the analysis of heavy atoms in a light matrix, the sensitivity being better, the larger the difference between the masses. A heavy element can be employed as an interface marker (e.g., Xe in silicide formation).[3] It is suited to determine diffusion coefficients down to 10^{-21} m^2·s^{-1}.[25]

8. *Secondary ion mass spectroscopy (SIMS)* instrument, also called an ion probe, ion microprobe, or ion microscope, offers the best and most reliable alternative to the classical radiotracer technique. This comprises mass spectroscopic analysis of the secondary ions emitted from the surface (2 to 10 atomic layers) of a specimen upon bombardment with an energetic ion beam (1 to 20 keV). Since the sputtering of specimen occurs inherently in the SIMS, additional sputtering equipment in the ultrahigh vacuum chamber for depth profile measurement is unnecessary. Its main advantage lies in the high detection sensitivity, i.e., impressive depth resolution possible, of the order of nanometers (simply by continuous sputtering of surface atoms from one area of about 250 μm in diameter);[25] its capability of isotopic analysis; and its well-proven ability to examine grain boundary and dislocation-enhanced diffusion as well as very slow diffusion processes or short-term experiments. The reader is referred to some reviews of Petuskey[36] and Kilner[37] on SIMS in diffusion studies.

2.3 ATOMISTIC DIFFUSION MECHANISMS

In crystalline solids the atoms are located at well-defined equilibrium positions (irrespective of thermal vibrations); they move by jumping successively from an equilibrium site to another site.[30] The various possible mechanisms are discussed below.

2.3.1 Direct Exchange and Ring Mechanisms

In the *direct exchange* (or *interchange*) *mechanism* (Fig. 2.11) an interchange of sites takes place between two nearest-neighbor atoms. This is not a likely mechanism for close-packed structures because it would involve large distortions and consequently require very large activation energies. The rotation of a triangle of three atoms in

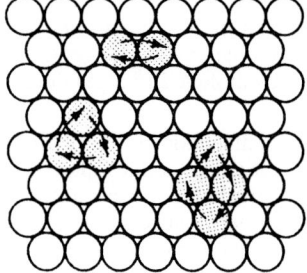

FIGURE 2.11 Mechanism of diffusion in crystals showing direct exchange (or two-ring), three-ring, and cyclic (or four-ring). (*P. A. Thornton and V. J. Colangelo, Fundamentals of Engineering Materials, Prentice-Hall, Inc., Englewood Cliffs, N.J.*)

a close-packed plane is another possible mechanism which requires a lower activation energy than that for the exchange of a pair of atoms (Fig. 2.11).[10]

In the *ring mechanism*, as proposed by Zener, a group of four atoms, forming a ring, jumps simultaneously to neighboring sites (Fig. 2.11). The activation energy required for this type of diffusion is quite low compared to that for the direct exchange, but the mechanisms remain unlikely due to the constraint imposed by a cooperative action involved in which all members of the ring move at once in close-packed structures.

So far, no experimental evidence has been found to support the above mechanisms in dense structures. In metallic liquids and probably in amorphous alloy, mechanisms involving cooperative motions are more likely.

2.3.2 Mechanisms Involving Point Defects

In thermal equilibrium, a crystal always contains a small concentration of point defects such as *vacancies, divacancies,* and *interstitials.* The presence of these defects in the crystals allows atoms to change sites most frequently, without producing a large lattice distortion.

2.3.2.1 Interstitial Mechanism. In the *(direct) interstitial mechanism* (Fig. 2.12),[10] interstitial atoms (or ions) jump (or move) from one interstitial site to an adjacent empty interstitial site. In general, small interstitial atoms such as H, N, O, B, or C in metals diffuse through the lattice by this mechanism. Likewise the mechanism is responsible for the diffusion of hydrogen in Si and Ge.[38] The possibility of interstitial movement decreases with the increasing size (say, >0.6 times the diameter of the solvent) of the interstitial atom. A large atom in a solvent can also move interstitially.

In the *interstitialcy (or indirect) mechanism* the normal substitutional atoms move through the lattice from the interstitial sites to normal substitutional sites and vice versa (Fig. 2.13). Such an atom in the interstitial position is called *self-interstitial.* This mechanism is the counterpart of the vacancy mechanism because the self-interstitial is the antidefect of vacancy.[12] This type of mechanism occurs in the high temperature range in certain classes of materials such as silver halide (AgBr) and many fluorite structures.[1] In close-packed metals and alloys the self-interstitial formation energy is so large that the concentration of these defects may be nonexistent at thermal equilibrium. In Si, this mechanism dominates self-diffusion which is responsible above about 1270 K and plays an important role in the diffusion of some

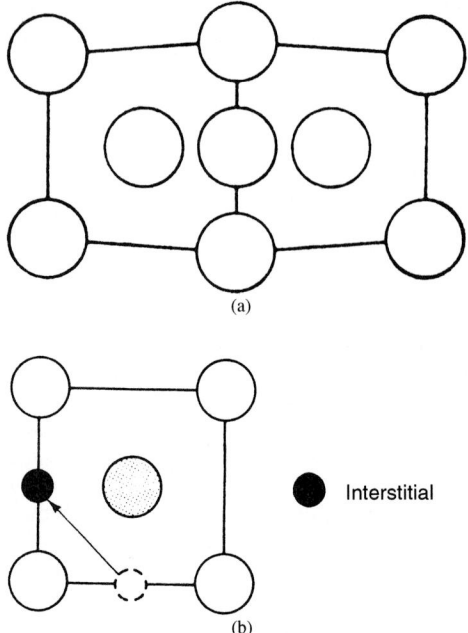

Interstitial

FIGURE 2.12   (*a*) The (100) plane in fcc lattice showing interstitial diffusion.[10] (*b*) Schematic example of an interstitial jump in bcc (center atom is shaded).

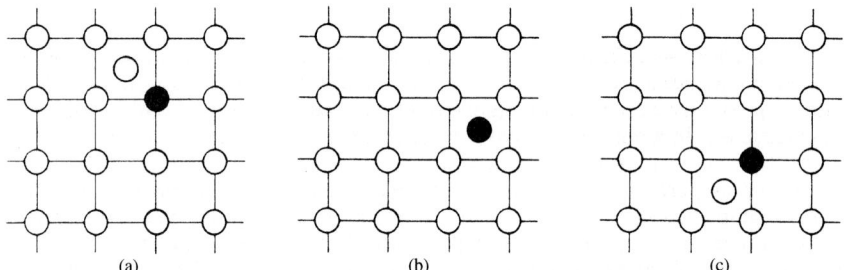

FIGURE 2.13   Interstitialcy mechanism. In (*a*) a self-interstitial (○ in the center of a lattice cell) has diffused to a radioactive self-atom or a substitutional foreign atom ●, respectively; in (*b*) the marked atom has exchanged its original position with the self-interstitial. In this manner the marked atom has temporarily become an interstitial, whereas the original self-interstitial atom has vanished by taking a regular lattice site. In (*c*) the marked atom has retaken a regular site by kicking a self-atom into an interstice.[38] (*Courtesy of W. Frank.*)

substitutional solutes, notably P, B, Al, and Ga.[39] Under nonequilibrium (e.g., irradiated or plastically deformed) conditions, Frenkel pairs (consisting of an equal number of self-interstitials and vacancies pairs) can be thermally created which will contribute simultaneously to the diffusion.[38,39]

In metals and alloys, a *self-interstitial atom* (SIA) is not centered on the interstitial position; rather, it has a *dumbbell (one solute + one solvent) split configuration*

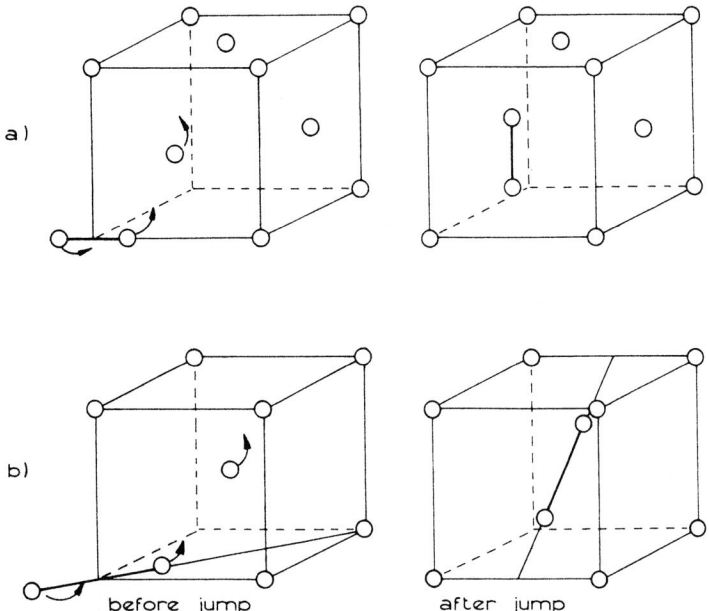

FIGURE 2.14 Elementary jumps of the split interstitials in (*a*) fcc metals and (*b*) bcc metals.[30] (*Reprinted by permission of Elsevier Science BV, Amsterdam.*)

around a stable site. Figure 2.14 shows the elementary jumps for the split interstitials in fcc and bcc metals.[40] It consists of two atoms arranged around a simple stable lattice site (i.e., two atoms centered at one site). It is generally believed that the self-interstitial is split along a $\langle 100 \rangle$ direction in fcc or a $\langle 110 \rangle$ axis in bcc materials,[40] and along $\langle 0001 \rangle$ (*c* axis) in the hcp structure.

At low temperature (annealing), after irradiation (damage), the interstitial atoms would occupy a *crowdion configuration* in which extra interstitial atoms are spread along the close-packed direction, as a result of a flux of high-energy particles knocking atoms out of their normal sites and into its interstitial position (Fig. 2.15).[41] At high temperatures, this crowdion would convert into a split interstitial.

2.3.2.2 *Vacancy Mechanism.*

Frenkel proposed that a certain fraction of lattice sites, known as vacancies, are empty at any temperature above absolute zero. Since the formation energy of vacancies is much smaller than that of interstitials, the concentration of vacancies in thermodynamic equilibrium is always much greater than that of interstials.[42] The extents of these vacancies depend on pressure, temperature, and composition. In metals and alloys near the melting point, the vacancy concentration is about 10^{-3} to 10^{-4} site fraction. Diffusion takes place easily by this mechanism when a "direct exchange" occurs between the atom and the vacancy (Fig. 2.16).[39] Vacancies move opposite to the movement of atoms. This is the well-established mechanism that is dominant in many metals and ionic compounds.[43] This mechanism controls self-diffusion in all metals, as studied by Seeger and Peterson.[39] In Si and Ge, this is the main contributor to self-diffusion below about 1270 K.

FIGURE 2.15 The (111) plane of fcc lattice showing a crowdion. (Note extra atom in the middle row.)[10]

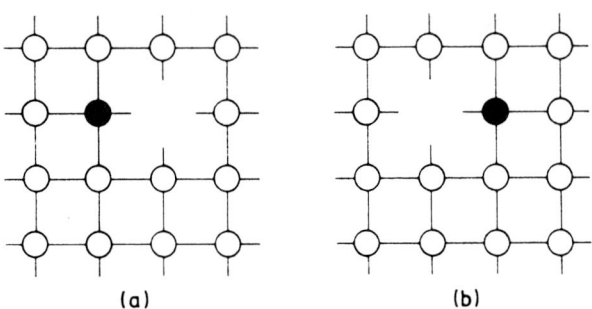

(a)   **(b)**

FIGURE 2.16 Vacancy mechanism: The radioactive self-atom in tracer self-diffusion or foreign atom in substitutional-solute diffusion (●) moves, by jumping into the vacancy on its right-hand side (*a*), to the right (*b*) by one nearest-neighbor distance of the regular lattice atoms.[38,39]

In addition to monovacancies, there are vacancy aggregates such as divacancies (or double vacancies), trivacancies, etc., which also contribute to the diffusion. The ratio of divacancies to monovacancies usually increases with temperature, thereby enhancing the divacancy contribution to diffusion. The observed curvatures of the Arrhenius plot at high temperatures are probably attributed to divacancies.

Molecular dynamic calculations by Da Fano and Jacucci[30] have demonstrated that at elevated temperatures, the atom jump frequency becomes large and a dynamic correlation between successive jumps can take place in such a way as to move a vacancy by more than one jump distance. These *vacancy double jumps* are another explanation for anomalies found in the Arrhenius plot at high temperatures.[30]

2.3.2.3 Mixed Mechanisms.   Complexes of intrinsic and extrinsic point defects may diffuse as entities. Examples are the *A center* (*vacancy-interstitial oxygen pair*), the *E center* (*vacancy-substitutional phosphorus pair*) in Si as reported by Watkins and Corbett[44] and Watkins et al.,[45] and the *mixed dumbbells* (*self-interstitial-substitutional*) *foreign-atom pairs* as observed by Dederichs et al.[46]

Foreign atoms (A) under thermal equilibrium conditions may occupy either as substitutional (A_s) or interstitial (A_i) atoms (such as Au or Ni in Si) and may diffuse *through* the *kick-out mechanism*[47] (Fig. 2.17) and/or *through* the *dissociative mechanism*[48] (Fig. 2.18). The two mechanisms are characterized by the much higher diffusivity of foreign atoms when they are located in the interstitial positions than when they are located in the substitutional sites. In these cases, the incorporation of A

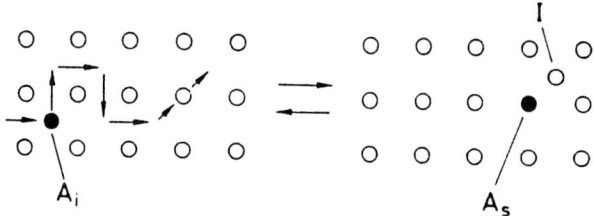

FIGURE 2.17 Kick-out mechanism: Foreign atom (full circle) interchange between interstitial sites A_i and substitutional sites A_s in cooperation with self-interstitial I.[38] (*Courtesy of W. Frank.*)

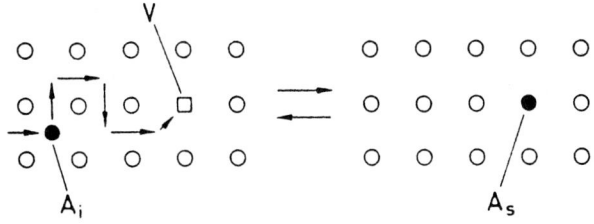

FIGURE 2.18 Dissociative mechanism: Foreign atoms (full circle) interchange between interstitial sites A_i and substitutional sites A_s in cooperation with vacancies V.[38] (*Courtesy of W. Frank.*)

atoms can take place by fast diffusion as A_i interstitials and their subsequent changeover to the substitutional sites. The difference between the two mechanisms depends on the type of intrinsic point defects which mediates this changeover.[12]

In the *kick-out mechanism*, as shown in Fig. 2.17, the interchange involves self-interstitial (I) according to the reaction

$$A_i \rightleftharpoons A_s + I \tag{2.54}$$

Examples of kick-out mechanisms have been established for some diffusing foreign atoms in Si, the diffusion of Zn in GaAs semiconductor, and the diffusion of Au in Si.[49,50] So far, no examples of kick-out mechanisms have been observed in metals.[12]

In the *dissociative, Frank-Turnbull, or Longini mechanism*, as shown in Fig. 2.18, the interchange involves vacancies (V) according to the reaction

$$A_i + V \rightleftharpoons A_s \tag{2.55}$$

The rapid diffusion of some foreign metals is said to be attributed to the dissociative mechanism. The backward reaction is the dissociation of a substitutional foreign atom A_s into a Frenkel pair with extrinsic interstitial partner A_i. The forward reaction is the recombination of such a pair. It is noted that reactions (2.54) and (2.55) are not symmetric under the exchange of I and V. That is why the theoretical prediction for the kick-out and the dissociative mechanisms may differ widely.[38]

In ordered compounds, diffusion by nearest-neighbor jumps of vacancies may cause interchange of atoms on different sublattices which, in turn, results in the formation or elimination of so-called antisite defects. An example is the diffusion in Group III-V semiconductor compounds.[51,52]

2.3.3 Short-Lived Frenkel Pairs

Numerical simulations have established that, at least in noncompact phases at elevated temperatures, short-lived Frenkel pairs can form *homogeneously* and generate rings of replacement of various sizes (four atoms and above). At the end of sequence, the pair recombines.[30,53]

2.3.4 Mechanisms Involving Extended Defects

Linear defects (dislocations) and planar defects (surfaces, interfaces, grain boundaries, etc.) are disordered regions where the atomic migration is easier than in the bulk. These preferential paths of diffusion are known as *short circuits*. In this area, the diffusion mechanisms have not yet been well established, but there is a great expectation of rapid theoretical advances as a result of the increasing power of computers.[30] For more details, see Sec. 2.11.

2.4 RANDOM WALK THEORY OF MICROSCOPIC DIFFUSION

Random walk theory has been formulated to explain the observed macroscopic diffusion from the atomic jumps. Since the directions of successive jumps of an atom through the cubic lattice are assumed to be random, the path followed by each atom during diffusion may be described as a random walk. According to Einstein, the diffusion coefficient of a given species for an isotropic material is given by

$$D = \frac{\langle \overline{R}_n^2 \rangle}{6t} \tag{2.56}$$

where R_n is the net displacement of an atom from the origin after making n jumps in time t, $\overline{R}_n^2$ is the mean squared displacement of an atom in time t, and $\langle \overline{R}_n^2 \rangle$ is defined by

$$\langle \overline{R}_n^2 \rangle = \sum_{i=1}^{n} \langle r_i \rangle^2 + 2 \sum_{i=1}^{n-1} \sum_{j=1}^{n-i} \langle r_i \cdot r_{i+j} \rangle \tag{2.57}$$

For a truly random walk, where each jump direction is unrelated to the direction of the preceding jumps, the second term in Eq. (2.57) reduces to zero, because for any product $r_i \cdot r_{i+j}$, there will always be equal probability of their positive and negative values. Consequently, we have the expression

$$\langle \overline{R}_n^2 \rangle = \sum_{i=1}^{n} \langle r_i \rangle^2 \tag{2.58}$$

or

$$\left(\overline{R}_n^2 \right)^{1/2} = \sqrt{n} \, r = (6Dt)^{1/2} = 2.45(Dt)^{1/2} \tag{2.59}$$

which states that root mean-square displacement of an atom in time t is proportional to the square root of the number of jumps, as well as to $(Dt)^{1/2}$. This equation is also similar to Eq. (2.42) used for carburizing case depth midway between surface and base concentrations.

There are two ways to derive the equation relating D to atomic jump frequency Γ and jump distance r. A simple approach considers atomic jumps occurring in both directions between adjacent crystal planes in the presence of a concentration gradient using Fick's law. In cubic crystals, if all atomic jumps involve the same distance r (usually the nearest-neighbor distance in the crystal), then

$$D = \frac{1}{6}\Gamma r^2 \qquad (2.60)$$

For example, for a fcc crystal, $r = a\sqrt{2}/2$ (a being the lattice parameter of the cubic cell) and $D = (a^2/12)\Gamma$, whereas for the bcc lattice, $r = a\sqrt{3}/2$ and $D = (a^2/8)\Gamma$.

These considerations of random walk of the diffusing atoms may not be fully correct when a defect mechanism is involved. For example, when a tracer atom interchanges positions with a vacancy, there will be a greater-than-random probability for that atom to make a return jump into the neighboring vacancy, thereby canceling the effect of the first jump. We, therefore, need to introduce an additional tracer correlation factor f that is written, in most cases, as

$$f = \frac{D_{\text{actual}}}{D_{\text{random}}} = \frac{D^*}{D} \qquad (2.61)$$

and

$$f = \frac{1 + \langle\cos\theta_1\rangle}{1 - \langle\cos\theta_1\rangle} \qquad (2.62)$$

where $\langle\cos\theta_1\rangle$ is the average value of the cosine of the angle between an initial atom jump and the first subsequent jump. The factor f is always less than 1, $\langle\cos\theta_1\rangle$ being negative due to the possibility of a return jump. In fact, $\langle\cos\theta_1\rangle$ is obtained from an infinite series that converges slowly, and methods have been devised to calculate it accurately. The most commonly used f values are 0.7815 and 0.7272 for self-diffusion via the vacancy mechanism in fcc and bcc crystals, respectively. Diffusion involving divacancies, however, provides quite different results; for example, $f = 0.475$ for fcc and hcp crystals.[30] In the case of bcc crystals, a pure divacancy mechanism is impossible, because the divacancy must dissociate during a nearest-neighbor atom jump. An extension of the concepts of correlation factors to noncubic crystals is somewhat more complex and will not be treated here.

2.5 VACANCY DIFFUSION

2.5.1 Vacancies in Thermodynamic Equilibrium

Among the several diffusion mechanisms discussed in Sec. 2.3, vacancy has been accepted as the most predominant mechanism. Vacancies always exist as an integral part of the structure in thermodynamic equilibrium. We will derive here the equilibrium concentration of vacancies in a pure metal as a function of temperature on the basis of a thermodynamic model. For a better understanding of this diffusion mechanism, the vital concept to be realized is that the entropy of a defect-containing crystal is larger than that for a perfect crystal and that the entropy change is of two types: (1) vibrational entropy change ΔS_V and (2) mixing or configurational entropy change ΔS_m as a result of the introduction of vacancies in a pure metal.

Let us assume that a large piece of pure metal contains n_a atoms and that n_v vacancies have been introduced and mixed throughout the metal.[10] The increase in mixing or configurational entropy ΔS_m, according to statistical mechanics, is given by

$$\Delta S_m = k \ln \Omega + k \ln \frac{(n_a + n_v)!}{n_a! n_v!} \tag{2.63}$$

where Ω is the configurational degeneracy and equals $(n_a + n_v)! / (n_a! n_v!)$. Applying Sterling's approximation and some algebraic calculations, we have

$$\Delta S_m = -k \left(n_a \ln \frac{n_a}{n_a + n_v} + n_v \ln \frac{n_v}{n_a + n_v} \right) \tag{2.64}$$

$$= k[(n_a + n_v) \ln(n_a + n_v) - n_v \ln n_v - n_a \ln n_a] \tag{2.65}$$

In Eq. (2.65), ΔS_m must be a positive large number to favor the process. Similarly an increase in vibrational entropy ΔS_v based on statistical mechanics can be given by[54]

$$\Delta S_v = k \sum_{i=1}^{N'} \ln \frac{v}{v'} \tag{2.66}$$

where N' is the number of degrees of freedom, v is the initial frequency of atom, and v' is the final frequency of the atom around the vacancy. The vacancy appears to increase vibrational amplitude and lower v' so that $v/v' > 1$. This means ΔS_v is positive.

In any isothermal isobaric system the free energy change for vacancies is

$$\Delta G = G_v^f - G = n \Delta H_v^f - T\left(n \Delta S_v^f + \Delta S_m^f\right) \tag{2.67}$$

where G_v^f and G are the free energy of a pure metal with and without vacancies, respectively; ΔH_v^f is the increase in enthalpy of the formation of a vacancy; and ΔS_v^f is the corresponding vibrational entropy change and ΔS_m^f is the mixing entropy change per vacancy.

Combining Eqs. (2.65) and (2.67), we obtain

$$\Delta G = n \Delta H_v^f - Tn \Delta S_v^f - kT[(n_a + n_v) \ln(n_a + n_v) - n_v \ln n_v - n_a \ln n_a] \tag{2.68}$$

ΔG must be a minimum at a given temperature when the equilibrium number of vacancies n_v^* is reached. That is, $(d\Delta G/dn_v)_T$ in Eq. (2.68) must be equal to zero, i.e.,

$$\left(\frac{d\Delta G}{dn_v} \right)_T = 0 = \Delta H_v^f - T \Delta S_v^f + kT \ln \frac{n_a + n_v}{n_v} \tag{2.69}$$

Rearranging Eq. (2.69) gives the equilibrium vacancy concentration C_v^*

$$C_v^* = \frac{n_v^*}{n_a + n_v^*} = \exp \frac{-\Delta H_v^f}{kT} \exp \frac{\Delta S_v^f}{k} \tag{2.70}$$

or
$$C_v^* = \exp\frac{-\Delta G_v^f}{kT} \tag{2.71}$$

where
$$\Delta G_v^f = \Delta G_{1v}^f = \Delta H_{1v}^f - T\Delta S_{1v}^f \tag{2.72}$$

where ΔG_v^f or ΔG_{1v}^f represents the free energy associated with the formation of a single or monovacancy. Since the vacancy formation enthalpy ΔH_v^f is quite larger than the ΔS_v^f term, ΔG_v^f is approximated to be equivalent to ΔH_v^f. In a pure metal, ΔH_v is the same for all sites. In a binary system, the ΔH_v^f term varies for different sites.

In addition to monovacancies, it is thermodynamically possible to have higher-order clusters of vacancies (such as divacancies and trivacancies) in equilibrium. The thermodynamic equilibrium concentration of divacancies C_{2v}^* can be given by the Arrhenius expression of the form[55]

$$C_{2v}^* = \frac{z}{2}\exp\frac{-\Delta G_{2v}^f}{kT} \tag{2.73}$$

where $z/2$ is the number of orientations of the divacancy, z the number of first or second nearest neighbors (based on the stable form of the divacancy), and ΔG_{2v}^f the Gibbs free energy of formation of a divacancy. An equation similar to Eq. (2.72) can be expressed for ΔG_{2v}^f. In addition to the existence of statistical divacancies, i.e., divacancies produced mainly due to random encounters of monovacancies, it is possible to have an attractive interaction between two monovacancies that increases C_{2v}^*. This attractive interaction, known as the Gibbs free energy of binding ΔG_{2v}^b, can be defined as

$$\Delta G_{2v}^b = 2\,\Delta G_{1v}^f - \Delta G_{2v}^f \tag{2.74}$$

where an attractive interaction $(2\,\Delta G_{1v}^f > \Delta G_{2v}^f)$ denotes a positive value of ΔG_{2v}^b. Therefore, a more useful equation for C_{2v}^* is given by[55]

$$C_{2v}^* = \frac{z}{2}\left(C_{1v}^*\right)^2 \exp\frac{\Delta G_{2v}^b}{kT} \tag{2.75}$$

A positive value of ΔG_{2v}^b means that C_{2v}^* increases more rapidly with temperature than when ΔG_{2v}^b becomes equal to zero. A combined field-ion microscopy and resistivity experiments on vacancies in quenched platinum performed by Seidman[56] suggests the direct evidence for the presence of divacancies and a positive value of ΔG_{2v}^b. However, these vacancies might correspond to divacancies at the surface or just below the surface and, therefore, do not prove their existence within the bulk of a crystal.[57]

In a similar way, we can express the equilibrium concentration of trivacancies C_{3v}^* as

$$C_{3v}^* = \eta\left(C_{1v}^*\right)^3 \exp\frac{\Delta G_{3v}^b}{kT} \tag{2.76}$$

where η is the number of physically distinct orientations of the most stable configuration of a trivacancy and ΔG_{3v}^b is the Gibbs free energy of binding of a trivacancy. If the vacancy remains the main point defect in thermodynamic equilibrium, the total equilibrium vacancy concentration C_{tv}^* is given by

$$C_{tv}^* = C_{1v}^* + 2C_{2v}^* + 3C_{3v}^* + \cdots + nC_{nv}^* + \cdots \tag{2.77}$$

where n is the number of vacancies contained in the cluster.[55] For equilibrium concentration measurements, higher aggregates than divacancies need not be considered because of their negligibly small concentration in the normal cubic metals.[57]

2.5.2 Migration Rates of Defects and Atoms

In moving from one site to another adjacent site, a defect has to overcome an activation barrier at the saddle point (midpoint position). The work done in this reversible, isothermal process, at constant pressure, is the change in Gibbs free energy of migration (ΔG_d^m) from a region when an atom moves from a normal site to the saddle-point position (the height of the barrier to be overcome), which again can be divided into an enthalpy of migration ΔH_d^m and a vibrational entropy ΔS_d^m term according to

$$\Delta G_d^m = \Delta H_d^m - T\,\Delta S_d^m \tag{2.78}$$

It is assumed that ΔG_d^m has all the properties held by ΔG_v^f of Eq. (2.72). The equilibrium concentration of activated complexes in the region of the saddle point C_d^* can be calculated using the same treatment as for C_v^* in Eq. (2.73). Instead of mixing into the lattice vacancies that increase the ΔG_v^f per mole of vacancies, we mix in activated complexes which increase the free energy by ΔG_d^m per mole of complexes. Since the ideal entropy of mixing remains the same for vacancies and activated complexes, the equilibrium concentration of activated complexes in the neighborhood of the saddle point at any instant can be given by

$$C_d^* = \frac{n_d^*}{n_a + n_d^*} = \exp\left(\frac{-\Delta H_d^m + T\Delta S_d^m}{kT}\right) = \exp\left(\frac{-\Delta G_d^m}{kT}\right) \tag{2.79}$$

where C_d^* is the equilibrium concentration of activated complexes, n_d^* is the equilibrium number of activated complexes, and n_a is the number of atoms. A statistical analysis, taking all degrees of freedom into consideration, leads to a simple final result for the probability per second w_d of a defect jumping into a particular neighboring site (i.e., the average jump frequency of defect per atom):

$$w_d = v C_d^* = v \exp\left(\frac{-\Delta G_d^m}{kT}\right) \tag{2.80}$$

where v is the oscillation (or vibrational) frequency at the bottom of the potential well of the atom that makes the jump, which is usually taken close to the Debye frequency v_D ($\sim 3 \times 10^{12}\,\text{s}^{-1}$).[28]

2.5.3 Experimental Measurement of Enthalpy (ΔH_v^f) and Entropy (ΔS_v^f) in Vacancy Formation

During quenching, three perturbing effects arise:[44] (1) clustering of vacancies, (2) loss of vacancies to sinks such as dislocations and grain boundaries, and (3) creation of additional defects by plastic deformation. Ideally, one would like to measure directly the individual concentration of various clusters C_{nv} which are present after quenching. But only the total quenched-in single or monovacancy C_{1v} has been approximately estimated to date.

Properties of vacancy-type defects are derived from results of measurements of the self-diffusion coefficient, the isotope effect, the lattice parameter, length expansion with temperature, the quenched-in resistivity, and positron annihilation.[42] Only the last three are briefly discussed here.

2.5.3.1 Direct Determination.

Individual concentrations of monovacancies C_{1v}, divacancies C_{2v}, and higher-order vacancies C_{nv} present at a quenched metal tip can be determined directly by field ion microscopy (FIM). However, the main problems involved in the measurement of C_{nv} are (1) existence of artifact defects, (2) the sampling problems linked with the observation of a statistically large defect concentration, and (3) stress-induced defect migration, because of the imposition of large electric field on the specimen at the imaging temperature (T); this causes a reduced concentration of defects.[58]

Differential Dilatometry. The second direct method for the determination of the total quenched-in equilibrium concentration of vacancies C_v^* is differential dilatometry. This involves the simultaneous measurement of macroscopic length expansion $\Delta L/L$ and microscopic lattice parameter expansion $\Delta a/a$ as a function of temperature. This gives an additional amount of vacancies if measured at higher temperature in a nonequilibrium condition because the mobility of vacancies increases with increasing temperature. Due to the lack of mobility of vacancies below a certain temperature, the specimen will always contain a nonequilibrium concentration $C_{v,0}$ of vacancies at ambient temperature. Hence, the total vacancy concentration at temperature T in isotropic media is[59]

$$C_v(T) = C_{v,0} + 3(1 + C_{v,0})\left(\frac{\Delta L}{L} - \frac{\Delta a}{a}\right) \tag{2.81}$$

If we assume $C_{v,0} = 0$, then Eq. (2.81) converts the well-known Simmons and Balluffi equation to

$$C_v(T) = 3\left(\frac{\Delta L}{L} - \frac{\Delta a}{a}\right) \tag{2.82}$$

This equation holds good for close-packed metals or cubic crystals with small concentration of vacancies. A plot of $3(\Delta L/L - \Delta a/a)$ against $1/T$ gives approximate values for ΔH_v^f and ΔS_v^f (Fig. 2.19). Table 2.3 lists ΔH_v^f and ΔS_v^f values for several metals. These values often imply the assumption that ΔH_v^f and ΔS_v^f are independent of T. However, this may not always be a sufficient approximation. For example, there appear to be large temperature variations in metals which exist in more than one crystal structure.[60] But in all cases, it is presumed that such variations arise from the normal lattice expansion with increasing T and the accompanying changes in interatomic forces.

In case both vacancies and interstitials form, the right-hand portion of Eq. (2.82) would be equal to the difference between vacancy and interstitial concentrations C_v and C_I, respectively.[10] That is,

$$C_v - C_I = 3\left(\frac{\Delta L}{L} - \frac{\Delta a}{a}\right) \tag{2.83}$$

This equation assumes that vacancies and interstitials occupy the same related volume; this is not true. It is also noted that interstitials are less likely to be quenched-in.

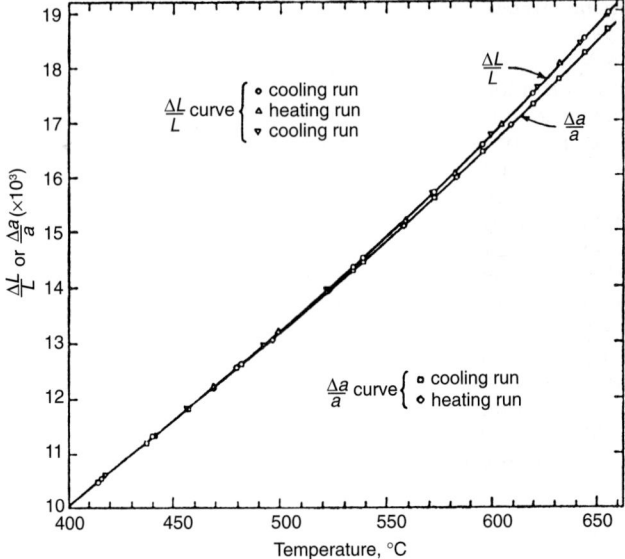

FIGURE 2.19 Differential dilatometry in vacancy equilibrium conditions for aluminum. Relative length change $\Delta L/L$ and relative lattice parameter change $\Delta a/a$ of a sample against $1/T$. The difference between the two lines is linearly proportional to the concentration of vacant atomic sites.[10] (*From R. Simmons and R. Balluffi, Phys. Rev., vol. 117, 1960, p. 52.*)

TABLE 2.3 Values of ΔH_V^f as Determined by Positron-Annihilation Spectroscopy (PAS) and Quenching Methods for a Number of FCC and BCC Metals[62]

Metal	H_{iv}^f (eV)	
	PAS	Quenching
Al	0.66 ± 0.02	0.66
Ag	1.16 ± 0.02	1.10
Au	0.97 ± 0.01	0.94
Cu	1.31 ± 0.05	1.30
Ni	1.7 ± 0.1	1.6
V	2.1 ± 0.2	
Nb	2.6 ± 0.3	
Mo	3.0 ± 0.2	3.2
Ta	2.8 ± 0.6	
W	4.0 ± 0.3	3.7

2.5.3.2 *Indirect Determination*

The Quenched-in Electrical Resistivity ρ_q Measurement.   A second and more accurate determination of total vacancy concentration can be obtained by measuring the change in quenched-in resistivity $\Delta\rho_q$ as a function of temperature and hydrostatic pressure. (The order of magnitude of pressure dependence is such that for gold, a 6-kbar pressure increment corresponds to a decrease of temperature of about 30 K around 900 K for constant C_v^{*}.[57]) The method consists of heating the metal (such as Au, Cu) samples to a high temperature T_q, followed by rapid quenching to a very low temperature T_1 (4.2 or 78 K) at which vacancies become immobilized and incapable of diffusing to sinks. This equilibrium concentration of vacancies corresponding to T_q is maintained at T_1. Subsequently, resistivity measurements at low temperatures are performed on the quenched as well as the unquenched (or well-annealed) samples for direct comparison. It is found that the quenched concentration C_v remains smaller than C_v^{*} (T_q) because of (1) the loss of migrating vacancies during quenching to sinks such as dislocations, grain boundaries, and surfaces and (2) vacancy clustering. The difference in these resistivities $\Delta\rho_q (= \Delta\rho_v)$ is then taken to be directly proportional to the total concentration of quenched-in vacancies. That is,

$$\Delta\rho_q = \Delta\rho_v = \alpha C_v \qquad (2.84)$$

where α is the resistivity per unit concentration of vacancies. Some deviations up to 10% may occur for vacancy defects at 78 K; however, these deviations become negligible at 4.2 K. Figure 2.20 shows an Arrhenius plot of the quenched-in residual resistivity of single and polycrystals of gold with different dislocation densities N_d.[57]

Positron-Annihilation   Spectroscopy.   Another   method   utilizes   *positron-annihilation spectroscopy* (PAS) to determine ΔH_v^f and thereby the vacancy concentration in thermal equilibrium in the medium temperature range $(T \approx 0.6T_m)$. High-energy positrons produced by a nuclear reaction injected into a metal specimen are rapidly thermalized within a picosecond by electron-hole excitations and interactions with phonons. The thermalized positrons diffuse through the lattice and end its life either by annihilation with electrons or by trapping with vacancies. The vacancy-trapped positrons cause a marked reduction in the local electron density and produce increased lifetimes by 20 to 80% with respect to the free positrons in the perfect lattice. It is noted that the average lifetime of a positron in the metal crystal is about 200 ps.[61] The increased lifetimes of captured positrons are proportional to the vacancy concentration $C_v(T)$. The temperature dependence of the lifetimes is also used for accurate ΔH_v^f determination. Figure 2.21 shows the Arrhenius plot of vacancy concentration as measured by the positron annihilation spectroscopy for Au and Cu. The range of measurement by differential dilatometry is indicated for comparison. The slope of the line is a measure of ΔH_v^f.[61] The PAS data extend to about two orders of magnitude lower vacancy concentration than the differential dilatometry data. Since monovacancies are predominant in this concentration range, PAS studies are very valuable as a complement to differential dilatometry on one side, and to resistivity measurements in quenched specimens on the other side. Table 2.3 lists values of ΔH_v^f obtained by PAS and resistivity methods for a number of fcc and bcc metals; usually a good agreement is found between these two methods.[62]

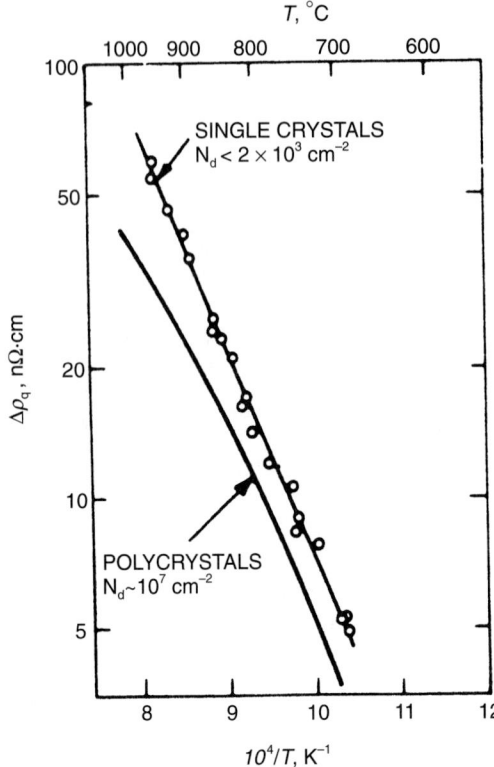

FIGURE 2.20 Arrhenius plot of the quenched-in residual resistivity of single crystals and polycrystals of gold, showing the effect of the high dislocation density N_d in causing the curved plot for quenched polycrystals.[57] (*Source: After B. Lengeler, Phil. Mag., vol. 34, 1976, p. 259.*)

2.6 EFFECT OF KEY VARIABLES ON DIFFUSIVITY

In this section we will study diffusion as a function of key variables such as temperature T, pressure P, and isotopic mass m.

2.6.1 Effect of Temperature

Normal self-diffusion (at elevated temperature) proceeds mainly through vacancies. Empirically it has been observed that the temperature dependence of the diffusivity of many systems can be expressed by the well-known Arrhenius-type equation

$$D = D_0 \exp\left(\frac{-Q}{RT}\right) \tag{2.85}$$

where D is the volume diffusivity; D_0 is the preexponential or frequency factor for a given diffusion couple which has a value within a range of $10^{-6}\,m^2 \cdot s^{-1}$ to $10^{-3}\,m^2 \cdot s^{-1}$; Q is the total activation energy, within a range of $50\,kJ \cdot mol^{-1}$ to $600\,kJ \cdot mol^{-1}$, for

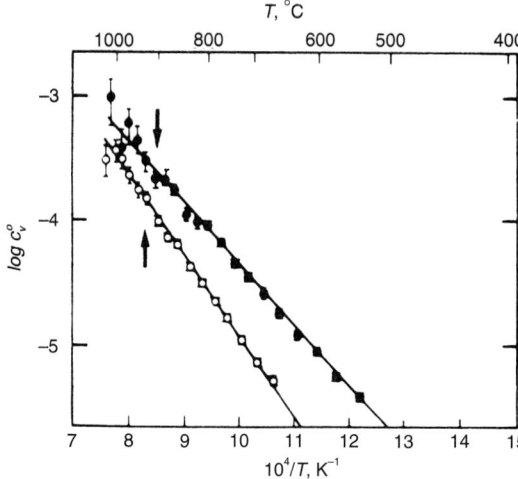

FIGURE 2.21 Arrhenius plot of the vacancy concentration as derived from positron annihilation spectroscopy for copper (open symbols) and gold (full symbols) according to Triftshauser and McGervey. The left-hand side of the arrow corresponds to the range covered by the differential dilatometry. (*Source: W. Triftshauser and J. D. McGervey, Appl. Phys., vol. 6, 1975, p. 177.*)

self-diffusion; and R is the molar gas law constant. Both D_0 ($= v_0 a^2 f$) [here v_0 is an attempt frequency of the order of magnitude of the Debye frequency, a is the cubic lattice parameter, and f is a geometric factor depending on the lattice structure and type of interstitial sites involved[12]], and Q ($= \Delta H^f + \Delta H^m$) will vary with composition, but are independent of temperature (provided the diffusion mechanism does not change) and are together called *Arrhenius parameters*. The above equation may also be written as

$$\log D = \log D_0 - \frac{Q}{2.302RT} \tag{2.86}$$

When $\log D$ is plotted against $1/T$, a straight line is obtained with a slope equal to $Q/2.302R$. The value of Q is observed to be a function of the melting temperature T_m by the expression $Q = 34T_m$ in calories per mole, or $Q = 141T_m$ J/mol and T_m (in kelvins). This behavior forms the basis of the Van Liempt relation and is well obeyed by compact metals (see Fig. 2.22a for fcc metals and Fig. 2.22b for bcc metals). It is noted here that bcc structures exhibit a wider dispersion around the Van Liempt straight line than that of the fcc structures. Table 2.4 lists diffusion parameters D_0 and Q for self-diffusion in various pure elements which can be employed to produce the Arrhenius diffusivity plots. Figure 2.23 shows the Arrhenius plot for certain metals. Figure 2.24 shows similar plots for the diffusion of impurity/tracer elements into silicon which are of importance in the fabrication of integrated circuits for the electronic industry. Figure 2.25 shows the Arrhenius plot of the different regimes of behavior which are possible for diffusion within the crystal lattice in a ceramic.

For most of the metals the Arrhenius plot is slightly curved, even if contributions from short-circuit diffusion can be excluded. This curvature can be explained with temperature-dependent enthalpies ΔH and entropy ΔS (one-defect model) or with the competition of two or more diffusion mechanisms with different diffusion energies (two-defect model). In this case, temperature-dependent self-diffusivity D^T near the melting point is given by

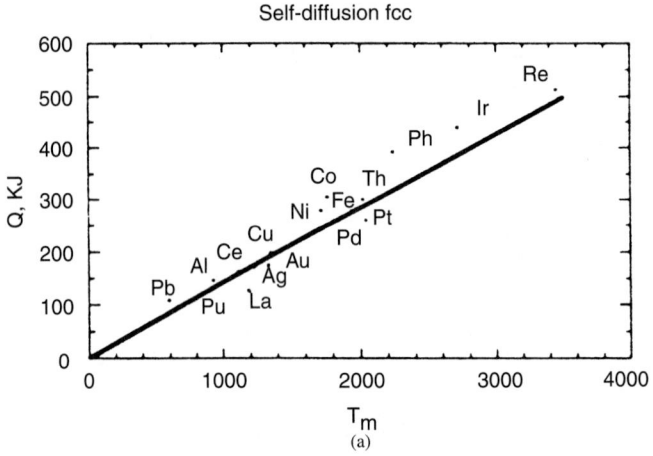

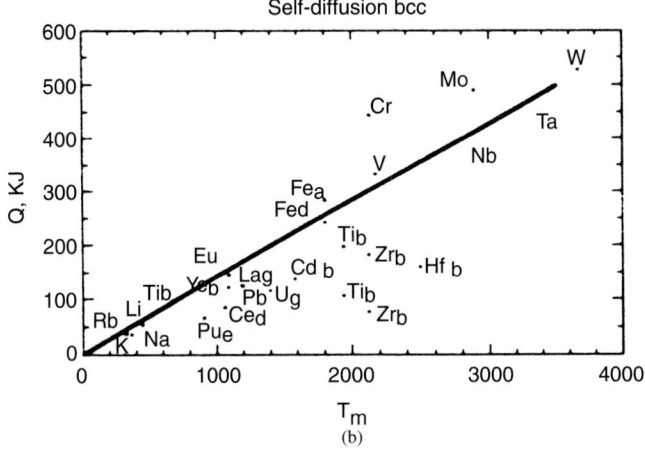

FIGURE 2.22 Van Liempt relation for compact metals: (*a*) fcc metals and (*b*) bcc metals. The straight lines represent the Van Liempt relation: $Q = 0.14T_M$ kJ/mol and illustrates a wide dispersion of bcc structures around the line compared to the fcc structures.[30]

$$D(T) = D^T = D_{01v} \exp\left(\frac{-Q_{1v}}{kT}\right) + D_{02v} \exp\left(\frac{-Q_{2v}}{kT}\right) \qquad (2.87)$$

where subscripts $1v$ and $2v$ denote the mono- and divacancy contributions, respectively.[28,63]

2.6.2 Effect of Pressure

The rate of diffusion process is generally given by the expression

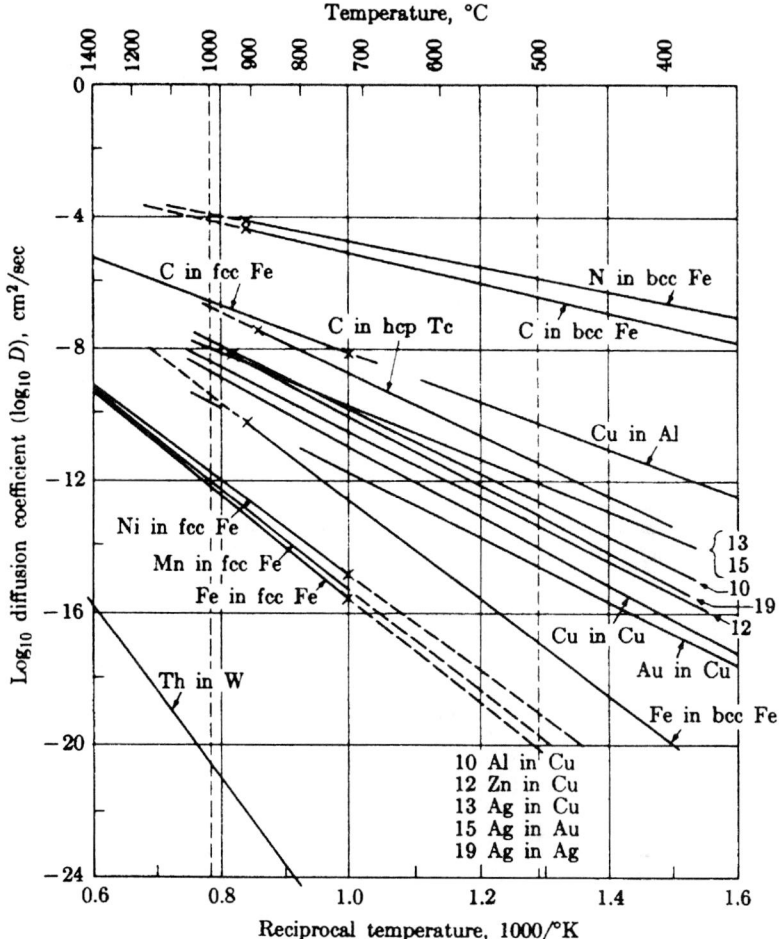

FIGURE 2.23 Arrhenius plots for certain metals.

$$D_{1v}^T = D = fv\alpha^2 \exp\left(\frac{-\Delta G_{1v}}{kT}\right) = fv\alpha^2 \exp\left[\frac{-\left(\Delta G_{1v}^f + \Delta G_{1v}^m\right)}{kT}\right]$$

$$= fv\alpha^2 \exp\left[\frac{-\left(\Delta H_{1v}^f + \Delta H_{1v}^m\right)}{kT}\right]\exp\left(\frac{\Delta S_{1v}^f + \Delta S_{1v}^m}{k}\right) \qquad (2.88)$$

where f is the geometric correlation factor (0.728 for Na),[64] α is the jumping distance, v is the average attempt frequency, and ΔG_{1v} is the Gibbs free energy change comprising ΔG_{1v}^f and ΔG_{1v}^m, corresponding to the formation and migration of defects, respectively. We can obtain a relation between D and the external hydrostatic pressure P by differentiating Eq. (2.88) with respect to pressure P at a constant temperature T as

TABLE 2.4 Self-diffusion Parameters for Pure Elements[30]

Element (see comments at end)[a]	C.S.[b]	T_m (K)[c]	D_0(m²s⁻¹) × 10⁴[d]	Q (kJ/mol)[e]	Temp. range (K)[f]	Q = 0.1422T_m Van Liempt[g]	$D(T_m)$(m²s⁻¹)[h]	D(ph. tr.)(m²s⁻¹)[i]	Reference in Ref. 30[j]
Ag	fcc	1234	$D_{01}=0.046$ $D_{02}=3.3$	$Q_1=169.8$ $Q_2=218.1$	594–994	175.5	4.9×10^{-13}		Rein and Mehrer (1982)
Al	fcc	933	2.25	144.4	673–883	132.7	1.85×10^{-12}		Beyeler and Adda (1968)
Au	fcc	1336	0.084	174.1	1031–1333	190	1.3×10^{-12}		Herzig et al. (1978)
Be	hex	1560	⊥c 0.52 //c 0.62	157.4 165	836–1342 841–1321	221.8 NA[k]	2.79×10^{-10} 1.85×10^{-10}		Dupouy et al. (1966)
Ca	bcc	1116	8.3	161.2	773–1073	158.7	2.36×10^{-11}		Pavlinov et al. (1968)
Cd	hex	594	⊥c 0.18 //c 0.12	82 77.9	420–587 NA	84.5 NA	1.11×10^{-12} 1.69×10^{-12}		Mao (1972)
Ceγ T < 999	fcc γ/δ 999		0.55	153.2	801–965	152.3*	4.9×10^{-11}	5.37×10^{-13} (999 K)	Dariel et al. (1971)
Ceδ T > 999	bcc	1071	0.007	84.7	1018–1064	152.3	2.65×10^{-13}	2.36×10^{-11} (999 K)	Languille et al. (1973)
Co	fcc	1768	2.54	304	944–1743	251.4	1.86×10^{-12}		Lee et al. (1988)
Cr	bcc	2130	1280	441.9	1073–1446	302.9	5.97×10^{-13}		Mundy et al. (1981)
Cu	fcc	1357	$D_{01}=0.13$ $D_{02}=4.6$	$Q_1=198.5$ $Q_2=238.6$	1010–1352	193			Bartdorff et al. (1978)
Er	hex	1795	⊥c 4.51 //c 3.71	302.6 301.6	1475–1685 NA	255.2	7.05×10^{-13} 6.2×10^{-13}		Spedding and Shiba (1972)
Eu	bcc	1099	1	144	771–1074	156.2	1.43×10^{-11}		Fromont and Marbach (1977)
Feα T < 1183	bcc	α/γ 1183	121	281.6	1067–1168	257.2*		4.45×10^{-15} (1183 K)	Geise and Herzig (1987)
Feγ 1183 < T < 1663	fcc	γ/δ 1663	0.49	284.1	1444–1634	257.2*		1.4×10^{-17} (1183 K) 5.83×10^{-14} (1663 K)	Heuman and Imm (1968)
Feδ T > 1663	bcc	1809	2.01	240.7	1701–1765	257.2	2.25×10^{-11}	5.5×10^{-12} (1663 K)	James and Leak (1966)
Gdβ	bcc	1585	0.01	136.9	1549–1581	225.4	3.07×10^{-11}		Fromont and Marbach (1977)
Hfα T < 2013	hex	α/β 2013	⊥c 0.28 //c 0.86	348.3 370.1	1538–1883 1470–1883	355.5*		2.56×10^{-14} 2.14×10^{-14} (2013 K)	Davis and McMullen (1972)

Hfβ $T > 2013$	bcc	2500	0.0011	159.2	355.5	5.19×10^{-11}	8.13×10^{-12} (2013 K)	Herzig et al. (1982)
In	tetr	430	$\perp c$ 3.7	78.5	61.1	1.08×10^{-13}		Dickey (1959)
			$//c$ 2.7	78.5	NA	7.85×10^{-14}		Arkhipova (1986)
Ir	fcc	2716	0.36	438.8	386.2	1.3×10^{-13}		Mundy et al. (1971)
K	bcc	336	$D_{01} = 0.05$	$Q_1 = 37.2$	47.8	1.32×10^{-11}		
			$D_{02} = 1$	$Q_2 = 47$				
Laβ $T < 1134$	fcc	β/γ 1134	1.5	188.8	169.6*		3×10^{-13} (1134 K)	Dariel et al. (1969)
Laγ $T > 1134$	bcc	1193	0.11	125.2	169.6	3.62×10^{-11}	1.88×10^{-11} (1134 K)	Languille and Calais (1974)
Li	bcc	454	$D_{01} = 0.19$	$Q_1 = 53$	64.5	3.13×10^{-11}		Heitjans et al. (1985)
			$D_{02} = 95$	$Q_2 = 76.2$				
Mg	hex	922	$\perp c$ 1.75	138.2	131.1	2.59×10^{-12}		Combronde and Brébec (1971)
			$//c$ 1.78	139		2.37×10^{-12}		
Mo	bcc	2893	8	488.2	411.4	1.22×10^{-12}		Maier et al. (1979)
Na	bcc	371	$D_{01} = 57$	$Q_1 = 35.7$	52.7	1.75×10^{-11}		Mundy (1971)
			$D_{02} = 0.72$	$Q_2 = 48.1$				
Nb	bcc	2740	0.524	395.6	389.6	1.5×10^{-12}		Einziger et al. (1978)
Ni	fcc	1726	$D_{01} = 0.92$	$Q_1 = 278$	245.4	9.35×10^{-13}		Maier et al. (1976)
			$D_{02} = 370$	$Q_2 = 357$				
Pb	fcc	601	0.887	106.8	85.4	4.63×10^{-14}		Miller (1969)
Pd	fcc	1825	0.205	266.3	259.5	4.9×10^{-13}		Peterson (1964)
Prβ $T > 1068$	bcc	1205	0.087	123.1	171.3	4×10^{-11}		Dariel et al. (1969)
Pt	fcc	2042	$D_{01} = 0.06$	$Q_1 = 259.7$	290.3	1.4×10^{-12}		Rein et al. (1978)
			$D_{02} = 0.6$	$Q_2 = 365$				
Puβ $395 < T < 480$	m	β/γ 480	0.0169	108	129.8*		2.98×10^{-18} (480K)	Wade et al. (1978)
Puγ $480 < T < 588$	ort	γ/δ 588	0.038	118.4	129.8*		4.95×10^{-19} (480K)	Wade et al. (1978)
							1.15×10^{-16} (588 K)	
Puδ $588 < T < 730$	fcc	δ/δ 730	0.0517	126.4	129.8*		3.05×10^{-17} (588 K)	Wade et al. (1978)
							4.66×10^{-15} (730 K)	
Puε $T > 753$	bcc	913	0.003	65.7	129.8	5.22×10^{-11}	8.3×10^{-12} (753 K)	Cornet (1971)
Rb	bcc	312	0.23	39.3	44.4	6.05×10^{-12}		Holcomb and Norberg (1955)

TABLE 2.4 Self-diffusion Parameters for Pure Elements[30] (Continued)

Element (see comments at end)[a]	C.S.[b]	T_m (K)[c]	D_0(m²s⁻¹) × 10⁴[d]	Q (kJ/mol)[e]	Temp. range (K)[f]	$Q = 0.1422T_m$ Van Liempt[g]	$D(T_m)$(m²s⁻¹)[h]	D(ph. tr.)(m²s⁻¹)[i]	Reference in Ref. 30[j]
Re	hex	3453	511.4		1520–1560	491			Noimann et al. (1964)
Rh	fcc	2239	391		903–2043	318.4			Shalayev et al. (1970)
Sb	trig	904	⊥c 0.1 //c 56	149.9 201	773–903	128.5	2.17×10^{-14} 1.36×10^{-14}		Cordes and Kim (1966)
Se	hex	494	⊥c 100 //c 0.2	135.1 115.8	425–488	70.2	5.18×10^{-17} 1.1×10^{-17}		Brätter and Gobrecht (1970)
Sn	tetr	505	⊥c 21 //c 12.8	108.4 108.9	455–500	71.8	1.29×10^{-14} 6.9×10^{-15}		Huang and Huntington (1974)
Ta	bcc	3288	0.21	423.6	1261–2993	467.5	3.9×10^{-12}		Werner et al. (1983)
Te	trig	723	⊥c 20 //c 0.6	166 147.6	496–640	102.8	2.03×10^{-15} 1.3×10^{-15}		Werner et al. (1983)
Thα T < 1636	fcc	α/β 1636	395	299.8	998–1140	287.7*		1.48×10^{-16} (1155 K)	Schmitz and Fock (1967)
Tiα T < 1155	hex	α/β 1155	6.6×10^{-5}	169.1	1013–1149	275.8*		5.4×10^{-14} (1155 K)	Dyment (1980)
Tiβ T > 1155	bcc	1940	D(m²s⁻¹) = 3.5 × 10⁻⁴ × exp(−328/RT) × exp{4.1(T_m/T)²}	275.8	1176–1893	275.8	3.11×10^{-11}		Köhler and Herzig (1987)
Tlα T < 507	hex	α/β 507	⊥c 0.4 //c 0.4	94.6 95.9	420–500	82*		7.16×10^{-15} 5.2×10^{-15} (507 K)	Shirn (1955)
Tlβ T > 507	bcc	577	0.42	80.2	513–573	82	2.3×10^{-12}	2.29×10^{-13} (507 K)	Chiron and Faivre (1985)
Uα T < 941	ort	α/β 941	0.002	167.5	853–923	199.8*		1×10^{-16} (941 K)	Adda and Kirianenko (1962)
Uβ 941 < T < 1048	tetr	β/γ 1048	0.0135	175.8	973–1028	199.8*		2.35×10^{-16} (941 K) 2.33×10^{-15} (1048 K)	Adda et al. (1959)
Uγ T > 108	bcc	1405	0.0018	115.1	1073–1323	199.8	9.46×10^{-12}	3.29×10^{-13} (1048 K)	Adda and Kirianenko (1959)
V	bcc	2175	1.79 26.81	331.9 372.4	1323–1823 1823–2147	309.3	3.05×10^{-12}		Ablitzer et al. (1983)
W	bcc	3673	$D_{01} = 0.04$ $D_{02} = 46$	$Q_1 = 525.8$ $Q_2 = 665.7$	1705–3409	522.3	1.7×10^{-12}		Mundy et al. (1978)

	a	b	c	d	e	f	g	h	i	j
Y α T < 1752	hex	α/β 1752	⊥c 5.2 //c 0.82	280.9 252.5	1173–1573	256.4			2.19 × 10⁻¹² 2.43 × 10⁻¹² (1752 K)	Gorny and Altovskii (1970)
Yb α T < 993 Yb β T > 993	hex bcc	α/β 993 1097	0.034 0.12	146.8 121	813–990 995–1086	156* 156			6.4 × 10⁻¹⁴ (993 K) 5.18 × 10⁻¹² (993 K)	Fromont et al. (1974) Fromont et al. (1974)
Zn	hex	693	⊥c 0.18 //c 0.13	96.3 91.7	513–691	98.5	2.08 × 10⁻¹² 9.92 × 10⁻¹³ 1.59 × 10⁻¹²			Peterson and Rothman (1967)
Zr α T < 1136	hex	α/β 1136	No value	Curved Arrh. plot	779–1128	302*			≈5 × 10⁻¹⁸ (1136 K)	Horvath et al. (1984)
Zr β T > 1136	bcc	2125	$D(\mathrm{m^2 s^{-1}}) = 3 \times 10^{-5} \times \exp(-3.01/RT) \times \exp[3.39(T_m/T)^2]$		1189–2000	302	1.37 × 10⁻¹¹		6.14 × 10⁻¹⁴ (1136 K)	Herzig and Eckseler (1979)

Source: These self-diffusion data have been extracted from the compilation by Mehrer et al. in *Diffusion in Solid Metals and Alloys*, Landolt-Bornstein New Series, vol. 26, ed. O. Madelung, Springer-Verlag, 1990. Reprinted by permission of Elsevier Science, Amsterdam, Netherlands.

a Symbol of the metal.

b Crystal stucture. bcc = body-centered cubic, fcc = face-centered cubic, hex = hexagonal, m = monoclinic, ort = orthorhombic, tetr = tetragonal, trig = trigonal.

c Melting temperature. For the phases which do not melt (for instance Ce γ, Fe, α etc.) we have given the temperature of the phase transition.

d Experimental D_0. The value in m²s⁻¹ is multiplied by 10⁴ (so that it is in cm²s⁻¹). For some of the metals the Arrhenius plot is curved and D has the form: $D = D_{01}\exp(-Q_1/RT) + D_{02}\exp(-Q_2/RT)$, in these cases D_{01} and D_{02} are given for D (in m²s⁻¹ without any multiplying factor). For Ti and Zr which have strongly curved Arrhenius plots, special expressions are given for D (they are also multiplied by 10⁴).

e Experimental Q in kJ mol⁻¹. Same remarks as for column 4.

f Temperature range of the experimental determination of D.

g Empirical value of Q according to the Van Liempt relation. For the phases which do not meet, this value is followed by an *.

h Value of D at the melting point.

i For metals which display several phases, the values of D are given at the temperature boundaries of the phase. For instance, U_β is stable between 941 and 1048 K, D values at these temperatures are given in column 9.

j References mentioned in Ref. 30.

k NA = not available.

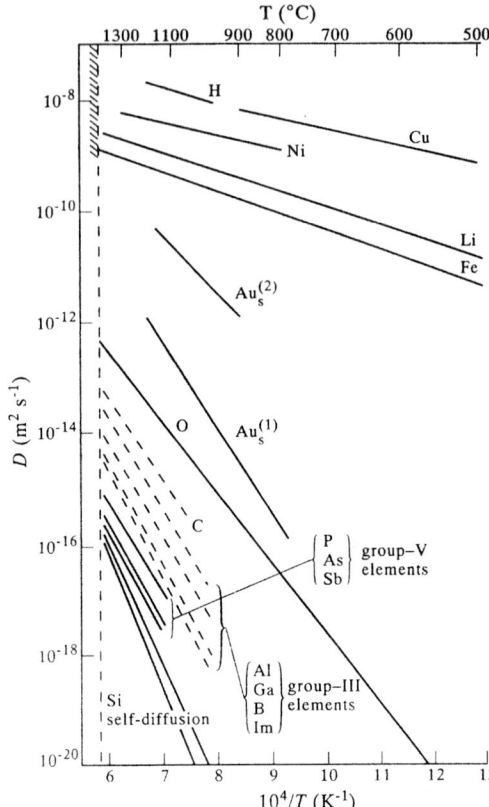

FIGURE 2.24 An Arrhenius plot ($\log_{10}D$ versus $10^4/T$) of the diffusion coefficients of a range of foreign/tracer solute elements in solid crystalline silicon and of silicon self-diffusion. The lines $Au_s^{(1)}$ and $Au_s^{(2)}$ correspond to different effective diffusivities of substitutional gold in silicon. (*After U. Gösele and T. Y. Tan, in Encyclopaedia of Materials Science and Technology, vol. 4, Verlagsgesellschaft Chemie, Weinheim, Germany, pp. 197–247.*)

$$\left(\frac{\partial \ln D}{\partial P}\right)_T = \left[\frac{\partial \ln(fv\alpha^2)}{\partial P}\right]_T - \left[\frac{\partial(\Delta G^f + \Delta G^m)/\partial P}{kT}\right] \qquad (2.89)$$

Using the thermodynamic relation $(\partial G/\partial P)_T = V$, we can define the activation volume ΔV of the diffusion process as

$$\Delta V = \frac{\partial(\Delta G^f + \Delta G^m)}{\partial P}\bigg|_T = -kT\frac{\partial \ln D}{\partial P}\bigg|_T + kT\frac{\partial \ln(fv\alpha^2)}{\partial P}\bigg|_T \qquad (2.90)$$

The second term in Eq. (2.90) can be estimated using the isotherm compressibility and the Gruneisen constant, but its value is only a few percent of the first term; thus, it can be neglected. Since the activation volume ΔV is the sum of the defect formation volume $\Delta V_d^f [= \partial(\Delta G^f)/\partial P|_T]$ and defect migration volume $\Delta V_d^m [= \partial(\Delta G^m)/\partial P|_T]$, we have

$$\Delta V = \Delta V_d^f + \Delta V_d^m \qquad (2.91)$$

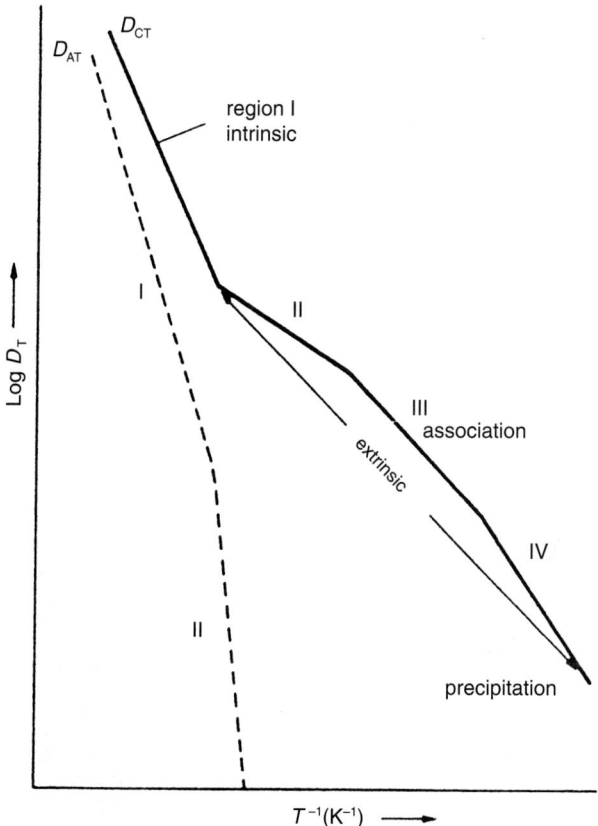

FIGURE 2.25 Schematic Arrhenius plot showing the typical temperature dependence of cation (D_{CT}) and anion (D_{AT}) tracer diffusion coefficients in a ceramic. (*Source: A. Atkinson, in Encyclopaedia of Materials Science and Engineering, Pergamon Press, Oxford, 1986, p. 1177.*)

Since the amount of ΔV_d^m associated with the jump process is much smaller than ΔV_d^f, we may assume $\Delta V_d^f \approx \Delta V$. Typically, ΔV varies between 0.5 and 1.3 Ω, at least in the case of monovacancy mechanism; however, in fcc metals, ΔV lies in the range of 0.7 to 1.1 atomic volume (Ω) and ΔV_d^m is about 0.15 ΔV. In some cases, ΔV is very small or even negative, which can be an indication of an interstitial-type mechanism.[30] For bcc sodium, a value of $\Delta V \sim 0.32$ atomic volume reflects a stronger relaxation into the vacancy, whereas a dependence of ΔV on temperature indicates that more than one defect mechanism is operative, as observed for the curvature of the Arrhenius plot. The thermodynamic equation for ΔV_d^f has been derived as[65]

$$\Delta V_d^f = -\left(\frac{\Omega}{B}\right)\left(\frac{dE_f}{d\Omega}\right) \qquad (2.92)$$

where Ω is the atomic volume, E_f is the vacancy formation energy under the constraint of constant volume, and B is the bulk modulus.[66]

2.6.3 The Isotopic Mass Effect

The isotopic mass effect of solute atoms is of large significance in volume diffusion because it provides information about the atomic mechanism of diffusion and experimental means of occurrence of the tracer correlation factor f_α of isotope α. It is customary to obtain the isotope effect E parameter (or the strength of the isotope effect) by measuring simultaneously the diffusion coefficients D_α and D_β of the two isotopes α and β of the same element with masses m_α and m_β, respectively, and using the relation

$$E = \frac{(D_\alpha/D_\beta)-1}{(m_\beta/m_\alpha)^{1/2}-1} = f_\alpha \Delta K \qquad (2.93)$$

where f_α is the correlation factor for isotope α and ΔK is the fraction of kinetic energy in the unstable mode (i.e., at the saddle point residing in the jumping atom) associated with motion in jump direction that belongs to the diffusing atom. Hence, ΔK is bound between unity and zero (of the order of 0.8 to 0.9 for self-diffusion in simple fcc metals).[67] The value of f_α suggests the specific mechanism and indicates whether it changes with temperature in a given system. Consequently, a measurement of E should enable one to identify the corresponding diffusion mechanism.

If the Cu, Ag, or Au diffuses fast in lead as a simple interstitial atom (like C in fcc iron), then $f_\alpha = 1$ and $f_\alpha \Delta K \approx 1$. However, the value of $f_\alpha \Delta K$ for Ag diffusing in lead is 0.25, suggesting that the mass of the activated complex is much greater than most of the Ag atom alone.[10]

Equation (2.93) is applied to a process involving one atom jump. The more general case of isotopic mass effect involving n atoms jumping simultaneously, e.g., the interstitialcy mechanism (where $n = 2$), is expressed by

$$E = f\Delta K = \frac{(D_\alpha/D_\beta)-1}{[m_\beta + (n-1)m_0/m_\alpha + (n-1)m_0]^{1/2}-1} \qquad (2.94)$$

where m_0 is the average mass of nontracers.

Assuming the same relationship to hold in the case of grain boundary diffusion, we can rewrite Eq. (2.93) as

$$E_b = \frac{(D_{b\alpha}/D_{b\beta})-1}{(m_\beta/m_\alpha)^{1/2}-1} = f_b \Delta K_b \qquad (2.95)$$

where the subscript on the various symbols suggests the same quantity for grain boundary diffusion.[68]

2.7 TYPES OF DIFFUSION COEFFICIENTS

There are various types of diffusion coefficients, which are briefly discussed in this section.

2.7.1 Self-diffusion or Tracer Self-Diffusion Coefficient

The self-diffusion or tracer diffusion coefficient D^* is measured experimentally by introducing a few radioactive isotopes (A*) into pure A and measuring D_A^* (or $D_A^{A^*}$) by the relation

$$D_A^{A^*}\left(=D_A^*\right) = f D_A^A (= D_A) = f\left(\frac{1}{6}\right)\alpha^2\Gamma = f\left(\frac{1}{6}\right)\alpha^2 z\Gamma \qquad (2.96a)$$

or more generally

$$D^* = f D_{\text{random}} \qquad (2.96b)$$

where Γ is the jump frequency for both A* and A atoms, z is the number of nearest neighbors, and f is the correlation factor (usually less than unity, which depends on the crystal structure and on the diffusion mechanism). Figure 2.26a shows the experimental situation for tracer self-diffusion coefficient $D_A^{A^*}(=D_A^*)$ when A* substitutes B*. The same idea can be extended to alloys and compounds. The tracer self-diffusion coefficients for both A* and B* tracer atoms in a homogeneous binary alloy AB (Fig. 2.26b) are represented by $D_{AB}^{A^*}$ and $D_{AB}^{B^*}$, respectively. The A* or B* concentration is always kept very small so that the alloy composition remains unaffected by the diffusant. These coefficients are functions of concentration.

According to Darken, simple thermodynamic considerations lead to the relationship between the intrinsic and the self-diffusion coefficients in binary solid solutions as

$$D_A = D_A^*\left(1 + N_A \frac{\partial \ln \gamma_A}{\partial N_A}\right) \qquad (2.97)$$

and

$$D_B = D_B^*\left(1 + N_B \frac{\partial \ln \gamma_B}{\partial N_B}\right) \qquad (2.98)$$

where $D_A \neq D_B$ because the solution is not ideal; D_A and D_B are the intrinsic diffusion coefficients, D_A^* and D_B^* are tracer self-diffusion (or tracer diffusion) coefficients, γ_A and γ_B are the activity coefficients of A and B, and N_A and N_B are the atom fractions (or fractional concentrations) of A and B, respectively. According to the Gibbs-Duhem equation, $N_A \partial \ln \gamma_A = -N_B \partial \ln \gamma_B$ and since $\partial N_A = -\partial N_B$, we have

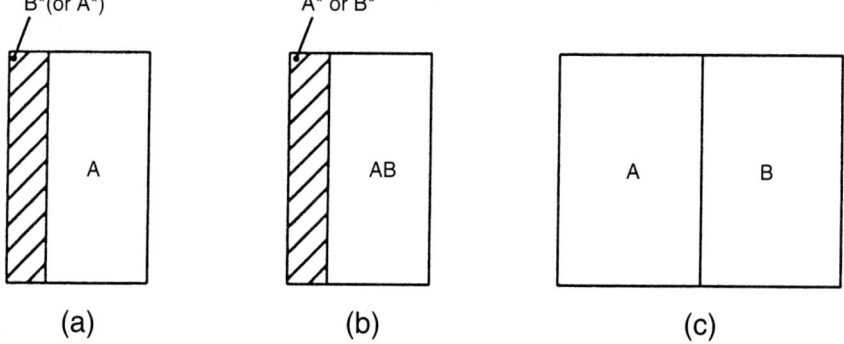

FIGURE 2.26 Three types of diffusion experiments to determine (a) tracer self-diffusion or impurity diffusion coefficients D_A^* and D_B^* in pure metal; (b) tracer self-diffusion coefficients $D_{AB}^{A^*}$ and $D_{AB}^{B^*}$ in homogeneous AB alloys; and (c) chemical diffusion coefficient $\bar{D}$ of two metals A and B.

$$1 + N_A \frac{\partial \ln \gamma_A}{\partial N_A} = 1 + N_B \frac{\partial \ln \gamma_B}{\partial N_B} \tag{2.99}$$

The two factors that give the intrinsic diffusivities when multiplied by the respective tracer diffusion coefficients are actually equal, and it is customary to call this quantity the thermodynamic factor.

Experimental values for lattice self-diffusivity are available for about 50 metallic and semimetallic elements. Among these available data, some tend to be unreliable due to extreme difficulty in obtaining results for highly reactive materials.[69]

Results on self-diffusion in homogeneous dilute *binary alloys* containing small atomic fraction N_B are usually denoted by

$$D_{AB}^{A*} = D_{AB}^{A*}(N_B) = D_A^{A*}(1 + b_1 N_B + b_2 N_B^2 + \cdots) \tag{2.100a}$$

Then D_{AB}^{A*} is called the *solvent self-diffusion coefficient* and D_{AB}^{B*} the *solute diffusion coefficient*. Experimental measurements of $D_{AB}^{A*}(N_B)$ are frequently given by Eq. (2.100a); and b_1, b_2, etc. are called *solvent enhancement factors*; D_A^{A*} is the tracer self-diffusion coefficient in the pure solvent. Mainly, b_1 is determined by perturbations due to isolated solute atoms, b_2 by pairs of solute atoms, and so forth.

Similarly, the *solute diffusion coefficient* D_{AB}^{B*} at low concentration can be denoted by a power series dependence

$$D_{AB}^{B*} = D_{AB}^{B*}(N_B) = D_A^{B*}(1 + B_1 N_B + B_2 N_B^2 + \cdots) \tag{2.100b}$$

where D_A^{B*} is represented by impurity diffusion coefficient of species B in solvent A and B_1 and B_2 are the *solute enhancement factors*.[12]

2.7.2 Impurity and Solute Diffusion Coefficients

Impurity diffusion denotes the diffusion of a solute element present in such low concentrations in a solvent (matrix) that the solute atoms may be considered as diffusing quite independently of one another, i.e., with no mutual interaction. It implies the simplest type of binary diffusion and is of special interest in furtherance of theoretical understanding.[69a]

In order to measure the impurity diffusion coefficient, the tracer B* acts now as the impurity that is chemically different from the host or solvent. However, the impurity concentration must be kept extremely low to eliminate a chemical composition gradient (Fig. 2.26a). In reality, the tracer impurity should be allowed to diffuse into the specimen already containing the same concentration of impurity. Since the impurity always remains in stable solid solution (unless implanted), it is often called the solute and the impurity diffusion coefficient is sometimes called the solute diffusion coefficient at infinite dilution. However, the term *solute diffusion coefficient* is often "reserved" for dilute alloys, where measurement of the solvent diffusion coefficient is often necessary. In these situations, both solute and solvent diffusion coefficients often depend on solute content; and, as in self-diffusion, the chemical composition of the specimen must remain essentially unchanged by the diffusion process—otherwise it turns out to be a chemical diffusion coefficient.

2.7.3 Chemical or Interdiffusion Coefficient

Unlike self- and impurity diffusion coefficients which are measured in the absence of chemical composition gradients, chemical, interdiffusion, collective, or mutual diffusion coefficient $\tilde{D}$ denotes the interdiffusion of A and B and is measured from the plot of C_A or C_B versus x. It is usually a function of concentration and depends markedly on temperature. Generally, the chemical diffusion coefficient does not equal the self-diffusion coefficient due to the effects resulting from the chemical composition gradients. Examples of this type of diffusion coefficient in a pseudo-one-component system include diffusion of an adsorbed monolayer onto a clean section of a surface, diffusion between two metal specimens with difference in their relative concentration of a highly mobile interstitial atom such as hydrogen, and diffusion between two nonstoichiometric compounds such as $Fe_{1-\partial}O$ and $Fe_{1-\partial'}O$.

Chemical diffusion in binary substitutional solid solutions is often known as interdiffusion and occurs when pure metal A is bonded to pure metal B and diffusion is allowed at elevated temperature (Fig. 2.26c). Although both A and B atoms move, only one independent concentration profile, say of A, is established. The resulting diffusion coefficient, which is derived from the profile by the Boltzmann-Matano analysis, is called the interdiffusion, chemical diffusion, or mutual diffusion coefficient D. For many practical purposes, $\tilde{D}$ is adequate to explain the diffusion behavior of a binary substitutional and is often quoted in the metals property databook. It is pointed out that the rates of diffusion of the two species relative to the local lattice planes are not equal in amount. For chemical diffusion in systems of more than two components, Eq. (2.5) or (2.6) and those following are insufficient.[5] In many cases, $\tilde{D}$ is simply given by an Arrhenius equation

$$\tilde{D} = \tilde{D}_0 \exp\left(\frac{-Q}{RT}\right) \qquad (2.101)$$

As the limit of concentration of one metal species approaches zero, the interdiffusion coefficient approaches the intrinsic diffusion coefficient.

2.7.4 Intrinsic Diffusion Coefficient

The intrinsic diffusion coefficients (or component diffusion coefficients) D_A and D_B of an AB alloy denote the diffusion of the two species A and B relative to the lattice planes. In general, the diffusion rates of A and B are not equal.[12]

2.8 SELF-DIFFUSION IN PURE METALS

Self-diffusion is the most basic diffusion process in solids by which substitutional host atoms move through the lattice to other substitutional sites. For the majority of metals, the sum of formation and migration energies of vacancies is equal to the activation energy for the self-diffusion measured for that metal. In this way, the single- or monovacancy mechanism plays a dominant role for self-diffusion in pure metals.[70] Experimentally, the diffusion coefficient is measured by employing radioactive pure metals and monitoring their movement through the metal lattice. Self-diffusion parameters for various pure elements are listed in Table 2.4. The pure metals are most extensively studied with respect to point defects and diffusion properties. Self-diffusion of metals is usually divided into *normal* and *anomalous* *self-diffusion.* The characteristics of normal self-diffusion (Fig. 2.27*a*) are as follows: (1) A single vacancy mechanism is operative, and D (or D^*) is represented by the Arrhenius relation: $D = D_0 \exp(-Q/kT)$ is obeyed which implies the occurrence of a straight-line diffusion data plot of $\ln D$ versus $1/T$. (2) The D_0 (preexponential factor) values lie in the range 5×10^{-6} and $5 \times 10^{-4} \, \text{m}^2 \cdot \text{s}^{-1}$. (3) A relationship between the value of Q (activation enthalpy) and the melting temperature T_m by the expression $Q = 34 T_m$ cal/mol or $Q = 141 T_m$ J/mol and T_m (in kelvins). Examples of metals in this class include the fcc metals such as Al, Ag, Au, Cu, Ni, and Pt and bcc metals such as Cr, Li, Na, Nb, Ta, and V. This behavior forms the basis of the Van Liempt relation and is well established in compact metals (see Fig. 2.22*a* for fcc metals and Fig. 2.22*b* for bcc metals). It is noted here that bcc structures exhibit a wider dispersion around the Van Liempt straight line than that of the fcc structures.

Metals whose properties do not conform to normal characteristics are called *anomalous.* These are usually specific phases of allotropic metals for which several distinct solid phases are present. Examples of the anomalous self-diffusion behavior occurring in bcc metals are V, Cr, β-Hf, β-Zr, β-Ti, γ-U, ε-Pu, γ-La, δ-Ce, β-Pr, γ-Yb, and β-Gd. All these metals are characterized by very low values of D_0 (and $D_{\text{melt.point}} \approx 10^{-10} \, \text{m}^2 \cdot \text{s}^{-1}$) and Q at low temperatures when compared to *normal* metals[22,71,72] (Fig. 2.27*a*); and a clearly visible (upward concave) curvature of the "Arrhenius plot."[73] The diffusion coefficient of most bcc metals at the melting temperature is $10^{-11} \, \text{m}^2 \cdot \text{s}^{-1}$, an order of magnitude higher than in most fcc metals. Various theories for this anomalous behavior have been advanced, and most of them have been discounted. The following explanation is presently the most favored one. The anomalous behavior is usually interpreted as resulting from (1) a strong temperature dependence of the vacancy migration energies, (2) the existence of allotropy or phase transformation associated with very high densities of short-circuiting paths such as dislocations and grain boundaries, (3) lowering of activation (or jump) barrier at lower temperatures, (4) unusually large divacancy binding energy, (5) two possible types of monovacancy jumps such as *nn* and *nnn*, and (6) a "dissociative" mechanism, in which mainly the substitutional atoms have a considerable probability of being excited into a highly mobile interstitial state. Among these, the most likely explanation of the anomalies is the lowering of the jump barrier at lower temperature.[74] These are linked to the fragments or embryos of lower-temperature ω phase occurring near the transition. This ω phase, which is structurally related to the bcc lattice when a migrating atom is at the saddle point, could enhance diffusion by providing easier jumps.[75]

If we assume, for simplification, temperature dependence of Q and D_0 values, then Eq. (2.102) is used to explain a curvature in the plot of $\ln D$ versus $1/T$

$$D = \sum A_i \exp\left(\frac{-Q_i}{kT}\right) \tag{2.102}$$

In such an equation, two or three terms are normally sufficient; each term represents a competing diffusion mechanism. One of the terms is possibly due to a single vacancy mechanism that is dominant in fcc metals. There are several possibilities for the other terms, and different mechanisms can actually work in different systems.

2.8.1 Self-diffusion in FCC and HCP Metals

The vacancy mechanism for the diffusion in these metals has now been widely recognized. However, there are still some controversies with respect to the origin of

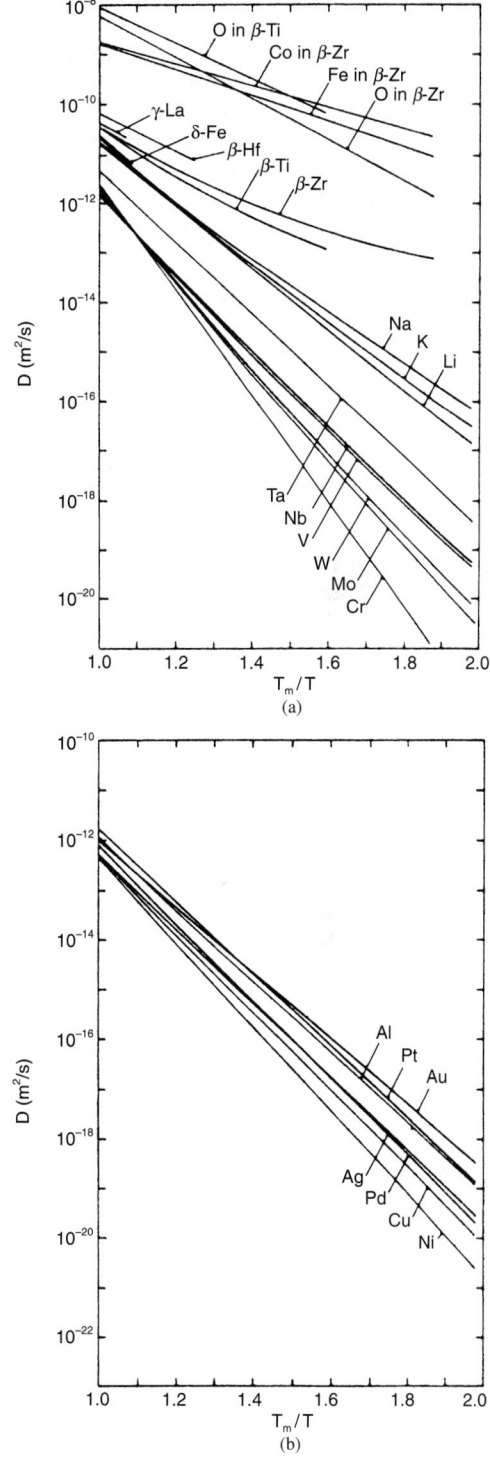

FIGURE 2.27 Self-diffusion coefficients D of (a) bcc metals and diffusion coefficients of the very fast impurity diffusion and (b) fcc metals as a function of their homologous temperature T_m/T.[70,71]

2.55

the possible curvature of the Arrhenius plots even in the so-called normal metals (Al, Cu, Ni, Ag, and Au) profoundly at high temperature and sporadic presence over the entire temperature range from T_m to $T_m/2$ (Fig. 2.27b). In order to elucidate this curvature, three hypotheses have been put forward:

1. A vacancy mechanism takes place over the whole temperature range. However, D_0 and Q increase with temperature due to strong thermal expansion coefficient for vacancies, or variation of elastic constants with the temperature. In these temperature ranges, it was found that the data did not follow an Arrhenius behavior; that is, ln D versus $1/T$ plots were curved.

2. Both vacancies and divacancies contribute to the diffusion; however, the single vacancy mechanism dominates at low temperatures, whereas there is an increasing contribution from divacancies at the higher temperatures approaching T_m.[70,76]

3. A vacancy mechanism takes place, and the curvature is attributed to the dynamic correlation between successive jumps (vacancy double jumps).[77]

Figure 2.27b also exhibits that, in Group IB metals, the D values in Cu became the lowest, followed by those of Ag and Cu, and in the Group VIII metals, the Ni values were the lowest followed by Pd and Pt.[70]

In hcp metals the limited number of available data are in agreement with a slight decrease of the ratio of the activation energies of the diffusion parallel to perpendicular to the c axis with increasing c/a ratio; the activation energy remains the same in the perfect lattice.[78]

2.8.2 Self-diffusion in BCC Metals

Self-diffusion in bcc metals exhibits three features: (1) There is much larger scatter of diffusivity in bcc metals than in the compact phases, and some exhibit an exceptionally large absolute value of D (Fig. 2.27a). (2) Their Arrhenius plots often display much larger curvature than those of the fcc or hcp metals, mostly accounted for by a divacancy mechanism. (3) They show an orderly variation of D with the position in the periodic table which needs to be elaborated; e.g., metals of the same column, like Ti, Zr, Hf in Group IV, have a very small activation energy and a large curvature (Fig. 2.27a). The self-diffusion in Group IV metals is much faster than in others; the lowest the temperature, the largest this difference.[79] The diffusivities of some transition impurities such as Fe, Co, and Ni are orders of magnitude faster than the already fast diffusion coefficient. Many explanations have been proposed to explain these anomalies based on strong contribution of short circuits, presence of extrinsic vacancies due to impurities, interstitial mechanisms, etc. All these hypotheses were excluded by experiments. The very origin of this behavior is now understood to be associated with the electronic structure of the metal and the structural properties of the bcc lattice.[80]

2.9 DIFFUSION IN DILUTE SUBSTITUTIONAL ALLOYS

In this case, dilute solute concentrations imply those below 1 at %, where one can avoid the complications of solute-solute interactions. Although fcc alloys have been mainly investigated in which defect mechanisms are needed, the same approach may be valid to all systems where the atomic migration involves motion of vacancies or interstitials or more complex defects formed from them (such as solute-vacancy pairs).[81] Analysis has been accomplished in terms of the five-jump-frequency model of fcc alloys involving the various vacancy jumps near an impurity atom, as shown in Fig. 2.28.[82] The jump frequencies w_0, w_1, w_3, and w_4 all involve solvent-vacancy exchanges, each in a different relationship to the solute atom. Only w_2 is the frequency of a solute (impurity)-vacancy exchange, and the solute diffusion coefficient is given by

$$D_2 = \alpha^2 w_2 f_2 \exp\left(-\frac{\Delta G_v^f + \Delta G_A}{kT}\right) \qquad (2.103)$$

where ΔG_A is the solute-vacancy association free energy. The major problem, however, is that the correlation factor f_2 is no more a constant but now is dependent in a complex fashion on all the various jump frequencies.[28]

For the normal dilute alloy systems, the experimental results provide an activation energy for D_2 within 25% of that of the value for solvent self-diffusion, and a preexponential D_{02} within one order of magnitude of that of the solvent. This similarity helps to advocate the vacancy mechanism for solute diffusion.

The (solvent) self-diffusion coefficient $D_s(c_2)$, in a dilute random alloy containing a low concentration c_2 of solute/impurity atoms, can be written in the form

$$D_s(c_2) = D_s(0)(1 + bc_2) \qquad (2.104)$$

where $D_s(0)$ is the self-diffusion coefficient of the pure solvent and b is a constant (for a given alloy system at a given temperature) called the *solvent enhancement factor* or *coefficient* (either positive or negative for particular alloy models). Positive b values may be attributed to (1) the effect of solute atoms in enhancing the total vacancy concentration in the crystal and (2) the effect of changed solvent jump frequencies (w_1, w_2, and w_4) near solvent atoms. Expressions for b can be very complex, based on the degree of approximation used. In entirety, the five-frequency model holds good in explaining data on both solute diffusion and solvent enhancement for a wide range of dilute alloy systems.[28]

Anomalous Dilute Systems. In some dilute alloys one observes striking departures from the just discussed behavior such as $D_2 \gg D_s$ by factors of 10^3 to 10^5 times,

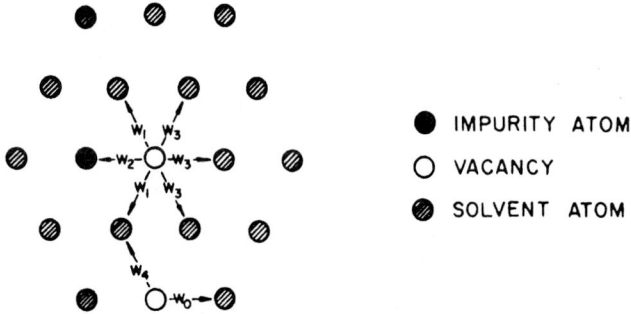

FIGURE 2.28 The five-jump frequency model for diffusion near an impurity atom in dilute fcc substitutional solid solutions.[82]

TABLE 2.5   Dilute Alloy Systems in which Anomalously Fast Solute Diffusion Occurs[80]

Solutes	Solvents
1. Low valency groups I and II	High valency groups III and IV
Cu, Ag, Au, Be, Zn, Cd, Hg, Pd	Pb, Sn, Tl, In
2. Later transition metals	Early members of d-transition groups
Ni, Co, Fe, Cr, Mn, Pd	β-Ti, β-Zr, β-Hf, Nb, La, Ce, Pr, Nd, γ-U, Pu
3. Noble metals	Li, Na, K

and an activation energy Q_2 by only approximately one-third to one-half of Q_s. Table 2.5 lists many of these (classified into three groups by Le Claire).[80] Such effects are most significantly found for dilute alloys of Pb, Na, and In containing the noble metals as solutes where interstitial solute diffusivity is often much larger than the solvent diffusivity. To elucidate these effects by a vacancy mechanism, it would need a very large solute-vacancy association/binding energy ΔG_A and both a rapid solute-vacancy exchange rate (w_2) and a rapid vacancy jump rate around the solute (w_1). These requirements can be written quantitatively. However, these requirements would impart a very high enhancement factor b [Eq. (2.104)], which is not observed. Alternatively, it was suggested that in such systems there is a fraction y_i of solute atoms that fills interstitial sites. The diffusion coefficient D_2 is then given by

$$D_2 = y_i D_i + (1 - y_i)D_s \qquad (2.105)$$

where D_i and D_s are the interstitial and substitutional diffusivities, respectively. Since $D_i \gg D_s$, the first term may dominate even if $y_i \ll 1$. The smaller enhancement factor b (than expected from the D_2/D_s ratio) coupled with the small activation volume ΔV discounts the involvement of vacancy, direct-exchange, and/or interstitial-vacancy pair mechanisms.[28]

One of the best investigated systems is Au in Pb, for which $Q_2 = 0.41\,\text{eV}$ (compared to $Q_s = 1.13\,\text{eV}$); others are transition metals in β-Ti and α-Ti.[84,85] While a vacancy mechanism has been ruled out, several measurements suggest that simple interstitial mechanism may be dominantly operative. In this case, one should expect an isotope effect $E = 1$; instead, for Ag and Cu in Pb, one measures $E \sim 0.25$. Such results have led to the belief that substitutional-interstitial dumbbell pairs called diplons are effective. Warburton and Turnbull have described such defects as Au-Pb and Au-Au diplons for the system Au in Pb. But they suggest that not all the anomalous systems can be explained in the same manner, and that other defects may dominate in some of these systems.[83]

2.10   DIFFUSION IN CONCENTRATED ALLOYS

2.10.1   Boltzmann-Matano Solution

In general, diffusion occurs due to a chemical potential gradient caused by a concentration gradient or the presence of a driving force. In commercial processes, the diffusion coefficient D depends on the chemical composition of the alloy; and since there is a concentration gradient, it implies that D changes with the frame of reference in the diffusion couple, which we indicate by the symbol $\tilde{D}$. In such a situation, Fick's second law or the diffusion equation in the one-dimension case may be written

$$\frac{\partial C}{\partial t} = \left(\frac{\partial}{\partial x}\right)\left(\tilde{D}\frac{\partial C}{\partial x}\right) \tag{2.106}$$

The term $\partial \tilde{D}/dx$ makes the equation inhomogeneous, and the solution in closed form then ranges from difficult (for special cases) to impossible. The solution to Eq. (2.106) is useful for obtaining $\tilde{D}$ over a composition range. However, it does not give a solution $C(x, t)$; rather it allows $\tilde{D}(C)$ to be determined from an experimental plot of $C(x)$. This method of analyzing experimental data is called a *Boltzmann-Matano solution*.

We assume C to be a function of new variable $\eta = x/\sqrt{t}$ (called the Boltzmann parameter) and transform the partial differential Eq. (2.106) into an ordinary differential equation as

$$\frac{\partial C}{\partial t} = \frac{\partial \eta}{\partial t}\frac{dC}{d\eta} = \frac{-x}{2t^{3/2}}\frac{dC}{d\eta} = \frac{-\eta}{2t}\frac{dC}{d\eta} \tag{2.107a}$$

and

$$\frac{\partial C}{\partial x} = \frac{\partial \eta}{\partial x}\frac{dC}{d\eta} = \frac{1}{t^{1/2}}\frac{\partial C}{\partial \eta} \tag{2.107b}$$

Substituting into Eq. (2.106), we obtain

$$\frac{-\eta}{2t}\frac{dC}{d\eta} = \frac{\partial}{\partial x}\frac{D}{t^{1/2}}\frac{dC}{d\eta} = \frac{1}{t}\frac{d}{d\eta}\left(D\frac{dC}{d\eta}\right) \tag{2.107c}$$

and finally,

$$\frac{-\eta}{2}\frac{dC}{d\eta} = \frac{d}{d\eta}\left(D\frac{dC}{d\eta}\right) \tag{2.107d}$$

Equation (2.107d) can be integrated for the boundary condition ($C = C_0$ for $\eta = \infty$ and $c = 0$ for $\eta = -\infty$), giving

$$D(C) = -\left(\frac{1}{2}\right)\left(\frac{d\eta}{dC}\right)_C \int_0^C \eta \, dC' = -\left(\frac{1}{2t}\right)\left(\frac{dx}{dC}\right)_C \int x \, dC' \tag{2.108}$$

And D can be obtained graphically from Fig. 2.29. First the Matano plane (at which $x = 0$) is fixed by the conservation condition $\int x \, dC = 0$, which is defined by the equality of the two hatched areas of Fig. 2.29. The double crosshatched area is the integral in Eq. (2.108) for $C = 0.2C_0$, and the tangent at this point is the corresponding multiplying factor. The product of these (slope and area under the diffusion profile) provides the diffusion coefficient at this concentration. In reality, the problem of mutual diffusion of two (mutually soluble) components is explained by one diffusivity, the so-called interdiffusion coefficient, which is designated $\tilde{D}$. This diffusion coefficient represents the combined effect of two individual diffusion coefficients D_A and D_B. A detailed use of this analysis is described by Borg and Dienes.[86]

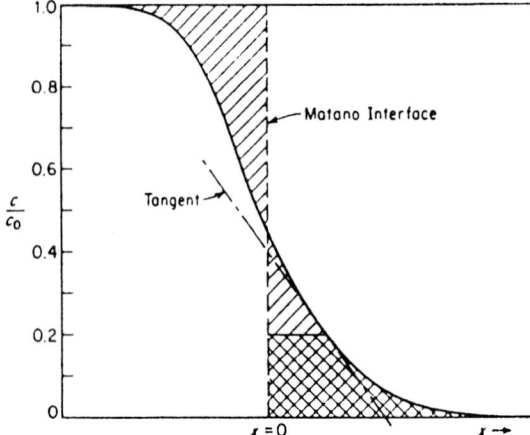

FIGURE 2.29 The Matano interface at $x = 0$ implies the equality of the hatched areas on either side of it. The Matano method is used for the calculation of $\tilde{D}$ at $C = 0.2C_0$.[10]

In recent years, least-square interpolation techniques were successfully used in conjunction with the Boltzmann-Matano treatment to compute the interdiffusion coefficient as a function of composition.[87]

2.10.2 Kirkendall Effect

The Kirkendall effect has served to explain a number of issues related to solid-state diffusion. The unambiguous identification of vacancy movement is recognized as the operative transport mechanism during interdiffusion in binary alloy systems.

If two semi-infinite bars of different concentrations of A and B are welded and diffusion-annealed, the Boltzmann-Matano solution provides only one diffusion coefficient $\tilde{D}$ (c) which completely explains the resulting homogenization. Thus, the problem involves a relation between a single diffusion coefficient and the self-diffusion coefficients at the same composition. To accomplish this, two new effects must be comprehended. The first deals with the type of material transport to be classified as diffusion. In a binary diffusion couple having a large concentration gradient, diffusion produces the transport of one type of material across the diffusion couple to another. The coordinate system used in the Boltzmann-Matano solution is fixed with respect to the end of the specimen, and the chemical diffusion coefficient is given by the equation

$$\tilde{D} = -\frac{J}{\partial c/\partial x} \tag{2.109}$$

Thus any movement of lattice planes with respect to the ends of the diffusion couple is recorded as an interdiffusion flux and influences $\tilde{D}$, although such translation does not reflect any jumping of atoms from one site to another.

In addition, a relation exists between the diffusion coefficient measured in a tracer experiment and the intrinsic diffusion coefficient of the separate elements in a binary diffusion couple. This requires a detailed treatment of the chemical forces, giving rise to solid-state diffusion. The development will depend mainly on thermodynamic or phenomenological considerations.[10]

2.10.2.1 Kirkendall Experiment.

The Kirkendall experiment involves welding together two pieces of pure metals A and B with small inert markers such as oxide particles or W wire on the weld interface and annealing for various periods of time to allow interdiffusion and determination of the concentration profile along the entire sample (Fig. 2.30a). In general, a shift of the initial welding interface (defined by inert markers) is obtained relative to the ends of the sample (Fig. 2.30b). This shift is attributed to the Kirkendall effect. This is based on the fact that (1) the chemical diffusion occurs by vacancies and (2) the intrinsic diffusivities D_A and D_B are unequal and unbalanced, say D_A (of low-melting-point species) > D_B. In this situation, across any lattice plane in the concentration gradient, more atoms of A migrate in one direction than do atoms of B in the opposite direction; i.e., a net mass transfer of atoms takes place across the plane and is associated with an equal flux of vacancies in the opposite direction. These unbalanced intrinsic fluxes result in (1) an increase in volume of the B-rich part of the sample to accommodate the net positive inward flux of matter, (2) shift of the lattice plane (or marker movement) toward the A-rich part (i.e., the shrinking side), thereby a state of compression in this region, and (3) generation or agglomeration of vacancies, voids, or pores on the A-rich side of the couple, commonly called Kirkendall voids or porosity (thereby causing the region in a state of tension). With continued exposure to high temperature, the voids increase and coalesce, producing porosity and loss of strength. This shift of crystal lattice (and related phenomena) during mutual diffusion was first observed by Smigelkas and Kirkendall on Cu-brass diffusion couple.[88] This experiment rejects the assumption of direct exchange of A $\leftrightarrow$ B mechanism which was formerly proposed and which would have implied equal diffusivity for both atoms. This phenomenon has been, subsequently, observed to occur in thin-foil couples of Ag-Au, Al-Au, Au-Sn, Au-Pb, Au-Sn/Pb, Cu-Al, Cu-Ni, Cu-Pt, Ni-Mo,[89] and Pt-Ir. The Kirkendall effect is time- and temperature-dependent but with some systems can even occur at ambient temperatures.

It is concluded, based on the effects mentioned in the previous paragraph, that a composition profile exists after interdiffusion of A and B (Fig. 2.31a); voids or pores form in the diffusion zone area from which there is a mass flow; and a net flow of vacancies occurs from the B-rich side of the bar toward the A-rich side (Fig. 2.31b). This movement increases the equilibrium number of vacancies on the A-rich side and reduces it on the B-rich side. The vacancies are thus created on the side of a couple that gains mass and are absorbed or destroyed on the side that loses mass (Fig. 2.31c).

Kirkendall porosity can be eliminated by careful selection of alloy composition. For example, Kirkendall voids observed in Pt-Cu couple can be avoided by select-

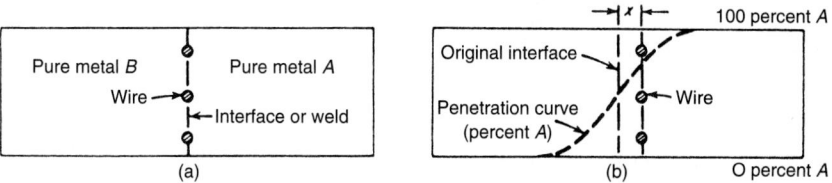

FIGURE 2.30 Marker shift in a Kirkendall diffusion couple.[10]

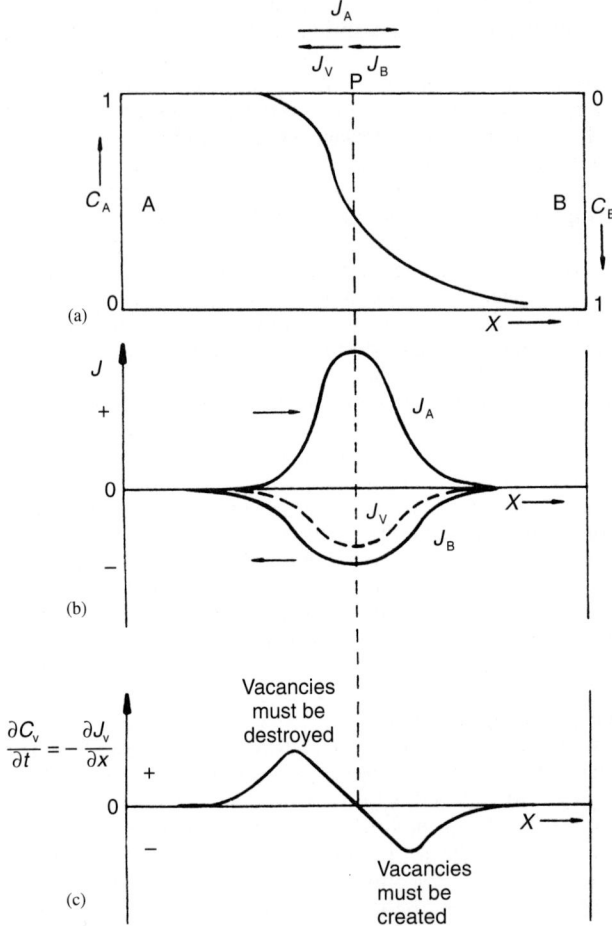

FIGURE 2.31 Schematic diagram showing interdiffusion and vacancy flow. (*a*) Concentration versus distance curve for A component. (*b*) The corresponding fluxes of A and B ($D_A > D_B$) and vacancies as a function of position x. (*c*) $\partial C_v/\partial T = -\partial J_v/\partial x$. That is, the rate of vacancy creation is equal to the rate of vacancy destruction or absorption.[14] [*Reprinted by permission of Van Nostrand Reinhold (U.K.) Ltd.*]

ing Pt and electrodeposited Ni after annealing for 8 hr at 700°C. Annealing in a hot isostatic press (HIP) has also been used to suppress Kirkendall void formation. One advantage of the Kirkendall effect is the development of deliberately controlled porosity.

2.10.2.2 Darken's Analysis. Fick's law contains the implicit assumption that the driving force for diffusion is the concentration gradient. However, a more basic viewpoint assumes that the driving force is a chemical free-energy gradient. There is a direct similarity between the two gradients, but it is seldom that the relationship becomes inverted and the so-called uphill diffusion, i.e., diffusion against a concentration gradient, occurs. Darken provided the following analysis to describe the Kirkendall effect.

Let us suppose that the vacancy mechanism of transport is operating. In every experiment, the inert markers are invariably made of insoluble materials with high melting temperature. The formation and migration energies of the vacancy in such materials are significantly larger than in the surrounding matrix. As a result, the markers are impermeable to the vacancy flux. In this situation, it can be shown that such a marker shifts along with the lattice planes,[90] irrespective of the type of its interface with the matrix (coherent or incoherent). Thus the measurement of Kirkendall shift is considered as the measurement of the lattice plane shift. The above formalism can be easily extended to incorporate the situation where the average atomic volume changes with the concentration of the alloy.[91]

Since the net intrinsic flux of atoms across a plane is associated with an equal flux of vacancies in the opposite direction, we can write

$$J_V = -(J_A + J_B) \qquad (2.110)$$

as shown in Fig. 2.31b. The flux can be maintained by the creation of new vacancies at sources upstream of the flux and by their destruction at sinks downstream (Fig. 2.31c), both processes striving to maintain equilibrium concentration of vacancies characteristic of the sample at each composition along the gradient. The most probable sources and sinks include dislocations generating and annihilating vacancies by climb processes. But such generation and annihilation involve, respectively, an increase and a decrease in the number of lattice planes and thus provide the mechanisms for exactly the amounts of expansion and shrinkage necessary to accommodate the net atom flux.

To express chemical interdiffusion coefficient $\tilde{D}$ in terms of the intrinsic (or component) diffusion coefficients D_A and D_B, we describe the fluxes J'_A and J'_B of A and B, respectively, with respect to the local lattice plane. Furthermore, $\tilde{D}$ elucidates the interdiffusion fluxes $\tilde{J}_A = -\tilde{J}_B$ measured with respect to laboratory fixed axes. In the simplest case, to analyze this problem, we require the flux of the two components relative to the fixed end of the sample and velocity of a lattice plane referred to these axes; hence, with

$$J'_A = -D_A \frac{\partial C_A}{\partial x}, \qquad \text{etc.} \qquad (2.111)$$

and
$$\frac{\partial C_A}{\partial x} = -\frac{\partial C_B}{\partial x} \qquad (2.112)$$

we can express the sum of the fluxes of both components J_{net} as

$$J_{net} = J'_A + J'_B = (D_B - D_A)\left(\frac{\partial C_A}{\partial x}\right) = J_A + J_B - (C_A + C_B)V \qquad (2.113)$$

If we assume that (1) the vacancies condense onto the lattice planes normal to the diffusion flow, (2) the volume per lattice sites is constant, i.e., atomic volume is constant and does not vary with composition, and (3) net flux J_{net} is zero by definition when the volume change on mixing is zero, then we find that the Kirkendall or marker velocity will equal the net flow of atoms, which is the vacancy flux times the atomic volume

$$V = (D_B - D_A)\left(\frac{\partial N_A}{\partial x}\right) = J_V\Omega = -(J_A + J_B)\Omega \tag{2.114}$$

and

$$J_A = (N_B D_A + N_A D_B)\left(\frac{\partial C_A}{\partial x}\right) = -J_B \tag{2.115}$$

where $\partial N_A/\partial x$ denotes the concentration gradient at the marker position.

The mutual, chemical, or interdiffusivity $\tilde{D}$ is the same for both species and equal to

$$\tilde{D} = N_B D_A + N_A D_B \tag{2.116}$$

where N_A and N_B are the molar (or atomic) fraction of species A and B, respectively. So D_A and D_B can be determined separately from measurements of $\tilde{D}$ and V. Equations for $\tilde{D}$ and V can be easily written that allow for volume changes and are required to be used in accurate work when these are appreciable, but it will not be considered here.[92] It is possible to express Darken's relation between tracer diffusion coefficients and chemical diffusion coefficients by using the Darken's equation (2.116), which is given by

$$\tilde{D} = N_B D_{A^*}\left(1 + \frac{N_A \partial \ln \gamma_A}{\partial N_A}\right) + N_A D_{B^*}\left(1 + \frac{N_B \partial \ln \gamma_B}{\partial N_B}\right) \tag{2.117a}$$

or

$$\tilde{D} = (N_B D_{A^*} + N_A D_{B^*})\left(1 + \frac{N_A \partial \ln \gamma_A}{\partial N_A}\right) \tag{2.117b}$$

or

$$D_A = D_{A^*}\left(1 + \frac{\partial \ln \gamma}{\partial \ln c}\right) \tag{2.118}$$

Because the two forms of the thermodynamic factors are equal and simple, it provides an obvious convenience. This is also called the *Darken equation*. Figure 2.32 shows how well this equation is obeyed in the Au-Ni system.

2.10.2.3 Vacancy Wind Effect.

A very important assumption in the above derivation is that the compensation of atom flux differences by bulk lattice motion is complete and achieved by motion only along the diffusion direction. In the absence of kinetic cross-interactions between diffusing components, intrinsic diffusion can be explained according to a simple atomic mobility model. For systems involving diffusional kinetic interactions among components, Manning expressed the interactions in terms of the vacancy wind effect where the net vacancy flux influences a component's intrinsic flux.[93] For a random concentrated alloy for which there is no solute-vacancy binding, the Manning approximation assumes that, at infinite dilution, the solid solution becomes ideal ($\varphi = 1$) and the intrinsic diffusion coefficient D_A must tend toward the tracer diffusion coefficient D_{A^*}. Hence,

$$D_A = D_{A^*}\varphi \quad \text{and} \quad D_B = D_{B^*}\varphi \quad \text{where } \varphi = 1 + \frac{\partial \ln \gamma}{\partial \ln c} \tag{2.119}$$

These relations are again called Darken's equations. Furthermore, from thermodynamics of irreversible processes, we know that the off-diagonal terms cannot be ignored. The more general expression can, therefore, be given by

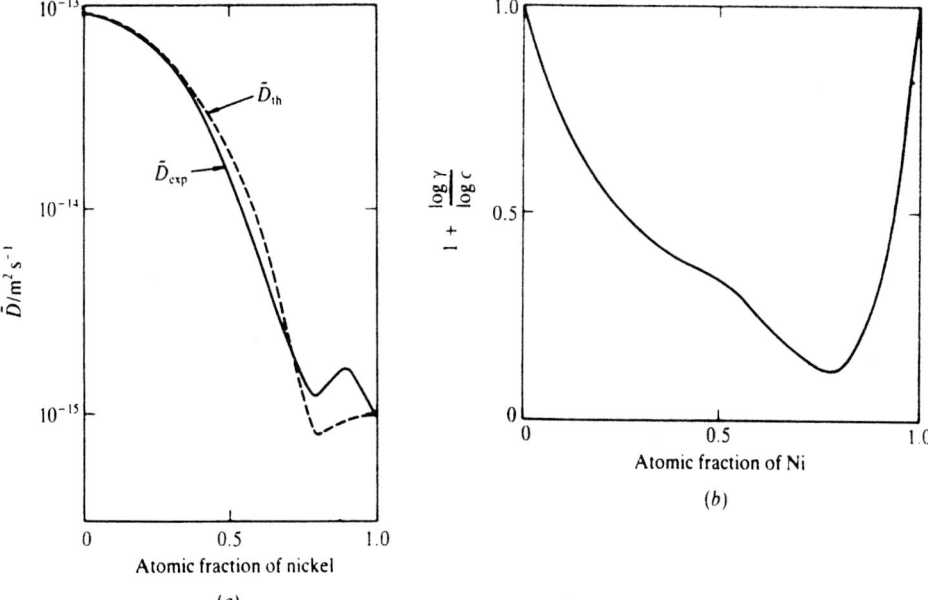

FIGURE 2.32 (*a*) Calculated and observed interdiffusion coefficients in gold-nickel alloys at 1173 K. (*b*) Corresponding thermodynamic factor for interdiffusion. (*Source: J. E. Reynolds, B. L. Averbach, and M. Cohen, Acta Met., 1957, p. 29.*)

$$D_A = \frac{kT}{N}\left(\frac{L_{AA}}{N_A} - \frac{L_{AB}}{N_B}\right)\varphi \quad \text{and} \quad D_B = \frac{kT}{N}\left(\frac{L_{BB}}{N_B} - \frac{L_{BA}}{N_A}\right)\varphi \quad (2.120)$$

where k is the Boltzmann constant; N is the total number of sites per unit volume; L_{AA}, L_{AB}, L_{BA}, and L_{BB} are the phenomenological coefficients which are functions of temperature, concentration, etc.; and according to the Onsager reciprocity relation, $L_{ij} = L_{ji}$ (i.e., $L_{AB} = L_{BA}$).

In a particular case of a simplified random-alloy model for which $\varphi = 1$, these expressions are assumed to hold even for a nonrandom alloy in which the thermodynamic factor φ is no longer unity. Therefore, the final expression for the intrinsic diffusivities is given by equations

$$D_A = D_{AB}^{A*}\varphi\, r_A \quad \text{and} \quad D_B = D_{AB}^{B*}\varphi\, r_B \quad (2.121)$$

where r_A and r_B denote the vacancy wind corrections. Hence

$$\tilde{D} = (N_A D_{B*} + N_B D_{A*}) \varphi \left[1 + \frac{2 N_A N_B (D_{A*} - D_{B*})^2}{M_0 D*(N_A D_{B*} + N_B D_{A*})} \right] \qquad (2.122)$$

Here, M_0 is a pure number that is based on the crystal structure. The rightmost term in the square brackets is known as the Manning vacancy wind term or correction because it represents the coupling between the transport of species A and B through the vacancy flux. It is noted that Manning's equations predict a chemical diffusion coefficient $\tilde{D}$ always greater than that given by Darken's equations. It is clear that the vacancy wind factor r_m depends on the physical effect that the isotopes of A and B differ. These vacancy wind factors and phenomenological coefficients can be found for dilute alloys by using a kinetic theory for the mobility of a substitutional tracer.[2] Thus the vacancy wind effect can improve or retard atomic flow, respectively, when it is opposite to or in the same direction as the net vacancy flux.[93]

2.10.3 Ternary Diffusion

Diffusion phenomena in multicomponent systems are complicated and encountered in a variety of materials and processes. The systems include steels, high-temperature alloys, coatings and composites, and processes ranging from diffusion bonding, cladding, and controlled heat treatment to surface modifications.[94] Diffusion against a concentration gradient in a ternary or higher-order couple can take place due to thermodynamic interactions, Onsager correlation cross-effects, vacancy winds and inhomogeneities, electrostatic neutralization interactions in ionic solids, and electron-solute interactions in semiconductors.[95]

One problem associated with multicomponent diffusion is the prediction of the diffusion path between two terminal alloys. In general, the diffusion path is defined as a line from one terminal alloy to another on the ternary isotherm, denoting the locus of the average composition in planes parallel to the original interface throughout the diffusion path. It also corresponds to the morphology of the diffusion area. It depends on the gradient of chemical potentials and mobilities of each species, taking into consideration the mass-balance requirements. If the phases are separated by planar interfaces, the diffusion path can cross the two-phase region along a tie line, and along the entire interface; here the same local equilibrium is assumed.

For ternary and multicomponent diffusion, the concept of zero-flux planes has been developed by Dayananda and co-workers.[96-98] These are defined as the positions where the interdiffusion flux of one component approaches zero, resulting in an interdiffusion flux reversal for that component.[99]

In a ternary system, Onsager's phenomenological diffusion coefficients forming an L (3×3) matrix have served as the basis for both theoretical and experimental diffusion studies. In substitutional alloys, the frequency of direct atom interchanges is small compared to atom-vacancy interchanges because it is assumed that metal diffusion occurs only by exchange with the neighboring vacancies. If vacancies are in equilibrium in the system and the contributions of vacancy wind and correlation effects are omitted, it is possible to eliminate the nondiagonal, or Onsager coefficients in the L_{ij} matrix. Based on the Kirkendall frame of reference, the intrinsic diffusion fluxes J_i (mol·cm^{-2}·s^{-1}) can be given by[100]

$$J_1 = -L_{11}\frac{\partial \mu_1}{\partial x} = -D_{11}\frac{\partial C_1}{\partial x} - D_{12}\frac{\partial C_2}{\partial x} \tag{2.123a}$$

$$J_2 = -L_{22}\frac{\partial \mu_2}{\partial x} = -D_{21}\frac{\partial C_1}{\partial x} - D_{22}\frac{\partial C_2}{\partial x} \tag{2.123b}$$

$$J_3 = -L_{33}\frac{\partial \mu_3}{\partial x} = -D_{31}\frac{\partial C_1}{\partial x} - D_{32}\frac{\partial C_2}{\partial x} \tag{2.123c}$$

where C_i is the concentration (mol·cm^{-3}), x the distance parameter (cm), μ_i the chemical potential (J·mol^{-1}), and D_{ij} the intrinsic diffusion coefficients (cm^2·s^{-1}). The relationship between Onsager diffusion coefficients L_{ij} and tracer diffusion coefficients D_i^*, according to Darken and Le Claire, is given by

$$L_{ij} = \frac{C_i D_i^*}{RT} \tag{2.124}$$

which yields

$$J_i = -\frac{C_i D_i^*}{RT}\frac{\partial \mu_i}{\partial x} = -\frac{N_i D_i^*}{RTV_m}\frac{\partial \mu_i}{\partial x} \tag{2.125}$$

where N_i is the atomic fraction of component i and V_m is the molar volume (cm^3/mol of atoms), assumed to be constant in this analysis. Since activity a_i is written as

$$\mu_i = \mu_i^0 + RT \ln a_i \tag{2.126}$$

this gives

$$J_i = -\frac{N_i D_i^*}{V_m a_i}\frac{\partial a_i}{\partial x} = -\frac{D_i^*}{\gamma_i V_m}\frac{\partial a_i}{\partial x} \tag{2.127}$$

where $\gamma_i = a_i/N_i$ is the activity coefficient of species i. In an ideal ternary system, the relationships among the *interdiffusion fluxes* $\tilde{J}_i$ and intrinsic diffusion fluxes J_i; interdiffusion coefficients $\tilde{D}_{ij}^n$, and concentration gradients $\partial N_i/\partial x$; and tracer diffusion coefficients D_i^*, and chemical potential gradients $\partial \mu_i/\partial x$ can be given by

$$\tilde{J}_i = J_i - N_i \sum_{j=1}^{3} J_j \tag{2.128}$$

$$\tilde{J}_1 = -\frac{\tilde{D}_{11}^3}{V_m}\frac{\partial N_1}{\partial x} - \frac{\tilde{D}_{12}^3}{V_m}\frac{\partial N_2}{\partial x} \tag{2.129a}$$

$$\tilde{J}_2 = -\frac{\tilde{D}_{21}^3}{V_m}\frac{\partial N_1}{\partial x} - \frac{\tilde{D}_{22}^3}{V_m}\frac{\partial N_2}{\partial x} \tag{2.129b}$$

$$\tilde{J}_3 = -\tilde{J}_1 - \tilde{J}_2 \tag{2.129c}$$

or through Eqs. (2.125) to (2.128)

$$\tilde{J}_1 = -\left(\frac{N_1 D_1^*}{RTV_m}\right)\left(\frac{\partial \mu_1}{\partial x}\right) + N_1\left[\left(\frac{N_1 D_1^*}{RTV_m}\right)\left(\frac{\partial \mu_1}{\partial x}\right)\right.$$
$$\left. + \left(\frac{N_2 D_2^*}{RTV_m}\right)\left(\frac{\partial \mu_2}{\partial x}\right) + \left(\frac{N_3 D_3^*}{RTV_m}\right)\left(\frac{\partial \mu_3}{\partial x}\right)\right] \tag{2.130a}$$

$$\tilde{J}_2 = -\left(\frac{N_2 D_2^*}{RTV_m}\right)\left(\frac{\partial \mu_2}{\partial x}\right) + N_2\left[\left(\frac{N_1 D_1^*}{RTV_m}\right)\left(\frac{\partial \mu_1}{\partial x}\right)\right.$$

$$\left. +\left(\frac{N_2 D_2^*}{RTV_m}\right)\left(\frac{\partial \mu_2}{\partial x}\right) + \left(\frac{N_3 D_3^*}{RTV_m}\right)\left(\frac{\partial \mu_3}{\partial x}\right)\right] \tag{2.130b}$$

$$\tilde{J}_3 = -\tilde{J}_1 - \tilde{J}_2 \tag{2.130c}$$

Using the known values of tracer diffusion coefficients, the vacancy flux J_V can be calculated from the equation

$$J_V = -\sum_{j=1}^{3} J_j = \frac{N_1 D_1^*}{RTV_m}\frac{\partial \mu_1}{\partial x} + \frac{N_2 D_2^*}{RTV_m}\frac{\partial \mu_2}{\partial x}$$

$$+ \frac{N_3 D_3^*}{RTV_m}\frac{\partial \mu_3}{\partial x} \tag{2.131}$$

Using the Boltzmann-Matano analysis for multicomponent systems, the interdiffusion fluxes $\tilde{J}_i$ can be calculated from the experimental diffusion profile as give by

$$\tilde{J}_i = \left(\frac{1}{2tV_m}\right)\int_{N_i(-\infty)}^{N_i'} x\,dN_i \tag{2.132}$$

where the origin ($x = 0$) is the Matano plane or interface. At the original (Kirkendall) plane, a relation between the vacancy flux and the marker shift Δx_m is given by the equation

$$\frac{\Delta x_m}{2t} = V_m J_V \tag{2.133}$$

In all analyses, note that $\Sigma_i N_i = 1$, $\Sigma_i \tilde{J}_i = 0$, and $\Sigma_i N_i \partial \mu_i = 0$, which implies that according to Eq. (2.125), $\Sigma(J_i/D_i^*) = 0$.

Since the interdiffusion and intrinsic fluxes are related by the lattice or marker velocity V, the interdiffusion coefficients can be represented in terms of the intrinsic coefficients by

$$\tilde{D}_{ij}^3 = D_{ij}^3 - N_i\sum_{k=1}^{3} D_{kj}^3 \qquad (i,j=1,2) \tag{2.134}$$

where N_i is the molar fraction of component i and the molar volume V_m is assumed to be constant. Usually, all the diffusion coefficients are dependent on composition. The D_{ij}^3 are interrelated by three atomic mobilities and thermodynamic data.[100a]

For a ternary system, a pair of solid/solid couples A/B and C/D with intersecting diffusion paths is required, as shown in Fig. 2.33. Here, the composition identified by the intersection point I is common to both couples, and it is only at this composition we can measure the four $\tilde{D}_{ij}^3$.

Ternary diffusion couple experiments elucidated by Darken involve welding a bar of γ-Fe–0.4% C with γ-Fe–0.4% C–4% Si and diffusion annealing (or austenitizing) at 1050°C. Figure 2.34 shows that initially a sudden rise in the carbon concentration develops on both sides of the weld plane because of the very rapid diffusion of interstitial carbon with respect to silicon which, although contrary to Fick's first law, decreases the μ_c (chemical potential of carbon) gradient caused by the difference in Si concentration. In the long term, after equilibrium of Si content

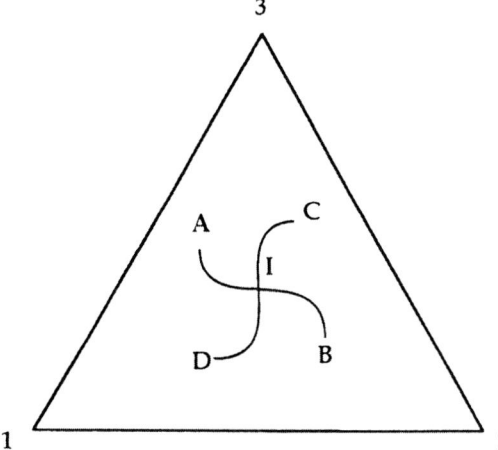

FIGURE 2.33 Schematic diffusion paths for a pair of ternary solid/solid diffusion couples A/B and C/D with a common composition point (C_1, C_2) of intersection I where four ternary interdiffusion coefficients $\tilde{D}^3_{11}$, $\tilde{D}^3_{12}$, $\tilde{D}^3_{21}$, and $\tilde{D}^3_{22}$ can be determined.[96] (*Reprinted by permission of Springer-Verlag, Berlin, Germany.*)

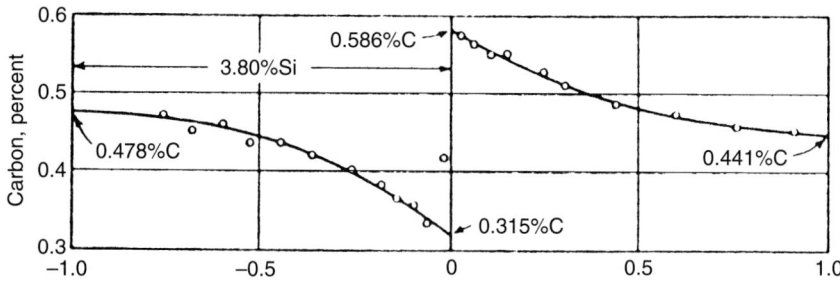

FIGURE 2.34 Distribution of carbon after a 13-day anneal at 1050°C in a diffusion couple between γ-Fe-0.44% C (right) and γ-Fe-0.4% C-4% Si (left). The chemical potential of carbon is continuous and monotonic across the couple throughout the diffusion anneal. (*Source: L. S. Darken, Trans. AIME, vol. 180, 1949, p. 430.*)

on either side of the weld plane, the carbon content is redistributed and maintains a constant chemical potential again. Figure 2.35 shows the concentration-time paths for two points to the left and to the right of the weld plane.

2.11 ELECTRO- AND THERMOMIGRATION

Electromigration is a diffusion-controlled mass transport phenomenon where the forced movement of individual atoms (or metal ions) occurs along the length of a conducting element under the direct influence of an applied electric field, current flow, or high current density. In metals and alloys, electrons carry practically all the electric current, and the ratio of electrons to atomic currents is high. It is clear that most of the atomic transport arises from the momentum transfer between large flux of electrons and lattice ions (*electron winds*) (which is attributed to the collision of electrons with atoms).[10] The direction of the drift movement in the electric field depends primarily on whether the alloy conducts by electrons or holes. In reality,

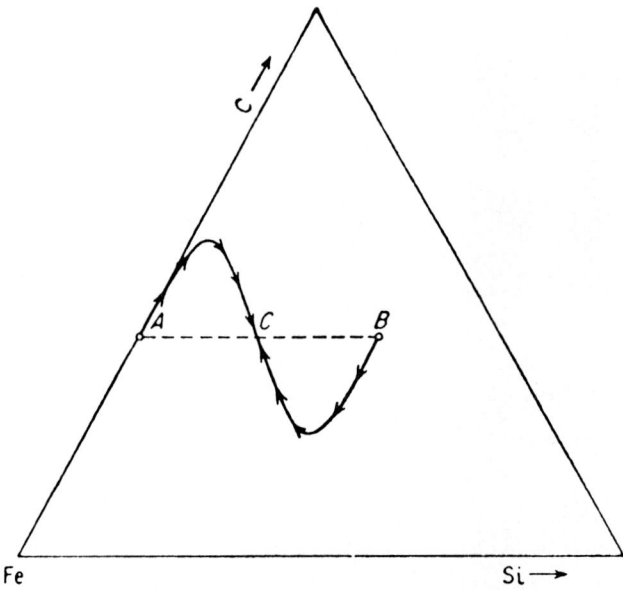

FIGURE 2.35 Schematic diagram showing the change in composition with time of two points *A* and *B* on opposite sides of the weld plane (diffusion couple) in Darken's diffusion couple of γ-Fe-0.4% C and γ-Fe-0.4% C-4% Si. C is the final equilibrium composition of the whole bar. (*Source: L. S. Darken, Trans. AIME, vol. 180, 1949, p. 430.*)

the *electron wind effect* tends to drag atoms which are changing places toward the anode. However, the greatest influence is observed in the case of α-iron where highly mobile interstitial solute carbon atom in metals moves toward the cathode.[43] This is an important mechanism responsible for several different kinds of failure in many electrical and microelectronics systems such as tungsten lightbulb filaments and computer chips. The most common are void failures along the length of narrow conducting lines (called internal failures) and diffusive displacements at the terminals of the line that destroy electrical contact. Recent research has determined that both of these failure modes are attributed to the microstructure of the conducting line and can, therefore, be inhibited by controlling or changing the microstructure.[101] This has assumed increasing importance in the performance and reliability of packaging systems that involve electrical contacts (Fig. 2.36). Thus, electromigration-induced damage in metallic conductors is a complex field which ranges from microscopic damage mechanisms to reliability modeling.[102]

Electromigration often occurs in the presence of uneven temperature distributions that develop at various sites within device structures, such as at locations near hot spots, which are caused by occasional nonadhesion of the stripe to the substrate; in areas of different thermal conductivity, such as metal-semiconductor contacts or interconnect-dielectric crossovers; at nonuniformly covered steps; and at terminals of increased cross section. In addition to the effect of microstructure and phase morphology, the temperature gradient plays a key role in adding complications.[103] Matter flow induced by a temperature gradient is called *thermotransport*.[104]

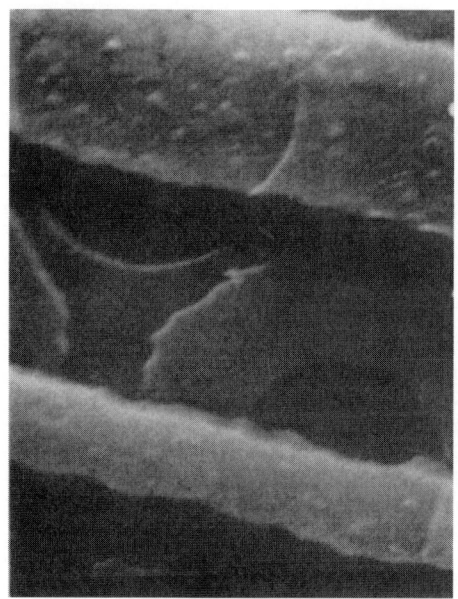 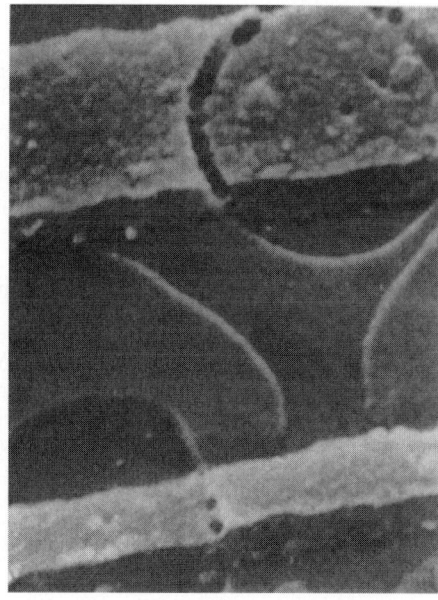

Before   After

FIGURE 2.36 SEM micrograph showing void formation after reliability test in thin-film metallization due to electromigration. (*Source: A. Christou and M. C. Peckerar, in Electromigration and Electronic Device Degradation, ed. A. Christou, Wiley, New York, 1994, pp. 105–137.*)

TABLE 2.6 Different Modes of Metallic Electromigration[105]

A. Electrolytic (ionic)
    Ambient temperatures (<100°C)
    Low current densities (<1 mA/cm^2)
    1. Normal conditions—no visible moisture ("humid" or silver electromigration)
    2. Visible moisture across conductors ("wet" electromigration)

B. Solid state (electron momentum transfer)
    High temperatures (>150°C)
    High current densities (>10^4 A/cm^2)

Thermotransport is usually a second-order effect with respect to electromigration and, therefore, can only occur when electromigration has been subdued as, for example, under an ac stressing or bidirectional pulsing.[104]

Table 2.6 shows the different modes of metallic electromigration. Electrolytic electromigration occurs primarily under normal ambient conditions when the local temperatures and current densities are low enough to allow water to be present on the surface. Solid-state electromigration has received much attention in microelectronics due to its role in producing failures in integrated circuits. For more details, see the article by Krumbein.[105]

Thermomigration, i.e., the diffusion of atoms under the influence of a temperature gradient, is also known to contribute to voids in thin-film metallization. In conclusion, the thin-film connecting stripes between the individual transistors, etc., are especially susceptible to failures due to diffusion-controlled processes.[104] Although the magnitude of thermomigration is usually small compared to electromigration, there can be occasions where temperature gradients are very steep to effect diffusion. Finally, it is a method to study the electronic structures of point defects (vacancies, impurity atoms) at elevated temperatures and its variation during a jump.

To understand and solve numerous diffusion-related problems such as electromigration, thermomigration, hillock growth, etc., the availability of diffusion data in thin films at low temperatures is of great significance.

2.11.1 Electromigration in Thin Films

Material transport phenomena on thin-film surfaces are of fundamental and technological interest. These processes include electromigration, thermomigration, and mechanodiffusion (diffusion in stress gradients). In multiphase materials, diffusion caused by a concentration gradient is often associated with these phenomena.

Thin metallic films are used for interconnects in integrated circuits, for superconducting devices, magnetic memories, etc., and their grain structure and preferred orientations can have important bearings on their performance. Thin-film metallic conductors are normally characterized by a polycrystalline microstructure comprising dense crystalline grains separated by an interconnected system of relatively open grain boundaries (GBs), as shown schematically in Fig. 2.37. These GBs usually provide short-circuit paths for interdiffusion, which permits the conductor to conform with external (driving) forces, thereby accelerating the degradation of requisite physical properties and reducing ultimate time to failure.[106]

The atomic or ionic flux J under usual drift conditions (i.e., induced by internal mechanical stress caused by nonequilibrium deposition and processing conditions,

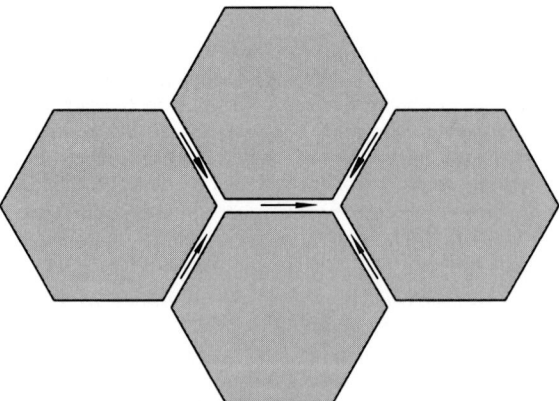

FIGURE 2.37 Schematic representation of a small segment of an idealized (two-dimensional) microstructure in thin-film conductor, comprising impenetrable (hexagonal) grains isolated by high-conductivity grain boundaries. Usually the corresponding mass flux (denoted by arrows) can be directed toward or away from individual grain boundary triple junctions, causing localized mass accumulation or depletion.

differential thermal expansion and contraction of the film, and substrate and/or chemical/interfacial reactions) is given by the Nernst-Einstein equation

$$J = NMF = \left(\frac{ND_b}{kT}\right)F = \frac{ND_b}{kT}\frac{d\mu}{dx} \tag{2.135}$$

where N is the number of atoms per unit volume; M, the atomic mobility of the rate controlling species ($= D_b/kT$); F, the electromigration induced force on the metal ions; D_b, the grain boundary diffusion coefficient; and $d\mu/dx$ the chemical potential gradient (force per atom) (where stress-dependent component of $\mu = \Omega\sigma$, where Ω is the atomic volume $= 1/N$ and σ the uniaxial stress).

The phenomenon of electromigration involves diffusion-controlled mass transport induced by current flow in a thin-film conductor, either through coulombic force applied on individual lattice ions or through momentum transfer between electrons and lattice ions (electron wind). The driving force on the ions due to applied electric field E ($= \rho j$, where ρ is the electrical resistivity and j the current density) is given by

$$F = -z^*qE = -qz^*\rho j \tag{2.136}$$

where q is normal coulombic (ionic or electronic) charge, z^* an empirical (dimensionless) parameter, qz^* the effective metal ion charge, as negative in sign, and E the electric field strength. Thus the electromigration force (chemical potential gradient) is directly related to the current density. The atomic flux induced by current flow along GB i is obtained by substituting $d\mu/dx$ with F so that

$$J_i = N\left(\frac{D_b}{kT}\right)qz^*\rho j\cos\varphi_i \tag{2.137}$$

where D_b is the GB diffusivity and φ_i is the angle subtended by the direction of current and the inclination of grain boundary i. The net atomic flux, due to both stress- and current-induced mass transport along GB, is given by

$$J_i = N\left(\frac{D_b}{kT}\right)\left(\frac{\Omega d\sigma}{dx} + qz^*\rho j\cos\varphi_i\right) \tag{2.138}$$

For low values of j, the corresponding driving force for electromigration is just balanced by the induced stress gradient so that the net atomic flux along the GB disappears. If we assume that l is the length of the thin-film conductor and $d\sigma/dx > 2\sigma_0/l$ (a critical value), stress at the GB extremities (adjacent grain boundary triple junctions) can no longer be supported. In that situation, matter is either absorbed or ejected, leading to the formation of either a hole or a hillock. Consequently, the threshold condition for the formation of either a hole or a hillock is given by

$$l_B j = 2\sigma_0/qz^*\rho\cos\varphi_i \tag{2.139}$$

where l_B is called the *Blech length*. When the threshold condition is exceeded (i.e., when $l > l_B$), the effective current density j is diminished by $2\sigma_0/l_B qz^*\rho\cos\varphi_i$. It may be noted that stress- and current-induced degradations are interrelated and depend strongly on the corresponding microstructure of thin-film conductors.[106]

Figure 2.38 shows three common examples of electromigration in which atomic transport occurs by (a) grain boundary movement, (b) saddle displacement, and (c) flux divergence at a grain boundary triple junction.[107] In the first case (Fig. 2.38a), grain boundary velocity V_{gb} ($=V_b$) is easily obtained from the condition for mass continuity between grains I and II and is given by

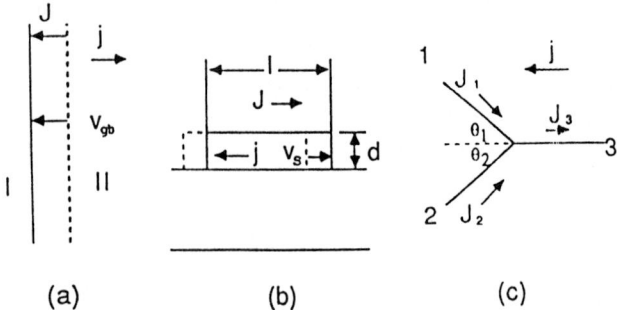

FIGURE 2.38 Three common examples of electromigration, in which atomic transport is manifested by (*a*) grain boundary movement, (*b*) saddle displacement, and (*c*) flux divergence at a grain boundary triple junction.[107]

$$V_b = \frac{D_b}{kT} F \qquad (2.140)$$

Thus, grain boundary migration arises from the net mass transport (net atomic flux across the boundary) attributed to electromigration. Although this effect is very small, it clearly illustrates the influence of a localized flux imbalance on a microscopically observable event (i.e., grain boundary migration). This specific mode is associated with the stability of bamboo structures in narrow conducting lines.[108]

In the second case (Fig. 2.38*b*), the velocity of a high conducting (thin-film) saddle, defined by length l ($l \gg d$), width w, and thickness d, is located on a low-conductivity (thin-film) rail, so that current is diverted by the saddle. The resultant saddle velocity V_s is obtained from the condition for mass continuity and is given by

$$V_s = \frac{A_i}{wd} \frac{D_i}{kT} F \qquad (2.141)$$

where D_i is the appropriate diffusivity for mass transport through cross-sectional area A_i. (Ideally, V_s is independent of saddle length.) If the principal mechanism for mass transport is controlled by volume diffusion of the film, $A_i = wd$ and Eq. (2.141) converts to that for the atomic drift velocity ($V = MF$). However, cross-sectional areas for surface and grain boundary diffusion are approximately equal to $2aw$ and $\delta wd/\lambda$, respectively, where a, δ, and λ are, respectively, nearest-atom distance, effective grain boundary thickness, and average grain size ($\lambda \gg \delta$). Thus, saddle velocity is a function of the principal mechanism for mass transport, thereby, of the characteristics of the rate-controlling species and saddle dimensions, because volume, surface, and grain boundary diffusions are the competing mechanisms for mass transport. This particular mode is related to the stability of multilevel interconnections in thin-film circuits.

In the third case (Fig. 2.38*c*), mass transport at the triple junction can be analyzed in a similar manner. If we assume the transport to be limited to the grain boundary, net flux at a given triple junction is given by

$$\Delta J_b = \frac{N_b \boldsymbol{F}}{kT} D_b = \frac{N_b \boldsymbol{F}}{kT} D_0 \exp\left(\frac{-Q}{kT}\right) \tag{2.142}$$

where N_b = number of atoms per unit volume at the grain boundaries, D_b is the GB diffusivity, D_0 is the preexponential diffusivity (0.1 cm^2·s^{-1} for Al self-diffusion), and $\boldsymbol{F}$ is the electromigration-induced force on the metal ions.[104] Equation (2.142) may also be written as

$$\Delta J_b = \frac{N_b \boldsymbol{F}}{kT} (D_1 \cos\theta_1 + D_2 \cos\theta_2 - D_3 \cos\theta_3) \tag{2.143}$$

where D_i and θ_i are, respectively, diffusivities of an angle formed by grain boundary i and the applied electric field ($\theta_3 = 0$ in Fig. 2.38c). Usually a flux divergence occurs at individual triple junctions, leading to preferential formation of microscopic holes and hillocks to various degrees at GB triple points or where the GBs intersect the edge of the metal stripe. Unlike the first two examples, however, a flux divergence favors growth or decay of voids, rather than displacement of a macroscopic entity. This particular mode is linked to the stability of thin polycrystalline conducting films.[107]

Formation of holes and hillocks is a function of a gradient in temperature, current density, grain size, geometric features, crystal orientation of the grains, etc. The growth of these holes provides ultimately the interruption of the continuity of a conducting metallic stripe, which is damaged by high current densities and high temperature.[101]

2.11.2 Microelectronic Device and Electromigration

In microelectronics, metallization, i.e., application of metals and metallike layers, plays an important role in controlling the device circuit performance. Basically metallization application can be divided into two groups, as illustrated in the schematic cross section of a completed metal-oxide semiconductor field-effect transistor (MOSFET). Metallization in contact with Si, such as on source and drain regions of Fig. 2.39, is termed *contact metal* whereas the one interconnecting various devices is termed the *interconnect metal*. Sometimes the metal on top of the gate of MOSFET can also act as the interconnect at that level. However, in a multilevel scheme of interconnects, the upper-level interconnects can be different from this gate-level interconnect.[109]

The effect of stress due to the ease of different processing temperatures of the metal and the dielectric on electromigration is of importance in fine multilayered aluminum interconnecting thin-film lines or stripes (typically, 1 μm or less in thickness and width) which are confined in quartz in VLSI or microelectronic devices.[110] These thin-film lines contain a high density of defects such as grain boundaries, dislocations, and normal vacancies. Also, multilevel thin-film metallizations are employed to produce steep composition gradients.[111]

Electromigration in integrated circuits, especially VLSI and ultralarge-scale integration (ULSI) circuits, describes the development of structural damage due to metal ion transport in thin metal films under high current density ($\geq 10^6$ A/cm^2). In integrated circuits, electromigration damage takes place either in the metallization or at the metal-semiconductor contacts (in GaAs MOSFET high-power electronic device). The former leads to an open-circuit localized mass depletion (holes) or short-circuit mass accumulation (hillocks), and the latter produces a poor perfor-

(a)

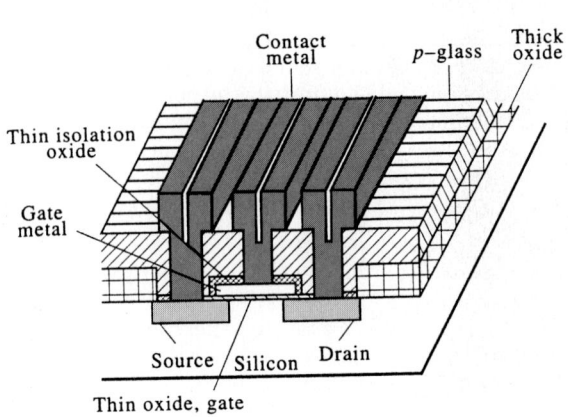

Contact metal   p–glass   Thick oxide

Thin isolation oxide

Gate metal

Source / Silicon   Drain

Thin oxide, gate

(b)

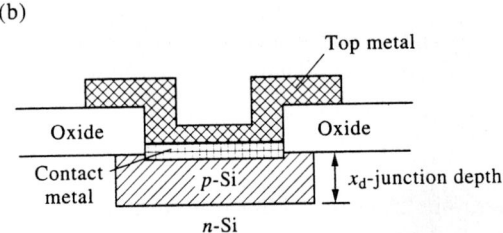

Top metal

Oxide   Oxide

Contact metal   p-Si x_d-junction depth

n-Si

FIGURE 2.39 Schematic illustration of a cross section of (*a*) MOSFET with same contact and interconnect metal and (*b*) a contact with different contact and interconnection metals.[109] (*Reprinted by permission of Pergamon Press Plc., Oxford, after S. P. Murarka.*)

mance or malfunction in semiconductor devices due to large power dissipation and high current density arising from the deterioration of ohmic and Shottky contacts.[112] (In ohmic contact, the current-voltage characteristics are linear; they are usually formed by alloying or annealing several metallization structures on the semiconductor.[113] High-power applications require low or ohmic barriers with low specific contact and parasitic resistance. Optimal Shottky barriers are the most demanding requirements of desirable metal-semiconductor interface properties.[114]) These accelerate degradation of requisite physical properties and reduce ultimate time to failure. Figure 2.40 shows examples of electromigration damage in thin films.

2.12 DIFFUSION ALONG SHORT CIRCUITS

It is generally recognized that polycrystals and even single crystals contain regions such as dislocations, grain boundaries, interfaces, and free surfaces where the diffusivity is much higher than that through the lattice. These areas interact chemically

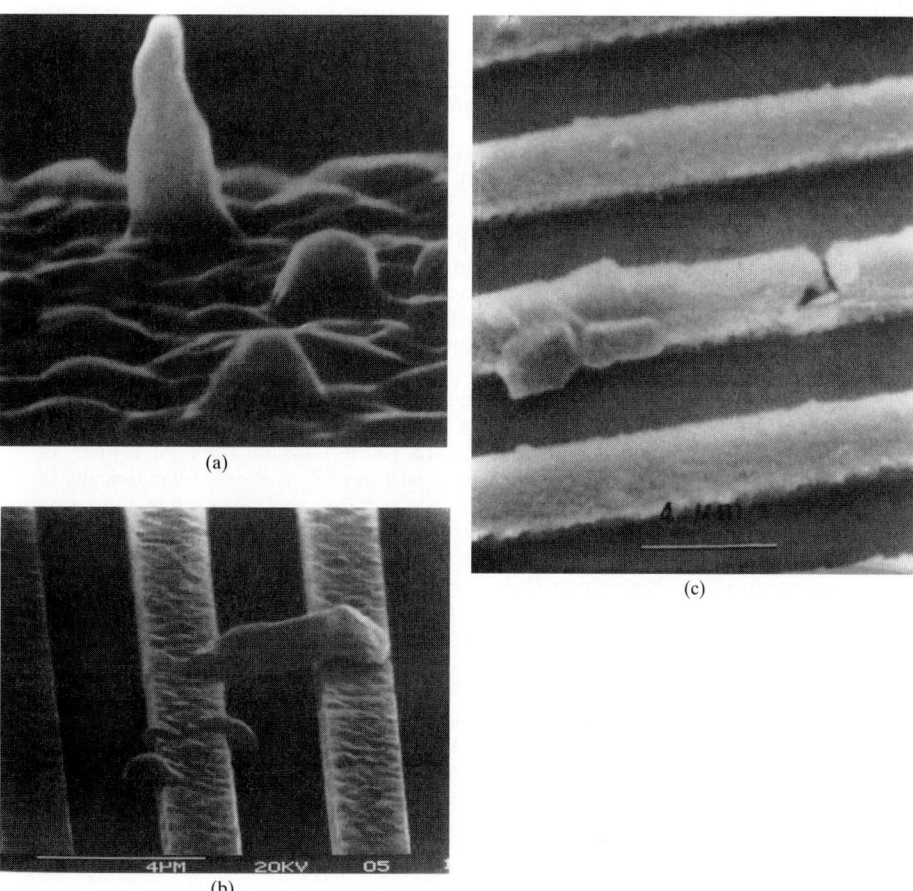

FIGURE 2.40 Examples of electromigration damage in Al films. This leads to preferential formation of microscopic (*a*) hillock growth and holes,[103] (*b*) whisker bridging two conductors,[108] and (*c*) nearby mass accumulation (hillocks) and depletion (cracks).[108] (*Reprinted with permission from Academic Press, Inc., Boston, Massachussets.*)

with the point defects, diffusing species, and the components of the alloy; the concentrations in the short circuits are quite different from those in the bulk. They can be modified by the diffusion process itself, which can produce changes in the ledge and kink densities on a surface, diffusion-induced migration of a grain boundary, etc. In the next section, we describe diffusion along dislocations, GB diffusion, diffusion-induced grain boundary migration (DIGM), and surface diffusion.

2.12.1 Dislocation Diffusion

Several methods have been used to measure dislocation diffusion rates (with single crystals to avoid the grain boundary effect); most provide directly the product $D_d a^2$, where a is the effective radius of dislocation "pipe" within which the mean effective diffusion coefficient is D_d. When segregation of the diffusant to the dislocation occurs, the product usually becomes $D_d a^2 s$, where s is the segregation coefficient (or factor). To obtain the dislocation diffusion coefficient D_d, it requires the knowledge of a and, if relevant, of s. Because these quantities are commonly unavailable, it is common practice to report directly the $D_d a^2$ or $D_d a^2 s$. (The value of $a = 5 \times 10^{-10}$ m is sometimes assumed to calculate D_d.[114a])

Let us consider now the effect of randomly dispersed dislocations or a fine grain size which is of great interest in the study of kinetics of diffusion-enhanced process at or below $(\frac{1}{2})T_M$ and is vital in determining the diffusion rate in fine-grained thin films.

A close similarity is observed between the results for diffusion enhanced by a three-dimensional array of dislocations and that resulting from a three-dimensional array of GBs. We first describe the findings for dislocations and then compare with those for boundaries. A proper mathematical description of penetration curves in dislocated crystals was provided by Le Claire and Rabinovitch, based on Smoluchowski's model, which is shown in Fig. 2.41a. Let us consider the regular array of dislocation pipes of radius a and average separation $2Z$, as shown schematically in Fig. 2.41b. They are perpendicular to the free surface, which has solute source on the top of it. The section can exhibit three different types of solute distributions based on the ratio of the mean diffusion distance in the lattice $(D_l t)^{1/2}$ to the separation Z, and remembering that D_d is always much higher than D_l.

A Kinetics. Case I: If $L = (D_l t)^{1/2} >> R$, an atom interacts with several dislocations in diffusing far enough in the lattice over the time t, and the effect of dislocations is to increase the jump frequency of the atoms in an isotropic manner. Thus, the dislocations enhance the effective diffusivity D_{eff} for the solid, and the resulting penetration depth becomes larger than it would be without dislocations. The advancing isoconcentration lines are relatively flat near the dislocations. To obtain a relation between D_{eff} and D_l, the concentration profiles obey generally the solution of Fick's equation for homogeneous medium (Gaussian profile for a thin-film source). However, we can measure an apparent (or effective) diffusion coefficient given by

$$D_{eff} = D_l\left[1 + \rho_D \pi a^2\left(\frac{D_d'}{D_l} - 1\right)\right] \tag{2.144}$$

where ρ_D is the (line) dislocation density, D_d' the diffusion coefficient inside the dislocation pipe, and D_l the diffusion coefficient in sound crystal. Alternatively, D_{eff} for a single crystal is given by, for no segregation of solute at dislocations,

$$D_{eff} = fD_d + D_l(1-f) = D_l\left[1 + f\left(\frac{D_d}{D_l}\right)\right] \tag{2.145}$$

For diffusion of solute segregation at dislocations

$$D_{eff} = fsD_d + D_l(1-fs) = D_l\left[1 + fs\left(\frac{D_d}{D_l}\right)\right] \tag{2.146}$$

where f is the fraction sites on dislocation. Since $D_d >> D_l$, it follows that

$$D_d a^2 s = \frac{D_{eff} - D_l}{\pi \rho_D} \tag{2.147}$$

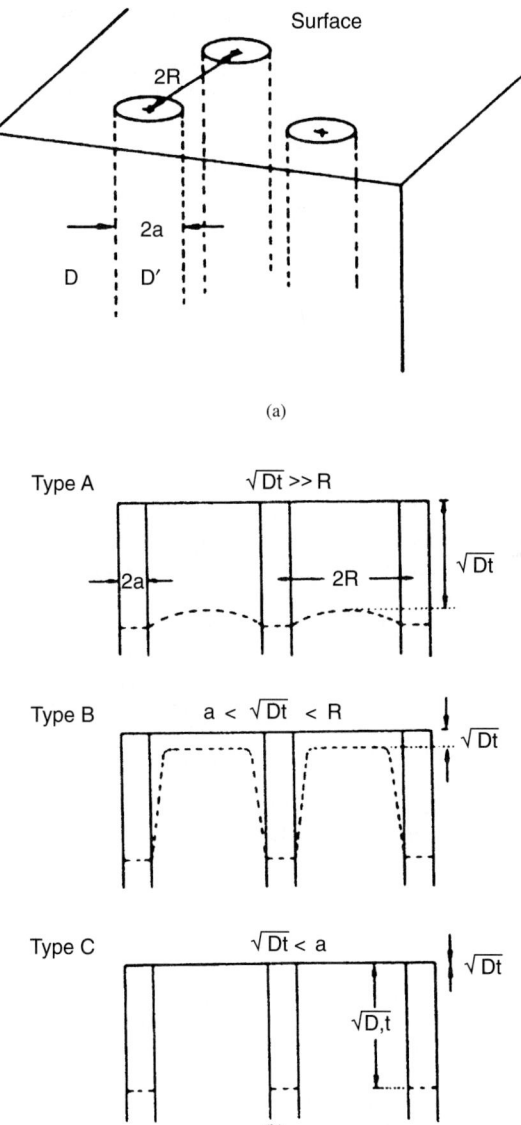

FIGURE 2.41 (*a*) Smoluchowski's model of a crystal with dislocation "pipes." (*b*) Schematic representation of Type A (top), Type B (middle), and Type C (bottom) diffusion kinetics; vertical lines indicate dislocation pipes; dashed lines are isoconcentration contours.[115]

If it is unknown, but constant, at least the activation energy for dislocation diffusion can be determined by this method.[114a]

B Kinetics. In this kinetic region, diffusion occurs simultaneously from the surface into the bulk and down and out of dislocations into the surrounding lattice. The solute field around each dislocation develops independently of its neighbors. If $(D_l t)^{1/2} \ll R$, but $(D_l t)^{1/2} \gg a$, the situation is similar to GB diffusion in bicrystals; the lateral diffusion zones surrounding the dislocations are not influenced by neighboring dislocations.[115]

C Kinetics. If $(D_l t)^{1/2} \ll a$, diffusion occurs wholly within the dislocation pipes with negligible loss into the surrounding lattice. This rarely occurs in bulk samples, but in a thin film at $T = 0.5 T_M$, D_d/D_l can be equal to 10^8 and diffusion through a 1-μm-thick film can take place with no loss of solute to the surrounding lattice.[8]

2.12.2 Grain Boundary Diffusion

It is usually recognized that diffusion in metals proceeds much faster along grain boundaries than through the corresponding crystal lattice. Ag, Au, and Ni, which diffuse 10^6 to 10^8 times faster than Pb in the bulk, were found to diffuse 10^5 to 10^6 times faster than Pb in grain boundaries, too.[116] Likewise, Fe, Co, and Ni have interstitial diffusivities in various polyvalent host metals such as in α-Zr that are about eight orders of magnitude larger than that of the host atoms.[117] However, their diffusivities along grain boundaries are much higher than those in the corresponding crystal lattice.

Thus, the atomic jump frequency in these planar defects, which are structurally only several atoms thick, is about a million times greater than the jump frequency of regular lattice atoms at $2T_m/3$. Because of the very high atomic mobility at grain boundaries relative to that in the bulk, GB diffusion plays a very important role in the kinetics of microstructural changes during physical and metallurgical processing and applications. Examples include electromigration, coble creep, sintering, *discontinuous precipitation* (DP), *discontinuous coarsening* (DC), *discontinuous dissolution* (DD), *eutectoid decomposition* (ED), *diffusion-induced grain boundary migration* (DIGM) investigation, cellular precipitation, solid-state reactions, and intermixing in thin-film technologies. A knowledge of diffusion characteristics of grain boundaries is, therefore, necessary for understanding the kinetics of these processes.[118]

Grain boundaries only exist as regions between two crystals. Since the GB diffusion coefficient D_b cannot be determined on a sample consisting entirely of boundaries, techniques have been developed for the measurement of D_b from determinations made on samples having a few grain boundaries.[119] An example is creep deformation where the deformation at low temperatures and stresses, known as coble creep, has been shown to occur as a result of diffusional flow along grain boundaries (Fig. 2.42a).[120] The mechanism during sintering at different temperatures, represented by sintering mechanism maps, also has been shown to involve low-temperature GB diffusion as the dominant mechanism (Fig. 2.42b).[121] The discontinuous precipitation reaction in the Co-Ti system suggests occurrence of the reaction through a GB diffusion mechanism.[122]

Measurement Techniques. Two main techniques have been used to obtain GB diffusion data. In the first, the *profiling technique*, concentration profile inside the specimen and in the second, the *accumulation technique*, change in surface concentration are monitored during the isothermal diffusion anneal.

Profiling Technique. In this technique, a layer of the radioactive tracer is deposited on the surface of a solid containing a GB. After diffusion annealing of the specimen, isoconcentration contours such as shown in Fig. 2.43a are obtained. These concentration profiles of radioactive species are determined by techniques such as chemical etching, autoradiography, or electron probe microanalyzer. Two different measurements, as given below, are normally made on the concentration profiles to obtain GB diffusivities:

1. The penetration depth y in the boundary at which a specific concentration occurs
2. The angle φ between the tangent to the concentration contour and the boundary

It is also possible from this type of experiment to obtain the total amount of material diffused into a number of thin slices below the surface. Successive layers from the surface are machined off using a lathe, an *electrochemical sectioning technique*, or *RF sputtering*. The composition of the material removed is then measured by chemical or radio tracer techniques. To obtain GB diffusion coefficients, a relation between any of the above measured experimental parameters and the GB diffusivity is required. Because of the wide variety and complexity of GB core structures, the GB is treated as a slab of uniform thickness, within which the diffusion coefficient is D_b and assumed to be larger than D_l. Mass-balance equations using Fick's law are then established for diffusion in the boundary and in the bulk material, and these are solved for different initial and final boundary conditions imposing the constraint that the composition be continuous between the grain boundary and the bulk material. It is assumed that the GB thickness is small and that the concentration in the GB does not vary across its thickness.[123]

Fisher and Whipple have obtained solutions assuming constant surface composition and infinite source at the surface, that is, $C = C_0$ at $y = 0$ for $t \geq 0$. For grain boundaries, each grain boundary region is considered as a uniform, thin slab of width δ with the average diffusion coefficient D', which is embedded in the lattice of diffusivity D_l.[122]

Accumulation Technique. In this method, atoms are allowed to diffuse along GBs from a source on one side of a foil to a free surface on the other side where they spread out and accumulate and are measured. It is especially suited to C-type kinetics at low temperatures where lattice diffusion is negligible. Figure 2.43b shows a schematic diffusion path geometry used for the accumulation method. Note that there exists a large scatter among the measurements of $D'\delta$ by this technique in different laboratories. The reason for such deviations is not fully understood.[123]

For a conventional diffusion experiment, Harrison's A-B-C classification (Fig. 2.44) is a good approach to describe and distinguish the three types of diffusion kinetics/regimes,[124] according to the bulk penetration depth $(D_l t)^{1/2}$ being larger than, equal to, or smaller than a characteristic length l of the network. For a GB network, l is the average diameter of the grains; and for a dislocation network, l is the average distance between two (dislocation) pinning points.

KINETIC REGIME A. When the volume diffusion distance or the bulk penetration depth $(D_l t)^{1/2} \gg$ average interboundary spacing l between the GBs, each diffusant atom diffuses along a large number of GBs as well as in the crystals in between or a multiple boundary diffusion zone. This condition, called Harrison's A regime, is met for small-grained materials or very long diffusion anneal times and/or a volume

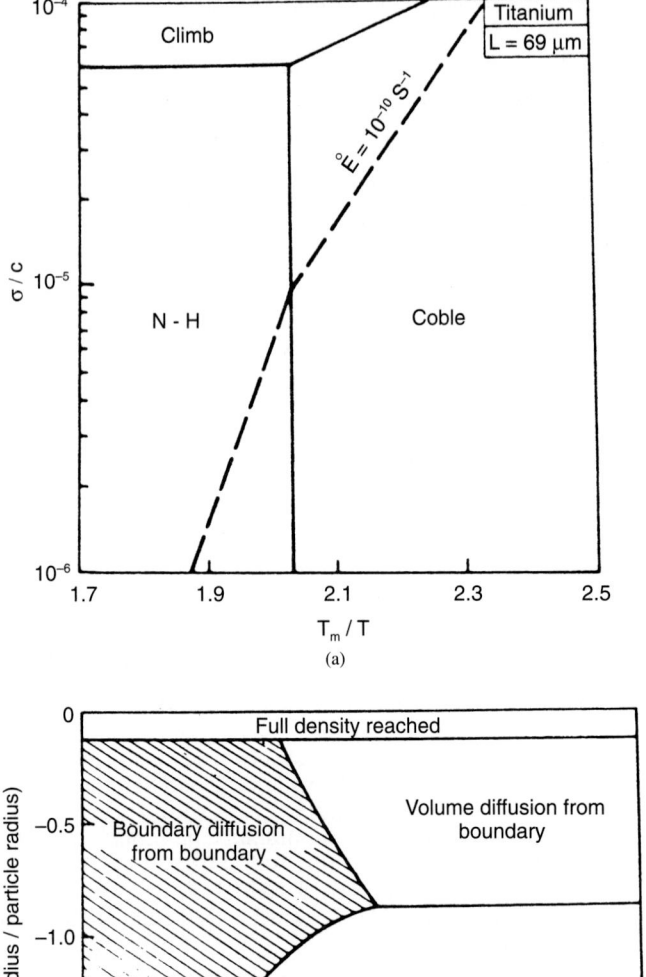

FIGURE 2.42 (*a*) Coble creep that occurs as a result of diffusional flow along grain boundaries for α-Ti of grain size 69 μm. A constant strain rate contour is exhibited by dashed lines.[120] (*Source: G. Malakonmdaiah and P. Rama Rao, Acta Metall., vol. 29, 1983, p. 1263.*) (*b*) Sintering map for aggregate of Cu spheres of radius 88 μm. Mechanism corresponding to diffusional flow along GBs dominates in the shaded area of the map.[121] (*Source: M. F. Ashby, Acta Metall., vol. 22, 1974, p. 275.*)

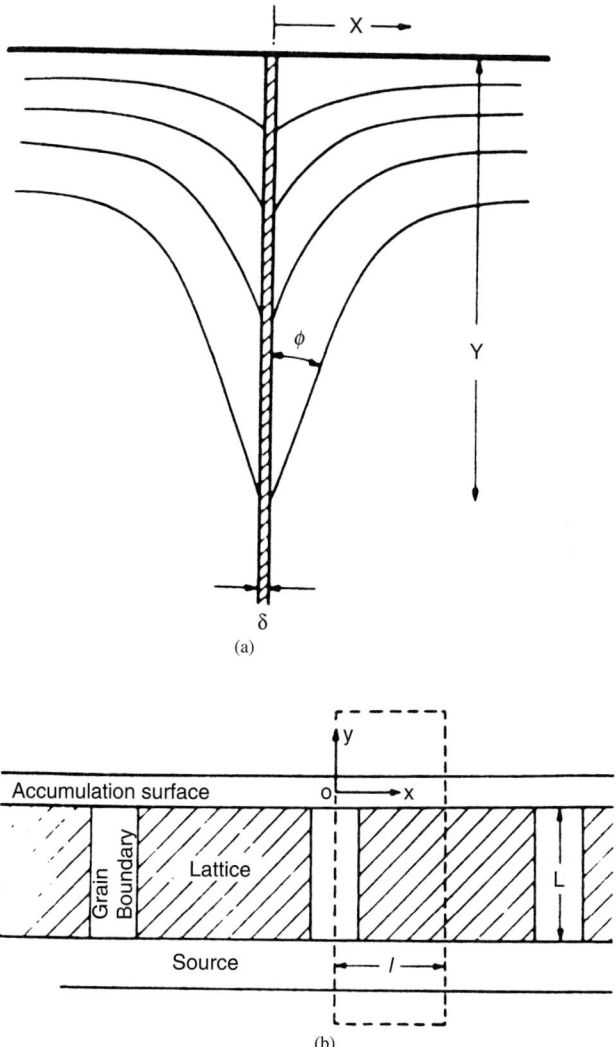

FIGURE 2.43 (*a*) Typical isoconcentration contours resulting from a rapid diffusion along a grain boundary slab of thickness δ.[123] (*b*) Diffusion path geometry used for the accumulation method.[123] (*Source: J. C. M. Hwang and R. W. Balluffi, J. Appl. Phys., vol. 50, 1979, p. 1339.*)

diffusion coefficient not much smaller than the grain boundary diffusion coefficient. A simple expression of the effective or true diffusion coefficient D_{eff} can be written, taking into consideration the fraction f of the lattice sites which belong to the grain boundaries (or dislocations as the case may be).[125]

$$D_{\text{eff}} = fD_b + (1-f)D_l \qquad (2.148)$$

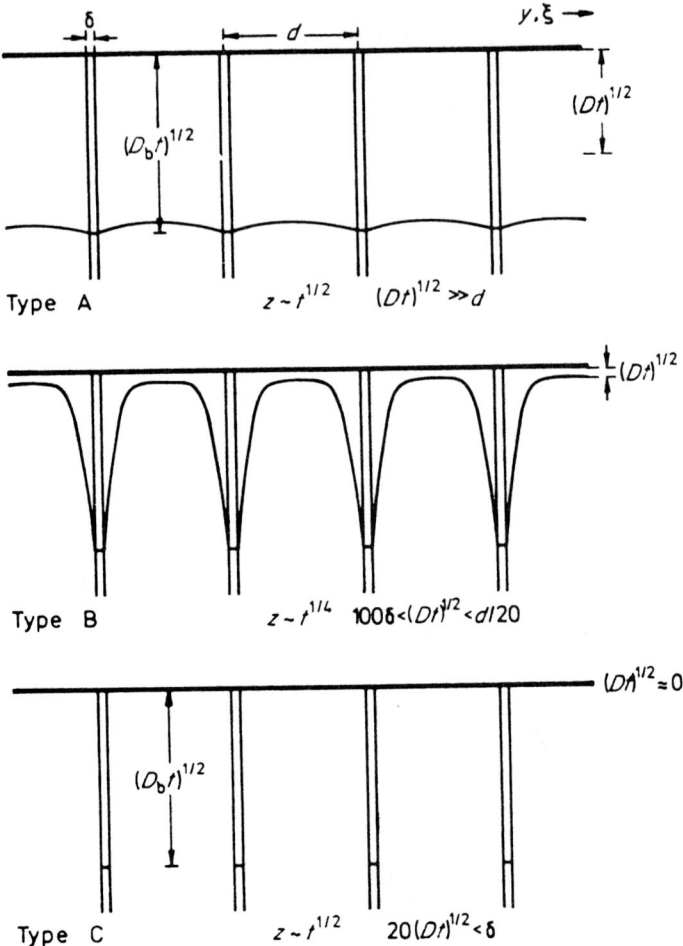

FIGURE 2.44   Harrison's A-B-C classification.[124]

where f is the volume fraction of grain boundaries in the specimen and D_l is the lattice diffusion coefficient. Equation (2.148) is generally known as the Hart-Mortlock equation. Le Claire has also given a rough calculation using this equation. Assuming an effective dislocation area of $\sim 10^{-14}\,cm^2$ and a typical dislocation density of annealed metals of $\sim 10^{-6}\,cm^{-2}$, we find that $f = 10^{-8}$. The dislocation contribution to a measured or effective D will then exceed $\sim 1\%$ (about the limit detectable), when $D_{eff}/D_l > 10^6$. From the activation energy Q (about $34T_M$ for volume diffusion) and the rough generalization that Q_d equals $0.5Q_v$, we estimate that $D_d/D_l > 10^6$ occurs for temperatures below about $0.5T_M$. That is why experiments where the lattice diffusion coefficient D_l only is of importance are always made at temperatures well above $0.5T_M$. At lower temperatures, D_{eff} may be enhanced and may be seen as a slight upward curvature of the Arrhenius plot.[13] For the diffusion of solute impurities that are bound or attracted to dislocations with a binding energy E_b, the Hart-Mortlock equation, Eq. (2.148), is written as

$$D' = fD_d \exp\frac{E_b}{kT} + (1-f)D_l \tag{2.149}$$

and the diffusion coefficients refer to impurities. The Hart equation can be shown to follow fairly accurately, if $l/(D_l t)^{1/2} \leq 0.3$ and also if $l/(D_l t)^{1/2} \geq 100$ (this represents the regime C kinetics as discussed below).[126,127] Dislocations may start to contribute to the measured diffusion coefficient in these cases at higher temperatures than for self-diffusion.[13]

In reality, the same Eq. (2.148) with D_b replacing D_d can be used for grain boundary-enhanced diffusion measurements in polycrystalline materials, provided grain size l is much less than L. According to the *Lavine-MacCallum model*, Hassner derived the following equation for the effective volume diffusion coefficient in the Type A kinetics regime:

$$D_{eff} = D_l + 4\delta D_b/3d \tag{2.150}$$

where d is the grain size in polycrystalline materials. Other situations require more detailed considerations, but a rough working rule is that in well-annealed polycrystals, contributions from GB diffusion are usually negligible at temperatures above about $0.75T_M$. For very fine-grained material, the limit may be greater.

KINETIC REGIME C. Under these situations, diffusion may be assumed to occur only within the grain boundaries with negligible sidewise leakage into the neighboring crystals (Fig. 2.44c). Short diffusion anneal times and/or negligibly small values of the volume diffusion coefficient compared to the grain boundary diffusion coefficient cause volume diffusion lengths much shorter than the grain boundary width $[(D_l t)^{1/2} << \delta)]$. It is convenient to discuss other limits out of sequence. If the volume diffusion distance l, that is, $(D_l t)^{1/2}$, is much smaller than the GB width, so that the material transport is limited to within the GBs, type C kinetics is said to prevail. In this instance, all material comes down the short-circuit paths, and the measured diffusion coefficient is given entirely by D'.

KINETIC REGIME B. In this intermediate and most often encountered situation, GB diffusion is accompanied by the sideways leakage of the diffusant into the nearby grains by volume diffusion, but, in contrast to type A kinetics, the grain boundary spacing is large for the boundaries to be assumed as isolated. In other words, it is assumed that $(D_l t)^{1/2}$ is equivalent to the interboundary spacing l in order for the material to be transported down a short-circuit path and the materials diffuse out into the lattice with less likelihood to reach another short-circuit path.[68]

There are various solutions available of Fick's second law to cope with tracer diffusion in the presence of short-circuit paths, but space does not allow us to discuss these in any detail. Suzuoka[128] and Le Claire[129] have given a solution for the GB problem for the usual case where there is a finite amount of tracer originally at the surface. It is found, among other things, that

$$\left(\frac{d\ln C}{dx^{6/5}}\right)^{5/3} = \left(\frac{4D_l}{t}\right)^{1/2} (D_b\delta)^{-1}(0.661) \tag{2.151}$$

or
$$KD_b\delta = 0.661\left(\frac{d\ln C}{dx^{6/5}}\right)^{-5/3}\left(\frac{4D_l}{t}\right)^{1/2} \tag{2.152}$$

where $KD_b\delta$ is the GB diffusion parameter and is ordinarily measured in the diffusion measurements,[130] K is the GB segregation factor, i.e., the ratio (≥ 1) between GB and lattice solute constants, and δ is the GB width. That is, the very common method of measuring D_b using either polycrystalline samples under conditions such that $(D_l t)^{1/2}$ is much less than grain size or specially prepared "bicrystals" with the boundary along the diffusion direction. It is pointed out that only the product $D_b\delta$ can be found experimentally by determining the slope of the linear region (penetrations reached by GB diffusion) in a plot of ln C versus $x^{6/5}$ (and not x^2) and with a knowledge of D_l itself. It can be difficult, however, to determine the two diffusion coefficients in a single experiment; the practicalities of this and alternatives are discussed by Rothman.[20] An example of a tracer penetration plot with a clear contribution from GB diffusion ($x^{6/5}$ dependence) is shown in Fig. 2.45a.

Le Claire and Rabinovitch[131] have addressed the dislocation pipe problem and provided near-exact solutions to Fick's second law for both an isolated dislocation pipe and arrays of dislocation pipes. The profiles generally are similar to the GB ones except that a linear region in a plot of ln C versus x is now found, i.e.,

$$\frac{d\ln C}{dx} = -\frac{-(A(a))}{\left[(D'/D_l - 1)a^2\right]^{1/2}} \tag{2.153}$$

where A is a slowly varying function of $a/(D_l t)^{1/2}$. An example of such a tracer penetration plot with a clear contribution from dislocation pipe diffusion (x dependence) is shown in Fig. 2.45b.

For further details on GB diffusion we refer to reviews by Kaur and Gust,[68] Peterson,[132] Mohan Rao and Ranganathan,[123] Balluffi,[133] Philibert,[134] and Mishin et al.[135] For dislocation pipe diffusion we refer to Le Claire and Rabinovitch,[131] Philibert,[134] and Fournelle.[136]

2.12.3 Diffusion-Induced Grain Boundary Migration

The phenomenon of diffusion-induced grain boundary migration, sometimes called chemically induced grain boundary migration (CIGM), was first noticed and explained for the Cr-W systems by den Broeder in 1972.[137] DIGM is a well-recognized phenomenon and occurs in a wide range of binary (interstitial and substitutional) alloy systems and a few ternary alloy and binary ceramic (or oxide) systems, in which a sideways, transverse, or lateral grain boundary migration in either direction in a pure metal or solid solution accompanies the diffusion of two chemically different elements at a temperature usually below $0.5T_m$ into or out of the material along an existing grain boundary and even inducement of nucleation and growth of new grains.[118,138-148] Accordingly, the area behind the migrating boundary is enriched by or depleted with the solute, thereby forming an alloyed zone (representing a better mixing of the alloy) in the former case and a dealloyed zone (or phase separation) in the latter (Fig. 2.46), usually with sharp solute discontinuity at the initial and final positions of the grain boundary.[149,150] The process is thus capable of producing an enhanced rate of mixing (diffusion) where little or none might be expected in the absence of grain boundary motion. This lateral displacement (normal to the grain boundary plane) is not necessarily uniform along the bound-

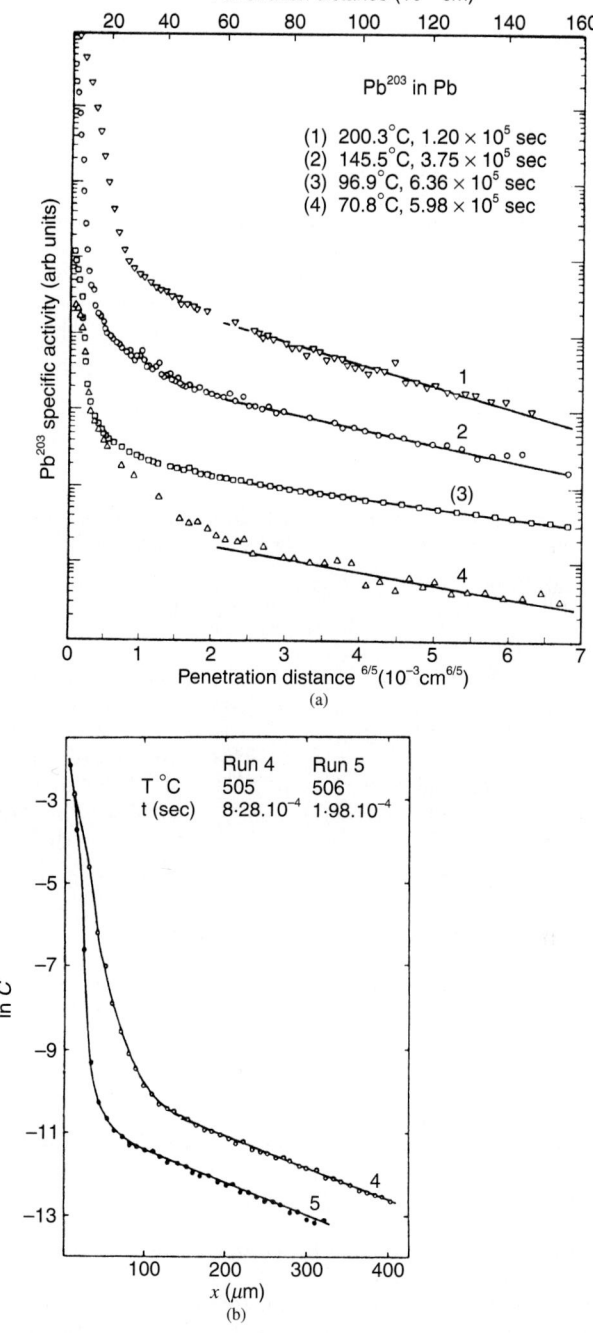

FIGURE 2.45 (*a*) Typical ^{203}Pb tracer concentration/penetration profiles into polycrystalline Pb along grain boundaries plotted in the 6/5th power of the penetration distance for extraction of GB diffusion coefficient equations.[120] (*b*) Tracer concentration profiles of ^{22}Na into single crystal of NaCl showing a contribution from dislocations. (*Source: Y. K. Ho, Thesis, Imperial College, London, 1982.*)

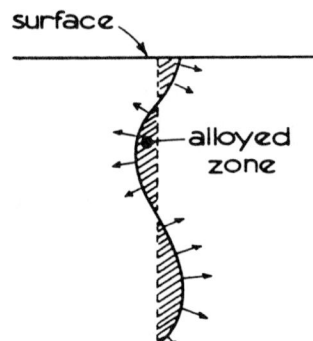

surface

alloyed zone

FIGURE 2.46 Lateral displacement of a grain boundary due to a Kirkendall effect along the boundary. The hatched regions have a composition different from that of the surrounding matrix. The dashed line is the initial position of the boundary.[30] (*Reprinted by permission of Pergamon Press Plc., Oxford.*)

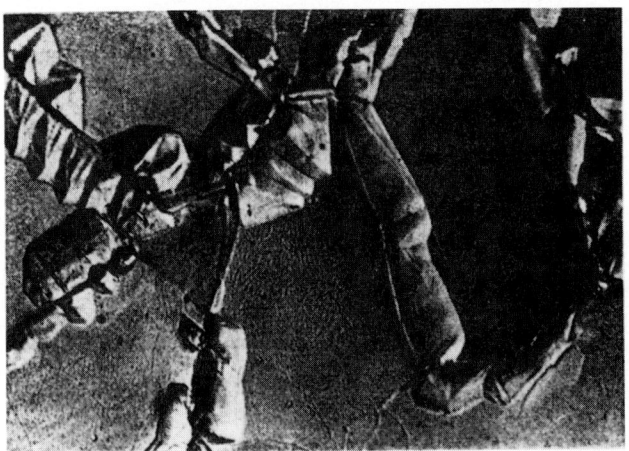

FIGURE 2.47 The surface structure of polycrystalline iron after heating in an atmosphere of zinc vapor at 873 K for 4 hr.[144] (*Courtesy of M. Hillert and G. R. Purdy.*)

ary, and as a result the latter is very often distorted. The principal difference between DIGM and the conventional stationary-boundary diffusion experiments is that in the former the transversed transport of diffusant occurs primarily by the GB migration and the effects most prevail at low temperatures, where the lattice diffusion is negligible or essentially frozen out; in the latter, volume diffusion is responsible for transversed transport of diffusant.[68,138–148]

Figure 2.47 shows numerous important microstructural features of DIGM where faster-diffusing Zn from a vapor source has penetrated into a thin α-Fe foil during dezincification at 873 K ($0.48T_m$ of Fe) for 4 hr.[139,144]

In contrast to the initial observations, the condition of a vanishingly small lattice diffusion is not a prerequisite, and the movement is observed in a fairly large temperature range up to almost the melting point. It produces a unique and thus characteristic microstructure. However, there is a controversy over the precise origin of the driving force for this process; consequently several models have been advanced.[141] According to the first model, the driving force for this process is the reduction in free energy of mixing caused by (the discontinuity in composition and) the formation of an alloyed (or dealloyed) region behind the moving GB.[151] The solute atoms enter the lattice at the grain boundaries as a result of trapping at lattice sites as the boundary migrates. Thus, enrichment of the lattice in solute takes place at the trailing edge of the migrating grain boundary.[16]

According to the second model, the driving force arises from the generation of elastic strain energy in the crystals adjoining the boundary due to outward diffusion of misfitting solute atoms from the boundary core.[140] Here, the elastic constants are usually anisotropic, the volume of the alloyed layer is changed, and the amount of elastic energy will essentially be different on the two sides of the boundary. The grains with the increased adjacent elastic energy will shrink and dissolve at the expense of the other by the sweeping movement of the boundary or of the liquid film.[152] In both situations, the free energy loss overcompensates the energy increase because of the increase of the grain boundary surface. Further evidence for the significance of solute misfit in the initiation of DIGM is found in the work of Rhee and Yoon[153] who systematically varied the misfit parameter in a ternary system and showed that the phenomenon is suppressed when the misfit parameter in a ternary system is brought to zero. This suggests that elastic-strain energy also plays a role in the grain boundary movement. However, DIGM could not be observed in an Ag-Mn system with lowest misfit parameter.[154]

Balluffi and Cahn[142] and Smith and King[155] have suggested that DIGM requires a direct coupling between the diffusive flux and the GB motion. This mechanism invokes the climb of grain boundary dislocations[142] and manifests no driving force itself but offers only a means to move the boundary laterally.

We have reasons to believe that more than one model such as both elastically derived and purely chemical terms can be important in maintaining DIGM.[156]

A wide range of experiments have shown that DIGM occurs for both alloying and dealloying, in alloy systems having positive and negative deviations from thermodynamic ideality for interstitial and substitutional alloys, metals, and oxide systems. It is also established that low-angle grain boundaries and some special grain boundaries, such as $\Sigma 3$ coherent twin boundaries in fcc metals, do not normally display DIGM, but all other high-angle grain boundaries appear to undergo DIGM. In some cases, namely, those with a large initial grain size and a large free energy of mixing, this leads to both the migration of the existing grain boundaries and nucleation of new grains and growth of these newly introduced or recrystallized grains into the sample, again leading to alloying. The process was introduced by, among others, Li and Hillert[157] and is described as *diffusion-induced recrystallization* (DIR). The general phenomena of DIGM and DIR have been reviewed in detail by Handwerker[158] and by Yoon.[159]

Discontinuous or *cellular precipitation* is the process by which a supersaturated solid solution releases the excess solute by the migration of a matrix grain boundary into a supersaturated solute, leaving a two-phase mixture of solute-depleted matrix and solute-rich precipitate. The conditions necessary for this process are similar to those of DIGM in that temperature is normally $0.5T_m$ or lower, and a composition change in the matrix is imposed which for discontinuous precipitation is by the temperature change. Generally, discontinuous precipitation provides a mechanically undesirable microstructure with coarse and usually brittle intermetallic plates; hence the usual solution is to avoid this phenomenon. DIGM is most pronounced in couples with a large concentration difference. It is most commonly observed in discontinuous precipitation, the interdiffusion of thin films, accelerated sintering, and developing surface alloyed layers. Several studies have been made in the Fe-Zn system with Fe-Zn diffusion couples. The initial Zn penetration appears to occur by DIGM. The mechanism of continued penetration is less clear.[143]

Liquid Film Migration. A phenomenon very similar to "classical" DIGM, called *liquid film migration* (LFM), is found in partially molten samples during either the introduction of the liquid or variation in chemical composition of the liquid, for example, by alloying or by varying the temperature. This type of behavior occurs easily during liquid-phase sintering, as was observed by Yoon and Huppmann.[160] When pure W powder is mixed with Ni and heated to above the melting point of Ni, the Ni-rich liquid promotes the easy atom transport of W by liquid-phase diffusion. The solid W, in equilibrium with liquid Ni, contained a small quantity of Ni (0.15 wt % at 1913 K), and it was noticed that the liquid films between W particles advanced into one of the particles, dissolving the pure W and reciprocating a W alloy containing that amount of solute. Similarly, Mo powder sintered with Ni exhibited the same phenomenon when the equilibrium solute content of the Ni-rich liquid and thus that of the Mo-rich solid were altered.[152]

2.12.4 Surface Diffusion

Surface diffusion deals with the motion of atoms or molecules over the surface of a substrate. Surface diffusion phenomena are of great importance in many areas such as creep (at low applied stress in a pure polycrystalline material through the lattice), sintering (of powders by lattice or boundary diffusion from regions of sharp curvature to low curvature), and thin-film (nucleation and growth process), crystal growth, chemisorption, and physisorption surface chemical reactions. The classical description of surface diffusion is the terrace-ledge-kink (TLK) model* for stepped surface, as shown in Fig. 2.48. Here, terraces are singular surfaces of lower Miller's indices (e.g., {100}, {110}, and {111} for cubic structures) which are separated by straight ledges (or steps) of atomic height or by zigzag ledges (i.e., ledges with kinks). Ledges and kinks have a double origin: (1) a geometric one, to yield the misorientation of the actual surface with respect to the dense planes of the terraces (θ and α angles in Fig. 2.48), and (2) a thermally activated one for entropy determinations. The TLK model is valid in a range of low temperature (between 0 K and $0.5 T_m$) due to surface roughening, reconstruction, and premelting transitions. As predicted by Burton et al.,[161] a drastic change in the surface topology takes place at a particular transition temperature T_R at which the formation free energy of the ledges disappears (or becomes very small); as a result, the surface becomes delocalized. This roughness transition is due to a large number of steps of increasing height which give rise to indistinguishable edges of the terraces. According to Leamy and Gilmer,[162] T_R is approximately given by $T_R = 0.5\varepsilon/k$, where ε is the strength of the first-neighbor bond. This transition has really been noticed on several metals employing He scattering spectroscopy.[68]

In a pure material, the point defects are adatoms and advacancies (or surface vacancies) (Fig. 2.48); they can be produced in pairs on the terraces, or they are created and destroyed at kinks. The latter case is energetically favored compared to others and is believed to be predominant. Multidefects can also form by cluster-

---

* The TLK model describes a surface of absolute zero. As the temperature increases, point defects appear on the terraces and kinks appear on the edges (created in pairs of opposite sign). No new ledges appear due to their very high energy of formation. The ledges, and particularly the kinks, play a role as sources and sinks for point defects.[3]

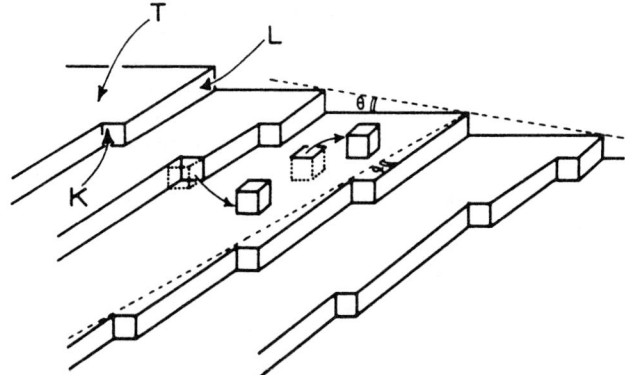

FIGURE 2.48   Terrace-ledge-kink (TLK) model for low-index surfaces. The formation of adatoms (the extra atoms bulging out from the plane of a low-index surface) and advacancies (the antidefects to adatoms) is represented.[30] (*Reprinted by permission of Elsevier Science BV, Amsterdam.*)

ing adatoms or advacancies. Impurity atom adsorption and diffusion on a metal substrate are called heterodiffusion and self-diffusion on the substrate surface. For both self- and heterodiffusion it is normal to call the movement of atoms for short distances, where there is one type of site, as in intrinsic diffusion. When the concentration of diffusing species is very low, the relevant diffusion coefficient is called a tracer diffusion coefficient. This is really a single particle diffusion coefficient. At higher concentrations, the relevant diffusion coefficient is called a *chemical diffusion coefficient.*[163]

A recent EAM (embedded atom method) study of surface self-diffusion of single adatoms of seven fcc metals demonstrated that the calculated activation energies for Ni using the EAM potentials were in excellent quantitative agreement with the experimental data.[164–167] This clearly suggests that the formation energies of point defects are dependent only on very fundamental and simple properties of the surfaces (such as the number of lateral neighbors or the packing): (1) The formation and migration energies for adatoms and advacancies are observed to be very sensitive to surface orientation. (2) The formation energies of both types of defect are comparable, with the exception of the (100) surface of an fcc lattice, where the formation energy of advacancy is significantly small with respect to the energy for the adatom. Thus, both defects appear to contribute significantly to mass transport. They will be created in roughly equal amounts, either separately at kinks or pairwise at terrace sites. (3) The migration energies have been primarily calculated for adatoms on fcc and bcc surfaces, and for the advacancy on Cu[164] and Ni[165] surfaces; the advacancy in most cases is found to be the slower-diffusing defect.

For fcc crystals, the migration energy of adatoms decreases roughly with decreasing surface roughness: $E_m(111) < E_m(113) \sim E_m(331) < E_m(001) < E_m(210)$. For bcc crystals, the migration energies of adatoms are roughly in the following order: $E_m(110) \sim E_m(211) \sim E_m(321) < E_m(310) < E_m(001) < E_m(111)$. Note that, because of the existence of the defect concentration term, the *surface self-diffusivities* are not essentially in the same order; a compensation effect takes place, which pairs a low migration to a large defect formation energy [say, on the (111) surface of the fcc lattice]. Consequently, the surface diffusivity of Ni is believed to be clearly larger than (113) and (133) surfaces than on any other.[167]

At low temperatures ($T < T_m$) and rough surfaces, exchange between an adatom and a surface atom is predominant. A particular low-temperature mechanism is tunneling. This has been found for adsorbed H atoms on W. At high temperatures ($T/T_m > 0.75$) several mechanisms have been reported. The first one is *nonlocal surface diffusion* of adsorbed atoms or complexes thereof. This mechanism describes a nonlinearity in the Arrhenius plot. The second mechanism depends on the order-disorder transition (below T_m) at the surface which results in a nonexponential increase in the number of diffusible species. This process, also called *surface melting*, leads to high activation energies of surface diffusion at high temperatures and, therefore, very large surface diffusion coefficients (which have been experimentally observed).[167a]

If mass transfer surface diffusion takes place by an adsorbed (adatom) or terrace vacancy mechanism, the macroscopic surface diffusivity D_s (for self-diffusion or diffusion of an impurity layer) can be expressed by[168,169]

$$D_s = \frac{1}{4}\sum_{i=1}^{n}\alpha_i^2 \Gamma_i \tag{2.154}$$

and
$$D_s = D_{s0}\exp\left(\frac{-Q_d}{kT}\right) \tag{2.155}$$

From an atomic approach, Eq. (2.154) relates to an atomic jump distance α_t between neighboring lattice sites, total number of jump types n, and a mean atomic jump frequency Γ_i. If the elementary diffusion process is viewed as the one where the atoms are activated over a free-energy barrier Q_d from one potential well to another, then according to absolute rate theory, Eq. (2.154) is converted to the Arrhenius equation, Eq. (2.155), in which the preexponential term $D_{s0} = \frac{1}{4}v_s\alpha^2 \exp(\Delta S_d/k)$. Here, v_s is the atomic vibration frequency, and ΔS_d and Q_d may include terms related to the formation of a suitable diffusible atom and its motion.[157a] However, this equation holds only when (1) all the diffusion mechanisms must contribute independently to mass transport and (2) all the surface sites are equivalent. The defect should be in equilibrium all over the surface, and the concentration should be uniform everywhere, without any preferential trapping or occupancy sites. This requirement can be fulfilled by the close-packed perfect surfaces without ledges or kinks, e.g., a (111) surface in the fcc lattice.[30]

Adsorption of surface contaminants, such as oxygen, sulfur, and halogen and low-melting-point metals, can enhance the value of D_s very greatly—by factors of up to 10^4 in some cases.[169] This effect is of great technical significance in sintering.

Surface free-energy mass transport can also be found in compounds (e.g., oxides), but it is more difficult to interpret this phenomenon if more than one chemical species is involved.

2.13 APPLICATION OF THIN FILMS TO DIFFUSION STUDY

Atomic diffusion in thin films is of great technical importance in the processing of electronic and electrooptic devices as well as from the standpoint of their performance and reliability.[170] Diffusion studies can be used to exploit the unique capability of control of structure and composition in thin films. This may consist of molecular beam epitaxial (MBE) growth of defect-free (dislocations and grain boundaries) single-crystal films and superlattices (SLs), ultrahigh-vacuum deposition of very pure thin films, the fine-tuning of composition in local regions by ion implantation, and the patterning (at submicrometer dimension) of sample by lithographic techniques employed in the electronic industry. In the following section we will only outline a few applications of thin films to diffusion study, such as irreversible processes, nonlinear diffusion, and diffusion in metastable phases.[110]

Irreversible Processes. This is the area where the effect of multiple driving forces (such as chemical affinity, stress, electric field, and temperature gradient) on diffusion in thin films and their interference and/or interactions can be studied. Examples include metallic thin-film deposition on inert substrate under stress due to the thermal mismatch between them; low-temperature reaction between Cu and Sn thin films leading to whisker and hillock growths in Sn films; and the effect of stress on electromigration in fine lines in microelectronic devices (see also Sec. 2.10.1).

Nonlinear Diffusion. Effect of nonlinear diffusion study is of importance due to the trend of miniaturization in the electronic industry. Very high gradients of concentration, temperature, and electrical field are present in localized regions in the structure of devices during operation. It is now possible to grow superlattices of Si/SiGe and GaAs/AlGaAs by MBE to a large degree of perfection with respect to the control of periodicity and defect density. They could be employed to study the nonlinear effect of large concentration gradients on diffusion, using better samples and high-intensity synchrotron radiation sources for X-ray diffraction.

Diffusion in Metastable Phases. Metastable phases such as amorphous alloys[171] and icosahedral crystals[172] can be prepared by thin-film deposition and preparation. Pseudoepitaxial growth can yield metastable crystalline phases. Self-diffusion and impurity diffusion in these metastable phases can be studied by using shallow-profiling techniques of thin films.[173] Structure-kinetics correlation should be set up by performing a parallel study of these phases by electron microscopy and X-ray diffraction. The effect of atomic movement on phase stability is interesting.

2.14 DIFFUSION IN IONIC SOLIDS

Extensive diffusion studies of ionic solids, involving the transport of electric charge, have contributed greatly to our understanding of defect solid-state physics. Various precise experimental techniques have emerged to probe the lattice defect structures, their transport, and defect interactions. In an ionic crystal, diffusion and electric (or ionic) conductivity σ occur by the same defect mechanism. The two are related by the Nernst-Einstein equation

$$D^* = \frac{\sigma f k T}{N Z^2 e^2} \quad \text{or} \quad D_\sigma = \frac{\sigma k T}{N Z^2 e^2} \tag{2.156}$$

where D^* is the diffusivity of the solute measured by radio tracer, D_σ is the "charge" diffusion coefficient, f is the correlation factor (usually between 0.5 and 1.00) accounting for the nonrandom motion of tracer ions, k is the Boltzmann constant, and T, N, and e are the absolute temperature, number of mobile ions (of charge Ze) per unit volume, and electronic charges, respectively.[174] A large number of ionic solids with exceptionally large ionic conductivity are called superionic conductors, fast-ion conductors, and solid electrolytes. Among these, *fast-ion conductor* is the term most used.[174]

In the next section, diffusion in ionic crystals, namely, oxides, will be briefly discussed, because of their technological importance in the oxidation process and in superionic materials, as used in fuel cells, etc. It is also important in the control of ceramics and, thereby, their properties.[176]

2.14.1 Defects in Ionic Crystals

Let us first consider stoichiometric crystals, say, the oxide MO, and "intrinsic" defect production. The *Shottky defect* (actually a pair of defects) contains a vacant anion site and a vacant cation site. It arises from thermal activation without undergoing interaction with the atmosphere. The Shottky defect production (assuming fully ionized defects) is expressed as a chemical reaction, i.e.,

$$0 \rightleftharpoons V_M'' + V_O^{\cdot\cdot} \tag{2.157}$$

where 0 refers to a perfect crystal, V_M'' is a vacant (V) metal (M) site, the primes denote effective negative charges (with respect to perfect crystal), $V_O^{\cdot\cdot}$ is a vacant (V) oxygen (O) site, and the over dots denote effective positive charges. This is a *Kröger-Vink defect* notation that is used as a shorthand for atomic point defects and electronic defects in compounds.[3]

We can express an equilibrium constant K_s for the reaction (also called the *Shottky product*) in the following form that is valid for low defect concentrations:

$$K_s = [V_M''][V_O^{\cdot\cdot}] = \exp\left(\frac{-\Delta G_s^f}{kT}\right) \tag{2.158}$$

where the brackets [ ] denote concentrations and ΔG_s^f is the Gibbs free energy of formation of the Shottky defect, which can be split into its enthalpy ΔH_s^f and entropy ΔS_s^f parts.

The other type of defects observed in the stoichiometric ionic crystal is the Frenkel defects. The Frenkel defect (actually a pair of defects) consists of a cation interstitial and a cation vacancy, or an anion interstitial and an anion vacancy. The latter is called an anti-Frenkel defect, although presently this nomenclature is seldom used. Like the Shottky defect, the Frenkel defect is thermally activated. The Frenkel defect production is also expressed as a chemical reaction; for example, for cationic disorder

$$M_M \rightleftharpoons M_i^{\cdot\cdot} + V_M'' \tag{2.159}$$

where M_M is a metal atom (M) on a metal site (M), $M_i^{\cdot\cdot}$ is an effectively doubly positively charged metal ion on interstitial i, site and V_M'' is an effectively doubly negatively charged metal ion vacancy. We have assumed double charges here solely for illustrative purposes.

The equilibrium constant for the cation Frenkel defect reaction K_{cF} is given by

$$K_{cF} = [V_M''][M_i^{\cdot\cdot}] = \exp\left(\frac{-\Delta G_{cF}^f}{kT}\right) \tag{2.160}$$

where ΔG_{cF}^f is the free energy of formation of the cation Frenkel defect. Equation (2.160) is valid for low defect concentration. This equation is often termed the *cation Frenkel product*. Likewise for anion Frenkel defects, we have

$$[V_{\ddot{O}}][O_i''] = K_{aF} = \exp\left(\frac{-\Delta G_{aF}^f}{kT}\right) = \exp\left(\frac{-\Delta H_{aF}^f}{kT}\right)\exp\left(\frac{-\Delta S_{aF}^f}{k}\right) \qquad (2.161)$$

where K_{aF} is the equilibrium constant for the anion Frenkel defect reaction, O_i'' represents doubly negatively charged anion interstitials, $V_{\ddot{O}}$ is a doubly positively charged oxygen vacancy, ΔG_{aF}^f is the free energy of formation of the anion Frenkel defect, and ΔH_{aF}^f and ΔS_{aF}^f are the corresponding enthalpy and entropy terms.[25,177]

2.14.2 Diffusion Theory in Ionic Crystals

In most diffusion mechanisms, except the interstitial mechanism, an atom has to "wait" for a defect to arrive at a nearest-neighbor site prior to a possible jump. Thus the jump frequency contains a defect concentration term such as the vacancy concentration C_V. Let us investigate an example case for diffusion comprising a Frenkel defect. Although both an interstitial and a vacancy are formed, in oxides one appears to be much more mobile, i.e., to possess a lower migration energy, than the other. In the case of stoichiometric UO_2, for example, theoretical calculation of migration energies indicates a much smaller migration energy for the O_2 vacancy than for the interstitial (by either interstitial or interstitialcy mechanisms).[25,176] At the stoichiometric composition,

$$[M_i^{\cdot}] = [V_M''] \qquad (2.162)$$

where $\quad [V_M''] = C_v = \exp\left(\dfrac{-\Delta G_{cF}^f}{2kT}\right) = \exp\left(\dfrac{-\Delta H_{cF}^f}{2kT}\right)\exp\left(\dfrac{\Delta S_{cF}^f}{2k}\right) \qquad (2.163)$

The measured activation enthalpy for diffusion consists of the migration enthalpy plus one-half the Frenkel defect formation enthalpy.

Another interesting process is the *intrinsic ionization process* in which an electron is advanced from the valence band to the conduction band, leaving behind a hole in the valence band. When the conduction electron and hole are localized at atoms, it is normal to distinguish the process with a self-ionization reaction such as $M^{2+} \rightarrow M^{3+} + M^+$. In the Krœger-Vink notation, we usually express the intrinsic ionization process by

$$0 \rightleftharpoons e' + \dot{h} \qquad (2.164)$$

Usually all ionic crystals are capable of becoming nonstoichiometric. The limit of nonstoichiometry is primarily determined by the ease with which the metal ion can change its valence and the ability of the structure to "absorb" defects without undergoing reversion to some other structure and, thereby, changing phase. Nonstoichiometry can occur by either (1) an anion deficiency, which is accommodated by either anion vacancies such as UO_{2-x} or metal interstitials such as $Nb_{1+y}O_2$ or (2) an anion excess, which is accommodated by either interstitials such as UO_{2+x} or metal vacancies such as $Mn_{1-y}O$.[25]

The extent of nonstoichiometry and the associated defect concentration are dependent on the temperature and partial pressure of the components. The defects generated in this manner are sometimes termed *extrinsic*; however, this nomenclature should be avoided because they are still intrinsic to the material. In an oxide, it is the usual practice to consider only the partial pressure of O_2 at the temperatures of interest. In carbides, where the temperatures of diffusion are quite high, the

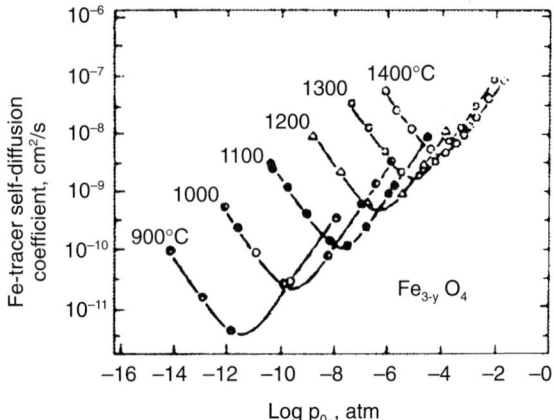

FIGURE 2.49 The iron tracer self-diffusion coefficient in magnetite as a function of oxygen activity.[16] (*Source: After Dieckmann and Schmalzried, Ber/Bunsenges, Phys. Chem., vol. 81, 1977, p. 344.*)

metal partial pressures can be comparable to the carbon partial pressure, and therefore either can be controlled.

Another important example is that of diffusion in magnetite, Fe_3O_4, where several diffusion mechanisms are operative. Magnetite has the inverse spinel structure, with Fe^{2+} on the octahedral sites and one-half of the Fe^{3+} on the tetrahedral sites at room temperature. At elevated temperatures the distribution of these two cations is random in the two types of sites. There exist four octahedral and eight tetrahedral sites and three cations. At high oxygen activities there is a metal-deficient oxide with cation vacancies as the dominant defect. At low oxygen activities, a cation excess is present with iron ions on interstitial sites as the dominant defect. The reactions controlling the species in magnetite are

$$8Fe_{Fe}^{2+} + 2O_2 = 8Fe_{Fe}^{3+} + 3V_{Fe} + 4O_0 \quad \text{(high oxygen activity)} \quad (2.165)$$

$$3Fe_{Fe} + 4O_0 = 3Fe_i + 2O_2 \quad \text{(low oxygen activity)} \quad (2.166)$$

In both reactions the stoichiometric ratio of lattice sites of the two species is conserved. Hence, the diffusivity should vary as the $\frac{2}{3}$ power of the oxygen partial pressure at high oxygen activity and as the minus $\frac{2}{3}$ power of the oxygen partial pressure at low oxygen activity. The data shown in Fig. 2.49 are consistent with these relations.[16]

2.15 DIFFUSION AND DIFFUSION-INDUCED DEFECTS IN SEMICONDUCTORS

Diffusion and diffusion-induced defects are important in the manufacture of solid-state circuits in semiconductor materials, and an appreciation of their importance is necessary to control the growth of single crystals and to regulate multilayer materials. Silicon and gallium arsenide are presently the most widely used semiconductors for fabricating microelectronic and optoelectronic devices.[178,179] A short introduction to diffusion in silicon comprising point defects, phenomenological description of diffusion processes, diffusion mechanisms, and diffusion in GaAs and AlAs-GaAs materials will be given in this section.

2.15.1 Intrinsic Point Defects

Equilibrium Condition. Intrinsic point defects are defects of atomic dimensions, free from any foreign atoms. The most basic and simple intrinsic point defects are vacancies (V) and self-interstitials (I) in Si. A Si crystal can achieve a thermodynamically more suitable state by introducing a specific concentration of intrinsic point defects under thermal equilibrium conditions. The thermal equilibrium concentration C_x^{eq} (in atomic fractions) of an electrically neutral intrinsic point defect x ($x = V, I, \ldots$) is expressed in terms of Gibbs free energy of formation ΔG_x^f by

$$C_x^{eq} = Z \exp\left(-\frac{\Delta G_x^f}{kT}\right) \tag{2.167}$$

where $\Delta G_x^f = \Delta H_x^f + \Delta S_x^f$ and Z is the dimensionless quantity characterizing the number of different configurations of the intrinsic point defects and is equal to 1 for vacancies.

Nonequilibrium Condition. Surface reactions such as thermal oxidation, nitridation, and silicidation of Si may also lead to the injection of nonequilibrium intrinsic point defects. In contrast to the case of implantation, where both vacancies and interstitials are produced in supersaturation, these surface reactions normally produce only one type of point defect (such as self-interstitials by the surface oxidation process and vacancies by the surface nitridation process). In this case, the resulting perturbed self-interstitial concentration C_I and vacancy concentration C_V are expressed by the local equilibrium relationship

$$C_I C_V = C_I^{eq} C_V^{eq} \tag{2.168}$$

which holds good for prolonged times and elevated temperatures. These nonequilibrium point defects influence both dopant diffusion and nucleation, growth or shrinkage of dislocation loops, and are therefore of great significance for modern process simulation programs for silicon devices, according to Antoniadis, Fichtner, Kump, and Dutton.[179]

Intrinsic point defects may also occur in a charged form x^r, where the superscript r denotes a positive or negative integer. In this case, the concentration of the charged intrinsic point defects is a function of the Fermi level, i.e., of the electron concentration n. Thus, a relation involving intrinsic electron concentration n_i, electron concentration n, and the corresponding thermal equilibrium concentrations $C_{x^r}^{eq}(n_i)$ and $C_{x^r}^{eq}(n)$, of charged intrinsic point defect and charged point defect x^r, is given by

$$\frac{C_{x^r}^{eq}(n)}{C_{x^r}^{eq}(n_i)} = \left(\frac{n}{n_i}\right)^{-r} \tag{2.169}$$

We can use the hole concentration p instead of the electron concentration n, which is related to n by

$$np = n_i^2 \tag{2.170}$$

2.15.2   Diffusion in Silicon

Silicon is the most important electronic material presently used. The dopant (impurity) diffusion technique is an important step in the fabrication of (doped) *p-n-* and hetero-junctions in semiconductor or microelectronic devices. These dopant atoms are usually substitutionally dissolved elements on Si lattice sites such as group III elements (B, Al, Ga) for *p* doping or group V elements (P, As, Sb) for *n* doping. Their (dopant) diffusion takes place mainly by processes involving lattice/intrinsic point defects such as monovacancies and self-interstitial atoms.[178] Hence, from the atomistic approach, diffusion involves the transport of atoms from one region of the Si crystal to another by the interaction of atoms with these point defects. Dopant diffusion from doped polysilicon or dopant-implanted silicides are used to avoid implantation-induced intrinsic point defects in the fabrication of ultra-shallow junctions in silicon devices. The diffusion of various dopants such as Zn or Si results in greatly enhanced disordering of GaAs/AlAs and related III–V compound superlattices or multiquantum wells.[180]

2.15.2.1   *Phenomenological Concepts.*   In the simplest case, Fick's first law, the Arrhenius relationship, and Fick's second law hold for diffusion in silicon. The diffusion coefficient of substitutionally dissolved elements can be shown to be primarily the product of the lattice vibration frequency Γ, the atomic jump distance α, the point defect fraction x_{pd}, and the fraction of atoms that are activated to make a jump x_m.[181]

$$D = \frac{1}{2}\alpha^2 x_{pd} x_m \Gamma \tag{2.171}$$

Thus, the activation energy for diffusion consists of the energy necessary to form the point defect and the migration energy of the diffusing atom. The concentration profile for a species diffusing with a constant diffusivity into Si from a surface source at a constant surface concentration C_s can be expressed by the traditional complementary error function, as shown in a normalized form in Fig. 2.50.[179,182] In many cases, impurity diffusion in Si and other semiconductors can be best described by a concentration-dependent diffusivity of the form

$$D = D_s \left(\frac{C}{C_s} \right)^{\gamma} \tag{2.172}$$

where C_s is the concentration at the surface, D_s is the diffusivity at the surface, and γ is the parameter describing the concentration dependence. Figure 2.50 is thus the resulting indiffusion normalized concentration profiles [in the log (C/C_s) versus $x/\sqrt{4D_s t}$ form] corresponding to the actually occurring cases of $\gamma = 0, 1, 2, 3$, and -2.[179] For $\gamma = 0$, the diffusion equation for constant diffusivity reduces to

$$C(x,t) = C_s \mathrm{erf}\left[\frac{x}{\sqrt{4Dt}} \right] \tag{2.173}$$

For $\gamma > 0$, the diffusivity decreases with decreasing concentration; $\gamma = 1$ is observed for high-concentration diffusion of B and As in Si; $\gamma = 2$ for high-concentration P

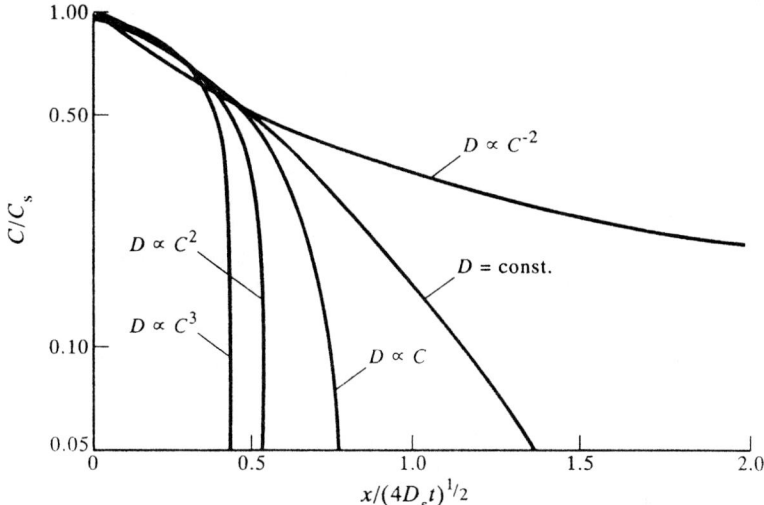

FIGURE 2.50 Normalized concentration profiles for different concentration dependencies of the diffusion coefficient as indicated.[179,182]

diffusion in Si and for Zn diffusion in GaAs. For $\gamma = -2$, the diffusivity increases with decreasing concentration which produces concave profile shape (in the semi-logarithmic plot of Fig. 2.50). Such concave concentration profile shapes have been noted for Au, Pt, and Zn in Si and for several elements in Group III–V compounds such as Cr in GaAs.[179] The concentration dependence of D can be determined from measured concentration profiles by Boltzmann-Matano analysis, as described in Sec. 2.10.1.

2.15.2.2 Diffusion Mechanisms. The dominant diffusion mechanisms in Si are vacancy diffusion, the direct interstitial diffusion mechanism, and the interstitialcy mechanism. In Si, self-diffusion via the interstitialcy mechanism predominates and occurs at higher temperature above about 1270 K. The interstitialcy mechanism also plays an important role in the diffusion of substitutional solutes such as P, B, Al, and Ge. Moreover, the ability to vary vacancy and self-interstitial concentrations by the addition of donors and acceptors can materially influence self-diffusivity. The open lattice of covalently bonded materials makes interstitial solute diffusion much more common.[16]

Interstitially dissolved foreign atoms can jump from one interstitial site to another. Intrinsic point defects do not play a role in this direct interstitial mechanism. However, the diffusion of substitutional foreign atoms such as p- and n-dopants and of host atoms (self-diffusion) requires intrinsic point defects. In the vacancy diffusion mechanism, the substitutional foreign atoms move by jumping into an adjacent vacant site. In the direct interstitial or interstitialcy diffusion mechanism, a self-interstitial drives out a substitutional atom, which then returns to a substitutional lattice site at a neighboring position. Generally, diffusion by the direct interstitial mechanism, which occurs for elements such as Cu, Ni, or H, is very fast compared to the diffusion of substitutionally dissolved elements such as Group III and Group V dopants via the interstitialcy or vacancy mechanism, as shown in Fig. 2.24.

Silicon Self-diffusion. Uncorrelated self-diffusion and tracer self-diffusion coefficients of Si atoms, D^{SD} and D^T, respectively, in polycrystalline silicon are given by

$$D^{SD} = D_I C_I^{eq} + D_V C_V^{eq} \tag{2.174a}$$

and
$$D^T = f_I D_I C_I^{eq} + f_V D_V C_V^{eq} \tag{2.174b}$$

where D_I and D_V are the diffusivities of self-interstitials and vacancies, respectively, and f_I and f_V are correlation factors for self-diffusion via self-interstitial (interstitialcy) and vacancy diffusion, respectively. The interstitial- and the vacancy-related parts have been measured predominantly by indirect methods comprising elements diffusing via interstitial-substitutional diffusion mechanisms, as shown in Fig. 2.51. However, it remains very difficult to determine the individual factors of these parts

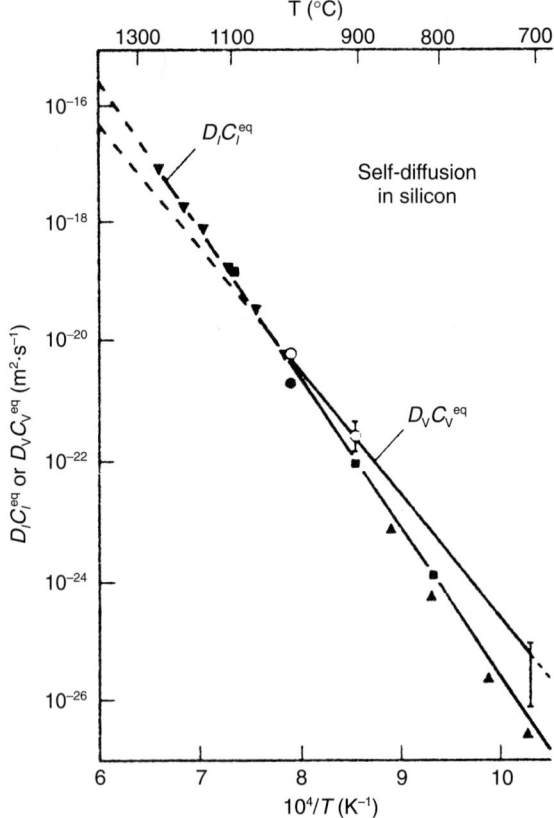

FIGURE 2.51 Self-interstitial contribution $D_I C_I^{eq}$ (full symbols) and vacancy contribution $D_V C_V^{eq}$ as a function of reciprocal absolute temperature.[178] (*Source: U. Gösele and T. Y. Tan, in Encyclopedia of Materials Science and Technology, vol. 4, Verlagsgesellschaft Chemie, Weinheim, Germany, 1991, pp. 197–247.*)

of the right side of Eq. (2.174) with good certainty, due to the dynamic interactions of vacancies and self-interstitials, which result in a single effective point defect diffusivity instead of two individual values of D_I and D_V, usually expected.[178]

Dopant Diffusion

INTERSTITIAL-RELATED CONTRIBUTION. The diffusivity D^s of substitutionally dissolved atoms such as dopants comprising an interstitial-related part D_I^s and a vacancy-related part D_V^s is given by

$$D^s = D_I^s + D_V^s \qquad (2.175)$$

The normalized interstitial-related fractional diffusion part $\varphi_I = D_I^s/D^s$ is close to unity for the small P and B atoms. The large-sized Sb atoms diffuse primarily by the vacancy mechanism, whereas the medium-size As atoms diffuse equally by both vacancy and self-interstitial mechanisms. The normalized φ_I part tends to decrease with decreasing temperature. For n-type dopants such as P, the φ_I part tends to decrease with increasing n-doping level. For perturbed intrinsic point defects such as during surface oxidation ($C_I > C_I^{eq}$) or surface nitridation ($C_V > C_V^{eq}$), the perturbed dopant diffusivity D_{per}^s, is given by

$$D_{per}^s = D_I^s \left(\frac{C_I}{C_I^{eq}} \right) + D_V^s \left(\frac{C_V}{C_V^{eq}} \right) \qquad (2.176)$$

In many instances, the local dynamic equilibrium of intrinsic point defects according to Eq. (2.170) is dominant and permits further simplification. Equation (2.176) may also be employed to simulate diffusion of dopants after ion implantation provided the nonequilibrium concentration of implanted-induced defects may be calculated.

FERMI LEVEL EFFECT. The diffusivities D^s of all dopants in Si depend on the Fermi level. The experimentally observed dopant-dependent diffusivity, resulting from the interaction of dopants with charged and neutral point defects, can be described by the relation

$$D^s(n) = D_0^s + D_+^s \left(\frac{n_i}{n} \right) + D_-^s \left(\frac{n}{n_i} \right) + D_=^s \left(\frac{n}{n_i} \right)^2 \qquad (2.177a)$$

which for intrinsic conditions $n = n_i$ converts to

$$D^s(n_i) = D_0^s + D_+^s + D_-^s + D_=^s \qquad (2.177b)$$

where the various subscripts denote the charge state of the intrinsic point defects forming a diffusing complex with the dopant atoms. Usually dopants tend to diffuse with neutral or oppositely charged intrinsic point defects. Table 2.7 lists the suitable diffusivities in the different charge states for B, P, As, and Sb dopants.

DOPANT DIFFUSION-INDUCED NONEQUILIBRIUM POINT DEFECTS. The nonequilibrium concentrations of supersaturated intrinsic point defects may also be induced by the in-diffusion of some dopants (e.g., P) starting from a high surface concentration. The highest supersaturation of self-interstitials is produced by high-concentration P in-diffusion. High-concentration B in-diffusion is accompanied by

TABLE 2.7 Diffusion of Various Dopants Fitted to Eq. (2.177a). Each term is fitted to $D_0^* \exp(-Q/kT)$; D_0^* values are $10^{-4}\,\mathrm{m^2s^{-1}}$ and Q values in electronvolts ($1\,\mathrm{eV} = 1.6 \times 10^{-19}\,\mathrm{J}$).[178]

Element	$D_0^\bullet$	Q_0	$D_+^\bullet$	Q_+	$D_-^\bullet$	Q_-	$D_=^\bullet$	$Q_=$
B	0.037	3.46	0.72	3.46				
P	3.85	3.66			4.44	4.00	44.20	4.37
As	0.066	3.44			12.0	4.05		
Sb	0.214	3.65			15.0	4.08		

Reprinted by permission of Pergamon Press Plc., Oxford.

a much smaller but still perceptible supersaturation. The self-interstitial supersaturation s_I may be predicted from the equation

$$s_I = \frac{C_I - C_I^{eq}}{C_I^{eq}} = \frac{h\phi_I D^s(n_s)C_D^s}{(\gamma+1)D_I C_I^{eq}} \tag{2.178}$$

where D^s and C_D^s are the diffusivity and the surface concentration of the dopants, respectively. Here it is assumed that the dopant diffusivity obeys a simple linear or quadratic relationship with the electron concentration n, resulting in $\gamma = 1$ or 2, respectively. For P, $\gamma = 2$ has to be used.[179] The supersaturated self-interstitials (SSI) can also condense into interstitial-type dislocation loops with or without a stacking fault or affecting oxygen or carbon agglomeration.

DIFFUSION-INDUCED MISFIT DISLOCATIONS. Both P and B (with smaller size than Si atoms) are dopants which can lead to the formation of an array of misfit dislocations formation provided they are present in very high concentrations. For most applications, it is preferred to avoid the formation of misfit dislocations.[178]

Interstitial-Substitutional Diffusion. As shown in Fig. 2.24, it seems that the much higher diffusivities of some elements are likely to be due to the presence of these solutes as interstitial atoms. A detailed discussion of these relative diffusivities is given in the article by Frank et al.[38,39] Note that Au and Pt (Fig. 2.24), used in power devices to decrease the minority carrier lifetime in a controlled fashion, diffuse in Si via the interstitial-substitutional diffusion mechanism. At high temperatures ($> \sim 800°C$) the kick-out diffusion prevails in their diffusion behavior with a strongly concentration-dependent effective diffusion coefficient $D_{eff}^{(I)}$ which is given by

$$D_{eff}^{(I)} = \frac{D_I C_I^{eq}}{C_s^{eq}} \left(\frac{C_s}{C_I}\right)^2 \tag{2.179}$$

yielding a fairly unusual concentration profile shown in Fig. 2.50 as the case of $\gamma = -2$. In Eq. (2.179), C_s and C_s^{eq} are the actual and equilibrium concentrations (solubilities), respectively, of substitutional Au or Pt, and C_I and C_I^{eq} are the actual and equilibrium concentrations, respectively, of self-interstitials. At lower temperatures, the dissociative diffusion mechanism dominates with a concentration-independent diffusivity $D_{eff}^{(V)}$

$$D_{\text{eff}}^{(V)} = \frac{D_V C_V^{\text{eq}}}{C_s^{\text{eq}}} \tag{2.180}$$

which represents a typical complementary error function profile (Fig. 2.50). In the event where the supersaturation of self-interstitials is canceled by a high concentration of internal sinks such as dislocations, for both the kick-out and the dissociative mechanisms, the in-diffusion of foreign interstitials prevails and leads to a concentration-independent diffusivity

$$D_{\text{eff}}^{(I)} = \frac{D_i C_i^{\text{eq}}}{C_s^{\text{eq}}} \tag{2.181}$$

where D_i is the diffusivity and C_i^{eq} the solubility, of interstitial Au or Pt.

2.15.3 Diffusion in GaAs and AlAs-GaAs Materials

Gallium arsenide is the most important compound semiconductor which has four principal advantages over silicon: (1) its higher electron mobility and saturated drift velocity, (2) ease of manufacture into semi-insulating substrate form, (3) the contribution of GaAs-based devices for much greater temperature tolerance and radiation hardness due to its larger band gap, and (4) use of GaAs as a direct band gap material. These properties make GaAs and related materials promising candidates to produce a wide range of high-frequency electronic and optoelectronic devices. To fabricate or even simply to operate the devices, self-diffusion and impurity diffusion of atoms are involved. Hence, a good understanding of both diffusion mechanisms and the behavior of point defect species controlling the diffusion processes is essential. For more details on this subject, see the article of Tan and others.[183]

Impurity diffusion profiles in GaAs usually exhibit complex, non-erfc function shapes. Laidig et al.[184] have shown that the diffusion of the p-type dopant Zn into a GaAs-AlAs superlattice (SL) dramatically increased the Al-Ga interdiffusion. Later, it was found that n-type dopants, e.g., Si, also give rise to an improved Al-Ga interdiffusion; however, the effect is small compared to that due to Zn. Because of the technological significance of using dopant-enhanced SL disordering for lateral patterning of device structures, the doping and As$_4$ pressure dependencies of Al-Ga interdiffusion have been studied in great detail.[185,186]

For the dominant native point defect species controlling the diffusion of self and impurity species on the Ga (such as gallium vacancy V_{Ga} and gallium-self interstitial species I_{Ga}) or group III sublattice, the following characteristics are very important:[185,186] (1) Both V_{Ga} and I_{Ga} species are contributors; (2) the point defects, V_{Ga} and I_{Ga}, species are charged; (3) the point defects can occur in nonequilibrium concentrations; and (4) point defect thermal equilibrium concentrations are functions of the GaAs crystal composition, which is readily determined by the pressure of a vapor species (e.g., that of As$_4$) in thermal equilibrium coexistence with the crystal.[183]

2.16 RADIATION EFFECTS AND DIFFUSION

Bombardment with energetic particles such as electrons, protons, neutrons, and light and heavy ions produces changes in the physical properties of metals and alloys.

These property changes include formation of defects such as point defects, defect clusters, voids, and bubbles; void swelling; enhanced diffusion; induced segregation or precipitation; inverse Kirkendall effects; induced phase transformation; embrittlement; enhanced creep; surface modifications; sputtering at surfaces; changes in thermal and electrical conductivity; and impurity atom production.[187–191] These physical property changes depend on the particular metal or alloy, its metallurgical history, the conditions of irradiations (namely, temperature and stress state), nature of irradiations, type of irradiation particles, particle energy or distribution of energies, instantaneous flux of particles, and the accumulated fluence of particles incident on the metal. Hence, an understanding of the various effects of irradiation is essential in several fields of materials science, namely, nuclear industry, surface treatment, semiconductors,[188] superconductors,[189] microelectronic devices, and so forth.[30]

This section briefly discusses only six areas of the physical property changes due to irradiation because of space limitations. However, readers should find the remainder elsewhere for further reading.

2.16.1   Types of Radiation

Electron and proton irradiations produce primarily isolated point defects and are used to study basic properties and interactions of point defects. Ion bombardment is most useful for studies of microstructural changes and is used for radiation creep and fatigue studies on thin specimens. Light-ion irradiations produce some combination of isolated Frenkel pairs and defect cascades over longer ranges. Heavy-ion irradiation is usually carried out to study the properties of defect cascades over a shorter range.

Neutron irradiations produced by fission of ^{235}U have an average energy of 1.98 MeV and can exhibit defect cascade damage in metals and alloys. In a thermal reactor, fission neutrons become slowed or moderated to energies typical of reactor temperatures (<1 eV). These thermal neutrons can produce defects in materials by the (n, γ) reaction comprising neutron absorption and atomic recoil from prompt γ emission. During this process, the atom recoil energies are quite low, and isolated and small clusters of Frenkel defects usually form. Additionally, each reaction provides one transmuted atom (impurity). An important aspect of fission or thermal neutron irradiation is that the damage events are evenly distributed in most materials at thicknesses on the order of centimeters. This fact, among others, makes the simulation of neutron damage by ion irradiation very difficult or impossible.[187]

Neutron damage in core components of a fast breeder reactor occurs primarily by fission neutrons, whereas the damage in core components of a thermal reactor is produced by both fission and thermal neutrons. Neutron irradiations produced by a fusion reactor have energies near 14 MeV which are capable of producing more *subcascade damage*. Neutron damage in a fusion reactor produces elastic recoils with higher energies and nonelastic events such as (n, p) or (n, α). These reactions lead to the production of H and He internally in a metal or alloy and can adversely affect its dimensional and mechanical properties.[187]

2.16.2   Defects Produced by Irradiation

Under irradiation by energetic particles, the atoms of an alloy suffer elastic and inelastic collisions with the projectiles. Except for very high densities of energy transfer, on the order of several keV/Å,[187] the electronic excitations normally do not tend to produce defects. In contrast, the part of the energy which is transferred elastically to a target atom (the primary "knock-on" atom, or PKA) will displace it, provided the energy imparted is greater than a threshold E_d on the order of 20 to 50 eV in metals. A simple defect called a Frenkel pair and defect clusters are then formed which are stable at low temperatures (<20 K) in metals. When the energy received by the PKA is satisfactorily high, it will act as a projectile and initiate a *cascade of displacements*, also known as a *displacement spike*, in the target.[30]

Frenkel pairs are grouped as freely migrated defects when they survive the cascade event and can freely migrate through the lattice. These mobile Frenkel pairs and defect clusters constitute the basic elements for radiation damage which adversely affect the mechanical properties of materials.[192]

If the recoil energy of the PKA is large (>10 keV), this atom begins a collision cascade involving many atoms. A *defect cascade* so produced consists of several hundred vacancies and interstitials in a very small region of the crystal lattice, on the order of 10 nm in diameter. Large defect cascades are the significant damage produced by fission and fusion neutron irradiation.

When defect cascades form at temperatures sufficiently high for defect migration, the vacancies and interstitials from a single cascade can cluster, annihilate each other, condense on a plane to form a dislocation loop, or freely migrate to react with other existing defects. It is usually expected that intra- and intercascade vacancy clusters or dislocation loops lead to irradiation hardening and embrittlement. The freely migrating vacancies and interstitials which are absorbed at sinks such as dislocations and cavities result in irradiation-enhanced creep and void growth, respectively.

Cascade Defects. The important events in the production and evolution of cascade defects are grouped into four stages: collisional, thermal spike, quenching, and annealing. In the collisional stage, the primary recoil atom initiates a cascade of displacive collisions that continue until its energy to create further displacements is drastically diminished. An energy of 10 to 100 eV is required to displace an atom, which depends on the material. At the end of this stage, which lasts for several tenths of a picosecond, damage comprises energetic displaced atoms and vacancies. Stable lattice defects do not form in this stage.

During the thermal-spike stage, the collisional energy of displaced atoms is shared among their neighboring atoms in a localized region of high deposited energy density. This stage lasts about 1 ps, and the spike may take the form of one or more molten zones where the energy density is very high.[193,194]

In the quenching stage, energy is imparted to the surrounding material, the molten zone returns to a condensed state, and the thermodynamic equilibrium is found. This cooling takes a few picoseconds more than the thermal-spike stage, depending on the cascade energy and energy transport properties of the material. The cascade regions return to the ambient crystal temperature, and stable lattice defects form as single-point defects or defect clusters. The total number of point defects present at this stage is very small compared to that in the collisional stage.

In the annealing stage of an individual cascade, further rearrangement and interaction of the remaining defects occur by normal, thermally activated diffusion of mobile lattice defects. This local or short-term annealing occurs until all mobile defects escape the cascade region or another cascade takes place within it. The time scale may range from microseconds to months, depending on the temperature and irradiation conditions. Vacancy-interstitial pairs recombine, defect clusters form, and mobile defects escape the local cascade region to interact with other components of the changing microstructure.

The specific actions and their results at each of these four stages depend mostly on the material being irradiated, especially with respect to the atomic mass and the nature of the atomic bonding.[193]

At very high PKA recoil energies > ~50 keV, defect cascades, instead of simply getting larger, split into distinct, well-separated damage regions or subcascades. These subcascades are closely spaced, and each exhibits typically a 25- to 30-keV primary recoil event. The frequent production of subcascades is typical of irradiations by fusion neutrons. *Planar channeling* of primary or secondary recoils also influences the subcascade formation when high-energy recoils discharge far from an earlier energetic collision.[195,196]

A small addition of impurities (<1 at %) has little effect on the defect production. However, they can significantly affect the fate of migrating defects. This is mostly through trapping of migrating vacancies or interstitials by certain impurities. This can alter the balance of steady-state concentrations of interstitials and vacancies and, therefore, change their effect on macroscopic properties.[196]

Defect Clusters. In irradiated materials, point defects are usually produced in much larger concentrations than the thermal equilibrium concentrations. Under irradiation, the main microstructural features can be grouped as defect clusters, voids, and bubbles.[197]

Defect cluster formation is a low-temperature phenomenon with limited atomic mobility. Defect clusters (in the range of 1.5 to 2.0 nm) denote the initial stage of agglomeration of the radiation-produced defects such as vacancies, interstitials, displacement spikes, and/or transmuted atoms. The vacancies and interstitials form a planar cluster and ultimately collapse and grow into a dislocation loop. Clusters can also be three-dimensional, which denotes the initial states of vacancy agglomeration or agglomeration of some specific elements in an alloy. Clusters can grow into more distinct structures with the increase of irradiation temperature. The interstitial clusters grow into large loops and finally a dislocation network. The three-dimensional vacancy clusters grow into voids and, in some cases, bubbles.[197]

Voids. In an irradiated material, voids are gas-free cavities and appear as light regions against a darker background in the electron microscope image (Fig. 2.52*a*). Smaller voids are usually imaged in an under-focus condition and can also produce strain contrast. Under irradiation up to large doses in a suitable temperature range, small roughly spherical voids can occur in certain materials such as W, Nb, and Mo. Such voids may form ordered arrangements of the same topological type as the atoms, i.e., body-centered or face-centered with a lattice parameter of several nanometers.[198] Large voids usually take a crystallographic shape (Fig. 2.52*a*).

Voids occur by agglomeration of radiation-created vacancies and are observed in materials at $\frac{1}{3}$ to $\frac{1}{2}$ of the absolute melting temperature T_m at which vacancies become relatively mobile to permit void growth. Voids can be homogeneously or heterogeneously distributed in a material. An example of homogeneous distribution is the void lattice where the voids arrange themselves in a periodic array like an atomic lattice (Fig. 2.52*b*). An example of the heterogeneous distribution is the denudation of voids near grain boundaries or free surfaces, and the attachment of voids to precipitate particles. The lattice parameter increases with the voids or bubble size.[197]

Bubbles. Bubbles are internal cavities that contain He and other inert gas atoms. An equilibrium bubble contains sufficient gas to balance the internal gas pressure with the surface tension of the cavity. Bubbles provide the same contrast in the

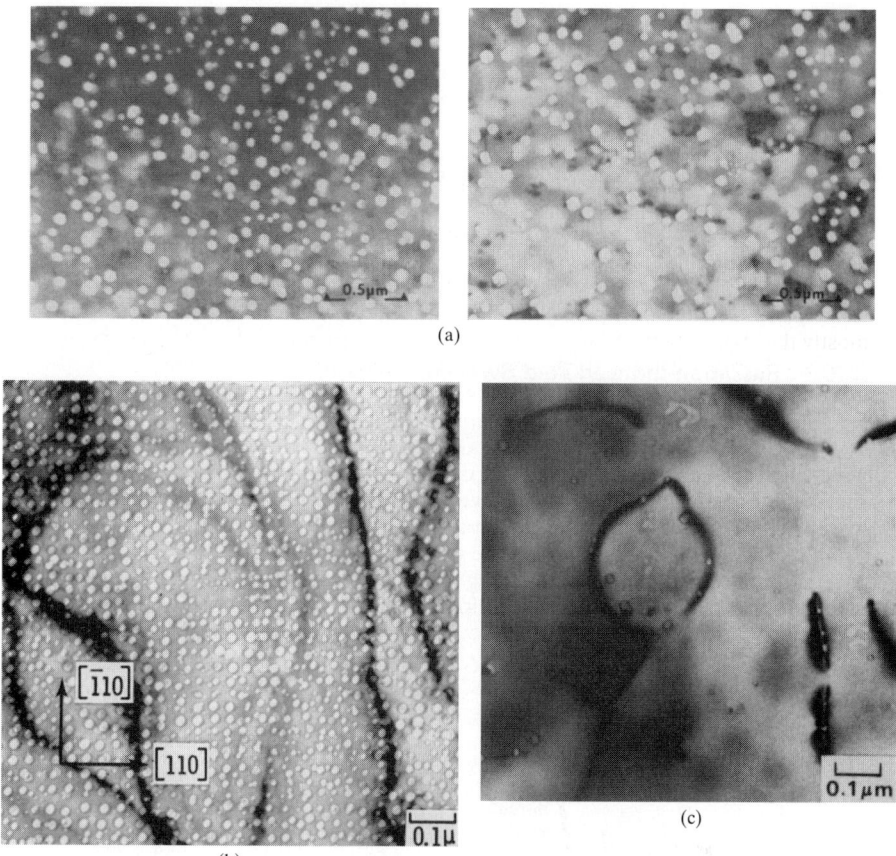

FIGURE 2.52 Transmission electron micrographs. (*a*) Faceted voids (left) in continuous irradiated (to 10 dpa) and (right) continuous irradiated (to 2 dpa) and pulsed irradiated (to 10 dpa) nickel at 775 K. Here the foil normal is near [110]. (*b*) A void lattice in a single crystal molybdenum irradiated with 7.5-MeV Ta^{3+} ions at 900°C to a dose of 150 dpa. (*c*) Preferential formation of helium gas bubbles at grain boundaries and dislocations in molybdenum bombarded with He^+ ions at 1275 K.[197] [*Source: (a) J. L. Brimhall, L. A. Charlot, and E. P. Simonen, J. Nucl. Mater., vols. 103 and 104, 1981, p. 1149; (b) courtesy of E. P. Simonen; (c) J. L. Brimhall and L. A. Charlot, J. Nucl. Mater., vol. 99, 1981, p. 17.*]

electron microscope as voids; hence, in this respect, they are indistinguishable. The main difference, however, is that the bubbles occur preferentially at dislocations, grain boundaries, and other internal defects, whereas voids do not. Preferential formation of He bubbles at dislocations in α-irradiated Mo is shown in Fig. 2.52c. Like voids, bubbles can also take a crystallographic form.

Bubbles are caused by coalescence of gas produced by transmutation reactions during neutron irradiation. Fission products (e.g., Kr and Xe in nuclear fuels), α-particles (He), and protons (H) are the primary sources of gas during irradiation. Since voids and bubbles form at about the same temperature, in any irradiation involving the formation of gaseous products, the voids will be at least partially filled with gas.

Bubbles filled with inert gas such as He are very resistant to high-temperature annealing (due to extreme lattice insolubility of inert gas atoms), whereas voids anneal out easily at temperatures $\geq 0.6 T_m$. Voids, partially filled with gas, have been found to shrink by vacancy emission during annealing to a size until they become equilibrium gas bubbles. No further shrinkage then takes place.

Bubble growth at high temperatures by migration and coalescence of smaller bubbles has been noticed directly in the electron microscope. In this way, large bubbles can occur both in the grain interior and in the grain boundaries. Grain boundary bubbles can lead to severe embrittlement of the material.[197]

2.16.3 Radiation-Induced Void Swelling

Swelling (or a macroscopic volume expansion) is very sensitive to microstructure, composition, and irradiation variables. On a microscopic scale, swelling is associated with the formation of high density of cavities with sizes up to 100 nm or more, in the range of $0.3 T_m$ to $0.55 T_m$.[199] Radiation-induced void swelling is caused by an imbalance of the steady-state concentration of interstitials and vacancies. This can cause density variations >50% in some grades of austenitic stainless steels.[200–202]

Figure 2.53 shows the schematic diagram relating swelling with dose dependence or fluence for different materials.[203] Alloying elements or impurities may have considerable influence on swelling.[202] The excellent swelling resistance of ferritic steels,

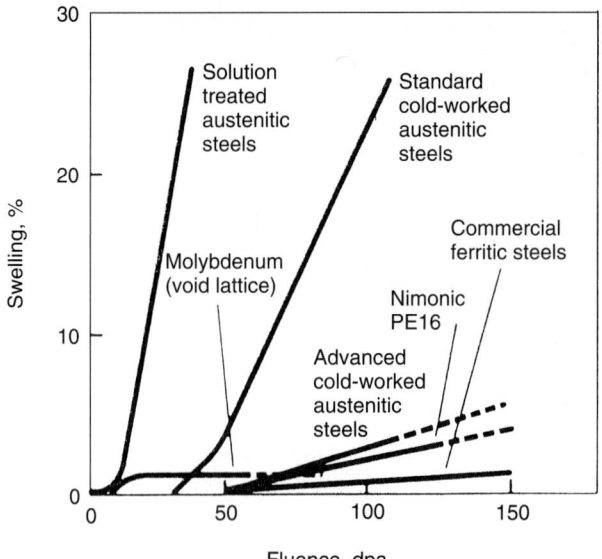

FIGURE 2.53   Swelling versus fluence for different materials and conditions (schematic).[203] (*After B. L. Eyre and J. R. Matthews, J. Nucl. Mater., vol. 205, 1993, p. 1.*)

and of other bcc metals and alloys, is attributed to the trapping of interstitials near dislocations and their availability for recombination.[204] It has been concluded, based on experimental results, that impurity trapping has greater effect on the vacancy supersaturation than the corresponding reactions occurring in fcc metals and alloys.[205] The interstitials C and N and the substitutional Si atoms are particularly found to be effective. The C-vacancy binding enthalpy appears to be as high as 0.85 eV and would, therefore, serve as an effective recombination catalyzer even at high temperatures. Also, Cottrell atmosphere around dislocations may screen the sinks and reduce climbing rates.

Transmutation-produced He affects both solute segregation and phase instability; the latter may exert strong effects on swelling.

2.16.4   Radiation-Enhanced Diffusion

Radiation-enhanced diffusion (RED) is responsible for inducing phase transformations in a wide range of alloys under various irradiation conditions. The concept of RED results from the creation of excess numbers of point defects and defect clusters by irradiation and their subsequent free migration. RED is useful to (1) describe thermodynamic force-driven microstructural changes and (2) determine the characteristics of phase diagrams in the temperature range where diffusion is usually too slow to achieve equilibrium in an experimentally viable time. The MeV electron and ~100-keV proton irradiations are commonly used to remove energetic cascade effects and to provide a large fraction of freely migrating defects.[206]

When defects annihilate primarily at sinks, the RED coefficient becomes temperature-independent and is linearly proportional to the dose rate. This implies that the number of jumps a defect makes between generation and annihilation at a fixed sink is constant irrespective of how long the defect takes to make each jump. When defects annihilate mainly by direct recombination, the RED coefficient displays an Arrhenius temperature dependence, with an apparent activation enthalpy equal to one-half of the migration enthalpy of the slowest-moving defect, and is proportional to the square root of the defect-production rate.[207]

At intermediate temperatures, irradiation-induced point defects form much faster than they can anneal (or migrate) to dislocations, grain boundaries, and other point defect sinks. The resulting vacancy and interstitial concentration are often orders of magnitude greater than their equilibrium values. RED coefficients can be many orders of magnitude enhanced by this random migration process. At high temperatures, the additional defects introduced by irradiation become trivial and have no effect on diffusivity, because of the very rapid production of equilibrium thermal vacancies. The activation energy for diffusion is the sum of the energies of vacancy formation and migration. As stated earlier, at lower temperatures, Frenkel pairs annihilate by direct recombination, and D has an activation energy of one-half that for vacancy migration.[30]

2.16.5   Irradiation-Induced Segregation and Precipitation

In general, an irradiation-induced flux of point defects to dislocations, grain boundaries, and voids usually produces a corresponding mass flux. In dilute alloys, the smaller and usually faster-diffusing atomic species is segregated toward defect sinks, whereas in concentrated alloys, the faster-diffusing species is segregated in the opposite direction. Radiation-induced segregation (RIS) toward a sink appears to cause precipitation in single-phase alloys whereas segregation away from sinks has produced precipitate dissolution. RIS is a nonequilibrium process, which induces nonequilibrium concentration gradients rather than nonequilibrium phases.[208]

The large concentrations of vacancies and interstitials so created exchange differently with the atoms of different elements and produce redistribution of the different atomic species in the regions. Strong segregation of solute atoms toward point defect sinks and irradiation-induced instabilities can lead to oscillations in the composition of irradiated alloys even when it appears that very few sinks are available. RIS can be attributed ultimately to the formation of mobile defect-solute complexes and/or inverse Kirkendall effects. Both mechanisms couple a net flux of solute to the defect fluxes. The former mechanism is especially predominant in dilute alloys, for example, the formation of Zn precipitates in an irradiated Al–1.9% Zn alloy[209] and occurrence of spinodal decomposition in an irradiated more complex Fe-Cr-Ni system.[210,211] The extent of segregation produced by quenching and annealing is usually minor, because of the formation of a smaller amount of point defects (mostly vacancies) and transient fluxes. In contrast, the incessant creation and subsequent migration of vacancies and interstitials during irradiation at high temperatures produce continuous defect fluxes. Preferential coupling of individual solute elements to these persistent fluxes favors large localized compositional changes that undergo a net influx or outflow of defects. For example, austenitic grains of stainless steel transform completely into ferrite during neutron irradiation because of Ni segregation toward, and Cr segregation away from, grain boundaries. Also, during neutron irradiation, coatings of brittle silicide phases form on grain boundaries in many alloy systems due to strong RIS of Si or tramp impurities. Such large microstructural changes significantly influence the in-service performance of irradiated components. This defect-flux-driven segregation during irradiation has the greatest technological importance.[207]

2.16.6 Inverse Kirkendall Effects

In the traditional Kirkendall diffusion mechanism, a gradient in composition can induce a net flux of defects (i.e., vacancies) across a diffusion couple with alloy components of unequal diffusion rates. In contrast, inverse Kirkendall segregation can occur in irradiated alloys where a vacancy (or interstitial) flux gives rise to a solute flux (i.e., a gradient in defect concentration can induce a solute concentration gradient). Both vacancy and interstitial defects can induce inverse Kirkendall effects. The inverse Kirkendall effect from a vacancy flux always results in the depletion of the faster-diffusing species at a sink, due to the atom flux associated with vacancies in the opposite direction to that of the defect flux. The inverse Kirkendall effect induced by an interstitial flux always causes enrichment of the faster-diffusing species at a sink, due to the same directions of both the interstitial and solute fluxes. However, the two inverse Kirkendall effects may assist or oppose each other in producing segregation at any specific sink.[207] During the irradiation of austenitic stainless steels, the vacancy flow to grain boundaries leads to Ni enrichment and Cr and Fe depletion at grain boundaries. These composition changes are in agreement with fast diffusion of Cr and slow diffusion of Ni in austenitic stainless steels.[211–213] Figure 2.54 is a schematic of the solute and defect concentration gradients near a vacancy sink.[214]

Heavy-ion and proton irradiation experiments have proved that the inverse Kirkendall segregation mechanism is quantitatively in agreement with several

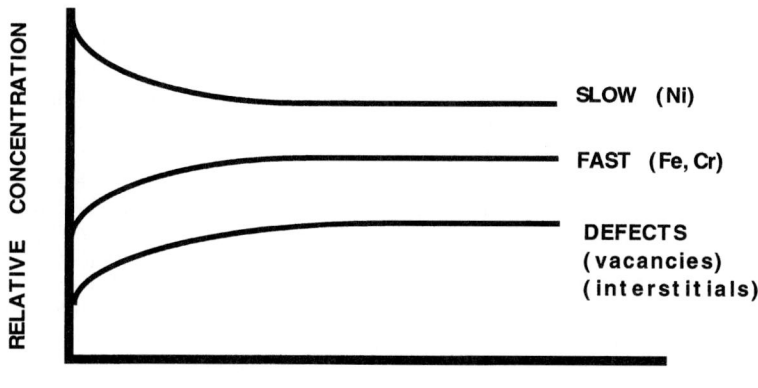

FIGURE 2.54 Inverse Kirkendall segregation: a schematic of concentration profiles for solute atoms and defects near irradiated grain boundaries. Fast-moving solutes move away from the boundary, and slow-moving solutes become segregated near the boundary.[214] (*Courtesy of E. P. Simonen.*)

hundred measurements of grain boundary composition in irradiated austenitic alloys.[214,215]

2.16.7   Radiation-Induced Phase Transformation

Irradiation of metals and alloys with energetic particles can introduce up to 10^8 J/mol of energy in the form of atomic displacements which is finally available to produce a range of phase and microstructural changes: These phenomena cannot be observed under thermal conditions.[216] The production of metastable phases during irradiation is an active research area, especially dealing with the ion implantation.[192]

Many theoretical and experimental studies have been carried out on radiation-induced phase transformations.[216] Russell has provided a brief review of phase stability under irradiation.[217] Experimental results have been briefly outlined here for various ferrous and nonferrous alloy systems.[216] Irradiation-induced amorphization is not discussed, but can be found elsewhere.[216]

2.16.7.1   *Experimental Results*

Ferrous Alloys.   Irradiation of martensitic stainless steel gives rise to the formation of Mo_6C, Mo_2C, χ, Laves, $M_{23}C_6$, and α' phases. Irradiation of low-alloy ferritic steels gives severe embrittlement of the steel due to irradiation-induced defect aggregates such as dislocation loops, voids, various kinds of precipitates, and point defect-solute complexes. Cu and P used as tramp elements are known to increase greatly the embrittling effect of irradiation.

Neutron irradiation of type 316 stainless steel provides γ' and G phases, which do not occur thermally; increases the formation of the η, Laves, and MC phases; and changes the composition of the Laves phases. The σ and Fe_2P phases may also be influenced by irradiation. The temperatures of appearance and abundances of these phases are very sensitive to minor changes in alloy composition.

Irradiation of types 316, 316L, 321, and 347 stainless steels has caused the formation of two kinds of magnetic phases. Some irradiations produce blocky ferrite particles, whereas others give a very high concentration of fine supermagnetic particles which are believed to be Ni- and Fe-rich.

Neutron, heavy-ion, and electron irradiations induce copious NbC precipitation in Nb-containing stainless steels. Irradiation of Fe-Ni or Fe-Ni-Cr Invar-type alloys provides a spinodallike decomposition of the matrix into Fe-rich and Ni-rich regions.

Nonferrous Alloy Systems. Irradiation of Al-Cu solid solutions appears to favor the formation of θ' phase and depress the formation of θ''. Irradiation leads to a large increase in precipitation nucleation rates in Al-Ge and Al-Si alloys. Electron irradiation raises the solvus temperature in Al-rich Al-Zn alloys by tens of degrees, based on the displacement rate.

Strong segregation effects take place in Cu-base alloys for Fe, Be, and Ag solutes. Irradiation usually augments GP zone formation and precipitation in Cu-Be alloys and leads to coherency loss in Cu-Co and Cu-Fe alloys. It also results in a decrease in resistivity of Cu-Ni alloys, which can be attributed to the decomposition of solid solution, perhaps by a spinodal-type transformation.

Strong solute segregation of Al, Be, Cr, Cu, Mo, Mn, Si, and Ti has been observed in irradiated Ni-base alloys. Precipitation has been noticed in irradiated, thermally single-phase Ni-Be and Ni-Ge alloys. A displacement rate-dependent threshold for precipitation is found in thermally single-phase Ni-Si alloys. Irradiation of Inconel 706 and 718 provides complex interplay between γ', γ'', and η phases.

Irradiation has been found to induce disorder in several Ni-base alloys. An interesting finding is the disordering of Ni_3Si and its replacement by Ni_5Si_2, which is not thermally stable, but appears to be more irradiation-resistant than Ni_3Si.

Irradiation induces strong segregation of Al, Mo, and V in Ti-base alloys. Segregation of Al in irradiated Ti-Al alloys produces redistribution of α_2 precipitates and formation of β phase.

Neutron irradiation of thermally single-phase W-rich W-Rh alloy causes the formation of χ, WRe_3 phase rather than the equiatomic σ phase. The effect of irradiation in ω-phase precipitation in Zr-Nb alloys is ambiguous, in some cases favoring ω-phase formation and in other cases exerting no effect.

Neutron irradiation greatly increases the rate of white-to-gray (tetragonal-to-bcc) tin transformation, perhaps by the vacancy-rich cores of displacement cascades serving as nucleation sites for the less dense gray phase.

High-fluence neutron irradiation of U-Mo and U-Nb alloys produces a stable α-U and causes γ' phases to revert to the metastable high-temperature γ phase. The reversion probably takes place by irradiation disordering of the γ' phase to yield the γ phase.

REFERENCES

1. D. Lazarus, in *Materials Science Forum*, vol. 1, 1984, *Diffusion in Solids*, guest eds. A. L. Laskar, G. P. Tiwari, E. C. Subba Rao, and R. Krishnan, *Trans Tech Publications*, Switzerland, 1984.

2. J. P. Stark, in *Treatise of Materials Science and Technology*, vol. 4, Academic, New York, 1974, pp. 59–111; *Solid State Diffusion*, Wiley, New York, 1976.

3. J. Philiburt, *Atom Movements: Diffusion and Mass Transport in Solids* (English translation by S. J. Rothman), Les Editions de Physique, Paris, 1991.

4. E. A. Brandes and G. B. Brook, eds., *Smithells Metals Reference Book*, 7th ed., Butterworth-Heinemann, Oxford, 1992, pp. 13.1–13.119.

5. J. Karger and D. M. Ruthven, *Diffusion in Zeolites and Other Microporous Solids*, Wiley, New York, 1992.

6. J. F. Shakelford and W. Alexander, *CRC Materials Science and Engineering Handbook*, CRC Press, Boca Raton, Florida, 1994, pp. 236–247.

7. C. P. Flynn, *Point Defects and Diffusion*, Clarendon Press, Oxford, 1972.

8. G. E. Murch, in *Materials Sc. Technolog.*, vol. 5, *Phase Transformations in Metals and Alloys*, chap. 2, VCH, Weinheim, 1991, pp. 75–142.

9. R. P. Smith, *Acta Met.*, vol. 1, 1953, p. 578; W. F. Smith, *Principles of Materials Science & Engineering*, McGraw-Hill, New York, 1986.

10. P. G. Shewmon, *Diffusion in Solids*, The Metallurgical Society, Warrendale, Pa., 1989.

11. G. H. Geiger and D. R. Poirier, *Transport Phenomena in Metallurgy*, Addison-Wesley, Reading, Mass., 1973.

12. H. Mehrer, in Landolt-Börnstein, vol. 26, *Diffusion in Solid Metals and Alloys*, ed. H. Mehrer, Springer-Verlag, Berlin, 1990, pp. 1–31.

13. A. D. LeClaire, in *Encyclopedia of Materials Science and Engineering*, Pergamon, Oxford, 1986, pp. 1186–1193.

14. R. A. Porter and K. E. Easterling, *Phase Transformations in Metals and Alloys*, Van Nostrand Reinhold (UK) Ltd., 1981.

15. M. Bishop and K. E. Fletcher, *Int'l. Met. Reviews*, vol. 17, 1972, pp. 203–225.

16. E. S. Machlin, *Thermodynamics and Kinetics/Materials Science*, Giro Press, New York, 1991.

17. *Decarburization*, ISI Publication133, Gresham Press, Old Working Surrey, England, 1970.

18. G. E. Totten and M. A. H. Howes, eds., *Steel Heat Treatment Handbook*, Marcel Dekker, New York, 1996.

19. M. C. Richman, *An Introduction to Science of Metals*, Blaisdell, Waltham, Mass., 1967.

20. S. J. Rothman, *Diffusion in Crystalline Solids*, eds. G. Murch and A. S. Nowick, Academic Press, Orlando, Fla., 1984, pp. 1–61.

21. H. Zabel, in *Nontraditional Methods in Diffusion*, eds. G. E. Murch, H. K. Birnbaum, and J. R. Cost, The Metallurgical Society, Warrendale, Pa., 1984, pp. 1–37.

22. W. Petry, A. Heiming, C. Herzig, and J. Trampenau, *Defect Diffusion Forum*, vol. 75, 1991, pp. 211–228.

23. H. T. Stokes, in *Nontraditional Methods in Diffusion*, eds. G. E. Murch, H. K. Birnbaum, and J. R. Cost, The Metallurgical Society, Warrendale, Pa., 1984, pp. 39–58.

24. H. T. Stokes, in *Conf. Proceed. AIME-TMS, 1984*, pp. 39–58.

25. G. E. Murch, in *The Encyclopedia of Advanced Materials*, Pergamon Press, Oxford, 1994, pp. 125–130; in *Phase Transformations in Materials*, chap. 2, vol. ed. P. Haasen, VCH, Weinheim, Germany, 1991, pp. 74–141.

26. G. K. Wertheim, *Mossbauer Effect: Principles and Applications*, Academic Press, New York, 1964.

27. J. C. Mullen, in *Nontraditional Methods in Diffusion*, eds. G. E. Murch, H. K. Birnbaum, and J. R. Cost, The Metallurgical Society, Warrendale, Pa., 1984, pp. 59–81.

28. A. S. Nowick, in *Encyclopedia of Materials Science and Engineering*, Pergamon Press, Oxford, 1986, pp. 1180–1186.

29. J. L. Snoek, *Physica*, vol. 6, 1939, p. 591.

30. J. L. Bocquet, G. Brebec, and Y. Limoge, in *Physical Metallurgy*, chap. 7, vol. 1, 4th ed., eds. R. W. Cahn and P. Haasen, Elsevier Science BV, Amsterdam, 1996, pp. 535–668.

31. C. Zener, *Trans. AIME*, vol. 152, 1943, p. 122; *Phys. Rev.*, vol. 71, 1947, p. 34.

32. J. Volkl, *Ber. Bunsen Gesell*, vol. 76, 1972, p. 797.

33. L. Neel, *J. Phys. Rad.*, vol. 12, 1951, p. 339; vol. 13, 1952, p. 249; vol. 14, 1954, p. 225.

34. S. Taniguchi, *Sci. Rep. Res. Inst. Tohoka Univ.*, vol. A7, 1955, p. 269.

35. W. A. Lanford, R. Benenson, C. Burman, and L. Weilunski, in *Nontraditional Methods in Diffusion*, eds. G. E. Murch, H. K. Birnbaum, and J. R. Cost, The Metallurgical Society, Warrendale, Pa., 1984, pp. 1–38.

36. W. T. Petuskey, in *Nontraditional Methods in Diffusion*, eds. G. E. Murch, H. K. Birnbaum, and J. R. Cost, The Metallurgical Society, Warrendale, Pa., 1984, pp. 179–202.

37. J. A. Kilner, *Mater. Sci. Forum*, vol. 7, 1986, pp. 205–222.

38. W. Frank, *Defect and Diffusion Forum*, vol. 75, 1991, pp. 121–148.

39. W. Frank, U. Gösele, H. Mehrer, and A. Seeger, *Diffusion in Crystalline Solids*, eds. G. Murch and A. S. Nowick, Academic Press, Orlando, Fla., 1984, pp. 63–142.

40. W. Schilling, *J. Nucl. Mater.*, vol. 69–70, 1978, p. 65.

41. A. Seeger, in *Proc. Conf. Fundamental Aspects of Radiation Damage in Metals*, Gathinburg, eds. M. T. Robinson and F. W. Young, Jr., ERDA Report Conf. 751006, P1 and P2, U.S. Energy and Research and Development Association, Washington, D.C.

42. W. Schule, *Defects and Diffusion Forum*, vols. 66–69, 1989, pp. 313–326.

43. P. Haasen, *Physical Metallurgy*, 3d ed., Cambridge University Press, London, 1993.

44. G. D. Watkins and J. W. Corbett, *Phys. Reviews*, vol. A134, 1964, p. 1359.

45. G. D. Watkins, J. W. Corbett, and R. M. Walker, *J. Appl. Phys.*, vol. 30, 1959, p. 1198.

46. P. H. Dederichs, C. Lehmann, H. R. Schober, A. Scholtz, and R. Zeller, *J. Nucl. Mater.*, vols. 69 and 70, 1978, p. 176.

47. U. Gösele, W. Frank, and A. Seeger, *Appl. Phys.*, vol. 23, 1980, p. 361.

48. F. C. Frank and D. Turnbull, *Phys. Rev.*, vol. 104, 1956, p. 617.

49. U. Gösele and F. Morehand, *J. Appl. Phys.*, vol. 52, 1981, p. 4617.

50. A. H. Van Ommen, *J. Appl. Phys.*, vol. 54, 1983, p. 5055.

51. V. M. Vorob'ev, V. A. Murav'ev, and V. A. Panteleev, *Sov. Phy. Solid State*, vol. 23, 1981, p. 2055.

52. D. Weiler and H. Mehrer, *Phil. Mag.*, vol. A49, 1983, p. 309.

53. N. V. Doan and Y. Adda, *Phil. Mag.*, vol. A56, 1987, p. 269.

54. H. Bakker et al., in *Diffusion in Solids: Recent Developments*, eds. M. A. Dayananda and G. E. Murch, TMS-AIME, Warrendale, Pa., 1984, pp. 39–65.

55. D. N. Seidman, in *Encyclopedia of Materials Science and Engineering*, Pergamon Press, Oxford, 1986, pp. 3591–3596.

56. D. N. Seidman, *J. Phys.*, vol. F3, 1973, pp. 393–421.

57. H. J. Wollenberger, in chap. 18, *Physical Metallurgy*, vol. 2, 4th ed., eds. R. W. Cahn and P. Haasen, Elsevier Science BV, Amsterdam, 1996, pp. 1621–1721.

58. R. W. Balluffi, K. H. Lie, D. N. Seidman, and R. W. Siegel, *Vacancies and Interstitials in Metals*, eds. A Seeger et al., North-Holland Publishing Co., Amsterdam, 1970, pp. 125–166.

59. H. Bakker et al., in *Diffusion in Solids: Recent Developments*, TMS-AIME, Warrendale, Pa., 1984, pp. 39–65.

60. C. Herzig and U. Kohler, *Mater. Sci. Forum*, vols. 15–18, 1987, p. 301.

61. A. Seeger, *J. Phys.*, vol. F3, 1973, p. 248.

62. R. W. Siegel, *Annu. Rev. Mat. Sci.*, vol. 10, 1980, pp. 393–425.

63. G. Neumann, in *Defects and Diffusion Forum*, vols. 66–69, 1989, pp. 43–64.

64. R. B. McLellan and Y. C. Argel, *Scripta Metall.*, vol. 23, no. 67, 1995, pp. 945–948.

65. M. W. Finnis and M. Sachdev, *J. Phys.*, vol. 6, 1976, p. 965.

66. F. J. Kedves and G. Edelyi, *Defect and Diffusion Forum*, vols. 66–69, 1989, pp. 175–188.

67. A. D. Le Claire, *Phil. Mag.*, vol. 14, 1966, p. 271.

68. I. Kaur, Y. Mishin, and W. Gust, *Fundamentals of Grain and Interphase Boundary Diffusion*, 3d ed., Wiley, Chichester, 1995.

69. J. R. Cahoon and O. D. Sherby, *Met. Trans.*, vol. 23A, 1992, pp. 2491–2500.

69a. A. D. Le Claire and G. Neumann, in Landolt-Börnstein, vol. 26, *Diffusion in Solid Metals and Alloys*, ed. H. Mehrer, Springer-Verlag, Berlin, 1990, pp. 85–212.

70. J. N. Mundy, *Defects and Diffusion Forum*, vol. 83, 1992, pp. 1–18.

71. N. L. Peterson, *J. Nucl. Mater.*, vol. 69–70, vol. 3, 1978; *Proc. Int. Conf. on the Properties of Atomic Defects in Metals*, Argonne, eds. N. L. Peterson and R. W. Siegel, 1976.

72. A. D. Le Claire, in *Diffusion in Body Centered Cubic Metals*, eds. J. A. Wheeler, Jr. and F. R. Winslow, ASM, Metals Park, Ohio, 1965, pp. 3–25.

73. J. N. Mundy, T. E. Miller, and R. J. Porte, *Phys. Rev.*, vol. B3, 1971, p. 2445.

74. L. Slifkin, *Met. Trans.*, vol. 20A, 1989, pp. 2577–2582.

75. J. M. Sanchez and D. De Fontaine, *Phys. Rev. Lett.*, vol. 35, 1975, p. 227.

76. A. Seeger and H. Mehrer, in *Vacancies and Interstitials in Metals*, eds. A. Seeger, D. Schumacher, W. Schilling, and J. Diehl, North-Holland Publishing Co., Amsterdam, 1970, p. 1.

77. A. Da Fano and G. Jacucci, *Phys. Rev. Lett.*, vol. 39, 1977, p. 950.

78. G. M. Hood, *Proc. Int. Conf. on Diffusion in Materials*, 1992, Kyoto, Japan, 1993, p. 755.

79. W. Petry, A. Heiming, J. Trampenau, and G. Vogl, *Defect and Diffusion Forum*, vols. 66–69, 1989, pp. 157–174.

80. A. D. LeClaire, *J. Nucl. Mater.*, vols. 69–70, 1978, p. 70.

81. A. R. Allnatt and A. B. Lidiard, *Atomic Transport in Solids*, Cambridge University Press, London, 1993.

82. N. L. Peterson, in *Diffusion in Solids: Recent Developments*, eds. A. S. Nowick and J. J. Burton, Academic Press, New York, 1975, pp. 115–170.

83. W. K. Warburton and D. Turnbull, in *Diffusion in Solids*, eds. A. S. Nowick and A. S. Burton, Academic Press, New York, 1975, pp. 171–229.

84. H. Nakajima, O. Ohshida, K. Nonaka, Y. Yoshida, and E. E. Fujita, *Scripta Metall.*, vol. 34, no. 6, 1996, pp. 949–953.

85. H. Araki et al., *Met. Trans.*, vol. 27A, 1996, pp. 1807–1814.

86. R. J. Borg and G. I. Dienes, *An Introduction to Solid State Diffusion*, Academic Press, New York, 1988.

87. R. K. Kapoor and T. W. Eager, *Met. Trans.*, vol. 21A, 1990, pp. 3039–3047.

88. A. Smigelskas and E. Kirkendall, *Trans. AIME*, vol. 171, 1947, p. 130.

89. T. C. Chou and L. Link, *Scripta Metall.*, vol. 34, no. 5, 1996, pp. 831–838.

90. M. A. Krivoglaz, *Theory of X-ray and Thermal Neutron Scattering by Real Crystals*, Plenum, New York, 1969.

91. R. W. Balluffi, *Acta Metall.*, vol. 8, 1960, p. 871.

92. J. Crank, *Mathematics of Diffusion*, 2d ed., Oxford University Press, London, 1975.

93. M. C. Petri and M. A. Dayananda, *Philosophical Magazine A*, vol. 76, no. 6, 1997, pp. 1169–1185; *Defect & Diffusion Forum*, vols. 143–147, 1997, pp. 575–579.

94. M. A. Dayananda, in *Defects and Diffusion Forum*, vol. 83, 1992, pp. 73–86.

95. D. J. Young, R. C. Doward, and J. C. Kirkaldy, in *Fundamentals and Applications of Ternary Diffusion, 29th Ann. Conf. of Metallurgists of CIM*, ed. C. Purdy, Pergamon, New York, 1990, pp. 3–14.

96. M. A. Dayananda, in *Diffusion in Metals and Alloys*, Landolt-Börnstein, Springer-Verlag, Berlin, 1991.

97. M. A. Dayananda, *Met. Trans.*, vol. 27A, 1996, pp. 2504–2509.

98. M. A. Dayananda, *Trans., TMS-AIME*, vol. 242, 1968, pp. 1369–1372.

99. F. J. J. Van Loo, G. F. Bastin, and J. C. Vrolijk, *Metal Trans.*, vol. 18A, 1987, pp. 801–809.

100. J. S. Kirkaldy and D. J. Young, on *Diffusion in Condensed State*, The Institute of Metals Society, London, 1987, pp. 172–272.

100a. M. A. Dayananda, in Landolt-Börnstein, vol. 26, Diffusion *in Solid Metals and Alloys*, ed. H. Mehrer, Springer-Verlag, Berlin, 1990, pp. 372–436.

101. J. W. Morris et al., *JOM*, May 1996, pp. 43–46.

102. O. Kraft and E. Artzt, *Acta Mater.*, vol. 46, 1998, pp. 3733–3743.

103. M. Ohring, *The Materials Science of Thin Films*, Academic Press, Boston, 1992.

104. R. E. Hummel, *International Materials Reviews*, vol. 39, no. 3, 1994, pp. 97–111.

105. S. J. Krumbein, in *Electromigration and Electronic Device Degradation*, ed. A. Christou, Wiley, New York, 1994, pp. 139–166.

106. C. L. Bauer, in *The Encyclopedia of Advanced Materials*, vol. 1, Pergamon, Oxford, 1994, pp. 711–717.

107. C. L. Bauer and P. F. Tang, in *Defect and Diffusion Forum*, vols. 66–69, 1989, pp. 1143–1152.

108. S. Vaidya, T. T. Sheng, and A. K. Sinha, *Appl. Phys. Lett.*, vol. 36, no. 6, 1980, p. 464.

109. S. P. Murarka, in *The Encyclopedia of Advanced Materials*, vol. 4, Pergamon, Oxford, 1994, pp. 2912–2919.

110. K. N. Tu, *Defect and Diffusion Forum*, vol. 83, 1992, pp. 141–143.

111. D. Gupta, in *Defect and Diffusion Forum*, vol. 59, 1988, pp. 137–150.

112. J. H. Zhao, in *Electromigration and Electronic Device Degradation*, ed. A. Christou, Wiley, New York, 1994, pp. 167–233.

113. C. J. Palmstrom, in T*he Encyclopedia of Advanced Materials*, vol. 1, Pergamon Press, Oxford, 1994, pp. 473–478.

114. L. J. Brillson, *The Encyclopedia of Advanced Materials*, vol. 1, Pergamon Press, Oxford, 1994, pp. 478–486.

114a. A. D. Le Claire, in Landolt-Börnstein, vol. 26, *Diffusion in Solid Metals and Alloys*, ed. H. Mehrer, Springer-Verlag, Berlin, 1990, pp. 626–629.

115. H. Mehrer and M. Lubbehusen, *Defect and Diffusion Forum*, vols. 66–69, 1989, pp. 591–604.

116. J. Bernardini, S. Bennis, and G. Moya, *Defect and Diffusion Forum*, vols. 66–69, 1989, pp. 801–810.

117. K. Viergge and C. Herzig, *Defect and Diffusion Forum*, vols. 66–69, 1989, pp. 811–818.

118. I. Kaur and W. Gust, *Defect and Diffusion Forum*, vols. 66–69, 1989, pp. 765–786.

119. N. L. Peterson, *Int'l. Metals Review*, vol. 29, no. 9, 1983, p. 65.

120. G. Melakondaiah and P. Rama Rao, *Acta Metall.*, vol. 29, 1983, p. 1263.

121. M. F. Ashby, *Acta Metall.*, vol. 22, 1974, p. 275.

122. J. Singh, S. Lele, and S. Ranganathan, *Zeitschrift für Metallkunde*, vol. 72, 1981, p. 469.

123. M. Mohan Rao and S. Ranganathan, *Materials Science Forum*, vol. 3, 1984, pp. 43–58; Trans Tech Publications, 1984, pp. 43–57.

124. L. G. Harrison, *Trans. Faraday Soc.*, vol. 57, 1961, p. 1191.

125. E. W. Hart, *Acta Metall.*, vol. 5, 1957, p. 597.

126. G. E. Murch and S. J. Rothmann, *Diffusion Defect Data*, vol. 42, 1985, p. 17.

127. D. Gupta, D. R. Campbell, and P. S. Ho, *Thin Films: Interdiffusion and Reactions*, eds. J. M. Poate, K. N. Tu, and J. W. Meyer, Wiley, New York, 1978, p. 161.

128. T. Suzuoka, *Trans. Jpn. Inst. Met.*, vol. 2, 1961, p. 25.

129. A. D. Le Claire, *Br. J. Appl. Phys.*, vol. 14, 1963, p. 351.

130. D. Gupta and J. Oberschmidt, *Diffusion in Solids: Recent Developments*, eds. M. A. Dayananda and G. E. Murch, TMS-AIME, Warrendale, Pa., 1984, p. 121.

131. A. D. Le Claire and A. Rabinovitch, in *Diffusion in Crystalline Solids*, eds. G. E. Murch and A. S. Nowick, Academic, New York, 1984, pp. 259–319.

132. N. L. Peterson, *Mater. Sci. Forum*, vol. 1, 1984, p. 85.

133. R. W. Balluffi, in *Diffusion in Crystalline Solids*, eds. G. E. Murch and A. S. Nowick, Academic, New York, 1984, pp. 320–378.

134. J. Philibert, *Diffusion et Transport de Matiere dans les Dolides*, Les Editions de Physique, Paris, 1985.

135. Y. Mishin, Chr. Herzig, J. Bernardini, and W. Gust, *Inter. Mater. Reviews*, vol. 42, no. 4, 1997, pp. 155–178.

136. R. A. Fournelle, *Materials Science and Engineering*, vol. A138, 1991, p. 133.

137. F. J. A. den Broeder, *Acta Metall.*, vol. 20, 1972, p. 319.

138. A. H. King, *Int'l. Mater. Review*, vol. 32, 1987, p. 173.

139. R. D. Doherty, *The Encyclopedia of Advanced Materials*, Pergamon Press, Oxford, 1994, pp. 1695–1698.

140. P. G. Shewmon and G. Meyrick, in *Diffusion in Solids: Recent Developments*, eds. M. A. Dayananda and G. E. Murch, The Metallurgical Society, Warrendale, Pa., 1985.

141. P. Shewmon, *Defect and Diffusion Forum*, vols. 66–69, 1989, pp. 727–734.

142. R. W. Balluffi and J. W. Cahn, *Acta Metall.*, vol. 29, 1981, p. 493.

143. P. G. Shewmon, M. Abbas, and G. Meyrick, *Met. Trans.*, vol. 17A, 1986, pp. 1523–1527.

144. M. Hillert and G. R. Purdy, *Acta Metall.*, vol. 26, 1978, p. 338.

145. D. B. Butrymowicz, J. W. Cahn, J. R. Manning, and D. E. Newbury, in *Advances in Ceramics*, vol. 6, *Grain Boundaries and Interfaces in Ceramics*, American Ceramic Society, Westerville, Ohio, 1983, p. 202.

146. R. S. Ahy and B. Evans, *Acta Metall.*, vol. 35, 1987, p. 2049.

147. L. Liang and A. H. King, *Acta Metall.*, vol. 44, no. 7, 1996, pp. 2893–2988.

148. J. Sommer, Y. M. Chiang, and R. W. Balluffi, *Scripta Metall.*, vol. 33, no. 1, 1995, pp. 7–12.

149. M. Hillert and R. Lagenborg, *J. Mater. Sci.*, vol. 6, 1971, p. 208.

150. I. G. Solorzano, G. R. Purdy, and G. C. Weatherly, *Acta Metall.*, vol. 32, 1984, p. 1709.

151. C. Y. Ma, W. Gust, R. A. Fournelle, and B. Predel, *Mater. Sc. Forum*, vol. 3/7, 1993, pp. 126–128.

152. Y.-J. Baik and D. N. Yoon, *Acta Metall.*, vol. 38, 1990, p. 1525.

153. W. H. Rhee and D. K. Yoon, *Acta Metall.*, vol. 37, 1989, pp. 221–228.

154. T. Tashiro and G. R. Purdy, *Scripta Metall.*, vol. 21, 1987, pp. 361–364.

155. D. A. Smith and A. H. King, *Phil. Mag.*, vol. A44, 1981, pp. 330–340.

156. G. R. Purdy, in *Phase Transformations in Materials*, vol. 5, vol. ed. P. Haasen, VCH Publishers, Weinheim, Germany, 1991.

157. C. Li and M. Hillert, *Acta Metall.*, vol. 29, 1981, pp. 1949–1960.

158. C. A. Handwerker, in *Diffusion Phenomena in Thin Films and Microelectronic Materials*, Noyes, Park Ridge, N.J., pp. 245–322.

159. D. N. Yoon, *Annu. Rev. Mater. Sci.*, vol. 19, 1989, pp. 43–58.

160. D. N. Yoon and W. J. Huppmann, *Acta Metall.*, vol. 27, 1979, pp. 973–979.

161. M. K. Burton, N. Cabrera, and F. C. Frank, *Phil. Trans. Roy. Soc.*, vol. A243, 1951, p. 299.

162. H. J. Leamy and G. H. Gilmer, *J. Crystal Growth*, vols. 24/25, 1974, p. 499.

163. J. P. Van der Eerden, P. Bennema, and T. A. Cherepanova, *Prog. Crystal Growth Charact.*, vol. 1, 1978, p. 219.

164. B. Perraillon, I. M. Torrens, and V. Levy, *Scripta Metall.*, vol. 6, 1972, p. 611.

165. P. G. Flahive and W. R. Graham, *Surface Sci.*, vol. 91, 1980, p. 449.

166. C. L. Liu, J. M. Cohen, J. B. Adams, and A. F. Voter, *Surface Sci.*, vol. 253, 1991, p. 334.

167. C. L. Liu and J. B. Adams, *Surface Sci.*, vol. 265, 1992, p. 262.

167a. H. P. Bonzel, in Landolt-Börnstein, vol. 26, *Diffusion in Solid Metals and Alloys*, ed. H. Mehrer, Springer-Verlag, Berlin, 1990, pp. 717–747.

168. J. Y. Choi and P. G. Shewmon, *TMS/AIME*, vol. 224, 1962, p. 589.

169. G. E. Rhead, in *Encyclopedia of Materials Science and Engineering*, Pergamon Press, Oxford, 1986, pp. 237–241.

170. J. M. Poate, K. N. Tu, and J. W. Mayor, eds., *Thin Films: Interdiffusion and Reactions*, Wiley-Interscience, New York, 1978.

171. S. R. Herd, K. N. Tu, and K. Y. Ahn, *Appl. Phys. Lett.*, vol. 42, 1983, p. 597.

172. D. M. Follstaed and J. A. Knapp, *Phys. Rev. Lett.*, vol. 56, 1986, p. 1827.

173. D. Gupta and P. S. Ho, eds., *Diffusion Phenomena in Thin Films and Microelectronic Materials*, Noyes, Park Ridge, N.J., 1988.

174. A. L. Laskar, *Defect and Diffusion Forum*, vol. 83, 1992, pp. 207–234.

175. A. V. Chadwick, *Defect and Diffusion Forum*, vol. 83, 1992, pp. 235–258.

176. C. R. A. Catlow, *Proc. Roy. Soc.*, London, Ser. A353, 1977, p. 533.

177. A. S. Nowick, *Defect and Diffusion Forum*, vols. 66–69, 1989, pp. 229–256.

178. U. Gösele, in *The encyclopedia of Advanced Materials*, Pergamon Press, Oxford, 1993, pp. 629–635.

179. U. M. Gosele and T. Y. Tan, in *Electronic Structure and Properties of Semiconductors*, vol. 4, vol. ed. W. Schroter, VCH, Weinheim, Germany, 1991, pp. 197–247.

180. U. Gösele and T. Y. Tan, *Defect and Diffusion Forum*, vol. 83, 1992, pp. 189–206.

181. R. B. Fair, in *Semiconductor Materials and Process Technology Handbook*, ed. G. E. McGuire, Noyes, Park Ridge, N.J., 1988.

182. U. Gösele, in *Microelectronic Materials and Processes*, Kluwer Academic Press, Dordrecht, The Netherlands, 1989, pp. 583–634.

183. T. Y. Tan, in *The Encyclopedia of Advanced Materials*, Pergamon Press, Oxford, 1993, pp. 629–635.

184. W. D. Laidig et al., *Appl. Phys. Lett.*, vol. 38, 1981, pp. 776–778.

185. D. G. Deppe and N. Holonyak, Jr., *J. Appl. Phys.*, vol. 64, 1988, pp. R93–113.

186. T. Y. Tan, U. Gosele, and S. Yu, *Critical Rev. Sol. State and Mater. Sci.*, vol. 17, 1991, pp. 47–106.

187. M. A. Kirk and R. C. Bircher, in *Encyclopedia of Materials Science and Engineering*, Pergamon Press, Oxford, 1986, pp. 4016–4022.

188. J. W. Corbett, in *Encyclopedia of Materials Science and Engineering*, Pergamon Press, Oxford, 1986, pp. 4034–4036.

189. B. S. Brown, in *Encyclopedia of Materials Science and Engineering*, Pergamon Press, Oxford, 1986, pp. 4036–4038.

190. K. Farrell and M. B. Lewis, in *Encyclopedia of Materials Science and Engineering*, Pergamon Press, Oxford, 1986, pp. 2407–2409.

191. A. Barbu, A. Dunlop, D. Lesueur, and R. S. Averback, *Europhys. Lett.*, vol. 15, 1991, p. 37.

192. W. Shilling and H. Ullmaier, in *Nuclear Materials*, vol. 10B, vol. ed. B. R. T. Frost, VCH, Weinheim, Germany, 1994, pp. 179–241.

193. H. L. Heinish, *JOM*, Dec. 1996, pp. 38–41.

194. R. S. Averback, T. Diaz de la Rubia, and R. Benedek, *Nucl. Inst., Meth.*, vol. B33, 1988, p. 693.

195. M. T. Robinson, in *Encyclopedia of Materials Science and Engineering*, Pergamon Press, Oxford, 1986, pp. 2516–2518.

196. L. E. Rehn, *Met. Trans.*, vol. 24A, 1993, pp. 1941–1945.

197. J. L. Brimhall, in *Encyclopedia of Materials Science and Engineering*, Pergamon Press, Oxford, 1986, pp. 1459–1462.

198. S. Amelinckx and D. Van Dyck, in *Encyclopedia of Materials Science and Engineering*, supp. vol. 1, Pergamon Press, Oxford, 1988, pp. 77–85.

199. L. K. Mansur, in *Encyclopedia of Materials Science and Engineering*, Pergamon Press, Oxford, 1986, pp. 4834–4838; *JOM*, Dec. 1996, pp. 28–32.

200. R. E. Stoller, *JOM*, Dec. 1996, pp. 23–27.

201. J. A. Wang, F. B. K. Kam, and F. W. Stallman, in *Effects of Radiations on Materials*, ASTM STP 1270, eds. D. S. Gelles et al., ASTM, Philadelphia, 1996, pp. 500–521.

202. R. E. Stoller, in *Effects of Radiations on Materials*, ASTM STP 1270, eds. D. S. Gelles et al., ASTM, Philadelphia, 1996, pp. 25–28.

203. H. J. Wollenberger, in *Physical Metallurgy*, 4th ed., Elsevier, 1996, pp. 1621–1723.

204. E. Kuramoto, *J. Nucl. Mater.*, vols. 192–194, 1992, p. 1297.

205. E. A. Little, *J. Nucl. Mater.*, vol. 206, 1993, p. 324.

206. L. E. Rehm, *Met. Trans.*, vol. 20A, 1989, pp. 2619–2626.

207. L. E. Rehm, in *Encyclopedia of Materials Science and Engineering*, supp. vol. 3, Pergamon Press, Oxford, 1993, pp. 1941–1945.

208. K. C. Russell, in *Encyclopedia of Materials Science and Engineering*, supp. vol. 1, Pergamon Press, Oxford, 1988, pp. 380–386.

209. R. Cauvin and G. Martin, *J. Nucl. Mater.*, vol. 83, no. 1, 1979, pp. 67–78.

210. H. R. Brager and F. A. Garner, *Effects of Radiations on Materials*, ASTM STP 870, eds. F. A. Garner and J. S. Perrin, ASTM, Philadelphia, 1985, pp. 139–150.

211. S. M. Murphy, *Met. Trans.*, vol. 20A, 1989, pp. 2599–2607.

212. E. P. Simonen and S. M. Bruemmer, *JOM*, Dec. 1998, pp. 52–55.

213. S. J. Rothman, L. J. Norwicki, and G. E. Murch, *J. Phys. F. Metal Phys.*, vol. 10, 1980, pp. 383–398.

214. E. P. Simonen, *JOM*, Dec. 1996, pp. 34–37.

215. R. Allen et al., *7th International Symp: Environmental Degradation of Materials in Nuclear Power Systems—Water Reactors*, NACE, Houston, 1995, pp. 997–1008.

216. K. C. Russell, in *Encyclopedia of Materials Science and Engineering*, supp. vol. 1, Pergamon Press, Oxford, 1988, pp. 380–386.

217. K. C. Russell, "Phase Stability under Irradiation", *Progr. Mater. Sc.*, vol. 28, 1984, pp. 229–434.

CHAPTER 3
SOLIDIFICATION

3.1 INTRODUCTION

Solidification is the most important method of material preparation and a very familiar phase transformation that is usually associated with the formation of crystalline
metals and alloys from liquid upon cooling.[1] The microstructure is determined
largely by the solidification process. Solidification and melting play important roles
in many processes used in the fields ranging from production engineering to
solid-state physics. Examples include ingot casting, foundry casting, single-crystal
growth, directionally solidified alloys, rapidly solidified alloys, and glasses.[2]

This chapter discusses heat transfer in solidification, thermodynamics of
solidification, nucleation, interface kinetics, solidification of alloys, cellular
and dendritic solidification, polyphase solidification, solidification processes and
casting structure, new solidification processes, and structure manipulation and
control.

3.2 HEAT TRANSFER IN SOLIDIFICATION

When hot metal is poured into a mold, the rate of heat extraction from the melt
is of special interest because it usually (but not always) constitutes the rate-
determining process for the progress of solidification.[3] Heat transfer may take
place in different portions of the metal-mold system by conduction, convection, and
radiation.[4] The solidification rate influences directly the coarseness or fineness
of dendritic structures and, therefore, controls the spacing and distribution of
microsegregates such as coring, second phases, and inclusions. Thermal gradient
during freezing is also of great importance because of its relation with the formation of microporosity in alloys.[5]

The analysis of heat transfer during solidification is complex due to continuous
evolution of latent heat of fusion at the moving solid-liquid (S-L) interface, the
behavior of S-L interface configuration, and the change in physical properties of the
metal-mold system with temperature.[5] In this section we briefly describe solidification in conducting molds, solidification limited by interface resistance, and solidification in insulating molds.

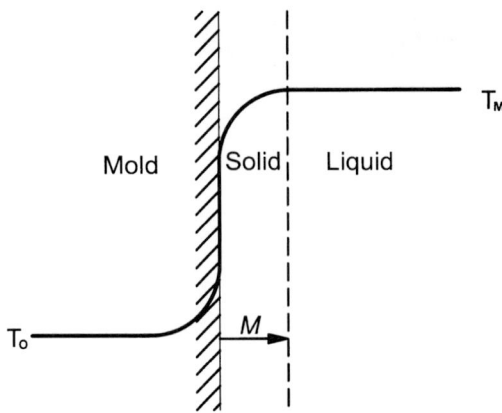

FIGURE 3.1 Temperature distribution during the solidification of a metal in a conductive mold.[5]

3.2.1 Solidification in Conducting Molds

If poured into metal molds, castings freeze rapidly and temperatures change fast in both the mold and the casting. An understanding of the factors influencing solidification in metal molds is of paramount importance in permanent and pressure diecasting molds. The analysis of heat transfer during pouring of metal against a chill wall is more complex than that when pouring into a sand mold (discussed later), because metal molds are superior heat conductors to sand molds and the size of the main thermal resistance, the solid itself, increases with the progress of solidification. As shown in Fig. 3.1, a temperature drop takes place at the solidified metal-mold interface, because of the thermal contact resistance. In this case, it is assumed that the mold is held at a temperature T_0 and the liquid metal is at its melting temperature T_m at the S-L interface. The surface temperature of the metal mold remains well below the melting point, because of high thermal conductivity of the metal being cast, while a considerable thermal gradient exists within the solidifying metal. More total heat extraction occurs during solidification due to the cooling of the solidified metal.

The solution to the heat-transfer equation, corresponding to the above boundary conditions, shows the amount of solid growth as the square root of time and is given, in simple form, by

$$M = 2B\sqrt{K_S/\rho_S C_S}\,\sqrt{t} \tag{3.1}$$

where M is the thickness of the solidified material, t is the time, and K_S, ρ_S, and C_S, respectively, are the thermal conductivity, density, and specific heat (capacity) of the solid metal. As in the case of insulating molds, the solidification time is proportional to the square of thickness or modulus. Here B, a pure number, is the integration constant, which has no specific algebraic form and must be numerically known for each particular condition. However, the values of B can be expressed, in good approximation, as

$$B = f\left[(T_m - T_0)\frac{C_S}{\Delta H_f}\right] \tag{3.2}$$

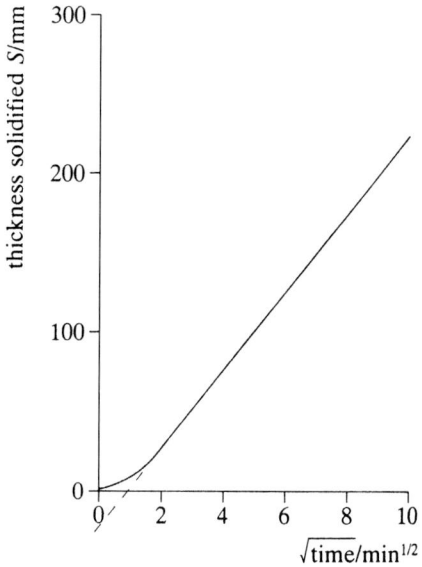

FIGURE 3.2 Measured solidified thickness and time for pure iron in a cast iron mold.[6] (*Reprinted by permission of Butterworths, London.*)

where ΔH_f is the latent heat of fusion. Here B increases with the increase of $T_m - T_0$ and C_S or with the decrease of ΔH_f. Since the function in Eq. (3.2) is not linear, it is difficult to calculate the precise effect of thermal properties on the rate of solidification in this situation. However, the parabolic law given by Eq. (3.1) holds well with experimental results, as seen from Fig. 3.2.

Metals or conductive molds are best suited to thin-walled castings because of yielding short cycle times. To overcome the tooling costs in a reasonably short time, short cycle time and larger production runs are required. Therefore, the metal-mold requirements for pressure die castings must include conductivity and durability. This analysis is also justified for injection molding where thick sections have long cycle times and are difficult to mold.[6]

3.2.2 Solidification Limited by Interface Resistance

In a large number of casting processes such as permanent mold casting, die casting, splat cooling, and powder manufacturing processes, there exists an insulating layer (air gap) between the solid and the mold due to shrinkage of the casting; then heat flow is controlled to a large extent by resistance at the mold-metal interface.[6,7]

Figure 3.3 shows the temperature distribution across the solidifying metal and mold, which illustrates that the temperature of the solid (effectively T_m) and that of mold (T_0) are both practically constant. At the S-L interface, heat is generated due to the evolution of latent heat of solidification. The rate of heat production is thus regulated by the volume of the material solidifying in a particular time. Thus the heat flow rate across this interface for metal poured at its melting temperature T_m is

$$\frac{Q}{A} = \Delta H_f \rho_s \frac{dM}{dt} \qquad (3.3)$$

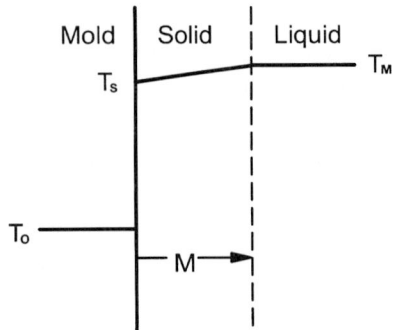

FIGURE 3.3 Temperature distribution across the solidifying metal and mold in which interface resistance is dominant.[5] (*Reprinted by permission of Addisson-Wesley Publishing, Inc.*)

where A is the area and dM/dt the velocity of the S-L interface. We can predict the rate of heat transfer across the solid-mold interface using a modified (form of) experimental law for heat flow

$$Q = hA(T_m - T_0) \tag{3.4}$$

where h is the heat-transfer coefficient of the interface and A the interface area. By combining Eqs. (3.3) and (3.4) for the case of large, flat mold wall and integrating from $M = 0$ at $t = 0$, we obtain the general solution for M as a function of t:

$$M = \frac{h(T_m - T_0)t}{\rho_s \Delta H_f} \tag{3.5}$$

Thus the solid thickness increases linearly with time, and its growth rate is a function of the interfacial heat-transfer coefficient. Since the shape of the casting does not affect in any way the heat transfer across the interface, Eq. (3.5) can be generalized to calculate the solidification time t_f for a simple-shaped casting in terms of its volume-to-surface-area ratio:

$$t_f = \frac{\rho_s \Delta H_f M}{h(T_m - T_0)} = \frac{\rho_s \Delta H_f}{h(T_m - T_0)} \frac{V}{A} \tag{3.6}$$

where V/A $(= M)$ is called the modulus of a casting and is widely employed in the foundry industry. It allows comparison of freezing times of castings with different shapes and sizes and assumes that castings with the same modulus will have the same solidification time. It is also beneficial in the determination of the feeding requirements of castings.[6]

In case of castings such as gravity die casting where heat transfer occurs as a result of interface resistance, the solidification time t_f is proportional to the casting thickness. Therefore, large castings are more economical than thin section castings. Solidification times are usually longer, and multiple molds are generally employed to increase the productivity. The large casting size implies the use of friable cores and increases of the range of shapes that can be produced.[6]

3.2.3 Solidification in Insulated Molds

Sand casting and investment casting are two important commercial processes for making shaped castings which employ relatively insulated molds.[7] The largest tonnage of metal is cast in sand molds, except the quantity of steel cast in ingot molds. The following analysis is used for solidification of melt in refractory mold such as sand molds, molds made of plaster, granulated zircon, or various other materials that have poor thermal conductivity. Hence mold itself is the main resistance to heat flow.

Consider pure liquid metal without superheat poured against a flat sand mold wall. Figure 3.4 depicts the temperature distribution during solidification of metal in a sand mold at some time. Since all the resistance to heat flow is entirely confined within the mold and the thermal conductivity of the casting is very much higher than that of its mold, mold surface temperature T_S is nearly equal to the melting point of the metal T_m. That is, the temperature drop during freezing through the solidified metal is small, and at the metal-mold interface a constant temperature $T_S \approx T_m$ is maintained.

We assume an infinitely thick mold whose outer surface is at T_0 and inner surface is immediately heated to T_m at time $t = 0$. Finally (at $t = \infty$) the temperature gradient in the mold will be linear, but the metal solidification takes a very long time before this happens.

The solution arising from this boundary condition is based on the assumption that the amount of heat which flows in the mold must equal the latent heat evolved during solidification. This solution can be finally expressed as

$$M = \frac{V}{A} = \frac{2}{\sqrt{\pi}}\left(\frac{T_m - T_0}{\rho_s \Delta H_f}\right)\sqrt{K_m \rho_m C_m}\sqrt{t} \tag{3.7}$$

where K_m, ρ_m, and C_m are the thermal conductivity, density, and specific heat capacity of the mold material, respectively. In this case the amount of solidification depends on certain metal characteristics $(T_m - T_0)/(\rho_s \Delta H_f)$ and the mold's heat diffusivity $K_m \rho_m C_m$ (which is a measure of the ability of mold to absorb heat at a certain rate) and is proportional to the square root of time. Note that since a high melting temperature favors solidification, steel castings solidify faster than similar cast iron castings. Similarly, low ΔH_f favors rapid solidification, so that, in spite of their similar melting temperatures, Mg alloy castings solidify faster than Al alloy castings.[6]

The thickness of the solid metal is a parabolic function of time, which implies the solidification rate is initially very rapid and decreases as the mold becomes heated.

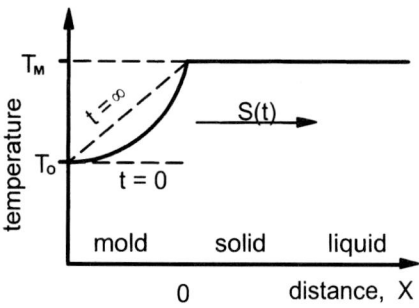

FIGURE 3.4 Temperature distribution during solidification of a metal in a sand mold.[5] (*Reprinted by permission of Addison-Wesley Publishing, Inc.*)

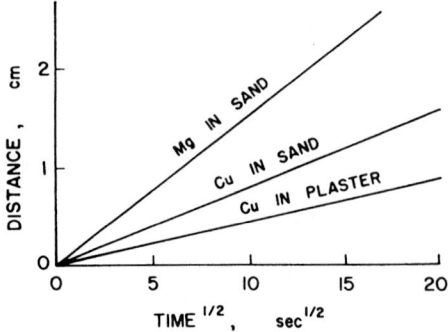

FIGURE 3.5 Distance solidified as a function of square root of time for several metals in insulating molds.[7] (*Reprinted by permission of McGraw-Hill, Inc.*)

Figure 3.5 shows the wide range of solidification rates achieved in practice, based on metal and mold and mold temperature.[7]

The product $K_m C_m$ is a useful parameter for evaluating the ability of mold material to absorb heat and is called *heat diffusivity*. If we rearrange to give solidification time t_f of a casting in terms of its volume-to-surface-area ratio (or modulus) V/A, we obtain the following equation:

$$t_f = C \left(\frac{V}{A} \right)^2 \tag{3.8}$$

where
$$C = \frac{\pi}{4} \left(\frac{\rho_s \Delta H_f}{T_m - T_0} \right)^2 \sqrt{K_m \rho_m C_m} \tag{3.9}$$

Equation (3.8) is the well-known Chvorinov's rule, and C is a Chvorinov constant for a given metal-mold material and mold temperature. This works well for casting configurations where none of the mold material becomes saturated with heat such as in internal cores or internal corners. The success of this relationship relies on the mold material absorbing the same amount of heat per unit area exposed to the metal. As said earlier, it holds for castings with similar shapes but different sizes.

For shapes such as cylinders and spheres, a more appropriate expression than Eq. (3.8) is derived without retaining the assumption of nondivergency of heat flow. In this situation, the partial differential equation for heat flow in the mold is given by

$$\frac{\partial T}{\partial t} = \alpha_m \frac{\partial^2 T}{\partial r^2} + \frac{n}{r} \frac{\partial T}{\partial r} \tag{3.10}$$

where α_m ($= K_m/\rho_m C_m$) is the thermal diffusivity of the mold, r is the casting radius, and n is 1 for cylinder and 2 for sphere. The resulting equivalent equation, similar to Eq. (3.7), is

$$\frac{V}{A} = \frac{T_m - T_0}{\rho_s \Delta H_f} \left(\frac{2\sqrt{K_m \rho_m C_m} \sqrt{t_f}}{\sqrt{\pi}} + \frac{n K_m t_f}{2r} \right) \tag{3.11}$$

By comparing Eqs. (3.7) and (3.11), it seems that the simple Chvorinov approximation becomes increasingly valid with the decrease in thermal diffusivity α_m

$(= K_m/\rho_m C_m)$. It holds more for cylinders than for spheres. For a certain V/A ratio, a sphere solidifies more rapidly than a cylinder and a cylinder more rapidly than a plate.[7]

3.3 NUCLEATION DURING SOLIDIFICATION

Nucleation during solidification is defined as the formation of a small crystal from the melt that is capable of continued growth.[4] Nucleation process plays a vital role in the solidification of castings by controlling to a large extent the initial structure type, size scale, and spatial distribution of the product phases. Nucleation effects in the solidification microstructure exert a great influence on the grain size, morphology, extent of segregation, and compositional homogeneity. The final microstructure is also modified by the crystal growth, fluid flow, and structural coarsening processes that dominate in the latter stages of ingot freezing.[8] These microstructural features have important bearings on the mechanical properties of an as-solidified material. Hence, it is necessary to understand and control the nucleation behavior.

Nucleation can occur homogeneously and/or heterogeneously. Homogeneous nucleation takes place in pure liquid without the aid of foreign particles. Heterogeneous nucleation implies that nucleation preferentially takes place on foreign substances.[9] (See Chap. 6 for nucleation in solids.)

3.3.1 Homogeneous Nucleation

The consideration of nucleation and the extent of undercooling introduce another type of deviation from full equilibrium, called *metastable equilibrium*. From the thermodynamic point of view, the solidification process cannot occur at full L-S equilibrium. Solidification is possible only when there is a departure from full equilibrium that causes the liquid to remain in a metastable, undercooled state.

For a pure metal at the thermodynamic melting point T_m, the free energy change per unit volume, which acts as the driving force, is given by

$$\Delta G_v = G_v^S - G_v^L = H_S - H_L - T_m(S_S - S_L) = 0 \tag{3.12}$$

or

$$\Delta G_v = \Delta H_f - T_m \Delta S = 0 \tag{3.13}$$

so that

$$\Delta H_f = \frac{\Delta S}{T_m} \tag{3.14}$$

where G_v^S and G_v^L, H_S and H_L, and S_S and S_L are free energy, enthalpy, and entropy per unit volume for the solid and liquid, respectively. The enthalpy change ΔH_f is the latent heat of fusion L_v per unit volume. At temperature T, other than T_m, $\Delta G_v \neq 0$, and Eqs. (3.12) and (3.14) can be combined to give

$$\Delta G_v = \frac{\Delta H_f(T_m - T)}{T_m} = \frac{\Delta H_f(\Delta T)}{T_m} = \frac{L_v \Delta T}{T_m} = \Delta S_f \Delta T \tag{3.15}$$

where ΔT is the undercooling or supercooling. The total free energy change per unit volume ΔG to form a solid embryo of spherical shape of radius r from pure liquid consists of the change of volume free energy (which has a negative contribution) ΔG_v and the interfacial free energy ΔG_i and is given by

$$\Delta G = \Delta G_v + \Delta G_i = -\frac{4\pi r^3 \Delta H_f (= L_v)\Delta T}{3T_m} + 4\pi r^2 \gamma_{SL} \qquad (3.16)$$

where γ_{SL} is the S-L interfacial free energy. The critical radius r^* occurs, when ΔG has a maximum given by the condition $[\partial \Delta G(r)/\partial r]_{r=r^*} = 0$ as

$$r^* = \frac{2\gamma_{SL}}{\Delta G_v} = \frac{2\gamma_{SL} T_m}{L_v \Delta T} \qquad (3.17)$$

Figure 3.6, due to Kurz and Fisher,[1] shows a comprehensive plot of free energy change of an embryo as a function of its radius and ΔT: (a) At $T > T_m$, both ΔG_v and ΔG_i increase with r. Hence, the total ΔG increases monotonically with r.(b) At T_m, $\Delta G_v = 0$, but ΔG_i still increases monotonically. (c) Below T_m, the sign of ΔG_v becomes negative due to the metastable liquid whereas the nature of ΔG_i remains unchanged, as in (a) and (b). At large values of r, the cubic dependence of ΔG_v dominates over ΔG_i, and ΔG passes through a maximum at the critical radius r^*. When a thermal fluctuation results in a growth of embryo larger than r^*, the total free energy of the system decreases if the growth of the solid occurs. Such clusters (or embryos) thus grow spontaneously and are termed *nuclei*, whereas clusters

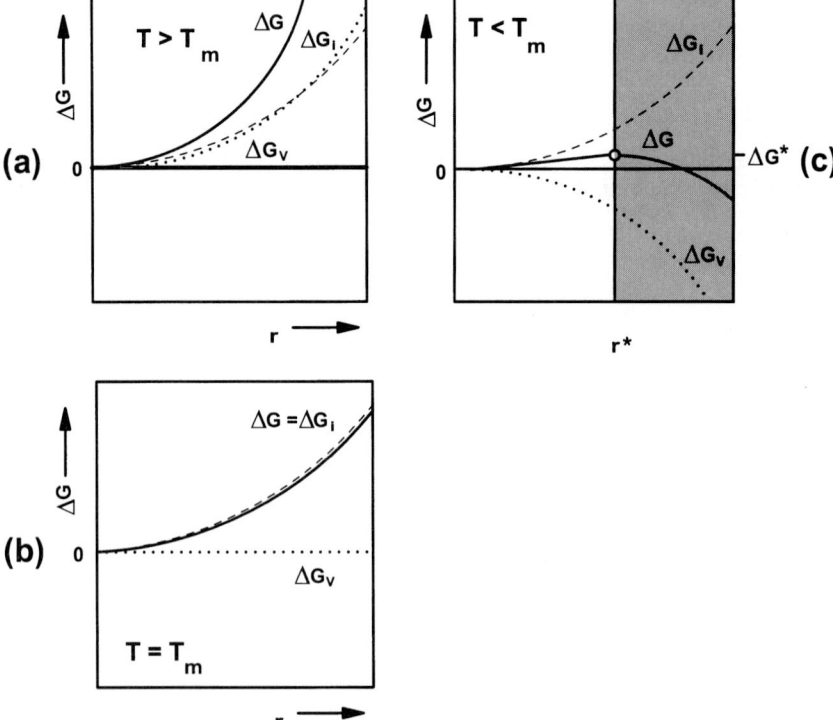

FIGURE 3.6 Free energy change of an embryo as a function of its radius at temperatures (a) $T > T_m$, (b) $T = T_m$, and (c) $T < T_m$.

smaller than r^* can lower its free energy by dissolution of the solid. Unstable solid particles with $r < r^*$ are called *clusters* or *embryos* whereas stable particles with $r > r^*$ are called *nuclei* and r^* is called the *critical radius*. Nucleation in a homogeneous melt (containing no solid phase) is called homogeneous nucleation, and, from Eq. (3.16), the critical free (or activation) energy for nucleation of a spherical nucleus of radius r^* in a pure melt is given by

$$\Delta G^* = \Delta G^*_{\text{hom}} = \frac{16\pi}{3} \frac{\gamma_{\text{SL}}^3}{\Delta G_v^2} = \frac{16\pi}{3} \frac{\gamma_{\text{SL}}^3 T_m^2}{L_v^2 \Delta T^2} \qquad (3.18)$$

Equation (3.16) shows effectively how the maximum cluster size r_{max} is present in the liquid and how it increases with the decrease in temperature, as shown schematically in Fig. 3.7. This indicates the undercooling ΔT_N at which nucleation occurs. The maximum cluster size exceeds r^* at an undercooling ΔT_N, but the probability of occurrence of clusters only slightly larger than r_{max} is very small. Also note that as undercooling ΔT increases, both ΔG^* ($\Delta G^* \propto \Delta T^{-2}$) and r^* ($r^* \propto \Delta T^{-1}$) are reduced.[2,10] A high value of γ_{SL} implies a high value of r^*, a high ΔG^*, and a nucleation problem. The value of γ_{SL} is considered independent of temperature.

It is noted that undercooling phenomena are of great importance in the nucleation of both equilibrium and nonequilibrium crystalline phases and in the formation of amorphous phases. The amount of undercooling is a critical factor in the determination of many practical aspects of solidification practice, including morphological evolution, final solidification structure, phase selection or manipulation, and grain refinement.[8]

3.3.2 Homogeneous Nucleation Rate in Liquid

To calculate the nucleation rate, we consider the entropy of mixing between a small number N_n of crystalline clusters, each of which contains n atoms, and N_L atoms per

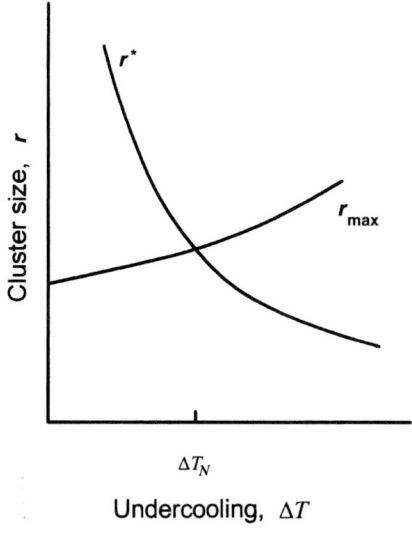

FIGURE 3.7 Schematic plot of critical radius r^* and the maximum cluster size radius in the liquid r_{max} as a function of undercooling ΔT. This indicates the undercooling ΔT_N at which homogeneous nucleation occurs.

unit volume in the liquid. The metastable equilibrium concentration of these spherical clusters or embryos N_n of a given size $(N_n \ll N_L)$ is given by

$$N_n = N_L \exp\left(-\frac{\Delta G}{kT}\right) \tag{3.19}$$

where N_L is the total number of atoms per unit volume in the liquid and ΔG is the total free energy change given by Eq. (3.16). Similarly, it can be shown, for an equilibrium number of critical embryos per unit volume, that

$$N_n^* = N_L \exp\left(-\frac{\Delta G^*}{kT}\right) \tag{3.20}$$

where ΔG^* is the activation energy for nucleation of the critical size clusters. The addition of one more atom to each of these critical embryos (or clusters) will make them stable nuclei. If we assume that an equilibrium number of critical nuclei can be maintained in the melt during the nucleation process, the homogeneous nucleation rate (number of nuclei/time/volume) $\dot{N}$ is then expressed by

$$\dot{N} = K_1 v_{SL} \frac{N_n^*}{N_L} \tag{3.21}$$

where K_1 is a constant involving the product of the number of atoms per unit volume and the number of atoms on a nucleus surface and v_{SL} is the frequency of attachment of atoms to the nucleus by diffusion and can be predicted from the liquid diffusivity D_L and jump distance a_0, as in D_L/a_0^2. This is called the *steady-state nucleation rate*.

For metals, this attachment rate is fairly independent of temperature, and so

$$\dot{N} = K_2 \exp\left[-\frac{K_3}{T(\Delta T)^2}\right] \tag{3.22}$$

where K_2 and K_3 are constants. Typically K_2 has a value of $10^{42}\,\mathrm{m^{-3}\,s^{-1}}$. It is clear from Eq. (3.22) that $\dot{N}$ increases sharply with an increase in supercooling at a critical supercooling ranging between $0.2T_m$ and $0.4T_m$. If it is assumed that the rate of cluster formation is so high or $\dot{N}$ is so small that the equilibrium concentration of critical clusters N_n^*/N_L, will not vary, i.e., the source of nucleation will not be exhausted, then the steady-state nucleation rate is given by

$$\dot{N} = K_4 N_L \exp\left(-\frac{\Delta G^*}{kT}\right) \tag{3.23}$$

where K_4 is a constant and represents a preexponential factor for diffusion. In nonmetallic systems, where the diffusivity in the liquid can depend strongly on temperature,

$$\dot{N} = K_4 N_L \exp\left[-\frac{(\Delta G^* + \Delta G_D)}{kT}\right] = \dot{N}_0 \exp\left[-\frac{(\Delta G^* + \Delta G_D)}{kT}\right] \tag{3.24}$$

where $\dot{N}_0$ is a preexponential factor and ΔG_D is the activation energy for diffusion in the liquid toward the nuclei across the S-L interface.

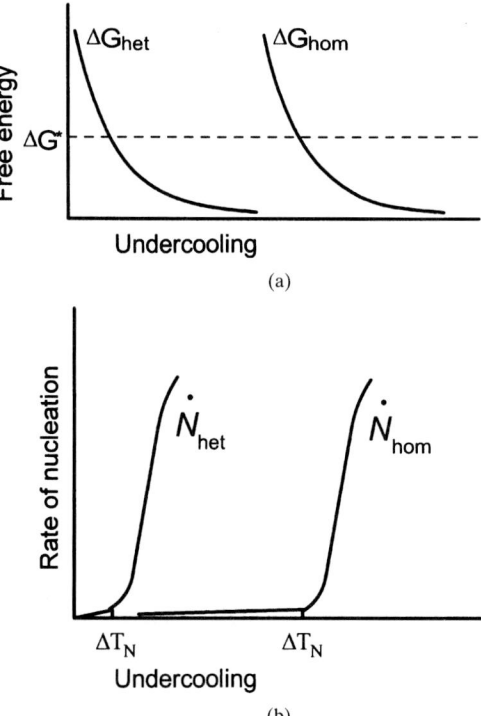

FIGURE 3.8 Variation of (*a*) critical free energy barrier ΔG^* and (*b*) nucleation rate $\dot{N}$ with melt supercooling for homogeneous and heterogeneous nucleation.

Figure 3.8 shows that virtually no nuclei are formed until ΔT_N is reached at which solid nuclei begin to form; after this point, there is a sharp increase of nucleation rate. The value of ΔT_N is approximately $0.33 T_m$ for homogeneous nucleation.[10]

3.3.3   Heterogeneous Nucleation

Homogeneous nucleation is rarely observed for crystal formation from the melt due to the relatively large activation barrier for nucleus development. To overwhelm this barrier, according to classical theory, the large undercooling values are required. However, in practice, a much lower undercooling of only a few degrees or less is required for heterogeneous nucleation (Fig. 3.8) in the majority of castings. It is achieved by reducing the free energy barrier, usually by effectively decreasing the interfacial energy. This behavior of the operation of heterogeneous nucleation can be based on the presence of preexisting surface or foreign substances such as impurity inclusions, oxide films, or mold walls which aid nucleation (or crystallization) by lowering ΔG^*.

The theory of heterogeneous nucleation of a crystal can be developed in terms of the nucleus volume that adopts the shape of a spherical cap (of the existing nucleant) rather than a sphere. If equilibrium conditions are assumed to prevail among the three isotropic interfacial energies at the edge of the spherical cap, the balance of forces can be depicted by the vector diagram shown schematically in Fig. 3.9 and can be expressed by

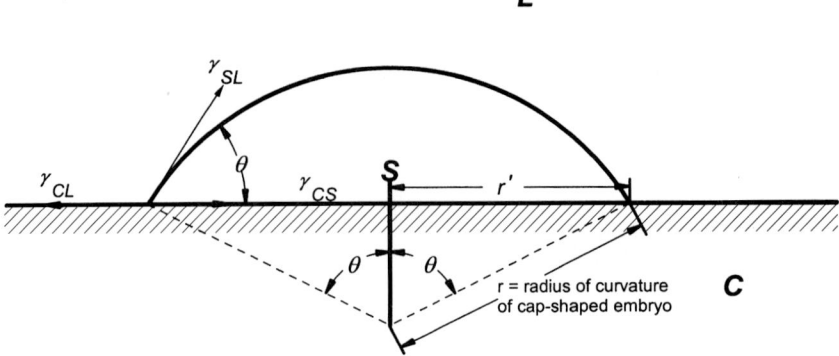

FIGURE 3.9 Geometric and surface energy balances for a cap-shaped nucleus on a foreign substrate involved in heterogeneous nucleation.

$$\gamma_{CL} - \gamma_{CS} = \gamma_{SL}\,\mathrm{Cos}\,\theta \tag{3.25}$$

where θ is the wetting angle with a flat nucleating particle and γ_{CL}, γ_{CS}, and γ_{SL} are the catalyst-liquid interfacial free energy, catalyst-solid interfacial free energy, and solid-liquid interfacial free energy, respectively. At a supercooling ΔT, the critical radius of the spherical cap is again given by Eq. (3.17), but the number of atoms in the critical nucleus is smaller than that for homogeneous nucleation due to the presence of the catalytic substrate. In reality, the thermodynamic barrier to ΔG_{het}^{*} is diminished by a shape factor $f(\theta)$. Thus, the free energy change during heterogeneous nucleation is expressed by

$$\Delta G_{het}^{*} = \frac{16\pi}{3}\,\frac{\gamma_{SL}^{3}T_{m}^{2}}{L_{v}^{2}\Delta T^{2}}\,f(\theta) \tag{3.26}$$

where $\qquad f(\theta) = \left[(2+\mathrm{Cos}\,\theta)(1-\mathrm{Cos}\,\theta)^{2}\right]/4 = (2-3\,\mathrm{Cos}\,\theta+\mathrm{Cos}^{3}\,\theta)/4 \tag{3.27}$

Combining Eqs. (3.18) and (3.26) gives

$$\Delta G_{het}^{*} = \Delta G_{hom}^{*}\,f(\theta) \tag{3.28}$$

If a nucleation occurs in a scratch or a cavity of the catalytic substrate, the number of atoms in a critical nucleus and the value of ΔG_{het}^{*} can be reduced significantly. For a planar catalytic surface, the reduction in the free energy barrier compared to that for homogeneous nucleation depends on the contact angle. Any value of θ between 0 and 180° represents a stable angle. If the wetting angle $\theta = 180°$, the catalyst is impotent as a nucleating agent and there is no interaction between the nucleating solid and the catalyst substrate (nonwetting situation); and as expected, $f(\theta) = 1$ and the homogeneous nucleation results. If the wetting angle $\theta = 0°$, the solid completely wets the substrate, $f(\theta) = 0$, and $\Delta G_{het}^{*} = 0$. That is, there is no free energy barrier to nucleation, and, therefore, nucleation occurs automatically with zero undercooling.[11] It implies that solidification can start immediately when the liquid cools to the freezing point.

For any other value of θ, $f(\theta) < 1$ and homogeneous nucleation becomes rapid only when the driving force is quite large, i.e., only at a sufficiently high undercooling. At low undercoolings, heterogeneous nucleation is dominant due to the catalytic surfaces; at high undercoolings, homogeneous nucleation is favored due to the greater number of potential homogeneous nucleation sites.[12,13]

From the viewpoint of classical heterogeneous nucleation, a good nucleant represents a small contact angle between the nucleating particle and the growing solid. According to Eq. (3.25), this means that γ_{CS} must be much smaller than γ_{CL}. However, usually the values of γ_{CS} and γ_{CL} are not known, and, therefore, it is rather difficult to predict the potential catalytic effectiveness of a nucleant. Tiller[14] suggested that there is no clear understanding of what determines θ and how it changes with (1) lattice disregistery between substrate and the stable phase, (2) topography of the catalytic substrate surface, (3) chemical nature of the catalytic surface, and (4) absorbed films on the catalytic substrate surface.

In commercial practice, nucleating agents (inoculants), which are added to many molten alloys to produce fine-grained materials, usually possess at least one crystal dimension that is similar to the solid being nucleated. Typical examples include high-melting-temperature nucleating agents such as TiB_2, TiC, and inoculants based on impure ferror-silicon, carbon, and zirconium which are used for aluminum alloys, steel castings, cast irons, certain magnesium alloys, and others. As a general rule, particles that are well wetted (small contact angle) by the liquid act as the good heterogeneous nucleation sites. (See Secs. 3.11.2, 3.11.3, and 3.11.4 for more details.)

3.3.4   Heterogeneous Nucleation Rate

Following an analysis similar to that developed for homogeneous nucleation, the heterogeneous nucleation rate expression $\dot{N}_{het}$ ($m^{-3} s^{-1}$) is given by

$$\dot{N}_{het} = K \exp\left[-\frac{f(\theta)\,\Delta G^*_{hom} + \Delta G_D}{kT}\right] \tag{3.29}$$

or

$$\dot{N}_{het} = \frac{D_L}{a_0^2}\frac{2\pi r^{*2}\,(1-\cos\theta)}{a_0^2}\,N_a \exp\left[\left(\frac{-\Delta G^*_{hom}}{kT}\right)f(\theta)\right] \tag{3.30}$$

where N_a is the number of surface atoms of the nucleation site per unit volume of liquid which is on the order of $10^{20}\,m^{-3}$. Since $\Delta G_D \ll \Delta G^*_{het}$ in liquids, $\dot{N}_{het}$ is dependent on ΔG^*_{het}.

For typical metals,

$$\dot{N}_{het} = 10^{30} \exp\left[-\frac{16\pi\,\gamma^3_{SL}\,T_m^2\,V_m^2\,f(\theta)}{3k\,\Delta H_f^2\,T\,\Delta T^2}\right] \tag{3.31}$$

The comparison between plots of heterogeneous and homogeneous nucleation rate with undercooling, assuming the same critical value of ΔG^*, is shown in Fig. 3.8. The form of this expression for heterogeneous nucleation is similar to that for homogeneous nucleation. Like homogeneous nucleation, many refinements have been made extensively in heterogeneous nucleation theory. One important factor is surface geometry. Surface roughness and local pits or grooves can change considerably the nucleation behavior. For example, it is possible to form and retain solid embryos inside the groove, pore, crevice, or cavity even at temperatures above its T_m,[13,15] i.e., at quite high supercooling ΔT. Here, the volume free energy destabilizes the embryos whereas the surface free energy serves as the stabilizing factor.[14] In this way a critical nucleus forms and grows in cylindrical cavities or reentrant cavities of sufficient size and sufficiently sharp angle.

3.4 GROWTH FROM MELT

Growth from a melt is widely used for the preparation of crystals, mass production of castings, and fabrication of high-quality materials essential for the modern technologies. For the growth of crystals, the S-L interface holds a key role, because it is this region in which the molecules are attached with the crystal lattice. Transformation of liquid into crystalline solid is basically a local change in structure, accompanied by a small density variation.[16] Here we first describe the interface structure and interface kinetics of pure materials and then discuss single-phase alloy effects on interface structure and interface growth.

3.4.1 Interface Structure

After nucleation in its melt, the solid phase grows by the addition of atoms to the S-L interface and/or by thermal and mass diffusion. The ease with which atoms attach themselves to the growing solid interface is determined by the atomic structure of the interface. Nonmetallic phases grow by the interface structure-controlled mechanisms whereas the metallic alloys usually grow by diffusion-controlled mechanisms where the flux of atoms controls the rate. Growth may also occur by a combination of these mechanisms.

Growth of nonmetallic phases is important in solidification in such systems as Al-Si alloys, cast iron, and diamond growth from metallic solutions. The interest focuses both primary nonmetallic phases, their growth and mechanism of instability, as well as in the certain class of irregular eutectics between nonfaceted and faceted phases.[17]

The S-L interface is a region between two condensed phases where interatomic cohesive energies are comparable.[8] They are grouped into (1) diffuse interface in which transition from liquid to solid occurs over a number of atomic layers and (2) atomically flat or sharp interface. Pure metals usually grow from its melt with a diffuse interface. For solidification of a pure substance, the interface supercooling ΔT_k, which is the difference between the thermodynamic melting temperature T_m and the interface temperature T_i, governs the atomic or molecular attachment kinetics, thereby the growth velocity of the interface. Different interfaces grow in different manners, producing different interface morphologies.[18]

Diffuse or rough interface moves forward more or less uniformly and is called continuous growth process without requiring steps. This results in the nonfaceted morphology of the interface and represents a gradual transition in atomic positions from the randomness associated with a liquid to perfect registry of the crystal. If the thicknesses of the transition layer of the faceted and nonfaceted interface are of the same order, the distinction between the diffuse and rough interfaces vanishes. The atomically flat or sharp interfaces advance by lateral (nonuniform) growth preferentially at steps which are typically interplanar distances in height. This lateral growth results in a faceted morphology of the interface. For a given substance and supercooling, it is necessary to determine the type of growth. The supercooling needed for lateral growth at a certain interface velocity is typically greater than that of continuous growth. Additionally, an interface that advances by continuous growth can propagate with a smooth curved interface on a microscopic scale. In fact, there is strong evidence that most metals freeze by continuous growth. A nonfaceted, diffuse, rough interface moves most easily while a faceted interface poses the most difficult situation for growth.[4]

In the case of faceted growth, since high-index faces grow more rapidly than low-index faces, it leads to the disappearance of high-index faces, leaving the low-index faces behind. For example, in Si, Ge, and other fcc systems, the low-index facets with {111} planes are developed.[18]

Jackson's Theory. Using a nearest-neighbor bond model in a simple statistical-mechanical treatment and assuming an extra population of N_A adatoms to be randomly deposited in any of the total number of atom sites on the surface N, Jackson developed an expression for the free energy change ΔG of the interface, with the concept of an α or roughness parameter, which is given by[19]

$$\frac{\Delta G}{RT_m} = \frac{\Delta G}{NkT_E} = \alpha x(1-x) + x\ln x + (1-x)\ln(1-x) \tag{3.32}$$

The value of α is given by

$$\alpha = \left(\frac{L_m}{RT_m}\right)\xi = \left(\frac{L_m}{NkT_E}\right)\xi = \left(\frac{\Delta S_f}{R}\right)\xi \tag{3.33}$$

where k is Boltzmann's constant, R is the gas constant, x ($= N_A/N$) is the fraction of nearest-neighbor sites which are filled, L_m is the latent heat of transformation, ξ is the crystallographic factor, and ΔS_f is the entropy of fusion. Figure 3.10 shows plots of Eq. (3.32) for different values of the parameter α. Note that for $\alpha \le 2$, the lowest free energy occurs when one-half the available surface sites (that is, $x = \frac{1}{2}$) are filled and implies that the interface is rough, because liquid molecules can transform to solid ones throughout the layer, so there should be an absence of preferential planes

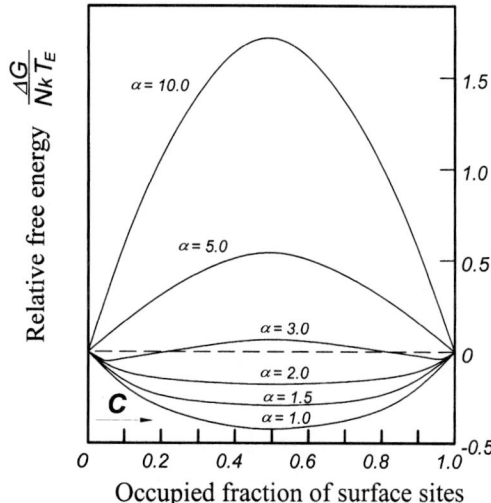

FIGURE 3.10 Relationship between the relative free energy change of the surface and the occupied fraction of the (monolayer) surface sites N_A/N for various values of Jackson's α parameter.

of growth and growth should be isotropic. For $\alpha > 2$, there are two minima in the ΔG-versus-x curve, implying that there will be segregation into two regions in the layer, one liquidlike and the other solidlike.[20] This represents a smooth (faceted) interface.[21] Thus the value of $\alpha = 2$ is the transition between the rough and smooth S-L interface. Mostly metals with $\alpha \approx 1$ grow fast and isotropically from the melt. At higher undercoolings, dendritic growth is predominant, as described in Sec. 3.7.2.1. Most organic compounds with $\alpha > 2$ grow facets. Polymers with large α values grow as spherulites having complex internal structures.

The parameter α that controls the equilibrium interface consists of two parts: (1) ξ, which is a function of the structure of the solid and the orientation of the interface, is always <1 and is closest to 1 for the most closely packed planes of structure; and (2) L_m/RT_m, which is the entropy change associated with the transformation, is about unity for metallic crystals, some ionic crystals, and few organic compounds growing from their melt. It is about 2 to 4 for a few near-metals and semiconductors, and for some ionic crystals growing from the melt. Usually, Jackson's theory has been successfully used to classify growth morphologies,[22] and the α factor of Eq. (3.33) is a roughness parameter to explain why crystals of some materials growing from their melt exhibit rough (or isotropic) and faceted (or smooth) S-L interfaces.[23]

Cahn's Theory. Although Jackson's theory provides good estimation in most cases, it does not explain the transition of some borderline materials from faceting to nonfaceting with the increase of interface undercooling, because this analysis is essentially an equilibrium one. Cahn has concluded that it is not appropriate to describe a surface as being singular or nonsingular or to describe whether the growth is by the lateral propagation of steps or the forward movement of the interface, without considering the effect of driving force on the nature of the interface.[24]

The essence of this theory is that a continuous range of possibilities occurs between a completely smooth and completely rough interface, based on both the criteria of the material and the driving force. For any interface, a driving force occurs below which growth is by lateral propagation of steps and above which normal propagation occurs. The value of the critical driving force depends on the diffusiveness of the interface.[18]

Other Models. Another model for predicting interface roughness arises from the consideration of how thermal vibrations affect the surface energy of a step on an otherwise faceted interface.[25] It is noticed that the step energy disappears when L_m/RT_m drops below a critical value of order unity. When the step energy reduces to zero, a barrier to surface roughening disappears. This analysis provides the same qualitative result as Jackson's model. Various statistical multilevel models of the interface structure and Monte Carlo simulations also suggest the importance of the L_m/RT_m ratio given in Eq. (3.33).[26,27] A common characteristic of all these models is that the interface roughness increases with the decrease of L_m/RT_m. This implies that when the surface is rough, a nucleation barrier to the formation of a new layer ceases.[19] Figure 3.11 illustrates simulations of an interface at different values of L_m/RT_m.[28]

Molecular dynamic (MD) simulations have also been performed to model the Lennard-Jones fcc (111) and (100) crystal-melt interface structure. Since each atomic position is computed as a function of time, the model permits an interface to be diffuse. In fact, simulations show that the liquid → solid transition occurs for

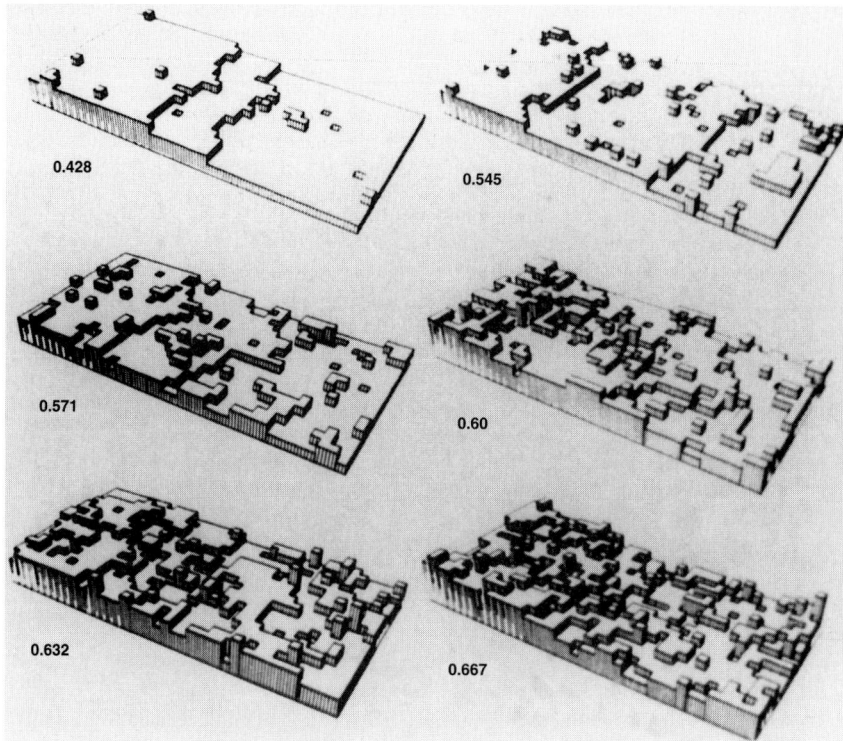

FIGURE 3.11 Computer simulation of a stepped surface showing that the steps gradually disappear into the background roughness at higher surface temperature.[19,28] (*After Leamy and Gilmer, 1974.*)

a material with a Lennard-Jones interatomic potential over several atomic layers.[29] Such potentials are believed to approximate nondirectional metalliclike bonding. Figure 3.12 shows the trajectories of the molecules in layers parallel to the interface (x-y plane), (111) interface. In another method, called the density functional theory, superposition of ordering waves is employed to show the local atomic density.[30] This method also depicts the interface to be several atom layers thick. The expansion relating the free energy to the local density uses order parameters that describe the amplitude to the ordering waves through the interfacial region. In reality, this is similar to a general statement of the gradient energy model of Cahn except that liquid structure factor data are employed to determine the interface thickness and the gradient energy coefficient.[4,24]

3.4.2 Interface Kinetics

As expected, different interfaces grow by different mechanisms. Ideally, diffuse interfaces should grow more easily than flat interfaces. The interface kinetics explains the relationship between the driving force, i.e., kinetic undercooling, ΔT_k and the growth rate of the interface V for small undercooling.

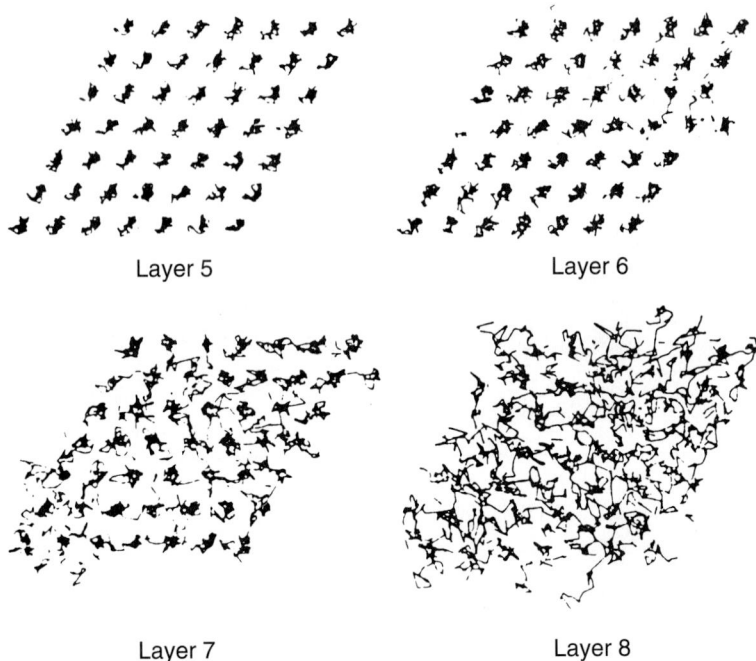

Layer 5   Layer 6

Layer 7   Layer 8

FIGURE 3.12 Trajectories of the molecules in layers parallel to the (111) interface (x–y plane) indicating the transition from rather sharp solid (layer 5) into the liquid (layer 8) that were obtained by molecular dynamics simulations.[29] (*J. Q. Broughton et al., J. Chem. Phys., vol. 74, 1981, p. 4029.*)

3.4.2.1 Continuous Growth.

The growth of an ideal diffuse S-L interface is called continuous or normal growth because the interface advances normal to itself by the continuous, random, and easy atom attachment. According to the analyses of Wilson[31] and Frenkel[32] for continuous growth, a linear relationship between T_k and V, for small undercooling, is given by

$$V = \mu \Delta T_k \tag{3.34}$$

where μ is the kinetic coefficient. Equation (3.34) has been derived by assuming atomic transfer between the liquid and solid which depends on the activation energy for self-diffusion in the liquid, and thus the liquid diffusion coefficient D_L. Flemings, for example, has obtained the equation[7]

$$V = h_m \frac{D_L}{akT_m^2} = \beta \frac{D_L}{D_{LM}} \Delta T_k \tag{3.35}$$

where h_m is the molecular (or atomic) heat of fusion, k is Boltzmann's constant, a is the interplanar spacing, β is a constant, and D_{LM} is the liquid diffusivity at the melting temperature. This equation shows that the solidification rate of a diffuse interface is linearly proportional to the interfacial undercooling.

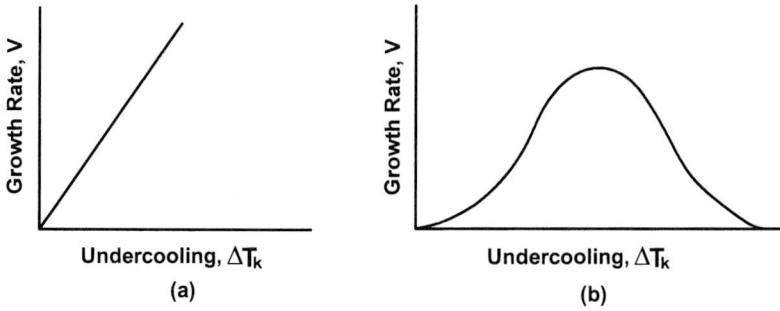

FIGURE 3.13 Schematic growth rate curves for (a) nonviscous liquids such as metals and (b) oxide glasses or polymers.[7] (*Reprinted with permission from McGraw-Hill, Inc., New York.*)

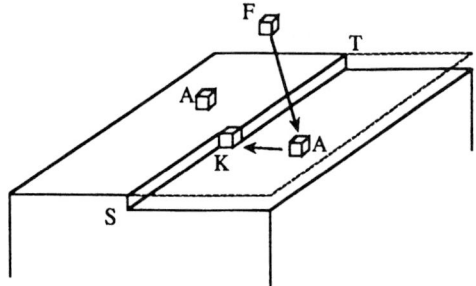

FIGURE 3.14 Atomically smooth flat interface with a step, a jog, or kink (K) in a step (S-T) and one free (F) and two adsorbed (A) atoms.[35] (*Courtesy of Materials Science and Engineering.*)

If D_L does not change appreciably with undercooling from D_{LM}, as in metals, for example, the relationship between V and ΔT_k remains linear (Fig. 3.13a). If D_L changes strongly with temperature, as in oxide glasses and polymers, V increases to a maximum at a certain undercooling ΔT_k and then decreases above this, as shown in Fig. 3.13b.[7]

Continuous growth kinetics in fcc materials have also been investigated by Broughton, Gilmer, and Jackson.[33] They determined that freezing in these situations, for small undercoolings, was athermal, which followed a velocity-undercooling relationship, as given by

$$V = \frac{a}{\lambda}\left(\frac{3kT}{m}\right)^{1/2}\left(\frac{f_0 h_m \Delta T_k}{k \Delta T_m^2}\right) \tag{3.36}$$

where m is the atomic mass, f_0 is the accommodation factor, and λ is the geometric factor ($= 0.4a$).[34]

3.4.2.2 Lateral Growth. Growth of a faceted, atomically smooth, or sharp interface can proceed only by the lateral motion of interfacial steps of a few atomic dimensions, called *lateral* or *layer growth.* Figure 3.14 shows an atomically smooth, flat, close-packed interface with a step where a transition from liquid to solid is

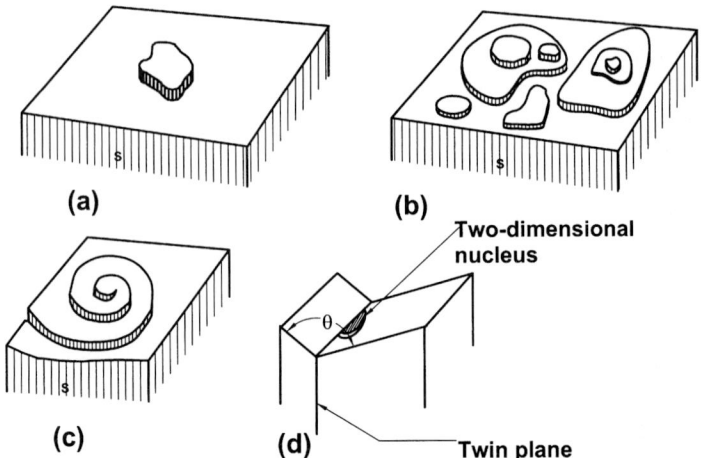

FIGURE 3.15 Schematic illustrations of interfacial processes for the lateral growth mechanisms: (*a*) two-dimensional mononuclear, (*b*) polynuclear, (*c*) spiral growth,[40] and (*d*) twin-plane reentrant corners.[13]

anticipated to occur across a single atomic or molecular layer.[35] Growth of the sharp interface takes place by such steps sweeping across it. The steps are created at the interface by two-dimensional nucleation (2DN), screw dislocations (SDs) intersecting the faceted interface, reentrant corner due to the presence of a twin boundary, and/or stacking faults intersecting the faceted interface. These different sources of steps result in different growth mechanisms and kinetic laws.[36–39]

Growth by Two-Dimensional Nucleation (2DN). In the case of 2DN and its growth (2DNG), the growth rate V is mostly explained in terms of the random nucleation rate N_S of 2D clusters of atoms, their lateral spreading velocity, and the S-L interfacial area A. The 2DNG kinetics are grouped into two regions, based on the relative time between nucleation events and layer completion (cluster spreading) face. The first, appearing at small supercoolings, is the mononuclear (monolayer) growth (MNG) (Fig. 3.15*a*), where a single critical nucleus forms and spreads over the entire surface prior to the occurrence of the next nucleation event. In this case, the growth velocity V is proportional to the interfacial area A and is exponentially dependent on the reciprocal of the interface supercooling, as given by

$$V = KA\exp\left(-\frac{B}{\Delta T_k}\right) \qquad (3.37)$$

where K and B are coefficients and are presumed to be independent of supercooling. Usually the MNG region is restricted, and a very close control of ΔT_k is required to observe it. In addition, its small growth rates and difficulty in its measurement lead to very rare observation of the MNG region.[40,41] The second mechanism is the polynuclear (multilayer) growth (PNG) region, which is found to appear at higher supercooling; here, the rate of 2DN becomes so high that a large number of (2DN) nuclei form at random locations at the interface prior to the completion of a layer

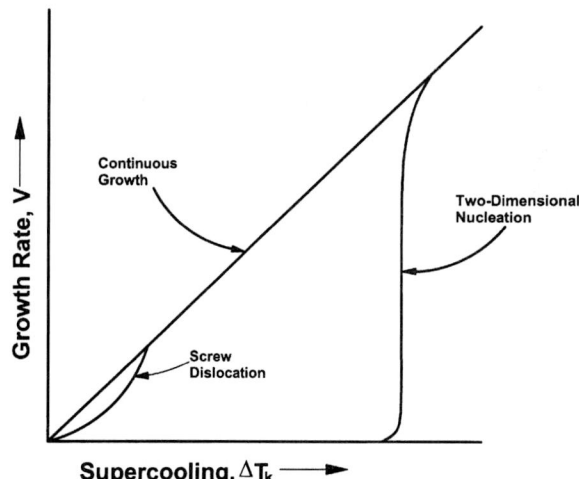

FIGURE 3.16 Growth rate V versus interface supercooling ΔT_k according to three classical laws of growth mechanisms.

or on top of the already growing 2D islands, as shown in Fig. 3.15b. It approaches the continuous growth law as an upper limit, as illustrated in Fig. 3.16. Here, the growth rate V is proportional to the exponential of the reciprocal of the super-cooling, but independent of the interface area as

$$V = K' \exp\left(-\frac{B'}{\Delta T_k}\right)$$

(3.38)

where K' and B' are constants. According to the classical theory, the step edge free energy is assumed to be constant relative to the supercooling, irrespective of the expected changes in interface structure upon increasing the supercooling.

Growth by Screw Dislocations (SDG). Frank[42] first proposed that each screw dislocation emerging at its L-S interface produces a ledge or step in the interface, allowing growth at much lower undercoolings than predicted by the 2DN theory. The step is self-perpetuating and persisting, irrespective of the deposition of many layers of atoms on the face. Since the step is anchored, it will rotate around the dislocation and lead to a spiral growth. Both rounded and nonrounded spirals have been observed during material growth (Fig. 3.15c).

For a single dislocation, the crystal grows up a spiral staircase by the continuous deposition of atoms at the exposed step. As the spiral ledge makes a full rotation about the screw dislocation, an extra plane is either added or depleted from the crystal, based on the fact of whether the crystal is growing or diminishing in size (Fig. 3.15c).[43] Support for spiral growth linked to screw dislocations is based on the fact that most crystals in bulk are understood to contain a significant number of growth dislocations.

Assuming the spiral to be Archimedean, which has been shown to give very nearly the correct final result, the growth rate is given by[7,44]

$$V = \beta \frac{D_L}{D_{LM}}(\Delta T_k)^2$$

(3.39)

where β is a constant. Thus we expect the squared dependence of the growth rate on undercooling. At high supersaturation, this growth rate by screw dislocations approaches the continuous growth law as an upper limit (Fig. 3.16).

The treatment resulting in Eq. (3.39) assumes only a single dislocation. However, real crystals usually have a dislocation density of 10^8 lines/cm^2 maximum. Another difficulty is that the dislocation emerging at the interface may be associated with different undercoolings, especially when there is a temperature gradient along the interface. A detailed analysis of the interactions between the growth spirals has demonstrated that the growth rate of the interface is the one which would result if the ΔT_k over the entire interface were equivalent to that of the maximum, i.e.,[44]

$$V \propto (\Delta T_k)^2_{\max} \tag{3.40}$$

where $(\Delta T_k)_{\max}$ is the maximum kinetic undercooling along the interface.

Unified Theory. Another mechanism proposed by Cahn[24] and Cahn et al.[45] predicts a smooth transition from lateral growth at low supercooling (classical regime) to continuous growth (or regime) at high supercooling (Fig. 3.17). Here the transition starts at a critical supercooling ΔT_k^* and ends at $\pi \Delta T_k^*$. The critical value depends on the extent of diffusiveness and the surface energy. For a material with highly diffuse interface, ΔT_k^* may be very small so that the lateral growth is difficult to find. An experiment showing the transition predicted by the theory has been

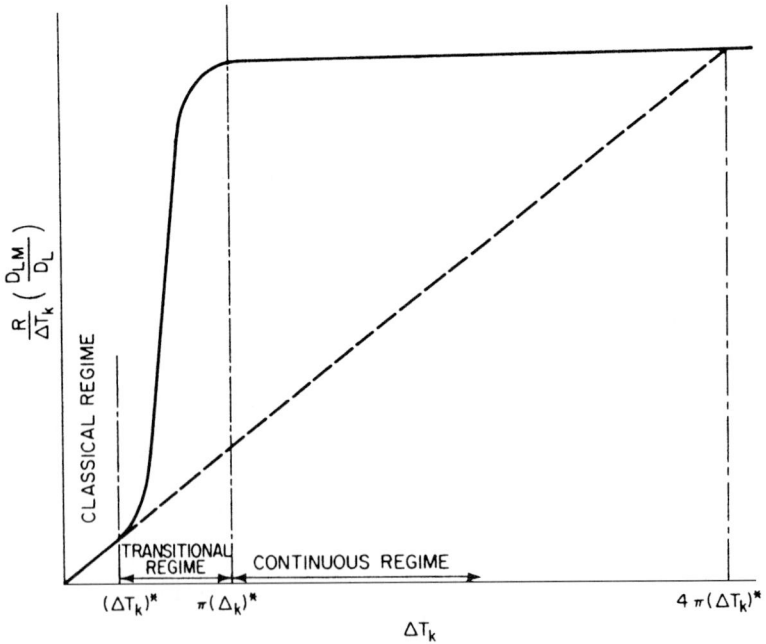

FIGURE 3.17 Predicted growth rate curve for surface with an emergent dislocation ordinate is interface velocity divided by undercooling, corrected for temperature dependence of the diffusion coefficient.[45] (*After Cahn et al.*)

accomplished by Peteves and Abbaschian on Ga.[36] The transition to continuous growth takes place at a supercooling of about 4 K and a growth rate of 5 mm/s for the (111) face.

With increasing supercooling, other models can also predict a transition to continuous growth. Chernov[25] suggests the nucleation of a disk on an otherwise faceted interface. For $L_m/RT_m > 2$, the edge energy becomes nonzero and the growth will be lateral with faceted interface. At some value of increased supercooling, the driving force is comparable to the work of creating a new disk, and the advancing interface becomes rough. The kinetic roughening then allows the occurrence of the continuous growth.

Growth by Propagation of Twin Plane. *Twin-plane reentrant corner* (TPRE) is another source of generation of layers which occurs at both low and high supersaturation due to low edge energy on the rough twinned area and the higher nucleation rate along the reentrant corner because of the reduction in the formation of the partial MNG at the reentrant corner (Fig. 3.15d).[38,39] This mechanism of step growth arises from the existence of reentrant grooves or angles on the surface which form when multiple twins terminate on the surface and which generate a source for the initial growth of steps.[46,47] This twinned crystal requires smaller undercooling for a given velocity than the 2DN and the screw dislocation mechanism. Twin boundary, rotational twin boundary, and twin planes are found as the operating growth mechanisms in silicon plate and wafer growth[48,49] and diamonds, eutectic Al-Si, and graphite growth.[50] Recently, it has been shown that stacking faults and twin lamellae, like screw dislocations, can also act as self-perpetuating step-generating sources during crystal growth.

3.4.3 Interface Structure of Single-Phase Binary Alloys

Kerr and Winegard[51] extended the Jackson model to include alloy melt, and they demonstrated that the entropy of solution $\Delta S^{\alpha L}$ should be employed instead of the entropy of fusion in the derivation, which is given by

$$\Delta S^{\alpha L} = \left(1 - x_e^\alpha\right)S_1^L + x_e^\alpha S_2^L - S^\alpha \tag{3.41}$$

where x_e^α is the concentration of component 2 in the liquid, S_1^L and S_2^L are the partial molar entropies of components 1 and 2, respectively, in the liquid alloy, and S^α is the molar entropy of the solid. Taylor et al.[52] and Croker et al.[53] have employed this approach to describe the various microstructures. More general approaches of Jackson[54,55] show that $\Delta S^{\alpha L}$ is really a significant parameter for the determination of the faceting behavior as well as the rate of crystal growth and its anisotropy. Mamta et al.[23] and Ramachandran and Srikanth[56] have confirmed the requirements of both a high value of entropy of solution and a strong temperature dependence of the entropy of solution for promoting faceting behavior in the primary phases during eutectic solidification of the Al-Sn, Ag-B, Ag-Pb, Ag-Pd, Sn-Zn, and In-Zn systems, but fails in the case of the Al-Ge system.[23,56] Usually Jackson's α criterion is useful in predicting the system for the occurrence of facet formation, whereas the criterion of strong temperature dependency of entropy of solution is advantageous in predicting the composition range for facet formation. However, Jackson's criterion describes most of the experimental results with a few exceptions. For example, crystals of Al_2Cu and white P exhibit faceting in spite of their α parameters being less than the critical value.[57]

3.5   *INTERFACE GROWTH*

Growth of the interface depends on the heat and solute flow and the conditions of the S-L interface. During crystal growth, the instability of the smooth and planar S-L interface will result in its breakdown and the formation of cellular or dendritic interface morphologies. Linear morphological stability theory deals with the conditions under which the interface becomes unstable.[58] Stability conditions describe the pattern formation of cellular and dendritic growth. Here we briefly describe the interface conditions and heat and solute transport during the unidirectional solidification of a single-phase alloy from its melt.

3.5.1   Interface Conditions

Under normal conditions, the growth rate of the interface is lower than the rate of atom transport; the local equilibrium is assumed to apply at the interface. When a liquid solidifies, the interface temperature T_I is related to the interface liquid composition C_I, the surface curvature, and the departure of the interface from the equilibrium, as given by[59]

$$\Delta T_I = T_0 - T_I = \Delta T_S + \Delta T_R + \Delta T_k \qquad (3.42)$$

where T_0 is the convenient reference temperature and $\Delta T_S, \Delta T_R, \Delta T_k$ are the undercoolings resulting, respectively, from solute, curvature, and kinetics.

For local equilibrium at the S-L interface, the coefficients obtained from the phase diagram are the liquidus slope m and the distribution coefficient k. If T_m is the melting temperature for an alloy of composition C_0, the solute undercooling ΔT_S resulting from the solute buildup around the planar interface or tip is given by

$$\Delta T_s = m(C_0 - C_I) \qquad (3.43)$$

The local liquid composition C_I deviates from the bulk composition due to the rejection of solute at the S-L interface. The distribution coefficient, for low-growth velocities, is defined by

$$C_S = kC_I \qquad (3.44)$$

where C_S is the solid solute concentration at the interface. During fast growth, such as in splat quenching, the usual assumption of local equilibrium at the S-L interface does not hold good, and k departs from the equilibrium value, that is, $k = f(V)$, leading to solid trapping at high growth rates.

The curvature or capillary undercooling ΔT_R is characterized by the equilibrium conditions at a curved interface which is associated with the depression in freezing temperature due to the Gibbs-Thomson curvature effect. It is given, for isotropic surface energies, by

$$\Delta T_R = A\left(\frac{1}{R_1} + \frac{1}{R_2}\right) \qquad (3.45)$$

where A is the Gibbs-Thomson coefficient and R_1 and R_2 are the principal radii of curvature. For a planar or faceted solidification, $\Delta T_R = 0$.

The kinetic undercooling at the interface ΔT_k, resulting from the nonequilibrium effects, is necessary to drive the growth process or interface reaction in which atoms are transferred from the liquid to the solid. ($\Delta T_k \to 0$ as the velocity $V \to 0$.) Kinetic undercooling ΔT_k has been predicted to be about 10^{-4} K for typical crystallization rates and can be totally neglected relative to other terms.[7] For nonfaceting materials, this term is assumed to be small and fairly isotropic, which is given by

$$\Delta T_k = BV \qquad (3.46)$$

where B is a constant. For a faceting material, the normal velocity of a facet depends on the rate of production of steps on the face and thus is dependent in a much more complex manner on the local undercooling kinetics.

In the case of nonfaceted growth, ΔT_k is normally omitted except for very fast growth rates. In nonfaceted growth, ΔT_k and ΔT_R should be fairly isotropic. However, certain crystallographic features can be described only by incorporating a slight anisotropy in these terms.

Both the solute and heat flow are described at the S-L interface, resulting in the final two interface equations:

For solute, $$V(C_S - C_T) = D_{LC}\left(\frac{\partial c}{\partial n}\right) - D_{SC}\left(\frac{\partial c}{\partial n}\right) \qquad (3.47)$$

For heat flow, $$VL = K_S\left(\frac{\partial T}{\partial n}\right) - K_L\left(\frac{\partial T}{\partial n}\right) = K_S G_S - K_L G_L \qquad (3.48)$$

where V is the growth rate; D_{LC} and D_{SC} are the liquid and solid diffusivities, respectively; n is normal to the interface; L is the latent heat of fusion per unit volume; K_L and K_S are the thermal conductivities of liquid and solid, respectively; and G_S and G_L are the normal components of the thermal gradient in solid and liquid, respectively.

If $k < 1$, the solute is rejected at the interface and thus concentrated in the liquid ahead of the interface. If $k > 1$, on the other hand, solute is depleted in the liquid ahead of the interface.

3.5.2 Heat and Solute Transport

In the simplest case, where fluid motion is neglected, heat and solute flow occurs by conduction and diffusion.[60] The equations

$$D_{iT} \nabla^2 T = \frac{\partial T}{\partial t} \qquad (3.49)$$

$$D_{iC} \nabla^2 C = \frac{\partial C}{\partial t} \qquad (3.50)$$

must be solved in both the solid and liquid phases (t is time, D_{iT} and D_{iC} are the thermal and solute diffusivities in the relevant phases, respectively).

For steady-state growth (i.e., planar front, cellular and lamellar eutectic growth, and dendrites in the vicinity of tip), these equations may be transformed to coordinates moving with the interface, which then become[61]

$$D_{iT} \nabla^2 T + V \frac{\partial T}{\partial x} = 0 \qquad (3.51)$$

$$D_{iC} \nabla^2 C + V \frac{\partial C}{\partial x} = 0 \qquad (3.52)$$

where x is the steady-state growth direction. A steady-state solution, therefore, involves solving Eqs. (3.51) and (3.52) for an interface shape that satisfies Eqs. (3.42), (3.47), and (3.48).

When fluid flow occurs in most solidification processes, the complex effect of convection should be incorporated in the transport equations. The effect of fluid flow is discussed later.

3.6 PLANE FRONT SOLIDIFICATION OF ALLOYS

The solidification rate in a pure metal is almost always a controlled heat flow problem; however, in single-phase and eutectic alloys, it is a heat and solute flow problem. The controlled solute diffusion plays an important role in the solidification of alloys. The assumption of local equilibrium or metastable equilibrium at the L-S interface also has been found useful in explaining solidification at rates normally encountered in commercial practices. This may be shown by considering the solidification of a single phase and a eutectic alloy.

Plane front solidification (or growth) is widely used to grow semiconductor single crystals, jewelry, and oxide crystals for laser systems and other applications; refine material (zone refining); obtain uniform or nonuniform composition within the solidified alloy by controlling solute redistribution; and understand the more complex interface morphologies involving cellular and dendritic growth. Planar growth of the S-L interface may be macroscopically or microscopically planar. The former is achieved by controlled directional solidification, good furnace design, and no convection in the melt. The latter is achieved by removing interface instabilities due to constitutional supercooling. This section addresses solute redistribution during unidirectional solidification under different conditions.

3.6.1 Solute Redistribution during Unidirectional Solidification

We will consider a one-dimensional quantitative treatment of the solute redistribution during the solidification of a single phase or a eutectic alloy with simple phase diagram in order to (1) describe the methods of purifying metals (based on this principle) and (2) understand the effects of this redistribution on the micro- and macrostructure of the solidifying alloy. Here it is essential to make the following simple assumptions:[62,63]

1. The S-L interface corresponds to the cross-sectional plane of a long, narrow crucible and advances at a constant velocity V.
2. In the idealized binary region of the phase diagram, the liquidus and solidus are straight lines. The ratio of their slopes is a constant, given by $C_S/C_L = k_0$ (Fig. 3.18), which implies that local equilibrium always exists at the S-L interface. The solute concentration gives rise to either a decrease (Fig. 3.18a) or an increase in melting temperature (Fig. 3.18b).
3. Solute diffusion is of vital importance for complete, limited, or partial mixing of the melt (rather than convection).
4. No solid-state diffusion occurs.

Figure 3.19 shows the solute distribution in the solid bar after unidirectional plane front solidification under four different limiting conditions, which are described below.

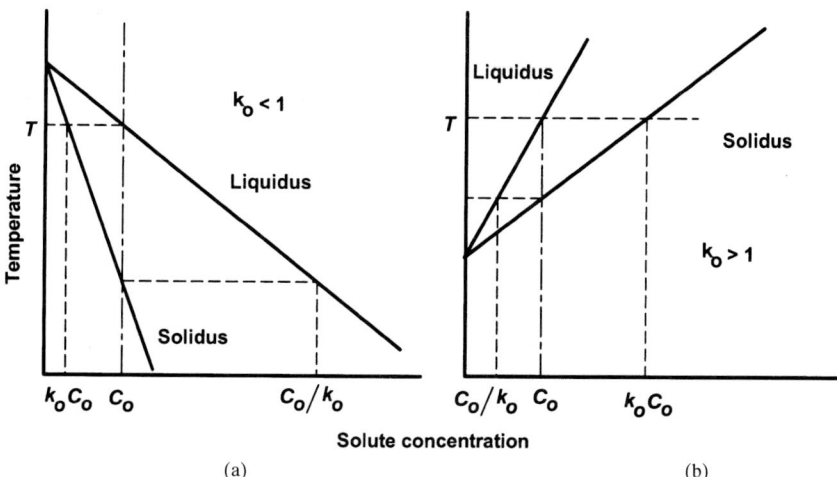

FIGURE 3.18 Schematic portion of binary-phase diagrams with linear solidus and liquidus curves and very small $T_m - T_i$ value where (*a*) solute lowers ($k_0 < 1$) or (*b*) raises ($k_0 > 1$) the melting point of the alloy relative to the pure solvent.

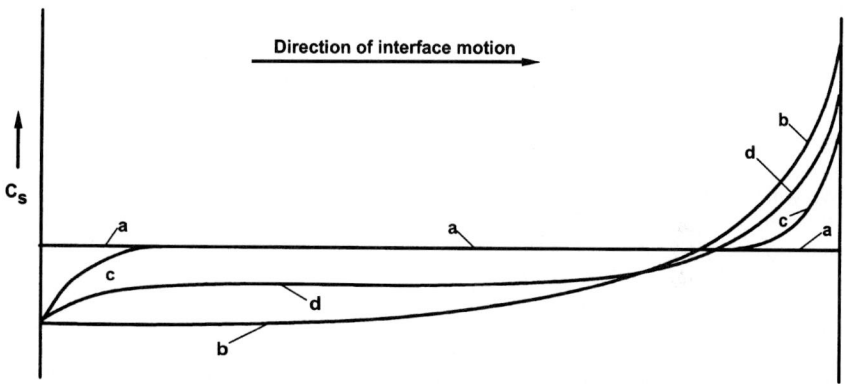

FIGURE 3.19 Solute distribution curves for a bar solidified unidimensionally under various conditions: (*a*) complete diffusion in solid and liquid; (*b*) complete mixing in the liquid, no solid diffusion; (*c*) diffusion in liquid only; and (*d*) partial mixing in the liquid, including convection.[4] (*Courtesy of Elsevier Science.*)

3.6.1.1 Equilibrium Solidification. Let us consider that the first solid of composition $k_0 C_0$ begins to form at the liquidus temperature T_L (Fig. 3.20*a*). The S-L plane front moves so slowly that equilibrium conditions at the interface are reached at temperature T^* ($<T_L$) at all stages of the solidification process. If the solidification occurs at temperature T^*, the solid composition C_S^* forms in equilibrium at the interface with liquid composition C_L^*. Since diffusion in the solid and liquid is complete, the entire solid and liquid become of uniform composition; that is, $C_S =$

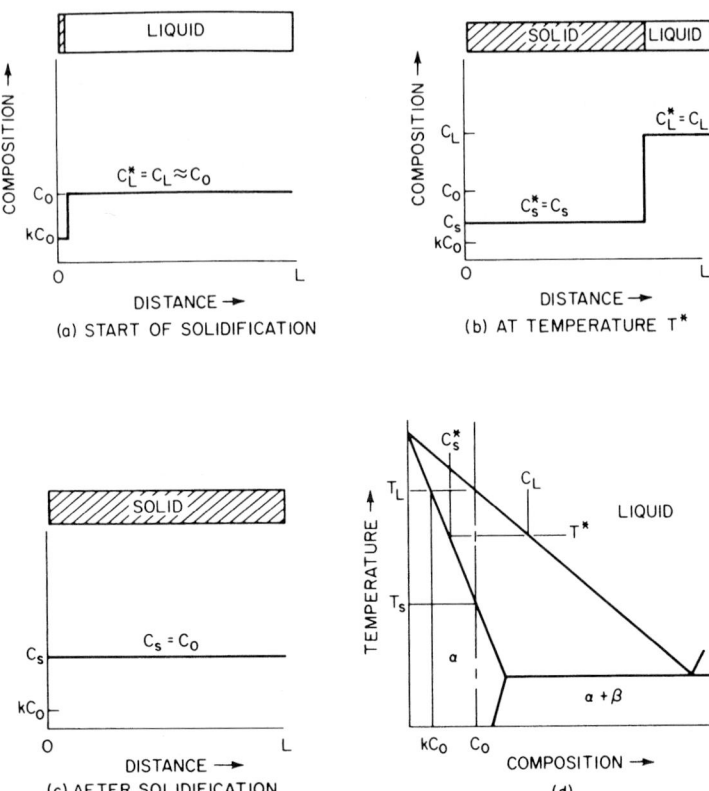

FIGURE 3.20 Solute redistribution in equilibrium solidification of an alloy of composition C_0. (a) At the beginning of solidification, (b) at temperature T^*, (c) after solidification, and (d) phase diagram.[7] (*Reprinted with permission from McGraw-Hill, Inc., New York.*)

C_S^* and $C_L = C_L^*$ (Fig. 3.20b). Any change in the solid and liquid composition occurring during the solidification process will vanish after the completion of solidification (Figs. 3.19a and 3.20c). In equilibrium solidification, the equilibrium distribution coefficient $k_0 = C_S^*/C_L^*$. At temperature T^*, a general materials balance (conserving solute atoms) yields

$$C_S f_S + C_L f_L = C_0 \qquad (3.53)$$

where f_S and f_L are weight fractions of solid and liquid, respectively, and $f_S + f_L = 1$ at a fixed temperature. This is simply the equilibrium lever rule applicable during solidification. One example is the growth of Fe-C at very low velocity due to high solid-state diffusion coefficient of C in Fe. In fact, equilibrium solidification does not take place in any practical casting or crystal growth conditions.[4] Rather, a reasonable solute redistribution occurs during solidification; the material is homogeneous only before and after solidification.[7]

3.6.1.2   Complete Liquid Mixing, No Solid Diffusion.   Let us consider the more realistic situations where (1) complete mixing (or diffusion) of the liquid is achieved (by vigorous convection or stirring); for $k_0 < 1$, all the solute rejected at the interface is immediately mixed uniformly throughout the liquid; (2) there is no or negligible solid diffusion during or after solidification; (3) there are equal mass densities of solid and liquid; and (4) local equilibrium at the S-L interface takes place, that is, $k_I = k_0$.[64–66] Thus, in the initial stages, where the volume of liquid is large, the solid formed is of composition $k_0 C_0$, where C_0 is the initial alloy composition and the overall composition of the melt remains constant (Fig. 3.21a). As solidification advances, the compositional changes in the liquid become more significant, and its concentration increases more progressively (Fig. 3.21b).

When a fractional distance of the rod f_S is solidified, the composition of the remaining liquid is found to conform to the following differential equation:

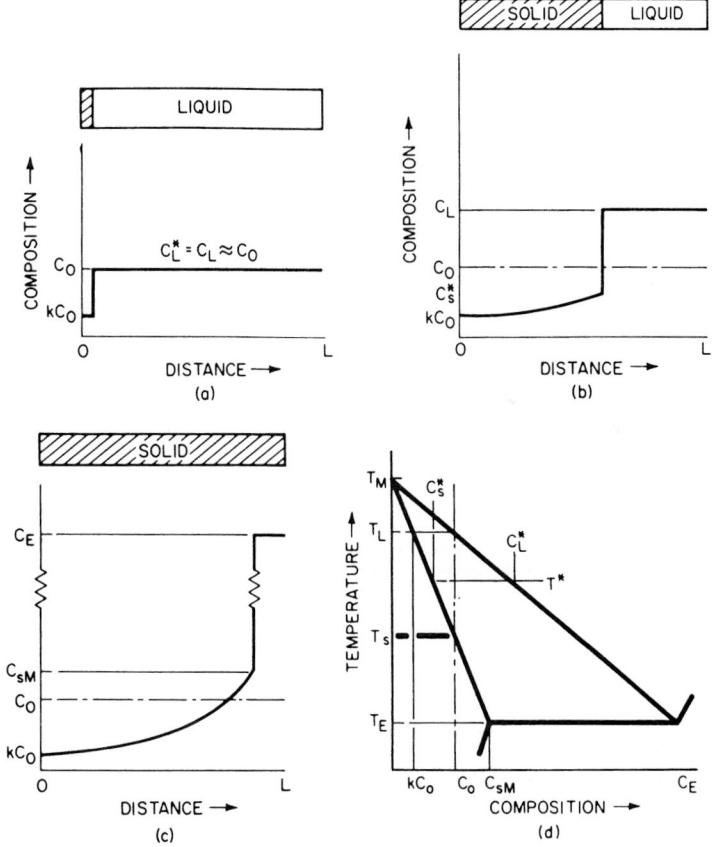

FIGURE 3.21   Solute redistribution in solidification with complete mixing (or diffusion) in the liquid and no solid diffusion. (a) At the beginning of solidification, (b) at temperature T^*, (c) after solidification; and (d) phase diagram.[7] (*Reprinted with permission from McGraw-Hill, Inc., New York.*)

$$\frac{dC_L}{C_L} = \frac{1-k_0}{1-f_S} df_S \tag{3.54}$$

which, on integration for constant k_0, gives the solute distribution in the liquid

$$C_L = C_0(1-f_S)^{k_0-1} = C_0 f_L^{k_0-1} \tag{3.55a}$$

or, for different densities of the liquid and solid ρ_L and ρ_S, respectively,

$$C_L = C_0\left(1 - \frac{\rho_S}{\rho_L} f_S\right)^{k_0-1} \tag{3.55b}$$

After the completion of solidification, the solute profile in the solid, as shown schematically in Figs. 3.19b and 3.21c, is given by the *normal freezing equation*

$$C_S = k_0 C_0(1-f_S)^{k_0-1} \tag{3.56}$$

where f_S is the solidified fraction. Equation (3.54) or (3.55) has been developed by Gulliver,[66] Hayes and Chipman,[67] Scheil,[68] and Pfann[69] and is called the *nonequilibrium lever rule*, or Scheil's law or equation, and (3.56) the normal freezing equation. The Scheil equation is applicable to growth of crystals as well as to describe microsegregation in dendritic solidification such as in ingot-casting practice.[70] In Sec. 3.6.1.5, the concentration profile of solute along the crystal for various values of k_0 is plotted in the normal freezing configuration (see Fig. 3.24);[71] for application purposes, a portion of the solidified product has to be discarded.[72]

It may be noted here that Eq. (3.56) cannot hold through the entire f_S range for $k_0 < 1$, since composition would approach infinity at high concentrations.[72] A complete mixing in the liquid implies $k_{eff} = k_0$ at very low solidification velocity, and there is no steady-state region of the curve.[66] Consequently, maximum segregation occurs in the solid, as shown in Fig. 3.19b.

These equations can only be considered a limited case of a more generalized experimental treatment which considers the effect of some limited liquid diffusion and incomplete liquid mixing. For alloys containing eutectic, some liquid composition will remain until a eutectic temperature is reached and the remainder liquid will solidify at eutectic composition. If k_0 is not constant, Eq. (3.54) can be numerically integrated. In this situation, the liquid is exhausted either when the eutectic is reached or when k_0 tends to unity.[7,73]

3.6.1.3 Limited Liquid Diffusion, No Convection.   Another practical important limiting case of normal freezing occurs when all the assumptions given in Sec. 3.6.1.2 apply except that mixing in the liquid is not complete, which implies that the diffusion is limited in the liquid and there is no convective transport. Under these conditions, steady-state diffusion-controlled planar solidification is approached if the bar is sufficiently long; this is sketched in Fig. 3.22. Here solidification of the melt starts with initial crystallization of solid of composition $k_0 C_0$. The solute rejected into the liquid is transported only by diffusion in the liquid, and thus an enriched *solute boundary layer* forms and progressively increases in solute concentration (Fig. 3.22a). At the steady state, solid forming and liquid have the same composition C_0. Equilibrium at the interface then requires the composition in the liquid at the S-L interface to be C_0/k_0 in the final transient stage and the solidification to occur at the solidus temperature T_S (Fig. 3.22b and c). During the steady-state region which

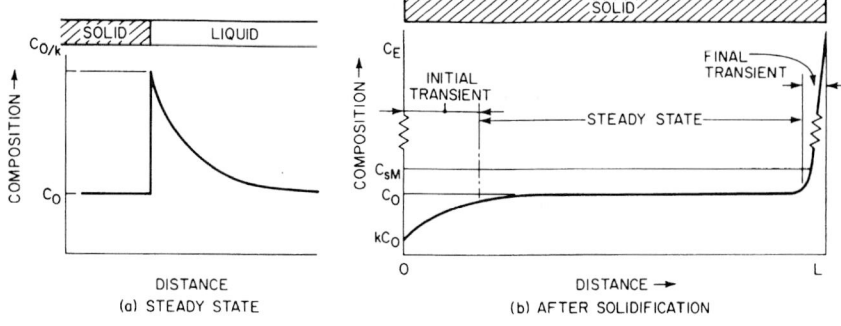

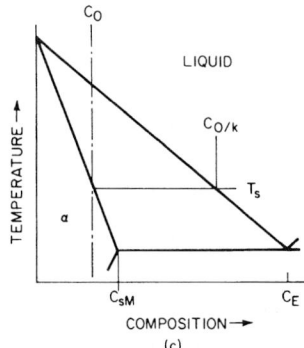

FIGURE 3.22 Solute redistribution in solidification with limited liquid diffusion and no convection: (*a*) composition profile during steady-state solidification, (*b*) composition profile after solidification, and (*c*) phase diagram.[7] (*Reprinted with permission from McGraw-Hill, Inc., New York.*)

occurs in the central portion of the bar, the sum of the amount flowing in and the amount flowing out must be zero, and the solute distribution is given by the differential equation

$$D_L \frac{\partial^2 C_L}{\partial x^2} + V \frac{\partial C_L}{\partial x} = 0 \tag{3.57}$$

where x is the distance from the interface, D_L is the diffusion coefficient in the liquid, C_L is the solute concentration in the liquid, and V is the velocity of the S-L interface. Also, the solute conservation requirement at the interface provides directly the composition gradient in the liquid at the interface:

$$\left(\frac{\partial C_L}{\partial x} \right)_{x=0} = -\frac{V}{D_L} C_L^* (1 - k_0) \tag{3.58}$$

Applying the correct boundary conditions at the steady state ($C_L = C_0/k_0$ at $x = 0$ and $C_L = C_0$ at $x = \infty$), the solution to Eq. (3.57), as developed by Tiller et al.,[61] for solute distribution in the liquid at any position x ahead of the S-L interface moving at a constant velocity V is given by

$$C_L = C_0 \left[1 + \frac{1-k_0}{k_0} \exp\left(-\frac{Vx}{D_L} \right) \right]$$ (3.59a)

The thickness of the solute-rich boundary layer is given by the characteristic distance D_L/V. Note that the liquid concentration is C_0/k_0 at the interface that produces a solid composition C_0. For small values of k_0, Eq. (3.58) reduces to

$$C_S = C_0 \left[1 - (1-k_0) \exp\left(-\frac{k_0 Vx}{D_L} \right) \right]$$ (3.59b)

Figures 3.19c and 3.22b show the final solute distribution along the rod after solidification.[4,7,73] In practice, the crystal conforming to the final transient has to be discarded to the very large concentration variation.[72]

3.6.1.4 Partial Liquid Mixing with Convection (Boundary Layer Approach).
The intermediate case (with boundary layer between zero and D_L/V), where the partial mixing is effected by single or combined effects of convection-free solute, thermal gradients in the liquid, and diffusion and forced convection by crystal rotation or electromagnetic stirring, strongly generates steady-state segregation. Numerous workers have extensively studied this subject.[71,74–79] According to the simple derivation made by Burton et al.,[79] an effective distribution coefficient is given by

$$k_{eff} = \frac{k_0}{k_0 + (1-k_0) \exp\left(-\dfrac{V\delta}{D_L} \right)}$$ (3.60)

The solute distribution (Fig. 3.19d) is then expressed by the normal segregation equation similar to Eqs. (3.55) and (3.56):

$$C_S = k_{eff} C_0 (1 - f_S)^{k_{eff}-1}$$ (3.61a)

and

$$C_L = C_0 f_L^{k_{eff}-1}$$ (3.61b)

Equation (3.60) is the famous Burton, Prim, and Schlichter (BPS) relationship. It shows that k_{eff} is a function of k_0 and dimensionless parameter $V\delta/D_L$. In this expression, the effect of convection is related to the boundary layer thickness δ. The stronger the convection is, the thinner the boundary layer becomes.[80] That is, for strong convection in the liquid, $\delta \to 0$, $k_{eff} \to k_0$, and Eq. (3.61a) reduces to the Scheil equation. For negligible convection, $\delta \to \infty$, $k_{eff} \to 1$, and Eq. (3.61a) gives a constant solute distribution profile.

The BPS boundary layer model is the improved version of pure diffusion transport models and gives some useful concepts. It has the following features:[76,79–81]

1. It deals only with the steady state.

2. It is only a one-dimensional approach and describes solute distribution along the growth direction (longitudinal segregation). Hence, it cannot estimate radial segregation due to lateral nonuniformities in the flow velocity and solute concentration which are present extensively in Czochralski, Bridgman, and zone refining crystal growth processes. It cannot explain variation in segregation

arising from the variation in flow intensity, other than to changes in the boundary layer thickness by convection.

3. It assumes no solute diffusion in the crystal.

4. It assumes no convective mixing inside the boundary layer, but a sudden, perfect mixing by convection outside the layer. In reality, the presence of the solute boundary layer results from the convection within the layer; otherwise the boundary layer will extend to infinity for pure diffusion.

5. The heat flow is only included indirectly by L-S interface velocity V.

Note that if the BPS analysis assumes velocity V_{eff} instead of $V\delta/D_L$, a lot of confusion could be avoided in terms of mathematical complexity.[72]

3.6.1.5 Zone Refining. Since zone refining (or melting) was first developed by Pfann[69] for more efficient purification process for high-purity metals and semiconductors, numerous attempts have been made to improve many aspects of this technique.[72,82–84] In this technique, a narrow zone of length l of the bar is melted, and this molten zone is moved from one end of the bar to the other, i.e., a single pass (Fig. 3.23a). The molten zone can be passed across the length of the bar numerous times in order to produce extremely high-purity crystals. The important variables in the zone refining process are as follows: (1) zone length, (2) specimen (or charge) length, (3) initial distribution of solute in the charge, (4) vapor pressure, (5) zone travel rate (constant or variable) (the higher travel rates permit more passes per unit time, but

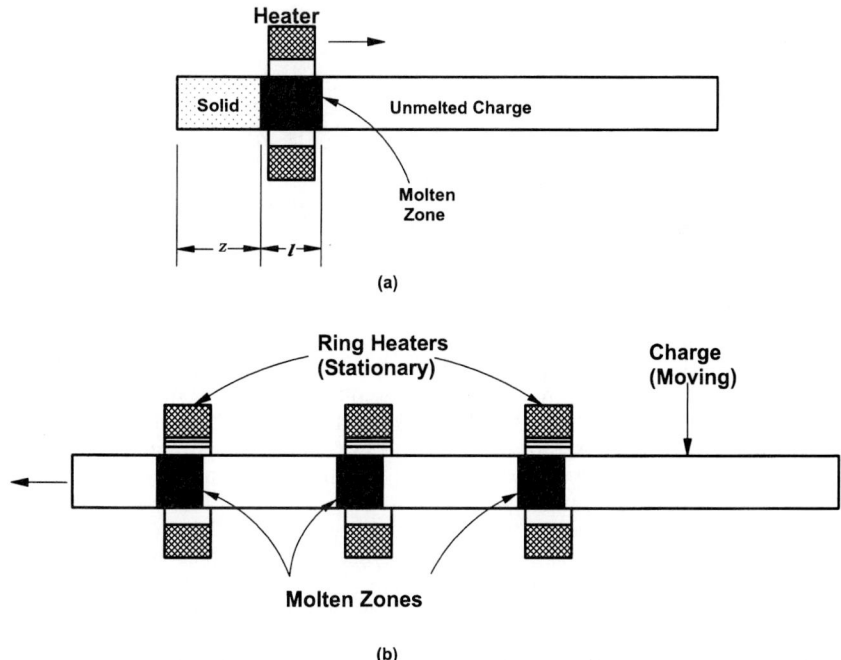

FIGURE 3.23 Zone refinement process based on (*a*) movement of a molten zone along a bar, (*b*) three molten zones traveling along an ingot.[4]

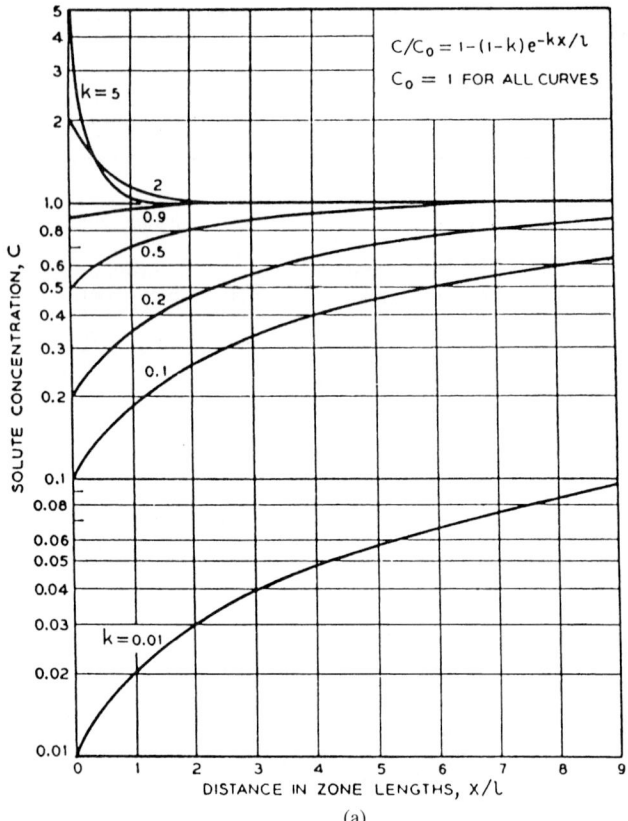

$$C/C_0 = 1 - (1-k)e^{-kx/l}$$

$C_0 = 1$ FOR ALL CURVES

(a)

FIGURE 3.24 (a) Solute distribution after one pass for various values of k_0 and (b) relative solute distributions after several passes with complete mixing in the zone.[69,82,89] (*Courtesy of John Wiley & Sons, New York.*)

low travel rates provide more advantageous k_{eff} values), and (6) k_{eff} values, which can be varied, for instance, by stirring the molten zone or imposing an electric field across it, resulting in electrodiffusion. The zone length of the molten zone relative to the length of the specimen affects both the ultimate solute distribution and the rate at which it is reached. Burris et al. suggested a long zone for the early passes, followed by a shorter length for the latter passes for effective purification.[83] On the other hand, Pfann suggested a longer zone in the first portion of the ingot, followed by a shorter zone in the second portion.

The molten zone can be produced by resistance element, high-frequency coil, or electron beam. The advantage of the process lies in the fact that the solidification and purification process can be repeated many times until an ultimate distribution is reached. Figure 3.23b shows a schematic view of a more efficient multipass zone refining device.

In the zone melting process, which is nonconservative, the transport of a molten zone along a charge of initially uniform composition C_0 leads to a distribution of the form

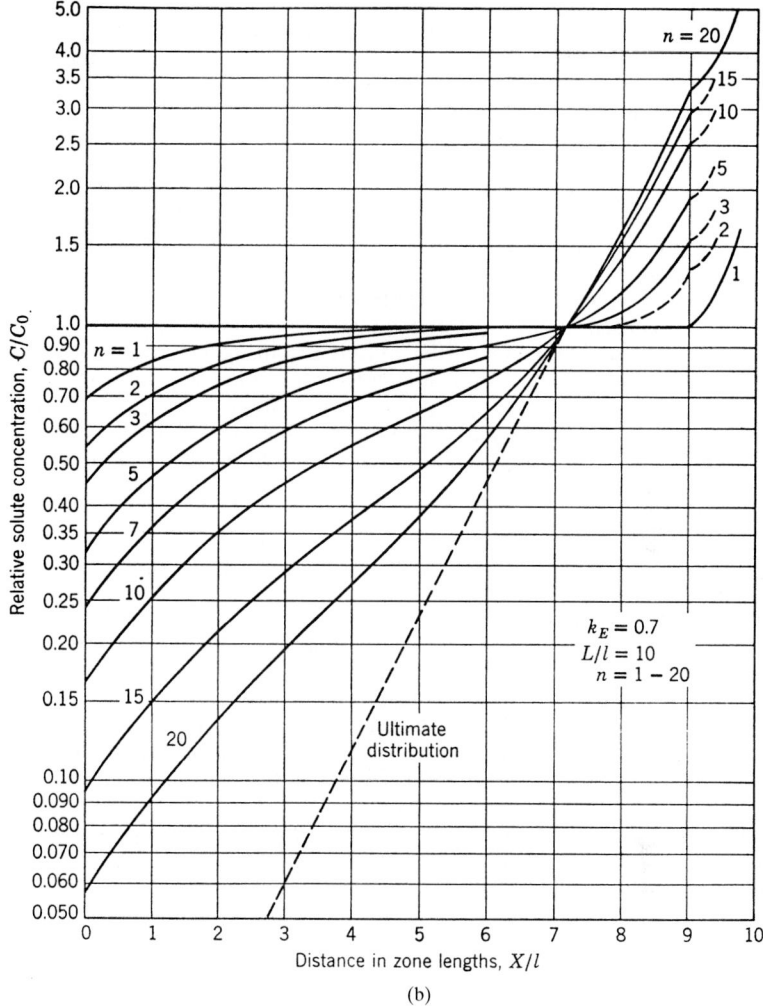

FIGURE 3.24   (*continued*).

$$C = C_S = C_0\left[1 - (1 - k_0)\exp\left(-\frac{k_{0}x}{l}\right)\right]$$ (3.62)

where $C = C_S$ is the solute concentration in the solid at any point x, C_0 is the original concentration, x is the distance from the starting end, and l is the zone length parallel to the direction of zone travel (x axis). Equation (3.62), often called *Pfann's law*, ceases to hold when the melt reaches the end of the ingot, since the subsequent solidification p occurs by normal freezing.

Figure 3.24a shows the solute distribution after one pass for various values of k_0 whereas Fig. 3.24b shows the progressive purification after n passes. It is clear that the smaller the distribution coefficient, the more efficient the purification. The closer

k_{eff} is to unity, the more passes (n) must be carried out to reach a certain degree of purification. If $k_0 > 1$, there is usually a solute enrichment at the front end of the specimen. The zone refining technique performed without a crucible on a vertical rod, which relies on the surface tension to support the molten zone, is of great significance in the production of pure semiconductors.

3.6.2 Lateral Segregation

The segregation in the lateral direction cannot be treated in a unidirectional manner, because of (1) the unavoidable presence of slightly curved S-L interface during steady-state unidirectional solidification; (2) the difference between the apparent distribution coefficients at the center and at the interface during the progress of solidification with a partially faceted front; and (3) a significant role of convective flow. The two-dimensional lateral segregation treatments present some difficulties in that such results do not give the information as to whether axially symmetric descriptions indicate reality. It is thus essential to study, first experimentally and then theoretically, what type of flow geometries appears under different thermal boundary conditions.[72,81]

3.6.3 Morphological Stability of a Planar Solidification Front

So far the solidification front of a single phase and a eutectic alloy has been assumed to be a microscopically planar S-L interface, coinciding with the solute redistribution during unidirectional solidification. However, actually, preferential rejection or incorporation of solute and local temperature fluctuation, mechanical shocks, and convection within the melt occur at the interface and may produce small morphological perturbations of the interface. This morphological instability of the front interface influences and leads to the development and pattern formation of cellular and dendritic growth. Thus morphological stability theory deals with the conditions under which the smooth and planar S-L interface becomes unstable. Such morphologies along with the associated microsegregation of solute or impurities determine the microstructure and defect structure of the growing phase. There are many reviews of the morphological stability; some of the recent ones are given.[1,55,85–87]

3.6.3.1 Constitutional Supercooling (CS). A thermodynamic criterion for interface instability on the principle of constitutional supercooling (Fig. 3.25) was proposed by Rutter and Chalmers,[88] Tiller et al.,[61] and Winegard.[89] This criterion considers steady-state solute distribution in the liquid C_L at any position x in front of an S-L interface (Fig. 3.25a), as given by Eq. (3.59), and the corresponding equilibrium liquidus temperature T_L (Fig. 3.25b), which is given by

$$T_L = T_m - mC_L = T_m - mC_0\left[1 + \frac{1-k_0}{k_0}\exp\left(-\frac{V_X}{D_L}\right)\right] \qquad (3.63)$$

where m is the slope of the liquidus line corresponding to the solidifying alloy and T_m is the melting temperature of the pure metal. If we assume that freezing occurs only at an advancing S-L interface, the interface temperature T_I for the flat front, i.e., for concentration C_0/k_0, is

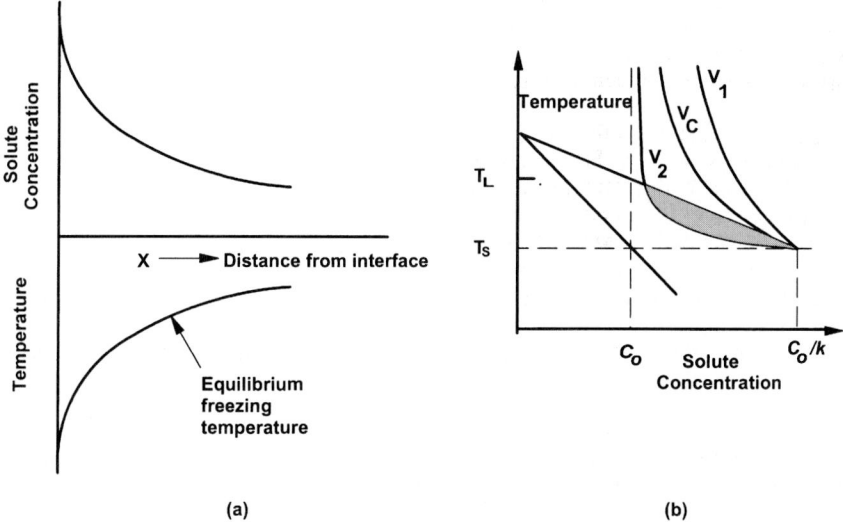

(a)   **(b)**

FIGURE 3.25 (*a*) Solute concentration C_L ahead of the L-S interface and the corresponding liquidus temperature T_L. (*b*) Building of constitutional supercooling showing actual temperature profile where part of the liquid ahead of the S-L interface is below its actual or normal freezing as the growth velocity is raised above the critical velocity V_C ($V_2 > V_C > V_1$).

$$T_I = T_0 - \frac{mC_0}{k_0} \tag{3.64}$$

The actual temperature profile T in the liquid is considered linear to a good approximation and is given by

$$T = T_I + G_L x \tag{3.65}$$

where G_L is the temperature gradient in the liquid ahead of the S-L interface. Combining Eqs. (3.64) and (3.65) yields

$$T = T_0 - \frac{mC_0}{k_0} + G_L x \tag{3.66}$$

When this temperature is superimposed on the curve for equilibrium freezing temperature, as in Fig. 3.25*c*, it is seen that the temperature and composition fields in the melt lie below the local liquidus temperature.[20] This is called *constitutional supercooling* because it develops by combined solute and temperature distribution in the liquid ahead of an advancing L-S interface, regardless of whether $k_0 > 1$ or $k_0 < 1$.

For onset of constitutional supercooling

$$\frac{G_L}{V} \leq m \frac{C_0}{D_L} \frac{1 - k_0}{k_0} \tag{3.67a}$$

or

$$\frac{G_L}{V} \leq m \frac{\Delta T_0}{D_L} \tag{3.67b}$$

or

$$G_L \leq m G_c \tag{3.67c}$$

where $\Delta T_0 = T_S - T_L$ = freezing range of the alloy = $mC_0(1 - k_0)/k_0$ and $G_c = C_0[(1 - k_0)/k_0]$ (V/D_L). According to the nature of Eq. (3.67), above a critical value of G_L/V, the S-L interface is stable, i.e., planar; below the critical value, instability occurs and the morphology changes. A planar interface under CS conditions becomes unstable because any bulge forming on the interface leads to transverse diffusion of solute atoms, their accumulation, and a localized decrease in the liquid temperature. This, in turn, will break up or result in the formation of cellular or dendritic microstructure.[90] Figure 3.26a is a schematic plot of the relation between G_L and V, called $G_L V$ *space*, showing its division by lines into different G_L/V values to differentiate between planar to cellular instability and cellular to dendritic instability.[91] Cellular and dendritic instabilities are shown in Fig. 3.26b. Equation (3.67) is applicable to both the presence and absence of convection because of the formation of laminar layer near the solidifying interface in every case.[7] Both theoretical and experimental results support the CS theory of predicting conditions of breakdown of the planar front in metals and semiconductors solidifying with a diffuse interface.

3.6.3.2 Mullins and Sekerka (MS) Instability Theory. The CS criterion has resulted in better understanding of many crystal growth, solidification, and casting processes. However, this concept does not yield any information about the size scale of instability. Also, for some solidification processes such as rapid solidification, the CS criterion can lead to erroneous results because this criterion ignores the temperature gradient in the solid and the S-L interfacial energy. In these situations, the theory of morphological instability was developed and analyzed by Mullins and Sekerka,[92] Sekerka,[93,94] and others[95,96] to fill in these gaps. They assumed a local equilibrium at the S-L interface, isotropy of the S-L surface tension (which is a very good approximation for many metals at low growth velocities), and initial occurrence of a very small amplitude of sinusoidally perturbed S-L interface with isotropic interface plane (x, y) moving in the pure melt in the z direction (Fig. 3.27), which is expressed by

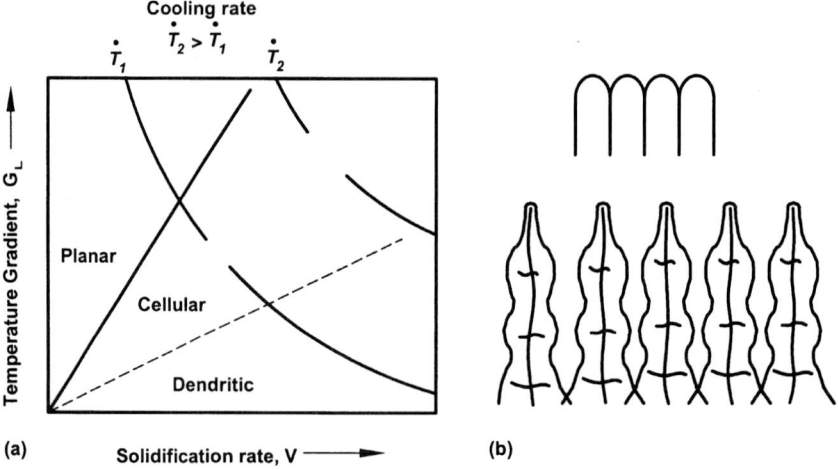

FIGURE 3.26 (*a*) Diagram of $G_L V$ space showing regions of cellular and dendritic growth as a function of temperature gradients and solidification rates. (*b*) Cellular and dendrite instability.

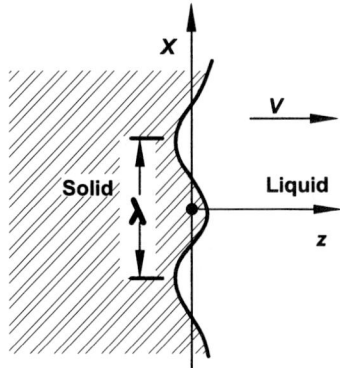

FIGURE 3.27 Sinusoidally perturbed S-L interface as observed from a reference frame that advances along with the unperturbed interface ($z = 0$).[94] (*Reprinted by permission of Pergamon Press, Plc.*)

$$z(x,t) = \delta(t)\cos\omega x = \delta(t)\exp(\sigma t + i\omega x) \qquad (3.68)$$

where $\delta(t)$ is the amplitude of a Fourier component of the interface shape, ω is the wave number of perturbation $= 2\pi/\lambda$, λ is the wavelength, and $\sigma\ (= \sigma_r + i\sigma_i)$ is the growth (or decay) rate of perturbation represented by a complex quantity. If σ_r is positive or negative, respectively, for all values of ω, the interface becomes unstable or stable with respect to perturbation of all Fourier components. Using the appropriate equations and boundary conditions, a relationship between σ and ω, called the *dispersion relation*, is required to determine the stability limit of the system and, therefore, derivation of the critical wavelength at the threshold $\lambda^* = 2\pi/\omega^*$. This is the stability exchange principle which is given by $\sigma_r(\omega) < 0; [\sigma_r(\omega)]_{max.} = 0$, and $\sigma_i(\omega) = 0$.[59]

The dispersion relation, calculated by the these authors by assuming the thermal conductivity weighted temperature gradient (G) on the liquid side and a (Sekerka's) stability function $S(\omega)$ or $S(A, k_0)$, dependent on the capillarity parameter A and k_0, is expressed by

$$\sigma(\omega) = \left[\frac{K_S G_S + K_L G_L}{K_S + K_L} + \frac{D_L}{V}S(A, k_0)\right]\frac{1}{\Delta T_0} \qquad (3.69)$$

$$A = \Gamma\left(\frac{T_m k_0}{mG_c}\right)\left(\frac{V}{D_L}\right)^2 = \Gamma\left(\frac{k_0^2}{1-k_0}\right)\left(\frac{T_m}{mC_0}\right)\left(\frac{V}{D_L}\right) \qquad (3.70)$$

$$G = \frac{K_S G_S + K_L G_L}{K_S + K_L} \qquad (3.71)$$

where K_S and K_L are the thermal conductivities of the solid and liquid, respectively, Γ is the Gibbs-Thompson parameter, and T_m is the melting point of pure metal. The final analysis of Eq. (3.69) leads to the instability criterion if

$$G < mG_c S(A, k_0) \qquad (3.72a)$$

or the stability criterion if

$$G > mG_c S(A, k_0) \qquad (3.72b)$$

The instability of the planar interface for dilute Al-0.1 wt% Cu alloy has been found to exist over a range between two critical velocities. Beyond this range, the growth (or decay) rate of perturbation $\sigma < 0$ for all wavelengths λ, and the interface remains always stable.[91]

The upper and lower critical velocities may be approached closely by two different limiting cases:

1. For normal solidification rates, i.e., at low growth rates of solidification (or interface velocity), the values of the capillary forces (or interfacial tensions) are so small that $A \ll 1$ and $S \approx 1$. In this case Eq. (3.72a) reduces to the instability criterion as

$$G \le mG_c \tag{3.73}$$

which is termed the *modified constitutional supercooling criterion*. If $K_S = K_L$, then G_L replaces G and Eq. (3.72) reduces to the constitutional supercooling criterion as given by Eq. (3.67).

2. At rapid rates of solidification, S decreases and A increases, which make stability easier for a given value of G in Eq (3.72b). For $A \ge 1$ and $S = 0$, the interface becomes stable for all values of G; this is called the *absolute stability phenomenon*. In this situation, stabilization due to the temperature gradient becomes negligibly small, and one finds the absolute stability condition as

$$mG_c \le k_0 T_m \Gamma \left(\frac{V}{D_L} \right)^2 \tag{3.74}$$

or

$$V \ge \frac{mC_0(k_0 - 1)D_L}{k_0^2 T_m \Gamma} \tag{3.75a}$$

$$\ge \frac{\Delta T_0 D_L}{k_0 \Gamma} \tag{3.75b}$$

The modified CS criterion and the absolute stability criterion act as asymptotes to the exact result at low and high velocities, respectively, as shown in Fig. 3.28.[97] Theoretically, the perturbation analysis appears to be more exact than constitutional supercooling. It has been shown again experimentally in unidirectional Al-0.1 wt%

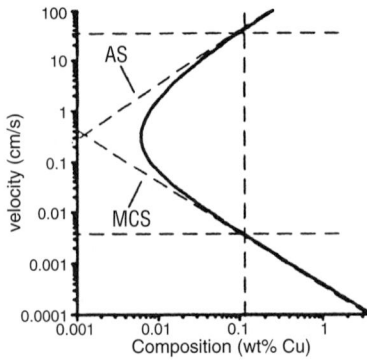

FIGURE 3.28 Critical velocity versus G_L for a fixed C_0 (0.1 wt% Cu), showing the modified CS and absolute stability criteria at low and high velocities.[4,97] (*Reprinted by permission of Elsevier Science B.V.*)

Cu alloy solidification front that the S-L interface becomes unstable at the conditions predicted by the perturbation analysis rather than by the constitutional supercooling theory, if $S \approx 1$.[98] This absolute stability criterion can be easily applied to rapid solidification conditions. Another effect of this criterion is the progressive variation of the k_0 value at the operative solidification front toward unity, called *solute trapping*, which is described elsewhere.[99,100]

3.7 CELLULAR AND DENDRITIC GROWTH

It has been established that when the degree of morphological instability is increased by varying the growth rate, one observes three regions of nonplanar solidification fronts. The first one comprises cell structure in which cooperative phenomena in the array are necessary. The second one is the dendritic structure, which evolves under two different (e.g., free and constrained or directional solidification) growth conditions according to the different mode of dissipation of latent heat of fusion away from the interface. The third one is the rapid solidification microstructures with special features where cells and dendrites may still be present.[101-103]

An understanding of cellular and dendritic growth is of great practical importance in (1) predicting intercellular and interdendritic spacings which, in turn, facilitates the determination of heat treatment cycles for the homogenization of ingots and mechanical properties of the as-cast materials, and (2) control of practical casting defects such as segregation, porosity, and hot tearing.[4]

This section deals with the cellular and dendritic growths, cell and dendritic spacings, and cell-dendrite transition.

3.7.1 Cellular Growth

Cellular growth is an important periodic pattern formation process of the S-L interface with strongly interacting tips, leading to a nonparabolic form of the interface and missing sidebranches. With this characteristic feature, when original melt composition C_0 and temperature gradient G_L are fixed, and growth velocity V is increased from a small threshold to a very large value, transitions occur from planar to cellular, from cellular to dendritic, from dendritic to cellular, and back from cellular to planar.[104] Thus cellular structures have been found to be stable at low and high L-S interface velocity, as shown recently by numerical analysis of Lu and Hunt.[105] The temperature gradient effect at low velocities and the capillarity effect at high velocities are responsible for the formation of cellular structure.

The origin of instability and evolution of morphologies during transition from initially smooth interface to cellular structure at low velocities have been studied by means of pulling off and autoradiography, decanting technique, or elaborate transverse sectioning metallography in Pb-Sn, Sn-Pb, Zn-Cd, and Al-Cu alloys.[106,107] Cellular structure at low velocities is divided into four subregions. Beginning with the threshold, these regions represent ordered microdepressions, pock marks, or nodes (Fig. 3.29a); rounded depletions of the S-L interface that can be either randomly or regularly distributed, resulting in the formation of grooves (Fig. 3.29b); elongated cells forming rambling bands (Fig. 3.29c); and regular or hexagonal cells when a honeycomb is formed (Fig. 3.29d). The alloy crystallography and crystal orientation exert profound effects on the morphologies of the interface after the breakdown of the planar interface.

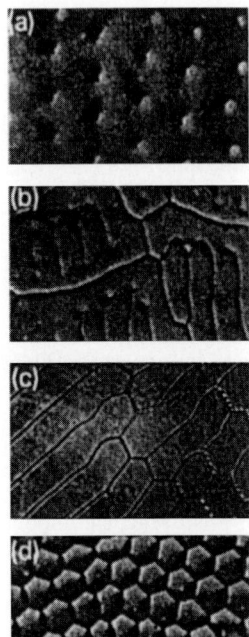

FIGURE 3.29 The succession of patterns usually found in 3-D samples when going above the threshold of morphological instability of planar S-L interfaces: (*a*) poxes, (*b*) irregular cells, (*c*) elongated cells, and (*d*) hexagonal cells in Sn-0.12 at % Pb.[105] (*Courtesy of Y. Malméjac.*)

Nonfaceted cellular growth, which occurs in the nonfaceting crystals, such as mostly metals and organic compounds, with relatively low entropy of fusion has been studied extensively.[108] The characteristic features of faceted cellular structure have been described first by Bardsley et al.[109] for Czochralski growth of Ga-doped Ge with several orientations. Shangguan and Hunt have reported occurrence of faceted cellular array growth in transparent organic compounds such as salol, thymol, and O-therphenyl.[18,108]

The most currently used models to interpret experimental data of steady-state cellular growth include the Bower-Brody-Flemings model,[110] its improvement by Hunt,[111] and the viscous-fingering approach of Pelce and Pumir,[112] which are not discussed here.

3.7.2 Dendrite Growth

The term *dendrite* derives from the Greek word *dendron*, which means tree; and like a tree, the dendrite is a branched structure with primary, secondary, tertiary, and higher-order branches.[113] Dendritic growth, which has attracted artists, physicists, engineers, and applied mathematicians alike, is the most frequently elegant form of crystallization process in nature and technology, where an instability of a simple system produces complex and highly structured patterns and the voids between the dendrite branches are filled by eutectic or intermetallic phases.[113–115] The development of various-length scales or spacings (e.g., dendrite tip radius R, the primary spacing λ_1, and secondary arm spacing λ_2) are controlled by the variation in growth conditions. These interarm spacings characterize the solidification morphology which constitutes the solute segregation pattern and, often, the formation of a second phase (e.g., precipitates or pores) in the interdendritic region and plays a larger role in the determination of as-cast mechanical properties which largely control the material.[1,7] Dendritic growth is commonly observed in metal ingots, alloy castings, and weldments, when metals and alloys solidify under small thermal or concentration gradients.[116] Reviews of dendritic growth of pure metals and alloy materials are given by Glicksman and Marsh[114] and other workers[116a,116b], and that of alloy dendrite growth is given by Trivedi and Kurz.[113]

In metallic alloys the coupled growth of two or more phases may also take place. Mostly dendritic growth tends to involve the coupling of two important processes: (1) steady-state propagation of the tip region, elucidating the formation of the main or primary, and (2) the time-dependent crystallization of the secondary and tertiary sidebranches, a process that constitutes the length scales of a dendrite, as shown in Fig. 3.30. Dendritic structures that develop under two different growth conditions are called *free growth* and *directional solidification* or *constrained growth*. In the case of dendritic growth from an undercooled melt, dendritic characteristics are controlled by (dendritic tip) composition, bath undercooling, and thermal diffusion in the liquid. The general characteristics of dendrite growth conditions are as follows:[101,113,114]

1. In free dendritic growth condition from the undercooled melt, mostly equiaxed dendritic crystals form, the temperature gradient in the liquid at the interface is negative and that in the solid is nearly zero, and the dendritic growth and heat

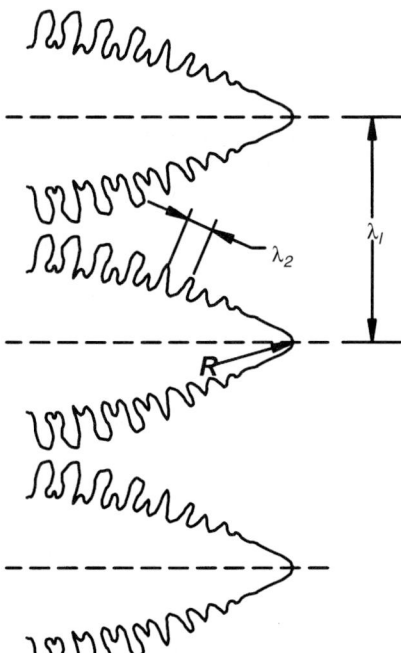

FIGURE 3.30 Two-dimensional representation of a dendritic structure.

flow directions are parallel for solidification. The presence of other dendrites in the adjacent grains significantly influences the thermal and solute fields ahead of any given dendrite.

2. In pure materials, thermal dendrites have smooth paraboloidal tip region and the presence of enthalpy microsegregation during crystal growth. After the completion of solidification, all vestiges of the prior dendritic structure vanish.

3. In pure materials, the growth rate of dendrite is controlled by the diffusion or dissipation of latent heat ahead of the interface through the supercooled liquid.

4. In alloys, free growth of dendrites also segregates solute and impurity contents. Furthermore, the latent heat of fusion is released away from the dendritic interface by transport processes such as thermal conduction or convection. Chemical diffusion, being normally slower than thermal transport, is often the rate-controlling process in alloy dendrites. The presence of the thermal and solutal diffusion fields in alloys and the temperature dependence of the equilibrium phase concentrations rather complicate the analysis of the process.

5. In directional solidification or constrained growth condition, mainly columnar dendritic crystals form, the dissipation of latent heat of fusion is through the solidified mass (and cold mold wall) to the external environment, the imposed temperature gradient in the liquid is positive, and dendritic growth and heat flow directions are antiparallel.

6. Usually in the final stages of solidification of a certain casting, freely growing alloy equiaxed dendrite crystals can occur due to the isometric development within the casting. Their formation, although technically important, is inadequately understood, and control of transition from columnar to equiaxed (CTE) dendrite is of great interest in shaped and continuous casting of alloys.[117]

3.7.2.1 Dendritic Growth Models in Alloy Melt. The growth behavior of the dendritic tip is important for microstructure formation because it allows one to determine the tip radius and tip composition as a function of supercooling or imposed velocity. This also enables one to estimate microstructural transition and to set up the microstructure selection maps. Dendritic growth models assume steady-state solutions of the thermal and solute diffusion processes which restrain the transport of latent heat of fusion and solute away from the interface.

Dendritic Growth in Undercooled Melt (Free Growth). Ivantsov[119] derived the relation of growth rate to the supercooling by assuming that the dendritic tip interface was nearly isothermal and isoconcentration in the tip region and the steady-state shape of the interface could be approximated to the constant shape of a paraboloid of revolution. The Ivantsov solution does not consider a unique dynamic operating state, rather the permissible combination of velocity and length scales which fulfill the steady-state equation, expressed in terms of the Peclet number.[120] For supercooled melt, Ivantsov analyzed the relation between dendritic growth rate and supercooling by assuming a supercooled alloy with initial composition C_0 (C_∞) and initial bath temperature (far away from the interface) T_∞. The Ivantsov solution for a paraboloid needle with constant interface (tip) composition C_t and isothermal interface (tip) temperature T_t, using solute and thermal flux at the tip interface, gives

$$C_t - C_0 = C_t(1 - k_0)\text{Iv}(P) \tag{3.76a}$$

or
$$C_t = \frac{C_0}{1 - (1 - k_0)\text{Iv}(P)} \tag{3.76b}$$

or
$$\Omega = \text{Iv}(P) \tag{3.76c}$$

and
$$T_t - T_\infty = \frac{\Delta H_m}{c_L}\text{Iv}(P_t) \tag{3.77a}$$

or
$$\Delta\theta = (T_t - T_\infty)\left(\frac{c_L}{\Delta H_m}\right) = \frac{c_L}{\Delta H_m}\Delta T = \text{Iv}(P_t) \tag{3.77b}$$

where $\text{Iv}(P) = P\exp(P)E_1(P)$, $E_1(P)$ is the first exponential integral function, $P = VR/2D_L$ and is the solute Peclet number of the dendrite tip, Ω is the dimensionless supersaturation, c_L and ΔH_m ($= L$) are the specific heat in the liquid and enthalpy (or latent heat) of fusion per unit volume, respectively, $\Delta\theta$ is the dimensionless supercooling (or Stefan number), $\text{Iv}(P_t) = P_t\exp(P_t)E_1(P_t)$, $P_t = VR/2\alpha_L$ and is the thermal (or growth) Peclet number, and α_L is the thermal diffusivity in liquid.

When capillarity and interface kinetics are absent, and for linear liquidus and solidus lines of the phase diagram, a relationship involving the dendrite tip temperature and composition is given by

$$T_t = T_m + mC_t \tag{3.78}$$

or
$$\Delta T = (T_t - T_\infty) - m(C_t - C_0) \tag{3.79}$$

where ΔT is the total bath undercooling and $\Delta T_t = T_m + mC_0 - T_\infty$. Replacing values of T_t from Eq. (3.77a) and C_t from Eq. (3.76b), we get

$$\Delta T = \frac{\Delta H_m}{c_L}\text{Iv}(P_t) + \frac{k_0\Delta T_0\text{Iv}(P)}{1 - (1 - k_0)\text{Iv}(P)} \tag{3.80}$$

For dendritic growth in a pure metal, the second term in Eq. (3.80) goes to zero since $\Delta T_0 = 0$. If capillary effects at the dendrite tip are assumed, then Eq. (3.79), which relates the interface temperature and interface composition, is modified. In this case, the modified Ivantsov result for the paraboloid needle is obtained.

$$\Delta T = \frac{\Delta H_m}{c_L}\text{Iv}(P_t) + \frac{k_0\Delta T_0\text{Iv}(P)}{1 - (1 - k_0)\text{Iv}(P)} + \frac{2\Gamma}{R} \tag{3.81}$$

Equations (3.80) and (3.81) show that the total bath undercooling consists of two and three terms, respectively. The first term yields the supercooling for thermal diffusion process ΔT_T, while the second term yields dendrite tip supercooling for solute diffusion ΔT_S. The last term in Eq. (3.81) is due to the capillarity effect ΔT_R. Note that for alloy systems as $k_0 \to 0$, $\Delta T_0 \to \infty$, but the product $k_0\Delta T_0$ becomes finite and is equal to mC_0. The above Ivantsov solutions hold only for isothermal interface and isotropic interface energy consideration. Other approximate solutions for non-isothermal interfaces have been developed by Temkin[121] and Trivedi.[122]

Figure 3.31 shows the relationship between velocity and tip radius for fixed undercooling in pure succinonitrile, illustrating Ivantsov solution and its modification due to capillary effects.[123] As mentioned before, the capillary term for metals

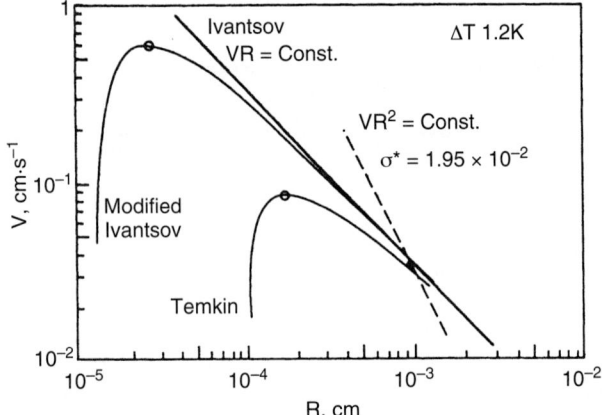

FIGURE 3.31 Relationship between the growth velocity V (log scale) and tip radius R (log scale) at a fixed undercooling in pure succinonitrile dendrite growth, illustrating Ivantsov solution (which is characterized by VR = constant). The nonisothermal solutions (Temkin and modified Ivantsov) show the gradual departure from the line VR = constant as the tip radius reduces. The broken curve, VR^2 = constant, corresponds to the conditions of marginal morphological stability with a separation constant $\sigma^* = 1.95 \times 10^{-2}$. (*Courtesy of M. E. Glicksman.*)

at low undercooling values is very small and, therefore, can be neglected from the calculations of thermal and solute fields; however, its effects remain important at rapid solidification conditions.

DENDRITE TIP UNDERCOOLING. Note that the second and third terms in Eq. (3.81) yield the dendrite tip undercooling ΔT_t so that $\Delta T = \Delta T_T + \Delta T_t (= \Delta T_S + \Delta T_R)$ (Fig. 3.32). That is, under the assumption of steady-state growth with a parabolic interface,

$$\Delta T_t = \frac{k_0 \, \Delta T_0 \, \mathrm{Iv}(P)}{1 - (1-k_0)\mathrm{Iv}(P)} + \frac{2\Gamma}{R} \qquad (3.82)$$

DENDRITE TIP RADIUS SELECTION CRITERION. A general expression for the dendrite tip radius R for an alloy can be written in terms of the corresponding temperature gradients relating the driving forces for stabilization/destabilization

$$mG_c^* \xi_c - G' = \frac{\Gamma}{\sigma^* R^2} \qquad (3.83)$$

where
$$G' = \frac{K_S G_S \xi_S + K_L G_L \xi_L}{K_S + K_L} \qquad (3.84)$$

Here G_c^* is the concentration gradient at the dendrite tip given by Eq. (3.84); ζ_c, ζ_S, and ζ_L are functions of P; and σ^* is the dendritic tip selection parameter, assumed to be constant. The terms in Eq. (3.82) relate the effective liquidus temperature gradient $mG_c^* \zeta_c$, the effective temperature gradient at the interface G', and the equilibrium temperature gradient due to curvature of the tip Γ/R^2.

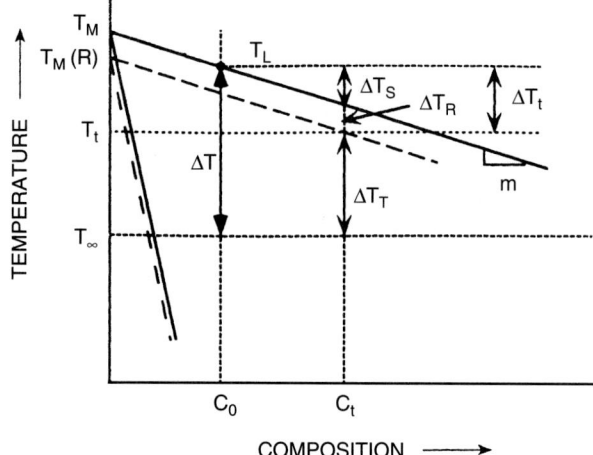

FIGURE 3.32 Division of bath undercooling ΔT in terms of undercoolings attributed to solute diffusion ΔT_S, thermal diffusion ΔT_T, and capillary effects ΔT_R for dendritic growth in undercooled melt; ΔT_t is the tip undercooling.[113] (*Courtesy of The Institute of Materials, London.*)

For dendritic growth, $mG_c^* \zeta_c$ in Eq. (3.83) is always positive because of the same signs of m and G_c^*, i.e., negative for $k_0 < 1$ and positive for $k_0 > 1$. The temperature gradient term G_L is always positive for an undercooled melt because of negative temperature gradient in the liquid. Hence, the driving force for destabilization, which is the difference between the two gradients in Eq. (3.83), representing the solute and thermal effects, is always positive in an undercooled alloy melt. This destabilizing effect is balanced by the stabilizing effect of the capillary term on the right-hand side.

For an Ivantsov dendrite which is assumed to be isothermal, the temperature gradient in the solid is zero, so that the effective temperature gradient reduces to $G' = [(K_S G_S + K_L G_L)/(K_S + K_L)]\zeta_L$. The values of composition gradient in the liquid at the tip G_c^* and the thermal gradient at the tip G_T^* are given from solute and thermal flux balance at the tip interface as

$$G_c^* = -\frac{V}{D_L}(1-k_0)C_t \tag{3.85}$$

$$G_T^* = -\left(\frac{V}{\alpha_L}\right)\left(\frac{\Delta H_m}{c_L}\right) = -\frac{2}{R}\frac{\Delta H_m}{c_L}P_t \tag{3.86}$$

Combining Eqs. (3.85), (3.86), and (3.83), one finds a general tip radius selection criterion in an undercooled alloy melt as

$$VR^2\left(\frac{k_0\Delta T_0}{\Gamma D_L}\right)\left(\frac{C_t}{C_0}\right)\xi_c + VR^2\left(\frac{\dfrac{\Delta H_m}{c_L}}{2\Gamma\alpha_L\beta}\right)\xi_L = \frac{1}{\sigma^*} \tag{3.87}$$

where $\beta = 0.5\,(1 + K_S/K_L)$, $K_L = \alpha_L c_L$, and $\Delta T_0 = mC_0(k_0 - 1)/k_0$. Note that $\beta = 1$ when $K_S = K_L$. The value of the composition at the tip is given by Eq (3.76b). Equations

(3.81) and (3.87) now explain completely the growth problem in an undercooled alloy. Equation (3.87) shows that the two terms on the left-hand side are quite similar. They differ only by the factor 2β due to the one-sided diffusion problem, that is, $D_S \ll D_L$ or the equivalent value of $2\beta = 1$ for the diffusion situation. On the other hand, the heat diffusion is a two-sided problem and becomes a symmetric problem when $K_S = K_L$ or $\beta = 1$.

For the Ivantsov model, the dendrite tip selection criterion in alloy melt can be simplified by combining Eqs. (3.81) and (3.87) into the following form:

$$R = \frac{\dfrac{\Gamma}{\sigma^*}}{\dfrac{2Pk_0\,\Delta T_0 \xi_c}{1-(1-k_0)\mathrm{Iv}(P)} + \dfrac{P_t \Delta H_m \xi_L}{c_L \beta}} \tag{3.88}$$

For a particular value of P or P_t, the value of R is known from Eq. (3.88) and that of V is determined from the definition of P_t as

$$V = \frac{2\alpha_L P_t}{R} \tag{3.89}$$

By using the values of P, R, and V, the corresponding dendrite tip composition and undercooling value are calculated from Eqs. (3.76b) and (3.81).

Dendritic Growth in Directional Solidification (Constrained Growth).   In directional solidification, an alloy of fixed composition is solidified at a desired velocity under the conditions of a constant positive temperature gradient in the liquid, and the dendrite tip temperature is mainly controlled by solute diffusion. However, the temperature gradient effect becomes important only at very low velocities where $V \le V_c$, and at very high velocities where $V \ge V_c$, where the latent heat contribution is substantial.

DENDRITE TIP UNDERCOOLING.   In directional solidification, the diffusion field of the solute is controlled by the Ivantsov equation (3.76a) so that the dendrite tip undercooling ΔT_t is given by

$$\Delta T_t = \frac{k_0\,\Delta T_0\,\mathrm{Iv}(P)}{1-(1-k_0)\mathrm{Iv}(P)} + \frac{2\Gamma}{R} \tag{3.90}$$

Comparing Eqs. (3.81) and (3.90), one finds a difference only in the thermal diffusion term in Eq. (3.81). The dendrite tip undercooling ΔT_t for both free and constrained growth as given by Eqs. (3.82) and (3.90) is the same, though for free growth one normally relates the velocity with bath undercooling ΔT rather than tip undercooling.

DENDRITE TIP RADIUS SELECTION CRITERION.   For directional solidification, the dendrite tip radius selection criterion given by Eq. (3.83) can be rewritten as

$$V\left(\frac{k_0\,\Delta T_0}{D_L}\right)\left(\frac{C_t}{C_0}\right)\xi_c - G' = \frac{\Gamma}{\sigma^* R^2} \tag{3.91}$$

Since temperature gradient G' is positive for directional solidification, the second term in Eq. (3.91) will be negative; hence, the thermal effect tends to stabilize the interface. The total driving force for stabilization is given by the difference in the two gradients representing solute and thermal fields. If this difference becomes zero or negative, a planar front arises.[92,124] The left side of Eq. (3.91) corresponds to the modified constitutional supercooling at the dendrite tip at low velocity when $\zeta_c = 1$. Moreover, at low velocities the latent heat effect is negligible, and if $K_S = K_L$, $G' = G_L$, and the constitutional supercooling criterion will be restored.

Note that if the temperature gradient effect is negligible, the dendritic tip radius criterion at low velocities becomes

$$VR^2 = \frac{\Gamma D_L}{\sigma^* k_0 \Delta T_0} \frac{C_0}{C_t} \frac{1}{\xi_c} \qquad (3.92)$$

For dendritic growth at low velocities, $C_t \approx C_0$, and $\zeta_c = 1$, which predicts a constant value of VR^2 and is given by the first expression of the right-hand side of Eq. (3.92). As the velocity increases, $C_t \to C_0/k_0$, and at the high velocity limit of the dendritic growth, $\zeta_c = 0$, and $VR^2 \to \infty$, that is, $R \to \infty$, for finite V which represents the absolute velocity for planar front stability.[113]

3.7.2.2 Dendrite Growth Models in Pure Supercooled Melt.
For dendrite growth in pure supercooled melt (free growth), only thermal diffusion and capillarity effects are considered, and the dendrite tip radius criterion, given by Eq. (3.87), reduces to

$$VR^2 = \frac{2\alpha_L \Gamma c_L}{\sigma^* \Delta H_m} \frac{\beta}{\xi_L} \qquad (3.93)$$

To calculate the dendrite tip radius or velocity as a function of bath undercooling, it is better to express the dendrite tip criterion as

$$R = \frac{1}{P_t} \frac{\Gamma c_L}{\sigma^* \Delta H_m} \frac{\beta}{\xi_L} \qquad (3.94)$$

so that, for a specific value of P_t, the value of R can be determined from the above equation, and V by using the relationship $V = 2\alpha_L P_t/R$. Inserting these values of P_t and R in the bath undercooling Eq. (3.81) for pure undercooled melt ($\Delta T_0 = 0$), tip undercooling ΔT_t is given by

$$\Delta T_t = \frac{\Delta H_m}{c_L} \mathrm{Iv}(P_t) + \frac{2\Gamma}{R} \qquad (3.95)$$

3.7.3 Anisotropy

Clearly anisotropy plays an important role in dendritic growth, as shown in Fig. 3.33. All dendrites grow with a strongly preferred growth direction such as (001) crystallographic axis in cubic materials which facilitate the main stem and sidebranches to propagate for long distances. This characteristic of anisotropy seems to be an important factor in establishing certain details of the dendritic microstructures such as crystallographic growth direction, tip morphology, and branching angles. This property has been effectively used to advantage in some crystal growth processes

SUCCINONITRILE   PIVALIC ACID

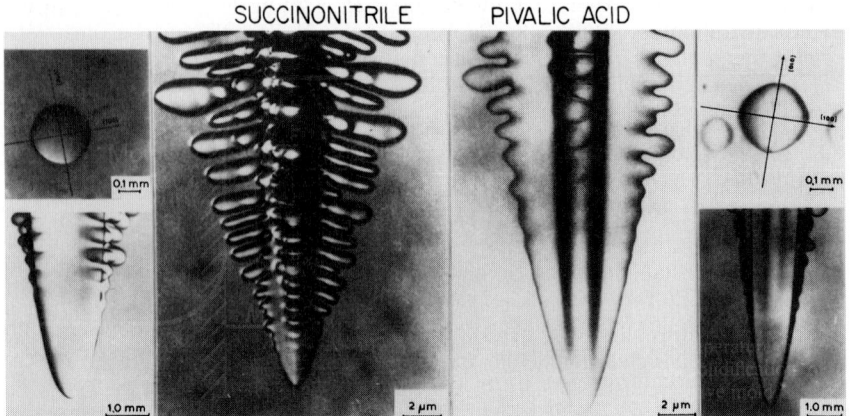

FIGURE 3.33 Anisotropy effects in dendritic growth. A comparison of microstructures of succi-nonitrile (SCN) dendrite (a bcc crystal with small anisotropy) and a pivalic acid (PVA) dendrite (an fcc crystal with large anisotropy) during growth from their melts at the same Stefan numbers. The inserts in the upper left and right corners illustrate droplets of liquid SCN and PVA, respectively, equilibrated within the solid phase. The droplets' shape exhibits the relative anisotropy of the inter-facial energies in the two crystals.[114] (*Courtesy of M. E. Glicksman.*)

such as continuous ribbon pulling of single crystals for the production of solar cell substrates, columnar region of cast alloys, and superior magnetic for high-temperature applications.[114,125]

More recent theoretical models have indicated that anisotropic interfacial prop-erties play a role in the morphological stability of planar interfaces and the cellular and dendritic structures.[125a] Experimental observations reported by Tiller and Rutter[125b] in Pb-Sn alloys as well as by Trivedi[125c] and Trivedi and his coworkers[125d] in transparent organic systems support these theoretical predictions.

3.7.4 Primary Dendrite Spacing

Several theoretical and experimental studies have shown that primary dendrite spacings (λ_1s) are functions of solidification parameters, and an increase in growth rate V decreases the λ_1 value. For a given value of temperature gradient in the liquid G_L, λ_1 goes through a maximum as a function of growth rate or velocity V.[126] For a given V, λ_1 decreases as G_L is increased at the L-S front. Since λ_1 increases linearly with the square root of the inverse cooling rate $1/|\sqrt{G_L V}|$, a relation between λ_1 and mass transport in the liquid is very likely.[127]

There are five important theoretical models which have been proposed to describe λ_1 as a function of V, G_L, $|G_L V|$, and the alloy characteristics.[128] According to the Hunt model,[111] theoretically λ_1 as a function of solidification processing para-meter and material properties can be predicted by

$$\lambda_1 = \left\{ \frac{64 D_L \Gamma [m C_0 (k_0 - 1) - k_0 G_L D_L V^{-1}]}{G_L^2 V} \right\}^{1/4} \qquad (3.96)$$

where Γ is the Gibbs-Thompson coefficient. According to Hunt, for the dendritic growth condition or if $V \ggg k_0 G_L D_L$, Eq. (3.96) converts to

$$\lambda_1 = 2\sqrt{2}\{D_L\Gamma[mC_0(k_0-1)]\}^{1/4}G_L^{-1/2}V^{-1/4} \qquad (3.97)$$

The Okamoto-Kishitake model[127] correlates λ_1 with solidification parameters and assumes the secondary arms to be plates which are thickened during the course of solidification, which is given by

$$\lambda_1 = 2\varepsilon[-D_L C_0 m(1-k_0)]^{1/2}(VG_L)^{-1/2} \qquad (3.98)$$

where ε is a constant (<1) and is about 0.5 for several aluminum alloys.

The Kurz-Fisher model,[129] developed on the marginal stability criterion, correlates λ_1 with solidification parameters. It assumes that (1) dendrite morphology can be approximated by an ellipsoid of revolution, (2) dendrites are located on the corners of a hexagon, and (3) a variation of growth rate exists.

For low growth rate or for $V < V_{tr}$, λ_1 is given by

$$\lambda_1 = \left\{\left[\frac{6\Delta T'}{G_L(1-k_0)}\right]\left(\frac{D_L}{V}-\frac{k_0\Delta T_0}{G_L}\right)\right\}^{1/2} \qquad (3.99)$$

where $V_{tr} = G_L D_L/k_0\Delta T_0$ (the cellular-dendritic transition growth rate), $\Delta T_0 = mC_0(k_0 - 1)/k_0$, and $\Delta T' =$ nonequilibrium solidification range and is the difference between the tip temperature and the nonequilibrium solidus temperature. So $\Delta T'$ is given by

$$\Delta T' = \left(1-\frac{G_L D_L}{V\Delta T_0}\right)\frac{\Delta T_0}{1-k_0} \qquad (3.100)$$

For high growth rate or for $V > V_{tr}$, λ_1 is given by

$$\lambda_1 = 4.3\Delta T^{1/2}\left(\frac{D_L\Gamma}{k_0\Delta T_0}\right)^{1/4}V^{-1/4}G_L^{-1/2} \qquad (3.101)$$

Both the Hunt and Kurz-Fisher models at high growth rates appear to be very similar except for the difference in the constant values. However, at low growth rate, the results obtained by applying these models are different.

The Trivedi model[130] is a modified Hunt model and assumes a marginal stability criterion which is given by

$$\lambda_1 = 2\sqrt{2}G_L^{-1/2}V^{-1/4}(Ak_0\Delta T_0\Gamma D_L)^{1/4} \qquad (3.102)$$

where A is a constant which depends on the harmonic perturbations and, for dendritic growth, is equal to 28.

According to the Hunt-Lu model[131] which is based on the numerical modeling of cellular/dendritic array growth, the dimensionless primary spacing λ' is given by

$$\lambda' = 0.7798 \times 10^{-1}V'(a-0.75)(V'-G')^{0.75}G'^{-0.6028} \qquad (3.103)$$

where $a = -1.131 - 0.1555\log G' - 0.7859 \times 10^{-2}(\log G')^2$. This yields a value of a between -0.34 and -0.58. Note that λ' is the radius rather than the diameter and is independent of k_0. Hence, the values obtained from such a model should be multiplied by 2 to 4 for comparing with the experimental data.

Among all the above five theoretical models, the Kurz-Fisher and Hunt-Lu models allow one to estimate reasonably λ_1 as a function of directional solidification parameters.

3.7.5   Secondary Dendrite Arm Spacing

A reduction in secondary arm spacing λ_2 obtained by the higher cooling rate of cast materials results in a reduction in microsegregation and will produce a more easily homogenized casting.[1,7,132] Secondary arms form initially very near the dendrite tip, presumably due to the presence of noise at the dendrite tip. The precise distance behind the tip where sidebranches are observed depends on the strength of the noise (due to thermal or hydrodynamic fluctuations in the system) which is yet to be established. For both free and constrained growth, the ratio of the initial secondary arm spacing to dendrite tip radius is approximately 2.5 (actually between 2.0 and 3.0) over a wide range of growth conditions in different systems. Theoretical work by Langer and Mueller-Krumbhaar[133] and two-dimensional numerical solution by Saito et al.[134] have confirmed this relationship. However, the actual secondary spacing developed in a fully solidified material is coarser than that of the initial spacing. The coarsening of the final secondary arm spacing arises first from melting or dissolution of the smaller arm at the expense of the larger arms and then from the minimization of surface energy during the coming closer of sidebranches to the neighboring dendrites.[101]

For all except the most dilute alloys, the time-dependent law of coarsening of secondary spacing λ_2 has been developed for directional solidification conditions, which is given by[135]

$$\lambda_2 = 5.5(Mt_f)^{1/3} \tag{3.104}$$

where t_f is the solidification time and

$$M = \frac{\Gamma D_L}{k_0 \, \Delta T_0} \frac{\ln\left(\dfrac{C_f}{C_0}\right)}{\dfrac{C_f}{C_0} - 1} \tag{3.105}$$

where C_f is the final composition of the liquid at the base of the dendrite. For directional solidification, a relation between the local solidification time and the magnitude of the cooling rate $|G_L V|$ can be expressed by

$$t_f = \frac{\Delta T_f}{|G_L V|} \tag{3.106}$$

where ΔT_f is the nonequilibrium temperature range of solidification, i.e., the temperature difference between the tip and the base of the dendrite. Bower et al.[110] obtained, within the experimental error, Eq. (3.92) to be valid in directionally solidified Al-4.5 wt% Cu alloy. The precise value of the parameter M differs slightly for different models; however, its effect on secondary arm spacing is small due to its relation through the cube root.

Recently, Li and Beckermann[116a] have investigated an evolution of the sidebranch spacings in three-dimensional dendritic growth.

3.7.6 Cell-Dendrite Transition

When the degree of morphological instability v becomes substantial at intermediate velocities, the cellular structure undergoes a transition sidebranching instability, thereby transforming to a dendritic morphology. On the basis of a sharp change of the primary spacing with velocity and approximation of the tip shape, Kurz and Fisher[129] proposed the condition for low-velocity cell-dendrite transition V_{C-D} as

$$V_{C-D} = \frac{G_L D_L}{k_0 \Delta T_0} \tag{3.107}$$

or

$$v = \frac{1}{k_0} \tag{3.108}$$

However, the correlation between theory and experimental results was very poor, as reported by Tewari and Laxmanan.[136]

The condition for low-velocity cell-dendrite transition based on the Ivantsov solution can be given by V_{C-D}, using the Ivantsov solution for solute diffusion given in Eq. (3.76b):

$$V_{C-D} = \frac{G_L D_L}{k_0 \Delta T_0} [1 - (1 - k_0) \mathrm{Iv}(P)] \tag{3.109}$$

This equation, at very small Peclet number, converts to the condition of Eq. (3.107) (for $k_0 < 1$), as proposed by Kurz and Fisher on the basis of a sharp change of primary spacing with velocity. In the low-velocity cellular regime, the tip undercooling is mainly given by the term $G_L D_L/V$. In reality, the cell-dendrite transition occurs when the solute diffusion length $\ll$ the effective thermal length or $k_0 \Delta T_0/G_L$.[101]

3.7.7 Twinned Dendrites

Recent *electron backscattering diffraction* (EBSD) experiments in combination with detailed microscopy examination have shown that, under certain solidification conditions (such as lack of, or improper, grain refining), feathery grains in some aluminum alloys exhibit a twinned dendrite morphology, also called *twinned columnar grains*.[†] According to these findings, a growth mechanism has been proposed: It involves a change in the surface tension anisotropy of aluminum alloys, a likely contribution of the attachment kinetics, and convection effects.[136a] Due to the deficiency of nucleation sites, the grains grow instantaneously into clusters. The commercially available titanium-boron grain refiners make it relatively simple to eliminate the formation of twinned columnar grains.[136b]

[†] Although these long, feathery grains cause minimal problem during casting, they pose problems in fabricated aluminum products. For example, ingots containing twinned columnar grains are prone to fracture during upset forging at the junction of two groups of grains with different orientations. The deleterious insoluble constituent concentrations also take place in the microstructure near the long, slender grains.

3.8 SOLIDIFICATION OF EUTECTICS, MONOTECTICS, AND PERITECTICS

In the previous sections, the solidification of single-phase metals and alloys as dendrites or cells leads to the rejection of one of the components from the solid into the liquid. In this section eutectic, monotectic, and peritectic solidification, each involving the formation of two or more phases from the melt, is discussed. In a eutectic system, the alloy melt of eutectic composition C_E solidifies simultaneously into a structure consisting of an intimate mixture of two α and β phases at a single eutectic temperature T_E. That is, the eutectic reaction is expressed as

$$L \underset{\text{heating}}{\overset{\text{cooling}}{\rightleftharpoons}} \alpha + \beta$$

(see Sec. 1.3 also). Below T_E the equilibrium exists between α and β solid phases.

In the monotectic system, the homogeneous liquid L_1 with a monotectic composition C_M transforms to one solid-phase α and a second B-rich liquid-phase L_2 simultaneously through the monotectic reaction at a constant monotectic temperature T_M, that is, $L_1 \rightleftharpoons \alpha + L_2$. Below T_M, nucleation and growth of a phase occur, leading to the supersaturation in L_1 with respect to L_2. After the nucleation of L_2, the growth morphology depends on the relative interfacial energies and densities of the three phases.[137]

In a peritectic reaction, one solid phase and one liquid phase react at a peritectic temperature T_P to form a peritectic composition C_P of the peritectic phase β on cooling, which is expressed by $\alpha + L \rightleftharpoons \beta$ (see Sec. 1.3 also). The liquid composition C_L is called the *peritectic point* or *peritectic limit*; however, peritectic point also has been referred to as C_p. Some peritectic reactions can occur at all compositions ranging from C_α to C_L, and they are known as *peritectic alloys*.[138] Alloys with solute content $C_0 > C_L$ are called nonperitectic alloys because they can form the peritectic phase β directly from the melt, and not through peritectic reaction.

3.8.1 Eutectic Solidification

The eutectic alloys are of great industrial importance because of their following characteristics: (1) lower melting points than those of the pure components; (2) zero or small freezing ranges which effectively remove the dendritic mushy zone, reduce segregation and shrinkage porosity, and provide excellent fluidity (valuable in casting, welding, and soldering processes);[138a] and (3) the prospect of forming fine in situ composites[139] which possess superior mechanical properties with a spacing which is an order of magnitude finer than the primary and final secondary arm spacing in dendritic alloys.[140] The most widely used eutectic alloys include low-temperature soldering (such as Pb-free, ternary eutectic Sn-3.5Ag-0.9Cu with eutectic temperature of 217°C)[140a] or brazing materials, wear-resistant alloys, and casting alloys such as cast irons and Al-Si alloys. Binary eutectics are usually classified, according to morphology, into two major classes, called normal or regular and anomalous or irregular eutectics. The factor that determines the formation of a particular type of microstructure in a given alloy system is the faceting tendency of the growing phases. When both phases grow with a nonfaceting S-L interface, the familiar regular eutectic microstructure is formed during solidification. The mechanism of formation of these microstructures is well established. Common examples are Al-

CuAl$_2$ and Pb-Sn eutectics. On the other hand, an irregular eutectic structure is formed when at least one of the phases solidifies with a faceted interface (e.g., binary Sn-Ag and Sn-Cu systems, where Sn is a nonfaceted phase and Ag$_3$Sn and Cu$_6$Sn$_5$ are faceted).[140a] The formation of these microstructures is not well understood and is sometimes classified, in order of increasing volume fraction of the faceting phase, as *broken lamellar, irregular, complex regular,* and *quasi-regular types.* Binary eutectics where both phases solidify with faceted interface have been little studied.[141]

Generally, normal binary eutectic systems show regular microstructures of lamellar or fibrous and rod form (Fig. 3.34) and phase diagrams with similar melting points of metals. They form by simultaneous movement of the two solid phases in a coupled fashion at a well-defined S-L interface. They exhibit well-defined grain structure and an orientation relationship between the lamellae within the grains. Anomalous eutectic alloys form in systems with metals of different melting temperatures. The anomalous eutectic systems exhibit an irregular and wider range of microstructures in which a single, isothermal growth front does not exist and one phase has the tendency to grow ahead of the other, leading to a more ragged

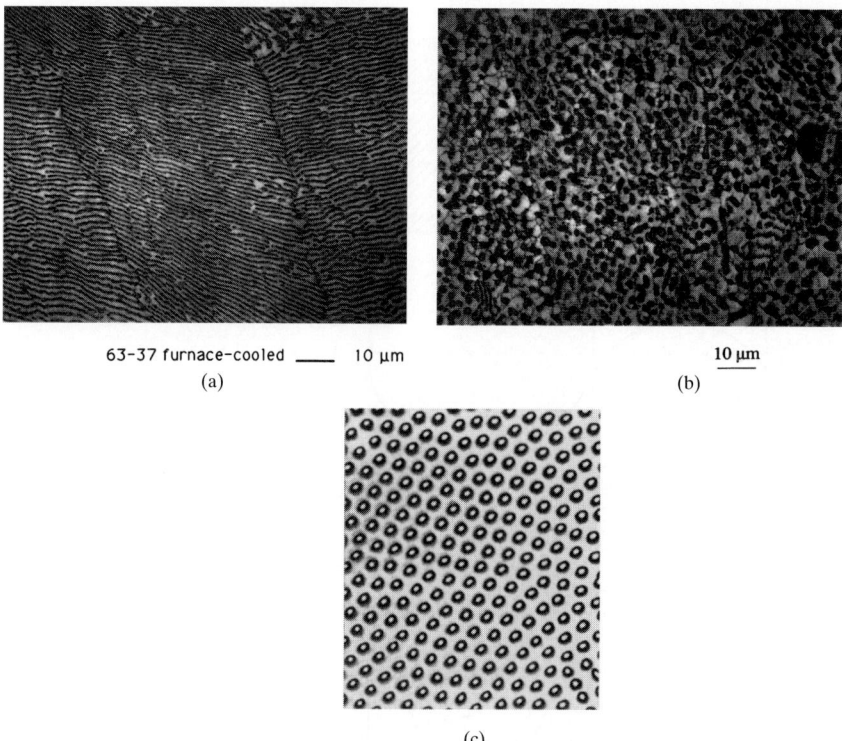

63-37 furnace-cooled ____ 10 μm

(a)

10 μm

(b)

(c)

FIGURE 3.34 Typical solidification microstructure of eutectic 60Sn-40Pb solder: (*a*) lamellar microstructure formed by furnace cooling and (*b*) fine equiaxed structure formed by rapid cooling. (*c*) Directionally solidified eutectic alloys showing a rod structure in the Sn-18 wt% Pb system. (*a* and *b: Courtesy of Z. Mei and J. W. Morris, Lawrence Berkeley National Laboratory, Report LBL 31240, Feb. 1992. c: after Trivedi and Kurz, 1988.*)

S-L interface.[142] That is, in anomalous structure, minor phase is crystallographically dominant, grains cannot be discerned, and preferred orientation relationship is absent.[143]

3.8.1.1 *Regular Lamellar Growth.* During eutectic growth, also called cooperative growth, the solute rejected from one phase is laterally transported across the interface to the other as the interface advances forward. This leads to a concentration profile across the eutectic interface and the associated variation in solute undercooling ΔT_S and the capillarity undercooling ΔT_R. Since thermal diffusivities are much higher than the solute diffusivities, the variation in the interface temperature T_I along the S-L interface, i.e., the total undercooling ΔT, is shown in Fig. 3.35. Simple lamellar or fibrous microstructures, also called *nonfaceted-nonfaceted* (nf-nf) *eutectics*, are produced when both phases have low entropy of melting. In the Jackson-Hunt model and most of the other models, it is assumed that heat flow or thermal diffusivity is very large relative to that of the solute diffusivity, and the simplified solution of the liquid diffusion problem is used for a planar rather than a curved front.

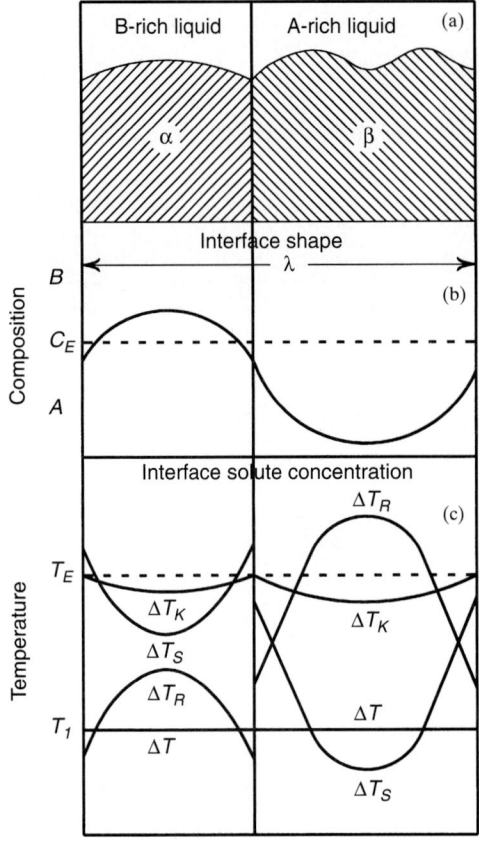

FIGURE 3.35 (a) Schematic representation of (a) lamellar α-β eutectic interface, (b) concentration profile (of liquid composition B) across an α-β interface, and (c) contribution of the total undercoolings ΔT present at the S-L interface, comprising undercoolings attributed to solute ΔT_S, solute curvature ΔT_R, and kinetic ΔT_K effects. (*After Hunt and Jackson, TMS-AIME, vol. 236, 1966, p. 843.*)

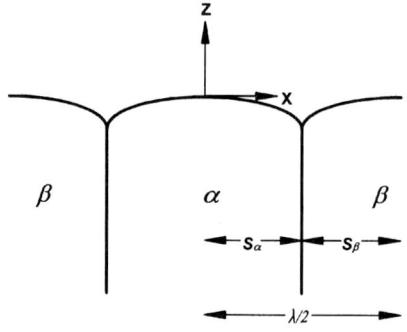

FIGURE 3.36 Schematic diagram of lamellar eutectic structure which defines coordinate system, contact angle at the triple junctions, x, z, λ, S_α, and S_β.

Let us consider the steady-state growth of a eutectic structure, in which the solute flow occurs in the liquid by diffusion and there is no convection. A schematic diagram of lamellar eutectic, showing the definition of relevant length scales and angles required for the theoretical model, is shown in Fig. 3.36. The steady-state diffusion transformed to coordinates x and z (Fig. 3.36) attached to the S-L interface and moving at a constant velocity V in the z direction is given by

$$\nabla^2 C + \frac{V}{D_L}\frac{\partial C}{\partial z} = 0 \tag{3.110a}$$

or

$$\frac{\partial^2 C}{\partial x^2} + \frac{\partial^2 C}{\partial z^2} + \frac{V}{D_L}\frac{\partial C}{\partial z} = 0 \tag{3.110b}$$

where C is the composition. The boundary conditions are

$$C = C_E + C_\infty \qquad \text{at } x = \infty \tag{3.111a}$$

$$\frac{\partial C}{\partial x} = 0 \qquad \text{at } x = 0 \text{ and } x = \frac{1}{2}\lambda = S_\alpha + S_\beta \tag{3.111b}$$

where C_E is the eutectic composition, λ is the eutectic lamellar spacing, and S_α and S_β are the half-width of the α and β phases, respectively (defined in Fig. 3.36).[145] For planar interface, the conservative equations at the boundary become

$$\left(\frac{\partial C}{\partial z}\right)_{z=0} = -\frac{V}{D_L}C_0^\alpha \qquad 0 < x < S_\alpha \tag{3.112a}$$

and

$$\left(\frac{\partial C}{\partial z}\right)_{z=0} = +\frac{V}{D_L}C_0^\beta \qquad S_\alpha < x < S_\alpha + S_\beta \tag{3.112b}$$

where $C_0^\alpha = C_{L\alpha} - C_{S\beta}$ and $C_0^\beta = C_{S\beta} - C_{L\beta}$ are both positive, are amounts of A and B rejected by the α and β phases as a unit volume solidifies, and depend on the local liquid composition [for a nonplanar interface the velocity and gradient in Eq. (3.112) would be normal to the interface].

At the S-L interface, it is essential to fulfill the undercooling equation as given by

$$\Delta T = T_E - T_I = \Delta T_S + \Delta T_R + \Delta T_k \tag{3.113}$$

where $\Delta T_S = m(C_E - C)$. The Jackson-Hunt model of eutectic growth which relates the eutectic spacing l with the growth rate V for directionally solidified alloys is effective at low velocities. This model assumes that, at low growth rates, ΔT_k is small and ΔT_R is the only major contributor to the total undercooling ΔT for nonfaceting materials. At very high growth rates, ΔT_k may be substantial and the C_0^α and C_0^β of Eq. (3.112) would be dependent on both the liquid composition and the growth rate.

The solution to diffusion Eq. (3.111) under these conditions is

$$C = C_E + C_\infty + \sum_{n=0}^{\infty} B_n \text{Cos}\frac{2n\pi x}{\lambda}\exp\left(-\left\{\frac{V}{2D_L}-\left[\left(\frac{V}{2D_L}\right)^2+\left(\frac{2n\pi}{\lambda}\right)^2\right]^{1/2}\right\}z\right) \quad (3.114)$$

Since $2n\pi/\lambda \gg V/2D_L$ for $n > 0$, i.e., when the spacing λ is smaller than the diffusion distance D_L/V, Eq. (3.114) reduces to

$$C = C_E + C_\infty + B_0\exp\left(-\frac{V^2}{2D_L}\right)+\sum_{n=1}^{n=\infty}B_n\text{Cos}\frac{2n\pi x}{\lambda}\exp\left(-\frac{2n\pi z}{\lambda}\right) \quad (3.115)$$

This assumption holds except at the highest velocities (typically > 10 mm/s). The Fourier coefficients B_n and B_0 may be evaluated using the continuity of matter equations at the interface and are given by

$$B_n = \frac{\lambda V C_0}{D_L(n\pi)^2}\text{Sin}\frac{2n\pi S_\alpha}{\lambda} \quad (3.116)$$

and

$$B_0 = \frac{2(C_0^\alpha S_\alpha - C_0^\beta S_\beta)}{\lambda} \quad (3.117)$$

where $C_0 = C_0^\alpha + C_0^\beta$ is shown in Fig. 3.37.

The average compositions in the liquid C_α and C_β, respectively, at the interface ahead of the α and β phases at $z = 0$ are given by

$$C_\alpha = C_E + C_\infty + B_0 + \frac{2(S_\alpha + S_\beta)^2 V C_0 P}{S_\alpha D_L} \quad (3.118a)$$

$$C_\beta = C_E + C_\infty + B_0 + \frac{2(S_\alpha + S_\beta)^2 V C_0 P}{S_\beta D_L} \quad (3.118b)$$

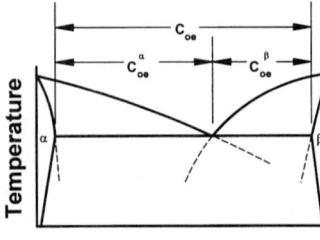

Composition, % B

FIGURE 3.37 A eutectic phase diagram showing the quantities used in equations for conservation of matter in two-phase α-β lamellar growth.[17] (*Courtesy of John Wiley & Sons, Ltd.*)

where

$$P = \left[\sum \left(\frac{1}{n\pi} \right)^3 \sin^2 \frac{n\pi S_\alpha}{S_\alpha + S_\beta} \right] \tag{3.119}$$

Here P depends on S_α/S_β and is usually tabulated. The average undercooling on each phase becomes

$$\Delta T_\alpha = m_\alpha \left[C_\infty + B_0 + \frac{2(S_\alpha + S_\beta)^2 V C_0 P}{S_\alpha D_L} \right] + \frac{a_\alpha^L}{S_\alpha} \tag{3.120a}$$

$$\Delta T_\beta = m_\beta \left[-C_\infty - B_0 + \frac{2(S_\alpha + S_\beta)^2 V C_0 P}{S_\beta D_L} \right] + \frac{a_\beta^L}{S_\beta} \tag{3.120b}$$

where m_α and m_β are the positive slopes of the liquidus lines and a_α^L and a_β^L are constants, obtained from the Gibbs-Thomson coefficient, and are defined elsewhere.[145]

By eliminating B_0 from Eq. (3.120) and assuming the volume ratio $\xi = S_\beta/S_\alpha$ to have its equilibrium value for the given alloy composition, the growth equation relating undercooling, growth velocity, and lamellar spacing, due to Jackson and Hunt,[144] is given by

$$\frac{\Delta T}{m} = V\lambda Q^L + \frac{a^L}{\lambda} \tag{3.121}$$

where

$$\frac{1}{m} = \frac{1}{m_\alpha} + \frac{1}{m_\beta} \tag{3.122}$$

$$Q^L = \frac{P(1+\xi)^2 C_0}{\xi D} \tag{3.123}$$

$$a^L = 2(1+\xi) \left(\frac{a_\alpha^L}{m_\alpha^L} + \frac{a_\beta^L}{\xi m_\beta} \right) \tag{3.124a}$$

$$a_\alpha^L = \left(\frac{T_E}{L} \right) \alpha \gamma_\alpha^L \tag{3.124b}$$

where γ_α^L is surface energy (α-liquid). Here the superscript L denotes the lamellar growth.

The Trivedi-Magnin and Kurz (TMK) model is a more general one to make theoretical predictions at high growth rate (large Peclet number) regime (under rapid solidification conditions). It consists of (1) the solution of the solute diffusion equation in the liquid with boundary conditions and nonequilibrium effect at the interface and (2) the use of a proper spacing selection criterion for lamellar eutectic for a given growth rate condition.[146]

Eutectic Spacing Selection. Assuming (1) planar and isothermal lamellar solidifying interface, (2) a coupling between the constitutional supercooling and the Gibbs-Thomson effect, and (3) the diffusion distance ahead of the interface much larger than the interlamellar spacing, Jackson and Hunt[144] derived the following expression for the average undercooling ΔT at the solidifying interface as a function of solidifying rate V and lamellar eutectic spacing λ:

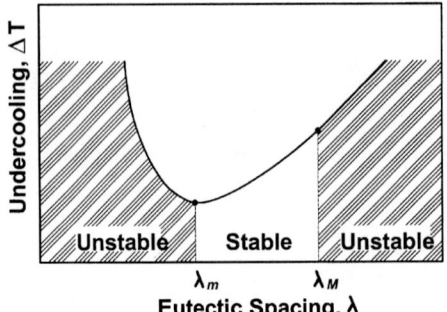

FIGURE 3.38 A schematic plot of undercooling versus eutectic interlamellar spacing at a given velocity showing the stable and unstable spacings (or regions), as predicted by Jackson-Hunt analysis.

$$\Delta T = K_1 \lambda V + \frac{K_2}{\lambda} \tag{3.125}$$

where K_1 and K_2, respectively, are the constant parameters represented by the constitutional supercooling effect and the Gibbs-Thomson effect of the solidifying interface. They used the theoretical values of the minimum and maximum stable spacings λ_m and λ_M, respectively, as shown in Fig. 3.38.[147] On the basis of this assumption, they were able to derive a simple relationship of λ_m and λ_M as a function of growth rate V as

$$\lambda_m^2 V = \frac{K_2}{K_1} \qquad \text{or} \qquad \lambda_m \propto V^{-1/2} \tag{3.126}$$

$$\Delta T = \left(2\sqrt{\frac{K_1}{K_2}}\right) V^{1/2} = A V^{1/2} \tag{3.127}$$

and

$$\lambda_M^2 V = K_3 \tag{3.128}$$

where K_3 is a constant factor and is dependent on properties of the system only.[147] Many experimental results in several alloy systems usually display a narrow band of spacings.

On the basis of Zener's maximum growth rate hypothesis for eutectic growth in an undercooled melt, Tiller proposed an equivalent criterion for directionally solidified eutectics. A more correct criterion should be based on the stability of the interface with respect to small variations in eutectic spacings. Jackson and Hunt concluded that of all the possible eutectic spacings predicted by Eq. (3.125), only a finite range of spacings will be stable with respect to fluctuations in the interface morphology. That is, the eutectic spacings $<\lambda_m$ will be unstable because any depression in the interface shape will produce a narrower lamella at the depression (center of the diagram), which will remove that lamella and increase the local lamella spacing, as shown in Fig. 3.39a. On the other hand, the eutectic spacings $>\lambda_m$ will be stable. For the eutectic spacings $\gg \lambda_m$, the steady-state interface shape develops a hollow or pocket at the center of the wider (or larger volume fraction) phase, as shown in Fig. 3.39b. At some large spacings, denoted by λ_M, the hollow interface gets deeper, i.e., the slope of the (hollow or pocket) interface becomes infinity so that all eutectic spacings above λ_M become unstable. Thus, Jackson and Hunt concluded that of all possible spacings based on diffusional growth considerations, only those spacings will be stable that lie in the range $\lambda_M < \lambda < \lambda_m$. Figure 3.40 exhibits

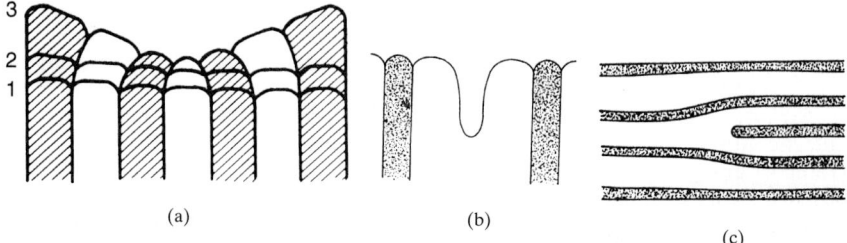

(a)   (b)

(c)

FIGURE 3.39   Schematic illustrations (*a*) of the instability of a lamella. Interface position 1 shows the originally regular lamella, position 2 the formation of narrower lamellae at the center, and position 3 the change in interface shape with time. (*b*) For a wider lamella spacing. The shape instability of the interface of one phase that occurs with very wide spacing. A new lamella forms in the hollow pocket. (*c*) A lamellar fault.[145] (*a: Courtesy of Elsevier Science, Amsterdam; b and c: Courtesy of McGraw-Hill, New York.*)

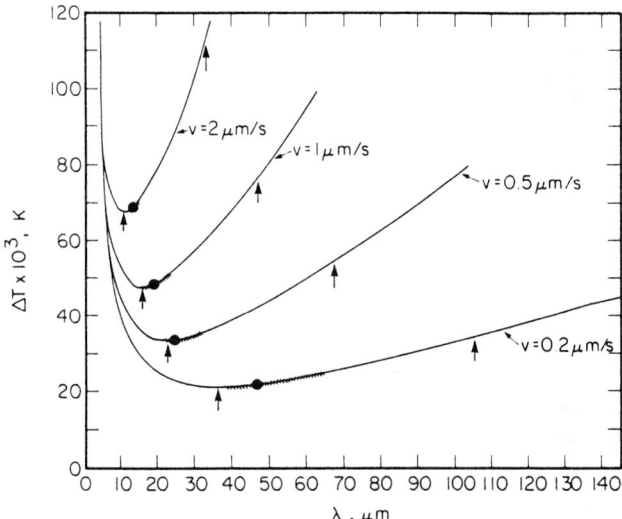

FIGURE 3.40   Theoretically predicted relationships between the interface undercooling ΔT and the lamellar spacing λ for different growth rates. The experimentally observed ranges of lamellar spacings at different velocities are denoted by the hatched regions, and the average spacings at the different velocities are shown by the solid circles.[139] (*Reprinted by permission of the Metallurgical Society, Warrendale, Pa.*)

the theoretically predicted minimum and maximum stable spacings λ_m and λ_M, given by Eqs. (3.126) and (3.128), respectively, and indicated by the pairs of vertical arrows. It is seen that ΔT is largest for large and small lamellar spacings, because diffusion is difficult in the former and curvature effects are dominant in the latter.

Criterion for Growth.   In the Jackson-Hunt analysis, the extremum condition was ignored by applying a criterion noted on an observed growth mechanism. It was suggested that the lamellar spacings are controlled by the stability of a lamellar fault (Fig. 3.39c). This figure shows that the average spacing decreases when the fault moves to the right. Stable lamellar growth occurs when the fault remains stationary.[144]

Extent of Eutectic Range.   Lamellar eutectics can grow from a liquid of both eutectic and off-eutectic compositions by solidifying these materials with varying gradients and growth rates. When an alloy having off-eutectic composition solidifies, first dendrites or cells of the primary phase will occur at some undercooling below the liquidus temperature, and then the eutectic phase will appear from the remaining liquid reaching the eutectic composition.[148] The extent of the range depends on the temperature gradient and the growth rate.

Mollard and Flemings[149] and Cline and Livingston[150] obtained the eutectic structure in the off-eutectic compositions by solidifying them at varying growth rates and temperature gradients to maintain a planar interface. The two-phase structures are achieved with different volume ratio of the phases from that given by the equilibrium eutectic composition.

According to the competitive growth mechanism, cells or dendrites cannot form unless they grow with a tip temperature higher than the eutectic growth temperature. Combining the Burden and Hunt dendritic growth model and the Jackson-Hunt eutectic model, the dendritic tip undercooling ΔT_t as a function of velocity and temperature gradient can be given by

$$\Delta T_t = T_0 - T_t = \frac{G_L D_L}{V} + B\sqrt{V} \qquad (3.129)$$

where T_0 is the alloy liquidus temperature and T_t is the tip temperature for eutectic; Eq. (3.127) is

$$\Delta T = T_E - T_I = A\sqrt{V} \qquad (3.130)$$

where A is a constant. At the critical condition $D_L = T_I$, the eutectic composition range on the α side of the eutectic becomes

$$\Delta C = C_E - C_0 = \frac{T_0 - T_E}{m_\alpha} = \frac{G_L D_L}{m_\alpha V} + [(B - A)m_\alpha]\sqrt{V} \qquad (3.131)$$

where B and A are constants for the dendrite and eutectic structures, respectively. At high gradients, the $G_L D_L$ term dominates and the eutectic range is proportional to G_L/V. At sufficiently higher velocities, the $(B - A)\sqrt{V}$ term dominates and the range is proportional to $\sqrt{V}$. It is thus clear that the eutectic can form and be stable at both very low and very high velocities.[151]

Figure 3.41 shows a schematic representation of this approach. When the composition is changed from A-rich to the B-rich alloy, the β liquidus temperature moves upward and α liquidus temperature downward with respect to the eutectic in order to change the transition velocities. Figure 3.42 shows the superimposition of results from Fig. 3.41 for different compositions on the phase diagram by plotting the eutectic growth temperature versus composition. This shows the composition-temperature zone in which eutectic microstructures can be found, and this region is usually called the *coupled zone of competitive growth*. Figure 3.42a shows an example for two different gradients and Fig. 3.42b for skewed eutectic due to very different growth kinetics of the primary phases. A comparison of theoreti-

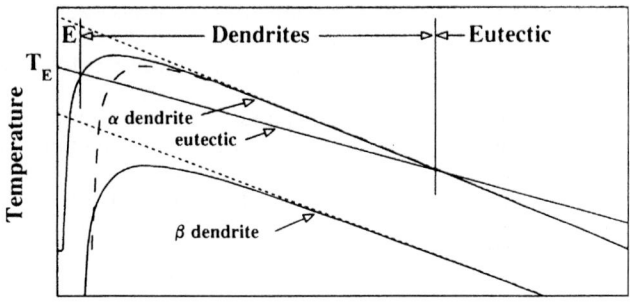

$$(\textbf{Velocity})^{0.5}$$

FIGURE 3.41 A schematic diagram of the α and β dendrite tip temperatures (the dotted lines represent plots for a zero gradient, the continuous lines are for a larger gradient) and the eutectic interface temperature plotted against square root of velocity for a given temperature gradient. It is assumed that the eutectic structure forms without dendrites when the eutectic has the higher growth temperature (i.e., at high and low velocities). Conversion from α-rich to β-rich phase will raise the β but lower the α line, resulting in a variation in the transition velocities.[145] (*Reprinted by permission of Elsevier Science, B.V., Amsterdam.*)

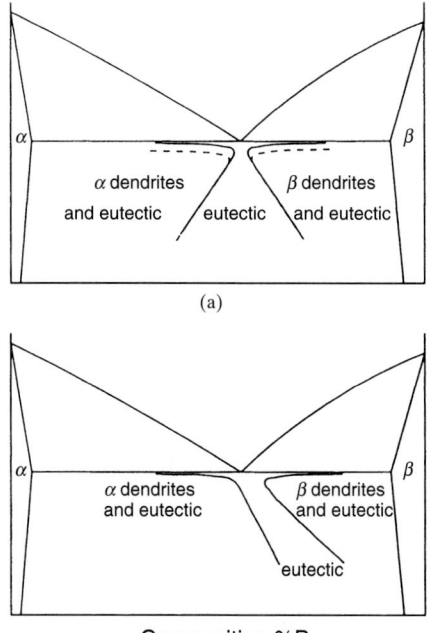

FIGURE 3.42 (*a*) The superimposition of results from Fig. 3.41 for different compositions on the phase diagram by plotting eutectic growth temperature against composition. The dashed line exhibits the boundary for a larger gradient. (*b*) The influence of a steeper β dendrite line on the eutectic range.[145] (*Reprinted by permission of Elsevier Science, B.V., Amsterdam.*)

cal and observed eutectic ranges shows a reasonably good agreement. In the case of rapid eutectic solidification, Trivedi and Kurz proposed a schematic plot similar to Fig. 3.41 except that the eutectic line has a maximum velocity and the dendrite line changes to cellular and then planar again at very high velocities (Fig. 3.43). The results again can be condensed by superimposing on the phase diagram (see Fig. 3.44a). Figure 3.44b is an example for calculated results showing metastable phases.[139]

3.8.1.2 Rod Eutectics. Like lamellar structures, two nonfaceted phases form normal rodlike structures, and rod growth has been analyzed. Figure 3.45 shows a schematic diagram defining the length scales for a rod eutectic. The rod growth can be given by

$$\frac{\Delta T}{m} = VRQ^R + \frac{a^R}{R} \tag{3.132}$$

where R is the rod spacing and

$$a^R = 2(1+\xi)^{1/2}\left(\frac{a_\alpha^R}{m_\alpha} + \frac{a_\beta^R}{\xi m_\beta}\right) \tag{3.133}$$

$$Q^R = \frac{4(1+\xi)C_0 M}{\xi D} \tag{3.134}$$

where a_α^R and a_β^R are surface energy terms and M is a sum of Bessel function terms similar to P, defined and tabulated elsewhere.[144]

Like lamellar growth, Eq. (3.132) requires an additional condition for a solution. In the absence of a detailed mechanism for variation of rod spacings, the structure can be expected to grow at the extremum spacing. In this situation

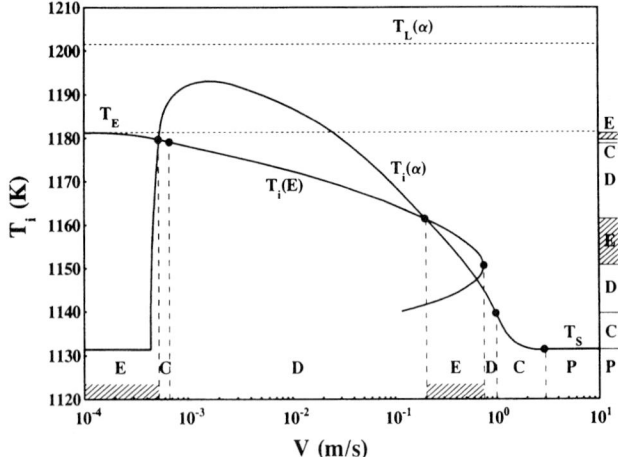

FIGURE 3.43 Schematic plot similar to Fig. 3.41 except that the eutectic line has a maximum velocity, as predicted from high-velocity eutectic analysis, and the dendrite line becomes cellular and then planar again at very high velocities.[139] (*Reprinted by permission of The Metallurgical Society, Warrendale, Pa.*)

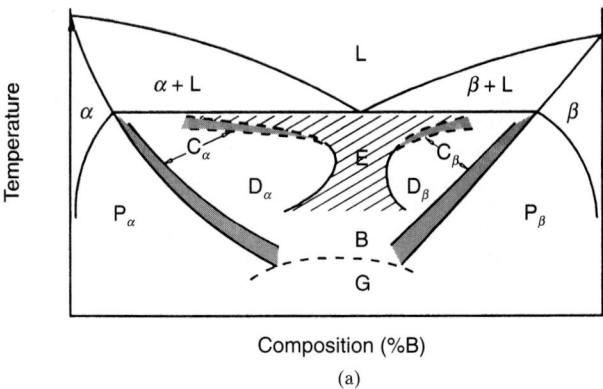

(a)

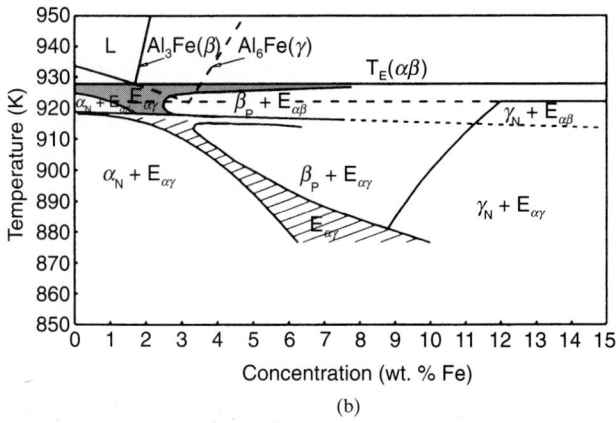

(b)

FIGURE 3.44 (*a*) Schematic phase selection produced from Fig. 3.41 for a given gradient. E: eutectic, P: planar, C: cellular, D: dendritic, B: banded, and G: glass. (*b*) A calculated phase selection map for Al-Fe system.[139] (*Reprinted by permission of The Metallurgical Society, Warrendale, Pa.*)

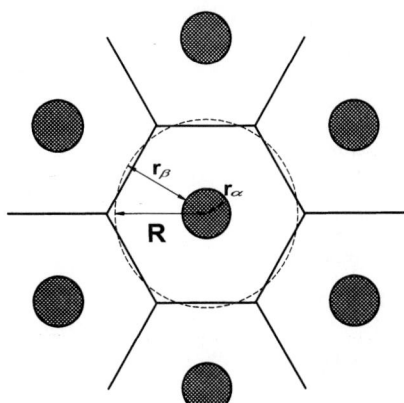

FIGURE 3.45 Schematic diagram defining a rod eutectic.

$$VR^2 = \frac{a^R}{Q^R} \tag{3.135}$$

and
$$\frac{\Delta T^2}{V} = 4m^2 a^R Q^R \tag{3.136}$$

According to Walter and Cline, VR^2 is a constant for NiAl-Cr rod eutectic.

Rodlike eutectics usually form when the volume fraction of one phase is much smaller than the other. Consequently, when the minimum undercooling for rod-eutectic formation becomes less than that for the lamellae, the rod eutectic structure will grow preferentially, i.e., when

$$\frac{\left(\dfrac{a_\alpha^L}{m_\alpha} + \dfrac{a_\beta^L}{\xi m_\beta} \right)}{\dfrac{a_\alpha^R}{m_\alpha} + \dfrac{a_\beta^R}{\xi m_\beta}} > \frac{4M}{P(1+\xi)^{3/2}} \tag{3.137}$$

For isotropic surface energies for rods and lamellae, the left-hand side of Eq. (3.137) becomes equal to unity. Table 3.1 shows the value of the right-hand side of this rela-

TABLE 3.1 Comparison of the Stability of Rods and Lamellae[†152]

$\dfrac{S_\alpha}{S_\alpha + S_\beta} = \dfrac{1}{1+\zeta}$	$\dfrac{4M}{P}\left(\dfrac{1}{1+\zeta}\right)^{3/2}$
0	—
0.01	$(0.8)^{\ddagger}$
0.02	0.75
0.05	0.74
0.1	0.806
0.2	0.946
0.3	0.994
0.4	1.054
0.5	1.088
0.6	1.116
0.7	1.152
0.8	1.142
0.9	1.374
1.0	—

† For isotropic interfacial free energies, rods grow with lower undercooling for values of ζ such that the number in the right-hand column is less than 1. A lamellar structure grows with lower undercoolings for the cases where it is greater than 1; see Eq. (3.138).

‡ The series defining M and P become difficult to sum for small values of $1/(1+\zeta)$.

Courtesy of the Institute of Materials, London.

tionship for different values of ξ. After the evaluation of M and P terms, it is found that the critical volume fraction for the minor phase to form rods is given by[152]

$$\frac{S_\alpha}{S_\alpha + S_\beta} = \frac{1}{1+\xi} < \frac{1}{\pi} \tag{3.138}$$

In reality, rodlike structures do not form for volume fractions greater than this value. Lamellar structures, however, are found to occur for volume fractions slightly less than this value, and it is suggested that this results from the generation of low-energy boundaries between the solid (lamellae) phases.[145]

3.8.1.3 Irregular Eutectic: Faceted-Nonfaceted Eutectics.

Faceted-nonfaceted (f-nf), irregular, or abnormal eutectic structures usually form when at least one of the solidified phases is faceted. The best examples of such eutectics are industrially important Al-Si alloys and cast irons which are also discussed in Secs. 3.11.4 and 3.11.6. The main characteristics of irregular eutectics which differentiate them from the regular eutectics are as follows:

1. The growth of two dissimilar phases is cooperative rather than the coupled in which the growing S-L interface is the nonisothermal.

2. The microstructure is irregular due to the nonparallelism of different lamellae arising from growth anisotropy of high entropy faceted phase.

3. It is found in practice that undercoolings and spacings are much larger than those in regular eutectics. This is described by indicating that lamellar eutectic grows, penetrating each other and thereby making the structure coarser. As a result of larger growth undercoolings, the nucleation of new eutectic grains often occurs ahead of the solidifying eutectic interface. These results do not follow the Jackson-Hunt lamellar theory of growth at the minimum undercooling condition.

4. Unlike regular eutectics, the irregular average eutectic spacing is a function of temperature gradient in the liquid, growth rate, and addition of very small amounts of a third element or impurities (e.g., Na to Al-Si and FeSi and Mg to cast iron). The final spacing is a measure of the ability of the structure to branch or form new phases.[59] The growth of each lamella of the faceted phase is determined mostly by its own crystallographic orientation, independent of heat flow direction. Consequently, irregular eutectics solidify with a wide range of spacings growing simultaneously according to the following mechanism (Fig. 3.46). According to Fisher and Kurz,[153] Magnin and Kurz,[154] and Magnin et al.,[155] for solidification of converging lamellae, when $\lambda < \lambda_{min}$, the growth of one lamella stops and the surface energy effects hinder further growth. In contrast, for solidification of diverging lamellae, when the local spacing $\lambda >$ critical spacing λ_{br}, the growth becomes unstable and one of them, especially the faceted phase, branches into two diverging lamellae, thereby decreasing the local spacing.[155]

5. The severity of faceting on the faceted phase is believed to reduce with the increase in growth velocity.[27,156]

The main characteristic required for the formation of irregular eutectic is the inability of the α-β boundary to change its position smoothly due to the presence of a facet in or adjacent to the boundary. Magnin et al. describe this departure of average eutectic spacing λ from the extremum eutectic spacing λ_{ex} (in the Jackson-Hunt analysis) and have introduced two dimensionless parameters, average value ϕ and its extent η, to describe the operating range for the irregular eutectic, which are defined as

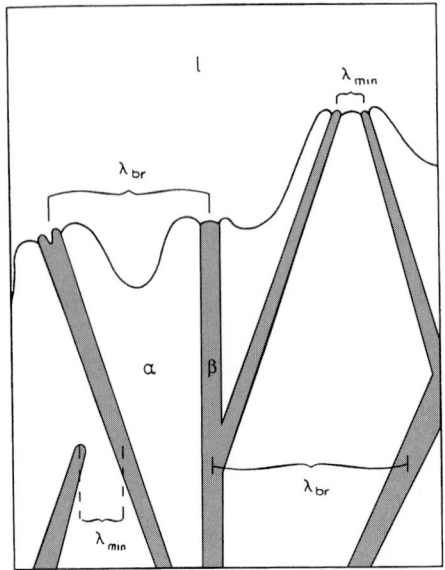

FIGURE 3.46 Growth mechanism of irregular eutectics, as proposed by Fisher and Kurz.[153-155] (*Courtesy of Acta Metall.*)

$$\lambda = \phi\lambda_{ex} \tag{3.139}$$

$$\eta = \frac{\lambda_{br} - \lambda_{min}}{\lambda} \tag{3.140}$$

Since the lamellar spacing varies between λ_m and λ_{br}, they showed that the average spacing for irregular eutectics will be larger than the average spacing by some operating factor ϕ. For gray cast iron, $\phi = 5.4$, and, for unmodified Al-Si alloy, $\phi = 3.2$. It appears that no criterion is available as yet to predict the ϕ value.

3.8.2 Monotectic Solidification

The monotectic alloys have unique features in that both A and B elements are almost insoluble in each other in the solid state, liquids L_1 and L_2 have different densities, and the alloy exhibits a liquid miscibility gap which appears to be very wide, as seen in the diagram in Fig. 3.47. Hence, it is very difficult to obtain a homogeneous hypermonotectic alloy melt, and alloys having compositions within the miscibility gap solidify with the associated heavy gravity segregation prior to the start of the monotectic solidification. These problems have restricted the application of monotectic alloys as industrial materials and systematic study on the solidification of monotectic alloys.[157] However, a few studies on the solidification of monotectic alloys reveal that the gravity segregation does not occur in alloys with a monotectic composition. It has been reported that in Al-Bi, Al-Pb, Al-In, Cu-Pb, and Bi-Si systems, fibrous or rodlike composite structures form in a similar manner as for eutectic alloys. Monotectic alloys having Al- or Cu-matrix with a fine particulate dispersion and/or fibrous arrangement of the second phases Bi, Pb, and In are considered as important materials for light and soft in situ superconductors, wear-resistant materials, etc.

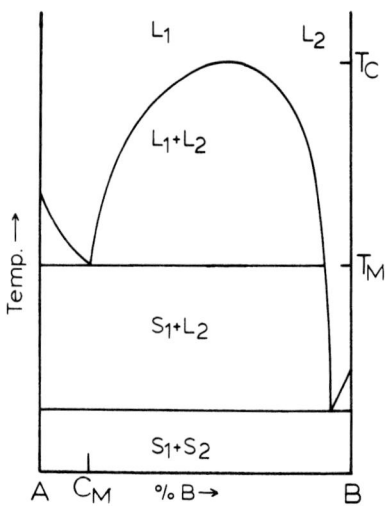

FIGURE 3.47 Idealized monotectic phase diagram.

3.8.2.1 Directional Solidification of Monotectic Alloys.

Two types of solidification processes have been distinguished during directional solidification of monotectic alloys. In type A solidification (exemplified by Al-In, Al-Pb, and Al-Bi alloys), at low growth velocities ($V \leq 5$ mm/s), well-aligned, regularly arrayed, close-packed fibrous composite structures are observed (Fig. 3.48a). At high growth velocity, the steady-state growth interface is replaced by an interface which is continually perturbed locally, the L_2 phase spheroidizing rapidly behind the interface. A transient stage precedes the breakdown or establishment of the steady-state growth in which the monotectic growth interface is prone to oscillatory perturbations over a wide area. This leads to the formation of rows of globular second-phase L_2 droplets aligned parallel and transverse to the growth direction in the solid matrix.[158] In type B solidification (exemplified by Cu-Pb system), no steady-state growth is noticed at any growth rate. Coarse, interconnected globules of L_2 are formed at low velocity, becoming more aligned as growth velocity is increased, wherein loose coupled diffusion occurs between α and L_2 (Fig. 3.48b). In this situation, a fibrous structure with heavy branching morphology and termination is noticed above a critical growth rate. The phase spacings in type B growth are approximately an order of magnitude greater than in type A. These morphological transformations of monotectic are strong functions of G_L/V ratio, the volume fraction of liquid L_2, and the interfacial energies between the solid and two liquids at the monotectic growth front.[157]

Figure 3.48c depicts the microstructure of a longitudinal section of a monotectic Zn-0.6 at% Bi alloy, showing two columnar grains of different morphology. The central region represents the perfectly aligned morphology, whereas the adjacent grains show cellular morphologies containing more randomly arranged, shorter Bi droplets. These latter morphologies are called monotectic cells, like eutectic cells.[158a]

Cahn extended Chadwick's hypothesis of relative surface energies of the solid and two liquid phases and pointed out that these types of growth can be related to the height of the miscibility gap. A high $T_C - T_m$ value promotes type A growth

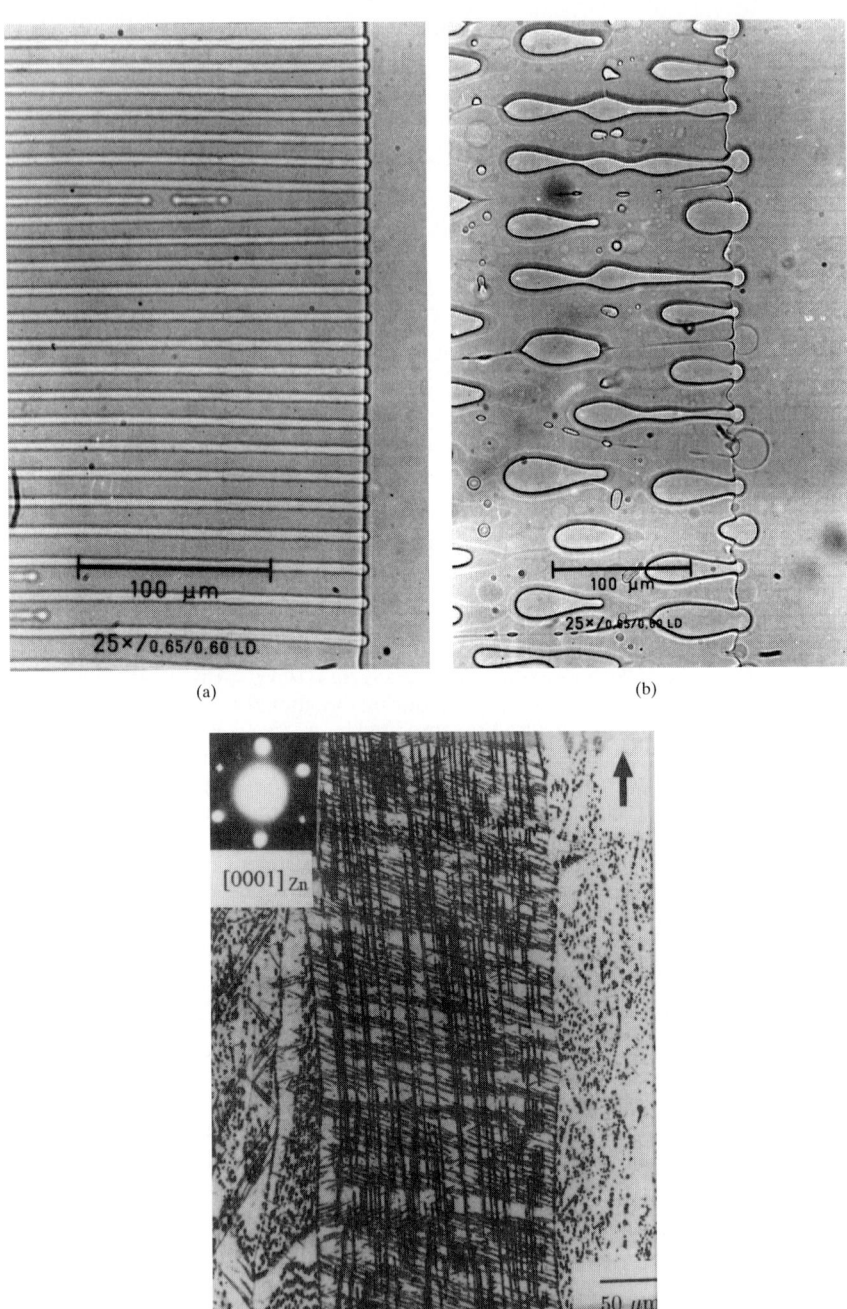

FIGURE 3.48 (*a*) Type A monotectic. Steady-state growth in succinonitrile-7.5 wt % glycerol.[137,145] (*Courtesy A. Hellawell*). (*b*) Type B monotectic. Non-steady-state growth front in succinonitrile 20 wt % ethanol.[137,145,153] (*Courtesy A. Hellawell*). (*c*) Microstructure of a longitudinal section of monotectic Zn-0.6 at % Bi alloy (grown at a temperature gradient of 30×10^3 K/m and velocity of 17×10^{-6} m/s) showing two columnar grains of different morphology. The central grain denotes the perfectly aligned morphology, while the other grain shows monotectic cells. The selected area diffraction pattern from the aligned grains at zero tilt reveals [0001] zone axis (inset).[158a] (*Courtesy of TMS, Warrendale Pa.*)

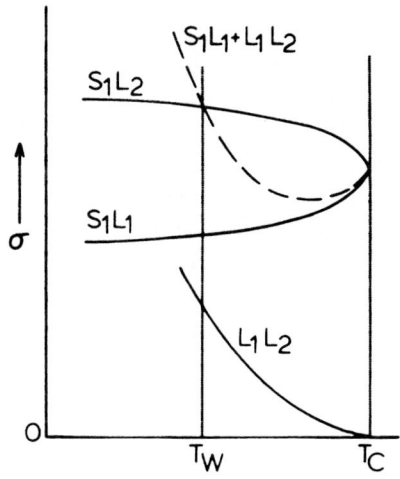

FIGURE 3.49 Variation with temperature of interfacial energies in monotectic system. (*Courtesy of Elsevier Science, Amsterdam.*)

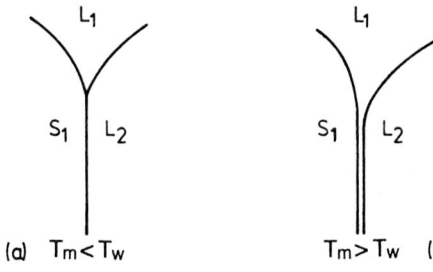

FIGURE 3.50 Three phase profiles at monotectic interface. (*a*) Balanced wetting of α by L_1 and L_2 (*J. W. Cahn, Met. Trans., vol. 10A, 1979, pp. 119–121*). (*b*) Perfect wetting of α by L_1.[131] (*Courtesy of TMS, Warrendale, Pa.*)

whereas a low value promotes type B growth. These findings can be described by Cahn's findings in terms of the relative surface energies between L_1 and L_2 and a third phase (α) having different temperature dependencies. Figure 3.49 illustrates this point. At T_C, $\gamma_{L_1 L_2} = 0$ and $\gamma_{\alpha L_1} = \gamma_{\alpha L_2}$. As the temperature falls, $\gamma_{L_1 L_2}$ increases (as nearly as $\Delta T^{1.2}$) and $\gamma_{\alpha L_1}$ and $\gamma_{\alpha L_2}$ diverge (as nearly as $\Delta T^{0.35}$). The stability of a three-phase junction, in terms of the parameter, can be given by

$$\Delta \gamma = \gamma_{\alpha L_2} - (\gamma_{\alpha L_1} + \gamma_{L_1 L_2}) \qquad (3.141)$$

Assuming a critical (or transition) wetting temperature T_W (above which the liquid L_1 will wet the third phase α) for which $\Delta \gamma = 0$, the two situations can be explained as follows. For $T_M < T_W < T_C$ (high-temperature region), a high miscibility gap occurs and L_2 perfectly wets α; this results in a stable three-phase junction representing type A solidification. For $T_W < T_M < T_C$ (low-temperature regime), a lower miscibility gap (i.e., large ratios of T_M/T_C) occurs, and with $\Delta \gamma < 0$, L_1 preferentially wets α which leads to the separation of L_2 away from the growth front (Fig. 3.50). This concept was supported by experimental studies by Grugel and Hellawell[159] and Grugel et al.[160] who showed that the transition from type A to type B composite structure occurred when the critical value of the T_M/T_C ratio was around 0.9 (Table 3.2). The fiber spacing in both types of growth was found to follow

TABLE 3.2 Summary of Directionally Solidified Monotectic Alloys Arranged in Order of Increasing T_M/T_C*

System	$\left(\dfrac{T_m}{T_c}\right)K$	Type
Ga-Pb	~0.5	Aligned "A"
Sb-Sb$_2$S$_3$	~0.5	Aligned "A"
Al-Bi	0.59	Aligned "A"
Al-In	0.75	Aligned "A"
(CH$_2$CN)$_2$-H$_2$O(S-W)	0.887	Aligned "A"
(CH$_2$CN)$_2$-C$_3$H$_5$(OH)$_3$ (S-G)	0.896	Aligned "A"
Cu-16 wt% Pb-3 wt% Al	~0.9 (est.)	Transition
(CH$_2$CN)$_2$-7.5 wt% E-6.9 wt% G	0.937	Transition
(CH$_2$CN)$_2$-C$_2$H$_5$OH (S-E)	0.94	Irregular "B"
Cu-Pb	0.97	Irregular "B"
Cd-Ga	0.98	Irregular "B"
C$_7$F$_{14}$-C$_3$H$_7$OH	0.857	Surface wetting (Schmidt and Moldover)

* A low ratio represents a high miscibility gap and vice versa.[160]
Courtesy of the TMS, Warrendale, Pa.

the relationship $\lambda^2 V$ = constant, where λ is the interfiber spacing. Since the fiber spacing is about 10 times greater for type B solidification, the constant for this type of growth is about 2 orders of magnitude greater than that for type A growth. Furthermore, the constant for type A growth lies between 3 and 10 times that for comparable eutectic systems.[137]

There are many common features between these two types of alloys with respect to the morphological transition of the S-L interface and the incorporation of L_2 phase into the monotectic solidification front. First, in unidirectional solidification of planar growth front at high G_L/V conditions, the liquid separated through the monotectic reaction is not incorporated into the growth front, which leads to the formation of an Al single-phase region without the L_2 perhaps in the Al-based alloys and the formation of L_2 phase banded segregation in the Cu-Pb alloy. Second, as the growth rate increases, the morphology of the monotectic growth front in both Al-based and Cu-Pb alloys becomes cellular, and the separated L_2 phases are incorporated into the intercellular grooves and composite growth of the solid and liquid L_2 advances.[157]

3.8.3 Peritectic Solidification

Many binary-phase diagrams display peritectic reactions. Perhaps the most well-known example is the Fe-C system. Other important examples include binary Fe-Ni, Cu-Zn, Cu-Sn, Cu-Al, and Ti-Al systems and ternary Fe-Ni-Cr systems, tool steels, magnetic materials such as Co-Sm-Cu and Nd-Fe-B alloys, and various inorganic materials such as $yBa_2Cu_3O_y$ ceramics (YBCO) which are produced via peritectic reactions.[138] Figure 3.51a resembles the Fe-C phase diagram and many Cu- and Ti-based systems, with the only exception in the latter cases of the formation of intermetallic peritectic phases. In these cases, $k_0 < 1$ in both the primary and peritectic phases. Hypo- and hyperperitectic alloys are those compositions in both the

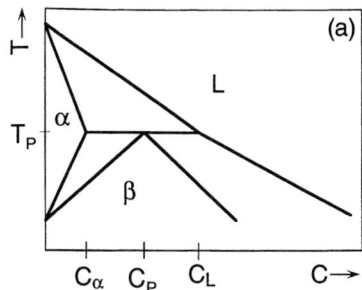
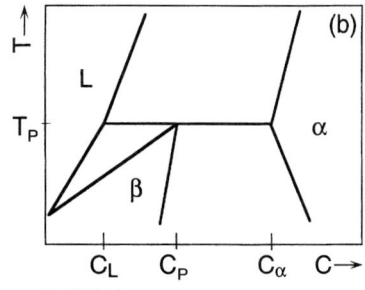

FIGURE 3.51 Schematic peritectic phase diagrams illustrating compositions of liquid C_L and primary phase C_α, which react to form peritectic β phase with composition C_P at peritectic temperature T_P. (a) $k_0 < 1$ and (b) $k_0 > 1$. (*Reprinted by permission of Elsevier Science, Amsterdam.*)

TABLE 3.3 Comparison of $k_0 < 1$ and $k_0 > 1$ Peritectic Systems

Definition	$k_0 < 1$	$k_0 > 1$
Nonperitectic	$C_o > C_L$	$C_o < C_L$
Peritectic alloys	$C_a < C_o < C_L$	$C_L < C_o < C_a$
Hypoperitectic	$C_a < C_o < C_P$	$C_L < C_o < C_P$

phase diagrams, at T_P, which lie between C_α and C_P and C_P and C_L, respectively. They constitute α and β phases below T_P under equilibrium conditions.

In contrast, many Al-based systems, notably Al-Ti, exhibit peritectic reactions in both phases for $k_0 > 1$, as shown in Fig. 3.51b. In these systems, the hypoperitectic term for peritectic alloys with composition $< C_P$ implies the range from C_L to C_P. These alloys constitute the liquid and β phases below T_P, in contrast to the α and β phases in hypoperitectic alloys with $k_0 < 1$ systems. Similarly, nonperitectic alloys with $k_0 > 1$ systems, which should form the peritectic phase directly from the liquid, are those compositions $C_0 < C_L$. These differences in peritectic systems between $k_0 < 1$ and $k_0 > 1$ are given in Table 3.3.

Despite their importance, limited studies have been done to comprehend the transformation behavior of peritectic alloys. Three short reviews[161–163] and one detailed review[138] are described elsewhere.

Peritectic growth consists of three stages: the peritectic reaction, peritectic transformation (solid-state diffusion for thickening of β phase), and direct solidification of the peritectic phase (β on the previously formed β layer). These are discussed in this section.

3.8.3.1 *Peritectic Reaction.*
Depending on the surface tension conditions, two types of peritectic reactions can take place: nucleation and growth of β phase in the liquid with or without contacting α phase. In the first case, which is most prevalent, secondary β phase nucleates at the interface between the primary α phase and the liquid. A lateral growth of the β phase over the α phase would require both dissolution and some resolidification of the α phase (Fig. 3.52a).

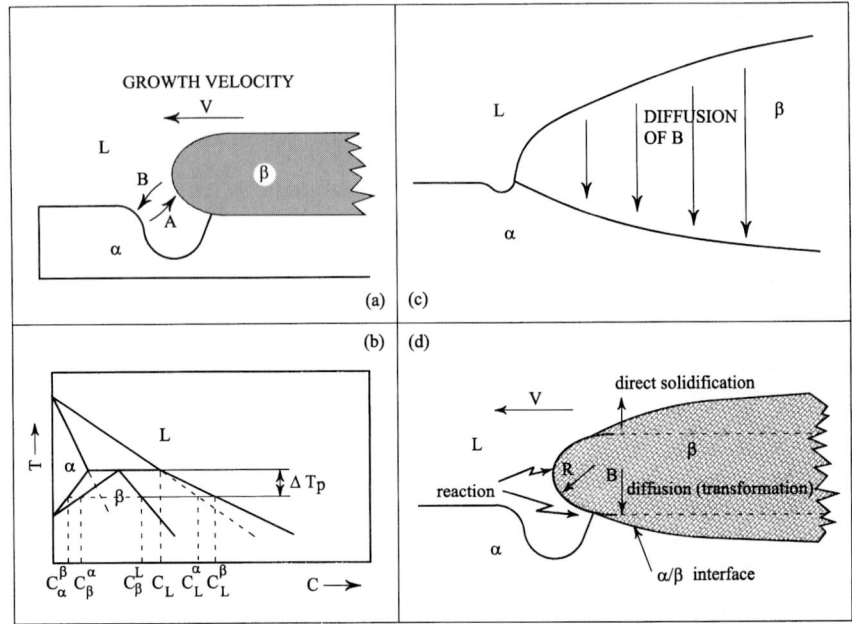

FIGURE 3.52 (a) Peritectic reaction $\alpha + L \rightleftharpoons \beta$ with all three phases in contact at the triple junction, showing the growth of the secondary solid phase β along the surface of the primary solid phase α by diffusion through the liquid. (b) Peritectic phase diagram exhibiting liquid compositions which would be in equilibrium with each of α and β phases at undercooling ΔT_P. (c) Peritectic transformation involving diffusion of B atoms through the already formed solid phase β. (d) Schematic representation of the three stages of peritectic transformation: the reaction, the transformation, and direct solidification of β on primary α.

In the second type of reaction, unhindered growth of the secondary phase in the liquid and the simultaneous dissolution of primary phase occur. This type of peritectic reaction has been found in the Al-Mn system. There has also been a tendency for the growth of a secondary phase around the primary phase at larger cooling rates, especially in Ni-Zn and Al-U systems.[163]

Based on the assumptions that (1) the diffusion of B in the liquid occurs in front of the advancing β phase due to the compositional difference $C_L^\beta - C_L^\alpha$, (2) the β phase is a platelike surface layer with a tip radius of curvature R which grows at its maximum velocity to provide a thickness $2R$, and (3) application of the Bosze and Trivedi model,[164] Frederiksson and Nylen[165] derived the velocity of plate tip as a function of supersaturation Ω. For low undercoolings, the β growth velocity is expressed by

$$V = \frac{9}{8\pi} \frac{D_L}{R} \left(\frac{\Omega}{1 - \frac{2\Omega}{\pi} - \frac{\Omega^2}{2\pi}} \right)^2 \qquad (3.142)$$

where

$$\Omega = \frac{C_L^\beta - C_L^\alpha}{C_L^\beta - C_\beta^L} \qquad (3.143)$$

and C_β^L is the composition of β in equilibrium with the liquid phase at the growth temperature (Fig. 3.52b).

According to Bosze and Trivedi, assuming maximum growth rate and small Peclet number ($VR/2D_L$), the ratio of critical radius for nucleation R^* to the actual radius R is given by

$$\frac{R^*}{R} = \frac{3}{32}\Omega \qquad (3.144)$$

For peritectic reaction, Frederiksson and Nylen predicted the critical thickness of the plate (perhaps twice the critical radius) as

$$R^* = \frac{\gamma V_m}{\Delta G_m} \qquad (3.145)$$

where $\gamma = \gamma_{\beta L} + \gamma_{\alpha\beta} - \gamma_{\alpha L}$ and $\gamma_{\beta L}$, $\gamma_{\alpha\beta}$, and $\gamma_{\alpha L}$ are the surface energies of the β-L, α-β, and α-L interfaces, respectively; ΔG_m is the driving force; and V_m is the molar volume of the liquid. This model was employed to compare the theoretical and experimentally found β-layer thickness due to the L-S peritectic reaction for unidirectionally solidified Ag-Sn and Cu-Sn alloys at various rates, which showed only limited success, partly due to difficulties experienced in separating it from other stages.[138]

3.8.3.2 Peritectic Transformation. The peritectic transformation involves solid-state diffusion and the movement of the α-β interface during cooling (Fig. 3.52c). As shown schematically in Fig. 3.52b and c, the solute diffusion in the solid state is driven by the compositional difference $C_\beta^L - C_\beta^\alpha$ through the β phase. Using a one-dimensional analysis, Hillert[161] and St. John and Hogan[166] estimated the thickness Δ of the β layer, for isotropic conditions, represented by the equation

$$\frac{\Delta^2}{2t} = D_\beta \frac{\left(C_\beta^L - C_\beta^\alpha\right)\left(C_L^\beta - C_\alpha^L\right)}{\left(C_L^\beta - C_\beta\right)\left(C_\beta - C_\alpha^\beta\right)} \qquad (3.146)$$

where t is the isothermal annealing time; D_β is the average interdiffusion coefficient in the β phase; C_β^α and C_α^β are the compositions of β phase in equilibrium with the α phase and that of α phase in equilibrium with the β phase, respectively; and C_β is the average composition of the β phase.

The peritectic transformation is of great significance when the solute diffusivity is large. For example, the most rapid conventional peritectic transformation in the Fe-C system allows a rapid interstitial carbon diffusion and consequent completion of transformation just a few degrees below T_P.[167] Based on the isothermal experiments, the time dependence of the thickening of austenite in Fe-C systems follows $n = \frac{1}{2}$, consistent with Eq. (3.146),[168] and is eclipsed by peritectic transformation.

Peritectic reaction in Fe-C system leads to some undesirable effects such as creation of tensile stress (leading particularly to the surface cracking of continuously cast slabs), precipitation of inclusions, segregation of alloying elements, and so forth.[168a]

3.8.3.3 Direct Solidification of Peritectic Phase. As the temperature decreases below T_P, the driving force for both the solid-state transformation of α to β and the solidification of β from the liquid increases. Hence, calculations of the total thickness or percentage of β formed below T_P consist of contributions from three stages of peritectic growth. However, Fredricksson and Nylen[165] have focused on the contribution of direct solidification in their studies on Ag-Sn and Cu-Sn systems and suggested the following equation for the formation of solid β fraction by direct solidification, which uses the Scheil equation and assumption of complete mixing in the liquid and no solid-state diffusion.

$$C_L^\beta = C_L^0 f_L^{k_0-1} \tag{3.147}$$

where k_0 is the distribution coefficient of B in the β phase and C_L^0 is the average composition of the liquid at the onset of direct solidification, which is nearly equal to the peritectic liquid composition C_L. According to their model, the temperature at which β forms depends on the surface energy factor γ in Eq. (3.145). A discussion of the agreement between theoretical and experimental values of β phase thickness in Cu-20 wt% Sn and Ag-Sn systems is given elsewhere.[138]

3.8.3.4 Solidification Growth Kinetics.

The breakdown of a steady-state planar front growth of initial composition C_0 can be predicted by the constitutional supercooling criterion as

$$\frac{G_L}{V} \geq \frac{mC_0(1-k_0)}{k_0 D_L} = \frac{\Delta T_0}{D_L} = \frac{\Delta T_{\text{planar}}}{D_L} \tag{3.148}$$

In peritectic systems, the equilibrium liquidus-solidus temperature interval ΔT_0 for composition C_0 can be very small in order to achieve planar front growth. In these cases, alloys with composition $C_0 < C_{P\alpha}$ grow as single-phase α without the formation of the β phase. Alloys with compositions $C_0 > C_{P\beta}$ can grow as single β phase without any α phase, if β-phase nuclei form. Initially α phase will be formed, followed by the decrease of interface temperature below T_P near the steady state; and finally, if β phase nucleates, it will grow very fast over the α phase.

In reality, planar front growth becomes unstable between the limits of constitutional supercooling, i.e., between

$$V_c = \frac{G_L D_L}{\Delta T_0} \tag{3.149}$$

and the limit of absolute stability

$$V_a = \frac{\Delta T_0^\gamma D_L}{k_v \Gamma} \tag{3.150}$$

where ΔT_0^γ is the nonequilibrium liquidus-solidus interval, Γ is the Gibbs-Thomson coefficient, and k_v is the nonequilibrium distribution coefficient. The critical velocity for nonequilibrium liquidus slope (m_v) and k_v to change drastically with velocity is given by the diffusional velocity (or diffusion rate) across the interface V_D (of the order of ms^{-1}).[138]

3.8.3.5 Layered Structure Formation.

Several theoretical and experimental studies have been made to gain greater understanding of the formation of layered or banded structure in peritectic systems, directionally solidified at low velocities in which two phases form alternate layers (or bands) that are aligned parallel to the interface. Boettinger[169] and Brody and David[170] have plotted the conditions of layer formation on the G_L/V versus C_0 plot, assuming the steady-state solidification condition. A recent model developed by Trivedi[171,172] takes into consideration the interaction between nucleation and growth for the two phases and non-steady-state conditions for the two phases for the formation of layered structure.

By using the equation developed by Tiller et al.[61,173] for the initial transient at a given velocity, the widths of each phase layer can be given by

$$\lambda_{ip} = \frac{D_L}{V} \ln \frac{\Lambda_i}{k_i}$$
(3.151)

where λ_{ip} is the spacing of the ith peritectic phase in the band ($i = \alpha$ or β), k_i is the equilibrium distribution coefficient for phase i, and Λ_i is a function of phase diagram parameters and the nucleation temperatures of the α and β phases. Note that, for a specific alloy system, the functions Λ_i are a function of composition only.[171] The total band spacing $\lambda = \lambda_{\alpha p} + \lambda_{\beta p}$, and

$$\lambda = \frac{D_L}{V} \left(\ln \Lambda_\alpha^{1/k_\alpha} + \ln \Lambda_\beta^{1/k_\beta} \right)$$
(3.152)

or $\lambda \propto D_L/V$, that is, λ is proportional to the diffusion length l_D.

From this model, some important restrictions for banding can be specified. The requisite criteria for the formation of non-layered morphology are given below:[138,172]

1. Sufficient convection is present because it leads to complete mixing in the liquid, destabilizes the layer formation, and prevents the interface to buildup the necessary boundary layer.

2. $C_0 < k_\alpha C_l^M$ or ΔT_N^β required for the nucleation of the β phase is large, i.e., below the α solidus temperature for the composition C_0. In the latter situation, the plane front α reaches C_0 and will grow in a steady-state manner without β forming in the steady-state regime.

3. In the situation that $C_0 > k_\beta C_l^M$ only one band forms, followed by steady-state growth of the plane front β. Hence, there is a narrow range of composition for the formation of peritectic bands (Fig. 3.53), given by the condition

$$k_\alpha C_l^M < C_0 < k_\beta C_l^M$$
(3.153)

This condition imposed on the phase diagram provides the following necessary criterion for band formation:

$$\frac{k_\alpha}{k_\beta} < \frac{\Delta T_M^{\alpha/\beta} - \Delta T_N^\alpha}{\Delta T_M^{\alpha/\beta} + \Delta T_N^\beta}$$
(3.154)

where $\Delta T_M^{\alpha/\beta}$ is the difference of the melting points of α and β. Phase diagrams which do not satisfy this condition will cease to form bands. Note that Eq. (3.154) is based on the nucleation undercoolings for the discrete phases. Hence, it depends on both the system and the nucleation state of the melt in contact with both phases, with the crucible and probable effects of presence of any inclusions or precipitates and convection.

This helps to describe the variation in results reported for different systems, as discussed earlier. Clearly, if both nucleation undercoolings are zero, the above inequality gives rise to $k_\alpha < k_\beta$, which is always the case in peritectics. In this situation, bands would not form.[138,172]

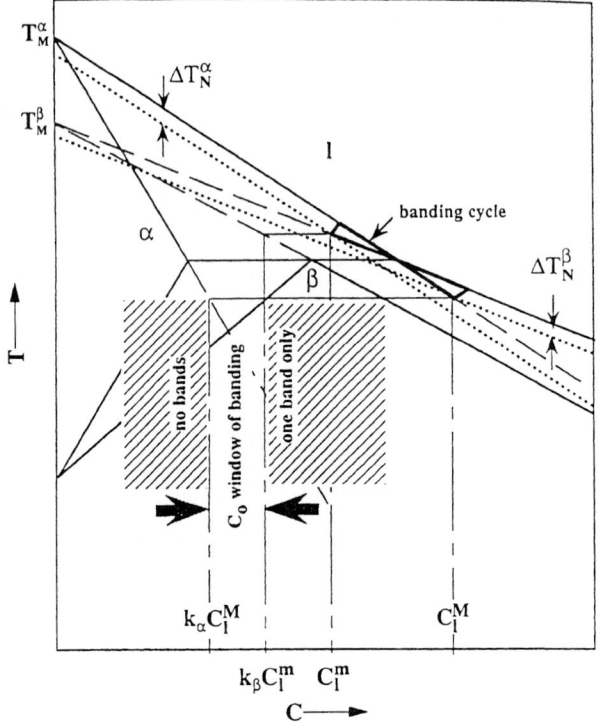

FIGURE 3.53 Peritectic phase diagram with banding cycle and concentration window for banding ($k_\alpha C_l^M < C_0 < k_\beta C_l^m$) which depends strongly on nucleation undercooling.[138,172] (*Courtesy of International Material Reviews.*)

3.9 SEGREGATION

Segregation occurs in solidified metals and alloys at two distinct length scales, macroscopic (macrosegregation) and microscopic (microsegregation). In general, segregation is the result of solute rejection at the solidification interface during solidification of a casting. Macrosegregation develops on the scale of entire casting product whereas microsegregation develops on the scale of dendrite. Macrosegregation depends on the type of morphology of the S-L interface: planar or columnar dendrite. Complete elimination of macrosegregation is very difficult.

3.9.1 Macrosegregation

Many early attempts to model macrosegregation have centered on the shrinkage effects as the only source of interdendritic liquid flow. Using a continuum model, Xu and Lu examined the shrinkage phenomenon in Cu-Al alloy and proposed that a large pressure gradient is required deep in the mushy zone to replenish liquid to feed associated volume changes. Consequently, shrinkage-induced flows might be predicted to dominate in the regions next to the solidus. They demonstrated macrosegregation patterns for the completely solidified ingot.

Mostly, macrosegregation in a solidified ingot is caused by a mechanism involving physical movement of liquid or solid phases through the interdendritic channels or regions in the L-S zone driven by solutal buoyancy and solidification contraction.[174] However, the settling of precipitated phase early in solidification or the deformation of solid skeleton of the mushy zone can also modify solute redistribution on the macroscopic scale. The motion of free liquid, which partially penetrates into the mushy region and presumably gives rise to the formation of segregated channels (freckles), has been simulated by the Boussinesq approximation method.[175] Figure 3.54 is a schematic representation of solidification of an ingot.

The macrosegregation (or solute redistribution) equation has been derived based on the following assumptions: (1) liquid composition is uniform in a sufficiently small volume element, which is considered as differentiable; (2) to perform a mass balance, solute enters or leaves the volume element only by liquid flow to feed shrinkage; (3) the liquid and solid phases have different densities; (4) there is complete diffusion of solute in the liquid, but no solid diffusion; (5) there is no solid movement; and (6) there is an absence of voids.

The character of (longitudinal) macrosegregation can be expressed quantitatively in three manners using necessary thermal relations and the Scheil equation [Eq. (3.55) or (3.56)].[7,176] The L-S region (or mushy zone) is treated as permeable (or porous media) to interdendritic liquid flow, as shown by the schematic volume element in Fig. 3.55. Using D'Arcy's law, the interdendritic liquid flow through the element with a fluid velocity v is given by

$$v = -\frac{K_p}{\eta f_L}(\nabla P + \rho_L g) \qquad (3.155)$$

where K_p is the permeability of a mushy zone; $\eta, f_L,$ and ρ_L are the viscosity, volume fraction, and density, respectively, of the interdendritic liquid; ∇P is the pressure gradient; and g is the gravitational acceleration.

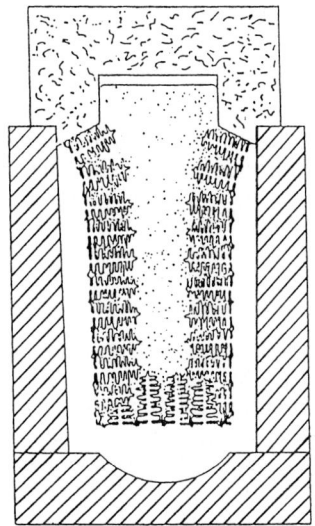

FIGURE 3.54 Sketch of an ingot solidifying in a metal mold (shaded) with a refractory "hot top." The center of the ingot (dotted) is fully liquid; the outer portion (white) is fully solid; and a semisolid region exists between the two.[73] (*Reprinted by permission of VCH, Weinheim.*)

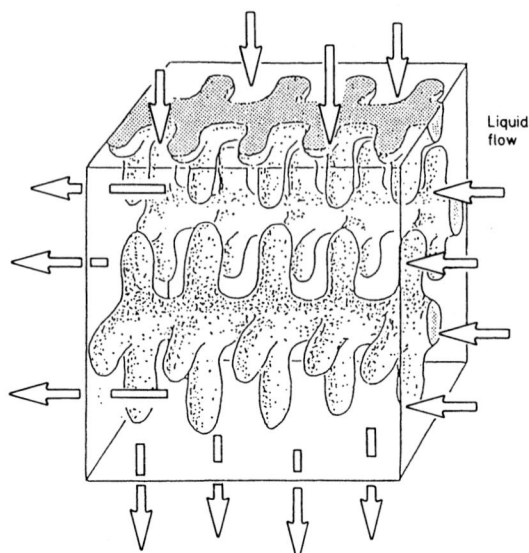

Liquid
flow

FIGURE 3.55 Fluid flow through a solidifying "volume element."[73] (*Reprinted by permission of VCH, Weinheim.*)

The solute flow may be gravity-driven, acting on a fluid of varying density within the L-S zone, solidification shrinkage-driven, convection-driven in the bulk liquid, or electromagnetic or centrifugal acceleration-driven. It may also be solid movement- or deformation-driven, as in bulging in continuous casting.[73] Since the liquid composition varies spatially within the L-S zone, any interdendritic flow except parallel to the isotherm must change the composition of the volume element. Conservation of solute leads to the modification of the Scheil equation into a new local solute redistribution equation, given by[175–177]

$$\frac{\partial f_L}{\partial C_L} = -\frac{1-\beta}{1-k_0}\left(1+\frac{v\nabla T}{\varepsilon}\right)\left(\frac{f_L}{C_L}\right) \tag{3.156}$$

where k_0 is the distribution coefficient, f_L is the volume fraction of liquid, $\beta = (\rho_S - \rho_L)/\rho_S$ = solidification shrinkage (or contraction), v is the velocity of interdendritic liquid flow, ∇T is the local temperature gradient vector, and ε is the local rate of temperature change. This expression considers steady-state solidification with planar isotherm moving with velocity V in the x direction, which becomes

$$\frac{\partial f_L}{\partial C_L} = -\frac{1-\beta}{1-k_0}\left(1+\frac{v_x}{V}\right)\left(\frac{f_L}{C_L}\right) \tag{3.157}$$

where v_x is the isotherm velocity normal to isotherms. Equation (3.157) converts to the Scheil equation when both v_x and β equal zero. It also reduces to the Scheil equation (written in terms of volume fraction) if the interdendritic flow velocity just equals the flow required to feed solidification shrinkage

$$v_x = -\frac{\beta}{1-\beta}V \tag{3.158}$$

In the steady-state solidification, macrosegregation is not observed. When flow is in the same direction and greater than (down the temperature gradient form) that of Eq. (3.158), it causes a negative segregation. When the speed is lower and in the opposite direction, it clearly produces a positive macrosegregation (for normal alloys where $\beta > 0$ and $k_0 < 1$). This is usually found in ingots and is called *inverse segregation*,[178] since it is reversed from what we would predict based on the initial transient of plane front growth.[4]

Figure 3.56 shows a schematic illustration of the application of the above principles to continuous casting. When interdendritic flow lines remain all vertical, no segregation results. When flow lines resemble those of Fig. 3.56b and c, respectively, negative segregation and positive segregation result. In the latter case, greater downward flow toward centerline promotes *centerline segregation*[73] (see also the next section for segregation in continuous casting).

Normal Segregation. Normal segregation, also called positive segregation, results in the rejection of a high concentration of the low-melting-point composition at the S-L interface and its accumulation (for $k < 1$) or lack (in the case of $k > 1$) at the last or central portion of the ingot or casting (Fig. 3.57a). Normal segregation is

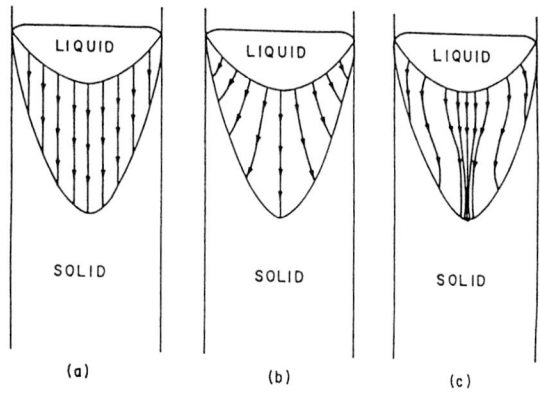

FIGURE   3.56 Interdendritic fluid flow in a continuous casting: (*a*) no segregation, (*b*) negative segregation at midradius, and (*c*) positive segregation at centerline.[7] (*Reprinted by permission of McGraw-Hill, New York.*)

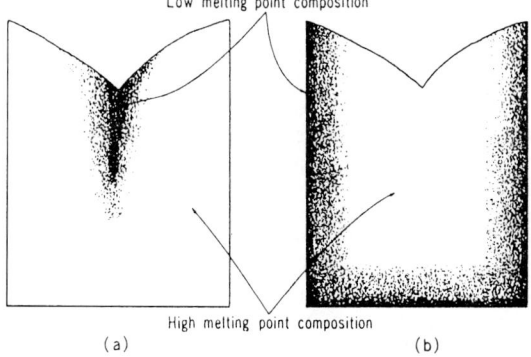

FIGURE 3.57 Normal segregation and inverse segregation of an alloy ingot.

often found in steel ingots, where higher concentrations of P and S are observed in the central region than in the outer one. Average concentrations are used to explain normal segregation, and microsegregation effects are assumed to be negligible and are ignored.[63] The solute distribution for $k < 1$ is predicted to conform broadly to one or the other of the types shown in Fig. 3.19b and c. The amount of segregation will be greater in the case of an alloy with a larger segregation coefficient value; however, normal segregation tends to lessen if the separation of crystals from the mold wall takes place by thermal convection or other crystal detachment mechanisms.[9] The principles of normal segregation are used in the zone refining of impure metals, which has been discussed previously.[69,82]

Inverse Segregation. In the inverse segregation, lower-melting-point composition is usually observed in the outer portion of the ingot, and the higher-melting-point region lies in the central region of the ingot (Fig. 3.57b). Alloys with a wide freezing range are especially prone to this type of segregation. A typical example is the zone of low-concentration negative segregation of impurities that often exists in the lower central portion of the steel ingots. In Al-Cu alloy castings, the outer surface contains low-melting (eutectic) alloy, and the central one the primary crystals of high melting point, which constitute the inverse segregation. Hence it is customary to remove about a 1-mm-thick surface by machining prior to subjecting to the subsequent plastic deformation. In Cu-10% Sn alloy casting, exudations of 20 to 25% Sn are often observed; these exudations are known as *tin sweat*.

The occurrence of these exudations or so-called inverse segregation results from the entrapping of low-melting-point composition of the interdendritic liquid during solidification. Its extent is a function of solidification time and the amount of contraction-stimulated flow.[63] As solidification progresses, the solid shell contracts and pulls away from the mold wall. Then the enriched low-melting interdendritic liquid in the center of the casting is captured by the roots of the crystals on the mold wall and is forced to the contracted region, where it begins to solidify.[9,179]

Banding Segregation. Banded microstructures consist of alternate structures or phases which develop mostly parallel to the transformation front in the unidirectional solidified alloy ingot (in contrast to the traditional solidified structures such as dendrites and eutectics which form perpendicular to the S-L interface). They form when the solute-rich liquid concentrated at the advancing S-L interface becomes unable to escape and is trapped among dendritic side arms of the crystals (because of the presence of greater undercooling by the rejected solutes). Usually they appear as two different bands which alternate in time. Different types and mechanisms of banding segregation are schematically represented in Fig. 3.58.[172]

Banding phenomenon has been found (1) at relatively low speeds, in welded or laser-treated materials; (2) in directionally solidified peritectic Ag-Zn, Sn-Sb, Sn-Cd, Zn-Cu, Ti-Al, and Ni-Al alloys; (3) in off-eutectic or off-monotectic compositions;[138a] and (4) in rapidly solidified Ag-Cu, Al-Cu, Al-Fe, Al-Pd, and Al-Zr alloys.[172] In eutectic cast irons, such a banding segregation occurs after passing gas bubbles through the melt. This is, however, eliminated by preventing any source of gas in the mold or within the flux used (see also Sec. 3.8.3.5).

Banding segregation in centrifugal castings occurs only with larger wall thickness (in excess of 2 to 3 in., or 50 to 75 mm) and not in thinner wall castings. They are characterized by a hard demarcation line at the outer edge of the band that usually disappears into the base metal of the casting. Most alloys susceptible to banding have wider solidification range and greater solidification shrinkage. It is believed to result from vibration, variations in gravitational force between the top

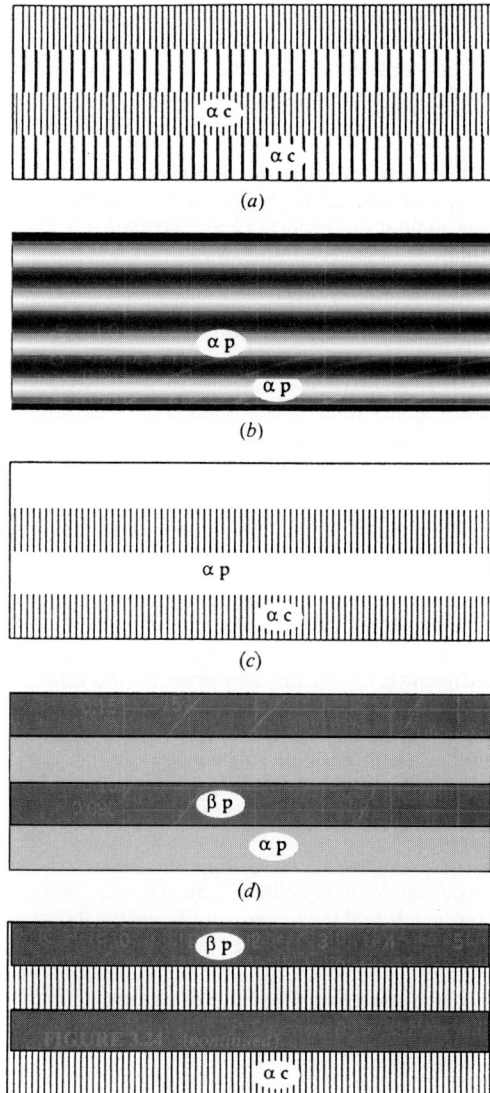

FIGURE 3.58 Schematic representation of different types and mechanisms of banding segregation: (*a*) different scales of same structure and phase α; (*b*) fluctuations of composition in plane front growth of same phase (usually do not form visible bands); (*c*) different microstructures of same phase; (*d*) different phases (α, β) with plane front growth morphology; and (*e*) different phases and microstructures (c = cells, p = plane front).[172] (*Reprinted by permission of The Metallurgical Society, Warrendale, Pa.*)

and bottom of the mold, and irregularities in the flow of the liquid metal during the entrance into the rotating mold.[180]

Gravity Segregation. Usually this occurs during early stages of solidification, before, or just after the initial growth nuclei have formed. It is caused by the density variation between the solid and liquid phases or between two nonmixing liquid phases (exhibiting miscibility gap) which leads to differential movements within the liquid. For example, Cu and Pb exist as two layers in the molten state, and unless they are stirred thoroughly during freezing, the lower-density Cu-rich layer remains on top of a Cu-Pb portion. Similarly, during the zone leveling of Ga-Si alloys, the Ga-rich portion settles at the bottom of a horizontal ingot whereas the Si-rich portion is at the top.

Gravity segregation in polyphase systems is of even greater significance and has been widely found in bearing alloys. During the solidification of the Sn-base bearing alloys containing Cu and Sb, two primary intermetallic phases form; for example, in isolation, cuboids of SbSn (lower density) float on the surface and needles of Cu_6Sn_5 (higher density) sink. In combination, they form an entangled network and remain unsegregated.

Negative cone segregation often occurring in steel ingots has been presumed to be formed by the settling of free primary dendrites or melted-off dendrites into the bottom of a steel casting, because of its density higher than the liquid (see the next section).

When an Al-Si alloy with hypoeutectic composition is cooled rapidly, only hypoeutectic primary crystals form throughout the entire casting; but when it is cooled slowly, both hypo- and hypereutectic sides of the alloy appear separately in the upper and lower part of the casting. The extent of gravity segregation depends on the amount of Fe, Mn, and Si and the thermal condition of the melt. Gravity segregation disappears when a melt with the recommended Fe, Mn, and Cr compositions is thoroughly stirred and mixed after heating to at least 720°C.[181] The reasons for this segregation have been attributed to (1) flotation of the primary Si particles to the top of the melt, (2) the nucleation of primary Si, especially at the sidewalls and the bottom of the molds, and (3) the localized growth of such primary crystals at the expense of the nucleation and growth of the primary Si in the bulk of the melt.[182] Figure 3.59 shows the segregation after the solidification of Sn-Pb alloy.

Likewise, during solidification of hypereutectic Al-Si alloys, clusters of primary Si particles have been found: (1) in slowly solidified castings, including block and bar, sand castings, and step castings in sand and graphite molds; (2) by rapid solidification such as in wedge castings, and in laser-treated surfaces; and (3) during isothermal holding in the freezing range.

The sink rate is controlled by a form of Stokes' equation which states that a Stokes or terminal velocity is the maximum velocity achieved by a solid particle of radius r falling through a convection-free liquid and corresponds to the point at which the relative weight of the particle just balances the viscous drag by the fluid; this is given by the relation

$$V = \frac{gr^2(\rho_S - \rho_L)}{9\eta} \tag{3.159}$$

where ρ_S and ρ_L are the densities of the solid and liquid, respectively. It is influenced by the shape of the particle, forced convective stirring, and centrifugal effect.

If lighter solids such as nonmetallic inclusions and kish or spheroidal graphite are formed early in solidification, they can float to the upper part of the casting, resulting in positive segregation areas.

In centrifugal casting, centrifugal force stimulates gravity and can result in the compositional variations between the internal and external parts of the casting. It is pointed out that gravity segregation does not occur in single-phase liquid.

Channel Segregation. Examples of channel segregation include "inverse-V" segregation; "A" segregate channels in large-scale steel billets and ingots with mostly

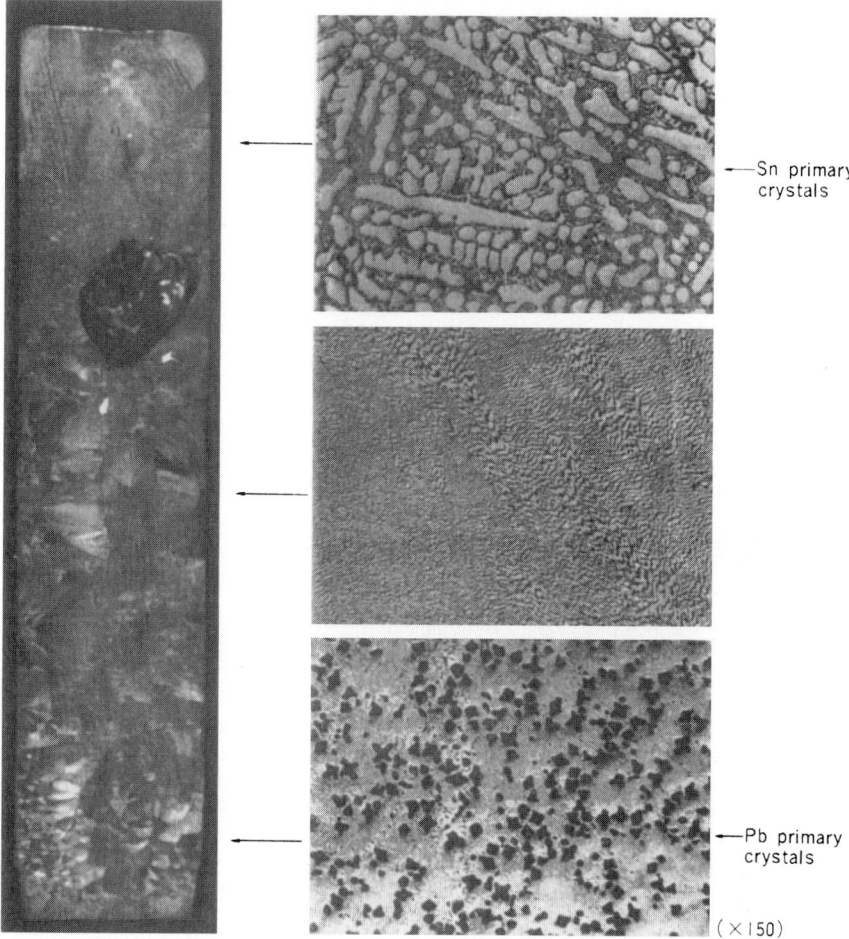

FIGURE 3.59   Gravity segregation after the solidification of Sn-Pb alloy.[179] (*Courtesy of A. Ohno.*)

horizontal heat flow; and "freckles" formation in unidirectional solidified Al-Mg and
Al-Mg-Cu, Pb-Sn, Pb-Sb, or Pb-Sn-Sb alloys and superalloys. This is caused by
convection due to the density variation between the bulk liquid and solute-rich
interdendritic liquid. Channel-type segregation occurs when the solidification
rate is below a critical value V_c, which is given by

$$V_c = c\left(\frac{\Delta\rho_L}{\Delta T_L}\right)^{0.6} (\text{Sin}\,\theta)^{1.3} \qquad (3.160)$$

where c is a constant (= 1.8×10^{-4} for Al alloys and 0.5×10^{-4} for steel ingots), $\Delta\rho_L$
is the density variation in the interdendritic liquid for a temperature change ΔT_L,
and θ is the inclination angle of the solidification interface to the horizontal plane.[183]

Freckles are formed mostly on the casting surface in single-crystal or direction-ally solidified castings; however, some may be found inside the castings. Metallur-gical examination reveals that freckles consist of chains of small equiaxed grains and eutectic constituents.[183a] Particularly, in directionally solidified single-crystal superalloy parts, freckles are usually a reason for rejection.[183b]

The Lutwig-Soret Effect. This type of segregation is an interesting behavior in which homogeneous melt of, say, Zn-Sn, Cu-Sn, and Pb-Sn alloys of uniform com-position, held in a temperature gradient, becomes nonuniform in composition[34] because solute transport related to temperature occurs in one direction and chem-ical diffusion in the opposite direction. The Soret coefficient is equal to D'/D, where D' and D are thermal and chemical diffusion coefficients, respectively.[17] That is, Sn always migrates to the higher-temperature end. A large concentration gradient could be generated in this manner, e.g., for a 36% Pb in Sn alloy the excess Pb content at the cooler end was 5.28%. Again, convection must result in a remark-able divergence from ideal behavior.[63]

3.9.2 Segregation Patterns in Steel Ingots

Ideally, an ingot with a homogeneous composition is desired, but during the solidi-fication of steel, solute elements such as C, Mn, S, and P become concentrated in the liquid ahead of the advancing dendrites, resulting in microsegregation between dendrites, called interdendritic microsegregation (due to settling of free-floating solid or flow of solute-rich liquid into or out of the volume elements during solidi-fication) and macrosegregation for dendritically solidified materials over large dimensions typically ranging from millimeters to the size of the entire ingot casting. Macrosegregation is of great concern because it may carry through the defect into the final product. In large killed steel ingots, three major types of macrosegregation such as A segregation, V segregation, and negative segregation are commonly found.[184] Figure 3.60 is a schematic representation of structure and segregation pattern in a killed big-end-up ingot.

The factors that are responsible for the formation of vertical macroscopic segregate channels called A segregates are the extent of solute enrichment, the dependence of steel density on solute concentration, and the morphology of the dendritic network. They are formed in the same way as freckles.

V bands, probably occurring during disturbances or fluid flow in the interden-dritic region, may also be observed running parallel to the solidification front. Since the V segregation forms in the center of the ingot and during the final stage of solid-ification process, it possesses the largest degree of solute (such as S, P, and C) enrich-ment. The inverse V segregation in steel is usually observed as string-shaped.[185]

The negative cone of segregation (implying lower concentration) normally appears as a zone in the bottom third of the ingot, and is associated with the pile of dendritic debris. This region also contains a relatively high concentration of inclu-sions, probably trapped by the equiaxed crystals in the core bottom of the ingot.

Other zones of positive segregation (implying a higher alloying content) also occur near the centerline, primarily at the top of the ingot. In fact, all three types of segregation arise from the unsolidified pool and are influenced by the ingot struc-ture (i.e., the relative amount of the columnar and equiaxed zones). Both positive and negative segregates can form by various mechanisms, the most common being interdendritic flow due to solidification shrinkage or density difference, generating a natural convection during the solidification process.[185] It is pointed out that any

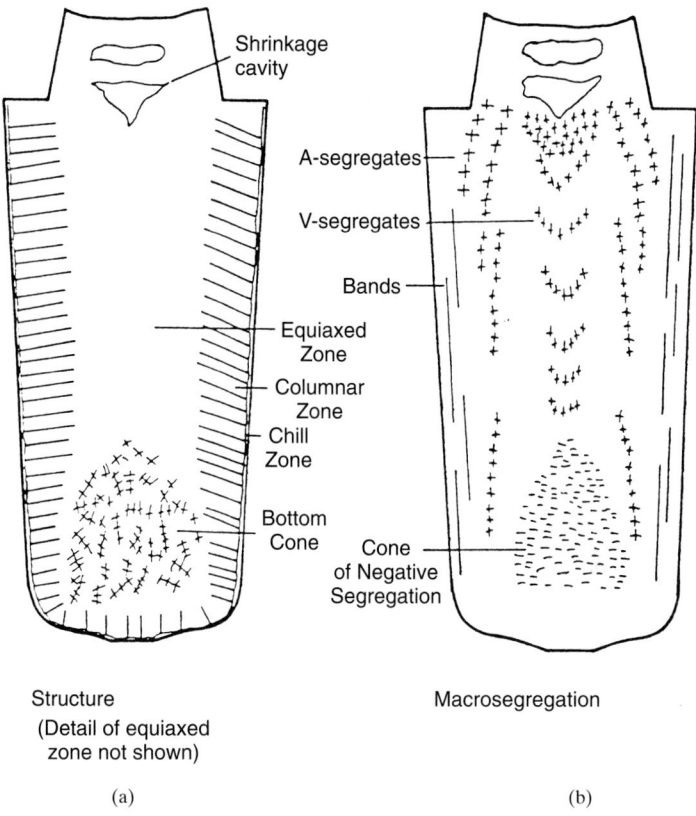

FIGURE 3.60 Schematic representation of (*a*) structure and (*b*) segregation pattern in a killed, big-end-up ingot. (*Reprinted by permission of Pergamon Press, Oxford.*)

factors that influence structure and fluid flow must also affect the segregation in ingots. More segregation is observed in rimmed ingots than in killed ingots. The rim zone displays negative segregation whereas positive segregation is measured at the center.[184]

3.9.3 Segregation Patterns in Continuous Steel Casting

Continuously cast billets solidify mainly in a dendritic structure that can lead to heavy centerline segregation and cavities. However, axial segregation, center unsoundness, and V segregates are less predominant in continuously cast steel. The axial segregation (Fig. 3.61*a*) and center unsoundness in billets are often found periodically along the strand that may pose serious quality problems. The factors responsible for severe axial segregation and porosity are minimal equiaxed structure, large superheat, predominant columnar structure, no electromagnetic stirring (EMS) application in the upper part of the strands, increased section size, higher

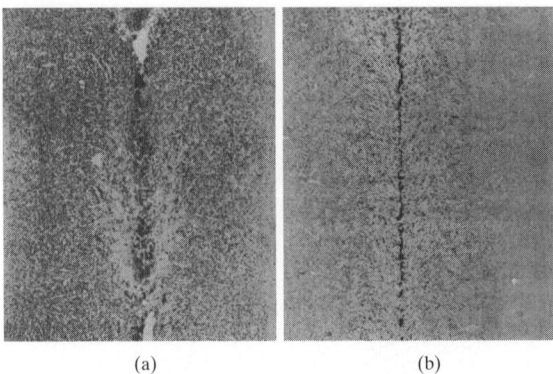

(a)   (b)

FIGURE 3.61   Sulfur prints in continuous casting showing axial
or longitudinal segregation in (*a*) a billet and (*b*) a slab. Notice
that segregation in the billet is periodic. (*After Alberny and Birat,
1976, The Metals Society, London. Reproduced with permission.*)

aspect ratio, and carbon content. The carbon content influences both the width of
the central segregation zone and the size of the axial voids: 0.3% < 0.1% < 0.6%
C.[186] For very high carbon contents, and in the case of Cr-bearing grades, the centerline consists of a tubular segregated region. The severity of V segregates increases
with carbon content. The mechanism of formation of V segregates involves the settling of dendrites, volume shrinkage, and interdendritic flow in the very long mushy
zone.

The axial segregation typically observed in slabs is shown in Fig. 3.61*b*. The main
factors responsible for segregation are the stability of the strand, bulging of the slab,
and superheat of the steel.

3.9.4 Microsegregation

Microsegregation is the segregation of solute elements over distances on the order
of cellular spacing, dendritic arm spacing, and grain boundaries. Microsegregation
due to the nonequilibrium solidification and consequent solute redistribution causes
nonequilibrium second phases, porosity, and crack formation.[187] Microsegregation
is observed when the solute diffusivity in the solid alloy is too low to homogenize
this second phase during its growth.[188] For melt alloy systems with $k_0 < 1$, the severity of segregation increases with the decrease in the k_0 value.[189] The extent of
microsegregates in the alloy structure is determined experimentally by measuring
(1) the amount of nonequilibrium eutectic, (2) the amount of nonequilibrium second
phases, (3) minimum solid composition, (4) ratio of minimum to maximum composition of primary phase, or (5) composition versus fraction solid profile.[190]

Microsegregation, due to time-dependent variations of the interface, convection
velocity, or unsteady convection in the melt, is often the major problem for the
crystal growth.[191] For example, rotation of the sample in an asymmetrical thermal
field, pulling device instability or vibration, or temperature fluctuation causes time-
dependent variations of the interface, thereby leading to striations. Another possible mechanism for microsegregation is associated with the unsteady convection flow
in the melt.[192] It can serve either directly on the solutal field, as in the case of g-jitter
during solidification in microgravity, or indirectly when the temperature fluctuations

linked with the flow modify the growth velocity and thereby the composition of the alloy.[193]

Microsegregation varies significantly with the history of the growth of the solid. For example, microsegregation frequently increases with the cooling rate for equiaxed dendritic solidification; however, it decreases for unidirectional solidification. Microsegregation can also change considerably if a phase transformation occurs during solidification due to the variation of k_0 values with phase.[194] In binary Al-4.5 wt% Cu alloy, the intercellular fluid flow has a small, but perceptible effect on microsegregation and cell morphology.[188]

During solidification of the melt when the uniform solute distribution is usually lost, it gives rise to a nonuniform cored solute distribution over distances of dendrite arm spacing, a spacing dependent on cooling rate. An extreme case of this segregation is the formation of insoluble inclusions with similar spacing. Although some nonuniformity can be eliminated by heat treatment after solidification, much segregation still persists and may cause a reduction in mechanical properties.[189] The reason is that, in the former case, the liquid composition is uniform in the interdendritic liquid whereas in the latter case there is a solute built up on the dendrite tip.

Cellular Microsegregation. During the early stages of growth of cellular solidification in single-phase alloys, low degrees of constitutional supercooling (CS) develop; the liquid near the advancing interface is richer in solute for $k_0 < 1$. This difference in solute concentration from the cell center to the cell boundaries is called *cellular segregation*, which is taken as the result of the cell thickening process and extends over distances on the order of the cell size of $\sim 5 \times 10^{-3}$ cm. Once the rounded cell projection is stabilized, solute will be rejected ($k_0 < 1$) from the sides of the projections as well as from the top, and the accumulation of solute will occur in the cell boundaries. The most severe segregation will be found at the junction points (nodes) in the hexagonal array. The situation then appears as shown in Fig. 3.62. For $k_0 > 1$, the cell boundary regions are depleted of solute.

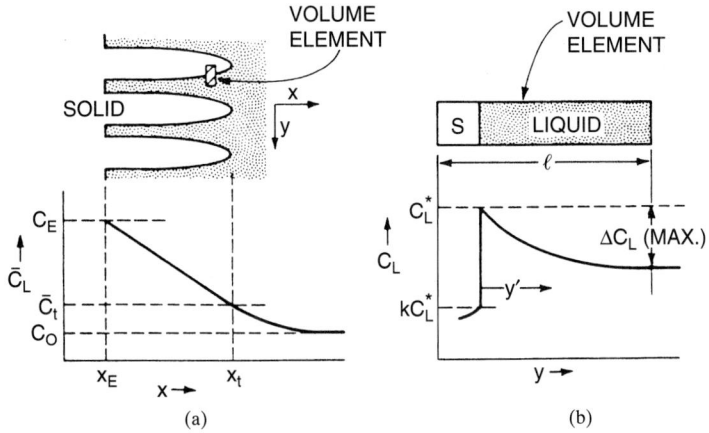

FIGURE 3.62 Cellular redistribution (microsegregation) during cellular growth. (*a*) Cellular growth and solute distribution in the growth direction. (*b*) An enlarged "volume element" and solute distribution transverse to the growth direction.[195] (*Reprinted by permission of Pergamon Press, Oxford.*)

Quantitative studies have shown that the solute concentrations at the cell boundary nodes could be up to two orders of magnitude greater than the melt average concentration. Annealing times of the order of 1 hr should be adequate to remove cellular segregation, assuming the solid-state diffusivity of $\sim 10^{-8}$ cm^2/s and a cell size of $\sim 5 \times 10^{-3}$ cm.

Ma and Sahm[195] have extended the Bower et al. model by taking into consideration the effects of cell geometry on the solute rejection in both longitudinal and lateral directions and, therefore, on the final segregation profile. It is suggested that the forward solute rejection at the advancing S-L interface could result in a reduction of solute enrichment in the intercellular region, and, thereby, a flatter segregation profile of alloying elements is anticipated. The application of the modified model holds well with the experimental data.[195]

Dendritic Microsegregation. In the interdendritic region, first solid to form has a concentration $<C_0$. For $k_0 < 1$, where solidification occurs first and last, the solute concentration is $<C_0$ or $\gg C_0$. Figure 3.63 shows the microsegregation in Sn-10 wt% Pb alloys according to equation

$$(C_L - C_S)df_S^* = (1 - f_S)dC \qquad (3.161)$$

where C_L [$= C_0(1 - f_S)^{k_0-1}$] and C_S are concentrations in the liquid and solid at the interface, respectively, in weight percent and f_S is the weight fraction of solid. When $C_L = C_E$ and $1 - f_S = f_E$, the liquid solidifies to eutectic mixture with a concentration C_E.[195] The microsegregation resulting from solute redistribution during dendritic solidification leads to coring in the primary phase, which is a variation in the solute concentration between the center and the outside of dendrite arms (Fig. 3.64). In extreme instances the accumulation of solute between the growing dendrite arms can result in the formation of second phases in the interdendritic region in amounts significantly higher than those predicted from the equilibrium diagram.[195a]

Brody and Flemings first considered the diffusion in the solid during dendritic solidification using the Scheil nonequilibrium solidification equation. In the simplest form, the equation for solute redistribution at the S-L interface during linear dendritic growth is

$$C_S^* = k_0 C_0 \left(1 - \frac{f_S}{1 + \alpha k}\right)^{k_0-1} \qquad (3.162)$$

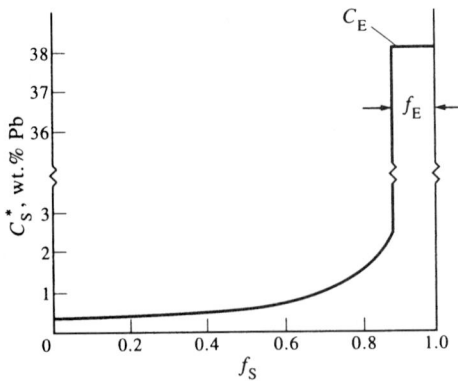

FIGURE 3.63 Dendritic microsegregation in Sn-10 wt% Pb alloys.[195] (*Reprinted by permission of Pergamon Press, Oxford.*)

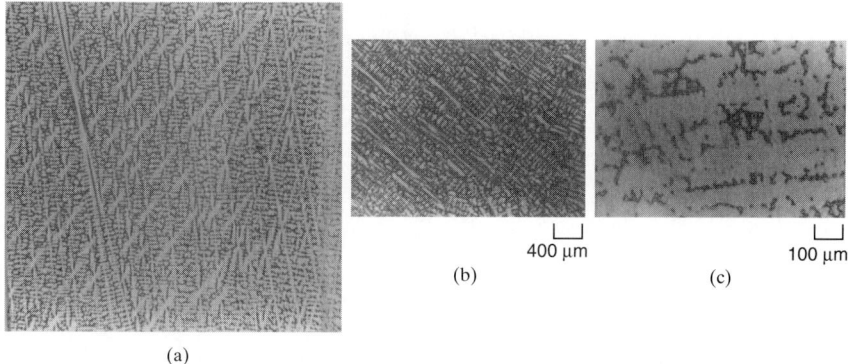

(b)

400 μm

(c)

100 μm

(a)

FIGURE 3.64 (*a*) Coring or microsegregation during directionally solidified Ni-60% Cu alloy. The centers of the dendrite are deficient in copper which segregates to the dark etching regions between the dendrite arms that are the last portions to solidify. (*b, c*) Dendritic microsegregation at the mid-radius location of the longitudinal cross section of the as-cast 718 (Ni-18.5Fe-18Cr-5.3Nb-3Mo-0.9Ti-0.5Al-0.03C) electroslag remelting (ESR) ingot. (*b*) The low-magnification microstructure reveals a dendritic solidification pattern with numerous primary dendrite spines and secondary arms. (*c*) The high-magnification microstructure reveals the enriched (in Nb, Mo, and Ti) interdendritic regions, which reflect the last metal to solidify.[195a] (*a: Reprinted by permission of Pergamon Press, Oxford, Plc; b, c: Reprinted by permission of TMS, Warrendale Pa.*)

and for parabolic growth of dendrite

$$C_S^* = k_0 C_0 [1 - (1 - 2\alpha k_0) f_S]^{(k_0-1)/(1-2\alpha k_0)}$$

where C_S^* is the solute concentration in the solid at the S-L interface, k_0 the equilibrium partition coefficient, C_0 the alloy composition, f_S the weight fraction of solid, and α the diffusion Fourier number, defined as $D_S t_f / \lambda^2$, where D_S is the solute diffusivity in the solid, t_f the local solidification time, and λ one-half of the characteristic *dendritic arm spacing* (DAS).[91,196] When $\alpha k_0 \ll 1$, there is practically no solid-state diffusion. When $\alpha k_0 \gg 1$, there is almost complete homogenization, for example, for the carbon distribution in Fe-C alloys, where very fast interstitial diffusion of carbon occurs.

Figure 3.65 illustrates the complex nature of isoconcentration profiles for a low-alloy steel columnar dendrite after complete solidification. Table 3.4 lists the microsegregation ratio k_0 (i.e., ratio of maximum solute concentration to the minimum solute concentration after solidification) found at 1.7, 2.5, and 5.75 in. from the chill, for Mn and Ni. However, the real microsegregation ratio is not easily estimated. Note that the microsegregation analysis also predicts the amount of non-equilibrium eutectic or other secondary phases.[4] Both these parameters are required to evaluate the extent of microsegregation. Microsegregation is more severe across and between primary dendrite arm spacings than secondary dendrite arm spacings. The microsegregation ratio is largest at low alloy contents, although there is no basic difference in the solidification mode. Both columnar and equiaxed dendrites display essentially the same microsegregation feature.[135]

The DAS defines the range of dendritic microsegregation. The smaller the DAS, the easier to homogenization by subsequent heat-treatment. For a given alloy heat-treated at a fixed temperature, the homogenization time varies as the square of the DAS. For coarse DAS ($\sim 10^{-2}$ cm) in steels, heat treatment of ~ 300 hr at 1200°C is required (in the absence of simultaneous working) to produce any appreciable

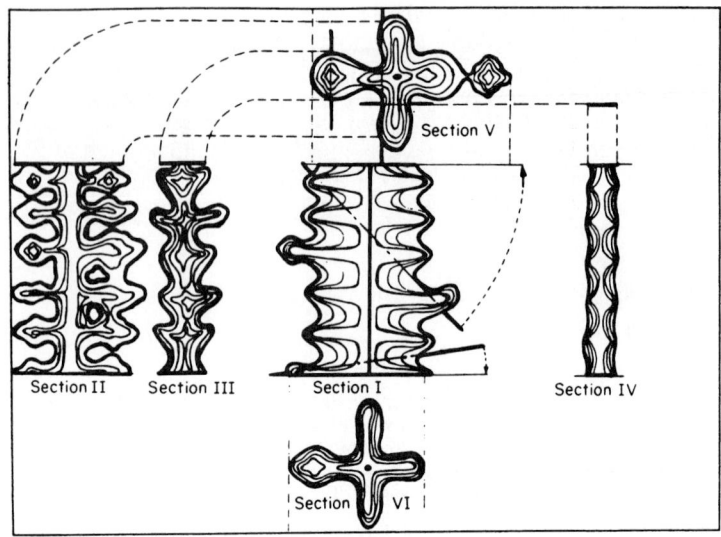

FIGURE 3.65 Isoconcentration profiles for a low-alloy steel columnar dendrite.[135] (*Reprinted by permission of McGraw-Hill, New York.*)

TABLE 3.4 Electron-Microprobe Measurements of Microsegregation[135]

Element	Distance from chill, in.	Solute concentration (max)	Solute concentration (min)	k_o
Mn	1.7	0.68	0.45	1.52
Mn	2.5	0.88	0.62	1.39
Mn	5.75	0.83	0.46	1.79
Ni	1.7	2.33	1.92	1.21
Ni	2.5	2.11	1.85	1.14
Ni	5.75	2.16	1.61	1.34

reduction in dendritic microsegregation. Simultaneous working improves the situation only slightly. Refining of DAS by increasing cooling rate can therefore have extremely advantageous effects.

Complete removal of dendritic microsegregation gives rise to improved mechanical properties. This is especially difficult if second-phase particles are formed. For example, Turkdogan and Grange[197] observed the formation of second-phase sulfide inclusions in the solute-rich interdendritic regions in steel toward the final stage of freezing. These inclusions were extremely stable, inhibited grain refinement, and along with the solute microsegregation were believed to be responsible for subsequent banding of the wrought steel. The importance of dendritic microsegregation to the occurrence of banding or fibering had been previously discussed by Flemings.[198]

Grain Boundary Segregation. Grain boundary (GB) segregation during solidification originates from two sources. First, if the GB lies parallel to the growth direction, surface energy requirements lead to a GB groove, where the boundary meets
the interface. The groove is typically about 10^{-3} cm deep.[199,200] In the presence of CS,
conditions are energetically favorable[201] for considerable segregation to GB groove.
The second type of GB segregation occurs due to the impingement of two interfaces moving with a growth component perpendicular to each other which may be
regarded as a form of microsegregation.[73]

3.10 SOLIDIFICATION PROCESSES AND CAST STRUCTURES

This section deals with three traditional solidification processes: ingot casting, continuous casting, and welding; however, the last two processes hold a key importance
in today's technology. Almost entire molten metal is cast into solid state by either
the traditional ingot casting or relatively modern continuous casting. The development of the ingot and continuously cast structure as well as fusion welding structure and weld cracking are briefly described.[59]

3.10.1 Ingot Casting

Ingots are cast shapes produced with a cross section that is suitable for rolling,
extrusion, and forging operations. They are commonly used in steel and copper
industries. In ingot casting, steel is poured into a top or bottom of cast iron molds.
The molds are often equipped with "hot tops" comprising insulating boards and
exothermic compounds to greatly reduce the depth of the shrinkage cavity formed
during the solidification of ingot (Fig. 3.54). After a required period, the molds are
removed from ingots and charged into soaking pits for rolling into finished products.[101] Unlike continuously cast semifinished products, ingots have larger thickness
in at least one transverse direction, the depth of liquid core is small, solidification
occurs in a stationary space, and time is measured from the viewpoint of a fixed
observer.[184]

3.10.1.1 Classification of Steel Ingot. Steel ingots can be classified into four
types according to the deoxidation practice used or, alternatively, by the amount
of gas evolved during solidification (Fig. 3.66). The gas is generated primarily by
the reaction between carbon and oxygen dissolved in the steel, which is favored
thermodynamically at lower temperatures.[184] These types are called killed, semikilled, capped, and rimmed steels. If practically no gas is evolved, the steel is termed
killed because it lies quietly in the molds. Increasing extents of gas evolution result
in semikilled, capped, or rimmed steels.[202,203]

Killed steel ingots are fully deoxidized so that there is only a slight or practically
no evolution of gas during solidification. These ingots are normally cast in "hot-
topped, big-end-up molds" to reduce the depth of the shrinkage cavity. The top of
the ingot solidifies much faster than in semikilled or rimmed steel. Killed steel is
characterized by a homogeneous structure (free of blowholes and segregation), even
distribution of chemical composition and properties, and formation of pipe in the
upper central portion of the ingot, which is later cut off and discarded. All steels
having more than 0.3% carbon are killed. Alloy steels, forging steels, and carburiz-

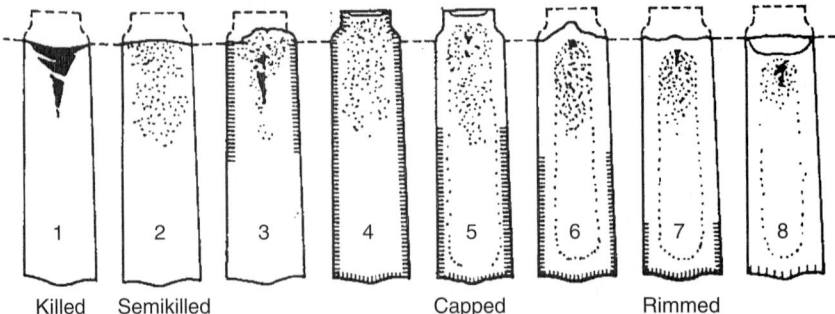

Killed   Semikilled   Capped   Rimmed

FIGURE 3.66   Classification of steel ingot according to the deoxidation practice used. These commercial ingots are cast in identical bottle-top molds, where the degree of suppression of gas evolution ranges from maximum for completely killed ingot (No. 1) to that of a minimum for violently rimmed ingot (No. 8).[204]

ing grades of steels are usually killed, where the essential emphasis is placed on soundness and homogeneity of structure.[184,204–206]

Killed steel is produced by various steel melting practices involving the use of certain deoxidizing elements which act with varying intensities. The most common deoxidizing elements are Al and ferroalloys of Mn and Si; however, calcium silicide and other special strong deoxidizers such as V, Ti, and Zr are sometimes used. Deoxidation practices in the manufacture of killed steels are generally left to the discretion of the producer.

Semikilled steel ingots are partially deoxidized so that gas evolution is not completely suppressed during solidification by deoxidizing additions. There is a greater degree of gas evolution than in the killed steel, but less than in the capped or rimmed steel. An ingot skin of considerable thickness is formed prior to the beginning of gas evolution. A correctly deoxidized semikilled steel ingot does not have a pipe, but does have well-scattered large blowholes in the top-center half of the ingot; however, they weld shut during rolling of the ingot. Semikilled steels, generally, have a carbon content in the range of 0.15 to 0.30%; they are commonly used in structural shapes, plates, merchant bars, skelp, and pipe applications.

The main features of semikilled steels are (1) variable degrees of uniformity in composition, which are intermediate between those of killed and rimmed steels, and less segregation than rimmed steel and (2) pronounced tendency for positive chemical segregation at the top center of the ingot (Fig. 3.66).

Rimmed steel ingots are characterized by only a small amount of deoxidation and a greater degree of gas evolution during solidification in the mold and a marked difference in chemical composition across the section and from top to bottom of the ingot (Fig. 3.66). This results in the formation of an outer ingot skin or rim of relatively pure iron (hence the name *rimming steel*) and an inner liquid (core) portion of the ingot with higher segregation/concentration of alloying elements, especially carbon, nitrogen, sulfur, and phosphorus, having lower melting temperature. The high-purity zone in the surface is preserved during rolling.[21] Most low-carbon steels (<0.25% C) are rimming steels. Rimmed ingots are best suited for the manufacture of flat-rolled products (plates, sheets) as well as wires, rods, and other products in the cold-rolled or subcritically annealed condition, where good surface or ductility is required.[206]

The technology of producing rimmed steels limits the maximum content of carbon and manganese, and these maximum contents vary among producers. Rimmed steels are less expensive to make than killed or semikilled steels (because they are tapped without deoxidizer addition in the furnace and with only small addition of deoxidizer made in the ladle) and have better ingot surfaces. Rimmed steels do not retain any significant amount of highly oxidizable elements such as aluminum, silicon, or titanium.

Capped steel ingots have similar characteristics to those of a rimmed steel ingot, but to a degree intermediate between those of rimmed and semikilled steels. Less deoxidizer is used to produce a capped ingot than to produce a semikilled ingot.[19] This induces a controlled rimming action when the ingot is cast, i.e., thinner zone and less segregation in the core. The gas entrapped during solidification is in excess of that required to counteract normal shrinkage, resulting in a tendency for the steel to rise in the mold.

Capping is a variation of rimmed steel practice. The capping operation confines the time of gas evolution and prevents the formation of an excessive number of gas voids within the ingot. The capped ingot practice is usually applied to steels with carbon contents greater than 0.15% that are used for the production of sheet, strip, tin plate, skelp, wire, and bars.

Mechanically capped steel is poured into bottle-top molds using a heavy cast iron cap to seal the top of the ingot and to stop the rimming action.[207]

Chemically capped steel is cast in open-top molds. The capping is accomplished by the addition of aluminum or ferrosilicon to the top of the ingot, causing the steel at the top surface to solidify rapidly. The top portion of the ingot is cropped and discarded.

3.10.1.2 Ingot Structures.   Generally, the chill castings are found to have complex structure comprising three distinct structural zones (Figs. 3.60 and 3.67):

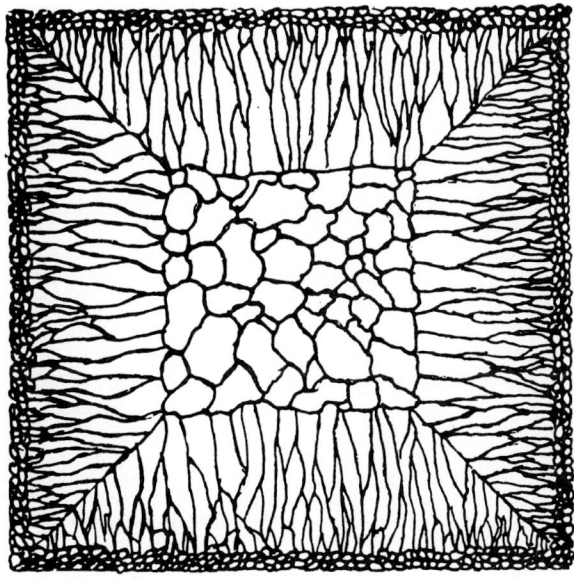

FIGURE 3.67 Chill casting showing three distinct structural zones. (*Reprinted by permission of McGraw-Hill, New York.*)

the chill zone which is composed of numerous fine, equiaxed grains with random orientations; the columnar (dendritic) zone, i.e., a band of elongated crystals aligned parallel to the direction of heat flow; and a central region of equiaxed (nondendritic) zone. Alloy castings may exhibit either entirely columnar or wholly equiaxed grain structures, depending on the alloy composition and solidification conditions. Within each grain a substructure of cells, dendrites, and/or eutectic prevails.

Chill Zone. The formation of a chill zone structure on the outer surface of the ingot involves complex interactions of liquid metal flow, metal-mold heat transfer, plenty of nucleants, and dendritic growth. The factors responsible for the formation of the chill zone at or near the mold walls by copious nucleation events and grain multiplication by fragmentation mechanisms include low pouring temperature (low superheat), fluid flow by convection current or vigorous stirring of the melt, a very cold mold (high rate of heat extraction), and a rough mold surface. These grains rapidly become dendritic.

Columnar Zone. Moving inward, the grain structure of the chill zone eventually evolves into a columnar structural zone, having grains with their preferred ⟨100⟩ crystallographic orientation (for the fcc and bcc alloys), relative to the maximum thermal gradients in the melt. The substructure in the columnar zone shows all stages of development through planar to cellular to cellular-dendritic.[208] Thus the dendrites are oriented perpendicular to the ingot surface along heat flow lines; in killed ingots the dendrites are usually inclined slightly upward to the mold wall, implying thereby that the liquid steel was flowing down the advancing solidification front, at least in the initial stages of solidification. If convection is reduced by magnetic fields or mold rotation, perpendicular columnar growth can be found.[209]

For a particular alloy, the degree of columnar region increases with the pouring temperature (Fig. 3.68a). For a given pouring condition, the columnar region decreases with alloying additions (Fig. 3.68b).[65] With pure metals, the as-cast structure exists normally as entirely columnar.

When columnar growth takes place in concentrated alloys having a low-temperature gradient, a substructure different from cellular-dendritic can some-

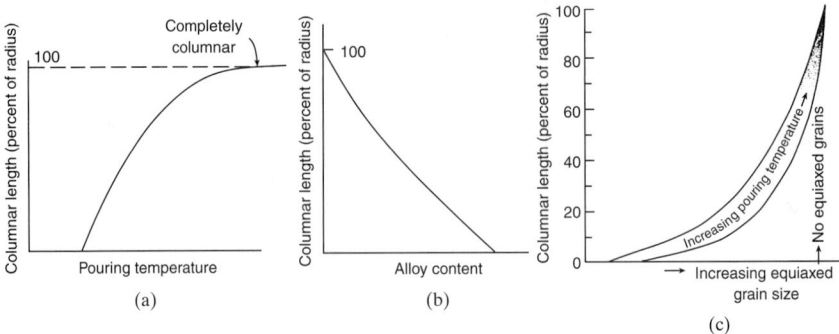

FIGURE 3.68 Variation of the length of the columnar zone with (*a*) pouring temperature for a particular alloy and (*b*) alloying additions for a given pouring temperature. (*c*) Correlation between columnar length, equiaxed grain size, and pouring temperature.[65] (*Reprinted by permission of John Wiley & Sons.*)

times be visible. Dendrite groups, often called *superdendrites*, originate out of a planar dendrite array (due to thermal instability) and grow cooperatively with a large sawtooth front. They evolve because equilibrium dendrite spacing becomes greater than the actual spacing, and then some particular dendrites will grow ahead of their nearest neighbors to minimize solutal interaction. These superdendrites have been found in various alloy systems such as Pb-base alloys, medium-carbon steels, nickel-base superalloys, and Al-Cu alloys. [210]

Under certain solidification conditions, namely, high thermal gradient, a minimal concentration of alloying elements, and convection associated with hot liquid feeding, Al alloys can develop a special, unusual microstructure called *feathery crystals* which is quite different from the conventional columnar or equiaxed grain morphologies.[211] This structure is composed of elongated grains which are observed to consist of twinned lamellae (or dendrites) that are parallel to the (111) twin plane. These structures are usually associated with continuous, direct-chill, or similar semi-continuous castings and welding processes.

Feathery grain growth is not desirable for both aluminum producers and users, because it leads to strong anisotropy, causing uneven deformation during hot rolling, heterogeneous recrystallization, nonuniform aspect after etching, and an abnormal microsegregation pattern. However, grain refinement methods can be employed to prevent it to a great extent.[212]

Equiaxed Zone.   Figure 3.68c is the correlation between columnar growth, equiaxed grain size, and pouring temperature. The tendency to form equiaxed grains decreases with the increase of pouring temperature; however, their occurrence results in a coarser grain size. The formation of an equiaxed zone in the center of the casting requires both the presence of active nuclei ahead of the columnar dendrites (or front) and favorable conditions promoting their growth.[213]

Columnar-to-Equiaxed Transition.   The columnar-to-equiaxed transition (CET) is a classical problem in the solidification behavior of cast alloys. CET in some as-cast structure may be sharp or gradual, depending on the mold wall and the size of the columnar grains. The CET occurs earlier when an alloy has higher solute contents, smaller thermal gradients, higher nuclei density present in the melt, and more vigorous melt convection.[213a] That is, the CET occurs when the equiaxed grain nuclei are sufficient in size and number to hinder the advance of the columnar front. The CET has also been reported to be sensitive to many casting parameters such as alloy composition, cooling rate, mold material, casting size, trace elements present in the melt, and so forth.[213b] The prediction of CET is of great significance for the evaluation and design of mechanical properties of solidified products. It is therefore essential to understand the CET mechanisms. Qualitatively, it can be expected that CET takes place earlier when an alloy has lower pouring temperature, larger solute content, smaller temperature gradient, higher nuclei density present in the melt, and more thorough melt convection.[214] Three theories have been proposed to explain the origin of equiaxed nuclei or CET. The first one, *constitutional supercooling (CS) theory*, predicts that the equiaxed crystals nucleate after the formation of columnar zone due to CS of the remainder liquid. The second one, known as the *big-bang mechanism* or *Chalmer's theory*, points out the possibility that all the crystals, equiaxed as well as columnar, originated during initial chilling of the liquid layer in contact with the mold walls.[215] Ohno has proposed a separation theory for the big-bang mechanism where the source of free equiaxed grains may be secondary dendrite arms that detach from the solidification front by a similar mechanism. The survival and growth of the equiaxed grains depend primarily on the superheat of the liquid and convection.[216] The third theory, proposed by Jackson et al. and O'Hara and Tiller, suggests that a remelting mechanism of dendritic arms is responsible for the formation of the equiaxed region. Alternatively, dendritic fragmentation occurs from the columnar grains or from dendritic crystals nucleated at the top of the ingot because of radiation cooling occurring in that region.[217]

Among these mechanisms, a combination of the last two theories appears to be responsible for equiaxed grains where convection (natural, reduced, or forced) is present. The CS mechanisms appear to be effective when convection is minimized or removed, which seems to be quite difficult. Mold-wall materials and coatings have a decisive effect on the structure and substructure of the ingot surface.[218]

3.10.2   Continuous Casting

Continuous casting is the established and relatively modern method of production of semifinished steel products such as billets, blooms, and slabs. More recently, single crystals have been developed. Various workers have reviewed and contributed in the fields of steel and nonferrous alloys which are described elsewhere.[17,219–227]

Continuous casting offers the following advantages over the ingot casting process: (1) It yields a semifinished shape (billet, bloom, or slab) or near-net shape casting in a nearly continuous fashion, unlike ingot casting, which produces unfinished shape of large cross-section. (2) It offers increased productivity and yield, reduced energy and manpower, improved product quality, and consistency of quality over the traditional ingot casting process. (3) It is the preferred casting process and provides lower emissions harmful to the environment and plant operator, reduces stock levels and enables shorter delivery times, and offers reductions in capital costs for new steel plants. Additional advantages of this process are the total elimination of primary mills, linking of the continuous caster directly with the reheating furnaces and rolling mills, and directing of the casting sections to the final product dimensions, thereby leading to further reduction of operations and investment costs.[228]

In contrast to ingot casting, in continuous casting, the depth of the liquid core may easily reach 20 m and above, which is far greater than in an ingot casting; as a result, the solidification of the center of a continuously cast section is largely removed from the mold. Solidification occurs in a moving section, and time is viewed from the standpoint of a reference frame that moves into the strand. In addition, the continuously cast strand usually changes its orientation from vertical to horizontal during its movement through the casting machine. Submerged pouring tubes and electromagnetic stirring are normally used in continuous casting rather than ingot casting. Most of these differences affect fluid flow conditions in the liquid core, which significantly affects the structures and segregation.

Historically, the use of shorter molds (700 to 1500 mm) in continuous casting has a clear cost advantage at speeds of up to 5 m/min. However, with the current trend toward high, thick slabs, longer molds seem to be required. Longer molds can improve operational safety during transient conditions and allow a longer heating period in case of sticker alarms.[229]

3.10.2.1   Continuous Casting of Steel.   In continuous process, the liquid steel is first poured from a ladle or melting furnace into a refractory-lined tundish, which then directs steadily its flow into a stationary, vertically oscillating water-cooled copper mold (contained by steel backing plates) controlled by stopper nozzle or a slide gate valve arrangement (Fig. 3.69*a*). Prior to start-up of a casting, a movable

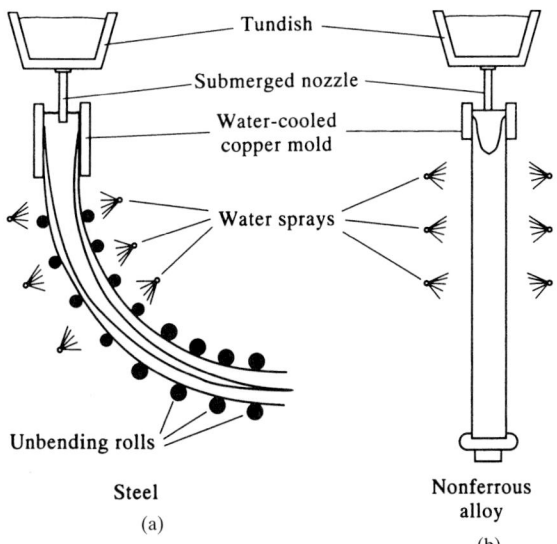

FIGURE   3.69 Schematic representation of the continuous casting process of (*a*) steel and (*b*) aluminum.

starter or dummy bar is introduced into the bottom of the mold and sealed in order to restrict the initial flow of steel from the mold and to form a solid skin. After commencement of pouring the mold, the dummy bar is gradually withdrawn at a rate proportional to the addition of molten steel into the mold. The strand of steel that follows immediately is semisolid (a liquid core encased in a solidified thick shell) over much of the length of the casting machine. Usually a reciprocating motion is superimposed toward the downward travel to prevent sticking and to eliminate breakout. After the completion of the solidification stage, the steel strand is cut by welding torch into the desired lengths.

Many of the solidification phenomena discussed earlier regarding the killed steel ingots also occur in continuous casting. In the continuous casting mold machine, fluid flow results from the momentum of the input stream(s), like ingot casting, during and just before teeming. With increasing time (or distance down the machine in continuous casting), natural convection (of fluid motion) takes over until the latter stages of solidification (near the end of the pool in continuous casting), whereupon solidification shrinkage and buoyancy driven by concentration become important.

The accumulation behavior of inclusions in a slab depends, to a large extent, on the fluid flow in the mold. To control the metal flow of molten steel in the mold and, therefore, the occurrence of inclusions in the slab, it is essential to control and prevent (1) the downward outlet stream of liquid steel from the immersion nozzle in order to avoid penetration too deeply into the mold and (2) entrainment of inclusions near meniscus and fluxes into the outlet stream. It is possible to reduce the number of inclusions and pinholes in slabs, especially at the inside accumulation zone of curved mold slab casters, by conducting downward longitudinal stirring (with a linear motor type of in-mold electromagnetic stirrer). It also reduces the penetration depth of the nozzle outlet stream and decelerates the molten steel flow rate at the meniscus, leading to the stabilization of the meniscus.[230]

Examples of a recent trend in continuous casting technologies of steel are beam blank casting, strip casting, and thin slab casting. Beam blank casting is employed as a starting material for hot-rolled H beams.

Continuous Cast Structure. The broad structural features of continuously cast billets, blooms, or slabs are similar to those of an ingot. Moving inward from the surface, usually, a chill zone, a columnar zone, and a central equiaxed zone, are observed. The origin of each of the zones appears to be same as in an ingot. The creation of equiaxed crystals that finally restricts the growth of the columnar dendrites, for example, almost certainly starts in the mold by a crystal multiplication mechanism comprising the remelting of secondary dendritic arms. The effect of superheats on the length of the columnar zone is the same as in ingot casting; increasing superheat increases the size of the columnar zone, presumably due to the small equiaxed crystal remelt.

By the application of electromagnetic stirring, nuclei are formed and distributed in the liquid core of the strand. These nuclei develop an equiaxed crystal core zone upon solidification. Application of EMS in the mold, however, can reduce the length of the columnar zone at a given level of superheat. This finding supports the theory that the free equiaxed crystals are created in the mold or just below it. The influence of superheat on continuous cast quality is of greater importance than for ingot casting because of smaller cross section and generation of predominantly smaller equiaxed structural zone as well as the fact that (1) segregation, central porosity, and internal crack formation are all minimized by a large equiaxed zone because of less frequent nozzle blockage and improved inclusion float-out; and (2) less reduction in area is imparted during rolling to the final product.

The solidifying shell in a continuous casting strand does not always form uniformly in the mold. Steel having 0.1% C characteristically displays a wavy shell with locally thin regions that are not observed in higher-carbon (>0.4%) steels.[231] This nonuniformity, which restricts heat transfer to the mold, is most likely connected with the $\delta \rightarrow \gamma$ phase transformations. Better roll containment, decrease in casting speed, or increase of spray water also overcomes these problems.

In continuous casting, the production of good-quality billets for low-carbon steels depends on a very low oxygen activity in the mold in order to avoid pinhole occurrence, and low total oxygen contents to minimize the amount of inclusions. This last requirement becomes important in the case of the billet casters with small radius where the inclusions accumulate near the upper side of the billets.[232]

Alloys made of Ca, Mg, and rare earth elements are used to remove, at least partly, nonmetallic trace elements in steels and to control the composition and morphology of residual inclusions.[233] Inclusion modification by CaSi injection (1) prevents the formation of Al_2O_3 inclusions and the resulting nozzle clogging and (2) produces cleaner steel with very good surface, improved mechanical properties, and anisotropy.[234] In the Al-killed continuous-cast steels, Ca treatment provides a steel with good fluidity if a critical ratio of Ca/Al is achieved; prevention of nozzle clogging; possibility of longer sequence; reduction in product rejects; and effective control of oxides and sulfides, particularly in low-silicon steel and in construction steel grades such as *electroslag refining* (ESR) quality.[235]

Selection of appropriate mold powder such as TIS ULC flux containing 46% (SiO_2 + Al_2O_3), 40.2% (CaO + MgO + SrO), 1.2% (Na_2O + K_2O + Li_2O), 2.3% (MnO + Fe_2O_3), 8.5% F, and 1.6% C (with higher surface tension and viscosity and low solidification temperature) has been reported to have considerable promise in reducing the entrainment and entrapment of the in-mold slag and achieving a higher surface quality of the cast slab and lower sliver occurrence in the coil.[236]

In a horizontally cast strand, crystal formations sink down whereas the upper part of the strand still solidifies with a dendritic structure. To avoid this, special stirring has to be applied to increase the initial homogeneity of the billet and to prevent deviation of the metallurgical center. Through the use of special stirring, a high-quality strand is attainable. Thus both surface and internal quality improvements are attainable with an optimum horizontal continuous casting process design.[237]

3.10.2.2 Continuous Casting of Nonferrous Alloys.

In continuous casting of nonferrous alloys, the molten metal is continuously poured within formed molds (rolls, belts, or a wheel and belt) which are constantly in motion and solidify as it moves along the water-cooled copper molds (Fig. 3.69b). The advantages of continuous casting include improved surface quality of the cast bar and more uniform structure. Rolls and belts are employed mainly in the casting of sheets, plates, foils, and strips for various thicknesses. In general, nonferrous bars and rods for the production of wire are cast using wheel belt, dip forming, and upcasting methods.[226]

The main advantages of a moldless vertical continuous casting process for the production of Al, Al-Cu, and Al-Si rods are (1) near-net shape material with small and complex cross-sectional shape; (2) full automation preventing breakout of molten metal and allowing easy start-up and easy shutdown; (3) unidirectionally solidified cast structures; and (4) geometries with variable cross-sectional configuration along their axes. An attractive casting process using a heated rather than a cooled mold has been developed by Ohno for single-crystal growth or unidirectional solidification of ingots, rods, and wires of unlimited length.[4]

Ohno Continuous Casting (OCC) Process. Figure 3.70 shows the schematic sketch of the integrated Ohno continuous casting (OCC) process for crystal growth and casting of alloy wires ~1.7 mm in diameter. Essentially, it consists of stainless steel or graphite crucible, a graphite mold, a cooling device, and pinch rolls for withdrawal of the cast product.[238] The process is based on the fact that the mold is heated just above solidification temperature of the metals to be cast to prevent the formation of new crystals on the mold surface. The cooling device is confined to a short distance away from the mold exit where solidification starts just before entering the cooling water, producing unidirectional cast structure. The process variables required in single-crystal growth are the mold exit temperature, the casting speed, and the cooling condition. The diameter of the single crystal or alloy wire decreases

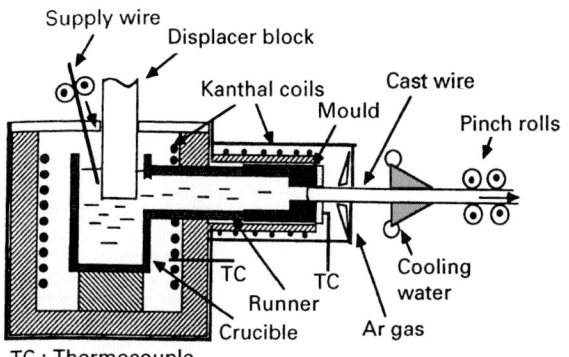

Supply wire
Displacer block
Kanthal coils   Cast wire
Mould
Pinch rolls
TC   TC   Cooling water
Runner
Crucible   Ar gas
TC : Thermocouple

FIGURE 3.70 Schematic illustration of the Ohno continuous casting equipment.[239] (*Courtesy of H. Soda, A. McLean, Z. Wang, and G. Motoyasu, J. Mater. Sci., vol. 30, 1995, p. 5438.*)

with the increased mold exit temperature. Single crystals of small diameter can be easily produced with excellent surface quality for use in vapor deposition applications.[239,239a] Another feature of this process is that the heated mold can be placed in a vertically upward, a vertically downward, or a horizontal position. Ohno has obtained encouraging results with Al, Pb, Sn, Cu, and their alloys.[238] More recently, Kim and Kou[238] and Soda et al.[239] investigated the experimental variables of the OCC process and accomplished numerical modeling of heat and fluid flow. Tada and Ohno extended the OCC principles for the production of Al strips using an open horizontal, heated mold and patented as the Ohno strip casting (OSC) process. This technology has the potential to offer an alternative route to traditional methods such as rolling and extruding for materials that are otherwise difficult to fabricate due to limitations over other physical properties.[239a]

Like most solidification processes, attempts have been made to mathematically model the freezing of a continuous cast nonferrous alloy ingot. For aluminum alloys, a short review has been made by Shercliff et al.[241]

Hot-Top Casting. The hot-top level-pour system has been developed for casting extrusion billets or slabs of normal-purity aluminum and aluminum alloys (containing higher Mg content) in order to achieve a superior cast surface quality, higher casting speed (due to a reduction in effective mold length), and a more uniform and finer subsurface structure that can be exploited to advantage in certain applications requiring a uniform surface finishing response.[241] The metal feeding into the mold occurs with spout and float (Fig. 3.71a).[241a]

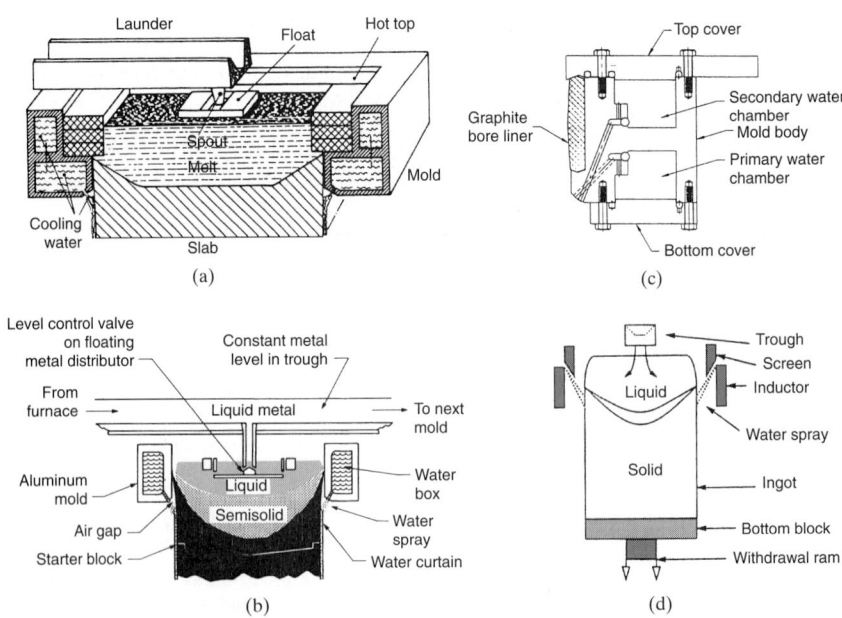

FIGURE 3.71 (*a*) Schematic view of VAW hot-top mold for casting slab.[241a] (*b*) The vertical direct chill (DC) casting process for Al and Cu alloy casting process.[225] (*c*) Schematic cross section of LHC mold.[245a] (*d*) The electromagnetic casting (EMC) process.[249] (*a and c: Courtesy of Light Metal Age. b: Courtesy of Academic Press, Boston. d: Courtesy of TMS, Warrendale, Pa.*)

To avoid molten metal bleedouts at start-up, due to thermal stresses and consequent curling developed in the first formed ingot, the metal level in the mold is established by start-up requirements. Other modifications have been introduced such as Isocast and Alcoa 729 processes that reduce the intensity of water cooling by promoting film boiling. This is achieved by passing CO_2 gas into the cooling water stream, thereby minimizing curl and allowing the ingot to be started with a low head.[242,243] Similarly, increasing casting speed will increase ingot surface temperature and promote film boiling.[243a] The Wagstaff Turbo process uses air bubbles while the Alcan pulsed water process uses a special rotary valve to turn the water on or off.

Direct Chill (DC) Casting. Important variations of direct chill casting process include vertical direct chill (VDC) casting, horizontal direct chill (HDC) casting, low-head composite (LHC) casting, electromagnetic casting (EMC), and so forth.[243b] More recently the trend has been in favor of VDC casting, which allows a far greater number of ingot or slab strands to be cast simultaneously and which permits the more advanced EMC and hot-top mold technologies to be applied.[244] The VDC casting process (Fig. 3.71*b*) is used widely to prepare rectangular rolling ingots or slabs and cylindrical billets of Al, Cu, Zn, and Mg and their alloys. The VDC casting process can directly produce billets for extrusion, blocks or ingots for rolling, and sheet for fabrication, thereby removing the intermediate mechanical processes by casting near-net shapes.[245] This offers the ability to cast the full range of alloys up to the highest aircraft alloys, but requires an expensive hot rolling. This allows a far greater number of ingot strands to be cast simultaneously and the application of more advanced EMC and hot-top mold technologies.

At the start-up, the open-end of the mold is closed off with the bottom block. As the mold begins to fill with degassed and filtered molten metal, the bottom block is lowered at such a rate that the molten metal level is controlled at a particular distance above the lower end of the mold, called the metal head. Thus, during casting, the mold exactly equals the metal exiting. As shown, solidification starts with the formation of a solid shell in the water-cooled mold. This shell shrinks away from the mold, and a gap forms which restricts the heat transfer to the mold. Most of the heat of solidification is eliminated from the ingot as it comes out from the mold into a water curtain which impinges directly onto the surface.

There are three factors that affect the separation of the ingot shell from the mold: (1) shrinkage at the ingot shell itself, (2) thermal strain within the ingot shell, and (3) the shrinkage in the block section under the mold and associated mechanical strains in the shell. The primary and secondary water cooling systems usually influence them and thereby the ingot structure, primarily at the ingot surface. The air gap formed when the solid shell contracts away from the mold can give rise to a rough surface due to a surface liquation phenomenon. This shell shrinks away from the mold, a gap forms which diminishes the heat removal to the mold, and the solidified metal (or skin) reheats, causing surface and subsurface defects such as micro- and macrosegregation. The shell then reheats the point in which a mushy zone extends to the outer surface, and droplets of interdendritic liquid bleed through that surface, to produce a rough surface ("liquation beads") on that surface as further contact is made with the mold.

Reheating produces micro- and macrosegregation exudations, runouts, retardation of the solidification in the subsurface zone, and variations in the cell/dendrite size of the outer surface of ingots. Zones of coarse dendritic substructure may extend 2 to 3 cm below the surface. Additionally, large particles of intermetallic constituents are formed by eutectic reactions, which may be exposed by surface machining that is usually performed before fabrication.[245] Several methods have been proposed to reduce surface defects (such as cold shuts or laps, liquation, butt curl, butt swell, folds, and major and minor bleed-outs) and associated hot tearing propensity. The most successful ones are those that reduce the heat extraction at the mold through the control of the microgeometry of the mold surface, e.g., by machining fine grooves in the face of the mold. The molten Al does not fill the grooves due to surface tension.

Highly automated DC casting is equipped with a programmable controller to control all or some of the parameters such as cast length, casting speed, cooling water, metal temperature, metal flow, and mold lubrication. Additional control capacity addition in the DC mold covers further requirements such as grain refiners wire rod feed rate, hot-top mold gas injection rate, and emergency shutdown due to loss of power.[245a]

Low-Head Composite Casting Technology. The low-head composite (LHC) casting technology is similar to traditional DC casting technology in several ways. This arises from casting with a low metallostatic head in a permeable graphite-lined aluminum mold surface with a strong direct chill. Figure 3.71c is a schematic cross section of an LHC mold.[245a] The additional advantages of LHC mold are:[245b] (1) reduction of scrap rates (up to 2% increase in overall mill recovery between casting and cold rolling); (2) increase of casting speeds by 25 to 40%; (3) reduction of oil usage to 5 gal per month (versus the previous use of 500 gal per month); (4) reduction of shell zones[†245c] or scalping to 50% (with respect to a traditional DC mold); (5) very good ingot geometry; (6) smooth surface, usually free from exudation; and (7) unique method of heat removal during the run portion of the cast.

Electromagnetic Casting. More recently, moldless electromagnetic casting (EMC), which has been developed by Getselev et al.,[246] is finding applications in the production of sheet ingots (i.e., of rectangular cross section) because of the improved metal surface quality. It is a semicontinuous casting operation in which solidification of metal occurs without contact between the liquid metal and the mold wall (Fig. 3.71d). For highly conductive metals, electromagnetic forces are concentrated only close to the surface, extending over only a few millimeters, that supports the liquid metal away from the mold and against gravity at the periphery of the pool and causes stirring.[247]

Although this method has been gaining increasing acceptance since the 1980s, it is confined to non-heat-treatable alloys, because of the difficulty in crack removal associated with the more crack-prone heat-treatable alloys.[247]

Advantages of EMC over DC castings include (1) smooth and segregation-free surface, thereby allowing hot rolling without scalping operation; (2) solidification of the cast slab that is more uniform; (3) edge trimming that is reduced or avoided; and (4) finer than usual cell size in the subsurface area, resulting in reduced or no edge cracking.[223]

Disadvantages of EMC over DC casting are as follows: (1) Higher capital, licensing, and operating costs of EMC do not seem to outweigh improved recoveries. (2) Additional energy consumption is due to dissipation (10 to 30 kWh/ton) of electric power as heat into the metal, screen, or inductor. (3) One of the horizontal (surface) defects observed on EMC ingots is a surface waviness with ~1-cm wavelength, presumably because of the variation of the metal head above the solidification line. This results in leaning of the meniscus outward or inward from its normal position to maintain a balance between electromagnetic and metallostatic pressures, with a

---

† Shell zone is considered as a combination of inverse segregation and depletion.

slight increase or decrease in ingot section. According to Dobatkin, the alloys exhibiting high hot brittleness or longer solidification time have susceptibility to develop surface cracks and coarse-grained structure in EMC. That is, EMC yields coarse-grained microstructure, leading to surface cracks.[248] Developments of EMC led to improvements in conventional DC casting such as control of metal level and development of automation systems.[243a]

There are two similarities between EMC and DC castings.[249] (1) Both are semi-continuous processes in that an ingot, supported on a descending bottom block, is withdrawn continuously from a molten metal pool (the sump) into a casting pit, at a speed of about 1 mm/s, whereas a fresh molten metal flows in the pool from above. (2) In DC casting, the solidifying alloy is in contact with the mold; in EMC, the liquid pool is constrained on the side by electromagnetic forces (see Fig. 3.71d) while it is chilled. These forces are moderated by a band of conducting material, usually stainless steel, partially interposed between the inductor and liquid pool and termed the *screen* or *shield*. Alternatively, water cooling of the screen, inductor, and metal is provided, with the last being by sprays that strike the periphery of ingot just below the solidification line.

3.10.3 Solidification and Structure of Fusion Welds

Weld metal solidification behavior controls the size and shape of grains, the extent of segregation, the distribution of inclusions, the extent of defects such as hot cracking and porosity, and the properties of weld metal. In the last three decades, several excellent reviews have been published emphasizing various aspects of weld solidification and weld microstructures.[4,250–256]

Figure 3.72 is a schematic diagram that describes an autogeneous welding process, exhibiting three distinct zones of a fusion weld. They are the fusion zone

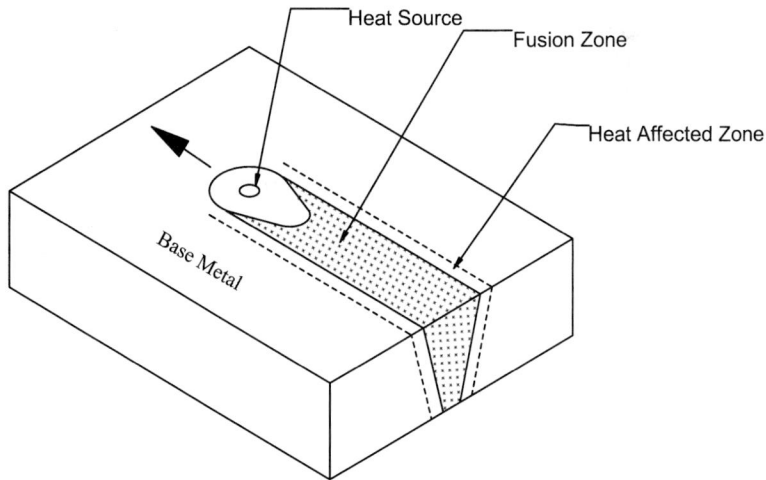

FIGURE 3.72 Schematic diagram showing three zones of a fusion weldment.

(FZ), also called the weld metal; the unmelted heat-affected zone (HAZ) near the FZ; and the unaffected base metal (BM). The characteristics of the FZ depend, to a great extent, on the solidification behavior of the weld pool. However, according to close metallographic evaluation, the FZ can be further divided into three sub-zones: the composite zone (CZ), the unmixed zone (UZ), and a partially melted zone (PMZ), present between the FZ and the HAZ[257] (Fig. 3.73). The UZ occurs in welds with filler metal additions and consists of molten base metal and a resolidi-fied zone without mixing with filler metal additions during the movement of the weld pool. This zone can act as initiation sites for microcracking as well as corro-sion susceptibility in stainless steel. In the PMZ region, the peak temperatures developed lie between the liquidus and solidus which leads to melting of low-melting-point inclusions and segregated zones. After cooling, these areas may serve as potential sites of microcracks.[258]

In this section, weld pool solidification, shape, macrostructure, microstructure, and cracking are briefly discussed.

3.10.3.1 Characteristics of Weld Pool Solidification.   Inherent to the welding process is the formation of a molten weld pool directly beneath the heat source which contains impurities. The weld metal shape is influenced by both the resultant heat and fluid (or metal) flow. A significant turbulence, i.e., good mixing, takes place in the molten metal. The heat input determines the volume of the molten metal and, therefore, dilution, weld metal composition, and the thermal condition. The molten metal volume is small when compared to the size of the base metal. The composi-tion of molten metal is very similar to that of the base metal. There are large tem-perature gradients across the melt. Since the heat source moves, weld solidification is considered as a dynamic process comprising very rapid localized melting and freezing. Crystal growth rate is geometrically related to weld travel speed and weld pool shape. Hence, weld pool shape, weld pool composition, cooling rate, and growth rate are all factors interrelated to heat input that will ultimately affect the solidifi-cation microstructure.

In the spectrum of solidification processes, cooling rates in conventional weld metal solidification lie between cooling rates for most castings (ranging from 10^{-2} to 10^{2} K/s) and those for rapid solidification process (ranging from 10^{4} to 10^{7} K/s). For conventional welding process such as shielded metal arc, gas tungsten arc (GTA), submerged arc, and electroslag welding, the cooling rates may vary from 10 to 10^{3} °C/s. For modern high-energy beam processes such as electron beam (EB) and laser welding (LW), cooling rate is typically of the order of 10^{3} to 10^{6} K/s within the welding pool. Weld pool solidification, therefore, involves features of both extremes of solidification, i.e., conventional casting as well as rapid solidification.

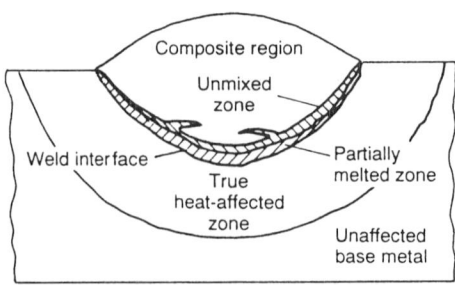

FIGURE 3.73   Schematic diagram of the different subzones of a fusion zone.[257] (*Courtesy of Welding J.*)

The local solidification conditions and cooling rates vary significantly within the weld pool.[250,254]

3.10.3.2   Weld Shape and Macrostructure.

In contrast to a casting, solidification in a weld pool is more transient, there is no nucleation barrier, and chill zone is absent. In fact, in Al- and Fe-base alloys, it has been established that the initial weld pool solidification occurs by epitaxial grain growth at the FZ-HAZ interface (i.e., partially melted solid grains).[251,258,259]

The development of weld pool geometry is influenced by the amount of heat transfer from the heat source to the workpiece, the welding speed, the nature of fluid flow in the weld pool, and the rate at which heat can be removed at the L-S interface. As shown in Fig. 3.74a, at low heat inputs and welding speeds, an elliptically shaped weld pool forms and the columnar grains curve along the welding direction. At high heat inputs and welding speeds, weld pool becomes teardrop-shaped and the columnar grains are straight (Fig. 3.74b). In both situations, grain growth begins from the substrate at the fusion boundary and advances toward the weld centerline. In this manner, the weld metal grain structures adjacent to the fusion boundary are eclipsed by the epitaxial growth process. An elliptical or spherical pool is usually observed in weldments of a higher thermal diffusivity (e.g., Al), whereas a tear-shaped weld pool is more likely in the weldments of a low thermal diffusivity

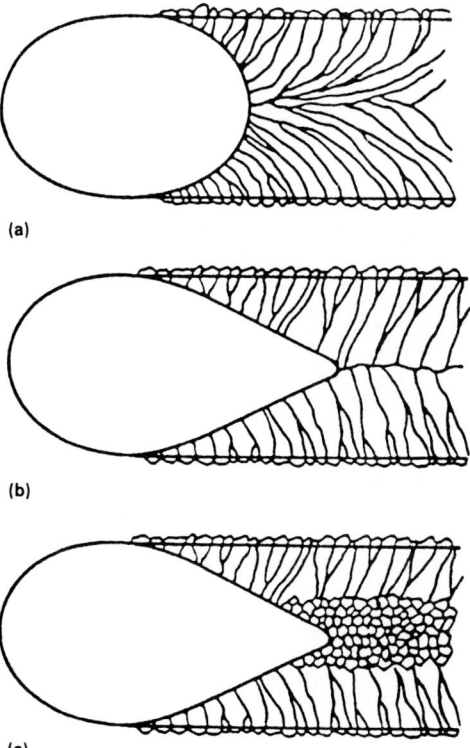

(a)

(b)

(c)

FIGURE 3.74 Schematic showing effect of welding parameters on a grain structure: (*a*) low heat input and low welding speed producing elliptical weld pool; (*b*) high heat input and high welding speed, in the absence of nucleation in the bulk weld metal; and (*c*) high heat input and welding speed, in the presence of nucleation in the bulk weld metal.[261] (*Reprinted by permission of The Metallurgical Society, Warrendale, Pa.*)

(e.g., austenitic stainless steels). In higher-carbon steels or in Al alloys such as 5083 Al alloy, the columnar grains have the tendency to multiply near the weld centerline with a corresponding increase in dendritic arm branching. This provides the formation of stray grain structure.[256]

In most commercial Al alloys, low-carbon steels, and 304 stainless steels (at still higher weld speed, say, >30 to 40 cm/min), equiaxed grains are found near the weld centerline (Fig. 3.74c). In this case, the competitive growth changes from epitaxial columnar grains to nucleation and growth of equiaxed grains in the bulk weld metal, similar to that noticed in ingots and castings.[251,260,261]

In addition to the factors mentioned above, the weld pool geometry is influenced by convectional and the resultant heat and fluid flow behavior due to three driving forces such as buoyancy effects, electromagnetic forces, and surface tension gradient forces. The interactions among these driving forces have been modeled by Wang and Kou,[262] illustrating the effects on the shape and weld penetration in GTA aluminum welds, and by Zacharia et al.[263] in GTA welding of type 304 stainless steel. It is important to mention that impurities in the weld metal, acting as surface active elements in the base metal, can change the surface tension of the liquid metal and its temperature dependence. In the absence of surface active elements such as O_2 and S (<150 ppm) in stainless steel, surface tension decreases with an increasing temperature, and the resultant flow becomes outward away from the center of the weld pool (or toward the fusion boundary). The result is a wider and shallower weld pool. In the presence of surface active elements, a positive temperature coefficient of the surface tension occurs, and the resultant inward flow leads to a deeper and narrower weld pool.[264] Recent work has shown that the controlling effect of surface active elements on the surface tension gradient flow is strongly dependent on the welding process and associated parameters.[265,266]

The ability of computational modeling to predict precisely the weld pool shape is illustrated in Fig. 3.75, where calculated temperature profiles are superimposed on a cross-section micrograph of an aluminum alloy GTA weld.[267] Although the

FIGURE 3.75   Calculated temperature profiles superimposed on a macrograph of a cross section of an aluminum alloy GTA weld.[4] (*After Zacharia et al., Weld J., vol. 67, 1988, p. 185.*)

correlation is rational, further refinement of the model and better thermophysical property data are required to more accurately simulate the real welds.[254]

3.10.3.3 Weld Microstructure.   The development of solidification structures within a weld is mostly dependent on the solidification conditions, which in turn depend on welding parameters. Hence, an understanding of the interrelationship between solidification conditions and welding parameters is essential to comprehend the relationship between welding parameters and microstructure.[251]

The constitutional solidification model is used to qualitatively describe the progress of different microstructures. Figure 3.76 shows the schematic weld microstructures for different solidification modes and the temperature gradients. At curve a, the temperature gradient is steeper than the liquidus temperature curve, and the S-L interface can move by planar front growth. At curve b, the temperature gradient is shallow and results in cellular growth. For curves c and d, large protrusion and cellular dendritic and columnar dendritic growth are likely. At curve e, the temperature gradient is very shallow and leads to the formation of equiaxed dendrites ahead of the S-L interface.[268,269]

Figure 3.77 shows experimentally observed relationships between the solidification parameter, alloy composition, and type of structure developed. The basis of this figure lies in Eq. (3.67). This equation predicts the occurrence of more nonplanar solidification at a smaller value of G_L/V. This figure shows that experimentally the correlation is with $G_L/\sqrt{V}$, rather than G_L/V.[268]

Since increase of travel speed changes weld pool shape from roughly oval to elliptical to teardrop-shaped, this change in weld shape causes a change in the solidification growth rate V_W. Assuming two-dimensional heat flow and isotropic crystal growth of the weld pool, the solidification rate V_W must be related to the travel or welding speed U_W by

$$V_W = U_W \cos \theta \qquad (3.163)$$

where θ is the angle between the normal to the S-L interface and the welding direction (Fig. 3.78a). At the back of the weld pool, when $\theta = 0$, $V_W = U_W$; and at the side of the weld pool, when $\theta = 90°$, $V_W = 0$. Equation (3.163) predicts the variation of V_W around the weld pool, and Fig. 3.79 shows the variation of microstructure with the variation of solidification rate V_W. That is, the results of these two effects lead to the variation of microstructure around the weld pool (Fig. 3.79).[254] A planar front solidification is expected near the sides of the weld pool, and an equiaxed dendritic growth is anticipated near the weld centerline. In between these, the extent of nonplanar solidification front increases from sides to the center of the weld pool.[270]

When anisotropic crystal growth is considered (Fig. 3.78b), Eq. (3.163) must be modified according to Nakagawa et al.[267] to become

$$V_W = \frac{U_W \cos\theta}{\cos(\theta' - \theta)} = \frac{V_n}{\cos(\theta' - \theta)} \qquad (3.164)$$

where V_n is the growth rate in the direction normal to the isotherm, the welding direction and direction of favored growth (Fig. 3.78b). Thus, in welding, the solidification rate is maximum on the weld centerline where $\theta = 0°$. At this point, the temperature gradients are shallow due to the large distance from the welding heat source.

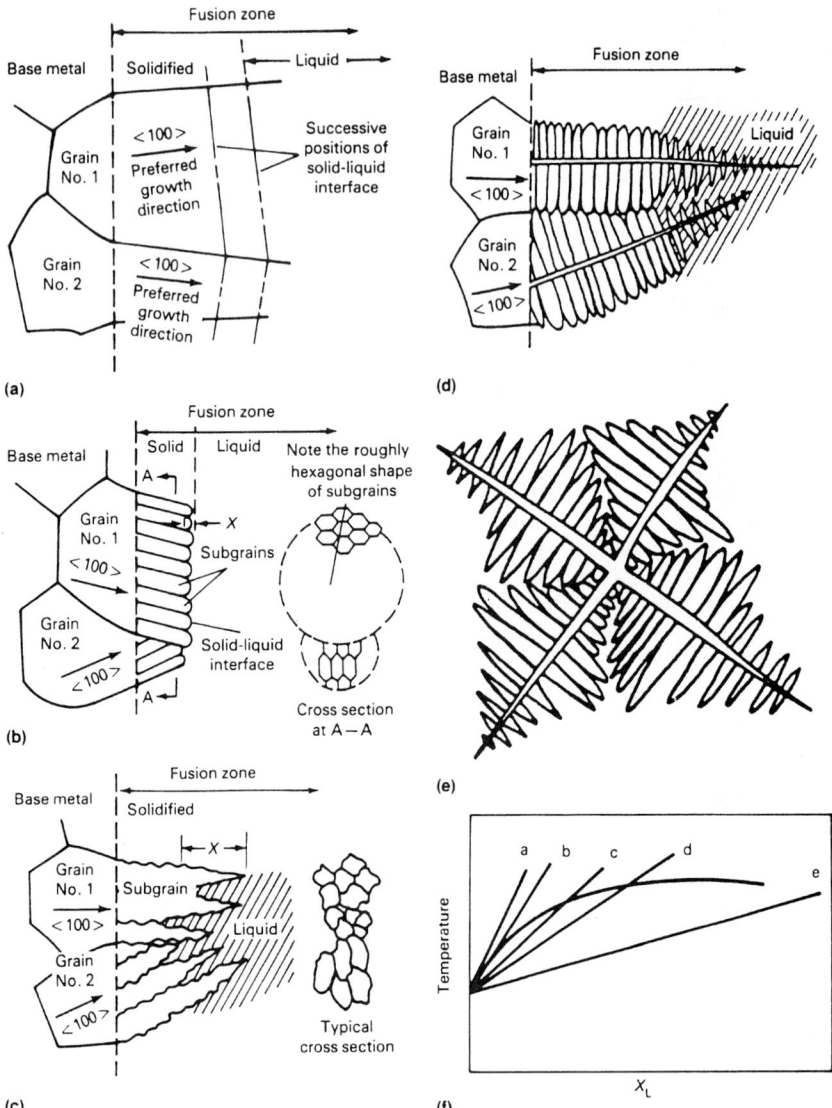

FIGURE 3.76 Schematic showing microstructure of S-L interface for different modes of solidification and the temperature gradients that generate each of the different modes: (*a*) planar growth, (*b*) cellular growth, (*c*) cellular-dendritic growth, (*d*) columnar-dendritic growth, (*e*) equiaxed dendritic growth and structure, and (*f*) five temperature gradients versus constitutional supercooling.[268,269] (*Reprinted by permission of ASM International, Materials Park, Ohio. After Savage et al., Weld. J., vol. 55, 1976.*)

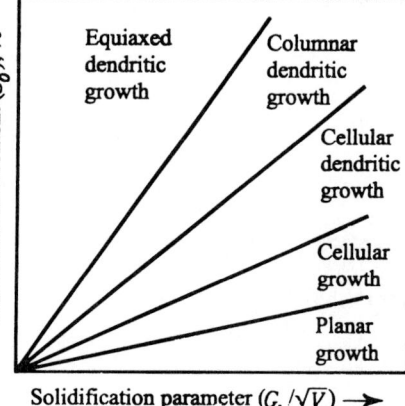

FIGURE 3.77 Schematic plot describing variation in solidification structure as a function of percent solute content and $G_L/\sqrt{V}$.

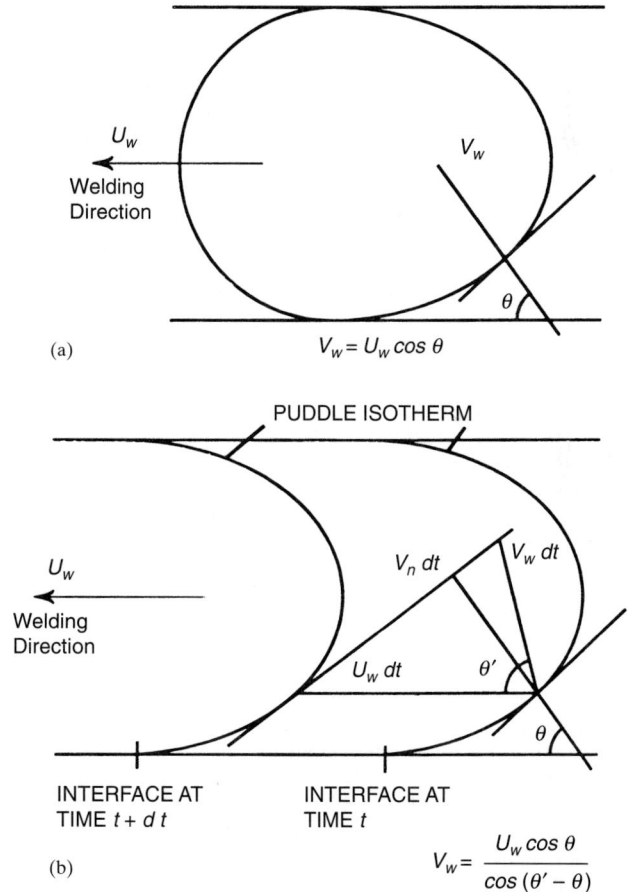

FIGURE 3.78 (a) Schematic solidification rates at different positions of the weld pool for isotropic growth rate. (b) Relationship between welding speed and actual growth rate for anisotropic growth rate.[267]

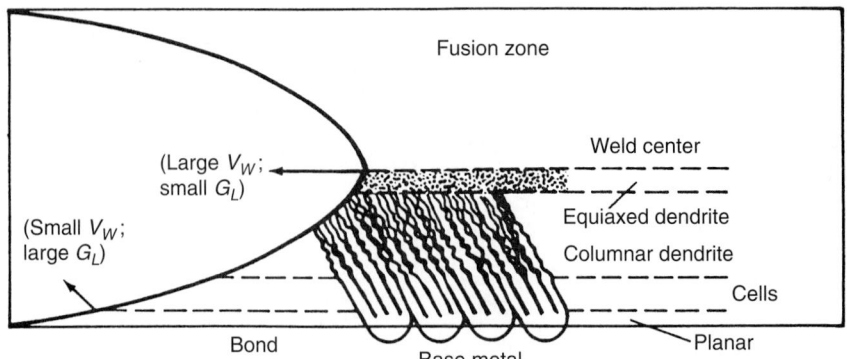

FIGURE 3.79 Schematic drawing of structural variation of weld microstructure in response to variations of the solidification rate around weld pool across the fusion zone.[254] (*After David and Vitek.*)

3.10.3.4 Weld Cracking. A crack located in the weld metal or heat-affected zone is called *weld crack*. There are four important types of welding cracks (defects) which occur at certain temperature intervals specific to a particular alloy. They are called hot or solidification cracks, heat-affected zone microcracking, cold crack, and lamellar tearing. Others include reheat cracking and chevron cracking.

Solidification Cracking or Hot Cracking. Cracking that occurs during solidification is termed *solidification cracking* or *hot cracking* (weld deposit). This occurs in the FZ during the last stage of solidification. It has been shown that a coarse columnar structure is susceptible to hot cracking (Fig. 3.80*a*), whereas a fine-grained structure, a distorted columnar structure, or an equiaxed grain in the weld metal promotes resistance to hot cracking. However, the FZ grain structure can be refined by controlling grain size by inoculation, arc oscillations, pulsing EMS, and mechanical and ultrasonic vibrations.[250]

Solidification cracking in austenitic steel welds is found in different areas of the weld with varying orientations in the weld zone such as transverse cracks, centerline cracks, and microcracks in the underlying weld metal or near HAZ. These cracks result from the segregation of specific low-melting alloying elements along grain boundaries during the last stages of solidification, eventually leading to grain boundary separation. Solidification cracking requires both an adequate amount of mechanical strain and a susceptible microstructure.[270] A simple device to reduce the restraint on a solidifying weld joint is to keep the joint gap to a minimum by designing hardware with good fit-up. The natural tendency to use high-speed welding to improve productivity can have a deleterious effect. Formation of a teardrop-shaped weld pool can result in the centerline solidification cracks. Alloys with a wide solidification temperature range are more susceptible to solidification cracking than alloys that solidify over a narrow temperature range.

HAZ Cracking. Cracks may form in the HAZ due to liquation of the low-melting constituents; this is termed *liquation cracking*, *HAZ burning*, or *hot tearing* (Fig. 3.80*b*). The reasons for this type of cracking are associated with grain boundary segregation aggravated by melting of boundaries near the fusion line. High tensile residual stresses occurring during the weld solidification tend to rupture

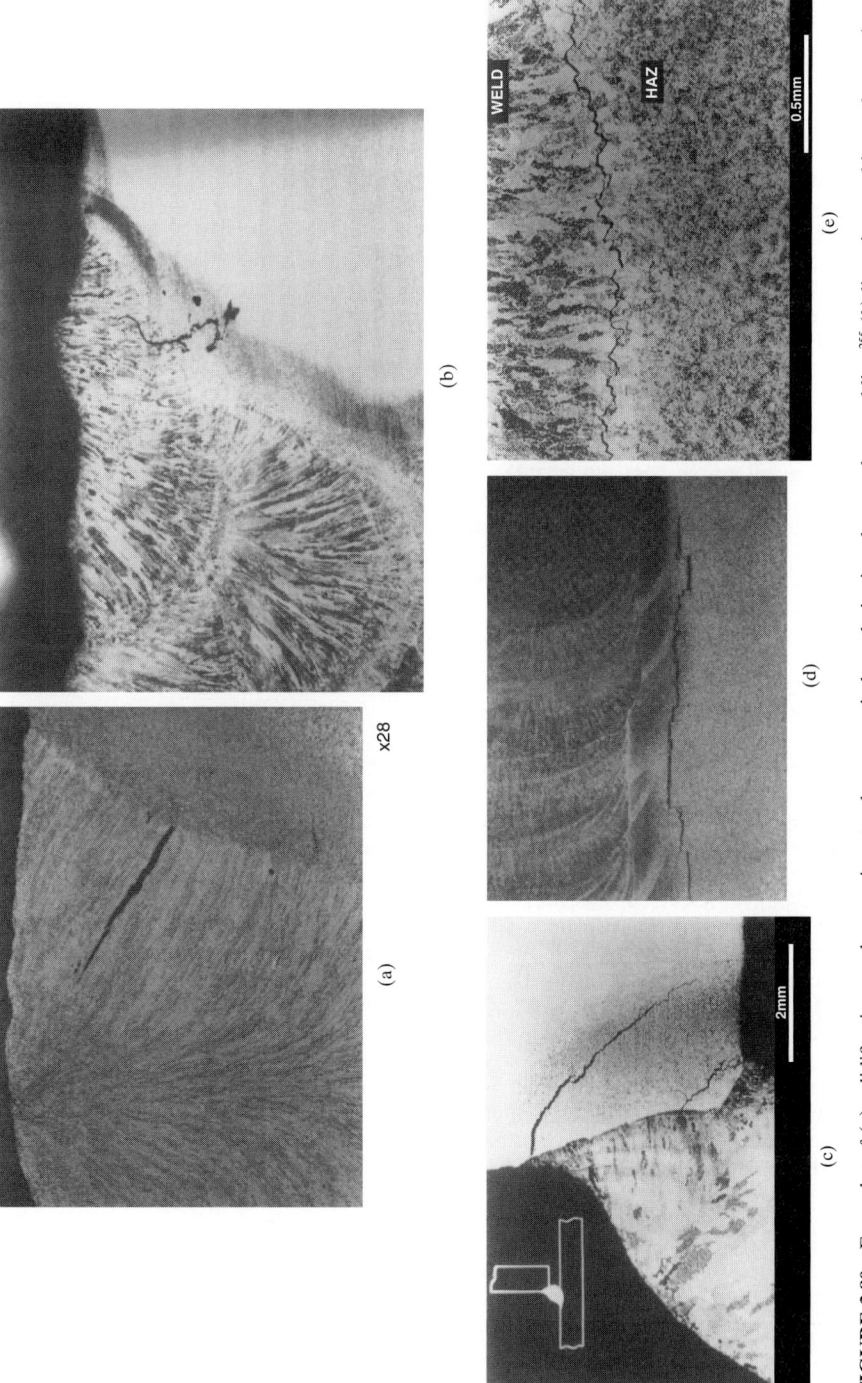

FIGURE 3.80 Examples of (*a*) solidification crack occurring at columnar grain boundaries in submerged arc welding,[255] (*b*) liquation cracking or hot tearing in MMA weld,[255] (*c*) HAZ hydrogen crack, (*d*) lamellar tear adjacent to a tee-butt weld in a structural steel,[255] and (*e*) reheat crack in a structural steel. (*a, b, d: Reprinted by permission of Butterworth-Heinemann, Oxford. c and e: Reprinted by permission of ASM International, Materials Park, Ohio.*)

these impurity-weakened boundaries. The HAZ cracks may also take place in alloys subjected to polymorphic transformations during the weld thermal cycle. Some researchers believe that the presence of liquid is not a necessary criterion for the formation of HAZ cracks, as they form at temperatures well below solidus, which has yet to be quantified.

Hydrogen-Induced Cracking. Also called cold cracking, delayed cracking, or HAZ hydrogen cracking, this occurs (Fig. 3.80c) when three factors are present, namely, a hardened microstructure, a stress beyond threshold level, and a critical amount of hydrogen (which is alloy- and microstructure-dependent). They appear in the HAZ (underbead, root crack), as a toe crack (from the surface), or in the weld metal. The use of preheating and postheating to minimize susceptibility to hydrogen-induced cracking is an accepted welding practice for many steels. The hazard of hydrogen-induced cold cracking is diminished by developing a non-sensitive microstructure, reducing the diffusible hydrogen generated under the welding arc at significantly elevated temperature, maintaining the temperature above the cold cracking range, or by avoiding excessive restraint, if required. The problem can be overcome in practice by using low-hydrogen electrodes or processes combined with minimum necessary preheat (in the 100 to 200°C range) and the minimum allowable heat input.[271] They usually occur with intergranular or trans-granular cleavage fracture, possibly weeks or even months after the welding operation. They tend to form in the temperature range of –50 to 150°C or –60 to 300°F in steels.

Lamellar Tearing. This is the cracking that appears between welds. It occurs in the rolled steel plate weldments (Fig. 3.80d). It always lies under the base metal (plate surface), usually outside the HAZ and parallel to the weld fusion boundary. It results from the combined effect of high localized stress and low through-thickness ductility, occasionally promoted by the presence of hydrogen and segre-gation of elongated (Mn, Fe)S inclusions. It can occur during flame cutting and cold shearing operations. It can occur shortly after welding or occasionally months later. Thicker, higher-strength materials appear to be more susceptible. The problem can be avoided by proper attention to joint details.

Reheat Cracking. Also called stress relief or strain age cracking, this is another defect type found in certain low-alloy steels during reheating of the weld metals in the temperature range of 500 to 650°C in order to relieve the residual stresses (Fig. 3.80e).[255] The creep-resisting steels containing Cr, Mo, and V as well as austenitic stainless and even microalloyed steels seem to be most at risk. It can be prevented by the removal of sharp notches and the use of welding methods to minimize grain coarsening in the HAZ (e.g., low heat input and refinement of weld beads by subsequent weld passes), and replacing more stable MnS with rare earth metal, Ca, Ti, or Zr additions to the steel.[272] The use of basic submerged arc fluxes may sometimes result in the so-called *chevron cracking* due to the cracking of the longitudinal section of the weld metal in the form of a chevron pattern.[273] Among the alloy steels, the 0.7 Ni-0.4Cr-0.6Mo steel used, for example, for the nuclear pressure vessels is prone to reheat cracking which appears in grain growth zones either as longitudinal microcracks or more usually as microcrack networks.[255]

3.11 SINGLE-CRYSTAL GROWTH, GRAIN REFINEMENT, AND EUTECTIC MODIFICATION

The control and the manipulation of microstructure have been used by metallurgists to achieve better physical and mechanical properties. However, this is closely influenced by the continued development of solidification processing and melt treatment. In this section we describe single-crystal growth from the melt, grain refinement, and eutectic modification.

3.11.1 Single-Crystal Growth from Melt

Single-crystal growth may be treated as a special case of solidification-front dynamics. Extensive applications for various optical, electronic, and magnetic devices and strong materials for the metal and aerospace industries have been attributed to the emergence and the production of single crystals of Si, GaAs, and other compounds. For successful crystal growth from melt, several criteria have to be met. These include one- or multi-component single-phase materials with controlled vapor pressure and congruent melting point (or stable phase of the crystal). A good understanding of phase diagram, heat and mass transfer, kinetics, and materials properties is necessary to prepare the good-quality crystals. The growth conditions can be predicted based on these parameters. We discuss here only the crystal growth techniques from the melt which are normally used in the industry to produce large crystals.[274,275]

Bridgman Technique. In the vertical Bridgman method, the crucible containing the melt is relatively moved to the axial temperature gradient through the heater (or furnace) where solidification starts either at the lowest point or on a seed placed at the bottom of the crucible.[276-278] The size of the crystal that can be grown by this technique is restricted by the size of the container and the ability to control heat flow. The intensity of natural melt convection plays a major role on the axial and radial segregation of the dopant and interface shape during the crystal growth.[276] Figure 3.81*a* shows a schematic of the process, and Fig. 3.81*b* shows the temperature distribution in a growing crystal. Many industrially important large-diameter binary, ternary, and multinary semiconductor and metallic crystals have been grown using this method, and extensive modeling has been made to control the bulk homogeneity. In the case of high refractory material growth such as cubic zirconia, a crucibleless modification of Bridgman growth, called *skull melting*, can be deployed where a solid skin of the refractory powder is placed around the molten charge to maintain its shape.[278] The shape of the S-L interface is determined by the imposed temperature gradient and translation velocity. The minimum thermal stresses can be obtained by regulating a planar interface at the S-L interface. The main advantage of this method arises from its relative simplicity, ease of automation, and ease of preparing compound from the elements as a single-crystal ingot which can be subsequently zone-refined in the same apparatus.[278] Horizontal Bridgman (HB) crystal growth assists in reducing stresses, but is not commonly used commercially due to the noncylindrical shape of the mold.[277]

    Gradient Freeze Method. Like the vertical Bridgman method, vertical gradient freeze (VGF) methods have been developed to find a cost-effective solution to the

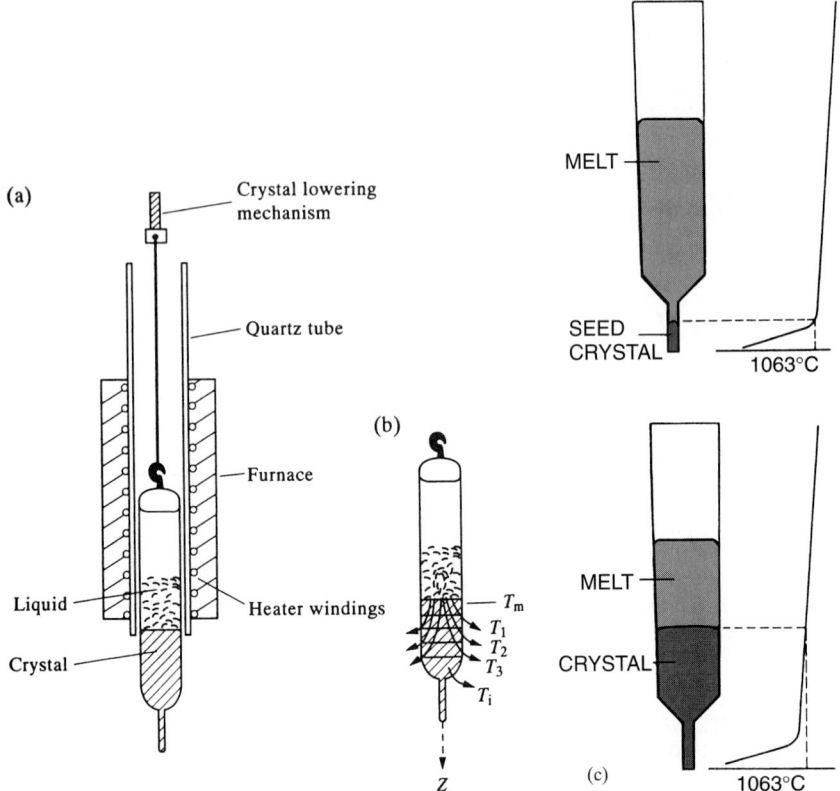

FIGURE 3.81 Crystal growth from melt by Bridgman technique: (*a*) crystal growth mechanism and (*b*) temperature distribution in a growing crystal.[277] (*c*) Vertical gradient freeze temperature profile for InP crystal growth.[278] (*a: Reprinted by permission of Pergamon Press, Plc., Oxford. b: Reprinted by permission of Elsevier Science, B.V., Amsterdam.*)

production of large and high-quality crystals of InP, GaP, InAs, GaAs, and other III-V compounds. The primary principles are the same, and the rate of crystal growth is dependent on the temperature gradient and material parameters.

The VGF technique comprises the controlled freezing from the bottom up of a molten charge of material held in a tube-shaped vertical container (Fig. 3.81*c*). The freezing is better brought about by the use of a furnace consisting of separate, independently controlled heating elements. Adjustment of the heating elements controls the position of the temperature gradients so that the movement of the L-S interface can be increased slowly to effect the crystallization of an ingot.[278]

This technique provides three beneficial effects: (1) low-temperature gradients which favor exceedingly low dislocation densities and absence of slip, (2) ideal shaped ingot of the required diameter, and (3) substrate with a lower thermoelastic stress level and higher degree of uniformity, particularly in InP growth.[278,279]

Chalmer Method. This is a variation of the Bridgman method using a horizontal boat. The advantage of this method is that the remaining liquid can be decanted by electromagnetic devices.

The Czochralski (CZ) Pulling Method. Usually this method consists of melting the material in a crucible by resistance or radio-frequency induction heaters under a controlled atmosphere to a temperature slightly above the melting temperature, into which a seed crystal attached to a seed holder or chuck is lowered into the melt and, after wetting, the seed is slowly lifted and rotated onto the end of the seed (without vibration). The aim of rotation of the seed crystal and melt container with respect to the melt is to force convective motion which facilitates maintenance of axial symmetry of the crystal and control of the segregation of solute or dopant during directional solidification from its melt. Ease of seeding provides very high yield, and absence of ampoule around the cooling crystal eliminates the problem of thermal stress during cooling. The visibility of the crystal through a window in the atmosphere containment vessel is an additional advantage.[280] The shape of the crystal depends on the shape of the meniscus under the seed. The entire process needs considerable operator skill and judgment.[279] Figure 3.82 is a schematic diagram showing the basic principles of the Czochralski pulling method.

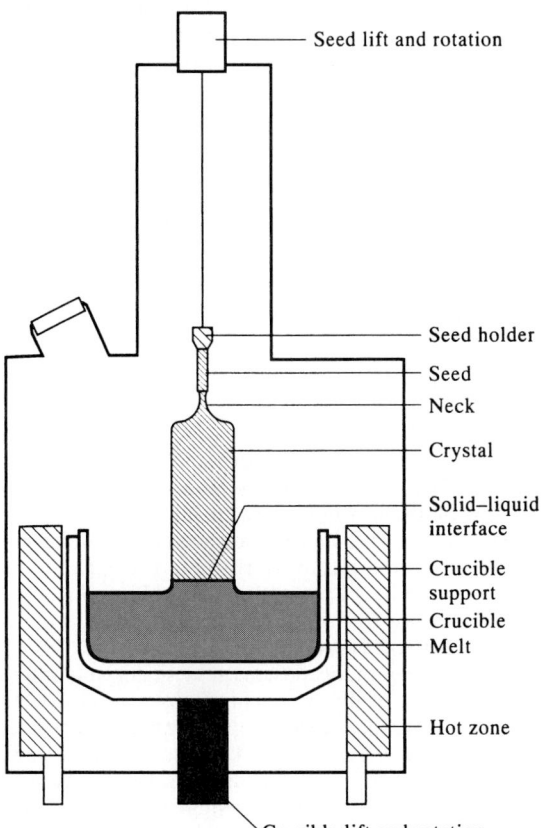

FIGURE 3.82 Schematic representation of basic principles of the Czochralski pulling method.[282] (*Reprinted by permission of Pergamon Press, Plc., Oxford.*)

The dimensional control is achieved by a fully automatic production method using a sensor to monitor the crystal diameter that provides feedback to the power control.[279] The important advantages of CZ method are (1) unconstrained crystal growth during cooling, (2) yield of more uniform distribution of dopant in the melt, (3) low concentration gradients at the interface, (4) monitoring of crystal growth, (5) ease of seeding, and (6) the formation of large cylinder-shaped high-quality single crystals.[276,280]

CZ pulling is usually preferred for large-scale growth of Bi, Cu, AgCu, Co, Cu_3Au, and rare earth metals[280] and semiconductor crystals such as silicon, gallium arsenide, InSb, and GaSb compounds. These compounds have virtually no vapor pressure at the melting point. The CZ silicon is almost exclusively used for worldwide manufacture of integrated circuit (IC) fabrication, large-scale integration (LSI), very large-scale integration (VLSI), ultra-large-scale integration (ULSI), etc., as well as for photovoltaic applications (solar cells). The CZ method is also employed for the production of bulk material for infrared optics, high-purity furnace parts, and the fabrication of micromachined components and devices.

Advanced CZ methods include magnetic field-induced CZ, termed the MCZ process, continuous growth (CCZ), semicontinuous growth (SCZ), hot wall, and liquid-encapsulated Czochralski (LEC) process. Czochralski growth under applied strong transverse magnetic field (MCZ) across an electrically conductive melt effectively increases the viscosity of the melt, thereby suppressing the thermal convection, dramatically reducing the melt turbulence, and yielding crystals approaching 100 kg in weight. This results in a reduction in temperature fluctuation-related growth rate variations, resulting in high-resistivity or low-impurity striation/microdefects in the crystal.[279,281] In continuous CZ method, the melt is filled up while the crystal is growing, allowing the melt size to remain small, thereby resulting in savings in crucible, hot zone, and chamber costs. In reality, the process is not precisely continuous, because growth must be stopped at some point due to equipment pull limitations. Semicontinuous CZ employs a conventional grower equipped with a vacuum-tight isolation chamber to allow retrieval of a grown crystal without permitting air into the growth chamber or cooling the furnace. After removal of the crystal, a hopper or other mechanical loading device is employed to refill the crucible for another growth cycle. In practice, only two or three crystals can be successfully grown from one crucible. Semicontinuous CZ growth is presently used commercially for solar cells.[282]

Hot-wall and LEC methods are used to overcome the vapor pressure problem and loss of group V components. The LEC method is simple and involves the use of an inert layer of transparent liquid B_2O_3, which floats on the melt surface, serves as a liquid seal, and prevents the loss of dissociating volatile components, provided the external gas pressure P_G is greater than the dissociation vapor pressure P_d of the volatile component. The additional required characteristics of the encapsulant include immiscibility with the melt, unreactivity toward the melt, ability to wet the crystal and the crucible, ease of pulling clear from the layer of encapsulant with its viscosity, and temperature dependence of its viscosity.[279]

Advances in the CZ method have greatly reduced the cost of silicon integrated circuits. CZ is currently capable of commercially producing 6- to 8-in. (152.4- to 203.2-mm) diameter, 100-kg dislocation-free silicon crystals.[282,283] Experimental sizes of 250-mm and 300-mm diameter have been reported to be produced, and there seems to be no immediate limit to crystal size.[282] To reduce undesirable convection in the melt, the liquid-encapsulated Czochralski (LEC) system is turned upside down. In the inverted CZ process, the crystal is under the melt and drawn downward. Heating is done from the top (through the melt), and the system is therefore stable and thermally stratified.[283]

Floating Zone (FZ) Melting Method. This does not use a crucible and keeps the molten material completely out of contact with the solid crucible.[282,283] This method controls impurity levels in materials, provides purifying effect by evaporation of impurities, especially for high-melting materials, and offers suitable conditions for doping from the gas phase.[284] In addition, crystal with less strain can be grown by this method. A simple schematic diagram for the vertical float zone process is shown in Fig. 3.83. In this technique, a polycrystalline source material (rod) is supported in the growth chamber only at its ends with a pulling and rotating mechanism. A small portion of the lower end of the rod is melted, the melt being held in place by the combination of high surface tension and a rf levitation from the induction coil heater. A seed crystal is inserted into the melt from below, using a second lifting and rotating device. Crystal growth is carried out by a combination of slowly lowering the seed and moving the zone passage upward or lowering the feed rate into the heating coil. The single crystals of silicon, metals, metallic and nonmetallic oxides, and II-VI and III-V compounds have been grown by the vertical FZ method. Very high-purity silicon crystals are commonly grown by this method.[282]

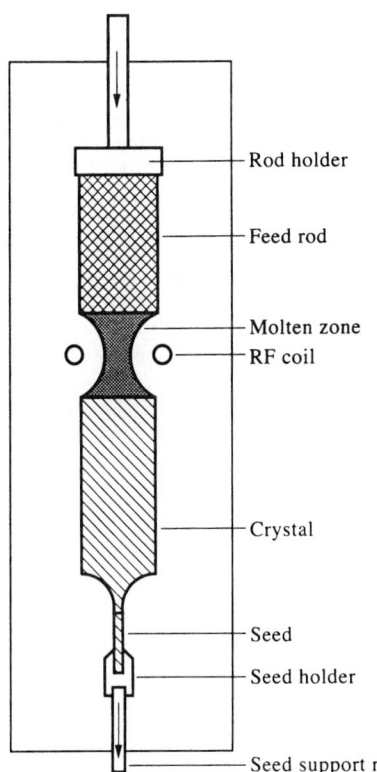

FIGURE 3.83 Schematic diagram of the vertical floating zone process for single-crystal growth.[282] (*Reprinted by permission of Pergamon Press, Plc., Oxford.*)

A variety of zone melting processes have been developed using vertical and horizontal configurations, single and multiple heating zones, and liquid-encapsulated molten zone (LEMZ). Usually the heating sources include special rf coils, arc image or halogen lamps, arc discharge, electron bombardment, and laser and conduction heating. In the case of rf heating, a single-turn coil is used to couple directly the material. Crystal growth methods such as the traveling heater method (THM) are developed based on zone melting and have been widely used to grow crystals of binary and multinary semiconductor compounds.

The FZ method is mostly used for high-power electronics, thyristors, and sensitive infrared radiation detectors, because it is capable of producing very high-purity crystals with high resistivity (silicon, in excess of $100\,\Omega m$) and high minority carrier lifetime and growing dislocation-free silicon with diameter of 150 mm and a length of 1.5 m. Some drawbacks of FZ silicon over CZ silicon are the smaller crystal diameter, tendency of crystal cracking during processing, preparation of smooth rods as starting materials, a more complicated and difficult method to control, and more complex and expensive equipment.[284] Silicon crystals grown by either the FZ or CZ method possess a cylindrical shape and are sliced into circular wafers, lapped, and polished prior to being processed into circuits or devices.[282]

The LEMZ technique was used to grow single crystals of InBi aboard Space Shuttle Endeavor (STS-51), which was characterized by better diameter control, surface appearance, and crystal quality than the unencapsulated one. It was also found that crystal quality improved with the increase of growth velocity.[284a]

3.11.2 Grain Refinement of Al-Si Alloy

Grain refinement has been a widely established practice in the aluminum casting industry for the last several decades. Currently, three types of grain refinement techniques are used: mechanical (or electromagnetic stirring); thermal method, such as chill effect, supercooling, and rapid solidification; and chemical additions (or inoculation). The grain refining inoculants commonly employed in the industry are master alloys of Al-Ti (consisting of $TiAl_3$ particles in an Al matrix) and Al-Ti-B[†284b] which are expected to enhance intensive heterogeneous nucleation by providing substrates in the form of aluminides and borides for the formation of primary aluminum phase grains during freezing.

A large variability of grain refiner performance has been noted which is attributed to several factors, such as slight chemical composition changes and nucleant particle morphology originated during master alloy preparation. The variability due to these factors is difficult to maintain under control during casting and solidification, which created a need for an adequate procedure to examine accurately the master alloy effectiveness.[285]

Three different grain morphologies, namely, equiaxed, columnar, and twinned columnar (TCG), can occur in the solidifying aluminum alloy castings. Using grain refiners, it is possible in practice (particularly with Al and Mg alloys) to achieve consistent grain sizes under about 0.12 mm. Grain refiners have been developed for ferrous and nickel-base alloys which are effective as mold coatings for thin-section castings, but their effective life is not long enough to refine sections greater than a few millimeters thick.[7]

---

† Boron is often added to master alloy which exerts an additive effect on the reduction of ultimate grain size in Al-Si alloys.

The AFS wall chart on *Microstructure Control in Hypoeutectic Al-Si Alloys* is a new tool that is useful for foundrymen to help control the grain size of their castings and to optimize mechanical properties.

Usually grain refinement can be expected to exert a positive effect on certain properties; particularly, less porosity and better shrinkage distribution and reduced hot-tearing tendency are achieved. However, the grain refinement may cause reduced fluidity and mold-filling capability, because the improved grain nucleation may limit the continued mass liquid flow during a casting process.[284b]

Thermal Methods.   The use of chills increases the solidification cooling rate and temperature gradient, promotes directional solidification, reduces DAS, and refines the microstructure. The shrinkage porosities formed are smaller, a more uniform pore distribution is obtained, and the susceptibility to form interconnected porosity is decreased.[286] The effects are more pronounced in thicker sections than in thinner sections. When the molten metal contacts the cold mold walls, the liquid superheat is removed and it becomes locally supercooled. The numbers of nucleation centers increase, and multiple nucleation occurs catastrophically in the liquid and results in a fine as-cast grain size. Special methods such as splat cooling and die casting as well as applications of chills use this approach with varying efficiencies according to the sample size.

Cast iron chills in Al-12%Si alloys produce lower porosity in a given chill-plate-feeder combination as compared to graphite chills.[286a]

Grain Refinement by Vibration (Ultrasonic and Sonic).   In dendritically solidifying materials this occurs by mechanical stresses in the dendrite roots, giving rise to failure by either (1) bending, causing formation of new grains or fragmentation of primary crystallites, or (2) shear, perhaps along a single slip plane, yielding an artificial source of more nuclei.[287,288]

From a practical point of view, introduction of convection by mechanical or electromagnetic device during the early stages of solidification promotes formation of fine, equiaxed grains. Similarly, mold oscillation is sometimes used to obtain fine grains in sand castings, and stirring with a cold rod is used to refine the ingot structure. Currently, electromagnetic stirring (EMS) or (the intensity and frequency of) vibration is widely used in continuous casting to achieve fine grain size.[289]

It is now generally recognized that vibration, low pouring temperature, and externally induced convection all favor grain refinement mainly by a dendrite fragmentation mechanism, although the basic mechanism has not yet been established.

Some possible dendrite fragmentation mechanisms include the following: (1) Dendrite arm fracture occurs due to the force on the arm from the fluid flow.[290] (2) Remelting of the arm at its roots occurs due to normal coarsening.[291] The function of the fluid flow in this case is simply to carry the dendrite arm away from its "mother grain" to where it can grow as a new grain. (3) Remelting as above is improved by thermal perturbation which occurs with turbulent convection in a liquid which is not at a uniform temperature. (4) Remelting occurs as above, but the melting at the root is accelerated by the stress introduced at the dendrite root as a result of the force of the fluid flow. (5) This occurs as in (3) above, but the melting at the root is further enhanced by a high-solute content in the solid at the dendrite root.[292] (6) Recrystallization due to the stress is introduced by the force of the fluid flow, with rapid liquid penetration along the new grain boundaries.[73,292–294]

The study of segregation substructure of the grains obtained at different supercooling appears to support the dendritic fragmentation mechanism.

Inoculation Methods. The grain refinement of cast aluminum alloys is most commonly effected by the addition of inoculants into the melt immediately before castings. These inoculants induce restricted diffusion time and instantaneous heterogeneous nucleation and produce fully, equiaxed fine-grain structure.

Grain refinement of aluminum alloy casting yields the following benefits: (1) uniform mechanical properties throughout the material; (2) distribution of secondary phases and shrinkage porosity on a finer scale; (3) improved machinability due to (2); (4) improved ability to facilitate the uniform anodization process; (5) better strength, toughness, and fatigue life; (6) better corrosion resistance; (7) attainment of better feeding of the casting; and (8) better surface finishes on both the basic casting and the machined part. Thus grain refinement improves resistance to hot tearing, decreases porosity, and increases mass feeding. A grain-refined cast part is more homogeneous, with better casting soundness and increased mechanical properties.[295]

Grain refiners for use in semicontinuous or continuous casting of ingot or strip are usually supplied as $\frac{3}{8}$-in.- (9.5-mm-) diameter rod which is produced by casting the master alloy followed by mechanical working. These alloys normally possess a microstructure in which the $TiAl_3$ phase (with an average of 40 to 60 μm size) has a rounded or blocky morphology whose surface is covered with small TiB_2 particles (with 0.5 to 3 μm size), which is quite different from the acicular aluminide morphology, commonly prevalent in master alloy waffle plate.[296]

The various grain refining inoculants employed in the aluminum industry are Al-Ti-B master alloys such as Al-10Ti-0.4B, Al-5Ti, Al-5Ti-1B, Al-5Ti-0.6B, Al-5Ti-0.2B, Al-3Ti-0.2B, Al-3Ti-1B, or Al-3Ti-3B, with Ti/B ratios ranging between 5:1 and 50:1. These master alloys are available in wrought (rolled or extruded) rod form, ingots or waffle, or as salt mixtures. Grain refiner rods made by rolling process are found to be superior (in effect) to rods made by the conform and extrusion processes.[296a] To be effective, they contain numerous small, controlled, predictable, and operative amounts of intermetallic crystals of $TiAl_3$ and TiB_2 in the correct form, size, and distribution for grain nucleation. The effectiveness of grain refiner in an alloy system will be determined by the number of potential nuclei, their potency, the alloy cooling rate, and alloy constituents. In addition, salts (usually in compacted form) that react with molten aluminum to form combinations of $TiAl_3$ and TiB_2 are used. Several mechanisms have been proposed for grain refinement. It is recognized that master alloy added to aluminum alloy melts results in its dissolution in the aluminum matrix and releases intermetallic phase, Al_3Ti and various borides and carbides.

The grain-refining ability of the master alloy will decrease with time. This process, called *fading time*, is very significantly prolonged with Al-Ti-B master alloys.

It is now a common practice to add Al-Ti-B grain-refining master alloy to molten aluminum alloy before casting, in order to produce an equiaxed grain structure in the solidified ingot or cast product.[296]

Grain-refining master alloy in the form of 16-lb (7.26-kg) notched waffle ingot is added to the holding furnace before the start of the casting; and the casting must wait until the waffle ingot reaches maximum effectiveness and must be complete before its effectiveness fades. Grain refiners supplied as $\frac{3}{8}$-in.- (9.5-mm-) diameter rod reaches maximum effectiveness within 1 min, is continuously introduced to a precise amount into the launder (or controlled molten metal stream) during the cast, close to the holding furnace, and prior to any in-line degassing and filtration devices. This time-dependent mechanism shows a limitation on the use of waffle for refining slow-casting alloys or continuous casting processes.[297]

It has been recently found that Al-4B has superior potency as a grain refiner. Figure 3.84 compares the effectiveness of various master alloy grain refiners in 356

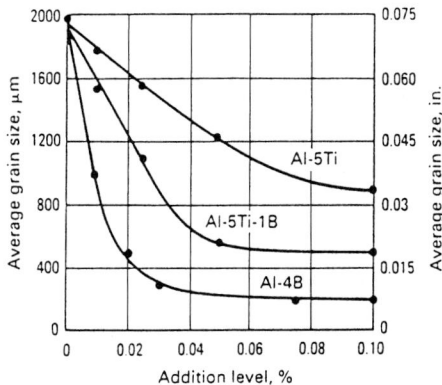

FIGURE 3.84 Comparison of the effectiveness of various master alloy grain refiners in A356 aluminum alloy.[298,299] (*Courtesy of AFS, Des Plaines, Ill.*)

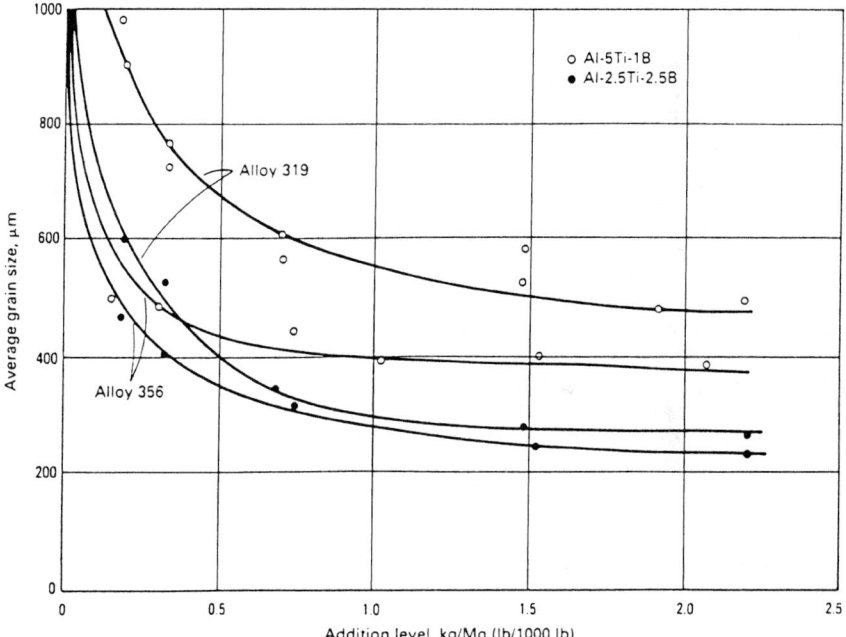

FIGURE 3.85 Substantial advantages of Al-2.5Ti-2.5B master alloys to 356 and 319 casting alloys over previously used Al-Ti-1B alloy.[300] (*Courtesy of ASM International, Materials Park, Ohio.*)

Al alloy[298,299] which clearly demonstrates the superior potency of Al-4B as a grain refiner compared to Al-5Ti-1B and Al-5Ti master alloys. Figure 3.85 shows the substantial advantage of Al-2.5Ti-2.5B master alloys to 356 and 319 casting alloys over previously used Al-5Ti-1B alloy.[300] Table 3.5 lists the sizes of the largest insoluble particles (TiB$_2$ or TiC) found in some master alloys investigated.

TABLE 3.5 Size of the Largest Insoluble Particles Found in Master Alloys Investigated

Master alloy	Largest insoluble particles
6Ti-0.02C	6 μm
Pure 6Ti	—
3Ti-0.2B	3 μm
6Ti-0.02B	2 μm

Source: G. Campbell, S. A. Danilak, S. R. Thistlethwaite, and P. Fisher, *Light Metals Alloy*, ed. R. R. Cotshall, TMS, 1991, pp. 831–836.

Al-Ti-C grain refiner with 5% Ti is reported to be superior to that of the 3% Ti refiner. For example, 5% Ti/0.19C is more than twice as effective as 3% Ti/0.15C and needs only one-half the addition level. The superior efficiency of Al-Ti-C product compared to the more conventional Al-Ti-B was attributed to the additional nuclei.

Grain Refinement Mechanisms. In principle, the grain refinement mechanism involves dispersion of copious potent heterogeneous nuclei in the melt where a large number of these sites become active during solidification and nucleate the solid.[300a] Three theories have been hypothesized for grain refining: the transition-element *carbide-boride particle theory*, *metastable boride theory*, and the *peritectic theory* based on the peritectic reaction in Al-Ti phase diagram: L + TiAl₃ ⇌ α-Al (solid solution). It has been concluded that a peritectic reaction comprising TiAl₃ is the most likely mechanism responsible for nucleation.[300a]

The *carbide-boride particle theory* was postulated by Cibula and others, based on the belief that nominal addition of Al-Ti master alloy favors TiC formation due to the reaction with the residual carbon present in the aluminum melt. In the case of Ti-Al-B master alloy addition, TiB_2 particles are dispersed in the melt; and TiC and TiB_2, being practically insoluble in the aluminum melt, act as heterogeneous nucleants. The observed fading nature is only linked to the particle settling and/or agglomeration.

The occurrence of complex $(Al,Ti)B_2$ phase has led researchers to propose another *metastable boride theory*. The reason is that this active phase may disappear with long holding time. It is suggested that the $(Al,Ti)B_2$ phase promotes grain refinement by acting directly as heterogeneous nucleation sites or by nucleating TiAl₃ crystals during cooling.

Although peritectic theory successfully describes the behavior of Al-Ti master alloys, it does not adequately explain the enhanced performance of commercial grain refiners containing Ti and B. The improved performance of B-containing master alloys is believed due to the displacement peritectic composition from 0.15% Ti toward the Al-rich end of the phase diagram (as experimentally asserted by Mondolfo and his coworkers), thereby ensuring the thermodynamic stability of TiAl₃ at low levels of Ti additions (~0.02%). This hypothesis, however, seems to be incompatible with the Al-Ti-B phase diagram predicted by various researchers as well as by the segregation experiments of Finch. It has also been indicated that when all the boron reacts to form TiB_2, its effect vanishes and fading of grain refinement follows.[300a]

The period of effectiveness after grain refiner addition and the potency of grain-refining action are improved by the existence of TiB_2. In the Al-Si system, AlB_2 and TiB_2 in the absence of excess Ti have been observed to yield effective grain refinement. However, the requirement of an excess Ti compared to the stoichiometric balance with B in TiB_2 is generally accepted to yield the best grain-refining results, and Ti or higher-ratio Ti-B master alloys are used exclusively for grain size control.[300]

The selection of a suitable grain refiner, practices for grain refiner addition, and procedures involving holding and pouring of castings after grain-refining addition are mostly developed by the foundry and cast house on the basis of casting and product requirements and after referral to the performance characteristics of commercial grain refiners furnished by suppliers. However, in most cases, grain refiners of 5Ti-1B and 5Ti-0.6B types, which are characterized by cleanliness and fine, uniform distribution of aluminide and boride phases when added to the melt at 0.01 to 0.03% Ti level, should be expected to give adequate grain refinement.[300]

More recently, Squarez and Perepezko have used the droplet emulsion technique (DET) and differential thermal analysis (DTA) for the isolation and observance of different potency levels of individual particles and study of microstructural development and active nuclei site evaluation in Al-5Ti-1B and Al-5Ti-0.2B alloys.

Despite the uncertainty about the details of nucleation mechanisms, improvements have been made in theory in order to predict the final grain size as a function of the dispersion and density of nucleant particles.

3.11.3 Grain Refinement of Mg Alloy

As discussed previously, the grain size of castings can be manipulated by changing various casting parameters, such as cooling rate, or by adding alloying elements and nucleants (grain refiner) prior to or during the casting process. However, unlike for Al alloys, a well-established, reliable grain refiner system is not commercially available for the range of Mg alloys.

Increasing the Al content in hypoeutectic Mg-Al alloys was reported to produce grain refinement up to 5 wt% Al, attaining a relatively consistent grain size for higher Al content (>5 wt%). Sr addition produced a significant grain refinement effect in pure Mg and low-Al containing alloys but a very small grain refinement effect in Mg-9Al alloy. The addition of a small amount of Zr, Si, and Ca to pure Mg produced an effective grain refinement effect. The grain refinement is mainly attributed to the growth restriction effects caused by constitutional undercooling, during solidification; but the influence of nucleant particles, either introduced with the alloying additions or as secondary phases formed due to these additions, may enhance the grain refinement. Readers are referred to an article by Lee et al. for more details on this topic.[300b]

3.11.4 Eutectic Modification of Al-Si Alloys

Among the most important foundry alloys are those on the Al-Si and Fe-C systems. The mechanical properties of these metal-nonmetal (nonfaceted-faceted) eutectics are mainly dependent on the morphologies where the nonmetals solidify.

The Al-Si system forms a simple eutectic with a composition of 12.5 wt% Si at 577°C (850°F) between an Al solid solution containing 1.65 wt% Si and virtually pure Si. Alloys with Si as the major alloying additions are very important Al casting alloys primarily due to their high fluidity, low casting shrinkage, high corrosion resistance, low coefficient of thermal expansion (CTE), good welding, and easy brazing.

In the eutectic Al-Si system, the Al phase is nonfaceted and in hypoeutectic alloy it is dendritic. The minor Si phase, in pure binary alloys, freezes as faceted flakes, either as primary phase or as the finer eutectic constituent; it is usually able to grow in only certain crystallographic directions, which plays a major role in modification. The modification is essentially a method of changing the shape/structure (morphology) in Al-Si alloys during solidification. In normal, unmodified Al-Si alloys, the Si crystals grow in a faceted manner, i.e., on the close-packed flat (111) faces of the diamond cubic structure, usually combined with a few twins across the $\langle 111 \rangle$ planes (Fig. 3.86).

Modification of eutectic phase is very desirable, particularly for slower rate solidification processes such as sand or permanent mold casting. However, recent work has shown that in pressure die casting, Sr modification also produces an improvement in feeding characteristics.[284b]

The modification is carried out by treating the melt with certain elements,[†] called modifiers, or by subjecting the melt to a fast cooling rate. The modification of eutectic structure in Al-Si casting alloys is obtained by the use of Na, Sr, and Sb, Ca, K, P, Li, B, Ti, and selected rare earth elements. Of all these, Na, Sr, and Sb are the most popular and effective modifiers at low concentration levels, typically 0.01 to 0.15 wt%.[301] The effectiveness of various elements as modifiers is clearly a function of the number of twins formed in the modified Si fibers.

Careful studies of TEM have revealed that modified Si fibers contain orders of magnitude more twins than do unmodified Si plates. The significant increase in twin density is caused by the addition of only a fraction of 1 wt% of modifier which becomes concentrated in the Si, and not in the Al phase.[295] The surface of the fibers is microfaceted and rough due to intersection of myriad twin planes with it. The modification features of silicon fibers in Al-Si alloys by Na addition are summarized in Fig. 3.87 and Table 3.6.[302]

The modification of Si in Al-Si alloys can improve its mechanical properties by changing the Si structure of the alloy (Table 3.7). Since Si is a major constituent of these hypoeutectic alloys—typically ranging between 5 and 12%—it plays a great role in the processing and final properties of the casting. The modification of the structure of these alloys and the resultant effect on the mechanical properties drive the foundrymen to manipulate the microstructure based on the casting alloys.[295]

The microstructural change from acicular to fibrous silicon is not a sharp one. Modification with Sr is often less uniform than with Na; and Sb will, of course, produce only a lamellar and never a fibrous structure. The formation of exact microstructure depends on five parameters: type and amount of modifier used, impurities present in the melt, Si content of the alloy, and freezing/solidification rate.

Type and Amount of Chemical Modification by Na, Sr, or Sb.   Sodium is the most efficient modifier, although the effect fades rapidly by evaporation and oxidation during holding in the liquid state. Additionally the use of Na modifier decreases the castability of Al-Si alloys. The addition of Na is associated with a violent reaction

---

[†] They are surface active agents with favored crystallographic orientations which facilitate nucleation of silicon grains.[284b]

(a)

(b)

(c)

FIGURE 3.86 Microstructure of Si flakes in unmodified Al-Si eutectic alloy. (*a*) Optical microstructure showing irregular growth front; (*b*) SEM, deep-etched removing metal matrix; and (*c*) TEM showing a few [111] twins and growth orientation.[301] (*Reprinted by permission of TMS, Warrendale, Pa.*)

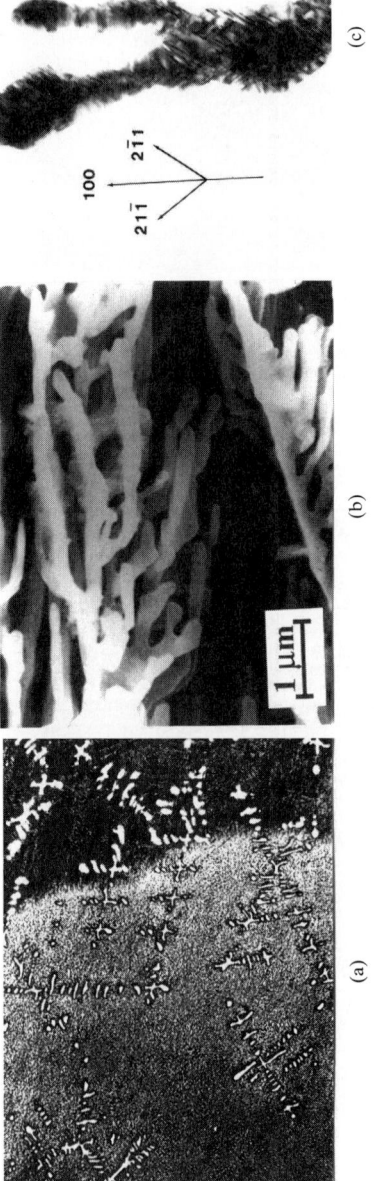

FIGURE 3.87 Microstructure of Si fibers in Na-modified Al-Si eutectic alloy. (*a*) Optical micrograph, showing a near planar growth front; (*b*) scanning electron micrograph of deep-etched removing metal matrix; and (*c*) transmission electron micrograph showing a high density of multiple twinning and growth orientation. (*Reprinted by permission of TMS, Warrendale, Pa.*)

TABLE 3.6 Summary of Experimental Evidence due to Modification[301]

	Unmodified	Modified minor additions	Chill
Microstructural shape	Faceted flakes	Microfaceted fibers	Smooth fibers (nonfaceted)
Internal structure (twin spacing)	Very few twins ($0.4–1\,\mu m$)	Heavily twinned ($0.005–0.1\,\mu m$)	Few twins (very large)
Growth undercooling	Relatively small	Large	Very large
Growth interface	Irregular	Near planar	Uncertain
Distribution of modifier	Not applicable	Within Si phase	Not applicable

Reprinted by permission of TMS, Warrendale, Pa.

TABLE 3.7 Tensile Properties of A356 Alloy Unmodified and Treated with Different Modifiers

Modifier	Silicon structure	As-cast			Heat treatment[†]		
		UTS, MPa	El., %	Q, MPa	UTS, MPa	El., %	Q, MPa
None	Acicular	180	6.8	305	304	11.8	465
Sb	Lamellar	201	11.9	362	293	16.5	476
Na	Fibrous	195	16.4	377	292	15.1	469
Sr	Fibrous	196	15.9	376	301	14.4	475

[†] Solution treatment at 450°C for 10 hr and aging at 160°C for 6 hr.
Source: *Modern Casting*, January 1990, pp. 24–27.

and results in the increase in H_2 levels. Sodium in excess of ~0.015% causes over-modification bands where the duplex eutectic front is briefly arrested and over-grown by aluminum film. The eutectic growth front is no more jagged and irregular but becomes more or less planar.

Na Modifier. Sodium has easy dissolution above 700°C, but poor and somewhat unpredictable recovery. Sodium is the more powerful modifier, easy to use, and the most beneficial to the eutectic structure (by producing more uniformly modified structures at low concentrations) and the mechanical properties; but it fades rapidly by evaporation and oxidizes during holding of the molten alloy. Prolonged holding (>>30 to 40 min) can require renewal of the treatment. The Na-modified melts have the tendency to gas pickup and increased porosity. Oxidation of melt reduces casta-bility and attacks aggressively against mold coatings or electrical resistance. Sodium interacts with P, if present in the alloy, necessitating increased additions of Na with the increased P content and decreased solidification rate. Typical retained Na levels sufficient to modify the Al-Si system are in the range of 0.005 to 0.015%. Remodi-fication is accomplished, if required, to maintain the desired modification level. Since sand castings cool more slowly than permanent mold castings, they will need higher Na content; the time delay for pouring is more critical in sand than in per-manent mold casting.[295]

Sodium is added as a flux, but many modern applications employ vacuum prepacked Na metal in small aluminum cans to minimize its oxidation and hydrogenation. Dissolution of Na is usually complete within 5 min.

Sr Modifier. Strontium has a similar effect as Na on the eutectic structure. The current trend is to use Sr as a modifier, which is introduced into Al-Si alloy melt as pure solid metal, salt, or master alloys such as Al-3.5Sr, Al-10Sr, Al-10Sr-14Si, Al-10Sr-15Si, and 90Sr-10Al, in the form of waffles, piglets, or rods. The rod form is reported to improve production-floor flexibility and effectiveness. This is attributed to a finer microscopic size of the contained Sr compound.[302] However, the problem of adding elemental Sr and its salts to liquid metal, in addition to its high cost, limits the commercial use of Sr as a modifier. Recently, less expensive master alloys have been put in market, making Sr addition practical from the cost standpoint. Additionally, Sr has less effect on the oxidizing behavior of the alloy; its action is more durable than that of Na, having a semipermanent effect. The Sr treatment does not introduce hydrogen into the liquid-metal state, which causes greater porosity in the casting.

The rate of loss of Sr is definitely less than that in Na, and the use of pre-modified alloy ingots is simple. Its sensitivity to gas pickup and porosity formation in part areas not well fed or cooled may be a drawback in day-to-day foundry practice and for mechanical properties in highly stressed parts. It is, however, easy for many types of castings. A range of 0.015 to 0.05% Sr is a standard industrial practice. Remodification through Sr additions may be required, although less frequently done than for Na. Its amounts of 0.02% are adequate to modify an A356 (Al-7Si-Mg) alloy, but up to 0.04% is required for a eutectic alloy such as A413.

Strontium additions can be made with ease in the form of Al-15Si-10Sr master alloy with optimum additions in the range of 0.04 to 0.1 wt%. The improvement in mechanical properties is similar to that found by Na modification without some of the undesirable features. Partial modification is obtained with additions of Ce, La, and Na with La, the last being the most effective in increasing the UTS by ~250%. An addition of 0.1% Ce also yields an improvement in machinability. A 0.2% addition of all three elements in a fluoride mixture is an effective modifier, and an addition of 1% mischmetal produces complete modification with a corresponding improvement in mechanical properties.

The addition of 150 to 300 ppm of Sr to the wrought 6061 alloy improves the formation of ductile intermetallic alpha phase (AlFeSi) which is more potent than the less desirable brittle intermetallic beta phase (AlFeSi) which forms during solidification of the alloy.[302]

Sb Modifier. The Sb modifier has some benefits with respect to its capacity as a permanent-type modifier, primarily through its extremely low oxidizability. However, Sb has some limitations. It can only be used in die casting, because the lower speed of solidification results in segregation. The degree of modification with Sb is less than that achieved by Na or Sr modification. Moreover, the addition of Sb itself into Al-Si alloy is associated with various problems due to the very long time needed for the Sb to be dissolved in the molten metal, even with continuous agitation of the melt.

Antimony yields a lamellar eutectic structure which is particularly sensitive to the freezing rate. In practice, Sb is used in the range of 0.06 to 0.50%. Its permanent effect is linked with a tendency to have less gas pickup and less porosity formation, both being beneficial for permanent mold castings that are subjected to high stresses and require consistent properties. Antimony-treated alloy is purchased as pre-modified ingot from primary Al suppliers and is simply remelted and cast. As

Sb is very stable in the melt, losses are virtually nil, and no extra additions are needed.[295] Antimony treatment is not recommended for sand casting, because of the occurrence of nonuniform lamellar structure due to the slow freezing rate. It is usually confined to permanent mold applications and is used mostly in Europe and Japan.[303]

Impurities and Contamination Present in the Melt. The presence of minor or impurity elements in Al-Si alloys may be either beneficial or detrimental to certain mechanical properties. Iron impurity leads to detrimental effect on the ductility and corrosion resistance of the alloy. Presence of Pb in Al-Mg-Si alloy yields a low-ductility intergranular fracture. However, mechanical properties of Al-Si alloys, especially elongation, depend on both the alloy composition and the eutectic-silicon size and morphology. It is pointed out that the presence of phosphorus makes modification difficult; that is, low P content in the alloy simplifies the modification. Since Sb interacts with both Na and Sr in a negative manner, Sb-containing melts need very high levels of either modifier to produce the desired structures. It is reported by some researchers that Mg makes modification easier and by others, more difficult.[295]

Silicon Content of the Alloy. Larger modifier content (up to 50%) is required to produce complete modification in higher (11%) Si concentration than in Al-7% Si alloy.

Solidification Rate and Modifying Efficiency. It is well recognized that Al-Si eutectic microstructures may be altered significantly by both the solidification conditions (cooling/growth rates) and minor additions of certain modifiers, which result in enhanced modification and finer structures. In the presence of an impurity- (or chemical-) modified eutectic Al-Si alloy, the Al phase is not markedly affected; but the modified Si phases/fibers become very heavily twinned and are actually microfaceted, and the angle of branching increases with solidification rate. In modified alloys these impurities are linked with the Si phase, and it is found that they are adsorbed upon the Si liquid growth front.[304] In contrast, quench-modified Si is essentially twin-free and nonfaceted. This observation is in agreement with the previous finding of twin formation in Si, grown epitaxially from Al solutions in the presence of Na.[305]

The impurity-induced modification is accomplished at much slower rates by minor additions of Na ($\sim$0.01 wt%), Sr ($\sim$0.1 wt%), and less certainly with other alkali, alkaline earth, and rare earth metals.[306] It is also noted that chemical modifiers are more effective at higher freezing (solidification) rates as in chill casting rather than in a heavy section sand casting. This structure is considered as a very fine scale of the unmodified eutectic. It is of little practical consequence, since commercial casting processes, with the possible exception of die casting, do not involve very high solidification rates to cause quench modification.

Like the Al-Si system, Si modification may be observed in other diamond cubic phases such as Ga and III-V compounds.[304]

Quench Modifier Fibers. These occur at growth rates of $\geq$1 mm/s and are quite smooth on external surfaces and more often twin-free than otherwise. The quench-modified Si fibers are characterized by isotropic growth, nonfaceted in appearance, very much finer than the slowly grown flakes, and also finer than the impurity-modified fibers which form at relatively low growth rates.

Impurity-Modified Fibers. Detailed examination of the Na-modified fibers shows them to be externally rough or microfaceted and that they contain a very high twin density on up to four {111} systems (Fig. 3.87); the preferred growth axis is then in a ⟨100⟩ direction, with symmetric branching in a ⟨211⟩ direction. In some cases, the average twin spacing may be as low as 5 nm, and when compared to the prevailing growth rates, this corresponds to up to 10^4 twinning events per second. The structure is therefore considerably imperfect crystallographically. Like quench-modified fibers, the external surfaces of impurity-modified fibers are not smooth, but are rough or microfaceted on a scale determined by twin density, and the occurrence of general fibrous morphology is a consequence of growth in many directions on that fine scale.

TEM studies of Sr-modified materials show a similar fibrous morphology although less heavily twinned or faceted for a comparable analyzed level of impurity concentration. Higher solidification rate accelerates the modification process; hence, lower modifier content is required in permanent mold casting than in heavy section sand castings. Additions of 0.02 to 0.03% Sr to die casting 380.0 alloy lead to a noticeably finer microstructure, thereby improving machining properties. Also, doubly modified alloys that are inoculated with Na as well as quenched showed a fine fibrous structure with a very high multiple-twin density, comparable with Na modification at slower growth rates; that is, the promotion of twinning by impurity is independent of the growth rate for medium to high rates.[304] A combination of grain refining and impurity modification seems to improve the mechanical properties, especially at low cooling rates.[307]

Overmodification. Overmodification with Na in excess of 0.018 to 0.02% produces coarsening of Si (due to the formation of AlSiNa compound) together with bands of primary aluminum in the final cast product. An overmodification with Sr is less critical than an overmodification with Na from the standpoint of mechanical properties. However, an overmodification may influence castability. Strontium overmodification leads to coarsening of the Si structure and the reversion of the fine fibrous Si to an interconnected platelike nature. The reasons for its occurrence are still unknown. Another evidence of Sr overmodification is the formation of Sr containing intermetallic phases in the microstructure such as Al_4SrSi_2 phase. It is evident that both of these effects will yield reduced properties of the alloy to the levels more typical of untreated alloy. Surprisingly, these two effects do not seem to occur simultaneously; that is, Al_4SrSi_2 can take place without remarkable Si coarsening, and vice versa.

Modifier Fading. Exact fading rates of Na depend on the circumstances of addition. Large melts are less susceptible to fading than small ones due to the lower ratio of melt surface area to melt volume. Stirring increases fading rapidly, and thus degassing, even with an inert gas, is not recommended after Na treatment. Usually, Sr fades considerably more slowly than Na; therefore, it is considered as a semi-permanent modifier. The major Sr losses from the melt are from oxidation because of the formation of stable strontium oxide.

The Effect of Phosphorus. As mentioned earlier, P interferes with modification by either Na, Sr, or Sb and alloys containing high P contents require higher retained modifier concentration in order to produce an acceptable cast structure. An Al-Si casting alloy with lower (<1 ppm) P concentration will solidify with a lamellar structure (class 2) without the addition of any modifier. It is only the higher P contents linked with commercial alloys that lead to acicular eutectic Si structure. Sometimes intentional P addition to the eutectic alloys is made for piston castings, which results in the nucleation of few primary Si particles and elimination of any traces of modification. These effects usually improve wear resistance and piston life.

In hypereutectic alloys, deliberate addition of P to Al melt to form AlP particles as nucleants for primary Si leads to a reaction with a fine dispersion of primary particles in contrast to Si coarsening in hypoeutectic alloys (with the P addition). This proves the different operative mechanisms for primary and eutectic crystallization/solidification of Al-Si alloy.

If the melt contains higher levels of P, higher (Na) modifier concentration is needed, which implies a lower holding time requirement, that is, shorter available fade time. On the other hand, for low P concentration, a low Na level and a correspondingly longer holding time are required (Fig. 3.88). The same reasoning holds for Sr modification.

Incubation Period.   An effect, probably related to P, is the presence of an incubation period with Sr modification. During the first 1 to 2 hr after Sr addition in the melt, an improvement in the extent of modification occurs. Figure 3.89 shows a typical incubation effect where the modification ratings (MRs) for Na- and Sr-treated samples are plotted against holding time after melt treatment. As Na fades, the microstructure and modification rating deteriorate. With Sr, an incubation period is observed; the extent of modification increases from MR = 2.5 after 10 min of Sr addition to MR = 4 after 1 hr in which the structure becomes completely fibrous.

The Mechanism of Eutectic Modification.   The most obvious mechanism that has been reported to be responsible for the modification of Al-Si alloy by Na or Sr

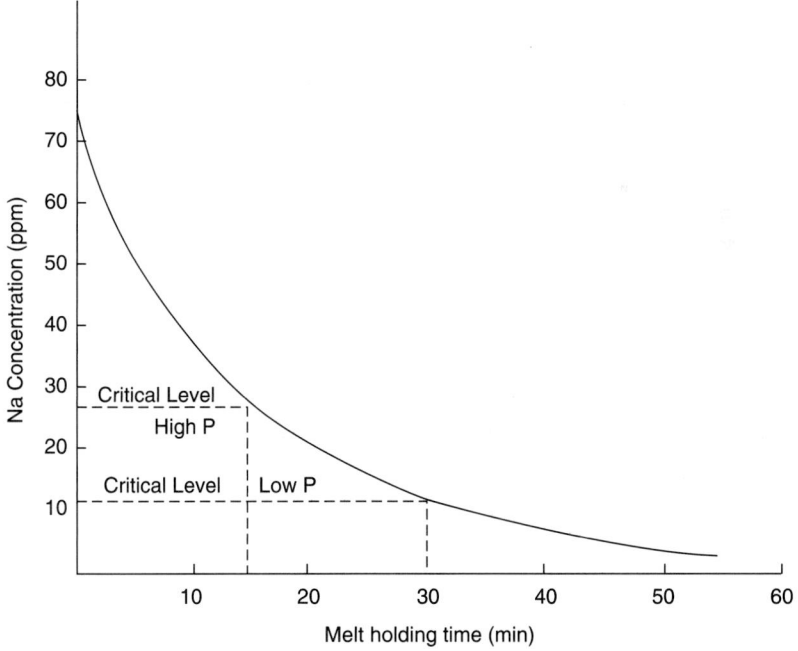

FIGURE 3.88   The effect of P level on the allowable melt holding time with Na modification.[295] For a high P level, only 15 min of holding is possible. For a low concentration, the melt can be held for 30 min. (*Reprinted by permission of American Foundry Society, Des Plaines, Ill.*)

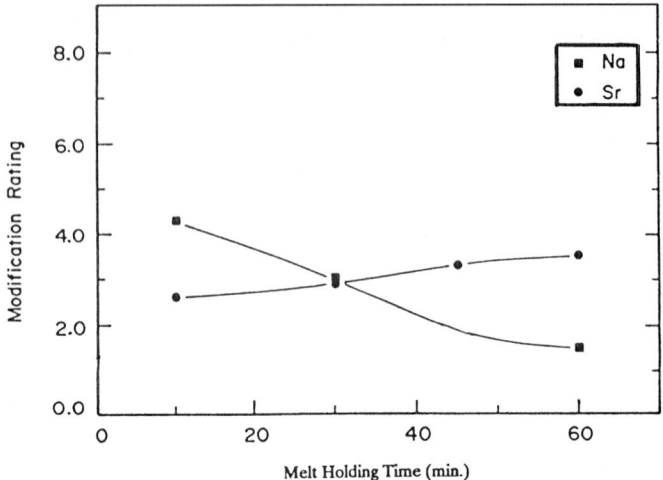

FIGURE 3.89 The evolution of microstructure with time as measured by modification rating for Na- and Sr-modified A356 alloy.[295] (*Reprinted by permission of American Foundry Society, Des Plaines, Ill.*)

is a twin plane reentrant edge (TPRE) mechanism by which a stable groove maintained at the surface of a twinned crystal favors rapid growth in a specific direction, which is $\langle 112 \rangle$ when the twinning plane is {111}. In this model, it is assumed that the impurity involved adsorbs selectively onto the reentrance edge sites and prevents, retards, or poisons the subsequent growth of Si and leads to an increase in undercooling of the Si phase/interface, producing more frequent outgrowth and more frequent twinning. The increase in undercooling in this case occurs due to the loss of TPRE mechanism in contrast to a simple increase in kinetic undercooling, as would be experienced during quench modification.[308]

In chill modification, the increased growth velocity would require an increase in kinetic undercooling of the Si phase, which also results in more frequent twinning. A higher twin population would produce more frequent branching and provide a more flexible growth front, which could adjust to the diffusion-controlled pattern of the normal fibrous structure. However, this hypothesis has been discounted recently. Very few twins are noticed in chill-modified alloys, and although multiple twins are found in impurity-modified alloys (Table 3.6), their occurrence appears to be due to the impurity effects rather than for the promotion of the growth of the [100] Si fibers.

Combined Effect of Grain Refinement and Modification. Application of a combination of AlSr and Al-Ti-1B treatment improves solidification, particularly in the case of low cooling rates.

Microporosity and Shrinkage. The effects of grain refinement, modification, and melt hydrogen on microporosity, shrinkage, and impact properties of A356 alloy were measured by LaOrchan and Gruzleski. Melt hydrogen was maintained at 0.1, 0.2, and 0.3 ml of $H_2/100$g of Al. The following major conclusions were drawn:[309]

1. Grain refinement, singly or in combination with modification, decreases microporosity by inducing mass feeding, particularly in slow-solidifying castings.

2. Varying melt treatments produce little differences in total shrinkage, and Sr-modified samples have less total shrinkage than non-Sr-modified samples.

3. Lower volumetric shrinkage in the presence of Sr causes less pipe, slumping, and contraction volume than in non-Sr-modified samples.

4. Microporosity displaces slumping and contraction volume more than it does pipe volume.

5. Silicon morphology plays an important role in controlling the A356 impact strength.

6. Increased H_2 content has a marginal effect on A356 impact strength.

7. Grain refinement with or without modification does not enhance A356 impact strength.

8. For optimum impact strength, a combination of modification and degassing to about 0.1 ml H_2/100 g Al is always recommended.

3.11.5 Primary Si Refinement of Hypereutectic Al-Si Alloys

Hypereutectic Al-Si alloys (containing more than 12.7% Si) are characterized by a discontinuous eutectic matrix containing primary Si particles of various sizes and shapes.[310] Effective refinement of primary Si particles is usually achieved by the addition of P to the hypereutectic alloy melt in the form of a brazing Cu–8% P master alloy rod or shot, or fluxes containing P compounds/salts at a level of 0.01 to 0.03% to introduce small, insoluble particles of AlP (with a crystal structure very similar to that of Si) which serve as effective heterogeneous nucleants. However, Cu is detrimental to corrosion properties of these alloys, especially in saltwater applications. Alternatively, fluxes containing red P as the active agent and salt additive to clean and degas the melt can be used. In this case, the recoveries of P range from 5 to 20% depending upon the type of additive and the exact addition technique. An increasingly popular form of P addition is by means of Ni_3P which offers excellent P recoveries. Practices recommended for melt refinement include these: (1) The casting temperature of P-refined alloy should be high enough to allow complete filling of the mold cavity prior to the onset of precipitation of the primary Si particles. (2) A rapid solidification rate produces a significant refining effect even in the absence of P. (3) Melting and holding temperatures and holding time should be held to a minimum. (4) The melt should be thoroughly chlorine or freon fluxed prior to refining to eliminate the P-scavenging impurities, namely, Na and Ca. (5) Brief inert gas fluxing (or degassing) after the addition of P is reported to remove the H_2 pickup during the addition and to produce more uniform distribution of AlP nuclei into the melt.

Effect of Refinement on Properties. The effect includes improvement in mechanical properties, castability, surface roughness, machinability, and probably increased wear resistance.

Modification and Refinement.   Current melt technology does not permit coexistence of a modified eutectic and the formation of a finer uniformly distributed primary Si phase by P treatment. The potential adverse effects of using modifying and refining additions in the melt are characterized by the interaction of P with Na and Sr to form the respective phosphides. Strontium has been reported to benefit hypoeutectic and hypereutectic structures, but this claim has not been well proven.

Refinement with Other Elements.   Good refinement can be found in alloys containing 0.05 to 0.01% S. Like P, S forms stable compounds with Na and Sr, and higher melt temperature produces better refinement. Also Se, As, Ce, La, and Nb have been reported to refine primary Si.

3.11.6   Eutectic Growth Morphology of Cast Iron

The growth morphologies and characteristics of eutectic exhibited by cast iron, whether stable or metastable, with or without modification, are briefly described and related to the classification in this section.

White Irons.   Growth of the metastable unalloyed γ-Fe_3C (ledeburite) eutectic, which is classified as quasi-regular, begins with the nucleation of a cementite plate followed by edgewise and sidewise growth of Fe_3C plates on which an austenitic dendrite nucleates and grows (Fig. 3.90).[311] As a result, two types of quasi-regular eutectic structures develop: (1) a lamellar or platelike eutectic structure with Fe_3C, as the leading edge, growing in the edgewise direction *a* more rapidly than the sidewise growth and (2) a rodlike eutectic structure growing cooperatively in the sidewise direction *c* and normal to the plates.
    Cooling rate is of importance on the morphology of γ-Fe_3C eutectic. At moderate undercoolings, ledeburite structure is expected. High cooling rates result in a degenerated eutectic structure characterized by Fe_3C plates. A coarse mixture of Fe_3C and γ takes up the spaces between the Fe_3C plates. In fact, this structure does not form from cooperative growth.[312] A platelike carbide structure associated with

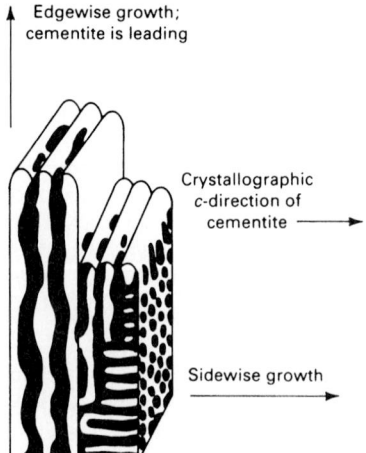

Edgewise growth;
cementite is leading

Crystallographic
c-direction of
cementite ⟶

Sidewise growth
⟶

FIGURE 3.90 Schematic showing edgewise and sidewise growth of ledeburite (austenite/cementite) eutectic.[311] (*Reprinted by permission of The Metals Society, London.*)

equiaxed eutectic grains can be achieved by increasing the undercooling by super-heating, by adding Cr or Mg, or by decreasing the Si level.

The gray or white solidification mode of cast iron depends on the relative nucleation events and the growth rates of graphite and Fe_3C phases. This, in turn, will depend on the cooling rate and chemistry of the alloy. Figure 3.91 shows the growth kinetics of gray and white iron eutectics, i.e., the range of existence of gray and white iron eutectic transition plotted as temperature versus growth rate.[312] The equilibrium temperatures for graphite-eutectic and Fe_3C-eutectic are 1153°C (2107°F) and 1148°C (2098°F), respectively. Between these eutectic temperatures, only the graphite eutectic can nucleate and grow. Below 1148°C (2098°F), both eutectics can be found. The growth rate of γ-Fe_3C eutectic rapidly exceeds that of the γ-Gr eutectic; and at a temperature of approximately 1140°C (2085°F), there should be change from gray to white solidification.[311,313,314]

The improved toughness of alloyed white iron is related to the different eutectic morphologies of the Fe-M_7C_3 eutectic. This is an anomalous structure with a broken lamellar structure and is composed of a mixture of blades and hollow faceted rods. The extent of rods in the structures increases with the cooling rate.[143]

Based on the solidification conditions, the white and gray iron eutectics can display either a columnar morphology or a completely equiaxed structure. More frequently, both eutectics occur in the casting, comprising a transition from an outer columnar region to a central equiaxed zone.[314a]

Gray Irons.   The growth of flake structure is well documented. Once graphite has nucleated, the γ-FG eutectic cell or colony grows in an approximately radial fashion, and each flake is in contact with the γ up to the growing edge. The leading phase during the eutectic growth is the graphite. The flakes can only grow in the close-packed, strong bonding *a* direction by a defect mechanism from the step of a rotation boundary. For a faceted crystal, which is the case for graphite, the defects in the crystals are in the form of rotation boundaries. One part of the crystal is rotated around the $\langle 0001 \rangle$ axis. According to Minkoff, this forms a step on the $\{10\bar{1}0\}$ planes due to screw dislocations on unstable $\{10\bar{1}0\}$ faces, which allows development into a branch and growth over the initiating crystal.[315] This imposes a constraint on growth in the graphite eutectic and sets up the slowly growing side-branched structures normally found in graphite eutectic cell. Figure 3.92 shows a schematic diagram of cooling curve and evolution of the microstructure and the associated cell

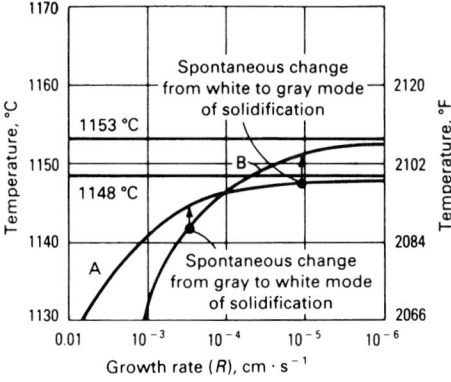

FIGURE 3.91 Gray-to-white eutectic transition as a function of temperature and growth rate.[311] (*Reprinted by permission of The Metals Society, London.*)

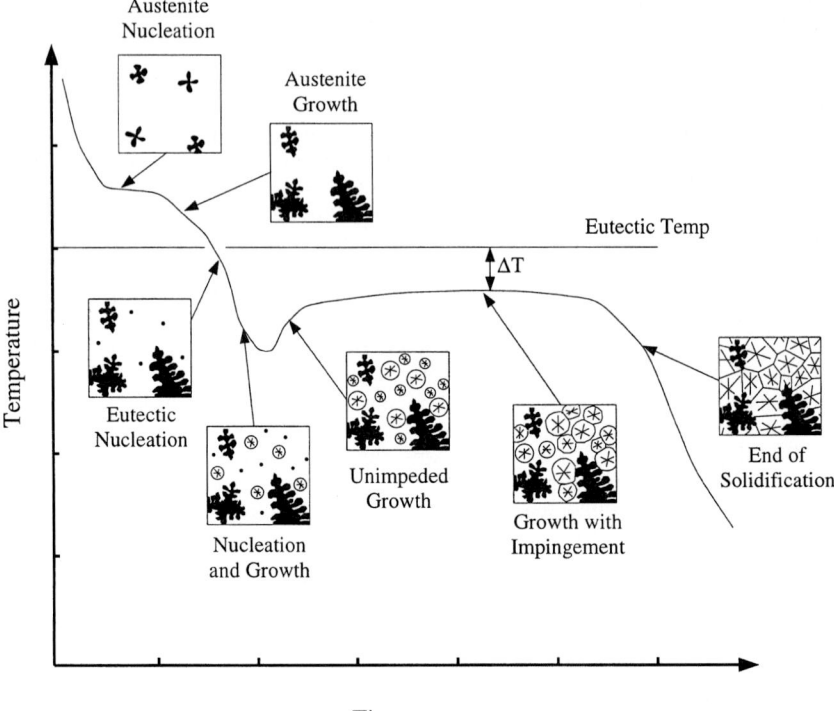

FIGURE 3.92 Microstructural development of a gray iron of eutectic composition and the associated evolution of the eutectic cell size distribution.[316] (*Courtesy of D. D. Goettsch and J. A. Dantzig.*)

size distribution during the different stages of solidification.[316] Coarse graphite eutectic structures are associated with small undercooling and existence of sulfur, whereas finer structures are attributed to larger undercooling and probably enhanced branching rate from the edges.

The graphite eutectic represents a typical, irregular eutectic structure which grows in a different manner than a regular eutectic growth; the crystal growth of the irregularly growing phase is nearly independent of the γ phase, and its branching mechanism is also independent of austenite.[17]

Nodular or Spheroidal Irons.   The spheroidal iron is considered a divorced or discontinued eutectic. It has been widely accepted until recently that the growth of γ-SG eutectic commences with nucleation and growth of the graphite in the liquid at higher undercooling than flakes, followed by early encapsulation of these graphite spheroids in the austenitic shells (envelopes), producing eutectic grains (often called *eutectic cells*). Thus, it is a common practice in the foundry industry to relate the number of nodules to the number of eutectic grains.

Sikora et al.[317] and Banerjee and Stefanescu[318] have proposed the presence of simultaneous nucleation of both the dendritic austenite and the spheroidal graphite. The interaction between both phases during solidification results in the formation of eutectic grains presenting several nodules. This fact should be considered when micromodeling of the structure is attempted. For the spheroidal growth of the graphite, several theories have been put forward in the literature which have been reviewed by Minkoff,[313] Elliott,[143] and Stefanescu.[319] According to Minkoff,[315] the relationship among undercooling (kinetic plus constitutional), melt chemistry, and crystalline defects determines the spheroidal growth of graphite. In this situation, with the increasing undercooling, the growth is determined mainly from screw dislocations, and the crystal grows to form a pyramid (Fig. 3.93). At small values of ΔT, the defect growth mechanism is operative. At higher values of ΔT (i.e., difference in temperature between the interface and the liquid), chemical composition becomes significant. In few cases of thin-section castings, the cooling rate may also exert some influence. The chemical composition controls primarily related to alloying elements result in constitutional and kinetic undercoolings. In the beginning, at small undercooling ($\sim 4°C$), growth of flake graphite occurs which depends on defect growth. This graphite morphology changes from plate to rod or coral graphite morphology (at undercooling of about 9°C), which is attributed to instability mech-

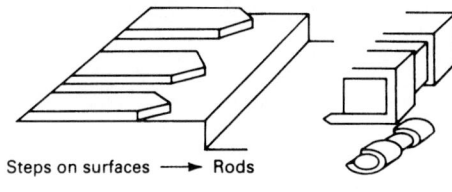

Steps on surfaces ⟶ Rods

(a)

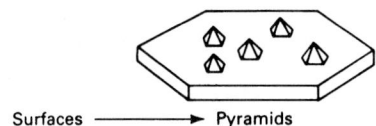

Surfaces ⟶ Pyramids

(b)

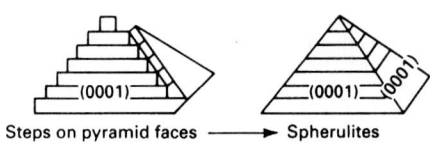

(0001)   (0001)   (0001)

Steps on pyramid faces ⟶ Spherulites

(c)

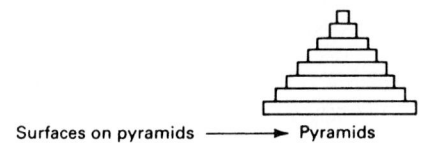

Surfaces on pyramids ⟶ Pyramids

(d)

FIGURE 3.93 Correlation among the different types of instability found in graphite growth and growth morphologies with increasing undercooling ΔT: (a) $\Delta T = 4°C$ (7°F), (b) $\Delta T = 9°C$ (16°F), (c) $\Delta T = 30°C$ (54°F), and (d) $\Delta T = 40°C$ (72°F).[314] (*Reprinted by permission of ASM International, Materials Park, Ohio.*)

anism from steps on surfaces. With an increase in undercoolings (>9°C), pyramidal instabilities occur on the faces of graphite crystal. At undercoolings of 29 to 35°C (50 to 65°F), instabilities occur on the $(10\bar{1}0)$ faces of the pyramid, and the theory predicts the occurrence of spherulite graphite at these undercoolings. Finally, at large undercoolings of 40°C (70°F), the growth form observed is a pyramidal one, bounded by $(10\bar{1}0)$ faces. These pyramidal crystals are part of the series of imperfect forms noticed especially in thick-wall SG iron castings.[314] More recently, it is shown that the solidification mechanism of SG iron is more complicated and that γ dendrites play an important role in eutectic solidification.[314]

Coral Graphite. This is an intermediate morphology between lamellar and spherulitic graphite and exists as cylindrical rods (Fig. 3.94)[320] in high-purity Fe-C-Si alloys with S < 0.001% and high Si.[321]

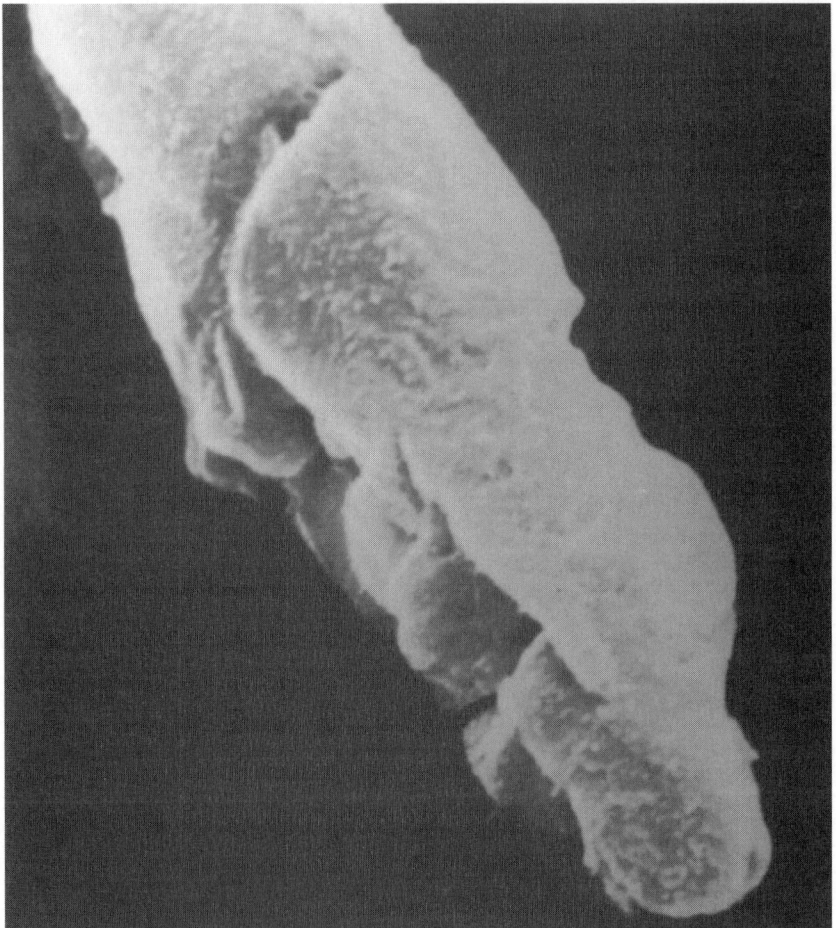

FIGURE 3.94 Cylindrical-shaped crystals called coral graphite in cast iron.[320] (*Reprinted by permission of Pergamon Press, Plc.*)

As the name suggests, the graphite occurs as fibers with roughly round cross section. It is an irregularly branched rod-shaped structure (in contrast to the branched flake structure of type D flake) and connected to form a highly convoluted and interconnected network in a manner similar to that of fibrous Al-Si eutectic.

TEM studies suggest that the internal structure of fibers comprises sheets of graphite wrapped around the fiber axis to provide irregular scrolls.[322] It is reported that coral graphite does not occur in commercial irons because of the reduction of interface undercooling and neutralization of the influence of silicon as a result of adsorbed sulfur.[143]

Compacted (Vermicular) Graphite Eutectic. Compacted graphite grows with at least one branch in direct contact with the liquid by an instability of flake surfaces leading to thick (chunky or compacted) graphite crystals as interconnected segments within an austenitic matrix. The pyramidal instabilities form on primarily a flake-type growth and may become rounded as recalescence occurs and the interface undercooling is decreased. Alternatively, the pyramidal promontories may flatten.[323] The scanning electron micrographs of chunky graphite extracted from an iron alloy are shown in Fig. 1.17*b*.

Compacted graphite occurs very frequently in heavy section spheroidal iron castings. It has been reported that even with the presence of sufficient spheroidizing agent, the very slow cooling rate can result in the formation of incomplete shell around spheroids and that thermal currents can dissolve a portion of the γ shell, exposing the graphite to the liquid. In reality, a balance between flake-promoting elements such as S and O, spheroidizing elements such as Mg, Ce, and La, and anti-spheroidizing elements such as Ti and Al is necessary for a successful growth of compacted graphite structure.

3.12 NEW SOLIDIFICATION PROCESSES

In this section we describe rapid solidification, squeeze casting, semisolid metal forming, metal-matrix composite fabrication, the Cosworth process, and improved low-pressure casting processes.

3.12.1 Rapid Solidification Process (RSP)

Rapid solidification is defined as the rapid cooling (in excess of 10^3 K/s) of the molten metal through the solidification temperature range which is well above the range achieved in conventional ingot metallurgy. In RSP the significant undercooling produces metastable effects which can be grouped into constitutional and microstructural categories.[324] Constitutional changes result in extension of solid solubility limits, formation of nonequilibrium or metastable crystalline or quasi-crystalline intermediate phases, production of metallic glasses, and retention of disordered crystalline structures in normally ordered materials and intermetallic compounds. The microstructural effects comprise morphology changes and refinement of scales of microstructural features (such as grain size, DAS, shape, and location of phases present) and a large reduction in scales of solute segregation effects.[324] Figure 3.95 shows the combination of molten metal stream and cooling medium used in the major rapid solidification processes.[325]

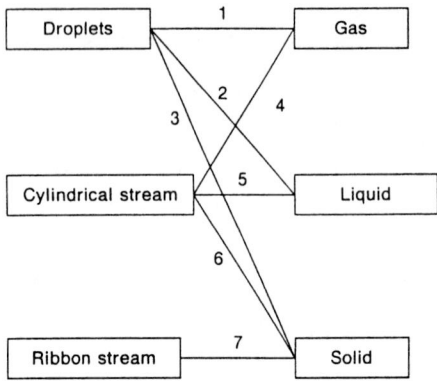

1. Atomization processes, rapid solidification rate (RSR), rotating electrode process (REP)

2. Water atomization, rapid spinning cup (RSC)

3. Duwez gun, piston-and-anvil, drum splat, electron beam splat quenching, controlled spray deposition, spray deposition, Osprey

4. Taylor wire, free-flight melt spinning

5. Free-flight melt spinning, in-rotating water spinning

6. Melt extraction, pendant drop melt extraction

7. Melt spinning, planar flow casting, melt drag, melt overflow

FIGURE 3.95 Combinations of molten metal stream and cooling used in the major rapid solidification processes.[325] (*Courtesy of VCH, Weinheim, Germany.*)

Important advantages of RSP are the chemical homogeneity of material (i.e., reduction in chemical and dendritic or microsegregation), microstructural refinement (i.e., small grain size, fine dendritic arm spacing, elimination of massive phases, and fine-scale dispersion of precipitates and dispersoids), large extension of solid-solubility limit of alloying elements (i.e., alloying flexibility), achievement of high-density point defects, formation of unique crystalline (amorphous or metallic glassy) metastable (or nonequilibrium) phases, and increased tolerance of tramp elements,[325–327] which have a significant bearing on the properties and structural engineering applications of alloys. Other advantages include better fabricability, the manufacture of net or near-net shapes, and the removal of highly textured products, especially in hcp Ti materials.[324]

A widespread application of RSP has been to develop new alloy systems (Al-, Mg-, Cu-, Fe-, Ni-, and Ti-based alloy systems) with greatly superior strength properties, improved corrosion resistance, and a highly desirable combination of magnetic properties which cannot be obtained in conventional materials.[328,329] Typical examples are ferromagnetic metallic glasses based on amorphous Fe-B-Si alloys for transformer applications; crystalline soft magnetic Fe-B, Fe-Si-Al, Fe-Al, and Fe-B-Si-Al alloys;[330] hard magnetic alloys based on crystalline Fe-Nd-B alloy; and Be-Cu thin strip for applications in electrical and electronic connectors, switches, relays, diaphragms, corrugated bellows, and miniaturization of many electronic devices.[331] In the last decade the RSP has been extended to the production of ceramic powders directed to the circumvention of toughness problems.

Rapid solidification methods for the production of metal powders may be classified into conduction and convection processes. Chill block melt spinning (CBMS), free-flight melt spinning (FFJM), free-jet melt spinning (FJMS), planar flow casting (PFC), crucible melt extraction (CME), and melt overflow (CMO) processes follow the conductive cooling. Spray deposition (or forming) and ultrasonic atomization follow the convective cooling.

RSP involves mostly rapid removal of latent heat of fusion of the melt by conductive and convective heat-transfer mechanisms. In both cases the solidification rate depends primarily on the heat-transfer coefficient of the liquid metal. Typically the cooling rate of conduction processes ranges between 10^6 and $10^8\,°C/s$ (1.8×10^6 and $1.8 \times 10^8\,°F/s$) whereas the convection processes may be limited to the range 10^4 to $10^6\,°C/s$ (1.8×10^4 to $1.8 \times 10^6\,°F/s$).

3.12.1.1   Conduction Processes

Chill Block Melt Spinning Process.   The CBMS, or simply melt spinning, process has become the most widely used technique to produce long and continuous ribbons. A wide variety of materials such as steel, Al, Cu, Ti-base alloys, superalloys, Fe- and Ni-based metallic glass alloys (for numerous applications including soft ferromagnetic lamination for power distribution transformers, cutting and forming tool materials, and wear- and corrosion-resistant hard facing coatings) have been successfully melt spun as filaments.[332,333] The CBMS process has been modified to manufacture helical glassy alloy ribbons,[334] composite alloys, and multilayer deposits.[335]

The thickness δ (μm) of ribbon filament in melt spinning is proportional to the jet diameter d and inversely proportional to the velocity v (m/s) of the spinning disk surface:

$$\delta = \frac{kd}{v} \tag{3.165}$$

In the CBMS process, a jet of liquid metal is ejected through a round orifice onto a rapidly moving highly thermally conductive, chill wheel substrate surface (Fig. 3.96a). Continuous, long, and uniform ribbons, typically about 5 mm wide and 15 to 100 μm thick, are formed. The important processing parameters involved in CBMS are selection of appropriate crucibles and wheels chamber atmosphere, substrate velocity, molten metal pressure, nozzle design, and nozzle-substrate distance.[324,336]

Free-Flight Melt Spinning Process.   Also called melt extrusion, this process consists of ejecting the liquid through an orifice and subsequently solidifying a stable liquid metal jet through a gaseous or liquid quenching medium (Fig. 3.96b). Typically, circular liquid jets ranging from about 50 to 1250 μm are used, and the pressure required to eject the molten metal from the orifice increases with the decrease of orifice area. The quenchants generally used to both stabilize the molten metal jet and accelerate its cooling rate include ambient air, air-water fog (for cast iron), liquid quenchant, liquid N_2, inert gases, and brine. The main advantage of this process is its simplicity and production of continuous, round filaments for wire-type applications such as Fe-base tire cord.[325]

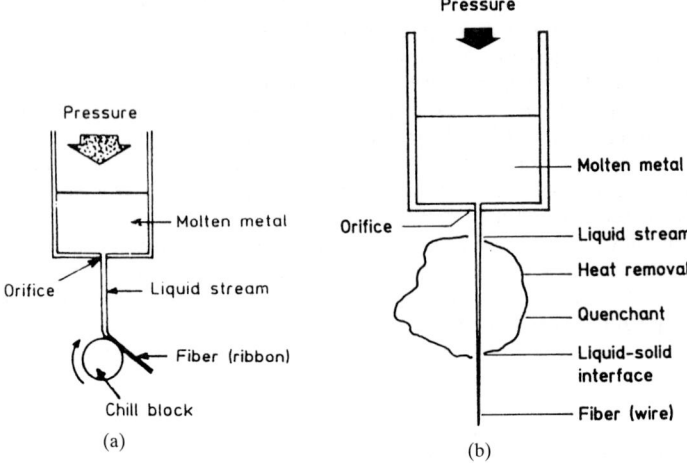

FIGURE 3.96 Schematic illustration of (*a*) chill block melt spinning (or melt spinning) and (*b*) free-flight melt spinning technique.[325] (*Courtesy of VCH, Weinheim, Germany.*)

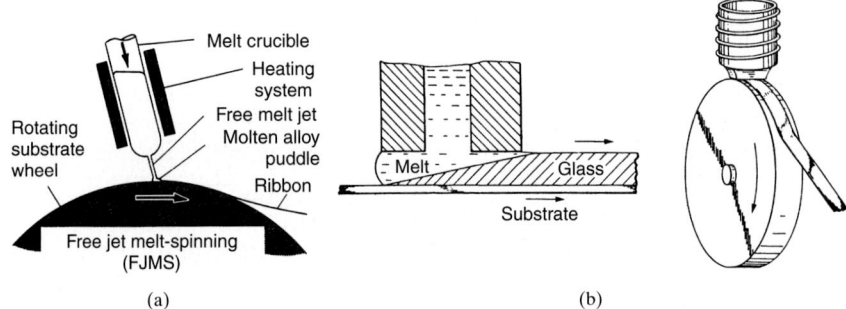

FIGURE 3.97 Schematic representation of (*a*) free jet melt spinning (FJMS) and (*b*) planar flow casting (PFC) processes.[337,339]

Free Jet Melt Spinning (or Jet Casting) Process. In the continuous FJMS process, a thin molten alloy is ejected under pressure through an orifice to form a free melt jet which strikes an external surface of a water-cooled copper wheel ~100 mm in diameter rotating with a speed of ~30 m/s where the jet forms a melt puddle (of thickness about equal to and length about double that of the jet) and solidified into ribbons, which is expelled from the surface of the wheel as shown in Fig. 3.97*a*.

This process is used to produce a wide range of amorphous alloys, but the melt-puddle does not remain considerably stable. As a result, the process is restricted to the production of ribbons <5 mm wide. This potentially serious production limitation was overcome by the development of planar flow casting (PFC) process with the production capability of 100- to 180-mm-wide strip of 20- to 30-μm thickness.[337,338]

Planar Flow Casting Process. This is a patented process (Fig. 3.97*b*) which forces the melt through a rectangular slotted orifice and onto a rapidly rotating chill surface, positioned in close proximity (~0.5 mm), so that increase in the uniformity of quench rate, stabilization (less perturbation) of the melt puddle, and a controlled thickness by the nozzle-chill surface gap are achieved.[339] Both FJMS and PFC processes have been used with a suitable protective atmosphere for a wide range of Fe-, Al-, Mg-, Ni-, and Ti-based alloys. The mechanical and corrosion properties of these alloys provide greater potential for a wide range of industrial applications.[340] It has been reported that ribbons up to 300 mm in width have been produced by this method, which can also be employed in an inert vacuum or atmosphere.[341] The thickness of the tapes ranges from 20 to 100 μm, and cooling rates achieved are about 10^6 K/s. However, care must be taken to maintain a fixed nozzle substrate gap because it significantly influences ribbon smoothness, dimensions, and quench rate. It has been calculated that a momentum and not a thermal boundary layer controls the ribbon.[342]

Crucible Melt Extraction Process. In the CME process, the periphery of multi-edged rotating water-cooled metal (Mo, Ni, Cu, Al, brass, stainless steel) disk is brought into contact with the surface of a molten pool (Fig. 3.98*a*) (by lowering the disk or raising the crucible) such that a thin melt film is extracted by momentum transfer onto the edge. The liquid solidifies on the edge wheel disk, clings there momentarily, then breaks away and is thrown free of the disk edge.

Materials with high surface tension (e.g., steels) can yield coarser fibers and large particulates, whereas materials with low surface tension (e.g., Pb-base alloy systems) give finer fibers and particulates. The solidification rate for stainless steel has been calculated as about 5×10^4 K/s.[325] Fibers in the form of ribbon, or with C-, D-, and L-shaped cross sections, can be cast simply by changing the shape of the disk periphery. Staple fibers of any specified length can be made by suitably interrupting the continuity of the casting (i.e., by providing suitable indentations in the periphery at appropriate intervals). The fiber thickness and quench rate are controlled mostly by the rotational speed of the disk. A 20-cm-diameter disk rotated at smooth speeds ranging from 200 to 2000 rpm can produce fibers at rates of 1.5 to 15 m/s. Melt-extracted fibers of Fe-Mn-C, Fe-Cr-C, Fe, and stainless steels are produced commercially as reinforcement of ceramics, concrete (flooring), and polymer safety

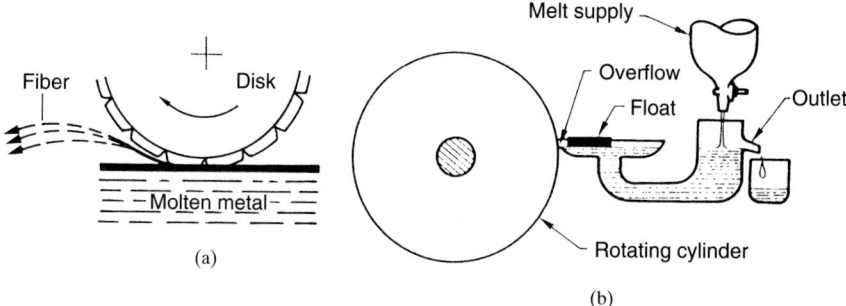

FIGURE 3.98 Schematic illustration of (*a*) crucible melt extraction[325] and (*b*) melt overflow process as described by E. H. Strange in U.S. patent 993904, 1911. (*a: Reprinted by permission of VCH, Weinheim, Germany.*)

installations and castable refractories.[340] This method is used for the laboratory-scale production of Ti alloy (such as Ti-6Al-4V) and other refractory and reactive metals and alloys.

The thickness of the ribbon filament in melt extraction depends inversely on the square root of the velocity v, implying that the filament thickness t is controlled by heat transfer through the solidified material.

$$t = \frac{K_2}{\sqrt{v}} \qquad\qquad (3.166)$$

The source of molten metal should have a clean, stable surface to produce a uniform rapid solidification processing; and for uniform product dimensions, it is necessary to employ a wiper to remove the metal debris from the disk edge.[325]

Crucible Melt Overflow Process. To overcome the inherent limitations of the CME process which is unable to produce fine-scale fibers (<100-μm diameter), an alternative process called crucible melt overflow has been developed (Fig. 3.98*b*). Here, a film of alloy melt is allowed to flow over a protruding lip of the crucible onto a rapidly rotating chilled wheel. In this way, ribbons, ~20 to 50 μm thick and a few mm wide can be formed (at cooling rates of about 10^6 K/s) for a wide variety of amorphous and microcrystalline materials ranging from lead through steels and ferroalloys. The rate of material removal is mainly governed by the melt flow rate, which, in turn, is usually adjusted by displacing the liquid metal in the crucible behind the overflow lip.[325]

3.12.1.2 Convection Processes. Spray deposition and ultrasonic atomization are examples of convection processes. Here the average cooling rate is linearly proportional to the heat-transfer coefficient and inversely proportional to the particle size. As a result, forced convective cooling as well as decreased particle size can be used to increase the cooling rate.

Spray Deposition Process. Spray deposition refers to both the spray forming process (to manufacture near-net-shape billets, tubes, and more complex shapes) and spray coating (to coat a metallic substrate with a thin coating of either metallic or ceramic materials).[343] Spray forming process was pioneered by Professor Singer in the 1960s and developed further by Osprey Metals, Ltd.[344,345] It is popularly called the Osprey process, which consists of three major steps. In the first stage, the alloy is induction-melted in a sealed crucible (Fig. 3.99). During melting the chamber is purged, and the crucible is pressurized using inert gas such as nitrogen and argon gas. In the second stage, the molten alloy is propelled away through a refractory nozzle into an atomizer where the melt stream with a chosen superheat in the range of 75°C (165°F) to 200°C (398°F) above the liquidus temperature is atomized into a spray of fine droplets by high-velocity atomizing gas jets. In the third stage, the resulting droplets impinge onto a suitably shaped collecting surface to directly solidify into a coherent, near-fully-dense preform or near-net-shape deposit. Nondeposited atomized powder which is entrained in the exhaust gas is removed by and collected from a cyclone. Typical flow rates are 6 to 10 kg/min for Al and 25 to 100 kg/min for steel and superalloys. Recovery efficiency is typically in the 60 to 90% range, depending on the product form.[337,344]

The main advantages of spray forming are the near-net-shaping capability in a single melt-heat operation, maintenance of fine-scale microstructural characteristics providing excellent workability, elimination of macrosegregation, decrease in the

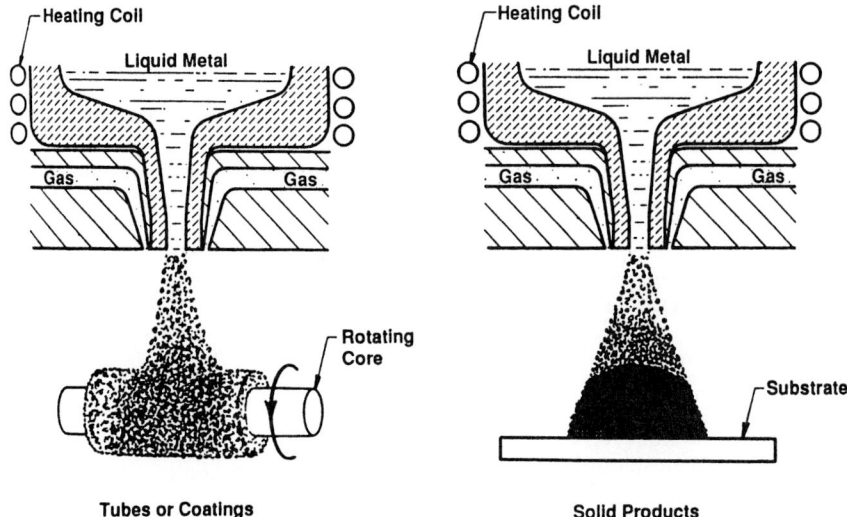

FIGURE 3.99 Schematic of the spray-forming process and the substrate motion required to produce billets, tubes, and strips.[346] (*Reprinted by permission of Butterworth-Heinemann, Boston, Mass.*)

oxygen level or surface oxide formation in the final product, production of metal-matrix composites (MMCs), and fabrication of various shapes such as solid flat disk preform or even thin- or thick-monolithic or bimetallic walled tubular structure (Fig. 3.99).[346]

Disadvantages of spray-formed preforms are porosity, which can be minimized by extrusion, hot/cold rolling, or hot isostatic pressing (HIPing). The efficiency of feedstock conversion to final product (yield) by spray forming is significantly less than 100%. Materials losses arise from (1) the preform top surface, (2) bouncing of impacting droplets or powders from the preform surface, (3) machining losses because of over-specification preforms and the removal of preform bases and crowns, and (4) preform rejection due to inadequate metallurgical quality.[347]

Spray-formed deposits are characterized by uniform, dense and refined (10- to 100-μm) equiaxed grains. In some cases, chemically homogeneous microstructure with refined low microsegregation allows alloy homogenization heat treatments to be eliminated or shortened.[347] The equipment can simply be modified to produce direct fabrication of a wide variety of simple shapes, such as thin- or thick-walled hollow tubes (stainless steel seamless tubes), hollow stainless strip, plate. Final geometry can be realized with near-net shape by thermomechanical processing such as hot forging, hot rolling, hot extrusion, and swaging. The other distinguishing features of this process are the uniform distribution of second-phase particles with a minimum oxygen content (typically 20 to 40 ppm for superalloys), good isotropic mechanical properties,[340,348] and elimination of thermal cracking.

It seems that no technical limitation exists for the production of larger preforms with fine-scale microstructures, for example, the manufacture of stainless tubes and clad tubes (up to 4.5 tons), roll rings (1 ton), clad rolls (500 kg), Cu extrusion preforms (2 tons), Al extrusion preforms (400 kg), Ni superalloy turbine ring blanks (500 kg), and special steel forging preforms (1 ton).[348] This method is now used for the production of a wide variety of materials such as Al alloys (notably, 2000 series, 7075, 7090, Al-Si, 6000 series, and Al-Cu-Pb alloys); discontinuously reinforced MMCs; Mg-, Pb-, Cu-, and Cu-Si-Pb bearing alloys; Ag- and Au-brazing alloys; Sn-Pb soldering alloys, zirconium, high-speed steels, die steels, stainless steels, and stellite; Ti alloys; Ni- and Co-based superalloys; alnico permanent magnetic alloys; etc.[348,350]

Plasma Spray Deposition Process.   Normally this process is used to melt and spray prealloyed powder in very hot, highly ionized, plasma jets (generated by heating an inert gas into an electric arc confined within a water-cooled nozzle) in a reduced-pressure environment. Subsequent consolidation occurs concurrently onto cold substrates. The particles injected into the plasma jet (temperature $\sim$10,000 K and higher) undergo rapid melting and are simultaneously accelerated toward the workpiece surface. Rapid quenching (10^5 to 10^7 K/s) of the molten particles occurs,[348] and the metal droplet velocities may reach up to 1000 m/s. In one pass of spray, a layer typically 100 μm thick may be deposited, but thicker deposits can be produced by continuing the deposition process.[325]

Another variation of conventional plasma spraying (CPS) is a low-pressure plasma spray deposition (LPPD) system where enclosed inert gas pressure is high, typically 2 to 3 Mach. Other advantages of LPPD over CPS are (1) broad spray pattern, leading to large regions of uniform deposition characteristics; (2) higher particle velocities, giving rise to denser deposits (often >98% of theoretical density); and (3) likelihood of automatic regulation to produce controlled deposits of complex designs at relatively high deposition rates (up to 50 kg/hr).

Figure 3.100*a* shows the main components of a typical plasma spray gun which comprises cathode, arc chamber, and exit nozzle (anode). Both cathode and anode are water-cooled to prevent heating by high gas temperature. The gas (Ar, Ar/He, or Ar/H$_2$) is introduced tangentially from behind the cathode to produce a vortex that stabilizes the electric arc. A shroud of inert gas is also injected around the periphery of the jet and/or around the substrate in order to provide sufficient protective atmospheres after particle melting. All these operating procedures are designed to eliminate environmental interactions and produce

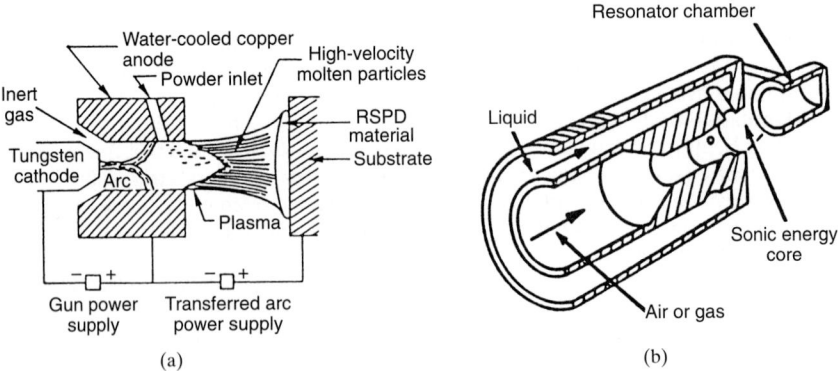

FIGURE 3.100   Schematic representation of (*a*) a low-pressure plasma deposition (LPPD)[348] and (*b*) ultrasonic gas atomizer.[343] (*a: Reprinted by permission of TMS, Warrendale, Pa. b: Reprinted by permission of Clarendon Press, Oxford.*)

high-quality metallurgically sound deposits forming a good bond with the substrate.

Plasma-sprayed deposits are characterized by homogeneous, fine-grained ($\sim$0.5-μm) microstructure with varying amounts of retained microporosity and oxide inclusions; these levels are sharply reduced with the use of protective atmosphere, i.e., under LPPD condition ($\sim$40 torr). A significant increase in thermal fatigue resistance has been found in LPPD superalloys when compared to the as-cast materials. However, the tensile strengths are comparable over the entire range of compositions of the Ni-base superalloys, especially U-700, while the elevated-temperature (1400°C) ductilities remain inferior to the wrought product. This arises from the onset of recrystallization in the wrought product. The stress rupture properties of LPPD René 80 are inferior to those of the cast product while the properties of LPPD U-700 and the cast products are comparable.[348,350]

CPS is used to produce powders including refractory metals (W, Ta, Mo), carbides (Cr_3C_2, WC, TaC), oxides (Al_2O_3, TiO_2, Y_2O_3), borides (TiB_2, ZrB_2), intermetallics (Cr_3Si_2, $MoSi_2$), mixed phases (WC-Co, Y_2O_3-ZrO_2), and special powders such as self-fluxing B-rich and Co- and Ni-based alloy powders, and Ni-Al composite powders. In addition, the LPPD process successfully produces advanced coatings of MCr-AlY type (M = Ni, Co, Fe, Ni-Co) and high-performance Co- and Ni-based superalloys.[340]

Low-pressure plasma-sprayed deposits find applications (1) as protective coatings in corrosion and wear applications; (2) as thermaflux barriers; (3) in building up warm bearing surfaces; (4) for making relatively thin-walled structures such as refractory rocket nozzles, ceramic bodies, industrial gas turbines, and jet engines; and (5) building up thick (>5-cm) deposit without sacrificing quality.[348]

Ultrasonic Gas Atomization (USGA) Process. An ultrasonic gas atomizer, also called a Hartmann whistle acoustic atomizer, involves the production of ultrafine and more uniform droplets from the molten metal stream and solidification into more uniform ultrafine powders ($\sim$20 μm) by impingement of ultrasonic-frequency (of 80 to 100 kHz), supersonic gas jets (often attaining 1.7 to 2.5 Mach speed of gas) by accelerating high-pressure gas (such as N, Ar, or He) through the resonant cavities (as the Hartmann tube).[343,346,351,352] Figure 3.100*b* shows a convergent-divergent nozzle to produce an ultrasonic-frequency (of 80 to 100 kHz), supersonic gas jet. It is claimed that droplet formation in the USGA is a one-step process compared to the three-stage mechanism of conventional gas and water atomization.

The advantages of the traditional USGA process include the more efficient breakup of molten stream, high relative velocity between gas and melt droplets leading to high average quench rates, finer and more uniform powder particles, and gas chilling factor due to expansion of high-pressure gas. The yields are >90%, and this process has been successfully used for the large-scale commercial production of low-melting alloys such as Al-Li alloys and for small-scale production of high-melting-temperature alloys such as Ti alloys, stainless steels, and Ni- and Co-based superalloys.[329] Because of the large specific areas, care must be taken in the safe handling of these powders.

3.12.2 Solidification Processing of Metal-Matrix Composites

Metal-matrix composites are defined as artificial or advanced material comprising any combination of fibers, whiskers, particles, and metal wires embedded in a metallic matrix.[353] They possess attractive properties such as enhanced specific strength, specific modulus, and damping properties; low specific gravity and coefficient of thermal expansion (CTE); superior wear and abrasion resistance and frictional properties; better elevated-temperature strengths; increased creep-rupture and fatigue properties; good microcreep performance;[353a] moderate to high toughness; improved corrosion resistance; high electrical and thermal conductivities; and in several cases, relatively moderate composite fabrication cost. MMCs also offer the opportunities to develop new materials with a unique set of properties that cannot be realized with conventional monolithic materials.[354–356] However, the main drawback with these materials is that they often suffer from poor tensile ductility, insufficient fracture toughness, and poor fracture resistance compared to the reinforced matrix alloy.[353a]

Recently there has been a widespread use of reinforced MMCs as materials of construction for high-performance structural, aerospace, automotive, marine, electronic, and sporting goods applications. A detailed review of the processes for producing MMCs has been available covering numerous metallic matrices such as light alloy matrices, high-temperature alloy matrices, high-thermal-conductivity matrices, and reinforcements in the form of a diversity of particulates, whiskers, and continuous fibers or short fibers.[1,357–361] Table 3.8 lists some of the potential applications of MMCs in automobile components and justification for their use.[357] Processing methods may be classified into (1) solidification, casting, or liquid-state processing; (2) solid-state processes; (3) deposition processes; and (4) deformation processes. Among these, solidification processing in which the molten metal matrix is combined with the reinforcing phase in the form of the particles, whiskers, or fibers in the final composite material is attractive and finding greater recognition in the production of complex-shaped components at high production rates due to simplicity, cost-effectiveness, and ease of handling liquid metal around the reinforcing phases. It also offers a wide selection of materials and processing conditions. Good wetting condition is a prime requirement for the formation of a satisfactory bond between solid ceramic phase and liquid-metal matrix during casting of composites.[353] Based on the mechanism for combination of reinforcement and molten-metal matrix, liquid-state processing of MMCs can be broadly grouped into four major categories: infiltration, dispersion, spraying, and in situ fabrication.[363]

Infiltration Casting Process. This involves injecting molten metal into a preheated preform of porous ceramic-reinforced skeleton in a metal mold in which open porosity is entirely filled, and a composite is produced with enhanced mechanical and thermal properties over conventional engineering materials. This is often accomplished with short alumina fibers such as Saffil or with very fine SiC whisker fibers. Figure 3.101 shows two examples of infiltration processes.

The main parameters common to infiltration processes are the morphology, initial composition, volume fraction, and temperature of the reinforcement; the initial composition and temperature of the infiltrating molten metal; and the nature and magnitude of the external force applied to the liquid metal, if any. Accordingly, infiltration processes are categorized into pressureless infiltration, vacuum-assisted infiltration, vibration-assisted infiltration, pressure infiltration, and electromagnetic body force-driven infiltration.[361,362] In the Lanxide pressureless infiltration, the metal or alloy ingot is placed on the ceramic preform in a graphite mold, and the assembly is heated in a nitrogen atmosphere above the liquid temperature, so that the molten metal spontaneously infiltrates the ceramic preform. The process variables are infiltration temperature, particle size, alloy composition, and nature of atmosphere. Here, infiltration occurs spontaneously, and thus the wettability of the dispersoid by the alloy is very important. In pressure infiltration the hydraulic pressure

TABLE 3.8 Some of Potential Applications of MMCs in Automobile Components[356]

Composite[+]	Components	Benefits	Manufacturers
Al-SiC (p)	Piston	Reduced weight, high strength and wear resistance	Duralcan, Toyota, Martin Marietta, Lanxide
	Brake rotor, caliper, cylinder liner	Higher wear resistance and reduced weight	Duralcan, Lanxide
	Propeller shaft	Reduction of weight and high specific stiffness	GKN, Duralcan
Al-SiC (w)	Connecting rods	Reduced reciprocating mass, high specific strength, stiffness and low CTE	Nissan
Mg-SiC (p)	Sprockets, pulleys, and covers	Reduced weight, high strength, and stiffness	Dow Chemical
Al-Al$_2$O$_3$ (sf)	Piston ring	Wear resistance, high running temperature	Toyota
	Piston crown (combustion bowl)	Reduced reciprocating mass, high creep and fatigue resistance	T&N, JPL, Mahle, etc.
	Driveshaft	Increase of driveshaft length for a constant cross-sectional area	Duralcan
Al-Al$_2$O$_3$ (lf)	Connecting rod	Reduced reciprocating mass, improved strength and stiffness	Dupont Chrysler
Cu-graphite	Electric contact strips, electronics packaging bearings	Low friction and wear, low CTE	Hitachi Ltd.
Al-graphite	Cylinder, liner piston, bearings	Low resistance, reduced friction, wear and weight	Associated Eng., CSIR
Al-TiC (p)	Piston, connecting rod	Reduced weight and wear	Martin Marietta
Al-fiber flax	Piston	Reduced weight and wear	Zollner
Al-Al$_2$O$_3$ (f) -C (f)	Engine block and cylinder liner	Reduced weight, improved strength and wear resistance	Honda

[+] p, particle; w, whiskers; sf, short fibers; lf, long fibers.
Reprinted by permission of Pergamon Press, Oxford.

3.151

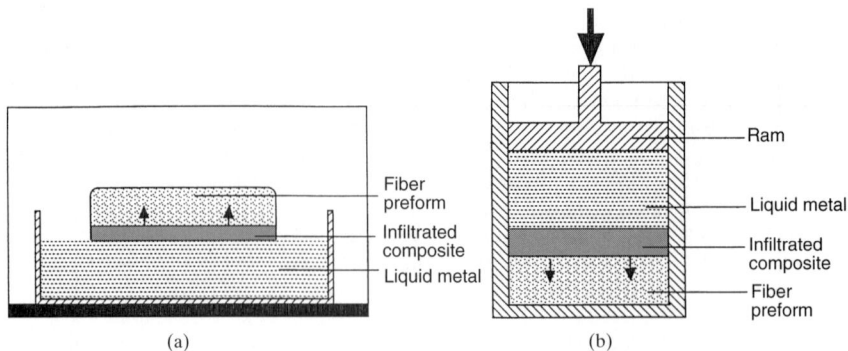

FIGURE 3.101 Schematic illustration of (*a*) pressureless (or spontaneous) infiltration and (*b*) pressure-driven infiltration.[365] (*Reprinted by permission of Butterworth-Heinemann, Boston, Mass.*)

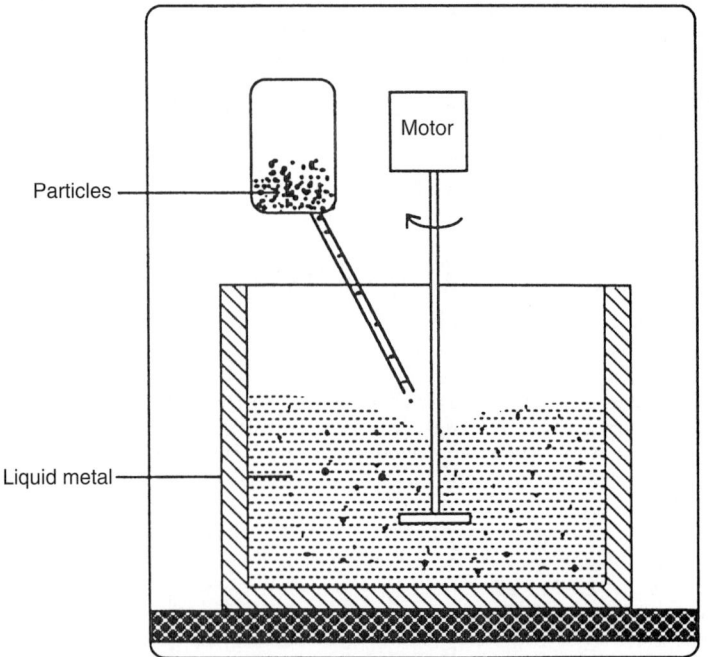

FIGURE 3.102 Schematic illustration of a dispersion process.[365] (*Reprinted by permission of Butterworth-Heinemann, Boston, Mass.*)

of squeeze casting is replaced by gas pressure. The process variables are similar to those in squeeze casting.[363]

Dispersion Process. In the dispersion process, shown schematically in Fig. 3.102, the reinforcement phase is introduced in loose form into molten or semisolid metal matrix and vigorously stirred. This method is now the most inexpensive one to produce particulate-reinforced MMCs, which can be further processed by casting or extrusion. This process is applied to mixing of SiC particulates in molten Al under vacuum, using a specially designed impeller which reduces impurities, oxides, or gases as a result of vacuum and limited vortexing. Other methods include the bottom mixing process, where a rotating blade is gradually lowered into an evacuated bed of particles covered with molten Al and the injection of particles is carried out below the melt surface using a carrier gas. The important parameters in dispersion processes are the reinforcement particle size, volume fractional content of the reinforcement, dispersion uniformity (by agitation of liquid slurry), the temperature and composition of the metal, and the applied shear rate.

Spray Casting Process. In this process (discussed earlier), droplets of liquid metal are sprayed together with the reinforcing phase and collected on a substrate, where they solidify completely. Alternatively, the reinforcement may be placed on the substrate, and liquid metal may be sprayed onto it. The important parameters in the spray processing are the initial temperature, size distribution, and velocity of metal drops; the velocity, temperature, and feeding rate of the reinforcement (provided it is simultaneously injected); and the position, nature, and temperature of the substrate collecting the material. According to Alcan International Limited, the particles may be injected within the droplet stream or between the liquid stream and the atomizing gas as in the Osprey process. Examples include SiC-, Al_2O_3-, or graphite-reinforced Al alloys.

Advantages of this technique include the matrix microstructure containing fine grain size and low segregation and minimal interfacial reaction, thereby allowing the production of thermodynamically metastable two-phase materials such as Fe particles in Al alloys.[364]

Disadvantages include the production of only simple forms such as ingots and tubes, extent of residual porosity, further processing requirements, and reduced economy compared to dispersion or infiltration processes, due to the high cost of gases used and the large amount of powders wasted.[365]

Reactive Processing (in situ MMCs). Recently, in situ processes of MMCs for nonferrous and intermetallic systems have been developed to produce a new class of naturally stable composites for advanced structural and wear applications due to their low production cost.[366] They result from the directional solidification of eutectic alloys where a binary alloy melt with sufficiently low volume fraction of one of the phases (such as carbides, nitrides, or borides) is allowed to solidify in the form of fibers (rather than the more commonly observed lamellae of the eutectic composition). The critical volume fraction depends on solute diffusion characteristics and interfacial energy but is typically about 5%. Additionally, by choosing the growth conditions, the fiber diameter and spacing can be monitored to some extent.[367] Some examples include eutectic systems based on Cr and Ta; Fe-TiC composites produced from solidification of Fe-Ti-C alloy melts; TiB rods in TiAl matrices produced from solidification of melts containing γ-TiAl, Ta, and B; and TiC/Ti composites from mixtures of Ti, C with Al additions.

The XD process is another in situ process which is also referred to as the self-propagating high-temperature synthesis (SHS) process. This involves reaction between liquid metal and reacting constituents or compounds to produce the required reinforced metallic alloy or suitable composites such as TiB_2-reinforced Al alloys, where Ti, B, and Al powders heated to 800°C react to form TiB_2; TiB whiskers obtained after laser melting of Ti and ZrB_2 powders; or TiAl matrix obtained after squeeze casting of molten aluminum into TiO_2 powders or short fibers. Reaction of molten metal with a gas also produces in situ composites such as Al_2O_3/Al composites by directional oxidation of molten Al and TiC particle reinforced Al-Cu alloys by injecting CH_4 and Ar gas through a melt of Al-Cu-Ti alloy (Fig. 3.103).[368]

The main advantages of this process are the homogeneous distribution of reinforcing phase and the control of spacing or size of the reinforcement. However, the choice of systems and the orientation of the reinforcements are restricted, and the kinetics of the processes (in the case of reactions) and the shape of the reinforcing phases are sometimes difficult to control.[365]

Metal-matrix composite solidification constituting the former two processes consists of three stages. The first stage corresponds to the interaction of the reinforcement material and the liquid matrix, which, in turn, is controlled by their wetting and adhesion or chemical bonding characteristics. In most instances, external positive pressure is provided to reach a favorable wetting. However, there are alternative methods to improve the wettability which include reinforcement pretreatment, alloy modification of matrix, and reinforcement coating. To accomplish this, a comprehensive understanding of the surface of metal matrix, the reinforcement surface, and the interface chemistry, influence of alloying additions, and reactive wettings is needed.[1]

The second stage involves fluid flow, heat-transfer, and solidification phenomena that take place during the infiltration and prior to complete solidification; they dictate to a large extent the microstructure of the cast composites.[361] The infiltration mechanism, thermal and solidification effects, and processing of MMC slurries (rheology and particle migration) must be given thorough consideration.

The third stage denotes the completion of the solidification process. The reinforcement particle size, its fractional content, and dispersion uniformity have a pronounced influence on nucleation, coarsening, microsegregation, and grain size during the solidification of composites, which, in turn, are very important in improving its mechanical properties.[1]

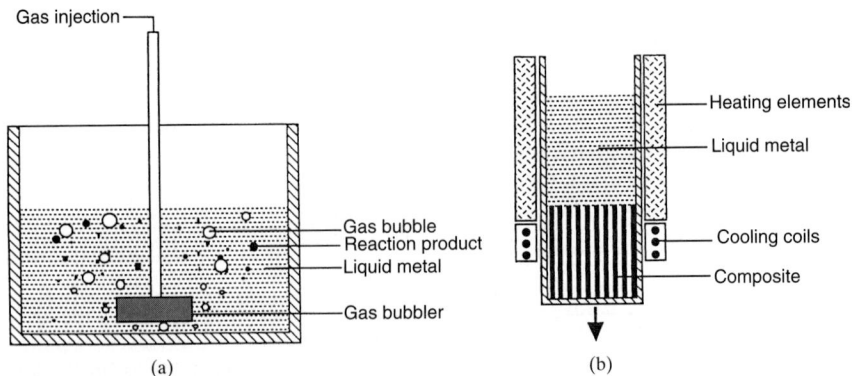

FIGURE 3.103 Schematic illustration of in situ processes. (*a*) Reaction between an injected gas and a liquid metal. (*b*) Directional solidification of a eutectic alloy.[365] (*a: From Koczak and Sahoo, 1991. b: Reprinted by permission of Butterworth-Heinemann, Boston, Mass.*)

3.12.3 Squeeze Casting

Squeeze casting (SC) is a casting process with a slow filling rate, minimum turbulence, and high pressure throughout the solidification to consistently produce high-integrity castings capable of solution heat treatment.[369] The applied pressure (ranging from 17.5 to 175 MPa, or 2.5 to 25 ksi) and the immediate contact of the liquid metal with the die surface produce a rapid heat-transfer condition that yields nearly defect (or pore) -free, fine-grain castings with improved mechanical properties. SC can be easily automated to produce near-net to net shape high-quality components with isotropic properties.[370]

The process was introduced in the United States in 1960 and has since gained widespread recognition within the nonferrous casting industry. The Al-, Mg-, and Cu-based alloy components are readily produced using this process. Several ferrous components with relatively simple shapes have also been produced by this process. The examples include aluminum automotive wheels, pistons, and dome; gear blanks made from brass and bronze; ductile iron morton shell; steel bevel gear; stainless steel blades; nickel hard-crusher wheel inserts; and superalloy disks. The process has been adapted to manufacture a wide range of MMCs using porous ceramic preform or fibers of SiC and Al_2O_3 at strategic site. Examples include automotive and diesel aluminum alloy matrix composite pistons and engine blocks that require increased mechanical properties, elevated temperature strength, and wear resistance.

There are two basic types of squeeze casting, direct (Fig. 3.104a) and indirect (Fig. 3.104b), each with its own history and development.[369] In direct squeeze casting method, a metered amount of liquid metal is poured into a preheated, lubricated mold, and the pressure (from a hydraulically activated source) is applied directly to the entire surface of liquid metal during solidification. In the indirect method, liquid metal is first poured into a shot sleeve and then injected vertically into the die cavity by a small-diameter piston, and pressure is applied through a runner system (during solidification). Of the two methods, the direct process is the more commonly used practice for the production of fully integrated castings such as automotive wheels, cylinder liners, suspension parts, hubs, hardware, drive train and steering components, and MMC components. Direct squeeze casting is mainly used for round, concentric shapes, but is not limited to these applications, and the use of inserts and slides of other materials is common.

The major advantages of SC over other casting processes are as follows:[369–380]

1. High-quality cast components without gas and shrinkage porosities are produced.

2. No feeders or risers are required, and, therefore, no metal wastage occurs. The cast components are produced with near-net shape (close tolerances), requiring only small machining and finishing allowances as well as offering exceptionally high strength-to-weight ratio of the product.

3. The high pressure involved in SC makes the process well suited to casting inserts in place and cast-in preforms that must be infiltered by the cast alloy to produce near-net-shape MMC components for engineering applications.

4. Both common (heat-treatable and non-heat-treatable) and presumably wrought alloys can be squeeze-cast to finished shape.

5. Because of the presence of very fine grain size and small primary and secondary dendritic arm spacing, and the absence of gas or shrinkage porosity, the mechanical and physical properties of squeeze cast parts are superior to those of conventional permanent mold castings.

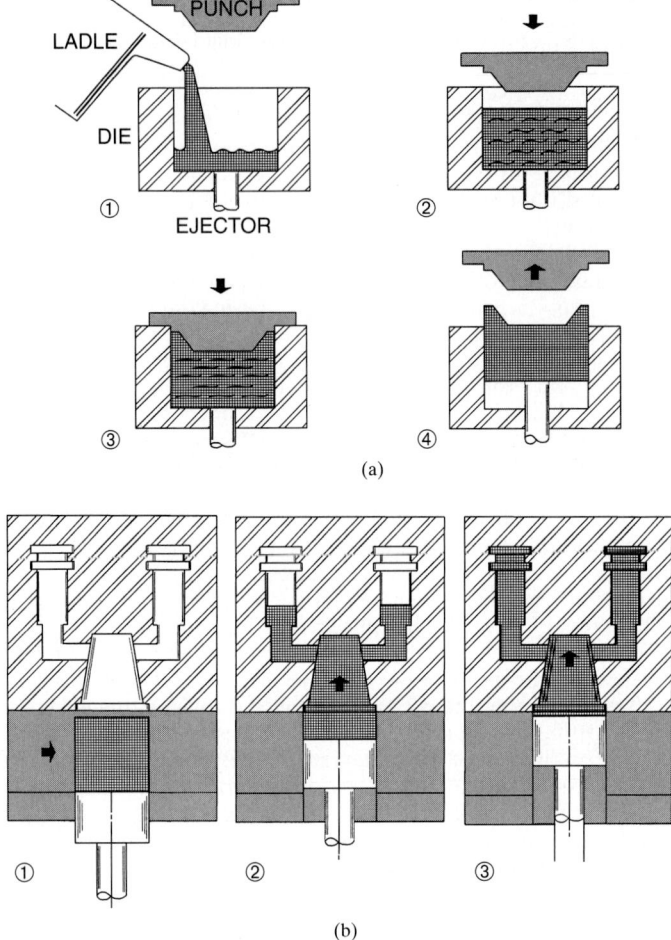

FIGURE 3.104   Schematic representation of (*a*) direct and (*b*) indirect squeeze casting processes. (*Courtesy of ELM Int., Inc., Mich.*)

6. Since squeeze casting produces sound castings, costly postsolidification examination by NDT may not be required.

The parameters that require close attention are the casting variables such as pouring temperature (6 to 55°C, or 10 to 100°F, above the liquidus temperature), die temperatures (between 190 and 315°C, or 375 and 600°F), pressure levels (between 50 and 175 MPa, or 7.5 and 25 ksi), pressure duration (30 to 120 s), lubrication (colloidal graphite spray for Al-, Mg-, and Cu-base alloys and ceramic-type coating for ferrous castings). Usually the optimization of process parameters is required for quality and reproducibility of each squeeze cast component. Failure to do so results in one or more of the following defects:[370]

1. Oxide inclusions can be minimized by cleaning melt handling and melt transfer systems, including filter in the melt-transfer system, filling the die cavity without turbulence of liquid metal, and preventing foreign objects from entering open dies.

2. Porosity and voids are removed usually by applying a sufficient pressure of at least 70 MPa or 10 ksi.

3. Extrusion segregation is avoided by increasing the die temperature, minimizing the die closure time, or selecting an alternate alloy. Blistering is avoided by degassing the melt and preheating the handling transfer equipment, using a slower die closing speed, increasing the die and punch venting, and reducing the pouring temperature.

4. Cold laps are alleviated by increasing the pouring temperature or die temperature and reducing the die closure time.

5. Hot tearing is overcome by reducing the pouring temperature and die temperature, increasing both the pressurizing time and draft angles on the casting.

6. Case debonding and extrusion debonding are eliminated by increasing the tooling or pouring temperature and decreasing the die closure time.

3.12.4 Semisolid Metal Forming Processes

Semisolid metal (SSM) processing is a relatively new process of forming metals and alloys in the semisolid state to near-net-shape products. It depends on the behavior of semisolid slurries in which stirring creates strong shear forces that break off original dendritic structure into a globular or spheroidal structure and produce very fine grain size without using grain-refining additions (Fig. 3.105).[381–383]

This process of stirring semisolid metallic alloys during solidification to produce nondendritic solid within a slurry and subsequently injecting this slurry directly into

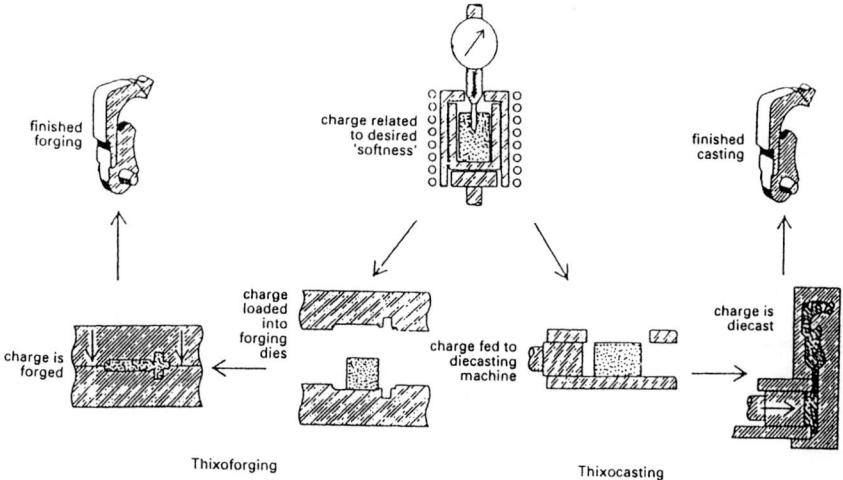

FIGURE 3.105 Semisolid metal forming processes.[381] (*Courtesy of The Institute of Materials, London.*)

dies, as in die casting, was initially called *rheocasting*. Alternatively, the slurries can be cast as a high-quality cylindrical billet, precisely cut to proper weight and sized slugs, and reheated rapidly to the desired semisolid (partially molten) condition, introduced into the diecasting machine or open die forging machine to make *thixocasting* or *thixoforging*.[1] Using this method, all types of casting can be produced, including continuous rheocasting with or without electromagnetic stirring (EMS). The lower heat content of the slurry, thixotropic characteristic, and much-reduced solidification shrinkage and cracking lead to the following advantages of SSM forming.[381-385]

1. There is smooth filling of the die with essentially laminar flow, no air entrapment, and low shrinkage porosity, producing parts of high quality or integrity to near-net shape rapidly and efficiently (Table 3.9).

2. There are substantial savings in risers and gating, closer tolerances and improved surface finish, pressure-tight components, reduced tendency toward hot tearing, and increased alloy range.

3. It is an energy-efficient process with the potential of easy automation, weight savings, minimal scrap, and increased die life and productivity.

4. There is ease of formation of composite materials by the addition of fibers or other solid particulates into the feedstock *compocasting*.

5. It is a cost-effective process due to finer and more uniform microstructure and reduction in component costs due to improved design.

6. There is faster formation of more intricate shapes, use of smaller presses, lower finishing costs, and significant reduction in forming stresses.

7. Capital investment and operating costs are lower than for conventional casting methods because of the containment of the entire process within one machine, easy maintenance of foundry cleanliness, and additional benefits, as in item 3.

8. Since the microstructure does not depend on the cooling rate, the mechanical properties were found to be independent of local sections.

Semisolid Metal Casting.   Its additional advantages include (1) control of metal quality at the source using filter, (2) no melt loss at the casting site, (3) elimination of liquid handling problems such as oxide and dross losses, (4) simple automation of reheating process because of the self-supporting slugs, and (5) energy saving because of the elimination of liquid metal-holding baths and melting of no more than 50% of the alloy at the SSM casting site.

The drawbacks of the process are (1) runners, overflows, etc., must be recycled through a complete melting cycle and cannot be reclaimed by semisolid processing, and (2) the slug heating furnace is more expensive than a liquid-metal furnace.

Unlike the liquid-metal casting process, SSM material flows under the action of applied pressure and does not follow gravity. This provides a degree of flexibility in die design not available elsewhere where liquid metal must always flow uphill and gating systems must adapt to this requirement.

The unique rheological properties of SSM materials require careful control of the injection cycle. A balance must be exercised between filling a cavity fast enough to avoid premature freezing and maintaining laminar flow and thereby excluding air entrapment.

Semisolid Al alloys exhibit only about one-half of the volumetric (solidification) shrinkage when compared to liquid alloys (3 versus 6%), and for some applications only moderate pressure is needed to produce acceptable parts. Since the parts do not contain any gas due to the laminar filling behavior, good surface and heat-treatable parts can be produced readily. For pressure-tight parts such as brake parts for automobiles, a final pressurization stage is essential to feed liquid metal from the biscuit to the shrinkage centers, and care must be taken in gating and risering design to ensure an adequate feeding path.

Raw Materials. For SSM forming, part production needs the special microstructure discussed above. When semisolid, this structure consists of spheroids or globule-shaped solid particles suspended in a matrix of low-melting alloy liquid. Preservation of this microstructure in materials heated from the solid state requires the retention of some residual microsegregation to provide differential melting between solid and liquid phases.[386]

SSM Components and Properties. SSM components are finding a wide range of applications in the automotive, aerospace, and electronic fields. Typical examples are multilink rear axle components for European automobile, fluid handling systems; Ford air conditioner front and rear compressor heads; and suspension parts, bicycle cranks, fuel system, electrical connectors, valve bodies, and threaded brass plumbing fittings. It is typically a competitor with wrought, permanent mold cast, or sometimes investment cast components. Mostly all these parts are being SSM cast from the well-established A356 and A357 (Al-7% Si-0.5% Mg) alloys and AZ91D Mg alloy. The mechanical and functional performance of SSM cast parts typically competes well with that of permanent mold cast parts and, in some instances, forged or machined parts.

A major advantage of SSM casting is the ability to provide near-net shape with minimum machining. Table 3.10 compares an SSM cast automotive master brake cylinder part with its permanent mold-cast counterpart which shows that the SSM part requires less machining.[381]

Slurry casting or compocasting is the simplest and most economically attractive method of MMC production, in which the liquid metal is mixed with solid ceramic particles and then the mixture is allowed to solidify. Being a variation of rheocasting, this is conducted through a rapid temperature increase (up to liquidus point) just before pouring.[382] It involves the development and production of MMCs containing nonmetallic particles, taking the benefits of rheological behavior and structure of partially solidified and agitated matrix. The particulate or fibrous nonmetals are introduced to the partially solid alloy slurry. The high viscosity of the slurry and the existence of a high volume fraction of primary solid in the alloy slurry prevent the floating, settling, or agglomeration of nonmetallic particles. As the mixing time, after addition, is increased, bonding increases due to the interaction between particles and the alloy matrix. The composites are then heated to the semisolid state in a second induction furnace and forged into shape with hydraulic press. Very promising wear-resistant alloys have been obtained by Sato and Mehrabian in Al alloys containing particulate additions of Al_2O_3 and SiC.[387] Matsumiya and Flemings[388] extended the application of SSM to strip casting, and the basic technology of SSM provides a potential method of metal purification.[1,389]

3.12.5 Cosworth Process

Conventional methods of aluminum castings exhibit turbulent liquid-metal transfer; dispersed, concentrated, and connected porosity; poor and inconsistent mechanical strength; dimensional inaccuracies (especially when extremely cored); and

TABLE 3.9 Comparison of Semisolid Forging (Thixoforging) and Permanent Mold Casting for Production of Aluminum Automobile Wheels

Process	Wt. from die/mold, kg	Finished part wt., kg	Production rate per die/mold, pieces/hr	Aluminum alloy	Heat treatment	Tensile strength, MPa	Yield strength, MPa	Elongation, %
Semisolid forging	7.5	6.1	90	357 (Al-7Si-0.3Mg)	T5	290	214	10
Permanent mold castings	11.1	8.6	12	356 (Al-7Si-0.5Mg)	T6	221	152	8

Source: M. P. Kenney et al., *ASM Metals Handbook*, 9th ed., vol. 15, *Casting*, ASM International Metals, Park, Ohio, 1988, pp. 327–338.

TABLE 3.10 Comparison of Semisolid Cast Aluminum Automotive Master Brake Cylinder with Its Permanent Mold Cast Equivalent (Al-7Si-0.5Mg)[381]

Process	Cast wt., kg	Finished wt., kg	Production rate, pieces/hr	Heat treatment	Tensile strength, MPa	Yield strength, MPa	Elongation, %
Semisolid cast	0.45	0.39	150	T5	303	228	8
Permanent mold casting	0.76	0.45	24	T6	290	214	8

Courtesy of International Materials Reviews, London.

dimensional instability during machining or service.[390] In addition, they may display considerable entrained oxide films; considerable contraction on the order of 7% by volume during solidification; blowholes from chills, cores, and adhesives; inaccurately-located cores or mold halves; metallurgical insufficiencies (especially poor hardness or strength); and considerable fettling required to remove large feeder heads and extensive gates.

Cosworth process is a unique low-pressure sand casting method, capable of producing a wide range of accurate, precise, completely reproducible, sound (free from porosity, oxide inclusions, and gas), high-integrity, pressure-tight, and quality-controlled components, including thin-walled section (0.16 in. or 4 mm). This process utilizes controlled melting, holding, and preparation of metal as well as liquid-metal transport through a programmable electromagnetic pump from the center of the bulk (or holding bath) to the final filling of the zircon sand mold cavity (located above the furnace) in a smooth, quiescent, nonturbulent fashion (Fig. 3.106). This approach enables one to dispense with fluxing, degassing, grain refinement, and modification procedures and reduce the subsequent heat treatment cycle to a very short time span compared to conventional practice. Extended holding (or dwell time) between the melting and casting under controlled atmosphere allows oxides and inclusions to separate by floating or sinking. The mold is permeable to allow air to escape from the cavity.[390] It is equally beneficial when applied to permanent molds with or without cores.

The Cosworth castings exhibit superior strength, ductility, and dimensional accuracy to gravity diecasting due to well-designed, high-quality tooling, a thermally stable high-purity zircon sand as mold material, programmable mold filling, consistent and predictable contraction, low casting temperature, high rate of solidification, positive metal feeding during solidification, need of minimal machining, and sound cast structure with consistent response to heat treatment. Additional advantages of the Cosworth process include rollover, weight savings, cost-effectiveness, recycling of zircon sand, excellent machining properties, and low tool wear rates.[391,392]

Applications include complex cylinder heads, engines of Formula 1 Grand Prix cars, MBA engines, cylinder blocks and Indianapolis CART racers, and high-pressure flight-refueling manifolds. Weight of Al castings ranges between 0.2 and 55 kg (0.44 and 121 lb). The maximum mold size used in the Cosworth process has been reported to be 36 × 24 in. (915 × 610 mm). Castings for aerospace and defense include gas-turbine front-end components; fuel system pumps and controls; flight-refueling manifolds; weapon mountings; lightweight undercarriage components, transmission housings, and manifolds and ductings. Automotive castings include high-performance cylinder heads, air-cooled cylinder heads, marine cylinder heads, engine blocks, engine sumps, and transmission cases for racing engines and passenger cars, rally cars, and upmarket domestic cars.

3.12.6 Improved Low-Pressure (ILP) Casting Process

The ILP process, developed in Australia, is a precision casting process that utilizes transfer of molten metal vertically through a riser tube to the bottom of the mold cavity. Degassed and filtered metal is delivered to the casting furnace which employs a pressurized N_2 atmosphere to enable nonturbulent, computer-controlled filling of the mold. Once filled, the mold is sealed and immediately removed so that solidification takes place away from the casting section. This permits another preassembled mold to be placed in the station which facilitates high productivity with cycle times of around one minute.

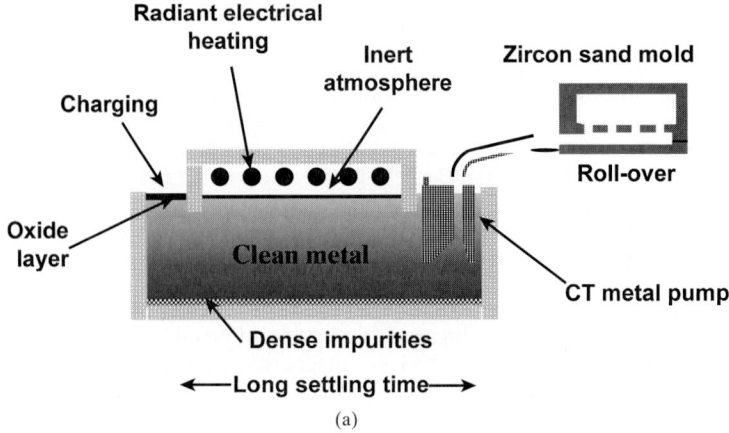

(a)

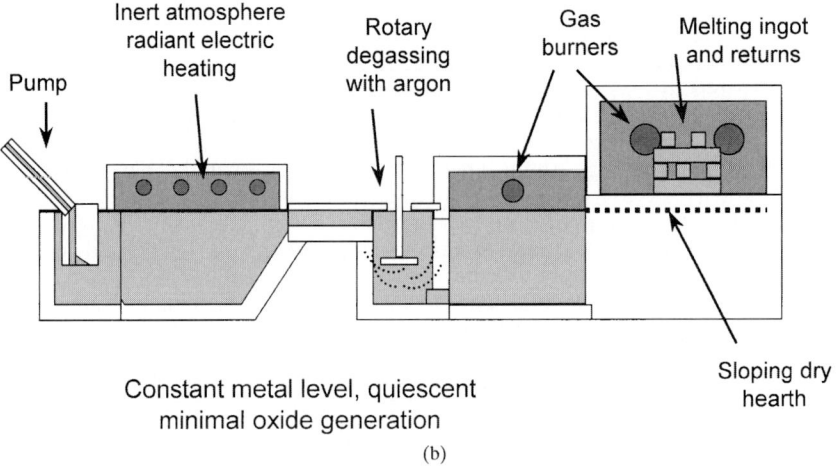

Constant metal level, quiescent
minimal oxide generation

(b)

- Closed loop control system for mold fill
  - precise, direct, real time feedback
  - repeatable, consistent mold filling

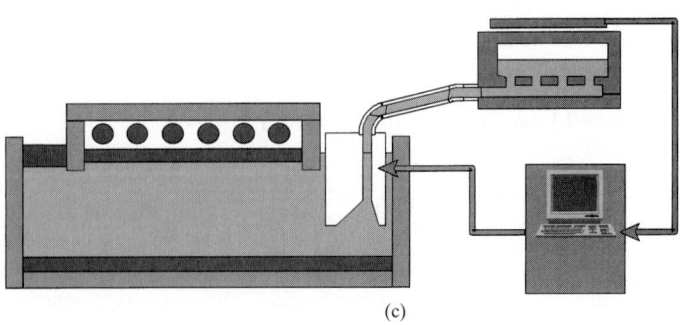

(c)

FIGURE 3.106 (*a*) The Cosworth technology process involving (*b*) metal preparation and (*c*) metal transport. (*Courtesy of Cosworth Technology Ltd., Northampton, England.*)

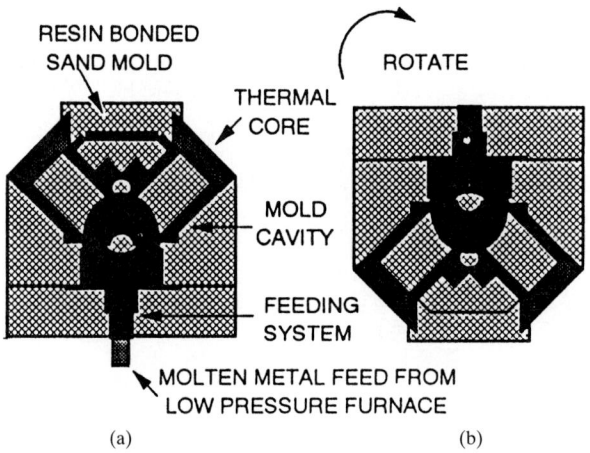

RESIN BONDED
SAND MOLD   ROTATE

THERMAL
CORE

MOLD
CAVITY

FEEDING
SYSTEM

MOLTEN METAL FEED FROM
LOW PRESSURE FURNACE

(a)   (b)

FIGURE 3.107 Schematic representation of ILP process: (*a*) cast and (*b*) solidify.[393] (*Courtesy of Comalco Aluminum Ltd.*)

As shown in Fig. 3.107, a special feature of the ILP process is the use of a combination of resin-bonded silica sand for the mold and metal cores, which promotes rapid unidirectional solidification in those areas of castings requiring optimum properties. Moreover, the mold can be inverted to facilitate this controlled solidification, which may yield DAS $<20\,\mu$m near the metal cores and $<0.5\%$ overall microshrinkage.

The ILP process has been adapted to permit robotic handling of the molds in order to precisely repeat the movements. It is being developed for mass production of automotive cylinder blocks.[393]

3.13 DIRECTIONAL SOLIDIFICATION PROCESSING

The directional solidification process can be applied extensively in the production of aligned eutectic structures, directional columnar structures, and single-crystal nickel-base superalloy turbine blades. In contrast to single crystals grown for electronic applications, single-crystal turbine blades (for aerospace applications) solidify with a dendritic structure, leading to improved ductility, creep rupture, and thermal and fatigue resistance.[393a] Consequently, the turbine blades contain microsegregation and frequently second-phase particles formed by eutectic reactions. Two conditions must be met in order to produce the desired grain structure: (1) the heat extraction from the casting in a unidirectional manner (i.e., from one end of the casting) and (2) maintenance of controlled movement of positive temperature gradient ahead of solidifying interface to prevent undesirable nucleation.[394,395]

In directional solidification, a mold standing on a water-cooled copper chill plate is heated in an induction-heated susceptor furnace, above the liquidus temperature

(a)

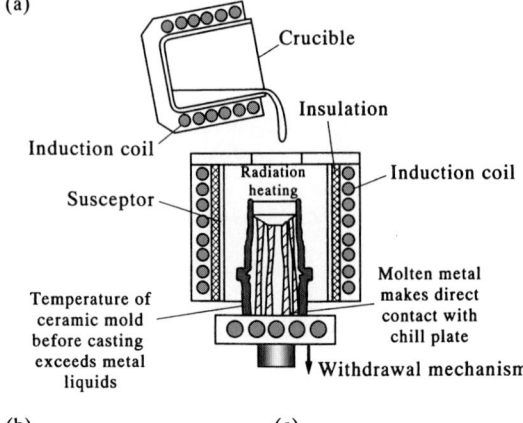

(b)　　　　　　　　　　(c)

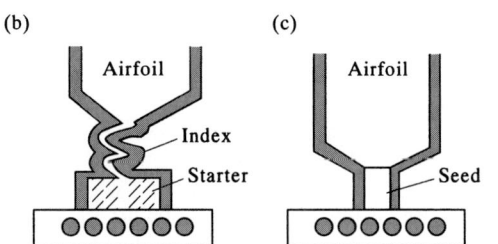

FIGURE 3.108　Schematic showing the directional solidification (DS) process. (*a*) Casting process, with heating and pouring accomplished under vacuum. (*b*) Use of helix for obtaining a single-crystal casting allowing primary orientation control only. (*c*) Use of a single-crystal seed in a DS setup to favor the production of a single-crystal component with primary and secondary orientation control.[396] (*Courtesy of T. Tom.*)

of the alloy to be poured. Subsequently, molten alloy is poured into the mold, which impacts the copper chill plate through starter holes in the bottom of the mold. The directional solidification casting process is started by simultaneously withdrawing the mold and chill plate from the susceptor furnace at a preselected rate to the cold zone, in order to produce a thermal gradient, as shown in Fig. 3.108a. A single-crystal casting is obtained by inserting a helix between the starter and the airfoil cavity so that only a single grain with a $\langle 001 \rangle$ orientation is selected and allowed to grow into the airfoil casting (Fig. 3.108b). Alternatively, the single-crystal grain is produced from the preferred oriented *seed* by epitaxial growth (Fig. 3.108c). This degree of control over the crystallographic direction provides the designer with further options to optimize part performance.[396]

Since grain boundaries in single-crystal turbine parts are nonexistent, the alloying elements such as carbon, boron, zirconium, and hafnium serving as grain boundary strengtheners can be omitted. Alloys without these elements have higher incipient melting temperatures, which allows increase of solutionizing temperature to provide more homogeneous castings. The improved homogeneity renders a more uniform distribution and higher percentage of γ' strengthening phase, which gives rise to an increase in elevated-temperature properties for a particular alloy. The benefits of single-crystal turbine airfoils include higher turbine operating temperatures (thereby increasing engine fuel efficiency) and greater durability and life at current operating temperatures. Additionally, monocrystal turbine blades have improved corrosion resistance because of the absence of grain boundaries.[394,395]

REFERENCES

1. W. Kurz and D. J. Fisher, *Fundamentals of Solidification*, Trans Tech Publications, Aedermannsdorf, Switzerland, 1999.

2. D. A. Porter and K. E. Easterling, *Phase Transformations in Metals and Alloys*, Van Nostrand Reinhold, Berkshire, England, 1992.

3. T. W. Clyne, *Materials Science and Engineering*, vol. 65, 1984, pp. 111–124.

4. H. Biloni and W. J. Boettinger, in *Physical Metallurgy*, 4th ed., eds. R. W. Cahn and P. Haasen, Elsevier Science, Amsterdam, 1996, pp. 669–842.

5. G. H. Geiger and D. R. Poirier, *Transport Phenomena in Metallurgy*, Addison-Wesley, Reading, Mass., 1973.

6. *Manufacturing with Materials*, eds. L. Edwards and M. Endean, Butterworths, London, 1990.

7. M. C. Flemings, *Solidification Processing*, McGraw Hill, New York, 1974.

8. J. H. Perepezko, in *Metals Handbook*, 9th ed., vol. 15, ASM International, pp. 101–108; *Materials Science and Engineering*, vol. 65, 1984, pp. 125–135.

9. A. Ohno, *The Solidification of Metals*, Chijin Shokan Co., Tokyo, 1976.

10. R. Elliott, *Cast Iron Technology*, Butterworths, London, 1988.

11. G. A. Chadwick, *Metallography of Phase Transformations*, Crane, Russak & Co., New York, 1972.

12. R. Cantor and R. D. Doherty, *Acta Metall.*, vol. 27, 1979, pp. 33–46.

13. W. A. Tiller, *The Science of Crystallization: Microscopic and Macroscopic Interfacial Phenomena*, Cambridge University Press, Cambridge, 1991.

14. W. A. Tiller, in *Physical Metallurgy*, 2d ed., ed. R. W. Cahn, North-Holland, Amsterdam, 1970, p. 403.

15. J. H. Hollomon and D. Turnbull, *Progr. Met. Phys.*, vol. 4, 1953, p. 333.

16. A. Bonissent, in *Modern Theory of Crystal Growth I*, eds. A. A. Chernov and H. M.-Krumbhaar, Springer-Verlag, Berlin, 1983, pp. 1–22.

17. I. Minkoff, *Solidification and Cast Structure*, Wiley, Chichester, England, 1986.

18. D. Shangguan, *Cellular Growth of Crystals*, Springer-Verlag, Berlin, 1991.

19. K. A. Jackson, in *Crystal Growth: A Tutorial Approach*, eds. W. Bardsley, D. T. J. Hurle, and J. B. Mullen, North-Holland, Amsterdam, 1979, pp. 139–154.

20. R. H. Doremus, *Rates of Phase Transformations*, Academic Press, Orlando, Fla., 1985.

21. K. A. Jackson, D. R. Uhlmann, and J. D. Hunt, *J. Cryst. Growth*, vol. 1, 1967, p. 1.

22. K. A. Jackson, in *Solidification*, ASM, Metals Park, Ohio, 1971, p. 121.

23. M. Saroch, K. S. Dubey, and P. Ramachandrarao, *J. Cryst. Growth*, vol. 126, 1993, pp. 701–706.

24. J. W. Cahn, *Acta Metall.*, vol. 8, 1960, p. 554.

25. A. A. Chernov, *Modern Crystallography III: Crystal Growth*, Springer-Verlag, Berlin, 1984.

26. D. E. Temkin, *Sov. Phys. Crystallogr.*, vol. 14, 1969, p. 344.

27. K. A. Jackson, *J. Cryst. Growth*, vols. 24/25, 1974, p. 130.

28. H. J. Leamy and G. H. Gilmer, *J. Crystal Growth*, vols. 24/25, 1974, p. 499.

29. J. Q. Broughton, A. Bonissent, and F. F. Abraham, *J. Chem. Phys.*, vol. 74, 1981, p. 4029.

30. D. W. Oxtoby and A. D. J. Haymet, *J. Chem. Phys.*, vol. 76, 1982, p. 6262.

31. H. A. Wilson, *Phil. Mag.*, vol. 50, 1900, p. 238.

32. J. Frenkel, *Physik. Z. Sowjet Union*, vol. 1, 1932, p. 4.

33. J. Q. Broughton, G. H. Gilmer, and K. A. Jackson, *Phys. Rev. Letters*, vol. 49, 1982, p. 1496.

34. G. H. Rodway and J. D. Hunt, *J. Cryst. Growth*, vol. 112, 1991, pp. 554–562.

35. Y. Shiohara and A. Endo, *Materials Science and Engineering*, vol. R19, nos. 1–2, March, 1997, pp. 1–86.

36. S. D. Peteves and G. J. Abbaschian, *J. Cryst. Growth*, vol. 79, 1986, pp. 775–782.

37. L. M. Fabietti and R. Trivedi, *Met. Trans.*, vol. 22A, 1991, pp. 1249–1258.

38. H. Li and N.-B. Ming, *J. Cryst. Growth*, vol. 152, 1995, pp. 228–234.

39. H. Li, X.-D. Pen, and N.-B. Ming, *J. Cryst. Growth*, vol. 139, 1994, pp. 129–133.

40. G. J. Abbaschian and S. F. Pavits, *J. Cryst. Growth*, vol. 44, 1978, p. 453.

41. V. Bostanov, W. Obretenov, G. Staikov, D. K. Roe, and E. Budovski, *J. Cryst. Growth*, vol. 52, 1981, p. 761.

42. F. C. Frank, *Discussions Faraday Soc.*, vol. 5, 1949, p. 48.

43. S. D. Peteves and R. Abbaschian, *Met. Trans.*, vol. 22A, 1991, pp. 1271–1286.

44. W. K. Burton, N. Cabrera, and F. C. Frank, *Phil. Trans.*, vol. A243, 1950, p. 299.

45. J. W. Cahn, W. B. Hillig, and G. W. Sears, *Acta Metall.*, vol. 12, 1964, p. 1421.

46. G. F. Bolling and W. A. Tiller, *J. Appl. Phys.*, vol. 132, 1961, pp. 2587–2605.

47. G. F. Bolling and W. A. Tiller, in *Metallurgy of Elemental and Compound Semiconductors*, ed. R. Grubel, Interscience, New York, 1963, pp. 97–105.

48. E. S. Wagner, *Acta Metall.*, vol. 8, 1960, p. 57.

49. D. R. Hamilton and R. G. Seidensticker, *J. Appl. Phys.*, vol. 31, 1960, p. 1165.

50. M. Oron and I. Minkoff, *Phil. Mag.*, vol. 9, 1964, p. 1059.

51. H. W. Kerr and W. C. Winegard, in *Crystal Growth*, ed. H. S. Peiser, Pergamon Press, Oxford, 1967, p. 179.

52. M. R. Taylor, R. S. Fidler, and R. W. Smith, *Met. Trans.*, vol. 2A, 1971, p. 1797.

53. M. N. Croker, R. S. Fidler, and R. W. Smith, *Proc. Roy. Soc. (London)*, vol. A335, 1973, p. 15.

54. K. A. Jackson, *J. Cryst. Growth*, vol. 3/4, 1968, p. 507.

55. K. A. Jackson, *J. Cryst. Growth*, vol. 5, 1969, p. 13.

56. P. Ramachadran and S. Srikanth, in *Advances in Physical Metallurgy*, eds., S. Banerjee and R. V. Ramanujan, Gordon and Breach Publishers, Canada, 1994.

57. N. Apaydin and R. W. Smith, in *Solidification Processing '87*, Institute of Metals, London, 1988.

58. S. R. Coriell and G. B. McFadden, in *Handbook of Crystal Growth*, vol. 1b, *Fundamentals: Transport and Stability*, ed. D. T. J. Hurle, North-Holland, Amsterdam, 1993, pp. 785–857.

59. J. J. Favier and D. Camel, in *Crystal Growth in Science and Technology*, eds. H. Arend and J. Hulliger, Plenum Press, New York, 1989, pp. 69–105.

60. H. S. Carslaw and J. C. Jaeger, *Conduction of Heat in Solids*, Oxford University Press, London, 1959.

61. W. A. Tiller, K. A. Jackson, J. W. Rutter, and B. Chalmers, *Acta Metall.*, vol. 1, 1953, pp. 428–437.

62. P. Haasen, *Physical Metallurgy*, 3d ed., Cambridge University Press, Cambridge, 1996.

63. G. J. Davies, *Solidification and Casting*, Wiley, New York, 1973.

64. J. D. Verhoeven, *Fundamentals of Physical Metallurgy*, Wiley, New York, 1975.

65. B. Chalmers, *Principles of Solidification*, Wiley, New York, 1964.

66. G. H. Gulliver, ed., *Metallic Alloys*, Charles Griffin, London, 1922, p. 397.

67. A. Hayes and J. Chipman, *Trans. Met. Soc., AIME*, vol. 135, 1938, p. 85.

68. E. Scheil, *Z. Matallk.*, vol. 34, 1942, p. 70.

69. W. G. Pfann, *Trans. Met. Soc., AIME*, 1952, vol. 194, p. 747.

70. D. Apelian, in *Encyclopedia of Materials Science and Engineering*, Pergamon Press, Oxford, 1986, pp. 4525–4529.

71. D. T. J. Hurle, in *Crystal Growth of Electronic Materials*, ed. E. Kaldis, North-Holland, Amsterdam, 1985, pp. 1–9.

72. J. P. Garandet, J. J. Favier, and D. Camel, in *Handbook of Crystal Growth*, vol. 2b, North-Holland, Amsterdam, 1994, pp. 659–707.

73. M. C. Flemings, in *Materials Science & Technology*, vol. 15, *Processing of Metals and Alloys*, vol. ed. R. W. Cahn, VCH, Weinheim, 1991, pp. 1–56.

74. J. R. Carruthers, in *Preparation and Properties of Solid State Materials*, vol. 2, Marcel Dekker, New York, 1976.

75. M. Pimputkar and S. Ostrach, *J. Cryst. Growth*, vol. 55, 1981, p. 614.

76. M. E. Glicksman, S. Coriell, and G. S. McFadden, *Ann. Rev. Fluid Mech.*, vol. 18, 1986, p. 307.

77. R. A. Brown, *J. AiChE*, vol. 34, 1988, p. 881.

78. J. J. Favier, *J. Cryst. Growth*, vol. 99, 1990, p. 18.

79. J. A. Burton, R. C. Prim, and W. P. Schlichter, *J. Chem. Phys.*, vol. 21, 1953, p. 1987.

80. R. Zuo and Z. Guo, *J. Cryst. Growth*, vol. 158, 1996, pp. 377–384.

81. G. Mueller, *Crystal Growth from the Melt*, Springer-Verlag, Berlin, 1988.

82. W. G. Pfann, *Zone Melting*, 2d ed., Wiley, New York, 1966.

83. L. Burris, C. H. Stockman, and I. G. Dillon, *Trans. AIME*, vol. 203, 1955, p. 1017.

84. G. H. Rodway and J. D. Hunt, *J. Cryst. Growth*, vol. 97, 1989, pp. 680–688.

85. S. R. Coriell, G. B. McFadden, and R. F. Sekerka, *Ann. Rev. Mats. Sc.*, vol. 15, 1985, p. 119.

86. S. de Cheveigne, C. Guthmann, P. Kurvaski, E. Vicente, and H. Biloni, *J. Cryst. Growth*, vol. 92, 1988, p. 616.

87. J. M. Flesslles, A. J. Simon, and A. J. Liebechaber, *Adv. Phys.*, vol. 40, 1991, p. 1.

88. J. N. Rutter and B. Chalmers, *Can. J. Phys.*, vol. 31, 1953, p. 15.

89. W. C. Winegard, *An Introduction to the Solidification of Metals and Alloys*, Institute of Metals, London, 1964.

90. R. Jansen and P. R. Sahm, *Materials Science and Engineering*, vol. 65, 1984, pp. 199–212.

91. W. W. Mullins and S. F. Sekerka, *J. Appl. Phys.*, vol. 34, 1963, p. 323.

92. W. W. Mullins and S. F. Sekerka, *J. Appl. Phys.*, vol. 35, 1964, p. 444.

93. S. F. Sekerka, in *Crystal Growth*, ed. H. S. Peiser, Pergamon Press, Oxford, 1967, p. 691.

94. R. F. Sekerka, in *Encyclopedia of Materials Science and Engineering*, Pergamon Press, Oxford, 1986, pp. 3486–3493.

95. V. V. Voronkov, *Soviet Phys.-Solid State*, vol. 6, 1965, p. 2378.

96. R. T. Delves, in *Crystal Growth*, ed. B. Pamplin, Pergamon, New York, 1975, p. 40.

97. S. R. Coriel and W. J. Boettinger, *NIST, unpublished research*, 1994.

98. T. Sato and G. Ohira, *J. Cryst. Growth*, vol. 40, 1977, pp. 78–89.

99. J. C. Baker and J. W. Cahn, *Acta Metall.*, vol. 17, 1969, p. 575.

100. H. Jones, *Materials Science and Engineering*, vol. A137, 1991, pp. 77–85.

101. B. Billia and R. Trivedi, in *Handbook of Crystal Growth*, vol. 1b, ed. D. T. J. Hurle, Elsevier Science, Amsterdam, 1993, pp. 899–1073.

102. R. Trivedi, J. A. Sekhar, and V. Seetharaman, *Metall. Trans.*, vol. 20A, 1989, p. 769.

103. M. Gremaul, M. Carrand, and W. Kurz, *Acta Metall. Mater.*, vol. 38, 1990, p. 2587.

104. R. Trivedi and W. Kurz, *Acta Mater.*, vol. 42, 1994, p. 15.

105. S.-Z. Lu and J. D. Hunt, *J. Cryst. Growth*, vol. 123, 1992, p. 17.

106. L. R. Morris and W. C. Winegard, *J. Cryst. Growth*, vol. 5, 1969, pp. 361–375.

107. H. Biloni, G. F. Bolling, and G. S. Cole, *Trans. AIME*, vol. 236, 1966, p. 930.

108. D. Shangguan and J. D. Hunt, *Metall. Trans.*, vol. 22A, 1991, pp. 1683–1687.

109. W. Bardsley, J. B. Mullin, and D. T. J. Hurle, in *The Solidification of Metals*, Iron & Steel Institute, London, 1968, p. 93.

110. T. F. Bower, H. D. Brody, and M. C. Flemings, *Trans. AIME*, vol. 236, 1966, p. 624.

111. J. D. Hunt, in *Solidification and Casting of Metals*, Metals Society, London, 1979, p. 3.

112. P. Pelcé and A. Pumir, *J. Cryst. Growth*, vol. 73, 1985, p. 337.

113. R. Trivedi and W. Kurz, *Int. Mat. Rev.*, vol. 39, 1994, pp. 49–74.

114. M. E. Glicksman and S. P. Marsh, in *Handbook of Crystal Growth*, vol. 1b, ed. D. T. J. Hurle, Elsevier Science, Amsterdam, 1993, pp. 1075–1122.

115. R. N. Grugel, *J. Mater. Sc.*, vol. 28, 1993, p. 677.

116. B. J. Spencer and H. E. Huppert, *J. Cryst. Growth*, vol. 148, 1995, pp. 305–323.

116a. Q. Li and C. Beckermann, in *Solidification 1999*, eds. W. H. Hofmeister, J. R. Rogers, N. B. Singh, S. P. Marsh, and P. W. Vorhees, The Metallurgical Society, Warrendale, Pa., 1999, pp. 111–120.

116b. V. Pines, A. Chait, and M. Zlatkovsky, in *Solidification 1999*, eds. W. H. Hofmeister, J. R. Rogers, N. B. Singh, S. P. Marsh, and P. W. Vorhees, The Metallurgical Society, Warrendale, Pa., 1999, pp. 143–149.

117. M. E. Glicksman, in *Crystal Growth of Electronic Materials*, ed. E. Kaldis, Elsevier Science, Amsterdam, 1985, pp. 57–69.

118. J. S. Langer and H. Mueller-Krumbhaar, *Acta Metall.*, vol. 26, 1978, p. 1681.

119. G. P. Ivantsov, *Dokl. Akad. Nauk S.S.S.R.*, vol. 58, 1947, p. 567.

120. L. A. Tennenhouse, M. B. Kross, J. C. La Combe, and M. E. Glicksman, *J. Crystal Growth*, vol. 174, 1997, pp. 82–89.

121. D. E. Temkin, *Dok. Akad. Nauk S.S.S.R.*, vol. 132, 1960, p. 1307.

122. R. Trivedi, *Acta Metall.*, vol. 18, 1970, p. 287.

123. H. C. Huang and M. E. Glicksman, *Acta Metall.*, vol. 29, 1981, p. 701.

124. W. Kurz, B. Giovanola, and R. Trivedi, *Acta Metall.*, vol. 34, 1986, p. 823.

125. P. W. Voorhees and M. E. Glicksman, *J. Electron. Mater.*, vol. 12, 1983, p. 161.

125a. R. Abbaschian, S. Coriell, and A. Chernov, in *Solidification 1999*, eds. W. H. Hofmeister, J. R. Rogers, N. B. Singh, S. P. Marsh, and P. W. Vorhees, The Metallurgical Society, Warrendale, Pa., 1999, pp. 219–228.

125b. W. A. Tiller and J. W. Rutter, *Can. J. Phys.*, vol. 34, 1956, p. 96.

125c. R. Trivedi, *Appl. Mech. Rev.*, vol. 43, no. 5, part 2, 1990, p. 579.

125d. R. Trivedi, V. Seetharaman, and M. A. Eselman, *Met. Trans.*, vol. 22A, 1991, p. 585.

126. R. Trivedi, in *Encyclopedia of Materials Science and Engineering*, Pergamon Press, Oxford, 1986, pp. 1044–1047.

127. T. Okamoto and M. Kishitake, *J. Cryst. Growth*, vol. 29, 1975, pp. 137–146.

128. C. T. Rios and R. Caram, *J. Cryst. Growth*, vol. 174, 1997, pp. 65–69.

129. W. Kurz and J. D. Fisher, *Acta Metall.*, vol. 29, 1981, p. 11.

130. R. Trivedi, *Met. Trans.*, vol. 15A, 1984, pp. 977–982.

131. J. D. Hunt and S.-Z. Lu, *Met. Trans.*, vol. 27A, 1996, pp. 611–623.

132. L. F. Shiau, W. G. Lo, and A. W. Cramb, *ISS Trans.*, vol. 13, 1992, pp. 53–58.

133. J. S. Langer and H. Mueller-Krumbhaar, *Acta Metall.*, vol. 29, 1980, p. 145.

134. Y. Saito, G. Goldbeck-Wood, and H. Mueller-Krumbhaar, *Phys. Rev.*, vol. A38, 1988, p. 2148.

135. T. Z. Kattamis and M. C. Flemings, *TMS-AIME*, vol. 233, 1965, p. 992.

136. S. N. Tewari and V. Laxmanan, *Metall. Trans.*, vol. 18A, 1987, p. 167.

136a. M. Rappaz and S. Henry, in *Solidification 1999*, eds. W. H. Hofmeister, J. R. Rogers, N. B. Singh, S. P. Marsh, and P. W. Vorhees, The Metallurgical Society, Warrendale, Pa., 1999, pp. 131–141.

136b. J. E. Jacoby, in *5th Australasian-Asian Pacific Conference on Aluminum Cast House Technology*, The Metallurgical Society, Warrendale, Pa., 1997, pp. 245–251.

137. D. M. Herlach et al., *Inter. Materials Rev.*, vol. 38, 1993, pp. 273–347.

138. H. W. Kerr and W. Kurz, *Inter. Metals Rev.*, vol. 41, 1996, pp. 129–164.

138a. S. H. Han and R. Trivedi, *Met. and Mats. Trans.*, vol. 31A, 2000, pp. 1819–1832.

139. R. Trivedi and W. Kurz, in *Solidification Processing of Eutectic Alloys*, eds. D. M. Stefanescu, G. J. Abbaschian, and J. J. Bayuzick, The Metallurgical Society, Warrendale, Pa., 1988, pp. 3–34.

140. A. Karma and A. Sarkissian, *Met. Trans.*, vol. 27A, 1996, pp. 635–656.

140a. M. E. Loomans and M. E. Fine, *Met. and Mats. Trans.*, vol. 31A, 2000, pp. 1155–1162.

141. M. A. Ruggiero and J. W. Rutter, *Mat. Sc. & Technol.*, vol. 13, 1997, pp. 5–11.

142. S. D. Bagheri and J. W. Ritter, *Mat. Sc. & Technol.*, vol. 13, 1997, pp. 541–550.

143. R. Elliott, *Eutectic Solidification Processing*, Butterworths, London, 1983.

144. K. A. Jackson and J. D. Hunt, *TMS-AIME*, vol. 236, 1966, p. 1129.

145. J. D. Hunt and S.-Z. Lu, in *Handbook of Crystal Growth*, vol. 2b, ed. D. T. J. Hurle, Elsevier Science, Amsterdam, 1994, pp. 1111–1166.

146. W. Kurz and R. Trivedi, *Met. Trans.*, vol. 22A, 1991, pp. 3051–3057.

147. V. Seetharaman and R. Trivedi, *Met. Trans.*, vol. 19A, 1988, pp. 2955–2964.

148. J. Glazer, *Intl. Materials Rev.*, vol. 40, no. 2, 1995, pp. 65–93.

149. F. Mollard and M. C. Flemings, *TMS-AIME*, vol. 239, 1967, p. 1526.

150. H. E. Cline and J. D. Livingston, *TMS-AIME*, vol. 245, 1969, p. 1987.

151. M. H. Burden and J. D. Hunt, *J. Cryst. Growth*, vol. 22, 1974, p. 328.

152. J. D. Hunt and J. P. Chilton, *J. Inst. Met.*, vol. 91, 1963, p. 338.

153. D. J. Fisher and W. Kurz, *Acta Metall.*, vol. 28, 1980, p. 777.

154. P. Magnin and W. Kurz, *Acta Metall.*, vol. 35, 1987, p. 1119.

155. P. Magnin, J. T. Mason, and R. Trivedi, *Acta Metall.*, vol. 39, 1991, pp. 469–480.

156. K. A. Jackson, *Materials Science and Engineering*, vol. 65, 1984, p. 7.

157. A. Kamio, S. Kumai, and H. Tezuka, *Materials Science and Engineering*, vol. A146, 1991, pp. 105–121.

158. J. Livingston and H. E. Cline, *TMS-AIME*, vol. 245, 1969, p. 351.

158a. B. Mazumdar and K. Chattopadhyay, *Met. and Mats. Trans.*, vol. 31A, 2000, pp. 1833–1842.

159. R. N. Grugel and A. Hellawell, *Met. Trans.*, vol. 12A, 1981, p. 669.

160. R. N. Grugel, T. A. Lograsso, and A. Hellawell, *Met. Trans.*, vol. 15A, 1984, p. 1003.

161. M. Hillert, in *Solidification and Casting of Metals*, The Metals Society, London, 1979, pp. 81–87.

162. H. E. Exner and G. Petzow, in *Metals Handbook*, 9th ed., vol. 9, ASM, Metals Park, Ohio, 1985, pp. 675–680.

163. H. Fredriksson, in *Metals Handbook*, 9th ed., vol. 15, ASM, Metals Park, Ohio, 1988, pp. 125–129.

164. W. P. Bosze and R. Trivedi, *Met. Trans.*, vol. 5A, 1974, pp. 511–512.

165. H. Fredriksson and T. Nylen, *Met. Sci.*, vol. 16, 1982, pp. 283–294.

166. D. H. St. John and L. M. Hogan, *Acta Metall.*, vol. 35, 1987, pp. 171–174.

167. Y. K. Chuang, D. Reinisch, and K. Schwerdtfeger, *Met. Trans.*, vol. 6A, 1975, pp. 235–238.

168. K. Matsura, H. Maruyama, Y. Itoh, M. Kudoh, and K. Ishii, *ISIJ Int.*, vol. 35, 1995, pp. 183–187.

168a. K. Matsura, H. Maruyama, M. Kudoh, and Y. Itoh, in *Solidification Science and Processing*, eds. I. Chandra and D. M. Stefanescu, The Metallurgical Society, Warrendale, Pa., 1996, pp. 11–18.

169. W. J. Boettinger, *Met. Trans.*, vol. 5A, 1974, p. 2023.

170. H. D. Brody and S. A. David, in *Solidification and Casting of Metals*, The Metals Society, London, 1979, p. 144.

171. R. Trivedi, *Met. Trans.*, vol. 26A, 1995, pp. 1583–1590.

172. W. Kurz and R. Trivedi, *Met. Trans.*, vol. 27A, 1996, pp. 625–634.

173. V. G. Smith, W. A. Tiller, and J. W. Rutter, *Can. J. Phys.*, vol. 33, 1955, pp. 723–745.

174. S. Chang and D. M. Stefanenscu, *Met. Trans.*, vol. 27A, 1996, pp. 2708–2721.

175. M. C. Flemings and G. E. Nereo, *Trans. AIME*, vol. 239, 1967, pp. 1449–1461.

176. M. C. Flemings, R. Mehrabian, and G. E. Nereo, *Trans. AIME*, vol. 242, 1968, p. 41.

177. R. Mehrabian, N. Keane, and M. C. Flemings, *Met. Trans.*, vol. 1A, 1970, p. 1209.

178. J. S. Kirkaldy and W. V. Youdelis, *TMS-AIME*, vol. 212, 1958, p. 833.

179. A. Ohno, *Solidification*, Springer-Verlag, Berlin, 1987.

180. R. L. Dobson, *Metals Handbook*, 9th ed., vol. 15, *Casting*, ASM International, Metals Park, Ohio, 1988, pp. 300–307.

181. S. G. Shabestari and J. E. Gruzleski, *Met. Trans.*, vol. 26A, 1995, pp. 999–1006.

182. D. Liang, Y. Bayrakter, and H. Jones, *Acta Metall.*, vol. 43, 1995, pp. 579–585.

183. N. Mori and K. Ogi, *Met. Trans.*, vol. 22A, 1991, pp. 1663–1672.

183a. W. Yang, W. Chen, K.-M. Chang, S. Mannan, and J. deBabadillo, *Met. and Mats. Trans.*, vol. 32A, 2001, pp. 397–406.

183b. C. Beckermann, J. P. Gu, and W. J. Boettinger, *Met. and Mats. Trans.*, vol. 31A, 2000, pp. 2545–2557.

184. J. K. Brimacombe, in *Encyclopedia of Materials Science and Engineering*, ed. M. B. Bever, Pergamon Press, Oxford, 1986, pp. 2312–2317.

185. A. Shahani, G. Amberg, and H. Fredriksson, *Met. Trans.*, vol. 23A, 1992, pp. 2301–2311.

186. H. Mori, N. Tanaka, N. Sato, and M. Hirai, *Trans. ISI Jpn.*, vol. 12, 1972, pp. 102–111.

187. J. H. Kim, J. W. Park, C. H. Lee, and E. P. Yoon, *J. Cryst. Growth*, vol. 173, 1997, pp. 550–560.

188. N. F. Dean, A. Mortensen, and M. C. Flemings, *Met. Trans.*, vol. 25A, 1994, pp. 2295–2301.

189. R. D. Doherty, in *Encyclopedia of Materials Science and Engineering*, ed. M. B. Bever, Pergamon, Oxford, 1986, pp. 3045–3049.

190. M. N. Gungor, *Met. Trans.*, vol. 20A, 1989, pp. 2529–2533.

191. J. I. D. Alexander, J. Ouazzani, and F. Rosenberger, *J. Cryst. Growth*, vol. 97, 1989, p. 285.

192. J. R. Carruthers and A. F. Witt, *J. Cryst. Growth*, vol. 131, 1993, pp. 431–438.

193. J. P. Garandet, *J. Cryst. Growth*, vol. 131, 1993, pp. 431–438.

194. I. Ohnaka, *Metals Handbook*, 9th ed., vol. 15, *Casting*, ASM International, Metals Park, Ohio, 1988, pp. 136–141.

195. D. R. Pourier and J. C. Heinrich, *The Encyclopedia of Advanced Materials*, eds. D. Bloor et al., Pergamon, Oxford, 1994, pp. 2362–2368.

195a. R. M. Forbes Jones and L. A. Jackman, *JOM*, Jan. 1999, pp. 27–31.

196. S. Ganesan and D. R. Pourier, *J. Cryst. Growth*, vol. 97, 1989, pp. 851–859.

197. E. T. Turkdogan and R. A. Grange, *JISI*, vol. 208, 1970, p. 482.

198. M. C. Flemings, in *Solidification of Metals*, Iron and Steel Institute, London, Publ. 110, 1968, p. 277.

199. G. F. Bolling and W. A. Tiller, *J. Appl. Phys.*, vol. 31, 1960, p. 1345.

200. G. F. Bolling and W. A. Tiller, *J. Appl. Phys.*, vol. 31, 1960, p. 2040.

201. W. A. Tiller, *J. Appl. Phys.*, vol. 33, 1962, p. 3106.

202. *ASM Handbook*, 10th ed., vol. 1, ASM International, Materials Park, Ohio, 1990, pp. 140–194.

203. *1997 SAE Handbook*, vol. 1, Society of Automotive Engineers, Warrendale, Pa., SAE J411, pp. 2.01–2.04.

204. W. D. Lankford and H. E. McGannon, eds., *The Making, Shaping, and Treating of Steel*, 10th ed., U.S. Steel, Pittsburgh, 1985.

205. *Steel Product Manual: Plates; Rolled Floor Plates: Carbon, High Strength Low Alloy, and Alloy Steel*, American Iron and Steel Institute, Washington, DC, August 1985.

206. E. Hornbogen, in *Physical Metallurgy*, eds. R. W. Cahn and P. Haasen, Elsevier, New York, 1983, pp. 1075–1138.

207. J. D. Smith, *Fundamentals of Ferrous Metallurgy*, Course 11, Lessons 2 and 5, Materials Engineering Institute, ASM International, Metals Park, Ohio, 1979.

208. H. Biloni, in *The Solidification of Metals*, ISI publ. 110, The Institute of Metals, London, 1968, p. 74.

209. G. S. Cole, in *Solidification*, ASM, Metals Park, Ohio, 1971, p. 201.

210. D. Fainstein-Pedraza and G. F. Bolling, *J. Cryst. Growth*, vol. 28, 1975, pp. 311–318 and 319–333.

211. H. Biloni, in *Aluminum Technology and Applications, Proc: Int'l. Symposium at Puerto Madryn Chubut, Argentina*, Aug. 21–25, 1978, eds. C. A. Pampillo et al., ASM, Metals Park, Ohio, 1980, pp. 1–79.

212. S. Henry, J. Jerry, P. H. Jounesu, and H. Rappaz, *Met. and Mats. Trans.*, vol. 28A, 1997, pp. 207–213.

213. S. C. Flood and J. D. Hunt, *J. Cryst. Growth*, vol. 82, 1987, pp. 552–560.

213a. C. Y. Wang and C. Beckermann, *Met. and Mats. Trans.*, vol. 27A, 1996, p. 2754.

213b. A. E. Ares and C. E. Schvezov, *Met. and Mats. Trans.*, vol. 31A, 2000, pp. 1611–1625.

214. C. Y. Wang and C. Beckermann, *Met. and Mats. Trans.*, vol. 25A, 1994, pp. 1081–1093.

215. B. Chalmers, *J. Aust. Inst. Metals*, vol. 8, 1963, p. 255.

216. A. Ohno, *J. Japan Inst. Metals*, vol. 34, 1970, p. 244.

217. R. T. Southin, *TMS-AIME*, vol. 236, 1967, p. 220.

218. H. Biloni and R. Morando, *Trans. AIME*, vol. 242, June 1968, pp. 1121–1125.

219. C. R. Taylor, *Met. Trans.*, vol. 6B, 1975, p. 359.

220. F. Weinberg, in *Solidification and Casting of Metals*, The Metals Society, London, 1979, p. 235; *Metals Technology*, February 1979, p. 48.

221. J. K. Brimcombe and I. V. Samarasekera, in *Principles of Solidification of Materials Processing*, Trans. Tech. Publ., Switzerland, 1990, p. 179.

222. W. Irving, *Continuous Casting of Steel*, The Institute of Metals, London, 1993.

223. E. F. Emley, *Intl. Met. Rev.*, vol. 21, 1976, pp. 75–175.

224. J. C. Baker and V. Subramanian, in *Aluminum Transformation Technology and Applications 1979*, eds. C. A. Pampillo, H. Biloni, and D. Embury, ASM, Metals Park, Ohio, 1980, p. 335.

225. D. A. Granger, in *Treatise on Materials Science and Technology*, vol. 31, *Aluminum Alloys—Contemporary Research and Applications*, eds. A. K. Vasudevan and R. D. Doherty, Academic Press, Boston, 1989, pp. 109–135.

226. E. H. Chia, in *Encyclopedia of Materials Science and Engineering*, Pergamon, Oxford, 1986, pp. 832–835.

227. R. D. Pehlke, in *Metals Handbook*, 9th ed., vol. 15, *Casting*, ASM International, Metals Park, Ohio, 1988, p. 308.

228. A. Etienne and W. R. Irving, in *Continuous Casting '85, Intl. Conf. Proc., 1985*, The Institute of Metals, London, 1985, pp. 1.1–1.9.

229. M. M. Wolf, *I&SM*, vol. 23, no. 2, 1996, pp. 47–51.

230. I. Hoskikawa, T. Saito, M. Kimura, Y. Kaihara, K. Tanikawa, H. Fukumoto, and K. Ayata, *I&SM*, April 1991, pp. 45–52.

231. Singh and Blazek, *Open Hearth Proc.*, vol. 57, 1974, p. 6.

232. A. Palmers, P. Dauby, P. Russe, and F. Anselin, *Int'l. Conf. on Clean Steel*, The Metals Society, London, 1983, p. 138.

233. M. Olette and C. Gatellier, *Int'l. Conf. on Clean Steel*, The Metals Society, London, 1983, pp. 165–186.

234. V. Presern and P. Bracun, in *Continuous Casting '85*, The Institute of Metals, London, 1985, pp. 6.1–6.8.

235. J. R. Bourguinon, J. M. Dixmier, and J. M. Henry, in *Continuous Casting '85*, The Institute of Metals, London, 1985, pp. 7.1–7.9.

236. R. Bommaraju, T. Jackson, J. Lucas, G. Skoczylas, and V. B. Clark, *I&SM*, April 1992, pp. 21–27.

237. P. Machner, E. Reist, T. Tarmann, G. Holleis, and W. Polanschuetz, *I&SM*, April 1986, pp. 15–18.

238. Y. J. Kim and S. Kou, *Met. Trans.*, vol. 19A, 1988, pp. 1849–1852.

239. M. Soda, A. McLean, Z. Wang, and G. Motoyasu, *J. Mater. Sc.*, vol. 30, 1995, pp. 5438–5448.

239a. S. Sengupta, H. Soda, A. McLean, and J. W. Rutter, *Met. and Mats. Trans.*, vol. 31A, 2000, pp. 239–248.

240. Y. H. Wang, Y. J. Kim, and S. Kou, *J. Cryst. Growth*, vol. 91, 1988, p. 50.

241. H. R. Shercliff, O. R. Nyhr, and S. T. J. Totta, *Mater. Res. Bull.*, vol. xix, no. 1, 1994, p. 25.

241a. A. I. (Ed) Nussbaum, *Light Metal Age*, April 1991, pp. 8–83.

242. R. E. Spear and H. Yu, *Aluminum*, vol. 60, 1984, pp. 440–442.

243. D. A. Granger and C. L. Jensen, *Light Metals*, AIME, Warrendale, Pa., 1984, pp. 1249–1263.

243a. J. F. Grandfield, in *5th Australasian-Asian Pacific Conference on Aluminum Cast House Technology*, eds. M. Nilmani et al., The Metallurgical Society, Warrendale, Pa., 1997, pp. 231–243.

243b. J. E. Jacoby, in *5th Australasian-Asian Pacific Conference on Aluminum Cast House Technology*, eds. M. Nilmani et al., The Metallurgical Society, Warrendale, Pa., 1997, pp. 245–252.

244. D. Altenpohl, *Proc. 5th Light Metals Congress*, Leoben, Austria, 1968, p. 372.

245. A. I. (Ed) Nussbaum, *Light Metal Age*, vol. 8, 1990.

245a. A. I. (Ed) Nussbaum, *Light Metal Age*, Feb. 1995, pp. 59–65.

245b. S. C. Servé, *Light Metal Age*, June 1995, pp. 30–34.

245c. J. L. Davis, B. Wagstaff, and C. Shaber, in *5th Australian Asian Pacific Conference on Aluminum Cast House Technology*, eds. M. Nilmani et al., The Metallurgical Society, Warrendale, Pa., 1997, pp. 253–269.

246. T. F. Bower, D. A. Granger, and J. Keverian, in *Solidification*, ASM, Metals Park, Ohio, 1971, p. 385.

247. Z. N. Getselev, *J. Metals*, vol. 23, 1971, p. 38.

247a. J. F. Grandfield, in *5th Australian Asian Pacific Conference on Aluminum Cast House Technology*, eds. M. Nilmani, P. Whiteley, and J. Grandfield, The Metallurgical Society, Warrendale, Pa., 1997, pp. 231–243.

248. V. I. Dobatkin, *Tsvet Met.*, vol. 2, 1980, pp. 54–59.

249. J. W. Evans, *J. Metals*, vol. 47, no. 5, 1995, pp. 38–41.

250. S. A. David and J. M. Vitek, in *The Metal Science of Joining*, eds. A. J. Cieslak, J. H. Perepezko, S. Kang, and M. E. Glicksman, The Metallurgical Society, Warrendale, Pa., 1992, pp. 1–9.

250a. D. J. Davies and J. G. Garland, *Int. Met. Rev.*, vol. 20, 1975, p. 83.

251. J. A. Brooks and K. W. Mahin, in *Welding: Theory and Practice*, eds. D. L. Olson, R. Dixon, and A. L. Liby, Elsevier Science, Amsterdam, 1990, pp. 35–78.

252. W. F. Savage, *Weld World*, vol. 18, 1980, p. 89.

253. K. E. Easterling, *Mats. Sc. & Engrg.*, vol. 65, 1984, p. 91.

254. S. A. David and J. M. Vitek, *Int. Met. Rev.*, vol. 34, No. 5, 1989, pp. 213–245.

255. K. E. Easterling, *Introduction to Physical Metallurgy of Welding*, 2d ed., Butterworth-Heinemann, Oxford, 1992.

256. J. F. Lancaster, *Metallurgy of Welding*, 5th ed., Chapman & Hall, London, 1993.

257. W. F. Savage and E. S. Szekeres, *Welding J.*, vol. 46, 1967, p. 94.

258. W. Savage and A. H. Aaronson, *Welding J.*, vol. 45, no. 2, 1966, p. 85.

259. M. S. Misra, D. L. Olson, and E. R. Edwards, in *Grain Refinement in Casting and Welds*, Conf. Proc. TMS-AIME, 1983, pp. 259, 274.

260. O. Grong, *Metallurgical Modelling of Welding*, The Institute of Materials, London, 1994.

261. S. Kou and Y. Le, *Met. Trans.*, vol. 19A, 1988, pp. 1075–1082.

262. Y. H. Wang and S. Kou, in *Advances in Welding Science and Technology*, ed. S. A. David, ASM, Metals Park, Ohio, 1987, p. 65.

263. T. Zacharia, S. A. David, and J. M. Vitek, in *The Metals Science of Joining*, eds. H. J. Cieslak, J. H. Perepezko, S. Kang, and M. E. Glicksman, The Metallurgical Society, Warrendale, Pa., 1992, p. 257.

264. C. R. Heiple, P. Burgardt, and J. R. Roper, in *Modeling of Casting and Welding Processes, III*, eds. J. A. Dantzing and J. T. Berry, TMS-AIME, 1984, p. 193.

265. T. Zhacharia, S. A. David, J. M. Vitek, and T. Debroy, Part I, Theoretical Analysis, *Welding J.*, vol. 68, no. 2, 1989, p. 499.

266. T. Zhacharia, S. A. David, J. M. Vitek, and T. Debroy, Part II, Experimental Correlation, *Welding J.*, vol. 68, no. 2, 1989, p. 510.

267. H. Nakagawa, H. Kato, F. Matsuda, and T. Senda, *J. Jpn. Weld. Soc.*, vol. 39, 1970, p. 94.

268. M. Rappaz, S. A. David, J. M. Vitek, and L. A. Boatner, *Met. Trans.*, vol. 20A, 1989, p. 1125.

269. H. D. Solomon, in *ASM Handbook*, vol. 6, 10th ed., *Welding, Brazing, and Soldering*, ASM Int., Materials Park, Ohio, 1993, pp. 45–54.

270. J. A. Brooks and A. W. Thompson, *Intl. Materials Rev.*, vol. 36, no. 1, 1991, pp. 16–44.

271. N. Yurioka and H. Suzuki, *Intl. Materials Rev.*, vol. 35, no. 4, 1990, pp. 217–249.

272. P. T. Houldcroft, in *Materials Science and Technology*, vol. 7, *Constitution and Properties of Steel*, VCH, Weinheim, 1992, pp. 739–776.

273. J. F. Lancaster, in *Encyclopedia of Materials Science and Engineering*, Pergamon, Oxford, 1986, pp. 5289–5292.

274. M. Piputkar and S. Ostrach, *J. Cryst. Growth*, vol. 55, 1981, pp. 614–646.

275. J. M. Robertson, in *Encyclopedia of Materials Science and Engineering*, Pergamon, Oxford, 1986, pp. 961–970.

276. S. Meyer and A. G. Ostrogorsky, *J. Cryst. Growth*, vol. 171, 1997, pp. 566–576.

277. N. B. Singh and R. Mazelsky, in *The Encyclopedia of Advanced Materials*, vol. 1, 1994, Pergamon Press, Oxford, pp. 519–530.

278. E. Monberg, in *Handbook of Crystal Growth*, vol. 2a, ed. D. T. J. Hurle, Elsevier Science, Amsterdam, 1994, pp. 51–97.

279. J. B. Mullin, in *Processing of Semiconductor*, vol. 16, ed. K. A. Jackson, chap. 2, VCH, Weinheim, 1996, pp. 63–105.

280. D. T. J. Hurle and B. Cockayne, in *Handbook of Crystal Growth*, vol. 2a, ed. D. T. J. Hurle, Elsevier Science, Amsterdam, 1994, p. 98; D. T. J. Hurle, *J. Cryst. Growth*, vol. 85, 1987, pp. 1–8.

281. K. E. Benson, in *The Encyclopedia of Materials Science and Engineering, Supplement*, vol. 2, 1992, Pergamon Press, Oxford, pp. 1261–1271.

282. R. L. Lane, in *The Encyclopedia of Advanced Materials*, vol. 4, Pergamon Press, Oxford, 1994, pp. 2528–2533.

283. W. J. Yang et al., *Transport Phenomena in Manufacturing and Materials Processing*, Amsterdam, 1994, pp. 136–145.

284. L. F. Mondolfo, in *Grain Refinement in Castings and Welds*, eds. G. J. Abbaschian and S. A. David, TMS-AIME, 1983, pp. 3–50.

284a. R. Abbaschian, R. C. Lopez, A. Gokhale, E. Jensen, and R. Raman, in *Solidification Science and Processing*, eds. I. Ohnaka and D. M. Stefanescu, The Metallurgical Society, Warrendale, Pa., 1996, pp. 319–329.

284b. D. V. Neff, in *5th Australian Asian Pacific Conference on Aluminum Cast House Technology*, eds. M. Nilmani et al., The Metallurgical Society, Warrendale, Pa., 1997, pp. 155–177.

285. M. Suarez and J. H. Perepezko, *Light Metals*, ed. E. R. Coshall, The Metallurgical Society, Warrendale, Pa., 1991, pp. 851–859.

286. E. N. Pan, C. S. Lin, and C. R. Loper, Jr., *AFS Trans.*, 1990, pp. 735–746.

286a. K. V. Prabhakar and M. R. Seshadri, *AFS Trans.*, 1979, pp. 377–386.

287. J. Campbell, in *Solidification Technology*, Proceeding of International Conference held at the University of Warwick, Coventry, 15–17 September 1980, The Metals Society, 1983, pp. 61–64.

288. D. G. McCartney, *Intl. Met. Rev.*, vol. 34, 1989, p. 247.

289. M. C. Flemings, *Met. Trans.*, vol. 22A, 1991, p. 957.

290. H. Garabedian and R. F. Strickland-Constable, *J. Cryst. Growth*, vol. 22, 1974, pp. 188–192.

291. T. Z. Kattamis and M. C. Flemings, *Modern Casting*, vol. 52, 1967, p. 97.

292. D. R. Uhlmann, T. P. Seward, III, and B. Chalmers, *Trans. AIME*, vol. 236, 1966, pp. 527–531.

293. A. Vogel, *Metal Sci.*, vol. 12, 1978, pp. 576–578.

294. N. Apaydin, K. V. Prabhakar, and R. D. Doherty, *Mat. Sc. Engrg.*, vol. 46, 1980, pp. 145–150.

295. J. E. Gruzleski and B. M. Closset, *The Treatment of Liquid Aluminum-Silicon Alloys*, American Foundry Society, Des Plaines, Ill., 1990.

296. C. D. Mayers, D. G. McCartney, and G. J. Tatlock, *Light Metals*, ed. E. R. Cotshall, The Metallurgical Society, Warrendale, Pa., 1991, pp. 813–819.

296a. S. C. Servé, *Light Metal Age*, August 1999, pp. 90–92.

297. R. A. P. Fielding and C. F. Kavanaugh, *Light Metal Age*, Oct. 1996, pp. 46–59.

298. G. K. Sigworth et al., *Trans. AFS*, 1985, p. 907.

299. D. Apelian et al., *Proc. Conf. Intl. Molten Alloy Aluminum Processing*, California, American Foundry Society, Des Plaines, Ill., 1986, p. 179.

300. *Aluminum Alloys Handbook*, ASM Intl., Materials Park, Ohio, 1994, pp. 199–230.

300a. P. S. Mohanty and J. E. Gruzleski, *Acta Metall.*, vol. 43, no. 5, 1995, pp. 2001–2015.

300b. Y. C. Lee, A. K. Dahle, and D. H. St. John, *Met. and Mats. Trans.*, vol. 31A, 2000, pp. 2895–2906.

301. S.-Z. Lu and A. Hellawell, *JOM*, February, 1995, pp. 38–43.

302. S. C. Servé, *Light Metal Age*, June 1995, pp. 36–43.

303. N. Handiak, J. E. Gruzleski, and D. Argo, *AFS Trans.*, 1987, pp. 31–38.

304. S.-Z. Lu and A. Hellawell, *Met. Trans.*, vol. 18A, 1987, pp. 1721–1733.

305. V. del Davies and J. M. West, *Inst. of Metals*, vol. 92, *1963–1964*, pp. 175–180.

306. A. Hellawell, *Progr. Mats. Sci.*, vol. 15, 1980, p. 1.

307. B. M. Closset and J. E. Gruzleski, *AFS Trans.*, 1982, pp. 453–464.

308. J. F. Major and J. W. Rutter, *Mats. Sci. & Technology*, vol. 5, no., 7, 1989, pp. 645–656.

309. *Foundry*, July 1992, pp. 35–38.

310. G. K. Sigworth, *AFS Trans.*, 1987, pp. 303–314.

311. M. Hillert and V. V. Subbarao, *The Solidification of Metals*, Iron and Steel Institute publ. 110, London, 1968, pp. 204–212.

312. J. S. Park and J. D. Verhoeven, *Met. and Mats. Trans.*, vol. 27A, 1996, pp. 2328–2337.

313. I. Minkoff, *The Physical Metallurgy of Cast Iron*, Wiley, New York, 1983.

314. *Cast Irons*, ASM Intl., Materials Park, Ohio, 1996, pp. 16–31.

314a. A. Jacot, D. Maijer, and S. Cockcroft, *Met. and Mats. Trans.*, vol. 31A, 2000, pp. 2059–2068.

315. I. Minkoff, in *F. Weinberg Int. Symposium on Solidification Processing*, eds. J. E. Lait and I. V. Samarasekera, Pergamon Press, New York, 1990, p. 225.

316. D. D. Goettsch and J. A. Dantzig, *Met. Trans.*, vol. 25A, 1994, pp. 1063–1079.

317. J. A. Sikora, G. L. Rivera, and H. Biloni, in *Proc: F. Weinberg Symposium on Solidification Processing*, eds. J. E. Lait and I. V. Samarasekera, Pergamon Press, New York, 1990, p. 255.

318. D. K. Banerjee and D. M. Stefanescu, *Trans. AFS*, vol. 99, 1991, p. 747.

319. D. M. Stefanescu, *Metals Handbook*, 9th ed., vol. 15, *Casting*, ASM, Metals Park, Ohio, 1988, p. 168.

320. B. Lux and M. Granges, *Practical Metallogr.*, vol. 5, 1968, p. 123.

321. I. Minkoff and B. Lux, *Nature*, vol. 225, 1979, p. 540.

322. J. S. Langer and H. Mueller-Krumbhaar, *Acta Met.*, vol. 26, 1978, p. 1697.

323. E. R. Evans and J. K. Dawson, *BCIRA J.*, vol. 46, Jan. 1974.

324. C. Suryanarayana, F. H. Froes, and R. G. Rowe, in *Inter. Materials Rev.*, vol. 36, no. 3, 1991, pp. 85–123.

325. C. Suryanarayana, in *Materials Science and Technology*, vol. 15, *Processing of Metals and Alloys*, vol. ed. R. W. Cahn, VCH, Weinheim, 1991, pp. 57–110.

326. A. Lawley, in *Mechanical Behavior of Rapidly Solidified Materials*, eds. S. M. L. Sastry and B. A. MacDonald, The Metallurgical Society, Warrendale, Pa., 1986, pp. 3–19.

327. R. Bowman and S. D. Antolovich, *Met. Trans.*, vol. 19A, 1988, pp. 93–103.

328. H. Jones, A. Joshi, R. G. Rowe, and F. H. Froes, *Int. J. of Powder Metall.*, vol. 23, 1987, pp. 13–24.

329. H. J. Fecht and J. H. Perepezko, *Met. Trans.*, vol. 20A, 1989, pp. 785–794.

330. N. C. Koon and R. Hasegawa, in *Rapidly Solidified Crystalline Alloys*, eds. S. K. Das et al., The Metallurgical Society, Warrendale, Pa., 1985, pp. 245–262.

331. R. L. Ashbrook, H. H. Liebermann, A. Guha, and J. C. Harkness, in *Rapidly Solidified Crystalline Alloys*, eds. S. K. Das et al., The Metallurgical Society, Warrendale, Pa., 1985, pp. 307–318.

332. R. Ray, in *Metals Handbook*, vol. 7, 9th ed., ASM International, Metals Park, Ohio, 1984, p. 47.

333. R. E. Maringer, C. E. Mobley, and E. W. Collins, in *Rapidly Quenched Metals*, eds. N. J. Grant and B. C. Giessen, *Proc. 2nd Int'l. Conf.: Rapidly Quenched Metals*, MIT Press, Cambridge, Mass., 1976, p. 29.

334. H. H. Lieberman, *Mats. Sc. & Engineering*, vol. 49, 1981, pp. 185–191.

335. M. J. Tenwick, H. A. Davies, H. Solderhjelm, and L. Mandal, *Mats. Sc. & Engng.*, vol. 63, 1984, pp. L.1–L.4.

336. H. H. Lieberman and R. L. Bye, Jr., in *Rapidly Solidified Crystalline Alloys*, eds. S. K. Das et al., The Metallurgical Society, Warrendale, Pa., 1985, pp. 307–318.

337. S. K. Das, C. G. Chang, and D. Raybould, *Light Metal Age*, vol. 44, 1986, pp. 5, 6, 8.

338. M. C. Narsimham, U.S. Patent 442571, Mar. 6, 1979.

339. E. J. Lavernia, J. D. Ayres, and T. S. Srivatsan, *Inter. Materials Rev.*, vol. 37, no. 1, 1992, pp. 1–44.

340. C. M. Adams, in *Mechanical Behavior of Rapidly Solidified Materials*, eds. S. M. L. Sastry and B. A. MacDonald, The Metallurgical Society, Warrendale, Pa., 1986, pp. 21–39.

341. W. A. Heineman, in *Rapidly Quenched Metals*, eds. V. S. Steeh and H. Warlimont, Elsevier Science, Amsterdam, 1985, pp. 27–34.

342. J. H. Vincent, J. G. Herbertson, and H. A. Davies, in *Rapidly Quenched Metals*, vol. I, eds. T. Masumoto and K. Suzuki, Sendai, Japan, The Japanese Institute of Metals, 1982, pp. 77–80.

343. A. J. Yule and J. D. Dunkley, *Atomization of Melts*, Clarendon Press, Oxford, 1994.

344. A. R. E. Singer, *Jn. Inst. Metals*, vol. 100, 1972, p. 185; *Int'l. J. of Powder Metall. and Powder Tech.*, vol. 21, 1985, pp. 219–234.

345. T. C. Willis, *Metals and Materials*, vol. 4, 1988, pp. 485–488.

346. A. Bose, *Advances in Particulate Materials*, Butterworth-Heinemann, Boston, 1995.

347. P. Grant, *Progress in Materials Sci.*, vol. 39, 1995, pp. 497–545.

348. D. Apelian, B. H. Kear, and W. H. Schadler, in *Rapidly Solidified Crystalline Alloys*, eds. S. K. Das et al., The Metallurgical Society, Warrendale, Pa., 1985, pp. 93–109.

349. A. G. Leatham, J. S. Coombs, J. B. Forrest, and S. Ahn, in *Proc: 2nd Pacific Rim Int. Conf. on Advanced Materials Processing*, Kyongju, Korea, 1995.

350. P. G. Brooks, A. G. Leatham, J. C. Combs, and C. Moore, *Metall. and Met. Forming*, vol. 4, 1977, p. 157.

351. N. M. Khokhlacheva et al., *Soviet Powder Metall. and Ceramics*, vol. 23, no. 4, 1984, pp. 145–149; vol. 25, no. 5, 1986, pp. 355–358.

352. L. Jin et al., *Powder Metall. Int.*, vol. 20, 1988, pp. 17–22.

353. C. Zweben, *Encyclopedia of Advanced Materials*, vol. 2, Pergamon Press, Oxford, 1994, pp. 1489–1493.

353a. T. S. Srivatsan and R. Annigeri, *Met. and Mats. Trans.*, vol. 31A, 2000, pp. 959–973.

354. S.-Y. Ohn, J. A. Cornie, and K. C. Russell, *Met. Trans.*, vol. 20A, 1989, pp. 527–532.

355. F. Rana et al., *AFS Trans.*, 1989, pp. 255–264.

356. A. Mortensen, in *Encyclopedia of Advanced Materials*, vol. 2, Pergamon Press, Oxford, 1994, pp. 1493–1497.

357. P. Rohatgi, *JOM*, April 1991, pp. 10–15.

358. A. K. Gosh, in *Principles of Solidification and Materials Processing*, Oxford and IBH Publications Co., 1990, p. 585.

359. P. Rohatgi, in *Metals Handbook*, 9th ed., vol. 15, *Casting*, ASM, Metals Park, Ohio, 1988, p. 840.

360. A. Mortensen, in *Proc of the 12th RISO Int. Symp. on Mat. Sc.: Metal Matrix Composites Processing, Microstructures and Properties*, ed. H. Hansen et al., Nat. Lab. RoIsklide, Denmark, 1991, p. 101.

361. A. Mortensen and I. Jin, *Int. Mat. Rev.*, vol. 37, no. 3, 1992, pp. 101–128.

362. V. J. Michaud, in *The Encyclopedia of Advanced Materials*, vol. 2, Pergamon Press, Oxford, 1994, pp. 1507–1513.

363. S. Ray, *J. Mater. Sci.*, vol. 28, 1993, pp. 5397–5413.

364. A. R. E. Singer, *Mater. Sc. Engrg.*, vol. A135, 1991, pp. 13–17.

365. V. J. Michaud, in *Fundamentals of Metal Matrix Composites*, eds. S. Suresh, A. Mortensen, and A. Needleman, Butterworth-Heinemann, 1993, pp. 3–22.

366. M. J. Kozak and M. K. Premkumar, *JOM*, vol. 45, no. 1, pp. 44–48.

367. T. W. Clyne and P. J. Withers, *An Introduction to Metal Matrix Composites*, Cambridge University Press, Cambridge, 1993.

368. D. J. Lloyd, *Int. Mats. Rev.*, vol. 39, no. 1, 1994, pp. 1–23.

369. *Product Design & Specifications for Squeeze Castings*, ELM Int., Inc., MI, 1994.

370. J. L. Dorcic and S. K. Verma, in *Metals Handbook*, 9th ed., vol. 15, *Casting*, ASM International, Metals Park, Ohio, 1988, pp. 323–327.

371. G. A. Chadwick and T. H. Yue, *Metals and Materials*, Jan. 1989, pp. 6–12.

372. R. F. Lynch et al., *AFS Trans.*, 1975, pp. 569–590.

373. S. Rajagopal and W. H. Altergott, *AFS Trans.*, 1985, pp. 145–154.

374. R. D. Maier et al., *Int'l. Congress and Exposition*, SAE Technical Paper #920456, Detroit, Mich., Feb. 1992, pp. 1–7.

375. N. Yamamoto et al., *AFS Casting Congress*, 1996, Paper # 92-087, pp. 1–15.

376. S. Suzuki, *Product Design and Specification for Squeeze Casting*, UBE Industries Ltd., Michigan.

377. S. Okada et al., *AFS Trans.*, 1982, pp. 135–146.

378. H. J. Heine, *Foundry*, July 1989, pp. 22–26.

379. M. Dixon, P. Heuler, P. E. Irvine, and B. Neuwmann, *The Use of Squeeze Cast Aluminum Alloys for Automotive Applications, COST 504, Final Report 1989, Report # TF-2536.*

380. P. R. Gibson, A. J. Clegg, and A. A. Das, *MST*, 1985.

381. D. H. Kirkwood, *Int. Mat. Rev.*, vol. 39, no. 5, 1994, pp. 173–189.

382. K. P. Young, *The Encyclopedia of Advanced Materials*, Pergamon, Oxford, 1994, pp. 2411–2417.

383. A. Madronero et al., *JOM*, Jan. 1997, pp. 46–49.

384. A. I. Nussabaum, *Light Metal Age*, vol. 54, no. 5/6, 1996, pp. 6–22.

385. M. P. Kenney, J. A. Courtois, R. D. Evans, G. M. Farrior, C. P. Kyonko, A. A. Koch, and K. P. Young, *ASM Handbook*, vol. 15, 9th ed., ASM Int., Metals Park, Ohio, 1988, pp. 327–338.

386. *Int'l. Conf.: Semisolid Alloys and Composites*, University of Sheffield, England, June 19–21, 1996.

387. A. Sato and R. Mehrabian, *Met. Trans.*, vol. 7B, 1976, p. 443.

388. T. Matsumiya and M. C. Flemings, *Met. Trans.*, vol. 12B, 1981, p. 17.

389. R. Mehrabian, R. G. Rick, and M. C. Flemings, *Met. Trans.*, vol. 5, 1974, pp. 1899–1905.

390. M. H. Lavington, *Metals and Materials*, Nov. 1986, pp. 713–719.

391. N. K. Graham and K. R. Irvine, Paper presented to the Annual Conf. on the Institute of British Foundrymen, 20 pages.

392. R. Worseley, *Metalworking Production*, Sept. 1986, pp. 80–84.

393. J. Polmear, *Light Alloys*, 3d ed., Arnold, London, 1995.

393a. N. D. Souza, M. G. Ardakani, M. McLean, and B. A. Shollock, *Met. and Mats. Trans.*, vol. 31A, 2000, pp. 2877–2886.

394. T. S. Piwonka, *ASM Handbook*, 9th ed., vol. 15, ASM Int'l., Metals Park, Ohio, 1988, pp. 319–323.

395. F. L. VernSnyder and A. F. Giamei, in *The Encyclopedia of Materials Science and Engineering*, Pergamon Press, Oxford, 1986, pp. 1202–1204.

396. T. Tom, in *Encyclopedia of Advanced Materials*, vol. 1, Pergamon Press, Oxford, 1994, pp. 645–649.

CHAPTER 4
PLASTIC DEFORMATION

4.1 INTRODUCTION

Tension testing is widely used to obtain vital information concerning the progress of deformation and to characterize the various mechanical properties. Slip (dislocation motion) and deformation twinning, respectively, are the chief and the second important mechanisms of plastic deformation. Slip is the more common mode of plastic deformation in materials with high crystal symmetry (e.g., fcc and bcc). Deformation twinning, which is a homogeneous shear process, becomes more predominant [in the plastic deformation of low-symmetry (e.g., hcp) materials, and also of high-symmetry materials] at room temperature under high strain rate and at very low temperatures under the normal strain rate and by high alloying in order to decrease the stacking fault energy. Plastic deformation is a sensitive function of many variables such as extent, rate, type, operative mechanism, temperature of deformation, past history, grain size, and stacking fault energy. On a microscopic scale, the dislocation distribution in plastically deformed material is nonuniform because of the formation of dislocation cell walls and cells or subgrains. Nonrandom orientation of grains occurs on extensive deformation at room temperature, and the materials are said to have a preferred orientation or deformation texture. Such a texture, which develops plastic anisotropic properties, may have an important technological advantage. In this chapter, an attempt has been made to present the tensile properties, yielding phenomena, work hardening, strain aging, twinning, and deformation texture in polycrystalline metals and alloys.

4.2 TENSILE PROPERTIES

In the standard tension testing, a test specimen of 2.000-in. (50.8-mm) gauge length and 0.500-in. (12.7-mm) width (or diameter) (Fig. 4.1a) is gripped into a tension testing machine, and then grips are pulled apart at a constant rate of crosshead speed or a constant rate of loading to fracture, as described in ASTM E8.[1] The applied load and extension are measured by means of a load cell and strain gauge extensometer. The load-elongation (or extension) relation is recorded from which both the engineering (or nominal) stress s—engineering (or nominal) strain e and the true stress σ—true strain ε curves are plotted (Fig. 4.1b)[2] using the equation

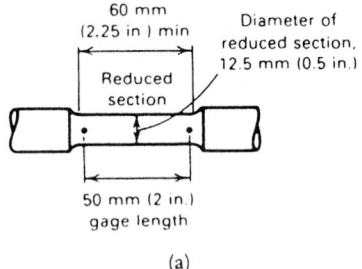

(a)

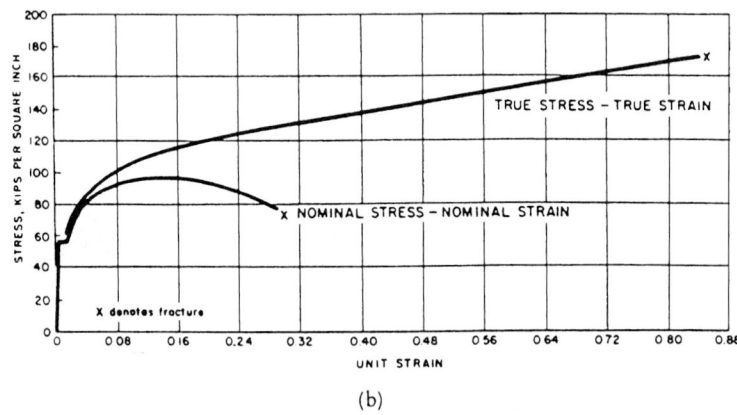

(b)

FIGURE 4.1 (*a*) Standard round tensile test specimen and 2-in. gauge length.[1] (*b*) Engineering stress–engineering strain curve and true stress–true strain curve.[2] [(*a*) *Reprinted by permission of ASM International, Materials Park, Ohio; (b) reprinted by permission of Iron and Steel Engineers, Pittsburgh, Pennsylvania.*]

$$\text{Engineering stress } s \text{ (psi or MPa)} = \frac{F}{A_o} \qquad (4.1)$$

where F is the applied load or force and A_o is the original cross-sectional area in gauge length, and

$$\text{Engineering strain } e = \frac{(l_f - l_o)}{l_o} = \frac{\Delta l}{l_o} \qquad (4.2)$$

where l_o is the original gauge length of the specimen, l_f is the instantaneous or final gauge length after the occurrence of deformation, and Δl is the change in length. *True strain ε* is defined as the sum of instantaneous length changes dl divided by the instantaneous length l, or

$$\varepsilon = \int_{l_o}^{l_f} d\varepsilon = \int_{l_o}^{l_f} \frac{dl}{l} = \ln\left(\frac{l_f}{l_o}\right) = \ln\frac{l_o + \Delta l}{l_o} = \ln(1 + e) \qquad (4.3)$$

During the plastic deformation, the volume of the specimen remains unchanged; that is, volume conservation is maintained:

$$A_f l_f = A_o l_o \tag{4.4}$$

where A_f is the true, final, or instantaneous uniform cross-sectional area in the gauge length after the occurrence of some deformation. We can rewrite Eq. (4.3) as

$$\varepsilon = \ln\frac{A_o}{A_f} \quad \text{or} \quad \frac{A_o}{A_f} = \exp\varepsilon \quad \text{or} \quad A_f = A_o \exp(-\varepsilon) \tag{4.5}$$

Equation (4.5) indicates that the cross-sectional area decreases exponentially with true strain. Now we are able to define *true stress* as

$$\sigma = \frac{F}{A_f} = \frac{F}{A_o \exp(-\varepsilon)} = s \exp\varepsilon \tag{4.6}$$

It is thus obvious that the engineering stress-strain system depends on the original dimensions of the test specimen, while the true stress-strain system depends on the instantaneous specimen dimensions. Moreover, the true stress is always greater than the engineering stress, and the ratio of the two increases with the increase in the plastic strain. At large strains (above ~10%) the difference between the two is quite large, whereas at low strains (<1%) there is no measurable difference. The stress-strain curves remain identical for both tension and compression tests when true stress and true strain values are taken into account. On the other hand, their data differ when engineering stress and engineering strain values are inserted, which does not seem to be logical. However, engineering stress-strain data are employed for convenience, particularly for small strain; the true stress-strain data provide a more rational approach and are successfully used in helping to understand the forming operations involving plastic deformation.

Equations (4.1) through (4.6) are normally only applicable during uniform deformation. After necking starts, the typical longitudinal extensometer output is no longer useful for strain calculation.

Table 4.1 lists typical tensile properties of selected thin (0.02- to 0.04-in., or 0.5- to 1.0-mm) sheet materials.[3] The following properties, usually obtained from a tensile test, are used in specifications:

Modulus of Elasticity (Young's Modulus) E.   This is the slope of the stress/strain curve in the elastic or linear region and is expressed as

$$E(\text{psi or MPa}) = \frac{s}{e} \tag{4.7}$$

It is used to calculate the deflection or stiffness of beams or other members under load, and it is an important design parameter. It is the structure-insensitive property of a material.[4] The magnitude of the elastic modulus varies widely for different metals (Table 4.1). For example, E for low-carbon steel is approximately 30×10^6 psi, and E for aluminum is 10×10^6 psi. This is only marginally affected by a small variation in material structure such as small alloying additions or the presence of defects, such as vacancies, dislocations, or grain boundaries. However, the variation becomes appreciable where the alloys either show complete solid solubility or form intermediate phases. The general rule is that the stronger the interatomic forces (curvature of the potential energy well), the higher the modulus. In addition to the variation in elastic modulus, crystallographic dependency in elastic modulus occurs; that is, if we determine E along different crystallographic directions in a single

TABLE 4.1 Typical Tensile Properties of Selected Sheet Metals[3]

Material	Young's modulus E		Yield strength		Tensile strength		Uniform elongation, %	Total elongation, %	Strain-hardening exponent n	Average normal anisotropy r_m	Planar anisotropy Δr	Strain rate sensitivity m
	GPa	10^6 psi	MPa	ksi	MPa	ksi						
Aluminum-killed drawing-quality steel	207	30	193	28	296	43	24	43	0.22	1.8	0.7	0.013
Interstitial-free steel	207	30	165	24	317	46	25	45	0.23	1.9	0.5	0.015
Rimmed steel	207	30	214	31	303	44	22	42	0.20	1.1	0.4	0.012
High-strength low-alloy steel	207	30	345	50	448	65	20	31	0.18	1.2	0.2	0.007
Dual-phase steel	207	30	414	60	621	90	14	20	0.16	1.0	0.1	0.008
301 stainless steel	193	28	276	40	690	100	58	60	0.48	1.0	0.0	0.012
409 stainless steel	207	30	262	38	469	68	23	30	0.20	1.2	0.1	0.012
3003-O aluminum	69	10	48	7	110	16	23	33	0.24	0.6	0.2	0.005
6009-T4 aluminum	69	10	131	19	234	34	21	26	0.23	0.6	0.1	-0.002
70-30 brass	110	16	110	16	331	48	54	61	0.56	0.9	0.2	0.001

4.4

crystal, we will obtain different results. The directional variation in properties is termed *anisotropy*. For example, E_{iron} varies between 41×10^6 psi (2.83×10^5 MPa) in the [111] direction and 19×10^6 psi (1.31×10^5 MPa) in the [100] direction.[5]

Last, E decreases approximately linearly as the temperature is increased up to about one-half the melting temperature; rapid decrease, however, occurs with further increase in temperature. For example, E at one-half the melting temperature becomes 0.8, and at near the melting temperature it reaches 0.4 of the value near absolute temperature.[5] The modulus is a major parameter in determining elastic springback. The lower-modulus metal has greater springback when unloaded from a given flow stress.

Tensile Strength s_{ut}. This is also referred to as the *ultimate tensile strength* (UTS) and is expressed in pounds per square inch or megapascals. It is a maximum stress on the engineering stress-strain curve. It is an index of quality of a material; that is, it is a good indication of defects, flaws, or harmful inclusions present in the material.

The maximum in the engineering stress-strain curve (that is, s_{ut}) is also significant because homogeneous and stable plastic deformation is obtained at strains less than e_{ut}, while inhomogeneous or localized deformation along the gauge section is obtained for strains longer than e_{ut}, leading finally to failure by ductile (cup-cone) fracture. However, the extent of the tendency toward necking is mostly governed by the strain rate sensitivity m of the material. The higher the value of n (i.e., closer to 1), the lower the tendency toward localization of strain. It should be pointed out that the amount of strain prior to the occurrence of necking is referred to as *uniform strain e_u*, which may be equivalent to uniform elongation with respect to the engineering strain. For steels without yield point elongation, the UTS/YS ratio is related to the n value.

Yield Strength σ_y. The applied stress required to induce the onset of plastic deformation (yield point) is called *yield stress*. This stress is the most important value for structural design because it avoids the practical difficulties encountered in measuring the exact stress for the proportional limit or elastic limit at which a material starts to deform plastically. It is extremely sensitive to the structure and prior history of a material. It is usual practice to use offset yield strength by a specific plastic strain because it is easily measurable and reproducible from one laboratory to another. Its measurement involves taking a measured stress-strain curve and drawing a line parallel to the elastic or straight-line portion of the curve that is offset in strain by 0.002 until it intersects the stress-strain curve. This intersection gives the value of yield strength σ_y, the 0.2% offset yield strength. This 0.2% set or offset is tolerable in most cases. In some cases, however, 0.1% or 0.5% offset is required. In materials showing sharp yield point (mild steel), the lower yield stress is considered as the yield strength.

Ductility. Ductility is a measure of the ability of a material to undergo larger plastic deformation under stress without failing. Failure may take place by fracture or by highly localized, perhaps unstable, deformation. Ductility contributes to the toughness of metals and alloys and is also an important factor in determining formability. Ductility of a material is a function of its chemistry, microstructure, and imposed stress and strain rate. Usually materials with high volume fractions of non-deformable second-phase particles seem to possess reduced ductility under any condition of deformation.[6] Commonly ductility, in a tension test, is measured by (1) engineering or plastic strain required after fracture e_f, usually called the *elongation* (or *tensile ductility*), or (2) the *reduction of area* at fracture RA(q). Both elongation e_f and reduction of cross-sectional area (at the fracture) q(RA) are expressed as percentages:

$$\text{Elongation} = \frac{l_{\text{break}} - l_o}{l_o} \times 100 = \frac{l_f - l_o}{l_o} \times 100 = e_f \qquad (4.8)$$

$$\text{RA} = \frac{A_o - A_f}{A_o} \times 100 = q \qquad (4.9)$$

where l_o, l_f, A_o, and A_f have the usual meanings. Percent elongation is a better indicator of quality than the tensile strength because this is drastically reduced in the presence of inclusions and porosity. Elongation times tensile strength is an index of toughness at low strain rates. The gauge length over which the percent elongation is measured must be quoted. The materials are called *ductile* when RA > 50%, *semi-brittle* when RA is small (<10%), and *notch-brittle* when RA is moderate in a simple tension test (say 30%) but very small or zero when a notch or crack is introduced into the tensile specimen before the testing.

Ductility may, in some instances, be determined in terms of the *uniform strain* (or *elongation*) e_u or the *zero gauge length strain* e_o. The former, being the strain prior to the onset (or occurrence) of necking, is an important value in the formability of sheet metal because its magnitude specifies the permissible strain to obtain uniform or homogeneous deformation, after which local thinning (i.e., inhomogeneous deformation) may occur. For metals that obey the power law curve (of strain hardening), e_u is related to the strain-hardening exponent n by

$$n = \ln(1 + e_u) \qquad (4.10)$$

This relationship holds good for steels; for other materials the actual e_u measured in tension should be employed. As shown in Fig. 4.2, both e_u and total elongation

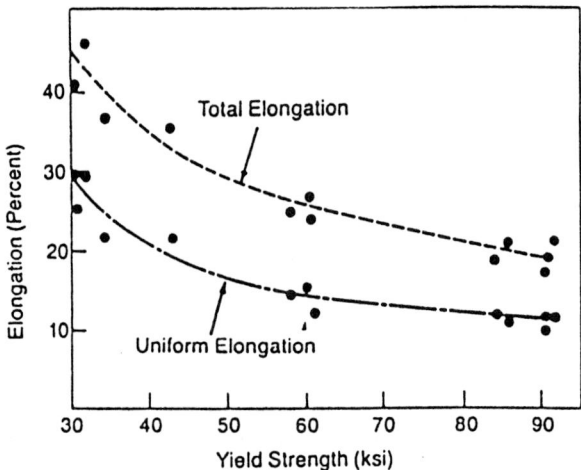

FIGURE 4.2   Effect of yield strength on both uniform elongation and total elongation.[7] (*Courtesy of AISI, Washington, D.C.*)

e_T decrease with an increase in yield strength.[7] The e_u, like the n value and UTS/YS ratio, is employed as a measure of maximum allowable stretchability. Zero gauge length strain e_0 is measured from the reduction of area at fracture by

$$e_0 = \frac{q}{1-q} \qquad (4.11)$$

This quantity is important in forming operations where gauge length is very short.

Aluminum alloy sheets tend to have a considerable low elongation compared to mild steel in contrast to the slightly low UTS. The uniform elongation of aluminum sheets is higher than that of steel sheets, although the local elongation of them is very small.

The relation between the reduction in area at fracture q (in tension test) and the minimum permissible bend radius R_m for a given thickness of sheet can be predicted fairly accurately (for less ductile materials) by[8]

$$\frac{R_m}{t} = \frac{1}{2q} - 1, \qquad \text{for } q < 0.2 \qquad (4.12)$$

and for ductile materials (due to the shift of the neutral radius in tight bends) by

$$\frac{R_m}{t} = \frac{1-q^2}{2q-q^2}, \qquad \text{for } q > 0.2 \qquad (4.13)$$

In addition to the significance of e_u on formability, e_T (e.g., in 2-in. gauge length) at fracture is being considered as an important indicator of formability (e.g., in dual-phase steels) and is a measure of cleanliness of the material. Note that total elongation e_T = uniform elongation e_u + postuniform (or necking) elongation e_{pu}; the first term is dependent on the strain-hardening exponent n, and the second term is dependent on the strain rate sensitivity m. These combined factors, therefore, contribute to the overall formability of the sheet material.[9] Thus, e_T is dependent on the length, width, and thickness of the gauge section employed for the measurement. More specifically, e_T is dependent on the sum of n and m variables. It also correlates well with the hole expansion, bending capacity, elongation of a blanked edge, and other forming operations.

Toughness. The toughness of a material is its resistance to crack propagation or its ability to absorb energy in the plastic deformation range. It is a measure of the energy to resist fracture. For qualitative determination it is a measure of the work per unit volume which can be done on the material without fracture. For a simple tension test, it can be expressed as follows:

$$\text{Work per unit volume } U = \int (\text{force } F)(\text{increment of extension } dl) \qquad (4.14a)$$

or $\qquad$ Work per unit volume $U = \int_{l_o}^{l_f} F \dfrac{dl}{Al} = \int_{l_o}^{l_f} \sigma \, d\varepsilon \qquad (4.14b)$

where A and l are simple dimensions and ε_f is the true strain at fracture. Toughness, therefore, represents the area under the true stress-true strain curve (Fig. 4.3). Materials with high strength and low ductility have low toughness values. They are called *brittle materials* which exhibit fast fracture and need a low absorbed energy for fracture. Materials with high ductility but lower strength also provide low toughness. Tough materials denote slow fracture and would need a large absorbed energy for

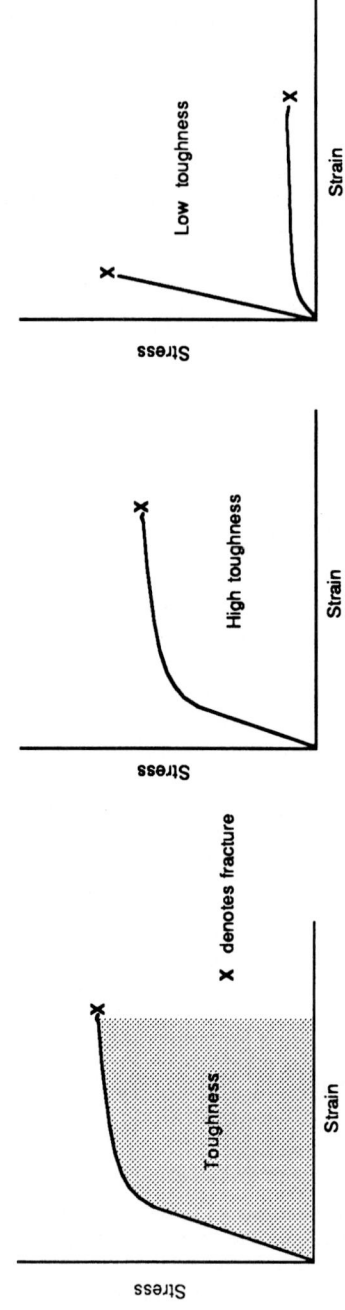

FIGURE 4.3 Schematic stress-strain curve representing qualitative measures of toughness.

fracture. Maximum toughness can be attained by an optimum combination of strength and ductility. Toughness is a strong function of the temperature, rate of loading (or strain rate), thickness of the material, and the presence of stress concentration. Improvement of toughness (i.e., resistance to brittle fracture) for structural materials has been attempted by various methods such as addition of alloying elements, special heat treatments, thermomechanical treatments, production methods which reduce harmful impurity elements, or nonmetallic inclusions.[10] It should be realized that an optimum level of toughness is very important because attainment of a certain high strength level may be of no significance unless coupled with sufficient toughness to meet service requirements in parts such as freight car couplings, chains, gears, and crane hooks.[1,2]

For quantitative measurements of the fracture toughness of a material, the procedure outlined in ASTM E399, Standard Test Method for Plane-Strain Fracture Toughness (K_{IC}) of Metallic Materials, should be followed. The parameter K_{IC}, often termed the *fracture toughness* of a material, is the critical value of stress intensity K_I at which fracture occurs; and it is given by

$$K_{IC} = s(\pi a)^{1/2} \qquad (4.15)$$

where s is the nominal tensile stress away from the stress concentration and a is the half-length of a through-thickness crack. Thus, K_{IC} denotes a concentration criterion that provides some knowledge of applied stress where the crack starts.[11]

Fracture toughness of steels is dependent on numerous factors which influence the mechanical properties such as test temperature, chemistry, melt practice, microstructure, grain size, strain rate, section thickness, and notch acuity. It has been established that the fracture toughness of high-strength alloy steels such as 4340 and 18% maraging steels increases with the decrease of their yield and tensile strengths.[11] Table 4.2 shows comparative data of K_{IC} for various materials.[12]

Brittle Behavior. The brittle behavior is usually described by the Griffith-Irwin criterion, whereas the ductile behavior needs an Orowan correction. In brittle materials, the stress at which fracture occurs is controlled by the propagation of small flaws, the size of the largest flaw present, and fracture toughness.[13] Irrespective of the manner of crack initiation, brittle fracture possesses a built-in instability—characteristics that can be triggered with little or no plastic deformation at applied stress below the conventional yield strength of the material and little or no evidence of local or microstructural-scale plastic strain (or deformation).[13a] Even the presence of the smallest internal or external defects or sharp notches can result in catastrophic failure (or brittle fracture) of large structural members.[11]

Studies of various brittle fractures have led to the development of the crack starter techniques and the concept of the *nil-ductility transition* (NDT) temperature. This measures the position of ductile-brittle transition temperature (DBTT) on the curve that can be defined as the point of no detectable ductility. The tests widely employed for the determinations of DBTT are Charpy V-notch (CVN), dynamic tear (DT) energy, drop weight tear (DWT), NRL explosion tear, and notched tensile tests. However, certain metals and alloys do not exhibit a DBTT.

Although the main task for the design engineer is to avoid brittle characteristics during the selection of structural materials and configurations, complete removal of defects is not feasible. Even with the best detection of any original defect, there can be at least two or three dozen material, environmental, and operational related degradation mechanisms that, over the lifetime of components or a structure, can aggravate to failure.[11]

TABLE 4.2 Typical Values of Plane Strain Fracture Toughness K_{IC}[12]

	Strength		Fracture toughness	
	MPa	ksi	$MPa \cdot m^{1/2}$	$ksi \cdot in.^{1/2}$
Steels				
Medium-carbon steel	260	37.7	54	49
Pressure-vessel steel A533B Q and T	500	72.5	200	182
Maraging steel	1390	202	110–176	100–160
	1700	247	99–165	90–150
	1930	280	85–143	80–130
300M	{ 2033	295	66	60
	{ 1895	275	83	75
AISI 4340	{ 1758	255	77	70
	{ 1930	280	61	55
AISI 4340				
Tempered 200°C, commercial purity	1650	239	40	36
Tempered 200°C, high purity	1630	236	80	73
Aluminium alloys				
2024-T4	346	50	55	50
2024-T851	414	60	24	22
7075-T6	463	67	38–66	35–60
7075-T651	482	70	31	28
7178-T6	560	81	23	21
7178-T651	517	75	24	22
Titanium alloys				
Ti-6Al-1V	830	120	50–60	45–55
Ti-6Al-5Zr-0.5Mo-0.2Si	877	127	60–70	55–65
Ti-4Al-4Mo-2Sn-0.5Si	960	139	40–50	36–45
Ti-11Sn-5Zr-2.25Al-1Mo-0.2Si	970	141	35–45	32–41
Ti-4Al-4Mo-4Sn-0.5Si	1095	159	30–40	27–36
Thermoplastics				
PMMA	30	4	1	0.9
GP polystyrene	—	—	1	0.9
Acrylic sheet	—	—	2	1.8
Polycarbonate	—	—	2.2	2.0
Others				
Concrete	—	—	0.3–1.3	0.3–1.2
Glass	—	—	0.3–0.6	0.3–0.5
Douglas fir	—	—	0.3	0.3

Courtesy of Butterworths, London.

4.2.1 Plastic Instability and Necking Formation

The true and engineering stress-strain curves obtained from a tensile test reveal important differences, as shown in Fig. 4.1*b*. In the engineering stress-strain curve, the maximum stress point is achieved at a maximum load, which decreases to fracture, whereas in the latter the curve rises continually to fracture. The inflection in the engineering curve is due to the onset of localized plastic flow and the way in which the engineering stress has been defined. In a ductile material with low strain rate sensitivity m (see Chap. 15 for more details), the condition for plastic stability is the well-known Considére criterion

$$\sigma \le \frac{d\sigma}{d\varepsilon} \tag{4.16}$$

where σ and ε are true stress and true strain, respectively. During the tensile deformation of a ductile material, the material resists progressive necking by work hardening, and necking begins at a point of maximum load, that is, $F = F_{max}$. If no strain hardening took place, the material would become unstable in tension and start to neck as soon as yielding occurred. However, in normal strain-hardening alloys, the load-carrying capacity of the specimen increases with the increase in deformation. That is, strain hardening raises the yield stress. This effect is offset, to some extent, by the gradual elongation and subsequent reduction in cross-sectional area of the specimen as the deformation proceeds. Since necking or instability of plastic deformation in tension commences at a maximum load at which the increase in stress due to reduction in the cross-sectional area of the specimen becomes greater than that due to strain hardening, we may write

$$dF = 0 \tag{4.17}$$

At any instant, we have

$$F = \sigma A_f \tag{4.18}$$

where all the terms have the same meaning. Hence, Eq. (4.17) becomes

$$dF = \sigma\, dA_f + A_f\, d\sigma \tag{4.19}$$

From constancy of volume during deformation, we obtain

$$\frac{dl_f}{l_f} = -\frac{dA_f}{A_f} = d\varepsilon \tag{4.20}$$

Hence it follows that

$$-\frac{dA_f}{A_f} = \frac{d\sigma}{\sigma} = d\varepsilon \tag{4.21}$$

That is, at the tensile instability, we have

$$\sigma = \frac{d\sigma}{d\varepsilon} \tag{4.22}$$

Equation (4.22) illustrates that necking will start at a strain at which the slope of the true stress-true strain curve equals the true stress at that strain. This necking criterion can also be easily expressed in terms of engineering strain as follows:

$$\frac{d\sigma}{d\varepsilon} = \frac{d\sigma}{de}\frac{de}{d\varepsilon} = \frac{d\sigma}{de}\frac{dl/l_o}{dl/l_f} = \frac{d\sigma}{de}\frac{l_f}{l_o}$$

$$= \frac{d\sigma}{de}(1+e) = \sigma \quad \text{or} \quad \frac{d\sigma}{de} = \frac{\sigma}{1+e} \tag{4.23}$$

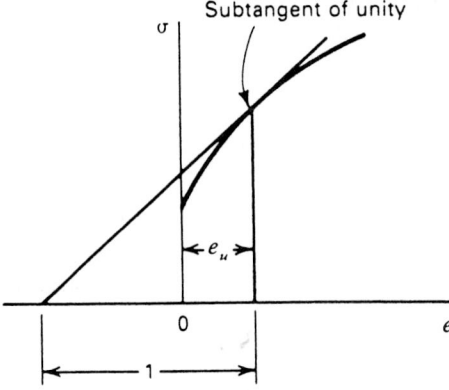

FIGURE 4.4 Considére's construction for the determination of the point of maximum load.[1] (*Reprinted by permission of ASM International, Materials Park, Ohio.*)

Figure 4.4 shows the plot of true stress versus engineering strain where the point of maximum load (or value of true stress at maximum load) can be measured geometrically by drawing a tangent to a curve from a true strain of −1. The slope of this tangent is $\sigma/(1 + e)$.

4.3 YIELDING AND PLASTIC FLOW

4.3.1 Slip

Body-centered cubic (bcc) iron deforms by the slip on the {110}, {112}, and {123} planes while the slip direction is always the <111> direction. Usually the slip plane is the plane of highest atomic density, and the slip direction is the closest-packed direction within the slip plane. The combination of slip plane and slip direction is termed the *slip system*. It may be shown that the bcc metals (irons) have 48 possible slip systems, although not all are operative. In contrast, γ-iron and other fcc metals deform by slip on the close-packed octahedral {111} planes in the <110> close-packed directions. There are four different orientations of {111} planes and each plane contains three <110> directions which constitute a total of 12 possible slip systems in the deformation process.

We can summarize the characteristics of slip formation:

1. Slip occurs when the shear stress exceeds a critical value.

2. There is a similar orientation of the slipped and unslipped regions.

3. The atomic movements are integral multiples of interatomic spacing (along the slip plane).

4. Slip lines or bands are removed after polishing.

5. This (shear) process is heterogeneous, and it occurs on widely spaced and well-defined individual planar crystallographic planes. However, the very wavy appearance of slip lines or bands observed in a number of α-iron and other bcc metals suggests that the dislocations producing slip are not confined to a unique slip plane; that is, multiplicity of possible slip systems is reflected.

6. Its formation is not very rapid—it takes milliseconds.

7. The slip systems in hcp structures are more complex than those in cubic structures due to anisotropy. The observed slip systems in the former are considerably less than in the latter.

The shortest lattice vector extending from an atom at the corner of a bcc iron unit cell to the atom at the center is the Burgers vector $(a/2)$ <111> of magnitude 2.48 Å. Since the edge dislocation moves in the direction of its Burgers vector, the slip plane of an edge dislocation is defined by the dislocation line and its Burgers vector. In contrast, the line of a screw dislocation is parallel to the Burgers vector, and it glides in a direction normal to the Burgers vector. Thus, no specific plane can be defined; that is, screw dislocations can move on any plane that contains its Burgers vector, <111> direction, and can cross-slip. In all instances, as said earlier, plastic deformation by slip occurs in a given slip system when the shear stress τ exceeds a maximum or critical value τ_{cr} called the *critical resolved shear stress*. Table 4.3 lists the prominent slip systems of metals and other crystalline solids.[14,15]

4.3.2 Micro- and Macroyielding

In high-purity polycrystalline annealed iron, the stress-strain curve in tension is observed as partly shown in the left side in Fig. 4.5[16] for a given strain rate and at room temperature. In a series of loading and unloading unidirectional tensile tests, the first elastic deformation occurs on the application of load, which is followed by a microscopic yielding, microyielding, or preyield microstrain at stresses well below the macroscopic upper yield stress,[16–18] during which free or mobile dislocations are produced (thereby showing the slip activity) in a few grains. The stress at which the first detectable movement of dislocations is found upon loading and reloading is called the *elastic limit* (where elastic behavior ends), as shown by σ_E in Fig. 4.5 (right side).[16] This denotes the stress required for reversible motion of edge dislocations. This represents an energy loss. The mobile edge dislocations available increase rapidly when the stress is increased beyond σ_E. With increasing stress, a stress σ_A is observed where a permanent (irreversible) strain or deformation occurs; that is, the hysteresis loop fails to close. This stage of yielding is called *anelastic*. This is the stress required for long-range irreversible motion of edge dislocations. It is slightly below the *proportional limit* σ_{pl} (Fig. 4.5), which is defined as the stress at which an initial deviation from the linearity occurs. Both σ_E and σ_A constitute different stages in the yielding process, and this preyield behavior is strongly influenced by temperature, prior strain (i.e., initial dislocation substructure), and concentration of impurities.[16,19]

At stresses higher than σ_A but lower than the yield stress, a high hardening stage occurs in which edge dislocations become highly mobile while screw dislocations become sessile, thereby hindering dislocation multiplication by Frank-Read sources.[17] It has also been noted that at stresses above σ_E, edge dislocations are continuously liberated and consequently become exhausted, until macroscopic yielding commences.

The speculation concerning exhaustion of mobile edge dislocations (*exhaustion hardening*) during the early stage of yield comes from B. Escaig (1966) and is further developed by R. J. Arsenault (1966) and B. Sestak (1970).[20] They all assume that the core structure of the screw dislocation is protruded on three planes, thereby leading to a high Peierls barrier. Unfortunately, such aspects of deformation of bcc metals do not agree with the direct observation but must be inferred from various

TABLE 4.3 Prominent Slip Systems of Metals and Some Other Crystalline Solids[14,15]

Crystal structure	Lattice type	Slip plane	Slip direction	Ref.
Cu, Au, Ag, Ni, CuAu α-CuZn, AlCu, AlZn	FCC	{111}	<110>	a
Al	FCC	{111} {100}	<110> <110>[1]	a
α-Fe	BCC	{110} {112} {123}	<111> <111> <111>	a
Mo, Nb, Ta, W, Cr, V	BCC	{110} {112}	<111> <111>[2]	a
Cd, Zn, ZnCd	HCP c/a > 1.85	(0001) ($10\bar{1}0$) ($11\bar{2}2$)	$<2\bar{1}\bar{1}0>$ $[11\bar{2}0]$[3] $[\bar{1}\bar{1}23]$[3]	a
Mg	HCP c/a = 1.623	(0001) {$10\bar{1}1$} {$10\bar{1}0$}	$<2\bar{1}\bar{1}0>$[4] $<2\bar{1}\bar{1}0>$[4] $<2\bar{1}10>$[5]	a
Be	HCP c/a = 1.568	(0001) {$10\bar{1}0$}	$<2\bar{1}\bar{1}0>$ $<2\bar{1}\bar{1}0>$	a
Ti	HCP c/a = 1.587	{$10\bar{1}0$} {$10\bar{1}1$} (0001)	$<2\bar{1}\bar{1}0>$[6] $<2\bar{1}\bar{1}0>$[6] $<2\bar{1}\bar{1}0>$[6]	a
Ge, Si, ZnS	Diamond cubic	{111}	<101>	a
As, Sb, Bi	Rhombohedral	(111) ($11\bar{1}$)	$[10\bar{1}]$ [101]	b
NaCl, KCl, KBr, KI, AgCl, LiF	Rocksalt structure	{110} {001}	<110> <110>[7]	b
MgTl, LiTl, AuZn, AuCd, NH₄Cl, NH₄Br, CsI, CsBr, TlCl-TlBr, CsCl	Cesium chloride structure	{110}	<100>	a c

1. Above 450°C.
2. Secondary system at higher temperature.
3. Above 250°C.
4. Above 225°C.
5. At RT and below.
6. Pyramidal and basal planes less active than prism planes.
7. Secondary system at higher temperature.
Sources: Reprinted by permission of Elsevier Science B.V., Amsterdam.
a. A. Seeger, in *Encyclopedia of Physics*, vol. 7/2 (Crystal Physics II), ed. S. Flugge, Springer, Berlin, 1958, p. 1.
b. E. Schmid and W. Boas, *Kristallplastizitat*, Springer, Berlin, 1935.
c. M. T. Sprackling, *The Plastic Deformation of Simple Ionic Crystals*, Academic Press, New York, 1976.

models of ambiguous validity, i.e., atomistic models using assumed interatomic potentials. Direct thin-foil deformation observations are subject to the artifacts introduced by image forces. W. C. Leslie, who supports the dislocation dynamics model for upper and lower yield point in steel, also employs the concept of exhaustion hardening for the microstrain stages. The dislocation dynamics model, as originally proposed by G. T. Hahn, does not advocate the effects of grain size on yield behavior. It appears that contribution to Fig. 4.5 results from interactions, in addi-

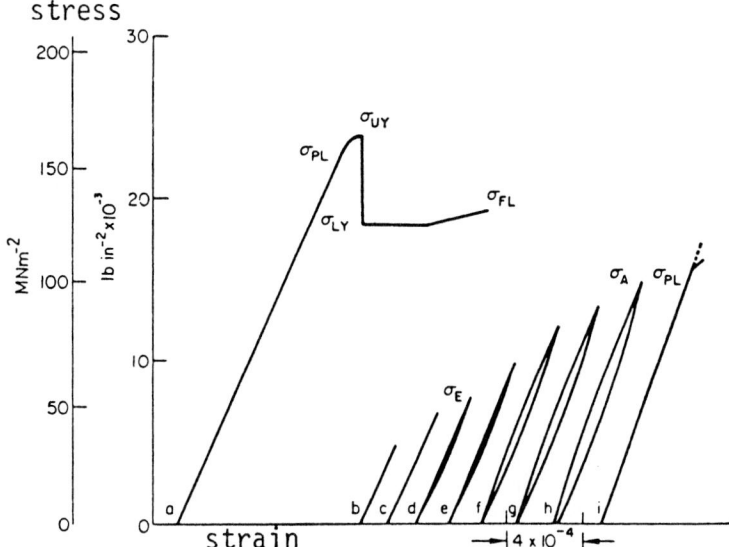

FIGURE 4.5 Stress-strain curve in polycrystalline annealed iron in tension (left) and during unidirectional loading and unloading (right) where interactions other than just dislocation motion are involved.[16] (*Reprinted by permission of Pergamon Press, Plc.*)

tion to just dislocation motions. The etch pit studies of Brentnahl and Rostoker on polycrystalline iron, where the effects of elastic interactions between elastically anisotropic grains prevail in the early stages of yield, do not subscribe to the edge dislocation exhaustion model. Keh and Weissman state that "most dislocations introduced into iron by slight deformation have screw orientations." Moreover, all observations clearly demonstrate that the early stages of yield are associated with activity at the grain boundary. It is thus clear that the exhaustion-hardening concept is inconclusive.[20]

During macroscopic yielding, the motion of screw dislocations is significant and is associated with dislocation multiplication through the double-cross-slip mechanism.[21] At the stress approaching the macroscopic yield point, which is measured by a sharp drop in the load extension or stress-strain curve, the slip activity has been shown to occur in essentially all the grains. Upon any further increase in load, fresh multiplication of dislocations occurs, and the instability sets in. This results in the drop from the upper to the lower yield point. Note that the difference between the microyield and macroyield is very narrow in coarse-grained materials but increases with finer grain size.[22]

4.3.3 Upper and Lower Yield Points

Two criteria have been found necessary for a sharp yield point: (1) an increase in the density of mobile dislocations (by dislocation multiplication) and (2) direct relationship between stress and average dislocation density.

When the dislocation density is studied in the early stages of deformation by etch pits or thin-foil microscopy, it is found, for $\varepsilon < 0.1$, that

$$\rho = \rho_0 + K\varepsilon_p^m \tag{4.24}$$

where ρ is the total dislocation density; ρ_0 is the average density of mobile, grown-in, or initial dislocations present before straining (between $10^3/\text{cm}^2$ and $10^7/\text{cm}^2$); K is a constant; and m is also a constant, usually $\cong 1 \pm 0.5$. We will see in Sec. 4.4.2 that this equation gives a relationship between immobile dislocations and the plastic strain for polycrystalline materials. The plastic strain rate $\dot{\varepsilon}_p$ can also be expressed by the well-known equations[19,22–25]

$$\dot{\varepsilon}_p = \rho b v \tag{4.25}$$

and
$$v = B\sigma^m \tag{4.26}$$

where ρ is the density of mobile dislocations (not the total dislocation density) contributing to the deformation at any instant,[23,24] b is the magnitude of their Burgers vector, σ is the average applied stress operating on them, and B and m are the temperature-dependent and perhaps purity-dependent constants designated, respectively, as the dislocation velocity at unit applied stress and the dislocation velocity stress exponent. The strain rates at the upper yield point $\dot{\varepsilon}_{pu}$ and the lower yield point $\dot{\varepsilon}_{pl}$ can be defined according to Eq. (4.25) as

$$\dot{\varepsilon}_{pu} = \rho_u v_u b \tag{4.27}$$

and
$$\dot{\varepsilon}_{pl} = \rho_l v_l b \tag{4.28}$$

where ρ_u and ρ_l are the density of *mobile dislocations* at the upper and lower yield points, respectively, and v_u and v_l are their velocities, so that

$$\frac{v_u}{v_l} = \frac{\rho_l}{\rho_u} \tag{4.29}$$

Combining Eqs. (4.26) and (4.29), we get

$$\frac{\sigma_u}{\sigma_l} = \left(\frac{\rho_l}{\rho_u}\right)^{1/m} \tag{4.30}$$

where σ_u is the upper yield stress and σ_l is the lower yield stress. It is obvious from Eq. (4.30) that the ratio σ_u/σ_l will be large; that is, a large yield drop will occur when m is small and $\rho_l \gg \rho_u$. This fact is supported by the experimental observation that the dislocation density immediately after the lower yield point is much greater than that at the upper yield stress.[26]

4.3.4   Yield Point Elongation and Lüder's Band Formation

Many metals, especially low-carbon steel, show a localized, heterogeneous type of transition from elastic to plastic deformation that produces a *yield point*. The load at which the sudden drop occurs in the load-elongation curve is called the *upper yield point*. The constant load is referred to as the *lower yield point*, and the elongation or stretching that occurs for a while without any increase in its flow stress (i.e., constant load and before the load starts to increase monotonically) is called the *yield point elongation* YPE, or e_y (Fig. 4.6).[27] The extent of YPE depends on the

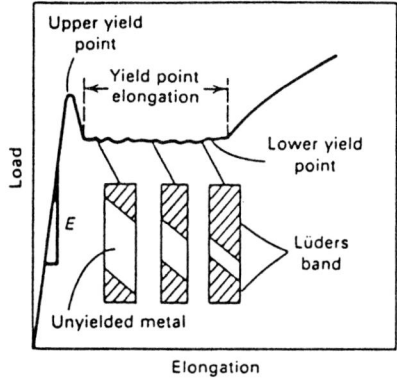

FIGURE 4.6 Schematic showing yield point elongation.[27]

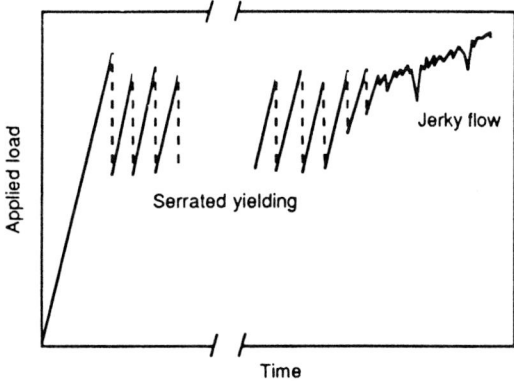

FIGURE 4.7 Schematic load-time curve showing two variations of Lüder's band.[28] (*Reprinted by permission of Pergamon Press, Plc.*)

grain size, strain rate, and temperature. The plastic deformation that occurs during YPE is inhomogeneous, discontinuous, small, and localized and is oriented approximately 55° to the tensile axis. This is called *Lüder's band*, *Lüder's strain*, or *stretcher strain*. Successive generation of Lüder's band continues over the entire length of the specimen at a comparatively low and constant applied stress (i.e., at the lower yield point) (Fig. 4.6). Lüder's band formation does not produce strain hardening until it is complete (i.e., when the YPE is complete). That is, when YPE zone or Lüder's zone ends, work hardening is registered on the stress-strain curve. Lüder's zone occurs in steels that are prone to strain aging, that is, strong interaction between solutes (e.g., C or N in iron and steel and Mg in Al), and dislocations that produce solute segregation to, and immobilization (pinning) of, dislocations.

There are two variations of distinguishable Lüder's band: (1) serrated yielding, that is, occurrence of plastic flow, where the leading edge of a single Lüder's band propagates intermittently along the entire gauge lengths—this is characterized by a very uniform sawtooth stress-time plot (Fig. 4.7)—and (2) jerky flow, that is, occurrence of plastic flow at random locations along the specimen. When load serrations stop, a single Lüder's band front passes from one end to the other end of the specimen. It is characterized by a randomly fluctuating stress-time plot (Fig. 4.7).[28]

Discontinuous or inhomogeneous yielding and flow by Lüder's band formation and propagation has long been a nuisance and is sometimes an unacceptable problem in the fabrication of complex shapes such as automobile doors and bumpers from low-carbon steels because of the surface appearance marred by Lüder's lines. However, discontinuous yielding and strain aging have sometimes been proved to be advantageous, as in machining and paint-baking of low-carbon steels.

With coarsening of grain size, the difference in upper and lower yield stress is lowered and the yield drop and Lüder's band fronts become less pronounced (also mentioned earlier) and finally become diffuse and difficult to observe. In this case, e_y decreases according to[29]

$$e_y = \text{YPE}(\%) = \frac{k_y d^{-1/2}}{\alpha}$$ (4.31)

where d is the grain size and k_y and α are constants depending on the material and tensile testing conditions, respectively. Usually, both the yield stress and the YPE are increased by the presence of fine precipitates.[30]

4.4 FLOW STRESS

The *flow stress* is defined as the stress required to maintain deformation (or cause the metal to flow plastically) at any given strain. Flow stress of many metals in the region of uniform plastic deformation can be described by the nth-power hardening equation (Ludwik-Hollomon)

$$\sigma = K\varepsilon^n$$ (4.32)

where σ is the true flow stress; ε is the true plastic strain; and K and n are the strength coefficient and strain-hardening exponent (or index), respectively, and both are independent of temperature at sufficiently lower temperatures. A $\log \sigma$ versus $\log \varepsilon$ plot for the true stress–true strain curve up to maximum load provides the n value, represented by the slope of a straight line for most polycrystalline materials, as shown in Fig. 4.8.[8] Tables 4.1 and 4.4 list the n values for several materials. Typically, the value of n, for most metals, varies between 0.1 and 0.56.

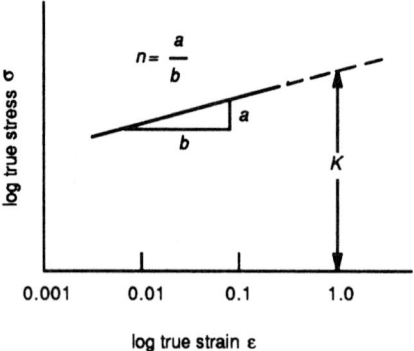

FIGURE 4.8 A $\log \sigma$ versus $\log \varepsilon$ plot for the true stress–true strain curve.[8] (*Reprinted by permission of McGraw-Hill, New York.*)

TABLE 4.4 Values of n and K for Metals at Room Temperature[1]

Metal	Condition	n	K MPa	K ksi
0.05% Carbon steel	Annealed	0.26	530	77
SAE 4340 Steel	Annealed	0.15	641	93
0.6% Carbon steel	Quenched and tempered at 540°C (1000°F)	0.10	1572	228
0.6% Carbon steel	Quenched and tempered at 705°C (1300°F)	0.19	1227	178
Copper	Annealed	0.54	320	46.4
70/30 Brass	Annealed	0.49	896	130

Reprinted by permission of ASM International, Materials Park, Ohio.

The alternative equation suggested for another group of metals including some steels with a temperature-dependent yield stress is expressed as

$$\sigma = \sigma_0 + K\varepsilon^n \qquad (4.33)$$

where σ_0 is the yield stress and K and n are the same constants as in Eq. (4.32). It can be shown that in a tensile test, the limit of true uniform strain is equal to the strain-hardening exponent, that is, $\varepsilon_u = n$. The specimen will begin to neck at a point of maximum load (stress) on the engineering stress-strain curve. Hence, the specimen will neck down and fail when $\varepsilon_u \geq n$.

The true stress–true strain curve of fcc metals having low stacking fault energy, such as austenitic stainless steel, can be expressed by Ludwigson as

$$\sigma = K\varepsilon^n + \exp(K_1)\exp(n_1\varepsilon) \qquad (4.34)$$

where K_1 and n_1 have been introduced as additional constants, $\exp K_1$ is nearly equal to the proportional limit, and n_1 is the slope of the deviation from Eq. (4.32) plotted against ε. The true strain term in Eqs. (4.32) and (4.33) precisely should be the plastic strain

$$\varepsilon_p = \varepsilon_{\text{total}} - \varepsilon_E = \varepsilon_{\text{total}} - \frac{\sigma}{E} \qquad (4.35)$$

4.4.1 Strain-Hardening Exponent

Wok hardening (or strain hardening) is a phenomenon whereby, during the deformation of metals at lower temperatures, the yield stress (required to continue) increases with increasing strain. Strain hardening is caused by the storage of dislocations within a metal and the resistance they offer to the passage of other dislocations. Strain hardening influences the cold formability of sheet. As a result of strain hardening, in many instances, an annealing treatment is required after each forming operation, to enhance the formability and obtain the desired deformation.[31] The strain-hardening behaviors of fcc, bcc, NaCl with an abundance of equivalent

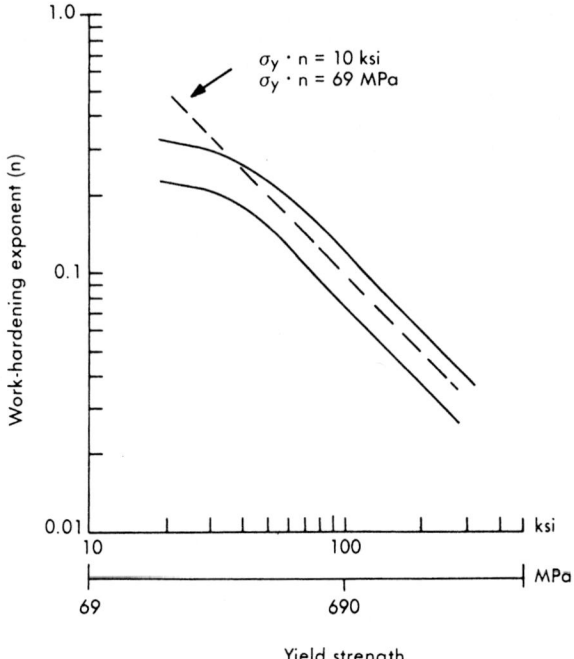

FIGURE 4.9 Relationship between work- (or strain-) hardening exponent n and the yield strength. (*National Steel Corporation.*)

slip systems are different from those of hcp metals that do not undergo intersecting slip until very late in their deformation.[15]

The strain-hardening exponent (n value) is a measure of the ability of a material to resist localized straining and thereby increase uniform elongation. It is a very significant parameter in sheet metal forming. The final strength of a cold-worked part can be determined from the n value. Figure 4.9 can be used to approximate the n value of low-carbon steel sheet.

The amount of e_u, the level of the *forming limit diagram* (FLD), the strain distribution, and many other forming variables are directly related to the n value. The n value is not influenced by texture, but is reduced by the addition of solid solution elements in steel and by the refinement of ferrite grain size.[32] The n values of aluminum alloy sheets decrease sharply with an increase in the tensile strain and are lower than those of steel sheets. The n value of commercial HSLA steels decreases with increasing strength, and its low n value makes it less formable than mild steel.[33] There are different strain-hardening behaviors for mild steels and high-strength steels. The strain distribution ability of steels increases with the increase in overall n value. The peak n value at a low strain level increases the strain distribution ability of the steel. Aging of rimmed steels causes the n value to decrease with time. Additionally, excessive temper rolling beyond that required to eliminate YPE (e_y) will also reduce the n value.

For some materials such as dual-phase steels, some aluminum alloys, and so forth, the n value is not constant, as given by Eq. (4.32). In such cases, two or three

n values may be required to be determined for initial (low), intermediate, and terminal (high) strain regions. The initial n value relates to the low-deformation region, where springback is often a problem. The terminal n value relates to the high-deformation region, where fracture may occur.[3]

Recrystallized structures produce low yield strength and high hardening capacities and are ideal candidates for forming applications. In addition, low stacking fault energy, such as in brass, is advantageous because the difficulty of cross-slip leads to a high hardening rate. In certain stainless and high-strength steels, the decomposition of metastable austenite to martensite during deformation results in a very high n value.

Because most engineering materials need high strength and good formability, a desirable means of strengthening them is by dispersion of hard, spheroidal phases in a soft, ductile matrix. This is also the basis for the dual-phase, ferrite-martensite (high-strength) steels used in the automotive industry. This type of microstructure provides a high initial hardening rate and allows improved shape fixability and less springback compared to high-yield-strength steels.[34]

4.4.2 Work Hardening and Dislocation Density

It has usually been agreed upon by many investigators that the increase of flow stress is proportional to the square root of total dislocation density $\sqrt{\rho}$.[35–38] Since the density of fresh dislocations is much smaller than that of the immobile dislocations, we can write $\rho = \rho_{\text{immobile}}$. Many work-hardening theories, therefore, predict the following interrelationship between initial flow stress and dislocation density produced (in the cell walls), which has been verified by several experiments in metals with varying grain sizes and temperatures.[39–41]

$$\sigma_f = \sigma_0 + \alpha G b \sqrt{\rho} \qquad (4.36)$$

where σ_f is the applied flow stress at a given strain rate, σ_0 is the friction stress on dislocations arising from all sources except dislocation-dislocation interaction (e.g., grain boundary strengthening, solid solution strengthening), α is a numerical constant measuring the efficiency of dislocation strengthening (usually between 0.2 and 0.4), G is the shear modulus, and b is the Burgers vector. The flow stress, therefore, increases with increasing density of dislocations. The work-hardening rate depends on the rate at which dislocations increase with strain.[40] Using Eq. (4.24) to represent the relationship between the density of immobile dislocations building up in the cell structure (to be discussed in the next section) of dislocation and plastic strain for the early stage, we finally get

$$\sigma_f = \sigma_0 + \alpha G b \sqrt{\rho_0 + K \varepsilon_p^m} \qquad (4.37)$$

Equations (4.36) and (4.37) predict a parabolic strain-hardening or stress-strain curve for a polycrystalline metal.

4.4.3 TEM Study of Dislocation Tangles and Cells after Work Hardening

Heavily deformed (or cold-worked) metal at room temperature raises the dislocation density from 10^7 to 10^{12} dislocations/cm^2. Both x-ray diffraction data and TEM studies suggest that the distribution of dislocation is not uniform throughout the deformed or cold-worked structure; but this leads to the formation of

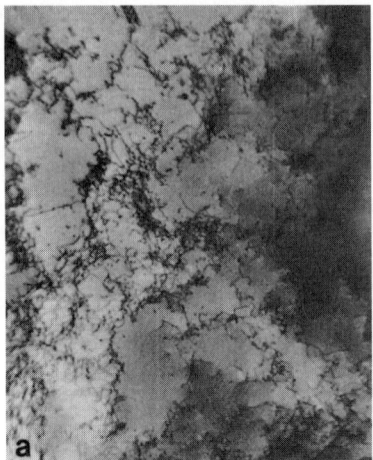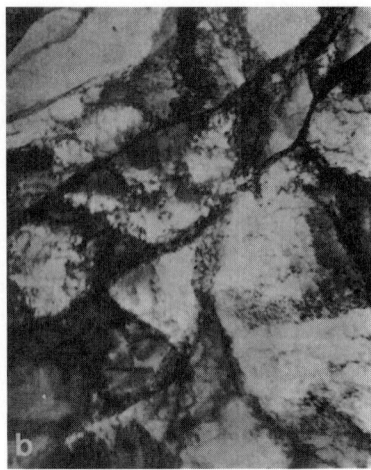

FIGURE 4.10   Cell structure of iron rolled at room temperature to a strain of (*a*) 9% and (*b*) 70%,[25] 6850X. (*Reprinted by permission of John Wiley & Sons, New York.*)

cell walls containing dense or tangled (immobile) dislocations (which represent the boundaries of a subgrain structure) that are surrounded by low-dislocation-density regions called *cells* or *subgrains* (Fig. 4.10). The misorientations across these boundaries increase with the extent of deformation (or cold work); values of 2 to 6° are typical. The cell size in pure metals usually varies between 1 and 3 μm. The dislocation structure comprising cell walls and cells varies from metal to metal depending on their stacking fault energy, the amount, the type, and the rate of plastic strain (deformation) and temperature of deformation. As the temperature of deformation decreases below room temperature, the dislocations develop an increasingly random distribution, and subgrains are not formed. Moreover, dense, tangled dislocations are not formed in the dislocation structure below a certain plastic strain (i.e., below 3%). In polycrystalline iron the cell size decreases with increasing deformation up to about 10% strain. Then they maintain a stable cell size; further increase in plastic strain or flow stress increases the immobile dislocation density in the cell walls (Fig. 4.10),[25,42] which makes these walls or tangles more effective obstacles to fresh dislocation motions. This increased flow stress causes the observed strain hardening and can be related to the average dislocation density by Eq. (4.36).[43] We, however, find slip lines, deformation (transition or micro-) bands, and overall structural features at low magnification by using conventional microscopy. It is pointed out that the deformation bands are linked with particularly well-developed subgrain structures and marked misorientations across the bands.

4.4.4   Work Hardening and Grain Size

The relatively large average grain size dependence of lower yield stress σ_l for polycrystalline metal is described by the classical Hall-Petch relationship as

$$\sigma_l = \sigma_0 + kd^{-1/2} \tag{4.38}$$

where σ_0 and k are constants for the metal (the latter is often termed the Hall-Petch slope and is material-dependent) and d is the average grain size (typically, $1 \mu m$ or larger). This equation has wide applicability and is valid for bcc metals (including iron) as well as fcc and hcp metals. Figure 4.11a and b illustrates such results for various alloys of iron, aluminum, copper, and brass.[44-46] This equation is also applicable to cellular substructures (i.e., subgrain boundary structures), which are normally produced by severe cold working. It may equally apply for small strains showing no yield and for strains well beyond the yield or beyond Lüder's extension, if present. In that case, this equation can be related to dislocation density, and it can also be extended to higher strains in analogous form, in which σ_l is replaced by the flow stress

$$\sigma_f = \sigma_0' + k'd^{-1/2} \tag{4.39}$$

where σ_0' is the frictional force required to move dislocations through the crystal and k' reflects the pinning effect of the grain boundaries.[47] A wide variation in slope of Fig. 4.11b is due to factors such as solute pinning of dislocations (for increased slope) and cross-slip at the head of the pile-up (for decreased slope).[48]

For very small grain sizes, Coble creep is expected to be active and the σ versus d relationship is given by

$$\sigma_c = \frac{A}{d} + Bd^3 \tag{4.40}$$

where B is both temperature- and strain-rate-dependent. The additional (threshold) term A/d seems to be large if d is in the nanometer scale. For intermediate grain sizes, both mechanisms might be operative for the specimen containing a range of grain size distribution.[49] A threshold of the form A/d has been suggested by S. M. Sastri.[50]

Note that a dislocation theory of the Hall-Petch effect only provides a linear relationship of σ with $d^{-1/2}$ when a large number of dislocations in a pile-up are present and plasticity is not source-limited. At sufficiently small grain sizes, the Hall-Petch model based upon dislocations may not be functional. In this case, Choksi et al.[51] have suggested room-temperature Coble creep as the mechanism to explain their results.

If it is assumed that a grain size d^* prevails at which value the classical Hall-Petch mechanism converts to the Coble creep mechanism, $\sigma_{th} = \sigma_c$ at $d = d^*$, as shown schematically in Fig. 4.12.[49]

4.4.5 Relationship between Dislocation Density and Grain Size

For a particular strain, the relationship between the dislocation density and grain size was first developed by Meakin and Petch[52] based on the assumptions that the average slip distance L^s is approximately equal to grain diameter d and that the strain in tension ε equals $\rho b L^s$. Thus, by inserting the dislocation density in the hardening equation (4.36), we get[46]

$$\sigma_f = \sigma_0 + \alpha Gb^{1/2}d^{-1/2} \tag{4.41}$$

This equation is different from the Hall-Petch equation, (4.38), in the sense that here σ_0 is strain-independent.

Ashby[53,54] has proposed that when metals or alloys are deformed, a deformed grain consists of a uniform deformation (by slip on one system) and a local nonuni-

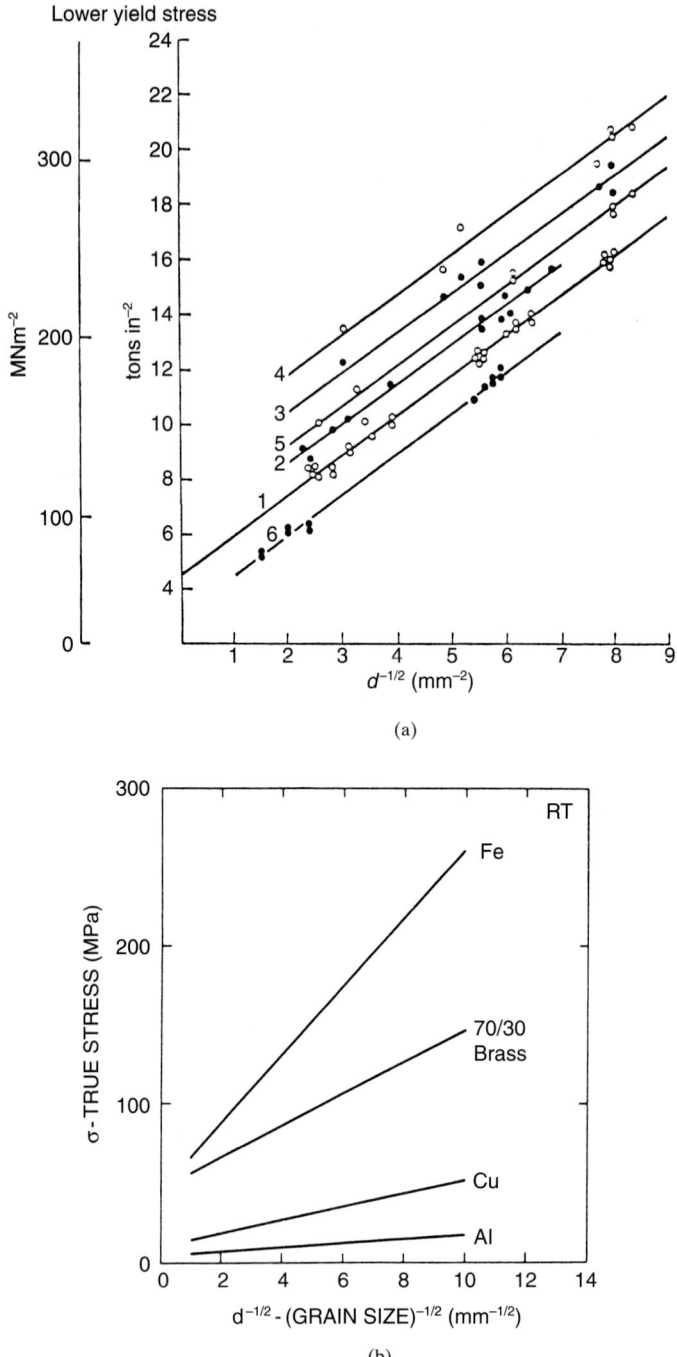

FIGURE 4.11 (*a*) Dependence of lower yield stress on grain size for (1) annealed mild steel (En2); (2) En2, nitrided; (3) En2, quenched from 650°C; (4) En2, quenched and aged for 1 hr at 150°C; (5) En2, quenched and aged for 100 hr at 200°C; (6) annealed Swedish iron.[44] (En2 steel contains 0.115% C and Swedish iron, 0.02% C.) (*b*) Yield stress—grain size relationship at room temperature for Al, Cu, 70-30 brass, and Fe.[46] [(*a*) *Reprinted by permission of Pergamon Press, Plc.* (*b*) *Reprinted by permission of The Metallurgical Society, Warrendale, Pa.*]

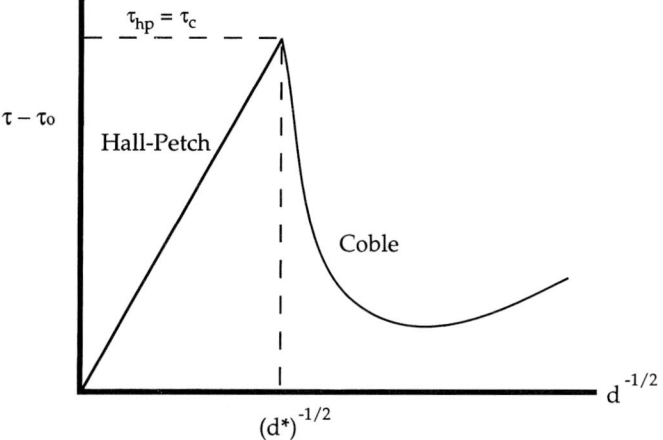

FIGURE 4.12 Combined Hall-Petch response and Coble creep. The critical d^* where transition occurs is represented by dashed line.[49]

form deformation in the grain boundary area. The uniform deformation causes the work hardening within the grain due to accumulation of dislocation density ρ^s of statistically stored dislocations. The accumulation of geometrically necessary dislocation density ρ^g causes the grain-size-dependent (part of stress-strain curve) work hardening (near the grain boundary). Accordingly, Ashby developed the following equations:[53,54]

$$\rho_{\text{total}} = \rho^s + \rho^g \tag{4.42}$$

$$\rho^s = \frac{C_1 \varepsilon}{bL^s} \tag{4.43}$$

$$\rho^g = \frac{C_2 \varepsilon}{bd} \tag{4.44}$$

where C_1 and C_2 in Eqs. (4.43) and (4.44), respectively, are constants. Equation (4.43) is independent of grain size contribution, while ρ^g is dependent on the grain size. The flow stress thus depends on the total dislocation density in the following form:

$$\sigma_f = \sigma_0 + \alpha G \sqrt{\varepsilon} \sqrt{\frac{C_1 b}{L^s} + \frac{C_2 b}{d}} \tag{4.45}$$

If grain boundary strengthening is predominant, we have

$$\sigma_f = \sigma_0 + \alpha G \sqrt{\varepsilon} \sqrt{\frac{C_2 b}{d}} \tag{4.46}$$

If the grain boundary contribution is small (e.g., in coarse-grained materials), Eq. (4.45) reduces to

$$\sigma_f = \sigma_0 + \alpha G \sqrt{\varepsilon} \frac{C_2 b}{2} \sqrt{\frac{L^s}{C_1 b}} \frac{1}{d} \qquad (4.47)$$

Equations (4.46) and (4.47) illustrate that the flow stress at a constant strain may be a function of $d^{-1/2}$ or d^{-1} according to the magnitude of the grain size contribution.[46]

4.4.6 Effect of Temperature

Temperature has a significant effect on the stress-strain curve of polycrystalline iron when compared with the stress-strain curve obtained at room temperature. When the temperature of deformation is decreased, the upper yield point rises considerably; the yield drop and yield point elongation region both become progressively larger. This has also been established in other bcc metals (e.g., Mo, Nb, and Ta).[55] At low temperatures the grown-in dislocations are firmly locked by interstitial impurities; consequently a quite high stress is required to produce new and mobile dislocations. Hence the temperature dependence results mainly from the frictional stress. It has also been experimentally evidenced that frictional stress increases substantially with interstitial solute (C, N) concentrations in the range of 0.001 to 0.3 wt%.

In contrast, when the temperature of deformation is raised above room temperature, the upper and lower yield points as well as the yield point elongation region slowly disappear and are replaced by serrated curves (Fig. 4.13) (see Sec. 4.6.1 for more details).[56]

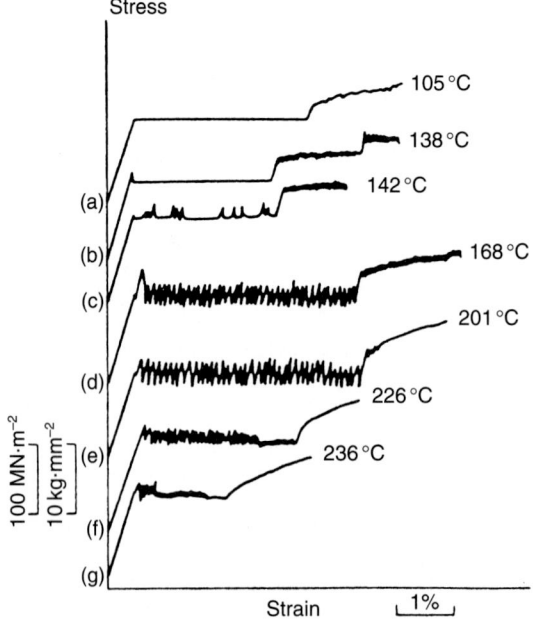

FIGURE 4.13 Stress-strain curve for polycrystalline mild steel at increasing temperature. (*Reprinted by permission of The Institute of Metals, England.*)

4.5　COLD WORKING

Cold working plays an important role in the latter stages of manufacture of steel products, as it offers greater control of dimensions, surface finish, and properties. The terms *cold working* and *annealing* are closely linked because the subtle sequence of these operations allows material properties to be closely controlled.

At the microscopic level, the original ferrite grains are retained during cold working but are prone to drastic shape changes, representing the macroscopic shape changes linked with the working operation. Thus, in cold rolling the grains assume elongated pancake shape and result in a deformation texture or preferred orientation (see Sec. 4.8).

The slip mechanism associated with the plastic deformation of the ferrite crystals in steel generates crystal defects in excess of the thermodynamic equilibrium concentration. The progressive increase of these defects, particularly dislocation density, differentiates cold working from hot working, because in the latter the dislocations are removed spontaneously during the operation. Deformation twinning also occurs during plastic deformation at low temperature and high strain rates, which is discussed in the next section. The dislocations created during cold working at room temperature and above form tangled networks, which represent the boundaries of a subgrain structure (Fig. 4.10).

The process of cold working is associated with the phenomenon of work hardening. As a result of cold working of iron and steel, the tensile strength, yield strength, hardness, and electrical resistivity, as well as the susceptibility to intergranular corrosion, are increased while the ductility, density, and magnetic permeability are decreased.

4.5.1　Stored Energy of Cold-Worked Materials

When a metal is plastically deformed by cold working, most of the mechanical energy is released as heat, but a small fraction of this energy is retained as stored energy inside the metal grains, thereby raising its internal energy. Thus the stored energy is the change in internal energy produced by plastic deformation.[57] The stored energy is mainly in the form of elastic energy in the strain fields of lattice defects (e.g., interstitials, vacancies, dislocations, and stacking faults). That is, this stored energy exists in the metallic grains (or crystals) as point defects, dislocations, and stacking faults. Thus, a heavily deformed metal, being in a stage of higher energy, is thermodynamically unstable.

It has been indicated by many workers that the ratio of stored energy E_s to the total energy of deformation E_w varies from 2 to 10%. In general, the amount of stored energy and the structure of a deformed material depend on many variables: impurity content or composition of the metal or alloy; type or process of deformation; deformation temperature; grain size; deformation (or strain) rate; and other factors.[56]

An increase in impurity content hinders the dislocation motion, which enhances the dislocation density (or multiplication) and the stored energy for the same amount of strain. The complex deformation process (e.g., extrusion and wire drawing) usually activates many slip planes when compared to the simple process (as in tension); with pronounced dislocation intersection, this leads to cross-slip and high dislocation density. Lower temperature of deformation or increase in strain rate produces an increased amount of stored energy and dislocation density. Fine-grained materials cause high levels of stored energy and dislocation density for a given state of strain.

4.5.2 Effect of Variables on the Amount of Stored Energy

The amount of stored energy depends on purity, extent, rate, type and temperature of deformation, and grain size, as described below:[57]

1. *Purity.* The increased impurity content in a metal increases the amount of stored energy. Moreover, it causes a significant change in the kinetics of its release during recovery and usually raises the recrystallization temperature.[57] These effects apparently arise from the pinning of dislocations by impurity atoms and subsequent dislocation multiplication.

2. *Extent, rate, type, and temperature of deformation.* The amount of stored energy increases with the extent of plastic deformation, for a particular metal and deformation process. It has been suggested that more energy is stored in the fast primary deformation process than in the slow one. This is attributed to a change in heat losses arising from varying time required for deformation. Simple deformation processes of tension, compression, and torsion produce lower stored energy. As the deformation becomes more complicated (e.g., wire drawing, explosive shock loading, etc.), the level of stored energy increases. This is presumably due to (enhanced) slip activity on all possible slip planes. The stored energy increases with decreasing temperature of deformation as a result of the formation of smaller cell diameter and larger dislocation density. That is, the rate of work hardening is larger at a lower temperature.[57]

3. *Grain size.* The fine-grained material stores more energy than the coarse-grained material. This is due to the increased grain boundary–dislocation interaction and multiplication of dislocations in the deformed fine-grained material.

4.6 STRAIN AGING

Strain aging is the term used to characterize the time-dependent strengthening process in plastically deformed metals and alloys by (1) strong elastic interactions (and high diffusivities) of interstitial solute atoms (e.g., C and N) with mobile dislocations and point defects and (2) allowing solute atmosphere (or cloud) formation around dislocations due to their high diffusivities at temperatures as low as ambient during or after straining or plastic deformation. Aging reactions that occur after plastic deformation (or press-forming operations) are grouped as *static strain aging* (where plastic deformation and aging are separated in time) and *dynamic strain aging* (where both plastic deformation and aging processes occur concurrently).[58] Static strain aging can take place when the concentration of solute atoms is low and deformation temperature is adequate to allow low-range diffusion of solute atoms. After aging, higher stress levels are needed to produce further straining of the material, either to nucleate fresh dislocations or to pull the dislocations free from these atmospheres. Dynamic strain aging can take place at deformation temperatures below ambient when the concentration of solute atoms is high, or at temperatures above ambient when the solute concentration is low.[59]

4.6.1 Static Strain Aging

In practice, strain aging is of great importance in press-formed high-strength low-carbon (or bake-hardening) steels during paint baking where strength anisotropy is strongly developed. The strain-aging rate also depends on the extent of deformation and increases when the deformation occurs at elevated temperatures or lower strain rates. Usually, about 15% reduction in thickness gives the maximum effect. Strain aging is common in sheet and plate steels and sheet-metal forming. It is undesirable in deep-drawing steels. Figure 4.14a schematically describes the effects of static strain aging on the flow stress of a low-carbon steel.[8]

In continuous wire-drawing operations, strain aging can develop to such a large extent that, in extreme cases, the stress generated by the reduction of the last die leads to splitting and cracking in the center of the wire. For small amounts of strain aging, a reduction in torsional properties of otherwise defect-free wire will occur. Strain aging can also develop erratic behavior in a forming operation because of the erratic variation of the yield strength of the wire at elevated temperatures.[60]

Static strain aging in drawn pearlitic steels greatly influences the drawability and properties of the finished products such as ropes and cables. In addition, higher susceptibility to central burst formation and to delamination is also observed. Static strain aging in a drawn pearlitic steel wire is controlled by the decomposition of cementite. This appears to occur in two distinct stages, each linked to different atomistic mechanisms. The first stage occurs at low aging temperature interval (between 60 and 100°C) and is characterized by a slight increase in yield strength and a decrease in electrical resistivity and background damping. The second stage of aging takes place at higher temperatures (between 120 and 200°C) or longer aging times and is exhibited by a considerably larger increase in yield strength and an increase in electrical resistivity, whereas background damping reaches very small values.[60a] Note that the second stage of aging is not reported in low-carbon steels.[60b]

Certain coating treatments such as hot-dip galvanizing can yield a greater extent of strain-aging embrittlement in areas that were cold-worked to the critical amount ε_C; this can lead to brittle fractures. This problem can be eliminated by annealing the part before it is galvanized. Alternatively, strain aging can be controlled by using interstitial-free steels which are Al-killed and contain <20 ppm C and stabilizing

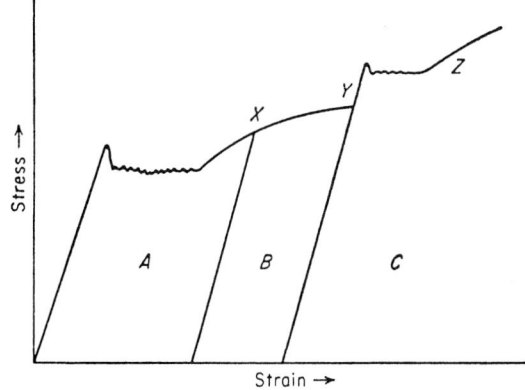

FIGURE 4.14a Strain aging for low-carbon steel. Region A, original material strained through yield point. Region B, load removed from specimen at point X and specimen immediately reloaded. Region C, load removed at point Y and specimen reloaded after aging at 300°F.[8]

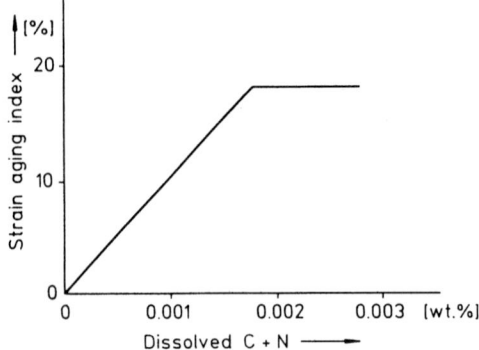

FIGURE 4.14*b*   Variation of strain-aging index with dissolved C and N contents.[30]

elements such as Nb and Ti.[61] The usual industrial solution to combat strain aging involves temper-rolling reduction of about 1% to flatten the cold-rolled and annealed low-carbon steel sheet and its immediate use without storage; but if the steel containing even small amounts of C and N is allowed to be stored for 1 to 3 days, subsequent forming operations at or slightly above room temperature may cause the return of the sharp yield point, increase in yield and tensile strengths, inhomogeneous or stretcher strain, YPE, decrease in ductility, unsatisfactory or spoiled surface appearance, and some loss in stretch formability.[58] Strain aging is also a possible hardening mechanism in AISI 4340 steels.[62]

The aging index AI, used to describe the strain-aging characteristics, is given by

$$AI = 100\frac{\sigma_a - \sigma_s}{\sigma_s} \tag{4.48}$$

where σ_a is the yield stress after aging for 1 hr at 100°C and σ_s is the flow stress for a nominal strain of 8 to 10%. Figure 4.14*b* is a plot of AI index against dissolved C and N contents.[30]

Quench Aging.   Strain aging is different from quench aging or *quench-aging embrittlement*, which occurs in low-carbon steel (containing 0.04 to 0.12% C). Quench aging, by producing increased hardness, reduced ductility, and enhanced yield strength, is deleterious to the formability of sheet steel, and especially of hot-dipped galvanized sheets that are readily cooled at the exit of the molten zinc bath. It is also deleterious to the magnetic properties of iron. In this case, the peak of magnetic hardening is attained much later than the peak of the mechanical ferrite hardening; i.e., the most effective particle size for magnetic hardening is much larger than that for mechanical hardening.[59]

Quench aging is a precipitation-hardening process that is caused by the precipitation of carbide and/or nitride from ferrite solid solution after quenching from the α or α + Fe$_3$C phase fields rather than from austenite.[59] Like age hardening of aluminum alloys, subsequent aging at ambient temperature or slightly above produces an increase in yield strength and hardness. Plastic deformation is not essential to produce quench aging.[63] Quench aging of steel can involve the formation of transition phases such as ε-carbide (Fe$_{2.4}$C) or α''-Fe$_{16}$N$_2$ because activation energy barriers to the formation of these phases at low temperatures are lower than those for the equilibrium phases.[59]

4.6.2 Dynamic Strain Aging

Dynamic strain aging (DSA) is the phenomenon of attractive interactions between diffusing solute atoms in the alloy and mobile dislocations, during plastic deformation and generation of more dislocations prior to the continuation of strain. It is also called the *Portevin-Le Chatelier (PLC) effect.*[64] DSA has been observed to occur in a broad range of alloys, where both interactions between interstitial (C and N) solutes and substitutional solutes (O, Si, Mn, Ni, Ru, Rh, Re, Ir, and Pt) and strong interactions of the resulting substitutional-interstitial complexes with dislocations over a greater temperature range have been believed to play dominant roles in iron-base alloys. Other alloys reported to produce DSA effects include alloys of Ni, (also H-charged Ni and C-doped Ni), Cu, Al (e.g., AlZnMg alloys),[64a] Au, and V. The effect occurs in both fcc (austenitic steels) and bcc iron alloys. Even hcp materials have been observed to deform with serrated (or jerky) flow to exhibit DSA effect.[65]

DSA can be expected to be more pronounced when the average velocity of moving dislocations $\bar{V}_d$ equals the diffusion rate or solute velocity V_s of C and N atoms in the stress field of the dislocations. The values of $\bar{V}_d$ and V_s depend on the strain rate $\dot{\varepsilon}$ (m/ms) and the temperature T(K), respectively. Hence, the condition for DSA is described by an optimum strain rate $\dot{\varepsilon}^*$ and temperature T^* for which $\bar{V}_d = V_s$. For example, DSA occurs in low-carbon steel when $\dot{\varepsilon}^* = 10^{-3}$ m/ms and $T^* = 500$ K ($\sim$230°C). At very high strain rates ($\dot{\varepsilon} > \dot{\varepsilon}^*$) and very low temperatures ($T < T^*$), C and N atoms are unable to segregate to the moving dislocations because $\bar{V}_d > V_s$. At very low strain rates ($\dot{\varepsilon} < \dot{\varepsilon}^*$) and very high temperatures ($T > T^*$), any solute atmosphere formed can easily move with the dislocations, and deformation occurs without DSA effects.[30,64]

DSA caused by interstitial elements is reported to be irregular and finds difficulties to be systematically analyzed. However, five types of serrated flow curves in the deformation of substitutional solid solution alloys have been observed which are attributed to DSA effects. These serration types are termed A, B, C, D, and E (Fig. 4.15*a*). Type A is characterized by periodic or repeated appearance of deformation bands and occurs after a low strain. This is observed in Al-Mg alloys deformed at room temperature or a mild steel deformed in blue brittle temperature range ($\sim$175°C). Type B refers to the stress oscillations around a constant level of the σ-ε curve. It is sometimes called Lüder's or PLC bands and occurs at a specific angle ($\sim$55°) to the specimen tensile axis, which may be influenced by the rolling texture.[65] Type C serration is recognized from load or yield drops below the prevalent level of the flow curve, and type D from peak or plateau in the σ-ε curve due to band propagation. Type E serrations usually appear at high strains or near the maximum load.[64,65]

The increase in flow stress and strain-hardening rate during DSA results partly from the increased migration and locking of solute atoms to dislocations (Fig. 4.15*b*); also a reduced dislocation annihilation occurs.[66] Precipitates and atmosphere drag also enhance the hardening. When the low-carbon steel is strained at a normal tensile strain rate of about 1.75×10^{-4} m/ms, DSA effects occur in the stress-strain curve at temperatures between 100 and 250°C, with very low total elongation and high tensile strength (Fig. 4.15*c*). If the interstitial solute content is substantial, DSA can be observed at room temperature. At very high strain rate, as in impact testing, these same limits take place between 350 and 650°C.[63,64] In steels, clustering of N atoms may occur in solution with substitutional atoms such as Mn; this decreases their diffusion rate and, therefore, shifts these DSA effects to higher temperatures.[66] DSA in steel wire is a function of steel composition, drawing strain, highest temperature attained during wire drawing, and the extent of time at that temperature.

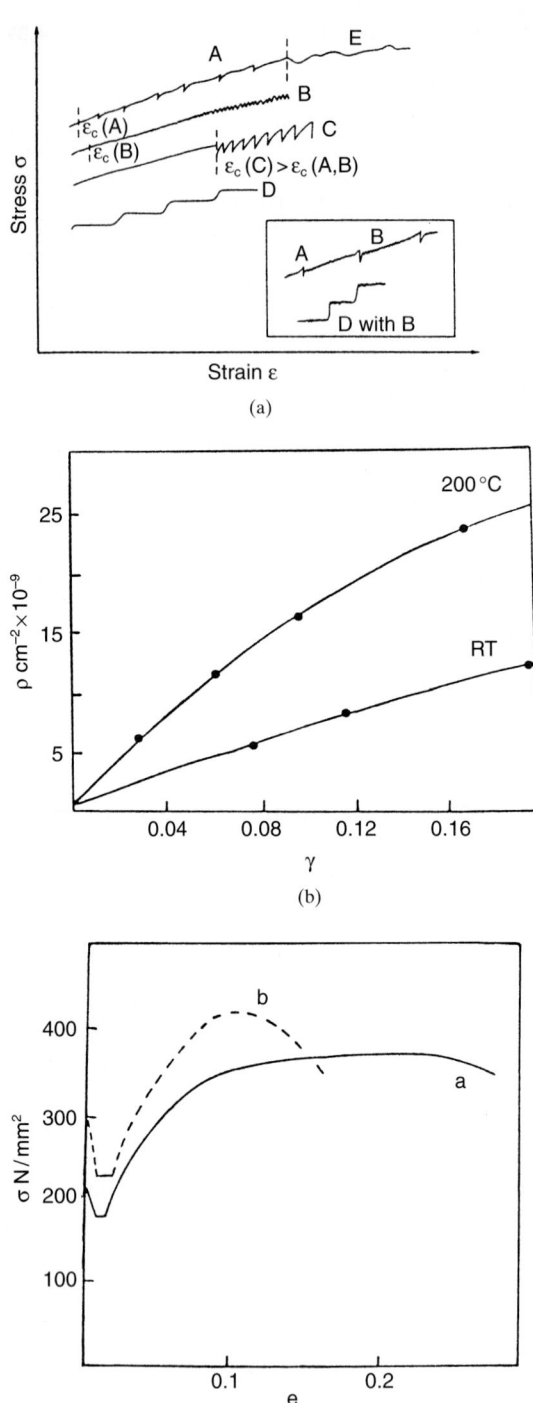

FIGURE 4.15 (*a*) Types of serrations due to DSA.[64] (*b*) Dislocation density ρ versus plastic shear strain γ for a 0.035% C steel deformed at room temperature and 200°C. (*c*) DSA of a mild steel. Nominal stress versus elongation at (*a*) room temperature and (*b*) 150°C.[66]

Table 4.5 summarizes the effects of DSA on the room-temperature properties of five steels, such as increased strength, YS/UTS ratio, fatigue life and the DBTT, and decreased ductility and notch impact toughness.[63]

In polycrystals, the DSA-like serrations of flow stress and work hardening become more pronounced with decrease in grain size. It has been suggested that grain boundary regions are preferred sites for DSA and for consequent increased rate of dislocation accumulation. DSA has been found in a large number of dilute interstitial and substitutional solid solutions as well as in most of the dilute and substitutional concentrated commercial alloys. It is also observed in specimens subjected to uniaxial tension, compression, and torsional loading. Whether due to solute atmosphere, ordered regions, or precipitates, it is clear that DSA can be a very effective method of strengthening[61] and can be used in combined forming and strengthening operations in the DSA temperature range.[59] It is of importance to study the effect of DSA on mechanical behavior such as negative *strain rate sensitivity* (SRS) and unstable plastic flow (or strain localization). These phenomena are interrelated with mechanical anisotropy and may influence the formability at room temperature.[64a]

DSA effects in drawn pearlitic steels (1) are characterized by lower intensities than those found in low-carbon steels, (2) start or attain their maximum values at higher temperatures, and (3) are controlled by the decomposition of cementite, as evidenced by different techniques such as Mössbauer spectroscopy and neutron diffraction.[60b]

The dependence of kinetics of DSA is expressed with the strain rate $\dot{\varepsilon}$ as

$$\dot{\varepsilon} = K\rho_m b \exp\frac{-Q}{RT} \tag{4.49}$$

where K is a constant, ρ_m is the density of mobile dislocations, b is the Burgers vector, Q is the activation energy involved in the process responsible for the PLC effect Q_{PLC} or for the maximum stress Q_{max}, R is the universal constant, and T is the absolute temperature.

Blue Brittleness.   One of the well-known manifestations of DSA is the blue brittleness phenomenon which occurs during plastic deformation at a strain rate of 10^{-3} m/ms in the *blue-heat range* of 230 to 370°C (or 450 to 700°F). This shows an extremely low tensile ductility and notch-impact resistance and high tensile strength. The onset of necking takes place when the rate of strain hardening equals the flow stress. Baird and Jamieson have observed that the loss of ductility does not occur with high N levels (>0.03%).[66] Carbide- and nitride-forming elements are usually added to prevent blue brittleness.[27,63]

4.7 DEFORMATION TWINNING

Mechanical or deformation twinning has long been accepted as a second important mode of plastic deformation in crystalline solids that have a deficient number of independent slip systems. Recently, it has been recognized that deformation twinning greatly influences the strength and ductility of some interesting materials.[67] The actual extents to which deformation twinning can influence plastic flow in polycrystalline solids appear to depend on the amount of total strain associated with twinning. Deformation twinning has been found to be formed in many bcc, hcp, and lower-symmetry metals and alloys; in many fcc metals and alloys; in ordered alloys

TABLE 4.5 Changes in Charpy V-notch Impact Properties Due to Dynamic Strain Aging[63]

Steel	Shelf energy		Transition temperature at 10.9 J (8 ft·lbf)		FATT, 50% ductile		Grain size, μm	Yield strength	
	J	ft·lb	°C	°F	°C	°F		MPa	ksi
1008[†]									
As-rolled	26	19	−65	−85	−75	−103	14	262	38
Prestrained 3%, 250°C (480°F)	21.5	16	−40	−40	−45	−49	14	372	54
1020[†]									
As-rolled	21.5	16	−55	−67	−50	−58	14	352	51
Prestrained 3%, 250°C (480°F)	18	13.5	−10	14	−25	−13	14	507	73.5
1035[†]									
As-rolled	17.5	13	−35	−31	−20	−4	13	379	55
Prestrained 3%, 250°C (480°F)	13.5	10	+3	37	−5	23	13	576	83.5
1522[†]									
As-rolled	26	19	−70	−94	−55	−67	10	386	56
Prestrained 3%, 250°C (480°F)	24	17.5	−50	−58	−39	−38	10	503	73
1010 renitrogenized[‡]									
As-rolled	24.5	18	−63	−81	...	...	10	331	48
Prestrained 3%, 250°C (480°F)	20.5	15	−52	−62	...	...	10	427	62

FATT: fracture appearance transition temperature.
† Specimens 5 × 5 × 55 mm (0.2 × 0.2 × 2.2 in).
‡ Specimens 2.5 × 10 × 55 mm (0.1 × 0.4 × 2.2 in).

and other intermetallic compounds; in elemental semiconductors and compounds; in nonmetallic compounds such as calcite and sodium nitrate; and even in complex minerals and crystalline polymers.[14,68–74] Table 4.6 lists the twinning systems in some important metals and alloys.[14,68] Twinning occurs when a portion of the crystal takes up an orientation that is related to the orientation of the untwinned lattice in a distinct symmetrical manner. The twinned portion of the crystal forms a mirror image across the twinning plane (also called *composition plane*) of the parent matrix. The following are the characteristics of the deformation twins:[15,27,74–76]

1. It is produced by a sudden localized shear process and involves a small, but well-defined, plastic deformation, in contrast to the apparently chaotic process of formation and growth of formation and growth of slip bands during glide deformation.[68]

2. In the twinned portion, realignment of atoms within a finite crystalline volume of the parent phase reproduces the original symmetry and crystal structure, but with a different orientation.

3. Deformation is a pure shear (Fig. 4.16)[71] in which a cooperative atomic movement occurs; that is, individual atoms move by a fraction of the lattice interatomic spacing relative to each other. It produces a change in shape in the form of a small lens or plate; the parallel (or nearly parallel) sides represent the low-index planes, called a *twin habit plane* or *twinning plane*. This distorts the surrounding matrix.

4. Twins are visible after repolishing and etching the sample.[77]

5. The shear strain γ associated with twinning is homogeneous (i.e., uniformly distributed), in which every adjacent atomic plane in the twinned portion is involved.

6. Its formation occurs very rapidly (in microseconds) (often with an audible click sound and with a sharp drop in yield) in materials with even moderate plastic slip resistance. Its formation consists of two parts: the nucleation of a twin nucleus and development or growth of the twin.[69] Twin initiation stress is much larger than the stress required to propagate a preexistent twin.

7. Deformation twinning is always accompanied (or preceded) by the formation of some microslip, even though this slip may be difficult to recognize.[68] Greater deformation (48 to 75%) in fcc alloys results in the formation of deformation twins which increase in density with cold working.

8. Twins usually form easily in large grains. Twin thickness is a function of grain size; for example, thicker twins form in coarse-grained specimens.[68] Deformation twins have imperfect structures comprising many stacking faults. In many cases, the twinning shear can only transform a sublattice relative to the exterior, and reposition all other atoms properly within the twinned region, requiring internal atom switches which are called *shuffles*.[68]

9. There are a number of important geometrical differences between twinning and slip. Like slip, the occurrence of twinning depends on the least stress to be initiated in the crystal undergoing shear. The flow stress levels required for slip or twinning behavior of a particular material are not constant, but depend on the crystal structure of the metal or alloy, temperature of deformation, strain rate, mode of deformation, microstructure (including the grain size, texture, and presence of second-phase precipitates or dispersoids), chemical composition, and prestrain (or prior deformation).[68]

10. The nucleation of a twin is accompanied by a sudden load drop (responsible for serrated stress-strain curve), while the growth of a twin displays smoother behavior.[77]

TABLE 4.6 Deformation Twining Elements in Some Important Metals and Alloys[5,14]

Metal	K_1	η_1	γ^r	Notes
		Face-centered cubic		
Cu	(111)	[11$\bar{2}$]	0.707	Only recrystallization twins (with these elements) occur in Ag, Al, Au, γ-Fe, and Co
		Body-centered cubic		
Cr, α-Fe, Mo, Na	(112) (441) (332)	[11$\bar{1}$]	0.707	
W	(112)	[11$\bar{1}$]	0.707	
		Diamond cubic		
Ge	(111)	[11$\bar{2}$]	0.707	
		Hexagonal close-packed		
Be		[$\bar{1}$011]	0.199	c/a = 1.568
Cd		[10$\bar{1}\bar{1}$]	0.171	c/a = 1.886
Mg	(10$\bar{1}$2)	[$\bar{1}$011]	0.129	c/a = 1.624
Ti		[$\bar{1}$011]	0.189	c/a = 1.587
Zn		[10$\bar{1}\bar{1}$]	0.139	c/a = 1.856
Be	(10$\bar{1}$1) (10$\bar{1}$3)			Additional forms
Mg	(10$\bar{1}$1)	[$\bar{1}$012]	1.066	Ditto
	(11$\bar{2}$1)	[$\bar{1}\bar{1}$26]	0.638	"
	(11$\bar{2}$2)	[11$\bar{2}$3]	0.957	"
Ti	(11$\bar{2}$3)	[$\bar{1}\bar{1}$22]	1.194	"
	(11$\bar{2}$4)	[$\bar{2}\bar{2}$43]	0.468	"
	(30$\bar{3}$4)	⟨20$\bar{2}$3⟩	?	"
				"
				"
	(10$\bar{1}$3)*	⟨30$\bar{3}$2⟩*	?	* 150 and 268°C
Bi	($\bar{1}$012)	[10$\bar{1}$1]	0.118	
Hg	($\bar{1}$012)	[10$\bar{1}$1]	0.447	
Sb	($\bar{1}$012)	[10$\bar{1}$1]	0.146	
	(10$\bar{1}$2)	?	?	
α-Zr	(11$\bar{2}$1)	"	"	
	(11$\bar{2}$2)	"	"	
	(11$\bar{2}$3)	"	"	
		Tetragonal		
β-Sn	(301)	[$\bar{1}$03]	0.119	c/a = 0.541
In	(101)	[10$\bar{1}$]	0.150	c/a = 1.078
		Orthorhombic		
	(1$\bar{7}$2)	[312]	0.228	
	(112)	X[3$\bar{7}$2]	0.228	X = irrational twin
α-U	(121)	X[100]	0.329	
	(111)	[12$\bar{3}$]	0.214	
	(130)	[3$\bar{1}$0]	0.299	

4.7.1 Crystallography of Twinning

Figure 4.16[71] represents an atomic rearrangement where twinning occurs along the twinning plane, called K_1 in the twinning direction. The open circles are the original atomic positions, whereas the black circles represent the new atomic positions after the twin displacement. The formation of mechanical twinning is shown schematically in Fig. 4.17 when a shear stress is applied to the single-crystal specimen as shown in Fig. 4.17a. The crystal deformed into the final form is shown in Fig. 4.17b. The nature of the atomic movement occurring in twinning is shown in Fig. 4.17c. The atomic rearrangement in the twinned portion occurs in such a way that it preserves the original crystal structure symmetry; that is, the size and shape of the unit cell remain unchanged.

According to crystallographic theory, it is possible to retain the size or shape of the unit cell only when three noncoplanar, rational lattice vectors in the original crystal having the particular length and mutual angles remain unchanged after shear

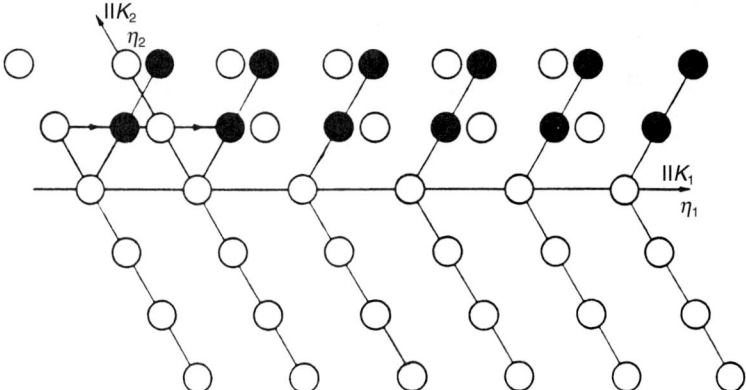

FIGURE 4.16 Atomic arrangement of twinning transformations.[71] (*Reprinted by permission of Butterworths, London.*)

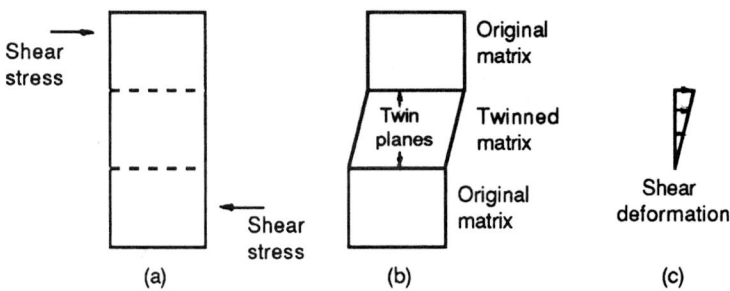

FIGURE 4.17 Twin formation by a shear stress. (*a*) Application of shear stress. (*b*) Twin formation. (*c*) Shear deformation.[75] (*Reprinted by permission of John Wiley & Sons, New York.*)

deformation. For the determination of the first criterion (i.e., undistorted planes upon transformation), let us consider a spherical single crystal that has been sheared to produce a twinned region in the upper hemisphere and an untwinned region in the lower hemisphere (Fig. 4.18).[78] Shear stress has transformed the hemisphere to an ellipsoid in which all planes joining the twinning plane and a point on the original matrix have moved to a position after twinning and have suffered a change in shape and length except for a single plane (shaded area in Fig. 4.18) above the twinning plane. Its length and shape remain undistorted upon twinning because it makes the same angle θ with the twinning plane. It is obvious, from the above, that only two crystallographic planes in a shear transformation have remained unchanged in size and shape. The first is called the *twinning plane* or the *first undistorted plane* (*equitorial plane*) and is designated by the symbol K_1. The other plane (shaded) intersects K_1 in a line that is perpendicular to the shear direction and makes equal angles with K_1 prior to, as well as after, the occurrence of shear transformation. This is called the *second undistorted plane* and is crystallographically designated as K_2. The plane of shear is the plane that contains both the perpendicular to the twinning plane K_1 and shear direction η_1. The intersection of the plane of shear with the second undistorted plane K_2 is the second shear direction η_2, which is represented by an arrow. Two shear directions η_2 and η_2' correspond to the positions of K_2 and K_2' before and after twinning, respectively.

The second criterion for twinning transformation is that the mutual angles between the second undistorted plane K_2 and first undistorted plane K_1 must be retained upon transformation. If we assume h to be the width of the twin, s the magnitude of shear displacement as represented in Fig. 4.18, and θ the angle between K_1 and K_2 (or K_2'), then the shear strain or shear γ can be related to the angle subtended between K_1 and K_2 by

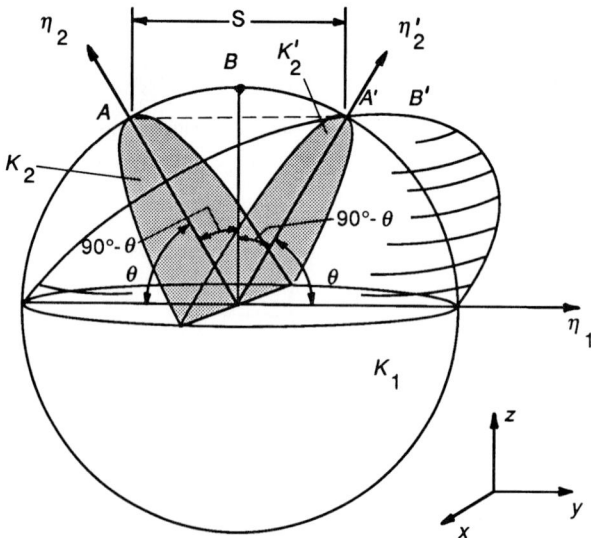

FIGURE 4.18 Schematic diagram showing the transformation of a sphere into an ellipsoid by twinning.

$$\gamma = \frac{s}{h} = 2\tan(90-\theta) = 2\cot\theta \qquad (4.50)$$

Thus the shear strain becomes fixed and can be easily determined after measuring the mutual angle between two undistorted planes.[76]

If we consider any vector ε in the plane K_1 (Fig. 4.19), there remains only one vector η_2 in plane K_2 which makes the same angle α with ε before and after twinning. This arises because η_2 is the only vector perpendicular to the intersection of two planes K_1 and K_2. This condition is applicable to ε and all other vectors lying in K_1. Finally, if it is assumed that K_1 is a rational plane containing rational directions and that η_2 is a rational direction, then the requirements of having these noncoplanar vectors unchanged in shape and size can be realized.

In this manner, it can be shown that η_1, being perpendicular to the intersection of K_1 and K_2, is the only vector in plane K_1 that makes the same angles with all other vectors lying in plane K_2 before and after twinning. It can, therefore, be concluded that another condition for preserving the crystal structure during twinning is that direction η_1 and plane K_2 be rational. There is also a third possibility that the two planes K_1 and K_2 and the two vectors η_1 and η_2 are all rational.

The classical theories of the crystallography of deformation twinning define a particular twinning mode. Therefore, in summary, there are three modes of lattice shear for retaining its symmetry and crystal structure, thereby satisfying the conditions for twinning:

1. A twin of the first kind (when K_1 and η_2 are rational plane and direction)
2. A twin of the second kind (when K_2 and η_1 are rational plane and direction)
3. A rational or compound twin (when all four elements K_1, K_2 planes and η_1, η_2 directions are rational)

It is apparent that the lattice rotation produced during twinning will depend upon whether twins of the first or second kind or compound twins are formed.[26,76]

The theory assumes that the twin interface is planar whereas, in reality, twin boundaries contain linear steps which have the characteristics of dislocations. Twins, therefore, are controlled by the properties of twinning dislocations.[79]

4.7.2 Twin Boundary Interface

Twin boundary may be classified as coherent, semicoherent, or incoherent. In the former case, there is a perfect one-to-one matching (i.e., registry) of the lattice

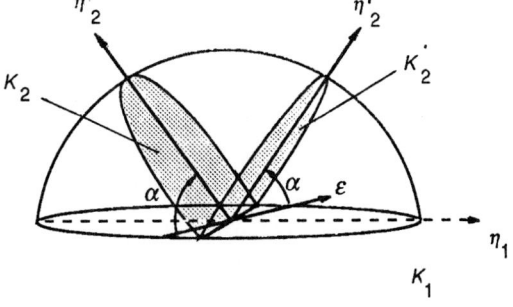

FIGURE 4.19 Twinning of the first kind.

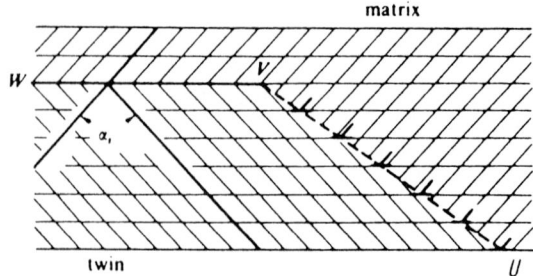

FIGURE 4.20 Coherent (*WV*) and semicoherent (*VU*) twin boundaries; the latter contains partial dislocations.[80] (*Reprinted by permission of Cambridge University Press, New York.*)

planes at or across the twin boundary *WV* (Fig. 4.20).[80] This is also a special low-energy grain boundary. In the semicoherent case, the twin boundary plane *VU* rotates off the symmetry plane (as shown in Fig. 4.20), which results in the disregistry δ of the boundary. This contains partial dislocations as shown in the figure. The disregistry or misfit is defined as

$$\delta = \frac{a_p - a_m}{a_m} \qquad (4.51)$$

where a_p and a_m are the lattice parameters of the twinned and matrix lattices, respectively. Semicoherent boundaries include low-angle grain boundaries and may be further divided into glissile and nonglissile. Incoherent boundaries will be discussed in Chap. 5.

4.7.3   Twinning in BCC Materials

In bcc metals such as Fe, Nb, or W, the twinning plane K_1 is usually [112] and the shear direction η_1 is <111> (or equivalent planes and directions).[68] Mechanical twins (or Neumann bands) are generally observed in polycrystalline pure iron (or α-iron), Fe-Si alloys, and Fe-Cr alloys[81] deformed at low temperatures. They are more predominant in coarse-grained than in fine-grained iron and other bcc metals. In α-iron (and other bcc metals) the twins form on {112} planes with a shear of 0.707 along the <111> direction.[43] The geometry of the process is shown in Fig. 4.21.[78] Figure 4.21*a* shows a (1$\bar{1}$0) section through a bcc structure. The twinning planes {112} packed in the stacking sequence of *ABCDEFAB* . . . is normal to the (1$\bar{1}$0) plane whose trace is shown on the diagram. The twinning shear is $1/\sqrt{2}$ in a <111> direction on a {112} plane. This can be produced by the shear displacement of $\frac{1}{6}$ <111> partial dislocation on every successive {112} plane. This indicates that the bcc twinning results from the propagation of partial dislocations $\frac{1}{6}$ <111> on every {112} plane.[70] Figure 4.21*b* shows the movement of a set of twinning dislocations, lying along *XY*, to a portion of the crystal which produces a twin-oriented region. Figure 4.21*c* illustrates the completely twinned crystal where the twinning dislocations have crossed the whole crystal. This type of dislocation has usually been observed in α-iron, Fe-Si alloys, and Fe-Cr alloys deformed at low temperatures.

    Mechanical twins form in bcc metals as long and thin plates—usually not more than 5 μm thick—because of the high shear strain energy involved;[26] bcc twinning may be induced by a high deformation rate (e.g., impact loading or explosive shock

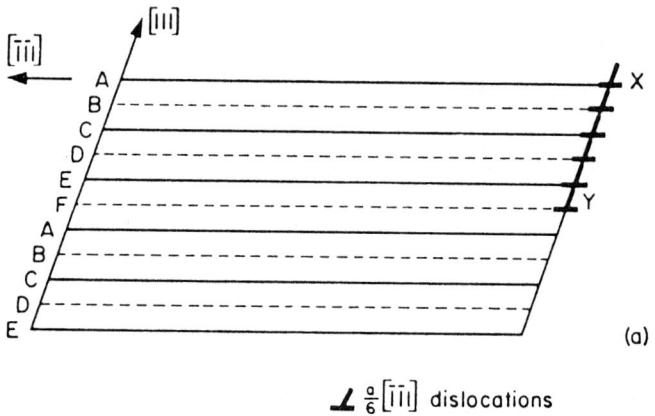

$$\perp \tfrac{a}{6}[\bar{1}\bar{1}1] \text{ dislocations}$$

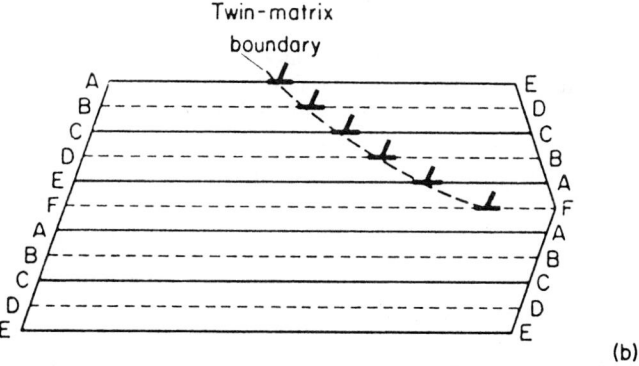

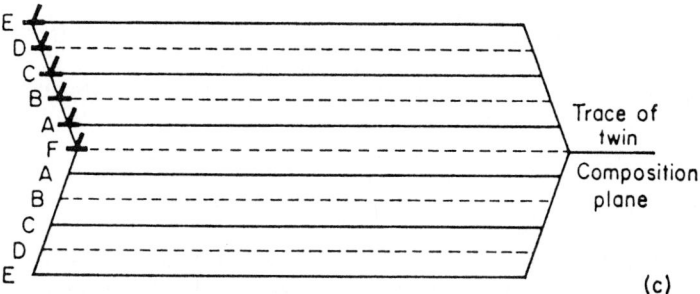

FIGURE 4.21 Schematic representation of twinning in a body-centered lattice. (a) Section parallel to ($1\bar{1}0$) showing the stacking sequence of (112) planes, X-Y is a row of $\tfrac{1}{6}$ [111] twinning dislocations, one dislocation on every (112) plane. (b) The twinning dislocations have moved halfway across the crystal to produce a twin-oriented region, changing the stacking sequence. (c) A twinned crystal.[78] (*Reprinted by permission of Pergamon Press, Plc.*)

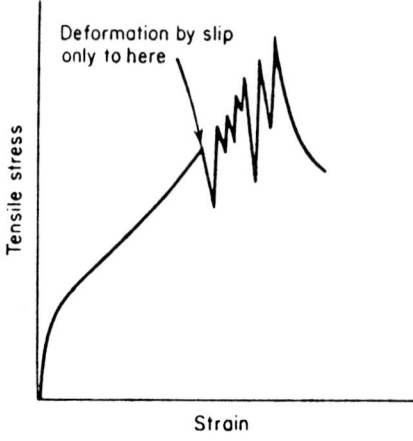

Deformation by slip
only to here

Tensile stress

Strain

FIGURE 4.22 Stress-strain curve in a single crystal showing load serrations due to twinning.[76] (*Reprinted by permission of PWS-Kent Publishing Co., Boston, Massachusetts; after Schmidt and Boas.*)

loading at room temperature). In the conventional (slow) deformation process, stress required to produce twinning is usually larger than that for slip at room temperature, and in this case slip precedes twinning. The stress-strain curve for iron may show the resultant serrated yielding due to twinning after the occurrence of plastic deformation by slip[74] (Fig. 4.22). However, the tendency to twin increases with falling temperature of deformation at normal or slow strain rate because critical shear stress for slip deformation increases, and then the deformation twinning will occur with ease at stresses much below that for slip. This is experimentally supported by the fact that twin formation has been found to occur at evidently lower stress than the flow stress for slip in Fe-Be alloys and pure iron at 4.2 K and Fe-10% Cr alloy at 112 and 147 K.[81] The addition of Be, Si, and Cr to iron causes deformation twinning to occur more prominently.

4.7.4 Twinning in FCC Materials

In fcc metals such as Ag, Cu, Al, and Ni, the twinning mode is [111]<112>. That is, the twin plane is {111} slip plane having the stacking sequence $ABCABC\ldots$, while the twinning direction is <112>. The twin formation changes the stacking sequence into

$$\downarrow$$
$$ABCABACBA$$
$$\uparrow$$

The arrows denote the twin boundary beyond which the crystal has opposite stacking to that in the untwinned portion.[24] In high-stacking fault energy (SFE) fcc pure metals such as pure Cu or Ni, deformation twinning can only occur at high stress levels (e.g., 150 MPa for Cu and 300 MPa for Ni), which are usually reached at low temperatures (or under heavy deformations). Many high-SFE metals such as Al-4.8% Mg, 6061-T6 Al alloy, and Cu-5 at% Ge crystals that do not display twinning during conventional low-strain deformation, easily twin under severe stress levels developed by shock deformation.[69] In these cases, a subdivision of a

FIGURE 4.23 Bundles in brass with 15% Zn rolled to 37% reduction viewed in longitudinal section (with rolling direction indicated).[83] (*Courtesy of E. Aernoudt, P. V. Houtte, and T. Leffers.*)

large fraction of grains by dense dislocation walls (DDWs)/first-generation microbands (MB1s) into cell blocks (CBs) has been observed. For Al, a specific macroscopic role of the DDWs/MB1s for the fraction of the mechanical anisotropy has been suggested.[82]

When fcc metals are highly alloyed (e.g., Fe-Ni-Co-Cr-Mo alloys), they develop low SFE, which gives rise to easy nucleation of fcc twins by the dissociation of perfect dislocations. Here the twins form in high density on a fine scale when heavy deformation is effected. These twins subdivide the matrix and are functionally equivalent to finer grain size in that they act as strong obstacles to dislocation motions.[80,82] In some cases, the microstructure (e.g., of brass with 15% Zn) is dominated by twin lamellae and bundles of twin lamellae, separated by matrix without twins (Fig. 4.23).[83] In Hadfield steel (12 wt% Mn, 1 wt% C steel), very high flow stress can be developed by the deformation at low temperatures. The occurrence of deformation twinning in compression, together with the continued subdivision of the original austenite matrix into finer domains by the twin lamellae and the high density of dislocations within the twin boundaries, gives rise to a high work-hardening rate.

For deformation (or annealing) twins, it has been shown and observed that coherent twin boundaries act as strong barriers to dislocations. Most of the extra hardening due to twinning is, therefore, attributed to the existence of coherent twin boundaries.[57]

4.7.5 Twinning in HCP Materials

In hcp metals, the pyramidal $[10\bar{1}2]$ planes are the most common twinning system. In hcp and other low-symmetry crystals, a simple shear will not satisfy the twin relationship, causing some atoms to be slightly out of atomic positions. In addition to the shear, it is necessary to *shuffle* local arrangements of some of the atomic positions or undergo secondary twinning, as in the case of hcp metals, to complete the twinning relationship.[69]

Research on several hcp metals has proved that the stress for twinning on $\{10\bar{1}2\}$, $\{11\bar{2}1\}$, and $\{11\bar{2}2\}$ planes increases with increasing temperature whereas that for $\{10\bar{1}1\}$ twinning decreases. The increase in stress for $\{11\bar{2}2\}$ twinning with increasing temperature may suggest that slip is a prerequisite for this twinning behavior in α-Ti.[69] In hcp metals and alloys such as Zr and Ti-6Al-4V, very high strain rates (by shock loading or explosive deformation) often lead to {11$\bar{2}$1} deformation twins.

4.8 DEFORMATION TEXTURE

When a polycrystalline metal is severely deformed by rolling, deep drawing, swaging, or wire drawing, certain crystallographic directions or planes of the majority of individual crystals or grains tend to rotate themselves in a non-random or preferred orientation with respect to the direction of deformation. This preferred (grain or crystal) orientation is usually called the *deformation texture* or *cold-worked texture*. Thus, texture introduces anisotropy in mechanical properties of metals and alloy sheet materials.[84] It produces peaking or earing (Fig. 4.24) and nonuniform wall thickness during deep drawing of aluminum alloy and steel sheets. Texture is of special importance in designing Ti and Zr alloys with reasonable drawability and makes a distinction between an acceptable and a completely unstable material.[85] The development of such structures is most conveniently characterized by the measurement of the *plastic strain ratio* or *plastic anisotropy factor r*, which is defined as the ratio of the true width (or lateral) strain ε_w to the true thickness strain ε_t (i.e., $r = \varepsilon_w/\varepsilon_t$) in a tensile test on the sheet-metal specimen. For isotropic sheet material tested in tension, $r = 1$; and for plastic anisotropy, $r > 1$. The r is a measure of capacity of a sheet to resist thinning. The higher the r value, the greater the resistance to thinning during deep drawing.[32] This behavior, due to plastic anisotropy in the rolling plane, is undesirable because it leads to frequent interruptions of production runs and causes material wastage. It is, therefore, necessary to control earing to tolerable minimum by introducing an appropriate production schedule.[86] In contrast, plastic anisotropy producing a high deformation resistance in thickness direction and less in the rolling plane is said to have a larger *limiting draw ratio* (LDR) and can be used advantageously in various aspects of formability.[87] Thus, LDR is defined as the largest ratio of blank-to-cup diameters that may be drawn successfully. The LDR has been reported to vary linearly with the r_m value, as shown in Fig. 4.25 for mild steel.[88,89]

In sheet metal forming technology, the drawability of the deep-drawn containers (e.g., oil filters and compressor housings)[32] can be predicted from the average normal anisotropy r_m (or $\bar{r}$) in the sheet, that is given by

$$r_m(\text{or } \bar{r}) = \frac{r_0 + 2r_{45} + r_{90}}{4} \qquad (4.52)$$

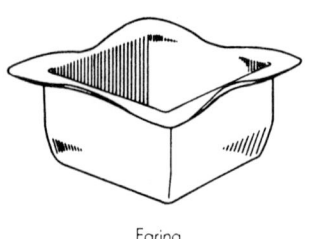

Earing   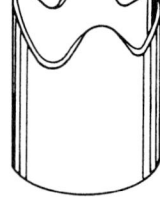 Directional earing

FIGURE 4.24 Typical earing.

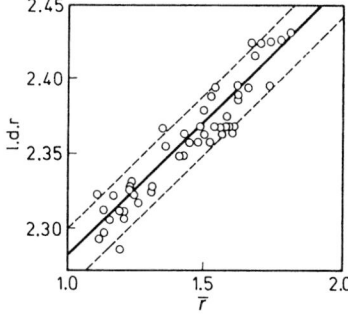

FIGURE 4.25 The deep drawability of mild steel, represented by the limiting draw ratio (LDR) as a function of the r_m (or $\bar{r}$) value for the material. (*After Atkinson and McLean, 1965.*)

The planar anisotropy which is responsible for the magnitude and position of ears in a deep-drawn cylindrical cup and to undesirable metal flow in the blank holder, in the general case, is usually defined by

$$\Delta r = \frac{r_0 + r_{90} - 2r_{45}}{2} \tag{4.53}$$

where r_0, r_{45}, and r_{90} are the experimentally measured plastic strain ratios in the rolling $0°$, $45°$, and transverse $90°$ directions in the sheet, respectively. When Δr is positive, the earing occurs in the $0°$ and $90°$ direction; when Δr is negative, the earing (i.e., uneven edges on cylindrical cups drawn from circular blanks)[32] occurs in the $45°$ direction. When $\Delta r = 0°$, the earing disappears. Thus, planar anisotropy Δr is bad for earing, particularly with beverage cans drawn from Al alloys because (1) it increases the machining or other processing to be done on the drawn product and (2) it means an azimuthal variation in the thickness distribution of the product, which necessitates a thicker gage than that would be necessary for isotropic strains. *Orientation distribution functions* (ODFs) can be used to predict earing behavior; however, they can also be determined by a deep-drawing test.

An r_m value of 1 denotes complete isotropy or equal flow strength, in the plane of the sheet and in the thickness direction. A large r_m value improves LDR, which indicates the high average depth (or wall height) and a high through-thickness strength relative to the strength in the plane of the sheet. A high r_m value also improves the dent resistance property of formed panels.[90–92] Rimmed and semikilled steels have r_m values close to unity (1.0 to 1.3) due to their weak texture and are, therefore, not suited for the most demanding deep-drawing applications. In reality, a low r_m value implies limited drawability. On the other hand, aluminum-killed steels with strong {111} <100> textures have r_m values in the range of 1.5 to 1.8. The Ti- or Nb-stabilized steels with strong {111} and {554} orientations have r_m values approaching or exceeding 2.0. There is a direct relationship between r_m value and the grain size of steel, as shown in Fig. 4.26.

It is important to point out that the press-forming processes used in the industry to form automobile bodies, domestic consumer durables, beverage cans, and so forth can be considered as specific forms of deep drawing; therefore, a close control of r values affects a large sector of manufacturing industry. A combination of a high r_m value and low Δr value gives optimum drawability. Table 4.7 lists the main texture components observed in the cold rolling and annealing textures of low-carbon and extra-low-carbon (ELC) steels, along with the calculated average r_m and Δr values, relating to each texture component.[93,94]

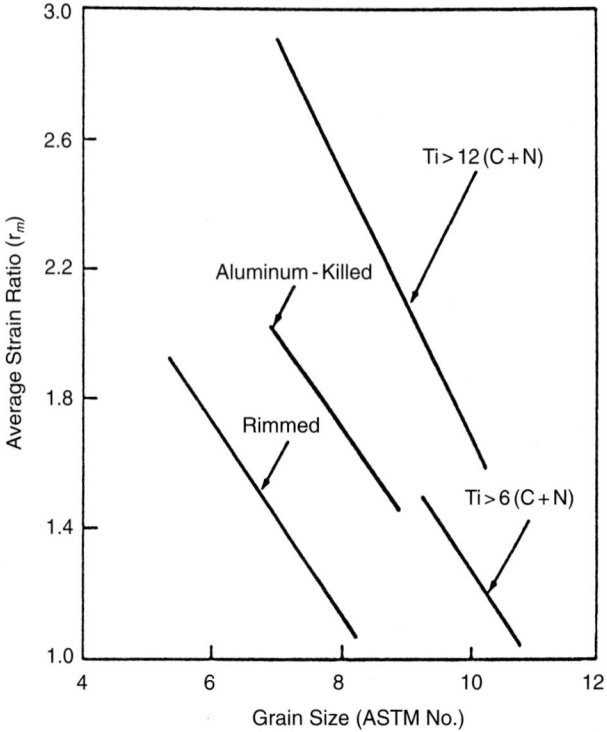

FIGURE 4.26 Variation of r_m with the grain size of steel.[7] The normal range of conventional rimmed and killed steel is shown. In addition to chemical composition, the value r_m is determined by coil processing. Two Ti-treated Al-killed steels are also shown. To produce an interstitial-free steel with a high r_m value, the amount of Ti present must exceed 6 times the amount of C + N.

TABLE 4.7 Major Components in Cold-Rolling and Annealing Textures of Low-Carbon Steels

Texture Component	r_m	Δr
{001} ⟨110⟩	0.4	−0.8
{112} ⟨110⟩	2.1	−2.7
{111} ⟨110⟩	2.6	0
{111} ⟨112⟩	2.6	0
{554} ⟨225⟩	2.6	1.1
{110} ⟨001⟩	5.1	8.9

Source: After D. Daniel and J. J. Jonas, *Met. Trans.*, vol. 21A, 1990, p. 331.

Preferred orientations are determined by X-ray methods. The X-ray pattern of a fine-grained randomly oriented material will show uniformly intense rings for all angles, whereas that of the preferred oriented grains will show breaking up of rings into short arcs or spots. The textures are usually represented on a stereographic projection by means of *pole figures*. Actually, they are two-dimensional plots of three-dimensional distributions of crystal orientations and do not, therefore, describe the true texture. In this two-dimensional plot, the angular relationships between planes in the crystal aggregate are maintained. There are two (e.g., normal and inverse) pole figure methods that provide a description of the texture of a sheet material.[95] In the former, X-ray data are plotted directly on the stereographic projection where the center of projection corresponds to the normal to the sheet plane in the rolling or in the wire axis. The latter represents the density distribution of an important direction in the polycrystalline specimen (such as wire axis or the sheet rolling direction) drawn on a stereographic projection of the lattice in standard orientation.[95] This usually illustrates the intensities of a certain set of planes such as {100}, {110}, or {111}. The plane of projection is the plane of sheet where the *rolling direction* (RD) and the *transverse direction* (TD) lie. Recently, mathematical methods have been developed for the calculation of the three-dimensional ODF from the numerical data obtained from several pole figures,[93] but they do not contribute much to the understanding of the theory of texture.[96] Precht et al. have developed a simple method for the prediction and control of texture and associated earing in rolled 3004 aluminum which is based on the measurement of the X-ray diffraction peak height ratios of the {110} planes.[97]

Deformation texture of a metal depends on the stacking fault energy, the extent, the type, the operative mechanisms, and the temperature of deformation processes. As the degree of deformation is increased, the texture improves toward perfection. In polycrystalline materials, deformation texture can be broadly grouped into fiber texture and rolling texture.

4.8.1 Classification of Texture

For practical purposes, polycrystalline texture can be grouped into two types, according to the symmetry of the preferred orientations in the specimen. When the crystals are aligned along one crystallographic direction <*uvw*>, parallel to the major axis of the specimen, and around this common axis the crystals are in random orientation, the texture is termed *fiber texture*. It is denoted by <uvw>, the fiber axis. In a rolled sheet or strip, the texture exhibits symmetry with the three orthogonal directions of the product such as the rolling direction (RD), the transverse direction (TD), and the normal direction (ND). Thus the rolling texture of such samples needs two parameters, {*hkl*}<*uvw*>, for its description.[98]

4.8.1.1 Fiber Texture. This type of texture occurs by uniaxial deformation methods (e.g., wire drawing, extrusion, etc.) in which the grains are elongated in the forming direction. This texture is characterized by a crystallographic direction <*uvw*> of low indices parallel to the wire axis, for example, fiber texture of wire-drawn bcc metals in the <110> direction parallel to the wire axis.[99] However, cold-drawn wires of fcc metals may have a double-fiber texture such as <111> + <100> directions, in varying degree, parallel to the wire axis. The <111> becomes predominant in metals with high SFE (which causes easy cross-slip), while those with low SFE (by the alloying additions) have a predominantly <100> texture containing some <111> direction. In hcp metals (e.g., Mg), basal planes with <10$\bar{1}$0> direction rotate and coincide with the wire axis.

4.8.1.2 Rolling Texture. A rolling texture {*hkl*}<*uvw*> describes that the crystallographic planes {*hkl*} are parallel to the rolling plane, and that the particular crystallographic direction <*uvw*> in that rolling plane is parallel to the rolling direction.[98] In fcc metals and alloys, two rolling textures predominate. Generally, {110} <112> texture (α-brass-type texture) is developed in the early stage of deformation, whereas [112] <111> texture (copper-type texture) becomes effective with

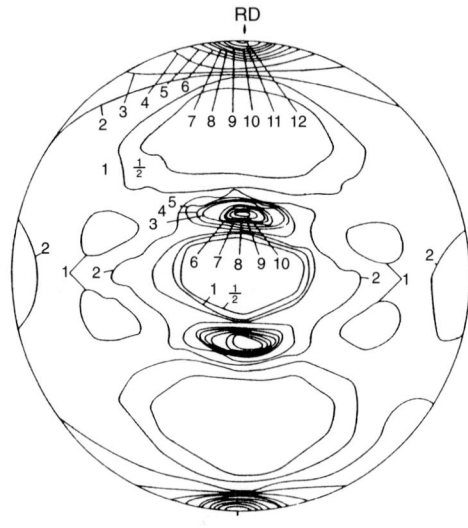

FIGURE 4.27 A {110} pole figure of rolling texture of iron (90% reduction in one direction only after quenching from 925°C). A pole density of unit represents randomness.[21] (*Reprinted with permission from Trans. AIME, vol. 221, 1961, p. 764, a publication of The Metallurgical Society, Warrendale, Pennsylvania.*)

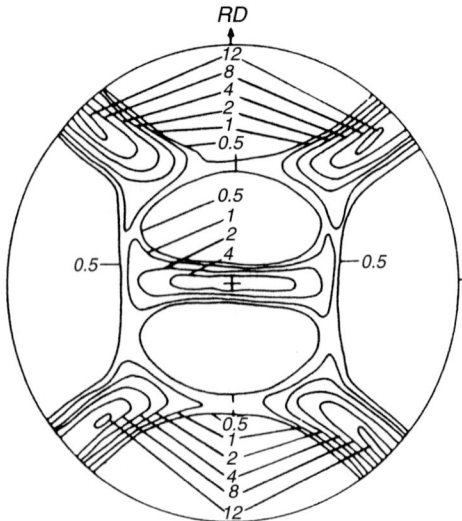

FIGURE 4.28 A {100} pole figure of the rolling texture of carbonyl iron (99.7% deformed).[96] (*Reprinted by permission of Dr. Riederer Verlag, GmbH, Stuttgart; after Liesner and Wahl.*)

heavy plastic deformation in which extensive cross-slip occurs. Type of texture is also dependent on the SFE, deformation temperature, and grain size; for example, high SFE and high temperature of deformation produce the copper-type texture {112} <111>, and lower-SFE material gives rise to α-brass-type texture.

In bcc metals the predominant texture is {001} <110> (cube-on-edge) with cube planes in the rolling plane; but other texture components such as {112} <110> and {111} <112> are also found (Table 4.7). Figure 4.27[20] shows the {110} pole figure of a 90% cold-rolled iron sheet, whereas Fig. 4.28[96] shows the {100} pole figure of the 99.7% (deformed) rolling texture of a carbonyl iron. HCP metals tend to have the basal plane oriented parallel to both the close-packed <11$\bar{2}$0> and the rolling directions.

Note that the texture of the tubes is a function of the relative reductions of the wall thicknesses and the tube diameter in their fabrication. When the wall thickness and the tube diameter are reduced proportionally, the texture resembles that of a wire or rod. If only the wall thickness is reduced, the texture is similar to that of a rolled sheet.[98]

REFERENCES

1. *Metals Handbook*, 9th ed., vol. 8, American Society for Metals, Metals Park, Ohio, 1985.

2. *Making, Shaping and Treating of Steels*, 9th ed., ed. H. E. McGannon, United States Steel Corp., Pittsburgh, Pa., 1971.

3. *Carbon and Alloy Steels*, ASM International, Materials Park, Ohio, 1996; B. Taylor, *Metals Handbook*, vol. 14, 9th ed., ASM, Metals Park, Ohio, 1988, pp. 877–899.

4. D. J. Mack, *Trans. AIME*, vol. 166, 1946, pp. 68–85.

5. C. S. Barrett, W. D. Nix, and A. S. Tetelman, *The Principles of Engineering Materials*, Prentice-Hall, Englewood Cliffs, N.J., 1973.

6. J. Gurland and R. J. Asaro, in *Encyclopedia of Materials Science and Engineering*, Pergamon Press, Oxford, 1986, pp. 1258–1260.

7. *Sheet Metal Formability*, American Iron and Steel Institute, Washington, D.C., 1984.

8. G. E. Dieter, *Mechanical Metallurgy*, 3d ed., McGraw-Hill, New York, 1986.

9. S. Kalpakjian, in *Encyclopedia of Materials Science and Engineering*, Pergamon Press, Oxford, 1986, pp. 4378–4381.

10. A. Hamano, *Met. Trans.*, vol. 24A, 1993, pp. 127–139.

11. A. Blake, *Practical Fracture Mechanics in Design*, Marcel Dekker, New York, 1996.

12. F. A. A. Crane and J. A. Charles, *Selection and Use of Engineering Materials*, Butterworths, London, 1987.

13. B. Derby, in *The Encyclopedia of Advanced Materials*, Pergamon Press, Oxford, 1994, pp. 295–299.

13a. A. W. Thompson and J. F. Knott, *Met. Trans.*, vol. 24A, 1993, pp. 523–534.

14. M. V. Klassen-Neklyudova, *Mechanical Twinning of Crystals*, Consultant Bureau, New York, 1964.

15. A. S. Argon, in *Physical Metallurgy*, 4th ed., eds. R. W. Cahn and P. Haasen, Elsevier Science B.V., Amsterdam, 1996, pp. 1877–1955.

16. N. Brown and R. A. Ekvall, *Acta Metall.*, vol. 10, 1962, pp. 1101–1107.

17. W. D. Brentnahl and W. Rostoker, *Acta Metall.*, vol. 13, 1965, pp. 187–198.

18. A. Abel and H. Munir, *Acta Metall.*, vol. 21, 1973, pp. 99–105.

19. L. P. Kubin, *Hardening of Metals*, ed. P. Feltham, Freund, Israel, 1980, pp. 67–112.

20. H. Jones, private communication, 1998.

21. W. C. Leslie, *Physical Metallurgy of Steels*, McGraw-Hill, New York, 1981.

22. M. Bernstein, *Acta Metall.*, vol. 17, 1969, pp. 249–259.

23. H. Conrad, in *Iron and Its Dilute Solid Solutions*, eds. C. W. Spencer and F. E. Werner, Interscience, New York, 1963, p. 315.

24. W. G. Johnston and J. J. Gilman, *J. Appl. Phys.*, vol. 30, 1959, p. 129.

25. W. C. Leslie, J. T. Michalak, and F. W. Aul, in *Iron and Its Dilute Solid Solutions*, eds. C. W. Spencer and F. E. Werner, Interscience, New York, 1963, p. 119–212.

26. R. W. K. Honeycombe, *The Plastic Deformation of Metals*, 2d ed., Edward Arnold, London, 1984.

27. G. E. Dieter, *Metals Handbook*, vol. 14, 9th ed., ASM International, Metals Park, Ohio, 1988, pp. 20–27.

28. B. W. Christ and M. L. Picklesimer, *Acta Metall.*, vol. 22, 1974, pp. 435–447.

29. H. Conrad, *Acta Metall.*, vol. 11, 1963, pp. 75–77.

30. M. Abe, in *Constitution and Properties of Steel*, vol. 7, vol. ed. F. B. Pickering, VCH, Weinheim, 1992, pp. 285–333.

31. *Handbook of Metal Forming*, ed. K. Lange, McGraw-Hill, New York, 1985.

32. B. L. Bramfitt, in *Encyclopedia of Materials Science and Engineering*, Pergamon Press, Oxford, 1986, pp. 4897–4899.

33. C-Y. Sa, in *Sheet Stamping Technology: Applications and Impact*, SP-779, Society of Automotive Engineers, Warrendale, Pa., 1989, pp. 37–45.

34. A. K. Ghosh, S. S. Hecker, and S. P. Keeler, in *Workability Testing Techniques*, ed. G. E. Dieter, American Society for Metals, Metals Park, Ohio, 1984, pp. 135–195.

35. G. I. Taylor, *Proc. R. Soc., London*, vol. 89, 1934, p. 660.

36. E. Orowan, *Dislocations in Metals*, American Institute of Mining, Metallurgical, and Petroleum Engineers, New York, 1954, p. 69.

37. S. J. Basinski and Z. S. Basinski, in *Dislocations in Solids*, vol. 4, North-Holland Physics Publishing, Amsterdam, 1979, p. 263.

38. V. I. Startsev, in *Dislocations in Solids*, vol. 6, ed. F. R. N. Nabarro, North-Holland Physics Publishing, Amsterdam, 1979, p. 145.

39. D. J. Dingley and D. McLean, *Acta Metall.*, vol. 15, 1967, pp. 885–901.

40. F. B. Pickering, *Physical Metallurgy and Design of Steels*, Applied Science Publishers, London, 1978.

41. D. Kuhlmann-Wilsdorf, *Met. Trans.*, vol. 16A, 1985, pp. 2091–2108.

42. H. J. McQueen, *Met. Trans.*, vol. 8A, 1977, pp. 807–824.

43. J. R. Low, in *Iron and Its Dilute Solid Solutions*, eds. C. W. Spencer and F. E. Werner, Interscience, New York, 1963, pp. 217–263.

44. A. Cracknell and N. J. Petch, *Acta Metall.*, vol. 3, 1955, pp. 186–189.

45. E. O. Hall, *Yield Point Phenomena in Metals and Alloys*, Macmillan, London, 1970.

46. N. Hansen, *Met. Trans.*, vol. 16A, 1985, pp. 2167–2189.

47. A. W. Thompson, M. Baskes, and W. F. Flangan, *Acta Metall.*, vol. 21, 1973, pp. 1017–1028.

48. R. W. Armstrong, in *Yield, Flow and Fracture of Polycrystals*, ed. T. N. Baker, Applied Science Publishers, London, 1983, pp. 1–31.

49. R. A. Masumura, P. M. Hazzledine, and C. S. Pande, *Acta Mater.*, vol. 46, no. 13, 1998, pp. 4527–4534.

50. Ref. 19 in ref. 49.

51. A. H. Choksi, A. Rosen, J. Rarch, and H. Gleiter, *Scripta Metall.*, vol. 23, 1989, p. 1679.

52. J. Meakin and N. J. Petch, *Role of Substructure in Mechanical Behavior of Metals*, Report ASD-TDR-63-324, U.S. Air Force Wright-Patterson AFB, Ohio, 1963, pp. 243–251.

53. M. F. Ashby, *Phil. Mag.*, vol. 21, 1970, pp. 399–424.

54. M. F. Ashby, *Strengthening Methods in Crystal*, eds. A. Kelly and R. B. Nicholson, Wiley, New York, 1971, pp. 137–192.

55. E. T. Wessell, *Trans. AIME*, vol. 209, 1957, p. 930.

56. J. S. Blackmore and E. O. Hall, *JISI*, vol. 204, 1966, p. 817.

57. M. V. Bever, D. L. Holt, and A. L. Titchener, *The Stored Energy of Cold Work*, in *Progress in Materials Science*, vol. 17, Pergamon Press, Elmsford, N.Y., 1973.

58. D. V. Wilson, in *Encyclopedia of Materials Science and Engineering*, suppl. vol. 1, Pergamon Press, Oxford, 1990, pp. 508–512.

59. W. C. Leslie, in *Encyclopedia of Materials Science and Engineering*, Pergamon Press, Oxford, 1986, pp. 4007–4011.

60. *Carbon Steel, Wire and Rod*, Iron and Steel Institute, Warrendale, Pa., 1993.

60a. V. T. L. Buono, M. S. Andrade, and B. M. Gonzalez, *Met. Trans.*, vol. 29A, 1998, pp. 1415–1423.

60b. R. H. U. Queiroz, E. J. Foneseca, V. T. L. Buono, M. S. Andrade, E. M. PeSilva, and B. M. Gonzalez, *Wire J. Int.*, June 1999, pp. 76–81.

61. W. C. Leslie and E. Hornbogen, in *Physical Metallurgy*, 4th ed., eds. R. W. Cahn and P. Haasen, Elsevier Science B.V., Amsterdam, 1996, pp. 1555–1620.

62. V. T. L. Buono, G. M. Gonzalez, and M. S. Andrade, *Scripta Metall.*, vol. 38, no. 2, 1998, pp. 185–190.

63. *Carbon and Alloy Steels*, ASM Specialty Handbook, ASM International, Materials Park, Ohio, 1996, pp. 308–328.

64. P. Rodriguez, in *Encyclopedia of Materials Science and Engineering*, suppl. vol. 1, ed. R. W. Cahn, Pergamon Press, Oxford, 1990, pp. 504–508.

64a. A. Fjeldly and H. J. Roven, *Met. and Mats. Trans.*, vol. 31A, 2000, pp. 669–678.

65. J. M. Robinson and M. P. Shaw, *Int. Mater. Rev.*, vol. 39, no. 3, 1994, pp. 113–122.

66. N. J. Petch, in *Advances in Physical Metallurgy*, eds. J. A. Charles and G. C. Smith, The Institute of Metals, London, 1990, pp. 11–25.

67. Q. H. Tang and T. C. Wang, *Acta Mater.*, vol. 46, no. 15, 1998, pp. 5313–5321.

68. J. W. Christian and S. Mahajan, *Progr. Mat. Sci.*, vol. 39, 1995, pp. 1–157.

69. G. T. Gray, III, in *Encyclopedia of Materials Science and Engineering*, suppl. vol. 2, ed. R. W. Cahn, Pergamon Press, Oxford, 1990, pp. 859–866.

70. L. Remy, *Met. Trans.*, vol. 12A, 1981, p. 847.

71. E. O. Hall, *Twinning*, Butterworths, London, 1954.

72. P. B. Partridge, *Int. Metall. Rev.*, vol. 12, 1967, p. 169.

73. F. J. Turner, in *Deformation Twinning*, eds. R. E. Reed-Hill, J. P. Hirth, and H. C. Rogers, Gordon and Breach, New York, 1964.

74. T. L. Altshuler and J. W. Christian, *Acta Metall.*, vol. 14, 1968, pp. 903–908.

75. J. D. Verhoeven, *Fundamentals of Physical Metallurgy*, Wiley, New York, 1975.

76. R. E. Reed-Hill and R. Abbaschian, *Physical Metallurgy Principles*, 3d ed., PWS-Kent Publishing Co., Boston, 1992.

77. R. W. Hertzberg, *Deformation and Fracture Mechanics of Engineering Materials*, 3d ed., Wiley, New York, 1989.

78. D. Hull, *Introduction to Dislocations*, Pergamon Press, Oxford, 1975.

79. A. Crocker, in *Twinning in Advanced Materials*, eds. M. R. Yoo and M. Wuttig, The Metallurgical Society, Warrendale, Pa., 1994, pp. 3–16.

80. P. Haasen, *Physical Metallurgy*, 2d ed., Cambridge University Press, New York, 1986.

81. M. J. Kelley and N. S. Stoloff, *Met. Trans.*, vol. 7A, 1976, pp. 331–333.

82. D. Juul Jensen and N. Hansen, *Acta Metall. Mater.*, vol. 38, 1990, p. 1369.

83. E. Aernoudt, P. V. Houtte, and T. Leffers, in *Plastic Deformation and Fracture of Materials*, vol. 6, ed. M. Mughrabi, VCH, Weinheim, 1993, pp. 89–136.

84. H. J. Prask and C. S. Choi, *Encyclopedia of Materials Science and Engineering*, Pergamon Press, Oxford, 1986, pp. 4895–4897.

85. C. Tome, A. Pochettino, and R. Penelle, in *ICOTOM—Eighth International Conf.: Texture of Materials*, The Metallurgical Society, Warrendale, Pa., 1988, pp. 985–990.

86. P. M. B. Rodrigues and P. S. Bate, in *Textures in Non-ferrous Metals and Alloys*, eds. H. D. Merchant and J. G. Morris, The Metallurgical Society-American Institute of Mining, Metallurgical, and Petroleum Engineers, Warrendale, Pa., 1985, pp. 173–187.

87. R. Sowerby, in *Textures in Non-ferrous Metals and Alloys*, eds. H. D. Merchant and J. G. Morris, The Metallurgical Society-American Institute of Mining, Metallurgical, and Petroleum Engineers, Warrendale, Pa., 1985, pp. 99–115.

88. S. P. Keeler, in *Automotive Sheet Metal Formability*, Final Report on AISI Project no. 1201-409, American Iron and Steel Institute, Washington, D.C., 1989.

89. R. W. Cahn, in *Processing of Metals and Alloys*, vol. 15, ed. R. W. Cahn, VCH, Weinheim, 1991, pp. 429–480.

90. W. B. Hutchinson, *Int. Metall. Rev.*, vol. 29, no. 1, 1984, pp. 25–42.

91. K. Yoshida et al., *Proceedings of the 8th Biennial IDDRG Congress*, Gothenburg, 1974, International Deep Drawing Research Group, pp. 258–268.

92. A. J. Klein and E. W. Hitchler, *Metall. Eng. Q.*, vol. 13, 1973, pp. 25–27.

93. R. K. Ray, J. J. Jonas, and R. E. Hook, *Int. Mats. Rev.*, vol. 39, no. 4, 1994, pp. 129–172.

94. D. Daniel and J. J. Jonas, *Met. Trans.*, vol. 21A, 1990, p. 331.

95. C. S. Barrett and T. B. Massalski, *Structure of Metals*, 3d ed., McGraw-Hill, New York, 1966.

96. J. Grewen and J. Huber, in *Recrystallization of Metallic Materials*, ed. F. Haessner, Dr. Riederer Verlag GmbH, Stuttgart, 1978, pp. 111–136.

97. W. Precht, L. Christodoulou, and F. Lockwood, in *Textures in Non-ferrous Metals and Alloys*, eds. H. D. Merchant and J. G. Morris, The Metallurgical Society-American Institute of Mining, Metallurgical, and Petroleum Engineers, Warrendale, Pa., 1985, pp. 223–227.

98. H. Hu, *The Encyclopedia of Advanced Materials*, vol. 4, Pergamon Press, Oxford, 1994, pp. 2798–2807.

99. D. V. Wilson, *J. Inst. Metall.*, vol. 94, 1966, pp. 84–93.

CHAPTER 5
RECOVERY, RECRYSTALLIZATION, AND GRAIN GROWTH

5.1 INTRODUCTION

Annealing behavior of cold-worked materials has been of considerable interest for many years from both a theoretical and a practical standpoint. This is a widely accepted industrial practice which is broadly divided, based on the effects of temperature, into three main stages of the process: recovery, recrystallization, and grain growth. Usually the term *recrystallization* is synonymous with *primary recrystallization*. *Grain growth* can be subdivided into *normal* or *continuous grain growth* and *abnormal* or *discontinuous grain growth*, also called *secondary recrystallization*. These classifications are approximate because overlapping of some stages is a common phenomenon. This chapter deals with the property changes, mechanisms, kinetics, and commercial importance of these procedures.

5.2 RELEASE OF STORED ENERGY

When the temperature of a cold-worked metal is raised sufficiently, the stored energy of cold work is released as heat[1] and is usually measured in a highly sensitive differential calorimeter as the difference in specific heat (usually called *power difference* ΔP) between the cold-worked and the undeformed (or annealed) specimens. Such measurements are carried out by anisothermal, isothermal, or isochronal annealing methods. These methods determine the total amount of stored energy as well as the kinetics of release of energy. A knowledge of kinetics of the energy release is vital in understanding recovery and recrystallization processes and in interpreting the mechanism of stored energy.

In the anisothermal annealing method, the temperatures of cold-worked and annealed specimens are raised at equal and constant rates, and the power difference ΔP between the two specimens is determined as function of temperature (Figs. 5.1 and 5.2). The area below the curve is proportional to the amount of stored energy released. Generally the power-difference curves exhibit single or multiple peaks (Fig. 5.1), each representing a surge in the stored energy released. The peak occurring at the higher and lower temperatures denotes the recrystallization and

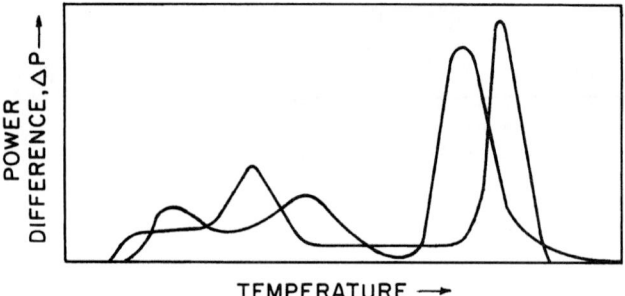

FIGURE 5.1   Schematic representation of the rate of energy release as a function of annealing time.

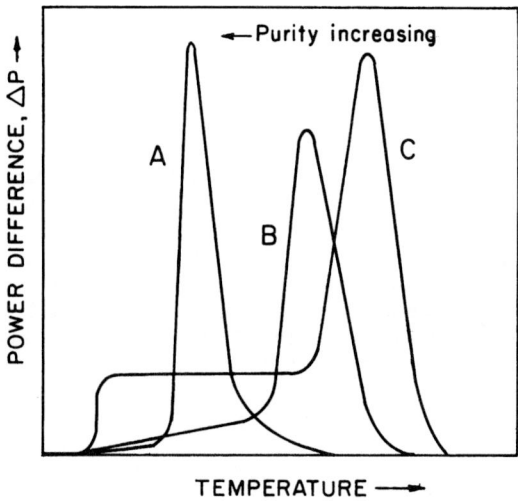

FIGURE 5.2   Schematic plot of the rate of energy release as a function of annealing temperature for three different purities.

recovery processes, respectively. However, there are some cases where recovery occurs without any surge in the energy release.

Figure 5.2 shows the schematic illustration of three different patterns of energy release curves as a function of temperature as obtained by the anisothermal annealing method. Curve A, which is for the highest-purity metal, represents (1) a very small amount of energy released for recovery and (2) the lowest recrystallization temperature, as measured by a main (i.e., large) power peak. A slightly impure metal is represented by curve B, where a small but appreciable amount of energy is released during recovery. This is obviously clear from the existence of a shoulder prior to the main peak. Metal with increased impurity is represented by curve C, in which it appears that nearly one-half of the stored energy is released preceding the main peak. These patterns of energy release curves have been experimentally observed for copper and silver with varying purity. Thus we can conclude here that

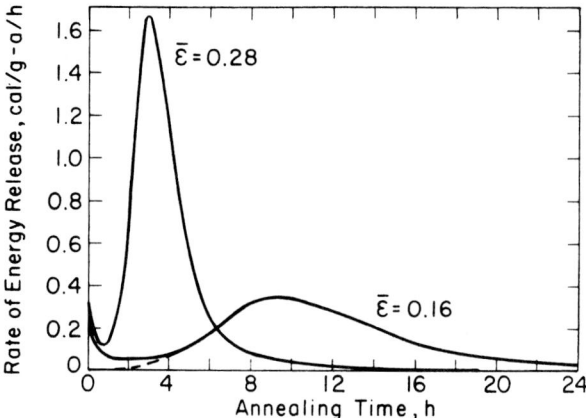

FIGURE 5.3 The rate of energy release as a function of annealing time for 99.999% copper deformed by extension of two strains. In each case the isothermal annealing temperature was 189.7°C.[1,2] (*Reprinted by permission of The Metallurgical Society, Warrendale, Pennsylvania.*)

increasing the impurity content in the metal raises not only the recrystallization temperature but also the fraction of the stored energy $E_{recovery}/E_{total}$ released during recovery.

As in the anisothermal annealing method, in the isothermal annealing method the power difference ΔP between the two specimens is plotted against the (isothermal) annealing time (Fig. 5.3).[2] The area under the curve is proportional to the amount of energy released. In the isochronal annealing method, the power difference is plotted against different annealing temperatures for equal times. This method provides the data for the amount of stored energy released as a function of temperature and does not render any information concerning the kinetics of the released energy during individual annealing treatment. Hence, the resolution obtained by this method is poor.

Figure 5.4a shows the schematic representation of a change in mechanical properties (e.g., hardness, strength, ductility, etc.) and other physical properties (e.g., electrical resistivity, density, cell size, etc.) occurring during the annealing process of plastically deformed metal at room temperature.

5.3 RECOVERY

Recovery embraces all structural changes that do not involve the sweeping of the deformed structure by migrating high-angle boundaries.[3] Alternatively, recovery is defined as the restoration of physical properties of the cold-worked metal without any observable change in microstructure. The driving force for this process is a decrease in the thermodynamic (or stored) energy differences between the deformed state and the recovered structure. It involves the annihilation of dislocations and the rearrangement of dislocations into lower-energy configurations.[4]

Recovery thus produces changes in structures and properties prior to the onset of new strain-free recrystallized grains. More strain-free regions that form during

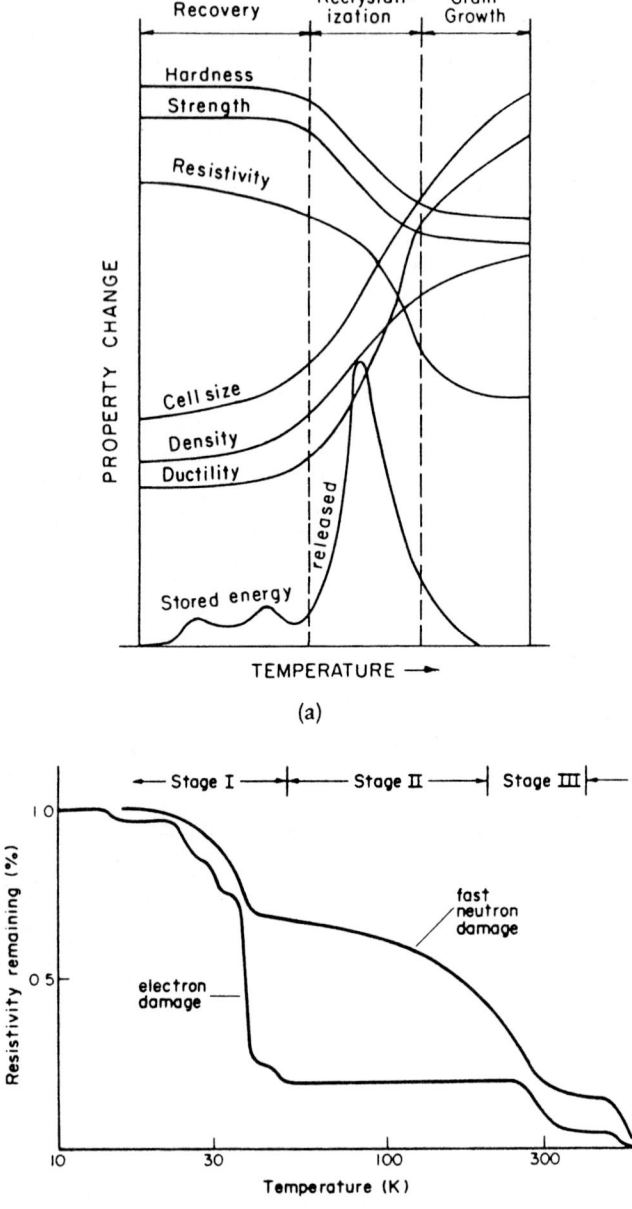

FIGURE 5.4 (a) Changes in several physical and mechanical properties of plastically deformed metal as a function of temperature. (b) Fractional recovery of defects introduced by electron or neutron bombardment of liquid helium temperature.[9] [(b) *Reprinted by permission of Pergamon Press, Plc.*]

5.4

recovery can act as the origin or nucleus for subsequent recrystallization annealing. Recovery is often measured directly by calorimetry from the changes in stored energy which are achieved by glide, climb, and cross-slip of dislocations because stored energy is directly related to the number and configuration of dislocations in the material. The calorimetric measurements can be made with no phase transformation over the temperature range of experiment. Recovery is measured indirectly by following the changes in mechanical or physical property.[4] The recovered substructure is usually weaker—but more ductile, more stable for high-temperature service, and more corrosion-resistant—than that formed by cold working.[5] A special type of recovery commonly encountered in heat treatment is called *stress relief annealing*, which is often desirable and is used to retain much of the increased strength resulting from cold work and to remove the residual stresses produced by cold-working operations.

5.3.1 Mechanical Properties

The extent of recovery depends on (1) the kinetics of recovery itself, (2) the temperature or time at which recrystallization occurs, and (3) the nature of metallic materials. In general, complete recovery can occur, particularly in lightly deformed materials. The larger the deformation, the smaller the fraction of the property (e.g., hardness) change on deformation which can be restored. This is presumably due to the greater ease of recrystallization at increased deformation[6] (see also Fig. 5.10). However, there is an exception: Large deformation in single crystals of hexagonal metals that is produced by easy glide can be completely recovered to the original structure and properties on annealing.[6]

Based on the changes in mechanical properties upon recovery, metallic materials are subdivided into two groups.[7] Metals with low *stacking fault energy* (SFE), denoted by γ_{SFE} (e.g., Cu, Ni, α-brass, and austenitic stainless steel), usually show very little recovery in mechanical properties (e.g., hardness and strength) prior to recrystallization. This is presumably due to the occurrence of little climb and consequently little recovery or rearrangement of dislocation structure. Since hardness and strength (or flow stress) are controlled by concentration and disposition of dislocations, they are not affected at the recovery temperature. However, metals with high SFE (e.g., α-iron, Al, etc.) usually show an appreciable amount of softening during a recovery anneal. This is attributed to the fact that the dislocation rearrangement can easily occur by the rapid climb process within the deformed metal during recovery anneal.[7,8] Solutes may influence recovery by their effect on the SFE, by pinning dislocations, or by altering the concentration and mobility of vacancies.[4] However, a large drop in hardness and strength occurs during recrystallization because of the presence of strain-free new recrystallized grains.

5.3.2 Physical Properties

During plastic deformation, concentration of lattice vacancies and dislocation density increase. As a result, the electrical resistivity is increased and density is decreased. During annealing, the resistivity continues to decrease and finally approach toward the original (annealed) value, together with the change in the stored energy released, increase in density, and considerable reduction of lattice strains. In contrast to normal deformation, deformation caused by high-energy electron or fast-neutron irradiation in nuclear reactors involves a complex mechanism;

when such a damaged material is annealed, the resistivity decreases in a distinct manner, different from that due to cold work (Fig. 5.4b).[3,8,9] The study of damage recovery provides the general understanding of lattice imperfections in metals and the physics of defect reactions occurring at higher-temperature irradiations. This knowledge is of technological significance to both fusion and fission reactor systems.[9]

The drop in resistivity, increase in density, and reduction in lattice strain during recovery anneal give an indication of a significant decrease of concentration of point defects. Thus the properties that are most influenced by recovery are those that are sensitive to point defects.

5.3.3 Microstructure (Cell Size)

During recovery anneal, the cell walls or boundaries sharpen, and then the cells gradually grow to a large size while the interior dislocations are further attracted into them. However, the sharpness of such cell walls occurring during annealing is influenced by the sharpness of the original cell walls formed during plastic deformation, which, in turn, depends on its SFE, that is, the extent of dissociation of dislocations and their ability to climb. Thus higher and lower SFE of the metal gives rise to well-defined and ill-defined cell structures, respectively, after moderate or heavy plastic deformation.[8]

5.3.4 Mechanism

Recovery annealing is a thermally activated process, which is reflected by the fact that its operative mechanism at low-temperature, intermediate-temperature, and high-temperature range involves vacancy motion, dislocation motion without climb, and dislocation motion with climb, respectively.

5.3.5 Rearrangement of Dislocations into Stable Arrays

5.3.5.1 Polygonization, Subgrain Formation, and Subgrain Growth. An especially simple form of recovery process occurring at higher (recovery) temperature is called *polygonization*. It is associated with single crystals that have been slightly (plastically) bent (about an axis parallel to their active slip planes) so that only one glide system operates and is subsequently annealed. After annealing, the curved crystal breaks up into small (single or perfect) crystal segments (called *subgrains*), each retaining the local orientation of the original bent crystal and separated by dislocation walls known as *subboundaries* (or *tilt boundaries*) that are right angles to the glide vector of the active slip plane.[8,10] These represent a simple type of low-angle grain boundary.

When an X-ray beam is diffracted by such bent crystals, Laue photographs with elongated or asterated spots are observed. When the X-ray beam is allowed to fall on the annealed specimen, elongated spots of the bent crystal are replaced by a series of tiny, sharp spots, which is a characteristic of the polygonized structure (Fig. 5.5).[11]

The sub-boundaries (or polygon walls) present in the polygonized structures can be revealed by etching; these sub-boundaries appear as dense rows of pits in the etched photograph (Fig. 5.6).[12] To understand this, let us consider the polygoniza-

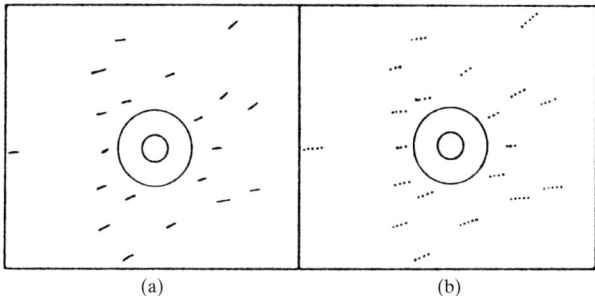

(a)   (b)

FIGURE 5.5   Schematic Laue pattern showing (*a*) elongated spot obtained from a bent single crystal and (*b*) breaking up of elongated spots into a series of small sharp spots after polygonization.[11] (*Reprinted by permission of PWS-Kent Publishing Co., Boston, Massachusetts.*)

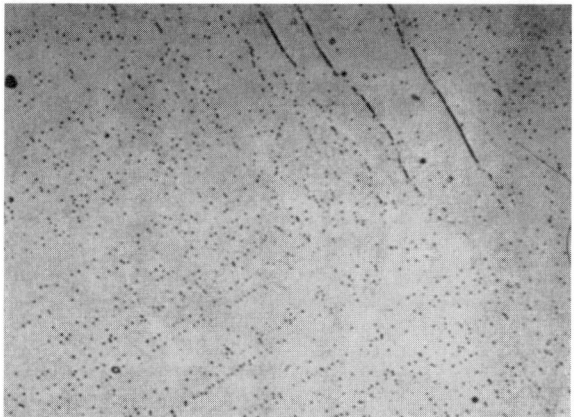

FIGURE 5.6   Complete polygonized structure in bent and annealed iron (single crystal) at 700°C is revealed by dense rows of etch pits.[12] 750X. (*Reprinted by permission of Pergamon Press, Plc.*)

tion process in terms of dislocation distribution. When a crystal is subjected to plane gliding, edge dislocations of both signs will emerge at the surface as a result of easy glide. However, in bend gliding, which is a prerequisite for polygonization, a large number of excess edge dislocations of one sign are generated in the crystal to accommodate the plastic curvature, especially near the outer surface[6] (Fig. 5.7*a*).[13] The density of these excess dislocations is given by $1/rb$, where r is the average radius of curvature and b is the Burger's vector. Annealing then causes these dislocations to rearrange themselves by lining up over one another to form the dislocation walls or tilt boundaries, normal to the Burgers vector, by the process as shown in Fig. 5.7*b*.[13] It is thus clear that alignment of edge dislocations into the low-angle tilt boundaries in the polygonized structure occurs by the dislocation glide and climb, which leads to a large relief of each other's elastic strain field (i.e., lowering of strain

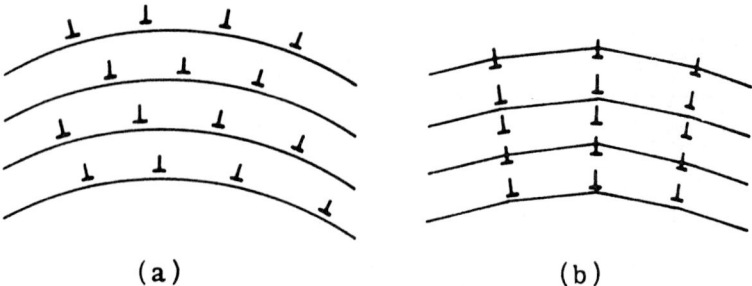

(a)   (b)

FIGURE 5.7 Schematic representation of the polygonization process: (a) Random arrangement of excess (positive) edge dislocations which remain on active slip planes after bend-gliding. (b) Regrouping of edge dislocations to form dislocation walls.[13] (*Reprinted by permission of Butterworths, London.*)

energy). The driving force for this process is, therefore, the decrease of strain energy achieved in forming a low-angle tilt boundary as a result of polygonization.

In general, the relationship between the pure tilt (or grain) boundary energy Eb and angle of misorientation across the boundary (or relative tilt angle of the two subgrains) θ [$\approx b$ (Burgers vector)/h (spacing of dislocations)] (Fig. 5.8a) is expressed by the Read-Shockley equation

$$E_b = E_0\theta(A - \ln\theta) \tag{5.1}$$

where $E_0 = Gb/4\pi(1 - v)$, G is the shear modulus, v is the Poisson ratio, $A = 1 + \ln(b/2\pi r_0)$, and r_0 is the radius (or size) of dislocation core (often between b and $5b$). According to Eq. (5.1), the energy of a tilt boundary and energy per dislocation increases and decreases, respectively, with increasing misorientation (decreasing h), as shown in Fig. 5.9.[4.14] This theory holds good for small values of θ ($< \sim 15°$). Equation (5.1) can be expressed in normalized forms of E_b and θ, i.e., in terms of E_m and θ_m, when the boundary becomes a high-angle boundary (that is, $\theta \sim 15°$).[4]

$$E = \frac{E_m\theta}{\theta_m}\left(1 - \ln\frac{\theta}{\theta_m}\right) \tag{5.2}$$

Initially, segments of tilt boundaries are produced (from the scattered dislocations) which are on the order of 10 dislocations high. Such boundaries then exert a small force on isolated dislocations some distance away. This causes these dislocations to glide into the neighborhood of the boundary and then to climb sufficiently to find a niche in the boundary. In the latter stage, polygonization progresses by the merger or coalescence of pairs of adjacent subboundaries or short-range boundaries, thereby providing a somewhat large boundary which is still of the low-angle (tilt) type. This is shown in Fig. 5.8b.[8] This part of the process is associated with the migration of vacancies, to or from the climbing edge dislocations. The occurrence of the removal of vacancies from the lattice can be explained by the fact that a significant decrease in electrical resistivity has been observed prior to recrystallization.

Work on iron has shown that polygonization occurs with great ease after annealing of moderately deformed high-purity iron at temperatures as low as 300°C. However, this tendency decreases with increasing impurity, especially carbon content.[15]

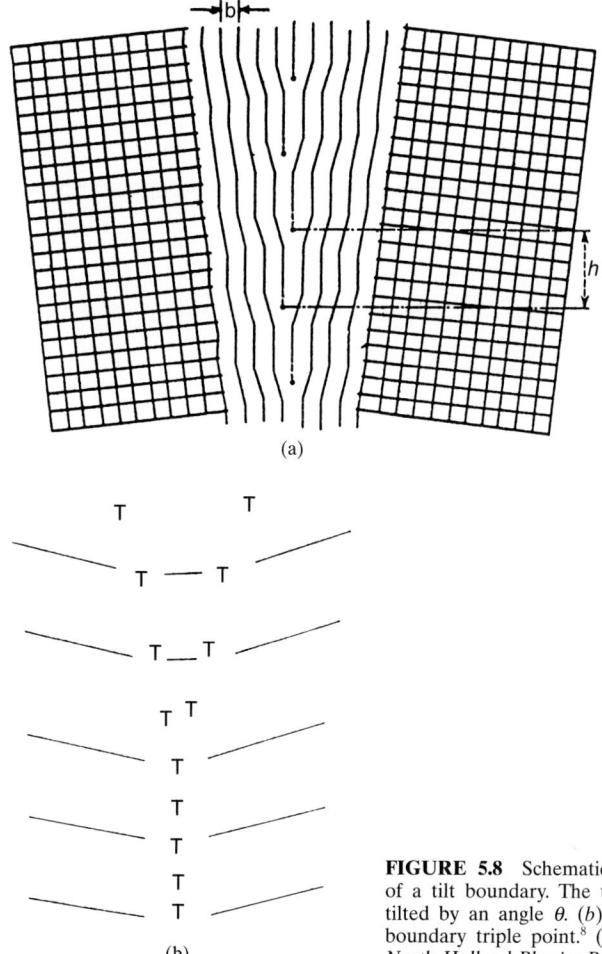

(a)

(b)

FIGURE 5.8 Schematic illustration. (*a*) Structure of a tilt boundary. The two subgrains are mutually tilted by an angle θ. (*b*) Edge dislocations at a tilt boundary triple point.[8] (*Reprinted by permission of North-Holland Physics Publishing, Amsterdam.*)

5.3.5.2 *Relationship between Subgrain Size and Flow Stress.*

The relationship between the subgrain structure formed by recovery after cold work and the flow stress of a material can be expressed in three ways by assuming negligible dislocation density within the subgrains.

1. If we assume that subgrains behave as subgrains, the flow stress σ will be given by a Hall-Petch equation:

$$\sigma = \sigma_0 + k_1 D^{-1/2} \tag{5.3}$$

where σ_0, the friction stress, and k_1 are constants and D is the subgrain diameter.

2. If we assume the operation of dislocation sources whose length will be closely related to their subgrain diameter, the recovered flow stress will be given by the Kuhlman-Wilsdorf[15a] equation

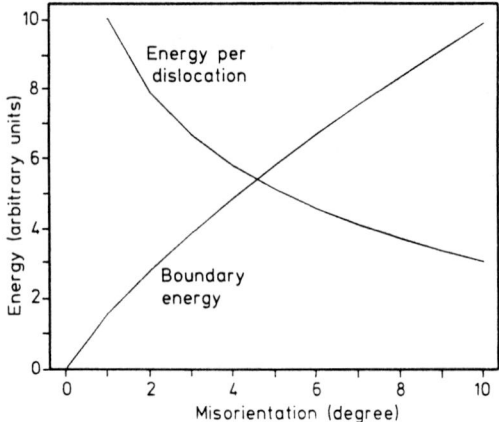

FIGURE 5.9   The energy of a tilt boundary and energy per dislocation as a function of the crystal misorientation.[4,14] (*Reprinted by permission of Elsevier Science Publishing, B.V., Amsterdam.*)

$$\sigma = \sigma_0 + k_2 D^{-1} \tag{5.4}$$

Equations (5.3) and (5.4) may both be expressed in the form

$$\sigma = \sigma_0 + k_3 D^{-m} \tag{5.5}$$

3. If we assume the cell boundary in terms of dislocation tilt boundary, the area of subgrain per unit volume A becomes $\sim 3/D$; for small orientations (angles of tilt boundary), $\theta = b/h$ and the length of dislocation per unit area of boundary $L = 1/h$. Hence

$$\rho = AL = \frac{3}{Dh} = \frac{3\theta b}{D} \tag{5.6}$$

Assuming the relationship between flow stress and dislocation density of Eq. (4.36), we can write

$$\rho = \sigma_0 + k_4 \left(\frac{D}{\theta}\right)^{-1/2} \tag{5.7}$$

which is similar to Eq. (5.5) with $m = \frac{1}{2}$ and $k_3 = k_4 \theta^{1/2}$. All these equations have been employed to analyze substructure strengthening; however, there is little agreement in their application. According to Thompson's findings, the well-developed subgrain structures tend to exhibit $m \sim 0.5$, and cell structures exhibit $m \sim 1$.[15b]

5.3.6   Kinetics

The recovery anneal has no incubation period. It is a homogeneous process because it occurs uniformly throughout the entire volume of the specimen.[4] During isothermal annealing after cold work, all properties (e.g., resistivity, flow stress) that change during recovery have a characteristic kinetic form: The rate of recovery (of the relevant property) is initially very rapid, and then it steadily diminishes with increasing time and eventually ceases after a long time.

In the most extensive study,[16,17] the recovery of the initial flow stress of polycrystalline zone-refined iron was evaluated as a function of the amount and temperature of deformation and time and temperature of the recovery anneal. The fraction of residual strain hardening, which is a measure of recovery, was plotted against time at various temperatures, as shown in Fig. 5.10.[18] The fraction of residual strain hardening was defined as

$$1 - R = \frac{\sigma_r - \sigma_0}{\sigma_m - \sigma_0} \tag{5.8}$$

where $\qquad R$ (fraction or extent of recovery) $= \dfrac{\sigma_m - \sigma_r}{\sigma_m - \sigma_0}$ $\qquad$ (5.9)

σ_r is the flow (or yield) stress after recovery, σ_m is the flow stress after deformation, and σ_0 is the flow stress after full annealing.

If we assume that the change of flow stress occurring during the recovery anneal is proportional to the concentration of defects, we can write the recovery process by the equation

$$\frac{d(\sigma_m - \sigma_r)}{dt} = K(\sigma_m - \sigma_r)e^{-Q/RT} \tag{5.10}$$

where n, the order of the reaction, is an integer; K is the rate constant for the reaction; Q is the activation energy; R is the universal gas constant; and T is the absolute temperature. Rearranging Eq. (5.10) gives

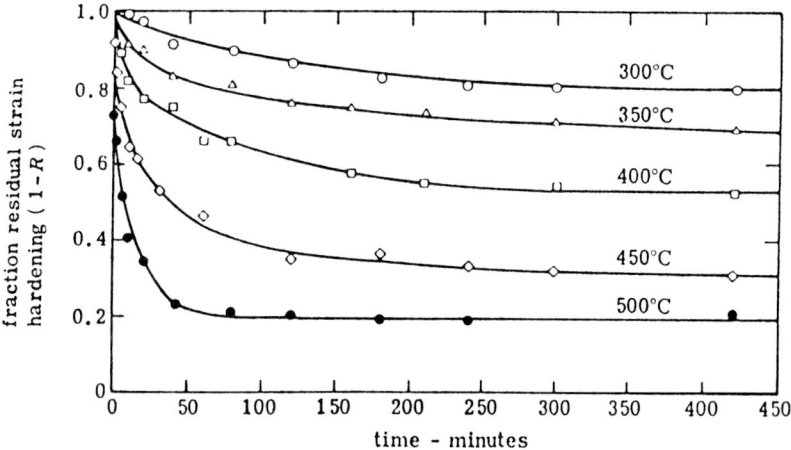

FIGURE 5.10 Recovery kinetics represented by the change in residual strain hardening as a function of time during isothermal annealing at various temperatures for iron strained 5% in tension at 0°C: $1 - R$ denotes a fraction of flow stress increment which remains after annealing.[18] (*Reprinted by permission of John Wiley & Sons, New York.*)

$$\int \frac{d(\sigma_m - \sigma_r)}{(\sigma_m - \sigma_r)^n} = \int K e^{-Q/RT} dt \tag{5.11}$$

which reduces to

$$A = kt e^{-Q/RT} \tag{5.12}$$

where A is an unknown function of $(\sigma_m - \sigma_r)$ and t is the annealing time. Equation (5.12) can also be represented in logarithmic form as

$$\ln t = \ln A + \frac{Q}{RT} \tag{5.13}$$

Equations (5.10) and (5.12) can be used to measure the time-dependent decay of any other physical property during recovery. We can determine the value of Q from the slope of the plot of $\ln t$ versus $1/T$ for a given value of $(1 - R)$. The activation energy for recovery Q is usually the same as that for self-diffusion, which may be represented by the equation

$$D = D_0 e^{-Q/RT} \tag{5.14}$$

where

$$Q (\text{self - diffusion}) = \Delta H^f + \Delta H^m \tag{5.15}$$

where ΔH^f and ΔH^m represent the enthalpy of vacancy formation and vacancy migration, respectively. The diffusion coefficient D can be measured from

$$x = \sqrt{2Dt} \tag{5.16}$$

where x is the diffusion distance in centimeters, D is the self-diffusion coefficient in square centimeters per second, and t is the time in seconds.

It has been shown by Leslie et al.[18] and Michalak and Paxton[16] that Q does not remain the same during recovery of the flow stress of iron but becomes closer to ΔH^m at small recovery time and closer to Q (self-diffusion) at the latter stage (i.e., large recovery time). This implies that vacancy migration is the controlling factor at short recovery times and that self-diffusion (i.e., dislocation climb) is the controlling factor at longer times.

5.4 RECRYSTALLIZATION

Recrystallization is the thermally activated microstructural evolution process in which a new set of comparatively strain-free grains nucleates at the expense of a deformed matrix until it is consumed.[19] Alternatively, recrystallization is defined as the reorientation of crystals in a solid material by the passage of a high-angle boundary. The process is akin to phase transformation in the sense that it can be described phenomenologically in terms of the constituent nucleation and growth rates. Nucleation may be either time-dependent or site-saturated. The driving force for this process is the reduction in free energy, which is accomplished by the reduction of the dislocation network remaining after the recovery stage.[20] Unlike recovery, recrystallization produces a drastic change of mechanical and other physical properties of the deformed metal to the level corresponding to those of the annealed condition. For example, during recrystallization over a small temperature range, hardness, yield strength, and tensile strength are generally reduced (Fig. 5.4a). Similarly, the ductility increases rapidly. Physical properties such as electrical resistivity decrease, and for iron and steel, the magnetic permeability increases markedly to its original level.

The rate of heating of the specimen to the annealing temperature plays a significant role. During rapid heating, recovery is less and the driving force for recrystallization is greater. Hence recrystallization occurs faster.[20]

5.4.1 Origin of Recrystallized Nuclei

A transmission electron microscopic (TEM) study of recrystallization in cold-rolled silicon-iron and single crystals by Hu has endorsed the hypothesis that the strain-free recrystallized nuclei are formed by *subgrain coalescence* via the "evaporation" of edge dislocations constituting the subboundaries between them.[21–23] The enlarged subgrains so formed, much larger than their neighbors, could act as nucleus. Figure 5.11 is a schematic model due to Jones et al.[24] which illustrates the nucleation of recrystallized grain based on subgrain coalescence at an original high-angle grain boundary (along *AB*) in a polygonized structure. They established that subboundary dislocations link continuously with dislocations in a neighboring high-angle grain boundary; the disappearance of subboundary involves a perceptible rearrangement of the dislocations (by climb and glide) in the grain boundary, and in reality, it was observed that grain boundaries serve more effectively as dislocation sinks than as dislocation sources.

Recently Doherty and Szupunar[25] modified Li's theoretical model of subgrain coalescence[26] by assuming the climb mobility of dislocation loops via pipe diffusion instead of that of discrete edge dislocations by lattice diffusion to predict a much faster rate of coalescence. They also employed a nonuniform array of dislocations

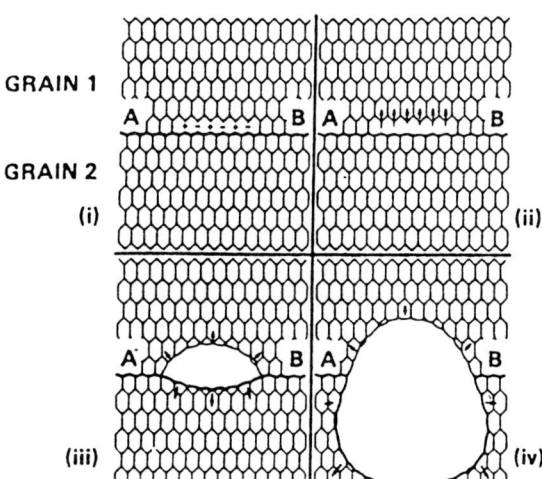

FIGURE 5.11 Schematic model illustrating the nucleation of recrystallized grain based on subgrain coalescence at an original high-angle grain boundary (along AB) in a polygonized structure.[22,24] (*Reprinted by permission of The Metallurgical Society, Warrendale, Pennsylvania.*)

FIGURE 5.12 An optical micrograph of aluminum compressed 40% and annealed for 1 hr at 328°C, showing strain-induced boundary migration into both grains. The direction of growth is not always parallel to the plane sectioned, as can be seen from the two white "bulges" that do not contact the lower parent grain in the section seen.[29] (*Courtesy of S. P. Bellier and R. D. Doherty.*)

to model the disappearance of a low-angle boundary adjacent to a high-angle boundary.

The second nucleation model, which was first reported by Beck and Sperry[27] and later studied by Bailey and Hirsch,[28] is *strain-induced boundary migration* (SIBM). In this model a subgrain within a deformed grain grows into its adjacent grain, forming a bulge which has the same orientation as the original grain, but is predominantly free of dislocations. Figure 5.12 is an example of SIBM in a compressed and annealed aluminum. Some of the tongues originating from above or below the plane of section were shown to have the same orientation with a nearby region of the deformed structure, below the central band limits.[29] The salient point of the micrograph is that it exhibits "two-way" SIBM, which cannot occur by one grain with a larger subgrain size than the other grain. Thus it suggests that a process such as subgrain coalescence has occurred at the grain boundary. In subsequent work with Paul Faivre, they showed that subgrain coalescence occurred at grain boundaries where a high-angle transition band met the grain boundary.[29a]

Note that the difference in nucleation models between coalescence and SIBM lies only in the initial stage, whereas the later stages of nucleus growth are indistinguishable in the two models. Based on the studies on Al, Cu, Fe, and Cu, it has been inferred that a moderate amount of deformation promotes the occurrence of subgrain coalescence in the early stages, whereas high strain and low SFE promote SIBM.[8]

Based on the TEM study of rolled brass by Huber and Hatherly[30] and further observation by Jones[31] in Cu alloys and stainless steels, another nucleation mechanism was proposed which involved nucleation from *recovery twins* (formed at a very early stage of subgrain growth) and growth by a SIBM mechanism.[8] Similarly, the high-vacuum electron microscopic (HVEM) studies of tensile deformed and annealed single crystals by Wilbrandt and Haasen,[32] and later those of Cu-Mn and Cu-P alloys with a very low SFE by Haasen,[33] have prompted them to propose another *multiple twinning mechanism* (i.e., successive twin generation mechanism), as shown in Fig. 5.13. They emphasize that most of the recrystallized orientations do not exist in the deformed microstructure, which become modified by multiple twinning on specific twinning variants after nucleation. That is, orientations of the recrystallized grains are related to multiple twinning.[32,33]

This multiple twinning nucleation may be valid in materials that do form extensive annealing twins, e.g., brass and even copper, but it seems very unlikely in aluminum.[33a] Humphrey and Ferry (1996) have reported that there is much more twinning at a free surface than away from the surface. These results on the effect of a single surface in samples annealed after polishing for observation in HVEM clearly indicate the probability that the multiple twinning, generating new orientations, as reported by Haasen for the case of thin-foil TEM annealing with a free surface, is valued for a process occurring in thin-section annealing and not for materials annealed in bulk.[33a]

The concept of *preexisting* or *preformed nuclei* which turn into a viable nucleus, as proposed by Hutchinson in 1992, is also gaining favor. It is thus clear that the orientation of new grains is clearly imprinted in the previous deformed microstructure.

The nucleation sites for recrystallization in the deformed microstructure may be either those present prior to deformation, such as second-phase particles or grain boundaries, or those induced by the deformation, such as transition bands and shear bands.[8]

5.4.2 Kinetics of Recrystallization

Like phase transformation and unlike recovery, the process of primary recrystallization proceeds by the nucleation and growth process in a sigmoidal manner with respect to time. *Recrystallization nucleus* may be defined as a crystallite of low internal energy growing into deformed material from which it is separated by a high-angle boundary. The basic characteristics of the kinetics of the primary recrystallization process are as follows:

1. An initial incubation period is necessary.
2. It is followed by slow rate of change at the initial stage, accelerated rate of change at the intermediate stage, and finally slow rate of change at the final stage. This is schematically shown in Fig. 5.14*a*.

5.4.2.1 The Johnson-Mehl-Avrami-Kolmogorov (JMAK) Model. The main features of a formal theory of recrystallization kinetics due to Johnson, Mehl, Avrami, and Kolmogorov, called the JMAK model, are discussed below.[34,35]

Let us consider that N is the nucleation rate (i.e., number of new grains formed per unit time per unit volume of unrecrystallized metal) and G is the linear growth rate of any new grain. During linear growth rate G, the radius of a growing nucleus r is given by

$$r = G(t - t_0) \tag{5.17}$$

where t is the time of recrystallization and t_0 is the incubation period. This relationship is represented in Fig. 5.14*b*. If we consider the nucleus to be a sphere, then

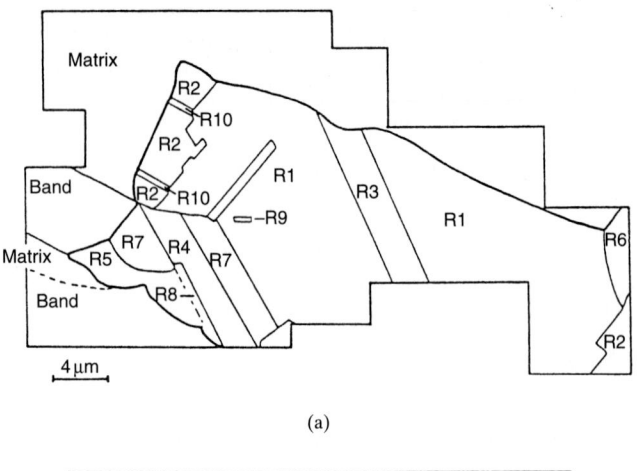

(a)

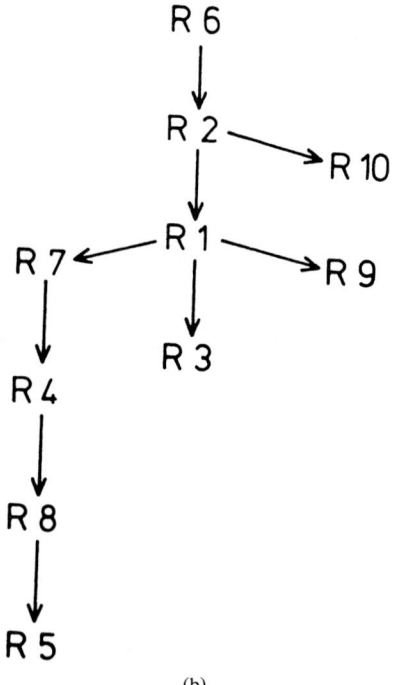

(b)

FIGURE 5.13 (*a*) Schematic of recrystallized multiple twinned grain in Cu-0.03% P showing microstructure after 80% strain and 20 min annealed at 500°C; (*b*) twin chain corresponding to the microstructure.[33] (*Reprinted by permission of The Metallurgical Society, Warrendale, Pa.; after P. Haasen.*)

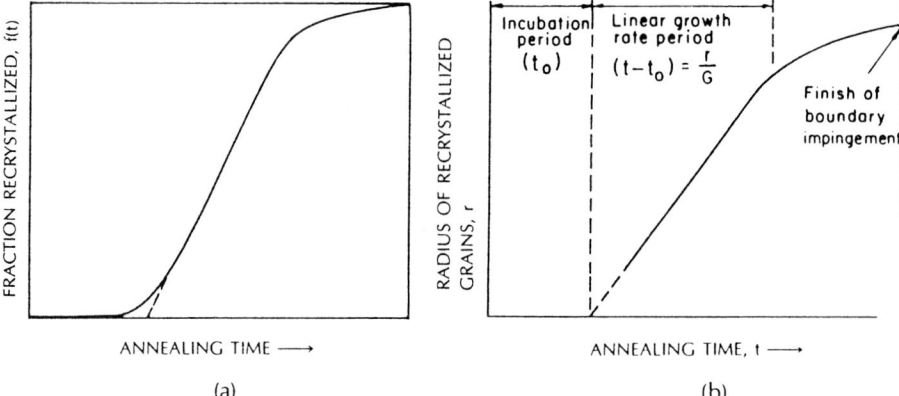

FIGURE 5.14 Schematic diagram illustrating (*a*) the basic characteristics of the kinetics of the primary recrystallization process and (*b*) the variation of the radius of the new recrystallized grain with isothermal annealing time during the growth stage of primary recrystallization.

the recrystallized volume per nuclei is $(4\pi/3)[G(t - t_0)]^3$. Since the number of nuclei increases at the same time as the growth occurs, we can express

$$dn = \dot{N}[1 - f(t)]dt \qquad (5.18)$$

where dn is the number of recrystallized nuclei formed in time interval dt and for the recrystallized volume fraction of the material at time t. Equation (5.18) is valid for the actual recrystallization process, and it does not take into consideration the boundary impingement taking place, in some portion of the material, even at the early stage. The boundary impingement causes a reduction of the actual number of nuclei formed at any instant by an amount represented by the ghost nuclei which would have formed if the recrystallization had not advanced to this stage. The extended number of nuclei dn_{ex} is the sum of the actual nuclei dn and the ghost nuclei dn' at any stage of recrystallization. Thus we can rewrite Eq. (5.17) in the following form:

$$dn_{ex} = dn + dn' = \dot{N}[1 - f(t)]dt + \dot{N}f(t)dt = \dot{N}dt \qquad (5.19)$$

The extended volume fraction transformed $f_{ex}(t)$ can thus be written as

$$f_{ex}(t) = \int_0^t \frac{4\pi}{3}[G(t - t_0)]^3 \dot{N}dt \qquad (5.20)$$

If we further assume $\dot{N}$ and G to be constant and $t_0 = 0$, we get

$$f_{ex}(t) = \frac{\pi}{3}\dot{N}G^3t^4 \qquad (5.21)$$

Since the actual and ghost nucleus formed in any time interval dt will occupy the same volume per nucleus, we can write, according to Johnson and Mehl,[36] the following relationship between extended volume fraction $f_{ex}(t)$ and actual recrystallized volume fraction, f or $f(t)$:

$$\frac{dn}{dn_{ex}} = \frac{\text{untransformed volume}}{\text{total volume}} = \frac{df(t)}{df_{ex}(t)} = 1 - f(t) \qquad (5.22)$$

This simple differential equation reduces to

$$f \text{ or } f(t) = 1 - \exp[-f_{ex}(t)] \qquad (5.23)$$

The above analysis assumes that

1. The grains grow isotropically in three dimensions until impingement occurs. The JMAK exponent n decreases if the grains are constrained to grow in one or two dimensions.
2. Nucleation sites are assumed to be randomly distributed.

Combining Eqs. (5.21) and (5.23), we get

$$f(t) = 1 - \exp\left(\frac{-\pi}{3}\right)\dot{N}G^3t^4 \qquad (5.24)$$

For the three most common forms of specimens (wire, sheet, and lump), the limiting equations for isothermal recrystallization, as derived by Johnson and Mehl,[36] are as follows:

For one-dimensional recrystallization,

$$f(t) = 1 - \exp\left(\frac{-\eta\dot{N}G\delta_w^2t^2}{2}\right) \qquad (5.25)$$

where δ_w represents the wire diameter. For two-dimensional recrystallization,

$$f(t) = 1 - \exp\left(\frac{-\eta\dot{N}\delta_sG^2t^3}{3}\right) \qquad (5.26)$$

where δ_s represents the sheet thickness. For three-dimensional recrystallization,

$$f(t) = 1 - \exp\left(\frac{-\eta\dot{N}G^3t^4}{4}\right) \qquad (5.27)$$

where η is the shape factor. Most investigators agree on the following fraction softening equation due to Johnson, Mehl, Avrami, and Kolmogorov (JMAK) who assumed $\dot{N}$ to decrease exponentially with increasing time and G to be the same for all grains:

$$f(t) = 1 - \exp(-kt^n) \qquad (5.28)$$

where $f(t)$ and t have the usual meanings; k is a kinetic parameter related to constant growth rate, nucleation rate, and a shape factor; and n is the annealing or JMAK exponent. The value of n lies between 1 and 2 for one-dimensional recrystallization (i.e., nucleation in wire); between 2 and 3 for two-dimensional recrystallization (i.e., nucleation in thin sheet), and between 3 and 4 for three-dimensional (i.e., nucleation in lump form) recrystallization. This equation can be used to obtain the isothermal annealing curves in Fig. 5.15. Both the Johnson-Mehl and the Avrami equations will be discussed again in Chap. 7.

Thus, the n value can provide useful fundamental information regarding the type and morphology of grain growth. Rearranging Eq. (5.28) and taking logarithms of both sides, we get

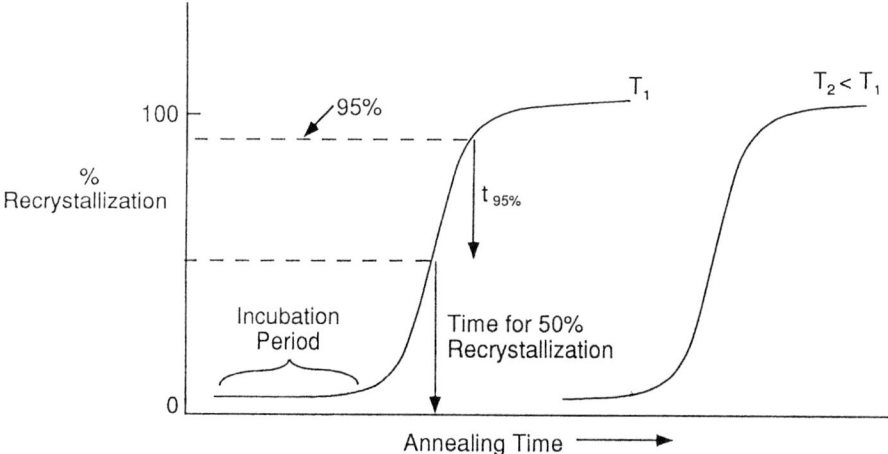

FIGURE 5.15   Percentage recrystallization as a function of annealing time at a fixed temperature. (*Courtesy of H. Pops.*)

$$\log\frac{1}{1-f(t)} = kt^n \tag{5.29}$$

or

$$\log\log\frac{1}{1-f(t)} = \log k + n\log t \tag{5.30}$$

which is a straight-line equation whenever the parameters $\log\{1/[(1-f(t)]\}$ and annealing time t are plotted on double-log paper. In this case the annealing exponent is defined as the slope of the line, as shown in Fig. 5.16 for a large range of oxygen concentration and annealing temperatures. Slopes for all compositions and annealing temperatures are nearly equal to 1, and calculated n values are listed in Table 5.1. The fact that n is less than 2 and independent of oxygen content gives significant insight into the nature of the nucleation rate.[37]

To measure the JMAK exponent, it is essential to correct the time axis of the JMAK plot to $t - t_0$, where t_0 is the incubation time. In some recent studies by Samajdar and Doherty[37a] on the recrystallization of warm deformed aluminum, there was an extensive incubation period for samples deformed 83%; but when deformed 96%, the failure to correct for incubation time yielded absurdly high values of JMAK exponent ($n = 25$) for the material deformed 83%. However, a more reasonable value of $n = 2.4$ was found for the material that showed no incubation time.[37a]

5.4.2.2   *Microstructural Path Methodology.*   A significant attempt to improve the JMAK model has recently been made by Vandermeer and Rath[38–40] using the *microstructural path methodology* (MPM). In this model, more realistic and more complex geometric approaches are employed by using additional microstructural properties in the analysis and, if required, relaxing the uniform grain impingement constraint. The essential features of MPM are (1) the extraction of more detailed information about nucleation and growth rates from experimental measurements than those obtained by the JMAK approach, (2) greater flexibility than in the

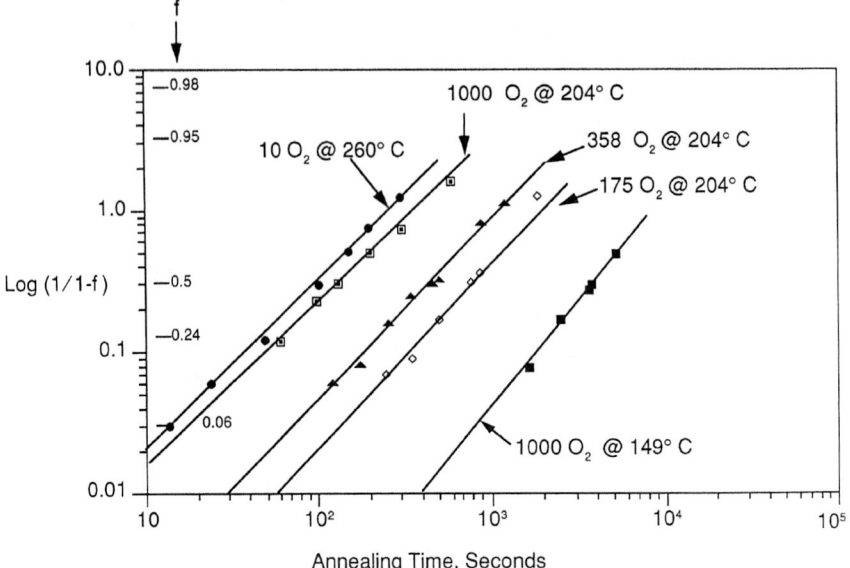

FIGURE 5.16 Typical isothermal transformation plots showing the effects of annealing time upon $\log\{1/[1 - f(t)]\}$. Recrystallization data obtained for different oxygen contents and annealing temperatures.[37] (*Courtesy of H. Pops.*)

TABLE 5.1 Values of Exponent n in Annealing Time-Transformation Equation.[†] All wires drawn 65% reduction area prior to annealing[37]

Oxygen, ppm	Annealing temperature, °C		
	149	204	260
10		0.88	1.2
175		1.23	
358		1.27	
642		1.41	
1000	1.50	1.13	

[†] $f(t) = 1 - \exp(-kt^n)$ = volume fraction transformed.

JMAK model, (3) spatial or random distribution of recrystallized grains, and (4) consideration of global microstructural properties such as extended volume fraction of recrystallized grains and extended interfacial area per unit volume.[4]

As in the JMAK model, it is simple to use the concept of extended volume fraction recrystallized $f_{ex}(t)$ or $f_{ex}(= \int V \, dn')$ [Eq. (5.23)] and the extended interfacial area per unit volume between recrystallized and unrecrystallized (deformed) grains by means of Eq. (5.31), due to Gokhale and DeHoff,[41]

$$S_{ex} = S_{ex}(t) = \frac{S}{1 - f} \tag{5.31}$$

where V is the volume of the recrystallized grains and S is the interfacial area per unit volume between recrystallized and unrecrystallized grains. The progress of isothermal recrystallization, which can be determined metallographically, is given by

$$f_{ex}(t) = \int_0^t \dot{N}_{(t)} V_{(t-t_0)} dt \tag{5.32}$$

and

$$S_{ex}(t) = \int_0^t \dot{N}_{(t)} S_{(t-t_0)} dt \tag{5.33}$$

where $V_{(t-t_0)}$ and $S_{(t-t_0)}$ are the volume and interfacial area, respectively, at time t, of a grain which has nucleated at time t_0'; $\dot{N}_{(t)} \, dt$ is the number of new nodules nucleated per unit volume in the time interval between t_0 and $t + t_0$. If it is further assumed that the grains have the same spheroidal shape which is preserved during growth, the volume $V_{(t-t_0)}$ and interfacial area $S_{(t-t_0)}$ of individual nucleated grains are given by

$$V_{(t-t_0)} = K_v a_{(t-t_0)}^3 \tag{5.34}$$

$$S_{(t-t_0)} = K_s a_{(t-t_0)}^2 \tag{5.35}$$

where K_v and K_s are shape factor constants and $a_{(t-t_0)}$ is a major semiaxis of the spheroid, which is related to the interface migration rate $G(t)$ by

$$a_{(t-t_0)} = \int_{t_0}^t G(t) dt \tag{5.36}$$

where $a_{(t-t_0)}$, being a local property, may be predicted by measuring the diameter of the largest unimpinged grain intercept D_L on a plate polished surface. By assuming D_L to be the diameter of the earliest nucleated grain, we get

$$a_{(t-t_0)} = \frac{D_L}{2} \tag{5.37}$$

For isothermal recrystallization, f_{ex}, S_{ex}, and D_L can be expressed by the time-dependent power law functions of the form

$$f_{ex} = \ln \frac{1}{1-f} = kt^n \tag{5.38}$$

$$S_{ex} = \frac{S}{1-f} = Kt^m \tag{5.39}$$

$$D_L = St^s \tag{5.40}$$

If the derived functions are also assumed to be expressed by power law functions, we have

$$\dot{N} = N_1 t^{\delta-1} \tag{5.41}$$

and
$$a_{(t-t_0)} = G_a(t - t_0)^r \tag{5.42}$$

where N_1, δ, G_a, and r are constants. Since shapes of spheroidal grain remain preserved during growth, $a_{(t)} = D_L/2$, and thereby $G_a = S/2$, and $r = s$.

Vandermeer and Rath show that, for spheroidal grains, which preserve their shape during growth, δ and r become

$$\delta = 3m - 2n \tag{5.43}$$

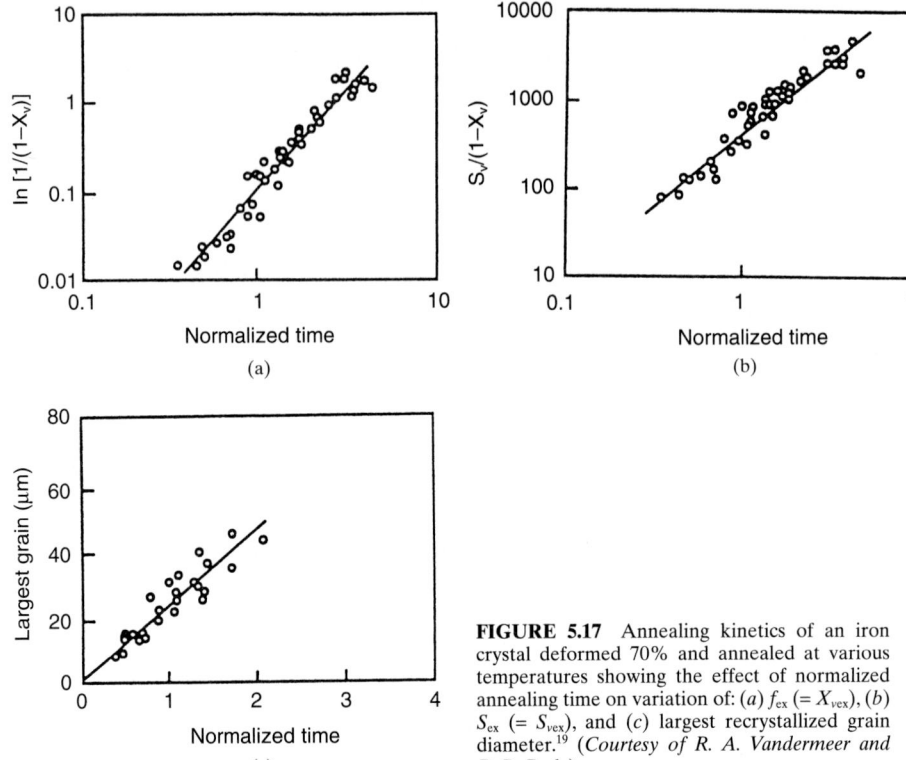

FIGURE 5.17 Annealing kinetics of an iron crystal deformed 70% and annealed at various temperatures showing the effect of normalized annealing time on variation of: (a) f_{ex} ($= X_{vex}$), (b) S_{ex} ($= S_{vex}$), and (c) largest recrystallized grain diameter.[19] (*Courtesy of R. A. Vandermeer and B. B. Rath.*)

$$r = s = n - m \qquad (5.44)$$

For Eqs. (5.38) to (5.42), it is evident that $\delta = 1$ represents the case of constant nucleation rate, $\delta = 0$ denotes the site saturation nucleation, and $r = s = 1$ represents a constant growth (or interface migration) rate. Hence, by measuring f, S, and D_L experimentally as a function of annealing time, we can determine the values of n, m, and s experimentally; calculate δ and r; and identify the form of the nucleation kinetics. In contrast, if the JMAK model is applied for constant nucleation and growth rates, we obtain $n = 4$, $m = 3$, $s = 1$, $\delta = 1$, and $r = 1$.

Experimental Application. Figure 5.17 represents the recrystallization kinetics of deformed (111)[112] iron crystals at various temperatures which illustrates the variations of $f (= X_v)$, $S (= S_v)$, and the largest unimpinged recrystallized grain diameter as a function of normalized annealing time. The slopes of these lines in Fig. 5.17a, b, and c, respectively, yield n, m, and s values. The values obtained are $n = 1.90$, $m = 1.28$, and $s = 0.60$ which, in turn, give $\delta = 0.04$ and $r = 0.62$. It is inferred from these data that nucleation is site-saturated (due to negligibly small δ value), the grains grow as spheroids (due to $r \sim s$), and growth rate decreases with time (due to low r value). However, for $s \neq r$, the grain shape varies during recrystallization.[19] The model has also been extended to incorporate the influence of recovery during the recrystallization anneal.[39]

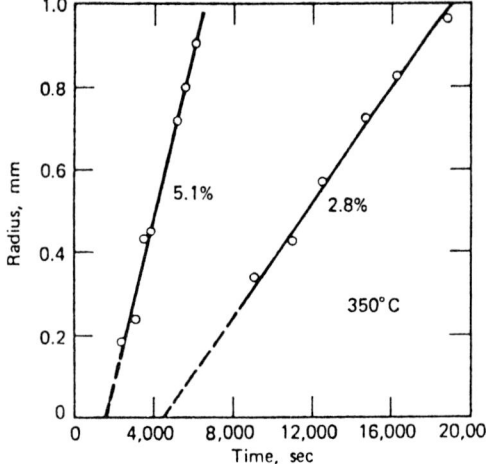

FIGURE 5.18 Radius of largest grain versus isothermal annealing time at 350°C for 2.8 and 5.1% elongation of aluminum.[42] (*Reprinted by permission of The Metallurgical Society, Warrendale, Pennsylvania.*)

5.4.3 Experimental Determination of *G* and *Ṅ*

For the measurement of G, a series of identical specimens are first given a specific amount of strain, then isothermally recrystallized for different lengths of time, and finally quenched to room temperature. The radius of the largest grain (which has not suffered impingement with another growing grain) observed individually in metallographically prepared specimens is plotted as a function of time of isothermal recrystallization, as shown in Fig. 5.18.[42] The slope of this curve determines the rate of growth G, while the intercept on the time axis determines the incubation period t_0.

For the determination of $\dot{N}$, first the number of new recrystallized grains per unit area N_s is plotted as a function of time, as shown in Fig. 5.19a. The slope of this curve yields the (two-dimensional) surface nucleation rate $\dot{N}_s$, as illustrated in Fig. 5.19b. The two-dimensional nucleation rate can be approximately related to the three-dimensional nucleation rate $\dot{N}$, which can be found in any book on quantitative metallography.

5.4.4 Effect of Process Variables on *G* and *Ṅ*

The concepts of $\dot{N}$ and G are useful in explaining the influence of different variables on the recrystallization process. Since G is invariant with time, its values can be easily compared. On the other hand, $\dot{N}$ rapidly increases with increasing temperature whereas $\dot{N}$ varies with time: Hence it is not directly compared. Figure 5.20 shows the effect of the extent of deformation and grain size, respectively, on G. The results represented in Figs. 5.20a and 5.18 illustrate that the growth rate increases up to about 10% strain; beyond this level, G still increases but at a slower rate up to about 15% strain. Figure 5.20a also demonstrates the plot of incubation time versus extent of deformation. The incubation time falls rapidly with strain to a small level at about 15% strain, beyond which it is constant. This implies ease of nucleation (i.e., accelerated recrystallization) at high strain.

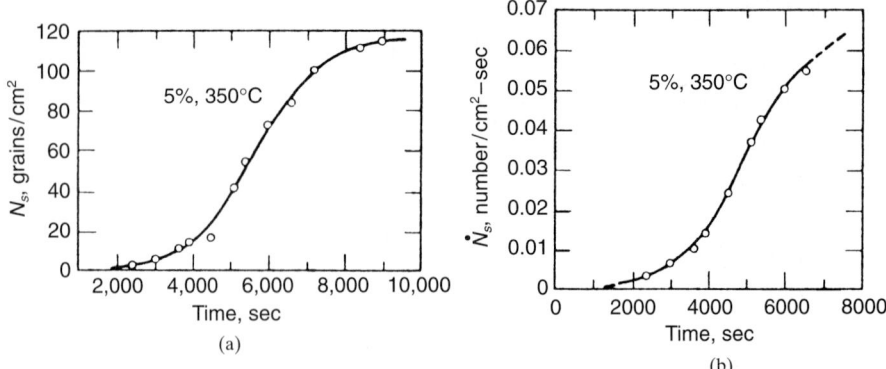

FIGURE 5.19 (*a*) Aluminum after 5.1% elongation: surface density of the recrystallized grains versus time.[42] (*b*) Aluminum after 5.1% elongation: surface nucleation rate N_s at the surface versus time.[42] (*Reprinted by permission of The Metallurgical Society, Warrendale, Pennsylvania.*)

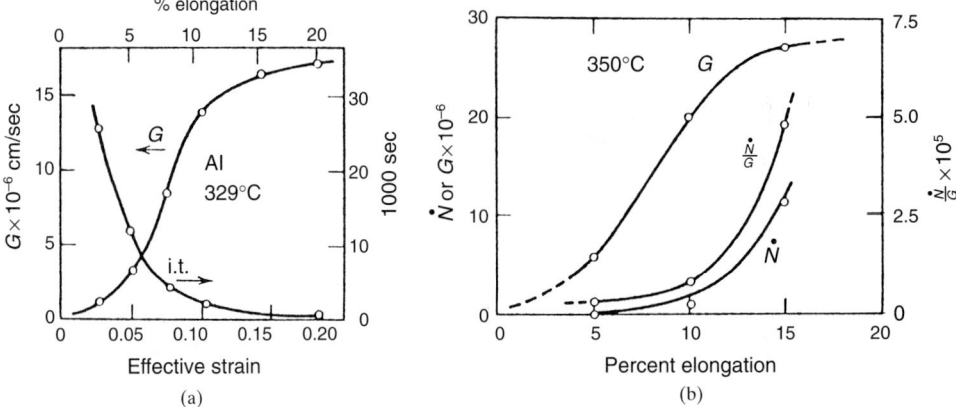

FIGURE 5.20 (*a*) Effect of prior strain in aluminum on growth rate G and induction period t_0 during recrystallization at 329°C.[42] (*b*) Effect of a prior strain in aluminum on growth rate G and nucleation rate N during recrystallization at 350°C.[42] (*Reprinted by permission of The Metallurgical Society, Warrendale, Pennsylvania.*)

5.4.5 Effect of Recrystallization Time and Grain Size

The important characteristic of a transformation (e.g., recrystallization) is the time needed to transform a given volume fraction. We do not experimentally determine the time for $f(t) = 1.00$ because it is very difficult. However, we often calculate the time to achieve a given volume fraction of transformation, say, $f(t) = 0.5$ or 0.95. We get the following relationship by using Eq. (5.24):

$$t_{0.5} = \left(\frac{0.662}{\dot{N}G^3}\right)^{1/4} \tag{5.45}$$

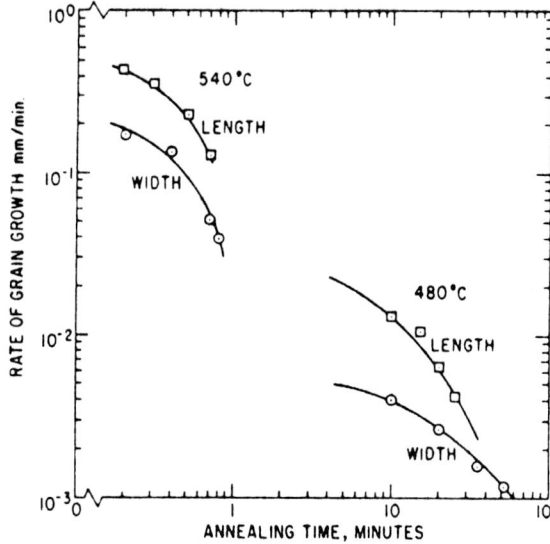

FIGURE 5.21 Effect of time and temperature on rate of linear growth in zone-melted iron (cold-rolled 60%) showing rapid decrease of linear growth rate with time at constant annealing temperature.[18] (*Reprinted by permission of John Wiley & Sons, New York.*)

where $t_{0.5}$ denotes the time required to obtain 50% volume transformed. We know that both $\dot{N}$ and G increase when the temperature is raised. With this in mind, Eq. (5.45) indicates that we can get a certain volume fraction of recrystallization accomplished in a shorter time by increasing the values of $\dot{N}$ and G, which, in turn, lowers the 1-hr recrystallization temperature. That is, there is an inverse relationship between T_{recry} and $\dot{N}$ and G. It can also be generalized from Eq. (5.45) that a change in G has a far-reaching influence on $t_{0.5}$ or any other annealing time corresponding to a given volume fraction of transformation. In other words, there is a sharp decrease of G with increasing annealing time compared to that of $\dot{N}$ which has been experimentally found in cold-worked and annealed zone-refined iron specimens (Fig. 5.21).[18]

The number of nuclei formed per unit volume is given by

$$\dot{N}t_{0.5} = 0.90\left(\frac{\dot{N}}{G}\right)^{3/4} \tag{5.46}$$

If d is the average diameter of the recrystallized nuclei, the number of nuclei per unit volume will be roughly $1/d^3$, so that

$$d_{f(t)=0.5} = 1.035\left(\frac{G}{\dot{N}}\right)^{1/4} \tag{5.47}$$

We can generalize Eq. (5.47) for the recrystallized grain size as

$$d = (\text{constant})\left(\frac{G}{\dot{N}}\right)^{1/4} \tag{5.48}$$

It is apparent from Eq. (5.48) that the ratio of $\dot{N}/G$ is useful to determine the recrystallized grain size. Fine recrystallized grain size can be obtained by maintaining a high $\dot{N}/G$ ratio, that is, high nucleation rate and a slow growth rate or large number of nuclei formation with very little growth. In contrast, a low ratio represents a slow nucleation rate with respect to the growth rate and thus represents coarse recrystallized grains.

Figure 5.20*b* shows a plot of $\dot{N}$ and G along with their ratio $\dot{N}/G$ versus the amount of strain for a typical metal (aluminum). These curves illustrate that as the degree of cold deformation (prior to annealing) decreases, $\dot{N}$ falls more rapidly than G; that is, the $\dot{N}/G$ ratio drops more sharply. When the amount of cold deformation decreases below a particular (about 3%) strain, called *critical strain* (which is different for different metals), the $\dot{N}/G$ ratio becomes practically zero and, consequently, it is not possible to form recrystallized grains in a reasonable length of time. Note that the amount of critical strain is very different for different metals.

Many factors restrict the number of recrystallized nuclei which successfully grow into fully formed grains. Such factors greatly reduce the final grain density. Other factors influencing the recrystallized grain size include annealing temperature and heating rates used in the annealing treatment. A high annealing temperature yields a high density of potent nuclei and a fine grain size if the grain growth and secondary recrystallization are absent. A high heating rate leads to a higher temperature for the start of recrystallization. Hence, the density of the active nucleation sites is increased, and the likelihood of a strongly preferred nucleus starting to grow at relatively low temperature is diminished. A slow heating rate produces excessive growth of a few nuclei, resulting in an extremely coarse and inhomogeneous structure.[43]

5.4.6 Strain-Anneal Technique

Well-developed single crystals are successfully prepared by the strain-anneal technique, in which the specimen is first strained just above the critical strain and then annealed at a slow heating rate and the lowest possible temperature so that the $\dot{N}/G$ ratio is drastically reduced.

5.4.7 Activation Energies Q_n and Q_g

The activation energy of recrystallization Q is a measure of the driving force, i.e., of the energy difference between the cold-worked and annealed states. Since the Arrhenius equation applies for this process, we can write[37]

$$\text{Recrystallization rate} = \frac{1}{t} = Ae^{-Q/RT} \tag{5.49}$$

Note that activation energy for pure metals is typically one-half that for self-diffusion.[44] Since activation energy for recrystallization increases with increasing impurity contents, oxygen concentration in the high-conductivity Cu wire leads to the formation of metal oxides upon annealing, thereby removing impurities from solid solution.[45] Moreover, $\dot{N}$ and G vary exponentially with temperature (that is, $\dot{N} \propto e^{-Q_n/RT}$ and $G \propto e^{-Q_g/RT}$), which indicates that nucleation and growth in recrystallization are processes that are controlled by their respective activation energies (for example, Q_n and Q_g decrease with increasing strain and decreasing initial grain size). Surprisingly, the values of both Q_n and Q_g have been found to be of the same order of magnitude, provided the strain is higher than about 5%.[42]

5.5   *RECRYSTALLIZATION OF TWO-PHASE ALLOYS*

As most commercial alloys contain more than one phase, an understanding of the recrystallization behavior of two-phase materials is of great importance and scientific interest. The second phase may be in the form of dispersed particles, which are present during the deformation; or, if the matrix is supersaturated solid solution, the particles may precipitate during subsequent annealing. In aluminum-treated low-carbon steel, it has been observed that a low annealing temperature results in the precipitation of AlN on sub-boundaries prior to recrystallization, thereby causing a considerable retardation of the process. However, high annealing temperature accelerates the recrystallization to such a degree that the steel may completely recrystallize before the onset of precipitation of AlN.[46] In steels, the presence of pearlite or spheroidized cementite has been found to accelerate recrystallization.[47]

5.5.1   Particle-Stimulated Nucleation

Coarse, widely dispersed second-phase particles can produce a local concentration of lattice distortion caused by the applied deformation which, in turn, accelerates recrystallization by altering N rather than G. This *partial stimulated nucleation* (PSN) has been observed in many Al, Cu, Fe, and Ni alloys and is usually only found adjacent to large ($\geq 1\,\mu m$) particles; however, a lower limit ($0.8\,\mu m$) was deduced from the indirect measurements of Gawne and Higgins on Fe-C alloys.[48] In this manner, rapid recrystallization leads to a final grain size of the order of interparticle spacing, in a relatively pure Al matrix. In most cases, a maximum of one grain is found to nucleate at any particle, although it seems that multiple nucleation occurs at very large particles, as seen in Fe-O alloy (0.33% O_2) (Fig. 5.22). Another example of PSN is the precipitation of ~1-μm $FeAl_3$ particles in Fe-Al alloy which enhances N.

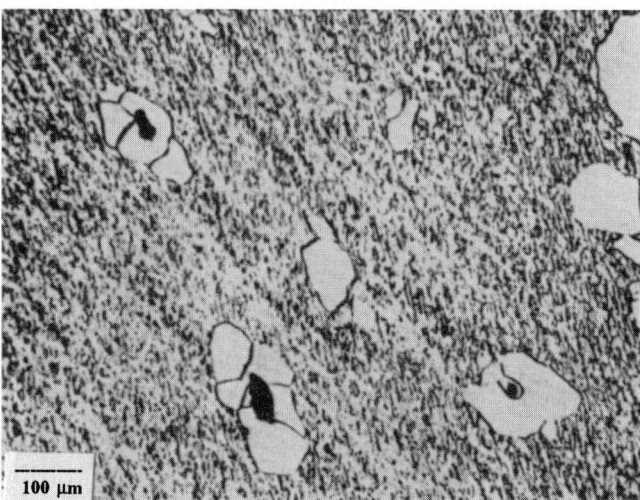

FIGURE 5.22   Particle-stimulated nucleation of recrystallization at oxide inclusions in iron; 60% rolling reduction, 2 min at 540°C.[18] (*Reprinted by permission of John Wiley & Sons, New York.*)

A dense distribution of finely dispersed second-phase particles has been found to hinder recrystallization by delaying both the nucleation and growth processes.[40] For example, dense distributions of fine (0.1- to 1-μm) particles have a retarding effect on the growth rate, which results in a fine grain structure. However, microstructural inhomogeneities often lead to the preferential growth of just a few grains. The effect of a fine and dense particle dispersion is thus an increase in T_{recry}, and also a coarse final structure.[43]

Very small (<<0.1-μm) particles do not considerably affect the deformation microstructure except for a preferred formation of deformation bands in alloys with coherent precipitates or very small incoherent particles. However, such precipitates may form by the decomposition of a supersaturated solid solution during the annealing process. The presence of solute atoms leads to processes which may precede or accompany and interact with recrystallization.[43]

5.5.2 Segregation and Precipitation

Segregation and precipitation processes affect both $\dot{N}$ and G. When these reactions take place in the deformed and/or recovered material, the substructure rearrangement necessary for the formation of viable nuclei is strongly impeded due to the pinning action of segregated atoms and preferential precipitation at the dislocations. Therefore, T_{recry} is substantially increased. However, at driving force locally exceeding the retarding force, anomalous growth of a few grains produces again a coarse final grain structure.

The retarding effects of segregation and precipitation are found, for example, in Al-Mn alloys, where they lead to intermittent grain boundary motion and simultaneous occurrence of discontinuous (cellular) precipitation and recrystallization, if the cold-worked alloy is annealed in a certain temperature range. The result of these partially hindered recrystallization processes is a highly inhomogeneous grain structure.

A fine grain structure in a deformed supersaturated solid solution can be best produced by a high-temperature annealing treatment where recrystallization precedes precipitation and the two processes do not interact (Fig. 5.23). In general, the large particles, the prior grain boundaries, and the deformation bands provide a very high density of viable nucleation sites in cold-worked commercial Al alloys.[43]

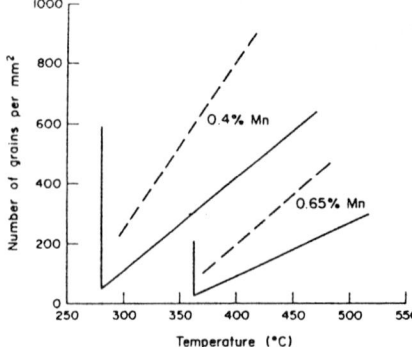

FIGURE 5.23 Grain density of a commercial Al-1% Mn alloy when recrystallization and precipitation processes interact (0.65% Mn in solid solution) and when precipitation occurs only after completion of recrystallization (0.4% Mn in solid solution): ——, lower limit; --- mean values.[43] (*Reprinted by permission of Pergamon Press, Plc.*)

5.6 RECRYSTALLIZATION TEMPERATURE

Recrystallization Temperature $\mathbf{T}_{recry}$. This temperature is defined as the temperature at which a given metal with a certain amount of cold deformation will completely recrystallize in a specific time, usually 1 hr. For practical reasons, 95% recrystallization is considered to be completion of recrystallization;[49] sometimes 50% recrystallization after a 1-hr anneal is also used. It is a function of purity, prior strain, initial grain size, deformation temperature, deformation texture, and thickness.

Purity. The T_{recry} of several metals and alloys is listed in Table 5.2.[20,50,51] Table 5.3[11,52] shows the influence of small amounts (0.01 at %) of solute atoms (in solid solution) which raise the T_{recry} of pure copper by an amount varying between 0 and 240°C. As a general rule, the purification of a metal results in an appreciable decrease of its T_{recry}, and the effect is very pronounced at low concentration range. Solute atoms form clouds or atmospheres around dislocations as well as around boundaries that drastically retard the interface mobility, thereby decreasing N. Similarly, it has been shown that finely dispersed second-phase particles have a strong influence in inhibiting recrystallization, particularly nucleation.[46]

TABLE 5.2 Approximate Recrystallization Temperature T_{recry} for Several Metals and Alloys[20,43,44]

Material		T_{recry} (°C)	Material		T_{recry} (°C)
Copper	Zone-refined	80	Iron	Zone-refined	300
	OFHC	200		Electrolytic	399
	5% Zn	315		Low-carbon steel	538
	2% Be	371	Tungsten	Zone-refined	1000
Aluminum	Zone-refined	−50		(5 ppm Mo, O_2)	
	99.999%	80		Impure	1500
	99.0+	288		(50–60 ppm Mo,	
	Alloys	316		30 ppm O_2)	
Nickel	Zone-refined	300	Zirconium	Zone-refined	170
	99.99%	371		Commercial	450
	Commercial	600	Magnesium	99.99%	66
	Monel alloy	593		Alloys	232
Silver		200	Zinc		10
Platinum		450	Tin		−4
			Lead		−4

TABLE 5.3 Increase in Recrystallization Temperature of Pure Copper by 0.01 at% of Elements[11,52]

Alloying elements	Increase in T_{recry} (°C)	Alloying elements	Increase in T_{recry} (°C)
Ni	0	Ag	80
Co	15	Sn	180
Fe	15	Te	240

Prior Strain.   As the prior strain is increased above a certain value, the activation energy for nucleation and growth Q_g decreases. This increases $\dot{N}$ and G within the deformed metal; consequently, T_{recry} and the associated grain size d are both decreased with the increase in strain[7] (Fig. 5.20b).

Initial Grain Size.   During cold working of a polycrystalline material, the grain boundaries tend to interrupt the slip process that occurs in the crystals. As a result, the lattice in the vicinity of the grain boundaries is, on average, much more distorted than that in the center of the grains. Thus, smaller grain size increases the grain boundary area; this, in turn, increases local deformation, thereby increasing $\dot{N}$ faster than G. Consequently, T_{recry} decreases. Thus, a fine-grained material will recrystallize more rapidly than a coarse-grained material.

Deformation Temperature.   It has been observed that a reduction in the initial deformation temperature produces (1) lowering of T_{recry} due to the decreased activation energies and, consequently, increased values of $\dot{N}$ and G and (2) higher $\dot{N}/G$ because of the increased dislocation density and the strain energy with a more uniform dislocation density distribution, which will also decrease d.[7]

Deformation Mode.   The effect of deformation mode on the recrystallization rate is complex. For example, in both iron[53] and copper,[54] it was noticed that, for the same reduction in thickness, straight rolling produced more rapid recrystallization than that of cross-rolling. Barto and Ebert found that, in deformed Mo to a true strain of 0.3 by tension, wire drawing, rolling, and compression, the subsequent recrystallization rate was highest for the tensile deformed metal and decreased in the above order for the other samples.[55] For the same equivalent strain in pure Al, deformation by multi-axis compression leads to slower recrystallization than the specimens deformed by monotonic compression.[56]

Heating Rate.   The heating rate of the material to the annealing temperature may be of importance because a rapid rate may reduce the amount of recovery occurring prior to recrystallization and may produce a large driving force for recrystallization, thereby accelerating the recrystallization in both single-phase and complex alloys. In the latter, these effects have been attributed to the role of solute and second-phase particles in hindering recovery. Conversely, it is also found that in steels, rapid heating may decelerate recrystallization, presumably attributed to carbon content present in solid solution.[57]

Pores and Bubbles.   Pores and bubbles act as second-phase particles or inclusions in pinning boundaries and tend to inhibit recrystallization and increase T_{recry} in many metals, such as Pt, Al, Cu, and W. Thomson-Russell,[58] Dillamore,[59] Bewlay et al.,[60] Briant,[61] Bewlay and Briant,[62,63] Walter and Briant,[64] and Pugh and Lasch[65] have shown that in blue tungsten oxide powder doped with aqueous solutions of potassium disilicate and aluminum chloride, called AKS (implying Al_2O_3, KCl, and SiO_2)—doped or nonsag (NS) doped—a stable compound that decomposes only at a latter stage of sintering produces potassium bubbles, 0.5 to 5 μm in diameter, that are aligned in rows in the direction of wire drawing. When this is drawn down, elongated grains have lengths several times the wire diameter (Fig. 5.24b), and fine strings of micropores form (Fig. 5.24c), act as barriers to grain boundary migration, and fix the positions of grain boundaries. That is, an elongated and well-interlocked secondary grain structure is generated in tungsten wire (by impeding the transverse grain growth across the wire),[65a] as shown in Fig. 5.24b,[63] that can prevent the for-

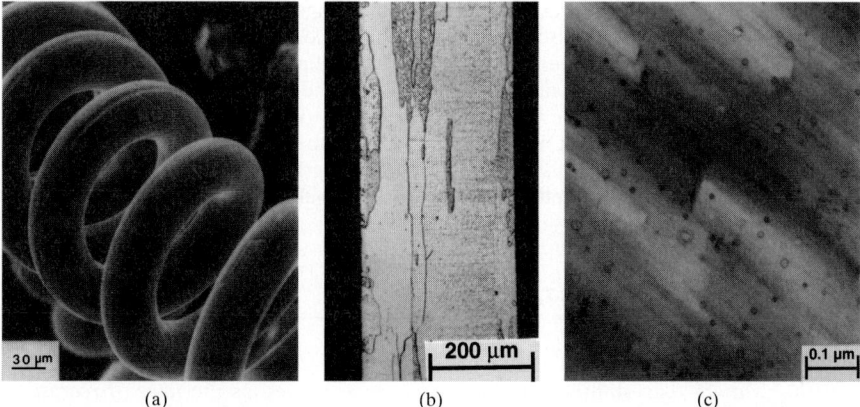

(a)   (b)   (c)

FIGURE 5.24 (*a*) Photograph of a filament and (*b*) micrograph of the cross section of NS-doped tungsten wire showing the interlocking secondary recrystallized grain structure. (*c*) Transmission electron micrograph of tungsten wire showing strings of potassium bubbles.[63] (*Courtesy of B. P. Bewlay and C. L. Briant.*)

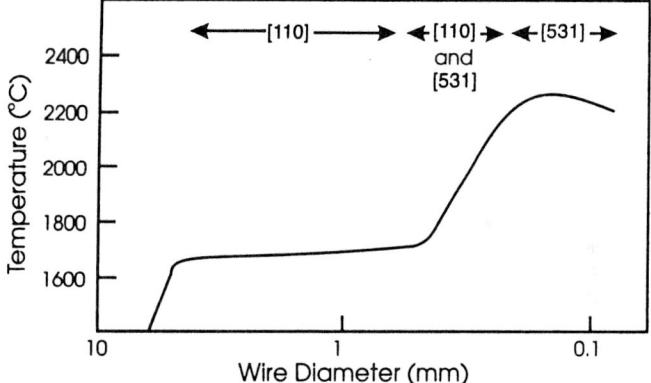

FIGURE 5.25 Increase in recrystallization temperature and change in texture, with increasing strain in drawn tungsten wire.[58]

mation of bamboo structure (which would otherwise cause premature sag and rapid failure of W filaments (Fig. 5.24*a*) during operation of incandescent lightbulbs due to grain boundary sliding), control the grain structure and sag resistance, increase the T_{recry} (Fig. 5.25),[58] reduce the creep strains (due to grain boundary sliding and cavitation),[65b] and increase the high-temperature strength and ductility of the as-drawn wire. This incandescent lamp wire structure is obtained by close control of the complete wire manufacturing process starting from ammonium paratungstate (APT) through to powder, pressing and sintering of ingot, wire drawing, and filament coiling.[63]

Mechanical deformation usually modifies the tungsten grain structure and both narrows and elongates the potassium bubbles into the ellipsoid in the rod and wire. When W wire or rod is annealed, the potassium ellipsoid breaks up into rows of bubbles due to Rayleigh instabilities. To understand the evolution of potassium bubbles from the initial doped powder of the ingot to sintering and their final formation in the lamp filament, readers should refer to the recent paper by Bewlay and Briant.[63] The most effective mechanism of strengthening the matrix by bubbles is the modulus-defect interaction which tends to block the climb of dislocations at high temperatures.[65b]

5.7   LAWS OF RECRYSTALLIZATION

It is very useful, for practical applications, to know the effect of important variables on the primary recrystallization process. These are listed as the *laws of recrystallization*,[34] as described below. These rules are followed in most cases and are easily explained in a rational manner if recrystallization is a nucleation and growth phenomenon, controlled by thermally activated processes, whose driving force is given by the stored energy of deformation.

1. A minimum amount of deformation is necessary to initiate recrystallization (Fig. 5.20*b*).

2. The smaller the degree of deformation, the higher the temperature necessary to cause recrystallization (Fig. 5.26).[66]

3. Increasing the annealing time decreases the temperature required for recrystallization[67] (Fig. 5.27).

4. The recrystallized grain size depends mainly on the degree of deformation (Fig. 5.28)[68] and, to a lesser degree, on the annealing temperature. Usually, the greater the degree of deformation and the lower the annealing temperature, the smaller the final grain size. Restated, the rate of recrystallization of a metal increases with increased amounts of cold deformation (i.e., with increased stored energy) and with increased recrystallization temperature.

5. The larger the initial grain size, the greater the amount of cold work required to achieve an equivalent recrystallization temperature and time. This general rule can be understood, if it is remembered that nuclei usually form at grain boundaries.

6. The amount of cold work required to give equivalent deformation hardening increases with increasing deformation temperature.

7. New grains do not grow into deformed grains of identical or slightly deviating orientation or into grains close to a twin orientation.[8,69]

8. Recrystallization is retarded if the temperature of deformation is increased for a given prestrain, but is accelerated if, at the higher temperature, the metal is strained at the same stress.[70]

9. Continued heating after the completion of recrystallization causes the grain size to increase.

10. For a given percentage deformation, the recrystallization (annealing) temperature is raised with slower heating rates, because recovery usually occurs during heating. This rule may become an important issue when batch annealing is compared with the high-speed resistance annealing processes.

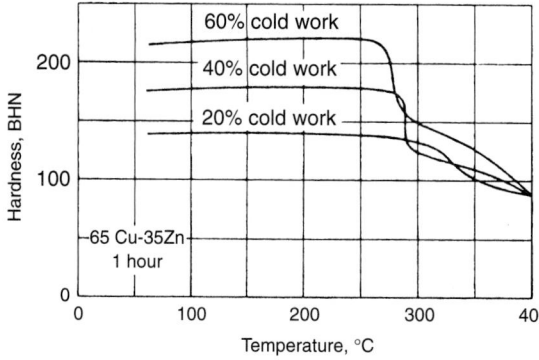

FIGURE 5.26 Variation of hardness of 65% Cu–35% Zn brass with percent cold reduction and annealing temperature. Each annealing time is for 1-hr duration. Highly deformed brass recrystallizes at a lower temperature.[66] (*Reprinted by permission of Addison-Wesley Publishing Co., Reading, Massachusetts.*)

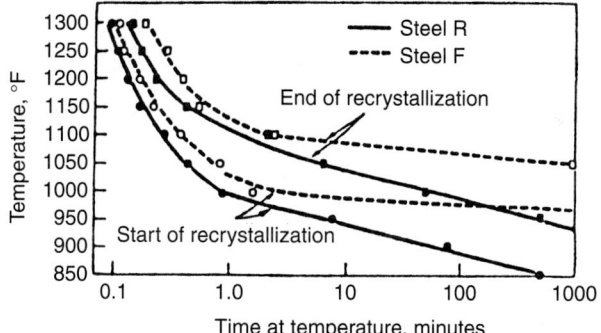

FIGURE 5.27 Time-temperature recrystallization diagrams for the rimmed steel (R) and the aluminum-killed steel (F) annealed in salt bath.[67] (*Reprinted by permission of ASM International, Materials Park, Ohio.*)

In view of the recent developments in the understanding of the recrystallization process, the seventh law is considered to be significant and appropriate. It emphasizes both the definition aspect for recrystallization accomplished by high-angle boundary migration and the orientation aspects of recrystallization which are very important for a complete understanding of the process.[22]

5.8 RECRYSTALLIZATION TEXTURE

The recrystallization of a cold-worked metal having preferred orientation or texture also produces a preferred orientation of the recrystallized grains which is generally different and more pronounced than that of the deformation texture. This is called the *annealing texture* or *recrystallization texture*.[71]

There are two theories of the origin of annealing texture: One is based on the formation of oriented nuclei (near the grain boundaries) with respect to the deformation texture, and the other is based on the growth of favorably oriented grains

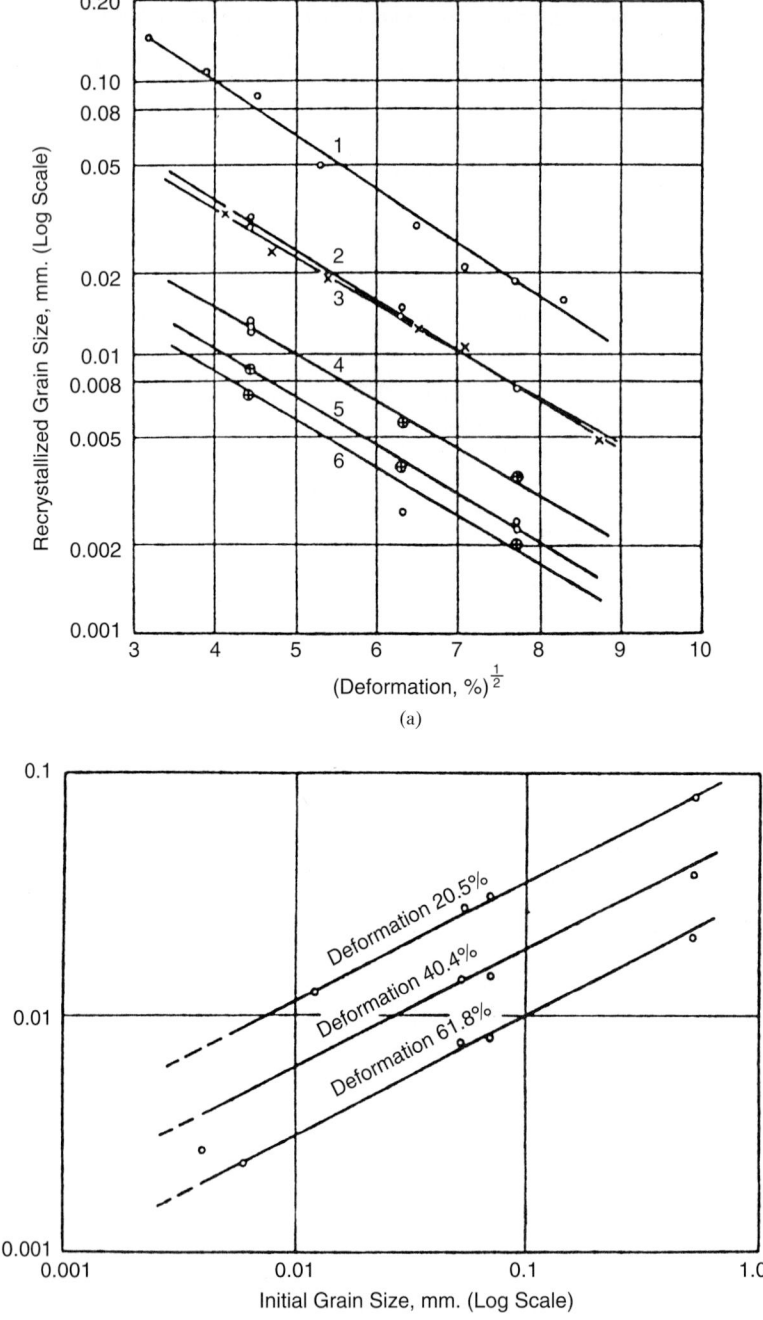

FIGURE 5.28 (*a*) Relation between recrystallized grain size *d* and percentage deformation *q* for pure 70-30 brass and initial grain sizes being rolled. The initial grain sizes corresponding to curves 1 through 6 are 0.53, 0.066, 0.053, 0.012, 0.006, and 0.004 (all in mm), respectively. The general equation for this relationship may be written as $\log d = n\sqrt{q} + \log m$, where *m* is a function of initial grain size.[68] (*b*) Relationship between recrystallized grain size *d* and initial grain size *i* in pure 70-30 brass for different values of prior deformation. The empirical relationship is given by $d = bi^a$, where *b* is a function of percentage deformation.[68] (*Reprinted by permission of ASM International, Materials Park, Ohio.*)

in relation to the deformation texture to a large size. The first is called the *oriented nucleation hypothesis*, while the second is called the *oriented growth hypothesis*. Like many controversial theories, later work has established that both mechanisms probably play a role in the development of recrystallization texture. Recent experimental evidence has suggested that orientation nucleation appears to be, in aluminum and copper, an important mechanism for the development of cube texture during recrystallization; however, final values might be dependent on the recovery behavior as well as factors influencing growth.[72] Dillamore et al.[73] and later Hutchinson[74] studied the annealing texture of rolled mild steel and showed that differently oriented deformed grains set up different amounts of strain energy, highest in (110)[001] and (111)[*uvw*] grains. However, if a fine critical dispersion of second-phase particles such as AlN is present, this greatly hinders the {100} component with lower stored energy from growing into large recrystallized grains, which allows the technologically desirable grains with a {111} component to grow. Thus these steels, after proper processing, possess a strong {111} <110> texture and a very weak {100} component. Annealing texture present in copper and α-brass consists of two or three components.[75]

A recent development in a microtexture method which entails a correlation of microstructure and *local* texture is now used for providing the three-dimensional orientation distribution to obtain a better understanding of the formation of annealing texture.[76]

The important process variables that affect the annealing texture are the boundary mobility, hot band texture, rolling reduction, preferential orientation of the nuclei of the recrystallized grains, chemical composition, initial grain size, hot band coiling temperature, heating rate during annealing, annealing temperature and time, and relative growth velocity of grains of different orientations.[4] Figure 5.29 summarizes the effects of coiling temperature and heating rate during final annealing on r_m values for different steel grades.[77] This demonstrates that low coiling temperatures ($\leq 600°C$) after hot-rolling, and slow heating rates represented by the batch annealing approach, impart high r_m values to Al-killed steels. This condition is reversed for high coiling temperatures ($\sim 700°C$), but the maximum value that can be achieved is lower. The r_m values of interstitial free (IF) steels stabilized with Ti, Nb, or both elements do not tend to be particularly sensitive to either coiling temperature or heating rates.[78] That is, these steels can be used to produce excellent r_m results with all types of annealing. The influence of coiling temperature is relatively weak for IF steels. Figure 5.30*a* and *b* shows the effect of C and Mn contents, respectively, on the r_m value of steel. Since small solute contents have an adverse effect on the development of desired strong {111} texture during annealing, Nb- or Ti-stabilized low-carbon steel is annealed at a temperature to dissolve a small amount of carbon into solid solution without affecting the texture during the completion of recrystallization.

In general, a well-defined recrystallization texture will be produced, provided the deformation texture is sharp. With small cold reduction, the final texture in annealed steel is weakly developed and often contains the Goss component {110} <001> (with shear band nucleation), together with a {111} <110> sheet plane texture. Shear band formation is promoted by coarse grain size and the presence of large C and N interstitial solute contents. The Goss component is weakened on increasing deformation, while the {111} intensity increases and the center of spread changes from {111} <110> to {111} <112> component.[71,79] A combination of high cold reduction and a small hot band grain size improves the (111) recrystallization texture. The (111)[112] component is especially enhanced by a large cold reduction, due to development of component by oriented growth into the (112)[110] matrix. Figure 5.31 summarizes the

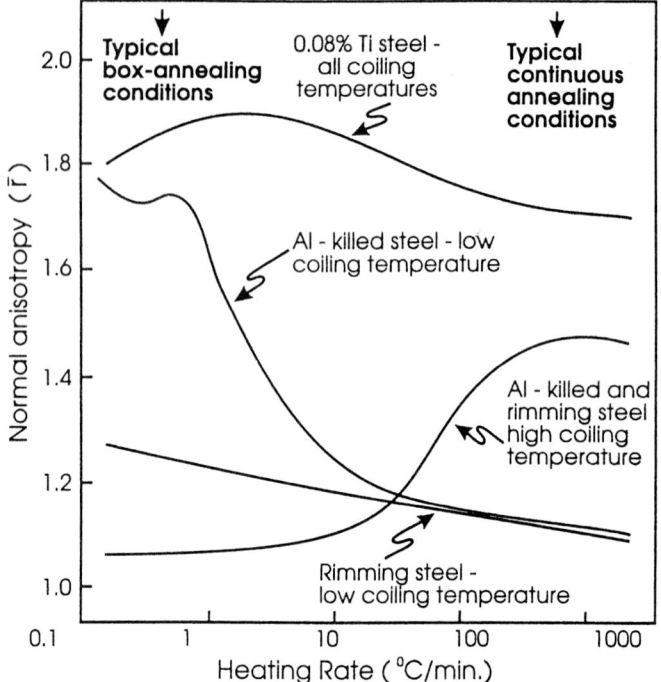

FIGURE 5.29 Summary of effect of coiling temperature (CT) and heating rate for final anneal on the r_m (or $\bar{r}$) values for different steel chemistries.[77,78a]

variation of important texture components during box annealing of rolled, low-carbon steel. This also shows that texture continues to change during grain growth.[77,80] For Al-killed steels, the hot band grain size affects both the recrystallized grain size and texture. It is usually necessary to ensure that the solutes C, N, and Mn are reduced to a low level in the hot band structure of Al-killed steel and during recrystallization itself, if high r_m values are to be obtained by cold rolling and annealing. Also, both Al and N present in the solid solution before the start of cold rolling should not precipitate during or after hot rolling. In fact, AlN precipitation is prevented by spraying water over the hot band after rolling and before coiling at ~580°C.

An outstanding example of a sharp recrystallization texture is the production of a strong cube texture {001} <100> in many fcc metals with medium to high SFE (notably Fe-Ni alloys as well as Cu, Au, and Al), after cold rolling and subsequent recrystallization of sheet material.[81] This texture is strengthened by high rolling reductions and annealing temperatures. If the SFE is quite low, minor components that correspond to the twins of this orientation may also be present. The strong cube texture of Cu is eliminated by small amounts of many alloying elements such as 5% Al, 1% Be, 0.2% Cd, 1.5% Mg, 4.2% Ni, 0.03% P, 0.3% Sb, 1% Sn, and 4% Zn.[81] Figure 5.32 shows such a sharp orientation of individual recrystallized grains toward the ideal orientation that the Fe-Ni alloy sheet material resembles the cube texture

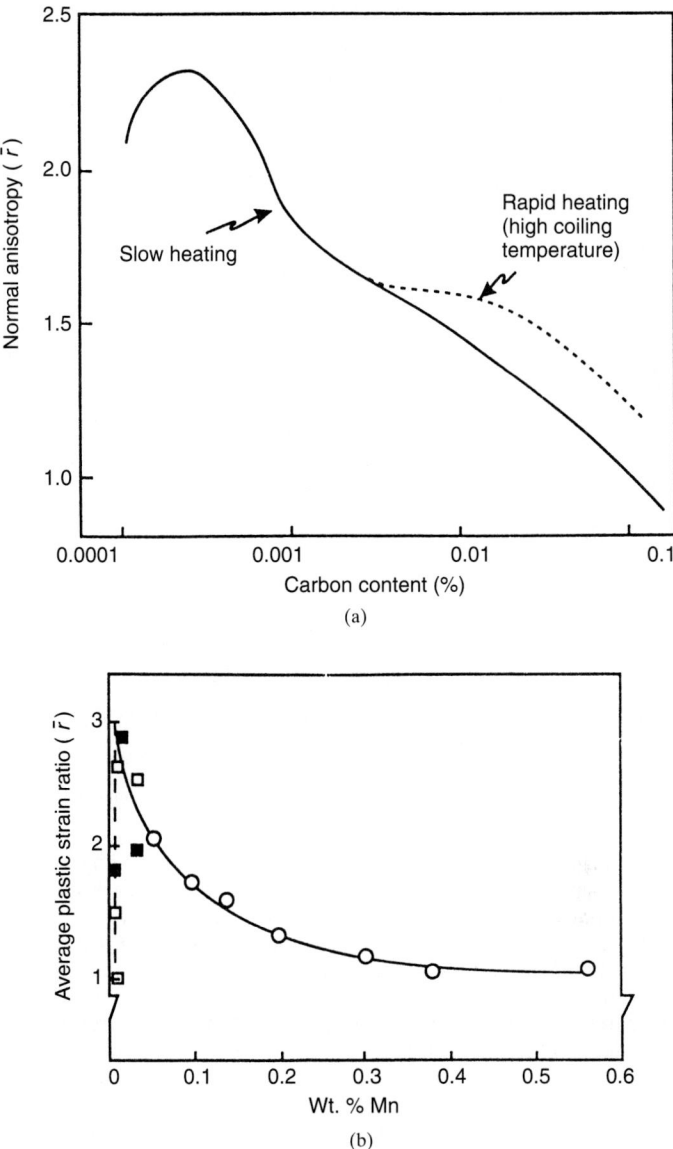

FIGURE 5.30 Effect of (*a*) carbon content on r_m (or $\bar{r}$) value of steel[77] and (*b*) manganese contents on r_m (or $\bar{r}$) value of steel. (*After Cline and Hu, quoted in Hu in 5th ICOTOM vol. 2, eds. G. Gottstein and K. Lücke, Springer-Verlag, Berlin, 1978, p. 3.*)

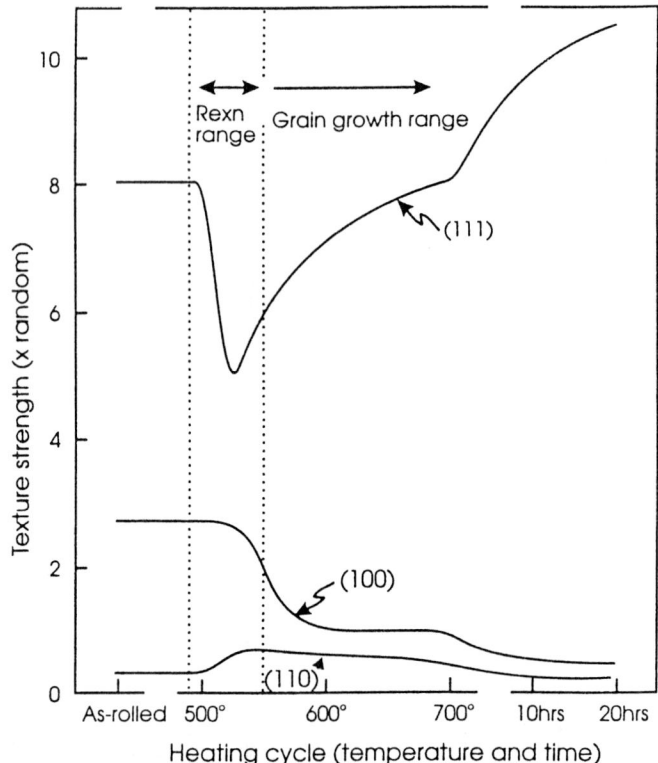

FIGURE 5.31 Variation of important texture components during box annealing of rolled, low-carbon steel.[77,80]

of a single crystal.[71] A dominant recrystallized cube texture with a minor Goss component is found by conventional processing of hot band Al-Mn-Mg (AA3004) alloy sheet for aluminum beverage can manufacturing. This includes annealing at 343 to 454°C, followed by cold rolling for an 85 to 90% reduction in thickness. The properties achieved by this processing are 276-MPa yield strength, 3 to 4% elongation, and ~1.6% earing. The modified processing consists of a high-temperature solution treatment instead of the conventional anneal, 70 to 75% cold-rolling reduction, and low-temperature aging at the final stage. This produces recrystallization texture with similar or even stronger cube component, better yield strength (~350 MPa), 4 to 6% elongation, and lower (~1%) earing values which are always desired for the aluminum beverage can manufacture.[82]

Based on the ODF study on rolled and recrystallized Ti, Inoue and Inakazu[83] showed that the major component after annealing within an α-phase field was $(02\bar{2}5)[2\bar{1}\bar{1}0]$ irrespective of the extent of rolling reduction and the annealing temperature. This orientation corresponds to the major component of the rolling texture $(\bar{1}2\bar{1}4)[10\bar{1}0]$ by a 30° [0001] rotation. As in bcc metals, the intensity increased greatly with the increase of cold reduction of annealing temperature. When annealing was accomplished within the β-phase field at 1000°C, a modified texture was reported due to variant selection associated with the phase transformation.[4]

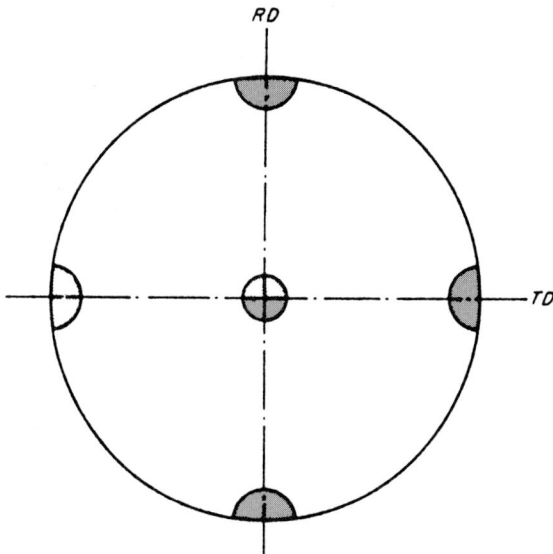

FIGURE 5.32 A {100} pole figure of the primary recrystallized texture of rolled and annealed Fe-50% Ni alloy at 1373K (cube texture).[71] (*After Burgers and Snoek.*)

Usually, the factors favoring the formation of fine recrystallized grains and that of randomly oriented recrystallized grains are similar. In steels, moderate plastic deformation and low annealing temperature are beneficial in developing the recrystallized texture.[84] In copper sheets, the change in the values of r (a measure of plastic anisotropy) remains the same after both moderate and heavy deformation, but in brass these changes are different. Planar anisotropy increases with increasing deformation and higher annealing temperature in copper sheets, whereas the reverse is found for brass sheets.[75]

Application.   Deep-drawing steels are obtained by using cold-rolling controlled composition with the aim of producing an essentially ferritic structure with a suitable recrystallization texture. An ideal texture is one with {111} crystallographic planes being strongly rotated parallel to the sheet plane. Its advantage is that it sets up the greatest through-thickness strength and consequently reduces wall thinning during drawing.[85]

5.9   GRAIN GROWTH

Further annealing after the completion of primary recrystallization leads to an increase in the average size of the grains. This process, called *grain growth*, is entirely controlled by grain boundaries (GBs)[86] and is carried out by the migration of high-angle grain boundaries into a fully recrystallized matrix. There are two types of grain growth: *normal* or *continuous grain growth* and *abnormal, exaggerated*, or *discontinuous grain growth*, also called *secondary recrystallization*. Normal grain growth is associated with the growth of all grains which, in turn, is associated with a decrease in GB energy. These types of grain growth together with annealing twins are discussed here.

5.9.1 Normal Grain Growth

5.9.1.1 Grain Boundary Migration and the Influence of Other Variables. *Grain boundary* (or *interface) migration* is defined as the displacement of the boundary normal to its tangent plane. Grain boundary mobility M is the characteristic feature explaining the migration behavior of a grain boundary and is defined as the migration rate or velocity V, divided by the driving force F_d arising from the stored energy of cold work or deformation (velocity per unit driving force). GB energy and mobility strongly influence the microstructural development and properties of polycrystals. GB mobility is a function of orientation in a complex manner.[86] The growth of a single recrystallized grain may be explained by this rate.[87]

5.9.1.2 Driving Force. Surface free energy (surface tension) of the grain boundaries is associated with the grain boundary movement. The driving force for grain growth is the decrease in the grain boundary energy of a material as the grain size is increased. Alternatively, the driving force for grain growth relates to GB energy and the kinetics of grain growth as determined by GB mobility.[86] In the soap bubble experiment, the higher pressure on the concave side of the soap film relative to that on the convex side of the soap film causes the gas molecules to move through the water-membrane film toward the convex side, so that cell boundaries move toward the concave side. This causes small bubbles finally to shrink and larger ones to grow. A similar effect occurs in metallic grains, where the atoms diffuse from the high-pressure side of the shrinking grain boundary toward the growing grain boundary.

The pressure difference caused by the curved boundary produces a difference in free energy ΔG (arising from the difference in stored energy between the new strain-free grains and the deformed grains) or in chemical potential $\Delta\mu$ (between the two adjoining grains), which, in turn, results in the atomic movement across the boundary according to the relation for a pure metal

$$\Delta G = \Delta\mu = \frac{-2\gamma V_m}{r} \tag{5.50}$$

where γ is the specific (interfacial) energy of the boundary, V_m is the atomic volume, and r is the radius of curvature (or spherical pinning particle). This expression is often called the *Gibbs-Thompson equation.* This free-energy difference exerts a force pulling the grain boundary toward the grain with higher free energy. The driving force F_d acting on the grain boundary is the *chemical potential gradient* across the boundary, which can be expressed by

$$F_d = \frac{\Delta\mu}{b} \tag{5.51a}$$

where b is the boundary width. Alternatively, the driving force due to curvature is expressed by the pulling force per unit area of the boundary as

$$F(N/m^2) = \frac{\Delta G}{V_m} = \frac{2\gamma}{r} \tag{5.51b}$$

To explain the behavior of grain boundary migration or instability of the grain structure, let us consider a two-dimensional network analogy of grain structure (Fig. 5.33).[88] For a stable grain structure the grains tend to appear as regular hexagons; that is, the sides of the grains would be flat, and every set of three (straight) grain boundaries would meet at a triple junction point at a dihedral angle of 120°. Grains that have less than six sides will necessarily have some of (if not all) the boundaries curved concave inward to give 120° intersections, whereas those having more than

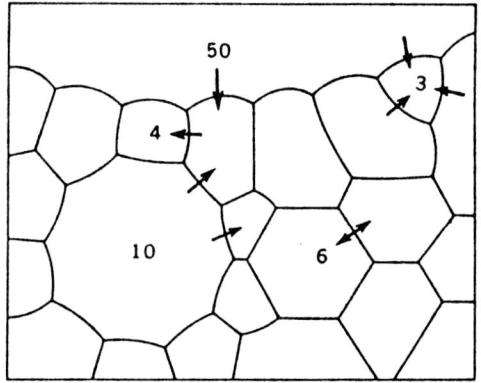

FIGURE 5.33 Schematic diagram showing two-dimensional array of grains in which all grain boundaries intersect at an angle of 120° with one another. Grains with more than six sides grow, while the grains with smaller than six sides disappear. The arrows indicate the direction of migration of grain boundary.[89] (*Courtesy of P. G. Shewmon; after Burke.*)

six sides will have some of (if not all) their boundaries curved concave outward to achieve included angles of 120°. As a result of these curvatures, the grains with less than six sides will tend to shrink or vanish, and those with more than six sides will tend to grow by the migration of the boundary toward its center of curvature. The smaller the number of sides, the sharper the curvature for a given grain size and the higher the probability for further shrinkage or absorption.

5.9.1.3 Grain Boundary Mobility. From the previous discussion we can conclude that grain boundary migration is associated with the transfer of atoms from one grain on one side of the grain boundary and joining the lattice of the other grains on the other side. This occurs by a series of diffusion jumps of individual atoms across the boundary. Figure 5.34 shows the free-energy-distance relationship that can be applied to the reaction rate theory given below. We assume here that (1) atomic jumps are independent of one another and (2) the free-energy barrier to be overcome in the jump process is similar for all atoms.[89]

In the presence of driving force, activation energy for the jump process ΔG^A is decreased in one direction and is increased in the reverse direction. According to various workers,[89–91] the following general expression has been developed for the migration velocity V and mobility M based on the absolute reaction rate theory and continuous transfer of atoms back and forth across the boundary.

The mean jump frequency of atomic transfer from grain A to B is $v \exp (-\Delta G^A/RT)$, where v is the atomic jump frequency, R is the universal gas constant, and T is the absolute temperature. The mean jump frequency in the reverse (from B to A) direction is $v \exp [-(\Delta G^A + \Delta\mu)/RT]$. The net jump frequency, therefore, becomes[92] $v \exp (-\Delta G^A/RT)[1 - \exp (-\Delta\mu/RT)]$. Expanding $\exp (-\Delta\mu/RT)$ and assuming $\Delta\mu \ll RT$, we get for the net jump frequency

$$\frac{v\Delta\mu}{RT}\exp\left(\frac{-\Delta G^A}{RT}\right)$$

The velocity of the grain boundary migration V can be represented as

$$V = bv\frac{\Delta\mu}{RT}\exp\left(\frac{-\Delta G^A}{RT}\right) = bv\frac{\Delta\mu}{RT}\exp\frac{\Delta S^A}{R}\exp\left(\frac{-Q^A}{RT}\right) \qquad (5.52)$$

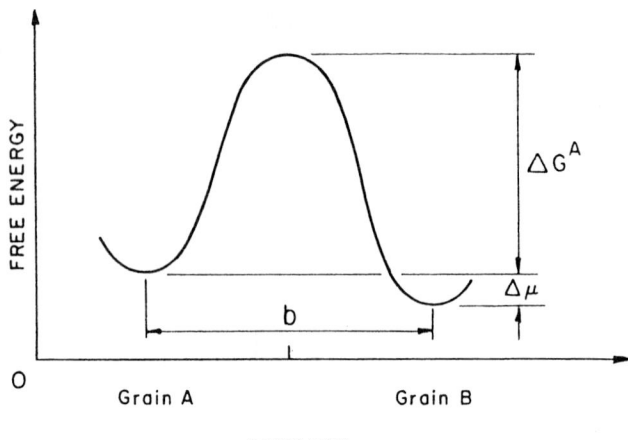

FIGURE 5.34 Free energy versus distance during the atomic movement across the grain boundary.[89] (*Courtesy of P. G. Shewmon.*)

Since

$$\Delta G^A = \Delta H^A - T\Delta S^A = Q^A - T\Delta S^A \tag{5.53}$$

where ΔH^A and ΔS^A are enthalpy and entropy, respectively, of activation for the migration process, Q^A is the activation energy of the migration process. Since the velocity of the grain boundary migration is regarded as the product of the mobility M and the driving force F_d, we can write Eq. (5.52) as

$$M = \frac{b^2 v}{RT} \exp\frac{\Delta S^A}{R} \cdot \exp\left(\frac{-Q^A}{RT}\right) \tag{5.54}$$

where M can be considered to be proportional to the diffusion coefficient of atoms across the boundary and where $(b^2 v/RT)$ exp $(\Delta S^A/R)$ is the preexponential parameter. Equations (5.53) and (5.54) both hold well for high-purity metals. That is, the boundary mobility M is influenced by impurity, second-phase particles, inclusions, orientation relationship between two grains, temperature, texture, and thickness.

Impurity (or Solute Drag) Effect. Recent (solute drag) theory[91] predicts that when the impurities' segregated boundary moves, the impurity atmosphere must be dragged by the moving boundary. This leads to a decrease in the mobility of the boundary at any given temperature by the limitation of the diffusing impurity atoms along with the moving boundary. Also, the boundary cannot be held back by solute atoms either at high temperatures or at low concentrations, and as a consequence, the moving grain boundary will pull free of the impurity atmosphere. In this case it is apparent that the grain boundary self-diffusion coefficient determines the rate of motion of the boundary. However, at low temperatures or at high impurity content, the moving boundaries are held back by the solute atoms. The rate of the boundary motion is then governed by the rate at which the impurity atoms diffuse behind the boundary. The activation energy for boundary motion and the volume diffusion of the solute atoms, therefore, should be nearly the same.

Second-Phase Particles. Finely dispersed second-phase particles have a strong inhibiting effect on recrystallization, particularly nucleation and boundary mobility.

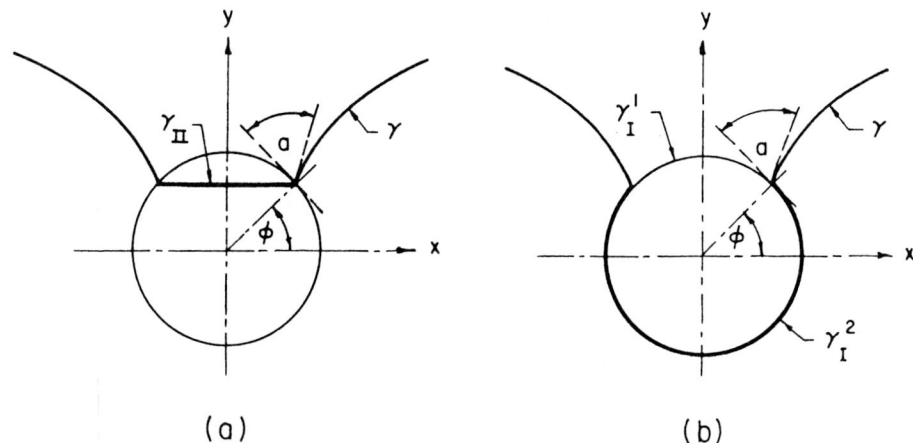

(a)   (b)

FIGURE 5.35 (*a*) Schematic diagram showing the condition for entry of the grain boundary inside the coherent inclusion. Specific energy of the grain boundary within the coherent inclusion γ_{II} is lower than that of outside the inclusion, γ.[94] (*b*) The schematic diagram showing the condition for the grain boundary to surround the incoherent inclusion. Specific energy of the boundary inside the inclusion is higher than that of the outside.[94] (*Reprinted by permission of Pergamon Press, Plc.*)

Hence they are used to control the grain size. Ashby et al.[93] and Gleiter and Chalmers,[94] after the revision of Zener's calculations,[95] have shown that when the grain boundary free energy within a coherent inclusion γ_{II} is lower than that of the outside γ, the boundary penetrates the inclusion (Fig. 5.35a). But when the inclusion is incoherent or when the boundary free energy inside the inclusion is higher than that of the outside, the boundary envelopes the inclusion (Fig. 5.35b). In the first case, when the migrating boundary enters inside a spherical coherent inclusion, local equilibrium of boundary tensions resolved parallel to the tangent of the inclusion surface yields α, which is given by

$$\cos\alpha = \frac{\gamma_{II}\cos\phi}{\gamma} \tag{5.55}$$

In the second case, after the boundary envelopes the incoherent inclusion, α is given by

$$\cos\alpha = \begin{cases} \dfrac{\gamma_i^2 - \gamma_i^1}{\gamma} & \text{when } \cos\alpha < 1 & (5.56a) \\[2ex] 1 & \text{when } \dfrac{\gamma_i^2 - \gamma_i^1}{\gamma} > 1 & (5.56b) \end{cases}$$

The restraining or particle pinning force F is the component of the boundary tension γ resolved parallel to the y axis (the vertical direction) (Fig. 5.35)[94] and is given by

$$F = 2\pi r \gamma \cos\phi \cos[90 - (90 + \phi - \alpha)] = 2\pi r \cos(\alpha - \phi) \tag{5.57}$$

where r is the radius of the spherical pinning particle, γ is the pinned grain boundary energy, and $2\pi r \cos\phi$ is the circumference. The maximum restraining force F_{max} exerted by a single particle can be obtained by using Eq. (5.55) or (5.56) for α and putting $\partial F/\partial\phi = 0$. For the boundary enveloping the inclusion, F_{max} is given by

$$F_{max} = \pi r \gamma (1 + \cos \alpha) \qquad (5.58)$$

where F_{max} varies from $\pi r \gamma$ to $2\pi r \gamma$ for incoherent inclusion, depending on the values of α. A good approximation is found when $F_{max} = \pi r \gamma$, which is equal to Zener's estimate for the particle pinning force.[95] In the case of the Zener model, the maximum force occurs when the boundary has moved $0.707r$ from the diametrical position. The coherent particles are twice as effective in pinning of a grain boundary as incoherent particles of the same size. It appears that the variables affecting the pinning force include the particle size, the particle distribution, and the nature of the particles, all of which affect the values of α. If the driving (or pulling) force for grain boundary migration [Eq. (5.51)] becomes larger than the pinning force exerted by the second-phase particles [Eq. (5.58)], the boundary is pulled free from the particles and moves through the lattice, unaffecting the particle distribution, as shown in Fig. 5.35b. If there are N of these second-phase (inclusion) particles per unit volume, all randomly distributed with an average particle radius r, their volume fraction f is $(4/3)\pi r^3 N$. A boundary of the unit area will intersect all inclusion particles within a volume of $2r$, that is, $2Nr$ particles. Hence, the average number n of inclusions intercepted by a unit area of the boundary is given by

$$n = \frac{2Nrf}{(4\pi/3)r^3 N} = \frac{3f}{2\pi r^2} \qquad (5.59)$$

The restraining force per unit area of grain boundary is given by

$$F_{res} = \frac{3f}{2\pi r^2} \pi r \gamma (1 + \cos \alpha) = \frac{3f\gamma(1 + \cos \alpha)}{2r} \qquad (5.60)$$

when this force counterbalances the driving force due to curvature $2\gamma/r$ and the radius of curvature $r \cong \overline{D}$, where $\overline{D}$ is the mean grain diameter of the particle. When $\overline{D} (\cong r)$ is small, F_{res} will be negligibly small. For continuous grain growth, the driving force $2\gamma/r$ decreases with increasing grain size. Grain growth will stop when the driving force becomes equal to the F_{res}. Equating both driving and restraining (or pinning) forces, we obtain

$$\frac{3f\gamma(1 + \cos \alpha)}{2r} = \frac{2\gamma}{\overline{D}} \qquad (5.61)$$

at the maximum (or limiting) grain size. We finally obtain a maximum grain size

$$\overline{D}_{crit} = \overline{D}_{lim} = \frac{4r}{3f(1 + \cos \alpha)} \qquad (5.62)$$

This equation is based on the assumptions that there is a uniform distribution of inclusion particles and that the particles are spherical in shape—conditions that are difficult to achieve. This equation permits the prediction of final grain sizes in commercial alloys.[96] It also explains the necessity of effectiveness of a large volume fraction of very small particles in retarding grain growth, thereby producing the fine-grained commercial alloys. Thus, for $f = 0.01$ and $r = 10^{-6}$ cm (just visible in the electron microscope), $\overline{D}_{crit} = \overline{D}_{lim} \approx 0.013$ cm; in effect, the particles exert effective grain size control.[8] Thus, a stable grain size is expected if the particle pinning forces exceed the driving force, and an unstable grain size or grain growth if the driving forces exceed the pinning forces.

The Zener limiting grain size $\overline{D}_{Zlim}$ is obtained when $\alpha = 90°$, that is,

$$\overline{D}_{Zlim} = \frac{4r}{3f} \qquad (5.63)$$

Gladman[97,98] developed a geometric model by (1) assuming tetrakaidecahedral grains, (2) considering two effects of second-phase particles, namely, coalescence and growth and partial or total dissolution of these particles, and (3) using Ostwald ripening in controlling the onset of grain growth. The solubility products of various grain-refining particles such as AlN, VC, VN, NbCN, and TiC in austenite as well as the commencement of austenite grain growth in steels without substantial dissolution of the second-phase particles are well documented and established. The particles, when small and freshly precipitated, are potent in inhibiting the grain growth processes. However, with extended holding, the particles coalesce and grow until they exceed the critical particle size $\overline{D}_{G\lim}$ as given by

$$\overline{D}_{G\lim} = \frac{\pi r}{3f}\left(\frac{3}{2} - \frac{2}{Z}\right) \tag{5.64}$$

where Z is the size advantage of a particular grain over that of its neighbors $= R$ (radius of the growing grain) / R_0 (radius of the matrix grains). Although this parameter is difficult to calculate, its value is expected to be between 1.33 and 2.[4]

The alloys that are resistant to grain growth are produced with the aim of achieving increased temperature range or increased heat treatment time over which the particles are effective. This is accomplished by increasing $\overline{D}_{G\lim}$ by increasing the volume fraction of particles or selecting specific types of particles which are resistant to coarsening.

Inclusion Shape. It is well known that oxide, sulfide, and nitride inclusions in steel sheets inhibit normal grain growth. When oxides in nonoriented electrical steels are elongated, they inhibit grain growth during consumer annealing, and average grain size becomes small, causing bad core loss. When oxides are spherical, they do not inhibit grain growth, and the average grain size becomes large, causing good core loss.[98a]

Temperature Effect. The relationship between mobility M and grain boundary diffusion coefficient D_b, for pure metal, can be given by[99]

$$M = \frac{D_b}{RT}\frac{V_m}{d} \tag{5.65}$$

where V_m is the atomic volume, d is the interatomic jump distance, R is the gas constant, and T is the absolute temperature. This equation shows that the grain boundary mobility is very temperature-dependent. As the temperature is increased, the difference between the mobility of special boundaries in materials of moderate purity and that of random boundaries diminishes.[93] This may be due to the evaporation of solute atmosphere at boundaries at high temperatures.

Orientation Effect. Orientation of the crystals across the boundary affects the mobility of the grain boundary. It has been concluded by many workers[22,100] that for fcc metals a [111] rotation of 30 to 40°, for bcc metals a [110] rotation of 20 to 30°, and for hcp metals a [0001] rotation of 30° correspond to the highly mobile boundaries.[101] As the orientation mismatch decreases to zero, the diffusion coefficient in the boundary falls from the higher grain boundary diffusion coefficient level to the low matrix diffusion coefficient level.[96] According to Gottstein and Shvindlerman, a gap within which orientation dependence of mobility is noticed may be determined by both the boundary structure and the purity level, as shown schematically in Fig. 5.36.[102]

Vacancy Effect. Unlike the retardation or pinning effects of solute atoms, inclusions, second-phase particles, and so on, on migration of the boundary, vacancies act to improve the boundary mobility. The vacancy-enhanced mobility causes (1) the

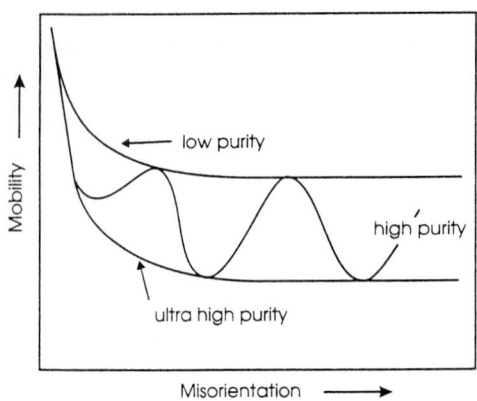

FIGURE 5.36 Schematic diagram showing the dependence of the activation energy of grain boundary mobility for different purity levels.[95]

vacancy absorption within the grain boundary and (2) a subsequent steady supply of excess vacancies. The diffusion of excess vacancies along the grain boundary facilitates the transfer of atoms across the boundary and, therefore, increases the grain boundary mobility.[15]

Thickness Effect. The rate of grain growth diminishes sharply once grain size becomes greater than the thickness of a sheet specimen. This is due to the reduced driving force associated with boundaries which are cylindrically curved in one direction rather than spherically curved, also with the drag caused by the formation of grooves on the sample surface by surface diffusion. This influence is called *thickness inhibition*.[8,14]

Texture Effect. Grain growth may also be seriously affected by the presence of a sharp or strong crystallographic texture. This arises from the much reduced low-angle grain boundary energy γ associated with grains having mutual misorientations of less than 15°; this, in turn, reduces the driving force for grain growth, which for a specific grain size is proportional to γ. This effect is called *texture inhibition*.[8]

5.9.1.4 Kinetics of Normal Grain Growth.
In a pure metal and single-phase alloys, $d\overline{D}$ depends on the mobility and the driving force of the grain boundary. If we assume that r [in Eq. (5.50)], the mean radius of curvature of all moving boundaries, is proportional to the mean grain diameter $\overline{D}$ (at any instant during growth), then using the equation

$$V = \frac{-M\,\Delta\mu}{b} \tag{5.66}$$

we can write

$$V \propto \frac{d\overline{D}}{dt} = \frac{M(2\gamma V_m)}{b\alpha\overline{D}} \tag{5.67}$$

where α is a proportionality constant on the order of unity. This equation shows that $d\overline{D}/dt$ increases with the decrease of $\overline{D}$ and increases with temperature for a given $\overline{D}$, because the mobility again increases with temperature. Integration of Eq. (5.67), assuming $\overline{D} = D_0$ at $t = 0$ (the initial pregrowth grain diameter), gives

$$\overline{D}^2 - \overline{D}_0^2 = Kt \tag{5.68a}$$

where $\overline{D}$ is the average grain size after time t at temperature T; $\overline{D}_0$ is the initial grain size; Q is the energy differential across the boundary; and $K = 4\gamma MV_m/b\alpha$. Since this process is temperature-dependent, it can be represented by the following Arrhenius-type relationship:

$$\overline{D}^2 - \overline{D}_0^2 = kt\exp\left(-\frac{Q}{RT}\right) \tag{5.68b}$$

When $\overline{D} \gg \overline{D}_0$, Eq. (5.68a) reduces to

$$\overline{D} = \sqrt{Kt} \tag{5.69}$$

Various models and theories of grain growth predict that at large times

$$\overline{D} = K't^{1/n} \tag{5.70}$$

where t is the annealing time, K' is a proportionality constant that depends on material composition and temperature, and n is the grain growth exponent. A grain growth exponent $n = 2$ (i.e., parabolic growth law) was predicted by the simple theory of Burke and Turnbull[103] and is mostly considered as an ideal value for n. Mullins and Venals[104] have predicted $n = 2$ for a single-phase polycrystal. However, experimental values of n are usually larger than 2 and may change with time. These are called *normal grain growth equations*.

5.9.1.5 Grain-Coarsening Behavior. Coarse grains are usually avoided during cold-working operations because of the reduction in strength, toughness, and hardness with increasing grain size. Moreover, the plastic flow under stress becomes uneven, which causes the smoothness of the metallic surface to be impaired; this produces surface roughness or orange-peel appearance. Hence, the annealing conditions are selected in such a way as to limit the secondary recrystallization.[105]

The coarsening or Ostwald ripening of the second-phase particles used to inhibit austenite grain growth in steel is usually found to follow the Wagner equation for diffusion-controlled growth

$$r^3 - r_0^3 = \frac{8\gamma D[M]V_m t}{9RT} \tag{5.71}$$

where r is the particle radius after a time t; r_0 is the initial particle radius; γ is the interfacial energy; D is the diffusion coefficient of the relevant (or rate-limiting) species; $[M]$ is the dissolved content of the solid; V_m is the molar volume; R is the gas constant; and T is the absolute temperature.

Plain carbon steels without grain-refining additions exhibit grain coarsening by normal grain growth (Fig. 5.37). When grain-refining particles such as AlN, VC, VN, NbCN, TiC, and TiO are added, fine grains persist up to a coarsening temperature, and the grain-coarsening temperature increases. In Al-killed steels, growth inhibition fails by abnormal grain growth process at temperatures between 1050 and ~1200°C (Fig. 5.37a). Above this temperature, normal grain growth occurs. However, if a coarser dispersion of more stable particles such as oxides or TiCN is present (Fig. 5.37b), all grain growth is hindered up to a very high temperature.

5.9.2 Formation of Annealing Twins

Annealing twins have been commonly observed in recrystallized structures, particularly in fcc metals (such as Cu-group metals and alloys, lead, nickel-base superalloys, and austenitic steels) and disordered alloys as well as in many intermetallic

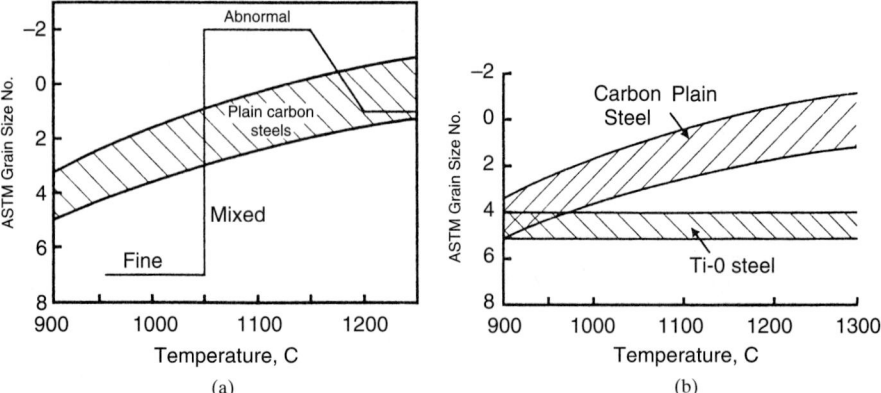

FIGURE 5.37 (*a*) Austenite grain sizes in plain carbon and aluminum-killed low-carbon steel at various tyemperatures showing normal and abnormal grain growth. (*b*) Austenite grain sizes at various temperatures in Ti-O grain-refined steels showing normal grain growth.[97,98]

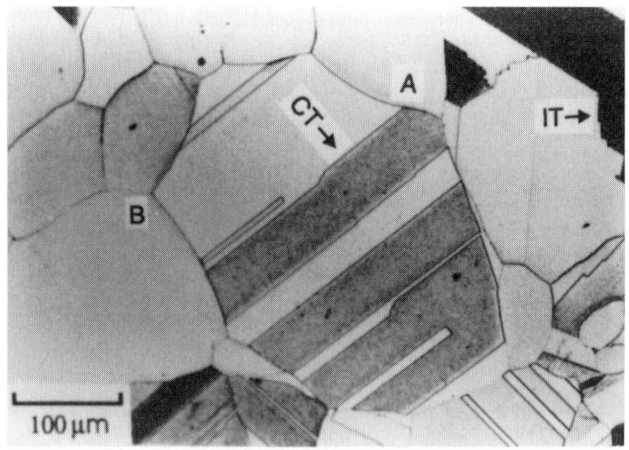

FIGURE 5.38 Annealing twins in annealed 70-30 brass. Some coherent twin (CT) and incoherent twin (IT) boundaries are marked.[4] (*Courtesy of M. Ferry.*)

compounds, ceramics, and minerals. These twins take the form of parallel-sided lamellae, bounded by {111} planes or *coherent twin* (*CT*) *boundaries* and at their ends, steps or terminations, by *incoherent twin* (*IT*) *boundaries*, as shown schematically in Fig. 5.38. Twins may form during recovery, recrystallization, and grain growth (see also Sec. 4.7).

Grindvaux and Form[106] have established, by direct observation, that most annealing twins occur during primary recrystallization and few additional twins form during subsequent grain growth. When coherent twin interfacial energy γ_T is greater than the ordinary grain boundary energy γ_B, as in aluminum, where $\gamma_T/\gamma_B = 0.2$, then annealing twins are rarely visible.

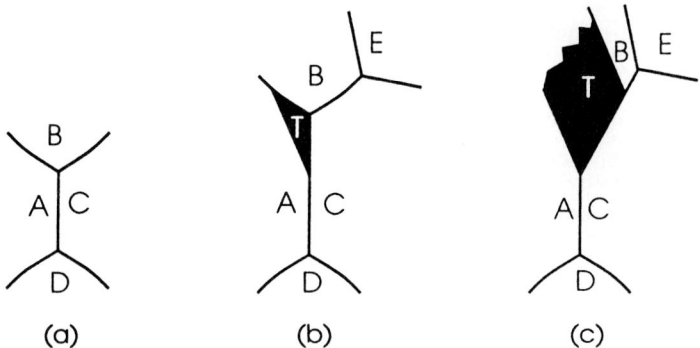

FIGURE 5.39 Mechanism for twin formation during grain growth following recrystallization.[4]

The important models for the formation of annealing twins can be classified into three groups: (1) growth accident model, (2) grain encounter model, and (3) model consisting of nucleation twins by stacking faults or fault packets. The first model assumes that a coherent twin boundary forms at a migrating grain boundary due to a stacking error during growth under energetically favorable conditions. The second model assumes that different grains initially separated "encounter" each other during grain growth. If these grains tend to be in twin orientation to each other, the boundary between them becomes a coherent twin boundary after reorienting itself. In the third model, a grain boundary during its migration nucleates a twin so that its incoherent segment lies at the grain boundary. The twin then grows presumably by the migration of the other noncoherent boundary.[107]

Figure 5.39 shows the postulated mechanism of twin formation during grain growth following recrystallization. As grain growth occurs, it is assumed that the triple point between grains A, B, and C moves vertically. As grain growth advances, a *growth fault* may be generated, and a twin will appear at the preceding B-A-C triple point, and such a fault will be stable in order to grow the twin (T), as shown in Fig. 5.39c. If the relative orientations of the grains are such that the energy of the boundary AT is lower than that of AC, then, because the energy of the coherent twin boundary AT is very low, there may be a reduction in total boundary energy albeit the extra boundary area is created, and thus the twin configuration will be stable and grow. This condition, in two-dimensions, is given by

$$\gamma_{AT}L_{13} + \gamma_{TC}L_{23} + \gamma_{TB}L_{12} < \gamma_{AC}L_{23} + \gamma_{AB}L_{12} \tag{5.72}$$

where γ_{ij} is the energy of the boundary between grains i and j and L_{xy} is the distance between points x and y.[4]

The growth will stop if the triple point ABC reacts with another triple point, say, BCE, leading to a grain configuration with a less favorable energy balance (Fig. 5.39c). On this model, the number of twin lamellae should vary linearly with the number of triple-point interactions; this evidence was observed by Hu and Smith.[108] The actual atomistic mechanisms of twin formation during grain growth and during recrystallization are likely to be similar.

Annealing twins have also been observed in bcc metals and alloys in which the crystallography is more complex in that one twin can possess three distinct coher-

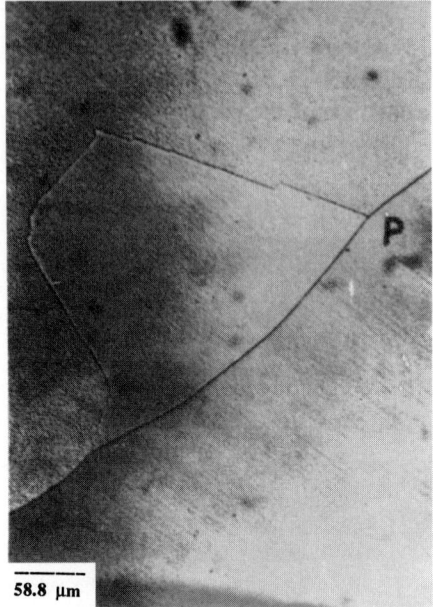

FIGURE 5.40   Annealing twins in an Fe-Al alloy after prolonged grain growth.[8] (*After R. W. Cahn and J. A. Coll, Acta Metall., vol. 4, 1961, p. 683.*)

58.8 μm

ent interfaces, as shown in Fig. 5.40. The low energy of the coherent boundary can be inferred from the fact that the normal grain boundary is only slightly diverted where a coherent boundary adjoins on it (at P).

Recently, the methods of mesotextural measurement have been used to study the twin formation during recrystallization and grain growth.[109] Twin formation has also been cited in one model of nucleation in primary recrystallization, namely in Cu; this model was fully established by Haasen.[33]

5.9.3 Secondary Recrystallization

When the annealing of an initially deformed material is continued for a long time (even) after the complete formation of a (primary) recrystallized structure, a few grains start growing preferentially and very rapidly until they impinge upon one another. Thus, they consume all small neighboring (recrystallized or normal) grains and produce a very coarse-grained structure, on the order of several centimeters. In extreme cases, single-crystal metals can be produced, for example, by annealing commercially pure Mo wire, drawn to a proper degree, at 2000°C.[110] This is called *secondary recrystallization* or *abnormal growth*. The requirement for secondary recrystallization is the strong impediment of normal grain growth, with the exception of a few grains that act as "nuclei" for secondary recrystallization; here the large grains are not freshly nucleated but are merely large-grown grains of the primary structure. The discontinuous growth of selected grains has similar kinetics to primary recrystallization and has some microstructural similarities, as shown in Fig. 5.41. Secondary recrystallization is promoted if one or more of the following inhibiting conditions for normal grain growth are realized:[105]

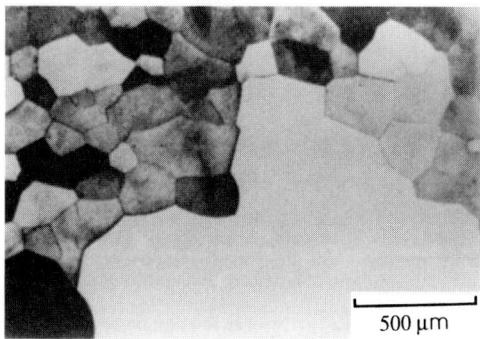

500 μm

FIGURE 5.41 Secondary recrystallization in Fe-3% Si during an anneal at 1373K.[105] (*Detert, 1978.*)

1. Introduction of an array of less stable particles capable of totally inhibiting grain growth even for a grain with an infinite size advantage favors secondary recrystallization.[97] Examples are the presence of finely dispersed second-phase particles such as MnS and AlN in Fe-3% Si alloys and at least 1% retained austenite in certain grades of 12% Cr steels used for the manufacture of turbine disks.[111] As the grain growth inhibitors coarsen and dissolve, abnormal grain growth occurs (Fig. 5.37). Abnormal grain growth is viable in alloys where normal grain growth has stagnated due to particle pinning.

2. The average grain diameter $\overline{D}$ after primary recrystallization has reached a limiting value which is twice the thickness of the sheet t or is equal to the diameter of the wire d. That is, grain coarsening stops when $\overline{D} = 2t$ or $\overline{D} = d$. Hence, abnormal growth may not be possible.

3. Presence of a strong single-orientation texture component in a fine-grained recrystallized material favors the formation of abnormal grain growth on further annealing at high temperatures. The grain boundaries within a highly textured volume have a lower misorientation and, therefore, a much lower energy and mobility than those within a normal grain structure.

4. Abnormal grain growth may result either from an abnormally high boundary mobility or from a higher driving force. Abnormal grain growth is possible if abnormal grain growth is faster than the growth in the average grain assembly.

5. It is obvious from condition 2 that sufficient thermal energy or Ostwald ripening is necessary to facilitate significant grain boundary displacement. That is, abnormal grain growth is likely to occur if the temperature is raised above the grain-coarsening temperature and as the particle dispersion becomes unstable. The *coarsening temperature* is defined as the temperature at which particles can grow to, or exceed, the critical radius (i.e., limiting size).[111]

Like primary recrystallization, secondary recrystallization is a nucleation and growth process. An incubation period is usually observed. A plot of the volume fraction of the secondary (recrystallized) grains versus isothermal annealing time produces a characteristic sigmoidal curve. Figure 5.42 shows such a curve which represents the kinetics of secondary recrystallization in Fe-3% Si alloy, upon isothermally annealing, producing the cube texture.[112]

Like primary recrystallization, the progress of secondary recrystallization can be described by the Avrami relationship

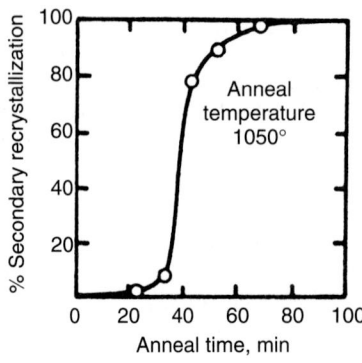

FIGURE 5.42 Sigmoidal curve for the secondary recrystallization for the formation of the cube texture in Fe-3%Si upon isothermal annealing.[22,112] (*Reprinted by permission of The Metallurgical Society, Warrendale, Pennsylvania.*)

$$f(t) = 1 - \exp(-kt^n) \tag{5.73}$$

where $f(t)$ is the fraction of secondary recrystallized grains, t is the annealing time, and k and n are constants. However, the driving force for secondary recrystallization is the large decrease in total interfacial energy of the grain boundary.

Dunn and Walter[113] have given the following equation expressing the relationship between relative contributions of grain boundary, surface energy, sheet thickness, and driving force of a secondary grain in a weakly textured matrix:

$$\frac{\text{Grain boundary energy term}}{\text{Surface energy term}} = \frac{\gamma_b/\bar{r} - \gamma_b/r}{2\,\Delta\gamma_s/t} \tag{5.74}$$

where γ_b is the grain boundary energy per unit area, $\Delta\gamma_s$ is the difference in surface energy, $\bar{r}$ is the average grain radius in a stable matrix, r is the radius of the secondary (or potential secondary) grain, and t is the sheet thickness. This ratio is maximum when $r \gg \bar{r}$, and it is zero when $r = \bar{r}$.

5.9.4 Commercial Application of Secondary Recrystallization Principles

A better solution for avoiding the intergranular fracture problems associated with the application of fine filamentary Mo and W wires for electron tubes, emitters of thermoionic electrons, incandescent lamps, and so forth is to use them as long, pure, single-crystal wires. This is achieved by exploiting the secondary recrystallization phenomenon.[110] A coarse-grained or secondary recrystallized structure is highly desired in the production of soft ferrous magnetic sheets (e.g., Fe-Si and Fe-Mn alloy sheets) to be used as transformer cores. Low coercive force (or magnetizing force H) is a prerequisite in the design of soft magnets used as transformer cores. The presence of (1) dispersed second-phase particles, (2) minimal nonmagnetic inclusion, and (3) the negligible orientation differences between the neighboring grains are characteristic features for the production of such soft magnetic materials, because all these reduce coercive force. Coercive force decreases with increasing grain size because the domain pattern across the grain boundary is simple as a result of the smaller orientation difference between the neighboring grains. A simple domain pattern can also be produced at the free surface when this is parallel to the direction of easy magnetization. Thus, coercive force can be reduced to a very low level by using rolling and anneal schedules to produce either a cube-textured or a cube-on-edge textured material.[114]

When a strong cube-texture is obtained, two easy <100> directions lie in the plane of the sheet parallel to the rolling and transverse directions. On the other hand, when a cube-on-edge texture is produced, it makes the easy <100> direction in all grains almost parallel to one another and to the rolling direction of the sheet. The former texture has superior magnetic properties relative to the latter texture.[13] This technique is very important because transformer material is applied in the form of thin sheets to achieve minimum hysteresis loss.

When strips made of Fe-3% Si alloy containing a fine dispersion of MnS particles are given a final treatment in the temperature range of 900 to 950°C, it produces a (110) [001] cube-on-edge (or Goss) texture of the secondary grains with a directionally preferred ("easy") magnetization along the strip-rolling direction, thereby requiring a smaller mass of transformer laminations, usually with a 0.30- to 0.35-mm thickness.[105]

Dunn and Walter[115] have produced the secondary recrystallization texture in high-purity irons and 0.6% Si-Fe. They have found that the surface energies control the preferred orientation, which, in turn, depends on the annealing atmosphere. If the annealing atmosphere is contaminated with high O_2 content, γ_{100} (surface energy of the {100} planes) becomes the lowest {hkl} surface energy that leads to the preferential grain growth in {100} <001> orientation. On the other hand, if O_2 is low or absent in the atmosphere, γ_{110} (surface energy of {110} planes) becomes the lowest {hkl} surface energy that permits the secondary grains to grow preferentially in the (110) [001] orientation. Thus secondary recrystallized texture can be either {100} <001> or {110} <001>, or a mixture of both, depending upon the O_2 content in the annealing atmosphere.[100]

Another variation to improve and obtain a sharp preferred (110) [001] texture is accomplished by the controlled dispersions of small AlN particles which precipitate in the deformed material. Minor additions of boron, sulfur, and nitrogen which segregate preferentially at grain boundaries have also been used by Grenoble and Fiedler to improve the preferred (110) [001] texture.[116]

In contrast to the findings of Dunn and Walter,[115] Taguchi and Sakakura[117] have observed that when single crystals of 3% Si-Fe alloy containing needlelike AlN particles and having (001) [100] orientation were rolled with 70 to 80% reduction and annealed between 800 and 1000°C, they noticed the secondary recrystallization texture with the initial (001) [100] orientation, irrespective of the purity and thickness of crystals, the purity of gas atmosphere, and the formation of surface scales. The driving energy for this texture is surprisingly not the surface energy. According to them, the effectiveness of AlN to cause (001) [100] secondary recrystallization texture should have the following characteristics:

1. Very fine needles (~1 μm in length) of AlN, which precipitate at 800 to 1000°C, are easily dissolved into solution within a short period by raising the temperature to, say, 1200°C, for 5 min. However, they reprecipitate on slow cooling.

2. These fine needles have an hcp structure with $a = 3.104$ Å, $c = 4.965$ Å, and $c/a = 1.60$.

3. The orientation relationship between needles and parent crystal, as obtained by analysis of the electron diffraction pattern, is given as follows:

$$\{10.1\}_{AlN} \,//\, \{120\}_{\alpha} \qquad\qquad (5.75a)$$

$$\{12.2\}_{AlN} \,//\, \{122\}_{\alpha} \qquad\qquad (5.75b)$$

Mostly, the needles tend to precipitate on $\{100\}_{\alpha}$ or $\{120\}_{\alpha}$.

Secondary recrystallization with either (100) [001] or (110) [001] texture has also been reported in Fe-3.25% Si rolled strips from sintered compacts by controlling the purity, particle size of the metal powder, temperature, and O_2 content of the annealing atmosphere.[118]

5.9.4.1 Production of Grain-Oriented Silicon Steel Sheets.

Two processes are used to manufacture *grain-oriented silicon steel* (GOSS) sheet as a transformer core material for electrical applications. The first is called the *Armco or two-stage cold-rolling process*, as summarized in Table 5.4. The essential requirements in the Armco process include (1) nucleation of {110}<001> grains, (2) ability of these grains to grow, and (3) grain growth inhibition of other orientations. The process consists of initial hot rolling, two light cold-rolling stages, and three annealing stages.

The necessary conditions for their abnormal growth are achieved by microstructural control. A fine dispersion of MnS precipitates produced by rapid cooling of the slab prior to hot rolling is resistant to rapid grain coarsening, which maintains the small matrix grain size during the early stages of the high-temperature annealing. As Ostwald ripening and dissolution progress, abnormal grain growth with a strong cube-on-edge or GOSS {110}<001> texture is produced. Abnormal grain growth is also favored by the presence of sharp texture. Preferential grain growth in desirable orientations at the surface is also achieved by the addition of S to the MgO surface coating. This prevents the growth of surface grains, and the sulfide formed is eventually eliminated by reaction with the H_2 atmosphere for steels containing ~0.25% Al and ~0.01% N.

The second method, called the *Nippon Steel process*, was developed by Taguchi et al.[119] and Sakakura.[120] This process consists of a single-stage cold-rolling reduction process and requires both MnS and AlN precipitates as grain growth inhibitors. This process involves rapid cooling after hot rolling step, large cold rolling, decarburization at 850°C in wet H_2 atmosphere, and addition of metal nitrides and sulfur in the final annealing treatment to control the decomposition of AlN particles. Although both processes are being continually refined, it is believed that the Nippon Steel process leads to a stronger GOSS texture but to a large grain size.

A cube-oriented sheet has been described by Arai and Yamashiro[121] for a 3.26% Si-Fe alloy directly cast to 0.37-mm thickness prior to final cold rolling. Here surface-energy-controlled grain growth occurred with a strong {100}<uvw> texture in the final sheet thickness of 0.15 mm without using any inhibitor.

TABLE 5.4 Processing of Grain-Oriented Silicon Steel Sheets[4]

Armco process	Nippon Steel process
a) Soak at ~1340°C; hot-roll to 2 mm.	a) Soak at ~1350°C; hot-roll to ~2 mm.
b) Cold-roll to 0.5 to 1 mm.	b) Hot-band anneal 1125°C, air-cool to 900°C, water-quench to 100°C
c) Anneal at 950°C in dry H_2/N_2 (80:20).	c) Cold-roll
d) Cold-roll to finished size.	d) Decarburize at 850°C in wet H_2; dew point 66°C.
e) Decarburize at 800°C in wet H_2; dew point 50°C.	e) Coat with MgO + 5% TiO_2.
f) Coat with MgO.	f) Texture anneal at 1200°C in H_2/N_2 (75:25).
g) Texture anneal at 1150°C in pure H_2.	

5.9.4.2 Surface-Controlled Secondary Recrystallization. A process called *surface-controlled secondary recrystallization* or alternatively *tertiary recrystallization*, was discovered by Detert[122] and Walter and Dunn[123] in thin (<0.6-mm) cold-rolled Fe-3% Si alloy sheet when recrystallized at a high temperature in the range of 1000 to 1200°C. Tertiary recrystallization has also been found in highly pure Pt—annealed at 1500°C, where the grain growth with (111) planes parallel to the surface was obtained.[124] Normal grain growth was retarded in the thin sheets, and large (secondary) grains with a (001) [100] (cube texture) appeared in contrast to normal (110) [001] GOSS texture. This only happens if the annealing atmosphere contains a slight trace of O_2 or S which results in the grain growth with a lower $\gamma_{s(100)}$ energy of (100) faces. If the atmosphere is changed to very dry H_2 or vacuum (i.e., the strip is clean), the GOSS texture grows because (110) faces have the lower specific surface energy than $\gamma_{s(100)}$. This change can be repeatedly performed by repeated changes of atmosphere. It has been well established that the development of these textures is due to the anisotropy in surface energy (surface tension) γ_s;[8] addition of small alloying elements in silicon steel sheet also affects the texture. Of the Group VI elements, the surface adsorption of Se gives rise to the preferred growth of {111} surface-oriented grains.[125]

This process has never been applied to produce cube-textured transformer laminations, because of probably small critical thickness of such laminations and the cost of necessary severe process control.

5.9.4.3 Secondary Recrystallization and Sintering. Secondary recrystallization is often observed in the final stage of sintering and aggravates the pore entrapment problem due to grain boundary separation from the pores. Here the pores remain stranded within their large grains, far away from the nearest boundary; vacancies then are unable to diffuse rapidly away from the pores, and sintering practically stops. Pore entrapment results when the driving force for grain boundary motion is substantially large and when the mobility of the grain boundary is greater than that of the pores, thereby resulting in grain boundary separation from the pores.[8,126]

For effective sintering to high densities, therefore, secondary recrystallization must be prevented; this is done by the addition of selected dopants and additives. For example, Al_2O_3 has been sintered to a high degree of translucency with the addition of small amounts of MgO and extended high-temperature sintering in H_2 or O_2 (these gases are soluble and mobile in Al_2O_3). Other examples are the sintering to full densification of Y_2O_3 doped with La_2O_3;[127,128] W and Mo doped with transition metals (Fe, Ni);[129] and Fe doped with P[115] at comparatively low temperatures without using an external pressure.[130]

REFERENCES

1. M. B. Bever, D. L. Holt, and A. L. Titchener, *Stored Energy of Cold Work, Progress in Materials Science*, vol. 17, Pergamon Press, Elmsford, N.Y., 1973.

2. P. Gordon, *Trans. AIME*, vol. 203, 1955, p. 1043.

3. B. R. Benerjee, *Met. Prog.*, November 1980, p. 59.

4. F. J. Humphreys and M. Hatherly, *Recrystallization and Related Annealing Phenomena*, Pergamon Press, Oxford, 1996.

5. H. J. McQueen, *Met. Trans.*, vol. 8A, June 1977, pp. 807–824.

6. F. J. Humphreys, in *Encyclopedia of Materials Science and Engineering*, Pergamon Press, Oxford, 1986, pp. 4101–4105.

7. P. Cotterill and P. R. Mould, *Recrystallization and Grain Growth in Metals*, Wiley, New York, 1976.

8. R. W. Cahn, in *Physical Metallurgy*, eds. R. W. Cahn and P. Haasen, chapter 28, Elsevier Science B. V., Amsterdam, 1996, pp. 2399–2501.

9. R. C. Bitchner and M. A. Kirk, in *Encyclopedia of Materials Science and Engineering*, Pergamon Press, Oxford, 1986, pp. 4099–4101.

10. R. W. Cahn, *J. Inst. Met.*, vol. 76, 1949, pp. 121–143.

11. R. E. Reed-Hill and R. Abbaschian, *Physical Metallurgy Principles*, 3d ed., PWS-Kent Publishing, Boston, 1992.

12. W. R. Hubbard, Jr., and G. C. Dunn, *Acta Metall.*, vol. 4, 1956, p. 306.

13. R. E. Smallman, *Modern Physical Metallurgy*, Butterworths, London, 1985.

14. F. J. Humphreys, in *Processing of Metals and Alloys*, ed. R. W. Cahn, vol. 15, VCH, Weinheim, 1991, pp. 371–428.

15. J. Talbot, in *Recovery and Recrystallization of Metals*, ed. L. Himmal, Interscience, New York, 1963, pp. 269–303.

15a. D. Kuhlmann-Wilsdorf, *Met. Trans.*, vol. 1, pp. 3173–3179.

15b. A. W. Thompson, *Met. Trans.*, vol. 8, 1977, pp. 833–842.

16. J. T. Michalak and H. W. Paxton, *Trans. AIME*, vol. 221, 1961, p. 850.

17. A. S. Keh, in *Direct Observations of Imperfections in Crystal*, eds. J. B. Newkirk and J. H. Wernick, Interscience, New York, 1962, p. 213.

18. W. C. Leslie, J. T. Michalak, and F. W. Aul, in *Iron and Its Dilute Solid Solutions*, eds. C. W. Spencer and F. E. Werner, Interscience, New York, 1963, pp. 119–212.

19. R. A. Vandermeer and B. B. Rath, in *Recrystallization '90*, ed. T. Chandra, The Metallurgical Society, Warrendale, Pa., 1990, pp. 49–58.

20. H. Pops, in *Non-Ferrous Wire Handbook*, vol. 3, Wire Association International, Guilford, Conn., 1995, pp. 7–22.

21. H. Hu, *Acta Metall.*, vol. 10, 1962, p. 112; *TMS-AIME*, vol. 224, 1962, p. 75.

22. H. Hu, in *Metallurgical Treatise*, eds. J. K. Tien and J. F. Elliott, TMS-AIME, Warrendale, Pa., 1983, pp. 385–407.

23. H. Hu, B. B. Rath, and R. A. Vandermeer, *Recrystallization '90*, The Metallurgical Society, Warrendale, Pa., 1990, pp. 3–16.

24. A. R. Jones, B. Ralph, and N. Hansen, *Proc. Roy. Soc., London, Ser. A*, vol. 368, 1979, p. 345.

25. R. D. Doherty and J. A. Szupunar, *Acta Metall.*, vol. 32, p. 1789.

26. J. C. Li, *J. Appl. Phys.*, vol. 33, 1962, p. 2558.

27. P. A. Beck and P. R. Sperry, *Trans. AIME*, vol. 180, 1949, p. 240; *J. Appl. Phys.*, vol. 21, 1950, p. 150.

28. J. E. Bailey and P. B. Hirsch, *Proc. Roy. Soc.*, vol. 267A, 1962, p. 11.

29. S. P. Bellier and R. D. Doherty, *Acta Met.*, vol. 25, 1977, p. 521.

29a. P. Faivre and R. D. Doherty, *J. Mater. Sci.*, vol. 14, 1979, pp. 897–919.

30. J. Huber and M. Hatherly, *Z. Metallk.*, vol. 71, 1980, p. 15.

31. A. R. Jones, *J. Mater. Sc.*, vol. 16, 1981, p. 1374.

32. P.-J. Wilbrandt and P. Haasen, *Z. Metallk.*, vol. 71, 1980, p. 273.

33. P. Haasen, *Met. Trans.*, vol. 24A, 1993, pp. 1001–1015.

33a. R. D. Doherty, private communication, 1999.

34. J. E. Burke and D. Turnbull, *Prog. Met. Phys.*, vol. 3, 1952, p. 220.

35. J. G. Byrne, in *Recovery, Recrystallization and Grain Growth*, Macmillan, London, 1965.

36. W. A. Johnson and R. F. Mehl, *TMS-AIME*, vol. 135, 1939, p. 416.

37. H. Pops and J. Hollomon, *Wire J. Int.*, May 1994, pp. 70–83.

37a. I. Samajdar and R. D. Doherty, *Acta Mater.*, vol. 46, 1998, pp. 3145–3158.

38. R. A. Vandermeer and B. B. Rath, *Met. Trans.*, vol. 20A, 1989, p. 391.

39. R. A. Vandermeer and B. B. Rath, *Proc. 10th Riso Int. Symp.*, eds. Bilde Sorensen et al., Riso, Denmark, 1989, p. 589.

40. R. A. Vandermeer, R. A. Masumura, and B. B. Rath, *Acta Metall.*, vol. 39, no. 3, 1991, pp. 383–389.

41. A. M. Gokhale and T. T. DeHoff, *Met. Trans.*, vol. 16A, 1985, pp. 559–564.

42. W. Anderson and R. F. Mehl, *Trans. AIME*, vol. 161, 1945, p. 140.

43. P. Furrer, in *Encyclopedia of Materials Science and Engineering*, Pergamon Press, Oxford, 1986, pp. 2051–2054.

44. F. Haessner and S. Hofmann, in *Recrystallization of Metallic Materials*, 2d ed., ed. F. Haessner, Dr. Riederer-Verlag, GmbH, Stuttgart, Germany, 1978, p. 63.

45. S. Murphy and C. J. Ball, *J. Inst. Met.*, vol. 100, 1972, pp. 225–232.

46. R. D. Doherty and J. W. Martin, *J. Inst. Met.*, vol. 91, 1963, p. 332.

47. T. Gladman, I. D. McIvor, and F. B. Pickering, *JISI*, May 1971, pp. 380–390.

48. D. T. Gawne and R. A. Higgins, *J. Mats. Sci.*, vol. 6, 1971, p. 403.

49. R. M. Brick, R. B. Gordon, and A. Phillips, *Structure and Properties of Alloys*, 3d ed., Eurasia Publishing, New Delhi, India, 1965.

50. D. Dimitrov, R. Fromageau, and C. Dimitrov, in *Recrystallization of Metallic Materials*, ed. F. Haessner, Dr. Riederer Verlag GmbH, Stuttgart, Germany, 1978, pp. 137–157.

51. A. G. Guy, *Elements of Physical Metallurgy*, 2d ed., Addison-Wesley, Reading, Mass., 1959.

52. J. S. Smart and A. A. Smith, *Trans. AIME*, vol. 144, 1942, p. 48; vol. 147, 1946, p. 166.

53. J. T. Michalak and W. R. Hibbard, *Trans. Met. Soc., AIME*, vol. 209, 1957, p. 101.

54. J. T. Michalak and W. R. Hibbard, *Trans. ASM*, vol. 53, 1961, p. 331.

55. R. L. Birto and L. J. Ebert, *Met. Trans.*, vol. 2, 1971, p. 1643.

56. J. D. Embury, W. J. Pool, and E. Koken, *Scripta Metall.*, vol. 27, 1992, p. 1465.

57. K. Ushioda, W. B. Hutchinson, J. Agren, and U. von Schlippenmbach, *Materials Science and Technology*, vol. 2, VCH, Weinheim, 1989, p. 807.

58. K. C. Thomson-Russell, *Planseebericht Pulvermetall.*, vol. 22, 1974, p. 264.

59. I. L. Dillamore, in *Recrystallization of Metallic Materials*, ed. F. Haessner, Dr. Riederer Verlag GmbH, Stuttgart, Germany, 1978, p. 223.

60. B. P. Bewlay, N. Lewis, and K. A. Lou, *Met. Trans.*, vol. 23A, 1992, pp. 121–133.

61. C. L. Briant, *Met. Trans.*, vol. 20A, 1989, pp. 179–184.

62. B. P. Bewlay and C. L. Briant, *Met. Trans.*, vol. 22A, 1991, pp. 2153–2155.

63. B. P. Bewlay and C. L. Briant, *Int. J. of Refractory Metals and Hard Materials*, vol. 13, 1995, pp. 137–159.

64. J. L. Walter and C. L. Briant, *J. Mater. Res.*, vol. 5, no. 9, 1990, pp. 2004–2021.

65. J. W. Pugh and W. A. Lasch, in *Tungsten and Tungsten Alloys: Recent Advances*, eds. A. Crowson and E. S. Chen, The Metallurgical Society, Warrendale, Pa., 1991, pp. 195–201.

65a. K. Tanoue, K. Watanabe, and H. Matsuda, *J. Jpn. Inst. Met.*, vol. 59, 1995, pp. 1230–1236.

65b. K. Tanoue, *Met. and Mats. Trans.*, vol. 29A, 1998, pp. 519–526.

66. L. H. VanVlack, *Elements of Materials Science and Engineering*, Addison-Wesley, Reading, Mass., 1989.

67. R. Goodenow, *Trans. ASM*, vol. 59, 1966, p. 804.

68. S. Channon and H. Walker, *Trans. ASM*, vol. 45, 1953, p. 200.

69. T. J. Tiedema, W. May, and W. G. Burgets, *Acta Crystallogr.*, vol. 2, 1949, p. 151.

70. G. Gottstein, P. Nagpal, and W. Kim, *Mat. Sci. & Eng.*, vol. A108, 1989, p. 165.

71. R. W. K. Honeycombe, *The Plastic Deformation of Metals*, 2d ed., Arnold, London, 1984.

72. P. M. B. Rodriguees, H. Bichsel, and P. Furrer, in *Texture in Non-Ferrous Metals and Alloys*, eds. H. D. Merchant and J. G. Morris, TMS-AIME, Warrendale, Pa., 1985, pp. 45–59.

73. I. L. Dillamore, C. J. Smith, and J. W. Watson, *Met. Sci.*, vol. 1, 1967, p. 49.

74. W. B. Hutchinson, *Met. Sci.*, vol. 8, 1974, p. 185.

75. B. Perovic and Z. Karastojkovic, *Met. Technol.*, February 1980, pp. 79–82.

76. V. Randle, *Microtexture Determination and Its Applications*, The Institute of Materials, London, 1992.

77. W. B. Hutchinson, *Int. Met. Rev.*, vol. 29, 1984, p. 25.

78. R. K. Ray, J. J. Jonas, and R. E. Hook, *Int. Mater. Rev.*, vol. 39, no. 4, 1994, pp. 129–172.

78a. B. Hutchinson and L. Rydes, in *Thermomechanical Processing, in Theory, Modelling, and Practice [TMP]²*, 4–6 Sept. 1996, The Swedish Society of Materials Technology, 1997, pp. 145–161.

79. M. Cook and T. L. Richards, *J. Inst. Met.*, vol. 66, 1940, p. 1.

80. J. D. Michalak and H. Hu, quoted in Hu, in *Texture of Materials, 5th ICOTOM*, vol. II, eds. G. Gottstein and K. Lucke, Springer-Verlag, Berlin, 1978, pp. 3–20.

81. C. S. Barrett and T. B. Massalski, *Structure of Metals*, 3d ed., Pergamon Press, Oxford, 1980.

82. S. Ding and J. G. Morris, *Met. Trans.*, vol. 28A, 1997, pp. 2715–2721.

83. H. Inoue and N. Inakazu, in *ICOTOM 8*, eds. J. S. Kallend and G. Gottstein, The Metallurgical Society, Warrendale, Pa., 1988, p. 997.

84. G. E. Dieter, *Mechanical Metallurgy*, 3d ed., McGraw-Hill, New York, 1986.

85. A. R. Morgridge and R. Priestner, *Heat Treatment '84*, The Metals Society, London, 1984, pp. 7.1–7.7.

86. G. Gottstein and F. Schwarzer, in *Materials Science Forum*, vols. 94–96, eds. G. Abruzzese and P. Brozzo, 1992, pp. 187–208.

87. R. A. Vandermeer and B. B. Rath, *Met. Trans.*, vol. 21A, 1990, pp. 1143–1149.

88. R. L. Coble and J. E. Burke, *Progress in Ceramic Science*, Pergamon Press, Oxford, 1963.

89. P. Shewmon, *Transformations in Metals*, McGraw-Hill, New York, 1969.

90. L. E. Murr, *Interfacial Phenomena in Metals and Alloys*, Addison-Wesley, Reading, Mass., 1975.

91. K. Lucke and H. P. Stuwe, in *Recovery and Recrystallization of Metals*, ed. L. Himmal, Interscience, New York, 1963, pp. 171–210.

92. J. W. Martin and R. D. Doherty, in *Stability of Microstructures in Metallic Systems*, Cambridge University Press, Cambridge, 1976.

93. M. F. Ashby, J. Harper, and J. Lewis, *Harvard University Report No. 547*, 1967.

94. H. Gleiter and B. Chalmers, *Progress in Materials Science*, vol. 16, Pergamon Press, 1972.

95. Ref. 24 in G. S. Smith, *Trans. AIME*, vol. 175, 1948, p. 15.

96. H. P. Stuwe, in *Recrystallization of Metallic Materials*, ed. F. Haessner, Dr. Riederer Verlag Gmbh, Stuttgart, Germany, 1978, pp. 11–21.

97. T. Gladman, *JOM*, Sept. 1992, pp. 21–24; in *Materials Science Forum*, vols. 94–96, eds. G. Abruzzese and P. Brozzo, 1992, pp. 113–127.

98. T. Gladman, *Encyclopedia of Materials Science & Engineering*, Pergamon Press, Oxford, 1986, pp. 2045–2051.

98a. Y. Kurosaki et al., *ISI J. Int.*, vol. 39, no. 6, 1999, pp. 607–613.

99. R. A. Vandermeer, D. J. Jensen, and E. Woldt, *Met. Trans.*, vol. 28A, 1997, pp. 749–754.

100. K. T. Aust and J. W. Rutter, in *Recovery and Recrystallization of Metals*, ed. L. Himmal, Interscience, New York, 1963, p. 131.

101. M. Feller-Kriecpmeier and K. Schwartzkopf, *Acta Metall.*, vol. 17, 1969, p. 497.

102. G. Gottstein and L. S. Shvindlerman, *Scripta Metall.*, vol. 27, 1992, p. 1521.

103. M. A. Fortes, *Materials Science Forum*, vols. 94–96, eds. G. Abruzzese and P. Brozzo, 1992, pp. 319–324.

104. W. W. Mullins and J. Venals, *Acta Metall.*, vol. 37, 1989, p. 991.

105. K. Detert, in *Recrystallization of Metallic Materials*, ed. F. Haessner, Dr. Riederer Verlag GmbH, Stuttgart, Germany, 1978, pp. 97–109.

106. G. Grindvaux and W. Form, *J. Inst. Met.*, vol. 101, 1973, p. 85.

107. C. S. Pande, M. A. Imam, and B. B. Rath, *Met. Trans.*, vol. 21A, 1990, pp. 2891–2896.

108. H. Hu and C. S. Smith, *Acta Metall.*, vol. 4, 1956, p. 638.

109. Y. Y. Gertsman and R. Birringer, *Scripta Metall. et Mater.*, vol. 30, 1994, p. 577.

110. Y. Ohba, *Acta Metall.*, vol. 34, 1986, pp. 1329–1334.

111. C. Musiol, *Met. Technol.*, April 1976, pp. 173–182.

112. F. Assumus, K. Detert, and G. Ibe, *Z. Metallik.*, vol. 48, 1957, p. 44.

113. C. G. Dunn and J. L. Walter, in *Recrystallization, Grain Growth and Texture*, ed. H. Mergolin, ASM, Metals Park, Ohio, 1966, p. 461.

114. B. D. Cullity, *Introduction to Magnetic Materials*, Addison-Wesley, Reading, Mass., 1969.

115. C. G. Dunn and J. L. Walter, *Trans. AIME*, vol. 224, 1962, p. 518.

116. A. Mager, *Ann. Phys.*, 6F, vol. 11, 1952, p. 16; T. Yamamoto, S. Taguchi, A. Sakakura, and T. Nozawa, *IEEE Trans. Mag.*, vol. MAG-8, 1972, p. 677.

117. S. Taguchi and A. Sakakura, *Acta Metall.*, vol. 14, 1966, pp. 405–423.

118. J. Howard, *Trans. AIME*, vol. 230, 1964, p. 588.

119. S. Taguchi, A. Sakakura, and H. Takashima, U.S. Patent 3287183, 1966.

120. A. Sakakura, *J. Appl. Phys.*, vol. 40, 1969, p. 1539.

121. K. J. Arai and Y. Yamashiro, in *MRS Int. Mtg. on Advanced Materials*, vol. 11, 1989, p. 187.

122. K. Detert, *Acta Metall.*, vol. 7, 1959, p. 589.

123. J. L. Walter and C. G. Dunn, *Acta Metall.*, vol. 7, 1959, p. 424; *J. Metals*, vol. 11, 1959, p. 599.

124. R. W. Cahn, in *Processing of Metals and Alloys*, Chapter 10, vol. 15, ed. R. W. Cahn, VCH, Weinheim, 1991, pp. 429–480.

125. J. G. Benford and E. B. Stanley, *J. Appl. Phys.*, vol. 40, 1969, p. 1583.

126. H. E. Exner and E. Arzt, in *Physical Metallurgy*, Chapter 31, 4th ed., eds. R. W. Cahn and P. Haasen, Elsevier Science B. V., Amsterdam, 1996, pp. 2627–2662.

127. W. H. Rhodes, *J. Am. Ceram. Soc.*, vol. 64, 1981, pp. 13–19.

128. D. L. Johnson, in *Encyclopedia of Materials Science and Engineering*, Pergamon Press, Oxford, 1986, pp. 4520–4525.

129. W. A. Kaysser, *Sintern mit Zusatzen (Sintering with Additions)*, in German, *Materialkund.-Technische Reihe*, vol. 11, Gebr. Borntraeger, Berlin, Stuttgart, 1992.

130. S. Gowri and J. A. Lund, in *Advances in Powder Metallurgy*, vol. 1, eds. F. Gasbarre and W. F. Jandeka, Metal Powder Industries Federation, Princeton, N.J., 1989, p. 139.

CHAPTER 6
NUCLEATION IN SOLIDS

6.1 INTRODUCTION

Nucleation is defined as the formation, through thermally activated fluctuations, of the smallest stable particles of a new phase.[1] J. Williard Gibbs[2] classified the fundamental modes of transformations occurring in solid systems into two categories. The first category involves those transformations that initiate by composition fluctuations which are large in magnitude but localized in extent (e.g., classical homogeneous and heterogeneous nucleation processes). This is usually described as a *nucleation and growth reaction*. The second category involves those transformations that initiate by small composition fluctuations over large distances (e.g., spinodal decomposition). In the same manner, J. W. Christian[3] grouped the precipitation reactions into two categories: heterogeneous and homogeneous transformations. Heterogeneous transformation, which includes both the classical homogeneous and heterogeneous nucleation processes, usually involves the spatial partitioning of the system into transformed and untransformed regions separated by a precipitate/matrix interface and is associated with changes in crystal structure. On the other hand, homogeneous transformations are those in which precipitation occurs uniformly throughout the entire matrix, and they involve the progressive increase in composition and/or the order parameter from initial small values to large values, characteristics of a more stable condition. This can be called *continuous transformation* because the entire system transforms by continuous enlargement of initially small fluctuations within the supersaturated or undercooled solid solution.[4]

Classical nucleation theory was formulated by Farkas[5] and Volmer and Weber[6] for the simplest nucleation process, namely, the condensation of a pure vapor to form a liquid. It was applied to other phase transformations, for example, vapor $\rightarrow$ solid (sublimation), liquid $\rightarrow$ solid (solidification), and solid $\rightarrow$ solid (precipitation, allotropic, and recrystallization) processes.[4] Among these applications of the classical theory of nucleation, the one in solids was first developed by Becker and Doring.[7] However, the degree of complexity is highest in the solid $\rightarrow$ solid phase transformations because it involves, in addition to structural changes, strain energy effects, the possibility of metastable phases, the existence of dislocations and grain boundaries, highly anisotropic interfacial energies, complex interfacial structures, sluggish mass transport, and the complicated solution thermodynamics.[8,9]

In summary, homogeneous nucleation occurs at random and with rapidity within the matrix, independent of lattice defects and inclusions. In practice, homogeneous nucleation is rarely observed in solids and in liquids because of the surmounting higher activation barriers in the absence of heterogeneities. Heterogeneous nucleation occurs at preexisting high-energy sites such as lattice defects, inclusions, and so on, whose available free energy helps to cause a large reduction of critical (activation) energy for nucleation. This chapter deals with classical homogeneous and heterogeneous nucleation at various nucleating sites and also deals with spinodal decomposition and precipitation hardening.

6.2 CLASSICAL HOMOGENEOUS NUCLEATION

For a normal phase transformation in solids, let us consider the precipitation of a new β phase from a single-phase α on cooling below the equilibrium transformation temperature T_E. Figure 6.1 shows the free energy (per unit volume)-temperature relationship. The free-energy–temperature curve for α (designated G^α) intersects that for β (called G^β) at T_E at which both α and β phases coexist. At this temperature, both phases have the same free energy, because they are in thermodynamic equilibrium. Thus the volume free-energy change ΔG_v at T_E is given as

$$\Delta G_v = G^\beta - G^\alpha = \Delta G^{\alpha \to \beta} = 0 = \Delta H_v - T_E \Delta S_v \quad \text{or} \quad \Delta S_v = \frac{\Delta H_v}{T_E} \tag{6.1}$$

where ΔH_v and ΔS_v are changes in enthalpy and entropy, respectively; G^β and G^α are greater for all temperatures above and below T_E, respectively. That is, the β phase is stable below T_E, while the α phase is stable above T_E. For a given amount of undercooling (that is, ΔT) below T_E, the volume free-energy change ΔG_v associated with the transformation is given as

$$\Delta G_v = \Delta H_v - T \Delta S_v \tag{6.2}$$

Since ΔH_v and ΔS_v are insensitive functions of temperature within a narrow temperature range, we may combine Eqs. (6.1) and (6.2), which gives

$$\Delta G_v = \Delta H_v - T \frac{\Delta H_v}{T_E} = \Delta H_v \frac{\Delta T}{T_E} \tag{6.3}$$

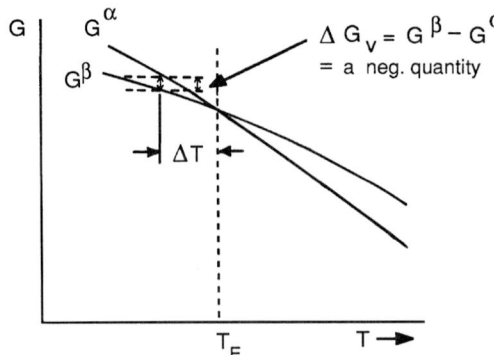

FIGURE 6.1 Schematic free energy versus temperature for single phases, α and β.

where ΔT is the amount of undercooling (or supercooling). Figure 6.1 and Eq. (6.3) indicate that ΔG_v is (1) a function of ΔT and (2) always negative below the transformation temperature T_E. It is thus clear that ΔG_v is the driving force for formation of the new phase below the transformation temperature. However, in practice, the nucleation of the new phase will not occur just below the transformation temperature, because as soon as a very small region of a few atoms' assembly of β phase, called an *embryo*, forms within the parent phase α (at the beginning of the transformation), it creates an interface between the two phases with associated interfacial (or surface) energy and distortion of the interface planes as a result of differences in size and/or shape of the unit cells of the α and β phases, thereby giving rise to misfit or elastic strain energy per unit volume ΔG_ε of β. Both misfit and new interface contribute to the positive free energy and act as energy barriers to nucleation.

The total free-energy change ΔG associated with the formation of a coherent embryo can now be written as

$$\Delta G = V\Delta G_v + \sum_i^n A_i \gamma_i + V\Delta G_\varepsilon \qquad (6.4)$$

where V is the volume of the embryo β produced, A_i is the area of the embryo-matrix boundary with the constant interface energy γ_i, and Σ_i denotes the summation over all surfaces of the embryo; ΔG_v and ΔG_ε have the usual meanings.

When we assume a spherically shaped embryo, Eq. (6.4) becomes

$$\Delta G = \frac{4}{3}\pi r^3 \Delta G_v + 4\pi r^2 \gamma_{\alpha\beta} + \frac{4}{3}\pi r^3 \Delta G_\varepsilon \qquad (6.5)$$

where r is the radius of the embryo. The plot of free energy versus r in Fig. 6.2 shows ΔG and its constituents, with the negative ΔG_v term causing the process to occur and the positive $\gamma_{\alpha\beta}$ and ΔG_ε terms providing, respectively, the principal and secondary barriers to nucleation.

As the embryos increase in size, the ΔG_v and ΔG_ε terms exert a continuously greater effect on ΔG, causing it to reach a maximum ΔG^*_{Hom} at the critical radius r^*. The critical radius r^* is obtained by solving Eq. (6.5) for $[\partial (\Delta G)/\partial r]_{r=r^*} = 0$:

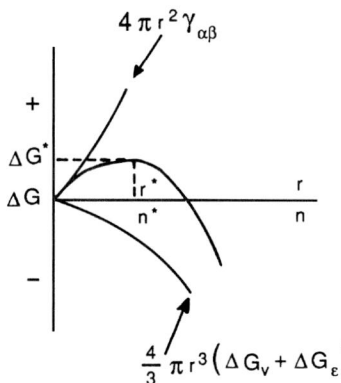

FIGURE 6.2 Schematic diagram representing the changes in interfacial free energy $\gamma_{\alpha\beta}$, strain energy ΔG_ε, volume free energy ΔG_v, and total free energy ΔG associated with embryo formation as a function of embryo radius r or number of atoms in the embryo n.

$$r^* = -\left[\frac{2\gamma_{\alpha\beta}}{\Delta G_v + \Delta G_\varepsilon}\right] \tag{6.6}$$

When $r < r^*$, the embryo will tend to shrink; when $r > r^*$, a critical embryo (called a *nucleus*) will grow in size. It is clear that further increase or decrease in size corresponding to the critical value reduces the ΔG value. The activation energy for nucleation or the free-energy change of formation of the critical nucleus, in the presence of elastic strain energy, is obtained after combining Eqs. (6.5) and (6.6) as

$$\Delta G^*_{\text{Hom}} = \frac{16\pi\gamma^3_{\alpha\beta}}{3(\Delta G_v + \Delta G_\varepsilon)^2} \tag{6.7}$$

Alternatively, Eq. (6.5) can be expressed in the following form because it is convenient to focus attention on the number of atoms in the embryo instead of its radius:

$$\Delta G = n\Delta G_v + \eta n^{2/3}\gamma_{\alpha\beta} + n\Delta G_\varepsilon \tag{6.8}$$

where n is the number of atoms in the embryo and η is the shape factor.

Similarly, the critical number of atoms and the critical free-energy change accompanying the formation of a potent coherent nucleus are given by

$$n^* = \left[\frac{2\eta\gamma_{\alpha\beta}}{3(\Delta G_v + \Delta G_\varepsilon)}\right]^3 \tag{6.9}$$

and
$$\Delta G^*_{\text{Hom}} = \frac{4}{27}\frac{16\eta^3\gamma^3_{\alpha\beta}}{(\Delta G_v + \Delta G_\varepsilon)^2} \tag{6.10}$$

Note: Equations (6.6), (6.7), (6.9), and (6.10) are applied for coherent nucleation with sharp interfaces.

6.2.1 Strain Energy

Strain energy plays an important role in the nucleation of solid-state phase changes. Strain energy developed by the formation of a new phase in a matrix phase is grouped into two types. The first one is caused by lattice mismatch or misfit strain due to different lattice parameters between two different crystalline phases. The second one is associated with the volume strain or dilatational strain due to the volume difference between two structures.[10] Thus the elastic energy is a strong function of morphologies of the two phases such as shapes, orientations, and mutual arrangements of the precipitates.[11]

The elastic strain energy associated with a coherent second-phase precipitate has been studied by Laszlo,[12] Robinson,[13] and Eshelby.[14] Eshelby's approach, based on the transformation strain, will be presented here.

Let us consider an infinite isotropic linear elastic matrix of shear modulus μ, modulus of elasticity E, and Poisson's ratio v which contains a fully coherent ellipsoid precipitate with the corresponding elastic constants μ^*, E^*, and v^* formed by transformation. If the transformation of the matrix into the coherent precipitate occurs without producing constraints of the surrounding matrix, the transformation is said to be a *stress-free* transformation and can be described by its uniform stress-free transformation strain e_{ij}^{T*}.[8,14]

In a pure dilatation transformation, a change in size without a change in shape of the precipitate occurs, which is represented by $e_{ij}^{T*} = \varepsilon \delta_{ij}$, where δ_{ij} is a Kronecker delta [a second-rank unit isotropic tensor (δ_{ij} is 1 if $i = j$ and is 0 if $i \neq j$)] and ε is the constant linear misfit strain. When the elastic constants are the same in the matrix and precipitate particle (that is, $\mu = \mu^*$, $E = E^*$, and $\nu = \nu^*$), the strain energy per unit volume of the precipitate is given by[8,14,15]

$$\Delta G_\varepsilon = \Delta G_\varepsilon^0 = \frac{E\left(\varepsilon_{11}^T\right)^2}{1-\nu} = \frac{2\mu(1+\nu)\left(\varepsilon_{11}^T\right)^2}{1-\nu} \tag{6.11}$$

which is independent of precipitate shape. However, when the matrix and precipitate have different elastic constants, ΔG_ε becomes shape-dependent (Fig. 6.3).[8,15] Figure 6.3 is the variation of $\Delta G_\varepsilon/\Delta G_\varepsilon^0$ with particle eccentricity (or ellipsoid aspect ratio) $\beta = c/r$, where c denotes semithickness and r is major radius of the particle for $\nu = 0.291$ and $\mu = 8.6 \times 10^{11}$ dyne/cm^2 (the elastic constants for α-iron), with $\nu^* = \nu$ and $\mu^*/\mu = 1/3, 1$, and 3.[15] For $c/r = \infty$, the shape of the particle becomes a needle; for $c/r = 1$, the shape becomes a sphere with the maximum strain energy existing; and for $c/r \ll 1$ (i.e., near zero), the shape approximates a thin plate or disk.

Many transformations involve both shear and dilatational transformation strains. In those cases, shear strain energy per unit volume of the precipitate depends on the shape factor c/r, and the shear strain ($\varepsilon_{13}^T = \varepsilon_{31}^T$) is given by[8,15]

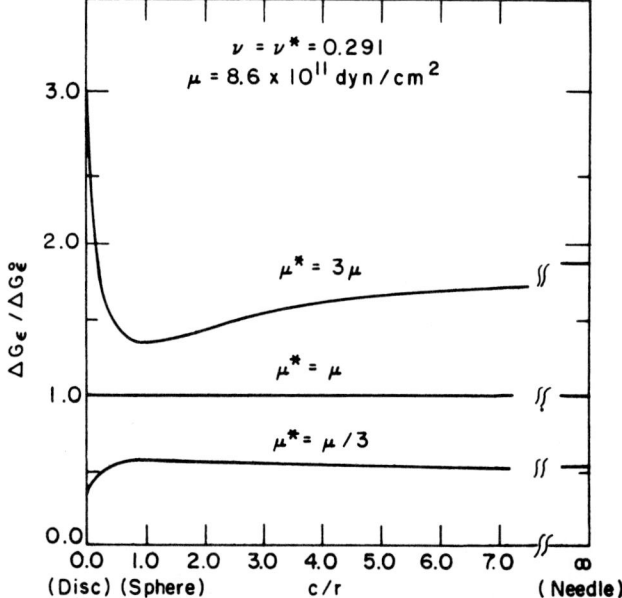

FIGURE 6.3 Variation of normalized strain energy of a coherent ellipsoid nucleus with its shape or eccentricity (c/r) for three combinations of elastic constants.[8,15] (*Reprinted by permission of North-Holland Physics Publishing, Amsterdam.*)

$$\Delta G_\varepsilon = \frac{\pi E(2-v)}{4(1-v^2)}\left(\varepsilon_{13}^T\right)^2\left(\frac{c}{r}\right) = \frac{\pi(2-v)}{2(1-v)}\mu\left(\varepsilon_{13}^T\right)^2\left(\frac{c}{r}\right) \tag{6.12}$$

The total strain energy for an ellipsoid will become

$$\Delta G_\varepsilon = \frac{E}{1-v}\left(\varepsilon_{11}^T\right)^2 + \frac{\pi E(2-v)}{4(1-v^2)}\left(\varepsilon_{13}^T\right)^2\left(\frac{c}{r}\right) \tag{6.13}$$

For the disk morphology ($c/r \to 0$), the strain energy due to shear strain goes to zero.

Let us consider a plate or disk precipitate whose strain energy is expressed by Eq. (6.11). We assume that the precipitate/matrix interface has a small disregistry or mismatch δ. Then ε_{11}^T becomes approximately equal to δ.

Note that ΔG_ε is largest, and γ is lowest, for coherent embryos. As the embryos increase in size, ΔG_ε decreases and γ increases. Thus the coherent interfaces corresponding to good lattice matching have low energy (10 to 30 ergs/cm²), while incoherent interfaces representing poor lattice matching have high energies ($\cong 500$ ergs/cm²).[16]

It is thus clear that for an incoherent precipitate, misfit between the precipitate particle and matrix does not exist. As a result, it becomes almost strain-free; the interphase boundary energy γ becomes isotropic and of very large magnitude. Consequently the barrier to incoherent nucleation is very high.[8]

Let us consider again a disk morphology with an aspect ratio $A = r$ (radius)/t (thickness). The barrier to nucleation for an incoherent precipitate is approximately equal to $\gamma\,[2\pi(At)^2 + 2\pi(At)t]$, and the barrier to nucleation for the coherent precipitate (interface) is given by Eq. (6.11) as $[E\delta^2/(1-v)]\pi(At)^2t$.

At critical thickness (t_{cr}) of the precipitate, the aspect ratio will be independent of shape; any decrease in strain energy is completely counterbalanced by an increase in surface energy. Equating these two expressions, we get

$$\frac{E\delta^2}{1-v}\pi(At)t = \gamma\left[2\pi(At)^2 + 2\pi(At)t\right]$$

which yields

$$t_{cr} = \frac{2\gamma(1-v)}{E\delta^2}\left(1+\frac{1}{A}\right) \tag{6.14}$$

Figure 6.4 shows a schematic representation of the barrier to nucleation versus plate thickness. When $t < t_{cr}$, the precipitate remains coherent, and the barrier to nucleation is small. When $t > t_{cr}$, it becomes incoherent. This represents a qualitative picture of the effect of precipitate size on the nature of the precipitate/matrix interface. In fact, the precipitate forms initially with a coherent interface, and as it grows in size, it becomes partially coherent (or semicoherent); finally, coherency is completely lost and the precipitate becomes incoherent. This has been experimentally observed in many solid → solid transformations. The coherent precipitates are invariably associated with the orientation–habit-plane relationship (e.g., formation of disk-shaped Guinier-Preston (GP) zones in Al-Cu alloys with orientation—habit-plane relationship $\{100\}_{zones}$ // $\{100\}_{matrix}$). Similarly, Widmanstätten precipitates have been initially found to be coherent with sharp interfaces and an orientation—habit-plane relationship.

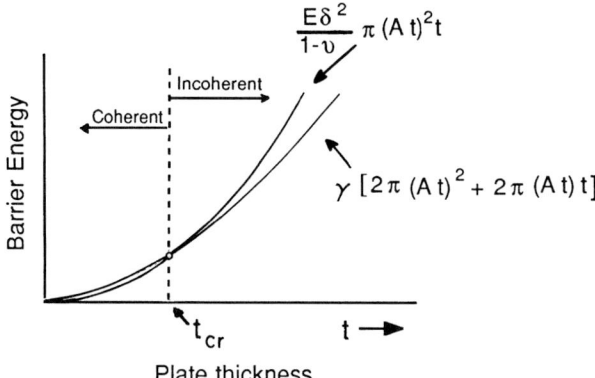

FIGURE 6.4 Schematic representation of the barrier to nucleation for coherent and incoherent precipitates versus its thickness.[16] (*Reprinted by permission of John Wiley & Sons, New York.*)

6.2.2 Vacancy Effect

In coherent nucleation, excess vacancies do not contribute to the driving force. For incoherent nucleation, thermal vacancy concentration relieves any transformation strain that results in a stress-free critical nucleus.[8] Hence the change in free energy of forming a spherical incoherent embryo is given by

$$\Delta G = \frac{4}{3}\pi r^{3}(\Delta G_{v} + \Delta G_{vac}) + 4\pi r^{2}\gamma_{\alpha\beta} \tag{6.15}$$

where ΔG_{vac} is the driving force due to excess vacancies; other terms have their usual meanings. Similarly, the critical free-energy change $\Delta G^{*}_{Hom,incoh}$ is given by

$$\Delta G^{*}_{Hom,incoh} = \frac{16\pi\gamma^{3}}{3(\Delta G_{v} + \Delta G_{vac})^{2}} \tag{6.16}$$

Note: For homogeneous nucleation in liquid $\rightarrow$ solid transformation, the strain energy barrier is nonexistent. Hence in this case, Eqs. (6.6), (6.7), (6.9), and (6.10) can be applied after deleting the ΔG_{ε} term. (See Chap. 3 for more details.)

6.3 CLASSICAL HOMOGENEOUS NUCLEATION RATE IN SOLIDS

Since the probability of observing any embryo of a given size with n atoms in the equilibrium state is shown to be proportional to $\exp(-\Delta G/kT)$, the number of embryos N_{n} containing n atoms can be expressed by the relation[8]

$$N_{n} = N\exp\!\left(-\frac{\Delta G}{kT}\right) \tag{6.17}$$

where N is the total number of (active) atomic sites per unit volume, k is the Boltzmann constant, and ΔG is the total free-energy change represented in Eq. (6.5).

Similarly, it can be shown, for an equilibrium number of critical-sized embryos per unit volume, that

$$N_n^* = N \exp\left(-\frac{\Delta G^*}{kT}\right) \tag{6.18}$$

The addition of one more atom to each of these critical embryos (or microclusters) will make them stable nuclei. These never decompose; instead, they continue to grow. If this occurs with a frequency β^*, the overall nucleation rate $\dot{N}$ of a new phase in the solid state is given by the general nucleation equation[15,16]

$$\dot{N} = Z\beta^* N \exp\left(-\frac{t_0}{t}\right)\exp\left(-\frac{\Delta G^*}{kT}\right) \tag{6.19}$$

where Z is the Zeldovich nonequilibrium factor (typically $Z \cong 0.1$); β^* is the frequency factor, which is the rate at which a single atom joins the critical nucleus; N is the density of atomic nucleation sites of the matrix phase to establish a steady-state nucleation condition; t_0 is the incubation time; t is the isothermal reaction time; and ΔG^* $(= \Delta G_{\text{Hom}}^*)$ is the critical free energy of activation for nucleation. This equation is not exact. For one thing, it does not take into account the effect of strain energy on nucleus shape. In fact, Lee et al.[19] have recently shown that strain energy does not influence the critical nucleus shape unless it constitutes a large fraction of the absolute value of the volume free-energy change.

In general, the values of Z, β^*, ΔG^*, and t_0 can be calculated from a specific model of critical nucleus shape and are all specific to the system and to the type of nucleation process considered.[8,20,21] Usually, the classical nucleation rate consists of transient $\dot{N}_t$ and steady-state $\dot{N}_{ss}$ periods, which can be in the form of equations such as

$$\dot{N}_{SS} = Z\beta^* N \exp\left(-\frac{\Delta G^*}{kT}\right) \tag{6.20}$$

and

$$\dot{N}_t = \dot{N}_{SS} \exp\left(-\frac{t_0}{t}\right) \tag{6.21}$$

In the transient period, the nucleation rate rises continuously until the incubation time is reached. In the steady-state period, the nucleation rate is constant.[20] Note that the steady-state period may not hold for long before a depleted matrix concentration causes a decrease.

6.4 HETEROGENEOUS NUCLEATION

In this section we will deal with the modified theory of nucleation in the presence of impurity particles in the assembly on the container wall or at grain boundaries, grain edges, grain corners, or on dislocations.

6.4.1 Nucleation at Container Wall

Let us consider the formation of the β embryo in the form of a spherical cap on a flat container wall, as shown in Fig. 6.5a. Assuming the surface energy between the α/β interface $\gamma_{\alpha\beta}$ to be isotropic (i.e., constant), the contact angle between the embryo and surface is θ; the condition for static equilibrium can be given by[3]

$$\gamma_{\alpha s} = \gamma_{\beta s} + \gamma_{\alpha\beta}\cos\theta \qquad (0 \le \theta \le \pi) \tag{6.22}$$

where $\gamma_{\alpha s}$, $\gamma_{\beta s}$, and $\gamma_{\alpha\beta}$ are the surface energies of the α/wall interface, β/wall interface, and α/β interface, respectively. When θ, the so-called wetting or dihedral angle, lies beyond the stated limits, Eq. (6.22) cannot remain valid; in that situation, either α (no wetting) or β phase will spread over the surface (complete wetting). The free energy associated with heterogeneous formation of a β embryo may be written as

$$\Delta G_{\text{Het}} = V_s(\Delta G_v + \Delta G_\varepsilon) + A_{\alpha\beta}\gamma_{\alpha\beta} + A_{\beta s}\gamma_{\beta s} - A_{\beta s}\gamma_{\alpha s} \tag{6.23}$$

where V_s is the volume of the cap-shaped embryo; $A_{\alpha\beta}$ and $A_{\beta s}$ are the areas of the α/β and β/wall interfaces; and $\gamma_{\alpha\beta}$, $\gamma_{\beta s}$, and $\gamma_{\alpha s}$ are the surface energies per unit area

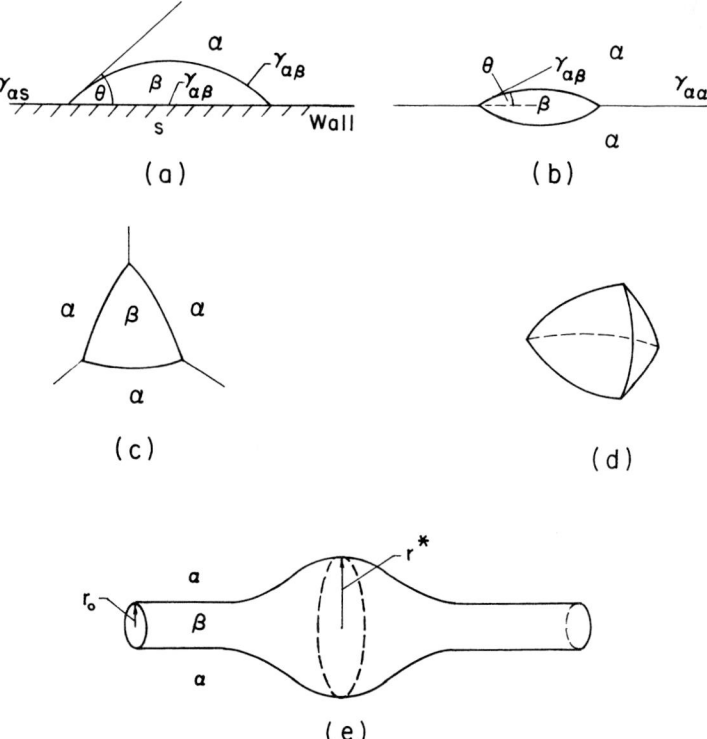

FIGURE 6.5 The formation of a β embryo on (a) a flat impurity surface s, (b) an α grain boundary surface, (c) a grain edge, (d) a grain corner, and (e) a dislocation. [(b–d) *After Clemm and Fisher*; (e) *after Cahn.*]

of the various interfaces, respectively. From the geometry shown in Fig. 6.5a, Eq. (6.23) can be expressed in terms of the contact angle θ and cap radius r:

$$\Delta G_{\text{Het}} = \left[\frac{4}{3} \pi r_{\alpha\beta}^3 (\Delta G_v + \Delta G_\varepsilon) + 4\pi r_{\alpha\beta}^2 \gamma_{\alpha\beta} \right] S(\theta) \tag{6.24}$$

where
$$S(\theta) = \frac{(2+\cos\theta)(1-\cos\theta)^2}{4} = \frac{2-3\cos\theta+\cos^3\theta}{4} \tag{6.25}$$

In Eq. (6.24) the term in the first set of brackets is the same as the free energy associated with the formation of a spherical embryo by homogeneous nucleation if the ΔG_ε term is not different from homogeneous nucleation. However, the second term after the brackets is a function only of the angle of contact between the container wall and embryo. Therefore, we can obtain $r_{\alpha\beta}^*$ in the same manner as we did for homogeneous nucleation:

$$r_{\alpha\beta}^* = -\frac{2\gamma_{\alpha\beta}}{\Delta G_v} \tag{6.26}$$

The relationship for the radius of the critical nucleus or the radius of curvature of the spherical cap, as expressed in this equation, is the same in all cases of nucleation (including homogeneous) within a single matrix phase, differing only in the values of interfacial energy used. Consequently, the ΔG^* for heterogeneous nucleation ΔG_{Het}^*, if ΔG_ε is the same, can be written as

$$\Delta G_{\text{Het}}^* = \Delta G_{\text{Hom}}^* S(\theta) \tag{6.27}$$

Comparing Eqs. (6.7) or (6.10) and (6.27), we see that the activation energy barrier for heterogeneous nucleation ΔG_{Het}^* is lower than ΔG_{Hom}^* by a shape factor $S(\theta)$. For example, when $\theta > 0$, this term is positive; when $\theta = 10°$, $S(\theta) \cong 10^{-4}$; when $\theta = 30°$, $S(\theta) \cong 0.02$; when $\theta = \pi/2$, $S(\theta) = 1/2$; and when $\theta = \pi$, $S(\theta) = 1$. When $\theta \to 0$, ΔG_{Het}^* decreases to zero; that is, the β phase wets the substrate S in the presence of the α phase.

6.4.2 Nucleation on Grain Boundaries, Grain Edges, and Grain Corners

The barrier to nucleation on grain boundary ΔG_B^*, is reduced by two mechanisms: (1) destruction or elimination of the portion of the planar α-α grain boundary, which is the surface separating the two matrix grains, as shown by a dashed line in Fig. 6.5b, and (2) a decrease in the critical nucleus size when compared to homogeneous nucleation.[3] When we consider that the embryo formed on the grain boundary is a symmetrical doubly spherical cap, the condition for static equilibrium can be given by

$$\gamma_{\alpha\alpha} = 2(\gamma_{\alpha\beta}\cos\theta) \tag{6.28}$$

where $\gamma_{\alpha\alpha}$ is the grain boundary energy and $\gamma_{\alpha\beta}$ is the particle-matrix boundary energy. If we ignore the strain energy term, the free energy associated with grain boundary nucleation is given by

$$\Delta G = V_B \Delta G_v + A_{\alpha\beta}\gamma_{\alpha\beta} - A_{\alpha\alpha}\gamma_{\alpha\alpha} \tag{6.29}$$

where V_B is the volume of embryo, $A_{\alpha\beta}$ is the area of the α/β interface of the energy $\gamma_{\alpha\beta}$ created, and $A_{\alpha\alpha}$ is the area of α-α grain boundary of energy $\gamma_{\alpha\alpha}$ destroyed during nucleation.

Alternatively, Eq. (6.29) can be written, using the geometry of Fig. 6.5b, in the form[3,22]

$$\Delta G = \frac{2}{3}\pi r^3 \Delta G_v (2-3\cos\theta+\cos^3\theta) + \gamma_{\alpha\beta} 4\pi r^2 (1-\cos\theta) - \gamma_{\alpha\alpha}\pi r^2 \sin^2\theta \quad (6.30)$$

Again it can be shown that the critical radius of curvature of the nucleus (i.e., of the spherical cap) for grain boundary nucleation is given by

$$r_B^* = -\frac{4\gamma_{\alpha\beta}}{\Delta G_v} \quad (6.31)$$

and the critical free energy for nucleation at the grain boundary ΔG_B^* is related to that of homogeneous nucleation ΔG_{Hom}^* by the equation

$$\Delta G_B^* = \Delta G_{Hom}^* (1/2)(2-3\cos\theta+\cos^3\theta) \quad (6.32)$$

$$= \Delta G_{Hom}^* 2[S(\theta)] \quad (6.33)$$

It should be pointed out that the ratio of $\Delta G_B^*/\Delta G_{Hom}^*$ decreases with the increasing ratio of $\gamma_{\alpha\alpha}/\gamma_{\alpha\beta}$. In a similar manner, for nucleation on grain edges, where three planar boundaries meet in a line at an angle of 120° and the shape of an embryo is bounded by three spherical surfaces (Fig. 6.5c), greater reduction in the barrier energy for nucleation (than that required for grain boundary nucleation) is found. For nucleation at grain corners, where four different grains and the four different grain edges meet (Fig. 6.5d), a much greater reduction in the barrier energy to nucleation occurs and, according to Clemm and Fisher, may even cause nucleation without any nucleation barrier.[23] The reason is as follows: At a grain edge or corner, the surface area of the grain boundary destroyed in comparison with the critically sized nucleus is larger than that for a planar boundary.

Cahn[24] has plotted the ratios of $\Delta G_B^*/\Delta G_{Hom}^*$, $\Delta G_E^*/\Delta G_{Hom}^*$, and $\Delta G_C^*/\Delta G_{Hom}^*$ as a function of $\cos\theta$ (Fig. 6.6), from which it is obvious that

$$\frac{\Delta G_C^*}{\Delta G_{Hom}^*} < \frac{\Delta G_E^*}{\Delta G_{Hom}^*} < \frac{\Delta G_B^*}{\Delta G_{Hom}^*}$$

That is, for all values of $\cos\theta$, we have

$$\Delta G_C^* < \Delta G_E^* < \Delta G_B^*$$

Such reduction in the barrier to nucleation does not cause a large increase in the nucleation rate, because the density of atomic sites for nucleation decreases sharply as the nucleation barrier changes in the order: homogeneous, grain boundary, grain edge, and grain corner. Cahn has approximately expressed the following equations for the number of atoms per unit volume on the various nucleation sites as N_B for grain boundary, N_E for grain edges, N_C for grain corners:

$$N_B = N_A\left(\frac{\delta}{D}\right) \qquad N_E = N_A\left(\frac{\delta}{D}\right)^2 \qquad N_C = N_A\left(\frac{\delta}{D}\right)^3 \quad (6.34)$$

where N_A is the number of atomic sites per unit volume, δ is the grain boundary thickness, and D is the mean grain diameter. Cahn has calculated the nucleation

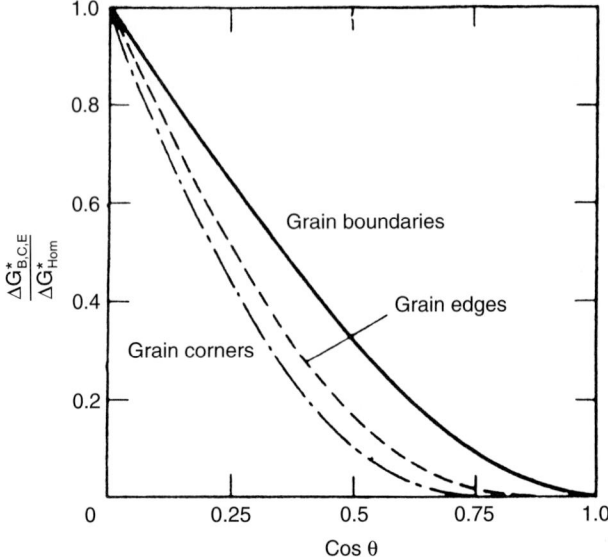

FIGURE 6.6 The ratio of $\Delta G_B^*/\Delta G_{Hom}^*$, $\Delta G_E^*/\Delta G_{Hom}^*$, and $\Delta G_C^*/\Delta G_{Hom}^*$ as a function of $\cos \theta$ ($= \gamma_{\alpha\alpha}/2\gamma_{\alpha\beta}$).[24] (*Reprinted by permission of Pergamon Press, Plc.*)

rates in all these conditions and has summarized his findings in Fig. 6.7. This is the plot of $[kT \ln (D/\delta)]/\Delta G_{Hom}^*$ (along the vertical axis) versus $\gamma_{\alpha\alpha}/\gamma_{\alpha\beta}$ (along the horizontal axis), which illustrates the highest nucleation rates for homogeneous nucleation; as we proceed from (a) grain boundary to (b) grain edge to (c) grain corner nucleation, the rates decrease.

6.4.3 Nucleation at Dislocations

Cottrell[25] and Koehler and Seitz[26] first proposed that dislocations can act as catalysts for nucleation of a new phase in solid. They suggested independently that the accommodation of the misfit strain by the strain field of dislocations would decrease the activation energy for the formation of a nucleus.[27]

Many electron-microscopic observations have also shown that dislocations act as excellent catalysts for precipitation processes and numerous alloy systems in the solid state,[28–30] especially at small values of ΔG_v (small undercooling) and if there is a modest elastic misfit strain energy due to the modest volume difference between the precipitate and the matrix. For small undercooling, this effectiveness of the nucleation site depends upon the magnitude of the Burgers vector, the extent of supersaturation, and the type of dislocations.[31] Note that edge dislocations are more effective nucleation sites than screw dislocations. Nucleation on dislocations may occur with the incoherent, semicoherent, or coherent embryo with the matrix. In this section, we will discuss only incoherent and coherent interfaces.

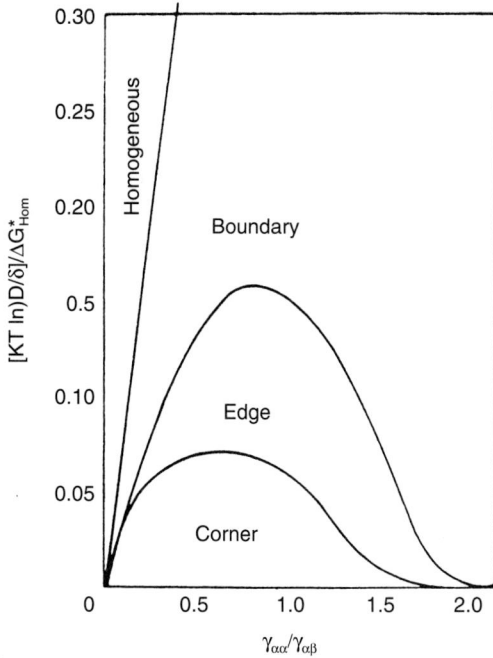

FIGURE 6.7 The plot of $[kT \ln (D/\delta)]/\Delta G^*_{\text{Hom}}$ versus $\gamma_{\alpha\alpha}/\gamma_{\alpha\beta}$ showing the highest nucleation rates for homogeneous nucleation and the lowest rates for grain corner nucleation.[24] (*Reprinted by permission of Pergamon Press, Plc.*)

6.4.3.1 Incoherent Nucleation on Dislocations. When the incoherent precipitate forms on a dislocation, it causes a decrease in the length of the dislocation core, a decrease in a portion of the dislocation strain field, and a decrease in strain energy. In other words, all the strain energy of the dislocation in the volume of the dislocation core, now occupied by the new phase (nucleus), is relieved. This strain reduction of dislocation core enhances the nucleation.[31–34]

Cahn[32] has proposed the incoherent nucleation model at dislocations by considering the nucleus as shown in Fig. 6.5e. The free-energy change produced by the formation of a small length of cylindrical nucleus of radius r around the dislocations is given by the relation

$$\Delta G = \pi r^2 \Delta G_v - A \ln r + 2\pi r\gamma + \text{constant} \qquad (6.35)$$

where A is a constant equal to $\mu b^2/4\pi (1 - \nu)$ for edge dislocation and to $\mu b^2/4\pi$ for screw dislocation, μ is the shear modulus, b is the Burgers vector, ν is Poisson's ratio, and γ and ΔG_v have the usual meanings. The first term is negative; the second term, being the elastic strain energy of dislocation within the radius r, is also negative; and the third term is positive.[32]

When both the first and second terms are not appreciable, the ΔG versus r curve has the form A (Fig. 6.8) with parameter $\alpha [= \mu b^2 \Delta G_v/2\pi^2 \gamma^2_{\alpha\beta}$ or $\mu b^2 \Delta G_v/2\pi^2 \gamma_{\alpha\beta}(1 - \nu)] < 1$. This parameter represents the catalytic power of dislocations. On the other hand, when the above terms are large compared to the surface-energy term, the free-energy curve has the form B (with $\alpha > 1$), demonstrating the nonexistence of barrier energy for the nucleation of embryos on dislocations. In this situation, growth-controlled transformation occurs.[3] Figure 6.9 illustrates the ratio of inco-

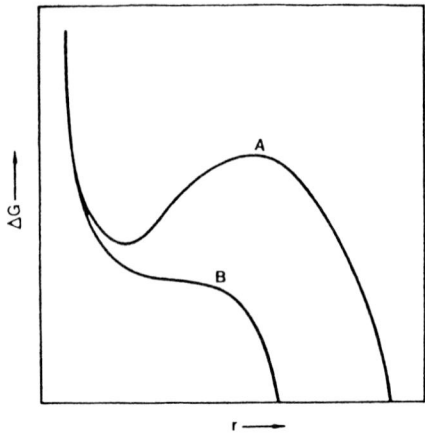

FIGURE 6.8 Schematic representation of the free energy of formation of a dislocation nucleus as a function of its radius.[32] Curve A, $\alpha < 1$; curve B, $\alpha > 1$. (*Reprinted by permission of Pergamon Press, Plc.*)

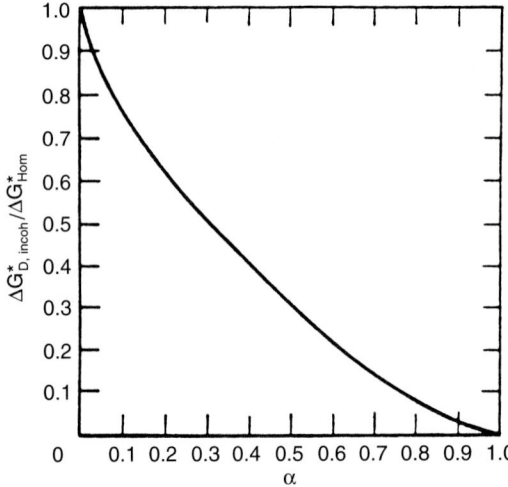

FIGURE 6.9 The ratio of the critical free energy for nucleation on a dislocation to that for homogeneous nucleation as a function of the parameter α.[32] (*Reprinted by permission of Pergamon Press, Plc.*)

herent nucleation energy on dislocations ($\Delta G^*_{D,\text{incoh}}$) and in the matrix as a function of the parameter α. If $\alpha < 0.4$, the dislocation does not act as a potent nucleation site. When $\alpha > 1$, precipitation occurs simultaneously without any nucleation barrier, resulting in a bulge of radius r^* over the stable cylindrical precipitate (Fig. 6.5e).[28,32]

Nucleation on dislocations may also be aided, to a small extent, by solute atmosphere. It has been shown that small precipitates are more stable with a coherent interface, whereas large precipitates are more stable with an incoherent interface.[27,35]

6.4.3.2 Coherent Nucleation on Dislocations. A coherent nucleus with the matrix does not form at the dislocation core. The strain energy and core energy persist in the presence of the formation of coherent particles as a result of continuity of the lattice planes. However, in some regions of the dislocation strain field, nucleation might occur in order to minimize ΔG_ε.[27] The critical free energy for coherent nucleation at nearby dislocation $\Delta G^*_{D,coh}$ can be expressed after replacing $\gamma_{\alpha\beta}$ in Eq. (6.7) by $[\gamma_{ab} - \mu b (1 + v) |\varepsilon^T_{11}|/9\pi(1 - v)]$, which takes into account the effect of elastic interactions (for a spherical nucleus) with a nearby edge dislocation[8,15,27,36]

$$\Delta G^*_{D,coh} = \frac{16\pi \left[\gamma_{\alpha\beta} - \mu b(1+v)|\varepsilon^T_{11}| \right]}{\frac{9\pi(1-v)^3}{3(\Delta G_v + \Delta G_\varepsilon)^2}} \qquad (6.36)$$

where μ is the shear modulus of the matrix, v is Poisson's ratio, ε^T_{11} is the misfit strain of the particle, b is the Burgers vector of dislocation, and $\gamma_{\alpha\beta}$, ΔG_v, and ΔG_ε have the usual meanings.

6.5 MECHANISM OF LOSS OF COHERENCY

It is thus apparent once again from the above discussions that the precipitate initially nucleates coherently and becomes incoherent during growth. The mechanism of loss of coherency (in order to become an incoherent precipitate) that is summarized below also holds in precipitation sequences occurring in a wide range of precipitation-hardening alloys (Sec. 6.7):

1. Creation or punching of prismatic dislocation loops at, or close to, the particle/matrix interface due to high shear stress arising from coherency strain[37,38]

2. Formation and growth of small dislocation loops inside the precipitate by the collapse of clusters of vacancies or interstitials[37,39]

3. The accumulation of point defects from the matrix

4. The climb or glide or absorption of dislocations from an exterior source (in the matrix), or from the "grown-in" dislocation network, to the particle/matrix interface[40]

6.6 SPINODAL DECOMPOSITION

Spinodal decomposition (i.e., continuous phase separation) has received wide theoretical attention since its current ideas were put forth by Hillert and Cahn in 1961 and 1962.[41–44] Spinodal decomposition represents a departure from (or a limiting case of) the conventional theories of diffusion-controlled phase transformation. Spinodal decomposition occurs in systems that exhibit a simple miscibility gap in an otherwise single-phase region and is usually associated with the ordering reaction.

Figure 6.10b is a simple hypothetical binary phase diagram with a miscibility gap. A spinodal line observed on the phase diagram represents the loci of $g'' = (\partial^2 G/\partial c^2)_{T,P} = 0$ (Fig. 6.10a) at different temperatures; this line has also been described thermodynamically as a limit of stability and represents the boundary between metastable and unstable regions of the phase diagram.[45,46] The region of negative curvature in the free energy versus composition diagram defines the region of spinodal decomposition, while the region lying between the points corresponding to $g'' = 0$ and g'

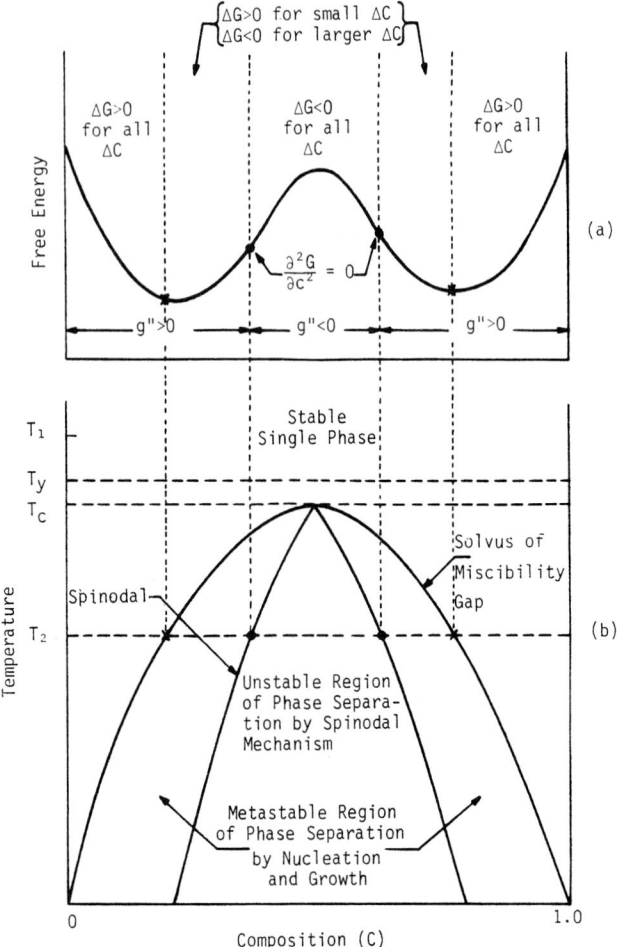

FIGURE 6.10 (*a*) Schematic free energy versus composition curve, at temperature T_2, showing the regions of stability; (*b*) phase diagram illustrating the equilibrium miscibility gap (the locus of the common tangent points ×) and the spinodal (the locus of the inflection points of the free energy versus composition curve, •).[45] (*Reprinted by permission of Academic Press, Orlando, Florida.*)

= 0 in the free-energy versus composition curve defines normal nucleation and growth processes.

In general, the heat treatment steps involved in the spinodal decomposition are (1) solution treatment of the alloy at a temperature T_1 above the miscibility (or solubility) gap to produce a single-phase material; and (2) quenching to an intermediate temperature within the spinodal region (for example, T_2) and holding (i.e., aging) at that temperature for a short time or continuously cooling the specimen from T_1

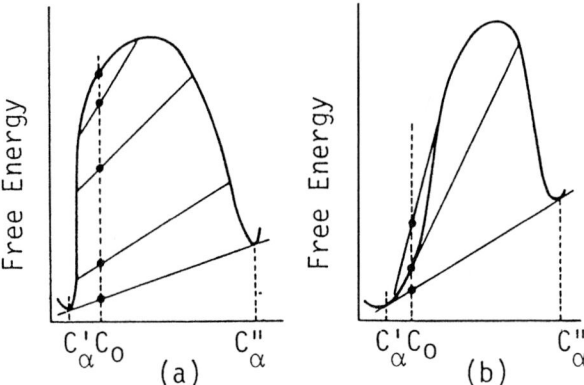

FIGURE 6.11 Enlarged schematic free energy versus composition curves. (a) The free-energy change of an unstable phase C_0; (b) free-energy changes during decomposition of a metastable phase of composition C_0.[45] (*Reprinted by the permission of The Metallurgical Society, Warrendale, Pennsylvania.*)

to room temperature within the spinodal region.[47] During this period the supercooled alloy becomes unstable because a small departure from composition C_0 will lower the total free energy (Fig. 6.11a). Hence the decomposition of unstable solid solution into two-phase mixtures with essentially the same crystal structure, but with composition different from that of the parent phase, proceeds with a lowering of successive lines of Fig. 6.11a until the line corresponding to the lowest free-energy state is reached. This state is defined by the common tangent of the mixture C_α' and C_α''.[48] Spinodal structures are, therefore, fine-scale, homogeneous, two-phase mixtures produced by phase separation under certain conditions of temperature and composition.

In contrast, if an alloy of composition C_0 is quenched at a small undercooling or supersaturation, that is, at a temperature T_1'' outside the spinodal region but within the solvus of the miscibility gap (where $g'' > 0$), then the infinitesimal fluctuation from C_0 increases the free energy and the solution becomes a metastable phase. However, if nuclei are formed with a large localized composition variation toward C_α'', a lowering of the free energy can be effected. The reason is that for a small composition change, a positive interfacial term is more pronounced. Thus the metastable region is represented by the portion of the phase diagram where $g'' > 0$, and the phase transformation proceeds by classical nucleation and growth process.

6.6.1 Nucleation and Growth Process versus Spinodal Decomposition

Unlike classical nucleation and growth process that occurs in metastable solutions, the spinodal reaction is a spontaneous unmixing or diffusional clustering process in unstable solutions. The number of features that differentiate nucleation and growth process from spinodal decomposition are listed in Table 6.1. Figure 6.12a shows the composition profile associated with the classical nucleation and growth process. A composition discontinuity occurs at the distinct precipitate/matrix interface. The

TABLE 6.1 Factors Distinguishing Nucleation and Growth Process from Spinodal Decomposition

Nucleation and Growth Processes	Spinodal Decomposition
1. Nucleation and growth occur within a metastable supersaturated solid solution during the early period.	1. Spinodal decomposition occurs within the supersaturated solid solution, which is inherently unstable to small fluctuations in compositions; therefore, the solution decomposes, spontaneously producing A-rich and B-rich regions.
2. The nucleus formed at a sufficient undercooling is a distinctly separated particle of new phase which may have the crystal structure, as well as orientation, different from the matrix.	2. The nucleus formed at a sufficient undercooling is not really a distinctly separated particle of the new phase but has the same crystal structure and orientation as the parent phase.
3. It is a large concentration fluctuation over a small volume, i.e., involving short-range composition change.	3. In the vicinity of spinodal, it is a small concentration fluctuation over a large volume, i.e., involving long-range composition change. The transformation thus occurs simultaneously throughout the matrix.
4. The classical nucleation process occurs in the region of the free energy versus composition plot where the curvature is positive, i.e., $g'' > 0$ (Fig. 6.10).	4. The spinodal decomposition occurs within the region where the curvature of the free energy versus composition plot is negative, i.e., $g'' < 0$ (Fig. 6.10).
5. This is associated with the establishment of a distinct precipitate-matrix interface with a positive surface-energy barrier. Consequently, a nucleation barrier exists which needs to be overcome prior to the start of transformation. This is a diffusion-controlled process.	5. The precipitate-matrix interface initially is not sharp but is diffuse in nature, without possessing a distinct structural discontinuity. Associated with this is a gradient energy which is overcome by g'' if the interface is largely diffused. In this extreme situation, there is no thermodynamic barrier to spinodal decomposition. However, this reaction is governed by the activation energy of diffusion.
6. Normally there is an incubation time.	6. There is no incubation time.
7. Within the nucleation and growth region, the interdiffusion coefficient, $\tilde{D} > 0$ (i.e., downhill diffusion) occurs, and the composition fluctuations increase nonexponentially with time.	7. Within the spinodal region, $\tilde{D} < 0$ (i.e., uphill diffusion) occurs, and composition fluctuations increase exponentially with time.

composition in the precipitate nucleus rises immediately to C''_{α} with the reduced matrix concentration in the vicinity of the nucleated particles. With the precipitate growth, normal downhill diffusion occurs (from C_0 to C'_{α}) in the depleted zone of the matrix toward the precipitate. Figure 6.12b shows the composition profile associated with the spinodal decomposition. The composition increase, which extends

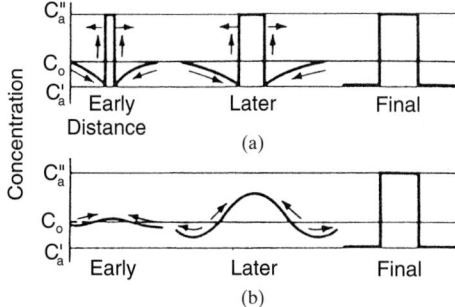

FIGURE 6.12 Schematic evolution of composition profiles to show the difference between (a) classical nucleation and growth and (b) spinodal decomposition.[48] (*Reprinted by the permission of The Metallurgical Society, Warrendale, Pennsylvania.*)

over a large area, is initially much less. During the precipitate growth, the solute concentration increases toward the precipitate region by uphill diffusion (i.e., diffusion up the concentration gradient).[48]

6.6.2 Coherent Spinodal Decomposition

There are two types of spinodal decomposition, namely, chemical spinodal decomposition (as discussed in the previous section) and coherent spinodal decomposition. In either case the decomposition occurs by a continuous process; that is, there must be an existence of continuity of free energy versus composition curve from one phase to another. Chemical spinodal decomposition also occurs in fluids and noncrystalline solids. In coherent spinodal decomposition, the coherent fluctuations in solids can lead to effective elastic strains across the diffuse interface; in chemical spinodal decomposition, however, the gradient energy contribution due to diffuse interface is more pronounced. The coherent phase diagram is always a metastable phase diagram and lies within the unstressed equilibrium phase diagram because it involves a reversible metastable equilibrium, which tends to be constrained because the lattice remains continuous.[45]

Figure 6.13a shows the coherent and incoherent free-energy composition curves for a binary solid solution at temperature T_2. It is clear from this diagram that coherent spinodal decomposition lies between the inflection points at α^{IV} and β^{IV} of the free energy versus composition curve. The coherent free-energy curve is higher than the incoherent equilibrium free-energy curve. Figure 6.13b shows both coherent and incoherent miscibility gaps and the respective coherent and chemical spinodals.[45] The coherent critical temperature occurs at a lower temperature that is depressed by ΔT from the incoherent critical temperature.[43]

6.6.3 Nonclassical Homogeneous Nucleation Theory

Cahn and Hilliard[49] have defined the classical nucleation regime as one which involves the constancy of composition near the center of the nucleus so that the volume free energy and the interfacial energy can be taken separately into account. In contrast, the nonclassical regime is one in which the presence of composition variation with position throughout the critical nucleus necessitates the use of both the energies together. Cahn and Hilliard[49,50] have developed a continuum nonclas-

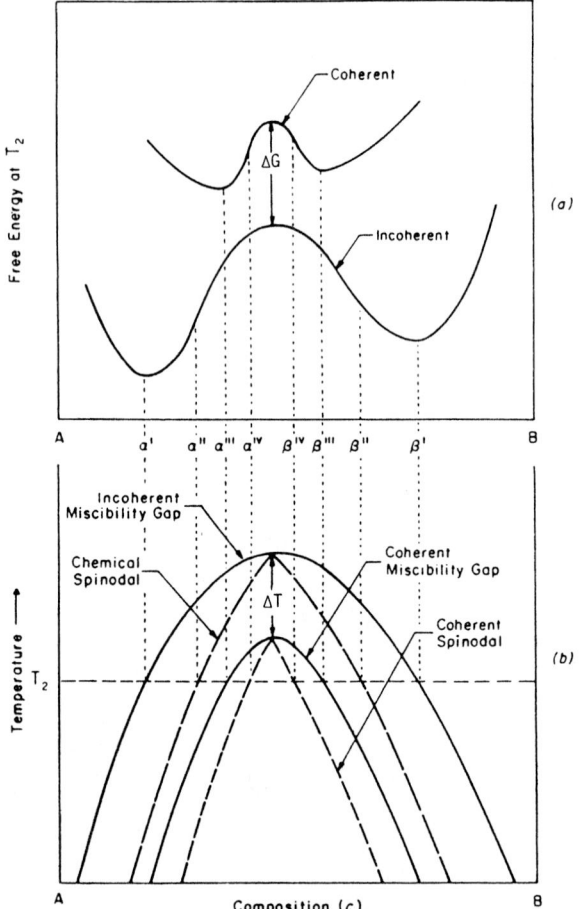

FIGURE 6.13 (*a*) Corresponding coherent and incoherent free-energy curves for a binary solid solution at temperature T_2. The coherent spinodal is represented by the inflection points on the free-energy curve at α^{IV} and β^{IV}. The coherent free-energy curve is higher than the incoherent equilibrium free-energy curve (after Hilliard). (*b*) Coherent and incoherent miscibility gaps and the respective coherent and chemical spinodal.[45] (*Reprinted by permission of the Academic Press, Orlando, Florida.*)

sical theory of homogeneous nucleation which depends upon the free energy of an inhomogeneous binary solution and does not involve a change in crystal structure. The assumptions included in this theory were that both phases were incompressible fluids of constant molar volume and that the range of fluctuations within the matrix fluid is large with respect to the interatomic or intermolecular spacing. The fluid flow assumption holds with reasonable accuracy in solid → solid reactions provided both phases have the same crystal structure and orientation, the same lattice parameters, and the same isotropic elastic properties. The Cahn-Hilliard theory is applied to high temperature (i.e., transformation temperature not much below the critical temperature T_c of the equilibrium or metastable equilibrium miscibility gap) and close to the spinodal decomposition, where the interface is diffuse. Recently, LeGoues et al.[51] have found that in many cases an alternative discrete lattice-point nonclassical model (i.e., sharp interface model) is more applicable at high temperatures.

For the continuum Cahn-Hilliard model, we first assume an incompressible binary solution of constant molar volume. The free energy G_0 of a homogeneous solid solution of composition C_0 is then given by

$$G_0 = \int_v g(C_0) dV \tag{6.37}$$

where $g(C_0)$ is the local free energy per unit volume and V is the volume. The free energy of a solution of composition C can be given by the Cahn-Hilliard relationship as

$$G = \int_v \left[g(C) + K(\nabla C)^2 + \left(\frac{\eta^2 E}{1-v} \right)(C - C_0)^2 \right] dV \tag{6.38}$$

where $g(C)$ is the free energy per unit volume of a homogeneous solid solution of composition C; K is the gradient energy coefficient, defined by

$$K = -\frac{\partial^2 G}{\partial C}(\partial \nabla^2 C) + \frac{1}{2}\frac{\partial^2 G}{(\partial |\nabla C|)^2} \tag{6.39}$$

which is a positive proportionality constant (for clustering system); $K(\nabla C)^2$ is the positive gradient energy; and, therefore, the barrier energy to nucleation (like interfacial energy), η is the linear change in the lattice parameter per unit composition change $(1/a_0)(\partial a/\partial c)$ (where a_0 is the lattice parameter at the average composition C of the solid solution), E is Young's modulus, v is Poisson's ratio, and $\int_v [\eta^2 E/(1-v)](C - C_0)^2 dV$ is the total elastic strain energy of an infinite isotropic solid solution across the diffuse interface arising from coherent composition fluctuations.

The total free-energy change resulting from a composition variation in an initially homogeneous solution is given by

$$\Delta G = G - G_0 = \int_v \left[g(C) - g(C_0) + K(\nabla C)^2 + \eta^2 Y(C - C_0)^2 \right] dV \tag{6.40}$$

where $Y = E/(1 - v)$. On expanding $g(C)$ about the average composition C_0 as

$$g(C) \cong g(C_0) + (C - C_0)g' + \frac{1}{2}(C - C_0)^2 g''$$

and inserting in Eq. (6.40) for a conservation system, we get

$$\Delta G = \int_v^0 \left[\frac{1}{2}(C - C_0)^2 g'' + K(\nabla C)^2 + \eta^2 Y(C - C_0)^2 \right] dV \tag{6.41}$$

If we further assume that the difference in composition between the initial homogeneous solution C_0 and one with a composition C varies sinusoidally with distance, then we may write

$$C - C_0 = A\cos\frac{2\pi X}{\lambda} = A\cos(\beta X) \tag{6.42}$$

where β is the wave number and λ is the wavelength. Combining Eqs. (6.41) and (6.42) yields

$$\Delta G = \int_v^0 \left[\frac{1}{2} A^2 \cos^2(\beta X) g'' + K(\nabla C)^2 + \eta^2 Y A^2 \cos^2(\beta X) \right] dV \tag{6.43}$$

that is,

$$\frac{\Delta G}{V} = \frac{1}{4} A^2 (g'' + 2K\beta^2 + 2\eta^2 Y) \tag{6.44}$$

This equation is the free-energy change per unit volume between the homogeneous solid solution and the solid solution with a composition fluctuation, as shown by Eq. (6.42).

It can be seen that the condition for a homogeneous solid solution to become unstable with respect to sinusoidal fluctuations of wavelength and $2\pi/\beta$ and thereby decompose spinodally is that ΔG should be negative,[41,43] that is,

$$g'' + 2K\beta^2 + 2\eta^2 Y < 0 \tag{6.45}$$

For coherent spinodal composition, β is negligibly small, that is, $\lambda \to \infty$ and

$$g'' + 2\eta^2 Y < 0 \tag{6.46}$$

The loci of $g'' + 2\eta^2 Y = 0$ at different temperatures represent the limit of stability; these loci form the line that is usually called the *coherent spinodal line* in the phase diagram. This lies within the chemical spinodal line (loci of $g'' = 0$), as shown in Fig. 6.13b.

It is clear from Eq. (6.43) or (6.44) that ΔG will have either a negative or a positive value depending on the magnitudes of various quantities. For example, for a particular value of g'', when β increases from a very low value (long wavelength) to a very high value (short wavelength), ΔG varies from negative (an unstable solid solution) through zero to a positive value (a stable solid solution). Thus a critical value of β (called β_c) exists, above which the solution is stable to composition fluctuations and below which the solution is unstable to infinitesimal composition fluctuations. Since $\Delta G = 0$ at the critical value β_c, we can write

$$g'' + 2K\beta_c^2 + 2\eta^2 Y = 0 \tag{6.47}$$

which yields

$$\beta_c = \left(-\frac{g'' + 2\eta^2 Y}{2K} \right)^{1/2} \tag{6.48}$$

Alternatively, the critical wavelength, $\lambda_c = 2\pi/\beta_c$, becomes[41,44]

$$\lambda_c = \left(-\frac{8\pi^2 K}{g'' + 2\eta^2 Y} \right)^{1/2} \tag{6.49}$$

Equation (6.47) asserts that the gradient or surface-energy term varies as $K\beta^2$ and limits the decomposition on a fine scale.[41]

6.6.4 Diffusion Equation

Let us consider the kinetics of spinodal decomposition in terms of a diffusion equation involving gradient in chemical potential. Cahn formulated an equation relating the spontaneous diffusional flux to the gradient chemical potential and demonstrated that for the flux to be spontaneous, it must lead to the lowering of free energy.[41,48] Thus Cahn's formulation of the diffusion equation for the initial-stage kinetics of spinodal decomposition in modified form is[48,52]

$$\frac{\partial C}{\partial t} = \frac{M}{N_v}[(g'' + 2\eta^2 Y)\nabla^2 C - 2K\nabla^4 C + \text{nonlinear terms}] \tag{6.50}$$

where M is the positive atomic mobility and N_v is the number of atoms (or molecules) per unit volume of homogeneous solution of composition C_0.

Let us compare Eq. (6.50) with the usual equation for Fick's second law, namely,

$$\frac{\partial C}{\partial t} = \tilde{D}(\nabla^2 C) \tag{6.51}$$

The term $(M/N_v)(g'' + 2\eta^2 Y)$ in Eq. (6.50) can be identified with the interdiffusion coefficient $\tilde{D}$ after including the strain energy term but neglecting the higher-order and nonlinear terms. It is thus clear that, in an unstable solid solution, $\tilde{D}$ is negative; that is, uphill diffusion takes place.

6.6.4.1 Solution of Diffusion Equation.
Assuming linear approximation and that g'', K, η, and M are independent of composition, the general solution of Eq. (6.50) can be expressed in the form

$$C - C_0 = A(\beta,t)\cos(\beta,r) \tag{6.52}$$

where A is the amplitude of the Fourier component of the composition wave of wave number β (vector) at time t and r is the position vector. The general solution of Eq. (6.52) over all values of β and r gives

$$C(r,t) - C_0 = \left(\frac{1}{2\pi}\right)^3 \int_\beta A(\beta,t)\exp(i\beta r)d\beta \tag{6.53}$$

The initial or as-quenched amplitude of the Fourier component of wave number β at $t = 0$ is

$$A(\beta,t) = A(\beta,0)\exp[R(\beta)t] \tag{6.54}$$

where $R(\beta)$ is the time-dependent amplitude factor and is defined by

$$R(\beta) = -\frac{M}{N_v}(g'' + 2\eta^2 Y + 2K\beta^2)\beta^2 \tag{6.55}$$

and $R(\beta)$ is positive if

$$g'' + 2\eta^2 Y + 2K\beta^2 \le 0 \tag{6.56}$$

Plots of $R(\beta)$ versus β and λ are shown by solid curves in Figs. 6.14 and 6.15, respectively. The dashed line is the solution [Eqs. (6.54) and (6.55)] to the classical diffusion equation [Eq. (6.50)] after neglecting the gradient energy term, the elastic

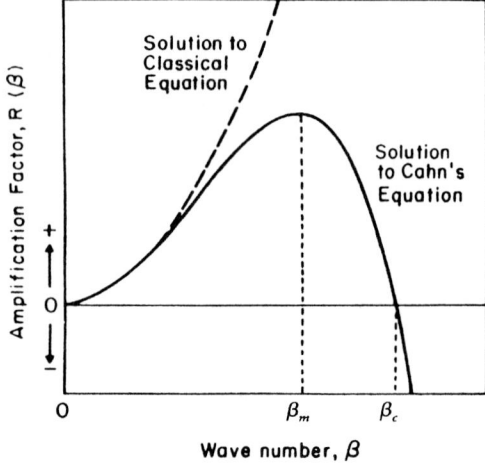

FIGURE 6.14 Amplification factor $R(\beta)$ versus wave number β. The dashed curve denotes the solution [Eqs. (6.54) and (6.55)] to classical diffusion equation [Eq. (6.50)] after neglecting the gradient energy term, elastic strain energy, and the nonlinear term. The solid curve represents the solution [Eqs. (6.54) and (6.55)] to Eq. (6.50) without the higher-order nonlinear β terms.[52] The β_m is the wave number at maximum amplification factor, and β_c is the critical wave number. (*Reprinted by permission of ASM International, Metals Park, Ohio.*)

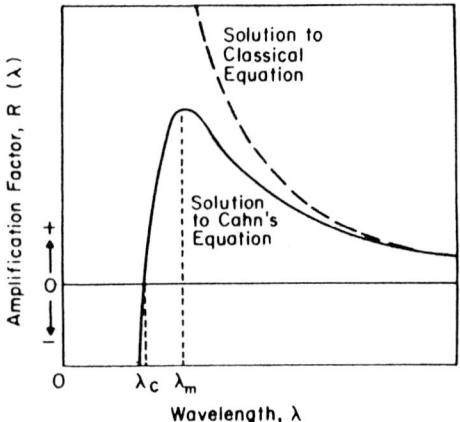

FIGURE 6.15 The amplification factor $R(\beta)$ plotted against wavelength λ (= $2\pi/\beta$) instead of the wave number β for the same solutions as given in Fig. 6.14.[52] (*Reprinted by permission of ASM International, Metals Park, Ohio.*)

strain energy term, and the nonlinear term. The solid curve represents the solution [Eqs. (6.54) and (6.55)] to Eq. (6.50) without the higher-order nonlinear β terms.

6.6.5 Early Stages of Spinodal Decomposition

Within the spinodal region $g'' + 2\eta^2 Y < 0$ and $R(\beta) > 0$ for all values of β, decomposition occurs spontaneously by the evolution of fluctuations with wave numbers around a specific growth rate $R(\beta_m)$. For low values of β (high λ), the gradient energy is negligibly small and the partitioning of the atomic components is slow; however, for high values of β (short λ), which is greater than β_m, the gradient energy term begins to dominate. For $\beta > \beta_c$, the decomposition cannot take place because $R(\beta)$

will decay as a result of shorter compositional fluctuations (i.e., shorter distances between clusters).

For a one-phase field or a metastable portion of a two-phase field of the phase diagram (Fig. 6.13b), we have $g'' + 2\eta^2 Y > 0$. In this case, $R(\beta) < 0$ and Eq. (6.54) predicts that any existing composition fluctuations will decay out.

For a particular temperature and composition within the spinodal region, there exists a critical wave number β_c ($= 2\pi/\lambda_c$) that satisfies

$$g'' + 2\eta^2 Y + 2K\beta_c^2 = 0 \tag{6.47}$$

for which $R(\beta)$ becomes zero. It follows from Eqs. (6.46) and (6.47) that coherent spinodal represents the locus of $\beta_c = 0$. We can thus rewrite Eq. (6.55) in terms of β_c by using Eq. (6.47):

$$R(\beta) = \frac{2KM}{N_v}(\beta_c^2 - \beta^2)\beta^2 \tag{6.57}$$

This function is shown by the solid curve in Fig. 6.14. The maximum growth rate is found at

$$\beta_m = \frac{\beta_c}{\sqrt{2}} \tag{6.58}$$

After equating the derivative of Eq. (6.57) to zero, β_m is called the *spinodal wave number*. Combining Eqs. (6.57) and (6.58), we get the maximum value of $R(\beta)$[52]

$$R(\beta_m) = \frac{2KM}{N_v}\beta_m^4 \tag{6.59}$$

$$= \frac{KM}{2N_v}\beta_c^4 \tag{6.60}$$

Thus $R(\beta_m)$ is strongly dependent on β_m.

6.6.6 Experimental Study of Spinodal Decomposition

The study of the kinetics of spinodal reactions is made by using small-angle X-ray and neutron scattering by monitoring the changes in the intensity distribution of the composition fluctuation around the direct beam. The lattice imaging technique by transmission electron microscopy has also been used to characterize the modulated and tweed structures occurring during the initial period of decomposition.[4]

The isothermal kinetics of an early stage of spinodal decomposition can be determined from a quenched and aged specimen of a spinodal alloy (e.g., Al-22 at% Zn-0.1 at% Mg alloy specimen aged at 125°C) by a plot of small-angle X-ray spectra intensity versus the scattering angle 2θ, where $\beta = 2\pi/\lambda \cong$ deg $2\theta/14$ (Fig. 6.16).[53] This angle is a characteristic of the wavelength of composition modulations corresponding to the peak position λ_m. Note that there is a steady change in spectral shape from a broad peak at low intensity for quenched specimen to a very sharp peak at much greater intensity for specimen aged for 80 min. The small-angle region representing small β (high λ) is growing in intensity, while the large-angle region is shrinking. This is predicted by assuming the occurrence of a critical wave number β_c.

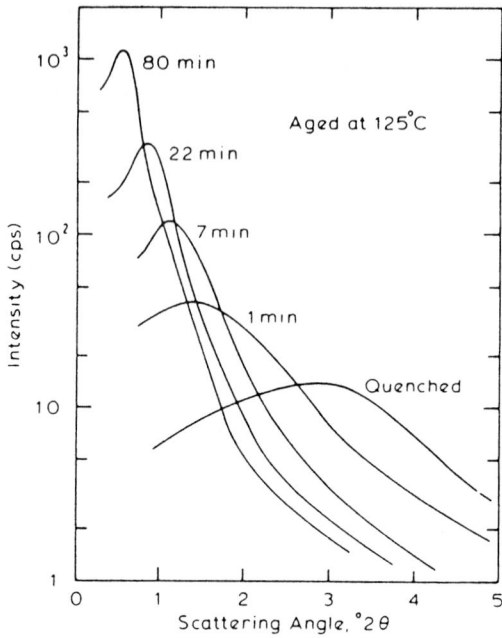

FIGURE 6.16 SAS spectra from an Al-22 at% Zn-0.1 at% Mg alloy specimen aged at 125°C using the correction for the effect of beam height. The radiation employed was CuK_α with a characteristic wavelength of 1.541 Å.[53] (*Reprinted by permission of The Metallurgical Society, Warrendale, Pennsylvania.*)

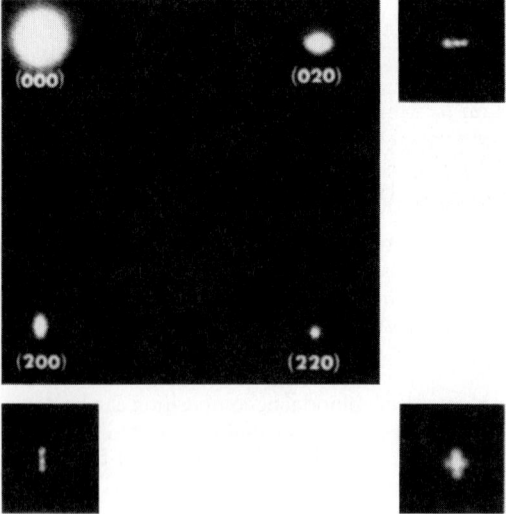

FIGURE 6.17 The [001] electron diffraction pattern from spinodally decomposed Cu-4 wt% Ti alloy aged at 400°C (750°F) for 100 min, illustrating satellites flanking the matrix reflection.[4] (*Reprinted by permission of The Metallurgical Society, Warrendale, Pennsylvania.*)

Figure 6.17 is a [001] electron diffraction pattern from a spinodally decomposed Cu-4 wt% Ti alloy showing the satellite configuration flanking the matrix reflection.[4] The appearance of "side bands" or satellite spots in the electron diffraction pattern is a characteristic feature of spinodal decomposition.[54]

If the strain energy term in the free-energy expression is very small (small misfit) or if the material (e.g., Fe-Cr-Co permanent magnetic alloy) is elastically isotropic, the microstructure developed will be isotropic (Fig. 6.18).[55] The two-phase mixture is interconnected in three dimensions and does not exhibit any directionality.[55] However, aligned modulated structure is developed; that is, preferential growth or a dominant concentration wave occurs in anisotropic materials (cubic crystals), for example, a Cu-Ni-Fe alloy, along the elastically soft <100> matrix directions (Fig. 6.19),[55] where minimum strain energy exists.[52] In several Cu-Ni-Fe spinodal alloys, a coherency loss (of modulated structures) has been found when $\lambda > 800\,\text{Å}$ (80 nm).[56] The tweed structure also appears to form in alloys where coherent tetragonal precipitates become embedded in a cubic matrix, being aligned along the <011> direction.[11]

6.6.7 Spinodal Alloys

Spinodal decomposition has been observed in a wide variety of metallic alloy systems, for example, Al-base alloys such as Al-Li, Al-Cu, Al-Zn, and Al-22 at% Zn-0.1 at% Mg; Cu-base alloys such as Cu-4 to 5 at% Ti, Cu-Zn, Cu-Co, Cu-Ni-Fe, Cu-Ni-Si, Cu-Ni-Sn, and Cu-Ni-Cr; Fe-base alloys such as Fe-Al, Fe-Be, Fe-Mo, Fe-Cr, Fe-Ni-Al, Fe-Ni-C, Fe-Cr-Co, and Fe-Al-Cr-Ni; Mn-base alloys such as

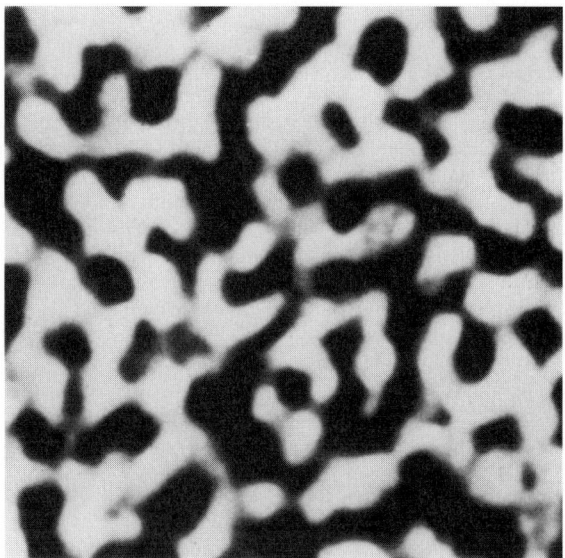

0.05 μm

FIGURE 6.18 Transmission electron micrograph of isotropic spinodal structure developed in permanent magnetic Fe-28.5Cr-10.6Co(wt%) alloy aged at 600°C (1100°F) for 4 hr.[55] (*Reprinted by permission of ASM International, Metals Park, Ohio; after A. Zettser.*)

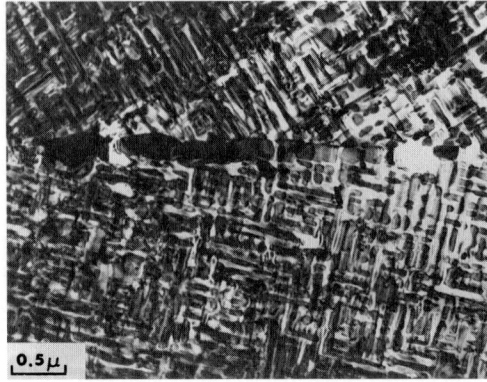

FIGURE 6.19 Transmission electron micrograph showing spinodal microstructure in a 51.5Cu-33.5Ni-15Fe (at%) alloy solution treated at 1050°C (1922°F) for 2 hr and aged at 775°C (1427°F) for 15 min. The dark regions are Ni-rich, and the light regions are Cu-rich. Foil normal is ~[001], and the alignment along elastically soft (100) matrix direction is apparent. The wavelength of the modulated structure is ~25 nm (250 Å). (*Courtesy of G. Thomas.*)

Mn-Cu; Ni-base alloys such as Ni-Al, Ni-12 wt% Ti, Ni-Cr, Ni-V, Ni-Mo, Ni-Be, and Ni-Si; Nb-base alloy such as Nb-Zr; Ti-base alloys such as βTi-Cr; and Au-base alloys such as Au-77 at% Ni and Au-Pt alloys.[4,57,58] Many of the splat-cooled alloys are also expected to exhibit spinodal decomposition, and in many systems, for example, Al-4 wt% Cu alloy below 130°C, Cu-15Ni-8Sn (wt%), and Cu-10Ni-6Sn (wt%) alloys, modulated structures precede the formation of GP zones or transition phases.

6.6.8 Applications

This mode of transformation is useful in improving the physical and mechanical properties of commercial alloys. For example, spinodal structure favors high coercivity in the production of permanent magnetic materials. This structure can be optimized by thermomechanical treatment, step aging, and magnetic aging. This structure is important in high-cobalt Alnico, Cu-Ni-Fe, and Fe-Cr-Co alloys.[55] Spinodal structure in Cu-Ni-Fe appears to possess good mechanical stability under fatigue conditions because sheared precipitates would withstand very severe distortion prior to the development of an effective wavelength.[59]

6.7 PRECIPITATION HARDENING

In Secs. 6.1 through 6.4, we have described homogeneous nucleation and heterogeneous nucleation on various nucleation sites. This knowledge is the foundation for understanding diffusional transformations. For example, let us begin our discussion with the *precipitation-hardenable alloys* (also called *age-hardenable alloys*). If an alloy system exhibits (1) moderate or extended solubility that decreases with decreasing temperature, for example, Al-rich region of Al-Cu system (Fig. 6.20),[60] and (2) the formation of one or several metastable transition phases prior to, or in addition to, the equilibrium phase, it fulfills the conditions necessary for precipitation hardening. These alloys are called precipitation-hardening alloys because their hardness or yield strength increases with time at a constant temperature

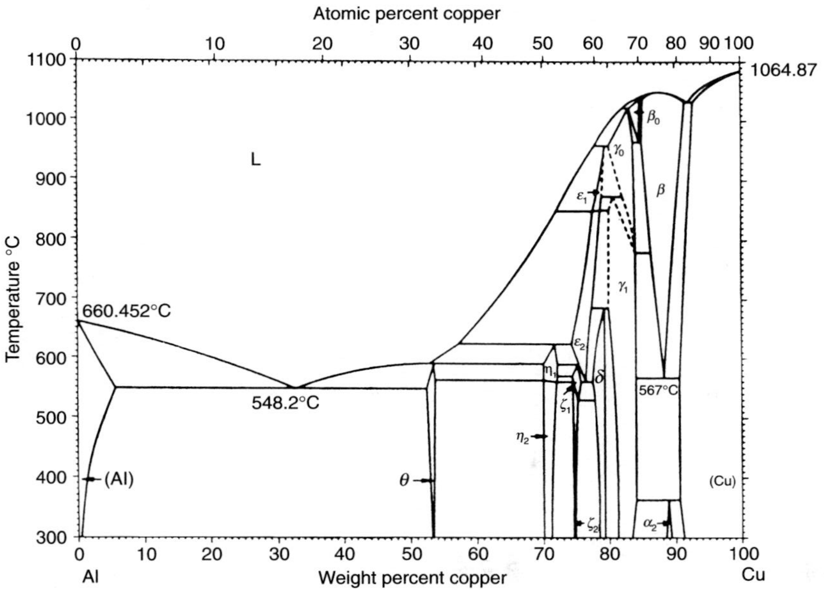

Phase	Composition, wt% Cu	Pearson symbol	Space group
(Al)	0 to 5.65	cF4	$Fm\bar{3}m$
θ	52.5 to 53.7	tI12	$I4/mcm$
η_1	70.0 to 72.2	oP16 or oC16	$Pban$ or $Cmmm$
η_2	70.0 to 72.1	mC20	$C2/m$
ζ_1	74.4 to 77.8	hP42	$P6/mmm$
ζ_2	74.4 to 75.2	(a)	...
ε_1	77.5 to 79.4	(b)	...
ε_2	72.2 to 78.7	hP4	$P6_3/mmc$
δ	77.4 to 78.3	(c)	$R\bar{3}m$
γ_0	77.8 to 84	(d)	...
γ_1	79.7 to 84	cP52	$P\bar{4}3m$
β_0	83.1 to 84.7	(d)	...
β	85.0 to 91.5	cI2	$Im\bar{3}m$
α_2	88.5 to 89	(e)	...
(Cu)	90.6 to 100	cF4	$Fm\bar{3}m$

Metastable phases

Phase	Composition, wt% Cu	Pearson symbol	Space group
θ'	...	tP6	...
β'	...	cF16	$Fm\bar{3}m$
Al_3Cu_2	61 to 70	hP5	$P\bar{3}m1$

(a) Monoclinic? (b) Cubic? (c) Rhombohedral. (d) Unknown. (e) DO_{22}-type long-period superlattice

FIGURE 6.20 Al-Cu phase diagram with description of phases.[60,60a] (*Reprinted by permission of ASM International, Materials Park, Ohio.*)

after quenching from a solution treatment temperature. These sequences of precipitate nucleation are of considerable interest both in the understanding of the mechanisms of phase transformations and in the controlled evolution of the mechanical properties of precipitation-hardening alloys.[61] The use of precipitation hardening as a strengthening technique is very important in certain alloy systems, notably Al-based alloys, Cu-Be, steels (including maraging steels), oxide systems (such as Ti-rare earth oxides), and amorphous alloys.[62] The precipitation nucleation sequences occurring in Al-, Cu-, Fe-, and Ni-based precipitation-hardening alloys have been given in Table 6.2.[63–65] Table 6.3 lists the mechanical properties and applications of some Al-, Cu-, and Ni-based precipitation-hardening alloys,[66,67] while Table 6.4 enlists only the mechanical properties of important precipitation-hardening stainless steels.[67–69]

Precipitation strengthening also includes dispersion strengthening, which involves the most commonly known mechanism of overaging, namely, the Orowan mechanism. For dispersion strengthening, the dispersoid particles are introduced by natural or artificial aging as well as by other methods such as powder metallurgy and mechanical alloying. Precipitation strengthening used widely in ferrous alloys is recognized in the quench aging of low-carbon steels, in microalloyed steels, and in tempering of martensite.[70] Readers will find some interesting discussions on these topics elsewhere in this book.

6.7.1 General Characteristics of the Precipitation-Hardening Process

1. The hardness (and tensile strength) measured at room temperature increases up to a maximum (called *peak hardness*) in the hardness versus aging-time curve (Fig. 6.21).[71,72] This peak hardness occurs at an optimum size distribution and degree of coherency.

2. As the aging temperature is increased (that is, $\dot{N}$ and G increase rapidly), the peak hardness value decreases but is readily attained.

3. A single-stage (peak) hardening takes place at higher aging temperatures for high supersaturation or low aging temperatures for lower solute content (or supersaturation). In contrast, a two-stage hardening occurs at lower aging temperatures for higher supersaturation exhibiting high peak-hardness level[69] (Fig. 6.21).

TABLE 6.2 Precipitation Sequence Observed in Common Precipitation-Hardening Systems

Base metal	Alloy	Precipitation sequence	Equilibrium precipitate
Al	Al-Cu (containing up to 1.7–4.5 wt% Cu)	Spinodal decomposition → GP1 zones rich in Cu (plates, fcc) on $\{100\}_\alpha$ → θ'' phase (plates, tetragonal) on $\{100\}_\alpha$ → θ' phase (plates, tetragonal)	θ (CuAl$_2$, bct)
	Al-Ag (containing up to~23 at% Ag)	GP zones rich in Ag (spheres, fcc) on $\{111\}_m$ → γ' (Ag$_2$Al, plates, hcp)	γ (Ag$_2$Al, hcp)
	Al-Mg	Ordered GP zones (plates or rods, fcc) → β' (plates, fcc)	β (Mg$_3$Al$_2$)
	Al-Zn	Spinodal decomposition → GP zones (spheres, fcc) → α' (plates, rhombohedral) → α' (cubic)	Zn

Base metal	Alloy	Precipitation sequence	Equilibrium precipitate
	Al-Cu-Mg (2xxx)	GPB zones rich in Mg and Cu (rods) on $\{110\}_m \to S'$ (plates) on (021) planes	S (Al$_2$CuMg, orthorhombic)
	Al-Mg-Si (6xxx)	GP zones rich in Mg and Si (needles, fcc) along $\{100\}_\alpha \to \beta''$ ordered zones rich in Mg and Si $\to \beta'$ phase (rodlike, hcp), Mg$_2$Si	β (Mg$_2$Si, plates)
	Al-Zn-Mg(-Cu) (7xxx)	GP zones rich in Zn and Mg (spheres) $\to \eta'$ (MgZn$_2$, plates, hcp) on $\{111\}_\alpha \to \eta$ (MgZn$_2$, rods or plates, hcp) along $\{111\}_\alpha$	T [(Al,Zn)$_{48}$Mg$_{32}$, cubic]
	Al-Li	δ' (Al$_3$Li)	δ (AlLi)
	Al-Li-Zr	$\delta' + \alpha'$ (Al$_3$Zr)	$\delta + \alpha'$
	Al-Li-Mg	δ' (Al$_3$Li)	Al$_2$MgLi, bcc, rod like
	Al-Cu-Li	GP1 zones $\to$ GP2 zones $\to \delta'$, θ', and/or T'_1	θ(CuAl$_2$) $\to$ T_1 (Al$_2$CuLi)
	Al-Li-Cu	T'_1 (Al$_2$CuLi, thin plates)	T_1 (Al$_2$CuLi, thin plates, hcp)
	Al-Li-Cu-Mg (high Mg/Cu ratio)	GP zones $\to S'$ (Al$_2$CuMg)	S (Al$_2$CuMg)
	Al-Li-Cu-Mg-Zr	$\delta' + S'$(Al$_2$CuMg) + T'_1 (Al$_2$CuLi) + α'	T_1 (Al$_2$CuLi) + δ + α' (Al$_3$Zr)
Cu	Cu-2Be	GP zones $\to \gamma'' \to \gamma'$ (bct) first forms with a $\{112\}_\alpha$ habit plane and later with a $\{113\}_\alpha$	γ (CuBe, ordered, bcc)
	Cu-1–3 at% Co	GP zones (spheres)	β (Co, plates)
	Cu-2.5% max Ti	β' phase (Cu$_4$Ti, D1$_a$ structure)	—
Fe	Fe-C	ε carbide (Fe$_{2.4}$C disks, hcp)	Fe$_3$C (plates, orthorhombic)
	Fe-N	α'' (disks, Fe$_8$N, bct) on (100)$_m$	γ' (Fe$_4$N, fcc)
	PH stainless steels		
	17-4 PH	ε phase, Cu-rich clusters (spheres, bcc)	ε (fcc)
	17-7 PH	NiAl (sphere, fcc)	β (NiAl, laths and plates)
	A286	γ' (Ni$_3$TiAl)	η (Ni$_3$Ti)
	17-10 P alloy	M$_{23}$(CP)$_6$ or (CrFeP)$_{23}$C$_6$	
Ni	Inconel X-750	Ordered γ' [Ni$_3$(Al,Ti), fcc, L1$_2$ structure, cuboidal]	η [Ni$_3$(AlTiNb) plate, hexagonal]
	Inconel 718	Ordered γ'' (Ni$_3$Nb, bct, DO$_{22}$ structure, disks) + γ' [Ni$_3$(Al,Ti), fcc, in small amounts]	δ Ni$_3$Nb (orthorhombic)
	Incoloy 901, 902	Ordered γ' Ni$_3$(Al,Ti)	Ni$_3$Ti, hcp, DO$_{24}$ structure
	Haynes 242	Ordered Ni$_2$(Mo,Cr)	

TABLE 6.3 Typical Mechanical Properties and Applications of Some Precipitation Hardening Al-, Cu-, and Ni-Based Alloys[66,67]

Base metal	Alloy number (UNS #)	Chemical composition, wt%	Temper	Tensile strength ksi	Tensile strength MPa	Yield strength ksi	Yield strength MPa	Elongation, (%)	Typical application
					Wrought alloys				
Al	2024 (92024)	Al-4.4Cu-1.5Mg-0.6Mn	T86	64	440	57	395	6	Aircraft structures, rivets, truckwheel hardware, screw machine products
	3004 (93004)	Al-1.2Mn-1.0Mg	H38	41	—	36	250	4–6	Rigid containers (cans), chemical handling and storage, sheet-metal works, builder's hardware, incandescent and fluorescent lamp bases
	6061 (96061)	Al-1.0Mg-0.6Si-1.3Cu-0.2Cr	T6	45	310	40	276	12	Trucks, towers, marine structures, canoes, railroad cars, pipelines, furniture
	7075 (97075)	Al-5.6Zn-2.5Mg-1.6Cu-0.23Cr	T6	83	572	73	503	11	Aircraft and other structural parts
	7178 (97178)	Al-6.8Zn-0.3Mn-2.7Mg-2.0Cr	T6, T65	88	605	78	540	10	Aircraft and aerospace structures
Cu	C17200	98Cu-2Be-0.6Mn		175	1207	140	965	7	Bourdon tubing, flexible metal hose, bellows, clips, washers, retaining rings, firing pins, springs, flexible metal hose, bushings, valves, pumps, shafts, diaphragms, contact bridges, bolts, screws, navigational instruments, nonsparking safety tools
	C17000	98Cu-1.7Be-0.3Co		180	1240	155	1070	4	Bellows, bourdon tubing, fuse clips, fasteners, lock washers, retaining rings, switch and relay parts, electric and electronic parts, roll pins, springs, spine shafts, rolling mill parts, welding equipment, valves, pumps, diaphragms, nonsparking safety tools
Ni	Monel K-500 (N05500)	66Ni-30Cu-2.7Al-0.6Ti		151	1041	111	765	30	Corrosion-resistant and high-strength parts such as pump shafts, oil well drill collars, tools and instruments, doctor blades and scrapers, impellers,

Alloy (UNS)	Composition						Applications
Nimonic 80 (N07080)	Ni-1.4Al-0.008B*-0.10C*-2.0Co*-19.5Cr-0.2Cu*-3.0Fe*-1.0Si*-2.25Ti						valves, gears, springs, valve trims, fasteners, marine propeller shaft, gyroscope applications
Nimonic 90 (N07090)	Ni-1.4Al-0.13C*-18.0Co-19.5Cr-3.0Fe*-1.0Mn*-1.5Si*-2.4Ti						
Udimet 500	53Ni-18.5Co-18Cr-4Mo-3Ti-3Al						
Inconel 718 (N07718)	52.5Ni-19Cr-3Mo-5.1Nb-0.9Ti-0.5Al-18.5Fe-0.15Cu*-0.07C*	180	1240	150	1036	22	
Inconel 718 SP (N07719)	Same as above with 0.05%C*						
Inconel X-725 (N07725)	57.0Ni-20.75Cr-8.25Mo-3.4Nb-1.4Ti-0.35Al-0.35Mn*-0.03C*						See Sec. 6.7.4
Inconel X-750 (N07750)	73.0Ni-15.5Cr-3Mo-1.0Nb-2.5Ti-0.7Al-7Fe-0.25Cu*-0.04C*	162	1120	92	635	24	See Sec. 6.7.4
Haynes 240	65Ni-25Mo-8Cr-2.5Co*-2Fe*-0.8Mn*-0.8Si*-0.5Al*-0.5Cu*-0.03C*-0.006B*						Aerospace and chemical processing industries, seal and container rings, duct segments, casings, fasteners, rocket nozzles, pumps, etc.
Incoloy 925 (N09925)	42Ni-0.3Al-0.03C*-0.5Nb*-21.5Cr-2.25Cu-22.0Fe min-1.0Mn*-3.0Mo-0.50Si*-2.15Ti						
Nimonic 105	Ni-0.12C*-1.0Si*-1.0Mn*-15Cr-20Co-5Mo-1Ti-4.7Al-0.2Cd*-1.0Fe*-0.0065B-0.15Zr*						

TABLE 6.3 Typical Mechanical Properties and Applications of Some Precipitation Hardening Al-, Cu-, and Ni-Based Alloys[66,67] (Continued)

Base metal	Alloy number (UNS #)	Chemical composition, wt%	Temper	Tensile strength		Yield strength		Elongation, (%)	Typical application
				ksi	MPa	ksi	MPa		
			Cast alloys						
Al	296.0 (A02960)	Al-4.5Cu-2.5Si	T6	40	275	26	179	5	Aircraft fittings, fuel pump bodies, wheels and gun control parts, railroad car seat frames, connecting rods, compressors, fuel pump bodies
	355.0 (A13550)	Al-5Si-1.3Cu-0.5Mg	T61	35	241	25	172	3	Aircraft supercharger covers, pump housings, aircraft fittings and engine crankcases, air compressor pistons, fuel pump bodies, liquid-cooled cylinder heads, water jackets, and blower housings
	356.0 (A13560)	Al-7.0Si-0.3Mg	T61	38	260	28	195	5	Aircraft pump and control parts, automotive transmission cases, aircraft fittings, water-cooled cylinder blocks, and wheels
	712.0 (A17120)	Al-5.8Zn-0.6Mg-0.5Cr-0.2Ti	T5	35	241	25	172	5	Machine parts
Cu	C82400	Cu-1.7Be-0.3Co		150	1034	140	965	1	Safety tools, cams, bushings, molds for forming plastics, pump parts, valves, bearings, gears, parts for submarine telephone cable repeater system and hydrophone, and plunger tips for die casting
	C95400	Cu-4Fe-11Al		105	725	54	373	8	Bearings, bushings, gears, worms, valve seats and guides, pickling hooks, pump impellers, rolling mill slippers, slides, and nonsparking hardware
Ni	Monel 505 (N05505)	63Ni-29Cu-4Si		127	876	97	669	3	Valve seats
	Inconel 705 (N09705)	68Ni-15Cr-9Fe-6Si		110	758	95	665	3	Exhaust manifolds

* Indicates maximum value.

TABLE 6.4 Minimum Mechanical Properties of Some Precipitation-Hardening Stainless Steels[67-69]

Trade name	AISI (UNS) designation (ASTM specification)	Composition	Product form[a]	Condition	UTS, MPa (ksi)	YS, MPa (ksi)	Elong. (%)	Reduction in area[b] (%)	Hardness[c] Rc Min	Hardness[c] Rc Max
				Martensitic						
17-4PH	630 (S17400) (A564, A693, A705)	0.09C-1.0Mn-1.0Si-17.0Cr-4.0Ni-3.6Cu-0.25(Nb+Ta)	B,F,P,Sh,St	H900	1310 (190)	1170 (170)	10[d]	40-35[e]	40	48
			B,F,P,Sh,St	H1025	1070 (155)	1000 (145)	12[d]	45[e]	35[d]	42[d]
			B,F,P,Sh,St	H1075	1000 (145)	860 (125)	13[d]	45[e]	32[d]	38[d]
			B,F,P,Sh,St	H1150	930 (105)	725 (105)	16[d]	50[e]	28[d]	36[d]
15-5PH	(S15500) (A564, A705)	0.07C-15Cr-1Mn-4.5Ni-3.5Cu-0.35(Nb+Ta)		(Similar to 17-4PH)						
PH13-8Mo	(S13800) (A564, A693, A705)	0.05C-0.02Mn-0.1Si-12.75Cr-8Ni-2.25Mo-0.01Nb	B,F,P,Sh,St P,Sh,St	H950	1520 (220)	1410 (205)	6-10[f]	45-35[f]	45	—
				H1000	1380 (200)	1310 (190)	6-10[e]	—	43	—
13-8 SuperTough		Fe-13Cr-8Ni-2Mo-1Al		H1000		1406 (204)				
Custom 450	(S45000) (A564, A705)	0.05C-1Si-15.5Cr-6Ni-0.75Mo-1.5Cu-(8 × C)Nb	B,F,P,Sh,St	H900	1240 (180)[g]	1170 (170)	6-10[f]	20-40[f]	40	—
			B,F,P,Sh,St	H1000	1100 (160)	1030 (150)	8-12	45	43	—
			B,F,P,Sh,St	H1150	860 (125)	515 (75)	15	50	28	—
Custom 455	(S45500) (A564, A705)	0.05C-0.5Mn-0.5Si-12Cr-8.5Ni-1.1Ti-0.3(Nb+Ta)-2Cu-0.5Mo	B,F, shapes	H900[h]	1620 (235)	1520 (220)	8	30	47	—
			B,P,Sh,St	H950	1586 (230)	1517 (220)	12	50	48	—
			B,P,Sh,St	H1000	1483 (215)	1345 (195)	14	55	44	—
Custom 465		0.02C-0.25Mn-0.25Si-12.0Cr-11.0Ni-1.65Ti-1.0Mo		H900	1758 (256)	1648 (239)	14	62	49	—
				H1000	1593 (231)	1503 (218)	16	66	47	—
				H1025	1565 (227)	1475 (214)	15	67	47	—
				H1050	1517 (220)	1413 (205)	17	66	46	—
				Semiaustenitic						
17-7PH	631 (S17700) (A693)	0.07C-1.0Mn-1.0Si-17Cr-7Ni-1.15Al	B,Sh,St	CH900	1650 (240)	1590 (230)	1	—	46	—
			B,Sh,St	RH950	1450 (210)	1310 (190)	1-6[d]	—	41[d]	44[d]
			B,Sh,St	TH1050	1240 (180)	1030 (150)	3-7[d]	—	38	—
PH15-7Mo	632 (S15700) (A693)	0.07C-1Mn-1Si-15Cr-7Ni-1.15Al	B,Sh,St	RH950	1550 (225)	1380 (200)	1-5[d]	—	43[d]	46[d]
			B,Sh,St	TH1050	1310 (190)	1170 (170)	2-5[d]	—	38[d]	46[d]
			B,Sh,St	Cold-rolled and aged	1650 (240)	1590 (230)	1	—	46	—

TABLE 6.4 Minimum Mechanical Properties of Some Precipitation-Hardening Stainless Steels[67-69] (*Continued*)

Trade name	AISI (UNS) designation (ASTM specification)	Composition	Product form[a]	Condition	UTS, MPa (ksi)	YS, MPa (ksi)	Elong. (%)	Reduction in area[b] (%)	Hardness[c] Rc Min	Max
Stainless W	635 (S17600)	0.08C-1Mn-1Si-16-17.5Cr-6.0-7.5Ni-0.4Al-0.4-1.2Ti	B,P,Sh,St B,P,Sh,St B,P,Sh,St	H950 H1000 H1050	1310 (190) 1240 (180) 1170 (170)	1170 (170) 1110 (160) 1070 (150)	8 8 10	25 30 40	39 37 35	— — —
AM 350	S35000 (A693)	0.07C-0.5Mn-0.5Si-16-17Cr-4-5Ni-2.5-3.25Mo-0.07-0.13N	P,Sh,St P,Sh,St	H850 H1000	1275 (185) 1140 (165)	1030 (150) 1000 (145)	2-8[d] 2-8[d]	— —	42 36	— —
AM 355	S35500 (A693, A705)	0.01C-0.5Mn-0.5Si-15-16Cr-4-5Ni-2.5-3.25Mo-0.07-0.13N	F P,Sh,St	H1000 H850	1170 (170) 1310 (190)	1070 (155) 1140 (165)	12 10	25 —	37 —	— —
Austenitic										
17-10P		0.5C-1Mn-1Si-17Cr-10.5Ni-0.28P	B B	Annealed Aged	615[i] (89)[i] 945[i] (137)[i]	255[i] (37)[i] 605[i] (88)[i]	70[i] 25[i]	76[i] 39[i]	82[i,j] 30[j,i]	— —
17-14Cu-Mo		0.12C-1Mn-16Cr-14Ni-2.5Mo-0.4Mo-0.3Ti-3Cu		Solution-annealed 1230°C, 0.5 hr, WQ, and aged 730°C, 5 hr	593 (86)	290 (42)	45	63	—	—
A286	600 (S66286)	0.08C-1.4Mn-0.4Si-15Cr-26Ni-1.3Mo-0.3V-2Ti-0.35Al-0.003B	P,Sh,St	Annealed 980°C (1800°F)	641 (93)	248 (36)	48	70	75[i]	—
JBK-75		0.015C-0.005Mn-0.02Si-14.5Cr-29.5Ni-1.25Mo-2.15Ti-0.25Al-0.21V-0.0015B	P,Sh,St	Aged 720°C (1330°F)	1000 (145)	690 (100)	24	37	29	—

[a] B, bar; F, forging; P, plate; Sh, sheet; St, strip.
[b] Values are for bar products.
[c] Where minimum value is also given, maximum value applies only to flat-rolled products. Both max and min values may vary with thickness for flat-rolled products.
[d] Value varies with thickness or diameter.
[e] Value is generally lower for flat-rolled products and varies with thickness.
[f] Higher value is longitudinal; lower value is transverse.
[g] Tensile strength only applicable up to sizes of 13mm (½in.).
[h] Up to and including 150mm (6in.).
[i] Values are typical.
[j] Rockwell B hardness.
Reprinted by permission of ASM International, Materials Park, Ohio.

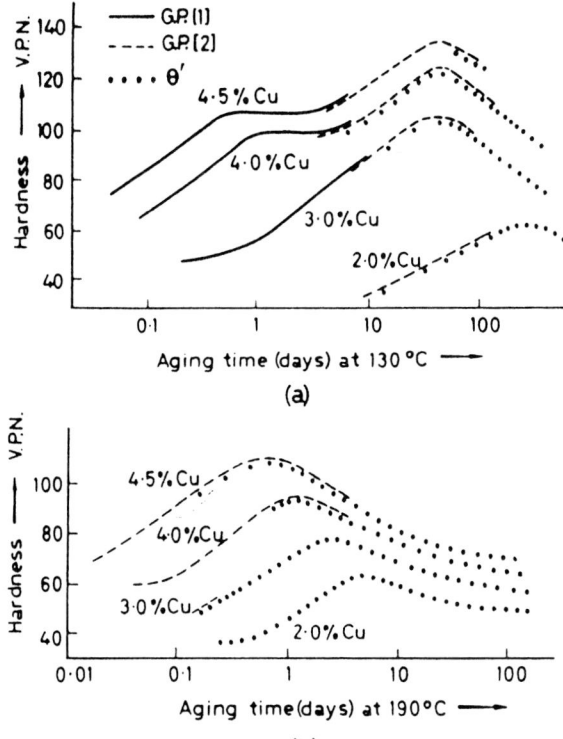

FIGURE 6.21 The hardness versus aging-time curves for Al-Cu alloy at (a) 130°C and (b) 190°C.[71,72]

4. The precipitation occurs by a nucleation and growth process. The precipitate particles that are nucleated during the early aging treatment are coherent with the matrix and are metastable phases that can form more rapidly than the equilibrium phase.

5. When aging is allowed to proceed beyond the peak hardness, the hardness decreases in a monotonic manner into the overaged condition. This is attributed to the loss of coherency of the precipitate particles.

Double (or *two-step*) *aging treatment* is sometimes used to produce initially a fine distribution of GP1 zones at a temperature below the zone solvus, followed by zone coarsening at an elevated temperature. Such treatments are sometimes necessary to reduce *precipitate-free zones* (PFZs). However, the choice between double-aging treatment and alloying in order to optimize the structures and properties should be exercised based on the economic factors.[70]

6.7.2 Al-Base Alloys

6.7.2.1 Al-Cu Alloys.   Let us focus our attention now on a typical Al-4.0 wt% Cu alloy. The basic steps involved in precipitation-hardening heat treatment are as follows:

1. *Solution heat treatment* (*solutionizing*) at a temperature of about 550°C (i.e., near the eutectic temperature and between the solvus and solidus temperatures) to produce a uniform and homogeneous single-phase solid solution.

2. *Quenching* in water to room temperature to retain an unstable, *supersaturated solid solution* (with copper) by suppression of the precipitation of equilibrium θ phase. This is also effective in retaining most of the concentration of vacancies present in thermal equilibrium at the solutionizing temperature.[73] Note that enhanced vacancy concentration formed by quenching from the solutionizing temperature gives rise to an increased rate of solute diffusion, thereby an enhanced rate of clustering. The vacancies may finally condense to form dislocation loops, which tend to act as nucleation sites for precipitation or may exit to various sinks such as grain boundaries. The loss of vacancies at these sinks results in precipitate-free zones.

3. Aging by holding for a sufficient length of time at room temperature (called *natural aging*) or at an intermediate temperature below about 190°C (called *artificial aging*) to precipitate a finely dispersed second phase within the matrix. In general, the (artificial) aging temperature for any precipitation-hardening alloy lies between ~15 and 25% of the temperature difference between the solutionizing temperature and room temperature.[74]

Metastable and Equilibrium Phases. The age-hardening sequence in Al-4% Cu alloy below 130°C is described as

$$\alpha_s \rightarrow \text{spinodal decomposition} \rightarrow \text{GP1 zones} \rightarrow \text{GP2 zones } (\theta'') \rightarrow \theta' \rightarrow \theta$$

The prefix GP is used to indicate the initials of the discoverers, Guinier[75] and Preston,[76] of these zones. Age-hardening of this alloy, however, at 130°C has been found to have the following precipitation sequence:

$$\alpha_s \rightarrow \text{GP1 zones} \rightarrow \text{GP2 zones } (\theta'') \rightarrow \theta' \text{ phase} \rightarrow \theta \text{ phase } (CuAl_2)$$

On the other hand, there is a direct precipitation of equilibrium phase at high temperatures, which is represented by

$$\alpha_s \rightarrow \theta$$

SPINODAL DECOMPOSITION. There is strong evidence that the formation of GP1 zones below 130°C initially proceeds by a spinodal decomposition. Rioja and Laughlin[77] made an electron diffraction study of an Al-4 wt% Cu alloy aged at room temperature for 5 hr and showed that the initial decomposition product was the modulated microstructure comprising the periodic and aligned two-phase mixture (i.e., solute-rich and solute-poor regions) which was attributed to spinodal decomposition. This is followed by GP1 zones arranged in a semiperiodic array. Hence they proposed that zone formation occurred continuously from the spinodal decomposition. Other workers have suggested that there is no nucleation barrier for a spinodal mechanism for GP1 zone formation or for direct GP1 zone formation.[28] A kinetic study of precipitation also supports a spinodal mechanism. In other important age-hardening alloys (e.g., Inconel 80 and the Nimonics), spinodal decomposition occurs initially.

GP1 ZONES. The existence of GP1 zones is detected as streaks in the X-ray or electron diffraction patterns arising from the fact that they are very thin; they are characterized by compositional segregation together with the strain field induced by coherent zones (Fig. 6.22).[8]

GP1 zones are formed directly from the solid solution as the first precipitate on aging at 130°C and consist of circular disk- or platelike clusters of Cu atoms on the

0.4 μm

FIGURE 6.22 Electron micrograph of an aged Al-4% Cu alloy. The fine equiaxed particles are either GP1 zones or θ''; the elongated particles are θ', which have formed on dislocations.[8] (*Reprinted by permission of North-Holland Physics Publishing, Amsterdam; after G. W. Lorimer.*)

{100} planes of the matrix. These disks have fcc crystal structure (Fig. 6.23)[28,78,79] and are about 2 atoms (0.4 to 0.6 nm) thick and 30 atoms (~10 nm) in diameter and are usually coherent with the matrix because the atoms just replace the Al atoms in the lattice. Figure 6.24 is Gerold's model of a GP1 zone in Al-Cu.[79] It shows the extension of the associated displacement field of the surrounding Al-rich matrix to approximately 15 layers on each side of the single Cu-rich plane.[28] This model is in good agreement with the investigation of (1) Al-17 at% Cu alloys aged at room temperature for 12 years and examined by diffuse X-ray scattering[80] and (2) Al-1.5 wt% and Al-4.0 wt% Cu alloys aged at 80, 100, and 130°C, studied by weak beam electron microscopy.[81] However, later results have confirmed that GP1 zones are really a mixture of single-layer and multilayer pure copper regions on {100}$_\alpha$ planes.[82] The compositional range of GP1 zones varies between ~100 at% Cu (according to Gerold's model) and 30 to 40 at% Cu.

The formation of GP1 zones is a homogeneous nucleation and growth process because this establishes the three conditions: random formation without any perceptible incubation time and in very high number densities (up to 10^{18}/cm^3), which is several orders of magnitude larger than the density of available heterogeneous nucleation sites. However, excess vacancies appear to play a vital role in their formation. On the other hand, θ'', θ', and θ phases usually form by a heterogeneous nucleation process. Lorimer and Nicholson[83] have indicated that GP1 zones act as nucleating sites for other metastable precipitates.

GP2 ZONES. (θ'' PHASE). On aging for a longer time or at higher temperatures (130 to 170°C), the diffraction streaks are resolved into a well-defined intensity maxima. The precipitate, formed homogeneously[82] in this stage, is a disk-shaped θ'' phase; the precipitate size is larger than that of GP1 zones, being 1 to 4 nm (10 to 40 Å) thick and 10 to 100 nm (100 to 1000 Å) in diameter. The crystal structure is

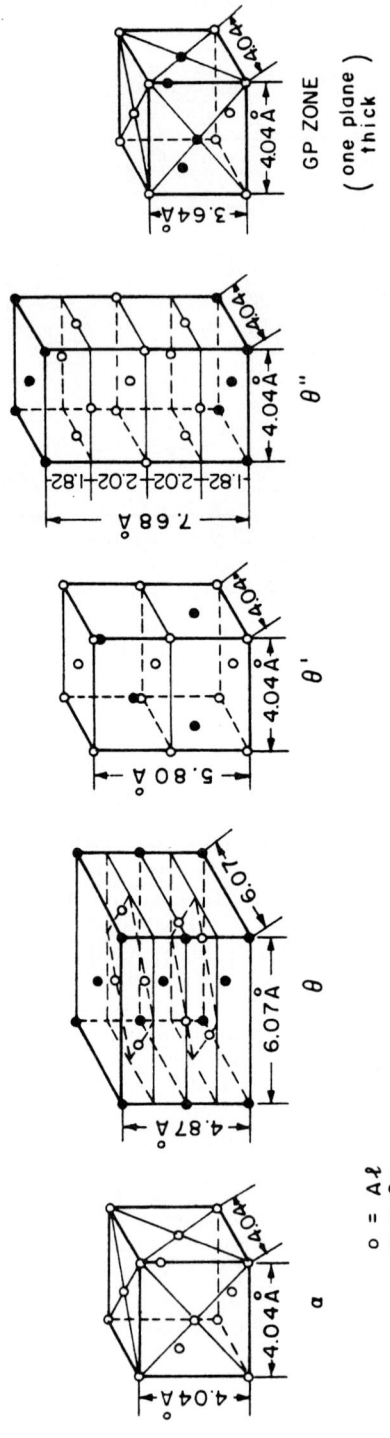

FIGURE 6.23 The crystal structures of the precipitates formed during aging in Al-4% Cu alloy. The matrix phase is α; GP1 zones, θ'', and θ' are transition phases of increasing stability; and the stable precipitate is θ.[1.28] (*Reprinted by permission of North-Holland Physics Publishing, Amsterdam; after Hornbogen[78] and Gerold.[79]*)

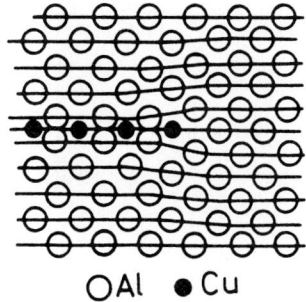

O Al ● Cu

FIGURE 6.24 Gerold's model of a GP1 zone in Al-Cu.[79] (*Reprinted by permission of Van Nostrand Reinhold International.*)

tetragonal (Fig. 6.23) with $a = 4.04$ Å and $c = 7.68$ Å, not due to ordering but due to coherency constraints with the matrix. They are fully coherent platelike precipitates with a $\{001\}_\alpha$ habit plane. The orientation relationship with the matrix is $(001)_{\theta''}$ // $(001)_\alpha$; $[100]_{\theta''}$ // $[100]_\alpha$. GP2 zones are the main contributors to the strength of Al-Cu alloys, which is nearly double when compared to the solid solution itself (Fig. 6.21). Like the GP1 zones, θ'' phases are visible as a result of coherency strain fields caused by misfit normal to the plates (Fig. 6.22). The composition of θ'' was deduced to be ~17 at% Cu by comparing the θ'' lattice parameter with that of splat-quenched metastable fcc Al alloy samples containing 17.3 at% Cu.

θ' PHASE.   After prolonged aging times, the θ'' phase yields an elongated disk-shaped θ' phase with a thickness of 10 to 150 nm (100 to 1500 Å). The θ' phase has a more ordered tetragonal structure than the θ'' (Fig. 6.23), with $a = 4.04$ Å and $c = 5.8$ Å. The approximate composition of θ' is $CuAl_2$, and it obeys the same orientation relationship with the matrix as θ'', while edges of the plate may be either semi-coherent or incoherent.[84] Hence, it is appropriate to say that θ' is semicoherent with the matrix. The density of the θ' phase in the matrix is lower than that of θ'', which nucleates on matrix dislocations (Fig. 6.25)[8] and grain boundaries.

θ PHASE.   After a very long aging time, the equilibrium θ, $CuAl_2$, forms, which is incoherent with the matrix phase and is bct with $a = 6.066$ Å and $c = 4.874$ Å. The θ forms at planar interfaces, dislocations, grain boundaries, or θ'/matrix boundaries. Figure 6.25 shows the nearly equiaxed particles which have nucleated on the θ'/matrix interface.

Free-Energy Relationship.   Let us consider the free-energy relationship in the nucleation of GP1 zones, the transition phases θ'' and θ', and the equilibrium phase θ. Figure 6.26 shows the schematic bulk free energy versus composition diagram at a given temperature T for the Al-Cu system with composition C_0. Since GP1 zones and the matrix have the same crystal structure, they lie on the same free-energy curve. The solubilities of θ, θ', θ'', and GP1 zones, respectively, in equilibrium with the matrix at a given aging temperature T are higher than those of the parent phase. From the common tangent rule, it is clear that GP1 zones have the highest solubility (lowest stability) and that the solubility decreases in the order θ'', θ', θ. That is, the solute concentration α_1 of the matrix in equilibrium with GP1 zones is higher than α_2 for θ''. Similarly, the solute concentration α_2 of the matrix in equilibrium with the θ'' phase is higher than α_3 for θ', and the solute concentration α_3 of the matrix in equilibrium with θ' is higher than α_4 for θ. Thus, only alloys of solute content greater than α_1 can form GP1 zones; alloys of solute content between α_2 and α_1 can form the θ'' phase; alloys in the composition range between α_3 and α_2

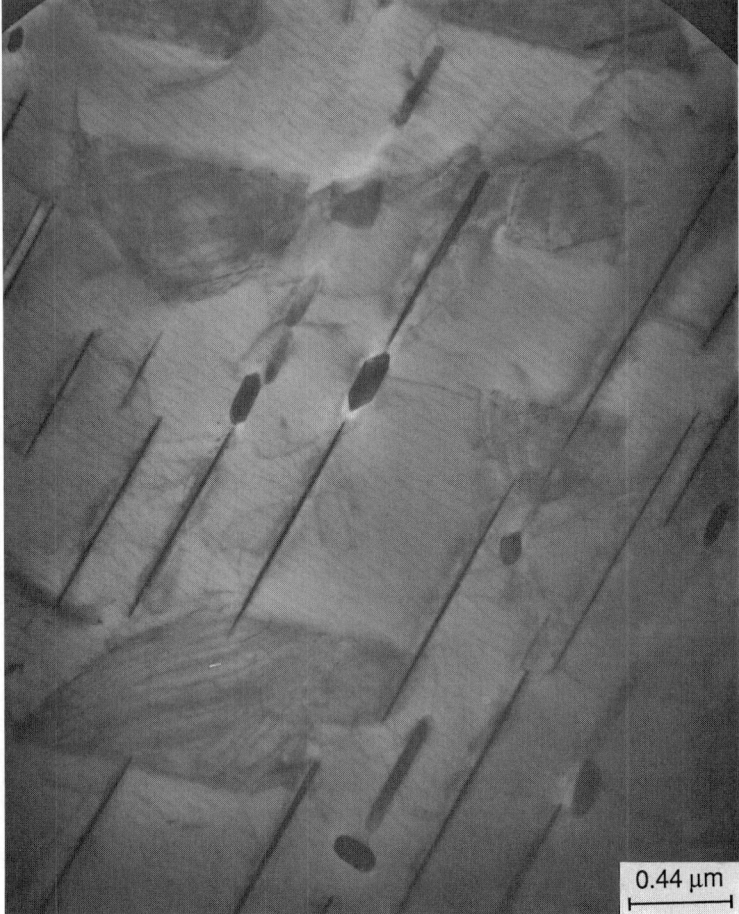

0.44 μm

FIGURE 6.25 Precipitation of θ' and θ phases in Al-Cu alloy. The equiaxed θ particles have nucleated at the θ' plate/matrix interfaces and have grown to consume the
metastable precipitate.[8] (*Reprinted by permission of North-Holland Physics Publishing,
Amsterdam; after R. Carpenter.*)

can form the θ' phase; and alloys in the composition range from α_4 and α_3 must
decompose directly to the stable θ phase.

We can thus conclude that the solubilities α_1, α_2, α_3, and α_4 represent the
maximum stability limits of GP1 zones, θ'', θ', and θ, respectively, at a given temperature T; and that the number of intermediate reaction stages decreases with the
decrease of supersaturation. These matrix compositions represent the points on the
solvus lines for GP1 zones, θ'', θ', and θ, at a particular aging temperature. By taking
into account similar free-energy relationships for all temperatures and the common
tangent rule, the appropriate phase diagram (Fig. 6.27)[8,78] may be constructed which
will include the extra solubility lines, that is, metastable solvus curves. It illustrates
that as the metastable equilibrium solvus curve drops to a lower temperature range,
the stability of the transition phase it represents is lowered. The complete precipi-

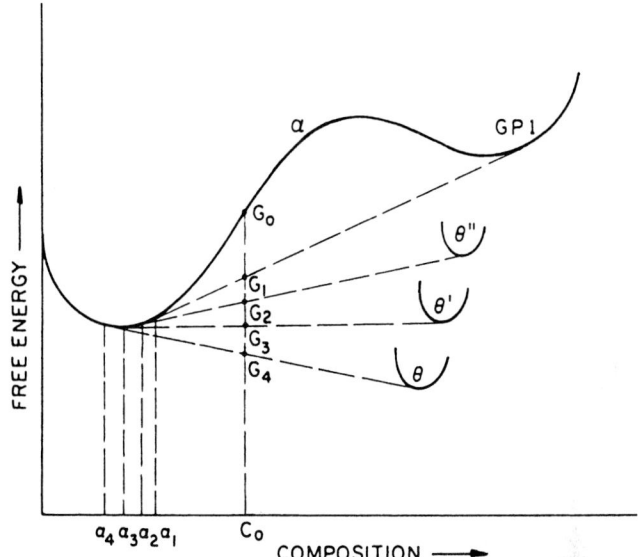

FIGURE 6.26 Schematic bulk free energy versus composition diagram at a given temperature T for the Al-Cu system with composition C_0. It illustrates the relative free energies of GP1 zones and of θ'', θ', and θ phases.

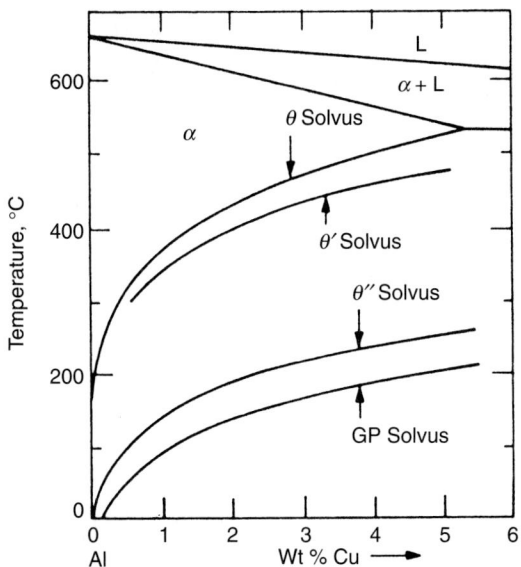

FIGURE 6.27 Al-rich portion of the Al-Cu phase diagram including the metastable solvus curves.[8] (*Reprinted by permission of North-Holland Physics Publishing, Amsterdam; after Hornbogen.*)

tation sequence comprising GP1 zones and transition phases can only be achieved if the alloy is aged below the GP1 zones solvus. When aging is carried out at a temperature between the θ'' solvus and θ' solvus, the precipitation sequence will comprise only θ' as the first metastable phase. Thus, with the increase of aging

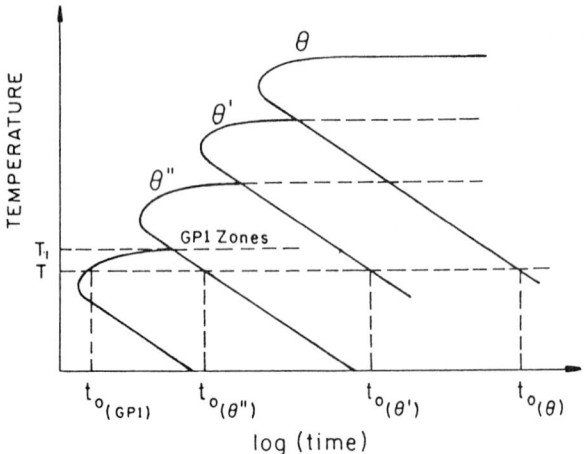

FIGURE 6.28 Schematic TTT diagram of stable and metastable phases in the Al-Cu system.

temperature, the number of transition phases formed diminishes; and above θ' solvus, no transition phase appears. However, in all situations the precipitation sequence is completed with the formation of a stable θ phase.

The effect of aging temperature on the time required to nucleate various phases is shown by a schematic TTT diagram in Fig. 6.28. Some important points may be elucidated here. Since the hardening sequence at low temperature comprises $\alpha_s \rightarrow$ GP1 zone $\rightarrow \theta'' \rightarrow \theta' \rightarrow \theta$ and given the fact that $\Delta G^*_{\text{GP1}} < \Delta G^*_{\theta''} < \Delta G^*_{\theta'} < \Delta G^*_{\theta}$ (where ΔG^* denotes the activation barrier energy for nucleation of the respective phase in the subscript), the time t_0 required to nucleate each phase is given by $t_{0(\text{GP1})} < t_{0(\theta'')} < t_{0(\theta')} < t_{0(\theta)}$. That is, GP1 zones nucleate more easily than θ'', θ', and θ; θ'' nucleates more easily than θ' and θ; and so on. Figure 6.26 illustrates the relative free energies of GP1 zones and those of θ'', θ', and θ phases. This is exemplified in Fig. 6.28, where the upper temperature corresponding to the asymptotic curve is the solvus temperature for a particular phase in an alloy of composition C_0. For example, GP1 zones will not form above T_1 (as shown in Fig. 6.28) as a result of its instability arising from the increased free energy of this alloy of composition C_0.

Effect of Trace Additions. Polmear and his coworkers[85,86] have noticed that small additions of Mg and Ag into Al-Cu alloys promote a precipitation of disk- or platelike Ω phase with a $\{111\}_\alpha$ habit plane. The chemistry of the Ω phase is close to the θ (Al_2Cu) phase but with trace amounts of Mg and Ag.[87,88]

6.7.2.2 Al-Cu-Mg (or 2xxx) Alloys.
Some of the Al-Cu-Mg alloys, notably 2014, 2024, 2025, 2124, 2219, and 2618 alloys, have been developed for aircraft construction. For high Cu/Mg ratio (= 8), θ phase is favored and the aging sequence is summarized as[89–91]

$$\alpha_s \rightarrow \text{GP1 zones} \rightarrow \text{GP2 zones } (\theta'') \rightarrow \theta' \rightarrow \theta$$

where GP1 zones are Cu-rich thin plates, θ'' is fully coherent intermediate precipitate probably nucleated at GP1 zones, and θ' is tetragonal with $a = 0.404$ nm, $c =$

0.580 nm; θ is the incoherent equilibrium phase nucleated at the surface of θ' and is bct: $a = 0.607$ nm, $c = 0.487$ nm.

For alloys with a low Cu/Mg ratio ($1.5 < $ Cu/Mg $ < 4$), the precipitation sequence can be described as

$\alpha_s \rightarrow$ GPB1 zones $\rightarrow$ GPB2 zones/$S'' \rightarrow S'$ (coherent/semicoherent,

orthorhombic, lath-shaped, Al_2CuMg) phase $\rightarrow S$ phase

(incoherent, orthorhombic, Al_2CuMg)

The GPB1 and GPB2 represent Mg- and Cu- rich (Bagaryatski) GP1 and GP2 zones (or Cu-Mg co-clusters) and form as thin rods on {110} matrix planes; and S' plays a key role in age hardening at higher temperatures.[90] The formation of these intermediate precipitates may be influenced by trace elements. For example, Si additions restrict the vacancy condensation and favor more homogeneous distribution of S' precipitates[89] and produce an increase in hardness. The equilibrium S phase forms either by loss of coherency of the S' or by heterogeneous nucleation. Like Al-Cu alloys, a small cold deformation after solution treatment is more potent in refining and in improving the precipitate distribution.

In an experimental Al-4.2Mg-0.6Cu alloy, Ratchev et al.[92] have reported the heterogeneous formation of S'' phase on dislocation loops and helices and proposed that this is the predominant cause of strengthening in the early stages of precipitation hardening. They also observed S'' precipitation at the later aging stage, which suggests its significance for precipitation hardening.

Chopra et al.[93] have observed cubic Z phase as the primary strengthening precipitate in Al-1.5Cu-4Mg-0.5Ag alloy aged at 473 and 513 K. This phase has two orientation relationships with the α-matrix phase, such as $(100)_Z$ // $(100)_\alpha$; $[010]_Z$ // $[010]_\alpha$ at shorter aging time, and $(011)_Z$ // $(1\bar{1}1)_\alpha$; $[01\bar{1}]_Z$ // $[01\bar{1}]_\alpha$.

Gao et al.[94] have found that the addition of > 0.2 wt% Si to the Al-4Cu-0.3Mg-0.4Ag alloy causes complete suppression of Ω phase that dominates the microstructure of the quaternary alloy. However, the hardness of the Si-modified alloys can still be enhanced with an appropriate amount of Si addition, even in the absence of Ω phase.

6.7.2.3 Al-Mg-Si (or 6xxx) Alloys.

In the Cu-free Al-Mg-Si alloys, the needle-like precipitate plays a main role of age hardening. The precipitation sequence in these alloys is given by

$\alpha_s \rightarrow$ Co-clusters of Si and Mg atoms $\rightarrow$ GP1 zones $\rightarrow$ GP2 zones /β''

(needle-shaped precipitates or zones along {100}$_\alpha$ containing a high

concentration of vacancies) $\rightarrow B'$ (lath-shaped precipitates) +

β'(rod-shaped, Mg_2Si) $\rightarrow \beta$ (plate-shaped, fluorite structure, Mg_2Si)

The Mg/Si ratio in the Mg-Si co-clusters, the small precipitates, the β'' precipitates, and B' precipitates are close to 1:1.[95] The zones are oriented parallel to the <100> direction of the Al matrix and contain coherency strain around the needles.[90]

When $\sim$1% Cu is added to the Al-Mg-Si alloy, θ ($CuAl_2$, bct), ϑ' (Mg_2Si), and a third hardening Q phase ($Al_4Cu_2Mg_8Si_7$, hexagonal with a unit cell: $a = 1.04$ nm and $c = 0.405$ nm), being possibly precursor to Q' phase ($Al_4CuMg_5Si_4$), appear. When present as very fine precipitates after age hardening, these phases combine to produce an alloy (e.g., 6013) with a higher yield strength than 6061, 6009, and 6010.[96]

Among the Al-Mg-Si(-Cu) system, 6011-T4 has become the material of choice as an autobody (outer panel) sheet alloy due to greater demands for lightweight vehicles to overall improve fuel efficiencies and reduce vehicle emissions. In 6011 alloy, the Mg/Si ratio in the Q phase is ~1.0.[96a] The strengthening of Al-Mg-Si(-Cu) alloys in automotive applications by precipitation hardening is achieved during the paint bake cycle with an average temperature of 175°C for about 20 min.

The 6016 Al alloys, aged under the automotive paint bake (180°C for 30 min) and peak-strength conditions (180°C for 11 hr), contain only one recognized precipitating phase, β'' with the monoclinic structure. In the 6111 Al alloy, aged at 180°C for 30 min and 11 hr, two phases β'' (monoclinic) and Q (hexagonal) appear together. In 6016 alloy, the Mg/Si ratio increases from 0.85 to 1.01 with extended aging time in the β'' phase and attains 1.35 in the Q phase.[96a]

In a 6022 alloy containing small amounts of Cu, the precipitation sequence involves $\alpha_s \rightarrow$ GP zones $\rightarrow$ needlelike $\beta'' \rightarrow$ rodlike β' + lath-shaped $Q' \rightarrow \beta$ + Si. However, with increased Cu content to 0.91%, the precipitation sequence becomes[96b]

$$\alpha_s \rightarrow \text{GP zones} \rightarrow \text{needlelike } \beta'' \rightarrow \text{lath-shaped } Q'' \rightarrow Q(\text{hexagonal}) + \text{Si}$$

6.7.2.4 Al-Zn-Mg(-Cu) (or 7xxx) Alloys.

The Al-Zn-Mg(-Cu) systems are widely used in the aerospace industry due to their high specific strength. These alloys attain their strength by precipitation through a complex sequence.[96c] In Al-Zn-Mg alloy, the role of GP zones and of vacancy-rich clusters in the nucleation of η' phase offers a greater response to precipitation-hardening treatments with substantially enhanced strength. In addition to the major alloying elements, and impurity and dispersoid additions, some alloys (e.g., 7010 and 7050) contain Zr for more efficient grain refinement, decreased quench sensitivity, and improved strength and toughness.[89] The precipitation sequence in these alloys is summarized as follows:

For higher Zn/Mg ratio, $\alpha_s \rightarrow$ GP zones $\rightarrow \eta'$ [semicoherent, hcp, $MgZn_2$ platelets nucleated on (111) plane] $\rightarrow \eta$ (incoherent, hcp, $MgZn_2$ plates or rods)[96d] and for lower Zn/Mg ratio, $\alpha_s \rightarrow$ GP zones $\rightarrow T'$ [semicoherent, metastable, hcp, irregular morphology, $Mg_{32}(Al, Zn)_{49}$ with high Mg/Zn ratio] $\rightarrow T$ (incoherent, cubic, $Al_2Zn_3Mg_3$, with high Mg/Zn ratio).

The additions of Fe and Si to 7050 Al alloy retard the formation of GP zones and η' precipitates and decrease the mechanical properties.[96e]

In 7075, strengthening occurs by η' precipitates which are regarded as partially ordered intermetallic compound with a formula $Mg_4Zn_{11}Al$, hcp structure.[96f] An addition of 0.7 wt% Li in 7075 alloy results in the following precipitation sequence:

$$\alpha_s \rightarrow \text{GP zones (vacancy-rich)} \rightarrow T' \rightarrow T(Al, Zn)_{49}Mg_{32}$$

According to this mechanism, the Li-vacancies (Li-v) aggregates act as nuclei for subsequent clustering of Zn and Mg atoms, result in the limited formation of vacancy-rich GP zones, and produce narrow PFZs. This results in an early inhibited nucleation and reduced hardening which can be improved by using increased solution heat treatment or two-step aging.[97]

6.7.2.5 Al-Li Base Alloys.

Al-Li alloys have been developed primarily to reduce the weight of aircraft and aerospace structures. Quaternary alloys are used in a wide range of demonstrator parts on civil and military aircraft; and Weldalite finds applications in the large welded cryogenic fuel tanks in the space launch systems for the NASA program and in the welded structure for the ESA space lab project.[89]

Like other age-hardenable alloys, Al-Li alloys achieve precipitation strengthening by artificial aging after a solution heat treatment. The precipitate structure is a function of quenching rate, the following solution heat treatment, the extent of plastic deformation before aging, and the aging temperature and time. Minor alloying additions have also a major effect on the aging process by altering the interfacial energy of the precipitate, by increasing the vacancy concentration, and/or by increasing the critical temperature for homogeneous precipitation. Furthermore, heterogeneous precipitation at interfaces and grain boundaries has a deleterious effect on fracture behavior.

In Al-Li alloys the homogeneous precipitation of coherent, ordered, $L1_2$, spherical δ' (Al_3Li) particles may advance via many transformation steps, including short- and long-range ordering and decomposition. Both ordering and decomposition may occur by nucleation and growth or by spinodal (decomposition) mechanism, increasing thereby the number of possible combinations of consecutive transformation steps.[98] In the presence of grain boundaries and dislocations, coarsening of δ' is enhanced.[99] The precipitation of equilibrium δ ($AlLi$) at the grain boundaries can lead to PFZs which can produce further strain localization and promote intergranular failure. Consequently, dispersoids (Mn, Zr) and semicoherent / incoherent precipitates such as T_1 (Al_2CuLi), θ' (Al_2Cu), or S (Al_2LiMg) introduced by Cu or Mg additions and thermomechanical processing have been used to minimize the formation of PFZs and optimize Al-Li microstructures, thereby achieving the best combinations of strength and toughness.[100]

The important Al-Li alloys are 1420 (Al-5Mg-2Li-0.5Mn), 2020, 2090, 2091, 8090, 8091, and Weldalite 049, which are grouped into two categories: those with small Li additions to Al-Cu to enhance strength (e.g., 2020 and Weldalite) and those with high Li contents such as 1420, 2090, 2091, and 8090 alloys to offer the maximum density reduction. Among the various precipitation sequences for Li-containing aluminum alloys, the one for Al-Li-Cu-Mg(-Zr), e.g., 8090 alloy, is very complicated. These are given as[89,90,99]

Al-Li alloys:	$\alpha_s \rightarrow \delta'$ (Al_3Li) $\rightarrow \delta$ ($AlLi$)
Al-Li-Zr:	$\alpha_s \rightarrow \delta' + \alpha'$ (Al_3Zr) $\rightarrow \delta + \alpha'$
Al-Li-Mg:	$\alpha_s \rightarrow \delta' \rightarrow Al_2MgLi$ (rod-like precipitate, cubic: $a = 1.99$ nm)
Al-Cu-Li (low Li/Cu ratio):	$\alpha_s \rightarrow$ GP1 zones $\rightarrow$ GP2 zones / $\theta'' \rightarrow \theta'$ $\rightarrow \theta$ ($CuAl_2$)
or	$\alpha_s \rightarrow$ GP zones $\rightarrow T_1$ [thin plates with $\{111\}_\alpha$ habit plane, hcp ($a = 0.497$ nm, $c = 0.934$ nm), Al_2CuLi]
Al-Li-Cu (high Li/Cu ratio):	$\alpha_s \rightarrow T_1'(Al_2CuLi) \rightarrow T_1(Al_2CuLi)$
Al-Li-Cu-Mg (high Mg/Cu ratio):	$\alpha_s \rightarrow$ GP zones $\rightarrow S' \rightarrow S$ (Al_2CuMg)
Al-Li-Cu-Mg-Zr:	$\alpha_s \rightarrow \delta' + S'$ (Al_2CuMg) $+ T_1(Al_2CuLi) + \delta + \alpha'$ (metastable or stable Al_3Zr)

The role of Zr is considerably more complex. The Al-Li-Zr alloys achieve a peak yield strength of about 360 MPa in 2 to 3 hr whereas corresponding Al-Li-Mn alloys exhibit a maximum strength of ~250 MPa after ~50 hr at 200°C.[99]

In Al-Li-Mg alloy, since Mg reduces the solubility of Li, for a given Li content, an increased volume of δ' forms. The precipitation of incoherent Al_2MgLi occurs as rods with a $<110>_\alpha$ growth direction, either by overaging or by heterogeneous precipitation, primarily at grain boundaries, and leading to significant deleterious effects upon ductility and toughness.[99]

Primarily, the Al-Cu-Li system provides the likelihood of three precipitates such as δ', θ', and T_1. The balance between them depends on the Cu content and stretching: High Cu content leads to increased θ' precipitation while stretching and low Cu content promotes T_1 at the expense of δ'.[99]

In the Al-Li-Cu system, the precipitation of the T_1 phase takes up a portion of the Li, and further δ' nucleation on θ' occurs so that the volume fraction of δ' available for strengthening is significantly decreased. In a typical composition with 2.5% Li, 1.4% Cu, and Mg > 0.5%, S' dominance prevails.[99]

In Al-Li-Cu-Mg alloys, the T_1 (Al_2CuLi) phase still forms at low Mg levels. However, as the Mg/Cu ratio increases, S (Al_2CuMg) phase progressively replaces the T_1 phase.[99]

Hirosawa et al.[101] have reported that a small addition of Mg to the Al-Li-Cu-Zr alloy (e.g., 8090 alloy) accelerates the formation of GP1 zone, not δ' Al_3Li phase, leading to improved age hardening; this is attributed to the decreased activation energy for the GP1 zone nucleation due to the Mg/Cu/vacancy complexes. It is believed that dislocations are essential for the nucleation of T_1 phase in Al-Li-Cu such as 2090 alloy and S' phase in Al-Li-Cu-Mg-Zr such as 8090 alloy.

The addition of Ag and Mg induces a very high age-hardening response in Al-Cu-Li alloys due to the promotion of a finer and more uniform dispersion of T_1 phase precipitate without recourse to cold work prior to aging.[102]

In the Al-Li-Cu-Mg-Zr system, Zr precipitates coherently and homogeneously as α', which controls the grain size and shape and acts as a nucleation site for δ'; consequently a composite α'/δ' precipitate containing a shell of δ' phase around α' precipitate forms, which is more resistant to planar slip than δ' and thereby improves the strength of the alloy. That is, these composite precipitates cannot be sheared by moving dislocations, thereby leading to dispersed slip band and subsequent high tensile ductility and fracture toughness, even at cryogenic temperatures.[103]

6.7.2.6 Al-Ag Alloys. The Al-rich Al-Ag system is useful for the study of heterogeneous nucleation of precipitates because of (1) small differences in size between Al and Ag atoms, resulting in an extremely small strain energy interaction between Ag atoms, vacancies, and dislocations; (2) very small vacancy-solute binding energy; and (3) simplicity of the crystallographic nature of the γ' precipitation.[104]

Figure 6.29 is the Al-Ag phase diagram including coherent miscibility gaps for GP zones.[105] If Al-rich Al-Ag alloys containing up to ~23 at% Ag are (1) solution-ized above the solvus temperature, (2) quenched, and (3) given a low-temperature aging treatment, decomposition of the metastable supersaturated solid solution occurs[104] in the sequence

$$\alpha_s \rightarrow \text{GP zones} \rightarrow \gamma' \rightarrow \gamma(Ag_2Al)$$

GP zones in this system are Ag-rich spherical clusters that tend to form by homogeneous nucleation and growth. On aging, the GP zones increase in size but decrease in number, which is revealed by the X-ray pattern which exhibits decrease in ring diameter. Faceting of GP zones in an Al-5.07 at% Ag alloy has been found predominantly at a lower aging temperature (160°C) than at a higher aging temperature (350°C), as shown in Fig. 6.30A and B, respectively.[106] These zones also contain ~68% Ag.[107]

On prolonged aging, GP zones gradually dissolve with the nucleation and growth of intermediate γ' platelets. However, GP zones do not take part in the nucleation

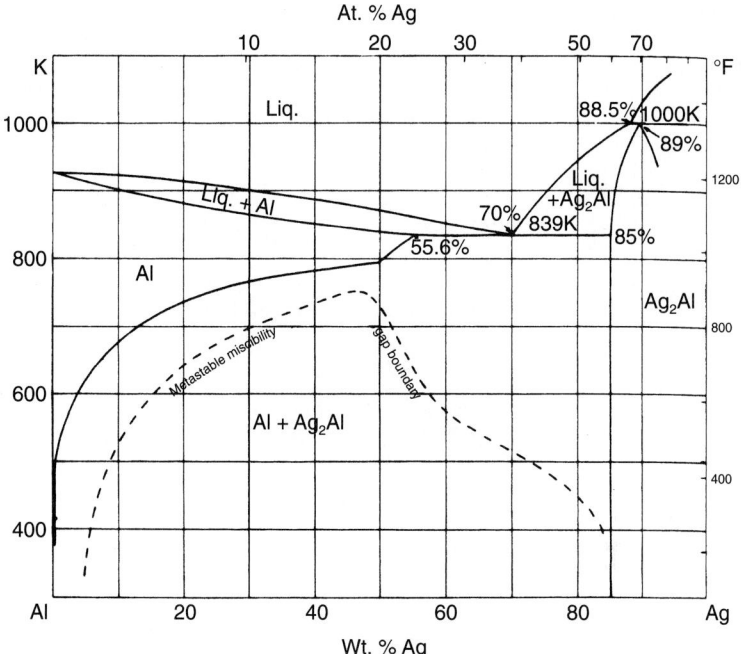

FIGURE 6.29 Al-Ag phase diagram.[102] (*Reprinted by permission of Butterworths, London.*)

of γ'. This fact has been established by the existence of short streaks in the X-ray pattern. The γ' has an hcp structure with $a = 2.858$ Å and $c = 4.607$ Å, and the orientation relationship with the matrix is

$$(0001)_{\gamma'} \, \| (111)_\alpha; \, [11\bar{2}0]_{\gamma'} \, \| [1\bar{1}0]_\alpha$$

where γ' forms as thin coherent platelets (Ag-rich). Initially they form as rod-shaped precipitates and later, by growth, become plates. Figure 6.30C shows an electron micrograph of an Al-Ag alloy aged at 160°C for 5 days, illustrating γ' precipitates with stacking fault contrast.[108]

The steps involved in the formation of Ag-enriched stacking faults are (1) Ag enrichment of helical dislocations by the Suzuki mechanism, thereby decreasing the stacking fault energy (SFE); (2) the climb of dislocation onto {111} planes; and (3) subsequent dissociation of the dislocation into two partials separated by a ribbon of a stacking fault (γ' precipitate) by the reaction[28]

$$\left(\frac{a}{2}\right)[\bar{1}10] \rightarrow \left(\frac{a}{3}\right)[\bar{1}11] + \left(\frac{a}{6}\right)[\bar{1}1\bar{2}]$$

The equilibrium γ phase Ag$_2$Al is hexagonal with $a = 2.879$ Å and $c = 4.573$ Å and has the same orientation relationship with the matrix as γ'. It forms either by discontinuous precipitation (i.e., cellular mechanism) at grain boundaries or from γ' into γ by acquiring misfit dislocations.

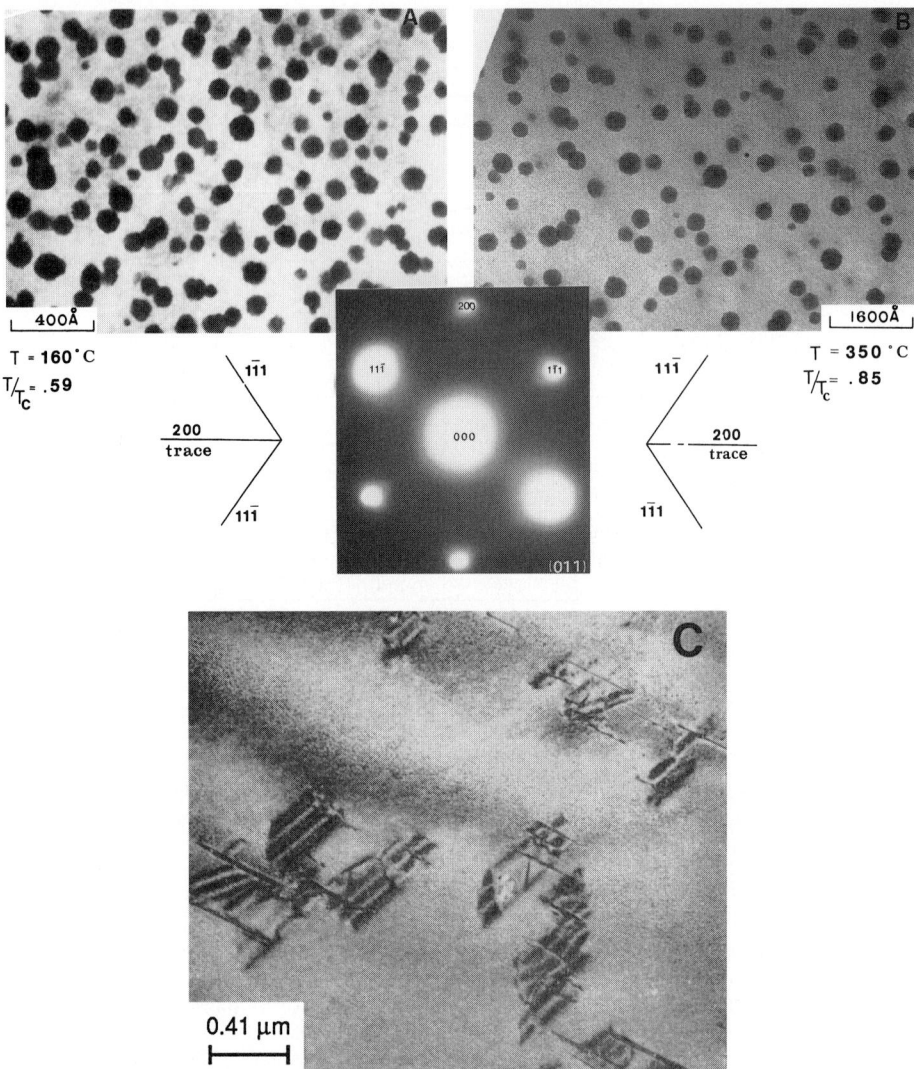

FIGURE 6.30 Electron micrographs of Al-5.07 at% Ag alloy illustrating (*A*) dominant faceted GP zones at lower aging temperature $T = 160°C$ ($T/T_c = 0.59$) and (*B*) less faceted GP zones at higher aging temperature $T = 350°C$ ($T/T_c = 0.85$). Note that the $T = 160°C$ specimen has more "angular" zones. Also note that the electron diffraction pattern exhibits diffuse scattering in the {111} and {100} directions around the fundamental reciprocal.[103] (*C*) Electron micrograph of AlAg alloy aged at 160°C for 5 days showing γ' precipitate within stacking fault contrast.[105] [(*A*) and (*B*) *Reprinted by permission of Pergamon Press, Plc*; (*C*) *reprinted by permission of The Institute of Metals, England.*]

6.7.3 Cu-Base Alloys

Precipitation-hardening wrought Cu-Be alloys find wide applications in electronic components, electrical equipment, control bearings, housings for magnetic sensing devices, and resistance welding systems. The wrought high-strength alloys (C17000 and C17200) contain 1.6 to 2.0% Be and nominal 0.25% Co. The traditional wrought high-conductivity alloys (C17500 and C17510) contain 0.2 to 0.7% Be and nominal 2.5% Co (or 2% Ni). The newer version of high-conductivity alloy is C17410, which contains <0.4% Be and 0.6% Co.

The precipitation sequence in C17200 is given by

$$\text{GP zones} \rightarrow \gamma'' \rightarrow \gamma' \rightarrow \gamma \text{ (stable intermetallic compound, CuBe)}$$

When the Cu-Be alloys contain a third element, either Co or Ni addition, it leads to the dispersion of fine insoluble BeCo and/or BeNi particles in the matrix and grain boundaries,[108a] improves the magnitude of age-hardening response, and retards the tendency to overage or soften at extended aging times and higher aging temperatures. In C17500 and C17200, the beryllides are (Cu,Co)Be with an ordered bct CsCl (B2) superlattice. The beryllides in C17510 are (Cu,Ni)Be; they also exhibit the B2 superlattice.[109]

The Cu-2Ni-1Si alloys can be age-hardened typically at ~450°C, after quenching from solution treatment temperature, which produces coherent precipitation of δ'-Ni$_2$Si as disks on {110} plates with a well-defined orientation relationship:

$$(100)_{Cu} // (001)_{\text{precipitates}} \quad \text{and} \quad [011]_{Cu} // [010]_{\text{precipitates}}$$

These alloys find applications as switch gears, relays, and circuit breakers which require high electrical conductivity, medium strength, and resistance to cyclic deformation.[109a]

6.7.4 Nickel-Base Superalloys

These alloys are essentially γ, Ni-Cr solid solution hardened by additions of Al, Ti, Nb, and Ta, to precipitate a coherent, ordered fcc metastable γ' Ni$_3$(Al,Ti) phase.[110] The unique characteristics of these alloys largely derive from the mechanical behavior of the γ' particles and from the unusual crystallographic relationship between the γ and γ' phases. The γ and γ' phases are both fcc and are almost totally coherent and have similar lattice parameters and identical lattice orientations.[111] An important precipitate that forms in many Ni-, Ni-Cr-, Ni-Cr-Co- (e.g., Udimet 115), and Ni-Fe-Cr-base alloys (e.g., B-1900, Udimet 500, Udimet 700, Udimet 720, René 41, Waspaloy, Unitemp AF2–1DA, Inconel-100, Inconel X-750, Inconel X-751, MAR-M alloy-Hf modified, M-252, etc.)[112,113] is the coherent, ordered fcc γ' phase with an L1$_2$ structure and a cube-cube orientation relationship with the matrix.[114] The γ' phase is usually based on Ni$_3$Al, Ni$_3$(Al,Ti), or (Ni,Co)$_3$(Al,Ti) composition.

Incoloy 901, Incoloy 902, and Incoloy 925 are Ni-Fe-Cr alloys which are hardened by the precipitation of Ni$_3$(Al,Ti) phase. Incoloy 903 is the Ni-Fe-Co alloy with Nb, Ti, Al for precipitation hardening; and Incoloy 909 is the Ni-Fe-Co alloy with an Si addition and containing Nb and Ti for precipitation hardening; in these alloys Cr has been removed to promote low expansion. Thus, the usual precipitate in these Incoloy alloys is metastable γ', Ni$_3$Ti, whereas the stable phase exhibits hcp, DO$_{24}$ structure.[115]

In Inconel 718 alloys, aging reaction produces the finely dispersed metastable γ'' Ni_3Nb (bct, DO_{22}-structure phase), coherent disk-shaped particles on the {100} planes which contribute a major strengthening effect. This is associated with a smaller proportion of metastable, ordered fcc precipitate ($L1_2$) γ', bearing a cube-to-cube orientation relationship with the γ matrix (A1).[116,116a] Both γ'' and γ' can exist at peak hardness. The stoichiometry of both can be given by $Ni_3(Nb,Ti,Al)$. The stable precipitate phase in overaged Inconel 718 has an acicular shape and δ Ni_3Nb orthorhombic structure.[115,116b] It forms as intergranular laths along (111) planes and/or at grain boundaries.[116a] Custom Age 625 Plus alloy is strengthened by the precipitation of γ'' $Ni_3(NbTiAl)$ during aging. In Ni-Mo-base alloys, for example, solid solution strengthened Hastelloy B, ordered bct, Ni_4Mo and fcc, Ni_3Mo form. The most advanced superalloys (e.g., SRR99) have ~70 vol% γ' and a γ' solvus temperature of ~1200°C. Among the most minor additions in nickel-base superalloys are B, Mg, Zr, Hf, Y, rare-earths, and Re.[111]

Haynes 242 alloy is an age-hardenable Ni-Mo-Cr alloy (Table 6.3) which derives its strength from a long-range ordering reaction upon aging, forming very fine, ordered $Ni_2(MoCr)$-type precipitates. It has tensile and creep to high-temperature fluorine and fluoride environment strength properties up to 705°C (1300°F), very good thermal stability and oxidation resistance up to 815°C (1500°F), and excellent low cycle fatigue properties.[116c]

6.7.4.1 Precipitate Morphology.

Mostly γ' phase forms by a continuous precipitation reaction that produces a uniform dispersion of coherent and ordered precipitate particles within the grains. The coherent and ordered precipitates in Ni-base superalloys can assume a sphere-, cube-, or disk-shaped morphology, depending upon the precipitate-matrix misfit. Spherical precipitates are usually observed with misfit less than ~0.3%; cube-shaped morphology forms when the misfit is large (0.3 to 1.0%); disk-shaped precipitates occur when the misfit is greater than ~1%. Examples of disk-shaped precipitates are γ'' in Inconel 718 alloy and Ni-14% Cr-25% Fe-6% Nb alloy.[37]

6.7.4.2 Applications.

Ni-base superalloys comprise a broad range of alloy compositions which find widespread applications in industries (e.g., aerospace, rocket, missile, and spacecraft; fossil-fueled power plants; nuclear power plants; aircraft and gas turbines, jet engines, and high-speed airframe parts; industrial and domestic heaters; thermal processing and heat treatment; chemical; and petroleum such as oil/gas well downholes and wellhead parts) that require low creep rates under high stresses, high stress-rupture strength, high fatigue strength, and high resistance to chemical corrosion and oxidation at elevated temperatures.[117,117a] Custom-Age 625 plus is an age-hardenable material which exhibits corrosion resistance similar to that of Alloy 625 but superior to that of Inconel 718, and is used in deep sour gas wells and in a variety of refinery and chemical process industry applications.[117] Nimonic 80A, Inconel X751, and in some instances Nimonic 90 (N07090) are used as engine exhaust valves for applications requiring superior performance due to higher operating temperatures and harsher environments. Inconel alloy MA 754 (N07754) is now specified in the X-33 prototype space shuttle. Incoloy 908 (N09908) is being used for sheathing and piping in the International Thermonuclear Experimental Reactor Program where special controlled expansion properties are essential.[117a] Furnace muffles made from N0 6600 and N0 6025 (Nicrofer alloy 6025 HT) can be easily fabricated from plate or sheet and can offer several years of uninterrupted life.[117a] Haynes 242 finds applications in aerospace and chemical processing industries and as rocket nozzles, seal and container rings, casings, fasteners, pumps, and so forth.[116c]

6.7.5 Cobalt-Base Alloys

The most widely used cobalt alloys—Haynes 188 (Co-Ni-Cr-W alloy), L-605, and X-40—are carbides plus solid solution strengthened alloys.[117b] Unlike the γ'-strengthened Ni-base superalloys containing Ti or Al, Co does not display a stable, ordered coherent precipitation similar to $Ni_3(Al,Ti)$ intermetallic compounds (Cu_3Au $L1_2$ type). Some engineering value has been derived, however, from metastable alloys involving Co_3Ti, $(Co,Fe)_3Ta$, or $(NiCo)_3Al(Ti)$ where predominant strengthening occurs up to ~800°C. Another class of cobalt alloys for applications in the cryogenic to 1000°C temperature range derives its strength from precipitation hardening closely coupled with the retention of cold work in a martensitic structure.

In a wear-resistant Co-Mo-Si-Cr alloy, precipitation hardening is due to 35 to 70 vol% CoMoSi or Co_3Mo_2Si Laves phase ($MgZn_2$-type hcp).[118]

In the solution-annealed condition, the carbides in Haynes alloy 188 are of the M_6C type. Aging in the temperature range of 650 to 1175°C (1200 to 2150°F) promotes secondary carbide precipitation (M_6C at the higher aging temperatures and $M_{23}C_6$ at the lower temperature). In the MAR-M alloy 509 (with higher C content), the active carbide-forming elements Ta, Ti, and Zr display a significant Chinese script MC carbide in the as-cast condition.[118a]

Haynes 188 exhibits high-temperature strength, very good resistance to oxidation up to 1095°C (2000°F) for prolonged exposure, and excellent resistance to sulfate deposit hot corrosion. Other characteristic features include excellent resistance to molten chloride salts and good resistance to gaseous sulfidation. It is widely used for fabricated parts in the aerospace industry, in the military, and commercially in gas-turbine engines for transition ducts, combustor cans, and after-burner liners. It is used together with 230 alloy, a Ni-Cr-W-Mo alloy, in newer engine programs.[118a]

6.7.6 Precipitation-Hardening (PH) Stainless Steels

The PH stainless steels offer improved formability in the manufacturing stage together with appreciable ductility, higher strength and hardness, and superior corrosion resistance to that of martensitic stainless steels in the final product which are developed by a simple heat treatment comprising solution treating, martensite formation, and low-temperature aging treatment; the latter heat treatment step may be applied after fabrication into wire and strip. Precipitation-hardening grades contain 11 to 18% Cr, 3 to 27% Ni, and small additions of Cu, Al, and/or Ti and occasionally Nb and Mo, to produce the precipitates as hard intermetallic compounds.[117] AM 350 and AM 355 are precipitation-strengthened, but by Cr_2N rather than intermetallic precipitates. The precipitation-hardening class is divided into three broad groups:[68] (1) martensitic type, (2) semiaustenitic type, and (3) austenitic type. Many of these steels are classified by a three-digit number in the AISI 600 series or by a five-digit UNS designation. However, most of them are better known by the trade name of their manufacturers, as shown in Table 6.4, which lists the important grades of PH stainless steels along with their mechanical properties.[69]

6.7.6.1 Martensitic PH Stainless Steels. These are also called *single-treatment alloys.*[67] These utilize low-carbon contents with Ni additions to limit δ-ferrite formation and enhance through-thickness properties in heavy sections. Other additions promote age hardening. The advantages of these steels include improved weldability, corrosion resistance, and toughness, particularly in the overaged condition.[113]

Martensitic grades also display dimensional stability (that contract 0.0005 in./in.) during age hardening, making them ideal candidates for formed parts requiring tight tolerance specifications. Important alloys are 17-4 PH (AISI 630), Stainless W(AISI 635), 15-5 PH, PH 13-8 Mo, 13-8 SuperTough, Custom 450, Custom 455, and Custom 465 alloys.[113] The chemical composition of these grades is listed in Table 6.4. These alloys are solution-annealed at an appropriate temperature (which varies with the grade) (e.g., at 1040°C, or 1900°F, for 17-4 PH alloy)[118b] to dissolve the main hardening precipitate phase, followed by air-cooling to room temperature to transform the austenite into martensite (being a supersaturated solid solution with respect to the hardening agent). The as-quenched hardness of the martensite of these steels is much lower than that of martensitic 410 stainless steel because of the substantially lower amount of carbon and because carbon is tied up as NbC in 17-4 PH and as TiC in Stainless W (AISI 635).[68] A single aging treatment in 17-4 PH alloy is carried out by reheating at temperatures ranging from 480 to 620°C (900 to 1150°F), depending upon the properties desired. This causes, in addition to stress relief, the precipitation of the hardening intermetallic second phase (such as copper-containing particles) within the bcc martensite, which results in an appreciable strengthening of the alloy. The final heat treatment condition is designated "Hxxx", where xxx refers to the tempering temperature in °F.[118b]

6.7.6.2 Semiaustenitic PH Stainless Steels.

The semiaustenitic PH stainless steels are modifications of standard 18-8 stainless steels. These materials contain lower Ni contents and other alloying elements such as Al, Cu, Mo, and Nb. These alloys are semiaustenitic because they are essentially austenite in the as-quenched condition after solution annealing, and eventually martensite can be formed by simple thermal or thermomechanical heat treatments. These steels are used at temperatures up to 315°C (600°F) for long-term exposure and at [480°C (900°F)] high temperatures for some applications. Commercial examples of semiaustenitic PH stainless steels are 17-7 PH, PH 15-7 Mo, PH 14-8 Mo, AM 350 and AM 355. The chemical composition of some semiaustenitic PH stainless steels is shown in Table 6.4.[67-69]

Semiaustenitic PH stainless steels (also called *double-treatment alloys*), for example, 17-7 PH, are solution-treated at about 1050°C (1922°F), followed by water-quenching to give a predominantly soft and metastable austenitic structure with some δ-ferrite. Then this austenite in the solution-treated condition is given one of three hardening sequences (to transform to martensite), as given below:[68,69]

1. In the first TH (temper-hard) sequence, the austenite is conditioned (or primary-aged)[119] at 760°C (1400°F) for 1.5 hr to allow the precipitation of $M_{23}C_6$, which causes depletion of Cr and C content in the austenite matrix, thereby making the austenite unstable and consequently raising the M_s temperature to between 65 and 95°C (150 and 200°F). Upon cooling to a temperature below 15°C (60°F) but above 0°C (32°F), the transformation of austenite into bcc martensite occurs. Finally, aging is carried out at 480 to 620°C (900 to 1150°F) for 1.5 hr, which results in further strengthening by an intermetallic hardening phase. This sequence provides higher ductility, but lower strength, than other sequences (see Table 6.4).

2. In the RH (refrigeration-hard) sequence, metastable austenite is conditioned (i.e., trigger-annealed) at a higher temperature [925°C (1750°F) for 10-min exposure], which allows more carbon to remain in solution in the austenite (than at the 1400°F temperature discussed above); thereby a small but significant amount of chromium carbides is precipitated, and consequently the M_s temperature becomes lower than that in the TH condition. Hence, cooling to subzero temperature at −75°C (−100°F) for 8 hr is required to bring about complete transformation into martensite. The final step is aging at 480 to 620°C for 1 hr, which produces the desired strengthening. The RH sequence produces a small degree of intergranular chromium depletion that slightly sensitizes the material.

3. In the CH (cold-hard) sequence, severe cold working (60 to 70% reduction) is applied, which results in substantial martensitic transformation above the M_s temperature. Heat treatment is, therefore, reduced to a single aging at 480 to 620°C for 1 hr to induce precipitation strengthening. The precipitation-hardening step in these hardening sequences has dual functions: (1) it provides additional hardening by precipitation of an intermetallic compound and (2) it stress-relieves the martensite in order to increase ductility, toughness, and corrosion resistance. This sequence lacks carbide precipitation and prevents sensitization. That is, CH condition displays the greatest corrosion resistance, followed by the RH condition, and then the TH condition, which displays the lowest corrosion resistance.[118b]

Mechanical properties developed by these heat treatment sequences for 17-7 PH stainless steels are listed in Table 6.4.[67–69]

PH 15-7 Mo is a high-strength modification of 17-7 PH and needs the same heat treatment procedures.

6.7.6.3 Austenitic PH Stainless Steels. These steels possess their austenite structures at all temperatures. The most important steels in this class include A-286 (AISI 660), 17-10 P, and 14–17 Cu-Mo alloys. Of these grades, A-286 is the most extensively used in aerospace applications. A-286 is solution-annealed at 900 or 980°C (1650 or 1800°F), followed by rapid cooling to room temperature to suppress the precipitation of the second phase. In this condition, A-286 can be fabricated by forming or machining. Aging treatment is done at 720°C (1325°F) for 16 hr. The choice of the solutionizing temperature is made, depending on the desired properties. The 900°C (1650°F) solutionizing treatment causes finer grain size and improved room-temperature properties, whereas the 980°C (1800°F) treatment produces coarse grain size and better elevated-temperature creep strength.[68]

The 17-10 P alloy is solution-annealed at 1149°C (2100°F), followed by rapid cooling to produce a supersaturated austenite matrix with exceptionally high ductility and formability. Subsequent aging in the temperature range of 649 to 760°C (1200 to 1400°F) produces strengthening by a fine dispersion of carbides. Higher P content of this alloy plays a vital role in promoting carbide precipitation within the grain. This aging treatment produces higher ductility, but lower tensile strength, than other precipitation-hardening alloys (Table 6.4). This alloy is seldom used now.

6.7.6.4 Precipitation Sequence

17-4 PH Stainless Steel. Unaged 17-4 PH stainless steel is a mixture of equiaxed α-martensite matrix and stringers of δ-ferrite. The precipitation sequence consists of the initial formation of coherent Cu-rich spherical clusters, which have the same crystal structure as the matrix (i.e., bcc). After prolonged aging, this transforms to an incoherent fcc ε-precipitate.[120] The other grades (e.g., 15-5 PH, Custom 450, PH 13-8 Mo, Custom 455, and Custom 465) are similar to 17-4 PH but contain little δ-ferrite.

17-7 PH Stainless Steel. During precipitation hardening of 17-7 PH stainless steel, the Al in the martensite initially combines with some amounts of Ni to produce a fine homogeneous precipitation of bcc, spherical Ni-Al particles, which are coherent with the bcc/bct martensitic matrix phase and are the main contributor to strengthening. In the overaged condition, β-Ni-Al forms as Widmanstatten laths and plates.[119,121]

A-286 and 17-10 P Alloys.   The precipitate responsible for age hardening in austenite A-286 alloy is a coherent fcc γ', $Ni_3(Ti, Al)$ phase, but secondary precipitates which form during aging are η or Ni_3Ti, alloy carbides such as $(Ti,Mo)C$, and a boride, M_3B_2. Another austenitic PH stainless steel is JBK-75, which is a minor modification of the A-286 alloy, developed for better weldability.[122] In 17-10 P alloy, the age-hardening precipitate phase is $M_{23}(CP)_6$ or $(Cr, Fe, P)_{23}C_6$.[123]

6.7.6.5   Application.   The precipitation-hardening (PH) martensitic stainless steels are the most widely used of this class of steels because they possess excellent mechanical properties; for example, Custom 465 can produce aged yield strength in the range of 965 to 1620 MPa (140 to 235 ksi), high toughness, and superior corrosion resistance. The combination of these properties makes them attractive materials[121] for forging and for fasteners in the aerospace industries and naval propulsion turbines.[124] With respect to the strength/weight viewpoint, 15-5 PH, PH 13-8 Mo, Custom 450, Custom 455, and Custom 465 are preferred to 17-4 PH in heavy sections because there is essentially no undesirable δ-ferrite which decreases the cross-sectional ductility. In all cases, martensitic PH stainless steels offer advantages over the conventional quenched and tempered martensitic stainless steels. They can be machined to accurate size because they undergo very little dimensional change (nominally a contraction of ~0.0005 in./in.) during aging of solution-treated material. As a result, they are cost-effective and can be cheaper for a given component than the traditional quenched and tempered low-alloy steels (e.g., AISI 4340).

Custom 450 has similar corrosion resistance to type 304 stainless steel but about 3 times the yield strength; hence it has been used for applications requiring higher strength. Custom 455 is used where simplicity of heat treatment, ease of fabrication, high strength, and corrosion resistance are required together. Custom 465 is designed to have about 260-ksi tensile strength, excellent notch tensile strength, and fracture toughness in the H900 condition.[117]

PH 13-8 Mo stainless steel has found applications as valve parts, fittings, cold-headed and machined fasteners, shafts, landing gear parts, pins, lockwashers, aircraft components, nuclear reactor components, and petrochemical applications requiring resistance to stress corrosion cracking. The 15-5 PH is used for applications requiring high transverse strength and toughness such as valve parts, fittings, fasteners, shafts, gears, engine parts, chemical process equipment, paper mill equipment, aircraft components, and nuclear reactor components.[117]

The semiaustenitic PH stainless steels find applications mostly in flat-rolled sheet or strip form in the aircraft and aerospace industries. Other applications include lightweight high-strength parts for use in corrosive media, for example, filters (wire mesh), and moving components in weaving machines.[106] The AM 350 alloy (UNS S35000) is exclusively a sheet product and finds applications where welding is required. The AM 355 (UNS S35500) alloy is the matching bar and plate products. The two alloys differ in that AM 350 contains δ-ferrite while AM 355 is controlled to prevent the presence of δ-ferrite in the finished product. Welded metal bellows are used in turbine fuel controllers, and there is one such bellows in each aircraft turbine engine (including helicopter and turboprop engines as well as jets). The absence of active, slag-forming elements (such as Al and Ti) improves the weldability of AM 350 compared to 17-7 PH or PH 15-7Mo.[124a] Both AM 350 and AM 355 are used for gas-turbine compressor components such as blades, disks, shafts, and rotors or for similar components requiring high strength at intermediate elevated temperatures. Typical applications for PH 15-7 Mo (UNS S15700) include aircraft bulkheads, retaining rings, diaphragms, welded and brazed honeycomb paneling, and other aircraft parts requiring high strength at elevated temperatures. Applications for 17-7 PH (S17700) are aerospace components and flat springs.[41]

The austenitic PH stainless steels are used for high-temperature applications due to their excellent oxidation-resistance properties. They are nonmagnetic and possess good cryogenic (low-temperature) properties. The A-286 is used in jet engines, superchargers, and various high-temperature applications such as turbine wheels and blades, frames, casings, afterburner components, and fasteners.[117]

6.8 STRENGTH OF PRECIPITATION-HARDENED ALLOYS

A change in strength with time at room temperature was first noticed in aluminum alloys by Wilm in 1911.[125] The first probable explanation for this "aging" process was put forward by Merica et al.,[126] who suggested that this hardening was associated with the precipitation of a new phase. Since that time, there have been many reviews of precipitation-hardening mechanisms, notably the famous review of Kelly and Nicholson[127] and the more recent reviews by Brown and Ham,[128] Gerold,[129] Lloyd,[130] Ardell,[131,132] and Nembach.[133]

The formation of metastable second-phase precipitate particles greatly increases the strength of the alloy. These particles represent effective obstacles to dislocation movement. The factors responsible for the increased yield stress are strength, structure, spacing, size, distribution and shape of the precipitate particles, degree of coherency or misfit with the matrix, and their orientation relationship with the matrix. To understand the precipitation hardening, we must study, in detail, the manner in which dislocations interact with a statistical precipitate particle distribution.

There are seven mechanisms of precipitation strengthening:[123,134]

1. Coherency strain strengthening
2. Orowan strengthening
3. Chemical strengthening
4. Order strengthening
5. Modulus strengthening
6. Stacking fault strengthening
7. Hardening by spinodal decomposition

This section discusses the theoretical yield strength or the *critical resolved shear stress* (CRSS) of single crystals attainable by various strengthening mechanisms and how these compare with the experimental observation in commercial alloys. The maximum interaction force F_m for the presumed spherical particle has been expressed here in a normalized form.

6.8.1 Yield Stress and Interaction Force

When a gliding dislocation moves, under the influence of applied stress, and encounters a row of strong and uncuttable obstacles (or particles), it bows around them. If the applied stress is sufficiently increased, semicircular loops of dislocations between particles are formed; subsequently, the two dislocation arms meet and thus annihi-

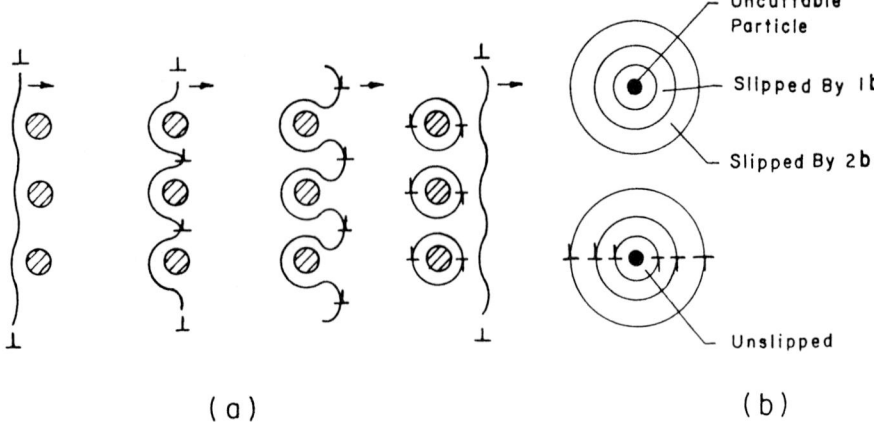

FIGURE 6.31 Schematic illustration of (*a*) different stages of a moving dislocation in bypassing a row of precipitate particles and (*b*) unslipped area and slipped area around a particle after bypassing of several dislocations.

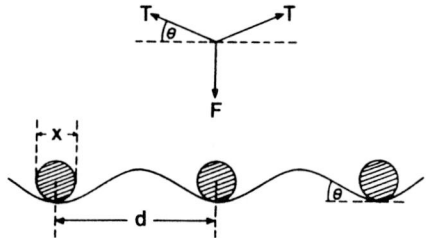

FIGURE 6.32 Interaction of a dislocation with a row of precipitate particles with a spacing *d*. The upper sketch represents the forces acting on the dislocation at a precipitate particle.[129] (*Reprinted by permission of North-Holland Physics Publishing, Amsterdam.*)

late each other, and the main dislocation moves on. However, there is a dislocation loop left around each particle which separates the slipped area of the matrix from the unslipped area on the glide plane. Every time the dislocation bypasses the precipitate particle, it leaves one dislocation loop around the particle. This situation (which may be observed experimentally by using electron microscopy) is schematically shown in Fig. 6.31. This mechanism was first suggested by Orowan and is therefore called the Orowan mechanism of hardening.

To estimate the critical shear stress for the Orowan process, let us consider a row of obstacles equidistant from one another with an intersection (i.e., particle) diameter *x* in the slip plane. The dislocation line is pressed against this row by an applied stress and bends between the particles with bending angle θ, as shown in Fig. 6.32.[129] The force balance on the dislocation line gives

$$\Delta\tau bd = 2T\sin\theta \qquad (6.61)$$

where $\Delta\tau = \tau - \tau_m$ is the increase in applied stress due to interaction with the particle, τ_m is the shear stress required to move the dislocation across the slip plane, *d* is the interparticle distance, *b* is the magnitude of the Burgers vector of the dislocations, and *T* is the dislocation line tension. The resistance force *F* of each particle is

$$F = 2T \sin \theta \qquad (6.62)$$

which can also be defined by $\Delta \tau b$ (force per unit length) times d (length of dislocation). The upper part of the figure has a maximum value of $2T$. The maximum resisting (or pinning) force of the second-phase particle is

$$F_m = 2T° \qquad \text{when } \theta = 90° \qquad (6.63)$$

where $T°$ is a particular value of the varying dislocation line tension. In reality, the obstacles on the glide plane will have a range of strength. If $F_m > 2T°$ (i.e., if θ becomes 90° prior to reaching F_m), the dislocation bypasses the particle by the Orowan process. If we assume that this is the onset of plastic deformation, the following equation for the increase in CRSS, denoted by τ_0, is obtained:

$$\Delta \tau_0 = \frac{2T°}{bd} \qquad \text{for } F_m > 2T° \qquad (6.64)$$

Since the applied stress, τ_0, acts only on the free distance $l = (d - x)$ between particles, which also represents the average separation between two particles in the slip plane, Eq. (6.64) should be written as

$$\Delta \tau_0 = \frac{2T°}{b(d - x)} = \frac{2T°}{bl} \qquad \text{for } F_m > 2T° \qquad (6.65)$$

Since the applied shear stress $\Delta \tau_0$ causes a dislocation to bow into a radius $r (= l/2)$ for it to extrude between the particles, we may write

$$\Delta \tau_0 = \frac{T°}{rb} \qquad (6.66)$$

where the line tension is assumed to be independent of dislocation character. Again it can be shown from Fig. 6.32 that

$$2r \sin \theta = l \qquad (6.67)$$

We can also rewrite Eq. (6.61) as

$$\Delta \tau = \frac{2T \sin \theta}{bl} \qquad (6.68)$$

6.8.2 Interparticle Spacing

Figure 6.33 shows the Friedel process for dislocation-point-obstacle interaction, which illustrates that each time a glide dislocation breaks free from the obstacle pinning it and encounters one new obstacle, it sweeps out an area A of the slip plane.[130] For a regular square array of obstacles, if there are n_s obstacles per unit

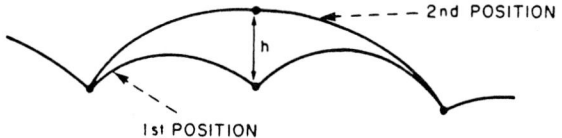

FIGURE 6.33 The Friedel process for dislocation-point-obstacle interaction.[130] (Reprinted by permission of Pergamon Press, Plc.)

area of the slip plane, the spacing between them, called *square lattice spacing* l_s, is defined as

$$l_s = n_s^{-1/2} \tag{6.69}$$

where

$$A = l_s^2 \tag{6.70}$$

From the geometry in Fig. 6.33, we have

$$A \cong hl \cong l_s^2 \tag{6.71}$$

and

$$r^2 = l^2 + (r-h)^2 \tag{6.72}$$

which yields

$$l^2 \cong 2hr \qquad \text{for } r \gg h \tag{6.73}$$

Combining Eqs. (6.67), (6.71), and (6.73), we get

$$l = l_f = \frac{l_s}{\sqrt{\sin\theta}} \tag{6.74}$$

where l_f is the effective Friedel obstacle spacing. This parameter is a function of square lattice spacing and obstacle strength. For strong obstacle strength, effective obstacle spacing becomes smaller.

Combining Eqs. (6.68) and (6.74) yields

$$\Delta\tau = \frac{2T(\sin\theta)^{3/2}}{bl_s} \tag{6.75}$$

Equation (6.75) is based upon the following assumptions: (1) a straight dislocation [i.e., a weak obstacle (or soft particle)] and (2) a square array of obstacles. However, for strong obstacles where Orowan looping occurs, Brown and Ham[128] have suggested the equation

$$\Delta\tau = \frac{0.8(2T\sin\theta)}{bl_s} \tag{6.76}$$

6.8.3 Effect of Finite Particle Size

The above approach deals with the idealized situation involving point obstacle or smaller particle size relative to the planar particle spacing (i.e., when the volume fraction is small). However, these assumptions do not hold good for real systems, where we have a large volume fraction and a large particle size. For the Orowan process, with finite particles, Ardell[131,132] has summarized the following relationships among average particle (or dispersion) radius $<r>$, the average planar radius $<r_s>$, and the volume fraction of precipitates f:

$$\frac{\pi <r_s>^2}{l_s^2} = f \tag{6.77a}$$

$$<r_s> = \frac{\pi <r>}{4} \tag{6.77b}$$

$$<r_s^2> = \frac{32<r_s>^2}{3\pi^2} \tag{6.77c}$$

Substituting Eq. (6.77c) into Eq. (6.77a) results in

$$l_s = \left(\frac{32}{3\pi f}\right)^{1/2} <r_s> = \left(\frac{2\pi}{3f}\right)^{1/2} <r> \tag{6.77d}$$

For cuttable particles, an allowance must be made for the finite-sized particles when the bow-out area is measured. This depends on the specific dislocation configuration at breakthrough.[130]

6.8.4 Coherency Strain Hardening

In most cases, coherent particles (formed particularly in the early stages of precipitation) are surrounded by an elastic strain field as a result of small differences in the average atomic volume of the precipitate and matrix. The precipitation of coherent particles—very strong and resistant to cutting and with slight misfit in the matrix—gives rise to strain fields, which hinder the dislocation movement in the matrix.[135] Recent theories of coherency strengthening have been developed by Gerold and Haberkorn,[136] Gleiter,[137] and Brown and Ham,[128] based upon treatments using isotropic linear elasticity theory. This theory considers an elastic interaction between a spherical coherent particle of radius r and misfit strain ε with an infinite straight-edge dislocation where

$$\varepsilon = \frac{1}{3}\left(\frac{1+v}{1-v}\right)\left(\frac{\Delta a}{a}\right) \cong \frac{2}{3}\delta \tag{6.78}$$

where δ is the fractional difference in lattice parameter between the particle and matrix. Based on this theory, Gerold and Haberkorn found that the interaction force F between a coherent precipitate and a straight-edge dislocation, gliding on a slip plane and located at a distance z from the center of the particle (see Fig. 6.34), is given by

$$F = \left(\frac{3}{2}\right)^{3/2}\frac{\mu b|\varepsilon|r^3}{z^2} \quad \text{for } \frac{z^2}{r^2} > \frac{3}{4} \tag{6.79}$$

and
$$F = 8\mu b|\varepsilon|z\left(1-\frac{z^2}{r^2}\right)^{1/2} \quad \text{for } \frac{z^2}{r^2} < \frac{3}{4} \tag{6.80}$$

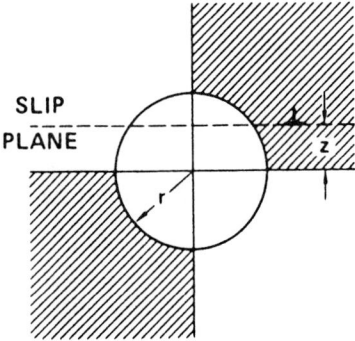

SLIP PLANE

FIGURE 6.34 Schematic illustration of the interaction between a spherical coherent precipitate, $\delta < 0$, and a positive edge dislocation on a slip plane at a distance z from the center of the particle. In the hatched regions, the elastic interactions are repulsive for this combination of δ and b. Reversing the sign of either also reverses the quadrant of repulsive and attractive interaction.[131] (*Reprinted by permission of The Metallurgical Society, Warrendale, Pennsylvania.*)

where μ is the shear modulus of the matrix. This expression yields maximum interaction force F_m

$$F_m = 4\mu b|\varepsilon|r \qquad \text{at} \qquad \frac{z^2}{r^2} = \frac{1}{2} \tag{6.81}$$

Note that for $z^2/r^2 > 3/4$, F_m occurs for an edge dislocation on a slip plane at a distance z from the center of the particle; for $z^2/r^2 < 3/4$, F_m lies at the matrix-particle interface, and the slip plane is located at the center plane of the particle; for $z = 0$ (i.e., slip plane located at the center of the particle), $F_m = 0$.

For coherency strain hardening, the flow stress is derived by substituting the average value of the interaction force F of all of the precipitates in the microstructure over a distance z of the central position of the particles with respect to the slip plane (Fig. 6.34)[131] into Eq. (6.62). Once a suitable method of averaging is chosen, an average value of F_m is measured. For underaged alloys (small coherent particles), the flow stress $\Delta\tau_{0,\varepsilon}$, which is equivalent to the CRSS, is given by

$$\Delta\tau_{0,\varepsilon} = 4.1\varepsilon^{3/2}\mu\left(\frac{f<r>}{b}\right)^{1/2} \tag{6.82}$$

For large coherent particles a different formula is used, namely,

$$\Delta\tau_{0,\varepsilon} = 0.7\mu f^{1/2}\left(\frac{\varepsilon b^3}{<r>^3}\right)^{1/4} \tag{6.83}$$

where μ, ε, and f have the usual meanings and $<r>$ is the averaged value of the radius of spherical precipitate particles. Thus for small particles, coherency strain hardening increases with an increase in average particle size $<r>$, whereas for larger particles, the coherency strain hardening decreases with an increase in average particle size $<r>$, leading to maximum in the strengthening at a critical particle size which is of the order of $<r>/b = 1/4\ \varepsilon^{-1}$, irrespective of the volume fraction.[70]

The systems in which the coherency strengthening mechanism is expected to predominate are Cu-Co, Cu_3Au-Co, Cu-Fe, and Cu-Mn. However, it has been concluded, on the basis of experimental results, that the theories of coherency strengthening are quite inadequate for estimation of the expected strengthening in the underaged, peak-aged, and overaged hardening alloy regimes.[131]

6.8.5   Orowan Strengthening

When (coherent or incoherent) precipitate particles or obstacles are impenetrable (i.e., nonshearable) by the glide dislocation and $F_m > 2T^\circ$, the contribution of the precipitates (i.e., the CRSS, $\Delta\tau_0$, required to make the dislocation pass between the impenetrable obstacles) is determined by the Orowan equation

$$\Delta\tau_0 = \frac{2T^\circ}{bl} \tag{6.65}$$

By assuming $T^\circ = 1/2\mu b^2$ in Eq. (6.65), this equation can be simplified to the form

$$\Delta\tau_0 = \frac{\mu b}{l} \tag{6.84}$$

This theory was modified by Ashby and others by introducing the relation between the dislocation line tension T and dislocation line energy E for both a long straight dislocation [(Eq. (6.85)] and a curved segment of dislocation [(Eq. 6.86)] lying in the slip plane of an isotropic crystal:[128]

$$E(\theta) = \frac{\mu b^2}{4\pi} \left[\frac{1 - v \cos^2 \theta}{1 - v} \right] \ln\left(\frac{r_1}{r_0}\right) \tag{6.85}$$

$$T(\theta) = E(\theta) + \frac{d^2 E(\theta)}{d\theta^2} \tag{6.86}$$

where θ is the angle between the dislocation line or its tangent and its Burgers vector for straight and curved segments of dislocations, respectively, and r_1 and r_0 are the outer and inner cutoff distances used in the calculation. Combining Eqs. (6.85) and (6.86), we obtain, for an isotropic medium, an effective dislocation line tension T

$$T = \frac{\mu b^2}{4\pi} \left(\frac{1 + v - 3v \sin^2 \theta}{1 - v} \right) \ln\left(\frac{2r}{r_0}\right) \tag{6.87}$$

where $2r$ is the averaged particle diameter. The dislocation under an applied stress no longer lies on a circular arc, but it takes up a position such that its radius of curvature at any position is

$$r = \frac{T}{\Delta \tau b} \tag{6.88}$$

Since for a pure screw $\theta = 0$, whereas for a pure edge $\theta = 90°$, the line tension is different for edges and screws. The line tension is lower for an edge dislocation. As a result, edge dislocations will be more flexible, bowing out with ease under an applied stress and encountering more obstacles than screws.

Note that Orowan stress is independent of the nature of dislocation due to smaller curvature of a screw dislocation (than that of an edge) associated with a larger average interparticle spacing for the screw dislocation; this causes a compensating effect of the difference in interparticle spacing for the difference in line tension.[118] In fact, the appropriate line tension is the geometric mean for edges and screws. Based on all these factors, the final expression for the Orowan stress for a random array of noncuttable, hard spherical particles can be given as[138–140]

$$\Delta \tau = \frac{0.84 \mu b}{2\pi(1 - v)^{1/2} l_s} \ln\left(\frac{2r}{r_0}\right) \tag{6.89}$$

where $2r$ is the average particle diameter intersecting the slip plane, the cutoff radius $r_0 \cong 4b$, and l_s is the effective interparticle spacing [i.e., Eq. (6.69)] due to the large size of the noncuttable particle.

For nondeformable precipitates (e.g., incoherent oxides such as silicon oxide and beryllium oxide particles in dispersion-strengthened copper and carbides and coherent γ' in both the overaged Co-based alloys and Ni-based superalloy Nimonic PE16), the strength is independent of the properties of the precipitates but is strongly dependent on their dispersion and size.[133,141] This can also be concluded from Eq. (6.89). For a constant f, coarsening of particle size is linked with the increase in effective interparticle spacing l_s, which leads to decreased yield stress and overaging (Fig. 6.35). This has also been experimentally observed in Co-based alloys containing γ' precipitates, where the lowering of strength at larger particle radii is consistent with the Orowan strengthening.[130]

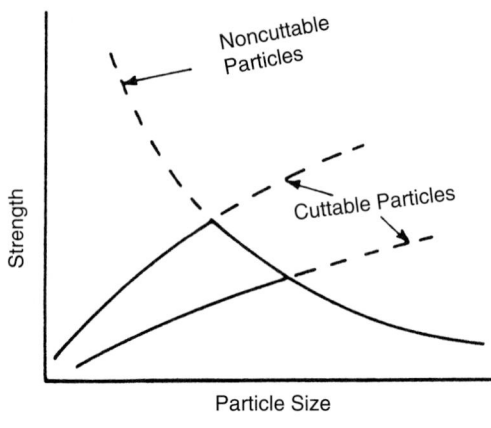

FIGURE 6.35 Variation of strength with particle size defining cuttable and noncuttable particle regimes.[72] (*Reprinted by permission of Butterworths, London.*)

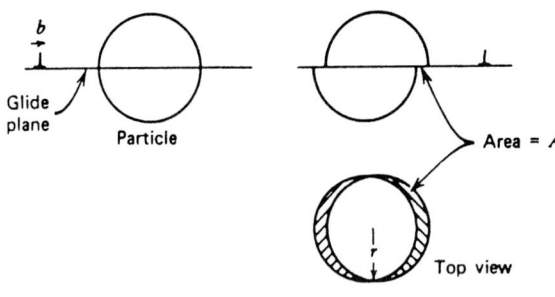

FIGURE 6.36 Shearing of a coherent and soft particle by dislocation across the glide plane.[16] (*Reprinted by permission of John Wiley & Sons, New York.*)

6.8.6 Chemical (or Surface) Strengthening

Chemical strengthening arises when a dislocation actually passes through the coherent precipitate particle, causing the particle to be sheared by one Burgers vector across the glide plane by breaking favorable bonds within the particle, thus creating a new lunar-shaped precipitate particle/matrix interface area (Fig. 6.36).[16]

Using the same schematic diagram (Fig. 6.32) as in particle looping, the CRSS, $\Delta\tau_0$, for the cutting mechanism is

$$\Delta\tau_0 = \frac{F_m}{bd} \qquad \text{for } F_m < 2T° \tag{6.90}$$

It is thus noted that the stress required to shear the obstacles is much less than that necessary to force dislocation loops between precipitate particles.[142] Since this process creates two ledges of new precipitate/matrix interface of specific energy γ_s, interaction force F_m is given by

$$F_m = 2\gamma_s b \tag{6.91}$$

Substituting into Eq. (6.62), we get

$$\sin\theta = \frac{\gamma_s b}{T} \tag{6.92}$$

Combining Eq. (6.92) and Eq. (6.75), we obtain the theoretically predicted CRSS due to chemical hardening $\Delta\tau_{0,c}$:

$$\Delta\tau_{0,c} = \frac{2\gamma_s^{3/2}b^{1/2}}{T^{1/2}l_s} \tag{6.93}$$

When the value of l_s is inserted in Eq. (6.93), the final expression becomes[128,131,132,143]

$$\Delta\tau_{0,c} = \left(\frac{6\gamma_s^3 bf}{\pi T}\right)^{1/2} <r>^{-1} \tag{6.94}$$

This theory predicts that for a given volume fraction f, the CRSS occurs at minimum particle size. In most cases, in which precipitate nucleation is relatively easy, strength increases with the increase in particle size in the initial stages of aging. This is opposite to Eq. (6.94).[130] It thus appears that the chemical strengthening mechanism will be of interest in the early stages of the precipitation processes. This is not a principal contributor to the strength in the aged alloys, except for the steadily increasing volume fraction of precipitates with aging and for very small-sized precipitates. This mechanism seems to be important in Al-Cu and Al-Cu-Mg alloys where GP1 zones, θ'', and S' are coherent platelike precipitates,[65] and in Al-Ag alloys containing Ag-rich spherical particles. Brown and Ham[128] interpreted strengthening of copper by Be-rich zones as being due to chemical strengthening. Figure 6.35 shows the variation of strength with particle size for cuttable and noncuttable (nondeformable) particles.[72]

6.8.7 Order Strengthening

Strengthening by ordered coherent precipitate particles takes place when a matrix dislocation shears (cuts) through an ordered precipitate and creates disorder, for example, *antiphase boundary* (APB) on the slip plane within the precipitate particle (Fig. 6.37).[72] In such materials, dislocations move in pairs to restore the perfect long-range order on the glide plane of the precipitate where the trailing dislocation is attracted to the leading dislocation and removes the antiphase boundaries created by the first dislocation (Fig. 6.38).[57,128,144,145] In contrast to chemical strengthening where the planar fault is limited to the periphery of the particle, the antiphase boundary extends over the entire glide-plane intersection.[133] The APB energy per unit area of the slip plane γ_{apb}, which is the force per unit length, opposes the motion

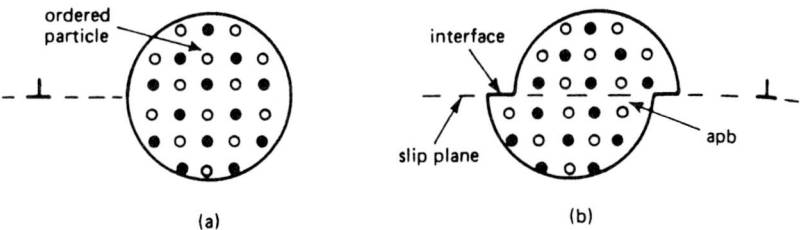

FIGURE 6.37 Ordered precipitate particle (*a*) sheared by a dislocation in (*b*) to produce an antiphase boundary (APB) on the slip plane.[72] (*Reprinted by permission of Butterworths, London.*)

(a)
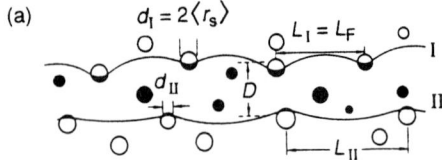

(b)
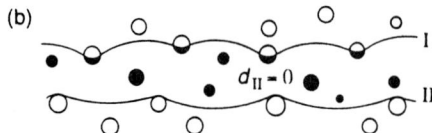

FIGURE 6.38 Schematic illustrations of the shearing of ordered precipitates by a pair of dislocations where d_{II} is finite in (a) and is equal to zero in (b). It is assumed here at the critical cutting configuration that the spacing L_I of the precipitates along the leading dislocation I is equal to L_F. The trailing dislocation II interacts with a different set of precipitates, which have been previously sheared by dislocation I. The obstacle spacings L_{II} and L_I are different because sheared, ordered precipitates are attractive obstacles to dislocation II. The distance D between the two dislocations changes along the length of the pair.[132] (*Courtesy of A. J. Ardell.*)

of dislocation as it enters the particle. Gleiter and Hornbogen[146] developed this theory based on the models of Castagné,[147] Ham,[148] Raynor and Silcock,[144] and Brown and Ham.[128] The maximum force of interaction for a single dislocation shearing an ordered precipitate is given by

$$F_m = 2\gamma_{apb}<r_s> = \frac{\pi\gamma_{apb}<r>}{2} \tag{6.95}$$

where $<r_s>$ is the average planar radius and $<r>$ is the avarage radius of a spherical precipitate particle.

The CRSS, $\Delta\tau_{0,0}$, which arises from the interaction of a single dislocation with a random row of ordered precipitate particles by creating antiphase boundaries within them, can be given by substituting Eq. (6.95) into Eqs. (6.62) and (6.75) and using Eq. (6.77).[131,132]

$$\Delta\tau_{0,O} = \frac{\gamma_{apb}}{b}\left(\frac{3\pi^2\gamma_{apb}f<r>}{32T}\right)^{1/2} \tag{6.96a}$$

For a pair of dislocations, the CRSS is one-half of the value given in Eq. (6.96a) which becomes

$$\Delta\tau_{0,O} = \frac{\gamma_{apb}}{2b}\left(\frac{3\pi^2\gamma_{apb}f<r>}{32T}\right)^{1/2} \tag{6.96b}$$

Figure 6.38 depicts the schematic illustration of the shearing of ordered, coherent precipitates by a pair of leading and trailing dislocations.[132] It is normally assumed that the trailing dislocation is straight. If the trailing dislocations cannot be entirely pulled through the sheared precipitates due to repulsive force exerted by the leading dislocations, Eq. (6.96b) becomes

$$\Delta\tau_{0,O} = \frac{\gamma_{apb}}{2b}\left[\left(\frac{3\pi^2\gamma_{apb}f<r>}{32T}\right)^{1/2} - f\right] \tag{6.97}$$

This equation is valid when the cutting mechanism operates. For alloys aged to peak strength, one obtains

$$\Delta \tau_{0,O} = \frac{\gamma_{\text{apb}}}{2b} \left(\frac{3\pi f}{8} \right)^{1/2} \tag{6.98}$$

where $\Delta \tau_{0,O}$ is constant for fixed f and Eq. (6.98) does not make complete allowance for particles of finite size.

In complex disordered particles, more than two dislocations may be needed to eliminate the disorder; consequently, several dislocations can be able to move as a coupled unit.[149]

Order strengthening is the relevant mechanism in Al-Li; Ni-base alloys such as Ni-Al, Nimonic PE16, Nimonic 105, Nimonic 80A, etc.; Co-Ni-Cr superalloys; and Fe-base alloys such as stainless steels and A-286 where strengthening occurs by γ' precipitates. In all these systems, the matrices are fcc, and the coherent precipitates have the $L1_2$ superlattices. Other alloys strengthened by $L1_2$ ordered precipitates are Fe_2TiSi in Fe-Ti-Si alloys: Al_3Sc in Al-Sc alloys, metastable Al_3Zr in Al-Zr alloys, and Pb_3Na in Pb-Na alloys.[132] Some other systems strengthened by long-range ordered precipitates with different crystallographic structures are NiAlTi, B2 in Fe-Ni-Al-Ti; Ni_4Mo, $D1_a$ in Ni-Mo; Cu_4Ti, $D1_a$ in Cu-Ti; Ni_3Nb, DO_{22} in Co-Ni-Cr-Nb-Fe; NiAl, B2 in Fe-Cr-Ni-Al; various intermetallic phases in maraging steels; and magnesium ferrite, spinel in magnesium oxide–iron oxide systems.[133] Unless they are very large, they are spherical. In Al-Li alloys and in superalloys, the lattice mismatch is very small, below $<\sim 0.002$. In these alloys the dislocations move in pairs due to the passage of a pair of matrix dislocations ($b = (a/2)<110>$) through a γ' particle which restores perfect order on the {111} slip plane. For more detailed discussion of this mechanism, readers are referred to an interesting review by Ardell.[132]

6.8.8 Modulus Strengthening

Modulus hardening arises from the different elastic modulus of the matrix and the precipitate phase.[150,151] Russell and Brown[150] have considered the fact that a dislocation is "refracted" when it crosses the boundary between two media of different elastic constants.[131] By assuming that (1) E_1^∞ is the energy per unit length of dislocation within the particle (of infinite volume) and E_2^∞ is the energy per unit length of dislocation in the matrix and (2) $E_2^\infty > E_1^\infty$, the force F can be written as

$$\frac{F}{2T} = \left(1 - \frac{E_1^2}{E_2^2} \right)^{1/2} \tag{6.99}$$

with

$$\frac{E_1}{E_2} = \left(1 - \frac{E_2^\infty - E_1^\infty}{E_1^\infty} \right) \frac{\ln(r/r_0)}{\ln(R/r_0)} \tag{6.100}$$

where E_1 and E_2 are the dislocation line energies in the particle and matrix, respectively,[130] and R and r_0 are the outer and inner cutoff radii, respectively, for the elastic strain field of the dislocation. As the particle radius r increases, its strength increases logarithmically. This equation suggests a maximum strength at very small particle sizes.

There is a clear evidence that modulus hardening may be an important mechanism of overaging of precipitates growing at a constant volume fraction in several alloy systems. It is unfortunate that the modulus of metastable precipitates are mostly unknown but can be measured indirectly, thus making comparison between theory and experiment, in underaged and peak-aged alloys, more difficult.[130]

6.8.9 Stacking-Fault Strengthening

When the stacking-fault energies γ_{sfp} and γ_{sfm} of the precipitate and matrix phases, respectively, are different, the separation of the partial dislocations will also be different. In addition, when they are both fcc or both hcp structures and when the dislocations move from the matrix into the particle, the glide of the dislocation is hindered and consequently stacking-fault strengthening occurs. A large difference in SFE (denoted by $\Delta\gamma = |\gamma_{sfm} - \gamma_{sfp}|$, where $\gamma_{sfp} \ll \gamma_{sfm}$) produces the large hindrance of dislocation motions because the degree of separation of partial dislocations depends on the phase in which the dislocations reside.

Hirsch and Kelly,[152] and later Gerold and Hartmann,[153] considered a variety of cases with $\gamma_{sfm} > \gamma_{sfp}$ involving different combinations of relative values of $<r_s>$ and partial dislocation separation (or ribbon width), W_m and W_p, in matrix and precipitate, respectively (Fig. 6.39); and they arrived at the following equation for maximum force exerted by the split dislocation:[131]

$$F_m = \Delta\gamma l \tag{6.101}$$

where $\Delta\gamma$ has the usual meaning and l is the length of the chord within the particle at the critical breaking condition, as shown in Fig. 6.39c.

In the underaged alloy when $2<r_s> <W_m$ and $l = 2<r_s>$, the increment in the CRSS can be predicted as

$$\Delta\tau_{0,\gamma} = \Delta\gamma^{3/2}\left(\frac{3\pi^2 f <r>}{32Tb^2}\right)^{1/2} \tag{6.102}$$

This equation agrees well with the strengthening by γ' in dilute Al-Ag alloys[151] and by GP zones in Al-Zn-Mg alloys.[154]

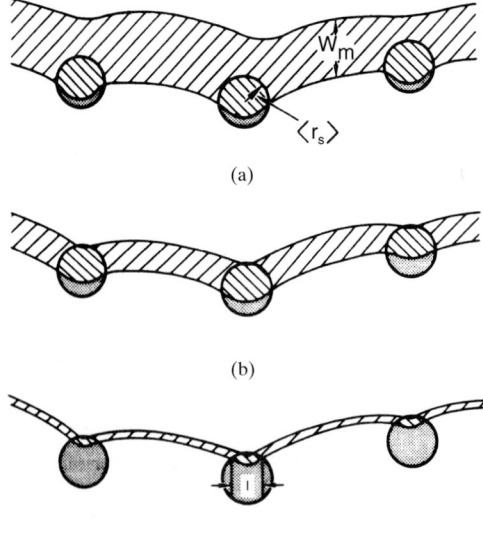

(a)

(b)

(c)

FIGURE 6.39 Representation of some of the possible configurations encountered in dislocation-precipitate interactions during stacking-fault strengthening when $\gamma_{sfm} > \gamma_{sfp}$ $(W_m < W_p)$. (a) $W_m > 2 <r_s>$, $l = 2 <r_s>$; (b) $W_m \cong 2 <r_s>$, $l = 2 <r_s>$; (c) $W_m < 2 <r_s>$, $l < 2<r_s>$.[131] *(Reprinted by permission of The Metallurgical Society, Warrendale Pennsylvania.)*

In the Al-Ag alloy system, stacking-fault strengthening is more dominant. The Ag content of the coherent precipitates was observed to be higher for the lower aging temperature. It is observed that SFE decreases significantly with increasing Ag content of the precipitate phase; this increases $\Delta\gamma$, thereby increasing the strength of the precipitate phase.[129] There are initially no coherency strain fields around these precipitates, and no strong ordering is found. For large particles, Hirsch and Kelly[152] found a decrease in stacking-fault strengthening, with an increase in particle size, which explains the overaging behavior.[130]

6.8.10 Hardening by Spinodal Decomposition

This mechanism occurs due to the periodic variation in composition that produces periodic variation in elastic strain, which can result in a significant increase in strength of the alloys. One theory of hardening by spinodal decomposition has been given by Cahn,[155] who predicted an increment in the CRSS by assuming the decomposition process to be perfectly periodic in nature:

$$\Delta\tau_{0,s} = \frac{(A\eta Y)^2 b}{3\sqrt{6}\beta T} \qquad \text{for screws} \qquad (6.103a)$$

$$= \frac{(A\eta Y)^2 b}{\sqrt{2}\beta T} \qquad \text{for edges} \qquad (6.103b)$$

where β is the wavenumber of the composition modulation, A is the amplitude of that modulation (in atom fraction), η is defined as $(1/a)(da/dc) = d(\ln a)/dc$, where da/dc is the variation of the lattice constant a with composition c (in atom fraction), Y is a function of the elastic constants of the alloy, and T is the dislocation line tension.

For elastically isotropic materials, we have $Y = E/(1 - v)$. However, for cubic crystals of normal anisotropy, we have

$$Y = \frac{(C_{11} + 2C_{12})(C_{11} - C_{12})}{C_{11}} \qquad (6.104)$$

where C_{11} and C_{12} are the elastic constants.

Another theory of spinodal decomposition was put forward by Kato et al.,[156] who found it much more difficult to move a mixed dislocation than either a pure edge or pure screw dislocation in a periodic stress field, and found this in good agreement with theory.[2] Their expression for the flow stress is

$$\Delta\tau_{0,s} = \frac{A\eta Y}{\sqrt{6}} \qquad (6.105)$$

The predictions involved in Eqs. (6.103) and (6.105) are quite different. In Cahn's theory, the increment in CRSS is linearly dependent on $\lambda = 2\pi/\beta$; however, in the theory of Kato et al., CRSS is independent of λ (i.e., independent of b and T).

In fcc spinodal alloys, it appears that the periodic distribution of coherent internal stress due to composition variation results in an increase in the yield stress;[155,156] however, in bcc spinodal alloys, the periodic variation of elastic moduli arising from very large amplitude of the composition variation plays an important role in the hardening.[157] Kato[158] has found that both the misfit strain effect due to coherent internal stress and the modulus effect due to periodic fluctuation of elastic modulus are the major contributors to the increased yield strength.

6.8.11 Conclusions

In spite of many attempts, there is no precise agreement between the theoretical and experimental results on precipitation-strengthening mechanisms. This situation arises partly due to the shortcoming of the theories and partly due to their experimental inaccuracies in the measurement of small-sized particles and their distribution and in accounting for various shaped particles deviating from the spherical shape. It is also very difficult to calculate the precipitate properties of metastable phases such as misfit, surface energy, modulus, and so forth, in order to measure the contributions of individual strengthening mechanisms. In the case of noncuttable particles, particularly in the overaged and dispersion-hardened conditions, experimental and theoretical Orowan stresses are in good agreement. In other cases any specific mechanism may operate entirely. Moreover, in many commercial alloys, a superposition of strengthening mechanisms prevails because of the formation of different types of precipitates, for example, a combination of coherent hardening and modulus hardening in the Cu-based alloys and a combination of coherency strengthening and order strengthening in Ni-base superalloys.[131] In some cases, an additional strengthening mechanism, for example, grain boundary strengthening, that is not associated with the particles may also operate. However, the strengthening mechanisms so far discussed provide only a semiquantitative picture in our understanding.

REFERENCES

1. K. C. Russell, in *Encyclopedia of Materials Science and Engineering*, Pergamon Press, Oxford, 1986, pp. 3263–3267; *The Encyclopedia of Advanced Materials*, Pergamon Press, Oxford, 1994, pp. 1807–1810.

2. J. W. Gibbs, *Scientific Papers*, vol. 1, Dover, New York, 1961.

3. J. W. Christian, *The Theory of Transformations in Metals and Alloys*, Part I, Pergamon Press, Oxford, 1975.

4. W. A. Soffa and D. E. Laughlin, in *Solid → Solid Phase Transformations*, eds. H. I. Aaronson et al., TMS-AIME, Warrendale, Pa., 1982, pp. 159–183.

5. L. Farkas, *Z. Phys. Chem.*, vol. 125, 1927, p. 236.

6. M. Volmer and A. Weber, *Phys. Chem*, vol. 119, 1926, p. 227.

7. R. Becker and W. Doring, *Ann. Phys.*, vol. 24, 1935, p. 719.

8. K. C. Russell, *Adv. Colloid Interface Sci.*, vol. 13, 1980, p. 205.

9. J. E. Burke and D. Turnbull, *Progress in Metal Physics*, vol. 3, Pergamon Press, Oxford, 1952, p. 220.

10. R. D. Doherty, in *Physical Metallurgy*, 4th ed., eds. R. W. Cahn and P. Haasen, Elsevier Science BV, Amsterdam, 1996, chap. 15, pp. 1363–1505.

11. Y. Wang, L. Q. Chen, and A. G. Khachaturyan, *Scripta Met.*, vol. 25, 1991, pp. 1969–1974.

12. F. Laszlo, *JISI*, vol. 164, 1950, p. 5.

13. K. Robinson, *J. Appl. Phys.*, vol. 22, 1951, p. 1045.

14. J. D. Eshelby, *Proc. Roy. Soc. London*, vol. A241, 1957, p. 376.

15. D. M. Barnett, J. K. Lee, H. I. Aaronson, and K. C. Russell, *Scripta Met.*, vol. 8, 1974, pp. 1447–1450.

16. J. D. Verhoeven, *Fundamentals of Physical Metallurgy*, Wiley, New York, 1975.

17. W. C. Johnson, C. L. White, P. R. Marth, S. M. Tuominen, K. D. Wade, K. C. Russell, and H. I. Aaronson, *Met. Trans.*, vol. 6A, 1975, p. 911.

18. K. C. Chan, J. K. Lee, G. J. Shiflet, K. C. Russell, and H. I. Aaronson, *Met. Trans.*, vol. 9A, 1978, pp. 1016–1017.

19. J. K. Lee, D. M. Barnett, and H. I. Aaronson, *Met. Trans.*, vol. 8, 1977, p. 963.

20. S. R. Speich, L. J. Cuddy, C. R. Gordon, and A. J. DeArdo, in *Proceedings of the International Conference on Phase Transformations in Ferrous Alloys*, eds. A. R. Marder and J. L. Goldstein, TMS-AIME, Warrendale, Pa., 1984, pp. 361–389.

21. H. I. Aaronson and K. C. Russell, in *Solid → Solid Phase Transformations*, eds. H. I. Aaronson et al., TMS-AIME, Warrendale, Pa., 1982, pp. 371–397.

22. P. J. Clemn and J. C. Fisher, *Acta Metall.*, vol. 3, 1955, p. 70.

23. W. Huang and M. Hillert, *Met. and Mats. Trans.*, vol. 27A, 1996, pp. 480–483.

24. J. W. Cahn, *Acta Metall.*, vol. 4, 1956, p. 449.

25. A. H. Cottrell, *Report on Strength of Solids*, The Physical Society, London, 1948.

26. J. S. Koehler and F. Seitz, *J. Appl. Mech.*, vol. 14, 1947, p. 217.

27. F. C. Larche, in *Dislocations in Solids*, vol. 4, ed. F. R. N. Nabarro, North-Holland Physics Publishing, Amsterdam, 1979, pp. 135–153.

28. G. W. Lorimer, in *Solid → Solid Phase Transformations*, eds. H. I. Aaronson et al., TMS-AIME, Warrendale, Pa., 1982, p. 613.

29. R. M. Allen, in *Solid → Solid Phase Transformations*, eds. H. I. Aaronson et al., TMS-AIME, Warrendale, Pa., 1982, p. 655.

30. H. I. Aaronson, in *Lectures on the Theory of Phase Transformations*, TMS-AIME, Warrendale, Pa., 1975.

31. D. M. Vanderwalter and J. B. Vander Sande, in *Solid → Solid Phase Transformations*, eds. H. I. Aaronson et al., TMS-AIME, Warrendale, Pa., 1982, pp. 371–397.

32. J. W. Cahn, *Acta Met.*, vol. 5, 1957, p. 169.

33. J. N. Hobstetter, in *Decomposition of Austenite by Diffusional Processes*, Interscience, New York, 1962, p. 138.

34. R. G. Ramirej and G. H. Pound, *Met. Trans.*, vol. 4, 1973, p. 1563.

35. J. Friedel, *Dislocations*, Pergamon Press, Oxford, 1964.

36. G. C. Dollins, *Acta Metall.*, vol. 18, 1970, p. 1209.

37. H. F. Merrick, in *International Symposium on Superalloys*, ASM, Metals Park, Ohio, 1980, pp. 161–190.

38. H. Brooks, *Metal Interfaces*, ASM, Metals Park, Ohio, 1982, p. 20.

39. R. G. Baker, D. G. Brandon, and T. Nutting, *Philos. Mag.*, vol. 4, 1959, p. 1339.

40. G. C. Weatherly, *Philos. Mag.*, vol. 17, 1968, pp. 791–799.

41. J. W. Cahn, *Acta Met.*, vol. 9, 1961, pp. 795–801.

42. J. W. Cahn, *Acta Met.*, vol. 10, 1962, pp. 179–183.

43. J. W. Cahn, *Acta Met.*, vol. 10, 1962, pp. 907–913.

44. M. Hillert, *Acta Met.*, vol. 9, 1961, pp. 525–535.

45. C. M. F. Jantzen and H. Herman, in *Phase Diagrams: Materials Science and Technology*, vol. 5, Academic Press, New York, 1978, pp. 128–184.

46. J. E. Epperson, in *Encyclopedia of Materials Science and Engineering*, ed. R. W. Cahn, suppl. vol. 1, Pergamon Press, Oxford, 1988, pp. 311–318.

47. K. B. Rundman, in *Metals Handbook*, vol. 8, ed. T. Lyman, ASM, Metals Park, Ohio, 1975, pp. 184–185.

48. J. W. Cahn, *TMS-AIME*, vol. 242, 1968, pp. 166–180.

49. J. W. Cahn and J. E. Hilliard, *J. Chem. Phys.*, vol. 31, 1959, p. 688.

50. J. W. Cahn and J. E. Hilliard, *J. Chem. Phys.*, vol. 28, 1958, p. 258.

51. F. K. LeGoues, Y. W. Lee, and H. I. Aaronson, *Acta Met.*, vol. 32, 1984, pp. 1837–1843.

52. J. E. Hilliard, in *Phase Transformations*, ASM, Metals Park, Ohio, 1970, pp. 497–560.

53. T. N. Bartel and K. B. Rundman, *Met. Trans.*, vol. 6, 1975, pp. 1887–1893.

54. E. L. Huston, J. W. Cahn, and J. E. Hilliard, *Acta Met.*, vol. 14, 1966, p. 1053.

55. D. E. Laughlin and W. A. Soffa, in *Metals Handbook*, vol. 9, 9th ed., ASM, Metals Park, Ohio, 1985, pp. 652–654.

56. K. G. Kubarych, M. Okada, and G. Thomas, *Met. Trans.*, vol. 9A, 1978, pp. 1265–1272.

57. K. Binder, in *Materials Science and Technology*, vol. 5: *Phase Transformations in Materials*, vol. ed. P. Haasen, VCH, Weinheim, 1991, pp. 405–471.

58. A. M. Mebed and T. Miyazaki, *Met. and Mats. Trans.*, vol. 29A, 1998, pp. 739–749.

59. R. K. Ham et al., *Acta Met.*, vol. 15, 1967, p. 861.

60. J. L. Murray, *Int. Met. Rev.*, vol. 30, no. 5, 1985, pp. 211–233.

60a. *ASM Handbook*, vol. 3, 10th ed., ASM International, Materials Park, Ohio, 1992.

61. K. C. Russell and H. I. Aaronson, *J. Met. Sci.*, vol. 10, 1975, pp. 1991–1999.

62. J. B. Cohen, in *Solid State Physics*, eds. H. Ehrenreich and D. Turnbull, Academic Press, Orlando, Fla., vol. 39, 1986, pp. 131–196.

63. J. W. Martin, *Precipitation Hardening*, 2d ed., Pergamon Press, Oxford, 1988.

64. R. J. Rioja and D. E. Laughlin, *Acta Met.*, vol. 20, 1980, pp. 1301–1313.

65. S. M. Copley and J. C. Williams, in *Alloy and Microstructural Design*, eds. J. K. Tien and G. S. Ansell, Academic Press, New York, 1976.

66. *Metals Handbook*, vol. 2, 10th ed., ASM International, Materials Park, Ohio, 1991.

67. *Stainless Steels*, ASM International, Materials Park, Ohio, 1994.

68. R. A. Lula, *Stainless Steel*, ASM, Metals Park, Ohio, 1986.

69. *Heat Treater's Guide*, 2d ed., ASM International, Materials Park, Ohio, 1995.

70. T. Gladman, *Materials Sc. and Tech.*, vol. 15, 1999, pp. 30–36.

71. J. M. Silcock, T. J. Heal, and H. K. Hardy, *J. Inst. Met.*, vol. 82, 1953–1954, p. 239.

72. R. E. Smallman, *Modern Physical Metallurgy*, 4th ed., Butterworths, London, 1985.

73. J. W. Martin and R. D. Doherty, *Stability of Microstructure in Metallic Systems*, Cambridge University Press, Cambridge, 1976.

74. W. F. Smith, *Principles of Materials Science and Engineering*, McGraw-Hill, New York, 1986.

75. A. Guinier, *Nature, London*, vol. 142, 1938, p. 142.

76. G. D. Preston, *Proc. Roy. Soc., London*, vol. A167, 1938, p. 526.

77. R. J. Rioja and D. E. Laughlin, *Met. Trans.*, vol. 8, 1977, pp. 1257–1261.

78. E. Hornbogen, *Aluminum*, vol. 43, 1967, p. 9.

79. V. Gerold, *Z. Metallk.*, vol. 45, 1954, pp. 593–599.

80. X. Auvray, P. Georgeopoulos, and J. B. Cohen, *Acta Metall.*, vol. 29, 1981, p. 1061.

81. H. Yoshida, D. J. H. Cokayne, and M. J. Whelan, *Philos. Mag.*, vol. 34, 1976, p. 89.

82. T. Matsukara and J. B. Cohen, *Acta Metall.*, vol. 33, 1985, pp. 1945–1955.

83. G. W. Lorimer and R. B. Nicholson, *The Mechanism of Phase Transformations in Crystalline Solids*, Institute of Metals, London, 1968, p. 36.

84. G. C. Weatherly and R. B. Nicholson, *Philos. Mag.*, vol. 17, no. 148, 1968, pp. 801–831.

85. I. J. Polmear, *Trans. TMS-AIME*, vol. 230, 1964, pp. 1331–1339.

86. B. C. Muddle and I. J. Polmear, *Acta Metall.*, vol. 37, 1989, pp. 777–789.

87. I. S. Suh and J. K. Park, in *Solid → Solid Phase Transformations*, eds. W. C. Johnson et al., The Metallurgical Society, Warrendale, Pa., 1994, pp. 159–164.

88. W. E. Bensen and J. M. Howe, in *Solid → Solid Phase Transformations*, eds. W. C. Johnson et al., The Metallurgical Society, Warrendale, Pa., 1994, pp. 1109–1114.

89. P. J. Gregson, in *High Performance Materials in Aerospace*, ed. H. M. Flower, Chapman & Hall, London, 1995, pp. 49–84.

90. Y. Murakami, *Materials Science and Technology*, vol. 8: *Structure and Properties of Alloys*, vol. ed. K. H. Matucha, VCH, Weinheim, 1996, pp. 213–276.

91. A. M. Zahra, C. Y. Zahra, C. Alfonso, and A. Charai, *Scripta Mater.*, vol. 39, no. 11, 1998, pp. 1555–1558.

92. P. Ratchev, B. Verlinden, P. De Smet, and P. Van Houtte, *Acta Mater.*, vol. 46, no. 10, 1998, pp. 3523–3533.

93. H. D. Chopra, B. C. Muddle, and I. J. Polmear, *Philos. Mag. Letters*, vol. 73, no. 6, 1996, pp. 351–357.

94. X. Gao, J. F. Nie, and B. C. Muddle, *Materials Sc. Forum*, vol. 217–222, pt. 2, 1996, pp. 1251–1256.

95. G. A. Edwards, K. Stiller, G. L. Dunlop, and M. J. Couper, *Acta Mater.*, vol. 46, no. 11, 1998, pp. 3893–3904.

96. R. A. Jeniski, Jr., B. Thanboonsombut, and T. H. Sanders, Jr., *Met. and Mater. Trans.*, vol. 27A, 1996, pp. 19–27.

96a. A. Perovic, D. D. Perovic, G. C. Weatherly, and D. J. Lloyd, *Scripta Mater.*, vol. 41, no. 7, 1999, pp. 703–708.

96b. W. F. Miao and D. E. Laughlin, *Met. and Mats. Trans.*, vol. 31A, 2000, pp. 361–371.

96c. A. Deschamps, X. Bréchet, and F. Livet, *Mat. Sc. & Technol.*, vol. 15, 1999, pp. 993–1000.

96d. S. K. Maloney, K. Hono, I. J. Polmear, and S. P. Ringer, *Scripta Mater.*, vol. 41, no. 10, 1999, pp. 1031–1038.

96e. H. Tsubakino, A. Yamamoto, and T. Ohnishi, *Thermec '97*, The Metallurgical Society, Warrendale, Pa., 1997, pp. 1059–1064.

96f. J. H. Auld and S. M. Cousland, *J. Aust. Inst. Met.*, vol. 19, 1974, p. 194.

97. Z. W. Huang, M. H. Loretto, R. E. Smallman, and J. W. White, *Acta Met. et Mater.*, vol. 42, no. 2, 1994, pp. 549–559.

98. J. Lendvai, *Mater. Sci. Forum*, vol. 217–222, pt. 1, 1996, pp. 43–46.

99. R. Grimes, in *The Encyclopedia of Advanced Materials*, eds. D. Bloor et al., Pergamon Press, Oxford, 1994.

100. R. S. James, *ASM Handbook*, vol. 2, 10th ed., ASM International, Materials Park, Ohio, 1990, pp. 178–199.

101. S. Hirosawa, T. Sato, A. Kamio, K. Kobayashi, and T. Sakamoto, *Materials Sc. Forum*, vol. 217–222, pt. 2, 1996, pp. 839–844; S. Hirosawa, T. Sato, and A. Kamio, *Materials Sc. and Engineering A*, no. 1–2, February 1998, pp. 195–201.

102. S. P. Ringer, K. Hono, I. J. Polmear, and T. Sakurai, in *Solid → Solid Phase Transformations*, eds. W. C. Johnson et al., TMS, Warrendale, Pa., 1994, pp. 165–170.

103. E. W. Gayle and J. B. Vander Sande, *Scripta Met.*, vol. 18, 1984, p. 473.

104. D. E. Passoja and G. S. Ansell, *Acta Metall.*, vol. 19, 1971, pp. 1253–1261.

105. L. F. Mondolfo, *Aluminum Alloys: Structure and Properties*, Butterworths, London, 1976.

106. K. B. Alexander, F. K. LeGoues, H. I. Aaronson, and D. E. Laughlin, *Acta Metall.*, vol. 32, 1984, p. 2241.

107. M. E. Fine, *Met. Trans.*, vol. 6, 1975, pp. 625–630.

108. R. B. Nicholson, G. Thomas, and A. Nutting, *J. Inst. Met.*, vol. 87, 1958–1959, p. 431.

108a. M. Miki, Y. Ogino, and S. Ishikawa, *Thermec '97: Int. Conf. on Thermomechanical Processing of Steels and Other Materials*, eds. T. Chandra and T. Saki, TMS, Warrendale, Pa., 1997.

109. J. H. Harkness, W. D. Spiegelberg, and W. R. Cribb, *ASM Handbook*, vol. 2, ASM International, Materials Park, Ohio, 1990, pp. 403–427.

109a. S. A. Lockyer and F. W. Noble, *Mat. Sc. and Technol.*, vol. 15, October 1999, pp. 1147–1153.

110. J. M. Oblak and B. H. Kear, *Trans. ASM*, vol. 61, 1961, pp. 519–527.

111. M. McLean, in *High Performance Materials in Aerospace*, ed. H. M. Flower, Chapman & Hall, London, 1995, pp. 135–154.

112. *Metals and Alloys in the Unified Numbering System*, 8th ed., SAE-ASTM, Warrendale, Pa., 1999.

113. E. F. Bradley, ed., *Superalloys*, ASM International, Materials Park, Ohio, 1988.

114. R. A. Ricks, A. J. Porter, and R. C. Ecob, *Acta Metall.*, vol. 31, 1983, pp. 43–53.

115. W. L. Mankins and S. Lamb, *ASM Handbook*, vol. 2, 10th ed., ASM International, Materials Park, Ohio, 1990, pp. 428–445.

116. I. Kirman and D. H. Warrington, *J. Inst. Met.*, vol. 94, 1971, pp. 197–199.

116a. R. M. Forbes Jones and L. A. Jackman, *JOM*, January 1999, pp. 27–31.

116b. Y. Rong, S. Chen, G. Hu, M. Gao, and R. P. Wei, *Met. and Mats. Trans.*, vol. 30A, 1999, pp. 2297–2303.

116c. *Adv. Mater. & Process.*, January 2001, pp. 51–54.

117. *Alloy Data: Carpenter Specialty Steels*, 1998.

117a. J. R. Crum, E. Hibner, N. C. Farr, and D. R. Muriasinghe, in *Casti Handbook of Stainless Steels and Nickel Alloys*, Casti Publishing Inc., Edmonton, Alberta, Canada, 1999, pp. 259–324.

117b. G. Aggen, *Phase Transformations and Heat Treatment Studies of a Controlled-Transformation Stainless-Steel Alloy*, Ph.D. Thesis, Renssalaer Polytechnic Institute, Troy, N.Y., August 1963, pp. 30–31.

118. A. M. Beltran, in *The Encyclopedia of Advanced Materials*, vol. 1, eds. D. Bloor et al., Pergamon Press, Oxford, 1994, pp. 437–441.

118a. *Haynes International, Inc. Publication*, 1996.

118b. J. F. Grubb, in *Uhlig's Corrosion Handbook*, 2d ed., ed. R. W. Revie, John Wiley, & Sons, Inc., New York, 2000, pp. 667–676.

119. S. Paetke and A. R. Waugh, in *Solid → Solid Phase Transformations*, eds. H. I. Aaronson et al., TMS-AIME, Warrendale, Pa., 1982, p. 769.

120. H. J. Rack and D. Kalish, *Met. Trans.*, vol. 5, 1974, pp. 1595–1605.

121. G. T. Murray, *Met. Trans.*, vol. 12A, 1981, pp. 2138–2141.

122. G. Krauss, *Steels: Heat Treatment and Processing Principles*, ASM International, Materials Park, Ohio, 1990.

123. F. B. Pickering, *Int. Met. Rev.*, 1976, pp. 227–268.

124. J. M. Wright and M. F. Jordan, *Met. Technol.*, vol. 7, 1980, pp. 473–482.

124a. J. F. Grubb, *private communication*, 1999.

125. A. Wilm, *Metallurgia*, vol. 8, 1911, p. 225.

126. P. D. Merica, R. G. Waltenberg, and H. Scott, *Trans. AIME*, vol. 64, 1920, p. 41.

127. A. Kelly and R. B. Nicholson, in *Strengthening Methods in Crystals*, eds. A. Kelly and R. B. Nicholson, Applied Science Publishers, London, 1971, pp. 1–8.

128. L. M. Brown and R. K. Ham, in *Strengthening Methods in Crystals*, eds. A. Kelly and R. B. Nicholson, Applied Science Publishers, London, 1971, p. 9.

129. V. Gerold, in *Dislocations in Solids*, vol. 4, North-Holland Physics Publishing, Amsterdam, 1979, pp. 221–260.

130. D. J. Lloyd, in *Strength of Metals and Alloys*, eds. H. J. McQueen, J. P. Bailon, J. I. Dickson, J. J. Jonas, and M. G. Akben, *Proceedings, 71st International Conference on the Strength of Metals and Alloys*, Montreal, Canada, August 12–16, 1985, pp. 1745–1778.

131. A. J. Ardell, *Met. Trans.*, vol. 16A, 1985, pp. 2131–2165.

132. A. J. Ardell, in *Intermetallic Compounds*, vol. 2, eds. J. H. Westbrook and R. L. Fleischer, John Wiley, & Sons, New York, 1994, p. 257.

133. E. Nembach, *Particle Strengthening of Metals and Alloys*, John Wiley, & Sons, Inc., New York, 1997.

134. F. B. Pickering, *Physical Metallurgy and the Design of Steels*, Applied Science Publishers, London, 1978.

135. N. F. Mott and F. R. N. Nabarro, *Bristol Conference on the Strength of Solids*, Physical Society, London, 1948.

136. V. Gerold and H. Haberkorn, *Phys. Status Solidi*, vol. 16, 1966, p. 675.

137. H. Gleiter, *Z. Angew. Phys.*, vol. 23, 1967, p. 108.

138. P. B. Hirsch and F. J. Humphreys, *Proc. Roy. Soc.*, *London*, vol. A318, 1970, p. 45.

139. D. J. Bacon, O. F. Kocks, and R. O. Scattergood, *Philos. Mag.*, vol. 28, 1973, p. 1241.

140. I. D. Embury, *Met. Trans.*, vol. 16A, 1985, pp. 2191–2200.

141. R. B. Nicholson, in *Strengthening Methods of Crystals*, eds. A. Kelly and R. B. Nicholson, Applied Science Publishers, London, 1971, pp. 535–613.

142. J. W. Martin, *Micromechanisms in Particle-Hardened Alloys*, Cambridge University Press, Cambridge, 1980.

143. A. Kelly and M. E. Fine, *Acta Metall.*, vol. 5, 1957, p. 365.

144. D. Raynor and J. M. Silcock, *Met. Sci. J.*, vol. 4, 1970, pp. 121–130.

145. H. Gleiter and E. Hornbogen, *Acta Metall.*, vol. 13, 1965, pp. 576–578.

146. H. Gleiter and E. Hornbogen, *Phys. Status Solidi*, vol. 12, 1965, p. 235.

147. J. L. Castagné, *J. Phys. (Paris)*, vol. 27, 1966, pp. C3–C33.

148. R. K. Ham, *Trans. Jpn. Inst. Met.*, vol. 9, Suppl., 1968, p. 52.

149. J. Greggi and W. A. Soffa, in *Strength of Metals and Alloys, Proceedings of the 5th International Conference on the Strength of Metals and Alloys*, eds. P. Haasen, V. Gerold, and G. Kestorz, Pergamon Press, Oxford, 1979, p. 651.

150. K. C. Russell and L. M. Brown, *Acta Metall.*, vol. 20, 1972, p. 969.

151. G. Knowles and P. M. Kelly, in *The Effect on 2nd Phase Particles on the Mechanical Properties of Steel*, BSC/ISI Conference, Scarborough, England, 1971.

152. P. B. Hirsch and A. Kelly, *Philos. Mag.*, vol. 12, 1965, p. 881.

153. V. Gerold and K. Hartmann, *Trans. Jpn. Inst. Met.*, vol. 9, Suppl., 1968, p. 509.

154. I. Kovas, J. Lendvai, T. Uangar, T. Urmezey, and G. Groma, *Acta Metall.*, vol. 25, 1977, p. 673.

155. J. W. Cahn, *Acta Metall.*, vol. 11, 1963, p. 1275.

156. M. Kato, T. Mori, and L. H. Schwartz, *Acta Metall.*, vol. 28, 1980, p. 285.

157. D. Chandra and L. H. Schwartz, *Met. Trans.*, vol. 2, 1971, p. 511.

158. M. Kato, *Acta Metall.*, vol. 29, 1981, pp. 79–87.

CHAPTER 7
PEARLITE AND
PROEUTECTOID PHASES

7.1 INTRODUCTION

We shall now consider the diffusion-controlled austenite decomposition that occurs in the eutectoid region of the phase diagram. Pearlite, ferrite, and cementite are the main constituents of the microstructure of slowly cooled and isothermally reacted steels, whereas interphase boundary carbides and fibrous carbides embedded in ferrite often occur in low- and high-alloy steels containing carbide formers in addition to the foregoing constituents.

In this chapter, important features of various diffusional transformations are discussed, such as nucleation and growth; interlamellar spacing; overall kinetics and mechanisms of pearlite transformation; interphase precipitation; fibrous carbides; proeutectoid phases; and structure-property relationships of ferrite, and medium- and high-carbon ferrite-pearlite, and fully pearlitic microstructures. Finally, the applications of some important medium- and high-carbon steel products are briefly described.

7.2 PEARLITE

Pearlite is a two-phase lamellar product of eutectoid decomposition which can form in steels and various nonferrous alloys (such as Ti-Al, Ag-Ga, Cu-Al, and Zn-Al alloys) during transformations under isothermal, continuous-cooling, or forced-velocity (directional) growth conditions below the eutectoid temperature due to cooperative and synchronized growth of two constituent phases from a parent phase.[1-3] A pearlite nodule is composed of multiple colonies; each colony has parallel lamellae which are oriented differently with respect to lamellae in adjacent colonies. Figure 7.1 shows the lamellar structure of pearlite formed isothermally from austenite in a plain carbon eutectoid steel.[4] This also exhibits a wide range of apparent interlamellar spacings in different colonies because of intersection of pearlite colonies at different angles with the plane of polish. Table 7.1 lists some ferrous and nonferrous alloy systems that contain eutectoid transformations and produce numerous pearlite morphologies.

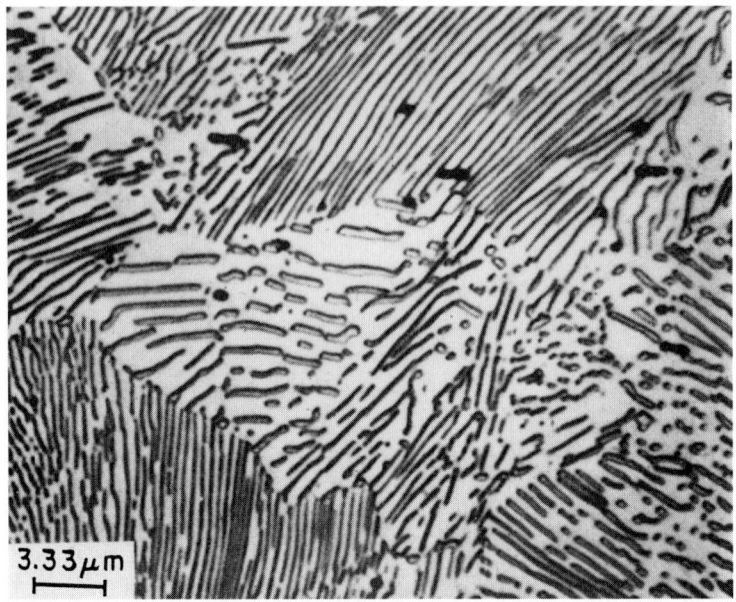

FIGURE 7.1 Lamellar pearlitic structure formed isothermally at 705°C in a plain carbon eutectoid steel.[4] (*Reprinted by permission of ASM International, Materials Park, Ohio.*)

7.2.1 Mechanism of Pearlite Reaction

According to Hillert,[5] the possible sites for nucleation of pearlite may be either ferrite or cementite on an austenite grain boundary, depending on the composition and reaction temperature. In hypereutectoid steel, cementite will usually nucleate first; in hypoeutectoid steel, the ferrite will nucleate first. Nicholson,[6] in analyzing kinetic data for the formation of pearlite, has also emphasized that ferrite can nucleate first in lower-carbon hypoeutectoid steels.

Mehl and Hagel[7] proposed the formation of pearlite nodules by *sidewise nucleation/growth and edgewise growth* into the austenite. Modin[8] and Hillert[5] first showed that the proeutectoid phase can be continuous with the same phase present in the pearlite. Later work by Dippenaar and Honeycombe[9] on 13% Mn-0.8% C (hypereutectoid) steel showed conclusively that nucleation of pearlite occurred on grain boundary cementite and there is a continuity of grain boundary and pearlitic cementite. They have, therefore, confirmed the earlier observation of Hillert[5] that in this alloy sidewise growth occurred as a result of a *branching mechanism* rather than by *repeated nucleation or multiple nucleation events* of cementite at the proeutectoid ferrite/austenite interface. That is, all the cementite lamellae were branches from one single lamellae or nucleus of cementite which had grown from the cementite network, and all the ferrite lamellae also joined together to form a continuous crystal. It is thus thought that a pearlite unit is regarded as a bicrystal comprising two interwoven crystals of ferrite and cementite.[10] During further growth, more branching occurred until the interlamellar spacing characteristic of the transformation temperature was obtained (Fig. 7.2).[5] Considerable evidence has been found from repeated sectioning studies favoring the branching mechanism.[5]

TABLE 7.1 Eutectoid Transformations in Nonferrous and Ferrous Alloys[51,54]

Alloy	Eutectoid composition, wt%	Eutectoid temperature °C	Eutectoid temperature °F	High-temperature phase and crystal structure	Low-temperature phases and crystal structures	Reactions observed
Ag-Ga	15.3 Ga	380	716	β-hcp structure type hP2 → α-fcc (cF4) + β' ordered hcp (hP9 crystal structure)	α-fcc β" metastable (9R or hR3 crystal structure)	Lamellar pearlite
Cu-Al	11.8 Al	565	1049	β-bcc	α-fcc γ2 (gamma brass)	Lamellar pearlite; granular pearlite
Cu-Be	6 Be	605	1121	β-bcc	α-fcc β' (bcc, CsCl)	Lamellar pearlite
Cu-In	31.4 In (20.15 at% In)	574	1065	β-bcc	α-fcc δ (deformed gamma brass)	Lamellar pearlite; granular pearlite
Cu-Si	5.2 Si	555	1031	κ-hcp	α-fcc γ-cubic (β-Mn)	Granular pearlite
Cu-Sn	22–26 Sn	520	968	β-bcc (e/a = 3.2, Cu$_5$Sn)	α-fcc γ (complex cubic with a DO$_3$ type superlattice)	Non-lamellar; needles of α about which γ precipitates
Cu-Sn	27.0 Sn	520	968	γ-bcc	α-fcc ε (e/a = 21/13, gamma brass)	Lamellar pearlite; needles of α about which δ precipitates
Cu-Sn	32.5 Sn	350	662	δ (gamma brass)	α-fcc ε-orthorhombic (Cu$_3$Sn)	Lamellar pearlite
Fe-C	0.8 C	723	1333	γ-fcc (interstitial C)	α-bcc Fe$_3$C (orthorhombic)	Lamellar pearlite
Fe-N	2.35 N	590	1094	γ-fcc (interstitial N)	α-bcc γ'-fcc (interstitial N)	Lamellar pearlite; granular pearlite
Fe-O	23.3 O	560	1040	Wüstite cubic (NaCl)	α-bcc Fe$_3$O$_4$ cubic (spinel)	Lamellar pearlite; granular pearlite
Ni-Zn	56 Zn	675	1247	β-cubic (CsCl)	β$_1$-tetragonal (Cu-Au) γ (gamma brass)	Lamellar pearlite
Ti-Cr	15 Cr	680	1256	β-bcc	α-hcp TiCr$_2$-fcc (MgCu$_2$)	Lamellar pearlite; granular pearlite
Zr alloy				β-phase	α-hcp + intermetallic phase	Lamellar pearlite

Source: After A. R. Marder and J. A. Kowalik, in *Metals Handbook*, vol. 9: *Metallography and Microstructures*, ASM, Metals Park, Ohio, 1985, p. 659.

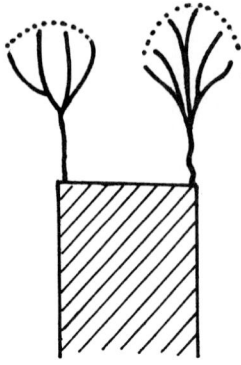 

FIGURE 7.2 Schematic branching mechanism for the growth of fine pearlite at a lower temperature formed by partial transformation from coarse pearlite at a higher temperature showing sideways growth of pearlite.[5] (*Reprinted by permission of The Metallurgical Society, Warrendale, Pennsylvania.*)

It has been observed that the formation of pearlite requires the establishment of cooperative growth of ferrite and cementite.[5] This cooperation gradually develops and increases with time, and finally lamellar pearlite forms. The necessary criterion for this cooperative growth is that both phases should have an incoherent interface with the austenite. In some hypereutectoid steels, a low degree of cooperation between the two phases resulted in nonlamellar growth of ferrite and cementite, thereby producing the so-called *degenerate pearlite structure*. Partially coherent interphase boundaries seriously inhibit cooperation. When the two phases form in a noncooperative mode, it is called a *divorced eutectoid transformation* (DET) in which divorced or spheroidal cementite particles grow directly from the austenite phase along a cellular γ/α reaction front, without encasement in ferrite shells. Use of DET permits the development of fine, equiaxed microstructure with spheroidized carbide particles.[11] DET mode occurs at lower undercoolings compared to the pearlite mode observed at higher undercoolings.[12]

Hackney and Shiflet[13,14] presented direct experimental evidence, in an Fe-C-Mn alloy, similar to that of Dippenaar and Honeycombe, for the existence of mobile growth ledges associated with both phases of pearlite at the advancing pearlite/austenite interface. This, in turn, implies that the advancing pearlite/austenite interface, in Fe-C-Mn alloy, is partially coherent with the austenite and migrates by the lateral movement of steps. Edgewise growth is attributed to ledgewise growth mechanism which takes place despite the lack of a reproducible orientation relationship between the constituents of pearlite and the austenite grain into which the growth is occurring. Later, Hackney[15] proposed a linear stability theory which supports Hillert's branching mechanism by assuming (1) morphological instability of each phase (say of ferrite which produces branching of cementite and vice versa) and (2) ledgewise growth.[10,13] This is in contrast to Hillert's mechanism of unhindered incoherent pearlite interface growth, i.e., uniform attachment. Based on the Hackney-Shiflet results, it may be inferred that the pearlite reaction occurs by ledgewise cooperative (or shared) growth, and the essential conditions to be encountered are $G_\alpha = G_{cm}$ and $h_\alpha/\lambda_\alpha = h_{cm}/\lambda_{cm}$ where G's are growth rates, h's are ledge heights, and λ's are interledge spacings for the respective phases.[16] Currently, the ledge growth mechanism has been experimentally confirmed by many researchers, and it is now widely accepted. This ledge mechanism of pearlite growth was also verified by Whiting and Tsakiropoulos[17] for the Cu-Al lamellar eutectoid growth. Recently, Lee and Park[10] have suggested a new model in which sidewise growth occurs by *time-*

sequential branching (and not random branching) via the lateral movement of growth ledges, which also causes edgewise growth.

7.2.2 Theories of Pearlitic Growth

Figure 7.3*a* shows a portion of the Fe-Fe$_3$C phase diagram. The region outlined by the extrapolated Ae_3 and Ae_{cm} phase boundaries corresponds to the area, or range,

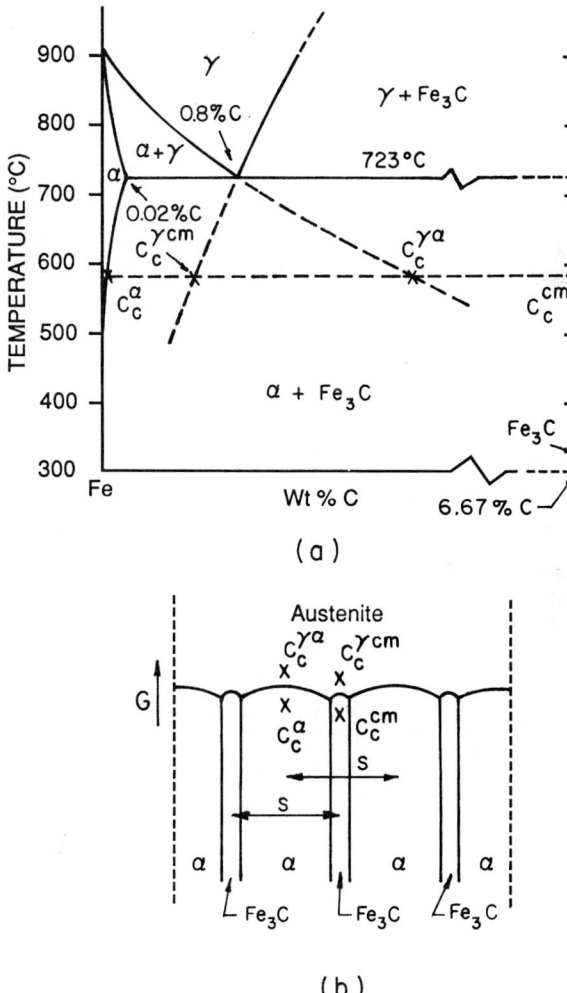

FIGURE 7.3 Schematic illustrations of (*a*) a portion of the Fe-Fe$_3$C phase diagram and (*b*) the austenite/pearlite growth front.[2] (*Reprinted by permission of The Metallurgical Society, Warrendale, Pennsylvania.*)

of compositions and temperatures for which the austenite compositions can be simultaneously saturated with respect to both ferrite and cementite. As the reaction temperature falls from just below Ae_1 toward the nose temperature (of the TTT diagram), the pearlite formed becomes finer in grain size with smaller interlamellar spacing.

Figure 7.3b shows a schematic representation of the austenite/pearlite reaction front. Since ferrite and Fe_3C have low (<0.02%) and high (~6.67%) carbon content, respectively (Fig. 7.3a), during pearlite growth an appreciable redistribution of carbon occurs at or near the reaction front. In the case of alloy steels, additional redistribution of alloying elements between the constituent phases takes place at the growth front.

For theories of pearlite growth, the important parameters are pearlite growth rate G (into the austenite) and its lamellar spacing S, which are assumed to be constant for a given temperature. During pearlite growth, the redistribution of solute (carbon) may take place in front of the reaction front by volume diffusion through austenite and/or by diffusion along the pearlite/austenite interface.[2,4]

The Zener[18]-Hillert[19] equation for pearlite growth G is based on two important assumptions: (1) The supersaturated austenite is in local equilibrium with the constituent phases of pearlite at the reaction front, and (2) volume diffusion of carbon in austenite is rate-controlling. This equation, arising from the flux balance of solute required for the reaction and from an accounting made for the influence of lamellar interfacial energy on the driving force for growth, is given by

$$G = \frac{D_c^\gamma}{k} \frac{S^2}{\lambda^\alpha \lambda^{cm}} \left(\frac{C_c^{\gamma\alpha} - C_c^{\gamma cm}}{C_c^{cm} - C_c^\alpha} \right) \frac{1}{S} \left(1 - \frac{S_c}{S} \right) \tag{7.1}$$

where D_c^γ is the volume diffusivity of carbon in austenite; k is the geometric parameter (~0.72 for Fe-C eutectoid alloy); $C_c^{\gamma\alpha}$ and $C_c^{\gamma cm}$ are the equilibrium carbon concentrations in austenite in front of the ferrite and cementite lamellae, respectively; C_c^α and C_c^{cm} are the respective equilibrium concentrations of carbon within the ferrite phase and cementite phase; λ^α and λ^{cm} are the respective thicknesses of ferrite and cementite lamellae; S_c is the critical interlamellar spacing for which the growth rate becomes zero because the entire driving force is consumed as interfacial energy; and S is the interlamellar spacing.

The pearlite interface (i.e., α/γ and Fe_3C/γ interfaces) must be curved. This curvature will, however, lower the concentration gradient due to the Gibbs-Thomson effect. In the foregoing equation the $[1 - (S_c/S)]$ term accounts for the curvature effect resulting from the Gibbs-Thomson relation, which reduces the maximum value of the concentration difference term $(C_c^{\gamma\alpha} - C_c^{\gamma cm})$ and thereby reduces the effective concentration gradient driving the growth process. However, later, the $[1 - (S_c/S)]$ term is believed to arise from the α/Fe_3C lamellar interphase boundary energy and not from curvature of the α/γ and Fe_3C/γ interfaces.

Since Eq. (7.1) contains two unknown parameters, G and S, for a given undercooling ΔT, it does not provide a unique solution. But it can be solved if we impose another condition through an optimization (or maximization) principle. Zener[18] proposed the maximum growth rate criterion to stabilize the system at a particular spacing, which can be given by

$$S = 2S_c \tag{7.2}$$

where S_c is the theoretical critical spacing at zero growth rate. Also, the equation relating the critical spacing and undercooling ΔT for isothermal and continuous cooling transformation is given by

$$S_c = \frac{2\gamma^{\alpha cm} T_E}{\Delta H_v \Delta T} \qquad (7.3)$$

where $\gamma^{\alpha cm}$ is the interfacial energy of the α/Fe_3C lamellar boundary in pearlite and ΔH_v is the change in enthalpy per unit volume accompanying the transformation of austenite to a mixture of ferrite and cementite. Combining Eqs. (7.2) and (7.3) gives an expression for S based on Zener's hypothesis:

$$S = 2S_c = \frac{4\gamma^{\alpha cm} T_E}{\Delta H_v \Delta T} \qquad (7.4)$$

This equation has been modified by Puls and Kirkaldy[20] using the hypothesis of *maximum rate of entropy production* (which gives somewhat larger critical spacings) and is written as

$$S = 3S_c = \frac{6\gamma^{\alpha cm} T_E}{\Delta H_v \Delta T} \qquad (7.5)$$

where, rearranged, these equations take the following form because ΔH_v and $\gamma^{\alpha cm}$ are relatively independent of temperature:

$$S\Delta T = \text{constant} \qquad (7.6)$$

Equation (7.6) predicts that interlamellar spacing and undercooling are inversely related. Zener[18] predicted that the reciprocal spacing should vary linearly with undercooling; this is in good agreement with experimental observations in commercial-purity plain carbon steels.[1]

Boundary diffusion-controlled pearlite growth has also been considered by several workers;[19,21,22] all of them found similar results. The relationship due to Hillert[19] is

$$G = 12KD_B\delta\left(\frac{S^2}{\lambda^\alpha \lambda^{cm}}\right)\left(\frac{C_c^{\gamma\alpha} - C_c^{\gamma cm}}{C_c^{cm} - C_c^\alpha}\right)\frac{1}{S^2}\left(1 - \frac{S_c}{S}\right) \qquad (7.7)$$

where K is the boundary segregation coefficient, D_B is the boundary diffusion coefficient, δ is the boundary thickness, and the other terms have the usual meanings. This equation assumes a major proportion of solute redistribution by boundary diffusion and a smaller one by volume diffusion. For boundary diffusion-controlled growth, the maximum growth rate and maximum rate of entropy production criteria give $S = (3/2)S_c$ and $S = 2S_c$, respectively. These relationships can be combined with Eq. (7.3) as before.

7.2.2.1 Effect of Alloying Elements.
The addition of small amounts of substitutional alloying elements such as Mn, Ni, Cr, Mo, W, Ti, and Nb significantly retards the pearlite reaction due to the change in A_1 temperature and slow diffusion of alloying elements. It is known that austenite stabilizers such as Mn and Ni depress the A_1 temperature and, therefore, are usually expected to slow down the pearlite reaction (Fig. 7.4)[23] because of the decreased thermodynamic driving force for the transformation at any given temperature below A_1 due to a reduced concentration difference (i.e., $C_x^{\gamma\alpha} - C_x^{\gamma cm}$). This may also result, probably by an average carbon content that is lower than in the pearlite than in the original alloy.[23] They form *unpartitioned pearlite*. The *no-partitioning* of ternary solute in eutectoids can be compared to the diffusionless precipitation of single-phase products, and the reduction in

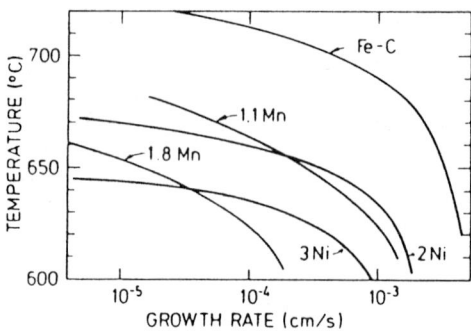

FIGURE 7.4 Comparison of growth rate curves for pearlite in nickel and manganese steels as a function of falling temperature.[23] (*Reprinted by permission of The Metallurgical Society, Warrendale, Pennsylvania; after N. A. Razik et al., Acta Metall., vol. 22, 1974, p. 1249; and D. Brown et al., JISI, vol. 207, 1969, p. 1232.*)

growth velocity of these ternary solute effects increases with the increase of the low-alloy solute additions.[24]

Other series of elements, called *carbide-forming elements* (which are also *ferrite stabilizers* such as Si, Mo, Co, W, Cr, V, and Ti), increase the A_1 temperature; but in their presence, at lower temperatures, the transformation rates are usually lower than that of a plain carbon eutectoid steel. These alloying additions partition between ferrite and cementite over the entire range of pearlite formation. In some instances, e.g., in steels containing ~5% Cr, pearlitic cementite can be substituted by chromium carbide, producing *alloy pearlite*.[25] The drastic reduction in the growth rate of pearlite at lower temperatures and the associated development of an austenite bay in the TTT diagram have been attributed to solute drag effect and partitioning of strong carbide formers.[26,27]

It is now well agreed that during pearlite growth, the partitioning of alloying elements occurs at low supersaturation, favoring alloying-element-boundary diffusion-controlled growth, while less partitioning occurs at high supersaturation. At high temperatures, strong solute partitioning occurs in low-alloy steel pearlites; with decreasing temperature, the partitioning becomes incomplete.[24] The actual temperature below which no partition occurs, if it exists, is a function of the alloy system and composition; and for a given alloying element this *transition (partition-no partition) temperature* decreases with an increase in alloy content. These substitutional alloying elements have low diffusivity, and therefore pearlite growth is controlled by the alloying element diffusion along the austenite-pearlite boundary; the following equation for this model in the Fe-C-Cr system has recently been developed by Sharma and coworkers[28,29] following Hillert[19] and Kirkaldy:[30]

$$G \cong 54 K D_X^B \delta \frac{C_X^{\gamma\alpha} - C_X^{\gamma cm}}{\overline{C}_X} \cdot \frac{1}{S^2} \qquad (7.8)$$

where K is the boundary segregation coefficient for alloy element X (= Cr); D_X^B is the interface diffusivity of the alloying element X; δ is the boundary thickness; $\overline{C}_X$ is the average composition of the alloying elements; $C_X^{\gamma\alpha}$ and $C_x^{\gamma cm}$ are the alloying element concentrations (in austenite in equilibrium with ferrite and cementite, respectively); and S is the interlamellar spacing (which is influenced by alloy additions[7]).

7.2.3 Interlamellar Spacing

Pearlitic interlamellar spacing is an important parameter controlling strain hardening and ductility of carbon steels. Fine pearlite is the desirable microstructure for drawing high-carbon steels with exponential strain-hardening rate, resulting in a high final tensile strength.[31] Reduction in interlamellar spacing in hypereutectoid steels reduces the carbide thickness. Note that the pearlitic constituent having a lower than equilibrium carbon content (or a lower proportion of cementite phase) in pearlite is called *dilute pearlite*. More dilute pearlites can be obtained by decreasing both the eutectoid carbon content by alloying addition and the transformation temperature in a hypoeutectoid steel which leads to the formation of more than the equilibrium amount of pearlite and decrease of the impact transition temperature by reducing the carbide plate thickness (i.e., $t < 0.14S$) at a specific interlamellar spacing. Thus, a dilute pearlite can provide a good combination of strength and toughness.[32]

A pearlite dilution factor D_p can be expressed by

$$D_p = \frac{0.8 f_p}{\text{wt\% C}} \tag{7.9}$$

where f_p is the volume fraction of pearlite and wt% C is the carbon concentration of the steel. For a fully pearlitic eutectoid steel, $D_p = 1$; and for a dilute pearlite having $t < 0.14S$, $D_p > 1$. The relation between t, D_p, and S is given by:[32]

$$t = \frac{0.12S}{D_p - 0.12} \tag{7.10}$$

In a pearlitic steel wire, the relation between f_p, initial interlamellar spacing S, and cementite thickness t can be expressed, according to O'Donnelly's results, by

$$t = \frac{0.15(\text{wt\% C})S}{f_p} \tag{7.11}$$

The increased interlamellar spacing leads to the large size globular cementite particles in colonies.[33]

7.2.4 The Overall Kinetics of Pearlite Formation

The pearlite reaction is a typical nucleation and growth process.[34] That is, the rate of pearlite transformation depends on (1) the nucleation rate of pearlite nodules N and (2) the growth rate of these nodules G. Studies by Mehl et al. have shown that N increases with time and with decreasing temperature down to the knee temperature of the TTT curve for a eutectoid steel, as shown in Fig. 7.5.[7] N is a structure-sensitive parameter and is influenced by austenite grain size, inhomogeneities, and impurities. Like N, G increases rapidly with decreasing reaction temperature or increasing degree of undercooling ΔT below the A_1 temperature, reaching a maximum at the nose of the TTT curve and decreasing again to a lower temperature (Fig. 7.5). Though G is structure-insensitive (independent of grain size), it is a strong function of both temperature and alloying elements. The external morphology of pearlite nodules strongly depends on the ratio N/G, which, in turn, increases with increase in supersaturation, as will now be described.

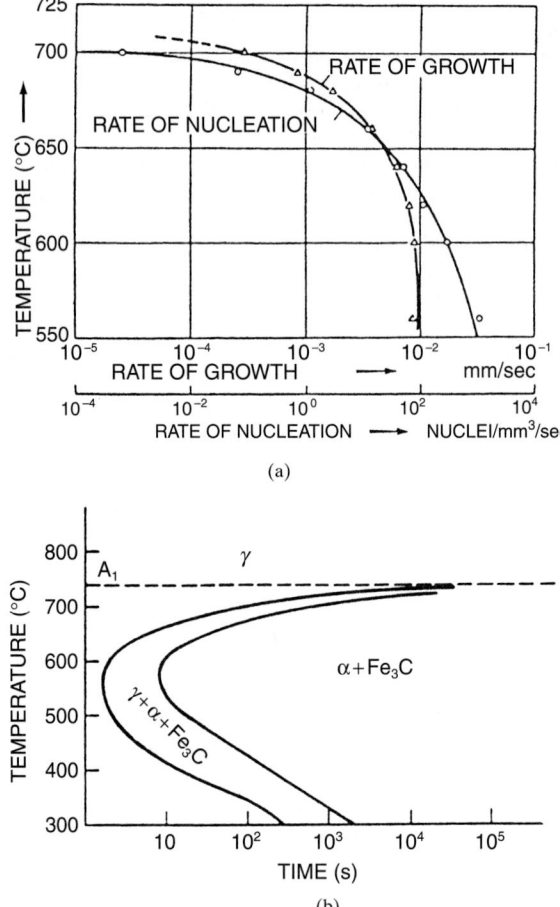

FIGURE 7.5 (*a*) Variation of nucleation rate and growth rate of pearlite with temperature.[7] (*b*) Isothermal transformation curve for plain carbon eutectoid steel. [(*a*) *Reprinted by permission of Pergamon Press, Plc.; (b) reprinted by permission from R. E. Smallman, Modern Physical Metallurgy, 4th edition, Butterworths, London, 1985.*]

Pearlite nodules usually nucleate on austenite grain boundaries and grow at a roughly constant radial velocity into the surrounding austenite grains. At temperatures immediately below A_1, where the ratio N/G is small, the reaction proceeds very slowly, and therefore relatively few pearlite nodules nucleate; these nodules grow as hemispheres or spheres without interfering with one another (i.e., before impinging on other growing nodules). Though the nodules nucleate at the austenite grain boundaries, the distribution of nuclei may be taken effectively as randomly distributed throughout the austenite because the ratio of the nodule diameter to the austenite grain diameter is large. "Accordingly, the following time-dependent reaction equation due to Johnson and Mehl[35] applies:

$$f(t) = 1 - \exp\left(-\frac{\pi \dot{N} G^3 t^4}{3}\right)$$ (7.12)

where $f(t)$ is the volume fraction of the pearlite formed isothermally at a given time t; $\dot{N}$ is the nucleation rate, assumed to be constant; and G is the growth rate, also assumed to be constant. As shown in Fig. 7.6a, this expression yields a typical sigmoidal curve for fixed values of $\dot{N}$ and G.[36] When $f(t)$ is plotted against $\sqrt[4]{\dot{N}G^3 t^4}$, a sigmoidal master curve is obtained which illustrates the fundamental kinetic behavior of a nucleation and growth process in a particular alloy (Fig. 7.6b).[7]

At large undercoolings (i.e., lower reaction temperatures), the nucleation rate becomes much higher (that is, $\dot{N}/G$ is quite large) and site saturation occurs;[37] that is, the boundaries become covered with pearlite nodules prior to the transformation of a significant fraction of austenite. Transformation simply proceeds by the thickening of these pearlite layers into the grains[37,38] (Fig. 7.7). Thus the morphology changes as the cooling rate increases, predominantly from a spherical nodule to a hemispherical one, and Eq. (7.12) no longer holds because:[9] (1) random nucleation does not prevail; instead, grain boundary nucleation predominates; (2) $\dot{N}$ is not a constant but a function of time; (3) G also varies from nodule to nodule with time;[39] and (4) the nodules are not really spheres. Consequently, Avrami applied the following time-dependent reaction equation:

$$f(t) = 1 - \exp(-kt^n)$$ (7.13)

and
$$\ln \ln \frac{1}{1 - f(t)} = \ln k + n \ln t$$ (7.14)

where k is a temperature-dependent parameter related to constant growth rate, nucleation frequency, and shape factor; and n, varying between 1 and 4, is a quan-

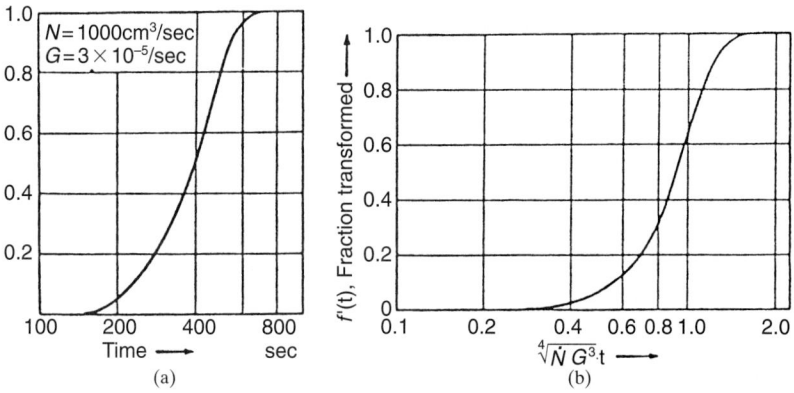

FIGURE 7.6 (a) Theoretical fraction of austenite transformed to pearlite as a function of time for specific $\dot{N}$ and G. (b) Master reaction curve for general nucleation, abscissa scale logarithmic.[7] (*Reprinted by permission of Pergamon Press, Plc.*)

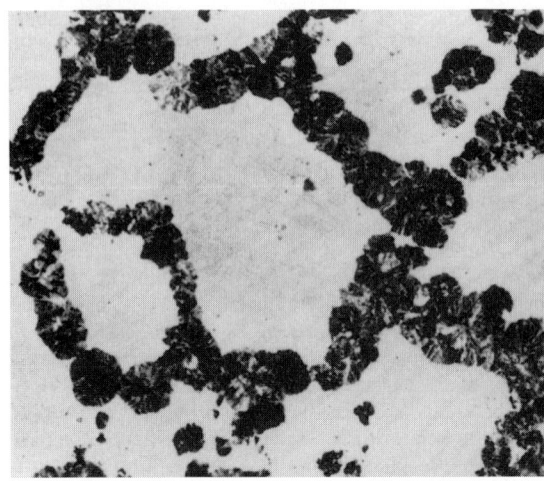

FIGURE 7.7 A partially transformed eutectoid steel at 550°C. Nucleation of pearlite on grain boundaries and subsequent thickening into grains, 120X. (*Courtesy of Paul G. Shewmon; after Aaronson.*)

tity dependent on the dimension of growth.[40] Christian has tabulated possible values for the n exponent [(Eqs. (7.13) and (7.14)] for both the interface-controlled growth including pearlite reaction and the diffusion-controlled growth (early stages of reaction only), which are listed in Table 7.2.[41]

Cahn and Hagel[37] have further developed the Avrami equation into the following forms, assuming site-saturated transformations on various nucleation sites, namely, grain faces, grain edges, and grain corners (see Fig. 6.5):

$$f(t) = 1 - \exp(-2AGt) \tag{7.15a}$$

$$f(t) = 1 - \exp(-\pi L G^2 t^2) \tag{7.15b}$$

$$f(t) = 1 - \exp\left(-\frac{4\pi}{3}\eta G^3 t^3\right) \tag{7.15c}$$

Equation (7.15a) corresponds to grain face nucleation, where A is the grain boundary area; Eq. (7.15b) applies to nucleation at grain edge where L is the grain edge length; and Eq. (7.15c) corresponds to grain corner nucleation,[37] where η is the number of grain corners per unit volume. [Equations (7.15a) to (7.15c) can be applied only as limiting cases; they have never been verified experimentally for the austenite → pearlite reaction.[42]] The nucleation rate N is irrelevant in all three situations because all grain boundaries at which pearlite can form have already been transformed. Transformation by site saturation is completed when the pearlite nodule migrates halfway across the grain. Hence the time for completion of the reaction is given by:[37]

$$t_f = 0.5\frac{d}{G} \tag{7.16}$$

where d = average grain diameter, G = average rate of pearlite growth, and d/G is the time taken for one nodule to transform one austenite grain. In eutectoid plain carbon steels, N would have no effect below about 660°C, while in hypereutectoid steels, the overall rate would be independent of N below 720°C.[37] At high temper-

TABLE 7.2 Possible Values of n in Kinetic Law[41] $f(t) = 1 - \exp(-kt^{-n})$

Conditions	n
(a) Polymorphic changes, discontinuous precipitation, eutectoid reaction, interface-controlled growth, etc.	
Increasing nucleation rate	>4
Constant nucleation rate	4
Decreasing nucleation rate	3–4
Zero nucleation rate (saturation of point sites)	3
Grain edge nucleation after saturation	2
Grain boundary nucleation after saturation	1
(b) Diffusion-controlled growth (early stages of reaction only)	
All shapes growing from small dimensions, increasing nucleation rate	>2½
All shapes growing from small dimensions, constant nucleation rate	2½
All shapes growing from small dimensions, decreasing nucleation rate	1½–2½
All shapes growing from small dimensions, zero nucleation rate	>1½
Growth of particles of appreciable initial volume	1–1½
Needles and plates of finite long dimensions, small in comparison with their separations	1
Thickening of long cylinders (needles), e.g., after complete edge impingement	1
Thickening of very large plates, e.g., after complete edge impingement	½
Segregation of dislocations (very early stage only)	2.3

Source: Courtesy of Pergamon Press, Plc., Oxford; after J. W. Christian.

atures, $\dot{N}$ remains very low and site saturation will not occur. In this situation, $\dot{N}$, where measured, appears to increase with time according to the following equation:

$$\dot{N} = at^m \qquad (7.17)$$

where a and m are constants. Cahn has found the time exponent n in the Avrami equation to be just $4 + m$ and k to be a function of grain boundary area for grain boundary nucleation.[7]

7.2.5 Measurement of Growth Rate of Pearlite

In the isothermal reaction technique, the rate of growth of pearlite nodules can be determined by allowing pearlite to grow isothermally for progressively increasing times, by measuring the radius of the largest pearlite nodule from each polished and etched specimen, and by plotting radii of these nodules versus time.[5,9] A straight line is obtained; its slope is taken as G, the growth rate of pearlite. This method provides the maximum growth rate rather than an average. It cannot be applied with ease to rapidly transforming specimens and where impingement of growing nodules has become considerable. Despite these limitations, this technique provides results similar to those of the average growth rate method developed by Cahn and Hagel.[43]

A hot-stage technique is an alternative method in which cinephotomicrography and thermal analysis are used under continuous cooling conditions, which enables us to perform in situ measurements of growth rate on several rapidly growing nodules.[44,45] In the forced-velocity growth method, a high-purity Fe-C specimen, in the form of rod, is moved at a constant velocity with respect to a high-temperature gradient, which causes the pearlite/austenite front to move unidirectionally along the rod in a manner similar to that observed in the directional solidification technique. It is based on the assumption that the pearlite/austenite interface temperature for growth at a given velocity corresponds to the temperature of isothermally transformed pearlite producing the same growth velocity.[46,47]

7.2.6   Orientation Relationships

Based on selected area diffraction (SAD) pattern analysis of pearlite nodules nucleated on the austenite grain boundary, it is found that pearlitic ferrite bears a K-S or N-W orientation relationship (OR) to that of the austenite grain into which it is not growing and has no simple OR with the austenite grain into which it is growing.[48] The ORs existing between pearlitic ferrite and cementite are represented by:[49]

Isaichev OR: $(\bar{1}03)_{cm}$ // $(110)_\alpha$; $[010]_{cm}$ // $[1\bar{1}\bar{1}]_\alpha$; $[311]_{cm}$ 0.91° from $[1\bar{1}1]_\alpha$

Bagaryatskii OR: $(001)_{cm}$ // $(11\bar{2})_\alpha$; $[100]_{cm}$ // $[0\bar{1}1]_\alpha$; $[010]_{cm}$ // $[1\bar{1}\bar{1}]_\alpha$

Pitsch-Petch OR: $(001)_{cm}$ // $(5\bar{2}\bar{1})_\alpha$; $[100]_{cm}$ 2.6° from $[13\bar{1}]_\alpha$; $[010]_{cm}$ 2.6° from $[113]_\alpha$

where the subscripts cm and α denote cementite and ferrite, respectively.

   In pearlite four new ORs between pearlitic ferrite and cementite have recently been identified by Zhang and Kelly,[49] employing the more accurate CBKLDP method. In pearlitic steel the pearlite follows the New-2, New-3, and New-4 ORs; in hypoeutectoid steels the Isaichev OR is observed. These authors were not able to find the two widely accepted orientation realtionships, namely, the Pitsch-Petch and Bagaryatskii ORs; and in hypereutectoid steels the New-4 OR is found. These are expressed as follows:[49]

New-2 OR: $(\bar{1}03)_{cm}$ // $(\bar{1}01)_\alpha$; $[010]_{cm}$ 8.5° from $[131]_\alpha$; $[3\bar{1}1]_{cm}$ // $[1\bar{1}1]_\alpha$

New-3 OR: $(0\bar{2}2)_{cm}$ // $(\bar{1}01)_\alpha$; $[\bar{1}01]_{cm}$ 2.4° from $[1\bar{3}1]_\alpha$; $[311]_{cm}$ // $[1\bar{1}1]_\alpha$

New-4 OR: $(210)_{cm}$ // $(101)_\alpha$; $[001]_{cm}$ // $[131]_\alpha$; $[\bar{1}21]_{cm}$ 5.95° from $[101]_\alpha$

New-4 OR can also be represented by $(100)_{cm}$ 1.4° from $(\bar{5}12)_\alpha$; $[010]_{cm}$ 5.4° from $[1\bar{1}3]_\alpha$; $[001]_{cm}$ // $[131]_\alpha$.

New-5 OR: $(\bar{1}03)_{cm}$ // $(\bar{1}01)_\alpha$; $[010]_{cm}$ // $[131]_\alpha$; $[311]_{cm}$ 8.5° from $[1\bar{1}1]_\alpha$

7.2.7   Pearlite in Nonferrous Alloys

Unlike the pearlite reaction in interstitial Fe-C alloys, the eutectoid transformation in a substitutional (nonferrous) solid solution involves diffusion of solute atoms either through the matrix or along the ledges or boundaries.[50] In the Cu-Al system, the austenite β phase (bcc) transforms into the α phase (fcc) and the γ_2 phase (complex bcc) at 565°C (838 K) at a composition of 11.8 wt% Al, 88.2 wt% Cu (Table 7.1). Recently, the eutectoid transformation in this system, previously reported to be volume-diffusion-controlled by Asundi and West (1970), has been recognized to undergo a ledge-growth mechanism assisted by boundary diffusion.[50] In the Cu-(22–26)Sn systems, the β phase (bcc electron compound with $e/a = 3/2$ at the eutectoid composition with Cu_5Sn formula) decomposes into the eutectoid product of α (fcc) and γ (complex cubic with a DO_3 type superlattice) phases by nucleation and growth process when held at temperatures between 520 and 586°C.[51] Although the equilibrium phases at temperatures between 480 and 520°C are α and δ, the β phase initially decomposes to α and γ phases in this temperature range, followed by δ precipitation at the α/γ interface. In the Cu-In system, the eutectoid transformation occurs as β matrix $\rightarrow \alpha + \delta$ at 20.15 at% In below $T_{eu} = 847$ K by volume-diffusion-controlled decomposition or interphase boundary diffusion-controlled process (Table 7.1). Here the eutectoid colonies nucleate at the β/β grain boundaries and grow into the untransformed β matrix.[52] In the fcc Al-78% Zn system, Cheetham and Ridley[53] have shown that the eutectoid transformation occurs by boundary diffusion.

The eutectoid products in many Ti- and Zr-base alloys consists of a mixture of α (hcp) phase and an appropriate intermetallic phase. In some of these so-called "active" eutectoid systems, the decomposition of the β (bcc) phase into a lamellar aggregate of two phases cannot be suppressed in alloys of near eutectoid composition, even by rapid quenching. Some of these alloys include Ti-Cu, Ti-Ni, Zr-Cu, Zr-Ni, and Zr-Fe. Active eutectoid decomposition in Zr-3 wt% Fe alloys occurs on water quenching from the β phase fields. Sympathetic nucleation and growth may be regarded as a possible mechanism for the straight to wavy morphological transition during this decomposition process.[54]

Livingston and Cahn (1974) have observed the discontinuous coarsening reaction in polycrystalline eutectoids in Co-Si, Cu-In, and Ni-In when the microstructures were annealed near the eutectoid temperature. Chuang et al. (1988) reported that the transport mechanism in both discontinuous precipitation and subsequent coarsening reaction in Ni-7.5 at% In alloy was attributed to the grain-boundary diffusion.[24] Doherty (1982) described the formation of similar coarsened lamellar structure by deformation-induced boundary migration (recrystallization) in Ni-based alloys containing coherent Ni_3 Al precipitates. In this case the driving force for boundary movement was the stored energy of the dislocations.[24]

In Ag-Ga systems, two different pearlite morphologies occur. The first one is the coarse pearlite occurring at high reaction temperatures near the eutectoid composition and constitutes equilibrium phases α and β', with fcc and hcp (hP2) crystal structures, respectively. The second one is blocky or fine pearlite occurring at a lower reaction temperature and constitutes α and metastable β'' with a 9R (hR3) crystal structure. Both types of pearlite undergo coarsening by a discontinuous or cellular reaction arising from the lower interfacial energy required to form this type of pearlite compared to traditional pearlite. The lower interfacial energy linked to blocky/fine pearlite results from the planar feature of the interlamellar boundaries and the similar structures of its product phases.[3]

The unusual presence of both α and β'' found within each lamella of blocky pearlite has not been fully understood, but may be attributed to the low concentration in the metastable parent phase in front of the β' (or β'') product lamellae.[3]

7.3 INTERPHASE PRECIPITATION

In 1964, Mannerkoski[55] and Relander[56] identified two new mechanisms of eutectoid decomposition of austenite containing appreciable amounts of strong carbide-forming elements such as Nb, Ti, V, Cr, Mo, and W with different alloy carbide morphologies.[57] These are called *interphase precipitation* (or sometimes *interphase boundary carbide precipitation*) and *fibrous carbide precipitation*.[58] In both struc-

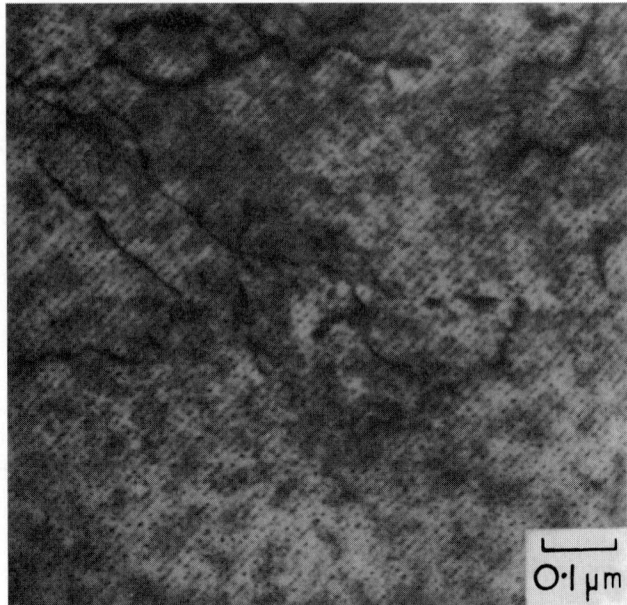

FIGURE 7.8   Fe-1.04% V-0.20% C after 5 min at 725°C. Thin-foil electron micrograph showing very fine banded dispersions of vanadium carbide within the ferrite.[64] (*Courtesy of A. D. Batte and R. W. K. Honeycombe.*)

tures, nucleation and growth of alloy carbides occur only on the partially coherent region of the γ/α boundaries; their spatial distribution and orientation tend to be strictly controlled; and the carbides are observed to be crystallographically related to the ferrite into which they are growing.[57] In the interphase boundary carbide structure, the precipitation of alloy carbides takes place periodically or repeatedly during the $\gamma \rightarrow \alpha$ transformation at the advancing allotriomorphic α/γ interface boundaries within a ferrite matrix. Further growth of allotriomorphic ferrite alternates with carbide precipitation; both the distribution and the crystallography of the carbides are influenced by the interphase boundary. This results in a finely banded structure comprising three-dimensional sheets or parallel layers of separate, small carbide particles (typically VC) embedded in a ferrite matrix[58–62] (Fig. 7.8). The bands are associated with either planar or curved interfaces which depend on the nature of the mobile γ/α interface. The interband spacing is dependent on the temperature of transformation and on the composition. As the temperature of transformation is decreased, the band spacing decreases and extremely fine-scale and rapid transformation occurs. Detailed investigations have indicated that the fine carbide precipitates are often much less than 100 Å (10 nm) in diameter and that the sheet and fibrous interband spacing may vary between 50 Å (5 nm) and 500 Å (50 nm).[58]

Classically, interphase precipitation reaction is observed in Ti, Nb, or V microalloyed steels (TiC_XN_{1-X}, NbC_XN_{1-X}, VC_XN_{1-X}) (see Chap. 15 also) and in alloy steels (such as Fe-12Cr-0.2C, Fe-10Cr-0.4C, and Fe-4Mo-0.2C alloy steels) with $Cr_{23}C_6$, $(FeMo)_{23}C_6$, or $(MoCrFe)_{23}C_6$ precipitation, as well as in medium- and high-carbon steels, with predominantly or entirely pearlitic microstructure.[31,57,63]

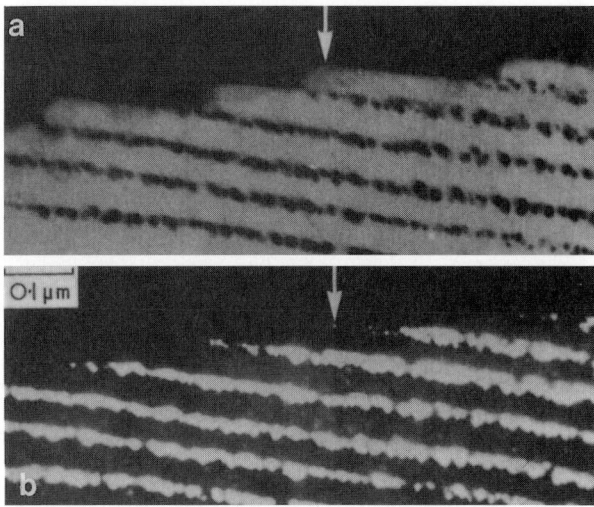

FIGURE 7.9 (*a*) Fe-12% Cr-0.2% C transformed after 30 min at 650°C. Interphase precipitation of $Cr_{23}C_6$ at γ/α interface. (*a*) Bright field showing ledges in γ/α interface. (*b*) Dark field showing white areas corresponding to precipitate particles.[9] (*Courtesy of K. Campbell and R. W. K. Honeycombe.*)

The planar interphase boundary carbide structure occurs usually at the lower transformation temperature and is produced by the thickening of ferrite allotriomorphs by the ledge mechanism.[62] This mechanism of interphase precipitation is operative on the γ/α interface, as clearly shown in electron micrographs of isothermally transformed thin foils of Fe-12Cr-0.2C alloy steel at 650°C in Fig. 7.9. Figure 7.9*a* is a bright field image showing the ledges on the interface (i.e., stepped region of the interface), and Fig. 7.9*b* is a dark field image showing white areas corresponding to the precipitate particles. For the most part, the nucleation and growth of carbide particles occur only on the immobile, low-energy, partially coherent broad faces of ledges, while the mobile, incoherent risers of the ledges are free of precipitate. The size of the carbide particles becomes smaller as each ledge riser is approached.[58,64] The process is repeated for each ledge. Figure 7.10 schematically represents the nucleation and growth of carbides on the γ/α interface. Davenport and Honeycombe showed that the relationship between the carbides and ferrite in the interphase precipitated condition (sheets of finely distributed carbides) corresponds to that of the Baker-Nutting orientation relationship: $(001)_{carbide}$ // $(001)_\alpha$; $[010]_{carbide}$ // $[110]_\alpha$. This is also valid for the case of $V(C_XN_{1-X})$.

The occurrence of curved type of interphase precipitation has been observed predominantly at the higher transformation temperature in Fe-C-Cr, Fe-C-V, Fe-C-Mo, and Fe-C-Ni alloys. Based on both the occurrence of this type of transformation mainly at grain boundaries and the absence of ledged planar arrays of carbide particles, some workers have proposed that it is the absence of a low-energy orientation relationship (and thus the ledge mechanism) which causes the γ/α interface to bow out between coarsely spaced carbide particles and move between them in the form of a regular sequence of curved interfaces. This produces curved rows of

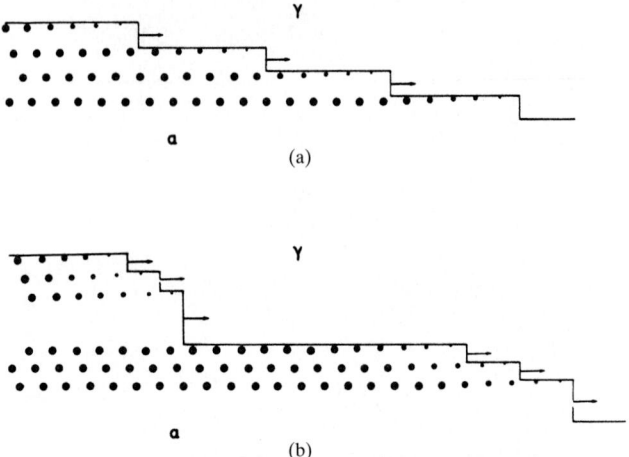

FIGURE 7.10 Schematic planar interphase precipitation showing (*a*) uniform steps and (*b*) irregular steps.[64] (*Reprinted by permission of The Metallurgical Society, Warrendale, Pennsylvania.*)

interphase precipitates. However, migration of the γ/α interface by ledge mechanisms has still been noticed in such structures, which indicates that partially coherent facets may not be completely absent.[65]

Based on direct observation of β/α interface in a Ti-Cr alloy, Furuhara and Aaronson[66] proposed that curved interphase interfaces are, in fact, partially coherent and can be explained by closely spaced growth ledges lying at an adequate angle to each other or terrace planes stepped up or down by structural ledges. Hence, the ledge mechanism theory can qualitatively describe the interphase boundary precipitation at both planar and curved interfaces.[66,67] The sheet and fibrous morphologies of erbium-rich particles in the Ti-1.7 at% Er alloy are surprisingly similar to those found in isothermally transformed alloy steels.[67]

Note that in low-carbon steel, the pearlitic transformation temperature range extends from 500 to 550°C up to A_1 temperature, and the interphase precipitation occurs usually in ferritic microstructure. However, during isothermal pearlitic transformation in vanadium alloyed medium- and high-carbon steels, interphase precipitation of linear or curved arrays of uniformly sized and shaped particles of VC or VCN occurs within both the proeutectoid ferrite and lamellar pearlitic ferrite in the entire temperature and composition ranges of their formations. Figure 7.11*a* shows a typical example of interlamellar interphase precipitation.[68] These interphase precipitates form curved arrays of uniformly sized and shaped VC (or VCN) particles (white) whose alignment is normal to the cementite long axis.[63,69] In a medium-carbon steel, dark field electron micrographs often exhibit interphase precipitation in both the ferrite and pearlitic ferrite (as shown in Fig. 7.11*b*). Note that the same vanadium carbonitride reflection will illuminate the interphase precipitates (under dark field conditions) in both the ferrite and contiguous pearlite regions, which implies the presence of some continuity in the ferrite orientation between the polygonal ferrite and interlamellar ferrite. However, VC does not precipitate within the pearlitic cementite; moreover, the pearlitic cementite is not supersaturated in vanadium.[63,68]

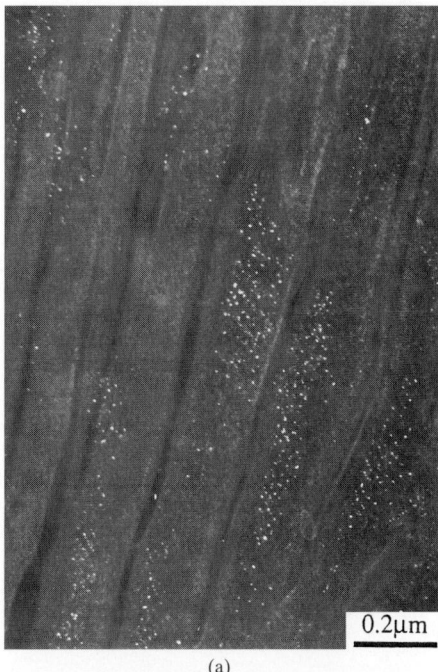

(a)

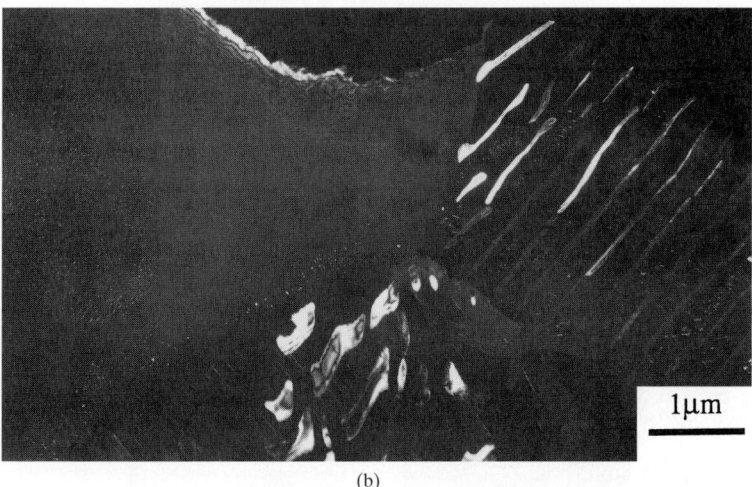

(b)

FIGURE 7.11 (*a*) Vanadium carbonitride particles having the interphase precipitate morphology and distribution in the interlamellar ferrite of the pearlite eutectoid structure (dark field TEM). (*b*) Vanadium carbonitride precipitation produced in a medium-carbon (Fe-0.1C-0.1SV) steel (by isothermal transformation at 600°C for 1 hr) having a ferrite-pearlite structure. Note that the VCN in both the ferrite and contiguous pearlite is illumined by the same diffracted beam, suggesting a common variant of the same orientation relationship in a common ferrite orientation dark field TEM.[63] [(*a*) *Courtesy of T. Gladman;* (*b*) *courtesy of G. Fourlaris.*]

7.4 FIBROUS CARBIDES

Unlike the interphase precipitations that usually form periodically in bands and have planar, irregular, curved γ/α interfaces, fibrous carbides involve colonies of fine (~20 to 100 nm in diameter and up to 10 μm in length), parallel, alloy carbide needles, rods, or fibers with little or no branching. The rare occurrence of branching, the fibrous morphology, and growth nearly normal to the reaction front are characteristic features which distinguish this mode of transformation from that of pearlite.

In heat-treated low-alloy steels containing carbide-forming alloying elements, fcc alloy carbides (e.g., NbC, TiC, VC, and $Cr_{23}C_6$), hcp carbides (e.g., Mo_2C and W_2C), and rhombohedral carbide Cr_7C_3 can form fine fibrous aggregates.

In isothermally transformed steels, fibrous carbide can form in association with interphase precipitation. For example, within one allotriomorphic ferrite crystal formed at an austenite grain boundary, the fibrous carbides (Mo_2C or VC) can form on one side and interphase precipitation on the other (Fig. 7.12a).[39,64] On the fibrous carbide side of the boundary, the ferrite does not show a low-energy orientation relationship with the austenite in which it is growing, and the γ/α interface does not correspond to a rational plane, thereby possibly resulting in a high-energy incoherent interface. In contrast, on the interphase precipitation side, the ferrite bears a *Kurdjumov-Sachs orientation relationship* with the austenite, and the bands are parallel to the reaction front such that $(111)_\gamma // (110)_\alpha$.

Figure 7.12b is a TEM micrograph of the Fe-5Cr-0.2C alloy specimen isothermally transformed at 650°C for 30 min. This figure illustrates both the carbide

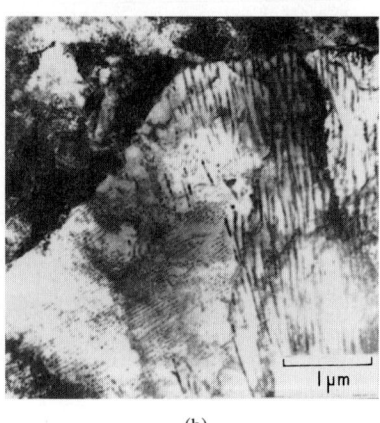

(a)   (b)

FIGURE 7.12   TEM micrographs. (a) The formation of fibrous carbide phase on one side and interphase precipitation on the other side of the allotriomorphic ferrite. Specimen Fe-4% Mo-0.2% C steel isothermally transformed at 700°C after 0.5 hr.[64] (b) The Fe-5% Cr-0.2% C alloy specimen isothermally transformed at 650°C for 0.5 hr, illustrating both the fibrous and interphase precipitation.[70] [(a) Courtesy of F. G. Berry and R. W. K. Honeycombe; (b) courtesy of K. Campbell and R. W. K. Honeycombe.]

morphology—fibrous—and interphase precipitation, exhibiting their characteristic features but incorporating the same Cr_7C_3 phase.[70]

The parameters that determine the relative proportion of fibrous and interphase carbides are alloying additions, transformation temperature, and the crystallography of the α/γ boundary from which these structures develop. However, studies made on Fe-C-Mo alloy partially reacted at 850°C and then further reacted at 725°C show that an interphase boundary carbide structure obtained at the higher temperature could be changed into a fibrous one at the lower temperature. Associated with this conversion was a change in the shape of the γ/α interface from ledged to smoothly curved, which suggested that there were significant influences of both lattice orientation and boundary orientation in the determination of the γ/α interfacial structure and mechanism of eutectoid reaction.[64] For isothermal transformation in the temperature range of 600 to 900°C, fibrous carbide does not occur in Fe-1V-0.2C steel, while it is predominant in Fe-4Mo-0.2C steel due to favorable reaction kinetics;[64] the extent of precipitation of fibrous Mo_2C increases when the transformation temperature falls from 850 to 600°C, which implies that the slower transformation enhances the growth of fibrous carbides.[71] It has been noted that when a sufficient proportion of an austenite-stabilizing element (e.g., Mn) is present in Fe-13Mn-2V-0.8C steel, the vanadium carbide phase occurs profusely in the fibrous mode.[72] It may be inferred again that the slow reaction favors the γ/α interface movement toward the fibrous mode, while rapid reaction causes abundant formation of interphase precipitation.

7.5 PROEUTECTOID PHASES

As discussed in Chap. 1, within the $\alpha + \gamma$ region or below the A_1 temperature, the separation of proeutectoid ferrite precedes pearlite formation in hypoeutectoid steels, and the separation of proeutectoid cementite precedes pearlite formation in hypereutectoid steels. The proportion of proeutectoid constituents decreases as the carbon content of steel approaches the eutectoid composition (0.8%C) and as the reaction temperature decreases. The proeutectoid constituents (especially ferrite) are identical in crystal structure to those present in pearlite, but their distribution in the microstructure is quite different from their lamellar arrangement in pearlite.

7.6 FERRITE MORPHOLOGY

Precipitation of proeutectoid ferrite forming from austenite, at the austenite grain boundary, occurs in shapes that have been classified as allotriomorphs, primary and secondary sideplates, and primary and secondary sawteeth. Ferrite morphologies forming in the interior of austenite grains include intragranular plates and equiaxed idiomorphs.[73-75] The so-called *massive structure* results from the impingement of crystals of other morphologies. Figure 7.13 shows the Dubé morphological classification system for proeutectoid ferrite in steels which includes six components.

As in the Fe-C system, Widmanstatten plates and needles have been found in nonferrous systems, for example, Cu-Zn, Cu-Cr, Al-Cu, Al-Mg-Zn, Al-Ag, and Al-Au systems. Most of these systems also exhibit other Dubé morphologies, particularly those of grain boundary allotriomorphs.

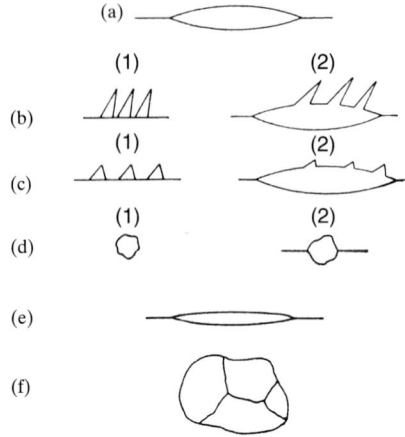

FIGURE 7.13 Dubé morphological classification system for ferrite crystals: (*a*) grain boundary allotriomorphs; (*b*) Widmanstätten side-plates, (1) primary, (2) secondary; (*c*) Widmanstatten sawteeth, (1) primary, (2) secondary; (*d*) idiomorphs; (*e*) intragranular Widmanstätten plate; and (*f*) massive structure.[73] (*Reprinted by permission of The Metallurgical Society, Warrendale, Pennsylvania.*)

7.6.1 Allotriomorphic Ferrite

The major ferrite morphology that forms at relatively high transformation temperature near A_3 in hypoeutectoid plain carbon and low-alloy steels is that of grain boundary allotriomorphs.[73] Allotriomorphic ferrite crystals nucleate first during the decomposition of austenite, at the prior austenite grain boundaries (Fig. 7.14*a* and *b*),[76,77] and grow preferentially along these boundaries (i.e., more rapidly than thickening), and have shapes approximating a double spherical cap or an oblate ellipsoid. The γ/α interface is thought to be of the disordered type; however, increasing evidence appears to indicate this to be in favor of an interface comprising coherent and disordered areas.[73] They form at small, as well as at large, undercoolings below A_3 and also below A_1. Their appearance at small undercooling requires the prolonged isothermal reaction or slow cooling through $\alpha + \gamma$ region.

7.6.1.1 Growth Kinetics
Thickening Kinetics. This section deals with the planar boundary growth problem. This approximates growth of a grain boundary allotriomorph in thickness direction, that is, migration of the broad faces of a grain boundary allotriomorph. During the formation of ferrite, carbon diffuses in the austenite ahead of the α/γ interface. Figure 7.15*a* shows a portion of the Fe-Fe$_3$C diagram, and Fig. 7.15*b* shows a plot of carbon concentration normal to the α/γ boundaries. If we assume complete local equilibrium at the α/γ interface, the carbon concentrations in α and γ at the α/γ interface, denoted by $C_\alpha^{\alpha\gamma}$ and $C_\gamma^{\gamma\alpha}$, respectively, correspond to those in the equilibrium phase diagram (Fig. 7.15*a*). If a unit area of the α/γ interface advances a distance ds, the amount of carbon removed from α and piled up in front of the advancing α/γ boundary is $ds\,(C_\gamma^{\gamma\alpha} - C_\alpha^{\alpha\gamma})$. In order to maintain equilibrium at the α/γ interface, this amount of carbon must be equal to the product $J\,dt$, where J is the flux of carbon atoms diffusing away from the α/γ interface in the austenite in a direction perpendicular to this interface. That is,

$$-J\,dt = D_C^\gamma \left(\frac{\partial C}{\partial S} \right)_{\gamma\alpha} dt = ds\left(C_\gamma^{\gamma\alpha} - C_\alpha^{\alpha\gamma}\right) \tag{7.18}$$

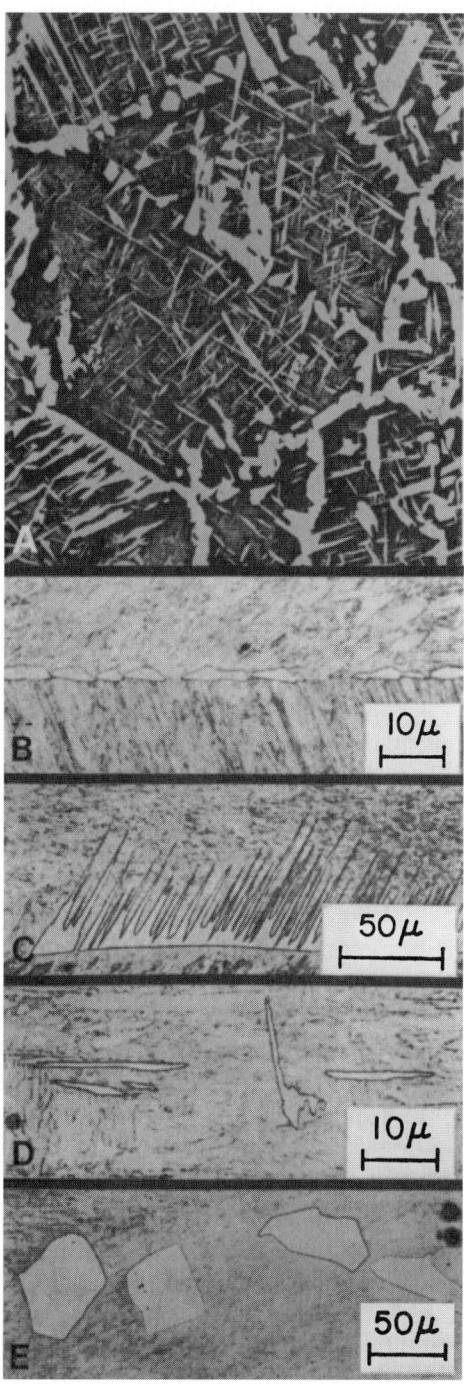

FIGURE 7.14 0.5% C-0.06% Si-0.7% Mn steel austenitized for 1 hr at 1350°C, cooled at 300°C/hr (etchant; Picral). (*A*) Different morphologies of ferrite. 100X (reduced 66%).[27] Observation of separate ferrite morphology in isothermally reacted 0.29%C steel at (*B*) 13 s at 750°C (grain boundary α); (*C*) 2 min at 700°C (sideplate); (*D*) 20 s at 600°C (intragranular plate); (*E*) 5 min at 725°C (intragranular idiomorph).[76] (*Reprinted by permission of ASM International, Materials Park, Ohio.*)

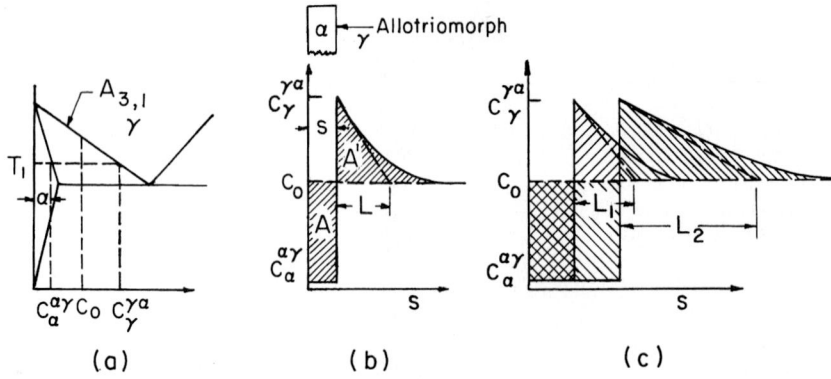

FIGURE 7.15 (*a*) A portion of the Fe-Fe$_3$C phase diagram. (*b*) Carbon concentration profile along a direction perpendicular to the γ/α interface. (*c*) Carbon concentration profile in front of growing ferrite at different time intervals or different volume fractions of ferrite formed. Shaded regions below and above C_0 in (*b*) and (*c*) are the same.

The growth rate G of ferrite is

$$\frac{ds}{dt} = G = \frac{-J}{C_\gamma^{\gamma\alpha} - C_\alpha^{\alpha\gamma}} = \frac{D_c^\gamma}{C_\gamma^{\gamma\alpha} - C_\alpha^{\alpha\gamma}} \left(\frac{\partial C}{\partial s}\right)_{\text{interface}} \tag{7.19}$$

The concentration gradient $\partial C/\partial s$ at the interface in Eq. (7.19) can be approximated, using Zener's linearized concentration gradient expression, as $\Delta C/L$, where ΔC is the difference of carbon concentration in austenite near to, and remote from, the γ/α interface and L is the effective diffusion distance:

$$\left(\frac{\partial C}{\partial s}\right)_{\text{interface}} = \frac{C_\gamma^{\gamma\alpha} - C_0}{L} \tag{7.20}$$

It is clear from Fig. 7.15*b* that the shaded region below C_0 represents the amount of carbon which diffused away from the ferrite as it grows a distance s. All of this carbon content piled up near the α/γ interface must diffuse further into the austenite. Assume the same cross-sectional area of the two phases; then conservation of solute requires that the two shaded areas (above and below C_0) must be equal, that is, $A = A'$, $A = s(C_0 - C_\alpha^{\alpha\gamma})$, and $A' = (L/2)(C_\gamma^{\gamma\alpha} - C_0)$. Substituting Eq. (7.20) into Eq. (7.19), we obtain

$$\frac{ds}{dt} = \frac{D_c^\gamma}{C_\gamma^{\gamma\alpha} - C_\alpha^{\alpha\gamma}} \left(\frac{C_\gamma^{\gamma\alpha} - C_0}{L}\right) = \frac{C_\gamma^{\gamma\alpha} - C_0}{2s(C_0 - C_\alpha^{\alpha\gamma})(C_\gamma^{\gamma\alpha} - C_\alpha^{\alpha\gamma})} \tag{7.21}$$

On integration of this equation, we get

$$s = \frac{(C_\gamma^{\gamma\alpha} - C_0)\sqrt{D_c^\gamma t}}{\sqrt{(C_0 - C_\alpha^{\alpha\gamma})(C_\gamma^{\gamma\alpha} - C_\alpha^{\alpha\gamma})}} \tag{7.22}$$

or

$$s = \alpha t^{1/2} \tag{7.23}$$

where

$$\alpha = \frac{\left(C_\gamma^{\gamma\alpha} - C_0\right)\sqrt{D_c^\gamma}}{\sqrt{\left(C_0 - C_\alpha^{\alpha\gamma}\right)\left(C_\gamma^{\gamma\alpha} - C_\alpha^{\alpha\gamma}\right)}} \qquad (7.24)$$

where s is the half-thickness of the allotriomorph and α is the parabolic rate constant for thickening.

This equation is the parabolic law of thickening and states that the thickening of ferrite layer s varies with the square root of the growth time t. Since the extent of the carbon diffusion field, that is, effective diffusion distance, increases with time, the growth rate of ferrite must decrease with time (Fig. 7.15c). That is, $G \propto t^{-1/2}$ because $L \cong \sqrt{D_c^\gamma t}$. Note that this equation is based on three assumptions: (1) The interface boundary is planar and disordered, (2) the migration kinetics are volume diffusion-controlled, and (3) D_c^γ is composition-invariant.[76]

As already noted, the foregoing derivation of allotriomorph thickening kinetics is based upon Zener's linearized gradient approximation. An exact expression for the parabolic rate constant for thickening α,[78] though still based on the assumption that D_c^γ is independent of carbon concentration, has been derived by Dubé[79] and Zener:[80]

$$\frac{C_\gamma^{\gamma\alpha} - C_0}{C_\gamma^{\gamma\alpha} - C_\alpha^{\alpha\gamma}} \left(\frac{\overline{D}_c^\gamma}{\pi}\right)^{1/2} = \frac{\alpha}{2} \exp\left(\frac{\alpha^2}{4\overline{D}_c^\gamma}\right) \mathrm{erfc}\left[\frac{\alpha}{2\left(\overline{D}_c^\gamma\right)^{1/2}}\right] \qquad (7.25)$$

where $\overline{D}_c^\gamma$ is the weighted average diffusivity of carbon in austenite[81] and t is the growth time.

7.6.1.2   *Experimental Study of Growth Kinetics of Allotriomorphs.*   An experimental study of the kinetics of lengthening and thickening of grain boundary allotriomorphs in the Fe-C system consists of reacting isothermally a number of similar samples for different times and then quenching. The half-thickness (or half-length) of the largest allotriomorph on the plane of polish in each specimen is measured and plotted as a function of the square root of the reaction time. Provided that the austenite grain boundaries are perpendicular to the plane of polish, the allotriomorph length and thickness should correspond to the dimensions occurring in three-dimensional space of those first nucleated. In this way, Bradley and Aaronson determined the values of α and β, the parabolic rate constants for thickening and lengthening of allotriomorphs, respectively. When these values of α and β were compared with the theoretical values [assuming allotriomorphs to be oblate ellipsoid with the experimentally observed aspect ratio (~1/3) and the variation of diffusivity with concentration[82]], they found experimental allotriomorph growth kinetics data to be considerably lower (i.e., less than an order of magnitude) than those calculated, although mainly at lower undercooling and carbon concentration. This was attributed to the existence of a variable proportion of partially coherent facets at the broad, disordered-type faces of the allotriomorphs, as proposed by Kinsman and Aaronson, which resulted from the low-energy orientation relationship of the allotriomorphs with at least one adjacent austenite grain.[83] These facets migrate (and grow) by a ledge mechanism at a slower rate compared to those of incoherent γ/α boundaries where migration occurs by essentially uniform attachment of Fe atoms and removal of carbon atoms.

Measurements of growth kinetics of ferrite allotriomorphs in Fe-C-X alloys (where X = Mn, Ni, Si, or Cr) reacted over the widest experimentally accessible temperature range were made by Bradley and Aaronson[84] utilizing the Hillert-

Staffansson[85] regular solution treatment of ternary-phase equilibrium, and they compared their results on the basis of three different models (hypotheses) of the growth process. These models included (1) full local equilibrium, also called *orthoequilibrium*, at α/γ boundaries with bulk partition of both X and C between γ and α above a critical temperature. This is called X *partition under local equilibrium* (P-LE); (2) full local equilibrium at α/γ boundaries with localized "pile-up" of X in front of the advancing α/γ boundaries below the critical temperature, called X *negligible partition under local equilibrium* (NP-LE); and (3) paraequilibrium at γ/α boundaries without partitioning of X in α and γ (i.e., with partitioning only of carbon in α and γ). The best overall agreement between experimental and calculated growth kinetics was achieved with the paraequilibrium hypothesis.

It has been demonstrated by Aaronson and Domian[86] that the bulk partitioning of alloying elements during ferrite growth in Fe-C-X alloys (X = Mn, Ni, Pt) occurs at low undercooling but not at higher undercooling. No partitioning was found for X = Si, Mo, Co, Al, Cr, and Cu. They also found, in some cases, a significant deviation of the ratio of the corrected experimental parabolic rate constant for allotriomorph thickening α_{corr} to the rate constant calculated from the paraequilibrium model α_{para} from unity. Shiflet et al.[87] suggested interphase boundary carbide precipitation and segregation of either Si or Ni to disordered γ/α boundaries to be responsible for this discrepancy in Fe-C-Si and Fe-C-Ni alloys. Mainly, however, these discrepancies are attributed to a *solute draglike effect* (SDLE). Elements that lower the activity of carbon in γ segregating to disordered areas of α/γ boundaries diminish the activity gradient of carbon in γ driving ferrite growth.[78,84]

In case of three quaternary Fe-C-Mn-X_2 alloys (where X_2 = Si, Ni, and Co), the parabolic rate constant α for thickening was observed to be an order of magnitude higher than the amount predicted by the P-LE mode in the alloying element diffusion-controlled regime in the applicable temperature range, whereas the opposite was found to be true in the carbon-diffusion-controlled regime. Likewise, the calculated paraequilibrium constant was normally quite larger than the experimentally determined value. Substantial discrepancies between the measured and calculated curves were attributed to the synergistic effects of Mn and X_2 upon growth.[88]

7.6.1.3 Interface Characteristics and Orientation Relations. Based on the foregoing discussion, the allotriomorph interface does not move uniformly at all elements of the interface. In fact, partially coherent facets present along the disordered-type interface tend to slow down the growth rate. At lower isothermal transformation temperatures, the disordered segments tend to decrease while the slowly moving partially coherent interface becomes more dominant. The ferrite allotriomorphs nucleate with a Kurdjumov-Sachs orientation relationship with (at least) one bounding austenite grain γ_1:

$$\{111\}_\gamma \; // \{110\}_\alpha ; < 110 >_\gamma // < 111 >_\alpha$$

But they also grow into the neighboring austenite grain γ_2 with which they have a random orientation relationship. However, recent TEM observations of the interfacial structure of the broad faces of α allotriomorphs precipitated from a hypoeutectoid Ti-Cr alloy—where retention of the β matrix is permitted by an M_s far below room temperature—have shown that even in the absence of a low-energy orientation relationship (Burgers relationship in this bcc-to-hcp transformation), both the rationally and the irrationally oriented interfaces of proeutectoid α allotriomorphs are partially coherent in β Ti-Cr alloys.[89] Presumably the same situation applies to ferrite allotriomorphs.

7.6.1.4   Relief Effects.   Using scanning tunneling microscopy, Bo and Fang[90] have reported the existence of surface relief effects associated with the formation of grain boundary allotriomorphs in Fe-C alloys. According to them, the interface between the grain boundary allotriomorph and austenite matrix at one side of the grain boundary is partially coherent, and the interface between the grain boundary allotriomorph and adjacent austenite matrix at another side of the same grain boundary may be disordered. The sympathetic nucleation-ledgewise growth mechanism is suggested to explain the formation of grain boundary allotriomorphs. Note that the formation of massive ferrite does not produce martensite-like surface relief effects. However, growth of both sideplates and intragranular plates is associated with relief effects at the free surface similar to those produced by bainite and martensite plates.

7.6.2 Widmanstätten Ferrite

At lower transformation temperature, the ferrite precipitates adopt a plate morphology called Widmanstätten ferrite α_w. Widmanstätten plates nucleate mainly from prior γ grain boundaries or develop from existing grain boundary allotriomorphs. The broad faces of α_w plates are believed to be partially or fully coherent with the matrix phase.[73]

Figure 7.16a shows a schematic TTT diagram for a plain carbon steel for the precipitation of grain boundary and Widmanstätten ferrite from austenite. An approximate temperature T_w, at which Widmanstätten ferrite appears from allotriomorphic ferrite, is also shown in the figure. Widmanstätten ferrite becomes dominant below this temperature, but ferrite allotriomorphs do not disappear completely (Fig. 7.14A).

Figure 7.16b[73] is a portion of the Fe-Fe$_3$C diagram showing temperature and composition regions in which various morphologies are dominant. Above the temperature T_w, the ferrite forms as grain boundary allotriomorphs, and below this temperature it is mainly in the form of Widmanstätten ferrite plates. It is apparent that T_w is roughly parallel to the A_3 temperature. Lowering the isothermal transformation temperature causes: (1) a decrease of the proportion of grain boundary allotriomorphs, (2) an increase in the number of Widmanstätten ferrite sideplates, (3) appearance of sideplates in groups into several plates growing from the same allotriomorph, (4) a decrease in an average spacing between parallel sideplates, and (5) physical impingement of Widmanstätten sideplates.[91] Recently, Enomoto has used finite difference techniques to develop a rigorous model for estimating T_w temperature in Fe-C alloys, based on the growth of the ledge mechanism.[92] In fact, increasing supersaturation causes changes in various Widmanstätten morphologies.[93]

At moderate undercooling below A_3, primary sideplates, the so-called *Widmanstätten ferrite primary sideplates*, grow directly from the preexisting grain boundary as platelets in the matrix phase. However, these morphologies are scarce. They become finer with increasing undercooling. The important factors which contribute to the fineness of these sideplates are the increasing aspect ratio (i.e., thickness/length) with high undercooling and the fact that the radius of curvature at plate tip is inversely proportional to the undercooling. This is encouraged by large austenite grain size. According to Aaronson, they form at low-angle austenite grain boundaries provided the matrix is fcc and the precipitate has another structure;[93,94] their rate of nucleation is large. It has been indicated by many workers that these sideplates lengthen by a ledge mechanism.[95] This morphology has also been found in Al-Mg-Zn-type alloys.[96]

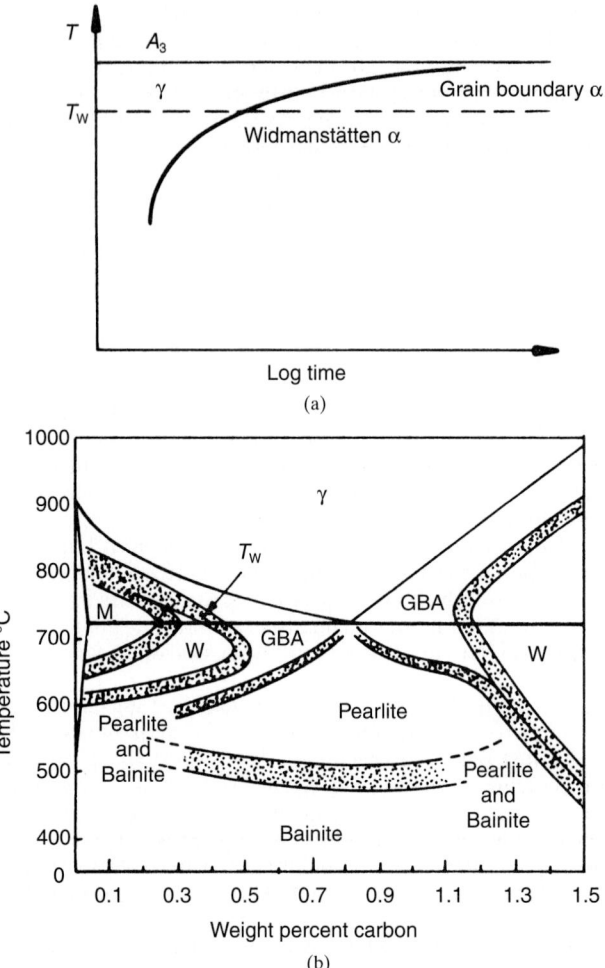

FIGURE 7.16 (*a*) Typical TTT curve for proeutectoid ferrite transformation. (*b*) Fe-Fe$_3$C diagram showing temperature versus composition regions in which various morphologies are dominant at late reaction times in specimens with ASTM grain size nos. 0 to 1. GBA, grain boundary allotriomorphs; W, Widmanstätten sideplates or intragranular plates; M, massive ferrite.[73] [(*a*) *Reprinted by permission of Van Nostrand Reinhold International;* (*b*) *reprinted by permission of The Metallurgical Society, Warrendale, Pennsylvania.*]

Secondary sideplates develop from grain boundary precipitates (usually allotriomorphs) formed earlier (Fig. 7.14*A* and *C*). These secondary ferrite sideplates (also called Widmanstätten ferrite secondary sideplates), have been reported to occur much more frequently than the primary variety.[93] They usually form in group morphologies which display a large degree of regularity, and consist of several evenly spaced plates growing such that their leading edges define an approximately planar front.[93] They form most readily in plain carbon steels containing less than 0.3 to 0.4% carbon under conditions of large austenite grain size and intermediate temperature range of transformation. Another mechanism of secondary sideplate formation involves a process in which primary sideplates initially nucleate directly from the grain boundary, followed by rapid lateral impingement along their bases to form apparent grain boundary allotriomorphic film; hence, at very early stages in their growth, they resemble secondary sideplates. Face-to-face sympathetic nucleation of additional primary sideplates may also contribute to this process.

On the second mechanism, sympathetic nucleation of ferrite sideplates takes place atop preexisting grain boundary ferrite allotriomorphs. Thus, small misorientations at the $\alpha:\alpha$ boundaries are noticed between the sideplates and the allotriomorphic regions with which they are associated.[97]

Primary and secondary sawteeth ferrite can be regarded as morphologies intermediate between grain boundary allotriomorphs and primary or secondary sideplates. Primary ferrite sawteeth are observed at only a few austenite grain boundaries. Although some impingement of primary sawteeth has occurred, it is evident that the majority form with a truncated rather than a sharp tip. (Examples may be observed along the left grain boundary above the center in Fig. 7.14A.) Aaronson and Wells have observed a ferrite/ferrite boundary between sawtooth and the allotriomorph, thereby suggesting sympathetic nucleation to be responsible for their formation.[98]

Intragranular Widmanstätten plates are needle- or platelike precipitates formed at large undercooling. Large austenite grain size favors the formation of intragranular plates which may form on isolated dislocations or at small-angle grain boundaries in austenite; direct determination of these nucleation sites, however, has not been achieved due to the destruction of the austenite matrix by martensite formation during quenching to room temperature. They are often grouped in starlike and more complex configurations at low transformation temperature in the proeutectoid ferrite region.[99] They have a lenslike shape and appear as double-ended isosceles triangles when formed during continuous cooling (Fig. 7.14A and D). At still lower temperatures the precipitation of nonlamellar carbides converts the ferrite plates to bainitic ferrites which often appear in sheaves,[100] probably arising from sympathetic nucleation.[99] This ferrite is found in Fe-C alloys and in carbon-free Fe-10% Cr alloy when transformed at 525°C (977°F).[75]

7.6.2.1 Mechanism of Widmanstätten Plate Formation. The atomistic mechanism of the formation of Widmanstätten plates has generated great interest. The important characteristic features of Widmanstätten plate formation are as follows:

1. Well-defined double-tilted tentlike or single-tilted surface-relief effects appear at T_w temperature (about 50°C higher than the calculated T_0 temperature), upper limiting temperature for the composition-invariant transformation in many ferrous alloys such as ternary Fe-C-X alloy (except in alloys having high Mn or Ni contents).

2. The growth of precipitate plates exhibiting surface relief is associated with the composition variation.

3. Plates exhibiting complex surface reliefs are actually monocrystals.

Recently, several studies have been performed on the thermodynamic and kinetic aspects of plate formation.[101,102] TEM studies have shown that α_w plates consist of a pair of two plates with a different orientation that forms in a mutually accom-

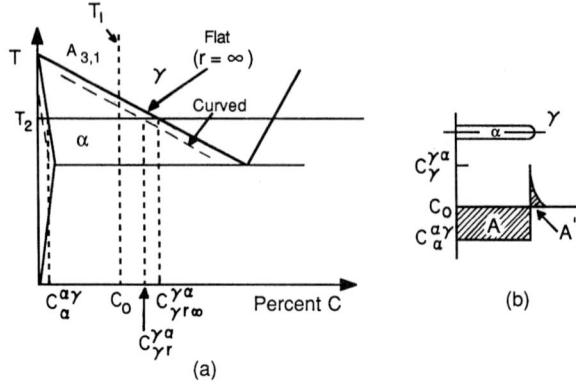

FIGURE 7.17 (*a*) Effect of curvature on the phase boundaries. (*b*) Composition profile along centerline of growing Widmanstätten ferrite plates/needles.

modating fashion. It has been reported that this pairing of plates gives rise to double-tilted surface relief. The subunit in α_w exhibits the same habits with $<111>_\alpha$ growth direction surrounded by two sets of parallel planes close to $\{451\}_\alpha$. This suggests that α_w forms via a displacive mechanism with respect to substitutional elements. If Widmanstätten plates are truly formed by displacive mechanism, the solute concentration has to be inherited from the matrix, at least initially. The ledgewise lengthening has been shown to provide a good account for reported diffusion-controlled lengthening kinetics of ferrite plates in Fe-C and Fe-Ni alloys in a wide range of reaction temperatures and supersaturation.[103]

7.6.2.2 Growth Kinetics of Widmanstätten Plates.

There are three important models for the growth kinetics of Widmanstätten plates. Two models for plate lengthening are called the *Zener-Hillert* treatment and the *Trivedi* treatment, while the third one for plate thickening is known as the *Cahn-Hillig-Sears* (CHS) model.

Lengthening Kinetics of Plates. Note that when the steel is austenitized at a high temperature T_1, the austenite may have a bulk composition C_0 (Fig. 7.17a). On cooling to a temperature $A_{3,1}$, the compositions of austenite and ferrite in equilibrium with each other are $C_\gamma^{\gamma\alpha}$ ($= C_{\gamma r\infty}^{\gamma\alpha}$) and $C_\alpha^{\alpha\gamma}$, respectively, assuming the planar, disordered interfaces (i.e., incoherent boundaries with infinite radii of curvature). As described earlier, the concentration difference, ΔC, within the austenite between a point near the planar growing ferrite interface and a distant point in the bulk austenite is given by

$$\Delta C = C_\gamma^{\gamma\alpha} - C_0 \qquad (7.26)$$

Let us assume a Widmanstätten ferrite (α_w) plate growing into austenite of composition C_0 (Fig. 7.17b) where the edge has a constant radius of curvature r. With the growth of a_w plate, most of the carbon rejected from the tip diffuses to the sides of the thin plate. As before, if we assume the same cross-sectional area of two phases, the increased amount of carbon atoms piled up ahead of the advancing tip (denoted by area A') becomes much smaller than the depleted amount of carbon behind the interface (represented by area A) (Fig. 7.17b).

For the curved growing plate interface, Zener[18] used the local equilibrium model and took into account the pressure difference between ferrite and austenite across the plate tip and derived the following relationship:

$$(\Delta C)_{\text{tip}} = \Delta C\left(1 - \frac{r^*}{r}\right) \tag{7.27a}$$

or

$$C_{\gamma r}^{\gamma\alpha} - C_0 = \left(C_\gamma^{\gamma\alpha} - C_0\right)\left(1 - \frac{r^*}{r}\right) \tag{7.27b}$$

where $C_{\gamma r}^{\gamma\alpha}$ is the actual composition of austenite at the tip interface (Fig. 7.17a) and r^* is the critical radius of curvature at which the concentration difference $(\Delta C)_{\text{tip}}$ is zero and growth ceases. The value of r^* can be approximately determined from the following Gibbs-Thomson equation

$$r^* = \frac{\gamma V_m^\alpha}{RT\left(C_\gamma^{\gamma\alpha} - C_0\right)} \tag{7.28}$$

where γ is the α/γ surface tension, V_m^α is the molar volume of the α phase, R is the universal gas constant, and T is the absolute temperature.

If we assume, in addition to the local equilibrium at the interface, that the effective diffusion distance L [in Eq. (7.20)], in front of the moving interface is proportional to the radius of curvature of the plate tip $(L = kr)$, then the equation for the concentration gradient at the tip, on comparing Eq. (7.20), becomes

$$\left(\frac{\partial C}{\partial r}\right)_{\text{tip}} = \frac{\left(C_\gamma^{\gamma\alpha} - C_0\right)(1 - r^*/r)}{kr} \tag{7.29}$$

This equation indicates the greater steepness of the concentration gradient for the edge of α plate (Fig. 7.17b). Substituting Eq. (7.29) in Eq. (7.19) yields the growth rate of the advancing plate G, given by

$$G = \frac{D\left(C_\gamma^{\gamma\alpha} - C_0\right)(1 - r^*/r)}{\left(C_\gamma^{\gamma\alpha} - C_\alpha^{\alpha\gamma}\right)kr} = \frac{D}{kr}\Omega_0 \tag{7.30}$$

where $\Omega_0 = \left(C_\gamma^{\gamma\alpha} - C_0\right)/\left(C_\gamma^{\gamma\alpha} - C_\alpha^{\alpha\gamma}\right)$ is the dimensionless supersaturation parameter. Equation (7.30) shows the variation of growth rate with tip radius. On differentiating with respect to r and setting $\partial G/\partial r = 0$, we obtain the maximum growth rate G^* at $r = 2r^*$:[18]

$$G^* = \frac{D}{4}\frac{\Omega_0}{kr^*} \tag{7.31}$$

Hillert[104] has extended this treatment further and found $k = 2$. Hence the equation of maximum (or steady-state) growth rate of a Widmanstätten plate becomes

$$G^* = \frac{D}{8r^*}\Omega_0 \tag{7.32}$$

This equation is called the *Zener-Hillert equation for plate lengthening*.

Hillert[105] further modified the above Zener-Hillert equation for the lengthening of a precipitate plate; this modification agrees well with Trivedi's results (discussed below) for medium and high values of Ω_0:

$$G^* = \frac{(D/4r^*)\Omega_0}{1 - \Omega_0}\exp[-5.756(1 - \Omega_0)] \tag{7.33}$$

The value of $r*$ depends on the composition of the alloy and can be computed from Eq. (7.28). Figure 7.18a shows a comparison between experimental lengthening rates of edges of Widmanstätten plates and calculated lengthening rates employing Hillert and Trivedi growth equations, which suggests the superior accuracy of the Trivedi model.[106]

The most advanced and rigorous treatment of the volume diffusion-controlled plate-lengthening kinetics is presented by Trivedi.[107] His solution is developed from Ivantsov's treatment[108] and includes the following assumptions:[106,109]

1. The growth of the plate maintains a constant shape, i.e., that of a parabolic cylinder. The steady-state shape of the interface near the growing tip corresponds to the parabolic cylinder for the plate morphology.

2. The elastic strain energy and anisotropy of interface properties may be ignored.

3. The concentration of solute in the parent phase is such that the theory of capillarity applicable to dilute solution can be employed.

4. The diffusivity is concentration-invariant.

5. This is valid provided the concentration variation along the interface is not very large so that the plate shape does not deviate much from that of a parabola.

Trivedi obtained the following result of plate thickening:

$$\Omega_0 = (\pi p)^{1/2} e^p \operatorname{erf}(p^{1/2}) \left[1 + \frac{G_l}{G_c} \Omega_0 S_1(p) + \frac{r*}{r} \Omega_0 S_2(p) \right] \qquad (7.34)$$

where $p = G_l r/2D$ (a dimensionless quantity, called the *Peclet number*), D is the diffusion coefficient of solute in the matrix, $S_1(p)$ and $S_2(p)$ are mathematically complicated functions of p and are presented graphically, G_l is the lengthening rate of a plate, G_c is the lengthening rate provided that growth was completely controlled by interfacial reaction, r is the radius of curvature at the tip of the plate, $r* =$ radius at which growth stops by capillarity, and the other terms have the usual meanings. The right-hand side of Eq. (7.34) has three terms: The first term is the result due to Ivantsov for the case of isoconcentrate interface, and the second and third terms are correction factors to Ivantsov's solution due to interface kinetics and capillarity effects, respectively, which grow larger with the decrease of supersaturation.[110]

Equation (7.34) holds for the case of isotropic energy but does not hold directly for the anisotropic case of Widmanstätten plate.[111] This equation, however, can be simplified to low-to-medium values of the supersaturation Ω_0 by the approximation due to Bosze and Trivedi.[112] The growth rate equation can now be obtained by using the definition of the Peclet number as

$$G = \frac{8}{9\pi} \frac{D}{r} \Omega_{BT}^2 = \frac{27D}{256\pi r*} \Omega_{BT}^3 \qquad (7.35)$$

and

$$p = \frac{9}{16\pi} \Omega_{BT}^2 \qquad (7.36)$$

The Bosze and Trivedi supersaturation Ω_{BT} is related to the supersaturation Ω_0 by

$$\Omega_{BT} = \frac{\Omega_0}{1 - (2\Omega_0/\pi) - (\Omega_0^2/2\pi)} \qquad (7.37)$$

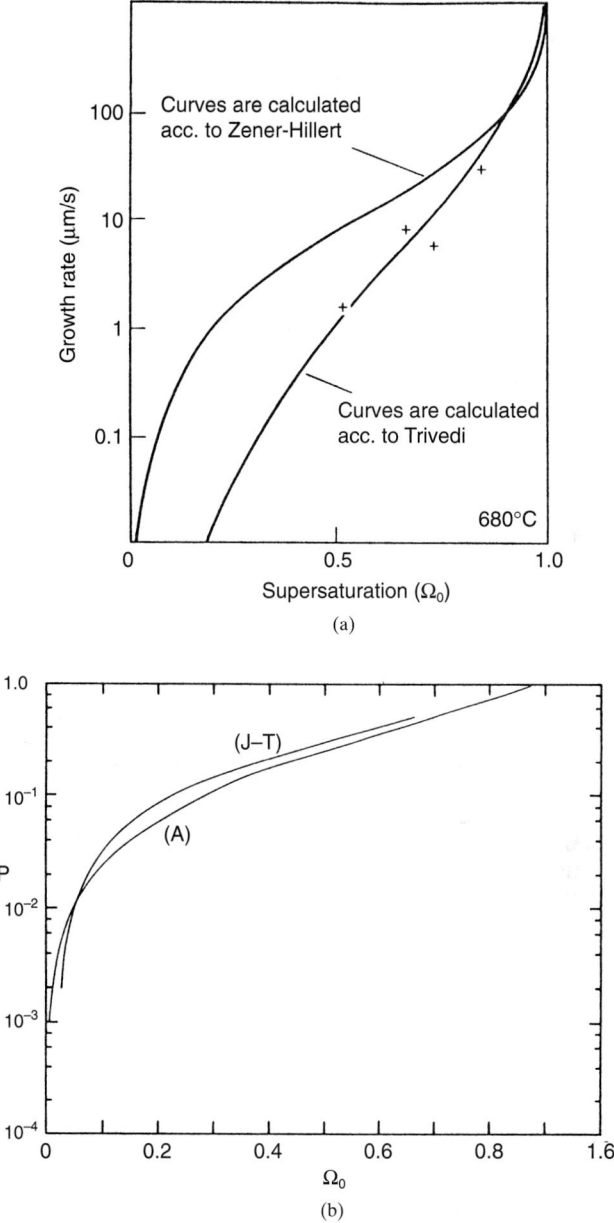

FIGURE 7.18   (*a*) Comparison between experimental lengthening rates of the edges of Widmanstätten plates and the calculated lengthening rates using the Hillert and Trivedi growth equations.[105,106] (*b*) Relationship between p and Ω_0 for the growth of an isolated ledge, as predicted by Jones and Trivedi (J-T) and Atkinson (A).[106, 115–116] (*Reprinted by permission of The Metallurgical Society, Warrendale, Pennsylvania.*)

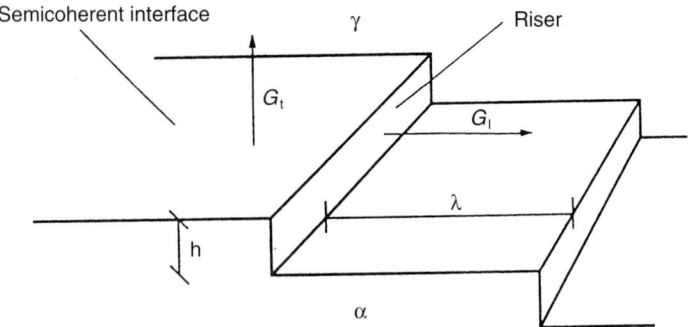

FIGURE 7.19. Schematic illustration of a growth ledge on a semicoherent interface.

where r^* is the critical radius of nucleation. Hillert[111] has shown that Eq. (7.35) holds well with the observed rate of linear lengthening of Widmanstätten α-iron plates in Fe-C alloys. Purdy[113] demonstrated similar agreement for the needlelike α-brass precipitates forming from β-brass in Cu-Zn alloys. In both situations, the precipitates grow from grain boundary nucleation sites because, for the small compositional differences between the two phases, inadequate driving force exists for nucleation within grains.

Thickening Kinetics of Plates. The growth ledge mechanism involving even kinks was introduced to describe the thickening rates or kinetics α_w plates. It is well agreed that the broad faces of a_w plates are partially coherent with the γ matrix. The migration of such interfaces must advance by a ledge mechanism, as shown in Fig. 7.19. The movement of growth ledges occurs by atom additions at the riser of the ledge. It is usually thought that the thickening of α_w plates arises from the movement of growth ledges across the partially coherent interface. The dependence of the thickening rate of Widmanstätten plates G_t on the lateral migration rate of ledges (i.e., broad faces of the ledges are considered entirely immobile and the risers of the ledges are mobile) was first proposed by Aaronson[73] and is expressed as:[114]

$$G_t = \frac{hG_l}{\lambda} \tag{7.38}$$

where h is the ledge height, G_l is the lateral velocity of a ledge, and λ is the interledge spacing. Jones and Trivedi[115] and Atkinson[116] used different assumptions and mathematical methods to explain the diffusion-controlled kinetics of ledge movement. There are important differences in the concentration profiles around the ledges as estimated by the two approaches.[106] Later they extended their treatments to incorporate the migration of multiple noninteracting and interacting ledges.[106,115,116] The following relation was obtained initially by Jones and Trivedi, using the flux balance equation

$$G_l = \frac{D\Omega_0}{\alpha\lambda} \tag{7.39}$$

where D is the averaged (carbon) diffusivity in the austenite matrix, and a is the effective diffusion distance ahead of the riser, which is given by the relationship

$$\Omega_0 = 2p\alpha \tag{7.40}$$

Atkinson's detailed and final treatment for the growth of a single ledge is represented by the following equation for $p < 0.1$:

$$\Omega_0 = \frac{2p}{\pi}(2.954 - \ln p) \tag{7.41}$$

Figure 7.18b compares the relationship between p and Ω_0 obtained by Jones and Trivedi (JT) and Atkinson (A). The differences between the two results are negligibly small; however, Atkinson's analysis provides faster (2 to 3 times) growth rates at low p values and remarkably smaller lengthening rate at high supersaturations than those obtained from the Jones-Trivedi treatment.[103,115]

Recent finite difference computer simulations of the migration of single ledges have provided more insights about the migration kinetics of the ledge growth mechanism during $\gamma \rightarrow \alpha + \gamma$ transformation which suggest that the ledge mechanism may also play a vital role in controlling the morphological evolution of ferrite plates during growth.[92]

Experimental evidence of diffusional growth by ledge mechanism has been found by many workers in nonferrous systems such as for γAl-Ag$_2$ plates in Al-Ag alloy, θ' Al-Cu plates, and α reactions in Ti-Cr and Ti-Fe alloys; and in ferrous systems such as α Fe-C, Fe-C-Si and Fe-C-Ni alloys in a wide range of reaction temperatures and supersaturation.[103,117–119]

7.6.2.3 Experimental Procedure to Determine Growth Rate. The experimental methods used to measure the lengthening kinetics of Widmanstätten plates are similar to those described earlier for pearlite and allotriomorphs. The thickening kinetics of Widmanstätten sideplates have been found to be highly variable.[120] The problem in measuring is much more difficult, especially when TEM is not used.

7.6.3 Massive Ferrite

7.6.3.1 General Aspects of Massive Ferrite. Massive transformation occurs during rapid cooling and involves a change in crystal structure, similar composition of product and parent phases, and lack of any definite crystallographic orientation relationship between them.[121,122] Usually, the massive transformation is observed if the competing equilibrium and other (notably, Widmanstätten) transformation modes are subdued. As a rule, it can occur in any system where: (1) the relevant phase fields overlap in composition and (2) a metastable two-phase field exists below an invariant reaction (e.g., eutectoid) temperature.[123–125] Growth is accomplished primarily by noncooperative (random) transfer of atoms across the relatively high-energy incoherent product/matrix interfaces. As a result of these features, once nucleated, the product phase grows at high velocities (up to 1 to 2 cm/s) with approximately the same rate in all directions,[126] consuming much of a parent grain, or sometimes several grains of the parent phase. This high growth rate puts the massive transformation in a position between diffusional transformation occurring at very slow rates (Å/sec) and the diffusionless martensitic transformation occurring at a velocity approaching the speed of sound ($\sim 10^5$ cm/s) (Fig. 7.20).[127,128]

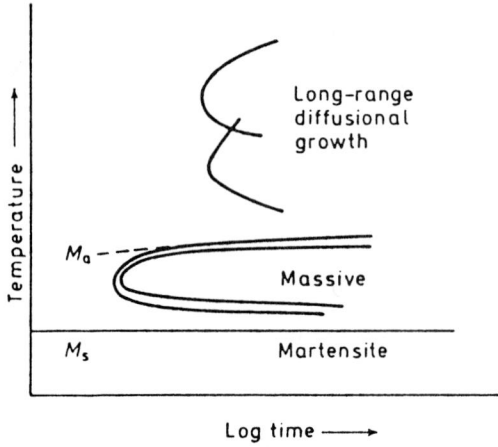

FIGURE 7.20 Schematic α TTT diagram illustrating the region of massive C-curve in between diffusional and martensitic transformations.[127,128] (*Reprinted by permission of Butterworths, London.*)

Another important feature of massive transformation is that transformation temperature displays a "plateau" behavior (Fig. 7.21) when initial transformation temperature (i.e., thermal arrest temperature measured in a continuous cooling experiment) is plotted against the different cooling rate for high-purity iron, mild steel, and nickel steel.[128,129] However, some researchers have discounted this as a fundamental feature.[121]

Evidence of both the presence and the absence of definite crystallographic relationships between the massive product and parent phase exists.[123,124] In these situations, it has been debated that massive transformation must always involve special crystallographic relations (between the product and parent), resulting in coherent nucleation followed by growth, where the interface changes from coherent to semicoherent and moves by a ledge mechanism.[124]

Figure 7.22 shows a free-energy-composition curve illustrating the thermodynamic condition for a massive transformation of β to α phase by quenching from a higher-temperature stable β phase. If the transformation temperature lies below the transition T_0 in the region II (i.e., with faster cooling rate than any reaction which produces long-range diffusion) and the matrix composition is lower than the critical concentration $C(T_0)$ corresponding to the temperature T_0, the free energy of the system can be lowered and the β phase may transform directly to α (that is, α_m) of the same composition. However, if the matrix composition exceeds the critical concentration $C(T_0)$ in the region I of the two-phase field, it will transform only to solute-depleted α phase.[130] However, Hillert has noted that massive transformations follow the line starting at the solvus point rather than at the line starting at the T_0 point.[126] That is, massive transformation requires undercooling below the solvus line.

Massive transformations are thermally activated processes that exhibit diffusional nucleation and growth characteristics. The kinetics of these transformations are controlled mainly by (1) interfacial diffusion rather than volume diffusion and (2) other interface features such as lack of coherency between the product and parent phases,[122,127] which permit the parent-phase grain boundary crossing. It applies to nonmartensitic composition-invariant reactions, which involve short-range diffusion at the disordered transformation interface. This transformation does not seem to involve invariant plane-strain surface-relief effects.[131] However, further

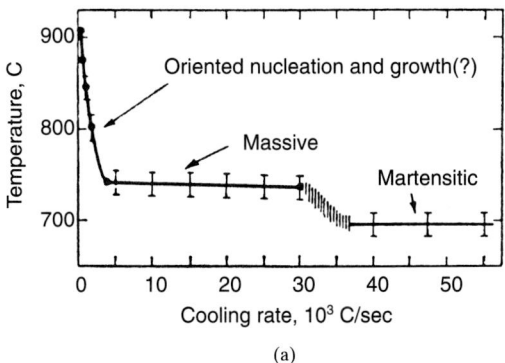

(a)

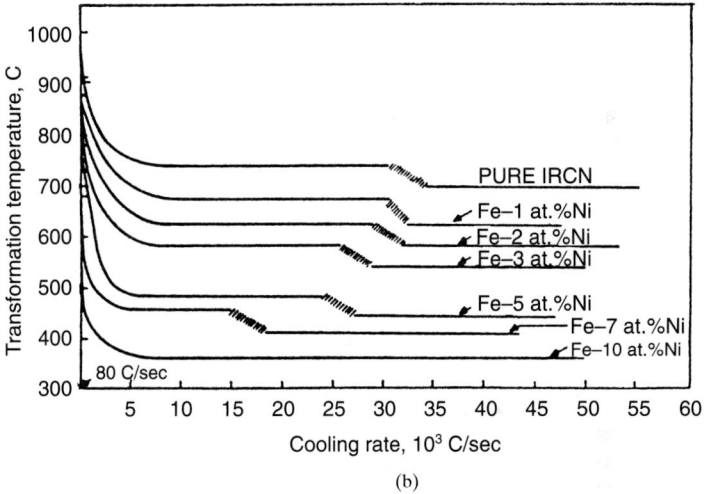

(b)

FIGURE 7.21 Variation of transformation temperature with cooling rates: (*a*) pure iron[128] and (*b*) iron and Fe-Ni alloys.[129] [(*a*) *Reprinted by permission of ASM International, Materials Park, Ohio;* (*b*) *reprinted by permission of The Institute of Metals, England.*]

study of surface-relief effects associated with the massive transformations is necessary in alloys, where it is possible to control the rapid growth rate and produce a pronounced anisotropic massive morphology.[132]

Massive transformations occur in many ferrous and nonferrous systems during quenching within the single-phase field or two-phase field; in the latter, at least one phase must be metastable and can form simultaneously, provided that it has the same composition as the parent phase but a different crystal structure.[131,133,134] Figure 7.23 shows schematic phase diagrams for pure metals and three types of alloys, illustrating phase relations which are essential for the occurrence of massive transformation.[122] Table 7.3 lists typical massive transformations for pure metals and binary alloys.[122] The composition in Fig. 7.23*a* corresponds to a high-purity metal and dilute

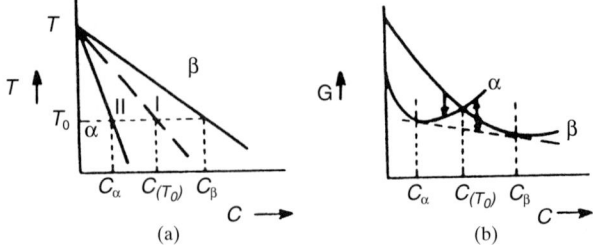

FIGURE 7.22 (*a*) Phase diagram illustrating transition temperature T_0 below equilibrium temperature T_E. (*b*) Free-energy–composition diagram showing the thermodynamic condition for a massive or a diffusionless transformation of β to α phase in region II. In region I only, a reaction to give solute-depleted α can decrease the free energy.[130] (*Reprinted by permission of North-Holland Physics Publishing, Amsterdam.*)

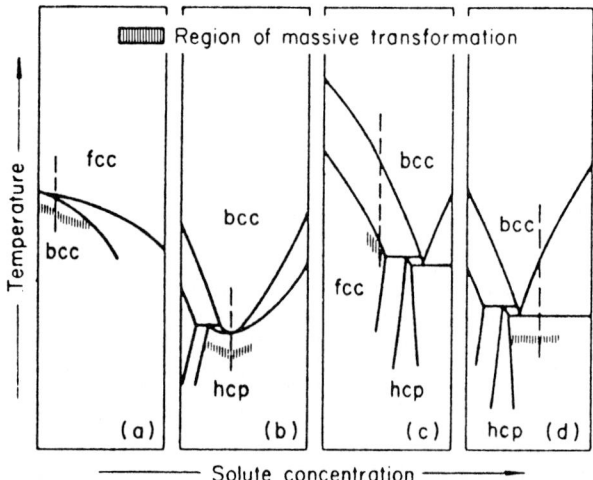

FIGURE 7.23 Schematic phase diagram for (*a*) pure metal and (*b–d*) three types of alloys, all of which may undergo massive transformations. Critical compositions are indicated by the dashed vertical lines.[122] (*Reprinted by permission of ASM International, Materials Park, Ohio.*)

binary alloy exhibiting a polymorphic transformation. In this case the long-range diffusion of the impurity and solute partitioning corresponding to the critical composition line do not occur during rapid cooling through the two-phase region, which facilitates the massive transformation. Examples are iron, low-carbon steels, and low-nickel steels. Figure 7.24*a* shows the resulting microstructure in Fe-0.002% C alloy. The critical composition in Fig. 7.23*b* corresponds to an alloy in which the bcc phase transforms to massive hcp phase on cooling through a congruent point (a junction of two-phase fields). It is similar, in several ways, to a polymorphic element. Examples are

TABLE 7.3 Typical Massive Transformations for Pure Metals and Binary Alloys[121]

Alloy system or metal	Amount of solute at which transformation occurs (at%)[†]	Temperature during quenching at which transformation occurs[†]		Change in crystal structure[‡]
		°C	°F	
Silver-aluminum	23–28	600	1110	bcc → hcp
Silver-cadmium	41–42	300–450	570–840	bcc → fcc
	50	300	570	bcc → hcp
Silver-zinc	37–40	250–350	480–660	bcc → fcc
Copper-aluminum	19	550	1020	bcc → fcc
Copper-zinc	37–38	400–500	750–930	bcc → fcc
Copper-gallium	21–27	580	1075	bcc ⇌ hcp
	20	600	1110	bcc → fcc
Iron	—	700	1290	fcc → bcc
Iron-cobalt	0–25	650–800	1200–1470	fcc → bcc
Iron-chromium	0–10	600–800	1110–1470	fcc → bcc
Iron-nickel	0–6	500–700	930–1290	fcc → bcc
Plutonium-zirconium	5–45	450	840	bcc → fcc

[†] Values listed are approximate.
[‡] bcc, body-centered cubic; fcc, face-centered cubic; hcp, hexagonal close-packed.
Reprinted by permission of ASM International, Materials Park, Ohio.

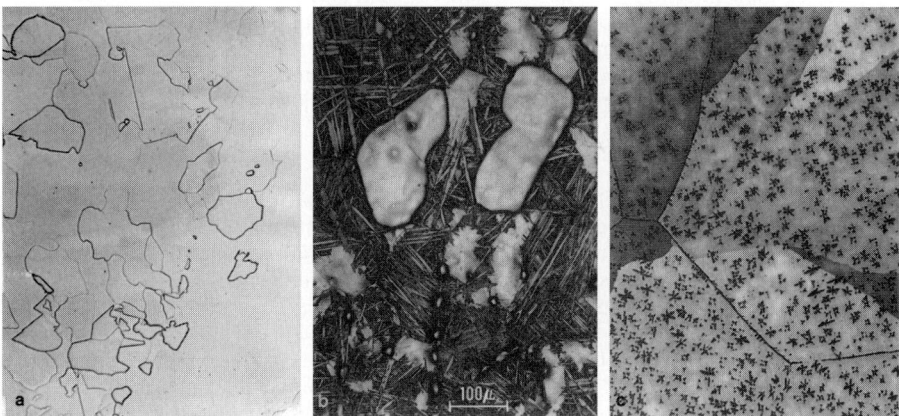

FIGURE 7.24 (a) Massive ferrite with irregular grains in Fe-0.002% C alloy after quenching in iced brine from an austenitizing temperature of 1000°C (1832°F), 385X (reduced 57%). (b) Partial $\beta \to \alpha$ massive transformation in Cu-19.3 at % Al which initially consisted of an equilibrium two-phase mixture of α and β phases. (c) Massive hcp phase grains in a quenched Cu-18.4 Ga-5Ge (at %) alloy which are able to cross prior β (bcc) boundaries. 60X (reduced 57%).[122] (*Courtesy of T. B. Massalski.*)

1. β (bcc) $\rightarrow \xi_m$ (hcp) transformations occurring massively in Al-24.5 at% Ag alloy and Cu-23.7 at% Ga alloy

2. $\xi \rightarrow \xi'$ in Ag-27 at% Ga alloy.

The composition in Fig. 7.23c corresponds to the decomposition of the high-temperature bcc phase on cooling through the two-phase field with an entire or partial transformation to the massive fcc phase. Examples are β (bcc) $\rightarrow$ massive ξ_m (fcc) transformation in Cu-Ga rich in Cu and bcc $\beta \rightarrow$ massive α_m (fcc) in the Cu-Zn, Cu-Al, Ag-Cd, and Ag-Zn systems. Figure 7.24b is a partial, $\beta \rightarrow$ massive transformation in Cu-19.3 at% Al which initially consisted of an equilibrium two-phase mixture of α and β phases. The composition in Fig. 7.23d corresponds to an alloy where bcc phase transforms to massive hcp phase on quenching through the two-phase field. Examples are (1) β (bcc) $\rightarrow \alpha_m$ (hcp) transformation in Ti-Ag, Ti-Au, and Ti-Si eutectoid systems[135] and (2) β (bcc) $\rightarrow \xi_m$ (hcp) transformation in Al-26 at% Ag, Cu-Ga, and Ag-Cd systems. Figure 7.24c shows the massive hcp phase grains in a quenched Cu-18.4Ga-5Ge (at%) alloy which are able to cross prior β (bcc) boundaries.[122]

Mechanism. Based on experimental evidence, in some cases, and theoretical consideration, it may be concluded that, during the nucleation stage of a massive phase, some form of a rational (i.e., with low Miller indices) or nearly rational orientation relationship with one of the matrix crystals seems necessary, in order to lower the activation free-energy barrier for critical nucleus formation ΔG^*. This greatly increases the formation rate of critical nuclei to the extent that critical nuclei enclosed by a smaller proportion of low-energy interface cannot successfully compete and therefore are not experimentally observed.[136] In this situation, small areas of their interfaces can be coherent with the matrix crystal which adopts the least coherent growth mode, while the remaining areas of their interfaces, being disordered and incoherent, have high mobility for rapid massive growth.[124,132]

It appears that the movement of the incoherent boundary is hindered by solute drag. That is, the boundary is pinned by solutes at various points along its length, giving rise to irregular grains with ragged boundaries.[137]

7.6.3.2 Massive (Ferrite) Transformation in Ferrous Alloys.

Ackert and Paar[138] have shown that the temperature of massive transformation is sensitive to the interstitial concentration. For massive transformation in ultrapure iron, solute drag does not appear; however, the strain energy ΔG_ε due to volume change will exist.

The $\gamma \rightarrow \alpha$ transformation in pure iron and in low-carbon (with ~0.002 wt% C) and low-nickel steels can occur massively. The austenite is quenched with a higher cooling rate to avoid the transformation with long-range diffusion (near the equilibrium), but with a slow enough rate to avoid the diffusionless martensite-type transformation.[129,139] Massive ferrite can only occur below T_0 ($Ae_3 < T_0 < Ae_1$) but above M_s temperature. (see Fig. 7.22, where β is considered as a parent phase). Figure 7.24a shows the characteristic microstructural appearance of massive patches of ferrite with irregular boundaries. Figure 7.21a shows the plot of temperature versus cooling rate for high-purity iron; the graph exhibits an arrest plateau at about 740°C, identified with massive transformation at cooling rates between 5000 and 35,000°C/s, depending upon composition. Such a distinct plateau may be considered as a manifestation of a special nucleation event.[123] Another plateau is found at a lower temperature (~690°C) when the cooling rate exceeds ~35 × 10³°C/s, corresponding to the onset of martensitic transformation.[128] Figure 7.21b represents the results obtained by Swanson and Parr,[129] which demonstrate that both the higher

FIGURE 7.25 Optical micrograph of carbon steel quenched at 913 K during cooling from 1523 K. The arrow indicates a typical intragranular ferrite idiomorph. (*Reprinted by permission of The Metallurgical Society, Warrendale, Pennsylvania.*)

plateau massive temperature and critical cooling rates required to suppress the massive transformation are strongly dependent on the nickel content.[105] Their data indicate that the massive transformation does not occur beyond the Fe-7 at% Ni alloy, but Massalski et al. have reported the observation of massive transformation in Fe-8.7 at% Ni alloy.[140,141]

Later Hayzeldon and Cantor[142] observed massive ferrite in melt-spun Fe(up to 25 wt%)-Ni. They attributed this to be due to increased nucleation rate at many austenite grain boundaries and grain corners in the very fine austenite grains of size $3\,\mu$m. Recently Chong et al.[143] have reported the formation of both the massive ferrite and equiaxed ferrite in the two-phase field in Fe-9.14 wt% Ni alloy when furnace (or continuously) cooled from 1000°C to the temperature range of 575 to 558°C (i.e., below T_0 and A_3 temperatures). They tend to nucleate at grain corners and grain boundaries with at least one coherent boundary. TEM examination showed occasional massive ferrite grains with bulged boundaries of heavy dislocation density and equiaxed grains associated with variable dislocation density. They attributed the presence of paraequilibrium during transformation to massive ferrite and equiaxed ferrite.[143]

7.6.4 Idiomorphic Equiaxed Ferrite

Idiomorphic ferrite can form intragranularly (within the austenite grains) or at austenite grain boundaries at small undercooling (Figs. 7.14C and 7.25). These idiomorphs are thought to be nucleated heterogeneously at nonmetallic inclusions and dislocations. They exhibit a roughly equiaxed morphology.

In medium-carbon vanadium steels, intragranular ferrite idiomorphs nucleate at vanadium nitride precipitates which, in turn, form at MnS particles during cooling in the austenite matrix. It is observed that intraganular ferrite idiomorph has the Baker-Nutting orientation relationship (B-N OR) with VN which precipitated at MnS. In the B–N OR, the lattice mismatch across the $(001)_\alpha$ // $(001)_{VN}$ atomic habit planes is probably very small.[144]

7.7 PROEUTECTOID CEMENTITE

Proeutectoid cementite morphologies obtained on isothermally transforming hypereutectoid steels are quite similar to those discussed above for proeutectoid ferrite, which include grain boundary precipitates; long, thin Widmanstätten sideplates; intragranular Widmanstätten sideplates; and intragranular idiomorphs.

7.7.1 Cementite Morphology

Grain boundary allotriomorphs usually form at high temperatures in hypereutectoid carbon and alloy steels. The allotriomorphic cementite phases nucleate with a Pitsch relationship to one austenite grain and grow into the adjacent grain with which they have no orientation relationship. It seems that the interfacial structures of the allotriomorphic carbide phases in steels are similar to allotriomorphic ferrite and other interfaces in both steels and some nonferrous alloys.[48,89]

The volume fraction of these thin-film carbides decreases with increasing grain size and increasing cooling rate. In high-purity Fe-C alloys and plain carbon steels, Heckel and Paxton[145] found the thickening growth rates of cementite allotriomorphs that were an order of magnitude lower than the expected equilibrium thickness based on carbon-diffusion-controlled growth. The thickening kinetics were observed to be insensitive to grain size and transformation temperature. It was suggested that in high-purity Fe-C alloys the thickening of cementite allotriomorphs occurred presumably by a diffusion-interface-controlled growth mechanism and that in plain carbon steels the thickening took place by silicon-diffusion-controlled growth.[145,146]

The morphology of cementite sideplates is usually similar to that of ferrite sideplates. The primary cementite sideplates nucleate at only a few austenite grain boundaries. Most cementite sideplates are of the secondary type rather than of the primary type. Both primary and secondary sideplates form at twin boundaries. Decreasing the transformation temperature (1) increases the tendency to form cementite sideplates, until interrupted at lower temperature by the pearlite or bainite reaction (7.16b), (2) increases the average length/width ratio, and (3) decreases the average sideplate spacing. Increasing the austenite grain size under given conditions of carbon content and reaction temperature increases the tendency to form more cementite sideplates.

Intragranular cementite plates usually nucleate shortly after cementite sideplates; and, in this morphology, apart from the increase in the average length/width ratio, there is not much variation with the transformation temperature. Intragranular Widmanstätten cementite plates also occur at lower heat treatment temperatures and are observed to follow the Pitsch orientation relationship with the austenite matrix.[48] The main difference between sideplate and intragranular Widmanstätten cementites which form at lower temperature is that the former can develop from the grain boundary allotriomorphs, while the latter precipitates within γ grain by sympathetic nucleation and growth.[73,147]

In alloy steels (such as Fe-0.96C-12.85Mn-0.78Si steel), a sawtooth morphology of fcc $M_{23}C_6$ (M = Fe, Mn) carbide forms at the grain boundaries and exhibits a cube-cube orientation relationship with respect to one of the γ grains (A_2) into which it grows. These findings resemble the recent work on high-Mn steels. The formation of facets and macroscopic steps at the growth interface of the grain boundary carbide was also reported. Planar faults noticed in the carbide are recognized as stacking faults lying along {111} planes, and these are associated with linear features at the planar interfaces.[48]

7.8 FERRITE-PEARLITE AND PEARLITIC STEELS

The most popular microstructure in the context of structural steel has been a mixture of ferrite and pearlite. These steels are usually characterized by their higher strength compared to low-carbon steels in both normalized and heat-treated conditions. Here formulations of quantitative relationships between microstructural parameters and mechanical properties, particularly yield or flow stress σ_y, tensile or ultimate tensile strength σ_T, *impact transition temperature* (ITT), and *Charpy shelf energy* (CSE) of steels containing ferrite and pearlite are described. In the following equations, quantities of dimensions of length are expressed in millimeters, stress in megapascals (meganewtons per square meter), energy in joules, and temperature in degrees Celsius. Important applications including rail steels, wire rod steels, wire ropes, bridge ropes, spring wires, tire bead wire and tire cord, wire for prestressed concrete, cold heading wire, and microalloyed forging steels are briefly described.

7.8.1 Structure-Property Relationships

7.8.1.1 Ferrite in Low-Carbon Ferrite-Pearlite (or C-Mn) Steels. The typical empirical strengths and impact (or ductile-brittle) transition temperature of C-Mn steels up to 0.25%C are given by[148–150]

$$\sigma_y(\text{MPa}) = K + 37(\%\text{Mn}) + 83.2(\%\text{Si}) + 2918(\%\,\text{N}_f) + 15.1d^{-1/2} \qquad (7.42)$$

$$\sigma_T(\text{MPa}) = 294.1 + 27.7(\%\text{Mn}) + 83.2(\%\text{Si}) + 3.9(\%\text{ pearlite}) + 7.7d^{-1/2} \quad (7.43)$$

$$ITT(°\text{C}) = -19 + 44(\%\text{Si}) + 700(\%\text{N}_f)^{1/2} + 2.2(\%\text{ pearlite}) - 11.5d^{-1/2} \quad (7.44)$$

Equation (7.42) can be used to plot the yield strength as a function of ferrite grain size, as seen in Fig. 7.26a.[148]

In the above equations, K is 88 MPa for normalized steel and 62 MPa for fully annealed (i.e., furnace-cooled) steel; Mn and Si are elements dissolved substitutionally in the ferrite solid solution; N_f is the free nitrogen content dissolved interstitially in the ferrite lattice (i.e., not present as a stable nitride); and d is the average linear intercept in polygonal ferrite. Both Mn and Si increase the yield and tensile strengths by solid solution strengthening of ferrite. Pearlite does not contribute significantly to the yield strength unless it is present in large amounts; but carbon lowers the transformation temperature, thereby decreasing the ferrite grain size. Grain refinement increases yield strength and toughness (i.e., lower ITT) while the strengthening without the grain refinement still reduces the toughness. Thus, an increase in pearlite content increases the tensile strength and ITT and decreases the CSE value. Figure 7.27 shows the exponential decrease of upper shelf energy C_v with the increase in pearlite fraction.[155] All solid solution strengtheners except Mn and Ni raise the ITT. Mn and Ni lower the ITT by 30 and 13°C per wt%, respectively. The behavior of Al is different. In small amounts, it lowers ITT by eliminating nitrogen from the solution.

Similar equations for yield and tensile strengths have been listed by Baird and Preston.[151] However, one of the more sophisticated equations for yield strength, relating to microstructure for a plain C-Mn steel, is[152]

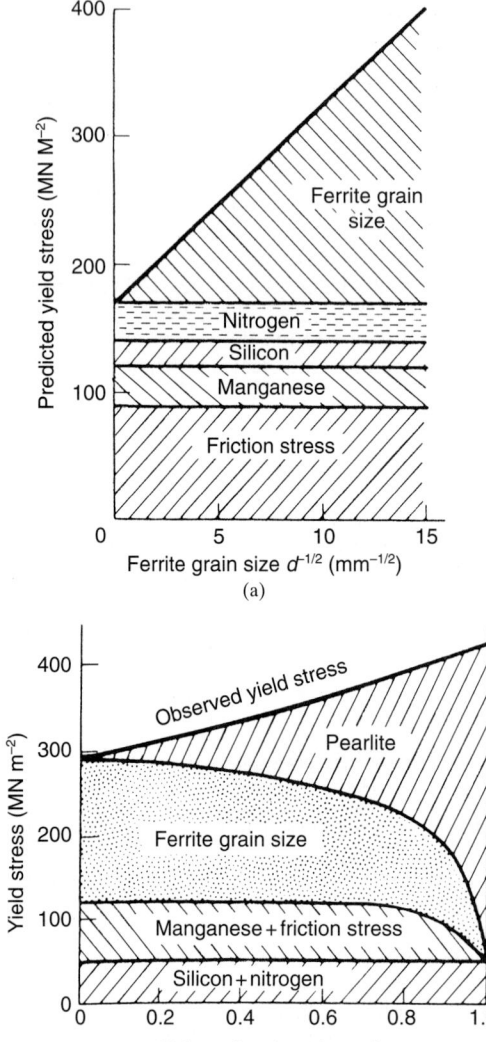

FIGURE 7.26 (*a*) Effect of ferrite grain size on the yield stress of normalized 0.2% C-1.0% Mn-0.2% Si-0.01% N steel.[148] (*b*) Contributions of various strengthening mechanisms to the yield stress of plain carbon steels.[148] (*Reprinted by permission of Pergamon Press, Plc.*)

$$\sigma_y(\text{MPa}) = 70 + 32(\%\,\text{Mn}) + 84(\%\,\text{Si}) + 680(\%\,\text{P}) - 30(\%\,\text{Cr}) + 33(\%\,\text{Ni})$$
$$+ 11(\%\,\text{Mo}) + 38(\%\,\text{Cu}) + 5000(\%\,\text{N}_f) + 18.1d^{-1/2} \qquad (7.45)$$

It is clear from the above expressions that the grain size of steel must be controlled if consistent mechanical properties are required.

7.8.1.2 Pearlite in Ferrite-Pearlite (C-Mn) Steels. Regression analysis on a wide range of medium- to high-carbon steels by Gladman et al.[72] has demonstrated that the yield strength σ_y, tensile strength σ_T, and impact transition temperature ITT can

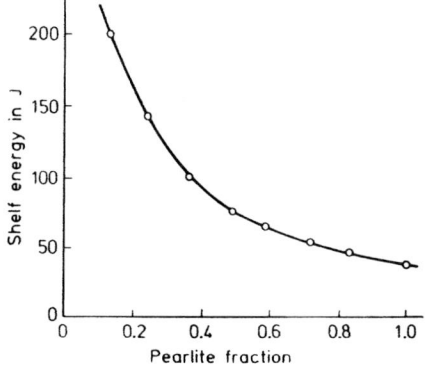

FIGURE 7.27   Effect of the pearlite fraction on the shelf energy in normalized ferrite-pearlite structures.[155] (*Reprinted by permission of VCH, Weinheim.*)

be related to the compositional and microstructural variation by $\sigma_y = f^n\sigma_\alpha + (1 - f^n)\sigma_p$, where σ_y is the yield stress of the aggregate (in megapascals), f is the volume fraction of ferrite, σ_α is the yield strength of the ferrite (in megapascals), σ_p is the yield strength of the pearlite (in megapascals), and n is an index representing the nonlinear contributions of the ferrite and pearlite.

Both yield strength and tensile strength showed an index of $n = \frac{1}{3}$, suggesting that the ferrite fraction contributes more to the yield and tensile strength than would be expected on a pro rata basis. On the other hand, the ITT showed an index of $n = 1$, suggesting a simple law of mixtures.[72]

The following yield strength σ_y and tensile strength σ_T relationships have been found to be applicable for ferrite-pearlite steels with pearlite content ranging between 20 and 100% (eutectoid composition):[32]

$$\sigma_y(\text{MPa}) = f_\alpha^{1/3}[35 + 58.5(\%\text{Mn}) + 17.4d^{-1/2}] + (1 - f_\alpha^{1/3})(178 + 3.8S^{-1/2})$$
$$+ 63.1(\%\text{Si}) + 425(\%\text{N}_f)^{1/2} \tag{7.46}$$

$$\sigma_T(\text{MPa}) = f_\alpha^{1/3}[246 + 1142(\%\text{N}_f)^{1/2} + 18.2d^{-1/2}] + (1 - f_\alpha^{1/3})$$
$$(719 + 3.56S^{-1/2}) + 97(\%\text{Si}) \tag{7.47}$$

where f_α is the volume fraction of ferrite, d is the average linear ferrite grain size intercept (in millimeters), and S is the pearlite interlamellar spacing (in millimeters). The index $n = \frac{1}{3}$ in the volume fraction σ_α in both expressions for σ_y and σ_T is used to show the nonlinear variation of yield and tensile strengths with ferrite and pearlite contents. This also suggests that the ferrite fraction contributes more to the yield and tensile strengths than would be expected on a pro rata proportion basis. As the volume fraction of pearlite increases, the influence of ferrite grain size on yield strength becomes progressively less significant (Fig. 7.26b). As the pearlite content approaches 100% (eutectoid composition), the pearlite becomes the main contributor to the strength; this is controlled by S. The contribution to the yield strength $\Delta\sigma_y$ from the interlamellar spacing can be predicted from the regression equation as:[153]

$$\Delta\sigma_y(\text{MPa}) = 3.86S^{-1/2}(1 - f_\alpha^{1/3}) \tag{7.48}$$

Increased yield strength can be achieved by an increased cooling rate from the austenitizing temperature so that pearlite forms at a lower temperature, resulting in smaller values of S. Also, for a fully pearlitic structure Eq. (7.46) reduces to

$$\sigma_{pe}(\text{MPa}) = 178 + 3.8 S^{-1/2} + 63(\%\text{Si}) + 425(\%\text{N}_f)^{1/2} \tag{7.49}$$

The contributions of other strengthening mechanisms may be added.

The notch impact transition temperature in ferrite-pearlite steels with increased pearlite volume fraction can be represented by the following regression equation:

$$\text{ITT}(°\text{C}) = f_\alpha(-46 - 11.5 d^{-1/2}) + (1 - f_\alpha)[-335 + 5.6 S^{-1/2} - 13.3 p^{-1/2}$$
$$+ (3.48 \times 10^6)t] + 49(\%\text{Si}) + 762(\%\text{N}_f)^{1/2} \tag{7.50}$$

where ITT is the Charpy V-notch 50% FATT ($°\text{C}$), p is the pearlite colony size (i.e., mean linear intercept) (in millimeters), and t is the pearlitic carbide lamellar thickness (in millimeters); other symbols remain the same as in Eq. (7.46). Since in most commercial steels $d^{-1/2}$ varies from 4 to 10 mm$^{-1/2}$, Eq. (7.50) illustrates that ITT decreases with the CSE (impact toughness) and decreases with the increased pearlite volume fraction. A small pearlite colony size lowers the ITT.[154] Again, other factors involving the influence of solid solution, precipitation strengthening, and dislocation strengthening on ITT may be added. Also, for a fully pearlitic structure Eq. (7.50) reduces to

$$\text{ITT}(°\text{C}) = -335 + 5.6 S^{-1/2} - 13.3 p^{-1/2} + (3.48 \times 10^6)t \tag{7.51}$$

again with requisite terms for other strengthening effects. Since the detailed pearlite morphology has a significant effect on *ITT*, we can incorporate a term for the prior γ grain size D. In this situation, the ITT for a pearlitic structure is given by Krauss, Hyzak, and Bernstein as

$$\text{ITT}(°\text{C}) = 218 - 0.83 p^{-1/2} - 2.98 D^{-1/2} \tag{7.52}$$

A decrease in interlamellar spacing adversely affects the CSE (impact toughness) because of the increased strength, but a decrease in pearlitic cementite plate thickness improves it. That is, S and t act in opposite directions (as shown in Fig. 7.28);[155] therefore, a compromise should be made for an optimum interlamellar spacing which will produce the best impact toughness (i.e., lowest ITT) properties.

7.8.2 Applications

7.8.2.1 Rail Steels. Since the advent of railways, rail steels have been challenged by the continuous increase of wheel loads on rails with the increasing speed and tonnage of traffic.[156] Rail steels should fulfill five major properties, viz., plastic deformation (including corrugations) resistance, wear resistance, fatigue (comprising both the rolling contact and the internally initiated type) resistance, residual stresses, and weldability.[157] Presently, high-volume rail steels have been based on fully pearlitic microstructures which are characterized by high strength, fatigue resistance, and fracture toughness. Failures in rail track are commonly linked to fatigue cracks of various types, while the rail life depends on the extent of wear and loss of the rail profile.[158]

The traditional methods for extended rail life include alloying and/or heat treatment and lubrication of the wheel/rail interface, especially the gauge face/flange

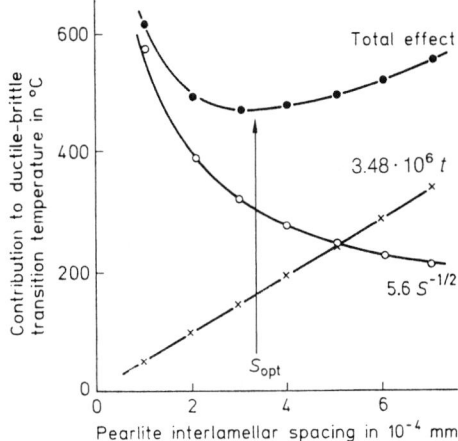

FIGURE 7.28 The contributions of the pearlite interlamellar spacing S and the carbide lamella thickness t to the ductile-brittle transition temperature, illustrating the occurrence of an optimum value of interphase lamellar spacing.[155] (*Courtesy of F. B. Pickering.*)

interface. Both strengthening and lubrication improve wear and deformation resistance while strengthening provides an additional advantage of enhancement in fatigue resistance.[159] However, rail life is determined in millions of gross tons (MGT) of traffic, and current rail life is in excess of 250 MGT. The average rail life is 70 years. It is extremely expensive to replace rails. The wear life of rail in curved track is so heavily dependent upon curvature and lubrication that it is meaningless to mention a life in MGT. However, in tangent track, rail life is mostly controlled by fatigue, and heavy-haul lines can yield rail service lives in excess of 1500 MGT.[160]

The standard axle load for most heavy-haul railroads is now 33 tons with a transition toward 39 tons. Additionally, significant roles of greater track utilization, better lubrication of the wheel/rail interface, and use of higher-strength rail steels have been emphasized. Consequently, the rail wear has decreased while surface-initiated *rolling contact fatigue* (RCF) has become more prominent. The combination of high axle loads and adequate lubrication of wheels and rails during curving is of special relevance.[161]

Wear resistance of a rail steel is directly related to both hardness and interlamellar spacing, as shown in Fig. 7.29. Thus, the pearlitic interlamellar spacing, which is a function solely of transformation temperature, becomes the most important microstructural parameter to control hardness and wear resistance.[162] A pearlitic microstructure with a finer interlamellar spacing and thinner cementite lamellae produces a higher wear resistance. Table 7.4 lists the values of the bulk average true interlamellar spacing S_t and hardness for four fully pearlitic rail steels. Clearly, the hardness increases with the decreasing S_t and, in fact, is a function of $S_t^{-1/2}$. In general, hardness is expressed in the form of the Hall-Petch relationship as:[163]

$$\text{Hardness}(\text{HV, kg/mm}^2) = 150(\text{kg/mm}^2) + 2.15(\text{kg/mm}^{3/2}) \times S_t^{-1/2}(\text{mm}^{-1/2}) \quad (7.53)$$

An increase in hardness increases both wear and fatigue resistance. Refinement of pearlitic interlamellar spacing is usually limited to a maximum hardness of about Rc 38 to 40. But fatigue resistance seems to improve with hardness as one changes from pearlitic microstructure below Rc 40 to bainitic microstructure above Rc 40.[160,161]

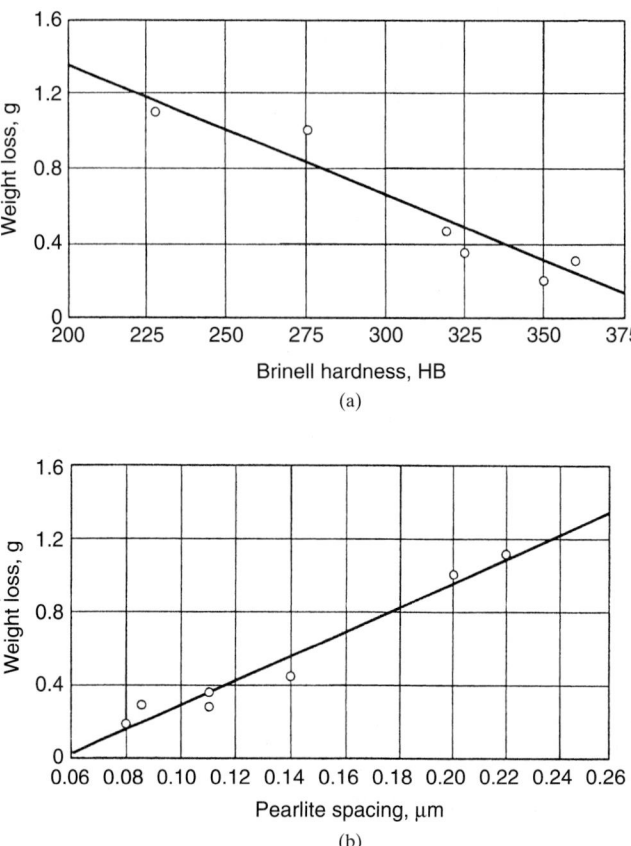

FIGURE 7.29 Relationships between (*a*) hardness and wear resistance (weight loss) for rail steels and (*b*) pearlite interlamellar spacing and wear resistance (wt% loss) for rail steels.[162] (*Courtesy of B. L. Bramfitt.*)

The desire to avoid premature failure, to increase resistance to surface damage, and to meet the need for improvement in straightness and line geometry has led to the requirement of tighter control for chemistry, inclusion types, segregation, blister, pipe, hydrogen, and surface condition of rails. Rails containing these improvements would be flatter, straighter, longer, and harder.[164]

Head Hardening of Rail.   To provide a rail steel with the highest hardness and wear resistance, a head-hardening process is commonly practiced which is off-line reheating process or in-line heat treatment. Off-line head hardening is a two-stage controlled induction heating process (comprising preheating and reaustenitizing the rails after rolling and cooling to room temperature) followed by simply an accelerated cooling process using forced-air cooling. This process is slow and costly (when compared to the in-line process). The in-line head hardening utilizes the rail exiting the rolling mill and controlled accelerated cooling based on water spray system, air blasting, or an aqueous polymer solution to produce a fine pearlitic microstructure

TABLE 7.4 Values of the Mean True Interlamellar Spacing and Hardness for Rail Steels[163]

Steel	Hardness (HV 10 ± 5%)	$\bar{S}_t$ (nm)[†]	$\bar{S}_t^{-1/2}$ (mm$^{-1/2}$)[‡]
1	276	226 ± 20	66
1	280	200 ± 20	71
1	285	204 ± 20	70
2	306	185 ± 13	74
2	316	163 ± 15	78
2	325	178 ± 15	75
2	330	123 ± 13	90
3	330	164 ± 10	78
2	335	145 ± 10	83
2	342	111 ± 12	95
3	342	145 ± 10	83
3	350	126 ± 10	89
2	360	109 ± 13	96
4	360	142 ± 10	84
3	365	118 ± 10	92
3	368	87 ± 12	107
4	370	92 ± 8	104
4	378	106 ± 5	97
3	380	77 ± 10	114
4	390	84 ± 5	109
4	398	77 ± 5	114
4	405	68 ± 5	121

[†] $\bar{S}_t = 0.5\bar{S}_t$ is the mean true interlamellar spacing.
[‡] $\bar{S}_t^{-1/2}$ is the reciprocal root of the true interlamellar spacing.

in the rail.[164–167] In this process, the head of the as-rolled rail is held (to about 1000°C) for several minutes and is stood head up. Only the head of the rail is allowed to develop a fully, fine pearlitic microstructure with the highest possible hardness (up to 425 HV at the near-surface position at the top and side of the head, down to 280 HV in the more slowly cooled parts of the head, web, and flange with coarse pearlitic microstructure) and wear resistance. This is accomplished, in both processes, by rapidly cooling the rail from the austenite condition to fully transform it to pearlite: (1) at 550 to 600°C (930 to 1100°F) for which a cooling rate up to 1100°C/min is needed or (2) by interrupted cooling at 570°C (i.e., at or near the pearlite transformation temperature), according to the continuous cooling transformation (CCT) diagram.[168,169] Figure 7.30a shows the CCT diagram of a rail steel with a cooling rate of 900°C/min (the heavy dashed cooling curve in Fig. 7.30b) with the resulting microstructure consisting of a mixture of pearlite, bainite, and martensite (which causes less than optimum wear properties). Figure 7.30b shows the same CCT diagram illustrating interrupted cooling at about 570°C producing 100% pearlitic microstructure.[166] The head-hardened rail thus has a fine pearlite structure, high toughness, and high-wear-resistance head, while the web and flange have the comparable toughness characteristics but lower strength than the standard grades of rail steel. The advantage of head hardening over full hardening is that the rail can be bent, roller-straightened, and drilled more readily.[160]

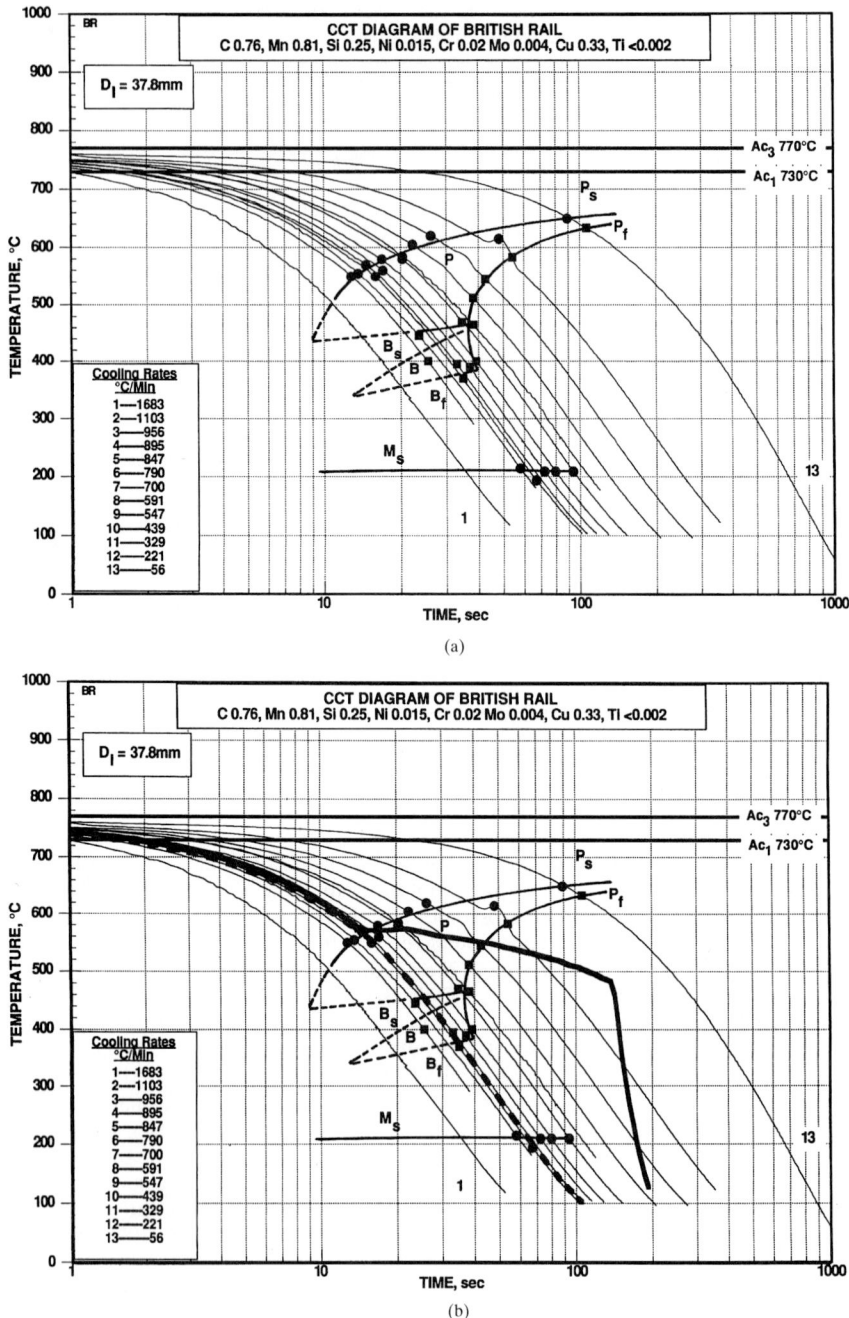

FIGURE 7.30 (*a*) Continuous cooling transformation (CCT) diagram of a rail steel and (*b*) same CCT diagram showing uninterrupted (dashed line) and interrupted (solid line) cooling paths.[166] (*Courtesy of B. L. Bramfitt, R. L. Cross, and D. P. Wirick.*)

TABLE 7.5 Rail Steel Grades—Chemical Compositions and Tensile Strengths[158]

Steel grades	Chemical composition (wt%)							Ultimate tensile strength (MPa)
	C	Mn	Si	S (max.)	P (max.)	Cr	V	
Standard								
UIC 700 (1986)	0.40–0.60	0.80–1.25	0.05–0.35	0.050	0.050	—		680–730
BS 11 (1978)	0.45–0.60	0.95–1.25	0.05–0.35	0.050	0.050	—	—	710 (min.)
Wear-resistant								
UIC 900 A (1986)	0.60–0.80	0.80–1.30	0.10–0.50	0.040	0.040	—	—	880–1030
UIC 900 B (1986)	0.55–0.75	1.30–1.70	0.10–0.50	0.040	0.040	—	—	880–1030
BS 11 A (1978)	0.65–0.75	0.80–1.30	0.05–0.50	0.050	0.050	—	—	880 (min.)
BS 11 B (1978)	0.50–0.70	1.30–1.70	0.05–0.50	0.050	0.050	—	—	880 (min.)
Highly wear-resistant— *special alloyed (typical analysis and strength)*								
S 1000	0.75	1.00	0.90	—	—	—	—	980–1120
S 1100	0.75	1.00	0.70	—	—	1.00	—	1080–1220
S 1200	0.75	1.00	0.80	—	—	1.00	0.10	1180–1280
Highly wear-resistant— *heat-treated (typical analysis and strength)*								
Heat-treated	0.75	0.90	0.25	—	—	—	—	1120–1320

Rail Steel Specifications.   Modern rail steels currently used in Europe and Asia are listed in Table 7.5, and broadly classified into four types according to tensile strength:[158, 170]

1. Standard grades, ~700-MPa minimum tensile strength

2. Wear-resistant grades, 880-MPa minimum tensile strength

3. Highly wear-resistant grades, 1080 to 1200 MPa

4. Other heat-treated high-strength grades, 1120 to 1320 MPa

The tensile strength of these steels can be related to the chemical composition by an empirical relation of the form, due to Kouwenhoven, which is different from Eq. (7.47):

$$\sigma_T(\text{MPa}) = 227 + 803(\%\text{C}) + 87(\%\text{Si}) + 115(\%\text{Mn}) + 133(\%\text{Cr})$$
$$+ 891(\%\text{P}) + 614(\%\text{V}) \tag{7.54}$$

Equation (7.54) is useful in predicting the effects of alloy additions on the tensile strength of typical rail steel products cooled within a specified range of natural air cooling rates.

1. Standard grades. Typical examples of standard grades include BS 11:1985, Normal and UIC Grade 70. These are high-tonnage grades which are employed in normal service conditions in traditional railways, including high-speed passenger traffic (200 km/hr) and medium-speed (100 km/hr), relatively heavy-axle-load (25-ton) freight.

2. Wear-resistant grades. The hardness and wear resistance of pearlitic steels are increased by refining the pearlite lamellae. This is carried out by increasing the C and Mn concentrations, both of which decrease the transformation temperature from austenite to pearlite. These wear-resisting grades are employed for heavy-axle loads, high-density traffic routes, or tightly curved track. However, the use of wear-resisting rails on conventional railways can also engender economic benefits.

3. High-strength grades. For extremely difficult service conditions faced in tightly curved track and under very high axle loads, even higher strengths and further refinement of the pearlitic structure with satisfactory level of hardenability and weldability are needed. This is achieved by head hardening of rails. Increasing the cooling rate of the railhead results in a refinement of the interlamellar spacing of the pearlite and an increase in tensile strength.

4. Other heat-treated grades. These grades include Fe-0.75C-0.90Mn-0.25Si, more alloyed steels such as Fe-0.7C-1.25Mn-0.1Si-1.0Cr, and CrMo steel (Fe-0.71C-0.59Mn-0.41Si-0.57Cr-0.21Mo), and microalloyed rail steels containing V or Nb to achieve 1035 to 1140 MPa (or 170 ksi minimum) of tensile strength.[170] Austenitic manganese (or Hadfield's) steel containing 1.2% C and 12% Mn is used in railway trackwork at frogs, switches, and crossings, where wheel impacts at intersections are very severe.[170]

American Rail Steel Specification. Table 7.6 lists the chemical composition of the standard and alloy (or premium) rail steels according to the AREA 1996 specification. The standard rail has the minimum 300 BHN hardness and minimum

TABLE 7.6   Chemical Analysis of Rail Steel (AREA: 1996)[171]

Element	Chemical analysis, weight percent		Product analysis, weight percent allowance beyond limits of specified chemical analysis	
	Minimum	Maximum	Under minimum	Over maximum
Carbon	0.72	0.82	0.04	0.04
Manganese	0.80	1.10	0.06	0.06
	(Note 1)	(Note 1)		
Phosphorus	—	0.035	—	0.008
Sulfur	—	0.037	—	0.008
Silicon	0.10	0.60	0.02	0.05
Nickel		(Note 1)		
Chromium		(Note 1)		
Molybdenum		(Note 1)		
Vanadium		(Note 1)		

(Note 1): The manganese and residual element limits may be varied by the manufacturer to meet the mechanical property requirements as follows:

Manganese		Nickel	Chromium	Molybdenum	Vanadium
Minimum	Maximum	Maximum	Maximum	Maximum	Maximum
0.60	0.79	0.25	0.50	0.10	0.03
1.11	1.25	0.25	0.25	0.10	0.05

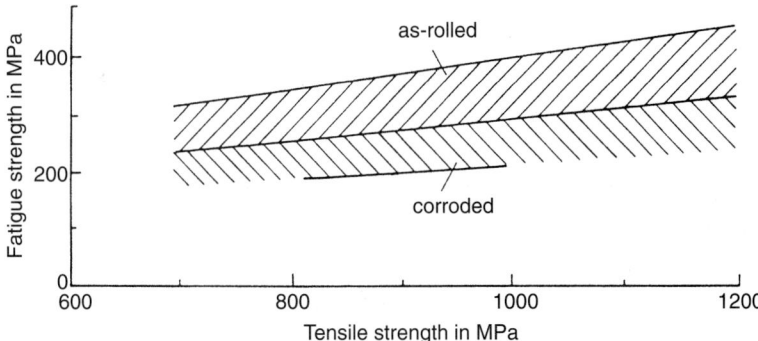

FIGURE 7.31 The relationship between bending fatigue strength and the tensile strength of a rail steel.[158] (*Courtesy of T. Gladman.*)

140 ksi tensile strength which lies between wear-resistant grade 2 and highly wear-resistant grade 3 of Table 7.5. The high-strength, Cr-Mo-V alloy and heat-treated rail steel lies between highly wear-resistant grade 3 and other heat-treated grade 4 of Table 7.5 and has the 341 to 388 BHN range.[171]

Recent tests of North American rail suggest that the ratio of endurance strength to tensile strength is greater than 0.4 at the high strength and contrary to the results shown in Fig. 7.31. Life tests indicate that fatigue life is proportional to hardness raised to about the fifth power. The endurance strength of electric flash butt welds approaches that of rail, and the North American welding practices would yield a much higher endurance strength of about 325 MPa (51 ksi) as observed by G. Fowler in 1976.[160]

To improve fatigue life, cleaner steels (fewer aluminum- or silicon-based oxide inclusions) are required but fracture toughness could not be improved sufficiently in a fully pearlitic steel even with finer interlamellar spacings. To accomplish this, a change from a fully pearlitic microstructure to a bainitic microstructure was made. Rail producers around the world are now studying high-strength bainitic rail steels. A German rail manufacturer has an experimental steel rail in track. Note that over the last 15 years, the introduction of clean steel practices has vastly improved the metallurgical cleanliness (both oxides and sulfides) of rail steels—and this has increased the fatigue (particularly internal fatigue) performance of rails. Usually oxide inclusions promote the development of shells (internally) which then cause a crack to turn from a transversal into a detail fracture, while sulfide inclusions favor dry wear and, in combination with oxide, play a vital role in the development of vertical split head cracks.[160]

In general, tougher rail being expensive, North American rail steels need not be very tough; 30 to 35 ksi√in. is usually adequate. Toughness is a concern if it involves roller-straightened chromium rails (German-made) where toughness is down near 25 ksi√in. These rails are prone to catastrophic web cracking.[160]

Rail Wear. A typical rail profile exhibiting the main regions of rail wear is shown in Fig. 7.32. The rail wear can be described in terms of lateral wear, vertical wear, or the area of rail removed. With both lateral and vertical wear, the extent of wear is directly dependent on the accumulated tonnage supported. Particularly, the lateral wear can be related to the local track curvature, a small radius of curvature inducing a higher wear rate. However, an adequate scatter of wear rate has been

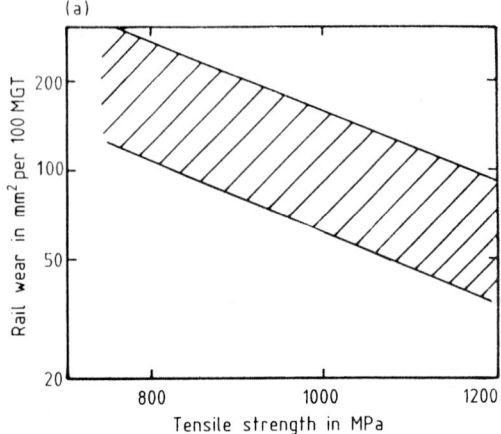

(a)

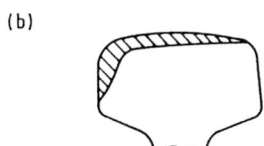

(b)

FIGURE 7.32 Rail wear exhibiting (*a*) the relationship between the tensile strength and rail wear in square millimeters per 100 million gross tons (MGT) of traffic and (*b*) a typical railhead wear profile.[158] (*Courtesy of T. Gladman; after C. Esvald.*)

observed from one rail position to another based on the local environmental and geometric features of the track, axle loads, and wheel flange lubrication.[172]

Dry wear rate of rail steel is a function of pearlitic interlamellar spacing, cementite lamellar fragmentation, MnS inclusion content, chemical composition, and strength parameters. Note that thin cementite lamellae are most ductile and remain relatively intact after deformation.[173] Sulfide inclusions are linked also to ductility exhaustion-related processes.[159]

The dry wear resistance capabilities of commercially available pearlitic rail steels have achieved a maximum limit with an interlamellar spacing near 1000 nm. But the wider use of effective lubrication and the introduction of bainitic steels in some cases can attain the wear demands placed on rail by heavier wheel loads.[174]

The wear rate can evidently be affected by the tensile strength (or hardness) of the rail material. For a given rail position, the rail head wear rate can decrease from 200 mm²/100 MGT for a rail with 900-MPa tensile strength to 40 mm²/100 MGT for a rail with 1200- to 1300-MPa tensile strength. It is a standard practice to use high-tensile-strength rails in tracks with a small radius of curvature (less than 200-m radius), e.g., in curves and turnouts and in track carrying high axle loads; in this manner, it is possible to acquire rail lives in these positions exceeding 500 MGT.[158]

Rail Corrugation. The formation of rail corrugations is considered to be a periodic wear process which is started by rail roughness and wheel/rail creepage. Current theories to describe the formation of corrugations fall into two categories: large creepage and small creepage. The large creepage is most desirable in heavily curved track; the small creepage is suitable for straight track and, therefore, is likely to be most important for future high-speed railways. Lateral creepage, which appears to be the most damaging form of creepage, is caused by the wheelset with an angle of yaw.[175]

Residual Stresses, Grinding, and Catastrophic Failure. Residual stresses in rails play a key role which affects the rail performance in modern railway service. Ground rails during service may develop transverse fatigue cracks (*detail fractures*) of unusual character.[176] [Note that not all shell cracks turn to detail fractures. Unground rails and rails ground to fit the wheel profiles (conformally ground) can also produce monoplanar shells, some of which turn into detail fractures.] Proper control of railhead profile at the contact interface by grinding is the most effective means to enhance the fatigue resistance of rail, but the optimum grinding rate tends to decrease with the increase of rail strength.[174] Residual stresses due to service and/or roller straightening establish crack propagation and fracture, leading to rail failure. Residual stresses due to wheel contact contribute to economic life limits (gauge corner spalling and subsurface shelling) when lubrication increases wear life and increased axle loads reduce fatigue life.[176]

The (thermal) residual stresses induced by rail welding act longitudinally and can reach 104 MPa (15 ksi) on cold days.[160] The use of welded track demands an additional longer stress cycle period (due to welding stresses).

Softer track (less stiff) support of the rail is beneficial, from a fatigue viewpoint, in that it increases base longitudinal tension stresses, but the improved compression stresses in the head are advantageous in preventing head-on-web bending stresses which lead to the development of bending tensile stresses at the bottom of the head when the wheel is directly over the stiff track. Additionally, the soft track minimizes the effects of dynamic loads. But the problem lies in the fact that one cannot produce the uniformity, consistency, and stability (or longevity) of, and track alignment with, the soft track; therefore, soft track is sought primarily to control track geometry.[160] Therefore, railroad attempts to attain modestly stiff track which is stable and uniform in character, even though impact loads will lead to higher stresses in the rail and the head on web bending, may cause the occurrence of longitudinal tensile stresses toward the bottom of the railhead directly under the wheel location.[160]

Fatigue Cracking. One of the major reasons for catastrophic rail failure is the development of *rolling contact fatigue* (RCF) cracks. The fatigue failure may be displayed in a number of different ways, such as *shelling* on the upper rail face (in a sharply curved track), *black spots*, *star cracking* from fish-plate bolt-holes, and *rail foot cracking*.

In Europe and for wheels in North America, shelling is a surface-initiated fatigue defect that usually develops in heavily deformed metal in the wheel loaded region of the running surface. In North America, for rails, shells are subsurface-initiated cracks that develop in undeformed metal—usually just at the boundary between the worked metal and the unworked metal; this is $1/8$ to $1/4$ in. beneath the surface.

Rail steels can (frequently but not always) turn to detail fractures which break rails. If internally initiated cracks (such as shells, vertical split heads, and horizontal split heads) are not considered as RCF defects, this catastrophic rail failure may not occur.

The problem of *rail-end bolt-hole cracking* (Fig. 7.33a) is usually eliminated, in Europe, by incorporation of residual stresses around the hole by introducing split sleeve cold expansion on existing tracks.[177] In North America, bolt-hole cracking is eliminated by replacing bolted rail with continuous welded rail. *Squat defect* (or *dark spot*) due to *surface initiated rolling contact fatigue* is prevalent in Europe and Japan and is characterized by transverse cracks (Fig. 7.33b) with the possibility of causing broken rails and train derailment. *Head checking* is another surface-

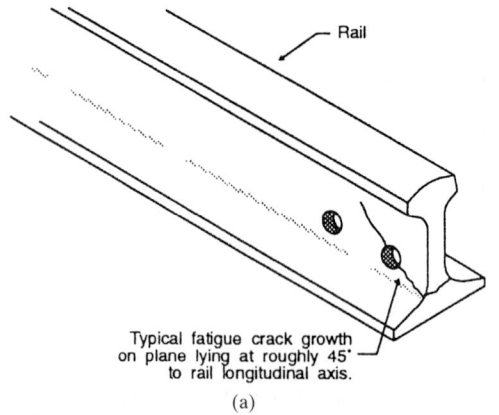

Typical fatigue crack growth
on plane lying at roughly 45°
to rail longitudinal axis.

(a)

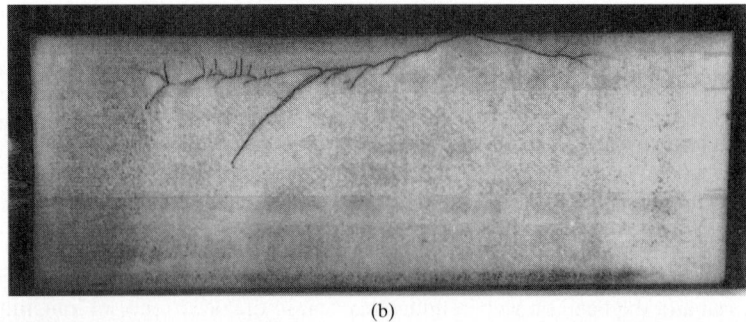

(b)

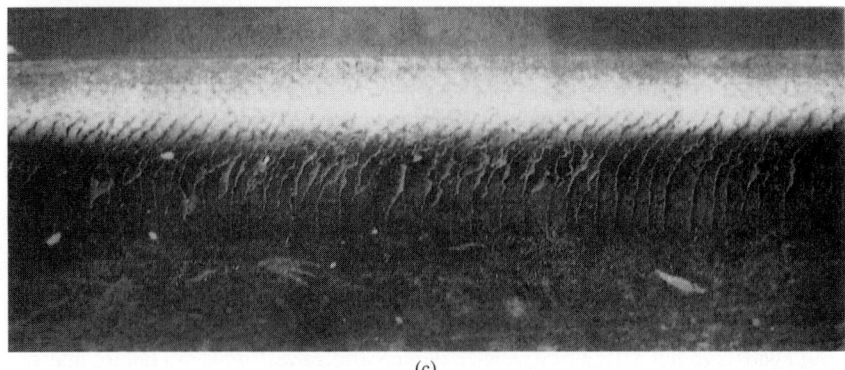

(c)

FIGURE 7.33 (*a*) Typical cracks originating at rail-end bolt-hole.[177](*b*) Vertical/longitudinal section through a squat-type rolling contact fatigue defect. Traffic from right to left. (*c*) Headcheck. Traffic from right to left. [(*b*) and (*c*) *After M. C. Dubourg and J. J. Kalker, in Rail Quality and Maintenance for Modern Railway Operation, International Conference, Delft, Netherlands, 1992, p. 374.*)

initiated RCF in steels which normally takes place on the rail gauge corner, especially on the outer rail in curved tracks (Fig. 7.33c). *Deep spalling* (to a depth of ~3 mm) due to crack-initiated surface fatigue from the white etching layers masks the details of fractures, and may cause the formation of potentially dangerous transverse cracks. *Crushed heads*, a defect due to the formation of extensive surface-initiated RCF, occur in older, softer rails with high nonmetallic inclusion contents which can lead to lack of support for the surface material, resulting in its lateral flow.[178]

The application of a cyclic stress pattern is an integral characteristic of railway track. The specific stress range in the stress cycle will clearly be affected by the axle loading and by residual stresses in the rail itself. It is, therefore, necessary to pay careful attention to correct ballasting of the track, because any local loss of support may result in a marked increase in the stress levels applied to the rail, effectively increasing the unsupported rail span and, thus, the applied stress.

The bending fatigue strength of rails for 2×10^6 cycles is a direct function of the tensile strength (Fig. 7.31), ranging from ~300 MPa at a tensile strength of 700 MPa to ~400 MPa at a tensile strength of 1100 MPa, in the as-rolled situation. However, as in all fatigue applications, the fatigue strength is strongly influenced by the presence of surface imperfections and damage, and by the presence of internal defects such as nonmetallic inclusions and hydrogen cracks. Corrosion effects on rails in service can decrease the fatigue strength level mentioned above by about 100 MPa,[158] as shown in Fig. 7.31. The fatigue strength of weldments can be lower than that of the parent rail, and the fatigue strength of both flash-butt or thermit welds is of the order of 200 MPa. Note that residual stress adds to the fatigue load, and in this respect, stresses around drilled fish-plate holes should be minimized by reaming.[158] Catastrophic web cracking in roller-straightened rail has been linked to high-residual-stress conditions.[168,169]

7.8.2.2 Wire Rod Steels.

Many of the applications of medium- to high-carbon steels (containing 0.30 to 0.85% C, 0.40 to 0.80% Mn, 0.10 to 0.35% Si, 0.01 to 0.045% S, and 0.008 to 0.045% P) in rod form include the conversion of hot-rolled rod to wire by a cold-drawing operation.[158] These hot-rolled rods should be free from surface defects due to casting and rolling of steel, segregation, surface imperfections, nonmetallic inclusions, and decarburization. Table 7.7 lists typical estimated tensile strength values for 5.6-mm ($^7/_{32}$-in.) medium-high carbon and high-carbon steel rods rolled on a mill using controlled cooling.[179] The microstructure of such a rod is near to that obtained by patenting. The strength usually lies between those obtained by air patenting and lead patenting. Most high-carbon steel wire is drawn from such rods without prior patenting.

The wire's properties depend on the steel chemistry and manufacturing processes such as surface treatment, heat treatment, and drawing, which, in turn, determine the quality of the finished wire rope.[180a]

7.8.2.3 Wire Ropes.

Rope wire is a commodity manufactured mainly with careful control for use in the construction of wire rope.[180] Wire ropes are made in various sizes and cover a wide diversity of applications, such as in suspension bridges, as the main load-carrying cables, and as suspension elements connecting the carriage-way with these cables; suspension elements for suspended roofs; in haulage applications such as lifting ropes or cranes, elevators, and mine hoists; load transmission ropes on excavators; and trawl ropes for fishing vessels.[158]

A standard wire rope consists of three basic components: the wires, the strands, and a core (Fig. 7.34a).[181] The selection of wire rope is made based on six factors: (1) strength resistance to breaking, (2) resistance to bending fatigue, (3) resistance

TABLE 7.7   Tensile Strengths of 5.6-mm ($^7/_{32}$-in.)-Diameter Hot-Rolled Medium-High-Carbon and High-Carbon Steel Rod[179]

Data produced from rod produced with controlled cooling

Carbon content of steel, %	Tensile strength for steel with manganese content of					
	0.60%		0.80%		1.00%	
	MPa	ksi	MPa	ksi	MPa	ksi
0.30	641	93	676	98	717	104
0.35	689	100	731	106	793	115
0.40	745	108	779	113	820	119
0.45	793	115	834	121	869	126
0.50	848	123	883	128	931	135
0.55	896	130	938	136	972	141
0.60	951	138	986	143	1020	148
0.65	1000	145	1041	151	1076	156
0.70	1055	153	1089	158	1124	163
0.75	1103	160	1138	165	1179	171
0.80	1151	167	1193	173	1227	178
0.85	1207	175	1241	180	1282	186

to vibrational fatigue, (4) abrasion resistance, (5) crushing resistance, and (6) reserve strength. There are six grades of wire rope referred to as *traction steel* (TS), *mild plow steel* (MPS), *plow steel* (PS), *improved plow steel* (IPS), *extra improved plow steel* (EIPS), and *extra extra improved plow steel* (EEIPS).[181] These steel grades denote the strength of a particular size and grade of rope. The plow steel strength curve is used as the basis for determining the strength of all steel rope wires; the tensile strength of any steel wire grade is a function of the diameter and is highest in the smallest wires. The steel wire has the "bright" or uncoated finish. Steel wires can be galvanized. *Drawn galvanized* wire has the same strength as the bright wire; however, wire *galvanized at finished size* has commonly 10% lower strength.[181] For more detailed discussion, readers are referred to a recent wire rope user's manual.[181] The strength of the rope depends on the number of individual wires and their arrangement in the rope. A wire rope is produced by helically laying wires about a central axis into strands, followed by helically laying the strands into a rope (Fig. 7.34). The helically wound strands may or may not be wound around an axial member called the core. Individual strands may be based on a centerless grouping principle or comprise a layer of wires wound around a center wire (single-layer principle) or of several layers wound around a center wire (multiple-operation principle). Wire rope is distinguished by its construction, i.e., by the way the wires have been laid to form strands, and by the way the strands have been laid around the core. For example, if the strand in the rope is wound in the opposite direction to the wires in the strand, it is said to be a left or *right regular lay* rope; and if the strands are wound in the same direction as that of the wires in the strand, it is said to be left or *right lang lay* rope.[158,182]

Among all types of wire rope, *right regular lay* (RRL) is widely used. Or alternate lay consists of alternating regular and lang lay strands. Figure 7.34*b* shows the cross sections of four basic strand patterns around which standard wire ropes are

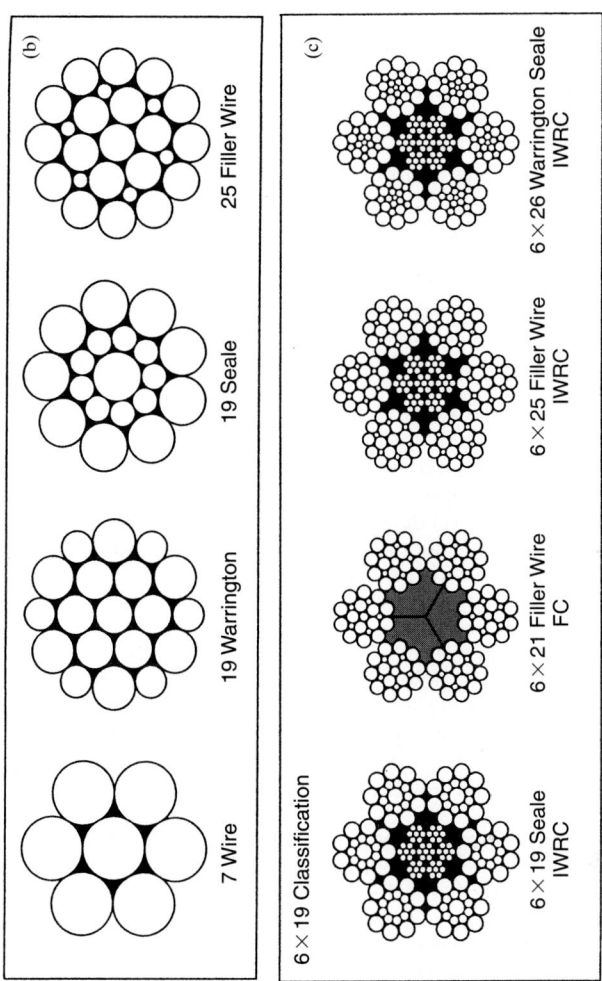

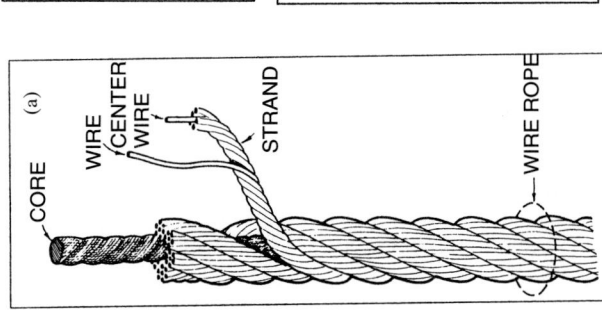

FIGURE 7.34 (*a*) Three basic components of a typical wire rope. (*b*) Four basic strand patterns. (*c*) Cross section of commonly used 6 × 19 classification wire rope construction.[181] (*Courtesy of Wire Rope Technical Board, Maryland.*)

built. The wire ropes are classified by the number of strands in the rope, the number and arrangement of wires in each strand, and a descriptive word or letter to recognize the type of construction or the geometric arrangement of wires. Some commonly used wire rope constructions are: 6×7 classification such as 6×7 FC (fiber core); 6×19 classification such as 6×19 Seale IWRC (independent wire rope core), 6×21 filler wire FC, 6×25 filler wire IWRC, 6×26 Warrington Seale IWRC (Fig. 7.34c); 6×37 classification such as 6×31 Warrington Seale IWRC, 6×36 Seale filler wire IWRC, 6×41 Warrington Seale (WS) IWRC, 6×46 Seale filler wire IWRC, 6×49 filler wire Seale (FWS) IWRC, etc.; 6×61 classification such as 6×55 (two-operation) filler wire Seale IWRC, 6×57 Seale filler wire IWRC, and 6×61 filler wire Warrington Seale IWRC. Figure 7.34c represents the commonly used cross sections of wire rope construction according to 6×19 classification.[181] Rotation-resistant ropes are a special class of wire rope designed to resist the likelihood of spinning or rotating under load. They are available as single-layer or multilayer strand types.[181]

The rope construction has important bearings on the application. The geometric arrangement of the wires can affect the rope strength, the internal fretting of the wires, the rope stiffness, and the untwisting propensities.

The ultimate break strength of a wire rope is by design less than the aggregate strength of all the wires and will depend on the construction of rope and grade of wire used. The proper design factor of a wire rope requires consideration of all loads. These loads should incorporate (if applicable) acceleration, deceleration, rope speed, rope attachments, number and arrangement of sheaves and drums, conditions producing corrosion and abrasion, and length of rope. Usually, the more flexible rope, which contains the largest number of wires and has fiber cores, will stretch more than all-metal ropes with fewer ropes and thus less flexibility.[183]

In the case of suspension bridge rope, galvanizing treatment is applied to the steel wire to improve corrosion resistance properties. In nonsevere conditions, a light oiling treatment serves the purpose of protective coating. However, in severe service conditions, stainless steel wires are used for very demanding applications.

7.8.2.4 Bridge Ropes.   Bridge rope is formed in a similar manner to a helical strand except that strands are formed helically around a center strand or rope, instead of wires. Bridge rope is usually made with a regular lay construction.

Bridge strands and ropes are not usually subjected to fluctuating bending stresses over sheaves. The loads are generally in direct tension and comprise mainly a large, constant dead load plus a relatively smaller pulsating live load. The fatigue characteristics of such tension members are, therefore, of interest.[182]

7.8.2.5 Spring Wires.   Mechanical spring wire can be grouped into seven types. Table 7.8 lists the ranges of chemical composition for seven types:[180,182,184]

   1. *Hard-drawn spring wire* (ASTM A227) is less costly and is used for the manufacture of mechanical springs in applications requiring infrequent stress repetitions or static loads. Its surface quality is relatively low with such imperfections as hairline seams. For hard-drawn spring steel wire (class 1 and class 2), the major requirement is the tensile strength. Class 2 is a higher-strength product.

   2. *High-tensile hard-drawn spring wire* (ASTM A629) is a special-quality, hard-drawn carbon steel spring wire with restricted size tolerances. This is used where such restricted dimensional requirements are essential for the production of highly stressed mechanical springs and wire forms. It is used for applications subject to static load and infrequent stress repetitions.[182]

TABLE 7.8 Mechanical Spring Wire (Chemical Composition, %)[180,182,184]

Alloying element	Uncoated, drawn galvanized at finish size hard-drawn wire	High-tensile hard-drawn wire	Oil-tempered wire	Wire for heat-treated components	Music spring steel wire	Valve spring-quality wire	Upholstery spring wire
Carbon	0.45–0.85	0.65–1.00	0.55–0.85	0.50–1.03	0.70–1.00	0.60–0.75	0.45–0.75
Manganese	0.30–1.30	0.20–1.30	0.30–1.20	0.30–1.30	0.20–0.60	0.60–0.90	0.60–1.20
Phosphorus	0.035 max.	0.035 max.	0.035 max.	0.035 max.	0.025 max.	0.025 max.	0.025 max.
Sulfur	0.045 max.	0.045 max.	0.045 max.	0.045 max.	0.030 max.	0.030 max.	0.030 max.
Silicon	0.15–0.35	0.15–0.35	0.15–0.35	0.15–0.35	0.15–0.35	0.15–0.35	0.15–0.35

3. *Oil-tempered spring wire* (ASTM A 229) is a general-purpose wire subjected to static loads or relatively infrequent stress repetitions. This is slightly more expensive and more susceptible to embrittling effects of plating than type 1; however, it is superior in surface smoothness. It is used in automotive and related industries.

4. *Spring wire for heat-treated components* (ASTM A713) is used for spring steel wire intended for heat-treated parts. The major requirement is a composition suitable for heat treatment.

5. *Music spring wire* (ASTM A228) is the least subjected to hydrogen embrittlement by electroplating and is similar to valve spring wire in surface quality. This is intended for applications requiring high stresses and good fatigue properties. Final cold drawing is usually accomplished by a wet white liquor method or by phosphating to develop a characteristic smooth bright surface. Specialized coiling tests, twist tests, torsion tests, and bend tests are employed to ensure that the exacting requirements of the uniformity and quality with exceptional high tensile strength of this type of wire are met.

6. *Valve spring quality wire* (ASTM A230) is used for the manufacture of engine valve springs and other springs requiring dynamic or high-stress repetitions with high-fatigue-strength properties. To meet this requirement, wire will have the highest degree of uniformity with respect to surface imperfections, internal soundness, and definite mechanical property values. The wire is available in both the oil-tempered and hard-drawn conditions or in one of the three conditions available for spring steel for heat-treated components.[185]

7. *Upholstery spring steel wire* (ASTM A407) is used in the manufacture of coil spring constructions for mattresses, furniture, beds, and automotive seats and cushions. It is available in sizes from 0.89 to 5.77 mm (0.035 to 0.225 in.) in diameter.[185] This is not used for the manufacture of other types of springs. This is drawn from thermally treated or controlled cooled wire rod or wire.

7.8.2.6 Tire Bead Wires and Tire Cords. Tire bead wire is wrapped and reinforced by the rubber-coated plies for pneumatic tires.[184] Wire for steel cord is drawn down from 5.5-mm-diameter hot-rolled, high-carbon steel (0.67 to 0.82% C, 0.4 to 0.6% Mn) rods to 0.94-mm (0.037-in.) diameter with a bronze-plated finish. Uniformity in chemical and mechanical properties and a good surface finish for rubber adhesion are essential for satisfactory performance. Mechanical property tests are performed on wire samples that have been heated for 1 hr at 150°C (300°F). Minimum breaking load for this wire is 129 kg (285 lb). In torsion tests, the wires must withstand 58 twists minimum in a 203-mm (8-in.) gauge length.

Tire cord in the form of strands is used for reinforcing the rubber matrix for automobile radial (ply) tires.[184] For tire cord produced from high-carbon hot-rolled steel rods (containing 0.67 to 0.82% C, 0.4 to 0.6% Mn) which is drawn to wire 0.15 to 0.25 mm (0.006 to 0.010 in.) in diameter. The deformation process involves a larger extent of strain ($\varepsilon > 1$). The deformability of inclusions is a critical parameter that affects the product performance.[184a] Elimination or minimization of the size of nondeformable inclusions such as spinel, calcium aluminates, and especially alumina in the steel rod is necessary because they are the most common cause of breakage of filament wire (and wear of the die) during final drawing and bunching into tire cord.[184b] This extremely fine, high-tensile-strength wire is produced from controlled cooled rods by drawing, followed by drawing, lead patenting (at 500 to 550°C), second drawing and lead patenting, brass plating, wet drawing, and forming into strands to be embedded into the tire rubber.[185] The patenting treatments restore an undeformed structure and ensure the necessary refinement of the pearlitic carbide lamellae for the subsequent large amounts of cold work, by shear cracking of the pearlite.

Stranding of wire involves laying several wires helically like rope wire; and, again, the strands can be laid to produce the tire cord. The number of filaments in a strand and number of strands in a cord can be varied based on the requirements of reinforcement. Additionally, the cord can include an outer spiral wrap, with a single filament, to stop strands from separating when cord is under an axial compressive load. The length of the lay, i.e., the length of the strand or cord required for one rotation of filament or strand, respectively, is important from the viewpoint of the contact bearing area and fretting, like wire rope.

The current conventional tensile strength of the tire cord steel is about 3600 MPa (522 ksi). Steelmakers are attempting to improve the strength to 5000 MPa (725 ksi) to satisfy the requirements of automobile and tire manufacturers. Higher-strength tire cord allows the automobile and tire manufacturers to reduce the weight and roll resistance of tires, leading to significant increase in vehicle fuel economy.[184b]

Accordingly, current developments include small alloy additions such as 0.20 to 0.25% Cr; maintaining ultra-low levels of Mg, Ca, and Al contents to prevent the formation of nondeformable inclusions; adding Wollastonite ladle flux to modify the inclusions during refining of tire cord steels;[184b] using 0.82 to 0.92% C, controlled cooling for rod coils in the hot mill to eliminate the need for lead patenting treatments; substitution of air patenting for lead patenting for certain products; elimination of a nonlamellar structure out of pearlite to improve the delamination resistance and work-hardening rate during drawing;[185a] and the use of low alloyed microalloyed medium- to high-carbon steels to provide added strength to the drawn product.[158,185a]

7.8.2.7 Wire for Prestressed Concrete.

7.8.2.7 Wire for Prestressed Concrete. There are two types of uncoated round high-carbon steel wire for prestressed concrete applications: cold-drawn and cold-drawn and suitably stress-relieved. The wire is used for linear or circular pretensioning or posttensioning structural members (ASTM A416 and ASTM A421). The stress-relieved product can be used as a single wire or as a strand that has been stress-relieved after stranding.[185]

Stress-relieved uncoated high-carbon steel wire is normally employed for the linear prestressing of concrete structures. It is produced in diameters of 4.88, 4.98, 6.35, and 7.01 mm (0.192, 0.196, 0.250, and 0.276 in.) to tensile strengths of 1620 to

1725 MPa (235 to 250 ksi). The wire can also be made in a low-relaxation mode which after drawing is subjected to a continuous thermomechanical treatment to produce the desired mechanical properties. The tensile strength of low-relaxation wire is the same as that of normal-relaxation stress-relieved wire; however, the minimum yield strength is also 90% of the minimum tensile strength (ASTM A421).

After stress-relieving the inside diameters of coils are usually greater than those of the drawn wire, in order to prevent stress set or reintroduction of coil stress. Coil inside diameters may be as large as 200 times the wire diameter.

High-carbon steel wire for mechanically tensioning is generally used for circular prestressing in the manufacture of concrete pressure pipe.[185]

7.8.2.8 Cold Heading Wires.
Cold heading is a cold forging process where the force, developed by one or more blows of a mechanical hammer or heading tool, is employed to displace or upset a portion of a blank to form a precise section of different contour or configuration than the original blank. Although the process is cold, heat is produced by the work performed.[182]

The manufacture of fasteners such as high-tensile-strength bolts is done with medium-carbon (0.30 to 0.45%) steels whereas lockwashers or screw drivers are made from high-carbon steel wire rod which is spheroidize annealed either in-process or after drawing finished sizes. Hot-rolled coiled rod is usually subcritically annealed to furnish low strength and ductility.

Cold heading steel wire rod is produced with carefully controlled manufacturing practices and rigid inspection practices to ensure the necessary degree of homogeneity, internal soundness, cleanliness, and freedom from surface imperfections. Decarburization must be held to a minimum for those products that are quenched and tempered. A fully killed fine-grain steel is usually required for the most difficult operations.

7.8.3 Microalloyed Ferrite-Pearlite Forging Steels

The concept of simultaneously increasing the pearlite content and making it more dilute has been exploited to advantage in microalloyed medium-carbon ferrite-pearlite forging steels which lead to an increase of both strength and toughness. This is achieved by increasing Mn content, decreasing C content, and/or increasing the cooling rate through the transformation range. An increase of Si content to 0.6 to 0.7% improves the toughness; microalloying with V and N raises the strength with some sacrifice of toughness; and reduction of γ grain size decreases the pearlite and ferrite grain size, which, in turn, improves the toughness without sacrifice of strength.[186]

Figure 9.21 shows the advantages of manufacturing automotive components such as connecting rods, crank shafts, steering knuckles, truck components, front axle beam, etc., from hot-rolled microalloyed medium-carbon forging steels over the hardened and tempered steels (see Sec. 9.5.2 also).[187,188]

REFERENCES

1. N. Ridley, *Met. Trans.*, vol. 16A, 1984, pp. 1019–1036.
2. N. Ridley, in *Proceedings of the International Conference on Phase Transformations in Ferrous Alloys*, eds. A. R. Marder and J. I. Goldstein, TMS–AIME, Warrendale, Pa., 1984, pp. 201–236.

3. J. K. Chen, S. W. Spencer, M. E. Ekstrand, D. Chen, and W. T. Reynolds, Jr., *Met. Trans.*, vol. 27A, 1996, pp. 1683–1689.

4. J. R. Vilella, G. E. Guellich, and E. C. Bain, *Trans. ASM*, vol. 24, 1936, pp. 225–252.

5. M. Hillert, in *The Decomposition of Austenite by Diffusional Processes*, eds. V. F. Zackay and H. I. Aaronson, Interscience, New York, 1961, pp. 197–237.

6. M. E. Nicholson, *J. Met.*, vol. 6, 1954, p. 1071.

7. R. F. Mehl and W. C. Hagel, *The Austenite-Pearlite Reaction, Progress in Metal Physics*, vol. 6, Pergamon Press, Oxford, 1956, p. 74.

8. S. Modin, *Jernkontorets Ann.*, vol. 1235, 1951, p. 169.

9. R. W. K. Honeycombe and H. K. D. H. Bhadeshia, *Steels: Microstructure and Properties*, 2d ed., Arnold, London, 1995.

10. D. L. Lee and C. G. Park, *Scripta Metall.*, vol. 32, no. 6, 1995, pp. 907–912.

11. E. M. Taleef, C. K. Syn, D. R. Lesuer, and O. D. Sherby, in *Thermomechanical Processing and Mechanical Properties of Hypereutectoid Steels and Cast Irons*, eds. D. R. Lesuer, C. K. Syn, and O. D. Sherby, TMS, Warrendale, Pa., 1997, pp. 127–141.

12. J. D. Verhoeven and E. D. Gibson, *Met. Trans.*, vol. 29A, 1998, pp. 1181–1189.

13. S. A. Hackney and G. J. Shiflet, *Scripta Metall.*, vol. 19, 1985, p. 757.

14. S. A. Hackney and G. J. Shiflet, *Acta Metall.*, vol. 35, 1987, pp. 1007–1017, 1019–1028.

15. S. A. Hackney, *Scripta Metall.*, vol. 25, 1991, p. 1453.

16. H. J. Lee, G. Spanos, G. J. Shiflet, and H. I. Aaronson, *Acta Metall.*, vol. 36, 1988, p. 1129.

17. M. J. Whiting and P. Tsakiropoulos, *Scripta Metall.*, vol. 29, 1993, p. 401.

18. C. Zener, *Trans. AIME*, vol. 167, 1946, p. 550.

19. M. Hillert, *Mechanism of Phase Transformations in Crystalline Solids*, Monograph no. 33, Institute of Metals, London, 1968, p. 231.

20. M. P. Puls and J. S. Kirkaldy, *Met. Trans.*, vol. 3, 1972, pp. 2777–2796.

21. J. M. Shapiro and J. S. Kirkaldy, *Acta Metall.*, vol. 16, 1968, pp. 579–585.

22. B. E. Sundquist, *Acta Metall.*, vol. 16, 1968, pp. 1413–1427.

23. M. Hillert, in *Proceedings of the International Conference on Solid $\rightarrow$ Solid Phase Transformations*, eds. H. I. Aaronson et al., TMS–AIME, Warrendale, Pa., 1982, pp. 789–806; J. W. Cahn and W. C. Hagel, *Acta Met.*, vol. 11, 1963, pp. 561–574.

24. R. D. Doherty, in *Physical Metallurgy*, 4th ed., eds. R. W. Cahn and P. Haasen, Elsevier Science BV, Amsterdam 1996, chap. 15, pp. 1363–1506.

25. K. Han, G. D. W. Smith, and D. V. Edmonds, *Met. Trans.*, vol. 26A, 1995, pp. 1617–1631.

26. Ref. 12 in A. R. Marder and B. L. Bramfitt, *Met. Trans.*, vol. 7A, 1976, pp. 902–905.

27. N. Ridley, in *Proceedings of the International Conference on Solid $\rightarrow$ Solid Phase Transformations*, eds. H. I. Aaronson et al., TMS–AIME, Warrendale, Pa., 1982, pp. 807–817.

28. R. C. Sharma, G. R. Purdy, and J. S. Kirkaldy, *Met. Trans.*, vol. 10A, 1979, pp. 1129–1139.

29. S. K. Tewary and R. C. Sharma, *Met. Trans.*, vol. 16A, 1985, pp. 597–603.

30. J. S. Kirkaldy, *Met. Trans.*, vol. 4A, 1973, pp. 2327–2333.

31. E. Lemaire, J. Copreaux, and F. Roch, *Scripta Metall.*, vol. 35, no. 1, 1996, pp. 83–89.

32. T. Gladman, I. D. McIvor, and F. B. Pickering, *JISI*, vol. 210, 1972, p. 916.

33. C. M. Bae, W. J. Nam, and C. S. Lee, *Scripta Metall.*, vol. 35, no. 5, 1996, pp. 641–646.

34. R. F. Mehl, C. S. Barrett, and D. W. Smith, *Trans. AIME*, vol. 105, 1933, p. 215.

35. W. A. Johnson and R. F. Mehl, *Trans. AIME*, vol. 135, 1939, p. 416.

36. R. F. Mehl and A. Dube, *Phase Transformations in Solids*, John Wiley & Sons, New York, 1951, p. 574.

37. J. W. Cahn and W. C. Hagel, in *The Decomposition of Austenite by Diffusional Processes*, eds. V. F. Zackay and H. I. Aaronson, Interscience, New York, 1962, pp. 131–192.

38. P. G. Shewmon, *Transformations in Metals*, McGraw-Hill, New York, 1969.

39. F. C. Hull, R. A. Colton, and R. F. Mehl, *Trans. AIME*, vol. 150, 1942, pp. 185–207.

40. B. B. Rath, in *Proceedings of the International Conference on Solid → Solid Phase Transformations*, eds. H. I. Aaronson et al., TMS–AIME, Warrendale, Pa., 1982, pp. 1097–1103.

41. J. W. Christian, *Theory of Transformations in Solids*, Pergamon Press, London, 1965.

42. P. R. Howell, *Materials Characterization*, vol. 40, 1998, pp. 227–260.

43. J. W. Cahn and W. C. Hagel, *Acta Metall.*, vol. 11, 1963, pp. 561–574.

44. A. R. Marder and B. L. Bramfitt, *Met. Trans.*, vol. 6A, 1975, pp. 2009–2014.

45. B. L. Bramfitt and A. R. Marder, *Met. Trans.*, vol. 7A, 1976, pp. 902–905.

46. D. D. Pearson and J. D. Verhoeven, *Met. Trans.*, vol. 15A, 1984, pp. 1037–1045.

47. B. L. Bramfitt and A. R. Marder, *Int. Metallogr. Soc. Proc.*, 1968, pp. 43–45.

48. F. A. Khalid and D. V. Edmonds, *Acta Metall.*, vol. 41, no. 12, 1993, pp. 3421–3434.

49. M.-X. Zhang and P. M. Kelly, *Scripta Mater.*, vol. 37, no. 12, 1997, pp. 2009–2015.

50. M. J. Whiting and P. Tsakiropoulos, *Scripta Metall. Mater.*, vol. 30, 1994, p. 1031; *Mater. Sci. and Tech.*, vol. 11, 1995, pp. 717–727.

51. M. B. Cortie and C. E. Mavrocordatos, *Met. Trans.*, vol. 22A, 1991, pp. 11–18.

52. A. Das, S. K. Pabi, I. Manna, and W. Gust, *J. Mater. Sci.*, vol. 34, 1999, pp. 1815–1821.

53. D. Cheetham and N. Ridley, *J. Inst. Metals*, vol. 99, 1971, p. 371.

54. L. Kumar, R. V. Ramanujan, R. Tewari, P. Mukhopadhyay, and S. Banerjee, *Scripta Mater.*, vol 40, no. 6, 1999, pp. 723–728.

55. M. Mannerkoski, *Acta Polytech. Scand.*, 1964, chap. 26, p. 7.

56. K. Relander, *Acta Polytech. Scand.*, 1964, chap. 34, p. 7.

57. J. V. Bee and D. V. Edmonds, *Metallography*, vol. 12, 1979, pp. 3–21.

58. R. W. K. Honeycombe, *Met. Trans.*, vol. 7A, 1976, pp. 915–936.

59. A. T. Davenport and P. C. Becker, *Met. Trans.*, vol. 2A, 1971, pp. 2962–2964.

60. P. R. Howell, in *Proceedings of the International Conference on Solid → Solid Phase Transformations*, eds. H. I. Aaronson et al., TMS–AIME, Warrendale, Pa., 1982, pp. 399–425.

61. R. W. Honeycombe and F. B. Pickering, *Met. Trans.*, vol. 3A, 1972, pp. 1099–1112.

62. A. T. Davenport and R. W. K. Honeycombe, *Proc. Roy. Soc., London*, vol. A322, 1971, p. 191.

63. T. Gladman, *The Physical Metallurgy of Microalloyed Steels*, The Institute of Materials, London, 1997.

64. R. W. K. Honeycombe, in *Proceedings of the International Conference on Phase Transformations in Ferrous Alloys*, eds. A. R. Marder and J. J. Goldstein, TMS–AIME, Warrendale, Pa., 1984, pp. 259–280.

65. T. Obara, G. J. Shiflet, and H. I. Aaronson, *Met. Trans.*, vol. 14A, 1983, pp. 1159–1161.

66. T. Furuhara and H. I. Aaronson, *Scripta Metall.*, vol. 22, 1988, pp. 1635–1637.

67. M. V. Kral, W. H. Hofmeister, and J. E. Wittig, *Met. Trans.*, vol. 28A, 1997, pp. 2485–2497.

68. G. Fourlaris, *Mater. Sc. Forum*, vol. 284–286, 1998, pp. 427–434.

69. G. Fourlaris, A. J. Baker, and G. D. Papadimitriou, *Solid → Solid Phase Transformations*, eds. W. C. Johnson, J. M. Howe, D. E. Laughlin, and W. A. Soffa, TMS, Warrendale, Pa., 1994, pp. 535–540.

70. K. Campbell and R. W. K. Honeycombe, *Met. Sci.*, vol. 8, 1974, p. 197.

71. P. G. Berry and R. W. K. Honeycombe, *Met. Trans.*, vol. 1A, 1970, p. 3279.

72. M. H. Ainsley, G. J. Cocks, and D. R. Miller, *Met. Sci.*, vol. 13, 1979, p. 20.

73. H. I. Aaronson, in *Decomposition of Austenite by Diffusional Processes*, Interscience, New York, 1962, pp. 387–546.

74. C. A. Dubé, H. I. Aaronson, and R. F. Mehl, *Rev. Met.*, vol. 55, 1958, p. 201.

75. H. S. Fong and S. G. Glover, *Met. Trans.*, vol. 15A, 1984, pp. 1643–1651.

76. H. I. Aaronson, C. Laird, and K. R. Kinsman, *Phase Transformations*, ASM, Metals Park, Ohio, 1970, pp. 313–396.

77. L. E. Samuels, *Optical Microscopy of Carbon Steels*, ASM, Metals Park, Ohio, 1979.

78. K. R. Kinsman and H. I. Aaronson, *Transformation and Hardenability in Steels*, Climax Molybdenum Co., Ann Arbor, Mich., 1967, pp. 39–53.

79. C. A. Dubé, Ph.D. Thesis, Carnegie Institute of Technology, 1948.

80. C. Zener, *J. Appl. Phys.*, vol. 20, 1949, pp. 950–953.

81. C. Atkinson, H. B. Aaron, K. R. Kinsman, and H. I. Aaronson, *Met. Trans.*, vol. 4A, 1973, pp. 783–792.

82. J. R. Bradley, G. J. Shiflet, and H. I. Aaronson, in *Proceedings of the International Conference on Solid → Solid Phase Transformations*, eds. H. I. Aaronson et al., TMS–AIME, Warrendale, Pa., 1982, pp. 819–824.

83. A. D. King and T. Bell, *Met. Trans.*, vol. 6A, 1975, pp. 1419–1429.

84. J. R. Bradley and H. I. Aaronson, *Met. Trans.*, vol. 12A, 1981, pp. 1729–1741.

85. M. Hillert and L. I. Staffansson, *Acta Chem. Scand.*, vol. 24, 1970, p. 3618.

86. H. I. Aaronson and H. A. Domian, *TMS–AIME*, vol. 236, 1966, p. 781.

87. G. J. Shiflet, H. I. Aaronson, and J. R. Bradley, *Met. Trans.*, vol. 12A, 1981, pp. 1743–1750.

88. T. Tanaka, H. I. Aaronson, and M. Enomoto, *Met. Trans.*, vol. 26A, 1995, pp. 561–580.

89. T. Furuhara and H. I. Aaronson, *Acta Metall. et Mater.* vol. 39, 1991, p. 2887.

90. X-Z. Bo and H-S. Fang, *Acta Mater.*, vol. 46, no. 8, 1998, pp. 2929–2936.

91. R. G. Kamat, E. B. Hawbolt, L. C. Brown, and J. K. Brimcombe, *Met. Trans.*, vol. 23A, 1992, pp. 2469–2480.

92. E. Enomoto, *Met. Trans.*, vol. 22A, 1991, pp. 1235–1245.

93. H. B. Aaron and H. I. Aaronson, *Met. Trans.*, vol. 2, 1971, pp. 23–37.

94. H. I. Aaronson, in *The Mechanism of Phase Transformations in Metals*, Institute of Metals, London, 1956, p. 47.

95. E. P. Simonen, H. I. Aaronson, and R. Trivedi, *Met. Trans.*, vol. 4, 1973, pp. 1239–1245.

96. P. N. T. Unwin and R. B. Nicholson, *Acta Metall.*, vol. 17, 1969, p. 139.

97. G. Spanos and M. G. Halls, *Met. Trans.*, vol. 27A, 1996, pp. 1519–1534.

98. H. I. Aaronson and C. Wells, *Trans. AIME*, vol. 206, 1956, pp. 1216–1223.

99. W. T. Reynolds, Jr., M. Enomoto, and H. I. Aaronson, in *Proceedings of International Conference on Phase Transformations in Ferrous Alloys*, eds. A. R. Marder and J. I. Goldstein, *TMS–AIME*, Warrendale, Pa., 1984, pp. 155–200.

100. H. K. D. H. Bhadeshia, *Progr. Mater. Sci.*, vol. 29, no. 4, 1985, pp. 321–386.

101. H. K. D. H. Bhadeshia and J. W. Christian, *Met. Trans.*, vol. 21A, 1990, pp. 776–797.

102. H. I. Aaronson, W. T. Reynolds, Jr., G. J. Shiflet, and G. Spanos, *Met. Trans.*, vol. 21A, 1990. pp. 1343–1380.

103. M. Enomoto, *Met. Trans.*, vol. 25A, 1994, pp. 1947–1955.

104. M. Hillert, *Jernkontorets Ann.*, vol. 14, 1957, p. 757.

105. M. Hillert, *Met. Trans.*, vol. 6A, 1975, pp. 5–19.

106. A. Vander Ven and L. Delaey, *Progr. Mater. Sci.*, vol. 40, 1996, pp. 181–264.

107. R. Trivedi, *Met. Trans.*, vol. 1, 1970, p. 921.

108. G. P. Ivantsov, in *Growth of Crystals*, vol. 3, eds. A. V. Shubnikov and N. N. Sheflat, Consultant Bureau, New York, 1962, p. 53.

109. M. M. Kostic, E. B. Hawbolt, and L. C. Brown, *Met. Trans.*, vol. 7A, 1976, pp. 1643–1653.

110. R. Trivedi, in *Proceedings of the International Conference on Solid → Solid Phase Transformations*, eds. H. I. Aaronson et al., TMS–AIME, Warrendale, Pa., 1982, pp. 477–502.

111. M. Hillert, *Met. Trans.*, vol. 25A, 1994, pp. 1957–1966.

112. W. R. Bosze and R. Trivedi, *Met. Trans.*, vol. 5, 1974, p. 511.

113. G. R. Purdy, *Met. Sci.*, vol. 5, 1971, p. 81.

114. J. W. Cahn, W. B. Hillig, and G. W. Sears, *Acta Metall.*, vol. 12, 1964, p. 1421.

115. G. J. Jones and R. Trivedi, *J. Appl. Phys.*, vol. 42, 1971, p. 4299; *J. Cryst. Growth*, vol. 29, 1975, p. 155.

116. C. Atkinson, *Proc. Roy. Soc.*, *London*, vol. 378A, 1981, p. 351; vol. 384A, 1982, p. 167.

117. C. Laird and H. I. Aaronson, *Acta Metall.*, vol. 17, 1969, p. 505.

118. H. I. Aaronson and C. Laird, *Trans. TMS–AIME*, vol. 242, 1968, p. 1437.

119. R. Sankaran and C. Laird, *Acta Metall.*, vol. 2, 1974, p. 957.

120. K. R. Kinsman, E. Eichen, and H. I. Aaronson, *Met. Trans.*, vol. 6A, 1975, pp. 303–317.

121. S. K. Bhattacharya, J. H. Perepezko, and T. B. Massalski, *Scripta Metall.*, vol. 7, 1973, pp. 485–488.

122. T. B. Massalski, in *Metals Handbook*, vol. 9, 9th ed., ASM, Metals Park, Ohio, 1985, pp. 655–657.

123. T. B. Massalski, *Met. Trans.*, vol. 15A, 1984, pp. 421–425.

124. M. R. Plichta, W. A. T. Clark, and H. I. Aaronson, *Met. Trans.*, vol. 15A, 1984, pp. 427–435.

125. P. Wang, G. B. Vishwanathan, and V. K. Vasudevan, *Met. Trans.*, vol. 23A, 1992, pp. 690–697.

126. M. Hillert, *Met. Trans.*, vol. 15A, 1984, pp. 411–419.

127. T. B. Massalski, *Massive Transformations in Phase Transformations*, ASM, Metals Park, Ohio, 1970, pp. 433–486.

128. G. A. Chadwick, *Metallography of Phase Transformations*, Crane, Russak and Company, New York, 1972.

129. D. Swanson and J. G. Parr, *JISI*, vol. 202, 1964, p. 104.

130. R. D. Doherty in *Physical Metallurgy*, eds. R. W. Cahn and P. Haasen, North-Holland Physics Publishing, Amsterdam, 1983, pp. 933–1030.

131. J. H. Perepezko, *Met. Trans.*, vol. 15A, 1984, pp. 437–447.

132. E. S. K. Menon, M. R. Plichta, and H. I. Aaronson, *Acta Metall.*, vol. 36, 1988, p. 321.

133. J. E. Kittl and T. B. Massalski, *Acta Metall.*, vol. 15, 1961, p. 161.

134. G. A. Sargent, L. Delay, and T. B. Massalski, *Acta Metall.*, vol. 16, 1968, p. 723.

135. M. R. Plichta, J. C. Williams, and H. I. Aaronson, *Met. Trans.*, vol. 8A, 1977, pp. 1885–1892.

136. J. C. Caretti, *Acta Metall.*, vol. 34, 1986, pp. 385–393.

137. E. A. Wilson, *ISIJ Int.*, vol. 34, no. 8, 1994, pp. 615–630.

138. R. J. Ackert and J. G. Paar, *JISI*, vol. 209, 1971, p. 912.

139. M. J. Bilby and J. G. Parr, *JISI*, vol. 202, 1964, p. 100.

140. Ref. 44 in M. Hillert, *Met. Trans.*, vol. 6A, 1975, pp. 5–19.

141. T. B. Massalski, J. H. Perepezko, and J. Jaklovsky, *Mater. Sc. Eng.*, vol. 18, 1975, p. 193.

142. C. Hayzeldon and B. Cantor, in *Int. J. Rapid Solidification*, 1984–1985, p. 237.

143. S. H. Chong, A. Sayles, R. Keyse, J. D. Atkinson, and E. A. Wilson, *Mater. Trans., JIM*, vol. 39, no. 1, 1998, pp. 179–188.

144. I. Ishikawa, T. Takahashi, and T. Ochi, *Met. Trans.*, vol. 25A, 1994, pp. 929–936.

145. W. Heckel and H. W. Paxton, *TMS–AIME*, vol. 218, 1960, p. 799.

146. T. Ando and G. Krauss, *Acta Metall.*, vol. 29, 1981, pp. 351–363.

147. W. Heckel and H. W. Paxton, *Trans. ASM*, vol. 53, 1961, p. 539.

148. F. B. Pickering, in *Encyclopedia of Materials Science and Engineering*, Pergamon Press, Oxford, 1986, pp. 4621–4632.

149. T. Gladman, D. Dulieu, and I. D. McIvor, in *Microalloying '75*, Union Carbide Corporation, New York, 1977, pp. 32–34.

150. F. B. Pickering, in *Towards Improved Toughness and Ductility*, Climax Molybdenum Co., Greenwich, Conn., 1971, p. 9.

151. J. D. Baird and R. R. Preston, *Processing and Properties of Low-Carbon Steels*, AIME, New York, 1973, p. 1.

152. B. Mintz, *Met. Technol.*, vol. 11, 1984, pp. 265–272.

153. B. Mintz, *Met. Technol.*, vol. 11, 1984, pp. 52–60.

154. A. R. Marder, in *Proceedings of the International Conference on Phase Transformations in Ferrous Alloys*, TMS–AIME, Warrendale, Pa., 1984, pp. 11–41.

155. F. B. Pickering, in *Constitution and Properties of Steels*, vol. ed. F. B. Pickering, VCH, Weinheim, 1992, pp. 41–94.

156. M. G. M. F. Gomes, L. H. Almeida, L. C. F. C. Gomes, and I. L. May, *Materials Characterization*, vol. 39, 1997, pp. 1–14.

157. J. H. Martens, and D. P. Wirick, *Proceedings of the International Symposium on Rail Steels for 21st Century*, Iron and Steel Society, Warrendale, Pa., 1995, pp. 1–3.

158. T. Gladman, in *Constitution and Properties of Steels*, vol. ed. F. B. Pickering, VCH, Weinheim, 1992, pp. 401–432.

159. R. K. Steele, *Rail Quality and Maintenance for Modern Railway Operation*, International Conference, Delft 1992, Kluwer Academic Publishers, Dordrecht, Netherlands, 1993, pp. 77–97.

160. R. K. Steele, *private communication*, 1999.

161. X. Su and P. Clayton, *Wear*, vol. 197, 1996, pp. 137–144.

162. B. L. Bramfitt, *Metals Handbook*, vol. 20: *Materials Selection and Design*, ASM International, Materials Park, Ohio, 1997, pp. 357–382.

163. A. J. Perez-Unzueta and J. H. Beynon, *Wear*, vol. 162–164, 1993, pp. 173–182.

164. W. H. Hodgson, *Rail Quality and Maintenance for Modern Railway Operation*, International Conference, Delft 1992, Kluwer Academic Publishers, Dordrecht, Netherlands, 1993, pp. 29–39.

165. *Railway Age*, September 1996, p. 55.

166. B. L. Bramfitt, R. L. Cross, and D. P. Wirick, *Proc. of the International Symp. on Rail Steels for 21st Century*, Iron and Steel Society, Inc., Warrendale, Pa., 1995, pp. 23–29.

167. B. L. Bramfitt, D. P. Wirick, and R. L. Cross, *Iron & Steel Engineer*, vol. 75, no. 4, 1996, pp. 33–36.

168. D. Utrata, *Rail Steels Symp. Proceedings*, Iron and Steel Society, Inc., Warrendale, Pa., 1994, pp. 131–135.

169. B. L. Bramfitt, *Mechanical Working and Steel Processing Proceedings Conference*, Cincinnati, Ohio, 1990, pp. 485–495.

170. *Carbon and Alloy Steels*, ASM International, Materials Park, Ohio, 1996, pp. 169–200, 684–687.

171. *AREA Manual for Railway Engineering*, 1996, pp. 4-2-6 to 4-2-10.

172. C. Esvald, *Modern Railway Track*, Thyssen Stahl, Duisburg, Federal Republic of Germany, 1989.

173. D. M. Fegredo and J. Kalousek, *Wear of Materials*, ed. K. C. Ludema, American Society of Mechanical Engineers, New York, 1987, pp. 121–132.

174. R. K. Steele, *1990 Mechanical Working and Steel Processing Proceedings*, 1990, pp. 131–142.

175. C. O. Frederick, *Rail Quality and Maintenance for Modern Railway Operation*, International Conference, Delft 1992, Kluwer Academic Publishers, Dordrecht, Netherlands, 1993, pp. 3–14.

176. O. Orringer, *Rail Quality and Maintenance for Modern Railway Operation*, International Conference, Delft 1992, Kluwer Academic Publishers, Dordrecht, Netherlands, 1993, pp. 253–271.

177. L. Reid, *Rail Quality and Maintenance for Modern Railway Operation*, International Conference, Delft 1992, Kluwer Academic Publishers, Dordrecht, Netherlands, 1993, pp. 337–347.

178. P. Clayton and X. Su, *Wear*, vol. 200, 1996, pp. 63–73.

179. R. J. Glodowski, *Metals Handbook*, vol. 1: *Properties and Selection: Iron, Steel, and High Performance Alloys*, 10th ed., ASM International, Materials Park, Ohio, 1990, pp. 272–276.

180. *Steel Products Manual: Carbon Steel Wire and Rods*, Iron and Steel Society, Inc., Warrendale, Pa., 1993.

180a. Z. Muskalski, J. W. Pilarczyk, H. Dijja, and B. Golis, *Wire J. Int.*, December 1999, pp. 108–113.

181. *Wire Rope User's Manual*, 3d ed., Wire Rope Technical Board, Baltimore, Md., 1993.

182. *Ferrous Wire*, vol. 2, The Wire Association International, Inc., Guilford, Conn., 1989.

183. F. L. Jamieson, *Metals Handbook*, vol. 11: *Failure Analysis and Prevention*, ASM, Metals Park, Ohio, 1986, pp. 514–528.

184. 1998 *SAE Handbook*, vol. 3: *On-Highway Vehicles and Off-Highway Machinery*, Society of Automotive Engineers, Warrendale, Pa., 1998, pp. 30.01–30.78.

184a. X. Zhang and S. V. Subramanian, *Wire J. Int.*, December 1999, pp. 102–107.

184b. G. M. Faulring, *I&SM*, July 1999, pp. 29–36.

185. A. B. Dove, *Metals Handbook*, vol. 1: *Properties and Selection: Iron, Steel, and High Performance Alloys*, 10th ed., ASM International, Materials Park, Ohio, 1990, pp. 276–288.

185a. I. Ochiai, S. Nishida, and H. Tashiro, *Wire J. Int.*, December 1990, pp. 50–61.

186. R. Lagneburg, O. Sandberg, and W. Roberts, in *Fundamentals of Microalloyed Forging Steels*, eds. G. Krauss and S. K. Banerji, TMS, Warrendale, Pa., 1987, pp. 39–54.

187. J. F. Held, in *Fundamentals of Microalloyed Forging Steels*, eds. G. Krauss and S. K. Banerji, TMS, Warrendale, Pa., 1987, pp. 39–54.

188. F. D. Gear, in *Fundamentals of Microalloyed Forging Steels*, eds. G. Krauss and S. K. Banerji, TMS, Warrendale, Pa., 1987, pp. 291–296.

CHAPTER 8
MARTENSITE

8.1 INTRODUCTION

The name *martensite* was first used by Osmond in 1895, in honor of German metallurgist Adolf Martens, to identify the very hard, platelike or acicular constituent
produced in many steels rapidly quenched from the austenite state. Martensite in
steel is a metastable body-centered tetragonal (bct) phase. Later, this name was
extended to include a number of other solid-state transformations in pure metals,
nonferrous alloys (e.g., Cu-Al-, Cu-Sn-, and Ti- and Zr-base alloys), semiconductors,
ceramics, minerals, superconducting compounds (such as V_3Si and Nb_3Sn just above
the superconducting transition temperature), solidified gases (such as oxygen and
helium), proteins, and polymeric materials which exhibit some common features,
notably diffusionless displacive (shear) transformation.[1,2] It is now common to
describe all such transformation processes as martensitic and the product phase as
martensite, irrespective of its crystal structures. Different types of martensitic transformation in both ferrous and nonferrous alloys (e.g., athermal or diffusionless;
isothermal or diffusional; interstitial; substitutional; etc.) have now been recognized
which may be distinguished from one another by kinetics, morphology, and crystallography or internal structure. Table 8.1 provides such a classification of diffusionless displacive transformation in metallic materials, where the alloy systems are
grouped into three categories.[3] The following definition has been given: A *martensitic transformation* involves the coherent formation of one phase from another
without change in composition by a diffusionless, homogeneous lattice shear.[4] In
this case, *diffusion* means long-range diffusion. Christian coined the term *military
transformation* for this type of reaction in which the most orderly and highly coordinated or disciplined atomic rearrangement occurs where every atom has the same
neighbors as in the parent but where the individual atoms move by a fraction of an
interatomic distance.[5,6] It has been defined by Cohen as a subset of diffusionless,
displacive phase transformation, involving sufficiently large lattice-distortive shear
displacements where strain energy dominates the transformation kinetics and
product morphology.[7] Martensitic transformation is classified as a nonequilibrium
phase transformation which always occurs far away from equilibrium and which
results in nonequilibrium products.[8] Martensitic transformation involves a cooperative motion of a set of atoms across an interface, causing a shape change and sound.[9]
Thus, martensitic transformation can be defined as diffusionless, lattice-distortive,
shear-dominant transformation occurring by nucleation and growth.[10] It is this class
of solid-state transformation and its products that constitutes the focus of this
chapter.

TABLE 8.1 Classification of Metallic Alloy Systems Showing Diffusionless Displacive Transformations[3]

1.	Martensite based on allotropic transformation of solvent atom		

1. Iron and iron-based alloys			
2. Shear transformation, close packed-to-close packed			
1. Cobalt and alloys	fcc → hcp, 126 R	SF[†]	
2. Rare earth and alloys	fcc, hcp, dhcp, 9R		
3. MnSi, TiCr$_2$	NaCl → NiAs, Laves		
3. Body centered cubic to close packed			
1. Titanium, zirconium and alloys	bcc → hcp, orth fcc	tw, d[†]	
2. Alkali and alloys (Li)	bcc → hcp		
3. Thallium	bcc → hcp		
4. Others: plutonium, uranium, mercury, etc., and alloys	Complex structures		

2.	β-bcc Hume-Rothery and Ni-based martensitic shape-memory alloys		

1. Copper-, silver-, gold-, β-alloys (disord., ord.) bcc	AB, ABABCBCAC, ABAC		
2. Ni-Ti-X β-alloys	bcc → 9R, AB	tw, SF[†]	
Nickel β-alloys (Ni-Al)	bcc → ABC	tw, SF[†]	
Ni$_{3-x}$M$_x$Sn (M = Cu, Mn)	bcc → AB	tw*	
(Cobalt β-alloys, Ni-Co-X)			

3.	Cubic to tetragonal, stress-relaxation twinning or martensite		

1. Indium-based alloys	fcc → fct, orth.	tw, tw$^{\sigma}$[†]	
2. Manganese-based alloys	fcc → fct, orth.	tw$^{\sigma}$[†]	
3. A 15 compounds, LaAg$_x$In$_{1+x}$	β-W → tetr.		
4. Others: Ru-Ta, Ru-Nb, YCu, LaCd			

[†] SF: stacking faults; tw: twins; tw$^{\sigma}$: (stress relaxation) twins; d: dislocated.
Courtesy of L. Delaey.

In steels, a large proportion of hardness and strength may be developed in martensite in which distortion produced by forcibly retaining the carbon atoms within the ferrite lattice is intense and unidirectional. In substitutional martensite (e.g., carbon-free Fe-30% Ni and nonferrous alloys), less hardening is developed[11] because of the small and omnidirectional distortion. It should be emphasized that the properties of some martensites are of vast technological importance. These include conventional hardening of steel, a type of age hardening (maraging); ductility increase in transformation-induced plasticity (TRIP) steels; partially stabilized zirconia (PSZ); rubberlike elastic ductility; shape memory effects; and high damping capacity.[12]

This chapter deals with the characteristics, types, and products of martensitic transformations, phenomenological theory of crystallography martensite, morphology, nucleation and growth of the transformation, shape memory alloys, and strengthening mechanisms. Finally, omega transformation, another type of diffusionless, displacive solid-state transformation, is briefly described.

8.2 GENERAL CHARACTERISTICS OF MARTENSITIC TRANSFORMATION

The main characteristics of martensitic transformation and of the product phase are summarized below:

1. It is a diffusionless transformation, which means that the chemical compositions of the parent and product phases are identical.

2. The transformation interface between the martensite and the parent phase depends greatly on the transformation growth process. Such an interface remains highly glissile and does not tend to require thermal activation for its movement, as verified from low-temperature experiments. This interface may be completely coherent or semicoherent, depending upon the crystallography of the particular material undergoing transformation. In the case of most ferrous martensites, the interface remains semicoherent and the product and parent lattices are coherently accommodated over a small portion of the boundary, producing an accumulating misfit. In contrast, in fcc $\rightarrow$ hcp transformations in Co and its alloys, the austenite/martensite interface is fully coherent.[1,2]

3. *Shape change and surface relief.* When martensite is formed, a macroscopic deformation (or shape change) is observed which results in surface upheavals, called *surface relief*, on a polished flat surface of the parent phase (Fig. 8.1), indicating that the transformation occurs by a displacive shear parallel to the *habit plane* (i.e., the interface plane between the parent and product phase), which is an invariant plane (i.e., an undistorted and unrotated plane like the K_1 plane in twinning). The habit plane is usually expressed as a plane in the parent phase and is of special importance in the crystallography of martensite. This observation of surface relief is one of the main experimental criteria of a martensitic transformation.

The surface in the transformed region remains plane but is tilted about its line of intersection with the habit plane. Straight lines inscribed on the surface are

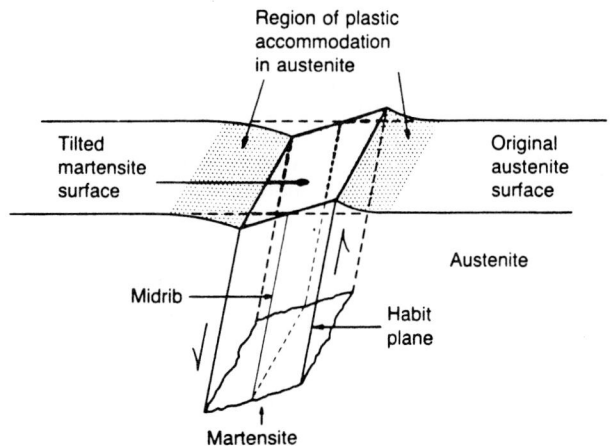

FIGURE 8.1 Schematic shape deformation produced during the formation of a martensite plate.[13] (*Courtesy of The Institute of Metals, England.*)

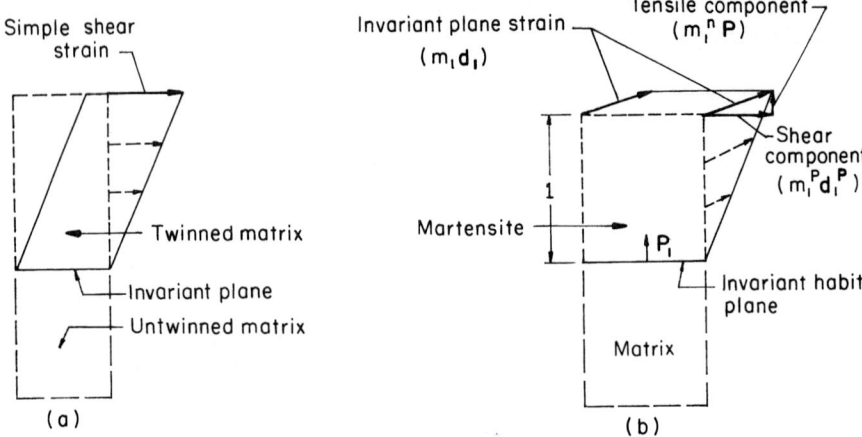

FIGURE 8.2 (*a*) A simple shear strain occurring in twinning; (*b*) invariant plane strain of martensite ($P_1 = 1 + m_1 d_1 P_1'$) comprising a simple shear component ($m_1^p d_1^p$) and tensile component ($m_1^n P_1$).

transformed into straight lines, and planes are transformed into planes.[13,14] Such a transformation is said to be homogeneous. Sometimes it is described mathematically as an *affine transformation*. The strain that produces a *net macroscopic* deformation associated with an invariant habit plane is termed an *invariant plane strain* (IPS) deformation. In this case the displacement of any point is proportional to its distance from the invariant plane. In martensitic transformations, the shear that produces the invariant plane strain is complex, consisting of (1) simple shear strain and (2) uniaxial tensile or compression strain normal to the habit plane (Fig. 8.2); the normal component of strain is attributed to the volume change produced during structural change in martensitic transformation. In contrast, merely the simple shear strain occurs in twinning. Table 8.2 illustrates two strain components for several martensites.[13,14]

4. Martensite occurs usually in the form of plates or laths which seem to be embedded in the matrix along certain well-defined planes (Fig. 8.3). The martensite (sometimes designated as α') plate, forming closer to room temperature, is usually of lenticular shape which is partially internally twinned, but has a core, called the *midrib*, within it [as a result of constraints imposed by the untransformed matrix (Fig. 8.1)]. On two-dimensional metallographic sections, the lenticular platelets lead to the characteristic microstructure. The plates forming at very low temperatures are usually thin and flat. Thin plate is fully internally twinned of the <111> {112} type, a common deformation twinning in bcc crystals. However, in low-carbon steels and dilute iron alloys, martensites form with lathlike morphology (i.e., with planar interface extending entirely across the parent grains). It has been suggested that α' crystals form initially at the midrib plane and grow laterally.[15]

5. The habit plane is usually irrational (i.e., the Miller indices are not very simple) in almost all cases. This fact is best illustrated by the experimentally determined habit planes in three steels on the stereographic triangle shown in Fig. 8.4. There is a considerable amount of scatter in the experimental data for the habit plane,[14] but martensite plates in a particular alloy possess a unique or definite habit plane.

TABLE 8.2 Crystallographic Strain Component Data for Several Martensites[13,14]

System	Structure change	Habit plane	Direction displacement	Shear strain component γ_T	Normal strain component ε_n
Fe-C (1.35% C)	fcc → bct	~{225}	$<\bar{1}\bar{1}2>$	0.19	0.09
Fe-C (1.8% C)	fcc → bct	~{259}	$<\bar{1}\bar{1}2>$		
Fe-Ni (30% Ni)	fcc → bcc	~{9, 22, 33}	~$<\bar{1}\bar{5}6>$	0.20	0.05
Fe-Ni-C (22% Ni, 0.8% C)	fcc → bct	~{3, 10, 15}	~$<\bar{1}\bar{3}2>$	0.19	
Pure Ti	bcc → hcp	~{8, 9, 12}	~$<11\bar{1}>$	0.22	
Au-Cd (47.5% Cd)	bcc → Orthorhombic	$\begin{bmatrix} 0.70 \\ -0.69 \\ 0.21 \end{bmatrix}$	$\begin{pmatrix} 0.66 \\ 0.73 \\ 0.18 \end{pmatrix}$	0.05	
In-Tl (20% Tl)	fcc → fct	{011}	~$<01\bar{1}>$	0.02	

Reprinted by permission of John Wiley & Sons, Inc, New York.

FIGURE 8.3 Optical micrographs. (*a*) Lath martensite formed in 1018 steel by water-quenching from 925°C (1697°F). (*b*) Plate martensite in 1060 steel by water-quenching from 815°C (1500°F).

It has been found that the habit plane is a function of composition and temperature rather than strain. In nearly pure iron or in lower-carbon steels, the crystals of martensite appear to be needle- or lath-shaped in cross section with habit planes of $\{111\}_\gamma$ or $\{112\}_\gamma$ type, for carbon steels containing 0.5 to 1.4% carbon,[16] and the usual plane observed is near $\{225\}_\gamma$ and is formed at higher temperatures. In carbon

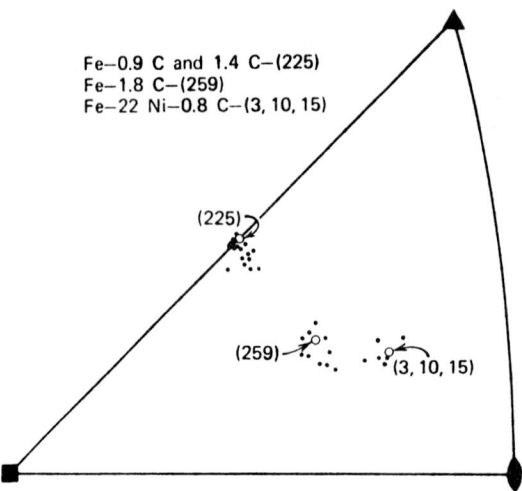

Fe–0.9 C and 1.4 C–(225)
Fe–1.8 C–(259)
Fe– 22 Ni–0.8 C–(3, 10, 15)

(225)

(259)

(3, 10, 15)

FIGURE 8.4 Experimentally measured habit planes of martensite plates on the stereographic triangle.[14] (*Reprinted by permission of John Wiley & Sons, New York; after Greninger and Troiano, and Dunne and Bowles.*)

steels of still higher carbon content (1.5 to 1.8% C), the habit plane may be approximately $\{259\}_\gamma$.

6. *Lattice orientation relationships.* Associated with the habit planes, there is always a precise orientation relationship between parent phase and martensite.[4]

The Kurdjumov-Sachs (K-S) relation[17] is $\{111\}_\gamma // \{011\}_{\alpha'}$ with $<011>_\gamma // <111>_{\alpha'}$ (α' = martensite) and is usually associated with $\{225\}_\gamma$ habit plane. This relationship has been found to occur in Fe-C alloys with carbon content in the range of 0.5 to 1.4 wt% (Table 8.3).[3]

The Nishiyama (N) relation, similar to the Greninger-Troiano (G-T) relation,[18,19] is again $\{111\}_\gamma // \{011\}_{\alpha'}$ but with $<112>_\gamma // <011>_{\alpha'}$ and is generally associated with an $\approx \{259\}_\gamma$ habit plane. The G-T orientation relation is an irrational relation between K-S and N and corresponds to the prediction of the IPS crystallographic theory. This relationship has been found to occur in Fe-C alloys with carbon content >1.4 wt%, which was observed by Greninger and Troiano in Fe-22 wt% Ni-0.8 wt% C alloy and by Nishiyama for Fe-Ni alloys containing 27 to 34% Ni[18,19] (Table 8.3).[3]

As regards the transformation of fcc (austenite) → hcp martensite and that of hcp → bcc martensite, the following relations are observed:[3]

$$(111)_\gamma // (0001)_\varepsilon // (101)_{\alpha'} \quad \text{and} \quad [110]_\gamma // [1210]_\varepsilon // [111]_{\alpha'} \quad (8.1)$$

It is evident from the K-S relationship that the close-packed planes and close-packed directions, respectively, of the γ lattice are parallel to those of the α' lattice. Moreover, this direction is parallel to the Burgers vector.

In the K-S relations, $\{111\}$ represents any four kinds of austenite planes, namely, (111), ($\bar{1}$11), (1$\bar{1}$1), or (11$\bar{1}$). In each plane, any one of six different directions can exist, as shown in Fig. 8.5a. These directions consist of three pairs in one direction and three pairs in the opposite direction. These pairs of crystals are twin-related.

TABLE 8.3 Crystallographic Observable Parameters of the Martensitic Transformations in Some Metals and Alloys[3]

Alloy system	Structural change	Composition, wt%	Orientation relationship	Habit plane
Fe-C	fcc ↓ bc tetr.	0–0.4% C	$(111)_P//(101)_M$ $[1\bar{1}0]_P//[11\bar{1}]_M$ K-S relationship	$(111)_P$
		0.55–1.4% C	K-S relationship	$(225)_P$
		1.4–1.8% C	Idem	
Fe-Ni	fcc ↓ bcc	27–34% Ni	$(111)_P//[101]_M$ $[1\bar{2}1]_P//[10\bar{1}]_M$ N relationship	$\approx(259)_P$
Fe-C-Ni	fcc ↓ bc tetr.	0.8% C, 22% Ni	$(111)_P \approx 1°$ of $(101)_M$ $(1\bar{2}1)_P \approx 2°$ of $[10\bar{1}]_M$ G-T relationship	$(3, 10, 15)_P$
Fe-Mn	fcc ↓ hcp (ε-phase)	13–25% Mn	$(111)_P//(0001)_\varepsilon$ $[1\bar{1}0]_P//[1\bar{2}10]_\varepsilon$	$(111)_P$
Fe-Cr-Ni	fcc ↓	18% Cr, 8% Ni	$(111)_P//(0001)_\varepsilon//(101)_{\alpha'}$ $[1\bar{1}0]_P//[1\bar{2}10]_\varepsilon//[11\bar{1}]_{\alpha'}$	$\varepsilon(111)_P$ $\alpha'(211)_P$
Austenite inox. iron	hcp (ε), bcc (α')			
Cu-Zn β	bcc → 9R	40% Zn	$(011)_P//?(\bar{1}\bar{1}4)_M$	$\approx(2, 11, 12)_P$
Cu-Sn	idem	25.6% Sn	$[1\bar{1}1]_P//?[\bar{1}10]_M$	$\approx(133)_P$
Cu-Al	bcc ↓	11–13.1% Al	$(10\bar{1})_P$ at 4° of $(0001)_M$ $[111]_P//[10\bar{1}0]_M$	2° of $(133)_P$
	hcp distorted	12.9–14.7% Al	$(10\bar{1})_P//[10\bar{1}1]_M$ $[111]_P//[10\bar{1}0]_M$	3° of $(122)_P$
Pure Co	fcc ↓ hcp		$(111)_P//(0001)_M$ $<110>_P//<11\bar{2}0>_M$	$(111)_P$
Pure Zr	bcc ↓		$(101)_P//(0001)_M$ $[111]_P//[11\bar{2}0]_M$	$(596)_P$ $(8, 12, 9)_P$
Pure Ti	hcp			$(334)_P$ $(441)_P$
Pure Li			Burgers relations	

Source: Courtesy of G. Guénin et al.; after P. F. Gobin, G. Guénin, M. Morin, and M. Robin, in *Transformations de Phases à l' État Solide-Transformations Martensitiques.* Lyon: Dep. Génie Phys. Mat., INSA, 1979.

Therefore, K-S relations yield α' crystals with $4 \times 6 = 24$ different orientations in a γ crystal. These different oriented martensite crystals are termed *variants*.

In the N relations there are four types of austenite planes, and three different directions exist in each plane, as shown in Fig. 8.5b. Thus N relations lead to $4 \times 3 = 12$ variants, i.e., half of those in the K-S relations. It should be pointed out that all these relationships are not precise and that they represent a deviation of 1° (or

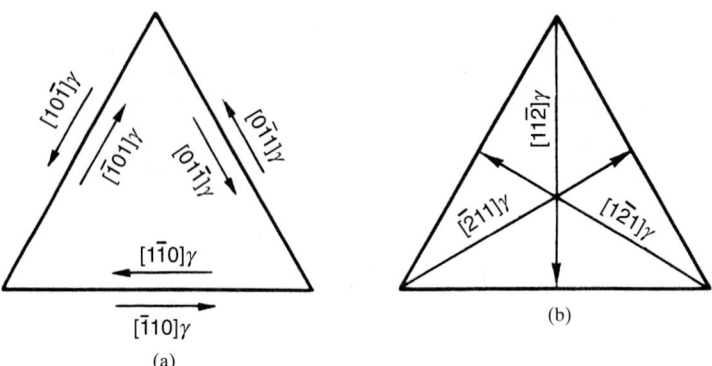

FIGURE 8.5   Directions of shears in $(111)_\gamma$ plane: (*a*) K-S relationship; (*b*) N relationship.[15] (*Reprinted by permission of Academic Press, Orlando, Florida.*)

TABLE 8.4   The M_s Temperature and Approximate Hardness of Martensite Product for a Number of Materials[2]

Composition	M_s/K	Hardness HV
ZrO_2	1200	1000
Fe-31Ni-0.23C wt%	83	300
Fe-34Ni-0.22C wt%	<4	250
Fe-3Mn-2Si-0.4C wt%	493	600
Cu-15Al	253	200
Ar-40N$_2$	30	

several degrees) from the ideal orientation.[15] Also the orientation relationships and the orientations of the habit plane alter from one alloy system to another, and within a particular alloy system even with alloy composition (Table 8.3).[3] A more detailed list of these and other crystallographic features of various martensite is presented by Nishiyama.[15]

7. Martensite transformation can propagate at temperatures approaching absolute zero, as shown by Kulin and Cohen for Fe-Ni and Fe-Ni-C alloys.[20] This does not imply that all martensitic transformations occur at low temperatures; in fact, many occur at comparatively high temperatures (Table 8.4).[2] It has been shown[21] that individual martensite plates (in Fe-Ni-C alloys, for example) form within 10^{-7} s, which means that the linear growth velocity of martensite plates approaches one-third the speed of sound in the solid.[21] However, in Fe-C alloys with very low carbon content, a much lower speed of platelet formation is observed.

The high velocity at low temperatures and the lack of temperature dependence support the view that the growth is not a thermally activated process; i.e., the transformation is athermal. Exceptions to this include the transformations in Au-Cd, Mn-Cu, In-Tl, Al-bronze, U-Cr, and possibly Fe-Ni alloys, for which slow growth has been observed under the microscope. The overall kinetics of a martensitic reaction are functions of both the nucleation and growth states and will be largely influenced by the slower of the two. For example, slow thermal nucleation may lead to isothermal transformation characteristics.[22]

8. Martensite structures are always harder than the parent phase, but extreme hardness is found only in steels containing interstitial solute elements.

9. The transformation on cooling proceeds by the formation of new plates rather than by the growth of preexisting plates.[23] Exceptions are isothermal growth of martensitic plates in U-Cr alloys and isothermal nucleation and slow interface migration in Cu-Al-Ni alloys.

10. *Evidence of fine inhomogeneous substructures in the martensite phase.* Experimental investigation with electron microscopy and X-ray line broadening has shown that the plate and lath morphologies of martensite have fine substructure. They may be composed of stacks of very fine, fully twinned or partially twinned regions or may contain fine-scale slip lines or faults which do not change the lattice. The twin spacing is usually in the range of 15 to 200 Å in ferrous martensite. Internal twins of $\{112\}\alpha'$ type have been found on a very fine scale in plates of high-carbon martensite and Fe-Ni-C martensite containing 20% Ni and 0 to 1% C. This has been confirmed as a result of observing twin-related Laue X-ray diffraction patterns. In low-carbon steels and 18-8 stainless steels, martensite is in the form of laths or needles with a fine-scale slip. Patterson and Wayman[24] have found that platelike martensite in Fe-Ni is always at least partially twinned and that the extent of twinning increases with decrease in the M_s temperature or increase in nickel concentration. Nonferrous martensites, such as those found in Cu-Zn alloys, consist of banded or internally twinned microstructure; in Cu-Al alloys, however, orthorhombic martensite possesses internally twinned microstructure, and cubic or tetragonal martensite contains arrays of stacking faults rather than twins.[25]

11. Depending on the alloy composition, fcc austenite in steels transforms either to bcc or bct martensite or to hcp ε-martensite;[3] even some systems exhibit more than one martensitic crystal structure. In some composition ranges, both types may coexist (e.g., bcc α- and hcp ε-martensite in Fe-Mn alloys and deformed 18-8 stainless steel). However, martensite formed by cooling may be quite different from that formed by deformation.

12. A comparison of three main types of martensitic transformation based on kinetic modes is listed in Table 8.5.[50] Table 8.6 summarizes a set of criteria selected to differentiate qualitatively the ferrous and nonferrous martensites.[26]

8.3 TYPICAL CHARACTERISTICS OF MARTENSITIC TRANSFORMATION

Martensitic transformations can be broadly classified into two groups, with respect to kinetics; isothermal and athermal (including "burst"). The former transformation displays explicitly time-dependent kinetics, but the latter does not. Other characteristics which are essential for a particular type of martensitic transformation are given below.

8.3.1 Athermal Martensitic Transformation

The athermal martensitic transformation without external stress may be either thermoelastic, as in the case of shape memory alloys, or nonthermoelastic, as in the case

TABLE 8.5 Comparison of Martensitic Transformation Characteristics in the Three Kinetic Modes[50]

Characteristics	Athermal	Burst	Isothermal
Fraction transformed at a given temperature	Increases with decreasing temperature	Increases with decreasing temperature	Increases with decreasing temperature (except below the nose of the C curve)
Grain size	Fine grain size depresses M_s	Fine grain size depresses M_b	Fine grain size decreases transformation rate
Morphology	Lath at higher temperatures Plate at lower temperatures	Midrib plates Lath or plate after the burst	Lath at higher temperatures Plate at lower temperatures
Autocatalysis	?	Extreme form	Important contribution
Thermal stabilization	Present	Present	Present
Superimposed magnetic field	Increases M_s	Increases M_b	Increases transformation rate
Superimposed stress field	Increases M_s	Increases M_b	Increases transformation rate
Plastic deformation	Increases M_s up to M_d	Increases M_b	Increases or decreases transformation rate
Small amounts of residual carbon	Sharply decreases M_s	Decreases M_b	Drastically lowers transformation rate

of the majority of iron-based alloys. The essential difference between thermoelastic and nonthermoelastic martensitic transformation is the manner in which the materials accommodate the transformation strain (transformation shape and volume changes). In either case, the martensite forms and grows continuously as the temperature is decreased.[27]

1. *Effect of temperature and time.* Martensite begins spontaneously forming after a very large undercooling (below Ae_1) at a well-defined temperature, designated M_s (martensite start), which is nearly constant for an alloy, irrespective of cooling rate up to 50,000°/s, but is dependent on the concentration of γ-stabilizing alloying elements in the steel, on prior thermal and mechanical treatment, and on grain size.[28,29] As the temperature is lowered below the M_s, further transformation occurs until the reaction ceases at the M_f temperature. If a plain carbon steel is quenched from the austenitic condition to a temperature between M_s and M_f, a definite proportion of martensite is formed instantaneously. This is called *athermal martensite.* This means that the reaction proceeds only while the temperature is changing. The amount of martensite formed depends only on the temperature to which the steel is cooled and is independent of holding time at that temperature; i.e., the nucleation process occurs without thermal activation. Figure 8.6 shows a curve for the amount

TABLE 8.6 A Qualitative Comparison between Ferrous and Nonferrous Martensites[3]

Ferrous martensite		Nonferrous martensite
Interstitial and/or substitutional	Nature of alloying	Substitutional
Martensitic state in interstitial ferrous alloys is much harder than the austenite state	Hardness	Martensitic state is not much harder and may even be softer than the austenite state
Large	Transformation hysteresis	Small to very small
Relatively large	Transformation strain	Relatively small
High values near the M_s	Elastic constants of the parent phase	Low values near the M_s
Negative near the M_s in most cases	Temperature coefficient of elastic shear constant	Positive near the M_s in many cases
High	Transformation enthalpy	Low to very low
Large	Transformation entropy	Small
Large	Chemical driving force	Small
Self-accommodation is not obvious	Growth character	Well-developed self-accommodating variants
High rate, "burst," athermal, and/or isothermal transformation	Kinetics	Slower rate, no "burst," no isothermal transformation, thermoelastic balance
No single interface transformation observed	Growth front	Single interface possible
Low and nonreversible	Interface mobility	High and reversible
Low	Damping capacity of martensite	High

Courtesy of L. Delaey.

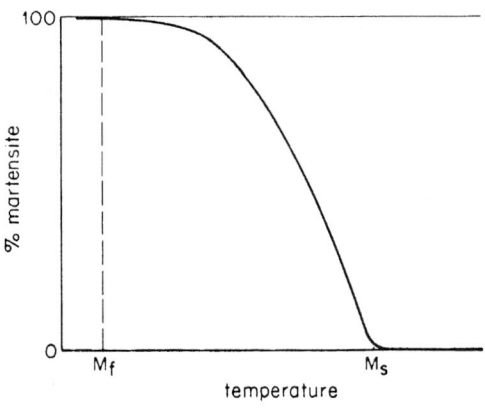

FIGURE 8.6 The amount of martensite as a function of temperature to which it is cooled. (*Courtesy of Hemisphere Publishing Corporation, New York; after Brooks.*)

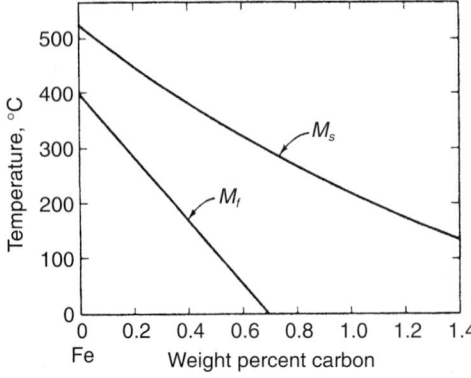

FIGURE 8.7 The effect of carbon content in steel on M_s and M_f temperatures.[18,28] (*Reprinted by permission of Kent Publishing Co., Boston Massachusetts.*)

of martensite formed as a function of temperature to which austenite has been cooled.

2. The M_s temperature is usually depressed by increasing the concentration of solute such as carbon (Fig. 8.7),[4] nickel, and manganese; but M_s increases with the addition of Co and Al.

An empirical relationship due to K. W. Andrews[30,31] has been given for the determination of M_s:

$$M_s(°C) = 539 - 423(\%C) - 30.4(\%Mn) - 17.7(\%Ni) - 12.1(\%Cr) - 7.5(\%Mo) + 10(\%Co) - 7.5(\%Si) \tag{8.2}$$

In high-carbon steels, the M_s temperature may decrease with increasing austenitizing temperature mainly due to the progressive solutionizing of undissolved carbides, which increases the carbon and alloy contents of the parent austenite. The M_f may be defined as the temperature below which no more martensite forms. It is approximately 215°C below the M_s temperature.

In practice, the martensite reaction can never be complete; i.e., a small amount of austenite always remains untransformed, which is called *retained austenite*. A large volume fraction of retained austenite is obtained in some highly alloyed steels where the M_f temperature is well below room temperature. The transformation of the last traces of austenite becomes more and more difficult as the amount of retained austenite becomes smaller. Figure 8.8 illustrates the amount of retained austenite produced by quenching carbon steels into water; most of the retained austenite may be transformed on subsequent cooling to liquid air temperature.[32]

3. Like M_s, the M_f temperature in steels is lowered by increasing carbon content. The $M_s - M_f$ range is extended by the increase in the carbon percentage and appears to widen with first increases of carbon content; but beyond a certain limit, further carbon addition does not have a significant effect. The room temperature lies between the M_s and M_f temperatures.

4. Martensite transformation in steels cannot be suppressed, even by extremely rapid cooling rate.[33] Although athermal kinetic behavior is not an essential feature of martensitic reactions, it is often observed.

5. *Reversibility of the transformation.* Martensite transformation can be reversible, which is usually associated with a considerable temperature hysteresis

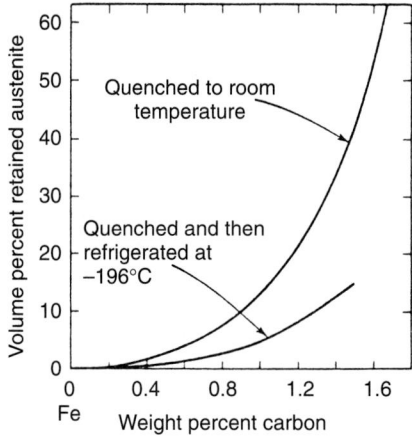

FIGURE 8.8 Variation of the amount of retained austenite, with plain carbon steels.[28,32] (*Reprinted by permission of PWS-Kent Publishing Co., Boston Massachusetts.*)

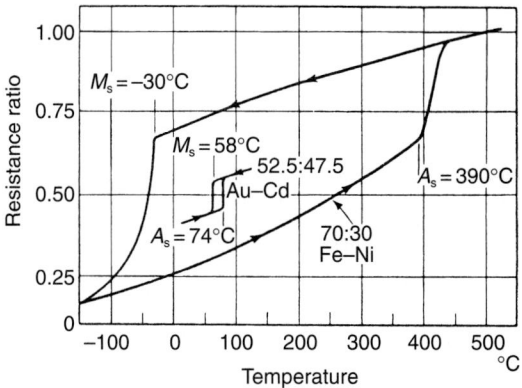

FIGURE 8.9 Variation of physical property (electrical resistivity, etc.) with temperature for an Fe-Ni and an Au-Cd alloy, illustrating martensitic transformations with different temperature hysteresis. (*Reprinted by permission of Pergamon Press, Plc.*)

between the martensitic transformation on cooling and the reverse transformation on heating. The reverse transformation begins at a temperature, called A_s, well above the M_s, provided no precipitation reactions occur in the martensite to alter its composition. Figure 8.9 shows such transformation hysteresis for Fe-Ni and Au-Cd alloys.[34] Similar phenomena are observed with Cu-Zn, Al-Cu, and Fe-Mn martensites. It is very difficult to study the reverse reaction from martensite to austenite in steel due to intervention of the tempering reaction. However, it is possible to study the reverse reaction by varying the pressure rather than the temperature.

   6. *Thermal stabilization* of the retained austenite (or parent phase) against further transformation is also a common feature of athermal martensitic transformation. This phenomenon occurs if the cooling is interrupted by holding at a temperature in the M_s–M_f range for a time interval Δt or if the cooling is slowed down in that temperature range (Fig. 8.10)[35] and is manifested as a retardation of transformation. When subsequent cooling is resumed, the athermal transformation does

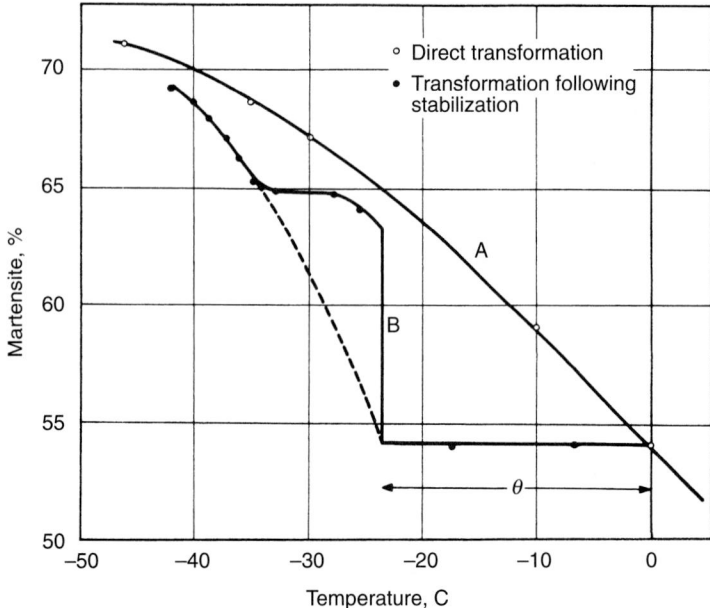

FIGURE 8.10 Percent martensite in a 0.96% C, 3% Mn, 0.5% C, 0.4% Si, 0.2% Ni steel for continuous cooling treatment, curve *A*; curve *B* corresponds to quenching to 0°C to give 54% martensite, aging at 18°C for 45 min, and then continuous cooling. The temperature difference θ is used as a measure of the extent of stabilization.[35] (*Reprinted by permission of ASM International, Materials Park, Ohio.*)

not start immediately, but only after an appreciable drop in temperature ΔT or further degree of supercooling.[36] At all subsequent temperatures, the amount of transformation is less than that produced by direct cooling. The extent of stabilization increases with time for which the article is held at that temperature as well as with the temperature approaching M_f (Fig. 8.11).[37] It appears, therefore, that minimum stabilization is found when a large proportion of martensite exists in the matrix.[38] It is a common observation that an oil-quenched specimen contains more retained γ than that of a water-quenched specimen.

A short classification of different types of stabilization of retained austenite is given in Tables 8.7 and 8.8.[39] Among various types, thermal stabilization is a common feature of athermal (or dynamically stabilized) martensite formation, which has been most extensively studied. There are three important theories (or mechanisms) of thermal stabilization; these involve the assumption that accumulation of interstitial carbon and nitrogen atoms leads to either the loss of mobility (or locking) of the γ/α' interface or the strengthening of the residual austenite by a strain-aging process or the destruction of the autocatalytic effect of previously nucleated martensite plates or phases.[6]

7. *Stabilization of retained austenite at low temperatures.* Studies conducted at subzero temperatures have shown that the stabilization of retained γ can also develop rapidly in Fe-Ni-C alloys at aging temperatures as low as −90°C. These

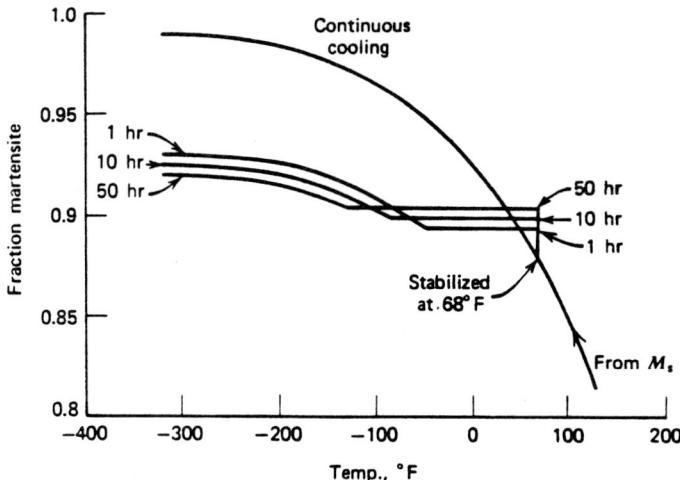

FIGURE 8.11 The amount of stabilization as a function of time and temperature in a W-2 type tool steel.[14,37] (*Reprinted by permission of John Wiley & Sons, New York.*)

results exhibit clearly that the mechanism of stabilization of retained γ proposed by early researchers is not complete because it cannot be used to describe the stabilization phenomenon at low temperatures. In fact, it has been observed in several studies that, for a particular aging treatment, the greater the amount of previously formed martensite, the larger will be the magnitude of stabilization of retained γ.[40]

8. *Effect of applied stress.* Uniaxial tensile or compressive stresses raise the transformation temperature, whereas hydrostatic stresses depress it.[41] Mechanically induced deformation (comprising the plastic deformation or the elastic stress) above M_s may frequently result in the formation of martensite; the highest temperature at which martensite can be nucleated is called M_d. Plastic deformation above the M_d temperature often lowers the M_s temperature on subsequent cooling and hence inhibits transformation; it leads to *mechanical stabilization* of the austenite against subsequent transformation when cooling is continued. This effect appears to result from an increase in difficulty of plate propagation into the austenite phase due to work hardening of the austenite.[14]

The stress-induced martensitic transformation is a complete analog of the martensitic transformation without external stress. In this case, the transformation proceeds continuously at a certain temperature with increasing applied stress. There will be one or more variants of martensite formed in any given direction where the transformation is favored. Usually, such martensite is expected to consist of two or more, but not all, of the many possible martensite variants. Unlike the martensite formed during the martensitic transformation without external stress, the associated strain will now be directional and will clearly be influenced by applied stress.[27]

9. *Effects of magnetic field and hydrostatic pressure.* The M_s temperatures for all ferrous alloys increase with an increase in magnetic field, but they remain unaffected

TABLE 8.7 Schematic Illustration of Stabilization of Austenite in Steels[39]

TABLE 8.8 Broad Composition of Steel Showing the Various Stabilization Types[39]

Type of stabilization (with reference to)	Composition of alloys (investigated Table 8.7)	Recent references (see ref. 39 for more)
A	Fe-Ni alloys	J. F. Breedis and W. D. Robertson, *Acta Metall.*, vol. 11, 1963, p. 547.
B	Fe-Ni alloys	L. F. Porter and G. T. Dienes, *Trans. AIME*, vol. 215, 1959, p. 854.
C	Fe-Ni-Al-C (0.013)	E. Hornbogen and W. A. Meyer, *Arch. Eisenhutt.*, vol. 1, 1968, p. 73.
D	Fe-Ni-P (0.7)	H. J. Neuhauser and W. Pitsch, *Arch. Eisenhutt.*, vol. 44, 1973, p. 235.
E	Fe-Ni alloy	B. Edmondson and T. Ko, *Acta Metall.*, vol. 2, 1954, p. 235.
F	1C-5Ni; 1C-0.1Ni, low alloys	G. S. Ansell et al., *Met. Trans.*, vol. 2, 1971, p. 2443.
G	1C-1.6Cr and other steels	M. Izumiyama and Y. Imai, *Nippon Kink. Gakk.*, vol. 24, 1960, p. 58.
H	Low-C, Ni steels; high-C, Cr steels	A. J. Baker et al., *Electron Microscopy and Structure of Crystals*, Interscience, New York, 1963, p. 903.
I	Steels with 1C and 0–12 Ni; steels with 1.4C, 5Ni, and Si, Mn, Cr, V; steels with ~1C, 1.6Cr, and ~2C, 12Cr, 0.12Ni	O. N. Mohanty, *Proc. Symp.: Phase Transitions and Phase Equilibria*, Bangalore, 1975, p. 38; Z. Nishiyama (ref. 15); R. Brook et al., *Met. Trans.*, vol. 8, 1977, p. 1449.

for Ti-Ni and Cu-Al-Ni shape memory alloys (SMAs). In contrast, M_s temperature decreases with increasing hydrostatic pressure in some ferrous alloys, but increases in Cu-Al-Ni alloys. Isothermal process in Fe-Ni-Mn alloys changes to the athermal one under magnetic field, and athermal process changes to the isothermal one under hydrostatic pressure.[42]

8.3.2 Isothermal Martensitic Transformation

Although a large number of martensitic transformations are athermal, some alloys show isothermal characteristics. It was first reported by Kurdjumov and Maksimova[43] in an Fe-0.6% C-6% Mn-2% Cu alloy. Other isothermal examples in ferrous alloys are Fe-Ni-Mn,[44] Fe-Ni-Cr,[45] Fe-Ni, Fe-Ni-C,[46] and Fe-Cr-C; in nonferrous metals and alloys they include U-Cr alloy and ordering transformations in Cu-Au, Co, Pt, and other close-packed lattices.[44] In some systems, isothermal transformation follows athermal transformations. In isothermal martensite transformations, slow nucleation but extremely rapid growth occurs. The number of nuclei of martensite formed depends on both temperature and time, increasing with time at a particular reaction temperature;[23] in athermal transformation, the nucle-

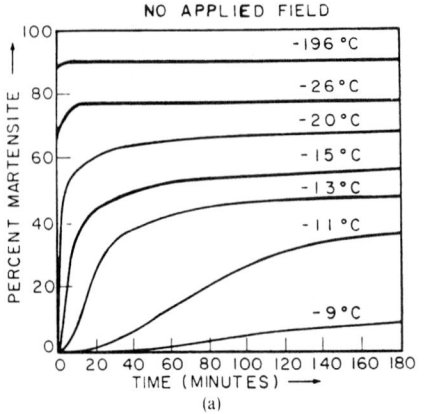
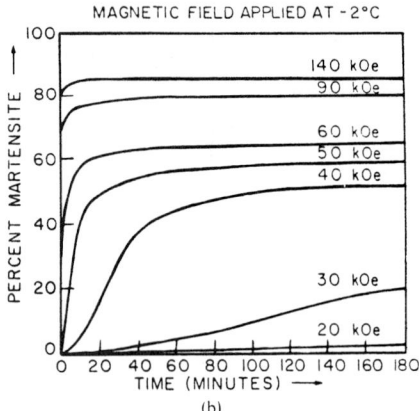

FIGURE 8.12   (*a*) Transformation kinetics for Fe-29.6%Ni alloy at varying temperatures. (*b*) Transformation kinetics for same alloy at –2°C and varying levels of applied magnetic field.[47] (*Reprinted by permission of Pergamon Press, Plc.; after M. K. Korenko, D. Sc. Thesis, MIT, Cambridge, Massachusetts, 1973.*)

ation rate is a function of temperature rather than time (Fig. 8.6). As a result, the slower volume transformation rate of isothermal martensite is produced, compared to that of athermal martensite. Figure 8.12*a* shows the transformation curves for an Fe-29.6% Ni alloy at different reaction temperatures.[47]

The isothermal reaction characteristics, first described by Kurdjumov, are as follows:[48]

1. The main requisite of this transformation is low reaction temperature. Whether the isothermal reaction occurs in the presence or absence of a fixed amount of athermal martensite, it is represented by a classical C-curve behavior with the maximum of isothermal transformation rate at some intermediate temperature (in the vicinity of –140°C in Fig. 8.13).[49]

2. Usually the isothermal reaction is observed to occur below both the M_s and the room temperature, for example, Fe-Ni-Mn (Fig. 8.13) and Fe-Ni-Cr alloys. But there are some examples where isothermal reaction can form above the M_s and the room temperature.

3. In some alloys, the transformation may be completely suppressed by rapid quenching to liquid nitrogen temperature (–180°C), but isothermal transformation occurs on reheating to temperature in the range of –80 to –160°C or to room temperature.[42]

4. The basic nature of isothermal reaction is best studied in the absence of athermal martensite, when the former reaction occurs above M_s.

5. Isothermal transformation curves have been found to be very sensitive to effects of magnetic fields and changes in grain size, as observed in Fe-29.6% Ni alloy (Fig. 8.12*b*) and Fe-26% Ni-2% Mn alloy (Fig. 8.14).[47]

6. The isothermal transformation curve depends on the prior plastic straining of the austenite (Fig. 8.15*a*) and the small variations in the residual carbon (Fig. 8.15*b*).[50]

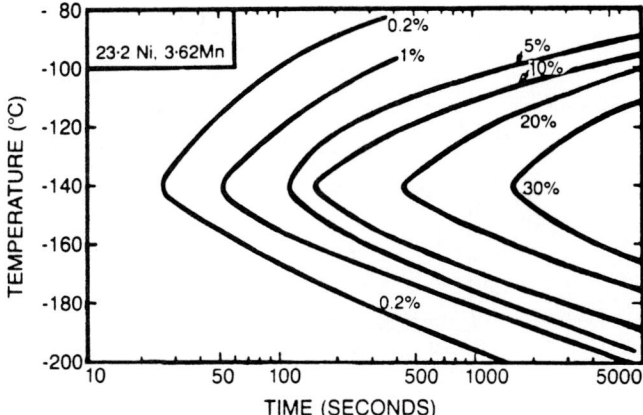

FIGURE 8.13 Isothermal transformation curves for different percentages of martensite formation in Fe-23.2Ni-3Mn alloy.[34] (*Reprinted with permission from Trans. AIME, vol. 203, 1955, pp. 183–187, a publication of The Metallurgical Society, Warrendale, Pennsylvania.*)

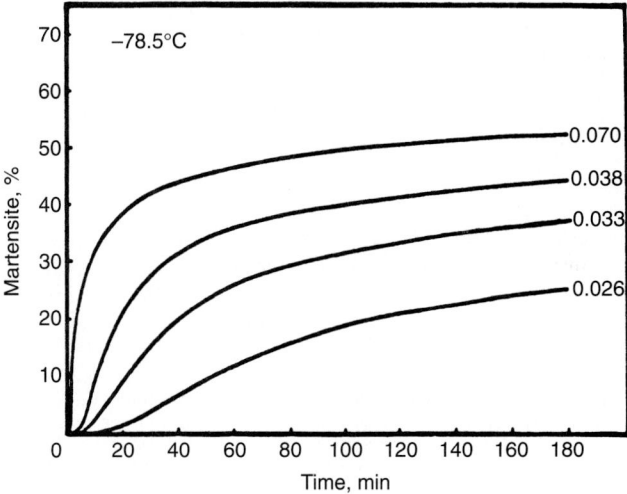

FIGURE 8.14 Isothermal transformation curves for Fe-26Ni-2Mn alloy at −78.5°C for different values of austenite grain size. Grain sizes in millimeters are shown in plot. (*Courtesy of The Institute of Metals, England; after V. Raghavan and A. R. Entwisle, Iron and Steel Institute, Special Report No. 93, 1965, p. 30.*)

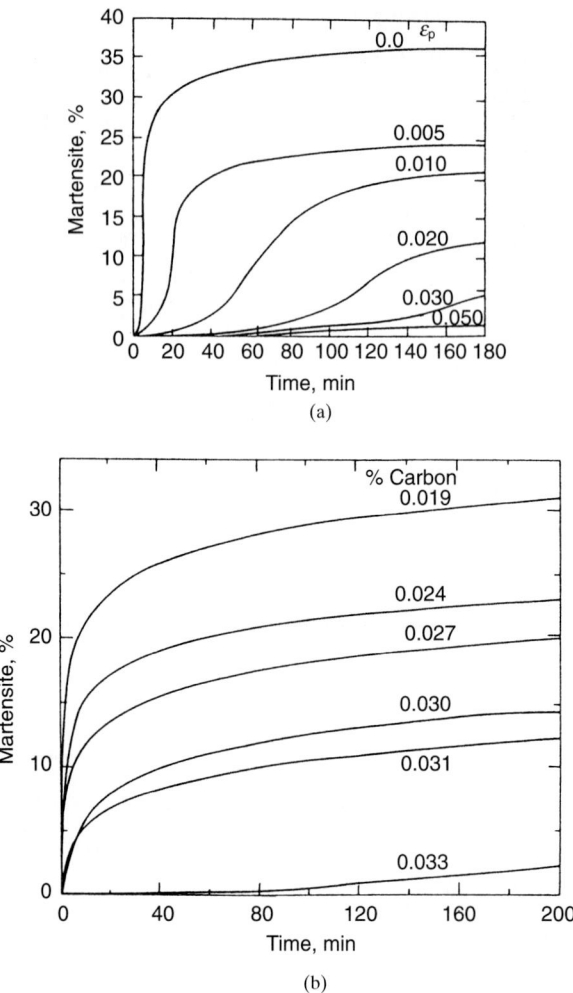

FIGURE 8.15   (*a*) Isothermal transformation curves for an Fe-23Ni-3Mn alloy tested at −100°C as a function of the prior plastic strain in austenite. (*After P. C. Chen and P. G. Winchel, Metall. Trans., vol. 11A, 1980, 1333.*) (*b*) Isothermal transformation curves for an Fe-22Ni-4Mn alloy tested at −196°C as a function of the residual carbon content in austenite. (*After V. I. Izotov and L. M. Uteskiy, Phys. Met. Metallogr., vol. 25, no. 1, 1968, p. 86.*)

7. To obtain the purely isothermal transformation kinetics of martensite below M_s, it is essential to begin with 100% supercooled austenite (without any prior athermal martensite) and monitor its decomposition isothermally.

8. Both lath martensite and twinned martensite could be obtained isothermally in one steel (say, Fe-30 Ni) at different temperatures, e.g., lath martensite formed at a higher temperature (140°C) and twinned martensite at a lower temperature (−30°C).[51]

56.5 μm

FIGURE 8.16 Structure of zigzag pattern of plate martensite by autocatalytic nucleation in an Fe-25.7Ni-0.48C alloy.[54] (*Reprinted by permission of The Metallurgical Society, Warrendale, Pennsylvania.*)

The readers are referred to the extensive reviews provided by Thadhani and Meyers[47] and Zhao and Notis[52] on isothermal transformation of martensite.

8.3.3 Burst Martensitic Transformation

In burst transformation, which has been observed in Fe-Ni and Fe-Ni-C alloys with $M_s < 0°C$, the mode of transformation is quite different from that of athermal or isothermal martensites discussed above (see also Table 8.5). Transformation starts suddenly (at M_B), and a comparatively large volume fraction of martensite plates may form simultaneously in a single event (within milliseconds)—a *burst*, perhaps triggered by a single plate or a few plates formed earlier. This burst transformation is characterized by the following:[49,53]

1. Zigzag morphologies of lenticular martensite plates tend to form by an autocatalytic nucleation and rapid growth of numerous plates (Fig. 8.16) (Table 8.5).

2. Each individual martensite plate is entirely formed with a speed greater than 10^5 cm/s, and the transformation advances by the generation of new plates.

3. This process is often associated with an abrupt release of heat of transformation, which causes a marked temperature rise of the specimen (sometimes 20 to 30°C).

4. The temperature of occurrence of the first burst, called the M_B temperature, is below M_s and is related to the grain size. For example, fine grain size is normally associated with a lower M_B.

5. This type of transformation often lasts over a temperature range of 20 to 30°C; hereafter, this temperature range is known as *the burst region*.

6. In some favorable conditions, the volume of transformation (also called the burst size) may exceed 70% (at a particular temperature), with a rise in temperature in excess of 30°C.

7. The transformation occurring before and after the burst region is associated with mainly single events, whereas the transformation within the burst region seems to be multievent in that many plates may overlap within a very short time interval.

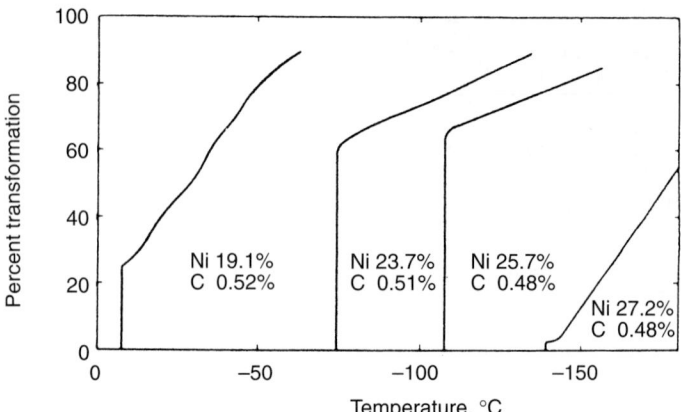

FIGURE 8.17   Transformation curves for Fe-Ni-C alloys. All the steels were austenitized at 1000°C.[54] (*Reprinted by permission of the Metallurgical Society, Warrendale, Pennsylvania.*)

8. The morphology of martensite plates exhibits a platelike or lenticular shape with a midrib and an irregular interface between the martensite and austenite. The midrib plane is considered to be the "habit plane" of the martensitic transformation.[53] A *tweaking* technique, which permits the burst temperature to be varied over a limited range in the same alloy, has been found to increase the usefulness of burst transformation studies.[50]

Figure 8.17 shows typical transformation curves for Fe-Ni-C alloys. With the increase in Ni content of the alloy, the burst phenomenon appears beginning with a few vol% transformation near 0°C, increasing steadily to 70 vol% at M_B −100°C. Beyond this temperature, further alloying causes a sharp reduction of burst size. However, in Fe-Ni alloys, such a decrease of burst size has not been observed; it stays at the same optimum value until the austenite becomes fully stable.

The habit plane of burst martensite is a function of the volume transformation. When the burst is small above 0°C, the habit plane of the so-called butterfly or chevron martensite is $(225)_\gamma$. When large bursts take place, the habit plane is near $\{259\}_\gamma$, and the lenticular martensite plates exhibit the characteristic midrib and internal twins (Fig. 8.16).[54]

8.3.4   Mechanically Induced Martensitic Transformation

The martensitic transformation induced by deformation (i.e., externally applied stress) has been classified into two modes, namely, stress-assisted martensite and strain-induced martensite transformations, based on the origin of the nucleation sites for the martensite plates.[2,55–58] When a stress or a deformation is applied during the phase transformation, a simultaneous change in the mechanical behavior occurs, called *transformation-induced plasticity* (TRIP) or phase transformation plasticity. TRIP occurs in some high-strength metastable austenitic steels exhibiting increased uniform ductility when plastically deformed.

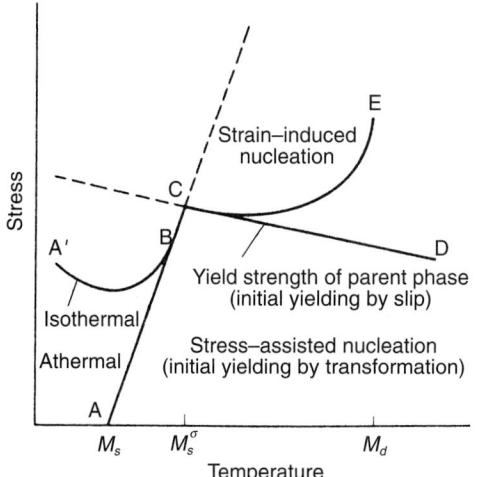

FIGURE 8.18 Schematic diagram showing the critical stress required for the stress-assisted and strain-induced martensitic transformations.[55] (*Reprinted by permission of Pergamon Press, Plc.*)

In stress-assisted martensite, nucleation on the preexisting nucleation sites is assisted thermodynamically by externally applied stress. Strain-induced martensite nucleation is based on the assumption that new nucleation sites or embryos are produced by the substantial plastic deformation. The conditions required for each mode of transformation can be shown in a stress versus temperature diagram (Fig. 8.18). The stress-assisted martensite is most effective when the stress is below the yield stress of the parent phase (point C); this defines the highest temperature M_s^{σ} for which transformation can solely be induced by elastic stresses. For an alloy exhibiting athermal transformation kinetics, the critical stress for stress-assisted martensite formation increases linearly with an increase in temperature in the range between M_s and M_s^{σ}; that is, stress-assisted athermal transformation follows the line ABC.

The strain-induced transformation becomes the dominant mode at a higher temperature ($\geq M_s^{\sigma}$) when the yield stress C is reached. Thus the strain-induced martensite transformation follows the curve CE. Because of the effectiveness of the strain-induced nucleation sites, this curve can initially follow the curve CD for the initiation of parent phase slip. Point E represents the amount of applied stress limited by fracture, which, in turn, defines the highest temperature M_d at which martensite can be mechanically induced. Above this temperature the chemical driving force is so small that it is practically impossible to nucleate martensite by mechanical inducement, even in the plastic strain regime.[2]

In the neighborhood of M_s^{σ}, both transformation modes operate; however, a single transformation mode can be observed only when we move farther away from the M_s^{σ}.

8.3.4.1 *Stress-Assisted Transformation.* For martensitic transformation with an invariant plane strain consisting of shear and normal strain components (γ_T and ε_n, respectively), applied stress increases the thermodynamic (or chemical) driving force by the mechanical driving force U, which can be expressed by[41]

$$U = \tau\gamma_T + \sigma_n\varepsilon_n \tag{8.3}$$

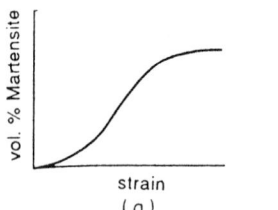 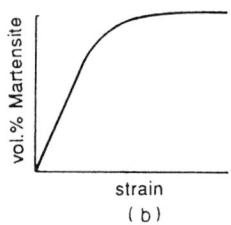

FIGURE 8.19 Schematic representation of transformation behavior showing an increase in the volume percent of mechanically induced martensite with strain at temperatures (*a*) above M_s^σ and (*b*) below M_s^σ.[55] (*Reprinted by permission of Pergamon Press, Plc.*)

where τ is the shear stress resolved along the transformation shear direction in the martensite habit plane and σ_n is the normal stress resolved normal to the habit plane. When transformation occurs at low temperature and well below the yield stress of the material, the volume fraction of martensite formed f is initially proportional to the strain ε_n. For a constant strain, a constant transformation rate f can be obtained. For an alloy exhibiting isothermal characteristics, application of Eq. (8.3) to the theory of isothermal martensitic transformation kinetics gives the observed stress-assisted transformation curve $A'BC$, as shown in Fig. 8.18.[55]

8.3.4.2 Strain-Induced Transformation. The transformation behavior observed in metastable austenitic steels and high-strength TRIP steels at temperatures above M_s^σ and below M_s^σ is schematically shown in Fig. 8.19, which illustrates an increase in the volume percent of mechanically induced martensite with strain.[55]

Figure 8.20 shows a model prediction of equivalent shear stress τ versus equivalent plastic shear strain $\bar{\gamma}^p$ curves during strain-induced transformation in high-strength austenitic steel for various stress conditions at a temperature just above M_s^σ for plane-strain tension. This strain-induced transformation yields a sigmoidal $\bar{\tau}$–$\bar{\gamma}$ curve where low-strain mode is dominated by the "dynamic softening" contribution of transformation strain and high-strain mode is dominated by the "static hardening" effect of the transformation product. These effects become more pronounced by a higher rate of transformation involving an increase in hydrostatic tensile component. Thus strain-induced transformation provides (1) a reverse curvature to the stress-strain curve (to exert a strong stabilizing influence on plastic flow behavior) and (2) an unusual property of strongly pressure-dependent strain hardening.[56]

8.4 FERROUS MARTENSITE

8.4.1 Crystal Structure

Essentially, three different crystal structures of ferrous martensites exist: the bcc or bct α'-martensite, the hcp ε-martensite, and the fct martensite.[3] The most popular ferrous martensite is α' formed in Fe-C and Fe-Ni alloys. The ε martensite forms only in ferrous alloys with a low stacking fault energy (SFE) of austenite such as

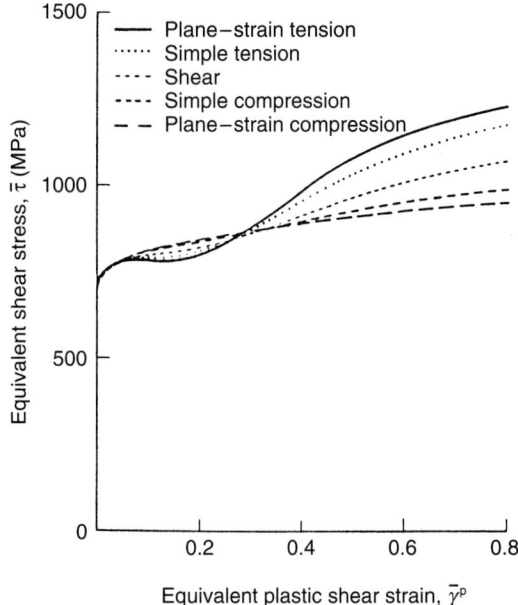

FIGURE 8.20 Model prediction of equivalent shear stress $\bar{\tau}$ versus equivalent plastic shear strain $\bar{\gamma}^p$ during strain-induced transformation in high-strength austenitic steel under various stress states.[56]

Fe-Cr-Ni and Fe-high-Mn alloys. Two types of martensite such as bcc α' and hcp ε form spontaneously on cooling austenitic stainless steels below room temperature. Fct martensite is rare and has been observed only in Fe-Pt and Fe-Pd alloys.[3,59]

In plain carbon steels, martensite is a supersaturated solid solution of carbon in iron which has a body-centered tetragonal structure. Since the carbon atoms are present in the octahedral interstitial positions in the γ phase, they find themselves automatically in the tetragonal c axis (= z axis) of the martensitic lattice (Fig. 8.21), which will result in expansion of the c axis.[60] Thus there is largely a unidirectional distortion which produces the tetragonal lattice. Figure 8.22 shows the effect of carbon content on the lattice parameter of austenite and martensite. This illustrates that in the case of martensite the c axis increases markedly, and the a axis decreases a little with an increase in carbon content above 0.25%.[15] Therefore, the axial ratio c/a increases with carbon content and is given approximately by the equation[48]

$$c/a = 1 + 0.467 \times \text{wt\% C} \tag{8.4}$$

In pure iron or in very low-carbon steel, the structure would be bcc, free of distortion. Bcc martensite is also favored when only substitutional solutes are present. However, tetragonal martensite forms when a large amount of substitutional elements is added to steel and ordering occurs (e.g., the addition of Ti to Fe-30% Ni alloy).[15]

8.4.2 Crystallography

Almost all the mechanisms for martensite formation are aimed at producing an invariant plane strain, and only a few amongst them are discussed below.

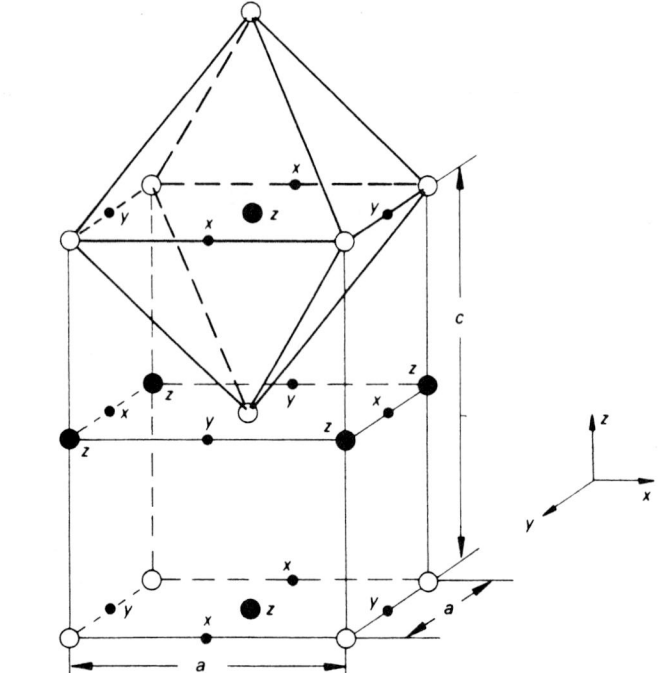

FIGURE 8.21 Body-centered tetragonal structuré of martensite in Fe-C alloys showing z sites fully occupied by carbon atoms.[41] (*Reprinted by permission of Pergamon Press, Plc.*)

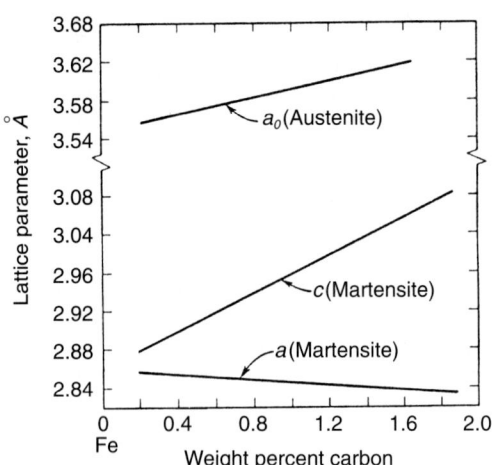

FIGURE 8.22 Effect of carbon on the lattice parameter of austenite and martensite in steel.[28,32] (*Reprinted by permission of PWS-Kent Publishing Co., Boston, Massachusetts.*)

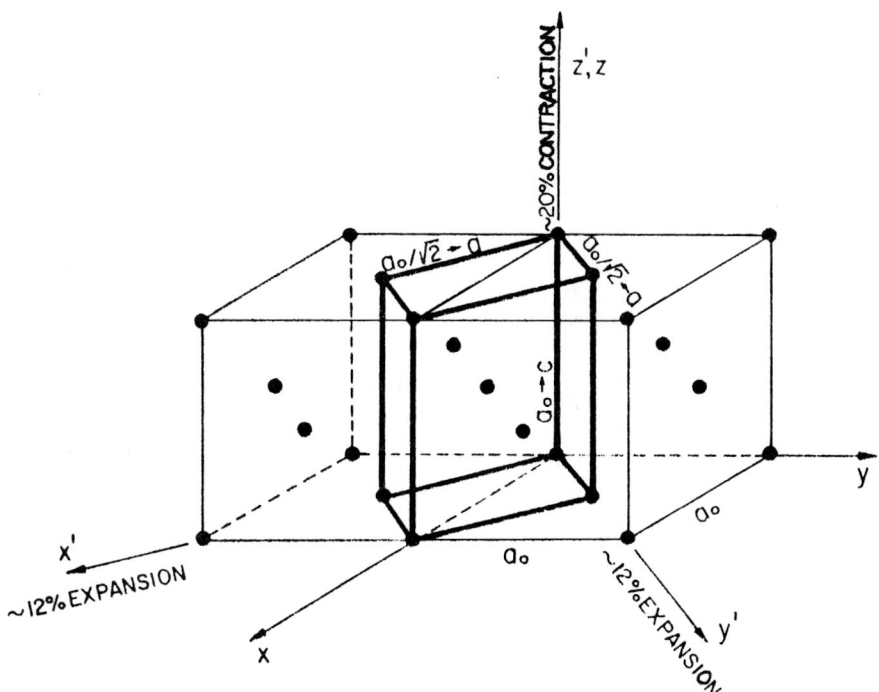

FIGURE 8.23 Bain distortion and Bain correspondence for the formation of bcc (bct) martensite from fcc austenite in ferrous alloys. The correspondence cell (heavy lines) in the austenite transforms to a unit cell of martensite after a homogeneous lattice deformation which involves a homogeneous expansion of about 12% along the x $[100]_\gamma$ and y $[010]_\gamma$ directions, respectively, and homogeneous contraction of about 20% along the z $[001]_\gamma$ direction. The principal directions along x' $[100]_\alpha$, y' $[010]_\alpha$, and z' $[001]_\alpha$ are indicated.[61] (*Reprinted by permission of The Metallurgical Society, Warrendale, Pennsylvania.*)

8.4.2.1 The Bain Mechanism. In 1921, Bain made a simple, but elegant, proposal of the deformation of the austenite lattice for the fcc, austenite → bcc (or bct), martensite transformation. He suggested that the fcc unit cell may be regarded as a unit cell with an axial ratio of $\sqrt{2}/1$ and showed how a bct unit cell could be outlined within a two-unit cell of austenite (Fig. 8.23).[61] According to the Bain mechanism, transformation to martensite is achieved by merely homogeneous contraction of about 20% along the $[001]_\gamma$ (corresponding to the c axis of the martensite unit cell) and a homogeneous expansion of about 12% along the $[100]_\gamma$ and $[010]_\gamma$ (corresponding to x and y directions, respectively); these amounts depend, of course, on the alloying elements. Such a homogeneous distortion is often called *Bain strain*, *Bain distortion*, or *lattice deformation* **B**.

The Bain distortion involves minimum displacement of individual atoms in generating the martensite lattice. This idea has been retained in subsequent analyses. The maximum c/a ratio of martensite is 1.08, and therefore the martensite lattice is much nearer to bcc than to fcc. Figure 8.22 shows the increase of tetragonality with increase in carbon content.

Although the presence of cubic martensite in carbon-free iron and ordered Fe-25 at% Pt alloy and tetragonal martensite in Fe-10% Al-1.5% C alloy supports his view, weaknesses of this are:[62]

1. The mechanism does not involve shear.
2. Bain transformation does not fulfill the requirement of occurrence of transformation with an undistorted plane.[63] That is, it does not predict the observed habit plane and orientation relationship between the martensite plates and the austenite matrix. The magnitude of the principal strains of the Bain distortion is inconsistent with the surface relief observations, which suggests that the $\gamma \rightarrow \alpha'$ habit plane is essentially undistorted.
3. Such a mechanism can give rise to only 3 orientations of martensite; in practice, however, 24 are found.
4. This alone cannot account for the observed invariant plane strain associated with martensite transformation.

However, this model is useful because it indicates the corresponding planes of atoms in the two systems. TEM and X-ray studies have confirmed the occurrence of Bain distortion and inhomogeneous shear in ferrous martensitic transformations;[64] the former is effected by coherency (or transformation) dislocations, while the latter is by anticoherency (or misfit) dislocations.[65] The Bain distortion can be applied to other types of martensitic transformation, provided that different lattice deformations are taken into consideration.[66]

8.4.2.2   Greninger and Troiano Model.   Greninger and Troiano[65] first observed that the shape change could not be produced only by lattice deformation, and they proposed that another deformation was needed to satisfy the requirement of invariant plane strain (IPS). Thus the homogeneous lattice deformation, when combined with fine-scale inhomogeneous deformation (lattice-invariant deformation) such as slip, twinning, or faulting, leads to the undistorted γ/α' interface. This additional deformation, being a shear, is also called the inhomogeneous (or complementary) shear of crystallographic theory. Figure 8.24 shows schematically how the lattice deformation combines with either the slip mode or the twinning mode of lattice-invariant deformation to produce martensite plates.[67,68]

As a result of inhomogeneous shear, one can expect to observe any type of substructure in the martensite. In fact, the study of the TEM micrograph has revealed the occurrence of heavy dislocated (i.e., slipped) lath martensite in Fe-0.16% C alloy (Fig. 8.25) and transformation twins (in the form of fine striations) across the plate martensite in 1.28% C steel (Fig. 8.26).[69] The term *transformation twin* is synonymous with *internal twins*, but a problem still persists. The habit plane produced by the incorporation of Bain distortion and inhomogeneous shear is still not rotated, as required from observation.

8.4.2.3   Phenomenological Crystallographic Theory of Martensitic Transformation.   The above approach, which was further developed by the incorporation of Wechsler, Lieberman, and Read[70] and Bowles and Mackenzie,[71] is called the *double shear theories, invariant plane strain theories,* or, better, the *phenomenological theories* of *martensitic crystallography (PTMC)*. The PTMC allows all the crystallographic characteristics of the parent and product phases to be mathematically related; however, it does not provide clear understanding about the transformation mechanisms which depend precisely on the structure of the transformation inter-

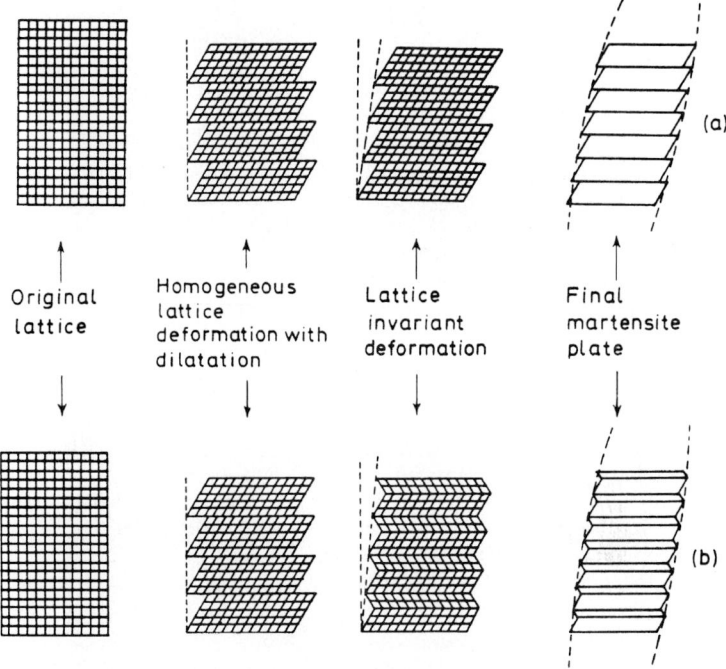

FIGURE 8.24 Schematic diagram illustrating the combined effect of lattice deformation and lattice invariant deformation by (*a*) slip and (*b*) twinning to produce martensite plates.[67,68] (*Reprinted by permission of Butterworths, London.*)

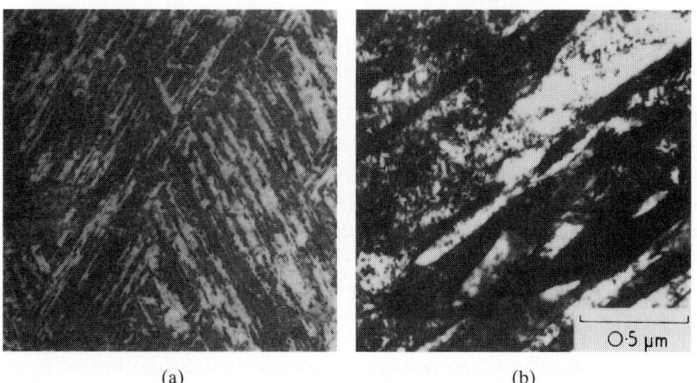

(a)   (b)

FIGURE 8.25 Martensite formation in Fe-0.16C alloy on quenching from 1050°C: (*a*) optical micrograph; (*b*) thin-film electron micrograph illustrating heavily dislocated laths.[38] (*After Y. Ohmori.*)

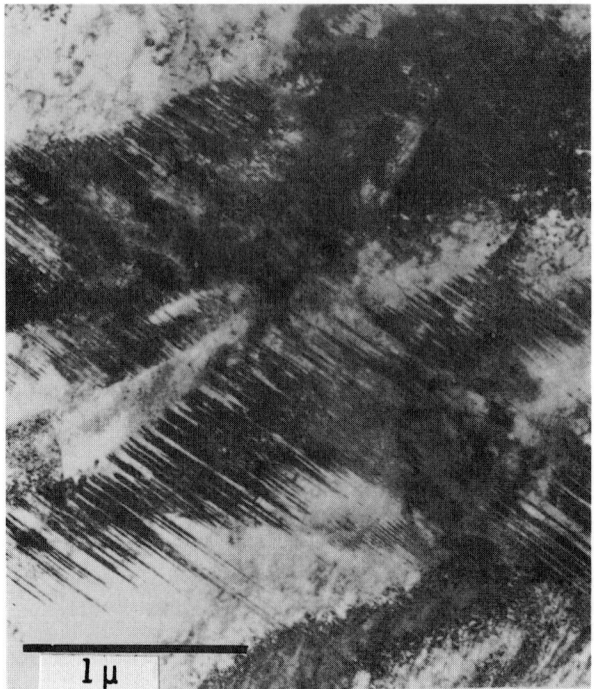

FIGURE 8.26 Electron micrograph of plate martensite in a quenched 1.28% C steel showing fine transformation twins.[69] (*Reprinted by permission of ASM International, Materials Park, Ohio.*)

face. The total transformation crystallography is described by three basic or phenomenological deformation steps **B**, **R**, and **P**:

1. *Homogeneous Bain distortion* **B** for the necessary lattice change from parent phase to martensite (or product) phase

2. An *inhomogeneous lattice-invariant shear* (or *deformation*) **P** which produces undistorted habit plane without altering the lattice (which is associated with faulting, slip, or twinning) and ensures that the martensite/parent interface is semicoherent and glissile

3. A *rigid body rotation* **R** to ensure that the undistorted habit plane becomes also unrotated (i.e., invariant habit plane results)

There is no time sequence implied as to which deformation step will occur when.[72] However, the net effect of these three operations must be equivalent to the total shape deformation and, hence, to the observed surface distortion.[64] These theories have gained general acceptance because of the agreement between theory and experiment in a number of cases. However, when they are applied to iron alloys and steels, it can be concluded that only a few specific martensitic transformations are adequately described.[73]

8.4.2.4 *Mathematical Expression of the Phenomenological Theory of Martensitic Crystallography.* The basic equation of the crystallographic theory of martensitic transformation just described can be written in matrix form by IPS shape deformation $\mathbf{P}_1$ as[61,74]

$$\mathbf{P}_1 = \mathbf{R}\overline{\mathbf{P}}\mathbf{B} \tag{8.5}$$

where $\mathbf{B}$, $\overline{\mathbf{P}}$, and $\mathbf{R}$ are all (3×3) matrices representing, respectively, Bain distortion, simple inhomogeneous lattice-invariant shear (twinning, slip, or faulting), and rigid-body rotation and $\mathbf{P}_1$ is the total strain due to transformation (called the shape strain). In other words, the matrix product $\mathbf{R}\overline{\mathbf{P}}\mathbf{B}$ is equivalent to the shape deformation $\mathbf{P}_1$ where rotation $\mathbf{R}$ rotates the plane left undistorted by $\overline{\mathbf{P}}\mathbf{B}$ to its original position. That is, $\mathbf{P}_1 = \mathbf{R}\overline{\mathbf{P}}\mathbf{B}$ is an IPS shape deformation.

Although Eq. (8.5) suggests the inhomogeneous shear $\mathbf{P}$ with Bain distortion, the same end result can be found by allowing the shear to take place in the parent phase prior to the Bain distortion. In this situation, the basic equation becomes

$$\mathbf{P}_1 = \mathbf{RBP} \tag{8.6}$$

where $\mathbf{P}$ is a simple shear as before. The IPS shape strain or deformation $\mathbf{P}_1$ can be written as

$$\mathbf{P}_1 = \mathbf{I} + m\mathbf{dp}' = \begin{pmatrix} 1 & 0 & 0 \\ 0 & 1 & 0 \\ 0 & 0 & 1 \end{pmatrix} + m[d_1\ d_2\ d_3](p_1\ p_2\ p_3) \tag{8.7}$$

$$= \begin{pmatrix} 1+md_1p_1 & md_1p_2 & md_1p_3 \\ md_2p_1 & 1+md_2p_2 & md_2p_3 \\ md_3p_1 & md_3p_2 & 1+md_3p_3 \end{pmatrix} \tag{8.8}$$

where $\mathbf{I}$ is a (3×3) identity matrix, m is the magnitude of the lattice-invariant deformation, and $\mathbf{p}'$ (prime implying transpose) being a plane normal is represented by a (1×3) single-row matrix, in contrast to $\mathbf{d}$, which is a vector represented by a (3×1) single-column matrix.

With respect to Fig. 8.2b, the Bain distortion for the fcc → bcc (or bct) transformation can be expressed as

$$\mathbf{B} = \begin{pmatrix} \sqrt{2}a/a_0 & 0 & 0 \\ 0 & \sqrt{2}a/a_0 & 0 \\ 0 & 0 & c/a_0 \end{pmatrix} \tag{8.9}$$

However, when typical values of lattice parameters a for the fcc lattice and a_0 and c for the bct lattice are inserted for martensitic transformation in steels, a representative $\mathbf{B}$ matrix becomes

$$\mathbf{B} = \begin{pmatrix} 1.12 & 0 & 0 \\ 0 & 1.12 & 0 \\ 0 & 0 & 0.80 \end{pmatrix} \tag{8.10}$$

which does not contain an undistorted plane and so does not conform to the IPS condition. Referring to Eq. (8.5), we see that $\mathbf{P}$ is a simple shear and therefore can be expressed as $\mathbf{I} + m d\mathbf{p}'$. Also the inverse of $\mathbf{P}$, $\mathbf{P}^{-1} = \mathbf{I} - m d\mathbf{p}'$, denotes a simple shear of the same amount, on the same plane, but in the opposite direction. This means that both $\mathbf{P}$ and $\mathbf{P}^{-1}$ are invariant-plane strains. It becomes now easy to rewrite Eq. (8.5) as

$$\mathbf{P}_1 \mathbf{P}_2 = \mathbf{RB} \qquad (8.11)$$

where $\mathbf{P}_2 = \mathbf{P}^{-1}$. Since $\mathbf{P}_1$ and $\mathbf{P}_2$ are invariant-plane strains, their product $\mathbf{RB}$ is also an invariant-line strain $\mathbf{S}$. Here *invariant line* is defined as the line of intersection of the *planes* which are invariant to $\mathbf{P}_1$ and $\mathbf{P}_2$. Once the invariant-line strain $\mathbf{S}$ is determined, all the crystallographic features of a specific martensite transformation can be predicted. Note that $\mathbf{RB} = \mathbf{S}$ develops the lattice deformation of the transformation.

The important findings of the invariant-line strain analysis, assuming the shape strain to be $\mathbf{P}_1 = \mathbf{I} + m_1 \mathbf{d}_1 \mathbf{p}'_1$ and the simple shear (mathematically prior to Bain distortion) to be $\mathbf{P}_2 = \mathbf{I} + m_2 \mathbf{d}_2 \mathbf{p}'_2$, are expressed below, where the magnitudes, directions, and planes of the component invariant-plane strains are given, respectively, in terms of m, $\mathbf{d}$, and $\mathbf{p}'$ as

$$\mathbf{d}_1 = \frac{\mathbf{S} Y_2 - Y_2}{\mathbf{p}'_1 Y_2} \qquad (8.12)$$

$$\mathbf{p}'_1 = \frac{\mathbf{q}'_2 - \mathbf{q}'_2 \mathbf{S}^{-1}}{\mathbf{q}'_2 \mathbf{S}^{-1} \mathbf{d}_1} \qquad (8.13)$$

where Y_2 is any vector lying in $\mathbf{p}'_2$ (except the invariant line $\mathbf{x}$) and $\mathbf{q}'_2$ is any normal (other than $\mathbf{n}'$, the row eigenvector of $\mathbf{S}^{-1}$, that is, $\mathbf{n}' \mathbf{S}^{-1} = \mathbf{n}'$) to a plane containing $\mathbf{d}_2$. The normalization parameter for $\mathbf{d}_1$ in Eq. (8.12) is $1/m_1$, and thus $\mathbf{P}_1$, m_1, $\mathbf{d}_1$, and $\mathbf{p}'_1$ all can be determined. Finally, $\mathbf{RB} = \mathbf{S}$ defines the elements of $\mathbf{P}_1$.

The above description of the crystallographic theory is analogous to the analysis presented by Bowles and Mackenzie, but the treatments of Wechsler et al. and Bullogh and Bilby are similar.

Some variations in the basic theory just mentioned include the introduction of a dilatation parameter, which relaxes slightly the requirement for the habit plane to be undistorted, and the incorporation of two inhomogeneous shear systems such that

$$\mathbf{P}_1 = \mathbf{RBS}_2 \mathbf{S}_1 \qquad (8.14)$$

where $\mathbf{S}_2$ and $\mathbf{S}_1$ are the two inhomogeneous shear systems involved. Neither of these modified methods is free of criticism, and it has been suggested that the double shear method loses generality.

Figure 8.2b shows the invariant plane strain of martensite comprising shear (or tensile) and dilatational (or compressive) components, i.e., one $m_1^P \mathbf{d}_1^P$ parallel to, and the other $m_1^n \mathbf{P}_1$ normal to, the invariant plane, which can be written as

$$m_1 d_1 = m_1^P \mathbf{d}_1^P + m_1^n \mathbf{P}_1 \qquad (8.15)$$

Since shear strain does not produce a volume change, m_1^n is simply the volume change associated with the transformation:

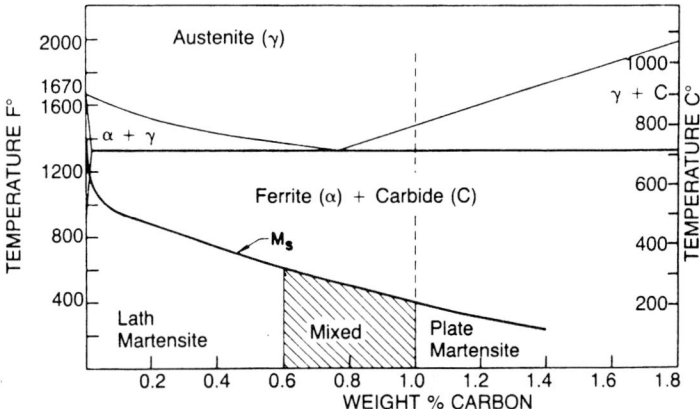

FIGURE 8.27 Effect of carbon concentration of the ranges of lath and plate martensite formation of iron-carbon alloys.[77] (*Reprinted by permission of The Metallurgical Society, Warrendale, Pennsylvania.*)

$$\frac{\Delta V}{V_\gamma} = \frac{V_{\alpha'} - V_\gamma}{V_\gamma} \qquad (8.16)$$

8.4.3 The Morphology of Martensite

Fe-C martensite can be broadly classified into two types: (1) lath martensite (Type I) with a high density of tangled dislocations with a few twins or without twinned regions within lath martensite[75] and (2) plate martensite (Type II) containing internal substructure of twins with or without dislocations.[76] These morphologies depend upon the carbon content of the alloy. The lath martensite forms in low-carbon and medium-carbon steels with up to 0.6% C; mixed lath and plate martensite takes place between 0.6 and 1.0% C; and plate-type martensite predominates above about 1.0% C (Fig. 8.27).[77]

8.4.3.1 Type I: Lath Martensite. Many workers have indicated the habit plane of lath martensite to be near $\{111\}_\gamma$ and the occurrence of the Kurdjumov-Sachs orientation relationship between the sets of laths and parent austenite grains.[77] Lath martensite is observed in low-carbon and medium-carbon steels; in Fe-V, Fe-W, Fe-Sn, Fe-Mn, and Fe-Mo alloys;[78] and in Fe-Ni (<25% Ni) alloys.[24] They are usually grouped together in differently oriented packets (or bundles)—the fundamental growth units—with relatively uniform lath sizes within them (Fig. 8.25*a*). Each adjacent parallel lath within a packet has the same habit plane, the same orientation relationship, and the same shape deformation and is separated mostly by low-angle boundaries,[79] although high-angle boundaries are sometimes observed.[80,81] In a few cases the misorientation corresponds to a near-twin orientation.[82] They are normally untwinned and contain a high density of internal dislocations (Fig. 8.25*b*) on the order of 10^{12} dislocations per square centimeter (in low-carbon lath martensites).

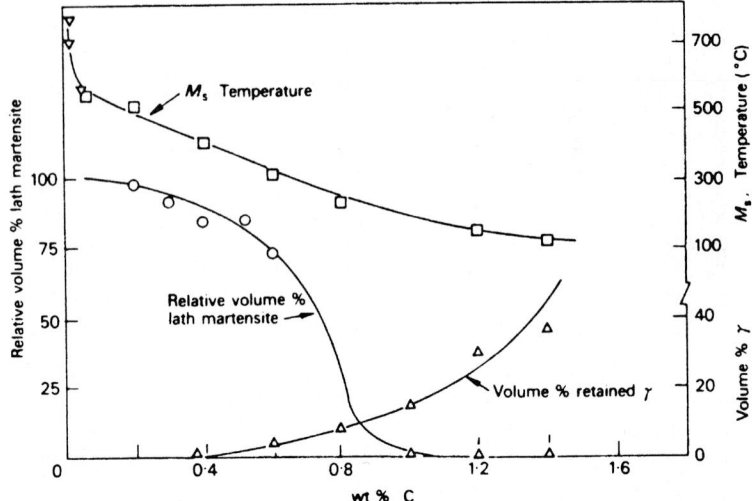

FIGURE 8.28 Effect of carbon content in steels on volume percentage of lath marten-site and retained austenite.[45] (*Courtesy of The Institute of Metals, England; after G. Speich.*)

Since transformation to lath martensite occurs well above room tempera-ture, there is a relatively small amount of retained austenite between the laths (Fig. 8.28).[45]

In plain carbon steels, high density of dislocation tangles is observed within laths as a result of high M_s temperature, while in Fe-Ni alloys, straight, uniformly dis-tributed dislocations are found within laths, consistent with low-temperature defor-mation of these alloys.

8.4.3.2 Mixed Lath and Plate Martensite. For mixed lath and plate martensite, the habit plane is $\{225\}_\gamma$ and there is a Kurdjumov-Sachs orientation relationship.

There is a transition of morphology from a lath- to a plate-type martensite between 0.6 and 1.0% C in Fe-C alloys (Fig. 8.28). With the increase in carbon content, the amount of lath martensite decreases and that of plate martensite increases. This is associated with the lowering of the M_s temperature and more retained austenite (Fig. 8.28).

8.4.3.3 Type II: Plate Martensite. The habit plane changes from $\{225\}_\gamma$ at 1.0 to 1.4% carbon concentration to $\{259\}_\gamma$ at carbon concentrations in excess of 1.5%. The addition of N, Ni, Pt, and Mn (austenite stabilizers) tends to favor the transition to plate martensite, but Si, Cr, W, V, and Mo (ferrite stabilizers) prevent its formation.

In high-carbon steels (0.8 to ~1.8% carbon content) and Fe-Ni (29 ~ 33%) steels, the martensite is found to be exclusively plate or lenticular martensite. This is asso-ciated with lower M_s temperature and more retained austenite (Fig. 8.28). In the medium- and high-carbon steels, martensite consists of lenticular plates with a core, called the *midrib*, within them. The midrib is planar and nearly parallel, occasion-ally to $\{225\}_\gamma$ or mostly to $\{259\}_\gamma$ habit.[15] The high-nickel $\{259\}_\gamma$ martensite also con-tains the midrib. In general, $\{225\}_\gamma$ martensites are partially twinned, while $\{259\}_\gamma$

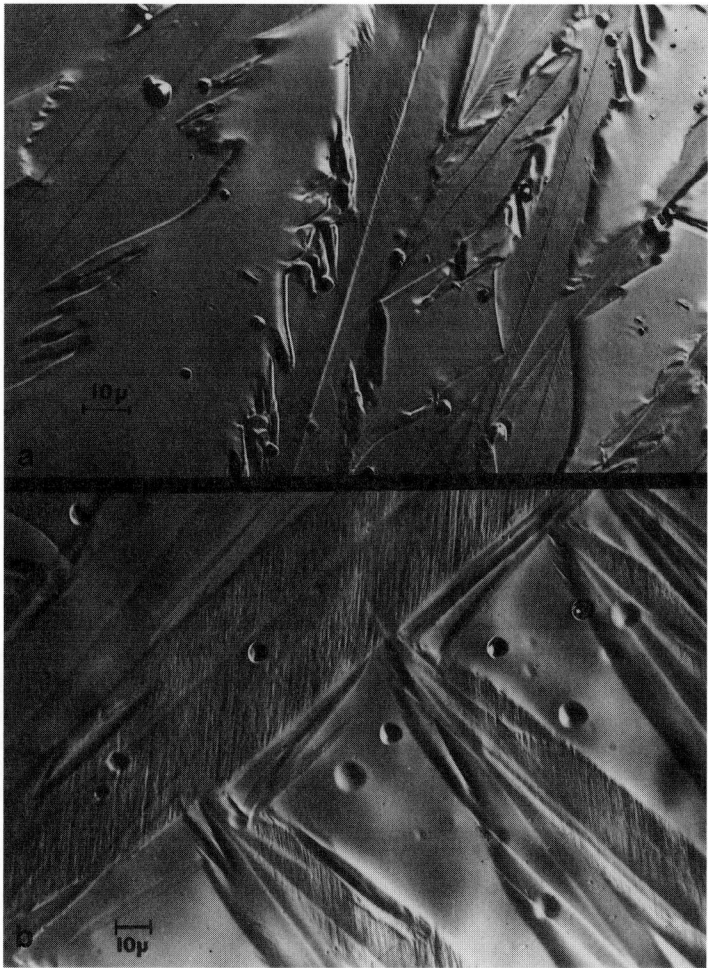

FIGURE 8.29 (*a,b*) Optical micrographs of lenticular plate martensite in Fe-33.2Ni alloy where internal twins are revealed by deep etching (*b*). (*Courtesy of C. M. Wayman.*)

martensites are completely twinned.[63,83] Note that the midrib is the plane where initiation of the transformation occurs and that the interface advances on either side in the opposite direction by means of ledges parallel to the midrib plane.[15] Figure 8.26 is a typical transmission electron micrograph of plate martensite in a 1.28% C steel showing variation in size and fine structure of parallel {112} type twins.[69] Figure 8.29 shows the optical micrograph of the lenticular plate martensite in Fe-33.2 Ni alloy where internal twins are revealed by deep etching (Fig. 8.29*b*).

Unlike lath α', plate-type ferrous α' usually forms in a self-accommodating manner consisting of several habit-plane variants. As a result, we observe nonparallel α' plates with relatively large size differences.[14]

8.4.3.4   Other Morphologies.   In Fe-Ni-C alloys, three distinct morphologies of martensites exist which arise either by quenching to below the M_s temperature or by strain ε at a temperature above M_s but below the thermodynamic equilibrium temperature T_0.[84] These are as follows:[85]

1. *Lenticular martensite* exhibits a $\{259\}_\gamma$ habit and is partially internally twinned with a midrib and forms at intermediate (or closer to room) temperature (Fig. 8.30a).

2. *Thin-plate martensite* exhibits a $\{3, 10, 15\}_\gamma$ habit and is fully twinned without a midrib and forms at the lowest temperature (below −150°C). When it is viewed under a microscope, the sides of the martensite plates appear parallel. Figure 8.30b is a transmission electron micrograph of a thin martensite plate formed at −40°C in an Fe-30 Ni-0.39 C alloy. The factors responsible for the formation of thin-plate martensite in ferrous alloys appear to be (1) a high strength of austenite, (2) a low M_s temperature, (3) a high tetragonality of martensite, (4) a small transformation volume change, and (5) a small magnitude of transformation shape strain. A high tetragonality of martensite represents a small twinning shear, a low twin boundary energy, and a low magnitude of shape strain.[86,87]

3. *Butterfly martensite* morphology (named after its appearance)—Types A and B— form at the highest temperature (between 0 and −60°C). Figure 8.31 shows the butterfly martensite morphology formed at −10°C in an Fe-18 Ni-0.7 Cr-0.5 C alloy. It is characterized by a substructure of (straight) dislocations and twins, $[225]_\gamma$ habit plane, and K-S orientation relationship.

In some ferrous alloys, a needlelike morphology, called *surface martensite*, forms under some conditions. It preferentially nucleates at the surface and grows into a needle shape due to relaxation of matrix constraints near the free surface[1,2] during polishing or thinning of specimens. This is not representative of the bulk transformation.

Another form is observed in certain steels, particularly austenitic stainless steels and Fe-Mn-C alloys in which the fcc → hcp transformation occurs in the banded morphology (Fig. 8.32). These bands delineate $\{111\}_\gamma$ planes.[2] This type of morphology is similar to that which is observed in cobalt and its alloys and involves a fully coherent interface without lattice-invariant deformation in the usual sense.[2]

8.5   NONFERROUS MARTENSITE

Table 8.1 provides a classification of the nonferrous martensitic transformation which has been reviewed by Delaey et al.[26] Examples of a first group are the Co and Ti alloys.[1] This fcc → hcp transformation is linked to the $\{111\}_\gamma$ habit plane. The Co-based martensites usually possess the hcp structure; however, more complex close-packed layered structures such as 48 R, 84 R, and 126 R types are observed in Co-Al alloys. The Ti-based martensite is hcp at low temperature and bcc at high temperature. Both lath and plate morphologies are found in Ti- and Zr-based alloys. Slip mode is observed as the lattice-invariant deformation in lath martensite, whereas the twinning mode is likely in the plate martensites.[3]

Examples of a second group are the Cu-, Ag-, and Au-based β alloys which have been reviewed by Delaey et al.[88] Depending on the composition, three types of close-packed martensite can form by quenching or stressing. The factors responsible for exact structures include the stacking sequence of the close-packed struc-

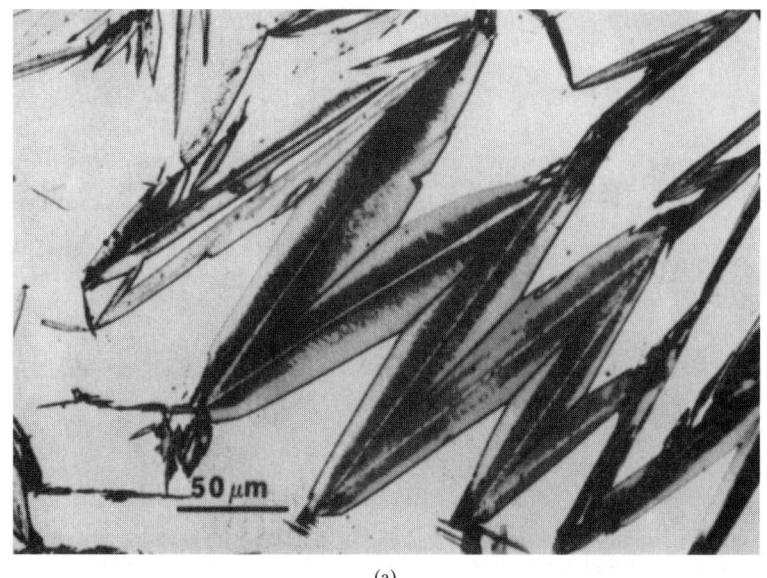

(a)

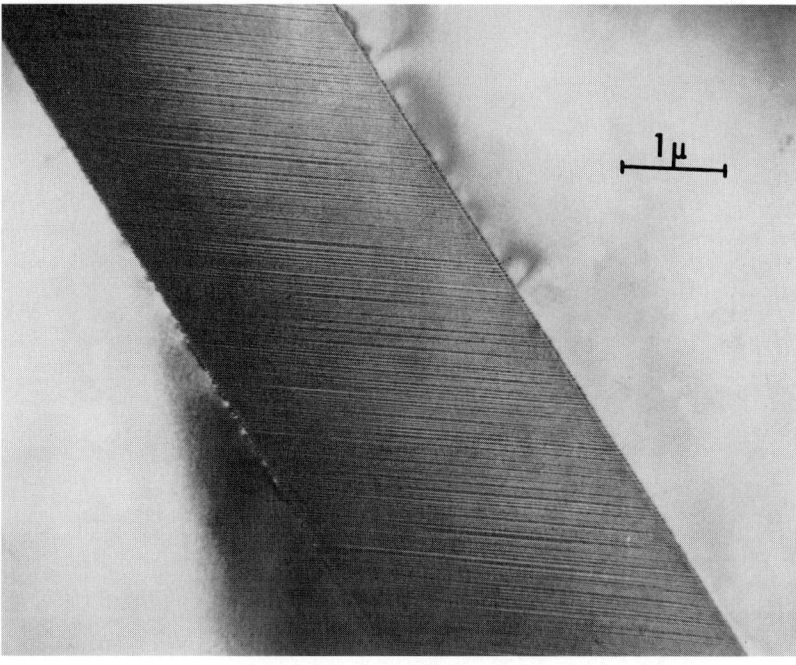

(b)

FIGURE 8.30 (*a*) Optical micrograph showing lenticular martensite formed at −79°C in an Fe-31Ni-0.23C wt% alloy. The more heavily etched linear central region of the plates is known as a midrib.[2] (*b*) Transmission electron micrograph showing a thin martensite plate and adjacent retained austenite formed at −40°C in an Fe-30Ni-0.39C alloy. The striations within the plate running from left to right are transformation twins. (*Courtesy of C. M. Wayman.*)

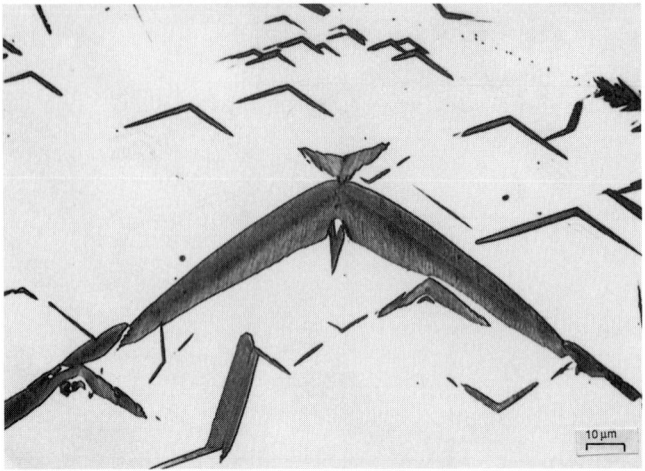

FIGURE 8.31 Optical micrograph showing butterfly martensite morphology formed at –10°C in an Fe-18Ni-0.7Cr-0.5C alloy. (*Courtesy of M. Umemoto.*)

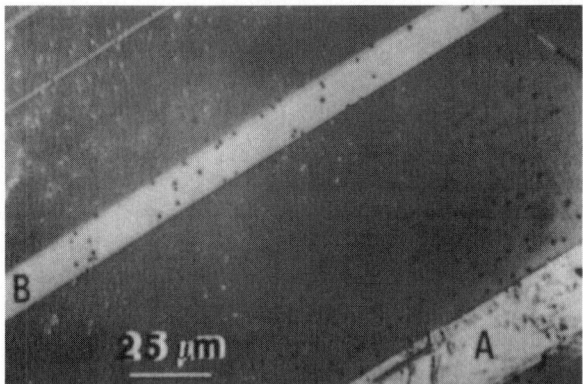

FIGURE 8.32 Optical micrograph showing bands of hcp martensite (A and B) formed in the fcc austenite matrix in an Fe-12Mn-0.48C wt% alloy. (*Courtesy of C. M. Wayman and H. K. D. H. Bhadeshia.*)

ture, the long-range order of the martensite as determined from the parent β-phase ordering, and the deviations from the regular hexagonal arrangements of martensites. The last factor arises from the difference in the sizes of the constituent atoms. The stacking sequences of the major three phases are ABC, ABCBCACAB, and AB, respectively.

One of the important findings in these β-phase alloys is the successive stress-induced martensitic transformations, as shown in Fig. 8.33.[89] When a single crystal

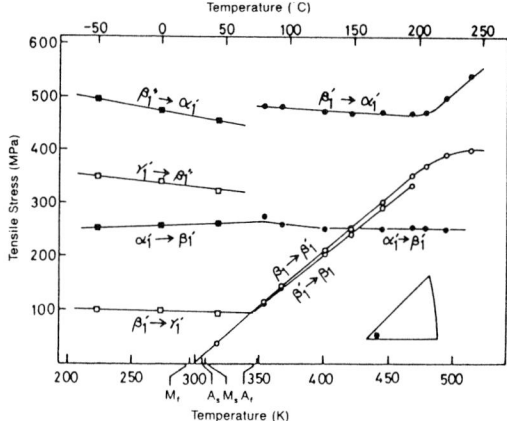

FIGURE 8.33 Critical stresses as a function of temperatures for the various stress-induced martensite transformations in a Cu-Al-Ni alloy.[89]

of Cu-Al-Ni is stressed above a critical value at, for example, 320 K, one obtains the parent $\beta_1 \rightarrow \beta_1'$, the $\beta_1' \rightarrow \gamma_1'$, the $\gamma_1' \rightarrow \beta_1''$, and finally the $\beta_1'' \rightarrow \alpha_1'$ martensite. Other examples of the second group are the Ni-Ti-based alloy systems which constitute the shape memory alloys (SMAs), which are extensively discussed later.

Regarding the third group, in only a few cases, as in In-Tl, has martensitic transformation been well established. In other systems, they are classified as quasi-martensitic.[3]

Martensite Morphologies. Martensite morphologies existing in Cu-Zn-Al SMA are plate, spear, bamboo, noose, round-spot, line, and dotlike martensites. During transformation and inverse transformation, the thermoelastic martensites are rising and falling, growing and shrinking, splitting and merging. The growth patterns of thermoelastic martensite are grouped into three types: fast growth, very slow expansion, and uniform automatic growth.[90]

8.6 NUCLEATION AND GROWTH IN MARTENSITIC TRANSFORMATION

8.6.1 Free-Energy Change of a Martensitic Transformation

Figure 8.34 is a schematic representation of the change in bulk free energies of martensite and austenite as a function of temperature. Temperature T_0 is the temperature at which these phases are in thermodynamic equilibrium. At any temperature T below T_0, it is thermodynamically possible to form martensite, since the change in bulk free energy given by $\Delta G_V^{\gamma \rightarrow \alpha'} = G^{\alpha'} - G^{\gamma}$ is negative and is therefore the driving force for the martensitic transformation.

The formation of a martensite plate produces a substantial amount of elastic strain due to the shape change and the constraints of the surrounding matrix. This strain energy ΔG_ε, which is a positive quantity, consumes a portion of the free energy released by the transformation. Also, the formation of a martensite nucleus creates a new interface with associated surface energy γ. Since both ΔG_ε and γ are the

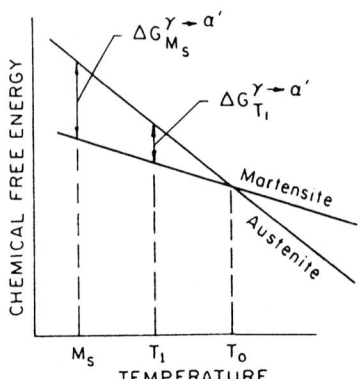

FIGURE 8.34 Schematic diagram showing the free-energy change for a martensitic transformation.

activation barriers to nucleation, further cooling to a temperature M_s much lower than T_0 is essential to give sufficient driving force to begin the martensitic transformation. Thus, at M_s temperature, we obtain

$$\Delta G_v^{\gamma \to \alpha'} = \Delta S(T_0 - M_s) \tag{8.17}$$

This is a negative number, where ΔS is the entropy change for the martensitic transformation. Experimental results show that $\Delta G_v^{\gamma \to \alpha'}$ at the M_s temperature in a typical iron-base alloy is approximately 300 cal/mol (1225 J/mol).[61]

8.6.2 Nucleation of Martensite

The martensitic transformation is a first-order solid-state reaction[2] that occurs by nucleation and growth. Here both parent and martensite phases coexist on cooling between M_s and M_f and on heating between A_s and A_f. In most cases (excluding thermoelasticity) and particularly in ferrous alloys, the growth of martensite plate proceeds so fast that the transformation kinetics are nucleation-controlled; this allows the direct measurement of nucleation kinetics from macroscopic observations.[91]

Various mechanisms of martensite nucleation have been proposed in the past; they can be classified into (1) localized nucleation models (using the concepts of diffusional kinetics) and (2) (static and dynamic) lattice instability models. All nucleation models can further be grouped into classical and non-classical. The former model comprises lattice perturbations of fixed amplitude and varying size.[91]

In the classical nucleation theory, martensite nuclei form along a path of constant composition and structure, and embryo size is preferably expressed in terms of particle volume or linear dimension. Let us assume that the embryos are in the form of thin oblate spheroids of radius r and semithickness c (Fig. 8.35) so that the volume is approximately $4\pi r^2 c/3$ and surface area is $2\pi r^2$ ($r \gg c$). The martensite shape deformation is IPS with a shear component γ_T and a normal (or dilatational) component ε_n. The total free-energy change associated with the formation of a classical martensitic embryo $\Delta G(r, c)$ can be expressed, according to Olson and Cohen, as

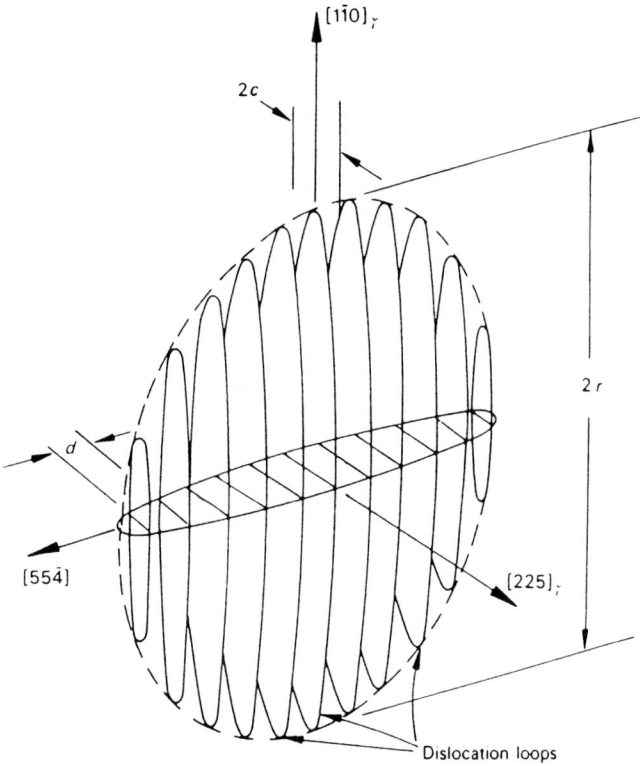

FIGURE 8.35 Martensite embryo.[34] (*Reprinted by permission of Pergamon Press, Plc; after Knapp and Dahlinger.*)

$$\Delta G(r, c) = \frac{4\pi r^2 c}{3}\Delta G_v + \frac{4}{3}\pi r^2 \frac{Ac}{r} + 2\pi r^2 \gamma + \Delta G_d + \Delta G_i \qquad (8.18)$$

where ΔG_v with negative sign is the chemical free energy per unit volume of transformation, γ is the martensitic nucleus/parent phase interfacial energy per unit volume, and Ac/r is the strain free energy per unit volume of martensite formed, arising from IPS strain or shape deformation, ΔG_d is the lattice defect free energy, and ΔG_i is the nucleus-defect interaction energy. And A, being the elastic strain energy factor, is a function of the elastic constants and of the shear and dilatational strains, which is expressed as

$$A = \frac{r}{c}\Delta G_\varepsilon = \frac{\pi(2-\nu)}{8(1-\nu)}\mu\gamma_T^2 + \frac{\pi}{4}(1-\nu)\mu\varepsilon_n^2 \qquad (8.19)$$

where ν (Poisson's ratio) and μ (isotropic shear modulus) are each considered to be the same for the parent and product phases; γ_T and ε_n are the transformation shear strain and expansion strain, respectively.

When the transformation strain is not an exact IPS, distortion in the habit plane gives rise to an additional shape-insensitive strain energy ΔG_p^{str}, such that the total strain energy ΔG_s^{str} is

$$\Delta G_s^{str} = \Delta G_\varepsilon + \Delta G_p^{str} \qquad (8.20)$$

Three cases are considered in calculating the critical free-energy barrier to nucleation ΔG^*, critical nucleus radius r^*, and critical semithickness c^*. In the case of homogeneous nucleation, ΔG_d and ΔG_i are zero and the total free-energy change of an isolated, thin oblate spheroidal coherent martensitic particle (Fig. 8.35) can be expressed by[92]

$$\Delta G(r,c) = \frac{4\pi r^2 c}{3}(\Delta G_v + \Delta G_p^{str}) + \frac{4}{3}\pi r c^2 A + 2\pi r^2 \gamma \qquad (8.21)$$

Equation (8.21) can be represented schematically as in Fig. 6.2 and can yield c^*, r^*, and ΔG^* by differentiation:

$$c^* = \frac{-2\gamma}{\Delta G_v + \Delta G_p^{str}} \qquad (8.22a)$$

$$r^* = \frac{4A\gamma}{\left(\Delta G_v + \Delta G_p^{str}\right)^2} \qquad (8.22b)$$

$$\Delta G^* = \frac{32\pi A^2 \gamma^3}{3\left(\Delta G_v + \Delta G_p^{str}\right)^4} \qquad (8.22c)$$

By inserting all the known quantities in these equations, we obtain a value of ΔG^*, several orders of magnitude larger than the experimentally found barrier energy for spontaneous martensitic transformation. Even assuming local compositional fluctuations or the presence of pre-existing embryos does not provide full satisfaction. Hence, it becomes clear that homogeneous nucleation of martensite in steels is practically impossible. In this case, Eq. (8.18) prevails.

Despite this fact, recently it has been suggested that the homogeneous nucleation of martensitic transformation in iron-based alloys such as Fe-Ni, Fe-Ni-Cr, and Fe-Ni-Mn alloys, which exhibit isothermal transformation characteristics, can be stimulated by thermal activation. The rate of nucleation $\dot{N}$ of isothermal martensite at operational nucleus size r, where rapid growth follows, may be given by

$$\dot{N} = N\beta^* \exp\frac{-Q}{kT} \qquad (8.23)$$

and is controlled completely by interfacial movement. Here N is the number of pre-existing nucleation sites per unit volume (taken in the range of 10^5 to $10^7/cm^3$ for iron-based alloys), Q is the activation energy for nucleation in the range of 10^{-18} to 10^{-19} J/event, β^* is a vibration frequency (lying between 10^5/s and 10^{13}/s), and k and T have the usual meanings.[61,93] The evaluation of activation energy for a homogeneous nucleation event is supported by the experimental results of numerous studies involving the kinetics of iron-based alloys.[94]

In fact, there is much experimental evidence which demonstrates that martensite nucleation is a heterogeneous process and the nucleation occurs at a lattice defect or inhomogeneity with barrier or without barrier similar to ones as shown schematically by curves A and B, respectively, in Fig. 6.8. The literature abounds with cases where martensite nucleates at dislocation clusters, twin boundaries, grain boundaries, etc., and quite specific models have appeared.[92,95]

It has been convincingly demonstrated that dislocations act as preferred sites for martensitic nucleation, and accordingly several dislocation-type nucleation models have been proposed. It is postulated that unspecified movements of groups of dislocations or arrays of dislocation loops[94] in the parent phase at elevated temperature are the potential sites for martensite nucleation. These groups of dislocations lead to the idea of the existence of "frozen-in" martensite embryo during cooling until the condition of embryo growth is attained.[34] At the M_s temperature, in the absence of a barrier at the preexisting group of arrays of dislocation loops (sometimes called *superdislocations*), martensite may form spontaneously due to a pronounced interaction between their strain fields and the strain field of the Bain distortion, causing the elimination of the energy barrier to nucleation. This feature of the model is applied to thermal martensite.[96] Olson and Cohen have provided an excellent and detailed review of the dislocation theory of martensitic transformation, elucidating the mechanism and kinetics of nucleation and growth.[92]

Kajiwara[97] has observed that for the martensite plate to nucleate from a potential nucleation site, three types of deformation steps must occur, namely, lattice deformation, lattice-invariant deformation, and the deformation accommodating the shape strain of the martensite plate. Of these steps, the "necessary plastic accommodation" in austenite has been considered to be the most important step in the nucleation event. However, the Kajiwara hypothesis of control by necessary plastic accommodation is discounted because this has no theoretical foundation and is irrelevant in thermoelastic systems.

Yu, Sanday, and Rath[98] have emphasized the model of preexisting martensite embryo in Kaufman and Cohen's theory, where the dislocation loops in the parent phase are themselves the so-called *preexisting embryos*. The strain energy stored in this distorted parent phase (the dislocation loop) offers the strain energy required to nucleate martensite embryo during transformation. The calculated critical characteristics of this embryo hold well with the model of Chen and Chiao and their experimental results.[98]

Recently, an alternative "quasi-martensitic" mechanism of microstructure evolution by a continuous strain modulation path characterizing the displacive analog of spinodal decomposition has been envisaged. Analysis indicates that the energy conservation can result in the selection of a fastest-growing wavelength controlling the dynamics of such a process. A quantitative description of the competition between nucleation and modulation would complete a unified prospect of first-order phase transformation mechanisms.[99] The reader is referred to a recent review by Olson and Roitburd for a fairly current assessment of martensitic nucleation.[99]

8.6.3 Methods of Identifying Nucleating Defects

It has been universally agreed that identification of the nucleating defect is necessary in order to understand the initiation of martensite transformation. The suggested methods include small-particle experiment (as accomplished on Fe in Cu and on zirconia in alumina); close control of defect distribution (involving, e.g., pseudoelastic or nonmetallic systems); surface nucleation; nucleation in bicrystals; short-impulse induced nucleation; study of martensite-to-martensite transformation; and careful study of autocatalytic nucleation (instead of primary nucleation) as a simplified alternative to the parent-to-martensite transformation.[100]

8.6.4 Growth of Martensite Plate

Once the nucleation barrier has been surmounted, the volume free-energy term becomes so large that an individual martensite plate grows at extremely rapid rates corresponding to the wave regime until it strikes another plate or a high-angle grain boundary. Very thin plates have been observed to be formed in the beginning with a very large r/c ratio; these plates later thicken only to a limited extent.[101] The final structure consists of isolated plates embedded in untransformed parent phase.

The growth velocity during athermal and isothermal transformation remains the same over a large temperature range; the difference between the two reactions is only in the frequency and temperature dependence of nucleation. This clearly indicates that the growth is not a thermally activated process. The different morphologies are attributed to the difference in the growth of martensite plates.[102]

The high growth rate may be interpreted as the movement of a glissile semicoherent interface consisting of an array of parallel dislocations, all having the same Burgers vector. The dislocations in the interface have to satisfy two requirements: (1) They must accommodate the misfit between the lattice planes of the parent and the martensite phases, and (2) they must produce the lattice-invariant shear transformation. Such type of movement of transformation dislocations causes the motion of the martensitic interface (habit plane) in a direction normal to itself.[38]

Grujicic, Ling, and Haezebrouck[103] have briefly reviewed the growth of martensite based on the dislocation models of the martensitic interfacial structure, interfacial mobility, and both high-velocity continuous and low-velocity thermally activated motion. They also investigated the growth behavior of thermoelastic martensite and growth dynamics of nonthermoelastic martensitic transformation.

Recently, Rao, Sengupta, and Sahu[104] have carried out simulations of the athermal martensite phase transformation and have shown that the growth process can be analyzed in terms of a renormalization group with two fixed points. The unstable heterogeneous fixed point of the normalization group controls the martensite grain size distribution or the behavior of a crossover function between the heterogeneous and homogeneous limits of the model. However, Crosby and Bradley[105] have used a mean-field theory of the martensite transformation to obtain approximately the heterogeneous limit of this model where the line segment (or size) distribution obtained is in good agreement with the numerical fit of the Rao et al. approach.

8.6.5 Nucleation Kinetics

The best quantitative understanding of the kinetics of martensitic transformation is gained from isothermal transformations, because they allow the determination of both the nucleation and the transformation rates. Olson and Cohen have developed a model for isothermal kinetics that is really related to the thermal components of the stress required to move a dislocation.[92]

For isothermal martensitic transformation, Thadhani and Meyers[47] have shown that the transformation initiates in the austenite and proceeds as a function of time at constant temperature (Fig. 8.12a). As a function of temperature, the transformation kinetics exhibit a typical C-curve characteristic (Fig. 8.13). Isothermal martensitic transformation kinetics involve two effects: an initial increase, which is due to autocatalytic nucleation of new martensite plates, followed by a decrease, due to the compartmentalization of austenite grains into smaller and smaller areas.[3,47]

Lin-Olson-Cohen Theory. This theory uses separate defect potency distributions for preexisting (initial) and autocatalytic nucleation sites (taken from nucleation experiments in an Fe-32.3 Ni alloy). The former follows an exponential form where the number density of preexisting defects continues to increase monotonically with decreasing defect potency. The latter obeys a Gaussian function where the cumulative number density of autocatalytic defects saturates with decreasing defect size. This potency distribution theory unifies all previous theory to provide a comprehensive framework describing the entire course of martensitic transformation curves (including isothermal, athermal, and burst kinetics).[106]

Two equations have been derived to explain the athermal transformation kinetics of martensite formation. The first empirical power relationship between the volume fraction of transformation f and the quenching temperature $T < M_s$ has been given by Harris and Cohen as:[107]

$$f = 1 - 6.956 \times 10^{-15}[455 - (M_s - T)]^{5.32} = 1 - 6.956 \times 10^{-15}(455 - \Delta T)^{5.32} \quad (8.24)$$

where the constant 455 is the M_s (K) for a 1.1% carbon steel and ΔT is the undercooling below M_s. This equation is valid for carbon and low-alloy steels containing ~1% carbon content and is mostly based on data from the later stages of transformation.

The second empirical "exponential relationship" has been provided, for a 0.37 to 1.1% carbon steel, by Koistinen and Marburger as

$$f = 1 - \exp[1 - 1.10 \times 10^{-2}(M_s - T)] \quad (8.25)$$

For stress-induced martensitic transformation under applied stress σ_{ik}^a with temperature $T > M_s$, the relationship becomes[27]

$$f = f_0[1 - \lambda_{ik}^\sigma(\sigma_{ik}^a - \sigma_{ik}^{M_s})] \quad (8.26)$$

where f_0 is the initial detectable amount of martensite formed at martensitic starting stress $\sigma_{ik}^{M_s}$ and λ_{ik}^σ is a material constant.

8.7 THERMOELASTIC MARTENSITIC TRANSFORMATION

We can also classify martensite into nonthermoelastic and thermoelastic varieties according to their difference in transformation hysteresis, defined by $A_f - M_s$ (Fig. 8.9). The magnitude of the hysteresis represents the transformation driving force. The nonthermoelastic martensites (for example, Fe-C and Fe-30% Ni alloys) are characterized by nonreversibility of transformation on cooling and heating, large hysteresis (>420°C for Fe-Ni), and large shear strain component of transformation (for example, $\gamma = 0.19$ for Fe-C and $\gamma = 0.20$ for Fe-30% Ni). Transformation strains are primarily relaxed by plastic accommodation.[27] In Fe-Ni alloys, the martensite/parent interface tends to become immobilized after thickening of a martensite plate to a certain extent upon cooling; further growth does not ensue after subsequent cooling. During heating the interface does not move backward, presumably due to the pinning effect by the damaged surroundings; instead, the nucleation of small platelets of the parent phase occurs within the immobilized martensite plates, and a given plate as a whole does not revert to the original parent-phase orientation.[108] In plain carbon steels, martensite on heating produces different stages of tempering by its diffusional decomposition.

On the other hand, in thermoelastic martensitic transformation, the martensite plates nucleate and grow continuously on lowering the temperature and disappear by the exact reverse path, i.e., on raising the temperature. These martensites are characterized by the following:

1. A small hysteresis [for example, 16°C for Au-Cd (Fig. 8.9)].
2. A small shear strain component (for example, $\gamma = 0.05$ for Au-Cd and $\gamma = 0.02$ for In-Tl). Note that this may be as large as in nonthermoelastic martensitic transformation, as commonly observed in ferrous alloys and steels.[109]
3. Inherently low elastic strains associated with the change of crystal structure so that the elastic limit of the parent-phase matrix is not exceeded.
4. Absence of irreversible plastic deformation events.
5. Largely or fully, accommodation of the transformation strain, elastically.[27]
6. Effective canceling out of the built-up strains during growth of martensite plates by the formation of groups of mutually accommodating plates.
7. The glissile martensite/parent interface, capable of "reversible" movement even at very low temperatures.

Other prerequisites for the occurrence of thermoelastic martensites are that (1) individual martensite plates be internally twinned (that can be easily detwinned by condensation of many twin variants in a single favored one)[110] or faulted to accommodate the transformation strains and (2) martensites contain an ordered structure that raises the elastic limit of the parent phase, thus inhibiting the generation of irreversible dislocation debris in the parent (or martensite) phase. That is, an ordered structure of martensite cannot be destroyed by slip.[15] Figure 8.36 shows the transformation characteristics of thermoelastic and nonthermoelastic martensite hysteresis loops observed in Fe-24.5 at% Pt alloy in the ordered (O) and the disordered (D) states, respectively.[61]

In the thermoelastic transformations, it is assumed that the chemical free-energy change between the two phases which drives the transformation is equal in magnitude to the nonchemical free-energy change comprising the reversible elastic strain energy $\Delta G_\varepsilon^{P \to M}$ (required to accommodate the transformational shape and volume changes) and the reversible frictional energy $V_m \tau_0 \gamma_T$ opposing the interfacial motion through 1 mol of the transformation phase in either direction. Because the interface is basically coherent, the surface energy remains very small and the elastic energy term is predominant, increasing the plate size.[109] The growth and shrinkage of martensite plates, therefore, take place under a balance between the chemical and elastic forces as expressed, for any geometry, by

$$\Delta G_{ch}^{P \to M} + \eta G_\varepsilon^{P \to M} = \overset{\text{cooling}}{\underset{\text{heating}}{\pm}} V_m \tau_0 \gamma_T \qquad (8.27)$$

where $\Delta G_{ch}^{P \to M}$ is the chemical free-energy change per mole (referred to as the *martensite phase*), which is negative below T_0; ΔG_ε is the stored elastic energy per mole; η is the geometric factor coming from the shape dependence of stored elastic energy; V_m is the molar volume; and τ_0 is the shear stress required to move an interface which produces the transformational shear strain γ_T.[111]

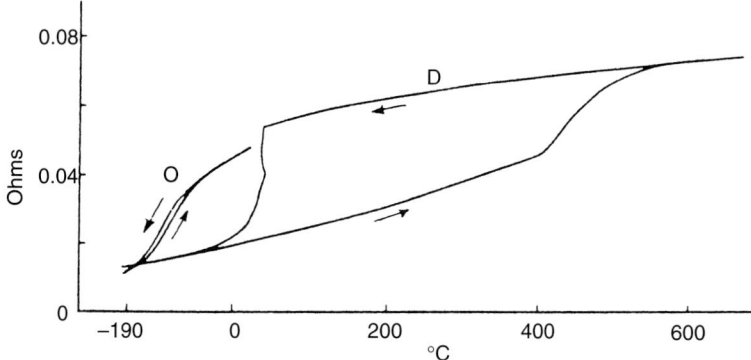

FIGURE 8.36 Changes in electrical resistivity due to martensitic transformation on cooling and its reverse on heating in an Fe-24.5 at% Pt alloy. In the disorderd state (D), the hysteresis is large and the transformation is nonthermoelastic; in the ordered state (O), the transformation characteristics change significantly and become thermoelastic.[61] (*Reprinted by permission of The Metallurgical Society, Warrendale, Pennsylvania.*)

Transformation proceeds by continuous growth of martensite on cooling (or application of external stress) below M_s. If cooling is stopped, growth ceases; but if it is resumed, further growth of martensite plates occurs until the plates encounter grain boundaries or other martensite plates. During heating (or unstressing), the reverse transformation takes place by the "reverse" movement of the interface, and the martensite plates shrink and crystallographically revert completely to the original phase.[108] The existence of such a reverse transformation provides a distinct and unique mechanical behavior, as shown below.[74]

The thermoelastic growth mode is characterized by the formation of thin, parallel-sided plates or a wedge- or spear-shaped pair of plates. The wedges consist of two sets of fine martensitic twins (alternating fine bands of differently oriented martensites) separated by a midrib. This microstructure provides an easy way for the initiation of transformation which is important for thermoelasticity and reversibility of transformation.[112]

8.7.1   Shape Memory Alloys

Shape memory alloys (SMAs) are mixtures of martensites and austenites in which the composition of mixtures varies and the martensite and austenite transform into one another. These transformations can be brought about by thermal or mechanical means. SMAs can be studied (1) at the microscopic level by describing the microstructures of constitutive crystals and (2) by using statistical thermodynamics of a lattice of particles.[112a]

SMAs have become an important class of functional materials due to their unique characteristics that are different from those of the traditional engineering alloys. The most fundamental descriptors of property are stress, recovery stress, and transformation temperatures rather than modulus, yield strength, and ductility for engineering alloys. Their properties are very sensitive to both intrinsic parameters such as alloy systems, minor alloying additions, lattice structures, and external conditions such as thermomechanical treatments, training history, processing, working environment temperatures, and so forth.[113] Thus SMAs exhibit a complex, nonlinear thermomechanical behavior with hysteresis which depends on both intrinsic and extrinsic parameters. Unlike martensite in steel which undergoes both a shape change and a volume change, SMAs such as Ni-Ti (or Ti-Ni) involve primarily only a shape change.[110]

Some of the best-known and well-developed SMAs are brass-type alloys; numerous other noble metal-based alloys based on copper group metals such as Cu-Al-Ni, Ag-Zn, and Au-Cd; Ti-Ni alloys; Mn-Cu-based alloys; Co-Cr-Mo alloys; and Fe-based alloys.[114]

The important advantages of SMAs in the temperature-activated applications are (1) simplicity, reliability, and compactness; (2) creation of clean, noise-free, and spark-free working environments; and (3) high power/weight (or power/volume) ratios.[115]

The main limitations of SMAs are as follows:[114]

1. For the most part, the transformation range involved is 20 to 100°C wide and is near the room temperature (±150°C). Hence the service temperature for SME devices should not be further away from the transformation range. Otherwise, the SMAs may undergo aging effects, and their characteristics may be altered or degraded.

2. Very high temperatures may give rise to creep or diffusional phase transformation of the alloy that adversely influences the shape memory properties.

3. Only around 6 to 10% strain can be recovered by unloading and heating; the strain beyond this limiting value will remain as permanent plastic deformation.

The nonferrous SMAs have a bcc parent phase which on cooling undergoes an ordering reaction (i.e., both parent and martensite phases are ordered) producing a B2 (cP2) CsCl structure, DO_3 (cF16), or $L2_1$ (cF16) Heusler symmetry. On further cooling, martensite begins to form on three types of close-packed planes with a common stacking sequence, using the Ramsdell notation of 2H (. . . A B A B A B . . .), 3H (. . . A B C A B C A B C . . .), 9R (. . . A B C B C A C A B . . .), or 18R (. . . A B A B C B C A C A B A B C B C A C . . .). [There are also six types of close-packed planes in martensites which are produced from an Fe_3Al β_1 (cF16) parent phase.] Figure 8.37 shows the stacking sequence of close-packed planes in martensite, using the Ramsdell and Zhdanov notations.[116] When viewed under the transmission electron microscope, the above martensites appear to contain a fine-scale deformation which is either internal twinning (2H, 3R) or internal faulting (9R, 18R). This represents the inhomogeneous shear of the phenomenological theory of martensitic crystallography.[55,56]

Tables 8.9[59] and 8.10[74] list the nonferrous and ferrous SMAs, respectively, exhibiting perfect or nearly perfect shape memory effect. Table 8.11 shows the mechanical and physical properties of three commercial SMAs.

Ti-Ni-Based Alloys. TiNi alloys are well established as SMAs due to their various applications based on the *shape memory effect* (SME), pseudoelasticity, and their superior properties in fatigue, corrosion resistance, biocompatibility, physiologic compatibility, ductility, elastic deployment (or large recoverable strain), kink resistance, constancy of stress, thermal deployment, dynamic interference, hysteresis, and MRI compatibility.[117] They also exhibit a high mechanical damping because of the easy movement of twin boundaries.[118]

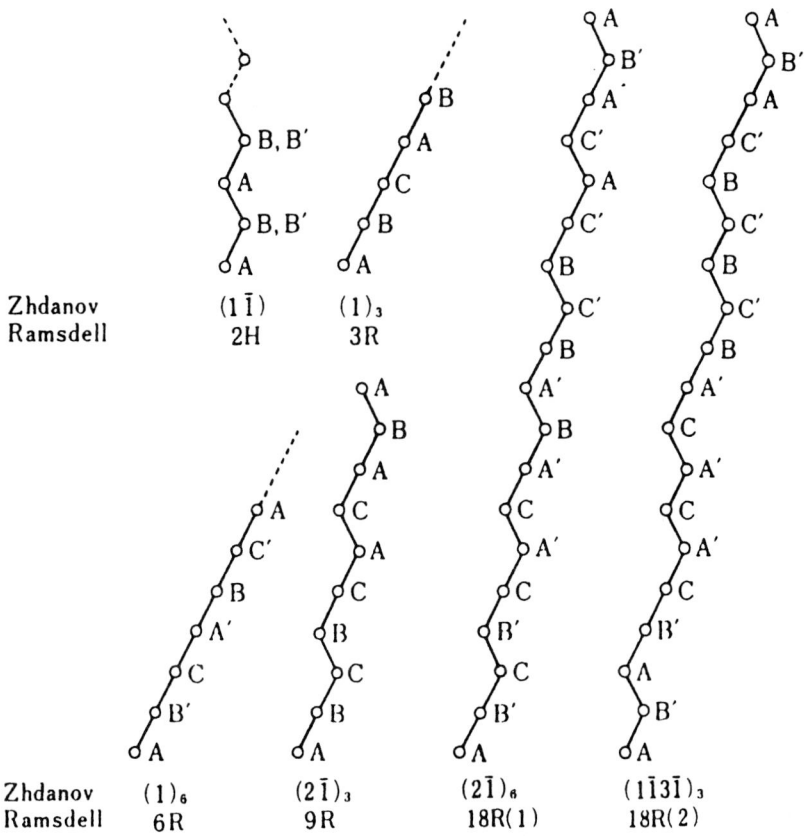

Zhdanov   (1 $\bar{1}$)   (1)₃
Ramsdell   2H   3R

Zhdanov   (1)₆   (2 $\bar{1}$)₃   (2 $\bar{1}$)₆   (1 $\bar{1}$3 $\bar{1}$)₃
Ramsdell   6R   9R   18R(1)   18R(2)

FIGURE 8.37 Periodic stacking structures with stacking sequence of close-packed planes in martensite produced from a CsCl-type B2 parent phase and from an Fe₃Al β_1 parent phase. The planes marked with a prime appear when the parent phase is an Fe₃Al superlattice.[116]

Usually, the martensitic transformation temperature of NiTi SMAs decreases with increasing Ni content and cold work and decreasing heat treatment temperature.[119] In TiNi alloys, the martensite can be either thermally induced or stress-induced. In thermally induced martensite, all martensitic variants nucleate and self-accommodate within the parent-phase matrix whereas in stress-induced martensite only the variant that best matches the applied stress system survives.[120]

The stress versus strain curves of polycrystalline martensitic NiTi SMAs are often different for loading under tension and compression. Under tension, a clear stress plateau occurs, while under compression, the material is rapidly strain-hardened and no stress-plateau can be noticed (Figs. 8.38 and 8.39).[121] This illustrates the difference in deformation mechanisms of martensitic NiTi SMA for tension and compression stress modes. These results suggest that the early stage of martensitic deformation under tension may be related mainly to the interfacial migration between two adjacent martensite plates, while under compression it is mainly related to the creation and migration of dislocations.[121]

TABLE 8.9 Nonferrous Alloys Exhibiting Perfect Shape Memory Effect and Pseudoelasticity[74]

Alloy	Composition (at%)	Structural change	Ordering	Transformation hysteresis		Transformation temperature range	
				°C	°F	°C	°F
Ag-Cd	44–49Cd	B2 → 2H	Ordered	~15	~25	−190 to −50	−310 to −60
Au-Cd	46.5–48.0Cd	B2 → 2H	Ordered	~15	~25	30 to 100	85 to 212
	49–50Cd	B2 → trigonal	Ordered	~2	~4	—	—
Cu-Zn	38.5–41.5Zn	B2 → R, rhombohedral M9R	Ordered	~10	~20	−180 to −10	−290 to 15
Cu-Zn-X (X = Si, Sn, Al, Ga)	A few at%	B2 (DO_3) → 9R, M9R (18R, M18R)	Ordered	~10	~20	−180 to 200	−290 to 390
Cu-Al-Ni	28–29Al 3–4.5Ni	DO_3 → 2H	Ordered	~35	~65	−140 to 100	−220 to 212
Cu-Sn	~15Sn	DO_3 → 2H, 18R	Ordered	~15	~25	−120 to 30	−185 to 85
Cu-Au-Zn	23–28Au 45–47Zn	Heusler → 18R	Ordered	~6	~11	—	—
Ni-Al	36–38Al	B2 → 3R, 7R	Ordered	~10	~20	−180 to 100	−290 to 212
Ti-Ni	49–51Ni	B2 → monoclinic	Ordered	~30	~55	−50 to 110	−60 to 230
		B2 → rhombohedral	Ordered	~2	~4	—	—
Ti-Ni-Cu	8–20Cu	B2 → orthorhombic → monoclinic	Ordererd	4–12	7–22	—	—
Ti-Pd-Ni[†]	0–40Ni	B2 → orthorhombic	Ordered	30–50	54–90	—	—
In-Tl	18–23Tl	fcc → fct	Disordered	~4	~7	60 to 100	140 to 212
In-Cd	4–5Cd	fcc → fct	Disordered	~3	~5	—	—
Mn-Cu	5–35Cu	fcc → fct	Disordered	—	—	—	—

† Ti-Pd-Ni alloys with high Pd content do not exhibit good SME unless specially thermomechanically treated.

TABLE 8.10 Ferrous Alloys Exhibiting Perfect or Nearly Perfect Shape Memory Effect[59]

Crystal structure of martensite	Alloy	Composition	Structural change	Nature of transformation†	Temperature hysteresis	M_s (°C)	A_s (°C)	A_f (°C)	Ordering
BCC or BCT (α')	Fe-Pt	~25 at% Pt	$L1_2 \to$ bct	TE	Small	131	—	148	Ordered
	Fe-Ni-Co-Ti (ausaged γ)	23Ni-10Co-10Ti	—	—	—	173	243	~443	—
		33Ni-10Co-4Ti	fcc $\to$ bct	TE	Small	146	122	219	Disordered
		31Ni-10Co-3Ti	—	Non-TE	—	193	343	508	—
	Fe-Ni-C (ausformed γ)	31Ni-0.4C	fcc $\to$ bct	Non-TE	Large	<77	—	~400	Disordered
	Fe-Ni-Nb (ausaged γ)	31Ni-7Nb	—	Non-TE	—	~160	—	—	—
HCP(ε)	Fe-Mn-Si	30Mn-1Si (single crystal)	fcc $\to$ hcp	Non-TE	—	~300	~410	—	—
		(28–33)Mn-(4–6)-Si	fcc $\to$ hcp	Non-TE	Large	~320	~390	~450	Disordered
	Fe-Cr-Ni-Mn-Si	9Cr-5Ni-14Mn-6Si	fcc $\to$ hcp	Non-TE	Large	~293	~343	~573	Disordered
		13Cr-6Ni-8Mn-6Si-12Co	fcc $\to$ hcp	—	—	—	—	—	—
		8Cr-5Ni-20Mn-5Si	fcc $\to$ hcp	Non-TE	—	~260	~370	<573	—
		12Cr-5Ni-16Mn-5Si	fcc $\to$ hcp	—	—	—	—	—	—
	Fe-Mn-Si-C	17Mn-6Si-0.3C	fcc $\to$ hcp	Non-TE	—	323	453	494	—
FCT	Fe-Pd	~30 at% Pd	fcc $\to$ fct	TE	Small	179	—	183	Disordered
	Fe-Pt	~25 at% Pt	fcc $\to$ fct	TE	—	—	—	300	—

† TE: Thermoelastic martensite. Non-TE: Nonthermoelastic martensite.

8.51

TABLE 8.11 Mechanical and Physical Properties of the Three Commercial Shape Memory Alloys[116]

Item	Ni-Ti (NiTi)	Cu-Zn-Al (CuZn)	Cu-Al-Ni (Cu$_3$Al)
Melting point (°C)	1250	1020	1050
Density (kg/m^3)	6450	7900	7150
Electrical resistivity ($10^{-6}\,\Omega\cdot$m)	0.5–1.1	0.07–0.12	0.1–0.14
Thermal conductivity, RT (W·m^{-1}K^{-1})	10–18	120	75
Thermal expansion coefficient (10^{-6}·K^{-1})	6.6–10	17	17
Specific heat (J·kg^{-1}·K^{-1})	490	39	440
Transformation enthalpy (J·kg^{-1})	28,000	7000	9000
E modulus (GPa)	95	70–100	80–100
Ultimate tensile strength, martensite (MPa)	800–1000	800–900	1000
Elongation at fracture, martensite (%)	30–50	15	8–10
Fatigue strength at $N = 10^6$ (MPa)	350	270	350
Grain size (10^{-6} m)	20–100	50–150	30–100
Transformation temperature range (°C)	−100–110	−200–110	−150–200
Hysteresis (K)	30	15	20
Maximum one-way memory (%)	8	4	6
Normal two-way memory (%)	1.2	0.8	1
Normal working stress (MPa)	100–130	40	70
Normal number of thermal cycles	100,000	10,000	5000
Maximum overheating temperature (°C)	400	150	300
Damping capacity, SDC (%)	20	85	20
Corrosion resistance	Excellent	Fair	Good
Biological compatibility	Excellent	Bad	Bad

The disadvantages of (NiTi) SMAs are (1) the high price of the alloying elements and the high requirements during fabrication and (2) the A_s temperature, which is limited to around 100°C.

Near-equiatomic Ti-Ni (or Ni-Ti, called *Nitinol*) and Ti-Pd alloys undergo thermoelastic martensitic transformation from B2 to B19′ and B19 structures upon cooling, respectively. The former is technologically important material having superior SME and superelasticity. The latter seems to be high-temperature SMA because M_s is above 800 K.[122]

There are several types of precipitates in near equiatomic Ni-rich Ti-Ni alloy; among them Ti$_3$Ni$_4$ precipitates have the strongest effect on the R-phase and martensitic transformation. Use of constrained aging can promote the all-round SME (ARSME or TWSME) because the stress field created by the Ti$_3$Ni$_4$ precipitates is believed to introduce univariant R-phase and martensite transformation.[123]

TiNi alloys of near-equiatomic composition are being applied to many medical fields, i.e., dental arch-wire, catheter guide wire, active endoscope, active catheter, etc. TiNi SMA tubes have been developed as active forceps for laparoscopic surgery.[124] The addition of third or fourth elements such as Cu, Nb, and precious metals in nearly equiatomic TiNi alloys provides a powerful tool for controlling the properties. It can be employed to control the transformation temperatures and the crystal structure of martensite, increase the stability of M_s with respect to thermal history, control the hysteresis width, increase or decrease martensitic strength,

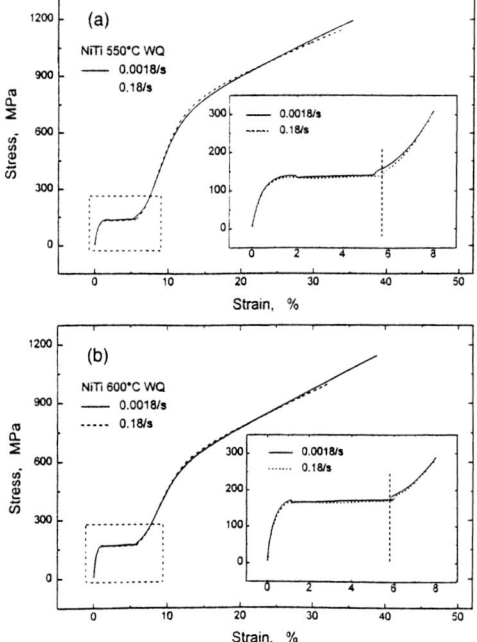

FIGURE 8.38   Stress-strain curves of a Ni-Ti SMA under tension as a function of strain rate for two annealing conditions showing that the stress-strain curve under tension is almost strain-rate-independent.[121]

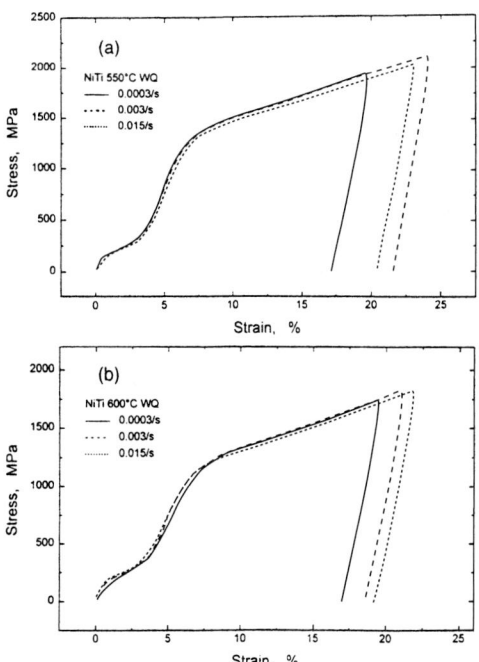

FIGURE 8.39   Stress-strain curves of a NiTi SMA under compression as a function of strain rate for two annealing conditions showing the insignificant effect of strain rate.[121]

increase austenitic strength, suppress the R phase, and improve corrosion strength.[125,126] The R phase forms in the alloys containing Fe, Cr, or Al, while the B19 phase forms in those containing Au, Cu, or Pd prior to the B2 → B19′ martensitic transformation. These intermediate phases do not form in solution-treated binary alloys.[125]

Two-stage martensitic transformation occurs in Ti-Ni alloys containing an excess of Ni or Ni partially replaced by Cu, Co, or Fe: parent B2 → intermediate phase (rhombohedral structure, called R phase) → martensite.[113] For example, $Ti_{50}Ni_{40}Cu_{10}$ SMAs usually exhibit a typical B2 ↔ B19 and B19 ↔ B19′ two-stage martensitic transformation, where B2, B19, and B19′ denote cubic, orthorhombic, and monoclinic structures, respectively. The exceptionally high plateau of damping capacity in $Ti_{50}Ni_{40}Cu_{10}$ alloy arises from the easy movement of twin boundaries of B19 martensite due to its inherently low elastic modulus and yield stress.[127] Rapidly quenched TiNiCu ribbons have been found to develop successful applications in various engineering fields such as actuators, thermal sensors, optical modulators, etc.

Recently, a pseudoelastic β Ti-alloy Ti-11Mo-3Al-2V-4Nb has been developed which possesses desirable properties of high springback, low stiffness, good corrosion resistance, and good formability for orthodontic arch-wire application for the intermediate stage of orthodontic treatment.[128] For more details on Ti-Ni SMAs, the reader is referred to reviews presented by Tang,[113] Melton,[126] and Saburi.[129]

Cu-Based Alloys.   Among the various SMAs, Cu-based SMAs (such as CuZnAl and CuAlNi) are of special importance due to their lower cost and easier formability and advantages in electrical and thermal conductivities.[130,131] They are developed for the transformation temperatures up to 200°C. The SME in Cu-Zn-Al alloys is associated with a reversible thermoelastic martensitic transformation. The internally faulted martensites in CuZnAl systems are characterized by a long period stacking order such as the 9R- or 18R-type structures, in accordance with the number of close-packed layers in the unit cell.[132] In reality, the 9R and 18R structures are very similar, except for a doubling of the unit cell along the b axis and c axis directions in the 18R case.[133] However, at elevated temperatures, the metastable β_1 (B2 or DO_3) and β_1' (M9R or M18R) phases transform into more stable structures of bainitic α_1, complex cubic γ_2, or fcc α phases by thermally activated processes, which leads to a thermal stability problem and poor mechanical strength, thereby placing an obstacle to the application of CuZnAl alloys.[130] Note that the main factor controlling the transformation temperatures is alloy composition; however, other factors such as heat treatment, quenching rate, grain size, and the number of transformation cycles also play important roles. For example, usually the smaller the grain size, the lower the M_s temperature.[15]

Cu-Zn-Al-Mn-Zr SMA exhibits a superior thermal stability to the Cu-Zn-Al SMA.[134] The coexistence of two types of martensites, 18R and 2H, has been found in Cu-Zn-Al-Mn alloys in which the contribution of a nonchemical term such as undercooling below T_0 to reach M_s and the need of continuous cooling for further growth of martensites are necessary. Additionally, the interaction between the coexisting two types of martensites, which can favor or impede the simultaneous growth, has to be taken into consideration.[135]

Three Cu-Al-Mn alloys—Cu-14 at% Al-13 at% Mn, Cu-16 at% Al-10 at% Mn, and Cu-17 at% Al-10 at% Mn—exhibit excellent ductility with ~15% elongation, 60 to 90% cold-workability due to the decrease in the degree of order in the β phase, and an SM recovery of 80 to 90%. Two types of martensitic transformation are found in these alloys: β_1 ($L2_1$) → β_1' (18R) martensite and β (A2) → α' (A1) martensite with a high degree of twins.[136]

The martensitic transformation in the Cu-15 at% Sn alloy is characterized by (1) the rapid aging effect in the parent phase and (2) the large transformation hysteresis (~150 K) of the thermal transformation. Khandros, Miura et al., and Stice and Wayman have observed that the reverse transformation in Cu-Sn alloys cannot be completed by a lower heating rate. However, with a higher heating rate the complete reverse transformation can be achieved, showing the lower A_f temperature. It suggests that the reverse transformation needs the driving force attained by the overheating in the rapid heating procedure.[137]

A TEM study of equiatomic Cu-Zr composition has shown that (1) above 715°C a B2 phase (CsCl type, bcc-based) exists as a line phase and (2) below this temperature this phase decomposes into two stable phases $Cu_{10}Zr_7$ and $CuZr_2$ with orthorhombic and tetragonal structures, respectively. By rapid cooling to below 140°C, the decomposition process is suppressed, and the B2 structure transforms into two metastable structures with monoclinic symmetry, martensitic characteristics, and shape memory behavior.[138]

Au-Cd and Au-Zn Alloys. Au-Cd alloys are the most interesting alloys among thermoelastic ones which have two different compositions: one is Au-47.5 at% Cd which is characterized by $\beta_2 \rightarrow \gamma_2'$ (orthorhombic) martensitic transformation and the presence of SME and rubberlike behavior; the other is Au-49.5 ~ 50.0 at% Cd which has the characteristics of the unique $\beta \rightarrow \zeta_2'$ (trigonal) martensitic transformation, showing a very small temperature hysteresis (~2 K) and small transformation strain.[139]

In the Au-Zn system, the β_1 phase with the V-shape is present near the equiatomic Au-Zn alloy and undergoes transformation. The reverse transformation has been observed in Au_3Zn alloy.[140]

Fe-Based Alloys. Although NiTi alloy and its ternary alloys are most widely used in the various fields of applications, they are very expensive. Among ferrous SMAs, Fe-Pt (near the composition Fe_3Pt) and Fe-31 at% Pd have been studied only for academic interest due to the expensive Pt and Pd elements. The other Fe-based SMAs are Fe-high Mn, Fe-Mn-Si, Fe-Mn-Co, Fe-Mn-C, Fe-Mn-Si-Cr-Ni, Fe-Ni-Co-Ti, and Fe-Ni-C alloys, which have attracted much attention due to their low costs, good SMEs, and excellent cold and hot workability. Table 8.10 summarizes the ferrous alloys exhibiting a perfect or nearly perfect SME.[59]

Fe-Mn alloys showing an fcc → hcp phase transformation have been extensively investigated to enhance the understanding of SMEs. The Fe-21Mn alloy exhibits an increase of the volume fraction of ε-martensite with an increase of grain size. Its highest damping capacity after heat treatment at 1000°C is attributed to the largest area of γ/ε boundaries.[141] Jun and Choi have observed the damping capacity of an Fe-23Mn alloy with respect to deformation at room temperature and described it in relation to the microstructural evaluation.[142] The shape memory in Fe-(28–33)Mn alloys is caused by the large amount of stress-induced fcc $\gamma \rightarrow$ hcp ε martensitic transformation to produce a shape change without accompanying slip deformation and $\varepsilon \rightarrow \gamma$ reverse transformation on heating by the backward movement of Shockley partial dislocations.

There have been some attempts to achieve a good shape memory effect in Fe-Mn-Co alloys, because like Si, Co reduces the Neel point and lowers the SFE of austenite. However, FeMnCo alloys display only a slight SME like Fe-Mn binary alloys, probably due to the fact that Co does not strengthen the austenite matrix, which is essential both for the suppression of slip deformation during the stress-induced ε transformation and for the $\varepsilon \rightarrow \gamma$ reverse transformation and thereby for the perfect shape memory effect.[59]

The damping capacity of the Fe-17Mn-0.06C has been found to have a maximum value around 10% cold-rolling reduction in thickness (due to an increase in ε-martensite) and decreases with further deformation due to dislocations and α' martensite formed during rolling, which hinder the operation of the damping sources.[143] The Fe-23Mn-0.015C alloy undergoing γ(fcc) $\rightarrow \varepsilon$(hcp) martensite transformation has received considerable attention due to its low cost and significant damping characteristics.[142]

In Fe-Mn-Si alloys, ordered structure or ordered coherent particles in austenite have not yet been reported. The composition range of Fe-(28–33)Mn-(5~6)Si exhibits nearly perfect shape recovery. The main advantage of this alloy is that M_s is around room temperature.[143a] Recently, Tsuzaki et al. have observed that carbon addition to Fe-Mn-Si alloy is very effective in improving the SME.[144]

Otsuka et al. have found that a training is effective in improving SME in Fe-Mn-Si alloys. Anisotropic shape memory has been predicted in a training-treated Fe-Mn-Si specimen where the stress-induced martensitic transformation depends on the Schmidt factor.[145]

The shape memory effect of Fe-Mn-Si and Fe-Mn-Si-Cr-Ni alloys with a decrease in a Neel temperature (T_N) and lower SFE is closely related to the stress-induced fcc (γ) $\rightarrow$ hcp (ε) martensitic transformations, usually associated with the $\gamma \rightarrow \alpha'$ (bcc) or $\varepsilon \rightarrow \alpha'$ martensitic transformations under applied stress and the subsequent reverse $\varepsilon \rightarrow \gamma$ transformation by recovery annealing.[146-148] The presence of α' martensite depends on the alloy chemistry and the transformation temperature.

Among the Fe-Mn-Si-Cr-Ni alloys, Fe-14Mn-5Si-9Cr-5Ni alloy is the most significant alloy which is a modification of Fe-Mn-Si alloy and has a corrosion-resistance property.[149] When it is compared to TiNi and Cu-based SMAs, they exhibit a lower degree of shape recovery, low yield elongation, and less net reversible strain. These are the main drawbacks which restrict the large-scale use of this type of alloy.[150]

In the Fe-32Mn-6Si-0.04C-0.05Nb alloy, a substantial SME has been found for which the following conditions are to be fulfilled:

1. A large amount of stress-induced ε hcp martensite must be generated by the stacking fault mechanism at the beginning without a permanent slip intrusion.

2. The reverse $\varepsilon \rightarrow \gamma$ transformation has to occur by the reverse movement of Shockley partial dislocations on $\{111\}_{fcc}$ planes which moved at the forward $\gamma \rightarrow \varepsilon$ transformation.[151]

Recently, Ogawa and Kajiwara and Kikuchi et al. have found that in a training treated Fe-Mn-Ni-Cr-Si SMA, a γ/ε lamellar structure with 1- to 10-mm-thick layers is formed. This lamellar structure has been considered to favor the backward motion of partial dislocations in an $\varepsilon \rightarrow \gamma$ reversion, due to easier nucleation of γ and/or due to unrelaxed backstress at the tip of an ε plate.

Like Fe-Mn-Si alloy, Fe-(7–15)Cr- < 10Ni- < 15Mn- < 7Si-(0–15)Co SME alloy shows the $\gamma \rightarrow \varepsilon'$ stress-induced transformation and complete shape recovery, if the deformation is up to 4% strain and the M_s temperature is between 173 and 323 K.[151a]

In the case of Fe-Ni-Ti alloys, lenticular martensite usually forms even in the ausaged condition. However, in Fe-Ni-Co-Ti alloys, thin plate martensites are easily produced by ausaging. This might be attributed to the decrease of transformation volume change, increase of austenite hardness, and increase in the amount of γ' precipitation.[59] The Fe-Ni-Co-Ti and Fe-Ni-C alloys represent γ(fcc) $\rightarrow \alpha'$(bct) transformation.[152] Several researchers have reported a SME and superelasticty in ausaged Fe-(23–33)Ni-(10–12)Co-(3–10)Ti alloy containing fine coherent ordered γ' Ni$_3$Ti (L1$_2$-type) precipitates.[153–155]

A perfect or nearly perfect SME has also been found in ausaged Fe-31Ni-7Nb alloy which contains the fine coherent plate-shaped γ'', Ni$_3$Nb, tetragonal ordered (DO$_{22}$-type structure) particles, despite the relatively small tetragonality of martensite phase and the relatively wide thermal hysteresis.[156]

The Fe-Ni-(TiNbTaAl)-based (FNB) alloys are potential SMAs due to the storage by the fine coherent particles of the elastic energy and finding applications, which require complex operations.

Ferrous SMAs are being used in China, but they have been abandoned in Japan because of low recovery stress. A potential very large field of application is as a tendon for pre- and poststressed concrete structures; but until the yield and recovery stress can be increased, they do not have a good future. In reality, the properties and pricing of NiTi alloys already make them candidates for large structural applications.[157] (See seismic application below.)

Mn-Cu-Based Alloys. The Mn-Cu-based alloys such as Mn-Cu and ones with small amounts of Cr, Ni, Ge, Si, and so forth (for example, 80Mn-19Cu-1Ni alloy) exhibit a pronounced two-way shape memory effect (TWSME).[158]

Ni-Based Alloys. The NiAlMn β SMAs are used as high-temperature SMAs.[159] Here the crystal structure of the parent phase always shows B2 [Ni(Al,Mn)] structure; and martensite phase is a 2M (L1$_0$), 10M, 12M, or 14M (7R) structure or a mixture of them. The ongoing studies of Heusler ferromagnetic Ni$_2$MnGa alloys are interesting and might provide a magnetically triggered martensitic transformation more valuable than the magnetic interaction found in the FeNiCoTi ferrous SMAs. Recently, Cherneko et al.[160] have observed that ferromagnetic NiGaMn SM alloys with an L2$_1$ (Ni$_2$GaMn: Heusler) structure are potential candidates as a new type of smart material whose SM characteristics can be controlled by temperature, stress, and magnetic field. Since the L2$_1$ Heusler structure is considered as stoichiometric Ni$_2$AlMn alloy,[161] it is expected that, like NiGaMn alloys, the NiAlMn Heusler alloys could also display unique magnetic and SM properties.[162,162a]

The ductile NiAlFe fcc γ-phase SMA exhibits martensitic transformation from B2 to L1$_0$ structure in the β phase near the $\beta + \gamma$ field. The crystallographic natures of martensitic transformation in NiAlFe ternary alloy are similar to those in the binary NiAl system.[163]

Co-Based Alloys. In Co-31.8% Ni alloy, fcc/hcp martensite transformation occurs by the passage of Shockley partial dislocation ledges [$\mathbf{b} = (1/6)<112>$] along alternate (111) plane in the fcc matrix. The hcp martensite thickens by the lateral motion of ledges across the fcc/hcp interface. Although superledges were noticed, the bulk of the growth ledges were two (0002) planes high, which tends to be the basic ledge height.[164]

Low-carbon wrought Co-Cr-Mo alloys have been widely used for improved hip and knee orthopedic implants due to their excellent mechanical properties, biocompatibility, and wear resistance.[165] They can exhibit an fcc $\rightarrow$ hcp martensitic phase change under athermal, isothermal, or strain-induced transformations (SIT). In the case of SIT, the mechanisms involved are rather similar to those exhibited by TRIP steels. In this case, a high level of alloy strength exists while the ductility is significantly increased (along with yield strength and volume fractions of SIT ε phase f_{hcp} with the amount of cold work) by a dynamic martensitic transformation.

The SIT phenomenon has been attributed to the spontaneous creation of new martensite nuclei or embryos via plastic deformation. It assumes that shear band intersections during plastic deformation act as likely plastic strain-induced nucleation sites. Accordingly, the rapid and localized strain hardening exhibited by Co alloys has been linked to numerous intrinsic stacking fault (ISF) intersections which act as tremendous obstacles for lattice dislocation motion, while promoting SIT martensite.[166]

Improvements of Shape Memory Alloys

Grain-Refined SMAs. Ti-doped grain-refined CuAlNi SMA offers great improvements in strength, strain to fracture, and fatigue life. For example, with grain refinement to $<20\,\mu m$, the number of cycles to failure for Ti-doped CuAlNi SMA increases to the level for Ti-Ni-based SMAs, i.e., 10^5.[167] Other methods of grain refinement include thermomechanical treatment (TMT), mechanical alloying, rapid solidification, sputtering, and powder metallurgy. For example, CuAlNi SMAs with grain sizes $\leq 5\,\mu m$ were obtained by TMT, which resulted in the fracture stress of 1200 MPa and fracture strain $\leq 10\%$.[168]

High-Temperature SMAs. The Cu-Al-Ni alloys possess better thermal stability and higher operating temperatures and are less prone to stabilization than other Cu-based ones;[131] however, they suffer from their lack of ductility. In the Cu-28 at% Al-4 at% Ni, $\beta_1 \rightarrow \gamma_1'$ martensitic transformation occurs upon cooling, where a stress-induced DO_3-type ordered martensite structure forms from the γ_1' phase, as reported by Otsuka et al.[169,170] Consequently, Cu-Al-Ni-Ti-Mn SMAs have been developed as high-temperature SMAs for use above 373 K, where Ti addition induces grain refinement and Mn addition affects the transformation temperature, thereby effectively improving the ductility.[131] It has been observed that a eutectoid Cu-Al-Ni-Mn-Ti SMA exhibits the best thermal resistance for the SME; no precipitation occurs upon aging at 623 K after 5 hr. On parent-phase aging at intermediate temperatures, however, the Cu-12Al-5Ni-2Mn-1Ti alloy becomes more susceptible to aging effects than ternary CuAlNi alloys.[171]

The Ni-Al-based SMAs are anticipated to be a good alternative to Ti-based alloys for use at temperatures greater than 373 K. The inherent poor ductility of the Ni-Al SMAs can be significantly increased by the addition of 3d elements, which produces two phase structures containing β (B2) and ductile γ (fcc) phases.[172] However, their application was limited to $<\sim$523 K because when the $L1_0$ martensite was heated above that temperature, Ni_5Al_3 precipitates were formed with a loss of ductility and degradation of the SME.[173]

Recently, it has been observed that Zr or Hf addition to a binary Ti-Ni SMA can increase the phase transformation temperatures. Both the NiTiHf and NiTiZr alloys exhibit SME and have better and worse stability, respectively, with respect to NiTi alloys during thermal cycling. NiTiZr alloys have relatively poor SM properties and ductility compared to NiTiHf and NiTiPd alloys. Also, the NiTiHf alloy exhibits clearly the TWSME after proper thermomechanical training.[174]

It seems that the ternary $Ti_{38}Ni_{50}Hf_{12}$ alloy exhibits an improved shape memory recovery when prestrained in the austenite condition (for $M_s < A_s$).[175] The NiTiHf SMA patented by Johnson Controls of Milwaukee (USA) can offer transition temperatures as high as 600°C.

Constitutive Relations of Shape Memory Alloys.

Liang and Rogers[176] have developed and presented a complete, unified, one-dimensional thermomechanical constitutive relation based on Tanaka's work, which provided a theoretical guide in the design of SMA-based intelligent materials and structures. Later, Tang and Sandstrom[113] provided an evaluation of the phase transformation tensor and limitations of the constitutive relations for TiNi SMAs. Readers interested in fundamental concepts that are essential for various intelligent material system applications such as SM hybrid composites should refer to these reviews.

8.7.2 Shape Memory Effect

When a plastically deformed specimen of martensite exhibiting thermoelastic behavior is heated, the (low-temperature, low-symmetry) martensite reverts to its original (high-temperature, high-symmetry) parent phase. This phenomenon of reversible martensitic transformation is called the *shape memory effect* (SME).[14] Here the term *shape memory* implies a material's capability of remembering its original shape as a result of reversible solid-state phase transformation. During the recovery of their shape, the alloys can produce a force or displacement, or a combination of the two, as a function of temperature.[177] The SME and related characteristics such as superelasticity, training, two-way shape memory effect (TWSME), damping behavior, and generation of high stresses during the reverse martensite-to-parent transformation are now well known to be observed in various alloy systems.[2,178]

Deformation processes associated with SMAs, in the form of single crystals or polycrystals, involve three stages. The first stage involves the transformation of the original single crystal of the high-temperature parent phase on cooling below M_f into martensite in a self-accommodating manner by forming groups of four habit plane variants with respect to each {110} pole of the parent crystal. Six such plate groups can form in a parent crystal, making a total of 24 variants of the habit plane. Such a combined grouping effect produces zero net macroscopic shape change in the given specimen when cooled below M_f. The second stage involves the application of stress in the heavily twinned martensite state (say, below M_f) in which the deformation proceeds by migration of the twin interface and variant coalescence. This produces a single crystal of martensite which is most favorably oriented with respect to the applied stress (i.e., largest extension in the stress direction). Finally, the deformed single-crystal martensite, when heated above the A_f temperature, transforms back into a single crystal of the parent phase. That is, the specimen as a whole reverts to the original shape or orientation. Hence, what is memorized in SME is the shape of the parent phase.[74] Figure 8.40 is a schematic representation of the martensite transformation and shape memory effect, which seem to be identical, irrespective of the alloy system or crystal structure of martensite.[2,177] Figure 8.41 is a typical transmission electron micrograph of thermoelastic martensite in polycrystalline Cu-Zn-Al shape memory alloy.[179]

Figure 8.42*a* illustrates a stress versus strain curve for a Cu-Zn shape memory alloy single crystal deformed below the M_f temperature.[2] It is apparent that martensite deformation occurs at a low flow stress a' of 35 MN/m^2 (5 ksi). During heating, the residual strain at point c is completely recovered.

Note that SME can also be realized when a specimen deformed at a temperature below A_s is heated to a temperature above A_f. At a temperature T ($M_f < T < A_s$), it is partly in a parent phase. The deformation in this regime takes place partly by the deformation and partly by the stress-induced martensite (SIM) transformations.

8.7.3 Two-Way Shape Memory Effect

The SMEs are of two types: the one-way SME and the two-way SME (TWSME). If a SMA is mechanically deformed at a temperature below M_s, its shape returns to

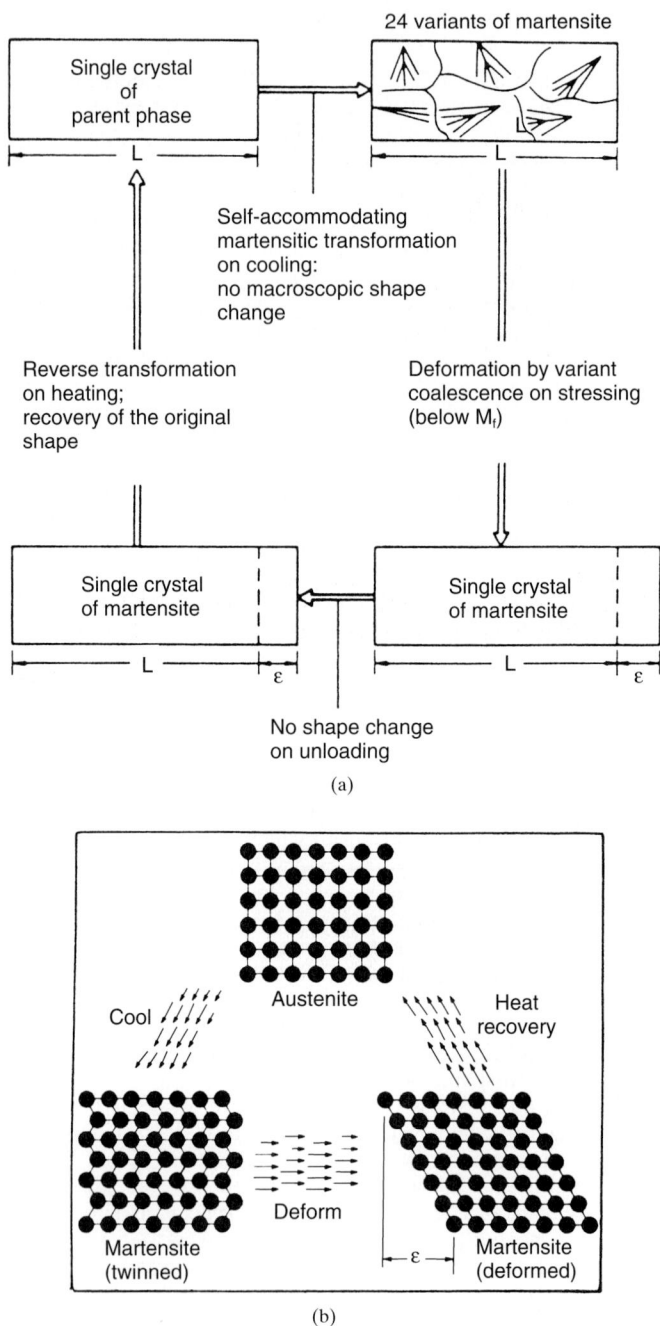

FIGURE 8.40 Shape memory process (*a*) by schematic diagram[2,178] and (*b*) at the crystal lattice level.[177] [*(a) Reprinted by permission of North-Holland Physics Publishing, Amsterdam; (b) reprinted by permission of Metallurgia, England.*]

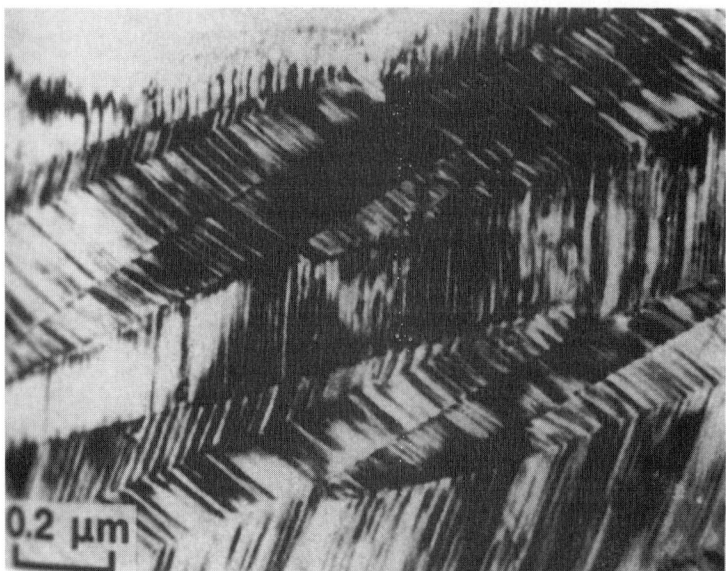

FIGURE 8.41 Typical transmission electron photomicrograph of thermoelastic martensite in a polycrystalline Cu-Zn-Al shape memory alloy.[114,179] (*Reprinted by permission of Pergamon Press, Plc.*)

the original form on heating above the A_f temperature. However, the deformed shape at low temperature cannot be recovered when the specimen is recooled below M_s temperature. Alternatively, if the SMA is cooled without applying stress, the shape of the alloy will not exhibit any macroscopic shape change. This process is called one-way SME because only the hot (initial) shape is memorized. That is, the specimens can spontaneously and reversibly alter their shape to two different configurations upon cooling down and heating up treatments without application of external stress/load.[115,180] However, in two-way shape memory (TWSM) effect or reversible SME, the shapes of both parent and martensite phases are memorized (recovered). In contrast to the one-way memory effect, no external forces are needed here to obtain the memorized cold shape.[115] TWSM effect can be achieved by subjecting a SMA specimen to one or several thermomechanical cycles or by a special training program during which the sample is either temperature- or stress-cycled via the martensitic transformation. Most commonly, TWSM behavior can be induced by using the following techniques:[181]

1. *Shape memory training.* Repeat the cycle several times of cooling the specimen to a temperature below M_f, then deform the martensite above the elastic limit, and subsequently heat it to a temperature above A_f under no stress.

2. *Strain-induced martensite (SIM) training.* Deform the parent phase accompanying SIM transformation, and then reverse the transformation by removing the load. Repeat the cycle several times.

3. *Combined mode I training.* Deform the parent phase accompanying SIM transformation and cool the specimen to a temperature below M_f with the constraint, then release the constraint. Repeat the cycle many times.

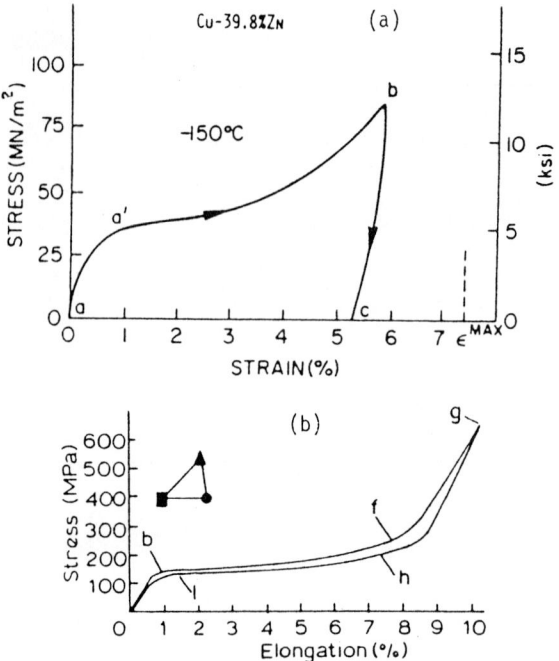

FIGURE 8.42   (*a*) Stress-strain curve for a Cu-Zn shape memory alloy single crystal deformed below its M_s temperature. The flow stress at point *c* was completely removed by heating.[2] (*b*) Stress-strain curve for a Cu-14.2Al-4.2Ni (wt%) alloy showing $\beta_1 \rightleftharpoons \beta_1'$ (18R) transformation pseudoelasticity.[74] [(*a*) Reprinted by permission of North-Holland Physics Publishing, Amsterdam; (b) reprinted by permission of ASM International, Metals Park, Ohio.]

4. *Combined mode II training.* Deform the specimen with martensite structure to reorient the martensitic variants, and heat the specimen to a temperature above A_f with the constraint to induce reverse transformation, then release the restriction.

5. *All-round SME training.* Age a NiTi specimen under a uniaxial stress to develop the preferred oriented precipitates in the matrix.

6. *Stabilized stress-induced martensite (SSIM) training.* Recently Guilamany and coworkers have developed the SSIM training to achieve TWSM effect in Cu-Zn-Al-Co SMA. A one-cycle training by constraint aging at parent-phase condition offers a TWSME in this material.

Recent experimental results have shown that in Cu-Zn-Al single crystals, it is not essential to apply this training to obtain a stable TWSME. Rather, it is adequate to stabilize the martensite by diffusion at room temperature, subsequently retransforming partially or completely to the parent phase by heating. In this way, it is possible to obtain a full TWSME, and induce martensite variant on cooling against an applied force.[182]

Recently, it has been reported by Stalmans, Humbeeck, and Delaey[183] that local martensite stabilization and true plastic deformations have to be considered as undesirable side effects of training. This finding rules out the usual explanation of the physical origins of TWSME, i.e., microscopic residual stress fields or locally retained martensite. Following Stalmans et al., the variants which are formed during cooling are those which possess the lowest energy.[184]

TWSME can also be obtained by stress-inducing a single variant above the M_s temperature, then stabilizing it and retransforming it on unloading to high-temperature β phase. Even when no retained martensite can be detected by optical microscopy, a perfect TWSME is present.[185]

There are three general markets where TWSM is used: engineering (industrial), toys, and biomedical. A typical potential use in biomedical applications is a cotter-pin type of fastener. It can be used to produce rigid anchoring of the prosthesis, e.g., inside the bone cavity or inside drilled holes in a bone.[186] Other applications include implanted devices such as intrauterine (contraceptive) devices (IUDs), aneurysm clips, and vena cava filters (to trap blood clots).[187]

8.7.4　Pseudoelastic Effects

In a more general term, pseudoelastic effect (PE) or pseudoelasticity refers to any non-linearity during unloading. Pseudoelasticity can be effected by either twinning or by stress-induced phase transformation.[188] In pseudoelastic effect, the martensitic transformation proceeds with an increase in stress well above the elastic limit and reverses completely along the same microstructural path, with shape memory, upon removal of the stress, usually above A_f temperature.[179] In these alloys the deformation produces both parent-to-martensite and martensite-to-martensite transformations. The SME and PE are actually complementary aspects in the deformation and reversion of the thermoelastic martensite in a SMA; when SMAs are deformed, some of the strain is recovered on unloading (PE) and the rest on heating (SME). In general, all SMAs exhibit PE above A_f temperature. However, there are exceptions, including Ti-Ni alloys where the PE is composition-dependent; that is, PE does not appear in alloys with a Ni content lower than 51 at% at any temperature, but the PE appears in alloys with a higher Ni content. Figure 8.42b shows a typical pseudoelasticity loop for a Cu-14.2Al-4.2Ni wt% alloy single crystal.[189] Clearly, during loading, elastic deformation occurs until point b on the stress-strain curve is reached where martensite plates begin to appear; plastic deformation occurs between the b and f points where the specimen becomes a single crystal of the β_1' martensite, through coalescence of variants; and elastic deformation proceeds from point f to point g. During unloading the reverse transformation from β_1' martensite $\rightarrow \beta_1$ parent phase occurs, first with the recovery of elastic strain from point g to point h, then with the rapid recovery of plastic strain between point h and l, and finally with the complete recovery of elastic strain from near point l.

Application of PE is based on its three characteristics:

1. The amount of recoverable strain varies from a few percent to 20%, which is very large compared to the ordinary elastic strain.

2. There is a constancy of stress during the transformation, which facilitates in making a spring that deforms without a change in stress.

3. There is a use of elasticity in the region prior to the onset of the stress-induced transformation.

PE is not as extensively used as the SME. Some applications of PE are (1) dental arch wires for orthodontics and (2) springs with very large energy storage capability for obtaining high fuel mileage in automobiles.

8.7.5 Pseudoelastic Effects: Superelasticity

There are two types of stress-induced pseudoelastic effects—superelasticity and rubberlike behavior.[189] Superelasticity is a mechanical-type shape memory as opposed to thermally induced (by heating) shape memory, as described above where certain metals with an exceptional ability undergo extensive elastic deformation.[117] Superelastic effect of SMA leads to a unique combination of high strength, high stiffness, and high pliability, which is unmatched by any other material or technology. Superelastic effect is isothermal in nature and involves the storage of potential energy.[115] Superelasticity is realized when the aging temperature is above A_f. In this case, the martensite is stress-induced and reverts to the parent phase upon unloading (i.e., on the removal of stress); thus the stress contributes the mechanical driving force for transformation. Superelastic alloys must exhibit an inflection point whereas pseudoelastic alloys show only nonlinear unloading behaviors. An inflection point in the unloading behavior suggests the presence of an unloading plateau, or a strain range with approximately constant stress. This is an important distinguishing feature in medical applications. Superelastic alloy (SEA) is increasing geometrically in recent years. Among SMAs, Ni-Ti-based alloys appear to be chemically and biologically compatible with the human body.[190]

A stress-strain curve showing typical superelastic behavior for a single-crystal sample is illustrated in Fig. 8.43a. The upper plateau region denotes the formation of a preferred variant of stress-induced martensite plates whose shape strain complies with the applied stress.[191] During unloading, the lower plateau region involves the formation of parallel plates of only one variant of the parent phase. The level of plateau stresses shown in Fig. 8.43a is a function of the test temperature. The upper plateau stress is naturally zero at the M_s temperature. Figure 8.43b shows the stress-strain behavior of superelastic Nitinol.[190]

8.7.6 Pseudoelasticity: Rubberlike Behavior

Martensites in several SMAs such as In-Tl (fct), Au-47.5 at% Cd (2H), Au-Cu-Zn (18R), Cu-Al-Ni (2H), Cu-Zn-Al (9R, 18R), and Cu-Al-Ni alloys, etc. have been found to exhibit rubberlike behavior which involves the deformation of existing single martensite phase without undergoing phase transformation. In Cu-Zn-Al SMA, a model to explain the rubberlike effect assumes that the development of short-range order (SRO) structure lowers the free energy of the martensite and results in an increase in the relative stabilization of martensite with respect to the parent phase. When an external stress is applied on a sample (after aging at room temperature for some time) which contains several twin-related variants of the martensite phase, deformation proceeds by reorientation and rearrangement of the martensite variants through movement of their twin boundaries. During unloading or releasing the stress, the boundaries move back to their starting positions, and the original shape of sample is recovered. Unlike the case of superelasticity, which occurs by stress-induced phase transformation, this behavior arises from a single martensite phase.[192,193]

The characteristics of rubberlike behavior (RLB) has been summarized based on experimental results in a Au-Cd system as follows:[139]

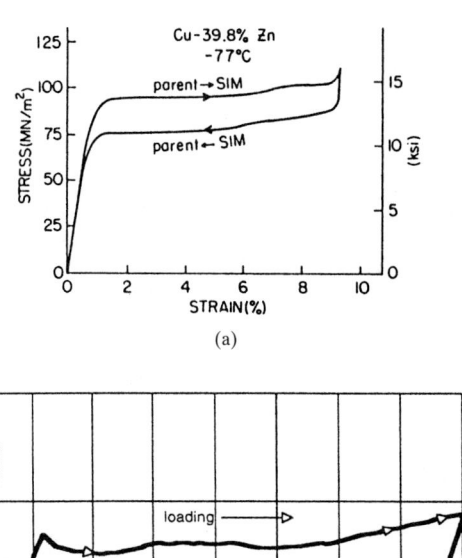

FIGURE 8.43 (*a*) Stress-strain curve for a thermoelastic Cu-Zn SMA deformed above the M_s temperature, showing superelastic behavior due to the formation of and reversion of a reversible, stress-induced martensite.[2] (*b*) Stress-strain curve for a Ti-50.8 at% Ni alloy tested at 10°C above the A_f temperature. A typical elastic loading-unloading curve for stainless steel is shown as straight line on extreme left. [(*b*) *Courtesy of Nitinol Devices and Components.*]

1. RLB arises from martensite aging.
2. The introduction of quenched-in vacancies or structural vacancies enhances the RLB.
3. RLB occurs by reversible motion of twin boundaries.
4. RLB is not a boundary effect, but a volume effect.
5. Development of SRO of martensite and rearrangement of the strain dipoles in martensite remain as viable candidates to explain the peculiar RLB.

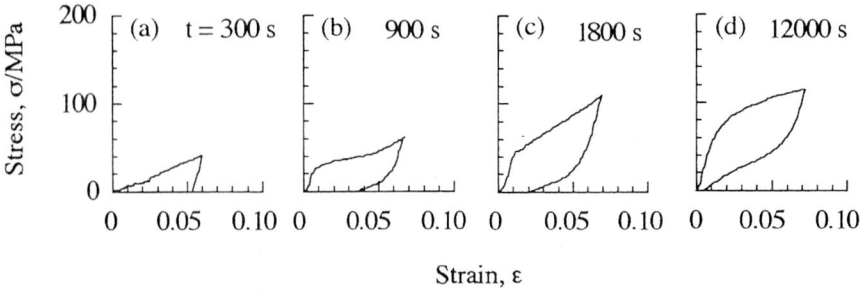

FIGURE 8.44 Tensile stress-strain curves obtained for Cu-14Zn-17Al martensite with different aging time at room temperature.[193]

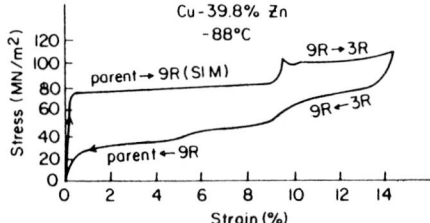

FIGURE 8.45 Stress-strain diagram showing the deformation behavior of a Cu-39.8% Zn alloy single-crystal specimen deformed above the M_s temperature. A double superelastic loop is observed which is the result of two successive stress-induced martensitic transformations.[2,195]

The origin of the restoring force in rubberlike behavior has not been established as yet. Figure 8.44 shows an example of rubberlike behavior for the case of 18R martensite in Cu-14 at% Zn-17 at% Al alloy.[193]

Among the nonferrous systems, the $\beta_2 \to \zeta_2'$ (trigonal) martensite transformation in Au-49.5 at% Cd alloy shows a quite significant aging effect in which RLB appears by room-temperature aging only for 1 hr. Nakajima et al. have shown that the martensite aging effect becomes more dominant with an increase of vacancy concentration and proposed the close relation between RLB and point defect (e.g., vacancy) in the martensite, which is consistent with the SRO model and the strain dipole model.[194]

8.7.7 Martensite-to-Martensite Transformations

When several Cu-based alloy martensites are continuously deformed above the single-crystal martensite stage, a new martensite phase is produced, i.e., a stress-induced martensite (9R structure)-to-martensite (3R structure) transformation occurs. This successive mode of martensite deformation permits the realization of recoverable strains of more than 17%. This occurs by shearing on the basal (close-packed) plane of the 9R structure such that the structural stacking sequence is changed from . . . ABCBCACAB . . . to ABCABCABC. . . . Figure 8.45 shows the deformation behavior of a Cu-39.8Zn single crystal strained at −88°C (~35°C above the M_s).[195] The first upper plateau represents the formation of SIM, as described earlier. The second plateau, which begins at about 9% strain, denotes a second martensitic transformation which is stress-induced from the first martensite "mother." The two lower plateaus arise from the reverse transformation occurring in an inverse sequence. This figure shows a double "superelastic loop." Note that the second stress-induced martensite in Cu-Zn alloy can be obtained from the first SIM, or from the normal thermally formed (upon cooling) martensite.[2]

8.7.8 Martensite Stabilization

The transformation temperature at which the martensite reverts to its original parent phase increases by martensite aging. This aging effect is called *martensite stabilization*, because the A_s temperature of some SMAs increases with aging time and finally aging suppresses the reverse transformation.[131] The stabilization of martensite in Cu-Zn-Al and Cu-Zn-Al-Mn alloys is related to reordering in martensite reaction B2 $\rightarrow$ DO$_3$, which explains the reason why Mn improves the stabilization of martensite.[196] In Cu-Zn-Al SMA, the martensite stabilization is due to pinning of interfaces between parent and martensite and between martensite variants by quenched-in vacancies and/or precipitates. However, the martensite stabilization occurs even in the single crystalline states of martensite. The pinning mechanism is thus not always considered as valid.[197] Severe martensite stabilization normally occurs when the alloy is aged in the martensitic phase, which may occur even at ambient temperature and which produces drastic changes in the SME characteristics. On the other hand, aging in the parent phase may result in a change in the degree of order and the formation of precipitate.[131]

Bidaux[198] has experimentally observed the following general features of stabilization of 18R martensite phase in Cu-Zn-Al alloys:

1. The stabilization of 18R phase does not affect the tensile stress for subsequent transformation to the 6R phase.
2. The stabilization of the 18R phase hinders further transformation to the 2H phase.
3. Stabilization can be accomplished in alloys with an electron concentration near e/a ratio = 1.48 and concerns the 18R phase. At higher e/a ratio, the hexagonal 2H, the fct 6R, and 18R phases can be obtained as well as advantageously studied in the same specimen.

Whatever the reason, the martensitic stabilization can be greatly suppressed by hot rolling of a Cu-Zn-Al alloy. It is postulated that the introduction of high density of dislocations at high temperature during rolling is a key player in providing high-mobility pathways and sinks for vacancies.[131]

It has been observed that minor additions of alloying elements such as B, V, Zr, and Ti can refine the grain size. Hence the aging effect becomes less severe in Cu-Zn-Al-Mn-Zr alloy. Vacancies pinning at the martensite boundaries can be decreased with suitable parent-phase aging after martensite-phase aging, which, in turn, can revitalize the stabilized martensite. Cu-Zn-Al-Mn-Zr SMA exhibits a better thermal stability than the Cu-Zn-Al SMA.[134]

8.7.9 Application of Shape Memory Alloys

SM alloys are used as an important class of functional materials because of the widespread applications of various heat-sensitive devices and even heat engines. A worldwide interest in SM alloys is evidenced by the fact that there are well in excess of 10,000 patents on SME actuators only in international records. A large focus on SMA commercialization has been on medical and dental applications, notably those with superelasticity. Applications, which are classified depending on their use, are briefly described here.

In Shape Recovery.   SMAs are used as heat-activated fasteners or as tight seals, plugs, clamps, and rivets, especially in inaccessible places in the nuclear industry, vacuum apparatus, deep-sea, and space systems which will eliminate welding. Another application is a self-erectable antenna developed by NASA and Russia for use in the MIR space station.

Couplings and Electrical Connectors.   SMA couplings are a classical example of restrained recovery where the forces created by preventing the SME from advancing to completion are employed to effect the joint and provide a permanent connection.[199] At the chilled low temperatures (which are attained by a blast of cold fluorocarbon gas from an aerosol can) and in the martensitic state, the Ti-Ni SM alloy electric connector ring becomes stretched and is opened, while at high temperature this ring recovers to its original shape (by the SME) and tightens and thus secures the electric connection. This device is suited for multiconnector electrical plugs and is used safely for high current in aerospace systems having high-vibration environment. Other applications are in the space assembly of the MIR space station; pivot connect with a NiTi prestrained wire in an antenna rib system; Ni-Ti type leakproof pipe couplings in hydraulic fluid lines in F-14 jet fighters and as connectors and for pipelines in the North Sea, power plant systems, and chemical and petroleum piping. The couplings are expanded ~4% in the martensitic state at liquid N temperature, placed around the tubes to be joined, and warmed to ambient temperature during which they contract and make a leakproof seal. Figure 8.46 shows the principles involved in the operating mode of tube or pipe coupling made from a Ni-Ti type SM alloy. Similar Ni-Ti types fixtures are widely used for plumbing on submarines and surface ships.[200] Advantages of cryogenic NiTi couplings for aircraft hydraulic tubings are light weight, easy craft-sensitive installation, and proven reliability. NiTiNb couplings have proved themselves as repair couplings and as a reinforcement of a pipe weld and are used as a substitute to welding in oil pipelines.[201]

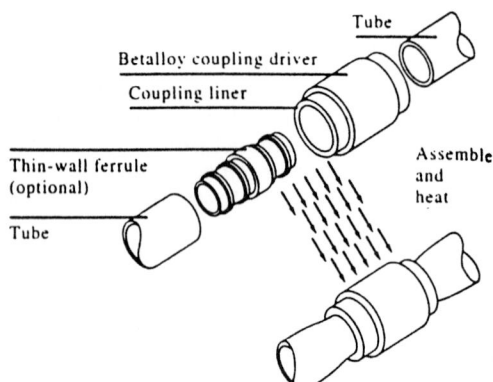

FIGURE 8.46 Operating mode of tube or pipe coupling made from a Ni-Ti shape memory alloy.[200] (*Reprinted by permission of Pergamon Press, Plc.*)

SM devices are also used for joining fiber optics with a hermetic seal to prevent intrusion of moisture.[116]

Fe-based pipe coupling is also presently being used for connecting the pipeline in the oil fields and chemical engineering plants which will increase the life of pipeline in service.[202] Fe-Mn-Si-based alloy has also received most attention for smooth tube and pipe couplings to retain their coupling force at liquid N temperature, unlike NiTi couplings.

SMA systems have been developed for nuclear power system piping couplings and to repair nuclear system and heat exchanger pipings which will eliminate welding. SMA is used as connectors for electronic interconnections in computers and other digital control systems and as zero insertion force (ZIF) connectors of the connecting board. The ZIF feature allows one to make the connector with little, if any, force.[203] Constrained recovery of SMA devices to connect signal wires to flat-plate liquid-crystal displays is now being made in large quantities for portable PCs.

Electrical Seals and Packaging. NiTiNb alloy is used in the fastening of metallic back-braid shielding on an electrical cable to the end connector and as a welded ring seal with polymer coat and prestraining.[204] The advantages of SMA fasteners are that (1) they can provide a hermetical good shock, vibration, and thermal cyclic properties; (2) they can be installed to precisely locate components with a controlled preload, e.g., bearings on a shaft;[201] and (3) they can result in reduced manufacturing cost.

Valves. SMA valve actuators have been developed using R phase transformation for various types of fluid gas valves, which have several advantages over other products, such as (1) an actuator has a large force/weight and force/volume output and (2) it shows very low hysteresis and has a temperature range of only -40 to $+50°C$.[204] Other applications include thermostatic valves for a domestic shower; NiTiCu actuators to prevent accidental scald burns in a tub, sink, or shower; SMA spring actuator to shut off the gas in the event of fire in domestic gas supply; SMA actuator to open a vent controlling a pneumatic control valve on a gas cylinder; SMA valve to control fluid flow in automatic transmission passenger cars; and SMA actuators for various automotive uses such as windshield wipers, fuel control valves, climate controls, mirror adjustment, door and hatch latches, retractable headlights, and so forth. Other applications of actuators are in the turbine axial compressor and load release system as well as in flexible robotic grippers.[205]

Electric Power Systems. SMA applications such as steam pipe hangers are expected to be used when there is a need of utility demand for implementation. Other applications include an overtemperature indicator device using a small flag attached to a bus bar bolt in a Ukrainian power system; shape memory valves to overcome the problem of balancing electrical demand and the best use of off-peak power. Ni-Fe-Co, Cu-Al-Ni, and NiTi SMAs are applied for best cavitation erosion resistance in hydroelectric utility such as Hydro Quebec.[204] Other applications in the electric power industry are as large electrical connectors; CuAlNi circuit breakers (to protect network equipment from overheating due to short circuits or overloading which are designed to switch off the current instantaneously in the case of short circuits or after specific time in the case of overload); fuses (to reduce or even eliminate maintenance costs); switches and safety devices such as fire detectors; control of transmission/distribution line sag caused by temperature changes; electrical contacts such as Belleville washers; thermal valves; deicing of transmission lines; energy converters—high-power heat engines, robotic devices, electrical/thermal actuators, thermomarkers, overcurrent protection, optical-fiber splices, contact-bounce dampers, and nuclear power applications.[206,207]

Aerospace Systems. Applications include the SM couplings in the space assembly of trusses used as masts on the MIR space station; high damping parts; hermetic seals; as replacements for explosive bolts; sprag-type coupling for composite tubing;[208] latching systems; ball-socket assemblies; antenna rib systems; CuAlNi SM actuators to provide control of the tip clearance in the turbine axial compressor; and SMA-actuated device in a load release system.[204,206] The use of intelligent structures incorporating sensors, processors, and actuators with enhanced damping characteristics is ongoing throughout the aircraft and aerospace industry.[116,207]

Thermal Sensors and Actuators. Thermal actuators fulfill two functions: (a) to detect a temperature change and (b) to actuate. They respond to changes in temperature by changing their shape and/or generating a force. Memory rings made from Cu-Zn-Al SM alloys in series with a bias spring, usually of steel, are used as thermal actuators or thermostats for its operation between the upper and lower temperature limits. Other applications include an air flow controller or vent controller for an air conditioner; a greenhouse fan wheel, window, ventilator, or fire door; switches for an automobile radiator and an automotive electric system; an automotive carburetor jet assembly; an automotive drive element for the choke; liquefied natural gas lead detectors; liquid gas switches; thermic safety valves for fire protection; thermostatic radiator valves for domestic hot-water heating systems; thermal protection device for domestic water filtration unit; steam trap for passenger train steam heating system; electric switch for diesel engine radiator fan; thermostatic mixing valve; SME element to switch off electric kettles upon boiling water; and NiTi alloy in servomechanism for driving recording pens and indicating-pointer assemblies. Many of these devices have been cycled over half a million times without any observable fatigue, "creep," or changes in deflection characteristics.[7,109,200,209]

The advantages of SMA thermal sensors and actuators are (1) large force output per unit weight, (2) large stroke/weight ratio, (3) rapid motion at a given temperature, (4) flexibility of design of actuator motion direction, (5) design that can be employed where cleanliness is critical, (6) insensitivity to a wide variety of environmental conditions, and (7) quiet operation.[209]

Electrical Actuators. In the case of electrically heated actuators, SMA element is heated by externally bypassing an electric current. Examples of applications include an air damper for a multifunction electric oven; fog lamp protective louver (mounted in front of a car); SM wire robotic hands, developed by Hitachi, which can flex and are driven by electric current; and robotic crab with six legs.[209]

High-Force Devices. High-force delivery of SMA actuators includes wheel-pulling devices to facilitate the disassembly of large gas and steam turbine wheels from their shaft with minimum possibility for damage; SMA high-force generators for rock breaking (even under water) and similar materials as a replacement for explosives in demolition.[116,204] Exploding bolts are used in aerospace vehicle launch procedures to separate the rocket from its hold-down or to separate various stages during the initial flight trajectory; they also find applications in underwater systems as safety devices in the petroleum industry, in petrochemical equipment, and in nuclear power plants.[116,207]

As Microactuators. SMA micromachine actuators can be realized by microma-chining and subsequent assembling of individual SMA devices or by integrated SMA thin-film NiTi microvalve actuation by sputter deposition or laser ablation.[210] The recovery stress and strain associated with martensitic transformation in TiNi and TiNiCu thin films reached up to 500 MPa and 6%, respectively.[211] Other exam-ples include the SM micron sized wires; pen drive mechanisms in recording; and industrial-control instruments. Sputter-deposited thin films of Ti-51.9 at% Ni are used for microactuators, due to their large deformation and strong recovery force.[212] Mollusk-type manipulator is used for transplanting saplings (in biotechnology studies), and miniature clean grippers are used in clean rooms or in vacuum.[209]

Conversion of Thermal Energy to Mechanical Energy. The principle involved in heat engine application of a SM alloy is shown in Fig. 8.47. These engines, being a third type of actuator, usually operate between two fixed temperatures (usually maintained by two water reservoirs) and have modest efficiencies of about 4 to 6% when operated at room temperature and above.[109,200] Consequently they are useful for the conversion of low-grade energy (such as solar-cell boom for aerospace appli-cations, locking/unlocking, and tracking mechanism for use in satellites), waste energy in factories, cold stores, geothermal energy in hot springs, and heat of water at the surface of the ocean into mechanical energy.

Biomedical Applications. Recently it has become clear that the largest commer-cial successes of SMAs are associated with the biomedical applications. The current developments of applications are grouped into four biomedical fields, which are pre-sented below.[115]

1. *Dental (or orthodontics) applications.* These include teeth braces; Smartini (standing for Smart Martensitic TiNi)-orthodontic arch wire (which produces the corrective forces to move and correct the tooth alignment); specific endosseous implants in blade or rod form; dental root prostheses or implants for a stable attach-ment in the bone; partial dentures; adjustable dental abutments with self-locking temperature-sensitive ball joints; adjustable telescopic head gears for reposition of the molars; small superelastic tension and compression springs to create space or

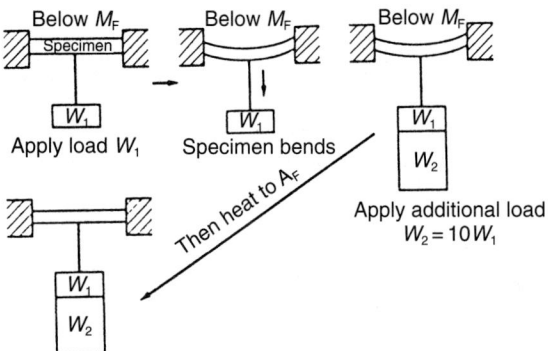

FIGURE 8.47 Principle involved in heat engine application of a shape memory alloy.[200] (*Reprinted by permission of Pergamon Press, Plc.*)

close gaps between the elements in the mouth; heavy suture expanders with super-elastic springs to alter the width of the mandible; special connectors with a wedge shape to hold the two teeth tightly together; martensitic, very adaptable wires and strips for retention of the teeth after correction; connectors for threadless anchoring of abutments to dental root implants; and adjustable orthodontic brackets with torque and angulation control.[115]

2. *Stents.* Intravascular scaffolding using various designs is a technique that is greatly increasing. The devices employed are usually called *stents*, named after a dentist, Charles Stent. Besides many types of endovascular stenting of blood vessels and tubular structures, stents for noncirculating conduits have been developed such as bronchial stenting, esophageal stenting, stenting of the ureter,[115,213] dynamic interference stents, and hysteresis stents for laparoscopic procedures.

NiTi stents are distinguished as permanent internal stents and removable (heat-shrinkable) stents. Self-expanding permanent stents are used to brace the inside circumference of a tubular passage such as esophagus, bile duct, or blood vessel; and cardiovascular stents. Intervention devices are used as a follow-up to balloon angioplasty.

Removable stents are devices, when deployed in the vessel after balloon expansion in the martensitic state, that can be removed by a specialty recovery catheter, coaxially placed in the stent in combination with an infusion of normal saline at 55°C to heat the stent for the occurrence of shape recovery. In addition to the advantage of being a removable foreign object after healing, other important applications include the prospect of using local drug or radiation therapy.[214]

3. *Orthopedic applications.*[115,215,216] These include bone-straightening plates; distraction osteogenesis device (to lengthen bones); external distraction devices with shape memory actuators; double ratcheting device for distraction of long bones or soft tissues, lengthening of a growing prosthesis, and correction option in, e.g., oral and maxillo-facial surgery; orthopedic external and internal fixations; Harlington (or superelastic) rods for the treatment or correction of scoliosis (abnormal curvature of the spine); prosthetic or artificial limb joints; osteosynthesis (a surgical treatment for bone fracture) compression plates; jaw plates; spacer for bone-chip arthodesis of a spinal column; straightening deformed spinal columns; separation of adjacent vertebra in a fusion process; superelastic cerclage/cartilage wires for a tight connection of bone parts; shape memory locking mechanism to adjust tension in cerclage/cartilage wires; shape memory locking rings to fix the liners in the cups of hip prostheses; intermedullary fixation pins or nails to adapt to the local diameter for the fixation of fractures in the major long bones; osteosynthesis staples for a stable reposition and fixation of two bone parts after fracture or in corrections of various osteotomies (or bone deformities in the legs and feet); tools for the removal of broken bone screws; straight superelastic tension wires for the fixation of hip implants to the bone; external fixators with a large extent of adjustability, locked by temperature-sensitive rings and ball joints; and connectors for modular orthopedic implants.[115,116]

4. *Medical instruments and tools.* The following long list of present applications illustrates the growing importance of the SMA alloys.[115]

Steerable instruments using superelastic NiTiNb such as catheters, guidewires, and other medical instruments; steerable lasers with superelastic, bending housing in an endoscope; superelastic torque wrench made of Cu-Zn-Al single crystal; superelastic guide wires often with optical system and a steerable tip; assembly of memory modular dental or medical devices or instruments and implants in various ways such as sliding stop, lockable ball joint in dental implants, hip implants, skull plates, etc.; catheters with a retrieval basket to catch stone fragments from gallbladder; cylindrical locking rings (to lock telescopically engaging parts of orthodontic face bows and orthopedic external and internal fixations); Homer Mammolok (curved locator wires) to pinpoint the location of nonpalpable breast lesions (after they have been identified by mammography); adjustable instruments and all types of adjustable clamps for spinal fixation; marrow needle for fixation of a broken thigh bone; needles and probes used in the repair of joints, typically the knee meniscus; self-fixating needles, electrodes, and sensors with a tip for shape variation during insertion into the body; locking device for cartilage wire or ligatures for use in complicated operations like spinal corrections;[215] superelastic springs for anterior cruciate ligament (ACL) prosthesis; microactuator for Braille systems of rheumatic joints; superelastic deployable umbrellas for atrial septal defect occlusion; intraocular lenses (for cataract patient) with fixation loops that are temporarily collapsed during insertion into the eye; vena cava filter (for the prevention of recurrent and fatal pulmonary embolism) to be inserted into veins in a straight shape through a catheter; clips for aneurisms and sterilization; arthroscopic procedures using fiber optics and superelastic suture needles for the repair of the damaged cartilage; pacemaker lead wires in strain-controlled fatigue environment; artificial sphincters for stress incontinence; MRI compatibility-instrument needles;[190] micropumps for artificial kidneys and artificial heart; application to laparascopic procedures to various cancer amelioration techniques and various areas of urology; temperature indicators to control the maximum temperature level of blood, organ, etc., during storage and transportation; hingeless instruments for minimal invasive surgery functions such as endoscopy, tomography, and radiology; hingeless gripper made of one single part and heat-treated beaks; miniature cutters and grippers; miniature endoscopic suturing instruments; NiTi needles, stylets, guidewires, catheters, stents, filters, tissue anchoring and connection, flow control devices, rhinosurgical instruments, etc.; very thin tubes (O.D. $\leq 100\,\mu m$) employed in arteries; instrument boxes with a heat-sensitive lock with opening at sterilization temperature only; self-locking instruments for reuse without sterilization; and endoscopic superelastic knives with continual variation of angle.[115]

Mechanical Dampers.   SMAs have a high damping (i.e., noise reduction) capacity in the martensitic state and two-phase conditions that is far greater than that of traditional alloys such as brass, steel, or aluminum; for example, SMAs and gray cast iron exhibit values of specific damping capacity (SDC), $\psi > 40\%$ and in the range of 5 to 19%, respectively. Incramute (58Cu-40Mn-2Al), on aging at 400°C, seems to have ψ up to 68%. Nitinol, Ni-(45–50)Ti provides high levels of damping (ψ 26%) due to stress-induced movements of twin boundaries in both $\beta19'$ martensite (M) and R phase.[217] Cu-Zn-Al and Cu-(10–14)Al-(2–4)Ni with minor additions are also available as damping alloys; the latter yield maximum damping with 50% martensite when both the transformation rate and the degree of easy-moving interphase boundaries are large.[218] This behavior is due to high hysteresis/mobility of interphase boundary surfaces or twin boundaries within the martensite plate variants, which under oscillating stress move back and forth, producing frictional energy dissipation. Other materials with high damping properties such as Fe-Co-Mo-Cr alloys and Co-Ni alloys have also been developed and partially used in the industry.[142,217–219]

Specific damping capacity ψ is expressed in percentage terms as $100(\Delta W/W)$, where W is the specific elastic energy stored at a certain stress amplitude σ_0. The damping intensity is also characterized by the logarithmic decrement $\delta = \ln(\varepsilon_m/\varepsilon_{m+1})$, where ε_m and ε_{m+1} are two subsequent strain amplitudes. In addition to these characteristics, often the inverse mechanical Q factor ($Q^{-1} = \Delta W/2\pi W$), called *internal friction*, is used. The relationship between the damping characteristics is given by:[219]

$$\psi = 1 - \exp(2\delta) \sim 2\delta = 2\pi Q^{-1} \qquad (8.28)$$

In Automotive Industry. Important automotive applications are found in fuel injection, cold starters, rattling-noise reduction (temperature compensating valve for), shock absorbers, temperature-sensitive boost-compensators, temperature-compensated valve lifter; Belleville- or wave-washers; door-locking mechanism; climate control, thermal valves, fog lamp louvres, and windshield wipers.[206]

As Structural Dampers. Damping can be used to control the vibration at the source or as a means to isolate the vibrating member from other adjacent structures. Cu-Zn-Al SMA is used to achieve a significant damping effect, when a centering force can be created to restore and isolate various structures to their original position from seismic disturbances after an earthquake.[220]

Consumer Products. SMAs have been used effectively in consumer products such as NiTi for eyeglass frames, ear and temple pieces, and nose and brow bridges; SMA wire to provide reinforcing elements in women's brassieres and in the heel of a shoe, as the head band of head phones; Cu-based SMAs for the nib on an ink pen; and the short superelastic NiTi stub in the antenna of the cellular phone.[204]

Smart Materials. *Smart, active, or intelligent materials*, also called *adaptive structures*, refer to various material systems that can combine the use of the sensors, some signal processing, and actuation functions into the structural system, i.e., typically a composite, which automatically or remotely changes their dynamic characteristics or their geometry to meet their intended performance.[207] These smart materials are characterized by their ability to change physically, in a predictable and measurable way, in response to the temperature, load, pressure, corrosion, fatigue, and other factors.[221]

For a linear relationship between the stimulus and response, some form of controlled feedback may be required through some type of signal conditioner, microelectronics being sufficiently inexpensive currently. However, in many cases, a step function is sufficient. For example, a resistance increase with temperature of several orders of magnitude can be observed with carbon black-filled polyethylene. This large, nonlinear effect can be employed to protect against overcurrent. Another similarity between SMA and many smart materials is their programmable "adaptiveness." Thus, for SMAs it is accomplished by the shape-setting process during annealing whereas for piezoelectric actuators it is done by alignment in an electric field. Ceramic smart materials find a great potential because of having an electrical charge as either input or output.[201]

A major development will probably be multilayer materials because of increased response, either to provide a much larger output signal for the same stimulus or to generate the same output from a much smaller stimulus. Ceramic piezoelectric devices possess relatively small field-induced strains (typically ~0.01 to 0.1% strain) at 1 MV/m applied field.[201] Multilayer constructions may allow a combination of several active functions in the same package by building composites with different ceramics. Thin-film techniques may also facilitate the integration of the control function with the sensor actuator, producing an integrated device, or the integration of the signal conditioner with the sensor such that the output is linear and proportional.[201]

Smart materials have already become commonplace in society, with such examples as SMA thin film devices, actuators (to provide control of the shape of masks and towers used in space structures), devices (to restore structural components to its original shape, stiffness, or orientation), and TRIP steel sensors; piezoelectric ceramics (lead zirconium titanate, PZT) (as piezoceramic copier devices, to implement high-speed data access on optical hard drives, and fiber-optic monitoring composite cure cycles) and electrostrictive devices (to implement essential adaptive optics on the Hubble telescope and correction factors on ground-based telescopes); silicon-based sensors (incorporated during the construction phases to provide internal load-bearing sensor capability and their own data transmission lines from these strategically placed sensor elements); polymer [polyvinylidene fluoride (PVDF), piezopolymer sensors, BST sensors, active implants, and actuators], fiber-optic and chemical sensors; ultraflat and conformal microantennas; microelectromechanical system (MEMS); and microsensors. In the medical field, applications range from SMA catheters to active prosthetic materials and piezoelectric controlled drug delivery systems. The automotive industries are now using smart materials in low-frequency electrorheological damping and high-frequency piezoelectric damping mechanisms, viscoelastic brake shoes with piezoelectric sensor modules, airbag sensors, and piezoceramically controlled mirror isolation. Smart materials also find applications in aircraft, spacecraft, medical, and automotive industries. These may be surface-mounted, as on an I beam, or embedded in structural elements, as in concrete or composite building, bridge components, roads, etc.[221] These devices can, for example measure instantaneous or peak strain in load-bearing members for later data retrieval; detect fatigue, corrosion, or cracking in metal rods and girders; and measure ice buildup on bridges, and weigh-in motion. All the data can be transmitted to remote monitoring sites. Table 8.12 provides a brief account of energy sensor materials and their applications based on device type whereas Table 8.13 summarizes some smart materials based on input and output.[201] Readers are referred to the recent articles provided by Melton,[201] Schetky,[207] and others.[221]

Incorporation of smart material devices and monitoring system will offer cost-effective solutions to maintain and monitor the structural integrity of the infrastructure comprising highways, communication, military installation, bridges, interstate pipeline, and storage tank system. These new materials will monitor the life cycle of roads, bridges, buildings, and pipelines in a noninvasive manner;[221] measure environmental conditions; provide data about materials (fatigue and cracking), about daily or total use (traffic flow including speed, volume, and loads), and about environment (icing and water flow rates).

8.8 STRENGTHENING MECHANISMS

Martensite is one of the most technologically interesting and complex cases of combined strengthening. The inherent brittleness of virgin martensite and the real danger of quench cracks associated with the high-carbon steels put the validity of strengthening mechanisms in jeopardy. The important factors that seem likely to contribute to the strength of martensite are (1) grain size, (2) substitutional solid-solution hardening, (3) interstitial solid-solution hardening, (4) precipitation hardening, and (5) substructure (including dislocation) hardening.

Grain Size. The martensite crystal grain size or the martensite "packet" size is determined to a first approximation by the prior austenite grain size, because a martensite plate, lath, or packet size can never exceed the austenite grain size within which it forms. It is expected that the yield strength increases with the decrease of the martensite plate size or packet size in accordance with the Hall-Petch equation. Note that the coefficient for the "packet" size is much smaller for carbon-free martensite than for carbon martensite.[222] The strong dependency of yield stress on

TABLE 8.12 Energy Sensor Materials[220]

Material	Device type	Applications
Piezo and Electrostrictive	Shear, pressure, load sensor. Local stiffening actuator.	Vibration sensing. Load sensing. Noise cancellation. Shear sensing. Cracking. Delamination.
Transducers	Acoustic emission. Surface wave.	Fatigue and corrosion. Cracking.
Silicon-based devices	Sensors. Data transmission. Communications medium.	Vibration sensing. Temperature sensor. Load sensor. Data transmission uplink. Automated travelers' advisories. IVHS data uplink.
Memory metals	Thin-film devices. Shape memory actuators and sensors. TRIP steel sensors.	Load sensing. Shape and position restoration. Local stiffening. Peak load storage. Out-of-plane deflections storage. Maximum load storage.
MEMS	Pressure, load sensors. Micromachines.	Load sensing. Microrepair.
Polymer	Piezopolymer sensors. BST sensors and actuators.	Load and shear sensing. Corrosion. Microcracking.
Active antenna	Integrated sensing. Microwave communication. Information networking.	IDT shear sensors. Corrosion sensors. Local wireless communication. Automated travelers' advisories. Satellite uplinks. IVHS data uplink.
Rare earth	Chemical sensing. Magnetostrictive actuation.	Corrosion. Local and global stiffeners.

"packet" size is due to carbon segregation at the "packet" boundaries and prior austenite grain boundaries, but tempering decreases the dependency of σ_y on the packet size due to the formation of carbides.[223]

The data in Fig. 8.48 shows that over the usual grain size ranges for commercial steels (i.e., grain diameters greater than ~10^{-3} cm) which are obtained by normal heat treatment, the contribution of grain size to martensite strength is relatively small.[224] With additional refinement to one order of magnitude smaller, this contribution may be substantial in steels given cyclic austenitization or thermomechanical treatment. The former technique is reaustenitizing and involves repetitive rapid heating and cooling to temperatures just above and just below the A_s temperature. The latter method is discussed in Chap. 15. Thus the development of an ultrafine grain size appears to be an important method of increasing the strength of martensitic steels.

TABLE 8.13 Smart Materials Based on Input and Output[201]

Input or stimulus	Output or response	Material	Application	Reference*
1. Overcurrent (or temperature increase)	Resistance increase	Ceramic, e.g., La-doped $BaTiO_3$, $Pb(ZrTi)O_3$	Thermistor	14
		Polymer, e.g., C-black filled polyethylene	Overcurrent protectors	15
2. Overvoltage	Resistance decrease	Varistor, e.g., Bi-doped ZnO	Overvoltage surge protector	16
3. Change in oxygen partial pressure	Electric signal	Y_2O_3-doped ZrO_2	Oxygen sensor; the electrical signal can be used as part of a feedback mechanism, e.g., for vehicle exhaust	14
4. Deformation or strain electric signal	Electric signal deformation or strain	Piezoelectric material	Active noise control devices	17, 18
		Ceramic, e.g., PZT Polymer, e.g., PVDF	Pressure and vibration sensitizing (intruder alarm)	21, 22
			Vibration control (road roughness ride)	23
			Acoustic emitters	24
5. pH change	Swelling or contracting	Polymeric gel	Artificial muscle	25
6. Electric signal	Viscosity change (increase with electric field on)	Electrorheological fluid (e.g., 35% cornstarch, 65% silicone oil)	Torsional steering system damper	20
7. Temperature	Electric signal	Pyroelectric material (ferroelectric), e.g., $Pb(Zr, Ti)O_3$	Personnel sensor (open supermarket door)	24
8. Humidity change	Capacity change	Polymer, e.g., thin-film cellulose esters Ceramic, e.g., Al_2O_3	Humidity sensors	26
	Resistance change	Polyelectrolyte, e.g., poly(styrene) sulfonate Ceramic, e.g., $MgCr_2O_4$-TiO_2		

*Reference in Melton.[201]

Substitutional Solid-Solution Hardening.   Substitutional elements in solid solution contribute little to the strength of ferrous martensite. However, it is difficult to isolate the indirect effect of substitutional alloying elements on grain size, M_s temperature, hardenability, and retained austenite. The solid-solution hardening effect by Mn is found to be relatively small (35 Pa per % Mn) which remains constant

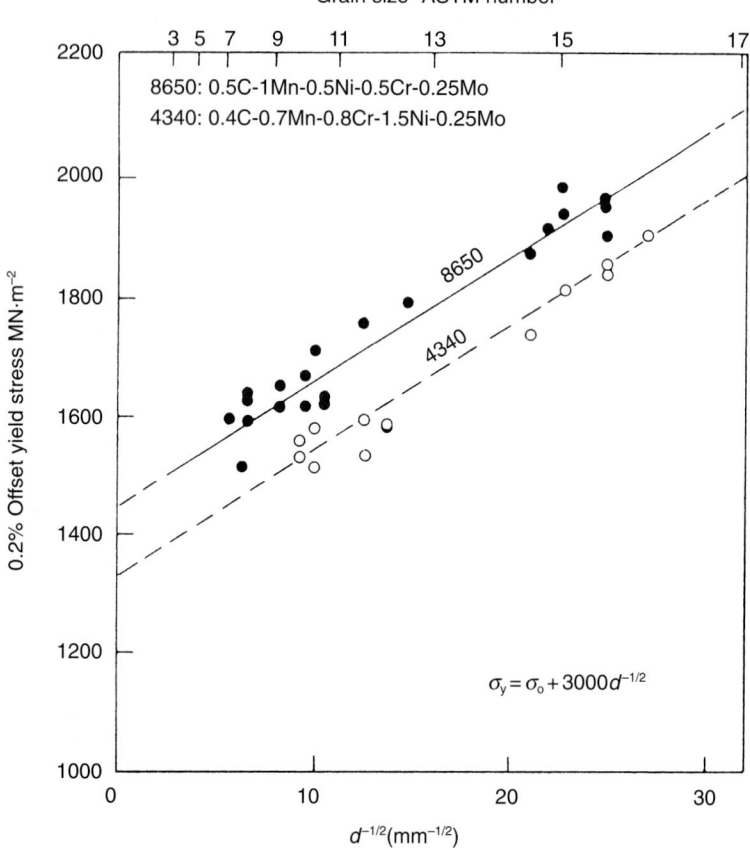

FIGURE 8.48 The effect of prior austenite grain size on the yield strength of two commercial martensitic steels.[224] (*Reprinted by permission of ASM International, Materials Park, Ohio.*)

above 5%.[222] Solid-solution hardening by Ni to carbon-free martensite is about 70 to 90 HV [that is, 232 to 309 MPa (15 to 20 tons/in.^{2})] at 10% Ni. When Ni content is increased from 10 to 30%, it does not produce any effect on the strength of Fe-Ni alloys quenched to −196°C.[225] The solid-solution hardening effect of Ni in Fe-20% Ni alloy accounts for only one-third of the measured flow stress; the remaining strengthening is due to the combined effect of substructure, grain size, and frictional stress of pure iron.[226]

Interstitial Solid-Solution Hardening. Fleischer[227] studied the effect of interstitial atoms on the strength of martensite and assumed a model of dislocation bending away from interstitial solute atoms with short-range interactions. He arrived at the following expression for the flow stress σ:

$$\sigma = \sigma_0 + \frac{2}{3}\mu\,\Delta\varepsilon C^{1/2} \tag{8.29}$$

where $\Delta\varepsilon$ = the difference between longitudinal and transverse strain caused by an interstitial atom. This equation thus predicts that the flow stress is a function of $C^{1/2}$ (the square root of carbon concentration in solution). The curve has a slope of $\mu/15$ to $\mu/20$.

Speich and Warlimont[228] observed that the yield strength of low-carbon martensite obeyed the following relationship:

$$\sigma_{0.2}(\text{MPa}) = 413.7 + 17.2 \times 10(\text{wt\% } C)^{1/2} \tag{8.30}$$

Chilton and Kelly[229] and Roberts and Owen[230,231] had also experimentally observed that the flow stress in both substructures (e.g., lath substructure and the partially twinned structure of martensite plates) varied as $C^{1/2}$ at room temperature, thus supporting the Fleischer model. However, the slopes of the curves of yield stress against $C^{1/2}$ are slightly different, being $\mu/13$ to $\mu/17$ for lath martensite and $\mu/12$ to $\mu/13$ for twinned martensite.

Precipitation Hardening. In plain carbon steels, M_s lies well above room temperature, and precipitation or clusters of carbon occur upon quenching, after the transformation to martensite. This phenomenon is called *autotempering* or *quench tempering*. It is thus difficult to determine the individual contribution of precipitation hardening in such (high-M_s) martensites.

In contrast, Winchel and Cohen[232] had studied the Fe-Ni-C alloys, which all had M_s temperatures well below 0°C (say, −35°C). If the martensite was formed on quenching to −196°C, the carbon became fully immobilized and elimination of autotempering (i.e., no precipitation of carbon) occurred (lower curve in Fig. 8.49). But if these martensites were aged at 0°C for 3 hr, a substantial increase of flow stress occurred because of clustering of carbon or precipitation of carbides. This resulted in the upper curve in Fig. 8.49.

Chilton and Kelly[229] showed similar results and concluded that precipitated carbon as carbide is more effective in raising the strength of martensite than the carbon in solution. That is, precipitation strengthening alone gives rise to a major increment in strength in low-M_s martensites, but solid-solution strengthening and strengthening from clusters or dislocation atmospheres are also important.

Substructure Strengthening. The proposal that the fine twin substructure characteristics of high-carbon martensites make a major contribution to strength has not been well received. However, in an untwinned densely dislocated lath martensite substructure, the dislocation density is similar to the structure obtained by heavy deformation. The dislocation density has also been found to increase to very large values with the increase in carbon content. Moreover, experimental data have suggested that most of the carbon present in lath α' segregates to dislocations and lath boundaries. This segregation of carbon and/or very fine carbides is very effective in increasing the dislocation density.

Norstrom[233] has related the yield strength of lath martensite with its packet diameter d and total dislocation density ρ_T by the following expression:

$$\sigma_y = \sigma_0 + \sigma_i + k_y d^{-1/2} + \alpha\mu b\sqrt{\rho_T} \tag{8.31}$$

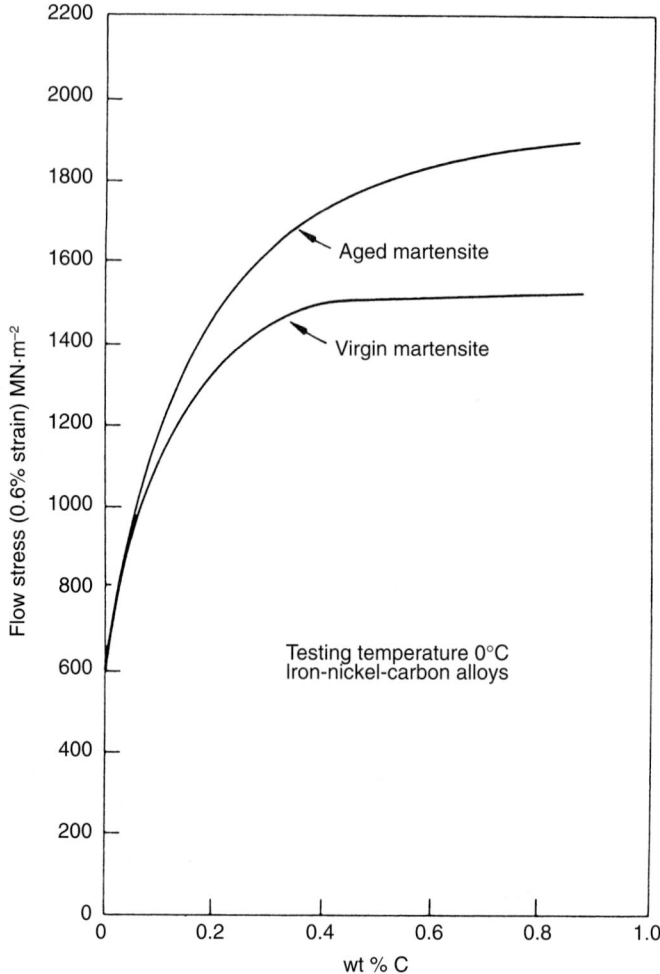

FIGURE 8.49   Flow stress of 100% martensite at 0°C versus carbon content. Aging treatment at 0°C for 3 hr.[232] (*Reprinted by permission of The Metallurgical Society, Warrendale, Pennsylvania.*)

where σ_0 is the friction stress for α-iron, σ_i is the solid-solution hardening effect, k_y is the Hall-Petch constant for the packet size d, α is a constant, μ is the shear modulus, b is the Burgers vector, and ρ_T may be given by

$$\rho_T = \rho_0 + K(\%C) + \frac{\theta}{b}\frac{2}{w}$$

(8.32)

where ρ_0 is the dislocation density present within the laths, K is a constant, $\%C$ is the carbon content, θ is the misorientation between laths, and w is the lath width. A major proportion of strength in lath martensite is, therefore, due to total dislo-

TABLE 8.14 Some Estimates of Contributions of Different Strengthening Mechanisms to the 0.2% Proof Stress of 0.4 Wt% Martensite[235]

Strengthening mechanism	Strength contribution (MPa)		
	Kelly and Nutting (1965)	Smith and Heheman (1971)	Williams and Thompson (1981)
Boundaries	—	450–600	620
Dislocations	150–300	—	270
Solid solution	350	—	400
Clustering, precipitation, etc.	750	—	1000

cation density comprising the individual contribution of dislocations within laths, and at lath boundaries and carbon-content-enhanced dislocation density.[234]

Table 8.14 lists some predictions of the contributions of different strengthening mechanisms to the 0.2% yield stress of a 0.4 wt% carbon steel martensite of $\sigma_y = 2200\,\text{MPa}$.[235]

8.9 TOUGHNESS OF MARTENSITE

The hardness and inherent brittleness of martensite, as well as the possibility of quench cracking leading to premature failure, make the determination of the ductile-brittle transition temperature T full of uncertainty. Furthermore, the shelf energy C_v can only be known at testing temperature at which some decomposition by tempering occurs, and the same problem can render uncertainty in the determination of the transition temperature.[235] It has been recognized that there exists a relationship between the prior austenite grain size and the martensite "packet size." Consequently, it appears that the martensite packet size controls T.[236,237] It has been further reported that a decrease in lath width and packet diameter in low-carbon steels and low-alloy martensitic/bainitic steels increases the toughness,[238] and it may be possible that some difficulties have been experienced in identifying the lath boundaries.

There are very few qualitative data available on the effect of grain diameter on C_v in martensitic structures.[235]

8.10 OMEGA TRANSFORMATION

The β (bcc) $\rightarrow \omega$ (hcp) phase transformation occurs both athermally (obtained by rapidly quenching from the β-phase field) and isothermally (produced on aging an alloy containing athermal ω phase or the untransformed β phase) in the group IV metals; Ti, Zr, and Hf alloys such as Zr-Ti, Zr-Nb, Zr-Mo, and so forth; and in β-phase alloys of the noble metals.[2,239] The reader is referred to reviews by Sass,[240] Williams et al.,[241] Sikka et al.,[242] and Srivastava et al.[239] for more details. The athermal $\beta \rightarrow \omega$ phase transformation is a displacive, diffusionless one so that the ω-phase composition is the same as the β phase; however, it forms as extremely fine, coher-

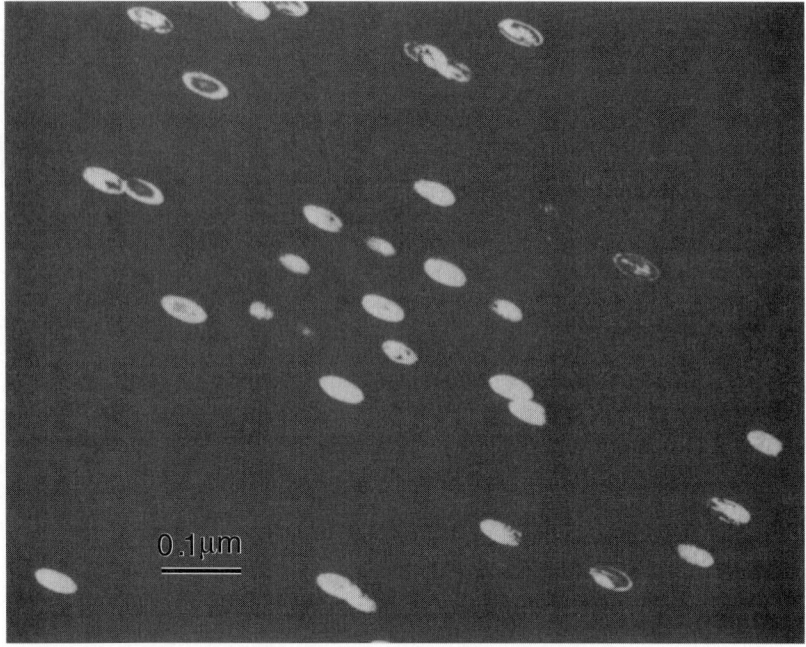

FIGURE 8.50 Transmission electron micrograph (dark field) illustrating ω particles formed at 480°C in a Ti-Mo-Sn-Zr alloy where only one of four variants of the ω phase is in contrast.[2] (*Courtesy of C. M. Wayman and H. K. D. H. Bhadeshia.*)

ent precipitates (<5 to 20 nm in size) in high number density with cuboidal or ellipsoidal morphology, depending on the degree of misfit between the ω precipitates and the β matrix. Figure 8.50 is an example of omega particles.[2,3]

A number of alloys displaying $\beta \rightarrow \omega$ transformation are characterized by the following orientation relationship:

$$(111)_\beta // (0001)_\omega \qquad [110]_\beta // [2110]_\omega$$

Significantly these same alloys at different compositions also experience $\beta \rightarrow \alpha'$ martensitic transformation where the orientation relationship follows the Burgers type:

$$(110)_\beta // (0001)_{\alpha'} \qquad [111]_\beta // [1120]_{\alpha'}$$

Note that the $\beta \rightarrow \omega$ orientation relationship depicts a multiplicity of only four when compared to much higher multiplicity of martensitic transformation; this implies that the respective lattice correspondences are different. Diffraction patterns often exhibit networks of diffuse intensity (sheets of intensity on $<111>_\beta$ planes) which conceal the description of diffraction maxima from the ω phase. These diffuse diffraction patterns are noticed at temperatures above those for which the ω phase are clearly recognized, and below those which are attributed to the pre-transformation "linear defects."[2]

Impurities play a major role on the $\beta \to \omega$ transformation; for example, the presence of 1200 ppm of oxygen can decrease the ω-start temperature by about 600 K, perhaps due to the stiffening of matrix and depression of the transformation start temperature by oxygen. It seems that the oxygen atoms someway interact with the pretransformation linear defects along $<111>_\beta$ and hinder their ordering.[2]

There are some important difficulties with ω-phase formation: these are the possible role of lattice vacancies in favoring linear defects and therefore nucleation, the high nucleation frequency and low particle growth rate, and the extreme embrittlement of the matrix β phase after the formation of ω phase.[2]

REFERENCES

1. C. M. Wayman, in *Physical Metallurgy*, 3d ed., eds. R. W. Cahn and P. Haasen, North-Holland Publishing, 1983, chap. 15, pp. 1031–1074; in *7th International Conference Proceedings on Strength of Metals and Alloys, 1985*, vol. 3, eds. H. J. McQueen et al., Pergamon Press, Oxford, 1986, pp. 1779–1805.

2. C. M. Wayman and H. K. D. H. Bhadeshia, in *Physical Metallurgy*, 4th ed., eds. R. W. Cahn and P. Haasen, Elsevier Science Publishers BV, Amsterdam, 1996, chap. 16, pp. 1507–1554.

3. L. Delaey, in *Materials Science and Technology*, vol. 5: *Phase Transformations in Materials*, vol. ed. P. Haasen, VCH, Weinheim, 1991, pp. 339–404.

4. E. R. Petty, in *Martensite*, ed. E. R. Petty, Longmans, New York, 1970, p. 5.

5. J. W. Christian, in *Physical Properties of Martensite and Bainite*, Iron and Steel Institute, London, 1965, pp. 1–19.

6. J. W. Christian, in *Martensite*, ed. E. R. Petty, Longmans, New York, 1970, pp. 11–41.

7. M. Cohen, G. B. Olson, and P. C. Clapp, in *Proceedings of International Conference "Martensitic Transformations," ICOMAT-79*, eds. C. M. Wayman et al., Cambridge, 1979, p. 1.

8. W. Quiming and K. Mokuanig, *Met. Trans.*, vol. 22A, 1991, pp. 1761–1765.

9. C. Clapp, *J. de Physique, ICOMAT 95*, vol. 5, 1995, pp. C8-11–19.

10. G. B. Olson, in *Martensite*, ASM International, Materials Park, Ohio, 1992, pp. 1–10.

11. R. D. Garwood, in *Martensite*, ed. E. R. Petty, Longmans, New York, 1970, pp. 95–118.

12. W. S. Owen, in *Encyclopedia of Materials Science and Engineering*, Pergamon Press, Oxford, 1986, p. 2736.

13. B. A. Bilby and J. W. Christian, in *The Mechanism of Phase Transformations in Metals*, The Institute of Metals, London, 1956, pp. 121–172.

14. J. D. Verhoeven, *Fundamentals of Physical Metallurgy*, Wiley, New York, 1975.

15. Z. Nishiyama, *Martensitic Transformation*, eds. M. Fine, M. Meshii, and C. M. Wayman, Academic Press, New York, 1978.

16. T. L. Richards and W. T. Roberts, in *The Mechanism of Phase Transformations in Metals*, The Institute of Metals, London, 1956, pp. 193–207.

17. G. V. Kurdjumov and G. Sachs, *Z. Phys.*, vol. 64, 1930, p. 325.

18. B. A. Greninger and A. R. Troiano, *Trans. AIME*, vol. 140, 1940, pp. 307–336.

19. Z. Nishiyama, *Sci. Rep.*, Tohoku Univ., Japan, vol. 23, 1934, p. 637.

20. S. A. Kulin and M. Cohen, *Metall. Trans. AIME*, vol. 188, 1950, pp. 1139–1141.

21. R. Bunshah and R. F. Mehl, *Metall. Soc. AIME*, vol. 230, 1954, p. 681.

22. J. W. Christian, in *Physical Metallurgy*, 2d ed., ed. R. W. Cahn, North-Holland, Amsterdam, 1970, chap. 10.

23. S. R. Pati and M. Cohen, *Acta Metall.*, vol. 29, 1971, pp. 1327–1339.

24. R. L. Patterson and C. M. Wayman, *Acta Metall.*, vol. 24, 1966, p. 347.

25. H. Warlimont, in *Physical Properties of Martensite and Bainite*, Special Report no. 93, Iron and Steel Institute, London, 1965, p. 58.

26. L. Delaey, K. Mukherjee, and M. Chandrasekaran, in *International Summer Course on Martensitic Transformation*, Leuven: Department of Metallurgy and Materials Engineering, KV Leuven, 1982, pp. 7.1–7.24.

27. H. Y. Yu, *Met. & Mat. Trans.*, vol. 28A, 1997, pp. 2499–2506.

28. R. E. Reed-Hill and R. Abbaschian, *Physical Metallurgy Principles*, 3d ed., PWS-Kent Publishing Co., Boston, 1992.

29. W. S. Owen and E. A. Wilson, in *Physical Properties of Martensite and Bainite*, Special Report no. 93, Iron and Steel Institute, London, 1965, pp. 53–57.

30. K. W. Andrews, *JISI*, July 1955, pp. 721–727.

31. C. Y. Kung and J. J. Rayment, *Metall. Trans.*, vol. 13A, 1982, pp. 328–331.

32. C. S. Roberts, *Trans. AIME*, vol. 197, 1953, p. 203.

33. M. Cohen, *Phase Transformations in Solids*, Wiley, New York, 1951, p. 591.

34. L. Kaufman and M. Cohen, in *Progress in Metal Physics*, vol. 7, Pergamon Press, Oxford, 1958, p. 165.

35. S. Glover and T. Smith, in *The Mechanism of Phase Transformations in Metals*, The Institute of Metals, London, 1956, p. 265.

36. J. Philiburt and C. Crussard, *JISI*, vol. 200, 1962, p. 102.

37. P. Payson, *The Metallurgy of Tool Steels*, Wiley, New York, 1962, p. 49.

38. R. W. K. Honeycombe and H. K. D. H. Bhadeshia, *Steels: Microstructure and Properties*, Arnold, London, 1996.

39. O. N. Mohanty, *Mat. Sc. & Engrg.*, B32, 1995, pp. 267–278.

40. Z. L. Xie, Y. Liu, and H. Hanninen, *J. De Physique, ICOMAT '95*, vol. 5, 1995, pp. C8-333–338.

41. J. R. Patel and M. Cohen, *Acta Metall.*, vol. 1, 1953, p. 531.

42. T. Kakeshita, T. Saburi, and K. Shimizu, *J. De Physique, ICOMAT '95*, vo. 5, 1995, pp. C8-367–372.

43. G. V. Kurdjumov and O. P. Maksimova, *Dokl. Akad. Nauk. SSR*, vol. 73, 1950, pp. 95–98.

44. R. E. Cech and J. H. Hollomon, *Trans. AIME*, vol. 197, 1953, p. 685.

45. K. C. Jones and A. R. Entwisle, *Met. Sci. J.*, vol. 5, 1971, pp. 190–195.

46. C. A. V. de Rodrigues, C. B. Prioul, and L. Hyspecka, *Metall. Trans.*, vol. 15A, 1984, pp. 2193–2203.

47. N. Thadhani and M. A. Meyers, in *Progress Mater. Sci.*, vol. 30, no. 1, 1986, pp. 1–37.

48. G. V. Kurdjumov, *JISI*, vol. 195, 1960, p. 26.

49. C. H. Shih, B. L. Averbach, and M. Cohen, *Trans. AIME*, vol. 203, 1955, pp. 183–187.

50. V. Raghavan, in *Martensite*, eds. G. B. Olson and W. S. Owen, ASM International, Materials Park, Ohio, 1992, pp. 197–225.

51. J. C. Zhao and M. R. Notis, *Mat. Sc. & Engrg.*, R15, nos. 4–5, November 1995, pp. 135–208.

52. J. C. Zhao and M. Notis, *J. Phase Equilibria*, vol. 14, 1993, p. 303.

53. Z.-Z. Yu and P. C. Clapp, *Met. Trans.*, vol. 20A, 1989, pp. 1601–1615.

54. A. R. Entwisle, *Metall. Trans.*, vol. 2, 1971, pp. 2395–2407.

55. G. B. Olson, in *Encyclopedia of Materials Science and Engineering*, Pergamon Press, Oxford, 1986, pp. 2929–2931.

56. G. B. Olson, in *Encyclopedia of Materials Science and Engineering*, suppl. vol. 3, Pergamon Press, Oxford, 1993, pp. 1787–1789.

57. A. Sato et al., *Acta Metall.*, vol. 28, 1980, pp. 367–376.

58. T. N. Durul, *Acta Metall.*, vol. 26, 1978, pp. 1855–1861.

59. T. Maki, in *Shape Memory Materials*, eds. K. Otsuka and C. M. Wayman, Cambridge University Press, Cambridge, 1998, pp. 117–132.

60. M. Cohen, *TMS-AIME*, vol. 224, 1962, pp. 638–656.

61. M. Cohen and C. M. Wayman, in *Metallurgical Treatise*, eds. J. K. Tien and J. F. Elliott, *TMS-AIME*, Warrendale, Pa., 1981, pp. 445–468.

62. R. Kumar, *Physical Metallurgy of Iron and Steel*, Asia Publishing House, Bombay, India, 1968.

63. D. A. Porter and K. E. Easterling, *Phase Transformations in Metals and Alloys*, Van Nostrand Reinhold, Berkshire, United Kingdom, 1981.

64. K. Shimizu and Z. Nishiyama, *Metall. Trans.*, vol. 3, 1972, pp. 1055–1062.

65. A. B. Greninger and A. R. Troiano, *Trans. AIME*, vol. 185, 1949, pp. 590–598.

66. H. Kubo and K. Hiranop, *J. Jpn. Inst. Met.*, vol. 37, 1973, pp. 400, 516.

67. G. A. Chadwick, *Metallography of Phase Transformations*, Crane, Russak and Co., New York, 1972.

68. C. M. Wayman, *J. Sheffield Univ. Metall. Soc.*, vol. 7, 1968, p. 19.

69. M. Oka and C. M. Wayman, *Trans. ASM*, vol. 62, 1969, pp. 370–379.

70. M. S. Wechsler, D. S. Lieberman, and T. A. Read, *Trans. AIME*, vol. 197, 1953, pp. 1503–1515.

71. J. S. Bowles and J. R. Mackenzie, *Acta Metall.*, vol. 2, 1954, pp. 129–137, 138–147, 224–234.

72. C. M. Wayman, in *Encyclopedia of Materials Science and Engineering*, Pergamon Press, Oxford, 1986, pp. 2736–2740.

73. D. P. Dunne and C. M. Wayman, *Metall. Trans.*, vol. 2A, 1971, pp. 2327–2341.

74. K. Otsuka and K. Shimizu, *Int. Met. Rev.*, vol. 31, no. 3, 1986, pp. 93–114.

75. A. Wirth and J. Bickerstaffe, *Metall. Trans.*, vol. 5A, 1974, pp. 799–808.

76. G. R. Speich and H. Warlimont, *JISI*, vol. 206, 1968, pp. 385–392.

77. A. R. Marder and G. Krauss, *Trans. ASM*, vol. 60, 1967, pp. 651–660.

78. Y. H. Tan, D. C. Zang, X. C. Dong, Y. H. He, and S. Q. Hu, *Met. Trans.*, vol. 23A, 1992, pp. 1413–1421.

79. G. Krauss and A. R. Marder, *Metall. Trans.*, vol. 2A, 1971, pp. 2343–2357.

80. C. S. Sharma, J. A. Whiteman, and J. H. Woodhead, *Met. Sci. J.*, 1976, pp. 391–395.

81. S. K. Das and G. Thomas, *Metall. Trans.*, vol. 1A, 1970, pp. 325–327.

82. G. R. Speich and L. P. R. Swanson, *JISI*, vol. 203, 1965, p. 480.

83. G. Thomas, *Metall. Trans.*, vol. 2A, 1971, pp. 2373–2385.

84. I. Kozelkova, C. Daghert, P. Grejoine, J. Gallard, and L. Hyspecka, *J. de Physique IV*, 1995, pp. C8-323–328; 1995, pp. 323–328.

85. M. Umemoto et al., *Acta Metall.*, vol. 32, 1984, pp. 1191–1203.

86. S. Kajiwara and W. S. Owen, *Scr. Metall.*, vol. 11, 1977, p. 137.

87. M. Umemoto and C. M. Wayman, *Met. Trans.*, vol. 9A, 1978, p. 891.

88. L. Delaey, R. V. Krishnan, H. Tas, and H. Warlimont, *J. Mater. Sc.*, vol. 9, 1974, p. 1521.

89. K. Otsuka and K. Shimizu, *Int. Met. Rev.*, vol. 31, no. 3, 1986, pp. 93–114.

90. X. M. Zhang, M. Z. Liu, and R. S. Liu, *Acta Metall. Sinica*, Series A: *Physical Metall. & Mater. Sci.*, vol. 10, no. 3, June 1997, pp. 157–165.

91. G. B. Olson and M. Cohen, in *Proceedings of the International Conference on Solid → Solid Phase Transformations*, eds. H. I. Aaronson et al., TMS-AIME, Warrendale, Pa., 1982, pp. 1145–1164.

92. G. B. Olson and M. Cohen, in *Dislocations in Solids*, vol. 7, ed. F. R. N. Nabarro, North-Holland Publishing, Amsterdam, 1986, chap. 37, p. 295.

93. X. Q. Zhao and Y. F. Han, *Met. and Mat. Trans.*, vol. 30A, 1999, pp. 884–887.

94. M. Suezawa and H. E. Cook, *Acta Metall.*, vol. 28, 1980, pp. 423–432.

95. C. M. Wayman and H. R. P. Inoue, in *Intermetallic Compounds*, vol. 1: *Principles*, eds. J. H. Westbrook and R. L. Fleischer, Wiley, New York, 1995, pp. 827–847.

96. K. E. Easterling and A. R. Tholen, *Acta Metall.*, vol. 24, 1976, pp. 333–341.

97. S. Kajiwara, *Metall. Trans.*, vol. 17A, 1986, pp. 1693–1702.

98. H. Y. Yu, S. C. Sanday, and B. B. Rath, *Mats. Sc. & Engrg.*, vol. B32, 1995, pp. 153–158.

99. G. B. Olson and A. C. Roitburd, in *Martensite*, eds. G. B. Olson and W. S. Owen, ASM International, Materials Park, Ohio, 1991, pp. 149–174.

100. R. Sinclair and K. Otsuka, in *Proceedings of the International Conference on Solid → Solid Phase Transformations*, eds. H. I. Aaronson et al., TMS-AIME, Warrendale, Pa., 1982, pp. 1583–1584.

101. K. E. Easterling and A. R. Tholen, *Acta Metall.*, vol. 28, 1980, pp. 1229–1234.

102. W. S. Owen, F. J. Schoen, and G. R. Srinivasan, in *Phase Transformations*, ASM, Metals Park, Ohio, 1979, pp. 157–180.

103. M. Grujicic, H. C. Ling, and D. M. Haezebrouck, in *Martensite*, eds. G. B. Olson and W. S. Owen, ASM International, Materials Park, Ohio, 1991, pp. 149–174.

104. M. Rao, S. Sengupta, and H. K. Sahu, *Phys. Rev. Lett.*, vol. 75, 1995, p. 2164.

105. K. M. Crosby and R. M. Bradley, *Phil. Mag. Lett.*, vol. 75, no. 3, 1997, pp. 131–135.

106. M. Lin, G. B. Olson, and M. Cohen, *Met. Trans.*, vol. 23A, 1992, pp. 2987–2998.

107. W. J. Harris and M. Cohen, *Trans. AIME*, vol. 180, 1949, pp. 447–470.

108. H. Kessler and W. Pitsch, *Acta Metall.*, vol. 15, 1967, p. 401.

109. C. M. Wayman, in *Proceedings of the International Conference on Solid → Solid Phase Transformations*, eds. H. I. Aaronson et al., TMS-AIME, Warrendale, Pa., 1982, p. 1119; *Encyclopedia of Materials Science and Engineering*, Pergamon Press, Oxford, 1986, pp. 4371–4374.

110. T. W. Duerig and C. M. Wayman, *Engineering Aspects of Shape Memory Alloys*, eds. T. W. Duerig et al., Butterworth-Heinemann, London, 1990, pp. 3–20.

111. R. J. Salzbrenner and M. Cohen, *Acta Metall.*, vol. 27, 1979, pp. 739–748.

112. K. Bhattacharya, *Acta Metall.*, vol. 39, no. 10, 1991, pp. 2431–2444.

112a. M. Fremond, in *Shape Memory Alloys*, eds. M. Fremond and S. Miyazaki, Springer Wien, New York, Italy, 1996, pp. 3–68.

113. W. Tang, *Analysis of Property Data for Ti-Ni Shape Memory Alloys*, doctoral thesis, Royal Institute of Technology, Report no. KTH/AMT-161, May 1996.

114. J. Perkins, in *The Encyclopedia of Advanced Materials*, Pergamon Press, Oxford, 1994, pp. 2439–2441.

115. J. V. Humbeeck, R. Stalmans, and P. A. Besselink, in *Metals as Biomaterials*, eds. J. A. Helsen and H. J. Breme, Wiley, Chichester, 1998, pp. 73–100.

116. L. M. Schetky, in *Intermetallic Compounds*, vol. 2: *Practice*, eds. J. H. Westbrook and R. L. Fleischer, Wiley, New York, 1995, pp. 529–558.

117. T. W. Duerig, A. R. Pelton, and D. Stockel, *Bio-Medical Materials and Engineering*, vol. 6, Amsterdam, Netherlands, 1996, pp. 255–266.

118. H. M. Liao, H. C. Lin, J. L. He, K. C. Chen, and K. M. Lin, *International Conference: Displacive Phase Transformations and Their Applications in Materials Engineering*, eds. K. Inoue et al., TMS, Warrendale, Pa., 1998, pp. 251–256.

119. C. Zhang, R. H. Lee, and P. E. Thoma, *International Conference: Displacive Phase Transformations and Their Applications in Materials Engineering*, eds. K. Inoue et al., TMS, Warrendale, Pa., 1998, pp. 267–274.

120. G. Airoldi, A. Corsi, and G. Riva, *Mat. Sc. & Engrg.*, vol. A241, 1998, pp. 233–240.

121. Y. Liu, Z. Xie, J. V. Humbeeck, and L. Delaey, *Acta Mater.*, vol. 46, no. 12, 1998, pp. 4325–4338.

122. M. Nishida, T. Hara, A. Chiba, and K. Hiraga, *International Conference: Displacive Phase Transformations and Their Applications in Materials Engineering*, eds. K. Inoue et al., TMS, Warrendale, Pa., 1998, pp. 257–266.

123. Q. Chen, Q. F. Wu, and T. Ko, *Scripta Met. Mater.*, vol. 29, 1993, pp. 49–53.

124. K. Yamauchi and Y. Nakamura, *International Conference: Displacive Phase Transformations and Their Applications in Materials Engineering*, eds. K. Inoue et al., TMS, Warrendale, Pa., 1998, pp. 281–284.

125. Y. Nakata, T. Tadaki, and K. Shimizu, *International Conference: Displacive Phase Transformations and Their Applications in Materials Engineering*, eds. K. Inoue et al., TMS, Warrendale, Pa., 1998, pp. 187–196.

126. K. N. Melton, *Engineering Aspects of Shape Memory Alloys*, eds. T. W. Duerig et al., Butterworth-Heinemann, London, 1990, pp. 21–35.

127. S. K. Wu and H. C. Lin, *International Conference: Displacive Phase Transformations and Their Applications in Materials Engineering*, eds. K. Inoue et al., TMS, Warrendale, Pa., 1998, pp. 197–206.

128. C. Y. Lei, M. C. Wu, L. McD. Schetky, and C. J. Burstone, *International Conference: Displacive Phase Transformations and Their Applications in Materials Engineering*, eds. K. Inoue et al., TMS, Warrendale, Pa., 1998, pp. 281–284.

129. T. Saburi, in *Shape Memory Alloys*, eds. K. Otsuka and C. M. Wayman, Cambridge University Press, Cambridge, 1998, pp. 49–96.

130. E.-S. Lee and Y. G. Kim, *Metall. Trans.*, vol. 21A, 1990, pp. 1681–1688.

131. T. Tadaki, in *Shape Memory Alloys*, eds. K. Otsuka and C. M. Wayman, Cambridge University Press, Cambridge, 1998, pp. 97–116.

132. D. W. Roh, E.-S. Lee, and Y. G. Kim, *Metall. Trans.*, vol. 23A, 1992, pp. 2753–2760.

133. E. S. Lee and S. Ahn, *Acta Mater.*, vol. 46, no. 12, 1998, pp. 4357–4368.

134. C. W. H. Lam, C. Y. Chung, W. H. Zou, and J. K. L. Lai, *Met. and Mat. Trans.*, vol. 28A, 1997, pp. 2765–2767.

135. C. Segui, E. Cesari, and J. V. Humbeeck, *Mats. Trans., JIM*, vol. 32, no. 10, 1991, pp. 898–904.

136. R. Kainuma, S. Takahashi, and K. Ishida, *Met. and Mat. Sc.*, vol. 27A, 1996, pp. 2187–2195.

137. H. Kato and S. Miura, *Acta Metall. Mater.*, vol. 43, no. 1, 1995, pp. 351–360.

138. J. W. Seo and D. Schryvers, *Acta Mater.*, vol. 46, no. 4, 1998, pp. 1165–1175.

139. K. Otsuka, T. Ohba, and Y. Murakami, *International Conference: Displacive Phase Transformations and Their Applications in Materials Engineering*, eds. K. Inoue et al., TMS, Warrendale, Pa., 1998, pp. 167–176.

140. K. Histasune, Y. Takuma, Y. Tanaka, K. Udoh, T. Morimara, and M. Hasaka, *Solid State Comm.*, vol. 106, no. 8, 1998, pp. 509–512.

141. K. K. Jee, W. Y. Jang, S. H. Baik, M. C. Shin, and C. S. Choi, *J. de Physique*, vol. 5, 1995, pp. C8-385–390.

142. J.-H. Jun and C.-S. Choi, *Met. and Mats. Trans.*, vol. 30A, 1999, pp. 667–670.

143. Y.-K. Lee, S.-H. Baik, and C.-S. Choi, *International Conference: Displacive Phase Transformations and Their Applications in Materials Engineering*, eds. K. Inoue et al., TMS, Warrendale, Pa., 1998, pp. 365–373.

143a. S. Miyazaki, in *Shape Memory Alloys*, eds. M. Fremond and S. Miyazaki, Springer Wien, New York, Italy, 1996, pp. 69–147.

144. K. Tsuzaki, Y. Natsume, Y. Kurokawa, and T. Maki, *Scripta Metall.*, vol. 27, 1992, p. 471.

145. K. Yamaguchi, Y. Morioka, and Y. Tomota, *Scripta Mater.*, vol. 35, no. 10, 1996, pp. 1147–1152.

146. C.-L. Li, X.-P. Ma, and Z. H. Jin, *Mats. Sc. & Engrg.*, vol. 13, September 1997, pp. 727–730.

147. A. Sato and Y. Yamazi, *Acta Metall.*, vol. 34, 1986, pp. 287–294.

148. Q. Liu, Z. Ma, and N. Zu, *Met. and Mats. Trans.*, vol. 29A, 1998, pp. 1579–1583.

149. T. Kikuchi, S. K. Kajiwara, and Y. Tomota, *Mats. Trans.*, vol. 36, no. 6, 1995, pp. 719–728.

150. *Mats. Sc. Bulletin*, vol. 33, no. 10, 1998, pp. 1433–1438.

151. L. Jian and C. M. Wayman, *Scripta Metall. Mater.*, vol. 27, 1992, pp. 279–284.

151a. T. Sampei and Y. Moriya, private communication.

152. T. Kikuchi, S. Kajiwara, and Y. Tomota, *J. de Physique IV, ICOMAT '95*, vol. 5, Lausanne, Switzerland, 1995, pp. C8-445–450.

153. N. Yu, V. Koval, V. Kokorin, and L. G. Khandros, *Phys. Met. Metall.*, vol. 48, no. 6, 1981, p. 162.

154. T. Maki, K. Kobayashi, M. Minato, and I. Tamura, *Scripta Metall.*, vol. 18, 1984, p. 1105.

155. N. Jost, in T. Maki, *Proceedings of International Conference of Martensitic Transformations, ICOMAT-89*, ed. B. C. Muddle, Trans. Tech. Publication, Switzerland, 1989, p. 157.

156. Y. N. Koval and G. E. Moriastyrsky, *Scripta Metall. Mater.*, vol. 28, 1993, pp. 41–46.

157. L. M. Schetky, private communications, 1999.

158. G. Nosova and E. Vintaikin, *Scripta Mater.*, vol. 40, no. 3, 1999, pp. 347–351.

159. R. Sainuma, N. Ono, and K. Ishida, *Mater. Res. Symp. Proc.*, vol. 310, 1995, pp. 467–478.

160. V. Cherneko, E. Cesari, V. V. Kokorin, and I. N. Vitenko, *Scripta Metall. Mater.*, vol. 33, 1995, pp. 1239–1244.

161. J. Soltys, *Phys., Status Solid: (a)*, vol. 66, 1981, pp. 485–491.

162. Y. Sutou, I. Ohnuma, R. Kainuma, and K. Ishida, *Met. and Mats. Trans.*, vol. 29A, 1998, pp. 2225–2227.

162a. F. Gejima, Y. Sutou, R. Kainuma, and K. Ishida, *Met. and Mats. Trans.*, vol. 30A, 1999, pp. 2721–2723.

163. R. Kainuma, K. Ishida, and T. Nishizawa, *Met. Trans.*, vol. 23A, 1992, pp. 1147–1153.

164. D. W. Bray and J. W. Howe, *Met. and Mats. Trans.*, vol. 27A, 1997, pp. 3362–3370.

165. D. C. Mears, *Int. Met. Rev.*, vol. 22, 1977, pp. 119–155.

166. P. Huang and H. F. Lopez, *Mats. Sc. & Tech.*, vol. 15, February 1999, pp. 157–164.

167. G. N. Sure and L. C. Brown, *Scripta Metall.*, vol. 19, 1985, p. 401.

168. K. Mukunthan and L. C. Brown, *Met. Trans.*, vol. 19A, 1988, pp. 2921–2929.

169. K. Otsuka, M. Tokonami, K. Shimizu, Y. Iwata, and I. Shibuya, *Acta Metall.*, vol. 27, 1979, pp. 965–972.

170. J. Ye, M. Tokonami, and K. Otsuka, *Met. Trans.*, vol. 21A, 1990, pp. 2669–2678.

171. Z. G. Wei, H. Y. Peng, W. H. Zou, D. Z. Yang, *Met. & Mats. Trans.*, vol. 28A, 1997, pp. 955–967.

172. R. Kainuma, H. Nakato, K. Oikawa, T. Ishida, and T. Nishizawa, *Mater. Res. Soc. Symp., Proc.*, vol. 246, 1992, p. 403.

173. J. H. Yang and C. M. Wayman, *Mater. Sc. Engrg.*, vol. A160, 1993, p. 241.

174. K. H. Wu, Z. J. Pu, Y. Gao, and C. M. Wayman, *International Conference: Displacive Phase Transformations and Their Applications in Materials Engineering*, eds. K. Inoue et al., TMS, Warrendale, Pa., 1998, pp. 285–290.

175. P. Olier, J. C. Brachet, J. L. Bechade, C. Foucher, and G. Guenin, *J. De Physique IV*, vol. 5, 1995, pp. C8-741–758.

176. C. Liang and C. A. Rogers, *J. Intell. Mater. Syst. and Struct.*, vol. 1, April 1990, pp. 207–234.

177. Shape Memory Alloys, *Metallurgia*, vol. 51, 1984, pp. 26–29.

178. T. Saburi and C. M. Wayman, *Acta Metall.*, vol. 27, 1979, p. 979.

179. J. Perkins, in *Encyclopedia of Materials Science and Engineering, Pergamon Press*, Oxford, 1986, pp. 4365–4368.

180. P. Da Silva, *Mats. Lett.*, vol. 38, 1999, pp. 341–343.

181. C.-T. Hu and T.-M. Chen, *International Conference: Displacive Phase Transformations and Their Applications in Materials Engineering*, eds. K. Inoue et al., TMS, Warrendale, Pa., 1998, pp. 407–412.

182. E. Cingolani, J. V. Humbeeck, and M. Ahlers, *Met. and Mats. Trans.*, vol. 30A, 1999, pp. 493–499.

183. R. Stalmans, J. V. Humbeeck, and L. Delaey, *Acta Metall.*, vol. 40, 1992, pp. 501–511, 2921–2931.

184. C. S. Thuillier, D. Favier, and G. Canova, *J. De Physique IV, ICOMAT '95*, vol. 5, 1995, pp. C8-587–592.

185. E. Cingolani, A. Yaway, and M. Ahlers, *J. De Physique IV, ICOMAT '95*, vol. 5, 1995, pp. C8-865–869.

186. K. Otsuko, C. Wayman, K. Nakai, H. Kakamoto, and K. Shimizu, *Acta Metall.*, vol. 24, 1976, pp. 207–226.

187. J. Perkins and D. Hodgson, *Engineering Aspects of Shape Memory Alloys*, eds. T. W. Duerig et al., Butterworth-Heinemann, London, 1990, pp. 195–206.

188. T. W. Duerig and R. Zadnos, *Engineering Aspects of Shape Memory Alloys*, eds. T. W. Duerig et al., Butterworth-Heinemann, London, 1990, pp. 3–20.

189. K. Otsuko and C. M. Wayman, *Reviews on the Deformation Behavior of Materials*, vol. 2, 1977, p. 81.

190. T. W. Duerig, A. R. Pelton, and D. Stockel, *International Conference: Displacive Phase Transformations and Their Applications in Materials Engineering*, eds. K. Inoue et al., TMS, Warrendale, Pa., 1998, pp. 141–147.

191. T. A. Schroeder and C. M. Wayman, *Acta Metall.*, vol. 27, 1979, p. 405.

192. K. Tsuchiya and K. Marukawa, *J. De Physique, ICOMAT '95*, vol. 5, 1995, pp. C8-901–905.

193. K. Tsuchiya and K. Marukawa, *International Conference: Displacive Phase Transformations and Their Applications in Materials Engineering*, eds. K. Inoue et al., TMS, Warrendale, Pa., 1998, pp. 177–186.

194. Y. Murakami, K. Otsuka, H. Muzubayashi, and T. Suzuki, *International Conference: Displacive Phase Transformations and Their Applications in Materials Engineering*, eds. K. Inoue et al., TMS, Warrendale, Pa., 1998, pp. 225–232.

195. T. A. Schroeder and C. M. Wayman, *Acta Metall.*, vol. 26, 1978, p. 1745.

196. H. Ming, *J. Functional Materials*, vol. 28, no. 6, 1997, pp. 580–582.

197. J. V. Humbeeck, J. Janssen, M. Ngoie, and L. Delaey, *Scripta Metall.*, vol. 18, 1984, p. 893.

198. J.-E. Bidaux, *Scripta Met. et Mater.*, vol. 25, 1991, pp. 1895–1899.

199. M. Kapgan and K. N. Melton, *Engineering Aspects of Shape Memory Alloys*, eds. T. W. Duerig et al., Butterworth-Heinemann, London, 1990, pp. 137–148.

200. C. M. Wayman, in *The Encyclopedia of Advanced Materials*, eds. D. Bloor et al., Pergamon, Oxford, 1994, pp. 2442–2444.

201. K. N. Melton, in *Shape Memory Materials*, eds. K. Otsuka and C. M. Wayman, Cambridge University Press, Cambridge, 1998, pp. 220–239.

202. D. Z. Liu, W. X. Liu, and F. Y. Gong, *J. De Physique IV*, vol. 5, 1995, pp. C8-1241–1245.

203. E. Cydzik, *Engineering Aspects of Shape Memory Alloys*, eds. T. W. Duerig et al., Butterworth-Heinemann, London, 1990, pp. 149–157.

204. L. M. Schetky, *International Conference: Displacive Phase Transformations and Their Applications in Materials Engineering*, eds K. Inoue et al., TMS, Warrendale, Pa., 1998, pp. 149–156.

205. M. Mertmann, A. Bracke, and E. Hornbogen, *J. De Physique IV*, vol. 5, 1995, pp. C8-1259–1264.

206. M. Braunovic, *Use of Shape Memory Materials in Distribution Power Systems*, Report for the Canadian Electrical Association, Montreal, Québec, May 1993.

207. L. M. Schetky, *The Role of Shape Memory Alloys in Smart/Adaptive Structures*, Materials Research Society Proceedings, winter meeting, 1992.

208. L. M. Schetky, *Engineering Aspects of Shape Memory Alloys*, eds. T. W. Duerig et al., Butterworth-Heinemann, London, 1990, pp. 170–177.

209. I. Ohtaka and Y. Suzuki, in *Shape Memory Materials*, eds. K. Otsuka and C. M. Wayman, Cambridge University Press, Cambridge, 1998, pp. 240–266.

210. M. Kohl, K. D. Skrobanek, E. Quandt, P. Schloßmacher, A. Schußler, and D. M. Allen, *J. De Physique IV*, vol. 5, 1995, pp. C8-1187–1192.

211. T. Hashinga, S. Miyazaki, T. Ueki, and H. Horikawa, *J. De Physique IV, ICOMAT '95*, vol. 5, 1995, pp. C8-689–694.

212. A. Ishida, M. Sato, A. Takei, K. Nomura, and S. Miyazaki, *Met. and Mats. Trans.*, vol. 27A, 1996, pp. 3753–3759.

213. U. Sigwart, ed., *Endoluminol Stenting*, W. B. Saunders, London, 1996, p. 601.

214. R. Makkar, N. Eigler, J. S. Forrester, and L. Litwach, in *Endoluminol Stenting*, W. B. Saunders, London, 1996, pp. 230–237.

215. P. A. Besselink and R. C. L. Sachdeva, *J. De Physique IV, ICOMAT '95*, vol. 5, 1995, pp. C8-111–116.

216. P. Filip, J. Musialek, and K. Mazanec, *J. De Physique IV, ICOMAT '95*, vol. 5, 1995, pp. C8-1211–1216.

217. H. C. Lin, S. K. Wu, and M. T. Yeh, *Met. Trans.*, vol. 23A, 1993, pp. 2189–2194.

218. I. B. Kekab, in *The Encyclopedia of Advanced Materials*, vol. 2, eds. D. Bloor et al., Pergamon Press, Oxford, 1994, pp. 986–991.

219. A. V. Srinivasan, D. G. Cutts, and L. M. Schetky, *Met. Trans.*, vol. 22A, 1991, pp. 623–627.

220. *Materials for Tomorrow's Infrastructure*, Technical Report, CERF Report no. 94-5011, chap. 10, pp. 109–126.

221. P. R. Witting and F. A. Cozzarelli, *Shape Memory Strcutural Dampers: Material Properties, Design and Seismic Testing*, Technical Report NCEER-92-0013, May 26, 1992.

222. M. J. Roberts, *Met. Trans.*, vol. 1, 1970, pp. 3287–3294.

223. T. Swarr and G. Krauss, *Met. Trans.*, vol. 7A, 1976, pp. 41–48.

224. R. A. Grange, *Trans. ASM*, vol. 59, 1966, p. 26.

225. W. S. Owen et al., in *The Structure and Properties of Iron Alloys, Proceedings of the 2d International Symposium on Materials*, University of California, June 1964.

226. M. J. Roberts, in *Martensite*, ed. E. R. Petty, Longmans, New York, 1970, pp. 119–136.

227. R. L. Fleischer, *Acta Metall.*, vol. 10, 1962, pp. 835–842.

228. G. R. Speich and H. Warlimont, *JISI*, vol. 206, 1968, p. 385.

229. J. M. Chilton and P. M. Kelly, *Acta Metall.*, vol. 16, 1968, p. 637.

230. M. J. Roberts and W. S. Owen, *JISI*, vol. 206, 1968, p. 375.

231. M. J. Roberts and W. S. Owen, in *Physical Properties of Martensite and Bainite*, Special Report 93, Iron and Steel Institute, London, 1965, pp. 171–178.

232. P. G. Winchel and M. Cohen, *Trans. Met. Soc.*, *AIME*, vol. 224, 1962, p. 638.

233. L. A. Norstrom, *Scand. J. Metall*, vol. 5, 1976, p. 159.

234. A. R. Marder, in *Proceedings of the International Conference on Phase Transformations in Ferrous Alloys*, eds. A. R. Marder and J. I. Goldstein, TMS-AIME, Warrendale, Pa., 1984, pp. 11–41.

235. F. B. Pickering, in *Materials Science & Technology*, vol. 7: *Constitution and Properties of Steels*, vol. ed. F. B. Pickering, VCH, Weinheim, 1991, pp. 41–94.

236. H. Ohtani, F. Terasaki, and T. Kunitake, *Trans. ISI, Jpn.*, vol. 12, 1972, p. 118.

237. J. P. Naylor and P. R. Krahe, *Met. Trans.*, vol. 5, 1974, pp. 1699–1701.

238. J. P. Naylor and R. Blondeau, *Met. Trans.*, vol. 7, 1976, pp. 891–894.

239. D. Srivastava, P. Mukhopadhyay, E. Ramadasdan, and S. Banerjee, *Met. Trans.*, vol. 24A, 1993, pp., 495–501.

240. S. L. Sass, *J. Less Common Met.*, vol. 28, 1972, p. 157.

241. J. C. D. Williams, D. De Fontaine, and N. E. Paton, *Met. Trans.*, vol. 4, 1973, p. 2701.

242. S. K. Sikka, Y. K. Vohra, and R. Chidambaram, *Progress Mater. Sci.*, vol. 27, 1982, p. 245.

CHAPTER 9
BAINITE

9.1 INTRODUCTION

The word *bainite* was coined in 1934 in honor of Dr. E. C. Bain to describe the non-lamellar aggregate of ferrite and carbide which formed in steel by the decomposition of austenite above that of martensite formation but below that of fine pearlite formation through either isothermal transformation or continuous cooling.[1] In both simple iron-carbon alloys and plain carbon steels, the pearlitic and bainitic temperature ranges overlap each other to a considerable extent, which considerably complicates the interpretation of microstructure and kinetics. Since alloying elements usually cause the retardation of proeutectoid ferrite precipitation and the pearlite reaction as well as depression of the bainite reaction to lower temperatures, the TTT curves, for most alloy steels, exhibit two distinct C curves—one for the pearlitic reaction and the other for the bainitic reaction. Hence the bainite transformation can be studied unambiguously in alloy steels of suitable composition (e.g., containing Mo, Cr, etc., or a combination of Mn and Si). However, isothermal bainite transformation in alloy steels (containing Mn, Ni, and Cr, singly or in combination) does not go to completion; rather, it terminates prior to completion of decomposition of austenite at any temperature within the bainite transformation range between B_s (bainite start) and B_f (bainite finish). This phenomenon is called *incomplete reaction phenomenon or transformation stasis*; the extent of reaction, which depends on the steel composition and the transformation temperature, increases with the lowering of temperature.[2] Although it is not always observed in plain carbon steels, the incomplete reaction phenomenon is considered an inherent property of the bainite reaction.[3]

Although there have been numerous studies, the mechanism of the bainite reaction has not yet been established. One of the most important controversies dealing with the bainitic reaction is the growth mechanism of the ferritic portion of bainite. Two entirely different models, such as diffusional or reconstructive and diffusionless or displacive mechanism, have been proposed in steels. The diffusional model assumes that the ferritic portion of bainite develops by a diffusional ledge mechanism like Widmanstätten proeutectoid ferrite.[4-7] On the other hand, the displacive model proposes that the bainitic reaction occurs diffusionlessly as far as substitutional atoms are concerned.[8-11] The complexities of its formation mechanism and kinetics, as well as a wide variety of microstructures, have led to the disagreement in ascertaining its correct definition.[12-14] An alternative definition for the bainitic transformation is proposed in which the bainitic mechanism involves (1) long-range diffusion of at least one atomic species, (2) maintenance of a lattice correspondence between the parent and product phases, and (3) a remarkable difference in crystal structure between the two phases.[15] The well-recognized bay in the TTT diagram, in the case of alloy steel, seems to give clear evidence in support of the displacive mechanism for bainite growth, as predicted by Olson et al.[16] However, the well-recognized difference in carbide distribution between bainite formed at high and low temperatures exists in most steels and makes the classical terminology of upper and lower bainite useful, both in explaining the microstructural appearance and in classifying the overall reaction mechanism.[17] To make it more explicit, the precipitation of carbides either at the γ/α interface (upper bainite) or partially within supersaturated ferrite for lower bainite is an essential feature (or component) of most, but not all bainitic reaction. Bainite does not grow with a purely martensitic type of interface but may become increasingly coherent at lower temperatures.[18]

Bainite differs from the Widmanstätten ferrite α_w formed in low-carbon and low-alloy steels mainly in the temperature range where it grows and in the growth events.[19,20] That is, both the Widmanstätten ferrite and bainite occur near the B_s temperature, but the growth of α_w clearly lags behind that of bainite. It is, therefore, proposed that while α_w and bainite develop from identical nuclei, their growth processes are different. The growth rate of a bainitic ferrite lath is faster and occurs in the paraequilibrium condition at the γ grain boundaries at temperatures below 580°C. The growth rate of a Widmanstätten ferrite α_w lath is slower, which occurs in the orthoequilibrium condition from the grain boundary ferrite allotriomorphs α_a below 600 to 625°C.[20-22] Both structures induce sharp surface relief, typical of invariant plane strain (IPS), and have laths (or plates) comprising several needle-like subunits. There is a likelihood that both structures form via a displacive mechanism with respect to substitutional elements.[21,22] However, α_w can grow continuously from upper bainite by increasing the temperature during isothermal transformation.

Bainite differs from martensite in that the nucleation of bainite involves the partitioning of carbon and because its morphology (i.e., sheaf) is greatly influenced by its formation at a relatively lower driving force and higher temperatures.[23]

This chapter deals with the three definitions of bainite and their disposition—mostly in ferrous and some in nonferrous systems—including morphologies, orientation relationships, habit planes, mechanisms of their nucleation, and growth. Finally, acicular ferrite, bainitic steels, and various parameters required to attain their optimum properties as well as weldability of bainitic steels are discussed.

9.2 DEFINITIONS OF BAINITE

Here three definitions of bainite in current use, each of which covers a different group of incompletely overlapping phase transformation phenomena, are discussed.[13,14,24]

9.2.1 Microstructural Definitions

The generalized microstructural definition of bainite now described as nonlamellar, noncooperative, competitive ledgewise formation of the two precipitate phases (ferrite and carbide, in case of steel) is again suggested as the definition of bainite.[4] Here the high-temperature parent phase is usually termed γ, and the low-temperature product phases are termed α and β, with α being assumed to be the active nucleus. A better perspective of the microstructural definition can be given in terms of internal and external morphologies of eutectoid reaction.

9.2.1.1 Ferrous Bainite. For steels, we change the nomenclature of phases to austenite (γ), ferrite (α), and cementite (Fe_3C or θ) or epsilon- (ε-) carbide ($Fe_{2.4}C$) (instead of β). A broader microstructural definition of bainite[13] would include all nonlamellar aggregates of ferrite and carbide (irrespective of the ferrite morphology) as well as divorced pearlite. Thus, the structures conforming to this class may also form well above the kinetic B_s (bainite start) temperature and represent the internal morphological forms of eutectoid reaction. The *microstructural B_s* is defined as the highest temperature at which bainite is observed. It has been suggested that a microstructural B_s temperature should, in principle, be the eutectoid temperature for all hypoeutectoid alloys and should follow the extrapolated $\gamma/(\gamma + \alpha)$ curve for hypereutectoid alloys.[12] A narrower microstructural definition depends on the formation of nonlamellar carbide together with Widmanstätten ferrite, but this also causes discrepancies due to the difference between the (high-temperature) proeutectoid Widmanstätten ferrite and bainitic ferrite; i.e., the latter occurs below the microstructural B_s (which is equivalent to kinetic B_s) with distinctly different growth mechanisms, substructure, and kinetics of formation of ferrite.[2] A strict application of the microstructural definition would, therefore, not consider carbide-free structures as bainite.[2] However, a completely ferritic structure (e.g., in steel containing high silicon) can be developed which is very similar in morphology, growth mechanisms, and kinetics to the formation of classical bainite; but it is still not bainite—it is just ferrite formed at low temperatures.

As in the case of pearlite, the external morphology of nonlamellar bainite is presumed to be (1) nearly spherical when formed intragranularly and (2) hemispherical when formed near or at the grain boundaries. However, such morphology is rarely observed because two-phase bainite deposits on the proeutectoid phase with more uniformity, thereby causing the composite morphology to resemble more closely that of the proeutectoid phase.[25,26] These morphologies have been observed in hypereutectoid Ti-Cr alloys.[27]

Classical Ferrous Bainite: Morphology and Crystallography. Two distinct forms of classical bainite, called *upper bainite* and *lower bainite*,[24,26–28] occur in steels in different temperature ranges. As the names imply, upper bainite forms at higher temperatures than lower bainite.

UPPER BAINITE. Upper bainite predominates at carbon concentrations less than 0.57 wt% and at temperatures between 350 and 550°C, and it is often characterized by a feathery structure. Upper bainitic ferrite laths are practically free of carbon (<0.03% C) (and do not themselves contain any carbide precipitates) which can nucleate coherently, in general, at the austenite grain boundaries. Consequently, with the growth or thickening of ferrite laths, the remaining austenite becomes enriched in carbon and finally reaches such a level that cementite begins to precipitate at the α (lath)/γ interfaces.[24,31] (According to some workers, enrichment of the remaining austenite is not directly relevant because the carbon concentration in austenite at the γ/α interface $C_\gamma^{\gamma\alpha}$ is not a function of time.) This secondary reaction, involving the formation and growth of cementite, depletes the surrounding austenite region of carbon so that it will transform to ferrite. The mechanism of this secondary reaction is not completely understood.[10,25,32] However, the amount of carbide precipitated is low. In this way, side-by-side nucleation of the bainitic ferrite and cementite is repeated several times in order to produce aggregates of bainitic plates, called *sheaves*, consisting of groupings of fine, parallel intragranular ferrite plates

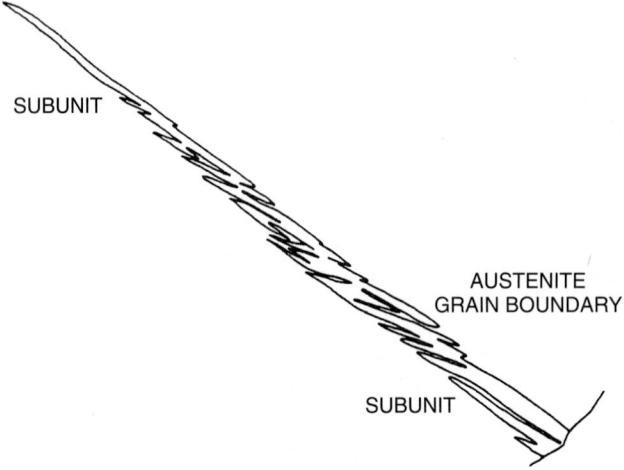

SUBUNIT

AUSTENITE
GRAIN BOUNDARY

SUBUNIT

FIGURE 9.1   Schematic illustration of the morphology of a sheaf of bainite.[10] (*Courtesy of H. K. D. H. Bhadeshia and J. W. Christian.*)

called *subunits*,[33,34] with cementite precipitated along the ferrite/austenite boundaries. The individual platelets or subunits within a given sheaf may not be entirely isolated from one another by the residual phases; in that case, they are generally separated by low-angle grain boundaries. This is due to the fact that within a sheaf, the subunits all appear to be in a common crystallographic orientation. Usually the subunit or the sheaf width increases with the increase of transformation temperature whereas the number of subunits per sheaf decreases with increasing temperature.[35] The cluster of platelets forming a sheaf is often called a *packet* of bainite because, on a microscopic scale, completely transformed γ grains tend to be grouped into distinct regions (i.e., packets) within which the traces of the habit planes of bainite are parallel. The overall morphology of a sheaf is shown schematically in Figs. 9.1[10] and 9.7c. Sheaf formation appears to have developed mainly, if not entirely, by sympathetic nucleation of new ferrite laths at the interphase boundaries of those previously formed.[34] (*Sympathetic nucleation* may be defined as the nucleation of a precipitate crystal whose composition is at all times different from that of the matrix at the interphase boundary of another crystal of the same phase.[34,36,37])

In high-carbon steels, transformation to upper bainite produces elongated carbide particles (i.e., continuous stringers) along the lath boundaries. It is presumed that very small carbides form at α/γ boundaries which then grow laterally into contact in higher-carbon steels. The source of this is that growth kinetics of ferrite at a given temperature decreases as carbon concentration increases.[38] In low-carbon steels, however, these carbides (formed from entrapped austenite) may be present as discontinuous stringers between the ferrite laths.

As the transformation temperature is decreased or the amount of carbon is increased in the upper bainite range, the carbon content of the enriched austenite increases more quickly and the laths increasingly form adjacent to each other in parallel groups by increased rates of sympathetic nucleation.[34] The resulting structure becomes much finer; i.e., ferrite laths become thinner and are generally accompanied by a significant increase in both the volume fraction of ferrite and

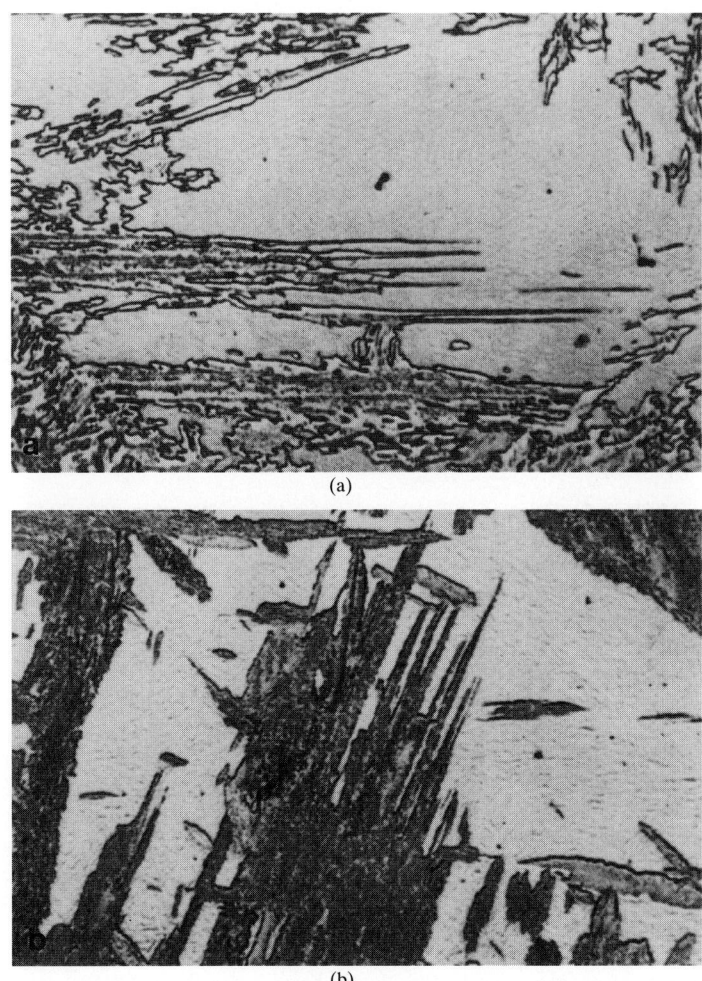

(a)

(b)

FIGURE 9.2   Upper bainite in 4360 steel transformed at (*a*) 495°C and (*b*) 415°C. Etchant: 5% picral. 750X.[24] (*Reprinted by permission of ASM International, Materials Park, Ohio.*)

the carbide particle density.[2,24] Such a structure forms presumably due to the restricted lateral growth of the ferrite laths by the slower rates of diffusion of carbon away from the α/γ interface as a result of the reduction of $C_\gamma^{\gamma\alpha} - C_0$ at constant temperature.[39]

Figure 9.2 shows the change of morphology of upper bainite with the lowering of reaction temperature in 4360 steel. In these optical micrographs, the light areas are ferrite regions, and the dark areas within these lighter patches are presumably fine dispersions of carbide particles in ferrite.[38] The feathery appearance of the upper bainite is clearly revealed in these microstructures. Usually, electron microscopy is required to resolve individual laths in fine bainitic structure. Figure

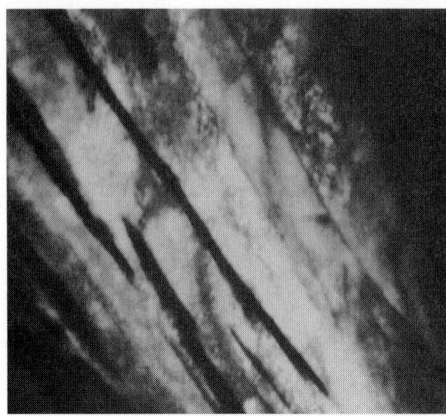

FIGURE 9.3 Electron micrograph of upper bainite formed at 495°C in 4360 steel. 15,000X.[24] (*Reprinted by permission of ASM International, Materials Park, Ohio.*)

TABLE 9.1 Orientation Relationships for Bainitic Structures

Kurdjumov-Sachs[41]	**Isaichev**[55]
$\{111\}_\gamma // \{110\}_\alpha; [110]_\gamma // [111]_\alpha$	$\{103\}_\theta // \{101\}_\alpha; <010>_\theta // <1\bar{1}\bar{1}>_\alpha$
Nishiyama-Wasserman[41]	**Jack**[59]
$\{111\}_\gamma // \{110\}_\alpha; [112]_\gamma // [110]_\alpha$	$(0001)_\varepsilon // (011)_\alpha; (10\bar{1}1)_\varepsilon // (101)_\alpha$
Pitsch[40]	**Bagaryatskii**[53]
$\{001\}_\theta // \{\bar{2}25\}_\gamma$	$\{001\}_\theta // \{211\}_\alpha; <100>_\theta // <0\bar{1}1>_\alpha$
$<100>_\theta$ within 2.6° of $<\bar{5}5\bar{4}>_\gamma$	
$<010>_\theta$ within 2.6° of $<\bar{1}10>_\gamma$	

9.3 is a thin-foil electron micrograph which shows the interlath cementite particles in the form of dark stringers in 4360 steel transformed to bainite at 495°C. The dark areas are far too thick to be composed entirely of carbides.[38]

The orientation relationship (OR) between ferrite and its parent austenite in upper bainite is always close to the classic Kurdjumov-Sachs (K-S)[19] relationship (Table 9.1). The K-S relationship is also observed in various microstructures such as Widmanstätten ferrite, allotriomorphic ferrite, and even pearlite.[19] The OR varies from K-S to N-W for sideplates from grain boundary ferrite allotriomorphs.[38] However, the proeutectoid cementite (θ) plate is related to the austenite by the Pitsch relationship (Table 9.1).[40] The α/θ relationship can be deduced from the γ/θ relationship by using the ferrite to be a variant of the K-S α/γ orientation relationship.[11] Note that the presence of a particular OR is not a prerequisite for a displacive transformation, but a certain relationship must exist for the occurrence of a reaction in a displacive manner.[19]

Upper bainite can also be distinguished from Widmanstätten ferrite by its morphology and dislocation density.[33] In general, dislocation density in bainite is observed to be higher than that in the allotriomorphic ferrite which forms at a similar transformation temperature. The dislocation density of bainitic ferrite appears to increase with the lowering of transformation temperature, although there are only few quantitative data to this effect.[10]

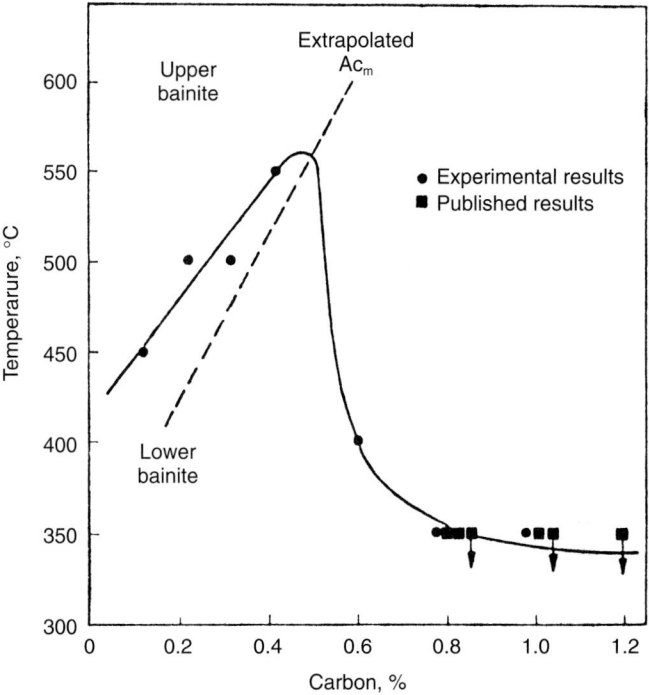

FIGURE 9.4 Effect of carbon concentration on the temperature of transformation separating upper and lower bainites.[45] (*Courtesy of the Institute of Metals, England.*)

It is difficult to determine the true habit of bainitic ferrite because of its morphological complexities. In upper bainite, no generally agreed-upon habit plane has been identified, and $\{111\}_\gamma$,[41] $\{232\}_\gamma \approx \{154\}_\alpha$,[42] $\{569\}_\gamma$,[43] and $(0.373, 0.663, 0.649)_\gamma$[44] have been reported. However, there is a good agreement that the longitudinal direction of ferrite laths is parallel to the close-packed direction $\langle 111 \rangle_\alpha$.[2] The habit plane was not observed to change appreciably with transformation temperature.[10]

LOWER BAINITE. Lower bainite also consists of a nonlamellar aggregate of ferrite and carbides and can occur at all carbon concentrations. The temperature of transformation separating the two types of bainite depends upon the carbon concentration, as shown in Fig. 9.4. This transition temperature is usually about 350°C in steels containing more than about 0.7% C, rising to a maximum of ~550°C at ~0.5% C and then decreasing again to ~450°C in low (0.1%) -carbon steels.[45] On the other hand, Oka and Okamoto have reported a constant transition temperature of 350°C for steels containing 0.54 to 1.1% C. Note that a transition temperature of 350°C represents the M_s temperature of a 0.55% C steel being the boundary of the martensite morphology between a lath and a plate (or needle) type. The transition temperatures of steel containing >1.1% C decrease along the T_0-composition line by 65°C which is attributed to the nonchemical free energy for the displacive growth of lower bainite.[46] Figure 9.5 shows that the transition temperature between upper and lower bainite (also called *lower bainite start temperature* LB$_s$) is paralleled by

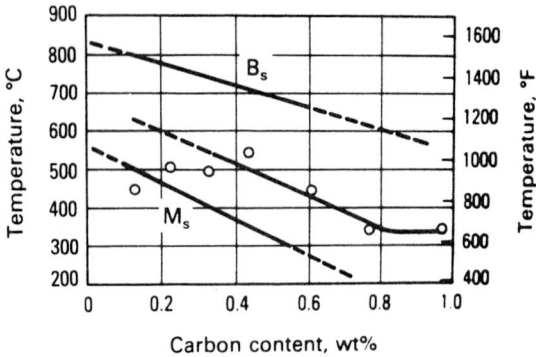

FIGURE 9.5 Lower bainite start temperature LB_s in relation to B_s and M_s temperatures.[47] Note that, here, LB_s differs from that in Fig. 9.4 with respect to the extrapolated LB_s line, though not of the data points. (*Reprinted by permission of ASM International, Materials Park, Ohio.*)

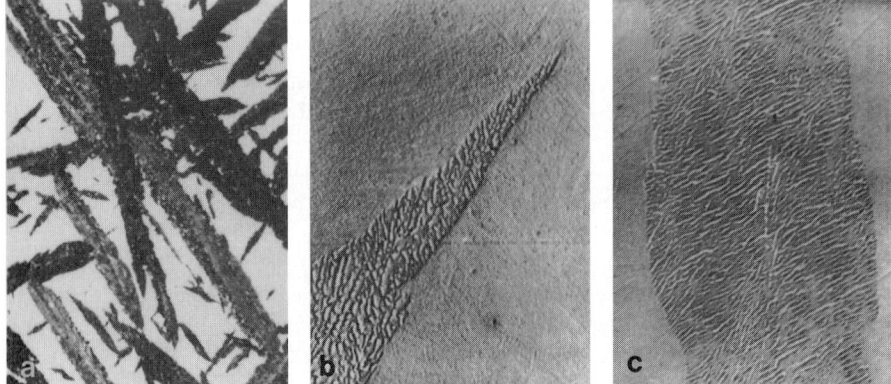

FIGURE 9.6 Lower bainite formed at 350°C in a 0.66% C-3.3% Cr steel. (*a*) Light micrograph, 1000X, (*b*, *c*) electron micrographs, 16,000X.[48] (Reduced 92%.) (*Courtesy of Paul G. Shewmon; after G. Speich.*)

similar variations with carbon content in the B_s and M_s temperatures; i.e., a decrease in LB_s occurs with increasing carbon content.[47] Note that LB_s in Fig. 9.5 differs from that in Fig. 9.4 in respect of the extrapolated LB_s line, though not of the data points.[38] Lower bainite is frequently characterized by a plate rather than a lath morphology (Fig. 9.6*a*),[48] as was determined by Srinivasan and Wayman[49] by careful two-surface examination. Here the carbide tends to precipitate as smaller particles, in parallel arrays which appear as cross-striations, making an angle of 55 to 60° with the longitudinal sheaf axis of the ferrite plate (Figs. 9.6*b* and *c* and 9.7*d*). This feature is in contrast to that of tempered martensite, where more than one variant is always found.

These ferrite plates nucleate at austenite grain boundaries as well as within the grains. They are thin because of the smaller diffusivity of carbon in austenite at the lower temperature corresponding to the formation of lower bainite. The carbide precipitation reactions for both upper and lower bainite are secondary events which take place after the growth of bainitic ferrite. Subsequently, according to the major-

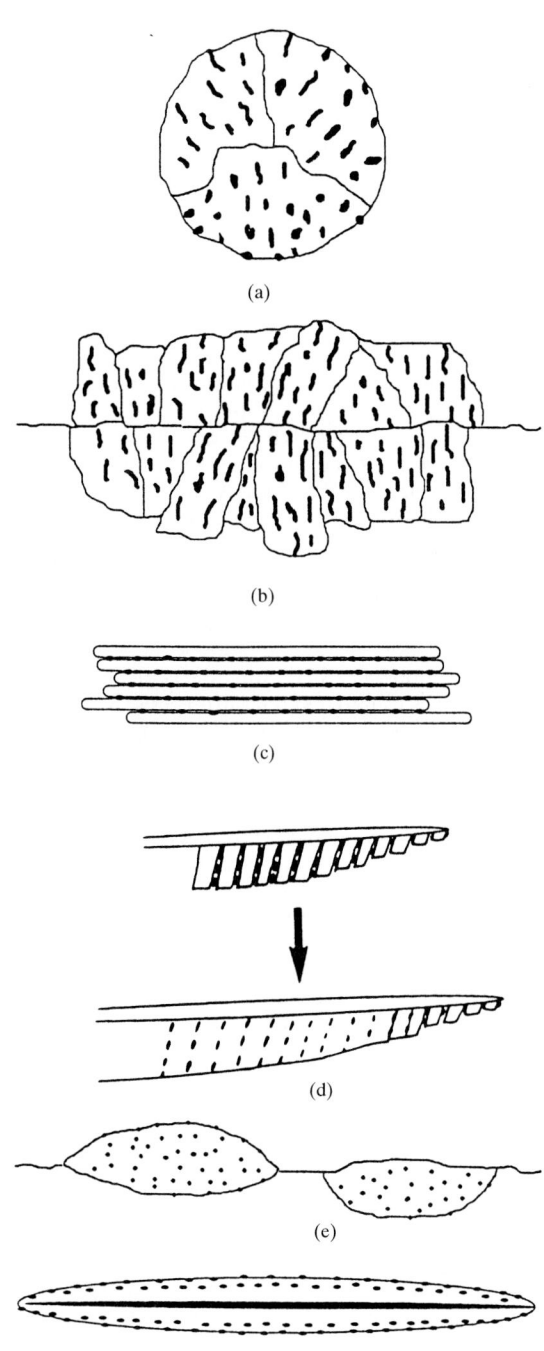

(a)

(b)

(c)

(d)

(e)

(f)

FIGURE 9.7 Schematic illustrations of various bainite morphologies: The white constituent represents the majority eutectoid phase (such as ferrite), and the dark constituent represents the minority phase (such as cementite). (a) Nodular bainite, (b) columnar bainite along a prior matrix grain boundary, (c) a sheaf of upper bainite laths, (d) lower bainite, (e) grain boundary allotriomorphic bainite, and (f) inverse bainite.[6] (*Courtesy of W. T. Reynolds, Jr., H. I. Aaronson, and G. Spanos.*)

ity of workers, the ferrite plates thicken most rapidly by the repeated precipitation of carbides at the advancing interface.[13,32] Some workers believe that carbide precipitation within the ferrite plates is very rapid and dense with increasing carbon content or decreasing reaction temperature.[17,50] Some of the carbides also precipitate from the austenite between platelets of bainitic ferrite; this quantity is very small when compared with upper bainite. Other researchers have proposed a diffusional mechanism for the formation of lower bainite sheaves (to be discussed later).[13]

Many measurements have been made of the orientation relations between bainitic ferrite and austenite that provide results close to either K-S or Nishiyama-Wasserman (N-W) relations (Table 9.1), indicating the true relations to be irrational.[2]

Study of the orientation relationship between bainitic carbide and ferrite has suggested that the initial ferrite is supersaturated with carbon and that the carbides precipitate from the ferrite plates at the advancing interface.[13,31,51,52] This is supported by the observation of the typical Bagaryatskii relationship (Table 9.1) between ferrite and cementite similar to that of the ferrite-cementite orientation relationship occurring in tempered martensite.[53] Later work by Ohmori[54] and Huang and Thomas[52] has suggested θ/α orientation relations to be those of Isaichev[55] (Table 9.1) for cementite within lower bainitic ferrite, which is slightly different from the Bagaryatskii relation by a rotation of $4°$ around $\langle 111 \rangle_\alpha$; the two are difficult to distinguish experimentally. In addition, Ohmori and Honeycombe have observed a unique variant of cementite habit plane in each lower bainite plate. Huang and Thomas[52] have also made similar observations on lower bainite formed in silicon-bearing steels. All these results strongly support the theory that the cementite forms by interphase precipitation at the γ/α interface, more likely from austenite than from supersaturated ferrite. However, a rational and uncommon orientation relationship between cementite (θ) and ferrite matrix, namely, $(011)_\theta // \{011\}_\alpha$ and $\{1\bar{2}2\}_\theta // \langle 100 \rangle_\alpha$, with up to four different variants of the cementite $\{011\}_\theta$ habit plane, has been reported in a single ferrite crystal of Si-Mn steel. Since this unique orientation relationship cannot be combined with the Kurdjumov-Sachs relations to provide the expected three-phase α-γ-θ relationship, it is consistent only with the hypothesis that the cementite precipitates directly from a supersaturated bainitic ferrite. Since the orientation relationship is uncommon, this result is unlikely to be general and may be discounted.

A TEM study of interfacial structure of the broad faces of ferrite/bainite plates in an Fe-C-Si alloy formed at 450 to 475°C has shown that the thickening of ferrite plates by shear is mechanistically impossible.[56] This is in agreement with the more recent high-vacuum electron microscope observation on migrational behavior of the α/γ interface in a similar steel.[57]

Carbides in lower bainite can be either ε-carbide ($Fe_{2.4}C$) or cementite (Fe_3C), depending on both the transformation temperature and the composition of steel. It has been reported that ε-carbide is the first carbide formed in low-temperature bainite which is subsequently replaced by cementite on further reaction.[58] This fact is supported by the formation of ε-carbide in steels containing silicon during isothermal transformation at 275°C, which eventually transforms to cementite upon further holding at this temperature.[52] Formation of ε-carbide is also promoted when the carbon concentration is sufficiently high (>0.55%) to overcome the energetically more favorable segregation of carbon atoms to dislocations.[47] However, this concept does not deal with the relevant competition between Fe_3C and $Fe_{2.4}C$ carbides.[38]

The ε-carbide formed in lower bainite exhibits an orientation relation with the bainitic ferrite close to that proposed by Jack[59] (Table 9.1), which is also found in the case of martensite tempered at low temperatures. As with cementite, this orientation relationship implies that ε-carbide in lower bainite is probably precipitated directly from a rather high-carbon supersaturated lower bainitic ferrite or both ferrite and ε-carbide precipitate from the austenite. In contrast, other researchers[52] have supported the theory, based on their experimental observations, that ε-carbide precipitates directly from austenite at α/γ boundaries instead of from ferrite.

A new transition carbide, termed κ-carbide, has been found by Deliry and Pomey to occur in high-carbon steels transformed to lower bainite which precipitates at a later stage of transformation, from the carbon-rich austenite between the ferrite platelets.[10] It is characterized by high solubility for silicon and its transformation to χ-carbide and eventually to more stable cementite on continuous holding at the isothermal transformation temperature. Similarly, the transition carbide, called η-carbide, with an orthorhombic crystal structure, discovered by Konoval et al. in high-Si transformer steel, has been reported to precipitate from lower bainitic ferrite in Fe-1.15C-3.9Si alloys.[10]

Using single-surface trace analysis, Shackleton and Kelly have reported the habit plane of cementite within lower bainitic ferrite in the neighborhood of the zone containing $\{1\bar{1}2\}_\alpha$ and $\{0\bar{1}1\}_\alpha$ (corresponding to $\{101\}_\theta$ and $\{100\}_\theta$, respectively);[31] approximately $\{001\}_\theta$ // $\{211\}_\alpha$; and $\{201\}_\theta$ due to faulting on the $\{001\}_\theta$ planes.[10,54]

Like the orientation relationships, the habit plane in lower bainite is found to be irrational. In a pure iron-carbon alloy[60] and Si-Mn plain carbon steel,[54] the habit plane is $\{456\}_\gamma$; while in an Fe-Cr-C alloy,[61] it is close to $\{254\}_\gamma$. An early postulation,[62] based partly on Smith and Mehl's observations, was that the crystallography of upper bainite should be similar to that of martensite corresponding to an appreciably lower-carbon steel than the specimen under study, while the crystallography of lower bainite should be very similar to that of martensite in the same steel. In fact, when the crystallographic theory of martensite is applied to bainite, the observed habit plane is not predicted on the basis of the hypothesis of fine-lattice-invariant deformation twinning on $\{112\}_\alpha$ but is consistent with different lattice-invariant shear (deformation). Srinivasan and Wayman also concluded in their study on lower bainite in 7.9% Cr-1.1% C steel that the orientation relations, habit planes, and shape deformations are quite different from those observed for martensite in the same steel.[61]

Nonclassical Ferrous Bainite Terminology

NODULAR BAINITE. *Nodular bainite* first becomes a major constituent at the 0.57% C content and dominates the microstructure at C concentrations of $\geq 0.95\%$ and in the temperature range of 275 to 400°C. (Sometimes it is considered as the low-temperature pearlite.) The growth of nodular bainite occurs in roughly isotropic manner, and the final shape of the eutectoid mixture is nearly a sphere, or a portion thereof, when growth is not restricted (Fig. 9.7a). Bainite nodules have been noticed in eutectoid and hypereutectoid Fe-C and Fe-C-Mn steels and in hypoeutectoid Ti-Cr alloys. Nodular bainite has also been found in hypoeutectoid Fe-C-Mo and Fe-C-Ni alloys. In the latter alloys, it occurs after large precipitation of proeutectoid ferrite allotriomorphs and sideplates. The nodular bainite adopts the shape of the pockets of remaining γ where it forms.[4,6,27]

GRANULAR BAINITE. In low- or medium-alloy steels, during continuous cooling rather than during isothermal treatment, enriched austenite does not transform to carbide films but either may remain as a film of retained austenite or may transform to high-carbon martensite between the bainitic ferrite laths at room temperature, depending upon the cooling rate.[51-61,63-65] This variety, frequently called *granular bainite*, usually associated with a large proportion of classical bainite,[66] adversely influences ductility. Note that, in a true sense, this structure is unlikely to fit any of the three definitions of bainite. It has been inferred that granular bainite consists of a ferrite matrix with a high dislocation density surrounding martensite-austenite (M-A) "islands."

INVERSE BAINITE.   In hypereutectoid steels, after appropriate isothermal treatment but at higher temperatures ($\geq 400°C$), proeutectoid cementite plates provide the "spine" of *inverse bainite*, originally recognized by Hillert.[67] The initial cementite is soon surrounded by a rim of ferrite laths followed by the rapid side-by-side nucleation of a nonlamellar, spearlike structure containing a mixture of ferrite and cementite around the ferrite rim (Fig. 9.7f).[6,27] Further study by Kinsman and Aaronson[68] has revealed that the formation of inverse bainite begins with the single spines of cementite; later, in the adjacent region, a normal bainitic structure develops with large ferrite laths and small carbide precipitates.

BLOCKY OR COLUMNAR BAINITE.   Columnar bainite, being a modification or derivative of the nodular bainite morphology, has been found to be formed in medium (0.44%) and high (0.82%) carbon steels when transformed in the bainitic temperature range under high hydrostatic pressure (for example, 3 GPa or 24 to 32 kbar). It consists of nonacicular ferrite nucleated preferentially at grain boundaries and fine dispersion of cementite arrayed parallel to the most rapid growth direction of the ferrite[69] (Fig. 9.7b). A similar structure can also be formed in hypereutectoid steels containing more than 1.4% C when transformed at atmospheric pressure.[4,27]

GRAIN BOUNDARY ALLOTRIOMORPHIC BAINITE.   In hypereutectoid steels, carbide precipitation often occurs along with grain boundary ferrite allotriomorphs formed below the eutectoid temperature. When these carbides are not formed by a cooperative growth process (as in pearlite), the composite microstructure forms as bainite, according to the microstructural definition. The fact that the allotriomorphic bainite takes over or supersedes the upper bainite with decreasing grain size at a particular transformation temperature underlines the significance of the proeutectoid component's morphology in determining the overall morphology of bainite. The external shape of this form of bainite is thus determined by the proeutectoid ferrite allotriomorphs from which it develops (Fig. 9.7e).[4,6]

LOWER MIDRIB BAINITE (LB_m).   Lower midrib bainite LB_m has been observed in ferrous alloys by several researchers who have put forward different hypotheses as to its origin. According to Okamoto and Oka, in hypereutectoid steels (0.85 to 1.8 wt% C), LB_m is characterized by[70–72] (1) its formation in the temperature range of 150 to 200°C; (2) a thin "spiny" lower bainitic ferrite plate (Fig. 9.8a) with a thin plate of isothermal martensite (TIM) as its midrib (Fig. 9.8b);[71] (3) a habit plane near $\{3\ 15\ 10\}_\gamma$, which is similar to that of TIM but different from that of LB; (4) the transition temperature range between LB and LB_m that lies between 210 and 170°C with a tendency toward lower temperatures as the carbon content is increased; and (5) the transition temperature between LB_m and TIM of ~127°C.

Ohmori[73] and Spanos et al.[74] have observed a single, nearly carbide-free ferrite thin "spine" plate with lower bainite sheaves containing a high density of carbides. This thin spine ferrite plate is similar to the LB_m, except that it lies mainly along one side of the sheaf, in contrast to the central part of the bainite sheaf. This spine was believed to be formed by purely diffusional ledgewise migration of partially coherent α/γ boundaries. Sun et al. have elucidated the nature of the midrib associated with the transformation mechanism of lower bainite in three silicon-bearing steels.[75]

9.2.1.2 Nonferrous Microstructural Bainite.   The morphological transition from upper to lower bainite (discussed above) does not occur in nonferrous alloys.

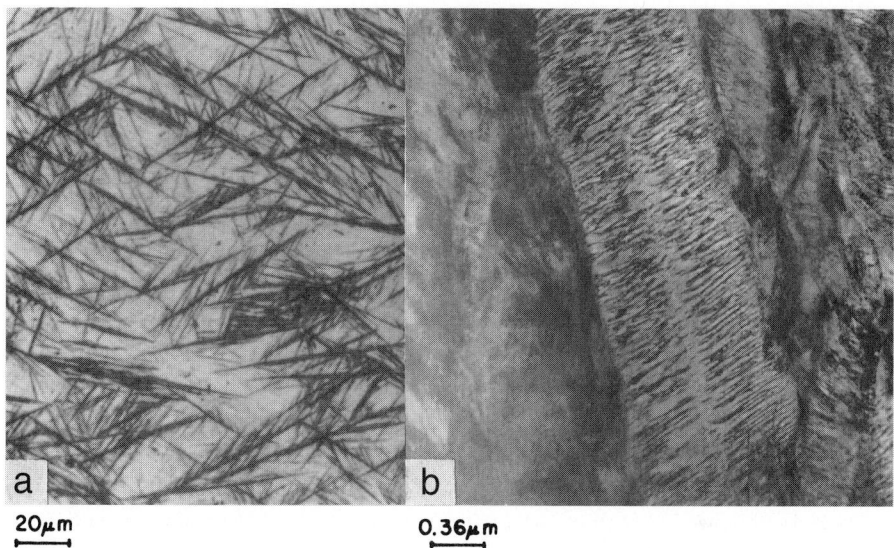

a   20μm

b   0.36μm

FIGURE 9.8 (*a*) Optical micrograph showing thin and spiny lower bainite formed at 190°C for 5 hr in a 1.1% C steel.[71] Etchant: 2% nital. (*b*) Transmission electron micrograph showing a lower bainite with midrib (LB$_m$) in the above steel. (*Courtesy of M. Oka.*)

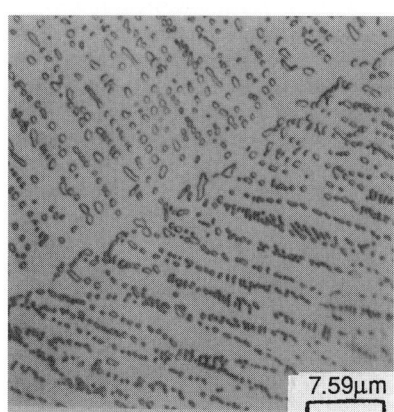

7.59μm

FIGURE 9.9 Fully reacted bainitic microstructure in Ti-4.6 at% Cu showing closely spaced proeutectoid α plates separated by rows of compound particles. Reacted at 775°C for 2 hr.[77] Etchant: 2% HF + 98% H$_2$O. (*Reprinted by permission of The Metallurgical Society, Warrendale, Pennsylvania.*)

Several nonferrous alloys exhibit two-phase nonlamellar microstructures similar to upper bainite in steels or, more frequently, only one lathlike phase precipitating from the high-temperature parent phase, analogous to the high-silicon steel mentioned above.[66]

In hypoeutectoid Ti-X (X = Bi, Co, Cr, Fe, Mn, Pd, and Pt) alloys, the microstructural bainite forms by eutectoid decomposition.[76] A fully reacted bainitic microstructure in a typical hypoeutectoid Ti-Cu binary alloy is shown in Fig. 9.9.[77] This was produced by the eutectoid reaction $\beta \rightarrow \alpha + Ti_2Cu$, below the eutectoid temperature. Its microstructure and morphology are similar to those of upper bainite in steel. This alloy exhibits rapid eutectoid decomposition and can be clearly resolved microstructurally.[77]

A TEM study of interfacial structure of microstructurally defined bainite in a hypereutectoid Ti-Cr alloy has shown the bainite to comprise hcp α + ordered (C15) $TiCr_2$ constituents. These constituents are partially coherent with their bcc parent matrix as determined by the existence of misfit dislocations in both interfaces.[25]

The bainite plates in the β-phase alloys, such as Cu-Zn, Ag-Cd, and Cu-Zn-Al alloys, exhibit, in common, a close-packed structure such as 3R or 9R structure. In contrast, complex bainitic structures with α and γ_2 phases begin to appear from the start in Cu-Al and Cu-Sn alloys. It is proposed that the bainitic reaction is just the martensitic one triggered by the atomic diffusion.[78]

It is well established that $\beta' \rightarrow \alpha_1$ in Cu-Zn and Cu-Zn-Al is undisturbed by diffusion of interstitial atoms.[79] Cliff et al. disproved the initial formation of bainitic α_1 plates in a Cu-39.3 wt% Zn and γ_1 plates in a Cr-33 wt% Ni by a mechanism only involving crystallography shear. Hsu and Zhou's thermodynamic calculations also showed that the bainitic transformations in Cu-26.77 wt% Zn-4 wt% Al could solely proceed with a diffusional mechanism. Wu et al. concluded that the formation and early growth of α_1 plates in a Cu-26.67 wt% Zn-4 wt% Al alloy involved an initial shear process and final classical diffusion-controlled growth of α_1 (α) plates. In a Cu-24.3 wt% Zn-4.7 wt% Al-0.8 wt% Mn alloy and a Cu-19.51 wt% Zn-5.88 wt% Al alloy, however, Liu and coworkers noticed the midribs in initial α_1 plates, the slabby stress field ahead of the tip of the plates, and the twisting of the intersected plates. These phenomena provided clear evidence for the shear mechanism of bainitic transformation.[80]

9.2.2   Surface Relief Definition

On this definition the bainite (1) consists simply of (ferrite) plates or laths, produced by a shear mode of phase transformation, usually occurring above the M_s or even the M_d temperatures, which differ in composition from its parent phase; (2) grows slowly; and (3) exhibits a shear or martensitic-like invariant plane strain (IPS) surface relief effect (Figs. 9.10c and 9.11) when formed at a free surface. In other words, bainite in this definition is the growth of a precipitate plate by a diffusion-controlled shear where the diffusion process may occur either before or after the formation of a single product phase.[76] This is now the most widely recognized and internationally used definition.[12-14] In the true sense, the term *relief effects* cannot be used to mean that the bainite formation is martensitic because of the difference in composition between the new bainitic ferrite and parent γ.[81] This definition, however, is not general. Some workers are of the opinion that upper bainite and lower bainite form by high-velocity shear; even that ferrite plates form in a similar manner.[11]

When such surface reliefs develop as a result of a precipitate plate formed by diffusional growth, it implies that there exists the coordinated atom movements at or near a semicoherent γ/α interface with a lattice correspondence between the ferrite and austenite to produce a sharp change (Fig. 9.10). Again, the interfacial structures so far observed at α/γ boundaries cannot support growth by shear—even slow growth. The relief effects for upper and lower bainite are different because in the respective structures the ferrite component forms as *packets* of laths and individual plates.[82] Accordingly, upper bainite exhibits multiple (and complex) surface relief, whereas lower bainite exhibits a single, uniform surface relief (or tilted planar surfaces).[24] In some instances, particularly lower bainite plates in an Fe-C-Cr alloy

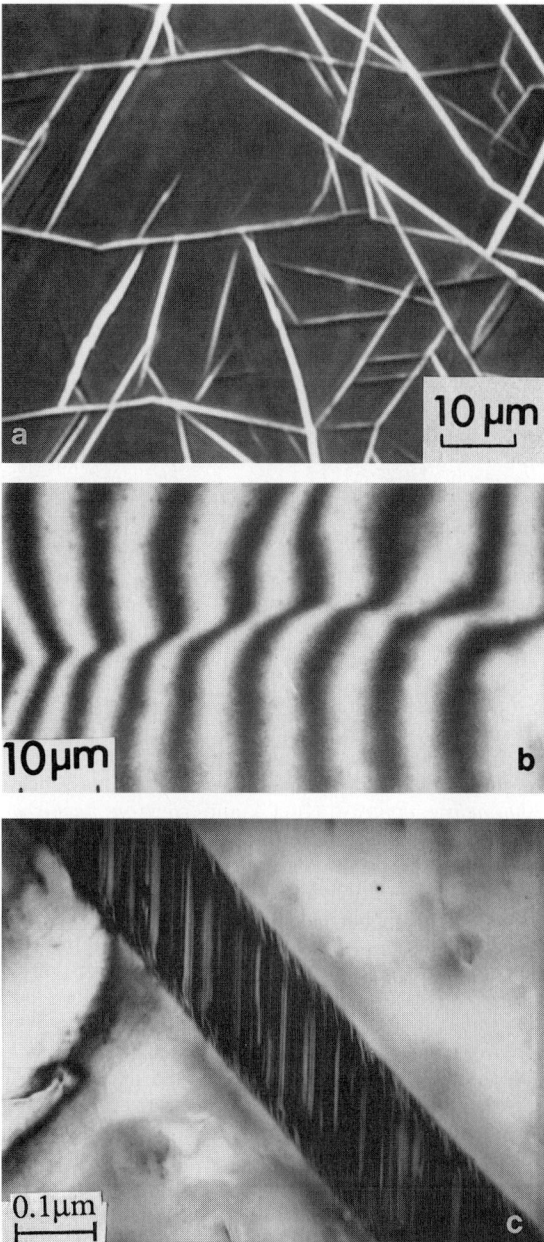

FIGURE 9.10 (*a*) Optical micrograph showing α plates in ternary Cu-9.3Au-40.1Zn (at%). (*b*) Interference micrograph illustrating single-tilt associated with plates. (*c*) Transmission electron micrograph of a faulted α plate in ternary 4.2 at% Au alloy.[90] (*Courtesy of P. E. J. Flewitt.*)

FIGURE 9.11 Photomicrograph showing surface relief associated with upper bainite formed at 450°C in 1.48% C steel. Unetched. (*Courtesy of N. F. Kennon.*)

can exhibit tent- or V-shaped relief tilt. Ferrite usually also forms as tent-shaped relief.[38]

The tent-shaped surface reliefs associated with single-crystal proeutectoid ferrite plates in low-carbon and medium-carbon, low-alloy steels have been demonstrated using *electron channeling contrast* (ECC) and *electron back-scattering patterns* (EBSPs), generated in an SEM. The tent-shaped surface relief is caused by the ledgewise growth of opposite broad faces in opposite directions.[4,5,83,84]

9.2.2.1 Nonferrous Surface Relief Bainite. Nonferrous surface relief bainite appears as sideplates (-laths) or intragranular plates (-laths). There need be no eutectoid decomposition possible in order for such morphologies to be termed *surface relief bainite*. In the past they were described as *Widmanstätten structure*.[38]

A classical upper bainite-type appearance has been observed in U-0.75Ti alloy,[85] binary Cu-base alloys such as Cu-Zn;[86,87] binary Ag-Cd; and ternary Cu-Zn-Al alloys and Cu-Zn-Au alloys. These alloys contain substitutional solute atoms in the matrix and precipitate lattices, in contrast to carbon steels, where movement of interstitial atoms plays a major role in bainite transformation.

Among these alloys, surface relief bainite formation in binary Cu-(~40at%) Zn and ternary Cu-(~40at%) Zn-(up to 8at%) Au systems has been extensively studied. Here metastable β' Cu-Zn phase transforms during isothermal heat treatment, to either α_1 plates or α rods based on the composition and aging temperature.[88,89] Both reactions follow their own C-curve kinetics, with the α_1 plates forming at lower temperatures and with characteristic obtuse-angled V-shaped pairs (Fig. 9.10a).[90] In the ternary alloy, the individual dimensions of α_1 plates increase by a factor of 4 compared to those of the binary alloys at a given temperature. An invariant plane strain surface relief is developed on a prepolished surface (Fig. 9.10b).[90] These plates show clear evidence of inhomogeneous shear by slip in the form of high-density arrays of stacking faults (Fig. 9.10c).[90] Moreover, they have an irrational orientation relationship with the parent β' phase and have irrational habit planes near $\{2\ 11\ 12\}_{\beta'}$. All these transformation characteristics are similar to those of low-temperature martensite formed in the same alloy.[66,89,90] Based on microanalysis of thin foils, their results appear to suggest that bainitic transformation should proceed in two stages: The first is lengthwise growth by the martensite shear mechanism, followed by the (second) sidewise growth by diffusion-controlled mechanism.[90,91] However, a further work by Tadaki et al.[92] reported the absence of meaningful difference in composition between plates and matrix phase, contrary to the previous one by Doig and Flewitt.[90] Consequently, a tentative model was proposed for the formation mechanism of bainite plates in the substitutional β-phase Cu-Zn-based alloys. This model involves diffusion of atoms only in the bainitic transformation in the substitutional alloy from the start, neither before nor after it.[92] High-temperature TEM in situ observation of bainite in Cu-Zn alloys by Yang et al.[93] supports the shear mechanism of bainite reaction based on the existence of (1) stacking-fault substructure in the entire bainite plate, especially just in the growing tip of the fresh-formed bainite plate; (2) shear field in the matrix around the tip; (3) movement of the superledges on the stacking fault plane in the direction of stacking fault growth, which is different than that predicted by the diffusional ledge mechanism; and (4) faster degeneration of bainite to equilibrium α phase. However, a TEM study of the interfacial boundary structure of isothermally formed α_1 plates in a β' Cu-(38–43 at%) Zn was shown to disprove the shear mechanism of thickening of α_1 plates but was shown to be consistent with the diffusion-controlled ledgewise growth mechanism.[79,94] The formation of hcp γ Ag_2Al precipitate plates from fcc α in Al-Ag represents an example of bainite reaction which occurs by a combination of diffusive flux of solute to the precipitate and a martensitic, fully displacive, interface motion.[95] Experimental evidence presented by Laird and Aaronson[96] suggests this transformation to be a classical diffusional type.

Examples of substitutional bainitic reaction exhibiting both changes in chemical composition and surface relief effects during precipitation include AuCuII, Cu-Zn, Cu-Al, Cu-Sn, Ag-Cd, Cr-Cu, Ti-Cr, and Cr-Ni, which suggests that the reactions are not fully martensitic when growth occurs with a compositional change. The bainitic growth involving precipitation of γ' (later γ) $AlAg_2$ plates from α Al-Ag solid solutions takes place by a combination of diffusive flux of solute to the precipitate and a displacive interface motion. Here, displacive reaction occurs by the glide of Shockley partial dislocations on every second $\{111\}$ plane, providing purely a displacive transformation of fcc matrix $\rightarrow$ hcp precipitate,[9,95] taking place at rates controlled by the long-range volume interdiffusion of Ag and Al.[96]

High-resolution TEM analyses by Wu et al.[97] have shown that α_1 plates in Cu-Zn-Al alloys precipitate at solute-depleted defects through a shear mechanism, and subsequent plate growth is then controlled by a diffusional process. However, other workers have shown that the α_1 plates in these alloys are initially free of stacking faults; mobile ledges are found to exist on the broad faces and the leading edge of α_1 plates. This suggests that the growth of α_1 plates is accomplished by a diffusion-controlled ledgewise mechanism.[98] Thermodynamic consideration also favors the latter mechanism.[99]

Recently, Muddle et al.[100] have shown that the surface relief and crystallography of a range of transformation products such as γ' $AlAg_2$ plates, ordered orthorhombic AuCuII plates in equiatomic Au-Cu, and 9R α_1 plates in ordered B2 Cu-Zn(-X) and Ag-Cd alloys, differing in composition throughout growth from the matrix phase, can be quantitatively described by the phenomenological theory of martensite crystallography (PTMC). Later, Aaronson and coworkers,[101] based on further examination of the experimental result in precipitate γ' AlAg$_2$ plates, Cu-Zn-X plates, α Fe-C laths/plates, and α Ti-Cr laths/plates developed above M_d during these model diffusional phase transformations, have concluded that their crystallography can be predicted by the PTMC.

9.2.3 Kinetic Definition

The kinetic definition is based on the fact that bainite reaction has its own (or a separate) C-curve (Fig. 9.12a)[102] on a TTT diagram whose upper limiting temperature, termed the *kinetic-B$_s$* (bainite start), in analogy with the M_s temperature for the martensitic reaction, lies much (100 to 300°C) below the eutectoid temperature. As the B_s temperature is approached, the bainitic reaction becomes increasingly

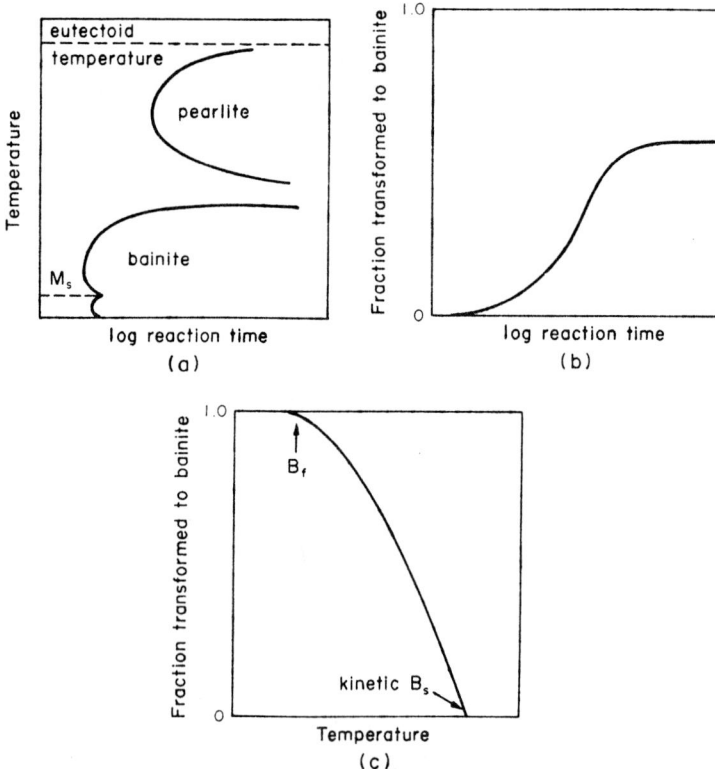

FIGURE 9.12 (*a*) Schematic TTT diagram with a separate bainite C-curve. (*b*) Schematic time-dependent isothermal bainite transformation curve showing incomplete transformation. (*c*) Schematic isothermal transformation diagram showing the amount of bainite formed as a function of temperature.[14,102] (*Reprinted by permission of ASM International, Materials Park, Ohio.*)

incomplete. The kinetic-B_s temperature is designated as the highest temperature at which bainite forms on the overall kinetics definition of bainite.[4] It is defined as one above which austenite will not transform to bainite. This characteristic is called the *transformation stasis* or the *incomplete-reaction phenomenon*. Bainite formed in this manner is often believed to form by shear. Between the kinetic-B_s and some lower temperature, incorrectly termed *bainite finish* or kinetic-B_f temperature, the proportion of austenite transformed to bainite increases from zero at the kinetic-B_s to unity at the kinetic-B_f.

Those who use the kinetic definition of bainite insist that the bainite C-curve is also present in plain carbon steels as well as in alloy steels in which distinct ferrite, pearlite, and bainite C-curves cannot be discerned. In such steels, the overlap between the bainite reaction and the higher-temperature reactions is said to obscure the separate C-curve. However, on the microstructural definition of bainite the view has been taken, with experimental support, that the proeutectoid ferrite and bainite C-curves are identical, since bainite is simply proeutectoid ferrite with which the carbide precipitation (in nonlamellar form) is associated at temperatures beginning at undercoolings below the eutectoid temperature which vary with carbon and alloying element concentrations and austenite grain size.[12] Further, the usual explanation for the absence of incomplete transformation is that this is only apparent and occurs because of overlap of the bainite and pearlite C-curves. However, the fact that the formation of pearlite obscures this phenomenon has been experimentally disproved in a number of alloy steels. This means that incomplete transformation of austenite to bainite is a characteristic of specific alloying elements.[3]

On the kinetic definition, the B_s temperature can be easily determined by the chemistry of the steel and by austenitizing variables such as temperature, holding time, and grain size. The effect of alloying additions on the B_s temperature may be represented by the following empirical relation (for a range of wrought low-alloy steels containing 0.1 to 0.55% C, 0.1 to 0.35% Si, 0.2 to 1.7% Mn, 0.0 to 5.0% Ni, 0.0 to 3.5% Cr, and 0.0 to 1.0% Mo) due to Steven and Haynes:[103]

$$B_s(^\circ C) = 830 - 270(\%C) - 90(\%Mn) - 37(\%Ni) - 70(\%Cr) - 83(\%Mo) \quad (9.1)$$

Recently Bodnar et al.[104] derived an equation for apparent B_s temperature at which the bainite starts to form during continuous cooling (which should lie below the true B_s temperature) for a series of Fe-C-Ni-Cr-Mo steels (containing 0.15 to 0.29% C, 0.01 to 0.23% Si, 0.02 to 0.77% Mn, 0.21 to 3.61% Ni, 1.13 to 2.33% Cr, and 0.44 to 1.37% Mo) used in the power generation industry:

$$``B_s"(^\circ C) = 844 - 597(\%C) - 63(\%Mn) - 16(\%Ni) - 78(\%Cr) \quad (9.2)$$

Another feature of bainitic reaction kinetics is that below B_s, as in any nucleation and growth process, bainite formation is time-dependent. A typical isothermal reaction curve is shown in Fig. 9.12b.[102] This figure also illustrates that the kinetically defined bainite reaction does not necessarily go to completion. In the absence of overlap with the pearlite reaction, the proportion of austenite transformed to bainite supposedly increases from zero at the kinetic-B_s to 100% at the kinetic-B_f (Fig. 9.12c); however, bainite can still form at temperatures below B_f.[102] Upon cessation of the bainitic reaction, the untransformed proportion of the austenite phase enters a period of stasis, during which further transformation does not occur, although redistribution of carbon continues throughout the remaining austenite. Decomposition of this (retained) austenite can resume after times varying from seconds to many hours, depending upon the alloy chemistry and reaction

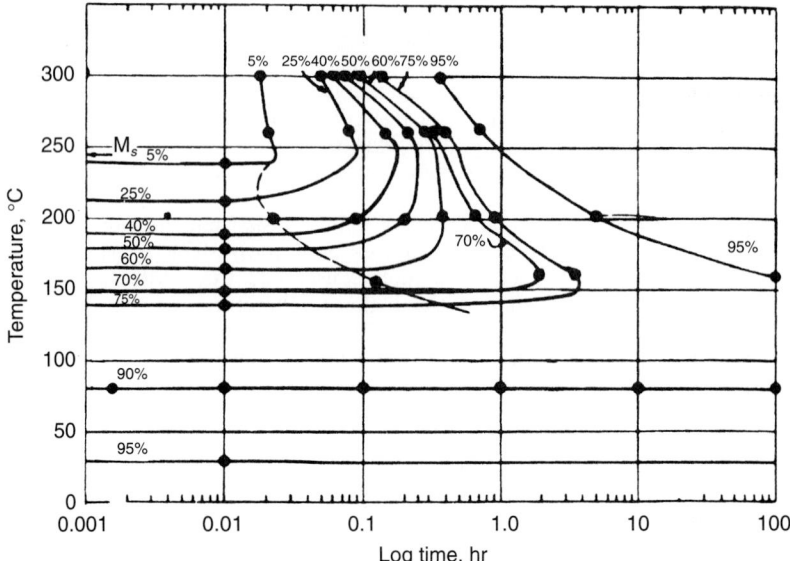

FIGURE 9.13 Experimentally determined TTT diagram for 0.75% C steel in the vicinity of M_s temperature.[105] (*Reprinted by permission of The Metallurgical Society, Warrendale, Pennsylvania.*)

temperature. The transformation product which completes isothermal decomposition of austenite (between 200 and 250°C) after the stasis interval can be either pearlite or some form of bainite.[3,102] When bainite forms at or below the M_s temperature, the prior formation of martensite catalyzes the bainite reaction, thereby leading to an abrupt deviation of the TTT curve for initiation of the bainite reaction to shorter times (Fig. 9.13).[105,106] When the B_f lies below the M_s, the B_f temperature cannot, of course, be experimentally determined.

9.2.3.1 Incomplete Transformation (or Stasis) Phenomenon in Fe-C-Mn-Si and Fe-C-Mo Steels.

When an intentionally prepared Fe-0.43C-3.0Mn-2.12Si alloy was studied, separate upper and lower bainitic curves (in addition to pearlite curves) were found to exist on the TTT diagram (Fig. 9.14),[17] which corresponded to two basically different reaction mechanisms separated by a very small temperature range near 350°C. [In steels where other reactions do not interfere (or occur simultaneously) with the growth of bainitic ferrite, the maximum volume fraction of bainitic ferrite that forms on prolonged holding at the isothermal transformation temperature is much lower than expected according to the equilibrium or paraequilibrium transformation.[107]] Bainitic formation, during isothermal transformation, ceased in a matter of minutes; but if the specimen was held at the transformation temperature for 32 hr, the residual γ initiated to transform to pearlite. Likewise, when an Fe-4.08Cr-0.3C alloy was isothermally transformed to lower bainite, the reaction was observed to cease within ~1/2 hr, but prolonged holding at the transformation temperature (43 days) resulted in the residual γ undergoing extremely sluggish reconstructive transformation to two different products such as pearlite at the γ grain boundaries and lower bainite with extremely irregular morphology at

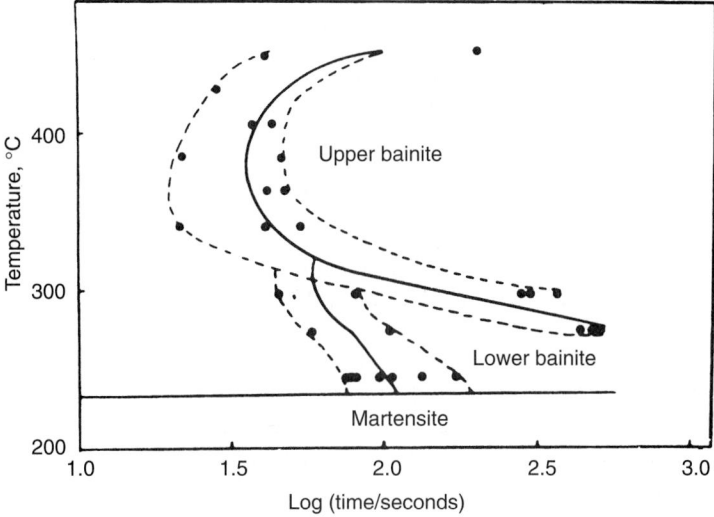

FIGURE 9.14 Dilatometrically determined 5% transformation TTT curve for Fe-0.43C-3.0Mn-2.12Si steel.[17] (*Reprinted by permission of The Metallurgical Society, Warrendale, Pennsylvania.*)

the lower bainite/γ interfaces.[108,109] This is in sharp contrast to the kinetic definition of bainite.

In high-purity Fe-C-Mo alloys containing 0.06 to 0.27 C and 0.27 to 4.28 Mo, the occurrence of incomplete transformation depends on both C and Mo contents, taking place only when critical proportions of C and Mo are exceeded. The Mo concentration needed to produce stasis increases as the C concentration of the alloy decreases.[110] Transformation stasis takes place when the sympathetic nucleation of carbide-free ferrite ceases because of the carbon enrichment of the remaining austenite.[110] The growth of the initially formed ferrite crystals and subsequent growth of sympathetic nucleated ferrite crystals are inhibited by a solute drag-like effect (SDLE).[110] However, it is suggested that stasis is not a general feature of bainitic reaction in Fe-C-X alloys, because of its absence in several other ferrous alloys, such as Fe-0.26C-2Mo, Fe-0.38C-1.73Si, Fe-0.38C-3.1Mn;[111] low-C or low-Mo, Fe-C-Mo alloys; low-Si, Fe-C-Si alloys; low-Ni, Fe-C-Ni alloys and Fe-C-Cu alloys; and nonferrous alloys.[76] However, Fe-0.1C-3Cr, Fe-0.1C-3Mn, and Fe-0.38C-3 at% Mn exhibit transformation stasis.[6,112]

9.3 OVERALL MECHANISMS

The mechanism of bainitic ferrite formation (i.e., primary stage) at the various temperatures of austenite decomposition, the factors controlling the growth rate, and the mechanism of bainitic carbide (i.e., secondary-stage) precipitation have not yet been unequivocally established. The main controversy exists between the proponents of an interface ledge-growth mechanism and a shear mechanism.

9.3.1 Upper Bainite

During TEM study, growth ledges were observed on bainite plates by many investigators, notably Purdy[113] in Fe-C-Mo, Nemoto[114] in Fe-0.51C-9.1Ni, Menon and Aaronson[115] in Ti-Cr (not microstructural bainite), Simonen et al.[116] in Cu-Zn (not microstructural bainite), and others in different systems.[117] Based on their data, it has been inferred that the bainite plates propagate by ledge mechanism along the broad faces with carbon volume-diffusion-controlled kinetics[117–119] and that the condition for microstructural bainite formation by the ledge-growth mechanism is $G_\alpha \neq G_\beta$ with $h_\alpha/\lambda_\alpha \neq h_\beta/\lambda_\beta$, where G, h, and λ are the growth rate, the average ledge height, and the average interledge spacing, respectively, of a particular precipitating phase, indicated as the subscript. The faster-growing phase (arbitrarily designated as α) inevitably terminates the access of the slower-growing β phase to the γ matrix phase. Hence, supersaturation of the γ matrix phase with respect to nucleation of the slower-growing β phase develops.[4] When such nucleation occurs, this cycle repeats several times to produce a fully developed bainite structure.[25] Variations in G_α/G_β and in renucleation kinetics of the slower-growing β phase at α/γ boundaries lead to the wide range of bainite morphologies developed in different alloy systems.[25]

On the other hand, one indirect deduction made by Oblak and Hehemann and supported by Bhadeshia and coworkers is that the initial growth to a limited size of individual ferrite subunits constituting the classical sheaf of upper bainite is essentially rapid or instantaneous, thereby suggesting the displacive mechanism of upper bainite to be operative in the primary stage. This may be interpreted that individual subunits cease growth for some other reasons, possibly due to strain-energy accumulation, the accumulating defects, or a loss of coherency at the interface.[2] Atom probe studies have provided limited and uncertain support for the belief that initial growth of (bainitic) ferrite occurs without carbon diffusion.[120] This is followed by a diffusion-controlled secondary stage where a slower rate with respect to the primary stage occurs.[47,121] However, growth-kinetics measurements with thermionic emission microscopy on individual subunits have shown that the subunits can grow by D_c^γ-controlled kinetics.[13]

Ohmori et al.[21,122–124] classified upper bainite formed in low-carbon low-alloy steels as well as in plain carbon (0.04 to 0.69%) steels into three types (B-I, B-II, and B-III) according to the carbide precipitation morphology in the lath-type bainite. The B-I type upper bainite is a carbide-free bainite (α_B) lath, formed at temperatures above ~550°C. The B-II type structure consists of α_B laths with interlath cementite which forms at temperatures between 500°C and just above M_s. The B-III type forms at temperatures just above the M_s and can grow quite continuously along the ferrite lath boundaries of a specific orientation. This suggests that bainite should be classified according to ferrite morphology and not by cementite dispersion.

9.3.2 Lower Bainite

As stated earlier, two types of carbide precipitation reactions are associated with lower bainite transformation. In the first, faster reaction type, it is believed that initially the supersaturated ferrite plate forms by a diffusionless shear process and that internal precipitation of carbide within this supersaturated bainitic ferrite is a subsequent stage of reaction. This is supported by the fact that oftentimes the bainite plates do not contain carbides at the primary stage of their formation.[17] However, growth kinetics measurements on bainite do not support this mechanism.[18,116] In the second, more sluggish reaction type, carbides are precipitated between the bainitic ferrite platelets by the decomposition of the carbon-rich residual austenite into the mixture of cementite and ferrite. This reaction is, therefore, similar to the precipitation of carbides during the upper bainite transformation.[10,11]

Recently, Ohmori[125] has proposed the mechanism of lower bainite in high-carbon Fe-C alloys as follows: Initially, bainitic ferrite with some supersaturation with respect to carbon nucleates by the diffusionless shear transformation of the carbon-depleted austenite adjacent to the cementite platelets, and then fine cementite platelets nucleate at the ferrite side on the ledges of α/γ interfaces and grow epitaxially within the supersaturated ferrite during the progress of transformation. The primary stage of this mechanism is analogous to that by Bhadeshia and Edmonds[17] and by Sandvik[44] based on their TEM investigations in silicon steels; the carbide-free bainite plates have been invariably found in these steels at the primary stage of their formation.

Based on experimental observations in Fe-C-2Mn alloys, Spanos et al.[27] have proposed a diffusional mechanism for the formation of lower bainitic sheaf which includes the following sequence of events:[126] (1) initial formation of a largely carbide-free ferrite plate or spine; (2) the formation of fine (secondary) ferrite plates/crystals mainly at one broad face of the initial plate or "spine" by edge-to-face sympathetic nucleation (the broad faces of the secondary plates lie at an acute angle with respect to the longitudinal axis of the spine); (3) the precipitation of carbides along these broad faces in the usually very narrow austenite gaps (or films) between adjacent (secondary) ferrite plates/crystals, probably at α/γ boundaries; and (4) continued growth of ferrite crystals which rapidly consumes the austenite gaps/films and causes the sheaf to appear as a single monolithic plate (Fig. 9.7d).[126]

9.3.3 Effects of Alloying Elements

A comparison of the lengthening growth rates of low-alloy nickel and plain carbon steels at the same temperature (400°C) has further supported the theory that the volume diffusion of carbon in austenite plays a vital role in the growth of bainite.[33,127,128] However, Rao and Winchell[129] found that the measured growth rates of bainite plates in Fe-C-Ni (up to 10 at%) alloys were lower by a factor of 100 than the values calculated from the carbon diffusion-controlled, paraequilibrium (no partitioning) model, thereby representing an anomaly. Reynolds and Aaronson[130] have reexamined this result by utilizing the highly developed Trivedi[131] analysis of plate-lengthening kinetics and found that the ratio of calculated to measured lengthening rates decreased from 100 to less than 5 after assuming the interfacial energy of 800 mJ/m^2 for disordered γ/α boundaries and an infinite interface kinetic coefficient μ_0. They attributed the remaining discrepancy to a possible effect of Ni in increasing the interledge spacing at the edges of the bainite plates.[130]

They subsequently concluded that their result is in agreement with the current views of the existence of solute draglike effect (SDLE). It should be kept in mind that SDLE occurs when an alloying element (X) increases the activity of carbon in bulk austenite (a_c^γ) in contact with the (disordered type) areas of γ/α boundaries below the level needed for paraequilibrium $\gamma/(\alpha + \gamma)$ phase boundaries. (Here Ni increases a_c^γ.) In this situation the driving force for, and hence the kinetics of, ferrite growth decreases.[107]

9.3.4 Lower Bainite Start Temperature LB$_s$

The effect of alloying elements on the LB$_s$ temperature, particularly in high-silicon steels (containing 0.095 to 0.46% C, 1.63 to 2.13% Si, 1.99 to 2.18% Mn with or without 2.0 to 2.07% Ni, and 0.5 to 1.97% Cr) may be given by the following regression equation developed by Chang,[20] Pickering,[50] and Llopis:[132]

$$LB_s(^\circ C) = 500 - (155 \pm 40)\%C - (38 \pm 14)\%Si - (17 \pm 13)\%Mn - (4 \pm 11)\%$$
$$Ni - (10 \pm 13)\%Cr - (5 \pm 20)\%Al - (4 \pm 56)\%Co \qquad (9.3)$$

A good correlation between experimental and calculated values has been observed. Here C and Si depress the LB$_s$ temperatures more effectively. The LB$_s$ temperatures of these Si-bearing steels were found to be very close to the M_s temperatures; thus the effect of Si in depressing the LB$_s$ temperature by retarding the formation of cementite is established. This result supports the displacive formation mechanism of bainite.[20]

9.3.5 Lower Midrib Bainite LB$_m$

The formation mechanism of LB$_m$ is a two-stage process,[71] namely, (1) initial formation of TIM in the first stage at the temperature range of 200 to 150°C, which constitutes a midrib of LB$_m$, and (2) austenite decomposition to bainite at the TIM/γ interfaces, acting as nucleation sites, in the same temperature range. Hence, LB$_m$ is a composite of TIM and lower bainite. This mechanism is supported by the fact that habit planes of TIM and LB$_m$ closely match, in contrast to the habit planes of conventional LB$_s$.

9.4 ACICULAR FERRITE

9.4.1 General Characteristics and Morphology

Acicular ferrite (AF) is the most exciting recent development in wrought and welded low-alloy steel technology because it provides a relatively tough and strong microstructure. These structures form the basis of one type of high-strength low-alloy steel (see also Chap. 15). The term *acicular* means shaped and pointed like a needle; but, in reality, acicular ferrite has, in three-dimensions, the morphology of thin, lenticular plates (Fig. 9.15). The true aspect ratio of these plates, in random planar sections, is smaller than 0.1, that is, typically about 10 μm long and 1 μm wide. Most commonly AF is observed due to the transformation of austenite during the cooling of low-alloy steel arc-weld deposits at relatively low temperatures, somewhat below those for upper bainite in steels of similar composition.[10,133–137] It is also found in wrought steels which have been intentionally inoculated with nonmetallic inclusions.[138,139]

AF plates nucleate heterogeneously and intragranularly at small indigenous non-metallic oxide inclusions (of the order of 0.4 μm) in the transformation temperature range between α_w and lower bainite, and the plates radiate from these inclusions (or "point" nucleation sites) within relatively large γ grains in many different directions, giving an overall morphology of clusters of nonparallel platelets dispersed throughout the austenite. Since this microstructure is far from being organized, it is

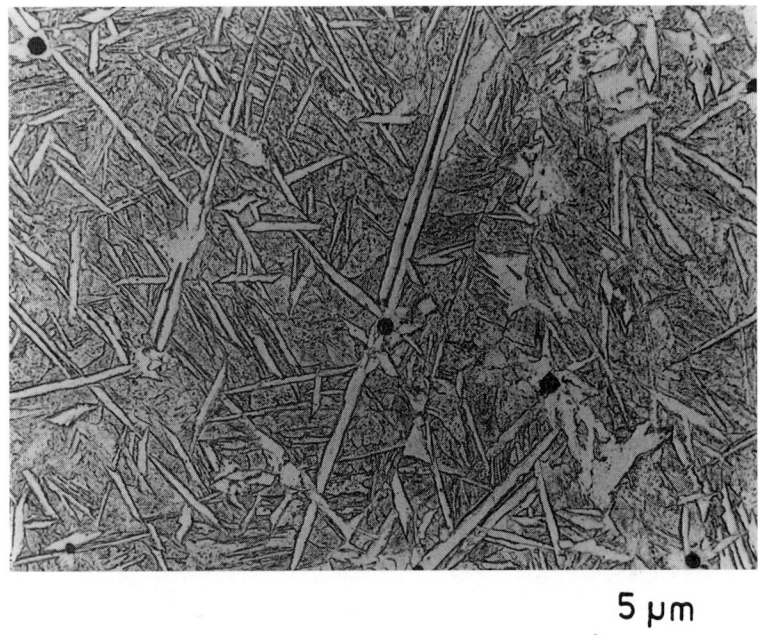

5 µm

FIGURE 9.15 Replica transmission electron micrograph of acicular ferrite plates in a steel weld deposit.[11] (*Courtesy of H. K. D. H. Bhadeshia; after Barritte.*)

better described as chaotic[11,140] or has randomness.[38] Only the distribution of inclusions seems a possible fit to this description.[38]

9.4.2 Growth (or Transformation) Mechanism

Acicular ferrite is shown to have all the essential features of bainitic transformation mechanisms: (1) AF plates exhibit the incomplete reaction phenomenon; (2) AF plates display surface relief and the K-S orientation relationship to the parent γ and the prior δ-ferrite columnar grains where they grow (thereby suggesting its occurrence either by a ledge mechanism or by a pure shear transformation[141]); (3) there is an IPS shape deformation; and so forth. Thus bainite and AF are essentially identical in transformation mechanism, except in morphology.[142] Some of the experimental results prove that AF is essentially intragranularly nucleated bainite, as summarized in Fig. 9.16.[136] When the wrought steels have small γ grain size and are practically free of nonmetallic inclusions, conventional bainite initially nucleates at the γ/γ grain surfaces and tends to grow by the repeated formation of subunits in organized packets to produce the classical sheaf morphology (Fig. 9.16a).[137] Usually AF does not grow in sheaves because of its development at higher temperatures and stifling of sheaves by hard impingement between plates nucleated freely at adjacent sites.[26,137]

In reality, both conventional bainite and AF can occur under similar isothermal transformation conditions in the same (inclusion-rich) steel. In the former

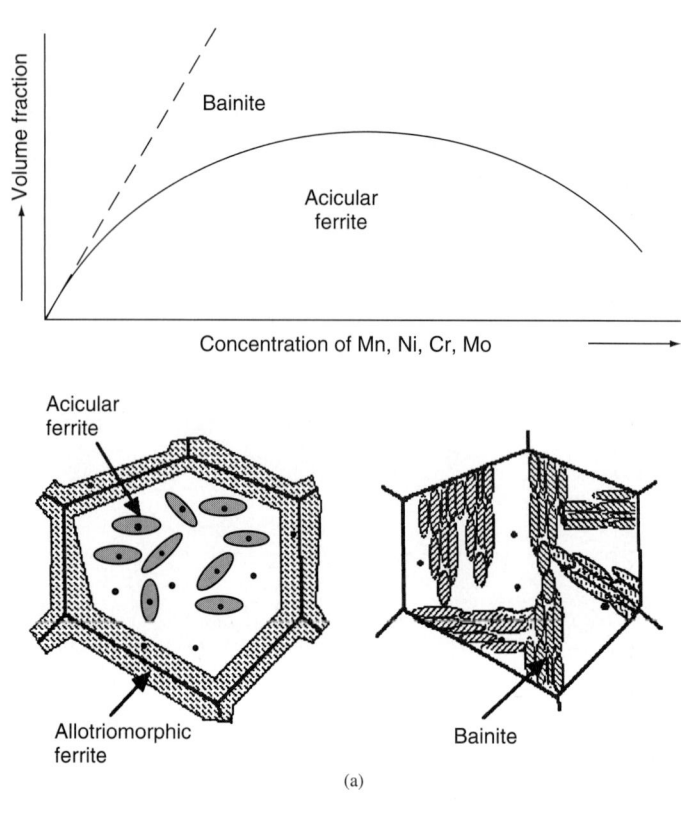

Acicular
ferrite

Allotriomorphic
ferrite

Bainite

(a)

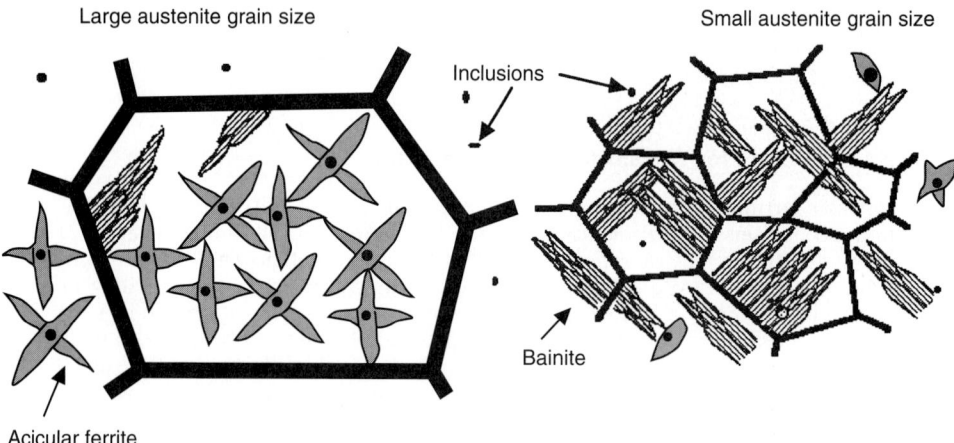

Large austenite grain size

Small austenite grain size

Inclusions

Bainite

Acicular ferrite

(b)

FIGURE 9.16 Schematic illustration of the mechanism by which: (*a*) the presence of allotriomor-phic ferrite at the austenite grain boundaries induces a transition from a bainitic to acicular ferrite microstructure and (*b*) large austenite grain size with a relatively large number density of intra-granular nucleation sites induces mostly acicular ferrite, and small grain-sized sample with large number density of grain boundary nucleation sites favors the bainitic structure.[11,64,136] (*Courtesy of H. K. D. H. Bhadeshia.*)

case, smaller γ grain size is needed to ensure dominant nucleation at the γ grain boundaries. In the latter case, larger γ grain-sized microstructure is required to ensure a large number density of intragranular nucleation sites (inclusions) to promote the formation of AF (in many directions) in preference to bainite (Fig. 9.16b).[11,64,136]

9.4.3 Effect of Allotriomorphic Ferrite

In weld deposits, AF occurs after the growth of allotriomorphic α_a and Widmanstätten ferrite α_w. Similarly, in wrought alloys containing mixed microstructure, the extent of AF decreases with decreasing γ grain size [as expressed by Eq. (9.4) for isothermal reaction] because grain boundary nucleated phases such as allotriomorphic ferrite (α_a) become more pronounced.

$$\ln(1-f) \propto S_v \qquad\qquad (9.4)$$

where f is the volume fraction of α_a divided by its equilibrium volume fraction at a particular temperature and S_v is the amount of γ grain surface per unit volume of the specimen. In some cases, Eq. (9.4) can also be applied to continuous cooling transformation in low-carbon, low-alloy steels, where $(1-f)$ is nearly equal to the volume fraction of AF, thereby relating the latter to the γ grain size. However, the dependence of the volume fraction of AF on the γ grain size is not dominant at lower cooling rates (from the γ phase field) because a larger proportion of the austenite is consumed during high-temperature formation of α_a.[38]

In the presence of relatively large solute concentrations (of Cr > 1.5% and/or Mo > 0.5% bearing steels), mixed microstructures, and nonmetallic inclusions, steel weld develops bainite instead of AF plates. This is probably due to the effect of large solute concentrations in reducing the volume fraction of α_a, which causes γ grain boundaries to become free (or less covered) to nucleate bainite[11] (or by increasing the hardenability of the steels, thereby allowing transformation to begin in the bainite region).[38] The observations of Sneider and Kerr[143] support this finding. Horri et al.[144] also found the large amount of bainite in a series of low-alloy steel welds containing increased concentrations (>1.5%) of Mn and (>2.9%) of Ni. In the latter case, however, the toughness also increased owing to the beneficial effects of Ni in solid solution on the toughness.[144] Furthermore, bainite was not observed when the α_a volume fraction was >0.08, probably because in their welds (1) that amount was adequate to entirely cover the γ grain surface, which hindered the grain boundary nucleation of bainite at a lower transformation temperature,[11] and/or (2) the cooling rate was then sufficiently slow to permit enough carbon to diffuse throughout the γ grains so that the remaining supersaturation at inclusions was insufficient to permit the nucleation of AF.

The AF formation in steels can also be enhanced by reducing the γ grain boundary nucleation sites by increasing γ grain size, by decorating the grain boundaries with thin layers of inert α_a, by increasing the inclusion content, or by rendering the grain boundaries impotent with the addition of small amounts of segregating elements such as 30 ppm of boron.[11] The boron addition causes nucleation of high-dislocation-density acicular ferrite at isolated boron-containing intermetallic compound particles at the γ grain boundaries, i.e., the boron hardenability effect.[38,145]

9.4.4 Nucleation of Acicular Ferrite and Role of Inclusions

It has been shown, assuming classical nucleation theory, that inclusions are less effective than the γ grain boundaries in nucleating ferrite.[64] Experiments have established the fact that ferrite formation initiates at the γ grain boundaries, and the larger inclusions are more effective. The most favorable size of inclusions was observed to be in the range of 0.3 to 0.9 μm and 0.35 to 0.75 μm for the high-oxygen-content and the lower-oxygen-level weld metals, respectively. A high [Al]/[O] ratio depresses the formation of AF by increasing the solute matrix Al content (which promotes sideplate formation).[146] Since Al increases the paraequilibrium $A_{e_{3,1}}$, the undercooling at a given temperature during the cooling process will be larger and the kinetics of ferrite formation will be more rapid.[147] The greater undercooling will also encourage sideplate formation.[148] Although the diffusion field associated with the edge of a ferrite plate is only the order of the tip radius,[149] the combination of the high sideplate lengthening rates and the thickening of the plate broad faces will expedite the diffusion of carbon through the remaining austenite.[38]

9.4.4.1 Lattice Matching Theory.
By applying the same principle of using inclusions to control the solidification of aluminum alloys to the solid-state transformation in steels, it is inferred that those inclusions which have the best crystallographic lattice matching with ferrite nuclei are most potent in providing preferential sites for the nucleation of ferrite.[64] The lattice matching is expressed as an average percentage planar misfit κ. To calculate κ, let us assume that (1) the inclusion is faceted on a plane $(hkl)_I$ and (2) the ferrite forms epitaxially with its plane $(hkl)_\alpha \, // \, (hkl)_I$ and the corresponding rational low-index directions $[uvw]_I$ and $[uvw]_\alpha$ make the angle ϕ with respect to each other. The interatomic spacings d along these three directions ($j = 1, 2, 3$) within the plane of epitaxy have been analyzed by Bramfitt[150] to give

$$\kappa = \frac{100}{3} \sum_{j=1}^{3} \frac{|d_j^I \cos\phi - d_j^\alpha|}{d_j^\alpha} \tag{9.5}$$

where d_j^I and d_j^α are the interatomic spacings of the inclusion and ferrite lattice, respectively, along each low-index direction j and ϕ is the angle between the low-index directions. Table 9.2 lists the calculated data for several inclusion phases. To

TABLE 9.2 Some Misfit Values between Different Substrates and Ferrite

The data are taken from a more detailed set published by Mills et al.[†] and include all cases where the misfit is observed to be less than 5%. The inclusions all have a cubic-F lattice, and the ferrite is body-centered cubic (cubic-I).[11,64]

Inclusion	Orientation	Plane of epitaxy	Misfit (%)
TiO	Bain	{100}	3.0
TiN	Bain	{100}	4.6
γ-Al$_2$O$_3$	Bain	{100}	3.2
Galaxite	Bain	{100}	1.8
CuS	Cube	{111}	2.8

[†] After A. R. Mills, G. Thewlis, and J. A. Whiteman, *Mater. Sc. and Technol.*, vol. 3, 1987, pp. 1051–1061.

TABLE 9.3 List of Ceramics Which Have Been Tested for Their Potency in Stimulating the Nucleation of Ferrite Plates[64]

Effective: Oxygen sources	Effective: Other mechanisms	Ineffective
TiO_2, SnO_2	Ti_2O_3	TiN, $CaTiO_3$
MnO_2, PbO_2	TiO	$SrTiO_3$, α-Al_2O_3
KNO_3		NbC

Source: *After Gregg.*

compare the lattice matching concept with experiments, it is essential to obtain both the right orientation relationship and the faceting of the inclusion on the appropriate epitaxy plane. Many compounds such as some of the titanium oxides exhibit good matching with ferrite and do tend to be very potent in nucleating ferrite. Also TiN seems to be effective in this respect, but is less stable at high temperatures when compared to Ti_2O_3.[11] On the other hand, some other compounds, such as γ-Al_2O_3, in spite of showing good fit, are impotent nucleants. Perhaps, several mechanisms are operative in making a nonmetallic phase an effective heterogeneous nucleation site.[64]

9.4.4.2 Other Possibilities. Other ways in which inclusions may prove to be beneficial in the formation of AF include stimulation by thermal strains, chemical heterogeneities in the neighborhood of the inclusion/matrix interface, acting as inert sites for heterogeneous nucleation, and/or chemical reactions at the inclusion/matrix interface (Table 9.3). Those minerals such as TiO_2 which are natural oxygen sources are observed to be powerful stimulants for nucleation, presumably by oxidizing and enhancing the decarburization of the adjacent steel at elevated temperatures. This effect does not appear to be dependent on the crystallographic nature of the mineral, except in the ability of the mineral to tolerate oxygen vacancy defects, or to decompose thermally. Ti_2O_3 appears to act as a sink for Mn in that it causes a large reduction in the Mn levels in the adjacent steel which, in turn, accelerates nucleation because Mn is an austenite stabilizer and is known to retard transformation of steel to ferrite[141] (Table 9.3). TiO is mystifying because it is an effective nucleant or stimulant; yet it does not produce any significant modification of the adjacent steel, and like TiN, it has good lattice matching with ferrite.[64]

9.5 BAINITIC STEELS

Bainitic steels usually display continuous yielding behavior with YS/UTS ratios <0.8. This gradual yielding arises from mobile dislocations, the presence of heterogeneities in the microstructure, and residual stresses due to transformation. Higher carbon contents lead to a reduction in ductility and toughness, mainly due to the void nucleating tendency of coarse cementite particles. The impact toughness of upper bainite decreases with increasing strength whereas that of lower bainite (associated with much finer carbides) is superior at the same strength level. There has been considerable experimental evidence that suggests that mixed microstructure of bainite and martensite can, in some cases, provide higher strength and toughness than fully martensitic alloys. The mechanism underlying this has not yet been established.[11] Figure 9.17 summarizes the range of currently used bainitic alloys.[11]

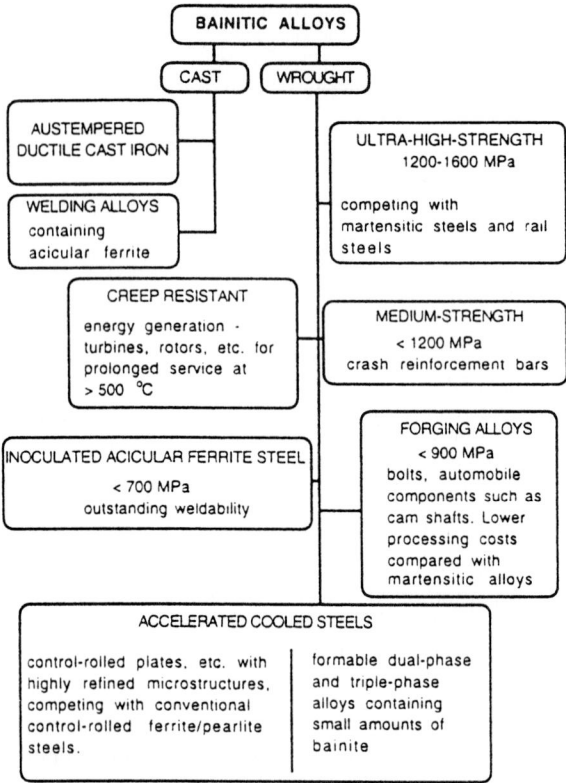

FIGURE 9.17 Bainitic alloys currently available on a commercial basis.[11,170] (*Courtesy of H. K. D. H. Bhadeshia.*)

Commercial bainitic steels usually possess 0.1 to 0.4% C, 0.3 to 0.9% Mn, 0.2 to 0.3% Si, and alloying elements such as Ni, Cr, Mo, B, and V. A TTT diagram for a bainitic steel is shown schematically in Fig. 9.18, which illustrates that bainite can form over a large range of cooling rates with relatively small change in transformation temperature. Table 9.4 lists typical compositions of recent commercial and experimental bainitic steels. Although the steels listed seem to be similar, they possess quite different mechanical properties due to the important role of trace element additions and differences in processing routes.[140] Small additions of Ti and Al are usually made to react with any free N because the latter can take the B out of solution, thereby severely restricting its hardenability effect. For more details on bainitic steels, readers should refer to the other recent source.[11]

9.5.1 Ultralow-Carbon and Low-Carbon Bainitic Steels

It has been made clear by classical experiments of Irvine and Pickering that good mechanical properties (with respect to martensitic steels) can be developed in bainitic steels containing low carbon concentrations. Otherwise, the coarse cementite particles and considerably large volume of untempered martensite (due to

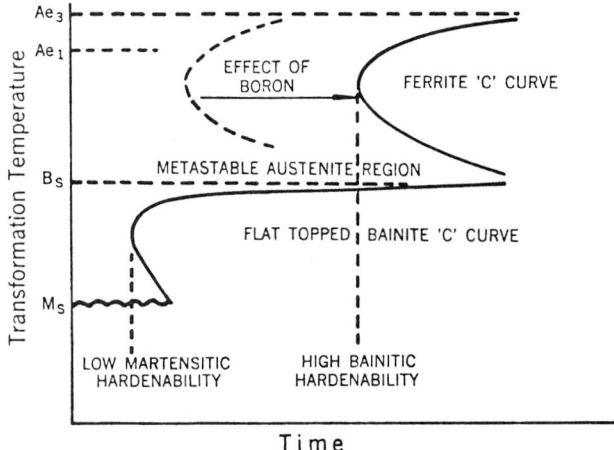

FIGURE 9.18 Schematic TTT diagram showing transformation requirements for low-carbon $^1/_2$Mo bainitic steel.[50] (*Courtesy of Climax Molybdenum Co., Ann Arbor, Michigan.*)

incomplete transformation to bainitic ferrite) will form which has a very adverse influence on the toughness of the steel. Very large concentrations of substitutional alloying additions can have detrimental effects as well, provided they hinder the possible extent of bainitic transformation. On the other hand, very low alloy concentration reduces the hardenability of the steel, making the production of bainitic microstructures more difficult.

Ultralow-carbon bainitic (ULCB) steels (see Table 9.4), with carbon concentrations of 0.01 to 0.03%, possess extremely good toughness, moderate strength (500 to 620 MPa), and improved weldability; they are used in high-strength line pipe in Arctic or submarine environments. The highest strength values are achieved by adopting low finishing temperatures and accelerated cooling through the transformation temperature range (see also Chap. 15). An additional strengthening can be achieved by retarding cooling below about 550°C (by coiling or stacking the hot product) to enhance the precipitation of NbC in the ferrite. The size of TiN particles which form in the melt should be sufficiently small; this can be achieved by controlling the cooling rates during the solidification process. Sulfide shape control using rare-earth or Ca additions is also a standard steelmaking practice, and the overall sulfur concentration must be kept below about 10 ppm for pipeline-grade steels.[11] Thus ULCB steels are similar to Irvine and Pickering alloys; they contain B, Mo, or Cr to improve bainite hardenability and Ti to getter N.

9.5.1.1 Strength. Hot-rolled plates of ULCB steels can be made possible with yield strength in the range of 450 to 900 MPa (which is much above that of ferrite-pearlite structures) by doing the following:

1. Decreasing the carbon content to about 0.1%. Although C is the most effective strengthening element in steel, it has an adverse effect on weldability, because the region of the heat-affected zone (HAZ) of welds then develops unacceptably high hardness levels immediately after welding. Improvement in weldability arising from reduction in carbon content can, therefore, result in a major

TABLE 9.4 Typical Compositions of Advanced Bainitic Steels (wt%)[11]

No.	C	Si	Mn	Ni	Mo	Cr	Nb	Ti	B	Al	N	Others	Steel type
1	0.100	0.25	0.50	—	0.55	—	—	—	0.003	—	—	—	Early bainitic
2	0.039	0.20	1.55	0.20	—	—	0.042	0.015	0.0013	0.024	0.003	—	Rapidly cooled bainitic
3	0.081	0.025	1.86	0.20	0.09	—	0.045	0.016	—	0.025	0.0028	—	Rapidly cooled bainitic
4	0.110	0.34	1.51	—	—	—	0.029	—	—	—	—	—	Rapidly cooled bainitic
5	0.100	0.25	<1.00	—	—	—	0.029	—	—	—	—	—	Bainitic dual phase
6	0.04	—	0.40	—	—	—	—	—	—	0.05	—	—	Bainitic dual-phase steel for building
7	0.150	0.35	1.40	—	—	—	0.022	0.011	—	0.035	—	—	Triple phase
8	0.02	0.20	2.00	0.30	0.30	—	0.050	0.020	0.0010	—	0.0025	—	ULCB
9	0.028	0.25	1.75	0.20	—	0.30	0.100	0.015	—	0.030	0.0035	Cu 0.3 Ca 0.004	ULCB
10	0.08	0.20	1.40	—	—	—	—	0.012	—	0.002	0.0020	—	Acicular ferritic TiO_x (0.0017 O)
11	0.08	0.20	1.40	—	—	—	—	0.008	0.0015	0.038	0.0028	—	Acicular ferritic, TiB
12	0.08	0.20	1.40	—	—	—	—	0.019	—	0.018	0.0050	—	Acicular ferritic, TiN
13	0.15	0.80	1.40	—	0.20	—	0.10	—	—	—	—	V 0.15	Forging steel (HS[†])
14	0.09	0.25	1.00	0.50	1.00	—	0.10	0.02	0.002	0.04	0.06	—	Forging steel, 100% bainite
15	0.09	0.40	1.40	—	—	—	0.07	—	—	0.04	0.010	V0.06	Forging steel (Nb+V)
16	0.09	0.25	1.40	—	—	—	0.07	0.02	0.002	0.04	0.006	—	Forging steel (Nb+V)
17	0.12	—	1.60	—	—	—	0.08	—	0.004	—	—	—	Cold-heading

[†] High-strength.

9.32

cost advantage in fabrication, particularly if the welding can be accomplished without any preheat.

2. Retarding the allotriomorphic ferrite α_a formation, thereby increasing the bainite reaction.

3. Adding elements such as Mo (~0.5%) and B (~0.02%) (to ensure that the hardenability with respect to bainitic transformation was good whereas that with respect to martensitic transformation was poor).

4. Controlling the grain size with Nb.[50]

The main purpose is to allow the formation of bainite over a wide range of cooling rates, as indicated in Fig. 9.18. Further control of the reaction is achieved by adding Ni, Mn, and Cr, which depress the temperature of the maximum rate of transformation to bainite.[50,152]

For a constant cooling rate, the strength of the steel can be increased significantly by lowering the B_s temperature [expressed by Eq. (9.1)], which can be achieved by the addition of carbon and other alloying elements. The B_s temperature is related to the temperature for 50% bainite transformation, denoted by B_{50}, and that for complete bainite transformation B_f by[153]

$$B_{50}(°C) = B_s - 60°C \tag{9.6}$$

$$B_f(°C) = B_s - 120°C \tag{9.7}$$

In this manner, in 0.5% Mo-B base steel, we can produce, after alloying, tensile strengths in the range of 530 to 1200 MPa and yield strength between 450 and 980 MPa. These strength levels can be achieved by austenitizing at 900 to 950°C and subsequent air cooling for a wide range of cooling rates or section thickness, without any appreciable variation.[154]

In the transformation temperature range of 650 to 450°C and for the carbon content between 0.05 and 0.20%, the strength of the bainitic structure has been found to increase linearly with the decrease of 50% transformation temperature (Fig. 9.19).[50,154] Note here that upper bainite structure changes to lower bainite at about 550°C without any visible discontinuity in strength.

An empirical linear relationship between the tensile strength σ_T and composition has been given by[154]

$$\sigma_T(\text{MPa}) = 15.4[16 + 125(\%C) + 15(\%Mn + \%Cr) + 12(\%Mo) + 6(\%W) \\ + 8(\%Ni) + 4(\%Cu) + 25(\%V + \%Ti)] \tag{9.8}$$

Depending upon the steel chemistry and transformation temperature, bainitic steels can be produced over a wide range of strength levels. The selection of the alloying elements and transformation temperature is usually made depending upon the cost (Mn and Cr being inexpensive), desired strength level, toughness, formability, and weldability. The last three factors are encountered when steels with low carbon concentration are chosen.

Strengthening Mechanisms. The operative strengthening mechanisms in low-carbon bainitic steels are

1. A fine bainitic ferrite grain (lath) size due to low transformation temperature which gives a Hall-Petch relationship with yield stress (Fig. 9.20)[155]

2. Increased dislocation density due to increased transformation strains associated with the decrease in transformation temperature and formation of dense carbides

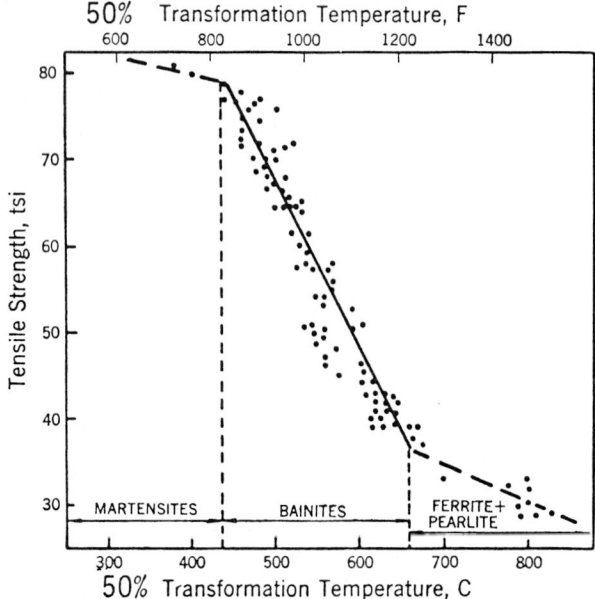

FIGURE 9.19 Effect of temperature for 50% transformation on tensile strength of low (0.05 to 0.15%) -carbon bainitic steels.[50] (*Courtesy of Climax Molybdenum Co., Ann Arbor, Michigan.*)

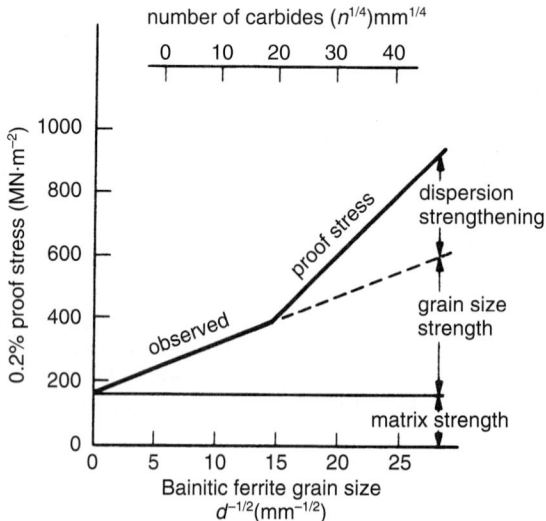

FIGURE 9.20 The contributions of the bainitic ferrite grain size and carbide dispersion strengthening to the 0.2% proof stress of low (0.05 to 0.15%) -carbon bainitic steels.[156] (*Courtesy of F. B. Pickering.*)

3. Increased carbide dispersion strengthening with the increasing carbon content and decreasing transformation temperature (Fig. 9.20) (this dispersion is denoted by the number of carbide particles per unit planar surface area)

4. Increased interstitial (carbon) and substitutional (such as Mn, Si, and Mo) solid-solution hardening of bainitic ferrite with decreasing transformation temperature

5. Strengthening by interaction of solute (carbon) atoms with dislocations

The interrelation between yield strength σ_y (0.2%) of bainite, bainitic ferrite lath size d (mm), and the number of carbide particles per square millimeter on a planar section of the structure is given by[155]

$$\sigma_y(0.2\%)(\text{MPa}) = -191 + 17.2d^{-1/2} + 14.9n^{1/4} \tag{9.9}$$

This empirical equation shows the negative constant, suggesting a threshold carbide distribution above which carbides are effective contributors to the strength; i.e., this equation applies to finely dispersed carbides, and the contribution due to carbide dispersion becomes significant provided the carbide spacing is less than the bainitic ferrite lath size. Thus in low-carbon upper bainite, bainitic ferrite lath size is the main contributor to strengthening because the carbides at the ferrite lath boundaries are not the strength contributors because their spacing is larger than that of the ferrite lath size. On the other hand, in lower bainite, carbide dispersion contributes greatly to the strengthening as a result of its occurrence within ferrite laths. Figure 9.20 shows the significant carbide strengthening only at the finer lath sizes, i.e., at lower transformation temperatures.

Other regression analyses include prior austenite grain size, dislocation density, and carbide interparticle spacing but exclude bainitic ferrite grain size because of a much smaller value of carbide interparticle spacing. This is expressed by[156]

$$\sigma_y(0.2\%)(\text{MPa}) = 9.8(60 + 1.25\mu\mathbf{b}n^{1/2} + 1.2 \times 10^{-4}\sqrt{\rho}) \tag{9.10}$$

where n is the number of carbides per square millimeter intersecting a planar section, μ is the shear modulus, $\mathbf{b}$ is the Burgers vector, and ρ is the dislocation density (lines/cm^2). This equation incorporates a constant term for austenite grain size which does not include the dislocation density, nor does it follow the $d^{-1/2}$ relationship. The third recent analysis is given by[157]

$$\sigma_y(0.2\%)(\text{MPa}) = 650 + 12.3 \times 10^{-2}w^{-1} + 41 \times 10^{-3}\lambda^{-1} \tag{9.11}$$

where w is the mean bainite plate width and λ is the mean planar interparticle carbide spacing, both in millimeters. A more recent yield strength equation for low-carbon bainite has included (1) the effect of common alloying elements present (i.e., solid-solution strengthening terms), (2) the effect of bainitic ferrite grain size d, and (3) precipitation and dislocation strengthening terms (that is, σ_p and σ_d, respectively):[155,158]

$$\sigma_y(0.2\%)(\text{MPa}) = 88 + 37(\%\text{Mn}) + 83(\%\text{Si}) + 2900(\%\text{N}_{\text{free}}) + 15.1d_L^{-1/2} + \sigma_d + \sigma_p \tag{9.12}$$

where d_L is the mean linear intercept of the bainitic ferrite lath size in millimeters; σ_d (the dislocation strengthening term) $= \alpha\mu\mathbf{b}\sqrt{\rho}$, where μ is the shear modulus, $\mathbf{b}$ is the Burgers vector of the dislocations, and α is a constant; and σ_p is the Ashby-

Orowan strengthening term from dispersion of carbide particles present only within the ferrite laths. The equation determined for σ_p is

$$\sigma_p = \left(\frac{5.9 f^{1/2}}{\bar{x}}\right) \ln\left(\frac{\bar{x}}{2.5 \times 10^{-4}}\right) \tag{9.13}$$

where f is the volume fraction and $\bar{x}$ is the mean planar intercept diameter of the precipitate.

9.5.1.2 Hardness. For fully bainitic microstructures, the hardness increases approximately linearly with carbon content, by about 190 HV per weight percent.[39] This is in contrast to a change of about 950 HV per weight percent for untempered martensite. The austenitizing temperature has no effect on the hardness unless it is not high enough to dissolve all the carbides.[39] The hardness of bainite is found to be independent of the prior γ grain size (i.e., insensitive to the austenitizing temperature), even though the latter has bearing on the bainite sheaf thickness.[159]

9.5.1.3 Ductility. The tensile ductility (as measured by elongation) of fully bainitic, low-carbon steel is observed to be superior to that of the quenched and tempered martensitic steels of equal strength,[39] but the situation reverses at high carbon concentrations. The ductility as measured by reduction of area, on the other hand, becomes always worse for bainitic steels compared to that of martensitic steels.[11]

9.5.1.4 Impact Toughness. In general, the transition temperature increases with increasing strength for both upper and lower bainite, and it is found that lower bainite has a lower impact (or ductile brittle) transition temperature (ITT or DBTT) and upper bainite has a higher ITT, as shown in Fig. 9.21; this is mainly because of the much greater transition temperature range in the lower bainite than in the upper bainite. This can be explained as follows.

Lower bainite exhibits a finer ferrite plate size and a higher dislocation density than upper bainite. The fine carbides within ferrite plates impede the cleavage-crack propagation because of their different orientations and high-angle ferrite boundaries; therefore, fracture toughness is raised (Fig. 9.21).[153] In upper bainite, easy cleavage-crack initiation occurs due to the cracking of the large or coarse interlath carbides and the inability of the low-angle bainitic ferrite lath boundaries to hinder cleavage-crack propagation;[158] hence ITT is raised, and a lower toughness value[155] is obtained.

A recent analysis of the factors influencing ITT of bainite or acicular ferrite has been given as follows:[155]

$$\begin{aligned}
\text{ITT(or DBTT)} = &-19 + 44(\%Si) + 700(\%N_{\text{free}})^{1/2} \\
&+ 0.26(\sigma_p + \sigma_d + \sigma_b) - 11.5 d^{-1/2}
\end{aligned} \tag{9.14}$$

where σ_p is the precipitation strengthening term given by Eq. (9.12); σ_d and σ_b, respectively, are the dislocation strengthening terms due to random forest dislocations and dislocations at low-angle bainitic ferrite lath boundaries; and d is the mean spacing between high-angle boundaries. However, in very low-carbon low-alloy steels, thermomechanical treatment prior to its transformation to upper bainite produces a considerable improvement in toughness level as a result of substantial reduction of average grain (or packet) size, thereby increasing the resistance to brittle crack propagation.[160] Low-temperature tempering of bainitic structures lowers the ITT owing to the decrease in σ_p and σ_d. However, at high tempering tem-

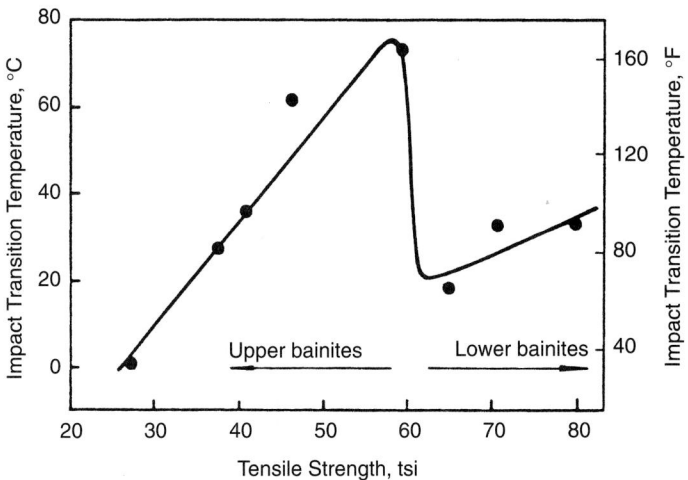

FIGURE 9.21   Effect of tensile strength on the impact (or ductile-brittle) transition temperature of low (0.05 to 0.15%) -carbon bainitic steels.[50] (*Courtesy of Climax Molybdenum Co., Ann Arbor, Michigan.*)

perature, ITT, in spite of decreased strength, increases as a result of the coarsening of ferrite grains.[155]

The main advantages of low-carbon and ultralow-carbon bainitic steels are that relatively high strength associated with improved ductility can be obtained without further heat treatment.

9.5.1.5 Fracture Toughness. Figure 9.22 shows a comparison of the tensile strength-fracture toughness relationship for various microstructures produced by heat treatment in commercially available AISI 4340 steel.[151] Typical properties of standard (300 HB minimum) and head-hardened (352 HB minimum) rail are superimposed on Tomita's data.[151,161] It is clear that a mixed microstructure of lower bainite and tempered lath martensites possesses the best strength-toughness combination, followed by tempered lath martensite, lower bainite, and head-hardened pearlitic rail. Standard rail with a coarse pearlite microstructure and AISI 4340 steel with upper bainite (either singly or together with tempered martensites) have the poorest strength-toughness combination.[151]

9.5.1.6 Wear Resistance. Several laboratory studies have been made to compare the wear properties of bainitic and fully pearlitic steels. According to Clayton and Jin,[162,163] low-(0.04–0.26%) -carbon bainitic steels containing Mn, Si, Cr, and Mo alloying elements have wear resistance comparable to that of fully pearlitic steels. Clayton observed that a bainitic structure comprising carbide-free bainite (mixed with lath martensite) had enhanced wear resistance over a fully pearlitic rail steel.[151]

9.5.1.7 Applications. Bainitic steels are usually produced via thermomechanical treatment, followed by various continuous cooling processes. Bainitic steels exhibit a favorable balance of strength, toughness, and elevated-temperature performance and a greater degree of uniformity in thick sections, and are often chosen

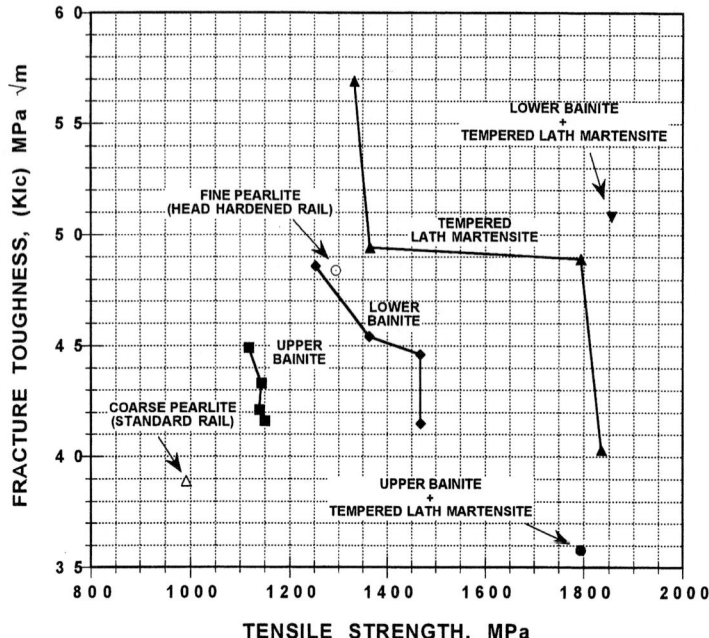

FIGURE 9.22   Fracture toughness versus tensile strength plot for AISI 4340 steel heat-treated to produce different microstructures.[103] The data points for standard and head-hardened rail are from Ref. 161.

for high-temperature applications due to their improved creep properties compared to other microstructures.[15,164]

Typical applications include Ni-Cr-Mo (for example, HY-80) and Cr-Mo steel plate and heavy forgings for nuclear reactors, and pressure vessel components; low-carbon Cu-Ni-Cr-Mo (for example, ASTM A710 grade) steel plate for ship construction; high-toughness ultralow-carbon bainitic (Mo-B) steels for use in the oil and gas industries;[165] Ni-Cr-Mo-V, Ni-Mo-V, and Cr-Mo-V heavy forgings for steam-turbine rotors; boilers; cranes; lifting equipment; structural members of light-weight bridges; mine supports; earthmoving vehicles; strong structural components in aircraft engineering; engine mounts;[154] and forged components (such as crank-shafts, connecting rods, piston shafts, bolts, axles, fasteners, etc.).[11,15,133] Other examples include Si-rich high-strength, directly transformed steels as crash reinforcement bars in the automobile industry.[11]

9.5.2   Bainitic Forging Steels

Currently modern forging steels (see also Table 9.4) are thermomechanically treated medium-carbon microalloyed steels with vanadium to manufacture highly refined ferrite-pearlite microstructure or are alloyed to produce fully bainitic, or mostly bainitic, microstructures. The main advantage in using direct-cooled forging steels having these microstructures is their production by direct transformation during continuous or controlled cooling from the forging temperature with the elimination or reduction of the number of processing steps associated with the production of

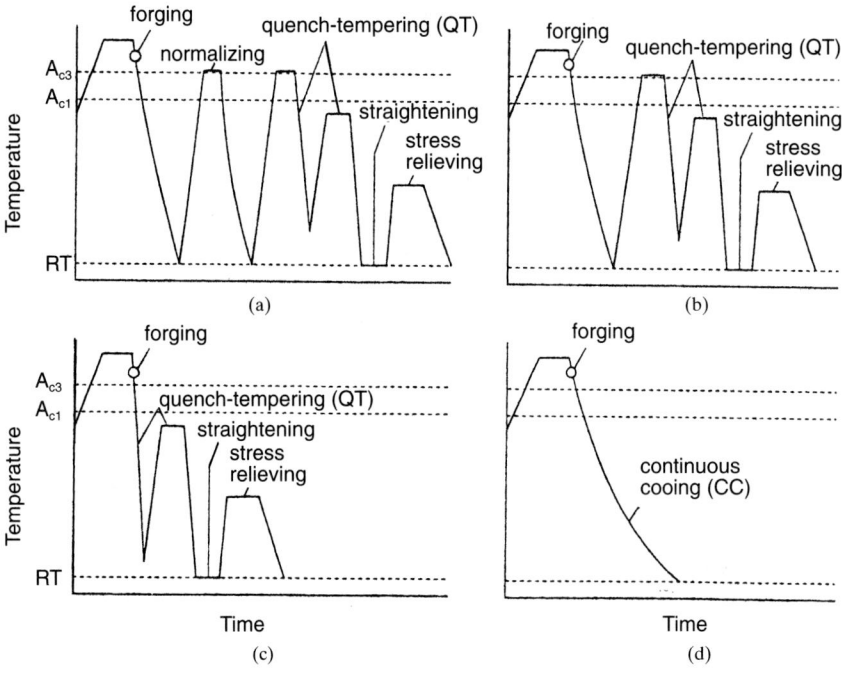

FIGURE 9.23 Various illustrations of heat treatment cycles for the manufacture of crankshafts by drop forging. (*a–c*) Conventional quenched and tempered martensitic steels requiring expensive straightening and stress-relieving cycles. (*d*) Controlled cooling of a microalloyed steel as a replacement for expensive quench and temper operations.[166a]

quenched and tempered microstructures. The composition for the loss of strength related to the reduction of carbon level, from 0.49% in the 1970s to 0.27% nowadays, required an increased amount of Mn and Si. A small amount of Ti provides an adequate austenite grain size control. This, in turn, provides savings in heat treatment, handling, and other fabrication costs (Fig. 9.23);[166,166a] strength level in the range of 500 to 750 MPa without losing toughness; improvements in weldability, machinability, and fatigue properties; and the possibility of controlled forging to refine the γ grain size prior to transformation (see also Chap. 15 for more applications of microalloyed forging steels).[11]

9.5.3 Carbide-Free, High-Strength Bainitic Steels

Typical compositions of experimental high-strength steels with good toughness include Fe-0.22C-2.0Si-3.0Mn and Fe-0.40C-4.0Ni. Figure 9.24 shows the comparison of the mechanical properties of these high-strength steels with quenched and tempered steels. It is clear that, in some instances, the properties are equivalent to those obtained from much more expensive maraging steels.[167] The mechanical properties data on these high-silicon steels look very promising. It is unlikely that the experimental steels correspond to the optimum compositions, and further development work could be required to achieve a further enhancement in properties.[11]

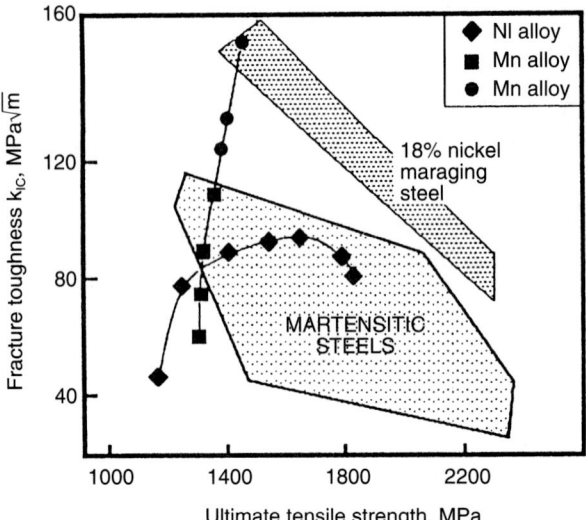

FIGURE 9.24 Comparison of mechanical properties of mixed microstructures of bainitic ferrite and austenite, versus those of quenched and tempered martensitic alloys.[167] (*Courtesy of V. T. T. Miihinen and D. V. Edmonds.*)

9.5.4 Bainitic Rail Steels

Low- to medium-carbon bainitic steels are emerging as viable candidates to replace conventional pearlitic (eutectoid) rail steels,[151] particularly in very severe loading environment. Their use becomes more widespread as wheel loads creep upward. For example, carbide-free molybdenum-boron, low-carbon bainitic steels have been used successfully in railway crossings and turnouts where impact wear ("batter") and fatigue of the crossing nose and turnouts were the severe wear problems with the traditional pearlitic rail steels.[162,168,169] Important advantages of using bainitic steels include improved weldability, comparable or superior wear resistance [for higher (>300 to 320 HV) hardness range], and much higher toughness and hardness (of 350 to 375 HB)[170] when compared to the traditional pearlitic steels. Table 9.5 lists the chemical composition of bainitic rail steels.

Clayton et al.[171] have observed empirically that Cr addition gives rise to improvements in fatigue strength, wear resistance, ductility, and toughness in bainitic rail steels. Figure 9.25 shows a comparison of wear rates and hardness range of 220 to 400 HB of conventional pearlitic rail steels versus the new carbide-free duplex microstructure of bainitic ferrite and retained austenite.[173] A typical carbide-free, low-carbon bainitic steel, coded J6 (with composition shown in Table 9.5), was found to have excellent wear resistance. This rail is being tested in track at the Facility for Accelerated Service Testing (FAST) in Pueblo, Colorado.[172]

Recently, researchers at British Steel[173] and Bhadeshia[11] have shown that the carbide-free bainite developed in C-Mn-Si steel can make a tough, wear-resistant rail steel. They reported 0.23C-1.30Mn-0.40Si steel (no. 17 in Table 9.5) to have a mixed microstructure of bainitic ferrite laths and retained austenite at the lath boundaries, a hardness of 400 HV, and better wear resistance than conventional rail

TABLE 9.5 Typical Chemical Compositions of Bainitic Rail Steels (with and without Hardness)[11,164]

No.	C	Si	Mn	Ni	Mo	Cr	Nb	Ti	B	Al	N	Steel grade[11]
1†	0.04	0.20	0.75	2.0	0.25	0.28	—	0.03	<0.01	0.03	—	Best available bainitic rail steel
2†	0.09	0.20	1.00	—	0.50	—	—	0.03	0.003	0.03	—	Experimental bainitic rail steel
3†	0.07	0.30	4.50	—	0.50	—	0.01	—	—	—	—	Experimental bainitic rail steel
4†	0.10	0.30	0.60	4.0	0.60	1.7	—	0.03	<0.01	0.03	—	Experimental bainitic rail steel
5†	0.30	0.20	2.00	—	0.50	1.0	—	0.03	0.003	0.03	—	Experimental bainitic rail steel
6†	0.30	1.00	0.70	—	0.20	2.7	0.1	—	—	—	—	Experimental bainitic rail steel
7†	0.52	0.25	0.35	1.5	0.25	1.7	—	—	<0.01	—	—	Experimental bainitic rail steel

No.	C	Si	Mn	Ni	Mo	Cr	Nb	Ti	B	Al	Hardness RB	Steel grade[179]
8‡	0.18	1.13	2.01	—	0.48	1.94	—	0.03	0.003	0.03	342	Experimental bainitic rail steel
9‡	0.12	0.27	3.97	—	0.47	0.02	—	0.04	0.003	0.03	327	Experimental bainitic rail steel
10‡	0.02	0.27	2.02	1.93	0.48	1.96	—	0.02	0.003	0.03	282	Experimental bainitic rail steel
11‡	0.26	1.81	2.00	—	0.49	1.93	—	0.04	0.003	0.04	367	Experimental bainitic rail steel

No.	C	Si	Mn	Ni	Mo	Cr	Nb	Ti	B	Al	Others	Steel producer[151]
12§	0.25	1.80	2.00	—	0.45	1.95	—	0.035	0.003	—	—	Association of American Railroads
13§	0.40	0.70	1.50	—	0.80	1.10	—	—	—	—	—	Thyssen Stahl
14§	0.07	—	4.50	—	0.50	—	—	—	—	—	V 0.10	Krupp
15§	0.30	—	—	—	0.20	2.70	—	—	—	—	—	Krupp

No.	C	Si	Mn	Ni	Mo	Cr	Nb	Ti	B	Al	Others	Steel producer
16§	0.50	1.50	2.00	—	—	—	—	0.022	0.002	—	V 0.04	British Steel
17§	0.23	0.40	1.30	—	0.30	0.30	—	—	—	—	—	British Steel
18§	0.15	0.51	1.41	3.89	0.41	0.95	—	—	—	—	—	Nippon Steel
19§	0.25	0.15	0.31	2.41	—	2.98	—	—	—	—	—	Nippon Steel
20§	0.28	0.36	1.21	—	—	1.65	0.04	—	—	—	V 0.08	Nippon Steel
21§	0.29	0.35	1.16	—	—	2.21	—	—	—	—	—	Nippon Steel
22§	0.31	0.31	1.31	—	0.26	1.32	—	—	—	—	—	Nippon Steel
23§	0.32	0.29	0.41	—	0.59	2.81	—	—	0.0015	—	—	Nippon Steel
24§	0.34	0.32	0.76	—	—	2.51	—	—	—	—	—	Nippon Steel
25§	0.35	1.98	0.74	—	—	2.41	—	0.032	—	—	—	Nippon Steel
26§	0.38	0.51	1.99	—	—	0.51	—	—	—	—	Cu 0.11	Nippon Steel
27§	0.45	0.31	0.64	—	—	2.21	—	—	—	—	—	Nippon Steel

† From Ref. 11.
‡ From Ref. 179.
§ From Ref. 151.

9.41

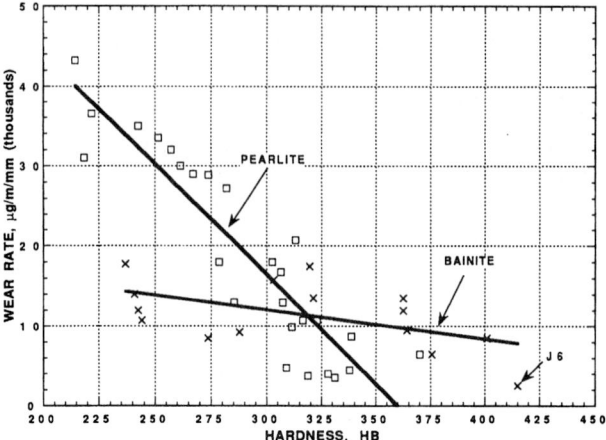

FIGURE 9.25 Wear rate versus hardness data for pearlitic (□) and bainitic (×) rail steels.[163] One steel, coded J6, is found to have excellent wear resistance. (*Courtesy of P. Clayton and N. Jin.*)

steels. Usually, carbide-free bainite is produced in low-carbon steels; however, increased Si content in the C-Mn-Si steel impedes the carbide formation at the lath boundaries, even with 0.4% C (no. 16 in Table 9.5). The rails having head hardnesses of 440 to 460 HB have been produced using an in-line head-hardening process with water sprays. The rails are "autotempered" during exit of the head-hardening unit in which the interior heat facilitates the tempering of the outer head surface. The characteristic features of the medium-carbon lath martensite rails are (1) wear resistance reasonably less than that of a pearlitic head-hardened rail and (2) higher fracture toughness [of 110 MPa $\sqrt{m}$ (100 ksi $\sqrt{in.}$)] than the toughness of [50 to 60 MPa $\sqrt{m}$ (45 to 55 ksi $\sqrt{in.}$)] fully pearlitic head-hardened rail. This is an outstanding attribute of low- to medium-carbon lath martensite toward improvement in rail properties.[151]

Nippon steel is developing bainitic rail steels (Table 9.5), using a forced-air, in-line head-hardening process[174] where the hardness range of 350 to 450 HV (331 to 425 HB) is achieved. NKK steel is also actively interested in the evaluation of rail steels.[175] Proprietary work is being done at Pennsylvania Steel Technologies (PST), where bainitic and tempered lath martensitic rails are being accelerated-cooled in their in-line head-hardening processes.[176,177]

When a conventional (medium to eutectoid) carbon rail steel such as Fe-0.71C-0.57Cr-0.88Mn-0.1Ni-0.2Mo-0.4Si was heat-treated to produce variations of pearlitic and bainitic microstructures, it was reported that the wear resistance of bainite was inferior and superior, respectively, to that of pearlite in the hardness range of 38 to 40 Rc and much above Rc 40 (450 to 500 BHN).[178] Clayton and Devanathan further showed that Cr-Mo rail steel heat-treated to upper and lower bainite microstructure with hardnesses near Rc 49 and Rc 59 offer significant wear resistance at very high contact pressures (1575 and 1645 MPa) with very large creepages (33 and 55%).[178–180] In spite of the attractive performance of the bainitic steels, their high alloying costs are likely to limit their use in comparison with in-line hardened pearlitic steels.[180]

Bainitic railway wheels are presently being investigated to produce an alloy that must resist microstructural changes due to "wheel spin burning." Otherwise brittle martensite can form when heated momentarily above the austenitizing temperature and rapidly cooled by heat dissipation into the underlying bulk wheel mass. Steels with a lower carbon content and a bainitic microstructure have been observed to outperform the traditional pearlitic steels, because of their low hardenability, lower carbon content, and high M_s temperature—these characteristics reduce the likelihood of formation of martensite and its consequent brittle behavior.[11]

9.5.5 High-Carbon Bainitic Steels

9.5.5.1 Strength. The increased strength obtained in high-carbon bainitic steels arises from (1) further refinement of the bainitic ferrite grain size and increased dislocation density due to the lower transformation temperature; (2) increased carbide dispersion strengthening; and (3) increased interstitial carbon content due to a decrease in transformation temperature.[51]

9.5.5.2 Composition. The base compositions used for high-carbon bainitic steels are the high-strength low-carbon steels containing 1% Cr (maximum), 1.5% Mn, and 0.5% Mo-B with carbon additions up to 1.0%. Tensile strength achieved in 0.6% C steel is >1500 MPa, which can be further increased to about 1750 MPa by increasing the austenitizing temperature to dissolve all the carbides.[39] Note that since eutectoid composition is reached with the increase in carbon content, the influence of boron on hardenability diminishes to zero.

For a 0.5% Mo-B base composition with carbon additions, up to 0.8 to 1.0% fully bainitic structures are developed after austenitizing and subsequent air cooling. The maximum tensile strength obtainable with 0.8% C in 10- to 300-mm-diameter specimen is about 1300 MPa.

For better design of these steels, the choice of alloying elements should be made such that (1) the B_s temperature is greatly depressed, but (2) the bainitic reaction is not appreciably retarded at higher carbon levels. Both these requirements lead to the highest strength, and the possibility of minimum carbon content to be used results in the improvement of ductility and impact strength. In addition, in the case where the steel is not hypereutectoid, the pearlitic reaction is retarded, and consequently the bainite forms over a wide range of cooling rates. Mn and Cr are the cheapest alloying elements that depress the B_s temperature. However, Mn is preferred to Cr. The carbon content should be kept at a low level consistent with the strength desired. Otherwise, the bainitic reaction is retarded and the M_s temperature is lowered, thereby running the risk of quench-cracking. Moreover, the impact properties and ductility are decreased.[154]

9.5.5.3 Advantages. Where weldability, formability, and toughness are less important, high-carbon bainitic steels may possess economic advantages over the quenched and tempered martensitic steels. These advantages include (1) use of large section thickness, which enables these steels to compete with quenched and tempered martensitic steels; (2) freedom from distortion, quench-cracking, and inferior impact strength, which are common for mixed bainite-martensite structures; (3) no need for expensive and elaborate quenching equipment; and (4) lower cost, owing to less substitutional alloying addition at the expense of higher carbon content.

In contrast, a quenched and tempered steel would require larger alloying additions to obtain the necessary martensitic hardenability.

9.5.5.4 Disadvantages.   Application of these steels is limited by the fact that the bainitic reaction is considerably retarded with both increased carbon and alloy contents. Martensite thus forms at slower cooling rates; i.e., martensite hardenability is increased. Also, these steels become hypereutectoid when the carbon content is higher; this causes the ineffectiveness of boron, thereby accelerating the pearlitic reaction.

9.5.5.5 Applications.   Typical applications include large die blocks, backup rolls, mandrel bars, and high-strength statically loaded machine components. Their use is, however, much more restricted than that of the low-carbon bainitic steels.[154]

9.6 WELDABILITY OF BAINITIC STEELS

For steels with a high hardenability and high carbon concentration, the heat-affected zone region may consist of unacceptable hardened microstructure such as untempered martensite after welding and subsequent cooling. These hardened microstructures are prone to cold cracking or HAZ hydrogen cracking due to hydrogen embrittlement and other impurity effects (see Sec. 3.10.3.4). That is why hardenable steels are difficult, if not impossible, to weld (Fig. 9.26).[11] Hardened areas of the HAZ can be avoided by preheating the workpiece prior to welding, but it is not cost-competitive.

Based on much empirical experience, it is now possible to correlate the cold-cracking susceptibility with a *carbon-equivalent* (CE) value of steel, which considers the effect of alloying elements on the hardenability. Two empirical formulas are popular. The first one, being a modified version of the equation initially developed by Dearden and O'Neill[181] and adopted by the International Institute of Welding (IIW), is commonly applied to steels containing <0.2% C concentration. The CE of the base metal is given by

$$\text{IIW CE} = \%C + \frac{\%Si + \%Mn}{6} + \frac{\%Cu + \%Ni}{15} + \frac{\%Cr + \%Mo + \%V}{5} \quad (9.15)$$

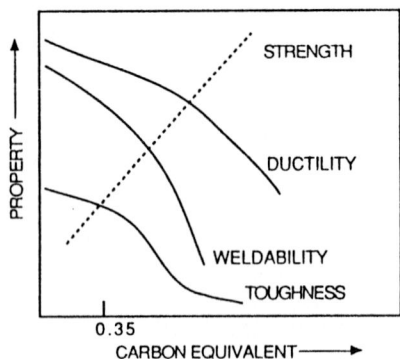

FIGURE 9.26   Variation in mechanical properties of the heat-affected zone as a function of the carbon equivalent.[11] (*Courtesy of H. K. D. H. Bhadeshia.*)

The second equation, recently developed by Ito and Bessyo[182] and adopted by the Japanese Welding Engineering Society (JWES), is sometimes preferred:

$$\text{JWES CE} = \%C + \frac{\%Si}{30} + \frac{\%Mn + \%Cu + \%Cr}{20} + \frac{\%Ni}{60} + \frac{\%Mo}{15}$$

$$+ \frac{\%V}{10} + 5(\%B) \tag{9.16}$$

Note that if the CE value is in the range of 0.35 to 0.55%, preheating of the part prior to welding is recommended (the preheating can be ≤400°C). For CE > 0.55%, both preheating and postheating are suggested to avoid cold cracking and other problems.[10] Of Eqs. (9.15) and (9.16), Eq. (9.16) is widely employed to appraise the weldability of ULCB and low-carbon, low-alloy steels. For these steels Eq. (9.15) provides overestimation of hardenability or gloomy estimation of weldability, whereas Eq. (9.16) is reliable. For low-carbon pipeline applications, a more realistic CE is expressed by Lorenz and Duren[183] as

$$CE = \%C + \frac{\%Si}{125} + \frac{\%Mn + \%Cu}{16} + \frac{\%Cr}{20} + \frac{\%Ni}{60} + \frac{\%Mo}{40}$$

$$+ \frac{\%V}{15} \tag{9.17}$$

for weld cooling times of 2 to 3 s over the temperature range of 800 to 500°C, corresponding to typical girth welds in pipelines.

REFERENCES

1. V. G. Paranjpe and D. D. Kaushal, *Trans. Indian Inst. Met., TP 55*, vol. 5, 1951, p. 147.

2. J. W. Christian and D. V. Edmonds, in *Proceedings of the International Conference on Phase Transformations in Ferrous Alloys*, eds. A. R. Marder and J. I. Goldstein, TMS-AIME, Warrendale, Pa., 1984, pp. 293–325.

3. W. T. Reynolds, Jr., S. K. Liu, F. J. Li, S. Hartfield, and H. I. Aaronson, *Met. Trans.*, vol. 21A, 1990, pp. 1479–1491.

4. H. I. Aaronson, W. T. Reynolds, Jr., G. J. Shiflet, and G. Spanos, *Met. Trans.*, vol. 21A, 1990, pp. 1343–1380.

5. H. I. Aaronson, T. Furuhara, J. M. Rigsbee, W. T. Reynolds, Jr., and J. M. Howe, *Met. Trans.*, vol. 21A, 1990, pp. 2369–2409.

6. W. T. Reynolds, Jr., H. I. Aaronson, and G. Spanos, *Mats. Trans.*, JIM, vol. 32, no. 8, 1991, pp. 737–746.

7. H. I. Aaronson, J. P. Hirth, B. B. Rath, and C. M. Wayman, *Met. Trans.*, vol. 25A, 1994, pp. 2655–2673.

8. J. W. Christian, *Met. Trans.*, vol. 21A, 1990, pp. 799–803.

9. J. W. Christian and D. V. Edmonds, *Scripta Metall.*, vol. 22, 1988, p. 273; vol. 23, 1989, p. 285.

10. H. K. D. H. Bhadeshia and J. W. Christian, *Met. Trans.*, vol. 21A, 1990, pp. 767–797.

11. H. K. D. H. Bhadeshia, *Bainite in Steels*, The Institute of Materials, London, 1992.

12. H. I. Aaronson, in *The Mechanisms of Phase Transformations in Crystalline Solids*, Proceedings of an International Symposium, Manchester, July 1968, Institute of Metals, London, Monograph no. 33, 1969, pp. 270–281.

13. R. F. Hehemann, K. R. Kinsman, and H. I. Aaronson, *Met. Trans.*, vol. 3A, 1972, pp. 1071–1094.

14. H. I. Aaronson, in *Encyclopedia of Materials Science and Engineering*, Pergamon Press, Oxford, 1986, pp. 263–266.

15. B. L. Bramfitt and J. G. Speer, *Met. Trans.*, vol. 21A, 1990, pp. 817–829.

16. G. B. Olson, H. K. D. H. Bhadeshia, and M. Cohen, *Met. Trans.*, vol. 21A, 1990, pp. 805–809.

17. H. K. D. H. Bhadeshia and D. V. Edmonds, *Met. Trans.*, vol. 10, 1979, p. 895–907.

18. M. Hillert, *Met. Trans.*, vol. 25A, 1994, pp. 1957–1966.

19. Y. Ohmori and T. Maki, *Mats. Trans., JIM*, vol. 32, no. 8, 1991, pp. 631–641.

20. L. C. Chang, *Met. and Mats. Trans.*, vol. 30A, 1999, pp. 909–916.

21. Y. Ohmori, H. Ohtsubo, Y. C. Jung, S. Okaguchi, and H. Ohtani, *Met. Trans.*, vol. 25A, 1994, pp. 1981–1989.

22. S. Okaguchi, H. Ohtani, and Y. Ohmori, *Mats. Trans., JIM*, vol. 32, no. 8, 1991, pp. 697–704.

23. H. K. D. H. Bhadeshia, in *Encyclopedia of Materials Science and Engineering*, suppl. vol. 2, Pergamon, Oxford, 1990, pp. 719–724.

24. R. F. Hehemann, in *Phase Transformations*, ASM, Metals Park, Ohio, 1970, pp. 402–406; and in *Metals Handbook*, 8th ed., vol. 8, ASM, Metals Park, Ohio, pp. 194–196.

25. H. J. Lee, G. Spanos, G. J. Shiflet, and H. I. Aaronson, *Acta Metall.*, vol. 36, 1988, p. 1129.

26. H. I. Aaronson and C. Wells, *Trans. AIME*, vol. 203, 1955, p. 1002.

27. G. Spanos, H. S. Fang, D. S. Sharma, and H. I. Aaronson, *Met. Trans.*, vol. 21A, 1990, pp. 1391–1411.

28. E. S. Davenport, *Trans. ASM*, vol. 27, 1939, p. 837.

29. A. Hultgren, *Trans. ASM*, vol. 39, 1949, p. 915.

30. S. M. Kaufman, G. M. Pound, and H. I. Aaronson, *Trans. AIME*, vol. 209, 1957, p. 855.

31. D. N. Shackleton and P. M. Kelly, *Physical Properties of Martensite and Bainite*, Special Report no. 93, Iron and Steel Institute, London, 1965, p. 126.

32. H. I. Aaronson, H. R. Plichta, G. W. Franti, and K. C. Russell, *Met. Trans.*, vol. 9A, 1978, pp. 363–371.

33. G. R. Purdy and M. Hillert, *Acta Metall.*, vol. 32, 1984, pp. 823–828.

34. H. I. Aaronson and C. Wells, *Trans. AIME*, vol. 206, 1956, pp. 1216–1223.

35. S. B. Singh and H. K. D. H. Bhadeshia, *Mater. Sc. and Engrg.*, vol. A245, 1998, pp. 72–79.

36. E. S. K. Menon and H. I. Aaronson, *Acta Metall.*, vol. 35, 1987, pp. 549–563.

37. H. I. Aaronson, G. Spanos, R. A. Masumura, R. G. Vardian, D. W. Moon, E. S. K. Menon, and M. G. Hall, *Mater. Sc. and Engrg.*, vol. B32, 1995, p. 107.

38. H. I. Aaronson, private communications, 1999.

39. K. J. Irvine and F. B. Pickering, *Physical Properties of Martensite and Bainite*, Special Report no. 93, Iron and Steel Institute, London, 1965, p. 110.

40. W. Pitsch, *Acta Metall.*, vol. 10, 1962, p. 897.

41. G. V. Smith and R. F. Mehl, *Trans. AIME*, vol. 150, 1942, pp. 211–226.

42. A. T. Davenport, *The Crystallography of Upper Bainite*, Republic Steel Corporation Research Center, Report on Project 12051, Cleveland, Ohio, February 1974, pp. 1–35.

43. S. Hoekstra, *Acta Metall.*, vol. 28, 1980, pp. 507–517.

44. E. P. J. Sandvik, *Met. Trans.*, vol. 13A, 1982, pp. 777–787, 789–800.

45. F. B. Pickering and B. R. Clark, *Physical Properties of Martensite and Bainite*, Special Report no. 93, Iron and Steel Institute, London, 1965, p. 143.

46. M. Oka and H. Okamoto, *J. de Physique*, vol. 4, 1995, pp. C8-503–508.

47. H. K. D. H. Bhadeshia, *Acta Metall.*, vol. 28, 1980, pp. 1103–1114.

48. P. G. Shewmon, *Transformations in Metals*, McGraw-Hill, New York, 1969, p. 238.

49. G. R. Srinivasan and C. M. Wayman, *Acta Metall.*, vol. 16, 1968, pp. 507–517.

50. F. B. Pickering, in *Transformations and Hardenability in Steels*, Climax Molybdenum Co., Ann Arbor, Mich., 1967, pp. 109–129.

51. P. Vasudevan, L. W. Graham, and H. I. Axon, *JISI, London*, vol. 190, 1958, p. 386.

52. D. H. Huang and G. Thomas, *Met. Trans.*, vol. 8A, 1977, p. 1661.

53. P. M. Kelly and J. Nutting, *JISI*, vol. 197, 1961, p. 199.

54. Y. Ohmori, *Trans. ISI, Japan*, vol. 11, 1971, pp. 95–101.

55. I. V. Isaichev, *Zh. Tekh. Fiz.*, vol. 17, 1947, pp. 835–838.

56. J. M. Rigsbee and H. I. Aaronson, *Acta Metall.*, vol. 27, 1979, pp. 365–376.

57. G. M. Purdy, *Scripta Metall.*, vol. 21, 1987, pp. 1035–1038.

58. J. M. Oblak and R. F. Hehemann, in *Transformations and Hardenability in Steels*, Climax Molybdenum Co., Ann Arbor, Mich., 1967, p. 15.

59. K. H. Jack, *JISI, London*, vol. 169, 1951, pp. 26–36.

60. Y. Ohmori and R. W. K. Honeycombe, *Trans. ISI, Japan*, suppl. vol. 11, 1971, pp. 1160–1164.

61. G. R. Srinivasan and C. M. Wayman, *Acta Metall.*, vol. 16, 1968, pp. 609–620, 621–636.

62. J. W. Christian, *The Theory of Transformations in Metals and Alloys*, Pergamon Press, Oxford, 1965, pp. 824–831.

63. R. W. K. Honeycombe and F. B. Pickering, *Met. Trans.*, vol. 3, 1972, p. 1099.

64. R. W. K. Honeycombe and H. K. D. H. Bhadeshia, *Steel: Microstructure and Properties*, 2d ed., Arnold, London, 1995.

65. L. J. Habraken and M. Economopoulos, in *Transformations and Hardenability in Steels*, Climax Molybdenum Co., Ann Arbor, Mich., 1967, p. 69.

66. D. V. Edmonds, *Metals Handbook*, 9th ed., vol. 9, ASM, Metals Park, Ohio, 1985, pp. 662–667.

67. M. Hillert, *Jernkontorets Ann.*, vol. 141, 1957, pp. 757–789.

68. K. R. Kinsman and H. I. Aaronson, *Met. Trans.*, vol. 1A, 1970, pp. 1485–1488.

69. T. G. Nilan, in *Transformations and Hardenability in Steels*, Climax Molybdenum Co., Ann Arbor, Mich., 1967, pp. 57–66.

70. N. F. Kennon and R. H. Edwards, *J. Aust. Inst. Met.*, vol. 15, 1970, pp. 195–200.

71. H. Okamoto and M. Oka, *Met. Trans.*, vol. 17A, 1986, pp. 1113–1120.

72. M. Oka and H. Okamoto, *Met. Trans.*, vol. 19A, 1988, pp. 447–452.

73. Y. Ohmori, *Mats. Trans., JIM*, vol. 30, 1989, p. 487.

74. G. Spanos, H. S. Fang, and H. I. Aaronson, *Met. Trans.*, vol. 21A, 1990, pp. 1381–1390.

75. J. Sun, H. Lu, and M. Kang, *Met. Trans.*, vol. 23A, 1992, pp. 2483–2490.

76. H. I. Aaronson and H. J. Lee, *Scripta Metall.*, vol. 21, 1987, pp. 1011–1016.

77. G. W. Franti, J. C. Williams, and H. I. Aaronson, *Met. Trans.*, vol. 9A, 1978, pp. 1641–1649.

78. A. Nagasawa, *Mats. Trans., JIM*, vol. 32, no. 8, 1991, pp. 774–777.

79. K. Chattopadhyay and H. I. Aaronson, *Acta Metall.*, vol. 34, 1986, pp. 695–711.

80. X. K. Meng, M. K. Kang, Y. Q. Yang, and D. H. Liu, *Met. Trans.*, vol. 25A, 1994, pp. 2601–2608.

81. G. R. Speich, in *Decomposition of Austenite by Diffusional Processes*, eds. V. F. Zackay and H. I. Aaronson, Interscience, New York, 1962, pp. 353–370.

82. N. F. Kennon, *Met. Trans.*, vol. 9A, 1978, pp. 57–66.
83. M. G. Hall and H. I. Aaronson, *Met. Trans.*, vol. 25A, 1994, pp. 1923–1931.
84. J. P. Hirth, G. Spanos, M. G. Hall, and H. I. Aaronson, *Acta Mater.*, vol. 46, 1998, p. 857.
85. Ref. 30 in Ref. 55.
86. C. W. Spencer and D. J. Mack, in *Decomposition of Austenite by Diffusional Processes*, eds. V. F. Zackay and H. I. Aaronson, Interscience, New York, 1962, pp. 549–603.
87. R. D. Garwood, *Physical Properties of Martensite and Bainite*, Special Report no. 93, Iron and Steel Institute, London, 1965, pp. 90–99.
88. G. W. Lorimer, G. Cliff, H. I. Aaronson, and K. R. Kinsman, *Scripta Metall.*, vol. 9, 1975, p. 271.
89. P. E. J. Flewitt and J. M. Towner, *J. Inst. Met.*, vol. 95, 1967, p. 273.
90. P. Doig and P. E. J. Flewitt, in *Proceedings of the International Conference on Solid → Solid Phase Transformations*, eds. H. I. Aaronson et al., TMS-AIME, Warrendale, Pa., 1982, pp. 983–987.
91. M. M. Kostic, E. B. Hawbolt, and L. C. Brown, *Met. Trans.*, vol. 7A, 1976, pp. 1543–1651.
92. T. Tadaki, C. J. Quiang, and K. Shimizu, *Mats. Trans., JIM*, vol. 32, no. 8, 1991, pp. 757–765.
93. Y. Q. Yang, D. H. Liu, X. K. Merg, and M. K. Kang, *Met. Trans.*, vol. 25A, 1994, pp. 2609–2614.
94. N. Ravishankar, H. I. Aaronson, and K. Chattopadhyay, *Met. Trans.*, vol. 25A, 1994, pp. 2631–2637.
95. R. D. Doherty, in *Physical Metallurgy*, 4th ed., eds. R. W. Cahn and P. Haasen, Elsevier, Amsterdam, 1996, pp. 1363–1506.
96. C. Laird and H. A. Aaronson, *Acta Metall.*, vol. 17, 1969, p. 505.
97. M. H. Wu, Y. Hamada, and C. M. Wayman, *Met. Trans.*, vol. 25A, 1994, pp. 2581–2599.
98. C. M. Li, G.-Z. Wang, Y. K. Chang, and H.-K. Fang, *Met. Trans.*, vol. 28A, 1997, pp. 1617–1623.
99. H. I. Aaronson, and J. P. Hirth, *Scripta Met.*, vol. 33, 1995, p. 347.
100. B. C. Muddle, J. F. Nie, and G. R. Hugo, *Met. Trans.*, vol. 25A, 1994, pp. 1845–1856.
101. H. I. Aaronson, B. C. Muddle, and J. F. Nie, *Scripta Mater.*, vol. 41, 1999, pp. 203–208.
102. R. F. Hehemann and A. R. Troiano, *Met. Progr.*, vol. 70, no. 2, 1956, p. 97.
103. W. Steven and A. G. Haynes, *JISI*, vol. 183, 1956, p. 349.
104. R. L. Bodnar, T. Ohhashi, and R. I. Jaffe, *Met. Trans.*, vol. 20A, 1989, pp. 1445–1460.
105. R. T. Howard and M. Cohen, *Trans. AIME*, vol. 176, 1948, p. 384.
106. R. Kumar, *Physical Metallurgy of Iron and Steel*, Asia Publishing House, Bombay, India, 1968.
107. S. K. Liu, W. T. Reynolds, Jr., H. Hu, G. J. Shiflet, and H. I. Aaronson, *Met. Trans.*, vol. 16A, 1985, pp. 457–466.
108. H. K. D. H. Bhadeshia, in *Solid → Solid Phase Transformations*, eds. H. I. Aaronson et al., TMS, Warrendale, Pa., 1981, pp. 1041–1048.
109. H. K. D. H. Bhadeshia, *J. De Physique*, vol. 43, 1982, C4-437–441.
110. W. T. Reynolds, Jr., F. J. Li, C. K. Shiu, and H. I. Aaronson, *Met. Trans.*, vol. 21A, 1990, pp. 1433–1463.
111. K. R. Kinsmann and H. I. Aaronson, in *Transformations and Hardenability in Steels*, Climax Molybdenum Co., Ann Arbor, Mich., 1967, pp. 39–53.
112. A. Ali, M. Ahmed, F. Hashmi, and A. Q. Khan, *Met. Trans.*, vol. 25A, 1994, pp. 2345–2350.
113. G. R. Purdy, *Acta Metall.*, vol. 26, 1978, pp. 477, 487.
114. N. Nemoto, *High Voltage Electron Microscopy*, Academic Press, New York, 1974, p. 230.

115. E. S. K. Menon and H. I. Aaronson, *Acta Metall.*, vol. 34, 1986, pp. 1975–1981.

116. E. P. Simonen, H. I. Aaronson, and R. Trivedi, *Acta Metall.*, vol. 23, 1977, p. 945.

117. H. I. Aaronson, J. M. Rigsbee, and R. K. Trivedi, *Scripta Metall.*, vol. 20, 1986, pp. 1299–1304.

118. E. P. Simonen, H. I. Aaronson, and R. Trivedi, *Met. Trans.*, vol. 4A, 1973, pp. 1239–1245.

119. L. Kaufman, S. V. Radcliffe, and M. Cohen, in *Decomposition of Austenite by Diffusional Processes*, eds. V. F. Zackay and H. I. Aaronson, Interscience, New York, 1962, p. 313.

120. H. K. D. H. Bhadeshia and A. R. Waugh, *Acta Metall.*, vol. 30, 1982, pp. 775–784.

121. H. K. D. H. Bhadeshia and D. V. Edmonds, *Acta Metall.*, vol. 28, 1980, pp. 1265–1273.

122. Y. Ohmori, H. Ohtani, and T. Kunitake, *Trans. ISI, Jpn.*, 1971, vol. 11, 1971, pp. 250–259.

123. Y. Ohmori, in *International Conference on Displacive Transformations and Their Applications in Materials Engineering*, eds. K. Inoue et al., TMS, Warrendale, Pa., 1998, pp. 85–92.

124. M. Oka and H. Okamoto, in *International Conference on Displacive Transformations and Their Applications in Materials Engineering*, eds. K. Inoue et al., TMS, Warrendale, Pa., 1998, pp. 79–84.

125. Y. Ohmori, in *Proceedings of the International Conference on Martensitic Transformation, 1986*, The Japan Institute of Metals, Sendai, Japan, 1986, pp. 587–594.

126. G. Spanos, *Met. Trans.*, vol. 25A, 1994, pp. 1967–1980.

127. M. Hillert, *Met. Trans.*, vol. 6A, 1975, p. 5.

128. T. Y. Hsu and M. Yiewen, *Acta Metall.*, vol. 32, 1984, pp. 1469–1481.

129. M. M. Rao and G. Winchell, *TMS-AIME*, vol. 239, 1967, p. 956.

130. W. T. Reynolds, Jr., and H. I. Aaronson, *Scripta Metall.*, vol. 19, 1985, pp. 1171–1176.

131. R. Trivedi, *Met. Trans.*, vol. 1A, 1970, p. 921.

132. A. M. Llopis, Ph.D. thesis, University of California, Berkeley, 1977.

133. O. Grong and D. Matlock, *Int. Met. Rev.*, vol. 31, 1986, pp. 27–48.

134. D. J. Abson and R. J. Pargeter, *Int. Met. Rev.*, vol. 31, 1986, pp. 141–194.

135. H. K. D. H. Bhadeshia, in *Recent Trends in Welding Science and Technology*, ed. S. A. David, ASM, Metals Park, Ohio, 1989, p. 189.

136. S. S. Babu and H. K. D. H. Bhadeshia, *Mater. Trans.*, *JIM*, vol. 32, no. 8, 1991, pp. 679–688.

137. J. R. Yang and H. K. D. H. Bhadeshia, in *Advances in Welding Science and Technology*, ed. S. A. David, ASM, Metals Park, Ohio, 1986, pp. 187–191.

138. K. Nishioka and H. Tamehiri, *Microalloying '88*, ASM International, Metals Park, Ohio, 1989, p. 1.

139. H. K. D. H. Bhadeshia, *Steel Technology International*, ed. P. H. Scholes, Stirling Publc. Int'l. Ltd., London, 1989, p. 289.

140. H. K. D. H. Bhadeshia, *Mater. Sci. Forum*, vols. 284–286, 1998, pp. 39–50.

141. A. O. Kluken, O. Grong, and J. Hjelen, *Met. Trans.*, vol. 22A, 1991, pp. 657–663.

142. J. M. Gregg and H. K. D. H. Bhadeshia, *Met. Trans.*, vol. 25A, 1994, pp. 1603–1611.

143. G. Sneider and H. W. Kerr, *Canadian Metall. Quart.*, vol. 23, 1984, pp. 313–325.

144. Y. Horri, M. Wakabayashi, S. Okhita, and M. Namura, *Nippon Steel Technical Report No. 37*, April 1988, Nippon Steel Co., Japan, pp. 1–9.

145. G. J. Sojjka, M. R. Krishnadev, and S. K. Banerji, *Boron in Steel, Conference Proceedings*, eds. S. K. Banerji and J. E. Morral, TMS-AIME, Warrendale, Pa., 1980, pp. 165–180.

146. Z. Zhang and R. A. Farrar, *Mater. Sci. and Technol.*, vol. 12, March 1996, p. 273.

147. K. R. Kinsman and H. I. Aaronson, *Met. Trans.*, vol. 4, 1973, p. 959.

148. C. A. Dubé, H. I. Aaronson, and R. F. Mehl, *Rev. de Met.*, vol. 55, 1958, p. 201.

149. C. Zener, *Trans. AIME*, vol. 167, 1946, p. 550.

150. B. L. Bramfitt, *Met. Trans.*, vol. 1, 1970, pp. 1987–1995.

151. B. L. Bramfitt, *39th MWSP Conference Proceedings*, Iron & Steel Society, vol. 35, 1998, pp. 989–996.

152. W. C. Leslie, *Physical Metallurgy of Steels*, McGraw-Hill, New York, 1981.

153. K. J. Irvine and F. B. Pickering, *JISI*, vol. 21, 1963, p. 201.

154. F. B. Pickering, *Physical Metallurgy and Design of Steels*, Applied Science Publishers, London, 1978.

155. F. B. Pickering, in *Encyclopedia of Materials Science and Engineering*, Pergamon Press, Oxford, 1986, pp. 4621–4632.

156. M. E. Bush and P. M. Kelly, *Acta Metall.*, vol. 19, 1971, p. 1363.

157. D. W. Smith and R. F. Hehemann, *JISIJ*, vol. 209, 1971, p. 476.

158. F. B. Pickering, in *Constitution and Properties of Steels*, vol. ed. F. B. Pickering, VCH, Weinheim, 1992, chap. 2, pp. 41–94.

159. A. Kamada, N. Koshizuka, and T. Funakoshi, *Trans. ISI*, vol. 16, 1976, p. 407.

160. N. P. Naylor and P. R. Krake, *Met. Trans.*, vol. 5A, 1974, pp. 1699–1701.

161. Y. Tomita, *Met. Trans.*, vol. 19A, 1988, pp. 2513–2521.

162. N. Jin and P. Clayton, *Wear*, vol. 202, 1997, pp. 202–207.

163. P. Clayton and N. Jin, *Wear*, vol. 202, 1996, pp. 74–82.

164. B. L. Bramfitt and J. G. Speer, *Mechanical Working and Steel Processing Proceedings*, Iron & Steel Society, 1989, Warrendale, Pa., pp. 443–454.

165. N. Nakasugi, H. Matsuda, and H. Tamehiro, *Alloys for 1980's*, Climax Molybdenum, Ann Arbor, Mich., 1980, p. 213.

166. P. H. Wright, T. L. Harrington, W. A. Szilva, and T. R. White, *Fundamentals of Microalloying Forging Steels*, eds. G. Krauss and S. K. Banerjee, TMS-AIME, Warrendale, Pa., 1987, pp. 541–566.

166a. R. Kaspar, I. G. Baquet, N. Schreiber, J. Richter, G. Nussaum, and A. Köthe, *Steel Research*, vol. 68 no. 1, 1997, pp. 27–31.

167. V. T. T. Miihinen and D. V. Edmonds, *Materials Sci. & Technol.*, vol. 3, 1983, pp. 441–449.

168. J. E. Garnham, University of Leicester, as quoted in ref. 11.

169. J. E. Garnham and J. H. Beynon, *Wear*, vol. 157, 1992, pp. 81–109.

170. H. K. D. H. Bhadeshia, in *International Conference on Displacive Transformations and Their Applications in Materials Engineering*, eds. K. Inoue et al., TMS, Warrendale, Pa., 1998, pp. 69–78.

171. P. Clayton, K. J. Sawley, P. J. Bolton, and G. M. Pell, *Wear of Materials*, 9th International Conference, Houston, Tex., 1987; *Publ. Wear*, vol. 120, 1987, pp. 199–200.

172. K. J. Sawley, *Bainitic Steels for Rails, Technology Digest*, TD97-001, Association of American Railroads, January 1997.

173. V. Jerath, K. Mistry, P. Bird, and R. R. Preston, *British Steel Report on Collaborative Research Project between The University of Cambridge*, Report SL/RS/R/S/1975/1/91A, 1991, pp. 1–43.

174. H. Kageyama, M. Ueda, and K. Sugino, U.S. Patent 5,383,307, Jan. 17, 1995.

175. Japanese Patent 7,090,494, Apr. 4, 1995.

176. B. L. Bramfitt, R. L. Cross, and D. P. Wirick, *Proceedings of International Symposium on Rail Steels*, Iron & Steel Society, 1994, pp. 23–29.

177. B. L. Bramfitt, D. P. Wirick, and R. L. Cross, *Iron and Steel Engineer*, vol. 75, no. 4, May 1996, pp. 33–36.

178. P. Clayton and R. Devanathan, *Wear*, vol. 156, 1992, pp. 121–131.

179. P. Clayton and X. Su, *Wear*, vol. 200, 1996, pp. 63–73.

180. R. Steele, private communication, 1999.

181. J. Dearden and H. O'Neill, *Trans. Inst. Weld*, vol. 3, 1940, p. 203.

182. Y. Ito and K. Bessyo, *Inter. Inst. Weld*, Document IX-576-68, 1968.

183. K. Lorenz and C. Duren, *Steels for Linepipe and Pipeline Fittings*, The Metals Society, London, 1983, p. 322.

CHAPTER 10
AUSTENITE

10.1 INTRODUCTION

A large majority of heat treatment processes of steels commence with the formation of austenite. Despite this fact, many fewer detailed studies have been carried out on the formation of austenite as compared to the vast amount of studies done on the decomposition of austenite, because the mechanical properties of steel are functions of transformation processes of austenite.[1-3] The occurrence of these final transformation products and the resulting properties depend considerably on the homogeneity and grain size of the austenite. The influence of austenite grain size on diffusional and diffusionless transformations of austenite and on hardenability is well documented in Chaps. 7 through 9 and in Chap. 13.

Although the austenite in carbon steels is stable only above the A_{e1} temperature, it greatly affects the transformation and deformation modes of heat-treated steels. Recent interest in austenite transformation has been focused on the study of mechanisms and kinetics of formation of various types of austenite. Since dislocation reactions are not entirely reversible, the austenite transformation leads to an increase in the number of lattice defects in the reversed austenite which hinders the reproduction of the original orientation of austenite and gives rise to phase hardening. An understanding of the austenitic transformation offers new possibilities of influencing the mechanical and physical properties of dual-phase, duplex stainless steels and austenitic alloys and the likelihood of developing the new grades of these alloys with useful properties.[4] This chapter deals with the austenite formation, austenite grain size, retained austenite, austenitic and superaustenitic stainless steels, duplex and superduplex stainless steels, and physical properties of stainless steels.

10.2 AUSTENITE FORMATION

The effect of the prior microstructure of steel comprising pearlite, ferrite, spheroidized cementite, bainite, and martensite on the rate of nucleation and growth—and, therefore, on the time requirement for austenitization and on the final grain size after completion of austenitization—has been studied by several workers.[1-3,5-16]

Figure 10.1 shows the process of formation of austenite in eutectoid carbon steel with a composition of 0.78% C, 0.35% Si, 0.55% Mn, 0.02% S, and 0.02% P when a series of steel specimens with starting annealed pearlitic microstructure are heated

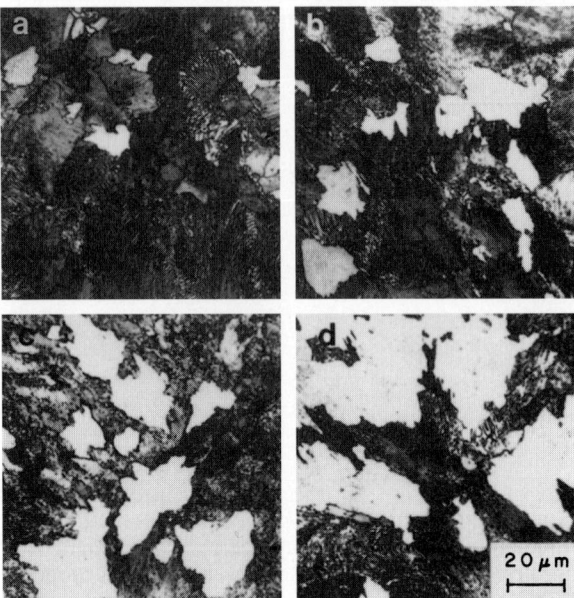

FIGURE 10.1   The process of formation of austenite (light patches) from pearlite at a temperature of 735°C as a function of time: (*a*) 40 s, (*b*) 59 s, (*c*) 80 s, and (*d*) 120 s. Prior heat treatment: annealed from 1050°C followed by furnace cooling.[3] (*Reprinted by permission of Pergamon Press, Plc.*)

at a temperature of 735°C, held for progressively increasing times, and subsequently quenched into ice water to transform the untransformed austenite to martensite.[3] This process may also be used for studying austenitization from the different initial microstructures. The white patches in these photomicrographs illustrate the regions of austenite phase formed. There are three favorable nucleation sites for the formation of austenite in a pearlitic microstructure (Fig. 10.2): (1) the ferrite/cementite interface, (2) the line of intersection between the platelets and surfaces of the pearlite colony, and (3) the points of intersection between the platelets and the edges of the pearlite colony.[3]

The growth of austenite into pearlite is a carbon diffusion-controlled process. A complete dissolution of cementite lamellae does not occur during the initial austenite formation, and some residual carbide remains for some periods (as revealed by dark spots in the white areas in the photomicrographs). However, these particles dissolve completely after a prolonged hold time at the austenitizing temperature.[1,17] The growth rate of austenite into pearlite is decreased by impurities present in plain carbon steel.[1]

The nucleation and growth process can be expressed by the Avrami equation [Eq. (7.10)], where the value of the exponent that characterizes the kinetics is usually found to be 4. A relationship between nucleation rate $\dot{N}$, the average true interlamellar spacing S_0, and the average pearlite colony edge length a^p has been found:

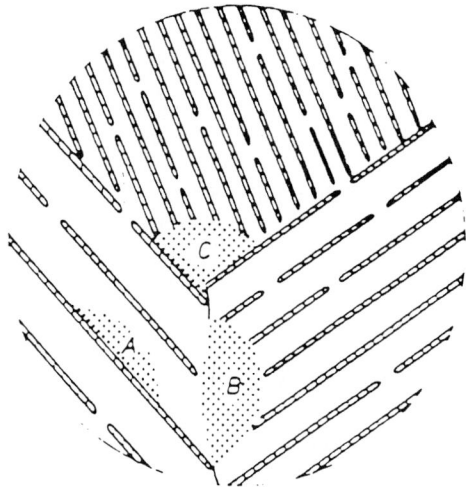

FIGURE 10.2 Three nucleation sites for austenite formation in a plane section of three pearlite colonies.[3] (*Reprinted by permission of Pergamon Press, Plc.*)

$$\dot{N} \propto \frac{1}{\left(a^p\right)^2 S_0} \tag{10.1}$$

On the other hand, the rate of grain growth G is

$$G \propto \frac{1}{S_0^2} \tag{10.2}$$

The time versus temperature relationship for austenitizing a normalized 1080 steel with fine pearlite microstructure is shown in Fig. 10.3a.[5] The first curve at the left illustrates the start of the disappearance of pearlite; the second curve, the final disappearance of carbide; and the final curve, the total disappearance of carbon concentration gradients. Figure 10.3b shows a new isothermal austenitizing diagram of annealed eutectoid carbon steel based on the structural factor $F = a^p S_0^2$ of $5 \times 10^{-9}\,\mathrm{mm}^3$. At 745°C, the times required to obtain 50% austenite [$f(t) = 0.5$] and complete austenite transformation [$f(t) = 0.99$] are 95 and 155 s, respectively. However, when the structural factor F remains lower or higher than $5 \times 10^{-9}\,\mathrm{mm}^3$, an auxiliary bar shown on top of the diagram should be taken into consideration. For example, when the value of F is $10^{-8}\,\mathrm{mm}^3$, the distance h between $F = 5 \times 10^{-9}$ and $F = 10^{-8}\,\mathrm{mm}^3$ is measured as the auxiliary bar and is added to the time of 155 s at 745°C to determine the time for complete austenitization. In this case, the time required for complete isothermal austenitization is 320 s.[3]

In a low-carbon steel with a ferritic microstructure, austenite transformation is also a nucleation and growth process, and the nucleation of austenite takes place mainly at ferrite grain boundaries (Fig. 10.4a)[17] and usually develops allotriomorphic austenite. Like allotriomorphic ferrite, an allotriomorphic austenite bears a Kurdjumov-Sachs orientation relationship with one ferrite grain and grows predominantly into adjacent grain only (Fig. 10.5a) by the migration of incoherent interfaces. The extraction replica electron micrograph also illustrates a region (arrowed) of coarse carbide dispersion around the growing austenite allotri-

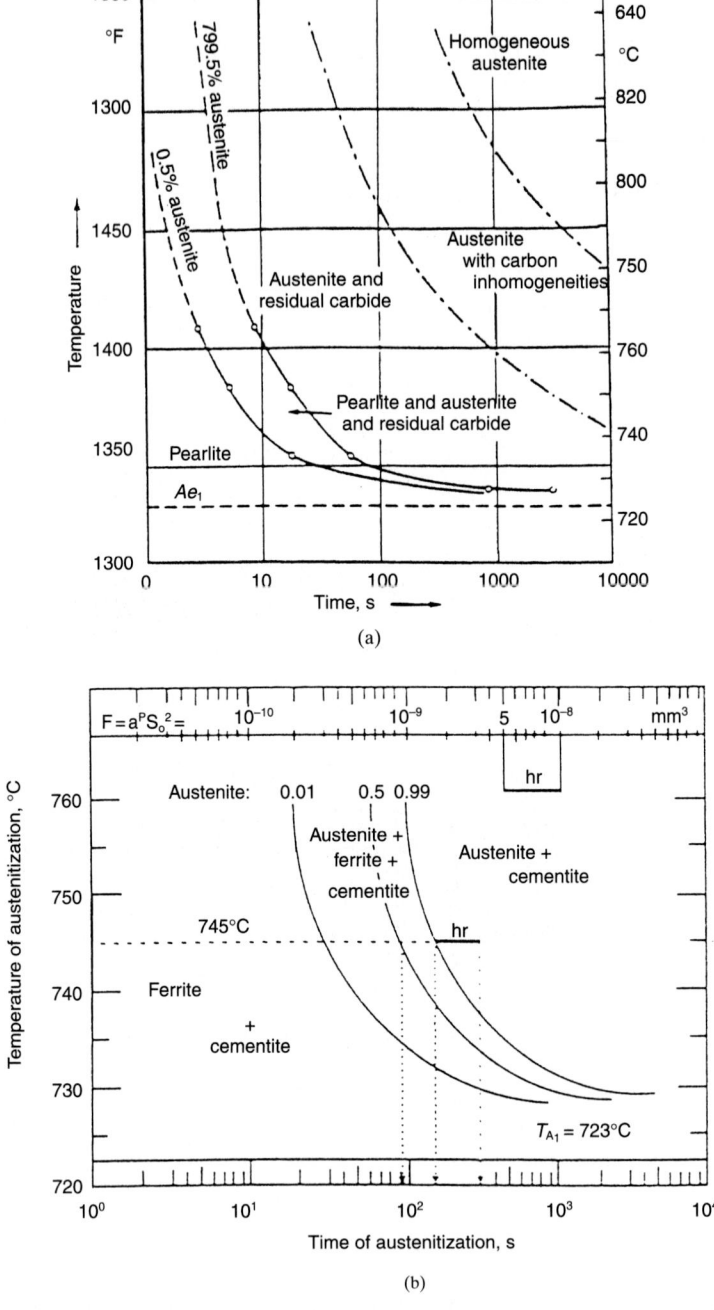

FIGURE 10.3 (*a*) The time versus temperature relationships for austenitizing a normalized 1080 steel from 875°C (1610°F) with fine pearlite structure.[5] (*b*) A new isothermal austenitizing diagram of annealed eutectoid carbon steel at 1050°C.[3] [(*a*) *Reprinted by permission of ASM International, Metals Park, Ohio;* (*b*) *reprinted by permission of Pergamon Press, Plc.*]

10.4

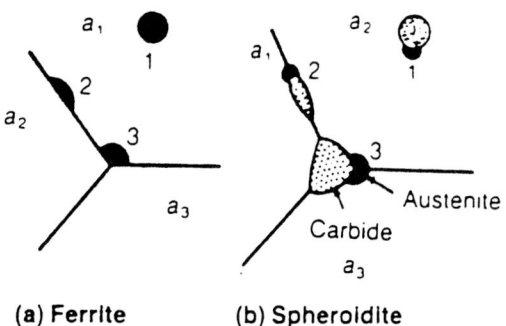

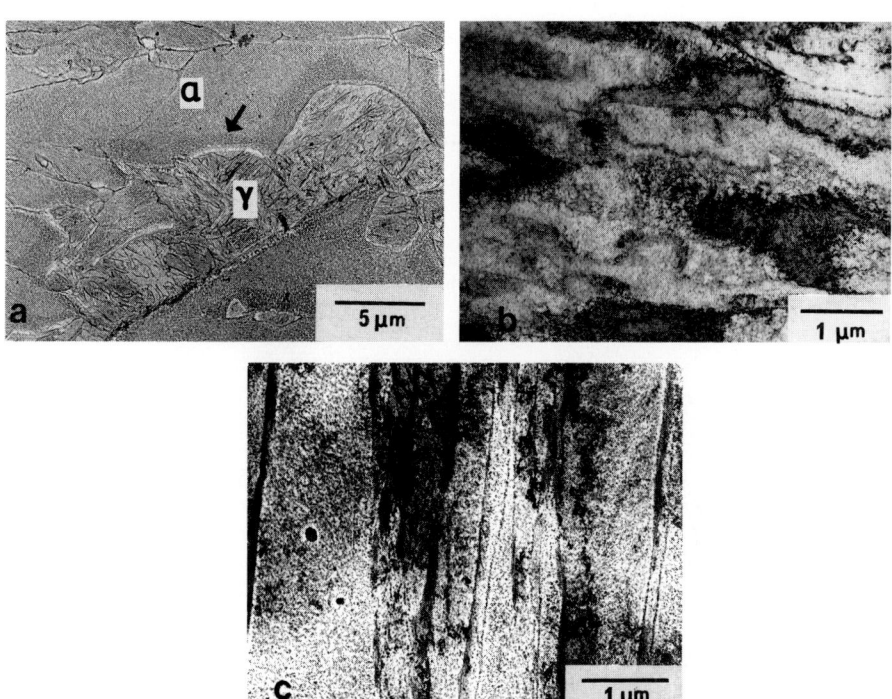

FIGURE 10.4 Nucleation sites for austenite formation in microstructures of (*a*) ferrite and (*b*) spheroidite.[17] (*Reprinted by permission of The Metallurgical Society, Warrendale, Pennsylvania.*)

FIGURE 10.5 (*a*) Ferritic microstructure heat-treated for 15 s at 850°C. Extraction replica electron micrograph. (*b*) and (*c*) Bright-field electron micrographs showing the acicular austenite formed by heat-treating the bainitic and martensitic structures, respectively, for 15 s at 840°C.[16] (*Courtesy of N. C. Law and D. V. Edmonds.*)

omorphs.[16] Figure 10.6 shows the TTT curves for austenitization from ferritic, bainitic, and martensitic structures of low-carbon (0.25%) low-alloy steel which illustrate a continuous increase of reaction rate with increasing temperature.[16]

In the ferrite-spheroidized cementite microstructures, austenite nucleates at the junctions between the cementite particles and the ferrite grain boundaries

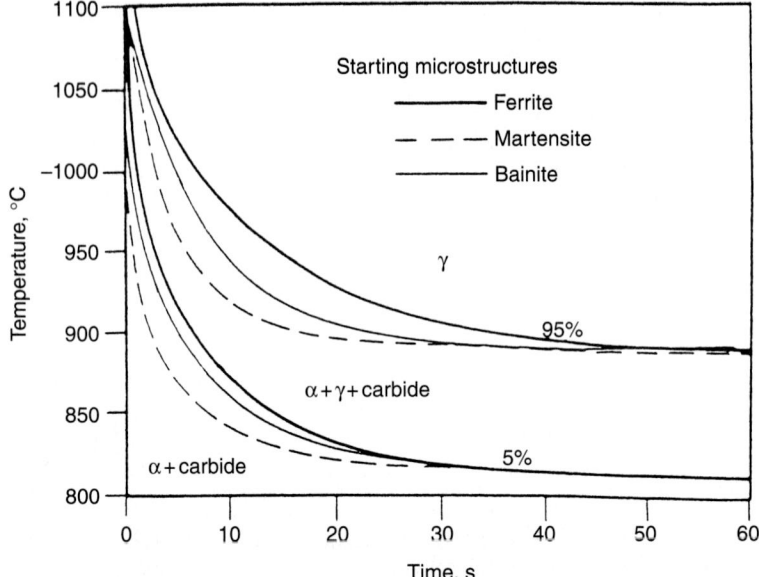

FIGURE 10.6   TTT curves for austenitization from ferritic, bainitic, and martensitic structures.[16] (*Reprinted by permission of The Metallurgical Society, Warrendale, Pennsylvania.*)

(Fig. 10.4*b*).[1,13] Then austenite film grows and completely envelopes the carbide
where further growth of austenite occurs by the continuous carbon diffusion from
the dissolution of the carbide toward the growing austenite envelope.[1,6]

The austenitization of annealed low-carbon steels (Fe-0.15%C-0.1%Si-0.01%S-
0.02%P alloys) with a fine distribution of pearlite in a ferrite matrix consists of two
stages. In the first stage, pearlite transformation to austenite occurs and becomes
complete in less than 1 s at temperatures greater than 800°C. In the second stage,
ferrite-to-austenite transformation occurs at temperatures higher than 910°C. If
the temperature is lower than 910°C, the second stage of further nucleation of
austenite does not occur, and the process is completed by the growth of the
remaining austenite particles within the ferrite matrix.[15]

In bainitic and martensitic microstructures of low-carbon low-alloy steels, austenite forms in two morphologies such as acicular and globular austenite. The acicular
austenite is promoted by acicular starting microstructure, low isothermal austenitizing temperature, and slow heating rates. The nucleation sites in this case are primarily the lath boundaries and prior austenite grain boundaries. These transitional
acicular nuclei grow rapidly and eventually coalesce to form a globular morphology. Globular austenite can also appear directly, which is promoted by conditions
opposite to those mentioned above.[16] The mechanism of austenite formation from
these structures tends also to be diffusion-controlled and is linked with fine dispersion of carbide particles.

Figures 10.5*b* and *c* are the bright-field electron micrographs showing the acicular austenite formed by heat-treating the bainitic and martensitic structures, respectively, at 840°C for 15 s.[16]

Rapid Austenitizing. Rapid austenitizing of several steels such as carbon steels with fine aggregate of martensite and tempered martensite and low-carbon Fe-20%Ni and Fe-30%Ni alloys results in appreciable grain refinement of the prior austenite grain size over conventional austenitizing and a large increase in austenite strength and toughness.[17-21] In Fe-30%Ni alloy, such an increase in strength is equal to the martensitic strength which is attributed to the combined effect of dislocation density and distribution in the austenite. However, Ansell and coworkers have correlated the increased austenite strength with lowering of M_s temperature.[17,22]

10.3 AUSTENITE GRAIN SIZE

Precise determination of austenite grain size is often essential in metallurgical studies of steel properties. For example, fine-grained austenite usually transforms to fine-grained ferrite, pearlite, bainite, or martensite, thereby providing increased strength and toughness. Coarse-grained austenite increases hardenability, as evidenced in Fig. 13.22, where the ideal diameter of the steel increases by about 8% with every decrement in ASTM grain size number (e.g., from 8 to 7). Most expressions for hardenability calculations use a term for austenite grain size [e.g., Eq. (13.13)]. For a good result in its final treated condition, the measured austenite grain size must be equal to the initial austenite grain size of the steel. However, the prior austenite grain size is related to the chemical composition and prior processing of the steel.[23]

10.3.1 Development of Austenite Grain Boundaries

The most commonly used methods to reveal the austenite grain boundaries are (1) the McQuaid-Ehn carburizing method, (2) etchant solutions based on picric acid, (3) delineation by ferrite or cementite, (4) delineation by pearlite, (5) oxidation etching method, and (6) thermal etching method. The choice of these methods is based on the chemical composition of the steel and its microstructure at room temperature.

The McQuaid-Ehn Test.[24] This is based on pack carburizing at a temperature of 925°C (1690°F) for 8 hr and slow cooling to allow the delineation of the austenite grain boundaries by the precipitation of proeutectoid cementite network. The specimen is cut in half, polished, and etched with nital, picral, or alkaline sodium picrate. While originally used for carburizing grades with reasonable accuracy, it has been subsequently applied to test most carbon and alloy steels. It is time-consuming, and in some steels it may cause coarsening of the original austenite grain size. After polishing and etching, the grain boundaries are clearly visible under the microscope.[23] This method is used for plain carbon steels up to 0.45% C and low-alloy steels up to 0.25% C.

Etchant Solution Based on Picric Acid. Etchants based on picric acid and a wetting agent can be used to reveal the true prior austenite grain boundaries in the as-received steels at room temperature having a martensitic or bainitic microstructure. More consistent results are obtained when the steel has been tempered in the embrittling temperature range (see Chap. 14 for more details). It is more suitable for hardenability studies and is applied to the quenched end of the Jominy bar. However, it has some drawbacks, which include these:

1. A standard etching solution cannot be employed for all grades; hence a variety of picric acid solutions, and even tempering treatments, should be tried to reveal clearly the austenite grain boundaries for a given steel.
2. Appearance of subgrains and twins in the austenite or in the final martensite may lead to ambiguity of the grain boundaries.

Delineation by Ferrite or Cementite. In this method, the steel specimen is heated into the austenite phase field and cooled at a controlled rate. In hypoeutectoid steels, the austenite grain boundaries are outlined by ferrite; in hypereutectoid steels, they are outlined by cementite.

Outlining the grains with ferrite is best done with medium-carbon (0.25 to 0.60% C) steels and low-alloy steels in the range of 0.25 to 0.50% C. In this case, the heat treatment step may not be needed because etching of the as-received material with nital depicts the well-delineated grain structures by ferrite. This not only results in a great reduction in time and labor but also gives a true measurement of the original austenite grain size.[23] With lower carbon content, the specimen is heated to the austenitizing temperature [say, 915°C (1679°F)], transferred to a furnace at intermediate temperature [740 to 790°C (1364 to 1454°F)], and allowed to cool down to this temperature, followed by water-quenching. After polishing and etching, microscopic examination will reveal the ferrite network around prior austenite (low-carbon martensite) grains.[23] This method has direct application to hardenability studies because of the appearance of ferrite-pearlite structures, in both shallow and medium hardening steels, at a distance away from the Jominy-quenched end. The precipitation of proeutectoid cementite is used to reveal prior austenite grain boundaries with greater than 1% C.

Delineation by Pearlite. In a eutectoid steel, the prior austenite grains can be revealed by outlining the martensite grains with fine pearlite. In this method the specimen is austenitized and is (1) brine-quenched so that the interior is not completely hardened or (2) gradient-quenched by immersing one end of the specimen into the brine, to be fully hardened, thereby leaving an incompletely hardened region a short distance away. In this region a prior austenite grain structure consists of martensite grains surrounded by a small amount of fine pearlite network. These methods are applied to steels with a near eutectoid composition. A Jominy bar usually exhibits this effect in the region of the drop-off in hardness.[25]

Oxidation Etching.[24] A polished face of hypoeutectoid steel (or interstitial-free steels with very low C content) is heated in air or in a controlled oxidizing atmosphere at the required temperature [say, 855°C (1575°F)] for 1 hr so that oxidation occurs at the grain boundaries. The steel is then quenched in water or brine to form martensite. It is next repolished to remove the oxide layer slightly and is finally etched. The grain boundaries are revealed by preferential oxidation. This method appears to be faster than the McQuaid-Ehn test. It is used for fine-grain-size carbon and alloy steels with which many other methods fail.[25]

Thermal Etching. Thermal etching or grain boundary grooving in vacuum is an established method for delineating austenite grain boundaries in a wide range of steels.[23,26] The method consists of heating the polished specimen to the desired temperature [say, 850°C (1562°F) for 1 hr] into the austenite range under a vacuum of 10^{-3} mm of mercury. Thermal grooving occurs at the austenite grain boundaries as a result of preferential evaporation and surface tension effects which are clearly revealed after cooling to room temperature.

10.3.2   Methods of Measuring Austenite Grain Size

In general, grain size can be measured in a wide range of metallic materials by several methods.[24] Their success is based on the clear delineation of (austenite) grain boundaries. However, their choice depends on many factors, such as (1) speed and accuracy requirements, (2) the appearance of (austenite) grains, and (3) the availability of the equipment.[23] Important methods are (1) comparison with standard charts, (2) lineal (Heyn) intercept method, (3) planimetric (Jeffries) method, (4) automated instruments, (5) Snyder-Graf intercept method, and (6) fracture grain-size method.

Just as for metallic materials, the first three methods can also be applied to nonmetallic materials having the structural appearances similar to those of the metallic structures shown in the comparison charts.[25]

Standard Chart Method.   The polished and etched specimen with delineated grain boundaries is viewed through the eyepiece of a microscope and is compared with eyepiece reticles having ASTM charts at a magnification of 100X. Alternatively, a photomicrograph of the structure at the same magnification is compared with the photograph of standard ASTM grain-size series 0 to 10 (ASTM E112). By trial and error, a close match of the photomicrograph with the image of the test specimen is found. The grain size of the metal is then reported by a number expressed by the index number of the matching chart. Grain-size estimates should be made on three or more representative areas of each specimen section. Metals with a mixed grain size are rated in a similar manner, and it is the normal practice to designate the grain size in terms of two numbers indicating the approximate percentage of each size present.

This is the most convenient and fastest method, and it is sufficiently accurate for most commercial equiaxed materials where an experienced operator can evaluate grain sizes to ±0.5 ASTM. As a consequence, the standard chart method is commonly used in quality control work.

The basic equation used to establish the ASTM grain-size number G for austenite grain-size measurement is

$$n = 2^{G-1} \qquad (10.3)$$

or

$$G = \frac{\log n}{\log 2} + 1 \qquad (10.4)$$

where n is the number of grains per square inch observed in a microstructure at a magnification of 100X. The relationship between G and the actual size (diameter) of the grains is listed in Table 10.1. Note that as the ASTM grain-size number increases, the grain size decreases. The common range encountered in a hardened medium-carbon steel is between (including) ASTM numbers 1 and 9.

TABLE 10.1 Micrograin Size Relationship Computed for Uniform Randomly Oriented Equiaxed Grains[24]

ASTM micrograin size number G	"Diameter" of average grain section		Average intercept distance $\bar{l}$ (mm)	Intercept count n/l (per mm)	Area of average grain section $\bar{a}$ (mm²)	Calculated number of grains[c] per mm³, n/v^c	Average	
	Nominal d_n (mm)	Feret's d_F (mm)					Grains[d] per mm² at 1X, n/a^d	Grains per in.² at 100X, n/a
00[e]	0.51	0.570	0.453	2.210	0.258	6.11	3.88	0.250
0	0.36	0.403	0.320	3.125	0.129	17.3	7.75	0.500
0.5	0.30	0.339	0.269	3.716	0.0912	29.0	11.0	0.707
1.0	0.25	0.285	0.226	4.42	0.0645	48.8	15.50	1.000
1.5	0.21	0.240	0.190	5.26	0.0456	82	21.9	1.414
(1.7)[f]	0.200	0.226	0.177	5.64	0.0400	100	25.0	1.613
2.0	0.18	0.202	0.160	6.25	0.0323	138	31.0	2.000
2.5	0.15	0.170	0.135	7.43	0.0228	232	43.8	2.828
	μm	μm	μm	μm	$mm^2 \times 10^{-3}$			
3.0	125	143	113	8.84	16.1	391	62.0	4.000
(3.2)[f]	120	135	106	9.41	14.4	463	69.4	4.480
3.5	105	120	95	10.51	11.4	657	87.7	5.657
(3.7)[f]	100	113	89	11.29	10.0	800	100	6.452
4.0	90	101	80.0	12.5	8.07	1,105	124	8.000
4.5	75	85	67.3	14.9	5.70	1,859	175	11.31
(4.7)[f]	70	79	62.0	16.1	4.90	2,331	204	13.17
5.0	65	71	56.6	17.7	4.03	3,126	248	16.00
(5.2)[f]	60	68	53.2	18.8	3.60	3,708	278	17.92
5.5	55	60	47.6	21.0	2.85	5,258	351	22.63
(5.7)[f]	50	56	44.3	22.6	2.50	6,400	400	25.81
6.0	45	50	40.0	25.0	2.02	8,842	496	32.00
(6.4)[f]	40	45	35.4	28.2	1.60	12,500	625	40.32
6.5	38	42	33.6	29.7	1.43	14,871	701	45.25
(6.7)[f]	35	39	31.0	32.2	1.23	18,659	816	52.67
7.0	32	36	28.3	35.4	1.008	25,010	992	64.00
(7.2)[f]	30	34	26.6	37.6	0.900	29,630	1,111	71.68
7.5	27	30	23.8	42.0	0.723	41,061	1,403	90.51
(7.7)[f]	25	28	22.2	45.1	0.625	51,200	1,600	103.23

	μm	μm	μm	μm	mm² × 10⁻⁶	×10⁶	×10³
8.0	22	25	20.0	50.0	504	0.0707	1.98	128.0
(8.4)f	20	23	17.7	56.4	400	0.1000	2.50	161.3
8.5	19	21	16.8	59.5	356	0.1190	2.81	181.0
9.0	16	18	14.1	70.7	252	0.200	3.97	256.0
(9.2)f	15	17	13.3	75.2	225	0.237	4.44	286.7
9.5	13	15	11.9	84.1	178	0.336	5.61	362.0
10.0	11	13	10.0	100	126	0.566	7.94	512.0
(10.3)f	10	11.3	8.86	119	100	0.800	10.00	645.2
10.5	9.4	10.6	8.41	110	89.1	0.952	11.22	724.1
(10.7)f	9.0	10.2	7.98	125	81.0	1.097	12.35	796.5
11.0	8.0	8.9	7.07	141	63.0	1.600	15.87	1,024
(11.4)f	7.0	7.9	6.20	161	49.0	2.332	20.41	1,317
11.5	6.7	7.5	5.95	168	44.6	2.692	22.45	1,448
(11.8)f	6.0	6.8	5.32	188	36.0	3.704	27.78	1,792
12.0	5.6	6.3	5.00	200	31.5	4.527	31.7	2,048
(12.3)f	5.0	5.6	4.43	226	25.0	6.40	40.0	2,581
12.5	4.7	5.3	4.20	238	22.3	7.61	44.9	2,896
13.0	4.0	4.5	3.54	283	15.8	12.80	63.5	4,096
13.5	3.3	3.7	2.97	336	11.1	21.54	89.8	5,793
(13.8)f	3.0	3.4	2.66	376	9.0	29.6	111.1	7,168
14.0	2.8	3.2	2.50	400	7.88	36.2	127	8,192
(14.3)f	2.5	2.8	2.22	451	6.25	51.2	160	10,323

a Feret's diameter = height between tangents; $d_F = a/l$. Values of d_n and d_F rounded to digits shown.
b Value of Heyn intercept or mean free path.
c Computation of n/v based on grains averaging to spherical shape for which $n/v = 0.5659(n/l)^3$.
d To obtain grains per mm² at 100x, multiply by 10^{-4}.
e The use of "00" is recommended instead of "-1" to avoid confusion.
f The G values shown in parentheses are calculated to one decimal place and correspond to some of the nominal "diameter" sizes (d_A) customarily used in reporting average grain size by the copper and brass industry.
Reprinted by permission of ASTM Standards, Philadelphia, Pennsylvania.

Lineal Intercept (Heyn) Method. In this method, the grain size is evaluated by counting (on a ground-glass projection screen, on a photomicrograph of a representative field of the specimen, on an eyepiece reticle, or on the specimen itself) the number of grains intersected by one or more straight lines (sufficiently long to give at least 50 intercepts). Grains touched by the end of the line are counted as half grains, while grains or grain boundaries intercepted by line elements are counted as full grains. Counts are made on at least three mutually perpendicular test directions to obtain a true average grain size. The product of the average number of intercepted grains n and magnification M, divided by the total line length l, gives the average number of intercepts per unit length n_l or grain diameter. That is,

$$n_l = \frac{n}{l/M} \tag{10.5}$$

and the mean intercept length $\bar{l}$ is given by

$$\bar{l} = \frac{1}{n_l} \tag{10.6}$$

The relationship between the mean lineal intercept $\bar{l}$ and the ASTM grain size G can be given by

$$G = -6.6353 \log \bar{l}(\text{in.}) - 12.6\bar{l}(\text{in.}) \tag{10.7}$$

$$G = -6.6457 \log \bar{l}(\text{mm}) - 3.928\bar{l}(\text{mm}) \tag{10.8}$$

For spheres and for uniform equiaxed grains, the relation between intercept size and planimetric average grain area $\bar{a}$ is

$$\bar{l} = \left(\frac{\pi\bar{a}}{4}\right)^{1/2} \tag{10.9}$$

The intercept method is also applied when measuring the grain size of two-phase structures, duplex grain structures, or all structures deviating from the uniform equiaxed form, such as elongated grains. It is more convenient to use than the planimetric method.

Other variations of the average mean lineal intercept method for oriented structures include the single circular test line suggested by Hilliard and the three concentric circular test lines developed by Abrams.

Planimetric (Jeffries) Method. In the planimetric method, a circle or a rectangle of known area (usually 79.8 mm in diameter or 5000 mm^2 in area) is inscribed on a photomicrograph or on the ground-glass screen of the metallograph. A magnification is then chosen which will give at least 50 grains in the field to be counted. The total number of equivalent whole grains within the total area is determined by the sum of (1) all the grains completely within the known area and (2) one-half the number of grains intersected by the circumference of the area. This equivalent number of whole grains is multiplied by Jeffries' multiplier, f, for the magnification of the specimen employed (Table 10.2) to obtain an estimate of the number of grains per square millimeter. A minimum of three fields is counted to secure a reasonable average. Thus, if the equivalent number of whole grains is observed to be 60 at a magnification of 100X, the number of grains per square millimeter is equal to 60×2, or 120. An accurate count requires marking of a photograph or overlay,

TABLE 10.2 Relationship between Magnification M Used and Jeffries' Multiplier f for an Area of 5000 mm² (a Circle of 79.8-mm Diameter) $(f = 0.0002 M^2)^{24}$

Magnification used M	Jeffries' multiplier f to obtain grains/mm²
1	0.0002
10	0.02
25	0.125
50	0.5
75[†]	1.125
100	2.0
150	4.5
200	8.0
250	12.5
300	18.0
500	50.0
750	112.5
1000	200.0

[†] At a 75-diameter magnification, Jeffries' multiplier f becomes unity if the area used is 5625 mm² (a circle of 84.5-mm diameter).
Reprinted by permission of ASTM Standards, Philadelphia, Pennsylvania.

which makes this method much slower than the intercept method. For obtaining equal accuracies the Jeffries method requires 60% more time than the intercept method.

In case of dispute, for all cases, the intercept method is preferred over the comparison and planimetric methods.

Automated Image-Analyzing Instruments. These have not been extensively employed to determine austenite grain size because they require a complete grain boundary delineation without any interior grain detail.

Snyder-Graff Intercept Method. This is a modified version of the intercept count method, developed by Snyder and Graff to measure the prior austenite grain size of high-speed tool steels which exists in the ASTM range of 9 to 12 where the fracture grain-size method, to be described later, is insensitive. In this method of grain-size rating, as-received or lightly tempered high-speed tool steels are polished and etched with an etchant (addition of 10% HCl to 3% nital). Subsequently, the etched surface is viewed on the ground-glass projection screen at 1000X. A number of 5-in. (127-mm) lines are drawn at random across the field, and the number of grains cut or interrupted by the lines is counted. The average of 10 such measurements is the Snyder-Graf (S-G) intercept grain-size number. In a high-speed steel, an intercept grain size of 6 to 8 is termed coarse, 8 to 12 is average, and 12 and above is fine-grained.

The relationship between the ASTM grain-size number G and the Snyder-Graff intercept count number S-G is given by

$$G = [6.635 \log(\text{S-G})] + 2.66 \qquad (10.10)$$

For conversion to mean lineal intercept, $\bar{l}$ is given by the following relationship:

$$\bar{l}(\text{mm}) = \frac{1}{n_l} = \frac{1}{[(\text{S-G}) \times 7.874]} \qquad (10.11)$$

$$\bar{l}(\text{in.}) = \frac{1}{n_l} = \frac{1}{[(\text{S-G}) \times 200]} \qquad (10.12)$$

Shepherd Fracture Grain Size. This test was introduced by Arpi in 1931 and later expanded by Shepherd. It consists of a standard set of 10 fracture grain-size specimens that have been specially prepared, ranging from specimen 1, an extremely coarse and pronounced intergranular-type fracture, to specimen 10, an extremely fine and fibrous or cleavage-type fracture. The grain-size measurement is accomplished by heat-treating (involving austenitizing for 1 hr at a suitable temperature and oil-quenching) the tool steel (which is about 1 in. thick and several inches long, usually without tempering) and fracturing by impact on a notched bar, usually at room temperature. The fracture plane should be transverse to the rolling direction. This is then given a fracture grain-size rating by comparing against a set of 10 fracture standards.

Note that the fracture grain-size numbers agree well with the corresponding ASTM grain size listed in Table 10.1. This coincidence makes the fracture grain sizes interchangeable with the austenite grain sizes measured microscopically for single-phase structures. This method of rating fracture specimens is only applied to high-carbon martensitic structures.[25]

10.4 RETAINED AUSTENITE

It was mentioned in Chaps. 8 and 9 that martensitic and bainitic transformation at sufficiently low temperature is often associated with some amount of retained (i.e., untransformed) austenite γ_R . During isothermal decomposition in the martensite range (between M_s and M_f temperatures), the thermal stabilization of austenite increases with time, and martensite transformation can commence only by a marked decrease in temperature.[27] The γ_R can occur in the quenched structure of steels at room temperature whether M_f lies above or below room temperature. However, in the former case, a small amount of γ_R will result. A significant amount of γ_R has been observed in plain carbon steels containing as low as 0.5% carbon, in low-alloy steels, and in Si-Mn TRIP steels with high (>1%) Si.[28,29] When the carbon content increases to 1.8%, the volume percent of γ_R rises to 70%. AISI 52100 ball-bearing steels contain 5 to 15% γ_R,[30] and Fe-(5–9) Ni steels used for cryogenic purposes contain 5 to 10 vol% γ_R, which is believed to contribute low-temperature toughness.[31] In other steels, particularly Fe-Cr-C, Fe->20%Ni-0.5%C, high-speed steel, and so forth, large amounts of γ_R occur, especially in large sections. The high-nickel gear steels such as 4620, 4820, and 9310 are likely to possess γ_R in their microstructure after heat treatment.

The isothermal austenite (retained) decomposition from the as-quenched structure proceeds immediately after the quenching operation but may continue at a diminishing rate for several months. However, the γ_R decomposes into carbide and ferrite by tempering the as-quenched steel at a relatively high temperature in the bainite range (see Chap. 14). Note that γ_R obtained by direct quenching contains more carbon and alloying elements in solution and may be characterized by chemical homogeneities.[32]

10.4.1 Role of Retained Austenite

10.4.1.1 Deleterious Effect. The deleterious effects of the presence of retained austenite in steel are reductions in mechanical properties, especially hardness, tensile strength, and dimensional stability, and greater susceptibility to the development of cracks at grinding.

Retained austenite is of the greatest concern when both high hardness ($\geq$60 Rc) and dimensional stability are desired. Heat-treated ball and roll bearings, dies, gauge blocks, and components for gauges and machine tools are examples of heat-treated products requiring high hardness and dimensional stability.[33] It is suspected by many that the significant growth (i.e., volume expansion) in the transformed regions produces additional concentration of stress and provides a site of crack nucleation and a path of crack propagation. It is thus clear that retained austenite, being an unstable phase, greatly influences steel properties. Its ultimate transformation to untempered martensite while the steel is in service may change mechanical behavior and cause unacceptable dimensional instability.[34]

The transformation of the γ_R to martensite in the gear case surface is a variable factor that does not offer beneficial effects. The first concern with gears that are to be shot-peened is that γ_R severely hinders the attainment of high compressive stress values (by shot-peening) in carburized and hardened gears. The second objection to the existence of high percentages of γ_R in gear surfaces is that it can result in premature wear of sliding and rolling gear teeth because of its soft microstructure. This, in turn, will cause tooth-surface pitting and case spalling, eventually leading to tooth failure. The third objection to the presence of γ_R in a case carburized part arises when it is subsequently ground because in some grinding situations, severe grinding burns and cracking are likely to be present.[35]

10.4.1.2 Beneficial Effect. Retained austenite tends to improve the fatigue properties of carburized and carbonitrided cases. For example, γ_R improves the contact fatigue resistance of case-hardened layer of carburizing Cr-Ni-W steel at low contact stresses and a large number of cycles.[32] The γ_R is advantageous in quenched tool steels and dual-phase steels. In the former it facilitates straightening of parts prior to complete transformation to martensite, and in the latter it increases the work-hardening rate and ductility at a particular strength level (see also Chap. 15). In M2 high-speed-steel, enhanced toughness is believed to occur from the presence of γ_R.[36]

Quarternary alloys based on Fe-Cr-C steel with Mn or Ni additions have been developed to reveal increasing amounts of γ_R with Mn contents up to 2% and Ni contents at 5% after quenching from 1100°C (2012°F). Figure 10.7 shows the extensive amount of interlath film of retained austenite found in the Fe-0.27%C-3.8%Cr-5%Ni steel upon high-temperature austenitization followed by interrupted quenching to room temperature.[37] A corresponding increase in strength and toughness properties has been found which is attributed to the production of dislocated lath martensite from a homogeneous austenite phase free from undissolved alloy carbides.[37]

The presence of γ_R in low-alloy ultrahigh-strength steels such as AISI 4340 and 300-M has been found to be beneficial for the plane strain fracture toughness (K_{IC}).[38] Two mechanisms have been postulated to explain this beneficial effect of γ_R on toughness.[39]

1. *Localized TRIP.* The mechanically produced phase transformation of γ_R to martensite in the plastic region absorbs additional energy, thereby effectively increasing the fracture toughness.

 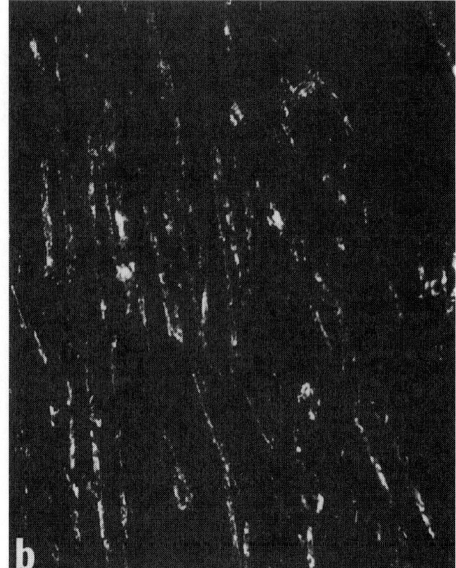

FIGURE 10.7 (*a*) Bright-field and (*b*) dark-field micrographs showing formation of extensive austenite in the Fe-0.27C-3.8Cr-5Ni steel (coarse-grained) upon high-temperature austenitization followed by interrupted quenching to room temperature.[37] (*Courtesy of B. V. Narsimha Rao and G. Thomas.*)

2. *Crack tip branching or blunting.* When an advancing crack encounters a retained austenite region, it is diverted and branched or blunted, causing enhanced energy absorption.

10.4.2 Factors Affecting the Formation of Retained Austenite

Alloy Content. Retained austenite may form from several causes, one of the most important being the alloy content. If sufficient alloying elements are added, the M_s and M_f temperatures are lowered to a marked degree. A 1% increase in C, Mn, Cr, Mo, and W content produces a 50%, 20%, 11%, 9%, and 8% increase, respectively, in as-quenched retained austenite. Thus the cold-work die steels containing high amounts of carbon and chromium and high-speed steel possess high percentages of retained austenite because of their high overall alloy content, while this constituent is low in low-carbon hot-work die steel.

Temperature. As the austenitizing temperature is raised, the amount of retained austenite is increased because of a greater degree of solution of carbon and other elements in the austenite acquired at the higher austenitizing temperatures. The effect of austenitizing temperature upon γ_R content is illustrated in Table 10.3 for a water-quenched steel containing 0.85% C and about 2% other alloying additions.

Thus, overheating tends to take more carbon into solution as a result of dissolution of more carbides and greater diffusion rates at higher temperature. With more carbon in solution, the M_s temperature decreases; this results in the reten-

TABLE 10.3 Effect of Austenitizing Temperature on Retained
Austenite[40]

Austenitizing temperature [°C (°F)]	Retained austenite (%)
843 (1550)	6.4
927 (1700)	12.4
1038 (1900)	47.0

tion of additional austenite at room temperature, to which the component is
quenched.[40]

Cooling Rate. The rate of cooling or severity of the quench also affects the
amount of γ_R in steel. If on quenching the rate of cooling in the M_s–M_f range is
slowed down, there is a greater amount of retained austenite formed than if the
cooling rate is continuous and rapid (see Chap. 8). This is the reason why there is
more γ_R in steels hardened by oil-quenching than in the same steels hardened by
water- or brine-quenching. The oil-quenching produces a slower cooling rate in the
lower temperature range.

Stresses. The amount and distribution of stress in a quenched steel also influence
the amount of γ_R.

10.4.3 Elimination

Several methods of eliminating or reducing transformation of retained austenite
such as cold treatment (or refrigeration), deep cryogenic treatment, tempering, cold-
working, shock, ultrasonic vibration, and correct carbon potential and austenitizing
temperature are available.[39,41]

Refrigeration (or Subzero Cold Treatment). Generally carried out at a temperature of −80 to −196°C (−112 to −321°F), this leads to transformation of all except
about 1% γ_R to martensite. Cold (or subzero) treatment, however, does not seem to
decrease bend ductility; rather it decreases impact fatigue resistance. One drawback
of refrigeration is that it forms coarse martensitic crystals. These large crystals tend
to develop subsurface microcracks unlike finer martensite crystals developed in a
properly carburized and hardened heat-treating cycles.[35] Refrigeration is most effective when done (prior to tempering) directly after the quench. When some quench-
hardened steels are refrigerated, further transformation to martensite progresses.
This results in greater stability (less change in the finished part size) and, in the case
of anticipated grinding, less likelihood of grinding burns and cracks.[35]
    Four methods are used for the subzero treatment:[42]

1. *Evaporation of dry ice (CO_2 in the Solid State).* This method provides a temperature of −75 to −78°C (−103 to −108°F) and is employed for a small number and
   weight of parts.
2. *Circulating air that has been cooled in a heat exchanger.* Such metal treating freezers have been set up with a capacity to cool 270 to 680 kg of parts to −85°C
   (−122°F) in about 2 hr.

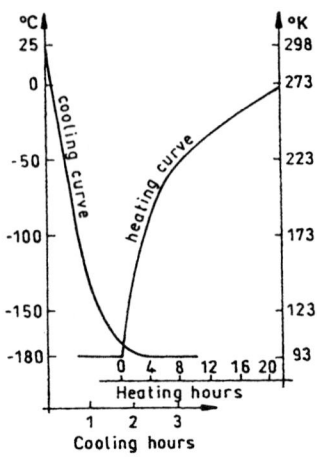

FIGURE 10.8 Cooling curve from room temperature to −180°C (−292°F) and the natural reheating curve from −180 to 0°C (−292 to 32°F) of an empty container of 100 dm³ space, connected to a cryogenerator.[42] (*Reprinted by permission of Marcel Dekker, Inc., New York.*)

3. *Evaporation of liquid N.* This subzero treatment equipment is used to cool small quantities of parts down to −180°C (−292°F) which can be reached in less than 10 min.

4. *In a container connected to a cryogenerator.* This method results in subzero treatment of large number of parts to a temperature up to −190°C (−310°F). The cryogenerator works on the principle of the Stirling cycle. By continuous circulation of air, the working space with parts is slowly cooled to the desired temperature. Figure 10.8 illustrates the cooling curve from room temperature to −180°C (−292°F) and the natural reheating curve from −180 to 0°C (−292 to 32°F) for an empty container of 100 dm³ space, connected to a cryogenerator. It is seen that a temperature of −120°C (−184°F) can be reached after cooling for 1 hr, but to reach a temperature of −180°C, an additional 1.5 hr is required. The natural reheating from −180 to 0°C (−292 to 32°F) takes nearly 20 hr.[42]

Advantages of deep cryogenic treatment on tool steels include improvements in hardness and/or wear resistance. There are two mechanisms involved in the occurrence of these changes: (1) The rapid transformation of γ_R causes the increase in hardness, and (2) "low-temperature conditioning" or partial decomposition of martensite, usually at liquid N temperature, results in a time-dependent distribution of fine alloy carbides during tempering in the microstructure, giving rise to improvement in wear resistance.[43]

Ultrasonic Vibration. Powerful ultrasonic waves significantly reduce the γ_R content in quenched steels and quenched carburized steels. This is attributed to the increased number of martensite nucleation sites during the ultrasonic treatment of steel above the M_s temperature.[44]

Correct Carbon Potential and Austenitizing Temperature. There are two methods to prevent the development of γ_R: (1) use of the correct carbon potential during the carburizing and diffusion portions of heat treatment cycles and (2) avoidance of quenching from a very high austenitizing temperature.[35]

10.4.4 Determination of Retained Austenite

In view of the fact that some product specifications for aircraft gears and precision bearings, gauges, and so forth quote quantitative limits for retained austenite content, it is necessary to determine accurately the amount of retained austenite in hardened steels for good quality control in production and to follow the process of transformation in research work. Industrial laboratories use X-ray diffraction techniques for measurement of retained austenite in amounts ranging from 16% to as little as 0.1%. It is very easy to find an accuracy of ±1% by this method. However, the smaller the quantity of γ_R to be measured, the poorer the precision of the measurement.[34] It takes a lot of time to get much better results. Other methods for measuring γ_R in heat-treated steels are metallographic, dilatometric, magnetic saturation analysis, Mössbauer effect spectroscopy (MES),[45] and dark-field electron microscopy methods.

10.5   AUSTENITIC STAINLESS STEELS

Stainless steels may be defined as complex alloy steels containing a minimum of 11 to 12% chromium in solid solution with or without other elements to ensure the formation of a continuous thin film of protective Cr-rich oxide (Cr_2O_3) on the surface, the so-called *passive layer*. The advantages of stainless steels over other steels and nonferrous metals for many applications are due to a combination of unique properties, such as resistance to corrosion in any environments, their superior resistance to oxidation and scaling at very high temperatures, and their good mechanical properties over an extremely wide temperature range.[46] Historically, stainless steels have been classified by microstructure and are described as ferritic, martensitic, austenitic, superaustenitic, duplex, superduplex, precipitation-hardening, and cast grades.

Ferritic stainless steels containing only Cr and no Ni, with the balance mostly Fe, are based upon the type 430 composition of 17% Cr. Type 409 with a nominal composition of 11% Cr is the most widely used ferritic grade of stainless steel. This nickel-free grade is the least expensive stainless steel. Type 446 (25% Cr) is the highest chromium grade in the conventional range of stainless steels and offers the best oxidation resistance. Ferritic stainless steels remain ferritic at all temperatures and never reach the austenitic state and are magnetic. They possess good ductility and formability, but relatively poor elevated-temperature strength. They cannot be hardened or strengthened by heat treatment and are generally less ductile than the austenitic types. In some cases, they can also be subjected to nitriding surface-hardening treatment.[47] They can be moderately hardened by cold-working and softened by annealing. They are more corrosion-resistant than the martensitic types, but generally inferior to austenitic grades. They have limited low-temperature and large-section toughness.[47] Their ductile-to-brittle transition is a function of grain size, carbon, nitrogen, and second phases. The σ-phase formation and *475°C embrittlement* occur in these alloys. The loss of toughness is attributed to the precipitation of a Cr-rich α' phase which becomes more predominant at higher Cr content. However, 475°C embrittlement can be eliminated by reheating to a temperature of about 600°C and rapid cooling to room temperature.[48] They are used as decorative (or highly polished) trim, sinks, and automotive applications such as exhaust systems; in food processing; and for consumer goods.[46]

Martensitic stainless steels (which also carry the 400 series number) are based upon the most popular grade type 410 composition with straight 12% Cr, 0.12% C, no Ni, and balance mostly Fe; however, Ni is used in some martensitic stainless steels such as type 431 to expand the γ loop and allow higher Cr in solid solution. They contain higher C plus N to expand the γ loop, and C and Cr are balanced to ensure austenitization. They are magnetic and are hardened by heating above the transformation temperature to produce a fully austenitic structure and subsequent rapid cooling in oil or air. Depending on the strength requirement, 12% Cr steels are tempered up to about 675°C where the tempering parameter can be given by $T(20\log t) \times 10^{-3}$, where T is the temperature in kelvins and t is the time in hours. Types 440A, 440B, and 440C are highly hardenable martensitic stainless steels with progressively increasing carbon content.

Intermediate tempering temperatures (450 to 550°C) are usually avoided because embrittlement and degradation of corrosion resistance are often encountered in this range. At the upper end of this range, the steepness of the tempering curve (hardness or strength versus temperature) makes it difficult to obtain consistent results. However, high Cr levels and high C content limit corrosion resistance of martensitic steels when compared to other stainless steels.[49,51]

Martensitic stainless steels possess maximum corrosion resistance in the hardened state. They can be annealed for best cold-working and machining characteristics. They are used where hardness, strength, and wear resistance are required. Notable examples are cutlery, knife blades, razor blades, and blades in turbine engines, aerospace equipment, petroleum production and refining, vegetable choppers, food processing equipment, surgical instruments, magnets, firearms, instrument bearings, wear-resistant parts, valves and stems, and so forth.[46,49]

Cast stainless steels are similar to equivalent wrought alloys. They are specified according to whether they are mainly intended for corrosion-resistance applications (and designated by the letter C) or for high-temperature applications (and designated by the letter H). Mostly the cast alloys are direct derivatives of one of the wrought grades, as CF-8 is equivalent to the wrought type 304 and CK and CH are equivalent to wrought type 310 with a difference in carbon content. These above three grades are not discussed in the following sections.

Among other grades, the structures, heat treatment, properties, and applications of precipitation-hardening (PH) stainless steels are discussed in Chap. 6. Here, austenitic and duplex stainless steels are treated. Between these two grades, greater emphasis has been given to the austenitic types because they constitute about 66 to 77% of the total world stainless steel production. Their growth rate of production and consumption in the world in recent years has exceeded that of most other classes of engineering materials. They have occupied a dominant position because of their higher corrosion and oxidation resistance and fabricability, their excellent cryogenic properties, their enhanced mechanical properties such as strength and toughness at both elevated and ambient temperatures, and their aesthetic appeal, and because of varied specific combinations of properties that can be obtained by different compositions within the group.[50] For this reason, they are used in a wide variety of applications such as consumer products, transportation, architecture, food and beverage, chemical and petrochemical, paper and pulp, pharmaceutical and biotechnology, semiconductor, energy, environmental, and aerospace.

In general, austenitic stainless steels have an fcc structure that is obtained by the liberal use of austenitizing elements such as Ni, Mn, and N. These steels are practically nonmagnetic in the annealed state and can be hardened only by cold-working. The 200 series (or Cr-Mn-Ni type) austenitic steels contain N, 4 to 15% Mn, and up to 7% Ni. The 300 series (or Cr-Ni type) steels contain about 16 to 30% Cr, 7 to

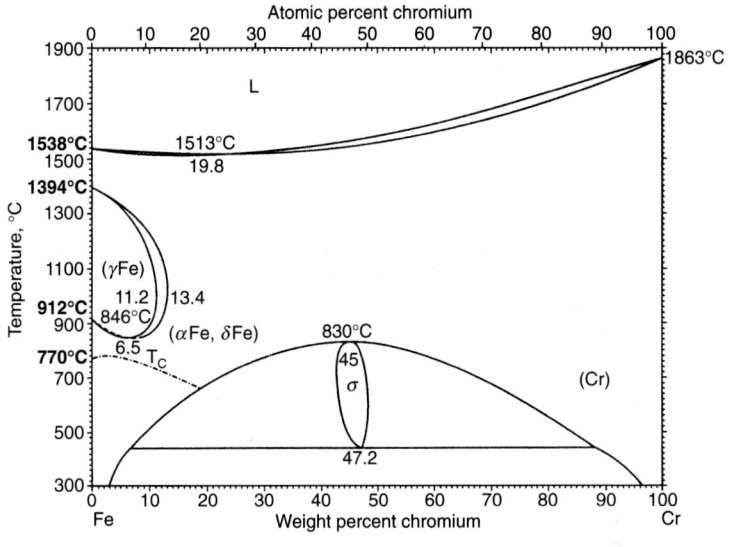

FIGURE 10.9 The binary iron-chromium equilibrium diagram with description of phases (from H. Okamoto, 1990) in the adjoining table.[55] (*Reprinted by permission of ASM International, Materials Park, Ohio.*)

Phase	Composition, wt% Cr	Pearson symbol	Space group
(αFe, Cr)	0 to 100	*cI*2	$Im\bar{3}m$
(γFe)	0 to 11.2	*cF*4	$Fm\bar{3}m$
σ	42.7 to 48.2	*tP*30	$P4_2/mnm$

35% Ni, up to 2% Mn, and up to 0.15% C. The alloying elements such as Cr, V, Ni, N, Cu, Mo, Si, Ag, and W are added to achieve beneficial effect on halide pitting resistance or oxidation resistance. Rare-earth metals may be used to enhance oxidation resistance.[51] S or Se may be added to certain grades to enhance machinability. They tend to become slightly magnetic by cold-working.

Superaustenitic stainless steels are alloys with large addition of Ni, Mo, or Cu and sometimes silicon or stabilizing elements. They are easily formed and welded and are made as bars or castings; but for the high-alloy products, most foundries prefer to cast duplex alloys, which are less prone to hot tearing.[52]

It is worthwhile to start with a brief discussion of individual constituent binary systems such as Fe-Cr, Cr-Ni, and Fe-Ni, followed by pseudobinary Fe-Cr-Ni systems.

10.5.1 Constituent Binary Phase Diagrams

Figure 10.9 shows the binary iron-chromium equilibrium diagram illustrating the chromium limits of the occurrence of the gamma loop up to 12% Cr. There is a narrow two-phase (fcc) austenite-(bcc) ferrite region between 11.2 and 13.4% Cr.

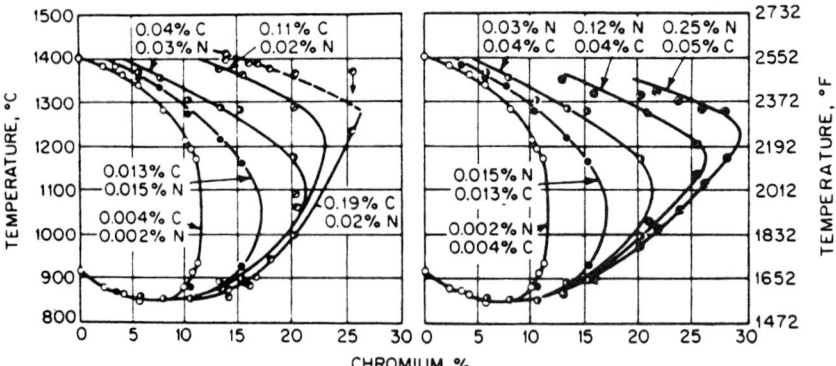

FIGURE 10.10   Shifting of the boundary line $(\gamma + \alpha)/\alpha$ in the Cr-Fe system through increasing additions of carbon or nitrogen.[54] (*Reprinted by permission of ASM International, Materials Park, Ohio.*)

Both above and below the gamma loop as well as above 13.4% Cr, an extensive single-phase ferrite field occurs. The ferrite is generally called *delta (δ) ferrite* because of its continuous existence right from the melting point down to room temperature. The miscibility gap occurs farther to the right where an intermetallic phase (Fe-Cr compound), referred to as the *sigma (σ) phase* with a tetragonal structure, forms below 830°C.[53–55] This will be discussed in a later section.

The dashed line in Fig. 10.9 represents the magnetic (Curie) transformation point above which iron is paramagnetic and below which it is ferromagnetic. Paramagnetic iron is essentially nonmagnetic, having a permeability only slightly above 1.00. Ferromagnetic iron is magnetic, having a permeability much greater than 1.00, the exact magnitude depending on the composition. When the fcc (γ) austenite structure is maintained at lower or ambient temperature, it will remain paramagnetic, except that Fe-Cr-Ni-N alloys become antiferromagnetic as the temperature approaches 0 K.[50,51]

The addition of other austenite stabilizers, especially carbon and nitrogen, shifts austenite plus ferrite fields, as well as the outside boundary of the austenite, to a higher chromium concentration, as shown in Fig. 10.10.[54,55]

The binary Ni-Cr equilibrium phase diagram is shown in Fig. 10.11a, illustrating the eutectic reaction and peritectic formation of the Ni_2Cr (γ) phase with an orthorhombic structure. The eutectic reaction occurs at 1345°C (2435°F) with eutectic composition at 56 at% Cr. This system is also characterized by the dramatic decrease in solubility of Ni in the α-Cr phase with temperature and the peritectoid decomposition of γ + α to γ′ (the ordered Ni_2Cr phase) at 590°C (1094°F).[55]

Figure 10.11b is a binary equilibrium diagram for the Fe-Ni system.[55] The peritectic reaction occurs at 1514°C (2257°F), where liquid and δ-ferrite react to give austenite; the eutectoid reaction into ferrite and γ′ (L1₂ ordered fcc Fe_3Ni) takes place at 347°C (657°F) and 53 wt% Ni.[55]

10.5.2   Pseudobinary Systems Based on Ternary Additions to Fe-C Alloys

Figure 10.12a and b shows the pseudobinary phase diagrams demonstrating the stability for various phases for Fe-18%Cr-4%Ni alloy and Fe-18%Cr-8%Ni alloy,

Phase	Composition, wt % Cr	Pearson symbol	Space group
(Ni)	0 to 47.0	cF4	Fm$\bar{3}$m
Ni$_2$Cr or γ'	21 to 37	oI6	Immm
(Cr)	65 to 100	cI2	Im$\bar{3}$m
Metastable phases			
σ	~28	tP30	P4$_2$/mnm
δ	100	cP8	Pm$\bar{3}$m

Phase	Composition, wt % Cr	Pearson symbol	Space group
(δFe)	0 to 3.7	cI2	Im$\bar{3}$m
(γFe, Ni)	0 to 100	cF4	Fm$\bar{3}$m
(αFe)	0 to 5.8	cI2	Im$\bar{3}$m
Fe$_3$Ni(a)	26	cP4	Pm$\bar{3}$m
FeNi(a)	51	tP2	P4/mmm
FeNi$_3$	64 to ~90	cP4	Pm$\bar{3}$m

(a) Metastable

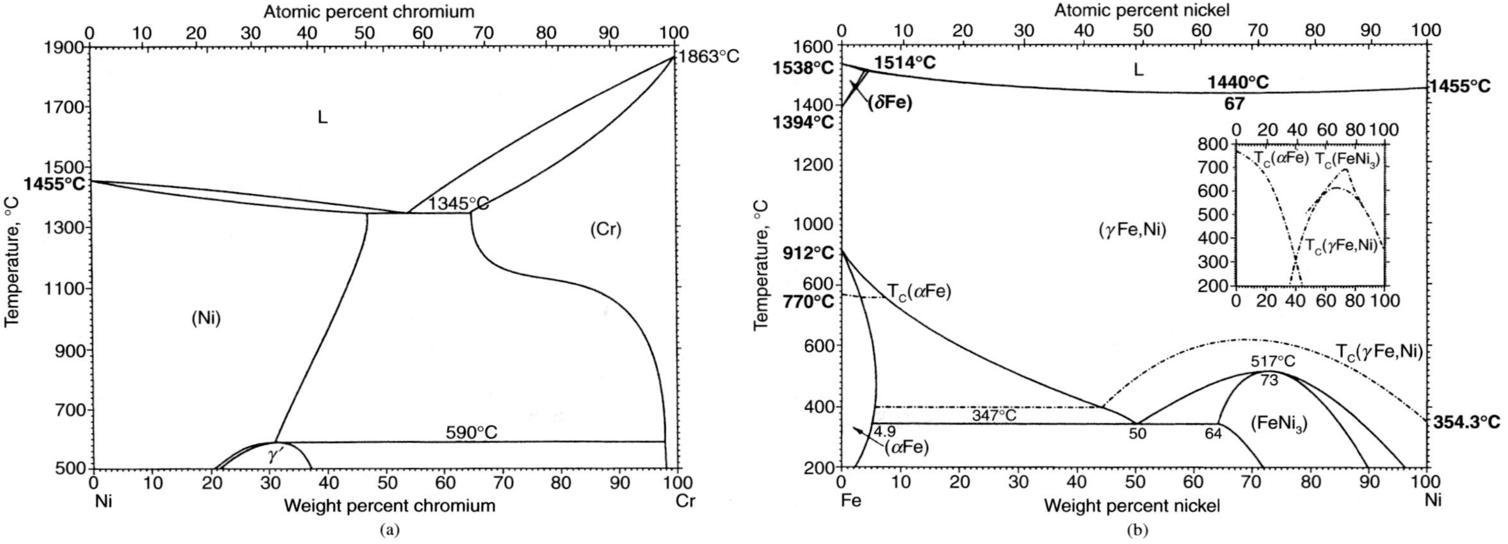

FIGURE 10.11 The binary (a) nickel-chromium equilibrium diagram (phase table from P. Nash, 1991) and (b) iron-nickel equilibrium diagram (phase table from L. J. Schwartzendruber et al., 1992).[55] (Reprinted by permission of ASM International, Materials Park, Ohio.)

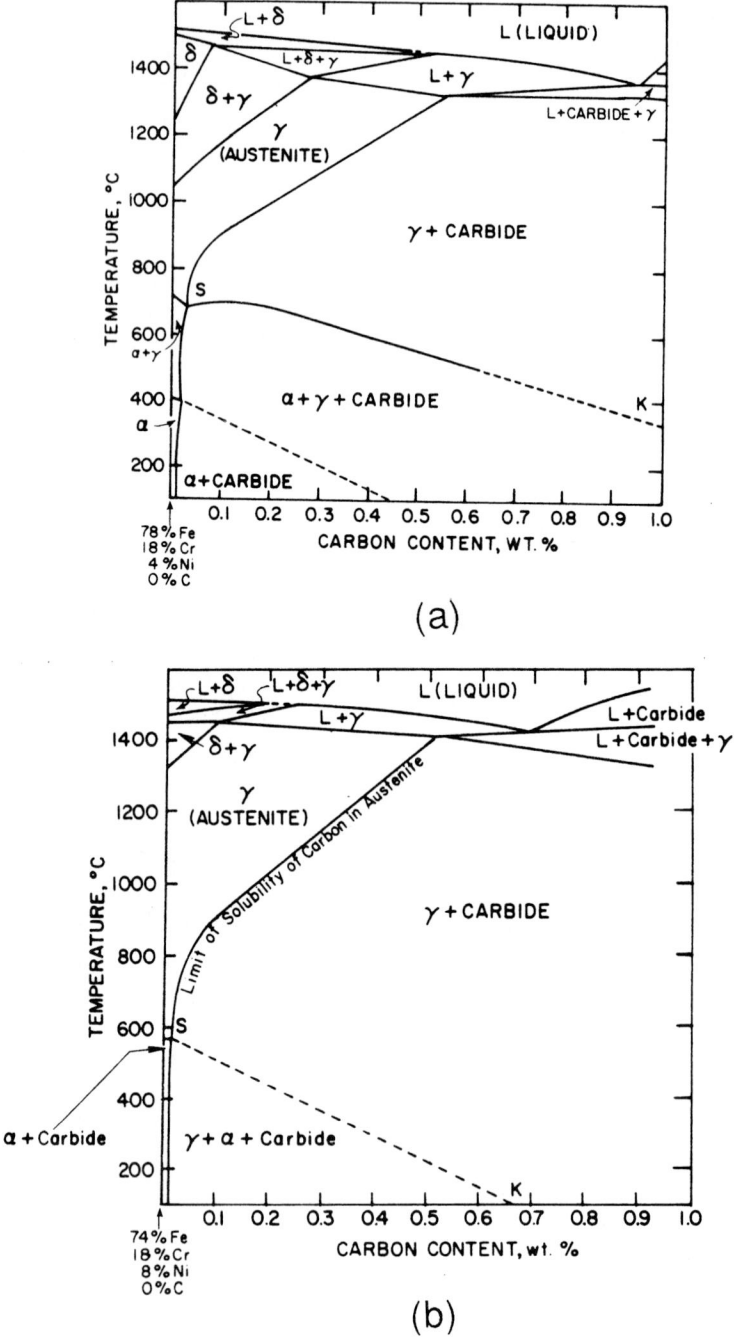

FIGURE 10.12 The pseudobinary phase diagrams showing the stability of various phases for (*a*) Fe-18%Cr-4%Ni alloy and (*b*) Fe-18%Cr-8%Ni alloy with varying carbon content.[56] (*Reprinted by the permission of John Wiley & Sons, New York; after Krivobok.*)

respectively, with varying carbon content.[56] The addition of C, N, Cu, and possibly Mn expands the range of Cr contents over which austenite can be formed at elevated temperatures.[57] The first step in the heat treatment of a steel containing low-carbon (0.1–0.2%)-18%Cr-4%Ni consists of solution annealing at 1100°C, followed by quenching, which produces an unstable austenitic structure; this structure can be readily transformed by tempering at 650°C or by cold-working. However, when the nickel content is sufficiently high, as in Fe-18Cr-8Ni-C systems, the $\gamma + \alpha$ phase field moves to a very high temperature range and to low carbon content, thereby extending the austenite phase field. Hence, for austenitic 18-8 stainless steel containing 0.03% or greater carbon content, a heat treatment operation comprising solution annealing at 1050 to 1100°C (1922 to 2012°F) and subsequent quenching produces a supersaturated austenite solid solution at room temperature and suppresses the carbide formation. However, if this alloy is slowly cooled after solution annealing or supersaturated austenite is thermally aged within the lower temperature range of 550 to 800°C, then carbide precipitation occurs, usually the chromium-rich $M_{23}C_6$ carbides at the grain boundaries, lattice defects, and so forth, even if the carbon content of the alloy is low (<0.005%).

For understanding δ-ferrite $\rightarrow$ austenite transformation in stainless steel weldments, Fe-Cr-Ni pseudobinary phase diagrams at three levels of iron such as 70, 60, and 55% are extremely useful (Fig. 10.13a, b, and c).[58,59] Each diagram has a eutectic triangle (labeled $\gamma + \delta + L$), which represents composition where austenite, ferrite, and liquid metal can all coexist. On the left (Ni-rich) side of each diagram are compositions which solidify as austenite and remain austenite at room temperature during the initial stages of solidification, but some ferrite usually appears during the last stage of solidification. This is called *primary austenite solidification*. Further cooling can result in ferrite transforming to austenite under equilibrium conditions; but at considerably rapid cooling conditions of welding, some of this ferrite can survive to room temperature.[57,58]

Close to, and within, the right side of the eutectic triangle are compositions that solidify as ferrite during the initial stages of solidification; however, some austenite can appear during the last stages of solidification. This is called *primary ferrite solidification*. Further cooling can result in ferrite transforming to austenite under equilibrium conditions, but at the considerably rapid cooling conditions of welding, some of this ferrite usually survive to room temperature.[57,58]

Further to the right in Fig. 10.13 are compositions that solidify fully as ferrite, but, upon further cooling under equilibrium conditions, transform either partly or fully to austenite. Finally, still farther to the right of the eutectic triangle in Fig. 10.13 are compositions that will never form austenite at any temperature and are called *ferritic stainless steels*.[57,58]

High-nickel alloys are very resistant to chloride stress corrosion cracking (SCC), as are nickel-free alloys. Fe-Cr-Ni alloys with ~8% Ni are the most susceptible to SCC.[51]

It is impractical to design a "wholly" austenitic weldment by simply using a high-Ni (or fully austenitic) filler metal because it would be prone to hot-cracking and/or catastrophic failure if such a weldment were used under conditions conducive to SCC.[60]

10.5.3 Classification, Properties, and Applications

The American Iron and Steel Institute (AISI) has designated the wrought standard stainless steels by three-digit numbers. The austenitic steels are designated by

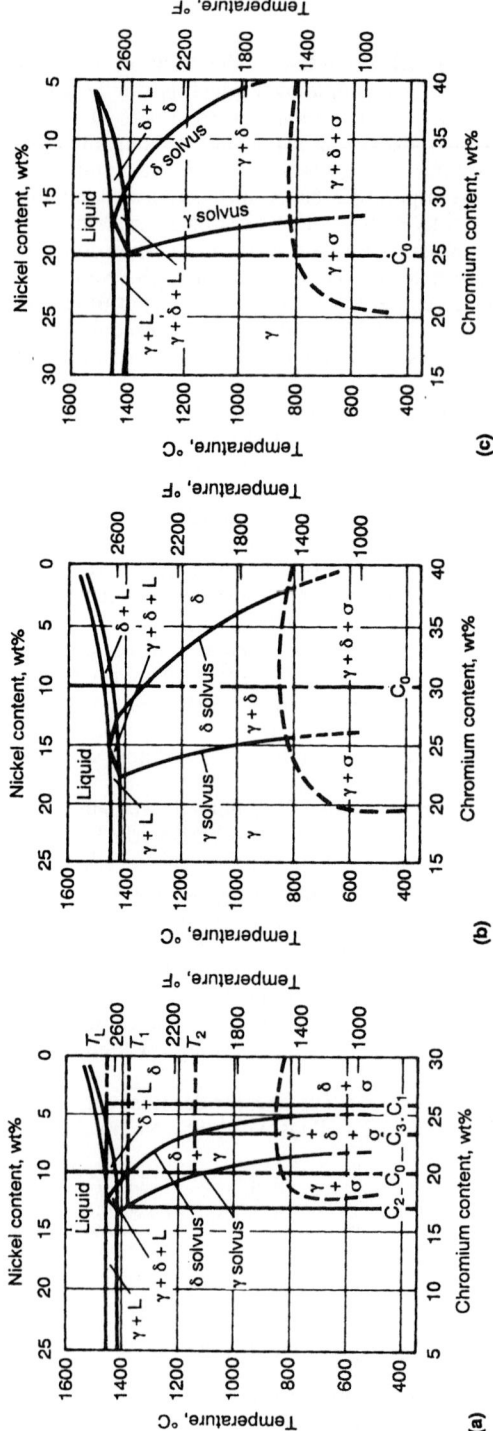

FIGURE 10.13 Iron-chromium-nickel pseudobinary diagram: (a) 70 wt% Fe, (b) 60 wt% Fe, and (c) 55 wt% Fe.[57] (*Reprinted by permission of ASM International Materials Park, Ohio.*)

numbers in the 200 and 300 series, whereas the ferritic and martensitic stainless steels are designated by numbers in the 400 series. For example, types 201, 301, 304, 310, and 316 represent the more common austenitic stainless steels; types 430 and 446, the ferritic stainless steels; and types 410, 420, and 440C, the martensitic stainless steels. In the Unified Numbering System (UNS) designation, stainless steels are identified by the letter "S" followed by five digits. A few stainless steels are given letter "N" in the UNS system because of their very high nickel content. Note that AISI stopped issuing new stainless steel designations two or three decades ago. For a time, ASTM assigned "XM" designations, but later stopped. Now the UNS designations are used by ASTM and ASME to describe various alloys.[51]

Tables 10.4 and 10.5 list the compositions of standard and nonstandard austenitic and superaustenitic stainless steels, respectively.[61,62] Not all the types in Table 10.4 are official AISI designations. Tables 10.6 and 10.7 list the mechanical properties of standard austenitic and high-nitrogen austenitic stainless steels, respectively.[62] Table 10.8 represents the suggested stress-relieving treatments for some standard austenitic stainless steels.[63] Note that most developments in the standard types have been made from the parent austenitic type 304, which is commonly called 18-8 stainless steel. Broadly, austenitic stainless steels can be grouped into.[53,54,61]

1. Manganese-substituted nonstabilized grades (types 201, 202, etc.)
2. The basic 18-8 stainless steels (types 302 and 304)
3. The metastable (lower-alloy) steels with high work-hardening rate (type 301)
4. The higher-nickel grade with suppressed rate of work-hardening and improved deep drawing (type 305)
5. The highly corrosion-resistant molybdenum-bearing steels (types 316, 317, 384)
6. The stabilized grades (types 321, 347, and 348)
7. The extra-low-carbon (ELC) steels (types 304L, 316L, and 317L)
8. The welding electrode grade (type 308)
9. The highly alloyed oxidation-resistant grades (types 308, 309, and 310)
10. The free-machining grades (types 303 and 303Se)
11. Dual-marked grades (types 304/304L and 316/316L)
12. The H grades

These classifications are not straightforward because of the overlapping effects, as will be discussed in the following sections.

10.5.3.1 Manganese-Substituted Grades (Types 201 and 202).
In the 200 series of austenitic grades, ~4% nickel is replaced by 7% manganese and 0.25% (maximum) nitrogen. Since these grades were introduced as a substitute, they are sometimes considered as substandard. However, their important features are as follows:

1. Their yield strengths are about 40% greater than those of the corresponding 300 series steels, and the densities are about 3% less.
2. These types are less expensive to produce than those of the 300 series alloys.
3. Like other austenitic grades, the notch ductility, measured by the impact test method, is higher, showing no transition temperature.

(Continues on page 10.47)

TABLE 10.4 Compositions of Standard Austenitic Stainless Steels[61]

Type	UNS no.	Composition (%)[†]							
		C	Mn	Si	Cr	Ni	P	S	Other
201	S20100	0.15	5.5–7.5	1.00	16.0–18.0	3.5–5.5	0.06	0.03	0.25 N
202	S20200	0.15	7.5–10.0	1.00	17.0–19.0	4.0–6.0	0.06	0.03	0.25 N
205	S20500	0.12–0.25	14.0–15.5	1.00	16.5–18.0	1.0–1.75	0.06	0.03	0.32–0.40 N
301	S30100	0.15	2.0	1.00	16.0–18.0	6.0–8.0	0.045	0.03	...
302	S30200	0.15	2.0	1.00	17.0–19.0	8.0–10.0	0.045	0.03	...
302B	S30215	0.15	2.0	2.0–3.0	17.0–19.0	8.0–10.0	0.045	0.03	...
303	S30300	0.15	2.0	1.00	17.0–19.0	8.0–10.0	0.20	0.15 min	0.6 Mo[‡]
303Se	S30323	0.15	2.0	1.00	17.0–19.0	8.0–10.0	0.20	0.06	0.15 min Se
304	S30400	0.08	2.0	1.00	18.0–20.0	8.0–10.5	0.045	0.03	...
304H	S30409	0.04–0.10	2.0	1.00	18.0–20.0	8.0–10.5	0.045	0.03	...
304L	S30403	0.03	2.0	1.00	18.0–20.0	8.0–12.0	0.045	0.03	...
304LN	S30453	0.03	2.0	1.00	18.0–20.0	8.0–10.5	0.045	0.03	0.10–0.16 N
302Cu	S30430	0.08	2.0	1.00	17.0–19.0	8.0–10.0	0.045	0.03	3.0–4.0 Cu
304N	S30451	0.08	2.0	1.00	18.0–20.0	8.0–10.5	0.045	0.03	0.10–0.16 N
305	S30500	0.12	2.0	1.00	17.0–19.0	10.5–13.0	0.045	0.03	...
308	S30800	0.08	2.0	1.00	19.0–21.0	10.0–12.0	0.045	0.03	...
309	S30900	0.20	2.0	1.00	22.0–24.0	12.0–15.0	0.045	0.03	...
309S	S30908	0.08	2.0	1.00	22.0–24.0	12.0–15.0	0.045	0.03	...
310	S31000	0.25	2.0	1.50	24.0–26.0	19.0–22.0	0.045	0.03	...
310S	S31008	0.08	2.0	1.50	24.0–26.0	19.0–22.0	0.045	0.03	...

314	S31400	0.25	2.0	1.5–3.0	23.0–26.0	19.0–22.0	0.045	0.03	...
316	S31600	0.08	2.0	1.00	16.0–18.0	10.0–14.0	0.045	0.03	2.0–3.0 Mo
316F	S31620	0.08	2.0	1.00	16.0–18.0	10.0–14.0	0.20	0.10 min	1.75–2.5 Mo
316H	S31609	0.04–0.10	2.0	1.00	16.0–18.0	10.0–14.0	0.045	0.03	2.0–3.0 Mo
316L	S31603	0.03	2.0	1.00	16.0–18.0	10.0–14.0	0.045	0.03	2.0–3.0 Mo
316LN	S31653	0.03	2.0	1.00	16.0–18.0	10.0–14.0	0.045	0.03	2.0–3.0 Mo; 0.10–0.16 N
316N	S31651	0.08	2.0	1.00	16.0–18.0	10.0–14.0	0.045	0.03	2.0–3.0 Mo; 0.10–0.16 N
317	S31700	0.08	2.0	1.00	18.0–20.0	11.0–15.0	0.045	0.03	3.0–4.0 Mo
317L	S31703	0.03	2.0	1.00	18.0–20.0	11.0–15.0	0.045	0.03	3.0–4.0 Mo
321	S32100	0.08	2.0	1.00	17.0–19.0	9.0–12.0	0.045	0.03	5 × %C min Ti
321H	S32109	0.04–0.10	2.0	1.00	17.0–19.0	9.0–12.0	0.045	0.03	5 × %C min Ti
330	N08330	0.08	2.0	0.75–1.5	17.0–20.0	34.0–37.0	0.04	0.03	...
347	S34700	0.08	2.0	1.00	17.0–19.0	9.0–13.0	0.045	0.03	10 × %C min Nb
347H	S34709	0.04–0.10	2.0	1.00	17.0–19.0	9.0–13.0	0.045	0.03	8 × %C min–10 max Nb
348	S34800	0.08	2.0	1.00	17.0–19.0	9.0–13.0	0.045	0.03	0.2 Cu; 10 × %C min Nb; 0.10 Ta
348H	S34809	0.04–0.10	2.0	1.00	17.0–19.0	9.0–13.0	0.045	0.03	0.2 Cu; 10 × %C min– 1.0 max Nb; 0.10 Ta
384	S38400	0.08	2.0	1.00	15.0–17.0	17.0–19.0	0.045	0.03	...

† Single values are maximum values unless otherwise indicated.
‡ Optional.
Reprinted by permission of ASM International, Materials Park, Ohio.

TABLE 10.5 Compositions of Nonstandard Austenitic and Superaustenitic Stainless Steels[61,62]

Designation*	UNS no.	Composition (%)†							
		C	Mn	Si	Cr	Ni	P	S	Other
Gall-Tough	S20161	0.15	4.00–6.00	3.00–4.00	15.0–18.0	4.00–6.00	0.040	0.040	0.08–0.20 N
203 EZ (XM-1)	S20300	0.08	5.0–6.5	1.00	16.0–18.0	5.0–6.5	0.040	0.18–0.35	0.5 Mo; 1.75–2.25 Cu
Nitronic 50 (XM-19)	S20910	0.06	4.0–6.0	1.00	20.5–23.5	11.5–13.5	0.040	0.030	1.5–3.0 Mo; 0.2–0.4 N; 0.1–0.3 Nb; 0.1–0.3 V
Tenelon (XM-31)	S21400	0.12	14.5–16.0	0.3–1.0	17.0–18.5	0.75	0.045	0.030	0.35 N
Cryogenic Tenelon (XM-14)	S21460	0.12	14.0–16.0	1.00	17.0–19.0	5.0–6.0	0.060	0.030	0.35–0.50 N
Esshete 1250	S21500	0.15	5.5–7.0	1.20	14.0–16.0	9.0–11.0	0.040	0.030	0.003–0.009 B; 0.75–1.25 Nb; 0.15–0.40 V
Type 216 (XM-17)	S21600	0.08	7.5–9.0	1.00	17.5–22.0	5.0–7.0	0.045	0.030	2.0–3.0 Mo; 0.25–0.50 N
Type 216 L (XM-18)	S21603	0.03	7.5–9.0	1.00	17.5–22.0	7.5–9.0	0.045	0.030	2.0–3.0 Mo; 0.25–0.50 N
Nitronic 60	S21800	0.10	7.0–9.0	3.5–4.5	16.0–18.0	8.0–9.0	0.040	0.030	0.08–0.18 N
Nitronic 40 (XM-10)	S21900	0.08	8.0–10.0	1.00	19.0–21.5	5.5–7.5	0.060	0.030	0.15–0.40 N
21-6-9 LC	S21904	0.04	8.00–10.00	1.00	19.00–21.50	5.50–7.50	0.060	0.030	0.15–0.40 N
Nitronic 33 (18-3 Mn)	S24000	0.08	11.50–14.50	1.00	17.0–19.00	2.50–3.75	0.060	0.030	0.20–0.40 N
Nitronic 32 (18-2 Mn)	S24100	0.15	11.00–14.00	1.00	16.50–19.50	0.50–2.50	0.060	0.030	0.20–0.40 N
18-18 Plus	S28200	0.15	17.0–19.0	1.00	17.5–19.5	…	0.045	0.030	0.5–1.5 Mo; 0.5–1.5 Cu; 0.4–0.6 N
303 Plus X (XM-5)	S30310	0.15	2.5–4.5	1.00	17.0–19.0	7.0–10.0	0.020	0.25 min	0.6 Mo
153MA‡	S30415	0.05	0.60	1.30	18.5	9.50	…	…	0.15 N; 0.04 Ce
304B1§	S30424	0.08	2.00	0.75	18.0–20.00	12.00–15.00	0.045	0.030	0.10 N; 1.00—1.25 B
304 HN (XM-21)	S30452	0.04–0.10	2.00	1.00	18.0–20.0	8.0–10.5	0.045	0.030	0.16–0.30 N
Cronifer 1815 LCSi	S30600	0.018	2.00	3.73–4.3	17.0–18.5	14.0–15.5	0.020	0.020	0.2 Mo
RA 85 H‡	S30615	0.20	0.80	3.50	18.5	14.50	…	…	1.0 Al
253 MA	S30815	0.05–0.10	0.80	1.4–2.0	20.0–22.0	10.0–12.0	0.040	0.030	0.14–0.20 N; 0.03–0.08 Ce; 1.0 Al

Alloy	UNS	C	Mn	Si	Cr	Ni	P	S	Others
Type 309 S Cb	S30940	0.08	2.00	1.00	22.0–24.0	12.0–15.0	0.045	0.030	10 × %C min to 1.10 max Nb
Type 310 Cb	S31040	0.08	2.00	1.50	24.0–26.0	19.0–22.0	0.045	0.030	10 × %C min to 1.10 max Nb + Ta
254 SMO	S31254	0.20	1.00	0.80	19.50–20.50	17.50–18.50	0.030	0.010	6.00–6.50 Mo; 0.50–1.00 Cu; 0.180–0.220 N
Type 316 Ti	S31635	0.08	2.00	1.00	16.0–18.0	10.0–14.0	0.045	0.030	5 × %(C + N) min to 0.70 max Ti; 2.0–3.0 Mo; 0.10 N
Type 316 Cb	S31640	0.08	2.00	1.00	16.0–18.0	10.0–14.0	0.045	0.030	10 × %C min to 1.10 max Nb + Ta; 2.0–3.0 Mo; 0.10 N
Type 316 HQ	...	0.030	2.00	1.00	16.00–18.25	10.00–14.00	0.030	0.015	3.00–4.00 Cu; 2.00–3.00 Mo
Type 317 LM	S31725	0.03	2.00	1.00	18.0–20.0	13.5–17.5	0.045	0.030	4.0–5.0 Mo; 0.10 N
17-14-4 LN	S31726	0.03	2.00	0.75	17.0–20.0	13.5–17.5	0.045	0.030	4.0–5.0 Mo; 0.10–0.20 N
Type 317 LN	S31753	0.03	2.00	1.00	18.0–21.0	11.0–15.0	0.030	0.030	0.10–0.22 N
Type 370	S37000	0.03–0.05	1.65–2.35	0.5–1.0	12.5–14.5	14.5–16.5	0.040	0.010	1.5–2.5 Mo; 0.1–0.4 Ti; 0.005 N; 0.05 Co
18-18-2 (XM-15)	S38100	0.08	2.00	1.5–2.5	17.0–19.0	17.5–18.5	0.030	0.030	...
19-9 DL	S63198	0.28–0.35	0.75–1.50	0.03–0.8	18.0–21.0	8.0–11.0	0.040	0.030	1.0–1.75 Mo; 0.1–0.35 Ti; 1.0–1.75 W; 0.25–0.60 Nb
20Cb-3[¶]	N08020	0.07	2.00	1.00	19.0–21.0	32.0–38.0	0.045	0.035	2.0–3.0 Mo; 3.0–4.0 Cu; (Nb + Ta) 8 × %C −1.00
20Mo-4[¶]	N08024	0.03	1.00	0.50	22.5–25.0	35.0–40.0	0.035	0.035	3.50–5.00 Mo; 0.50–1.50 Cu; 0.15–0.35 Nb
20Mo-6[¶]	N08026	0.03	1.00	0.50	22.0–26.00	33.0–37.20	0.03	0.03	0.10–0.16 N; 5.00–6.70 Mo; 2.00–4.00 Cu
Sanicro 28[¶]	N08028	0.03	2.50	1.00	26.0–28.0	30.0–34.0	0.03	0.03	3.0–4.0 Mo; 0.6–1.4 Cu
AL-6X[¶]	N08366	0.035	2.00	1.00	20.0–22.0	23.5–25.5	0.040	0.030	6.0–7.0 Mo
AL-6XN[¶]	N08367	0.030	2.00	1.00	20.0–22.0	23.50–25.50	0.040	0.030	6.0–7.0 Mo; 0.18–0.25 N
JS-700[¶]	N08700	0.04	2.00	1.00	19.0–23.0	24.0–26.0	0.040	0.030	4.3–5.0 Mo; (Nb + Ta) 8 × %C −0.4 max Nb; 0.5 Cu
Type 332[¶]	N08800	0.01	1.50	1.00	19.0–23.0	30.0–35.0	0.045	0.015	0.15–0.60 Ti; 0.15–0.60 Al
904L	N08904	0.02	2.00	1.00	19.0–23.0	23.0–28.0	0.045	0.035	0.1 max N; 4.0–5.0 Mo; 1.0–2.0 Cu

TABLE 10.5 Compositions of Nonstandard Austenitic and Superaustenitic Stainless Steels[61,62] (*Continued*)

Designation*	UNS no.	C	Mn	Si	Composition (%)†				Other
					Cr	Ni	P	S	
Cronifer 1925 hMo¶	N08925	0.02	1.00	0.50	24.0–26.0	19.0–21.0	0.045	0.030	6.0–7.0 Mo; 0.8–1.5 Cu; 0.10–0.20 N; ×10
Cronifer 2328¶	...	0.04	0.75	0.75	22.0–24.0	26.0–28.0	0.030	0.015	2.5–3.5 Cu; 0.4–0.7 Ti; 2.5–3.0 Mo
310MoLN¶	S31050	0.03	2.00	0.50	24.0–26.0	21.0–23.0	0.030	0.010	0.10–0.16 N; 2.0–3.0 Mo
254 SMo	S31254	0.02	1.00	0.80	19.5–20.5	17.5–18.5	0.030	0.010	0.18–0.22 N; 6.0–6.5 Mo; 0.5–1.0 Cu
B-66¶	S31266	0.03	2.0–4.0	1.00	23.0–25.0	21.0–24.0	0.035	0.020	0.35–0.60 N; 5.0–7.0 Mo; 0.5–3.0 Cu; 1.0–3.0 W
SR 50A¶	S32050	0.03	1.5	1.0	22.0–24.0	20.0–22.0	0.035	0.020	0.24–0.34 N; 6.0–6.8 Mo; 0.4 Cu
654 SMo¶	S32654	0.02	2.0–4.0	0.50	24.0–25.0	21.0–23.0	0.030	0.005	0.45–0.55 N; 7.0–8.0 Mo; 0.3–0.6 Cu
4565 S¶	S34565	0.03	5.0–7.0	1.0	23.0–25.0	16.0–18.0	0.030	0.010	0.4–0.6 N; 4.0–5.0 Mo; 0.10 max Nb
Nicrofer¶ 3127h Mo	N08031	0.015	2.00	0.3	26.0–28.0	30.0–32.0	0.020	0.010	0.15–0.25 N; 6.0–7.0 Mo; 1.0–1.4 Cu
20Modified¶ 825¶	N08320	0.05	2.5	1.0	21.0–23.0	25.0–27.0	0.040	0.030	4.0–6.0 Mo; (4 × %C min) Ti
	N08825	0.05	1.0	0.50	19.5–23.5	38.0–46.0	0.030	0.030	2.5–3.5 Mo; 1.5–3.0 Cu; 0.6–1.2 Ti
Inco25-6Mo	N08925	0.02	1.0	0.50	19.0–21.0	24.0–26.0	0.045	0.030	0.10–0.20 N; 6.0–7.0 Mo; 0.8–1.5 Cu
Inco25-6Mo¶ or Cronifer 1926hMo¶	N08926	0.02	2.00	0.50	19.0–21.0	24.0–26.0	0.030	0.010	0.15–0.25 N; 6.0–7.0 Mo; 0.5–1.5 Cu
UR SB-8¶	N08932	0.02	2.00	0.40	24.0–26.0	24.0–26.0	0.025	0.010	0.5–1.5 Cu; 0.15–0.25 N; 4.5–6.5 Mo; 1.0–2.0 Cu

* XM designations in this column are ASTM designations for the listed alloy.
† Single values are maximum values unless otherwise indicated.
‡ Nominal compositions.
§ UNS designation has not been specified; this designation appears in ASTM A887 and merely indicates the form to be used.
¶ Superaustenitic stainless steel grades.
Reprinted by permission of ASM International, Materials Park, Ohio.

TABLE 10.6 Minimum Room-Temperature Mechanical Properties of Austenitic and Superaustenitic Stainless Steels[61]

Product form[a]	Condition	Tensile strength		0.2% yield strength		Elongation %	Reduction in area, %	Hardness, HRB	ASTM specification
		MPa	ksi	MPa	ksi				
		Type 301 (UNS S30100)							
B	Annealed	620	90	205	30	40	...	95 max	A 666
B, P, Sh, St	Annealed	515	75	205	30	40	...	95 max	A 167
B, P, Sh, St	1/4 hard	860	125	515	75	25	...	...	A 666
B, P, Sh, St	1/2 hard	1030	150	760	110	18	...	...	A 666
B, P, Sh, St	3/4 hard	1210	175	930	135	12	...	...	A 666
B, P, Sh, St	Full hard	1280	185	965	140	9	...	...	A 666
		Type 302 (UNS S30200)							
B, F	Hot finished and annealed	515	75	205	30	40	50	...	A 276, A 473
B	Cold finished[b] and annealed	620	90	310	45	30	40	...	A 276
B	Cold finished[c] and annealed	515	75	205	30	30	40	...	A 276
W	Annealed	515	75	205	30	35[d]	50[d]	...	A 580
W	Cold finished	620	90	310	45	30[d]	40	...	A 580
P, Sh, St	Annealed	515	75	205	30	40	...	92 max	A 167, A 240, A 666
B, P, Sh, St	High tensile, 1/4 hard	860	125	515	75	10	...	...	A 666
B, P, Sh, St	High tensile, 1/2 hard	1030	150	760	110	10	...	...	A 666
B, P, Sh, St	High tensile, 3/4 hard	1205	175	930	135	6	...	...	A 666
B, P, Sh, St	Full hard	1275	185	965	140	4	...	...	A 666

TABLE 10.6 Minimum Room-Temperature Mechanical Properties of Austenitic and Superaustenitic Stainless Steels[61] (*Continued*)

Product form[a]	Condition	Tensile strength		0.2% yield strength		Elongation %	Reduction in area, %	Hardness, HRB	ASTM specification
		MPa	ksi	MPa	ksi				
		Type 302B (UNS S30215)							
B, F	Hot finished and annealed	515	75	205	30	40	50	...	A 276, A 473
B	Cold finished[b] and annealed	620	90	310	45	30	40	...	A 276
B	Cold finished[c] and annealed	515	75	205	30	30	40	...	A 276
W	Annealed	515	75	205	30	35[d]	50[d]	...	A 580
W	Cold finished	620	90	310	45	30[d]	40	...	A 580
P, Sh, St	Annealed	515	75	205	30	40	...	95 max	A 167
		Type 302Cu (UNS S30430)							
W[e]	Annealed	550	80	...	...	...	...	...	A 493
W[e]	Lightly drafted	585	85	...	...	...	...	...	A 493
		Types 303 (UNS S30300) and 303Se (UNS S30323)							
F	Annealed	515	75	205	30	40	50	...	A 473
W	Annealed	585–860	85–125	...	...	...	...	...	A 581
W	Cold worked	790–1000	115–145	...	...	...	...	...	A 581

Type 304 (UNS S30400)									
B, F^f	Hot finished and annealed	515	75	205	30	40	50	...	A 276, A 473
B	Cold finished^b and annealed	620	90	310	45	30	40	...	A 276
B	Cold finished^c and annealed	515	75	205	30	30	40	...	A 276
W	Annealed	515	75	205	30	35^d	50^d	...	A 580
W	Cold finished	620	90	310	45	30^d	40	...	A 580
P, Sh, St	Annealed	515	75	205	30	40	...	92 max	A 167
B, P, Sh, St	1/8 hard	690	100	380	55	35	...	...	A 666
B, P, Sh, St	1/4 hard	860	125	515	75	10	...	...	A 666
B, P, Sh, St	1/2 hard	1035	150	760	110	7	...	...	A 666
Type 304L (UNS S30403)									
F	Annealed	450	65	170	25	40	50	...	A 473
B	Hot finished and annealed	480	70	170	25	40	50	...	A 276
B	Cold finished^b and annealed	620	90	310	45	30	40	...	A 276
B	Cold finished^c and annealed	480	70	170	25	30	40	...	A 276
W	Annealed	480	70	170	25	35^d	50^d	...	A 580
W	Cold finished	620	90	310	45	30^d	40	...	A 580
P, Sh, St	Annealed	480	70	170	25	40	...	88 max	A 167, A 240
Type 304B4 (UNS S30424)									
P, Sh, St grade A	Annealed	515	75	205	30	27	...	95 max	A 887
P, Sh, St grade B	Annealed	515	75	205	30	16	...	95 max	A 887

TABLE 10.6 Minimum Room-Temperature Mechanical Properties of Austenitic and Superaustenitic Stainless Steels[61] (*Continued*)

Product form[a]	Condition	Tensile strength		0.2% yield strength		Elongation %	Reduction in area, %	Hardness, HRB	ASTM specification
		MPa	ksi	MPa	ksi				
Type 305 (UNS S30500)									
B, F	Hot finished and annealed	515	75	205	30	40	50	...	A 276, A 473
B	Cold finished[b] and annealed	260	90	310	45	30	40	...	A 276
B	Cold finished[c] and annealed	515	75	205	30	30	40	...	A 276
W	Annealed	515	75	205	30	35[d]	50[d]	...	A 580
W	Cold finished	620	90	310	45	30[d]	40	...	A 580
P, Sh, St	Annealed	480	70	170	25	40	...	88 max	A 167
B, W	High tensile[d]	1690	245	...	...	...	...	...	...
Cronifer 18-15 LCSi (UNS S30600)									
P, Sh, St	Annealed	540	78	240	35	40	...	...	A 167, A 240
Type 308 (UNS S30800)									
B, F	Hot finished and annealed	515	75	205	30	40	50	...	A 276, A 473
B	Cold finished[b] and annealed	620	90	310	45	30	40	...	A 276
B	Cold finished[c] and annealed	515	75	205	30	30	40	...	A 276
W	Annealed	515	75	205	30	35[d]	50[d]	...	A 580
W	Cold finished	620	90	310	45	30[d]	40	...	A 580
P, Sh, St	Annealed	515	75	205	30	40	...	88 max	A 167

Types 309 (UNS S30900), 309S (UNS S30908), 310 (UNS S31000) and 310S (UNS S31008)

B, F	Hot finished and annealed	515	75	205	30	40	50	...	A 276, A 473
B	Cold finished[b] and annealed	620	90	310	45	30	40	...	A 276
B	Cold finished[c] and annealed	515	75	205	30	30	40	...	A 276
W	Annealed	515	75	205	30	35[d]	50[d]	...	A 580
W	Cold finished	620	90	310	45	30[d]	40	...	A 580
P, Sh, St	Annealed	515	75	205	30	40	...	95 max	A 167

310Cb (UNS S31040)

P, Sh, St	Annealed	515	75	205	30	40	...	95	A 167, A 240
B, shapes	Hot finished and annealed	515	75	205	30	40	50	...	A 276
B, shapes	Cold finished[b] and annealed	620	90	310	45	30	40	...	A 276
B, shapes	Cold finished[c] and annealed	515	75	205	30	30	40	...	A 276
W	Annealed	515	75	205	30	35[d]	50[d]	...	A 580
W	Cold finished	620	90	310	45	30[d]	40	...	A 580

Type 314 (UNS S31400)

B, F	Hot finished and annealed	515	75	205	30	40	50	...	A 276, A 473
B	Cold finished[b] and annealed	620	90	310	45	30	40	...	A 276
B	Cold finished[c] and annealed	515	75	205	30	30	40	...	A 276
W	Annealed	515	75	205	30	35[d]	50[d]	...	A 580
W	Cold finished	620	90	310	45	30[d]	40	...	A 580

TABLE 10.6 Minimum Room-Temperature Mechanical Properties of Austenitic and Superaustenitic Stainless Steels[61] (*Continued*)

Product form[a]	Condition	Tensile strength		0.2% yield strength		Elongation %	Reduction in area, %	Hardness, HRB	ASTM specification
		MPa	ksi	MPa	ksi				
		Type 316 (UNS S31600)							
B, F[f]	Hot finished and annealed	515	75	205	30	40	50	...	A 276, A 473
B	Cold finished[b] and annealed	620	90	310	45	30	40	...	A 276
B	Cold finished[c] and annealed	515	75	205	30	30	40	...	A 276
W	Annealed	515	75	205	30	35[d]	50[d]	...	...
W	Cold finished	620	90	310	45	40[d]	40	...	A 580
P, Sh, St	Annealed	515	75	205	30	40	...	95 max	A 167, A 240
		Type 316L (UNS S31603)							
F	Annealed	450	65	170	25	40	50	...	A 473
B	Hot finished and annealed	480	70	170	25	40	50	...	A 276
B	Cold finished[b] and annealed	620	90	310	45	30	40	...	A 276
B	Cold finished[c] and annealed	480	70	170	25	30	40	...	A 276
W	Annealed	480	70	170	25	35[d]	50[d]	...	A 580
W	Cold finished	620	90	310	45	30[d]	40	...	A 580
P, Sh, St	Annealed	485	70	170	25	40	...	95 max	A 167, A 240

Product form	Condition							
Type 316Cb (UNS S31640)								
P, Sh, St	Annealed	515	75	205	30	30	...	95	A 167, A 240
B, shapes	Hot finished and annealed	515	75	205	30	40	50	...	A 276
B, shapes	Cold finished[b] and annealed	620	90	310	45	30	40	...	A 276
B, shapes	Cold finished[c] and annealed	515	75	205	30	30	40	...	A 276
W	Annealed	515	75	205	30	35[d]	50[d]	...	A 580
W	Cold finished	620	90	310	45	30[d]	40	...	A 580
Type 317 (UNS S31700)								
B, F	Hot finished and annealed	515	75	205	30	40	50	...	A 276, A 473
B	Cold finished[b] and annealed	620	90	310	45	30	40	...	A 276
B	Cold finished[c] and annealed	515	75	205	30	30	40	...	A 276
W	Annealed	515	75	205	30	35[d]	50[d]	...	A 580
W	Cold finished	620	90	310	45	30[d]	40	...	A 580
P, Sh, St	Annealed	515	75	205	30	35	...	95 max	A 167, A 240
Type 317L (UNS S31703)								
B	Annealed	585[g]	85[g]	240[g]	35[g]	55[g]	65[g]	85 max[g]	...
P, Sh, St	Annealed	515	75	205	30	40	...	95 max	A 167
Type 317LM (UNS S31725)								
B, P	Annealed	515	75	205	30	40	40	...	A 276
P, Sh, St	Annealed	515	75	205	30	40	40	96 max	A 167

TABLE 10.6 Minimum Room-Temperature Mechanical Properties of Austenitic and Superaustenitic Stainless Steels[61] (Continued)

Product form[a]	Condition	Tensile strength		0.2% yield strength		Elongation %	Reduction in area, %	Hardness, HRB	ASTM specification
		MPa	ksi	MPa	ksi				
Types 321 (UNS S32100) and 321H (UNS 32109)									
B, F	Hot finished and annealed	515	75	205	30	40	50	...	A 276, A 473
B	Cold finished[b] and annealed	620	90	310	45	30	40	...	A 276
B	Cold finished[c] and annealed	515	75	205	30	30	40	...	A 276
W	Annealed	515	75	205	30	35[d]	50[d]	...	A 580
W	Cold finished	620	90	310	45	30[d]	40	...	A 580
P, Sh, St	Annealed	515	75	205	30	40	...	95 max	A 167, A 240
Types 347 (UNS S34700) and 348 (UNS S34800)									
B, F	Hot finished and annealed	515	75	205	30	40	50	...	A 276, A 473
B	Cold finished[b] and annealed	620	90	310	45	30	40	...	A 276
B	Cold finished[c] and annealed	515	75	205	30	30	40	...	A 276
W	Annealed	515	75	205	30	35[d]	50[d]	...	A 580
W	Cold finished	620	90	310	45	30[d]	40	...	A 580
P, Sh, St	Annealed	515	75	205	30	40	...	92 max	A 167, A 240
18-18-2 (UNS S38100)									
P, Sh, St	Annealed	515	75	205	30	40	...	95 max	A 167, A 240

Type 384 (UNS S38400)

W[e]	Annealed	550	80	...	...	...	...	...	A 493
W[e]	Lightly drafted	585	85	...	...	...	...	...	A 493
20Cb-3 (UNS N08020), 20Mo-4 (UNS N08024), and 20Mo-6 (UNS N08026)									
B, W	Annealed	550	80	240	35	30	50	...	B 473
Shapes	Annealed	550	80	240	35	15	50	...	B 473
B, W	Annealed and strain hardened	620	90	415	60	15	40	...	B 473
W	Annealed and cold finished	620–830	90–120	...	...	...	...	...	B 473
P, Sh, St	Annealed	550	80	240	35	30	...	95 max	B 463
Pi, T	Annealed	550	80	240	35	30	...	...	B 464, B 468, B 474, B 729
Sanicro 28 (UNS N08028)									
P, Sh, St	Annealed	500	73	215	31	40	...	70–90[g]	B 709
Seamless tube	Annealed	500	73	215	31	40	...	...	B 668
Type 330 (UNS N08330)									
B	Annealed	485	70	210	30	30	...	...	B 511
P, Sh, St	Annealed	485	70	210	30	30	...	70–90[g]	B 536
Pi	Annealed	485	70	210	30	30	...	70–90[g]	B 535, B546
AL-6X (UNS N08366)									
B, W	Annealed	515	75	210	30	30	...	...	B 691
P, Sh, St	Annealed	515	75	240	35	30	...	95 max	B 688
Pi, T	Annealed	515	75	210	30	30	...	...	B 675, B 676, B 690
Welded T	Cold worked	515	75	210	30	10	...	...	B 676

TABLE 10.6 Minimum Room-Temperature Mechanical Properties of Austenitic and Superaustenitic Stainless Steels[61] (*Continued*)

Product form[a]	Condition	Tensile strength MPa	Tensile strength ksi	0.2% yield strength MPa	0.2% yield strength ksi	Elongation %	Reduction in area, %	Hardness, HRB	ASTM specification
				JS-700 (UNS N08700)					
B, W	Annealed	550	80	240	35	30	50	...	B 672
P, Sh, St	Annealed	550	80	240	35	30	50	75–90[g]	B 599
				Type 332 (UNS N08800)					
Pi, T	Annealed	515	75	210	30	30	...	...	B 163 B 407, B 514, B 515
Seamless Pi, T	Hot finished	450	65	170	25	30	...	...	B 407
B	Hot worked	550	80	240	35	25	...	...	B 408
B	Annealed	515	75	210	30	30	...	...	B 408
P	Hot rolled	550	80	240	35	25	...	...	B 409
P, Sh, St	Annealed	515	75	210	30	30	...	...	B 409
				Type 904L (UNS N08904)					
B	Annealed	490	71	220	31	35	...	...	B 649
W	Cold Finished	620–830	90–120	...	...	...	...	...	B 649
Pi, T	Annealed	490	71	220	31	35	...	...	B 673, B 674, B 677
P, Sh, St	Annealed	490	71	220	31	35	...	70–90[g]	B 625

[a] B, bar; F, forgings; P, plate; Pi, pipe; Sh, sheet; St, strip; T, tube; W, wire.
[b] Up to 13 mm (0.5 in.) thick.
[c] Over 13 mm (0.5 in.) thick.
[d] For wire 3.96 mm ($\frac{5}{32}$ in.) and under, elongation and reduction in area shall be 25 and 40%, respectively.
[e] 4 mm (0.156 in.) in diameter and over.
[f] For forged sections 127 mm (5 in.) and over, the tensile strength shall be 485 Mpa (70 ksi).
[g] For information only; not a basis for acceptance or rejection.
Reprinted by permission of ASM International, Materials Park, Ohio.

TABLE 10.7 Minimum Mechanical Properties of High-Nitrogen Austenitic Stainless Steels[61]

Product form*	Condition	Tensile strength		0.2% yield strength		Elongation, %	Reduction in area, %	Hardness, HRB	ASTM specification
		MPa	ksi	MPa	ksi				
				Type 201 (UNS S20100)					
B	Annealed	515	75	275	40	40	45	...	A 276
P, Sh, St	Annealed	655	95	310	45	40	...	100 max	A 276, A 666
Sh, St	1/4 hard	860	125	515	75	25	...	...	A 666
Sh, St	1/2 hard	1030	150	760	110	18	...	...	A 666
Sh, St	3/4 hard	1210	175	930	135	12	...	...	A 666
Sh, St	Full hard	1280	185	965	140	9	...	...	A 666
				Type 202 (UNS S20200)					
B	Annealed	515	75	275	40	40	45	...	A 276
P, Sh, St	Annealed	620	90	260	38	40	...	...	A 666
Sh, St	1/4 hard	860	125	515	75	12	...	...	A 666
				Type 205 (UNS S20500)					
B, P, Sh, St	Annealed	790	115	450	65	40	...	100 max	A 666
				Nitronic 50 (UNS S20910)					
B	Annealed	690	100	380	55	35	55	...	A 276
W	Annealed	690	100	380	55	35	55	...	A 580
Sh, St	Annealed	725	105	415	60	30	...	100 max	A 240
P	Annealed	690	100	380	55	35	...	100 max	A 240
				Cryogenic Tenelon (UNS S21460)					
B, P, Sh, St	Annealed	725	105	380	55	40	...	...	A 666

10.43

TABLE 10.7 Minimum Mechanical Properties of High-Nitrogen Austenitic Stainless Steels[61] (*Continued*)

Product form*	Condition	Tensile strength		0.2% yield strength		Elongation, %	Reduction in area, %	Hardness, HRB	ASTM specification
		MPa	ksi	MPa	ksi				
Types 216 (UNS S21600) and 216L (UNS S21603)									
Sh, St	Annealed	690	100	415	60	40	...	100 max	A 240
P	Annealed	620	90	345	50	40	...	100 max	A 240
Nitronic 40 (UNS S21900)									
B, W	Annealed	620	90	345	50	45	60	...	A 276, A 580
21-6-9 LC (XM-11) (UNS S21904)									
B, W, shapes	Annealed	620	90	345	50	45	60	...	A 276, A 580
Sh, St	Annealed	690	100	415	60	40	...	...	A 666
P	Annealed	620	90	345	50	45	...	...	A 666
Nitronic 33 (UNS S24000)									
B, W	Annealed	690	100	380	55	30	50	...	A 276, A 580
Sh, St	Annealed	690	100	415	60	40	...	100 max	A 240
P	Annealed	690	100	380	55	40	...	100 max	A 240
Nitronic 32 (UNS S24100)									
B, W	Annealed	690	100	380	55	30	50	...	A 276, A 580
Type 304N (UNS S30451)									
B	Annealed	550	80	240	35	30	...	...	A 276
P, Sh, St	Annealed	550	80	240	35	30	...	92 max	A 240

Form	Condition								Specification
Type 340HN (UNS S30452)									
B	Annealed	620	90	345	50	30	50	...	A 276
Sh, St	Annealed	620	90	345	50	30	...	100 max	A 240
P	Annealed	585	85	275	40	30	...	100 max	A 240
Type 304LN (UNS 30453)									
B	Annealed	515	75	205	30	...	...	...	A 276
P, Sh, St	Annealed	515	75	205	30	40	...	92 max	A 167, A 240
253 MA (UNS S30815)									
P, Sh, St	Annealed	600	87	310	45	40	...	95 max	A 167, A 240
B, shapes	Annealed	600	87	310	45	40	50	...	A 276
254 SMO (UNS S31254)									
P, Sh, St	Annealed	650	94	300	44	35	...	96	A 167, A 240
B, shapes	Annealed	650	95	300	44	35	50	...	A 276
Type 316N (UNS S31651)									
B	Annealed	550	80	240	35	30	...	...	A 276
P, Sh, St	Annealed	550	80	240	35	35	...	95 max	A 240
17-14-4 LN (UNS S31726)									
P, Sh, St	Annealed	550	80	240	35	40	...	96	A 167, A 240
B, shapes	Annealed	550	80	240	35	40	...	...	A 276
Type 317LN (UNS S31753)									
P, Sh, St	Annealed	550	80	240	35	40	...	95	A 167, A 240

TABLE 10.7 Minimum Mechanical Properties of High-Nitrogen Austenitic Stainless Steels[61] (*Continued*)

Product form*	Condition	Tensile strength		0.2% yield strength		Elongation, %	Reduction in area, %	Hardness, HRB	ASTM specification
		MPa	ksi	MPa	ksi				
				AL 6XN (UNS N08367)					
B, W	Annealed	715	104	315	46	30	...	...	B691
P, Sh, St	Annealed	715	104	315	46	30	...	100	B688
Flanges, fittings, valves, and so on	Annealed	715	104	315	46	30	50	...	B462
Seamless Pi, T	Annealed	715	104	315	46	30	...	...	B690
Welded Pi	Annealed	715	104	315	46	30	...	...	B676
Welded T	Solution treated and annealed	715	104	315	46	30	...	...	B676
Welded T	Cold worked	...	...	...	...	10	...	...	B676
				Cronifer 1925 hMO (UNS N08925)					
B, W	Annealed	600	87	300	43	40	...	...	B649
Seamless Pi, T	Annealed	600	87	300	43	40	...	...	B677
Welded Pi	Annealed	600	87	300	43	40	...	...	B673
Welded T	Annealed	600	87	300	43	40	...	...	B674

* B, bar; P, plate; Pi, pipe; Sh, sheet; St, strip; T, tube; W, wire.
Reprinted by permission of ASM International, Materials Park, Ohio.

TABLE 10.8 Stress-Relieving Treatments for Austenitic Stainless Steels[63]

Application or desired characteristics	Suggested thermal treatment*		
	Extralow-carbon grades, such as 304L and 316L	Stabilized grades, such as 318, 321, and 347	Unstabilized grades, such as 304 and 316
Severe stress corrosion	A, B	B, A	†
Moderate stress corrosion	A, B, C	B, A, C	C†
Mild stress corrosion	A, B, C, E, F	B, A, C, E, F	C, F
Remove peak stresses only	F	F	F
No stress corrosion	None required	None required	None required
Intergranular corrosion	A, C‡	A, C, B‡	C
Stress relief after severe forming	A, C	A, C	C
Relief between forming operations	A, B, C	B, A, C	C§
Structural soundness¶	A, C, B	A, C, B	C
Dimensional stability	G	G	G

* Thermal treatments are listed in order of decreasing preference. A: Anneal at 1065 to 1120°C (1950 to 2050°F) and then slow-cool. B: Stress relieve at 900°C (1650°F) and then slow-cool. C: Anneal at 1065 to 1120°C (1950 to 2050°F) and then quench or cool rapidly. D: Stress relieve at 900°C (1650°F) and then quench or cool rapidly. E: Stress relieve at 480 to 650°C (900 to 1200°E) and then slow-cool. F: Stress relieve at below 480°C (900°F) and then slow-cool. G: Stress relieve at 205 to 480°C (400 to 900°F) and then slow-cool (usual time, 4 hr per inch of section).
† To allow the optimum stress-relieving treatment, the use of stabilized or extralow-carbon grades is recommended.
‡ In most instances, no heat treatment is required, but where fabrication procedures may have sensitized the stainless steel, the heat treatments noted may be employed.
§ Treatment A, B, or D also may be used, if followed by treatment C when forming is completed.
¶ Where severe fabricating stresses coupled with high service loading may cause cracking. Also, after welding heavy sections.
Reprinted by permission of ASM International, Materials Park, Ohio.

4. Their corrosion properties have been less extensively studied because of the limited use of the 200 series alloys.

5. For a given carbon content, it is claimed that the 200 series alloys are less susceptible to sensitization (discussed in a later section) than the 300 series.

The Mn-N stainless steels may exhibit a ductile-brittle transition temperature (DBTT) at very low temperatures. This is certainly true for Nitronic 40 (21Cr-6Ni-9Mn, type 219, UNS S21904) which has a dramatic loss of toughness at ~10 K. Actually, this loss of toughness is observed in slow-strain-rate K_{IC}-type tests, since low heat capacity at ~4 K means that adiabatic heating during impact testing raises the specimen temperature to 77 K. For this reason, it has been proposed that ASTM restrict the E-23 Charpy impact test to 77 K and above. The loss of toughness at low temperatures is not caused by a phase change (martensite formation, etc.), but is related to the strong thermal component of strengthening by N in solid solutions in austenite. Incidentally, the Mn-N alloys display a transition near 10 K, going from paramagnetic (>>10 K) to antiferromagnetic (below ~10 K).[51]

In recent years, new proprietary high-manganese stainless steels have been developed (Tables 10.4 and 10.5) in which higher manganese content not only saves nickel but also increases the nitrogen solubility in the austenitic phase (as shown in Fig. 10.14),[56] thereby making them useful for structural control, strengthening, and

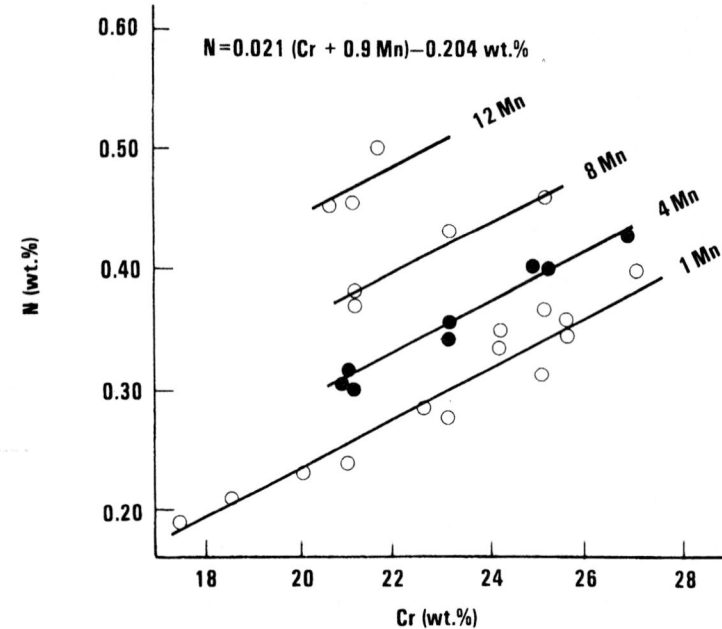

FIGURE 10.14 The solubility of nitrogen in austenitic stainless steels containing 14% nickel as a function of chromium and manganese contents.[56] (*After H. I. McHenry, ONR Far East Scientific Bulletin, vol. 10, no. 2, p. 122, 1985, reporting Nippon Steel Company data.*)

increased corrosion resistance. Nitronic 32 and Tenelon contain very low amounts of nickel. [However, Cryogenic Tenelon contains a higher nickel content (5 to 6%) in order to increase low-temperature toughness.[56]] All these steels possess higher room- and elevated-temperature strengths than the steels of the 300 series and can be rolled or cold-drawn to high strength without undergoing transformation into martensite, thereby producing the retention of very low magnetic permeability. Another important characteristic of these alloys is their good wear resistance. Nitronic 50 with 1.5 to 3.0% Mo is developed as a substitute for 304 and 316 types in applications where a combination of strength, corrosion resistance, and competitive price is of greater importance. Unlike many austenitic stainless steels, Nitronic 50 does not become magnetic upon cold-working. The grades Rex 734, Remanit 465S, and alloy 24 (Cronifer 2418 MoN) represent new high-strength, highly pitting-resistant, and crevice corrosion-resistant offshoots of the traditional Nitronic 50. Type 216 with 2.0 to 3.0% Mo also has both excellent (pitting and crevice) corrosion resistance and high yield strength. Nitronic 60, with 4% silicon, gives wear resistance in unlubricated metal-to-metal sliding.[54]

In the annealed condition, type 201 is stronger and slightly less ductile than type 301, while type 202 is stronger and somewhat more ductile than type 302.[56] Type 201 experiences martensite formation during cold-working. It is normally employed as cold-rolled sheet. Applications of type 201 and 201L stainless steels include airbag containers; automobile wheel covers; automotive trim; kitchen pot, pans, and lids; flatware; piston rings; thermal window spacers; hose clamps; transit car roofing and siding; transit car structural members; truck trailer posts and door frames; and washing machine baskets.[46,64] Type 201L is also used in cryogenic applications. With approximately the same chemistry but with 3 to 4% Si addition, a grade with the trade name *Gall-Tough* is developed for abrasion resistance.[54]

Applications of type 202, the basic stainless steel grade, include kitchen pots, pans, and lids; children's slide; hub cap; electrical range part stamping; railcar, truck trailer; milk handling, storage, and pasteurizing equipment; sink, countertop; and window flashing.

Like type 305, type 205 has a lower rate of work-hardening, which makes it beneficial for deep-drawn or spun parts.

Nitronic 40 (21Cr-6Ni-9Mn, 219, UNS 21904) is used for aircraft engine parts; steam and autoclave applications; and chemical processing and pollution control equipment. It is the most commonly produced among the Nitronic alloys.[46,51]

The new additions of high-Mn-N stainless steels include 7% Mo superaustenitic, S32654 (Avesta) (to compete with Ni-based corrosion-resistant alloys); 5% Mo superaustenitic, S34565 (VDM) (with comparable corrosion resistance to the 6% Mo superaustenitic stainless steels and higher strength); S21000 (with high strength, low magnetic permeability, developed for drill collars); and S20430 (Carpenter) (with low work-hardening and comparable formability to S30400 type).[64a]

10.5.3.2 The Basic 18-8 Grades (Types 302 and 304).

Austenitic 18Cr-8Ni stainless steels have better strength and oxidation resistance at high temperatures than low-alloy and carbon steels and are cheaper than Ni-base alloys. That is why they find a wide variety of applications, as given below. In addition, they enjoy a service life of several (sometimes over 10) years.[65]

Types 302 and 304 represent the "bread and butter" alloys, which possess somewhat greater stability and improved corrosion resistance. In these grades, particularly type 304, strain-hardening takes place throughout the application of stress, but the amount of strain hardening for a given stress increment decreases with the increase in stress. Hence, these grades are infrequently used in the cold-worked condition. The properties of cold-rolled materials are illustrated in Fig. 10.15. The cold-working temperature influences markedly the mechanical properties; e.g., cold-working at subzero temperatures gives rise to an increased strength and ductility.[54]

These stable steels, especially type 304, are more popular. Type 304 is the general-purpose grade which accounts for more than 50% of the total stainless steel production in the United States. It is nonmagnetic in the annealed condition and becomes slightly magnetic when cold-worked. It is widely used in applications requiring a combination of higher corrosion resistance, formability, deep drawability, and weldability. Some typical applications of type 304 are as follows.

Architectural Applications. Because of its durability, its adaptability to a wide variety of surface finishes, and its nearly maintenance-free surface, more people specify type 304 for architectural work. For example, the high lobby walls of World Trade Center in New York City[†] were made of type 304 with a number 8 mirror finish surface; the Bermuda-type building roof in Benjamin F. Yak Recreation Center in Wyandotte, Mich., is made of type 304 with a mirror finish surface; and the interior and exterior of Cleveland Federal Office Building in Cleveland, Ohio, are extensively covered with type 304. Another use of type 304 with high reflectivity is the famed CN Tower in Canada, the Chunnel Terminal in London, Canary

---

[†] Destroyed in a terrorist attack on September 11, 2001.

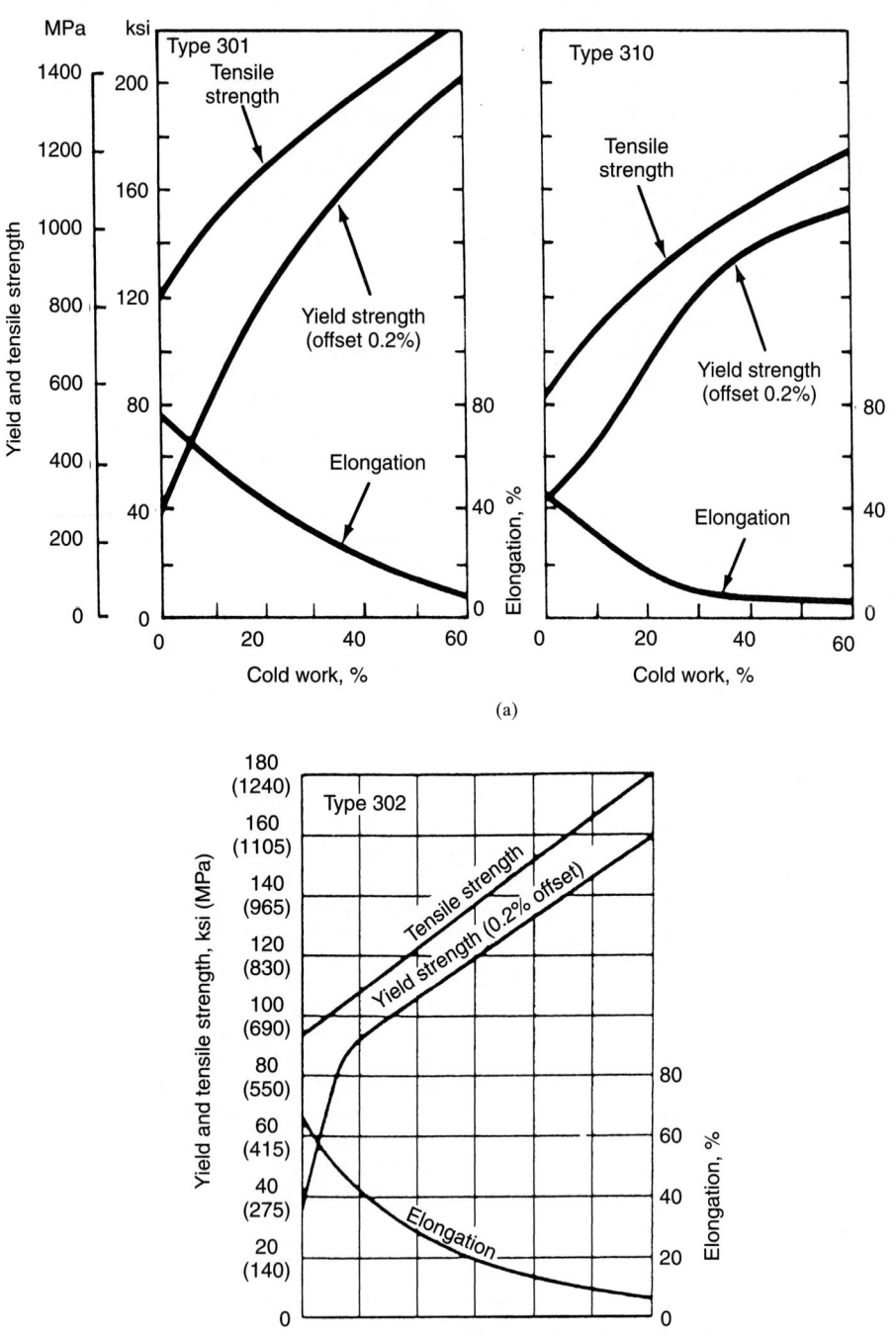

FIGURE 10.15 Effect of cold-working on tensile properties of the following types of stainless steels: (a) 301 and 310; (b) 302 (Fe-0.09C-1.39Mn-0.50Si-17.56Cr-9.69Ni); (c) 304 (Fe-0.056C-0.87Mn-0.43Si-18.6Cr-10.25Ni); (d) 305 steel.[54] [(a) *Stainless steel industry data. (b, c, and d) Reprinted by permission of ASM International, Materials Park, Ohio.*)]

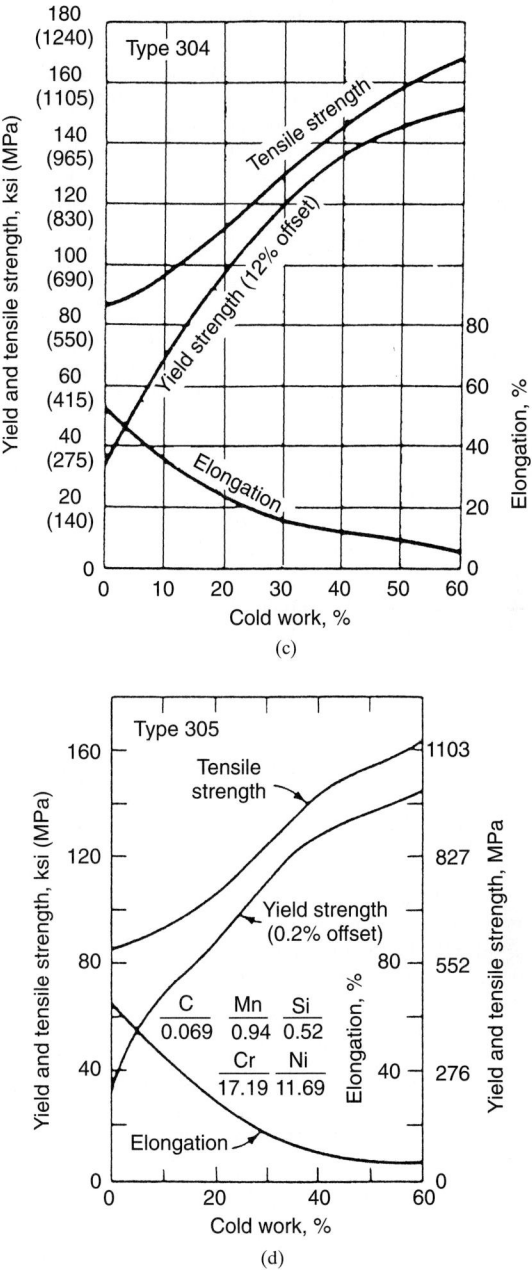

FIGURE 10.15 Effect of cold-working on tensile properties of the following types of stainless steels: (*a*) 301 and 310; (*b*) 302 (Fe-0.09C-1.39Mn-0.50Si-17.56Cr-9.69Ni); (*c*) 304 (Fe-0.056C-0.87Mn-0.43Si-18.6Cr-10.25Ni); (*d*) 305 steel.[54] [(*a*) *Stainless steel industry data.* (*b, c, and d*) *Reprinted by permission of ASM International, Materials Park, Ohio.*) (*Continued*)

Wharf Tower, the Petronas Towers in Kuala Lumpur, and the 630-ft-high "Gateway to the West" arch in Jefferson National Expansion Memorial, St. Louis, Mo.[54,66] Other applications include anchoring systems (Pittsburgh's Fort Tunnel and the Amoco Building, Chicago, Ill.), curtain walls (Center for Women and Newborn, Boston, Mass.), roofs and roofing systems (Plant Hall, University of Tampa, Fla.; Society Tower, Cleveland, Ohio; Pittsburgh International Airport with 80-ft-radius roof; North Terminal of Washington Airport; Kansai International Airport Terminals; The Fresno City Hall, Fresno, Calif.; Ohio Institute at NASA, Cleveland, Ohio; St. John Berbeuf Church, Niles, Ill.; Great American Pyramid, Memphis, Tenn., and Monticello, Charlottesville, Va.), revolving and balanced doors (Annapolis Office Plaza, Annapolis, Md.), column covers, storefronts, and/or wall panels (Miami International Airport; Siskin Hospital, Birmingham, Ala.; IBM Building in Kansas City, Kan.; and Landerbrook Place, Cleveland, Ohio), elevator and/or entrance doors (Miami International Airport; Skylon Tower; and private residence, Miami Beach, Fla.), railings and barriers (Canadian Embassy, Washington, D.C., and Cable-Rail product), highway and traffic control (Toll Collection Facilities, Fla.), and interior and exterior hardware components.[54,67,67a]

Marine Applications.[68] O-ring seals; sonar domes. Limited application is due to its low resistance to chloride crevice corrosion.[51]

Chemical and Process Industries.[51,69,70] All types of equipment for handling (1) nitric acid (all concentrations), (2) organic acids including acetic acid (up to glacial strength) at moderate temperatures, (3) liquid fertilizers, and (4) very dilute or highly concentrated sulfuric acid at room or slightly elevated temperatures (up to about 50°C). This is not suitable for handling 5 to 95% (approximate limits) sulfuric acid.

Transportation. Automobile parts such as radiator cap springs and hose clamps; milk-tank trailer; ancillary equipment such as pumps, valves, and connectors; sheathing of front and rear bumpers of the metro-type buses;[64] automotive trim and emission control applications; and truck components.

Food Industry. Dairy automatic milkers; walls and stalls of milking parlors; milk-tank trucks with number 4 polish finish; equipment for making milk products and for pasteurization of milk; equipment for processing and storage of beer, wine, orange juice, tomato juice, and so forth; tanks for brewing, fermenting, and thermal cooling; desalination and water purification plants; processing and containing vegetables, fruit, meat, cereal, flour, sugar, and so on;[71] restaurant counters; and cooking stoves and ovens.

Alcoholic Beverage Industry.[72] Pumping of crushed grapes; transferring of wines and brandy; beer barrels; tanks for mixing, brewing, storage, fermenting, and thermal cooling; filtration equipment; distillation vats; brewing plants.

Petrochemical Industry. Sheathing for transfer lines carrying high-temperature gas; cryogenic vessels; lining of chemical reactors; towers and drums containing internals (screens, trays, and demisting pads).

Fabricated Components. Storage tanks and vessels for storing liquid nitrogen, oxygen, and hydrogen; columns and pillars, walkways; railings and ladders; structural supports; and wall and ceiling panels.[67a]

Semiconductor Industry.[73] Vacuum equipment.

Nuclear Applications. Type 304 has been used for the American Fast Flux Test Facility (FFTF), for the German SNR 300, and for the Clinch River Breeder Reactor Plant (CRBRP). It was used widely in the EBR2 Reactor Plant.[74]

Energy Sector.[73] Piping and heat exchanger applications.

Medical Equipment. Many applications, but not good where edge retention is critical. Used for disposable hypodermic needles.[51]

Miscellaneous.   Coal-handling equipment such as bunkers and chutes, bins, discharge hoppers, screening, and lined hopper cars (improving life and discharging three times faster than the unlined cars due to slideability) and components in dyeing industry. Other applications include bridges (Toronto's Humber River), canopies (Garden Mausoleum/Archdiocese of Chicago), and fasteners and fastening systems (Dan-d-lok product); roofings, copings, gutters, downspout and leaders, and flashings; hospital equipment; bins and boxes; water handling and delivery systems; concrete reinforcements; wire cloth and screens (industrial); floor plates; bars and rods.[46,47a]

Type 304 HN is a high-nitrogen version of type 304 stainless steel. In this case, solid-solution strengthening of N offers quite higher yield strength and tensile strength in the annealed condition than the traditional type 304, without adversely affecting ductility, nonmagnetic, or corrosion resistance properties. In some instances, this removes the need for cold-drawing to attain higher strength. However, cold working can provide higher strength levels with slight loss of ductility. It finds applications as aircraft and aerospace components, marine shafting and hardware, pump parts, and so forth.[75]

Boron modified type 304 is similar to the conventional type 304 except that it contains a boron addition which imparts a much higher thermal neutron absorption capability, and increased hardness, tensile strength, and yield strength. However, decreased tensile ductility, impact toughness, and corrosion resistance have been observed. These grades have been used in the nuclear industry for spent-fuel storage racks and cask baskets, control rods, burnable poison, and shielding.[75]

Type 302[†] is a higher-carbon version of type 304 which can be hardened and readily fabricated by cold-working (Fig. 10.15) and has most often been used for wire applications. It is often used in the annealed state due to its nonmagnetic properties (with extreme toughness and ductility) which are of great importance to certain instruments. Because of its high red-hardness, more power for a certain reduction during hot working is required than with mild steel. It is easily fabricated by cold-working and responds well to bending, deep drawing, forming, and upsetting.[75] Typical applications of type 302 are as follows.

Architectural Applications.   Architectural paneling on New York's Chrysler Building, the Empire State Building, and others; in building construction as trim, frames and moldings for storefronts, banks, and curtain wall construction.

Food Industry.   Equipment for baking, candy making, cheese manufacturing, dairy refrigeration, soda fountain, beverage, and brewing; baby bottle warmers and sterilizers; broilers; bulk milk coolers; cooking utensils; dairy equipment; deep-drawn sugar bowls, creamers, and shallow-drawn trays; food processing and packing equipment; food serving and handling equipment; ice cream freezers; ice cream molds; meat packing equipment; and milk dispensers.[46]

Miscellaneous.   Antennas; springs; sheet metal products and stampings; moldings; woven screens; hospital equipment; cables; aircraft cowling, jewelry, oil refinery equipment, signs, saltwater fishing tackle, nitric acid vessels, animal cages; camera parts, thermometers, aerosol springs;[75] gasoline tanks; hinges, hosiery forms, household appliances; institution equipment; laundry and dry cleaning equipment; oil refining equipment; pharmaceutical equipment; press plates; plumbing fixtures and fittings; racks to hold concrete building blocks during curing; sports equipment; and switchboard panels.[46]

[†] Actually, type 302 was the earlier alloy, and type 304 was the lower-carbon version of type 302 that could be welded with less risk of sensitization.

Type 302HQ (also called SM 2000), with a nominal composition of 8.1Ni-17.2Cr-3.2Cu-0.01C-0.025N-1.8Mn-0.4Si, has been developed to draw up to a true strain of about 6 without intermediate annealing (i.e., drawing from 5.5 to 0.25 mm). This produces a considerable amount of martensite, thereby an increased level of tensile strength. Other advantages of this new grade include: (1) precipitation of ε-copper phase with fcc structure in the bcc martensite matrix, thereby competing with precipitation hardening 17-7 PH and 15-5 PH alloys and fully stimulating the antimicrobial properties of copper; (2) higher than or equal corrosion resistance to type 302 grade; (3) resistance to oxidizing nitric acid; (4) higher wire tensile strength with a lower production cost; and (5) improved fatigue resistance. It is used for spring wire and severe cold heading process.[76]

10.5.3.3   Metastable Type 301.   The metastable 301 stainless steel exhibits an accelerated rate of strain hardening with consequent greater increase in yield strength after an initial 10 to 15% plastic strain as a function of cold deformation owing to the strain-induced martensitic transformation.[61] It is clear from Fig. 10.15 that both the strength and the elongation of cold-rolled and annealed type 301 are greater than those of types 302, 304, and 310. As a result, it finds decorative structural applications where high strength and high elongation are required at ambient temperature. Examples are railway and subway cars; rapid transit railroad equipment; truck and trailer bodies; automobile molding and trim, wheel covers, hose clamps, and pole line hardware; endless belts in large bakeries for extensive distribution; forks, spoons, pots, pans, and covers for the household; door plates; fasteners; roof drainage parts; and large liquid launch vehicles and high-strength cryoformed pressure vessels in aircraft and aerospace.[46,54,64,77,78] Its diverse structural applications put it third in the U.S. tonnage production of austenitic stainless steels, slightly behind type 316.[54]

A new type of stainless steel, *type 301L*, which contains 0.03% carbon or less and 0.04 to 0.2% nitrogen, has been developed in which the intergranular stress corrosion cracking (IGSCC) in welds is completely removed.[79] This has resulted in its use for railroad vehicles with great success due to higher strength, workability, and improved IGSCC in welds.

10.5.3.4   Higher-Nickel Steel (Type 305).   The higher nickel content of this grade (10 to 13%) stabilizes the austenite to prevent the formation of martensite. It exhibits low work-hardening rates (Fig. 10.15) and is used where severe cold formability (or deep drawability) is important and where the finished part must remain nonmagnetic after severe cold-working. Typical applications include complex pots, sinks, pans, fountain pens, cold heading fasteners,[54] cold drawing in two drawing stages, automatic eyelet machines, electrical instrumentation, coffee urn tops, expanded metal parts, mixing bowls, and reflectors.[46,75]

10.5.3.5   High-Corrosion-Resistant Steels (Types 316, 317, and 384).   Types 316 and 317 are molybdenum-bearing austenitic stainless steels which contain 2 to 3% and 3 to 4% molybdenum, respectively, and are more resistant to general corrosion in reducing media and pitting/crevice corrosion in chloride solutions than the conventional Cr-Ni austenitic steels such as type 304. These alloys also offer increased elevated-temperature creep and stress-rupture and tensile strengths.[54] They are widely employed in handling many of the chemicals used by chemical process industries.[75]

Applications of type 316 include the following.

Architectural Applications. It is used particularly where the atmosphere is relatively high in chloride ions. Examples include wall cladding on accommodation modules, ventilation louvres, and other architectural features on more than one-half of the oil fields in the North Sea.[48]

Marine Atmospheres.[61,68] In some seawater environments, type 316 is superior to types 301, 302, and 304. These applications are boat rails and hardware and facades of buildings near the ocean, structures, heat exchangers, O-ring seals, piping or pump parts, and special items such as propellers, shafts, impeller wear rings, and yacht rigging and deck fittings (involving erosion, corrosion, and fatigue damage).[61] Type 316 is not sufficiently resistant to chloride crevice corrosion to survive long-term corrosion in live (unsterilized) seawater, unless it is cathodically or galvanically protected.[51]

Chemical and Process Industry.[70] Equipment for handling dilute sulfuric acid is one application. Both types 316 and 317 are used for handling and transporting concentrated sulfuric acid[51] as well as in all concentrations of acetic acid at temperatures up to the boiling point.[61] Type 317 is more resistant than type 316 to corrosive attack in hot acetic acid and is employed extensively in distillation equipment.

Food Industry.[61,71] Impellers for handling corrosive fumes and vapors from sauces and pickle liquor tanks and kettles for cooking ketchup; tanks for brewing, fermenting, or cooling equipment involving extremely corrosive areas; desalination and water purification plants; and yeast tubes.[46]

Alcoholic Beverage Industry.[72] Pumps for producing wine and brandy and vinegar distillation vats (because of the type 316's high corrosion resistance to alkaline solutions, acids, sterilants, and various fermentation by-products and aesthetic image).

Pulp and Paper Industry. Like type 304, type 316 has been used in traditional paper making for washers, evaporators, head boxes, green- and white-liquor equipment, save-alls, pipe, and tubing. Types 316 and 317 have been employed in pulp washers and rotary-drum filter equipment. They are also used as fasteners in the paper and pulp industry.[80]

High-Pressure Synthesis. Autoclaves and pressure-vessel linings, valves, pipelines, and other equipment dealing with a variety of products, often at quite high temperatures.[81]

Nuclear Industry. Steam pipes and superheater tubing in modern power generating plants, especially nuclear power plants such as the Advanced Gas-Cooled Reactor and Commercial Fast Breeder Reactor (CFBR)[82] [because of the type 316's higher creep strength than the type 304 and better resistance to heat-affected zone (HAZ) cracking during welding than the types 347 and 321[83]] and in fossil-fuel-fired power plants.[84] Type 316 is also being used in the construction of Civil Demonstration Fast Reactors (CDFRs) and the French Super Phoenix Fast Reactor (SPFR).

Pharmaceutical and Biotechnology.[73] Equipment used to produce and process pharmaceutical products, such as penicillin, etc. take advantage of its hygienic and cleanability properties. Its use helps minimize metallic contamination.

Other Applications. Process equipment for producing photographic chemicals, inks, rubber, rayon, textile bleaches and dyestuffs and high-temperature equipment, components of manual metal arc (MMA) welding equipment, and plant equipment dealing with hydrogenation or other treatment of stearic, oleic, and similar acids, in soap making involving glycerin recovery from soap liquor containing sodium chloride, phosphate industry parts, smoke stacks, textile finishing equipment, and textile mill vats, and water handling and delivery systems.[67a]

Typical applications of type 317 are pulp and paper processing equipment, ink manufacturing equipment, dyeing equipment, and flue gas desulfurization units.

Type 384 containing 17 to 19% Ni, has been found to give satisfactory performance for use in pumps for handling seawater because of its good resistance to cavitation—erosion in seawater. Type 384 has been used for fasteners, cold headed bolts, rivets, and screws; upset and punched nuts; instrument parts; thread rolling; and parts involving severe extrusion, coining, and swaging.

10.5.3.6 Stabilized Stainless Steels (Types 321, 347, and 348).

Sensitization or intergranular corrosion can be prevented by the presence of strong carbide formers in the basic stainless steel composition. Type 321 has a composition similar to that of type 304, except the titanium addition which is added at least 5 times the carbon content. Types 321, 347, and 348 should be used for applications requiring intermittent heating in the carbide precipitation range from 427°C (800°F) to 816°C (1500°F) such as aircraft collector rings, exhaust stacks and exhaust manifolds, expansion joints, fasteners, firewalls, flash boilers, cabin heaters, flexible couplings, jet engine parts, high-temperature chemical process equipment, pressure vessels, stack liners, large mufflers for stationary diesel engines, and for welded construction and parts undergoing heating in the carbide precipitation range. They also exhibit good mechanical properties in high-temperature service.

Type 347 is the (Nb + Ta)-stabilized grade where the Nb + Ta added at least 10 times the carbon content. This grade derives considerable creep strength from niobium carbide or carbonitride precipitation on dislocations. Like type 304, this grade can be used for a wide variety of acetic acid equipment, including stills, base heaters, holding tanks, pipelines, heat exchangers, valves, and pumps, at concentrations up to 50% and at temperatures up to the boiling point of the solution. Other applications are collector rings, expansion joints, fasteners, heat resistors, jet engine parts, wire cloth and screens (industrial), welded construction, and parts subjected to heating in the carbide precipitation range.[46]

Type 348, a nuclear grade, is similar to type 347 in composition, except that the tantalum and cobalt contents are restricted to 0.1 and 0.2%, respectively; this type is, therefore, basically niobium-stabilized. It is used for welded and seamless tubes and welded and seamless pipes for radioactive system service.

These steels show mechanical properties similar to those of type 304 but provide (1) greater resistance to creep and rupture at elevated temperatures and (2) better corrosion performance in some aspects of fabricability when compared to type 304.[54]

10.5.3.7 Extralow-Carbon and -Nitrogen Steels (Types 304L, 304LN, 310L, 316L, 316LN, 316LVM, 317L, 317LN, 317LM, 317LX, and 317LXN).

These steels contain 0.03% maximum carbon to avoid sensitization in welds or during stress-relieving treatments. They are used much more than stabilized steels because of their cheaper production by the AOD process. Except for improved weldability and corrosion resistance, type 304L shows properties similar to those of type 304. Since the strengths of the L-grades are lower than those of their 304 and 316 counterparts, their design allowances should be set at lower values. To enhance the strength at room temperature and reduce the susceptibility to sensitization, the modified low-carbon stainless steels with nitrogen addition, 304LN and 316LN, have been produced which contain up to 0.18% nitrogen. Nitrogen in solid solution increases the yield strength to the levels equivalent to that of standard 304 and 316 grades. Moreover, nitrogen hinders sensitization. Nitrogen addition to Mo-containing alloys improves resistance to chloride-induced pitting and crevice corrosion.[53]

Another modified version is the low-carbon, nitrogen, vanadium-bearing type 316LNV which offers considerable strength, toughness, and ductility, and nonmagnetism at cryogenic temperatures either in the solution-treated or in the solution-treated and aged condition.[85] As a result of these features, this grade is used in superconducting magnets, particularly for their applications to nuclear fusion, superconducting generators, and so forth.

Type 316LS is a vacuum arc melted (VAR) low-carbon, high-nickel and -molybdenum version of type 316 stainless steel. This alloy is nonmagnetic, even after severe cold-forming operations, and has a ferrite-free microstructure.[75]

Type 304L is used as bulk tankage for special bulk cargo containers for liquefied natural gas (LNG), chemicals, and beverages; coal hopper linings; tanks for liquid fertilizers, tomato paste storage, and for all types of equipment handling nitric acid up to about 50% concentration.[70] It also finds applications in the brewing industry, in the aircraft and aerospace industry (as pressure vessels), in the paper and pulp industry, in water handling and delivery systems; and for hydrogen storage. Types 304L and 316L steels are used extensively in the paper and pulp industry for numerous items.[86]

Type 310L is a modification of 310S alloy for enhanced corrosion resistance to concentrated nitric acids and other strong oxidizing environments.

Types 316L and 316 have equivalent resistance to cold, very dilute sulfuric acid, but type 316 provides protection against intergranular corrosion (IGC) in the as-welded condition. Type 317L is also slightly more corrosion-resistant because of its higher Mo content. Type 316L is resistant to acetic acid up to atmospheric boiling point.[70] In pure phosphoric acid solution, both types 316L and 317L are resistant to hot solutions up to about 40% concentration. These two grades also find extensive applications in sulfate and sulfite pulping industries.[86] In chemical tankages for sea transport applications, the applications include valves, cargo pumps, piping, and heating coils.[68] Applications of 316LVM steels with minimum nonmetallic inclusions include surgical implants and internal fixation devices such as bone plate fabrication.[87] Type 316L is used in hospital equipment and furniture and in orthopedic applications. Type 316LS finds applications in fracture fixation devices such as bone plates, screws, and intermedullary nails; surgical implant devices; and surgical instruments requiring no high hardness.[75] Types 317LX or 317LM (UNS S31725, containing 4 to 5% Mo), 317LXN or 317LMN (UNS S31726, containing 4 to 5% Mo and 0.1 to 0.2%N), and AL-6XN (containing 6 to 7% Mo and 0.18 to 0.25% N) have also been developed for use in high-chloride environments.

10.5.3.8 Welding Electrode Grade (Type 308 or 308L). Type 308 or 308L steel is used extensively in industrial furnaces, for sulfite liquor at high temperatures, and as welding rods for GMAW, GTAW, and SAW processes for joining 302, 304, 304L, or 305 base metals.[53] The composition of type 308L is balanced to provide the formation of enough δ-ferrite in the weld deposit in order to avoid hot cracking and intergranular corrosion in the heat-affected zone (weld decay). It is also designed to avoid the possibility of martensite formation under the conditions of solidification-induced segregation.[51] Many austenitic steel base metals have matching, or near matching, welding filler metals. Usually, a nearly matching filler metal is a good choice. Types 304, 309S, and 310S, being the lower-carbon versions of their counterparts, are also finding applications requiring welding.

10.5.3.9 Oxidation-Resistant Grades (Types 302B, 309, 309S, 310, 310S, 314, and 330). Types 302B and 314 contain 2 to 3% and 1.5 to 3% silicon, respectively, and are responsible not only for improved oxidation resistance but also for retardation, in many cases, or complete prevention of carburization at high tempera-tures.[54] Typical applications of type 302B (UNS S30215) are heat-resisting parts such as annealing boxes, tube supports, heat-treating fixtures, and radiant tube heating elements, for improved oxidation resistance at temperatures up to 926°C (1700°F) for intermittent service involving rapid cooling, and for service at both intermittent and continuous temperatures up to 871°C such as burner sections, annealing covers, and furnace parts.

Types 309, 310, 314, and 330 possess high amounts of chromium and nickel, which impart good oxidation resistance up to 1093°C (2000°F) and creep strength at ele-vated temperatures. Type 309 has been used for air heaters, annealing boxes, baffle plates, boiler baffles, dryers, kilns and oven linings, oil refining equipment, paper mill equipment, pump parts, rolls for roller hearth furnaces, tube hangers, still tube sup-ports, wire cloth and screens (industrial), furnace arch supports, furnace skids, heat exchangers, heat-treating trays, oil burner parts, firebox sheets, and weld wire.[46,75] The higher-alloy addition of type 310 leads to its superior elevated-temperature strength and scale resistance when compared to that of type 309. Types 310 and 310S have resistance to oxidation up to 1149°C (2100°F) in continuous service, and have offered good resistance to both carburizing and reducing atmospheres.[75] It is, there-fore, used for the following applications: lining for high-temperature applications or more oxidizing conditions in the petrochemical industry;[74] heat exchangers; recu-perators; construction chambers; furnace parts such as annealing boxes, carburizing boxes, baking oven equipment, furnace linings, firebox sheets, furnace arch supports, furnace castings, furnace supports and conveyors, furnace stacks and dampers, indus-trial oven linings, kiln linings, oil burner parts, rolls for roller hearth furnaces, and soot blower tubing; gas turbine parts; and in applications involving sulfur-bearing gases at elevated temperature, dye house equipment, incinerators, nitrate crystal-lizing pans, nozzle diaphragm assemblies for turbojet engines, paper mill equipment, oil refinery equipment, tube hangers, and wire cloth and screen (industrial).[46,75]

It is important to note that types 309 and 310 (0.25% C maximum) have no minimum carbon content, and types 309S and 310S (0.08% C maximum) qualify as types 309 and 310, respectively. Types 309H and 310H (0.04% C minimum) also exist.[51,88]

Types 309S and 310S, having lower carbon content, are used to restrict carbide precipitation resulting from welding or other fabrication operations and when the completed structure is subjected to moist corrosive environments.[46] This minimum required carbon content of 0.04% allows type 310S to be used for improved creep resistance compared to lower-carbon material.

Type 314 is rare in the United States. The German (DIN) version of type 310 has higher Si than the U.S. (ASTM) version. Standard equivalence tables (such as Stählschüssel) indicate that type 310 is equivalent to type 314. Type RA 85H alloy is more readily available in North America than type 314, and was developed specif-ically to exhibit improved carburization resistance.[51] Like type 302B, types 314 and RA 85H are used in annealing boxes, carburizing boxes, radiant tubes, and heat-treating fixtures.

Type 330 has high resistance to carburization and thermal shock.[56]

10.5.3.10   *Free-Machining Grades (Types 303 and 303Se).*   Types 303 and 303Se have high sulfur and selenium contents, respectively, and are therefore called *free-machining grades*. These steels have high machinability and enhanced surface finish. That is why they are produced mostly in bar and rod forms and are employed mainly in mass production operations carried out on screw machines. However, the corro-sion resistance of these steels is reduced by the existence of sulfides and selenides,

and therefore care has to be taken in selecting the proper environments in which they have sufficient corrosion resistance.[54] They find applications as shafts, bolts, bushings, nuts, rivets, screws, studs, saltwater spinning reels for surf casting, forged or swaged parts, valves, valve bodies, valve trim, and parts produced on automatic screw machines and other machine tool equipment.[46] They possess nongalling properties that make disassembly of parts easy and tend to avoid scratching or galling in moving parts. They are not recommended for vessels containing liquids or gas under high pressure.[75]

10.5.3.11 Dual-Marked Grades (Types 304/304L and 316/316L). These grades are characterized by the carbon content of the L grade and the strength of a straight grade by adding nitrogen (to offset the loss). ASME International now has formally stated that dual-marked grades may be employed at straight grade stresses for all product forms (tubing and piping) to 540°C (1000°F).[89]

10.5.3.12 The H Grades. Type H grades, comprising 304H (S30409), 316H (S31609), 321H (S32109), 347H (S34709), and 348H (S34809), all have carbon contents greater than 0.04% and are employed in solution-annealed conditions. These restrictions facilitate high and reproducible creep rupture strengths at elevated temperatures.[56]

10.5.3.13 High-Silicon Grades. These Fe-Cr-Ni-Si steels are UNS S30600 and S30601 (for handling hot nitric acid); UNS S32615 alloy (for handling hot, concentrated sulfuric acid); UNS S70003, with 7% Si; and UNS S38815, with 6% Si (for handling hot, concentrated sulfuric acid).[64a]

10.5.4 Sensitization

A major pitfall of austenitic or duplex stainless steel is the tendency for sensitization to occur. When nonstabilized austenitic stainless steels such as 304 and 316 or duplex stainless steels are slowly cooled from 1050°C (1922°F) through the critical temperature range of 871 to 427°C (1600 to 800°F) for a critical period of time (as a result of either heat treatment or welding), precipitation of chromium-rich $M_{23}C_6$ (as well as nitrides and intermetallic phase) occurs continuously, along the grain boundaries. Time at temperature determines the amount of carbide precipitation. This precipitation is associated with the depletion of chromium content below the critical 12% in the regions adjacent to the grain boundaries needed for corrosion resistance. As a result, the regions alongside these grain boundaries become more prone to intergranular corrosion (especially in oxidizing solutions) and intergranular fracture (Fig. 10.16)[90] than the remainder of the material. This phenomenon of breakdown in corrosion resistance is referred to as *sensitization*, and the heat treatment which produces this susceptibility is called the *sensitization process* or *heat treatment*. Sensitization also decreases resistance to other types of corrosion such as pitting, crevice corrosion, and stress corrosion cracking. Sensitization has caused concern because of failure of types 304 and 316 stainless steel parts by IGC and SCC in light-water reactors and coal conversion plants. The most critical temperature is about 650°C (1200°F), where holding for only a few seconds may be adequate to allow subsequent degradation.[91] Sensitization may occur during heat treatment, welding, reheating (or aging) through the sensitizing temperature range, or operating at these temperatures. However, sensitization typically takes several minutes. Type 304 can be welded in most cases without suffering sensitization. The

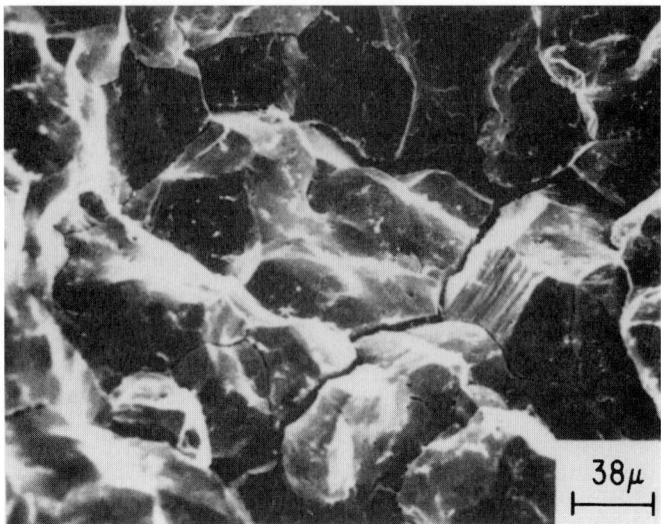

FIGURE 10.16 Intergranular corrosion of a sample of the type 316 stainless steel in the modified Strauss test (A262E) after sensitizing heat treatment.[90] (*Courtesy of C. L. Briant and A. M. Mitter.*)

real importance of L-grade stainless steels is that they resist sensitization when subjected to 1200 to 1250°F for 1-hr stress relief treatment specified by some welding and/or construction codes.[51]

Sensitized steels usually fail more rapidly in a chloride environment and may fail at lower temperatures and/or chloride content than the correctly heat-treated steel. Cracking may also occur as a result of stress-accelerated, intercrystalline corrosion. A good example is in catalytic reformers used in the petroleum industry, where heating during service causes sensitization of the steel which later cracks from the action of corrosive $H_2S_xO_6$ atmosphere (often termed *polythionic acid cracking*).[92]

Sensitized stainless steels (types 304 and 316) fail by SCC in high-temperature water (at 288°C) containing 10 ppm oxygen. However, no cracking was observed in solution-annealed type 304.[93] This SCC occurs when the materials are synergistically subjected to both high tensile stresses and corrosive environments to penetrate the protective oxide layer and proceed through the material at a rapid rate. For stainless steels, aqueous solutions of chlorides, fluorides, hydroxide ions, dissolved oxygen, or a very small amount of lead is considered as the SCC environments. The critical level of tensile stress for SCC may result from primary loading, thermally induced stresses, or residual stresses due to welding or fabrication.[94]

Figure 10.17*a* and *b* shows the temperature-time-sensitization (TTS) curves for type 304 steels with varying carbon and nitrogen contents, respectively.[92] The curves shown in Fig. 10.17*a* suggest that a type 304 with 0.062% C would have to cool below 595°C (1100°F) within 5 min to avoid sensitization, but a type 304L with ~0.030% C could require about 20 hr to cool down below 480°C (900°F) without becoming sensitized. These curves are general guidelines and should be verified prior to their applications to various types of stainless steels.[53]

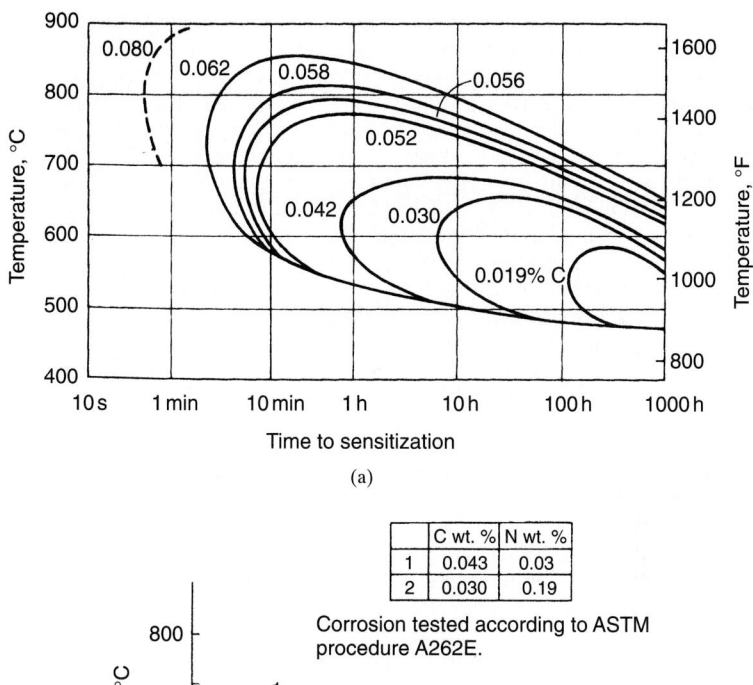

FIGURE 10.17 Temperature-time-sensitization curves for: (*a*) Type 304 stainless steels, solution treated at 1050°C, showing the effect of carbon content on carbide precipitation which forms in the areas to the right of the various carbon-content curves. Corrosion test used was based on ASTM A262E.[53] (*b*) Type 304 stainless steel with 0.03 and 0.19 wt% nitrogen, solution treated at 1050°C. The corrosion test used was based on ASTM A262E.[92] (*a. Reprinted by permission of ASM International, Materials Park, Ohio. b. Reprinted by permission of VCH, Weinheim, Germany.*)

Weld Decay. The term *weld decay* implies that the sensitization phenomenon is observed in the heat-affected zone of weldments in austenitic (types 304 and 316) stainless steels. This zone lies at a distance about ⅛ in. (3.175 mm) from the welded joint where a lower heating or cooling rate prevails during welding in a narrow temperature range of 649 to 871°C (1200 to 1600°F).[57]

10.5.4.1   Theories of Sensitization.   There are three models that are able to explain sensitization and intergranular corrosion of austenitic stainless steels: chromium depletion theory, thermodynamic theory, and solute segregation theory.

Chromium depletion theory, proposed by Bain and coworkers[95] more than 60 years ago, is the most well-recognized explanation because it nicely elucidates the effect of carbon content on sensitization and chromium depletion profile in sensitized steels, which can be easily measured using modern analytical electron microscopy. During sensitizing heat treatments, chromium carbides nucleate and grow along the austenite grain boundaries. This causes the local chromium concentration in the continuous region very near the growing carbides to fall below the critical 12 wt% (13 at%). As a result, the alloy becomes more prone to intergranular corrosion.[96] The major deficiencies of this theory are as follows: (1) This theory does not fully explain the occurrence of corrosion attack preferentially on one side of the carbide while depletion of chromium takes place on both sides of the carbide. (2) There is also a strong effect of cast structure on the extent of SCC in sensitized steels. (3) Failure of sensitized austenitic stainless steels has been reported by transgranular SCC in hot chloride solution rather than by intergranular SCC. This implies that the rate of sensitization is not solely a function of the chromium and carbon content of the alloy.[96]

Thermodynamic theory is based on (1) the simultaneous and independent diffusion of carbon and chromium to carbide nucleation sites, (2) constant ratio of the fluxes of carbon and chromium at the carbide/matrix interface, and (3) the thermodynamic equilibrium of elements at the carbide/matrix interface. Since the diffusion coefficient of carbon is several orders of magnitude larger than that of chromium, the chromium diffusion to the carbide site becomes the rate-controlling process. This model showed a good agreement with the experimental precipitation data.[97]

Segregation theory comprises segregation of residual impurity deleterious to the grain boundary and change of mechanical and chemical properties of the steel.[98] These impurities, which are usually phosphorus and sulfur, are segregated to grain boundaries of type 304 stainless steel in the form of $(Fe,Cr)NiP$ or $(Fe,Ni)CrP$ and increase the susceptibility of IGC or IGSCC[98–101] by independently accelerating corrosion or passivation or by affecting the chromium-carbon equilibrium at the grain boundary and thereby changing the kinetics and thermodynamics of $M_{23}C_6$ precipitation.[98–101] Note that the formation of phosphides of the type $(Fe,Cr)_3P$ also has been reported in other austenitic stainless steel welds which appear both at the ferrite/austenite interface and along solidification hot cracks.[102,103]

10.5.4.2   Control of Sensitization.   Harmful sensitization can be controlled by the use of the following:[94,104]

1. Solution annealing of the sensitized steel at 1038°C (1900°F) minimum to put carbides in solution, followed by rapid cooling in water through the sensitizing temperature range. This process is seldom practiced because of the distortion problem or the size of the component.[104]

2. Postweld heat treatment to rediffuse chromium back into the depleted solution.

3. Extra-low-carbon (0.03% C maximum) version of the alloys such as types 304L, 316L, and 317L, to minimize the amount of $M_{23}C_6$ formation and the corresponding chromium depletion during the relatively short exposure of welding. However, it will be sensitized by long exposure. Nitrogen addition up to 0.15% does not seem to adversely affect the usually excellent sensitization resistance of these steels and their weld metals.[105]

4. Stabilized stainless steels such as types 318, 321, 347, and 348, in which carbide-forming elements such as Nb+Ta, Ti, or Zr tie up the carbide preferentially (i.e., form stable MC-type carbides). Stabilization is not always sufficient to provide austenitic stainless steel immunity to sensitization and intergranular corrosion.[51,52] To obtain full benefits of stabilization, stabilization heat treatments, accomplished at 840 to 890°C (1544 to 1652°F), are carried out to give the most effective intragranular distribution of the alloy carbides.[50] However, at very high temperature in the vicinity of welds, these TiC and NbC carbides may redissolve and probably may lead to the precipitation of $M_{23}C_6$ if the weldments are held or slowly cooled in the sensitizing range. This may cause localized corrosion attack, called knife-line attack, and may be prevented by subjecting the weldments to a final stabilizing heat treatment.[50,106]

5. Duplex structures of ferrite in an austenite matrix (5% ferrite minimum) such as usually takes place in austenite stainless steel castings and weld metals such as type 308.[94]

10.5.5 Effect and Measurement of δ-Ferrite

Although austenitic stainless steels are predominantly austenitic, they often contain small amounts of δ-ferrite as a principal weld metal precipitate; this is desired because it has a higher solubility for sulfur and phosphorus, which avoids fissuring and reduces susceptibility to hot-cracking or hot-tearing.[†] The amount of δ-ferrite depends more on the composition and less on the cooling rate and subsequent heat treatment.

The amount of ferrite present can be determined by a standard magnetic measuring method in terms of ferrite number (FN) or by metallographic examination as a percentage value. For low ferrite levels, the ferrite number and percentage of ferrite are identical.[107] The Schaeffler diagram (Fig. 10.18a) and its modified DeLong diagram (Fig. 10.18b) have been used to predict weld microstructure and determine the approximate amount of δ-ferrite in the austenitic steel weld metal of a given composition, using, respectively, metallographically and magnetically determined percent ferrite.[53] The Schaeffler diagram, known as the "road map" of welding stainless steel, plots the composition ranges (representing the area of occurrence) at room temperature of austenitic, ferritic, martensitic, and duplex alloys in terms of percentage of chromium and nickel equivalents (Fig. 10.18a), considering the effects of prominent ferrite-forming elements as well as of austenite-forming alloying elements. The nickel and chromium equivalents forming the two axes of the Schaeffler diagram can also be calculated from the general formulas, due to Schneider,[108]

$$\text{Ni equivalent} = \%\text{Ni} + \%\text{Co} + 30(\%\text{C}) + 25(\%\text{N}) + 0.5(\%\text{Mn}) + 0.3(\%\text{Cu}) \qquad (10.13)$$

$$\text{Cr equivalent} = \%\text{Cr} + 2(\%\text{Si}) + 1.5(\%\text{Mo}) + 5(\%\text{V}) + 5.5(\%\text{Al}) + 1.75(\%\text{Nb}) \\ + 1.5(\%\text{Ti}) + 0.75(\%\text{W}) \qquad (10.14)$$

The constitution diagram includes nitrogen in the nickel equivalent.[54,61] The Welding Research Council (WRC) introduced further improvement of the DeLong

---

[†] Interdendritic cracking in the weld area that occurs prior to cooling of weld to room temperature is called *hot cracking* or *microfissuring*.[53] Hot cracking has been reported in several commercial grades when ferrite in the microstructure falls below ~4%. However, hot cracking also depends on the location of ferrite in the microstructure.[59]

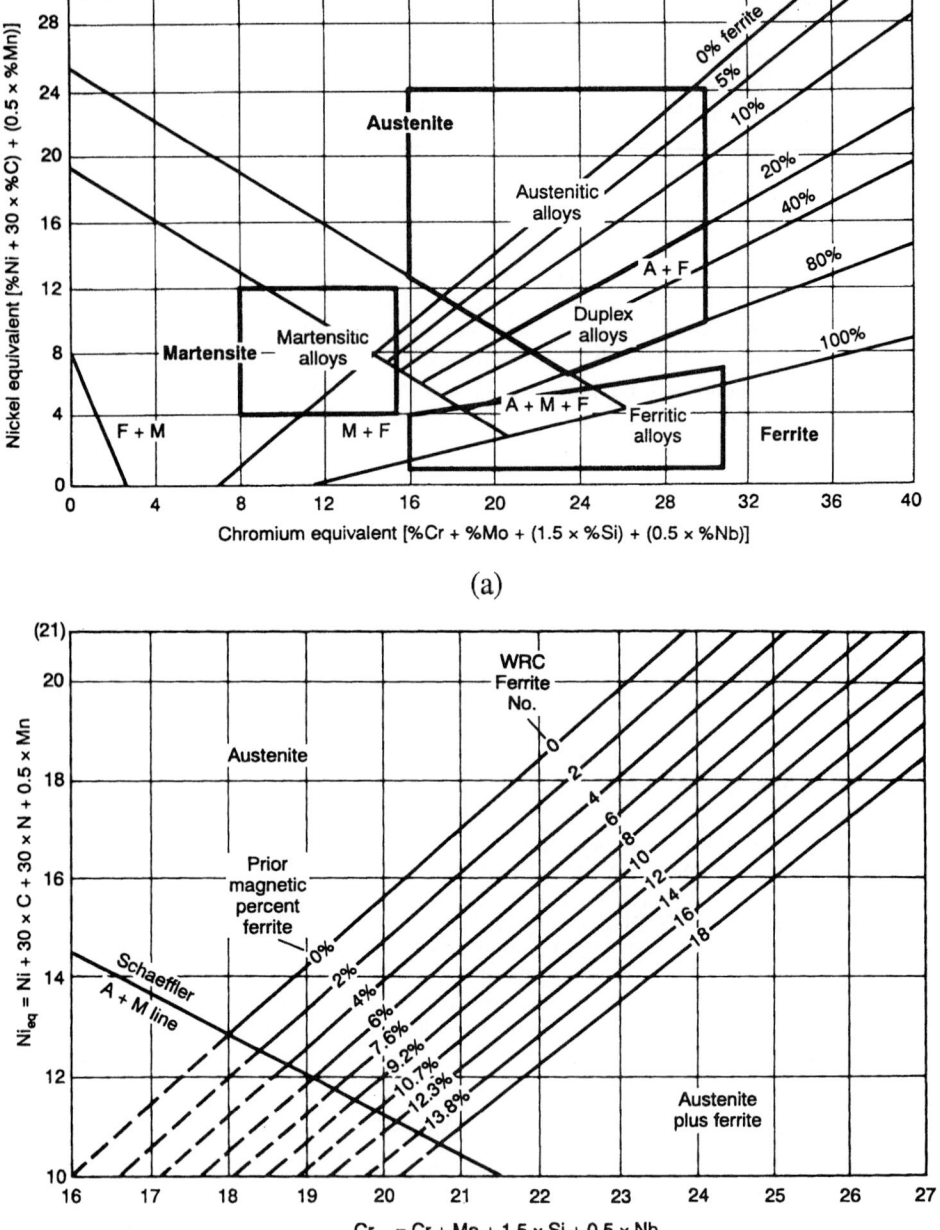

FIGURE 10.18 (*a*) Schaeffler constitution diagram for stainless steel weld metal and (*b*) DeLong constitution diagram for stainless steel weld metal where the Schaeffler austenite-martensite boundary is included for reference.[53,57,109] (*Reprinted by permission of ASM International, Materials Park, Ohio.*)

diagram which seriously (1) underestimated the FN of weld metals with high Mn content and (2) overestimated the FN of highly alloyed weld metals, such as type 309.[57] This one, called the WRC-1988 (or Siewert, McCowan, and Olson) diagram (Fig. 10.19a), is a more accurate diagram which uses computer mapping techniques, covers a larger range of compositions than does the DeLong diagram, and eliminates two serious drawbacks (of the DeLong diagram). The WRC-1988 diagram can be employed to predict both the weld metal ferrite content and solidification behavior. This diagram was again modified and called the WRC-1992 diagram (Fig. 10.19b), which involved addition of copper to the Ni-equivalent formula and extension of nickel-equivalent and chromium-equivalent axes to allow graphical prediction of the FN of weld metal comprising very different base metal(s) and filler metal, as was done less precisely with the Schaeffler diagram. Note that the WRC-1988 and WRC-1992 diagrams contain four solidification regimes, designated as (total) A, AF (primary austenite solidification), FA (primary ferrite solidification), and (total) F. Here solidification and transformation behavior is a function of both composition and cooling rate.[53,57,109]

From a steelmaker's point of view, δ-ferrite is undesirable because it leads to problems in hot-working, if present in excess amounts.[110] Its presence tends to reduce pitting corrosion resistance. When present as isolated ferrite grains in small amounts (4 to 8%) and in sufficiently high ferrite levels, as in duplex stainless steels, it significantly improves resistance to sensitization and SCC as well as avoids microfissuring and hot tearing. However, when present as a continuous grain boundary network, it can lower resistance to sensitization. Prolonged exposure of δ-ferrite at high temperatures (500 to 900°C) can transform it into the σ phase—a hard, brittle, and nonmagnetic phase that decreases ductility, toughness (i.e., induces low-ductility rupture), pitting corrosion resistance, and crevice corrosion resistance.[56,111] However, the range of stability of σ phase increases with Mo content; for AL-6XN alloy with 6% Mo, σ is stable up to ~1040°C (1900°F). The 254 SMo alloy (UNS S31254) with 6% Mo and 17.5% Ni has an even higher σ solvus. Also, it is reported that σ is the high-temperature phase with χ being the lower-temperature phase.[51] (See also the next section for more details.)

In view of the foregoing facts, the 300 series of austenitic stainless steels generally contain adequate nickel or its equivalent to limit the δ-ferrite content. However, in these steels, some δ-ferrite is intentionally allowed to be present in weld metal and castings to avoid microfissuring, hot tearing, or intergranular attack, or in some duplex stainless steels (e.g., type 329) where intentional heat-treating transforms δ-ferrite into brittle σ phase, thereby increasing the wear resistance.[56]

10.5.6 Diffusion-Controlled Precipitation in Cr-Ni Austenitic Steels

Austenite decomposition in austenitic stainless steels at higher temperatures is complex and can be known by referring to relevant equilibrium diagrams. Major precipitating phases include carbides, nitrides, borides, silicides, and a number of intermetallic phases; some of these are transient phases which occur only in the early stages of precipitation and do not exist in the equilibrium structure.[112] Techniques used to characterize and identify these precipitate phases include optical and electron microscopy; electron diffraction and energy dispersive X-ray (EDX) microprobe analysis; and X-ray crystallography.[74] The crystal structure and composition of different constituents in austenitic alloys (such as iron-based, nickel-based, and cobalt-based alloys) are listed in Table 10.9.[110,113–115]

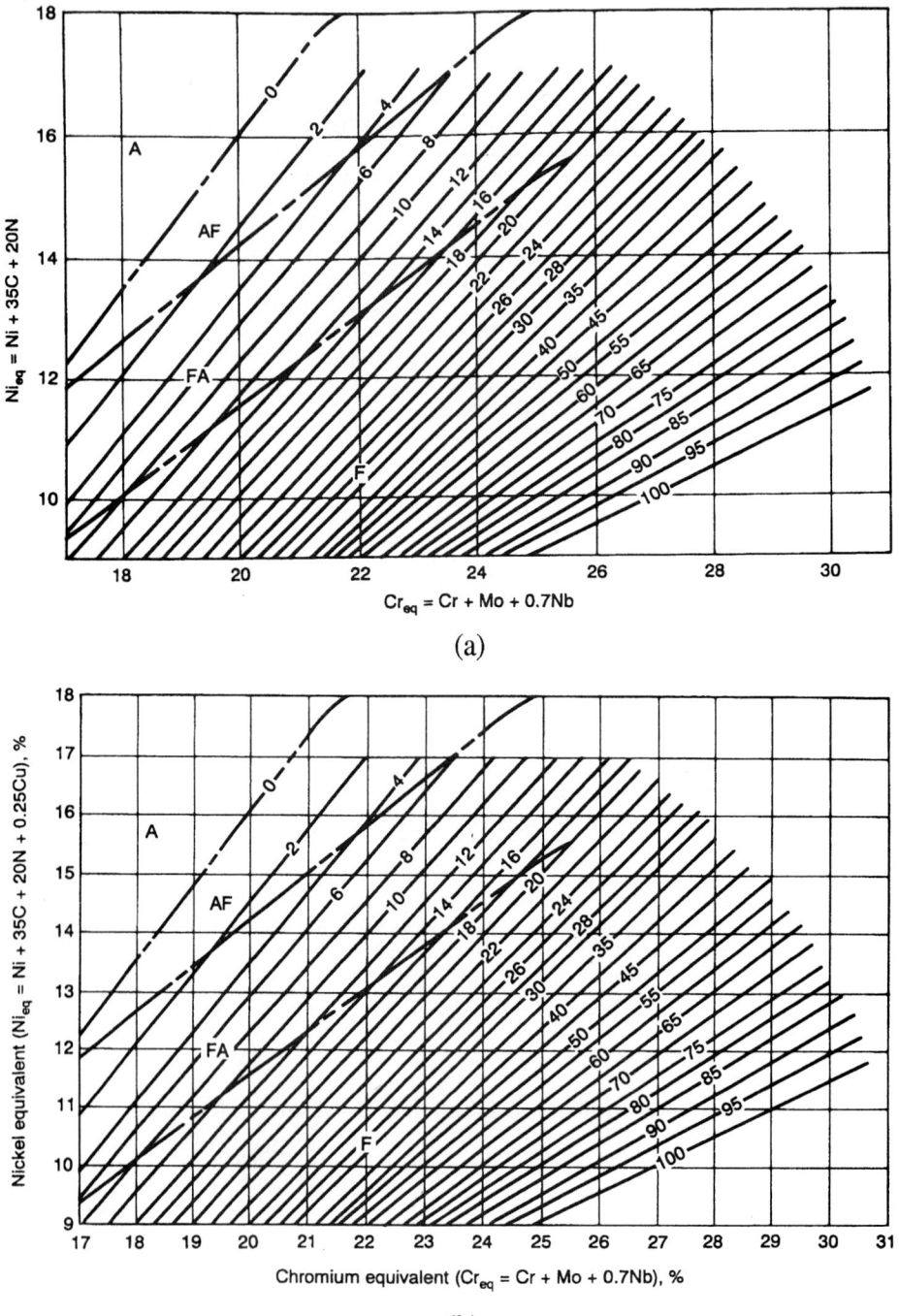

FIGURE 10.19 (*a*) WRC-1988 diagram, including solidification-mode boundaries.[57] (*b*) WRC-1992 diagram for predicting weld ferrite content and solidification mode.[57,109] [(*a*) *Reprinted by permission of ASM International, Materials Park, Ohio.* (*b*) *After D. Kotecki and T. A. Siewert, Weld J., vol. 71, 1992, pp. 171s–179s.*]

TABLE 10.9 Crystal Structure and Composition of Different Constituents in Austenitic Alloys (Such as Iron-Based, Nickel-Based, and Cobalt-Based Alloys)[110,113–115]

Phase	Crystal structure	Composition
	Carbides	
MC	fcc	TiC; NbC
M_7C_3	Pseudohexagonal	Cr_7C_3; $(Fe, Cr)_7C_3$
$M_{23}C_6$	fcc	$(Cr_{16}Fe_5Mo_2)C_6$; $(FeCr)_{23}C_6$
M_6C	fcc (diamond type)	$(Cr, Co, Mo, Ni)_6C$; $(Ti, Ni)_6C$; $(Nb, Ni)_6C$
	Nitrides	
M (C, N)	fcc	Nb (C, N); Ti (C, N)
M_2N	fcc	Cr_2N
	Borocarbides	
$M_{23}(C, B)_6$	fcc	$(Fe, Cr, Mo)_{23} (C, B)_6$
	Geometrically close-packed phases	
γ'	fcc	$Ni_3(Ti, Al)$
γ''	bct	Ni_3Nb
δ	Orthorhombic	Ni_3Nb; Ni_3Ti
β	bcc	$Ni(Al, Ti)$
	Topologically closed-packed phases	
σ	bct	$(Fe, Ni)_x(Cr, Mo)_y$
Laves	Hexagonal	Fe_2Mo; Fe_2Ti; Fe_2Nb
χ	bcc	$(Fe, Ni)_{36}Cr_{18}Mo_4$; $Cr_6Fe_{18}Mo_5$
μ	Rhombohedral	$(Co, Ti)_7(CrW)_6$
G	fcc	$(Ti, Zr, V, Nb, Ta, Mn)_6(Ni, Co)_{16}Si_7$
R	Hexagonal	Fe-Cr-Mo
	Others	
Carbosulfides	hcp	$Ti_4S_2C_2$
Cr-rich ferrite (α')	bcc	
α-Manganese sulfide	fcc	MnS

Reprinted by permission of the Institute of Metals, England.

10.5.6.1 Carbide Precipitation. It is suggested that at carbon contents in the ranges present in AISI austenitic stainless steels, the predominant carbide is $M_{23}C_6$. At very high carbon concentration, a second carbide, M_7C_3, may form. In the presence of strong carbide-forming elements, other carbides such as MC and M_6C form in these steels as well as in nickel-based and cobalt-based superalloys.

$M_{23}C_6$ Carbide. Cr-rich $M_{23}C_6$ has a complex face-centered cubic structure which contains 92 metal atoms and 24 carbon atoms (i.e., four "molecules")[93] per unit cell and has a lattice parameter of 10.65 Å. Iron, nickel, and molybdenum can partially substitute for chromium, with the amount of substitution usually being a function of aging temperature, time, and alloy composition. Its formation, along with

that of γ' (in only γ'-forming alloys), occurs at the lower aging temperature by the following reaction between MC and the matrix γ:

$$MC + \gamma \rightarrow M_{23}C_6 + \gamma' \qquad (10.15)$$

The $M_{23}C_6$ precipitation occurs in the temperature range of 550 to 950°C (932 to 1742°F); however, the upper temperature limit of its precipitation depends on the carbon content, as shown by the time-temperature-precipitation curves in Fig. 10.20a.[53,116] As a result, low-carbon L grades of stainless steels are preferred for corrosive service where as-welded parts must be used. The specific kinetics of precipitation of carbides are functions of the alloy chemistry, the prior condition, and the nucleation sites. In general, the kinetics are studied by solution-annealing of the alloy, water-quenching, to suppress precipitation, and reheating (i.e., aging) to the desired temperature.[50]

Note that $M_{23}C_6$ carbide precipitation can occur in relatively short times or at relatively fast cooling rates compared to other precipitates (Fig. 10.20b and c).[56] When $M_{23}C_6$ occurs in the presence of more stable MC carbides, it tends to be delayed and is obtained at the lower temperatures. In general, carbides with higher energy of formation form first and at the higher temperatures (Fig. 10.20b).[114,117]

The two precipitation kinetics of $M_{23}C_6$ carbides in type 304 stainless steel shown in Fig. 10.21[116,118] do not agree precisely because these curves were derived from X-ray and electron microscopy examination on two alloys with some variations in composition.[50,116,118] As in other austenitic alloys such as nickel- and cobalt-based superalloys, precipitation of this carbide takes place very rapidly on the

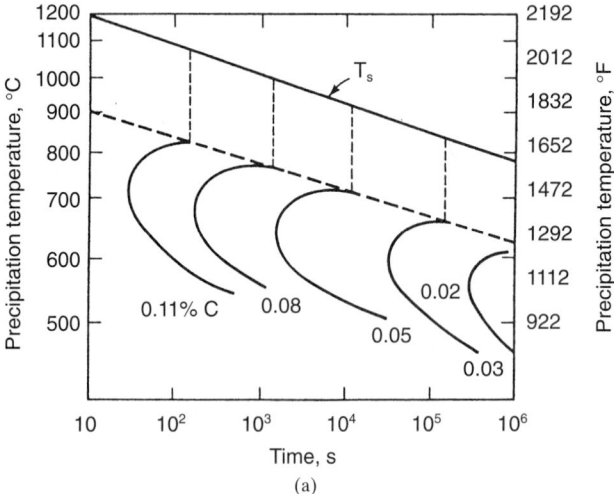

(a)

FIGURE 10.20 (a) Time-temperature-precipitation curves for an 18Cr-9Ni alloy with various carbon contents.[53,116] (*Reprinted by permission of ASM International, Materials Park, Ohio.*) (b) Temperature-time growth curves for $M_{23}C_6$, and (Nb,Ti)C in Cr-Ni austenitic steels.[117] (*Courtesy of R. W. K. Honeycombe.*) (c) Time-temperature-precipitation curve for type 316 stainless steel containing 0.066% carbon.[56] (*Reprinted by permission of John Wiley & Sons, New York. After B. Weiss and R. Stickler, Met. Trans., vol. 3, 1972, p. 851.*)

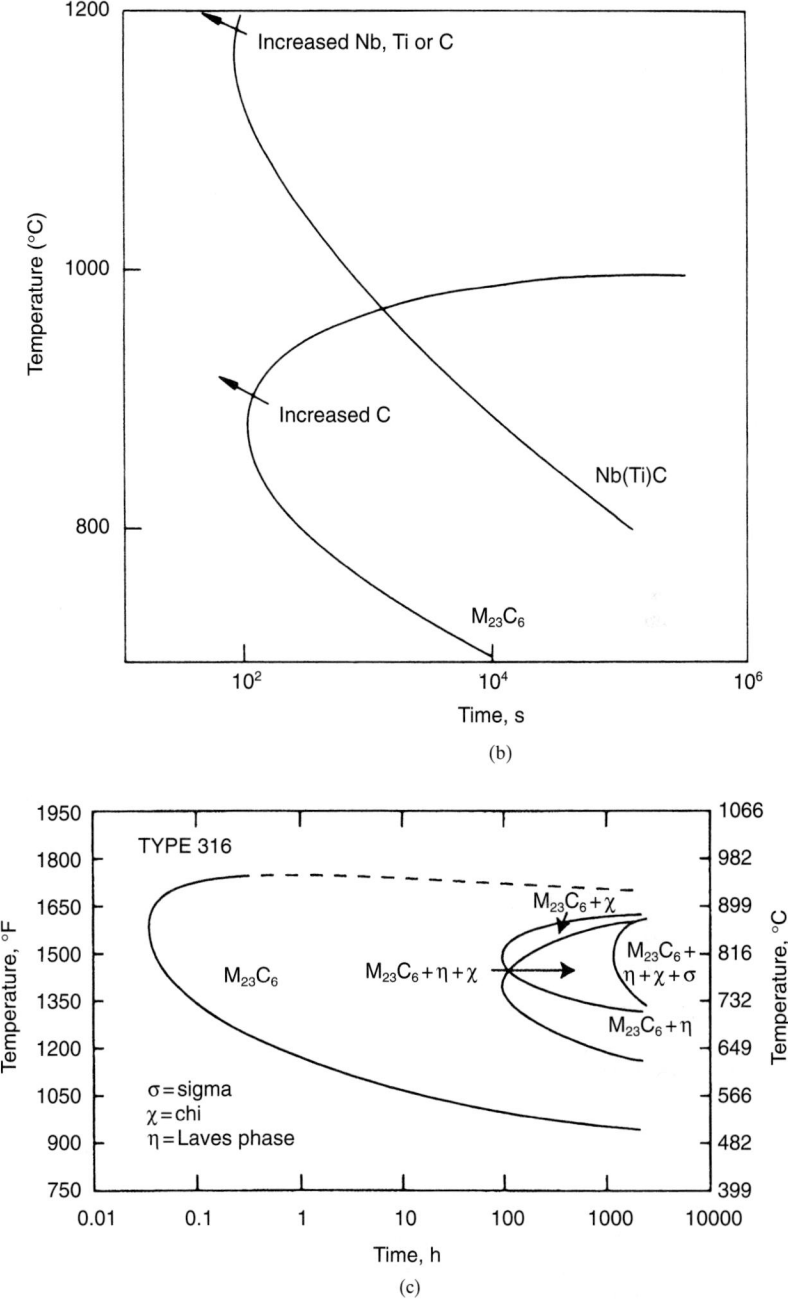

FIGURE 10.20 (*a*) Time-temperature-precipitation curves for an 18Cr-9Ni alloy with various carbon contents.[53,116] (*Reprinted by permission of ASM International, Materials Park, Ohio.*) (*b*) Temperature-time growth curves for $M_{23}C_6$, and (Nb,Ti)C in Cr-Ni austenitic steels.[117] (*Courtesy of R. W. K. Honeycombe.*) (*c*) Time-temperature-precipitation curve for type 316 stainless steel containing 0.066% carbon.[56] (*Reprinted by permission of John Wiley & Sons, New York. After B. Weiss and R. Stickler, Met. Trans., vol. 3, 1972, p. 851.*) (*Continued*)

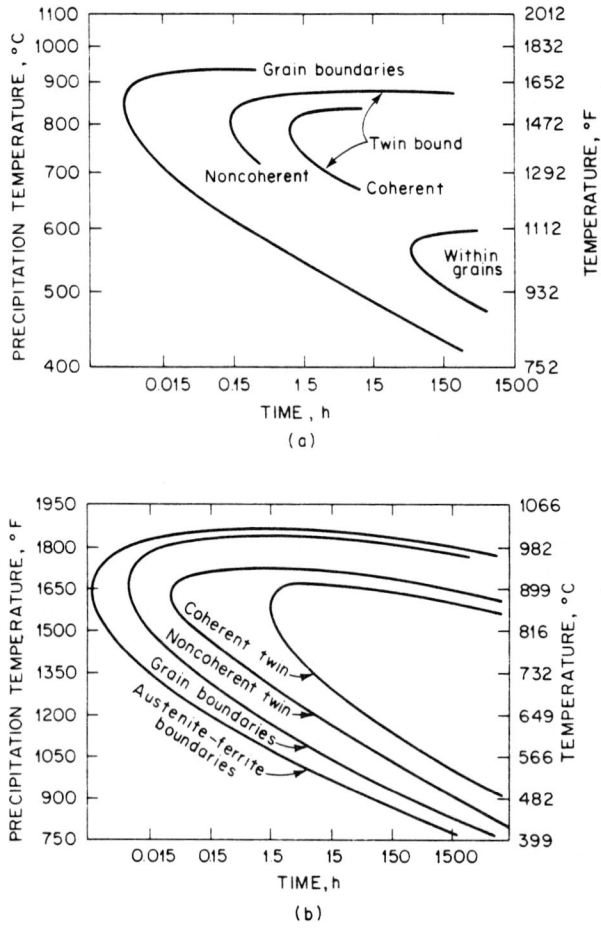

FIGURE 10.21   The precipitation kinetics of $M_{23}C_6$ carbides in type 304 (18Cr-9Ni) stainless steel. (*a*) Alloy containing 0.05% carbon originally quenched from 1250°C;[116] (*b*) alloy containing 0.038% carbon originally quenched from 1260°C (grain size ASTM no. 1).[118]

austenite/ferrite interface, followed very quickly by precipitation on other incoherent boundaries (grain boundaries and twin boundaries). Delayed precipitation occurs on coherent boundaries followed by intragranular precipitation.

Grain Boundary Precipitation.   Adamson and Martin[119] have concluded that ledges or dislocations can promote grain boundary precipitation of $M_{23}C_6$ which involves the development of small {111} facets on the surface of one of the grains and nucleation of carbide as a small plate on this facet. When the quenched steel is aged to precipitate grain boundary particles of $M_{23}C_6$, these linear defects disappear. However, Singhal and Martin[120] favor the grain boundary $M_{23}C_6$ precipitation by a process involving the migration of an austenite grain boundary. This produces, in some cases, regular arrays of very similar ($M_{23}C_6$) particle on the grain boundary so

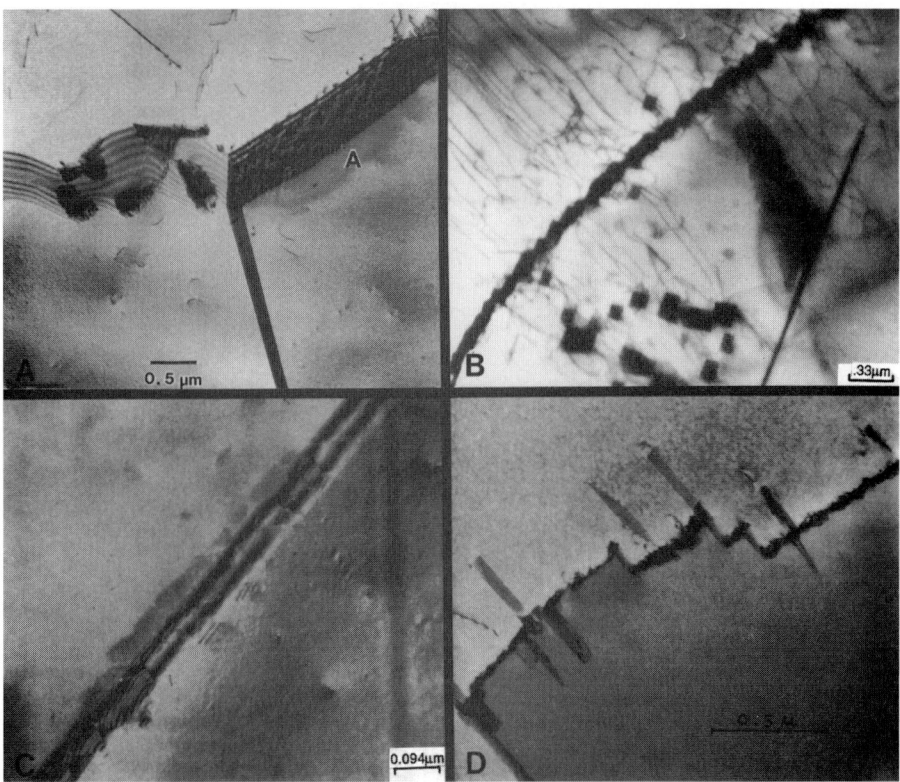

FIGURE 10.22   Thin-foil image from (A) the aged 316LN-3 steel specimen for 100 hr at 700°C illustrating the different carbide distribution at the three grain boundaries;[121] (B) the aged alloy (composition: Fe-0.191C-0.65Mn-0.27Si-9.85Ni-18.55Cr-0.010N) for 200 hr at 750°C showing the intragranular precipitation, growth, and coalescence of $M_{23}C_6$ carbide on dislocations with continued aging (this produces jagged primary stringers);[124] (C) the aged Fe-0.1C-15.16Cr-25.5Ni-0.23Si-0.6Mn alloy specimen for 0.5 hr at 750°C showing $M_{23}C_6$ plates on coherent twin boundaries;[126] and (D) aged Fe-0.031C-0.027N-24.1Ni-25.5Cr-0.27Ti-0.23Si-0.018Mn stainless steel for 3 hr at 750°C showing $M_{23}C_6$ precipitation at incoherent twin boundaries.[120] *[(A) Reprinted by permission of The Metallurgical Society, Warrendale, Pennsylvania; (B, C, D) reprinted by permission of Pergamon Press, Plc.]*

that each $M_{23}C_6$ particle is in parallel orientation with one or the other austenite grain with which it is in contact, in accordance with

$$\{111\}_{M_{23}C_6}//\{111\}_\gamma \qquad <110>_{M_{23}C_6}//<110>_\gamma$$

Figure 10.22*a* is a thin-foil image from the aged 316LN-3 steel specimen illustrating the different carbide distribution at the three grain boundaries.[121]

At the lowest $M_{23}C_6$ precipitation temperature, the grain boundary carbide assumes a continuous, thin sheetlike morphology. As the temperature increases through 600 to 700°C (1112 to 1292°F), the particles become feathery dendrites. With increasing time, the feathery dendrite structure progressively thickens and coarsens.[118] At higher temperatures the grain boundary carbides become discrete

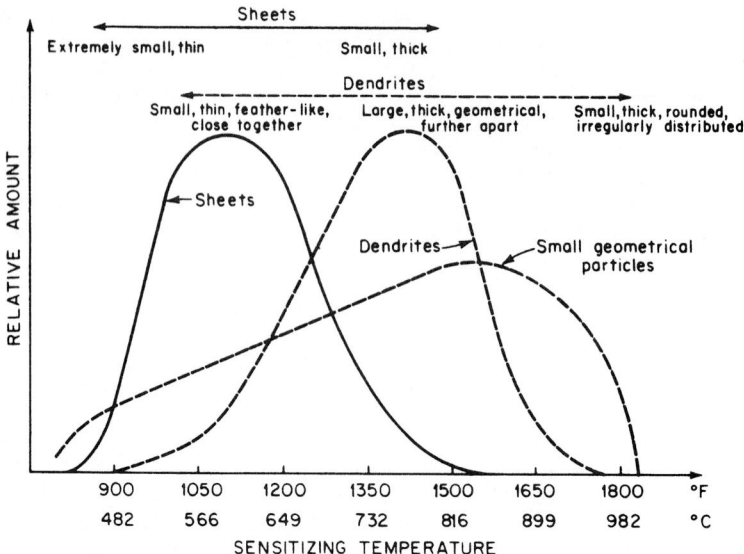

FIGURE 10.23   The morphology of grain boundary $M_{23}C_6$ as a function of sensitizing temperature.[116] (*Reprinted by permission of ASM International, Materials Park, Ohio.*)

geometric particles; their shape depends on the boundary orientation, the extent of misfit, and the particular temperature. Stickler and Vinckier have summarized the formation modes as shown in Fig. 10.23.[118]

Intragranular Precipitation.   Intragranular precipitation of $M_{23}C_6$ has been studied extensively.[122,123] Precipitation occurs on dislocations within the matrix grains as many evenly spaced angular particles. The dislocations are usually straight and close to $\langle 112 \rangle$ directions, i.e., edge dislocations having Burgers vector $\mathbf{b} = (a/2)$ $\langle 110 \rangle$ on $\{111\}$ planes. In some cases, stringers form on curved dislocations. These precipitates grow and coalesce with continued aging and produce jagged primary stringers (Fig. 10.22*b*).[124] Eventually, branches are found emanating from primary stringer particles growing in linear arrays along $\langle 110 \rangle$ directions. Lewis and Hattersley[125] have demonstrated that stringers of carbides form on dislocations generated by the prismatic punching mechanism around growing carbide particles. The orientation relationship between the austenite and carbide is given by

$$\{100\}_{M_{23}C_6} /\!/ \{100\}_\gamma \qquad <100>_{M_{23}C_6} /\!/ <100>_\gamma$$

and the carbide is bound by a low-energy $\{111\}$ interface.

During aging, coherent twin boundary precipitation of $M_{23}C_6$ occurs on preexisting pile-up of dislocations. The preferential growth of the precipitate along the twin plane is favored by the availability of vacancies along this interface. On continued aging, the particles grow as triangular plates with $\langle 110 \rangle$ edges along the boundary, and finally the precipitates link up and form a long continuous sheet of $M_{23}C_6$ (Fig. 10.22*c*).[126] The $M_{23}C_6$ precipitates exhibit Moiré fringes within which dislocations are occasionally seen.[126]

Nucleation of $M_{23}C_6$ occurs at incoherent twin boundaries at approximately the same time as that on grain boundaries on aging supersaturated solid solutions of austenite. Eventually, very thin ribbons or plates of $M_{23}C_6$ form, growing in directions parallel to the twin plane.[125] First, precipitates form upon matrix dislocations and upon an incoherent twin boundary. Subsequent loops of Shockley partial dislocations of type $\mathbf{b} = (a/6)\langle 211\rangle$ bow out from the incoherent twin boundary on {111} planes containing an intrinsic stacking fault. Carbide precipitation then proceeds along the bounding partial dislocation with growth across the stacking fault to form a sheet of $M_{23}C_6$ on {111} (Fig. 10.22d).[120] Further Shockley partial dislocations lead to repetition of the process.

MC Carbides.   The MC carbides such as NbC, TiC, and (Nb,Ti)C are thermodynamically stable and possess the B1 NaCl-type fcc structure. This structure consists of metal atoms occupying fcc positions and carbon atoms in the octahedral interstices. For the formation of MC carbides, usually sufficient amounts of niobium and titanium are added to exceed the stoichiometric (i.e., atomic weight) ratios of 4:1 and 8:1, respectively.[117]

The MC carbides are found to precipitate mostly intragranularly on dislocations and stacking faults and sparingly on the grain boundaries. When these carbides are growing, they show faceted structure; i.e., they adopt octagonal shapes. Moreover, carbides always have a cube-cube Widmanstätten orientation relationship with the matrix.

In the very early stage, MC particles appear to be coherent. At longer aging times, arrays of dislocations occur at the edges and around the growing particles. Since the specific volumes of MC particles are much greater than those of the austenite, a flux of vacancies into the precipitating carbide, which is required for its growth, is supplied by the dislocations and grain boundaries. No significant coarsening of MC is found at increased aging times and temperatures.

When austenitic stainless steel is quenched and aged at low aging temperatures (~700°C), thick disks of TiC, NbC, and TaC precipitate on dislocations and induce dislocation multiplication, usually in the form of loops. However, a small proportion of existing (or growth) dislocations are able to dissociate into partial dislocations and form stacking faults before growing into large particles.[126]

When types 321 (titanium-stabilized) and 347 (niobium-stabilized) steels are quenched and given stabilizing treatments at 842 to 898°C (1550 to 1650°F) for several hours, TiC and NbC particles form, respectively. Alternatively, during high-temperature [1100 to 1300°C (2012 to 2372°F)] solution-annealing treatment or welding, these carbides may dissolve to a greater extent in austenite and reprecipitate at some lower temperatures within the precipitation range, which produces a fine dispersion of intragranular or grain boundary MC particles, causing a greater strengthening effect at room and elevated temperatures. Such an occurrence of finer creep-induced Nb(C,N) precipitation (and large undissolved intermetallic Fe_2Nb particles) has been observed on dislocations within the grains (Fig. 10.24a) and on grain boundaries (Fig. 10.24b) when the high-niobium type 347 steel is given a prior solution treatment at lower temperature (1050°C).[127]

M_6C Carbide. M_6C carbide has a fcc structure having 96 metal atoms per unit cell. It is a ternary phase and is often represented as A_3B_3C such as Fe_3Mo_3C, Fe_3Nb_3C, and $(FeCr)_3Nb_3C$ or A_4B_2C. It may be observed in stainless steels containing >6% molybdenum or 0.8 to 2% niobium. They usually precipitate intragranularly.[53] In types 316 and 316L, M_6C appears to form from $M_{23}C_6$ after a prolonged aging time (>1500 hr) in a limited temperature range around 650°C according to the following reaction:[122,128]

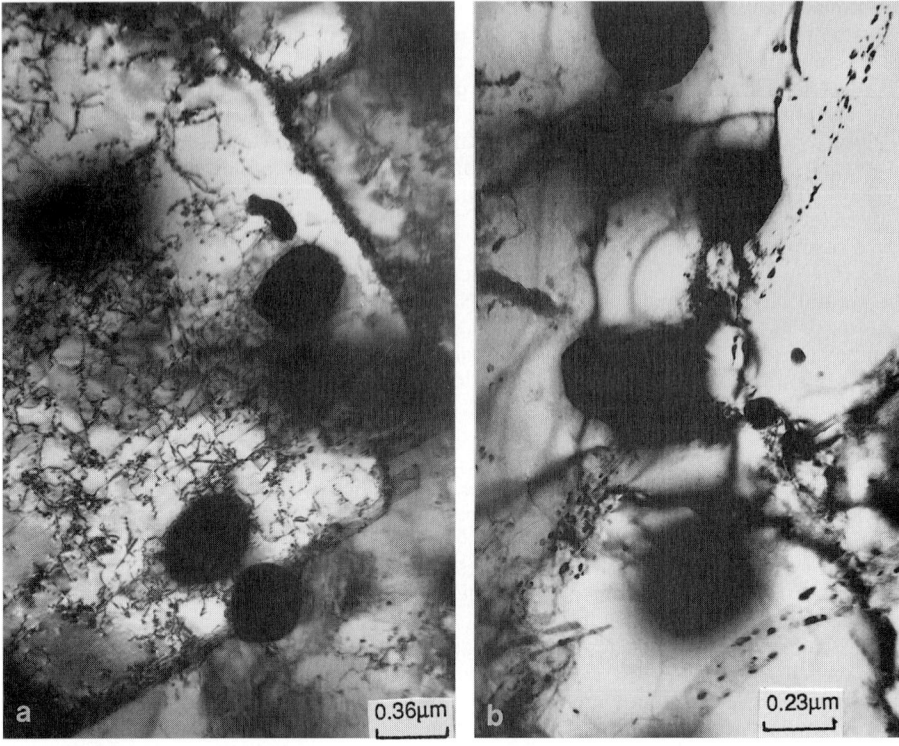

FIGURE 10.24 Type 347H (high-niobium steel) solution-treated at 1050°C showing large undissolved Laves phase, Fe₂Nb particles, and fine Nb(C,N) precipitation nucleated during creep on (*a*) dislocations and (*b*) grain boundaries.[127] (*Reprinted by permission of Elsevier Sequoia, Lausanne, Switzerland; after H. J. Kestenbach and L. O. Bueno.*)

$$M_{23}C_6 \xrightarrow[650°C\,(1202°F)]{+Mo} (Fe,Cr)_{21}Mo_2C_6 \xrightarrow{+Mo} M_6C \qquad (10.16)$$

This also forms in neutron-irradiated type 316.

M₇C₃ Carbide. Chromium-rich M_7C_3 probably forms in Fe-Cr-C or Fe-Cr-Ni-C alloys where carbon concentrations are considerably larger than those specified for the 300 series. Thus a major proportion of M_7C_3 has been observed to exist near surface regions of the steel.[74]

10.5.6.2 Boride Precipitation. Boron greatly reduces the solidus of stainless steel and is detrimental to weldability and at higher levels detrimental to hot workability.[51] However, boron additions up to 0.002 to 0.004% are often accomplished in austenitic stainless steels to enhance their hot workability and creep-rupture properties such as rupture-ductility by delaying grain boundary fissuring.[129] However, during reactor irradiation, helium forms as small bubble and induces sudden intergranular failure and severe embrittlement of austenitic stainless steels in elevated-temperature tensile and creep-rupture tests.[114,130] In austenitic stainless

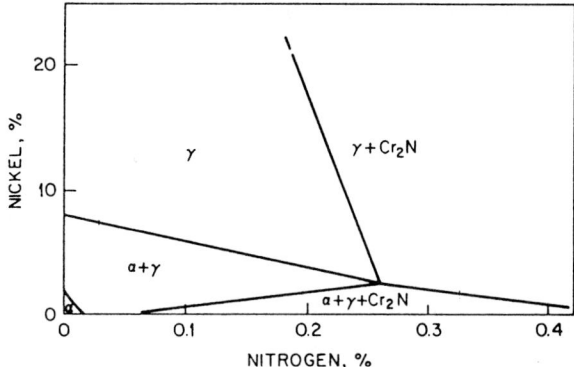

FIGURE 10.25   Equilibrium diagram for the Fe-18Cr-Ni-N system at 900°C (1652°F).[50]

steel, the solubility of boron is about 40 ppm at solution temperatures of about 1050°C and much less at the lower temperatures of 650 to 750°C. It has been observed that boron can replace some carbon, resulting in the intergranular $M_{23}(CB)_6$-type precipitation.[74,117]

Borated stainless steels (ASTM Specification A887) are austenitic stainless steels containing 0.20 to 0.25 wt% of natural or enriched boron which causes thermal neutron absorption due to the presence of ^{10}B isotope which occurs as a dispersion of Cr_2B-type boride precipitates in an austenitic stainless steel matrix. These steels have good mechanical properties, such as yield strength = 205 MPa minimum, tensile strength = 515 MPa, and elongation minima = 6 to 40%, and are primarily used as simple "strap-on" neutron shielding materials rather than structural alloys.[131] These steels also find applications in the nuclear industry for neutron critical control, spent-fuel storage tanks, and transportation parts.[131]

10.5.6.3 Nitride Precipitation.   N has attracted great interest as an alloying element in that its presence may retard the precipitation of secondary phases in austenitic stainless steels, provided nitride formation can be avoided. N tends to delay slightly the onset of sensitization and lower the maximum temperature of precipitation. In contrast, both Mo and Mn have the opposite effect, i.e., an acceleration of the onset of precipitation and an increase in the maximum precipitation temperature, while Ni decreases the maximum temperature.[112]

Figure 10.25 shows the equilibrium diagram for the Fe-18Cr-Ni-N system at 900°C (1652°F) illustrating the equilibrium between hexagonal Cr-rich Cr_2N (epsilon nitride) and austenite.[50] This holds for alloys containing no significant amounts of strong nitride formers. In the AISI 300 series, lower nitrogen content (<0.15%) produces complete dissolution of nitrogen in solid solution, whereas larger concentration (0.2 to 0.3%) results in the intragranular precipitation of Cr_2N. However, as the nitrogen level is increased and the material is exposed to air at higher temperature (>600°C), coarse Cr_2N precipitation occurs within the grain together with discontinuous lamellar precipitation at grain boundaries.[117]

In addition to the first precipitating Cr_2N phase, the nickel-rich pi nitride (as a transient phase) and silicon-containing eta-phase M_5SiN have also been observed in highly alloyed austenitic stainless steels. The latter was observed in a number of cases as a stable nitrogen-containing phase.

Higher nitrogen concentration, however, leads to large reduction of ductility and impact values in alloys aged between 700 and 800°C. The best mechanical properties are found to exist in alloys aged at 700°C.

Note that nitrogen content is important, especially in nonstabilized and stabilized austenitic steels with relatively low carbon concentrations. The nitrogen can substitute for carbon in NbC, TiC, $M_{23}C_6$, and M_6C precipitates. Hence these phases can be accurately described as carbonitrides having the general formula Ti(C,N), Nb(C,N), $M_{23}(C,N)_6$, and $M_6(C,N)$.[117] In general, increasing carbon content promotes general precipitation, whereas increasing nitrogen content and higher aging temperatures promote the formation of lamellar grain boundary precipitation.[132]

TiC and TIN may have complete intersolubility, but in normal steelmaking practice, Ti(C,N) is not usually observed. This is a kinetic effect. The much higher stability of TiN causes it to be formed at high temperatures. As the temperature is lowered, TiN precipitation depletes N. By the time the TiC solvus is reached, N is essentially gone. Metallography often reveals TiC (gray) precipitate growing, perhaps epitaxially on TiN (gold-colored) precipitate.[51]

10.5.6.4 Precipitation of Intermetallic Phase.
There are 10 different intermetallic phases commonly observed in austenitic stainless steels, as listed in Table 10.9. Many of them comprise the elements nickel, aluminum, and titanium.[133]

We will discuss here a majority of them.

Gamma Prime (γ') Precipitate.   γ' is principally an intermetallic $L1_2$ ordered fcc structure or Cu_3Au structure. It is based on Ni_3Al, in which nickel atoms occupy the face-centered positions while aluminum atoms occupy the cube corner sites; titanium, tantalum, or niobium may substitute for aluminum up to 60% without altering the basic structure. If the solubility limit becomes larger, η-Ni_3Ti or orthorhombic ε-Ni_3Nb phase will form instead of, or in combination with, γ'.[134]

It is generally recognized that γ' precipitate phase particles, being coherent and ordered, contribute the major age-hardening effects, and their large volume fraction (30 to 50%) is the major constituent of strength in nickel-based alloys and austenitic stainless steels as well as in iron-based superalloys (e.g., A286 alloy, an austenitic PH stainless steel) in the underaged and peak-aged conditions. This intermetallic phase has been briefly discussed also in Chap. 6 for Ni-based superalloys and austenitic PH stainless steels.

In austenitic stainless steels [for example, 12 to 25% Ni with up to 20% Cr and 1 to 5% Al (or Al + Ti)], quenching from a solution temperature of 1100 to 1250°C followed by aging in the range of 700 to 800°C produces ordered fcc γ' as either Ni_3Al or $Ni_3(Al,Ti)$ precipitates. These precipitates have the same crystal structure and lattice parameter as the parent phase and exhibit a cube-cube orientation relationship with the austenite matrix and therefore form coherently due to low precipitate/matrix interface energy. Coarsening rate of this phase follows a Lifshitz-Slazov-Wagner (LSW) theory of diffusion-controlled growth. This LSW theory has also been observed to hold invariably for austenitic nickel-base superalloys with γ' as the major strengthening phase.[135–138]

In general, initially fine, spherical γ' precipitates are usually observed when the misfit is small, but γ' precipitates assume a complex or different morphology on prolonged aging at 750°C in which either a larger misfit or the loss of coherency with the parent matrix occurs.

Figure 10.26 shows the formation of metastable γ' particles between the stable, lamellar η precipitates in a proprietary austenitic Fe-Ni-Cr-Al-Ti alloy specimen

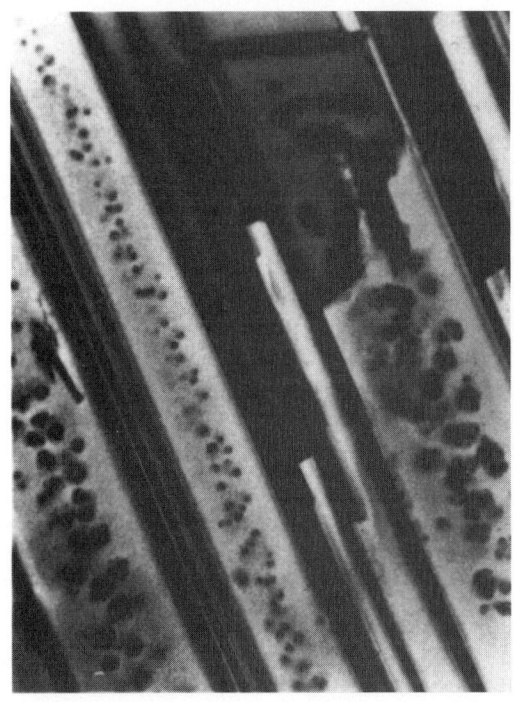

FIGURE 10.26 Formation of metastable γ' particles between stable lamellar η precipitate in a proprietary austenitic (Fe-Ni-Cr-Al-Ti) alloy specimen aged for 2 hr at 790°C following water-quenching from 1093°C (2000°F).[137] (*Reprinted by permission of Pergamon Press, Plc.*)

| 0.5μm |

aged at 790°C (1450°F) for 2 hr following water-quenching from 1093°C (2000°F). Note that three phases (γ, γ', and η) are found in the transformed region, and that the nucleation and growth of η precipitate are associated with the segmented motion of the twin boundaries.[139]

Note that the shape and arrangement of the γ' particles are dependent on the precipitate/matrix misfit strain and particle size.[140] The arrangement of γ' particles into rows usually occurs after much longer aging times.

By varying the Ti/Al ratio in γ', the coarsening characteristics can be considerably modified. Addition of Al to γ' Ni$_3$Ti reduces the lattice parameter from about 3.590 Å to around 3.582 Å, which is similar to that of the matrix of 25%Ni-15%Cr alloy and therefore leads to the precipitation of a fine coherent dispersion. However, complete substitution for titanium decreases the lattice parameter to 3.559 Å, which causes an increase in mismatch parameter. An optimum value for high-strength and corrosion-resistant austenitic steel has been found when the Al + Ti content of 1 to 1.5% Al and 3 to 3.5% Ti is present.[117]

Gamma Double Prime (γ'') Precipitate.   See Chap. 6 for details

Eta (η) and Beta (β) Precipitates.   In an austenitic steel essentially free from Al but containing high levels of Ti, ordered and incoherent hexagonal, DO$_{24}$ phase forms, from either intermediate or metastable γ' Ni$_3$Ti on overaging or at higher temperature, as a Widmanstätten intragranular precipitate or as a grain boundary cellular precipitate, or as both. The orientation relationship between lamellar η and the austenite matrix is

$$(0001)_{Ni_3Ti} // (111)_\gamma \qquad [1210]_{Ni_3Ti} // [110]_\gamma$$

The stacking sequence for η phase is ABACABAC, while that for γ' is ABCABC.

In steels containing a high Al/Ti ratio, the equilibrium intermetallic phase is ordered bcc β-NiAl or Ni(AlTi), which forms as massive plates after dissolving γ' phase. The β-NiTi forms at a relatively low Ni/Ti ratio and has a cube-cube orientation relationship. The β-NiTi exists as a small spherical precipitate with misfits larger than those for β-NiAl, due to the larger atomic diameter of titanium.

Both η- and β-phase transformations almost invariably lead to a deterioration of mechanical and physical properties.[141]

Sigma (σ) Phase.   The σ phase has a complex tetragonal (D8b) crystal structure with 30 atoms per unit cell and is generally found in composition ranges from AB_4 to A_4B, where A is a transition metal in Groups IIIA through VIA and B is from Group VIIA or VIIIA. The σ phase having a nominal composition FeCr is probably the important minor precipitate phase, particularly in Cr-Ni austenitic steel containing >18% (equivalent) Cr as well as in Fe-Cr system over a composition range between 25 and 60% Cr.[133] In addition to high chromium content, its formation is favored by Mo, Ti (to a lesser extent), Si (up to 2 to 3%), and cold work, whereas nickel, nitrogen, and carbon retard its formation. For example, AISI type 302B (containing 2 to 3% silicon), type 310 (25%Cr-20%Ni steel), and types 316 and 317 (molybdenum-bearing alloys) are susceptible to σ-phase formation (Fig. 10.20c).[50]

The occurrence of the σ phase in stainless steels and other high-alloy steels has been reviewed by Hall and Algie.[142] The mechanism of σ-phase precipitation in Cr-Ni austenitic steels has been shown to be a function of the chemical composition of the austenite after carbide precipitation whose formation always precedes nucleation of the σ phase. Depending on the chemical composition of the austenite, the σ phase may precipitate directly from the austenite (in the unstabilized austenitic steels, partly at the expense of $M_{23}C_6$) and partially through the $M_{23}C_6$ carbide or through the ferrite, or entirely through the $M_{23}C_6$ carbide. In the latter case, the amount of σ-phase precipitate does not exceed 0 to 5%.[143]

Cold deformation accelerates the start of σ-phase precipitation, but the number of nuclei increases only when the material has already been recrystallized to a fine grain size. The σ phase can occur in the temperature range of 540 to 900°C (1000 to 1650°F). Like sensitization, it can be corrected by (1) solution-annealing at 900°C or above it where σ phase redissolves[53] and (2) subsequent rapid cooling so that both σ-phase precipitation and 475°C embrittlement are prevented. Figure 10.27 shows an example of the effect of cold work on the amount of σ phase in type 310 stainless steel on aging at 800°C.[56]

The σ-phase precipitation occurs on triple points, incoherent grain boundaries, and incoherent twin boundaries and high-energy interfaces of intragranular inclusions. The orientation relationship between σ and the austenitic matrix often varies; according to Lewis,[144] this relationship is

$$(001)_\sigma // (111)_\gamma \qquad (140)_\sigma // (0\bar{1}1)_\gamma$$

and according to Beckitt,[145] it is

$$(001)_\sigma // (111)_\gamma \qquad (\bar{1}10)_\sigma // (\bar{1}10)_\gamma$$

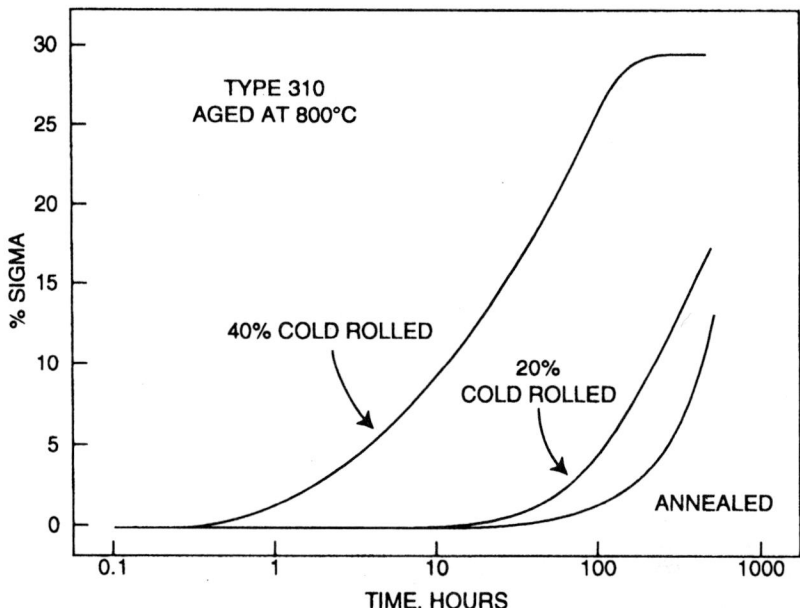

FIGURE 10.27   Effect of cold work on the formation of the σ phase in type 310 stainless steel on aging at 800°C.[56] (*After A. J. Lena and W. E. Curry, Trans. ASM, vol. 47, 1955, p. 193.*)

These are equivalent to within 1° about the $(001)_\alpha/(111)_\gamma$ axis.

Precipitation of the σ phase can lead to a significant embrittlement at room-temperature and elevated-temperature service, which reduces ductility, notch toughness, and (intergranular, pitting, and crevice) corrosion resistance. The extent of embrittlement is a function of the amount and distribution of the phase. A continuous grain boundary σ precipitation can be very damaging, but intragranular σ precipitation with a blocky morphology can even increase creep ductility.[146]

Laves (η) Phases. Beattie and Hagel[147] have observed that Laves phases form predominantly only in iron-based alloys or alloys containing approximately equal amounts of iron, nickel, and cobalt, but not to any large extent in nickel- or cobalt-based alloys. Laves phases can form from fcc matrices as type C15 ($MgCu_2$) cubic and as type C14 ($MgZn_2$) hexagonal. Readers should refer to the general reviews on Laves phases elsewhere.[148–150]

The Laves phase in type 316 steel occurs as hexagonal Fe_2Mo, isomorphous with $MgZn_2$, containing 12 atoms per unit cell. It remains as a minor constituent at a temperature below 600°C, whereas it predominates above 600°C (Fig. 10.20c). Silicon addition and cold work promote its formation, while nickel and carbon retard it. Similarly, the existence of Nb, Mo, or W in preference to Ti also suppresses the formation of Laves phases.

Laves phase formation is a significant problem for superalloy 718. It forms during ingot solidification or welding. It is $Fe_2(Nb,Mo)$ and is stabilized by Si. One of the functions of C addition (>0.12%) in superalloy 718 is to promote NbC formation during final solidification, thereby preventing the formation of Fe_2Nb eutectic phase.[51,148] Alternatively, the problem of Laves phases can be reduced by relatively lowering the Nb content in the weld or filler metal.[149] In Inconel 718, the final solidification event has been considered as a ternary eutectic of $L \rightarrow \gamma + NbC + Laves$ phases as well as a combination of two eutectic type reactions such as $L \rightarrow \gamma + NbC$ and $L \rightarrow \gamma + Laves$.[150,151] Recent reviews on Laves phases in alloy 718 are available elsewhere.[151,152]

Fe_2Mo Laves phase has been observed in alloy 904L (UNS N08904)[51] and 20Cr-25Ni-4.5Mo stainless steels.[153]

Laves phases usually form intragranularly (Fig. 10.24a), but sometimes grain boundary precipitation (Fig. 10.24b) has also been found. The orientation relationship between Fe_2Mo and austenite, according to Nutting and Parsons, is

$$\left(10\bar{1}3\right)_{Fe_2Mo} // (111)_{\gamma} \qquad \left[\bar{1}2\bar{1}0\right]_{Fe_2Mo} // \left[\bar{1}2\bar{1}\right]_{\gamma}$$

But Denham and Silcock found the orientation relations for Fe_2Nb in Nb-containing alloys to be[154]

$$\left(0001\right)_{Fe_2Mo} // (111)_{\gamma} \qquad \left[10\bar{1}0\right]_{Fe_2Mo} // \left[\bar{1}10\right]_{\gamma}$$

For the formation of binary Laves phases (AB_2) between elements, first, the atomic size difference should be neither too great nor too small. On the basis of a hard sphere packing model, the ideal ratio of the atomic diameters for the formation of the Laves phase is 1.225,[155] and, second, the average electron concentration of the AB_2 combination should be less than 8.[156] For example, average electron concentrations for Fe_2Mo and Fe_2Nb phases are 7.33 and 7.00, respectively. As a result of these requirements, the homogeneity ranges in the Laves phases are usually small.

Silicon decreases the effective electron concentration and therefore stabilizes the Laves phase in transition metal systems with iron, nickel, cobalt, and manganese, which do not normally form binary AB_2 Laves phases.[157] Silicon-stabilized ternary Laves phases include Ti_2Ni_3Si, Nb_2Ni_3Si, Mo_2Co_3Si, W_2Co_3Si, $W_{40}Fe_{50}Si_{10}$, $Ti_{30}Mn_{50}Si_{14}$, and so on.[133]

Chi (χ) Phase. The χ phase has a bcc α-Mn-type crystal structure (type A12) and has been found to occur in 17 binary systems of transition metals. The general formula of the phase is $A_{10}B_{48}$ (that is, 58 atoms per unit cell). The occurrence of the χ phase has been observed when the electron/atom ratio falls between 6.3 and 7.6 and the atomic diameter ratios fall between 1.017 and 1.207.[158]

The typical formula of ternary χ phase is $Fe_{36}Cr_{12}M_{10}$. Cold work, Mo (5 to 10%), Ti + Nb < 3%, and W (up to 6%) have been found to promote χ formation. Beattie and Hagel have observed that χ phase invariably tends to precipitate from iron-based alloys rather than from nickel- or cobalt-based alloys.[147]

As shown in Fig. 10.20c for type 316 stainless steel, the precipitation of σ, χ, and η (Laves) phases requires long-term exposure at elevated temperatures. The compositions of χ (52%Fe-21%Cr-22%Mo-5%Ni) and Laves phase (38%Fe-11%Cr-45%Mo-6%Ni) formed in the type 316 stainless steel show that they possess much higher Mo contents than the type 316 matrix. Hence, pitting and crevice corrosion due to depletion of Mo in the surrounding matrix are likely to be prevalent. Like σ phase, χ and η (Laves) phases can be redissolved by heating to 1050°C or above.[56]

Formation of the χ phase occurs at grain boundaries, incoherent twin boundaries, coherent twin boundaries, and intragranular dislocations. The χ phase that formed intragranularly as rods has the following orientation relationship with the matrix:[159]

$$(111)_\gamma // (110)_\chi \qquad [01\overline{1}]_\gamma // [\overline{1}10]_\chi \qquad [\overline{2}11]_\gamma // [001]_\chi$$

The χ phase contributes little to the hardening of the alloy.[146]

Mu (μ) Phase.   The B_7A_6 phase was referred to as a μ phase by Das et al.[160] It has type $D8_5$ crystal structure. It forms several binary phases with iron and cobalt such as Mo_6Co_7, Mo_6Fe_7, W_6Co_7, W_6Fe_7, and so forth. It has 13 atoms per unit cell in terms of hexagonal coordinates. The preferred electron/atom ratio lies in the range of 7.1 to 8.0.[146] This phase forms in iron-based alloys, iron-nickel-based alloys (e.g., Pyromet 860), nickel-based alloys (e.g., René 41), and iron-nickel-cobalt-based alloys.[133]

The μ phase ordinarily precipitates as long thin plates initially at grain boundaries and extends from one grain boundary to another. Its formation adversely affects the low-temperature ductility, high-temperature rupture strength, and stress-rupture life.[134,161]

G Phase.   The ternary silicides referred to as *G* phase were first reported by Beattie and his coworkers in an Fe-Ni superalloy A286.[162,163] Elliott and Rostoker have discussed the occurrence of Laves phase among transition elements.[164] It is often found in austenitic stainless steels after irradiation,[165] during low-temperature aging in cast duplex steels of ASME SA351 CF series,[163] and in the ferritic phase of AISI 329 duplex stainless steels.[166]

The *G* phase has an fcc structure with ideal composition as $Ti_6Ni_{16}Si_7$ or $Cr_6Ni_{16}Si_7$ and 116 atoms per unit cell. It is better described as $A_6B_{16}C_7$, where C is silicon or germanium. Cr or Ti can be stabilized by replacing V, Mn, or Ta.

The *G* phase forms as a massive globular grain boundary phase with embrittling effect when simultaneous addition of Ti and Si is made to 25Ni-(15–20) Cr austenitic steels.[162] In general, grain boundary *G*-phase precipitation readily occurs in nickel-based and cobalt-based alloys as compared to iron-based alloys.

10.5.6.5 Martensite Formation.   In austenitic stainless steels, martensite may form thermally during cooling below room temperature or form mechanically by cold working. The M_s temperature in austenitic stainless steel can be represented by the equation[167]

$$M_s(°F) = 75(14.6 - Cr) + 110(8.9 - Ni) + 60(1.33 - Mn)$$
$$+ 50(.047 - Si) + 3000[0.068 - (C + N)] \qquad (10.17)$$

This equation shows that C and N have a very strong effect whereas Cr and Ni have a moderate effect on the M_s. Usually residual N content of 0.03 to 0.07% present in austenitic stainless steels together with its carbon content strongly stabilizes the austenite relative to martensite formation. When $M_{23}C_6$ carbides form at austenite grain boundaries, both C and Cr are depleted from the neighboring austenite. This locally raises the M_s, thereby causing the formation of martensite at grain boundaries.[168,169] In practice, this phenomenon is employed as a means to develop martensitic structure in semiaustenitic PH stainless steel (see Chap. 6).

Two types of martensites, bcc α' and hcp ε, form simultaneously in austenitic stainless steel during cooling below room temperature. The α'-martensite forms as plates with (225) habit planes in groups delineated by faulted sheets of $(111)_\gamma$ planes. It is favored in Mn-free alloys. The ε-martensite forms on $(111)_\gamma$ planes, and is morphologically very similar, except for size, to deformation twins or stacking fault clusters forming on $(111)_\gamma$ planes.[170,171] Probably, α'-martensite forms directly from austenite while ε-martensite forms as an intermediate phase;[50] α' is ferromagnetic, whereas ε is not.[51]

Deformation- (or strain-) induced martensite formation, which strongly contributes to the increase of tensile strength, is another unique feature of austenitic stainless steels. The amount of martensite produced is a function of the chemical composition of the grade. The highest temperature at which strain-induced martensite forms under defined formation conditions, called M_d, is employed to characterize austenite stability relative to deformation. Angel[172] and Pickering,[173] respectively, have given the following equations for M_{d30} as a function of composition of austenitic steels:

$$M_{d30}(°C) = 413 - 462(C+N) - 9.2Si - 8.1Mn - 13.7Cr - 9.5Ni - 18.5Mo \quad (10.18a)$$

$$M_{d30}(°C) = 497 - 462(C+N) - 9.2Si - 8.1Mn - 13.7Cr - 20Ni - 18.5Mo \quad (10.18b)$$

where M_{d30} denotes the temperature at which 50% martensite forms by 30% true strain (reduction of area) in tension. Thus M_{d30} denotes the instability of an austenite grade. To achieve a large percentage of martensite transformation during cold drawing, the temperature of the wire entering the die must be lower than the M_{d30} temperature.[76] These equations also show (1) a very strong effect of C and N on austenite stability and (2) a large sensitivity of the extralow-carbon grades such as 304L to strain-induced martensite transformation, a feature that may provide them susceptibility to reduced performance in high-pressure hydrogen.[174] Deformation-induced martensite, however, significantly increases the strength produced by cold work. For example, types 301 and 302 stainless steels are designed to possess lower Cr and Ni contents in order to exploit this strengthening mechanism. However, type 301, with lower nickel content, produces more martensite than type 302 for a given deformation percentage (Fig. 10.15a and b). The effectiveness of this method is exemplified in the comparison of stress-strain curves of types 301 and 304 stainless steels, as shown in Fig. 10.15a and c, respectively, where the strain-hardening capability of type 301 over type 304 is evidently revealed.

The extent of strain-induced transformation of austenite to martensite is a function of alloy composition, temperature, strain, and strain rate. Martensite formation is thus favored by low temperatures, low strain rates, low amounts of alloying elements (such as Mn, Ni, and N), and large deformation.[175] Figure 10.28 shows the strain-induced martensite formation as a function of strain at various temperatures. A large amount of transformation occurs at low strain during low-temperature deformation, whereas the negligible amount of strain-induced martensite forms at high strain above room temperature.[55]

10.5.7 Structure-Property Relationships

Strengths. Pickering[176] and Irvine et al.[177] have derived empirical relationships for yield strength $\sigma_{y(0.2\%)}$ and tensile strength σ_{UTS} of austenitic stainless steels based on compositional and microstructural parameters which are given by the following equations:

$$\sigma_{y(0.2\%)}(MPa) = 68 + 354(\%C) + 493(\%N) + 20(\%Si) + 3.7(\%Cr)$$

$$+ 14.5(\%Mo) + 18.5(\%V) + 4.5(\%W) + 40(\%Nb)$$

$$+ 26(\%Ti) + 12.6(\%Al) + 2.5(\%\delta\text{-ferrite}) + 7.1d^{-1/2} \quad (10.19)$$

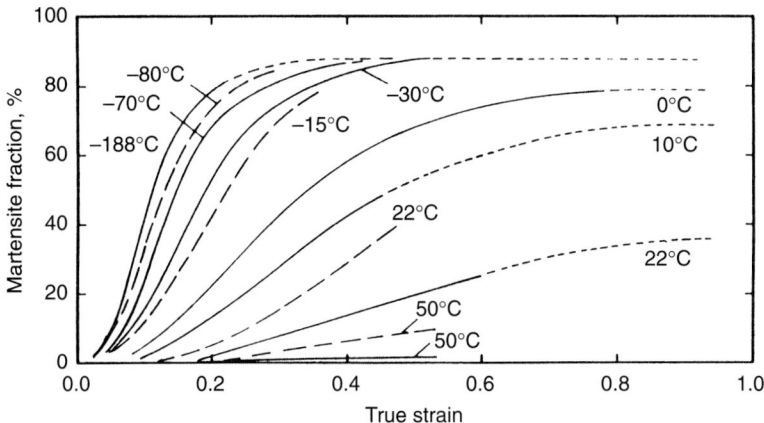

FIGURE 10.28 Strain-induced martensite formation in austenitic stainless steels as a function of strain at various temperatures.[a] Solid lines are original data of Angel,[b] dashed lines are data of Hecker et al.,[c] and dotted extrapolations are due to Olson's analysis.[d 53] (*Sources: a = Ref. 19; b = Ref. 15; c = Ref. 20; and d = Ref. 21 in Ref. 50.*)

$$\sigma_{UTS}(MPa) = 447 + 540(\%C) + 847(\%N) + 37(\%Si) + 1.7(\%Ni)$$
$$+ 18.5(\%Mo) + 77(\%Nb) + 46(\%Ti) + 18.5(\%Al)$$
$$+ 2.2(\%\delta\text{-ferrite}) + 12.6t^{-1/2} \qquad (10.20)$$

where d is the mean linear intercept of the grain diameter (i.e., austenite grain size) in millimeters and t is the annealing twin spacing in millimeters. Using these equations, the effect of a specific alloying addition such as chromium can be indicated in terms of various operative strengthening mechanisms (Fig. 10.29).[132,176,178]

The above relationships suggest that high-carbon and nitrogen specification stainless steels such as types 201, 202, and 301 have high yield and tensile strengths (see Table 10.6). Type 316 steel with high molybdenum content and types 321 and 347 with high amounts of titanium and niobium, respectively, also possess high tensile properties. The increase in tensile properties of niobium-stabilized steels has been attributed to the grain-strengthening effect induced by the intragranular precipitation of niobium carbide.[179]

It can be again seen from Eqs. (10.19) and (10.20) that the twin spacing does not contribute to the yield stress because the stacking fault energy (SFE), which controls the work-hardening rate, has negligible effect at the low strains at which the yield stress is measured. On the other hand, a small twin spacing (or high twinning frequency) resulting from the low SFE[180] is more important than the grain size effect in enhancing the work-hardening rate and hence the tensile strength. A large value of $t^{-1/2}$ denotes a low SFE and high work-hardening rate.[181]

Delta-ferrite increases both the yield and tensile strength values by a dispersion-strengthening mechanism. δ-ferrite produces a refinement of the austenite grain size which results in further strengthening effect over and above the ferrite itself, as shown in Fig. 10.29, which clearly illustrates how the strengthening due to grain refining increases sharply at the chromium level where δ-ferrite appears.[181] In addition, δ-ferrite tends to cause strain concentration in the softer austenite phase,

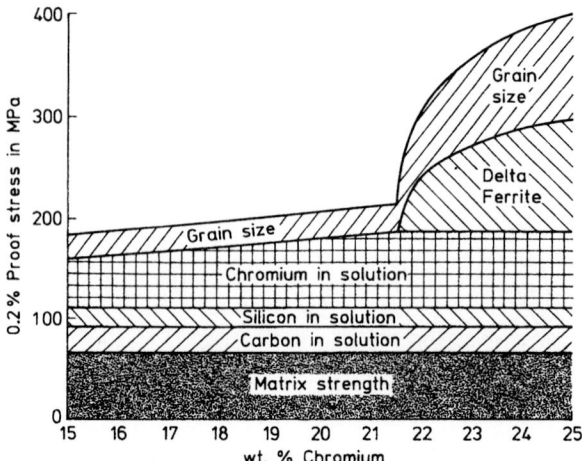

FIGURE 10.29 Effect of chromium on 0.2% yield stress of 8 to 10% Ni steel in terms of operative strengthening mechanisms.[132,176,178,181] (*Courtesy of F. B. Pickering.*)

which, in turn, results in work-hardening of austenite to a strain greater than the nominal 0.2% and thus gives a higher yield strength value.[132,176,182] In the case of tensile strength, about 80% of the strengthening due to δ-ferrite occurs by partitioning of carbon and nitrogen to the austenite, which increases the work-hardening rate. The interstitial solutes, carbon and nitrogen, have the greatest solid-solution strengthening effects, followed by substitutional ferrite-forming solutes, whereas the austenite-forming solutes have the least solid-solution strengthening effect (Fig. 10.30a).[176,180,183,184] A change in lattice parameter of austenite due to alloying additions induces lattice strains, which are effective in increasing the yield strength (Fig. 10.30b).[176,185]

An empirical equation for relating the yield stress of austenitic steels to nitrogen, grain size, and temperature has been developed by Norstrom[178]

$$\sigma_{y(0.2\%)} (\text{MPa}) = 15 + \frac{33{,}000}{T} + 65\frac{(1690-T)}{T}(\text{wt}\%\,\text{N})^{1/2}$$

$$+\left[7 + 78(\text{wt}\%\,\text{N})^{1/2}\right]d^{-1/2} \tag{10.21}$$

where T is in kelvins.

In metastable austenitic steel, martensite may form both before and after straining, which influences strength and ductility. Less than 35% martensite (M) prior to straining has no effect on the yield strength, but it increases linearly with the tensile strength according to[176]

$$\sigma_{\text{UTS}}(\text{MPa}) = 15.4T_c + 185 + 12.6(\%\text{M}) \tag{10.22}$$

where T_c is the tensile strength calculated from Eq. (10.20). When martensite formed prior to straining exceeds 35%, the yield stress increases with increasing martensite content. This may be due, in part, to the austenite being highly stressed,

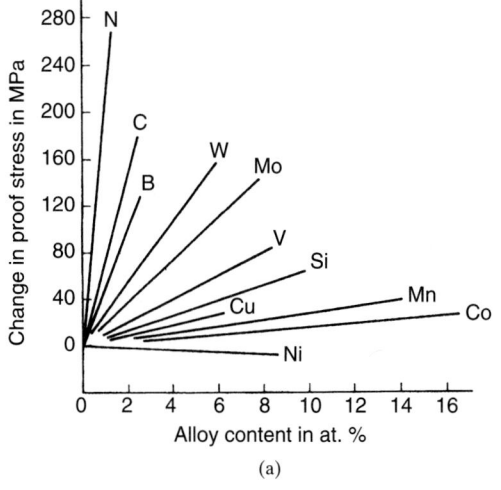

(a)

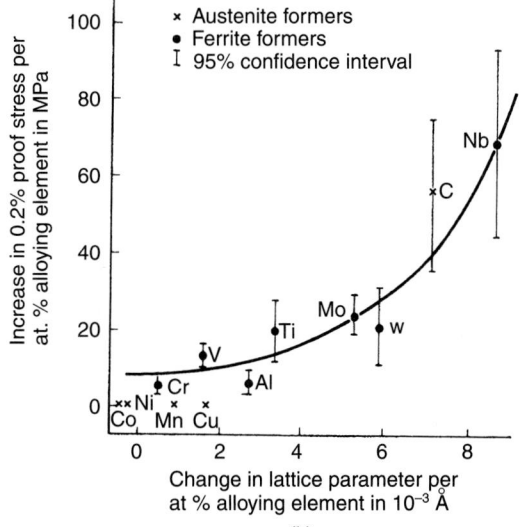

(b)

FIGURE 10.30 (*a*) Effect of solid-solution strengthening in austenite.[176,180,183,184] (*b*) Relationship between change in lattice parameter due to various solid solution strengthening alloying elements and increase in 0.2% yield (or proof) stress of an austenitic stainless steel.[176,185] (*Courtesy of F. B. Pickering.*)

and the austenite and the martensite may be differentially stressed during tensile testing. Equation (10.22) can be rewritten in the form

$$\Delta\sigma_{UTS}(\text{MPa}) = 185 + 9.84(\%M) \qquad (10.23)$$

where $\Delta\sigma_{UTS}$ is the difference between the experimentally observed and theoretically calculated tensile strengths. The intercept of 185 MPa at 0% prior martensite presumably corresponds to the strengthening induced by martensite which is transformed during straining up to the maximum uniform strain, i.e., the strain at which the tensile strength is measured. The general ductility, as determined by the percentage area reduction in a tensile test, decreases with increasing martensite content in the structure prior to straining.

Toughness.   Austenitic stainless steels possess good toughness down to very low subzero temperatures, provided the austenite remains stable and does not transform to martensite. That is why they are used for cryogenic applications. Stable austenitic steels do not exhibit a ductile-brittle transition temperature (DBTT) in the traditional manner, because the austenite does not undergo cleavage fracture. However, in case of austenite transformation to martensite, the toughness drastically decreases and there is a possibility of the appearance of a transition temperature. Usually, the influence of microstructure on toughness is small.[181]

10.6   SUPERAUSTENITIC STAINLESS STEELS

Superaustenitic stainless steels with UNS designation are listed in Tables 10.5 and 10.6. These are Fe-Cr-Ni-Mo alloys which contain 19 to 28% Cr, 17 to 40% Ni, 2 to 8% Mo, 0.1 to 0.6% N, and up to 4% Cu. Superaustenitic alloys can be refined to low C levels (≤0.03%) or can contain higher C content with Ti or Nb addition to stabilize against sensitization.

The *pitting resistance equivalent* (PREN)[†] (= Cr + 3.3Mo + 16N), being an index of pitting resistance in ferric chloride-containing environments, is a function of the stainless steel's composition and is considered as the "defining characteristic" for superaustenitic stainless steel.[62]

Superaustenitic stainless steels provide greater resistance to general corrosion, SCC in the presence of chlorides, and pitting and crevice corrosion resistance than those of the 300 series alloys, and, in some cases, Ti- or Ni-based alloys. UNS N08026 and other superaustenitic alloys with PREN > 44 have found broad application in paper and pulp industry for use in C- and D-stage bleach washers, paper machine head boxes, black liquor recovery boilers, effluent coolers, piping, reheaters, filter drums, and scrubbers. Superaustenitic grades, e.g., with PREN ~ 57 for S32654 and with PREN ~ 52 for S31266, are also used in pulp and bleach plant environments. They find applications in many industries including chemical processing, oil and gas production and refining, paper and pulp, food, pharmaceutical processing, electric power, and others exposed to severe adverse environments.[62] In the broad composition limits of superaustenitic alloys, the resistance to mineral and organic acids is extraordinary, often competing with and providing the cost-effective alternative to the Ni-based alloys.[62]

Carpenter 20Cb-3 alloy (UNS N08020) contains 32.5 to 35% Ni, 2 to 3% Mo, 3 to 4% Cu and Nb + Ta addition with at least 8 times the carbon content or 1% maximum. The 20Cb-3 alloy offers superior corrosion resistance to type 316. As a result, this alloy provides excellent resistance to (1) stress corrosion cracking in boiling 20 to 40% sulfuric acid, (2) all concentrations of sulfuric acid at temperatures up to 60°C (140°F), (3) concentrations as high as 10% sulfuric acid up to the boiling point, (4) all concentrations of formic acid up to the boiling point, (5) very dilute aerated hydrochloric acid solutions, and (6) the higher metal temperatures encountered with heat exchanger equipment. It is also stabilized to limit intergranular attack even in the sensitized condition. It is used extensively in the processing

---

[†] *Note:*There is a second PREN formula with higher multipliers, Cr + 3.3 Mo + 30N, which is often quoted.

of synthetic rubber, explosives, high-octane gasoline solvents, plastics, synthetic fibers, heavy chemicals, organic chemicals, pharmaceuticals, and agrichemicals. Other applications are in fans, heat exchangers, agitators, mixing tanks, distillation towers, process piping, bubble caps, pump shafts and rods, metal cleaning and pickling tanks, spray pickling equipment, valve stems, bolts, nuts, washers, tie rods, continuous-line pickling equipment including racks, etc.

The 20Mo-6 alloy (UNS N08026) has a higher strength than those of 300-series alloys such as 316L and is resistant to corrosion in hot chloride environments with low pH. It offers good resistance to pitting, crevice corrosion, and stress corrosion cracking in chloride environments as well as to oxidizing media and microbiologically active fresh and salt waters due to good *microbiologically influenced corrosion* (MIC) resistance. The alloy is designed for applications where better pitting and crevice corrosion resistance than that of 20Cb-3 alloy is required, e.g., in crystallizers, pressure vessels, mixing vessels, heat exchangers, tanks, columns, evaporators, piping, pumps, and valves. Other applications are seawater-cooled condensers, service water piping for nuclear power plants, and flue-gas desulfurization for scrubber's part (including ducting, internals, and absorbers).[53] They are also used in reverse-osmosis and flash distillation desalination plants, mainly as pipes.

The superaustenitic stainless steel grade, called 6545 Mo, which contains high N, Cr, and Mo contents and increased level of Mn, has high pitting and crevice corrosion resistance due to high levels of Mo + N + very low S contents.[56]

UNS N08028 is used as the primary pressure boundary material in water walls and heat exchanger bundles in syngas coolers. UNS N08904 or N08028 can be used in brackish water with chloride content in the range of 5000 to 10,000 ppm.

The Allegheny Ludlum AL-6XN alloy (UNS N08367)[186] is a low-carbon, high-nickel and -molybdenum, high-purity, nitrogen-bearing superaustenitic alloy. Its high strength and greater corrosion resistance make it a better choice than the traditional duplex stainless steels and a cost-effective alternative to more expensive nickel-base alloys in applications requiring excellent strength, formability, weldability, and chloride SCC, crevice, and pitting corrosion resistance in oxidizing chloride solutions to a degree previously achieved only by titanium- and nickel-base alloys. Exposure of N08367 at temperatures in the range of 538 to 1038°C (1000 to 1900°F) for times up to 10,000 hr has established that the current practice of 427°C (800°F) maximum use temperature lies well within safe limits for the alloy.[187,188] It is available in a wide range of product forms such as sheet, plate, tube, pipe, bar, billet, and forgings as well as components such as valves, pumps, fasteners, fittings, and castings. Its applications are as follows:[186]

Chemical and Process Industry. It is used in chemical process tanks and pipelines (as alternative to type 316, type 317, alloy 904L, alloy 20, alloy 276, alloy 625, and alloy 825) and tall oil distillation columns and packing.

Food Industry. Food processing equipment for use in salt (sodium chloride), chlorides such as sodium hypochlorite, and hot and cold environments are some applications. It is successfully used to replace types 304 and 316 that have suffered due to SCC, crevice, and pitting corrosion in many food processes including meat cookers, cereal cookers, baby food tanks, corn syrup refineries, and brewery piping.

Pulp and Paper Industry. It is used as filter washers (replacing types 316L, 317L, and 317LX and alloy 904L), vats and press rolls; white-water and kraft black liquor environments; and process changes involving recycling of wash water and high temperatures in piping at pulp and paper bleaching plants.

Marine and Offshore. Alloy N08367 has found wide applications in seawater piping systems (e.g., fire protection); process piping systems; condensers, heat exchangers and piping containing seawater or crude oil; splash zones and support structures; offshore drilling platforms; and desalination equipment and pumps.

Power Industry. It is an excellent choice for condenser tubing, service water piping systems for nuclear power plants, transformer cases exposed to marine environments, storage tanks, deaerator heaters, flue-gas reheaters and scrubber environments; fire protection; and emergency heat extraction.

Air Pollution Control. It is an effective alternative to type 317 and alloy 276 in absorber vessels, spray systems, and flue-gas cleaning.

Other Applications. These include Space Biosphere II in the Arizona desert with a planned 100-year life span; fuel-efficient residential heating furnaces incorporating secondary heat exchangers that condense flue gases before exhausting them; and pharmaceutical equipment (for product purity).

The new, higher-PREN, 6% alloys are VDM (N08031), Uranus B66 (S31266), and SR50A (S32050) alloys.[64a]

10.7 DUPLEX STAINLESS STEELS

Duplex stainless steel is a distinct class of stainless steel with a microstructure containing both the austenite and δ-ferrite phases. Duplex stainless steels are characterized by a favorable combination of strength, ductility, toughness, and corrosion properties. This duplex microstructure is produced by a combination of hot working and/or annealing in the temperature range of 1000 to 1150°C (1830 to 2100°F). Usually the steel is quenched in water or oil after solutionizing in this temperature range to avoid the formation of σ phase during cooling. The volume fractions of ferrite (V_α) and austenite (V_γ) vary between 0.25 and 0.75 in a duplex structure. The exact amount of each phase present in the microstructure can be deduced in terms of the major amount of alloying elements (Cr and Ni). The concentrations of these elements are adjusted using the ternary Fe-Ni-Cr equilibrium diagram to achieve a microstructure consisting of roughly equal amounts of ferrite and austenite, with ferrite comprising the matrix. Other alloying elements such as N, Mn, Cu, Mo, Si, and W are added to control structural balance and to enhance corrosion resistance characteristics.[189,190] The ratio of the ferrite and austenite phases (also called *phase balance*) determines the properties of the duplex steels. When compared to austenitic grades, they can offer improved strength, better pitting corrosion resistance (i.e., more resistant to sensitization), and greater resistance to chloride SCC; when compared to ferritic grades, they can provide improved toughness, formability, and weldability. These steels have been commercially available for many years but have not been used extensively due to hot-workability problems, their low toughness, and susceptibility to intergranular corrosion after welding or heat treatment.[191–193] However, the use of *argon-oxygen decarburization* (AOD) steelmaking refining process and continuous casting of slabs have now enabled us to achieve the closer control of chemical composition required to produce duplex steels with low carbon and proper nitrogen level in order to improve strength and corrosion resistance and modify some of the metallurgical reactions causing reduced weld toughness.[194] Very fine microduplex strcuture can be obtained by a suitable

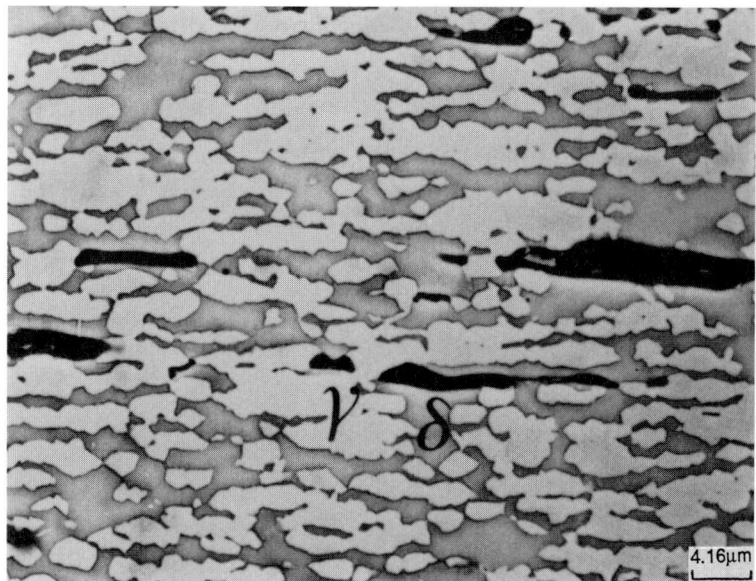

FIGURE 10.31 Microstructure of a longitudinal section of a 25 Cr-6.5 Ni-3 Mo-0.14 N duplex stainless steel specimen after isothermal deformation at 1050°C (1922°F) (at $\dot{\varepsilon} = 2 \times 10^{-1}\,\text{s}^{-1}$. The ductility obtained in this alloy exceeds 2500%. (Courtesy of Y. Ohmori.)

thermomechanical treatment using hot-working in the δ (or α) + γ phase region (between 900 and 1100°C)—or at even lower temperatures—which can exhibit superplastic properties during isothermal deformation at low strain rate $\dot{\varepsilon}$ in the range between ~4×10^{-4} and $2 \times 10^{-1}\,\text{s}^{-1}$, and greater resistance to SCC than the standard steels. Figure 10.31 shows the microstructure of a longitudinal section of 25Cr-6.5Ni-3Mo-0.14N duplex stainless steel specimen after isothermal deformation at 1050°C at $\dot{\varepsilon} = 2 \times 10^{-1}\,\text{s}^{-1}$. The ductility obtained in this alloy exceeds 2500%.[195] (See Chap. 15 for more discussion on superplasticity.) Thus, duplex steels combine some of the merits and demerits of austenitic and ferritic steels; however, for certain applications they are considered to be the optimum selection.[54]

Table 10.10 provides the compositions and PREN of selected wrought duplex grades. These grades are loosely divided into four generic types of increasing corrosion resistance: (1) type 2304 or S32304 (Fe-23Cr-4Ni-0.1N); (2) type 2205, S31803, or more restrictive S32205 (Fe-22Cr-5.5Ni-3Mo-0.15N); (3) type 2505 or S32550 and S31260 (Fe-25Cr-5Ni-2.5Mo-0.17N-Cu); and (4) type 2507 or S32750, S32760, and S32550 (Fe-25Cr-7Ni-3.5Mo-0.25N-W-Cu). The S32750 and S32760 alloys are often called *superduplex grades* (see Table 10.10) with PREN exceeding 38; they are characterized by more alloying with Cr, Mo, and N than ordinary stainless steels and have superior corrosion resistance to conventional stainless steel. According to PREN, duplex alloys can be grouped into low alloy or lean duplex (LD) with PREN < 32 (such as S32900, S32001, S31500, S32304, and S32404), moderately alloyed duplex (MD) with PREN between 32 and 38 (such as S31803, S32205, S32950, and S31200), and superduplex (SD) with PREN > 38 (such as S31260, S39274, S32550, S32750, and S32760).[196] The recommended annealing temperatures for selected duplex stainless steels are given in Table 10.11.[62]

TABLE 10.10 Composition and Pitting Resistance Equivalent Numbers of Some Wrought Duplex Stainless Steels[53]

UNS no.	Common name	Group	Composition, %†										PRE range‡
			C	Ma	S	P	Si	Cr	Ni	Mo	N	Other	
S31200	44LN	MD	0.03	2.00	0.03	0.045	1.00	24.0–26.0	5.5–6.5	1.2–2.0	0.14–0.20	...	30.2–35.8
S31260	DP3	SD	0.03	1.00	0.030	0.030	0.75	24.0–26.0	5.5–7.5	2.5–3.5	0.10–0.30	0.10–0.50 W, 0.20–0.80 Cu	33.9–42.4
S31500	3RE60	LD	0.03	1.2–2.0	0.03	0.03	1.4–2.0	18.0–19.0	4.25–5.25	2.5–3.0	0.05–0.10	...	27.1–30.5
S31803	2205	MD	0.03	2.00	0.02	0.03	1.00	21.0–23.0	4.5–6.5	2.5–3.5	0.08–0.20	...	30.5–37.8
S32304	2304	LD	0.03	2.5	0.04	0.04	1.0	21.5–24.5	3.0–5.5	0.05–0.60	0.05–0.20	0.05–0.60 Cu	22.5–29.7
S32550	255	SD	0.03	1.5	0.03	0.04	1.0	24.0–27.0	4.5–6.5	2.9–3.9	0.10–0.25	1.5–2.5 Cu	35.2–43.9
S32750	2507	SD	0.03	1.2	0.02	0.035	1.0	24.0–26.0	6.0–8.0	3.0–5.0	0.24–0.32	0.5 Cu	37.7–47.6
S32760	Zeron 100	SD	0.03	1.0	0.01	0.03	1.0	24.0–26.0	6.0–8.0	3.0–4.0	0.30	0.5–1.0 Cu, 0.5–1.0 W	40§
S32900	329	LD	0.06	1.00	0.03	0.04	0.75	23.0–28.0	2.5–5.0	1.0–2.0	¶	...	26.3–34.6
S32950	7-Mo Plus	MD	0.03	2.00	0.01	0.035	0.60	26.0–29.0	3.5–5.2	1.0–2.5	0.15–0.35	...	34.7

† Single values are maximum.
‡ PRE = %Cr + 3.3 (%Mo) + 16 (%N).
§ Minimum value.
¶ Not specified.

TABLE 10.11 Recommended Annealing Temperatures for
Some Duplex Stainless Steels[62]

UNS no.	Common name	Annealing temperature,[†] °C (°F)
S32900	329	925–955 (1700–1750)
S32950	7-Mo-plus	995–1025 (1825–1875)
S31500	3RE60	975–1025 (1785–1875)
S31803	SAF 2205	1020–1100 (1870–2010)
S31260	DP-3	1065–1175 (1950–2150)
S32550	Ferralium 255	1065–1175 (1950–2150)

[†] Cooling from the annealing temperature must be rapid and consistent with limitations of distortion.

Table 10.12 lists compositions of some cast duplex stainless steels. The microstructure of cast duplex stainless steels is coarser and exhibits a different morphology of austenite from that found in the wrought product.[53] In cast alloys, the proportion of ferrite and austenite can be calculated using the Schaeffler and DeLong (constitution) diagrams. However, these diagrams do not hold, in a strict sense, for wrought alloys because of the partial transformation of in situ ferrite during hot working at 1000 to 1200°C (1830 to 2190°F). An average correct equivalent diagram for wrought duplex steels has not yet been developed.

10.7.1 Precipitation of Phases in Duplex Steels

Annealing of duplex steels in the range of 925 to 1175°C (1700 to 2150°F) produces two phases, comprising elongated islands of austenite in a ferrite matrix. The high alloying addition and presence of ferrite matrix cause duplex stainless steels to be susceptible to embrittlement and deterioration of mechanical properties such as toughness, through prolonged exposure to elevated temperatures. This arises from the precipitation of intermetallic phases such as σ phase, α', χ phase, and/or η phase (Laves phase). Consequently, the upper temperature of their applications is typically ~280°C (535°F) for nonwelded material and 250°C (480°F) for welded products. The more highly alloyed steels such as 2505 and 2507 are the most susceptible to the formation of these deleterious phases.

The IGC or sensitization of the duplex steels depends primarily on the carbon content and the proportion of ferrite and austenite phases in the microstructure. Alloys that contain high proportions of carbon and a greater proportion of ferrite are prone to IGC and need annealing after welding. The duplex structure containing low carbon content (<0.03%) and about equal amounts of ferrite and austenite possesses good IGC resistance, i.e., greater resistance to sensitization.

Nucleation and precipitation of σ phase at δ/γ interfaces is much easier than at δ/δ or γ/γ. N helps even the Cr contents of δ and γ, thereby improving the corrosion resistance and slowing or minimizing the σ formation. Also N improves weldability. Thus N is a vital component of all new duplex stainless steels.[51]

In many aspects, σ-phase precipitation is similar to sensitization in that a chromium-depleted zone occurs at almost the same temperature range (450 to 1000°C, or 840 to 1830°F). Small amounts of σ phase (~1%) can lower Charpy V-notch (CVN) energy by 50%. Also σ phase can dramatically increase the corrosion rate and improve wear resistance in special duplex stainless steels (e.g., type 329).[56]

TABLE 10.12 Typical Compositions of Some Cast Duplex Stainless Steels[62]

UNS no.	ACI name	Other names	Composition,[†] wt%					
			C	Cr	Ni	Mo	N	Others
J93370	CD-4MCu	1A	0.04	24.5–26.5	4.75–6.0	1.75–2.25	—	2.75–3.25Cu, 1Mn
J93372	CD-4MCuN	1B	0.04	24.5–26.5	4.7–6.0	1.7–2.3	0.1–0.25	2.7–3.3Cu, 1Mn, 1Si
J92205	CD-3MN	2205, 4A	0.03	21.0–23.5	4.5–6.5	2.5–3.5	0.1–0.3	1Cu, 1.5Mn, 1Si
J93345	CE8MN	45D, 2A	0.08	22.5–25.5	8.0–11.0	3.0–4.5	0.1–0.3	1Mn, 1.5Si
J93371	CD6MN	3A	0.06	24.0–27.0	4.0–6.0	1.75–2.5	0.15–0.25	1Mn, 1Si
J93380	CD-3MWCuN	Z-100, 6A	0.03	24.0–26.0	6.5–8.5	3.0–4.0	0.2–0.3	0.5–1Cu, 0.5–1W, 1Mn, 1Si
J93404	CE3MN	A958, 5A	0.03	24.0–26.0	6.0–8.0	4.5–5.0	0.1–0.3	1.5Mn, 1Si

[†] Single values are maximum; all compositions contain balance of iron.

The 475°C (885°F) embrittlement occurs in ferrite phase at a temperature range of 300 to 500°C (570 to 930°F), corresponding to the nose of the curve (Fig. 10.32a). It is characterized by the formation of hard, brittle Cr-rich α' phase, which causes a precipitation-hardening effect in the ferritic phase and consequent reduction of toughness and corrosion resistance of the steel. The highest temperature for α' embrittlement in S31803 (2205) duplex is about 525°C (975°F). Figure 10.32b shows

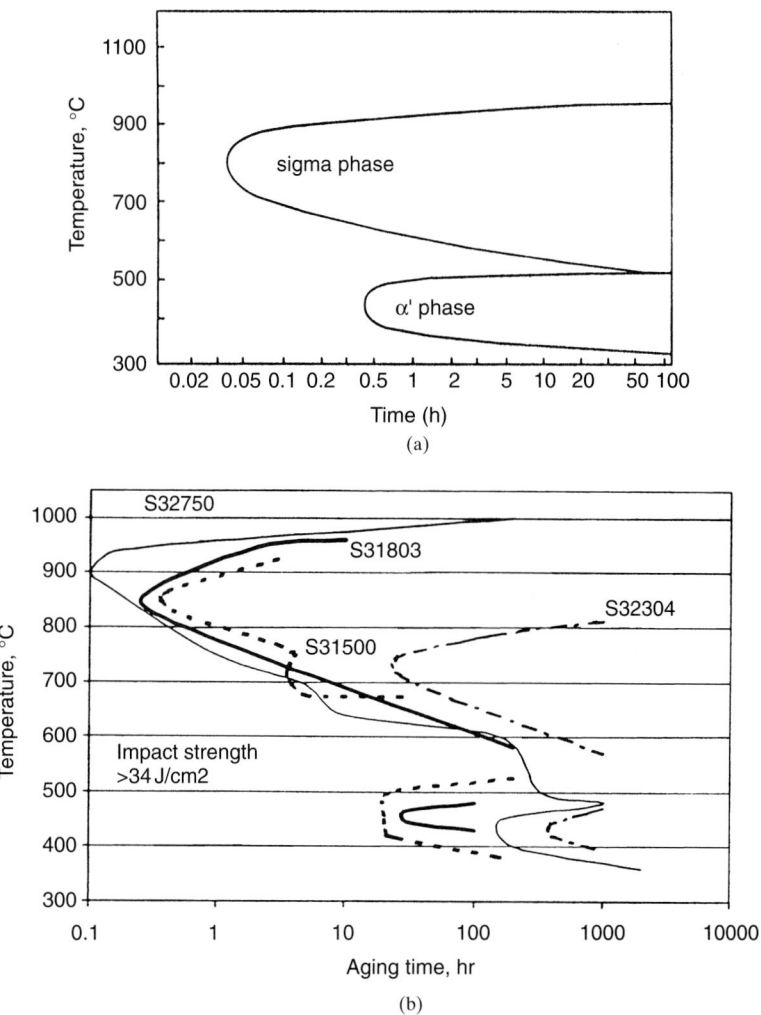

FIGURE 10.32   Time-temperature-transformation diagrams (a) for S31803 (2205) showing both σ and α' phases, and (b) for some duplex grades showing impact strength lower than 34 J/cm² (278-J full-size specimen),[196a] and (c) showing the effect of alloying additions on precipitation reactions in duplex stainless steel.[53] [(a): *After Lacombe et al., eds., Stainless Steels, Chapter 18, p. 624, Les Editions de Physiques, Les Ulis Cede A, France, 1993.*]

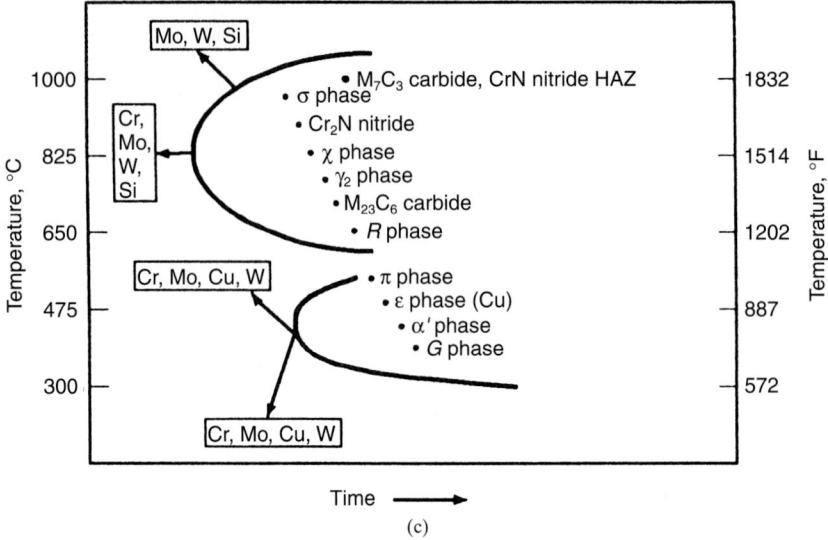

(c)

FIGURE 10.32 (*Continued*)   Time-temperature-transformation diagrams (*a*) for S31803 (2205) showing both σ and α' phases, and (*b*) for some duplex grades showing impact strength lower than 34 J/cm^2 (278-J full-size specimen),[196a] and (*c*) showing the effect of alloying additions on precipitation reactions in duplex stainless steel.[53] [(*a*): *After Lacombe et al., eds., Stainless Steels, Chapter 18, p. 624, Les Editions de Physiques, Les Ulis Cede A, France, 1993.*]

the TTT diagram for some duplex grades which have led to impact strength less than 34 J/cm^2 (27-J full-size specimen). Because of this risk of embrittlement, duplex grades are not provided in equipment with design temperatures above about 300°C (575°F).[196a]

In addition to α' phase, G phase has been implicated in the embrittlement of duplex alloys, as reported by Miller and Alexander. There is also a localized R-phase formation at an early stage of aging which causes extreme loss of Charpy impact toughness before the embrittlement attributed to the σ-phase formation.[197]

Figure 10.32c shows the TTT diagram illustrating the phases formed, approximate temperature ranges of their formation, and the effect of alloying elements on transformation kinetics.[53]

10.7.2   Mechanical Properties

In general, the high yield strength of duplex stainless steels (about 2 to 3 times greater than that of the austenitic steels—400 to 500 versus 200 to 250 MPa, or 58 to 80 versus 29 to 36 ksi) provides designers with the use of thin-wall material with sufficient load-bearing and pressure-containing capacity. This high yield strength factor can cause marked reduction in weight and welding time.[†] The elongation of duplex steels, although adequate for most service conditions and for fabrication, is lower than that of the austenitic steels.

[†] *Note*: But the ASME pressure vessel code is based on the tensile strength. This reduces the advantages of duplex alloys in North America. Many European design codes are yield strength-based. That is one reason why duplex alloys are more popular in Europe.[51]

TABLE 10.13 Room-Temperature Mechanical Properties for Some Duplex Stainless Steels per ASTM A 790[53]

UNS no.	Minimum yield strength		Minimum tensile strength		Elongation (minimum), %	Hardness	
	MPa	ksi	MPa	ksi		HB	HRC
S31200	450	65	690	100	25	280	...
S31500	440	64	630	92	30	290	30.5
S31803	450	65	620	90	25	290	30.5
S32304	400	58	600	87	25	290	30.5
S32550	550	80	760	110	15	297	31.5
S32750	550	80	800	116	15	310	32
S32760[†]	550	80	750	109	25	200–270	...
S32900	485	70	620	90	20	271	28
S32950	480	70	690	100	20	290	30.5

[†] Not listed in ASTM A790.

Because of their high yield strengths, duplex steels experience severe difficulties in cold forming when compared to the austenitic stainless steels. Further, since many high-temperature embrittlement phenomena can occur in these alloys, forging and other hot working operations need more attention than those with the austenitic stainless steels using the same processing operations.[53] Table 10.13 lists the room-temperature tensile properties and hardnesses of selected duplex stainless grades per ASTM A790.

The nonwelded base metal toughness, as expressed by the ductile-to-brittle transition temperature of the duplex steel, lies between the austenitic and ferritic grades; the extent of toughness is a function of ferrite content, the orientation of austenite-ferrite band, and the cooling rate from the annealing temperature. These alloys are not suitable for cryogenic applications. Most alloys contain about 50% ferrite to maintain a fairly good toughness because ferrite content exceeding 60 to 70% decreases the CVN energy sharply. Maximum toughness is achieved if impact tests are run with the crack growth transverse to the banded structure. Rapid cooling from the annealing temperature produces greater toughness, while slower cooling or holding at intermediate temperature, in the range of 400 to 500°C (750 to 930°F) and above 700°C (1290°F), gives rise to a variation in embrittlement due to precipitation of α' phase (475°C or 885°F embrittlement) and σ-phase particles.

10.7.3 Applications

The total tonnage of duplex stainless steels currently produced in the world is estimated to be <1% of the total stainless steel market.[196] Duplex stainless steels are widely used in various industries, including the oil and gas, petrochemical, pulp and paper, plastics, and pollution control industries. In the petrochemical industry, these alloys are used as welded pipe products for handling wet and dry CO_2, and sour gas and oil products. They are also employed in chemical, electric power, and other industries as welded tubing for heat exchangers, for handling chloride-containing coolants, and for handling hot brines and various organic chemicals. They also find applications as liners for ocean-going tankers and chemical transport barges; and as tanks and piping for breweries. The U.S. Navy has employed the duplex stainless steels for catapult trough covers on aircraft carriers and for retractable bow plane systems on submarines.[56]

TABLE 10.14 Typical Applications of Duplex Stainless Steels[53]

Grade(s)	Common name	Application
S31803, S32205, S32760, S32750, S32550	2205, 2507 Zeron 100, 255	Tubing for heat exchangers in refineries, chemical industries, process industries, and other industries using water as a coolant; kraft paper digesters
S32304	2304	Domestic hot water heaters and where pitting resistance is not of major importance
S32550	255	Application in marine corrosion areas, equipment for phosphoric acid and fertilizer industry, pollution control equipment, pulp and paper industry, and petrochemical industry
S32750, S32760	2507, Zeron 100	Pipe for seawater handling and firefighting system, oil and gas separators, salt evaporation equipment, desalination plants, geothermal well heat exchangers, human body implants (S32550 may suffer slight corrosion pitting and crevice in seawater service)
S32950	7-Mo-plus	Heat exchangers in petroleum refining, petrochemical, pulp and paper, and allied processing industries
All		Heat exchangers, chemical tankers, chemical reactor vessels, flue gas filters, acetic acid and phosphoric acid handling systems, oil and gas industry equipment (multiphase flow lines, downhole production tubulars, commonly cold worked)

As "problem-solver" alloys, they find applications to replace standard S30403 (type 304L) and S31603 (type 316L) or where total life-cycle concepts are used and the installed or long-term cost of the duplex alloy is still lower.[196]

Types 329 (S32900) and 7-Mo-plus (S32950) duplex stainless steels have superior corrosion resistance in boiling 65% nitric acid (ASTM Practice C), reducing mineral acid mixtures containing nitric acid, phosphoric acid medium, stronger organic acids, and sulfuric acid, as compared to other standard stainless steels.

Superduplex stainless steels such as Zeron 100 (S32760) and type 2507 (S32550) are resistant to oxygenated or chlorinated seawater. The 7-Mo-plus finds applications as heat exchangers in chemical, petroleum refining, pulp and paper, and allied industries. Other applications include food processing equipment, caustic evaporators, piping systems for cooling water in power generating plants, and as a possible substitute for type 316 or 316L in marine environments.[46,198] The leaner alloy grades are not resistant to live seawater and will corrode by pitting. Table 10.14 provides a list of duplex steels and their typical applications.[53,75]

10.7.4 Structure-Property Relationships

Yield Stress. Over a wide range of δ-ferrite phase field, the 0.2% yield stress increases linearly with an increase in ferrite content[199,200] (Fig. 10.33a).[181] This figure

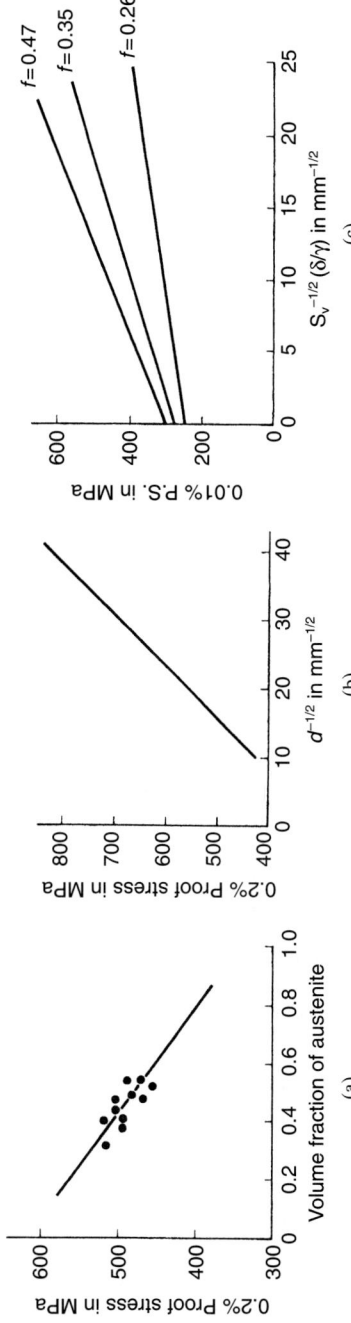

FIGURE 10.33 The effect of (a) the volume fraction of austenite on the 0.2% yield stress, (b) the austenite grain size d on the 0.2% yield stress in a 50% δ-ferrite matrix, and (c) the δ-ferrite/austenite interfacial area S_v and the δ-ferrite volume fraction f on the 0.01% yield stress of a duplex stainless steel.[181] (Courtesy of F. B. Pickering.)

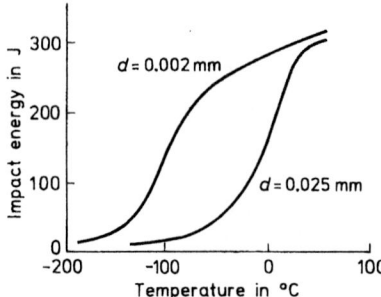

FIGURE 10.34 The effect of grain size d on the ductile-brittle transition curve for a duplex stainless steel.[181] (*Courtesy of F. B. Pickering.*)

shows an increase of yield stress by 27.5 MPa with an increase in ferrite volume fraction of 0.1. Figure 10.33b illustrates a Petch relationship for the dependence of austenite grain size on the 0.2% yield stress. Figure 10.33c shows the effect of δ-ferrite/austenite interfacial area S_v and the δ-ferrite volume fraction on the 0.01% yield stress. Here an increasing interfacial area implies a decreasing ferrite particle size (and a consequent refinement of the austenite grain size), and for a given interfacial area, the yield stress increases with the ferrite volume fraction.[181]

As a first approximation, the yield stress of a duplex stainless steel may be described by the equation[181]

$$\sigma_y = f_\delta\left(\sigma_{i\delta} - k_{y\delta}d_\delta^{-1/2}\right) + (1 - f_\delta)\left(\sigma_{i\gamma} + k_{y\gamma}d_\gamma^{-1/2}\right) \qquad (10.24)$$

where f_δ is the volume fraction of ferrite; $\sigma_{i\delta}$ and $\sigma_{i\gamma}$ are the friction stresses for the δ-ferrite and austenite, respectively; $k_{y\delta}$ and $k_{y\gamma}$ are the corresponding grain size coefficients; and d_δ and d_γ are the mean linear intercepts of grain size for δ-ferrite and austenite, respectively.

In reality, $\sigma_{i\delta}$ and $\sigma_{i\gamma}$ are not constants; rather, they are functions of solid-solution strengthening attributed to the solute partitioning between the two phases, which, in turn, depends on the composition of steel.[181]

Toughness. There has not been much systematic study on the toughness of duplex stainless steels; however, the DBTT, denoted by ITT, increases with an increase of ferrite content. Furthermore, for a given volume fraction of ferrite, ITT decreases with the improvement in grain refinement, as shown in Fig. 10.34. It is interesting to note that in the case of austenite transforming to martensite during impact testing, ITT does not appear to be adversely affected, probably because (1) the interface still tends to hinder cleavage crack propagation and (2) the carbon content of martensite is very low.[181]

There is a need to avoid the occurrence of intermetallic phases such as σ phase during exposure in the temperature range of 320 to 955°C (600 to 1700°F) which is a function of composition and cooling rate of the alloy, because they are detrimental to toughness and corrosion resistance (see ASTM A 923).

10.8 SUPERDUPLEX STAINLESS STEELS

Superduplex stainless steels are distinguished from duplex stainless steels mainly by the addition of higher levels of Cr, Mo, Ni, N, and W. They incorporate a minimum PREN of 40 as part of their specification to guarantee consistent localized corrosion performance.[201]

10.9 PHYSICAL PROPERTIES OF STAINLESS STEELS

The physical properties important for the selection of stainless steels are density and elastic modulus; thermal properties such as melting range, specific heat, thermal conductivity, heat-transfer coefficient, and coefficient of thermal expansion (CTE); electrical resistivity; and magnetic properties such as magnetic permeability. Table 10.15 provides typical physical properties of wrought stainless steels in the annealed condition.[54] This section briefly describes only five physical properties which differentiate stainless steels from other engineering materials. Table 10.16 summarizes the physical properties of duplex stainless steels. When compared to austenitic stainless steels, duplex grades possess a higher thermal conductivity and lower coefficient of thermal expansion.[202] Readers interested in more information on physical properties can find it elsewhere.[53,54,203–206]

Thermal Conductivity. Figure 10.35a illustrates a large variation in thermal conductivity among several materials, namely, aluminum alloy, copper alloy, 1080 steels, and stainless steels. Austenitic and PH stainless steels have the least thermal conductivity whereas type 6061 aluminum alloy has the highest thermal conductivity.[53] Table 10.17 lists room-temperature thermal conductivity values for some selected stainless steels,[202] whereas Table 10.15 gives the thermal conductivity values for a large number of stainless steels at 100 and 500°C.[54]

Coefficient of Thermal Expansion. Figure 10.35b shows a comparison of CTE for aluminum, copper alloys, carbon steels, and stainless steels which illustrates a large CTE value for type 6061 alloy and low CTE values for PH, ferritic, and martensitic stainless steel and carbon steels.[53] Tables 10.18 and 10.19 provide CTE values for some selected stainless steels at low and elevated temperatures, respectively,[202] whereas Table 10.15 gives the CTE values for a large number of stainless steels at intermediate temperatures.[54]

Electrical Resistivity. Composition variations through alloying additions can significantly increase the electrical resistivity of metallic alloys. Thus, austenitic, duplex, and PH stainless steels have higher electrical resistivity than ferritic and martensitic stainless steels, followed by 1080 carbon steels, aluminum bronze, and type 6061 aluminum alloys (Fig. 10.35c).[53] Table 10.20 illustrates the effect of temperature on the electrical resistivity of stainless steels,[202] whereas Table 10.15 provides the electrical resistivity values at room temperature.[54]

Heat Transfer. The overall heat-transfer rate of stainless steel is found to be superior to that of Admiralty brass (because of its ability to remain clean), as shown in Fig. 10.36, which illustrates a comparison of two condenser tubes exposed simultaneously to the same operating conditions. That is why stainless steels are used extensively for heat exchangers.[205]

Magnetic Property. The magnetic behavior of stainless steels varies significantly from paramagnetic (nonmagnetic) behavior in the annealed, fully austenitic grades (with dc magnetic permeabilities of ~1.003 to ~1.005), to hard, or permanent magnetic behavior in the hardened martensitic grades, to soft magnetic properties in ferritic stainless steels (Table 10.15). Certain austenitic steel grades such as types 302 and 304 develop increased magnetic permeability with cold work due to

TABLE 10.15 Typical Physical Properties of Wrought Stainless Steels in the Annealed Condition[54]

Type	UNS number	Density, g/cm³ (lb/in.³)	Elastic modulus, GPa (10⁶ psi)	Mean CTE† from 0°C to: 100°C, μm/m·°C (μin./in.·°F)	315°C, μm/m·°C (μin./in.·°F)	538°C, μm/m·°C (μin./in.·°F)
201	S20100	7.8 (0.28)	197 (28.6)	15.7 (8.7)	17.5 (9.7)	18.4 (10.2)
202	S20200	7.8 (0.28)	. . .	17.5 (9.7)	18.4 (10.2)	19.2 (10.2)
205	S20500	7.8 (0.28)	197 (28.6)	. . .	17.9 (9.9)	19.1 (10.6)
301	S30100	8.0 (0.29)	193 (28.0)	17.0 (9.4)	17.2 (9.6)	18.2 (10.1)
302	S30200	8.0 (0.29)	193 (28.0)	17.2 (9.6)	17.8 (9.9)	18.4 (10.2)
302B	S30215	8.0 (0.29)	193 (28.0)	16.2 (9.0)	18.0 (10.0)	19.4 (10.8)
303	S30300	8.0 (0.29)	193 (28.0)	17.2 (9.6)	17.8 (9.9)	18.4 (10.2)
304	S30400	8.0 (0.29)	193 (28.0)	17.2 (9.6)	17.8 (9.9)	18.4 (10.2)
304L	S30403	8.0 (0.29)	. . .	. . .	. . .	. . .
302Cu	S30430	8.0 (0.29)	913 (28.0)	17.2 (9.6)	17.8 (9.9)	. . .
304N	S30451	8.0 (0.29)	196 (28.5)	. . .	. . .	. . .
305	S30500	8.0 (0.29)	193 (28.0)	17.2 (9.6)	17.8 (9.9)	18.4 (10.2)
308	S30800	8.0 (0.29)	193 (28.0)	17.2 (9.6)	17.8 (9.9)	18.4 (10.2)
309	S30900	8.0 (0.29)	200 (29.0)	15.0 (8.3)	16.6 (9.2)	17.2 (9.6)
310	S31000	8.0 (0.29)	200 (29.0)	15.9 (8.8)	16.2 (9.0)	17.0 (9.4)
314	S31400	7.8 (0.28)	200 (29.0)	. . .	15.1 (8.4)	. . .
316	S31600	8.0 (0.29)	193 (28.0)	15.9 (8.8)	16.2 (9.0)	17.5 (9.7)
316L	S31603	8.0 (0.29)	. . .	. . .	. . .	. . .
316N	S31651	8.0 (0.29)	196 (28.5)	. . .	. . .	. . .
317	S31700	8.0 (0.29)	193 (28.0)	15.9 (8.8)	16.2 (9.0)	17.5 (9.7)
317L	S31703	8.0 (0.29)	200 (29.0)	16.5 (9.2)	. . .	18.1 (10.1)
321	S32100	8.0 (0.29)	193 (28.0)	16.6 (9.2)	17.2 (9.6)	18.6 (10.3)
329	S32900	7.8 (0.28)	. . .	. . .	. . .	. . .
330	N08330	8.0 (0.29)	196 (28.5)	14.4 (8.0)	16.0 (8.9)	16.7 (9.3)
347	S34700	8.0 (0.29)	193 (28.0)	16.6 (9.2)	17.2 (9.6)	18.6 (10.3)
384	S38400	8.0 (0.29)	193 (28.0)	17.2 (9.6)	17.8 (9.9)	18.4 (10.2)
405	S40500	7.8 (0.28)	200 (29.0)	10.8 (6.0)	11.6 (6.4)	21.1 (6.7)
409	S40900	7.8 (0.28)	. . .	11.7 (6.5)	. . .	. . .
410	S41000	7.8 (0.28)	200 (29.0)	9.9 (5.5)	11.4 (6.3)	11.6 (6.4)
414	S41400	7.8 (0.28)	200 (29.0)	10.4 (5.8)	11.0 (6.1)	12.1 (6.7)
416	S41600	7.8 (0.28)	200 (29.0)	9.9 (5.5)	11.0 (6.1)	11.6 (6.4)
420	S42000	7.8 (0.28)	200 (29.0)	10.3 (5.7)	10.8 (6.0)	11.7 (6.5)
422	S42200	7.8 (0.28)	. . .	11.2 (6.2)	11.4 (6.3)	11.9 (6.6)
429	S42900	7.8 (0.28)	200 (29.0)	10.3 (5.7)	. . .	. . .
430	S43000	7.8 (0.28)	200 (29.0)	10.4 (5.8)	11.0 (6.1)	11.4 (6.3)
430F	S43020	7.8 (0.28)	200 (29.0)	10.4 (5.8)	11.0 (6.1)	11.4 (6.3)
431	S43100	7.8 (0.28)	200 (29.0)	10.2 (5.7)	12.1 (6.7)	. . .
434	S43400	7.8 (0.28)	200 (29.0)	10.4 (5.8)	11.0 (6.1)	11.4 (6.3)
436	S43600	7.8 (0.28)	200 (29.0)	9.3 (5.2)	. . .	. . .
439	S43900	7.7 (0.28)	200 (29.0)	10.4 (5.8)	11.0 (6.1)	11.4 (6.3)
440A	S44002	7.8 (0.28)	200 (29.0)	10.2 (5.7)	. . .	. . .
440C	S44004	7.8 (0.28)	200 (29.0)	10.2 (5.7)	. . .	. . .
444	S44400	7.8 (0.28)	200 (29.0)	10.0 (5.6)	11.4 (6.3)	11.4 (6.3)
446	S44600	7.5 (0.27)	200 (29.0)	10.4 (5.8)	10.8 (6.0)	11.2 (6.2)
PH 13-8 Mo	S13800	7.8 (0.28)	203 (29.4)	10.6 (5.9)	11.2 (6.2)	11.9 (6.6)
15-5 PH	S15500	7.8 (0.28)	196 (28.5)	10.8 (6.0)	11.4 (6.3)	. . .
17-4 PH	S17400	7.8 (0.28)	196 (28.5)	10.8 (6.0)	11.6 (6.4)	. . .
17-7 PH	S17700	7.8 (0.28)	204 (29.5)	11.0 (6.1)	11.6 (6.4)	. . .

† CTE, coefficient of thermal expansion.
‡ At 0 to 100°C (32 to 212°F).
§ Approximate values.
Reprinted by permission of ASM International, Materials Park, Ohio.

Thermal conductivity					
100°C, W/m·K (Btu/ft·hr·°F)	500°C, W/m·K (Btu/ft·hr·°F)	Specific heat, J/kg·K (Btu/lb·hr·°F)	Electrical resistivity, nΩ·m	Magnetic permeability	Melting range, °C (°F)
16.2 (9.4)	21.5 (12.4)	500 (0.12)	690	1.02	1400–1450 (2550–2650)
16.2 (9.4)	21.6 (12.5)	500 (0.12)	690	1.02	1400–1450 (2550–2650)
. . .	. . .	500 (0.12)	. . .	. . .	. . .
16.2 (9.4)	21.5 (12.4)	500 (0.12)	720	1.02	1400–1420 (2550–2590)
16.2 (9.4)	21.5 (12.4)	500 (0.12)	720	1.02	1400–1420 (2550–2590)
15.9 (9.2)	21.6 (12.5)	500 (0.12)	720	1.02	1375–1400 (2500–2550)
16.2 (9.4)	21.5 (12.4)	500 (0.12)	720	1.02	1400–1420 (2550–2590)
16.2 (9.4)	21.5 (12.4)	500 (0.12)	720	1.02	1400–1450 (2550–2650)
. . .	. . .	. . .	. . .	1.02	1400–1450 (2550–2650)
11.2 (6.5)	21.5 (12.4)	500 (0.12)	720	1.02	1400–1450 (2550–2650)
. . .	. . .	500 (0.12)	720	1.02	1400–1450 (2550–2650)
16.2 (9.4)	21.5 (12.4)	500 (0.12)	720	1.02	1400–1450 (2550–2650)
15.2 (8.8)	21.6 (12.5)	500 (0.12)	720	. . .	1400–1420 (2550–2590)
15.6 (9.0)	18.7 (10.8)	500 (0.12)	780	1.02	1400–1450 (2550–2650)
14.2 (8.2)	18.7 (10.8)	500 (0.12)	780	1.02	1400–1450 (2550–2650)
17.5 (10.1)	20.9 (12.1)	500 (0.12)	770	1.02	. . .
16.2 (9.4)	21.5 (12.4)	500 (0.12)	740	1.02	1375–1400 (2500–2550)
. . .	. . .	. . .	. . .	1.02	1375–1400 (2500–2550)
. . .	. . .	500 (0.12)	740	1.02	1375–1400 (2500–2550)
16.2 (9.4)	21.5 (12.4)	500 (0.12)	740	1.02	1375–1400 (2500–2550)
14.4 (8.3)	. . .	500 (0.12)	790	. . .	1375–1400 (2500–2550)
16.1 (9.3)	22.2 (12.8)	500 (0.12)	720	1.02	1400–1425 (2550–2600)
. . .	. . .	460 (0.11)	750	. . .	. . .
. . .	. . .	460 (0.11)	1020	1.02	1400–1425 (2550–2600)
16.1 (9.3)	22.2 (12.8)	500 (0.12)	730	1.02	1400–1425 (2550–2600)
16.2 (9.4)	21.5 (12.4)	500 (0.12)	790	1.02	1400–1450 (2550–2650)
27.0 (15.6)	. . .	460 (0.11)	600	. . .	1480–1530 (2700–2790)
. . .	. . .	. . .	. . .	. . .	1480–1530 (2700–2790)
24.9 (14.4)	28.7 (16.6)	460 (0.11)	570	700–1000	1480–1530 (2700–2790)
24.9 (14.4)	28.7 (16.6)	460 (0.11)	700	. . .	1425–1480 (2600–2700)
24.9 (14.4)	28.7 (16.6)	460 (0.11)	570	700–1000	1480–1530 (2700–2790)
24.9 (14.4)	. . .	460 (0.11)	550	. . .	1450–1510 (2650–2750)
23.9 (13.8)	27.3 (15.8)	460 (0.11)	. . .	. . .	1470–1480 (2675–2700)
25.6 (14.8)	. . .	460 (0.11)	590	. . .	1450–1510 (2650–2750)
26.1 (15.1)	26.3 (15.2)	460 (0.11)	600	600–1100	1425–1510 (2600–2750)
26.1 (15.1)	26.3 (15.2)	460 (0.11)	600	. . .	1425–1510 (2600–2750)
20.2 (11.7)	. . .	460 (0.11)	720	. . .	. . .
. . .	26.3 (15.2)	460 (0.11)	600	600–1100	1425–1510 (2600–2750)
23.9 (13.8)	26.0 (15.0)	460 (0.11)	600	600–1100	1425–1510 (2600–2750)
24.2 (14.0)	. . .	460 (0.11)	630	. . .	. . .
24.2 (14.0)	. . .	460 (0.11)	600	. . .	1370–1480 (2500–2700)
24.2 (14.0)	. . .	460 (0.11)	600	. . .	1370–1480 (2500–2700)
26.8 (15.5)	. . .	420 (0.10)	620	. . .	. . .
20.9 (12.1)	24.4 (14.1)	500 (0.12)	670	400–700	1425–1510 (2600–2750)
14.0 (8.1)	22.0 (12.7)	460 (0.11)	1020	. . .	1400–1440 (2560–2625)
17.8 (10.3)	23.0 (13.1)	420 (0.10)	770	95	1400–1440 (2560–2625)
18.3 (10.6)	23.0 (13.1)	460 (0.11)	800	95	1400–1440 (2560–2625)
16.4 (9.5)	21.8 (12.6)	460 (0.11)	830	. . .	1400–1440 (2560–2625)

TABLE 10.16 Physical Properties of Duplex Stainless Steels according to EN 10088[206]

Property	20°C	100°C	200°C	300°C	Units
Density	7.8				kg/dm^3
Modulus of elasticity	200	194	186	180	kN/mm^2
Mean coefficient of thermal expansion between 20°C and T			13.0	13.5	14.0	10^{-6}/°C
Thermal conductivity	15				W/m·K
Specific thermal capacity	500				J/kg·K
Electrical resistivity	0.8				Ω·mm^2/m

Courtesy of National Association of Corrosion Engineers.

TABLE 10.17 Room-Temperature Thermal Conductivity Values for Selected Stainless Steels[203]

Type	UNS no.	Thermal conductivity at 20°C (68°F)	
		W/m·K	Btu/ft·hr·°F
201	S20100	14.6	8.4
304	S30400	14.6	8.4
310	S31000	14.6	8.4
316	S31600	14.6	8.4
316Cb	S31640	14.6	8.4
321	S32100	14.6	8.4
347	S34700	14.6	8.4
410	S41000	25.1	14.5
420	S42000	25.1	14.5
430	S43000	20.9	12.1
...	S31803	16.7	9.6

TABLE 10.18 Effect of Low Temperature on Mean Coefficient of Thermal Expansion Values for Selected Austenitic Stainless Steels[203]

Type	UNS no.	Mean CTE, μm/m·°C(μin./in.·°F) at			
		−184 to 21°C (−299 to 70°F)	−129 to 21°C (−200 to 70°F)	−73 to 21°C (−99 to 70°F)	−18 to 21°C (0 to 70°F)
301	S30100	13.7 (7.6)	14.1 (7.8)	14.8 (8.2)	15.7 (8.7)
304	S30400	13.3 (7.4)	13.9 (7.7)	14.8 (8.2)	15.7 (8.7)
310	S31000	12.6 (7)	13.5 (7.5)	14.1 (7.8)	14.4 (8)
316	S31600	12.8 (7.1)	13.3 (7.4)	14.1 (7.8)	14.8 (8.2)
347	S34700	13.5 (7.5)	14.6 (8.1)	15.3 (8.5)	15.7 (8.7)

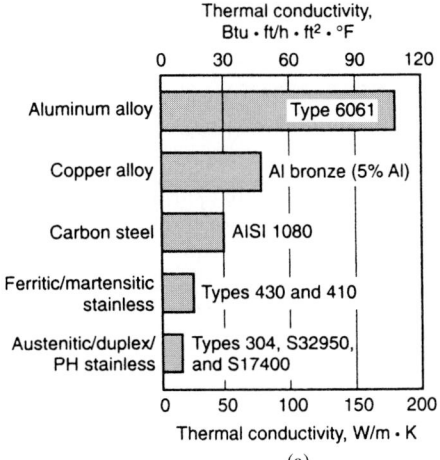

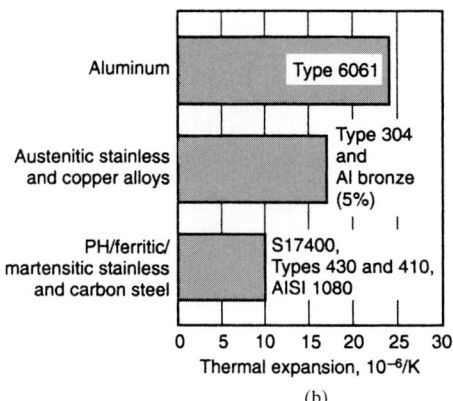

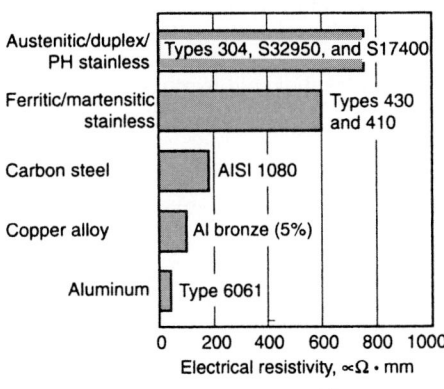

FIGURE 10.35 Comparison of (*a*) thermal conductivity for aluminum, copper alloy, carbon steel, and stainless steels; (*b*) thermal expansion for 6061 aluminum alloy, copper alloys, carbon steels, and stainless steels; and (*c*) electrical resistivity for stainless steels, 1080 steel, copper alloy, and aluminum alloy.[53] (*Reprinted by permission of ASM International, Materials Park, Ohio.*)

TABLE 10.19 Mean Coefficient of Thermal Expansion as a Function of Temperature for Selected Stainless Steels[203]

		Mean CTE, µm/m·°C(µin/in.·°F) at				
Type	UNS no.	20–200°C (70–390°F)	20–400°C (70–750°F)	20–600°C (70–1100°F)	20–800°C (70–1470°F)	20–1000°C (70–1830°F)
304	S30400	17 (9.4)	18 (10)	19 (10.6)	19.5 (10.8)	20.0 (11.1)
316	S31600	16.5 (9.2)	17.5 (9.7)	18.5 (10.3)	19.0 (10.6)	19.5 (10.8)
314	S31400	15 (8.3)	16 (8.9)	17 (9.4)	18 (10)	19 (10.6)
403	S40300	11 (6.1)	11.7 (6.5)	. . .	. . .	. . .
430	S43000	10.5 (5.8)	11.2 (6.2)	. . .	. . .	. . .
446	S44600	10.3 (5.7)	11 (6.1)	11.7 (6.5)	12.4 (6.9)	13.1 (7.3)
. . .	S31803	13 (7.2)	14 (7.8)	15 (8.3)	. . .	. . .

TABLE 10.20 Electrical Resistivity as a Function of Temperature for Selected Stainless Steels[203]

		Electrical resistivity, µΩ·cm, at							
Type	UNS no.	−196°C (−321°F)	−78°C (−108°F)	20°C (68°F)	200°C (390°F)	400°C (750°F)	600°C (1110°F)	800°C (1470°F)	1000°C (1830°F)
301	S30100	. . .	. . .	72	83	94	105	114	. . .
302	S30200	. . .	. . .	72	84	96	106	115	119
304	S30400	55	65	72	85	98	111	120	. . .
310	S31000	. . .	. . .	90	100	110	120	125	130
316	S31600	60	68	74	87	98	108	. . .	. . .
321	S32100	. . .	. . .	72	90	103	115	123	. . .
347	S34700	52	60	72	88	97	110	119	. . .
420	S42000	. . .	. . .	55	72	87	108	. . .	. . .
430	S43000	. . .	. . .	60	76	91	111	. . .	. . .

deformation-induced martensite formation, leading to weakly ferromagnetic behavior in heavily cold-worked condition (also see Sec. 10.5.6.5). Among austenitic stainless steels, higher-nickel grades exhibit lower magnetic permeabilities than the lower-nickel grades for equivalent level of cold working, as shown in Fig. 10.37.[204]

10.10 MACHINABILITY OF STAINLESS STEELS

Figure 10.38a shows the comparative machinability of stainless steels and other common metals, whereas Fig. 10.38b shows a comparative machinability between frequently used stainless steels and their free-machining counterparts. Mostly, stainless steels without composition modification are tough and gummy, and they have the tendency to seize and gall.

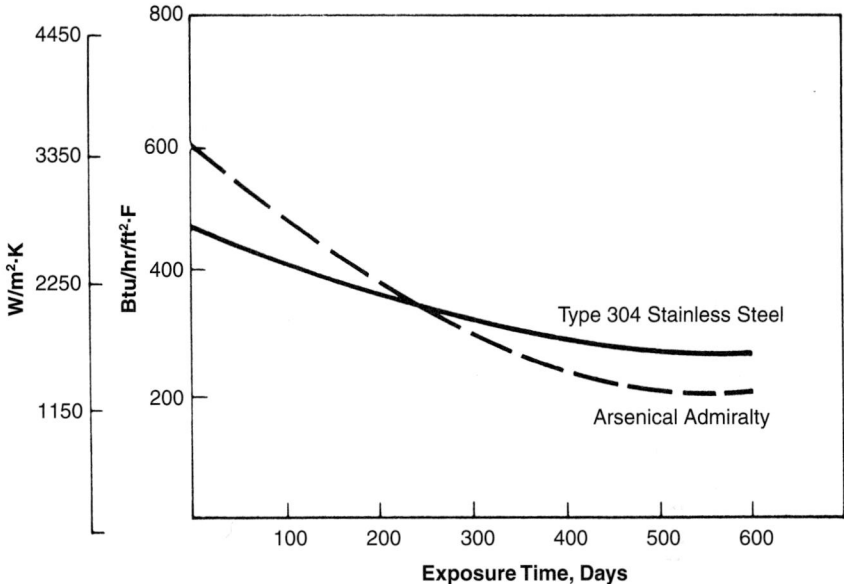

FIGURE 10.36 Overall heat transfer versus exposure time.[67a] (*Courtesy of Specialty Steel Industry of North America, Washington, D.C.*)

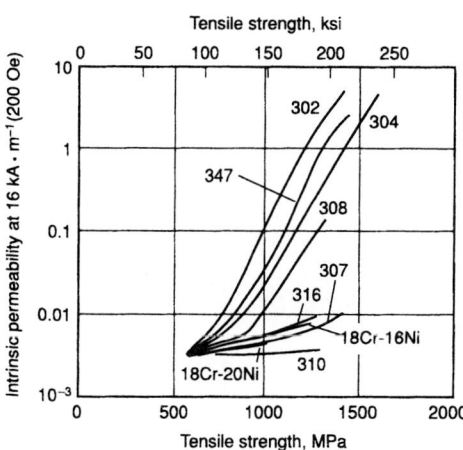

FIGURE 10.37 Correlation of increased tensile strength from cold working and the permeability of cold-worked austenitic steels. Annealed hot-rolled strips 2.4 to 3.2 mm (0.095 to 0.125 in.) thick before cold reduction. For normal permeability values, add unity to the number given on vertical scale.[204] (*Reprinted by permission of McGraw-Hill, Inc., New York.*)

Although the 400 series stainless steels are easily machinable, a stringy chip created during machining can reduce the productivity. In contrast, the 200 and 300 series have the most difficult machining properties, primarily due to their tendency to work-harden at a very rapid rate.

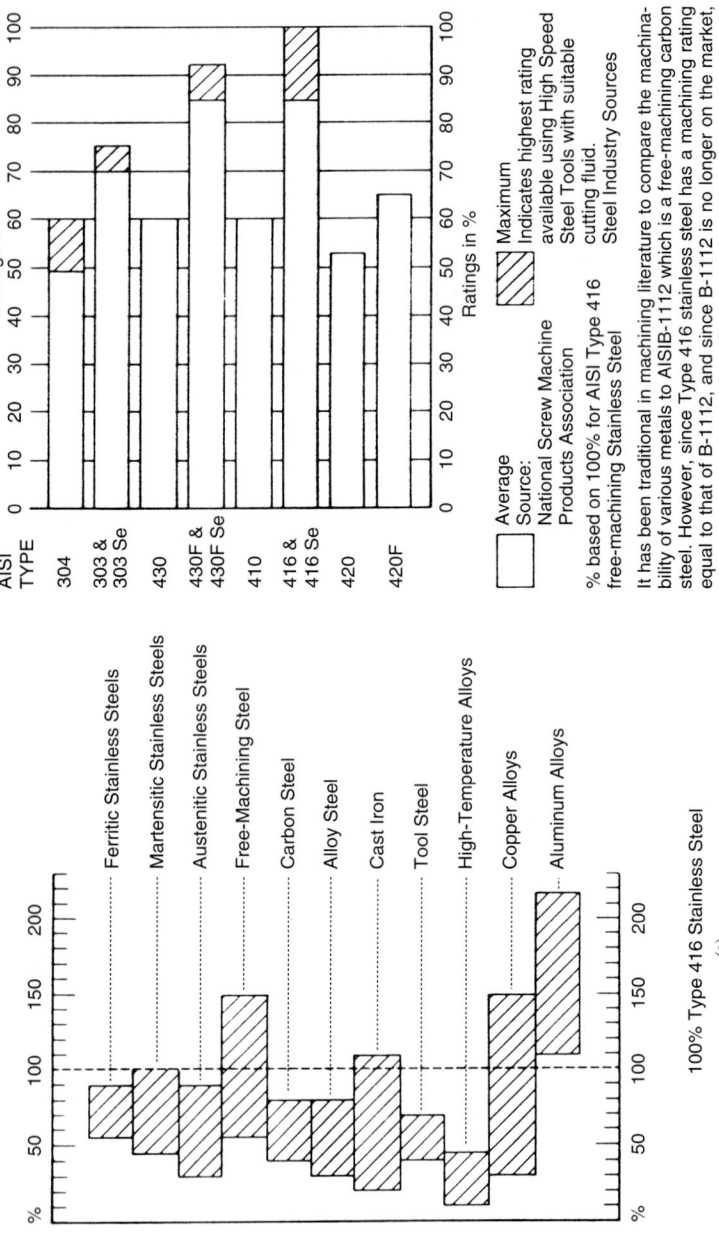

FIGURE 10.38 (*a*) Comparative machinability of common metals. (*b*) Comparative machinability of commonly used stainless steels and their free-machining counterparts.[67a] (*Source: Design Guidelines for Selection and Use of Stainless Steels, Specialty Steel Industry of North America, Washington, D.C.*)

The design/metallurgical engineers can help minimize machining problems and maximize productivity by specifying (1) a free-machining stainless steel, (2) a specific analysis stainless steel that is more suited for machining, or (3) a stainless steel bar in a somewhat hardened condition.[67a]

REFERENCES

1. G. R. Speich and A. Szirmae, *Trans. TMS-AIME*, vol. 245, 1969, p. 1063.

2. M. Nemoto, *Met. Trans.*, vol. 8, 1977, pp. 431–437.

3. A. Roosz, Z. Gacsi, and E. G. Fuchs, *Acta Metall.*, vol. 31, 1983, pp. 509–517.

4. V. G. Gorbach, J. Jelenkowski, and J. Filipiuk, *Mater. Sci. Technol.*, vol. 5, no. 1, 1989, pp. 36–39.

5. G. A. Roberts and R. F. Mehl, *Trans. ASM*, vol. 31, 1943, pp. 613–650.

6. R. R. Judd and H. W. Paxton, *Trans. TMS-AIME*, vol. 242, 1968, p. 206.

7. S. Kinoshita and R. Ueda, *Trans. ISI (Japan)*, vol. 14, 1974, p. 411.

8. G. Molonder, *Acta Metall.*, vol. 4, 1956, p. 565.

9. M. Hillert, K. Nilsson, and L. E. Torndahl, *JISI*, vol. 209, 1971, p. 49.

10. A. F. Nehenberg, *Trans. AIME*, vol. 188, 1950, p. 162.

11. M. Grossman and E. C. Bain, *Principles of Heat Treatment*, 5th ed., ASM, Metals Park, Ohio, 1964.

12. H. W. Paxton, in *Transformations and Hardenability in Steels*, Climax Molybdenum Co., Ann Arbor, Mich., 1967, pp. 3–12.

13. S. F. Dirnfeld, B. M. Horevaar, and F. Van't Spijker, *Met. Trans.*, vol. 5, 1974, pp. 1437–1444.

14. C. I. Garcia and A. J. DeArdo, *Met. Trans.*, vol. 12A, 1981, pp. 521–530.

15. D. P. Datta and A. M. Gokhale, *Met. Trans.*, vol. 12A, 1981, pp. 443–450.

16. N. C. Law and D. V. Edmonds, *Met. Trans.*, vol. 11A, 1980, pp. 33–46.

17. M. Tokizane, N. Matsumura, T. Maki, and I. Tamura, *Met. Trans.*, vol. 13A, 1982, pp. 1379–1388.

18. A. S. Sastry and D. R. F. West, *JISI*, vol. 203, 1965, p. 138.

19. O. A. Ankara and D. R. F. West, *in Physical Properties of Martensite and Bainite*, The Metals Society, London, *Special Report No. 93*, 1965, p. 183.

20. O. A. Ankara, A. S. Sastry, and D. R. F. West, *JISI*, vol. 204, 1966, p. 509.

21. O. A. Ankara, *JISI*, vol. 208, 1970, p. 819.

22. T. J. Nichol, G. Judd, and G. S. Ansell, *Met. Trans.*, vol. 8, 1977, pp. 1877–1883.

23. R. Millsop, in *Hardenability Concepts with Applications to Steel*, eds. D. V. Doane and J. S. Kirkaldy, *TMS-AIME*, Warrendale, Pa., 1978, pp. 316–333.

24. *ASM Standard E112-98*, vol. 01.05, Philadelphia, Pa., 1998.

25. G. F. Vandervoort, *Metallography Principles and Practice*, McGraw-Hill, New York, 1984.

26. W. I. Holiday, *ISI Special Report '81*, The Metals Society, London, 1982, p. 65.

27. G. Parrish, *Heat Treatment of Metals*, vol. 3, 1976, pp. 101–109.

28. A. Z. Hanzaki, P. D. Hodgson, and S. Yue, *Met. and Mats. Trans.*, vol. 28A, 1997, pp. 2405–2414.

29. B. L. Averbach and M. Cohen, *Trans. AIME*, vol. 176, 1948, pp. 401–415.

30. C. A. Stickels, *Met. Trans.*, vol. 8, 1977, pp. 63–70.

31. C. K. Syn, B. Fultz, and J. W. Morris, Jr., *Met. Trans.*, vol. 9A, 1978, pp. 1635–1640.

32. B. B. Vinokur and A. L. Geller, *JOM*, vol. 49, no. 9, 1997, pp. 69–71, 83.

33. P. K. Pearson, *Heat Treating*, vol. 15, no. 11, 1983, p. 18.

34. R. L. Banerjee, *J. Heat Treating*, vol. 3, no. 1, 1983, pp. 48–50.

35. J. Parrish, *Adv. Mater. & Processes*, vol. 145, no. 1, 1994, pp. 25–28.

36. S. C. Lee and F. J. Wazala, *Met. Trans.*, vol. 12A, 1981, pp. 1477–1484.

37. B. V. Narsimha Rao and G. Thomas, *Met. Trans.*, vol. 11A, 1980, pp. 441–457.

38. C. N. Sastry, M. H. Khan, and W. E. Wood, *Met. Trans.*, vol. 13A, 1982, pp. 676–680.

39. G. Y. Lai, W. E. Wood, R. A. Clark, Z. F. Zackay, and E. R. Parker, *Met. Trans.*, vol. 5, 1974, pp. 1663–1670.

40. D. D. Huffman, *Met. Prog.*, vol. 106, no. 7, 1974, pp. 79–85.

41. J. Y. Riedel, *Met. Prog.*, vol. 88, no. 3, 1965, pp. 78–82.

42. B. Liscic, in *Steel Heat Treatment Handbook*, eds. G. E. Totten and M. A. H. Howes, Marcel Dekker, New York, 1997, pp. 527–662.

43. D. N. Collins and G. O'Rourke, *Proceedings: 18th Conference on Heat Treating*, 12–15 Oct. 1998, ASM International, Materials Park, Ohio, 1998, pp. 229–247.

44. K. J. L. Iyer, *Ultrasonics*, vol. 27, July 1989, pp. 245–247.

45. D. L. Williamson, R. G. Schupmann, J. P. Materowski, and G. Krauss, *Met. Trans.*, vol. 10A, 1979, pp. 379–382.

46. *Steel Products Manual: Stainless Steels*, Iron and Steel Society, Warrendale, Pa., 1999.

47. E. Dodrill, *Metal Heat Treating*, August 1995, pp. 29–31.

48. D. T. Llewellyn and R. C. Hudd, *Steels: Metallurgy and Applications*, 3d ed., Butterworth-Heinemann, Oxford, England, 1998.

49. J. F. Grubb, in *Uhlig's Corrosion Handbook*, ed. R. W. Revie, Wiley, New York, 2000, pp. 667–676.

50. C. J. Novak, in *Handbook of Stainless Steels*, McGraw-Hill, New York, 1977, pp. 4.1–4.78.

51. J. F. Grubb, private communication, 1999.

52. J. F. Grubb and J. D. Fritz, in *Corrosion '97*, Paper no. 185, pp. 1–13.

53. *ASM Specialty Handbook: Stainless Steels*, ed. J. R. Davis, ASM International, Materials Park, Ohio, 1994.

54. J. Beddoes and J. G. Parr, *Introduction to Stainless Steels*, 3d ed, ASM International, Materials Park, Ohio, 1999; R. A. Lula, *Stainless Steel*, ASM, Metals Park, Ohio, 1986.

55. *ASM Handbook*, vol. 3: *Alloy Phase Diagrams*, ASM International, Materials Park, Ohio, 1992.

56. A. J. Sedriks, *Corrosion of Stainless Steels*, Wiley, New York, 1996 and 1979.

57. D. J. Kotecki, *ASM Handbook*, vol. 6: *Welding, Brazing, and Soldering*, 10th ed., ASM International, Materials Park, Ohio, 1993, pp. 677–707.

58. J. C. Lippold and W. F. Savage, *Welding J.*, vol. 61, no. 12, 1982, pp. 388s–396s.

59. M. J. Cieslak, A. M. Ritter, and W. F. Savage, *Welding J.*, *Welding Research Supplement*, 1982, pp. 1s–8s.

60. W. A. Baeslack, III, J. C. Lippold, and W. F. Savage, *Welding J.*, *Welding Research Supplement*, vol. 59, June 1979, pp. 168s–176s.

61. S. D. Washko and G. Aggen, *Metals Handbook*, vol. 1, 10th ed., ASM International, Materials Park, Ohio, 1990, pp. 841–907.

62. I. A. Franson and J. F. Grubb, in *Casti Handbook of Stainless Steels and Nickel Alloys*, tech. ed. S. Lamb, Casti Publishing, Alberta, Canada, 1999, pp. 215–257.

63. J. Douthett, *ASM Handbook*, vol. 4: *Heat Treatment*, ASM International, Materials Park, Ohio, 1991, pp. 769–792.

64. C. W. Vigor, J. N. Johnson, and J. E. Hunter, in *Handbook of Stainless Steels*, eds. D. Peckner and I. M. Bernstein, McGraw-Hill, New York, 1977, pp. 39.1–39.11.

64a. J. D. Fritz, J. F. Grubb, and R. E. Polinski, *Adv. Mater. & Processes*, vol. 159, no. 6, 2001, pp. 36–38.

65. Y. Minami, H. Kimura, and Y. Ihara, *Mater. Sci. & Technol.*, vol. 2, August 1986, pp. 795–806.

66. A. L. Doverspike, in *Handbook of Stainless Steels*, eds. D. Peckner and I. M. Bernstein, McGraw-Hill, New York, 1977, pp. 36.1–36.7.

67. *Bulletin*, Specialty Steel Industry of North America, Washington, D.C., 2000.

67a. *Designer Handbooks*, Specialty Steel Industry of North America, Washington, D.C., 2000.

68. B. F. Brown, in *Handbook of Stainless Steels*, eds. D. Peckner and I. M. Bernstein, McGraw-Hill, New York, 1977, pp. 37.1–37.8.

69. B. D. Craig, ed., *Handbook of Corrosion Data*, ASM International, Materials Park, Ohio, 1989, pp. 585–586.

70. M. H. Brown, in *Handbook of Stainless Steels*, eds. D. Peckner and I. M. Bernstein, McGraw-Hill, New York, 1977, pp. 38.1–38.10.

71. J. D. Spragg, in *Handbook of Stainless Steels*, eds. D. Peckner and I. M. Bernstein, McGraw-Hill, New York, 1977, pp. 40.1–40.5.

72. D. H. Detwiler and T. I. Williams, in *Handbook of Stainless Steels*, eds. D. Peckner and I. M. Bernstein, McGraw-Hill, New York, 1977, pp. 41.1–41.7.

73. C. W. Kovach and J. D. Redmond, in *Casti Handbook of Stainless Steels and Nickel Alloys*, tech. ed. S. Lamb, Casti Publishing, Alberta, Canada, 1999, pp. 159–178.

74. P. Marshall, *Austenitic Stainless Steels*, Elsevier, New York, 1984.

75. *Alloy Data Carpenter Specialty Steels*, Carpenter Steel, Reading, Pa., 2000.

76. M. E. Laverroux and J. Marandel, *Wire J. Int.*, vol. 32, no. 10, 1999, pp. 82–85.

77. W. Schumacher, in *New Developments in Stainless Steel Technology*, *Conference Proceedings*, ed. R. A. Lula, ASM, Metals Park, Ohio, 1985, pp. 107–116.

78. L. J. Korb, in *Handbook of Stainless Steels*, eds. D. Peckner and I. M. Bernstein, McGraw-Hill, New York, 1977, pp. 47.1–47.11.

79. T. Tanaka and K. Hoshino, in *New Developments in Stainless Steel Technology, Conference Proceedings*, ed. R. A. Lula, ASM, Metals Park, Ohio, 1985, pp. 129–137.

80. J. Mengel, in *Handbook of Stainless Steels*, McGraw-Hill, New York, 1977, chap. 48.

81. J. E. Truman, in *Corrosion*, vol. 1, ed. L. L. Shrier, Newnes-Butterworths, London, 1977, pp. 3.31–3.63.

82. J. K. L. Lai, D. J. Chastell, and P. E. J. Flewitt, *Mater. Sci. Eng.*, vol. 49, 1981, pp. 19–29.

83. J. K. L. Lai, *Mater. Sci. Eng.*, vol. 61, 1983, pp. 101–109.

84. J. K. L. Lai, *Mater. Sci. Eng.*, vol. 52, 1981, pp. 285–289.

85. K. Nohara, T. Kato, and A. Ejima, in *New Developments in Stainless Steel Technology, Conference Proceedings*, ed. R. A. Lula, ASM, Metals Park, Ohio, 1985, pp. 243–251.

86. K. E. Jonsson, in *Handbook of Stainless Steels*, eds. D. Peckner and I. M. Bernstein, McGraw-Hill, New York, 1977, pp. 43.1–43.12.

87. D. E. Bardos, in *Handbook of Stainless Steels*, eds. D. Peckner and I. M. Bernstein, McGraw-Hill, New York, 1977, pp. 42.1–42.10.

88. *Metals and Alloys in the Unified Numbering System*, 9th ed., SAE/ASTM, Warrendale, Pa., 2001.

89. G. Korbin, J. Lilly, J. MacDiarmid, and B. Moniz, *Materials Performance*, vol. 38, no. 10, 1999, pp. 58–59.

90. C. L. Briant and A. M. Ritter, *Met. Trans.*, vol. 11A, 1980, pp. 2009–2017.

91. S. Danyluk, G. M. Dragel, and D. Dubis, *J. Eng. Mater. Technol.*, vol. 101, 1979, p. 105; S. Danyluk and J. Y. Park, *Scr. Met.*, vol. 16, 1982, pp. 769–774.

92. J. E. Truman, in *Materials Science and Technology*, vol. 7: *Constitution and Properties of Steels*, vol. ed. F. B. Pickering, VCH, Weinheim, 1992, pp. 527–582.

93. W. E. Berry, E. L. White, and W. K. Boyd, *Corrosion*, vol. 29, 1973, pp. 451–469.

94. A. Taboada and L. Frank, *Intergranular Corrosion of Stainless Alloys*, ASTM, Philadelphia, Pa., 1978, pp. 85–98.

95. E. C. Bain, R. H. Aborn, and J. J. B. Rutherford, *Trans. Am. Soc. Steel Treat.*, vol. 21, 1933, p. 481.

96. C. S. Tedmon, D. A. Vermilyia, and J. H. Rosolowski, *J. Electrochem. Soc.*, vol. 118, 1971, p. 192.

97. G. Skyrme and J. Norbury, *CEGB Report, RD/B/N, 4943, 1980*, Ref. 167 in Ref. 45.

98. J. S. Armijo, *Corrosion*, vol. 24, 1968, p. 24.

99. C. L. Briant, *Corrosion*, NACE, vol. 38, no. 4, 1982, pp. 230–232.

100. R. M. Latanision and H. Opperhauser, *Met. Trans.*, vol. 5A, 1974, p. 483.

101. C. L. Briant, *Corrosion*, vol. 36, no. 9, 1980, pp. 497–509.

102. Y. Arata, F. Matsuda, and S. Katayama, *Trans.*, *JWRI*, vol. 5, no. 2, 1976, p. 135.

103. M. J. Cieslak and A. M. Ritter, *Scr. Met.*, vol. 19, 1985, pp. 165–168.

104. G. F. Vander Voort, *ASM Handbook*, vol. 1, 10th ed., ASM International, Materials Park, Ohio, 1990, pp. 689–736.

105. J. J. Eckenrod and C. W. Kovach, in *Properties of Austenitic Stainless Steels and Their Weld Metals*, eds. C. R. Brinkman and H. W. Garvin, ASTM STP 679, ASTM, Philadelphia, Pa., 1979, pp. 17–40.

106. G. Krauss, *Steels: Heat Treatment and Processing Principles*, ASM International, Materials Park, Ohio, 1990.

107. R. A. Walker and T. G. Gooch, *Met. Mater.*, vol. 2, no. 1, 1986, pp. 18–24.

108. H. Schneider, *Foundry Trade J.*, vol. 108, 1960, p. 562.

109. J. A. Brooks and J. C. Lippold, *ASM Handbook*, vol. 6: *Welding, Brazing, and Soldering*, 10th ed., ASM International, Materials Park, Ohio, 1993, pp. 677–707.

110. R. Stickler, *High Temperature Materials in Gas Turbines*, eds. P. R. Sahu and M. O. Speidel, Elsevier, Amsterdam, 1974, p. 115.

111. D. P. Edmonds, R. T. King, and G. H. Goodwin, in *Properties of Austenitic Stainless Steels and Their Weld Metals*, eds. C. R. Brinkman and H. W. Garvin, ASTM STP 679, ASTM, Philadelphia, Pa., 1979, pp. 56–67.

112. R. F. A. Jargelius-Patterson, *Z. Metallkd.*, vol. 89, no. 3, 1998, pp. 177–183.

113. K. W. Andrews, D. J. Dyson, and S. R. Keown, *Interpretation of Electron Diffraction Patterns*, Adam Hilger Ltd., London, 1971.

114. D. R. Harries, in *Mechanical Behavior and Nuclear Applications of Stainless Steels at Elevated Temperatures, Proceedings of the International Conference at Verese, 1981*, The Metals Society, London, 1982, pp. 1–14.

115. J. M. Leitnaker and J. Bentley, *Met. Trans.*, vol. 8, 1977, p. 1605.

116. V. Cihal, *Prot. Met. (USSR)*, vol. 4, no. 6, 1968, p. 563.

117. R. W. K. Honeycombe and H. K. D. H. Bhadeshia, *Steels: Microstructure and Properties*, Arnold, London, 1995.

118. R. Stickler and A. Vinckier, *Trans. ASM*, vol. 54, 1961, p. 362.

119. J. P. Adamson and J. W. Martin, *Acta Metall.*, vol. 19, 1971, pp. 1015–1018.

120. L. K. Singhal, and J. W. Martin, *Acta Metall.*, vol. 15, 1967, pp. 1603–1610.

121. E. L. Hall and C. L. Braint, *Met. Trans.*, vol. 15A, 1984, pp. 793–811.

122. B. Weiss and R. Stickler, *Met. Trans.*, vol. 3, 1972, p. 851.

123. T. Thorvaldson and G. L. Dunlop, *Met. Sci.*, vol. 16, 1982, pp. 184–190.

124. F. R. Beckitt and B. R. Clark, *Acta Metall.*, vol. 15, 1967, pp. 113–129.

125. M. H. Lewis and B. Hattersley, *Acta Metall.*, vol. 13, 1965, p. 1159.

126. L. K. Singhal, *Met. Trans.*, vol. 2, 1971, pp. 1267–1271.

127. H. J. Kestenbach and L. O. Bueno, *Mater. Sci. Eng.*, vol. 66, 1984, pp. L19–L23.

128. H. Goldschmidt, *The Phase Constitutions of Some Niobium Bearing and Associated Transition Metal Systems, Interstitial Alloys*, Plenum Press, New York, 1967.

129. S. Yamamoto and Y. Kobayashi, *Tetsu-to-Hagane*, vol. 78, no. 10, October 1992, pp. 1609–1616.

130. D. R. Harries, *J. Nucl. Mater.*, vol. 82, 1979, p. 2.

131. C. V. Robino and M. J. Cieslak, *Met. and Mats. Trans.*, vol. 26A, 1995, pp. 1673–1685.

132. K. J. Irvine, D. T. Llwellyn, and F. B. Pickering, *JISI*, vol. 199, 1961, p. 153.

133. G. Wallwork and J. Croll, *Rev. High Temp. Mater.*, vol. 3, no. 3, 1976, pp. 69–138.

134. L. R. Woodyatt, C. T. Sim, and H. J. Beattie, Jr., *Trans. AIME*, vol. 236, 1966, pp. 519–527.

135. A. K. Sinha, Ph.D. thesis, University of Minnesota, Minneapolis, 1984.

136. T. Y. Shimaniki, M. Masui, and H. Doi, *Met. Sci*, vol. 10, 1976, pp. 805–808.

137. M. C. Chaturvedi, D. J. Lloyd, and D. W. Chung, *Met. Sci.*, vol. 10, 1976, pp. 373–378.

138. D. W. Chung and M. C. Chaturvedi, *Metallography*, vol. 8, 1975, pp. 329–336.

139. R. W. Jones, *Scr. Met.*, vol. 10, 1976, pp. 1053–1058.

140. D. S. Duvall and M. J. Donachie, Jr., *Inst. Met. Monogr. Rep.*, vol. 100, 1972, pp. 6–12.

141. F. B. Pickering, *Heat Treatment '73*, Metals Society, London, 1975.

142. E. O. Hall and S. H. Algie, *Met. Reviews*, no. 104, 1966, pp. 61–88.

143. J. Barcik, *Mater. Sci. and Technol.*, vol. 4, Jan. 1988, pp. 5–15.

144. M. H. Lewis, *Acta Metall.*, vol. 14, 1966, p. 1421.

145. F. R. Beckitt, *JISI*, vol. 207, 1969, p. 632.

146. R. F. Decker and S. Floreen, in *Metals Society Conference on Precipitation from Iron Base Alloys at Cleveland, Ohio, 1963*, eds. G. R. Speich and J. B. Clark, Gordon and Breach, New York, 1965, p. 69.

147. H. J. Beattie, Jr., and W. H. Hagel, *Trans. AIME*, vol. 233, 1965, p. 277.

148. J. N. Dupont, C. V. Robino, and A. R. Marder, *Welding Research Suppl.*, October 1998, pp. 417s–431s.

149. C. Radhakrishna, K. P. Rao, and S. Srinivas, *J. of Mats. Sci. Lett.*, vol. 14, 1995, pp. 1810–1812.

150. B. Radhakrishnan and R. G. Thompson, *Met. Trans.*, vol. 22A, 1991, pp. 887–902.

151. J. N. Dupont, C. V. Robino, J. R. Michael, M. R. Notis, and A. R. Marder, *Met. and Mats. Trans.*, vol. 29A, 1998, pp. 2785–2796.

152. J. N. Dupont, C. V. Robino, A. R. Marder, and M. R. Notis, *Met. and Mats. Trans.*, vol. 29A, 1998, pp. 2797–2806.

153. R. F. A. Jargelius-Pettersson, *Scandinavian J. of Met.*, vol. 24, 1995, pp. 188–193.

154. A. N. Denham and J. M. Silcock, *JISI*, vol. 207, 1969, p. 585.

155. W. Hume-Rothery and G. Rayner, *The Structure of Metals and Alloys*, Institute of Metals, London, 1962.

156. A. Dwight, *Trans. ASM*, vol. 53, 1960, p. 477.

157. D. Bardos, K. Gupta, and P. Beck, *Trans. AIME*, vol. 221, 1961, p. 1087.

158. M. V. Nevitt, *Electronic Structure and Alloy Chemistry of the Transistor Elements*, Interscience, New York, 1963.

159. P. Duhej and J. Ivan, *JISI*, vol. 206, 1968, p. 1014.

160. D. K. Das, S. P. Rieout, and P. A. Beck, *Trans. AIME*, vol. 194, 1952, p. 1071.

161. C. T. Sims, *J. of Met.*, vol. 94, 1966, p. 1119.

162. H. J. Beattie, Jr., and W. C. Hagel, *Trans. AIME*, vol. 209, 1957, p. 911.

163. H. J. Beattie and F. L. Ver Snyder, *Nature*, vol. 178, 1956, p. 208.

164. R. P. Elliott and W. Rostoker, *ASM Trans.*, vol. 50, 1958, pp. 617–633.

165. E. H. Lee, P. J. Maziasz, and A. F. Rowcliffe, in *Phase Stability during Irradiation*, eds. J. R. Holland, L. K. Mansur, and D. I. Potter, TMS-AIME, Warrendale, Pa., 1981, p. 191.

166. A. Mateo, L. Llanes, M. Anglada, A. Redjaimia, and G. Metauer, *J. of Mats. Sci.*, vol. 32, 1997, pp. 4533–4540.

167. A. H. Eichelman, Jr., and F. C. Hall, *Trans. ASM*, vol. 45, 1953, pp. 77–104.

168. E. P. Butler and M. G. Burke, in *Solid → Solid Phase Transformations*, eds. H. I. Aaronson et al., TMS-AIME, Warrendale, Pa., 1982, pp. 1403–1407.

169. S. R. Thomas and G. Krauss, *Trans. TMS-AIME*, vol. 239, 1967, pp. 1136–1142.

170. R. P. Reed, *Acta Metall.*, vol. 10, 1962, pp. 865–877.

171. M. C. Mataya, M. G. Carr, and G. Krauss, *Mater. Sci. Eng.*, vol. 57, no. 2, 1983, pp. 205–222.

172. T. Angel, *JISI*, vol. 177, 1954, pp. 165–174.

173. F. B. Pickering, *Physical Metallurgy and Design of Steels*, Applied Science Publishers, London, 1978.

174. R. M. Vennett and G. B. Ansell, *Trans. ASM*, vol. 60, 1967, pp. 242–251.

175. S. F. Peterson, M. C. Mataya, and D. K. Matlock, *JOM*, September 1997, pp. 54–58.

176. F. B. Pickering, *Int. Met. Rev.*, vol. 21, 1976, pp. 227–277.

177. K. J. Irvine et al., *JISI*, vol. 207, 1969, p. 1017.

178. F. B. Pickering, in *Encyclopedia of Materials Science and Engineering*, Pergamon Press, Oxford, 1986, p. 4630.

179. J. Barford and J. Myers, *JISI*, vol. 12, 1963, pp. 1025–1031.

180. J. Nutting, *JISI*, vol. 207, 1969, p. 872.

181. F. B. Pickering, in *Materials Science and Technology*, vol. 7: *Constitution and Properties of Steels*, vol. ed. F. B. Pickering, VCH, Weinheim, 1992, pp. 41–94.

182. F. B. Pickering, in *Proceedings of the Symposium Towards Improved Ductility and Toughness*, Climax Molybdenum Co., Ann Arbor, Mich., 1971.

183. F. B. Pickering, *Metallurgical Achievements*, vol. 109, Pergamon Press, Oxford, 1965.

184. K. J. Irvine et al., *JISI*, vol. 196, 1960, p. 166.

185. B. Holmes and D. J. Dyson, *JISI*, vol. 208, 1970, p. 469.

186. *AL-6XN Alloy Bulletin*, Allegheny Ludlum Corporation, Brackenridge, Pa., 1995.

187. J. F. Grubb, *Corrosion '96*, Paper no. 426, 1996, pp. 1–15.

188. J. F. Grubb, in *Proceedings International Conference: Stainless Steels '96*, Düsseldorf, June 3–5, 1996, pp. 367–368.

189. K. Forch, C. Gillessen, I. von Hagen, and W. Wessling, *Stahl-Eisen*, vol. 112, no. 4, 1992, pp. 53–62.

190. J. A. Jimenez, F. Carreno, O. A. Ruano, and M. Carsi, *Mater. Sci and Technol.*, February 1999, pp. 127–131.

191. L. Colombier and J. Hochman, *Stainless and Heat Resisting Steels*, St. Martin's Press, New York, 1968, pp. 107–113.

192. P. Combrade, A. Desestret, P. Jolly, and R. Mayoud, in *Predictive Methods for Assessing Corrosion Damage to BWR Piping and PWR Steam Generators, Proceedings Japan, 1978*, National Association of Corrosion Engineers, Houston, Texas, 1982, pp. 153–172.

193. W. Wessling and H. E. Bock, *Stainless 77*, Climax Molybdenum Co., Ann Arbor, Mich., 1977, pp. 217–225.

194. J. J. Eckenrod and K. E. Pinnow, in *New Developments in Stainless Steel Technology, Conference Proceedings*, ed. R. A. Lula, ASM, Metals Park, Ohio, 1985, pp. 77–87.

195. Y. Maehara and Y. Ohmori, *Met. Trans.*, vol. 18A, 1987, pp. 663–672.

196. G. E. Coates, in *Casti Handbook of Stainless Steels & Nickel Alloys*, tech. ed. S. Lamb, Casti Publishing, Alberta, Canada, 1999, pp. 181–214.

196a. M. L. E. Falkland, in *Uhlig's Corrosion Handbook*, 2d ed. R. W. Revie, Wiley, New York, 2000, pp. 651–666.

197. Y. Shimoide, J. Cui, C. Y. Kang, and K. Miyahara, *ISJI*, vol. 39, no. 2, 1999, pp. 191–194.

198. T. A. DeBold, J. W. Martin, and D. A. Englehart, in *New Developments in Stainless Steel Technology, Conference Proceedings*, ed. R. A. Lula, ASM, Metals Park, Ohio, 1985, pp. 89–97.

199. R. C. Gibson, H. W. Hayden, and J. H. Brophy, *Trans. ASM*, 1968, vol. 61, p. 85.

200. S. Floreen and H. W. Hayden, *Trans. ASM*, vol. 61, 1968, p. 489.

201. C. V. Roscoe, K. J. Gradwell, M. Watts, and W. J. Nisbet, *Stainless Steels '87*, The Institute of Metals, London, 1988, pp. 325–333.

202. M. L. E. Falkland, in *Uhlig's Corrosion Handbook*, 2d ed. R. W. Revie, Wiley, New York, 2000, pp. 651–666.

203. M. Rouby and P. Blanchard, in *Stainless Steels*, Les Editions de Physique, 1993, pp. 111–158.

204. J. R. Lewis, in *Handbook of Stainless Steels*, eds. D. Peckner and I. M. Bernstein, McGraw-Hill, New York, 1977, pp. 19.1–19.36.

205. D. W. Dietrich, in *ASM Handbook*, vol. 2, 10th ed., ASM Interantional, Materials Park, Ohio, 1990, pp. 761–781.

206. R. A. McAllister, D. H. Eastham, N. A. Dougharty, and M. Hollier, *Corrosion*, vol. 7, 1961, pp. 579t–588t.

CHAPTER 11

ISOTHERMAL AND CONTINUOUS COOLING TRANSFORMATION DIAGRAMS

11.1 INTRODUCTION

There are two forms of transformation diagrams, such as the time-temperature trans-formation (TTT) and the continuous cooling transformation (CCT) diagrams, which are commonly used by metallurgists to assess the microstructure produced during heat treatment, and provide a judgment of the hardenability of a steel. The basic difference in interpretation of these transformation diagrams for a particular steel lies in the fact that the TTT diagram is read from left to right whereas the CCT diagram is read along the cooling curves from top to bottom right. This chapter describes methods of construction, significance in, and application to, heat-treating operations; the effect of alloying elements on shape and position; and the limita-tions of transformation diagrams. References to several atlases of transformation diagrams published so far for a wide variety of steels are provided, which allow us to make the decisive selection of a particular steel and the design of the heat treat-ment to achieve the desired microstructures and mechanical properties to meet the proper service conditions.

11.2 ISOTHERMAL TRANSFORMATION DIAGRAMS

The isothermal transformation (IT) diagram, alternatively known as the time-temperature-transformation diagram, represents the time-temperature relationship of austenite decomposition. It is a sort of kinetic phase diagram, where tempera-ture is plotted as the ordinate on a linear scale and concentration, as the abscissa, is replaced by the time on a logarithmic scale. It thus denotes the times required to start and complete austenite transformation into various microstructures under constant-temperature conditions.

The initial approach to study the progress of transformation of austenite at con-stant subcritical temperature was made by Davenport and Bain,[1] who determined the first IT diagram. Later, Cohen[2] modified this diagram by including the concept of M_s and M_f temperatures.

The following steps are adopted in order to construct an isothermal transformation diagram for a particular steel, e.g., a eutectoid steel.

1. A large number of thin specimens of steel under investigation, cut from the same wire, tape, or machined bar, are used so that their temperature can be quickly changed.

2. These specimens are austenitized at a proper temperature for sufficient time in a molten salt bath.

3. The specimens are then rapidly transferred and quenched into another liquid bath, e.g., lead bath or molten salt bath, held at the desired subcritical transformation temperature. When molten lead baths are used, changes occur very rapidly on cooling (e.g., in plain carbon steels).

4. After a lapse of progressively increasing times (from seconds to several days), each specimen is subsequently quenched in cold water or iced brine solution to transform any untransformed austenite to martensite.

5. Specimens are examined by optical metallography. Their hardness results are also ascertained. This provides the extent of transformation during the time the austenitized specimen is held in the molten salt (or lead) bath.

6. The above series of experiments are repeated at several subcritical temperatures until sufficient points are obtained.

An isothermal transformation diagram is then plotted by joining all the "start" points, all the 20%, all the 50%, all the 90%, . . . , and all the "finish" transformation points. Figure 11.1a illustrates the method of summarizing isothermal transformation in a simple TTT diagram.[3] It is general practice to draw the principal curves on this diagram by broad lines, which emphasize that their exact location on the time scale is not very precise. Some portion of these lines is frequently shown as dashed lines to indicate a still higher degree of uncertainty.

This method is expensive and demands a great skill for examination of more than 100 microstructures to achieve a reasonable accuracy. Any other property measurement such as dilatometry,[1] electrical resistivity,[4,5] or magnetic permeability[6,7] that changes with the proportion of austenite can also be employed as a variation of the method to determine the progress of transformation. However, note that the use of these alternative methods does not ensure a marked degree of accuracy obtained by microscopic examination. Hence, the results obtained by the various direct methods must be checked by performing a sufficient number of microscopic examinations to avoid errors.

Figure 11.1b shows the TTT curve for a plain carbon eutectoid (1080) steel.[3] Above Ae_1 the austenite is stable. Below Ae_1 and to the left of the start curve, an unstable austenite exists which represents austenite getting ready to begin

---

FIGURE 11.1 (a) Comparison of an isothermal volume fraction transformation curve to the TTT diagram at the corresponding temperature.[3] (b) TTT diagram for AISI 1080 steel (0.79% C, 0.76% Mn) austenitized at 900°C (1650°F).[3] Grain size: 6. A, austenite; F, ferrite; C, carbide; M, martensite; B, bainite; and P, pearlite.[3] (*Reprinted by permission of The United States Steel Corporation, Pittsburgh, Pennsylvania.*)

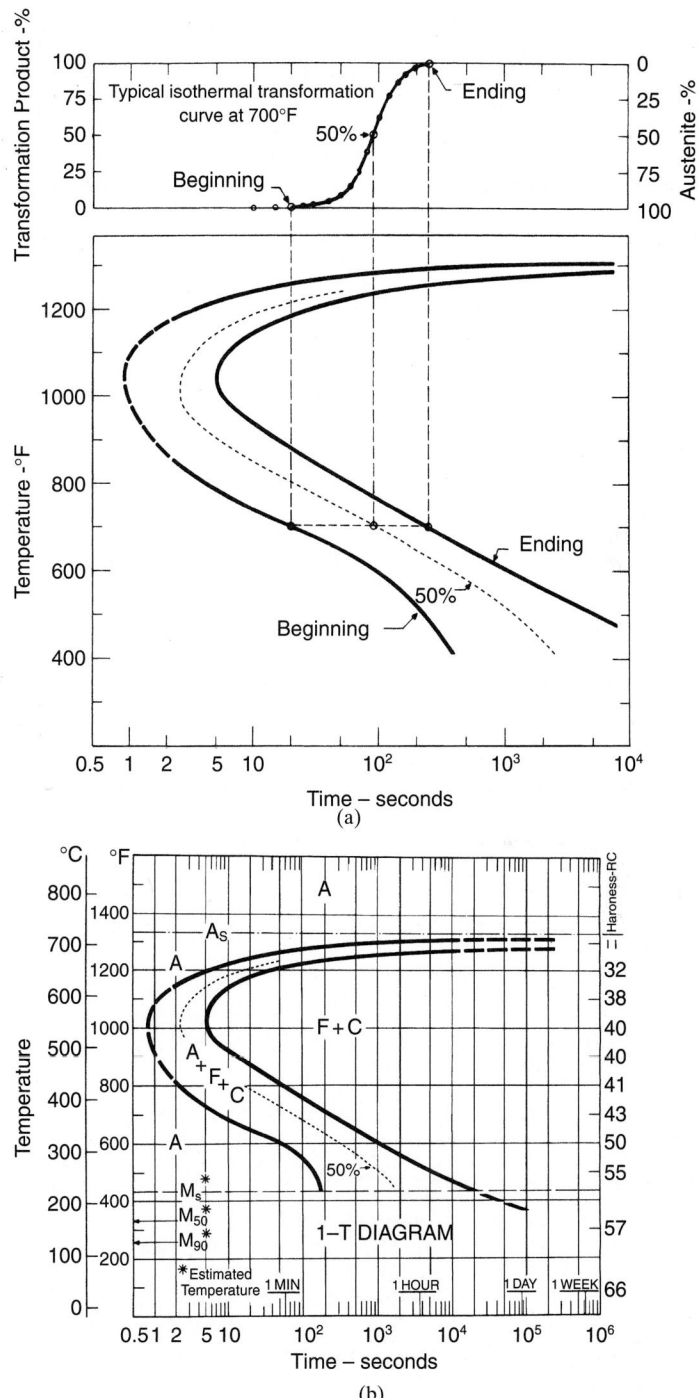

11.3

transformation. Between the two curves is austenite in the process of transformation along with the transformation products, ferrite and carbide. Finally, the section to the right of the finish curve represents the transformation products formed, after completion of the transformation. The point on the start curve farthest to the left (i.e., the point of minimum nucleation time) is termed the *nose* or *knee* of the diagram. At this temperature the reaction proceeds more rapidly, and finer pearlite lamellae are formed. However, this rate of transformation slows down both at high temperature below the Ae_1 line forming coarse pearlite and at lower temperature from about 550 to 250°C, where another transformation product, bainite, forms. In plain carbon steels, this transformation becomes more sluggish with the decrease in transformation temperature because of decreased diffusivity of carbon. This bainitic transformation was discussed in Chap. 9.

The austenite starts transforming to martensite below a critical temperature, called the M_s temperature. Below M_s, the percentage of austenite transformed to martensite is continuously increased with the falling temperature; the transformation becomes complete when the M_f temperature is reached. Between M_s and M_f the proportion of martensite formed is represented by a series of horizontal lines. We discussed martensitic transformation in Chap. 8.

It is clear from Fig. 11.1 that eutectoid steel must be rapidly cooled from the austenitizing temperature past 540°C (1000°F) and to M_s temperature in less than about 0.8s to produce an entirely martensitic structure and to M_s in about 2.5s to produce a 50% martensitic structure. Similarly, a bainitic structure is produced by fast-cooling past 540°C in less than 1s and subsequently holding the specimen for a time between 540°C and the M_s temperature corresponding to the time required to complete the transformation curve. Slower cooling rates (e.g., furnace- and air-cooling) produce coarse and fine pearlitic structure, respectively.

The effects of elements such as Mn, Cr, Mo, V, and Nb on the isothermal transformation behavior of eutectoid steels have been investigated by many researchers. Their results have shown that Mo up to 0.34% is effective in suppressing austenite decomposition at high temperatures (500 to 650°C), whereas Cr up to 0.76% is effective at intermediate temperatures (500 to 550°C) and Mn between 0.4 and 0.91% is effective at low temperatures (300 to 350°C). These findings have resulted in a growing interest in the development of high-strength alloyed rail steels, alloyed railroad wheels, and alloyed wire rods of eutectoid composition.[8]

11.2.1 Recent Study of Isothermal Transformation of Austenite to Pearlite and Upper Bainite in Eutectoid Steel

In an extensive study made by some investigators, using optical metallography and electron microscopy, both pearlite and upper bainite were found to be formed during isothermal transformation at temperatures between 350 and 600°C in eutectoid steel.[9–11] This established that they might form simultaneously, and probably independently, from austenite by different transformation mechanisms. Figure 11.2 shows a partial IT diagram illustrating the separate C curves for the start of pearlitic and upper bainitic transformation from point-counting measurements. However, the two overlapping curves were closer to the conventional start-of-transformation curve (the broken line in Fig. 11.2). This overlapping diagram is of academic, rather than of practical, interest; therefore we will use the conventional start-of-transformation curve throughout the text.

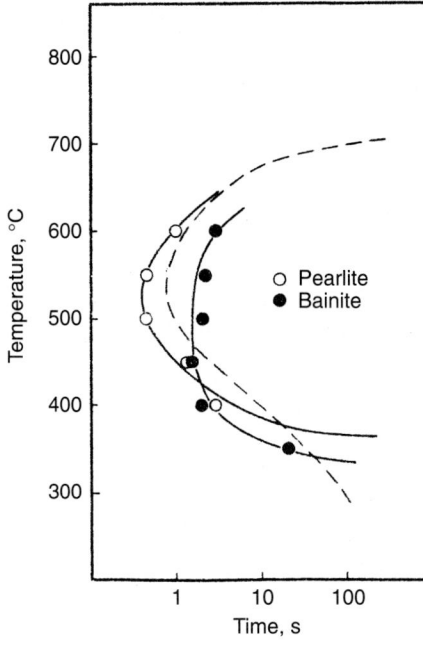

FIGURE 11.2 Partial isothermal transformation diagram illustrating C curves representing the start of austenite decomposition to pearlite and upper bainite.[11] (*Reprinted by permission of The Metallurgical Society, Warrendale, Pennsylvania.*)

11.2.2 Pearlite Nose Temperature

The pearlite nose temperature[12] is seldom measured accurately in experimental investigation. For commercial use, this method of precision does not constitute a considerable disadvantage because transformation times do not change significantly within ±20°C of the nose. This temperature usually occurs between 50 and 100°C below the Ac_1 temperature and can be expressed by either Eq. (11.1), which is based on Andrew's Ac_1 temperature measurements, or by Eq. (11.2), which is used for alloy steels containing a maximum limit of 2% Mn, 5% Ni, 3% Cr, and 0.5% Mo. Isothermal pearlite nose temperature (°C) is thus given by

$$650 - 11\%(Mn) - 17\%(Ni) + 17\%(Cr) \tag{11.1}$$

or

$$660 - 17\%(Ni + Mn - Cr - Mo) \tag{11.2}$$

11.2.3 Isothermal Transformation Diagrams for Noneutectoid Plain Carbon and Low-Alloy Steels

In hypoeutectoid plain carbon steels, proeutectoid ferrite forms at, as well as from, the austenite grain boundaries between the Ae_3 and Ae_1 temperatures and between Ae_1 and 550°C, which precedes the pearlite reaction. The occurrence of the ferrite reaction is represented by another curve which is often merged near the nose region.

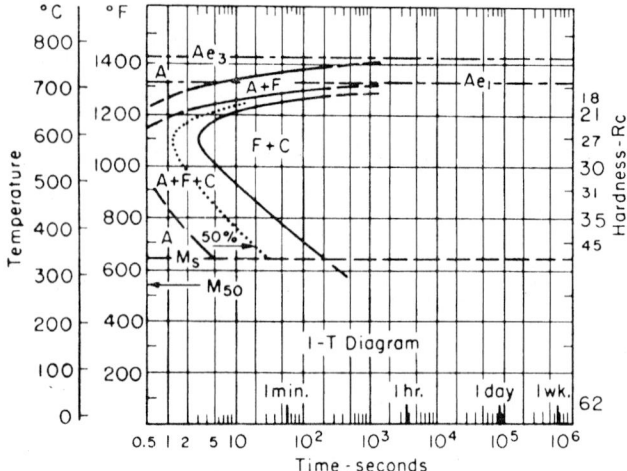

FIGURE 11.3   Isothermal transformation diagram for a hypoeutectoid steel containing 0.47% C and 0.57% Mn; austenitizing temperature 843°C (1550°F).[13] (*Reprinted by permission of ASM International, Materials Park, Ohio.*)

Figure 11.3 shows the TTT diagram for a steel with 0.47% C. Let us take small pieces of this material which have been austenitized at 843°C (1550°F) and then quenched and isothermally reacted at the different subcritical temperatures. At a higher isothermal reaction temperature [say, 690°C (1275°F)], an almost equilibrium structure containing proeutectoid ferrite and coarse pearlite is obtained (Fig. 11.4a).[12] At a lower isothermal transformation temperature [649°C (1200°F)], the amount of proeutectoid ferrite decreases and that of pearlite increases (Fig. 11.4b). However, quenching to a lower temperature [538°C (1000°F)] and subsequent partial transformation at 538°C produce pearlite nodules with some upper bainite (Fig. 11.4c). Quenching to a still lower temperature [427°C (800°F)] and partial isothermal transformation predominantly produce a lower bainitic structure (Fig. 11.4d).

Note that the lower the carbon content of hypoeutectoid steels, the higher the Ac_3 temperature and consequently the more expanded the region of proeutectoid ferrite coexisting with austenite. In addition, the M_s temperature will be raised, and such a diagram will be found to be shifted to the left when compared to that of eutectoid steel. Thus it is not possible to obtain the entire martensitic structure by quenching these steels from the austenitic phase fields. In this situation, the martensitic structure with small amounts of ferrite and pearlite is produced.

The isothermal diagram for a hypereutectoid steel is similar to that of hypoeutectoid steel, except that proeutectoid cementite, rather than ferrite, forms first.

Several factors determine the position and shape of the TTT diagram, the principal one being the amount of carbon and alloying elements present and the second one being the prior austenite grain size. Both have a marked effect on hardenability. An increase in the amount of carbon and alloying elements, particularly carbide formers such as Cr, Mo, V, and so forth, or coarsening of the (i.e., larger) austenite grain size, retards the (ferritic and) pearlitic transformations to a considerable

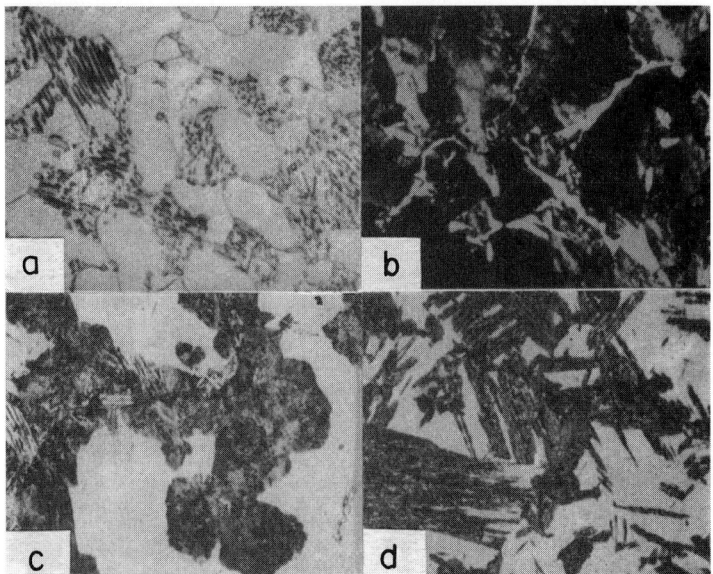

FIGURE 11.4   Microstructure of a hypoeutectoid steel containing 0.47% C after austenitizing at 843°C (1550°F) followed by isothermal transformation as follows: (*a*) After complete transformation at 690°C. (*b*) After complete transformation at 650°C. Structures show proeutectoid ferrite white and pearlite black in both (*a*) and (*b*). (*c*) After partial transformation at 538°C. (Pearlite nodules appear black with some upper bainite needles, and martensite appears white.) (*d*) After partial transformation at 427°C. (Lower bainite appears black, and martensite appears white.[13]) (*Reprinted by permission of ASM International, Materials Park, Ohio.*)

extent[3] and moves the TTT diagram to slightly longer times. However, these carbide-forming elements have a limited effect in delaying the bainitic reaction. As a consequence, a pearlite nose is shifted farther to the right than the bainite nose (Fig. 11.5*a* and *b*). Austenite stabilizing elements (e.g., Cu and Ni) exert a modest effect on the transformation and displace the whole diagram to a small degree toward the right (Fig. 11.5*c*). However, a simultaneous presence of several alloying elements, even if the combined concentration is not high, produces a greater influence in retarding the transformation rates, with the result that the two distinct knees are formed with or without separation (Fig. 11.5*d* and *e*).[3,14] The influence of boron in concentrations as low as 0.002 to 0.005% on the ferrite and pearlite is noteworthy, particularly in low-carbon steels. It has been found that Al, Co, and Si displace the TTT curve for the start of the proeutectoid ferrite reaction to shorter times compared to Fe-C alloys of similar carbon content. At lower temperatures, however, all the TTT curves appeared to converge.[15]

11.2.4   Application

The IT diagrams are very useful in (1) providing a better understanding of the transformation of austenite; (2) determining a suitable temperature for isothermal heat treatment of steel as isothermal annealing, martempering, and austempering are

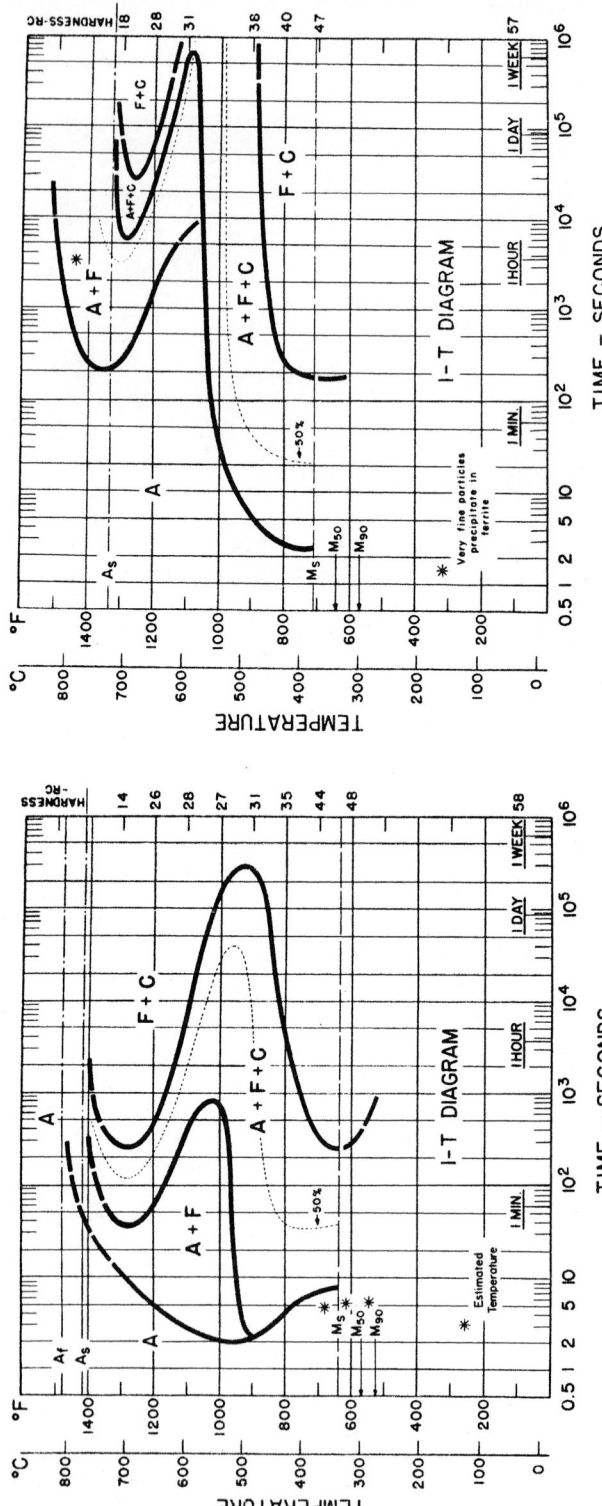

FIGURE 11.5 TTT diagrams for five commercial low-alloy steels. (a) 2Cr–0.3C steel containing 0.33% C, 0.45% Mn, and 1.86% Cr, austenitized at 871°C (1600°F). Grain size: 6 to 7.[3] (b) 2Mo–0.3C steel containing 0.33% C, 0.41% Mn, and 1.96% Mo, austenitized at 1038°C (1900°F). Grain size: 3 to 4.[3] (c) Nickel steel (AISI 2340) containing 0.37% C, 0.68% Mn, and 3.41% Ni, austenitized at 788°C (1450°F). Grain size: 7 to 8.[3] (d) AISI 8630 steel containing 0.3% C, 0.8% Mn, 0.54% Ni, 0.55% Cr, and 0.21% Mo, austenitized at 871°C (1600°F). Grain size: 9.[3] (e) High-strength steel (Air Steel X200) containing 0.44% C, 0.79% Mn, 1.63% Si, 2.10% Cr, 0.54% Mo, and 0.06% V, austenitized at 954°C (1750°F).[14] A, austenite; F, ferrite; C, carbide; P, pearlite; B, bainite; M, martensite; M_s, start of martensite formation. ((a–d) Reprinted by permission of the United States Steel Corporation, Pittsburgh, Pennsylvania; (e) reprinted by permission of ASM International, Materials Park, Ohio.)

11.8

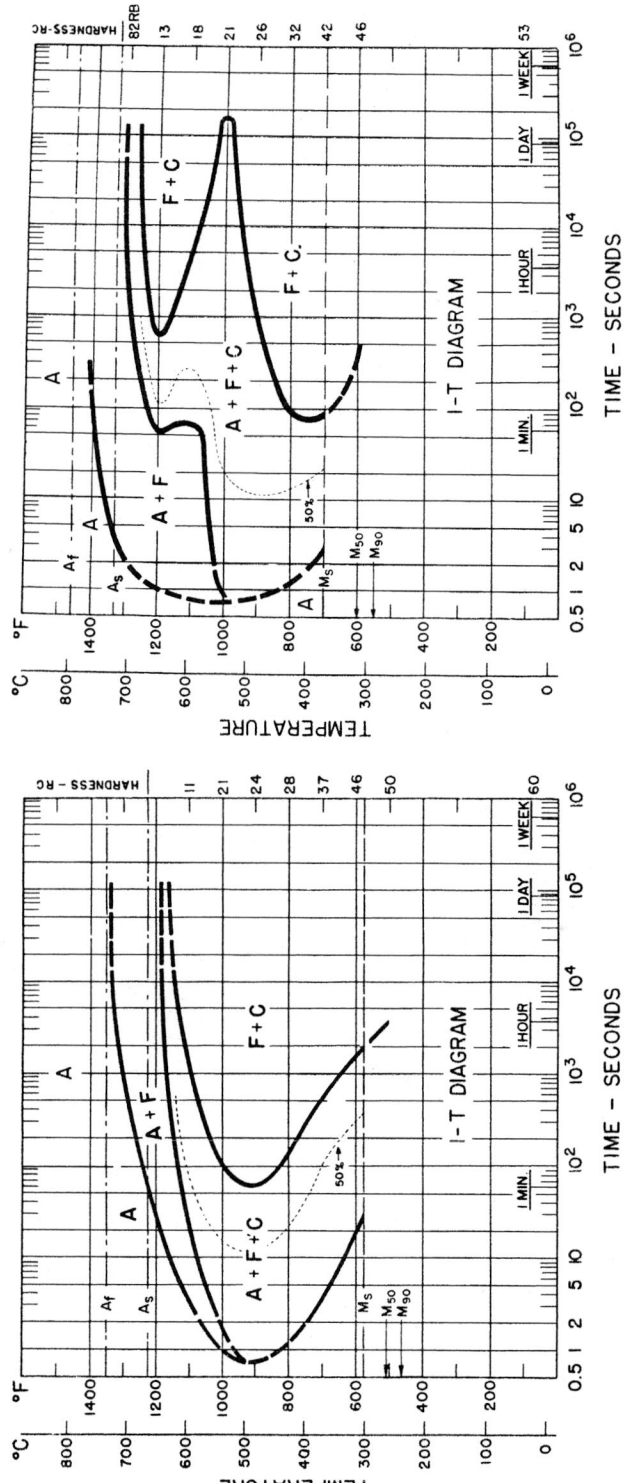

FIGURE 11.5 *(continued)* TTT diagrams for five commercial low-alloy steels. (*a*) 2Cr-0.3C steel containing 0.33% C, 0.45% Mn, and 1.86% Cr, austenitized at 871°C (1600°F). Grain size: 6 to 7.[3] (*b*) 2Mo-0.3C steel containing 0.33% C, 0.41% Mn, and 1.96% Mo, austenitized at 1038°C (1900°F). Grain size: 3 to 4.[3] (*c*) Nickel steel (AISI 2340) containing 0.37% C, 0.68% Mn, and 3.41% Ni, austenitized at 788°C (1450°F). Grain size: 7 to 8.[3] (*d*) AISI 8630 steel containing 0.3% C, 0.8% Mn, 0.54% Ni, 0.55% Cr, and 0.21% Mo, austenitized at 871°C (1600°F). Grain size: 9.[3] (*e*) High-strength steel (Air Steel X200) containing 0.44% C, 0.79% Mn, 1.63% Si, 2.10% Cr, 0.54% Mo, and 0.06% V, austenitized at 954°C (1750°F).[14] A, austenite; F, ferrite; C, carbide; P, pearlite; B, bainite; M, martensite; M_s, start of martensite formation. ((*a*–*d*) *Reprinted by permission of the United States Steel Corporation, Pittsburgh, Pennsylvania; (e) reprinted by permission of ASM International, Materials Park, Ohio.)*

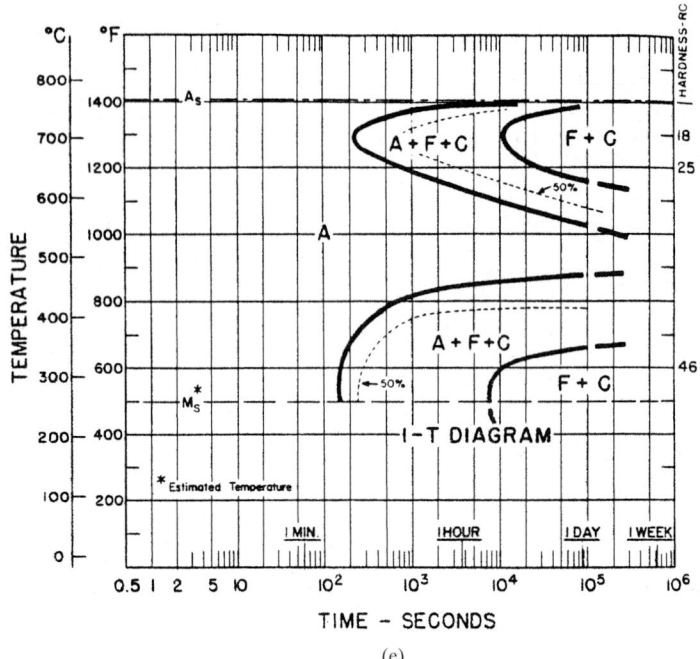

(e)

FIGURE 11.5 (*continued*) TTT diagrams for five commercial low-alloy steels. (*a*) 2Cr-0.3C steel containing 0.33% C, 0.45% Mn, and 1.86% Cr, austenitized at 871°C (1600°F). Grain size: 6 to 7.[3] (*b*) 2Mo-0.3C steel containing 0.33% C, 0.41% Mn, and 1.96% Mo, austenitized at 1038°C (1900°F). Grain size: 3 to 4.[3] (*c*) Nickel steel (AISI 2340) containing 0.37% C, 0.68% Mn, and 3.41% Ni, austenitized at 788°C (1450°F). Grain size: 7 to 8.[3] (*d*) AISI 8630 steel containing 0.3% C, 0.8% Mn, 0.54% Ni, 0.55% Cr, and 0.21% Mo, austenitized at 871°C (1600°F). Grain size: 9.[3] (*e*) High-strength steel (Air Steel X200) containing 0.44% C, 0.79% Mn, 1.63% Si, 2.10% Cr, 0.54% Mo, and 0.06% V, austenitized at 954°C (1750°F).[14] A, austenite; F, ferrite; C, carbide; P, pearlite; B, bainite; M, martensite; M_s, start of martensite formation. ((*a–d*) *Reprinted by permission of the United States Steel Corporation, Pittsburgh, Pennsylvania; (e) reprinted by permission of ASM International, Materials Park, Ohio.*)

desired; (3) ausforming; and (4) providing an approximate indication of the hardenability. Usually, the longer the time taken for the start of transformation, the greater the hardenability.

In later chapters we will discuss, in detail, isothermal heat treatment, ausforming, and hardenability.

11.3 CONTINUOUS COOLING TRANSFORMATION DIAGRAMS

Most heat treatment of steels is carried out by continuous cooling rather than by isothermal transformation from the austenitizing temperature to room temperature.

That is, IT diagrams are not directly related to normal thermal treatments. It is, therefore, necessary to consider the continuous cooling transformation diagram, which represents the progress of austenite transformation as a function of cooling rate. This allows us to predict the microstructure and associated mechanical properties of quenched, normalized, and annealed steels obtained by continuous cooling from the austenitizing temperature.

11.3.1 Derived CCT Diagrams

Figure 11.6 shows a CCT diagram derived from its IT counterpart for a eutectoid steel. Four cooling curves for a given austenitizing temperature corresponding to several positions along the Jominy end-quench bar are superimposed on the CCT diagram. The cooling rate at point D is slow enough to allow complete transformation of austenite into coarse pearlite. This slow-cooling curve is usually obtained by cooling the specimens, after austenitizing, in a furnace by shutting off the power supply. This rate of cooling brings the steel to room temperature in about a day. This transformation is associated with decreased hardness, as shown in the top part of Fig. 11.6.[7] Curve C represents a comparatively rapid cooling rate for which the transformation of austenite occurs in the temperature range of 550 to 600°C (1022 to 1112°F) and corresponds to the normalizing heat treatment (air cooling). The

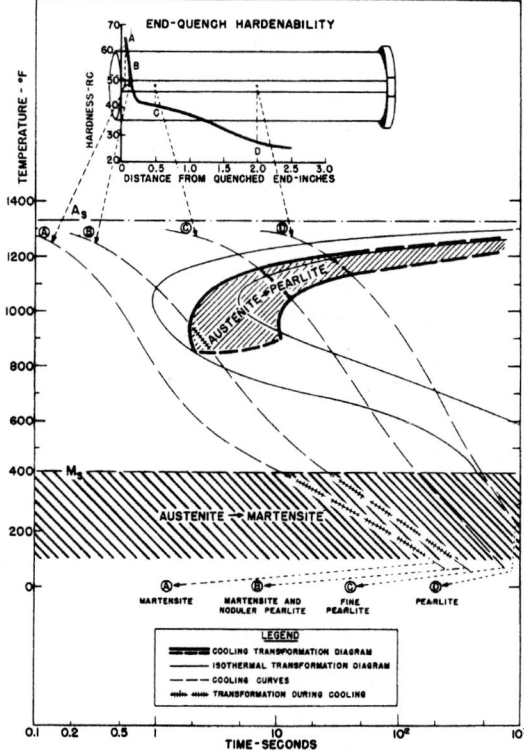

FIGURE 11.6 Correlation of CCT and IT diagrams with end-quench hardenability test data for eutectoid steel.[3] (*Reprinted by permission of The United States Steel Corporation, Pittsburgh, Pennsylvania.*)

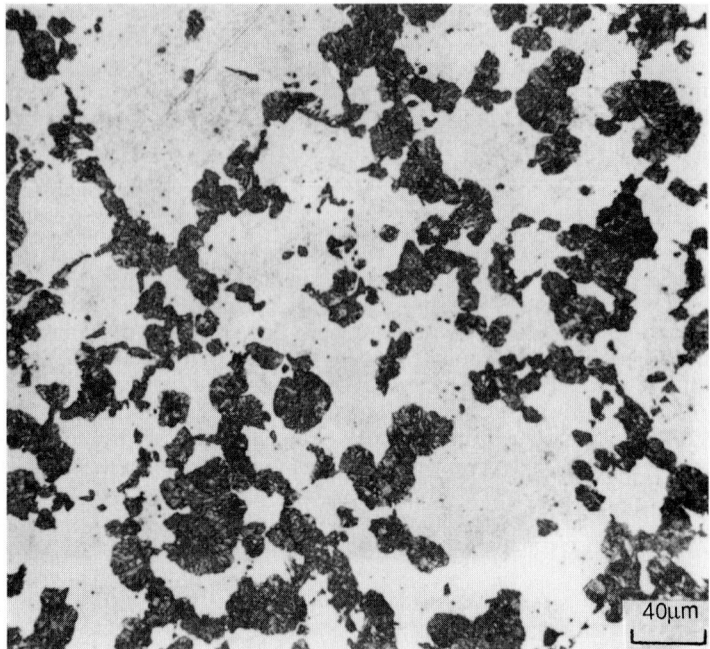

40μm

FIGURE 11.7 Microstructure with fine pearlite (large nodules) and martensite (white) produced by split transformation.[16] (*Reprinted by permission of Wadsworth, Inc., Belmont, California.*)

structure in this case is fine pearlite. Cooling curve *B* represents a still faster cooling rate which intersects the pearlite zone, and austenite partly transforms into fine pearlite; the remaining austenite transforms to martensite during cooling through a much lower temperature range. Since the transformation in this case occurs in two stages, it is called a *split transformation*. Figure 11.7 shows the microstructure of a eutectoid steel which has undergone such a transformation.[16] Finally, curve *A*, denoting the quenched end of the Jominy specimen, corresponds to the very rapid cooling (as by water-quenching) of austenite that produces entirely hard martensitic structure.

The essential points to be noted here are that (1) austenite transformation at a given cooling rate occurs only when the cooling curve passes across the shaded region of the CCT diagram, (2) the CCT diagram shifts the start and finish of transformation with respect to the corresponding IT diagram to lower temperature and longer time, and (3) the bainite transformation does not occur. As a result of these prominent features, the incubation time at the pearlite nose cannot be directly employed to predict the hardenability because the predicted hardenability is lower than is actually determined.[17]

Figure 11.8 shows a correlation of CCT and IT diagrams for 4140 steel. In this alloy steel, unlike the plain carbon eutectoid steel just discussed, the pearlite field has moved farther (lower temperatures) to the right (longer times) when compared to the bainitic region. In this hypoeutectoid alloy steel, the ferrite field appears on both the IT and CCT diagrams. Of course, the M_s and M_f temperatures remain the

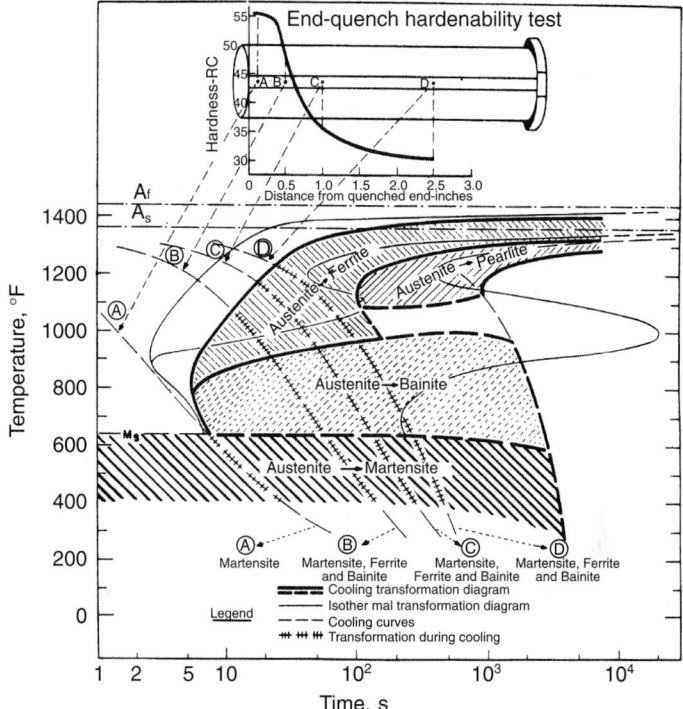

FIGURE 11.8   Correlation of CCT and TTT diagrams with end-quench hardenability test data for 4140 steel.[3] (*Reprinted by permission of The United States Steel Corporation, Pittsburgh, Pennsylvania.*)

same for the two diagrams. As in the previous example, several representative positions along the end-quench bar are related to a hardness curve and to the CCT diagram by means of cooling curves at each position.

The important features of CCT diagrams that have no counterpart in IT diagrams, particularly in alloy steels, are[18,19] (1) a depression of M_s temperature at slow cooling rates, (2) the tempering of martensite that occurs on cooling from M_s temperature to about 200°C, and (3) a large variety of microstructures.

It can be concluded, based on the available literature and the above examples, that the method of correlating IT and derived CCT diagrams did not give satisfactory results in plain carbon or alloy steels.[20] Hence the derived CCT diagrams from their counterpart IT diagrams should be abandoned as a valid technique; instead, the experimentally determined CCT diagrams should be used.

11.3.2   Experimentally Determined CCT Diagrams

The well-established method of direct experimental determination of CCT diagrams involves the use of a high-speed dilatometer in which small specimens are heated and cooled at controlled rates so that both temperature and changes in length can be simultaneously recorded.[17] This dilatometric method is sensitive because of the

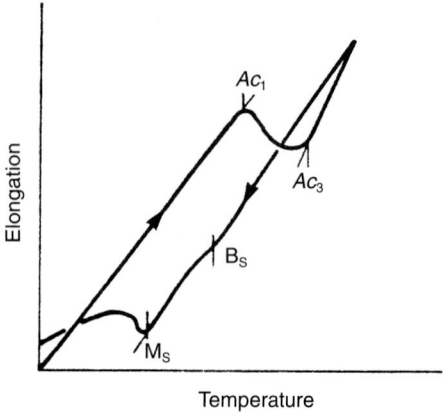

FIGURE 11.9 Schematic elongation-temperature curve on heating and cooling, illustrating the determination of Ac_1, Ac_3, B_s, and M_s temperatures.[23] (*Reprinted by permission of ASM International, Materials Park, Ohio.*)

fact that fcc austenite has a quite different thermal expansion coefficient compared to that of the bcc ferrite or other transformation product of austenite. First the thermal expansion characteristics of austenite are established, and the Ac_1 and Ac_3 (or Ac_{cm}) temperatures are determined. When the specimen is cooled at a controlled rate from the austenitizing temperature, the contraction of the specimen conforms to the contraction characteristics of austenite until transformation starts. A sharp change in the slope of the elongation-temperature curve takes place during transformation; and when transformation is arrested, further contraction proceeds at a different slope of the curve, characteristic of either a completely ferritic structure or a mixed structure containing untransformed austenites.[17,21,22] Figure 11.9 shows a schematic elongation-temperature curve on heating and cooling, illustrating the determination of Ac_1, Ac_3, B_s, and M_s temperatures.[23] The results are presented in such a manner that the individual cooling curves are shown on the temperature versus logarithmic time diagram. The hardness for each cooling rate is represented in a circle at the end of each cooling curve. However, this technique must be supplemented by a metallographic examination and hardness measurement of the as-cooled specimen.[22] More experimental details and explanations on the dilatometric method can be found in the atlases of CCT diagrams[17,20–22,24] and in the review article by Eldis.[25]

Figure 11.10 presents the experimentally determined CCT diagrams that show the effect of increasing carbon content in a Cr-Mo steel from 0.39 to 0.53% C. A substantial increase in the time interval for complete martensitic transformation (at the expense of the bainitic transformation) is clearly seen in addition to a delay in the start of the diffusion-controlled transformation and a decrease in the maximum B_s and M_s temperatures with the increased amount of carbon.[22]

Figure 11.11 shows the experimentally determined CCT diagrams for 0.4% C-0.75% Cr-0.5% Mo steels with 0.78% Ni and 4.53% Ni, respectively. These diagrams clearly demonstrate that increasing the amount of Ni, a strong austenite stabilizer, depresses the Ac_3 and Ac_1 temperatures and delays all transformations. It should be borne in mind that when Cr and Ni are present in a 0.4% carbon steel, the effectiveness of Mo is enhanced in retarding the phase transformation kinetics (Figs. 11.11a and 11.12). Mo additions greatly delay the ferrite transformation interval in the Ni-Cr alloy steels; the time intervals for bainite, bainite-martensite, and martensite transformations are significantly increased. The maximum B_s temperatures

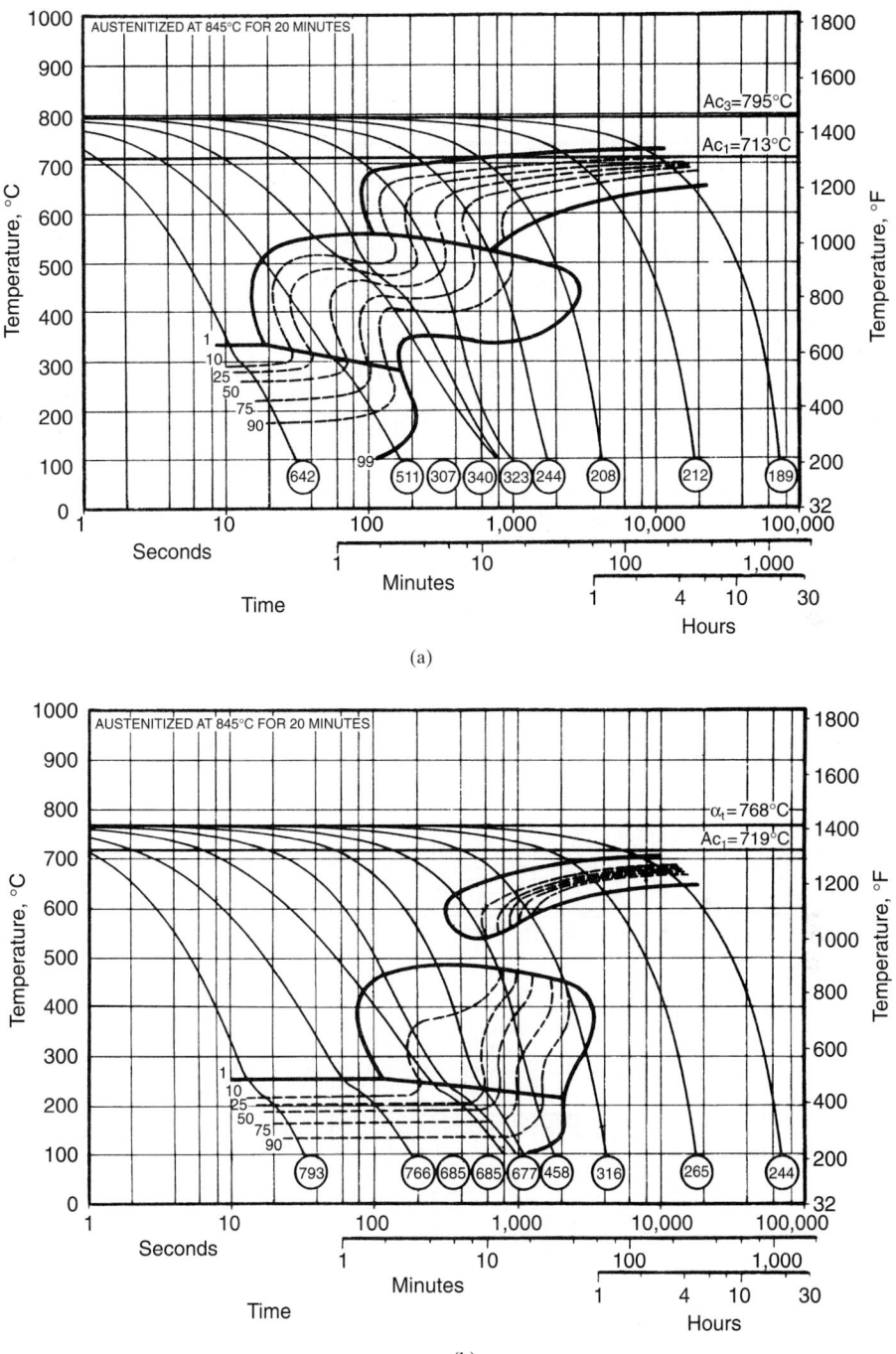

FIGURE 11.10 Experimentally determined CCT diagrams for (*a*) SAE 4140 steel (0.39% C, 0.26% Si, 0.82% Mn, 1.00% Cr, and 0.21% Mo) and (*b*) SAE 4150 steel (0.53% C, 0.34% Si, 0.83% Mn, 0.92% Cr, and 0.21% Mo). Both are austenitized at 845°C (1550°F) for 20 min.[22] (*Courtesy of Climax Molybdenum Co., Greenwich, Connecticut.*)

11.15

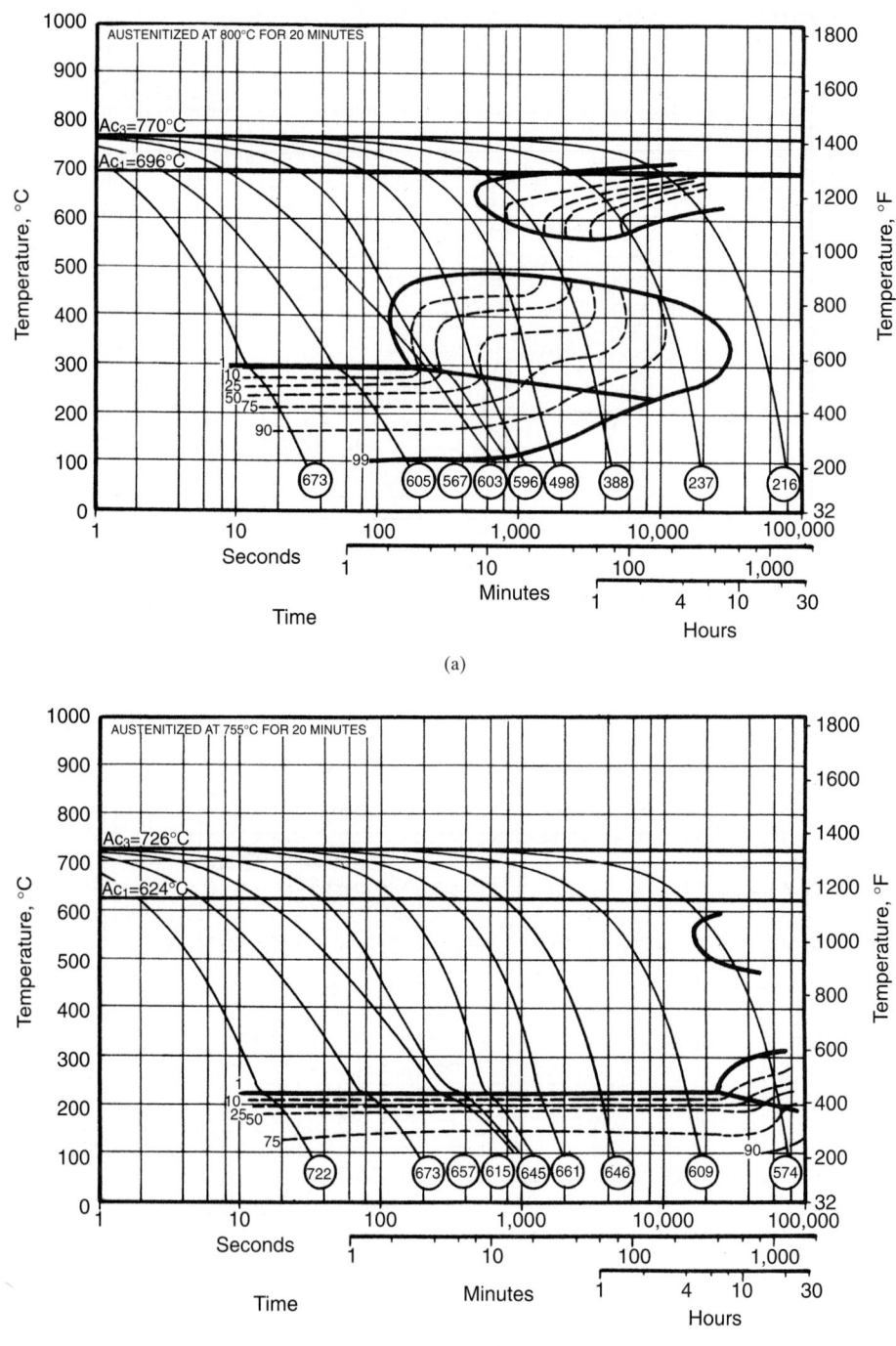

FIGURE 11.11 Experimentally determined CCT diagrams for 0.4% C, 0.4% Si, 0.72% Mn, 0.75% Cr, and 0.5% Mo steel with (a) 0.78% Ni and (b) 4.53% Ni. Both are austenitized at $Ac_3 + 30°C$ (54°F) temperature for 20 min.[22] (*Courtesy of Climax Molybdenum Co., Greenwich, Connecticut.*)

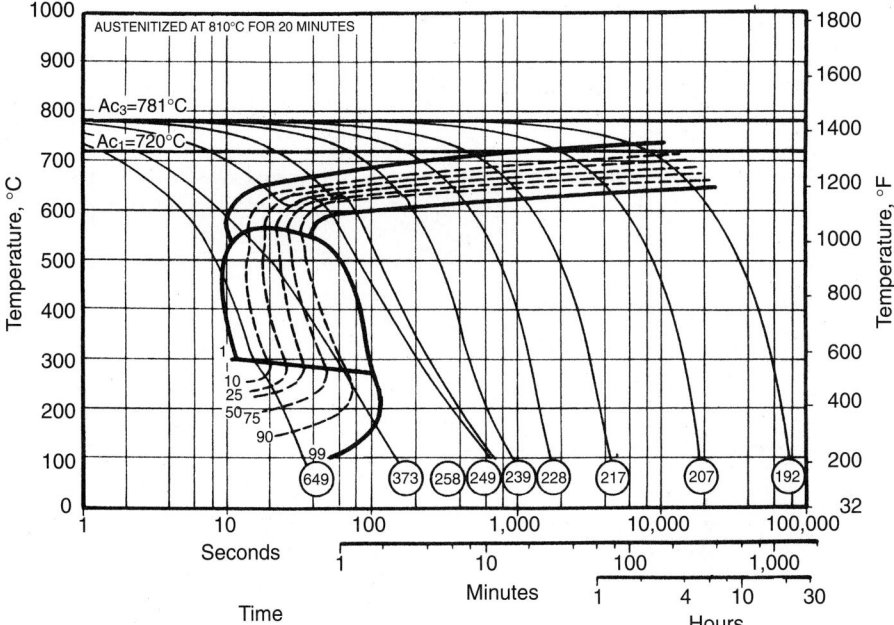

FIGURE 11.12 Experimentally determined CCT diagram of 0.4% C, 0.4% Si, 0.74% Mn, 0.75% Cr, 0.78% Ni, and 0.03% Mo steel, austenitized at $Ac_3 + 30°C$ (54°F) temperature for 20 min.[22] (*Courtesy of Climax Molybdenum Co., Greenwich, Connecticut.*)

decrease when compared to those of plain carbon, or of chromium or nickel steels. All these features produce much greater hardenability in the Ni-Cr-Mo alloy steels.

Note that the austenite-ferrite and the austenite-pearlite regions are not distinguishable in these five CCT diagrams (Figs. 11.10 to 11.12); however, the ferrite microstructure has definitely been observed for these continuously cooled specimens.

11.3.3 Modified CCT Diagrams

Modified continuous cooling transformation diagrams with individual bar diameter on the abscissa instead of transformation times have recently been adopted by the British Steel Corporation.[26] These diagrams provide an estimate of the microstructure comprising major phases that will form at the centers of bars of the given diameters for a wide variety of engineering steels in air-cooled, oil-quenched, and water-quenched conditions.[26] Moreover, the microstructure at positions different from the center can be inferred. These diagrams show the hardness of the microstructure so produced. Sometimes hardness values after tempering are also indicated.

The extent of hardenability of the steel can be known at a glance from these diagrams. For example, low-hardenability steels will exhibit ferrite, pearlite, or bainite as the early transformation products in the upper left side of the diagram. High-hardenability steels, on the other hand, will show predominantly martensite as the

TRANSFORMATION TEMPERATURE, °C

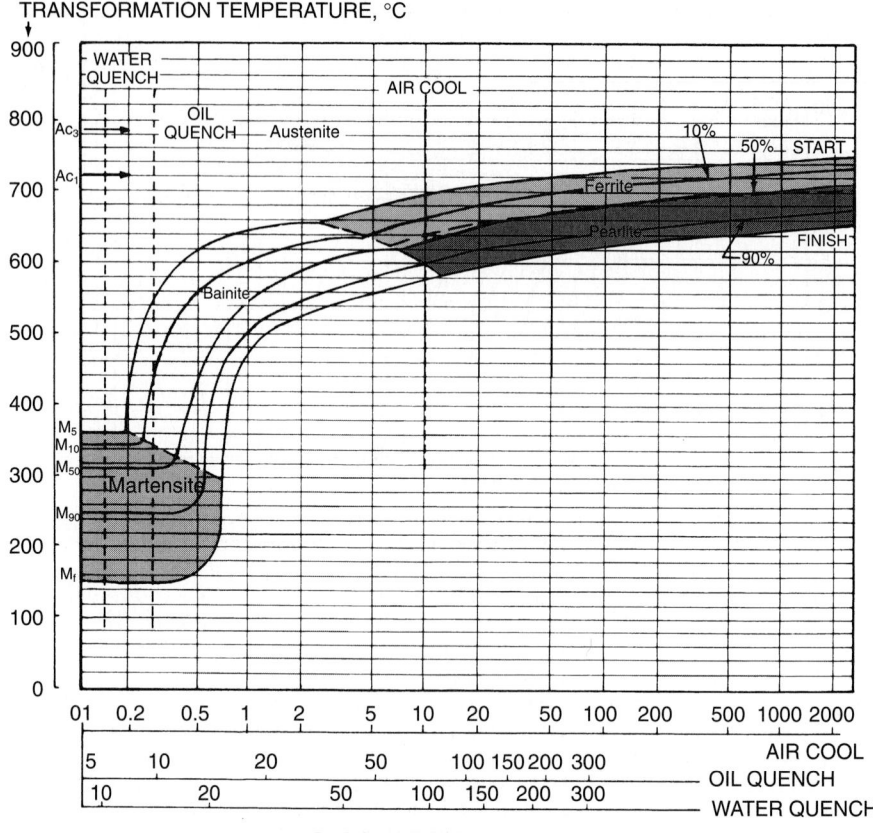

FIGURE 11.13 Modified CCT diagram plotted as a function of bar diameter for AISI 1035 to 1040 steel illustrating transformation behavior under different cooling rates.[26] (*Reprinted by permission of ASM International, Materials Park, Ohio.*)

transformation product in the lower right side of the diagram over an extended range of bar diameter and quenching values.

Figure 11.13 shows the CCT diagram for a 0.38% C steel (1035 to 1040 grade). It is apparent that ferrite and pearlite are produced above 660°C (1220°F), bainite is formed between 660°C and the M_s temperature, and fully martensitic structure will be obtained below the M_s temperature. Also, the resulting structures are changed from ferrite and pearlite through bainite to martensite with the decrease in bar diameter. In the case of air cooling, it is seen that increasing amounts of ferrite and pearlite are formed with progressively decreasing amounts of bainite in bars above 2 mm; bainite is formed at diameters up to 2 mm, whereas martensite is formed in diameters smaller than 0.18 mm. A vertical dashed line associated with a 10-mm-diameter air-cooled bar shows that continuous cooling from the austenitizing temperature produces a start of the ferrite transformation at 700°C (1292°F), 50% ferrite and a start of pearlite transformation at 640°C (1184°F), and a small amount of bainite at 580°C (1076°F) prior to the completion of transformation.

Similarly, the vertical dashed line corresponding to the 10-mm-diameter oil-quenched bar indicates that, in this case, the first bainite phase appears from austenite at 580°C, about 40% transformation at 330°C (626°F), and the remainder transformation to martensite until the completion of transformation occurs at 150°C (302°F). The vertical dashed line for a 10-mm-diameter water-quenched bar illustrates that martensite starts at 360°C (680°F) and finishes at 150°C.

Closer examination of the left portion of the curve reveals that martensite will form on water-quenching with bars up to 13-mm diameter, on oil-quenching up to 8-mm diameter, and on air-cooling up to 0.18-mm diameter.

Figure 11.14 illustrates the CCT diagram for a steel containing 0.15% C, 0.25% Si, 0.4% Mn, 1.15% Cr, 0.2% Mo, and 4.1% Ni. This figure shows that a fully martensitic structure could be developed in a 10-mm-diameter air-cooled bar and in a 100-mm-diameter oil-quenched bar. Also shown in the bottom of the figure is the hardness after transformation and the hardenability band.

11.3.3.1   Effect of Variation in Composition and Bar Diameter.   These CCT diagrams usually refer to an average composition within a particular steel grade. Variations in chemical composition within a specification range can sometimes result in a marked variation in microstructure and accompanying mechanical properties. Moreover, critical ranges of bar diameter are usually observed where slightly faster or slower cooling rates produce a drastic change in the predominant microstructure. This can be seen in Fig. 11.13, for example, where a very small increase in bar diameter would lead to the structural change from martensite commencing at 360°C to bainite starting at, say, 580°C.

11.3.3.2   Application.   From the correlation between Jominy hardenability and bar diameter produced by the American Society for Automotive Engineers (SAE), it is possible to determine the position of Jominy end-quench test results (in terms of hardness versus equivalent bar diameter) in relation to the center position of an oil- and water-quenched bar accompanying each CCT diagram. The additional applications of this modified CCT diagram include (1) the study of continuous heat treatment cycles for thin sections such as wire or strip; (2) the design of safe heat treatment cycles for sections that require retarded cooling after hot-working in order to avoid low-temperature reaction products; (3) prediction of machinability from expected microstructures and hardness values; (4) investigation of mass effect; and (5) assistance in the calculation of critical quenching rates for complex-shaped and varying sections.

Note that the microstructures indicated by the diagram correspond to the center of the bar diameter. In practice, the surface of the quenched bar may contain smaller amounts of ferrite and pearlite than its center, due to more rapid cooling.

11.3.3.3   Limitations.   Accuracy of the experimental determination of CCT diagrams may be affected by the following: (1) prior heat treatment and undissolved carbides and (2) noninclusion of the effect of agitation in the air-cooling, oil-, and water-quenching media.

11.4   CONCLUSION

There are many atlases of isothermal and continuous cooling transformation diagrams available for numerous steels[21-24,26-29] which have provided substantial

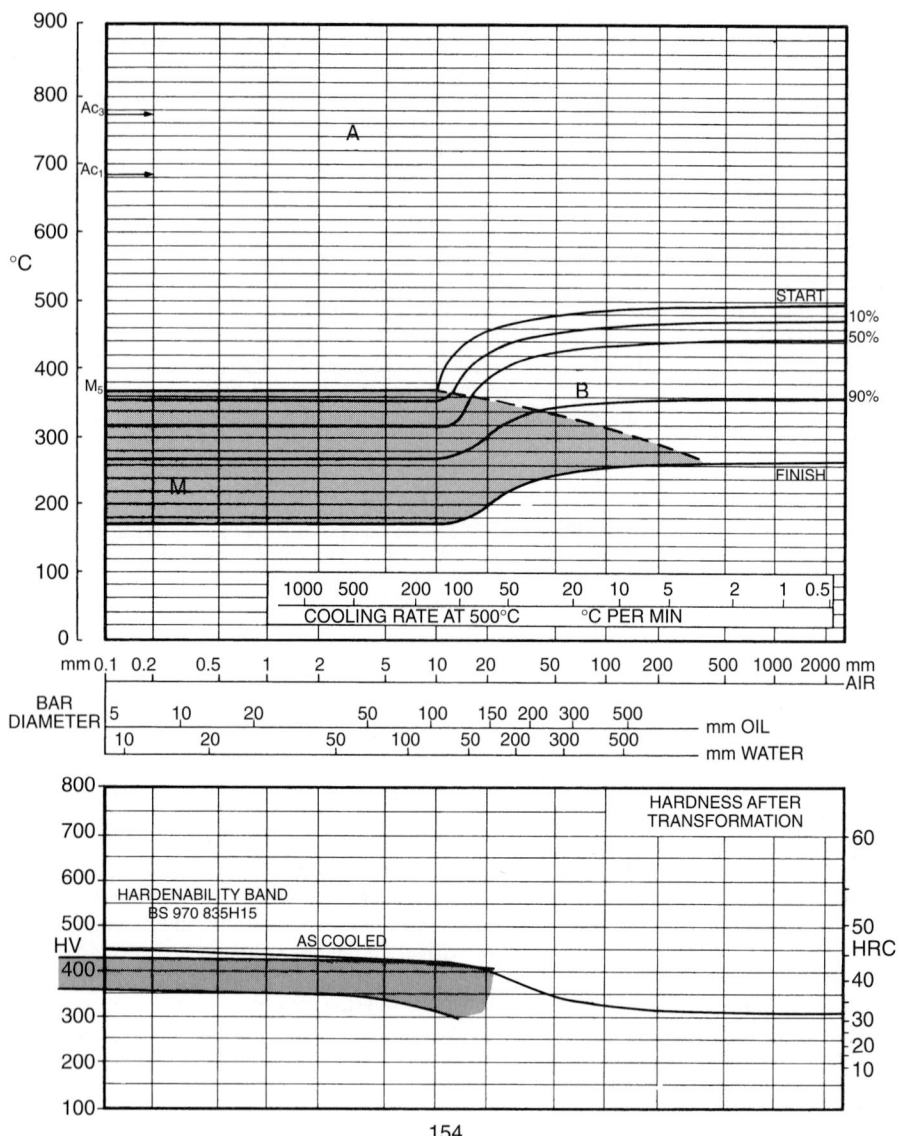

154

FIGURE 11.14 Modified CCT diagram plotted as a function of bar diameter for 4 Ni-Cr-Mo steel (analysis wt%: 0.15% C, 0.25% Si, 0.40% Mn, 1.15% Cr, 0.20% Mo, 4.10% Ni, 0.020% P, and 0.018% S) under different cooling rates. Also shown at the bottom is the hardness after transformation curve austenitized at 820°C (1508°F). Previous treatment: blank carburized at 900°C for 4 hr followed by air cooling.[26] (*Reprinted by permission of ASM International, Materials Park, Ohio.*)

contribution to our understanding of the influence of alloying elements on transformation kinetics. There is a widespread use of CCT diagrams because the majority of heat treatment is carried out by cooling continuously from the austenitizing temperature, and accurate, reliable, and experimentally determined CCT diagrams for all engineering steels are now accessible.

REFERENCES

1. E. S. Davenport and E. C. Bain, *Trans. AIME*, vol. 90, 1930, p. 117.
2. M. Cohen, *Trans. ASM*, vol. 28, 1940, p. 537.
3. *Isothermal Transformation Diagrams of Austenite in a Wide Variety of Steels*, United States Steel Corporation, Pittsburgh, Pa., 1963.
4. N. F. Kennon, *Metall. Trans. A*, vol. 9, 1978, pp. 57–66.
5. P. W. Brown and D. J. Mack, *Metall. Trans. A.*, vol. 4, 1973, pp. 2850–2851.
6. B. N. P. Babu, M. S. Bhat, E. R. Parker, and V. F. Zackay, *Metall. Trans. A.*, vol. 7, 1976, pp. 17–22.
7. C. E. Erickson, M. S. Bhat, E. R. Parker, and V. F. Zackay, *Met. Trans. A.*, vol. 7, 1976, pp. 1800–1803.
8. Y. J. Park and F. B. Fletcher, *J. Heat Treating*, June 1986, pp. 247–252.
9. H. Modin and S. Modin, *Jernkontorets, Ann.*, vol. 139, 1955, p. 481.
10. M. Hillert, in *Decomposition of Austenite by Diffusional Processes*, eds. V. F. Zackay and H. I. Aaronson, Interscience, New York, 1962, p. 197.
11. N. F. Kennon and N. A. Kaye, *Metall. Trans. A*, vol. 13, 1982, pp. 975–978.
12. A. G. Haynes, *Heat Treatment of Metals*, Iron and Steel Institute, London, 1966, pp. 13–23.
13. R. A. Grange, V. E. Lambert, and J. J. Harrington, *Trans. ASM*, vol. 51, 1959, p. 377.
14. *Atlas of Isothermal Transformations and Cooling Transformation Diagrams*, American Society for Metals, Metals Park, Ohio, 1977.
15. K. R. Kinsman and H. I. Aaronson, *Metall. Trans.*, vol. 4A, 1973, pp. 959–967.
16. D. S. Clark and W. R. Varney, *Physical Metallurgy for Engineers*, 2d edition, Litton Educational Publishing, New York, 1968.
17. C. A. Siebert, D. V. Doanne, and D. H. Breen, *The Hardenability of Steels*, ASM, Metals Park, Ohio, 1977.
18. R. Kumar, *Physical Metallurgy of Iron and Steel*, Asia Publishing House, New York, 1968.
19. D. A. Porter and K. E. Easterling, *Phase Transformations in Metals and Alloys*, Van Nostrand Reinhold, London, 1981.
20. A. K. Cavanagh, *Metall. Trans.*, vol. 10A, 1979, pp. 129–131.
21. W. W. Cias, *Phase Transformation Kinetics and Hardenability of Medium-Carbon Alloy Steel*, Climax Molybdenum Co., Greenwich, Conn., 1972.
22. W. W. Cias, *Austenite-Transformation Kinetics of Ferrous Alloy*, Climax Molybdenum Co., Greenwich, Conn., 1979.
23. M. Melander and J. Nicolov, *J. Heat Treating*, vol. 4, no. 1, 1985, pp. 32–38.
24. *Atlas for the Heat Treatment of Steel*, vols. 1–4, Vereins Deutscher Eisenhütten Leute in conjunction with the Max Planck Institute Verlag Stahleisen M.B.H., Dusseldörf, 1954–1976.
25. G. T. Eldis, in *A Hardenability Concept with Applications to Steel*, AIME, Warrendale, Pa., 1978, pp. 126–153.

26. M. Atkins, *Atlas of Continuous Cooling Transformation Diagram for Engineering Steels*, British Steel Corporation and American Society for Metals, Metals Parks, Ohio, 1980.

27. *Transformation Diagrams of Steels Made in France*, vols. 1–4, I.R.S.I.D., St. Germaine-en-Laye, 1953–1960.

28. M. Economopoulos, L. Harbraken, and N. Lambert, *Transformation Diagrams of Steels Made in the Benelux Countries*, C.N.R.M., Brussels, 1967.

29. A. A. Popov and L. E. Popova, *Isothermal and Thermokinetic Diagrams of the Breakdown of Supercooled Austenite*, Metallurgiya, Moscow, 1965.

CHAPTER 12
BASIC HEAT TREATMENT

12.1 INTRODUCTION

Heat treatment is the term that describes an operation, or a combination of operations, which involves the controlled heating and/or cooling of a metal or alloy in the solid state in order to modify the existing structure deliberately and/or bring about a change in its properties. However, the International Federation for the Heat Treatment of Materials (IFHT) has defined this term as "a process in which the entire object, or a portion thereof, is intentionally submitted to thermal cycles and, if required, to chemical or additional physical actions in order to achieve desired (change in the) structures and properties." *Thermal cycle* implies the change of temperature with time during a heat treatment process.[1] These changes may be essential to improve processings (e.g., machinability) or to meet proper service conditions [e.g., increased (or controlled) surface and core hardnesses, wear, or fatigue resistance; dimensional accuracy; desired microstructure and residual stress profile; and improvements of tensile and yield strength, toughness, ductility, impact strength, weld integrity, and magnetic and electrical properties].

In this chapter the importance, selection, and classification of heat treatment processes, in general, are presented. The basic heat treatments, such as annealing and normalizing, applied to steel are described in terms of Fe-Fe$_3$C and other transformation diagrams to produce a wide range of combinations of microstructures and mechanical properties, including uniform, soft, and refined microstructures; improved ductility, hardness, and strength; elimination of residual stresses; and improved machinability. Decarburization and secondary graphitization of steels are discussed. Furthermore, some basic principles and applications of annealing, stress relieving, and normalizing heat treatments to four basic types of iron casting (namely, gray, ductile, compacted graphite, and malleable irons) have been discussed. Finally the important engineering properties and applications of these processed irons are elaborated upon, and the structure-property relationship in gray iron is outlined.

12.2 IMPORTANCE, SELECTION, AND CLASSIFICATION OF HEAT TREATMENT

The importance of heat treatment in modern technology can be estimated by the fact that almost all metal engineering components, except for some cast irons, are subjected to at least one heat treatment process, usually as the final operation in, or close to, the final step of their production cycle. For example, large volumes of forgings, castings, pressings, and fabrications (forming and joining) are heat-treated prior to machining.[2]

By far, the greatest number of heat treatment processes carried out today in industry is done on ferrous metals; particularly, steels are subjected to a greater proportion of such treatments. The structure and accompanying mechanical properties of steels can be significantly altered by using a specific heat treatment, a fact well recognized by the early metalworkers of several-centuries-old civilizations. If the heat treatment operation is discarded or done incorrectly, serious consequences may result.[3]

Many factors have to be taken into consideration when a choice is made between different heat treatment processes that are accomplished to do a particular job on a particular component. Commonly, a process that will do one thing for a given part may not be at all suitable to do the same thing to a different part. Incorrect heat treatment will affect the service performance of the components more severely than any other interstage processes carried out during their manufacture. It is thus essential to pay equal attention to both correctly specified heat treatment and its inspection.[4,5] Moreover, proper selection of material for heat treatment is also essential, and one must be concerned with, in addition to the mechanical properties of the material, other factors as well, such as cost, availability, and fabricating characteristics.

There are various forms of heat-treating processes, each having a different term. Some of these processes are employed only for steels, whereas others are applied to certain metals and alloys. These processes can best be justified in terms of what they are expected to attain. The main objectives of these processes are

1. *To soften*—to improve plasticity (or slip band formation) by modifying the size, shape, and distribution of microconstituents.

2. *To stress-relieve*—to allow relaxation of residual stresses by raising the temperature in order to lower the yield strength and enhance recovery.

3. *To homogenize*—to obtain compositional uniformity in a phase by the diffusion of alloying elements at higher temperature, for example, austenitizing and solutionizing.

4. *To toughen*—to develop the ability to absorb energy or withstand occasional stresses in the plastic range without fracturing (to increase the total area under the stress-strain curve).

5. *To harden*—to provide slip interference by altering the size, shape, and distribution of microconstituents by grain refinement, quench-hardening, or age-hardening.

6. *To add chemical elements through the surface*—to improve wear and fatigue resistance caused by the development of surface compressive residual stresses arising from the absorption of interstitial solute atoms under a proper thermal cycle (carburizing, nitriding, etc.).

7. *To accomplish special treatments*—such as to develop magnetic properties by producing coarse-grained structure by high-temperature treatment, to remove chemical elements from the surface (by heat expulsion or chemical method at elevated temperature), or to improve electrical properties and cold, subzero, deep cryogenic treatment.[6]

TABLE 12.1 Some Heat Treatment Processes

1. Annealing processes:
   (*a*) full annealing; (*b*) homogenize annealing, solution treating, or austenitizing; (*c*) spheroidizing, critical range annealing, or subcritical annealing; (*d*) stress-relief annealing; (*e*) process annealing; (*f*) recrystallization annealing; (*g*) isothermal annealing; (*h*) dehydrogenation annealing, (*i*) blueing and bright annealing; (*j*) box annealing; (*k*) continuous annealing.

2. Normalizing.

3. Through hardening and tempering processes:
   (*a*) water-, oil-, or air-quenching and tempering; (*b*) time-quenching and tempering; (*c*) isothermal quenching and tempering (e.g., patenting); (*d*) austempering; (*e*) martempering.

4. Other through-hardening processes:
   (*a*) precipitation (age) hardening; (*b*) dispersion hardening; (*c*) maraging; (*d*) thermomechanical treatment; (*e*) order-disorder reactions.

5. Thermal surface hardening treatment (i.e., without compositional change):
   (*a*) flame hardening; (*b*) induction hardening; (*c*) laser hardening; (*d*) electron-beam hardening.

6. Thermochemical surface hardening treatment (i.e., with compositional change):
   (*a*) austenitic thermochemical treatment:
      (1) carburizing, solid, liquid, gas, vacuum, plasma, fluidized bed
      (2) carbonitriding, liquid, gas, plasma
      (3) cyaniding
   (*b*) ferritic thermochemical treatment
      (1) nitriding, liquid, gas, plasma
      (2) nitrocarburizing, liquid, gas

7. Other surface diffusion treatments:
   (*a*) siliconizing; (*b*) chromizing; (*c*) boronizing; (*d*) ion implantation; etc.

8. Miscellaneous heat treatments:
   (*a*) patenting; (*b*) stiffening temper; (*c*) malleabilizing; (*d*) cold, subzero, or deep cryogenic treatments; (*e*) restoring stainless characteristics; (*f*) developing magnetic characteristics; (*g*) improving electrical properties; (*h*) relieving embrittlement.

    The main classifications of heat-treating processes and their subdivisions are listed in Table 12.1.

12.3 ANNEALING OF STEELS

12.3.1 Annealing

Annealing treatment, being a very general term, has been defined as any heating operation above a certain temperature, holding at this temperature for a given length of time, followed by cooling at a predetermined rate (usually in the furnace) to room temperature in order to achieve or restore the desirable properties such as maximum ductility, low strength, and better machinability and cold-workability. This definition indicates the aim of annealing treatment. However, the purpose of annealing may be broadened, and the original definition of annealing should include other effects as well. Thus annealing process accomplishes one or more of the following:[7-9] (1) It induces softness and reduces hardness; (2) it facilitates machinability as well as the progress of subsequent manufacturing operations; (3) it produces a desired microstructure in order to improve or effect required mechanical, physical, electrical, and magnetic properties; (4) it eliminates any internal or residual stresses; (5) it refines the grain and produces grain orientation; (6) it homogenizes the microstructure; (7) it promotes dimensional stability; and (8) it removes gases from the steel. As a result, the following terms have been used for annealing: *full annealing, soaking, isothermal annealing, spheroidizing, dehydrogenation annealing, stress relief annealing, blueing, box annealing, soft annealing, finish or bright annealing,* and so forth.

In plain carbon steels, annealing usually produces a ferrite-pearlite microstructure. The exact annealing temperatures of steel parts are governed by their applications as well as their carbon and alloy content. For example, hypoeutectoid and hypereutectoid plain carbon steel parts are annealed to just above the upper critical temperature Ac_3 and the lower critical temperature Ac_1, respectively. Most high-carbon tool steels are annealed between 760 and 780°C.

12.3.1.1 Full or Supercritical Annealing.

The term *annealing* usually implies full annealing, which involves solution treatment in the austenite phase field, followed by cooling in the furnace. In *full annealing*, the steel is heated to about 40°C either above the Ac_3 temperature for hypoeutectoid steel or above the Ac_1 temperature for hypereutectoid steel and is held at this temperature for the desired length of time (for 1 hr/in. of maximum thickness) to dissolve all the carbides present in steel and to form a homogeneous austenite, followed by slow cooling in the shutoff furnace. The slow cooling causes the temperature of the hypoeutectoid carbon steel to remain at the temperature between A_3 and well above the pearlite nose temperature of 540°C (1000°F) for an extended period of time. As a result, the transformation product produced consists of equiaxed and relatively coarse-grained proeutectoid ferrite and pearlite with a coarse interlamellar spacing. Figure 12.1

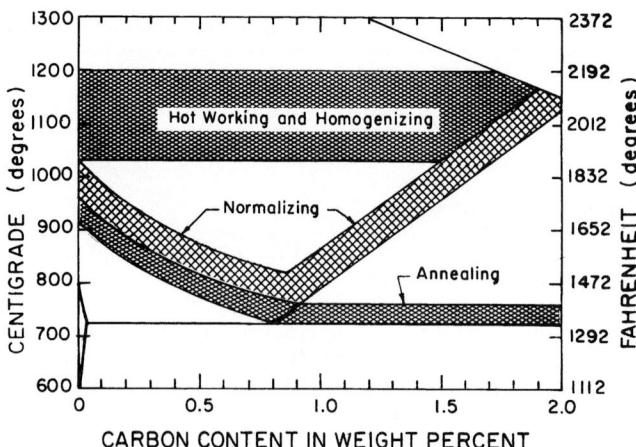

FIGURE 12.1 A portion of Fe-Fe$_3$C diagram representing the temperature ranges used for annealing, normalizing, hot working, and homogenizing treatments.[13]

shows a portion of the Fe-Fe$_3$C diagram with temperature ranges which are commonly used for some heat-treating processes, including full annealing.

The optimum cooling rate to obtain these structures can be predicted from continuous cooling transformation (CCT) diagrams.[8] Full annealing treatment can also be represented by the isothermal transformation diagram. Both procedures lead to similar hardness; however, isothermal transformation takes considerably less time. Figure 12.2 shows the use of isothermal transformation diagram in full annealing.

The purpose of full annealing includes one or more of the following: (1) to produce softening, (2) to remove the stress, (3) to change the mechanical properties, (4) to induce the desired and homogenized microstructure, and (5) to produce grain orientation or improve the grain structure.

If the hypereutectoid steel is heated above A_{cm} (i.e., in the single γ phase field), a brittle proeutectoid cementite phase will precipitate at the austenite grain boundaries on slow cooling, eventually producing a brittle carbide network in the structure which will render an easy fracture path. Consequently, the properties of this hypereutectoid steel in the slow-cooled condition become very poor.

Temperatures and associated Brinell hardnesses for simple annealing of carbon steels and alloy steels are given in Tables 12.2 and 12.3, respectively. Heating cycles employed in the upper austenitizing temperature ranges (for a longer time) given in Tables 12.2 and 12.3 generally form pearlitic structures. When lower temperatures are used for a prolonged time, the resulting structures are predominantly spheroidized.[9]

12.3.1.2 Homogenizing, Soaking, or Homogenize-Annealing. This is often carried out on ferrous and nonferrous ingot castings to remove or even out the inhomogeneous structure produced by segregation which occurs during solidification. It is accomplished by reheating the ingot at a high temperature, just below the solidus line (for example, ~1100°C for steel) for a longer time to increase the diffusion of solute atoms and to eliminate segregation. Sufficient care should be taken in this treatment not to cross the solidus line; otherwise, liquation of the grain boundaries will occur, adversely affecting the shape and properties of the casting. Figure 12.1

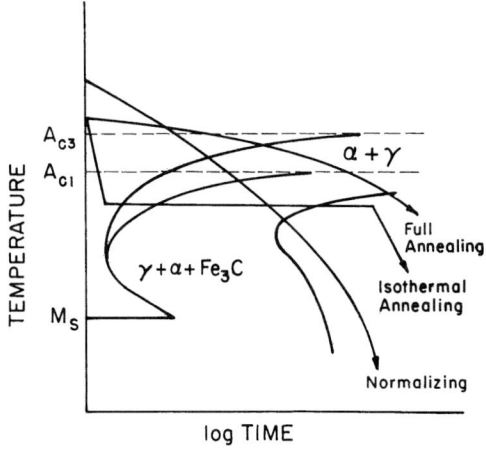

FIGURE 12.2 Schematic cycle for normalizing, full annealing, and isothermal annealing.

TABLE 12.2 Recommended Temperatures and Cooling Cycles for Full Annealing of Small Carbon Steel Forgings.[9]

Data are for forgings up to 75 mm (3 in.) in section thickness. Time at temperature usually is a minimum of 1 hr for sections up to 25 mm (1 in.) thick; $\frac{1}{2}$ hr is added for each additional 25 mm (1 in.) of thickness.

	Annealing temperature		Cooling cycle[†]				Hardness range, HB
			°C		°F		
Steel	°C	°F	From	To	From	To	
1018	855–900	1575–1650	855	705	1575	1300	111–149
1020	855–900	1575–1650	855	700	1575	1290	111–149
1022	855–900	1575–1650	855	700	1575	1290	111–149
1025	855–900	1575–1650	855	700	1575	1290	111–187
1030	845–885	1550–1625	845	650	1550	1200	126–197
1035	845–885	1550–1625	845	650	1550	1200	137–207
1040	790–870	1450–1600	790	650	1450	1200	137–207
1045	790–870	1450–1600	790	650	1450	1200	156–217
1050	790–870	1450–1600	790	650	1450	1200	156–217
1060	790–845	1450–1550	790	650	1450	1200	156–217
1070	790–845	1450–1550	790	650	1450	1200	167–229
1080	790–845	1450–1550	790	650	1450	1200	167–229
1090	790–830	1450–1525	790	650	1450	1200	167–229
1095	790–830	1450–1525	790	655	1450	1215	167–229

[†] Furnace cooling at 28°C/hr (50°F/hr).
Reprinted by permission of ASM International, Materials Park, Ohio.

illustrates the temperature range for homogenize-annealing of plain carbon steel. This treatment is applied prior to hot working operations (e.g., hot rolling or forging of steels). The best method of homogenization is to use a slow heating rate from room temperature to the homogenization temperature or step heating and the selection and control of the heating temperature based on the phase diagram. The final microstructure after soaking will usually exhibit higher strength than those without soaking due to homogeneous solution of the alloying elements and the precipitates such as carbides and nitrides.[8]

The advantages of the Hertwich continuous homogenizing process with step cooling/heating for aluminum extrusion billets or logs over conventional batch process are (1) superior billet quality by uniform homogenization of logs, leading to dependable high-level extrudability; (2) straightened logs emerging from the furnace; (3) lower energy consumption for both electric and gas furnaces than that for batch-type homogenizing: (4) very little manpower required, after integrating into the cast house production system; (5) low maintenance costs due to the reduced stress to the furnace operating at constant temperature in contrast to the batch furnaces involving rapid cooling and immediate reheating thereby concurrent heat expansions and thermal stress; (6) high-performance dependability to adapt to the particular production requirements of the billet production process due to its great flexibility; and (7) operation of the system with practically no downtime.[9a]

12.3.1.3 Spheroidize-Annealing or Spheroidizing. This treatment is usually selected to improve cold formability due to lower flow stress of the materials. Steels

TABLE 12.3 Recommended Annealing Temperatures for Alloy Steels (Furnace Cooling)[9]

AISI/SAE steel	Annealing temperature °C	°F	Hardness (max), HB	AISI/SAE steel	Annealing temperature °C	°F	Hardness (max), HB
1330	845–900	1550–1650	179	5140	815–870	1500–1600	187
1335	845–900	1550–1650	187	5145	815–870	1500–1600	197
1340	845–900	1550–1650	192	5147	815–870	1500–1600	197
1345	845–900	1550–1650	...	5150	815–870	1500–1600	201
3140	815–870	1500–1600	187	5155	815–870	1500–1600	217
4037	815–855	1500–1575	183	5160	815–870	1500–1600	223
4042	815–855	1500–1575	192	51B60	815–870	1500–1600	223
4047	790–845	1450–1550	201	50100	730–790	1350–1450	197
4063	790–845	1450–1550	223	51100	730–790	1350–1450	197
4130	790–845	1450–1550	174	52100	730–790	1350–1450	207
4135	790–845	1450–1550	...	6150	845–900	1550–1650	201
4137	790–845	1450–1550	192	81B45	845–900	1550–1650	192
4140	790–845	1450–1550	197	8627	815–870	1500–1600	174
4145	790–845	1450–1550	207	8630	790–845	1450–1550	179
4147	790–845	1450–1550	...	8637	815–870	1500–1600	192
4150	790–845	1450–1550	212	8640	815–870	1500–1600	197
4161	790–845	1450–1550	...	8642	815–870	1500–1600	201
4337	790–845	1450–1550	...	8645	815–870	1500–1600	207
4340	790–845	1450–1550	223	86B45	815–870	1500–1600	207
50B40	815–870	1500–1600	187	8650	815–870	1500–1600	212
50B44	815–870	1500–1600	197	8655	815–870	1500–1600	223
5046	815–870	1500–1600	192	8660	815–870	1500–1600	229
50B46	815–870	1500–1600	192	8740	815–870	1500–1600	202
50B50	815–870	1500–1600	201	8742	815–870	1500–1600	...
50B60	815–870	1500–1600	217	9260	815–870	1500–1600	229
5130	790–845	1450–1550	170	94B30	790–845	1450–1550	174
5132	790–845	1450–1550	170	94B40	790–845	1450–1550	192
5135	815–870	1500–1600	174	9840	790–845	1450–1550	207

Reprinted by permission of ASM International, Materials Park, Ohio.

may be spheroidized by any of the following methods,[8,9] as shown schematically in Fig. 12.3.

1. The simplest spheroidizing is achieved by using a prolonged subcritical annealing, that is, prolonged holding of steels at a temperature just below A_1. At this temperature, larger and globular cementite particles form, in time, at the expense of smaller ones. This involves slow action and normally requires from 20 to 40 hr of effective spheroidization. The actual temperature at which the transformation occurs has an important effect on the appearance of the resulting structure. The closer the transformation temperature lies to A_1, the coarser and softer the spheroidized structure. If the transformation takes place at a temperature further below A_1, the finer, more lamellar, and harder constituents will be obtained.

2. A second method used, a modified one, consists of heating and cooling alternately between temperatures just above Ac_1 and just below Ar_1. The number of cycles needed for effective spheroidization may be up to four.[8]

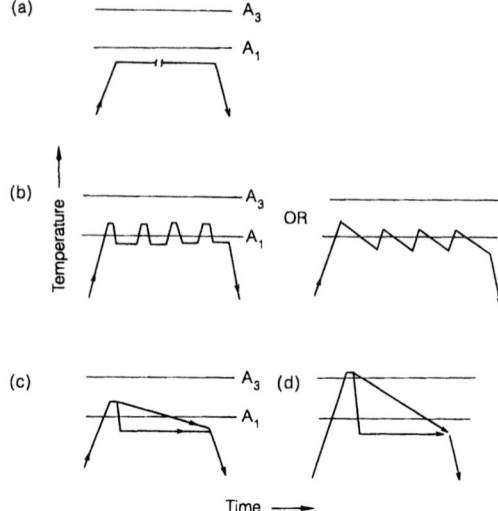

FIGURE 12.3 Schematic diagrams of spheroidize-annealing.

3. A more frequently used method of spheroidizing consists of complete or partial austenitizing (heating to a temperature above Ac_3 or Ac_1) and then either cooling very slowly in the furnace or holding the work at a temperature just below the Ar_1 point. During cooling, the carbon dissolved in the austenite will separate out as carbide spheroids. The higher the annealing temperature (over a few tens of degrees above Ac_1), the greater the amount of carbide dissolved, and the cementite will precipitate as lamellae.

4. The fourth procedure consists of cooling at a suitable rate from the minimum temperature at which all the carbides dissolve, such that the reformation of a carbide network is prevented; and subsequent reheating is done in accordance with the first or second method. This method is used for hypereutectoid steel containing a carbide network. Figure 12.4 shows the temperature range used for spheroidizing. It may be pointed out that if the steel is held too long at a spheroidizing temperature, the cementite particles become coalesced and elongated, thereby reducing machinability. When more inhomogeneous austenite (i.e., containing more undissolved carbides) is present or lower austenitizing temperature is used, the occurrence of the spheroidized structure is more complete. In contrast, fully homogenized austenite or higher austenitizing temperature produces a more lamellar structure (Fig. 12.5).[7,9]

The spheroidized microstructures are the most stable microstructures of the steel and are desirable when (1) maximum ductility and minimum hardness on low- and medium-carbon steels or (2) maximum machinability as in high-carbon steels and tool steels is required.

A large percentage of steels, including high-carbon, high-alloy tool steels and spring steels, are generally supplied by the manufacturer in the spheroidized condition to produce finely distributed spheroidized cementite particles in a ferrite matrix. This serves as the standard practice in the steel mills. It facilitates good machining associated with a reduction in labor cost (i.e., finish grinding and cutting

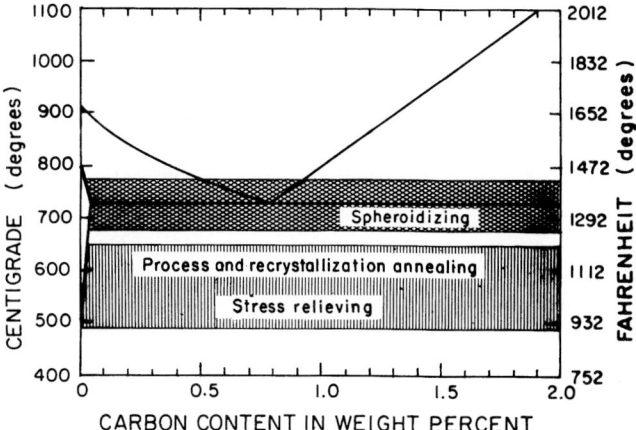

FIGURE 12.4 A portion of Fe-Fe₃C diagram with the temperature ranges used for spheroidizing, process, stress relief, and recrystallization annealing of plain carbon steels.

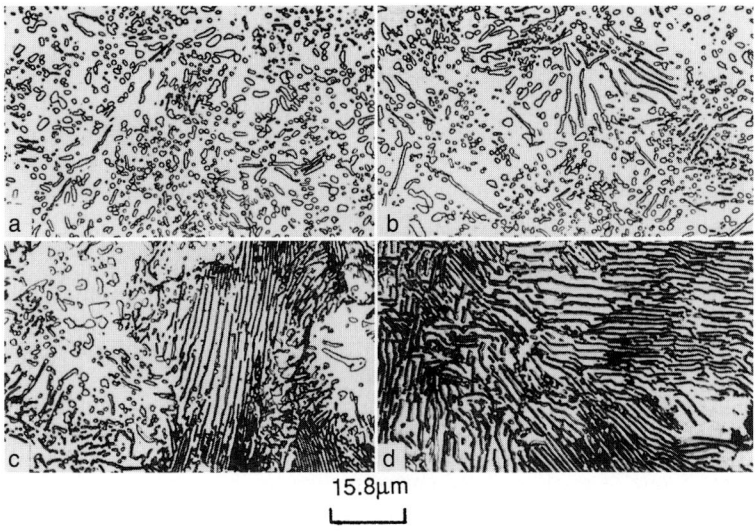

15.8μm

FIGURE 12.5 Effect of austenitizing temperature on the transformation product in 0.73% carbon steel. All samples were transformed at the same temperature 721°C (1330°F) after austenitizing at different temperatures: (*a*) 749°C (1380°F); (*b*) 760°C (1400°F); (*c*) 788°C (1450°F); and (*d*) 871°C (1600°F). 3% Picral.[7] (*Reprinted by permission of ASM International, Materials Park, Ohio.*)

cost) and an increase in tool life, and it ensures a better response to subsequent hardening operation due to a relatively lesser degree of dangerous hardening cracks and distortion. Moreover, uniformity in properties of hardened material is obtained. It should, however, be emphasized that spheroidized structure should not be employed for applications that will be flame- or induction-hardened on the surface unless the components are first normalized or quenched and tempered. The reason is that the globules of cementite do not readily go into the solution in the short heating time involved in these operations.[10]

For plain carbon steels the temperature range for spheroidization may be used from the Fe-Fe₃C diagram, as shown in Fig. 12.4. For alloyed steels, the spheroidization annealing temperature may be calculated according to the empirical formula:

$$T(^\circ C) = 705 + 20(\%\text{Si} - \%\text{Mn} + \%\text{Cr} - \%\text{Mo} - \%\text{Ni} + \%\text{W}) + 100\%\text{V} \quad (12.1)$$

This formula holds well up to the following values of the alloying elements: 0.9% C, 1.8% Si, 1.1% Mn, 1.8% Cr, 0.5% Mo, 5% Ni, 0.5% W, and 0.25% V. If the steel contained higher alloying contents, only those indicated maximum values are to be taken into consideration.[11]

Spheroidization of steels of much lower carbon content is seldom practiced for better machinability because in this condition they are soft and "gummy." However, this treatment is carried out for a number of cold-working operations.

The rate of spheroidizing reaction (1) increases with a prior cold work in a subcritical spheroidizing treatment and (2) decreases greatly when carbon-forming elements such as Cr, Mo, and V are present in plain carbon steels. A preliminary normalizing treatment is often employed to condition the steel for rapid spheroidization. However, this rapidity increases when the prior structure is bainite and attains a maximum when the prior structure is martensite. However, the carbide size is small. To obtain large globular carbides, the prior microstructure should be pearlite.

Carbon and the alloy content of steel greatly affect the hardness after spheroidization. An increase of carbon or alloy content or both gives rise to an increase in as-spheroidized hardness, which usually ranges from 163 to 212 HB. During prolonged spheroidizing treatment at high temperatures in the ferrite region, decarburization usually occurs, especially in high Si- and Mo-bearing steels. To prevent this decarburization, controlled atmosphere is regulated. Note that prevention of decarburization is more easily accomplished in the ferrite region than in the austenite region.[8]

The mechanism of pearlite spheroidization by heating the pearlitic structure to a temperature just below A_1 and holding for a prolonged period involves (1) breaking up the cementite lamellae into smaller particles, (2) formation and subsequent dissolution of smaller spherical shaped particles, and (3) growth of larger spherical particles due to the reduction in interfacial energy. The growth (or coarsening) rate of a spheroidized microstructure has been given by the following expression:[12,13]

$$\frac{dr}{dt} = \frac{2\gamma V_{cm}^2 C_c^\alpha D_c^\alpha}{V_\alpha RT r_1}\left(\frac{1}{\bar{r}} - \frac{1}{r_1}\right) \quad (12.2)$$

where γ is the interfacial energy, V_{cm} and V_α are the molar volumes of cementite and ferrite, respectively, C_c^α is the mole fraction of carbon in equilibrium with cementite in ferrite, D_c^α is the effective carbon diffusivity in ferrite, R is the gas constant, T is the absolute temperature, $\bar{r}$ is the average size of the already spheroidized particles, and r_1 is the radius of the newly formed particles. This equation indicates that the rate of spheroidization is directly proportional to the diffusivity of carbon in ferrite and decreases with the increase in mean size of particles in a spheroidized microstructure. When alloying elements are present, they reduce D_c^α to a considerable extent and hence reduce the rate of the spheroidization process.

12.3.1.4 Stress-Relief Annealing.
When steel is subjected to a cold-working, heat-treating, welding, or heavy machining process, residual stress is introduced into these processed parts. To reduce or eliminate these residual stresses to some safe value and to avoid fatigue or brittle fracture without intentionally changing its microstructure and mechanical properties, the usual practice is to provide stress-relief annealing for 1 to 2 hr in the plain carbon or low-alloy steels at a temperature of 550 to 650°C (1020 to 1200°F) (Fig. 12.4); in low-alloy ferritic steels at 595 to 675°C (1100 to 1250°F); in the hot-worked and high-speed steels at 600 to 750°C (1110 to 1380°F); in high-alloy steels at 900 to 1065°C (1650 to 1950°F); and in high-alloy austenitic steels at 480 to 925°C (900 to 1700°F). These temperatures are well above the recrystallization temperature of the respective steels, but below the lower transformation temperature Ac_1. As a result of this treatment, the yield strength of the materials is sometimes reduced. However, note that cold-extruded low-carbon steel (extrusion ratio 4:1) is completely stress-relieved at 500°C in 1 hr, with little change in hardness from that of the as-extruded rod.[14]

Stress-relief treatment is applied by the purchaser of semiprocessed electrical steel and (cold-rolled) magnetic lamination steel to optimize its magnetic properties.

Postweld heat treatment used after welding can fulfill many objectives such as stress relief of weldments, dimensional stability, stress corrosion resistance, and sometimes even elimination of fabrication cracking and improvements in toughness and mechanical properties.[14a]

For best results, it is customary to allow heating to, and cooling from, the stress-relief temperature to be relatively slow in the furnace, particularly in large welded assemblies and heavy sections. For example, large weldments, boilers of up to 30-ton weight, after fabrication; large diesel-engine crankshafts; nuclear steam generator vessels up to 600-ton weight; complex mild steel fabrications; and welded disks of turbine rotors are stress-relieved in many large furnaces by heating slowly in with subsequent slow cooling on the order of a few degrees Celsius per hour until the temperature reaches 300°C. Below this temperature, an increased cooling rate can be used. This practice limits the induced new thermal stresses from exceeding the lower yield point of steel.[15] If these conditions were not met, the resulting high-residual stresses would immediately pull the piece out of shape and cause possible cracking, and sometimes it would take a month or a year, or more, for the occurrence of the same problems.

This practice is also applied to large forgings with varying section thicknesses as well as to those forgings which have been subjected to severe straightening operations. These parts are given stress relieving at a temperature near the recrystallization temperature to relieve the elastic strain. Stress-relief annealing is also sometimes used to reduce distortion and high stresses from welding and castings of alloy steels that can affect service performance.[15] Similarly, stress relief of hardened steels must be done at the very low tempering temperature to avoid loss of hardness and strength.

Stress-relief treatment of copper alloys is usually accomplished at relatively low temperatures in the 200 to 400°C (390 to 750°F) range.

The conclusion drawn here is that mostly stress relieving is performed by recovery mechanisms rather than by recrystallization mechanisms, which are linked with distinct kinetics.[13] (See Chap. 5 for more details.)

12.3.1.5 Process Annealing. Process annealing refers to an in-process subcritical anneal and is akin to surface relief annealing. It involves heating the steel to a temperature between 538 and 677°C (1000 and 1250°F), soaking for a proper time (not longer than 30 min), and then cooling, usually in air, to near room temperature. In most cases, heating to within 11 to 22°C (20 to 40°F) below Ac_1 produces the best combination of hardness, microstructure, and mechanical properties.[9] Figure 12.4 shows the temperature range employed for process annealing.

This process eliminates the stiffening effects of cold working, relieves the stress induced by cold working, restores the ductility of the material, and softens the steel by recrystallization for further cold-working operations.

Process annealing is used as an intermediate treatment between manufacturing operations in the sheet, strip, and wire industries to soften the material for subsequent cold-forming operations. It can also be applied to hot-worked high-carbon and alloy steels in order to prevent cracking and soften for turning, shearing, or straightening.

12.3.1.6 Recrystallization Annealing. Recrystallization annealing is accomplished by holding the material for $\frac{1}{2}$ to 1 hr at a temperature above the recrystallization temperature (Fig. 12.4). In the case of continuous annealing, a short time will be adequate. Extremely large volumes of materials are heat-treated using this technique. Tin plate, cold-rolled steel sheets or strips, austenitic stainless steels (such as 18/8 stainless steels), and the 13% Mn (Hadfield) steel are processed in this way. This process is generally applied after cold-working of material. (See Chap. 5 for more elaboration.)

12.3.1.7 Isothermal Annealing. Isothermal annealing consists of heating into the austenite field; holding long enough to complete solutionizing; subsequent rapid cooling by transferring to a furnace held to a constant temperature within the pearlite range of the TTT diagram; and allowing complete transformation to occur isothermally into ferrite and pearlite, followed by rapid cooling, if desired, to room temperature (Fig. 12.2).[16,17]

Advantages over the conventional full annealing method are (1) shorter processing time for the similar final structure—in many cases the time required is 4 to 8 hr, compared to as much as 30 hr for conventional full annealing; (2) cost savings; and (3) consistency of microstructure and hardness over the normal full annealing and normalizing treatments when applied to case-hardening steels and low-alloy medium-carbon steels. The microstructure of isothermal annealed steels comprises the more uniform structure of ferrite and pearlite which is entirely suitable for machining operations.[18]

Isothermal annealing is used extensively for plain carbon and alloy steels, including bearing races made from high-carbon high-chromium bearing steels.[13] Both batch and continuous furnaces can be employed for this process. For certain applications, a succession of salt bath treatments also may be used advantageously and with success.[18]

12.3.1.8 Dehydrogenation Annealing. Dehydrogenation annealing is successfully carried out for large 9Ni-Cr-Mo steel forgings in the ferrite region after hot-working due to several orders of magnitude greater diffusivity of hydrogen in ferrite than in austenite.[8]

12.3.1.9 Decarburization Annealing. Decarburization annealing is usually accomplished on semiprocessed electrical and magnetic lamination steels at temperatures of 730 to 845°C (1350 to 1550°F) in a highly decarburizing atmosphere containing sufficient moisture to promote decarburization without excessively oxidizing the metal; this treatment leads to optimization of the magnetic properties [i.e., decreased core loss (hysteresis loss) and increased permeability] by relieving the stresses produced during cold forming (rolling, bending, shearing, and punching), increasing the grain size to ASTM 2 to 3 (because core loss depends on the grain size) and reducing the carbon level from about 0.08 to 0.02% to less than 50 ppm, because even small carbon contents interfere with the easy movement of magnetic domains and increase hysteresis loss.[†] The decarburization rate of iron becomes highest at annealing temperatures of 770 to 855°C (1420 to 1570°F).[18a‡] Time at temperature varies with lamination/strip dimensions, charge size, surface finish, and annealing furnace characteristics. An atmosphere containing about 20% H_2, 80% N_2, and a dew point of 15°C (55°F) often meets these conditions. Care must be taken to maintain the strips flat in the anneal and to allow easy access of the atmosphere to the edges of the metal strips. The roughened surface textures of the magnetic lamination steels promote decarburization and minimize sticking during lamination anneal (ASTM A683-99; A726/A726M-99).

The low-H_2, N_2-H_2-H_2O atmosphere is more commonly employed for decarburization annealing than the high-H_2 type. However, the high-H_2 type may offer better magnetic properties, depending on the material. It has been reported that N_2-H_2-H_2O atmospheres are intrinsically faster than exothermically generated atmospheres due to the absence of carbonaceous gases. Consequently, a higher-H_2-content atmosphere such as 20% H_2, 80% N_2, and a dew point of 15°C (55°F) permits more water vapor content in the furnace atmosphere, which increases both the decarburizing rate and the decarburizing potential. Hence, atmospheres containing a higher H_2 content offer better decarburizing of extra-low-carbon (ELC) steels.

12.3.1.10 *Annealing of Sheet and Strip.*

In large-scale industrial practice, both *batch* and *continuous annealing processes* are used for softening cold-reduced steel sheets and wires; however, certain types of steel products may only be processed by continuous annealing. Figure 12.6 shows a comparison of these processes with respect to the Fe-Fe₃C equilibrium diagram.[19] In the batch (also called box annealing) process, tightly wound steel coils are placed on the annealing base with a fan, diffuser, and charge (or load) plate with their axes vertical; separator or convector plates are placed between each coil to achieve better distribution of hot gas to the coils during the annealing treatment. This is covered with inner cover [to enshroud the coils that are sealed by using O-ring type base seals[19a] to hold and contain the appropriate protective atmosphere such as nitrogen, nitrogen plus up to 6% hydrogen (HNX gas) or 100% hydrogen which is recirculated by the use of fans in the annealing base and convector plate to promote heat transfer from the inner cover to the steel coils during both heating and cooling] and a surrounding furnace to heat the inner cover either through the use of radiant tubes or by direct firing. The charge

---

[†] Low-carbon cold-rolled motor lamination (CRML) steels are traditionally annealed at about 788°C due to the maximization of both kinetics of solid-state diffusion and carbon removal as well as minimization of lamination sticking at this temperature.[18a]

[‡] Ultralow-carbon (ULC) steels with 15 to 50 ppm carbon do not require decarburizing. Low-H_2, H_2-N_2-H_2O, or dry N_2/H_2 annealing atmospheres are not adequate to avoid carbon pickup and to maintain low carbon level in these steels without oxidation. However, a steam addition will be needed for subsequent blueing operation. Exothermically produced atmospheres are not recommended for ULC steels because of the existence of large amounts of CO, CO_2, and water vapors.[18a]

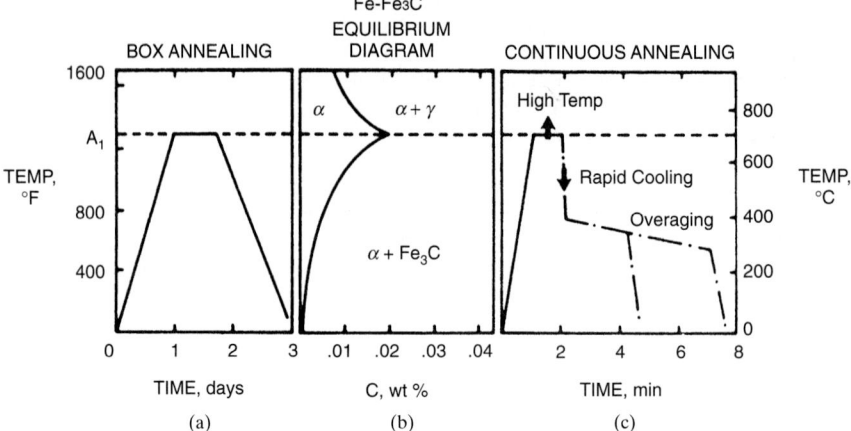

FIGURE 12.6 Comparison of (*a*) box annealing and (*c*) continuous annealing cycles with respect to the Fe-Fe₃C equilibrium diagram.[19] (*Reprinted by permission of the Metallurgical Society, Warrendale, Pennsylvania.*)

is heated to the desired temperature and held for an appropriate period that will produce the required properties. The outer furnace is then removed after completion of heating, and the coils are allowed to cool under the inner covers. When the temperature falls to the point where oxidation of the steel will not occur, the inner cover is removed and the steel is moved for further processing.[20–22]

The gas and the volatilized rolling oil (that is applied during the cold reduction process) escape from the inner cover through purge pipes during the initial burn-off period of the annealing cycle. The early removal of oil significantly improves the surface cleanliness of the steel because of the much reduced oil to crack and form carbon soot on the steel during the later stages of heating. A positive pressure of the protective gas is maintained under the inner cover during both heating and cooling. The oxygen concentration in the furnace atmosphere, held with an inner cover, should remain less than a few dozen ppm, irrespective of the type of metals being annealed, to prevent surface oxidation.[19a] Property variations can be obtained between the outside and inside laps of the coiled sheet due to much slower heating and cooling rates experienced at the coil center. The complete cycle may take several days. A schematic diagram of a conventional one-stool annealing base with associated equipment is shown in Fig. 12.7*a*.[20]

Rapidly recirculated hydrogen gas has a higher heat-transfer coefficient than HNX gas, and this leads to smaller temperature variations within the coil (i.e., no overheating) and enables one to attain faster heating and cooling rates. Hydrogen annealer with a 100% hydrogen atmosphere has become the quality standard in the coiled-sheet annealing industry for both steel and brass mills because it improves the metallurgical and physical properties and overall steel (or brass) product quality and provides a good combination of cleanliness and uniformity (because of the high atmosphere convection in the workload space) together with the highest productivity because of quicker heat-treat cycle (i.e., shortened overall cycle times by 40 to 60%) and low operating and energy costs.[23] Figure 12.7*b* shows a general assembly of a hydrogen annealer. Its basic design is similar to that of a conventional

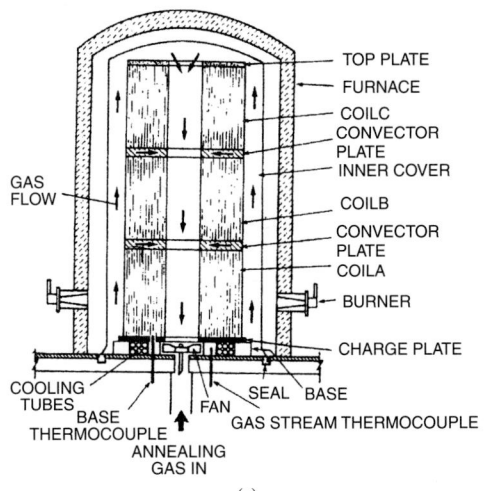

(a)

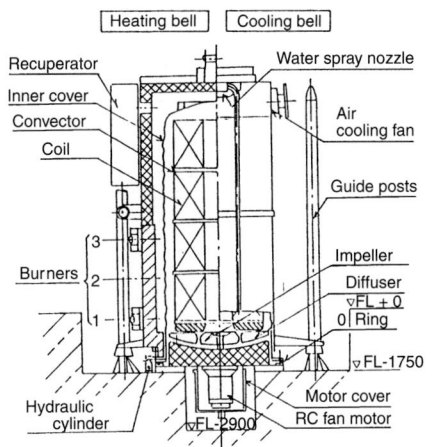

(b)

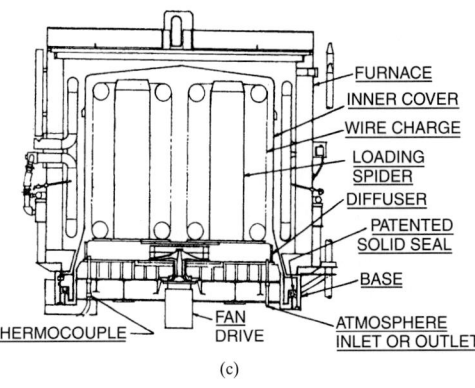

(c)

FIGURE 12.7 (*a*) Schematic sketch of a box annealing furnace components and coils.[20] (*b*) General assembly of hydrogen annealer.[24] (*c*) General arrangement of a batch-type furnace for wire annealing.[29]

12.15

annealer; however, the steel on the workbase is improved by employing a circular seal element and welded stainless steel structure.[24]

In the continuous annealing process, steel sheet coils are uncoiled and rapidly passed through a two-stage furnace where they are subjected to the annealing cycle on the order of a few minutes under a protective atmosphere.[25] After the sheet or strip has been rapidly cooled and exited the furnace, further in-line processing such as *zinc coatings* can be accomplished. These zinc coatings include *hot-dip galvanizing (HDG), zinc-aluminum hot-dip and batch coatings, galvannealing,* and *electrogalvanizing processes.* The more conventional HDG process may be carried out at a temperature of 450 to 470°C [in a zinc bath saturated with iron (~0.04%) and aluminum (0.14 to 0.2%)] for a few seconds to 10 minute and air-cooled. For other zinc coating processes, the interested reader is referred to a large number of papers published in a recent proceeding.[26] Alternatively, the steel may be cut into strips. In general, however, the steel is recoiled and then transferred/moved as in the batch process.

Figure 12.8 shows a typical arrangement of equipment in a continuous annealing production line.[27] The line comprises three main sections: the entry section, the furnace section, and the exit section. The first section consists of an uncoiler, shears, a welder, an electrolytic cleaning device, a water rinse, and an entry looper. The furnace section includes heating, soaking, and cooling zones. Mostly, the vertical-type furnace is used with a mixture of N + H atmosphere and separated radiant tube burners or direct burner heating. In the soaking zone, the homogeneity of temperature along the width and length of the strip is maintained. The mechanism of the cooling zone is of vital importance for realizing the precise control of the cooling rate of strip which directly affects the final product properties.[25,27]

The cooling methods include gas jet cooling with HNX gas or high concentration of hydrogen in nitrogen, hot-water quench, cold-water quench, roll contact, gas-assisted roll cooling, and water-mist.[28] The selection of cooling methods depends primarily on the product mix, which has to be treated in the line. In Fig. 12.9, the cooling rates, together with the metallurgical effects which can be obtained by these cooling methods, are shown. In the majority of the existing annealing lines, an over-

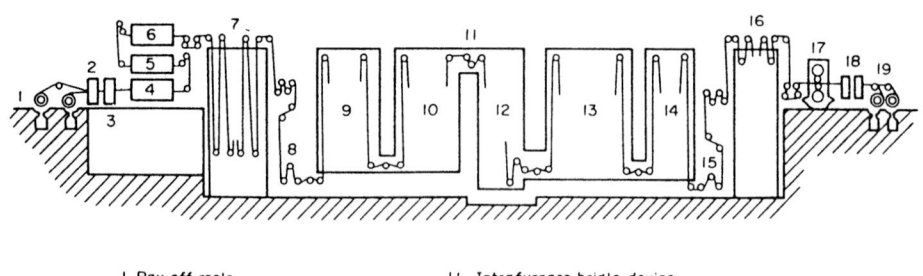

1 Pay off reels	11 Intrafurnace bridle device
2 Shear	12 Rapid cooling section
3 Welder	13 Overaging section
4,5 Electrolytic cleaning tanks	14 Final cooling section
6 Rinse and pretreatment tanks	15 No. 2 tension control unit
7 No. 1 looper	16 No. 2 looper
8 No. 1 tension control unit	17 Temper mill
9 Heating section	18 Shear
10 Soaking section	19 Tension reels

FIGURE 12.8 Typical arrangement of equipment in a continuous annealing production line.[27] (*Reprinted by permission of Pergamon Press, Plc.*)

TEMP, °F

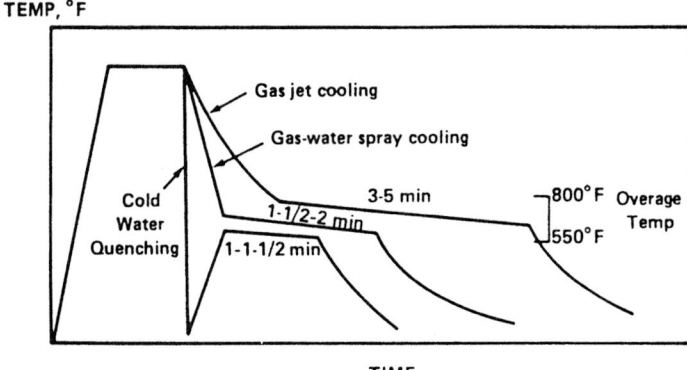

TIME

FIGURE 12.9 Various continuous annealing thermal profiles to produce over-aging in a continuous annealed sheet steel.[19] Anneal/soak temperature is ~750 to 850°C (1380 to 1560°F); overage temperature is ~350 to 400°C (660 to 750°F). (*Reprinted by permission of the Metallurgical Society, Warrendale, Pennsylvania.*)

aging treatment zone, where the temperature is held at 350 to 400°C, is put in front of the final cooling zone. The temperature and pass time in the overaging zone are determined by the chemical composition and the initial cooling rate of steel prior to the overaging treatment.

In the exit section, a looper, a temper mill, a side trimmer, an oiler, shears, and a tension reel are incorporated. The type and number of these features depend on the product mix, the size range, the line speed, and so forth. The temper mill, single or two-tandem stands are usually provided with a rapid roll exchange facility in order to maintain constant line speed in the furnace section without line stoppage.

The overall length of the existing lines varies between 150 and 350 m, and the length of the steel strip in the furnace section is between 1500 and 3000 m. The line speed is a function of product size and the expected production capacity. Usually the production capacity lies in the range of 350,000 to 650,000 tons/yr, and the line speed of the section is 150 to 300 m/min. In some lines, a tandem cold-rolling is directly connected, in front of the annealing line, and the line speed in this situation increases up to 450 m/min, capable of treating 1,000,000 tons/yr of strip. In the light-gauge (such as tin plate) annealing line, the production capacity is smaller, but the line speed is greater than "sheet-gauge" lines, where the highest speed reaches up to 700 m/min.[27]

Continuous annealing cycles are of shorter periods and are carried out at higher temperatures than batch annealing cycles. In some applications, the annealing temperature may exceed A_1. Typical cycles are 40 s at 700°C (1290°F) for cold-rolled commercial-quality steel and 60 s at 800°C (1470°F) for quality special killed steel sheet. Most continuous annealing of cold-rolled sheet includes an overaging treatment designed to precipitate carbon and nitrogen from solution in the ferrite and to reduce the likelihood of strain aging. Overaging for 3 to 5 min at 300 to 450°C (570 to 840°F) completes the required precipitation of carbon and nitrogen. It is widely used for the production of such sheet steel products as hot-dip galvanized steels, tin plates, nonoriented electrical steels, stainless steels, bake-hardening steels, interstitial free steels, and high-strength steels, using lower alloying additions.

Batch annealing and continuous annealing differ slightly in the properties they produce. Typical average properties of batch-annealed and continuous-annealed commercial-quality plain carbon steel can be summarized as follows:

Batch annealing	Continuous annealing
1. Typical yield strength and elongation are 210 MPa (30.4 ksi) and 43.0%.	1. Typical yield strength and elongation are 228 MPa (33 ksi) and 41.7%.
2. Poor productivity is due to a very long (several days) annealing time.	2. It is carried out in about 5 min which gives higher hardness levels and yield strengths.
3. It is conducted at lower temperatures.	
4. It is difficult to ensure uniformity of temperature throughout the charge because of the very large mass of steel. This results in a considerable scatter of properties along the coil length.	3. It is conducted at higher temperatures.
	4. It offers the potential of more uniform properties along the coil length, especially in high-strength low-alloy (HSLA) steels.[25]
5. Annealed grain size is coarse and results in excellent ductility.	5. The charge is passed through a two-stage furnace: the first stage is for heating above the recrystallization temperature, and the second one at a lower temperature is for overaging the steel and effectively removing carbon from solution.[25]
6. Inferiority of surface quality and product shape is due to annealing in the coil.	
7. A 100% H annealer offers superior performance for cold-rolled sheet which is superior to CAL provided the production tonnage is not very high or is applied to special steels.[23]	6. The investment and running cost of continuous annealing line (CAL) are very expensive for moderate tonnage of steels, and it is not used for special steels including high-carbon steels.[25] applied to special steels.[23]
	7. Further improvement in economies includes combining CAL with other processes such as galvanizing, tinning cycle, and so forth

12.3.1.11 Annealing of Bar, Rod, and Wire. For low-carbon steels up to 0.20% C, short-time subcritical annealing is adequate for preparing the material for further cold-working operation. For higher-carbon and alloy steel coiled products, spheroidizing treatment is needed to provide maximum ductility.

In batch annealing, it is customary to use higher-than-normal temperature (say, 650°C or 1200°F) during initial heating for purging because it promotes (1) the lower-temperature gradient in the charge during subsequent heating into the temperature between A_1 and A_3 and (2) the agglomeration of the carbides in the steel, which makes them more resistant to dissolution in the austenite when the charge temperature is finally raised. These undissolved carbides will facilitate the formation of a spheroidal rather than a lamellar structure when transformation is complete.[9]

Temperature distribution and control in the furnace and in the load are very critical in batch and vacuum furnaces, which can control loads up to 30 tons (27 Mg), than in continuous furnaces, where loads of only 2000 to 4000 lb (900 to 1800 kg) may be transferred from zone to zone. In spheroidizing, it is essential that no part of the charge be permitted to approach A_3, to minimize the formation of pearlite on cooling.

Prior cold working increases the degree of spheroidization and offers even greater ductility. Note that unless a reduction of at least 20% is applied, severe grain coarsening may be obtained after spheroidizing, which may severely impair subsequent performance.

Low-temperature annealing of steel wire in lead bath is widely used to lower the strength and increase elongation in a galvanizing operation, or to provide a workable dry-finished wire. Figure 12.7c is a general arrangement of batch-type furnace for wire annealing which is similar to that for sheet and strip annealing.

The wire annealing furnaces are limited in height by the allowable weight of the charge (of 40,000 lb) at the bottom of the stem to prevent sticking of the wire during the annealing cycle. Consequently, the wire charge height should not exceed 108 in.

Continuous annealing of wire, in strand form, may be performed in tube furnaces, muffle furnaces, molten salt, fluidized beds, or lead; this has the benefit of more precise heat treatment, cooling, and finishing when compared to batch annealing operations.[29]

12.3.1.12 Annealing of Forgings. Forgings are frequently annealed to facilitate subsequent operations such as machining or cold forming. The annealing treatment is determined by the type and amount of machining and cold forming to be carried out, as well as the type of material being processed.[9]

12.3.1.13 Annealing of Plate. Plate products are normally annealed to facilitate machining or forming operations. Plate is generally annealed at subcritical temperatures, and long annealing times are usually avoided. Maintaining flatness during annealing of large plate can pose a serious problem.[9]

12.3.1.14 Annealing of Tubular Products. Annealing is a standard treatment for tubular products which are often machined or formed. In most cases, subcritical temperatures and short annealing times are employed to reduce the hardness. Higher-carbon grades such as 52100 usually are spheroidized before machining. Tubular products produced in pipe mills are rarely annealed; rather, these products are used in the as-rolled, normalized, or quenched and tempered conditions.[9]

12.3.1.15 Blueing Annealing. Blueing annealing is a controlled-heat and controlled-humidity process to form a thin dark-blue, electrically insulated, tightly adherent, anticorrosive oxide (Fe_3O_4) layer with aesthetic appearance on the steel laminations in order to resist interlaminar eddy currents. This treatment involves subjecting the scale-free steel surface to the action of air, steam, wet exothermally generated atmosphere, or N_2-H_2-H_2O mixtures between 400 and 540°C (750 and 1000°F).[18a] Alternatively, in the initial stage, the steel parts are typically heated to 600 to 650°F for 1 hr and then in the final stage to 900 to 1200°F for $\frac{1}{2}$ hr. Steam is injected in the furnace in the final stage of the cycle. Here, consistent temperature and reproducible atmosphere are particularly important to form a uniform and adherent oxide surface on the steel laminations with good visual appearance and consistent magnetic properties. After the blueing treatment, parts may be given coatings or rustproof oils or waxes to further enhance corrosion resistance. This is applied to sheet, strip, or finished parts as well as to springs to improve their properties.[29a]

The blueing reaction is $3Fe + 4H_2O \rightleftharpoons Fe_3O_4 + 4H_2$. This reaction is reversible and can be controlled by maintaining the correct H_2/H_2O ratio and temperature.[29a] Using N_2-H_2-H_2O mixtures, this ratio can be controlled, leading to a uniform and adherent oxide film. Note that blueing of steel laminations is cheaper than the commercial coating alternative.[18a]

12.3.1.16 Bright Annealing. Bright annealing should be accomplished in a protective atmosphere such as hydrogen, vacuum, cracked (or dissociated) NH_3, a mixture of hydrogen and nitrogen, rich exothermic atmosphere, or inert gases with dew points of $-40°F$ ($-40°C$) or less to prevent discoloration and oxidation of the bright surface for processing stainless steel, brass and bronze alloys, copper and copper alloys, beryllium copper and beryllium nickel, nickel and nickel alloys, and silicon iron in the form of strip, tubing, wire, and stamped and formed parts.[29b]

If annealing occurs in a continuous furnace, air tightness is required which can be accomplished by the use of laminar flow curtains at the furnace inlet and outlet.[29c]

12.4 DECARBURIZATION

Decarburization is the loss of carbon from the surface layer of hot-worked steel products (during heating for hot rolling, forging, extruding, spinning, and fabrication) or other steel products such as bar, rod, wire, strip, tube, and so forth. The depth of decarburization is a function of the hot-working temperature, time at temperature, furnace atmosphere (such as slightly oxidizing atmosphere and protective gases containing small amounts of air or moisture), reduction in area from bloom to finished size, and type of steel. Decarburization softens and weakens the surface layers, which results in decrease of wear resistance and a serious drop in fatigue resistance. Hardening of the decarburized surface layer can introduce residual tensile stresses, which reduce the fatigue limits of the material. However, fatigue properties lost through decarburization can be regained by recarburization.

The presence of oxygen, water vapor, and carbon dioxide in the annealing atmosphere can cause decarburization, according to the reactions

$$2Fe\,(C)+O_2 = 2Fe+2CO \qquad (12.3)$$

$$Fe\,(C)+H_2O = Fe+CO+H_2 \qquad (12.4)$$

$$Fe\,(C)+CO_2 = Fe+CO \qquad (12.5)$$

where Fe (C) denotes carbon in solution in austenite.

Decarburization increases with increased (1) rate of carbon diffusion, (2) carbon activity, and (3) α-γ transformation temperature. Mn restricts the decarburization through its effect on carbon activity and the diffusion coefficient. Ni restricts carbon diffusion outward and reduces the depth of decarburization. Si tends to increase decarburization by increasing the carbon activity and the rate of carbon diffusion. Cr causes the reduction of carbon activity and would tend to reduce decarburization.

Decarburization control can be achieved by the use of (1) copper plating; (2) controlled atmosphere such as endothermic ($-40°C$ dew point), vacuum, inert N + 5% H_2, or H_2 gas; and (3) salt bath heating. Decarburization can be corrected by carbon restoration, i.e., by adequate carburizing to restore the carbon level equal to the original or intended content. The atmospheres used for carbon restoration are nitrogen-dissociated methanol; nitrogen enriched with hydrocarbon (usually, methane or propane); dry, purified exothermic gas; and endothermic gas. Successful carbon restoration depends on (in order of importance) (1) composition of the furnace atmosphere, (2) steel composition, (3) processing temperature, (4) surface condition of the steel encountering carbon restoration, (5) furnace type, (6) furnace zone separation, (7) atmosphere tightness of the furnace, (8) load distribution, and (9) flow rate and extent of circulation of the furnace atmosphere.[30,31]

Decarburization can be determined by optical metallography, arrest quench method, microhardness measurement (surface versus core hardness test), chemical analysis, grinding wheel visual test (from experience in making such examination), and carbon determination with a vacuum spectrograph. Among these methods, optical metallography and microhardness measurement are the most useful and convenient methods.[11,32]

For quality control and research studies, the average depth of decarburization—*complete*, *partial*, or *total*—which is easily observable under the microscope, may be necessary. *Complete decarburization* is the loss of carbon content at the surface of a steel specimen to a level below the solubility limit of carbon in ferrite to produce entirely ferritic surface layers. The *depth of complete decarburization* is the thickness of the ferrite layer, i.e., the distance from the surface to the first particle of a second phase. *Partial decarburization* is the loss of carbon at the surface of the steel specimen to a level below the bulk carbon content of the unaffected interior but greater than the room-temperature solubility limit of carbon in ferrite. In the partial decarburization zone, the carbon content increases progressively from the ferrite layer toward the interior. *Total depth of decarburization* is the perpendicular distance from the specimen surface to that location in the interior where the bulk carbon content is reached, i.e., the sum of the depths of complete and partial decarburization.

Total decarburization is not acceptable. However, a very small partial decarburization to a depth of 0.0015 in. maximum does not affect the carbon content of the case/surface, fatigue/mechanical properties, and heat-treatability performance of the final product. Therefore, most industries accept 0.0015-in. maximum depth of partial surface decarburization as a quality control measure.

12.5 (SECONDARY) GRAPHITIZATION OF STEELS

When carbon steels and low-alloy steels are annealed or treated for long times at moderate temperatures [425 to 680°C (800 to 1256°F)], decomposition of the pearlite (ferrite + iron carbide) phase into the equilibrium structure of iron and graphite is most likely; however, this becomes more pronounced in the higher temperature range (630 to 680°C, or 1166 to 1256°F) for both hypoeutectoid and hypereutectoid steels. This phenomenon is called *secondary graphitization*, in true sense, because of its formation during solid-state processing. However, it is usually called graphitization. This can severely embrittle the steel when the graphite particles or nodules form in a planar, band-continuous fashion. Graphitization has caused sudden failure of pressure boundary components such as high-energy piping and boiler tubes. This is also of concern in long-term aged carbon and carbon-molybdenum steels, both in base metal and in weldments. Graphitization is characterized by the factors that are summarized below:[33,34]

1. Graphitization is reported at subcritical temperature mostly in special carbon steels having large amounts of Ni, Si, or Al and in steels of extremely low P and S contents.

2. High levels of aluminum deoxidation present in fine-grained aluminum-killed steels strongly promote graphitization. This effect was thought to be nucleated by fine dispersion of Al_2O_3 in the steels acting as nucleants. Silicon-killed or low-aluminum killed steels are not prone to graphitization. Nitrogen appears to retard graphitization; high levels of aluminum remove or reduce nitrogen content and thus promote graphitization.

3. Prior deformation or cold working in both high-carbon (0.6 to 0.8%) and low-carbon (0.06%) steels accelerates the graphitization rate; tensile deformation favors graphitization, while compression does not. It is also reported that cold working was effective in accelerating graphitization in steels containing 0.6 to 0.8% C but not in carbon steels with 0.20 to 0.30% C.

4. Graphitization seems to be dependent on the microstructure of the material. For instance, pearlitic structures are much less susceptible to graphitization than martensitic structures. Thus quenched and tempered steels graphitize faster than the normalized ones. Similarly, cold-rolled microstructure is the most favorable for graphitization (as mentioned in item 3).[35]

5. Slightly oxidizing atmospheres tend to favor graphitization compared to reducing and neutral atmospheres containing CO, H_2, and vacuum. However, Inokuti has observed maximum graphitization with vacuum annealing, less with the use of nitrogen and argon, none with hydrogen.

6. Usually, the graphite exhibits nodular morphology, but sometimes graphite is distributed in a chainlike array, banded or stringer form aligned in the rolling direction (Fig. 12.10). In the latter form, it can embrittle steel parts. Thus, high-carbon steels become soft and ductile as low-carbon steels by graphitization.

7. Carbide-forming elements such as Cr (>0.5%) and 0.5% Mo as well as stress-relieving at 650°C (1200°F) appear to prevent or reduce the tendency of graphitization.

8. Presence or formation of small voids, during rolling, at the cementite particles present in the ferrite matrix plays a vital role in nucleating graphitization. Inokuti has proposed the mechanism of graphite nucleation at voids, augmented by the formation of CO gas in the voids. This was supported by the experimental observations made by Berge et al.[33]

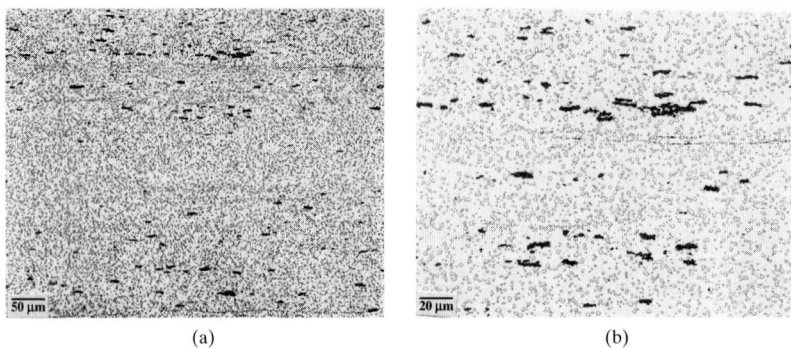

      (a)   (b)

FIGURE 12.10 Optical photomicrograph of flat stock for 0.035-in. cold-rolled 1074 spring pin showing graphitization produced at subcritical temperature. The dark-inclusionlike features are almost pure carbon as confirmed by scanning electron microscopy/energy disperse x-ray. Picral etch. (*a*) Low magnification and (*b*) high magnification.

9. Localized graphitization near a welded joint of steam piping or superheated tube appears to be much more damaging than general uniform graphitization because the former apparently produces notches that concentrate stress, provides a weak continuous path for fracture, and reduces load-bearing capability.[36]

12.6 NORMALIZING

Normalizing consists of heating the steel to a temperature 40 to 50°C above the upper critical points (that is, the Ac_3 or Ac_{cm} temperatures), holding it there for a period depending on the dimensions of the part and type of steel being treated until it is completely austenitized, followed by cooling, in still air, free of drafts, to room temperature. (Because the microstructure is a function of cooling rate, accelerated cooling such as air blowing or mist cooling is sometimes used to achieve a finer ferrite-pearlite or bainite microstructure to improve the mechanical properties of thick sections of steel products.) This term originated from the wrought steel industry, where air cooling was the normal practice. The temperature range representing the normalizing of a plain carbon steel is shown in Fig. 12.1. The minimum period for holding (or soaking) at the austenitizing temperature is 15 min, necessitating longer periods with larger sections. Table 12.4 lists the typical normalizing temperatures for standard carbon and low-alloy steels.[36a] Figure 12.11 illustrates the microstructure of a normalized plain carbon steel.

The main purpose of this operation is to (1) refine the grains of a steel bar, forging, welding, or casting that have become coarse grained because of being heated to a high-temperature treatment; (2) homogenize, that is, improve the uniformity of microstructure, by breaking up the coarse nonuniform structure (e.g., cast dendritic structure), although at a lower temperature and for shorter periods than those used for homogenizing; (3) allow smoother machining, thereby producing superior surface finish of the normalized product compared to that of the annealed one; and (4) produce harder and stronger steel than that of full annealing.

The majority of ordinary engineering steels do not form martensite when normalized. However, martensite can be produced in the case of highly alloyed steels; such steels are called *air-hardening steels*, and, in the true sense, it is not proper to call the air-hardening operation *normalizing* because it does not put the steel in the "normal" pearlitic conditions.

Since the cooling rate lies between that used for quenching and annealing, respectively, the hardness and strength resulting from this treatment will be somewhat less than if quenched and somewhat higher than if annealed. The hardness of the normalized product depends on the composition and dimensions of the steel. The difference in the cooling rate between the surface and center during this treatment is small for light sections and large for heavy sections.

Since air cooling does not represent the equilibrium condition, the Fe-Fe$_3$C phase diagram cannot be employed to predict the proportions of proeutectoid ferrite and pearlite or proeutectoid cementite and pearlite which will be present at room temperature. In fact, less time is available for the formation of proeutectoid constituents; consequently, a lesser proportion of the proeutectoid constituent will form in normalized steels than in annealed ones. Moreover, the faster cooling rate in normalizing will, in general, depress the temperature of austenite decomposition and will produce a finer pearlite.[37] Figure 12.2 shows the schematic time-temperature cycle for normalizing superimposed on the TTT diagram.

TABLE 12.4 Typical Normalizing Temperatures for Standard Carbon and Low-Alloy Steels.[36a]

Grade	Temperature[†] °C	Temperature[†] °F	Grade	Temperature[†] °C	Temperature[†] °F
Plain carbon steels			4815	925	1700
			4817	925	1700
1015	915	1675	4820	925	1700
1020	915	1675	5046	870	1600
1022	915	1675	5120	925	1700
1025	900	1650	5130	900	1650
1030	900	1650	5132	900	1650
1035	885	1625	5135	870	1600
1040	860	1575	5140	870	1600
1045	860	1575	5145	870	1600
1050	860	1575	5147	870	1600
1060	830	1525	5150	870	1600
1080	830	1525	5155	870	1600
1090	830	1525	5160	870	1600
1095	845	1550	6118	925	1700
1117	900	1650	6120	925	1700
1137	885	1625	6150	900	1650
1141	860	1575	8617	925	1700
1144	860	1575	8620	925	1700
Standard alloy steels			8622	925	1700
			8625	900	1650
1330	900	1650	8627	900	1650
1335	870	1600	8630	900	1650
1340	870	1600	8637	870	1600
3135	870	1600	8640	870	1600
3140	870	1600	8642	870	1600
3310	925	1700	8645	870	1600
4027	900	1650	8650	870	1600
4028	900	1650	8655	870	1600
4032	900	1650	8660	870	1600
4037	870	1600	8720	925	1700
4042	870	1600	8740	925	1700
4047	870	1600	8742	870	1600
4063	870	1600	8822	925	1700
4118	925	1700	9255	900	1650
4130	900	1650	9260	900	1650
4135	870	1600	9262	900	1650
4137	870	1600	9310	925	1700
4140	870	1600	9840	870	1600
4142	870	1600	9850	870	1600
4145	870	1600	50B40	870	1600
4147	870	1600	50B44	870	1600
4150	870	1600	50B46	870	1600
4320	925	1700	50B50	870	1600
4337	870	1600	60B60	870	1600
4340	870	1600	81B45	870	1600
4520	925	1700	86B45	870	1600
4620	925	1700	94B15	925	1700
4621	925	1700	94B17	925	1700
4718	925	1700	94B30	900	1650
4720	925	1700	94B40	900	1650

[†] Based on production experience, normalizing temperature may vary from as much as 27°C (50°F) below, to as much as 55°C (100°F) above, indicated temperature. The steel should be cooled in still air from indicated temperature.

Reprinted by permission of ASM International, Materials Park, Ohio.

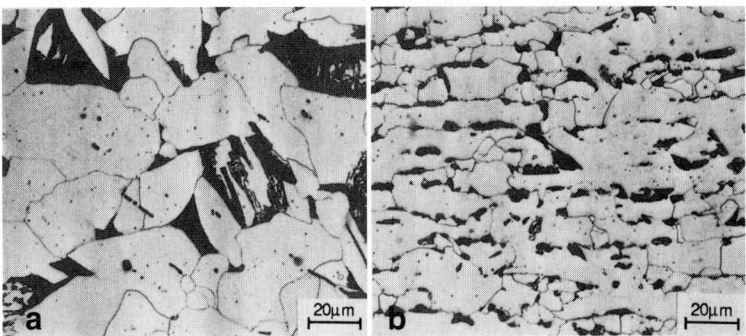

FIGURE 12.11 Microstructure of a normalized 1025 steel. (*a*) Produced by austenitizing at 1095°C (2000°F) and then air-cooling. Coarse grain structure; pearlite (black areas) in ferrite matrix (white areas). (*b*) Produced by austenitizing at 925°C (1700°F) followed by air cooling. Fine-grain structure. Picral etchant. (*Reprinted by permission of ASM International, Materials Park, Ohio.*)

The best practice of cooling in air is to suspend the steel parts in air, placing them on a special cooling bed in order for them to be surrounded by the cooling action of the air. Careless practice, such as dumping the material into a pile on the shop floor, may cause very poor treatment and very nonuniform structures.

Normalizing is frequently applied prior to spheroidize-annealing to disperse carbides homogeneously. It is also done as a prior heat treatment before quenching and tempering of thick sections. Normalizing often precedes hardening and annealing operations of cast steels of various cross-sectional sizes. Normalizing the part prior to hardening tends to minimize cracking upon quenching.[38] Spheroidization annealing and normalizing may be used to improve machinability; the method to be chosen depends on the carbon content. Normalizing is not usually employed for hypereutectoid steels, except as a pretreatment prior to subsequent annealing. The treatment is generally confined to *in-process treatments* performed by steel suppliers and is seldom required by tool manufacturers.

Effect of Hot Working.   The normalized microstructure is a function of the prior hot-working conditions. The finer the microstructure of the as-rolled microstructure, the finer the austenite grain size after reheating for normalizing. Hot-rolling at lower temperature corresponding to the nonrecrystallized austenite leads to a finer normalized austenite grain size. The reduction ratio during hot rolling has also a significant bearing on the austenite grain size during normalizing. Figure 12.12 shows that the reduction ratio is more important in improving toughness than is the cooling rate during normalizing. A large hot-rolling reduction and slow cooling during normalizing lower the Charpy ductile-brittle transition temperature [fracture appearance transition temperature (FATT)] more than a small reduction ratio and fast cooling during normalizing do.[39]

Double Normalizing.   In some cases a double normalizing treatment is used. This treatment consists of first heating to a temperature some 50 to 100°C above the usual temperature in order to produce complete dissolution of the constituents and homogeneous austenite grain size. The second normalizing treatment is accom-

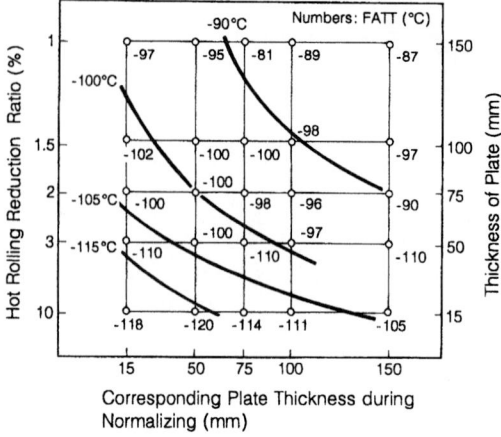

FIGURE 12.12 Effect of cooling rate during normalizing and the hot-rolling reduction ratio of the (0.05C-3.5Ni-0.1Mo) steel slab on the ductile-brittle transition temperature (FATT) measured by the Charpy impact test. The cooling rate is represented by the corresponding plate thickness. The heat treatment involved normalizing from 840°C, tempering at 600°C followed by stress-relieving at 580°C and cooling to 300°C at a rate of 100°C/hr.[8]

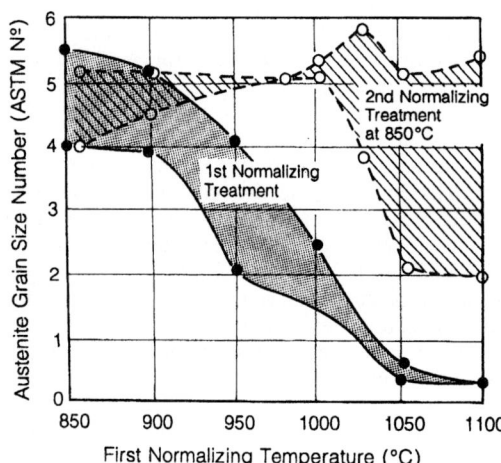

FIGURE 12.13 Variation of the austenite grain size with double normalizing in forged carbon steel containing 0.55% C.[8]

plished near the lower limit of the temperature range to produce a fine-grained structure. It is usually applied to carbon and low-alloy steels of large dimensions where too-high forging temperatures have been used. Figure 12.13 illustrates the change in austenite grain size during double normalizing treatment in forged medium-carbon (0.55% C) steels where the first normalizing lies between 850 and 1100°C and the second one at 850°C.[8]

Repeated normalizing (e.g., double normalizing) treatment is necessary for very coarse-grained forgings because it may be difficult to obtain a fine-grained structure in a single normalizing treatment.

12.7 ANNEALING AND NORMALIZING OF CAST IRONS

12.7.1 Gray Iron

12.7.1.1 Annealing.   Next to stress relieving, annealing treatment is occasionally applied to gray iron. Annealing of gray iron involves heating it to a temperature that is high enough to soften it and/or to minimize or remove eutectic carbides, thereby improving its machinability. Annealing significantly decreases the mechanical properties. In fact, it reduces the grade level approximately to the next-lower grade. Figure 12.14 illustrates the effect of annealing on the tensile strength of class 30 gray iron arbitration bars. The extent of decrease in properties is a function of the annealing temperature, the time at temperature, and the chemical composition of the gray iron.

Annealing treatments commonly applied to gray iron include *high-temperature (graphitizing) annealing, medium-temperature* ("full") *annealing,* and *low-temperature (ferritizing) annealing.*[40,41]

High-Temperature (Graphitizing) Annealing.   Graphitizing annealing produces the ultimate decomposition of chilled iron, massive, primary or free cementite, and reduction in strength and hardness. This annealing treatment consists of (1) heating to a temperature of 900 to 954°C (1650 to 1750°F) [however, at 925°C (1700°F) and above, the phosphide eutectic present in irons containing 0.10% P or more may melt]; (2) holding at this temperature for a period ranging from a few minutes to several hours; and (3) cooling at a rate depending on the final use of the iron. If it is desired to break down primary carbides and retain a pearlitic structure, and therefore a maximum strength and wear resistance, the casting should be air-cooled from the furnace to about 538°C (1000°F) to produce pearlite-graphite microstructure. If maximum machinability is the main concern, the casting should be furnace-cooled through the critical range to 538°C (1000°F) to obtain a ferrite-graphite structure. In both cases, cooling from 538°C (1000°F) to about 290°C (550°F) at no more than 110°C/hr (200°F/hr) should be employed to avoid the formation of residual stresses.

Medium-Temperature (Full) Annealing.   Full annealing in the absence of massive carbides or in the presence of small amounts of well-dispersed carbides can be performed by heating to just above the critical range between 790 and 900°C (1450 and 1650°F), depending primarily on the silicon content arising from the increase of critical temperature with the silicon content.[41] The soaking time in the furnace is the same as for ferritizing annealing but is shorter than for graphitizing

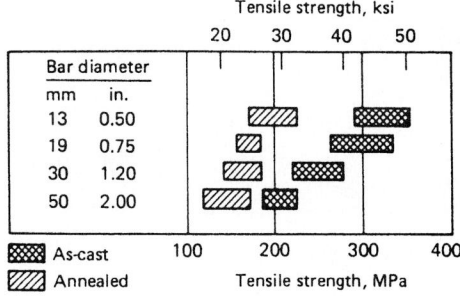

FIGURE 12.14   Effect of annealing on tensile strength of class 30 gray iron. Specimens were arbitration bars from 31 heats. Bars were annealed at 925°C (1700°F) for 2 hr plus 1 hr/in. (25 mm) of section over 1 in., then cooled at a maximum rate of 160°C (285°F)/hr from 925 to 565°C (1700 to 1050°F). Cooling then continued from that level at a maximum rate of 130°C (230°F)/hr to 200°C (390°F); subsequently the bars were air-cooled to room temperature.[40]

annealing. This is followed by slow cooling through the critical temperature range from about 790 to 675°C (1450 to 1250°F) to allow the combined carbon to precipitate as graphite.

Full annealing is used when a ferritizing annealing would be ineffective due to high alloy content of a particular iron. This treatment can also be used for pearlitic irons containing moderate amounts of Cr, V, Mo, or a higher than standard Mn content, but without free cementite, to obtain an entirely ferritic matrix.[41]

Low-Temperature (Ferritizing) Annealing. For unalloyed or low-alloy gray irons of normal composition with no free cementite, when the only requirement is to convert the pearlitic carbide to ferrite and graphite for improved machinability, it is advisable to ferritize anneal these castings by heating near the lower transformation temperature [i.e., between 704 and 760°C (1300°F and 1400°F)] and holding for approximately 1 hr/in. of section thickness, followed by slow cooling [that is, 55°C/hr (100°F/hr)].[41] There is no significant influence of temperature, up to 595°C (1100°F), on the structure and hardness of gray iron. However, the rate of decomposition of iron carbide into ferrite plus graphite increases markedly with the increase in temperature above 595°C (1100°F), reaching a maximum of 760°C (1400°F) for unalloyed or low-alloy gray iron.[40,42] Table 12.5 lists the effect of ferritizing anneal on tensile strength and hardness due to varying alloying addition.[40,42] Figure 12.15 shows the conversion of an as-cast pearlite structure of unalloyed gray iron (with 980 BHN hardness) to ferrite and graphite (with 120 BHN hardness) by ferritizing annealing at 760°C for 1 hr.[40,42]

For unalloyed irons, the rate of ferritization depends on the silicon content and the temperature employed. For example, when the unalloyed iron contains ~2% Si and is annealed at 760°C (1400°F), the rate of ferritization becomes maximum, reaching 90% of conversion within 20 to 30 min in light sections. However, this reaction is retarded by the presence of alloying elements such as Cr, Ni, Cu, and Mo; however, Mn in the range of 0.3 to 0.98% at a constant sulfur level is not a matrix strengthener in gray iron.[43] As the temperature is lowered, the rate of ferritization decreases drastically; and below 650°C (1200°F), the rate of pearlite conversion to ferrite is so slow as to require an excessively long holding time for the completion of the process.[41]

12.7.1.2 Stress-Relieving.
Invariably, as-cast gray irons contain residual stresses, the magnitude of which is a function of different parameters such as casting methods employed, composition, and properties of the cast material, as well as shape and

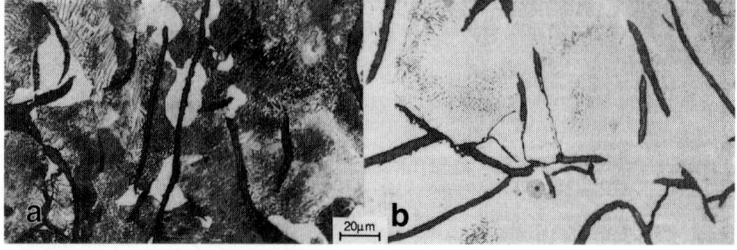

FIGURE 12.15 (*a*) As-cast unalloyed pearlitic gray iron structure (180 BHN). (*b*) Ferrite and graphite structures (120 BHN) produced by ferritizing annealing of (*a*) at 760°C (1400°F) for 1 hr.[40] (*Reprinted by permission of ASM International, Materials Park, Ohio.*)

TABLE 12.5 Effect of Ferritizing Annealing at 760°C (1400°F) on Hardness and Tensile Strength of Alloyed Gray Irons[40]

Gray iron no.	Alloy additions to base iron, %[†]					As-cast			After annealing		
	Cr	Mo	Cu	Ni	V	Hardness, HB	Tensile strength psi	MPa	Hardness, HB	Tensile strength psi	MPa
	Base iron without alloy additions[†]										
1	...	...	...	...	...	217	37,400	258	143	27,700	191
2	0.61	0.56	...	...	...	262	46,200	319	217	44,500	307
3	0.47	0.43	0.52	...	...	248	50,600	349	207	43,100	297
4	0.56	...	...	...	...	241	41,700	288	207	40,000	276
5	0.50	...	0.52	...	...	241	43,000	297	201	38,200	263
6	0.49	0.43	...	1.45	...	285	55,500	383	197	42,600	294
7	...	...	0.65	...	...	269	52,400	361	187	37,400	258
8	...	0.54	...	...	0.13	255	48,400	334	179	40,000	276
9	0.49	...	...	1.45	...	255	45,400	313	156	34,800	240
10	...	0.54	...	0.66	...	269	50,000	345	156	33,000	228
11	...	...	...	...	0.12	229	41,000	283	156	31,200	215
12	...	...	...	1.72	...	235	41,700	288	149	29,900	206
13	...	0.47	...	...	...	241	44,000	303	146	31,600	218
14	...	...	1.80	...	...	235	43,500	300	143	29,900	206

[†] Base iron analysis: 3.26% total C, 1.92% Si, 0.94% Mn, 0.03% S, 0.11% P.

Source: G. A. Timmons and V. A. Crosby, *Foundry*, vol. 69, October 1941, pp. 64–66, 142–147; November 1941, pp. 64–66, 145–147.

Reprinted by permission of ASM International, Materials Park, Ohio.

section size of the casting. Stress-relieving does not influence tensile strength, hardness, or ductility.

The temperature used for stress relieving lies well below the transformation range of pearlite to austenite. When maximum stress-relief (up to 75 to 85%) together with minimum decomposition of carbide in unalloyed irons is desired, stress-relief treatment at a temperature range of 540 to 565°C (1000 to 1050°F) for 1 hr is recommended. To attain more than 85% stress relief in unalloyed irons, a minimum temperature of 595°C (1100°F) can be used, but some sacrifices in hardness, strength, and wear resistance are inevitable.[40,42]

Low-alloy gray irons usually require a higher stress-relieving temperature of 595 to 650°C (1100 to 1200°F), which depends on the alloy content. Similarly for high-alloy gray irons, a temperature of 620 to 675°C (1150 to 1250°F) may be needed to eliminate most of the internal stresses. Figure 12.16 shows the effect of stress-relieving time and temperature on the extent of stress relief for seven low-alloy gray irons, and the accompanying table indicates that the stress relieving at 620°C (1150°F) for 8 hr does not have an adverse effect on hardness.[40] For austenitic (Ni-resist) cast irons, stress-relieving is done at 620 to 675°C (1150 to 1250°F) to remove residual stresses due to casting and/or machining. Stress-relieving should follow rough machining, especially for castings that require close dimensional tolerances, that have been extensively welded, or that are exposed to high stresses in service. In these alloys, stress-relieving at 675°C (1250°F) will eliminate about 95% of the stress. It is the usual practice to cool castings in air at a rate of 1 to 2 hr/in. (25 mm) of section thickness.

It is of particular importance in stress relieving to ensure that the rate of heating and rate of cooling are slow, to avoid the imposition of additional thermal stresses. It is normal practice to load the furnace at a temperature not exceeding 95°C (200°F), and the rate of heating should be such that it would take 3 hr to attain 620°C (1150°F), be held there for 1 hr, and be furnace-cooled to 315°C (600°F) or lower in about 4 hr before being allowed to cool in air. For castings of intricate shapes, it is recommended to continue furnace cooling until a temperature of about 95°C (200°F) has been attained.[40]

12.7.1.3 Normalizing.

As in steels, normalizing in gray iron castings is accomplished by heating about 50°C (100°F) above the critical temperature range or by heating typically to 885 to 920°C (1625 to 1700°F). The holding time at the normalizing temperature in the heat-treating furnace is about 1 hr/in. of maximum thickness, followed by cooling in still air to room temperature.

Iron castings may be normalized in three different ways, and the choice among them solely depends on the existing condition and the final hardness desired:[41]

1. Castings from the sand molds are usually given a separate normalizing treatment to obtain the increased hardness of castings.

2. Air cooling, just after the completion of solidification of castings, is done to remove excess (or free) cementite.

3. After the solidification of castings, and its subsequent fall of temperature to a region above the critical temperature range, they are stripped off from the mold, freed of sand, and cooled in still air to achieve an improvement in hardness.

Normalizing may be used to improve mechanical properties such as tensile strength and hardness or to restore as-cast properties that have been modified by

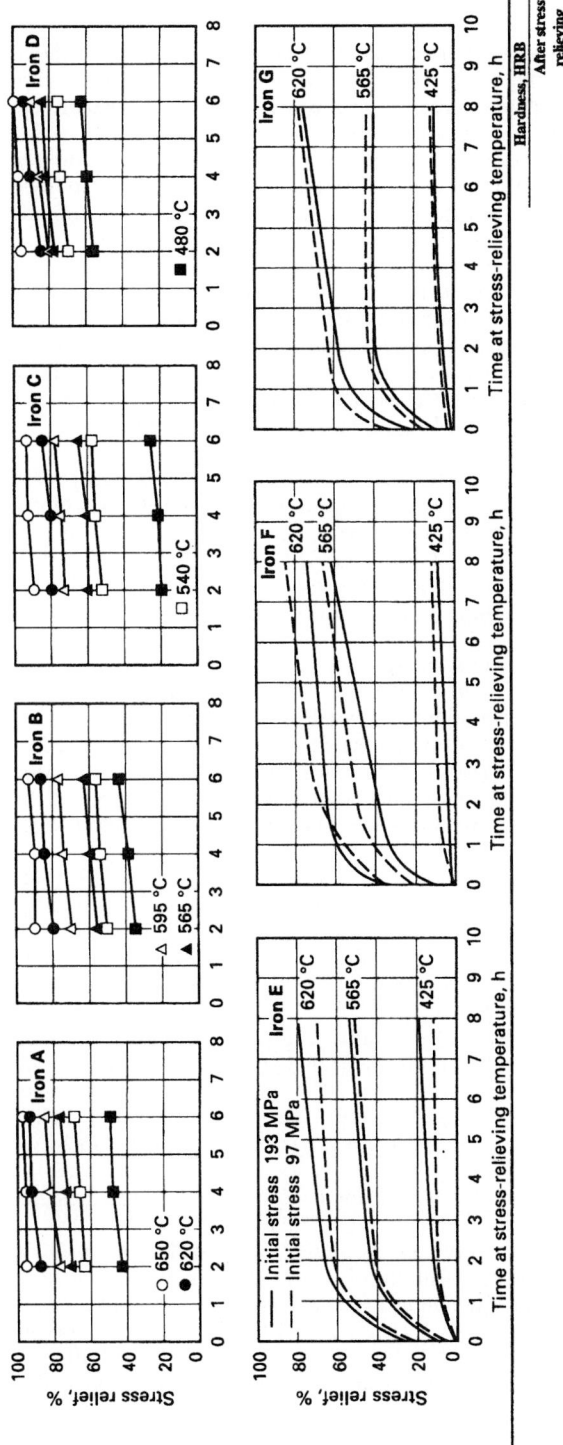

FIGURE 12.16 Effect of stress-relieving time and temperature on extent of stress relief obtained in low-alloy gray irons. Table lists compositions and negligible effect of maximum stress-relieving conditions on hardness.[40]

Iron	Composition, %										Hardness, HRB	
	C	Si	P	S	Mn	Ni	Cr	Mo	Cu	V	Before stress relieving	After stress relieving for 8 h at 620 °C (1150 °F)
A	2.93	2.14	0.110	0.57	0.47	0.35	0.10	...	...	...	98	94
B	3.43	2.12	0.104	0.70	0.81	0.34	0.18	0.23	...	...	98	94
C	3.24	2.55	0.107	0.62	0.87	0.51	0.20	0.22	...	...	95	95
D	3.91	1.43	0.54	0.25	0.32	1.56	0.06	...	...	...	82	80
E	3.18	2.13	0.73	0.125	0.70	1.03	0.33	0.65	...	...	98	98
F	3.12	1.76	0.075	0.097	0.78	1.02	0.41	0.58	...	...	94	95
G	2.78	1.77	0.065	0.135	0.55	0.36	0.10	0.33	0.46	0.04	96	96

another heat-treating process, such as graphitizing or the preheating and postheating associated with repair welding. The tensile strength and hardness of normalized gray iron castings are a function of combined carbon content, pearlite spacing, and graphite morphology.

Normalizing produces fine pearlite matrix, the fineness depending on the maximum (normalizing) temperature and alloy content. This structure exhibits good wear resistance with reasonable machinability and an excellent response to flame hardening.

For unalloyed gray irons, unless fans are used, normalizing produces structures softer than the as-cast material irrespective of the temperature used. For alloy irons, however, harder and stronger normalized structure is produced with higher normalizing temperature. It is thus apparent that normalizing retains as-cast properties to gray irons, and if the carbon equivalent is significantly small, normalizing even results in an improvement of these properties. Additionally, the alloying elements Cr, Ni, and Mo increase the strengthening effect of normalizing.[40]

Normalizing of alloy irons is often followed by tempering at 500 to 625°C (950 to 1150°F) to reduce hardness and to relieve some of the residual stresses developed when parts have variable section thicknesses.

12.7.2   Ductile (or Nodular or Spheroidal Graphite) Cast Iron

The principal types of annealing and normalizing treatments are discussed here.

12.7.2.1   Annealing.   Castings containing higher levels of C and Si are readily annealed due to accelerated decomposition of pearlite and carbides, thereby reducing the time at annealing temperature. This increases ductility and produces good machinability. Alternately, the alloying elements such as Mn, P, and Ni and the carbide stabilizers such as Cr and Mo retard the annealing process. This, in turn, reduces machinability on annealing.

Annealing is practiced to remove pearlite and eliminate any eutectic carbide which may form during solidification, especially in light sections.[44] Satisfactory annealing can be performed in four different ways:[45–47]

Single-Stage (Subcritical) Annealing.   The casting is heated to a subcritical temperature of 705 to 730°C (1300 to 1350°F), held for 1 hr/in. of section thickness, and then furnace-cooled at 55°C (100°F)/hr to 345°C (650°F), followed by air cooling to obtain grades 65-45-12 and 60-40-18. This treatment does not eliminate carbides, and therefore it is used when maximum impact strengths are not required. Figure 12.17 shows the effect of subcritical annealing at 705°C (1300°F) for various times on hardness of four ductile irons. For alloyed ductile irons, controlled cooling below 55°C (100°F)/hr through the critical temperature range down to 400°C (750°F) is recommended.[46,47]

Full (Two-Stage) Annealing with Carbides Present.   The iron casting is austenitized first by heating to temperatures in the range of 900 to 925°C (1650 to 1700°F), held there for 2 to 4 hr, then furnace-cooled at 95°C (200°F)/hr to 680 to 705°C (1256 to 1300°F); it is usually held for 2 to 6 hr to allow the graphitization process to proceed to completion with the production of a ferrite matrix. The material is finally furnace-cooled at 55°C (100°F)/hr to 345°C (650°F) prior to removal for air cooling to obtain grades 65-45-12 and 60-40-18.[45–47] Figure 12.18 shows the distribution of graphite nodules within a ferrite matrix produced by this two-stage annealing process.[46]

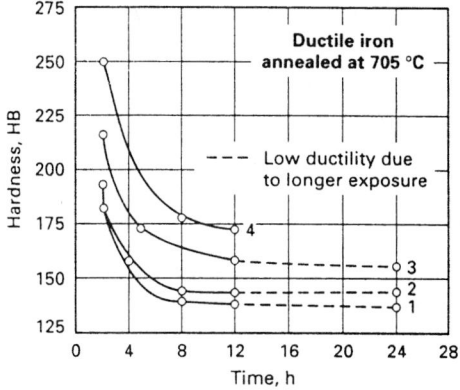

FIGURE 12.17 Effect of time at subcritical annealing temperature on hardness.[46,47] (*Reprinted by permission of ASM International, Materials Park, Ohio.*)

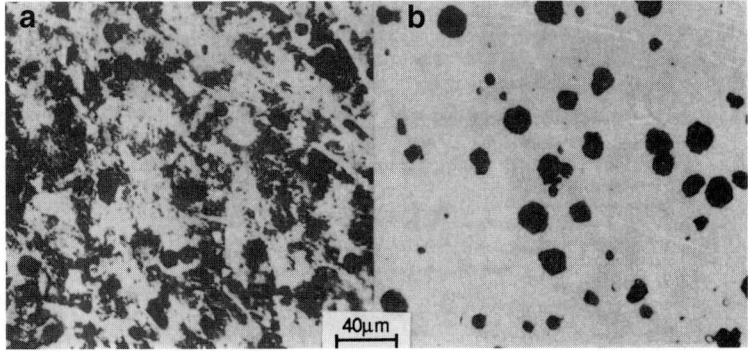

FIGURE 12.18 Microstructure of a 4-mm ($^5/_{32}$-in.) section of ductile iron: (*a*) As-cast. (*b*) After two-stage annealing treatment consisting of holding at 900°C (1650°F) for 4 hr, furnace-cooling to 690°C (1275°F), holding for 5 hr, then furnace-cooling to room temperature.[46] Picral etchant. (*Reprinted by permission of ASM International, Materials Park, Ohio.*)

Modified Two-Stage Annealing. The modified two-stage annealing cycles have been developed to produce the same or even better properties in remarkably shorter times. In this method, castings of 0.5-in. section thickness are first austenitized at 870 to 900°C (1600 to 1650°F) for 20 min to dissolve any free carbides and are then cooled rapidly (in 15 min) to 675°C (1250°F) to transform austenite to pearlite. In the second subcritical stage, the temperature is raised to 760°C (1400°F) and held there for 10 min, during which 90% ferritization of pearlitic carbide is achieved. The total time required after heating to 900°C (1650°F) for this two-stage process is 45 min.[41]

Full Annealing (for Unalloyed Casting in Absence of Carbides). The lower silicon iron casting is heated and held at 870 to 900°C (1600 to 1650°F) for 1 hr/in. of section thickness and then furnace-cooled at 55°C (100°F)/hr to 345°C (650°F), followed by air cooling to obtain grade 60-40-18 with maximum low-temperature impact strength.[46,47]

12.7.2.2   Stress Relieving.   This treatment removes residual stresses in (1) unalloyed castings by heating at 510 to 565°C (950 to 1050°F); (2) alloy castings by heating at 565 to 595°C (1050 to 1100°F); (3) high-alloy castings by heating at 595 to 650°C (1100 to 1200°F); and (4) austenitic castings by heating at 620 to 675°C (1150 to 1250°F). The required time at temperature is normally 1 hr plus 1 hr/in. section thickness. Castings should be furnace-cooled to 290°C (550°F), followed by air cooling. In majority of cases, austenitic ductile iron can be uniformly air-cooled from the stress-relieving temperature.[46,47] Cooling should be uniform to avoid reimposition of thermal stresses.

12.7.2.3   Normalizing.   This treatment can be carried out by holding to a temperature between 870 and 940°C (1600 and 1725°F), typically 100°C (180°F) above the critical temperature range, for 1 hr/in. section thickness in the heat-treating furnace, followed by air cooling. This treatment can be used to break down carbides, increase strength and hardness, and produce more uniform properties, but with a large decrease in ductility. It produces a fine pearlitic structure together with spheroidal graphite, provided that Si is not too high and that it contains a moderate Mn content.

Normalizing should be followed by tempering (or reheating) at 425 to 650°C (800 to 1200°F) and holding at that temperature for 1 hr/in. section thickness, which renders uniform hardness and mechanical properties (including high toughness and impact resistance) and relieves residual stresses arising from the uneven air cooling of different section thicknesses. The resulting microstructure depends on the composition of the castings and the cooling rate; the former, in turn, depends on the alloy content, while the latter depends on the mass and temperature of castings. *Step normalizing*, using a second lower-temperature stage prior to air cooling, can be employed to give the improved matrix control needed to produce the pearlitic/ferritic grades of ductile iron.[48] The effect of tempering on tensile properties and hardness depends on both the composition of the iron and the hardness level attained in normalizing. Figure 12.19 illustrates the effect of tempering temperature on the hardness of normalized ductile iron.[46]

12.7.3   Compacted Graphite (CG) Iron

Like gray and ductile irons, annealing and normalizing of CG iron can produce a variety of matrix structures such as ferrite, ferrite-pearlite mixture, and pearlite. However, CG irons are not heat-treated, due to their applications in the as-cast condition.

12.7.4   Malleable Iron

12.7.4.1   Annealing.   Annealing, as well as other heat treatment, is done in a controlled atmosphere to avoid decarburization, scaling, and loss of silicon during the extended periods at high temperature (955°C). This atmosphere is generated by packing the castings in a mixture of sand (or gravel) and carbonaceous material when heated in an open-fired furnace. Alternatively, castings are carefully sealed from air or the usual furnace atmosphere, and annealing is continued in an atmosphere created by the castings themselves. However, in recent years, the most effective atmosphere to prevent decarburization has been (1) dry nitrogen mixed with 1.5% hydrogen and 1.5% CO or (2) vaporized liquid nitrogen with the addition of

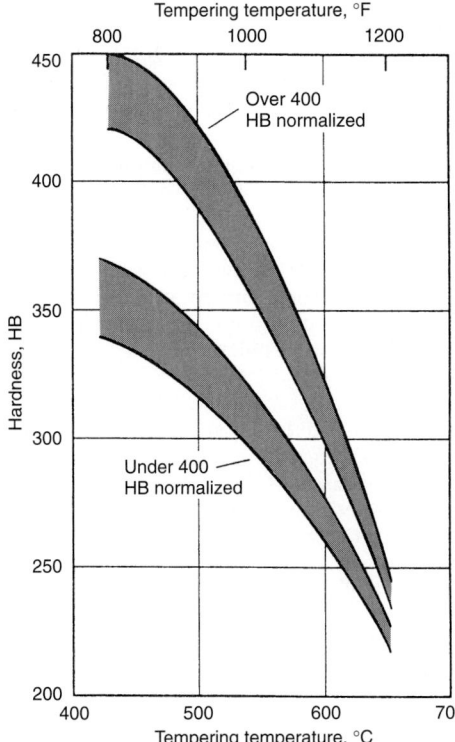

FIGURE 12.19 Hardness of normalized ductile iron tempered at various temperatures.[46] (*Reprinted by permission of ASM International, Materials Park, Ohio.*)

some methane.[49] The dew points of these mixtures should lie between −40 and −7°C (−40 and −20°F).

Ferritic, pearlitic, and martensitic malleable irons are produced through variations of controlled annealing of white cast iron of suitable composition. Thus, annealing is an integral part of the manufacturing process for these irons. During the annealing cycle, carbon that is present in the combined form, either as a microconstituent in pearlite or as massive carbides, is converted into free graphite in the form of irregular clumplike nodules called *temper carbon*. It occurs early during the holding or soaking period.

Ferritic Malleable Cast Iron.   The annealing heat treatment for ferritic malleable cast iron consists of the following two stages.

FIRST-STAGE GRAPHITIZATION.   The first stage converts primary carbides to temper-carbon nodules. In this stage, white iron castings are heated at a rate such that the temperature of 900 to 970°C (1650 to 1780°F) is reached in 4 hr. The soaking time varies from ~2 to 36 hr depending on the composition of the iron castings and the temperature of first-stage annealing. For example, longer soaking times are necessary if the silicon content of iron castings or the temperature employed is low.

During holding at the first-stage temperature, iron carbide dissolves in the austenite matrix, and the excess carbon diffuses to nucleate temper-carbon nodules at preferred sites such as boron nitride or at the (primary) carbide/(saturated) austenite interface. Thus the growth around these nuclei occurs by a reaction involving carbide decomposition and the diffusion rate of carbon. The rates of nucleation and graphitization are increased by high silicon and carbon contents. However, to solidify it as a white iron, these elements must be restricted to a certain maximum level. The first-stage annealing temperature also influences the rate of annealing and the number of temper-carbon nodules produced.

SECOND-STAGE GRAPHITIZATION. Castings are cooled as fast as practical to 740 to 760°C (1360 to 1400°F). This rapid cooling step requires 1 to 6 hr, depending on the equipment being used. Castings are then cooled slowly (through the transformation range of iron) at a rate of about 3 to 11°C (5 to 20°F)/hr to 649°C (1200°F); during this period, the remaining carbon dissolved in the austenite phase is converted to graphite and transferred to the existing temper carbon formed at the higher temperatures, and the austenite transforms to low-carbon ferrite. The final microstructure consists of uniformly dispersed temper-carbon nodules within a fully ferritic matrix[50,51] (Fig. 12.20a).

Pearlitic Malleable Iron. In the production of pearlitic malleable iron, the first-stage heat treatment includes annealing for about 13 hr at 970°C (1780°F). The second stage involves air cooling (Fig. 12.20b). Faster air cooling by an air blast avoids the formation of films of ferrite around the temper-carbon particles (bull's-eye structure) and produces less ferrite and a finer pearlitic structure[50,51] (Fig. 12.20c). The castings are then either tempered or reheated in a furnace to 870°C (1600°F), oil-quenched, and tempered.

Martensitic Malleable Iron. Uniformly high quality irons are produced by agitated oil-quenching of the castings directly from the first-stage annealing after stabilizing the temperature at 845 to 870°C (1550 to 1600°F) for 15 to 30 min. The martensitic matrix produced has a hardness of 415 to 601 HB. Finally, the castings are tempered at 590 to 725°C (1100 to 1340°F) to develop the desired microstructure (of tempered martensite plus temper-carbon nodules) and mechanical properties. If the hardness decreases by prolonged tempering, the resulting microstructure may not possess a good response to selective hardening.[50,51]

Note that during the heat treatment process (such as annealing), expansion or growth of castings occurs due to the nucleation and growth of temper-carbon nodules. Any restriction of the free growth such as the weight of castings placed on top of each other leads to distortion. This may, however, be corrected by die-pressing. The design of the molding pattern should reflect this dimensional change.[49]

12.8 ENGINEERING PROPERTIES AND APPLICATIONS OF CAST IRONS

12.8.1 Gray Iron

Mechanical Properties. Different graphite flake types, the amount of prior austenite dendrites, and the matrix structure present in the iron structure affect the mechanical properties.

Tensile strength is an important criterion in selecting a gray iron for parts that are subjected to static loads in direct tension or bending. Such parts include pressure vessels, housings, autoclaves, valves, fittings, and levers. Based on the uncertainty of loading, safety factors of 2 to 8 are used to estimate allowable design stresses. The tensile strength of the plain and lower-alloy gray irons varies from 69 to 414 MPa. The carbon equivalent and rate of solidification or section size play an

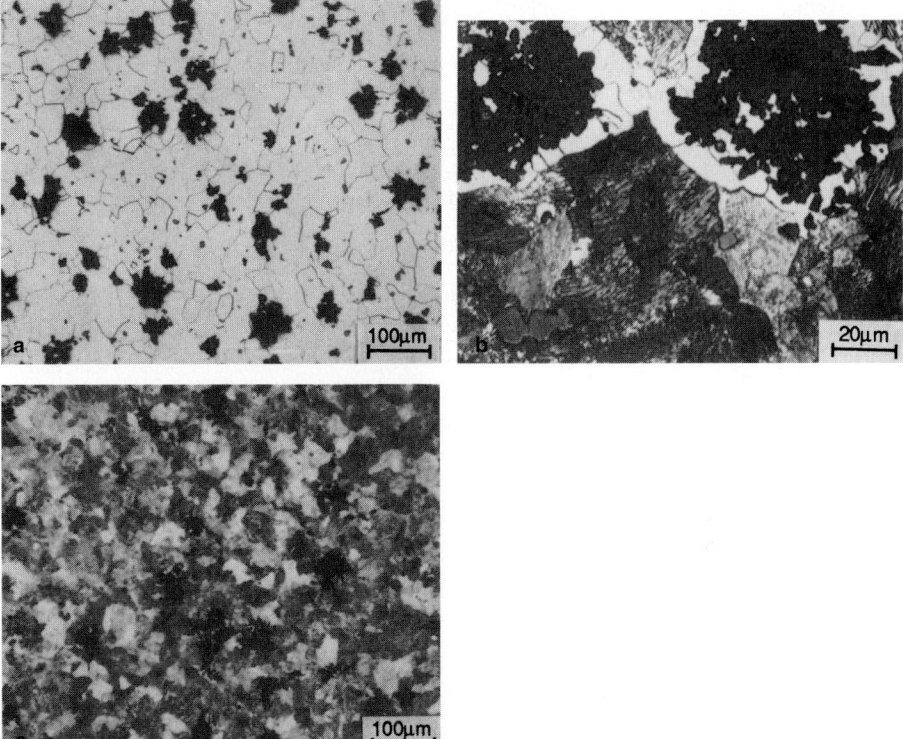

FIGURE 12.20 Microstructure. (*a*) ASTM A602, grade 3210, ferritic malleable iron with temper-
carbon (type III graphite) nodules (black constituents) in a matrix of granular ferrite produced
after two-stage annealing by holding at 954°C (1750°F) for 4 hr, cooling to 704°C (1300°F) in 6 hr,
followed by air cooling. Small gray particles are MnS (2% nital etch). (*b*) Pearlitic malleable iron
with the formation of free ferrite films (white constituent) around temper carbon (bull's-eye structure) obtained after first-stage annealing by austenitizing at 971°C (1780°F) for 13.5 hr and then air-
cooling slowly (2% nital etch). (*c*) Malleable iron containing 2.5% C and 1.5% Si with temper-carbon
(type III graphite) nodules in a matrix of fine pearlite (variegated gray), produced after first-stage
annealing by holding at 954°C (1750°F) for 2 hr and air-blast-cooling (1% nital etch). (*Reprinted by
permission from Metals Handbook, vol. 7, 8th ed., American Society for Metals, Metals Park, Ohio,
1972.*)

important role on this strength level. Higher strength is achieved by a combination
of small proportions of graphite, decreasing carbon equivalent, finer flakes, more
austenitic dendrites, and stronger matrices (obtained with decreasing section sizes
or faster solidification and cooling times).[49]

Gray irons do not show straight stress-strain curves (Fig. 12.21) due to the development of localized plastic deformation, during loading, which results from the
stress concentration (notch) effect at the edges of graphite flakes.[52]

The amounts of graphite, graphite flake types, and matrix structure also influence the modulus of elasticity. For example, increasing graphite content, longer
graphite flake length, increasing carbon equivalent, increasing section size of the
gray iron castings, and softer matrix structures (produced by annealing of unalloyed

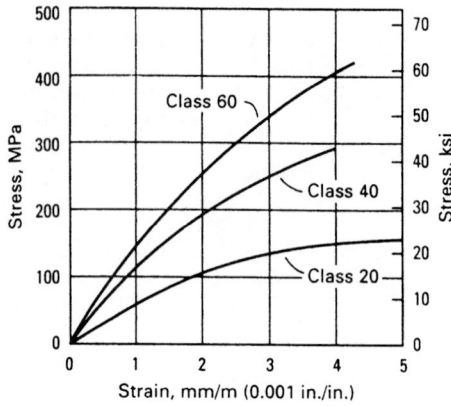

FIGURE 12.21 Typical stress-strain curves for three classes of gray iron in tension.[52] (*Reprinted by permission of ASM International, Materials Park, Ohio.*)

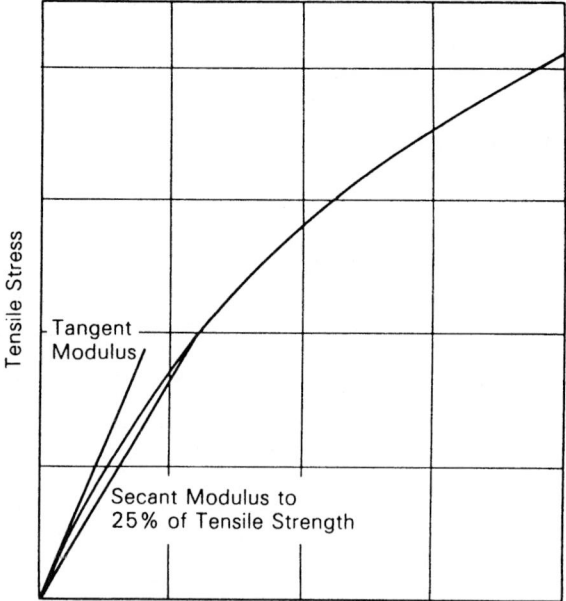

Strain, Inches Per Inch or Percent

FIGURE 12.22 Two methods for determining the modulus of elasticity for gray iron. The secant modulus to 25% of the tensile strength is the more commonly used.[41] (*Courtesy of Iron Castings Society, Inc., Des Plaines, Illinois.*)

iron) reduce the modulus.[41] Figure 12.22 shows the two methods for determining the modulus of elasticity for gray iron castings.[42] The more commonly used method of measuring the modulus of elasticity is the secant modulus, which corresponds to the slope of a straight line from the origin (i.e., at no load) to a point on the stress-strain curve at one-fourth of the ultimate tensile strength; and the deviation of the stress-strain curve from linearity is usually <0.01% at these loads. This is useful for most engineering work because design loads rarely exceed one-fourth of the ultimate strength. Another method is the tangent modulus, which represents the stress-strain curve at its origin. This is advantageous in the design of precision machinery where applied stresses are low.[41] The modulus of elasticity ranges from about 69 to 138 GPa. Poisson's ratio varies from about 0.24 for the soft iron to 0.27 for the high-strength irons at the start of loading. The values decrease progressively with increasing stress and higher carbon equivalent.

Compression strength is important when gray iron is employed for machinery foundation or supports. The graphite flakes exert much less influence on the compression properties than on the tensile properties. As a result, the compression strength of gray irons is 3 to 4 times its tensile strength. (See Table 12.6 and Fig. 12.23.)

Unlike steel and ductile and malleable irons, *the ratio of the tensile strength to hardness* in gray iron varies with carbon equivalent, being greater for the higher-

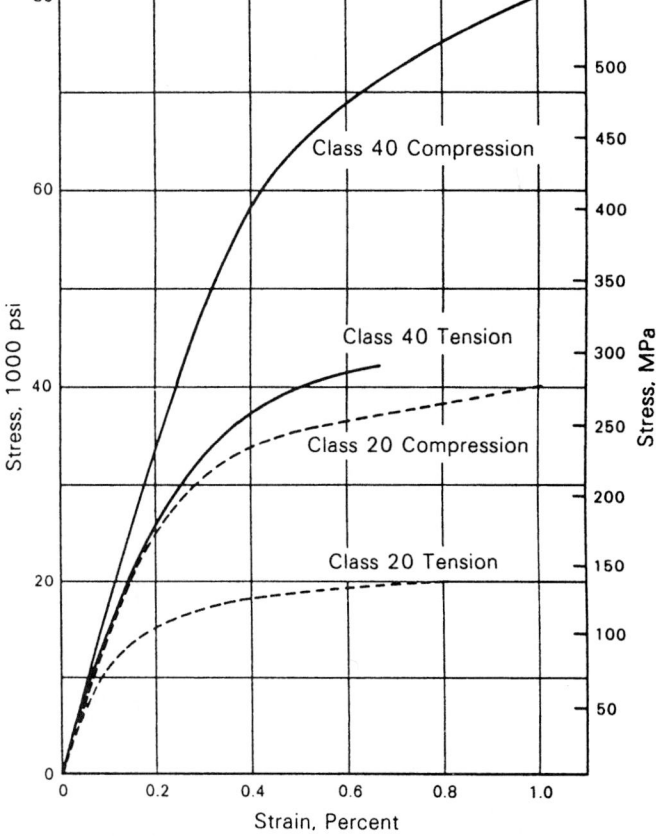

FIGURE 12.23 A comparison of stress-strain curves in tension and compression for a class 20 and a class 40 gray iron.[41] (*Courtesy of Iron Castings Society, Inc., Des Plaines, Illinois.*)

TABLE 12.6 Typical Mechanical Properties of As-Cast Standard Gray Iron Test Bars[52]

ASTM class	Tensile strength		Torsional shear strength		Compressive strength		Reversed bending fatigue limit		Transverse load on test bar B		Hardness, HB	Tensile modulus		Torsional modulus	
	MPa	ksi	MPa	ksi	MPa	ksi	MPa	ksi	kg	lb		GPa	10^6 psi	GPa	10^6 psi
20	152	22	179	26	572	83	69	10	839	1850	156	66–97	9.6–14.0	27–39	3.9–5.6
25	179	26	220	32	669	97	79	11.5	987	2175	174	79–102	11.5–14.8	32–41	4.6–6.0
30	214	31	276	40	752	109	97	14	1145	2525	210	90–113	13.0–16.4	36–45	5.2–6.6
35	252	36.5	334	48.5	855	124	110	16	1293	2850	212	100–119	14.5–17.2	40–48	5.8–6.9
40	293	42.5	393	57	965	140	128	18.5	1440	3175	235	110–138	16.0–20.0	44–54	6.4–7.8
50	362	52.5	503	73	1130	164	148	21.5	1638	3600	262	130–157	18.8–22.8	50–55	7.2–8.0
60	431	62.5	610	88.5	1293	187.5	169	24.5	1678	3700	302	141–162	20.4–23.5	54–59	7.8–8.5

Reprinted by permission of ASM International, Materials Park, Ohio.

strength low-carbon equivalent gray iron (Fig. 12.24). This implies that this ratio is influenced by the amount and shape of the flake graphite present.

Many grades of gray iron have *torsional shear strength* greater than that of some grades of steel. This characteristic, together with low notch sensitivity of gray iron, makes it a suitable material for various types of shafting, particularly in the higher-strength grades.[52] Most shafts are subjected to dynamic torsional stresses, and the designer should examine the true nature of the loads to be encountered. For high-strength grades, stress concentration factors associated with shape change in the part are vital both for torque loads and for bending and tension loads.[52]

Hardness of gray iron is measured by Brinell or Rockwell testers. To compare hardness, the type and amount of graphite in the irons must remain the same. Rockwell and Brinell hardness testers are used for hardened castings (such as camshafts). For unhardened castings, Brinell tests are preferred and used for strength correlations. Figure 12.25 shows the relationship between Brinell hardness and tensile strength.[52]

Fatigue properties of gray iron depend on the graphite and matrix structures, tensile strength, and mode of stressing. Since gray irons have higher strengths in bending, compression, and torsion tests as compared to their tensile strength, the endurance ratio or fatigue properties under these types of stresses are much higher than for tensile loading. Gray irons containing more austenitic dendrites or a Type D graphite in their structures possess a higher endurance ratio than those containing less dendrites or Type A graphite. The endurance limit of iron shows very little improvement with the increase in tensile strength because it is also influenced by graphite structure and mode of stressing, as mentioned before.[53]

Table 12.6 lists the typical mechanical properties of as-cast standard gray iron test bars.

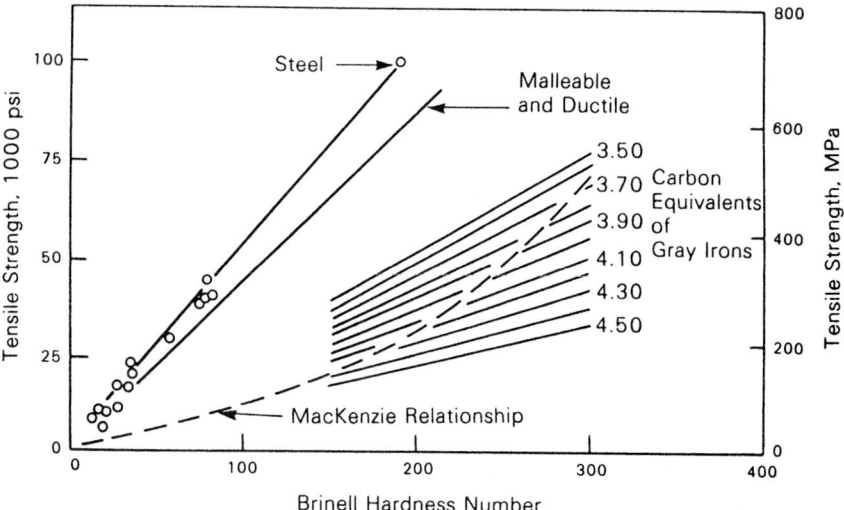

FIGURE 12.24 Tensile strength to hardness relationships for gray, ductile, and malleable irons and steels.[41] (*Courtesy of American Cast Metals Association, Des Plaines, Illinois; after L. G. Carmack. The Mackenzie relationship is from his paper in ASTM Proceedings, vol. 46, 1946, p. 1025.*)

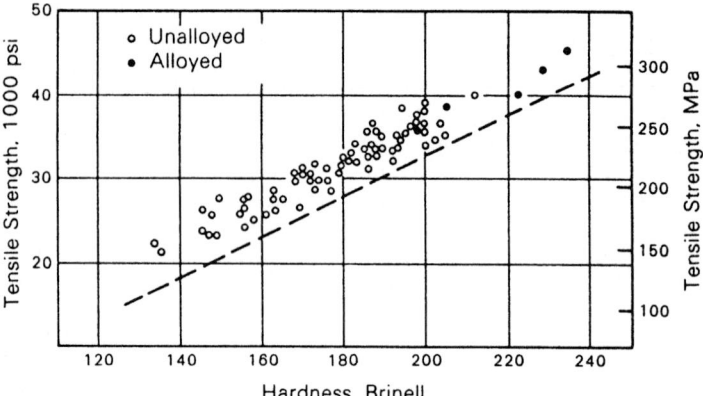

FIGURE 12.25 Relationship between tensile strength and Brinell hardness for a series of inoculated gray irons from a single foundry.[41] (*After D. E. Krause, "Gray Iron—A Unique Engineering Material," ASTM, STP 455, 1969, pp. 3–28.*)

Fatigue Limit. The fatigue strength of a material, as denoted by its fatigue limit or endurance limit, is the magnitude of the cyclic stress at which the fatigue life exceeds a certain number of cycles, usually 10^7. The fatigue strength of a material is related to its tensile strength by the endurance ratio—the ratio of the fatigue limit to tensile strength. The effect of a stress raiser on the fatigue limit is defined by the *notch sensitivity ratio*, also called the *fatigue strength reduction factor*. The notch sensitivity ratio is the ratio of the unnotched fatigue limit to the notched fatigue limit.

Figure 12.26 shows fatigue life curves at room temperature for a gray iron under completely reversed cycles of bending stress, where each point denotes the data from one specimen. This diagram primarily determines whether a certain condition of mean stress and cyclic stress offers a safe design for infinite life. The designer can also determine whether modifications in the mean stress and expected alternating stress will result in a design for the unsafe zone. Normally the data needed to analyze a given set of conditions are obtained experimentally. Note that the number of cycles of alternating stress implied in Fig. 12.26 is the number usually employed to determine fatigue limits, i.e., approximately 10^7. Fewer cycles, as encountered in infrequent overloads, will be safer than suggested by a particular point plotted on a diagram for infinite life. Too few data may be available to draw a diagram for less than infinite life.

Fatigue Notch Sensitivity. Usually, very small allowances must be made for a reduction in fatigue strength arising from notches or abrupt section variations in gray iron castings. The low-strength grades exhibit only a small reduction in strength in the presence of holes and fillets.[52] This implies that the notch sensitivity index approaches zero; i.e., the effective stress concentration factor for these notches approaches 1. High-strength grades usually display greater notch sensitivity, but perhaps not the full theoretical value represented by the stress concentration factor. Normal stress concentration factors are perhaps suitable for high-strength gray irons.[54]

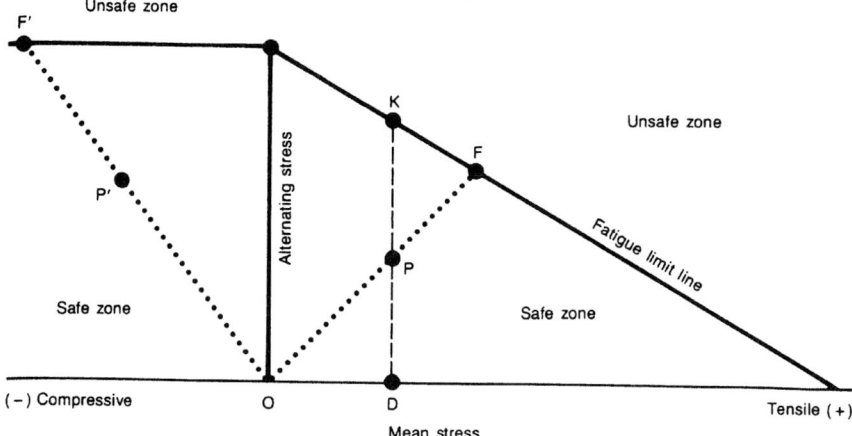

FIGURE 12.26 Diagram showing safe and unsafe fatigue zones for cast iron subject to ranges of alternating stress superimposed on a mean stress. Example point P illustrates conditions of tensile (positive) mean stress: P' illustrates compressive (negative) mean stress. The safety factor is represented by the ratio of OF to OP or OF' to OP'. For conditions of constant mean tensile stress, DK/DP is the safety factor.[52] (*Reprinted by permission of ASM International, Materials Park, Ohio.*)

Impact Resistance. Gray iron is not used for applications requiring high impact resistance because it has a considerably lower impact strength than ductile iron, malleable iron, and cast carbon steels. However, many gray iron castings require some impact strength to avoid breakage in shipment or service use.

Impact testing methods of cast iron given by ASTM A327 for notched and unnotched Charpy tests are employed as a research tool. Most producers of cast iron pressure pipe employ a routine pipe impact test as a quality control which helps prevent breakage in shipping and handling.[52]

Wear Resistance. *Wear* may be defined as the loss of material from a surface by its mechanical disintegration, caused by the relative movement between the surface and another material.[41] There are four types of wear: namely adhesive or frictional wear, abrasive wear, cutting wear, and corrosive wear. Among them, the first type, which is caused by metal-to-metal contact and production and breakage of friction welds, is important for gray irons. Gray irons have excellent resistance to the dry-sliding-friction type of wear because they do not friction-weld; they have good thermal conductivity and a low modulus of elasticity to cause minimum thermal stresses. That is why gray irons find applications as clutch plates and disk brake rotors. Gray iron is considered as the ideal material also for lubricated sliding wear because the exposed graphite flakes act as a reservoir for oil and displace the oil film under loads. This makes it resistant to lubricated sliding wear, galling, and scuffing.

It has been observed that hardened gray iron has 5 times greater wear resistance than pearlitic as-cast iron of the same composition. Hence in normal wearing applications such as valve guides, lathways, and various sliding components in machines and engines, as-cast gray iron is used; and in severe wearing applications such as high-speed diesel engine cylinder sleeves, gears, camshafts, and similar heavily loaded wearing surfaces where maximum wear resistance is the requirement, hardened gray iron is used.[52]

Machinability.   The majority of gray irons have superior machinability (Fig. 12.27) when compared to those of most other cast irons of equivalent hardness and to all grades of steel. The flake graphite interrupts continuities in the metallic matrix, which then act as chip breakers. In addition, graphite flakes act as a lubricant for a cutting tool.[52] It can be seen in Fig. 12.27*a* that the fully annealed soft ferrite matrix machines at higher speeds; however, a small amount (5%) of carbide, an extremely hard constituent in the medium pearlite matrix, has a significant detrimental effect, causing it to machine at slower speeds (Fig. 12.27*b*).[41] These carbides may exist at the edge of the casting due to rapid solidification of iron at this point.[52]

Damping Capacity.   This is defined as the ability of a material to absorb and dampen vibrations. (See also Chap. 8.) An accumulation of vibrational energy without adequate dissipation produces an increasing amplitude of vibration. Excessive vibration can cause (1) excessive wear on gear teeth and bearings, (2) fretting on mating surfaces, and (3) inaccuracies in precision machinery.

The damping capacity of gray irons, particularly of lower strength and with higher amounts of flake graphite, is considerably greater than that of steel or ductile and malleable irons (Fig. 12.28). This behavior is attributed to the flake graphite structure, together with its unique stress-strain characteristics.[41] This exceptional behavior of gray iron is advantageously used to prevent objectional vibrations and noise generation in machine bases and supports, gear covers, engine cylinder blocks and heads, and brake components.

Table 12.7 lists the relative damping capacity for several different materials. Here, the relative damping capacity δ is represented as a percentage of total energy or amplitude that is lost in a complete stress-strain cycle.[41]

Applications.   Gray iron is the most widely used cast iron because of its excellent wear resistance, machinability, high damping capacity, and ability to cast into complex configurations at the lowest possible cost.[41] Low cost has become a very vital criterion for its applications in different gear housings, pump housings, steam-turbine housings, guards and motor frames, fire hydrants, sewer covers, elevator counterweights, enclosures for electrical equipment, and industrial furnace doors.[41,52]

Gray iron is also used in more critical applications where mechanical and physical property requirements demand iron selection in pressure-sensitive castings, automotive castings, and process furnace parts. Table 12.8 lists typical applications for gray iron castings.[52]

12.8.2   Ductile Iron

Mechanical Properties.   Figure 12.29 shows the typical stress-strain curves for ductile iron.[55] Unlike gray iron, ductile irons show elastic behavior over a considerable stress range. Yield strength is determined by the usual 0.2% offset method from stress-strain curves as applied to plain carbon steels.

An increase in tensile and yield strengths can be attained in the ferritic grade by the addition of Si, Mn, or Ni but at the sacrifice of impact ductility, particularly with

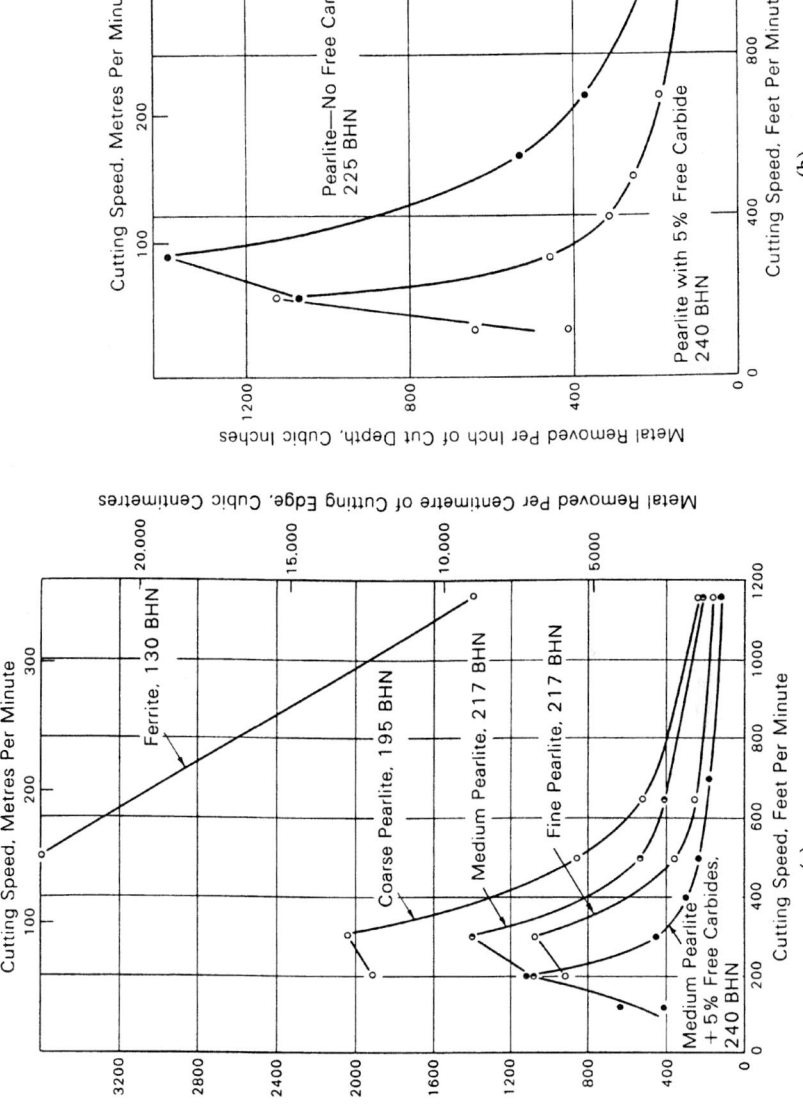

FIGURE 12.27 (*a*) Effect of matrix microstructure of iron on tool life during machining.[42] (*b*) Detrimental effect of as little as 5% of free carbides in the matrix structure on tool life which results in only a small increase in hardness.[41]

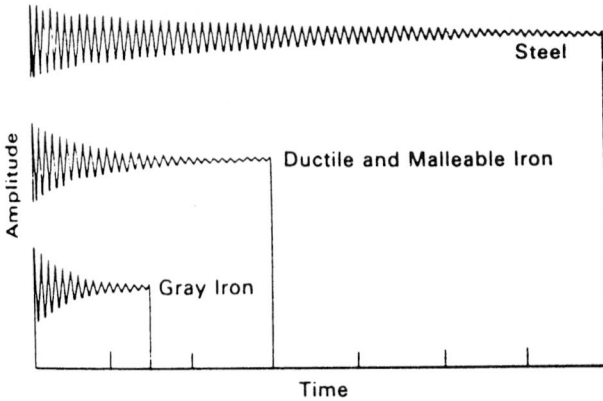

FIGURE 12.28 The relative ability of ferrous metals to dampen vibrations. The energy absorbed per cycle, or the specific damping capacity of these, can differ by more than 10 times.[41] (*Courtesy of American Cast Metals Association, Des Plaines, Illinois.*)

TABLE 12.7 Relative Damping Capacity of Different Materials[41]

Material	$\delta \times 10^{-4}$[†]
White iron	2–4
Malleable iron	8–15
Ductile iron	5–20
Gray iron, fine flake	20–100
Gray iron, coarse flake	100–500
Eutectoid steel	4
Armco iron	5
Aluminum	0.4

[†] Natural log of the ratio of successive amplitude.
Source: After Litovka et al.

Mn and Si.[56] The modulus of elasticity can be easily determined from the curve and is ~169 GPa for fully ferritic ductile irons and ~176 GPa for fully pearlitic ones.[44] Poisson's ratio does not vary much with grade and is about 0.275 for most ductile irons.

Figure 12.30 shows the correlation between dynamic elastic modulus (DEM) and nodularity, which emphasizes both the strong effect of nodularity on the DEM values obtained by sonic testing and the use of DEM values determined by sonic testing to measure nodularity (nodule count and graphite volume should be relatively constant).[48]

Table 12.9 provides the average tensile properties of one heat of ductile iron heat-treated to four different strength levels. As-cast tensile and elongation values are greater than the heat-treated ones. However, the yield strength values may be lower. Within each grade, strength and ductility values vary to some extent with

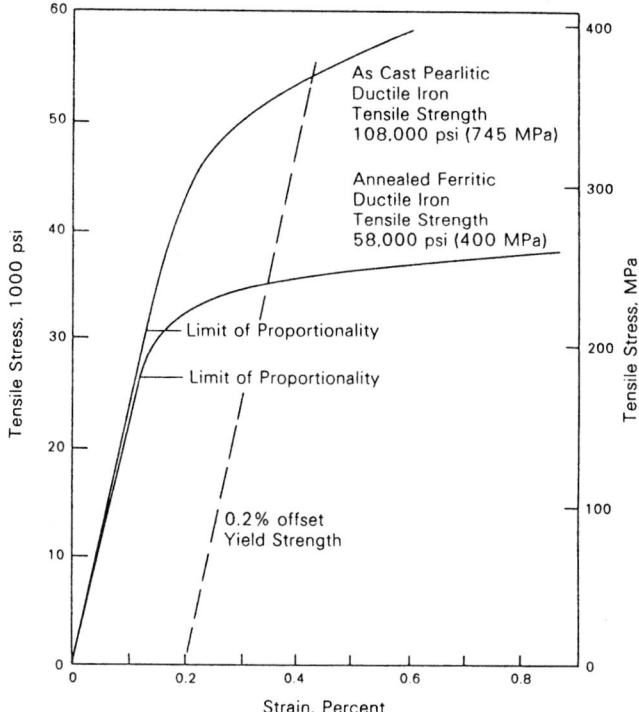

FIGURE 12.29 Stress-strain curves for two types of ductile iron.[41] (*Courtesy of American Cast Metals Association, Des Plaines, Illinois; after G. N. J. Gilbert.*)

hardness.[57] The modulus of elasticity in tension lies in the range of 164 to 176 GPa (23.8×10^6 to 25.5×10^6 psi) and remains more or less the same with grade. This value in tension cannot be employed in design for cantilever or three-point beam or torsion loading due to greater deflection. In this situation, a value of 142 GPa (20.5×10^6 psi) should be employed.

Relationships between Tensile Properties. The strong dependence of graphite morphology and matrix structure on different tensile properties of ductile iron provides important correlations between these properties. Figure 12.31 shows the nonlinear least-square relationships between yield and tensile strengths and dynamic elastic modulus. In 1970 Siefer and Orths[58] established a relationship between tensile strength and elongation in ductile iron samples as

$$(\text{Tensile strength})^2 \times (\text{Elongation}) = Q \qquad (12.6)$$

where Q is a constant which was defined by Crews[59] as the *quality index* (QI) of ductile iron. A large value of QI denotes a combination of higher strength and elongation and, therefore, higher material performance. High QI values, associated with high-quality castings, usually result from high nodularity (high percentage of

TABLE 12.8 Typical Applications for Gray Iron Castings[52]

Specification	Grade or class	Typical applications
ASTM A48	20, 25	Small or thin-sectioned castings requiring good appearance, good machinability, and close dimensional tolerances
	30, 35	General machinery, municipal and waterworks, light compressors, automotive applications
	40, 45	Machine tools, medium-duty gear blanks, heavy compressors, heavy motor blocks
	50, 55, 60	Dies, crankshafts, high-pressure cylinders, heavy-duty machine tool parts, large gears, press frames
ASTM A159, SAE J431	G1800	Miscellaneous soft iron castings (as-cast or annealed) in which strength is not of primary consideration. Exhaust manifolds may be made of this grade of iron alloyed or unalloyed. These may be annealed to avoid growth cracking due to heat.
	G1800h	Brake drums and disks where very high damping capacity is required
	G2500	Small cylinder blocks, cylinder heads, air-cooled cylinders, pistons, clutch plates, oil pump bodies, transmission cases, gearboxes, clutch housings, light-duty brake drums
	G2500a	Brake drums and clutch plates for moderate service requirements, where high-carbon iron is desired to minimize heat checking
	G3000	Automobile and diesel cylinder blocks, cylinder heads, flywheels, differential carrier castings, pistons, medium-duty brake drums, clutch plates
	G3500	Diesel engine blocks, truck and tractor cylinder blocks and heads, heavy flywheels, tractor transmission cases, heavy gearboxes

	G3500b	Brake drums and clutch plates for heavy-duty service where both resistance to heat checking and higher strength are definite requirements
	G3500c	Extra heavy-duty service brake drums
	G4000	Diesel engine castings, liners, cylinders, pistons
	G4000d	Heavy-duty camshafts
ASTM A126	A, B, C	Valve pressure-retaining parts, pipe fittings, flanges
ASTM A278	40, 50, 60, 70, 80	Valve bodies, paper mill dryer rollers, chemical process equipment, pressure vessel castings
ASTM A319	I, II, III	Stoker and firebox parts, grate bars, process furnace parts, ingot molds, glass molds, caustic pots, metal melting pots
ASTM A823	…	Automobile, truck, appliance, and machinery castings in quantity
ASTM A436[†]	1	Valve guides, insecticide pumps, floodgates, piston ring bands
	1b	Seawater valve and pump bodies, pump section belts
	2	Fertilizer applicator parts, pump impellers, pump casings, plug valves
	2b	Caustic pump casings, valves, pump impellers
	3	Turbocharger housings, pumps and liners, stove tops, steam piston valve rings, caustic pumps and valves
	4	Range tops
	5	Glass rolls and molds, machine tools, gauges, optical parts requiring minimal expansion and good damping qualities, solder rails and pots
	6	Valves

† Nickel-alloyed (13.5 to 36% Ni) austenitic gray iron.
Reprinted by permission of ASM International, Materials Park, Ohio.

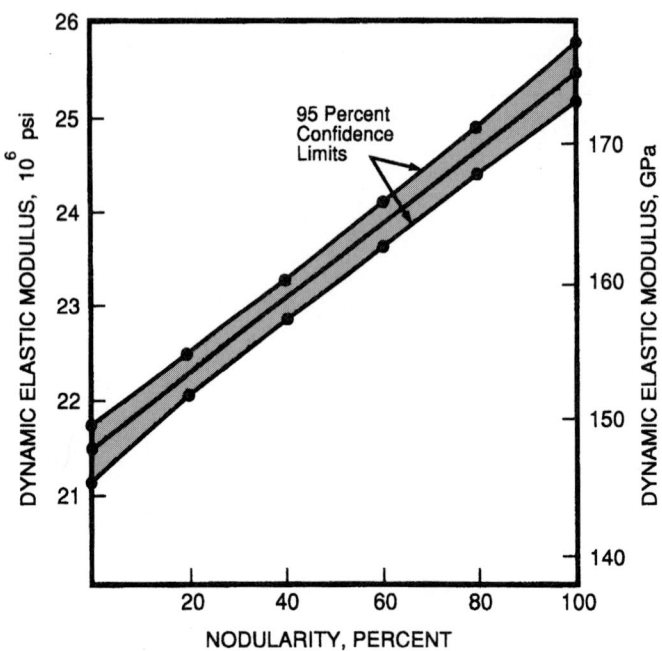

FIGURE 12.30 Relationship between dynamic elastic modulus (DEM) and nodularity.[48] (*Courtesy of QIT-Fer et Titane Inc.*)

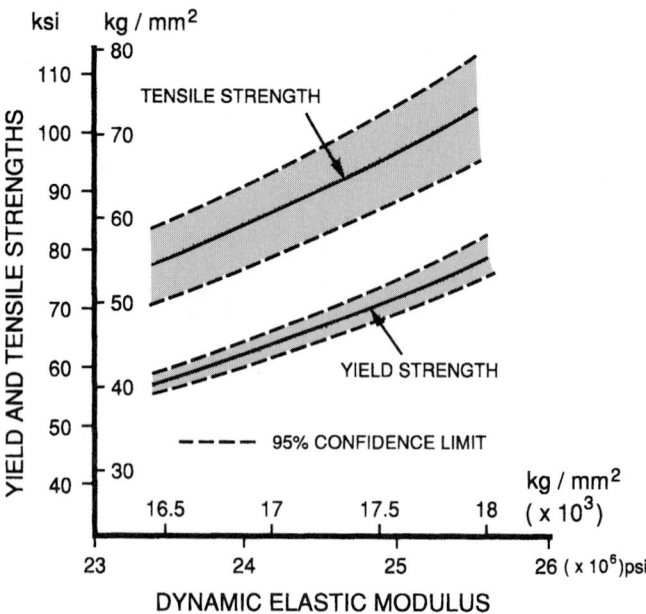

FIGURE 12.31 Relationships between yield and tensile strengths and dynamic elastic modulus for ductile iron.[48] (*Courtesy of QIT-Fer et Titane Inc.*)

TABLE 12.9 Average Mechanical Properties of Ductile Irons Heat-treated to Four Strength Levels[57]

Determined for a single heat of ductile iron, heat-treated to approximate standard grades. Properties were obtained using test bars machined from 25-mm (1-in.) keel blocks.

Nearest standard grade	Hardness, HB	Ultimate strength		Yield strength		Elongation in 50mm (2in.), %	Modulus		Poisson's ratio
		MPa	ksi	MPa	ksi		GPa	10⁶ psi	
Tension									
60-40-18	167	461	66.9	329†	47.7†	15.0	169	24.5	0.29
65-45-12	167	464	67.3	332†	48.2†	15.0	168	24.4	0.29
80-55-06	192	559	81.1	362†	52.5†	11.2	168	24.4	0.31
120-90-02	331	974	141.3	864†	125.3†	1.5	164	23.8	0.28
Compression									
60-40-18	167	...	...	359†	52.0†	...	164	23.8	0.26
65-45-12	167	...	...	362†	52.5†	...	163	23.6	0.31
80-55-06	192	...	...	386†	56.0†	...	165	23.9	0.31
120-90-02	331	...	...	920†	133.5†	...	164	23.8	0.27
Torsion									
60-40-18	167	472	68.5	195‡	28.3‡	...	63	9.1	...
							65.5§	9.5§	
65-45-12	167	475	68.9	207‡	30.0‡	...	64	9.3	...
							65§	9.4§	
80-55-06	192	504	73.1	193‡	28.0‡	...	62	9.0	...
							64§	9.3§	
120-90-02	331	875	126.9	492‡	71.3‡	...	63.4	9.2	...
							64§	9.3§	

† 0.2% offset.
‡ 0.0375% offset.
§ Calculated from tensile modulus and Poisson's ratio in tension.
Reprinted by permission of ASM International, Materials Park, Ohio.

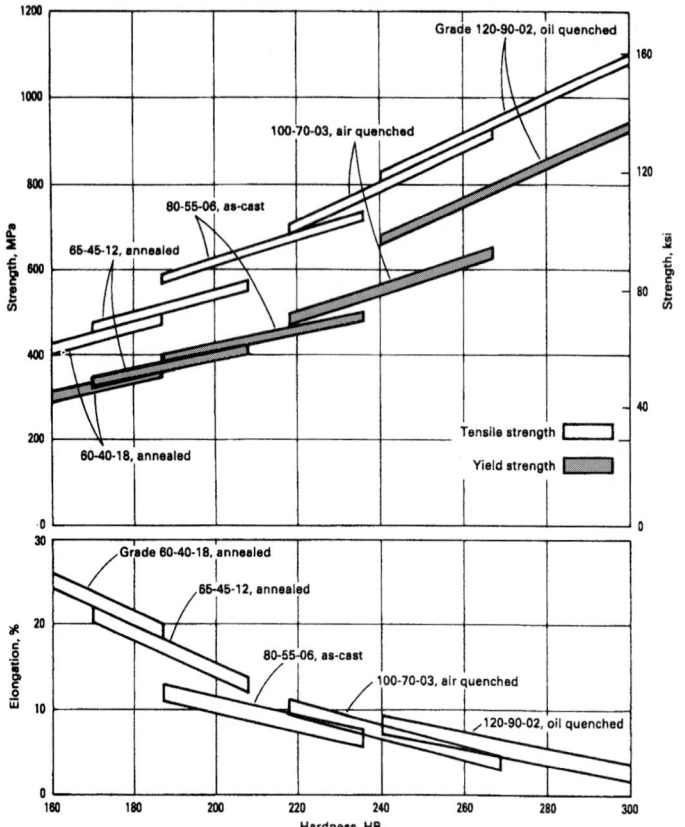

FIGURE 12.32 Tensile properties versus hardness values of ductile iron conforming to different grades of ASTM Specification A536.[48]

spherical or nonspherical particles), absence of intercellular degenerate graphite, high nodule count, a low volume fraction of carbides, low phosphorus (<0.03%) content, and freedom from internal porosity.[48]

Figure 12.32 shows the overlapping of tensile properties with hardness for different grades of ductile iron, according to ASTM A536 specification.

Tensile Properties versus Hardness. Microhardness data for the individual microstructural constituents can be used to quantitatively predict the tensile properties of as-cast, annealed, and normalized ductile irons. Figure 12.33, due to Venugopalan and Alagarsamy,[60] compares strengths and elongation data with the following linear progression curves:

$$\text{Tensile strength (ksi)} = 0.10 + 0.36 \times \text{CMMH} \tag{12.7}$$

$$\text{Yield strength (ksi)} = 12 + 0.18 \times \text{CMMH} \tag{12.8}$$

$$\text{Elongation (\%)} = 37.85 - 0.093 \times \text{CMMH} \tag{12.9}$$

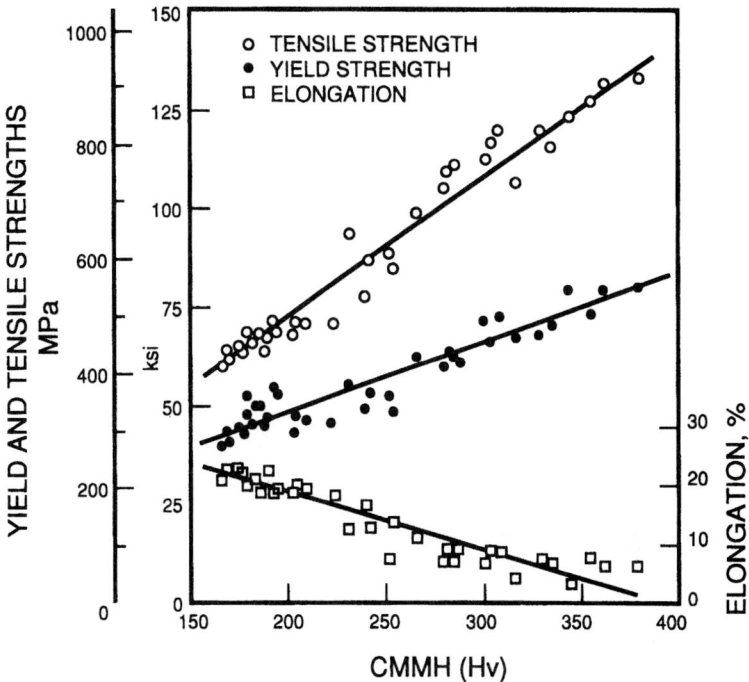

FIGURE 12.33 Relationships between actual tensile properties and properties calculated from CMMH values.[60] (*Courtesy of American Foundry Society, Des Plaines, Illinois.*)

where the *composite matrix microhardness* (CMMH) can be defined by the rule of mixtures as

$$CMMH = \frac{HF \times \%F + HP \times \%P}{100}$$ (12.10)

where HF and %F, HP and %P are the respective hardness and volume percentages of ferrite and pearlite contents.

Impact Properties.   Impact property of ductile iron depends strongly on the matrix microstructure. Like nonaustenitic steels, the unalloyed ductile irons exhibit a transition from ductile-to-brittle fracture behavior as the temperature of testing is lowered. As shown in Fig. 12.34, ferritic ductile iron is represented by the lowest ductile-to-brittle transition temperature and highest upper-shelf energy, which are influenced by both nodularity (Fig. 12.34a) and nodule count (Fig. 12.34b).[48] In Fig. 12.34a, notched Charpy energy in the upper-shelf region decreases drastically with a decrease in nodularity. Transition temperature and lower-shelf energy, however, remain unaffected by nodularity. In contrast, nodule count has a significant effect on both upper-shelf energy and transition temperature (Fig. 12.34b). Increasing the nodule count from 180/mm² to 310/mm² causes a decrease in transition temperature of 40°C (70°F) and a 25% decrease in upper-shelf energy.[48]

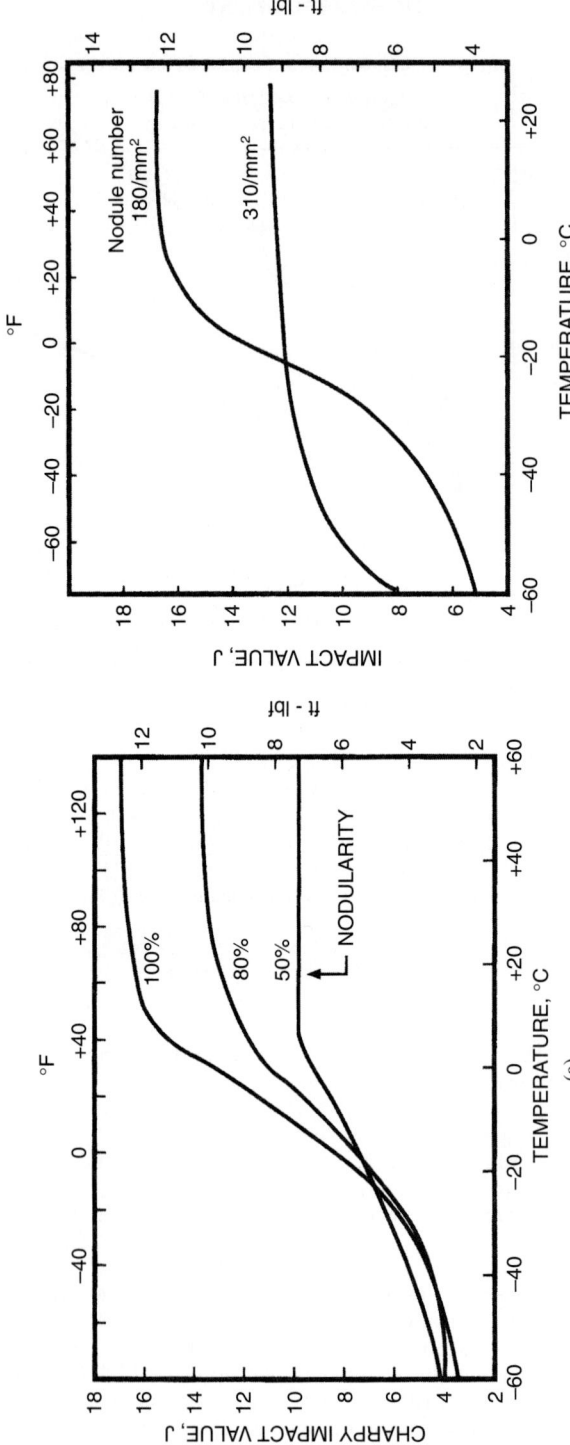

FIGURE 12.34 Effect of (a) graphite nodularity and (b) nodule count on the Charpy fracture properties of V-notched samples of ferritic ductile iron.[48]

12.54

Ductile irons have V-notch Charpy impact transition temperatures in the range of 0 to −60°C (32 to −76°F), depending on the composition, heat treatment, and graphite properties; they usually exhibit upper-shelf energy of 16 to 24 J (12 to 18 ft·lb) with room-temperature values in excess of 16 J (12 ft·lb), as seen in Fig. 12.34. Increasing the pearlite content in the matrix structure and raising the level of elements such as phosphorus and/or silicon raise the ITT and decrease the upper-shelf energies. The pearlitic grades are therefore used where high strengths with limited ductility and toughness are required and are usually not recommended for use in low-temperature applications requiring impact resistance. Quenched and tempered martensitic ductile irons usually display a combination of high strength and low-temperature toughness that is superior to those of pearlitic grades.[48]

Fracture Toughness. Some lower-strength ductile iron grades do not fracture in a brittle manner under nominal plane-strain conditions in a standard fracture toughness test, which is attributed to localized deformation in the ferrite film surrounding each graphite nodule.

Fatigue Limit. The fatigue limit of a ductile iron part depends on tensile strength; the size, shape, and distribution of graphite nodules; the volume fractions of inclusions; carbides and dross; the presence of stress raisers; the amount and location of porosity; and the condition of the part surface.[48]

Figure 12.35*a* shows *S-N* curves for notched and unnotched annealed ferritic ductile iron with a tensile strength of 454 MPa (65.8 ksi). This iron has a notch sensitivity factor of 1.65 [i.e., the ratio of unnotched and notched fatigue limits of 193 MPa (28 ksi) and 117 MPa (17 ksi), respectively] and an endurance ratio of 0.43. The endurance ratio of ductile iron is a function of the tensile strength and matrix. Figure 12.35*b* illustrates the similarity of endurance ratios of ferritic and pearlitic grades, decreasing from 0.5 to 0.4 with increasing strength within each grade. For tempered martensite matrices, the endurance ratio falls from 0.5 at a tensile strength of 415 MPa (60 ksi) to 0.3 at a tensile strength of 1035 MPa (150 ksi).[48] The fatigue limit of ductile iron can be increased, especially in bending and torsion tests, by surface rolling, shot peening, and flame and induction hardening.

Strain Rate and Notch Sensitivity. Like many steels, ductile iron is strain-rate-sensitive. In a forming operation such as coining to a close dimension, rolling an idler arm ball socket, or thread rolling, the rate of material movement should be lower to avoid cracking. In coining, a hydraulic press should be employed when the operation is performed at room temperature. In the case of a stroke press, the parts must be heated adequately to avoid cracking during forming. To tolerate the high strain rates, the section size should be increased to reduce the extent of strain.

The notch sensitivity of ductile iron must be looked into when one is designing a crankshaft. The fillet in the crankpin must be a minimum of 0.065 in. (1.65 mm) thick (radius), in contrast to 0.045 in. (1.14 mm) for the pearlitic malleable iron it replaces. It will possess significantly higher fatigue strength than the part it replaces.[57]

The mechanical properties and typical applications for most of the ductile irons are given in Table 12.10.[57]

Wear Resistance. The wear resistance of ductile irons is determined mainly by their microstructures. The presence of 8 to 11 vol% graphite gives both the graphitic lubrication and oil retention essential to some wear applications. In the

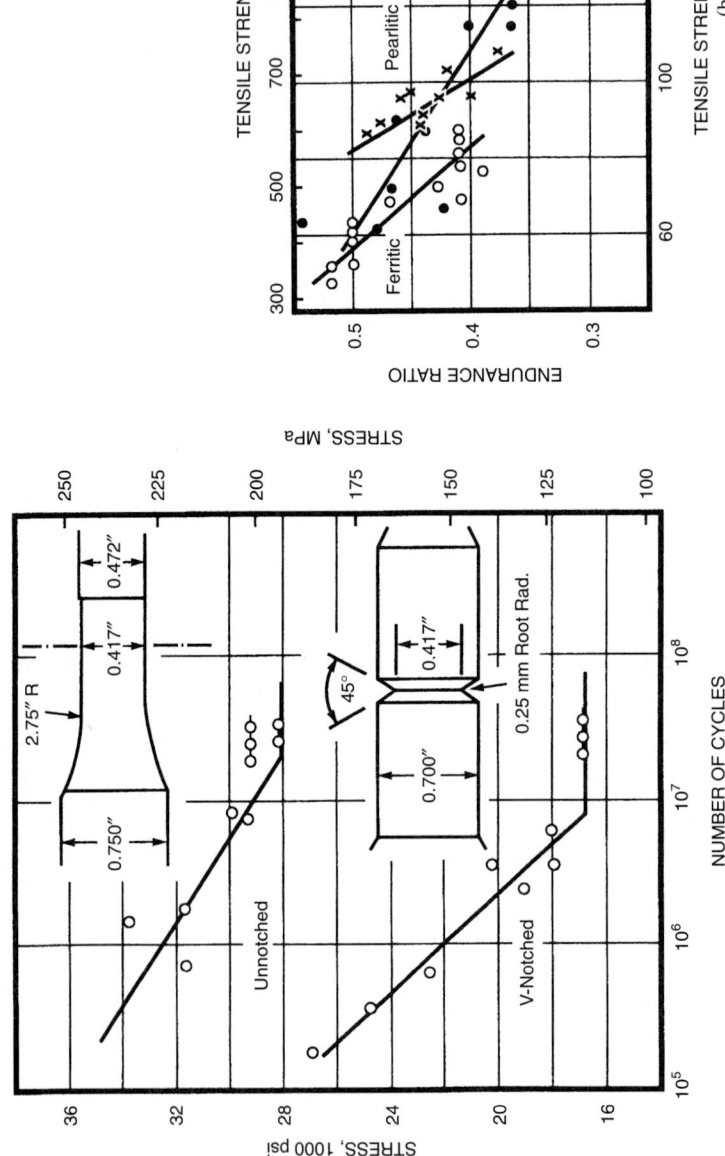

FIGURE 12.35 (*a*) Typical *S-N* curves for notched and unnotched ferritic ductile irons. (*b*) Relationships between endurance ratio, tensile strength, and matrix microstructure for ductile iron.[48] (*Courtesy of QIT-Fer et Titane Inc.*)

TABLE 12.10 Mechanical Properties and Typical Applications for Standard Grades of Ductile Iron[57]

Specification no.	Grade or class	Hardness, HB[†]	Tensile strength, min.[‡]		Yield strength, min.[‡]		Elongation in 50mm (2in.) (min), %[‡]	Typical applications
			MPa	ksi	MPa	ksi		
ASTM A395; ASME SA395	60-40-18	143–187	414	60	276	40	18	Valves and fittings for steam and chemical-plant equipment
ASTM A476[§]; SAE AMS 5316	80-60-03	201 min.	552	80	414	60	3	Paper mill dryer rolls
ASTM A536	60-40-18	…	414	60	276	40	18	Pressure-containing parts such as valve and pump bodies
	65-45-12	…	448	65	310	45	12	Machine components subject to shock and fatigue loads
	80-55-06	…	552	80	379	55	6	Crankshafts, gears, and rollers
	100-70-03	…	689	100	483	70	3	High-strength gears and machine components
	120-90-02	…	827	120	621	90	2	Pinions, gears, rollers, and slides
SAE J434	D4018	170 max.	414	60	276	40	18	Steering knuckles
	D4512	156–217	448	65	310	45	12	Disk brake calipers
	D5506	187–255	552	80	379	55	6	Crankshafts
	D7003	241–302	689	100	483	70	3	Gears
	DQ&T	§	¶	¶	¶	¶	¶	Rocker arms
SAE AMS 5315C	Class A	190 max.	414	60	310	45	15	Electric equipment, engine blocks, pumps, housings, gears, valve bodies, clamps, and cylinders

Note:
[†] Measured at a predetermined location on the casting.
[‡] Determined using a standard specimen taken from a separately cast test block, as set forth in the applicable specification.
[§] Range specified by mutual agreement between producer and purchaser.
[¶] Value must be compatible with minimum hardness specified for production castings.
Reprinted by permission of ASM International, Materials Park, Ohio.

dry-sliding-friction type of wear condition, the wear resistance increases with increasing pearlitic structures in ductile iron. In the lubricated condition, ductile iron provides a satisfactory running-in performance equivalent to that of gray cast iron.

Further improvements in abrasive wear resistance can be obtained by heat treatment and/or alloying to produce a harder bainitic or martensitic matrix.[48]

Machinability. Machinability is determined by both microstructure and hardness. Ductile iron has approximately the same machinability as gray iron of equivalent hardness. At low hardnesses (ductile iron with tensile strength up to ~550 MPa, or 80 ksi), the machinability of ductile iron is better than that of the cast mild steel; however, at high hardness levels, the difference in machinability becomes narrow.[48]

Damping Capacity. The average damping capacity of ductile iron in the hardness range of 156 to 241 HB is about 0.12 times that of class 30 gray iron and about 6.6 times that of 1018 steel (see also Table 12.7). Unalloyed ferritic ductile iron has a greater damping capacity than austenitic ductile iron. Thus, the damping capacity usually decreases with increased matrix hardness and increases with carbon content. At low stresses, the softer ferrite matrix has higher damping capacity, while at high stresses the pearlitic matrix has higher damping capacity.[48]

Applications. Ductile iron has attained an important position in the cast metals industry because it exhibits some of the engineering properties of steels and has the processing advantages of gray iron. Structural applications of ductile irons are based on their higher strengths, modulus, and toughness along with good machinability and lower manufacturing costs. They are mainly used in the automotive, agricultural implement, and pipe industries. For example, they are used for critical automotive/trucking parts such as crankshafts, pistons, gears and rollers, front-wheel spindle supports, complex shapes of steering knuckles, engine connecting rods, idler and rocker and follower arms, wheel hubs, truck axles, suspension system parts, power transmission yokes, disk brake calipers, exhaust manifolds, high-temperature applications for turbo housings and manifolds, and high-security valves for numerous applications.[57] (See Table 12.10.)

They also find applications in farm equipment, construction machinery, papermaking machinery, oilfield equipment, power transmission components (e.g., gears), mining, steel mill, and tool and die industries. Austempered ductile iron has also resulted in many new applications[57] (see Chap. 13).

12.8.3 Compacted Graphite (CG) Iron

Mechanical Properties. Figure 12.36a shows the typical stress-strain curves in tension for ferritic and pearlitic CG irons exhibiting linear elasticity to a lower limit of proportionality than does ductile iron. Poisson's ratio varies from 0.27 to 0.28.[42] Figure 12.36b shows the stress-strain curves for tension and compression for CG iron with 4.35 CE,[61] illustrating the higher-compression (76 MPa or 11 ksi), 0.1% proof stress than the tension ones.[41]

Table 12.11 shows some of the mechanical and physical properties of the three irons.[62] This list clearly demonstrates that tensile and yield strength, moduli of elasticity, hardness, impact and fatigue properties, thermal conductivity, and damping capacity of CG iron lie between the ranges for gray and ductile irons.[62,63]

The 35% increase in elastic modulus and >90% increase in tensile and fatigue strengths of CG iron with respect to conventional gray iron have led several auto-

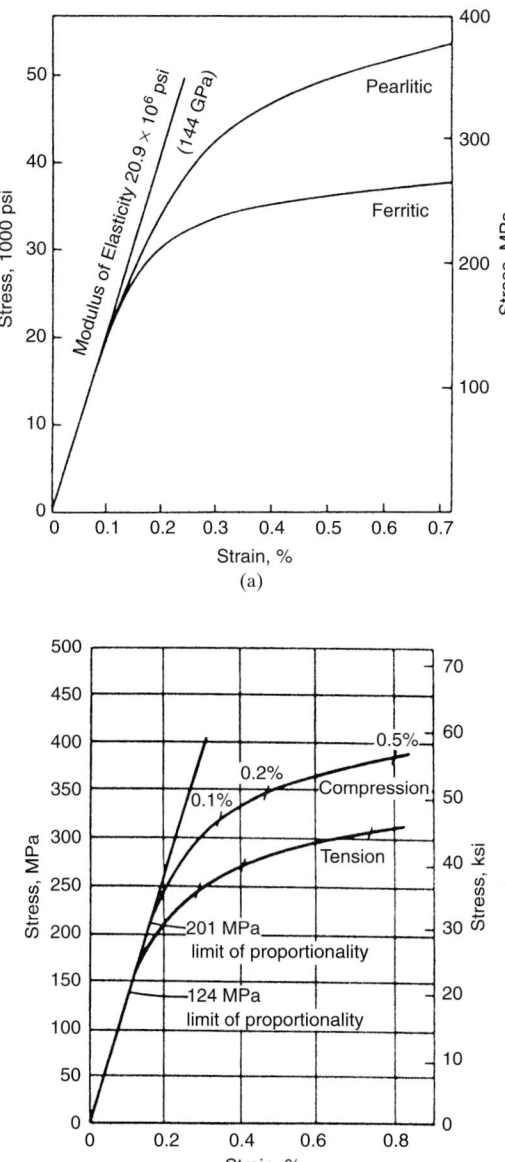

FIGURE 12.36 (*a*) Typical stress-strain curves for CG irons. The ferritic irons had a tensile strength of 320 MPa (46.4 ksi) and 3.5% elongation. The pearlitic iron had a tensile strength of 410 MPa (59.5 ksi) and 1.0% elongation.[41] (*b*) Stress-strain curves in tension and compression for a 4.35 carbon equivalent (CE) CG iron.[61] [(*a*) Courtesy of American Cast Metals Association, Des Plaines, Illinois. (b) Courtesy of British Foundryman.]

mobile manufacturers to redesign their existing aluminum and/or gray iron gasoline/diesel engine block in favor of CG iron blocks.[64]

Impact Toughness. Unlike gray irons, CG irons exhibit a ductile-to-brittle transition temperature between −20 and +20°C. Above the transition point, the impact

TABLE 12.11 General Comparison of Mechanical and Physical Properties of Compacted, Graphite, Gray, and Ductile Irons[62]

Property	CG iron	Gray iron	Ductile iron
Tensile strength (MPa)	276–586	138–414	414–428
0.20% Offset yield strength (MPa)	207–435		276–621
Modulus of elasticity (GPa)	121–152	76–131	152–172
Elongation (%)	1–4	<1.0	2–25
Hardness BHN (3000 N)	140–260	110–270	149–300
Impact breakage energy (notched bar, 20°C) (J)	2–9.5	<1.4	2.7–24.5
Fatigue endurance ratio (rotating beam, unnotched)	0.45–0.55	0.35–0.45	0.4–0.5
Thermal conductivity (W m^{-1} K^{-1})	33–50	44–57	31–38
Damping capacity (%)	3–6	4–10	2–3

Reprinted by permission of Pergamon Press, Plc.

energy progressively increases with an increase in temperature for both pearlitic and ferritic grades of CG iron, as shown in Fig. 12.37.[41] Between 20 and 100°C, gray iron usually shows no increment in impact energy, in contrast to CG irons and ductile irons which provide approximately 25 and 100% increase, respectively, over the same temperature range. Below room temperature, the impact strength of unnotched CG iron specimens is nearly 4 times higher than that of gray iron and only 5 to 20% lower than that of ductile irons with a similar matrix structure.[64]

As shown in Table 12.12, Ce-treated CG irons tend to exhibit higher impact energy than Mg-Ti-treated irons. This may be due to the presence of TiC and TiCN inclusions in the Mg-Ti-treated CG iron matrices.[65]

Fatigue Properties. The fatigue strength of unalloyed pearlitic CG irons at room temperature varies between 220 and 260 MPa, which is more than twice the strength observed in pearlitic gray irons used for engine blocks and cylinder head applications. The reduction in fatigue limit as a function of increasing temperature is shown for a pearlitic CG iron in Fig. 12.38.[66]

The 2 to 4 times higher thermal fatigue property of CG irons over conventional gray irons has made them superior in applications such as ingot molds, brake rotors or drums, diesel engine heads, cylinder liners, and a wide range of automotive parts.[64,67]

Wear Resistance. Unlike gray iron, CG iron does not suffer from the *graphite pull-out phenomenon* and does not possess stress raising "knife-edges" of graphite flake particles. Therefore, compacted graphite particles remain intact to offer the advantages of lubrication and surface heat removal, leading to substantially better wear characteristics than those of gray iron.

Like gray and ductile irons, the wear resistance of CG iron can by improved by alloying with Cr, Mo, or ~0.05% Ti. The improved wear resistance of CG iron bore was the sole impressive parameter in the performance of CG iron engine block compared to gray iron block.[64]

Machinability. CG iron can be machined on existing gray iron transfer lines with existing tools and cutting parameters. The machining of CG iron offers several indi-

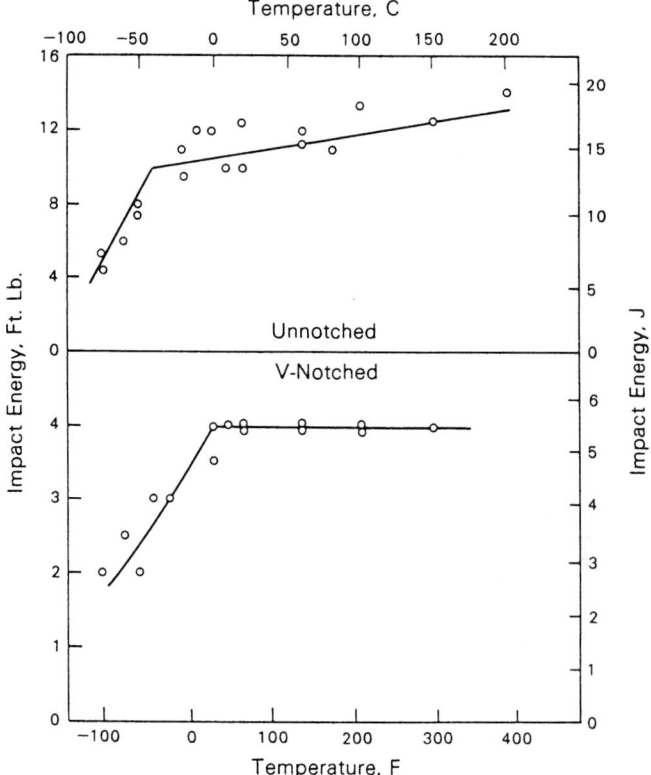

FIGURE 12.37 Impact transition curves for an annealed compacted graphite iron. (*Courtesy of American Cast Metals Association, Des Plaines, Illinois; after K. B. Palmer, BCIRA Journal, Report 1213, January 1976, p. 31.*)

rect advantages such as improved surface finish with respect to gray iron. In some cases, it is reported to reduce the number of milling passes required to obtain the desired surface finish, thereby offering considerable savings in machining time and cost. The improved surface finish and bore roundness of engine blocks is also significant. Other advantages include its ability to withstand higher broaching loads while providing flatter surfaces, which, in turn, reduces machining cost, increases productivity, and provides design opportunity to reduce wall thickness in the bulkhead area. It is also noted that machining of CG iron is linked to increased tool wear compared to gray iron, although in some cases CG iron is found to be easier to machine than gray iron.[68]

Damping Capacity.   While CG iron has less inherent damping capacity than gray iron, it is 3 times more effective than ductile iron.[64]

Applications.[67,69]   CG iron is suitable for noncritical applications such as flywheels and exhaust manifolds, or high-quality CG iron can be used for passenger car engine blocks. CG irons can be substituted for gray irons in bed plates for large diesel

TABLE 12.12 Impact Toughness of a Cerium-Treated CG Iron with Two Magnesium-Titanium-Treated CG Irons[65]

Iron	Structural condition and graphite type	Test temperature		Impact bend toughness[†]		Notched-bar[‡] impact toughness	
		°C	°F	J	ft·lbf	J	ft·lbf
Cerium-treated	>95% ferrite (as-cast); 95% CG, 5% SG	20	68	32.1	23.7	6.5	4.8
		−20	−4	26.5	19.5	4.6	3.4
		−40	−40	26.7	19.7	5.0	3.7
Magnesium (0.018%) and titanium (0.089%) treated	100% ferrite (annealed); CG	20	68	13.5–19	10–14	5.4	4.0
Magnesium (0.017%) and titanium (0.062%) treated	Ferritic (as-cast); CG	20	68	6.8–10.2	5–7.5	3.4	2.5

† Unnotched 10 × 10 mm (Charpy) testpiece.
‡ V-notched 10 × 10 mm (Charpy) testpiece.
Source: Ref 65.

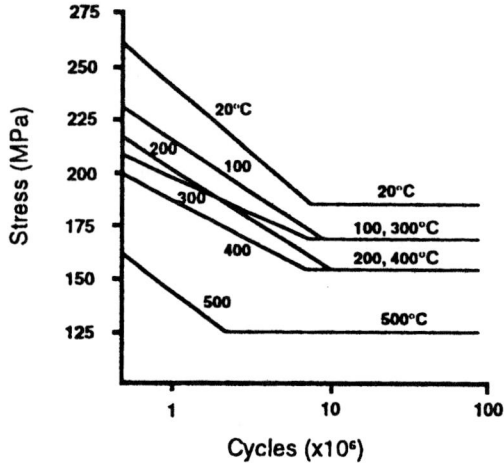

FIGURE 12.38 Effect of increasing temperature on the fatigue limit of pearlitic CG iron (CE = 4.39).[66]

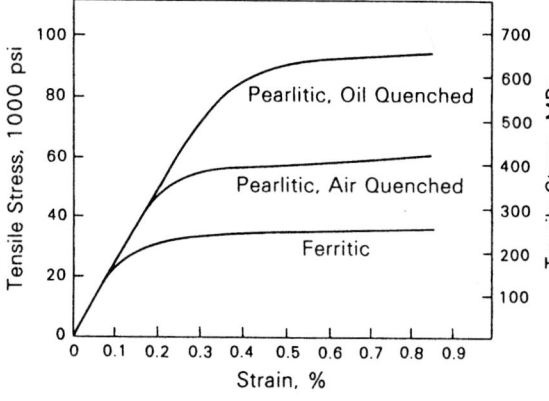

FIGURE 12.39 Typical stress-strain curves for three types of malleable iron.[41] (*After L. W. L. Smith and G. N. J. Gilbert, BCIRA Report 1363, January 1980, pp. 49–62.*)

engines, crankcases, turbocharger housings, gearbox housings, bearing brackets, connecting forks, sprocket wheels, pulleys for truck servo drives, and eccentric gears. It finds applications at elevated temperatures and/or thermal fatigue conditions, for crankcases, ingot molds, exhaust manifolds, cylinder heads, and brake disks.

12.8.4 Malleable Iron

Mechanical Properties.   The typical stress-strain curves for three types of malleable iron are shown in Fig. 12.39.[41]

The tensile properties of malleable irons are usually determined on machined test bars. The compressive strength of malleable irons is seldom measured due to nonoccurrence of failure in compression mode. For a given hardness, the compressive yield strengths are slightly higher than the tensile yield strengths.

The modulus of elasticity usually is lower with an increasing amount of graphite and less compact graphite nodule shape; its moduli vary from 175 GPa for the ferritic malleable iron to 193 GPa for the pearlitic variety.[49] Poisson's ratio is about 0.26.

Fatigue Properties. The unnotched fatigue limit of ferritic malleable iron is about 50 to 60% of the tensile strength. The unnotched fatigue limits of tempered pearlitic (air-cooled or oil-quenched) and tempered martensitic malleable irons (oil-quenched), respectively, are about 40 to 50% and 35 to 40% of tensile strength. Oil-quenched and tempered martensitic iron usually has a higher fatigue ratio than pearlitic iron made by the arrested anneal method.[70]

Based on 10^7 cycles, the ferritic iron has a fatigue strength of 185 MPa, and that of pearlitic iron is 195 MPa (at 217 BHN).[45] The typical mechanical properties and applications of malleable irons are shown in Table 12.13.[49]

Impact Properties. Malleable irons exhibit a smaller notch sensitivity effect on the impact toughness than does ductile iron.[49] The impact properties of malleable irons increase with a finer nodule size and higher nodule count at the same carbon level. Higher amounts of carbon, as well as silicon and phosphorus above 0.15%, decrease the impact strength but increase the ITT. Ferritic malleable iron has a high impact toughness and a lower ITT than pearlitic malleable iron.

Wear Resistance. Pearlitic and martensitic malleable irons have excellent wear resistance because of their structure, hardness, and high degree of structural uniformity.

Machinability. For the same hardness value, the machinability of the oil-quenched and tempered pearlitic iron is superior to that of air-quenched and tempered pearlitic malleable iron.

Damping Capacity. The higher damping capacity of ferritic malleable irons compared to pearlitic ones is attributed to the higher energy losses at the temper-carbon nodules in the softer matrix. (See also Fig. 12.28 and Table 12.7.)

Application. Ferritic malleable irons have been widely used for steering gear housings, mounting brackets, differential carriers, differential cases, wheel hubs, automotive components, agricultural implements, and railroads. Industrial applications of pearlitic malleable irons include camshafts and crankshafts in automobiles; pumps, rolls, cams, and rocker arms as machine parts; gears, sprockets, chain links, and elevator brackets in conveyor equipment; pistol parts, gun mounts, and tank parts in ordnance; and various tools such as clamps, hammers, wrenches, and shears. Martensitic malleable irons are used as connecting rods, universal joint yokes, and gears. (See also Table 12.13.)

Choice of Malleable Iron over Ductile Iron. Since malleable and ductile irons have comparatively similar engineering properties, the choice between the two is made depending upon the initial casting and final casting processing costs. The necessary condition for the production of malleable iron castings is that it must solidify as a white iron structure, which limits the maximum section thickness of the castings that can be produced. High-production foundries usually produce castings up to about 1.5 in. (40 mm) thick; however, in some cases, they go up to 4 in. (100 mm) thick.

It is thus apparent that lower casting yields and the necessity of heat treatment requirements for eutectic carbide decomposition are restrictive in making a favorable decision toward malleable iron. However, malleable iron castings are often selected due to their excellent machinability, better toughness, and shock resistance property.

Currently, malleable iron is less and less produced, and ductile iron production is increasing by 4% per year.

TABLE 12.13 Typical Mechanical Properties and Applications of Malleable Irons[49]

Specification no.	Use	Class or grade	Minimum tensile strength (MPa)	Minimum yield strength (MPa)	Elongation in tension (%)	Microstructure	Typical applications
ASTM A-47-77 ASME SA47	Ferritic, malleable, iron castings	32510 35018	345 365	220 240	10 18	Temper carbon and ferrite	General engineering service at normal and elevated temperatures for good machinability and excellent shock resistance
ASTM A197-79 ASME SA197	Cupola, malleable iron		275	205	6	Free of primary graphite	Pipe fittings and valve parts for pressure service
ASTM A220-76	Pearlitic, malleable, iron castings	40010 45008 45006 50005 60004 70003 80002 90001	415 450 450 485 550 585 655 725	275 310 310 345 415 485 550 620	10 8 6 5 4 3 2 1	Temper carbon in necessary matrix without primary cementite or graphite	General engineering service at normal and elevated temperatures; dimensional tolerance range for castings is stipulated
ASTM A338-61 (1977)	Flanges, pipe fittings, and valve parts		Property requirements as specified for each application				For railroad, marine, and other heavy-duty service at temperatures of up to 345°C

TABLE 12.13 Typical Mechanical Properties and Applications of Malleable Irons[49] (*Continued*)

Specification no.	Use	Class or grade	Minimum tensile strength (MPa) / Brinell hardness	Minimum yield strength (MPa)	Elongation in tension (%) / Heat treatment	Microstructure	Typical applications
ASTM A602-70 (1976) SAE J158a	Automotive, malleable, iron castings	M3210	156 max		Annealed	Ferritic	For good machinability, steering gear housings, carriers, and mounting brackets
		M4504	163–217		Air-quenched and tempered	Ferrite and tempered pearlite	Compressor crankshafts and hubs
		M5003	187–241		Air-quenched and tempered	Ferrite and tempered pearlite	For selective hardening, planet carriers, transmission gear, differential cases
		M5503	187–241		Liquid-quenched and tempered	Tempered martensite	For machinability and improved response to induction hardening
		M7002	229–269		Liquid-quenched and tempered	Tempered martensite	For strength as in connecting rods and universal joint yokes
		M8501	269–302		Liquid-quenched and tempered	Tempered martensite	For high strength and wear resistance as in gears

Reprinted by permission of Pergamon Press, Plc.

12.9 STRUCTURE-PROPERTY RELATIONS IN GRAY IRON

In recent years a considerable effort has been made to correlate the microstructural variations to the strength of gray iron. It has been found by many investigators[71-74] that both plane strain fracture toughness K_{IC} and tensile strength (equivalent to brittle fracture stress) of gray iron increase with decreasing proportions of ferrite in the matrix structure, carbon equivalent value, and eutectic cell size but increasing hardness. For a given eutectic cell size, plane strain fracture toughness K_{IC} is a linear function of tensile strength (representing critical stress) and is given by[75]

$$K_{IC} = 0.70\sigma_T \sqrt{\pi a_e}$$ (12.11)

where σ_T is the tensile strength, representing critical or brittle fracture stress, and a_e is the eutectic cell size, corresponding to the critical crack (or defect) size, measurable as the mean diameter from a metallographically prepared flat specimen.

REFERENCES

1. E. Tyrkiel, *J. Heat Treating*, vol. 1, no. 2, 1979, pp. 52–54.

2. B. Harrison, *Metallurgia*, March 1978, pp. 145–146.

3. D. J. Grieve, *Metall. Mater. Technol.*, vol. 7, no. 8, 1975, pp. 397–403.

4. T. L. Elliott, *Metallurgia*, December 1974, pp. 549–550.

5. T. L. Elliott, *Metall. Met. Form.*, April 1977, pp. 173–175.

6. L. J. Haga, *Practical Heat Treating*, L. J. Haga Co., Kenwood, Mich., 1981.

7. G. F. Melloy, *Annealing of Steel*, Metals Engineering Institute, ASM, Metals Park, Ohio.

8. H. Ohtani, in *Materials Science and Technology*, vol. 7: *Constitution and Properties of Steels*, vol. ed. F. B. Pickering, VCH, Weinheim, 1992, pp. 147–181.

9. B. L. Bramfitt and A. K. Hingwe, in *ASM Handbook*, vol. 4: *Heat Treating*, 10th ed., ASM International, Materials Park, Ohio, 1991, pp. 42–55.

9a. D. J. Roth, *Light Metal Age*, April 1991, pp. 86–89.

10. F. S. C. Brandous, *Met. Prog.*, July 1970, pp. 103–105.

11. B. Liscic, in *Steel Heat Treatment Handbook*, eds. G. E. Totten and M. A. H. Howes, Marcel Dekker, New York, 1997, pp. 527–662.

12. S. Chattopadhyay and C. M. Sellers, *Metallography*, vol. 10, 1977, pp. 89–105.

13. G. Krauss, *Steels: Heat Treatment and Processing Principles*, ASM International, Materials Park, Ohio, 1990.

14. M. B. Adeyeme, R. A. Stark, and G. F. Modlen, *Heat Treatment '79*, The Metals Society, London, 1979, pp. 122–125.

14a. B. Irving, *Welding J.*, February 1999, pp. 41–45.

15. K. E. Thelning, *Steel and Its Heat Treatment*, Butterworths, London, 1984.

16. V. H. Erickson, *Met. Treat.* (Rocky Mount, N.C.), vol. VII, no. 4, 1956, pp. 2–4.

17. E. C. Rollason, *Fundamental Aspects of Molybdenum on Transformation of Steel*, Climax Molybdenum Co., London.

18. R. G. Bretherton, *Metallurgia*, October 1961, pp. 179–184.

18a. P. F. Stratton and M. S. Stanescu, *Adv. Mats. Process.*, October 1999, pp. 224–227.

19. P. R. Mould, in *Metallurgy of Continuous Annealed Sheet Steel*, eds. B. L. Bramfitt and P. L. Mangonon, Jr., TMS–AIME, Warrendale, Pa., 1982.

20. K. G. Brickner and M. P. George, *Industrial Heating*, February 1993, pp. 25–27.

21. A. R. Perrin and B. F. Johnston, *Iron and Steel Engineer*, June 1983, pp. 39–45.

22. A. R. Perrin, R. I. L. Guthrie, and B. C. Stonehill, *I&SM*, October 1988, pp. 27–33.

23. T. M. Erdman, *33 Metal Producing*, February 1997, pp. 26–28, 35.

24. S. Tajima and M. Shrouzu, in *Developments in the Annealing of Sheet Steel*, eds. R. Pradhan and I. Gupta, TMS, Warrendale, Pa., 1992, pp. 529–537.

25. R. Pradhan, in *ASM Handbook*, vol. 4: *Heat Treating*, 10th ed., ASM International, Materials Park, Ohio, 1991, pp. 56–66.

26. A. R. Marder, ed., *The Physical Metallurgy of Zinc Coated Steel, Proceedings: International Conference:1994*, TMS/AIME, Warrendale, Pa., 1994.

27. N. Ohashi, in *Encyclopedia of Materials Science and Engineering*, Pergamon, Oxford, 1986, pp. 85–92.

28. M. Imose, *Trans. ISI, Japan*, vol. 25, 1985, p. 111.

29. *Ferrous Wire*, vol. 1, Wire Association International, Inc., Guilford Conn., 1989.

29a. R. A. Andreas, *Ind. Heating*, October 1991, pp. 59–61.

29b. R. G. O'Neill and D. H. Herring, *Ind. Heating*, August 1998, pp. 61–64.

29c. Y. Rancon, E. Duchateav, I. Latoumarie, and M. Phillips, *Ind. Heating*, May 1991, pp. 28–30, 32.

30. M. Stanescu et al., in *ASM Handbook*, vol. 4: *Heat Treating*, 10th ed., ASM International, Materials Park, Ohio, 1991, pp. 573–586.

31. B. A. Bercherer and T. J. Witheford, *ASM Handbook*, vol. 4: *Heat Treating*, 10th ed., ASM International, Materials Park, Ohio, 1991, pp. 207–218.

32. M. S. Stanescu, *Industrial Heating*, November 1993, pp. 35–40.

33. P. M. Berge, J. D. Verhoeven, D. Peterson, and A. H. Pendray, *I&SM*, March 1995, pp. 67–72.

34. J. R. Founds and R. Vishwanathan, in *First International Conference on Microstructures and Mechanical Properties of Aging Materials*, eds. P. K. Liaw, R. Vishwanathan, K. L. Murty, and E. P. Simonen, TMS, Warrendale, Pa., 1993, pp. 61–69.

35. M. A. Neri, R. Colas, and S. Valtierra, *Jn. Materials Engr., and Performance*, vol. 7, no. 4, 1998, pp. 467–473.

36. *ASM Specialty Handbook: Carbon and Alloy Steels*, ASM International, Materials Park, Ohio, 1996, pp. 308–328.

36a. T. Ruglic, in *ASM Handbook*, vol. 4: *Heat Treating*, 10th ed., ASM International, Materials Park, Ohio, 1991, pp. 35–41.

37. S. H. Avner, *Introduction to Physical Metallurgy*, 2d ed., McGraw Hill, New York, 1974.

38. R. T. Hook, *Met. Progr.*, October 1963, pp. 106–107.

39. H. Ohtani, Y. Kawaguchi, M. Nakanishi, N. Katsumoto, I. Seta, and Y. Ito, *The Sumitomo Search*, vol. 26, 1981, p. 62.

40. *ASM Specialty Handbook: Cast Irons*, ASM International, Materials Park, Ohio, 1996, pp. 179–191.

41. C. F. Walton, ed., *Iron Castings Handbook*, Iron Castings Society, Inc., Des Plaines, Ill., 1981.

42. B. V. Kovacs, *ASM Handbook*, vol. 4: *Heat Treating*, 10th ed., ASM International, Materials Park, Ohio, 1991, pp. 670–681.

43. C. E. Bates, *AFS Trans.*, vol. 94, 1986, pp. 889–912.

44. H. Morrough, in *Encyclopedia of Materials Science and Engineering*, Pergamon Press, Oxford, 1986, pp. 4539–4543.

45. A. Waluszewski, *Metallurgia*, vol. 53, no. 2, 1986.

46. K. B. Rundman, in *ASM Handbook*, vol. 4: *Heat Treating*, 10th ed., ASM International, Materials Park, Ohio, 1991, pp. 682–692; *Metals Handbook*, vol. 4, ASM, Metals Park, Ohio, 1981.

47. *ASM Specialty Handbook: Cast Irons*, ASM International, Materials Park, Ohio, 1996, pp. 192-204.

48. *Ductile Iron Data for Design Engineers*, QIT-Fer et Titane Inc., Chicago 1990.

49. L. Jenkins, in *Encyclopedia of Materials Science and Engineering*, Pergamon Press, Oxford, 1986, pp. 2725–2729.

50. L. R. Jenkins, in *ASM Handbook*, vol. 4: *Heat Treating*, ASM International, Materials Park, 1991, pp. 693–696.

51. *ASM Specialty Handbook: Cast Irons*, ASM International, Materials Park, Ohio, 1996, pp. 205-208.

52. *ASM Specialty Handbook: Cast Irons*, ASM International, Materials Park, Ohio, 1996, pp. 32–53; *Metals Handbook*, vol. 1, ASM, Metals Park, Ohio, 1978.

53. J. F. Wallace, in *Encyclopedia of Materials Science and Engineering*, Pergamon Press, Oxford, 1986, pp. 2059–2062.

54. R. E. Peterson, *Stress Concentration Factors*, Wiley, New York, 1974.

55. V. Patierno et al., in *Microstructural Characterization and Mechanical Properties of Various Grades of Nodularity in SG Iron*, 44th International Foundry Congress, September 1977.

56. H. T. Angus, *Cast Iron: Physical and Engineering Properties*, Butterworths, London, 1976.

57. *ASM Specialty Handbook: Cast Irons*, ASM International, Materials Park, Ohio, 1996, pp. 54–79.

58. W. Siefer and K. Orths, *AFS Trans.*, vol. 78, 1970, pp. 382–387.

59. D. L. Crews, *AFS Trans.*, vol. 82, 1974, pp. 223–228.

60. D. Venugopalan and A. Alagarsamy, *AFS Trans.*, vol. 98, 1990, pp. 395–400.

61. G. F. Sergeant and E. R. Evans, *British Foundryman*, May 1978, p. 115.

62. J. F. Wallace, in *Encyclopedia of Materials Science and Engineering*, Pergamon Press, Oxford, 1986, pp. 743–746.

63. C. F. Walton, in *Encyclopedia of Materials Science and Engineering*, Pergamon Press, Oxford, 1986, pp. 529–537.

64. S. Dawson, Publications by SinterCast SA, Switzerland, Feb/June/August 1994, and November 1995.

65. E. Nechtelberger, *Properties of Cast Irons up to 500°C*, Technology Ltd., 1980.

66. N. Nechtelberger, H. Puhr, J. B. von Nesselrode, and A. Nakayasu, *Cast Iron with Vermicular/Compacted Graphite—State of the Art*, Development, Properties, and Applications, World Foundry Congress, 1984.

67. *ASM Specialty Handbook: Cast Irons*, ASM International, Materials Park, Ohio, 1996, pp. 80–93.

68. S. Dawson, *The Machinability of Compacted Graphite Iron Engine Blocks*, SinterCast SA, Switzerland, 1995.

69. M. Tholi, A. Magata, and S. Dawson, *Sinter SA, Switzerland*, November 1995.

70. *ASM Specialty Handbook: Cast Irons*, ASM International, Materials Park, Ohio, 1996, pp. 94–106.

71. G. F. Ruff and J. F. Wallace, *AFT Trans.*, vol. 83, 1977, p. 179.

72. A. G. Glover and G. Pollard, *Fracture*, Chapman & Hall, London, 1969, p. 350.

73. T. V. Venkatesubramanian and T. J. Baker, *Met. Technol.*, vol. 5, 1978, p. 57.

74. D. K. Verma and J. T. Berry, *AFS Trans.*, vol. 89, 1981, p. 849.

75. R. N. Castillo and T. J. Baker, in *Physical Metallurgy of Cast Iron, Materials Research Society, Symposium Proceedings*, vol. 34, eds. H. Fredriksson and M. Hillert, North-Holland, New York, 1985, pp. 467–495.

CHAPTER 13
HARDENING AND HARDENABILITY

13.1 INTRODUCTION

Quenching of a steel is usually accomplished by immersion in water, brine, oils, polymer solution, or molten salts, although forced gas quenching has also been used extensively, especially after austenitizing in vacuum furnaces. The effectiveness of quenching may be varied by changing the steel composition and the type, concentration, temperature, and agitation of quenching media.

Different quenching methods such as quench-hardening, martempering, and austempering of steel as well as cast iron, in conventional and modified forms, will be discussed using a variety of quenching media.

We shall discuss the influence of carbon content on martensite hardness. Because the main purpose of quenching is to produce adequate depth of hardening, more emphasis will be given on various facets of hardenability such as definition, importance, and limitations of hardenability; in addition, several methods of determining hardenability (as developed by Grossman, DeRetana and Doane, Kramer et al., Moser and Legat, Grange, and Jatczak and Jominy), hardenability curves and their use, as well as computerized on-line applications of hardenability calculations for a wide range of steel compositions will be presented. Finally, hardenability band (or H-) steels, methods for specifying them on the basis of hardenability band, and an alloy steel selection guide for oil-quenched and highly and moderately stressed parts will be elaborated.

13.2 QUENCHING

Quenching process is accomplished to harden or provide desired microstructure in the heat-treated materials without crack or with minimum distortion. Quenching refers to rapid cooling of steel from high-temperature austenite to avoid changing completely to intermediate transformation products (e.g., ferrite, carbide, pearlite, or bainite) and to ensure formation of martensite at least on its surface, depending on the cooling rate, quenching media, section size and grade, or hardenability of steel. This condition produces certain desired properties, such as increased hardness and strength as well as grain refinement. Unalloyed steels are usually quenched in water, and alloy steels in oil, and highly alloyed steels are martempered, austempered, or even cooled in air in order to attain a desired depth of this hardened structure.

13.2.1 Heat Extraction during Quenching

There are three phases or stages of the quenching (i.e., heat extraction) cycle when steel is quenched in liquid media.[1-3] A, the vapor blanket (VB) or film boiling (FB) stage; B, the vapor transport (VT) or nucleate boiling (NB) stage; and C, the liquid cooling or convection cooling (CC) stage (Fig. 13.1).

Tamura identified the existence of a stage prior to the VB stage during oil quenching, where the heat removal by liquid-probe contact occurs in a very short time just after the immersion of a probe.[4] On the other hand, Narazaki et al.[5] reported either four or five stages of cooling during water quenching (Fig. 13.2) based on water temperature and probe geometry. In the five-stage type cooling, the additional stage (stage IIIa) between the VB stage (stage II) and the NB stage (stage IIIb) was named the *transition stage*. However, to simplify the subject matter, we will discuss henceforth the conventional three stages of quenching throughout the text.

Figure 13.1 shows a typical temperature-time or cooling curve of an oil, obtained from a thermocouple embedded in the center of a steel specimen. Stage A is characterized by the formation of a continuous, thin, insulating vapor blanket film around the entire part. In this stage severe boiling occurs due to a temperature exceeding the *critical overheat temperature* (COHT) of the liquid. Heat transfer to

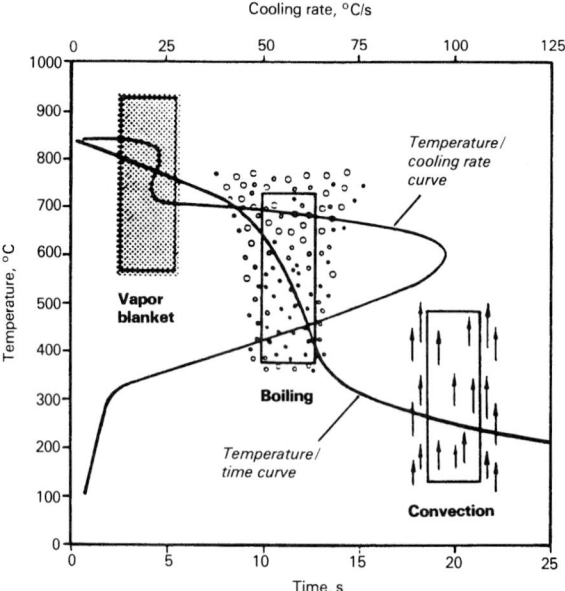

FIGURE 13.1 Three stages of the cooling curve for an oil. (*Courtesy of Wolfson Heat Treatment Center.*)

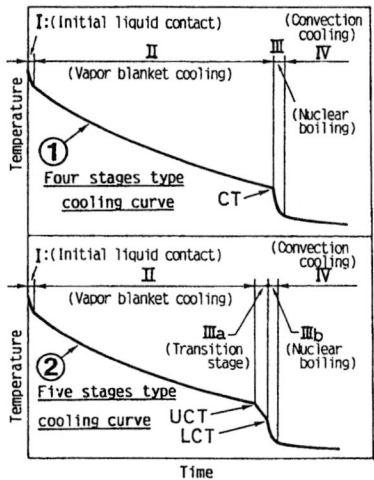

FIGURE 13.2 Typical cooling curves in water quenching: (1) Four-stage cooling curve: I, initial solid-liquid contact stage; II, vapor blanket stage; III, nucleate boiling stage; IV, convection cooling stage; CT, characteristic temperature. (2) Five-stage cooling curve; IIIa, transition stage; IIIb, nucleate boiling stage; UCT, upper characteristic temperature; LCT, lower characteristic temperature.[5]

the surrounding oil occurs by radiation and conduction through the vapor film; hence the heat extraction is very slow, which is represented by the flat appearance in this stage. This stage completes at the COHT of the liquid and is called the *characteristic temperature* (CT) or *Leidenfrost temperature*. During Stage B, the unbroken vapor blanket in contact with the surface of the component boils, collapses, and disperses, giving rise to the highest or most intensive cooling rate. Cooling in this region depends on the boiling point of the liquid. Stage C (the final stage) starts when the temperature of the metal surface falls below the boiling point of the quenchant. During this stage, the heat extraction or cooling rate is the slowest and is mainly affected by the difference of the viscosity and the temperature difference of the boiling point and the temperature of the liquid. Like Stage A, the curve is again flat here. The brine solution has the highest cooling rate, followed by water, synthetic polymer solutions, oils, molten salt baths, and air.[1]

The A-stage cooling is not observed when parts are quenched in aqueous solutions containing >5 wt% of an ionic material such as KCl, LiCl, NaOH, or H_2SO_4. In these solutions, cooling curves begin with stage B. The presence of the salts at the hot metal-quenchant interface initiates the nucleate boiling stage almost instantly.

A-stage cooling is not present during quenching in nonvolatile quenchant media such as molten salts. Conversely, heat transfer in gas quenchants such as air or inert gases takes place entirely by a vapor blanket mechanism.[6]

Many factors determine the actual cooling rate achieved in a steel part during quenching; the important ones are mass or section thickness; thermal diffusivity; thermal conductivity and volume specific heat of the steel part; and the type, temperature, velocity (or agitation), density, viscosity, and wetting characteristics of the quenchant.[1,7]

According to Newton's law of cooling, the rate of heat transfer from the steel part to the quenching medium through the surface is a function of (1) the temperature difference between the surface of a steel part and the quenching medium and (2) the coefficient of heat transfer at the metal/quenchant interface. That is,

$$q = hA(T_1 - T_2) \tag{13.1}$$

where q is the amount of heat transfer from the part to the quenchant (heat input) [or heat flux density (J/m^2)], h is the heat-transfer coefficient at the surface of the part [$J/(s·m^2·K)$], A is the surface area in contact with quenchant, T_1 is the part surface temperature (K), and T_2 is the bulk quenchant (bath) temperature (K).

The actual heat flow from the interior of a metallic part being quenched to the surface can also be written in terms of the temperature gradient dT/dx set up within the part (K/m)

$$q = -kA\left(\frac{dT}{dx}\right)_{surface} \tag{13.2}$$

where k is the thermal conductivity of the part [$J/(s·m·K)$], and q and A have the usual meanings. Since the heat removal is equal to the heat flow from the interior of a part to its surface, we have

$$-k\frac{dT}{dx} = h(T_1 - T_2) \tag{13.3a}$$

$$-\frac{dT}{dx} = \frac{h}{k}(T_1 - T_2) \tag{13.3b}$$

This is a one-dimensional equation for heat flow and transfer. Figure 13.3 is a schematic plot of the typical large variation of heat-transfer coefficient and the three regimes of heat transfer.[7a]

A ratio, more widely used for quenching media, is the quench severity factor, severity of quench, Grossman's number, or Grossman hardenability constant, expressed as $H = h/2k$. This quench severity factor H is ~1 for a 1-in. steel section quenched in still water.

Note that in Eq. (13.3), temperature T_1 at the surface of the steel part changes with time, which necessitates the use of a non-steady-state equation for heat conduction in the steel. Heat transfer in a solid metal, where the temperature varies

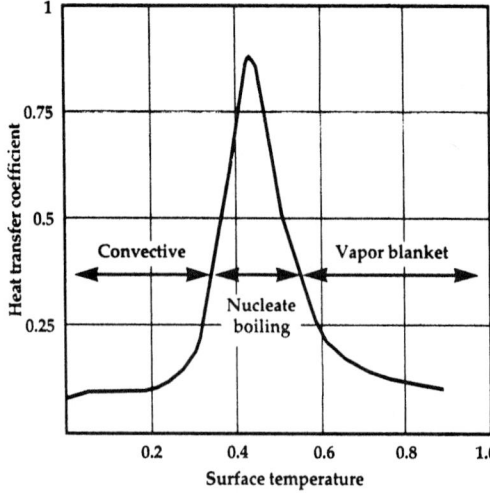

FIGURE 13.3 Schematic plot showing a typical variation of heat-transfer coefficient and the three regimes of heat transfer.[7a] (*Courtesy of ASM International, Materials Park, Ohio.*)

with time and heat sources are absent within the part, can be represented by the non-steady-state equation, in rectangular coordinates, as

$$\frac{\partial T}{\partial t} = \alpha \left(\frac{\partial^2 T}{\partial x^2} + \frac{\partial^2 T}{\partial y^2} + \frac{\partial^2 T}{\partial z^2} \right) \tag{13.4}$$

where α is the thermal diffusivity (m^2/s) of the steel and x, y, and z are local coordinates. Equation (13.4) is a general differential equation, in the absence of a heat source or sink, for the heat flow within an isotropic solid (metal). According to Eq. (13.4), the temperature distribution within a body depends on both the local temperature gradients and the thermal diffusivity, which include all thermodynamic parameters of the material. Note that a precise measurement of thermal conductivity is usually made using steady-state methods. However, non-steady-state methods are simpler and are easily performed, but with less precision.

Thermal diffusivity α can be defined as

$$\alpha = \frac{k}{\rho C_p} \tag{13.5}$$

where ρ is the density (kg/m^3) and C_p is the specific heat capacity at constant pressure [$J/(kg \cdot K)$]. In the simple form of Fourier equation [Eq. (13.4)], thermal diffusivity is assumed to be constant; but in reality, k, ρ, and C_p, and therefore α, are related to temperature. The larger the value of α, the more rapidly the heat will diffuse through the material. A high value of α can be obtained with either high thermal conductivity k or low heat capacity ρC_p, which suggests that less of the heat moving through the material is absorbed and thereby increases the temperature of the material. The fact is shown by faster cooling of silver with respect to austenitic steels. The thermodynamic data for silver, nickel-base alloys such as Inconel and austenitic stainless steel are given in Table 13.1.

To determine the H value of a quenchant steel specimen of simple geometry, for example, the cylindrical bar is cooled in the quenchant, and the cooling curve at the center (or surface) of the specimen is calculated. The cooling rate at a given temperature (say, 700°C) is determined from the cooling curve so obtained, and it is substituted in Eq. (13.4) and combined with Eq. (13.3) to measure the H.

The above method of measuring H does not yield a constant value. It is not surprising, because three assumptions inherent in this method do not hold;

TABLE 13.1 Approximate Values of Thermal Conductivity k and Thermal Diffusivity α of Selected Materials at Room Temperature

Material	Thermal conductivity, $J/(s \cdot K)$	Thermal diffusivity, 10^6 m^2/s
Silver	407	165
Inconel alloy 600	13	4.1
Austenitic stainless steel	15	3.8
Ferritic steel	19	5.1
1040 steel	55	14.3
Iron	75	21

namely, *both* the thermal conductivity k and the thermal diffusivity α of the steel are constant, and the quench severity H remains constant during the entire cooling period of the metal. In fact, both the thermal diffusivity and the thermal conductivity of steels are temperature-dependent, and both decrease with decreasing temperature (or increasing alloying elements) for austenite; however, both increase with decreasing temperature for the transformation products (e.g., ferrite-carbide mixture and martensite) (Fig. 13.4a and b). There is also a variation of volume specific heat ($= \alpha k$) with transformation temperature. Thus H is a function not only of the quenching medium but also of the temperature and section size of the part.

13.2.2   Methods of Evaluating the Cooling Power of Quenching Media

Cooling curve tests are the most useful means of testing the cooling power of liquid quenchants such as oil, water, polymer solution, and so forth. The actual shapes of the cooling curves are functions of both the quenchant properties (types of quenchant, quenching temperature, flow velocity and pattern, and so on), the probe shape or geometry, and its material.[5]

Probes used in cooling curve acquisition have been produced in various shapes such as spheres, cylinders, square bars, plates, coils and rings, wires, round disks, and production parts. These probes have been made from silver, nickel, copper, 5145 alloy, stainless steel, Alloy (Inconel) 600 (UNS N06600), gold, platinum, and aluminum.[5,8] The use of an instrumented silver ball probe has been relatively popular because silver has excellent heat-transfer properties; symmetrical shape of the spherical probe provides ease of heat-transfer calculations from the time versus temperature data. However, some limitations, such as (1) difficulty to manufacture a spherical probe with a consistent surface finish and with the thermocouple assembly located precisely at the geometric center and (2) the relatively high thermal conductivity of silver with respect to steel, restrict the realistic assessment of actual quenching conditions.[9,10] Cooling curves obtained using this probe are shown in Fig. 13.5. It is clear from this figure that brine has the highest cooling rate (power) and that oil-emulsion has the lowest.

Tagaya and Tamura[11] developed a Japanese Industrial Standard (JIS K2242) method for evaluating the cooling power that employs a cylindrical silver probe with a thermocouple assembly specially constructed to measure the surface temperature change with time during quenching. This probe has several advantages: (1) higher sensitivity of the probe than the ISO probe, (2) excellent reproducibility, (3) no effect of the phase transformation, and (4) no effect of the surface oxidation.[5]

Liscic[9] later developed a steel probe assembly[12] that utilizes a proprietary surface-temperature system which has several advantages, namely, (1) very small thermocouple response time (10^{-5} s), making it sensitive to small and rapid temperature changes; (2) ease in the maintenance of surface condition of the surface-temperature sensor before each measurement; (3) the body of the probe made from an austenitic stainless steel which is not subject to volume changes; and (4) the dimensions of the probe that allow it to be mathematically modeled as an infinite cylinder.

Currently, Inconel 600 alloy (12.5×60 mm, or 0.5×2.4 in.) has been accepted as a worldwide standard probe (Fig. 13.6a) for the ISO 9950-1995E method for testing the cooling power of nonagitated and agitated quenching oils, mainly

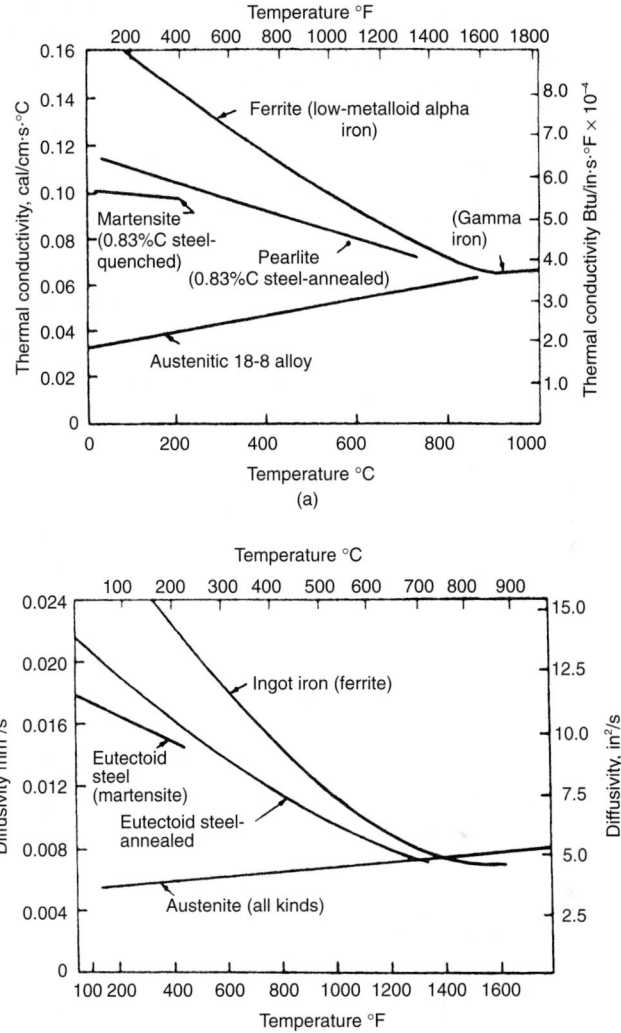

FIGURE 13.4 (*a*) The thermal conductivity of ferrite, pearlite, martensite, and austenite. (*b*) Thermal diffusivity of various structures. (*From J. B. Austin, Quenching Flow of Heat in Metals, ASM, Metals Park, Ohio, 1942.*)

because of its excellent mechanical properties.[13] Before use, this probe requires conditioning (passivation in an unprotected furnace atmosphere) to form a uniform surface oxide in order to provide reproducible quenching results, thereby eliminating the major problem associated with silver probes. Additional advantages of this probe include[14] (1) interlaboratory reproducibility, (2) a thermal conductivity much

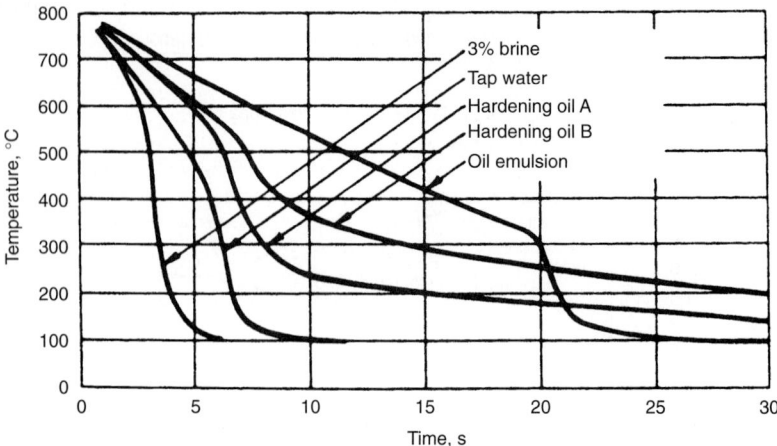

FIGURE 13.5 Center-cooling curves for various quenching media, determined by using a silver ball. Testing temperature is 40°C.[2]

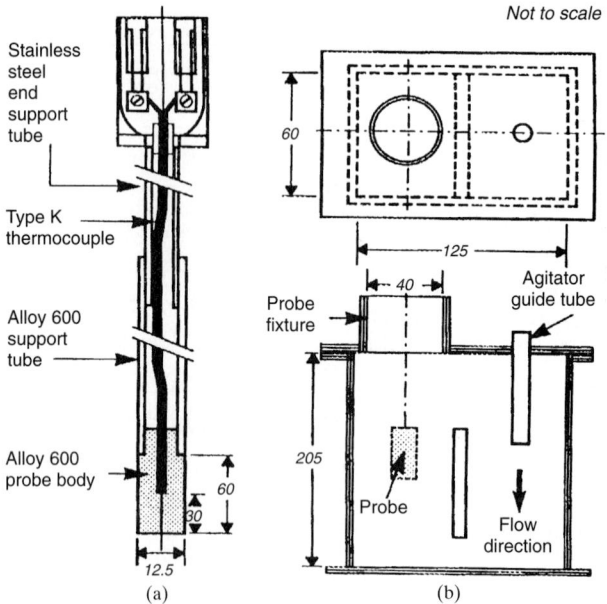

FIGURE 13.6 (*a*) The Alloy 600 (Wolfson) probe is said to provide improved data stability and heat-transfer properties more similar to those of steel. It is specified in ISO 9950-1995E and the proposed ASTM standard for nonagitated oils. It will also be used in standards for agitated quenchants being drafted by the International Federation of Heat Treatment and Surface Engineering and the ASM Heat Treating Society. (*b*) The IFHT-developed agitation system ensures uniform and stable quenchant flow. All dimensions are in millimeters.[13] (*Courtesy of ASM International, Materials Park, Ohio.*)

closer to that of steel than silver, (3) no phase transformation of Inconel 600 during quenching, and (4) the need of small quantities of quenchant solution (~2 L or 0.5 gal) to routinely evaluate the performance of quenching media due to a small probe size.

The agitation system (Fig. 13.6*b*) uses glass or a transparent plate and holds 1.5 L of quenchant. Agitation is effected by a rotating plastic stirrer. The main advantage of this system is high uniformity/stability of turbulence and flow throughout the quench zone. That is why the accuracy of the measurement is relatively insensitive to the location of the probe in the quench zone.[14] Totten et al.[15] have given an overview of currently existing national and international cooling curve analysis standards which have been developed so far.

13.2.3 Quenching Media

In general, quenching is carried out by immersion in water, brine solutions (aqueous, usually 8 to 10%), caustic solutions (aqueous, usually 3%), oils, polymer solutions (aqueous), molten salts, or molten metals. In some cases, still air or forced gas (or air), fog-quenching, fluidized beds, spray quenching, and water-cooled dry dies are used.

The main requirements of any quenching medium are the following.[16] The quenching medium should have a proper rate of cooling to produce the required properties. The ideal quenchant would have initial high quenching speeds during both the vapor-blanket and vapor-boiling phase periods to prevent the formation of pearlite and bainite, but would cool slowly during the final liquid-cooling (convection) stage to minimize distortion and cracking. Neither water nor oil is an ideal quenchant, because water causes fast cooling down to a temperature well below the M_s temperature, whereas oil lacks the high quenching speed of water above the M_s temperature but cools slowly below about 316°C (600°F).[17] The quenching medium should also possess good heat absorption properties; have reasonable oxidation and aging stability; not lose its physical properties during contact with hot solid mass; not distill, break up, or evaporate (i.e., have poor volatility); have, in case of an oil, a high flash point (a minimum of 200°C to avoid necessary fire hazards); and have a low viscosity so that it can flow freely around the part during quenching. For mineral oils, a low viscosity range of 70 to 300 SUS (Saybolt universal seconds) is most widely employed, which reduces drag-out. Finally, the quenching medium should be cheap and easily accessible.

13.2.3.1 Water.   Water is the most convenient, inexpensive, readily available, and disposable (easily stored, nontoxic, nonsmoking and noninflammable, and easy to filter and pump) of all quenching media. In addition, water can be used for easy scale removal and to produce a high cooling rate (in quench-hardened steels). It is an effective means of breaking scale from the surface of the steel part when it is quenched from furnaces operating without protective atmospheres. Water is also commonly used for quenching austenitic stainless steels and nonferrous metals after a solutionizing treatment at elevated temperatures.

On the other hand, water has several drawbacks:

1. The third stage in water quenching begins at a much lower temperature range than with oil, and during this state the cooling rate is rapid. This tends to cause a high residual stress, distortion, and cracking. Consequently, water usually is limited to the quenching of simple, symmetrical parts with no sharp stress raisers and to shallower hardening grades of steel (plain carbon and low-alloy steels and low-alloy carburizing steels).

2. Water has an extended vapor-blanket stage; this prolongation increases with the complexity of the carbon steel part being quenched and with the temperature of the quench water. This may result in uneven hardness and undesirable stress distribution, leading to excessive distortion, cracking, and/or soft spots.[17]

3. Water has a corrosive and oxidizing nature, requiring inhibitors to protect the quench tank and equipment.

4. Water also has an affinity for bacterial attack and fungus growth, which can produce unhygienic conditions.

5. Its quenching power is a strong function of temperature, intensity of agitation, and contamination.[17a]

To ensure reproducible results by water quenching, the temperature, agitation, and contaminants must be controlled. For efficient quenching in water, brine solutions, or other quenching media, sufficient volume of (water) quenchant at a temperature of 15 to 25°C (55 to 75°F) with circulation or agitation of water [to produce velocities greater than 0.25 m/s (50 ft/min)] and addition of soluble heat-treating salts to water (which raises the boiling point) must be provided, since most of them enhance the cooling power of water by restricting the duration of the first stage. On the other hand, insufficient quantity of water, higher water temperature, no agitation, and addition of soap, algae, slimes, or an emulsifier usually retard the rate of heat removal by trapping the steam or vapor film around the part, thereby extending the period of the vapor-blanket phase. Since insoluble salts either are carried over from activated salt baths into the water quench tank or are produced in the brine solution tank, periodic desludging or replacement of the quenchant is always recommended; otherwise, smooth functioning of a closed pumping system is disrupted, and reduced volume of available water will result in the quench tank.[17] Figure 13.7 shows a comparison of cooling rates of water, oil, and emulsion of water and soluble oil.[18]

The effects of water temperature on cooling curves and cooling rate curves in an Inconel 600 probe are shown in Fig. 13.8a and b, respectively. This figure also demonstrates that quenching into water of increasing temperature increases the period of A-stage cooling and decreases both the A-stage and B-stage cooling rates. Both of these effects will enhance the difficulty in obtaining desirable as-quenched hardness values.[6]

13.2.3.2 Brine and Caustic Solutions.

Brine solutions are aqueous solutions containing various proportions of common salt, frequently a 10% NaCl solution, whereas caustic solutions are aqueous solutions up to 10% NaOH. The advantages of brine solutions as well as caustic solutions over water or oil, for quenching, are the faster cooling rate; reduction or elimination of vapor-blanket stage around the heated part; improved uniformity of as-quenched hardness;[17a] reduced soft spots from steam pockets; less severe distortion and cracking; and less critical effect of small variations in operating temperatures.

The disadvantages of brine solutions are (1) the corrosive nature of brine, requiring the use of either protective coating, plating, or sheathing or corrosion-resistant metals such as copper-bearing or stainless steels for quench tank, pumping system,

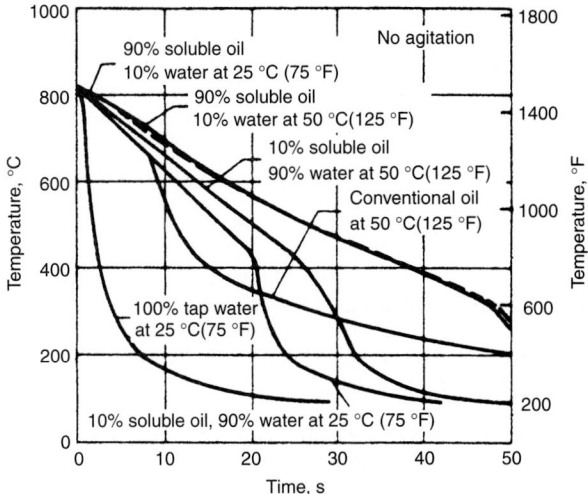

FIGURE 13.7 Center-cooling curves for still-quenched 18/8 stainless steel specimens 0.5 in. (13 mm) in diameter × 2.5 in. (64 mm) long, showing comparative cooling characteristics of plain water and soluble oil, at temperatures of 25 and 50°C (75 and 125°F).[18]

and so on; (2) requirement of a hood over the quenching tank to carry off the emanating (harmful) brine vapors; (3) corrosion inhibitors for cleaned quenched parts; (4) control and inspection for a safe environment; (5) higher cost than for water due to requirements of using special pumps, quenching facilities, and corrosion inhibitors; and (6) with certain solutions such as NaOH, nitrate salts or cyanide addition, fire hazards and toxication increase, and neutralization of wastes is essential.[17a] The main drawback of caustic solution is the harmful effect on human skin due to its high alkalinity.

The temperature of brine and caustic solutions should be at about 20°C (70°F) for maximum cooling power. Brine is favored for small tool hardening applications. On the other hand, caustic solutions are preferred for mass production applications.

Now proprietary salt mixtures having the quenching power similar to that of brine or caustic solutions are commercially available; these provide rust protection for the unwashed parts and are not harmful to human skin.[17]

13.2.3.3 Oils. Oils are probably the most extensively used quenchant, and they find applications from a wide range of low-, medium-, and high-alloy steels and for a wide cross section of sizes. Quenching oils can be classified in several ways. Based on their composition, quenching power, and operating temperatures, they are grouped as slow and conventional, fast, martempering, or hot-quenching.[17] Based on inherent major hydrocarbon compounds contained in the (mineral or base) oil, they are divided into aromatic, naphthenic, or paraffinic.[18] Aromatic oils exhibit a poor thermal resistance and a low viscosity, improve the wettability of the quenched surface, and prolong the boiling stage. The viscosity of naphthenic oils is smaller than that of paraffinic oils at the same temperature; however, the thermokinetic properties of these two groups of oils are quite similar.[17a]

All modern quenching oils are based primarily on mineral oils, particularly highly refined paraffinic oils of high thermal stability with additives to improve performance and increase tank life. Sometimes they contain naphthene fractions as well in their makeup.[17a] These additives are a combination of specially selected ingredients compatible with the base oil and carefully chosen and tested antioxidants, which minimize the aging process.

Other types of oil-based quenchants are oil-water emulsions and vegetable-based quenchants.

Slow and Conventional Quenching Oils. Also sometimes called *normal speed* or *nonaccelerated cold-quench oils*, these are the simplest and the most economical (paraffinic and naphthenic) mineral quenching oils, usually with a viscosity of 100 to 110 SUS at 40°C (100°F), but it may reach about 200 SUS at 40°C (100°F). These quench oils may contain antioxidants to impart oxidation stability, reduce thermal degradation, and prevent corrosion; but they are free from additives that would enhance their quenching rate. They exhibit a slow cooling rate during the long vapor-blanket stage, a higher cooling rate during short vapor-boiling stage, and a very slow cooling in the convection stage. Thus the quenching power of conventional quenching oils is very low compared to that of water and is often inadequate for steels of low hardenability. However, these oils can be employed for the quenching of high-carbon and alloy steels with good hardening response characteristics.

Fast-Quenching Oils (or Accelerated Cold-Quench Oils). These are blended from solvent-refined paraffinic mineral oils, accelerators, and antioxidants[19] that usually have better stability and viscosity [viscosity lies between 50 and 110 SUS at 40°C (100°F), but for the most part, it stays between 85 and 105 SUS at 40°C (100°F)], longer life, and higher flash and fire points.[20] They contain proprietary additives for increased quenching speed at the higher temperatures (i.e., during the vapor phase and boiling range periods) with the maintenance of a slower rate at lower temperatures (especially through the martensitic transformation region), which causes reduced distortion and cracking.[18] In some cases they contain very effective speed improver additives for faster quenching speeds in the convection stage, to cause a much better effect through hardening than that obtained by the usual fast-quenching oils. In addition, they may contain antioxidants, wetting agents, and other additives (to minimize viscosity increases and sludge formation).[18] The additives used in these oils are certain animal fats, polymers, and high-molecular-weight hydrocarbons, which reduce the duration of the vapor-blanket stage and increase the rate of heat extraction in the boiling stage.

Fast oils with lower viscosity are more suited for quenching parts of greater cross section than the denser oils, because of the lower boiling temperature of the former oils than that of the latter. The surfaces of the parts quenched in these oils normally darken during quenching.

Hot-Quenching (or Martempering) Oils. These are solvent-refined paraffin-base mineral oils with good oxidation and thermal (or aging) stability. They are commonly used at temperatures between about 95 and 230°C (200 and 450°F) for conventional and modified martempering treatment of ferrous parts. In some cases, they can be used even up to 250°C (482°F).[17a] Viscosities range from 250 to 3000 SUS at 40°C (100°F).[21] They contain inhibitors to slow down the rate of oxidation and thermal degradation, thereby prolonging their useful service life.[18] Sometimes they are available with additives which provide the high quenching rate, even at high operating temperatures. Figure 13.9a shows the effect of additives (such as calcium naphthenate and alkenyl succinimide) in increasing the wettability of an oil on steel, leading to higher cooling rates.[22] The ability of these additives to improve quenching rates is illustrated by the curves in Fig. 13.9b.[23]

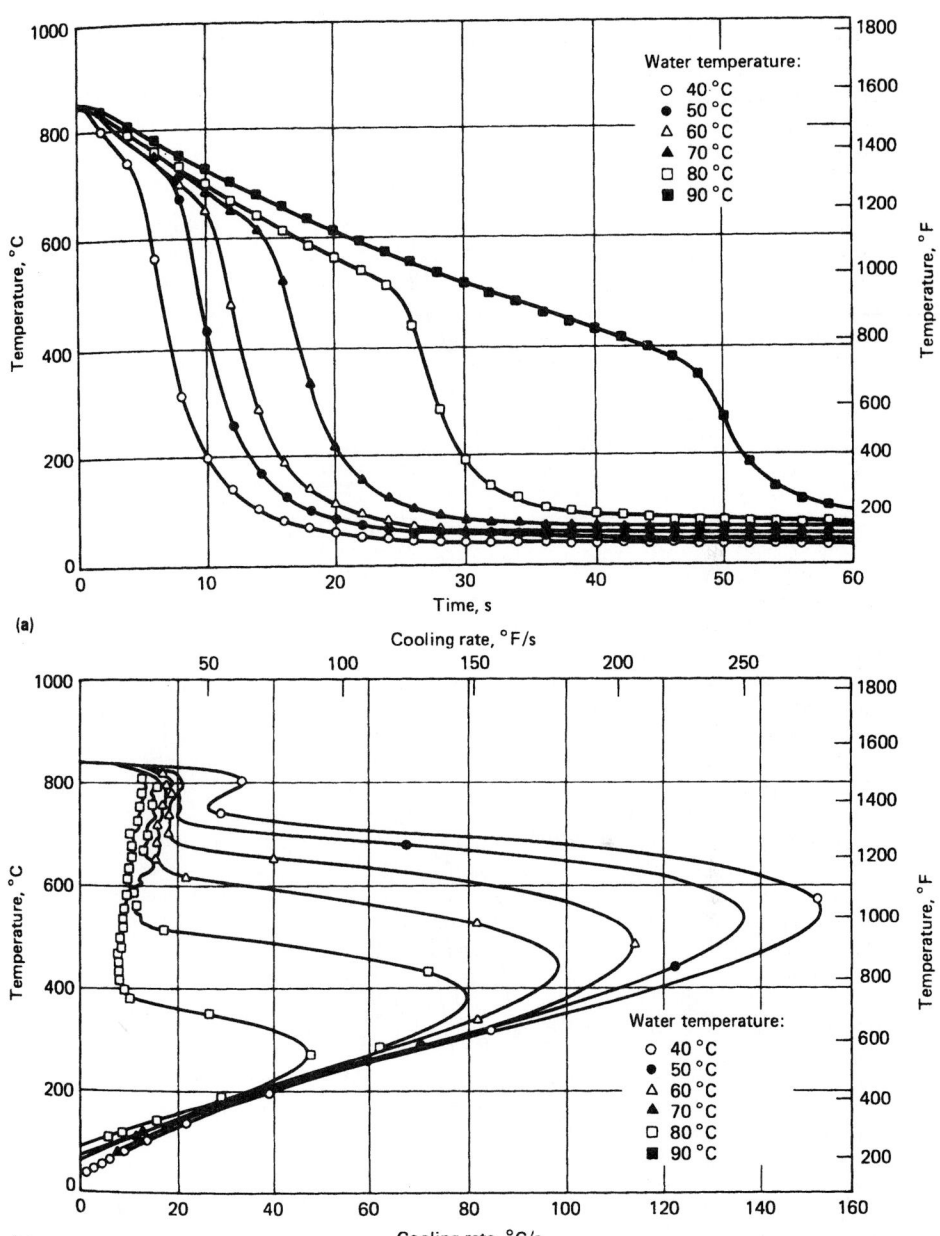

(a)

(b)

FIGURE 13.8 Effect of bath temperature on heat removal in a Wolfson probe. (*a*) Cooling curves. (*b*) Cooling rate curves. Quenchant is water having 0.25 m/s (50 ft/min) velocity.[6] (*Courtesy of ASM International, Materials Park, Ohio.*)

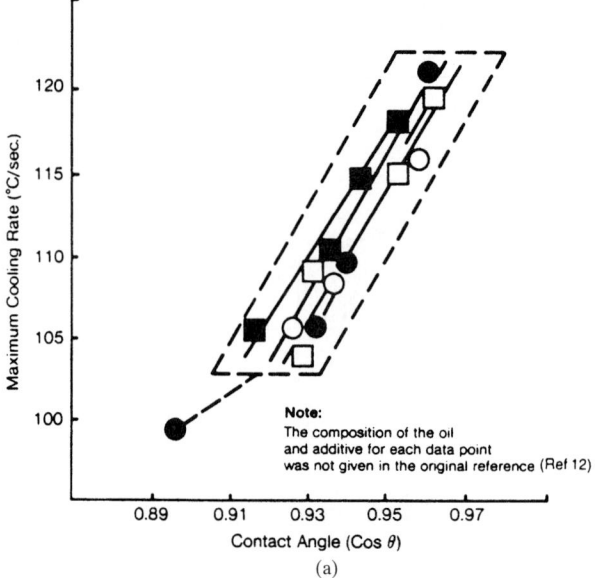

FIGURE 13.9a Relationship between cooling rate and wettability of oils with additives.[21]

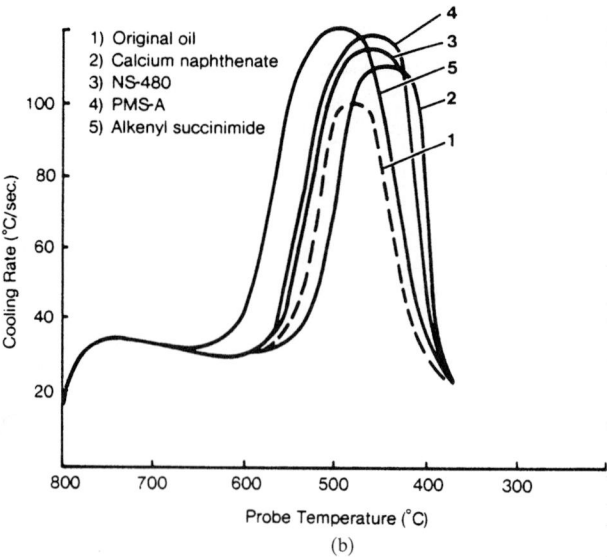

FIGURE 13.9b Cooling rates of a silver sphere quenched in oil containing different additives.[22]

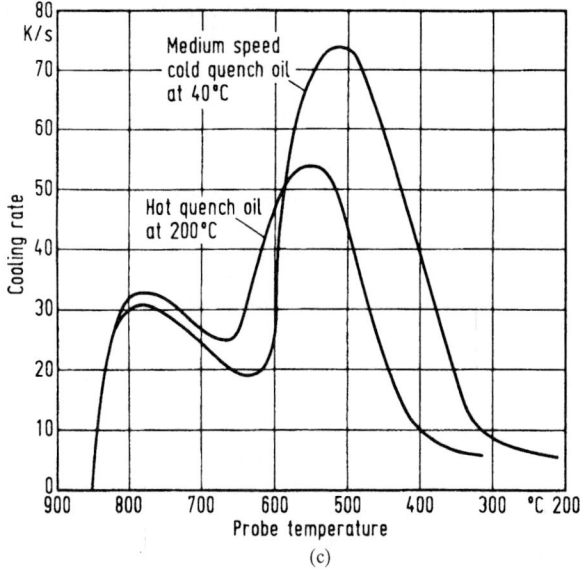

FIGURE 13.9c   Cooling rate curves for a hot oil at 200°C and a medium-speed cold oil at 40°C.[17a] (*After I. M. Hampshire, 1984.*)

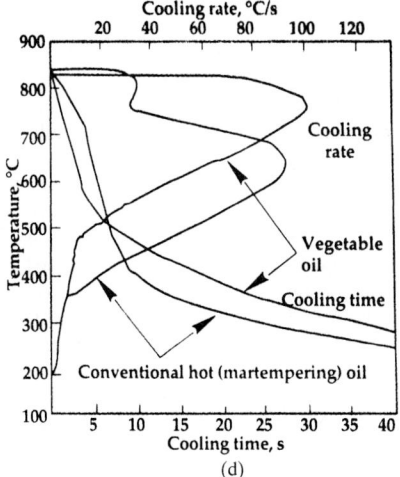

FIGURE 13.9d Comparison of the cooling characteristics of a bioquench vegetable oil and a conventional hot oil at 120°C (250°F).[25a] (*Courtesy of ASM International, Materials Park, Ohio.*)

The quenching power of hot oils is generally lower than that of low-temperature oils (Fig. 13.9c), but is much higher than that of nitrate baths. This facilitates hardening parts of increased section size, where good mechanical properties at the core are essential. Like low-temperature oils, all hot oils may contain emulgators, which are added to enhance washability without affecting the quenching power.[17a]

Martempering oils are used in a majority of heat-treating equipment such as batch-type and continuous furnaces, integral (sealed) quench furnaces, open quench tank, plug and die type press fixtures, vacuum oil quench furnace, and facilities for immersion-type quenching of induction heat-treating parts. These heat-treated parts include gears, shafts, bearing races, flat tools such as hand saws, high-alloy wear plates, and other distortion-prone parts.[24]

Table 13.2 shows the typical physical and chemical characteristics of several commercially available quenching oils, along with the corresponding ASTM test methods for their measurements.[17]

Demerits of Quenching Oils. First, all quenching oils have reduced quenching powers compared to those of water or aqueous inorganic salt solutions. However, heat removal is more uniform and particularly slow at the convection range, which causes the diminution of dangers of distortion or cracking. Second, water contamination, oxidation product contamination, smoke, fumes, spills, fire hazards, toxicity, aging, and disposal of spent oils require a provision of safety measures which are quite often ineffective. Third, quenched parts need washing in alkali solution, which increases the waste management problem.[17a,20,25]

Water-Oil Emulsion. Emulsions of soluble oils with concentrations of 3% to about 15% are used for spray-quenching in flame and induction hardening operations. They cannot be employed for immersion hardening, except in special cases. They are sometimes used for cooling steel parts, susceptible to temper brittleness, after the tempering of these parts at a high temperature.[17a] They also find applications as coolants for metalworking such as cutting, grinding, and sometimes forming. These quenching emulsions have characteristics similar to those of water-based quenchants.

Vegetable Oil-Based Quenchants. They have the following advantages over conventional oils: (1) They are readily biodegradable; (2) they have lower toxicological hazard potential and dragout and high flash and boiling points; and (3) they are easily replenished with consistent supply. The *demerits* of vegetable oil quenchants include (1) hydrolytic stability, (2) oxidative instability, (3) low-temperature characteristics, (4) narrow viscosity range, and (5) higher cost than mineral oil. Figure 13.9*d* shows a comparison of the quenching characteristics of a vegetable oil quenchant and a traditional hot quenching oil.[25a]

Measurement of Heat-Removal Properties. The *cooling rate, quenching speed,* or *quench-severity* of the mineral-oil base quenchants determines the ability to harden parts and is considered as a quality check for correct oil. For a more detailed discussion on this subject, readers may refer elsewhere.[8,26] It is often determined by the *GM quenchometer* or *chromized nickel ball test* (ASTM D3520).[27,28] In this test, a 22-mm- (7/8-in.-) diameter nickel ball is heated to 885°C (1625°F) and then dropped into a wire basket suspended in a beaker containing 200 ml of quenchant oil at 21 to 27°C (70 to 80°F);[29] and the cooling time (also called the *heat extraction rate* or *quenchometer time*) is the time for the nickel ball to reach its Curie or magnetic point (354°C, or 670°F) (Fig. 13.10) at which the ball becomes magnetic (thereby activating a sensor to stop the timer). These test results are employed to classify quench oils as slow (20- to 30-s oils), medium (14- to 20-s oils), or fast (8- to 12-s oils), and so forth (Table 13.3).[30] The main shortcoming of this method is that no information is obtained about the characteristics of the quenchant between the start of quench and the completion of the Curie point. The GM Quenchometer is not acceptable for evaluating polymer quenchants.[31]

The hot-wire test is another method for describing the heat-removal properties of an oil. In this test, a fine resistance wire is suspended between two terminals. The wire is lowered in 100 ml of oil and is heated by an electric current passing

TABLE 13.2 Typical Physical and Chemical Properties of Commercially Available Quenching Oils, along with the Corresponding Test Methods for Their Measurements[17]

Type of quenching oil	No.	API gravity	Flash point °C	Flash point °F	Pour point °C	Pour point °F	Viscosity at 40°C (100°F), SUS	Saponification	Ash, %	Water, %
Conventional, no additives	1	33	155	315	−12	10	107	0.0	0.01	0.0
	2	27	185	365	−9	15	111	0.0	0.03	0.0
Fast, with speed improvers	3	33.5	190	370	−12	10	95	0.0	0.05	0.0
	4	35	160	320	−4	25	60	0.0	0.20	0.0
Martempering, without speed improvers	5	31.1	235	455	−9	15	329	0.0	0.02	0.0
	6	28.4	245	475	−9	15	719	0.0	0.05	0.0
	7	26.6	300	575	−7	20	2550	0.0	0.10	0.0
Martempering, with speed improvers	8	28.4	230	450	−9	15	337	2.0	1.1	0.0
	9	27.8	245	475	−9	15	713	2.2	1.1	0.0
	10	25.5	300	570	−7	20	2450	2.5	1.4	0.0
ASTM test		D287	D92	D92	D97	D97	D445, D2161	D94	D482	D95, D1533

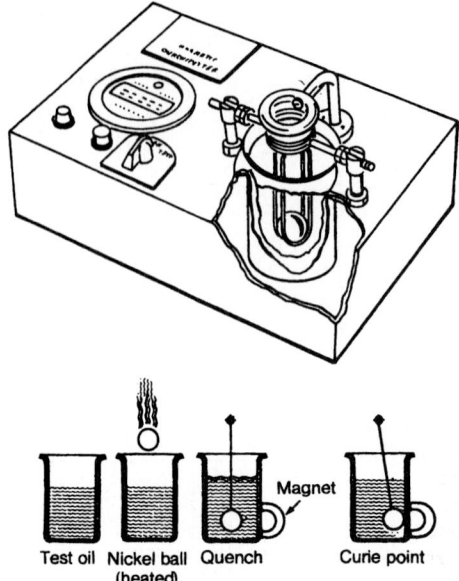

Test oil Nickel ball Quench Curie point
(heated)

FIGURE 13.10 GM Quenchometer and principle of operation.[29]

TABLE 13.3 Classification of Quench Oils by GM Quenchometer Times[30]

Classification	GM quenchometer time, s
Fast oil	8–10
Medium oil	11–14
Slow oil	15–20
Martempering oil	18–25
Deionized water	2.0
9% aqueous solution of NaCl	1.5

through it. During the immersion of wire, the current is stepped 5 A every 5 s, until the wire burns through. The (oil) quenching speed varies directly with the average amperage readings of the wire at burnout.[21,32,33] Its main drawbacks are that comparison of quenchants is only made in the higher temperature range and that no facts about the vapor-phase characteristics and the transition temperature to boiling can be obtained.[26] (A stable vapor phase cannot be achieved with a wire.) A correlation between hot-wire test and GM Quenchometer results is shown in Fig. 13.11.[34]

Alternative methods for the determination of quenching speed or the cooling curve test are *silver cylinder* (16-mm-diameter × 48 mm) *probe* (AFNOR NFT-60778-France), *Inconel 600 probe* (IFHT[†] and ISO 9950-1995E specification), *IVF quenchotest, Houghton quench test apparatus (system Meinhardt)*, and *interval test.*[26]

[†] International Federation of Heat Treatment and Surface Engineering.

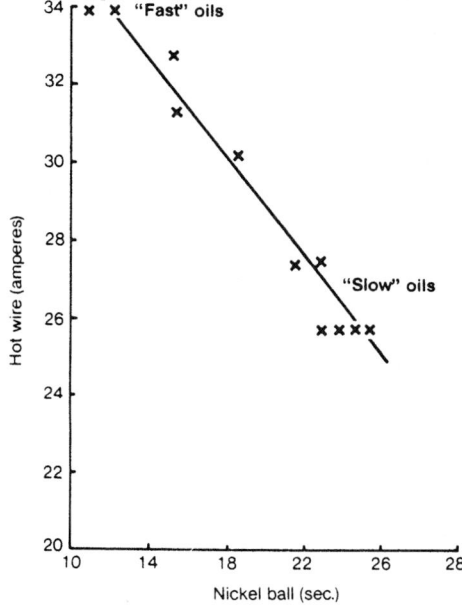

FIGURE 13.11 Correlation between hot-wire and GM Quenchometer (nickel ball) tests.[34]

The ability to computerize test data simplifies both comparison with standard reference curves and preparation of process control charts.[35]

The IVF Hardening Power (HP). Hardening power (HP) or quench severity of an oil quenchant, developed by IVF Quenchotest and using the Wolfson probe, has been calculated by measuring the three transition temperatures from the cooling rate curves of the quench oil: T_{VP}, CR, and T_{CP} (Fig. 13.12).[26,36] These parameters can be defined as follows:

- T_{VP} is the transition temperature (°C) where the vapor phase is broken. It is also called the *characteristic temperature* (CT), denoted by T_c.
- CR is the cooling rate (°C/s) over the temperature range of 600 to 500°C (1100 to 930°F).
- T_{CP} is the transition temperature (°C) where the convection phase starts. It is also called the *convection start temperature*, denoted by T_d.

For plain (or unalloyed) carbon steels, the IVF hardening power for quenching oils can then be calculated from the regression equation as a single value:

$$HP = 91.5 + 1.34T_{VP} + 10.88CR - 3.85T_{CP} \qquad (13.6a)$$

This algorithm was successfully used to rate the relative quench severities of various quench oils (Fig. 13.13).

In most cases, the quenching oil with the best quenching (hardening) power will have the highest T_{VP}, the lowest T_{CP}, and the maximum CR at a low temperature. By choosing a base oil with these properties, less additive will be needed to obtain the desired cooling curve.

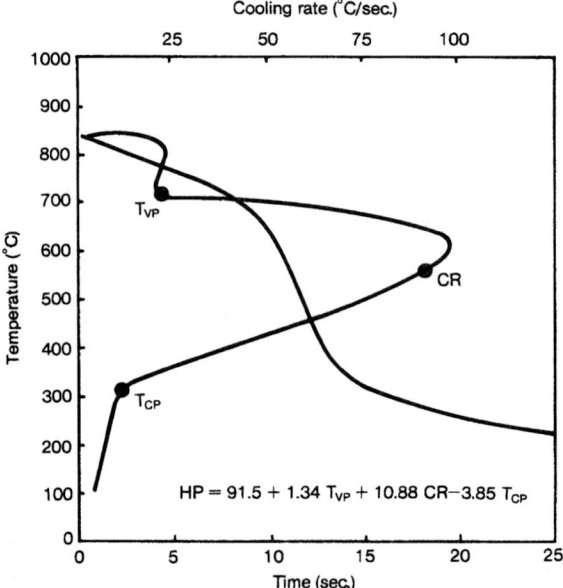

FIGURE 13.12 Calculation of hardening power of oil from cooling curves.[8]

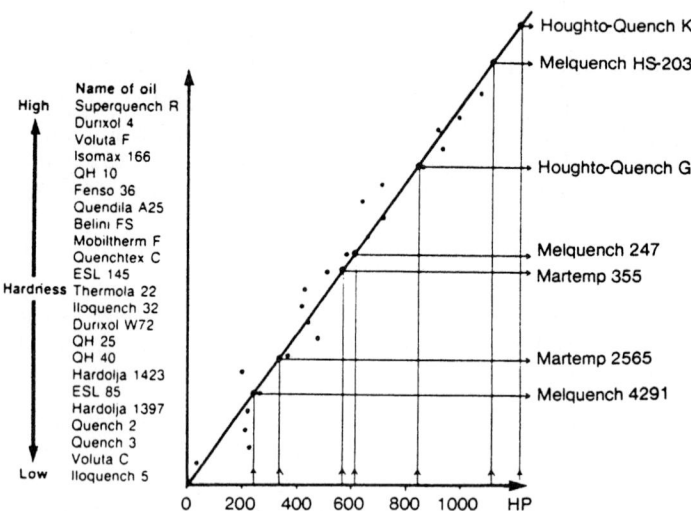

FIGURE 13.13 Correlation of calculated hardening power values for various quench oils when hardening unalloyed steels.[7]

The Castrol/Renault Hardening Power and the Castrol Index (CI). Another equation to predict the hardening power of the quenchant is the *Castrol index* (CI) which is given by

$$CI = \frac{K'V_{max}}{T_{max} - T_{bath}} \qquad (13.6b)$$

where K' is an apparatus constant, V_{max} is the maximum cooling rate (°C/s), T_{max} is the temperature corresponding to the maximum cooling rate (°C), and T_{bath} is the temperature of the quenchant (°C). The correlation between calculated HP values and hardness measured in steel test pieces holds well for both test probes.[26]

The major factors determining the quenching power of a given type of oil are (1) the fraction ratio of the so-called base oil according to the major hydrocarbon compounds contained in them (such as aromatic, naphthenic, and paraffinic oils) and according to their distillation temperature range [300 to 350°C (No. 1), 350 to 400°C (No. 2), 400 to 450°C (No. 3), and 450°C (No. 4)] and (2) the use of additives, called *accelerators*, such as metal sulfonates, nickel-zinc dithiophosphorate, various petrochemical compounds, and so forth.[17a] Since the quench severity is related to the wetting characteristics of the quenchant, additives which decrease the surface tension (i.e., increase the ability to wet the quenched part) should also increase the quench severity.

Modern oils also contain other types of additives such as oxidation inhibitors, sludge inhibitors, detergents, defoamers, and emulgators to facilitate the removal of oil residues.[17a]

Detergent/dispersant additives are ideal quenchant additives; their use increases the quench severity to a greater extent than the use of specialized sulfonates. Other additives to increase the quench severity are polyisobutylene polymers, where the quench rate depends on the polymer length.[19]

Oil Temperature. The conventional and fast-quenching oils are usually maintained at temperatures between 40 and 95°C (100 and 200°F), most frequently at 50 to 70°C (120 to 160°F). High temperature produces faster aging of the oil and can result in copious fuming. As a safety measure, the maximum oil use-temperature should be maintained at least 50°C (120°F) below the oil's flash point. Lower oil temperature may induce more distortion due to the faster quenching effect of the oil in the convection stage, and it may run the risk of fire hazard as a result of the higher viscosity of the cold oil, which does not effectively extract and rapidly distribute the heat, i.e., thereby leading to localized heating above the oil flash point.

The typical use-temperature ranges for commercially available covered and unprotected martempering oils are given in Table 13.4, which shows the higher use-temperature (near the flash points) for the covered and protective (inert, neutral, or reducing) atmosphere.[17]

Oil Laundering. The detrimental effect of higher-oil-temperature operation can be offset by methods of laundering (or cleaning) the oil which prolong its useful life and can help to trim a plant's production and operational cost.[37,38]

Dragout Loss. Dragout loss of a quenching oil depends primarily on the viscosity of the oil and the interface temperature during the part removal from the quench bath. It may occur due to heavy oxidation of oil, insufficient drainage, and complex part configuration.

Quench Loads. For effective quenching of the hot part, the requirements are lower quenching oil temperature (i.e., the maximum temperature rise that can be tolerated), higher flash points, and larger oil volumes. The oil volumes flowing per

TABLE 13.4 Typical Use Temperature Ranges for Commercially Available Covered and Unprotected Martempering Oils[6]

Viscosity at 40°C (100°F), SUS	Minimum flash point		Use temperature			
			Open air		Protective atmosphere	
	°C	°F	°C	°F	°C	°F
250–550	220	430	95–150	200–300	95–175	200–350
700–1500	250	480	120–175	250–350	120–205	250–400
2000–2800	290	550	150–205	300–400	150–230	300–450

minute which will satisfactorily quench a given weight of quench loads in a quench tank can be predicted at once by the following generalized rules regarding oil volumes, weight of the quench load, oil/air interface, and agitation:

1. Parts with a higher surface-to-volume ratio need lower quenching oil temperatures, larger oil volumes, or higher-flash-point oils. For massive parts, the oil volume may be lower because of its comparatively slow heat transfer to the oil. However, the installation of coolers and the provision of oil circulation from tank to cooler are still required to maintain the oil temperature at a safe value. A common rule is that oil volumes (liters) = weight of the quench load in kilograms/10, or volume of an oil quench tank (gallons) = weight of the quench load in pounds. This rule includes the weight of fixtures, trays, and so forth that are at the temperature of the parts and enter the quenching oil.

2. All quenching oils should have as small a contact with air as practical in order to minimize oxidation. Hence, the surface of open-air quench baths must be kept as small as possible.

3. Only mechanical agitation (pumps and impellers)—and no agitation by compressed air through the oil—should be employed to maintain a uniform bath temperature and to distribute heat rapidly throughout the quench cycle.

4. Preheating of quenching oils by heating elements should be minimized [$<1.5 \times 10^4$ W/m^2 (10.0 W/in.^{2})] to prevent local overheating and thereby prevent unnecessary aging of the oil and the formation of an insulating (carbonaceous) film of oil-coke on the heater elements and the possible burnout of heating elements.

5. Quenching oil systems comprising heating elements and coolers should be made of steel, ferrous castings, nickel-plated or tin-plated materials, or stainless steel, and should not be composed of copper or copper alloys (because the latter act as catalysts to accelerate the oxidation and polymerization of mineral oils).[6]

6. Brass or copper equipment should be tin-plated.

7. The precautionary measures adopted in the use of quenching oils include (*a*) use of inert-gas covers (curtains), (*b*) provision of snuffer plates to ensure the rapid entry of parts into the quench tank, and (*c*) possibly the installation of a vapor-degreasing or alkali-cleaning plant.[39]

8. The rate of immersion of load should not be less than 14 to 22 m/min (46 to 72 ft/min), and the depth at which parts are submerged should be at least 20 to 30 cm (8 to 12 in.) below the bath surface.[17a]

Quenching Oil Selection. Proper selection and a maintenance program of quenching oils are effective tools to reduce costs, improve quality and profits, reduce the consumption of natural resources, reduce waste treatment costs, minimize environmental pollution (by optimizing the tank-life of oils), and safeguard the health and safety of workers.[38] The deciding criteria or factors in the selection of quenching oils (media) are as follows:[8,19,40]

1. Selection of an oil with a relatively slow cooling rate for products with thick cross sections

2. Adjustment of oil temperature to provide a viscosity of 5 to 15 cSt; viscosity of 17 to 24 cSt at 40°C usually the most suitable for cold-quenching[19]

3. The flash point of the oil at least 50°C above the operating temperature

4. In case of heating of the products in air furnace, a quenching oil that offers maximum brightness

5. Cost effectiveness of the quenching oil

6. Utilization of available quench facilities

7. Freedom from excessive distortion and quench cracks

8. Production of desired residual-stress pattern

9. Chemical composition

10. Evaporation characteristics

11. Oxidation stability

12. Cooling curve characteristics

13. Availability

Quenching Oil Maintenance. Quenching and martempering oils deteriorate during use, although most premium quenching oils can last more than 10 years when reasonable use and care are provided. Quench oil degradation is usually associated with a change in cooling rate, decrease in flash point, insoluble sludge formation, part staining, and other physical and chemical properties.[41] Regular testing and analysis of a production quench oil is considered as a proactive maintenance policy and should be carried out every 3 months to determine its condition, extent of degradation, its suitability for continued use, and any corrective action to be taken. Used quenching oils are normally checked for their physical and chemical properties, which are briefly described below.

QUENCH OIL DEGRADATION. Quenching oils are subjected to two types of deterioration: oxidation and thermal cracking (or degradation). Mineral oils at high temperature when exposed to air or oxygen are oxidized at comparatively high rates. Thermal cracking leads to the formation of new materials, where some are light, relatively volatile products and others are heavy, less volatile products. The more volatile products decrease the flash point of the oil; the heavier products increase the oil's viscosity.[38]

Quench oil degradation is also associated with a change in cooling rate, buildup of organic acid and the formation of insoluble material or sludge, part staining, and other physical and chemical property changes. A general mechanism for molecular degradation of an oil has been provided by Igarashi.[42,43]

To determine the oxidation stability of mineral oil, the *panel coke test* and *IP 48/89 (BS 200: Part 48:1993)* have been developed. In the former test, the relative oxidation stability of the oil is measured by the amount of decomposition products adhering to the test panel and the amount of oil consumed during the test. In the latter test, the level of oxidation is determined by the increase of viscosity and carbon residue.[19]

The oxidation stability of quenchant can be increased by the addition of one or more antioxidants. Some examples of antioxidants are butylated hydroxy toluene, phenolic, amine, and amine/phenolic mixtures.[19]

VISCOSITY.   Change in viscosity of the tested oils indicates oxidation and thermal degradation, or the presence of contaminants. Usually, increased viscosity occurs with oil degradation, decreases heat-transfer rates,[11] and causes higher dragout rates, thereby leading to an increase in the operating cost.[35] A low-viscosity base oil is preferred due to low consumption (low dragout for parts with high surface-to-volume ratio), increased flow rate, and maximum agitation to maximize the severity of quench.[19] The degree of aging of a used quench oil can be evaluated by measuring its viscosity and flash point.[17a] Viscosity (SUS at 100 and 210°F) is determined by ASTM D445 and D2161 methods.

Kinetic viscosity η [centistokes (cS)] is calculated by multiplying the measured flow time t by the viscometer constant C:

$$\eta = Ct \tag{13.7}$$

Another unit of viscosity, called the centipoise (cP), is related to the centistoke viscosity η by the density ρ of the fluid at the same temperature.

$$cP = cS \times \rho \tag{13.8}$$

FLASH AND FIRE POINTS.   The higher the flash point, the greater the flexibility with respect to the operating temperatures. Ideally, flash point should be in excess of 180°C for cold-quenching oils. The Cleveland Open Cup Test (IP 36/89, BS 4689, ASTM D92-85) is regarded as the most suitable test method for evaluating mineral-oil-based quenchants.[19] The fire point is always higher than the flash point. Decrease in flash (and fire) point(s)with use normally represents contamination, the chain-scission degradation process of the oil, or the presence of dissolved gases. The oil operating temperature should be at least 30 to 50°C below the fire point and appropriately lower than the flash point.[17a]

WATER CONTAMINATION.   The presence of water in an oil quenchant, arising from a degradation process or by external contamination (such as a leaking cooling oil), poses a potentially serious problem. The water content should not exceed 0.05%.[17a] Even low (0.1%) water concentration can give rise to a foaming bath during quenching, greatly enhancing the risk of fire. Overflowing oil from the foaming bath can cause a more serious fire. When a sufficient amount of water accumulates in a hot bath, there is likelihood of an explosion due to steam generation.[30]

Water contamination in quenching oils may produce variable cooling rates, depending on the nature and extent of rate-accelerating additives present in the oil. For example, an increase in water content in normal-speed and accelerated (or martempering) quenching oils, respectively, produces increased and decreased cooling rates. Hence, martempering oils used at temperatures above 100°C (212°F) must be controlled regularly, because adverse effects of water contamination include uneven hardness, soft spots, worst distortion or cracking, and foaming of the quenching oil with consequent fumes and fire hazard. Figure 13.14a shows the variation in cooling rate of a martempering oil as a function of time in use, and Fig. 13.14b shows the effect of (0.06 to 2.0%) water addition in quenching oils on cooling curves. Small water contamination (<0.12%) increases the initial quenching speed, whereas higher water concentration causes a lower initial quenching speed due to a more stable vapor-blanket around the probe.[6]

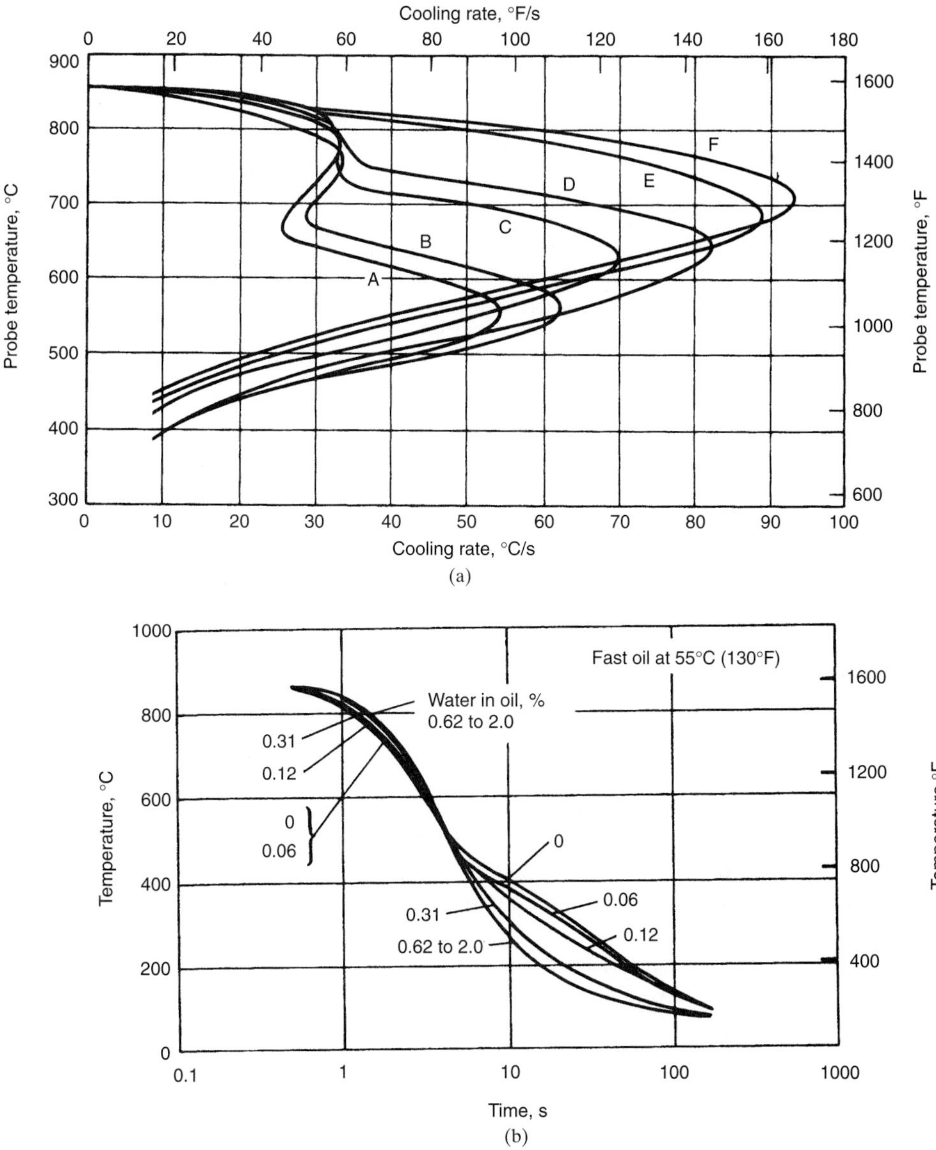

FIGURE 13.14 (*a*) Variation in cooling rate of a martempering oil as a function of time in use: A, new oil; B, 3-month-old oil; C, 7-month-old oil; D, 25-month-old oil. (*After G. M. Hampshire, Heat Treat. Met., vol. 1, 1984, pp. 15–20.*) (*b*) Effect of water contamination on the quenching power of fast oil at 55°C (130°F) in quenching type 304 stainless steel. Specimen measured 13 mm (0.5 in.) in diameter by 100 mm (4 in.) long. Oil was not circulated, and specimens were not agitated. Thermocouples were located at the geometric center of the specimen.[6]

The typical symptoms suggesting the presence of water in a quenching oil are (1) sputtering and characteristic sizzling (as during frying); (2) intensive "boiling" of the oil during cooling down of the load; (3) intensive foaming, particularly during cooling down of the load; (4) nonextinguishing fire, even after the complete immersion of the load; (5) oil turbidity and change of its color to milky-brown; and (6) considerable increase of pressure in the hardening aggregate.[17a]

Moisture analysis is done by the Karl Fisher method (ASTM D1744) which determines the dissolved water content in ppm.[37] Water content is determined by the ASTM D95 method. Usually bulk water is removed by settling to the bottom of the tank and then draining out; suspended water is removed by agitating and then slowly heating to 120°C (250°F). The water may continuously be considered to evaporate completely when foaming stops.[17a]

QUENCHING SPEED.   It was discussed earlier.

NEUTRALIZATION NUMBER.   The total acid number (TAN) or neutralization number is an indication of the extent of oxidation and measures the amount of free organic acid formed (or contained) in a quenching oil. Thus, increasing neutralization numbers usually signify that some aging processes have been proceeding in the oil and that the amount of sludge is increasing. The acid number of the quenching oil is calculated as

$$\text{Acid number (mg of KOH/g of oil sample)} = \frac{(A-B)N \times 56.12}{W} \qquad (13.9)$$

where A is the KOH solution needed to titrate the sample (mL), B is the KOH solution needed to titrate the blank (mL), N is the normality (meq of base/g of solution) of the KOH solution, and W is the weight of the sample (g) (ASTM D664). Usually, the neutralization number of 1 mg KOH/g is considered as adequate.

SAPONIFICATION NUMBER.   Saponification number is an indication of the presence of unstable or unsaturated hydrocarbons in the oil which may oxidize to form sludge. Like neutralization number, it is an indication of the degree of oxidation in the oil but quantifies the amount of free acids, esters, or fatty materials present in the oil. The saponification number of the quenching oil is calculated as

$$\text{Saponification number} = \frac{56.1N(V-V_1)}{W} \qquad (13.10)$$

where N is the normality of HCl, V is the volume of acid used to titrate the sample (mL), V_1 is the volume of acid used to titrate the blank (mL), W is the weight of the sample (g), and 56.1 is the molecular weight of KOH. The saponification number should not exceed 2 to 3 mg KOH/g.[17a] Both the saponification number and the neutralization number are required to quantify both ester and acidic by-products.

PRECIPITATION NUMBER OR SLUDGE.   Precipitation number in the test is an indication of the presence of high-molecular-weight compounds in the oil which can form sludge under operating conditions. This number indicates the extent of oxidation or degradation of the oil.[38] Quenching or martempering oils with a high precipitation number may cause staining of quenched parts.[35] Sludge, due to oxidation, thermal degradation or breakdown, and polymerization of the oil, can affect quenching properties, reduce the efficiency of heaters and coolers (by clogging filters and fouling heat-exchange surfaces), and adsorb on a part, resulting in reduced and nonuniform film heat transfer during quenching. The sludge formation also affects the viscosity, heat transfer, and quench severity of a quench bath.

Sludging (carbonaceous deposit or carbon buildup from atmosphere applications) is objectionable because it causes soft spots and its cleaning is difficult and expensive. Sludge is considered as a leading cause of problems in oil quenching baths, next only to water.[44]

It is important to maintain particulate contamination $\leq 1\,\mu m$ to optimize quench performance.[45] Sludge levels of $\geq 0.2\%$ may cause staining of normally bright metal surface.[44] It must be filtered at $\leq 1\,\mu m$. The amount of sludge can be determined by adding naphtha solvent to the oil sample and determining the volume of precipitate (sludge) after centrifuging (ASTM D91 and D2773).

ASH CONTENT.   Ash is an indication of the presence of incombustible material in the oil. Usually straight mineral oils are nearly ashless; many of the formulated quenching oils contain metallic additives that contribute to ash after combustion. The ash content of the oil is calculated as

$$\text{Ash}\% = \frac{w}{W} \times 100 \tag{13.11}$$

where w is the ash weight (g) and W is the sample weight (g).

ADVANCED ANALYTICAL METHOD.   Advanced analytical methods such as infrared (IR) spectroscopy and gel permeation chromatography (GPC) are widely used to provide an indication of oxidation in the oil and the presence of fatty additives, antioxidants, water contamination, and emulsifying agents. Figure 13.15 is a typical infrared analysis for a used quenching oil.[35]

13.2.3.4 Polymer Quenchants.   Polymer quenchants are dilute aqueous solutions of organic polymers and various additives to provide corrosion and foam control. These polymers are of high molecular weight or high degree of polymerization (10,000 to 20,000 amu). Additives are primarily corrosion inhibitors such as $NaNO_2$, mixtures of aromatic carboxyl acids, and amine salts of various aliphatic carboxyl acids.[46]

Advantages of polymer quenchants over the oil quenching include (1) more uniform heat transfer due to the so-called *explosive* wetting process that they exhibit;[47] (2) reduced quenchant cost and dragout losses; (3) their biodegradable nature; (4) lower capital equipment and maintenance costs; (5) easy concentration controls and excellent aging stability;[48] (6) virtual elimination of water contamination and smoke, fume, and fire hazards, thereby rendering improved environmental and easier and cleaner working conditions and greater operational safety; (7) flexibility of quenching speed possible to quench a broad range of alloys with different cross-section thicknesses by varying the concentration, temperature, and agitation of the polymer solution; (8) elimination of soft spots, distortion, and cracking problems associated with the trace water contamination in quenching oils; (9) low volatility and viscosity; (10) useful applications for both bulk quenching and induction hardening; (11) no cleaning required prior to tempering, thereby eliminating solvent and alkali degreasing;[49] (12) high specific heat, leading to reduced temperature rise during quenching; and (13) higher productivity and lower cooling-rate requirements.[25a]

A number of different polymer quenchants are available, but their quenching characteristics and field applications vary to a large extent. The notable polymers used in quenchant formulations are polyvinyl alcohol (PVA), produced by polymerization of vinyl alcohol (Fig. 13.16a); polyalkylene glycol (PAG), by copolymerization of ethylene oxide and propylene oxide (Fig. 13.16b); polyvinyl pyrrolidone (PVP), by polymerization of n-vinyl-2-pyrrolidone (Fig. 13.16c); poly

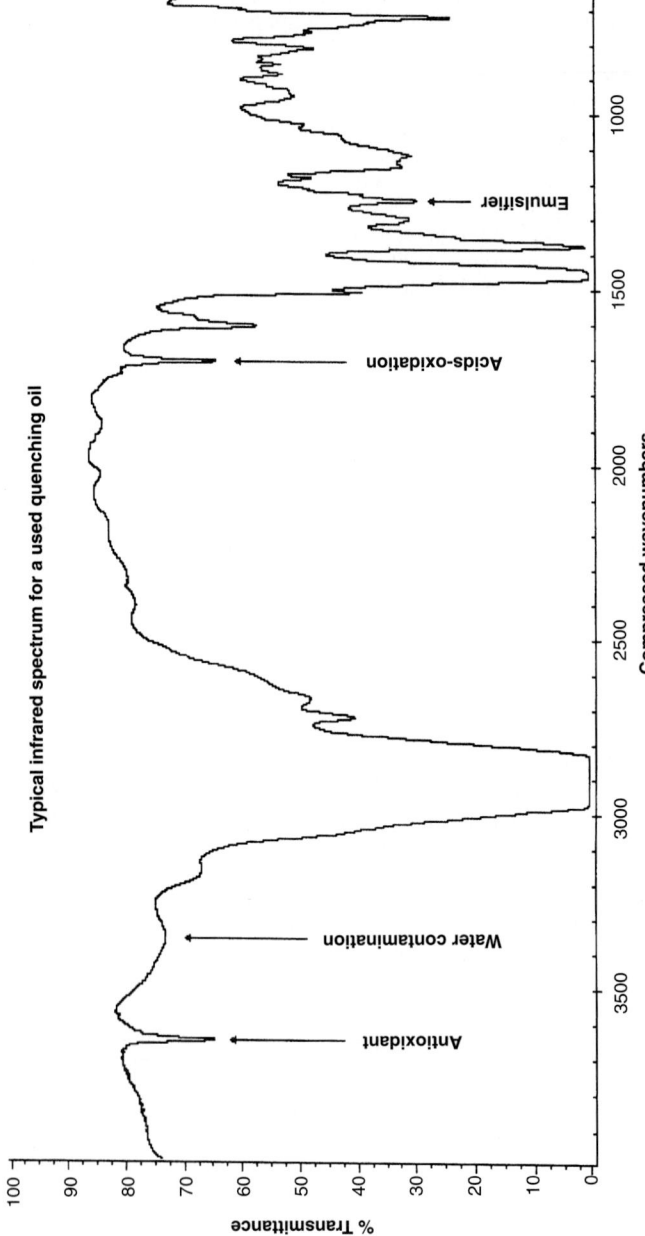

FIGURE 13.15 Infrared analysis of a used quenching oil.[35] (*Courtesy of E. F. Houghton & Co. Inc., England.*)

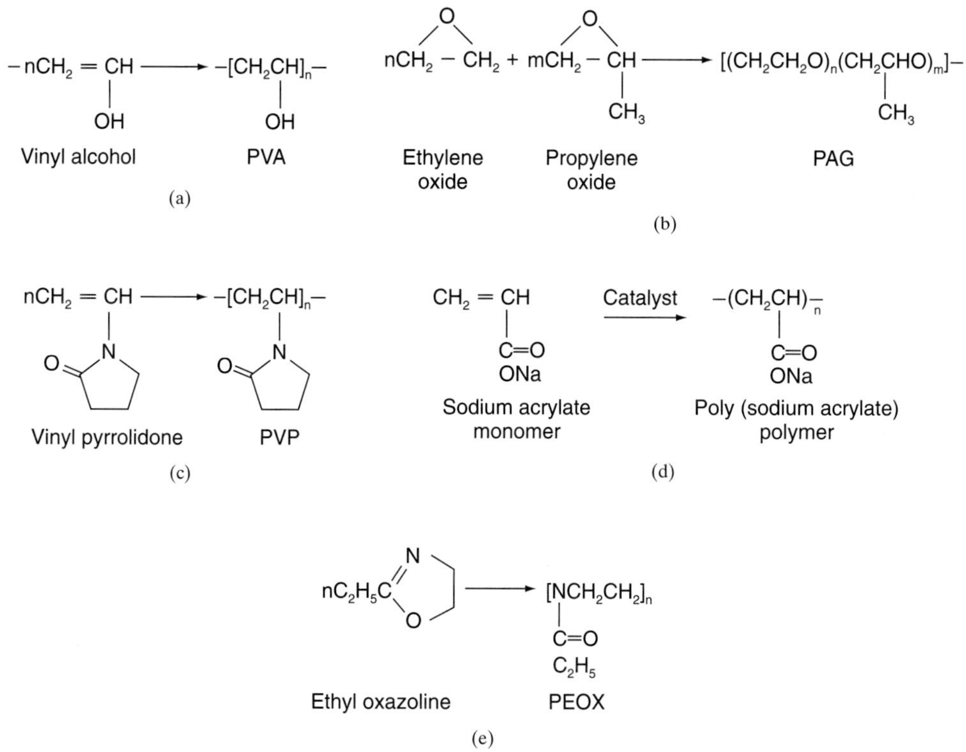

FIGURE 13.16 Synthesis of various polymer quenchants: (*a*) polyvinyl alcohol (PVA); (*b*) polyalkylene glycols (PAG); (*c*) polyvinyl pyrrolidone (PVP); (*d*) polysodium acrylates (PSA); and (*e*) polyethyl oxazoline (PEOX).

(sodium) acrylate (PSA), by polymerization/homopolarization of sodium acrylate (Fig. 13.16*d*) or by the alkaline hydrolysis of some polyacrylate ester; and polyethyl oxazoline (PEOX) by polymerization of ethyl oxazoline (Fig. 13.16*e*). These are linear, synthetic, water-soluble, and ionic or nonionic. The film strength of the polymers may be strong (in the case of PVP and PSA), weak (PAG), or intermediate (PEOX). The character of the film changes with the distance from the metal interface to the bulk solution. It implies that the heat transfer also changes under these conditions.[8]

Polymer quenchants have found widespread well-established use in the quenching of solution heat-treated aluminum alloys, hardening of plain carbon steels with less than 0.6% C, spring steels, boron steels, hardenable stainless steels, and all carburizing and alloy steels with section thickness greater than about 50 mm, through-hardening and carburizing steel parts, and induction-hardening treatments.

Polyvinyl Alcohol (PVA). The extent of hydrolysis, which controls commercial applications, may vary from partial (87 to 89%) through fully hydrolyzed (95 to 99%). The water solubility and quenching characteristics of PVA resins vary with the molecular weight of the polymer.

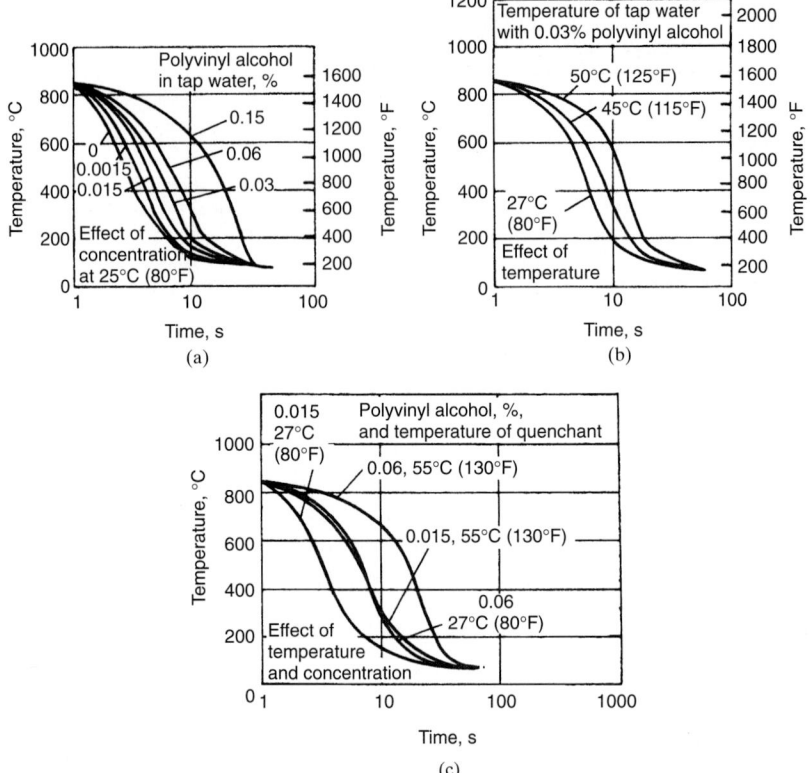

FIGURE 13.17 Effect of temperature and concentration on polyvinyl alcohol solutions. Center cooling curves are for type 304 stainless steel specimens 13 mm ($\frac{1}{2}$ in.) in diameter by 100 mm (4 in.) long, quenched in still tap water at 25°C (80°F) (or at other temperatures shown) containing various concentrations of polyvinyl alcohol. Thermocouple was placed in geometric center of each specimen. Water hardness was 130 ppm.[6]

COOLING CHARACTERISTICS. Small changes in solution concentrations are required to produce variations in cooling characteristics of PVA solution, as shown in Fig. 13.17. Since small (say, 0.10%) concentration variations have substantial effects, close control of PVA solution is essential.

CONCENTRATION CONTROL. This is cumbersome because quenched parts can become coated with an insoluble layer of resin, which lowers the bath concentration. To maintain an effective concentration requires specific control measures. Currently, very little PVA is used in quenching, mainly due to the difficulty of maintaining the correct bath concentration.[6] In fact, PVA-type quenchants have been superseded by PAG due to their ease of control and maintenance and greater flexibility.[49]

There are two other disadvantages of PVA-type quenchants: (1) A film of rubberlike compounds is formed on the surface of the quenched parts, pipes, and quench tanks which is difficult to remove. (2) Acetic acid is produced due to the thermal decomposition of these quenchants which must be neutralized because of its corrosive action and uncontrolled effect upon the cooling power.[17a]

Polyalkylene Glycol (PAG). PAG quenchants are characterized by stability to mechanodegradation, fairly low relative drag-out rates, thermal/oxidation stability, and good chemical stability. Small changes in polymer concentration display small effect on cooling rates, offering excellent process flexibility.

Unlike PVA, PVP, and PSA, PAG exhibits the unique behavior of inverse solubility in water, i.e., water solubility at room temperature and water insolubility at elevated temperatures above its cloud point.

Quenching of a hot steel part in a PAG solution also occurs in three stages. In stage 1, when the hot part is immersed into the conventional polymer solution, the solution in contact with it is heated to above the cloud point (i.e., the inverse solubility temperature). This causes the polymer to become insoluble in water, and a thin, noncontinuous polymer coating on the part surface associated with the vapor stage forms and renders a slow cooling period. Film thickness and the resulting cooling rates are functions of polymer concentration. Like oil quenching, this represents a slow quenching stage. The nature of the film (such as strength, thickness, and viscosity) controls cooling in this stage. In stage 2, polymer coating activates, partially ruptures, boils, and wets the part surface which increases the cooling power, similar to the boiling stage in oil quenching. In stage 3, the polymer layer redissolves because the local temperature falls below the inversion temperature of the polymer, and a heat extraction from the part is achieved by conduction through the remaining film and/or by convection through the solution.[40]

The viscosity of aqueous polymer solution is important; the effectiveness of heat transfer would be anticipated to decrease with an increase in the solution viscosity. Of course, very high solution viscosities, >10 to 12 cST, lead to considerable polymer dragout during the part removal from the bath after the quenching is completed.[8]

Like other polymer solutions, the cooling power of PAG depends on three main parameters, namely, concentration, temperature, and agitation of the PAG quenchant.[17] The effect of PAG concentration on cooling rates in a 25-mm- (1-in.-) diameter stainless steel probe is shown in Fig. 13.18*a*. The slower cooling rates obtained at the higher concentrations denote an increase in polymer film thickness that covers the part during quenching. The curves illustrated in Fig. 13.18*b* show the normal trends that occur with variations in bath temperature. Figure 13.18*c* shows the rapid increase of cooling rates with an increase in agitation.[8]

The performance of predicted characteristics of representative PAG quenchant under varying conditions of maximum cooling rate CR_{max} (°F/s), cooling rate at 650°F (°F/s), and Grossman H factor is shown in Fig. 13.19.[8,50] The data are based on the cooling curve apparatus described by Hines and Mueller.[50]

PAG has been applied successfully for immersion quenching of SAE 4140 valve parts in sealed quench furnaces, AISI 5140 crankshaft production, SAE 5160 coil springs and saw blades, AISI 52100 ball bearings, leaf spring production, small- and large- (>>24-in.-) diameter carburized automotive parts, automotive, truck and naval crankshaft production, alloy-steel forgings, tube stock for oil drilling applications, large alloy castings for the oil industry;[47] and induction hardening and spray quenching.[49]

PAG solutions, now referred to as "Type-1" polymer quenchants, have been used in the aluminum heat-treating industry for about 35 years as an excellent alternative to hot-water quenching for residual stress and distortion reduction and crack prevention of both forged and cast aluminum parts.[51,52]

Polyvinyl Pyrrolidone (PVP). PVP solution (aqueous) is characterized by its complexing and colloidal properties, by its physiological inertness, and by not exhibiting a thermal separation temperature. This is available in four molecular weight grades and may be used up to the boiling point of water. A 10% aqueous

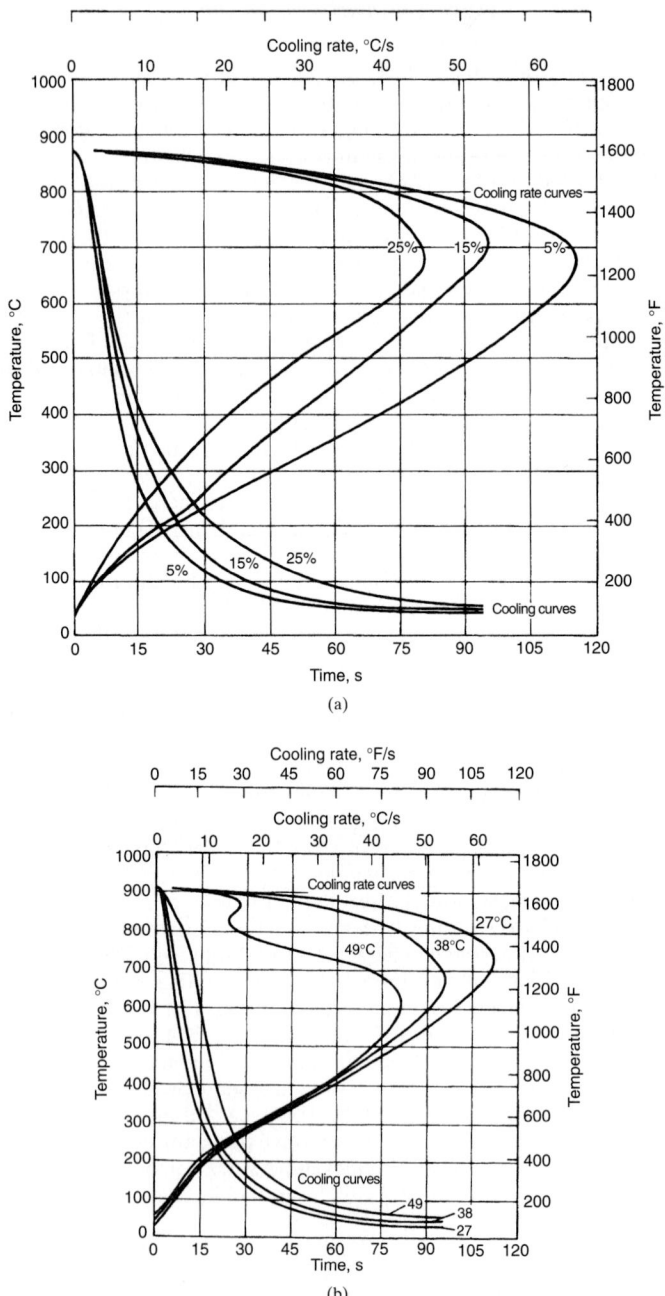

FIGURE 13.18 Cooling curves and cooling rate curves for a 25-mm- (1-in.-) diameter stainless steel probe: (*a*) quenched in 5, 15, and 25% PAG at 45°C (110°F) that is flowing at 0.25 m/s (50 ft/min); (*b*) quenched in 10% PAG at 27, 38, and 49°C (81, 100, and 120°F) that is flowing at 0.25 m/s (50 ft/min); and (*c*) quenched in 20% PAG at 45°C (110°F) that is flowing at 0, 0.25, and 0.50 m/s (0, 50, and 100 ft/min).[6]

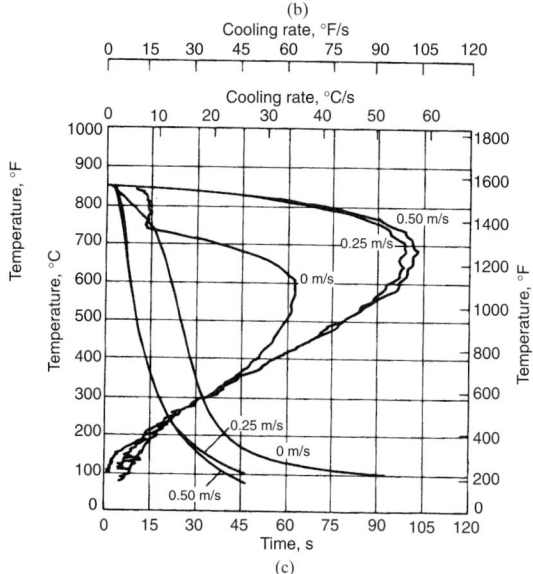

FIGURE 13.18 (*Continued*) Cooling curves and cooling rate curves for a 25-mm- (1-in.-) diameter stainless steel probe: (*a*) quenched in 5, 15, and 25% PAG at 45°C (110°F) that is flowing at 0.25 m/s (50 ft/min); (*b*) quenched in 10% PAG at 27, 38, and 49°C (81, 100, and 120°F) that is flowing at 0.25 m/s (50 ft/min); and (*c*) quenched in 20% PAG at 45°C (110°F) that is flowing at 0, 0.25, and 0.50 m/s (0, 50, and 100 ft/min).[6]

PVP solution should be used with a preferred rust inhibitor and a bactericidal preservative.

Like other polymer solutions, the quenching performance of PVP polymer solutions is relatively sensitive to small changes in concentration, bath temperature, and agitation. The quenching rates seem to be rapid with PVP quenchants in the upper temperature range (during the stable film and nucleate and boiling stages), but slower during the convection stage. Contour plot relationships showing the effect of quenchant concentration, agitation, and bath temperature on the maximum cooling rate (CR_{max}), cooling rate at 365°C (650°F), and Grossman H-factor under laboratory quenching conditions are illustrated in Fig. 13.19.[50] Since PVP is soluble up to the boiling point of water, a wider working range of temperatures for quenching can be used.

Optical refractometer readings can provide initial control of concentration, but backup with viscosity measurements is strongly suggested.[6]

Poly(sodium) Acrylate (PSA). Characteristic features of polysodium acrylate solutions are as follows:

1. They are ionic quenchants which separate them from the class of other nonionic polymer quenchants. This peculiar feature imparts strong polarity, which, in turn, produces water solubility and probably causes a different operating mechanism of heat extraction, as given below.

2. They neither split on heating nor form plastic films on the surface of the hot part.

3. By changing the molecular weights of the polymer, a complete family of quenchants can be developed, covering a wide spectrum of applications from the fast quenching of water to the slow quenching of oils.

4. Figure 13.20 shows the effect of concentration and temperature on cooling rates of polyacrylate quenchant. The unique property of the polyacrylate

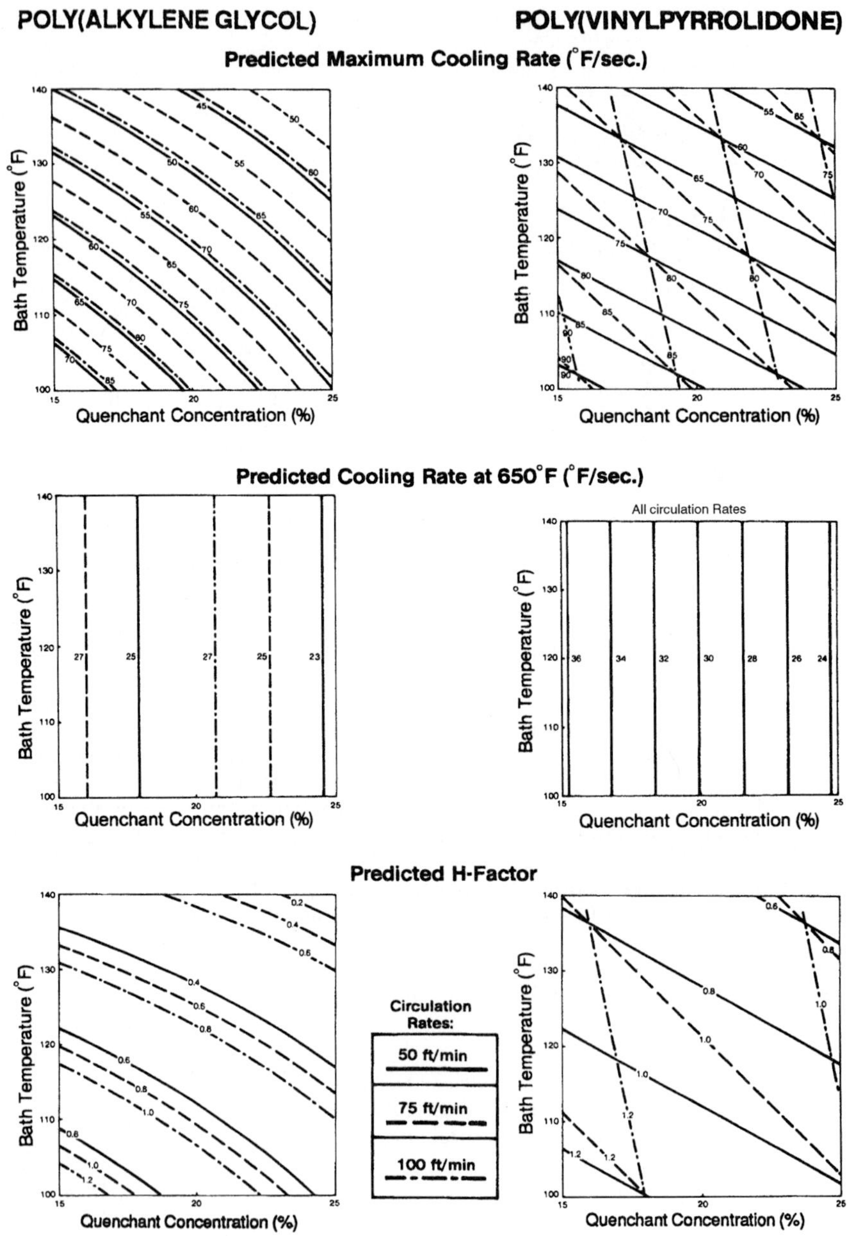

POLY(ALKYLENE GLYCOL)   POLY(VINYLPYRROLIDONE)

Predicted Maximum Cooling Rate (°F/sec.)

Predicted Cooling Rate at 650°F (°F/sec.)

Predicted H-Factor

FIGURE 13.19 Comparison of the predicted quenching characteristics of PAG and PVP polymers under varying conditions: (*a*) maximum cooling rate CR$_{max}$, (*b*) cooling rate at 345°C (650°F), and (*c*) Grossman H factor.[50]

quenchants allows its cooling curve to be nearly straight (Fig. 13.20) due to prolonged vapor-blanket stage and reduced nucleate boiling stage period.[6] This property facilitates their use in hardening crack-prone parts made of high-hardenability steels.

5. With increasing PSA concentration and bath temperature, the cooling rate can be decreased to such a level that many ferrous alloys do not transform to martensite at all, but rather form bainite or fine pearlite, thereby achieving lower hardness values.

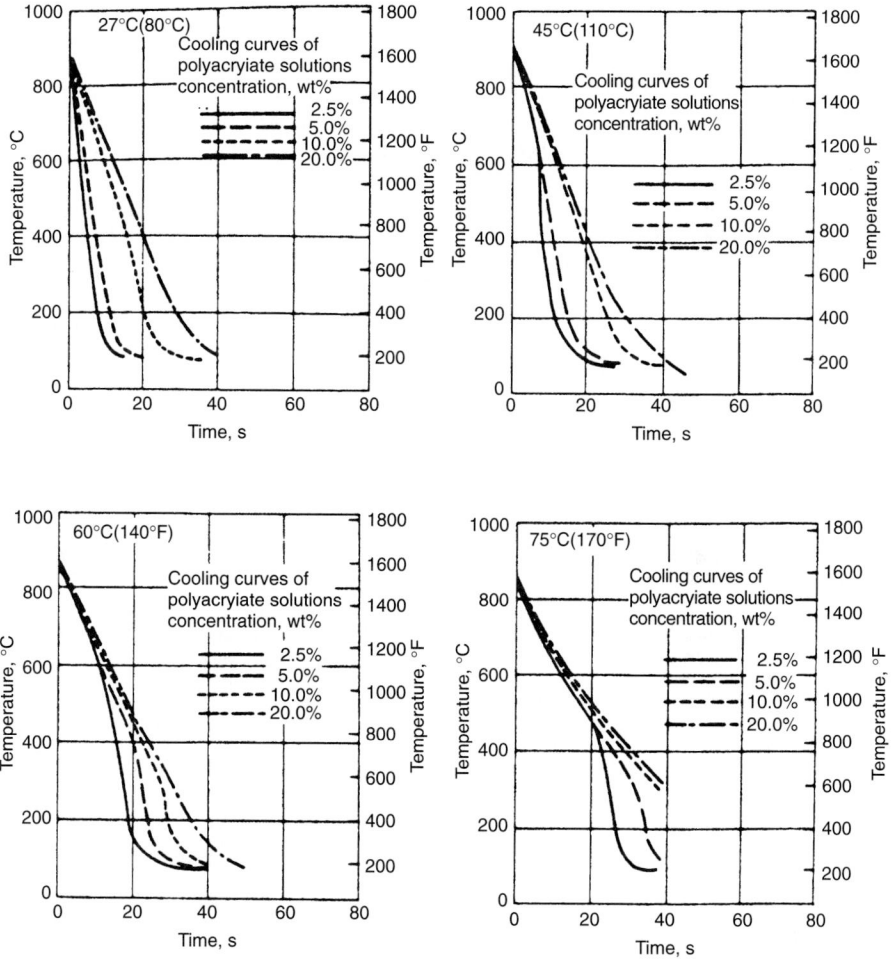

FIGURE 13.20 Cooling rate of an agitated polyacrylate quenchant as a function of concentration and temperature. Test specimen: 0.4-in. diameter × 2.4-in. (10-mm diameter × 60-mm) austenitic steel.[6] (*Reprinted by permission of ASM International, Materials Park, Ohio.*)

6. In general, higher agitation is recommended for hardening treatments, whereas minimal agitation is recommended for nonmartensitic quenching to form bainite or pearlite.

Figure 13.21 illustrates a comparison of cooling curves of a PSA quenchant with those of water, conventional oil, and a few typical polymer quenchants. This provides benefits in applications requiring slow quenching.[6]

APPLICATIONS. PSA solution is not widely used in the heat treatment industry. However, it is employed to harden high-hardenability steels that are particularly susceptible to quench cracking. Applications include

1. Use of polyacrylate solution as the first quenchant in the commonly practiced double-hardening treatment in oil for deep-carburized parts such as bearing races, balls, and rollers.

2. Direct quenching of high-carbon steels (in polyacrylate solutions) to produce similar mechanical properties as obtained by quench and temper or austempering (e.g., railroad or automotive forgings or sway bars, and rod and wire patenting made of SAE 1070 or 1090) as well as spring plates, torsion rolls, and gears of certain types.[17a]

3. The relatively long vapor-blanket stages encountered with PSA quenchants which have promoted its application in the quenching of railway rails, where pearlite and fine bainitic transformation products are required.

4. Quenching of hot-worked parts (in polyacrylate solutions) to obtain the same microstructure as by air cooling, without inducing excessive scaling and decarburization.

5. Replacement of the commonly used salt or lead bath process by polyacrylate solutions for patenting of wire at 510 to 565°C (950 to 1050°F).

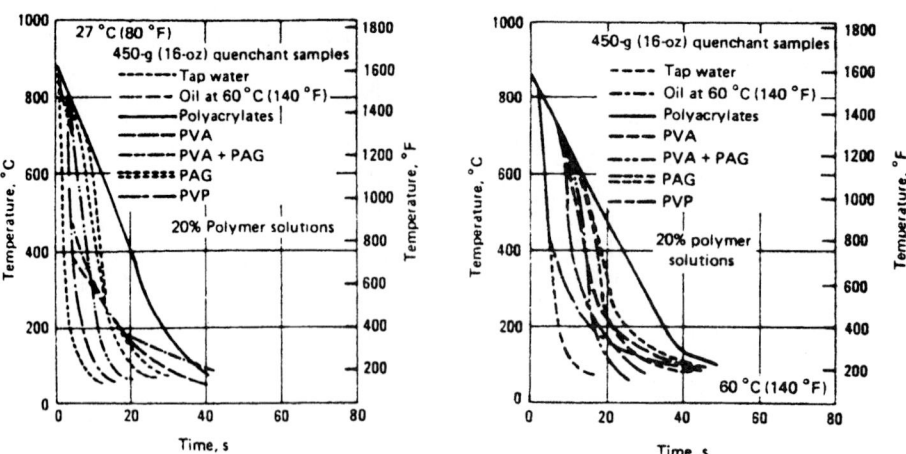

FIGURE 13.21 Cooling rates of various quenching media (including different polymer quenchants at 20% concentration) at 25 and 60°C (80 and 140°F). Test specimen: 10-mm diameter × 60 mm (0.4-in. diameter × 2.4 in.). Scaleproof austenitic steel, medium agitation.[6] (*Reprinted by permission of ASM International, Materials Park, Ohio.*)

6. Direct quenching of hot-worked parts (in polyacrylate solutions) to obtain good machinability without undergoing the conventional hardening and tempering process, especially for low-hardenability, low-alloy carburizing steel parts.

7. Quenching of large continuously cast steel slabs made of low- and high-carbon steel and alloy steels (in polyacrylate solutions) to allow inspection shortly after casting. This causes faster cooling without any tendency for quench crack.[6,17]

Aqueous PSA quenchants have some drawbacks that preclude their broader use in the heat treatment industry. However, with the probable exception of sensitivity to mechanodegradation, possibly the most important drawback is the propensity to form "polyelectrolyte complexes" with polyvalent cations that may form precipitates with polyvalent metals. Sources of polyvalent metal ions are Ca^{2+} and Mg^{2+} from hard water or solubilized Fe^{3+} ions from iron oxide quenching scale or quench tank corrosion products.[8]

Polyacrylate solutions can also be employed for quenching solution-treated aluminum alloys to minimize distortion and warpage.

Polyethyl Oxazoline (PEOX). Recently, PEOX products developed maintain the main properties of oil together with additional benefit of "nontacky" behavior of the residue for applications in induction heat treatment systems where the potential formation of residual tacky residues after quenching may interfere with the robotics operations.[12,51]

Although aqueous solutions of PEOX, like PAG-based quenchants, display inverse solubility temperatures, PEOX is not applied to purify quench bath. This may be due to a possibility of susceptibility to hydrolytic degradation of the amide linkages present in the PEOX polymer.[8]

The PEOX quenchant is also subjected to a film rupture process, but the rupture seems to be associated with a partial dissolution of PEOX. The haziness of the film is attributed to polymer instability, because the film temperature lies above the cloud point.[8]

Selection of Polymer Quenchant. Selection of polymer quenchant compared to other quenching media can be made based on a sensitivity band of steel hardenability and section complexity, as shown in Fig. 13.22a, which illustrates the more extended applications of PSA, PVP, and PEOX to steels of higher hardenability and thinner sections than PAGs, while still maintaining the environmental advantages of polymer quenchants.[25a]

Polymer Quenchant Stability. Polymer quenchant stability can be influenced by several factors such as mechanodegradation, thermal/oxidation stability, dragout, and hydrolytic stability. The extent of these effects is a function of the polymer structure and bath and quenching conditions.[8]

MECHANODEGRADATION. The polymer chains coil around each other in solution to provide a relatively long-range order, which increases the viscosity (thickness effect) with concentration (Fig. 13.22b).[8] The extent of polymer chain enlargement increases with an increase in molecular weight (chain length). Thus, for a given concentration, the viscosity of a polymer solution increases with the increase in molecular weight.

When mechanical energy (shear) is applied to a polymer solution by pump or impeller agitation, a shear-induced thinning of the solution takes place. With adequate energy application, covalent C—C bonds in entangled polymer chains easily break apart. With the occurrence of C—C bond scission, the average molecular weight of the polymer and the corresponding viscosity of the polymer solution are reduced. This is called mechanodegradation or shear degradation, which can be expected to produce a proportional increase in cooling rate.[53]

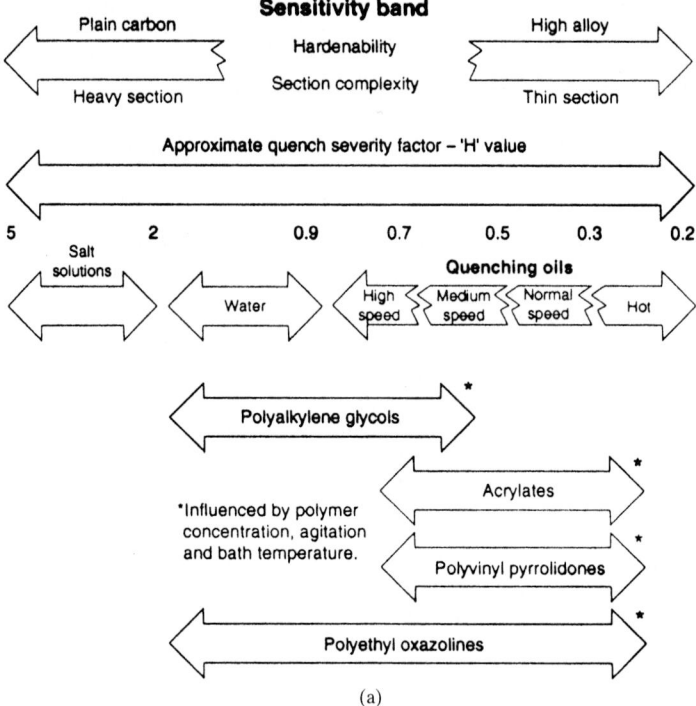

FIGURE 13.22 (*a*) Sensitivity band for quenchant selection which is influenced by concentration, agitation, and temperature of the quenchant bath.[25a] (*b*) Variations of aqueous solution viscosity with polymer molecular weight.[8] (*Reprinted by permission of ASM International, Materials Park, Ohio.*)

Variations in solution viscosity reflect both the polymer stability and potential quench severity. Polymer can display different shear stabilities when subjected to the same shear field, and a variety of results can be available with different polymers. The heat treater should require data on the relative shear stability of a polymer quenchant prior to using it.[8]

POLYMER DRAGOUT.   Polymer dragout occurs when the polymer after phase separation does not redissolve prior to the withdrawal of the part from the quench tank. In reality, polymer dragout always occurs because of the solution-wetting, whether polymer is completely or partially redissolved. However, dragout is expected to increase with the concentration (Fig. 13.23*a*),[14] solution viscosity (Fig. 13.23*b*),[14] and molecular weight of the polymer.

THERMAL/OXIDATIVE STABILITY.   Figure 13.24*a* shows the schematic illustration of the impact of polymer chain oxidation where polymer chain scission occurs in the middle. The impact of this chain scission causes higher viscosity reduction with higher-molecular-weight (longer) polymer (II) with respect to shorter polymer (I) (Fig. 13.24*b*). The ultimate effect of extensive oxidative/thermal degradation results in quench severity closer to water itself.[43]

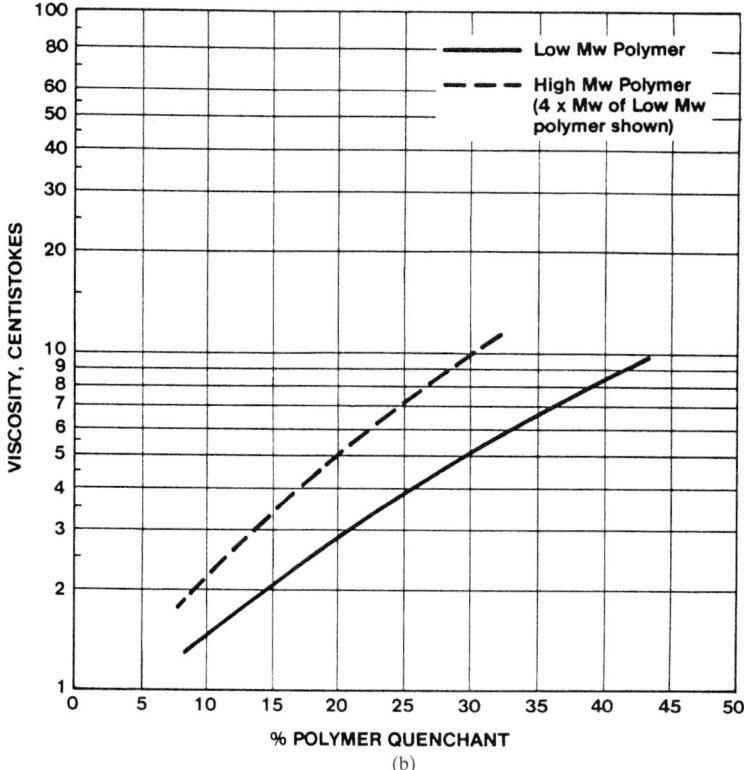

FIGURE 13.22 (*Continued*) (*a*) Sensitivity band for quenchant selection which is influenced by concentration, agitation, and temperature of the quenchant bath.[25a] (*b*) Variations of aqueous solution viscosity with polymer molecular weight.[8] (*Reprinted by permission of ASM International, Materials Park, Ohio.*)

CHEMICAL STABILITY. It is essential that the polymer quenchant selected be hydrolytically stable under quench bath conditions (e.g., elevated temperature and alkaline pH). This factor has limited the widespread use of polyacrylamide, PVA, and PEOX quenchants, because they are prone to hydrolytic degradation under some situations. Possible hydrolytic degradation reactions are shown in Fig. 13.25 and are based on the well-established organic chemistry reactions. A major purpose of quenchant development is to minimize degradation reactions and to increase long-term quenchant stability.

Polymer Quenchant Bath Maintenance. Various authors have used the cooling curve analysis to show that polymer quenchants are susceptible to oxidative/thermal degradation processes.[14,48,54–56] To ensure maximum performance, all precautionary measures should be taken so that the quenchants remain free of contaminants such as scale. It is, therefore, essential that a filter, preferably 5-μm, be used in the quenchant recirculation system.[57] The most common methods used for polymer quenchant analysis include concentration, viscosity, refractive index, conductance,

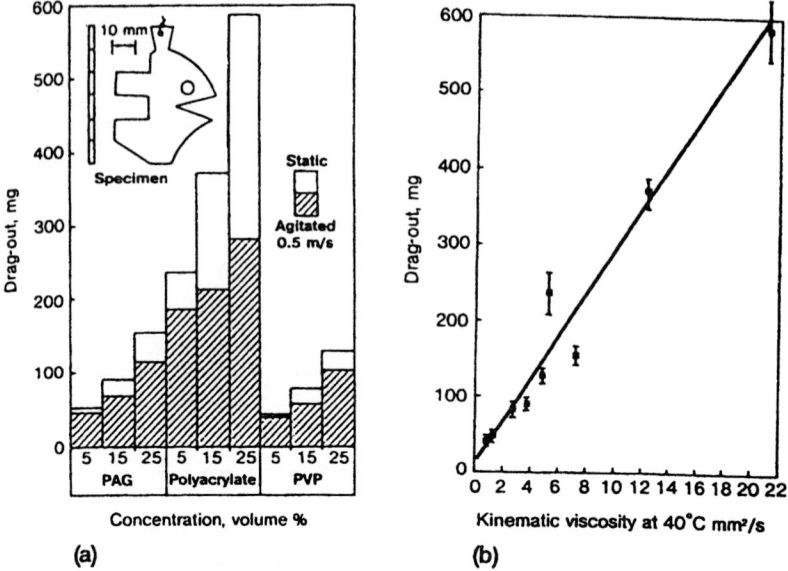

FIGURE 13.23   (*a*) Dragout specimen shape and results for polymer quenchants of various concentrations in both the static and agitated conditions. (*b*) Effect of polymer solution viscosity on the level of dragout in static tests.[14] (*Courtesy of N. A. Hilder.*)

separation temperature, and gel permeation chromatography (GPC) (Table 13.5). Only the most common testing methods are outlined here.

POLYMER CONCENTRATION.   The cooling rate of an aqueous polymer solution critically depends on the polymer concentration (Fig. 13.26*a*) and the condition (molecular weight and oxidation) (Fig. 13.26*b*). Polymer loss also occurs much more by dragout and less by polymer degradation.[55,56,58] It is, therefore, important that the quenchant concentration be maintained within approximately ±1.0%.

REFRACTIVE INDEX.   Refractive index n_D, being a linear function of polymer or quenchant concentration, is the most common method to measure polymer concentration (Fig. 13.27*a*). The subscript D implies that the refractive index is measured at the wavelength of the D line of sodium (5893 Å). This is determined using ASTM Standard D1747 and D1218 methods and (properly calibrated) handheld refractometer (Fig. 13.27*b*). A droplet of the solution, when put on the prism of this refractometer and directed against light, shows a related value to reading with the naked eye. The check should be made every week. This method is primarily restricted to quenchants derived from PAG and not applicable to dilute solutions of high-molecular-weight polymer such as PVP and PSA quenchants because of its unacceptable sensitivity to refractometry.[8]

Since refractive index measurements are also influenced by polymer degradation and the presence of contaminants in the quenchant system, its sensitivity decreases with the decrease in concentration.[55,59] Hence, it is essential to check the periodic validation of the quenchant concentration by complementary viscosity measurement. If carried out together, these measurements provide a good understanding of the condition of a polymer quenchant.

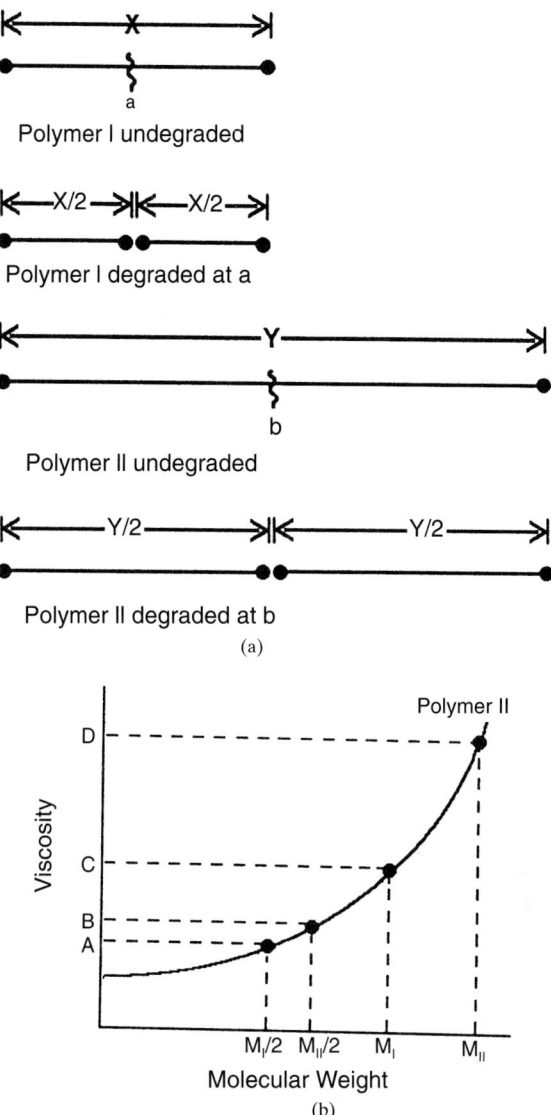

FIGURE 13.24 (*a*) Schematic illustration of the polymer chain scission process for a polymer chain with a low (polymer I) and a high (polymer II) molecular weight. (*b*) Relative effects of the polymer chain degradation process shown in Fig. 13.23*b*.[43]

Possible contaminants include the following: (1) Solid contaminants such as scale or carbon/soot have little effect on the quench rate but can adversely affect the concentration control by refractive index. (2) Tramp fluids, e.g., cutting fluids and hydraulic oils, can act as nutrients for biological growth and can increase the

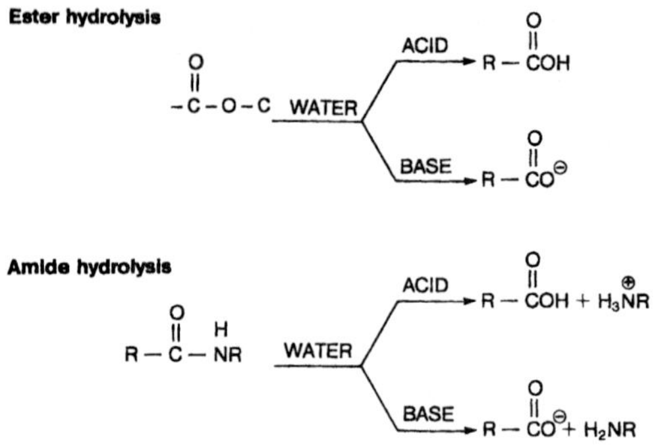

Ester hydrolysis

Amide hydrolysis

FIGURE 13.25 Hydrolysis of esters and amides.

TABLE 13.5 Methods for Polymer Quenchant Analysis

Test	Procedure
Concentration viscosity	ASTM D445
Refractive index	ASTM D1749, D1218
Salt content	Conductance
Corrosion protection	pH additive analysis
Biological activity	Dipstick test

vapor-phase stage of the quenching process. (3) Biological contamination such as fungi and bacteria results in foul smell and leads to corrosion-inhibitor depletion. Accumulation of fungus deposits can block nozzles and filters on spray systems. (4) Dissolved materials such as heat treatment salts, water hardness salts, and gases such as NH_3 for carbonitriding operations can be contaminants.[54]

VISCOSITY. Since viscosity η (cSt) of a polymer solution is dependent on concentration, molecular weight (or size), and temperature (Fig. 13.28), it is essential to compare the viscosities of fresh and used polymer quenchants at the same temperature by capillary viscometry. Low levels of salt contamination usually produce a minimal effect on viscosities of most quenchant solutions. As with quench oil, the viscosity of a polymer solution can be measured by using the ASTM Standard D445 procedure.

Since viscosity is sensitive to low concentration of polymer in solution, this method can be readily used to measure low concentrations of relatively high-molecular-weight polymers such as PSA.

SALT CONTENT. The salt content of a polymer quenchant, which is introduced either from hard water or by contamination from salt baths, usually increases during use. Salt contamination of polymer quenchants can greatly increase the quenchant cooling rates and decrease the cloud point. Hence, the makeup water for the quench bath should be distilled or deionized.

Perhaps the simplest and most inexpensive method to continuously check salt contamination is by variation in conductance, which has become an important quality control tool for some heat-treating products.[59] Such increases in conductance reflect the occurrence of salt contamination (Fig. 13.28b), and

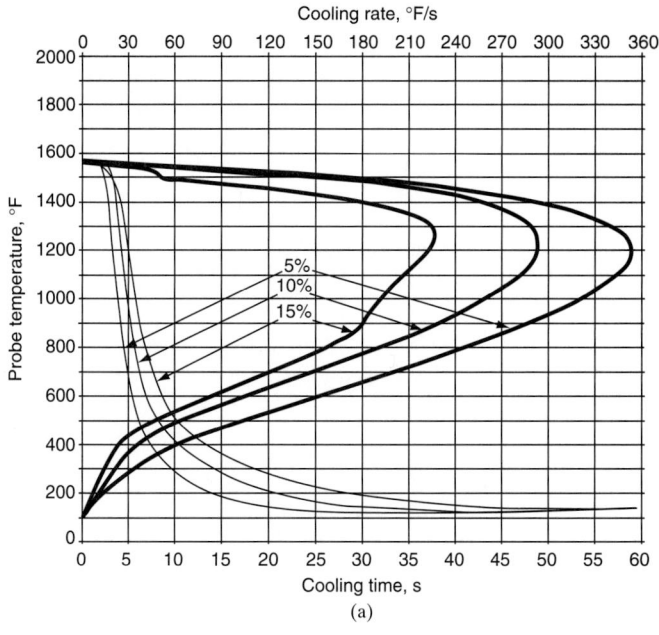

(a)

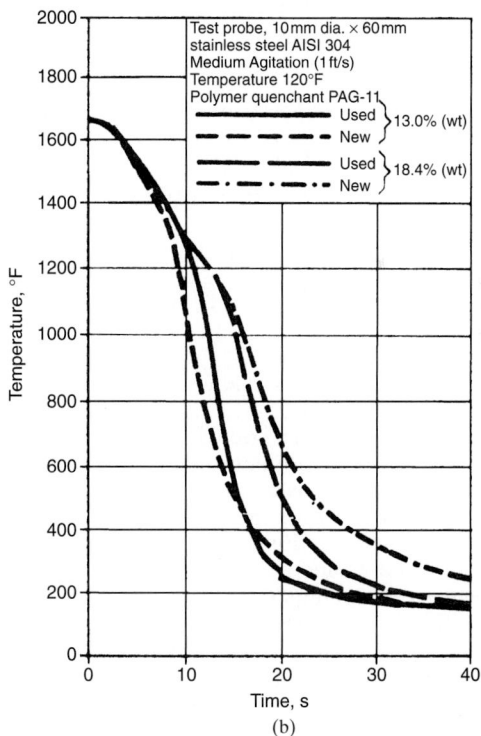

(b)

FIGURE 13.26 (*a*) Effect of polymer concentration on cooling rates. (*b*) Comparison of fresh and used PAG quenchant.[8]

13.43

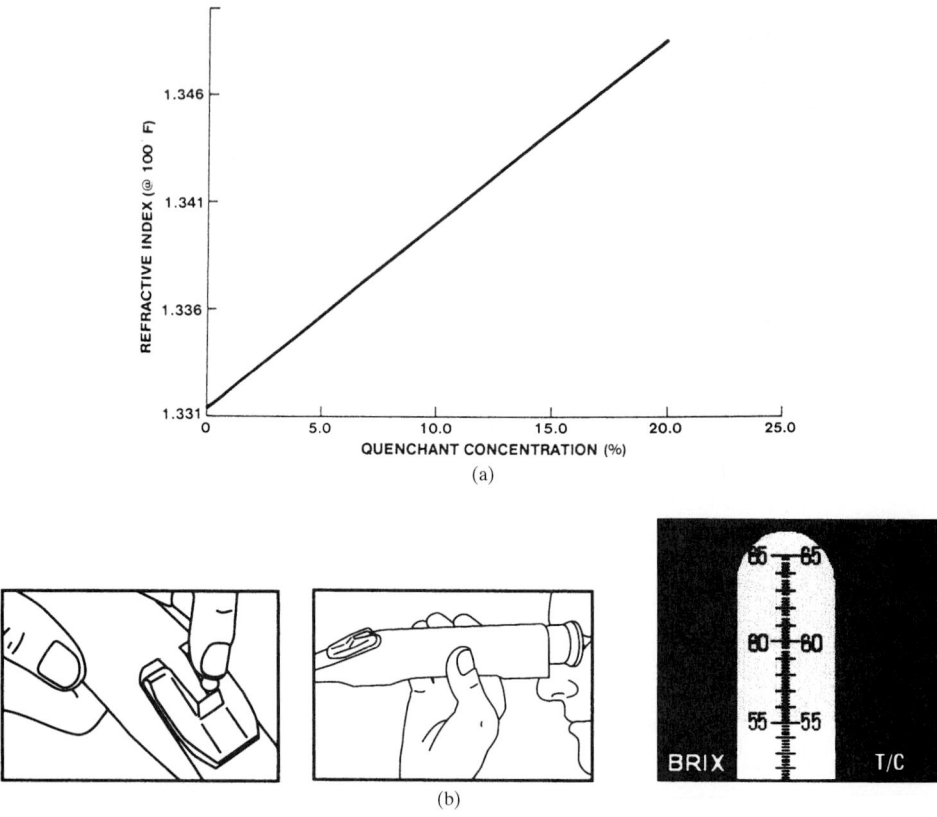

FIGURE 13.27 (*a*) Refractive index versus quenchant concentration.[8] (*b*) Use of a handheld refractrometer to measure polymer quenchant concentration. (*Courtesy of Leica Microsystems Inc.*)

decreases in conductance suggest the depletion of the corrosion inhibitor, which is mostly ionic.

Membrane Separation.   The principles of microporous membrane separation, such as reverse osmosis (RO) and ultrafiltration (UF), can be employed to change the concentration of a polymer quenchant, to recycle it (e.g., from a rinse tank), or to separate organic and additives (and salt contaminant) from an aqueous quenchant solution for subsequent disposal. The main differences between the two techniques are system pressure and molecular size that can be separated. Reverse osmosis is preferred because it offers the most complete separation.[60]

The beneficial effect of membrane separation is that it is a low-energy method which can be used on a continuous basis and can help to solve waste disposal problems. Similar systems can be used for quench oil separation and regeneration.

Thermal (Cloud Point) Separation and pH Measurement.   To obtain a homogeneous aqueous solution, it is necessary that the polymer undergoes a hydrogen bond

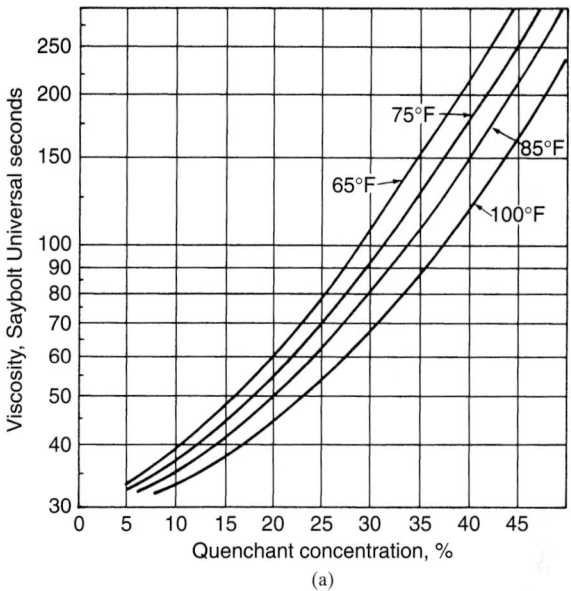

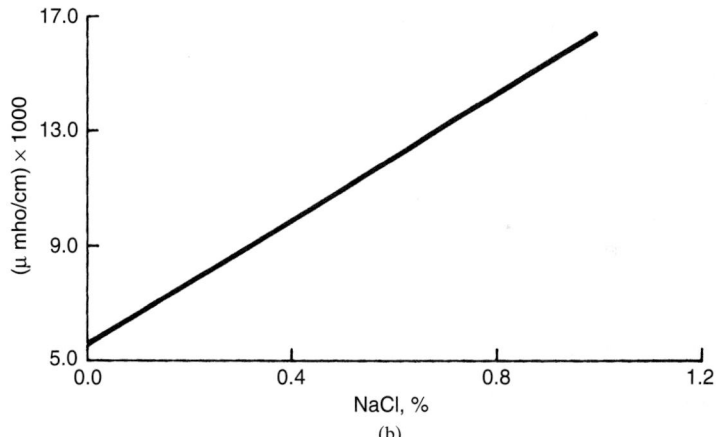

FIGURE 13.28 (*a*) Viscosity/concentration relationship for PAG polymer quenchant.[8] (*b*) Conductance versus salt (NaCl) concentration in 20% polymer quenchant solution.[8]

interaction with water.[8] Some polymers experience a phase separation with an increase in solution temperature at the hot metal interface during quenching. As a result, a two-phase system forms, one phase being polymer-rich and the other water-rich. The temperature corresponding to the occurrence of this phase separation is called the *separation temperature* or *cloud point*. It is this phenomenon that is employed to control the severity of quenching. Two currently used polymers, polyethyl oxazoline (PEOX) and polyalkylene glycol (PAG), display a separation temperature which depends on the ratio of monomers in these copolymers. For example, the separation temperature is reduced with an increase in the number of propylene oxide monomer units in the PAG copolymer.[8]

It has also been shown that the cloud point of PAG copolymers is a function of the salt concentration, particular polymer composition, and pH of the aqueous solution. The cloud point of a PAG polymer solution characteristically increases with the increase of polymer oxidation. Variations in the cloud point affect both the cooling rate and quench severity as well as the viscosity versus temperature properties of the solution. Hence, it is essential to avoid substantial pH variations in order to minimize variations in quench severity. The cloud point behavior of quenchant solutions is used advantageously to purify the salt-contaminated (PAG and PEOX) quenchant bath and should be part of any quality control program.

The pH measurement should be made at least once a week (to prevent corrosion of quenched parts and installation) and, if required, corrected by adding inhibitors.

Corrosion Protection. Polymer quenchants must contain (either inorganic or organic) corrosion inhibitor to ensure corrosion protection of the system; the most common inorganic corrosion inhibitor being $NaNO_2$ (or KNO_2). $NaNO_2$ may be easily determined by readily available colorometric reagents. The analysis of other commonly used corrosion inhibitors, such as amine/carboxyl acid salt or the sodium salts of mixed organofunctional aromatic acids, is relatively more complex and should be provided as a service by the polymer quenchant supplier.[8]

Biological Degradation. In closed systems such as automatic induction-hardening equipment, the bath is frequently infested with bacteria. Hence, a new bath must be prepared as a remedial step, after a few months of its operation. In immersion-hardening systems, the service life of the baths is longer, up to several years. One way to counteract the bacterial infection of the bath is to agitate it periodically or blow through with the compressed air.[17a]

Problems arising from microbial growth include foul odors, slime, staining, and depletion of nitrite corrosion inhibitors. In some cases, microorganisms (to be detected by simple dipstick test) develop that can degrade the polymer. When this occurs, the performance of the polymer quenchant is severely affected.[8]

Problems associated with biological degradation can generally be controlled by the addition of the proper biocide. Users faced with such problems should contact a biocide supplier or the polymer supplier.

Oil Contamination. The performance of aqueous polymer quenchants is influenced by the presence of oil contamination such as hydraulic, metalworking, and quench oils. The effect of oil contamination on cooling curve performance is shown in Fig. 13.29. As oil and aqueous polymer solutions do not provide friendly mixtures, they will form a discontinuous film around the hot metal part, which will produce substantial thermal gradients and may, in turn, lead to increased thermal and transformational stresses, probably producing cracking or increasing distortion.[8]

Changes in Polymer Quenchant with Use. Some degradation of polymer quenchant is inevitable.[61] Table 13.6 compares the physical properties of the "used" quenchant and a "fresh" quenchant comprising the same total polymer concentration. The cooling curves for these quenchant solutions are illustrated in Fig. 13.30. A large difference in the viscosities of used (1.93 cSt) and fresh (5.65 cSt) quenchants reveals

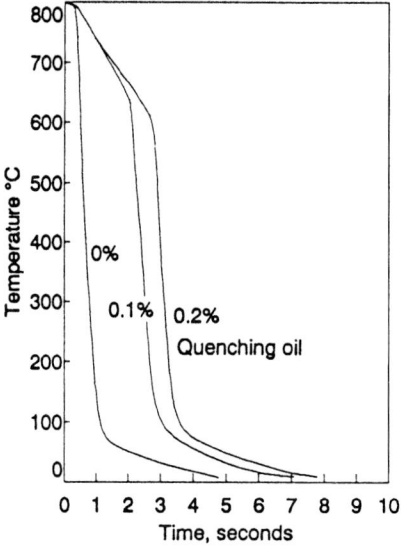

FIGURE 13.29 Effect of quench oil contamination on the cooling curve behavior of an aqueous polymer quenchant.[8]

TABLE 13.6 Characterization of a Severely Degraded and Contaminated Polymer Quenchant[8]

	Fresh quenchant	Contaminated quenchant	Water
Physical properties			
Concentration, %	20.0	20.0	. . .
Viscosity, cSt at 40°C (100°F)	5.65	1.93	. . .
Conductance, μmho/μm	5000	>20,000	. . .
Cloud point, °C (°F)	88.0 (190.0)	>100 (212)	. . .
GPC area shift, %	. . .	−40	. . .
Cooling curve results (bath temperature of 40°C, or 100°F)			
Time to cool from 732 to 260°C (1350 to 500°F), s	15.4	10.8	10.4
Maximum cooling rate, °C/s (°F/s)	39.2 (70.6)	69.3 (124.7)	61.8 (111.2)
Temperature at maximum cooling rate, °C (°F)	534 (993)	650 (1200)	610 (1130)
Cooling rate, °C/s (°F/s), at:			
704°C (1300°F)	20.6 (37.1)	65.0 (117.0)	56.8 (102.2)
343°C (650°F)	29.6 (53.3)	29.6 (53.3)	33.0 (59.4)

that extensive polymer degradation has taken place. This observation is supported by the GPC peak area shift of 40%. The amount of polymer degradation shown in Fig. 13.30 is clearly in excess of a permissible level.

Loss of the characteristic cloud point (>100°C or 212°F, versus 88°C or 190°F) for the used polymer quenchant further substantiates its large polymer degradation.

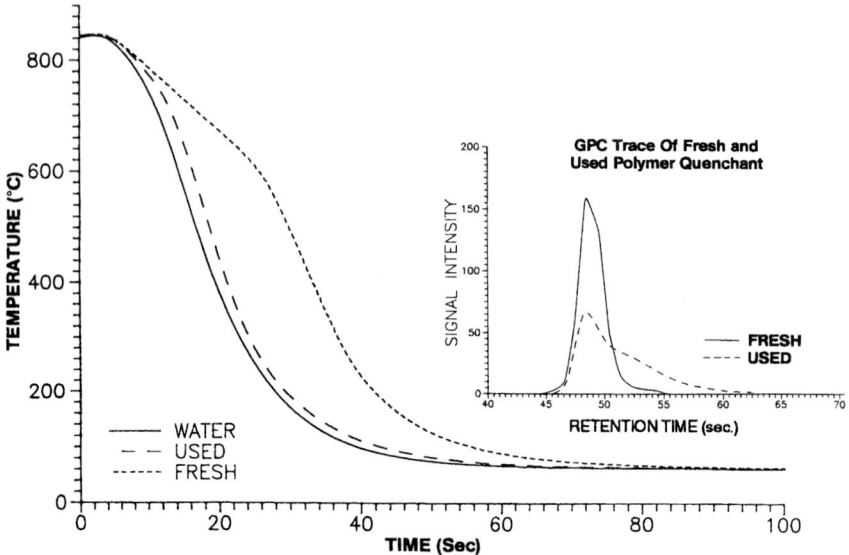

FIGURE 13.30   Cooling curves for an excessively degraded aqueous PAG quenchant and a fresh quenchant.[8]

Generally, used polymer should not display greater than a 1 or 2°C (2 to 4°F) cloud point elevation during normal use.

Table 13.6 also indicates that the used quenchant is contaminated with salt, as evidenced by higher conductance than that of the fresh quenchant. Although higher conductance can result from excess addition of corrosion inhibitor (in this case, $NaNO_2$), perhaps the high conductance suggests hard water contamination. It is, therefore, recommended to use deionized or distilled water to obtain maximum lifetimes for a polymer quenchant.

Cooling curve analysis shows the effects of both polymer degradation and salt contamination (Table 13.6). Here, the quench severity of the used quenchant was very close to water compared to that of the fresh quenchant. The degradation level of a polymer quenchant rarely approaches this point. In reality, if the physical properties and the quench severity of the bath had been regularly checked, proper corrective action presumably could have been taken.[8]

Gaseous Contamination.   Gaseous contaminants, such as NH_3, arising from nitriding or nitrocarburizing may produce a significant extension of vapor-blanket cooling behavior, shown in Fig. 13.31.[62] Similarly, other gases such as CO or CO_2 may extend A-stage cooling times, which results in inadequate hardening. Although a technical solution is possible, this explains why aqueous polymer quenchants are not widely used in nitriding and nitrocarburizing processes.

13.2.3.5 Molten Salt Baths.   Molten salt baths of various compositions are available to cover the temperature range from 150 to 1320°C (302 to 2408°F). They are made by heating selected mixtures of chemical compounds until they melt. The melt is then raised to the temperature at which the parts are to be processed. Molten salt baths are prepared in pot-type furnaces which are heated externally with gas, oil,

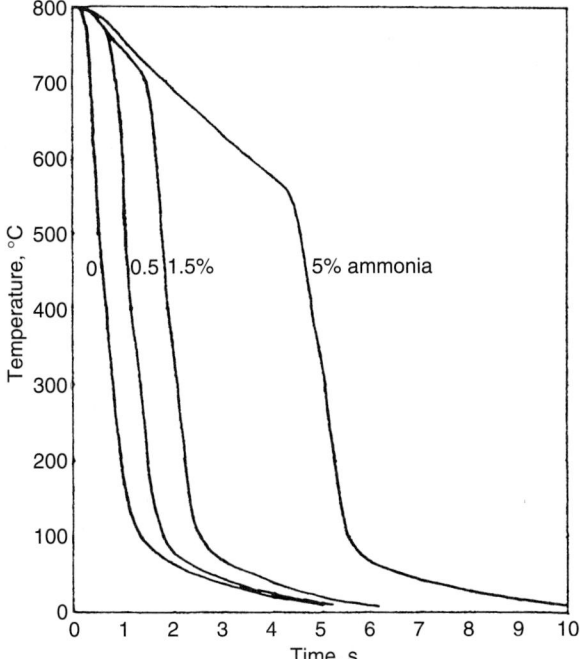

FIGURE 13.31 Effect of ammonia contamination on the cooling curve performance of an aqueous polymer quenchant.[8]

or electricity. Internal heating can also be done either with immersed electrodes or suitably protected electric resistors dipping into a salt bath contained in a metal (or ceramic) pot or with gas-fired immersion heaters. These are commonly employed together with a thermocouple which dips into the molten salt bath and is adjacent to the part being treated.[63]

Salt baths can be used for either heating or cooling. Agitation of the bath is important to improve uniform heating and/or quenching severity. This is effected by centrifugal pumps, air bubbles, or a belt-driven propeller; the last one is widely used.[64] To ensure consistent results, salt bath compositions are analyzed at regular intervals to maintain strict quality control. Bailing of the bath, coupled with replenishment with a suitable additive, is regularly done for surface-hardening molten salt baths. Sludge removal at regular intervals is also necessary.

Advantages and Disadvantages of Salt Bath Process. The salt bath process offers a number of well-established advantages over alternative processes as follow:[65–76]

1. There are a wide range of operating temperatures from 150 to 595°C (300 to 1100°F) for one composition salt compared to the inability of oil to be used above 230°C (450°F). Hence, the latter is restricted to low-temperature processes.

2. It is simple to operate, requiring only semiskilled labor.

3. Circulation of the molten salt bath by convection and/or stirring of the melt (a) causes rapid and uniform heating for all parts at the same time, (b) maintains a constant temperature, which, in turn, provides (c) a precise temperature control within ±1°C and (d) extremely uniform and reliable results.

4. It is energy-efficient because an extremely rapid rate of heat transfer is obtained by the very excellent thermal conductivity of the bath. Salt bath provides greater productivity due to rapid attainment of equalization temperature of parts compared to oil. Hence, it is well suited for quenching large parts and heavier sections without cracks by varying the temperature, agitation, and water content. This allows high-production systems to occupy a relatively small space (i.e., more compact furnace).

5. A greater range of quench severities (or rates) than in oil are possible.

6. In addition to the rapid heat-up capability, the automatic salt bath has a shutoff feature for turning the bath completely off or to a very low idle and is equipped with covers, which can yield a considerable energy saving at all temperatures. As a result of these features, the salt bath process can be regarded as one of the most cost-effective methods for high production.

7. The vapor stage (if present) in salt bath quenching is very little, and hence one can quench very rapidly past the pearlitic nose.

8. It has quite a different quenching mechanism from that of oil. Mostly the heat extraction during salt bath quenching is through the third (i.e., convection) stage liquid cooling, resulting in very low distortion and more uniform and consistent hardness.

9. There is superior thermal and chemical stability to oil with relative insensitivity to contaminants. Hence it enables one to go to a temperature for austempering where oils cannot compete. Also, it offers adequate quenching performance for many years. In contrast, oil deteriorates with use, necessitating close control and even partial or complete replacement.

10. Washing of adhering salts by water is easy, and salt recovery for reuse is simple, if desired. In contrast, washing of oil requires special cleaners and equipment, and its recovery is not simple. The salt recovery option eliminates disposal and reduces operating cost.

11. Being inflammable, salt bath does not pose a fire hazard compared to oil, which poses a serious fire hazard at a comparable temperature.

12. The capital cost of salt bath equipment is less than that for competitive methods of handling the same volume of work. The production costs with salt bath are lower due to rapid heating/cooling, cleaner surfaces, reduced cracking and distortion leading to uniform heating, and less furnace equipment.

13. Another distinct characteristic of salt baths is the ability to suspend parts from the top of the pot rather than laying them flat in a basket, and there is a buoyancy effect on the work immersed in them. The parts in salt baths weigh only about two-thirds to three-fourths of their weight in air. Thus the salt bath provides some degree of support to the components it surrounds and therefore offers better dimensional control and less distortion. Usually a distortion within 0.003 to 0.004 in. on automotive parts may be attained. (A little distortion can be corrected by post-heat-treatment straightening, the amount of which is far less than would be necessary with atmospheric or vacuum processing.)[74]

14. The salt bath process is extremely versatile and much more flexible than other processes. This process is carried out for a variety of treatments, such as liquid carburizing, liquid nitriding, liquid nitrocarburizing, quenching or quench-hardening, tempering, martempering, austempering, annealing, normalizing, stress-relieving of ferrous alloys, and aluminum dip brazing. It provides the treatment of small and large quantities of volume production as well as small individual piece weights.

15. There is virtual elimination of scaling, oxidation, and decarburization of the treated parts because of their full immersion in the neutral salt baths, without air contact. After processing through a chemical washing plant, parts have clean, safe, and good-quality finishes. After continued use, some salts exhibit the tendency to decarburize, which can be overcome by rectification, and self-rectifying hardening salts are available.

16. It produces a cocoon-type effect when a cold steel is immersed into a molten salt by freezing a large gob of molten salt around the part. This minimizes distortion due to uniform melting of this gob.

17. The salt bath offers speedy, reproducible microstructure, and uniform production by proper installation and operation of a salt bath.

18. Quench severity can be controlled to a large extent by varying the agitation, temperature, and water content of the salt bath.

The disadvantages of molten salt bath include the following:[65]

1. Since salt freezes at about 135°C (275°F), it cannot be used for lower-temperature quenching applications. Rather, it has to be used above its melting point of about 150°C (300°F).[67]

2. Freezing can lead to intermittent use, which is complicated and energy-inefficient.

3. Since nitrates are oxidizers, greater precautions are needed to avoid the introduction of combustible/oxidizable materials such as soot, oil, or cyanides and thereby the possibility of any violent reactions. Excessive overheating must be avoided; otherwise, salt can even oxidize steel or cast iron.

4. Since salt is a costlier quenchant on a volume basis, greater care must be exercised to avoid excessive *dragout*. Efficient washing and recycling can overcome this problem.

5. Removal of certain contaminants can be more complicated than that for conventional quenchants, although methods for removal are well-proven. In some cases, salt has the advantage over conventional quenchants with respect to contaminants.

6. Salt may seem to pose safety and environmental problems, but the newer technology removes these problems.

Advancement in Salt Bath Technology. The following advancement and undergoing developments make salt bath quenching more attractive in the future:[65]

1. Computer-controlled robot handling systems for moving parts are much faster and more convenient from austenitizing to quench.

2. More energy-efficient austenitizing and salt bath furnaces, respectively, use superior electrode designs and better insulation.

3. Integrated agitation and heater systems in the quench baths provide more uniform flow and temperature control together with safe water addition.

4. Water addition monitoring systems use fast quench rate determination. Continuous monitoring is also expected in the near future.

5. Development is underway of quench salt reclaim and treating systems that allow economic recycling and reduce disposal problems.

6. Data of quantitative salt quench rate-quench variables can be used to correlate with steel hardenability.

7. There is an established family of austempered ductile irons (ADI) that are superior to steel casting and forgings and other cast irons for both wear and structural applications.

Types of Salt Bath.   It has been established that one salt bath will never perform the entire range of heat-treating operations. Special types of salt baths have, therefore, been used to suit different heat treatment operations and metallurgical conditions.[69]

As a rule, manufacturers of heat-treating salts rarely specify the composition of the salts; instead, they provide detailed information regarding their properties and uses. The salt baths can be classified into four main groups, as discussed herein.[70]

LOW-TEMPERATURE SALT BATHS.   Having an operational range of 150 to 620°C, these baths are used chiefly for isothermal treatments such as austempering, martempering, low-temperature tempering of steels, and the solution treatment of Al alloys. Baths operating in the temperature range of 150 to 500°C consist of equal parts of KNO_3 and $NaNO_2$; those operating in the range of 260 to 620°C contain mixtures of KNO_3 and $NaNO_3$. The phase diagram shown in Fig. 13.32 illustrates the dependence of the melting point on the composition of the mixture.[77] Like all quenchants, the quench severity of molten $KNO_2/NaNO_2/NaNO_3$ salt mixture depends on the agitation rate and temperature, as shown in Fig. 13.33.[78]

These baths are readily water-soluble and strongly oxidizing, especially in the higher temperature range. Salt bath agitation can be used in several ways. One

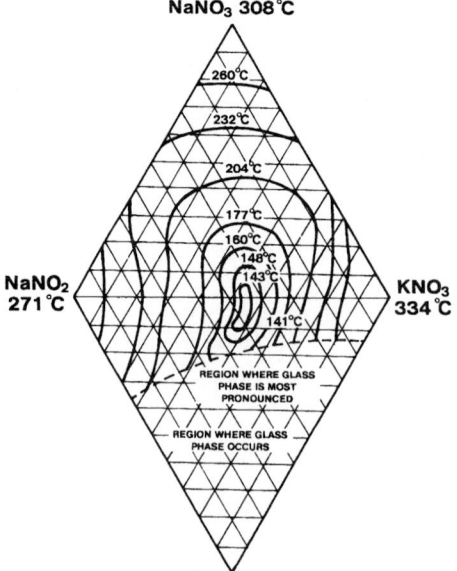

FIGURE 13.32 Freezing points of ternary alkali nitrate-nitrite mixtures with composition (%).[77]

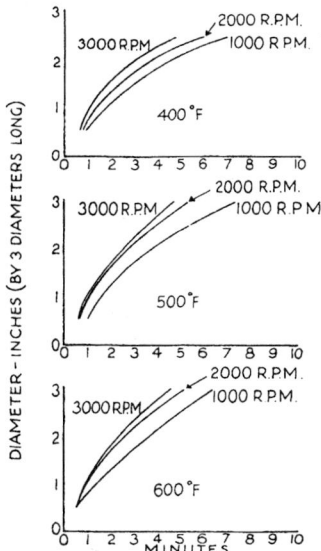

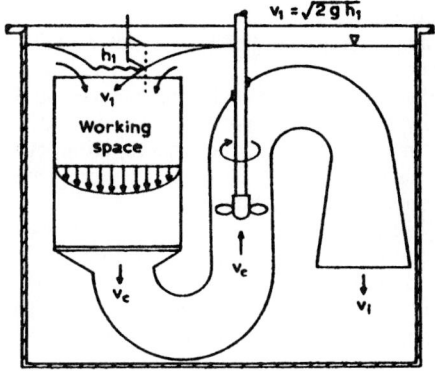

FIGURE 13.33 Effect of agitation on the quench severity of molten salt.[78]

FIGURE 13.34 Salt circulation system in a molten salt bath for martempering.[79] (*Courtesy of the Institute of Materials, London.*)

method of effective salt circulation with violent downward flow of liquid salt within the working space is described by Liscic and shown in Fig. 13.34.[79] By using a two-speed electromotor-driven propeller pump, it is possible to achieve agitation rates of 0.3 and 0.6 m/s in the working space.[79] Another method uses a dual-impeller bath agitation system (Fig. 13.35).[80]

Figure 13.36*a* illustrates a newly developed device for automatic continuous addition of small quantities (0.1 to 2.7%) of water into the hot salt bath, which improves its quenching speed.[79,81] The device comprises three essential portions: the probe for temperature measurement, the electronic control, and the system for water addition. To evaluate the effect of water addition and the agitation rate on the quenching intensity of the hot salt bath (Degussa AS-140) at 200°C, tempera-

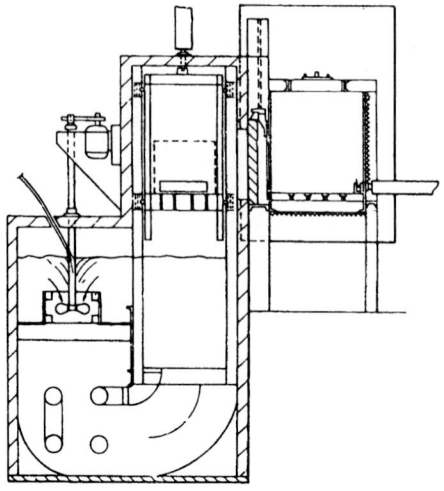

FIGURE 13.35 Dual-impeller salt bath agitation system.[80]

ture versus time curves have been determined in the center of a 25-mm-diameter steel specimen, and relevant curves have been calculated.[79] For safe working conditions without explosion hazard, the recommended working temperature of 180 to 250°C (355 to 480°F) should not be exceeded.[82]

Figure 13.36b shows the effect of the agitation rate and of 1 vol% water addition on the maximum cooling rate of a hot salt bath.

It is vital to note that all cyanides and all organic chemicals must be kept out of these baths. The baths, when satisfactorily maintained, have low viscosity and are excellent for the precise control needed in martempering and austempering. They can also be used for tempering, nonferrous annealing, steel blueing, general treatment of aluminum, and heat-treating copper and beryllium.

The cooling rates of nitrate salt baths are similar to, or even faster than, those of quenching oils. In addition, nitrate quenching baths have longer operating life.

Thermal decomposition of nitrate and subsequent reaction with atmospheric CO_2 form carbonates which decrease the quenching severity. However, reestablishment of the bath to a neutral condition can be achieved by adding a rectifier to the nitrate salt.[64]

One advantage of the use of nitrate/nitrite-type salts is their 90% recovery or recycling from the spent salts in washing waters. It saves not only the original cost of salt but also the problems caused by putting the nitrate down the drain and the cost of shipping it out to a safe disposal site.[72]

MEDIUM-TEMPERATURE NEUTRAL BATHS (650–1000°C). These baths are generally binary or ternary chloride baths comprising KCl, NaCl, $BaCl_2$, or $CaCl_2$. The typical compositions and working temperatures for these types of salt baths are 45% NaCl + 55% KCl (675 to 900°C) and 20% NaCl + 80% $BaCl_2$ (675 to 1060°C), which show that the composition of the salt bath mixtures is based on the required operating temperature range. The melting point of the bath is around 550°C, and these baths can be operated up to a maximum of 1000°C. When freshly prepared, these baths are inert to the steel surface so that neither decarburization nor carburization of the surface occurs. However, with prolonged use, atmospheric oxygen becomes absorbed by the molten salt, resulting in diminution of passivity and subsequent

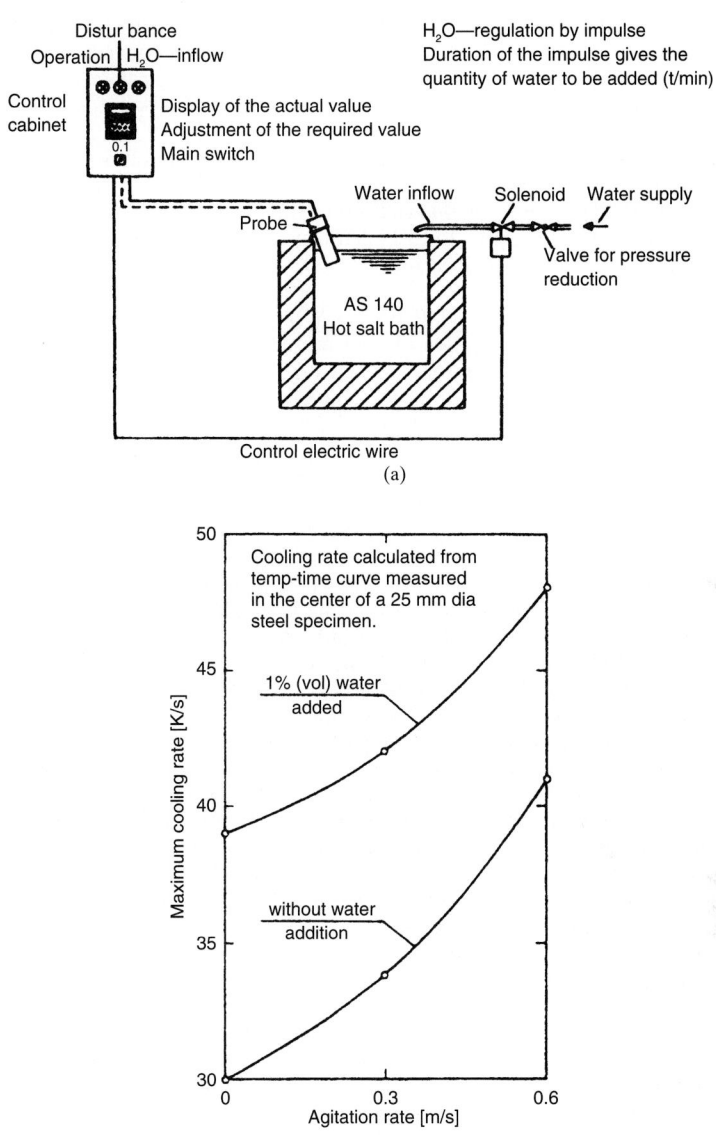

FIGURE 13.36 (*a*) Automatic system developed by Degussa for continuous addition of water to molten salt.[82] (*b*) Maximum cooling rate of a hot salt bath (Degussa AS-140) of 200°C as a function of the agitation rate and percentage of water addition.[82]

decarburization. To avoid this difficulty, a rectifying agent such as borax or boric oxide is introduced into the bath at regular intervals, which converts the oxychlorides to a sludge; this sludge is then periodically removed. Note that these rectifiers are not suitable for baths using mixtures of chlorides and carbonates. In that case, methyl chloride gas should be used as a salt bath rectifier.

This type of bath is employed for austenitizing carbon-, medium-, and high-alloy steels operating at temperatures up to a maximum of 1000°C. It can also be used for annealing stainless steel, ferrous and nonferrous annealing, cyclic annealing, and normalizing. It is usual practice to carry out tests at intervals for hardening characteristics and the presence of decarburization of the tool steel surface to retain a neutral condition of the bath.

HIGH-TEMPERATURE SALT BATHS (1000–1300°C). This type of bath is employed for austenitizing the high-speed steels (T and M series) and the hot-work steels (H series). These baths contain mixtures of barium chloride ($BaCl_2$), borax or sodium tetraborate (NaB_4O_7), sodium fluoride (NaF), and silicates. These salt baths are heated by utilizing the electric resistance of the salt mixture itself. The proprietary baths usually require introduction of a rectifying agent such as graphite or ferrosilicon, to prevent decarburization of high-speed steels as the bath ages. The melting range of the salt baths is usually between 870 and 1040°C. The high-melting-point salts do not drain well from the part and leave a layer that is somewhat difficult to remove.

Since the bath develops the decarburizing tendency after aging, tests are carried out periodically by quenching the test specimens in salt bath and by simply checking the surface by file. A "file soft" surface indicates the need for more rectification. This test can be combined with chemical analysis for barium oxide (BaO) content, which indicates the alkalinity of the bath. Another test for the decarburizing condition is performed with a *double-edge carbon steel razor blade* dipped in the bath. By changing the immersion time or temperature, a good prediction of bath conditions can be designed. If upon bending, the blade breaks after immersion, it means an absence of decarburization. On the other hand, if it bends, then the bath is decarburizing. Of course, stainless steel blades will not hold for this test.[66]

CYANIDE- AND NONCYANIDE-BEARING SALT BATHS FOR THERMOCHEMICAL SURFACE-HARDENING TREATMENTS. These baths are used for liquid carburizing and carbonitriding, nitriding, and nitrocarburizing of steel surfaces, which are treated in detail in Chap. 16.

Safety Precautions of Salt Baths. The following recommendations should be followed to protect the operating personnel from any hazards:[63,67]

1. Good exhaust around the salt bath is necessary, especially during charging of fresh salt and during the quenching operation.

2. Operators should always stand behind the furnace shields and wear face masks and gloves while salt or work is either introduced into or removed from the furnace.

3. To prevent salt from spurting or sudden expulsion from the molten bath, never introduce wet salt, wet work, or water into the high-temperature molten bath. That is, parts, fixtures, and conveyor entering a quench bath must be dry and free of any moisture, oil, or other liquid.

4. Before switching off furnaces using an externally heated pot, reduce the salt level so that the pot is no more than two-thirds full (depending upon the position of the burners), to avoid the risk of spurting upon remelting. Place the cover on the pot during remelting or cooling. Neutral salts, baled out to comply with the above, are normally reused when the furnace is restarted, provided that it has not been allowed to become damp in the meanwhile.

5. Dry out electrically heated brick-lined furnaces with an electric heater prior to restarting after use, to avoid spurts owing to entrapped moisture in the brickwork.

6. Never heat salts containing nitrate above 550°C.

7. Never mix salts containing cyanide with salts containing nitrates; otherwise, an explosion will take place upon heating.

8. Salts containing cyanide or barium are very poisonous. Do not take food where these salts are being used, and always wash the hands thoroughly prior to handling food.

9. Keep salts containing cyanides away from acids.

10. Handling and disposal of cyanide products are done by using special clothing, special handling, and state-of-the-art methods.

11. No water sprinkler should be installed near a molten salt system. No water or liquid extinguisher should be used in case of fire.

12. The salt bath protection from accidental overheating should be provided by installing audio/video alarms which go on when temperature is in excess of a set limit. If temperature rises above 595°C (1100°F), thermal breakdown of salt and the resultant reactions with the salt container may lead to salt leakout.

13. Salt should be stored in well-marked, sealed containers in a dry place and away from incompatible materials such as cyanide salts.

14. When parts are quenched from an atmosphere furnace into a salt bath, splashing of salt into the furnace should be prevented via a salt curtain, an intermediate chamber, or a vestibule.

Maintenance of Salt Baths. Molten salt baths are usually maintenance-free, primarily due to their excellent thermal stability and their tolerance for the contaminants. They offer consistently satisfactory performance for many years by the addition of new or recovered salt to replenish the dragout salt. The quantity of dragout salt is a function of the mass and the configuration of the parts and fixtures or conveying system. However, this is in the range of 50 to $100 g/m^2$ (~1 to $2 lb/100 ft^2$).

Contaminants such as metallic debris from the parts and carbonate formation may build up over a prolonged time. Agitation provides suspension of fine metallics, and when such suspension exceeds 0.5%, the quench severity is influenced. Similarly, when carbonate exceeds its solubility limit, it starts to slow down the quenching speed. Both metallics and carbonates can be eliminated by desludging the bath. The easy way to do this is to lower its temperature as far as possible, shut off agitation and heating, and wait for some period. This permits contaminants to settle to the bottom, where they are then scooped out as sludge.

If water addition is the usual practice, the water content needs to be checked periodically. Water addition can be decreased or increased based on the desired quench severity.

The melting point of the bath should be checked only occasionally, because it rarely changes over time. However, if any significant increase is observed, it may suggest thermal breakdown of the salt, perhaps due to unintentional overheating. The reason for the overheating should be examined and rectified.[67]

13.2.3.6 Lead Bath. With its low melting point (327°C, or 620°F), lead can be employed at temperatures up to 1200°C; however, it is commonly used up to a maximum temperature of 900°C. Since lead oxidizes readily, particularly with increasing temperature, sludge so produced forms a coating or scum on the surface of the steel being treated and is a problem, especially at higher temperatures. Therefore it is necessary to cover the surface of the bath with charcoal or a similar reducing agent above about 480°C (900°F), to minimize the oxidation.

The advantages of a lead bath over a salt bath are as follows:

1. A lead bath neither contains nor picks up moisture. Hence scaling or oxidation of the steel surface does not occur.
2. There is no danger of an explosion resulting from the hidden formation of steam as new lead is added to replenish the bath.
3. There is an absence of chemical attack on steel.
4. There is rapid heating and cooling.
5. It possesses a high stability (excluding its oxidizing tendency).

The disadvantages are as follows:

1. In view of the lower density of steel compared to that of lead, the steel parts must be held down by fixtures in the bath.
2. The rapid oxidation of the molten lead surface and the retention of lead oxide, mainly at the surface, are an adverse factor.
3. Being extremely toxic and health-hazardous when small traces are either inhaled or ingested, lead requires elaborate precautions and permission by the factory inspector for its use.
4. All parts and fixtures must be dry when immersed in the bath, to avoid the steam formation and resultant violent expulsion of the molten lead.

The method is used for specialized applications, such as files, and rapid local hardening and selective tempering where only a portion of the tool needs to be fully hardened.[16]

13.2.3.7 Pressure Gas-Quenching. Currently high-pressure gas-quenching has become the fastest-growing technology in heat-treating of components made from hardenable stainless steels, tool and die steels, and oil-hardening tool steels.[18,20,83] In the quench loop of a typical vacuum furnace, cold nitrogen, argon, helium, or hydrogen gas flows, under pressure (>5 bar), from the top to the bottom through the hot zone of the vacuum furnace, where it picks up heat from the hot workload. The heated gas is then passed out of the hot zone into a heat exchanger (water-cooled or through refrigerated coils) and is recirculated into the furnace chamber through the blower and directed at the work in a continuous fashion (Fig. 13.37).[84]

The cooling rate of the metal depends on the surface area and mass of the part and the type, velocity V, and pressure P of the cooling gas, according to the formula $h_g = C(VP)^m$, where h_g is the heat-transfer coefficient for a given gas and m and C are constants depending on the furnace design, part size, and load configuration ($m = 0.6$ to 0.8).[79] Thus, the cooling rate can be adjusted and controlled by varying the type, velocity, and pressure of the gas, thereby offering a significant flexibility.

The most important factors required in the design of a gas-quenching system are (1) the design of the gas circulation system to minimize distortion by using turbu-

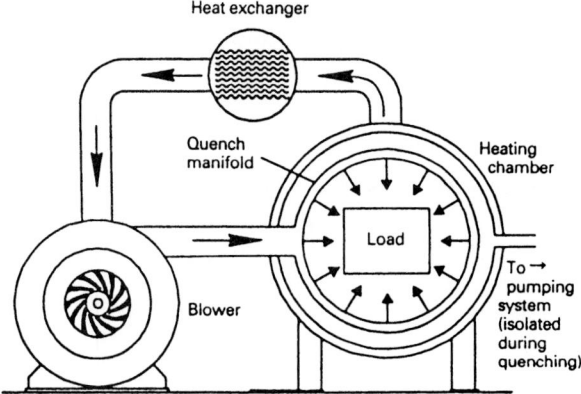

FIGURE 13.37 Vacuum furnace quench loop.[84] (*Courtesy of Wolfson Heat Treatment Center, England.*)

lent flow normal to all component surfaces, (2) the type of quenching gas and its pressure, in order to achieve the appropriate cooling rate, and (3) the specification of any gas recycle system to reduce the cost.[85]

Gas-quenching units are designed for either batch or continuous processing. Usually hydrogen or helium gas is used, both to produce clean, bright surface and to increase the heat-transfer rate between the gas and the workload. However, hydrogen has the greatest heat-transfer properties, as shown in Fig. 13.38a.[83] Compared to nitrogen (at 6 bar), hydrogen has 30% shorter cooling times and 40 to 50% higher heat-transfer coefficient. Hence, more heat treaters are using H_2-N_2, He-Ar, and He-N_2 mixtures.[84] The high cost of argon permits its use in situations where chemical inertness is required, such as aerospace parts, made of titanium, tantalum, or niobium.[86]

High-pressure gas (5-bar helium) of solution-treated alpha-beta Ti-6-Al-4V alloy castings in a vacuum furnace, together with improved circulation through the use of turboblower, has proved successful in achieving the very rapid cooling rates required for subsequent age hardening to obtain high strength. These titanium alloys with optimum mechanical properties (high strength and adequate ductility) are finding numerous and increasing applications in aircraft and aerospace parts.[87]

However, high-pressure gas-quenching technology (e.g., helium at 20 to 40 bar and nitrogen up to 10 bar)[88] can provide quench severities comparable to those of conventional recirculated oil. Very high pressure (e.g., hydrogen at 50-bar pressure) can produce heat-transfer coefficients even greater than that for water (3000 to 3500 W/m²·K, or 530 to 620 Btu/ft²·hr·°F), as illustrated in Fig. 13.38b.

In addition to single-chamber furnaces with high-pressure gas-quench systems for horizontal and vertical loading, double-, triple-, or multiple-chamber furnaces are now available with oil-quench or gas-quench chambers or both.[89] Multichamber systems and directed gas flows also can be employed for selective, partial, or directional hardening of parts such as a soft-shank drill.[90]

Advantages of high-pressure gas-quenching in vacuum furnaces over liquid quenching include[91–93]

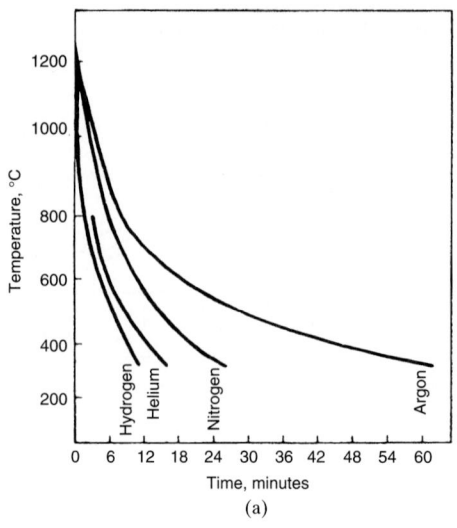

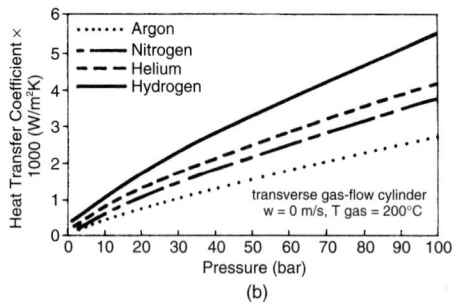

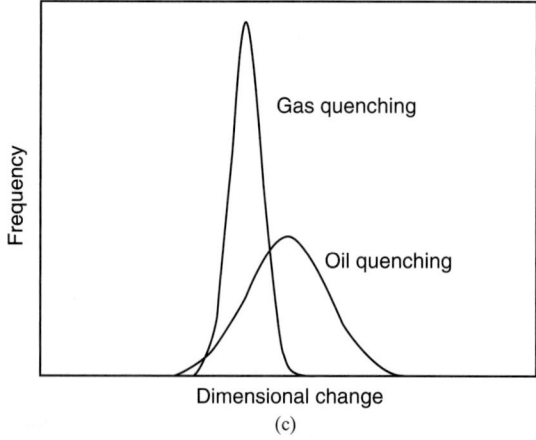

FIGURE 13.38 Comparisons of (*a*) cooling properties of common gases,[83] (*b*) various high-pressure gas quenchants, and (*c*) distortion (or dimensional change) in quenched parts between gas-quenching and oil-quenching. [(*a*) and (*b*): *Courtesy of Leybold Durferrit GmbH.*]

1. Optimum hardening
2. Dramatic reduction of distortion (Fig. 13.38c) and crack-free hardening of the workload due to more uniform cooling rates
3. Absence of oxidation, decarburization, or carburization on the surfaces of the cooled parts
4. Pleasant working operations due to clean operation without heat radiation; fume, fire, and explosion hazards; ventilation requirement; and environmental pollution
5. Reduced labor and total production costs, and excellent reliability and reproducibility by accurate microprocessor control
6. Reduced or no posthardening and finishing costs
7. Use in quenching a wide variety of materials by changing the cooling rate during a single cycle
8. Ability to provide cleaner parts, without posttreatment cleaning and attainment of reproducible results from batch to batch

The shortcomings of gas-pressure quenching over conventional liquid quenching are that[94]

1. Larger cross sections of some oil-hardening grades produce lower tensile properties, ductility or fracture toughness, and hardenability.
2. Certain carbon and alloy-steel grades must be liquid-quenched irrespective of their cross sections (such as 1045, 1075, 4130).

Applications. The following parts made from various high-speed steels, cold- and hot-worked steels, and air-hardening steels have been successfully hardened with high-pressure gas-quenching.[92,95]

High-speed steel: For drills, taps and reamers, hobs, shapers, shavers, milling cutters, flat and circular broaches, dovetail and circular form tools, and end mills having high-red hardness, good toughness, and excellent wear-resistance.

Cold-work steels: For progressive die stamping and forming, motor lamination production, can manufacturing, powder compaction, recycling and slitting (knives), and fine blanking and rollforming requiring wear-resistant tool steels with good toughness and high hardness. These materials include 12% Cr steels; X155CrVMo12 steel; X210Cr12 steel; 90MnCrV8 steel; A2, D2, and M2; and high-vanadium P/M tool steels.

Plastics-process steels: For mold/holder blocks, equipment for compounding extrusion injection molding, and pelletizing/granulating knives from P20 types, H13, S7, NAK55 (Ni + Al PH), 17-4(or 15-5)PH, maraging steels; A2, D2, 440C; high-vanadium P/M steels (9V, 10V, 440V, M390, etc.).

Hot-work steels: For aluminum die-casting molds, compression-molding dies, and hot-forging dies.

Air-hardening steels: For dies and tools from X37CrMoV51; X45NiCrMo4; A2, D2, and M2 grades; and high-vanadium P/M tool steels.

Other applications are for solution annealing/treating and hardening of aerospace/industrial turbine materials which include 300 series, 400 series, and PH stainless steels; maraging steels; iron-base superalloys; nickel-base superalloys; and high-strength (300M, 4340, etc.) alloys.[94]

Note that when endo gas atmosphere is already available, the cost of a vacuum furnace is very high.

13.2.3.8 Fluidized-Bed Quenching. A typical fluidized bed consists of a furnace system (shell, heaters, and insulation) and a quench system (gas diffusion assembly, fluidized-bed support, and gas) (Fig. 13.39). The fluidized bed is produced by blowing a gas such as nitrogen, argon, helium, hydrogen, or carbon dioxide through a solid support such as alumina, iron, copper, and molecular sieves.[8,96,97]

Heat transfer within a fluidized bed depends on the particle size of solid support (or medium), volumetric heat capacity, thermal conductivity of the gas, and gas flow rate through the bed. The effect of fluidized-bed variables on the heat-transfer coefficient is summarized in Table 13.7.[71]

The cooling curve behavior during fluidized-bed quenching can be described by the equation:[78]

$$\frac{T - T_f}{T_i - T_f} = \exp\left(-\frac{Aht}{C_p V \rho}\right) \tag{13.12}$$

where T_f is the temperature of the fluidized bed, T_i is the initial temperature of the part being quenched, A is the cooling surface area, V is the volume of the metal

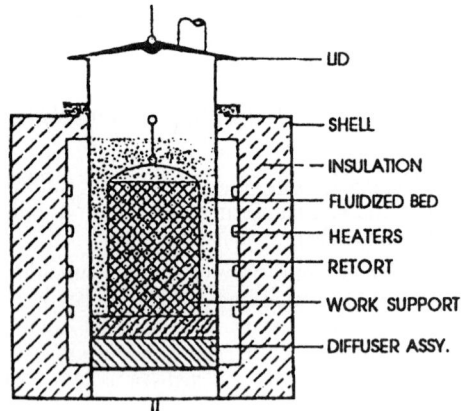

- LID
- SHELL
- INSULATION
- FLUIDIZED BED
- HEATERS
- RETORT
- WORK SUPPORT
- DIFFUSER ASSY.

FIGURE 13.39 Cross section of an electric fluidized-bed furnace.[8,46]

TABLE 13.7 Effect of Fluidized-Bed Variables on Heat Transfer[71]

Parameter	Effect[†]	Comment
Particle size d	$d \uparrow, \downarrow$	Valid for $d > d_{crit}$
Volumetric heat capacity C_v	$C_v \uparrow, \uparrow$	Al_2O_3
Gas conductivity k	$k \uparrow, \uparrow$	He, H_2
Gas flow rate U_g	α_{max} for $U_g \approx 5U_{mff}$	

[†] U_{mff} is minimum gas (fluid) flow rate.

(probe), ρ is the density of the metal, C_p is the heat-capacity of the metal, t is the cooling time, and h is the heat-transfer coefficient.

The important variables in the selection of a fluidized bed are the medium, particle size, temperature, and gas. The two variables that strongly affect heat transfer are the thermal conductivity and the flow rate of the gas, as shown in Fig. 13.40.[79] Clearly, helium and hydrogen provide much greater potential for cooling rate variation compared to the most commonly used nitrogen (and argon) gases utilized for fluidized beds.

Reynoldson has shown that air-fluidized beds produce cooling rates intermediate between those displayed by a mineral oil and by a low-pressure vacuum quench.[98] Other workers have shown that air-fluidized beds can produce the same hardness as oil and salt for several steels.[99]

A recent development in the fluidized-bed furnace includes computerized control of fluidization to optimize heat transfer/uniformity while minimizing gas usage as well as use of a specially-designed nitrogen-purged transfer hood to protect workload during transport from fluidized bed to quench facility, in order to achieve good surface finish.[100]

The advantages of fluidized-bed quenching over molten metal and molten salt bath quenching are the improved process control, improved process safety, flexibility, cleanliness, and reduced pollution and fire hazards. Cost is a significant shortcoming because the fluidizing gas is usually not recycled.

13.2.3.9 Spray Quenching. The term *spray quenching* refers to various quenching processes that allow a high rate of heat extraction from the hot metal surface by the impingement of a fast-moving stream of a quenchant medium, in order to produce a metastable structure with the required physical properties and development of the desired distribution and level of stress.[101] Hence, the time-temperature history of the part must be precisely controlled during quenching. In spray quenching, the heat-transfer coefficient from the part to the quenching medium is a direct function of the flow rate, turbulence, and impingement pressures of the quenchant at the hot surface. By adjusting these parameters during the quench, a cooling profile (or a controlled cooling operation) can be obtained which is not possible by

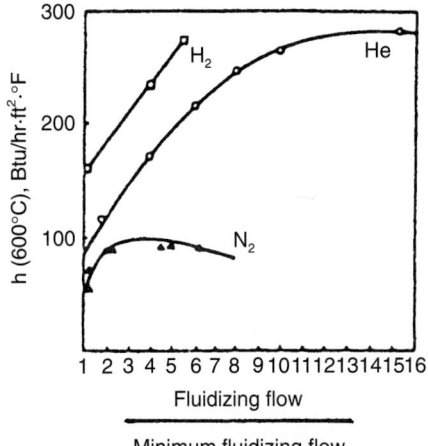

FIGURE 13.40 Effect of fluidizing gas on heat-transfer properties.[79] (*Courtesy of the Institute of Materials, London.*)

other methods (such as immersion quenching).[8,102] The rate of heat extraction can be varied over a wide range by changing the amount of sprayed liquid (e.g., by varying water and air pressure). Examples of sprayed liquid include (1) a mixture of water (or other volatile liquid) droplets to a gas-quenching stream (fog quenching), (2) water, or water/air streams, (3) sprays of other volatile liquid quenchants other than water, and (4) high-pressure jets of oil, water, or aqueous polymer solution under the liquid level in a bath.[8] By employing a CNC control system, it is possible to change continuously the heat flux extracted and to follow a predetermined cooling curve.[79]

Figure 13.41 shows the schematic heat transfer in the vapor-blanket stage for water-spray cooling of a heated metallic surface. The heat-transfer coefficient to water h_w comprises an internal heat-transfer resistance inside the water layer and a capacitative transport resistance for the quantity of water:

$$h_w = \left(\frac{1}{h_{wi}} + \frac{1}{\dot{m}_s C_w} \right)^{-1}$$
(13.13)

where h_{wi} is the internal heat-transfer coefficient of water ($W\,m^{-2}\,K^{-1}$), $\dot{m}_s$ is the flux of sprayed water ($kg\,m^{-2}\,s^{-1}$), and C_w is the specific heat capacity ($J\,kg^{-1}\,K^{-1}$).[79]

A well-established application of spray cooling, comprising 19 banks of water sprays covering a distance of 88 m, is the runout table of the accelerated cooling of hot strip mill in the integrated works of British Steel Strip Products, Port Talbot.[103]

13.2.3.10   Cold Die Quenching.   Thin flat disks, long slender rods, thrust washers, and thin blades (with a large surface area and relatively small weight) are usually not quenched in conventional liquid quenchants owing to the occurrence of excessive distortion. Rather, they are often quenched by squeezing between a pair of water-cooled, cold copper or beryllium-copper die blocks to avoid any distortion, where die blocks provide necessary quenching action together with maintenance of

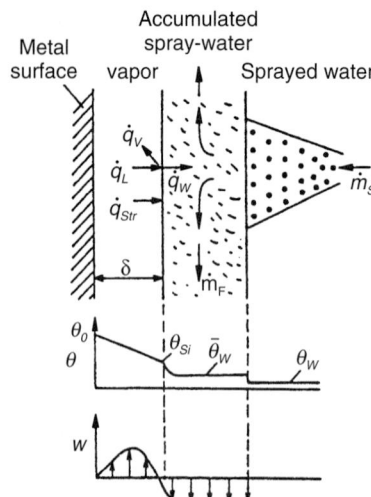

FIGURE 13.41   Heat transfer in the vapor-blanket stage for spray cooling, according to Jeschar et al. $\dot{q}_v$ = boiling heat flux density; $\dot{q}_L$ = conduction heat flux density; $\dot{q}_{str}$ = radiation heat flux density; $\dot{q}_w$ = heat flux density transferred to water; $\dot{m}_s$ = flux of sprayed water; $\dot{m}_F$ = flux of water flowing off; δ = thickness of vapor blanket; θ_0 = surface temperature; θ_{si} = boiling temperature; θ_w = water temperature; and w = velocity of fluid. (*After R. Jeschar, R. Maass, and C. Kohler, in Proceedings of the AWT–Tagung Induktives Randschichtharten, 23–25, March 1988, Darmstadt, pp. 69–81.*)

flatness. In this method, various forms of cold, flat, or shaped dies are used in order to suit the shape of the part being quenched.[6]

13.2.3.11 Press Quenching. Quenching presses are used for controlled quenching of ring gears, bearing races, and other round, flat, or cylindrical parts, to accomplish heat treating with minimal distortion.

In press quenching, the die contacts the hot and plastic part, and the pressure of the press aligns the part mechanically. The machine and dies then force the quenchant into contact with the part in a controlled manner. The amount of quench oil and its flow rate are controlled by the press. A uniform and high volume of oil flow around all parts is directed by the dies. Since quenching speed is a function of the quenching medium and the rate of oil flow, the cooling rate can be controlled by adjusting the rate of oil flow through the die.[104]

Although some basic types of dies are available for certain workload shapes, additional dies are usually needed to accommodate specific parts. Quenching to a tolerance of 0.025 to 0.050 mm (0.001 to 0.002 in.) and 0.025 to 0.125 mm (0.001 to 0.005 in.), respectively, for roundness and flatness is common practice for ring gears and bearing races when proper equipment and correctly designed dies are used.[6,104]

The use of press-quenching processes together with intense quenching using water has been developed. This significantly reduces cost and applied loading to reduce distortion compared to oil-quenching.[105]

13.3 QUENCH-HARDENING

13.3.1 Quench-Hardening of Steel

Quench-hardening consists of heating to a temperature 30 to 50°C above Ac_3 for hypoeutectoid (plain carbon and low-alloy) steels and above Ac_1 for hypereutectoid (plain carbon, low-alloy, and numerous tool) steels, soaking at that temperature for about 1 hr/in. section thickness, followed by quenching in a liquid quenchant (such as water, brine, oil, or polymer solution) to produce a martensitic and/or bainitic structure throughout the entire component. The cooling rate necessary to enable austenite to transform fully into martensite must be equal to, or greater than, the minimum rate of cooling, usually designated as the *critical cooling rate* to ensure maximum hardening.[106] Figure 13.42*a* shows the conventional quenching and tempering superimposed on the TTT diagram.[17]

In some of the very-high-alloy tool steels such as M1- and T1-type high-speed steels, austenitizing is carried out at a temperature greatly in excess of Ac_3 for such a time (typically 2 to 5 min) that a maximum solution of the alloy carbide occurs, without undue coarsening of the matrix.[107] But preheating of these parts for a longer time is required prior to high-temperature austenitization (see also Chap. 14).

In some more highly alloyed steels, where the position of the pearlite nose of the TTT diagram is far to the right, full hardening can be obtained by a slow cooling rate, such as would be encountered in air cooling. These grades of steels are usually called *air-hardening steels* (see also Chap. 12).

Advantages of conventional hardening of steel by rapid quenching in water or brine are that (1) the method is very simple, (2) minimum equipment and an easy control are needed, and (3) the quenchants are inexpensive.

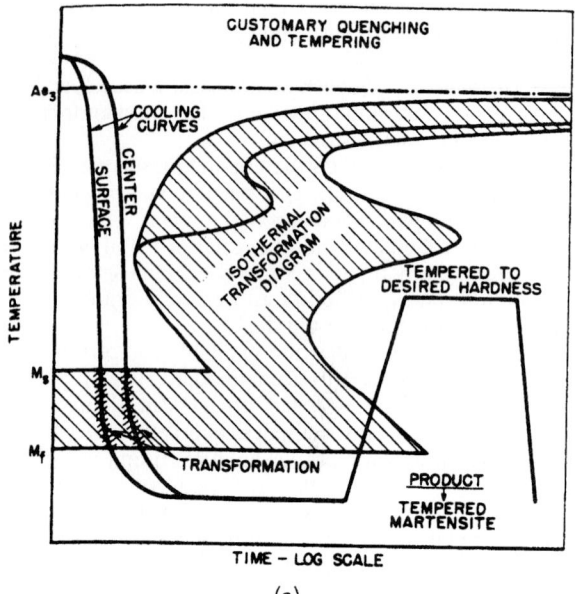

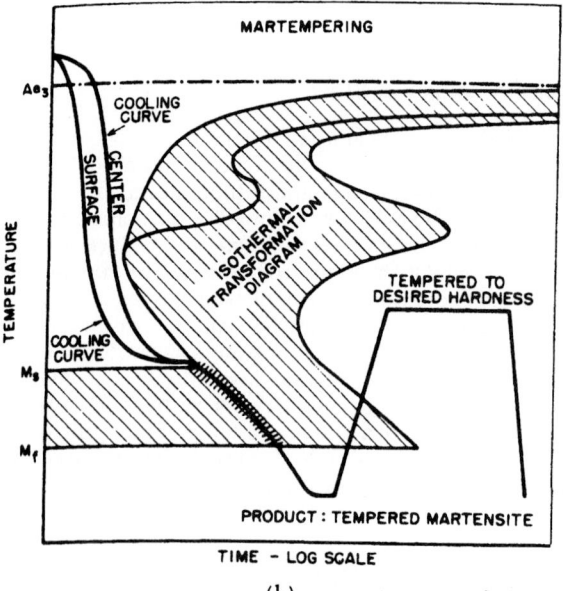

FIGURE 13.42 Typical curves for (*a*) conventional quenching and tempering, (*b*) martempering, and (*c*) modified martempering superimposed on TTT diagrams.[17] [(*a*) and (*b*): Courtesy of Association of Iron and Steel Engineers, Pittsburgh, Pennsylvania; The Making, Shaping and Treating of Steel, eds. W. D. Lankford and H. E. McGannon, United States Steel, 10th ed., 1985.]

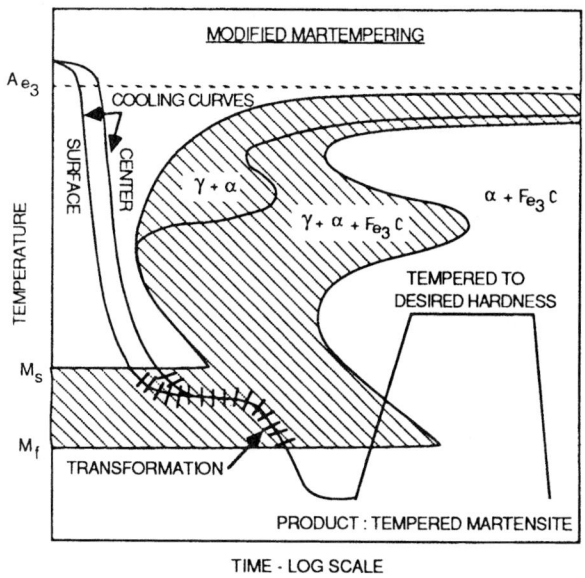

(c)

FIGURE 13.42 (*Continued*) Typical curves for (*a*) conventional quenching and tempering, (*b*) martempering, and (*c*) modified martempering superimposed on TTT diagrams.[17] [(*a*) *and* (*b*): *Courtesy of Association of Iron and Steel Engineers, Pittsburgh, Pennsylvania; The Making, Shaping and Treating of Steel, eds. W. D. Lankford and H. E. McGannon, United States Steel, 10th ed., 1985.*]

One serious disadvantage of this method, often observed in large-sized parts, is that after the conventional hardening of steel parts by rapid quenching, severe residual stresses occur, as a result of large variations in cooling rates across its entire cross section. This arises from a steep thermal gradient between the surface and center of the material as well as from nonuniformity in transformation. These stresses can produce an excessive distortion and can even cause cracking of the materials. (See also Chap. 17 for more elaborate discussion on residual stress, distortion, and cracking.)

The factors responsible for successful hardening of a specific part are (1) the heat-transfer characteristics of the quenchant, (2) the quenchant use conditions (such as bath temperature and flow velocity), (3) bath loading, (4) section thickness of the part, and (5) transformation characteristics of the particular alloy being quenched. Successful hardening normally means attainment of the required hardness, strength, and toughness with the minimum residual stress, distortion, and likelihood of cracking.[108]

13.3.2 Quench-Hardening of Iron Castings

Gray iron castings to be hardened are austenitized in a gas- or oil-fired furnace or in a salt bath at temperatures normally 30 to 55°C or sometimes as much as 95°C

(175°F) above the calculated A_1 temperature for 2 min to 1 hr per inch of section thickness (depending on their size and shape), followed by quenching in an oil, molten salt, or polymer solution. However, for high-alloyed irons, forced-air quenching is often the choice. Immediately after quenching, both gray and ductile castings are normally tempered at temperatures well below the transformation range for about 1 hr/in. section thickness.[109]

The calculated A_1 temperature of unalloyed gray iron is given by[17]

$$A_1(°C) = 730 + 28.0(\%Si) - 25.0(\%Mn) \tag{13.14a}$$

$$A_1(°F) = 1345 + 50.4(\%Si) - 45.0(\%Mn) \tag{13.14b}$$

Ductile irons are usually austenitized at a temperature of 845 to 925°C (1550 to 1700°F) and quenched in oil to minimize stresses. Nevertheless, water or brine can be employed for simple shapes.[110]

13.4 INVERSE QUENCH-HARDENING OF STEEL

The pattern of hardness distribution in round oil-quenched bearing-grade steel parts with relatively small mass and high quenchability has been found to exhibit lower hardness values at the surface than at the core. This phenomenon, called *inverse quench hardening* by Shimizu and Tamura, is attributed to the transformation of pearlite at the surface.[110a–e]

Liscic and his coworkers[110d,110e] have also observed the inverse quench-hardening phenomenon in their controllable delayed quenching (CDQ) technology polymer quench tests on AISI 4140 cylindrical specimens. Their explanation for the inverse hardening phenomenon agreed well with that of Shimizu and Tamura.[110a]

Such inverse hardness distribution obtained after quenching, when tempered at an appropriate tempering temperature, offers uniform microstructure of tempered martensite in the entire crosssection of the workpiece, providing the maximum ductility and toughness. The results of the bending fatigue tests and the impact loading behavior obtained in this investigation have shown an increase in the bending fatigue life by a factor of about 7, and an impact energy increase of about 7% in specimens with inverse hardness distribution, compared to specimens with normal hardness distribution.[110d]

13.5 DIRECT QUENCHING

For some applications such as coil springs, roll-formed balls, axle beams, steering and suspension parts, and track shoes, direct quenching after hot forging (or from high austenitizing temperature) provides a reduction in heat treatment cost by removing reheating for hardening. Direct quenching can be applied to carbon, boron-treated, and low-alloy steels and is mostly followed by tempering.[111] It is also used for rapid cooling of metals from the elevated solutionizing temperature.

13.6 INTENSE QUENCHING OF STEEL

Intense quenching (IQ) may be defined as a very rapid, uniform quenching process with very high agitation rate obtained by high-pressure quenchant jets or rapid impeller agitation systems, to produce a maximum surface compressive stress and very hard as-quenched surface. This occurs when the Biot number $\geq$18. The IQ should be stopped at the point of formation of maximum surface compressive stresses, which corresponds to the attainment of maximum tensile stresses in the core (which, according to some researchers, is at 450 to 500°C core temperature).[79] Successful implementation of IQ practices is associated with the emergence of a huge amount of dislocations (according to Ivanova) and the resulting physical property improvements.[79,112]

Pressurized jets of water, oil, 5% caustic soda, 3% sodium carbonate, and various concentrations of polymers can serve as quenchants. Here, *H* values of water can reach in the range of 2.5 to >5.0 and for oil >0.70.[113] Intense steel quenching technologies have become an increasingly interesting alternative to heat treatment of steel. It will be accomplished for autoparts (semiaxles, engine crankshafts, gears, spherical journals, bearing rings, springs, moil points); tools (dies, punches, etc.); fasteners (nuts, bolts, washers, etc.); large steel parts for machine, building, and power industries; mining equipment parts; various types of forge quenching; and cable wire quenching.

IQ has a great future owing to numerous advantages derived by using this technology. The advantages of IQ over conventional water-, water-polymer solution, and oil-quenching are[112,113]

1. Optimum quench uniformity achieved throughout the section thickness of the part.

2. Formation of nearly 100% martensite in the surface layers of a part and in critical areas of cheaper, shallow-hardenable steel parts (as in root fillets of a gear) (i.e., cost savings by allowing the replacement of alloy and high-alloy steels with cheaper 1040 steel, while attaining greatly enhanced ductility).

3. Development of maximum residual compressive stress value of about 1200 MPa (173 ksi) and the associated transformation of austenite to martensite, producing *additional strengthening* or *superstrengthening* of the material.

4. Increased hardening depth and high surface hardness for a given steel grade, which provide a reduced production cost of steel (Fig. 13.43).[113]

5. Minimum heat-treating distortion (e.g., 0.003 to 0.006 in./in. distortion is produced in an intensely oil-quenched AISI 1141 steel part against 0.015 to 0.022 in./in. in a normal oil-quenched part). This offers the elimination of many posthardening, straightening, or postquenching operations on steel parts.[79]

6. Increased safe life of machine steel parts by 3 to 4 times over oil-quenching.[112]

7. Effective and economical prevention of quench cracking in both batch processing of small parts and continuous conveyor lines.[112]

8. Maximum long-life bending and/or torsional fatigue properties of hardened parts.

9. Attainment of highest yield strength/tensile strength ratios with good toughness in the intensely quenched and tempered condition.

10. High abrasion resistance.

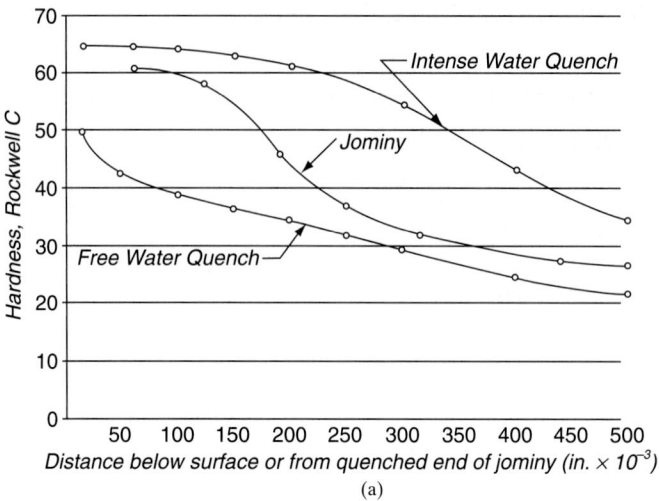

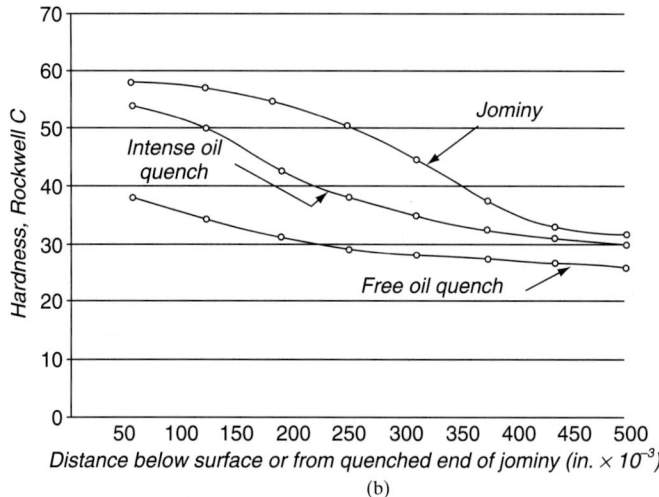

FIGURE 13.43   Typical advantages of (*a*) intense water-quenching for 3-in.-(76.2-mm-) diameter bars of AISI 1045 steel, *H* > 5.0, and (*b*) intense oil-quenching for 2.5-in.- (63.5-mm-) diameter AISI 1141 steel, *H* > 0.70.[113] The term *free* denotes mildly agitated. (*Courtesy of R. Kern.*)

11. Use of water or water-polymer solution instead of expensive oil quenchants.
12. Increased productivity during manufacturing.
13. Protection from dangerous fires.
14. Ease in automation of stable technological process.
15. Possibility of combining turnery work with heat treating when using induction heating.

Table 13.8 lists the surface hardness and bending fatigue life of a 2.25-in.- (57.1-mm-) diameter steel shaft [at applied stress of 50 ksi (345 MPa)] of varying compositions in the differently heat-treated condition.[113]

The disadvantages of intense quenching include the following:[113]

1. Intense quenching equipment is expensive.
2. Quenching of individual parts—or, at the most, 2 to 3 parts at a time—is possible compared to those of 6 to 12 parts (or more) quenched simultaneously by conventional method (Lowerator mechanism).

For successful operations of intense water-quenching, the following guidelines should be adopted:

1. Parts to be intensely quenched should be finish-machined except for final grinding.
2. Proper austenitization should be carried out in a gaseous atmosphere furnace or by induction heating. In the former case, the carbon potential of the furnace atmosphere should match closely (±0.05%) the carbon content of the part.
3. Part and holding fixtures should be preheated before loading in the austenitizing furnace.
4. After austenitization, transfer of the part to the appropriate quenching fixture should be rapid. Figure 13.44 is a schematic section drawing of a scanning-type intense quenching fixture for a round bar. A typical intense oil-quenching fixture for a gear is shown in Chap. 17 (Fig. 17.11).[113]
5. The quenchant should be flowing prior to loading of the hot part into the quenching fixture.
6. Once an intense quenching is set up with precise control of the quench pressure, scanning rate, and so forth, a start button is pressed. (See Chap. 17 for more discussion on better quenching fixture design.)
7. Vapor pockets should not be allowed to form.
8. The maximum carbon contents of steel to be used successfully with IQ are as follows:

Water-quenching: Complex shapes (e.g., splined shafts)	0.38%
Simple shapes (e.g., round pins)	No limit

TABLE 13.8 Results of Hardness Values and Full-Size Bending Fatigue Tests on a 2.25-in.- (57.1-mm-) Diameter Shaft at an Applied Stress of 50 ksi (345 MPa)[113]

Material	Heat treatment	Hardness, Rc	Life, cycles
8650H	Oil quench and temper	32–34	3.5–4.0×10^5
81B40	Oil quench and temper	30	2.9–4.8×10^5
1045	Intense quench and temper	43–45[†]	2.3–4.0×10^6
8620H	Carburize and harden (intense quench)	63	1.8–2.3×10^6
1045[‡]	Intense quench and temper	42	3.8×10^5

Note: The hardened depth to 40 Rc on the 1045 shafts was approximately 0.13 in.
[†] Decarburized 0.0015 in. deep.
[‡] Reprinted by permission of Fairchild Publications, New York.

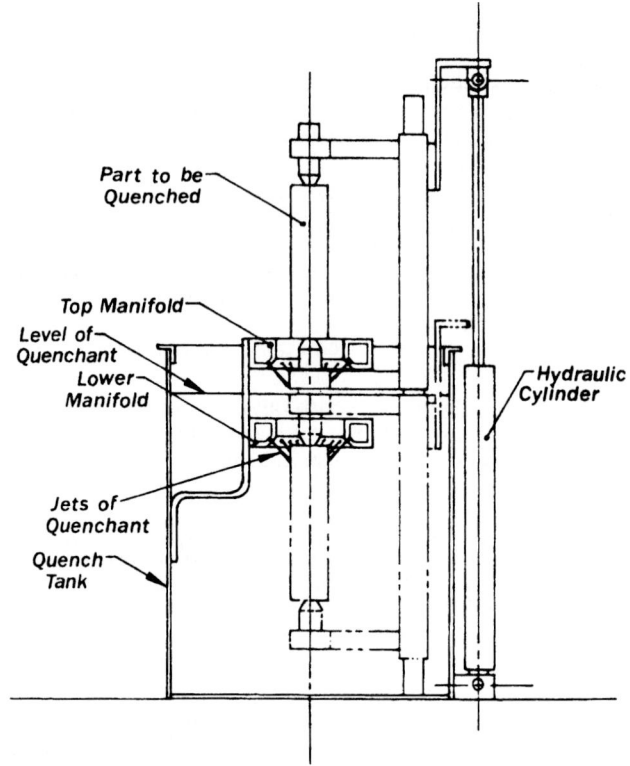

FIGURE 13.44 Schematic section drawing of an intense quenching fixture of the scanning type showing the completion of quenching of part.[113] (*Courtesy of R. Kern.*)

Oil-quenching:	Complex shapes (as above)	0.44%
	Simple shapes (as above)	No limit

9. Correct design of the part is also necessary to achieve its superior engineering properties after heat treatment. (See Chap. 17 for more details.)

13.7 MARTEMPERING OF STEEL

Martempering is a form of delayed or interrupted quenching. Martempering of steel consists of (1) quenching steels from the austenitizing temperature into a hot bath (such as hot quench oil, molten low-temperature nitrate/nitrite salt bath, molten lead bath, or a fluidized particle bed) maintained at a temperature slightly higher ($\sim$10°C) than the M_s temperature; (2) holding at this temperature in the quenching medium for a sufficient period of time (but within the incubation period for bainite at this temperature) to reach the equalization of the temperature of a section (i.e., within the part); and (3) subsequent cooling relatively slowly, usually in air, through the M_s–M_f range to room temperature. This enables the martensitic transformation to occur at one and the same time over the entire section, since the difference in temperature between the exterior and interior of the part during air cooling is small.[74] The term *martempering* is misleading since the martensite resulting from such treatment is not a tempered martensite in any sense of the word *tempering*.

Since there is less risk of differential martensitic transformations within the section, distortion, cracking, and residual stresses are minimized. However, an increase in volume and loss of ductility result, accompanying the transformation which set up a high degree of microstresses. The final structure and properties obtained by this method are similar to those obtained from conventional quench-hardening, but the proportion of retained austenite is usually larger after martempering. Straightening or forming can be easily performed after removal from the martempering bath while the part is still hot.[114]

After the martempering operation, tempering of the component must be carried out in the same manner as in the conventional quench-hardening to obtain the desired hardness level. The time versus temperature relationship of martempering and tempering is illustrated in Fig. 13.42*b*.[17]

Usually, the quenching temperature maintained in this process is 175 to 230°C (350 to 450°F) when hot-quench oils are used. Molten nitrate/nitrite salts (with water addition and agitation) are effective and used in the range of 160 to 400°C (320 to 750°F) which offers more metallurgical and operational advantages, due to their higher heat-transfer coefficients.[6]

When parts are immersed into the quenching bath, the oil or salt circulation around them must be uniform, and the bath temperature should be accurately stabilized.[115] For rapid cooling, the quenchant must have good cooling capacity and must be properly agitated, and a sludge-free bath must be maintained. Selection between salts and oil depends on several factors such as operating temperature, composition, and cooling power.

Most installations for martempering are equipped with batch-type furnaces utilizing manual operation for the transfer of steel parts from the austenitizing furnace to the molten salt quench. However, mechanized plants with nitrate/nitrite salt bath furnace are now employed where the parts are normally quenched in the temperature range of 250 to 300°C (482 to 572°F). At this temperature, austenite may remain for a maximum period prior to its transformation to bainite.

Martempering may also be feasible in vacuum, which reduces the thermal gradient, like the modified martempering process. The part held under vacuum at the austenitizing temperature is gas-quenched in the usual manner, but cooling is interrupted above the M_s to achieve temperature equalization in the part, this being followed again by gas-quenching to room temperature.

Advantages

1. Martempering is applicable to a greater range of steel grades than the normal hardening process.
2. Low-temperature quenchant (particularly oil) often allows the utilization of simpler and less expensive quenching equipment.
3. It greatly reduces the problems of pollution and fire hazard when fluidized beds or nitrate/nitrite salt bath (with recovery of the salts from wash water) is used.

4. It eliminates the need for quenching fixtures, thereby reducing the cost of tooling and handling.

5. Uniform and reproducible results are secured by both the normal and modified martempering process. However, more severe distortion of sensitive parts is likely with the modified process, which requires grinding or other finishing allowances.[17]

The advantage over the austempering process is that martempering can be used to harden those steels that cannot be hardened by austempering because of the very long time involved in complete transformation to bainite. A harder transformation product (martensite) is produced instead of bainite.

The advantages over through-hardening steels for carburized parts (especially splined shafts, cams, and gears) are realized because these parts do not require grinding or costly fabrication, due to fewer dimensional changes.[114]

Applications. This treatment is applied to oil-hardening and air-hardening (i.e., higher-alloy deep-hardening) steels. It is, however, not applicable to water-hardening tool steels or carburized carbon case-hardening steels unless these are of thin sections (to make them capable of hardening in oil), because the cooling rate in liquid bath is inadequate to prevent pearlitic transformation at higher temperatures of 500 to 600°C (932 to 1112°F). It is especially suitable for the treatment of irregular-shaped parts made from medium- or high-alloy steels. Steel castings are martempered in molten salts or hot oil baths at temperatures between 150 and 300°C (300 and 572°F).[116,117]

Typical applications of martempering steel parts in salt and oil baths are listed in Tables 13.9 and 13.10, respectively, which describe steel parts normally treated and provide details of martempering methods and hardness requirements.[114]

Modified Martempering. Figure 13.42c[17] shows the modified form of martempering in that the temperature of the quenching bath is held just below the M_s temperature, usually in the 150 to 175°C (300 to 350°F) range. The lower-temperature bath increases quench severity. Hence, a section thickness that is larger than that for normal martempering can be hardened to the required hardness level. This process can be applied to a wide range of steel grades.

Carbomartempering. Carbomartempering involves both carburizing or carbonitriding and martempering processes. It exhibits minimum distortion. If parts are carburized in a gas-fired furnace, they can be quenched directly into a salt bath. If they are carburized in a salt bath, however, they must pass through a neutral salt bath to prevent any violent reactions. The carbomartempering temperature is a compromise between the M_s temperature of the case and that of the core, and is selected not to jeopardize case hardness. It is usually in the 175 to 260°C (350 to 500°F) temperature range, and the holding time is only a few minutes.[67]

13.8 MARTEMPERING OF GRAY CAST IRON

Gray iron castings are quenched from above the A_1 temperature in a salt, oil, lead bath at temperatures slightly above the M_s [200 to 260°C (400 to 500°F) for unalloyed irons] until the entire section has reached the bath temperature, and then

TABLE 13.9 Typical Applications of Martempering Steel Parts in Salt[114]

Part	Grade	Maximum section thickness		Weight		Martempering conditions			Required hardness, HRC
						Temperature of salt		Minimum time in salt, min	
		mm	in.	kg	lb	°C	°F		
Compliant tube	4130	0.8	0.03	0.11	0.25	160[a]	320[a]	5	50[b]
Thrust washer	8740	5.1	0.20	0.05	0.1	230	450	1	52 mm[b]
Chain link	1045	5.6	0.22	0.11	0.25	205[c]	400[c]	1	45–50[b]
Cotton-picker spindle[d]	Type 410	6.4	0.25	0.05	0.12	315	600[e]	1½	44–48[b]
Accessory driveshaft	9310[e]	6.4	0.25	0.45	1.0	190	375	2½	90 (15N scale)
Clutch-adjustment nut	8740	7.6	0.30	0.14	0.3	230	450	2	52 min.[b]
Seal ring	52100	7.6	0.30	0.18	0.4	190	375	10	65[b]
Spur pinion	3312[e]	7.6	0.30	0.23	0.5	175	350	1½	90 (15N scale)
Internal gear	4350	8.9	0.35	0.36	0.8	245	475	2	54 min.[b]
Dual gear[f]	4815[e]	9.4	0.37	2.13	4.7	260	500	2	62–63[b]
Drive coupling	4340	10.2	0.40	0.27	0.6	230	450	2½	52 min.[b]
Spline shaft	8720[e]	10.2	0.40	0.50	1.1	190	375	2½	90 (15N scale)
Arbor sleeve	1117L[e]	10.2	0.40	0.59	1.3	205	400	3	...
Screw-machine spindle	8620[e]	10.2	0.40	6.35	14.0	205	400	3	...
Driving barrel	4350	12.7	0.50	0.45	1.0	245	475	3	48–52[g]
Bearing race[h]	52100	12.7	0.50	13.2	29.2	220	425	2½	63–64[b]
Hog knife	9260	15.2	0.60	8.16	18.0	175[j]	350[j]	15	62[b]
Landing-gear spring	6150	19.1	0.75	14.7	32.5	260	500	2¾	56–57[b]
Internal gear	1117L[e]	25.4	1.00	1.36	3.0	205	400	3	...
Spur pinion gear[k]	4047	25.4	1.00	16.4	36.2	230[c]	450[c]	3	50–52[m]
Screw-machine sprocket	8620[e]	38.1	1.50	9.07	20.0	205	400	3	...

Notes: OD, outside diameter; ID, inside diameter.

[a] Salt contained 1½% water. [b] As-quenched. [c] Salt contained water. [d] 6.4-mm (¼-in.) diameter by 203 mm (8 in.) long. [e] Carburized. [f] 124-mm (4⅞-in.) OD by 32-mm (1¼-in.) ID by 102 mm (4 in.). [g] Final. [h] 224-mm (8¹³⁄₁₆-in.) ID by 251-mm (9⅞-in.) OD. [j] Salt contained 1% water. [k] 19-mm (¾-in.) OD by 92-mm (3⅝-in.) ID by 140 mm (5½-in.). [m] As-quenched hardness of teeth.

TABLE 13.10 Typical Applications of Martempering Steel Parts in Oil[114]

Part	Grade	Maximum section thickness		Outside diameter		Weight		Carburizing temperature, °C	Depth of case		Quenching temperature, °C	Temperature of martempering oil,[†] °C	Surface hardness, HRC
		mm	in.	mm	in.	kg	lb		μm	0.001 in.			
Sleeve	52100	3.2	0.125	...	...	0.1	¼	...	...	...	790	165	58–59
Spacer plate	1065	3.2	0.125	...	...	0.1	¼	...	...	...	790	165	56–57
Bushing	1117	4.8	0.1875	51.0	2.009	0.2	½	910	1015–1220	40–48	910	190	58–62
	1117	6.4	0.25	76.3	3.0034	0.6	1¼	910	1015–1220	40–48	910	190	55–60
Shifter rail	1018	9.5	0.375	...	...	1.0	2⅛	845[‡]	255–455	10–18	845	165	55–60
	1018	9.5	0.375	...	...	1.4	3⅛	845[‡]	355–610	14–24	845	165	55–60
Spur gear	8620	12.7	0.5	320.6	12.620	12.7	28	925	1145–1525	45–60	845	150	55–60
Helical gear	4620H	19.1	0.75	331.5	13.050	16.9	37.2	925	760–1015	30–40	845	150	58–63
Herringbone gear	4820	19.1	0.75	283.2	11.150	16.3	36	925	1145–1525	45–60	845	150	55–61
Shifter rail	1141	25.4	1.0	25.4	1.0	0.9	1⅞	885[‡]	455–660	18–26	885	165	45–50
Spiral bevel gear	4620	25.4	1.0	210.6	8.29	5.1	11.25	925	1015–1270	40–50	845	150	55 min.
Helical pinion	8617H	25.4	1.0	35.8	1.409	0.4	0.9	925	510–710	20–28	845	150	58–63
Spur gear	8625	31.8	1.250	83.8	3.300	4.3	9⅜	925	1525–1725	60–68	925	190	58–62
	4817H	34.0	1.340	186.7	7.350	8.6	19	925	1400–1780	55–70	845	150	58–63
	8625	38.1	1.500	165.1	6.500	2.5	5½	925	1525–1725	60–68	925	190	58–62
Splined shaft	8625	39.7	1.564	39.7	1.564	2.7	5⅞	925	1400–1980	70–78	925	190	58–62
Spur gear	8625	44.4	1.750	108.0	4.250	3.5	7¾	925	1525–1725	60–68	925	165	58–62
Splined shaft	8625	44.4	1.750	44.4	1.750	2.0	4½	925	1525–1725	60–68	925	190	58–62
	8620	50.8	2.000	50.8	2.000	5.1	11¼	925	1525–1725	60–68	925	165	58–62
	8625	65.0	2.559	65.0	2.559	6.8	15	925	1525–1725	60–68	925	165	58–62
Spur gear	8625	84.7	3.3343	245.5	9.667	11.9	26¼	925	1525–1725	60–68	925	190	58–62

[†] Minimum time in oil, 5 min.
[‡] Carbonitriding temperature.

they are air-cooled to room temperature. A typical application includes cylinder liners (sleeves) for diesel and heavy-duty gasoline engines.[109]

13.9 AUSTEMPERING OF STEEL

Austempering, a hardening process, is the isothermal transformation of austenite into bainite. This process was first introduced by Bain and Davenport in the 1930s[118] and consists essentially of

1. Heating the steel part to the hardening or austenitizing temperature, usually between 790 and 915°C (1450 and 1675°F) in a molten nitrate-nitrite salt bath or in a controlled-atmosphere furnace and holding there for a proper length of time
2. Quenching in a molten lead or salt bath, vigorously agitated and maintained at an appropriate isothermal temperature, usually in the range of 260 to 400°C (500 to 752°F), and holding at bath temperature for a sufficient time to transform to fully bainitic structure
3. Cooling down to room temperature, usually in still air

The procedure is shown schematically in Fig. 13.45. The actual temperature of isothermal transformation is dependent upon the hardness and properties desired and the transformation characteristics of the steel being processed, as indicated in the TTT diagrams. One example of the influence of the low temperature of a salt bath is that it produces a high hardness of the resulting microstructure. The time in the quenching bath, usually varying from 20 min to several hours, depends entirely on the transformation temperature used, the steel grade, and the degree of hardness desired. Bath agitation, accomplished by mechanical stirring, pumping, and air agitation, can be a vital factor in austempering due to its increased quenching speed. The quenching severity of nitrate-nitrite salt can also be increased by careful and

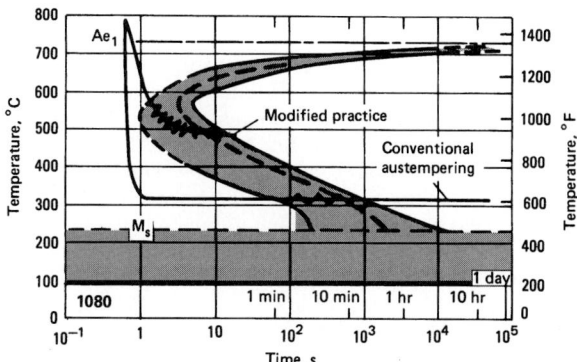

FIGURE 13.45 TTT diagram for 1080 steel showing a difference between conventional and modified austempering.[17] When applied to wire, the modification shown is called *patenting*. (*Reprinted by permission of ASM International, Materials Park, Ohio.*)

TABLE 13.11 Quench Severity Comparison for Salt Quenches[119]

Numbers given are estimated Grossman H–values.

Agitation	At temperature:	
	180°C (360°F)	370°C (700°F)
Still and dry	0.15–0.20	0.15
Agitated and dry	0.25–0.35	0.20–0.25
Agitated with 0.5% water	0.40–0.50	0.30–0.40
Agitated with 2% water	0.50–0.60	0.50–0.6[†]
Agitated with 10% water	0.90–1.30[†]	Not possible

[†] Requires special enclosed quenching apparatus

TABLE 13.12 Effect of Transformation Temperature on Hardness and Strength of Austempered 0.8% C Steel[120]

Rockwell C hardness	Approximate tensile strength, kpsi	Approximate transformation temperature, °C (°F)
40	180	400 (750)
45	210	343 (650)
50	240	316 (600)
55	280	271 (520)
58	300+	260 (400)

Reprinted by permission of Fairchild Publications, New York.

periodic water addition, which involves agitation of salt to disperse water uniformly (Table 13.11).[119] The rate of the last cooling stage (item 3) is not significant. Tempering after austempering may or may not be necessary because much depends on the properties desired in service conditions. However, if it is carried out, further improvement in ductility occurs by removal of carbon from any retained austenite.

Table 13.12 shows the transformation temperature of 0.8%C steel austempered to various hardness or strength levels.[120] Note that where high hardness and ductility are the prime considerations, it is best to austenitize at the highest temperature possible to dissolve all the carbides in the solution. Where both wear resistance and minimum distortion are required, such as in 1095 steel cutting edges of electric razors, a just-sufficient amount of carbides should be in solution during austenitizing, to meet the hardness by a bainitic reaction but to leave some undissolved carbides in it for wear resistance. Where minimum distortion is desired, austenitization is carried out at the lowest temperature possible and quenching into a bath at the highest temperature possible to meet the required hardness level.[120]

The mechanical properties of sway bars made of 1090 steel and hardened by austempering and quenching and tempering process are listed in Table 13.13. It is important to note that the austempered parts have desired mechanical properties in that they have a 100% bainitic structure. Table 13.14 lists section sizes of austempered parts made of various steels in which the formation of some pearlite or martensite is allowed in the microstructure.

TABLE 13.13 Comparison of Typical Mechanical Properties of Austempered and of Oil-Quenched and Tempered Sway Bars of AISI 1090 Steel[119]

Property[a]	Austempered at 400°C (750°F)[b]	Quenched and tempered[c]
Tensile strength, MPa (ksi)	1415 (205)	1380 (200)
Yield strength, MPa (ksi)	1020 (148)	895 (130)
Elongation, %	11.5	6.0
Reduction of area, %	30	10.2
Hardness, HB	415	388
Fatigue cycles[d]	105,000[e]	58,600[f]

  [a] Average values.
  [b] Six tests.
  [c] Two tests.
  [d] Fatigue specimens 21 mm (0.812 in.) in diameter.
  [e] Seven tests; range, 69,050 to 137,000.
  [f] Eight tests; range, 43,120 to 95,220.

TABLE 13.14 Hardness of Various Steels and Section Thicknesses of Austempered Parts[119]

Steel	Section size		Salt temperature		M_s temperature[a]		Hardness, HRC
	mm	in.	°C	°F	°C	°F	
1050	3[b]	0.125[b]	345	655	320	610	41–47
1065	5[c]	0.187[c]	[d]	[d]	275	525	53–56
1066	7[c]	0.281[c]	[d]	[d]	260	500	53–56
1084	6[c]	0.218[c]	[d]	[d]	200	395	55–58
1086	13[c]	0.516[c]	[d]	[d]	215	420	55–58
1090	5[c]	0.187[c]	[d]	[d]	...	...	57–60
1090[e]	20[c]	0.820[c]	315[f]	600[f]	...	...	44.5 (avg)
1095	4[c]	0.148[c]	[d]	[d]	210[g]	410[g]	57–60
1350	16[c]	0.625[c]	[d]	[d]	235	450	53–56
4063	16[c]	0.625[c]	[d]	[d]	245	475	53–56
4150	13[c]	0.500[c]	[d]	[d]	285	545	52 max.
4365	25[c]	1.000[c]	[d]	[d]	210	410	54 max.
5140	3[b]	0.125[b]	345	655	330	630	43–48
5160[e]	26[c]	1.035[c]	315[f]	600[f]	255	490	46.7 (avg)
8750	3[b]	0.125[b]	315	600	285	545	47–48
50100	8[c]	0.312[c]	[d]	[d]	...	...	57–60

  [a] Calculated.
  [b] Sheet thickness.
  [c] Diameter of section.
  [d] Salt temperature adjusted to give maximum hardness and 100% bainite.
  [e] Modified austempering; microstructure contained pearlite as well as bainite.
  [f] Salt with water additions.
  [g] Experimental value.

Advantages. The advantages include the following:

1. There is reduced distortion and cracking. The danger of retained austenite is also reduced. Hence this is a process of considerable importance where it is desired to harden intricate-shaped components successfully and safely.

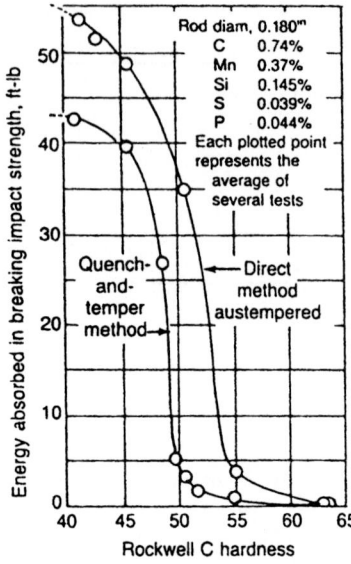

FIGURE 13.46 Comparison of impact toughness of a carbon steel after conventional hardening and tempering and after austempering, as a function of hardness.[124] (*Courtesy of Marcel Dekker, Inc., New York.*)

2. Ductility, toughness, and strength are increased at a given hardness value in the hardness range of Rc 47 to Rc 55. Alternatively, the material can be given an austempering treatment to achieve the highest hardness and tensile strength but a toughness similar to that of tempered martensite.[121] Figure 13.46 shows a comparison of the impact toughness produced by conventional quenching and tempering and austempering for 0.74% carbon steel and SAE 6150 steel.

3. Uniformity and consistency of properties throughout the section component are obtained.

4. There is a less heat treatment breakage.

5. These parts have high fatigue resistance; they often have a life 3 to 4 times longer than those of comparable quenched and tempered parts. Figure 13.47 shows the fatigue diagram of DIN 30SiMnCr4 steel after conventional hardening and tempering and after austempering. Note that the increase in fatigue strength is significant for notched specimens.

6. An austempered carbon steel wire has superior bendability to that hardened and tempered to Rc 50.

7. There are claims of less hydrogen embrittlement.

8. Much cleaner quench and less surface contamination are found.

9. Austempering in a protective atmosphere furnace with integral salt quenches has been designed to prevent air contact throughout the processing, which produces a clean and shiny attractive surface.[122] In this situation, austempered parts can, therefore, be immediately painted, phosphated, or zinc-plated. However, oil-quenched and tempered parts need pickling treatment prior to these finishing operations.[123]

10. It allows the use of inexpensive, lower-grade materials without sacrificing mechanical properties, thereby reducing production costs. For example, high-

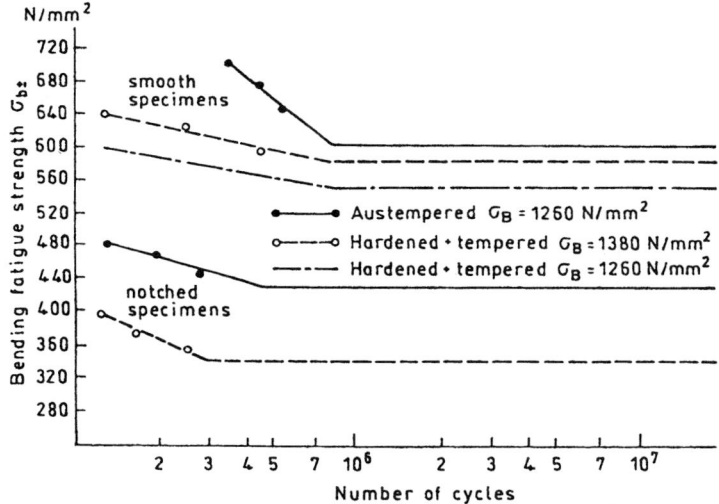

FIGURE 13.47 Bending fatigue strength of DIN 30SiMnCr4 steel after conventional hardening and tempering and after austempering. [*Source: F. W. Eysell, Die Zwischenstufenvergutung und ihre betriebliche Anwendung, Z. TZ Prakt, Metallbearb., 66:94–99, 1972 (in German).*]

tensile-strength bolts made of quenched and tempered 8640 steel can be replaced with austempered lower-grade 10B39 steel, resulting in a 40% saving on material costs alone while maintaining the 1724-MPa (250-ksi) minimum tensile strength and 17% elongation.[122]

11. It is a single-operation process and requires no tempering.

Limitations

1. Since tensile strength and impact values are lower than those of the hardened and tempered product, this process is not used in the case of low- and medium-carbon steels, except when risk of cracking and distortion is to be avoided. It appears to be more beneficial in steels with carbon content varying between 0.5 and 1.2%.

2. It is restricted to thin sections for plain carbon and low-alloy steels (Table 13.14), i.e., limited to those cross sections of steel which can be cooled from austenitizing temperature to austempering bath temperature with sufficient rapidity to prevent fine pearlite formation. It is common practice to austemper plain carbon steels up to a maximum diameter of 5 mm (0.2 in.) and certain low-alloy structural steels [e.g., spring steels up to 10-mm (0.4-in.) section thickness]; this is evidently dependent on the dissolved alloy content and grain size. Lower-carbon, boron-bearing steels, however, can be successfully austempered in heavier sections. In some alloy steels, section thickness up to about 25 mm (1 in.) can be satisfactorily austempered to fully bainitic structures.[119] The range of suitable sizes is wider for medium-alloy steels. With high-alloy steels, heavier sections (up to 25-mm, or 1-in., cross section) can be austempered, but an inordinately long time (sometimes several hours, days, or weeks) would be necessary to produce completely the bainitic structure. The austempering of such steel would be wholly uneconomical and impractical. Hence this process is limited to small tools where exceptional toughness together with reasonable hardness is desired.

3. This is a slower and somewhat more expensive process than conventional quenching and tempering and requires closer supervision to maintain a salt bath at a suitable temperature in the range of 260 to 400°C. However, when fully automated, it is capable of being cost-competitive with "quench and temper" systems as a result of labor savings of 50%.

Since most austempering is accomplished in molten nitrate-nitrite salt, the salt bath used for austenitizing must be compatible with the austempering salt. Hence, a chloride salt bath for austenitizing should have the following composition and characteristics: 40 to 55% NaCl + 45 to 55% KCl, melting temperature 650 to 675°C (1200 to 1250°F), working temperature range of 705 to 900°C (1300 to 1650°F).[119]

Recently a "supersaturation" method has been developed for salt bath quench on a high-production batch furnace. In this process, agitation—when combined with water addition up to 12% of the quench at 182°C (360°F)—raises the quench severity of the molten salt to a level equal to that of a brine solution. This allows austempering of parts up to 6 in. thick.[122]

Plain carbon steels containing (1) 0.5 to 1% carbon and at least 0.6% Mn, (2) more than 0.9% C and a little less than 0.6% Mn, and (3) less than 0.5% C and 1 to 1.65% Mn as well as low-alloy steels, such as 5140, 4140, 6145, and 9440, are well suited to austempering due to their high hardenability.[123]

It is not feasible to austemper AISI 1034 steel due to the extremely fast pearlite reaction at 540 to 595°C (1004 to 1103°F). The AISI 9261 steel is unsuitable to austempering treatment due to very slow bainite formation at 260 to 400°C (550 to 752°F).[124]

Comparison of Processes and Products. Table 13.15 shows the scheme of operation of austempering and of the oil-hardening and tempering process.[125] Although no tempering operation is involved in the austempering process, some gas input is necessary to maintain the quench bath at a temperature in the region of 260 to 400°C. However, it is considered good practice to temper austempered parts at

TABLE 13.15 Austempering versus Oil-Hardening and Tempering: Cost Compared[125]

Harden and temper		Austempering	
Process	Cost input	Process	Cost input
1. Heat components to hardening temperature	Fuel to raise temperature	Heat components to hardening temperature	Fuel to raise temperature
2. Quench (to room temperature)	Cost of oil + dragout + circulation + cooling	Quench (to 320°C)	Cost of salt + dragout + circulation + fuel to maintain temperature
3. Wash	Water heating	Wash	Water heating
4. Tempering	Fuel to raise temperature		
5. Cleaning (option)	Shot blast/oil dip		

TABLE 13.16 Comparison of Properties of Retaining Rings for 1080 Steel[126]

Property	Oil-quenching and tempering	Austempering
Tensile strength	254 ksi	260 ksi
0.2% yield stress	237 ksi	246 ksi
Percent elongation	3.0	4.6
Impact strength	0.4 ft·lb	1.9 ft·lb
Fatigue life, bend test	60,000 cycles	98,000 cycles

the quench temperature to transform any retained austenite present. Thus some saving in fuel can be made here. It should be pointed out that workpieces are going into an austempering bath at a higher temperature than the bath itself, whereas in tempering the work is heated from cold.

In a tempering treatment, the heat remains on full for 1 hr per 1.5-hr cycle; in austempering, however, the heat remains on full for 0.25 hr per 1.5-hr cycle. In the case of gaseous fuel, a savings of 25% on gas consumed is expected, and savings of as much as 75% can be made on the cost of the quenching medium if austempering operations are practiced instead of oil-quenching and tempering.

In a quenching operation, the costs involved are the cost of the quenching medium and dragout since the circulation cost is common in both. Another real savings in cost can be made in the salt bath austempering operation in the area of dragout.[125]

Table 13.16 shows a comparison of mechanical properties of retaining rings for 1080 steel with the composition range of 0.75 to 0.85% C, 0.65 to 0.85% Mn, 0.03% S and P maximum, and 0.35% Si maximum; with ASTM grain size Nos. 5 to 8.[126]

Application. Table 13.17 provides processing data for numerous parts made of various plain carbon, alloy, and carburized steels. These data are for representative austempering practices used by more than a dozen manufacturing plants.[119]

Modified Austempering. Modified austempering, also called *patenting*, is the well-accepted practice in the wire industry to produce spring wire, rope wire, piano wire, and so forth. The main purpose of this process is to convert the coarse, nonuniform ferrite/pearlite or carbide/pearlite structure of hot-rolled wire rod into a uniform structure of varying amounts of fine pearlite and bainite, to facilitate drawing of high tensile strengths accompanying large reduction of area without fracture. The production process consists in heating the wire or rod to a temperature above the Ac_3 point (850 to 1100°C) to enable it to be completely austenitized, followed by continuous quenching of the wire by passing it in air or lead or salt bath maintained at a temperature between 500 and 550°C (932 and 1022°F) according to the composition; holding in this bath for periods varying from 10 s (for small wire) to 90 s (for rod); and then cooling in air to ambient temperature. The process is described by the curve superimposed on the typical TTT diagram (Fig. 13.45).

The modified process can be applied to plain carbon steel and low-alloy steels having section thickness larger than normally considered practicable for austempering.[127]

During drawing, C atoms must abandon cementite to find favorable sites at dislocation. This dissolved the C pins dislocations, which act as a barrier to further

TABLE 13.17 Typical Production Applications of Austempering[119]

Parts listed in order of increasing section thickness.

Part	Steel	Maximum section thickness		Weight or parts per unit weight		Salt temperature		Immersion time, min	Hardness HRC
		mm	in.	kg or kg^{-1}	lb or lb^{-1}	°C	°F		
Plain carbon steel parts									
Clevis	1050	0.75	0.030	770/kg	350/lb	360	680	15	42
Follower arm	1050	0.75	0.030	412/kg	187/lb	355	675	15	42
Spring	1080	0.79	0.031	220/kg	100/lb	330	625	15	48
Plate	1060	0.81	0.032	88/kg	40/lb	330	630	6	45–50
Cam lever	1065	1.0	0.040	62/kg	28/lb	370	700	15	42
Plate	1050	1.0	0.040	0.5kg	1/4 lb	360	675	15	42
Type bar	1065	1.0	0.040	141/kg	64/lb	370	700	15	42
Tabulator stop	1065	1.22	0.048	440/kg	200/lb	360	680	15	45
Lever	1050	1.25	0.050	...	...	345	650	15	45–50
Chain link	1050	1.5	0.060	573/kg	260/lb	345	650	15	45
Shoe-last link	1065	1.5	0.060	86/kg	39/lb	290	550	30	52
Shoe-toe cap	1070	1.5	0.060	18/kg	8/lb	315	600	60	50
Lawn mower blade	1065	3.18	0.125	1.5kg	2/3 lb	315	600	15	50
Lever	1075	3.18	0.125	24/kg	11/lb	385	725	5	30–35
Fastener	1060	6.35	0.250	110/kg	50/lb	310	590	25	50
Stabilizer bar	1090	19	0.750	22kg	10lb	370	700	6–9	40–45
Boron steel bolt	10B20	6.35	0.250	100/kg	45/lb	420	790	5	38–43

Alloy steel parts

Part	Steel								
Socket wrench	6150	...	...	0.3 kg	$^1/_8$ lb	365	690	15	45
Chain link	Cr-Ni-V[†]	1.60	0.063	110/kg	50/lb	290	550	25	53
Pin	3140	1.60	0.063	5500/kg	2500/lb	325	620	45	48
Cylinder liner	4140	2.54	0.100	15 kg	7 lb	260	500	14	40
Anvil	8640	3.18	0.125	1.65 kg	$^3/_4$ lb	370	700	30	37
Shovel blade	4068	3.18	0.125	...	...	370	700	15	45
Pin	3140	6.35	0.250	100/kg	45/lb	370	700	45	40
Shaft	4140[‡]	9.53	0.375	0.5 kg	$^1/_4$ lb	385	725	15	35–40
Gear	6150	12.7	0.500	4.4 kg	2 lb	305	580	30	45

Carburized steel parts

Part	Steel								
Lever	1010	3.96	0.156	33 kg	15 lb	385	725	5	30–35[§]
Shaft	1117	6.35	0.250	66/kg	30/lb	385	725	5	30–35[§]
Block	8620	11.13	0.438	132/kg	60/lb	290–315	550–600	30	50[§]

[†] Contains 0.65 to 0.75% C.
[‡] Leaded grade.
[§] Case hardness.

dislocation motion. A high rate of dislocation pinning causes a very high rate of work-hardening.[128]

Carboaustempering.   Carboaustempering involves both carburizing or carbonitriding and austempering processes. Austempering of low-carbon-content steel after carburizing produces a high-carbon bainitic case and either a bainitic or martensitic core, depending on quench severity and steel composition. Carboaustempered parts possess increased fatigue strength and wear resistance, and they are dimensionally and functionally superior to carburized and conventionally quenched components.[67]

13.10 AUSTEMPERING OF DUCTILE IRON

Austempered ductile iron (ADI) is currently the subject of extensive study worldwide because of the development of the best combinations of low cost, design flexibility, good machinability, high strength-to-weight ratio, and good toughness, fatigue resistance, and wear resistance.[129] Table 13.18 lists the five standard ADI grades with minimum property requirements, according to the ASTM-A897M standard. Austempering of ductile iron consists of austenitizing components at a temperature usually between 815 and 955°C (1500 and 1750°F) for nearly 1.5 hr in salt bath, controlled endothermic atmosphere, or fluidized bed, followed by fast-quenching into a low-temperature quenchant such as molten salt bath, hot oil, or fluidized bed maintained isothermally in the bainitic transformation (or austempering temperature) range [205 to 450°C (400 to 842°F)], holding them isothermally for a specified time between 0.5 and 4 hr to complete the transformation; subsequently they are aircooled.[130-134] Figure 13.48a and b shows a schematic diagram of the austempering heat treatment cycle (A through H) and the typical austempering cycles for different grades of ADI, respectively.[129,134a]

Note that where limited facilities (with respect to the above-mentioned furnaces) are available, proprietary stop-off compounds can be used when heat-treating in air because these compounds effectively prevent surface degradation and can be easily removed after isothermal treatment.

Maximum thicknesses for complete austempering of unalloyed ductile irons are 15 mm at a temperature of 450°C and 30 mm at a temperature of 250°C. As section thickness increases beyond these values, the amount of pearlitic transformation increases.[131]

When quench severity is improved by the addition of 0.2 to 2 wt% water in a salt bath (called a *saturated salt bath*), the section thickness can be increased to ~1 in. in unalloyed irons and up to 2.5 in. (64 mm) thick in alloyed ductile irons. Subsequently it was found that the water content of the salt bath could be increased

TABLE 13.18   The Five Standard ADI Grades (ASTM A897M-1990)

Grade	Tensile strength, MPa	Yield strength, MPa	Elongation, %	Impact energy, J	Typical hardness, BHN
1	850	550	10	100	269–321
2	1050	700	7	80	302–363
1	1200	850	4	60	341–444
1	1400	1100	1	35	388–477
1	1600	1300	N/A	N/A	444–555

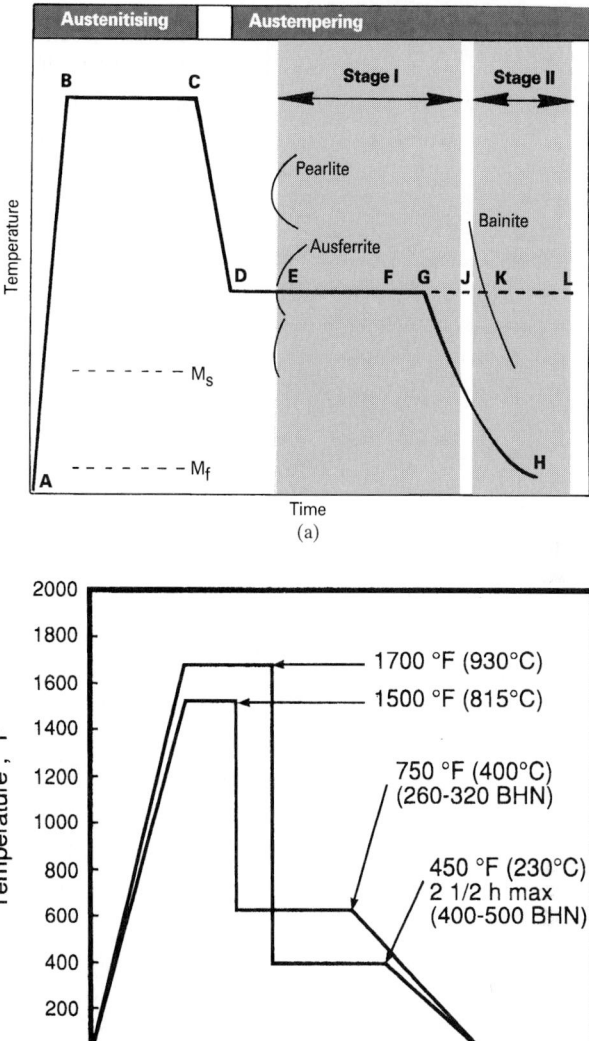

FIGURE 13.48 (*a*) Schematic diagram of the austempering process for a cast iron.[134a] (*b*) Typical austempering cycles for different grades of ADI.[129] [(*a*) *Courtesy of Wolfson Heat Treatment Center, England. (b) Courtesy of QIT, Fer et Titane, Inc., Chicago, Illinois.*]

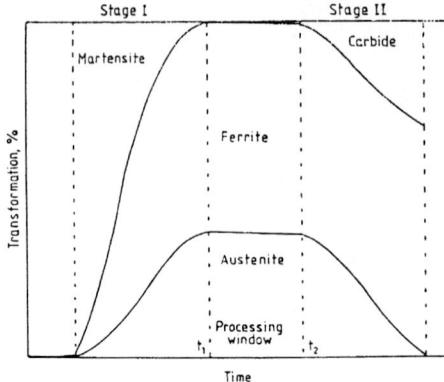

Stage I   Stage II

FIGURE 13.49 Percentage transformation and microstructural changes during stages I and II of austempering and processing window in unalloyed iron for higher austempering temperatures. The amount of high-C, retained austenite during the stage I reaction reaches a plateau representing the end of time t_1. The amount of retained γ starts to fall at time t_2 when the high-C austenite begins to undergo the stage II reaction. This behavior exposes a well-defined processing window $t_2 - t_1$, during which the metastable structure of bainitic ferrite and high-C austenite is relatively stable and the austempered iron exhibits optimum properties.[136]

further from 2 to 12% with the application of high rates of recirculation of the bath. The use of this bath, called *supersaturated salt bath*, increases the quench severity to the extent that 2-in. section thickness can be completely austempered in unalloyed irons. Note that water addition is never made in an open quench tank.[134]

The austempering reaction in ductile iron, over the austempering range of 205 to 450°C, occurs in two stages, as shown in Figs. 13.48a and 13.49. The first-stage transformation starts by the nucleation of bainitic ferrite (at high temperature) at interphase and grain boundaries, and its growth into austenite. This is associated with the carbon rejection from growing ferrite platelets into the surrounding austenite.[135,136] [The high Si content of ductile iron retards the iron carbide precipitation, which results in the carbon enrichment (up to 2%) of austenite, particularly between the growing ferrite plates, and the M_s temperature below −120°C (−180°F).] The carbon enrichment of the reacted austenite occurs (in the time period of EF) up to 1.2 to 1.6%. If the austempering period is extended further (FG), acicular ferrite continues together with additional enrichment of austenite up to 1.8 to 2.2%. This austenite is thermally and mechanically stable and is called *reacted stable austenite*, i.e., the stabilization of (retained) austenite after cooling to ambient temperature. This form of austenite is desirable in the ausferrite structures of grades 1 and 2.[134a] (Unreacted austenite is thermally unstable[†] and transforms to martensite on quenching to room temperature.) The structure of the transformation product corresponding to the first stage, nominally called *upper bainite*, consists of relatively coarse bainitic ferrite plates plus retained austenite (called *carbon-enriched stable austenite*) in the matrix (Fig. 13.50a and b). The reaction can be described as $\gamma \rightarrow \alpha_{bainite}$ (acicular) + $\gamma_{high\ carbon}$. This product is also known as *upper ausferrite*. The ADI transformed at a high austempering temperature (say, 400°C, or 750°F) exhibits high ductility, good fatigue and impact strength, and a tensile strength of 125 to 150 ksi (860 to 1030 MPa).[129]

In stage II with extended austempering period beyond about 2 hr (JKL), high-carbon (reacted stable) austenite further decomposes into the thermodynamically more stable fine acicular structure of bainitic ferrite and carbide and nominally

---

[†] Thermally unstable austenite can be distinguished in a microstructure using a double etchant of nital and sodium metabisulfite. However, it is more difficult to recognize between reacted metastable and reacted stable austenite. It can be differentiated only by a special technique such as heat tinting.[134a]

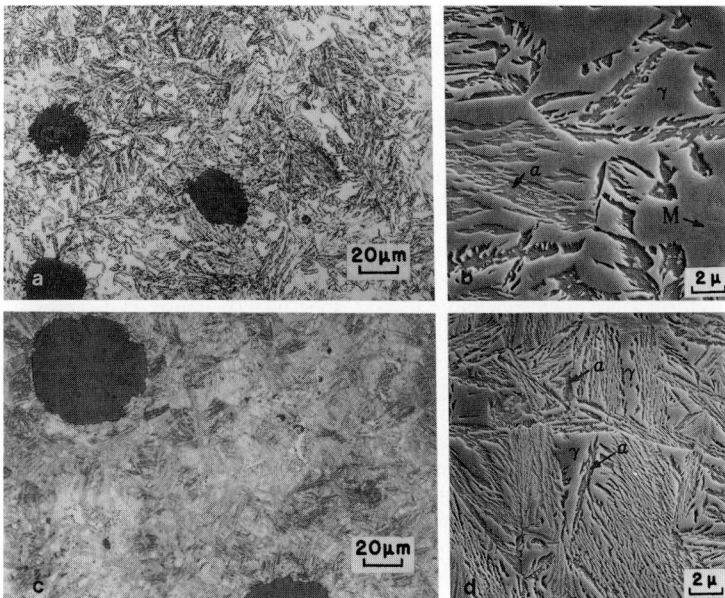

FIGURE 13.50 Light photomicrographs of austempered ductile iron at 370°C (700°F) (*a, b*) and at 315°C (600°F) (*c, d*). Both are austenitized at 900°C (1650°F) but for different times, namely, 125 and 90 min, respectively, corresponding to the higher and lower austempering temperatures. Microconstituents observed are austenite (γ), bainitic ferrite (α_bainite), and martensite (M). (*Courtesy of R. Gundlach and J. Janowak.*[132])

called *lower bainite* or *lower ausferrite*. The reaction can be represented by $\gamma_{\text{high carbon}}$ → α_{bainite} (acicular) + carbide. (See Fig. 13.50.[136]) (A transitional ε-carbide mostly occurs in the initial stages followed by an incoherent Fe_3C carbide. Usually this transformation is deleterious to ADI properties; however, the optimum properties can be achieved if the iron is cooled from the austempering temperature at a time between the two stages.[134a]) The ADI structure, transformed at around 260°C (500°F), has therefore a much finer structure that exhibits high hardness and excellent wear resistance equivalent to the case-hardened steel, a tensile strength >230 ksi (1600 MPa), and reduced ductility and impact strength.[129,137] The structure of this transformation product, nominally called *lower bainite*, contains a smaller amount (up to 15%) of stabilized austenite (Fig. 13.50*c* and *d*). In some cases (as seen in the above microstructures), complete stabilization of austenite is not found to occur, and the austempered structures invariably contain up to 12% martensite.

Untransformed low-carbon austenite from the stage I reaction, which is thermally unstable and readily transforms into martensite in the austempered structure, and the brittle carbide from the stage II reaction have deleterious effects on mechanical properties, especially ductility and toughness and fatigue strength.[136,137a] The important mechanical properties of ADI are realized when the quantity of these phases in the microstructure is small or inadequate to be deleterious to the mechanical properties.[138] Increasing the carbon content of austenite favors the stability of retained austenite, thereby eventually diminishing the martensite formation.

A marked reduction in ductility and toughness results when incomplete transformation occurs, i.e., when parts are quenched too early from the austempering temperature or when the parts are held for prolonged periods because the former produces varying amounts of martensite and retained (stabilized) austenite after cooling to ambient temperature and the latter induces undesirable transformation/decomposition of the stabilized austenite to form additional ferrite and carbide.[139]

It is clear that the two structures obtained from austempering ductile cast iron and steel and the accompanying properties are quite different; therefore it is appropriate to term them *ausferrite* and *bainite*, respectively, because ADI consists of two phase mixtures comprising acicular ferrite and high-carbon austenite rather than ferrite and carbide, in steel. Moreover, the ausferritic reaction in ductile iron is remarkably slower than that of bainitic reaction in steel (Fig. 13.51).[132] The presence of Si in ductile iron suppresses the formation of carbides. Austempering reaction in ductile iron is more sensitive to variations in the metal chemistries, times, and temperatures. It is the morphology of this austenite-ferrite structure that offers ADI its remarkable properties (Fig. 13.52).[140]

Based on the microstructure, transformation characteristics, and properties of the transformation product, ADI can be classified into the following grades, presently used commercially.

Grades 4 and 5. Ductile irons transformed from high austenitizing temperature to low austempering temperatures (that is, 205 to 235°C) represent lower bainite (ausferrite). These irons, containing lower amounts of retained austenite and higher volume fraction of finer acicular ferrite, have high hardness (400 to 500 BHN or 45 to 50 Rc) and high tensile and fatigue strengths and wear resistance, but limited ductility and fracture toughness.[129,134,137a]

Grades 1 and 2. Ductile irons transformed at higher austempering temperatures in the range of 330 to 450°C exhibit upper bainite (ausferrite) and contain large amounts of retained austenite (~20 to 50%) and ferrite. These irons possess hardness in the range of 260 to 300 BHN (29 to 35 Rc). They have excellent toughness, ductility, and fatigue strength, and good machinability. Typical grade 2 irons are used for applications requiring torsion stress, high impact loading, and high cycle fatigue life.

Process Control. Process control includes the consistent production of high-quality ductile irons and heat treatment process control. Austempering cannot be expected to improve the properties of poor-quality castings consistently. ADI castings should therefore be produced free from surface defects and segregation and should possess the microstructural features such as freedom from carbides and porosity, high nodule count (100 nodules/mm² or higher), good nodularity (80% type I and II nodules), uniform distribution of nodules, minimum inclusion content, and controlled pearlite/ferrite ratio, if specified.[129] Ductile irons require strict control of not only the austempering treatment parameters such as austenitizing temperature and time, austempering temperature and time, and transformation cooling rate, but also the materials-related parameters such as composition, hardenability, and dimensional control.[132,139]

When selecting the composition, the first consideration includes the limiting elements which adversely affect the casting quality through the production of non-spheroidal graphite, the improvement of shrinkage, and the formation of carbides

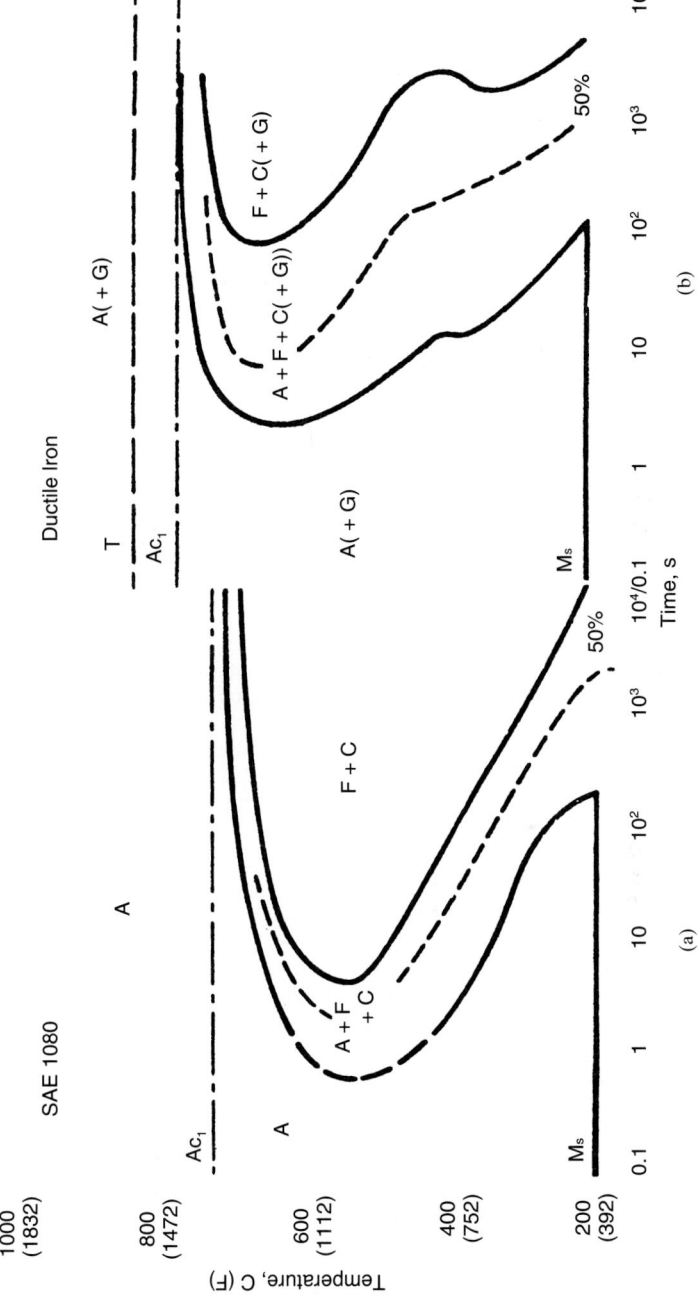

FIGURE 13.51 TTT diagrams for (*a*) SAE 1080 steel (0.79C–0.76Mn) and (*b*) ductile iron (3.3C–2.5Si–0.29Mn) showing the difference in bainitic reaction between the two. Both were austenitized at 900°C (1650°F).[132] (*Reprinted by permission of ASM International, Materials Park, Ohio.*)

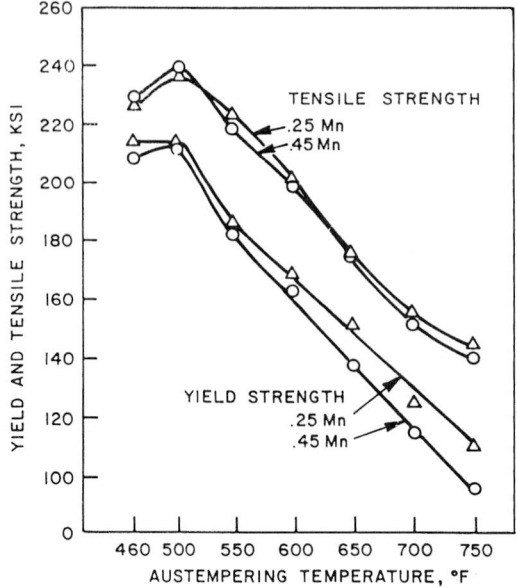

FIGURE 13.52 Yield and tensile strengths in ADI castings.[140]

and inclusions. The second consideration should be given to the control of carbon, silicon, and major alloying elements that control the hardenability of the iron and the properties of the transformed microstructure. When determining alloy requirements, both the section thickness and the severity of austempering quench must be given importance. For heavy section castings, selective alloying is needed to prevent pearlite/ferrite formation. For light section castings (up to $\sim\frac{3}{8}$ in., or 10 mm), a very rapid quench is normally adequate to avoid the pearlite/ferrite formation in even an unalloyed iron.[129]

Advantages of ADI

1. They have excellent mechanical properties, such as high strength, toughness, and wear resistance and through-hardenability, which are superior to those of quenched and tempered structures.[139] The new tougher material allows iron foundries to compete with heat-treated or high-alloy steels where greater reliability is in demand. In this case, the ADI is used as an upgrade for standard ductile iron parts and can also replace Ni hard in some applications.[141]

In the dry-sliding wear test mode, ADI wear resistance is about 4 times greater than that of pearlitic ductile iron (PDI) grade 100-70-03, more than 12 times that of leaded-tin bronze, and nearly 14 times that of aluminum bronze.[141a]

In abrasive wear tests, ADI exhibits equivalent wear resistance to that of AISI 4340 steel, nearly twice less than that of hardened and tempered AISI 1050 carbon steel, and remarkably greater than that of white and alloyed cast irons.[141a]

The higher hardness grades of ADI are used in wear applications while lower hardnesses are used in structural applications. When used as a substitute for forged/cast steel parts, the ADI parts exhibit fatigue strengths greater than those of forged steels and require an increased stiffness and larger fillet radii than do steel parts, to prevent stress concentration.[141]

2. No other single engineering materials available today show the range of properties illustrated by ADI (Fig. 13.52), because of its energy effectiveness, material conservation, better castability to near net-shape, reliability, and increasingly competitive world market.

3. ADI can be subjected to (a) work-hardening treatment of retained austenite under stress in the surface layers and (b) strain-induced transformation hardening of retained (enriched) austenite into martensite, producing a localized increase in volume and compressive stresses in the transformed material by surface layer deformation[142] by using surface treatments such as shot-peening and rolling,[129] which have been shown to produce inhibition of crack formation and an increase in surface hardness and, therefore, very high fatigue (bending) strength, endurance ratio, and rolling and sliding wear resistance. As a result of this hardening feature, care must be exercised in the sequence of operations on machined parts. The higher-hardness grades should be machined prior to heat treating whereas lower-hardness grades are best machined after heat treatment.[141]

4. ADI gears provide better noise damping, higher thermal conductivity, and far superior machinability compared to austempered forged steel gears.[143]

5. The excellent wear resistance of ADI (i.e., twofold improvement over steel at the same hardness level) is due to the presence of retained γ in the microstructure, transformation of high-carbon γ to martensite occurring in the surface layers during the wear tests, and the lubrication provided by the graphite content of the material.[141a,144] However, this property certainly causes problems with final machining operation of the heat-treated product.

6. ADI parts specified by ASTM A897 outperform steel castings, forgings, and fabrications (weldments) in both structural and wear applications.[141] They can replace wear-resistant steels and cast irons.

7. There have been significant cost and weight savings (10% lighter than steel), ease of machining, ready availability, and inexpensive raw materials.[144]

8. Austempered chilled ductile iron containing increased Ni and Mo contents is used as gears and pinions, crankshafts, driveshaft yokes, and related components to replace forged and case-hardened steels.[145]

9. ADI is 3 times stronger and weighs 2.6 times more than the best cast or forged aluminum. Since it has twice the stiffness, a properly designed ADI is a potential candidate to replace an aluminum part as a weight savings.[137]

10. Typically, an ADI part consumes 50% less energy than a steel casting and about 80% less energy than a steel forging.[137]

Applications. *Automotive applications* include camshafts, crankshafts (for high-speed diesel engines), transmission shafts, connecting rods and pistons (for diesel engines), timing gears, steering knuckles, CV joints, differential gears, differential housings, mounting brackets, chassis components, suspension components, spring hangers, spring stops, U-bolt stop plates, pintle hooks, pump rings, rocker arms, wheel hubs, axle structures, snowplow shoes, ring and pinion gears, bevel gear (with hypoid gearing) in rear-axle drives of light- and medium-duty trucks, and forklift truck parts.[144]

Railroad applications include car wheels, bogie and rail wheels, top caps, wear shoes, nipper and gauging hooks, shock absorbers, track plates and hardware, latches, covers, tie bars, engine parts, railway line components, suspension parts, and so forth.

Agricultural applications include plow and till points, eyebolts, wheel hubs, wear plates, chisels, steering shafts, tow hooks, plow shears and tips, slip clutch parts, fertilizer knives, soil turners, and so forth.

Military applications include projectiles, armor, track shoes and plates, track guides, engine rotors, suspension and engine components, end connectors, struts, projectiles, and so forth.

Construction and mining applications include digger teeth, grader blades, pavement breakers, carrier ring, guides and rollers, gears, housings, structural parts, slides, yokes, connectors, structural members, collars, drag conveyors, highway and pole line hardware, snowplow shoes, pitman arms and differential cases, mining drilling heads, track wheels, draw rolls, cam tracks, shift forks, and crusher components (in excavation).[144]

Industrial applications include gears (face, spur, bevel, and bull), brackets, grates, tool holders, jackhammer housings, paper mill components, parts for paper cutting industry, parts for printing machinery, water pump components, textile mill components, textile machinery-cam guides, steel mill rollers, steel rolling mill guides, rolls for pressing dies, bogie pins, cams, cover plates, shredder knives, high-volume repetitive punch for canning industry, parts for bottle-mold industry, mixer blades and worm wheels for food stuffs, crane support wheels, conveyor chain links, links for chain manufacturers, power handling parts, air compressor pump parts, shot-blast parts, gear wheels and gearbox parts for building equipment and stationary transmission units, differential spiders and spring seats, high-speed punching dies, gear wheels of stationary power units made of low-alloy (Cu-Mo) ductile iron weighing 30 kg to 2 tonnes and 300 to 2000 mm in diameter, planetary gear units as a substitute for case-hardened steel, joint sleeves for locomotives, cranes, and other hoisting equipment, on-site drill/milling components, segmented girth rings up to a diameter of 12 m, dolly wheels, driveshaft yokes, stub shafts, starter clutches, and others.[122,137,144,146]

13.11 QUENCH CRACKING

Usually, quench cracking follows austenite grain boundaries; i.e., it is intergranular, but it does not appear to be related to the austenite grain size. Although a majority of the quench cracking is associated with the quenchants, notably their excessive quench severity, many of the cases can be produced by material or mechanical flaws. To avoid quench cracking of steel, the proper quenching media should be selected for the range of alloys and type of heat treatment to be accomplished. It is also essential to control the bath temperature and flow characteristics of the quenching media. It is thus recommended to accomplish metallurgical analysis to establish the true cause of the problems.[147] (See also Sec. 17.4.) The potential sources contributing to quench cracking are given below.[147]

1. *Nonuniform quenching* is due to poor system design, inappropriate racking of parts to be quenched (which restricts the quenchant flow and uniform heat extraction), or incompatible quenchant contamination (water-in-oil or oil-in-water), air entrainment and excessive foaming, etc. All these factors have the ability to produce increased thermal gradients, leading to quench cracking.

2. *Prior steel structure*, e.g., cast, cold-formed, forged, extruded, etc., may increase the likelihood of cracking during the quench. Each as-formed structure needs a certain time and temperature cycle to condition the material for proper hardening.

For example, homogeneous cast structures, normalized and annealed cold-formed structures, and grain-refined forged structures (by normalizing) reduce the potential for cracking during quench.

3. *Nonuniform heating, localized overheating, exceptionally high austenitizing temperature*, and *grain coarsening* have the potential of cracking due to the presence of quench stresses. Overheating for hardening provides coarse grain size that leads to cracking due mostly to lower toughness.

4. *Excessive heating rate* to the austenitizing temperature causes surface oxidation and/or decarburization which, in turn, can result in quench cracking.

5. *Rapid cooling* through the M_s-M_f range causes internal stresses.[147a]

6. *Steel transformation temperature range* (M_s-M_f) may show effects on cracking propensity. The higher the carbon and alloy contents, the lower the M_s and M_f temperatures. The cracking tendency typically decreases with the increase of M_s temperature. Steels containing coarse austenite grains and with low M_s temperature are especially susceptible to quench cracking. Usually, the depth of transformation to martensite producing transformation stresses increases with the increase of steel hardenability. Excessively high austenitizing temperature increases the surface-to-core temperature differentials, which, in turn, causes a corresponding increase in residual stress and cracking potential. Cracking can thus result from surface carburization or decarburization, which can affect the transformation characteristics of the surface layers and particularly change the M_s temperature.

7. *Stress risers* created by improper design of keyways and holes; sharp changes in sections (such as fillets, threads, and gear roots from machining operations); notches and machining or grinding marks; and materials-related defects such as non-metallic inclusions, lap or seams, surface discontinuities, alloy nonuniformity such as *chemical segregation-banding* and *alloy depletion*, and porosity (due to trapped gases in the casting) promote quench cracking. Examples of alloys prone to alloy depletion are AISI 4100, 4300, and 8600 series.

8. *Improper selection of steel.* Hardenable alloy steel with more than 0.25% C may suffer from quench cracking due to martensite transformation. If the steel chemistry is in excess of the required hardenability limits for the products being processed, it results in cracking. Usually, quench cracks prevail in steels when the carbon equivalent (CE) [given in Eq. (13.15)][148] is in excess of 0.52 to 0.55, as shown in Fig. 13.53

$$CE = C + \frac{Mn}{5} + \frac{Mo}{5} + \frac{Cr}{10} + \frac{Ni}{50} \qquad (13.15)$$

where elemental concentrations are in wt%. Additions of small amounts of Nb and B reduce the tendency to quench cracking. Grain refinement of the austenite and high-purity steels, made by vacuum melting, are less sensitive to quench cracking.[149] High-carbon and tool steels that are inherently hard and brittle are invariably prone to quench cracking. In carbon tool steels, the fracture grain size should not be less than ASTM 8 (Fig. 13.54) to avoid quench cracking. It appears from Fig. 13.54 that an increasing austenitizing temperature causes less quench cracking for a particular grain size, presumably due to the formation of more retained austenite at higher austenitizing temperature. It is likely that retained austenite inhibits quench cracking. In high-speed steels, fine undissolved V, Nb, and Ti carbides are most effective in resisting grain coarsening.[149]

9. *Time delays between quenching and tempering* may also be a contributing factor.

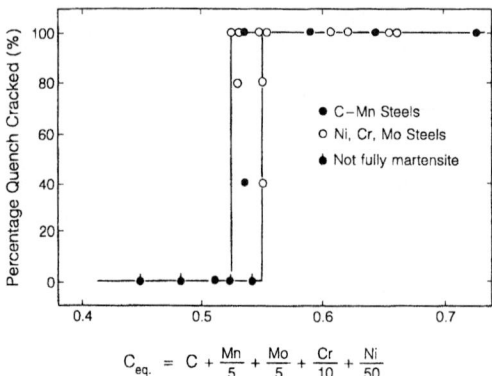

$$C_{eq.} = C + \frac{Mn}{5} + \frac{Mo}{5} + \frac{Cr}{10} + \frac{Ni}{50}$$

FIGURE 13.53 The relationship between the carbon equivalent and the percentage of specimens which quench-crack.[148]

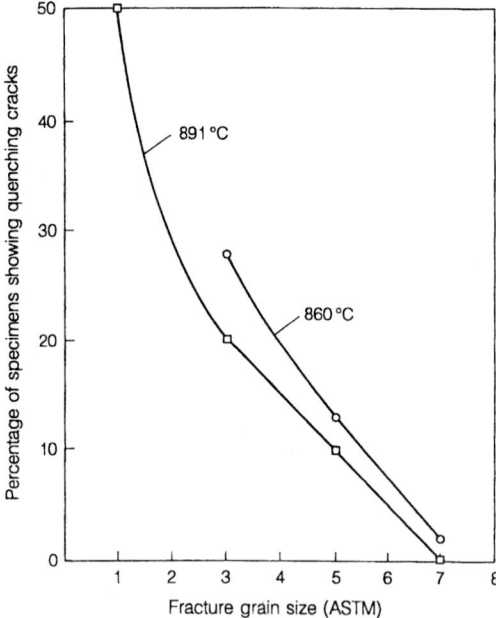

FIGURE 13.54 The effect of fracture grain size (ASTM) on the tendency to quench-crack in carbon tool steels, using different austenitizing temperature.[150]

13.12 EFFECT OF CARBON ON HARDNESS IN HARDENED STEEL

Hardness is one of the important indicators of the suitability of a steel for a specific application. Figures 13.55 and 13.56 are the plots of hardness of Fe-C martensites as a function of carbon content.[151,152] In the former case, the hardness is plotted against the Vicker's pyramid hardness numbers, while in the latter case it is plotted against the Rockwell C-scale. The essential feature of the Rockwell hardness plot

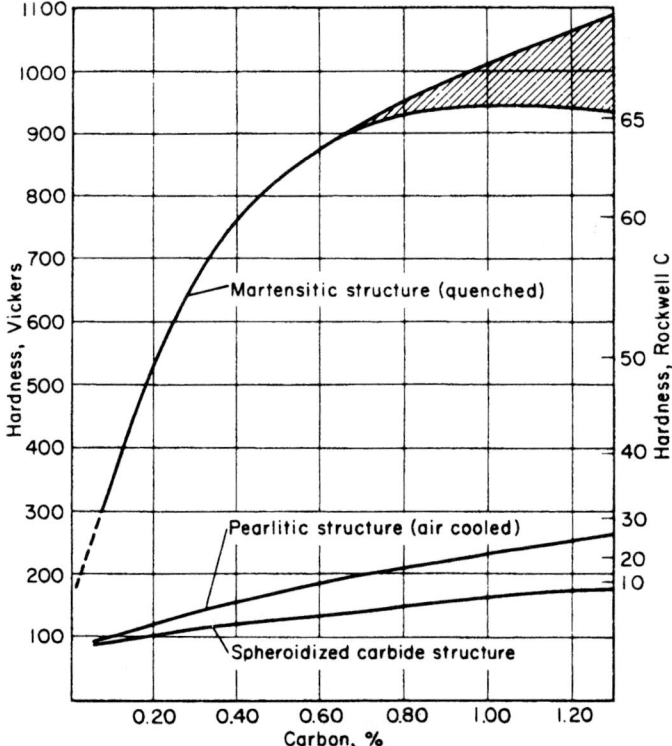

FIGURE 13.55 Relationship between Vickers hardness and carbon content for martensitic, ferrite-pearlite, and spheroidized microstructures in steels. Hatched area shows the influence of retained austenite.[151] (*Reprinted by permission of ASM International, Materials Park, Ohio.*)

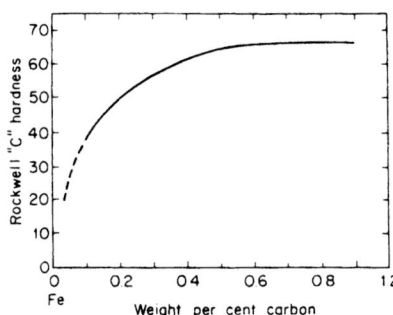

FIGURE 13.56 Rockwell hardness of martensite as a function of carbon content.[152] (*Reprinted by permission of PWS-Kent Publishing Co., Boston, Mass.; after J. L. Burns, T. L. Moore, and R. S. Archer, Trans. ASM, vol. 26, 1938, p. 1.*)

is that the hardness curve reaches about 60 Rc at 0.40% C, but it levels off at about 0.6% C, where the hardness attains about 65 Rc. Further increase in the hardness level, therefore, does not occur on exceeding the carbon content above 0.6% because the Rockwell hardness tester is insensitive in the hardness readings found

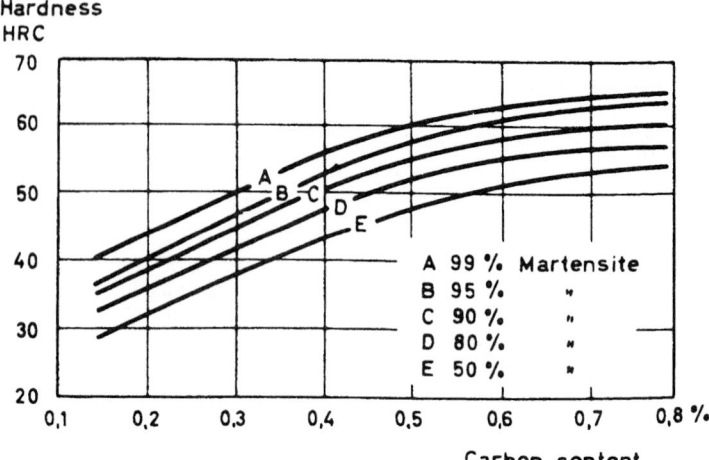

FIGURE 13.57 Average relationship between carbon content and hardness for steels containing different amounts of martensite in their microstructure.[153] (*Reprinted by permission of Butterworths, London; after Hodge and Orehoski.*)

in the hardened high-carbon steel. The reason for this is that a cone-shaped indentor with a round point in the Rockwell C-scale becomes so blunted with its use that it does not appreciably move (i.e., depth of penetration remains the same) under a fixed load to account for any observable increase in hardness. In the Vicker's hardness tester, on the other hand, since this indentor does not get blunted at its tip and the hardness measurement is based on load/surface area of an indentation formed, under a given load, by pressing into the metal a square-base diamond pyramid indentor, the DPH values (higher hardness branch of curve) provide a large effect of (increasing) carbon content on the hardness of martensite as compared to that of ferrite-pearlite or spheroidized microstructure in steel. However, a decrease in hardness at higher carbon contents in Fe-C alloys usually results (as indicated by the lower branch of the hardness curve) due to the M_f temperature lying below the room temperature, which causes the formation of a significant proportion of retained austenite in the microstructure.

Alloy martensites are usually harder at a given carbon level, due mainly to the fact that the M_s temperature is depressed and less auto-tempering occurs.

Figure 13.57 shows the usual relationship between carbon content and hardness for quenched microstructures containing various proportions of untempered martensite. Some of the variations observed in the maximum hardness at various carbon contents may be due to differences in the austenite grain size.

13.13 HARDENABILITY

13.13.1 Definition

The hardenability of steel can be defined in several ways. It is defined as the ability of a steel to be hardened by quenching (under given conditions) or as the property

TABLE 13.19 Normalizing and Austenitizing Temperatures[†155]

Steel series	Orderd carbon content, max., %	Normalizing temperature, °F (°C)	Austenitizing temperature, °F (°C)
1000, 1300, 1500, 3100, 4000, 4100	0.25 and under	1700 (925)	1700 (925)
4300, 4400, 4500, 4600, 4700, 5000, 5100, 6100,[‡] 8100, 8600, 8700, 8800, 9400, 9700, 9800	0.26 to 0.36, incl. 0.37 and over	1650 (900) 1600 (870)	1600 (870) 1500 (845)
2300, 2500, 3300, 4800, 9300	0.25 and under 0.26 to 0.36, incl. 0.37 and over	1700 (925) 1650 (900) 1600 (870)	1550 (845) 1500 (815) 1475 (800)
9200	0.50 and over	1650 (900)	1600 (870)

[†] A variation of ±10°F (6°C) from the temperatures in this table is permissible.
[‡] Normalizing and austenitizing temperatures are 50°F (30°C) higher for the 6100 series.

of a steel that determines the depth and distribution of hardness produced by quenching from the austenitizing temperature (Table 13.19).[153–155] Essentially depth and distribution of hardness are measured by the depth and distribution of martensite structure. The ability or ease to form martensite depends on the ability to avoid or suppress (partially or completely) the formation of diffusion-controlled transformation products of austenite, such as proeutectoid ferrite and cementite, pearlite, and bainite. Hence the microstructural definition of hardenability is the depth to which steel can be transformed from austenite to structure with 50%, 90%, or full martensite at a given temperature.[156] The 50% hardenability (martensite criterion) is widely used because of the ease with which it can be metallographically determined, and it is this position across which the hardness variations are most rapid and prominent. In some cases, the 90% martensite criterion is also employed with success (e.g., in shallow-hardenable steel).

Hardness and hardenability refer to distinct factors, and a steel capable of developing a high martensite hardness (with a high carbon content) can have a low hardenability with the resulting development of potential maximum hardness in very small sections. Conversely, a steel with a high hardenability will have the capability of hardening larger sections. However, if the carbon content is low, the maximum hardness developed will be also low. This effect is shown in Fig. 13.56.

It is clear that steels exhibiting deep-hardness penetration are known as *high-hardenable steels*, while those with shallow hardness penetration are called *shallow-hardenable steels*.

13.13.2 Relation between Nucleation Theory and Hardenability

It has been agreed that hardenability of a steel is governed mainly by the nucleation and growth rates of the microstructural components ferrite, pearlite, and/or bainite. Thus a theory of hardenability must include the chemical and structural factors which inhibit or suppress the nucleation and/or growth of these microconstituents.

Phosphorus is next to boron as a promoter of hardenability in hypoeutectoid steel; and like boron, phosphorus is highly surface-active and tends to retard the nucleation of ferrite at austenite grain boundaries.[157]

According to a current theory of hardenability, nucleation of ferrite takes place preferentially at specific locations of grain boundaries where these are poisoned by the segregation of boron, phosphorus, or other alloying elements which retard their growth. However, the use of phosphorus should be limited to 0.07 to 0.08% in low-alloy, and that of boron to 0.0015 to 0.002% in low- and medium-carbon steels and microalloyed steels because of the problem of temper embrittlement.[158]

13.13.3 Importance of Hardenability

Hardenability is an important property for any steel used in the heat-treated (i.e., quenched or quenched and tempered) condition. It is this concept that determines the formation of particular microconstituents during quenching of a steel component of a given size in a specific quenchant and, therefore, determines the final mechanical properties of the heat-treated material.[159]

The greater the hardenability of the steel, the slower the rate of cooling required to form a fully martensitic structure. Hardenability is primarily increased by the addition of alloying elements which help to stabilize austenite, resulting in slug-gishness of transformation of austenite upon cooling, thereby promoting the response to heat treatment.[160] The hardenability imparted by specific alloying elements is a very important consideration in the selection of steels for high-strength complex parts. The reason is that a certain depth of hardening is primarily controlled by the proportion of martensite formed in continuously cooled steels. If the hard-enability is low, fast cooling is needed to achieve a large amount of martensite. However, fast cooling causes the development of residual stress, distortion, and even cracking. That is why parts must be cooled as slowly as practical from the austenite phase, and thus the addition of certain alloying elements (such as Mn, Cr, B, or Ni) is necessary either in single addition or more often in combinations, to impart increased hardenability. Low Al and N contents enable austenite grain growth to occur, thereby leading to high hardenability in sheet steel.[161] The hardenability of cast and wrought steels is controlled mainly by the chemical composition, austenitizing temperature, grain size, and cooling rate from the austenitizing temperature.

Excess hardenability usually represents excess cost. Consequently, it is expedient to optimize the least expensive and most efficient alloy system. In this way, less expensive alternatives are sought to substitute highly alloyed steels, while retaining somewhat the same hardenability.[162]

In the context of hardenable steels, boron acts as a unique alloying element. Less expensive boron addition (with optimum boron content of 0.0015 to 0.002%) allows an increase of hardenability in certain grades of microalloyed steels with low carbon content, which, together with rapid quenching, produce greater hardened depth and notch toughness without the risk of quench-cracking. Another high-hardenable boron steel is quenched and tempered grade (or martensitic steel) containing <0.4%C.[163–165]

Variation (or nonuniformity) in hardenability leads to unpredictable behavior in heat treatment, which causes variable dimensional changes and variations in properties.[166]

In general, high hardenability and especially high carbon are not desirable for successful welding operations.

The ability to predict hardenability renders a great benefit to the materials engineer, who likes to know whether a proposed new steel or a particular heat of steel can meet the hardenability requirements of a specific application.[159]

13.13.4 Methods of Calculating Hardenability

The depth of hardness penetration or as-quenched structure obtained across a section is dependent upon (1) the quenching medium and quenching method which determine quench severity, (2) section size, (3) composition of the steel and method of manufacture, (4) austenite grain size, and (5) homogeneity of the austenite (or structure of the steel before quenching).

13.13.4.1 Grossman Method. In the Grossman method of calculating hardenability, a series of round steel bars of similar composition with increasing diameter from 0.5 to 2.5 in. (12.7 to 63.5 mm) are quenched from austenitic temperature to room temperature in an identical quenchant. A metallographic examination or hardness variation measurement of transverse sections is then carried out to determine the particular bar which has 50% martensite structure at the center. The diameter of this bar, called the *critical diameter D* (in inches or millimeters), refers to that diameter of the bar which can be fully hardened (in terms of 50% martensite at its center). Its value is a function of the composition of steel and the efficiency of quenching medium, and its importance lies in the fact that it gives a measure of the steel to respond to a hardening heat treatment.[152]

Ideal Critical Diameter. In order to eliminate the efficiency (or severity) of a quenching-medium variable, all hardenability measurements are expressed in terms of an *ideal quench.* (See above for definition and functions of ideal quench.) Under an ideal quench condition, the critical diameter, which has 50% martensite at the center of the bar when the surface is cooled at an infinitely rapid rate, that is, when $H = \infty$, is called the *ideal critical diameter.* Although an ideal quench does not exist, its cooling action on steel may be computed and compared with those of commercial quenching media.[167]

Figure 13.58[154] shows the correlation between ideal critical diameter D_I and the critical diameter D for given H values which are listed in Table 13.20[168,169] for a number of commercial quenching media. In practice, the D_I values are determined by first obtaining D and appropriate H values and then using Fig. 13.58.

The Effect of Grain Size and Chemical Composition on Hardenability. Since the hardenability of steel depends on the suppression of nonmartensite formations which nucleate heterogeneously at the austenite grain boundaries, coarse austenite grains (i.e., low ASTM grain-size number) with less grain boundary area will exhibit higher hardenability than fine-grained ones, all other factors remaining the same.[167,170] However, the use of coarse grain size in improving hardenability is not usually practiced because it is accompanied by undesirable changes such as decrease of strength, loss of ductility, increased notch sensitivity, and quench crack susceptibility.[167]

In addition to its influence on hardness, increasing amounts of carbon in a steel increases its hardenability. This fact is illustrated in Fig. 13.59, where the ideal critical diameter D_I is a function of carbon content and austenite grain size for iron-carbon alloys. These curves also demonstrate the very low hardenability of simple Fe-C alloys. For example, Fe-0.8%C eutectoid steel with an ASTM grain size of 8 possesses an ideal critical diameter of 0.278 in. (7 mm); an ordinary quench will not harden even this size bar to its center. Fortunately, commercial carbon steels

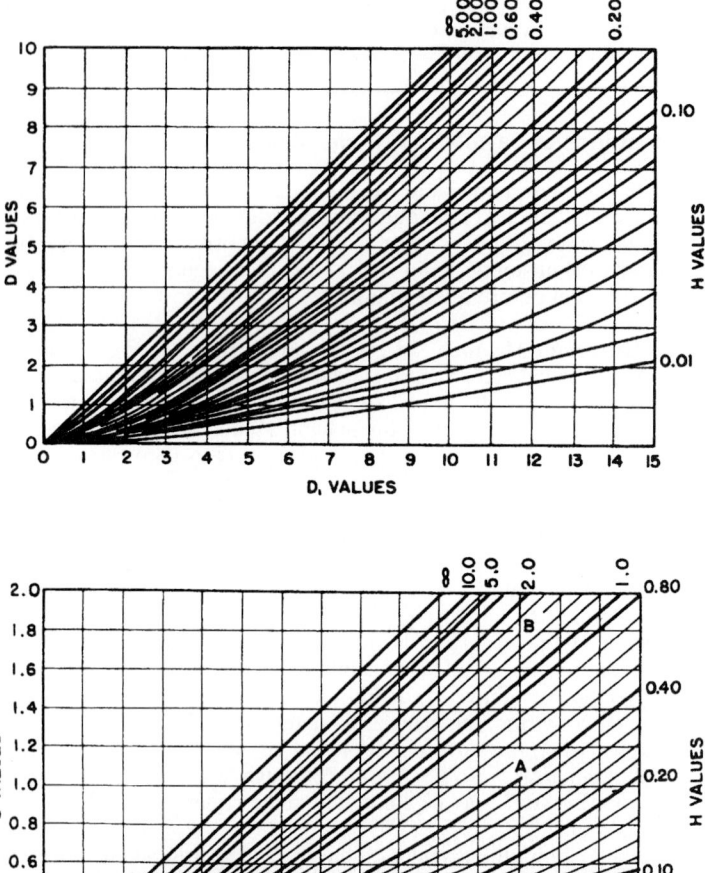

FIGURE 13.58   Relationship between actual critical diameter (D) and ideal critical diameter (D_I) for various severity of quench (H) values (two scales).[154] (*Reprinted by permission of ASM International, Materials Park, Ohio.*)

always contain some Mn, Si, and small amounts of other elements which increase their hardenability. A small amount of Mn in plain carbon steels strongly influences hardenability and therefore contributes to the economic manufacture of these steels.

Each alloying element in steel has some influence on its hardenability. All common alloying elements except cobalt increase the hardenability of steel.[171] The relative effects of some common alloying elements on hardenability are illustrated in Fig. 13.60. This plot gives the multiplying factors for various concentrations of individual alloying elements present in steel.

TABLE 13.20 Cooling Intensities H for Different Quenching Media and Quenching Conditions[169]

Quenchant agitation	Typical flow rates		Typical H values			
	m/min	sfm	Air	Mineral oil	Water	Brine
None	0	0	0.02	0.20–0.30	0.9–1.0	2.0
Mild	15	50	. . .	0.20–0.35	1.0–1.1	2.1
Moderate	30	100	. . .	0.35–0.40	1.2–1.3	. . .
Good	61	200	0.05	0.40–0.60	1.4–2.0	. . .
Strong	230	750	. . .	0.60–0.80	1.6–2.0	4.0

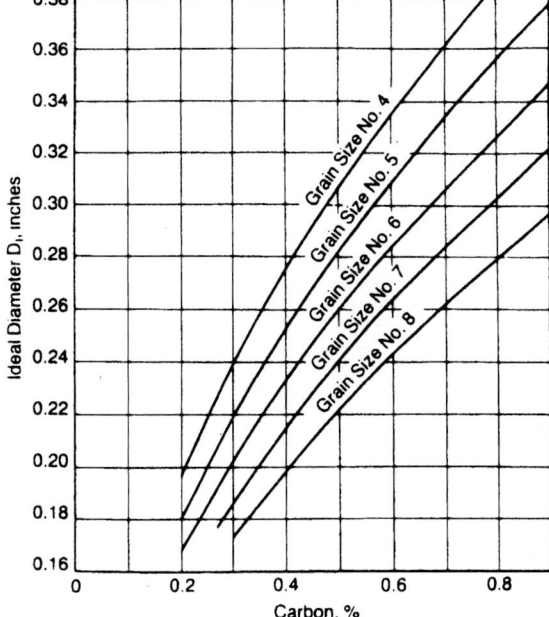

FIGURE 13.59 Hardenability, in terms of ideal critical diameter D_I, as a function of austenite grain size and carbon content of iron-carbon alloys.[154] (*Reprinted by permission of ASM International, Materials Park, Ohio.*)

An approximate hardenability of steel can be calculated if the chemical composition of the steel and the multiplying factors for alloying elements are known. An empirical relationship for ideal critical diameter D_I is expressed as

$$D_I = D \times F_{Mn} \times F_{Si} \times F_{Ni} \times F_{Cr} \times F_{Mo} \times \text{etc.} \qquad (13.16)$$

where F is the multiplying factor corresponding to the weight percentage alloying element present. This gives a diameter of maximum value which would be through-hardened under ideal quench conditions.

Example. The hardenability of 4340 steel containing 0.40 C, 0.80 Mn, 0.20 Si, 1.80 Ni, 0.90 Cr, 0.30 Mo, and an austenite grain size of ASTM 7 can be calculated as follows: First the base diameter D is determined for the carbon content of 0.40% and grain size of 7 from Fig. 13.59, which is found to be 0.213 in. Next the multi-

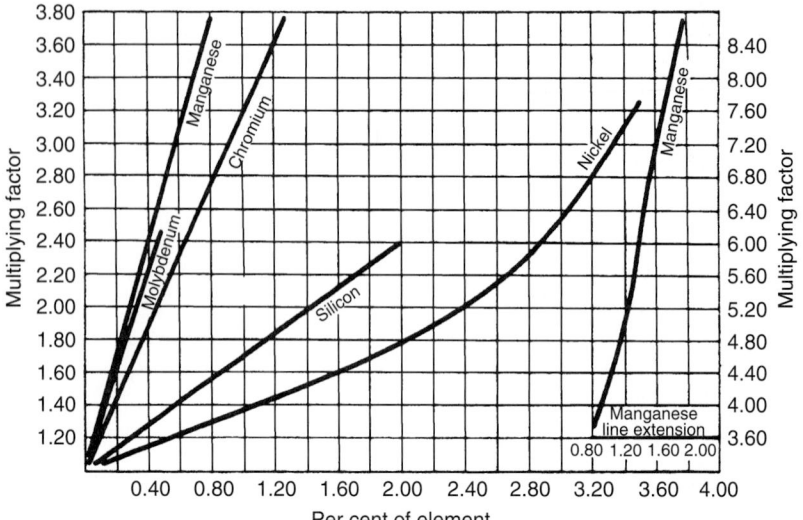

FIGURE 13.60 Multiplying factors for different alloying elements for hardenability calculations. (*Courtesy of Association of Iron and Steel Engineers, Pittsburgh, Pa.; after W. D. Lankford
and H. E. McGannon, The Making, Shaping and Treating of Steel, 10th ed., United States Steel
Corporation, 1985.*)

plying factors for each alloying element are determined from Fig. 13.60, using the
specified concentration. This gives $F_{Mn} = 3.667$, $F_{Si} = 1.14$, $F_{Ni} = 1.68$, $F_{Cr} = 2.944$, and
$F_{Mo} = 1.9$. Substituting all these values in Eq. (13.16), we obtain

$$D_I = 0.213 \times 3.667 \times 1.14 \times 1.68 \times 2.944 \times 1.9 = 8.367 \text{ in.}$$

The above numerical value reveals the contributions due solely to alloying elements present in solution at the austenitizing temperature. Thus, the alloying addition in 4340 alloy steel, which amounts to a total of 4.0%, produces a steel with an
ideal critical diameter of 8.367 in. (Here the alloy factor is the product of all the
multiplying factors, excluding the carbon multiplying factor, that is, $3.667 \times 1.14 \times$
$1.68 \times 2.944 \times 1.9 = 39.28$.) Even in a mild water quench ($H = 1.0$), the steel would
have a critical diameter of 7.5 in.

In comparison, an ideal critical diameter found in a 1040 plain carbon steel with
0.8% Mn is 0.781 in., and the corresponding critical diameter, using mild water
quench, is 0.28 in.[152,172]

Note that different alloying elements have greatly varying influence on hardenability, as seen in Fig. 13.60. Table 13.21 lists the ideal base critical diameter of
Fe-C alloys for different carbon content and a No. 7 austenite grain size as well as
the multiplying factors for common alloying elements in the range of 0.01 to 2.0%.[173]
To simplify the calculation of hardenability, this table may be used in place of Figs.
13.59 and 13.60. The multiplying factors for unspecified alloying elements, as well
as for phosphorus and sulfur present as impurities in steels, should be considered to
be unity. It means the missing elements and P and S have no effect on the D_I value.

Table 13.22 gives a range of D_I values for a variety of steels which are obtained
as a result of acceptable compositional variations of a given grade, grain size, and
concentration of residual elements.[154,174]

TABLE 13.21 Hardenability Multiplying Factors versus Element (Non-Boron Steels), in.[155]

% Alloy	Carbon, grain size	Mn	Si	Ni	Cr	Mo	Cu	V
0.01	0.005	1.033	1.007	1.004	1.022	1.03	1.00	1.02
0.02	0.011	1.067	1.014	1.007	1.043	1.06	1.01	1.03
0.03	0.016	1.100	1.021	1.011	1.065	1.09	1.01	1.05
0.04	0.021	1.133	1.028	1.015	1.086	1.12	1.02	1.07
0.05	0.026	1.167	1.035	1.018	1.108	1.15	1.02	1.09
0.06	0.032	1.200	1.042	1.022	1.130	1.18	1.02	1.11
0.07	0.038	1.233	1.049	1.026	1.151	1.21	1.03	1.12
0.08	0.043	1.267	1.056	1.029	1.173	1.24	1.03	1.14
0.09	0.049	1.300	1.063	1.033	1.194	1.27	1.03	1.16
0.10	0.054	1.333	1.070	1.036	1.216	1.30	1.04	1.17
0.11	0.059	1.367	1.077	1.040	1.238	1.33	1.04	1.19
0.12	0.065	1.400	1.084	1.044	1.259	1.36	1.05	1.21
0.13	0.070	1.433	1.091	1.047	1.281	1.39	1.05	1.22
0.14	0.076	1.467	1.098	1.051	1.302	1.42	1.05	1.24
0.15	0.081	1.500	1.105	1.055	1.324	1.45	1.06	1.26
0.16	0.086	1.533	1.112	1.058	1.346	1.48	1.06	1.28
0.17	0.092	1.567	1.119	1.062	1.367	1.51	1.06	1.29
0.18	0.097	1.600	1.126	1.066	1.389	1.54	1.07	1.31
0.19	0.103	1.633	1.133	1.069	1.410	1.57	1.07	1.33
0.20	0.108	1.667	1.140	1.073	1.432	1.60	1.07	1.35
0.21	0.113	1.700	1.147	1.077	1.454	1.63	1.08	...
0.22	0.119	1.733	1.154	1.080	1.475	1.66	1.08	...
0.23	0.124	1.767	1.161	1.084	1.497	1.69	1.09	...
0.24	0.130	1.800	1.168	1.088	1.518	1.72	1.09	...
0.25	0.135	1.833	1.175	1.091	1.540	1.75	1.09	...
0.26	0.140	1.867	1.182	1.095	1.562	1.78	1.10	...
0.27	0.146	1.900	1.189	1.098	1.583	1.81	1.10	...
0.28	0.151	1.933	1.196	1.102	1.605	1.84	1.10	...
0.29	0.157	1.967	1.203	1.106	1.626	1.87	1.11	...
0.30	0.162	2.000	1.210	1.109	1.648	1.90	1.11	...
0.31	0.167	2.033	1.217	1.113	1.670	1.93	1.11	...
0.32	0.173	2.067	1.224	1.117	1.691	1.96	1.12	...
0.33	0.178	2.100	1.231	1.120	1.713	1.99	1.12	...
0.34	0.184	2.133	1.238	1.124	1.734	2.02	1.12	...
0.35	0.189	2.167	1.245	1.128	1.756	2.05	1.13	...
0.36	0.194	2.200	1.252	1.131	1.776	2.08	1.13	...
0.37	0.200	2.233	1.259	1.135	1.799	2.11	1.14	...
0.38	0.205	2.267	1.266	1.139	1.821	2.14	1.14	...
0.39	0.211	2.300	1.273	1.142	1.842	2.17	1.14	...
0.40	0.213	2.333	1.280	1.146	1.864	2.20	1.15	...
0.41	0.216	2.367	1.287	1.150	1.886	2.23	1.15	...
0.42	0.218	2.400	1.294	1.153	1.907	2.26	1.15	...
0.43	0.221	2.433	1.301	1.157	1.929	2.29	1.16	...
0.44	0.223	2.467	1.308	1.160	1.950	2.32	1.16	...
0.45	0.226	2.500	1.315	1.164	1.972	2.35	1.16	...

TABLE 13.21 Hardenability Multiplying Factors versus Element (Non-Boron Steels), in.[155] (*Continued*)

% Alloy	Carbon, grain size	Mn	Si	Ni	Cr	Mo	Cu	V
0.46	0.228	2.533	1.322	1.168	1.994	2.38	1.17	...
0.47	0.230	2.567	1.329	1.171	2.015	2.41	1.17	...
0.48	0.233	2.600	1.336	1.175	2.037	2.44	1.18	...
0.49	0.235	2.633	1.343	1.179	2.058	2.47	1.18	...
0.50	0.238	2.667	1.350	1.182	2.080	2.50	1.18	...
0.51	0.242	2.700	1.357	1.186	2.102	2.53	1.19	...
0.52	0.244	2.733	1.364	1.190	2.123	2.56	1.19	...
0.53	0.246	2.767	1.371	1.193	2.145	2.59	1.19	...
0.54	0.249	2.800	1.378	1.197	2.166	2.62	1.20	...
0.55	0.251	2.833	1.385	1.201	2.188	2.65	1.20	...
0.56	0.253	2.867	1.392	1.204	2.210	...	...	...
0.57	0.256	2.900	1.399	1.208	2.231	...	...	...
0.58	0.258	2.933	1.406	1.212	2.253	...	...	...
0.59	0.260	2.967	1.413	1.215	2.274	...	...	...
0.60	0.262	3.000	1.420	1.219	2.296	...	...	...
0.61	0.264	3.033	1.427	1.222	2.318	...	...	...
0.62	0.267	3.067	1.434	1.226	2.339	...	...	...
0.63	0.269	3.100	1.441	1.230	2.361	...	...	...
0.64	0.271	3.133	1.448	1.233	2.382	...	...	...
0.65	0.273	3.167	1.455	1.237	2.404	...	...	...
0.66	0.275	3.200	1.462	1.241	2.426	...	...	...
0.67	0.277	3.233	1.469	1.244	2.447	...	...	...
0.68	0.279	3.267	1.476	1.248	2.469	...	...	...
0.69	0.281	3.300	1.483	1.252	2.490	...	...	...
0.70	0.283	3.333	1.490	1.256	2.512	...	...	...
0.71	0.285	3.367	1.497	1.259	2.534	...	...	...
0.72	0.287	3.400	1.504	1.262	2.555	...	...	...
0.73	0.289	3.433	1.511	1.266	2.577	...	...	...
0.74	0.291	3.467	1.518	1.270	2.596	...	...	...
0.75	0.293	3.500	1.525	1.273	2.620	...	...	...
0.76	0.295	3.533	1.532	1.276	2.642	...	...	...
0.77	0.297	3.567	1.539	1.280	2.663	...	...	...
0.78	0.299	3.600	1.546	1.284	2.685	...	...	...
0.79	0.301	3.633	1.553	1.287	2.706	...	...	...
0.80	0.303	3.667	1.560	1.291	2.728	...	...	...
0.81	0.305	3.700	1.567	1.294	2.750	...	...	...
0.82	0.307	3.733	1.574	1.298	2.771	...	...	...
0.83	0.309	3.767	1.581	1.301	2.793	...	...	...
0.84	0.310	3.800	1.588	1.306	2.814	...	...	...
0.85	0.312	3.833	1.595	1.309	2.836	...	...	...
0.86	0.314	3.867	1.602	1.313	2.858	...	...	...
0.87	0.316	3.900	1.609	1.317	2.879	...	...	...
0.88	0.318	3.933	1.616	1.320	2.900	...	...	...
0.89	0.319	3.967	1.623	1.324	2.922	...	...	...
0.90	0.321	4.000	1.630	1.327	2.944	...	...	...

TABLE 13.21 Hardenability Multiplying Factors versus Element (Non-Boron Steels), in.[155] (*Continued*)

% Alloy	Carbon, grain size	Mn	Si	Ni	Cr	Mo	Cu	V
0.91	...	4.033	1.637	1.331	2.966	...	...	...
0.92	...	4.067	1.644	1.334	2.987	...	...	...
0.93	...	4.100	1.651	1.338	3.009	...	...	...
0.94	...	4.133	1.658	1.343	3.030	...	...	...
0.95	...	4.167	1.665	1.345	3.052	...	...	...
0.96	...	4.200	1.672	1.349	3.074	...	...	...
0.97	...	4.233	1.679	1.352	3.095	...	...	...
0.98	...	4.267	1.686	1.356	3.117	...	...	...
0.99	...	4.300	1.693	1.360	3.138	...	...	...
1.00	...	4.333	1.700	1.364	3.160	...	...	...
1.01	...	4.367	1.707	1.367	3.182	...	...	...
1.02	...	4.400	1.714	1.370	3.203	...	...	...
1.03	...	4.433	1.721	1.375	3.225	...	...	...
1.04	...	4.467	1.728	1.378	3.246	...	...	...
1.05	...	4.500	1.735	1.382	3.268	...	...	...
1.06	...	4.533	1.742	1.386	3.290	...	...	...
1.07	...	4.567	1.749	1.389	3.311	...	...	...
1.08	...	4.600	1.756	1.393	3.333	...	...	...
1.09	...	4.633	1.763	1.396	3.354	...	...	...
1.10	...	4.667	1.770	1.400	3.376	...	...	...
1.11	...	4.700	1.777	1.403	3.398	...	...	...
1.12	...	4.733	1.784	1.406	3.419	...	...	...
1.13	...	4.767	1.791	1.411	3.441	...	...	...
1.14	...	4.800	1.798	1.414	3.462	...	...	...
1.15	...	4.833	1.805	1.418	3.484	...	...	...
1.16	...	4.867	1.812	1.422	3.506	...	...	...
1.17	...	4.900	1.819	1.426	3.527	...	...	...
1.18	...	4.933	1.826	1.429	3.549	...	...	...
1.19	...	4.967	1.833	1.433	3.570	...	...	...
1.20	...	5.000	1.840	1.437	3.592	...	...	...
1.21	...	5.051	1.847	1.440	3.614	...	...	...
1.22	...	5.102	1.854	1.444	3.635	...	...	...
1.23	...	5.153	1.861	1.447	3.657	...	...	...
1.24	...	5.204	1.868	1.450	3.678	...	...	...
1.25	...	5.255	1.875	1.454	3.700	...	...	...
1.26	...	5.306	1.882	1.458	3.722	...	...	...
1.27	...	5.357	1.889	1.461	3.743	...	...	...
1.28	...	5.408	1.896	1.465	3.765	...	...	...
1.29	...	5.459	1.903	1.470	3.786	...	...	...
1.30	...	5.510	1.910	1.473	3.808	...	...	...
1.31	...	5.561	1.917	1.476	3.830	...	...	...
1.32	...	5.612	1.924	1.481	3.851	...	...	...
1.33	...	5.663	1.931	1.484	3.873	...	...	...
1.34	...	5.714	1.938	1.487	3.894	...	...	...
1.35	...	5.765	1.945	1.491	3.916	...	...	...

TABLE 13.21   Hardenability Multiplying Factors versus Element (Non-Boron Steels), in.[155] (*Continued*)

% Alloy	Carbon, grain size	Mn	Si	Ni	Cr	Mo	Cu	V
1.36	...	5.816	1.952	1.495	3.938	...	...	...
1.37	...	5.867	1.959	1.498	3.959	...	...	...
1.38	...	5.918	1.966	1.501	3.981	...	...	...
1.39	...	5.969	1.973	1.506	4.002	...	...	...
1.40	...	6.020	1.980	1.509	4.024	...	...	...
1.41	...	6.071	1.987	1.512	4.046	...	...	...
1.42	...	6.122	1.994	1.517	4.067	...	...	...
1.43	...	6.173	2.001	1.520	4.089	...	...	...
1.44	...	6.224	2.008	1.523	4.110	...	...	...
1.45	...	6.275	2.015	1.527	4.132	...	...	...
1.46	...	6.326	2.022	1.531	4.154	...	...	...
1.47	...	6.377	2.029	1.535	4.175	...	...	...
1.48	...	6.428	2.036	1.538	4.197	...	...	...
1.49	...	6.479	2.043	1.541	4.217	...	...	...
1.50	...	6.530	2.050	1.545	4.239	...	...	
1.51	...	6.581	2.057	1.556	4.262	...	...	...
1.52	...	6.632	2.064	1.561	4.283	...	...	...
1.53	...	6.683	2.071	1.565	4.305	...	...	...
1.54	...	6.734	2.078	1.569	4.326	...	...	...
1.55	...	6.785	2.085	1.574	4.348	...	...	...
1.56	...	6.836	2.092	1.578	4.369	...	...	...
1.57	...	6.887	2.099	1.582	4.391	...	...	...
1.58	...	6.938	2.106	1.586	4.413	...	...	...
1.59	...	6.989	2.113	1.591	4.434	...	...	...
1.60	...	7.040	2.120	1.595	4.456	...	...	...
1.61	...	7.091	2.127	1.600	4.478	...	...	...
1.62	...	7.142	2.134	1.604	4.499	...	...	...
1.63	...	7.193	2.141	1.609	4.521	...	...	...
1.64	...	7.224	2.148	1.613	4.542	...	...	...
1.65	...	7.295	2.155	1.618	4.564	...	...	...
1.66	...	7.346	2.162	1.622	4.586	...	...	...
1.67	...	7.397	2.169	1.627	4.607	...	...	...
1.68	...	7.448	2.176	1.631	4.629	...	...	...
1.69	...	7.499	2.183	1.636	4.650	...	...	...
1.70	...	7.550	2.190	1.640	4.672	...	...	...
1.71	...	7.601	2.197	1.644	4.694	...	...	...
1.72	...	7.652	2.204	1.648	4.715	...	...	...
1.73	...	7.703	2.211	1.652	4.737	...	...	...
1.74	...	7.754	2.218	1.656	4.759	...	...	...
1.75	...	7.805	2.225	1.660	4.780	...	...	...
1.76	...	7.856	2.232	1.664	...	...	...	...
1.77	...	7.907	2.239	1.668	...	...	...	...
1.78	...	7.958	2.246	1.672	...	...	...	...
1.79	...	8.009	2.253	1.676	...	...	...	...
1.80	...	8.060	2.260	1.680	...	...	...	...
1.81	...	8.111	2.267	1.687	...	...	...	...
1.82	...	8.162	2.274	1.694	...	...	...	...
1.83	...	8.213	2.281	1.701	...	...	...	...
1.84	...	8.264	2.288	1.708	...	...	...	...
1.85	...	8.315	2.295	1.715	...	...	...	...

TABLE 13.21 Hardenability Multiplying Factors versus Element (Non-Boron Steels), in.[155] (*Continued*)

% Alloy	Carbon, grain size	Mn	Si	Ni	Cr	Mo	Cu	V
1.86	...	8.366	2.302	1.722	...	...	...	...
1.87	...	8.417	2.309	1.729	...	...	...	...
1.88	...	8.468	2.316	1.736	...	...	...	...
1.89	...	8.519	2.323	1.743	...	...	...	...
1.90	...	8.570	2.330	1.750	...	...	...	...
1.91	...	8.671	2.337	1.753	...	...	...	...
1.92	...	8.672	2.344	1.756	...	...	...	...
1.93	...	8.723	2.351	1.759	...	...	...	...
1.94	...	8.774	2.358	1.761	...	...	...	...
1.95	...	8.825	2.364	1.765	...	...	...	...
1.96	...	...	2.372	1.767	...	...	...	...
1.97	...	...	2.379	1.770	...	...	...	...
1.98	...	...	2.386	1.773	...	...	...	...
1.99	...	...	2.393	1.776	...	...	...	...
2.00	...	...	2.400	1.779	...	...	...	...

TABLE 13.22 Hardenability Expressed as a Range of D_I Values for Various Steels[154,174]

Steel	D_I	Steel	D_I	Steel	D_I
1045	0.9–1.3	4135H	2.5–3.3	8625H	1.6–2.4
1090	1.2–1.6	4140 H	3.1–4.7	8627H	1.7–2.7
1320H	1.4–2.5	4317H	1.7–2.4	8630H	2.1–2.8
1330H	1.9–2.7	4320H	1.8–2.6	8632H	2.2–2.9
1335H	2.0–2.8	4340H	4.6–6.0	8635H	2.4–3.4
1340H	2.3–3.2	4620H	1.4–2.2	8637H	2.6–3.6
2330H	2.3–3.2	4620H	1.5–2.2	8640H	2.7–3.7
2345	2.5–3.2	4621H	1.9–2.6	8641H	2.7–3.7
2512H	1.5–2.5	4640H	2.6–3.4	8642H	2.8–3.9
2515H	1.8–2.9	4812H	1.7–2.7	8645H	3.1–4.1
2517H	2.0–3.0	4815H	1.8–2.8	8647H	3.0–4.1
3120H	1.5–2.3	4817H	2.2–2.9	8650H	3.3–4.5
3130H	2.0–2.8	4820H	2.2–3.2	8720H	1.8–2.4
3135H	2.2–3.1	5120H	1.2–1.9	8735H	2.7–3.6
3140H	2.6–3.4	5130H	2.1–2.9	8740H	2.7–3.7
3340	8.0–10.0	5132H	2.2–2.9	8742H	3.0–4.0
4032H	1.6–2.2	5135H	2.2–2.9	8745H	3.2–4.3
4037H	1.7–2.4	5140H	2.2–3.1	8747H	3.5–4.6
4042H	1.7–2.4	5145H	2.3–3.5	8750H	3.8–4.9
4047H	1.8–2.7	5150H	2.5–3.7	9260H	2.0–3.3
4047H	1.7–2.4	5152H	3.3–4.7	9261H	2.6–3.7
4053H	2.1–2.9	5160H	2.8–4.0	9262H	2.8–4.2
4063H	2.2–3.5	6150H	2.8–3.9	9437H	2.4–3.7
4068H	2.3–3.6	8617H	1.3–2.3	9440H	2.4–3.8
4130H	1.8–2.6	8620H	1.6–2.3	9442H	2.8–4.2
4132H	1.8–2.5	8622H	1.6–2.3	9445H	2.8–4.4

Limitations of Grossman's Principle

1. The method of determining the ideal critical diameter is very time-consuming.

2. The application of the principle is complex.

3. A multiplying factor for a given alloying element is not always a linear function of its concentration.

4. Interaction effects are not taken into consideration.

5. When a hardness measurement is used instead of a microstructural criterion for the determination of 50% martensite location in the hardened zone of the test specimen, the accuracy of data remains very doubtful (usually ±15% of the actual).

6. The actual quenching intensity during the entire quenching process is described by a single H-value for a given quenching bath. In reality, the heat-transfer coefficient at the interface between the metal surface and the surrounding quenchant changes drastically during different stages of the quenching process.[175]

13.13.4.2 Reexamination of Grossman's Method. The hardenability factors developed by Grossman have been further studied by various workers; and a more precise relationship between base hardenability, grain size, and carbon content is shown by DeRetana and Doane,[176] after the extension of Kramer's factor for lower-carbon steels (0.15 to 0.25% C) (Fig. 13.61). Using the same model, these authors developed average multiplying factors for alloying elements in the lower-carbon steels which are shown in Fig. 13.62. Likewise, Moser and Legat[177] have also pre-

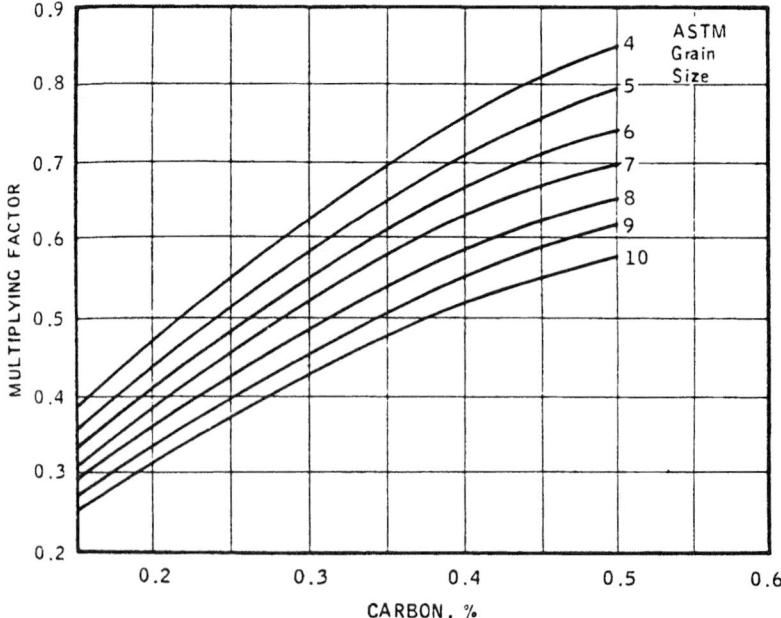

FIGURE 13.61 Multiplying factors for base hardenability for carbon content plus grain size. Empirical extension of Kramer factors by DeRetana and Doane.[176] (*Reprinted by permission of The Metallurgical Society, Warrendale, Pa.*)

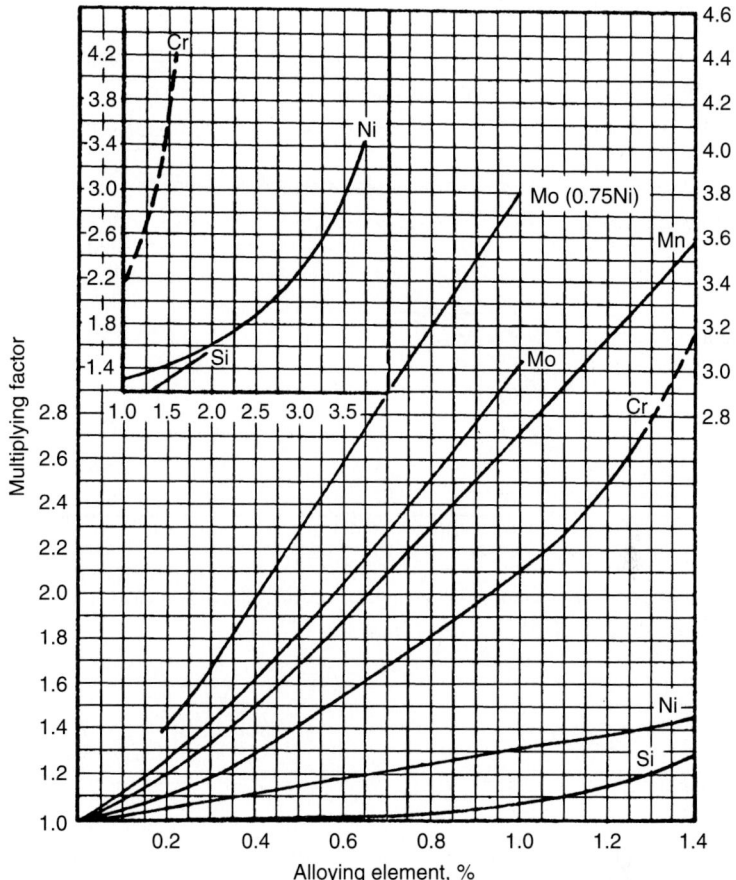

FIGURE 13.62 Average multiplying factor for several elements in alloy steels containing 0.15 to 0.25% carbon.[181] (*Reprinted by permission of The Metallurgical Society.*)

sented a reliable relationship between base hardenability, carbon content, and grain size as found in actual practice for medium- and high-carbon steels (Fig. 13.63). The corresponding multiplying factors for determining hardenability in the presence of common alloying elements are shown in Fig. 13.64.[177] The ideal critical diameter D_I, based on Moser and Legat diagrams, can be more accurately found from the empirical relationship

$$D_I = D \times 2.21(\%Mn) \times 1.40(\%Si) \times 1.47(\%Ni) \times 2.13(\%Cr) \times 3.27(\%Mo) \quad (13.17)$$

This relationship is quite effective in evaluating the actual hardening behavior. It is also possible to calculate the ideal critical diameter of 4340 steel with the above composition by using Figs. 13.63 and 13.64.

$$D_I = D \times F_{Mn} \times F_{Si} \times F_{Ni} \times F_{Cr} \times F_{Mo}$$
$$= 0.60 \times 1.862 \times 1.069 \times 2 \times 1.965 \times 1.41 = 6.66 \, in.$$

This value is lower than that of Grossman's value (found above).

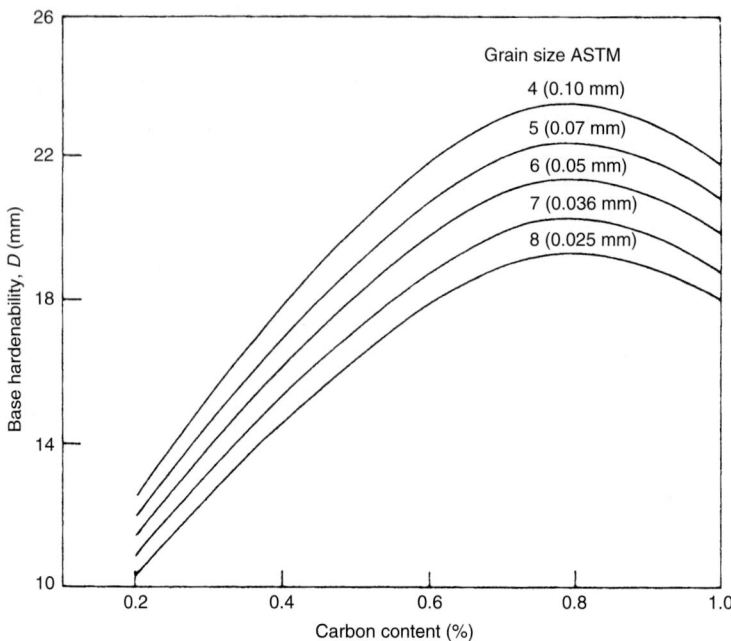

FIGURE 13.63 Relationship between base hardenability, carbon content, and grain size as found in actual practice.[177]

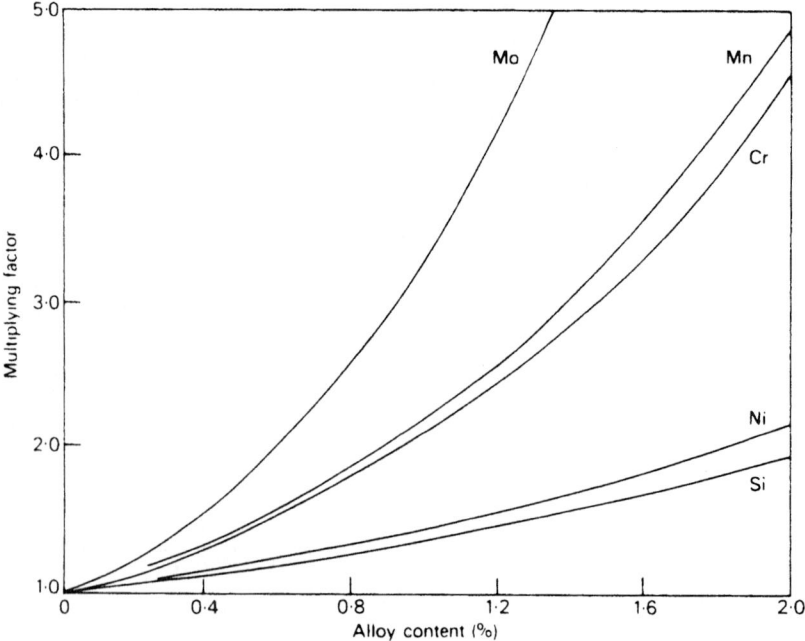

FIGURE 13.64 Multiplying factors for determining hardenability in the presence of Mo, Mn, Cr, Si, and Ni.[177]

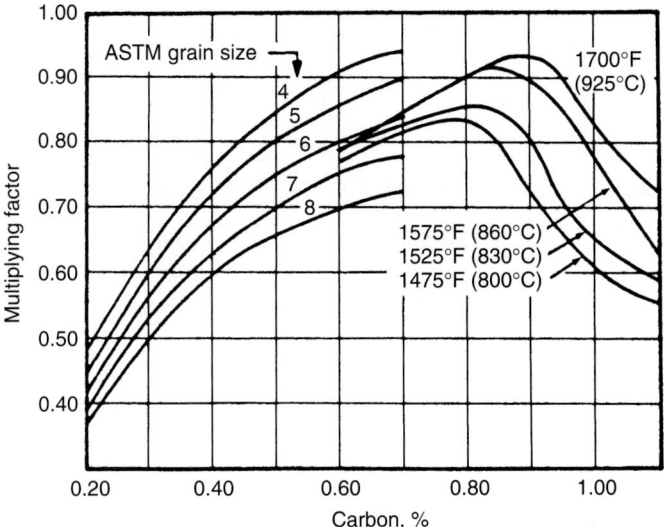

FIGURE 13.65 Multiplying factors for carbon at each austenitizing condition. Data are plotted on background of original Kramer data for medium-carbon steels with grain size variations from 4 to 8 ASTM.[178] (*Reprinted by permission of The Metallurgical Society, Warrendale, Pa.*)

Later, Jatczak[178] has extended and revised the Kramer et al.[179] data to higher-carbon contents (between 0.7% and 1.0%) (Fig. 13.65) and included the effect of austenitizing conditions (e.g., austenitizing temperature and prior microstructure—normalized or spheroidize-annealed) and multiplying factors for alloying elements at different temperatures (Fig. 13.66). Jatczak has further found the economic advantage of Mo, a carbide former, due to its strong influence on the austenitizing condition; that is, its effect at high temperature [927°C (1700°F)] in high-carbon steels is significantly greater than that in lower-carbon steels. His approach can be used to determine (1) core hardenability from compositions of homogeneous high-carbon steels, and (2) case-hardenability of high-carbon position in carburizing grades. The accuracy of hardenability prediction using his method is believed to lie within ±10% of the measured hardenability at D_I values up to 26 in. (660 mm).[178]

13.13.4.3 Hot Brine Hardenability Test/Method. In order to avoid the difficulty of obtaining fully martensitic structure at the quenched end of the Jominy bar made from unalloyed shallow-hardening steels, Grange devised a more sensitive, precise, and reproducible hardenability test, called *the hot-wire brine test*. This method consists of austenitizing several small specimens of fixed diameter between 2.5 mm (0.1 in.) and 22.5 mm (0.9 in.) and quenching in brine of increasing temperatures to produce a range of cooling rates of rounds when quenched in water or oil.[179,180] He defined hardenability in terms of *hardenable diameter* D_H of cylinders by water-quenching, which would exhibit 90% martensitic structure rather than a 50% martensitic structure at the center of the bar in Grossman's method.

Figure 13.67a[180] shows a plot of hardenable diameter D_H versus carbon content; D_H, which can be known from the curve, is the base hardenability of Fe-C alloys, free from alloying elements and impurities. This method is used successfully to

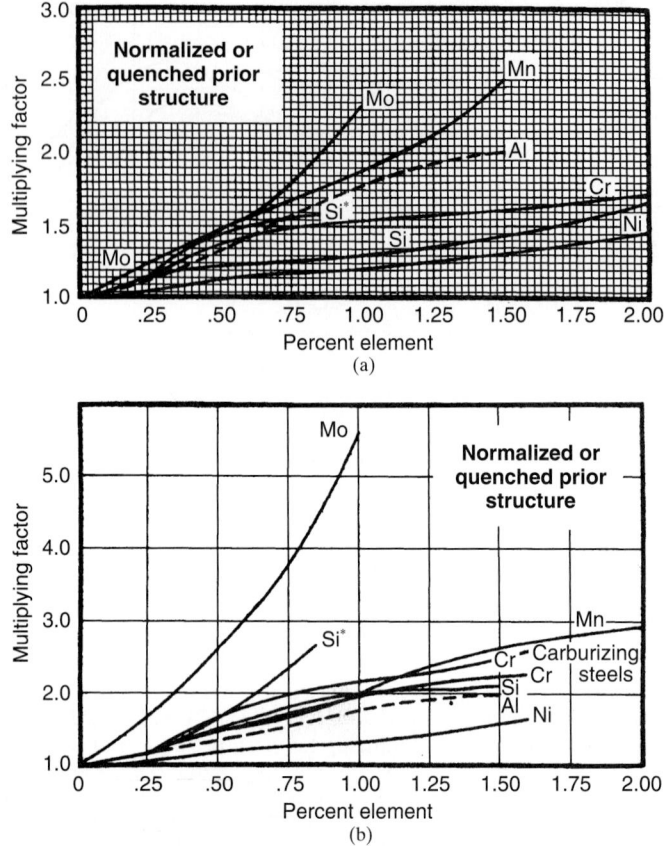

FIGURE 13.66 Revised multiplying factors for alloying elements for calculation of hardenability at high carbon levels when quenched from (a) 830°C (1525°F) and (b) 927°C (1700°F).[178] (*Reprinted by permission of The Metallurgical Society, Warrendale, Pa.*)

determine the additive influence of minor elements on the hardenable diameter of the shallow-hardening steels. The prediction of hardenability by this method is given by the equation[180,181]

$$D_H = \Delta D_C + \Delta D_{Mn} + \Delta D_P + \Delta D_{Si} + \Delta D_{Cu} + \Delta D_{Ni} + \Delta D_{Cr} + \Delta D_{Mo} + \Delta D_V \quad (13.18)$$

where ΔD's are the increase in bar diameter as a result of the presence of each individual element (in wt%) and are determined by using charts of ΔD values versus weight percentage of each element (Fig. 13.67b and c). These data give a hardenable diameter in terms of ASTM austenite grain size No. 4, which is convertible to any other grain size by the D_H versus $d_\gamma^{-1/2}$ relationship. An important feature of these data is the avoidance of the multiplying factor concept.[180]

Note that hardenability can be defined in lower-carbon steel by the DeRetana-Doane method, in alloy-free carbon steel by the Grange method, in low-alloy

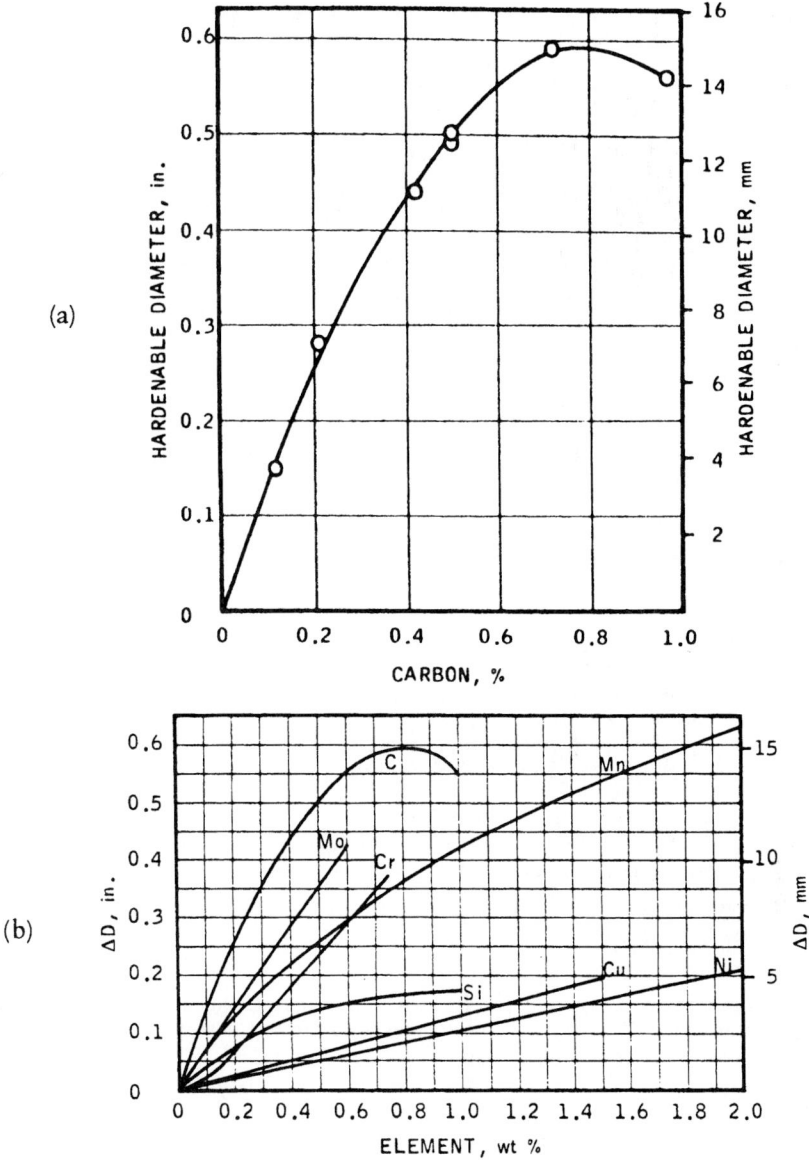

FIGURE 13.67 (*a*) Hardenability of Fe-C alloys (90% martensite, water-quenched, ASTM grain size No. 4). (*b*) Change in hardenable diameter (ΔD) with C, Mn, Si, Cu, Ni, Cr, and Mo. (*c*) Change in hardenable diameter (ΔD) with V and P.[180] (*Reprinted by permission of The Metallurgical Society, Warrendale, Pa.*)

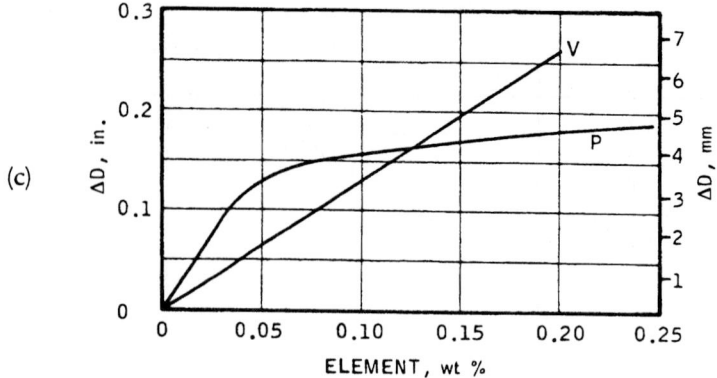

(c)

FIGURE 13.67 (*Continued*) (*a*) Hardenability of Fe-C alloys (90% martensite, water-quenched, ASTM grain size No. 4). (*b*) Change in hardenable diameter (ΔD) with C, Mn, Si, Cu, Ni, Cr, and Mo. (*c*) Change in hardenable diameter (ΔD) with V and P.[180] (*Reprinted by permission of The Metallurgical Society, Warrendale, Pa.*)

medium-carbon heat-treatable steels by the Kramer approach, in medium- and high-carbon steels by the Moser and Legat approach, and in carburizing and high-carbon steels by the Jatczak approach.

13.13.5 Methods of Determining Hardenability

13.13.5.1 Jominy Method.   The end-quench test proposed by Jominy and Boege-hold 64 years ago[182] is the most widely used method of determining hardenability. This test is popularly known as the *Jominy test*. The advantages of this method are as follows:[163]

1. This method is straightforward due to its operation, easy comparison of one result with another, and use of a simple test specimen.
2. A single test specimen is sufficient, rather than a series of round bars, to evaluate the hardenability of a particular steel.
3. It is capable of good reproducibility.
4. A high degree of accuracy can be achieved, provided that the appropriate precautions are taken during the performance of the test. The use of Vicker's hardness is more beneficial in testing low-hardenability steel.

There are two important shortcomings of this method:[183]

1. Since the cooling rates up to $\frac{5}{16}$ in. from the quenched end decrease drastically, and rates at a distance greater than $\frac{6}{16}$ in. decrease very slowly, it is difficult to use it on relatively shallow-hardenable steels. For such steels, the surface area center (SAC) test seems to be more appropriate.
2. Most of the important engineering characteristics of heat-treated steel parts are dependent on microstructure rather than hardness.

The methods that need to be controlled during heat treatment of the Jominy test bar are as follows:[163,184]

1. To ensure austenitizing of the test bar:

   a. Thermocouples and recorders should be calibrated periodically, and Jominy samples should be placed near the thermocouple.

   b. A scale-free quenched end-face must be achieved by the use of atmosphere-controlled furnaces; carbon block or fine graphite powder or gray iron chips should surround the specimen, or salt bath should be used during austenitizing.

2. To ensure proper quenching:

   a. Quenching should be started by removing the intervening plate rather than turning on a valve or starting a pump.

   b. A gravity-controlled constant-pressure water supply should be used rather than a valve-regulated water supply or pump, to avoid quenching-surges or fluctuations.

   c. The orifice or the umbrella should be uniform.

3. No burning or overheating should take place when the specimen is ground after hardening.

Procedure. The test has been standardized by the ASTM Standard A255-99[155] and SAE Standard J406.[173] In carrying out this test, a standard Jominy end-quench specimen is machined from normalized bar stock with some means of hanging it in a vertical position for end quenching. The normalizing temperature listed in Table 13.19 should be used to control the variations in prior structure. This is heated uniformly to the appropriate austenitizing temperature (Table 13.19) for $\frac{1}{2}$ hr. It is then withdrawn from the furnace and placed on a quenching fixture so designed that the specimen stands vertically, and the lower end of the specimen is at a distance of $\frac{1}{2}$ in. (12.7 mm) above the orifice where a jet of water immediately (within 5 s) impinges on the bottom end (Fig. 13.68a).[174] The fixture should be dry at the start of each test. The water-quenching device is adjusted so that the jet of water rises to a free height of 2.5 in. (63.5 mm) above the orifice, without the specimen in position. The water jet is directed at a temperature of 5 to 30°C (40 to 85°F) against the bottom face of the specimen for not less than 10 min. During quenching, a still-air environment should be maintained around the specimen. The entire specimen experiences a range of cooling rates varying from rapid cooling at the bottom end to the comparatively slow cooling at the upper end of the specimen. After quenching, two opposite faces (i.e., 180° apart) are flat-ground longitudinally to a depth of 0.015 in. (0.38 mm) to provide a stable base for Rockwell C hardness measurements and to eliminate the decarburized layer, if any. Rockwell hardness readings are taken at $\frac{1}{16}$-in. (1.59-mm) intervals along these surfaces, starting from the water-quenched end. A plot of these hardness values and their locations on the bar is shown in Fig. 13.69 for an 8650 low-alloy steel. This curve, called the *Jominy hardenability curve* of the steel, also represents the relation between hardness and cooling rate of the designated test.[172]

When using the end-quenching method, the hardenability represents the depth of the quenched zone. In most cases, the depth of the quenched zone, i.e., hardenability, is represented by the distance from the cooled end to the layer with a semi-martensitic structure (50% martensite) for carbon or bainite in alloyed steels.[185]

Figure 13.70[186] shows a comparison of hardenability curves for six grades of steel with indicated compositions and grain sizes. The high hardenability of 4340 steel is indicated by the retention of higher hardness level up to larger distances (40 Rc at 1 in.) from the quenched end. Among these steels, 1040 grade has the lowest hardenability where 12 Rc hardness is found at 2 in. (50.8 mm) from the quenched end. C-type curves for shallow-hardening steels may start high, say at Rc 60 or so, and

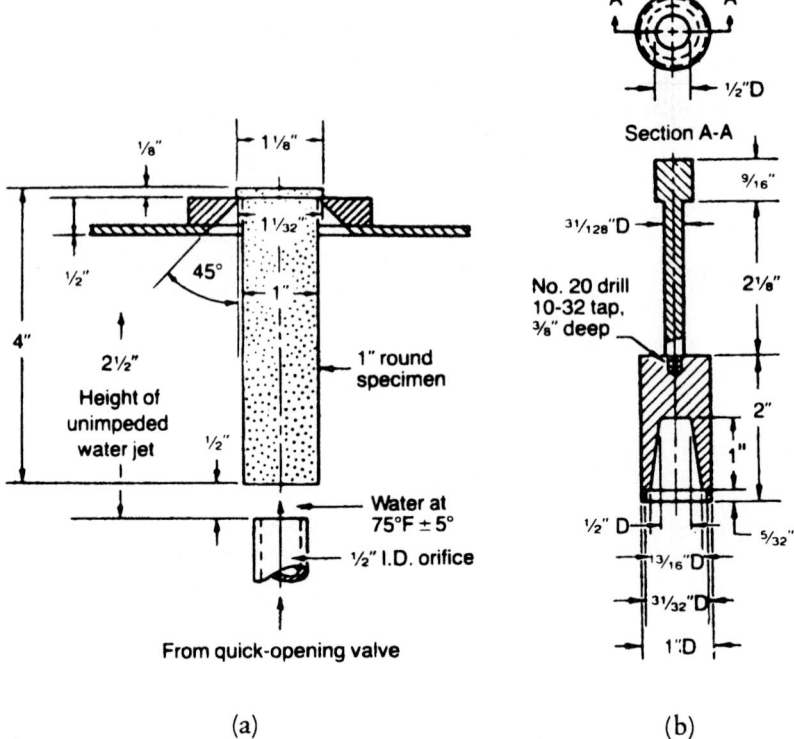

(a) **(b)**

FIGURE 13.68 (*a*) Jominy-Boegehold specimen for end-quench hardenability test. (*b*) The modified (called *hollow*) Jominy hardenability test specimen for shallow-hardening steel. (*Reprinted by permission of ASM International, Materials Park, Ohio.*)

drop very quickly as little as Rc 20 or so just at a minor fraction of the distance from the quenched end.

High-hardenability steels have nearly horizontal Jominy curves irrespective of how far they may be up the hardness scale (i.e., horizontally aligned curves correspond to strongly alloyed steel, and the height of the curves relates to the carbon levels).[187]

In general, steels exhibit two Jominy curves, where the upper curve corresponds to the maximum hardness values representing the upper limit of the composition range of the steel and the lower curve corresponds to the minimum hardness values representing the lower composition limit. Both the curves together are called a *Jominy band* or a *hardenability band*.

13.13.5.2 Shallow-Hardenability Tests

Modified Jominy Test. A modified test bar (Fig. 13.68*b*) is used to evaluate the performance of shallow-hardenability steels. A sufficiently accurate hardenability curve may be drawn under the following conditions:

1. Vicker's pyramid readings should be taken rather than Rockwell hardness readings.

2. Hardness impressions should be more closely spaced.

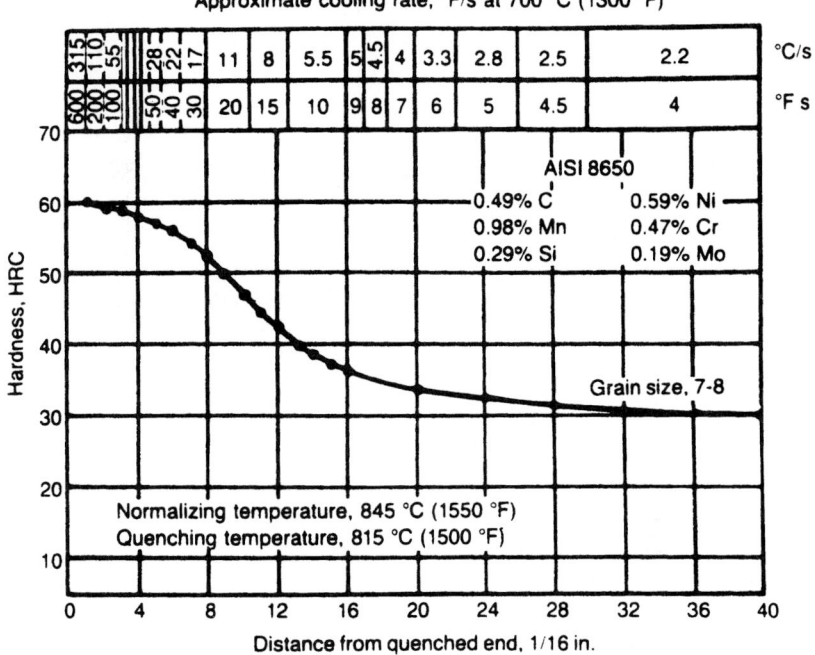

FIGURE 13.69 Procedure for presenting end-quench hardenability data for AISI 8650 steel showing the relationship between cooling rate (top) and Jominy distance from the quenched end.[153] (*Reprinted by permission of ASM International, Materials Park, Ohio; after Jatczak.*)

3. There should be a means to hold the specimen firmly during the measurement of hardness impressions.

The SAC Hardenability Test. The SAC hardenability test, also known as the *Rockwell inch test*, is another hardenability test for shallow-hardening steels. The acronym SAC denotes *surface area center* and is shown in Fig. 13.71.[169] The specimen is a cylinder measuring 1 in. (25.4 mm) diameter × 5.5 in. (140 mm) long. After normalizing for 1 hr and austenitizing for $\frac{1}{2}$ hr at a suitable temperature above Ac_3, the specimen is quenched overall in water at 24 ± 5°C. The 1-in.- (25.4-mm-) long section is cut from its center, and the cut faces are ground very carefully to avoid any burning or tempering effects that might have been introduced during cutting. Rockwell (Rc) measurements are then made at four positions 90° apart on the cylinder surface, and the average hardness of these readings becomes the surface reading. Next, a series of Rc readings are taken along the cross section at $\frac{1}{16}$-in. (1.59-mm) intervals from the surface to the center of the specimen and provide the type of hardness profile shown in Fig. 13.71. The total area under the curve gives the area *A* value in units of Rockwell-inch (using the original imperial unit), and the hardness at the center provides the C value. For example, SAC 65-51-39 indicates a surface hardness of 65 Rc, a Rockwell-inch area of 51, and a center hardness of 39 Rc. One important feature of the SAC test is that it can reveal central segregation in a bar, as shown by a hardness peak at the center position.[169] (See also Sec. 13.13.4.3.)

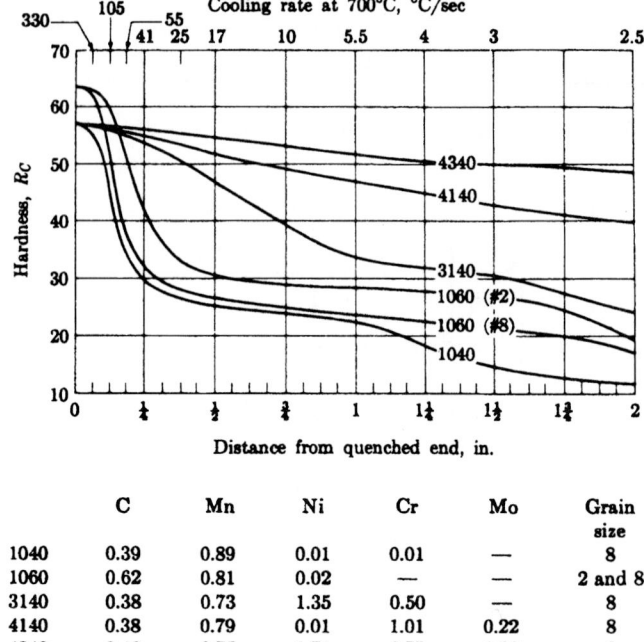

	C	Mn	Ni	Cr	Mo	Grain size
1040	0.39	0.89	0.01	0.01	—	8
1060	0.62	0.81	0.02	—	—	2 and 8
3140	0.38	0.73	1.35	0.50	—	8
4140	0.38	0.79	0.01	1.01	0.22	8
4340	0.40	0.75	1.71	0.77	0.32	8

FIGURE 13.70   Jominy hardenability curves for six steels with the indicated compositions and grain size.[186] (*Adapted from U.S. Steel data.*)

13.13.5.3   Air-Hardenability Tests

Modified Jominy Test.   Jominy test for air-hardening steels consists of the incorporation of a conical stainless steel cap with a hole through which the specimen can be fixed with the cap (Fig. 13.72).[188] During austenitizing, a "leg" is put on the lower end of the specimen to equalize heating, in order to provide the same austenitizing conditions along the entire test specimen. The total heating time is 1 hr, including the 20-min soaking time at the austenitizing temperature. The cap and leg are removed prior to quenching. Figure 13.73a shows the cooling time curves between austenitizing temperature and 500°C for both standard Jominy specimen and the modified Jominy specimen with added cap. Figure 13.73b shows two Jominy hardenability curves—one denoting the standard specimen and the other denoting the modified one, for the hot-working tool steel DIN 45CrMoV67 (containing 0.43 C, 1.3 Cr, 0.7 Mo, 0.23 V). Here both curves are nearly identical except at greater distances from the quenched end. The modified specimen shows slower cooling with the more decreased hardness reading starting at 23 mm compared to 45 mm from the quenched end for the standard specimen.[188]

    (Timken) Air-Hardenability Test.   In this air-hardenability test, a machined and partially threaded round test specimen, 1 in. (25.4 mm) in diameter × 10 in. (254 mm) long, is screwed into a cylindrical bar 6 in. (152 mm) in a hole drilled in a cylindrical bar 6 in. (152 mm) in diameter and 15 in. (383 mm) long, thereby exposing only 4 in. (102 mm) of the test bar length (Fig. 13.74). A second test specimen is screwed at the opposite end of the bar holder to act as a duplicate. The total assembly is

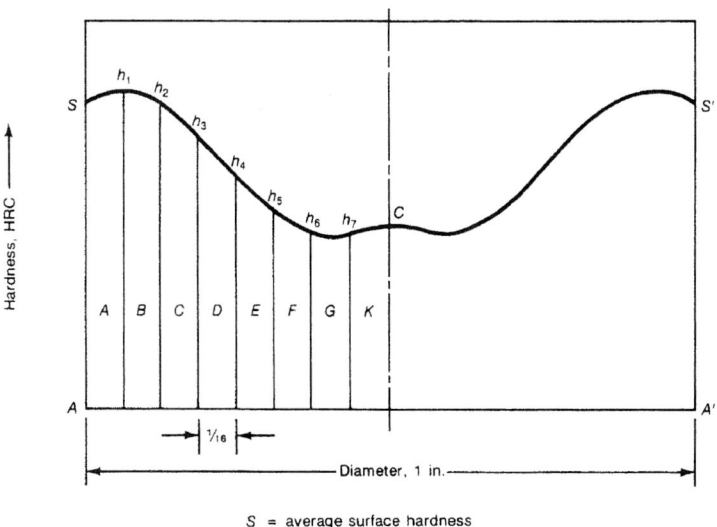

$$S = \text{average surface hardness}$$
$$h_1, h_2, h_3, \text{ and so on} = \text{average hardness at depths indicated}$$
$$C = \text{average center hardness}$$

Then area of $A = \dfrac{S + h_1}{2} \times \frac{1}{16}$

Area of $B = \dfrac{h_1 + h_2}{2} 4 \times \frac{1}{16}$

Total area $= 2(A + B + C + D + E + F + G + K)$

$\qquad\qquad = \frac{1}{8}\left(\dfrac{S}{2} + h_1 + h_2 + h_3 + h_4 + h_5 + h_6 + h_7 + \dfrac{C}{2} \right)$

FIGURE 13.71   Estimation of area according to SAC method.[169]

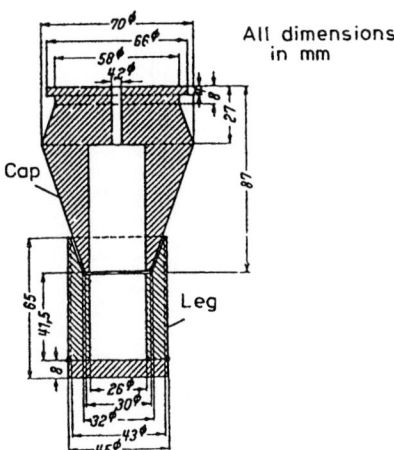

All dimensions in mm

FIGURE 13.72   Modification of the standard Jominy test by the addition of a cap to the specimen for testing the hardenability of air hardening steels.[188]

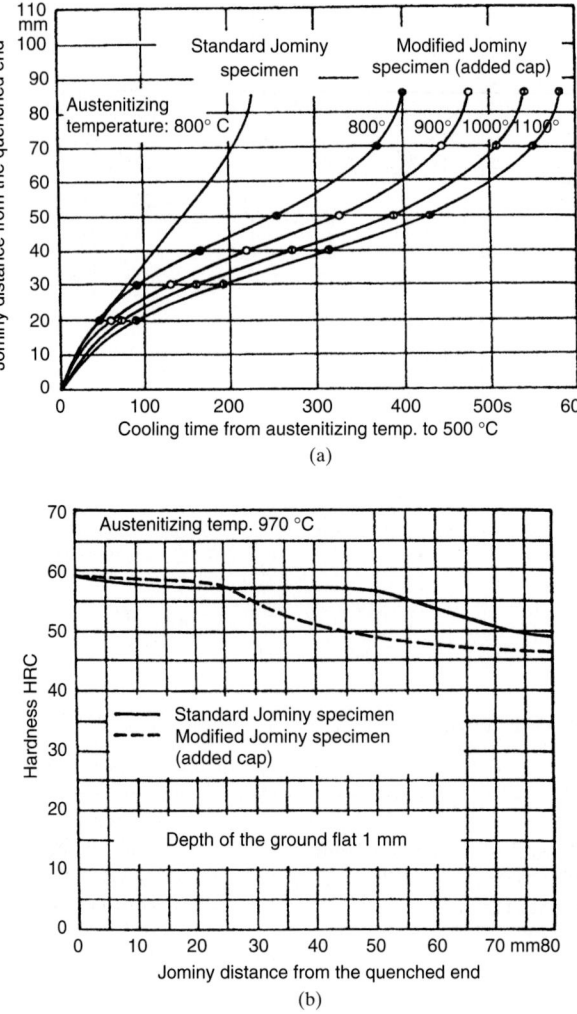

FIGURE 13.73 (*a*) Cooling times between austenitizing temperatures and 500°C for the standard Jominy specimen and for a specimen modified by the addition of a cap.[188] (*b*) Jominy hardenability curves of grade DIN 45CrMoV67 steel for a standard specimen and for a specimen modified by the addition of a cap.[188]

heated to the proper austenitizing temperature for 4 hr; it is then cooled in still air. The cylindrical bar restricts the cooling of the exposed section of each test bar, resulting in numerous cooling conditions along the bar length. Hardness is measured at the desired intervals along each test bar and plotted against distance from the exposed end on charts especially made for this purpose.[189]

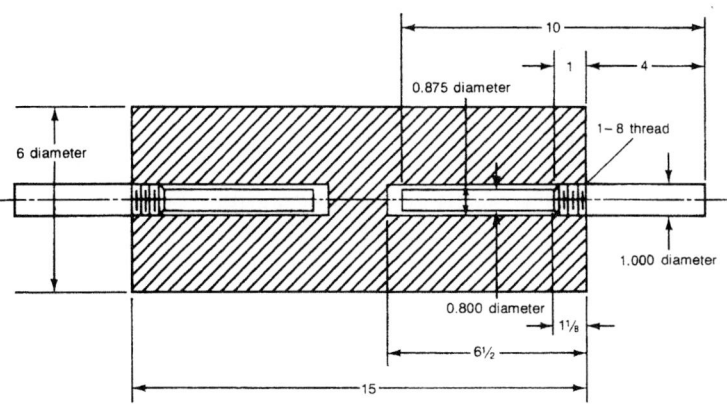

FIGURE 13.74   Timken air hardenability test setup.[189]

13.13.5.4   Boron-Hardenability.   It is well known that the hardenability of boron-treated steels is a function of the amount of boron present in its chemically free or atomic form relative to the total boron contained in the steel. Boron combines readily with oxygen and nitrogen because of its high affinity for them. Hence, during steel-making, sufficient deoxidation by the addition of Al (0.03%) and Si practice also requires the addition of strong nitride formers such as Ti (0.03%) and Zr to tie up the nitrogen and avoid the formation of boron nitride in steel, thereby providing soluble and metallurgically active boron. Without these additions, boron would be present as oxides and nitrides and act as noncontributors to hardenability.

Irvine et al. observed that the optimum acid-soluble boron content for maximum hardenability in 0.2 C-0.73 Mn-0.29 Si-0.55 Mo steel is ~0.002% (Fig. 13.75a),[164] compared to 0.0008% found by Kapadia et al.[189a] and Llwellyn and Cook (Fig. 13.75b).[190] However, a steady-state effect is achieved at boron contents in excess of about 0.0015%. Hence, soluble 0.0025% boron content is typically used commercially to produce a consistent effect by sacrificing a slight loss of harden-ability.[190a] Other workers have summarized that the concentration increase of "effec-tive" B up to 0.002% at simultaneous increase of austenitization temperature and cooling intensity at hardening and also duration reduction of a prehardening inter-val promote hardenability and increase the structural homogeneity.[191]

13.13.5.5   Calculation of Jominy Curves

Method of Calculating Hardenability from Chemical Composition

D_I CALCULATION FOR NON-BORON STEELS FROM GRAPHS.   The Jominy hardenabil-ity curve can be characterized in terms of the ideal critical diameter D_I. Figure 13.76 shows the relationship between an ideal critical diameter D_I which has the same cooling rate at its center as a given location along the surface of a Jominy bar. This relationship is of importance because it is possible, within limits, to interconvert the two methods of determining hardenability. This calculation depends on the fact that steel has a base hardenability predicated (or based) entirely upon carbon content and austenite grain size (Fig. 13.61). When this base hardenability is multiplied pro-gressively by the multiplying factors for each element present (Fig. 13.62), the product gives the hardenability value in terms of ideal diameter [Eq. (13.16)].

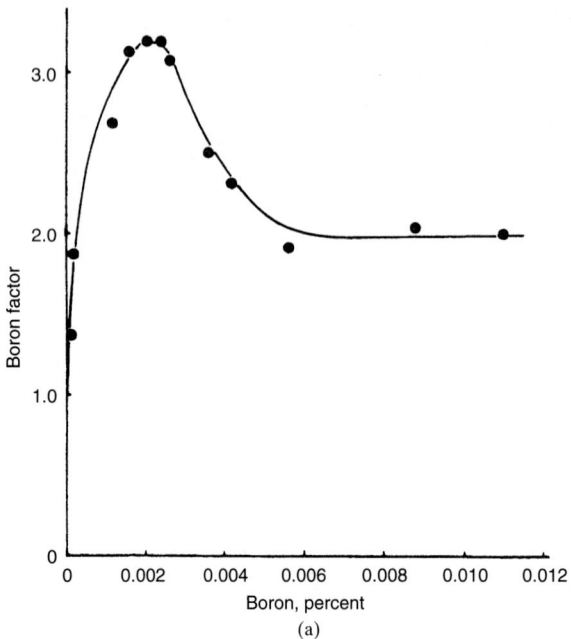

(a)

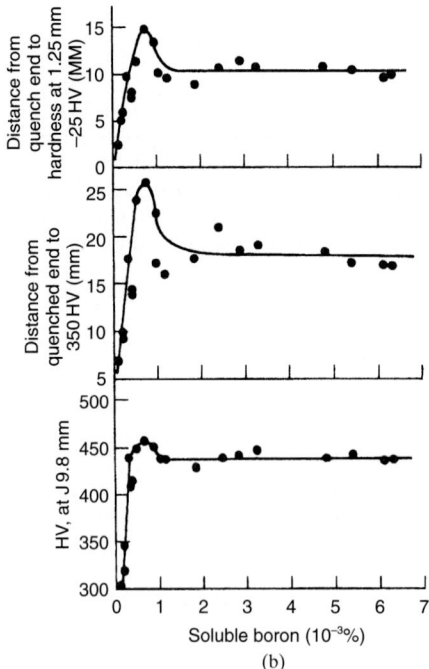

(b)

FIGURE 13.75 (*a*) Boron hardenability factor as a function of boron content for A514-J steel (composition: 0.2 C, 0.73 Mn, 0.29 Si, 0.55 Mo).[164] (*Reprinted by permission of The Metallurgical Society, Warrendale, Pa.*) (*b*) Effect of boron on hardenability.[190] (*After D. T. Llewellyn and W. T. Cook, Metals Technology, December 1974, p. 517.*)

DISTANCE FROM QUENCHED END, mm

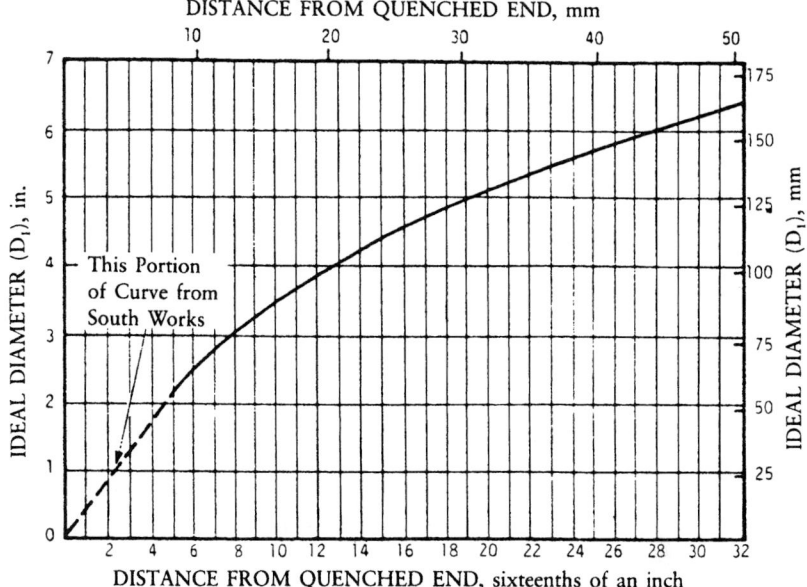

DISTANCE FROM QUENCHED END, sixteenths of an inch

FIGURE 13.76   Relationship between ideal critical diameter D_I and end-quenched distance.[181] (*Reprinted by permission of The Metallurgical Society, Warrendale, Pa.*)

EXAMPLE 13.1   *Calculate* D_I *based on Eq. (13.16), using the DeRetana-Doane approach (Figs. 13.61 and 13.62) for SAE 8620H steel with ASTM grain size No. 7 (composition: 0.2 C, 0.30 Si, 0.8 Mn, 0.5 Cr, 0.6 Ni, and 0.2 Mo).*

Answer. $D = 0.384$, $F_{Si} = 1.0$, $F_{Mn} = 2.30$, $F_{Cr} = 1.413$, $F_{Ni} = 1.18$, $F_{Mo} = 1.2$.

Hence, $D_I = 0.384 \times 1.0 \times 2.3 \times 1.413 \times 1.18 \times 1.2 = 1.767$ in.

These calculations are effective solely for core hardenability of boron-free carburizing steels.[159]

D_I CALCULATION FOR NON-BORON STEELS FROM TABLE.   For modified SAE 4118 steel containing 0.22 C, 0.80 Mn, 0.18 Si, 0.10 Ni, 0.43 Cr, 0.25 Mo, and 0.10 Cu with austenite grain size of ASTM 7, D_I can be calculated using only Table 13.20.[155,173]

$$D_I = (0.119)(3.667)(1.126)(1.036)(1.929)(1.75)(1.04) = 1.79 \text{ in. } (45.5 \text{ mm})$$

D_I *Calculation for Boron Steels.*   With an effective steelmaking process, the boron factor is an inverse function of the carbon and alloy content. The lower the carbon and/or alloy content, the higher the boron factor.[155] This contribution of boron to increased hardenability is called the *boron factor* B_F. The actual boron factor is defined as

$$B_F = \frac{D_I \text{ actual (measured from Jominy data and carbon content)}}{D_I \text{ calculated (from composition excluding boron)}}$$

$$= \frac{\text{hardenability of base (steel + boron)}}{\text{hardenability of base steel}}$$

The following table lists the end-quench test results, calculated D_I value without boron, and chemical composition for an SAE 15B30 modified steel.

EXAMPLE 13.2 *For SAE 15B30 modified steel (composition: 0.29 C, 1.25 Mn, 0.20 Si, 0.13 Ni, 0.07 Cr, 0.03 Mo, and 0.0015 B), the end-quench test results are as follows:*[155,173]

J_D position ($\frac{1}{16}$ in.)	1	2	3	4	5	6	7	8	9	10	12	14	16
Hardness (Rc)	50	50	49	48	47	45	41	38	33	28	25	22	20

J_D position (mm)	1.5	3	5	7	9	11	13	15	20	25
Hardness (Rc)	50	50	49	48	46	41	37	30	24	20

Answer. *First, the nearest position in the end-quench curve (or data) is determined where a hardness representing 50% martensite occurs as a result of only the carbon content. This can be found by the use of Fig. 13.57 or Table 13.23, which gives a hardness of 37 Rc. This hardness occurs at nearly $J_D = \frac{8}{16}$ in. from the quenched end. Next, D_I is calculated from the carbon content and the multiplying factor without boron, which gives 1.24 in. (31.5 mm) (using either Figs. 13.59 and 13.60 or Table 13.21). Using Fig. 13.76 [or Table 13.24 (in.) or Table 13.25 (mm)], the D_I value corresponding to a distance of $J_D = 13$ mm or $\frac{8}{16}$ in. equates to 2.97 in. or 76.4 mm (perhaps requiring interpolation). Thus*

$$B_F = \frac{2.97\,\text{in.}}{1.24\,\text{in.}} = 2.4$$

or

$$B_F = \frac{76.4\,\text{mm}}{31.5\,\text{mm}} = 2.43$$

Note that the ideal diameter with boron D_{IB} is given by

$$D_{IB} = D_I(\text{without boron}) \times B_F$$

The actual calculation of the Jominy curve consists of the following steps:

1. Determine the D_I value by using Eq. (13.16) and Figs. 13.61 and 13.62.
2. Determine the ratio of the initial (or quenched-end) hardness of the Jominy bar, I_H to the hardness at each location of the bar D_H (using the D_I value calculated in step 1), and plot or tabulate this I_H/D_H ratio as a function of the calculated D_I values for various steels.
3. Determine the I_H value from the carbon content, using the hardness versus carbon content plot or the equation involving regression coefficients or Table 13.22.
4. Divide the I_H value by the appropriate I_H/D_H ratio obtained in step 2 to obtain D_H for each position along the Jominy bar.

This method is valid only for the core-hardenability steels from which they are derived. These hardenability calculations (or curve determinations) can be found in the articles by Field,[192] Boyd and Field,[193] and Tartaglia and Eldis.[159]

TABLE 13.23 Initial 100% Martensite Hardness and 50% Martensite Hardness versus % Carbon[155]

% Carbon content	Hardness, HRC		% Carbon content	Hardness, HRC		% Carbon content	Hardness, HRC	
	Initial 100% Martensite	50% Martensite		Initial 100% Martensite	50% Martensite		Initial 100% Martensite	50% Martensite
0.10	38	26	0.30	50	37	0.50	61	47
0.11	39	27	0.31	51	38	0.51	61	47
0.12	40	27	0.32	51	38	0.52	62	48
0.13	40	28	0.33	52	39	0.53	62	48
0.14	41	28	0.34	53	40	0.54	63	48
0.15	41	29	0.35	53	40	0.55	63	49
0.16	42	30	0.36	54	41	0.56	63	49
0.17	42	30	0.37	55	41	0.57	64	50
0.18	43	31	0.38	55	42	0.58	64	50
0.19	44	31	0.39	56	42	0.59	64	51
0.20	44	32	0.40	56	43	0.60	64	51
0.21	45	32	0.41	57	43	0.61	64	51
0.22	45	33	0.42	57	43	0.62	65	51
0.23	46	34	0.43	58	44	0.63	65	52
0.24	46	34	0.44	58	44	0.64	65	52
0.25	47	35	0.45	59	45	0.65	65	52
0.26	48	35	0.46	59	45	0.66	65	52
0.27	49	36	0.47	59	45	0.67	65	53
0.28	49	36	0.48	59	46	0.68	65	53
0.29	50	37	0.49	60	46	0.69	65	53

13.127

TABLE 13.24   Jominy Distance for 50% Martensite versus D_I (in.)[155]

$J = {}^1/_{16}$ in.	D_I, in.	$J = {}^1/_{16}$ in.	D_I, in.	$J = {}^1/_{16}$ in.	D_I, in.
0.5	0.27	11.5	3.74	22.5	5.46
1.0	0.50	12.0	3.83	23.0	5.51
1.5	0.73	12.5	3.94	23.5	5.57
2.0	0.95	13.0	4.04	24.0	5.63
2.5	1.16	13.5	4.13	24.5	5.69
3.0	1.37	14.0	4.22	25.0	5.74
3.5	1.57	14.5	4.32	25.5	5.80
4.0	1.75	15.0	4.40	26.0	5.86
4.5	1.93	15.5	4.48	26.5	5.91
5.0	2.12	16.0	4.57	27.0	5.96
5.5	2.29	16.5	4.64	27.5	6.02
6.0	2.45	17.0	4.72	28.0	6.06
6.5	2.58	17.5	4.80	28.5	6.12
7.0	2.72	18.0	4.87	29.0	6.16
7.5	2.86	18.5	4.94	29.5	6.20
8.0	2.97	19.0	5.02	30.0	6.25
8.5	3.07	19.5	5.08	30.5	6.29
9.0	3.20	20.0	5.15	31.0	6.33
9.5	3.32	20.5	5.22	31.5	6.37
10.0	3.43	21.0	5.28	32.0	6.42
10.5	3.54	21.5	5.33		
11.0	3.64	22.0	5.39		

TABLE 13.25   Jominy Distance for 50% Martensite versus D_I (mm)[155]

J, mm	D_I, mm	J, mm	D_I, mm	J, mm	D_I, mm
1.0	8.4	18.0	94.2	35.0	137.1
2.0	15.7	19.0	97.1	36.0	139.1
3.0	22.9	20.0	100.6	37.0	140.9
4.0	29.7	21.0	103.7	38.0	142.8
5.0	36.3	22.0	106.5	39.0	144.7
6.0	42.9	23.0	109.7	40.0	146.4
7.0	48.2	24.0	112.2	41.0	148.3
8.0	54.2	25.0	114.9	42.0	150.1
9.0	59.5	26.0	117.4	43.0	151.7
10.0	64.2	27.0	119.9	44.0	153.4
11.0	68.6	28.0	122.4	45.0	154.1
12.0	72.1	29.0	124.7	46.0	156.5
13.0	76.4	30.0	127.1	47.0	157.8
14.0	80.1	31.0	129.0	48.0	159.2
15.0	84.0	32.0	131.4	49.0	160.5
16.0	87.6	33.0	133.5	50.0	161.8
17.0	90.1	34.0	135.2		

Just[194] has developed numerous equations based on linear, nonlinear, and interaction models as well as grain size for calculating the Jominy curve (i.e., determining the hardness at different Jominy distances, J_Ds) by using the multiple regression analysis technique to determine the quantitative influence of the alloying elements. This method is based separately on SAE bands, USS-Atlas, and MPI-Atlas. The usual compositional ranges of AISI-SAE H steels based on three models are: 0.10 to 0.64 C, 0.45 to 1.75 Mn, 0.15 to 1.95 Si, 0 to 5.0 Ni, 0 to 1.55 Cr, 0 to 0.52 Mo, and 0.02 V. He observed the increasing influence of all alloying elements below $J_D = J_4$,[†] but beyond this position their effects stay nearly the same.

The hardness in the quenched end (i.e., Jominy distance zero J_0) is only contributed by the carbon content. When his formulae are used, a hardenability expression for each Jominy position can be known. Since Jominy factors vary slightly between J_4 and J_{25}, a single formula may be used for this entire distance. For different grades of steel, different expressions for hardenability are used. Using the nonlinear hardenability model, the Rc hardness values at J_0 and J_1 are given by

$$J_0(Rc) = 60\sqrt{C} + 20\,Rc \qquad C < 0.6\% \tag{13.19}$$

$$J_1(Rc) = 60\sqrt{C} + 1.6\,Cr + 1.5\,Mn + 16\,Rc \tag{13.20}$$

Just[194] divided the steel into two groups based primarily on the interaction model (i.e., interaction between carbon and alloying elements), namely, case-hardenable (C < 0.28%) and hardenable (C > 0.29%) grades.

For case-hardening steels, we have the following equation:

$$J_{4-25}(Rc) = 87\,C + 14\,Cr + 5.3\,Ni + 29\,Mo + 16\,Mn - 21.2\sqrt{E} + 2.21\,E + 22\,Rc \tag{13.21}$$

where E is another factor, defined as the *Jominy depth* (in $\frac{1}{16}$ in.) from the quenched end.

For hardenable steels, we have

$$J_{4-25}(Rc) = 78\,C + 22\,Cr + 21\,Mn + 6.9\,Ni + 33\,Mo - 2.03\sqrt{E} + 1.86\,E + 18\,Rc \tag{13.22}$$

As expected, the equation for the hardenable steel possesses higher coefficients.

When both the composition and grain size of the 37 steels listed in USS Atlas were considered, Just[194] derived two different formulae:

$$J_{4-40}(Rc) = 88\sqrt{C} - 0.0135E^2\sqrt{C} + 16\,Mn + 5\,Si + 19\,Cr + 6.3\,Ni$$
$$+ 35\,Mo - 0.82\,G_{ASTM} - 20\sqrt{E} + 2.11\,E - 2\,Rc \tag{13.23}$$

and

$$J_{4-32}(Rc) = 98\sqrt{C} - 0.025E^2\sqrt{C} + 19\,Mn + 5\,Si + 20\,Cr + 6.4\,Ni$$
$$+ 34\,Mo + 28\,V - 0.82\,G_{ASTM} - 24\sqrt{E} + 2.86\,E - 1\,Rc \tag{13.24}$$

The compositional ranges applicable to these formulae are 0.08 to 0.56 C, 0.20 to 1.88 Mn, 0 to 3.8 Si, 0 to 8.94 Ni, 0 to 1.97 Cr, 0 to 0.53 Mo, and 1.5 to 11 ASTM grain size G.

[†] J_D is the Jominy distance; when expressed with a numerical subscript instead of the subscript D (Jominy position), it means that number multiplied by $\frac{1}{16}$ in. from the quenched end of the bar. For example, $J_2 = 2 \times \frac{1}{16}$ in. (3.175 mm).

These formulae do not predict the hardenability of steel precisely. However, these formulae are intended primarily to (1) aid the designer in determining the selection of steel and (2) assist the metallurgist in correcting the melt.

Kirkaldy[195,196] has proposed a method for calculating a Jominy curve by means of computers. He assumed initially that the nucleation of ferrite, pearlite, and bainite occurs immediately; however, the growth rate $G(T)$ needs calculation. He has formulated analytically an expression for the cooling rates at various Jominy distances. He further assumed that the cooling rate which produces 0.1% of transformed product at the temperature representing the maximum $G(T)$ also represents the Jominy inflection point where the martensite content is 50% (Fig. 13.77).

He was able to plot the whole Jominy curve by determining, first, the carbon content corresponding to the maximum hardness; second, the D_I value; and third, the position of the inflection point. Figure 13.78 shows the results due to Kirkaldy and his coworkers.

13.13.5.6 Use of Jominy Hardenability Curves. Jominy hardenability curves are the required method to characterize the steel. They are employed to compare the hardenability of different heats of the same steel grade as a quality control method in the steel manufacture and to compare the hardenability of different steel grades when selecting steel for a specific application. In the latter situation, Jominy curves are used to predict the hardening depth. Such predictions are usually dependent on the assumptions that the rates of cooling prevailing at different distances from the quenched end of the Jominy specimen may be compared with the cooling rates predominating at different locations on the cross sections of bars of different diameters. If the cooling rates are equal, it is assumed that equivalent microstructure and hardness can be expected after quenching.

Figure 13.79 shows the correlation between equivalent cooling rates along the Jominy specimen and at four locations in actual round bars up to 4 in. (102 mm) in diameter when quenched in mildly agitated water and oil at 60 m/min (200 ft/min). The cooling rates at both the surface and interior points decrease with the increase in bar diameter. The Jominy data found from Fig. 13.79 can be used directly in steel selection.

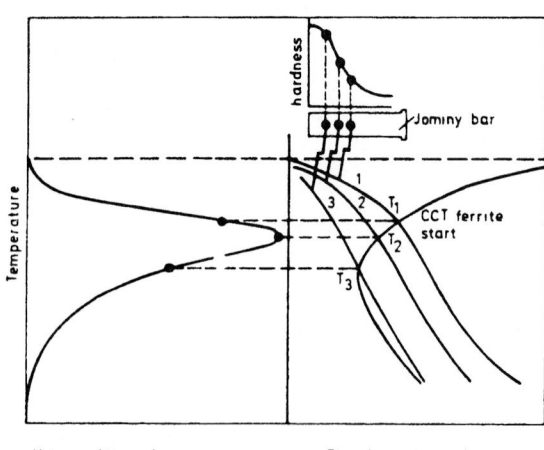

FIGURE 13.77 Construction for locating the inflection point of a Jominy curve.[195,196] (*Reprinted by permission of The Metallurgical Society, Warrendale, Pa.*)

Hardness
HRC

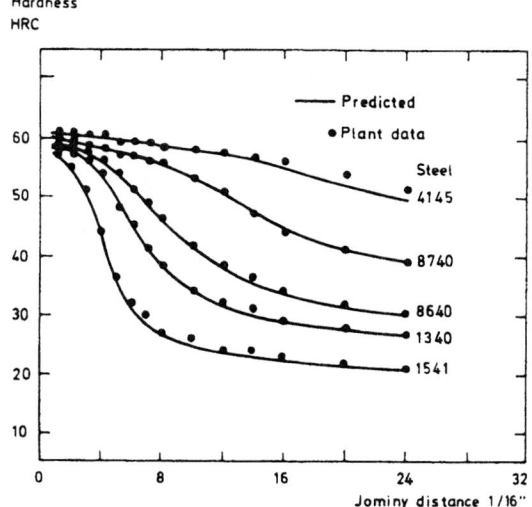

FIGURE 13.78 Comparison of predicted and observed Jominy curves for five steels.[195,196] (*Reprinted by permission of The Metallurgical Society, Warrendale, Pa.*)

13.13.5.7 Computer Calculations of Jominy Hardenability.

More reliable formulae were developed in the late 1970s to be employed in calculations facilitating alloy development, quality control, and control of steelmaking and heat-treating processes.[197]

Several computerized on-line systems for calculating hardenability curves and numerous properties have been proposed and used. The International Harvester Company has developed the Computer Harmonized Application Tailored (CHAT) Process. This system consists of two basic concepts: the first procedure, termed CH, is designed to quantitatively determine the hardenability requirement in terms of D_I by using a modified linear programming technique known as *separable programming*. The second procedure, termed AT, is employed to determine minimum and maximum hardenability levels required to design or select the most economical steel with composition sufficient to develop the desired engineering properties in a part for a given application.[198]

In the Creusot-Loire method, the general features of the kinetic properties of low-alloy steels are described by mathematical models fitted to experimental data by regression analysis. A set of 341 carefully selected experimental CCT diagrams have been employed to obtain regression equations for the critical cooling rate at 700°C for onset, specific percent transformation, or completion of different decomposition reactions of austenite in the quenched and quenched and tempered conditions, and to calculate the mechanical properties such as hardness, yield and tensile strengths, making use of a sum rule over volume fractions. The equations are valid in the following composition range: 0.2 to 0.5 C, 0 to 2 Mn, 0 to 1 Si, 0 to 4 Ni, 0 to 3 Cr, 0 to 1 Mo, and 0 to 0.2 V. This process is increasingly used for rapid calculation of hardness, Jominy curves, mechanical properties of welded steels, as-rolled steel products (plates and bars), and forgings and steels for nuclear application. The use of the personal computer software package HAZ CALCULATOR actually facilitates the application of this method with satisfactory accuracy.[199,200]

There are also several versions on the market which are sold as a compact disk. For best results, companies that use these should calibrate to their own

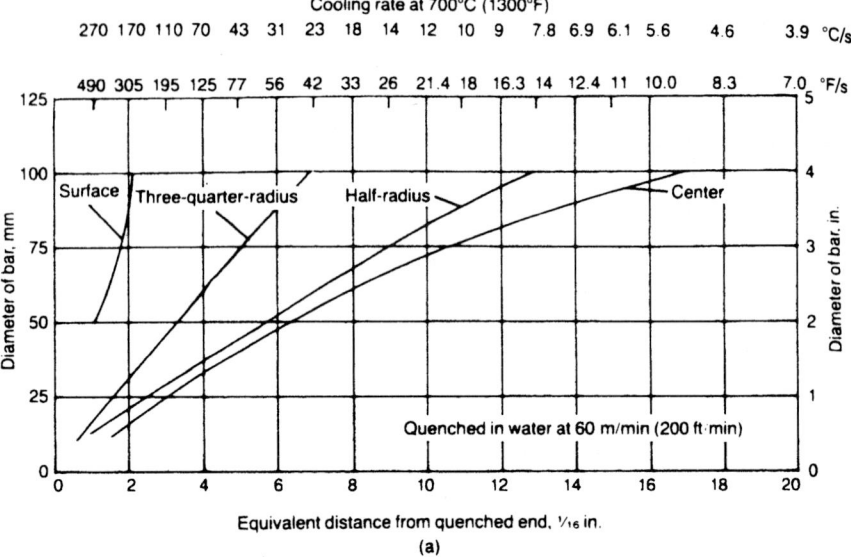

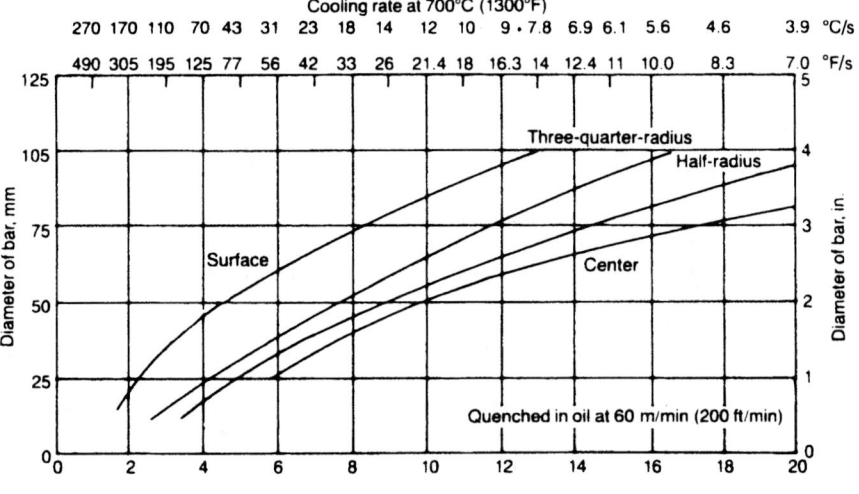

FIGURE 13.79 Equivalent cooling rates for round bars quenched in (*a*) water and (*b*) oil.[153] Data for surface hardness are for "mild agitation"; other data are for 60 m/min (200 ft/min). (*Reprinted by permission of ASM International, Materials Park, Ohio; after Jatczak.*)

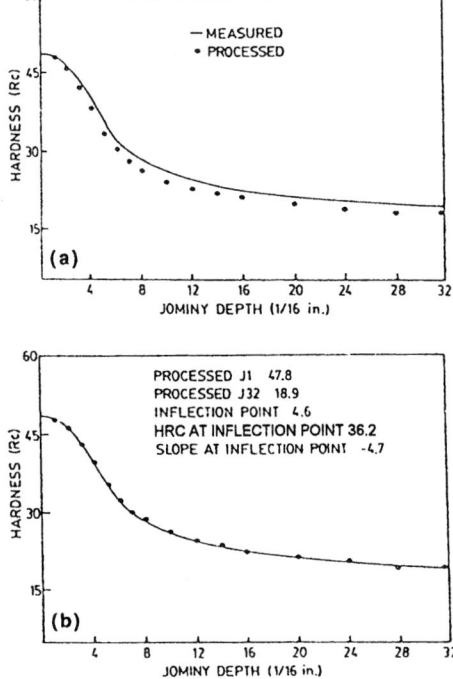

FIGURE 13.80 Outputs from Minitech Predictor data processing program for best fit to measured Jominy data. (*a*) Initial trial; (*b*) final trial.[202]

historical data. Current clean steels have a larger grain size and, therefore, higher hardenability.

Another system, the Minitech Computerized Alloy Steel Information System, involves 12 computer programs based on fundamental theory, alloy composition, and other input parameters; and it provides entire Jominy curves (based on the theoretical developments of the inflection point principle of Kirkaldy and co-workers), mechanical properties of hot-rolled products, quenched and tempered products, carburized products, hardness distribution, and weldability parameters with good accuracy for boron and non-boron steels containing 0.07 to 1.3% C and alloy ranges of AISI-SAE hardenability steels.[201] This system is strongly calibrated to plant data.

Figure 13.80 shows a typical output of the Minitech Predictor operating in the data processing mode for best fit to measured Jominy data at the initial and final trials, where the input values include chemical composition, Jominy hardness values, and predicted grain size.[202]

13.14 HARDENABILITY (OR H-) STEELS

Hardenability steels exhibiting hardenability limits have chemical composition ranges slightly broader than the equivalent normal AISI-SAE grades, to allow the manufacturers greater flexibility in achieving hardenability curves within the specified limits. Consequently, these hardenability steels offer a wide range of mechanical properties that are based on the maximum as-quenched hardness and the degree of tempering of martensite after quenching.[203] These hardenability limits are also called *hardenability bands*. These steel grades are designated by the letter H after the composition code or preceding UNS designation.

The width of the H-band for a particular steel is agreed upon by a compromise between the interests of the steel producer, who would prefer a larger width to increase the volume of heats that meet the specification, and the user, who would prefer a narrow band to obtain uniform mechanical properties of the finished products.

13.14.1 Shallow-Hardenable Steels (H-Band Carbon Steels)

Tables 13.26 and 13.27 list composition ranges of H-band AISI-SAE carbon steels and alloy steels, respectively. Carbon steels with very low Mn (say, 0.3%) and practically free of residual Ni, Cr, and Mo have the lowest hardenability. Most of the H-band 1xxx series carbon steels have 0.6 to 0.9% Mn content, although there are many exceptions. Those steels containing Mn in the range of 1.00 to 1.65% constitute the 15xx series, where higher Mn level contributes a significant increase in hardenability.

High-carbon steels (between 0.55 and 1.00% carbon) are more restricted in their applications because they are more costly and more difficult to fabricate due to decreased formability, machinability, and weldability. At the same time, they are less ductile in the quenched and tempered condition. Higher-carbon steels such as 1070 and 1095 are especially suitable for springs where resistance to fatigue and permanent set is a necessary requirement. Most of the parts made from steels in this

TABLE 13.26 Composition of Carbon and Carbon-Boron H Steels[204]

| UNS No. | SAE or AISI No. | Ladle chemical composition, wt% | | | | |
		C	Mn	Si	P, maximum[‡]	S, maximum[‡]
H10380	1038H	0.34/0.43	0.50/1.00	0.15/0.35	0.040	0.050
H10450	1045H	0.42/0.51	0.50/1.00	0.15/0.35	0.040	0.050
H15220	1522H	0.17/0.25	1.00/1.50	0.15/0.35	0.040	0.050
H15240	1524H	0.18/0.26	1.25/1.75	0.15/0.35	0.040	0.050
H15260	1526H	0.21/0.30	1.00/1.50	0.15/0.35	0.040	0.050
H15410	1541H	0.35/0.45	1.25/1.75	0.15/0.35	0.040	0.050
H15211	15B21H[†]	0.17/0.24	0.70/1.20	0.15/0.35	0.040	0.050
H15281	15B28H[†]	0.25/0.34	1.00/1.50	0.15/0.35	0.040	0.050
H15301	15B30H[†]	0.27/0.35	0.70/1.20	0.15/0.35	0.040	0.050
H15351	15B35H[†]	0.31/0.39	0.70/1.20	0.15/0.35	0.040	0.050
H15371	15B37H[†]	0.30/0.39	1.00/1.50	0.15/0.35	0.040	0.050
H15411	15B41H[†]	0.35/0.45	1.25/1.75	0.15/0.35	0.040	0.050
H15481	15B48H[†]	0.43/0.53	1.00/1.50	0.15/0.35	0.040	0.050
H15621	15B62H[†]	0.54/0.67	1.00/1.50	0.40/0.60	0.040	0.050

[†] These steels contain 0.0005 to 0.003% B.

[‡] If electric furnace practice is specified or required, the limit for both phosphorus and sulfur is 0.025%, and the prefix E is added to the SAE or AISI number.

Source: *2001 SAE Handbook*, vol. 1, *Materials, Fuels, Emissions, and Noise*, Society of Automotive Engineers, Warrendale, Pa., 2001.

TABLE 13.27 Composition of Standard Alloy H Steel[204]

The ranges and limits on this table apply only to material not exceeding $1.3 \times 10^5 \, mm^2$ ($200 \, in.^2$) in cross-sectional area, 460 mm (18 in.) in width, or 4.5 Mg (10,000 lb) in weight per piece. Ranges and limits are subject to the permissible variations for product analysis shown in Table 4 of SAE J409.

UNS No.	SAE or AISI No.	Ladle chemical composition, wt%						
		C	Mn	Si	Ni	Cr	Mo	V
H13300	1330H	0.27/0.33	1.45/2.05	0.15/0.35	...	...	...	...
H13350	1335H	0.32/0.38	1.45/2.05	0.15/0.35	...	...	...	...
H13400	1340H	0.37/0.44	1.45/2.05	0.15/0.35	...	...	...	...
H13450	1345H	0.42/0.49	1.45/2.05	0.15/0.35	...	...	...	...
H40270	4027H	0.24/0.30	0.60/1.00	0.15/0.35	...	...	0.20/0.30	...
H40280	4028H[c]	0.24/0.30	0.60/1.00	0.15/0.35	...	...	0.20/0.30	...
H40320	4032H	0.29/0.35	0.60/1.00	0.15/0.35	...	...	0.20/0.30	...
H40370	4037H	0.34/0.41	0.60/1.00	0.15/0.35	...	...	0.20/0.30	...
H40420	4042H	0.39/0.46	0.60/1.00	0.15/0.35	...	...	0.20/0.30	...
H40470	4047H	0.44/0.51	0.60/1.00	0.15/0.35	...	...	0.20/0.30	...
H41180	4118H	0.17/0.23	0.60/1.00	0.15/0.35	...	0.30/0.70	0.08/0.15	...
H41300	4130H	0.27/0.33	0.30/0.70	0.15/0.35	...	0.75/1.20	0.15/0.25	...
H41350	4135H	0.32/0.38	0.60/1.00	0.15/0.35	...	0.75/1.20	0.15/0.25	...
H41370	4137H	0.34/0.41	0.60/1.00	0.15/0.35	...	0.75/1.20	0.15/0.25	...
H41400	4140H	0.37/0.44	0.65/1.10	0.15/0.35	...	0.75/1.20	0.15/0.25	...
H41420	4142H	0.39/0.46	0.65/1.10	0.15/0.35	...	0.75/1.20	0.15/0.25	...
H41450	4145H	0.42/0.49	0.65/1.10	0.15/0.35	...	0.75/1.20	0.15/0.25	...
H41470	4147H	0.44/0.51	0.65/1.10	0.15/0.35	...	0.75/1.20	0.15/0.25	...
H41500	4150H	0.47/0.54	0.65/1.10	0.15/0.35	...	0.75/1.20	0.15/0.25	...
H41610	4161H	0.55/0.65	0.65/1.10	0.15/0.35	...	0.65/0.95	0.25/0.35	...
H43200	4320H	0.17/0.23	0.40/0.70	0.15/0.35	1.55/2.00	0.35/0.65	0.20/0.30	...
H43400	4340H	0.37/0.44	0.55/0.90	0.15/0.35	1.55/2.00	0.65/0.95	0.20/0.30	...
H43406	E4340H[d]	0.37/0.44	0.60/0.95	0.15/0.35	1.55/2.00	0.65/0.95	0.20/0.30	...
H46200	4620H	0.17/0.23	0.35/0.75	0.15/0.35	1.55/2.00	...	0.20/0.30	...

ABLE 13.27 Composition of Standard Alloy H Steel[204] (Continued)

The ranges and limits on this table apply only to material not exceeding $1.3 \times 10^5 \, mm^2$ ($200 \, in.^2$) in cross-sectional area, $460 \, mm$ ($18 \, in.$) in width, or $4.5 \, Mg$ ($10,000 \, lb$) in weight per piece. Ranges and limits are subject to the permissible variations for product analysis shown in Table 4 of SAE J409.

UNS No.	SAE or AISI No.	Ladle chemical composition, wt%						
		C	Mn	Si	Ni	Cr	Mo	V
H47180	4718H	0.15/0.21	0.60/0.95	0.15/0.35	0.85/1.25	0.30/0.60	0.30/0.40	...
H47200	4720H	0.17/0.23	0.45/0.75	0.15/0.35	0.85/1.25	0.30/0.60	0.15/0.25	...
H48150	4815H	0.12/0.18	0.30/0.70	0.15/0.35	3.20/3.80	...	0.20/0.30	...
H48170	4817H	0.14/0.20	0.30/0.70	0.15/0.35	3.20/3.80	...	0.20/0.30	...
H48200	4820H	0.17/0.23	0.40/0.80	0.15/0.35	3.20/3.80	...	0.20/0.30	...
H50401	50B40H^e	0.37/0.44	0.65/1.10	0.15/0.35	...	0.30/0.70	...	...
H50441	50B44H^e	0.42/0.49	0.65/1.10	0.15/0.35	...	0.30/0.70	...	...
H50460	5046H	0.43/0.50	0.65/1.10	0.15/0.35	...	0.13/0.43	...	...
H50461	50B46H^e	0.43/0.50	0.65/1.10	0.15/0.35	...	0.13/0.43	...	...
H50501	50B50H^e	0.47/0.54	0.65/1.10	0.15/0.35	...	0.30/0.70	...	...
H50601	50B60H^e	0.55/0.65	0.65/1.10	0.15/0.35	...	0.30/0.70	...	...
H51200	5120H	0.17/0.23	0.60/1.00	0.15/0.35	...	0.60/1.00	...	...
H51300	5130H	0.27/0.33	0.60/1.00	0.15/0.35	...	0.75/1.20	...	...
H51320	5132H	0.29/0.35	0.50/0.90	0.15/0.35	...	0.65/1.10	...	...
H51350	5135H	0.32/0.38	0.50/0.90	0.15/0.35	...	0.70/1.15	...	...
H51400	5140H	0.37/0.44	0.60/1.00	0.15/0.35	...	0.60/1.00	...	...
H51470	5147H	0.45/0.52	0.60/1.05	0.15/0.35	...	0.80/1.25	...	...
H51500	5150H	0.47/0.54	0.60/1.00	0.15/0.35	...	0.60/1.00	...	...
H51550	5155H	0.50/0.60	0.60/1.00	0.15/0.35	...	0.60/1.00	...	...
H51600	5160H	0.55/0.65	0.65/1.10	0.15/0.35	...	0.60/1.00	...	...
H51601	51B60H^e	0.55/0.65	0.65/1.10	0.15/0.35	...	0.60/1.00	...	...
H61180	6118H	0.15/0.21	0.40/0.80	0.15/0.35	...	0.40/0.80	...	0.10/0.15
H61500	6150H	0.47/0.54	0.60/1.00	0.15/0.35	...	0.75/1.20	...	0.15

13.136

UNS No.	SAE No.	C	Mn	Si	Ni	Cr	Mo	
H81451[e]	81B4S[e]	0.42/0.49	0.70/1.05	0.15/0.35	0.15/0.45	0.30/0.60	0.08/0.15	...
H86170	8617H	0.14/0.20	0.60/0.95	0.15/0.35	0.35/0.75	0.35/0.65	0.15/0.25	...
H86200	8620H	0.17/0.23	0.60/0.95	0.15/0.35	0.35/0.75	0.35/0.65	0.15/0.25	...
H86220	8622H	0.19/0.25	0.60/0.95	0.15/0.35	0.35/0.75	0.35/0.65	0.15/0.25	...
H86250	8625H	0.22/0.28	0.60/0.95	0.15/0.35	0.35/0.75	0.35/0.65	0.15/0.25	...
H86270	8627H	0.24/0.30	0.60/0.95	0.15/0.35	0.35/0.75	0.35/0.65	0.15/0.25	...
H86300	8630H	0.27/0.33	0.60/0.95	0.15/0.35	0.35/0.75	0.35/0.65	0.15/0.25	...
H86301[e]	86B30H[e]	0.27/0.33	0.60/0.95	0.15/0.35	0.35/0.75	0.35/0.65	0.15/0.25	...
H86370	8637H	0.34/0.41	0.70/1.05	0.15/0.35	0.35/0.75	0.35/0.65	0.15/0.25	...
H86400	8640H	0.37/0.44	0.70/1.05	0.15/0.35	0.35/0.75	0.35/0.65	0.15/0.25	...
H86420	8642H	0.39/0.46	0.70/1.05	0.15/0.35	0.35/0.75	0.35/0.65	0.15/0.25	...
H86450	8645H	0.42/0.49	0.70/1.05	0.15/0.35	0.35/0.75	0.35/0.65	0.15/0.25	...
H86451[e]	86B45H[e]	0.42/0.49	0.70/1.05	0.15/0.35	0.35/0.75	0.35/0.65	0.15/0.25	...
H86500	8650H	0.47/0.54	0.70/1.05	0.15/0.35	0.35/0.75	0.35/0.65	0.15/0.25	...
H86550	8655H	0.50/0.60	0.70/1.05	0.15/0.35	0.35/0.75	0.35/0.65	0.15/0.25	...
H86600	8660H	0.55/0.65	0.70/1.05	0.15/0.35	0.35/0.75	0.35/0.65	0.15/0.25	...
H87200	8720H	0.17/0.23	0.60/0.95	0.15/0.35	0.35/0.75	0.35/0.65	0.20/0.30	...
H87400	8740H	0.37/0.44	0.70/1.05	0.15/0.35	0.35/0.75	0.35/0.65	0.20/0.30	...
H88220	8822H	0.19/0.25	0.70/1.05	0.15/0.35	0.35/0.75	0.35/0.65	0.30/0.40	...
H92600	9260H	0.55/0.65	0.65/1.10	1.70/2.20	...	...	...	...
H93100[d]	9310H[d]	0.07/0.13	0.40/0.70	0.15/0.35	2.95/3.55	1.00/1.45	0.08/0.15	...
H94151[e]	94B15H[e]	0.12/0.18	0.70/1.05	0.15/0.35	0.25/0.65	0.25/0.55	0.08/0.15	...
H94171[e]	94B17H[e]	0.14/0.20	0.70/1.05	0.15/0.35	0.25/0.65	0.25/0.55	0.08/0.15	...
H94301[e]	94B30H[e]	0.27/0.33	0.70/1.05	0.15/0.35	0.25/0.65	0.25/0.55	0.08/0.15	...

[a] Small quantities of certain elements may be found in alloy steel that are not specified or required. These elements are to be considered incidental and acceptable to the following maximum amounts: copper to 0.35%, nickel to 0.25%, chromium to 0.20%, and molybdenum to 0.06%.

[b] For open hearth and basic oxygen steels, maximum sulfur content is to be 0.040%, and maximum phosphorus content is to be 0.035%. Maximum phosphorus and sulfur in basic electric furnace steels are to be 0.025% each.

[c] Sulfur content range is 0.035/0.050%.

[d] Electric furnace steel.

[e] These steels contain 0.0005 to 0.003% B.

Source: 2001 SAE Handbook, vol. 1, Materials, Fuels, Emissions, and Noise, Society of Automotive Engineers, Warrendale, Pa., 2001.

category are hardened by conventional quenching. Water-quenching is employed for heavy sections of the lower-carbon steels and for cutting edges, whereas oil-quenching is employed for general purposes. Austempering and martempering often successfully take advantage of the considerably reduced distortion, negligible breakage, and greater toughness at high hardness.

In spite of all favorable hardenability factors and the use of a drastic quench, these steels are much shallower hardening than are alloy steels because carbon singly, or in combination with Mn, in amounts even up to 1.65% does not promote deep hardening. The dangers of quench-cracking in these steel parts, especially in the nonuniform section, are very high, which requires closer supervision and control.

Screwdrivers, pliers, and wrenches (except the Stillson type) are generally oil-hardened, followed by tempering to the required hardness range. Even when reduction in as-quenched hardness is not desired, stress-relieving at 150 to 190°C (300 to 375°F) should be accomplished to help prevent sudden service failures. In Stillson-type wrenches, the jaw teeth are actually edges and therefore are always quenched in water or brine to produce a hardness of 50 to 60 Rc. To obtain considerable structural strength, either the jaws should be locally heated and quenched, or the parts should be heated all over and the jaws locally time-quenched in water or brine. The entire part is then oil-quenched for partial hardening of the remainder.

Since hammers must possess high hardness on the striking surface and lower hardness on the claws, they are usually locally hardened and tempered on each end. The striking face is always quenched in water or brine. Stress-relieving is done at about 175°C (350°F). The final hardness on the striking face may range from 50 to 58 Rc, and that on claws may range from 40 to 47 Rc.

Hand cutting tools (such as axes and hatchets) and mower blades must have high hardness and high relative toughness in their cutting edges, together with the ability to hold a sharp edge.

For hardening of hand cutting tools, their cutting edges are usually heated fast in liquid (salt) baths to the lowest temperature at which the pieces can be hardened and are then quenched in brine. Here, all the spheroidal carbides don't go into solution during austenitization. As a result, the cutting edge of the tool consists of martensite with less carbon than the actual composition of the steel and cementite particles. In this situation, the tool attains its maximum toughness with respect to its hardness, and particles of cementite provide a long service life of the cutting edge. The final hardness at the cutting edge is usually 55 to 60 Rc.[204]

It is more economical to use carbon steels whenever possible. The higher-Mn grades are more expensive than the lower-Mn grades, but are less expensive than the least expensive alloy grades.[204]

Agricultural machinery parts such as plowshares, moldboards, coulters, cultivator shovels, mower and binder knives, disks for harrows and plows, ledger plates, and band knives used for cutting or turning soil are made of high-carbon steels in the heat-treated condition. Grass-cutting and grain-cutting tools are usually made of 1090 or 1095 steel for the required edge to give a prolonged service life. Local hardening of cutting edges with the provision of fixture and subsequent oil quenching and tempering at low temperature produces a final hardness of 55 to 60 Rc on the cutting edges.[204]

13.14.2 Deep-Hardening Steels (H-Band Alloy Steels)

The common alloying elements dissolved in austenite prior to quenching increase hardenability in the following ascending order: Ni, Si, Mn, Cr, Mo, V, and B. The addition of several alloying elements in small amounts is more potent in increasing hardenability than the addition of much larger amounts of one or two alloying elements. To increase hardenability, alloying elements should go into the solution in the austenite during austenitization. Steels having carbide formers such as Cr, Mo, and V need higher temperatures to dissolve their (alloy) carbides. This dissolution proceeds more slowly than for the usual iron carbide. Hence, the heating schedule should be selected in such a way that a sufficient amount of these elements should dissolve in austenite. Otherwise, carbide-forming elements will remain undissolved, causing nonattainment of requisite hardenability.

Advantages of Alloy H-Steels

1. The presence of adequate alloying additions allows the use of a lower carbon content for a given application. The decrease in hardenability due to decrease in carbon content may be readily compensated by the increasing hardenability of the alloying addition.

2. The lower carbon content will induce a much lower susceptibility to quench-cracking which arises from the increased M_S temperature and from greater plasticity of low-carbon martensite.

3. It is possible to allow slower rates of cooling during quenching for a given section thickness due to increased hardenability, which, in turn, produces a decrease in thermal gradient and cooling stress. In fact, less drastic quenching used for deep-hardenable steels causes lower distortion and negligible cracking.

4. Increased hardenability of these alloy steels may allow austempering and martempering to be successfully performed, which decreases the residual stress to a very low level prior to tempering, thereby minimizing distortion and danger of cracking.

5. For the same hardness levels: (a) reduction in area is greater for alloy steels than for plain carbon steels, and (b) fully quenched alloy steels need higher tempering temperatures than carbon steels. This higher tempering temperature reduces the stress level in the finished components without sacrificing mechanical properties.

Specifying Hardenability Band. Figure 13.81 summarizes the alternative method used to specify a steel based on the hardenability band.[205] There are many methods to designate the hardness limits in the Jominy curve: (1) minimum and maximum hardness values at designated distance, such as A-A in the Jominy curve representing the section size used by the purchaser; (2) particular hardness value at the minimum and maximum Jominy distances, as at points B-B; (3) two maximum hardness values at two designated Jominy distances, as at points C-C; and (4) two minimum hardness values at two desired Jominy distances, as at points D-D; or (5) any minimum hardness plus any maximum hardness, as at E-E. When the full hardenability band is specified for alloy steel, it should be described by hardness values at the following distances from the quenched end of the Jominy bar: 1, 2, 3, 8, 12, 16, 20, 24, 28, and 32, in sixteenths of an inch.[205]

For carbon H steels, hardness values at $1/32$ in. should be reported through the distance 8 ($\times 1/16$ in.) from the quenched end as well as the distances listed for alloy H steel.

For each H steel, the limits of hardenability band at each Jominy position are presented in tabular form and graphically (Fig. 13.82), and specifications are written for these tabulations and graphical plots.[205]

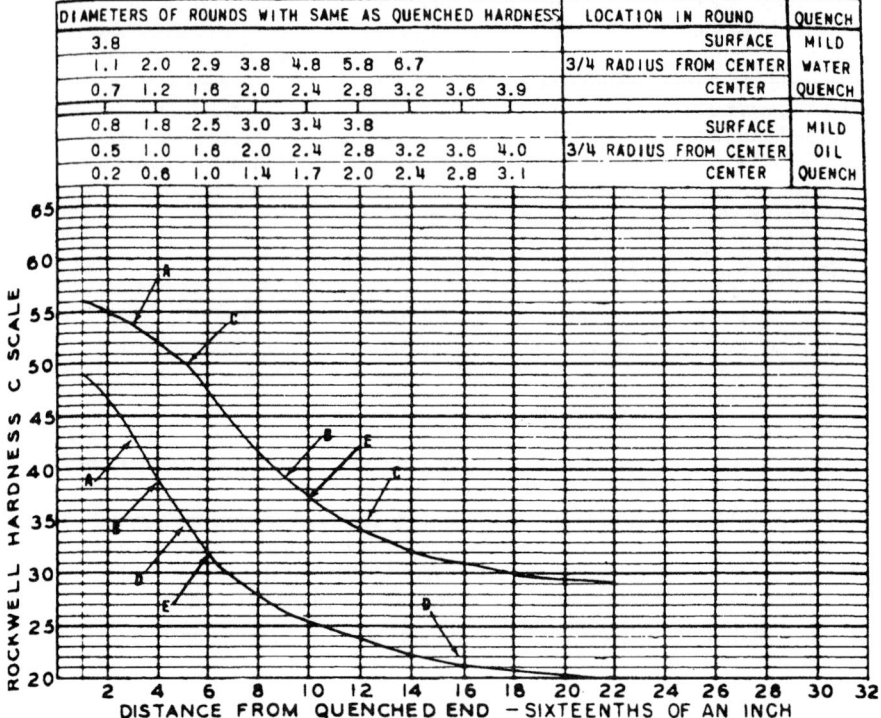

DIAMETERS OF ROUNDS WITH SAME AS QUENCHED HARDNESS									LOCATION IN ROUND	QUENCH
3.8									SURFACE	MILD
1.1	2.0	2.9	3.8	4.8	5.8	6.7			3/4 RADIUS FROM CENTER	WATER
0.7	1.2	1.6	2.0	2.4	2.8	3.2	3.6	3.9	CENTER	QUENCH
0.8	1.8	2.5	3.0	3.4	3.8				SURFACE	MILD
0.5	1.0	1.6	2.0	2.4	2.8	3.2	3.6	4.0	3/4 RADIUS FROM CENTER	OIL
0.2	0.6	1.0	1.4	1.7	2.0	2.4	2.8	3.1	CENTER	QUENCH

FIGURE 13.81   Examples showing alternative methods of specifying hardenability requirements.[205] (*Reprinted by permission of ASTM, Philadelphia, Pa.*)

It is necessary to determine the minimum hardenability in order to attain the proper functioning of the heat-treated steel part; this also serves as a conservative basis for hardenability calculations. More than requisite hardenability should not be specified because it increases costs and gives rise to tensile residual stresses on the quenched surface. There are some approximate guidelines that can be adopted in the selection of an appropriate alloy steel.[205] For highly stressed parts, a minimum of 80% martensite at the center of the largest round section is usually required. For moderately stressed parts, a 50% martensite structure at the center is frequently adequate. For most automotive parts, 80% martensite at the $\frac{3}{4}$ radius position is sufficient. The state of stress in each component should be known. If stress applied in bending is such that the outer layers are the most highly stressed in tension, the steel should have lower hardenability.

13.15   ALLOY STEEL SELECTION GUIDE BASED ON HARDENABILITY

This information is used as a guide in the selection of steel bars based on section thickness and mechanical properties desired in the final production part.

C	Mn	Si	Ni	Cr	Mo
0.37 / 0.44	0.55 / 0.90	0.15 / 0.35	1.55 / 2.00	0.65 / 0.95	0.20 / 0.30

DIAMETERS OF ROUNDS WITH SAME AS QUENCHED HARDNESS

LOCATION IN ROUND									QUENCH
SURFACE	3.8								MILD WATER QUENCH
3/4 RADIUS FROM CENTER	1.1	2.0	2.9	3.8	4.8	5.8	6.7		
CENTER	0.7	1.2	1.6	2.0	2.4	2.8	3.2	3.6	3.9
SURFACE	0.8	1.8	2.5	3.0	3.4	3.8			MILD OIL QUENCH
3/4 RADIUS FROM CENTER	0.5	1.0	1.6	2.0	2.4	2.8	3.2	3.6	4.0
CENTER	0.2	0.5	1.0	1.4	1.7	2.0	2.4	2.8	3.1

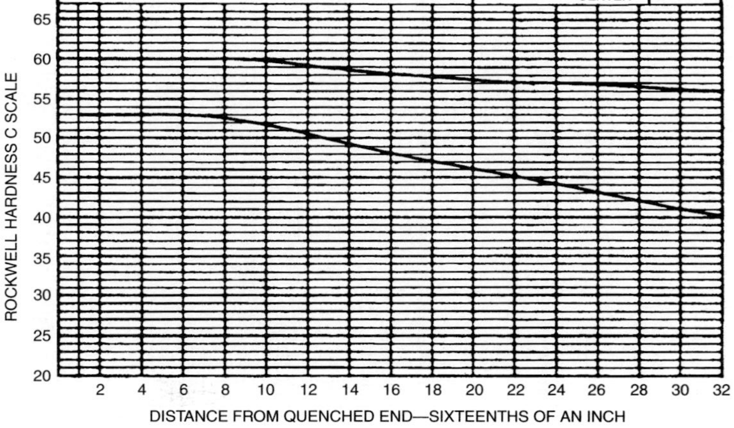

ROCKWELL HARDNESS C SCALE

DISTANCE FROM QUENCHED END—SIXTEENTHS OF AN INCH

NOTE – 1 in. = 25.4 mm.

HARDNESS LIMITS FOR SPECIFICATION PURPOSES

"J" DISTANCE SIXTEENTHS OF AN INCH	0300 N	
	MAX	MIN
1	60	53
2	60	53
3	60	53
4	60	53
5	60	53
6	60	53
7	60	53
8	60	52
9	60	52
10	60	52
11	59	51
12	59	51
13	59	50
14	58	49
15	58	49
16	58	48
18	58	47
20	57	46
22	57	45
24	57	44
26	57	43
28	56	42
30	56	41
32	56	40

HEAT TREATING TEMPERATURES RECOMMENDED BY SAE
*NORMALIZE   1600°F
*AUSTENITIZE   1550°F
*For forged or rolled specimens only.

FIGURE 13.82 Tabulation and graphical representation of the hardenability band of AISI-SAE 4340 H steel.[205] *(Reprinted by permission of ASTM, Philadelphia, Pa.)*

13.141

TABLE 13.28 Alloy Steel Selection Guide for Highly Stressed Parts[204,206]

Unless otherwise indicated in the footnotes, any steel in this table may be considered for a lower strength level or a smaller section, or both.

| Required yield strength | | As-tempered hardness | | Steels to give 80% martensite, minimum, for indicated location in a round section of indicated diameter | | | | | | | |
| | | | | At center | | At midradius | | At ¾ radius | | | |
MPa	ksi	HRC	HB	≤13 mm (½ in.)	13–25 mm (½–1 in.)	25–38 mm (1–1½ in.)	38–50 mm (1½–2 in.)	50–63 mm (2–2½ in.)	63–75 mm (2½–3 in.)	75–89 mm (3–3½ in.)	89–102 mm (3½–4 in.)
Oil-quenched and tempered											
620–860[a]	90–125[b]	23–30	241–285	1330H 5132H 4130H 8630H		94B30H					
860–1030[c]	125–150	30–36[d]	285–341	1335H 5135H	4135H 8640H 94B30H 8740H	4137H		4142H	9840H		4337H 4340H
1030–1170[e]	150–170	36–41[f]	331–375	1340H 5140H 4135H 8637H 94B30H 3140H	50B40H 4137H 8642H 8645H 8742H	4140H 94B40H		4145H 9840H	86B45H 4337H	4340H	
1170–1275[g]	170–185	41–46[h]	375–429	50B46H 5145H 50B44H 4140H 8640H 8642H 8645H 8740H 8742H	5155H 50B44H 5147H 94B40H 6150H	81B45H 4142H 4145H 8650H 8655H 4337H	86B45H 9840H	4147H 4340H	4150H		9805H E4340H
>1275[i]	>185	46 min.[j]	429 min.	5150H 5155H 50B44H 5147H 9260H	5160H 50B50H 9262H 4147H 8655H	50B60H 51B60H 8660H	4150H			9850H	

Water-quenched and tempered[k]

Tensile strength, MPa	620–860[a]	860–1030[c]	1030–1170[e]	1170–1275[g]	>1275[i]
Tensile strength, ksi	90–125	125–150	150–170	170–185	>185
Hardness, HRC	23–30[b]	30–36[d]	36–41[f]	41–46[h]	46 min.[j]
Hardness, HB	241–285	285–341	331–375	375–429	429 min.
Suitable H-steels	81B45H, 8650H, 86B45H, 6150H			4135H	94B40H
				94B30H	94B30H
	5130H, 5132H, 4130H, 8630H	5135H	5135H, 1335H	4137H, 4140H	4137H, 4140H
		1330H, 1335H, 5130H, 5132H, 5135H, 4130H, 8630H	5132H, 5135H, 4135H, 8637H	81B45H, 4337H	81B45H, 4142H, 4337H
			4135H[l], 8640H[l], 8740H[l], 3140H[l]	4142H	4145H, 4147H, 86B45H, 9840H, 4340H, E4340H
		5140H, 4037H, 4042H, 4137H, 8637H	50B40H[l], 4137H[l], 8642H[l], 8745H[l]	4140H[m], 8645H[m], 8742H[m]	8640H[m], 8740H[m]
			1340H, 50B46H, 5140H, 4135H, 8637H, 94B30H, 3140H; 1340H[m], 8637H[m]	5145H, 50B40H, 8640H, 8642H, 8740H; 50B44H, 5147H, 81B45H[l], 94B40H[l]	50B44H, 5147H, 4140H, 8645H, 8742H; 81B45H[m]
	5046H, 50B46H, 5145H, 4047H, 4142H, 8642H	5147H, 4145H, 8645H, 86B45H	1340H, 50B46H, 3140H	1340H, 50B46H, 3140H	50B44H

[a] Tensile strength, 790 to 940 MPa (115 to 138 ksi). [b] As-quenched hardness, 42 HRC, or 388 HB. [c] Tensile strength, 940 to 1100 MPa (136 to 160 ksi). [d] As-quenched hardness, 44 HRC, or 415 HB. [e] Tensile strength, 1100 to 1300 MPa (160 to 188 ksi). [f] As-quenched hardness, 48 HRC, or 461 HB. [g] Tensile strength, 1300 to 1530 MPa (188 to 222 ksi). [h] As-quenched hardness, 51 HRC, or 495 HB. [i] Tensile strength, over 1530 MPa (222 ksi). [j] As-quenched hardness, 55 HRC, or 555 HB. [k] Through steels with 0.47% C nominal. [l] May be substituted for steels listed under the 50 to 63 mm (2 to 2½ in.) column at same strength level or less. [m] Not recommended for applications requiring 80% martensite at midradius in sections 38 to 50 mm (1½ to 2 in.) in diameter because of insufficient hardenability.

Source: Republic Alloy Steels, Republic Steel Corporation, 1961.

TABLE 13.29 Alloy Steel Selection Guide for Moderately Stressed Parts[204,206]

Unless otherwise indicated in the footnotes, any steel in this table may be considered for a lower strength level or a smaller section, or both.

Required yield strength		As-tempered hardness		Steels to give 50% martensite, minimum, for indicated location in a round section of indicated diameter							
				At center		At midradius				At ¾ radius	
MPa	ksi	HRC	HB	≤13 mm (½ in.)	13–25 mm (½–1 in.)	25–38 mm (1–1½ in.)	38–50 mm (1½–2 in.)	50–63 mm (2–2½ in.)	63–75 mm (2½–3 in.)	75–89 mm (3–3½ in.)	89–102 mm (3½–4 in.)
Oil-quenched and tempered											
620–860[a]	90–125	23–30[b]	241–285	1330H 5132H 4130H 8630H	8737H	50B40H 8642H 94B30H 8740H 3140H	4140H 94B40H		4142H		
860–1030[c]	125–150	30–36[d]	285–341	1335H 4042H 4047H 5135H	4135H 8640H 94B30H 8740H 3140H	50B44H 5147H 4137H 8645H 8742H		4142H	4145H	4147H 86B45H 9840H	4337H 4340H
1030–1170[e]	150–170	36–41[f]	331–375	1340H 5140H 4135H 8637H 94B30H 3140H	5150H 50B40H 4137H 8642H 8645H 8742H	5160H 50B50H 4140H 94B40H 6150H	51B60H 8655H	4145H 9840H	4147H 86B45H 4337H	4150H 4340H	
1170–1275[g]	170–185	41–46[h]	375–429	5145H 50B40H 50B46H 4063H 4140H 8640H	5155H 50B44H 5147H 94B40H 6150H	81B45H 4142H 4145H 8650H 8655H 4337H	86B45H 9840H	4147H 8660H 4340H	4150H		9850H E4340H

>1275[i]	>185	46 min.[j]	429 min.	8642H 8745H 8740H 8742H 5150H 5155H 50B44H 5147H 9260H 81B45H 8650H 86B45H 6150H	5160H 50B50H 9262H 4147H 8655H	50B60H 51B60H 8660H	4150H			9850H	
Water-quenched and tempered[k]											
620–860[a]	90–125	23–30[b]	241–285		4037H 5130H 5132H 4130H 8630H	5135H	8637H[l]	5140H[m]	4135H	50B40H 8642H 94B30H 3140H	4137H
860–1030[c]	125–150	30–36[d]	285–341	1330H 5135H		1335H	4135H[l]	1340H[m] 8637H[m]	50B40H 8640H 8642H 94B30H 8740H 3140H	50B44H 5147H 4137H 8645H 8742H	4140H 94B40H
1030–1170[e]	150–170	36–41[f]	331–375	1330H 1335H 5130H 5132H 5135H 4130H 8620H	4042H 4047H	1340H 50B46H 5140H 4135H 8637H 94B30H 3140H	50B40H[l] 4137H[l] 8642H[l]	5145H[m] 8640H[m] 8740H[m]	50B44H 5147H 4140H 8645H 8742H	94B40H	81B45H 4142H 4337H

TABLE 13.29 Alloy Steel Selection Guide for Moderately Stressed Parts[204,206] (*Continued*)

Unless otherwise indicated in the footnotes, any steel in this table may be considered for a lower strength level or a smaller section, or both.

Required yield strength		As-tempered hardness		Steels to give 50% martensite, minimum, for indicated location in a round section of indicated diameter							
				At center		At midradius			At ¾ radius		
MPa	ksi	HRC	HB	≤13 mm (½ in.)	13–25 mm (½–1 in.)	25–38 mm (1–1½ in.)	38–50 mm (1½–2 in.)	50–63 mm (2–2½ in.)	63–75 mm (2½–3 in.)	75–89 mm (3–3½ in.)	89–102 mm (3½–4 in.)
1170–1275[g]	170–185	41–46[h]	375–429	5140H 4037H 4042H 4137H 8637H	1340H 50B46H 3140H	5145H 50B40H 8640H 8642H 8740H	50B44H[l] 5147H[l] 94B40H[l]	4140H[m] 8645H[m] 8742H[m]	4142H	81B45H 4337H	4145H 4147H 86B45H 9840H 4340H E4340H
>1275[i]	>185	46 min.[j]	429 min.	5046H 50B46H 5145H 4047H 4142H 8742H	5147H 4145H 8645H 86B45H	50B44H		81B45H[m]	4147H		

[a] Tensile strength, 790 to 940 MPa (115 to 138 ksi). [b] As-quenched hardness, 42 HRC, or 388 HB. [c] Tensile strength, 940 to 1100 MPa (136 to 160 ksi). [d] As-quenched hardness, 44 HRC, or 415 HB. [e] Tensile strength, 1100 to 1300 MPa (160 to 188 ksi). [f] As-quenched hardness, 48 HRC, or 461 HB. [g] Tensile strength, 1300 to 1530 MPa (188 to 222 ksi). [h] As-quenched hardness, 51 HRC, or 495 HB. [i] Tensile strength, over 1530 MPa (222 ksi). [j] As-quenched hardness, 55 HRC, or 555 HB. [k] Through steels with 0.47% C nominal. [l] May be substituted for steels listed under the 50 to 63 mm (2 to 2½ in.) column at same strength level or less. [m] Not recommended for applications requiring 50% martensite at midradius in sections 38 to 50 mm (1½ to 2 in.) in diameter because of insufficient hardenability.

Source: Republic Alloy Steels, Republic Steel Corporation, 1961.

Table 13.28 shows the H-band alloy steel selection guide for highly stressed parts, oil- or water-quenched to produce a minimum 80% martensite for round sections up to 4 in. (102 mm) in diameter. Table 13.29 provides the H-band alloy steel selection guide for moderately stressed parts, oil- or water- quenched to produce a minimum 50% martensite at the indicated locations for round sections up to 4 in. (102 mm) in diameter. An H-steel with the same grade specification as a standard SAE-AISI steel is capable of meeting the same section and strength requirements as the standard steel and is the preferred method of specification. These are ranked approximately in order of increasing cost. When one steel is substituted for another, it should be shifted to the left in the table, to a smaller diameter, or upward to a lower strength.[204,206]

REFERENCES

1. R. W. Monroe and C. E. Bates, *J. Heat Treat.*, vol. 3, no. 2, 1983, pp. 83–94.

2. K. E. Thelning, *J. Heat Treat.*, vol. 3, no. 2, 1983, pp. 100–107.

3. *Heat Treatment of Steels*, B. P., England.

4. M. Tagaya and I. Tamura, *KHG*, vol. 6, no. 1, 1955, pp. 7–12.

5. M. Narazaki, A. Asada, and K. Fukahara, in *Proceedings: Second International Conference on Quenching and Control of Distortion*, 4–7 Nov. 1996, eds. G. E. Totten et al., ASM International, Cleveland, Ohio, 1996, pp. 37–46.

6. C. E. Bates, G. E. Totten, and R. L. Brennan, in *ASM Handbook*, vol. 4: *Heat Treating*, 10th ed., ASM International, Materials Park, Ohio, 1991, pp. 67–120.

7. W. Leslie, *Physical Metallurgy of Steel*, McGraw-Hill, New York, 1981.

7a. A. V. Reddy, D. A. Akers, L. Chuzhoy, M. A. Pershing, and R. A. Wildow, *Heat Treating Progress*, June/July 2001, pp. 40–42.

8. G. E. Totten, C. E. Bates, and M. A. Clinton, *Handbook of Quenchants and Quenching Technology*, ASM International, Materials Park, Ohio, 1993.

9. B. Liscic, *Hart-Tech. Mitt.*, vol., 33, 1978, pp. 179–191.

10. V. Paschkis and G. Slotz, *Iron Age*, vol. 22, 1956, pp. 95–97.

11. M. Tagaya and I. Tamura, *Technol. Rep. Osaka Univ.*, vol. 7, 1957, pp. 403–424.

12. B. Liscic and T. Filetin, *J. Heat Treat.*, vol. 5, no. 2, 1988, pp. 115–124.

13. G. E. Totten, G. M. Webster, H. M. Tensi, and B. Liscic, *Advanced Mats. & Process.*, June 1997, pp. 68LL–68OO.

14. N. A. Hilder, Ph.D. thesis, University of Birmingham, Aston, U.K., January 1988.

15. G. E. Totten, G. M. Webster, H. M. Tensi, and L. M. Jarvis, in *Proceedings: Second International Conference on Quenching and Control of Distortion*, 4–7 Nov. 1996, eds. G. E. Totten et al., ASM International, Materials Park, Ohio, 1996, pp. 585–593.

16. W. F. Walker, *Tooling*, December 1968, pp. 13–25.

17. *Metals Handbook*, vol. 4, 9th ed., ASM, Metals Park, Ohio, 1981.

17a. W. Luty, in *Theory and Technology of Quenching*, eds. B. Liscic, H. M. Tensi, and W. Luty, Springer-Verlag, Berlin, 1992, pp. 248–340.

18. T. W. Dicken, *Heat Treat. Met.*, vol. 1, 1986, pp. 6–8.

19. D. Paddle, *Ind. Heat.*, January 1998, pp. 40–43.

20. J. A. Hasson, *Met. Progr.*, vol. 128, no. 4, 1985, pp. 67–72.

21. G. R. Weymueller, *Met. Progr.*, vol. 102, no. 1, 1972, pp. 38–50.

22. T. I. Tkachuk, N. Ya Rudakova, B. K. Sheremeta, and M. A. Al'tshuler, *Metalloved Term. Obrab. Met.*, October 1986, pp. 45–47.

23. T. I. Tkachuk, N. Ya Rudakova, B. K. Sheremeta, and R. D. Novoded, *Metalloved Term. Obrab. Met.*, October 1986, pp. 42–45.

24. R. J. Brennan, *Ind. Heat.*, January 1993, pp. 29–31.

25. I. F. Moreaux, J. M. Naul, G. Beck, and J. Oliver, *Heat Treatment '84*, The Metals Society, London, 1984, pp. 18.1–18.5.

25a. D. Moore, *Heat Treating Progress*, June/July 2001, pp. 29–33.

26. J. Bodin and S. Segerberg, *Heat Treat. Met.*, vol. 20, no. 1, 1993, pp. 15–23; in *Quenching and Carburizing*, The Institute of Materials, London, 1993, pp. 33–54.

27. H. J. Gilliland, *Met. Progr.*, October 1960, pp. 111–114.

28. E. A. Bender and H. J. Gilliland, *Steel*, December 1957, pp. 56–59.

29. C. A. Barley and J. S. Aarons, eds., *The Lubrication Engineer Manual*, U.S. Steel Corporation, Pittsburgh, Pa., 1971, pp. 56–57.

30. G. R. Furman, *Lubrication*, vol. 57, 1971, pp. 25–36.

31. T. Croucher, *Heat Treating*, December 1990, pp. 17–19.

32. W. H. Naylor, *Met. Progr.*, vol. 92, no. 6, 1967, pp. 70–73.

33. P. E. Cary, *Met. Progr.*, vol. 89, no. 2, 1966, pp. 90–92.

34. R. W. Foreman, Paper presented at ASM Short Course on Quenching Media, American Society for Metals, Metals Park, Ohio, November 1985.

35. *Houghton on Quenching*, products brochure, E. F. Houghton & Co., Inc., England, 1994.

36. S. O. Segerberg, *Heat Treat.*, December 1988, pp. 30–33.

37. V. Srimongkolkul, *Ind. Heat.*, September 1998, pp. 81–84.

38. J. Hasson, *Ind. Heat.*, November 1995, pp. 45–46.

39. R. K. Evans, *Met. Mater.*, vol. 12, 1978, p. 28.

40. R. F. Kern and M. E. Suess, *Steel Selection*, Wiley, New York, 1979.

41. K. Funfani and G. E. Totten, in *Proceedings: Second International Conference on Quenching and Control of Distortion*, 4–7 Nov. 1996, eds. G. E. Totten et al., ASM International, Materials Park, Ohio, 1996, pp. 3–15.

42. J. Igarashi, *Jpn. J. Tribol.*, vol. 35, 1990, pp. 1095–1105.

43. G. E. Totten, G. M. Webster, L. M. Jarvis, S. H. Kang, and S. W. Han, in *17th ASM Heat Treating Society Conference Proceedings*, 15–18 Sept. 1997, eds. D. L. Milam et al., ASM International, Materials Park, Ohio, 1997, pp. 443–448.

44. J. A. Hasson, *Ind. Heat.*, September 1981, pp. 21–23.

45. V. Srimongkolkul, *Heat Treat.*, December 1990, pp. 27–28.

46. H. M. Tensi, A. Stich, and G. E. Totten, *Steel Heat Treatment Handbook*, Marcel Dekker, New York, 1997, pp. 157–249.

47. Z. H. Ou, G. E. Totten, G. M. Webster, and Y. H. Sun, in *Proceedings: Second International Conference on Quenching and Control of Distortion*, 4–7 Nov. 1996, eds. G. E. Totten et al., ASM International, Materials Park, Ohio, 1996, pp. 517–523.

48. K. H. Kopietz, *Heat Treat.*, vol. 16, no. 9, 1984, pp. 20–26.

49. D. Moore, *Heat Treat. Met.*, vol. 26, no. 3, 1999, pp. 68–71.

50. R. W. Rhines and E. R. Mueller, *Met. Progr.*, vol. 122, no. 4, 1982, pp. 33–39.

51. G. E. Totten, G. M. Webster, and C. E. Bates, in *17th ASM Heat Treating Society Conference Proceedings*, 15–18 Sept. 1997, eds. D. L. Milam et al., ASM International, Materials Park, Ohio, 1997, pp. 247–255.

52. R. R. Blackwood, L. M. Jarvis, G. E. Totten, G. M. Webster, and T. Narumi, *Metal Heat Treat.*, May/June 1996, pp. 28–31.

53. L. S. Zarkhin, S. V. Sheberstov, N. V. Panfilovich, and L. I. Manevich, *Russ. Chem. Rev.*, vol. 122, no. 4, 1982, pp. 33–39.

54. R. T. von Bergen, *Heat Treat. Met.*, vol. 2, 1991, pp. 37–42.

55. S. Segerberg, *4th Int. Cong. Heat Treat. Mater.*, Berlin, 3–7 June 1985, vol. II, pp. 1252–1265.

56. T. Hibi, *Netsu Shori*, vol. 25, no. 1, 1985, pp. 46–50.

57. D. Diaz, H. Garcia, and B. Bautista, *Ind. Heat.*, June 1993, pp. 43–45.

58. N. Kobayashi, *Netsu Shori*, vol. 25, no. 1, 1985, pp. 51–54.

59. G. E. Totten, R. R. Blackwood, and L. M. Jarvis, *Heat Treating*, March 1991, pp. 17–18.

60. R. D. Howard and G. E. Totten, *Metal Heat Treat.*, September/October 1994, pp. 22–24.

61. *Heat Treat.*, October 1983, pp. 40–41.

62. E. H. Burgdorf, *Ind. Heat.*, October 1981, pp. 18–25.

63. *Cassels Salts for Heat Treatment and Case Hardening*, Birmingham, England.

64. *Heat Treat.*, vol. 11, no. 3, 1979, pp. 22–28; vol. 11, no. 4, 1979, pp. 42–44.

65. R. W. Foreman, *Ind. Heat.*, March 1993, pp. 41–47.

66. *Liquid Salt Baths*, product bulletin, E. F. Houghton & Co., Valley Forge, Pa., December 1981.

67. G. P. Dubal, *Proceedings: 18th Conference, 12–15 Oct. 1998, Heat Treating*, eds. R. A. Wallis and H. W. Walton, ASM International, Materials Park, Ohio, pp. 192–198; *Adv. Mats. & Process.*, December 1999, pp. H23–H28.

68. G. P. Dubal, M. T. Ives, A. G. Meszaros, and J. Recker, *1st International Automotive Heat Treating Conference*, 13–15 July 1998, eds. R. Colas, K. Funatani, and C. A. Stickels, ASM International, Materials Park, Ohio, pp. 90–95.

69. *Metall. Met. Form*, vol. 38, 1971, pp. 322–323.

70. R. Wilson, *Metallurgy and Heat Treatment of Tool Steels*, McGraw-Hill, London, 1975.

71. R. W. Foreman, *Heat Treat.*, vol. 12, no. 10, 1980, pp. 26–29.

72. *Heat Treat.*, vol. 15, no. 10, 1985, pp. 22–32.

73. B. Harrison, *Metallurgia*, vol. 45, no. 3, 1978, pp. 145–146.

74. D. Grieve, *Metall. Mater. Technol.*, vol. 7, no. 8, 1975, pp. 397–403.

75. J. A. Schliessman, *Meta. Treat.*, vol. 22, no. 6, 1971, pp. 3–8.

76. A. Creal, *Heat Treat.*, vol. 18, no. 10, 1986, pp. 26–28.

77. R. W. Foreman, Paper presented at ASM National Heat Treating Conference, Chicago, September 1988.

78. L. Rosseau, *Metallurgia*, vol. 49, 1954, pp. 27–33.

79. B. Liscic, in *Quenching and Carburizing, 3d International Seminar*, International Federation for Heat Treatment of Surface Engineering, Melbourne, Australia, 1991; The Institute of Materials, 1993, pp. 1–32.

80. J. A. Lincoln and A. Keogh, U.S. Patent 4,431,464, February 1984.

81. C. Skidmore, *Heat Treat. Met.*, vol. 2, 1986, pp. 34–38.

82. G. Wahl, *Carburizing—Processing and Performance*, ASM International, Materials Park, Ohio, 1989, pp. 41–56.

83. T. W. Ruffle and E. R. Byrnes, *Heat Treat. Met.*, vol. 6, no. 4, 1979, pp. 81–87.

84. J. E. Pritchard, G. Nurnberg, and M. Shoukri, *Heat Treat Met.*, vol. 23, no. 4, 1996, pp. 79–83.

85. P. F. Stratton, N. Saxena, and R. Jain, *Heat Treat. Met.*, vol. 24, no. 3, 1997, pp. 60–63.

86. W. R. Jones, *Heat Treat.*, September 1985, pp. 34–35.

87. J. M. Neidermann, *Heat Treat.*, vol. 16, no. 12, 1984, pp. 20–23.

88. S. Segerberg and E. Troell, *Heat Treat. Met.*, vol. 24, no. 1, 1997, pp. 21–24.

89. E. Edenhofer, *Heat Treat. Met.*, vol. 26, no. 1, 1999, pp. 1–5.

90. J. G. Conybear, *Adv. Mats. & Processes*, Feb. 2, 1993, pp. 20–21.

91. R. E. Andrews, J. Grooves, and M. H. Jacobs, in *Heat Treatment '84*, The Metals Society, London, 1984, pp. 41.1–41.8.

92. J. W. Bouwman, *Metallurgia*, vol. 52, no. 2, 1985, pp. 41–45.

93. J. Kowaleski and J. Olejnik, *Ind. Heat.*, October 1998, pp. 39–44.

94. R. Hill, Jr., *High Pressure Gas Quenching Typical Oil Hardening Grades of Steel*, SME 1998 Technical Paper, CM98-207, pp. 1–18.

95. R. B. Dixon, *Heat Treat. Met.*, vol. 24, no. 2, 1997, pp. 37–42.

96. Product brochure, Fluidtherm Technology Ltd., Madras, India.

97. M. A. Delano and J. Van den Sype, *Heat Treat.*, December 1988, pp. 34–37.

98. R. W. Reynoldson, *Heat Treatment in Fluidized Bed Furnaces*, ASM International, Materials Park, Ohio, 1993, p. 47.

99. Z. Kulin and H. Genlian, *Proc. 4th Int. Cong. Heat Treatment of Materials*, vol. 2, 3–7 June 1985, Berlin, pp. 1293–1320.

100. A. J. Hick, *Heat Treat. Met.*, vol. 21, no. 3, 1994, pp. 53–64.

101. G. Beck, in *Heat and Mass Transfer in Metallurgical Systems*, eds. D. B. Spalding and N. H. Afgan, Hemisphere Publishing, 1981, pp. 509–525.

102. G. Li, *Proc. 4th Ann. Conf. Heat Treat.*, Nanjing, 25–31 May 1987, Chinese Mechanical Engineering Society, pp. 171–175.

103. A. J. Hick, *Heat Treat. Met.*, vol. 26, no. 3, 1999, pp. 53–62.

104. L. Marshall and J. Canner, *Ind. Heat.*, June 1994, pp. 51–52.

105. N. I. Kobasko, M. A. Aronov, G. E. Totten, and A. V. Sverdin, *Proc. 18th Conference: Heat Treating 1998*, eds. R. A. Wallis and H. W. Walton, ASM International, Materials Park, Ohio, 1998, pp. 616–621.

106. G. F. Malloy, *Hardening of Steel*, Metals Engineering Institute, ASM, Metals Park, Ohio.

107. T. D. Atterbury, *Metall. Met. Form.*, August 1971, pp. 210–215.

108. C. E. Bates and G. E. Totten, *Heat Treat. Met.*, vol. 19, no. 2, 1992, pp. 45–48.

109. B. Kovacs, in *ASM Handbook*, vol. 4: *Heat Treating*, 10th ed., ASM International, Materials Park, Ohio, 1991, pp. 670–681.

110. K. B. Rundman, in *ASM Handbook*, vol. 4: *Heat Treating*, 10th ed., ASM International, Materials Park, Ohio, 1991, pp. 682–696.

110a. N. Shimizu and I. Tamura, *Trans. ISIJ*, vol. 16, 1976, pp. 655–663; *Trans. ISIJ*, vol. 17, 1977, pp. 469–476.

110b. N. Shimizu, Ph.D. thesis, Kyoto University, 1985 (in Japanese).

110c. K. Arimoto, D. Huang, D. Lambert, and W. T. Wu, *Heat Treating Conference and Exposition*, Oct. 9–12, 2000, St. Louis, Missouri.

110d. B. Liscic, G. Grubisic, and G. E. Totten, *Proc. 2d International Conference on Quenching and Control of Distortion*, Cleveland, Ohio, 1996, pp. 47–54.

110e. B. Liscic and G. E. Totten, *Advanced Materials and Processes*, vol. 9, 1997, pp. 180–184.

111. P. E. Reynolds, *Heat Treat. Met.*, vol. 17, no. 3, 1990, pp. 69–72.

112. N. I. Kobasko, *ASM 18th International Conference: 12–15 Oct. 1998*, eds. R. A. Wallis and H. W. Walton, ASM International, Materials Park, Ohio, 1998, pp. 613–616; *Adv. Mats. & Process.*, December 1999, pp. H31–H33.

112a. M. A. Aronov, N. I. Kobasko, J. F. Wallace, D. Schwam, and J. A. Powell, *Ind. Heat.*, April 1999, pp. 59–63.

113. R. F. Kern, *Heat Treat.*, vol. 18, no. 9, 1986, pp. 19–23.

114. H. Webster and W. J. Laird, Jr., in *ASM Handbook*, vol. 4: *Heat Treating*, 10th ed., ASM International, Materials Park, Ohio, 1991, pp. 137–151.

115. J. L. Yarne, *Met. Progr.*, vol. 84, no. 4, 1963, p. 105.

116. W. D. Lankford and H. E. McGannon, eds., *Making, Shaping, and Treating of Steel*, 10th ed., USS Corporation, Pittsburgh, 1985.

117. L. F. Spencer, *Met. Treat.*, vol. 11, no. 3, 1960, p. 55.

118. E. C. Bain and E. S. Davenport, *Trans. ASM*, 1930, p. 289.

119. J. R. Keough, W. J. Laird, Jr., and A. D. Godding, *ASM Handbook*, vol. 4: *Heat Treating*, 10th ed., ASM International, Materials Park, Ohio, 1991, pp. 152–163.

120. R. L. Siffredini, *Heat Treat.*, vol. 12, no. 1, 1980, pp. 14–19.

121. E. C. Harwood, *Metall. Met. Form.*, vol. 31, February 1964, pp. 82–83.

122. R. Creal, *Heat Treat.*, vol. 18, no. 8, 1986, pp. 24–27.

123. B. J. Hart, *Heat Treat.*, vol. 15, no. 10, 1983, pp. 36–38.

124. B. Liscic, in *Steel Heat Treatment Handbook*, Marcel Dekker, New York, 1997, pp. 527–662.

125. *Metallurgia*, vol. 47, no. 3, 1980, pp. 136–138.

126. B. J. Waterhouse, *Met. Progr.*, vol. 106, no. 3, 1974, p. 71.

127. D. Nicholson, S. Ruhamann, and R. J. Wingrove, *Heat Treatment of Metals, Special Report 95*, Iron and Steel Institute, London, 1966, p. 180.

128. V. K. Chandhok, A. Kasak, and J. P. Hirth, *Trans. ASM*, vol. 59, 1966, p. 288.

129. *Ductile Iron Data for Design Engineers*, QIT-Fer et Titane Inc., Chicago, Illinois, 1990.

130. R. A. Harding, *Met. Mater.*, vol. 2, no. 2, 1986, pp. 65–72.

131. R. A. Blackmore and R. A. Harding, *J. Heat Treat.*, vol. 3, no. 4, 1984, pp. 310–325.

132. R. B. Gundlach and J. F. Janowak, *Met. Progr.*, vol. 128, no. 2, 1985, pp. 19–25.

133. V. K. Sharma, *J. Heat Treat.*, vol. 3, no. 4, 1984, pp. 326–334.

134. J. A. Lincoln, *Heat Treat.*, vol. 16, no. 12, 1984, pp. 30–34.

134a. R. Elliott, *Heat Treat. Met.*, vol. 24, no. 3, 1997, pp. 55–59.

135. J. Mallia and M. Grech, *Mats. Sc. & Tech*, vol. 13, May 1997, pp. 408–414.

136. B. T. Sims and R. Elliott, *Mats. Sc. & Tech.*, vol. 14, February 1998, pp. 89–96.

137. *The New Bench Mark Material, Applied Process, Inc. Bulletin*, 1999.

137a. M. Bahmani, R. Elliott, and N. Varahram, *J. Mats. Sc.*, vol. 32, 1997, pp. 5383–5388.

138. H. Bayati and R. Elliott, *Mats. Sc. & Tech.*, vol. 13, April 1997, pp. 319–326.

139. R. C. Voigt and C. R. Loper, Jr., *J. Heat Treat.*, vol. 3, no. 4, 1984, pp. 291–309.

140. B. V. Kovacs, Sr., *Modern Casting*, March 1990, pp. 38–41; J. R. Keough, *Foundry, Manage., and Tech.*, October/November 1995.

141. J. R. Laub, *Heat Treat.*, March 1992, pp. 18–23.

141a. Y. S. Lerner and G. R. Kingsbury, *Jn. Mats. Engrg. and Performance*, vol. 7, no. 1, 1998, pp. 48–52.

142. A. Owhadi, J. Hedjozi, and P. Davami, *Mats. Sc. & Tech.*, vol. 14, March 1998, pp. 245–250.

143. K. Boiko, *Heat Treat.*, vol. 17, no. 9, 1985, pp. 22–24.

144. J. Race and L. Stott, *Heat Treat. Met.*, vol. 18, no. 4, 1991, pp. 105–109.

145. J. Hemanth, *Mats. Sc. & Tech.*, vol. 15, August 1999, pp. 878–884.

146. E. Dorazil, *High Strength Austempered Ductile Cast Iron*, Ellis Harwood, New York, 1991.

147. R. R. Blackwood and L. M. Jarvis, *Ind. Heat.*, March 1991, pp. 28–31.

147a. H. W. Rayson, *Constitution and Properties of Steels*, vol. ed. F. B. Pickering, VCH, Weinheim, 1992, pp. 583–640.

148. T. Kunitake and S. Sugisawa, *The Sumitomo Search* 5, May 16, 1971.

149. H. Ohtani, in *Materials Science and Technology*, vol. 7: *Constitution and Properties of Steel*, vol. ed. F. B. Pickering, VCH, Weinheim, 1992, pp. 147–181.

150. R. Arpi, *Jernkont. Ann.*, no. 115, 1931, p. 75.

151. E. C. Bain and H. W. Paxton, *Alloying Elements in Steel*, 2d ed., ASM, Metals Park, Ohio, 1966.

152. R. E. Reed-Hill and G. J. Abbaschian, *Physical Metallurgy Principles*, PWS-Kent Publishing, Boston, 1992.

153. *Metals Handbook*, vol. 1, 9th ed., ASM, Metals Park, Ohio, 1978.

154. M. A. Grossman and E. C. Bain, *Principles of Heat Treatment*, 5th ed., ASM, Metals Park, Ohio, 1964.

155. ASTM A255–99, ASTM, Pittsburgh, Pa., 1999.

156. *Metals Handbook*, vol. 1, 8th ed., ASM, Metals Park, Ohio, 1961.

157. R. M. Grange and T. M. Garvey, *Trans. ASM*, vol. 37, 1946, pp. 136–174; R. Grange and T. Mitchel, *Trans. ASM*, vol. 53, 1961, pp. 157–185.

158. C. J. McMahon, Jr., *Met. Trans.*, vol. 11A, 1980, pp. 531–535.

159. J. M. Tartaglia and G. T. Eldis, *Met. Trans.*, vol. 15A, 1984, pp. 1573–1583.

160. C. Skena, T. Prucher, R. Czarnek, and J. M. Jo, *The Int. Powder Met.*, vol. 33, no. 7, 1997, pp. 25–35.

161. A. Jones and P. E. Evans, *Heat Treat. Met.*, vol. 20, no. 4, 1993, pp. 99–100.

162. W. Hewitt, *Heat Treat. Met.*, vol. 8, no. 2, 1981, pp. 33–38.

163. C. T. Kunze, in *Hardenability Concepts with Applications to Steel*, eds. D. V. Doane and J. S. Kirkaldy, *Proceedings Symposium 1977*, TMS-AIME, Warrendale, Pa., 1978, pp. 290–305.

164. G. F. Malloy, P. R. Slimomon, and P. Kvaale, *Met. Trans.*, vol. 4, 1973, pp. 2279–2289.

165. O. M. Askelsen, O. Grong, and P. E. Kvaale, *Met. Trans.*, vol. 17A, 1986, pp. 1529–1536.

166. T. Lund, *Scand. J. Met.*, 1990, pp. 227–235.

167. R. Kumar, *Physical Metallurgy of Iron and Steel*, Asia Publishing House, Bombay, 1968.

168. K. E. Thelning, *Steel and Its Heat Treatment*, Butterworths, London, 1984.

169. H. Burrier, Jr., in *ASM Handbook*, vol. 1, 10th ed., ASM International, Materials Park, Ohio, 1990, pp. 464–484.

170. C. A. Siebert, D. V. Doane, and D. H. Breen, *The Hardenability of Steels*, ASM, Metals Park, Ohio, 1977.

171. R. F. Mehl and W. C. Hagel, *Prog. Met. Phys.*, vol. 6, 1956, p. 74.

172. W. F. Smith, *Structure and Properties of Engineering Materials*, McGraw-Hill, New York, 1981.

173. *Methods of Determining Hardenability of Steels*, SAE J406, February 1995, 2000 *SAE Handbook*, vol. 1, Society of Automotive Engineers, Warrendale, Pa., pp. 1.25–1.46.

174. G. Krauss, *Steels: Heat Treatment and Processing Principles*, ASM International, Materials Park, Ohio, 1990.

175. B. Liscic, in *Steel Heat Treatment Handbook*, Marcel Dekker, New York, 1997, pp. 93–156.

176. A. F. DeRetana and D. V. Doane, *Met. Prog.*, vol. 100, 1971, p. 65.

177. A. Moser and A. Legat, *Harterei-Techn. Mitt.*, vol. 24, no. 2, 1969, pp. 100–105.

178. C. F. Jatczak, *Met. Trans.*, vol. 4, 1973, pp. 2267–2277.

179. I. R. Kramer, S. Siegel, and J. G. Brooks, *Trans. AIME*, vol. 167, 1946, p. 670.

180. R. A. Grange, *Met. Trans.*, vol. 4, 1973, pp. 2231–2244.

181. D. V. Doane, in *Hardenability Concepts with Applications to Steel, Proceedings Symposium 1977*, TMS-AIME, Warrendale, Pa., 1978, pp. 351–378.

182. W. E. Jominy and A. L. Boegehold, *Trans. ASM*, vol. 26, 1938, p. 574.

183. R. F. Kern and M. E. Suess, *Steel Selection*, Wiley-Interscience, New York, 1979.

184. T. Brown, in *Hardenability Concepts with Applications to Steel, Proceedings Symposium 1977*, TMS-AIME, Warrendale, Pa., 1978, pp. 273–288.

185. S. I. Arkhangel'skii, E. S. Miroshnik, and I. V. Tikhonova, *Ind. Laboratory*, vol. 11, 1991, pp. 1349–1351 (English translation).

186. L. H. Van Vlack, *Materials Science for Engineers*, Addison-Wesley, Reading, Mass., 1971.

187. S. B. Lasday, *Ind. Heat.*, 1992, pp. 25–27.

188. A Rose and L. Rademacher, *Weiterentwicklung des stirnabschreckversuches zur Prufung der Hartbarkeit von tiefer einhartenden stahlen*, *Stahl Eisen*, vol. 76, no. 23, 1956, pp. 1570–1573 (in German).

189. C. F. Jatczak, *Trans. ASM*, vol. 58, 1965, p. 195.

189a. B. M. Kapadia et al., *Trans. AIME*, vol. 242, 1969, p. 1689.

190. D. T. Llewellyn and W. T. Cook, *Metals Technol.*, December 1974, p. 517.

190a. T. T. Llewellyn, *Ironmaking and Steelmaking*, vol. 20, no. 5, 1993, pp. 338–343.

191. E. M. Grinberg, in *21st Century Steel Industry 1994*, pp. 170–173.

192. J. Field, *Met. Prog.*, vol. 43, no. 3, 1943, p. 402.

193. L. J. Boyd and J. Field, *Contributions to the Metallurgy of Steel*, no. 12, American Iron and Steel Institute, New York, 1946.

194. E. Just, *Met. Prog.*, November 1969, pp. 87–88.

195. J. S. Kirkaldy, G. Pazionis, and S. E. Feldman, *Heat Treatment, 1976*, The Metals Society, London, 1976.

196. J. S. Kirkaldy, *Met. Trans.*, vol. 4, 1973, pp. 2327–2333.

197. J. S. Kirkaldy, in *Hardenability Concepts with Applications to Steel, Proceedings Symposium 1977*, eds. D. V. Doane and J. S. Kirkaldy, TMS-AIME, Warrendale, Pa., 1978, p. 491.

198. D. J. Keith, J. T. Sponzilli, V. K. Sharma, and C. H. Walter, in *Hardenability Concepts with Applications to Steel, Proceedings Symposium 1977*, eds. D. V. Doane and J. S. Kirkaldy, TMS-AIME, Warrendale, Pa., 1978, pp. 493–516.

199. P. Maynier, B. Jungmann, and J. Dollet, in *Hardenability Concepts with Applications to Steel, Proceedings Symposium 1977*, eds. D. V. Doane and J. S. Kirkaldy, TMS-AIME, Warrendale, Pa., 1978, pp. 518–544.

200. M. Larsson, B. Jansson, R. Blom, and A. Melander, *Scand. J. of Metall.*, vol. 19, 1990, pp. 51–63.

201. S. E. Feldman, in *Hardenability Concepts with Applications to Steel, Proceedings Symposium 1977*, eds. D. V. Doane and J. S. Kirkaldy, TMS-AIME, Warrendale, Pa., 1978, pp. 546–567.

202. J. S. Kirkaldy and S. E. Feldman, *J. Heat Treat.*, vol. 7, 1989, pp. 57–64.

203. J. S. Kirkaldy, in *Metals Handbook*, vol. 4, 10th ed., ASM International, Materials Park, Ohio, 1991, pp. 20–34.

204. E. R. Kuch, in *ASM Handbook*, vol. 1, 10th ed., ASM International, Materials Park, Ohio, 1990, pp. 451–463.

205. ASTM A304-96, ASTM, Pittsburgh, Pa.

206. ASTM A400-1995, ASTM, Pittsburgh, Pa.

CHAPTER 14
TEMPERING

14.1 INTRODUCTION

An attractive combination of strength, toughness, and ductility for a given application can be produced in steel when fully "as-quenched" martensite is reheated to a
suitable temperature below the lower critical temperature A_1. Such a heat treatment, following quench-hardening, is termed *tempering*. Tempering involves redistribution of carbon atoms; precipitation of fine, coherent carbides; decomposition
of retained austenite; coarsening and spheroidization of cementite (associated with
recovery and recrystallization of the matrix); and precipitation of alloy carbides, in
the presence of strong carbide-forming elements, from a supersturated as-quenched
martensite.[1] This ability of tempering to produce a wide range of mechanical properties, relieve quenching and/or other residual stresses, and ensure dimensional
stability was the principal reason for the lasting interest of the metallurgist in the
structures of iron-carbon martensites and their various decomposition products.
Crystallographic and kinetic theories have been developed to account for the formation of decomposition products of martensite, and numerous studies have identified several stages of decomposition during the tempering of martensite and the
effect of undesirable decomposition products in causing various types of embrittling
phenomena.

In this chapter we start our discussion with the structural and mechanical property changes involved in tempering and the role of alloying elements upon the
early-stage behavior and that of precipitation and growth of carbides. Then we
present secondary hardening, tempering parameter, strengthening mechanisms
of tempered martensite, and various types of embrittlement phenomena such as
tempered martensite embrittlement, temper embrittlement, secondary hardening
embrittlement, aluminum nitride embrittlement, hydrogen damage, and metal-
induced embrittlement associated with tempering. Finally, we describe maraging
steels in depth.

14.2 STRUCTURAL CHANGES ON TEMPERING

As-quenched steels are usually mixtures of martensite and retained austenite, with
the former constituent in abundance. Both these constituents are thermodynamically unstable and slowly transform, at least in part, if left at ambient temperature;
the retained austenite (γ_R) may convert to martensite (α'), and the martensite (α')
undergoes transformations which are described here.

Note at this point that as-quenched low-carbon steels at room temperature may be considered to have undergone the pre-precipitation (tempering) process during quenching because of its high M_s temperature and the high mobility of interstitial atoms. This is called *autotempering* and is evidenced by the precipitation of fine carbides or the carbon atom clustering, which precedes such precipitation.

In general, structural changes occurring during conventional tempering of martensitic steel can be divided into five distinct stages; but the temperature ranges are approximate, resulting in the overlapping stages of tempering and concurrent development of other transformation products. This fact, together with the occurrence of reactions on a very fine scale, produces complex structures of the tempered martensite.

The first stage (T1), occurring in the temperature range of 100 to 250°C (212 to 482°F), corresponds to the decomposition of supersaturated martensite to transition carbide, epsilon(ε)-carbide [or eta(η)-carbide, as described below], and low-carbon martensite (α''). The second stage (T2), occurring between 200 and 300°C (392 and 572°F), represents the decomposition of retained austenite to bainite. The third stage (T3) occurs in the temperature range of 250 to 350°C (482 to 662°F) and constitutes the transformation of the reaction products of first and second stages into ferrite and [chi (χ) and/or] cementite [theta (θ)] constituents. The fourth stage (T4), occurring above 350°C, comprises the growth and spheroidization of cementite. The fifth stage (T5) (see Sec. 14.5) holds primarily to alloy steels. In this stage, intermetallic precipitates and alloy carbides are formed.[2,3]

14.2.1 Aging Reaction Stage

Recent electrical resistivity,[4] x-ray diffraction,[5,6] transmission electron microscopy (TEM),[7] atom probe field ion microscopy (APFIM),[8] Mössbauer spectroscopy, and ^{13}C NMR studies have resulted in the incorporation of an additional stage prior to the conventional first stage of tempering;[9–13] this is called a *preliminary stage*, *aging reaction stage*, or *pre-precipitation process*, and it refers to a phenomenon that precedes T1, for example, during storage of the martensite specimen at room temperature.[4,7,9] The kinetics of this stage are such that aging is already accomplished in most "as-quenched" commercial steels with M_s temperature above room temperature.[10] These studies have recognized the development of four steps of structural arrangements of carbon atoms during the aging process:[10–12] (1) the formation of small clusters (of *carbon doublets*) into *isolated carbon multiplets* along $<101>_{\alpha'}$ directions; (2) coarsening by ordering of these *carbon multiplets* in $(100)_{\alpha'}$ monolayers; (3) thickening along $[100]_{\alpha}$ into a multilayer structure, called *the extended multiplet*; and (4) appearance of a superperiod of 12 lattice parameters along $[001]_{\alpha'}$, associated with antiphase domains by long-range interaction. Table 14.1 shows a sequence of precipitation of Fe_9C_4 (ε- or η-) carbide during the four-step aging of Fe-C martensite.[12]

The TEM was carried out by Nagakura et al.,[7] Genin et al.,[11–13] and Ohmori and Tamura[14] on high-carbon Fe-C alloys, while other studies were carried out by Taylor et al.[15] and other workers[4–6,8] on Fe-Ni-C alloys with subzero M_s temperatures.

A1 Stage. During the first substage of aging reaction A1 of ferrous martensites, a clustering reaction of carbon atoms occurs in the c-oriented octahedral sites, associated with the carbon trapping by lattice defects (or Bain correspondence). This is in good agreement with the foregoing experimental techniques used.

TABLE 14.1 Sequence of Precipitation of Fe_6C_4 ε- or η-Carbide during the Aging of Fe-C Martensite[12]

	Steps of aging			
	Short-range order: clustering of C atoms into isolated multiplets	Ordering of multiplets in $(100)_\alpha$ monolayers	Thickening along $[100]_\alpha$ into the multilayer B2 monoclinic structure	Long-range interaction: 12 a_0 superperiod along $[001]_{\alpha'}$ and antiphase domains
Resistivity	——————— Clustering and coarsening ———————			
Mössbauer spectroscopy	Clustering	——— Coarsening into ordered multilayer ———		
^{13}C NMR	Clustering	Ordering	Thickening	
Electron diffraction	———— Diffuse Scattering Streaks ————			Satellites
TEM	———————— Modulations of composition ————————			
APFIM	Clustering (Atomic static displacement of Fe in matrix)			
X-ray analysis	———————— Carbon depletion in the matrix ————————			C phase or κ phase

Each experimental technique is sensitive only to some specific aspects of the process and can reveal only some of the steps.
Courtesy of the Metallurgical Society, Warrendale, Pa.

The electrical resistivity measurements of Fe-Ni-C martensites containing 18 to 24% Ni and 0.003 to 0.62% C were first made following quenching in liquid nitrogen to form virgin martensite and then after up-quenching to temperatures between 80 and 350°C for tempering. Figure 14.1a shows a schematic resistivity versus aging time/temperature curve with the structural changes and regimes produced during the aging and first-stage tempering of virgin Fe-Ni-C martensite.[4] At subambient temperatures (in regime I), the first small drop in resistivity is attributed to isothermal transformation of a small amount of retained austenite to martensite; this may be important for "cryogenically" treated steels. In regime II, resistivity increase to peak and its subsequent decrease are attributed to the martensite aging where the resistivity increase to peak value corresponds to the carbon atom rearrangement prior to transition carbide formation. An analysis of integrated intensity changes of Fe-Ni-C martensite (002) x-ray diffraction peak profiles during subambient aging stage (Fig. 14.2) has shown that clusters are associated with two to four carbon atoms and leads to the formation of isolated multiplets. This figure also shows the dramatic changes of martensite (002) peak intensity for early and later stages of tempering up to 450°C. Mössbauer spectroscopy, ^{13}C NMR, and atom probe analysis also support the development of carbon atom clustering (and the depletion of carbon in the martensite matrix). Evidence of carbon atom clustering has been shown in TEM by the presence of diffuse intensity spikes around the fundamental electron diffraction spots. Nagakura et al.[7] have demonstrated from dark-field contrasts that the clusters were about 1 nm in size, which agreed well with the results obtained from x-ray diffraction studies.[5]

On the basis of APFIM study of Fe-Ni-C alloy, the formation of α''-Fe$_{16}$C$_2$ structure (similar to α''-Fe$_{16}$N$_2$ in Fe-N alloy) has been suggested as the pre-precipitation stage.[16] However, according to x-ray diffraction experiments using high-intensity synchrotron radiation, α''-like structure does not seem to be formed either upon aging at room temperature or upon tempering at higher temperatures.[17]

Taylor and his coworkers have proposed spinodal decomposition mechanism as the preprecipitation stage.[16] The Taylor-Cohen review paper[18] discusses the aging in terms of development of structural modulations (possibly spinodal decomposition mechanism). Figure 14.1b presents a pictorial summary of the aging of initially virgin martensites, as developed in this review, showing increasing amplitude in A1 and increasing wavelength in A2 (A3).[18]

A2 Stage. The second stage of aging, A2, occurring between subzero temperature and about 70°C, involves a fine modulated tweed microstructure containing carbon-deficient regions and carbon cluster regions (varying from 0.2 to 11 at%).[4,7,18a,19] Figure 14.3a shows the dark-field micrograph of a modulated A2 structure in Fe-1.31%C martensite tempered for 1 hr at 70°C, taken with the (002) fundamental spot and its satellites. This illustrates the fringes parallel to the (102) and ($\bar{1}$02) planes. Figure 14.3b is a high-resolution electron micrograph of an Fe-1.39%C martensite tempered for 1 hr at 70°C, taken with a beam parallel to [0$\bar{1}$0], which illustrates that the vertical fringes are ($\bar{1}$01) lattice fringes and that the white patches are due to the modulated A2 structure.[7] These observations suggest that the carbon atom clusters about 1 nm in size concentrate at random in the {102} planes spaced periodically at about 1 nm apart with the intervening carbon-depleted regions. Such a modulated structure formation was also provided by Tanaka and Shimizu[20] and Taylor et al.[15,16,18]

The x-ray diffraction and electrical resistivity measurements are unable to distinguish the A2 structure; the former, however, shows peak shifts in the tempering temperature range corresponding to the formation of A2 structures (Fig. 14.2),

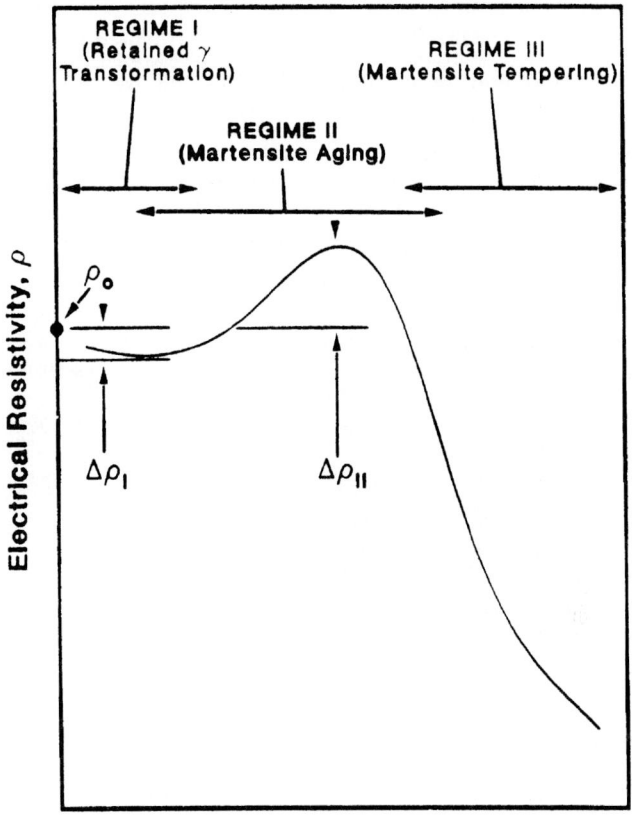

Time or Temperature

(a)

FIGURE 14.1 (*a*) Schematic resistivity versus aging time/temperature curve identifying the regimes which occur during the aging and tempering stages of initially virgin martensite.[4] (*b*) Summary of the structural changes that occur during aging and tempering stages of initially virgin martensite, with respect to the trends in electrical resistivity ρ and flow stress σ. Aging is identified with spinodal decomposition of the martensite, which produces the regime II resistivity peak and an increase in strength. The respective increases in amplitude Δc and wavelength λ of the carbon-concentration modulations that develop coherently are designated as substages A1 and A2; possible secondary ordering of the carbon atoms constitutes substage A3. Alternatively, substages A1 and A2 have also been associated with the formation of randomly distributed carbon clusters and subsequent evolution of the modulated structure by stress-induced alignment of these clusters. Carbide precipitation—ε-carbide in Stage T1 and cementite in Stage T3—marks the completion of aging and the subsequent sequence of tempering. Mechanical strength continues through a maximum near the onset of T1, while resistivity continues to decrease in Regime III. Structural changes are illustrated by electron micrographs obtained during the aging of initially virgin Fe-Ni-C martensites (magnifications for these micrographs are all the same).[18] (*Source:* K. A. Taylor, "Aging Phenomena in Ferrous Martensites," ScD thesis, Massachusetts Institute of Technology, Cambridge, 1985.) (*Courtesy of K. A. Taylor.*)

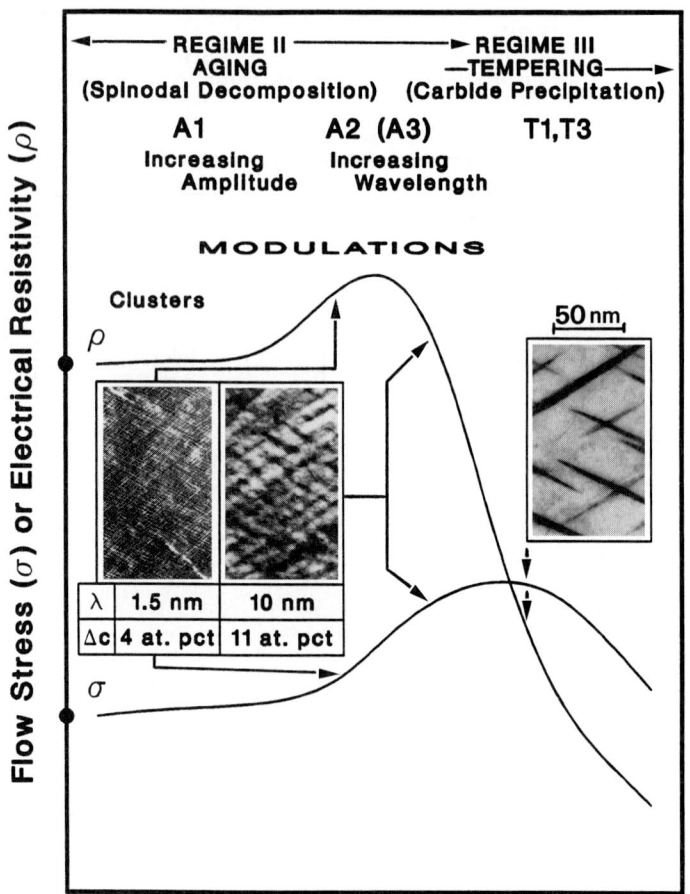

Time or Temperature

(b)

FIGURE 14.1 (*Continued*) (*a*) Schematic resistivity versus aging time/temperature curve identifying the regimes which occur during the aging and tempering stages of initially virgin martensite.[4] (*b*) Summary of the structural changes that occur during aging and tempering stages of initially virgin martensite, with respect to the trends in electrical resistivity ρ and flow stress σ. Aging is identified with spinodal decomposition of the martensite, which produces the regime II resistivity peak and an increase in strength. The respective increases in amplitude Δc and wavelength λ of the carbon-concentration modulations that develop coherently are designated as substages A1 and A2; possible secondary ordering of the carbon atoms constitutes substage A3. Alternatively, substages A1 and A2 have also been associated with the formation of randomly distributed carbon clusters and subsequent evolution of the modulated structure by stress-induced alignment of these clusters. Carbide precipitation—ε-carbide in Stage T1 and cementite in Stage T3—marks the completion of aging and the subsequent sequence of tempering. Mechanical strength continues through a maximum near the onset of T1, while resistivity continues to decrease in Regime III. Structural changes are illustrated by electron micrographs obtained during the aging of initially virgin Fe-Ni-C martensites (magnifications for these micrographs are all the same).[18] (*Source: K. A. Taylor, "Aging Phenomena in Ferrous Martensites," ScD thesis, Massachusetts Institute of Technology, Cambridge, 1985.*) (*Courtesy of K. A. Taylor.*)

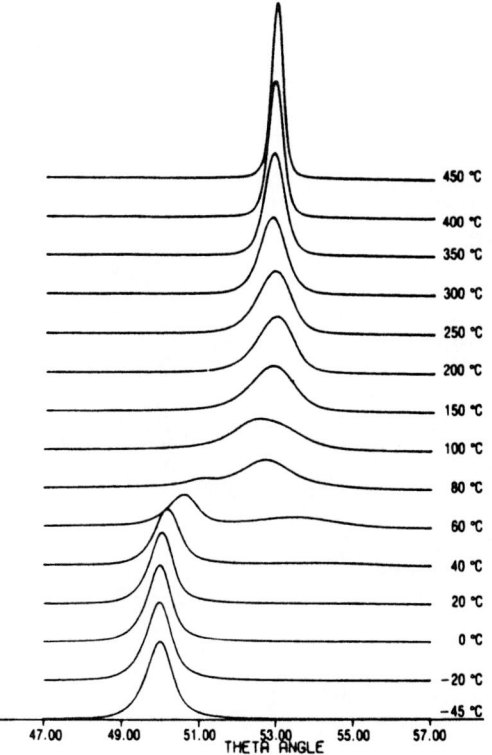

FIGURE 14.2 The (002) x-ray diffraction peak profiles for martensite in an 18Ni-0.98C-Fe alloy at successively higher temperatures, as indicated for times between 1 and 2.5 hr. (*Reprinted from P. C. Chen and P. G. Winchell, Met. Trans, vol. 11A, 1980, p. 1333; reprinted from a publication of The Metallurgical Society, Warrendale, Pa.*)

thereby illustrating a gradual decrease in tetragonality of the martensite, whereas the latter exhibits carbon depletion in the matrix.[12]

The entire aging process is diffusion-controlled and produces a depletion of carbon in the martensite lattice with the attendant conversion to low tetragonal martensite. Carbon clustering, the formation of isolated multiplets, and the development of modulated structure are insensitive to martensite morphology (provided the effect of "autotempering" during the quenching of alloys with high M_s temperature is ignored).[20a]

A3 Stage. During the third substage of aging reaction A3, an ordering reaction, perhaps intimately related to the prior clustering, may occur,[18] leading to tetragonal Fe_4C carbide particles. The A3 structure may be identified by superstructure spots or satellites appearing in the electron diffraction pattern of martensite tempered at about 60 to 80°C. This structure is also exhibited by the carbon-free regions between the antiphase domains, called κ phase, C-phase, or low-carbon martensite, which is perhaps coherently strained martensite adjacent to the transition carbide particles.[12] However, the recent TEM and APFIM results indicate that A1 and A2 may be the early and advanced stages, respectively, in the same overall process of spinodal decomposition. Furthermore, if A3 is associated with spinodal ordering, then A1, A2, and A3 stages could each represent different aspects of a single decom-

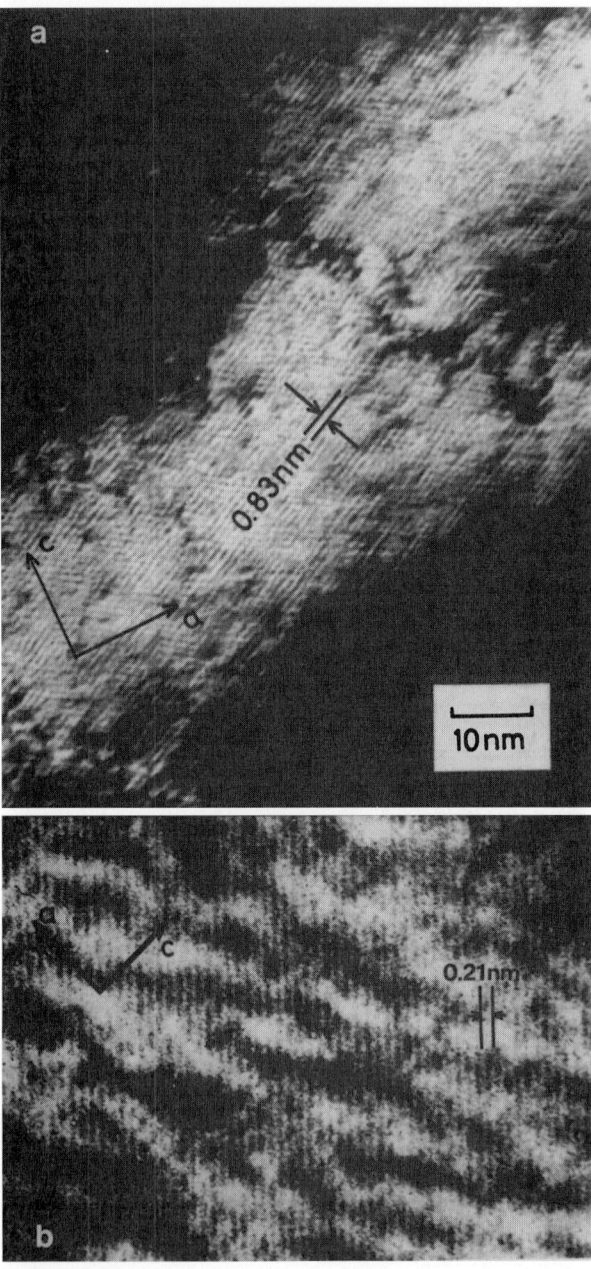

FIGURE 14.3 (*a*) Dark-field electron micrograph of an Fe-1.31C alloy martensite tempered for 1 hr at 70°C taken with the (002) fundamental spots and its satellites showing modulated A2 structure. (*b*) High-resolution electron micrograph of an Fe-1.39C martensite tempered for 1 hr at 70°C taken with beam parallel to [0$\bar{1}$0]. The vertical fringes are ($\bar{1}$01) lattice fringes, and the white patches are due to the modulated structure.[7] (*Reprinted by permission of The Metallurgical Society, Warrendale, Pa.*)

position process.[10] However, other workers[19] have discounted the A3 stage and hence the A3 structure.[21]

14.2.2   The First Stage of Tempering (T1)

The tempering reactions of martensite formed in low-carbon steels (<0.2% carbon) differ from those of medium- and high-carbon steels. On quenching and during the first stage of tempering (T1) of martensite containing up to 0.2% carbon concentration, the diffusion and segregation of carbon occur completely around the dislocations and lath boundaries which are limited to 0.2% carbon content. When supersaturated martensite containing more than 0.2% carbon is tempered at about 100 to 250°C (212 to 482°F), only a small proportion of carbon is bound with the dislocations and lath boundaries while a large proportion of carbon atoms remains in the normal interstitial positions and participates in the formation of a transition carbide. (This questions the segregation of carbon in 0.2% carbon or lower-carbon alloys. Data show such low-carbon alloys also exhibit clustering, as shown by sensitivity data; hence not all carbon contents get segregated to dislocations or lath boundaries.[20a])

Jack[22] first identified the transition carbide to be hexagonal close-packed ε-carbide with lattice parameter $a = 0.237$ nm, $c = 0.433$ nm, and $c/a = 1.58$. Its chemical composition was thought to be $Fe_{2.4}C$, and it contained about 8 wt% carbon. Steels with lower carbon content are unlikely to precipitate ε-carbide by nucleation and growth.

It is thus apparent that in the first stage of tempering (T1), high-carbon martensite decomposes into a two-phase mixture comprising fine transition ε- or η-carbides and low-carbon martensite (α''). The former constituent forms in linear clusters of particles 2 to 5 nm in size,[23,24] which increases linearly in amounts with increasing interstitial carbon content (above 0.2%) of the steel, while the latter results in an attendant decrease in tetragonality ($c/a = 1.008$). The carbon content of this low-carbon martensite (α'') is approximately 0.25 wt% and is independent of the carbon content of the original martensite. Tempering of an Fe-0.7%C alloy martensite at 150°C produces coherent and intragranular ε-carbide precipitation in a finely dispersed form on cubic planes of the matrix. They may also form on the dislocation lines, within the matrix, introduced by martensitic transformation. Two striking metallographic features, revealed only under the electron microscope, are shown in Fig. 14.4A and B;[25] the former micrograph exhibits the granular appearance, while the latter one is the fine crosshatched pattern similar to that observed in the aging process.

Tempering up to 200°C causes the growth of ε-carbide particles into well-developed arrays of needlelike morphology which are elongated in the <001>$_{\alpha''}$ direction with a {100} habit plane (Fig. 14.5A and B).[25] The orientation relationship between the ε-carbide and the matrix is

$$\left(10\bar{1}1\right)_{\varepsilon} \,/\!/\, (101)_{\alpha''}; (0001)_{\varepsilon} \,/\!/\, (011)_{\alpha''}; \left[12\bar{1}0\right]_{\varepsilon} \,/\!/\, [111]_{\alpha''}$$

according to Jack[22] and

$$(0001)_{\varepsilon} \,/\!/\, (011)_{\alpha''}; \, \left[\bar{1}2\bar{1}0\right]_{\varepsilon} \,/\!/\, [100]_{\alpha''}$$

according to Pitsh-Schrader. It is thus clear that such a precipitation occurs because of the similarity of $(10\bar{1}1)$ planes of ε-carbide (0.206 nm) with that of (101) planes

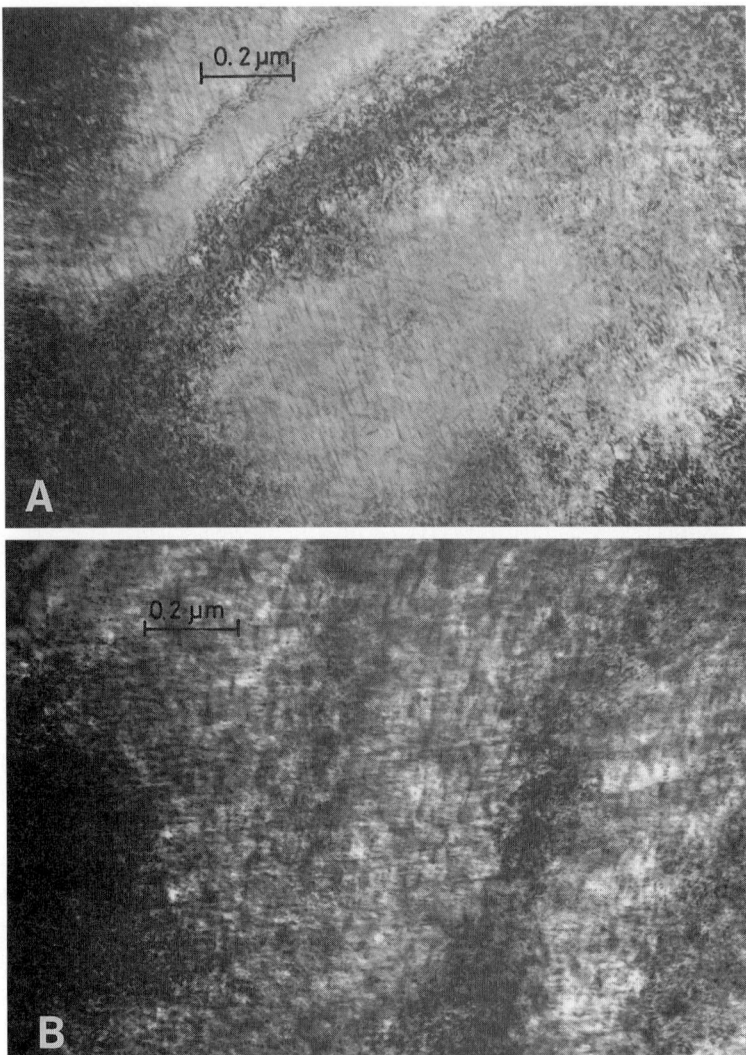

FIGURE 14.4 Thin-foil electron micrographs of Fe-0.7C alloy austenitized at 1100°C for 10 min, iced-brine quenched, and tempered at 150°C with a constant heating rate of 5°C/min illustrating ε-carbide precipitation in a finely dispersed form with (*A*) a granular appearance and (*B*) a crosshatched pattern. (*Courtesy: Y. Ohmori.*)

of martensite (0.206 nm) or the similarity of $(0001)_\varepsilon$ planes with $(001)_{\alpha''}$ planes. This gives rise to creation of coherency strains across these matching planes, which can account for an increase in hardness observed during the first stage of tempering.

When the synthetic diffraction pattern considering different variants of ε-carbide/matrix orientation relationships is compared with the experimental data, a good agreement can be found between them. Figure 14.6 shows such a comparison where the reflections from Fe_3O_4 formed on the thin film are also superimposed.[25]

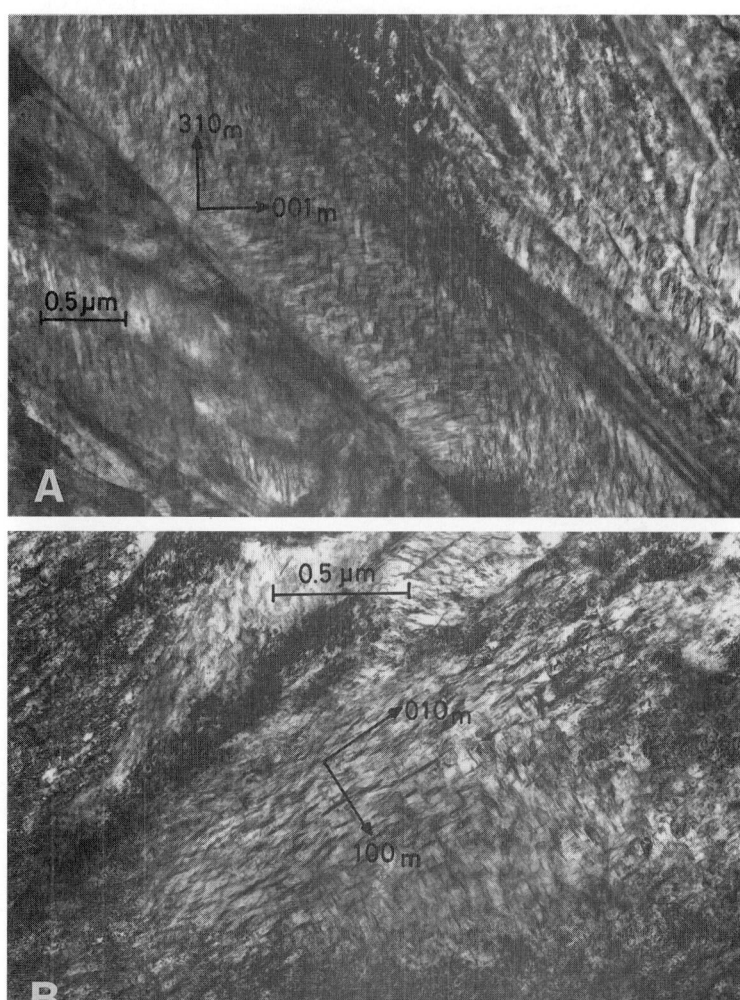

FIGURE 14.5(A,B) Dark-field electron micrographs of an Fe-0.7C martensite after tempering at 200°C taken with the different fundamental spots of martensite showing the growth of ε-carbide particles into well-developed arrays of needlelike morphology. (*Courtesy: Y. Ohmori.*)

The presence of ε-carbide during the first stage of tempering has also been found in subsequent x-ray and TEM studies.[26-29] However, Hirotsu and Nagakura[23,30] demonstrated that the transition carbide had an orthorhombic structure that was later designated as ordered eta (η)-Fe$_2$C phase, which is isomorphous with Co$_2$C and Co$_2$N precipitates. The structure of η-Fe$_2$C has lattice parameter $a = 0.4$ nm, $b = 0.433$ nm, and $c = 0.284$ nm. Carbon atoms regularly occupy one-half of the octahedral interstitial sites.[7] The structures of ε- and η-carbides are very similar, but the latter are distinguished by the presence of weak carbon superlattice spots in electron diffraction patterns. Taylor has suggested that η-carbide can be considered as

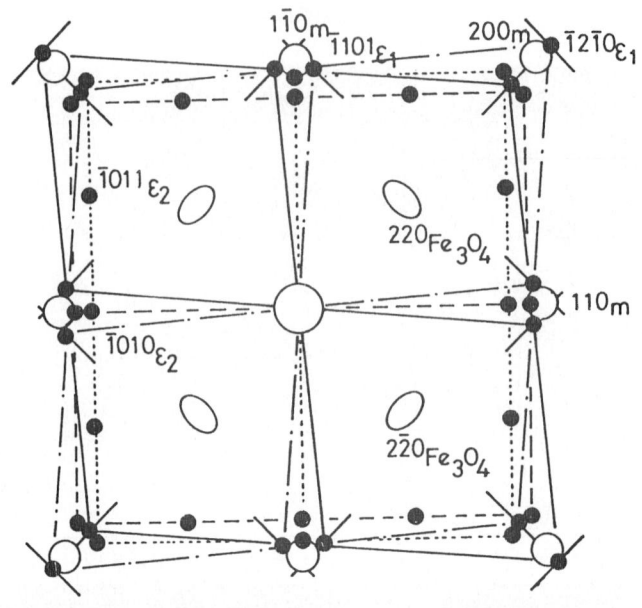

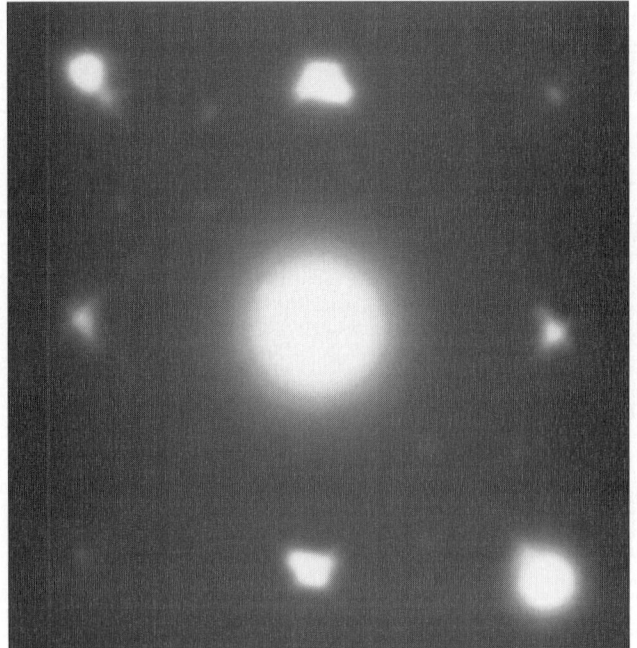

FIGURE 14.6 Correlation of synthetic diffraction pattern considering the different variants of ε-carbide/matrix orientation relationships with the observed electron diffraction pattern. Here the reflections from Fe_3O_4 formed on thin films are also superimposed. (Courtesy: Y. Ohmori.)

FIGURE 14.7 Bright-field image of Fe-1.22C alloy martensite tempered for 16 hr at 150°C, showing the distribution of η-carbide.[31] (*Reprinted by permission of The Metallurgical Society, Warrendale, Pa.*)

a derivative of the ε-carbide structure or ε'. Like the ε-carbide, the η-carbide contains substantially higher carbon content than that of the cementite. The apparently linear and fine platelets of η-carbides usually seen in bright-field TEM (Fig. 14.7) in the martensite of Fe-1.22%C alloy tempered for 16 hr at 150°C (302°F) is, in fact, composed of fine spherical particles about 2 nm in diameter; this is revealed in a dark-field micrograph taken with illumination from carbide diffraction spots.[31,32] In contrast, Fig. 14.8, an optical micrograph of Fe-1.22%C alloy after quenching and tempering at 150°C (302°F) for 1 hr, illustrates only the characteristic plate-type martensite and the presence of 21 vol% retained austenite in the microstructure based on the point-counting method. The η-carbide precipitate has also been found to form in low- and high-nickel steel[33] and manganese steel.[34]

The orientation relationship between the η-carbide and the martensite is given by Hirotsu and Nagakura[23] as $(110)_\eta // (010)_{\alpha''}$; $[001]_\eta // [100]_{\alpha''}$.

Kinetic data show that the first stage of tempering is dependent on the (clustering and) diffusion of carbon through the martensite with an activation energy of 60 to 80 kJ/mol, which, in turn, is a function of the carbon content in steel.[32] There is an appreciable reduction in specific volume of the metal specimen during the first stage of tempering.

14.2.3 The Second Stage of Tempering (T2)

During the second stage of tempering (T2), retained austenite formed during quenching decomposes, usually in the temperature range of 200 to 300°C, into

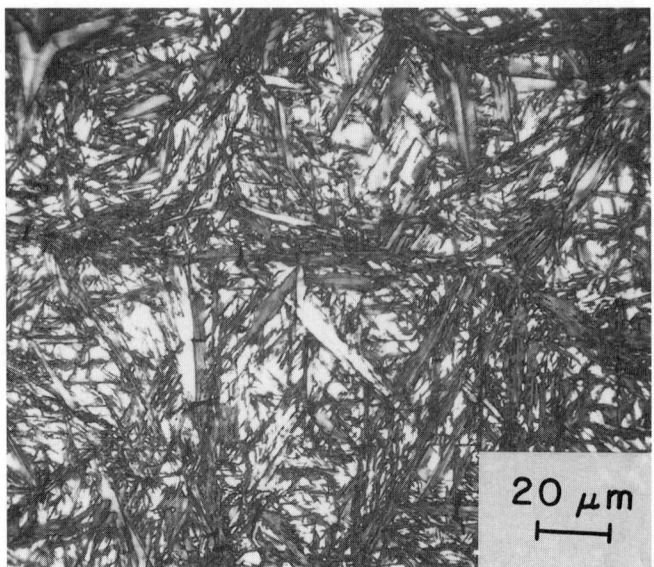

FIGURE 14.8   Optical micrograph of Fe-1.22C alloy martensite tempered for 1 hr at 150°C. Nital etch.[31] (*Reprinted by permission of The Metallurgical Society, Warrendale, Pa.*)

bainitic ferrite and carbide. At higher temperature the bainitic microstructure consists of ferrite and cementite, while at a lower temperature the bainitic microstructure may consist of ferrite and ε- or η-carbide. For example, ultrahigh-strength low-alloy steels such as AISI 4340 can be transformed fully into lower bainitic structure with beneficial effects on ambient temperature mechanical properties. The lattice relationship between cementite and ferrite formed at a higher temperature range of this T2 stage is

$$(100)_\theta \; // \; (011)_\alpha \; // \; (111)_\gamma \qquad \text{and} \qquad [010]_\theta \; // \; [11\bar{1}]_\alpha \; // \; [10\bar{1}]_\gamma$$

in accordance with Bagaryatski.[35]

Figure 14.9 shows the percent transformation of retained austenite in an Fe-1.22%C alloy as a function of time and temperature between 180 and 225°C.[32,36] This illustrates the complete transformation of retained austenite into ferrite and cementite aggregates when held for prolonged periods. Analysis of the kinetic data on the transformation of retained austenite in Fig. 14.9 provides an activation energy of 1.15×10^5 J/mol (27.5 kcal/mol), in agreement with the activation energy for the diffusion of carbon in austenite and the activation energy for the second stage of tempering as reported by Wells et al.[37] and Roberts et al.[38] Figure 14.10 shows the variations of retained austenite and cementite (dashed lines) as a function of tempering temperature for low-alloy medium-carbon steels such as 4130 and 4340 grades. As seen in the figure, the as-quenched austenite in 4130 and 4340 is 2% and 4%, respectively; and the transformation of retained austenite starts only above 200°C (392°F) for tempering times of 1 hr. The transformation is complete at about 300°C (572°F), and the cementite becomes a dominant phase in the microstructure after tempering at 300°C (572°F) and higher.[32,39]

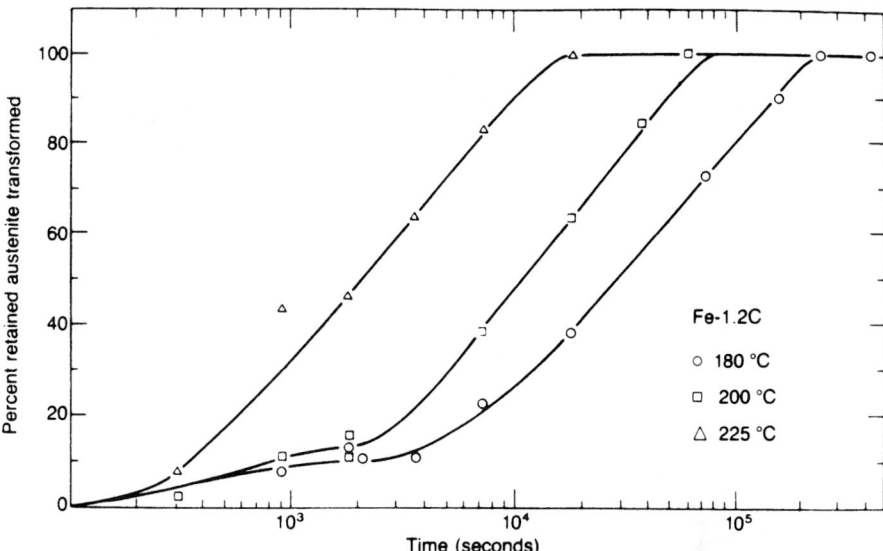

FIGURE 14.9 Transformation of retained austenite in an Fe-1.22C alloy as a function of time and temperature between 180 and 225°C.[36] (*Reprinted by permission of The Metallurgical Society, Warrendale, Pa.*)

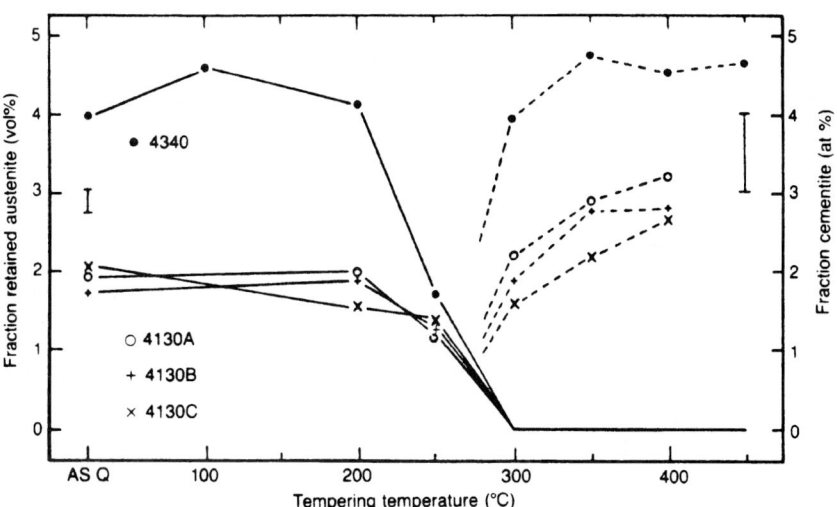

FIGURE 14.10 Variation of retained austenite and cementite (dashed lines) as a function of tempering temperature for low-alloy medium-carbon steels such as 4130 and 4340 grades.[39] (*Reprinted by permission of The Metallurgical Society, Warrendale, Pa.*)

A large positive dimensional change accompanies this transformation.[2]

Note that retained austenite exists in most low-alloy commercial steels with high M_s temperatures, and the transformation of retained austenite in these steels constitutes an important part of the tempering sequence. However, T2 stage may not be present in high-alloy steels, where the stabilization of austenite may produce no observable decomposition during the conventional tempering treatments.[10]

14.2.4   The Third Stage of Tempering (T3)

The third stage of tempering (T3) eventually develops ferrite and cementite at a temperature as low as 250°C.[40] According to Speich,[41] rodlike carbides dissolve and transform to spheroidal cementite by an Ostwald ripening type of process around 400°C. Transformation of ε/η transition carbide to cementite occurs somewhat later than the decomposition of retained austenite.[9] In high-carbon martensite, the most favored site for the formation of cementite is the ε- (or η-) carbide/matrix interface where the transition carbide gradually dissolves with the appearance of Widmanstätten cementite particles. The second potent sites for nucleation are the twins present in the high-carbon martensites where cementite appears to grow along the twin boundaries, forming colonies of similarly oriented lathlike particles of $\{112\}_\alpha$ habit which are quite different from the usual Widmanstätten habit. However, the Bagaryatski orientation relationship with the matrix is observed in both these cases, which is $(211)_{\alpha'} // (001)_\theta; [01\bar{1}]_{\alpha'} // [110]_\theta; [\bar{1}11]_{\alpha'} // [010]_\theta$. A third site for the nucleation of cementite is the grain boundary region, at both the original austenite grain boundaries and the interlath boundaries of the martensite. The cementite can form as very thin film which gradually becomes spheroidized. These grain boundary cementite films may be deleterious to mechanical properties, particularly ductility.[41,42] However, they can be modified by alloy additions.[10]

In low-carbon steels, during both T2 and T3 stages, cementite nucleates as fine platelets on $\{110\}_{\alpha'}$ habit planes within the martensite laths.[43] This results in a dramatic reduction in the martensite lath boundaries per unit volume and pinning of high-angle lath boundaries, thereby stabilizing the lath morphology of as-quenched martensite.[44]

Some investigators[45,46] have given evidence that at the commencement of the third stage, particularly in high-carbon steel, the carbide structure that forms is Hägg or chi (χ) carbide. This carbide has been identified to be monoclinic with lattice parameter $a = 1.1562$ nm, $b = 0.4573$ nm, $c = 0.5060$ nm, and $\beta = 97°44'$, based on x-ray powder pattern by Jack and Wild.[47] Its chemical composition is M_5C_2, where M stands for any combination of Fe and Mn atoms.[45,46]

There are three nucleation sites for the formation of χ-carbides within the martensitic structure of an Fe-1.22C alloy at the beginning of the third stage of tempering: along the martensite plate interface, along the twins within the martensite plates, and within the martensite matrix in plates without twins. Figure 14.11a shows both the matrix and interface χ-carbides formed on tempering of martensite in an Fe-1.22C alloy specimen at 350°C. Figure 14.11b shows the nucleation and growth of χ-carbides in a typical twin array in the same alloy specimen tempered at 300°C.[48]

This χ-carbide is progressively replaced by cementite at higher temperatures.[10]

As in the first stage of tempering, a negative dimensional change accompanies the third stage of tempering. The total shrinkage in length obtained in a 1.0% carbon steel during the first and third stages of tempering amounts to about 0.25%.[2]

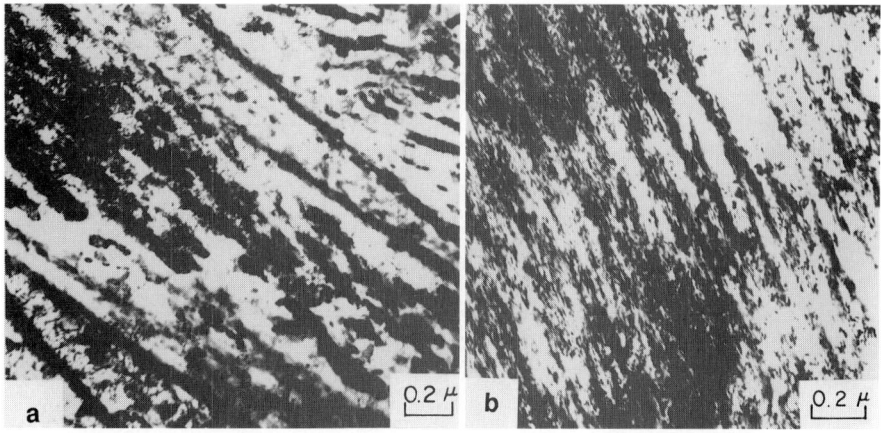

FIGURE 14.11 Bright fields. (*a*) The χ-carbide precipitation both at the interface and within the matrix upon tempering of martensite in Fe-1.22C alloy specimen at 350°C. (*b*) Nucleation and growth of χ-carbide in a typical twin array in the same alloy specimen tempered at 300°C.[48] (*Reprinted by permission of The Metallurgical Society, Warrendale, Pa.*)

14.2.5 Fine Structures of Faulted Cementite (θ') Platelets

Electron diffraction pattern analyses of cementite precipitate particles containing stacking faults and occurring during the tempering of Fe-0.7C martensite above 225°C have been made by various workers.[49–51] They have demonstrated that the faulted cementite (referred to as θ') platelets consisting of θ- and χ-carbides occur by *microsyntactic intergrowth mechanism*,[51] which involves local readjustments of stoichiometry and structure on lattice planes.[52] Both carbides precipitate on the prior austenite grain boundaries, on transformation twins, and within the matrix. The occurrence of stacking disorder during the early stages of cementite formation and growth may well be the general rule rather than an exceptional phenomenon. Figure 14.12 shows the detailed TEM structure.[25] The structural difference of faulted cementite in the tempered martensite in Fe-0.7C alloy has been determined by dark-field lattice imaging observation employing an extraction replica technique. Figure 14.13 shows these results. In both cases, fringe images are produced from $(001)_\theta$ faulted cementite planes, and two different lattice spacings 0.67 nm and 0.57 nm have been found for $(001)_\theta$ and $(200)_\chi$, respectively. The fringes resulting from χ-carbide planes exhibit a slightly different contrast, as shown by arrow χ in Fig. 14.13. Note that the density of χ-carbide lattice planes within a carbide precipitated on the twin interface in the tempered martensite is quite high (~35%). The boundaries comprising discontinuities of either $(001)_\theta$ or $(200)_\chi$ lattice planes within a carbide are often observed in tempered martensite (near A in Fig. 14.13). This difference of the faulted cementite structure suggests that the cementite platelets in tempered martensite grow in size by the coalescence of separately nucleated carbide particles on the twin interfaces.[25,51]

Nakamura et al.[50] examined the growth of T3 carbides with a hot-stage-equipped electron microscope and found that the θ' particles were not stable above 470°C; at higher temperatures only unfaulted cementite particles were observed. Surprisingly, their observations suggest that the θ' particles dissolve in favor of the cementite phase instead of undergoing an in situ transformation.[49]

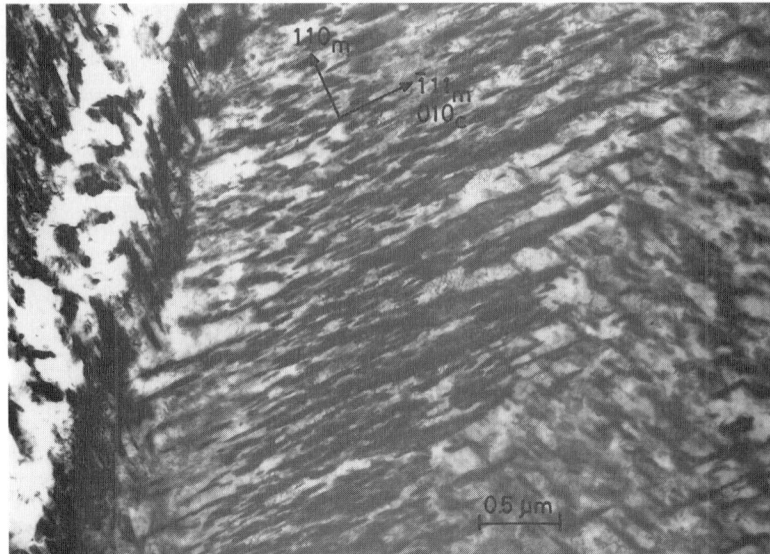

FIGURE 14.12 Detailed TEM structure: Faulted cementite platelets consisting of θ- and χ-carbides which occur by microsyntactic intergrowth mechanism on the prior austenite grain boundaries, transformation twins, and within the matrix. (*Courtesy: Y. Ohmori.*)

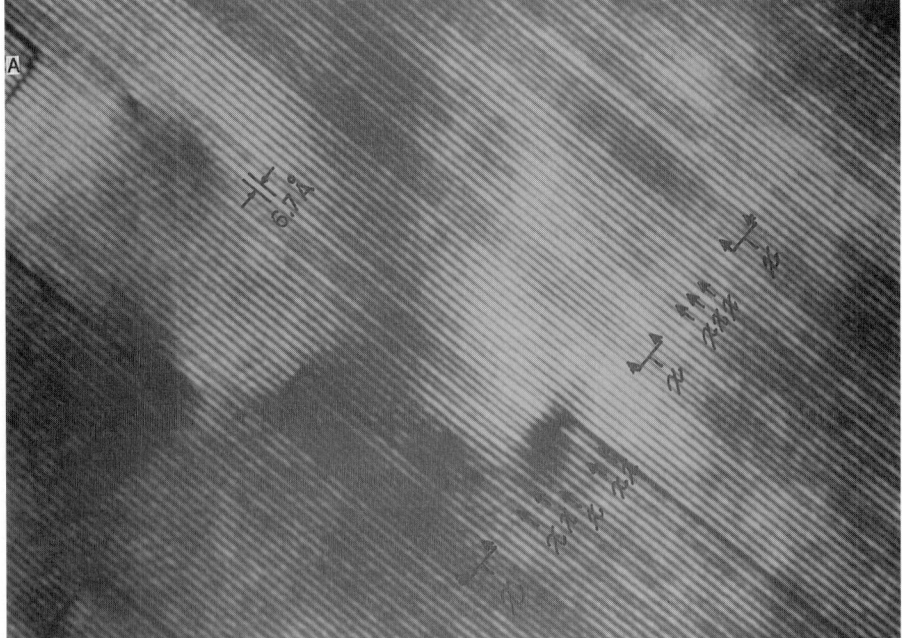

FIGURE 14.13 Dark-field lattice image for faulted cementite particles showing carbide formed on the twin interface in tempered martensite.[25,51] (*Courtesy: Y. Ohmori.*)

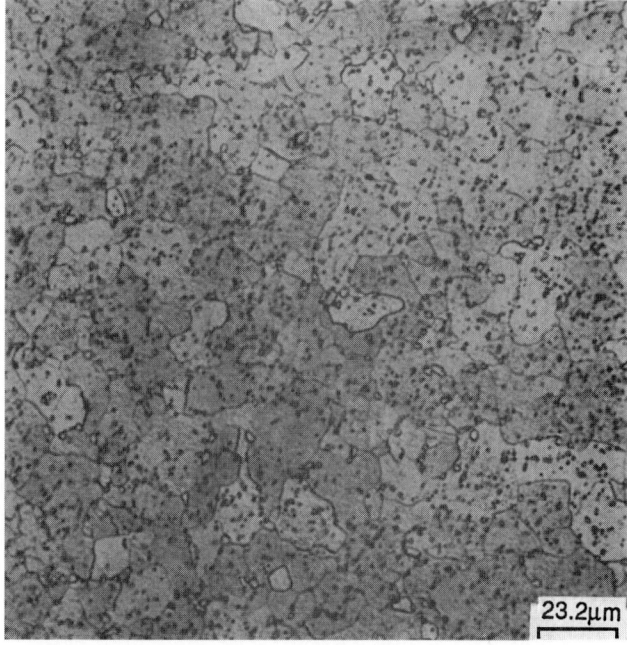

FIGURE 14.14 Optical micrograph: Fe-0.17C alloy specimen water-quenched from 900°C and tempered for 5 hr at 650°C. Spheroidized Fe_3C in equiaxed ferrite. (*Courtesy: R. W. Honeycombe; after Lenel.*)

14.2.6 The Fourth Stage of Tempering (T4)

The fourth stage of tempering (T4) is primarily a coarsening process of cementite particles and its spheroidization. For a lath martensite in a low-carbon steel, two distinct stages of structural change occur in the temperature range of 350 to 600°C.[32,42,44,53,54] The first stage is influenced by a recovery mechanism in which considerable rearrangement of the dislocations occurs within the laths and at the lath boundaries, which are essentially low-angle boundaries. This results in a significant reduction in the dislocation density, and the lathlike ferrite grains are closely related to the packets of similarly oriented as-quenched lath martensite. Simultaneously, the fine carbide precipitates formed help in retaining the elongated lath morphology. In the second stage, recrystallization is suppressed due to the pinning action of carbide particles on the grain boundaries. There is a considerable rearrangement of the remainder dislocations within the laths, which results in further elimination of low-angle lath boundaries and subsequent replacement by more equiaxed ferrite grains. The equiaxed grains possess subboundaries containing regular dislocation arrays. Such a breakup of large grains into smaller ones by dislocation boundaries is called *polygonization*. Further heating between 600 and 700°C involves recrystallization, which causes the aligned lath (or acicular) ferrite structure to recrystallize as an aggregate of equiaxed ferrite grains containing coarse dispersion of spheroidal cementite particles (Fig. 14.14).[27,42] The latter, being present both at the grain boundaries and within the grains, occurs by Ostwald ripening. In plate marten-site in high-carbon steels, a large density of cementite precipitation induces more sluggish recrystallization by pinning the ferrite boundaries. The final stage involves the continued coarsening of spheroidal cementite particles and progressive growth of equiaxed ferrite grains.

14.3 MECHANICAL PROPERTY CHANGES ON TEMPERING

The structural changes during tempering greatly influence the mechanical proper-ties of steel. Both tempering temperature and time affect the property changes. However, we will consider first the effect of tempering temperature (for a given tempering time) on hardness of plain carbon steels. Figures 14.15[41] and 14.16[55] show the influence of increasing tempering temperature (for 1 hr) on decreasing the hard-ness of quenched plain carbon steels. As seen in Fig. 14.15, tempering of Fe-C martensites containing 0.2% carbon or less up to 200°C (392°F) produces a very small change in hardness values because neither precipitation of transition carbide nor tetragonality in the martensite has been observed. Above 200°C (392°F) the hardness gradually decreases as the tempering temperature is raised to the eutec-toid temperature. When the carbon concentration is increased to 0.4%, the hard-ness decreases steadily with increase in tempering temperature from 150°C to the eutectoid temperature.

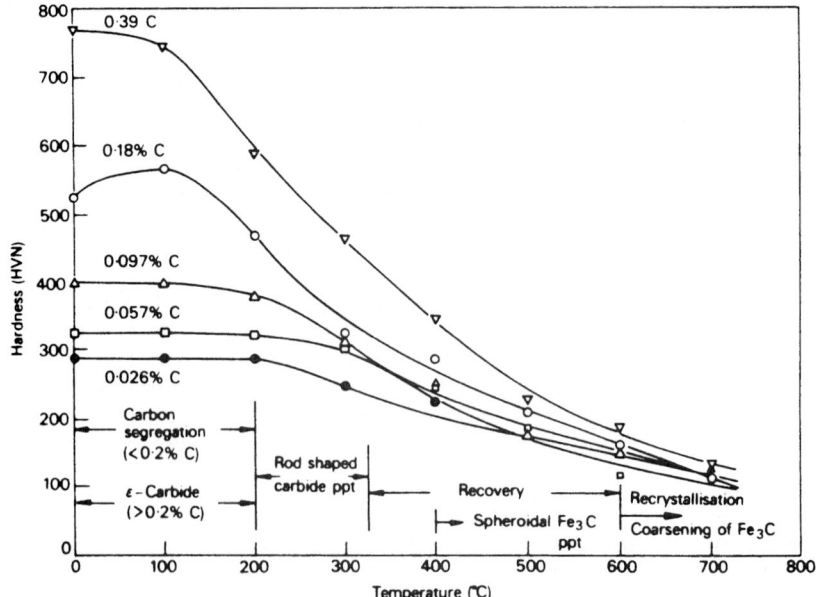

FIGURE 14.15   Hardness of iron-carbon martensites tempered for 1 hr at 100 to 700°C (212 to 1292°F).[41] (*Reprinted by permission of The Metallurgical Society, Warrendale, Pa.*)

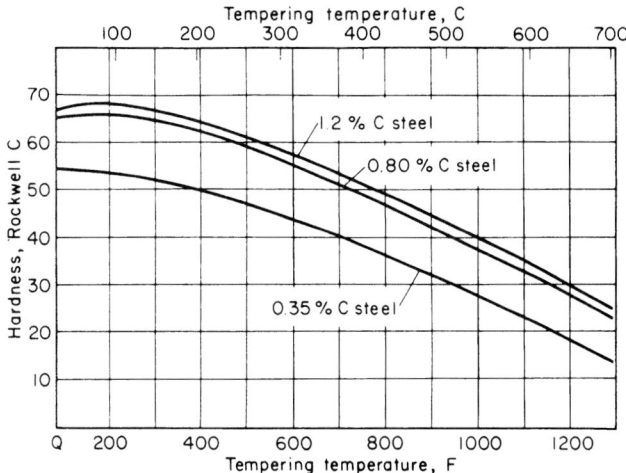

FIGURE 14.16 Influence of increasing temperature (for 1 hr) on decreasing the hardness of quenched carbon steels. Data taken from various sources by Grossman and Bain.[55] (*Reprinted by permission of ASM International, Materials Park, Ohio.*)

A small increase in hardness is observed when the martensite in high-carbon steels is given the first-stage tempering in the temperature range of 100 to 150°C (212 to 302°F) (Fig. 14.16). The reason is that low-temperature tempering produces a sufficient quantity of fine transition carbides, which, in turn, provides adequate dispersion strengthening to compensate for the softening effect due to the depletion of carbon from the martensite matrix. Hence, if maximum hardness is required, a choice should be centered on a high-carbon steel and a low-temperature tempering. Martensitic steels tempered in the first stage maintain high strength and possess very good fatigue and wear resistance.[52]

Above 200°C, and particularly during the third and fourth stages of tempering, a marked decrease in hardness and strength occurs as a result of (1) the pronounced softening of the matrix and (2) growth and coalescence of cementite particles. The hardness of tempered martensite reaches its lowest near the eutectoid temperature.

Figures 14.17[56] and 14.18 (right side of the hatched region, identified by solid points)[57] show the typical mechanical properties as a function of tempering temperature (for 1 hr) for water-quenched AISI 1050 steel and oil-quenched AISI 4340 steel, respectively. It is seen that tempering at higher temperatures reduces strength (like hardness, as noted above) but usually improves the ductility and reduction of area (ROA). When high toughness and ductility are of major concern and strength and hardness are of secondary concern, high-temperature tempering is used. On the other hand, when high-carbon (0.5 to 1.0%) steels are used for applications where high strength, hardness, wear resistance, and edge retention are of major importance and toughness and ductility are of secondary importance (e.g., axes, hammers, cutting tools, knives), lower-temperature tempering is the best choice.

Low-temperature tempering (LTT) [around 175 to 250°C (347 to 482°F)] process is also used in modified 4130 grade (with 1.96 Mn and 0.43 Mo contents) to achieve a better combination of improved toughness, hardness, and strength levels and

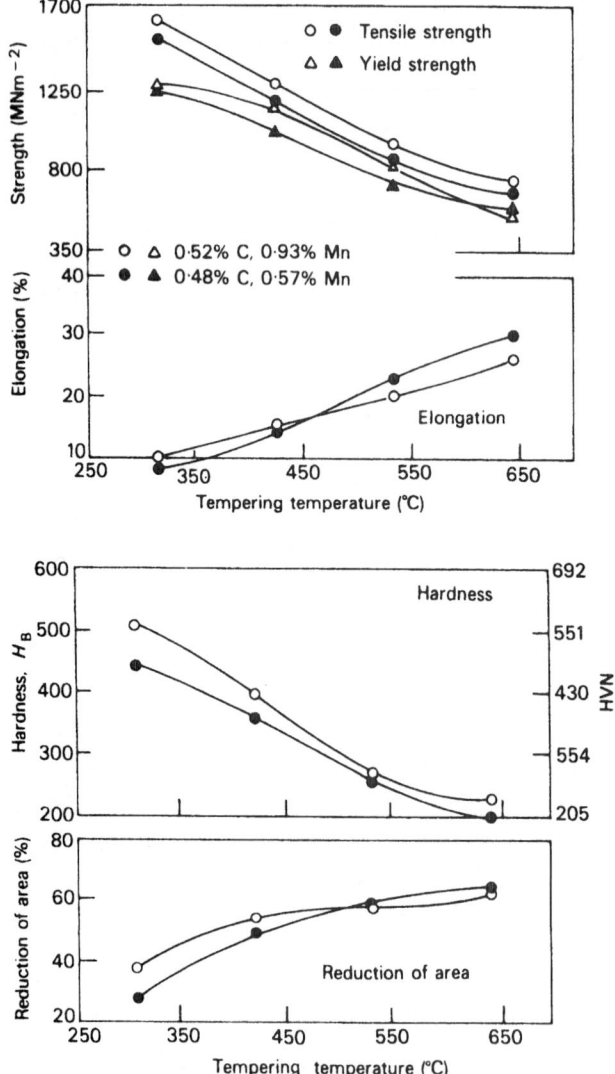

FIGURE 14.17 Mechanical properties of water-quenched and tempered 1050 steel (0.45 to 0.55 C, 0.6 to 1.07 Mn).[56] (*Reprinted by permission of ASM International, Materials Park, Ohio.*)

equivalent fatigue strengths compared to the high-carbon 4140 steel and high-temperature tempering.[57a]

The as-quenched (AQ) and as-quenched and low-temperature (150 to 200°C) tempered (AQLTT) martensite data (shown in the left portion of the Fig. 14.18),[58] when compared to the data for specimens tempered at higher temperatures (shown

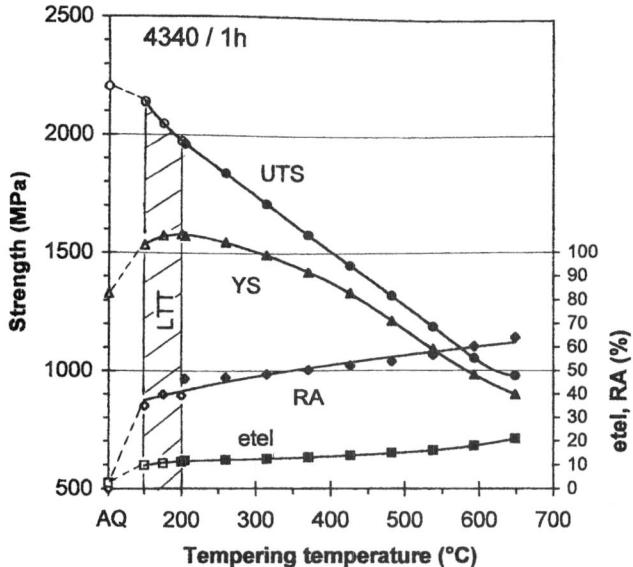

FIGURE 14.18 Change in mechanical properties with tempering temperature for oil-quenched 4340 steel.[57-59] (*Source:* For tempering temperature data above 200°C, Bethlehem Steel Corporation, Bethlehem, Pa., and for tempering temperature data below 200°C, M. Saeglitz and G. Krauss.)

in the right portion of the hatched area), illustrate that a large amount of strain hardening develops during uniaxial tensile testing, which is proportional to the difference between the yield strength and UTS. Figure 14.18 (in the left portion) thus shows that strain hardening is very high in the AQ and LTT martensitic specimens, which continuously decreases with the increase in tempering temperature. The high strain-hardening rates of LTT martensitic microstructures are partly due to low yield strength.[58]

Figure 14.19 shows the UTS versus hardness of all LTT martensitic 43xx steels, which illustrates that both UTS and hardness values increase with increasing low-temperature tempering temperature between 150 and 200°C.[59] The open circles, representing the highest strength and hardness measured in the 4350 steel, denote a tempered condition. (When Hollomon-Jaffe parameters are used, hardness and strength decrease with increasing temperature between 150 and 200°C.[58]) The linear equation relating UTS to hardness (HRC) for martensitic specimens tempered in the first stage of tempering can be given by[58]

$$UTS\,(MPa) = 74.04(HRC) - 1970$$

It is seen from Fig. 14.20 that the impact toughness decreases if 4340 steel is tempered in the range of 200 to 400°C (392 to 750°F), while hardness (or strength) also continues to decrease.[60] This decrease in toughness occurs in many AISI steels and is often called the *tempered martensite embrittlement* (TME) or *350°C (or 500°F)*

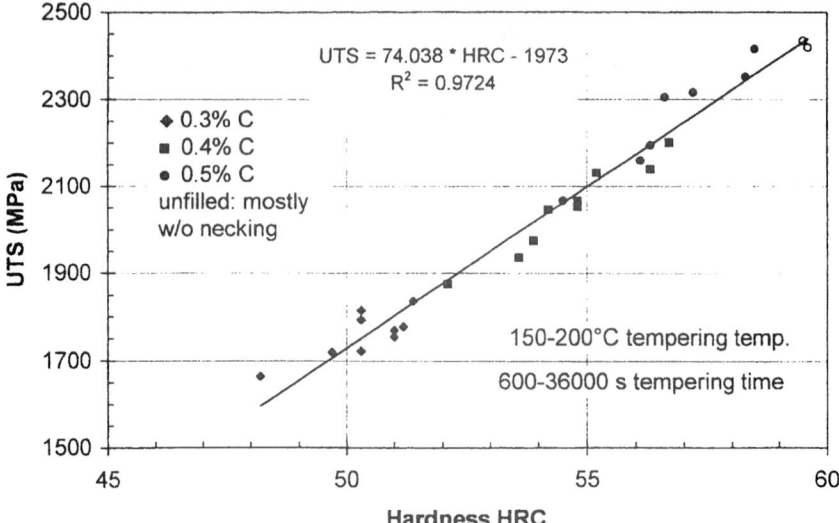

FIGURE 14.19   UTS versus hardness for 43xx steels tempered for various times between 150 and 200°C.[58]

embrittlement, and it is described in Sec. 14.11.1. Another variation of embrittlement is *temper embrittlement*, which may develop in martensitic steels during tempering above 425°C (800°F) or holding on slow-cooling through a certain tempering temperature range. This is also described in Sec. 14.12.1.

14.4   EFFECTS OF ALLOYING ELEMENTS ON TEMPERING

Like hardenability, alloying elements have pronounced influence on the tempering characteristics and the nature and amount of carbides present.[60a] They retard the rate of softening during tempering by stabilizing both the transition carbides and the supersaturated martensitic structure to higher tempering temperature, by delaying the precipitation and growth of cementite. For example, ε-carbide exists in the microstructure even after tempering at 400°C in steels containing 1 to 2.5% Si, and its further transformation to cementite is considerably retarded. The maximum effect of Si on the hardness of tempered martensite occurs at 316°C (600°F). The Ni content does not significantly affect the carbide conversion during tempering of 0.6C-2.5Si-yNi-0.2V steels. A nickel content of 3.25% does not offer any appreciable resistance to softening.[61] On the other hand, 0.1 to 0.2% V addition in 0.6% C steel causes an increase in hardness and impact toughness values by refining the microstructure.[62] The maximum effect of Mn on the hardness of tempered martensite occurs at 427 to 649 °C (800 to 1200 °F) by retarding the coalescence of carbides, thereby providing a resistance to grain growth in the ferrite matrix.[63]

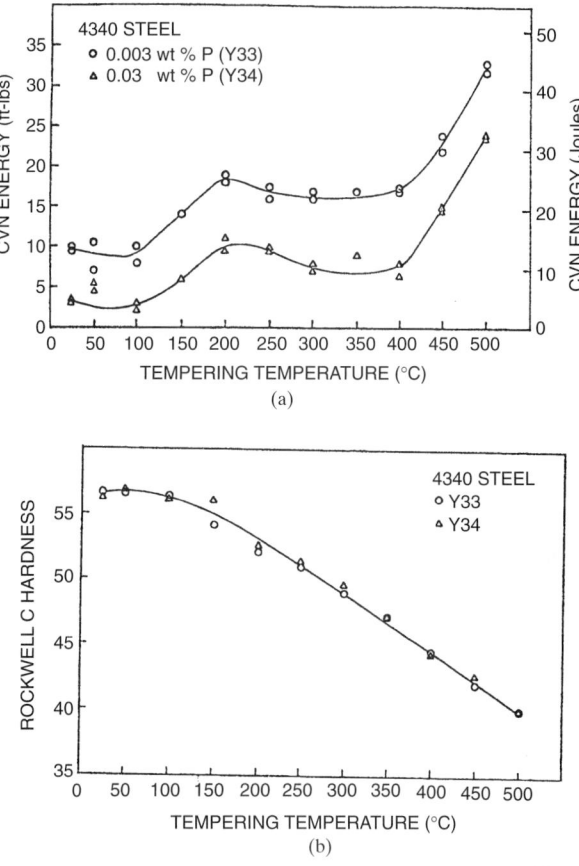

FIGURE 14.20 Change in room-temperature Charpy impact energy (*a*) and hardness (*b*) for 4340 steel austenitized at 870°C (1143 K), oil-quenched, and tempered for 1 hr at temperatures shown.[60] (*Courtesy of The Metallurgical Society, Warrendale, Pa.*)

Tetragonality of the matrix is revealed after tempering at 450 to 500°C (842 to 932°F) in steels containing certain alloying elements such as Cr, Mo, W, V, and Ti. For the strong carbide-forming elements, the greatest increases in hardness of tempered martensite have been found to occur at 427°C (800°F) for Cr, 538 to 592°C (1000 to 1100°F) for Mo, and 649°C (1200°F) for V. Mo partitions the carbide phase at higher temperatures and maintains the fine dispersion of carbide particles. The large effect of V is presumably due to the formation of V_4C_3 or VC carbides which replace cementite-type carbides at high tempering temperature and persist as fine dispersions up to A_1 temperature.[63] The Nb- and Ti-containing steels are more resistant than V-containing steels for tempering temperatures between 500 and 650°C.

Coarsening of cementite in the tempering temperature range of 400 to 700°C (752 to 1292°F), representing the fourth stage of tempering, can be hindered effectively when alloying elements such as Si, Cr, Mo, and W are present in steels. These alloying elements retain the fine Widmanstätten precipitation of cementite up to higher temperatures either by segregating at the carbide/ferrite interface or by entering into the cementite structure.[42] (See Table 1.10 for the effects of various alloying elements on some specific properties.)

14.5 SECONDARY HARDENING

A number of alloying elements in steels form stable carbides, nitrides, and borides. Among alloy carbides, cementite is the least stable compound with the largest solubility in the matrix. When steels containing a sufficient quantity of strong carbide-forming elements are tempered at temperatures below 538°C (1000°F), the tempering reactions preferentially lead to the formation of $(Fe,M)_3C$ cementite particles (where M denotes any substitutional alloying elements in steel). Because the diffusivities of substitutional elements are several orders of magnitude lower in iron than those of the interstitial elements (e.g., C, N, and B) at low temperatures, the alloying elements present in the cementite particles have the same ratio as that present in the matrix. However, when the tempering temperature exceeds 538°C (1000°F), substitutional diffusivities of carbide-forming alloying elements (e.g., Mo, W, V, Nb, Ti, and Cr) exceed those of the interstitial atoms; this results in the precipitation of appreciable amounts of more-stable alloy carbides such as V_4C_3, Mo_2C, W_2C, and so forth (in preference to cementite) within the tempered martensite matrix. This formation of fine alloy carbide dispersion in the range of 500 to 650°C is termed the *fifth stage of tempering* (T5), which is associated with (1) a marked increase in hardness, equivalent to or greater than the hardness in the as-quenched condition (Fig. 14.21),[42,64] and (2) improved toughness and wear resistance. This new form of precipitation hardening is called *secondary hardening*. Alloying additions of Mo, W, V, Nb, and Ti, in ascending order of effectiveness, are used to produce this secondary hardening effect.[60a] The fine precipitates which may form initially as coherent precipitates or zones, pin dislocations, remain very fine even after prolonged tempering and thus are quite resistant to coarsening of the matrix, thereby maintaining high hardness and strength levels. This characteristic of alloy carbides has been exploited to advantage in tool steels such as hot-work tool steels (type H), high-speed steels (types T and M), and Cr-, W-, Mo-, V-, and Co-Ni-bearing steels. As the tempering temperature is increased, secondary hardening alloy carbides grow and usually convert to one or more highly alloyed and stable carbides. Since the growth of alloy carbides is slow, the tempering resistance of these alloy steels is greater than that of plain carbon steels—with recovery, recrystallization, and grain growth being greatly retarded. Such a precipitation sequence of carbide in molybdenum- or tungsten-bearing steel (containing 4 to 6% of the element) is[65]

$$Fe_3C \rightarrow M_2C \rightarrow M_6C$$

The dislocation-nucleated dispersion of carbides responsible for the secondary hardening in both the steels containing Mo- and W-bearing steels, respectively,

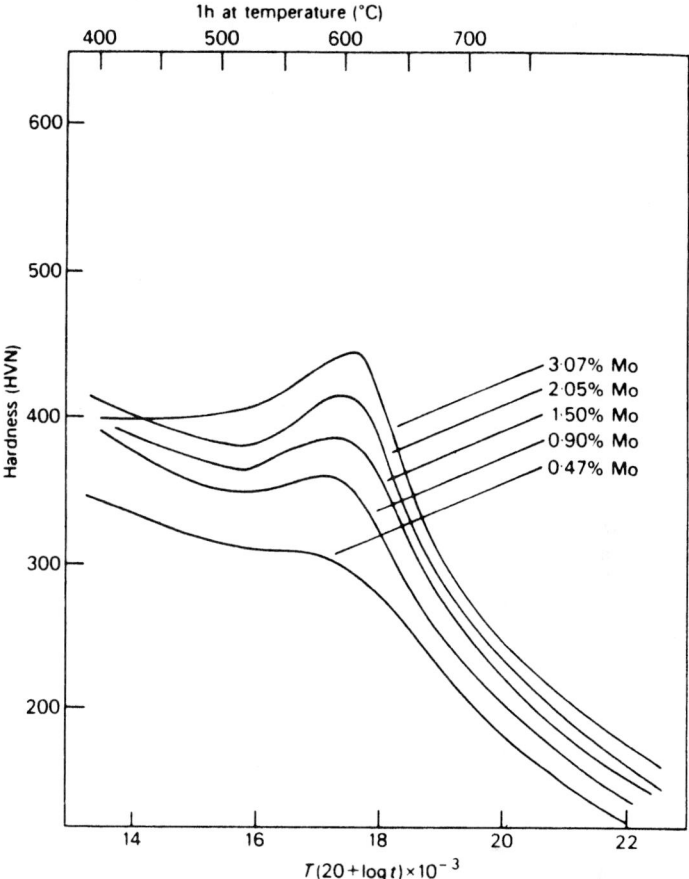

FIGURE 14.21   The influence of molybdenum on the tempering of quenched 0.1 C steel.[42,64] (*Courtesy of the Institute of Metals, England.*)

are the isomorphous hexagonal carbides Mo_2C and W_2C which exhibit rodlet morphology (Fig. 14.22a).[42]

For similar atomic concentrations, the secondary hardening response in the case of molybdenum steel is greater than that of tungsten steels. The M_2C dispersion in the latter is coarser, perhaps because slower diffusivity of tungsten produces coarsening of the dislocation network prior to being pinned by the M_2C precipitate particles.[42]

At lower concentrations of Mo and W (0.5 to 2 wt%), two other alloy carbides appear in the precipitation sequences as intermediate precipitates between M_2C and M_6C, namely, the complex cubic $M_{23}C_6$ and, perhaps, the orthorhombic Fe_2MoC.[42] Figure 14.22b shows the equilibrium structure of massive, coarse, faceted complex cubic M_6C carbide precipitates in the equiaxed ferrite obtained on prolonged tempering at 700°C.[42]

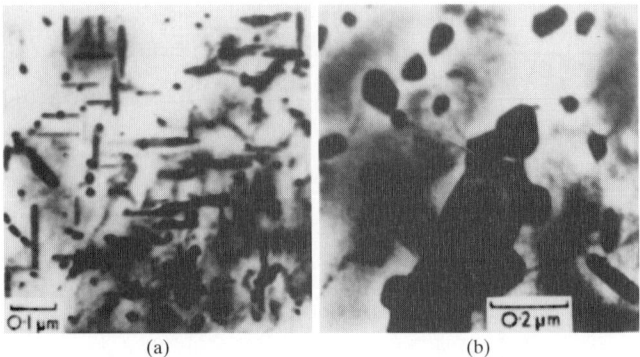

(a)   (b)

FIGURE 14.22 Fe-6W-0.23C quenched and tempered (Davenport). Thin-foil electron micrographs. (*a*) 100 hr at 600°C. W_2C needles along $(001)_\alpha$, some irregular particles of M_6C; (*b*) 26 hr at 700°C, massive M_6C. (*Courtesy of R. W. K. Honeycombe.*)

14.6 SECONDARY HARDENING STEELS

14.6.1 High-Speed Steels

High-speed steels are the most important alloy tool steels because of their very high hardness (in the range of Rc 63 to 69), good wear resistance in the heat-treated condition, and their ability to maintain high hardness at elevated temperatures often encountered during the operation of the tool at high cutting speeds. This property of resisting softening red-heat range (reaching 1000°F and more) is called *red* or *hot hardness* and is an important feature of a high-speed steel.[66,67] In addition to these three basic characteristics, high-speed steel must be capable of being fabricated, hot-worked, ground, machined, and whatever is essential to produce a given cutting tool.[68]

High-speed steels have been developed for various applications and contain tungsten and/or molybdenum for carbide formation and red hardness, vanadium for improved abrasion resistance, chromium for both increased hardness and oxidation resistance, and sometimes cobalt for excellent red hardness in very heavy work at high speeds.[69]

There are now 29 grades of high-speed steels on the AISI list. These are grouped into tungsten type T and molybdenum type M. For more details, see Sec. 1.8.2.6. The chemical composition and typical applications for selected high-speed steels are given in Tables 1.14 and 14.2, respectively.[70,71]

The properties of high-speed and other tool steels are dependent upon the types and hardness of primary and secondary carbides that precipitate from supersaturated martensite during tempering in the 530 to 590°C range, depending on the alloy. Figure 14.23a shows the hardness of as-quenched martensite and various carbides which are commonly present in tool steels.[72] Figure 14.23b illustrates a comparison of relative wear resistance at a typical hardness level for selected high-speed tool steels. Figure 14.23c compares the relative unnotched impact toughness for representative high-speed tool steels. Within a grade, a modest increase in toughness can be achieved by decreasing the tempered hardness. Lower austenitizing temperature increases the toughness for a particular hardness and grade.[73]

TABLE 14.2 Typical Applications for Selected High-Speed Steels[70,71]

AISI type	UNS no.	Typical applications
		High-speed tungsten
T1	T12001	Drills, taps, reamers, hobs, lathe and planer tools, broaches, crowners, burnishing dies, cold-extrusion dies, cold-heading die inserts, lamination dies, chasers, cutters, taps, end mills, milling cutters
T2	—	Lathe and planer tools, milling cutters, form tools, broaches, reamers, chasers
T4	—	Lathe and planer tools, drills, boring tools, broaches, roll-turning tools, milling cutters, shaper tools, form tools, hobs, single-point cutting tools
T5	—	Lathe and planer tools, form tools, cutoff tools, heavy-duty tools requiring high red hardness
T6	—	Heavy-duty lathe and planer tools, drills, checking tools, cutoff tools, milling cutters, hobs
T8	—	Boring tools, lathe tools, heavy-duty planer tools, tool bits, single-point cutting tools for stainless steel
T15	—	Form tools, end mills, lathe and planer tools, screw machine tools, gear hobs, cold-work tools, broaches, milling cutters, spade drills, shaper cutters, blanking dies, punches, heavy-duty tools requiring good wear resistance
		High-speed molybdenum
M1	T11301	Drills, taps, end mills, reamers, milling cutters, hobs, punches, lathe and planer tools, form tools, slitting saws, chasers, thread rolling dies, blanking dies, trimming dies, tool bits, broaches, routers, and woodworking tools
M2	T11302	Drills, spade drills, taps, end mills, reamers, milling and shaving cutters, tool bits, counterbores, hobs, form tools, saws, lathe and planer tools, chasers, broaches, boring tools, punches, gear thread, and wood knives
M3-1	MT11323	Drills, spade drills, taps, end mills, reamers and counterbores, broaches, hobs, chasers, form tools, lathe and planer tools, cheeking tools, milling cutters, slitting saws, punches, drawing dies, and woodworking tools
M3-2	—	Drills, spade drills, taps, end mills, reamers and counterbores, broaches, hobs, form tools, lathe and planer tools, cheeking tools, slitting saws, punches, drawing dies, and woodworking tools
M4—	T11304	Broaches, reamers, hobs, broach inserts, spade drills, taps, milling and shaper shaving cutters, thread chasers, form tools, counterbores, lathe and planer tools, cheeking tools, rolls, blanking dies, and punches for abrasive materials, inserts, heading dies, and swaging dies.

TABLE 14.2 Typical Applications for Selected High-Speed Steels[70,71] (*Continued*)

M4 mod.	—	Hobs, broaches, end mills, milling cutters, form tools, counterbores, taps, tool bits, punches, dies
M6	—	Lathe tools, boring tools, planer tools, form tools, and milling cutters
M7	T11307	Drills, taps, end mills, reamers, routers, slitting saws, thread rolling dies, blanking and trimming dies, lathe and planer tools, chasers, punches, shearing blades, borers, woodworking tools, hobs, counterbores, form tools, and punches
M10	T11310	Drills, taps, reamers, chasers, end mills, lathe and planer tools, blanking and trimming dies, shear blades, woodworking tools, routers, saws, counterbores, milling cutters, hobs, form tools, punches, and broaches
M30	—	Lathe tools, form tools, milling cutters, chasers
M33	—	Drills, taps, end mills, lathe tools, milling cutters, form tools, chasers
M34	—	Drills, taps, end mills, lathe tools, milling cutters, form tools, chasers
M36	—	Heavy-duty lathe and planer tools, boring tools, milling cutters, drills, cutoff tools, tool-holder bits
M41	—	} Twist drills, end mills, reamers, form cutters, turning and lathe tools, dovetail tools, gear hobs, taps, broaches, router bits, tool bits, milling cutters, gear cutters, keyway cutters, twist drills, counterbores, end mills. Hardenable to Rockwell C67 to C70.
M42	T11342	
M43	—	
M44	—	
M46	—	
M47	—	
M48	—	End mills, shaper cutters, form tools, broaching tools, gear hobs, tool bits, milling cutters, special taps, etc.
M35	—	Gear hobs, milling cutters, shaving cutters, milling cutters, shaper cutters
M62	—	End mills, shaper cutters, form tools, gear hobs, milling cutters, shaper cutters, broaching tools, spade drills, special taps, etc.
CPM	Rex 121(HS)	Hobs, broaches, milling cutters, end mills, form tools, punches and dies, guide rolls, wear parts
M50	K88165	Bearing and missile industry, used for bearings in aircraft and gas-turbine engines, used for tooling applications

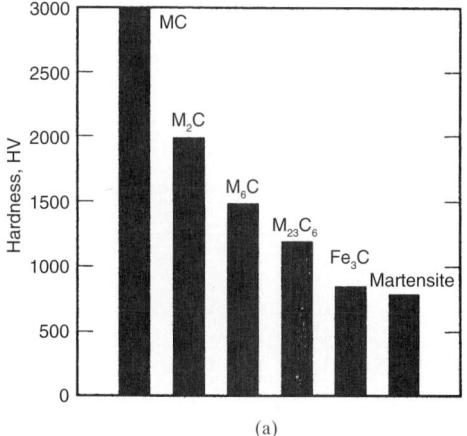

(a)

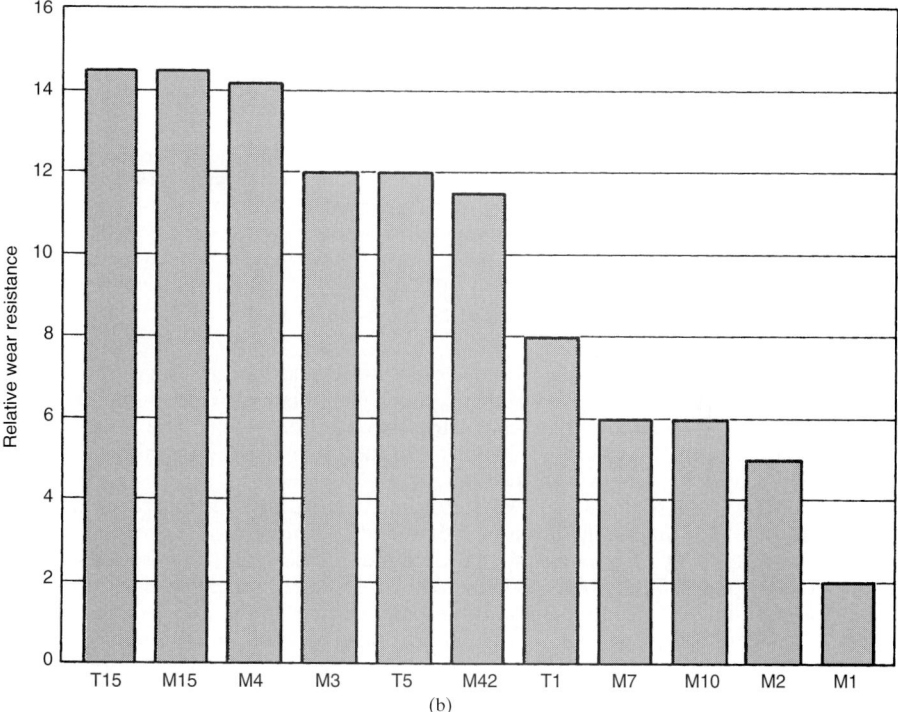

(b)

FIGURE 14.23 (*a*) Hardness of as-quenched martensite and of various carbides (HV: 20-g load).[72] (*b*) Comparison of relative abrasion wear resistance at a typical working hardness for high-speed tool steels.[73] (*c*) Relative toughness of high-speed tool steels at a typical working hardness.[73]

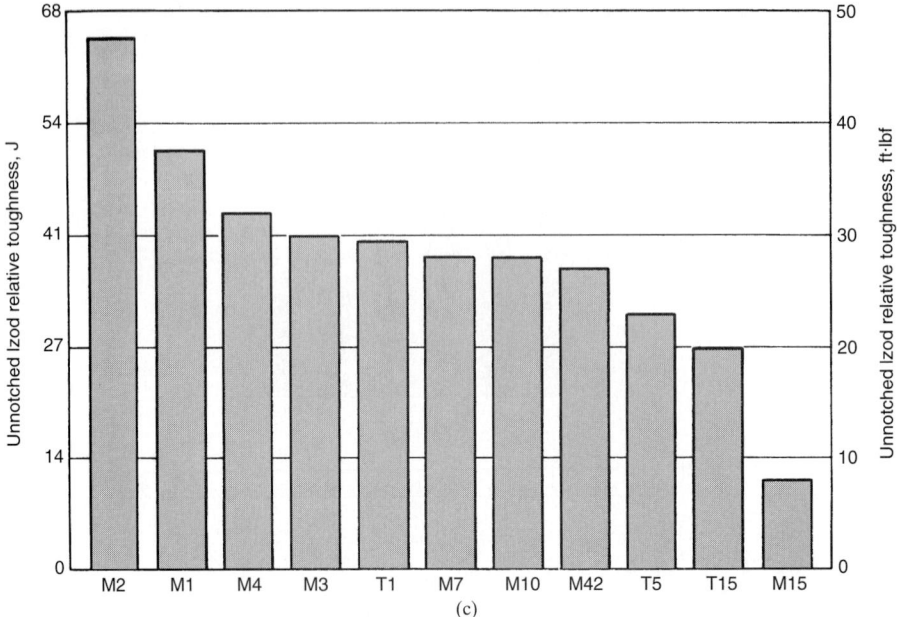

FIGURE 14.23 (*Continued*) (*a*) Hardness of as-quenched martensite and of various carbides (HV: 20-g load).[72] (*b*) Comparison of relative abrasion wear resistance at a typical working hardness for high-speed tool steels.[73] (*c*) Relative toughness of high-speed tool steels at a typical working hardness.[73]

14.6.1.1 Quenching and Tempering of T1 High-Speed Steel

Austenitizing. Austenitizing is the most critical heat-treating operation performed on high-speed steel. For example, excessively high austenitizing temperature or prolonged soaking times may cause excessive distortion, abnormal grain growth, low strength, and loss of ductility. On the other hand, underheating may cause low hardness and low wear resistance.[70]

In general, T series steels are always preheated before austenitizing to minimize thermal stresses that might develop due to the transformation to austenite at approximately 760°C. If a single preheat is employed, the T-type high-speed steels are preheated at 815 to 870°C (1500 to 1600°F). Usually, preheating time at temperature should be twice the length of the time required at the austenitizing temperature. Accordingly, to maintain a uniform flow of work, the capacity of the preheating installation is usually twice that of the austenitizing installation. If a double preheating is used, a first preheating is done in one furnace at 540 to 650°C (1000 to 1200°F) to equalize and remove any moisture present, and a second preheat is carried out in another furnace at 845 to 870°C (1550 to 1600°F) to minimize thermal shock, distortion, and grain growth and to reduce the holding time required at the higher temperature. It is then transferred to a high-temperature salt bath, where it is rapidly heated to the austenitizing temperature (just below the melting point) at 1260 to 1300°C (2300 to 2375°F) and held for 2 to 5 min (depending on the steel grade, tool configuration, and cross-sectional size), to dissolve completely

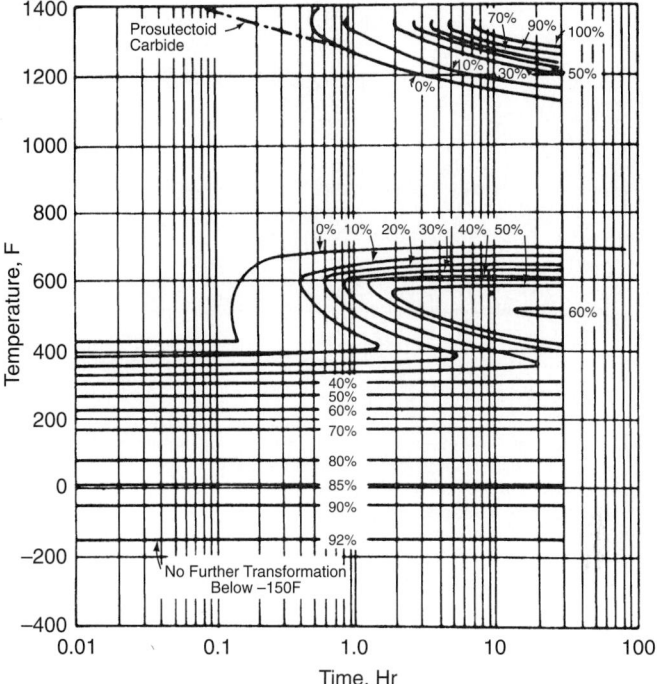

FIGURE 14.24 TTT diagram for a T1 (18-4-1) high-speed steel.[74] (*Reprinted by permission of ASM International, Materials Park, Ohio.*)

all secondary alloy carbides and partially the primary, stable carbides. Short time is used for small sections and longer time for large sections.

Although preheating is recommended for all high-speed steels and tool steels, small tools and those that do not possess complex design, abrupt changes in section, and sharp notches or corners, such as small bits and solid drill blanks, may be introduced directly into the austenitizing furnace with reasonable safety. If consumable carbonaceous muffles are employed, the preheating temperature should not exceed 650°C (1200°F), because of the possibility of decarburization at higher temperatures.

Quenching. It is the common practice to hot-quench in a salt bath in the range of 540 to 650°C (1004 to 1200°F) long enough to equalize the steel at the quench temperature, followed by air or oil cooling to about 50°C (122°F) to minimize distortion and cracking. This quenching temperature range also produces minimum grain boundary carbide precipitation, which normally forms in considerable amounts at high temperatures, as seen in the TTT diagram of T1 steel (Fig. 14.24).[74] Usually the quenched structure consists of 60 to 80% highly tetragonal martensite, 15 to 30% retained austenite, and 5 to 10% undissolved M_6C and VC carbides (Fig. 14.25a).[75]

Tempering. It is a good practice to transfer the quenched part from the quenchant to a tempering furnace prior to cooling below 65°C (150°F). Tempering

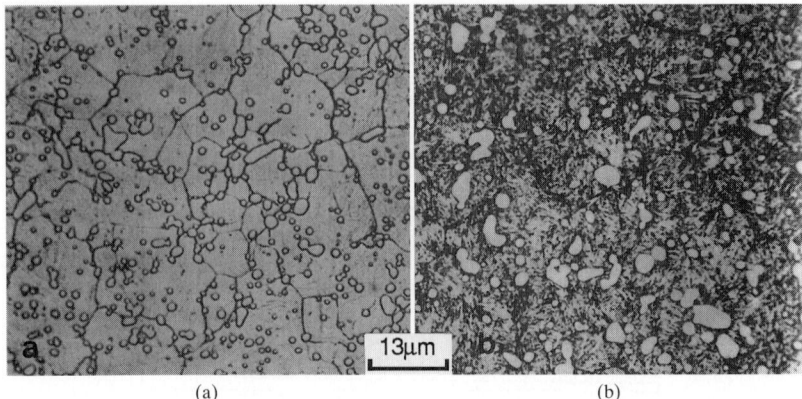

(a)   (b)

FIGURE 14.25   Optical micrographs of T1 high-speed steel. (*a*) Quenched at 607°C (1125°F) condition: austenitized for 3 to 4 min at 1279°C (2335°F); salt-bath-quenched to 607°C (1125°F), air-cooled. Structure consists of undissolved carbide particles in untempered martensite matrix. 10% Nital. (*b*) Quenched and tempered condition: austenitized and salt-bath-quenched the same as for (*a*), then double-tempered at 538°C (1000°F). Structure consists of undissolved carbide particles in a tempered martensite matrix. 4% Nital.[75] (*Reprinted by permission of ASM International, Materials Park, Ohio.*)

is done for 0.5 to 4 hr in the range of 550 to 600°C (1022 to 1112°F), depending on the composition of steel. During tempering of AISI T1 and M2 alloys, the following precipitation sequence of carbides occurs:

$$Fe_3C \rightarrow M_3C \rightarrow M_2C \rightarrow M_6C$$

$$M_{23}C_6 \rightarrow M_6C$$

(within the matrix)

and

$$Fe_3C \rightarrow W_2C + M_{23}C_6 \rightarrow M_6C + M_{23}C_6 \quad \text{(on the grain boundaries)}$$

Since M_6C precipitate particles coarsen rapidly, they cannot be responsible for the significant sustained strength exhibited by the steels at 600°C. The fine, homogeneous MC precipitates survive much longer, and the second wave of precipitation comprising coherent precipitates of M_2C and MC carbides, which are of the order of a few up to some tens of nanometers in size, is responsible for peak-hardened condition and for the continued hot strength of the tool material during cutting tool service; the stability of high dislocation density is presumed to provide nucleation sites for the continued precipitation process.[76]

During first tempering, some retained austenite is conditioned while much of the retained austenite transforms to martensite on cooling to room temperature, causing the introduction of internal microstresses. Upon second tempering for 2 hr or more, the tempering reaction proceeds further toward completion as an additional amount of retained austenite transforms to martensite and relieves the internal stresses induced during the first tempering.[67] In order to obtain components

Tempering temperature, °C

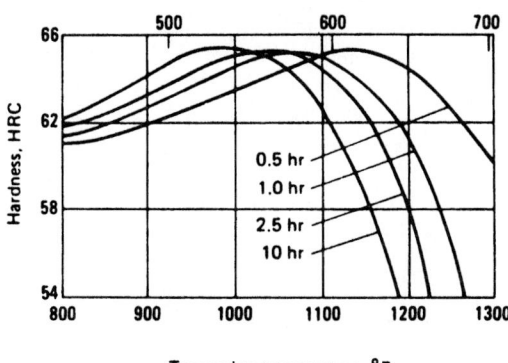

FIGURE 14.26 Hardness of T1 high-speed steel versus tempering time and tempering temperature.[78] (*Reprinted by permission of ASM International, Materials Park, Ohio.*)

of optimum mechanical properties, a compromise should be made between hardness and toughness. Approximate hardness, which is related to tempering temperature, varies between 65 and 60 Rc.[77] Figure 14.25*b* shows the final quenched and double-tempered microstructure of T1 alloy consisting of isolated globules of complex alloy carbides within a tempered martensite matrix.[75] Figure 14.26 shows the effect of tempering temperature and tempering time on hardness of T1 high-speed steel, illustrating the occurrence of secondary hardening peaks for tempering times of 10 hr at 515°C, 2.5 hr at 560°C, 1 hr at 570°C, and 0.5 hr at 615°C.[78]

Subzero treatments are sometimes used along with tempering in order to continue the austenite-to-martensite transformation. It is reported that cold treatments used after quenching and first tempering improve the martensite transformation, in a manner similar to the repeated tempering on martensite transformation. Cold treatment conducted after quenching may cause cracking or distortion, because the associated size changes are not accommodated by the freshly formed, brittle martensite. It is usually agreed that subzero treatments are not essential if the steel is properly hardened and tempered.[73]

14.6.1.2 Quenching and Tempering of M2 High-Speed Steel. The M series steels have lower peritectic temperature, which implies that austenitizing must be performed at lower temperature. For example, if a single preheat is used, M2-type alloys are preferably heated at 730 to 845°C (1350 to 1550°F). Double preheating in one furnace at 540 to 650°C (1000 to 1200°F) and in another furnace at 845 to 870°C (1555 to 1600°F) minimizes thermal shock. Preheating is then followed by further rapid heating to the austenitizing temperature at 1190 to 1230°C (2175 to 2250°F) for 2 to 5 min. Shorter time is used for small sections and longer time for larger sections. It is the usual practice to preheat at temperature twice the length of time used at the final austenitizing temperature.

The TTT diagram for the M2 high-speed steel is shown in Fig. 14.27.[74] Like the T1 alloy, the M2 alloy is hot-quenched at 540 to 650°C (1000 to 1200°F), followed by air-cooling or oil-quenching to avoid the transformation at the higher

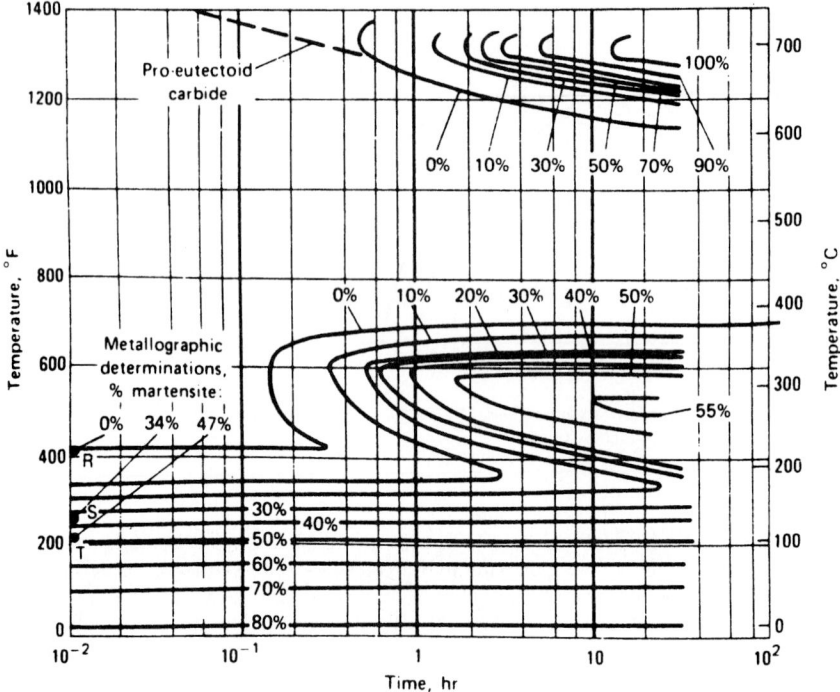

FIGURE 14.27   TTT diagram for the M2 high-speed steel.[74] (*Reprinted by permission of ASM International, Materials Park, Ohio.*)

temperature (i.e., above 700°C). As-quenched hardness lies in the range of 64 to 66 Rc.

Stabilizing (Optional).   For intricate shapes, stress-relief temper is accomplished at 150 to 160°C (300 to 320°F) briefly. It is then refrigerated at −100 to −195°C (−150 to −320°F), followed by immediate tempering after the part reaches room temperature.[68]

Tempering.   A minimum of two separate tempering treatments are accomplished at 540 to 595°C (1004 to 1103°F), with duration of 2 hr or more at each treatment temperature to ensure attainment of consistent tempered martensitic structures and to avoid uncertainties arising from variations in the amount of retained austenite in the as-quenched condition. Approximate tempered hardness, as related to tempering temperature, is 65 to 60 Rc. Figure 14.28 shows the tempering curves for M2-type steel austenitized at two different temperatures, illustrating the higher secondary hardening peak hardness at the higher austenitizing temperature. This high secondary hardening effect is due to the dissolution of more carbon and alloying elements in the austenite which, upon quenching, remain in martensite and later, during tempering, cause more precipitation hardening.[66]

The microstructures of the M2 used in the quenched and tempered condition are shown in Fig. 14.29, which are similar to those of T1 steel.[75]

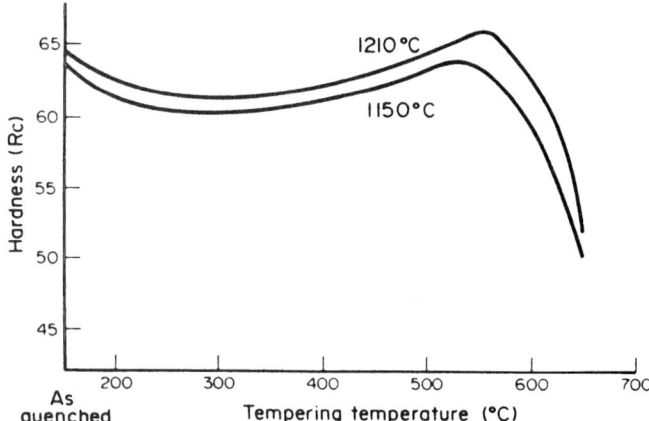

FIGURE 14.28   Tempering curves for M2 high-speed steel.[66] (*Reprinted by permission of Pergamon Press, Plc.*)

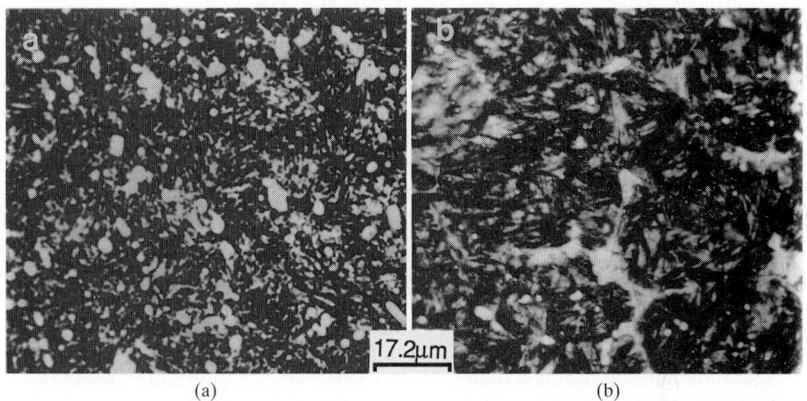

(a)   (b)

FIGURE 14.29   Optical micrographs of M2 high-speed steel. (*a*) A 50.8-mm- (2-in.-) diameter bar, austenitized at 1218°C (2225°F), oil-quenched, and tempered for 1 hr at 552°C (1025°F). Structure consists of spheroidal carbide particles in a tempered martensite matrix. Some small areas of retained austenite are apparent. 3% Nital. (*b*) A 22.2-mm- ($\frac{7}{8}$-in.-) diameter bar, austenitized at 1260°C (2300°F), oil-quenched, and double-tempered at 566°C (1050°F). Structure is tempered martensite containing a few spheroidal carbide particles. Incipient melting at grain boundaries occurred due to overheating. 6% Nital.[75]

14.6.2   Chromium-Bearing Secondary Hardening Steels

The low-alloy 2.25Cr-1Mo steel has been widely used in the production of steam and nuclear power plants and chemical, oil, and petrochemical industries as structural components, because of increased creep strength and resistance to stress corrosion cracking. The general trend in the tempering behavior of this steel in the tempering range of 823 to 1023 K can be schematically shown in Fig. 14.30, which can be divided into four distinct stages.[79] Stage I is characterized by the initial continuous softening due to rapid dissolution of bainite, releasing carbon into the surrounding ferrite. The

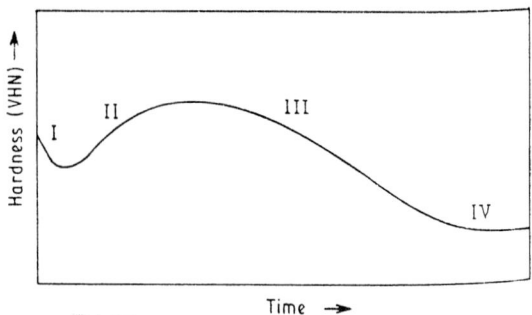

FIGURE 14.30 Schematic representation of four stages of tempering behavior in 2.25Cr-1Mo steel.[79]

relatively higher affinity of carbon for molybdenum causes the formation of Mo-C clusters and subsequent precipitation of fine, coherent, needlelike M_2C carbides, producing secondary hardening of the steel in stage II. As M_2C precipitate particles grow, there is a loss of coherency and precipitation of secondary alloy carbides, such as M_7C_3 and $M_{23}C_6$, which gives rise to softening in stage III. The slow overaging kinetics of coarse carbides leads to the saturation in hardness in stage IV.

Steels containing at least 9 wt% Cr exhibit secondary hardening peaks in tempering curves.[42] A typical standard 9Cr-1Mo steel (composition: Fe-0.1C-0.75Si-0.52Mn-9.04Cr-1.02Mo-0.27Ni-0.019N-0.022P-0.02S) is extensively used in the U.K. nuclear power industry as heat-exchanger parts in CARGS and in the fast reactor system.[80] This grade is austenitized for 1 hr at 1250°C in the $\gamma + \delta$ phase field, water-quenched to reveal lath martensite, autotempered to form M_3C carbides and a high density of transformation-induced dislocations, and tempered for 1 hr at ~550°C to produce secondary hardening peak hardness.

The tempering reaction proceeds by formation and growth of M_2X precipitate particles, decrease in dislocation density, and $M_{23}C_6$ precipitation at grain boundaries and tempered lath martensite boundaries. The main contribution to secondary hardening, in this alloy, seems to be from M_2X (where X = C, N, and/or CN) precipitation on dislocations within the matrix, which provides some dispersion strengthening and maintains a high dislocation density by a pinning mechanism. A small contribution of secondary hardening is from coherency strain.

In contrast, the modified 9Cr-1Mo steel with small (0.06%) Nb and (0.2%) V additions was developed for use as advanced nuclear steam generator and fusion reactor materials in the United States. Its high strength and thermal conductivity, good resistance to corrosion, and low thermal expansion make it an important candidate for many steam generator components such as tubing, piping, and headers for use in the temperature range of 450 to 600°C. This is austenitized at 1045°C for 1 hr, air-cooled to precipitate fine $(Fe,Cr)_3C$ particles within the martensite laths, and tempered for 1 hr at 700 to 760°C to precipitate initially Cr_2C to be replaced later by arrays of fine $(NbV)C$ carbides and V_4C_3 particles along the lath interface and enriched $(FeCrVMo)_{23}C_6$ carbides at lath interfaces and high-density dislocations. The improved martensite lath stability and consequent high-temperature fatigue and creep resistance of the modified alloy are attributed primarily to the interfacial pinning by V_4C_3 precipitates, which coarsen very slowly.[81,82]

The Fe-12Cr-1Ni-0.2C stainless steel is an important secondary hardening steel which can be quenched to martensite and tempered to provide a fine dispersion of chromium carbides in a ferrite matrix. The strength is well maintained up to the secondary hardening peak at 500°C with a reasonable amount of ductility (Fig. 14.31);

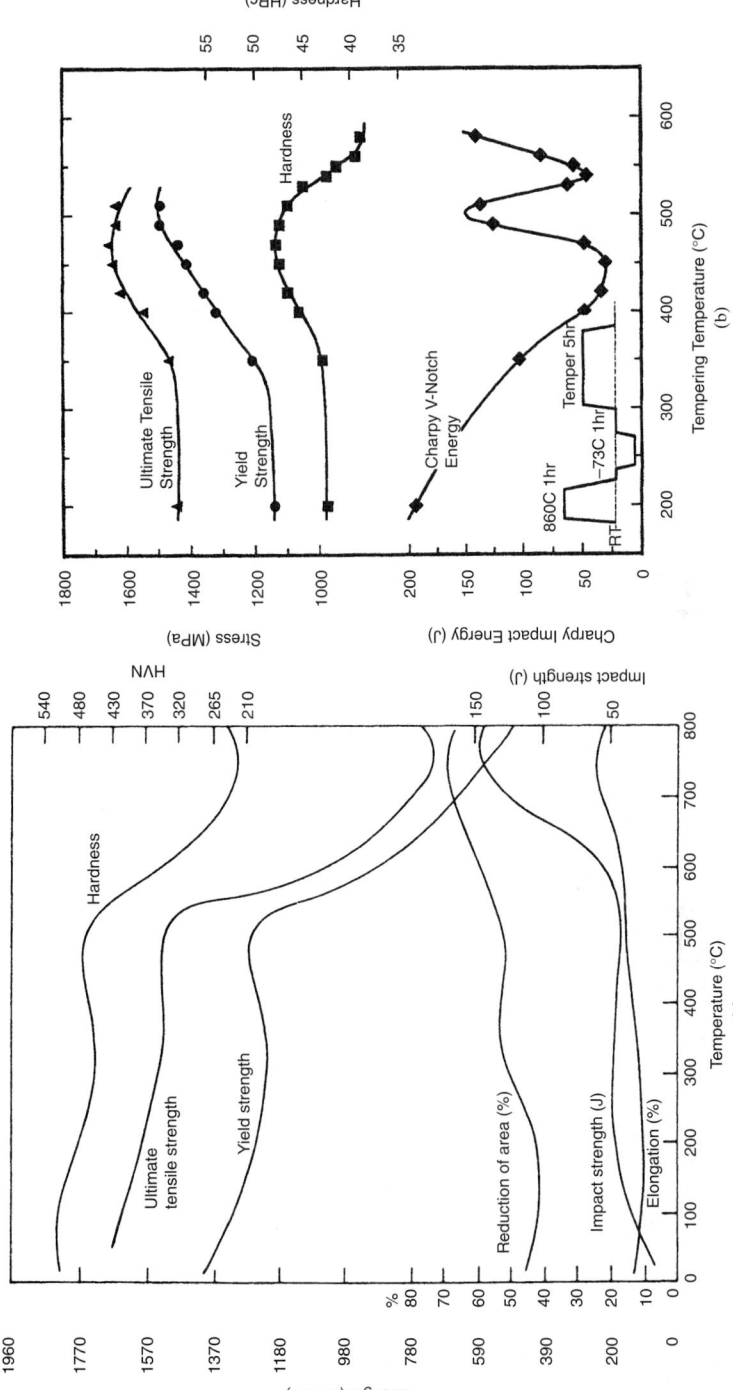

FIGURE 14.31 (*a*) Effect of tempering for 1hr on the mechanical properties of a 12Cr-1Ni-0.2C stainless steel. Typical results for 50-mm-diameter bars, oil-quenched.[42] (*Source: Steel and Its Heat Treatment*, Bofors Handbook, Butterworths, 1975.) (*b*) Effect of tempering for 5 hr on mechanical properties of an AF-1410 steel.[85a]

this is used in gas and steam turbines as well as for constructional purposes involving lower temperatures.[42]

14.6.3 High-Cobalt-Nickel Secondary Hardening Steels

Secondary hardening steels containing high Co-Ni contents, such as AF1410 (14Co-10Ni-2Cr-1Mo-0.16C), AerMet 310 (15Co-11Ni-2.4Cr-1.4Mo-0.25C), and AerMet 100 (13.4Co-11.1Ni-3.1Cr-1.2Mo-0.23C) (Table 1.10), have been developed for aerospace applications; these steels have high strength, high hardness, and superior fracture toughness among high-strength aerospace steels, even at ultrahigh strength levels (Fig. 14.32a), and are based on HY 180 (8Co-10Ni-2Cr-1Mo-0.1C).[83,83a] Figure 14.32b shows the superior fatigue strength of AerMet 100 alloy compared to five other high-strength alloys.[83b]

As shown in Tables 14.3 and 14.4, respectively, AF 1410 steel can be heat-treated to K_{IC} values up to 174 MPa$\sqrt{m}$ (158 ksi $\sqrt{in}$.), and AerMet 100 can be heat-treated to 1930 to 2070 MPa (280 to 300 ksi) tensile strength and K_{IC} in excess of 110 MPa$\sqrt{m}$ at 1930 MPa (100 ksi $\sqrt{in}$. at 280 ksi), in addition to the exceptional resistance to stress corrosion cracking and fatigue when compared to other steels.[84] AerMet 310 can be heat-treated to 2170-MPa (315-ksi) tensile strength and K_{IC}/YS ratio of 71 MPa$\sqrt{m}$ (65 ksi $\sqrt{in}$.). Its strength-to-weight ratio of 27.5 km (1.10×10^6 in.) is higher than that of Ti-6Al-4V alloy with 26.2 km (1.03×10^6 in. maximum).[83a]

The unique combination of strength and toughness of these steels up to secondary hardening peak is attributed to (1) retardation of dislocation recovery, (2) hardening through the precipitation of a fine dispersion of strengthening alloy carbides on lattice dislocations, (3) improvement of cleavage fracture resistance, and (4) an inclusion-free matrix.[85,85a] Cobalt plays a vital role in that it hinders dislocation recovery and increases the nucleation rate of the alloy precipitates.[85a]

Figure 14.31b shows the effect of tempering for 5 hr on the mechanical properties of AF 1410 steel. In AF 1410, the secondary hardening peak appears at 5 hr of tempering at 482°C, compared with about 1 hr at the same temperature in AerMet 100. This implies that the precipitation kinetics of M_2C particles in AerMet 100 is faster than that of AF 1410. This difference is due to the higher levels of C, Cr, and Mo in AerMet 100.[86] AerMet 100 alloy is used in applications such as armor, fasteners, landing gear, actuators, ordnance, ballistic-tolerant parts, jet engine shafts, structural members, driveshafts, and structural tubing.[71] Other applications include shanks for arrestor hooks and various structural parts where it has usually replaced 300M, AISI 4340, AF1410, and other low-alloy steels.[83a]

14.6.4 Chromium-Molybdenum-Vanadium-Bearing Secondary Hardening Steels

Typical examples are Fe-0.5Cr-0.5Mo-0.25V, Fe-1Cr-0.25V, Fe-3Cr-1Mo-0.25V, and Fe-1Cr-1Mo-0.75V. During their tempering treatments in the range of 550 to 650°C, the formation of fine vanadium carbide dispersion within the matrix on dislocations leads to the development of a significant secondary hardening peak; this condition remains unaffected even at elevated-temperature service (approaching 700°C).[42]

The Fe-1Cr-1Mo-0.5V with boron and zirconium additions possesses properties that make it a suitable material for elevated temperature applications. The boron and zirconium additions enhance notch ductility.

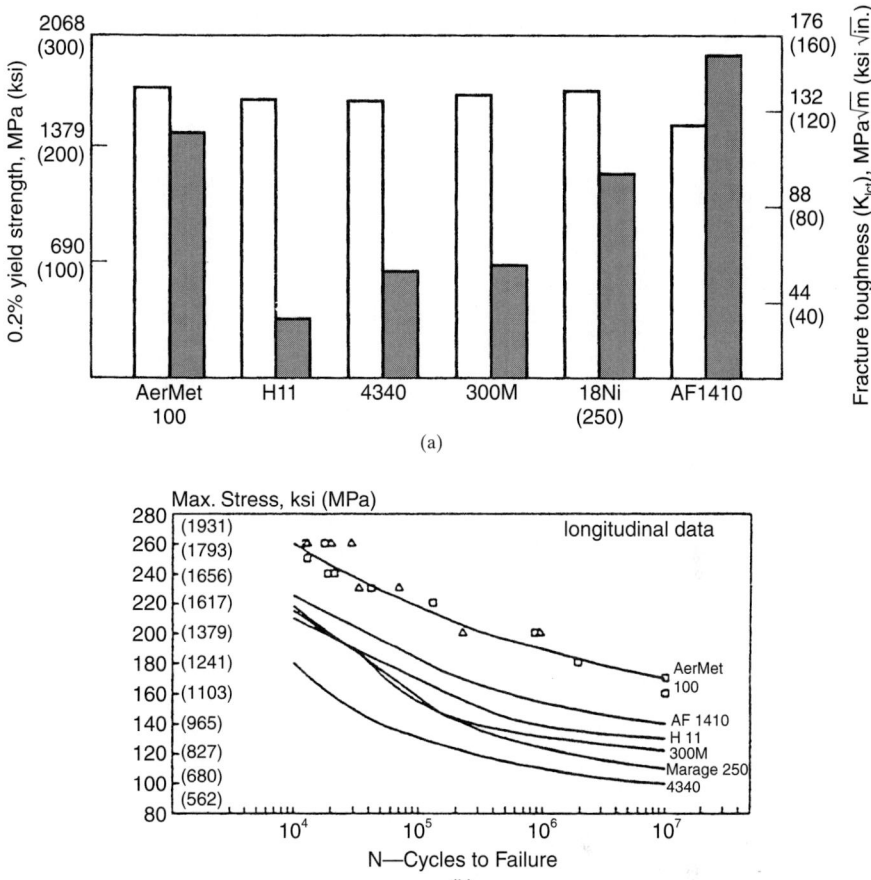

FIGURE 14.32   (*a*) Yield strength and fracture toughness (dark bars) data for ultrahigh-strength aerospace steels. (*Source: T. J. McCauffrey, Advanced Materials Processing, September 1992, pp. 47–50.*) (*b*) Axial fatigue resistance comparison of high-strength aerospace steels ($K_t = 1, R = 0$, test temperature = room temperature).[83b] (*Source: P. M. Novotny and T. J. McCauffrey, An Advanced Alloy for Landing Gear and Aircraft Structural Applications—AerMet 100 Alloy, SAE Technical Paper 922040, Aerotech '92, Anaheim, Calif., October 5–8, 1992, pp. 1–7.*)

14.6.5   Nitrogen-Bearing Secondary Hardening Steels

Secondary hardening during the tempering of nitrogen-bearing martensite has been observed at 450 to 500°C, in Fe-4.22Mn-0.7N[87] and Fe-Cr-N alloys.[88] In both these alloys, secondary hardening is attributed to the precipitation of thin segregates of nitrogen and substitutional solute atoms in {100} planes. The tempering reactions proceed in the former alloy as

$$\alpha''(Fe_{16}N_2) \rightarrow \gamma'(Fe_4N) \rightarrow Mn/N \text{ zones} \rightarrow \eta(Mn_3N_2)$$

and in the latter alloy as

TABLE 14.3 Mechanical Properties of AF1410 Steel in Various Quenching Media

Test specimens were 50-mm (2-in.) plate from VIM/VAR melt with the heat treatment: 675°C (1250°F) for 8 hr with air-cooling, 900°C (1650°F) for 1 hr, quenching, 830°C (1525°F) for 1 hr, quenching, refrigeration at −73°C (−100°F) for 1 hr, 510°C (950°F) for 5 hr, and air-cooling.

Quench medium	Ultimate strength		Yield strength		Elongation, %	Reduction in area, %	Charpy V-notch		Plane-strain fracture toughness (K_{IC})	
	MPa	ksi	MPa	ksi	%		J	ft · lbf	MPa√m	ksi√in.
Air	1680	244	1475	214	16	69	69	51	174	158
Oil	1750	254	1545	224	16	69	65	48	154	140
Water	1710	248	1570	228	16	70	65	48	160	146

Source: W. B. Brown, Jr., *Aerospace Structural Metals Handbook*, Code 1224, Metals and Ceramics Information Center, 1989, pp. 1–30.

TABLE 14.4 Typical Mechanical Properties of AerMet 100 Steel[71]

Solution heat-treated at 885°C (1625°F) for 1 hr, air-cooled, refrigerated at −73°C (−100°F) for 1 hr, and aged at 480°C (900°F) for 5 hr

Yield strength		Ultimate tensile strength		Elongation, %	Reduction of area, %	Charpy V-notch impact energy		Fracture toughness (K_{IC})	
MPa	ksi	MPa	ksi			J	ft · lbf	MPa√m	ksi√in.
Longitudinal orientation									
1724	250	1965	285	14	65	41	30	126	115
Transverse orientation									
1724	250	1965	285	13	55	34	25	110	100

$$\alpha''(Fe_{16}N_2) \text{ type zones} \rightarrow Cr\text{-}N \text{ monolayer plates}^{89}$$

14.6.6 Nitrogen- and/or Carbon-Bearing Secondary Hardening Martensitic Stainless Steels

Berns and Gavriljuk[90] have drawn the following conclusions on the tempering behavior of martensitic stainless steel containing 15% Cr, 1% Mo, and 0.6% C (Steel SC), N (Steel SN), and C + N (Steel SNC), respectively.

1. After quenching from 1100°C, the amount of retained austenite increases in the order of SC, SN, and SNC.

2. The transformation of retained austenite during tempering is retarded in the same order.

3. Secondary hardening is improved by nitrogen, which enhances both the amount of retained austenite and its resistance to tempering, especially in SNC.

4. Upon tempering, secondary hardening is more significant at about 450°C in high-nitrogen-bearing than in carbon-bearing grades.

5. The precipitation sequence during tempering between 200 and 650°C is represented by

SC: $\quad \varepsilon = (Fe,Cr)_2C \rightarrow \theta\,(Fe,Cr)_3C \rightarrow (Cr,Fe)_7C_3$

SN: $\quad \varepsilon = (Fe,Cr)_2N \rightarrow \zeta\,(Fe,Cr)_2N \rightarrow (Cr,Fe)_2N$

SNC: $\quad \varepsilon = (Fe,Cr)_2C \rightarrow \theta\,(Fe,Cr)_3C$

$\quad\quad\quad\quad \varepsilon = (Fe,Cr)_2N \rightarrow \theta\,(Cr,Fe)N \rightarrow (Cr,Fe)_2N$

As shown in Fig. 14.33a, the lower Cr content and finer distribution of the ζ-nitrides compared to the M_7C_3 carbides appear to contribute higher hardness and corrosion resistance after tempering in the regime of secondary hardening at

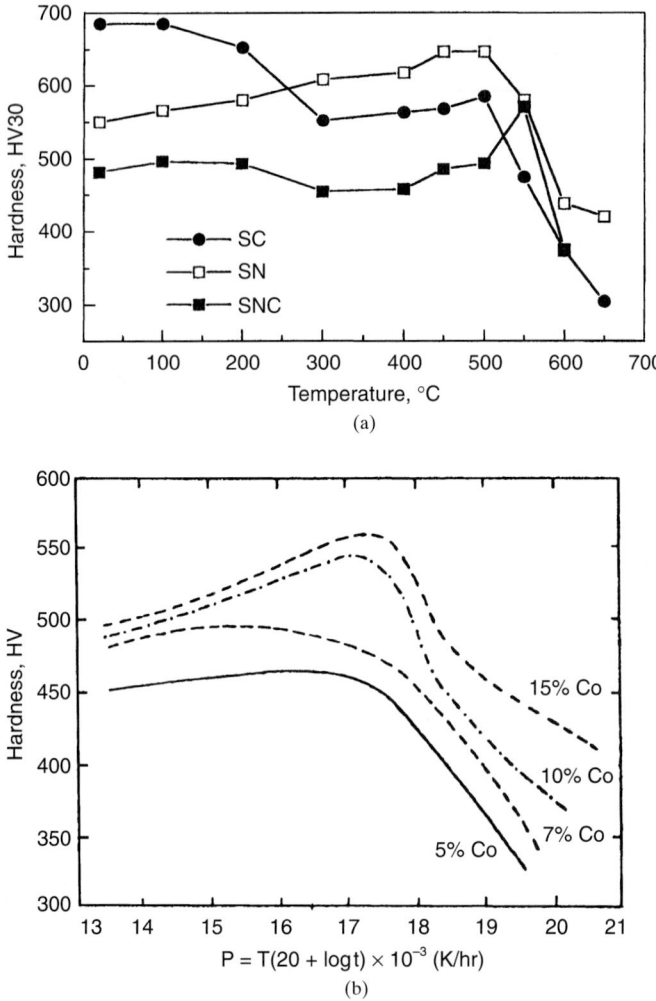

FIGURE 14.33   (*a*) Effect of tempering on the hardness of martensitic stainless steel with 0.6% N and/or C after quenching from 1100°C.[90] (*b*) Effect of
cobalt on tempering resistance of Fe-0.1C-12Cr-4Mo martensitic stainless steel.
(*Source: Cobalt Facts*, The Cobalt Development Institute, England.)

~450°C. Important aerospace applications of high-nitrogen steels include high-
strength parts in corrosive environment, tools, ball screws, and bearings.

14.6.7   Secondary Hardening Martensitic Stainless Steels

Increasing cobalt contents in Fe-0.1C-12Cr-4Mo martensitic stainless steel have
been observed to produce secondary hardening effect with increased peak hardness, as shown in Fig. 14.33*b*.

14.7 NUCLEATION AND GROWTH OF ALLOY CARBIDES

There are three ways of formation of alloy carbides:[36,42]

1. By in situ nucleation. The alloy carbides nucleate at several locations at the cementite/ferrite interfaces and grow by diffusion of carbon through the adjacent cementite until the cementite particles disappear.
2. By separate nucleation within the ferrite matrix on dislocations inherited from the martensite and subsequent growth with gradual disappearance of cementite.
3. At grain boundaries and prior austenite grain boundaries, the original martensite lath boundaries (now ferrite), and the new ferrite boundaries formed by coalescence of subboundaries, or by recrystallization. The growth is completed by the replacement of the intragranular Widmanstätten nonequilibrium carbide.

14.8 EFFECT OF TIME AND TEMPERATURE ON TEMPERING (OR TEMPERING PARAMETER)

Tempering is a thermally activated process. Temperature and time are interdependent variables in tempering of steels, and one can obtain the same result such as tempered hardness either by decreasing temperature and increasing time or by raising temperature and decreasing time. The relation between time and temperature can be represented by a simple rate equation

$$\frac{1}{t} = A \exp\left(\frac{-Q}{RT}\right) \qquad (14.1)$$

where t is the time to attain a given hardness, Q is an activation energy for the process, A is a constant, R is the universal gas constant, and T is the absolute temperature in Kelvin. Here the tempering treatment schedule may incorporate times other than 1 hr to achieve the final hardness.

14.8.1 Modified Hollomon-Jaffe Correlation

A temperature-time correlation was first developed by Hollomon and Jaffe,[91] and was later extended by Grange and Baughman,[92] to ensure a proper tempering cycle from chemical composition, provided that the tempering temperature is held in the range of 343 to 649°C (650 to 1200°F), that is, especially in the third and fourth stages of tempering, where aggregates of cementite and ferrite are involved. This method yields a reasonably reliable prediction of hardness of the final tempered martensite (usually about ±1 Rc or ±10 DPH) for AISI plain carbon and alloy steels containing 0.2 to 1.0% C and less than 5% total alloying elements, irrespective of the initial structure.[93] However, caution must be used in applying the method to secondary hardening steels.

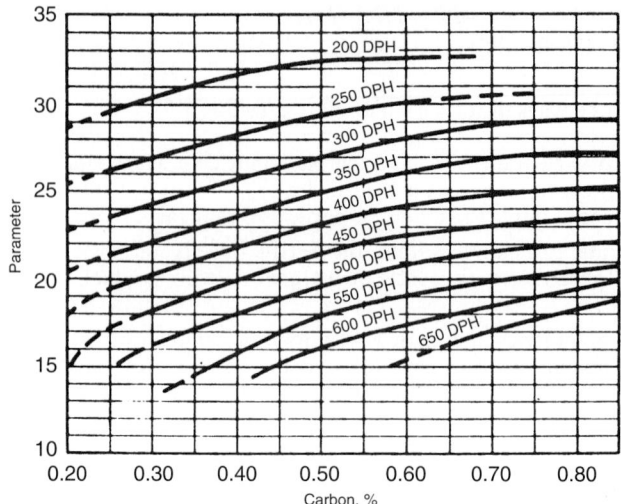

FIGURE 14.34 Variation in parameter value with carbon concentration at each indicated hardness level.[92,93] (*Reprinted by permission of ASM International, Materials Park, Ohio.*)

The Hollomon-Jaffe parameter P is defined by

$$\text{Hardness} = f(P) \tag{14.2}$$

based on experimental results where

$$P = T(C + \log t) \times 10^{-3} \tag{14.3}$$

where t is the time in hours and C is a material constant which is usually given a value of 18 for all steels by Grange and Baughman, to facilitate the direct comparison of tempering curves for different steels. Figure 14.34 shows the variation in parameters with carbon content in plain carbon steels at a number of hardness levels. Table 14.5 lists factors for predicting the incremental increase of hardness to carbon steel by each alloying element. Figure 14.35 shows the relationship of time-temperature with the Hollomon-Jaffe parameter $P = T(18 + \log t) \times 10^{-3}$.

EXAMPLE PROBLEM 14.1 *Construct an estimated tempered hardness versus tempering parameter curve for 4340 steel [composition (wt%): 0.42 C, 0.78 Mn, 0.24 Si, 1.85 Ni, 0.81 Cr, and 0.27 Mo].*

Answer. *Step 1. Construct on a parameter chart (Fig. 14.36) a tempering curve for a base plain carbon steel with similar carbon percentage (i.e., 1042 steel). Use data from Fig. 14.34.*

Step 2. Adjust the 1042 curve for effects of alloying elements, using values in Table 14.5 for a number of parameters, e.g., parameter 24:

$$6 \times \%\text{Ni} + 55 \times \%\text{Cr} + 80 \times \%\text{Mo} = 77.25\,\text{DPH}$$

TABLE 14.5 Factors for Predicting the Hardness of Tempered Martensite[92]

Element	Range, %	Factor at indicated parameter value					
		20	22	24	26	28	30
Manganese	0.85–2.1	35	25	30	30	30	25
Silicon	0.3–2.2	65	60	30	30	30	30
Nickel	Up to 4	5	3	6	8	8	6
Chromium	Up to 1.2	50	55	55	55	55	55
Molybdenum	Up to 0.35	40	90	160	220	240	210
		20[†]	45[†]	80[†]	110[†]	120[†]	105[†]
Vanadium[‡]	Up to 0.2	0	30	85	150	210	150

[†] If 0.5 to 1.2% Cr is also present, use this factor.
[‡] For AISI-SAE chromium-vanadium steels; may not apply when vanadium is the only carbide former present.
Note: Boron factor is 0.
Reprinted by permission of ASM International, Materials Park, Ohio.

Repeat for other parameters between 20 and 30 and develop the estimated curve.

If we want to estimate the hardness of a fully hardened and tempered 4340 steel for 5 hr at 800°F, the parameter for this tempering cycle would be 23.6 (Fig. 14.35); according to the estimated curve in Fig. 14.36, the required hardness would be 431 DPH or 43.8 Rc. Similarly, a tempering treatment for 3 hr at 1000°F would yield a parameter of 27.0 and a hardness of 372 DPH or 37.9 Rc.

EXAMPLE PROBLEM 14.2 *Determine the tempering time needed at 1100°F to give a similar hardness value to 37.9 Rc, obtained in 3 hr at 1000°F.*

Answer. *The time indicated in Fig. 14.35 by the intersection of the 1100°F curve and parameter 27.0 is ~0.2 hr, or 12 min.*

A similar result can be found by using the parameter relation $T_1(18 + \log t_1) = T_2 (18 + \log t_2)$, where $T_1 = 1460$ K, $t_1 = 3$ hr, $T_2 = 1560$ K, and t_2 is the time to be found. Substituting given values, we get $t_2 = 0.196$ hr (that is, ~12 min) as above.

The tempering schedule in the temperature range of 246 to 399°C (475 to 750°F) should be avoided, whenever possible, to prevent the occurrence of temper embrittlement.[93]

14.8.2 Grange-Hribal-Porter Correlation

The hardness of tempered martensite can also be estimated by the Grange-Hribal-Porter correlation, which is in agreement with the experimental results provided that the quenched martensite is free of any considerable amounts of ferrite and pearlite. Figure 14.37 shows the softening of as-quenched martensite for a series of Fe-C alloys with carbon content varying between 0.1 and 1% with increasing tempering temperature (for 1 hr) in the range of 204 to 705°C (400 to 1300°F). The hardness variation with carbon content in the Fe-C alloys was found to be significant at low tempering temperatures as compared to that at high tempering temperatures.[63]

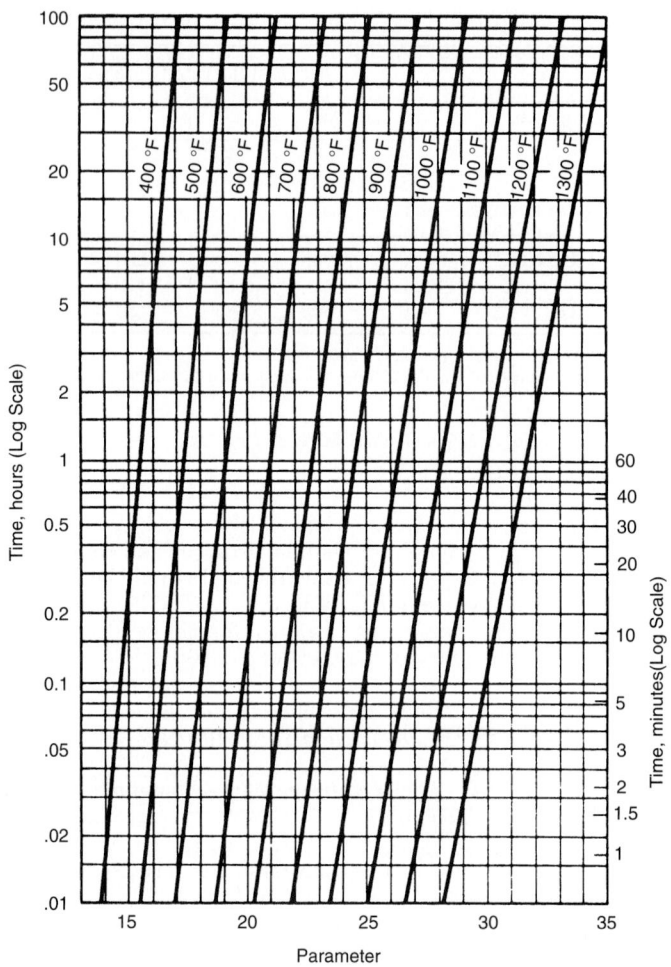

FIGURE 14.35   *Time versus temperature parameter chart with $C = 18$.[92,93] (*Reprinted by permission of ASM International, Materials Park, Ohio.*)

To estimate the increase in hardness (ΔHV) produced from alloying elements, a set of charts has been made at a series of tempering temperatures in the manner presented in Fig. 14.38. A semilogarithmic scale is used to allow accurate readings for small concentrations of alloying elements. The hardness of a given steel composition after tempering at any selected temperature is estimated by reading off the base hardness values of the Fe-C alloy of the same carbon content plus the ΔHV values for each of the alloying elements present in the steel at the selected temperature. Thus

$$\text{Estimated HV } [538°\text{C}(1000°\text{F})] = \text{HV(base value from Fig. 14.37)} + \Delta\text{HV}_{\text{Mn}}$$
$$+ \Delta\text{HV}_{\text{Si}} + \Delta\text{HV}_{\text{P}} + \Delta\text{HV}_{\text{Ni}} + \Delta\text{HV}_{\text{Cr}} + \Delta\text{HV}_{\text{Mo}} + \Delta\text{HV}_{\text{V}} \quad \text{(all from Fig. 14.38}g)$$

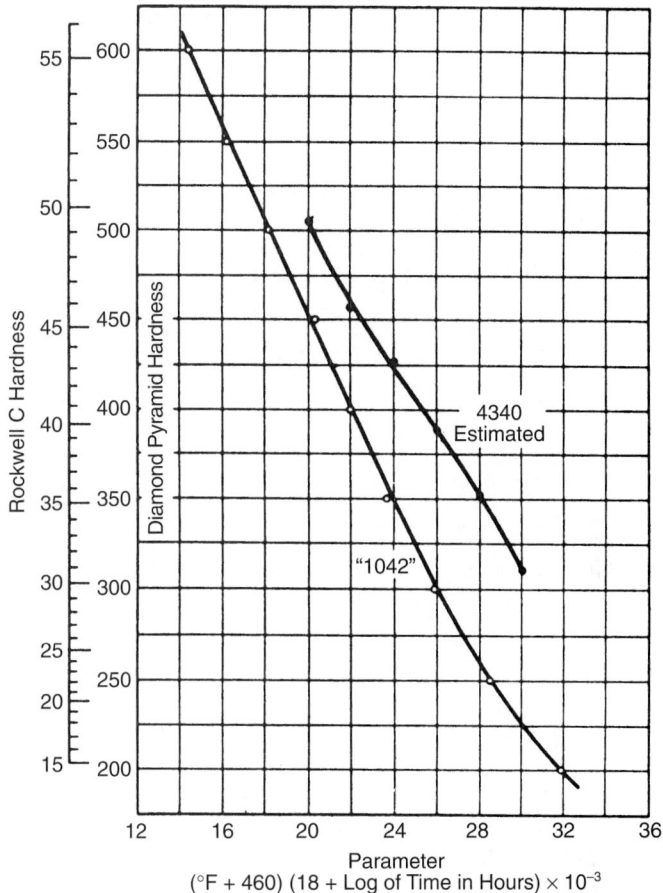

FIGURE 14.36 Estimated hardness of tempered martensite of a 4340 steel.[92,93] (*Reprinted by permission of ASM International, Materials Park, Ohio.*)

Grange et al. have provided a set of charts, obtained at intervals of 56°C (100°F) over the range of 205 to 705°C (400 to 1300°F). Three assumptions are made: (1) The initial structure is martensite. (2) The prior austenite grain size varies between sizes 2 and 4, which are insensitive to tempered hardness. (3) Alloying element interaction effects are not significant because they are present in limited amounts in most AISI low-alloy steels.

Note that the above estimation uses a tempering time of 1 hr. These data can be easily converted to a hardness value for any other tempering time by using Fig. 14.35. This is based on the fact that all combinations of tempering temperature and tempering time with the same parameter will have essentially the same hardness. For example, first the 1-hr tempering treatment is located on the chart, and then the constant tempering parameter ordinate is read vertically down for times less than 1 hr and is read vertically up the same ordinate for times greater than 1 hr.[63]

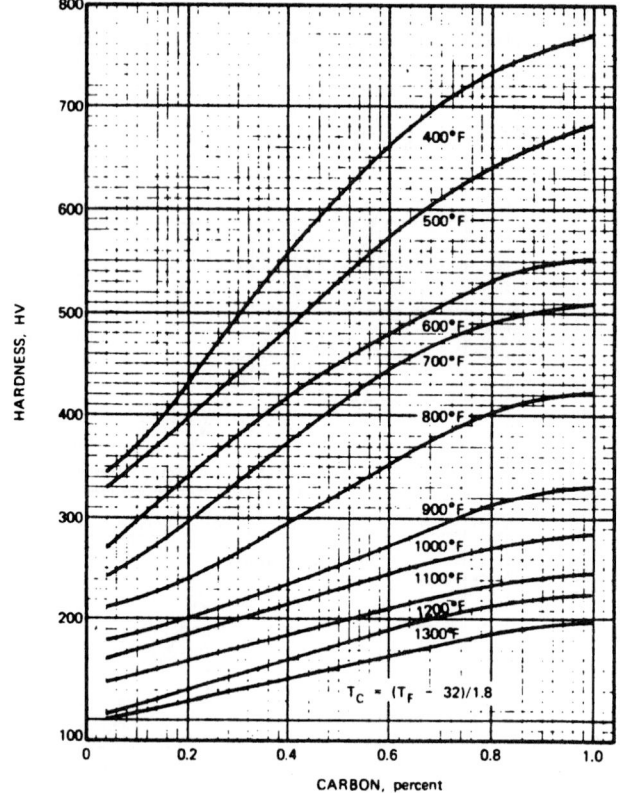

FIGURE 14.37 Hardness of tempered martensite in Fe-C alloys.[63] (*Reprinted by permission of the Metallurgical Society, Warrendale, Pa.*)

14.9 METHODS OF TEMPERING

Tempering can be performed by holding entire parts in the furnace for a sufficient period so that the tempering mechanism prevails to the desired point of completion, or by selective heating of certain regions of the parts to achieve toughness or plasticity in those areas. In general, bulk tempering processes may be accomplished in recirculating or forced-air convection furnaces, hot oil baths, molten salt baths, molten metal baths, and in vacuum. Flame tempering and induction tempering are the most commonly used selective methods in high-volume production applications, because of their controllable local heating capabilities. Selection of the types of furnaces is based mainly on the number and size of parts and on the required temperature. Other factors such as part configuration, ease of fixturing, heater frequency, and cost also affect the choice in selecting tempering equipment. Table 14.6 lists temperature ranges and general conditions of use for four types of tempering equipment.[70]

    Recirculating or *forced-air convection furnaces* are used in the temperature range of 150 to 750°C (302 to 1382°F) and are favored for large-scale batch or

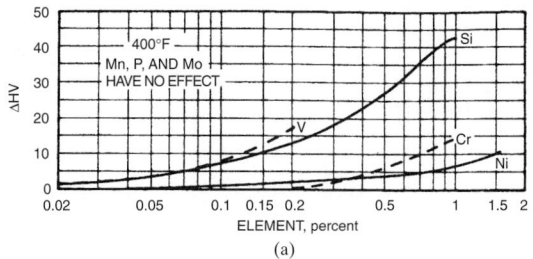

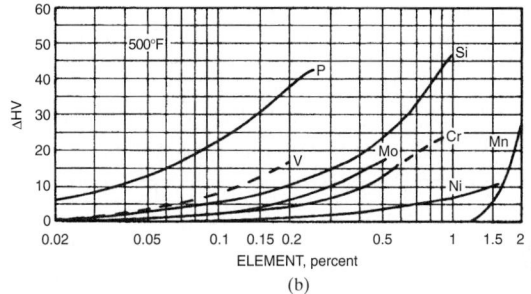

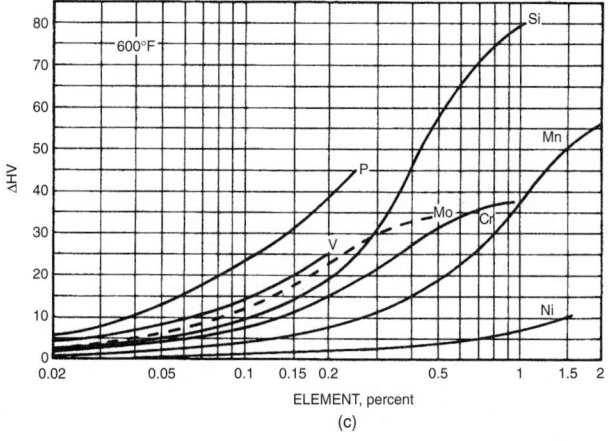

FIGURE 14.38 Effects of alloying elements on the hardness of martensite tempered for 1 hr at (*a*) 400°F (204°C), (*b*) 500°F (260°C), (*c*) 600°F (316°C), (*d*) 700°F (371°C), (*e*) 800°F (427°C), (*f*) 900°F (482°C), (*g*) 1000°F (538°C), (*h*) 1100°F (593°C), (*i*) 1200°F (649°C), and (*j*) 1300°F (704°C).[63] (*Reprinted by permission of ASM International, Materials Park, Ohio.*)

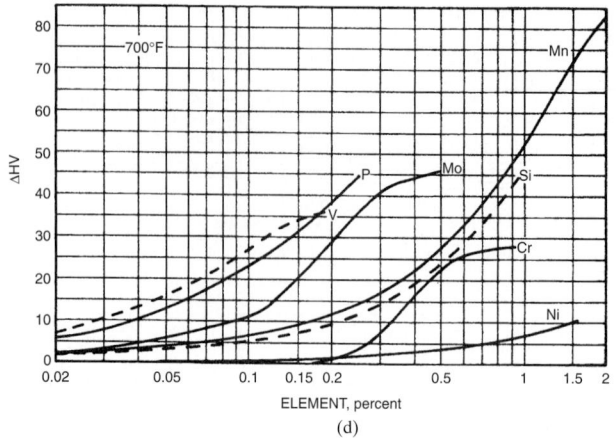

(d)

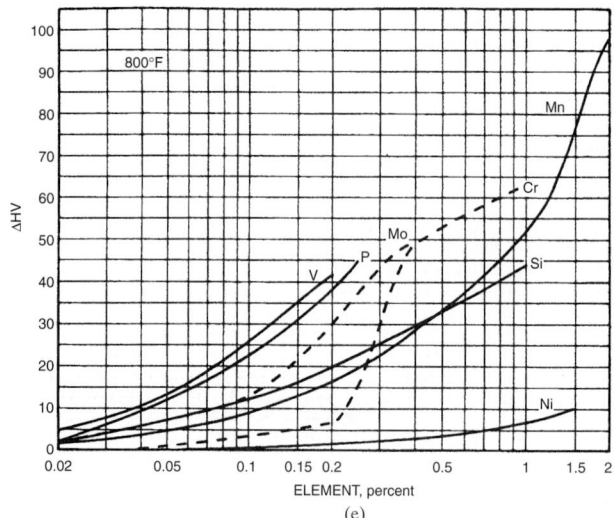

(e)

FIGURE 14.38 (*Continued*) Effects of alloying elements on the hardness of martensite tempered for 1 hr at (*a*) 400°F (204°C), (*b*) 500°F (260°C), (*c*) 600°F (316°C), (*d*) 700°F (371°C), (*e*) 800°F (427°C), (*f*) 900°F (482°C), (*g*) 1000°F (538°C), (*h*) 1100°F (593°C), (*i*) 1200°F (649°C), and (*j*) 1300°F (704°C).[63] (*Reprinted by permission of ASM International, Materials Park, Ohio.*)

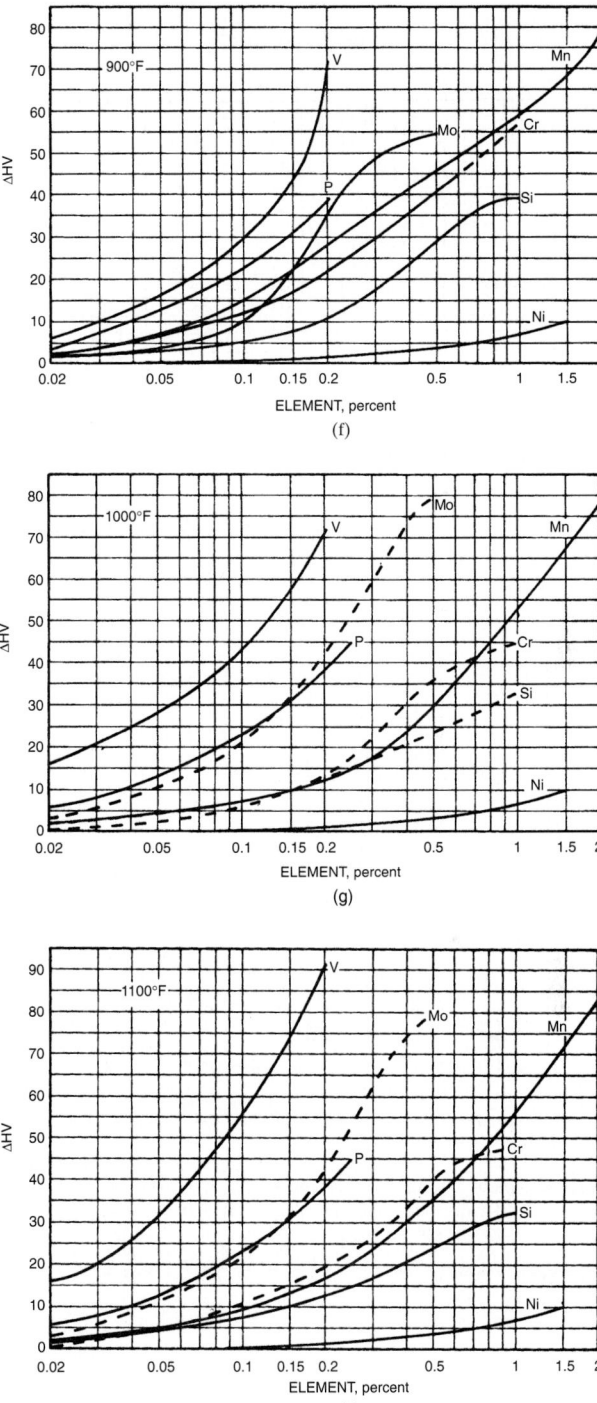

FIGURE 14.38 (*Continued*) Effects of alloying elements on the hardness of martensite tempered for 1 hr at (*a*) 400°F (204°C), (*b*) 500°F (260°C), (*c*) 600°F (316°C), (*d*) 700°F (371°C), (*e*) 800°F (427°C), (*f*) 900°F (482°C), (*g*) 1000°F (538°C), (*h*) 1100°F (593°C), (*i*) 1200°F (649°C), and (*j*) 1300°F (704°C).[63] (*Reprinted by permission of ASM International, Materials Park, Ohio.*)

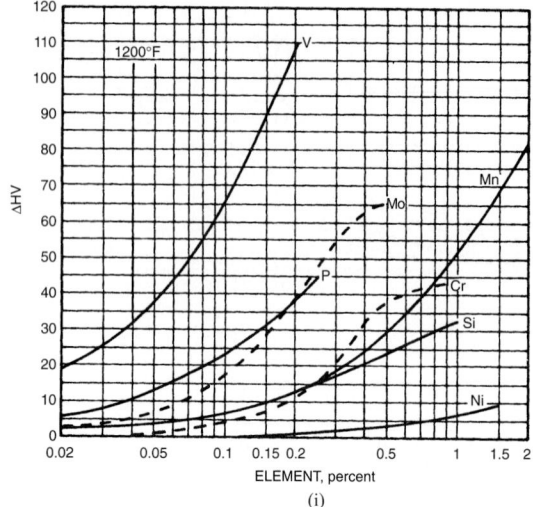

(i)

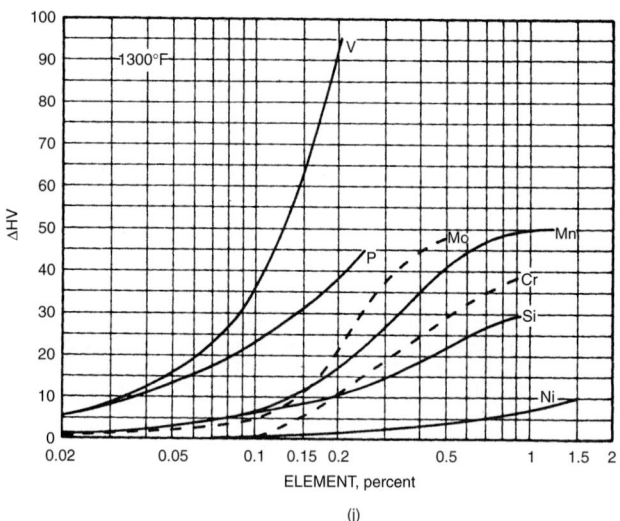

(j)

FIGURE 14.38 (*Continued*) Effects of alloying elements on the hardness of martensite tempered for 1 hr at (*a*) 400°F (204°C), (*b*) 500°F (260°C), (*c*) 600°F (316°C), (*d*) 700°F (371°C), (*e*) 800°F (427°C), (*f*) 900°F (482°C), (*g*) 1000°F (538°C), (*h*) 1100°F (593°C), (*i*) 1200°F (649°C), and (*j*) 1300°F (704°C).[63] (*Reprinted by permission of ASM International, Materials Park, Ohio.*)

TABLE 14.6 Temperature Ranges and General Conditions of Use for Four Types of Tempering Equipment[70]

Type of equipment	Temperature range		Service conditions
	°C	°F	
Convection furnace	50–750	120–1380	For large volumes of nearly common parts; variable loads make control of temperature more difficult
Salt bath	160–750	320–1380	Rapid, uniform heating; low to medium volume; should not be used for parts whose configurations make them hard to clean
Oil bath	≤250	≤480	Good if long exposure is desired; special ventilation and fire control are required
Molten metal bath	>390	>735	Very rapid heating; special fixturing is required (high density)

Reprinted by permission of ASM International, Materials Park, Ohio.

continuous operation of nearly similar parts. Large volumes of air at a controlled temperature are passed at high velocities over the workload. For temperatures of 550 to 750°C (1022 to 1382°F), alternatively radiant heating may be employed because of the greater transfer of radiant heat (or efficiency).

Oil baths are used in the temperature range of 120 to 250°C (250 to 480°F) and are preferred where long exposure time is desired. (Typical tempering time varies between 0.5 and 4 hr.) However, it is necessary to utilize stirring for temperature uniformity and longer oil life, use special ventilation for fume extraction, and avoid overheating for preventing fire hazard and rapid decomposition of the oil. For tempering temperatures in excess of 204°C (400°F), a salt bath is preferred over an oil bath.[54,70]

Oils for tempering are characterized by oxidation resistance with a flash-point much above the operating temperature. The preferred oils include high-flash-point paraffin oils containing antioxidant additives. [See also "Martempering of Steel" (Sec. 13.7).]

Salt baths for tempering are normally used in the temperature range of 160 to 750°C (320 to 1380°F) and are favored where a rapid and greater uniform heating and low- to medium-volume production are desired. They should not be employed for complex parts or parts with small or blind holes, since there are difficulties in cleaning them.

It is necessary to remove all moisture from parts prior to immersion in the molten salt to prevent violent reaction of hot salt with moisture. The parts must be clean and oil-free; otherwise, salt contamination will take place which will require more frequent rectification with chemicals or gaseous compounds. The introduction of cyanide salts or other reducing agents into nitrite tempering baths must be avoided to prevent violent explosion.[70] Thorough cleaning of tempered parts exiting the bath is also important because the adhering salts are hygroscopic and may result in severe corrosion. Table 14.7 lists the compositions and operating temperature ranges for salt baths used in tempering.[70] The reader is referred to the military specification MIL-S-10699A (Ordnance), mentioned above, for the chemical and other control

TABLE 14.7 Compositions and Operating Temperatures for Salt Baths Used in Tempering[70]

Class	Composition of bath, %								Operating temperature		Fuming temperature	
	NaNO$_2$	NaNO$_3$	KNO$_3$	Na$_2$CO$_3$	NaCl	KCl	BaCl$_2$	CaCl$_2$	°C	°F	°C	°F
1	37–50	0–10	50–60	…	…	…	…	…	165–595	325–1100	635	1175
2	…	45–57	45–57	…	…	…	…	…	290–595	550–1100	650	1200
3	…	…	…	45–55	…	45–55	…	…	620–925	1150–1700	935	1720
4	…	…	…	…	15–25	20–32	50–60	…	595–900	1100–1650	940	1725
4A	…	…	…	…	10–15	25–30	40–45	15–20	550–760	1025–1400	790	1450

Source: Military Specification MIL-S-10699A (Ordnance). Reprinted by permission of ASM International, Materials Park, Ohio.

procedures applicable to the different bath compositions. (See Chapter 13 for more details on molten salts.)

Molten lead bath for tempering is used above 390°C (735°F) and is useful for rapid local heating and selective tempering. Because of its high density, special fixtures are required to hold down the parts in the bath during tempering. Above 480°C (900°F), granulated charcoal may be used as a protective cover. For certain applications, lead-base alloys with lower melting point are used.

Vacuum tempering. Although a high-temperature vacuum tempering process can be carried out (along with a vacuum hardening), this practice proves to be inefficient and costly and provides less productivity than the lower-temperature tempering furnace design. When processed in a separate low-temperature tempering furnace, significant savings are achieved in production rates and utility consumption.[94]

Selective or *localized tempering methods* are employed either to temper specific areas of fully hardened parts or to temper areas that were locally hardened previously. The objective of this treatment is to increase the toughness and machinability in the selected area, while maintaining high yield strength for resistance to localized deformation and high hardness for wear resistance in the remaining area. Selective tempering is also used in preheating and postheating of weld areas when a decrease of hardness in the heat-affected zone (HAZ) is needed.

Special processes such as steam treatment, induction heating coils, special flame heads, protective atmospheres, defocused laser, or electron beam heating are often used to achieve certain desired properties. For some steels, the tempering mechanism is augmented by cyclic heating and cooling. A special significant method uses cycles between subzero temperatures and the tempering temperatures to increase the transformation of retained austenite. *Multiple tempering* is primarily used for the following reasons: (1) to relieve residual stress induced during quenching and straightening in complex-shaped carbon and alloy steel parts, thereby minimizing distortion; (2) to bring the retained austenite content to an acceptable level and improve dimensional stability; and (3) to increase both the yield strength and toughness (impact strength) without sacrificing hardness.[70]

14.9.1 Induction Tempering

Induction tempering has been widely adapted to automation in an inline manufacturing system such as for pipe, tube, chain, ball screw, shaft, and bar to produce specific mechanical properties. For induction tempering, mostly a different induction coil from hardening is used. The reason is that in induction hardening, heating and hardness pattern are different because of the component's shape, and the energy density in the hardening process is much higher than with tempering. In induction tempering the surface is heated at a much slower rate to obtain low-temperature gradient from surface to case depth.[95]

Advantages of induction tempering include the following:

1. There is a possibility of integration with production lines to avoid excessive handling of the workpieces, thereby minimizing labor cost.

2. The tempering cycle is very short compared to the furnace tempering cycle (i.e., a minute or less instead of 1 hr) which often requires an increase in induction tempering temperature to achieve similar hardness with a considerable reduction in residual stresses. Local induction tempering in high-stress areas on hardened parts can result in doubling of the component strength.

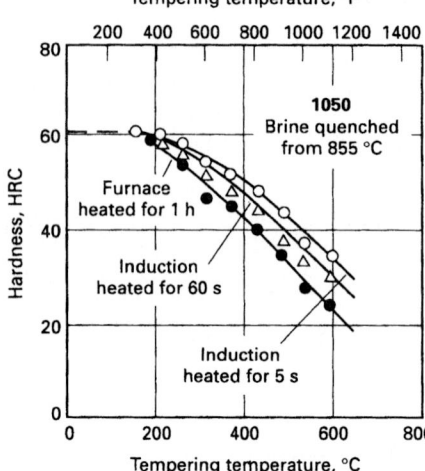

Tempering temperature, °F

FIGURE 14.39 Variation of room-temperature hardness with tempering temperature for furnace and induction heating.[70] (*Reprinted by permission of ASM International, Materials Park, Ohio.*)

3. There is greater energy efficiency. Energy is not needed to heat up and hold at temperature, as with a conventional furnace.

4. Much less floor space is required than for a conventional tempering furnace.

5. Material handling is minimized. In some instances, the same mechanism can load the green part and unload the finished hardened and tempered product.

6. There is precise control of power, monitoring of the final temperature of individual part, and enhanced operator environment.

7. There is a possibility of incorporation of options to track the parts through the hardening and tempering stations and cool down of the tempered parts before exiting the system.[96]

Figure 14.39 shows the increase in tempering temperature required to produce a specific hardness with the decrease in tempering time from 1 hr (furnace tempering) to 60 s and 5 s (induction tempering) in AISI 1050 steel quenched in brine from 855°C (1575°F). Small-section parts should be air-cooled immediately after reaching the tempering temperature, whereas large-section parts need slower heating rates or slow periods of time at temperature (5 to 60 s) before cooling to allow heat penetration. In scanning, however, the time of tempering is determined by the power density, the travel speed, and the length of the inductor.[70]

Equation (14.3) can be employed for short-time induction tempering to calculate the high temperature required with induction to produce the similar hardness to that in furnace tempering.[97]

EXAMPLE PROBLEM 14.3 *To illustrate the application of tempering parameter P, let us consider the required Rc 58 hardness in the root fillet, where P (for conventional tempering for 1 hr at 400°F) = (400 + 460)(18 + log 1) × 10^{-3} = 15.48; calculate the corresponding induction temperature T′ (°F) to achieve the above hardness level.*

Answer. *15.48 = (T′ + 460) [18 + log (5/3600)] × 10^{-3}, or 15,480 / 15.142668 = T′ + 460, or T′ = (1022.3 − 460)°F = 562.3°F.*

14.10   STRENGTHENING MECHANISMS OF TEMPERED MARTENSITE AND BAINITE

The factors contributing to the strengthening of tempered martensite (and bainite) in ferrous systems are the following:[98,99] (1) grain size (i.e., cell or lath size) hardening, (2) Peierls stress, (3) interstitial and substitutional solid-solution hardening, (4) substructure (twins and dislocations) hardening, and (5) precipitation hardening.

Grain Size (Cell or Lath Size) Hardening.   The Langford-Cohen[100] model for cell size hardening is based on the assumption that the dislocation sources in the cell or lath walls are activated more easily than in grain boundaries, and it ascribes the hardening to the stress needed to expand dislocation loops across the cell or lath. This can be represented by the following yield stress $\sigma_{y(0.2)}$ expression

$$\sigma_{y(0.2)} = \sigma_0 + KM^{-1} \tag{14.4}$$

where σ_0 is a friction stress, K is the slope coefficient of the yield strength versus inverse of the average cell/lath width plot, and M is the average lath width or its transverse thickness. This equation is applicable to tempered lath martensitic 0.4% C steel (e.g., AISI 4340) with prior heavy cold-work treatment to cause nonequiaxed and finer structures where yield strength is usually independent of packet size.

However, a Hall-Petch type relation is observed to exist, which can be expressed by

$$\sigma_{y(0.2)} = \sigma_0 + k_y d^{-1/2} + k_s M^{-1/2} \tag{14.5}$$

for packet size and for lath width or thickness, and has been found to be operative for tempered low-carbon martensite in low-carbon 9% Mn steel[101] and 5% Ni and 9% Ni steels[102] where the lath thickness remains remarkably constant. In the above expression, k_y and k_s are grain size coefficients for high-angle and low-angle boundaries, respectively.

Peierls Stress.   The stress required to move a dislocation through an otherwise perfect lattice is called the *Peierls stress*. This stress, based on a sinusoidal force relationship, is estimated to have a value of $10^{-4}\mu$, where μ is the shear modulus. Smith and Hehemann[99] have adopted a Peierls stress of 41.2 MPa (4.2 kg/mm²) based on the estimate of Speich and Swann[103] for AISI 4340 steel.

Interstitial Solid-Solution Strengthening.   According to Smith and Hehemann, in 4340 steels where complete precipitation of cementite occurs after tempering at 300°C or above, the equilibrium carbon content in the matrix remains at about 0.01%, which is estimated to contribute 137.3 MPa (14.0 kg/mm²) to σ_0. However, according to Cox,[104] a considerable amount of interstitially dissolved carbon was retained in a 0.3 to 0.38% carbon martensite even after tempering above 550°C, and he therefore noted that 60 to 70% of the strength of tempered martensite was attributed to interstitial strengthening over the entire tempering temperature range.

Substitutional Solid-Solution Strengthening.   According to Lacy and Gensamer, the assumption of additive effects of various substitutional elements on the yield strength of a 4340 alloy amounts to 165.7 MPa (16.9 kg/mm²). This contribution remains nearly the same up to a tempering temperature of 540°C, provided that the solute distribution is not markedly changed.[98]

Substructure Strengthening.   In a tempered carbon-free Fe-Ni martensite, Speich and Swann[103] have predicted the total contribution of twin substructure formation to the yield strength to be about 137.3 MPa (14.0 kg/mm^2). On the other hand, Kelly and Nutting[105] and Cox[104] predicted the substructure contribution of 206 MPa (21 kg/mm^2) and 68.7 to 206 MPa (7 to 21 kg/mm^2), respectively, to the flow stress in low-carbon and 0.3 to 0.38% carbon martensites. The contribution $\Delta\sigma$ of substructure dislocation strengthening in 0.42% carbon steel is given by[106]

$$\Delta\sigma = 2\Delta\tau = 2\alpha\mu b\rho^{1/2} \tag{14.6}$$

where $\Delta\tau$ is a flow stress contribution (consistent with the usual terminology), ρ is the dislocation density per square centimeter, and α is the dislocation strengthening coefficient, usually between 0.33 and 0.4. Since both the densities of internal twinning and of dislocations fall substantially with an increase in tempering temperature, their strengthening contributions to the yield strength must be reduced at elevated temperatures. This is supported by the observations made by Cox[104] and Malik and Lund.[106]

The work-hardening effect that occurs in attaining a 0.2% offset yield strength has been assumed to contribute a constant value of 68.6 MPa (7 kg/mm^2) to σ_0 for all structures in 4340 steel.

Precipitation Hardening.   The contribution of aging of lath martensite in low-carbon steel is negligibly small. However, the contribution of aging after the quench in twinned martensite is significantly large and can cause an increment in hardness, compared to the virgin martensite. The most well-recognized general expression for calculating the dispersion hardening strength $\Delta\sigma$ (= σ_p) due to the presence of carbide particles (dispersed phase) has been given by Kelly and Nicholson and Ashby and Orowan[107] as

$$\Delta\sigma = \sigma_p = \frac{\mu b}{\pi}\left[\frac{1}{2}\left(1+\frac{1}{1-v}\right)\right]\ln\frac{D}{2b} \tag{14.7}$$

and

$$\sigma_p = \frac{0.015}{l_s}\ln\frac{D}{2b} = 5.9 f^{1/2}\ln\frac{D}{2.5\times 10^{-4}} \tag{14.8}$$

where μ is the matrix shear modulus, b is the Burgers vector of a dislocation in the matrix, D is the average precipitate particle diameter intersecting the slip plane, l_s is the effective carbide (interparticle) spacing (μm), f is the volume fraction of carbide precipitate, and the term within the parentheses is an average Poisson's ratio v for edge and screw dislocations.

Onel and Nutting[107a] have estimated the contribution of carbide precipitation in a heavily tempered plain carbon steel occurring (1) solely at the ferrite grain boundaries and (2) also within the tempered ferrite grains. In the former case, the yield stress, without exhibiting additional strengthening contribution, is given by

$$\sigma_y = 108 + 18.2d^{-1/2} \tag{14.9}$$

In the latter case, the yield stress, with additional strengthening effect σ_p, is given by

$$\sigma_y = 77 + 23.9d^{-1/2} + \sigma_p \tag{14.10}$$

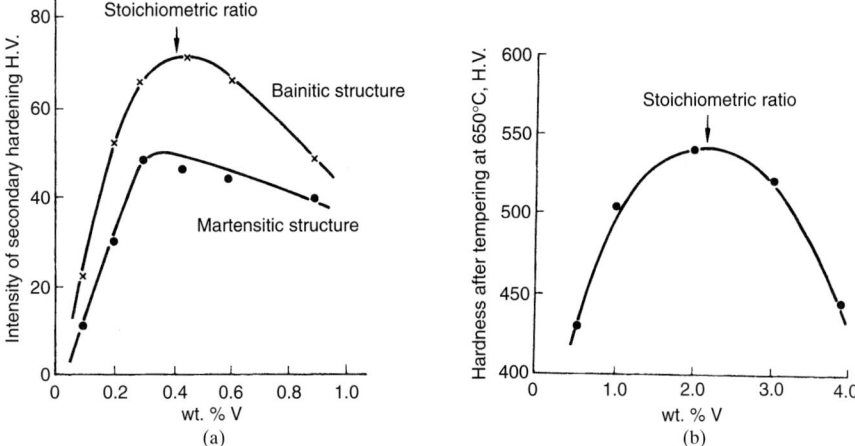

FIGURE 14.40 The effect of vanadium on (*a*) the intensity of secondary hardening in tempered 0.1% C steels consisting of prior martensitic or bainitic microstructures and (*b*) the overaged hardness of 0.6% C martensitic microstructures tempered for 1 hr at 650°C, both showing the maximum effect at the stoichiometric ratio.[108] (*Courtesy of F. B. Pickering.*)

where

$$\sigma_p = \frac{0.015}{l_s} \ln \frac{D}{2b} \qquad (14.11)$$

and all terms have the same meaning as mentioned above.

A significant aspect of secondary hardening alloy steels is that the 0.2% yield strength is a maximum at steel compositions corresponding to the stoichiometric metal/carbon ratio (wt%) for the precipitating alloy carbides. This takes place for both prior martensite and prior bainite microstructures, as shown by hardness peaks in Fig. 14.40*a*; this effect continues even into overaged microstructures (Fig. 14.40*b*).[108] The reason for this phenomenon is that at the stoichiometric ratio, the temperature dependence of the solubility is a maximum[109] so that an increased volume fraction of alloy carbides provides increased precipitation strengthening according to Eq. (14.8) or (14.11) (see Fig. 14.41).[110]

Final Expression of Yield Strength. In general, it is considered that the contributions from the carbide precipitates, from the grain or lath boundaries, and from the dislocation substructures are additive. For example, the yield strength of tempered martensite in an Fe-0.42C-1.1Mn-0.22Si-0.13Mo-0.07S-0.011P steel can be expressed as[106]

$$\sigma_{y(0.2)} = \sigma_0 + 2\alpha\mu b\rho^{1/2} + \frac{\mu b}{2\pi D}\left(1 + \frac{1}{1-v}\right)\ln\frac{D}{2b} \qquad (14.12)$$

where $\sigma_0 = 343\,\text{MPa}$ (35 kg/mm^2) and $\alpha \cong 0.4$.

On the other hand, some researchers are of the view that contributions from the dislocation substructure and the carbide precipitates are mutually dependent

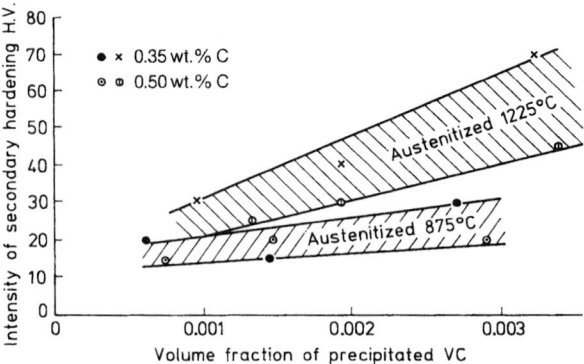

FIGURE 14.41 The dependence of the intensity of secondary hardening on the volume fraction of VC carbides precipitated during tempering of 0.35 to 0.50% C martensitic microstructures.[110] (*Courtesy of F. B. Pickering.*)

because both carbide dispersion and dislocation substructures (1) are encountered simultaneously by the mobile dislocations in a slip plane, (2) undergo considerable changes with increasing tempering temperature in the range of 250 to 550°C, and (3) account for 80% of the steel strength on tempering at 250°C and ~70% of the steel strength on tempering at 550°C in AISI 4340 steel. In this situation, Smith and Hehemann[99] and Daigne et al.[111] derived the yield strength of tempered martensite (MPa) in 4340 steel in the form

$$\sigma_{y(0.2)} = \sigma_0(=572) + 126M^{-1} + 68D^{-1} \tag{14.13}$$

where σ_0 is the sum of (1) the Peierls stress, (2) interstitial and substitutional hardening, (3) work hardening, and (4) the internal dislocation substructure hardening.

In order to account for the intergranular carbide particles, it was assumed that the intrinsic average resistance of lath boundaries to shear propagation increased from a value of K per unit area of clean boundary to a large value of σ_p per unit area of boundary covered with carbides. This assumption leads to an expression

$$\sigma_{y(0.2)} = \sigma_0 + 2KM^{-1} + 1.28(\sigma_p - K)f^{1/2}D^{-1} \tag{14.14}$$

where f is the volume fraction of carbides and the other terms have the usual meanings as stated above.

14.10.1 Toughness of Tempered Martensite and Bainite

There is sufficient evidence in the literature which suggests that the transition temperature T of tempered martensite and bainite increases with an increase in carbon content. Figure 14.42 shows the effect of tempering on the relationship between 0.2% yield stress and the ductile-brittle transition temperature (DBTT) T for low-carbon bainitic microstructures, illustrating the effect of different strengthening

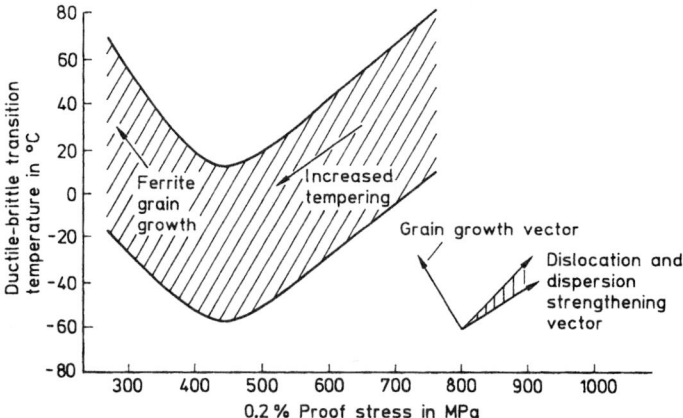

FIGURE 14.42 Effect of tempering on the relationship between the 0.2% yield stress and the DBTT for low-carbon bainitic steels, illustrating the effect of several strengthening mechanisms.[108] (*Courtesy of F. B. Pickering.*)

mechanisms. This observation also holds for tempered martensite. In both cases, at lower tempering temperature, T decreases with the decrease in the yield stress. However, at higher tempering temperatures, below A_1, the recrystallization and grain growth of the ferrite matrix cause an increase in T, even with a decrease in σ_y (as evidenced in Fig. 14.42).[108]

Like martensite and bainite, experimental evidence suggests that an increase in prior austenite grain size of tempered martensite increases T.[112] Tomita and Okabayashi have observed that, in martensitic and lower bainitic (low-alloy structural) steels, the packets are the principal constituent within the prior austenite grains, and the strength and toughness of the steels are increased with decrease in the packet diameter. Likewise, in upper bainitic microstructures, the well-defined blocks (i.e., matrix of laths having low-angle boundaries) rather than packets, control the yield stress and DBTT.[113]

These observations, however, indicate that, provided no embrittlement occurs during tempering, an analysis based on Eq. (7.44) might be employed to relate the microstructure to the transition temperature in tempered martensite. However, modifications pertaining to the effects of precipitation and dislocation strengthening, and omission of the pearlite term would be essential, as in Eq. (9.13), in conjunction with the fracture facet or martensite *packet* size. It would also be essential to consider a factor for the thickness of the carbide particles t. Thus, finally we can use the modified Eq. (14.15) for DBTT:

$$T(°C) = 46 + 0.45\sigma_p + 131t^{1/2} - 12.7d^{-1/2} \qquad (14.15)$$

14.10.2 Effect of Embrittlement on Toughness

Figure 14.43*a* and *b* shows the secondary hardening embrittlement during tempering of a 0.4C-5Cr-Mo-V steel, exhibiting minimum impact energy and K_{IC} values,

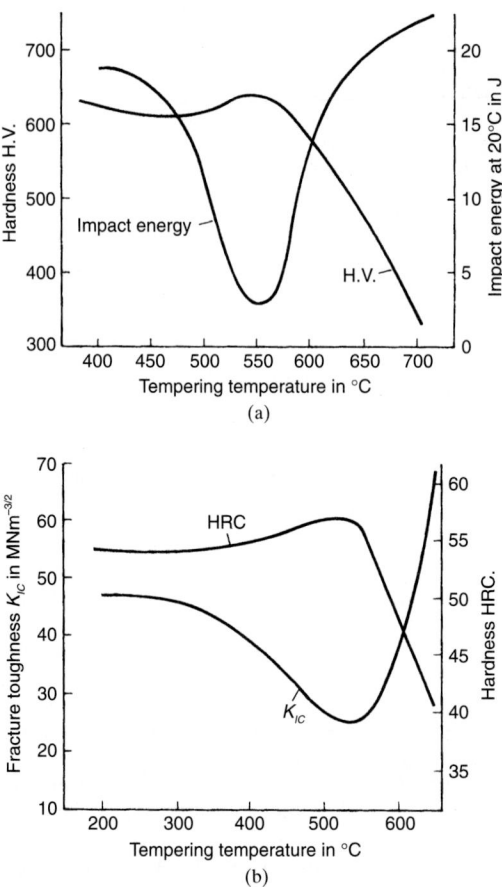

FIGURE 14.43 Effect of secondary hardening embrittlement during the tempering of a 0.4C-5Cr-Mo-V steel on (*a*) impact energy and (*b*) fracture toughness parameter K_{IC}.[108] (*Courtesy of F. B. Pickering.*)

respectively. Similar effects are also found in tempered martensite embrittlement, reversible temper embrittlement, and upper nose embrittlement, which depend on the tempering time.[32,108, 110]

Note that embrittlement always accompanies an increase in T where the brittle fracture mechanism usually changes from cleavage to various types of intergranular cracking, which again operates against an overall quantitative analysis.[108]

Some information on the effect of microstructure on the fracture toughness K_{IC} value suggests that a decrease in strength by carbide coarsening and a reduction in dislocation density result in about the same increase in K_{IC}.[110,114] However, a general relationship between microstructure and K_{IC} parameter is lacking, because K_{IC} is not a unique material property; rather, it is a function of both testing temperature

TABLE 14.8 Effect of Sulfur and Volume Fraction of MnS on the K_{IC} Value for a Quenched and Tempered Medium-Carbon Steel[108,115]

Wt%, S	Volume fraction, MnS	K_{IC}, MNm$^{-3/2}$
0.008	4.3×10^{-4}	72
0.016	8.6×10^{-4}	62
0.025	1.3×10^{-3}	56
0.049	2.6×10^{-3}	47

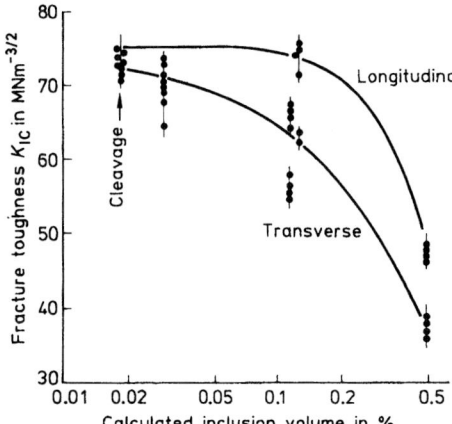

FIGURE 14.44 Effect of nonmetallic inclusion content (vol%) on the fracture toughness parameter K_{IC} for a quenched and tempered high-strength steel, showing the anisotropy of fracture toughness.[108,116]

and fracture mode. As expected, an increase in carbon content usually decreases K_{IC}.[108,114]

Table 14.8 shows that an increase in sulfur content and volume fraction of MnS in quenched and tempered high-strength steels lowers the K_{IC} value.[115] On the other hand, other workers have shown that the basic features of the inclusion distribution which have a bearing on ductility and C_v values also control K_{IC} in high-strength steels when a ductile fracture mode occurs. Figure 14.44 shows the effect of non-metallic inclusion volume fraction on the anisotropy of K_{IC}; at low volume fraction, there is a little anisotropy because of its small effect on K_{IC} for the cleavage fracture. Note that C_v decreases exponentially with increasing strength in tempered martensite, because perhaps this represents a decrease in dispersion strengthening σ_p rather than coarsening of grain size.[108]

14.11 THERMALLY INDUCED EMBRITTLEMENT PHENOMENA

Low-alloy quenched and tempered steels are prone to different types of embrittle-ment phenomena, which usually fall into two categories: (1) thermally induced embrittlement phenomena, resulting from structural changes produced during thermal treatment (including tempering or slow cooling) or elevated-temperature service in a specific temperature range, and (2) environmentally induced embrittlement phenomena, resulting from the interaction of the environment with the quenched and tempered structures. The former types of embrittlement comprise tempered martensite embrittlement, temper embrittlement, secondary hardening embrittlement, thermal embrittlement, and embrittlement due to aluminum nitride formation; and these are discussed mostly in this section, and one in Sec. 14.14.5. The latter types include hydrogen damage and metal-induced embrittlement, comprising liquid-metal embrittlement and solid-metal embrittlement and are discussed in subsequent sections. In many cases, an overlap between the two types of embrittlement occurs.

14.11.1 Tempered Martensite Embrittlement

A sudden loss of room-temperature toughness of quenched commercial high-strength low-alloy steels during tempering in the temperature range of 250 to 400°C (482 to 752°F), in spite of a loss in strength, has been variously referred to as *tempered martensite embrittlement* (TME), *350°C or 500°F embrittlement*, and *one-step temper embrittlement*. This has produced severe difficulties in the development of ultrahigh-strength steels. As a result of this embrittlement, this tempering range is usually avoided in commercial practice. Steels with tempered lower bainitic microstructures are also prone to TME, whereas steels with pearlite/ferrite and upper bainite microstructures are not embrittled by tempering in this region.[117–119]

14.11.1.1 Characteristics of TME[32,120–122]

1. There is a minimum in absorbed fracture energy (usually measured by room-temperature Charpy impact test), its magnitude indicating the extent or severity of embrittlement. The drop in CVN impact energy associated with TME can be explained by a change in three energy components associated with initiation zone, stable growth (or fibrous fracture) zone, and post-unstable (or termination) zone with tempering temperature.[117]

2. The ductile-brittle transition temperature (DBTT) increases through a maximum representing the minimum fracture energy (or Charpy toughness). In other words, it can be characterized by a trough in the plot of Charpy impact energy as a function of tempering temperature for a particular test temperature. Figure 14.20a shows room-temperature Charpy V-notch impact energy for AISI 4130 steels at two different phosphorus levels, 0.003 and 0.03%, as a function of tempering temperature.

3. Compared to temper embrittlement (discussed later), TME is a rapid process; the amounts of impurity segregation needed for intergranular fracture are much lower, and the intergranular fracture required to produce the toughness-minimum is relatively small.[122a]

4. TME is induced during tempering in the critical temperature range within the usual 1-hr time period and is independent of section dimension and/or cooling rate after tempering.

5. Fractures of test samples within the critical temperature range do vary, with transgranular fractures usually observed in the tempering temperature range of 200 to 300°C (390 to 570°F) and intergranular fracture mostly observed at 350°C (660°F) ± 50°C (122°F). These differences may be related to the differences in carbon, alloy, and impurity content, as well as in strength level, nature of the test, test temperature, and grain size. Impurities seem to affect coarse-grained steels to a larger extent than fine-grained steels.[119]

14.11.1.2   Types of TME Based on Fracture Mode

According to the fracture mode, the TME can be classified into two types, namely, transgranular TME and intergranular TME, both of which depend on the microstructure (mainly carbide formation) and variation in strain- (or work-) hardening rates induced by tempering.[117] Both ductile and brittle microcracks commence and develop at carbide particles, depending on the size, morphology, and distribution of carbides and on the strain-hardening rate. High work-hardening rates and high volume fraction of carbides (such as in 4150 steel) favor the brittle or intergranular fracture modes of TME. The transgranular TME or the ductile mode of TME is observed in 4130 steel as a result of the decrease in both the work-hardening rates and carbide distribution.[117]

The interlath (i.e., the area between the parallel martensite laths), intralath (i.e., the area across the martensite laths), and the grain boundary carbides formed by the decomposition of retained austenite or during the third stage of tempering, as well as the undissolved carbides, take part in fracture initiation. The transition carbides do not play any role in crack initiation; however, they influence the strain-hardening rate.[117]

The transgranular TME occurring at interlath or translath has been associated with (or ascribed to) the following:

1. The decomposition of interlath-retained austenite to interlath carbide films (or stringers) during tempering.[43,60,123] This microstructure is similar to that observed for upper bainite.[124]

2. The subsequent deformation-induced transformation on loading of remaining interlath-retained austenite which has become mechanically unstable due to the carbon depletion caused by this carbide precipitation.[125]

3. The formation of coarse (or thick) interlath cementite film as a result of the third stage of tempering in steels containing a very small volume fraction of retained austenite,[126] and/or the formation of coarse interlath Fe_2N. This is usually observed in high purity or lower-carbon quenched and tempered steels such as 4130 (Figs. 14.20a and 14.45a). As usual, crack initiation and crack growth take place by microvoid coalescence around carbide particles retained after austenitizing and formed during the second stage of tempering (Fig. 14.45a).

Transgranular TME has been found to develop upon tempering underhardened M2 (containing 0.8% C, 6% W, 5% Mo, 4% Cr, 2% V) and Vasco-MA tool steels (containing 0.5% C, 2% W, 2.75% Mo, 4.5% Cr, 1% V) austenitized at 1040°C (1900°F) or 980°C (1800°F) with a simultaneous decrease in hardness.[128]

The intergranular TME is ascribed to the decomposition of lath-boundary-retained austenite and the subsequent formation of long and platelike (or interlath) carbide (M_3C or Fe_xC) precipitates at prior austenite grain boundaries which are already weakened by segregation of impurity elements such as P and S, which occur during austenitization[120,129] (Fig. 14.45b), and which facilitate the alternate crack nucleation sites or easy crack path to follow. This is also linked with stress concentration susceptibility at the grain boundaries, which is associated with the matrix toughness. In most steels (e.g., 4340 and silicon stabilized M-300), the lowest K_{IC} and Charpy V-notch impact energy at room temperature is related to the increasing amount of intergranular fracture along prior austenite grain boundaries. However, in some cases, it has been suggested that the minimum in fracture energy may be due to cleavage fracture favored by the formation of carbides during tempering.[126]

14.11.1.3 Influence of Variables. In general, the level of toughness, the severity of TME (as indicated by the depth of the trough), the temperature range of embrittlement, and the mode of failure mechanisms depend on the complex interactions between steel composition, grain size, heat-treating condition, test temperature, and the testing methods.[117]

Composition. An increase in carbon content from 0.3 to 0.5% in 41xx series steels decreases the CVN absorbed energy and the apparent severity of TME (Fig. 14.46).[117] This increase in carbon content causes the crack initiation, ahead of the notch, to change from ductile to brittle mode; however, this depends on test temperature also.[129a]

Capus and Mayer[130] were the first to report the influence of impurities on TME. They found the absence of TME in the high-purity steels, and they showed the embrittlement in the steel which contained P, N, Sb, Sn, Mn, or Si alloying addition. Briant and Banerjee[131] found the occurrence of intergranular rupture of some heat-treating conditions in steels doped with 0.01% N which was believed to be due to precipitation of Cr_2N along prior austenite grain boundaries. The severity of TME developed by the addition of P remains the same (i.e., uniform decrease in impact toughness occurs) in both low (0.002% P) and nominal (0.02% P) 41xx series steels (Fig. 14.46).[117]

It has been seen that Mo addition reduces the P-induced embrittlement, whereas Mn and Cr are potent embrittlers in both P-free and P-doped (0.03%) alloy samples.[132] These elements probably change the ability of P to cause embrittlement by changing the chemical bond it can form at the grain boundaries.[131]

For the low-phosphorus steels (containing 0.003%), interlath cementite is the contributor in initiating cleavage fracture across the martensite laths (Fig. 14.45a), while for high-phosphorus steels (containing 0.03%), a combined effect of carbide formation and P segregation is responsible for intergranular fracture (Fig. 14.45b). However, the plane strain fracture toughness test data exhibited TME only in the high-phosphorus steels, while the Charpy V-notch data displayed TME in both steels.

Sulfur is a more potent embrittler (Fig. 14.47);[122] in many steels, however, it is precipitated as a sulfide, which does not segregate to the grain boundaries. It has been observed that if S is precipitated as chromium sulfide, embrittlement does not occur in the quenched and tempered steel with 3.5 Ni, 1.7 Cr, 0.3 C, 0.004 S, provided that the specimen is austenitized at or below 1000°C (1832°F). However, above an austenitizing treatment of ~1050°C (1920°F), sulfides dissolve and embrittlement is observed.

It is apparent that alloying elements such as Ni (up to 2%), Si (up to 2.5%), and Al, which retard the conversion of ε-carbide to cementite within martensite laths[132a] and/or the precipitation and growth of cementite, can delay the onset of TME to a higher temperature. On the other hand, Mn and Cr promote the decomposition of

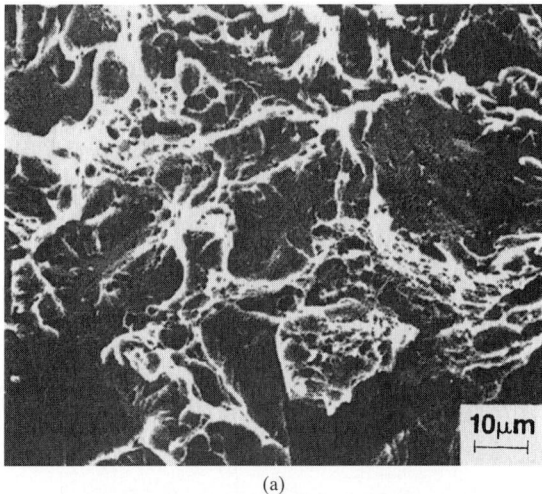

(a)

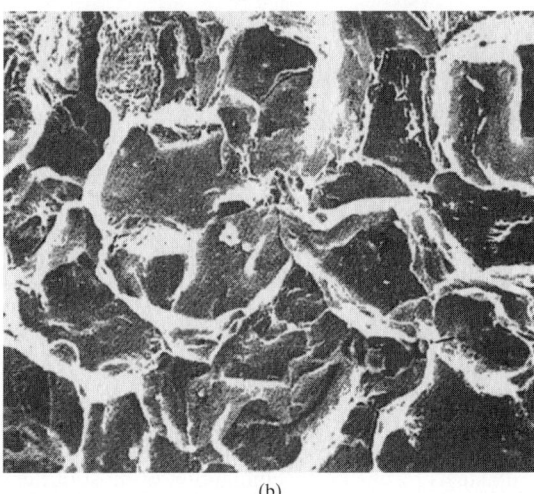

(b)

FIGURE 14.45 Fractography of AISI 4340 steel specimens that failed due to tempered martensite embrittlement. Specimens were broken by impact loading at room temperature. (*a*) Flat cleavage facets in a specimen containing 0.003% P after tempering at 350°C (662°F). (*b*) Intergranular fracture in a specimen containing 0.03% P after tempering at 400°C (752°F).[60] (*Reprinted by permission of the Metallurgical Society, Warrendale, Pa.*)

retained austenite at a lower temperature; therefore, steels containing Cr or Mn are highly susceptible to TME, and this is probably due to the precipitation of chromium or manganese nitride at the grain boundaries.[124] For example, the decomposition of retained austenite is completed for 1 hr at 300°C (572°F) in Mn-containing steels and

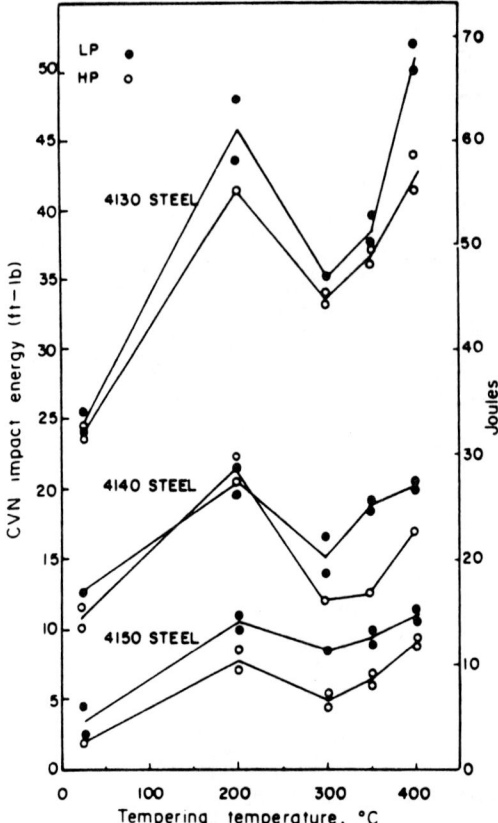

FIGURE 14.46 Room-temperature CVN impact energy versus tempering temperature for 4130, 4140, and 4150 steels austenitized at 900°C and tempered for 1 hr at the temperatures shown.[117] (*Reprinted by permission of Pergamon Press, Plc.*)

for 1 hr at 400°C in Ni-containing steels.[43] The extent of TME increases with the increase of Mn and Si contents corresponding to 4340 steel. Elimination of Mn and Si from high-purity steels results in the elimination of most of the susceptibility to TME. This may be interpreted in terms of the effects of these elements on impurity segregation.

Grain Size Effect and Test Variables. Reduced grain size reduces the TME only when the embrittlement is less severe. Fine-grained 0.01% P-doped steels do not show any embrittlement at room temperature but exhibit embrittlement at a lower test temperature (Fig. 14.48).[122] The embrittlement trough becomes more pronounced when the test temperature is below the DBTT of samples near 350°C.

The four-point bend test method with the Griffith-Owen elastic-plastic stress analysis is more sensitive to TME than is the Charpy test, and the data obtained have a clear physical interpretation.[120]

14.11.1.4 Control of TME. TME in low-alloy high-strength steels can be minimized by[118]

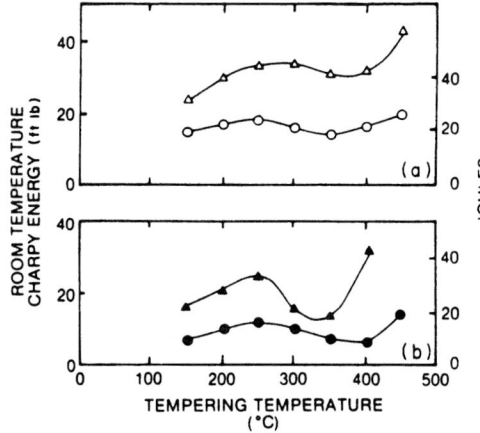

FIGURE 14.47 A comparison of TME in steels doped with either 0.01 wt% phosphorus or 0.01 wt% sulfur. The two different graphs represent two different austenitizing treatments carried out for samples with the base composition 3.5 Ni, 1.7 Cr, 0.3 C steel. (*a*) △, 0.01 P; ○, 0.01 S (at 850°C/1 hr). (*b*) ▲, 0.01 P; ●, 0.01 S (at 1200°C/3 hr). (*Source: Met. Trans., vol. 12A, 1981, p. 317.*)

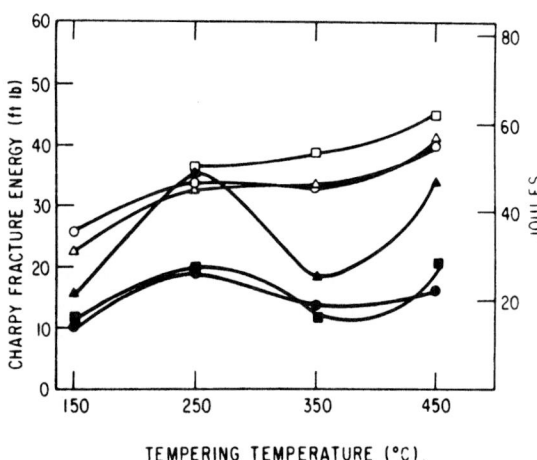

FIGURE 14.48 The fracture energy at different test temperatures plotted as a function of tempering temperature. Composition of steel: 3.5 Ni, 1.7 Cr, 0.3 C, 0.01 P. ●, −80°C; ■, −40°C; ▲, −5°C; △, 23°C; ○, 50°C; □, 90°C. (*Source: Met. Trans., vol. 10A, 1979, p. 1732.*)

1. Development of special steels with delayed onset of TME
2. Development of steels with faster rates of martensite tempering
3. Use of steels capable of transforming fully to upper bainite at the required strength level and section thickness
4. Avoiding tempering in the critical temperature region
5. Use of the lowest possible carbon concentration consistent with the required strength level and reduction of Mn and Si contents[129a]

14.11.2 Temper Embrittlement

When a quenched alloy steel is tempered or slowly cooled through or isothermally heated in the critical temperature range, usually 300 to 600°C (572 to 1112°F), it suffers an upward shift in DBTT,[†] a progressive reduction in impact toughness, and an increasing tendency toward intergranular (brittle) fracture;[133,134] this metallurgical phenomenon is termed *temper embrittlement* (TE), *reversible temper embrittlement* (RTE), *500°C embrittlement*, or *two-step temper embrittlement*.[135–137] TE is reversible; reheating an embrittled steel beyond 600°C (1112°F) and fast cooling eliminate most of the embrittlement.[122a] Note that TE occurs not only after the tempering in the embrittling temperature range (e.g., 550°C) but also on slow cooling (extending several hours) from high temperatures.[138] Although plain carbon steels are not embrittled by P[138a] and immune to TE, the substantial addition of Mn causes susceptibility to this problem.

This phenomenon is an extraordinarily complex metallurgical problem which had been known since the 19th century, and it was long considered as a metallurgical mystery. It has caused a potential concern in heavy-section components such as heavy armor plate, turbine rotors, and pressure vessels, which are slowly cooled through the embrittling temperature range following the tempering treatment during fabrication or experience service temperature operating within this embrittling range. Another area of concern is the prolonged use of such steels in the embrittling temperature range such as low-pressure rotors and disks of large steam turbines where concern over temper embrittlement has restricted the maximum use temperature, to the detriment of thermodynamic efficiency.[136,137] TE can also substantially promote the susceptibility of a steel to: hydrogen-induced embrittlement;[139] stress corrosion cracking;[140] fatigue crack propagation, especially at high maximum stress intensity;[141] and stress-relief cracking in heat-affected zones of a constrained weldment, particularly in creep-resistant low-alloy steels.[137,142] Table 14.9 lists a variety of TE studies.[143–147]

14.11.2.1 Metallurgical Variables.   TE studies on numerous steels have shown that the extent of TE depends primarily on the following metallurgical variables: intergranular concentration of embrittling metalloid impurities (also Mn); composition; grain size;[148] hardness;[143] and thermal treatment, including the rate of cooling.

Impurity Grain Boundary Segregation.   The general conclusions drawn from TE studies on impurity grain boundary segregation are as follows:

1. This phenomenon (i.e., grain boundary decohesion) is attributed to the grain boundary cosegregation, during tempering, of certain metalloid impurities (P, Sn, Sb, As) at prior austenite grain boundaries (Fig. 14.49a) or Fe_3C/martensite interfaces with alloying elements such as nickel, chromium, silicon, or manganese. TE will not occur in the absence of impurity segregation.

2. The experimental data on 3340 (Ni-Cr) steel doped individually with P, Sn, Sb, and As indicate that the relative potency of these impurities are in the order Sb > Sn > P > As.[149‡]

3. The embrittled grain boundary in a quenched and tempered steel provides a continuous fracture path (Fig. 14.49b). Embrittlement occurring at the Fe_3C/martensite interface boundaries, in the tempered martensite, provides a discontinuous fracture path.[137]

[†] The DBTT can be assessed in three ways: (1) the temperature for 50% ductile and 50% brittle fracture (50% fracture appearance transition temperature, or 50% FATT), (2) the lowest temperature for 100% ductile fracture (100% FATT), and (3) transition temperature based on absorbed energy values. Among these, the first one is most common, and the last one is normally not used.

[‡] Later work by McMahon, Jr., has not found any evidence of As as an embrittler.[129a]

TABLE 14.9 Examples of a Two-Step Temper Embrittlement

Type of steel	Reference	Composition, wt%	Embrittling heat treatment	Change in transition temperature, °C	Segregating impurities
HY130	144	0.11C-0.80Mn-0.003P-0.006S-0.0355Si-4.95Ni-0.53Cr-0.50Mo-0.08V	480°C for 1000hr	210	Si, P, Sn, N
3340 + 0.06 P	143	3.5Ni-1.7Cr-0.06P-0.4C	480°C for 100hr	175	P
Plain carbon	145	0.2C-2.0Mn-0.05Si-0.0815Sb-0.006P-0.011S	500°C for 480hr	Transition temperature not measured but fracture surface was 90% intergranular	Sb
A533B with high Cu	146	0.18C-0.25Si-1.39Mn-0.007P-0.01S-0.37Cu-0.63Ni-0.19Cr-0.55Mo-0.016Al-0.0108N-0.009Sb-0.008Sn-0.013As	500°C for 100hr	120	P, Cu(?)
Cr-Mo base	147	0.34C-3.1Cr-0.59Mo-0.89Mn-0.33Si-0.18Ni-0.036S-0.03P-0.015Sn-0.029As-0.008Sb	Furnace cool after temper	130	?

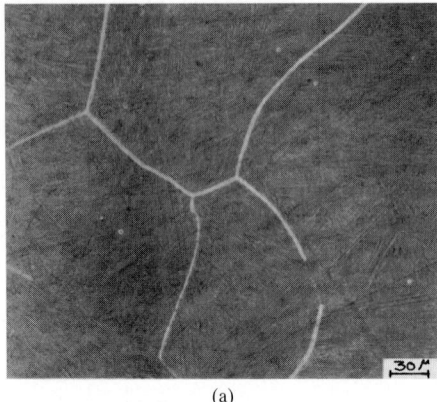 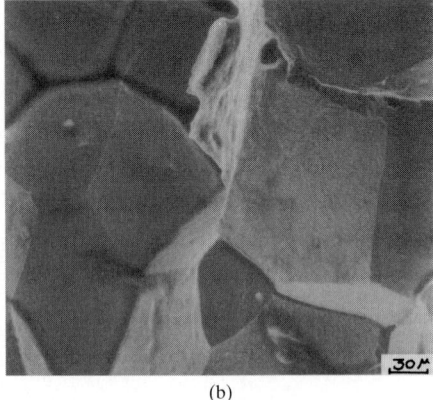

(a)   (b)

FIGURE 14.49 (*a*) Quenched and tempered Ni-Cr steel containing segregated phosphorus at the prior austenite grain boundaries, revealed by a picric acid etchant with a wetting agent. (*b*) Fracture surface of the same specimen showing the occurrence of brittle fracture along the phosphorus-contaminated prior austenite grain boundaries.[137] (*Courtesy of C. J. McMahon, Jr.*)

4. The impurity segregating at the grain boundaries appears to be an equilibrium and a reversible phenomenon. However, the results of Faulkner et al.[150] on combined quenching and tempering induced P segregation to grain boundaries in a 0.077% P doped 2.25Cr-1Mo steel indicate that combined equilibrium and nonequilibrium segregation plays a significant role in TE of the steel caused by direct tempering after quenching. Nonequilibrium segregation needs the formation of adequate amounts of vacancy-impurity complexes, and their migration to grain boundaries is of vital importance in the segregation.

5. The equilibrium grain boundary concentration of impurities increases with decreasing aging temperatures. However, aging time at the low aging temperature plays a vital role. For example, in a study made on 3.5Ni-1.7Cr-0.4C steel, it has been found that segregation of P is rapid and reaches equilibrium within 50 to 100 hr at 480 to 560°C; Sn and Sb require much longer to attain equilibrium. The rate of segregation falls off sharply below 400°C because of diffusional kinetics and above 560°C because of entropy effect.[132]

6. Among the impurity elements, P (as well as Mn) is the most common grain boundary embrittler in commercial alloy steels because of (*a*) its segregation during austenitization, tempering, and aging; (*b*) its rapid segregation even at low aging temperature; and (*c*) its larger concentration than those of other embrittling impurity elements in commercial steel. Sb is rarely present at the grain boundaries, and sulfur is usually precipitated as manganese or chromium sulfides. The next very important impurity elements are Sn and As. It is also evident that the complex sequence of precipitation of carbides in steels has an important influence on the embrittling process by affecting the impurity segregation.[122]

Composition: Alloying Elements. The embrittling impurity elements produce increased TE in Ni-Cr steels compared to the Ni and Cr steels. Mo additions are effective in decreasing or eliminating TE when impurity elements are present and retard the kinetics of TE of Fe-Mo-P alloy steels. However, to be more effective, Mo must be dissolved in the ferrite matrix and not tied up in carbides, with a maximum effect at a concentration of around 0.7%. The very strong interaction between Mo and P results in the precipitation of $(Mo,Fe)_3P$ or Mo-P atom cluster which prevents the segregation of P to grain boundaries.[32] It has been shown that C and P compete with each other in grain boundary segregation, in that a strong repulsive interaction between C and P may cause hindrance to P segregation. When more Mo is tied up in carbides, its beneficial effect diminishes. An increase in Cr concentration increases the embrittlement resulting from a fixed amount of P in Ni-Cr and other steels.[143,151,152] This influence of Cr on embrittling behavior is attributed to the reduction of carbon activity. The similar effect is found with any carbide formers such as Nb.[151†] Cr addition also raises the potency of Sn and Sb.[152–154] When the addition of Ni and Cr is made alone, the grain boundary segregation becomes large and small, respectively. However, the addition of both Ni and Cr in Sb-doped Ni-Cr steel leads to a very large Sb and Ni segregation compared to their cumulative individual effects.[122]

The interaction between Mo and C seems to be the dominant factor that has a bearing on the grain boundary cohesion in the Fe-Mo-P alloys.[154a]

Mn and Si also increase the susceptibility to embrittlement. The addition of 0.3% Mn to a NiCrMoV rotor-type steel with 0.02% P was found to produce a large increase in susceptibility of TE, compared to a Mn-free steel. However, when only 30 ppm P was present in a NiCrMoV steel, the presence of Mn did not cause TE. A study of the effect of 1% Mn in a high-purity, decarburized iron showed the segregation of Mn to the grain boundaries, associated with a large reduction in the intergranular fracture stress at 133 K. When P was added to the Fe-1%Mn alloy, it raised the amount of Mn segregation due to their attractive interaction and vice versa (with a Mn-free Fe-P alloy). Based on these findings, it has been concluded that Mn addition has two effects with respect to TE. One effect is attributed to its interaction (i.e., cosegregation effect) with P which increases the segregation level of P, thereby enhancing the embrittling potency of P. The other is an intrinsic embrittling effect due to segregation of Mn to the grain boundaries, which causes reduced intergranular fracture strength.[155] However, some workers have not observed the enrichment of Mn to the austenite grain boundaries in 10Mn-P steels.[156]

However, the presence of both Mn and S in the steel leads to the scavenging of S from solid solution by incorporation into existing sulfides, thereby its inability or less availability to segregate.[132,157] Similarly, Ti has a beneficial effect in the low-carbon Fe-Ni-Cr-Sb alloy, which is a scavenger of Sb.[132]

Steels containing Mo, W, and/or V exhibit delayed temper embrittlement due to the slow precipitation of alloy carbides of increasing stability.[137] The mechanisms of delayed TE in Mo-bearing steels can be explained in the following way:[137,158] The reduction of Mo (or C) activity in the ferrite matrix causes reduced segregation to grain boundaries, presumably due to the formation of Mo_2C carbides and the release of locked up impurities such as P into solid solution, which allows increased P segregation, finally reaching the level expected in Mo-free steels.

In Cr-Mo steel containing P as an impurity element, the kinetics of P segregation and hence the embrittlement is quite different from that in Ni-Cr steel as a result of the initial interaction of P and Mo. One mechanism to explain the influence of Mo is that it serves as a scavenger for P but that the resulting Mo-P compound is less stable than Mo-rich carbides. The other mechanism is Mo-P cosegregation which tends either to reduce greatly the embrittling potency of P at the grain boundaries or to exert an additional intrinsic strengthening effect due to the increased grain bound-

---

† According to McMahon, Jr., and Yu-Qing, Cr is mainly a scavenger of carbon which indirectly influences in reducing carbon segregation and allowing greater P segregation, both of which provide increased intergranular embrittlement.[154a]

ary cohesion. Experimental evidence favors both mechanisms.[132,158,159] The addition of Mn and Si in the Cr-Mo base alloy increases the potency of P.

The presence of V retards the rate of TE in Mo-bearing steel by a factor of 10; the mechanism, whether it is scavenging element or whether it retards the formation of Mo-rich carbides, has not yet been made clear.

The P-doped NiCrMoV steel has been found to show the least TE and the commercial Ni-Cr steel the most. The P-doped NiCrV steel is more prone to TE than P-doped NiCrMoV steel. This difference is due to Mo-enhanced grain boundary cohesion;[160] however, it is a questionable proposition.[129a]

Microstructure. Matrix microstructures are also very important because they control the toughness of both embrittled and nonembrittled steels. In general, tempered martensite is more susceptible than tempered bainite to TE; however, tempered bainite is more susceptible than ferrite-pearlite structures.[161]

Grain Size and Hardness. It has been shown in a quenched and tempered steel that for a fixed grain size and hardness, the extent of TE is a function of the metalloid element concentration in the grain boundaries. Similarly, for a fixed metalloid impurity concentration in the grain boundaries, the extent of embrittlement is a function of the alloying element, hardness, and grain size of the steel where the DBTT increases with either hardness or grain size or both. Thus, for a fixed level of impurities and constant embrittling temperature and time, there is a greater coverage of the grain boundaries in a coarse-grained steel than in a fine-grained steel. However, the distance over which the impurities must diffuse increases with the increase of grain size.

The degree of embrittlement increases with an increase in prior austenite grain size, and in steels with duplex grain structures, the size of the largest grains controls the deterioration in toughness. The reason is that less energy is needed to initiate and propagate cracking on the embrittled boundaries of a coarse-grained steel.[162]

In the case of CrMoV steels with P as the impurity element, susceptibility to TE during service is practically eliminated when the prior austenite grain size is ASTM number 9 or above. However, a decrease in ASTM grain size from 4 to 0 increases the shift in FATT by 61°C (110°F).[161]

Figure 14.50 shows the effect of prior-austenite grain size on the TE of a Ni-Cr steel (containing 0.33 C, 0.59 Mn, 0.03 P, 0.031 S, 0.27 Si, 2.92 Ni, and 0.87 Cr) that was heat-treated to produce coarse- and fine-grain size. It is quite clear that coarse-grained specimens were severely embrittled compared to fine-grained specimens.[163]

It is thus inferred that fine-grain size leads to a decrease in the amount of embrittlement for any given grain boundary segregation. The improvement induced in this case is much larger than in the TME. In general, the grain size effect increases with the potency and concentration of the embrittling element and with hardness.[144] Increasing hardness (or yield strength) increases the extent of TE for a fixed amount of grain boundary segregation. This is quite evident in plain carbon steels which are not embrittled by P because of being too soft.[138a]

In an experimental study on Ni-Cr steels doped with P and Sn, it was found that the transition temperature could be expressed as a Taylor series involving as variables the grain size, hardness, and average metalloid impurity concentration on the fracture surface. Figure 14.51 shows the variation of transition temperature for various prior austenite grain size and hardness values in these steels when doped with P and Sn as a function of Auger peak-height ratio (PHR) with respect to Fe in a Ni-Cr steel.[164] In more complex steel, other factors must be taken into consideration, such as a real fraction of hard carbide or nitride particles on the grain boundaries and presence of alloying elements, such as Mn, Ti, Mo, and so forth, cosegregated with the metalloid impurity elements.[137]

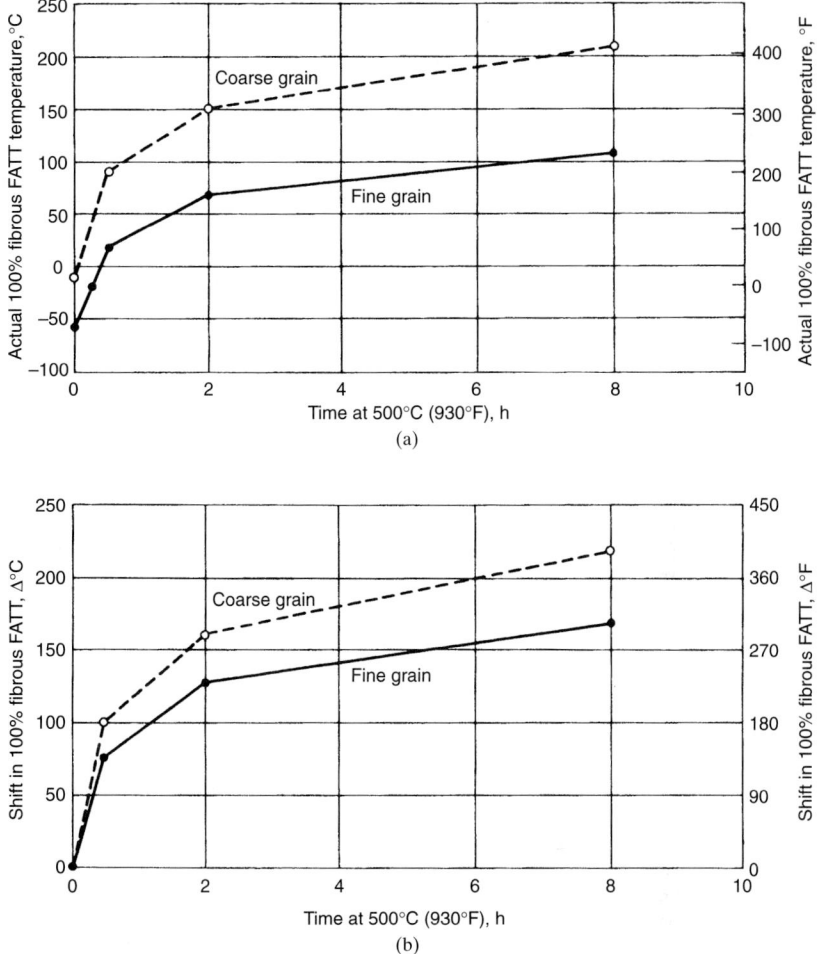

FIGURE 14.50 Effect of prior-austenite grain size on the temper embrittlement of a Ni-Cr alloy steel that was heat-treated to produce coarse- and fine-grain sizes. The alloy was tempered at 650°C (1200°F) and aged for various times at 500°C (930°F). (*a*) Actual 100% fibrous FATT. (*b*) Change in 100% fibrous FATT.[163]

14.11.2.2 Detection and Measurement of TE. Auger electron spectroscopy (AES) and DBTT have been the widely accepted tools used to detect and measure the embrittlement susceptibility. AES application to embrittlement studies allows the direct chemical analysis of impurity segregants on the intergranular fracture surfaces of embrittled specimens and alloy element segregation such as Ni at these boundaries, which act as stimulant for impurity element segregation to the prior austenite grain boundaries. The degree of enrichment of impurity elements may be 100 to 1000 times the bulk concentration, while the concentration of alloying elements may be only 2 to 3 times that of the bulk concentration, and the concen-

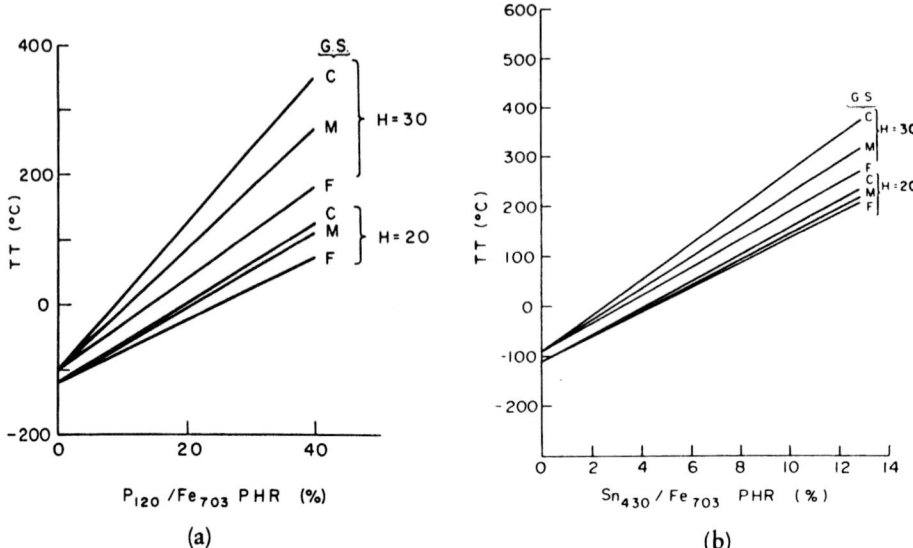

FIGURE 14.51   Variation of transition temperature with intergranular concentration of (*a*) phosphorus and (*b*) tin, expressed in terms of Auger peak-height ratio with respect to iron in Ni-Cr steel. The prior austenite grain size (C, coarse; M, medium; F, fine) and the Rockwell C hardness were also varied.[164] (*Reprinted by permission of The Metallurgical Society, Warrendale, Pa.; after C. J. McMahon, Jr.*)

tration profile from the grain boundary into the grain interior is usually much shallower than for the impurity element. Figure 14.52*a* shows an example of Auger analysis of Sb, S, and P segregated to either fracture grain boundaries or free surfaces.[165] These results have been accomplished by alternate argon ion sputtering (depth profiling) and analysis.[165]

An alternative method is the measurement of area fraction of intergranular facets on the fracture surface; all these measurements are directly proportional to the concentration of segregated impurities on prior austenite grain boundaries.[166] Monitoring acoustic emission activity may also be a very sensitive method of detecting the occurrence of TE in A 533B, which is a nuclear-grade MnMoNi low-carbon low-alloy steel.[167]

To identify RTE in the in-service embrittled CrMoV steel turbine bolts, Bulloch and Hickey[168] have developed an embrittlement estimative diagram (EED) by plotting grain size and grain boundary area S_v versus percent P (Fig. 14.52*b*). This exhibited two distinct regimes, i.e., the embrittled regime and nonembrittled regime, which were separated by a critical embrittled-nonembrittled interface which would be represented by the following equations:

$$d(\%P) = C_{RTE}$$

and $$S_v = 21.4 \times (\%P)$$

where d is the grain size in micrometers and S_v is the grain boundary per unit volume in mm^{-1}. C_{RTE} is called the RTE constant and is in the range of 0.12 to 0.59; an increased value of C_{RTE} represents an increase in the resistance of a series of bolts

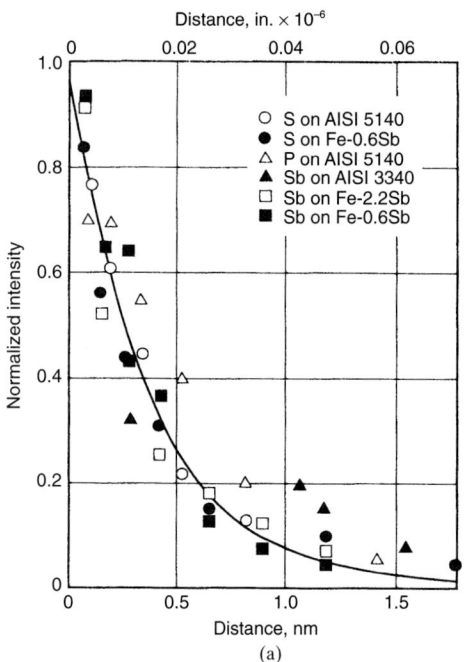

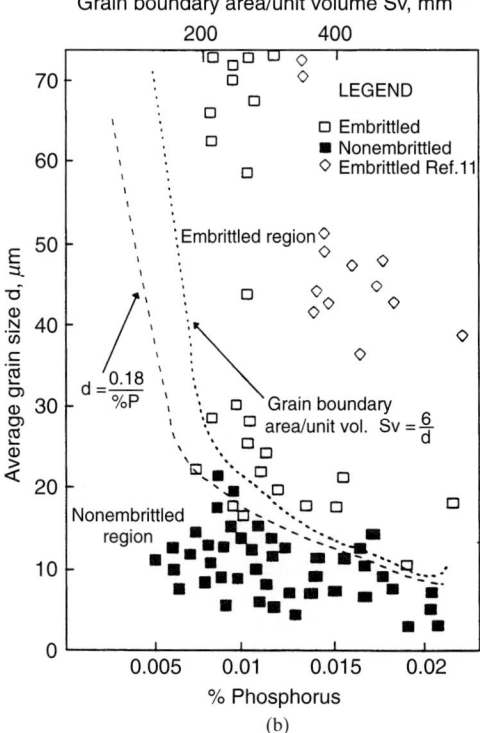

FIGURE 14.52 (*a*) Normalized intensities of Auger peaks (as a function of depth below the surface) from Sb, S, and P segregated to grain boundaries or free surfaces (depth profiling by argon ion sputtering).[165] (*b*) Relationship between grain size and bulk phosphorus level for the IP and HP turbine bolts with C_{RTE} value of 0.28.[168] (*Note*: Ref. 11 after J. J. Hickey and J. H. Bulloch, unpublished data, Electricity Supply Board, Dublin, Ireland, 1986–1987.) (*Courtesy of J. H. Bulloch.*)

to RTE during service at elevated temperatures. That is, RTE only occurs in bolts with larger grain sizes and bulk P levels. Essentially, it was illustrated that the extent of grain boundary area available for P segregation was the principal factor controlling turbine bolt embrittlement. Figure 14.52*b* shows the relationship between grain size and bulk phosphorus level for the intermediate-pressure (IP) and high-pressure (HP) turbine bolts.

Later work by Bulloch on steam-turbine casing steel bolts [composition: (0.36–0.44)C-(1.24–1.36)Cr-(0.56–0.81)Cr and (0.23–0.24)C-(1.28–1.37)Cr-(0.80–0.84)Mo-(0.25–0.28)V], undergoing operating temperatures of 450 to 490°C (842 to 914°F) for service times from 60,000 to over 200,000 hr, led to the development of a general embrittlement law for the occurrence of RTE in the form

$$d \times (\%P) \times (\%\varepsilon)^b = \alpha$$

where ε is the accumulated service strain and b and α are scaling constants equal to 0.64 and 0.0772, respectively. When the left-hand side product becomes greater than α, bolt embrittlement is envisaged; and when less, no service embrittlement is expected.[168a]

14.11.2.3 Control and Prediction of TE.   TE can be controlled by reducing susceptibility, which is primarily achieved by maintaining small concentration of embrittling impurities by control of raw materials and melting practice.[118] However, it is not possible, from both technical and economical grounds, to lower the concentrations of some metalloid impurities to completely harmless levels. An alternative method is given in the next section.

To predict the extent of TE of a particular steel such as a Ni-Cr steel, first the TE equation of a P-doped Ni-Cr steel and the McLean equation for equilibrium intergranular segregation of P in the same steel are derived experimentally and then used to construct the two-dimensional TE diagram.[169]

An embrittling treatment referred to as *step cooling* is sometimes employed to estimate the influence of extensive exposures. Step cooling comprises cooling the sample through the embrittling range in a series of steps from about 600 to 300°C (1100 to 572°F), with the time increasing progressively with the lowering temperature.[118] The extended step-cooling treatment (ESCT) provides more accurate indication of the degree of RTE for new 1CrMoV rotor steels during their service lifetime.[168]

14.11.2.4 The Kinetics of TE.   The embrittling kinetics follows the C-curve behavior with tempering time and temperature, with a minimum time for embrittlement at about 550°C (1022°F); these curves are plotted by using the Auger analysis of monolayers of P segregation at the prior austenite grain boundaries.

14.11.2.5 Segregation Theory of TE.   The use of AES, a semiquantitative analytical tool for measuring segregated elements at the embrittled grain boundaries, has confirmed the simultaneous segregation (or cosegregation) of metalloid impurities and alloying elements (Mn and/or Cr) to high-angle austenite grain boundaries in the temper-embrittled condition.[170] This *cosegregation theory* can be employed satisfactorily to elucidate the embrittlement after long exposure times of steels at 450 to 500°C, but it does not interpret well the two main features of TE, namely, (1) susceptibility to cooling rate after tempering and (2) the capacity to return to the ductile (i.e., deembrittled) state from the embrittled state on reheating the steel to 600 to 700°C (1112 to 1292°F) followed by rapid (water) quenching.

Site competition theory addressed a lack of evidence for the cosegregation model in the ternary Fe-based alloys. According to this mechanism, a site competition between C and P at the grain boundary is assumed. Consequently, activities of carbon in the ferrite matrix, precipitation of carbides and phosphides, partitioning of alloying elements between matrix and carbides, and presence of dislocation around precipitates should be the primary factors that influence the P segregation in the Fe-C-P-M system. This theory agrees well with the experimental results observed by Janovec et al.[171] in CrMoV steels.

Lei et al.[172] have concluded that En steels such as 40 Mn2Mo, 40Cr, and 40 CrNi steels are susceptible to TE and that this embrittlement is always associated with lowering of the Köster peak heights of the internal friction curves. The Mo addition to the steels as in 40 CrNiMo and 40 Mn2Mo, and NiCrMoV significantly improves grain boundary cohesion by interacting with carbon,[167] which inhibits the TE and simultaneously decreases the Köster peak heights in slowly cooled (brittled) or aged (embrittled) states. The reversibility of TE in En steels such as 30 CrMnSiNi 2 and 40 CrNi steels is closely associated with the reversibility of the internal friction behavior. Room-temperature impact toughness values are linearly related to the Köster peak height. This suggests the aging mechanism of α-solid solution to be the cause of TE, which causes the dead pinning of dislocations by ultrafine $Fe_3C(N)$ precipitate particles on slow cooling after tempering or on isothermal embrittling treatment. The dissolution of these ultrafine particles into the α-phase occurs on reheating and holding just a few minutes at about 600°C or above. Subsequent rapid cooling to below about 300°C (570°F) eliminates the TE; that is, ductility and impact toughness values of the embrittled steels are restored. The aging mechanism can be employed to elucidate the short-time TE, while McMahon's segregation mechanism is adequate to explain the TE after long-time exposure of boiler or steam engine parts at temperatures higher than 450 to 500°C (842 to 932°F).[172]

14.11.3 Secondary Hardening Embrittlement

The embrittlement occurring after tempering in the secondary hardening range is referred to as *secondary hardening embrittlement* (SHE). SHE in tungsten- and molybdenum steels is of two types, intergranular and transgranular. Intergranular SHE is associated with the impurity segregation and leads to easy intergranular fracture while the transgranular SHE is caused by coarse boundary carbides, resulting in easy transgranular fracture.[173]

14.11.4 Aluminum Nitride Embrittlement

When an Al-killed plain carbon steel (or Al-stabilized low-alloy steel) containing increased levels of aluminum and nitrogen is slowly cooled from high austenitizing temperature ~1300°C or from solidification, precipitation of long and dendritic- or plate-shaped AlN phase occurs along grain boundaries covering a large fraction of the grain boundary area. AlN dendrites form from the liquid near the completion of solidification and may act as nucleation sites for plate-like AlN that precipitates after solidification and appears as small, shiny fracture surface facets. This leads to

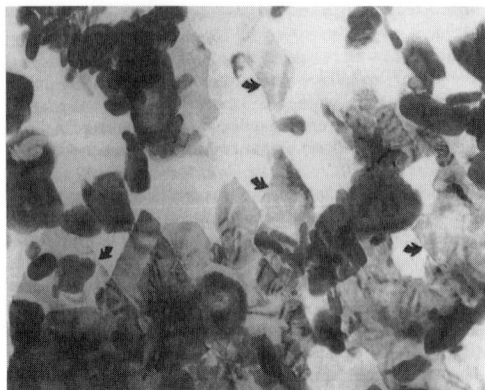

FIGURE 14.53 Thin aluminum nitride particles (arrows) extracted from the intergranular fracture surface of a medium-carbon steel casting. Thick/dark particles are carbides. Extraction replica electron micrograph. 82,500X; reduced to 75%. (*Courtesy of G. Krauss.*)

a drastic reduction in toughness and can cause intergranular fracture along the prior austenite grain boundaries which have been weakened by the existence of AlN. The resulting intergranular fracture is sometime called *rock-candy fracture* because the coarse intergranular facets of the castings produce a macroscopically crystalline appearance.[119,174] This can result in catastrophic failures in castings, panel cracking in ingots, and reduced hot ductility. Figure 14.53 shows the thin aluminum nitride and thick carbide particles extracted from the intergranular fracture surface of a medium-carbon steel casting.[32]

The minimum amounts of AlN necessary to produce intergranular fracture for plain carbon and alloy steels are 0.004 and 0.002%, respectively. AlN embrittlement in castings, in both the as-cast and heat-treated conditions, can be minimized or eliminated by[119,175–177]

1. Additions of Ti, Zr, B, S, Mo, Ni, or Cu

2. Use of the lowest possible amount of nitrogen (0.005%) and minimum requisite amount of aluminum (0.015 to 0.030%) for deoxidation

3. Increased cooling rate after solidification

4. Faster cooling rate in the range of 1150 to 700°C (2100 to 1290°F) after solutionizing at high austenitizing temperature to control the amount and size of AlN precipitation

Panel Cracking in Ingots. Panel cracks are longitudinal surface cracks on the side face of an Al-killed (0.4 to 0.7%) carbon steel ingot (or a low-carbon alloy steel ingot) that usually form (possibly below 850°C or 1560°F) near the center of the face and extend up to the midradius of the ingot.[178] These carbon levels produce ferrite grain boundary network film containing mostly pearlitic matrix structure. The extent of susceptibility to panel cracking is a function of the melt practice and aluminum and nitrogen contents. Thus, the severity of panel cracking decreases in the following order: electric arc furnace steel, basic open-hearth steel, basic oxygen furnace steel, and acid open-hearth steel. Panel cracking does not occur with less than 0.015% Al and 0.005% N. Small ingots are less prone than large ingots. Stripping of the ingot at a permissible high temperature perhaps reduces the susceptibility.[119]

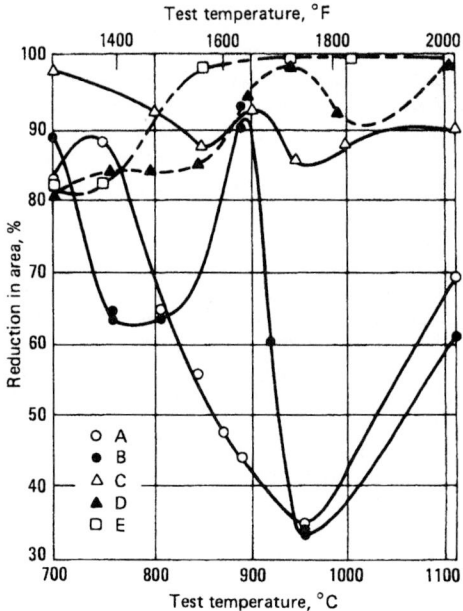

Low-carbon steel	Composition		Particle size	
	N, ppm(a)	Ti, %	nm	Å
Curve A	80	...	90	900
Curve B	70	...	80	800
Curve C	72	...	210	2100
Curve D	2	...	(b)	(b)
Curve E	1	0.06	(b)	(b)

(a) As AlN. (b) No data

FIGURE 14.54 Elevated-temperature tensile test results for five plain carbon steels with various levels of aluminum nitride. The nitrogen level (in ppm) of the steels in the form of aluminum nitride was: A, 80; B, 70; C, 72; D, 2; E, 1. (*Source: Ref. 181.*)

Reduced Hot Ductility.[179–183]   Increased levels of Al (>0.03%) and N (~ > 0.01%) have been found to degrade hot ductility in low-carbon steels, En 36 alloy steels, and Cr-Mo-V turbine rotor steels in the temperature range where the volume fraction and size of the AlN precipitates are maximum. These trends were improved with the decrease of AlN particle size.[119] Figure 14.54 shows the reduction in area for hot tensile tests over a temperature range for steels with different levels of soluble AlN and high levels of N present as AlN.[181] The presence of high residual impurity contents, mainly Cu and Sn, played vital roles in decreasing the hot ductility in medium- and high-carbon steels.[183] These impurities tend to segregate at the grain boundary ferrite networks. However, the addition of Ti and/or B in rotor steels has been observed to improve the hot ductility in the test temperature range of 800 to 1000°C (1470 to 1830°F) where high N contents are deleterious.[179]

It has been shown that Sn reduces the solubility of copper in austenite by a factor of 3; hence, in the presence of Sn, molten Cu can form at the surface at lower bulk Cu contents. Sn and Sb are extremely harmful to Cu-induced hot shortness. Ni reduces Cu-induced hot shortness, Mn and Cr slightly increase hot shortness, and As is slightly more harmful than Mn.[119,182]

14.12   HYDROGEN DAMAGE OF STEELS

Hydrogen damage is a term that describes a number of processes in metals by which the load-carrying capacity of the metal is reduced because of the presence of hydrogen, usually in combinations with residual or applied tensile stresses. Hydrogen damage can develop in a wide variety of environments and circumstances and, in one form or another, can largely limit the use of certain materials. Although it is more predominant in carbon and low-alloy steels, many metals and alloys can exhibit this phenomenon. Hydrogen can be retained in steels and other metals internally as a result of melting and casting practices (supersaturated) and/or present externally in the atmosphere around the alloy material as a gas or a constituent of gas as a result of pickling, electroplating, cathodic process, sour environment, contact with water or other hydrogen-containing liquids or gases, and so forth.

Hydrogen damage can be classified into:[184] (1) hydrogen embrittlement, (*a*) hydrogen-assisted cracking, (*b*) delayed failure, (*c*) sulfide stress cracking, and (*d*) hydrogen-induced ductility loss; (2) hydrogen attack; (3) shatter, cracks, flakes, and fish eyes; (4) blistering; (5) metal hydride formation; (6) microperforation; and (7) degradation in flow properties. These are discussed below.

14.12.1   Hydrogen Embrittlement

Hydrogen embrittlement has been found in various metallic materials. It is of great concern for the use of advanced high-strength materials in a wide range of high-technology applications.

Hydrogen embrittlement (HE) is a process resulting in the degradation in any one or more of a number of mechanical properties such as ductility, work-hardening rate, tensile and yield strengths, fracture toughness, and so on, depending on its application.[185,186] In general, a degradation of these mechanical properties occurs through a hydrogen-induced change in either the plastic behavior or the fracture behavior of the alloy, primarily the latter. The effect of hydrogen on the plastic behavior of an alloy is direct and is due to some hydrogen-dislocation interactions. On the other hand, the effect of hydrogen on the fracture behavior of the alloy is far less direct and can comprise any one or more of the hydrogen-metal interaction mechanisms (discussed later).

Hydrogen is capable of influencing all three stages of fracture, namely, initiation, slow crack growth, and the onset of rapid, unstable fracture. All the classical modes of fracture, such as intergranular fracture, transgranular ductile fracture, cleavage fracture, or a mixed fracture, can occur in the presence of hydrogen. The extent of a particular mode depends on the specific alloy and its application. For example, mixed modes of fracture can be produced in a single specimen of high-strength steels. The segregation of impurity elements such as S, As, and Sb at the grain boundaries in the nickel alloys and steels can promote intergranular HE by changing their fracture mode and the stress intensity necessary for the occurrence of HE.[185] Ti-, Nb-, or Zr-based alloys, which can form stable hydrides, tend to fracture by HE, exhibiting a stress-induced hydride formation and cleavage mechanism. HE in α-β Ti-alloy has been associated with slow tensile strain embrittlement and sustained load cracking.[187] HE in the Ti-24Al-11Nb alloy is a function of microstructure, H content, and type of hydrides formed in the microstructure.[188] Hydrogen-induced embrittlement in tensile tests has been reported for $L1_2$ intermetallic compounds such as those based on Ni_3Al, Co_3Ti, Ni_3Si, and $(Fe,Co,Ni)_3V$.[189]

HE of high-carbon austenitic stainless steels is mostly attributed to phase instability of austenitic phase with respect to cathodic hydrogen charging.[190] HE susceptibility in stainless steels has been associated with stacking fault energy (SFE) through its effect on planarity of slip and deformation twinning.[191]

HE is more likely a combination of several elementary steps of hydrogen, namely, surface entry or absorption, transport through the structure, accumulation or trapping, and decohesion, each being described by a different mechanism.

14.12.1.1 Hydrogen Entry, Transport, Trapping, and Resultant Embrittlement.

Hydrogen may be derived from hydrogen gas molecules, dissociated hydrogen molecules or atoms, or hydrogen-containing molecules such as H_2S, H_2O, or methanol. The slow crack growth behavior or severity of embrittlement observed in a high-strength martensitic steel is quite different when exposed to these three hydrogen-containing environments.

When hydrogen originates in the bulk of the alloy, hydrogen transport is a simple process and is most often controlled by the lattice diffusion process, which takes place by the movement of a screened proton that has given up its electrons to the electron gas of the metal. When hydrogen originates from an external environment, it is required to adsorb on an external surface, chemisorb, and enter the metal lattice as a screened proton.

Of all interstitial elements, hydrogen migrates fastest in metals and alloys, particularly in iron and steel. Hydrogen transport in dislocation cores or as associated Cottrell atmosphere may be several orders of magnitude faster than lattice diffusion. The transport process is important in certain phenomena such as the development of nonequilibrium internal gas pressure.[192] Grain boundary diffusion has also been suggested in the modeling of H transport during hydrogen-induced cracking of Ni and Ni-based alloys.[193]

Hydrogen trapping occurs at various depths and at a wide variety of locations in a microstructure, such as grain boundaries, carbide/matrix interfaces, inclusions, precipitates, voids, dislocations, dislocation arrays, and solute atoms.

These traps can be either reversible or irreversible according to their binding energy for H atoms. Microadditions of alloy elements have a vital effect on H diffusion in steels through trapping mechanisms. A large volume fraction of coherent ε-Ti(C,N) precipitates in microalloyed steels is the most appropriate state for the formation of irreversible trapping sites of H atoms.[194] According to Takahashi et al.,[195] the fine coherent TiC particles <100 Å are the most effective trapping sites. On the other hand, Pressouyre and Bernstein[196] observed that the incoherent pre­cipitates formed irreversible trapping sites of high binding energy ($E_B > 85$ kJ/mol).

Any of these locations which are most sensitive to fracture probably controls the magnitude of the hydrogen-induced changes. In many instances, trapping can be beneficial due to reduction of local concentration potential crack nuclei.[192]

A summary of the generalized processes which occur during HE is shown in Fig. 14.55.[192] These processes are influenced by temperature, stress state, and microstructure.[197]

14.12.1.2 Theories of Hydrogen Embrittlement.

Hydrogen embrittlement theories abound. Most theories involve the generation of atomic H at the crack tip with its subsequent absorption into the material and volume diffusion to specific microstructural sites. The accumulation of H at these sites results in a reduction in the work to fracture.[198] Hydrogen-metal interaction mechanisms can be grouped into the following categories:[186,192,199,200]

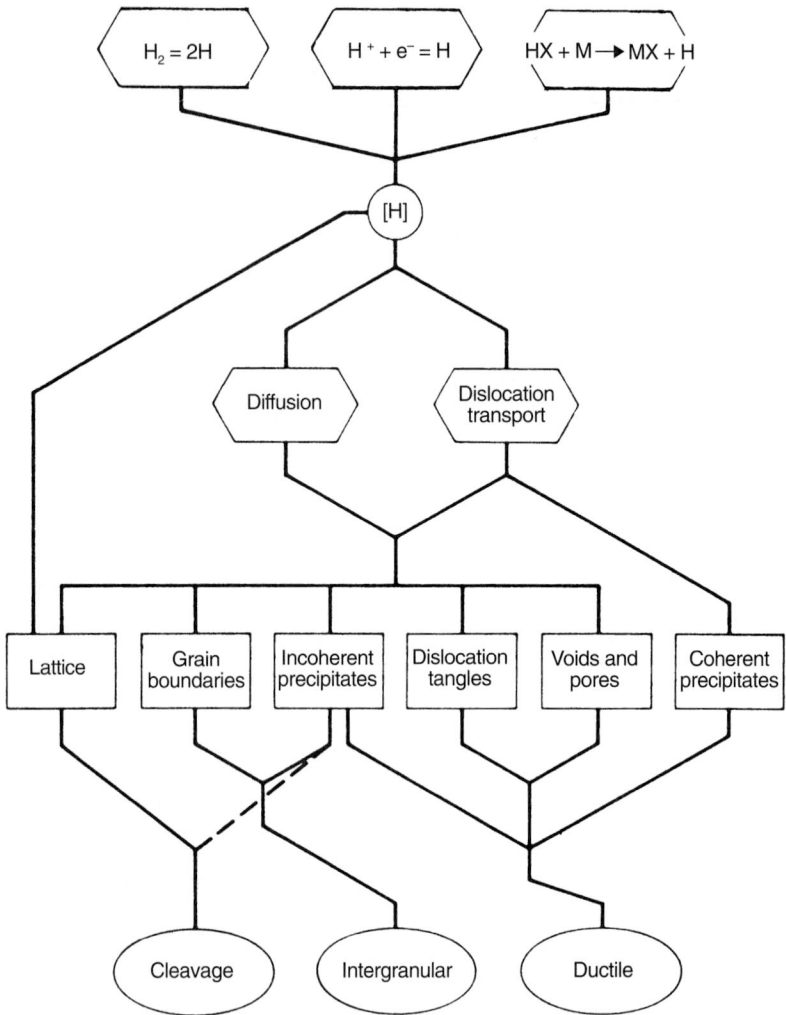

FIGURE 14.55   Summary of hydrogen sources, transport, and microstructural locations with corresponding end processes. The dashed line at the bottom refers to cleavage of hydrides.[192] (*Reprinted by permission of Pergamon Press, Plc.*)

1. *Local hydrogen pressure mechanism.* Hydrogen-dislocation interaction either hinders dislocation movement or provides localized hydrogen pressure at grain boundaries, thereby embrittling the lattice (by reducing the stress required to initiate voids or increasing the growth rate of voids). Initially it was suggested by Bastien and Azou[201] and was later modified by Tien et al.[202]

2. *Hydrogen-enhanced local plasticity (HELP) mechanism or slip softening mechanism.* Absorption of hydrogen is used to improve the generation of dislocations, mobility of dislocations, or both. This model depends on a decrease in general plasticity (or flow stress) due to the localization of H at the crack tip, where it is absorbed or drawn (or migrated) into the crack tip region by the hydrostatic stress field or perhaps by dislocations moving inward from the crack tip. This HELP mechanism of embrittlement, initially reported by Beachem,[203] and later expanded by Lynch[204] and Birnbaum and coworkers,[205,206] differs, in general, from the previous mechanisms, in which hydrogen has been assumed to improve deformation behavior locally rather than actually embrittling the lattice.[207]

3. *Decohesion mechanism.*[208–210] Interaction of dissolved hydrogen reduces the cohesive force between the atoms at high-stress region near the crack tip, leading to a decrease in stress required for fracture and eventually causing breaking of the atomic bond of the lattice at a crack tip, matrix/particle interface, or grain boundary. This model was proposed by Troiano[208] and later extended by Oriani.[209] Although observations of plasticity at the crack tip are consistent with this mechanism, the amount of plastic zone is assumed to decrease with the cohesive strength. The decohesion mechanism also holds for interfacial fracture. For the microvoid coalescence process, the decohesion mechanism could be employed to characterize the reduction in cohesive strength of the carbide/matrix interface.

According to this theory, the embrittlement effect of hydrogen is related to its concentration in the region of the steel where brittle cracking takes place. The equilibrium concentration of hydrogen C_H in the stressed crystal lattice depends on the hydrostatic tensile stress $\sigma_h [= (\sigma_{11} + \sigma_{22} + \sigma_{33})/3]$, according to the thermodynamic relation[211]

$$C_H = C_0 \exp \frac{\sigma_h \overline{V}}{RT} \tag{14.16}$$

where σ_{ii} denotes the three principal stresses; $\overline{V}$ is the partial molar volume of hydrogen in the solid solution, assumed to be $2.1\,cm^3/mol$ for iron and steel,[212] $1.72\,cm^3/mol$ for alloy 690, and $1.8\,cm^3/mol$ for X-750 alloy;[213] R is the universal gas constant; T is the absolute temperature in Kelvin; and C_0 is the equilibrium hydrogen concentration in the unstressed lattice (region), which is related to the square root of hydrogen pressure p_{H2}, for a gaseous atmosphere, in the form[214]

$$C_0 = 0.00185(p_{H_2})^{1/2} \exp\left(\frac{-Q_s}{RT}\right) \tag{14.17}$$

where Q_s is the heat of solution of hydrogen in iron, assumed to be $28.6\,kJ/mol$.[215] To estimate σ_h in front of a precrack in terms of tensile yield stress σ_y of the material, the following relation based on the elastic-plastic stress analysis of Rice and Johnson[216] should be used:

$$\sigma_h = 2.42\sigma_y \tag{14.18}$$

This theory predicts that in a given steel the K_{th} value for detectable cracking is a function of C_H, which, in turn, is a function of hydrogen pressure p_{H_2} or yield stress σ_y. This dependence is analogous to the metalloid impurities effect observed in many steels. The metalloid impurities, however, segregate during heat treatment rather than during the application of stress at room temperature.[217]

4. *Surface energy mechanism.* Reduction in surface energy due to hydrogen adsorption needed to form a crack is due to Petch.[218]

5. *Internal pressure mechanism.* There is recombination or precipitation of molecular hydrogen as a gas bubble and associated development of pressure at internal defects such as microvoids at the inclusion/matrix interface. This pressure, when added to the applied stress, leads to a decrease in apparent fracture stress. This mechanism, first suggested by Zapffe and Sims,[219] favors premature stress-assisted fracture.

6. *Hydride formation.* There is precipitation of brittle, less dense, metal hydride and its subsequent cracking near the crack tip.

In actual practice, a considerable overlap exists between these mechanisms; more than one or different combinations of mechanisms may dominate for different mechanical or environmental conditions.[192] For example, hydride can form along dislocations in the same manner as Cottrell-type hydrogen atmosphere and can change the ease and character of deformation in a structural alloy. Similarly, hydrides can preferentially form in front of a crack tip or a notch at the point of increased stress.

Several authors have provided brief reviews of the theories of HE and have shown that no one simple theory is able to explain the hydrogen degradation. In a broader sense, the decohesion theory encounters most of the observed phenomena.

14.12.1.3 *Forms of Hydrogen Embrittlement.*

1. Hydrogen-Assisted (or -Induced) Cracking, Hydrogen-Stress Cracking, and Static Fatigue. In general, hydrogen-induced cracking (HIC) occurs when the hydrogen atoms which are produced on the steel surface penetrate into the steel and precipitate at the inclusion/matrix interfaces. It has been suggested that large inclusions such as elongated MnS and clusters of oxide promote the HIC susceptibility. This type of cracking has the following characteristics:[220] (1) Cracks occur at sustained loads below the yield strength of the material and mostly in low-strength, ductile steels (or alloys). (2) All cracks are planar-oriented and occur with exhibiting tensile stress. (3) They are associated with brittle fracture and produce sharp singular cracks when compared to the extensive branching developed in stress corrosion cracking.[184]

When a steel structure containing dissolved hydrogen and/or exposed to hydrogen environment is subjected to a sustained load in service applications, it may fail at a stress level far below its tensile strength, as measured in a short-time notch tension test. This behavior is variously termed *delayed failure, delayed low-stress brittle failure, hydrogen-stress failure, hydrogen-induced delayed failure* (or *cracking*), and *hydrogen-induced cracking under static loading.* It is a manifestation of decreased fracture stress (or strain at fracture) in the presence of hydrogen, as noted above. This involves the fracture initiation in regions of highly localized stress where hydrogen is concentrated due to increased diffusion of hydrogen into triaxially stressed sites.

HIC under static loading has been recognized as the single most important design constraint for such structural applications as pressure vessels, pipelines, fasteners, and power equipment. This is also observed in many weldments. In reality, HIC (or HAC) in steel weldments is grouped into (1) *weld metal (HAZ) cracking,* from the location of initiation; (2) *macro-* (or *micro-*) *cracking,* from its size; (3) *longitudinal* (or *transverse*) *cracking,* from the direction of its propagation to the welding direction; (4) *root cracking, heel cracking, toe cracking,* or *underbead cracking,* from the weld location of initiation and propagation; and (5) *restraint* (or *distortion*) *cracking,* from the restraint condition.[220a]

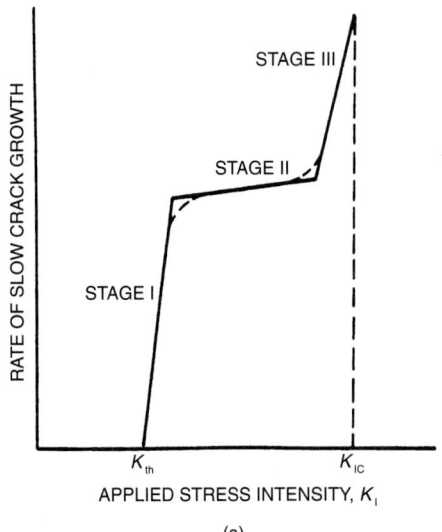

(a)

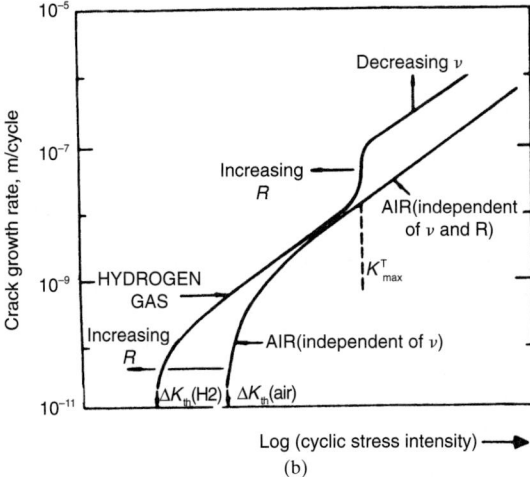

Log (cyclic stress intensity) ⟶
(b)

FIGURE 14.56 (*a*) Usual form of the hydrogen-induced slow crack growth rate for medium-to-high-strength steels as a function of applied static stress intensity.[186] (*b*) Crack growth rate behavior in gaseous hydrogen and in air as a function of applied cyclic stress intensity.[223] [(*a*) *Reprinted by permission of Academic Press, Inc., New York;* (*b*) *reprinted by permission of The Institute of Metals, England.*]

HIC can be controlled by reducing the amount of segregation of impurity elements at grain boundaries and inclusions and by accomplishing inclusion shape control.[220,221]

Hydrogen-induced slow crack growth under static loading in steels and other alloy systems exhibits the classical form of stress-intensity dependence, as shown in Fig. 14.56*a*,[186] where applied stress intensity for mode I (tensile) loading, represented by K_I, the stress intensity factor, is given by

$$K_I = \text{constant} \times \sigma \sqrt{a} \qquad (14.19)$$

where a is the effective crack length for a given applied stress σ. This plot consists of four regions. First, there is the threshold stress intensity level K_{th} below which crack growth is either very slow or nonexistent. The K_{th} value is believed to represent the maximum hydrogen degradation attainable in a material for a specific set of conditions. In high-strength steels, K_{th} is a function of the strength level of steel, temperature, and hydrogen pressure or fugacity, which indicates an equilibrium relationship between the hydrogen about or around the crack tip and that in the bulk of the alloy. In an α-phase titanium, however, K_{th} is independent of temperature, which indicates a maximum degree of degradation attainable by another hydrogen-interaction mechanism [category (6) above "hydride formation"].[222] Second, there is a region where slow crack growth rate increases rapidly with K_I and competes with the time-dependent hydrogen transport process (Stage I). Third, there is a region where the crack growth rate is nearly constant over a substantial range of K_I (Stage II). This is the result of the hydrogen transport process where slow crack growth is controlled by the rate of hydrogen transport to the area near the crack tip. Fourth, there is a region where crack growth rate increases rapidly as K_{IC} is reached (Stage III). In this regime, a failure criterion holds because of the increased mechanical contributions of the applied stress, and unstable fracture follows.[186,192]

Lower-strength steels were originally considered to be resistant to hydrogen-induced brittle crack growth under static loading.[221] However, both low-strength and high-strength steels have been found to exhibit crack growth under cyclic loading ratio $K_{min}/K_{max} = R$. Figure 14.56b[223] shows the general trends in the fatigue behavior of lower-strength steels in hydrogen gas when compared with air. Hydrogen tends to influence two important regions of crack growth behavior: (1) threshold-reducing ΔK_{th} and (2) growth measured by the Paris law, causing a steep rise in growth rate above a critical K_{max} level. The latter effect is usually associated with the onset of intergranular fracture.[223,224]

Yurioka and Suzuki[224a] have provided a review on hydrogen-assisted cracking in C-Mn and low-alloy steel weldments.

2. Delayed Failure. Above a tensile strength level of 1241 MPa (180 ksi), most high-strength low-alloy steels, notably AISI 4130 and 4340, and precipitation-hardening stainless steels are prone to hydrogen stress cracking in marine environments, when the applied or residual tensile stresses are quite high, and cracking normally takes place as a form of delayed failure.[184]

The delayed-failure behavior of steel has been extensively studied using static loading of precharged notched sample. Figure 14.57[199] shows a schematic representation of delayed failure in a high-strength steel, illustrating two stress-time curves, one for crack initiation and another for failure. These curves are bounded by two stress levels, an upper critical stress (UCS) and a lower critical stress (LCS), which correspond to the maximum and minimum stress levels, respectively, causing delayed fracture. At stresses above the UCS, failure takes place without a time delay; and at stresses below the LCS, hydrogen is not harmful. At intermediate stress level, failure occurs as a series of events consisting of crack initiation following an incubation period and crack propagation.

The important factors responsible for the delayed-fracture behavior of precharged steels are the hydrogen concentration (or potential or fugacity), the strength level or composition of steel, notch acuity, temperature, and grain boundary impurity concentration. An increase in either hydrogen concentration or the strength level usually causes a decrease in both the UCS and the LCS and leads to shorter delay times for failure.[199] Delayed failure of high-strength ferritic steels in hydrogen is believed to occur by diffusion of internal or external hydrogen to a region of high local constraint ahead of the crack tip.

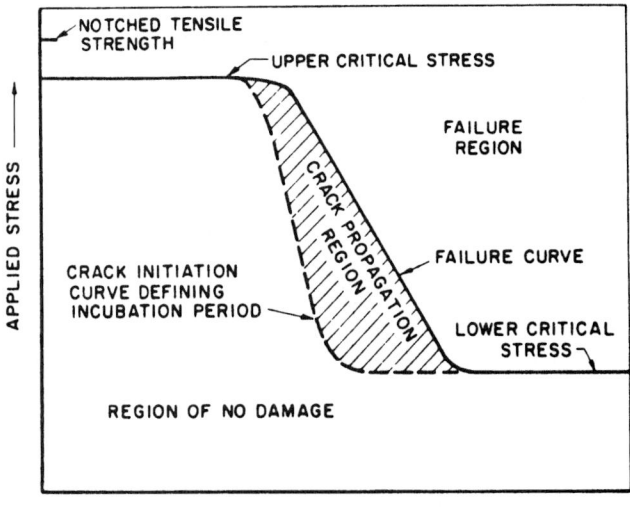

FIGURE 14.57 Schematic illustration of delayed-failure behavior of notched tension high-strength steel specimens containing hydrogen.[199] (*Reprinted by permission of ASM International, Materials Park, Ohio.*)

Delayed hydrogen cracking behavior has also been observed in zircaloy-2 pressure tubing during reactor service, where H is generated by corrosion processes and absorbed by these materials and eventually leads to precipitation of zirconium hydride platelets.[225]

GRAIN BOUNDARY IMPURITY CONCENTRATION. It is generally recognized that hydrogen-induced delayed failure (or cracking) occurs in quenched and tempered medium- and high-strength steels when stressed in a relatively low-hydrogen fugacity. It has been found that the stress intensity K_{th} required to produce a detectable amount of crack extension decreases with the increasing grain boundary impurity (such as P, S, and Sb) concentration, which lowers the intergranular cohesion. This is accompanied by a shift in cracking mode from displacement-controlled transgranular fracture at a high K_{th} level (in the unembrittled condition) to the stress-controlled intergranular fracture mode at a low K_{th} level (in the embrittled condition).[226,227] For a given yield strength and hydrogen pressure (or concentration), this type of hydrogen-induced delayed failure is believed to be mainly an impurity effect. The hydrogen and impurity effects tend to be simply additive. This is the minimum effect which can be expected, but no such details have yet been established.[228,229]

EXPERIMENTAL RESULTS. In a study of HY130-type steel (composition: Fe-5Ni-0.5Cr-0.5Mo-0.1V-0.1C) using static loaded edge-notched and precracked cantilever beam specimens, Yoshino and McMahon[226] demonstrated that TE by step cooling reduced drastically the K_{th} for crack growth from a high level, ~104.5 MNm$^{-3/2}$ (95 ksi$\sqrt{\text{in.}}$), to a quite low level, ~22 MNm$^{-3/2}$ (20 ksi$\sqrt{\text{in.}}$) (in a cathodically charged hydrogen atmosphere), in a 0.1 N H$_2$SO$_4$ solution and shifted the mode of cracking from transgranular fracture to a completely intergranular mode. They interpreted their results, in terms of Oriani's theory of HE, to occur from an interaction between

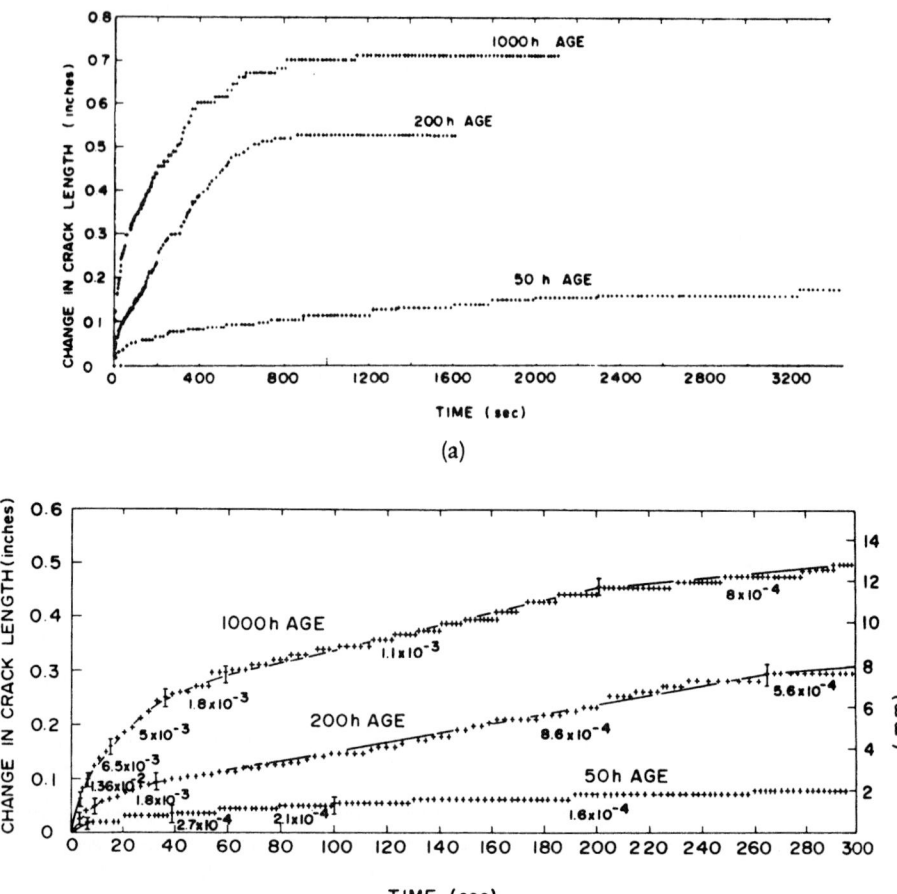

FIGURE 14.58 (*a*) Examples of crack growth data for samples aged for 50, 200, and 1000 hr, showing the decrease in crack growth rate as K_{th} is approached. (*b*) Crack growth rate data plotted on an expanded time scale show clearly the discontinuous nature of crack growth.[228] (*Reprinted by permission of The Metallurgical Society, Warrendale, Pa.*)

hydrogen-induced (delayed) cracking and weakening of grain boundaries by embrittling impurities. In another study, Briant et al.[228] used precracked bolt-loaded, wedge-opening-loaded (WOL) HY130 steel samples (1T-WOL) at fixed displacements, to determine the crack length near the K_{th} level. They measured the variation in K_{th} and crack length with time in these specimens aged for various times (i.e., grain boundary concentrations) and stressed to various initial K levels at a fixed temperature and hydrogen pressure (Fig. 14.58). These data show that hydrogen-induced (delayed) cracking proceeded in a stepwise fashion. All these results support the hypothesis that hydrogen-induced (delayed) cracking at low K_{th} levels in these quenched and tempered steels is stress-controlled intergranular decohesion

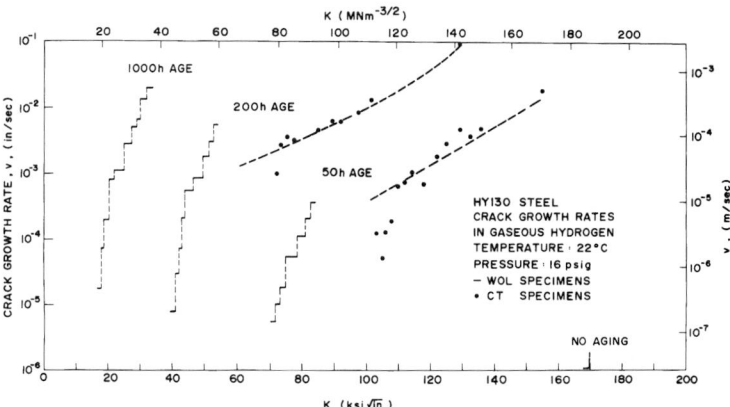

FIGURE 14.59 Variation of crack growth rate with stress intensity obtained from bolt-loaded WOL specimens (segmented curves) and from CT specimens (individual data) points.[228] (*Reprinted by permission of The Metallurgical Society, Warrendale, Pa.*)

which is essentially related to the presence of embrittling impurities in the grain boundaries. In the absence of the impurity effect, cracking would occur by displacement-controlled transgranular mode at high K_{th} levels, often referred to as *quasi-cleavage*. The hydrogen-plus-impurity effect can be rationalized in terms of the Oriani theory. To measure the crack velocity at $K > K_{th}$, compact test (CT) specimens were used at a fixed load, and the stress intensity was increased with an increase in crack length. Figure 14.59 shows macroscopic average crack velocity V versus stress intensity, illustrating well-defined K_{th} values below which the crack growth rate dropped below 10^{-6} m/s. Thus K_{th} values definitely decreased with the increased aging times (i.e., increased grain boundary impurity concentration). At K values just above K_{th}, the region of steeply rising crack velocity V is called the Stage I crack growth, while the regions where crack velocity appears to level off represent the Stage II crack growth.[228]

In 4340 and 300 M steels, following TME, similar studies have shown that high-purity Ni-Cr-Mo-C steel exhibits transgranular fracture mode at a higher K_{th} value, representing the intrinsic effect of hydrogen in these steels; the commercial grades, however, show greater susceptibility to hydrogen-induced delayed cracking at low K_{th} ($<30\,\mathrm{MNm}^{-3/2}$), representing the impurity effect. It is presumed that the addition of Mn and Si has caused the HE due to their effect on the intergranular segregation of metalloid impurities in these steels.[230]

3. Sulfide Stress Cracking. High-strength steels and alloys are necessary in the oil industry for deep sour gas production operations. However, the hydrogen-induced embrittlement or cracking of these materials occurs by externally applied stress and externally entered hydrogen through the corrosion reaction in wet H_2S-bearing sour environments and is designated as *sulfide stress-cracking* (SSC). This is a special case of hydrogen-stress (or -induced) cracking. The fracture mode of SSC consists of intergranular and/or transgranular. Hydrogen entering the steel recombines at type II manganese sulfide inclusion/matrix interfaces, generates hydrogen gas-pressurized cavities, and causes SSC.[231]

SSC occurs in steels exposed to a wet H_2S environment. In this case, a sulfide film, FeS_x, and nascent H develop at the steel surface according to the reaction

$H_2S + Fe \rightleftharpoons FeS_x + 2H$. The presence of adsorbed sulfur inhibits the surface recombination of atomic H, which leads to enhanced H permeation. The combination of internal or applied stresses and a susceptible microstructure are the determining factors for the critical H buildup necessary for crack initiation and growth.[232]

Steels with UTS < 690 MPa (100 ksi) seem to be resistant to hydrogen stress cracking, and the structures made of such steels have been employed in service without facing serious problems in various H_2S-free environments.

The susceptibility to SSC increases with the increase of H_2S concentration or partial pressure and decreases with the increase of pH. The ability of environment-causing SSC decreases substantially above pH 8 and below 101 Pa (0.001 atm) partial pressure of H_2S. The SSC susceptibility is maximum at room temperature and decreases with the increase of temperature. It is usually increased by (1) the presence of oxide and sulfide inclusions, (2) untempered martensite or bainite, and (3) impurities such as C, P, and Mn which produce segregation bands. It is well recognized that a uniform microstructure of fully tempered martensite is desirable for SSC resistance.

The effect of alloying elements on the SSC resistance of carbon and low-alloy steels is controversial, except that Ni is deleterious to SSC resistance. Steels with Ni > 1% are not suggested for service in sour environments.[232]

The SSC susceptibility of weldments seems to be greater than that of the base metal, and the high hardness and residual stresses arising from welding are believed to enhance the susceptibility. Recently, National Association of Corrosion Engineers (NACE) has issued a series of guidelines for safe operations in sour environments (NACE MR-0175-93). Accordingly, the strength of corrosion-resistant alloys (CRAs) for sour service has been restricted to maximum yield strength of 690 MPa and maximum hardness of 22 Rc.[232] Guidelines for dealing with hydrogen stress cracking that occurs in refineries and petrochemical plants have also been developed by the American Petroleum Institute (API).[184]

High-strength low-alloy steels such as C-90, C-100, and C-120 grades with improved SSC resistance have been used for the last two decades. These steels have minimum yield strengths of 620 MPa (90 ksi), 690 MPa (100 ksi), and 830 MPa (120 ksi), respectively; and they are used as oil- and gas-well tubing, casing and coupling, and advanced tool joints for deep sour-well drilling and production.[233]

Recently, the SSC resistance of API X-80 steel has been improved by microstructural modification involving water quenching from annealing temperature (852°C) and tempering for 1 hr at 600°C. This improvement is attributed to the increased number of H traps provided by the resultant fine distribution of (Nb,Ti)-containing carbides, which causes a tortuous crack path along ferrite interlath grain boundaries.[234] The development of X-100 linepipe steel has also been reported by Okatsu et al., based on optimum microstructural control by thermomechnical treatment.[234a]

4. Hydrogen-Induced Ductility Losses. The presence of dissolved hydrogen can often result in a loss of ductility, as measured by a decrease in the reduction of area (RA) value in a smooth tensile test specimen. Such ductility losses may take place with or without a change in the fracture mode, compared to a hydrogen-free test. This mode of failure is mostly observed in lower-strength alloys, and the embrittlement index, often quoted by the percent RA loss, is given by

$$RA \text{ loss} = \frac{RA - RA_H}{RA} \times 100 \qquad (14.20)$$

where RA and RA_H denote the values for uncharged and hydrogen charged specimens, respectively.

In addition to the effect of the hydrogen concentration, the measured tensile properties are affected by the temperature and strain rate of the test, the stress concentration of the specimen, the extent of prior cold work or work-hardening rate, and the strength level of the steel. Loss in tensile ductile behavior is more pronounced when strain rate decreases, from a normal strain rate (i.e., at $10^{-4}\,s^{-1}$) to a slow strain rate (i.e., on the order of $10^{-7}\,s^{-1}$ and below) testing. Hence, this behavior can pose the potential service problems under static loading and quasi-creep conditions,[192] but not during impact tests, such as the Charpy V-notch test. The temperature of minimum ductility has been found to increase with the increase in strain rate. This sensitivity to strain rate and temperature clearly demonstrates that the mechanism which leads to the ductility changes involves transport and trapping processes.[199]

As the notch acuity increases, the reduction in notch strength at a given value of the other factors will be greater due to increased strain localization.

Although a function of temperature, the work-hardening rate increases in the presence of hydrogen, with a maximum influence near ambient temperature.[192] In higher-strength steels, comparatively small hydrogen concentrations can produce large property changes, while in lower-strength steels the effect of hydrogen decreases.

14.12.1.4 *Control of Hydrogen Embrittlement.*
Hydrogen embrittlement (HE) of structural alloys is very complex and very specific, and it can be influenced by a large number of variables. To control HE in any structural applications, the HE process must be understood from the start to the end.[186] Numerous steps are used to control the HE:

1. Melting and casting practices and subsequent finishing operations such as pickling, plating, heat-treating, and welding should be ensured so that steel is free of residual internally dissolved hydrogen. For example, flaking can be eliminated by using vacuum stream degassing in steelmaking where the hydrogen content is held below 2 ppm.

2. To keep steel surface away from the accelerated entry of hydrogen, all hydrogen produced on the surface should be prevented from entering the material in deep sour oil and gas wells, which produce improved resistance to SSC.[235]

3. To control HE in applications involving aqueous pit and crevice corrosion, hydrogen concentration should be reduced by means of electrochemical reactions.[217]

4. Embrittlement becomes more severe with an increase in strength level, and strengths below 700 MPa do not show any marked embrittlement. Improved alloy design (rather than relying only on external hydrogen control) such as C-90, C-100, and C-120 high-strength steels should be used in tempered condition in petrochemical and ammonia industries, which will substantially decrease the susceptibility of steel to SSC.

5. C, Mn, and Si shift the crack-tip metal solution potential toward cathodic values so they increase steel sensitivity to HE. Hence, unnecessary increase of C, which induces low K_{th} values, should be avoided, and Mn and Si should be used with caution.[122a] Cr addition, which promotes grain boundary segregation of P, should be reduced. Mo and Ti should be used, which scavenge P and Al, which scavenges N. Increased Ni content, which promotes inherent toughness of the material, should be used.[230] Concentrations of sulfur and trace impurities in high- and low-strength steels should be reduced, and the addition of rare earth elements, such as La and

Ce, in amounts above 0.15 wt% should be used to improve substantially the resistance to HE.

6. *Microstructural control.* The microstructure and phase distribution plays a key role in determining the kinetics of HIC.[236] Fine and uniform structures of quenched and tempered steel and bainite are the most resistant to HE; spheroidized structure, intermediate resistant; ferritic or pearlitic (normalized) structure, somewhat less resistant; and untempered martensitic structure, the least resistant. The banded microstructure in ferrite-pearlite steel, being rich in Mn, is susceptible to HE. For example, carbon steels in the quenched and tempered conditions are more resistant to hydrogen attack (HA) than those in the annealed condition. This suggests that the presence of discrete carbide particles rather than continuous carbide films along the boundaries decreases the susceptibility to HA.[237]

Among the quenched, ferritic, and austenitic structures, the austenitic structures are just a little susceptible to HE. This is due to their higher H solubility, low H permeability, usual low sensitivity to notch effect, and low yield strength.[238]

SSC resistance is increased by (1) fine prior austenite grain size, (2) fully as-quenched martensitic structure, (3) tempering at high temperature, (4) rapid cooling from tempering temperature,[239] and (5) thermomechanical treatment.

Shape control of sulfide inclusions from crack-like flat to globular form increases substantially the steel's resistance to hydrogen-induced blistering and SSC.[239]

7. *Metallurgical modification.* Methods for mitigating HE by metallurgical modification include alloy additions which (*a*) favor adhesion at the solid/solid interfaces [e.g., by segregation of elements which tend to improve cohesion (B, C, or N) by displacing decohesion-improving elements (P, S, Sn) and consequently increase the critical H level needed to produce interfacial decohesion, or by interfacial segregation of elements which displace H from interfaces],[240] (*b*) reduce H entrance, thereby reducing the dissolved H concentration in the lattice,[241] or (*c*) force a redistribution of H trap that is partitioned between lattice sites and trapping sites, so that the critical H content required to induce interfacial decohesion is not easily reached at the same H fugacity.[242] In this way, Pd addition up to 1 wt% is shown to significantly change the hydrogen-assisted cracking phenomenon of PH 13-8 Mo stainless steel by suppressing intergranular hydrogen cracking.[243] Other workers have also obtained similar results in other steels (e.g., quenched and tempered AISI 4130 steel).[244]

8. Adsorption of O_2, SO_2, and CO, as well as organic inhibitors, appears to be effective in some metallic systems to hinder gaseous HE.

9. *Stresses.* A change in deformation mode, extent of cold-work, and increased applied stress can have a large influence on HE.[245]

10. *Passive oxide coating* consisting of TiO_2 and Al_2O_3 on Ti-45 at% Al [two-phase titanium aluminide, Ti_3Al (α_2) + TiAl (γ)] and Ti-50 at% Al alloys has been observed to be effective in preventing H penetration (or occlusion).[246] In the case of iron aluminide, small addition of Cr is effective in minimizing HE and improving ductility.

11. *Thermomechanical treatment* (*TMT*). High-temperature thermomechanical treatment (HTMT) reduces the susceptibility of steel to HE by favoring a change from intergranular to intragranular fracture mechanism. Presumably, impurity concentration on grain boundaries falls during the hot deformation step of HTMT due to its redistribution between boundaries and the substructure, which improves the resistance to development of grain-boundary crack.[247] It has also been shown that TMT plays a significant role in improving the mechanical properties of Fe-25Al-1B intermetallic alloy.[248] Agarwal et al.[249] demonstrated that partially recrystallized microstructure achieved by TMT in Fe-25Al iron aluminide also prevented H ingress through grain boundaries, thereby minimizing HE.

12. *Reversibility of HE.* HE, attributed to plating and cleaning, is reversible in that damage can be eliminated by baking treatment which disperses, diffuses out, or removes H, and consequently the time to failure and the lower critical stress limit increase. Appropriate baking time is a function of material hardness, plating processes, coating type, and coating thickness.[250]

14.12.2 Hydrogen Attack

Prolonged exposure of pressure vessel or piping steels to high-pressure hydrogen at elevated temperatures (>200°C) such as in petrochemical plant equipment[†] and hydrosulfurization reactors can develop a network of internal cracks and result in external surface decarburization by contact with H and internal decarburization by H permeation within the steel and consequent reaction with iron carbides to form methane gas bubbles, mainly along the grain boundaries. Driven by the (internal) methane pressure, cavities grow due to grain boundary diffusion and dislocation creep. This results in a progressive development or linking up of cavities to form intergranular *fissures* (or *cracks*) at grain boundaries or enlarged pores in the metal matrix. In cold-worked steels, bubbles can form within grains, probably initiating at voids formed at carbide/ferrite interfaces.[250a,251] This link-up of bubbles or cavities ultimately produces a substantial amount of swelling (breakaway stage) or blistering and premature failure at elevated temperature. This mode of material degradation, occurring at elevated temperatures, is called *hydrogen attack* (HA) and produces irreversible damage. The severity of hydrogen attack is a function of temperature, hydrogen partial pressure, exposure time, stress level, steel composition, and microstructure of the steel, especially its alloy carbides M_xC_y.

Internal decarburization and surface decarburization both result in marked reduction in strength; however, the former will tend to decrease the ductility of a steel, while the latter will tend to increase the ductility. Thus hydrogen attack can be a limiting design problem in both petroleum and synthetic ammonia industries.[192] This process of decarburization may accelerate cavity growth.

Fissures by HA are initially microscopic, and in advanced stage a large number of fissures lead to a large reduction in mechanical properties. A complete decarburized and fissured steel component may have its UTS of 41.3 MPa (60 ksi) reduced to 172.4 MPa (25 ksi) and its test bar elongation reduced from 30% in 2-in. (51-mm) sample to nil. Figure 14.60a is a microstructure of a C-0.5Mo steel sample damaged by HA comprising initial decarburization and cracking when exposed to service conditions of 421°C (790°F) at hydrogen partial pressure of 2.93 MPa (0.425 ksi) for approximately 65,000 hr in a catalytic reformer.[252]

The conditions under which different steels can be employed in high-temperature hydrogen service are listed in the American Petroleum Institute document 941, the January 1997 edition. Figure 14.61 shows the recent and modified Nelson curves developed by API. The main data are presented by a set of empirical operating curves on a T-P plot, known as *Nelson curves* which have been used by the industries for over 66 years[252,252a] to keep the hydrogen attack under control. These solid curves[253] provide the operating limits for plain carbon and low-alloy steels for various combinations of temperature and hydrogen partial pressure, above

---

[†] They handle hydrogen-hydrocarbon streams at pressures and temperatures up to 21 MPa (3 ksi) and 540°C or 1000°F, respectively.

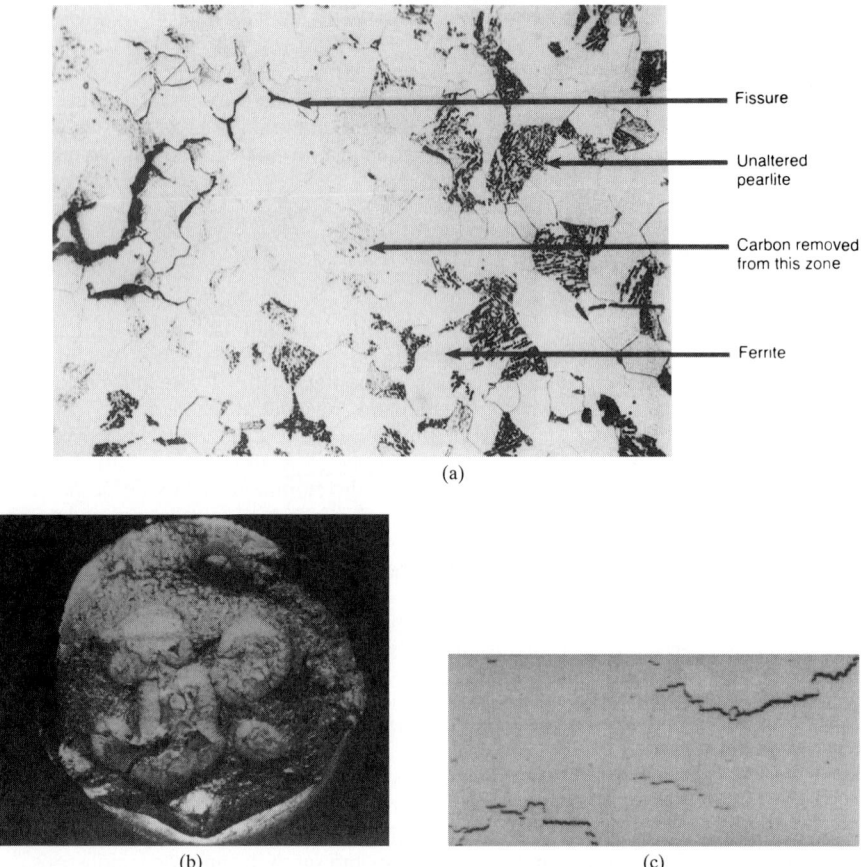

(a)

(b)   (c)

FIGURE 14.60 (*a*) Microstructure of a C-0.5Mo steel (ASTM A 204–A) specimen showing internal decarburization and fissuring in hydrogen attack. Service conditions: 65,000 hr in a catalytic reformer at 421°C (790°F) and 2.95 MPa (425 psi) absolute hydrogen partial pressure; nital etch. 520X.[252] (*Courtesy of American Petroleum Institute.*) (*b*) Fisheyes in as-welded E7018 tensile specimen tested at room temperature (4X).[184] (*c*) Stepwise cracking of a low-strength pipeline steel exposed to hydrogen sulfide (6X, shown here at 65%).[184]

which HA associated with internal decarburization and internal cracking will be observed and below which operations may be safely conducted in periods of plant operations. Although the curves establish a useful guideline, a safer approach uses alloys with (1) the addition of carbide stabilizers such as Cr, Mo, W, Ti, V, and Nb, thereby making methane formation thermodynamically and kinetically more difficult; (2) decreased carbon content to increase its resistance to hydrogen attack; (3) elimination of slags, segregated impurities, stringer-type inclusions, or laminations; and (4) neither inclusion in welds nor HAZs since they are more prone to hydrogen attack than the base or weld metal. However, HAZ susceptibility to HA can be minimized after tempering at 690°C for 1 hr for 2.25Cr-1Mo steel.[254]

Additional methods to minimize hydrogen attack are as follows:

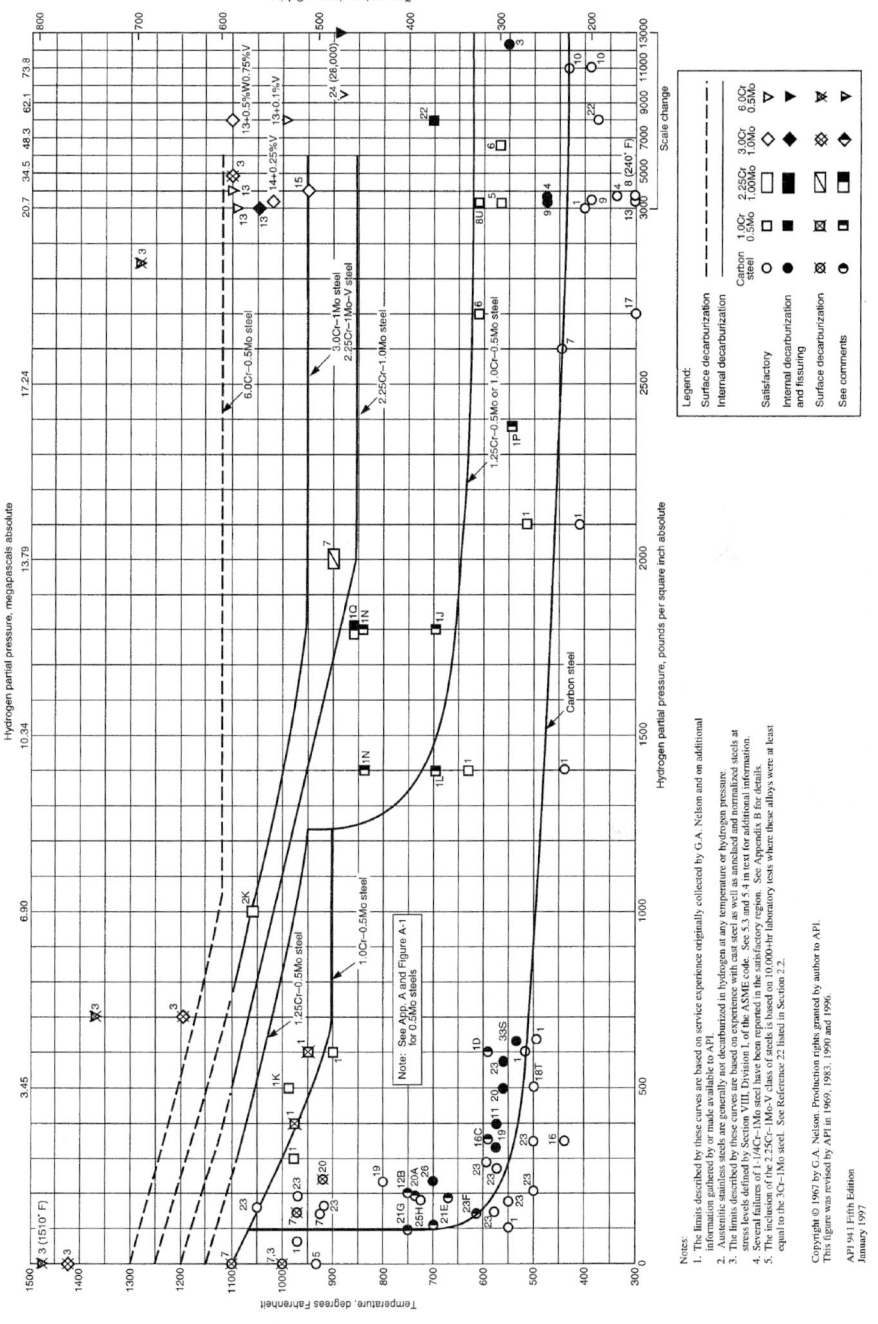

Notes:

1. The limits described by these curves are based on service experience originally collected by G.A. Nelson and on additional information gathered by or made available to API.

2. Austenitic stainless steels are generally not decarburized in hydrogen at any temperature or hydrogen pressure.

3. The limits described by these curves are based on experience with cast steel as well as annealed and normalized steels at stress levels defined by Section VIII, Division 1, of the ASME code. See 5.3 and 5.4 in text for additional information.

4. Several failures of 1-1/4Cr–1Mo steel have been reported in the satisfactory region. See Appendix B for details.

5. The inclusion of the 2.25Cr–1Mo-V class of steels is based on 10,000-hr laboratory tests where these alloys were at least equal to the 3Cr–1Mo steel. See Reference 22 listed in Section 2.2.

Copyright © 1967 by G.A. Nelson. Production rights granted by author to API.
This figure was revised by API in 1969, 1983, 1990 and 1996.

API 941 Fifth Edition
January 1997

FIGURE 14.61 Operating limits developed from industrial experience for various steels exposed to hydrogen-containing environments at elevated temperatures to avoid decarburization and fissuring.[252] (*Courtesy of American Petroleum Institute.*)

1. Addition of ethylene (up to 0.5%) and ethane decreases hydrogen permeation at high temperature (>200°C) in iron by a factor of 50 according to a reversible process.

2. The 1-2 dibromoethylene is even better, because it decreases the permeation rate by a factor of 90.

3. Use of low-alloy Cr-Mo-containing steels, for example, 2.25Cr-1Mo steels under hydrogen pressure of 9.8 to 24.9 MPa (100 to 300 kg/cm^2) at temperatures below 450°C, which is the critical temperature of a Nelson curve, provided that the weld joints have been completely stress-relieved.[255]

In the case of carbon steel, Shewmon and Xue have experimentally observed that (1) high-pressure H significantly increases the rate of crack growth at elevated temperature near the Nelson curve; (2) when K_1 is reduced to zero, the crack growth ceases; (3) a spheroidized structure is much more resistant to hydrogen-assisted crack growth than is a normalized (ferrite + pearlite) structure; and (4) metallographic examination reveals that the fracture comprises a mixed grain boundary and transgranular mode and displays little branching.[251]

14.12.3 Shatter Cracks, Flakes, and Fisheyes

Shatter cracks, *flakes*, *fisheyes*, *hairline cracks*, and *white spots* are common features of hydrogen damage in forgings, castings, and weldments. These internal defects in heavy sections are attributed to hydrogen pickup during steelmaking where moisture is present in the atmosphere and additives. These defects appear when steel cools below ~200°C (~392°F), at which temperature the amount of hydrogen degradation is operative.[199]

A thermal flake is a tight crack formed by the combined action of H and stress, always fully contained within a steel section, usually appearing as a disk or hairline crack in the central portion unless directional stresses or localized weaknesses change the shape. Flakes usually form during cooling after the first rolling or forging and not during cooling following solidification. Flakes are normally oriented within the forging grains or segregated bands. Flake susceptibility increases with the increase of H content. Medium and heavy flaking can be identified by ultrasonic pattern, using longitudinal waves.[256]

It is believed that hairline cracks appear in those steels melted in nonvacuum furnaces when measures such as slow cooling or isothermal annealing of slabs after rolling have not been carried out. It has been reported that HSLA steels with (1) higher Ni and Mn levels and/or (2) segregation of Mn, V, and Si are usually flake-sensitive at low-carbon levels. The flakes observed in such HSLA plates are present along grain boundary and other banded low-temperature transformation microstructures.[256a]

Fisheyes, being another form of localized HE, are described as small shiny spots frequently found on the fracture surface of tension specimens taken from steel forgings, plates, or welds (Fig. 14.60*b*) containing a high H level, which have the propensity to reduce tensile ductility. Fractographic examination generally exhibits fracture-initiation sites such as pores or inclusions, related to fisheyes. Baking or extended room-temperature aging of tension specimens mostly removes fisheyes and restores tensile ductility.

When this type of hydrogen damage takes place in welding, it is termed *underbead cracking* or *delayed cracking*, which develops in the HAZ region of the base metal (after several hours or days following welding) and runs nearly parallel to the fusion line. (See also Secs. 3.10.3.4 and 14.12.1.3 for more details.) The factors responsible for this type of cracking are dissolved H (arising from the arc atmosphere by the shielding gas, flux, or surface contamination), low-ductility (martensite) structure, and tensile stresses. The stresses produced by external restraint and by volume changes during transformation can readily develop cracks in this region.[184]

14.12.4 Blistering

Hydrogen-induced blistering is more common in low-strength steels, and it is observed in metals exposed to sufficiently higher hydrogen-charging conditions such as acid pickling, electroplating, cathodic processes, or in-service corrosive environments comprising H_2S.

If these damage processes occur at the surface or just below it in the interior, the hydrogen gas pressure in the cracks can lift up and bulge-out in blistering. When these blisters are on a line following plane precipitation, it may lead to step cracking, an opening of the wall, and a lower mechanical behavior of the steel part.

These irreversible damage processes were a common problem two decades ago in such applications as rails, plated parts, and enameling steels. Proper outgassing methods and compositional control have reduced its occurrence. However, sometimes they occur in line-pipe steels exposed to gases such as H_2S and moisture, which can cause high hydrogen fugacities.[192] In the refinery, H-induced blistering has been observed most often in vessels handling sour (H_2S-containing) light hydrocarbons and in alkylation units where HF acid is employed as a catalyst. Storage vessels containing sour gasoline and propane are most susceptible to blistering; however, sour crude storage tanks are less susceptible to blistering, perhaps due to the corrosion-inhibiting effect of oil film of the heavier hydrocarbon.

In storage vessels, blistering usually occurs at the bottom or in the vapor space containing water. Gas-plant vessels in catalytic hydrocarbon cracking units are very susceptible to blistering due to the generation of cyanides by the cracking reaction.

Hydrogen-induced blistering also takes place on steel plates as cathodes in industrial electrolysis.

Hydrogen blistering is often associated with HE in low-strength steel subjected to H_2S-containing environments in the unstressed condition. Internal hydrogen blistering on a microscopic scale along grain boundaries (fissures) can result in hydrogen-induced stepwise cracking (Fig. 14.60c). Cracking advances with the cracking of metal ligaments between adjacent fissures. In general, the severity of hydrogen blister is a function of the severity of corrosion, but even low corrosion rates can generate adequate hydrogen to cause extensive damage. In some instances, hydrogen blistering is restricted to dirty steel with highly oriented slag inclusions or laminations. Vapor/liquid interface areas in equipment frequently exhibit most of the damage, perhaps because NH_3, H_2S, and HCN concentrate in the thin water films or in water droplets that collect at these areas.[184]

14.12.5 Other Forms of Hydrogen Damage

Microperforation by high-pressure hydrogen takes place in steels at very high hydrogen pressure near ambient temperature. This form of hydrogen damage exhibits a network of small fissures that favor permeation of the alloy by gases and liquids.[184] *Degradation in flow properties* occurs in iron and steel in hydrogen environment at room temperature.[184] *Metal hydride formation* usually occurs in hydride-forming metals such as Ti, Nb, and Zr.[184]

14.13   METAL-INDUCED EMBRITTLEMENT

Metal-induced embrittlement is grouped into (1) liquid-metal embrittlement, where contact with liquid-metal embrittles a solid metal, and (2) solid-metal embrittlement, where contact with a solid metal just below its melting temperature embrittles a solid metal.[257]

14.13.1   Liquid-Metal Embrittlement

The exposure of a normally ductile solid metal or an alloy to a liquid-metal environment results in a thin film of liquid-metal coating, and subsequent stressing in tension may cause a reduction in the ductility or fracture stress together with a catastrophic failure by brittle intergranular mode or transgranular (cleavage) mode.[257–259] This phenomenon is referred to as *liquid-metal embrittlement* (LME). LME has received much less attention than HE. LME was first recognized in α-brass (embrittled) by mercury in 1914 by Huntington.[260] Since then, LME has been observed in various failure analyses.[118]

There are four distinct forms of LME:[118]

1. A sudden fracture of a certain metal occurs under an applied or residual tensile stress as a result of contact with a certain liquid metal. This is the common type of LME.

2. Delayed failure of a certain metal in contact with a particular liquid metal occurs after a fixed time period under static loading below the tensile strength of the metal. This form of LME is due to the liquid-metal penetration along the grain boundary and is not as common as the previous one.

3. Grain boundary penetration of a specific solid metal in the unstressed condition by a specific liquid metal causes the final disintegration of the solid metal.

4. Elevated-temperature corrosion of a solid metal by a liquid metal leads to embrittlement. This form is quite different from the others.

LME occurs only in combinations of specific liquid-metal and specific solid metal. For example, liquid mercury embrittles Zn but not Cd; liquid Ga embrittles Al but not Mg; liquid Li, Al, Cd, Cu, brass, bronze, Sb, Te, Ga, In, Zn, and Hg embrittle steel but not Na, Se, and Th.

The prerequisites for the occurrence of LME are:[118,258] (1) a good intimate contact or wetting between the surface of the solid metal and the liquid metal, i.e., complete coverage by the liquid metal which is usually difficult to remove; (2) an applied or residual tensile stress; (3) some measure of plastic flow and some stable obstacle to dislocation motion (or plastic flow) at the solid/liquid interface; and (4) little or no mutual solubility and absence of intermetallic compound formation. However, there are some exceptions.[257]

Additional factors that promote LME in solid metal are the presence of a sharp notch or stress raiser, high strain rate, coarse grain size, and the test temperature. LME does not depend on the purity of the liquid or its presaturation with the solid or on the time of exposure to the liquid metal.[257]

14.13.1.1   LME of Steel. This section provides the information on the embrittlement of a wide range of steels by various liquid embrittlers.

Aluminum Embrittlement. Tensile and stress rupture tests on steels exposed to liquid aluminum at 690°C (1275°F) for a short time revealed a selective attack with matrix corrosion, as reported by Radekar and others.[257,258]

Antimony Embrittlement. Bend tests were conducted by Shottkey et al. on plain carbon and silicon and chromium steels, and it was found that exposure to liquid antimony in the temperature range of 540 to 649°C (1000 to 1200°F) produced embrittlement. Fatigue testing of 4340 steel in the liquid Pb-35Sb at 540°C (1000°F) and in antimony at 675°C (1250°F) exhibited very severe embrittlement.[257]

Bismuth Embrittlement. Bismuth embrittlement of mild steel was observed by Tanaka and Fukunaga, on testing only at higher temperatures with maximum embrittlement and DBTT occurring at 350 and 550°C (662 and 1022°F), respectively.

Embrittlement by Brazing Alloy. Embrittlement of mild steel by brazing alloy was observed by Genders during bend tests at 900°C (1652°F). Riede reported failures in thin-walled steel tubing during dip-brazing operations.[258]

Cadmium Embrittlement. Delayed failures were observed by Iwata and Asayama in a range of cadmium-plated high-strength steels such as 4130, 4140, 4340, and 18 Ni-maraging steels down to 232°C (450°F), which is about 90°C (160°F) below the T_m of Cd. Radekar reported embrittlement along grain boundaries in a series of steels which were produced by pure cadmium at 350°C (662°F). He noted that the addition of 8 and 36% Zn to the cadmium enhanced the sensitivity to embrittlement at 400°C (752°F), while additions of 0.55% Al or 2% Ni did not produce any significant effect.

In the case of electroplated and vacuum-deposited cadmium, fatigue limits of 10 and 60% of room-temperature strength, respectively, have been reported at 300°C (570°F), and the catastrophic failure was associated with a transgranular ductile fracture.[257] Cadmium has been recognized as a more potent solid-liquid embrittler than lead, tin, zinc, or indium.[257]

Copper Embrittlement. During the hot-working of some steels, embrittlement by copper was noticed by several workers[258] in the range of 1100 to 1300°C (2010 to 2370°F). This occurred by diffusion-controlled grain boundary permeation of copper and associated dissolved alloying elements such as Ni, Mo, Sn, and As during oxidation. In another study made by Hough and Rolls, copper remarkably changed the creep behavior of notched pure iron specimen and caused intergranular fracture. Embrittlement was of the delayed type and occurred by penetration of copper along the prior austenite grain boundaries. The liquid copper appeared in front of the advancing surface cracks and was believed to favor the initiation and growth by reducing the cohesive strength of the boundaries and promoting grain boundary sliding.[118,258] Hot tensile testing in a Gleeble testing machine at high strain rates exhibited severe copper embrittlement in AISI 4340 steel (Fig. 14.62).[257]

Gallium Embrittlement. Severe embrittlement of iron alloy, Fe-3Si and 4130 steels by liquid gallium has been observed.[257]

Indium Embrittlement. Liquid indium embrittles pure iron (only above 310°C, or 590°F), carbon steel, and 4130 steel. Embrittlement depends on both the microstructure and the strength level.

Lead Embrittlement. Lead embrittlements of steel were found to be of two types: (1) external LME (i.e., embrittlement by molten lead and lead alloys) and (2) internal LME (i.e., the embrittlement of leaded steel, where lead is present as second phase or inclusions). Exposure of both pure lead and lead alloys induced external embrittlement in 4140 and 4145 steels. Additions of Zn, Sb, Sn, Bi, and Cu increase the extent of lead embrittlement. Additions of up to 0.5% Zn, 2% Sb, or 9% Sn increase the embrittlement of AISI 4145 steel; the embrittlement potency

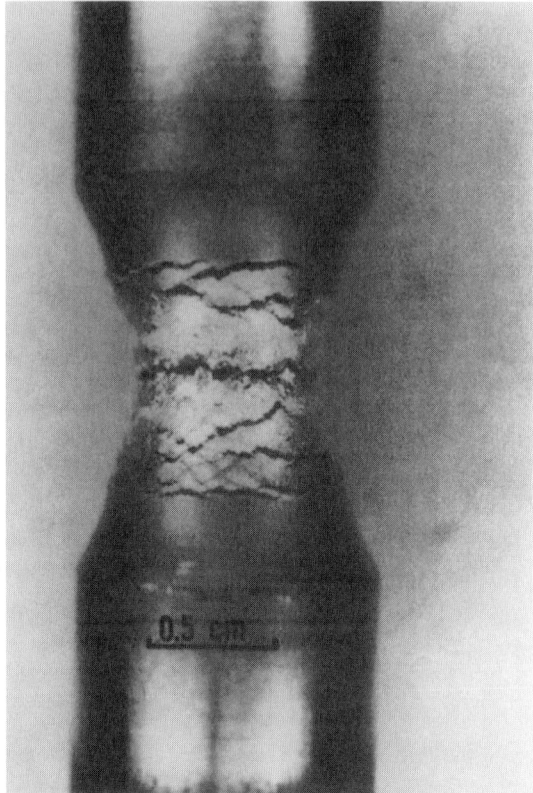

FIGURE 14.62 Copper-plated specimen that was pulled at 1100°C (2010°F) in a Gleeble hot tensile machine, showing liquid-copper embrittlement of 4340 steel.[257]

increases with the increase of impurity level. The severity of embrittlement or embrittling susceptibility and DBTT were found to increase with an increase in both surface roughness and grain size and a decrease in the amount of cold work.[258]

Failures of leaded steel parts such as shafts, gear teeth, die block, and compressor disks in jet aircraft and helicopter engines were reported by Breyer and Gordon, who noted that such failures were due to the presence of lead, a temperature range of 200 to 800°C (392 to 1472°F), and lowering of tensile stress toward the yield stress. However, the extent of lead embrittlement can be either eliminated or substantially decreased by controlling sulfide morphology and composition (by the addition of rare-earth elements) and cold-working of steel.

Lithium Embrittlement.   Tensile tests on lithium exposure were conducted by Cordwell on mild steel and Fe-2.25Cr-1Mo and Fe-9Cr-1Mo steels in the temperature range of 200 to 250°C (392 to 480°F) and at a strain rate of $2 \times 10^{-5} s^{-1}$. The tensile ductility of mild steel exposed to lithium at 200°C (392°F) showed drastic reduction of tensile ductility in lithium, with intergranular fracture after 2 to 3% elongation, without affecting its yield stress or the initial work-hardening behavior.[257]

Mercury Embrittlement.   The fracture toughness of notched 1Cr-0.2Mo steel was drastically decreased upon testing in mercury. The additions of solutes such as

Co, Si, Al, and Ni to iron reduced the tendency for cross-slip by a decrease in the number of active slip systems and changed the slip mechanism from wavy to planar glide, which enhanced the susceptibility to embrittlement. Iron alloys (such as Fe-2Si, Fe-4Al, Fe-8Ni, Fe-20V, and Fe-49Co-2V alloys) have been reported to be embrittled by mercury in unnotched tensile tests.[257]

Selenium Embrittlement. Selenium did not exhibit embrittling influence on the mechanical properties of a quenched and tempered steel (UTS ~1460 MPa or 212 ksi) that was subjected to bend test at 250°C (480°F).[257]

Silver Embrittlement. Silver showed little effect on the mechanical properties of plain carbon steels, silicon steels, and chromium steels that were subjected to bend test at 1000 to 1200°C (1830 to 2190°F). However, a silver base filler metal (composition: 45Ag-25Cd-15Sn) has been found to embrittle A-286 heat-resistant steel in static-load tests above and below 580°C (1076°F), the T_m of the alloy.[257]

Sodium Embrittlement. Unnotched tensile properties of low-carbon steels were unaffected when tested in sodium in the temperature range of 150 to 250°C (300 to 480°F). Similarly, no embrittlement by sodium was observed in Armco iron, low-carbon steel, and type 316 stainless steel in the temperature range of 150 to 1600°C (300 to 2910°F).[257]

Embrittlement by Solders and Bearing Metals. A wide variety of steels have the propensity to embrittlement by molten solders and bearing alloys at temperatures below 450°C (840°F). The extent of embrittlement increased with the grain size and the strength value of the steel, except in the temper-embrittled steels. In general, the embrittlement was associated with a change to a brittle intergranular fracture mode and penetration along prior austenite grain boundaries. Intercrystalline penetration of solder was not observed in (0.14 and 0.77%) carbon steel at 950°C (1740°F).[257]

Tellurium Embrittlement. Embrittlement by tellurium has been observed in both plain carbon and alloy steels. Hot-shortness has been reported in AISI 12L14 + Te steel, associated with drastic loss in ductility in the temperature range of 810 to 1150°C (1490 to 2100°F) and the most severe embrittlement at 980°C (1795°F) due to the formation of lead-tellurium compound film at the grain boundary, which has T_m = 923°C (1693°F).[257]

Thallium Embrittlement. Thallium did not exhibit embrittling effect on the properties of a quenched and tempered steel (UTS ~1460 MPa or 212 ksi) which was bend-tested at 325°C (615°F).[257]

Tin Embrittlement. The embrittlement behavior of tin on a range of steels can be represented by an embrittlement trough in the temperature range of 110 to 400°C (230 to 752°F); the location and extent of the trough were a function of the steel composition and the strain rate, as reported by Tanaka and Fukunaga.[258] In another study made on the fatigue properties of tin-embrittled mild steel, 13% Cr steel, and 18-8 stainless steel, it was indicated that in the case of the unnotched specimens the lifetime was crack-initiation-controlled whereas in the notched specimens the lifetime was crack-propagation-controlled.

Zinc Embrittlement. In many instances, zinc embrittlement cracks contain zinc-rich precipitates on fracture surfaces and at the crack-tip and are irregular in nature. Zinc-embrittled austenitic stainless steels are of two types. (1) Type I involves the metal penetration/erosion in the unstressed materials such as 18-8 austenitic stainless steel at 419 to 507°C (786 to 945°F) and above, type 316 stainless steel at 750°C (1380°F), and type 321 steel at 515°C (959°F). (2) Type II involves LME in the stressed materials at temperatures above 750°C (1380°F). According to Radekar, certain ferritic steels having greater thermal stability exhibited the maximum resistance to molten zinc embrittlement.[258]

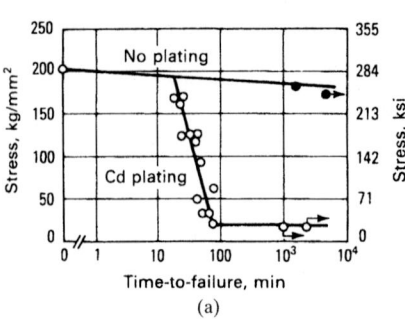
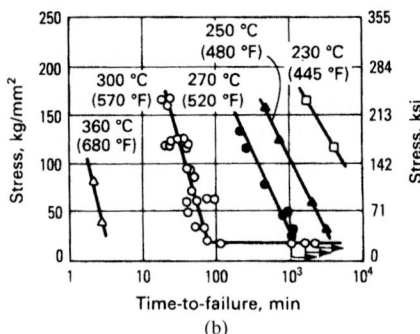

FIGURE 14.63 Embrittlement behavior of cadmium-plated 4340 steel. Specimens were tested in delayed failure (*a*) at 300°C (570°F) and unplated steel in air at 300°C (570°F) and (*b*) at temperatures in the range of 230 to 360°C (445 to 680°F).[262]

Zinc embrittlement was also observed in ferritic steels and Armco iron in the temperature range of 400 to 620°C (750 to 1150°F) and in AISI 4140 steel at 431°C (808°F).[257]

14.13.2   Solid-Metal Embrittlement

Solid-metal embrittlement (SME), also called *solid-metal-induced embrittlement* (SMIE), occurs below T_m of the solid when the embrittling solid is an internal environment, such as inclusion. Although SME has not been suggested or accepted as an embrittlement phenomenon in industrial processes, numerous examples of loss in ductility, strength, and brittle fracture have been noted for electroplated metals and coatings or inclusions of low-melting-point metals below their T_m.[261] Asayama has reported delayed failure of cadmium-plated high-strength steel bolt below the T_m of cadmium (Fig. 14.63); consequently, the use of cadmium-plated steel bolts above 230°C (450°F) is not recommended, despite their excellent corrosion resistance.[262] Solid-lead embrittlement has been reported in notched tensile specimens of various steels, resulting in considerable loss in ductility below the T_m of lead. This behavior seems to be responsible for several elevated-temperature failures of leaded steel, namely, failure of steel shafts during straightening at elevated temperatures, radial cracking of gear teeth during induction heating, and heat treatment failures of jet-engine compressor disks.[263]

14.13.2.1   Characteristics of SME.   To date, SME has been reported only in LME couples. However, SME can occur without LME. Table 14.10 lists the occurrence of SME, showing that all solid-metal embrittlers are also liquid-metal embrittlers. Since both SME and LME have similar behavior, the prerequisites for SME are the same as those for LME. Thus the prerequisites of SME are (1) intimate contact between the solid and the embrittler, (2) the presence of tensile residual or applied stresses, (3) the presence of the embrittler at the growing crack tip, and (4) crack initiation at the solid/embrittler interface from a barrier such as a grain boundary.

Furthermore, the metallurgical variables that increase the brittleness in steels (e.g., grain size, strain rate, yield strength, solute strengthening, and the presence of

TABLE 14.10 Occurrence of Solid-Metal Embrittlement in Steels[257]

Base metal	Embrittler (melting point)	Temperature at onset of embrittlement		Test type[†]	Specimen type[‡]
		°C	°F		
1041	Pb (327°C, or 621°F)	288	550	ST	S
1041 leaded	Pb	204	399	ST	S
1095	In (156°C, or 313°F)	100	212	ST	S
3340	Sn (232°C, or 450°F)	204	399	ST	N
	Pb	316	601	ST	N
4130	Cd (321°C, or 610°F)	300	572	DF	N
4140	Cd	300	572	DF	N
	Pb	204	399	ST	S
	Pb-Bi (NA)[§]	Below solidus		ST	S
	Pb-Zn (NA)	Below solidus		ST	S
	Zn (419°C, or 786°F)	254	489	DF	N
	Sn	218	424	DF	N
	Cd	188	370	DF	N
	Pb	160	320	DF	N
	In	Room temperature		DF	N
	Pb-Sn-Bi (NA)	Below solidus		ST	S
	In	80	176	DF	S
	Sn	204	399	ST	S
	Sn-Bi (NA)	Below solidus		ST	S
	Sn-Sb (NA)	Below solidus		ST	S
	In	110	230	DF	S
	In	93	199	DF	S
	In-Sn (118°C, or 244°F)	93	199	DF	S
4145	Sn	204	399	ST	S
	In	121	250	ST	S
	Pb-4Sn (NA)	204	399	ST	S
	Pb-Sn (NA)	204	399	ST	S
	Pb-Sb (NA)	204	399	ST	S
	Pb	288	550	ST	S
4145 leaded	Pb	204	399	ST	S
4340	Cd	260	500	DF	N
	Cd	300	572	DF	N
	Cd	38	100	DF	S
	Zn	400	752	DF	N
4340M	Cd	38	100	DF	S
8620	Pb	288	550	ST	S
8620 leaded	Pb	204	399	ST	S
A-4	Pb	288	550	ST	S
A-4 leaded	Pb	204	399	ST	S
D6ac	Cd	149	300	DF	N

[†] St, standard tensile test; DF, delayed-failure tensile test.
[‡] S, smooth specimen; N, notched specimen.
[§] NA, data not available.

stress raisers or notches) all seem to increase embrittlement. Susceptibility to SME is a function of the stress and temperature and does not take place below a certain threshold value.

Delayed-failure type embrittlement has also been noticed for both SME and LME.

The essential differences between SME and LME are

1. Formation of multiple cracks in SME, rather than the propagation of a single crack to failure in LME
2. Propagation-controlled fracture in SME, but their crack propagation rates being about 2 to 3 orders of magnitude slower than in LME
3. Possibility of a change from brittle intergranular fracture to ductile shear mode due to the inability of the embrittler to continue with the propagating crack tip
4. Presence of incubation periods, thereby implying that the crack nucleation process may not be the same as in LME
5. Nucleation and growth as two distinct stages of fracture in SME
6. Both SME and LME attributed to the reduction in the cohesive strength of the atomic bonds at the tip
7. Transport of the embrittler being the rate-controlling variable in SME
8. Crack-initiation attributed to the stress-assisted penetration of the embrittler in the grain boundaries, but crack growth controlled by the surface self-diffusion of the embrittling species, similar to those suggested for LME.[257]

The study of SME is of importance in eliminating the likelihood of LME that a crack, once formed, may propagate in a brittle fashion in the absence of embrittling species at the crack tip.[257]

14.14   MARAGING STEELS

Maraging steels are a special class of highly alloyed low-carbon iron-nickel martensitic steels which derive their ultrahigh strength not from carbon but from the precipitation of a uniform and dense distribution of various fine intermetallic compounds and Laves phases during aging (or age-hardening) treatment.[264–266] The term *maraging* is adopted because it involves martensite age-hardening, i.e., aging in the martensitic form.[267] Furthermore, the presence of σ-phase, μ-phase, and R'-phase has been reported in the literature, although these appear to be rare. The type and structure of the precipitates are functions of the composition of the material, aging temperature, and time.[268] In addition to the extremely high strength, the maraging steels have excellent fracture toughness and ductility.

The commonly available maraging steels contain 17–19% Ni, 7.5–12.5% Co, 3–5% Mo, 0.2–1.8% Ti, and 0.1–0.15% Al. Like other martensitic alloys strengthened by intermetallics, maraging steels show the transformational behavior involving austenite formation on heating; martensitic reaction on air-cooling from the solution-annealing (austenitizing) to room temperature to form a soft, ductile, heavily dislocated low-carbon, iron-nickel, bcc lath martensite (Fig. 14.64a) (with no twinning); and decomposition of martensite on aging below the austenite start temperature A_s (Fig. 14.64b).[266,269,270]

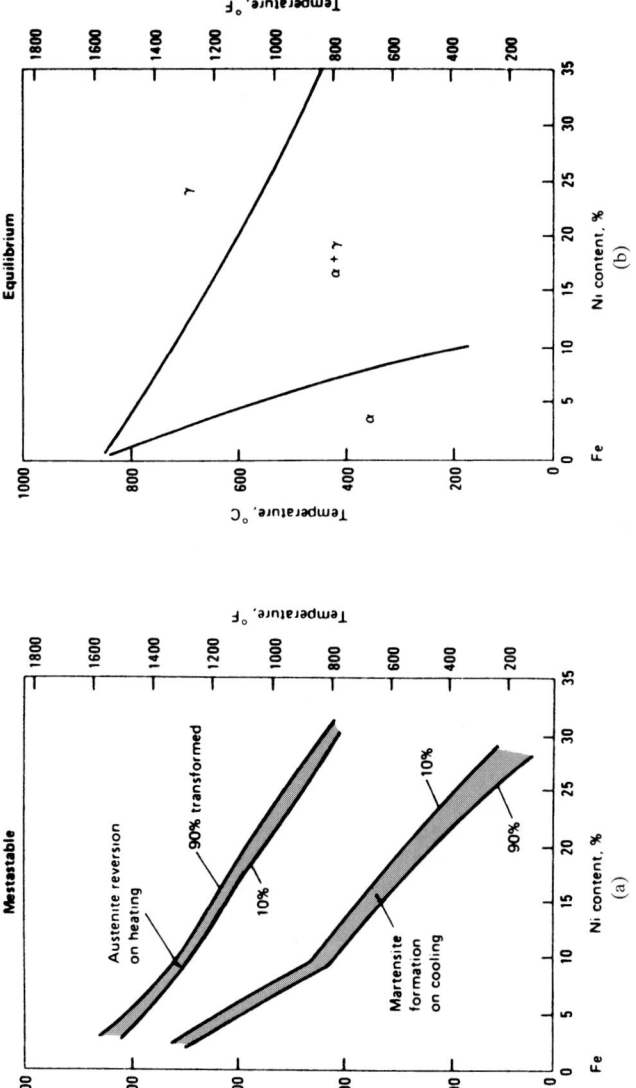

FIGURE 14.64 (*a*) Metastable phase diagram for the iron-rich end of the iron-nickel binary system showing the fcc austenite-to-bcc lath martensite transformation upon cooling and the martensite-to-austenite reversion upon heating.[266] (*b*) Equilibrium phase diagram showing, for higher nickel content, the austenite and ferrite equilibrium phases at low temperatures.[266] (*Reprinted by permission of ASM International, Materials Park, Ohio.*)

TABLE 14.11 Nominal Composition of Commercial Maraging Steels[273]

Grade	Composition,[†] wt%									
	Ni	Mo	Co	Ti	Al	Nb	Cr	Si	Cu	Mn
Standard grades										
18 Ni (200)	18	3.3	8.5	0.2	0.1	—	—	—	—	—
18 Ni (250)	18	5.0	7.75	0.4	0.1	—	—	—	—	—
18 Ni (300)	18	5.0	9.0	0.65	0.1	—	—	—	—	—
18 Ni (350)	18	4.2[‡]	12.5	1.6[‡]	0.1	—	—	—	—	—
18 Ni (cast)	17	4.6	10.0	0.3	0.1	—	—	—	—	—
13 Ni (400)[§]	13	10.0	15.0	0.2	—	—	—	—	—	—
12-5-3 (180)	12	3.0	—	0.2	0.3	—	5.0	—	—	6.0
Cobalt-free and low-cobalt-bearing grades										
Cobalt-free 18 Ni Marage 200	18.5	3.0	—	0.7	0.1	—	—	—	—	—
Cobalt-free 18 Ni Marage 250	18.5	3.0	—	1.4	0.1	—	—	—	—	—
Low-cobalt 18 Ni Marage 250	18.5	2.6	2.0	1.2	0.1	0.1	—	—	—	—
Cobalt-free 18 Ni Marage 300	18.5	4.0	—	1.85	0.1	—	—	—	—	—
Cobalt-free 18 Ni Marage 300	18.0	2.4	—	2.2	0.2	—	—	0.2	—	—
Cobalt-free Sandvik 1RK91	9.0	4.0	—	0.9	0.3	—	12.0	0.15	2.0	—
Cobalt-free C455	9.0	<0.3	—	1.3	0.14	—	12.0	0.18	2.0	—

[†] The carbon content for all grades is restricted to 0.03% max, Si to 0.10% max, S to 0.01% max, P to 0.01% max, Zr to 0.01% max, and B to 0.003% max.
[‡] Some producers use a combination of 4.8% Mo and 1.4% Ti, nominal.
[§] *Cobalt Facts*, The Cobalt Development Institute, England.

Ni-maraging steels were initially developed by C. B. Bieber at the International Nickel Company in the late 1950s, who used the age-hardening influence of 1.4% titanium, 0.3% aluminum, and 0.4% niobium on the Fe-(20–25%) Ni alloys. Later work led to the emergence of practically carbon-free 18% Ni-Fe alloys containing high molybdenum and cobalt content but lesser titanium and aluminum than those in the original steels which became of great commercial importance of fabrication (i.e., for cold-working and welding).[271] These later alloys exhibited an improved combination of strength and toughness compared to their earlier counterparts.[271a]

14.14.1 Types of Maraging Steels

Commercial maraging steels are grouped into standard grades and cobalt-free and low-cobalt-bearing grades; the latter grades have been developed as a result of the cobalt shortage and resultant price escalation of cobalt during the late 1970s and 1980s (Table 14.11). The former ones have six wrought grades, namely, 18 Ni (200), 18 Ni (250), 18 Ni (300), 18 Ni (350), 13 Ni (400), 12-5-3 (180), and one cast 18 Ni (cast) grade; the number in parentheses after the designation refers to the nominal yield strength in ksi (kpsi). Among these wrought grades, 18 Ni (200), 18 Ni (250), and 18 Ni (300) are the most widely employed and available grades. The 18 Ni (350) grade, available in two compositions, is an ultrahigh-strength type, produced in limited quantities for special applications. Some experimental maraging steels have yield strengths up to 500 ksi (3450 MPa). The 18 Ni (cast) grade was especially developed as a cast composition.

Two cobalt-free maraging steels with compositions 9Ni-12Cr-2Cu-4Mo (Sandvik 1RK91) and 9Ni-12Cr-2Cu (C455) are similar in composition; however, the hardening behavior of the material has been observed to be different after 2 hr of aging. The presence of Mo is responsible for the superior properties of 1RK91 with respect to C455 after long heat treatment.[268] The extremely high strength in 1RK91 can be ascribed to the precipitation of a quasicrystalline phase termed R' containing typically 30%Fe-15%Cr-5%Mo-5%Si, which impeded particle shearing.[272]

The yield strength of maraging steels is based on the titanium content, which varies between 0.2 and 1.6%. A high Ti level produces a larger volume fraction of Ni_3Ti-type precipitate, and Co's effect is to lower the solubility of Mo, which in turn enhances the precipitation of metastable Ni_3Mo. This effect is referred to as the *Co/Mo interaction*. Also, the presence of Fe_2Mo signifies the onset of overaging and austenite reversion, since the Ni present in the Ni_3Mo reverts back to the matrix and causes local Ni enrichment as Fe_2Mo replaces Ni_3Mo. Likewise, as the Fe_2Mo forms, there is a localized depletion in Fe, which further enhances the austenite reversion reaction.[271a] It is clear from Table 14.11 that the carbon content is maintained at a very low level (<0.03%) (to minimize the formation of TiC, which can deteriorate strength, ductility, and toughness);[273] the silicon and manganese contents are lower (0.1% maximum each), and phosphorus and sulfur contents are also very small (0.01% maximum each). The addition of aluminum up to 0.2% slightly increases the M_s temperature. However, this effect combined with its deoxidizing potency limits its use of about 0.1% in most maraging steels. In some cases, 0.01% zirconium and 0.003% boron additions are also made to improve the properties. Because of the high alloy content, especially the cobalt addition, maraging steels are very expensive.

14.14.2 Salient Features

The maraging steels possess a number of characteristic features as given below, which represent special advantages over conventional tempered martensitic steels that have led to their use in many demanding applications.

1. They offer the best available combination of (*a*) ultrahigh yield and tensile strength, ductility, and fracture toughness of any ferrous materials including quenched and tempered alloy steels and high-strength aircraft/aerospace steels (Fig. 14.65) with the exception of AF 1410 and AerMet 100 alloys, which offer a superior combination of these mechanical properties (Fig. 14.32);[83,83a,83b] (*b*) high notched tensile strength—up to 3089 MPa (400 ksi, 310 kg/mm^2); (*c*) exceptionally high resistance to crack propagation with K_{IC} values up to 240 MPa$\sqrt{m}$ (220 ksi$\sqrt{in}$.); and (*d*) retention of yield strength at least up to 350°C.[274]

2. Being a very low-carbon martensite, this structure is soft enough to be readily machinable and may be cold-rolled to as much as 80 to 90% without cracking prior to the aging treatment. Hot deformation is also possible.

3. There is consistently very little dimensional change following the aging treatment so that it is possible to finish-machine in the soft condition before aging.[274]

4. The heat treatment cycle is simple: solution treatment, followed by air-cooling and subsequent aging, followed by air-cooling. Quenching is thus not required, and decarburization is no problem. Parts uniformly harden throughout the entire section because they have high hardenability. Most manufacturers of maraging steel supply material in the solution-annealed condition. Thus, the heat treatment is further simplified in that aging is all that is required after machining.[271a]

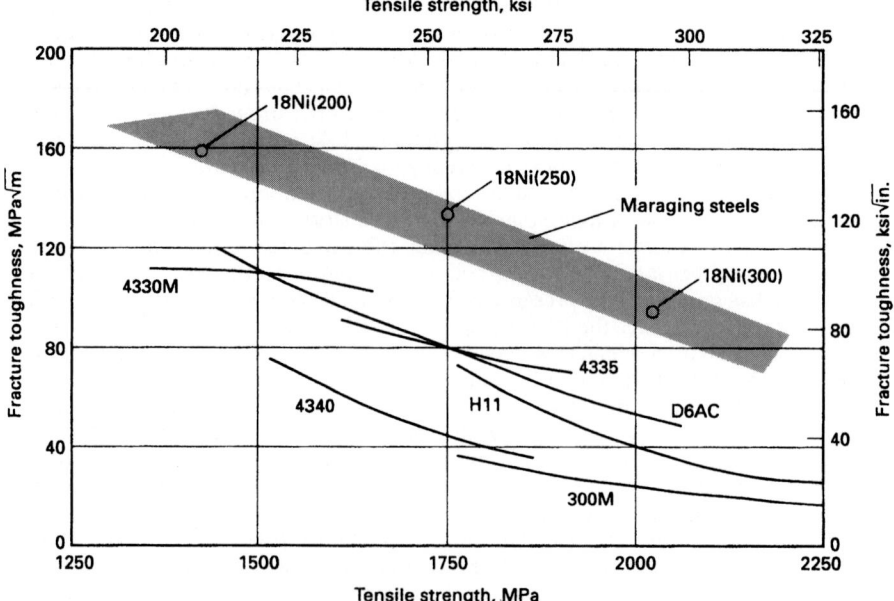

FIGURE 14.65 Plane-strain fracture toughness/tensile strength combinations of 18 Ni maraging steels compared to conventional ultrahigh-strength steels. (*Source: A. Magnee, J. M. Drapier, J. Dumont, D. Coutsouradis, and L. Habraken, Cobalt-Containing High-Strength Steels, Cobalt Information Center, England, 1974; J. C. Hamaker and A. M. Bayer, Cobalt, no. 38, 1968, p. 3.*)

5. They exhibit excellent weldability characteristics even in the fully heat-treated condition. This is a considerable advantage when compared to low-alloy steels containing typically 0.4% C. Preheating is not required, and the properties of the heat-affected zone can be restored by a postwelding treatment.

6. Since a ductile Fe-Ni martensite is produced by air-cooling from the austenitizing range, cracks and distortion are nonexistent or negligible.[275]

7. The fatigue properties of maraging steels, like those of conventional tempered martensite steels, can be improved by nitriding and by shot-peening.

8. The corrosion, stress corrosion, and hydrogen embrittlement characteristics of maraging steels are considered to be good compared to those of carbon-bearing steels of similar strength levels. However, these steels require protection, which is usually provided by cadmium plating or phosphating.[276]

9. There is no problem in maintaining the properties during and after low-temperature brazing.

14.14.3 Heat Treatment

The conventional heat treatment cycle of the maraging steels is simple. The typical heat treatment of these alloys consists of a solution-annealing treatment, usually at 820°C [which is significantly above the austenite finish (A_f) temperature] for a minimum period of 0.25 to 0.5 hr for a 1.3-mm-thick section and for 1 hr per 25 mm for a heavy section, to ensure the dissolution of alloying elements in solid solution and formation of a fully austenitic structure. This is followed by air-cooling to room temperature to form a soft (28 to 35 Rc), but heavily dislocated, massive or lath-type iron-nickel martensite throughout the section.[277,278] Most grades of maraging steels have an M_s of about 200 to 300°C (390 to 570°F) and are fully martensite at room temperature.[273] The orientation relationship between the lath martensite in 18 Ni maraging steels and the parent austenite has been identified as the Kurdjumov-Sachs type: $(110)_\gamma // (011)_M, [110]_\gamma // [111]_M$.[265,279] Subsequent aging for several hours (usually 3 to 6 hr) in the temperature range of 480 to 500°C, followed by air-cooling, produces a fine and uniform dispersion of coherent Ni_3Mo and Ni_3Ti or $Ni_3(Mo,Ti)$ intermetallic phases on dislocations within the lath martensite as well as at the lath boundaries. It is thus clear that such improved age-hardening response in maraging steels is promoted by the high dislocation density and uniform distribution of dislocations, which provide a large number of heterogeneous nucleation sites for the intermetallic precipitates. Furthermore, these same dislocations (and lath boundaries) also augment the diffusion rates of substitutional alloying elements. Both of these factors increase a uniform distribution of fine precipitates.[280] The final hardness in the maraged condition is 45, 50/51, and 54/55 Rc, respectively, for Marage 200, 250, and 300 grades.[271a]

Typical mechanical properties and heat treatment data for conventional maraging steels are given in Table 14.12, and Table 14.13 compares the mechanical properties of 18 Ni (250), low-cobalt-bearing 18 Ni (250), and cobalt-free 18 Ni (250) grades.

The standard aging treatments given in Table 14.12 produce contractions in length of 0.04% in 18 Ni (200), 0.06% in 18 Ni (250), and 0.08% in both 18 Ni (300) and 18 Ni (350) grades. These very small dimensional contractions during aging treatment permit many maraging steel parts to be finish-machined in the annealed condition, then hardened. When precise dimensional control is required, an allowance for contraction can be made, and finish machining should be accomplished after aging.

Role of Alloying Elements and Precipitation Reaction. The alloying elements that are involved in the precipitation reactions in maraging steels can be grouped into three categories:[280]

1. Ti and Be, called strong hardeners
2. Al, Nb, Mn, Mo, Si, Ta, W, and V, called moderate hardeners
3. Co, Cu, and Zr, called weak hardeners

Among these 13 elements, only three important age-hardening constituents in the above heat treatment are molybdenum, and titanium, and cobalt. The molybdenum forms metastable orthorhombic Ni_3Mo, which precipitates as rodlike (25 Å wide × 500 Å long in size in the peak-aged condition) and tends to prevent the formation of deleterious grain boundary precipitates that would otherwise degrade the toughness. Eventually, the metastable precipitates grow and dissolve and are replaced by Fe_2Mo which represents an overaged condition.

Titanium forms a stable η-Ni_3Ti (DO_{24} ordered hexagonal structure), $Ni_3(Ti,Al)$, or $Ni_3(Mo,Ti)$ and renders additional hardening. Titanium also aids in eliminating residual carbon and nitrogen by forming $Ti(C,N)$ precipitate particles. The intermetallics precipitate particles range in size from 10 to 50 nm (100 to 500 Å) and are uniformly precipitated.

TABLE 14.12 Heat Treatment and Typical Mechanical Properties of Conventional 18 Ni Maraging Steels[273]

Grade	Heat treatment[†]	Tensile strength		Yield strength		Elongation in 50mm (2 in.), %	Reduction in area, %	Fracture toughness	
		MPa	ksi	MPa	ksi			MPa√m	ksi√in.
18 Ni (200)	A	1500	218	1400	203	10	60	155–240	140–220
18 Ni (250)	A	1800	260	1700	247	8	55	120	110
18 Ni (300)	A	2050	297	2000	290	7	40	80	73
18 Ni (350)	B	2450	355	2400	348	6	25	35–50	32–45
18 Ni (cast)	C	1750	255	1650	240	8	35	105	95

† Treatment A: solution treat 1 hr at 820°C (1500°F), then age 3 hr at 480°C (900°F). Treatment B: solution treat 1 hr at 820°C (1500°F), then age 12 hr at 480°C (900°F). Treatment C: anneal 1 hr at 1150°C (2100°F), age 1 hr at 595°C (1100°F), solution treat 1 hr at 820°C (1500°F) and age 3 hr at 480°C (900°F).

TABLE 14.13 Comparison of the Longitudinal, Room-Temperature Mechanical Properties of Standard, Cobalt-Free, and Low-Cobalt-Bearing 18 Ni (250) Maraging Steels[273]

Heat treatment: solution heat 1 hr at 815°C (1500°F), then age 5 hr at 480°C (900°F). Testing was conducted on 63.5 × 88.9 mm (2.5 × 3.5 in.) billets produced from 200-mm (8-in.) diameter vacuum induction melted/vacuum arc remelted ingots.

Grade	Ultimate tensile strength		0.2% offset yield strength		Elongation in 25 mm (1 in.), %	Reduction in area, %	Charpy V-notch impact toughness[†]		Plane-strain fracture toughness[†]	
	MPa	ksi	MPa	ksi			J	ft · lbf	MPa√m	ksi√in.
18 Ni (250)	1870	271	1825	265	12	64.5	37	27	138	125
Cobalt-free 18 Ni (250)	1895	275	1825	265	11.5	58.5	34	25	127	115
Low-cobalt 18 Ni (250)	1835	266	1780	258	11	63.5	43	32	149	135

† Longitudinal-short transverse orientation tested (L-S orientation).

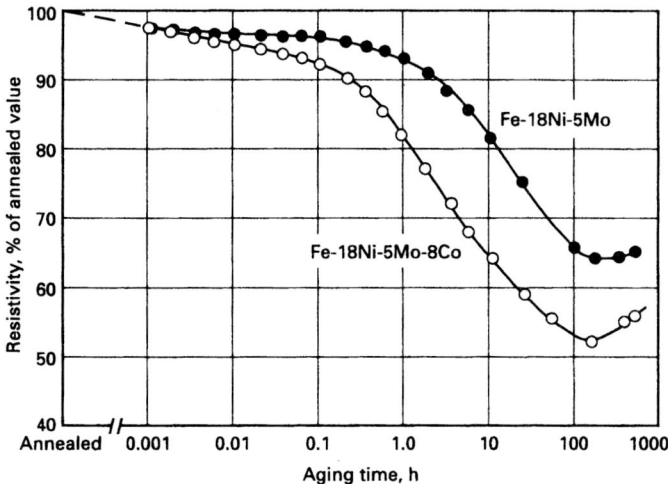

FIGURE 14.66 Plots of electrical resistivity (percent of annealed value) versus aging time at 455°C (850°F) for Fe-18Ni-5Mo and Fe-18Ni-5Mo-8Co maraging steels, showing cobalt/molybdenum interaction.[281]

Nickel forms the carbon-free martensite matrix as well as the intermetallics during age-hardening.

Cobalt/Molybdenum Interaction. Although cobalt does not participate directly to form a precipitate, it does participate indirectly into the age-hardening reaction through a so-called *cobalt/molybdenum interaction*, as described here. The addition of Co raises the M_s and M_f temperatures of 18% Ni-maraging steels so that greater levels of other alloying elements (such as Ti and Mo, which lower the M_s temperature) can be added while still allowing complete martensitic transformation prior to cooling of steel to room temperatures. Moderate cobalt contents (6 to 8%) are beneficial in the design of a maraging steel with strength above 2100 MPa (300 ksi), because that facilitates a high enough M_s temperature in the presence of a high alloy content. Cobalt is also very useful, because it causes a decrease in solubility of molybdenum in the bcc Fe-Ni martensite matrix, thereby promoting a finer, more uniform distribution of Ni_3Mo precipitate particles in the martensite during the aging treatment, which, in turn, will increase the yield strength of the alloy system.[280] This increase in strength is greater than the individual contributions of Mo and Co, and has thus been called a *synergistic reaction*. Fe_2Mo will eventually form whether Co is present or not.[271a]

This cobalt/molybdenum interaction theory, initially proposed by Peters and Cupp,[281] is supported by electrical resistivity measurements[281] and TEM studies by Miller and Mitchell[282] and Floreen and Speich.[283] Figure 14.66 shows that the amount of Mo in solid solution during the age-hardening reaction is less for the alloy system containing 8% Co.[281]

Figure 14.67 shows hardness versus aging curves for an 18 Ni 250 maraging steel. Very rapid hardening occurs at 482°C; this is due, in part, to the highly faulted martensite matrix which provides both a large number of nucleation sites for precipitation and rapid diffusion kinetics. With increasing aging times and/or higher

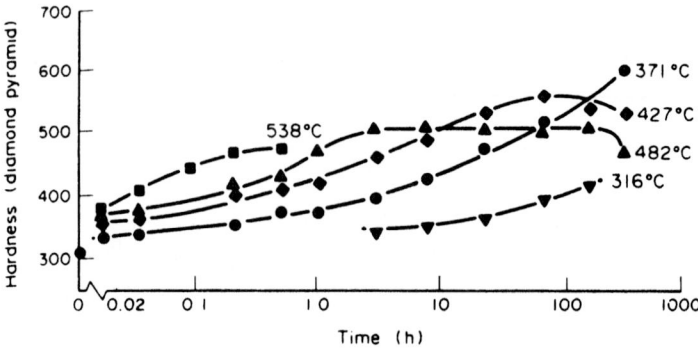

FIGURE 14.67 Hardness versus aging curves for 18 Ni (250) maraging steels at various aging temperatures.[266] (*Reprinted by permission of ASM International, Materials Park, Ohio.*)

temperatures, the hardness peak and decay occur early. This hardness drop is usually linked with precipitate particle-coarsening and austenite reversion.

Ordering Reactions. The high degree of short-range ordering phenomenon observed in 18 Ni marage (350) steel after aging at 510°C (950°F) for 3 hr has been attributed to the formation of iron- and cobalt-rich regions and nickel-rich regions. According to Rack and Kalish,[284] this phenomenon influences the subsequent precipitation-hardening reaction.

Austenite Reversion. Another basic aim during the aging of maraging steels is to minimize the reversion of metastable martensite into the equilibrium austenite and ferrite compositions, which occurs during reheating to a temperature below A_s. The rate at which this reversion occurs is a function of the temperature to which this alloy is heated. The rate of this reversion reaction in the temperature range of 455 to 510°C (850 to 950°F) is slow enough to allow adequate precipitation hardening to take place prior to where the reversion reaction begins to predominate. This behavior of sluggish austenite reversion in combination with rapid age hardening constitutes the basis for the aging treatment of maraging steels.[280] Martensite reversion in 1RK91 maraging steel was promoted by the accommodation of Mo and Cr in the precipitates.[272]

Based on the TEM work and microprobe analyses by Miller and Mitchell,[282] Banerjee and Hauser,[285] and Fleetwood et al.,[286] the austenite reversion behavior in maraging steels is attributed to an increase in Ni_3Mo, which, in turn, facilitates increased formation of Fe_2Mo and subsequent nickel enrichment of the matrix. Later, it was revealed that the austenite reversion reaction is not attributed to an increase in Ni_3Mo. The formation of this phase causes Ni depletion in the matrix. Initially, Ni_3Mo precipitation is favored due to the lower coherency strains associated with this metastable phase. As the precipitates grow, so do the coherency strains. Eventually the coherency strains reach a point where the stable Fe_2Mo phase is favored. As Fe_2Mo replaces the metastable Ni_3Mo, Ni is returned to the matrix and Fe is depleted. This combination alters the phase balance, and reversion to austenite occurs.[271a] On the other hand, cobalt additions up to about 8% limit the reversion reaction; however, extra additions of cobalt appear to augment the

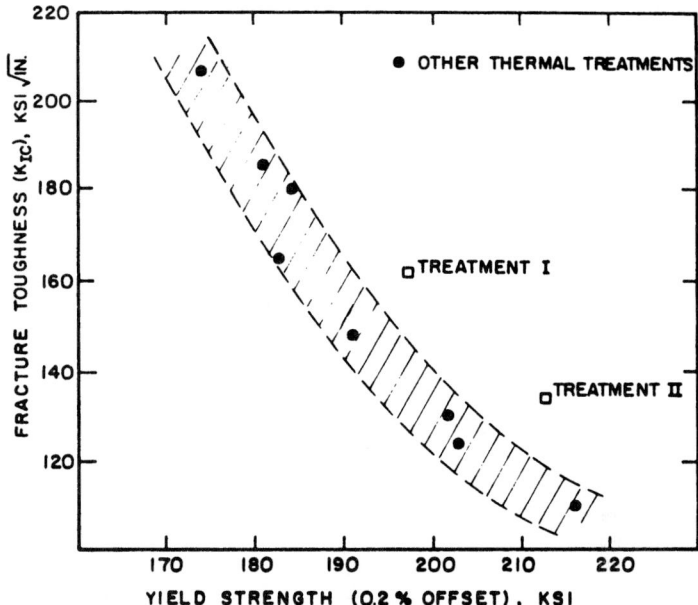

FIGURE 14.68 Comparison of strength-toughness characteristic of 250-grade maraging steels given reversion treatment (I, II) with that of the same steel after conventional thermal treatment.[290]

formation of reverted austenite.[287] Martensite reversion in 1RK91 maraging steel was promoted by the accommodation of Mo and Cr in the precipitates.[272]

Note that austenite reversion plays an important role in the magnetic properties of 18 Ni (350) grade maraging steel. It has been reported that remanence, coercive field, and saturation magnetization are functions of the reverted austenite content in the martensite matrix. The reverted austenite has Nishiyama-Wasserman and Kurdjumov-Sachs orientation relationship with the martensite matrix.[288] In contrast, reverted austenite aggravates the mechanical properties of the steel. Farooque et al.[289] have suggested that intralath reverted austenite is formed on or by the local dissolution of Ni_3Ti precipitates.

An inversion heat treatment has been developed by Jin and Cohen and Antolovich et al.[290] to achieve a further improvement in the strength-toughness combinations in maraging steels. This treatment involves the following steps: (1) the establishment of a maraging treatment to achieve the usual precipitate distribution, (2) rapid reversion (or heating) to austenite phase field, (3) subsequent transformation to martensite to achieve an extremely high dislocation density associated with a slight decrease in yield strength, and (4) finally reaging at low temperature (380°C) to pin and stabilize the high dislocation density, thereby producing an increment of 30 to 50 ksi in the yield strength for a given toughness level. Figure 14.68 shows a comparison of strength versus toughness results for 250-grade maraging steels produced by a combination of maraging, reverse transformation, and secondary aging treatment with that of the same steel processed by normal maraging treatment.[290]

14.14.4   Disadvantages

Unlike carbon-hardened steels, it is not practical to heat-treat maraging steels to a wide range of strength levels; usually, optimum properties cannot be attained by either underaging or overaging treatments. These restrictions have resulted in the development of several different compositions, each with a specific strength level. Whenever a different yield strength level is needed, it is better to use a different composition designed for that strength rather than manipulate the heat treatment by trial and error.

The wear resistance and fatigue resistance are inferior to those of low-alloy and high-carbon steels. On smooth test specimens, the fatigue ratio is typically 0.3 to 0.4, but the fatigue crack growth rate is similar to that in low-alloy steels.

The cost of maraging steels is high, due mainly to (a) the cost of the high-alloying elements involved and (b) the high purity required which is attainable by single- or double-vacuum melting.

14.14.5   Thermal Embrittlement

Maraging steels fracture intergranularly at low impact energies and tensile ductilities if improperly heat-treated or exposed to certain heat treatment schedules after hot-working. This problem of severe toughness degradation, termed *thermal embrittlement*, arises when maraging steels, after being heated to very high solution-annealing temperature, 1175°C (2150°F), are slowly cooled (or hot-worked) through, or held within the critical temperature range of 750 to 1090°C (1380 to 2000°F), resulting in the segregation and/or precipitation of TiC and/or Ti(C,N) film on the prior austenite grain boundaries or dislocations.[273,280,291] The extent of embrittlement increases with time within the critical temperature range. Increased carbon and nitrogen contents in maraging steels lead to enhanced susceptibility to thermal embrittlement. According to Auger analysis, embrittlement commences with the diffusion of Ti, C, and N to the grain boundaries, and the appearance of TiC or Ti(C,N) precipitates corresponds to an advanced stage of embrittlement.[280] Figure 14.69*a* shows the fracture of thermally embrittled cobalt-free high-titanium maraging steel. Light micrography (Fig. 14.69*b* and *c*) reveals the fracture profile (or edge) and internal cracks, respectively.[119]

Thermal embrittlement can be eliminated if the material, after being heated above 1175°C (2150°F), is rapidly cooled to room temperature or the thermally embrittled material is reheated to intermediate annealing temperature range of 750 to 1090°C (1380 to 2000°F), because the previously precipitated discrete, stable carbides cannot reprecipitate as a film on reheating. In addition, uniform precipitation of Ti(CN) occurs throughout the matrix on dislocations resulting from the martensite → austenite shear transformation during reheating.[280,292]

14.14.6   Applications

The wrought maraging steel product forms include plate, sheet, forgings, and bar stock. Most applications employ forgings or bars. Maraging steels have found a wide range of applications where lightweight structures with ultrahigh strength, high fracture toughness, hardness, and good magnetic properties are required and cost is not a major concern.

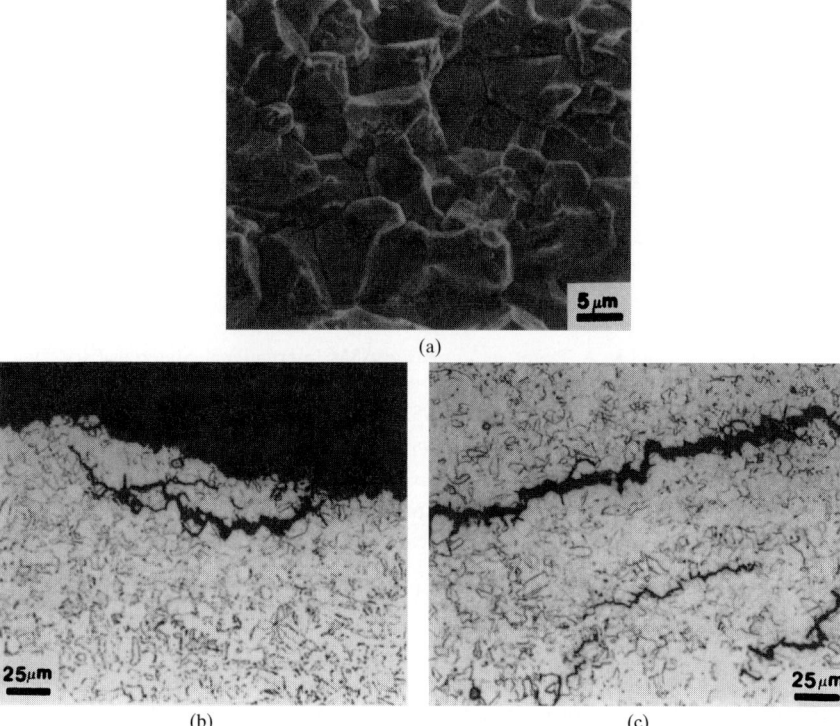

FIGURE 14.69 Fracture in a thermally embrittled cobalt-free high-titanium maraging steel. (*a*) SEM image of a fracture surface. (*b*) Optical micrograph of fracture edge. (*c*) Optical micrograph of internal cracks.[119] (*Reprinted by permission of ASM International, Materials Park, Ohio.*)

Typical applications are in the aerospace and aircraft industries for critical components such as rocket-motor casings, load cells, helicopter flexible drive shafts, transmission shafts, jet-engine drive (or fan) shafts, shock absorber for the Lunar Rover, couplings, hydraulic hoses, fasteners, bolts, punches, landing gear, wing hinge for swing-wing planes, aircraft forgings, and so forth; in the tool manufacturing industries for stub shafts, flexible drive shafts, splined shafts, pistons, cannon recoil springs, Belleville springs, bearings, plastic molds, extrusion rams, arrestor hooks, hot-forging dies, zinc-die casting dies, cold-heading dies and cases, diesel fuel-pump pins, clutch disks, gears in the machine tools, carbide die holders, autofrettage equipment, and so on. Other applications include parts for hydrospace (such as deep quest pressure hull), military, and auto racing cars, and surgical needles in surgical applications.[272]

REFERENCES

1. H. Ohtani, in *Materials Science and Technology*, vol. 7: *Constitution and Properties of Steels*, vol. ed. F. B. Pickering, VCH, Weinheim, 1992, chap. 4, pp. 147–181.

2. R. E. Reed-Hill and R. Abbaschian, *Physical Metallurgy Principles*, 3d ed., PWS-Kent Publishing, Boston, 1992.

3. G. Krauss and T. A. Balliett, *Met. Trans.*, vol. 7A, 1976, pp. 318–320.

4. A. M. Sherman, G. T. Eldis, and M. Cohen, *Met. Trans.*, vol. 14A, 1983, pp. 995–1005.

5. P. C. Chen, B. O. Hall, and P. G. Winchell, *Met. Trans.*, vol. 11A, 1980, pp. 1323–1331.

6. P. C. Chen and P. G. Winchell, *Met. Trans.*, vol. 11A, 1980, pp. 1333–1339.

7. S. Nagakura, Y. Hirotsu, M. Kusnoki, T. Suzuki, and Y. Nakamura, *Met. Trans.*, vol. 14A, 1983, pp. 1025–1031.

8. M. K. Miller, P. A. Weaver, S. S. Brenner, and G. D. W. Smith, *Met. Trans*, vol. 14A, 1983, pp. 1021–1024.

9. M. J. Van Genderen, M. Isac, A. Bottger, and E. J. Mittemeijer, *Met. and Mats. Trans.*, vol. 28A, 1997, pp. 545–561.

10. G. R. Speich and K. A. Taylor, in *Martensite*, ASM International, Materials Park, Ohio, 1992, pp. 243–275.

11. J. M. R. Genin, *Met. Trans.*, vol. 18A, 1987, pp. 1371–1388.

12. J. M. R. Genin, *Met. Trans.*, vol. 19A, 1988, pp. 2901–2909.

13. I. Fall and J. M. R. Genin, *Met. and Mats. Trans.*, vol. 27A, 1996, pp. 2901–2909.

14. Y. Ohmori and I. Tamura, *Met. Trans.*, vol. 23A, 1992, pp. 2147–2158.

15. K. A. Taylor, J. B. Vander Sande, and M. Cohen, *Met. Trans.*, vol. 24A, 1993, pp. 2585–2589.

16. K. A. Taylor, L. Chang, G. B. Olson, G. D. W. Smith, M. Cohen, and J. B. Vander Sande, *Met. Trans.*, vol. 20A, 1989, pp. 2717–2737.

17. M. J. Van Genderen, A. Bottger, R. J. Cernik, and E. J. Mittemeijer, *Met. Trans.*, vol. 20A, 1993, pp. 1965–1973.

18. K. A. Taylor and M. Cohen, *Progress in Materials Science*, vol. 36, 1992, pp. 225–272.

18a. L. Chang, Tempering of Martensite, D. Phil. thesis, University of Oxford, England, 1985.

19. G. B. Olson and M. Cohen, *Met. Trans.*, vol. 14A, 1983, pp. 1057–1065.

20. Y. Tanaka and K. Shimizu, *Trans. Jpn. Inst. Met.*, vol. 22, 1981, pp. 779–788.

20a. A. M. Sherman, Ford Motor Company, private communication, 2000.

21. G. Krauss, *Phase Transformations in Ferrous Alloys, International Conference Proceedings*, eds. A. R. Marder and J. I. Goldstein, TMS-AIME, New York, 1984, pp. 101–123.

22. K. H. Jack, *JISI*, vol. 169, 1956, p. 26.

23. Y. Hirotsu and S. Nagakura, *Acta Metall.*, vol. 20, 1972, p. 645.

24. H. C. Lee and G. Krauss, in *Gilbert R. Speich Symp. Proc.*, eds. G. Krauss and P. E. Repas, Iron and Steel Society, Warrendale, Pa., 1992, pp. 39–43.

25. Y. Ohmori, private communication, November 1987.

26. M. G. H. Wells, *Acta Metall.*, vol. 12, 1964, p. 389.

27. S. Murphy and J. A. Whiteman, *Met. Trans.*, vol. 1, 1970, p. 843.

28. C. J. Barton, *Acta Metall.*, vol. 17, 1969, p. 1085.

29. R. Padmanabhan and W. E. Wood, *Met. Sci. & Eng.*, vol. 66, 1984, pp. 125–143.

30. Y. Hirotsu and S. Nagakura, *Trans. Jpn. Inst. Met.*, vol. 15, 1974, p. 129.

31. D. L. Williamson, K. Nakazawa, and G. Krauss, *Met. Trans.*, vol. 10A, 1979, pp. 1351–1363.

32. G. Krauss, *Steels: Heat Treatment and Processing Principles*, ASM International, Materials Park, Ohio, 1990.

33. K. Shimizu and H. Okamoto, *Trans. Jpn. Inst. Met.*, vol. 15, 1974, pp. 193–199.

34. Y. Hirotsu, Y. Ilakura, K. C. Su, and S. Nagakura, *Trans. Jpn. Inst. Met.*, vol. 17, 1976, pp. 503–513.
35. Yu A. Bagaryatski, *Dokl. Akad. Nauk SSSR*, vol. 73, 1950, pp. 1161–1170.
36. T. A. Balliett and G. Krauss, *Met. Trans.*, vol. 7A, 1976, pp. 81–86.
37. C. Wells, W. Bartz, and R. F. Mehl, *Trans. AIME*, vol. 188, 1950, pp. 553–560.
38. C. S. Roberts, B. L. Averbach, and M. Cohen, *Trans. ASM*, vol. 45, 1953, pp. 576–604.
39. D. L. Williamson, R. G. Schupman, J. P. Materkowski, and G. Krauss, *Met. Trans.*, vol. 10A, 1979, pp. 379–382.
40. E. Tekin and P. M. Kelly, *Proceedings of a Conference on Precipitation from Iron-Based Alloys*, Metallurgical Society Conference at Cleveland, October 1963, eds. G. R. Speich and J. B. Clark, Gordon and Breach, New York, 1965, p. 173.
41. G. R. Speich, *Trans. Met. Soc., AIME*, vol. 245, 1969, p. 2553.
42. R. W. K. Honeycombe and H. K. D. H. Bhadeshia, *Steels: Microstructure and Properties*, 2d ed., Arnold, London, 1995.
43. M. Sarikaya, A. K. Jhingan, and G. Thomas, *Met. Trans.*, vol. 14A, 1983, pp. 1121–1133.
44. R. Caron and G. Krauss, *Met. Trans.*, vol. 3, 1972, pp. 2381–2389.
45. Y. Ohmori, *Trans. Jpn. Inst. Met.*, vol. 13, 1972, p. 119.
46. Y. Imai, *Trans. Jpn. Inst. Met.*, vol. 16, 1975, p. 721.
47. K. H. Jack and S. Wild, *Nature*, vol. 212, 1966, p. 248.
48. C. B. Ma, T. Ando, D. L. Williamson, and G. Krauss, *Met. Trans.*, vol. 14A, 1983, pp. 1033–1045.
49. S. Nagakura, T. Suzuki, and M. Kusunoki, *Trans. Jpn. Inst. Met.*, vol. 22, 1981, p. 699.
50. Y. Nakamura, T. Mikami, and S. Nagakura, *Trans. Jpn. Inst. Met.*, vol. 26, 1985, p. 876.
51. Y. Ohmori, in *International Conference Proceedings on Martensitic Transformations*, The Japan Institute of Metals, 1986, pp. 587–594.
52. G. Krauss, in *Constitution and Properties of Steels*, vol. ed. F. B. Pickering, VCH, Weinheim, 1992, pp. 1–40.
53. G. R. Speich, in *Metals Handbook*, vol. 8, 8th ed., ASM, Metals Park, Ohio, 1973, p. 202.
54. N. J. Nelson, *Heat Treatment '79*, The Metals Society, London, 1979, pp. 110–115.
55. M. A. Grossman and E. C. Bain, *Principles of Heat Treatment*, 5th ed., ASM, Metals Park, Ohio, 1964.
56. *Metals Handbook*, vol. 1, 8th ed., ASM, Metals Park, Ohio, 1961.
57. J. D. Murray, in *High Strength Steels, Report 76*, Iron and Steel Institute, London, 1962, pp. 41–50; *Modern Steels and Their Properties, Handbook*, 7th ed., Bethlehem Steel Corp., Bethlehem, Pa., 1972, p. 2757.
57a. J.-H. Feng and G. M. Michal, *38th MWSP Conference Proc.*, ISS-AIME, vol. 34, 1997, pp. 171–180.
58. M. Saeglitz and G. Krauss, *Met. and Mats. Trans.*, vol. 28A, 1997, pp. 377–387.
59. *Metals Handbook*, desk edition, ASM, Metals Park, Ohio, 1985, pp. 1-60–1-61.
60. J. P. Materkowski and G. Krauss, *Met. Trans.*, vol. 10A, 1979, pp. 1643–1651.
60a. H. W. Rayson, in *Constitution and Properties of Steels*, vol. ed. F. B. Pickering, VCH, Weinheim, 1992, pp. 583–640.
61. S. H. Samuel and A. A. Hussein, *Mater. Sci. Eng.*, vol. 58, 1983, pp. 113–120.
62. W.-J. Nam and H.-C. Choi, *Mats. Sci. and Technol.*, vol. 13, July 1997, pp. 568–574.
63. R. A. Grange, C. R. Hribal, and L. F. Porter, *Met. Trans.*, vol. 8A, 1977, pp. 1775–1785.
64. K. J. Irvine and F. B. Pickering, *JISI*, vol. 194, 1960, p. 137.

65. F. B. Pickering, in *Encyclopedia of Materials Science and Engineering*, Pergamon Press, Oxford, 1986, pp. 4605–4621.

66. M. G. H. Wells, in *Encyclopedia of Materials Science and Engineering*, Pergamon Press, Oxford, 1986, pp. 5115–5120.

67. R. A. Higgins, *Engineering Metallurgy*, 5th ed., Krieger, Malabar, Fla., 1983.

68. *Crucible Tool Steel and Specialty Alloy Selector*, Camillus, N.Y., 2000.

69. W. F. Smith, *Structure and Properties of Engineering Alloys*, McGraw-Hill, New York, 1981.

70. M. Wisti and M. Hingwe, in *Metals Handbook*, vol. 4, 10th ed., ASM International, Materials Park, Ohio, 1991, pp. 121–136.

71. *Alloy Data: Carpenter Specialty Steels*, Reading, Pa., 1999.

72. H. Brandis, E. Haberling, and H. H. Weigard, in *Processing and Properties of High Speed Tool Steels*, eds. M. G. H. Wells and L. W. Lherbier, TMS-AIME, New York, 1980, pp. 1–18.

73. *Tool Materials: ASM Specialty Handbook*, ASM International, Materials Park, Ohio, 1995, pp. 10–20.

74. P. Gordon, M. Cohen, and R. S. Rose, *Trans. ASM*, vol. 31, no. 1, 1943, p. 161.

75. *Metals Handbook*, vol. 7, 8th ed., ASM, Metals Park, Ohio, 1972.

76. S. Karagoz, H. F. Fischmeister, H.-O. Andren, and C. G. Jun, *Met Trans.*, vol. 23A, 1992, pp. 1631–1634.

77. N. Sarafianoz, *Met. and Mats. Trans.*, vol. 28A, 1997, pp. 2089–2099.

78. *Metals Handbook*, vol. 2, 8th ed., ASM, Metals Park, Ohio, 1964.

79. P. Parameshwaran, M. Vijayalakshmi, P. Shankar, and V. S. Raghunathan, *J. Mats. Sc.*, vol. 27, 1992, pp. 5426–5434.

80. S. J. Sanderson, *Met. Sci.*, vol. 11, 1977, pp. 490–492.

81. W. B. Jones, C. R. Hills, and D. H. Polonis, *Met. Trans.*, vol. 22A, 1991, pp. 1049–1058.

82. B. G. Gieseke, C. R. Brinkman, and P. J. Maziasz, in Microstructures and Mechanical Properties of Aging Materials, eds. P. K. Law et al., TMS, Warrendale, Pa., 1993, pp. 197–205.

83. H. Kwon, C. M. Kim, K. B. Lee, H. R. Yang, and J. H. Lee, *Met. and Mats. Trans.*, vol. 28A, 1997, pp. 621–627.

83a. J. M. Dahl, *Adv. Mater. & Proc.*, vol. 157, no. 5, 2000, pp. 33–36.

83b. P. M. Novotny and T. J. McCauffrey, *An Advanced Alloy for Landing Gear and Aircraft Structural Applications—AerMet 100 Alloy*, SAE Technical Paper 922040, Aerotech '92, Anaheim, Calif., October 5–8, 1992, pp. 1–7.

84. *Carbon and Alloy Steels: ASM Specialty Handbook*, ASM International, Materials Park, Ohio, 1996, pp. 637–643.

85. R. Ayer and P. Machmeier, *Met. and Mats. Trans.*, vol. 29A, 1998, pp. 903–905.

85a. R. P. Foley, V. Frainor, G. Baozhu, and G. Krauss, *38th MWSP Conf. Proc.*, ISS-AIME, vol. 32, 1995, pp. 425–442.

86. C. H. Yoo, H. M. Lee, J. W. Chan, and J. W. Morris, Jr., *Met. and Mats. Trans.*, vol. 27A, 1996, pp. 3466–3472.

87. B. A. Fuller and R. D. Garwood, *JISI*, vol. 210, 1972, p. 206.

88. L. K. V. Lou and D. H. Jack, *Met. Sci*, vol. 11, 1977, p. 46.

89. R. D. Garwood and C. A. Waine, in *Conference Proceedings on Solid → Solid Phase Transformations*, eds. H. I. Aaronson et al., TMS-AIME, Warrendale, Pa., 1982, pp. 875–879.

90. H. Berns and V. Gavriljuk, *J. de Physique IV*, vol. 7, 1997, pp. C5-263–C5-268.

91. J. H. Hollomon and L. D. Jaffe, *Trans. AIME*, vol. 162, 1945, p. 2023.

92. R. A. Grange and R. W. Baughman, *ASM Trans.*, vol. 48, 1956, p. 165.

93. R. F. Kern and M. E. Suess, *Steel Selection*, Wiley-Interscience, New York, 1979.

94. N. J. Orzechowski and D. L. Hughes, *Ind. Heating*, September 1999, pp. 45–49.

95. K. Weiss, in *8th International Induction Heating Seminar by Inductoheat, Inc.*, Kissimore, Fla., Nov. 3–6, 1998,

96. G. Welch, in *16th ASM Heat Treating Society Conference Proceedings*, 19–21 Mar. 1996, Cincinnati, Ohio, eds. J. L. Dossett and R. E. Luetze, ASM International, Materials Park, Ohio, 1996, pp. 89–93.

97. G. D. Pfaffmann, in *17th ASM Heat Treating Society Conference Proceedings*, 15–18 Sept. 1997, eds. D. L. Milan et al., ASM International, Materials Park, Ohio, 1998, pp. 923–926.

98. K. J. Irvine and F. B. Pickering, in *Physical Properties of Martensite and Bainite*, Iron and Steel Institute, London, Report 93, 1965, p. 126.

99. D. W. Smith and R. F. Hehemann, *JISI*, vol. 209, no. 6, 1971, pp. 476–481.

100. G. Langford and M. Cohen, *Trans. ASM*, vol. 62, 1969, p. 623.

101. H. J. Roberts, *Met. Trans.*, vol. 1, 1970, p. 3287.

102. L. A. Nordstrom, *Met. Sci.*, vol. 10, no. 12, 1976, pp. 429–436.

103. G. R. Speich and R. R. Swann, *JISI*, vol. 203, 1965, pp. 480–485.

104. A. R. Cox, *J. Inst. Met.*, vol. 9, 1968, p. 118.

105. P. M. Kelly and J. Nutting, *Proc. R. Soc. London*, vol. 259A, 1960, p. 45.

106. L. Malik and J. A. Lund, *Met. Trans.*, vol. 3, 1972, pp. 1403–1406.

107. M. F. Ashby, in *Oxide Dispersion Strengthening*, AIME, Warrendale, Pa., 1966, p. 143.

107a. K. Onel and J. Nutting, *Met. Sci.*, vol. 13, no. 10, 1979, p. 573.

108. F. B. Pickering, in *Materials Science and Technology*, vol. 7: *Constitution and Properties of Steels*, vol. ed. F. B. Pickering, VCH, Weinheim, 1992, pp. 41–94.

109. J. Wadsworth, J. H. Woodhead, and S. R. Keown, *Met. Sci.*, vol. 10, 1976, p. 342.

110. F. B. Pickering, in *Tool Materials for Molds and Dies*, eds. G. Krauss and H. Nordberg, Colorado School of Mines Pub. Dept., Golden, 1987, p. 3.

111. I. Daigne, M. Guttmann, and J. P. Naylor, *Mater. Sci. Eng.*, vol. 56, 1982, pp. 1–10.

112. E. A. Little, D. R. Harries, F. B. Pickering, and S. R. Keown, *Met. Tech.*, vol. 4, 1977, p. 205.

113. Y. Tomita and K. Okabayashi, *Met. Trans.*, vol. 17A, 1986, pp. 1203–1209.

114. H. Nordberg, in *Tools for Die Casting*, Uddeholm/Swedish Institute for Metals Research, Stockholm, 1983, p. 1.

115. J. F. Knott, in *Inclusions and Their Effects on Steel Properties*, British Steel Corporation, Rotherham, 1974, paper 8.

116. D. Dulieu and A. J. Gouch, in *Inclusions and Their Effects on Steel Properties*, British Steel Corporation, Rotherham, 1974, paper 16.

117. E. Zia-Ebrahimi and G. Krauss, *Met. Trans.*, vol. 14A, 1983, pp. 1109–1119; *Acta Metall.*, vol. 32, 1984, pp. 1767–1777.

118. *Metals Handbook*, vol. 11, 9th ed., ASM, Metals Park, Ohio, 1986.

119. *Carbon and Alloy Steels: ASM Specialty Handbook*, ASM International, Materials Park, Ohio, 1996, pp. 308–328.

120. N. Bandopadhyay and C. J. McMahon, Jr., *Met. Trans.*, vol. 14A, 1983, pp. 1313–1325.

121. H. Kwon and C. H. Kim, *Met. Trans.*, vol. 15A, 1984, pp. 393–395.

122. C. L. Briant and S. K. Banerji, in *Treatise of Materials Science and Technology*, vol. 25, eds. C. L. Briant and S. K. Banerji, Academic Press, New York, 1983, pp. 22–68.

122a. C. J. McMahon, Jr., *Materials Characterization*, vol. 26, 1991, pp. 269–287.

123. J. P. Materkowski and G. Krauss, *Met. Trans.*, vol. 10A, 1979, p. 1643.

124. G. Thomas, *Met. Trans.*, vol. 9A, 1978, pp. 439–450.

125. D. H. Huang and G. Thomas, *Met. Trans.*, vol. 2, 1971, pp. 1587–1598.

126. R. M. Horn and R. O. Ritchie, *Met. Trans.*, vol. 9A, 1978, pp. 1039–1053.

127. H. K. D. H. Bhadeshia and D. V. Edmonds, *Met. Sci.*, vol. 13, 1979, p. 325.

128. C. Kim, A. R. Johnson, and Y. F. Hosford, *Met. Trans.*, vol. 13A, 1982, pp. 1595–1605.

129. T. Ogura, C. J. McMahon, Jr., H. C. Feng, and V. Vitek, *Acta Metall.*, vol. 26, 1978, p. 1317.

129a. C. J. McMahon, Jr., private communication, 2000.

130. J. M. Capus and G. Mayer, *Metallurgia*, vol. 62, 1960, p. 133; *JISI*, vol. 201, 1963, p. 53.

131. C. L. Briant and S. K. Banerjee, *Scr. Met.*, vol. 13, 1979, p. 813; *Met. Trans.*, vol. 10A, 1979, p. 1151; *Met. Trans.*, vol. 12A, 1981, p. 309.

132. C. J. McMahon, Jr., W. Yu-qing, M. J. Morgan, and M. Manyhard, *Alloy Element Effects in Temper Embrittlement*, 4th Japan Institute of Metals Symposium: Grain Boundary Structure and Related Phenomena, Minakami, Spa Japan, November 25–29, 1985.

132a. W. J. Nam and H. C. Choi, *Mater. Sc. and Technol.*, vol. 13, July 1997, pp. 568–574.

133. I. Olefjord, *Int. Met. Rev.*, vol. 23, 1978, pp. 149–163.

134. J. R. Low, Jr., *Fracture of Engineering Materials*, ASM, Cleveland, Ohio, 1964.

135. J. M. Capus, in *Temper Embrittlement in Steel*, STP 407, ASTM, Philadelphia, Pa., 1968, pp. 3–19.

136. C. J. McMahon, Jr., in *Temper Embrittlement in Steel*, STP 407, ASTM, Philadelphia, Pa., 1968, pp. 127–176.

137. C. J. McMahon, Jr., in *Encyclopaedia of Materials Science and Engineering*, Pergamon Press, Oxford, 1986, pp. 4862–4866.

138. J. I. Ustinovschikov, *Acta Metall.*, vol. 31, 1983, pp. 355–364.

138a. C. J. McMahon, Jr., *Materials Sc. Forum*, vol. 46, 1989, pp. 61–76.

139. R. B. Clough and H. N. G. Wadley, *Met. Trans.*, vol. 13A, 1982, pp. 1965–1975; B. J. Schultz and C. J. McMahon, Jr., STP 499, ASTM, Philaldelphia, Pa., 1971, p. 104.

140. K. L. Moloznik, C. L. Briant, and C. J. McMahon, Jr., *Corrosion*, vol. 35, 1979, pp. 331–332.

141. R. O. Ritchie and J. F. Knott, *Acta Metall.*, vol. 21, 1973, pp. 639–648.

142. C. J. McMahon, Jr., *Z. Metallk.*, vol. 12, 1984, pp. 496–509.

143. R. A. Mulford, C. J. McMahon, Jr., D. P. Pope, and H. C. Feng, *Met. Trans.*, vol. 7, 1976, pp. 1183–1195.

144. C. L. Briant, C. J. McMahon, Jr., and H. C. Feng, *Met. Trans.*, vol. 9A, 1978, p. 625.

145. M. Guttmann, P. R. Krahe, F. Amel, G. Amsel, M. Brunaux, and C. Cohen, *Met. Trans.*, vol. 5, 1974, p. 167.

146. M. Hasegawa, N. Nakijama, N. Kusunoki, and K. Suzuki, *Trans. Jpn. Inst. Met.*, vol. 16, 1975, p. 641.

147. B. L. King and G. Wigmore, *Met. Trans.*, vol. 7, 1976, p. 1761.

148. J. M. Capus, *JISI*, vol. 200, 1962, p. 922.

149. C. J. McMahon, Jr., *Mater. Sci. Eng.*, vol. 25, 1976, p. 233.

150. R. G. Faulkner, S.-H. Song, and P. E. J. Flewitt, *Mater. Sci. and Technol.*, vol. 12, 1996, pp. 818–822.

151. H. Erhart and H. J. Grabke, *Met. Sci.*, vol. 15, 1981, p. 401.

152. J. R. Low, Jr., D. F. Stein, A. M. Turkalo, and R. P. Laforce, *TMS-AIME*, vol. 242, 1968, p. 14.

153. R. A. Mulford, C. J. McMahon, Jr., D. P. Pope, and H. C. Feng, *Met. Trans.*, vol. 7, 1976, p. 1269.

154. J. F. Smith, J. H. Reynolds, and H. N. Southworth, *Acta Metall.*, vol. 28, 1980, p. 1555.

154a. W. Yu-Qing and C. J. McMahon, Jr., *Mater. Sci. and Technol.*, vol. 3, 1987, pp. 207–216.

155. W. Yu-Qing and C. J. McMahon, Jr., in *The Effect of Mn on Intergranular Embrittlement of Iron and Steel*, 4th Japan Institute of Metals International Symposium: Grain Boundary Structure and Related Phenomena, Minakami, Spa Japan, Nov. 25–29, 1985.

156. M. Paju and H. J. Grabke, *Mater. Sci. and Technol.*, vol. 5, 1989, pp. 148–154.

157. I. A. Vatter, C. E. Lane, and C. A. Hippsley, *Mater. Sci. and Technol.*, vol. 9, 1993, pp. 915–922.

158. J. Yu and C. J. McMahon, Jr., *Met. Trans.*, vol. 11A, 1980, pp. 277–289.

159. J. Yu and C. J. McMahon, Jr., *Met. Trans.*, vol. 11A, 1980, p. 291.

160. N. Bandopadhyay, C. L. Briant, and E. Hall, *Met. Trans.*, vol. 16A, 1985, pp. 721–736.

161. N. S. Cheruvu and B. B. Seth, *Met. Trans.*, vol. 20A, 1989, pp. 2345–2354.

162. S. R. Holdsworth and D. V. Thornton, in *Microstructure and Mechanical Properties of Aging Materials*, eds. P. K. Laiw et al., TMS, Warrendale, Pa., 1993, pp. 83–89.

163. B. C. Woodfine, *JISI*, vol. 173, March 1953, pp. 240–255.

164. S. Takayama, T. Ogura, S. C. Fu, and C. J. McMahon, Jr., *Met. Trans.*, vol. 11A, 1980, pp. 1513–1530.

165. P. W. Palmberg and H. L. Marcus, *Trans. ASM*, vol. 62, 1969, pp. 1016–1018.

166. J. Yu and C. J. McMahon, Jr., *Met. Trans.*, vol. 16A, 1985, pp. 1325–1331.

167. M. Guttmann, Ph. Dumoulin, and M. Wayman, *Met. Trans.*, vol. 13A, 1982, pp. 1693–1711.

168. J. H. Bulloch and J. J. Hickey, in *Microstructure and Mechanical Properties of Aging Materials*, eds. P. K. Laiw et al., TMS, Warrendale, Pa., 1993, pp. 41–48.

168a. J. H. Bulloch, *Int. Jn. of Pressure Vessels and Piping*, vol. 76, 1999, pp. 63–78.

169. T. Ogura, *Met. Trans.*, vol. 13A, 1982, pp. 2205–2207.

170. E. C. Edwards, H. E. Bishop, J. C. Riviere, and B. L. Eyre, *Acta Metall.*, vol. 24, 1976, pp. 957–967.

171. J. Janovec, P. Sevc, and M. Koutnik, *Kovine Zlitine Technologije*, vol. 29, 1995, nos. 1–2, pp. 40–44.

172. L. T. Lei, C. H. Tang, and M. Su, *Heat Treatment '81*, The Metals Society, London, 1981, pp. 5.1–5.10.

173. H. Kwon, *Met. Trans.*, vol. 22A, 1991, pp. 1119–1122.

174. C. L. Briant and S. K. Banerji, *Int. Met. Rev.*, no. 4, 1978, pp. 164–199.

175. C. H. Lorig and A. R. Elsea, *Trans. AFS*, vol. 55, 1947, pp. 160–174.

176. B. C. Woodfine and A. G. Quarrell, *JISI*, vol. 195, 1960, pp. 409–414; J. A. Wright and A. G. Quarrell, *JISI*, vol. 200, 1962, pp. 299–307.

177. N. H. Croft, *Met. Technol.*, vol. 10, 1983, pp. 285–290; N. H. Croft et al., *Met. Technol.*, vol. 10, 1983, pp. 125–129; N. H. Croft et al., *Advances in the Physical Metallurgy and Applications of Steels*, Book 284, The Metals Society, London, 1982, pp. 286–295.

178. S. C. Desai, *JISI*, vol. 191, 1959, pp. 250–256.

179. R. Harris and L. Barnard, *Deformation under Hot Working Conditions*, S. R. 108, Iron and Steel Institute, 1968, pp. 167–177.

180. L. A. Erasmus, *JISI*, vol. 202, 1964, pp. 32–41.

181. G. D. Funnell and R. J. Davies, *Met. Technol.*, vol. 5, 1978, pp. 150–153.

182. W. J. M. Salter, *JISI*, vol. 207, 1969, pp. 1619–1623.

183. L. Gertsman and H. P. Tardif, *Iron Age*, vol. 169, no. 7, Feb. 14, 1952, pp. 136–140.

184. *Carbon and Alloy Steels*: ASM Specialty Handbook, ASM International, Materials Park, Ohio, 1996, pp. 477–488.

185. H. K. Birnbaum, in *Encyclopedia of Materials Science and Engineering*, Pergamon Press, Oxford, 1986, p. 2240.

186. H. G. Nelson, *Treatise of Materials Science and Technology*, vol. 25: *Embrittlement of Engineering Alloys*, eds. C. L. Briant and S. K. Banerji, Academic Press, New York, 1983, pp. 275–359.

187. D. A. Meyn, *Met. Trans.*, vol. 5, 1974, pp. 2405–2414.

188. K. S. Chan, *Met. Trans.*, vol. 24A, 1993, pp. 1095–1105.

189. M. Nakamura and T. Kumagai, *Met. and Mats. Trans.*, vol. 30A, 1999, p. 3089.

190. P. Rozenak and D. Eliezer, *Met. Trans.*, vol. 19A, 1988, pp. 723–730.

191. B. C. Odegard, J. A. Brooks, and A. J. West, in *Effects of Hydrogen Behavior of Metals*, eds. A. W. Thompson and I. M. Bernstein, TMS-AIME, New York, 1976, pp. 116–128.

192. I. M. Bernstein and A. W. Thompson, in *Encyclopedia of Materials Science and Engineering*, Pergamon Press, Oxford, 1986, p. 2241–2245.

193. T. M. Harris and R. M. Latanision, *Met. Trans.*, vol. 22A, 1991, pp. 351–355.

194. R. Valentini, A. Solina, C. Matera, and D. De Gregorio, *Met. and Mats. Trans.*, vol. 27A, 1996, pp. 3773–3780.

195. I. Takahashi, Y. Matsumoto, and T. Tanada, *Proc. JIMIS-2*, Minakami, Tokyo, Japan, 1979, pp. 285–289.

196. G. M. Pressouyre and I. M. Bernstein, *Met. Trans.*, vol. 9A, 1978, pp. 1571–1580.

197. X. Chen and W. W. Gerberich, *Met. Trans.*, vol. 22A, 1991, pp. 59–70.

198. R. G. Kelly, A. J. Frost, T. Shaharabi, and R. C. Newman, *Met. Trans.*, vol. 22A, 1991, pp. 531–541.

199. C. G. Interrrante, in *International Conference Proceedings on Hydrogen Problems in Steels*, ASM, Metals Park, Ohio, 1982, pp. 3–17.

200. J. P. Hirth, in *Hydrogen Embrittlement and Stress Corrosion Cracking*, eds. R. Gibala and R. F. Hehemann, ASM, Metals Park, Ohio, 1984, pp. 29–41.

201. P. Bastien and P. Azou, Effect of Hydrogen on the Deformation and Fracture of Iron and Steel in Simple Tension, in *Proceedings of the First World Metals Congress*, ASM, Cleveland, Ohio, 1951, pp. 535–552.

202. J. K. Tien, S. V. Nair, and R. R. Jensen, in *Hydrogen Effects in Metals*, eds. I. M. Bernstein and A. W. Thompson, TMS-AIM, Warrendale, Pa., 1980, pp. 37–56.

203. C. D. Beachem, *Met. Trans.*, vol. 3, 1972, pp. 437–451.

204. S. P. Lynch, *Metal Forum*, vol. 2, no. 3, 1979, pp. 189–200; *Acta Metall.*, vol. 36, 1988, pp. 2639–2661.

205. J. Eastman, F. Heubaum, T. Matsumoto, and H. K. Birnbaum, *Acta Metall.*, vol. 30, 1982, pp. 1579–1586.

206. D. S. Shih, I. M. Robertson, and H. K. Birnbaum, *Scripta Metall.*, vol. 36, 1988, pp. 111–116.

207. S. Sun, K. Shiozawa, J. Gu, and N. Chen, *Met. and Mats. Trans*, vol. 26A, 1995, pp. 731–739.

208. A. R. Troiano, *Trans. ASM*, vol. 52, 1960, pp. 54–80.

209. R. A. Oriani, in *Hydrogen Embrittlement and Stress Corrosion Cracking*, eds. R. Gibala and R. F. Hehemann, ASM, Metals Park, Ohio, 1984, pp. 43–59.

210. H. H. Johnson, *Met. Trans.*, vol. 19A, 1988, pp. 2371–2387.

211. C. M. Li, R. A. Oriani, and L. W. Darken, *Z. Phys. Chem.*, vol. 49, 1966, p. 271.

212. H. Wagenblast and H. A. Wriedt, *Met. Trans.*, vol. 2, 1971, p. 1393.

213. B. Baranowski, S. Majchrzak, and T. B. Flangan, *J. Phys. F. Met. Phys.*, vol. 1, 1971, pp. 258–261.

214. J. P. Hirth, *Met. Trans.*, vol. 11A, 1980, p. 861.

215. N. R. Quick and M. A. Johnson, *Acta Metall.*, vol. 26, 1978, p. 906.

216. J. R. Rice and M. A. Johnson, *Inelastic Behavior of Solids*, ed. M. F. Kanninen, McGraw-Hill, New York, 1970, p. 164.

217. N. Bandopadhyay, J. Kameda, and C. J. McMahon, Jr., *Met. Trans.*, vol. 14A, 1983, pp. 881–888.

218. N. J. Petch, *Philos. Mag.*, vol. 1, 1956, pp. 331–335.

219. C. A. Zapffe and C. E. Sims, *Trans. AIME*, vol. 145, 1941, pp. 225–259.

220. T. Taira et al., in *International Conference Proceedings on Hydrogen Problems in Steels*, eds. C. G. Interrante and G. M. Pressouyre, ASM, Metals Park, Ohio, 1982, pp. 173–185.

220a. H. Suzuki, *Jpn. Weld. Soc. Bulletin*, no. 1, 1977.

221. C. A. Hippsley, *Mater. Sci. Technol.*, vol. 3, 1987, pp. 912–922.

222. D. P. Williams and H. G. Nelson, *Met. Trans.*, vol. 3A, 1972, pp. 2107–2113.

223. R. O. Ritchie, in *Proceedings of the Conference on Analytical and Experimental Fracture Mechanics*, Rome, June 1980, pp. 81–108.

224. S. Suresh, C. M. Moss, and R. O. Ritchie, *Trans. Jpn. Inst. Met.*, 1980, vol. 21, 1980, p. 481.

224a. N. Yurioka and H. Suzuki, *Int'l. Mats. Reviews*, vol. 35, no. 4, 1990, pp. 217–249.

225. F. H. Huang and W. J. Mills, *Met. Trans.*, vol. 22A, 1991, pp. 2049–2060.

226. K. Yoshino and C. J. McMahon, Jr., *Met. Trans.*, vol. 5, 1974, pp. 363–370.

227. Y. Takeda and C. J. McMahon, Jr., *Met. Trans.*, vol. 12A, 1981, pp. 1255–1266.

228. C. L. Briant, H. C. Feng, and C. J. McMahon, Jr., *Met. Trans.*, vol. 9A, 1978, pp. 625–633.

229. J. Kameda and C. J. McMahon, Jr., *Met. Trans.*, vol. 14A, 1983, pp. 903–911.

230. S. K. Banerji, C. J. McMahon, Jr., and H. C. Feng, *Met. Trans.*, vol. 9A, 1978, pp. 237–246.

231. R. T. Hill and M. Iinoe, in *International Conference Proceedings on Hydrogen Problems in Steels*, eds. C. G. Interrante and G. M. Pressouyre, ASM, Metals Park, Ohio, 1982, pp. 196–199.

232. H. F. López, R. Raghunath, J. L. Albarran, and L. Martinez, *Met. and Mats. Trans.*, vol. 27A, 1996, pp. 3601–3611.

233. D. L. Sponseller, R. Garber, and T. B. Cox, in *International Conference Proceedings on Hydrogen Problems in Steels*, eds. C. G. Interrante and G. M. Pressouyre, ASM, Metals Park, Ohio, 1982, pp. 200–211.

234. J. L. Abarran, L. Martinez, and H. F. López, *British Corr. Jn.*, vol. 33, no. 3, 1998, pp. 202–205.

234a. M. Okatsu, F. Kawabata, and K. Amano, pp. 119–124, in *Proceedings of 16th International Conference on Offshore Mechanics and Arctic Engineering*, part III, Yokohama, Japan, 1997.

235. R. H. Hull and M. Lino, in *International Conference Proceedings on Hydrogen Problems in Steels*, eds. C. G. Interrante and G. M. Pressouyre, ASM, Metals Park, Ohio, 1982, pp. 196–199.

236. I.-F. Tsu and T.-P. Perng, *Met. Trans.*, vol. 22A, 1991, pp. 215–224.

237. M. Hasegawa and S. Nomura, *JISI, Japan*, vol. 7, 1977, p. 187.

238. J. Chene, in *International Conference Proceedings on Hydrogen Problems in Steels*, eds. C. G. Interrante and G. M. Pressouyre, ASM, Metals Park, Ohio, 1982, pp. 263–271.

239. H. Moikawa, T. Murata, and T. Inoue, in *International Conference Proceedings on Hydrogen Problems in Steels*, eds. C. G. Interrante and G. M. Pressouyre, ASM, Metals Park, Ohio, 1982, pp. 219–224.

240. H. J. Grabke, *Chemistry and Physics of Fracture*, Martinus Nijhoff Publishers/NATO Scientific Affairs, Boston, 1987, pp. 388–419.

241. J. R. Rice and J. S. Wang, *Presented at Symp. on Interfacial Phenomena in Composites*, Newport, Rhode Island, June 1988.

242. B. E. Wilde, C. D. Kim, and J. C. Turn, Jr., *Corrosion*, vol. 38, no. 10, 1982, pp. 515–524.

243. J. R. Scully, J. A. Van Den Avyle, M. J. Cieslak, and A. D. Romig, Jr., *Met. Trans.*, vol. 22A, 1991, pp. 2429–2444.

244. B. E. Wilde, I. Chattoraj, and T. A. Mozhi, *Scripta Metall.*, vol. 21, 1987, pp. 1369–1373.

245. S. Talbot-Besnard, in *International Conference Proceedings on Hydrogen Problems in Steels*, eds. C. G. Interrante and G. M. Pressouyre, ASM, Metals Park, Ohio, 1982, pp. 37–41.

246. A. Takasaki, Y. Furuya, and Y. Tanada, *Met. and Mats. Trans.*, vol. 29A, 1998, pp. 307–314.

247. A. Yu Kazanskaya, M. A. Smirnov, and V. V. Zabil'skiy, *Phys. Met. Metall.*, vol. 70, no. 2, 1990, pp. 191–193.

248. S. Suwas, Master's thesis, Indian Institute of Technology, Kanpur, India, 1993, p. 249.

249. A. Agarwal, R. Baalsubramaniam, and S. Bhargava, *Met. and Mats. Trans.*, vol. 27A, 1996, pp. 2985–2993.

250. E. D. McCarty, D. Wetzel, and B. S. Kloberdanz, in *Steel Products and Processing for Automotive Applications*, SP-1172, SAE, Warrendale, Pa., 1996, pp. 117–129.

250a. S. M. Schlogl, Y. V. Leeuwen, and E. V. D. Giessen, *Met. and Mats. Trans.*, vol. 31A, 2000, pp. 125–137.

251. P. G. Shewmon and Y. H. Xue, *Met. Trans.*, vol. 22A, 1991, pp. 2703–2707.

252. *Steels for Hydrogen Service at Elevated Temperatures and Pressures in Petroleum, Refineries, and Petrochemical Plants*, 5th ed., Publication 941, American Petroleum Institute, January 1997.

252a. F. H. Vitovec, in *International Conference Proceedings on Hydrogen Problems in Steels*, eds. C. G. Interrante and G. M. Pressouyre, ASM, Metals Park, Ohio, 1982, pp. 236–241.

253. C. G. Interrante, G. A. Nelson, and C. M. Hugens, *Weld. Res. Council Bull.*, vol. 145, 1969, pp. 33–42.

254. L. W. Tsway and W. L. Lin, *Corrosion Science*, vol. 40, no. 415, 1998, pp. 577–591.

255. I. Masoka et al., in *International Conference Proceedings on Hydrogen Problems in Steels*, eds. C. G. Interrante and G. M. Pressouyre, ASM, Metals Park, Ohio, 1982, pp. 242–248.

256. J. E. Steiner, in *International Conference Proceedings on Hydrogen Problems in Steels*, eds. C. G. Interrante and G. M. Pressouyre, ASM, Metals Park, Ohio, 1982, pp. 55–62.

256a. G. Nong and Y. Weixum, *HSLA Steels: Processing, Properties, and Applications*, eds. G. Tither and Z. Shauhua, TMS, Warrendale., Pa., 1992, pp. 245–249.

257. *Carbon and Alloy Steels: ASM Specialty Handbook*, ASM International, Materials Park, Ohio, 1996, pp. 489–499.

258. M. H. Kamdar, in *Treatise on Materials Science and Technology*, vol. 25: *Embrittlement of Engineering Alloys*, eds. C. L. Briant and S. K. Banerjee, Academic Press, New York, 1983, pp. 361–459.

259. S. P. Lynch, *Acta Metall.*, vol. 32, 1984, pp. 79–90.

260. A. R. Huntington, *J. Inst. Met.*, vol. 11, 1914, pp. 108–109.

261. S. Moslovoy and N. N. Breyer, *Trans. ASM*, vol. 61, no. 2, 1968, pp. 219–232.

262. Y. Asayama, in *Embrittlement by Liquid and Solid*, ed. M. H. Kamdar, AIME, Warrendale, Pa., 1984.

263. J. C. Lynn et al., *Mater. Sc. Engineer*, vol. 18, March 1975, pp. 51–62.

264. R. F. Decker, J. T. Each, and A. J. Goldman, *Trans. ASM*, vol. 55, 1962, p. 58.

265. S. Floreen, *Met. Rev.*, vol. 13, 1968, pp. 115–128.

266. S. Floreen, in *Encyclopedia of Materials Science and Engineering*, Pergamon Press, Oxford, 1986, pp. 5171–5177; *Metals Handbook*, vol. 4, 9th ed., ASM, Metals Park, Ohio, 1978, pp. 130–132.

267. T. Morrison, *Metall. Mater. Technol.*, vol. 8, 1976, pp. 80–85.

268. K. Stiller, M. Hattestrand, and F. Danoix, *Acta Mater.*, vol. 46, no. 17, 1998, pp. 6063–6073.

269. W. W. Cias, *Metall. Met. Form.*, vol. 38, no. 12, 1971, pp. 356–359.

270. F. H. Lang and N. Kenyon, *Weld Res. Council Bull.*, vol. 159, 1971.

271. B. R. Banerjee and J. J. Houser, *Symposium on Transformation and Hardenabilty in Steel*, Climax Molybdenum Co., Ann Arbor, Mich., 1967, pp. 133–150.

271a. M. S. Schmidt, Carpenter Technology Corp., private communication, 2000.

272. P. Liu, A. H. Stigenberg, and J.-O. Nilsson, *Fourth European Conference on Advanced Materials and Processes*, Padua–Venice, Italy, 25–28 Sept. 1995.

273. K. Rohrbach and M. Schmidt, *ASM Handbook*, vol. 1, 10th ed., ASM International, Materials Park, Ohio, 1990, pp. 793–800.

274. R. F. Decker, in *Source Book on Maraging Steel*, ASM, Metals Park, Ohio, 1979.

275. J. N. Pennington, *Modern Metals*, January 1994, pp. 106J–106M.

276. B. N. Schaf, *Met. Prog.*, vol. 95, no. 2, 1969, pp. 74–76.

277. G. Mayer, *Metall. Mater. Technol.*, vol. 9, no. 5, 1977, pp. 255–261.

278. S. Floreen and R. F. Decker, *Trans. ASM*, vol. 55, 1962, pp. 518–530.

279. M. D. Parker, *Met. Sci. Heat Treat.*, no. 7, 1970, p. 558.

280. M. Schmidt and K. Rohrbach, *ASM Handbook*, vol. 4, 10th ed., ASM International, Materials Park, Ohio, 1991, pp. 219–228.

281. D. T. Peters and C. R. Cupp, *Trans. AIME*, vol. 236, 1966, p. 1420.

282. G. P. Miller and W. I. Mitchell, *JISI*, vol. 203, 1965, p. 899.

283. S. Floreen and G. R. Speich, *Trans. ASM*, vol. 57, 1964, p. 714.

284. H. J. Rack and D. Kalish, *Met. Trans.*, vol. 2, 1971, p. 3011.

285. B. R. Banerjee and J. J. Hauser, *Technical Report 66-166*, Air Force Laboratory, 1966.

286. M. J. Fleetwood, G. M. Higginson, and G. P. Miller, *J. Appl. Phys.*, vol. 16, 1965, p. 645.

287. D. T. Peters, *Trans. ASM*, vol. 61, 1968, p. 62.

288. M. Ahmed, A. Ali, S. K. Hasnain, F. H. Hashmi, and A. Q. Khan, *Acta Metall. Mater.*, vol. 42, 1994, p. 631.

289. M. Farooque, H. Ayub, A. U. Haq, and A. Q. Khan, *Jn. of Mater. Sc.*, vol. 33, 1998, pp. 2927–2930.

290. J. W. Morris, Jr., J. I. Kim, and C. K. Syn, in *Advances in Metal Processing*, vol. 25: *Sagamore Army Material Research Conference Proceedings, 1978*, eds. J. J. Burke, R. Mehrabian, and V. Weiss, 1981, p. 173.

291. E. Nes and G. Thomas, *Met. Trans.*, vol. 7, 1976, p. 967.

292. P. P. Sinha, D. Sivakumar, T. Tharian, K. V. Nagarajan, and D. S. Sarma, *Mater. Sci. and Technol.*, vol. 12, 1996, pp. 945–954.

CHAPTER 15
THERMOMECHANICAL TREATMENT

15.1 INTRODUCTION

Thermomechanical treatment (TMT) or thermomechanical processing (TMP) involves a combination of heat treatment, plastic deformation, and phase transformation within a single hot deformation process (e.g., rolling, forging, extrusion), to control the structure, morphology, and grain size of parent phase (e.g., austenite) and subsequent grain and subgrain structure in the transformation product phases (such as ferrite) and/or interactions between dislocations and fine precipitates (e.g., alloy carbides and/or carbonitrides). Its main objective is to achieve microstructural refinement to produce ultrahigh strength, improved ductility, and greater toughness in a wide variety of finished and semifinished steel or nonferrous products.[1-5] Ferrous TMT has been extensively studied as evidenced by numerous detailed review papers by a number of authors[6-10] and several international conferences on TMT and microalloying in steels in late 1990s.[11-14] Nonferrous TMT (or TMP), though not as extensively studied as in steel, has useful potential applications in nickel-base superalloys and aluminum-, zinc-, titanium-, and copper-base alloys. This chapter describes both the ferrous and nonferrous thermomechanical treatments as well as applications of TMT to steels such as transformation-induced plasticity (TRIP) steels, dual-phase (DP) steels, ultrahigh-strength steels (UHSS), and ultrahigh carbon (UHC) steels, and to ductile irons. The former two steels are based, respectively, on the transformation of metastable austenite into martensite by plastic deformation and on intercritical annealing. In view of the great technological importance of superplastic forming, the final section of this chapter deals with superplastic materials and processing and their aerospace applications.

15.2 FERROUS TMT

15.2.1 Classification of TMT

Radcliffe and Kula[6] and, later, Greyday[10] classified the ferrous TMT process into three broad categories, depending on the introduction of the deformation process before, during, or after the phase transformation[7,8] (Fig. 15.1):[10]

Deformation temperature and metal phase	Name	Lay-out of the treatment	Treatment suitable to
Before transformation	(a) HTMT	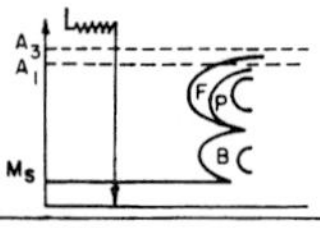	high or low alloy steels
Beyond A$_3$ stable austenite	(b) Controlled rolling	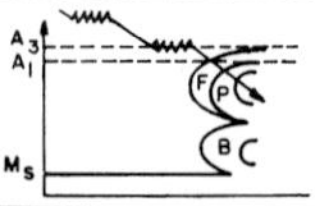	structural steels
Below A$_3$ metastable austenite	(c) LTMT ausforming	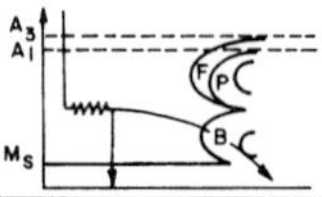	high or low alloy steels
During transformation between A$_1$ and M$_s$ austenite and ferrite pearlite or pearlite	(d) isoforming	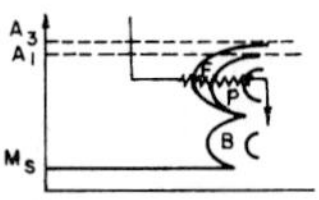	pearlite-forming steels
near M$_s$ austenite and martensite	(e) * treatment of TRIP steels ** zerolling	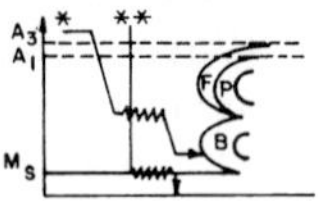	stainless or semi-stainless steels
below M$_s$ martensite	(f) marstraining	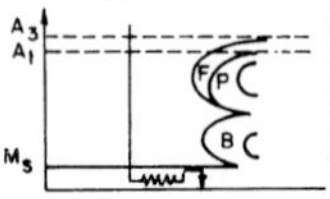	martensitic steels

FIGURE 15.1   Classification of thermomechanical treatment of steels.[10] (*Reprinted by permission of Pergamon Press, Plc.*)

1. Class I: Deformation is completed prior to the transformation of austenite. Examples are high-temperature thermomechanical treatment (HTMT) (Fig. 15.1a), controlled rolling (Fig. 15.1b), and low-temperature thermomechanical treatment (LTMT), also called *ausforming* (Fig. 15.1c). HTMT and LTMT are applied to martensitic steels (i.e., high- and low-alloy steels with increased hardenability). Controlled rolling and thermomechanical controlled processing have been developed for microalloyed ferrite-pearlite steels, acicular ferritic (i.e., low-carbon bainitic) steels, as-rolled pearlitic steels, and steels to be quenched and tempered.[15] Prominent among these are microalloyed (low-carbon low-alloy C-Mn) steels, which are extensively used as hot-rolled plate, sheet, and strip.

2. Class II: Deformation occurs during the transformation of austenite. Examples are the isoforming process, where spheroidal carbides within a ferrite matrix form during deformation of metastable austenite (Fig. 15.1d); and treatment of TRIP steels, where metastable austenite is initially deformed above the M_d temperature and finally cold-worked (i.e., zerolled) at room temperature to produce (strain-induced) martensite. The isoforming technique is applied to low-alloy or pearlite-forming steels, while treatment of TRIP steels is confined to stainless or semistainless steels.[10]

3. Class III: Deformation occurs after the transformation of austenite. Examples are marstraining (Fig. 15.1f), (b) marforming, and (c) strain tempering or warmworking. In marforming, the martensite is cold-worked prior to tempering to induce a dislocation substructure which improves the distribution of temper carbides.

15.2.2   Controlled Rolling and Thermomechanical Controlled Processing

Controlled rolling (CR) has been practiced for about five decades, in the manufacture of structural plate steels, and this technology falls under the broad category of TMT. An important difference between *hot rolling* and *controlled rolling* is that roughing and finishing steps are continuous and the nucleation of ferrite occurs primarily at austenite grain boundaries in the former, whereas there is a delay between the roughing and finishing stages and the nucleation of ferrite occurs in both the strain-hardened grain interiors and at grain boundaries in the latter, resulting in more refined grain structure.[16] During hot deformation in TMT operations, the temperature and strain rate may change continually; and in rolling and forging, the deformation occurs in a series of passes separated by intervals of time.[16a]

In controlled rolling, careful control of the time-temperature-deformation sequence is carried out. The main purpose of controlled rolling is to refine grain structure and thereby improve both the strength and toughness of steel in the hot-rolled condition to a level equivalent to, or better than, those of highly alloyed quenched and tempered steels.[15] Controlled rolling is performed on strip, coil plate, bar mills, and heavy plate production and offers a very attractive combination of properties with yield strength up to 700 MPa.

Controlled rolling involves three processes:[17] (1) *recrystallization controlled rolling* (RCR); (2) *dynamic recrystallization controlled rolling* (DRCR), which is used in cases where insufficient time for recrystallization between rolling passes is present; and (3) *conventional controlled rolling* (CCR).

The most important hot deformation parameters are the extent of deformation (or total reduction) below the recrystallization temperature and the temperature of the finishing pass. In general, the more severe the low-temperature deformation rolling, the better the final characteristics, provided the rolling mill capacity is not exceeded.[18]

On-line *accelerated cooling* (AC) following hot/controlled rolling has been recently incorporated for the additional benefits of mechanical properties over the CR. This integrated technology of combined CR and AC, referred to as *thermomechanical controlled processing* (TMCP), where all manufacturing parameters such as chemical composition and processing conditions are optimized and γ/α transformation is controlled to achieve excellent strength and fracture toughness, is now an established practice for flat products such as strip, coil plate, and heavy plate as well as bar products, sections, and rails.[19,20]

TMCP can be used to increase strength at a particular carbon equivalent (CE), to maintain a given strength with a lower CE value, or to increase strength and decrease CE simultaneously. Decreasing C content without advanced processing simply decreases strength. However, the addition of *microalloying elements* (MAEs) together with AC often is used to compensate for reduced strength. The benefits of TMCP decrease with increasing steel strength and thickness, but despite these limitations, the advantages of TMCP can markedly improve steel properties, fabrication, and structural performance.[21] Steels with lower C content and CE values exhibit less degradation of heat-affected zone (HAZ) fracture toughness. Similarly, steels, particularly microalloying steels containing lower C content and lower CE values, can be welded with ease using high-heat-input processes without significant deterioration in HAZ properties.[21]

15.2.2.1 Functions of Microalloying Elements. The main purposes of the addition of MAEs such as Nb, Ti, and V (≤0.1%) in steels are (1) refinement of (*a*) γ-grain size during rolling and (*b*) ferrite grain size after transformation and (2) production of precipitation strengthening in the ferrite lattice and lowering of the transformation temperature, thereby refining again the ferrite grain size with increased dislocation density. Additional benefits of MAEs include (3) inhibition of γ-grain growth, (4) retardation of γ recrystallization, (5) retardation of γ/α transformation, (6) promotion of bainite formation, and (7) removal of N from solid solution through the formation of nitrides.[22,23]

Effective thermomechanical rolling requires the assistance of MAEs that can form an extremely fine dispersion of very small and stable microalloy carbide (NbC, VC, TiC), nitrides (NbN, VN, TiN), and/or carbonitride [NbCN, VCN, TiCN, (Nb,V)CN, (Nb,Ti)CN][24] precipitates with C and N present in steel. It is thus essential for the MAEs to dissolve in the upper range of the austenite and tend to precipitate again in the lower-temperature range of austenite or in the ferrite. These stable second-phase particles form (i) during reheating to the austenite temperature; (ii) during hot rolling (or deformation); (iii) during and after transformation of ferrite; (iv) during and after accelerated cooling; (v) during texture formation; and (vi) in transformed martensite during tempering to provide secondary hardening.[25]

Nb is more effective in increasing strength with small additions, whereas large amounts of Ti (higher than 0.01 to 0.02%) are required, for optimum grain refinement, due to limited solubility of TiC in austenite, and to achieve high strength levels. Hence Ti additions are not extensively used for precipitation strengthening in many steels.[26]

Nb has a very powerful effect in that a normal (0.02 to 0.03%) addition increases the recrystallization stop temperature T_5 to 950 to 1000°C (Fig. 15.2).[27] This effect of MAEs is probably due to a combination of boundary pinning of carbonitride and solute drag effects.

The strengthening of Nb- and Ti-bearing steel plate is attributed to the precipitation of Nb(CN) and TiC, ferrite grain refining, and increased bainite volume fraction. The main factor of strengthening in Ti- and Nb-bearing accelerated cooled steel plate is the increase in bainite volume fraction.[28]

From hot torsion testing it has been observed that the Ti addition to Nb-microalloyed steels leads to a decrease in recrystallization limit temperature T_{nr} during TMT; the decrease becomes profound with larger Ti addition. This effect can be explained by the binding of Nb in stable Ti(C,N), thereby reducing the amount of Nb available for strain-induced precipitation, which tends to prevent recrystallization during hot deformation.[29]

Ti-treated microalloyed steels have been found to exhibit higher σ_p, ε_p, and Q values for deformation than Al-killed steels due to precipitation of TiN particles which retard recrystallization.[30] The addition of Ti to C-Mn steels reduces the austenite grain size up to the slab reheating temperature of 1250°C.[31]

Due to greater solubility of VC, V additions are mostly used for precipitation strengthening. However, additional strengthening by VC can be achieved by increasing N contents, which is attributed to (1) increased stability of V(CN) and

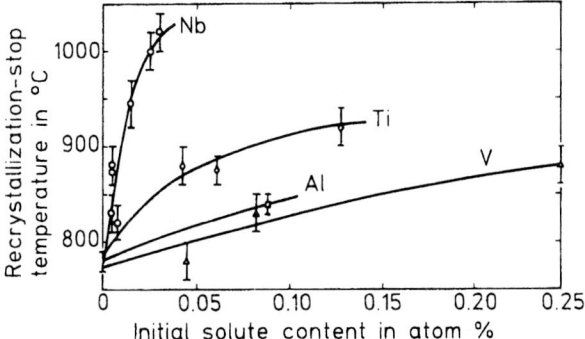

FIGURE 15.2 Increase in the recrystallization stop temperature with increase in the level of microalloy additions in a 0.07C-1.40Mn-0.25Si HSLA steel. Multipass rolling, quenched 10 to 18s after the final pass.[27] (*After L. J. Cuddy, Proc. Conf.: Thermomechanical Processing of Microalloyed Austenite, Pittsburgh, Pa., 1982, TMS-AIME, p. 129.*)

consequent reduced tendency for coarsening, (2) increased degree of dispersion of stable V(CN), and (3) increased volume fraction of V(CN) precipitates resulting from faster precipitation kinetics of VN or N-rich V(CN) compared to VC. The highest volume fraction of precipitates is obtained at a stoichiometric V/N ratio of $3.65:1$.[26,32]

N addition changes the transformation kinetics and structure of the V-microalloyed steels by (1) increasing the Ar_3 temperature, (2) increasing the rate of $\gamma \rightarrow \alpha$ transformation, and (3) decreasing the as-transformed ferrite grain size.[33]

15.2.2.2 Controlled Rolling Procedure.

The present controlled rolling practice for plate or slab steels of suitable composition consists of the following five steps, as shown schematically in Fig. 15.3.[2]

1. Reheating the steel plate or slab to the austenitizing temperature and soaking at this temperature to produce small and uniform γ grains.
2. A series of high-temperature rolling (roughing) passes to continually refine the γ grains by sequential recrystallization.
3. Holding (a time delay) the steel components (between the passes) to a lower temperature to allow a temperature drop below recrystallization to produce a static recrystallization of γ grains.
4. Incorporating a series of low-temperature finish-rolling passes is in the γ nonrecrystallization region to break up and flatten or pancake the recrystallized γ grains, completed above or below the Ar_3 temperature (intercritical rolling in the two-phase γ-α region).
5. Finally, air cooling to form ultrafine ferrite grains from such deformed (pancaked) γ grains (Fig. 15.4).[34]

The TMCP involves all the four steps followed by interrupted accelerated cooling of unrecrystallized austenite, where accelerated cooling is stopped at the optimum temperature, followed by air cooling. AC of unrecrystallized austenite activates

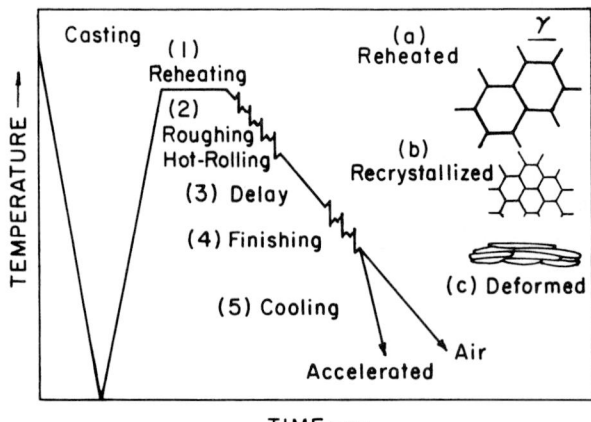

FIGURE 15.3 Controlled rolling (CR) and thermomechanical control process (TMCP).[2] (*Reprinted by permission of The Metallurgical Society, Warrendale.*)

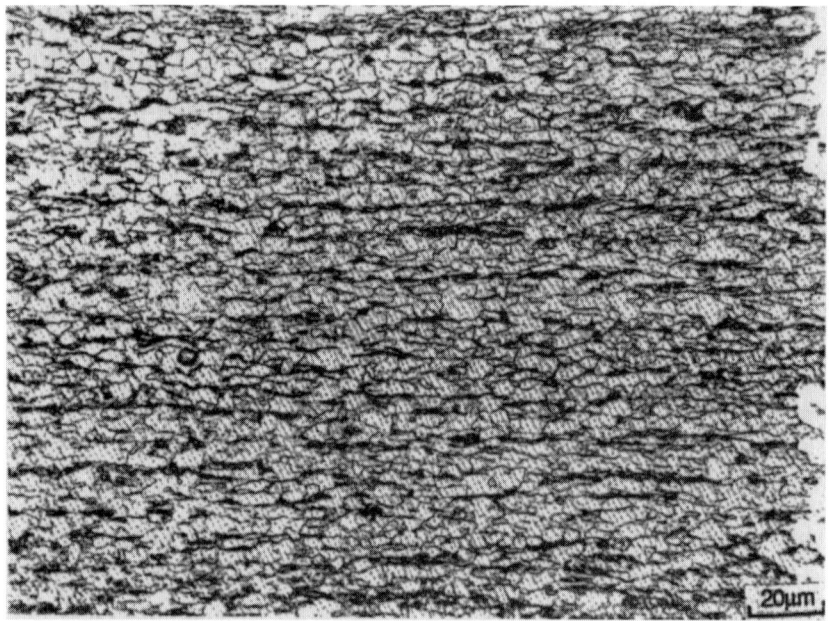

FIGURE 15.4 Ultrafine-grained ferrite produced from flattened unrecrystallized austenite.[34] (*Courtesy of L. J. Cuddy.*)

numerous ferrite nuclei within the austenite grains in addition to those activated by deformation bands. Both these cumulative factors add to the ferrite grain refinement with uniform grain size. In addition, AC modifies the transformed structure by replacing pearlite with an increased volume fraction of finely dispersed bainite as well as produces smaller and more effective microalloy precipitates.[19] All these factors enhance the strength of the product.[35]

Other variations of this process include use of RCR (high reheating temperature to produce coarse γ grain size), one pass heavy-reduction rolling, finish-rolling at low temperature (at or just above Ar_3), followed by AC for plate and strip production or heavy quenching of the strip surface by the work rolls;[36] high slab reheating temperature, large final rolling reduction, high finish-rolling temperature together with accelerated water cooling;[37] application of heavy-reduction (high-accuracy and high-efficiency) technology or rolling metallurgy with a plate pair cross mill;[38] direct quenching as an alternative route to reheat quenching; and water spray or direct quenching following the formation of deformed γ grains.[2,39,40]

In traditional controlled rolling, rolling is accomplished with lower than normal reheating and rolling temperatures (Fig. 15.3),[2] which are varied according to the requirements governed by the desired mechanical properties.[19] The roughing passes are usually separated by far longer interpass times to allow the plate or slab to cool to the specified starting temperature for finish-rolling, while the finishing passes are separated by very short times. Controlled rolled steel comprises α grains and pearlite bands and exhibits slightly large variations in α grain size.[41]

The *slab reheating temperature* (SRT) is an indicator of the initial grain sizes and dissolution of most MAEs. Usually, high reheat temperature produces high strength and poor toughness, while low reheat temperature produces the opposite properties. Hence, thick sections should be rolled at low reheat temperature to obtain optimum toughness. In contrast, thin sections, where high strength is of prime concern, should be rolled from high reheat temperature.

The serious shortcomings of the CCR process are the following:

1. The hold period necessary to reach the T_R temperature after the high-temperature roughing passes causes excessive processing time, which decreases mill productivity. However, application of accelerated cooling during the delay/hold period at the roughing stand exit temperature for 10 to 15 s has resulted in the improvement of productivity up to 300% in API X65 and API X70 linepipe steel plates and upgrading of yield strength range and notch toughness, when compared to conventional control-rolled plates of equivalent thickness.[42]

2. High roll separating forces (due to low finish-roll temperature) are often above the design loads of the mill. That is, CR is difficult to use with older, less powerful plate and strip mills, as well as modern high-speed bar mills because of the incorporation of a low finish-rolling temperature region. Moreover, low-temperature controlled rolling is hardly suitable for fast mills designed for rolling long products.[27]

15.2.2.3   *Stages of Controlled Rolling*

1. Reheating. The reheating conditions determine the amount of MAEs taken into solid solution and the initial γ grain size for thermomechanical processing. Solution of MAEs depends on two factors: thermodynamic stability of carbonitrides and the kinetics of dissolution for the size distribution of particles present.[43] *Reheated austenite* refers to the austenite at the solution temperature (above Ac_3) after casting and ready for the hot-rolling operation. It consists of annealed, equiaxed grains associated with high-angle boundaries, with some grains containing annealing twins.

Solubility of microalloyed precipitates. It should be noted that the solubility of carbide particles in the austenite increases in the order NbC, TiC, VC, while the nitrides with normally lower solubility increase in solubility in the order TiN, NbN,

AlN, VN. The same general order follows for the solubility in ferrite where the solubility is about two orders of magnitude smaller than in austenite.[26] It is thus apparent that NbC and TiN have the lowest solubility in austenite, are the most stable particles and the most effective grain size refiners, and are more resistant to coarsening by effective particle pinning of grain boundaries and dislocation arrays. However, Al, V, and Ti are more effective in high-nitrogen steels, by forming comparatively stable AlN, V(CN), and Ti(CN) in austenite,[44] which may be potent in preventing grain coarsening on reheating, but not effective in preventing recrystallization.[45] Nb is not used with increased N contents due to the low solubility of NbN, which is less soluble than TiC, and therefore has limited precipitation strengthening potency.[26]

Solubility data for the carbides and nitrides of Nb, V, and Ti in γ provide a guide to precipitation behavior during CCR. Precipitation of carbides and carbonitrides occurs during the γ/α transformation in several modes: planar and nonplanar interphase precipitation and fibrous carbide growth.[46]

Grain-coarsening temperature (GCT). During reheating of the slab austenite, grain growth occurs either continuously or discontinuously with time, temperature, and heating rate, depending on the type, size, and volume fraction of the second-phase microalloy carbides and/or nitride particles. Grain coarsening in C-Mn steels during reheating between 800 and 1300°C obeys a well-established continuous grain growth law, expressed by

$$\overline{D} = Kt^a \exp\left(\frac{-b}{T}\right) \qquad (15.1)$$

where $\overline{D}$ is the average grain diameter, t is the time, T is the temperature in absolute scale, and K, a, and b are constants.[47]

Grain coarsening in microalloyed and Al-killed steels in the temperature range of 900 to 1300°C is, however, characterized by continuous growth at low and high temperatures; between these temperature regions is a narrow temperature range where abnormal grain growth comprising a duplex distribution of austenite grain size occurs (Fig. 15.5).[45,48] This figure illustrates the grain-coarsening behavior during reheating in C-Mn steels with microalloyed additions that form with a volume fraction of ~0.0005 of alloy carbides and/or nitrides. It also illustrates that γ grains grow very little with increasing temperature during continuous grain growth, while grain growth becomes very rapid during abnormal or discontinuous grain growth. It is thus apparent that steels containing microalloyed second phases are susceptible to grain coarsening in a particular temperature range, where the undissolved precipitates can no longer suppress grain growth. The GCT may be defined as the temperature beyond which grain growth is very rapid; this depends on composition, initial γ grain size, and particle size distribution. (Sometimes the GCT has been defined as the temperature at which the material contains a volume fraction of ~3% of austenite grains derived from an abnormal growth process.) Abnormal grain growth is largely controlled by grain boundary-particle interactions.[49] When a critical area fraction of grain boundary is covered by particles, boundary pinning commences. Boundary pinning is promoted by high volume fraction and finer-sized particles. If the reheating temperature is too high, the particles tend to coarsen or dissolve. As a result, some boundaries can be released which would lead to abnormal grain growth.

Increased amounts of vanadium above 0.14% (in HSLA steel) refine the γ grain size at temperatures up to ~1000°C and do not contribute any marked effect above 1100°C[5] due to appreciable solubility of V(CN) in austenite.

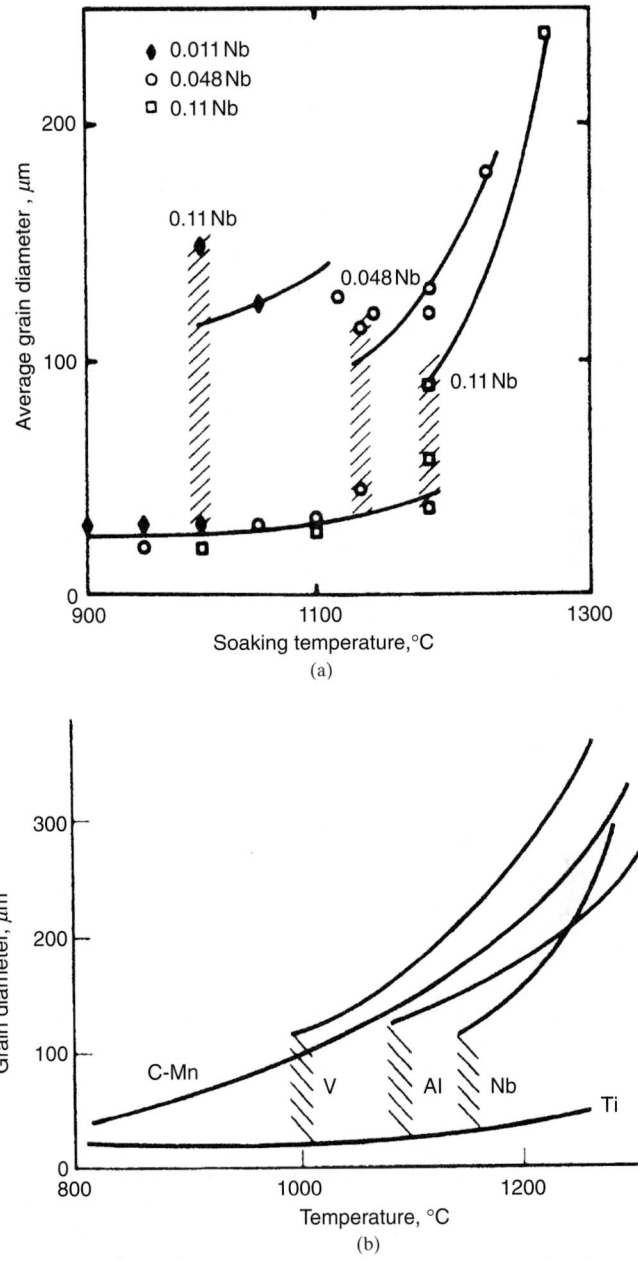

FIGURE 15.5 (*a*) Austenite grain growth characteristics in various Nb steels. Hatched bars represent the range of duplex grains produced at the coarsening temperature.[48] (*b*) Austenite grain growth characteristics in C-Mn steels with microalloyed additions containing a volume fraction of 0.0005 of alloy carbide or nitride.[45] (*Reprinted by permission of The Metallurgical Society, Warrendale, Pa.*)

15.9

The AlN particles prevent grain growth in normalized steels and in, some cases, in bars during forging.[24] The grain-coarsening temperature in Al-killed steel is above 1050°C.[50]

The effective use of niobium as grain-refining and precipitation-strengthening agents depends on the use of a quite high temperature (up to 1200°C) during reheating for rolling.[51] Nb-bearing steels have sufficiently higher grain-coarsening temperature than the V-bearing steels and Al-killed steels. The grain refinement of austenite by Nb addition is more effectively accomplished (relative to the Ti, V, and Al addition) on both the reheated grain size and recrystallized grain size when the Nb content is up to 0.1 wt% and the temperature is up to 1200°C. The precipitate formed during soaking and responsible for grain boundary pinning in high-temperature austenite is $Ti_xNb_{1-x}N$, where $x \cong 0.6$ and increases with the increase in solution temperature. The precipitate formed over a range of temperatures during hot rolling and responsible for grain boundary pinning is $Nb_x(C_yN_{1-y})$.[52] Nb addition can decrease the stability of TiN and produces a lower GCT. Rapid austenite grain growth arises from the solution of fine (<~15 nm) TiN or TiNbN particles and the consequent increase in mean grain size and decrease in volume fraction of these particles.

The starting austenite grain size on reheating after solution treatment or hot rolling is also a factor which can affect the GCT.[53] Ti addition up to 0.04% causes the formation of stable nitride, which raises the GCT to about 1300°C during both isothermal soaking and hot deformation. The stable precipitates formed during soaking and responsible for grain boundary pinning in high-temperature austenite, thereby restricting austenite grain growth, are TiN and TiNbN. Carbides or carbonitrides of titanium can form and result in precipitation strengthening in strip and thin plate steel during fast cooling. In Ti-V-N microalloyed steels, the GCT increases with increasing Ti/N ratio up to the stoichiometric value of 3.42:1. Titanium is also the most effective element for scavenging nitrogen from solution in ferrite in strip steels.[24]

For thinner plates and low-strength grades, where finish-rolling temperatures are low, Nb can be used to provide strength by grain refinement. However, in thicker plates and higher-strength grades, the strength from grain refinement in controlled rolled steels often has to be supplemented by the V addition to provide some precipitation strengthening. HSLA steels produced with Nb addition alone or together with V are normally cheaper than steels with larger V addition alone.

Since thicker plates cool more slowly and the finishing temperature is higher, some strengthening can take place through precipitation of Nb(CN). However, there is a likelihood of bainite formation with a high Nb level; both of these tend to cause a loss in toughness, but the bainite content increases the tensile strength. Carbonitrides of niobium precipitate at temperatures below about 1000°C when the steel is austenitic. In rolled steels, niobium prevents recrystallization of austenite that results in the formation of pancake-shaped grains and fine-grained ferrite, which, in turn, leads to increased strength and toughness. Niobium carbide precipitates also prevent grain coarsening of austenite during normalizing, thereby increasing its strength and toughness. Niobium also serves as a ferrite strengthener.

In plain carbon steels, recrystallization is sluggish at low temperatures in the γ region (about 900°C $\approx Ar_3$); hence, a finishing temperature should be above 750°C and preferably 800°C. Nb has a more powerful retarding influence on γ-recrystallization. It has been observed that the retardation of recrystallization due to Nb is more effective when an appreciable amount of Nb remains in solution in austenite prior to the start of rolling.[54] In Nb steels, the finishing temperature is above the temperature range of 850 to 900°C.

When more than one MAE are present in a steel, the cumulative benefits can be achieved in combination with some interactions which, in some conditions or compositions, can decrease the individual effectiveness of each element. When two MAEs are used in a steel, the element which usually precipitates at lower temperature will always have a tendency to precipitate together with the element that precipitates at higher temperature. For example, in steels containing Ti and V or Ti and Nb, V and Nb will be associated with the precipitates of TiN which form above 1300°C. This may influence the function both of the Ti and of the V or Nb. This precipitation of V or Nb with the Ti in the nitride can increase the TiN particle size, making them potent in preventing grain growth of the austenite. Conversely, when V and Ti are present in steels, the latter removes nitrogen from the system which reduces the precipitation of V(CN) in ferrite.[51]

2. Rough rolling. The roughing stage, involving repeated deformation and recrystallization of austenite and inhibition of grain coarsening, is referred to as *recrystallization controlled rolling* (RCR), where reductions are accomplished above the no-recrystallization temperature T_{nr}, as initially proposed by Sekine et al. for moderate grain refinement.[55] Steels designated for RCR conditions must possess a low recrystallization stop temperature T_5 (denoting 5% recrystallization), or T_{nr}, and a preexisting grain-coarsening inhibition system such as via (~0.01%) Ti additions to form fine, stable TiN precipitates to restrict growth of the recrystallized austenite grains.[56]

Recrystallized austenite. *Recrystallized grains* refer to the austenite formed during the hot-rolling operation (roughness pass). A certain minimum strain rate is necessary, if recrystallization is to result in grain refinement, as shown in Fig. 15.6.[43]

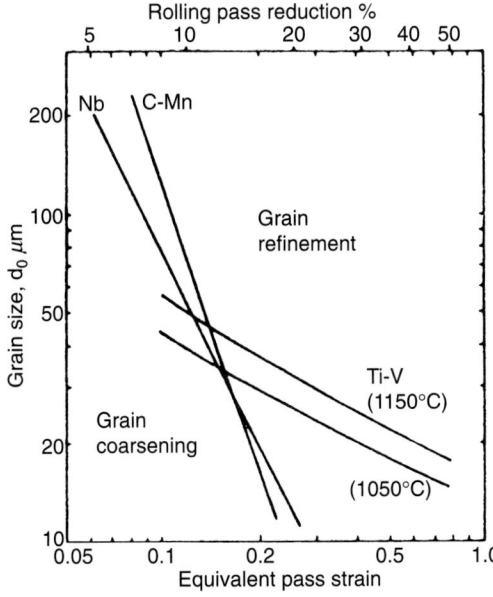

FIGURE 15.6 Critical strains for grain refinement by recrystallization calculated from the relationship for recrystallized grain size in C-Mn and Nb steels and in Ti-V steels.[43] (*Reprinted by permission of ASM International, Materials Park, Ohio.*)

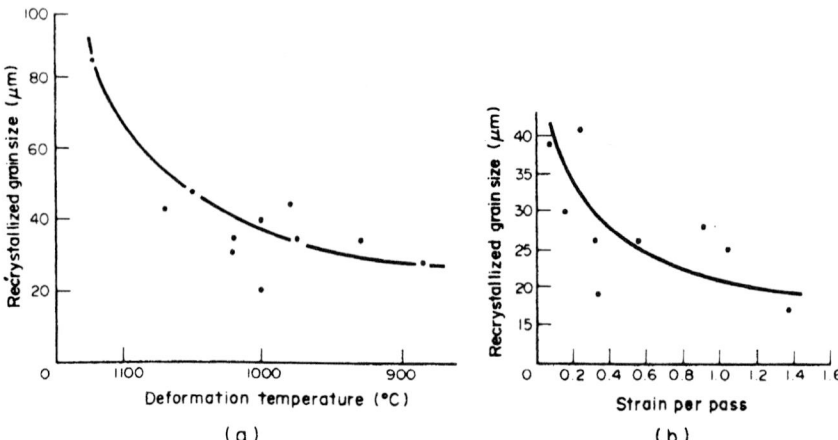

FIGURE 15.7   Effect on recrystallized grain size in 0.05 Nb steel of (*a*) deformation temperature and (*b*) percent reduction per pass. Total deformation is about 60% in multipass deformation.[45] (*Reprinted by permission of Pergamon Press, Plc.*)

The lines show the limit of rolling strain required for recrystallization. Strains below these lines produce grain coarsening, whereas strains above these lines produce the grain refinement. The extent of refinement of austenite grain size by sequential recrystallization, during hot deformation, increases with the increase in total reduction and percent reduction per pass, reaching a limiting value, and to a lower extent with a lowering in rolling temperature (Fig. 15.7).[45]

The effectiveness of common alloying elements (normalized to 0.1 wt% additions) in retarding γ recrystallization, in increasing order, is Si, Cr, Mn, Ni, V, Mo, Cu, Ti, Nb. Multiple alloying additions lead to greater retardation of recrystallization than the arithmetic sums of the individual effects.[57]

The recrystallization process occurring during the hot-rolling operation is called *dynamic recrystallization.* Likewise, the recrystallization of fine-grained austenite occurring in the delay or hold period between roughing passes is termed *static recrystallization.*

Dynamic processes. Plastic deformation of austenite at elevated temperature (during metalworking operations such as hot rolling, forging, or extrusion) leads to the occurrence of dynamic softening (or restoration) processes that comprise dynamic recovery and dynamic recrystallization. The term *dynamic* is employed to distinguish them from the *static* recovery and *static* recrystallization processes that occur either between intervals of hot-working or on heating, after the completion of cold-working. Both these dynamic softening processes are thermally activated and cause a strong temperature and strain rate dependence of flow stress, but they lead to different characteristic microstructural changes.[58]

These processes are important because they (1) decrease the flow stress of the material, (2) lead to deformation with ease, and (3) have an effect on the grain size and the texture of the worked material.[59]

Dynamic recovery associated with high-temperature deformation is rapid and extensive and appears to lower the dislocation density by the cross-slip of screw dis-

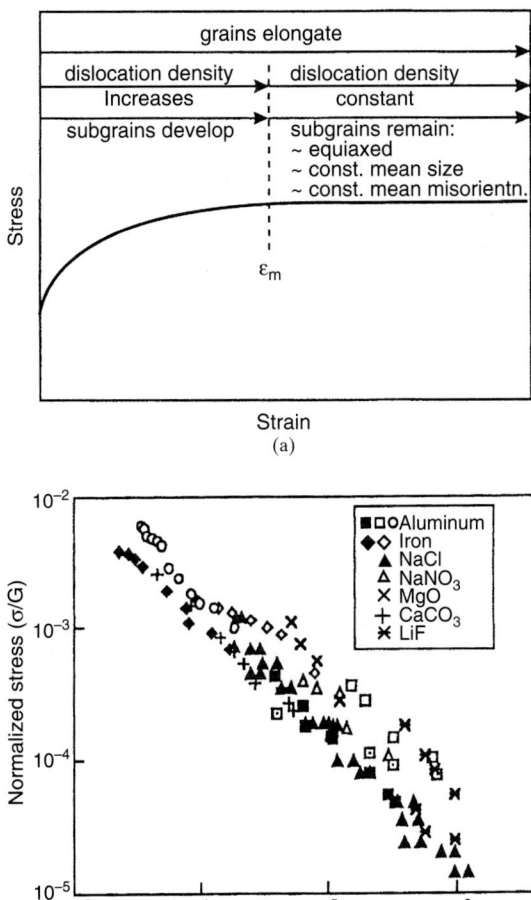

FIGURE 15.8 (a) Summary of microstructural changes that occur during dynamic recovery.[43] (b) The relationship between subgrain size and high-temperature flow stress.[64]

locations and the climb of edge dislocations, which lead to the annihilation of dislocations of opposite sign and the rearrangement of dislocations to form well-developed and delineated cells or subgrains containing low dislocation densities in their interiors.[58] These substructures are equiaxed in nature, even at large strains (because of their continuous breakup and reformation during hot deformation), and are present within the considerably elongated and flattened grains.[60] It occurs usually in the homologous temperature scale of 0.4 to 0.6 and is thus relevant to warm-working of materials.[60a]

Figure 15.8a is a schematic representation of a summary of the microstructural changes that take place during dynamic recovery. The stress-strain curve representing the restoration of dynamic recovery is fully characterized by a rise to a plateau followed by a steady-state flow stress, as shown in Fig. 15.8a.[61] A subgrain size d_{sg}, characteristic of the high-temperature and low-strain-rate conditions, can be related to the creep stress or to the steady-state flow stress σ_s by an equation of the form

$$\sigma_s = Kd_{sg}^{-n} \tag{15.2}$$

where K and n are constants and n, according to Twiss, has a value of about 1 for many materials.[58]

The high-temperature flow stress, for steels and Al, can also be related to the density of dislocations within the subgrains ρ_i by the equation of the form:[62]

$$\sigma_s = k_1 + k_2 + Gb\rho_i^{1/2} \tag{15.3}$$

which is somewhat similar to Eq. (4.36), except α relating the flow stress and overall dislocation density found at low temperatures. In the case of subgrain formation, high-temperature flow stress is inversely proportional to the mean subgrain diameter d_{sg} [Eq. (5.4)]. Takeuchi and Argon[63] have shown that if Eq. (5.4) is expressed in terms of normalized stress and subgrain size and a dimensional constant K, then K remains constant for a given class of material:

$$\frac{\sigma}{G}\frac{d_{sg}}{b} = K \tag{15.4}$$

Derby[64] has examined subgrain data for a range of metals and minerals, plotted them in Fig. 15.8b, according to Eq. (15.4), and found that the value of $K \sim 10$ for fcc metals and 25 to 80 for ionic crystals of the NaCl structure. At a constant flow stress (or Z), Eqs. (15.2) and (5.4) can be combined to form a unique relationship between the subgrain size and the dislocation density within the subgrains as

$$\rho_i^{1/2} = k_3 d_{sg}^{-1} \tag{15.5}$$

Such a relationship has been obtained in theoretical models of subgrain formation during hot deformation,[65,66] and experimental studies in aluminum,[62] copper,[67] and ferritic steels[68,69] have reported values of k_3 between 10 and 20.[59]

Dynamic recrystallization (DRX) is observed during the actual hot-working in the fcc metals and alloys (except commercial-purity Al) (such as Cu, Ni, and γ-Fe) containing moderate to low stacking-fault energy and is associated with relatively low rates of (i.e., sluggish) dynamic recovery and higher than average rates of work-hardening, so that the driving force for recrystallization is maintained.[70] In fact, Gottstein and Lee[71] have demonstrated in studies with copper single crystals that increased deformation temperature favors recrystallization when deformation is accomplished at an equivalent flow stress. Dynamic recrystallization may also occur during creep deformation,[72,73] which occurs at strain rates below $10^{-5}\,s^{-1}$ compared to high strain rates (in the range of 1 to $100\,s^{-1}$) obtained during hot-working. Dynamic recrystallization is also of interest to structural geologists because of its occurrence during the natural deformation of minerals in the earth's crust and mantle.[59]

The general characteristics of dynamic recrystallization are given below:

1. The *evolution of dynamically recrystallized* (DRX) *microstructure*[74] is quite different from the microstructure developed during static recrystallization after cold-working. The DRX process refers to the occurrence of simultaneous recrystallization during deformation by nucleation and growth processes. In static recrystallization, on the other hand, a certain amount of stored energy (depending on the cold work) is released by thermally activated dislocation

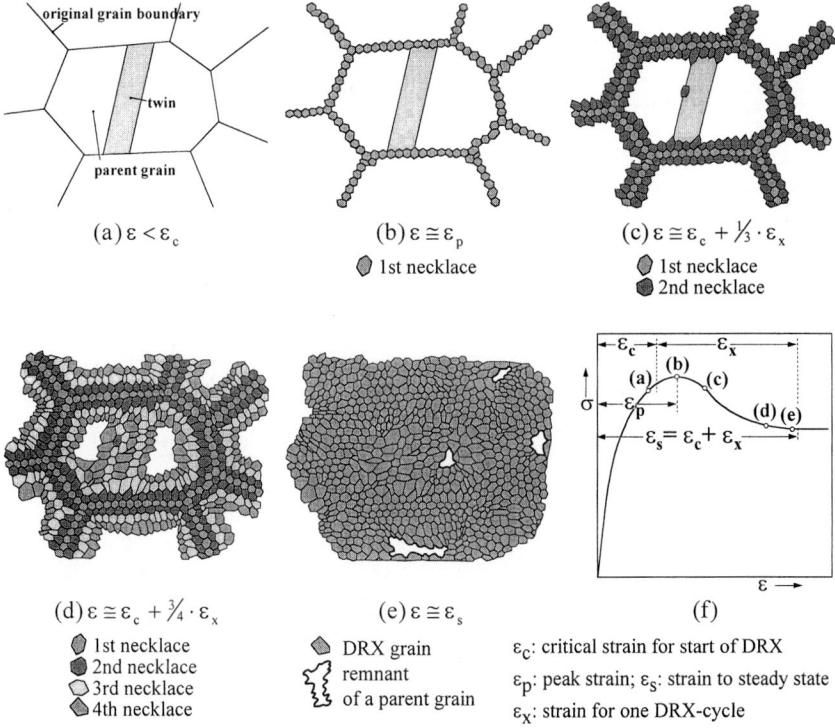

(a) $\varepsilon < \varepsilon_c$

(b) $\varepsilon \cong \varepsilon_p$

- ⬡ 1st necklace

(c) $\varepsilon \cong \varepsilon_c + \frac{1}{3} \cdot \varepsilon_x$

- ⬡ 1st necklace
- ⬡ 2nd necklace

(d) $\varepsilon \cong \varepsilon_c + \frac{3}{4} \cdot \varepsilon_x$

- ⬡ 1st necklace
- ⬡ 2nd necklace
- ⬡ 3rd necklace
- ⬡ 4th necklace

(e) $\varepsilon \cong \varepsilon_s$

- ⬡ DRX grain
- ⬡ remnant of a parent grain

(f)

ε_c: critical strain for start of DRX

ε_p: peak strain; ε_s: strain to steady state

ε_x: strain for one DRX-cycle

FIGURE 15.9 Microstructural evolution by consecutive necklace formation (schematically, according to Sellars). (*a*) For strains below ε_c the microstructure consists only of parent grains with dynamic recrystallized grains. (*b*) A first necklace has formed at prior grain boundaries. (*c*) A second necklace expands the DRX volume into the grain interior. (*d*), (*e*) Expansion of the DRX volume to consume the grain interior. (*f*) Corresponding flow curve.[74] (*Courtesy of D. Ponge and G. Gottstein.*)

recovery and grain boundary migration and may be called a *kinetic process*. Typically, during DRX, first necklace grains form at the prior grain boundary (Fig. 15.9*b*) by bulging mechanism, and then this *necklace structure* expands into the unrecrystallized volume to become fully recrystallized (Fig. 15.9*d* and *e*).

2. Dynamic recrystallization commences only when a *critical strain* (also called *incubation strain*) ε_c is reached during hot deformation. This is somewhat less than the maximum (or peak) strain ε_p (Fig. 15.9*f*).[74] The recrystallization start time t_{rs} for dynamic recrystallization may be determined from the relation:

$$t_{rs} = \frac{\varepsilon_p}{\varepsilon} \qquad (15.6)$$

where ε is the testing strain rate.[57]

3. This process leads to an initial increase of flow stress to the peak value and then a decrease of flow stress to a steady state. These flow curves are either periodic (multiple-peak) or continuous (single-peak)[75,76] (Fig. 15.10).[76] The former curves appear at low strain rates and high temperatures. This indicates that discrete cycles of grain coarsening are occurring. The latter curves occur at relatively high strain rates and low temperatures and are associated with the grain refine-

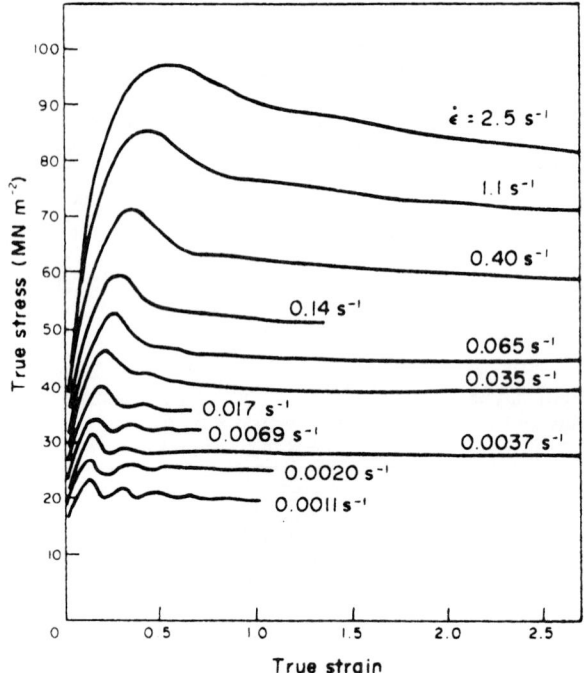

FIGURE 15.10   Flow curves derived from torsion data for a plain 0.25% C in the austenitic condition at 1100°C $(0.76T_M)$, showing the strong influence of strain rate. At higher strain rates they exhibit dynamic recrystallization with a single-peak flow curve corresponding to the occurrence of grain refinement. At low strain rates, the flow curve is periodic (or cyclic) and represents the occurrence of grain coarsening.[76] (*Reprinted by permission of Pergamon Press, Plc; after Rossard and Blain.*)

ment and with the operation of the *necklace mechanism* of dynamic recrystallization.[57] This is a beneficial process because it offers stable flow and good workability to the material by simultaneously softening it and reconstituting the microstructure.[60a] This is of commercial significance, as in extrusion and planetary hot-rolling, where large strains are given in a single operation.[77]

4. The size of the dynamically recrystallized grains D_R increases monotonically with a decrease in flow stress, according to the empirical relationship

$$\sigma = k_1 + k_2 D_R^{-q} \tag{15.7}$$

where q lies between 0.5 and 0.8.[78] Grain growth does not take place, and the grain size remains unaffected during the deformation.

5. The flow stress σ and D_R are virtually independent of the initial grain size D_0; however, the kinetics of dynamic recrystallization is enhanced in specimens containing smaller initial grain sizes.

6. In dynamic recrystallization, nuclei form preferentially at or near the preexisting grain boundaries (as mentioned earlier) but may also occur at deformation

bands, twins, or inclusions within the grains (i.e., intragranularly), particularly in coarse-grained materials and at high strain rates.

7. Dynamic recrystallization is a continuous process involving deformation, nucleation of grains, and subsequent migration of grain boundaries (i.e., growth), resulting in new dislocation-free grains which then deform further.[59]

8. In the case of dynamic recrystallization during constant strain rate deformation, the recrystallization curve can be represented by the Avrami expression

$$f(t) = 1 - \exp\left[-K(\varepsilon - \varepsilon_c)^n\right] \qquad (15.8)$$

where $f(t)$ is the volume fraction recrystallized; ε is the strain; ε_c is the critical strain for the onset of dynamic recrystallization; K is a constant, dependent on composition, grain size, temperature, and strain rate; and n is a constant (usually 1.2 to 2).

9. The dynamically recrystallized grain size D_R is uniquely related to the temperature-corrected (or -compensated) strain rate (also called the *Zener-Hollomon parameter*) Z and is given by an equation of the form[79]

$$D_R \text{ (or } D) = BZ^{-p} \qquad (15.9)$$

where
$$Z = \dot{\varepsilon} \exp \frac{Q}{RT} \qquad (15.10)$$

and where Q is activation energy, R is a gas constant, T is temperature in absolute scale, B is a constant, mostly dependent on materials' composition, and p is a constant of about 0.3 to 0.4 for all steels.[79]

10. Dynamic recrystallization produces both texture and grain size distribution that is different from that of static recrystallization.[80]

11. The occurrence of dynamic recrystallization during controlled rolling is favored by the following conditions: relatively high temperature and low strain (deformation) rate; large accumulated strains; reduced concentration of solute elements; and decreased rate of dynamic precipitation of carbide, nitride, and carbonitride particles.

12. Dynamic recrystallization is likely to produce very fine grain sizes, particularly in the presence of fine second-phase particles. This has been exploited commercially to produce superplastic alloys, which deform by Herring-Nabarro diffusional creep.[70]

13. Repeated dynamic recrystallization is apt to counteract incipient grain boundary cracks in low- or limited-ductility materials, by the repeated separation from their associated boundaries, thereby improving hot workability (which is similarly improved by the repeated softening of the grain interiors).[70]

14. DRX breaks down the as-cast microstructure to produce wrought microstructure, globularizes the acicular preform microstructure as in Ti and Zr alloys, redistributes the prior boundary defects in P/M components to promote further processing, or removes discrete particle effects by transferring mechanical energy across the hard particle interface to refine them. Thus, DRX is a desirable regime to maximize hot workability and control the microstructure and is a "safe" regime for bulk metalworking.[60a]

Critical strain model for dynamic recrystallization. The classical explanation for the transition from periodic to single-peak recrystallization was proposed by Luton

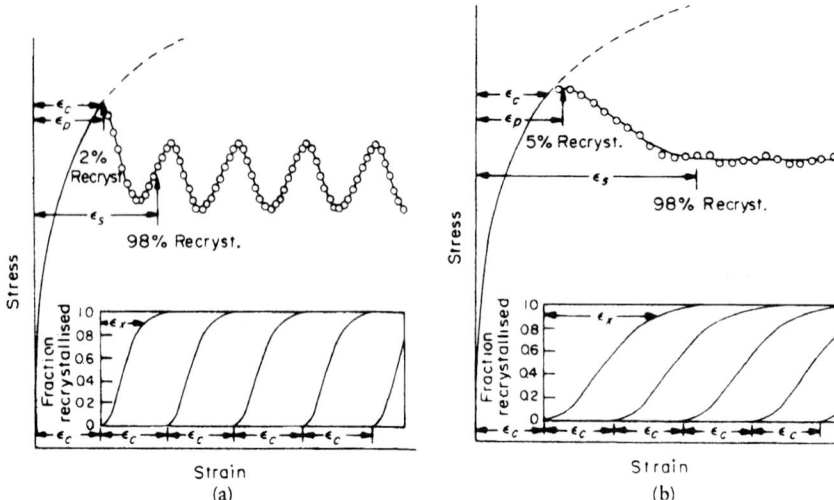

FIGURE 15.11 Predicted stress-strain curves for dynamic recrystallization. (*a*) A cyclic flow curve when $\varepsilon_x \ll \varepsilon_c$. (*b*) A single flow curve when $\varepsilon_x > \varepsilon_c$.[81] In both (*a*) and (*b*), ε_p and ε_s are the strains to peak flow stress and to the onset of steady-state stress, respectively. The inset illustrates the recrystallization kinetics. (*Reprinted by permission of Pergamon Press, Plc.*)

and Sellars in 1969.[81] On the basis of torsion testing results obtained on Ni and Ni-Fe alloys, they showed the occurrence of cyclic recrystallization when the peak strain ε_p was greater than recrystallized strain ε_x, where $\varepsilon_x = \varepsilon - \varepsilon_c$. At low strain rates (i.e., under creep conditions) and high temperature, $\varepsilon_x \ll \varepsilon_c$; that is, the recrystallization cycle is completed at a strain relatively lower than that of the critical strain. Further work-hardening of the recrystallized grains is needed for a second nucleation event of recrystallized grains. In this way, more than one cycle of recrystallization is occurring in the material at the same time, each cycle of which is at a different stage of the recrystallization process. The overlapping periods of recrystallization lead to the initial rise to the peak value and subsequent drop to a constant steady-state strain through intermediate periodicity (Fig. 15.11*a*).[81] Conversely, when $\varepsilon_p < \varepsilon_x$, that is, $\varepsilon_x > \varepsilon_c$, at relatively high strain rates and low temperatures, several recrystallization cycles can occur simultaneously and cause smooth $\sigma(\varepsilon)$ curves with only one initial peak and subsequent drop to a constant steady-state level (Fig. 15.11*b*). This model does not apply to tension or compression, or other homogeneous modes of deformation such as upset or continuous forging and rolling, due to the presence of strain and strain rate gradients during the torsion testing of the solid bar.[76]

Sah-Richardson-Sellars model for dynamic recrystallization. The nuclei initially grow rapidly, but the associated deformation causes work-hardening (i.e., an increase in the dislocation density within the dynamically growing grains), so that the driving force for boundary migration is gradually reduced until new dynamic grains stop growing. The dynamic grains therefore reach a size that is dependent on the deformation conditions. Recrystallization then progresses by further nucleation, frequently near the previously recrystallized grains.[82] When strain is increased, the recrystallization spreads by a process of repeated nucleation and limited grain growth rather than by limited nucleation and continuous grain growth until impingement of the new grains, as in classical static recrystallization.[58]

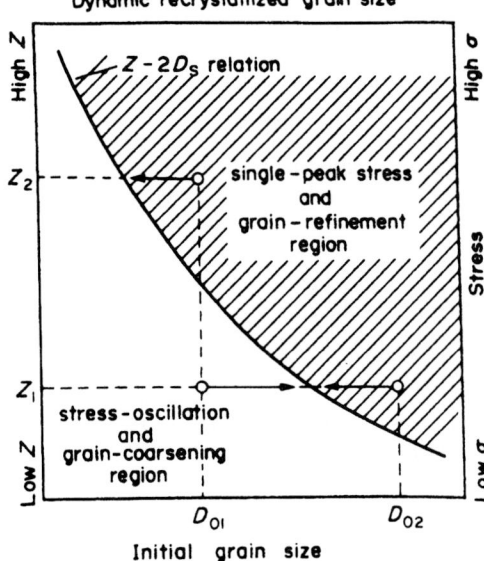

Dynamic recrystallized grain size

FIGURE 15.12 A microstructural mechanism map for distinguishing between the occurrence of two types of dynamic recrystallization. The curve describing the Z-$2D_s$ and Z_c-D_0 relations separates the single-peak (grain refinement) from the multiple-peak (grain-coarsening) region when straining is carried out at a given temperature and strain rate (fixed Z). Grain coarsening and repeated stress oscillations occur as long as each successive $D_0 < 2D_s$ (i.e., at D_{01} and Z_1). Conversely, when $D_0 > 2D_s$, grain refinement and a single-stress peak (i.e., at D_{02} and Z_2) occur. When a fixed initial grain size is used (e.g., D_{01}), stress oscillations and grain coarsening are associated with $Z < Z_c$ (e.g., $Z_1 < Z_c$), and a single-stress peak and grain refinement are associated with $Z > Z_c$ (e.g., $Z_2 > Z_c$).[76] (*Reprinted by permission of Pergamon Press, Plc.*)

Critical grain size (i.e., microstructural) model for dynamic recrystallization. Numerous workers have shown that the characteristic flow curve changes gradually from the multipeak to the single-peak type with increasing Z, or strain rate, and decreasing temperature. These two types of behavior can be distinguished from the critical grain size mechanism map of Fig. 15.12.[75] The vertical coordinate is Z, or its equivalent σ_p or σ_s. The horizontal coordinate is the initial grain size D_0 or the equivalent stable recrystallized grain size D_s. The full line dividing the crosshatched region from the plain region separates the initial conditions into two types, either multiple-peak or single-peak flow. The full line also indicates the experimentally observed locus of $D_0 = 2D_s$ or the Z-$2D_s$ relation. The same curve also denotes the dependence of Z_c on D_0, where Z_c is the critical value of Z. Thus, when the point (D_0, Z) lies above the line, $D_s < D_0/2$ and grain refinement takes place. Conversely, when (D_0, Z) lies below the line, $D_s > D_0/2$ and grain coarsening occurs.[77,80]

When the critical grain size model was verified in a 0.115% V, a 0.035% Nb, and a 0.040%Nb-0.30%Mo microalloyed steel under both solute retardation and precipitation retardation conditions, these results clearly supported the belief that necklace recrystallization (where grain growth is halted by concurrent deformation) is linked with the single-peak behavior. Conversely, termination of grain growth, under grain-coarsening conditions, by boundary impingement is associated with the cyclic flow curve.[75,80]

Progressive Lattice Rotation Mechanism. It has been reported in certain materials that new grains with high-angle boundaries may be formed during straining, by the progressive subgrain-rotation with *small associated boundary migration*. This *strain-induced* phenomenon comprises gradual rotation of subgrains near the preexisting grain boundaries with the straining of the material. The occurrence of this phenomenon has been noted in Mg[83] and in Al containing particular solute additions such as Al-Mg alloys[84,85] and Al-Zn alloys.[84]

In this type of dynamic recrystallization, it is usually observed that the size of grains formed at the old boundaries is only slightly greater than that of the subgrains, and the grain size is, therefore, expressed approximately by Eq. (15.4).

The mechanism by which this progressive subgrain rotation takes place is not yet fully understood. However, it is likely that it is associated with (1) inhomogeneous plasticity and accelerated dynamic recovery in the grain boundary regions and (2) grain boundary sliding. Although this phenomenon usually leads to partially recrystallized necklace microstructure, at large strains a fully recrystallized structure may be formed (Fig. 15.9d and e).[59,74,84]

3. Delay or hold period. Since the structural changes obtained by dynamic restoration are thermodynamically unstable, holding at temperature in the interstand period, particularly during roughing passes, modifies them by the static restoration process. These structural changes play an important role in determining the final microstructures and properties of HSLA steels.[86]

Static processes. Since the structures obtained by dynamic restoration are thermodynamically unstable, unloading the stress at strains less than ε_p and/or holding at temperature modifies them by the static restoration process. The static restoration process involves three different types of softening; namely, mode I is static recovery; mode II, classical (static) recrystallization; and mode III, metadynamic (i.e., post-dynamic) recrystallization, with the amount of each depending on the prestrain.[87,88] Static recovery occurs at low strains and causes a loss of dislocation density, which, in turn, leads to a small decrease in yield strength or flow stress. This does not produce any detectable microstructural changes under optical microscope.

In the case of classical (static) recrystallization, it is clear that recrystallized nuclei start to form after the completion of straining and an apparent incubation period. Classical static recrystallization progresses in the same manner as in cold-worked material. The rate of static recrystallization increases with the strain until the initiation of dynamic recrystallization.[89] The distribution of recrystallized nuclei is highly localized and inhomogeneous, occurring predominantly at triple junctions of grains and deformed grain boundaries[87,90] and less frequently at the intragranular sites such as twin boundaries and deformation bands.

Hence, the initial grain size has an important bearing on the progress of recrystallization and the recrystallized grain size. Like annealing after cold-rolling, the rate of static recrystallization is a strong function of the extent of deformation and the temperature and is a weak function of strain rate. Figure 15.13 illustrates the effects of initial grain size and deformation temperature on the critical deformation required for recrystallization.[22] The statically recrystallized grain size is determined primarily by the initial grain size and the extent of deformation, whereas the deformation temperature mainly affects the progress of recrystallization.[19]

The relationships among the statically recrystallized austenite grain size d_{sr} (μm), prior (initial) grain size (μm), and the applied strain, for (a) both austenitic stainless steel and transformable C-Mn steel and (b) Nb steels, are given by

$$d_{sr} = K d_0^{2/3} \varepsilon^{-1} \qquad \text{C-Mn steels} \qquad (15.11)$$

and
$$d_{sr} = K' d_0^{2/3} \varepsilon^{-0.67} \qquad \text{Nb steels, } T > 950°C \qquad (15.12)$$

where d_0 is the initial austenite grain size (prior to deformation, μm), ε is the equivalent true (rolling) strain, and K and K' are constants. For C-Mn steels, the values

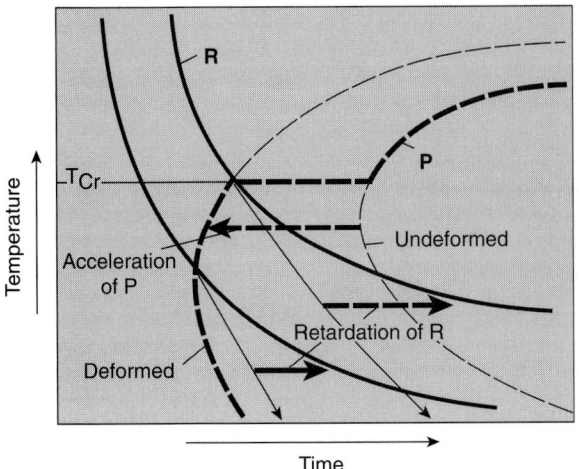

FIGURE 15.13 Schematic diagram representing the interaction of recrystallization R and strain-induced precipitation P.[22]

of K have been reported to be in the range of 0.35 to 0.83 $(\mu m)^{1/3}$, and a mean value of 0.5 $(\mu m)^{1/3}$ may be appropriately used.[91]

After the establishment of a statically recrystallized structure with an initial grain size d_{sr}, at some interval within the interpass period, then, in the absence of any second-phase pinning particles, grain growth will occur. As stated earlier, in this case, the basic form of the grain growth equation can be given by

$$d^n - d_{sr}^n = kt \qquad (15.13)$$

or
$$d = Kt^m \qquad (15.14)$$

where d is the growing grain size after a time t (s); K and k are constants; m is a temperature-dependent exponential; and n is the grain growth time exponential and in Eq. (15.13) is usually 2; i.e., classical parabolic growth follows. Note that in industrial situations, grain growth is not significant for the interpass time of 5 to 10s in a plate mill and 0.1 to 10s in a hot strip mill. However, after the completion of recrystallization at the end of the rolling pass, the grain growth rate within 20 to 40s is found to be quite high, and, in that situation, Eqs. (15.13) and (15.14) require further modifications.[92,93]

The time for dynamic or static recrystallization to take place is dependent on the microalloy content due to (1) solute drag effect in retarding recrystallization and (2) the pinning action of stable microalloy carbide, nitride, and/or carbonitride precipitated on migrating grain boundaries.

Metadynamic recrystallization, immediately after the termination of hot deformation, depends strongly on the strain rate; and it was reported that this condition occurred by transformation of partially recrystallized grain structure observed just after the deformation process to a more fully recrystallized structure by growth of recrystallized nuclei (grains) formed during deformation[93a] and, therefore, did not require an incubation period essential for conventional static recrystallization. It was also found that the rate of softening was a very weak function of temperature, composition, and strain.[89] Consequently, it is about an order of magnitude faster than classical static recrystallization. It also produces finer recrystallized grain structures compared with the classical mechanism, primarily due to the formation of higher density of nuclei by dynamic nucleation.[77,88,89]

4. Finish rolling. The deformation (rolling) of microalloyed austenite in the non-recrystallized temperature range above or below Ar_3 (intercritical rolling) leads to the breaks-up and formation of flattened (pancaked) *unrecrystallized austenite* grains. The process involving repeated flattening of the grains by repeated deformation below T_5 or T_{nr} is referred to as *classical* or *conventional controlled rolling* (CCR). This is usually achieved by Nb additions. The conditions of finish-rolling differ extensively for plate, TRIP, and rod; and the resulting microstructures depend on the relative kinetics of recrystallization and precipitation. This practice is well-suited for production conditions that require a high recrystallization temperature.

Retardation of recrystallization. In controlled rolling of HSLA steel plate, *strain-induced precipitation* of very fine Nb(CN) particles in austenite during recrystallization and at finishing temperatures plays a vital role of effectively stopping or retarding recrystallization of the austenite, which leads to increased ferrite nucleation on cooling due to grain elongation and accumulated strain (or flow stress).[94]

During hot-rolling, the spontaneous precipitation of MAEs in the austenite occurs with the falling temperature which results in the retardation of recovery and recrystallization. The effect of MAEs on recrystallization can also be observed with respect to the increased recrystallization temperature that takes place after a particular rolling deformation. Figure 15.2 shows the effects of MAEs on the critical temperature for γ recrystallization. Under the basic conditions of hot deformation, recrystallization occurs in the region above the curves in each case, whereas below the curves it is strongly retarded. It is thus clear that Nb is very potent in raising the recrystallization temperature.

The TMT is characterized by a complex interaction between the retardation of recovery and recrystallization occurring after rolling and a marked acceleration of strain-induced precipitation, if adequate MAEs and C and/or N are present, as shown schematically in Fig. 15.13, occurring below a critical temperature.[22]

Effect of deformation on ferrite formation. Among the austenitic conditions, the austenite grain structure is of vital importance since it determines the density of sites for ferrite nucleation during subsequent transformation. The nucleation sites are the internal defect structures, namely, grain boundaries, subboundaries, deformation bands, and incoherent annealing twin boundaries. Researchers have found that the austenite grain boundary per unit volume S_v, which is inversely proportional to the austenite grain diameter D, is an effective grain size parameter. Austenite grains with an increased S_v, obtained by producing large rolling reduction and highly elongated grains, favor large ferrite nucleation, while a small S_v causes low nucleation rates and coarse ferrite grains (Fig. 15.14a).[18]

Although RCR and CCR offer different approaches to austenite conditioning, they both have the same objective of achieving grain refinement in the final plate, beam, coil, or forging. The difference between RCR and CCR can be realized by using Fig. 15.14b which shows the different path employed by each to achieve high S_v values.[96]

The increase in S_v for the RCR practice is due solely to an increase in grain boundary area per unit volume owing to a decrease in average grain volume, whereas the increase in S_v for the CCR practice is due to the increase in grain

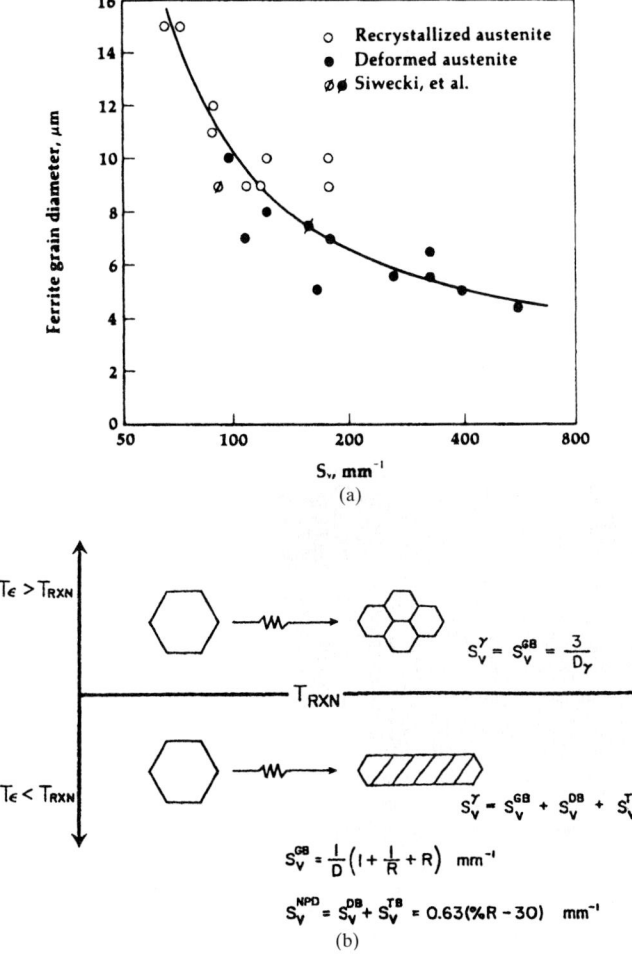

FIGURE 15.14 (*a*) A relationship between ferrite grain size and S_v.[18] (*Reprinted by permission of ASM International, Materials Park, Ohio.*) (*b*) Schematic illustration of austenite microstructures resulting from deformation above or below the recrystallization stop temperature of austenite, with corresponding description of S_v. The superscripts GB, DB, TB, and NPD are the contributions to the total S_v from grain boundaries, deformation bands, twin boundaries, and near-planar defects.[96] (*After E. J. Palmiere, University of Pittsburgh, unpublished research, 1990.*)

boundary area per unit volume resulting from a change in grain shape and through the addition of the transgranular twins and deformation bands.[95]

Precipitation kinetics of microalloying during hot-rolling. In microalloyed steels, the strain-induced precipitation (of carbonitrides) during hot deformation decreases the onset of dynamic recrystallization and the kinetics of static recrystallization when compared to C-Mn steels. Hence, the classical Avrami equation, usually employed for C-Mn steel, has to be modified to incorporate this retardation effect. Empirically, the strain-induced carbonitride precipitation in microalloyed Nb-bearing HSLA steel can be described by the Avrami equation of the form

$$f_p = 1 - \exp(-kt^n) \tag{15.15}$$

where f_p is the fraction of strain-induced precipitation at the holding time t (s), K is a constant, dependent on nucleation and growth rates, and n is a constant, near 0.6. Finally, the precipitation kinetics in austenite can be described by the equation

$$f_p = 1 - \exp(-0.38\varepsilon^{0.265}t^{0.6}) \tag{15.16}$$

in unrecrystallized austenite at 950°C, or

$$f_p = 1 - \exp(-Kt^{0.6}) \tag{15.17}$$

in recrystallized austenite where $K = 0.005$ to 0.006 or 0.004, respectively, at 900 to 950°C or 1050°C.[97]

5. Cooling. It has long been understood that the transformation temperature and cooling rate have profound effects on the transformation structures. A simple relationship between ferrite grain size (d_α, mm), austenite grain size (d_γ, mm), and some of the process variables has been given by:[98]

$$d_\alpha = a + b\dot{T}^{-1/2} + c[1 + \exp(-1.5 \times 10^{-2} d\gamma)](1 - 0.45\varepsilon^{1/2}) \tag{15.18}$$

where $\dot{T}$ is the cooling rate (°C/s); ε is the strain applied during finish-rolling below the recrystallization temperature of γ; and a, b, and c are constants, dependent on the composition of steel and, perhaps, on the processing parameters determining the degree of strained-induced precipitation.

Accelerated cooling lowers the transformation temperature and increases the density of effective nucleation sites within the unrecrystallized austenite grains in addition to those activated by deformation bands, which provides both the ferrite grain refinement with uniform grain size and finer carbonitride precipitation during or after transformation to ferrite, thereby further improving room-temperature strength at the expense of toughness.[43] In addition, AC modifies the transformed structure by replacing pearlite with an increased volume fraction of finely dispersed bainite as well as produces smaller and more effective microalloy precipitates.[19] All these factors enhance the strength of the product.[35]

15.2.2.4 Recrystallization-Controlled Rolling.

To overcome the above short-comings of the CCR method, attention was directed toward the RCR method of producing very fine ferrite grain size on the order of 5 to 10 μm (0.2 to 0.4 mil). This process involves a relatively low reheating temperature, repeated high-temperature finish-rolling above T_R (i.e., up to 1050°C), and retardation of austenite grain growth during and after rolling. In the RCR operation, recrystallization must go to completion within times available between rolling passes (i.e., occurrence of full recrystallization after each pass or after alternate passes to control grain refinement). After the last rolling pass, the fine recrystallized austenite grains transform to very fine ferrite grains. This approach is well-suited for production conditions that must have high finishing temperatures, e.g., in underpowered rolling mills and forging. The RCR process in combination with controlled accelerated cooling is often used to optimize the ferrite nucleation rate and higher precipitation-strengthening increment.

Advantages of the RCR processes are their high productivity, rapid processing, lower mill loading, energy savings, and easy implementation within existing plants,[99] compared to CCR. It also enhances the through-thickness properties of TMCP steel.[100]

To successfully utilize the RCR approach, it is necessary to determine the relationship between recrystallization behavior (time for static recrystallization and recrystallized grain size) and rolling parameters (strain, temperature, and interpass time).[101] The RCR process is especially suited to high-speed, high-finish-temperature bar and section mills, as well as plate and strip mills which are not designed for high mill loads produced at low-temperature CCR.

The RCR is well-suited to V-N and V-N-Ti steels. V steels with increased N contents (0.01 to 0.02% N) exhibit austenite grain refinement during high-temperature rolling, as well as increased VN precipitation strengthening when compared to low-nitrogen vanadium steels. The negative influence of Nb at higher finish-rolling temperature and the need for a low T_R prevent its use in RCR steels.

A small Ti addition (0.015%) to V-N steels causes a fine distribution of stable TiN particles, which effectively impede austenite grain growth during reheating and rolling, enhances the uniformity of grain size upon transformation, and contributes to a higher final S_v by hindering growth of recrystallized grains (Fig. 15.15). Figure 15.15 shows the microstructural evolution of austenite grain size in C-Mn and V-Ti-N steels during RCR.[102] This illustrates the occurrence of a significant grain growth in C-Mn steel after the last rolling pass with respect to V-N-Ti steel.[102] Properties of low-carbon V-Ti-N steels, processed by RCR in combination with accelerated cooling, are considered as an available alternative to steels processed by low-temperature CR. The concept of the RCR – AC approach has been successfully applied to produce heavy plates of Ti-V-N microalloyed HSLA steels which is found to be more economical than low-finish-temperature CR. Mechanical properties of the present 0.01Ti-0.08V-N steels in the RCR + AC condition (finish-rolling tem-

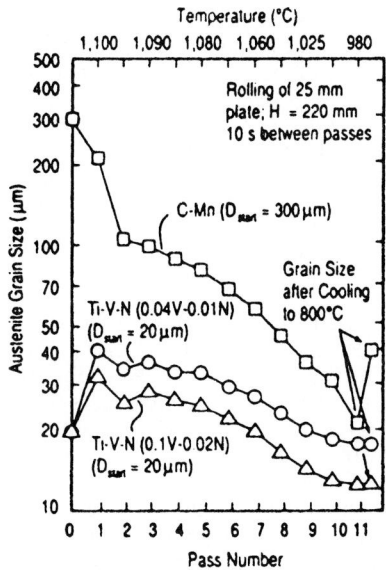

FIGURE 15.15 Microstructural evaluation of C-Mn and Ti-V-N austenites.[102]

perature ~1050°C) were reported to be superior to those after CR with finish-rolling temperature of ~800°C.[103]

15.2.2.5 *Some Applications of Microalloying and TMT.* A very wide variety of microalloying applications are clear from the diverse range of carbon in microalloyed steels from 0.02% C in interstitial-free steels to eutectoid composition for wire rod employed for prestressed concrete stands. In these applications, MAEs are used as getters for interstitials, as precipitation strengthener, or as microstructural modifier, depending on the property requirements. Different functions fulfilled by microalloying are described below as some examples.[104]

1. *High-strength linepipe steels.* The high-strength linepipe steels (API X60 to API X80) are produced via a judicious choice of (micro)alloy composition and optimization of TMT and subsequent accelerated cooling conditions to obtain a fully bainitic structure, a high level of toughness with a low DBTT.[22,105] The economic benefits of these higher-strength pipelines, such as lower pipe procurement cost, lower cost for transport to the site, reduced gas transportation, and reduced welding costs due to smaller-diameter and thinner walls, have led to their increasing use throughout the world for laying pipelines over long distances across the national and international borders and counteracting the consequent rise in operating pressure and exploitation of deposits with extremely sour media. From the weldability point of view, improvement in weldability provides a challenging problem arising from combinations of weld joint, welding procedure and technology, and alloy composition and properties. However, from a metallurgical point of view, weldability is associated with hardenability, HAZ properties, HAZ and weld metal cracking, and ability of steel to respond to post-weld treatments.[106] It is possible to achieve an improvement in weldability by using several metallurgical processes, such as inclusion shape control (CaSi-rare earth injection treatment), reduced P and S levels, improvement of steelmaking and continuous casting practices,[105] effective exploitation of TiN technology (Ti content between 0.010 and 0.015% and Ti/N ratio ≤3.42),[104] and a reduction of C, Mn, and Si concentrations to minimize segregation of these elements to the centerline during the continuous casting and welding.[106] It has been shown by William et al.[105] that lower Mn content reduces centerline microstructural banding, leading to low-segregation-ratio hardenability enhancing Mn, which lowers considerably the likelihood of martensite formation in the hot-rolled strip and plate. Taillard et al.[107] reported that Si content <0.1% increased the toughness of intercritical and coarse-grained HAZ (CGHAZ) in Ti-microalloyed steels. Lee and Pan[108] have proposed that reduced Si content in Ti-microalloyed steels increases the toughness of HAZ by allowing the Ti to form TiO, which serves as the most effective nucleants to form acicular ferrite microstructure which possesses a higher toughness than either allotriomorphic or Widmanstätten ferrite.[109]

2. *Ultra-low carbon steel or interstitial-free (IF) steels.* These are used where excellent formability is important.[32] Further increase of strength of these newer steels has been achieved by either incorporating P addition or by understabilizing of C and N followed by bake-hardening. The principal texture components in the hot band microstructure of these steels are {111}<110> and {111}<112>, independent of the steel composition and TMT path.[110] Recently, the demand for high-strength galvanized sheet for automotive body panels has increased tremendously, to achieve improved overall corrosion resistance of the vehicle. This development led to the use of continuous annealing lines of thin sheet of ULC Ti- or Nb/Ti-stabilized IF grades containing low (near 20–30 ppm) C and N contents for an improvement of the final texture (r_m value) and superior formability and nonaging properties, thereby creating its growth market.[111] (See Sec. 1.8.2.2 for more details.)

It is usually assumed that Ti forms TiN and Nb forms NbC. Recently, it was reported that Ti addition is also advantageous in C removal by first forming TiS phase and then transforming it into carbosulfide, $Ti_4C_2S_2$. Since Nb remains mostly in solid solution, its probable role in the formation of a favorable texture is questionable.[104]

3. *Medium- and high-carbon microalloyed steels.* Medium-carbon microalloyed steels in the as-rolled condition have also been used for various shaft, cylinder, steering, and piston rod components and springs. In addition, carburized low-alloy steels have been substituted by air-cooled microalloyed steels.[112]

TMT of 50CrV4 steel and (0.60–0.64)C-(1.75–1.96)Si-(0.70–0.85)Mn spring steels have been reported to develop exceptional steel properties which are not possible by conventional heat treatment.[113,114]

A new microalloy spring steel [composition: Fe-(0.4–0.5)C-(1.1–1.4)Mn-(1.15–1.5)Si-0.45Cr-(0.01–0.03)Nb-(0.1–0.2)V-(0.01–0.02)N] has been developed for applications as high-stress coil springs in automotive suspension systems. The lower carbon content of this microalloyed steel compared to conventional spring steels increases the fracture toughness. Hence a higher UTS coupled with better fatigue properties and sag resistance can be used to achieve mass savings which will depend on spring design and steel properties.[115]

4. *Weldable rebars.* In sharp contrast to ASTM standards for rebars, the European standard has recommended weldable high-strength rebars, containing 0.22%C max. and exhibiting good weldability. It is possible to attain specified yield strength of 500 MPa with good ductility and bendability in the rebars with lean chemistry and low CE, by applying heat treatment from the rolling temperature, or by (VN) precipitation strengthening.[104]

5. *Microalloyed forging steels.* For many automotive applications, requiring less demanding or critical impact toughness values, air-cooled microalloyed steels (containing V and/or Nb) can effectively replace more expensive quenched, tempered, straightened, and stress-relieved previously specified low-alloy steels, where direct quenching, air cooling, or controlled cooling from the forging temperature produces ferrite containing two-phase or even multiphase microstructures (without changing forging parameters). This ever-growing range of applications in the automotive industry for these steels includes crankshafts, connecting rods, steering and suspension components, wheel hubs, axle beams and shafts, steering and support arms, spindles and knuckles, suspension yokes and support shafts, bearing blocks, sun wheels, counterweights, shift forks, and piston crowns.[112]

The trend in the development of microalloyed ferritic steels designed for direct continuous cooling (or quenching) is shown in Fig. 9.23.[116] (See also Chaps. 7 and 9 for more details.)

6. *High-carbon steel wire rod.* Usually, strands for prestressed concrete are made of wire of eutectoid composition. (See Sec. 7.7.2.7 for some details on conventional wire for prestressed concrete.) Continuously cast billets often show centerline segregation of carbon, exceeding eutectoid composition. The resulting cementite adversely affects the wire drawing operations. This problem can be overcome by slightly decreasing the carbon content and compensating the loss in strength by precipitation-strengthening of ferrite constituent of pearlite. The addition of small amounts of V causes the precipitation of V(C,N) in the ferrite lamellae of pearlite, thereby leading to an increase in tensile strength of wire rod. The same strength increment is maintained in the drawn wire. Note that a partitioning of V between ferrite and cementite occurs; however, in the presence of N, the tendency of V to dissolve in cementite recedes, which, in turn, offers greater availability of V for precipitation-strengthening of ferrite lamellae.[104]

7. *Cr-Ni-Mn austenitic stainless steels.* TMCP applied to hot-forged products of 20Cr-4Ni-15Mn-2Mo-0.64N-0.06C austenitic stainless steel has been reported to develop a high strength (YS > 1000 MPa), nonmagnetic properties with a slight reduction in ductility and toughness. In this case the strengthening mechanism is due to a combination of the grain size hardening, substructure hardening, work-hardening, and solid-solution hardening, all of which are increased by higher N level.[117]

8. *AISI 4130 steel.* An optimum combination of strength and toughness (UTS of 660 MPa, K_{IC} of 70 MPa·m$^{1/2}$) and microstructure of fine equiaxed (with mean 8-μm diameter) ferrite grains in the finished product has been reported using a TMT involving a reheat temperature of 900°C, and 50% deformation above Ar_3, and reducing the finish-rolling temperature to 620°C (i.e., in the γ-α region).[118]

15.2.3   High-Temperature Thermomechanical Treatment (HTMT)

This is a highly effective nonconventional heat treatment operation which was developed by Kula and his associate, later followed up in the United States and Russia.[1] The usual method, schematically shown in Fig. 15.1a, consists of deforming the stable austenite at temperatures just above Ac_3, then direct quenching (to form martensite), followed by tempering. Alternative treatments include deforming the steel on a falling temperature between Ac_3 and Ar_3 or Ac_1 (i.e., between 1000 and 800°C), but in this case the accelerated decomposition into ferrite/carbide structure must be avoided.[119,120] As in other TMT processes, the increased strength achieved in this case is due mainly to γ grain size refinement (say, from 40 to 60 μm to ~3 μm) and suppression of recrystallization by alloy carbide precipitation.

Advantages over Conventional Heat Treatment.[3,48]   First, HTMT is used to provide superior strength, higher ductility, and improvement in fracture toughness by markedly refining both the initial austenite grains and the subsequently formed martensitic platelets. It improves the low-temperature mechanical properties of ultrahigh-strength steels (such as 4340) when appropriate combinations of deformation temperature with tempering conditions are applied to the steels.[120a] Second, the martensitic structure produced is less susceptible to quench-induced cracking and is less liable to premature or delayed fracture. It greatly hinders the initiation of defects in the martensitic structure by suppressing the dynamic influence of martensitic phase transformation. Third, tools and dies made of HTMT materials have considerably increased useful life. Fourth, it is simple to apply in industry.[121]

Advantages over Ausforming Process.   First, HTMT produces higher impact ductility, a substantial increase in fatigue limit but slight reduction in strength. Second, it is most practicable because there is greater flexibility in temperature control and reduction in the forming loads. Also, the optimum properties can be attained at a moderate deformation (~40%).

Application.   It is applied to highly alloyed steels, especially suited to those steels which recrystallize slowly, such as silicon-containing steels and microalloyed 38CrSi steel (containing 0.37% C, 0.62% Mn, 1.33% Si, 1.58% Cr, 0.02% S, 0.02% P) with

0.18% Ti, 0.11% Nb, or 0.15% V.[122] It can be applied to low-alloy high-carbon steels which are unsuitable for the lower-temperature treatment such as ausforming with significant success in improved strength and toughness. The fatigue limit is also found to improve in a wide variety of steels provided the deformation is limited to 25 to 30%.

15.2.4 Low-Temperature Thermomechanical Treatment (LTMT) or Ausforming

Ausforming, first suggested in 1954, now consists of quenching the steel from the austenite phase field to a temperature in the metastable bay, usually in the range of 450 to 600°C (i.e., below the pearlite nose), where it is deformed to a considerable amount (up to 80% reduction in cross-sectional area) before its transformation, and then quenching to develop a martensitic structure followed by tempering (Fig. 15.1c). For ausforming, the steel must have a TTT diagram with a deep metastable austenite region and a large bay between the pearlite and bainite noses,[123] which can be produced by carbon content of 0.3 to 0.4% and a high concentration of alloying elements (e.g., Cr, Mo, Ni, and Mn) in the ausformed steel. Other alloying elements such as Ti, Nb, and V may be advantageously added to enhance the coarsening resistance during tempering, thereby increasing the strength of the steel.[124] A schematic time-temperature-deformation diagram of this process is shown in Fig. 15.1c. The retained austenite is completely removed after double tempering of ausformed steel specimens.[125]

The important process variables which affect the overall properties of ausformed steels are the austenitizing temperature, amount (total percentage) and temperature of deformation, and deformation schedule. Improvement of mechanical properties due to ausforming is proportional to the amount of deformation. An increase in strength is observed with the decreasing deformation temperature, due, probably, to the greater strain-hardening; however, the temperature selected should be appropriate so that neither recovery and recrystallization nor transformation occurs during the deformation.[124]

Characteristic Features of Ausforming

1. Strength obtained by this process is usually independent of the prior γ grain size and carbon content.[126]

2. Improved strength (without adverse effect on ductility) and toughness can be obtained. Alternatively, for a given strength, ausformed steels have greater toughness than conventionally heat-treated steels. Large improvements in strength, during ausforming treatment, are brought about by the combined effect of precipitation-hardening due to fine dispersion of alloy carbide precipitates, very high dislocation density (up to $10^{13}/cm^2$), and martensite transformation associated with a large number of smaller martensite plates and inherent fine dislocation substructures from the deformed metastable austenite.[124] Dislocations are partly formed as uniformly distributed pile-up dislocations during deformation and are partly inherited as dislocation cell structure during transformation into martensite. The contribution of inherited dislocation structure is more effective at high deformation temperatures. The explanation is that as the ausforming temperature is increased, the dislocation configuration changes from uniformly distributed pile-up dislocations into dislocation cells.[127]

3. Steels required for ausforming exhibit serrated yielding during the warm deformation process as a result of dynamic strain aging (see Chap. 4) involving the formation of a fine dispersion of alloy carbides on the dislocation networks.[124]

4. Higher ductility is presumably related to the formation of a fine untwinned (i.e., twin-free) martensite plate.[8] The martensite plate size is considerably smaller than that in similar steels, given a conventional quenching and tempering treatment. This occurs due to the pinning effect of fine alloy carbide precipitates on the prior dislocation configuration (arrays or tangles) which serve as barriers to martensite plate propagation. Subsequent tempering further improves the impact properties because of the precipitation of carbides in a fine spheroidal form.[128]

5. It is beneficial for those parts which can be shaped to produce simple shapes by the deformation process, e.g., punches,[129] leaf springs, and bolts.[8]

6. The ausformed parts have excellent fatigue strength under severe stress conditions. They also offer better performance at high temperature because of their high strength at elevated temperature and greater ability to withstand heat fatigue.[123]

Disadvantages of Ausforming

1. Ausforming requires the use of expensive, high-alloying additions (e.g., Cr, Ni, Mo, etc.).

2. It has not been successful in practical use for steels with high concentration of carbon and alloying elements because the deformation resistance of such steels is usually high in ausforming compared to warm-forging, due to the precipitation of a substantial amount of alloy carbides during deformation.

3. High loads on the rolling mill or forming equipment are necessary to produce large deformation; this also limits the treatment to simple shaped parts.[8]

As a result of these shortcomings, ausforming is not a widely accepted industrial practice.

Application.   Ausforming is applied to steels with relatively high hardenability. It is especially suitable for high-alloy steels such as high-speed tool steels, AISI H13 tool steels (modified with Nb),[130] and 17-4 PH stainless steel[131]; and for low-alloy steels, especially containing Cr and Mo. The process can be applied to certain martensitic steels (with advantages)—examples are cold heading, cold piercing, extrusion, riveting, and hot piercing of steels.[132] Note that the ausforming of 17-4 PH stainless steel does not increase the strength of the product, but increases the ductility. Moreover, ausformed and aged 17-4 PH stainless steels offer a dramatic improvement in the impact value over the temperature range of -100 to $0°C$, due to the refined microstructure.[131]

15.2.5   Isoforming

Isoforming is a class II TMT which involves austenitizing the steel, quenching to an intermediate temperature in the pearlite nose region, deforming the metastable austenite during the isothermal transformation to pearlite, followed by air-cooling (Fig. 15.1*d*). Tempering is not needed at all. It is applied to any low-alloy steel with a suitable TTT diagram.

Irani et al.[133-135] have shown that the isoforming process can significantly improve the toughness of several low-alloy steels relative to the conventionally rolled products. It has been demonstrated that the lamellar morphology of pearlite lowers the toughness of ferrite-pearlite steel, the DBTT increasing with increasing pearlite content. However, isoforming, involving high deformation (>60%) during the transformation of a suitable low-alloy steel between 600 and 700°C, produces the change into a satisfactory morphology and distribution of ferrite and carbide from the prior ferrite-pearlite structure. The final structure of the isoformed steels is characterized by the formation of very fine ferrite subgrains (<1 μm across) together with fine, small, spheroidal carbide particles (~250 Å diameter) which are either mainly located at subgrain triple points[124] or uniformly dispersed throughout the ferrite matrix. This substructure is caused by a polygonization process at the isoforming temperature.[133] The misorientation across the subgrain boundaries lies between 1 and 10°.[124] Improved toughness of ferrite-pearlite steels is thus achieved by isoforming which is attributed to both the presence of substructure in a ferrite matrix (i.e., refinement of ferrite grain size) and the elimination of pearlite by replacement of lamellar cementite with spheroidized particles.[124,133-135]

15.3 HIGH-STRENGTH LOW-ALLOY STEELS

High-strength low-alloy (HSLA) low-carbon steels constitute a classical metallurgical innovation where a combination of MAEs and TMT provides improved mechanical properties through microstructural control. A general description of HSLA steel is one that contains (1) a low carbon (0.03 to 0.15%) content to obtain good toughness, formability, and weldability; (2) one or more of the strong carbide-forming microalloying elements (e.g., Nb, Ti, or V) (not usually in excess of 0.2%); (3) a group of solid-solution strengthening elements (e.g., Mn, Si); and (4) one or more of the additional microalloying elements (e.g., Ca, Zr) and the rare earth elements, particularly Ce and La for sulfide inclusion shape control.[136] In many other HSLA steels, small amounts of Ni, Cr, Cu, B, and particularly Mo are also present, which increase the atmospheric corrosion resistance and hardenability. The unique combination of strength, toughness, and decrease of the ductile-to-brittle transition temperature (DBTT) (to as low as −70°C), in microalloyed HSLA steels, arises from the precipitation and interaction of these microalloy precipitates with the process of recrystallization and grain growth of γ and of transformation from γ to a fine ferrite grain structure. Typical stress-strain curves for C-Mn (or mild) steel, HSLA steel, and DP steel in Fig. 15.16 show that the yield strength of HSLA steel in the range of 350 to 700 MPa (50 to 100 ksi) is double that of a C-Mn steel.[7,50,137,138] HSLA steels have sometimes better corrosion resistance than as-rolled carbon steels; however, their weldability is similar to or better than that of low-carbon steels. Carbides (NbC, TiC), nitrides (AlN, VN), and carbonitrides [e.g., Nb(CN), V(CN), Ti(CN), (Nb,V)CN, (Nb,Ti)CN] are the dispersed second-phase particles which act as grain size refiners and precipitation strengtheners in HSLA steels.[44]

HSLA steels are mainly hot-rolled into the usual wrought product forms (such as sheet, strip, plates, bars, rods, sections, and structural shapes) and are usually supplied in the as-hot-rolled condition. However, the production of hot-rolled HSLA products may also use special hot mill processing methods that further increase the mechanical properties of HSLA steels and product forms. The processing methods of these steels include the following:[17]

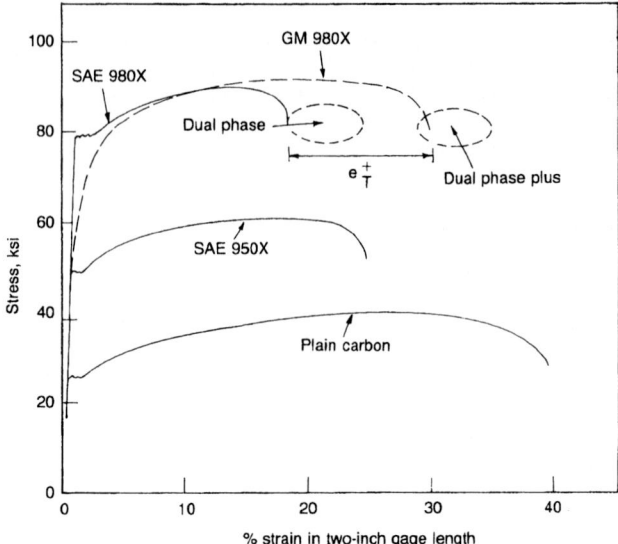

FIGURE 15.16   Stress-strain curves for plain carbon, HSLA, and dual-phase steels. The two dashed-line ellipses indicate reported ranges of elongation for dual-phase steels. (*After M. Rashid, "GM 980X—Potential Applications and Review," Research Publication GMR-232, General Motors Corporation, Warren, Mich., 1977.*)

1. *Controlled rolling* to produce fine austenite grains and/or highly deformed (pancaked) austenite grains, which transform into fine ferrite grains during cooling and provide improved strength-toughness combinations compared to normalized steels, thereby significantly improving toughness and enhancing yield strength. Commercially, it is possible to obtain both the yield strength greater than 550 MPa and the subzero ITT in plate thicknesses ≤40 mm.[139] (See earlier sections for more details.)

2. *Thermomechanical controlled processing* to obtain fine ferrite grain size during the transformation of austenite to eliminate or optimize the use of expensive alloying elements. Accelerated cooling increases the yield strength of controlled rolled plate up to 100 MPa to over 600 MPa. The improvement in strength is also associated with an increase in ductility and shelf energy.[139] (See earlier section for more details.)

3. *Direct quenching or accelerated air or water cooling and tempering* of low-carbon steels with sufficient hardenability to transform into low-carbon bainite (or acicular ferrite). They possess high yield strength (275 to 690 MPa, or 60 to 100 ksi) and tempering resistance, excellent formability, and higher toughness than those of quenched and tempered steels. These improvements in properties are attributed to both the presence of a finer precipitate distribution and more mobile dislocations in the direct quenched steels than in the reaustenitized-and-quenched steels.[140]

4. *Normalizing* of vanadium- and niobium-vanadium-bearing HSLA steels for grain refinement, thereby increasing yield strength and toughness.

5. *Intercritical annealing* of HSLA steels to produce a dual-phase microstructure.

The usefulness and cost-effectiveness of these processing methods mostly depend on alloy content and product form. Furthermore, HSLA steels are supplied as cold-rolled-sheet, -coated sheet, and -forgings. The principal benefit of HSLA forgings, like that of as-hot-rolled HSLA products, is its yield strength in the range of 275 to 485 MPa (40 to 70 ksi) or even higher without subjecting to any heat treatment. (See earlier section for more details.)

15.3.1 Classification of HSLA Steels

Although over 600 standard and proprietary grades of HSLA steels are available, they can be broadly described in seven categories:[136,141,142]

1. *Weathering steels* are steels with ~0.1% C, 0.15 to 0.30% Cu, 0.5 to 1.0% Mn, 0.05 to 0.15% P, 0.15 to 0.90% Si and MAEs, which exhibit superior atmospheric corrosion resistance and improved solid-solution strengthening. They are of two types: (a) weathering steels with normal low-P and multiple-alloy additions for solid-solution strengthening and improved corrosion resistance; and (b) proprietary weathering steels with high-P (0.05 to 0.15%) and multiple alloy additions for solid-solution strengthening and corrosion resistance. Typical applications include railroad cars, bridges, and unpainted buildings.

2. *Microalloyed ferrite-pearlite steels* are used for conventional HSLA steels, as mentioned earlier. They exhibit discontinuous yielding behavior. They are subdivided into nine types: V-microalloyed steels, Nb-microalloyed steels, Nb-Mo-microalloyed steels, V-Nb-microalloyed steels, V-N-microalloyed steels, Ti-microalloyed steels, Nb-Ti-microalloyed steels, V-Ti-microalloyed steels, and V-Ti-N-microalloyed steels. These steels find a wide variety of applications in automotive, agricultural, linepipe, construction, and pressure vessel industries.[138]

3. *As-rolled pearlitic steels* are C-Mn steels containing small amounts of other alloying elements to improve strength, formability, toughness, and weldability. The main drawbacks of these steels are an increase of DBTT, no enhancement of yield strength, and reduction in weldability.

4. *Acicular ferrite steels* are very low-carbon (typically, 0.03 to 0.06%) steels with enough hardenability (by Mn, Mo, and B additions) to transform on cooling to a very fine, high-strength acicular-ferrite (also called *low-carbon bainite*) structure rather than the usual polygonal ferrite structure. Nb can also be added for grain refinement and precipitation-strengthening. These steels have an excellent combination of high yield strength, toughness, formability, and weldability.[142] (See Sec. 9.4 for more details.) A major application of these steels is in linepipe in Antarctic conditions. Three readily available popular grades are X-65, 6–70, and X-80. A typical X-70 grade has a base composition of 0.03% C, 0.25% Si, 1.91% Mn, 0.008% P, 0.001% S, and 0.048% N with small additions of Ti, B, and Ca.

5. *Dual-phase steels* are steels having a microstructure of martensite dispersed in a ferrite matrix, providing a combination of high tensile strength and ductility and exhibiting continuous yielding behavior[142] (discussed later separately).

6. *Inclusion-shape-controlled steels*† are steels exhibiting improved ductility and through-thickness toughness by the small additions of Ca, Zr, Ti, and/or rare earth elements to change the sulfide inclusion shape from elongated stringers to small, dispersed, nearly spherical globules. This sulfide inclusion shape control in HSLA

---

† Currently, use of very low-sulfur-bearing steels has eliminated the need for sulfide shape control.

steels produces an improvement in transverse impact energy and can minimize lamellar tearing in welded structures.

7. *Hydrogen-induced cracking resistant steels* are steels with low carbon, low sulfur, >0.26% copper, inclusion shape control, and minimum Mn segregation.

15.3.2 Mechanical Properties

Hot-rolled steels with ferrite-pearlite microstructures are the most widely used HSLA steels. Commercially available microalloyed ferrite-pearlite HSLA steels have yield strengths in the range up to 700 MPa (100 ksi), which is about 3.5 times greater than 200 MPa (30-ksi) yield strength of conventional hot-rolled plain carbon steels. Tables 15.1 and 15.2 summarize tensile properties of some hot-rolled ferrite-pearlite HSLA steels.[142] These properties depend on the alloying additions and production methods.

15.3.3 Structure-Property Relationships

Strengthening Mechanisms. The major strengthening mechanisms in controlled rolled HSLA steels include grain refinement; precipitation-hardening by strain-enhanced precipitation of microalloyed carbonitrides in ferrite or at the interphase; solid-solution strengthening from Mn, Si, and uncombined N; and dislocation substructure (including dislocation tangles and cell walls) strengthening. When controlled rolling is used with a finishing temperature well below the Ar_3 temperature, an additional contributor to the overall strength comes from preferred orientation or texture hardening.

The observed yield stress of a polycrystalline controlled rolled microalloyed HSLA steel can be expressed by the following Hall-Petch relationship:

$$\sigma_y = \sigma_0 + k_y d^{-1/2} \tag{15.19}$$

$$= (\sigma_i + \sigma_{ss} + \sigma_d + \sigma_p + \sigma_{\text{tex}}) + k_y d^{-1/2} + k' d_s^{-1/2} \tag{15.20}$$

where σ_0 is the friction stress and is a function of steel composition; k_y is a material constant; d is mean ferrite grain size (diameter); d_s is the subgrain diameter; and σ_0 is divided into several terms: internal friction stress σ_i, solid-solution strengthening σ_{ss} [as given by Eq. (7.42)], dislocation strengthening σ_d [as given by Eq. (14.6)], precipitation strengthening σ_p [as given by Eq. (14.8)], and texture strengthening σ_{tex}; $k_y d^{-1/2}$ is the contribution to the strength by the ferrite grain size; and $k' d_s^{-1/2}$ is the subgrain or substructure strengthening term. However, the relative contribution of any individual mechanism presumably varies with the change in steel composition and rolling practice.[143]

A typical effect of $\gamma \rightarrow \alpha$ transformation temperature for an Fe-0.1C-1Mn-0.2Si steel with a prior γ grain size of ASTM 8 (or 20 μm) is shown in Fig. 15.17a. Evidently a decrease in the γ grain size and an increase in cooling rate associated with decrease in the transformation temperature will lead to a lower ferrite grain size, as shown by[26]

$$d_\alpha = a d_\gamma + b \dot{T}^{-1/2} + c \tag{15.21}$$

where d_α and d_γ are the ferrite and austenite grain sizes, respectively; $\dot{T}$ is the cooling rate; and a, b, and c are constants.

TABLE 15.1 Tensile Properties of HSLA Steel Grades Specified in ASTM Standards[142]

ASTM specification[†]	Type, grade, or condition	Product thickness[‡]		Minimum tensile strength[§]		Minimum yield strength[§]		Minimum elongation,[§] %		Bend radius[§]	
		mm	in.	MPa	ksi	MPa	ksi	in 200 mm (8 in.)	in 50 mm (2 in.)	Longitudinal	Transverse
A 242	Type 1	20	¾	480	70	345	50	18	...		
		20–40	¾–1½	460	67	315	46	18	21		
		40–100	1½–4	435	63	290	42	18	21		
A 572	Grade 42	150	6	415	60	290	42	20	24	¶	...
	Grade 50	100	4	450	65	345	50	18	21	¶	...
	Grade 60	32	1¼	520	75	415	60	16	18	¶	...
	Grade 65	32	1¼	550	80	450	65	15	17	¶	...
A 588	Grades A–K	100	4	485	70	345	50	18	21	¶	...
		100–125	4–5	460	67	315	46	...	21	¶	...
		125–200	5–8	435	63	290	42	...	21	¶	...
A 606	Hot-rolled	Sheet		480	70	345	50	...	22	t	2t–3t
	Hot-rolled and annealed or normalized	Sheet		450	65	310	45	...	22	t	2t–3t
A 607	Cold-rolled	Sheet		450	65	310	45	...	22	t	2t–3t
	Grade 45	Sheet		410	60	310	45	...	22–25	t	1.5t
	Grade 50	Sheet		450	65	345	50	...	20–22	t	1.5t
	Grade 55	Sheet		480	70	380	55	...	18–20	1.5t	2t
	Grade 60	Sheet		520	75	415	60	...	16–18	2t	3t
	Grade 65	Sheet		550	80	450	65	...	15–16	2.5t	3.5t
	Grade 70	Sheet		590	85	485	70	...	14	3t	4t
A 618	Ia, Ib, II	19	¾	485	70	345	50	19	22	t–2t	...
	Ia, Ib, II, III	19–38	¾–1½	460	67	315	46	18	22	t–2t	...
A 633	A	100	4	430–570	63–83	290	42	18	23	¶	...
	C, D	65	2.5	485–620	70–90	345	50	18	23	¶	...
	C, D	65–100	2.5–4	450–590	65–85	315	46	18	23	¶	...

TABLE 15.1 Tensile Properties of HSLA Steel Grades Specified in ASTM Standards[142] (Continued)

ASTM specification[†]	Type, grade, or condition	Product thickness[‡] mm	Product thickness[‡] in.	Minimum tensile strength[§] MPa	Minimum tensile strength[§] ksi	Minimum yield strength[§] MPa	Minimum yield strength[§] ksi	Minimum elongation,[§] % in 200 mm (8 in.)	Minimum elongation,[§] % in 50 mm (2 in.)	Bend radius[§] Longitudinal	Bend radius[§] Transverse
A 656	E	100	4	550–690	80–100	415	60	18	23	¶	...
	E	100–150	4–6	515–655	75–95	380	55	18	23	¶	...
	50	50	2	415	60	345	50	20	...	¶	...
	60	40	1½	485	70	415	60	17	...	¶	...
	70	25	1	550	80	485	70	14	...	¶	...
	80	20	¾	620	90	550	80	12	...	¶	...
A 690	...	100	4	485	70	345	50	18	...	2t	...
A 709	50	100	4	450	65	345	50	18	21	...	...
	50 W	100	4	485	70	345	50	18	21	...	...
A 715	Grade 50	Sheet		415	60	345	50	...	22–24	0	t
	Grade 60	Sheet		485	70	415	60	...	20–22	0	t
	Grade 70	Sheet		550	80	485	70	...	18–20	t	1.5t
	Grade 80	Sheet		620	90	550	80	...	16–18	t	1.5t
A 808	...	40	1½	450	65	345	50	18	22	...	...
		40–50	1½–2	450	65	315	46	18	22	...	...
		50–65	2–2½	415	60	290	42	18	22	...	...
A 812	65	Sheet		585	85	450	65	...	13–15	...	...
	80	Sheet		690	100	550	80	...	11–13	...	...
A 841	...	65	2.5	485–620	70–90	345	50	18	22	...	...
		65–100	2.5–4	450–585	65–85	310	45	18	22	...	...
A 871	60, as hot-rolled	5–35	3/16–1 3/8	520	75	415	60	16	18	...	...
	65, as hot-rolled	5–20	3/16–3/4	550	80	450	65	15	17	...	...

† For compositions, available mill forms, and special characteristics, see Table 1.13.
‡ Maximum product thickness except when a range is given. No thicknesses are specified for sheet products.
§ May vary with product size and mill form.
¶ Optional supplementary requirement given in ASTM A 6.
Reprinted by permission of ASM International, Materials Park, Ohio.

TABLE 15.2 Mechanical Properties of HSLA Steel Grades Described in SAE J410[142]

Grade[†]	Minimum tensile strength[‡]		Minimum yield strength[‡§]		Minimum elongation,[‡] %		Bend diameter[‡¶]
	MPa	ksi	MPa	ksi	in 200 mm (8 in.)	in 50 mm (2 in.)	
942X	415	60	290	42	20	24	t–3t
945A	415–450	60–65	275–310	40–45	18–19	22–24	t–3t
945C	415–450	60–65	275–310	40–45	18–19	22–24	t–3t
945X	415	60	310	45	19	22–25	t–2.5t
950A	430–483	63–70	290–345	42–50	18–19	22–24	t–3t
950B	430–483	63–70	290–345	42–50	18–19	22–24	t–3t
950C	430–483	63–70	290–345	42–50	18–19	22–24	t–3t
950D	430–483	63–70	290–345	42–50	18–19	22–24	t–3t
950X	450	65	345	50	18	22	t–3t
955X	483	70	380	55	17	20	t–3.5t
960X	520	75	415	60	16	18	1.5t–3t
965X	550	80	450	65	15	16	2t–3t
970X	590	85	485	70	14	14	3t
980X	655	95	550	80	10	12	3t

† For compositions, available mill forms, and special characteristics of these steels, see Table 1.13. ‡ May vary with product size and mill form; for specific limits, refer to SAE J410.
§ 0.2% offset.
¶ 180° bend test at room temperature. Used for mill acceptance purposes only; not to be used as a basis for specifying fabricating procedures.
Reprinted by permission of ASM International, Materials Park, Ohio.

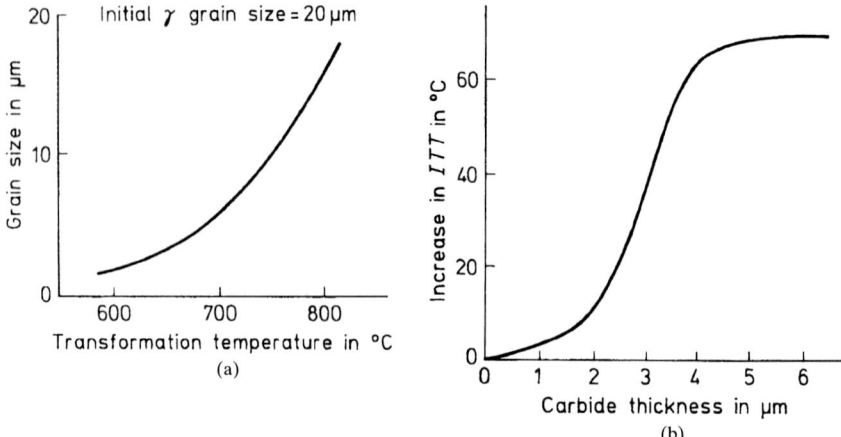

FIGURE 15.17   The effect of (a) $\gamma \rightarrow \alpha$ transformation temperature on the transformed polygonal ferrite grain size and (b) carbide thickness on the ITT of a ferrite-pearlite HSLA steel.[26] (*Courtesy of F. B. Pickering.*)

Toughness.   Both aspects of toughness, such as the ductile-brittle transition temperature (DBTT) or impact transition temperature (ITT) associated with the resistance to brittle cleavage fracture and the Charpy shelf energy (CSE) associated with the resistance to low-energy ductile fractures, must be considered for overall evaluation. The ITT of low-carbon pearlitic microstructures given by Eq. (7.44) can hold for ferrite-pearlite HSLA steels. In addition, both precipitation strengthening σ_p and dislocation strengthening σ_d increase the ITT by about 0.25°C per MPa increase in σ_p and 0.4°C per MPa increase in σ_d. Similarly, an increase in $\{111\}_\alpha$ texture in the rolling plane also raises the ITT; however, this is very small in hot-rolled or normalized plate or sheet.[26]

The ITT decreases with the increase of $d^{-1/2}$, as expressed by

$$ITT = T_o - k_y d^{-1/2} = f(\sigma_0) - k_y d^{-1/2} \tag{15.22}$$

where T_o is a function of the lattice friction stress σ_0 and k_y is a measure of the effectiveness of refining the ferrite grain size in order to lower ITT. For a given design stress, the beneficial effect of controlled rolling for improved toughness is attributed to the combined effects of lattice friction stress and grain size refinement; the latter factor affecting the most. For example, for each unit increment in $d^{-1/2}$ mm$^{-1/2}$, ITT drops by about 11.5°C. [144]

For 0.01Ti-0.08V-N microalloyed HSLA steels in the RCR + AC condition, the ITT may be expressed in the form of an inverse relationship between toughness and yield strength as

$$ITT(°C) = -217 + 0.33(YS) \tag{15.23}$$

This singular relationship is based virtually on constant γ grain size and the fact that increased N and V contents and increased cooling rate and decreased finish-cooling temperature all seem to simultaneously increase precipitation hardening while refining the ferrite grain size.[103]

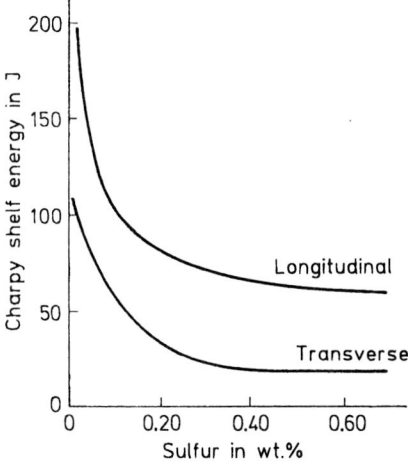

FIGURE 15.18 Effect of sulfur content on the Charpy shelf energy of a ferrite-pearlite HSLA steel, showing the anisotropy of the Charpy shelf energy. The sulfur content is proportional to the volume fraction of MnS.[26] (*Courtesy of F. B. Pickering.*)

Finally, if any coarse carbides form in the hot-rolled or normalized HSLA steels, which is unusual, ITT will increase with carbide thickness >2 μm, as shown in Fig. 15.17*b*. This effect is of great interest in evaluating the toughness of the acicular ferrite microstructure.

A high CSE is essential to avoid the likelihood of low-energy ductile fracture as well as to overcome *lamellar tearing* in welds, particularly in the short transverse or through-thickness direction. There is a significant decrease of CSE with increasing pearlite content (as shown in Fig. 7.27) and increasing nonmetallic sulfide inclusion content in ferrite-pearlite HSLA steels (Fig. 15.18). The remarkable anisotropy in CSE, as evidenced in the short transverse or through-thickness direction, has been eliminated/minimized by inclusion shape control. However, note that inclusion shape control is not advantageous for ITT, which really increases. An equation to describe the transverse CSE is given by[145]

$$CSE(J) = 112 - 2.8d^{-1/2} - 0.18\sigma_p(MPa) - 832(\%S) - 43(\%P)$$

$$-0.76(\%pearlite) + 107(\%Zr) \tag{15.24}$$

It is clear that increases in pearlite (i.e., C%) and S contents have deleterious effects. Zr offers beneficial effect through inclusion shape control. The deleterious effect of σ_p and grain refinement denote the lower CSE due to overall strength increase. Al is deleterious because aluminous inclusions formed by Al deoxidation decrease the CSE value.

15.4 TRIP OR MULTIPHASE STEELS

Martensite formed by plastic deformation of metastable (or retained) austenite γ_R is called *strain-induced martensite* (SIM). This irreversible *transformation-induced plasticity* (TRIP)[146] or *strain-induced martensite transformation* (SIMT) causes superior mechanical properties such as increased tensile strength (σ_T), resistance to local inhomogeneous necking (or strain inhomogenization), work-hardening rate, uniform elongation e_u up to extremely high strain rates, and formability (or $\sigma_T \cdot e_T\%$). The strain- (or stress-) induced transformation to martensite has become the basis to create a new class of ultrahigh-strength metastable austenitic steels, known as TRIP steels. In these steels, deformation of metastable austenite between the M_d and M_s temperatures little by little causes formation of martensite up to fracture, elongation well over 100%, and yield strength often greater than 1379 MPa (200 ksi).[147,148] The presence of γ_R in the microstructure also causes higher toughness via localized TRIP leading to crack-tip blunting. Fatigue crack propagation (FCP) studies (controlled ΔK) have further suggested that the deformation-induced martensitic transformation retards the propagation of a crack in the lower-strength metastable austenite, especially at low ΔK, and increases the fatigue strength of TRIP steels in smooth bar tests under stress-control situations.[149,150]

This phenomenon in delaying and stabilizing the necking was first reported by Banerjee.[151] Later, Zackay et al.[152] found it in highly alloyed austenitic Fe-Cr-Ni stainless steels. However, the large amount of expensive alloying elements imposed severe practical problems for their widespread use, notably, for automotive applications.[153] This deformation-induced α' transformation method of improving toughness and ductility is also used in various other materials such as in cast iron,[153a] $(\alpha + \beta)$ type Ti-6Al-2Sn-4Zr-6Mo alloy,[153b] and stabilized zirconia.[153c]

TRIP steels are based on multiphase microstructures comprising initially three phases, ferrite, bainite, and γ_R, with a fourth one being martensite transformation from γ_R during deformation such as deep drawing of sheet steel. A typical phase distribution of microalloyed Si-Mn TRIP steel [composition: 0.17C-1.4Mn-1.5Si-(0.02-0.04)Nb] in the as-shipped condition comprises about 50 vol% ferrite, 40 to 45 vol% bainite, and 5 to 10 vol% γ_R, but γ_R transforms to martensite during deformation. In this, Nb retards the isothermal bainite transformation kinetics, which causes an enhanced amount of γ_R during proper annealing. Nb also enhances the yield and tensile strengths by ~15 MPa per 0.01% addition due to grain refinement.[154]

The TRIP steels can be classified into (1) high-alloy austenitic TRIP steels, comprising Fe-Cr-Ni-Mo[150,155] and Fe-Mn-Si-Al alloys;[156] (2) low-alloy Si-Mn TRIP steels with varying amounts of Si and Mn; (3) TRIP-assisted steels (with increased γ_R stability);[153] and (4) microalloyed Si-Mn TRIP steel processed by a TMP route.[157] Note that TRIP-assisted multiphase steels are of two types: (a) cold-rolled TRIP-assisted multiphase steels by combination of intercritical annealing of cold-rolled product and isothermal holding at bainite transformation temperatures during a continuous annealing process and (b) hot-rolled TRIP-assisted multiphase steels by austenitic hot-rolling, holding in the ferrite field, followed by coiling in the bainitic range. Unfortunately both these TMT methods need a high Si content to suppress cementite precipitation in order to avoid a loss of γ_R stability. Also, high Si content causes red scale surface defects, leading to moderate hot-dip galvanizability.[158] Table 15.3 lists the typical composition of TRIP steels.[150,153–159,161,162,164–168]

15.4.1 Advantages of TRIP Steels

1. Compared to conventional high-strength steels, TRIP steels offer higher formability (i.e., strength-ductility combinations) and expand the range of cold-formable steels at high-strength values.[154] The product of $\sigma_T \cdot e_T$ percent was reported by Kawano et al.[159] in hot-rolled Fe-0.2C-1.5Si-(1.5-1.7)Mn steel in the range of 25,000 to 30,000 MPa·% and by Sakuma et al.[153] in cold-rolled 0.2C-2.1Si-0.94Mn and 0.2C-1.24Si-1.8Mn steels (held at 425°C in the bainite range) in excess of 27,500

TABLE 15.3 Typical Compositions, wt%, of TRIP Steels[150,153–159,161,162,164,166–168]

No.	C	Si	Mn	Ni	Mo	Cr	Other	γ_R%	Steel type
1	0.25	2.0	2.5	9.0	5.0	9.0	—	—	Austenitic TRIP steel[150,155]
2	0.30	2.0	2.0	8.0	4.0	9.0	—	—	Austenitic TRIP steel[150,155]
3	0.25	2.0	1.7	8.8	5.5	9.0	—	—	Austenitic TRIP steel[150,155]
4	0.25	2.0	2.0	8.8	3.0	9.0	—	—	Austenitic TRIP steel[150,155]
5	0.2–0.4	3.0	15–30	—	—	—	3.0 Al	—	Austenitic TRIP steel[156]
6	0.16	—	—	1.5	—	1.6	—	10	15CrNi6 TRIP steel[164]
7	0.2	2.0	1.5	—	—	—	—	—	Si-Mn TRIP steel[167]
8	0.2	1.5	1.5–1.7	—	—	—	—	15	Si-Mn TRIP steel[159]
9	0.39	1.37	1.45	—	—	—	—	—	Si-Mn TRIP steel[166]
10	0.18	0.39	1.33	—	—	—	0.018P, 0.012S, 0.029Al, 0.0069N	—	Si-Mn TRIP steel[161]
11	0.2	2.0	1.0	1.0	—	—	1.2Cu	—	Mod. Si-Mn TRIP steel[167]
12	0.2	2.1	0.94	—	—	—	0.0058S, 0.003P, 0.041Al, 0.0027N	—	Mod. Si-Mn TRIP steel[153]
13	0.20	1.24	1.8	—	—	—	0.006S, 0.003P, 0.041Al, 0.0028N	—	Mod. Si-Mn TRIP steel[153]
14	0.4	1.5	1.5	—	—	—	0.018S, 0.015P, 0.036Al	—	TRIP-aided DP steel[168]
15	0.17	1.5	1.4	—	—	—	0.002–0.04Nb	—	Microalloyed Si-Mn TRIP steel[154]
16	0.16	0.45	1.4	—	—	—	0.01P, 0.035Al, 0.045N, 0.005Ti, 0.015Nb, 0.03V	—	Microalloyed Si-Mn TRIP steel[158]
17	0.22	1.55	1.55	—	—	—	0.035Nb, 0.028Al, 0.002–0.004N	11	Hot-rolled TRIP-aided steel[157]
18	0.2	1.5	1.5	—	—	—	0.035Nb	10	Microalloyed Si-Mn TRIP steel[162]

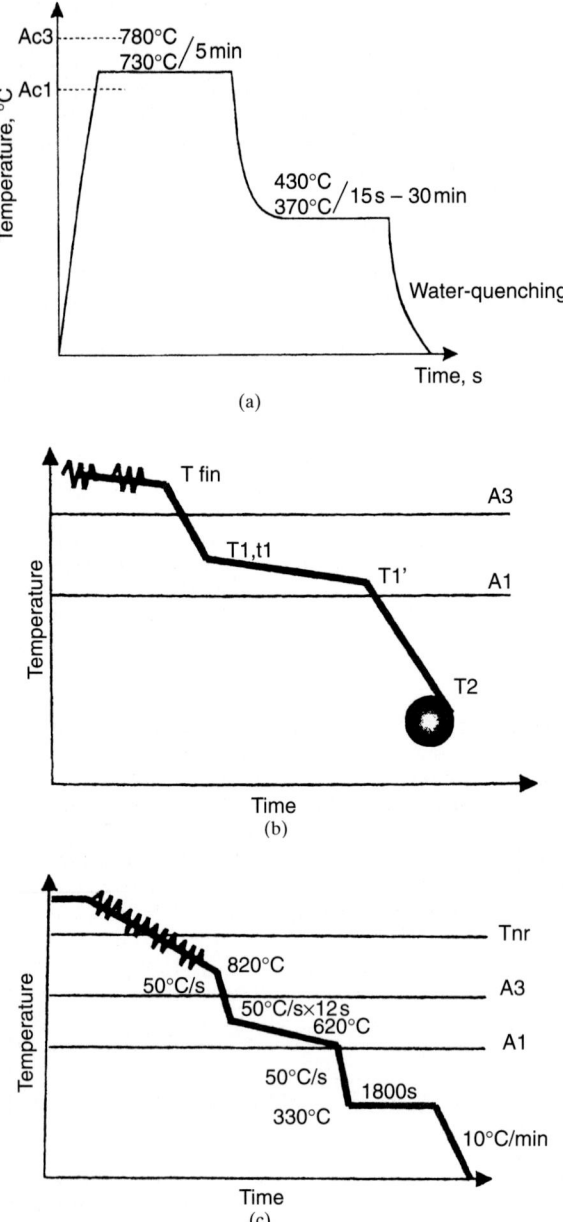

FIGURE 15.19 Schematic representation of (*a*) heat treatment condition for multiphase treatment consisting of an intercritical annealing followed by a bainite tempering stage,[161] (*b*) thermomechanical path for TRIP-assisted multiphase steels, and (*c*) new thermomechanical path for TRIP-assisted multiphase steels.[158]

MPa·%. However, the formability increases with increasing volume fraction of γ_R, C%, and plastic stability of γ_R.[159,160]

2. Through heat treatment (Fig. 15.19a)[161] or various thermomechanical paths (Fig. 15.19b and c),[158] it is now possible to stabilize considerable amounts of γ_R at room temperature in low-alloy ferritic [e.g., Fe-(0.1–0.2)C-(1–3)Mn-(1–2.5)Si] steels and microalloyed (e.g., Fe-0.16C-1.4Mn-0.45Si-0.015P-0.035Al-0.0045N-0.005Ti-0.015Nb-0.003V) steels, respectively.

15.4.2 Stability of Retained Austenite

As a major phase, the stability of γ_R is very important for enhancing TRIP steels. Large-sized γ_R is unstable and transforms to martensite at low strains, while small-sized (<1 µm) γ_R, with high-carbon enrichment, is more stable and can be retained to larger strains. It is believed that this latter type of γ_R is very potent for the TRIP effect.[162] These structural characteristics are also affected by the TMT schedule to which the steel is subjected.[160]

A low M_s temperature denotes a high γ stability, i.e., high resistance to martensitic transformation. Furthermore, γ stability is a function of γ composition, γ particle size, and partitioning of C and N during cooling. The presence of defects (micron and submicron sizes) in the microstructure after TMT can delay the martensite transformation and thus lead to more γ_R.[163]

Experiments have revealed that γ_R is less stable under tensile loading than under compressive loading. SIMT under isothermal conditions offers a larger volume fraction of martensite than under adiabatic conditions at the same applied stress levels. The difference between the two is dependent on the type of loading.[164]

It is shown that the control of microstructure by TMT produces the optimum conditions (size, distribution, amount, composition, and morphology) of γ_R in Si-Mn TRIP steels. The results show that deformation during dynamic recrystallization improves ferrite grain refinement and increases the volume fraction of γ_R.[165] Deformation in the intercritical region increases the γ_R stability due to: increased dislocation density, grain refinement, and carbon enrichment and the production of a more granular and a less interlath needlelike shape.[166] Granular-type γ_R mostly located along the ferrite grain boundaries is susceptible to martensite transformation by deformation, whereas interlath film-type γ_R is quite stable and would not transform to martensite by deformation.[167] The γ_R produced by heat treatment of Si-Mn multiphase steels, involving intercritical annealing followed by bainitic transformation, is large and very fine and becomes stabilized by carbon rejection from the bainite to avoid its transformation to martensite during the final stage.[161]

Increased Si improves the stability γ_R and makes e_u and $\sigma_T \cdot e_T$ % product larger. This enhancement is pronounced at higher holding temperatures and/or longer holding times in the bainite transformation region.[153]

Since Si and Mn have a beneficial effect on the γ_R, conventional TRIP-assisted steels usually contain large concentrations of these elements (1.5 to 2.5% for both). Mn is well known as a γ stabilizer. However, Si is considered as the γ stabilizer by inhibiting carbide precipitation and, thereby, promoting carbon enrichment of residual γ during the bainite transformation. Consequently, studies dealing with TRIP-assisted steels mostly focus on alloys containing more than 1% Si.[169] In contrast, Jacques et al.[161] have demonstrated the good mechanical properties of cold-rolled low-carbon, low-silicon steel (for example, 0.18C-0.39Si-1.33Mn-0.018P-0.012S-0.029Al-0.069N steel), heat-treated in the same conditions as those for conventional TRIP-assisted multiphase steels. These steels exhibit up to 10 vol% of γ_R stabilized by precise control of the bainitic transformation time and temperature (to hinder carbide precipitation).[161]

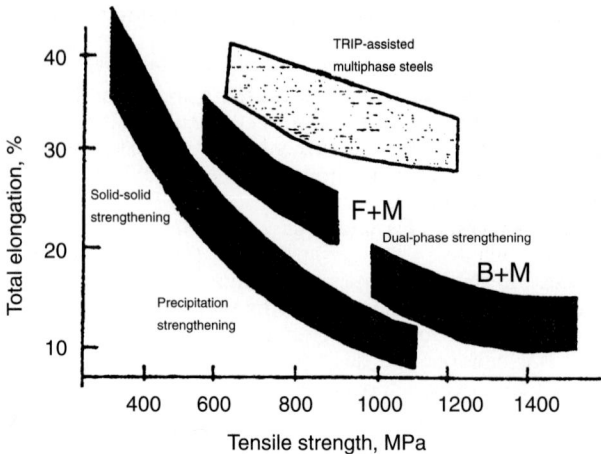

FIGURE 15.20   Tensile strength and total elongation of TRIP-assisted multiphase steels.[158]

15.4.3   Mechanical Properties

Mechanical properties of Nb-treated 0.2C-1.5Mn-1.5Si-0.035Nb TRIP steels depend on the processing conditions in the bainite transformation range (i.e., temperature and time). The σ_T is sensitive to microstructure, whereas the e_T is sensitive to both microstructure and volume fraction of γ_R.[170] The mechanical properties depend on the grain size, volume fraction of γ_R, the stability of γ_R, local transformation stresses after bainite formation, and the carbide precipitates.[154] Figure 15.20 shows a comparison of the tensile strength-ductility relationship of TRIP-assisted multiphase steels with respect to HSLA steels.[158]

15.4.4   Applications

The highly alloyed TRIP steel requires excellent metallurgical control and is very expensive to prepare. It is, therefore, used in special applications of flat-rolled products or wires, where very high mechanical properties are in demand. Other applications include shear-spun rings for containment of jet-engine rotor fragments in the case of a catastrophic failure and projectiles for the radically new design of a gun barrel.[9] The Si-Mn-Nb TRIP steels are used for: (a) cold-forming of wire intended for high-strength fasteners[169] and (b) automotive parts such as wheel disks, where light weight, high formability, and good hot-dip galvanizability are required.[158]

15.4.5   Theories

The unique behavior of TRIP steels is attributed to a combination of classical plastic flow by dislocation motion and inelastic transformation strain related to the martensite phase change. The latter creates a large additional plastic flow within austenite and martensite. Diani et al. have proposed a kinematic description of such a phenomenon depending on a local decomposition of strain rate into an elastoplastic portion and a given lattice inelastic strain rate field.[171]

Fischer et al. have proposed that an externally stressed specimen, during phase transformation, may exhibit a considerable nonlinear TRIP behavior. The TRIP strain may be irreversible for steels or reversible with a certain hysteresis for shape memory alloys. The basic mechanisms for this nonlinear phenomenon are the accommodation process of the transformation strain and the orientation process of the transforming micrograins. Fischer et al. presented a thermodynamic model to find a TRIP strain which deals with both phase transformation and microplasticity. The cross-coupling effect of the transformation and plasticity is well understood in the TRIP strain rate and the transformation kinetics.[172]

15.5   DUAL-PHASE STEELS

The high-strength low-alloy steels described earlier have improved strength-to-weight ratio over conventional low-carbon C-Mn steel grades, but they do not provide increased formability characteristics over a wider range of strength levels. *Formability* is a generic term relating to a number of properties such as bendability and stretch-forming and deep-drawing characteristics.[173] The reduced formability of HSLA steels has prevented them from being used in more complex cold-formed parts. The high strength allows the use of thinner-gauge sheets, which results in lighter and more fuel-efficient automobiles; furthermore, high formability is required when complex-shape automobile parts are to be pressed or stamped.

In the mid-1970s, an increased ductility at a particular tensile strength was found for low-carbon low-alloy steels (that is, Fe-C-Mn-Si steels) by an intercritical annealing treatment in the $\alpha + \gamma$ temperature range (between Ac_1 and Ac_3) followed by rapid cooling.[174] These steels, called *dual-phase steels*, had microstructures generally containing two phases, i.e., islands of fine dispersion of martensite (10 to 20%), $\geq 5\,\mu m$ in diameter, in a low-carbon ferrite matrix (80 to 90%) (Fig. 15.21).[175] Subsequent tempering at a low temperature further improved the ductility.

In addition to applications as flat-rolled products such as wheels, bumper reinforcement beams, door intrusion beams, and seating components for automobile usage, dual-phase steel has the potential for linepipe applications, particularly at low temperature;[155] and for producing high-strength wire as a replacement of the more expensive drawn and patented high-carbon steel wire,[176] wires for weaving wear-resistant screens,[177] and high-strength rebars.[178]

15.5.1   Process Routes

The dual-phase microstructures can be produced by intercritical heat treatment using either batch annealing or continuous annealing techniques and by hot-rolling followed by controlled/continuous cooling. Table 15.4 lists the typical dual-phase steel compositions.[179,180]

15.5.1.1   Intercritical Annealing.   The simplest approach to produce a dual-phase structure by intercritical annealing consists of (1) heating a low-carbon steel (e.g., AISI 1010) to the $\alpha + \gamma$ region (between Ac_1 and Ac_3) and holding, typically at 790°C for several minutes, with the aim to achieve 10 to 30% austenite; and (2)

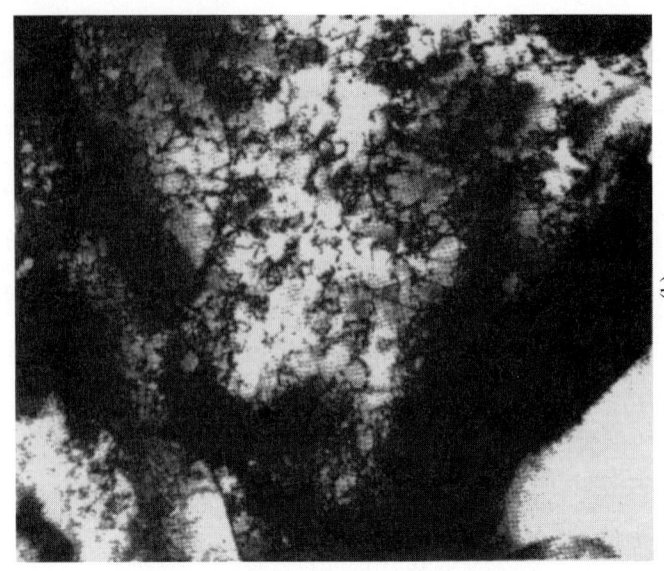

(b)

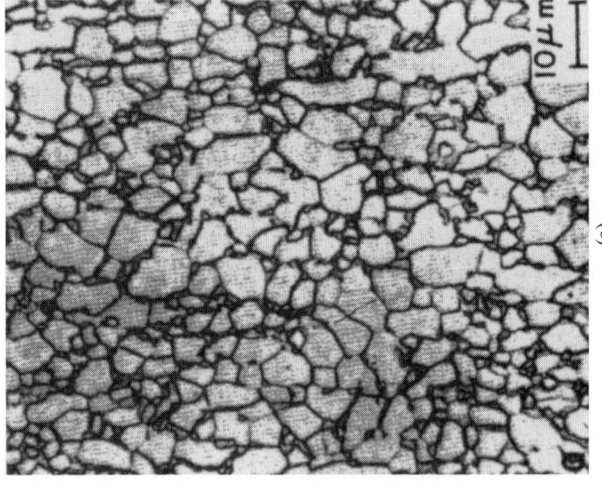

(a)

FIGURE 15.21 (a) (Optical) photomicrograph of a 0.06C-2Si dual-phase steel obtained after intercritical annealing at 780°C and subsequent quenching. The structure shows martensite regions of the grain boundaries and the fine-grained ferrite.[175] (b) Transmission electron photomicrograph of a dual-phase steel.[173] [(a) Reprinted by permission of The Metallurgical Society, Warrendale, Pa.; (b) courtesy of the Institute of Metals, England.]

TABLE 15.4 Typical Dual-Phase Steel Compositions[179,180]

Production method	Composition, wt%									
	C	Mn	Si	Cr	Mo	V	B	N	S	P
Continuous annealing, hot-rolled gauge	0.11	1.43	0.61	0.12	0.08	0.06	—	0.01	—	—
Continuous annealing, cold-rolled gauge	0.11	1.20	0.40	—	—	—	—	—	—	—
Batch annealing	0.12	2.10	1.40	—	—	—	—	—	—	—
Batch annealing, hot-rolled B-V microalloyed plate[180]	0.16	1.32	0.44	0.03	0.09	0.056	0.0019	0.4	0.02	0.013
As rolled	0.06	0.90	1.35	0.50	0.35	—	—	—	—	—

subsequent rapid cooling to transform austenite to martensite. However, some retained austenite usually forms, which is also termed the *martensite-austenite* (M-A) *constituent*. In some instances, dual-phase structures may contain a small quantity of bainite.

The manner in which the ferrite and austenite phases form on intercritical annealing has a bearing on the steel composition and the prior treatment. The cooling rate necessary to produce M-A constituent must be matched to the austenite composition, which, in turn, depends on the steel composition, the intercritical annealing temperature, and the amount of austenite formed. High alloying elements are needed for materials undergoing slow cooling (10°C/hr or 20°F/hr) after batch intercritical annealing, while much lower alloying additions, typical of HSLA steels, can be employed for rapid cooling after a continuous annealing process. In batch annealing, the heating times are longer (~3 hr).[179]

The relationship between the critical cooling rate $\dot{T}_{cr}$ (°C/s) for martensite formation and the chemical composition of the steel is given by Abe as[91]

$$\log \dot{T}_{cr} \geq 5.36 - 2.36(\%Mn) - 1.06(\%Si) - 2.71(\%C) - 4.72(\%P) \qquad (15.25)$$

The dual-phase structures will form at any cooling rate $\geq \dot{T}_{cr}$.[91]

On slow cooling from the $\alpha + \gamma$ region, a higher amount of M-A constituent is formed which may have a better strength-ductility combination than the water-quenched type.[181] To allow air cooling after intercritical annealing, alloy additions of Mo (between 0.2% and 0.4%), Cr, and V to a low C-Mn-Si steel is made. These elements increase the hardenability in steels.

Intercritical Austenitization. With the development of dual-phase steels, intercritical austenitization has become of technical importance. However, this reaction is a complex sequential process.

Partial austenitization,[†] during intercritical annealing of an Fe-0.04C-2.2Si-1.8Mn steel, has been studied on different types of initial microstructures, and the following results have been found:[182] (1) Formation of austenite occurs preferentially along the ferrite grain boundaries through which carbon atoms rapidly migrate from the carbide particle to the growing austenite particle. (2) Initially the austenite growth along the grain boundaries is rapid but slows down when grain boundaries become

---

[†] This was also discussed in detail by Bain and Paxton several decades ago.

site-saturated with γ particles. (3) In the final stage, γ particles can grow as Widmanstätten sideplates, which increases the austenite volume fraction.[182]

The intercritical austenitization of a fully homogenized ferrite-pearlite steel (with 0.11% C and 1.6% Mn) consists of three distinct stages, as proposed by Speich et al.[183] (after Bain and Paxton), and later observed by Souza et al.[184] and Cai et al.:[185] (1) a rapid transformation of the pearlite to austenite; (2) a slow growth of γ into untransformed ferrite; and finally (3) a slow homogenization of both the γ and α phases.

The preferred nucleation of γ tends to occur at the carbide particles on the α/α boundaries and in the pearlite colonies,[186] and this accelerates the γ growth and the onset of stage (3). The growth of γ into ferrite is a diffusion-controlled process, and either carbon or substitutional (Mn) element partitioning between α and γ has been suggested to be the critical step in the transformation.[184]

The stages involved in the austenite formation at low-temperature intercritical annealing (just above Ac_1) in a 0.08C-1.45Mn-0.21Si steel specimen with normalized ferrite-pearlite microstructures are:[187] (See also Bain effect of alloying elements in steel.) (1) spheroidization of the cementite within the pearlite colonies and coarsening of cementite particles present at the α/α boundaries; (2) nucleation of γ particles on and around the cementite particles located on α/α boundaries and within the spheroidized pearlite colony; and (3) growth of γ on α/α boundaries—initially by dissolution of carbide particles and later by diffusion of carbon from the particles, through the γ to the γ/α interfaces. This produces a fine distribution of γ which can result in the fine dispersion of martensite on quenching.

In an Fe-0.11C-1.6Mn steel with the prior structure of lath martensite with uniform Mn distribution, on intercritical annealing, γ forms at prior γ grain boundaries and later grows along a majority of the interfaces between the martensite laths. In this case, complete partitioning occurs in a very short time.[185]

Hot-Rolling. A dual-phase structure may also be produced both in low-C-Mn and microalloyed (or HSLA) steels by TMT or hot-rolling. Typically, these steels contain strong ferrite formers such as Si and transformation-delaying elements such as Cr, Mn, and/or Mo, and the rolling and coiling temperatures on the hot mill are so adjusted that no pearlite phase appears; instead, a large proportion of ferrite, together with some M-A constituents, is produced.[173] In the absence of MAE additions, the only way for improving strength is by increasing the volume fraction of martensite, which may be detrimental to formability. That is why 0.07 to 0.13% Ti is used as a MAE, because it has less effect on the increase of T_{nr} temperature, compared to Nb, and it leads to a more homogeneous microstructure.[188]

To be more specific, after completion of hot-rolling of low-carbon steel strip (containing 0.5% Cr and 0.4% Mo) around 870°C (i.e., at a temperature between A_3 and A_1), the steel is control-cooled on the water-cooled runout table to form about 80% ferrite. The material is then coiled in the metastable region (510 to 620°C) below the pearlite/ferrite transformation temperature; on subsequent cooling, the austenite region converts to martensite. The steel produced by this processing step is called *intercritically rolled dual-phase* (IRDP) *steel*. A good combination of strength (e.g., yield strength 575 MPa, tensile strength 690 MPa) and ductility (uniform elongation 14%) has been recently reported for AISI 1010 plain low-carbon DP steel, when 50% deformation by single-pass rolling is used after 10 min of intercritical annealing at 1010°C, followed by rapid quenching in iced brine.[174,189] Another grade produced nowadays by hot-rolling is called *austenitic rolled dual-phase* (ARDP) *steel*, where strips are rolled in the austenitic temperature range followed by rapid cooling to the $\alpha + \gamma$ field before final quenching to M_f.[10,190] The coiled

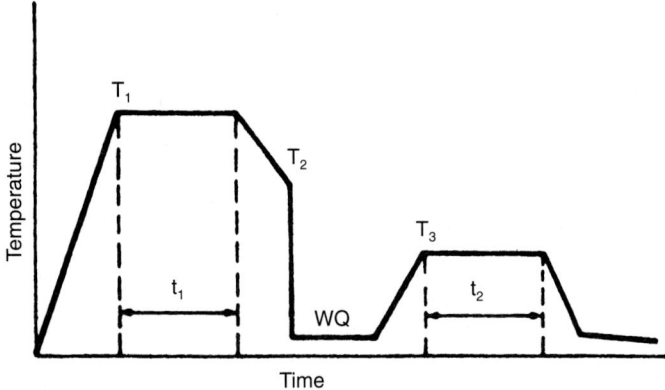

FIGURE 15.22 Typical thermal cycle for processing steel strips by a continuous annealing line method using water-quenching.[193] (*Reprinted by permission of The Metallurgical Society, Warrendale, Pa.*)

strips of ARDP steels are characterized by a dispersion of martensite-bainite pools in a ferrite matrix. An increase in bainite content in such pools changes the YS/UTS ratio from 0.5 to 0.75.

Continuous Annealing Method. The third method of producing dual-phase steel, usually called *intercritically annealed dual-phase* (IADP) *steel*, consists of continuous annealing of cold-rolled low-C-Mn steels above A_1, followed by a quenching and tempering treatment.[191] A continuous annealing line (CAL) is more suitable than batch annealing (BA) for the production of dual-phase steels (DPS) because of (1) the use of leaner-alloy steels, (2) better product uniformity, and (3) greater productivity. Several different CA lines have been developed, each using a different cooling method from the annealing temperature.[192]

The typical thermal cycle for processing steel strips with a nominal composition of 0.04 to 0.17% C, 0.4 to 1.6% Mn, and 0.1 to 0.3% Si by the continuous annealing method is schematically illustrated in Fig. 15.22.[193] The process involves (1) annealing of a cold-rolled steel strip in a molten salt bath for a short time (1 to 2 min) at a temperature T_1 above the A_1 temperature, (2) gas-jet cooling to temperature T_2, and (3) water-quenching. The minimum value of quenching temperature T_2, at which the dual-phase microstructure forms, is a function of Mn content in the steel. Gupta and coworkers[193,194] have found the minimum quenching temperature for 0.5%, 1.0%, and 1.5% Mn steels to be approximately 706, 650, and 595°C (1300, 1200, and 1100°F), respectively, in order to produce the dual-phase steels by the CAL process, irrespective of the carbon content in steel. Quenching from a lower temperature results in a ferrite + pearlite (F + P) or a ferrite + pearlite + martensite (F + P + M) structure. After quenching, the steels are usually overaged at a low temperature T_3 [~232 to 400°C (~450 to 750°F)] for 1.5 min so that part or all of the quenched-in solute carbon precipitates out, to increase the ductility and stabilize the properties. The annealing-quenching and overaging temperatures in a commercial CAL process may vary to within ±10°C (18°F). Since commercial CAL is designed on the basis of a constant mass throughput per unit time, the line speed is invariably proportional to the strip thickness. Thus the processing of a four times-thicker strip will need for the anneal soak, gas jet cool, and overaging a four times-longer time.[193]

15.5.2 Microstructures

In dual-phase structures, the second phase is either a lath or twinned martensite or a combination of the two (Fig. 15.21), inevitably with varying amounts of retained austenite.[193,195] The morphologies of retained austenite have been classified into three types:[181,196–199]

1. Isolated retained austenite particles were observed in 0.073%C-1.3%Mn-0.08%Nb-0.08%V; 0.12%C-2.04%Mn; 0.10%C-1.5%Mn-0.05%Si-0.1%V; 0.07%C-1.44%Mn-0.5%Si-0.13%V; and 0.10%C-2.19%Mn-0.51%Si-0.0072%P-0.0047%S-0.08%V steels.

2. Capsulated retained austenite was reported in 0.07%C-1.78%Mn-1.36%Si steel inside martensite particles without a definite shape.

3. Interlath or interfacial retained austenite was reported at the interface between martensite colonies and ferrite in brine-quenched 0.08%C-0.96%Mn and 0.12%C-1.44%Mn-0.5%Si-0.13%V steels.

The presence of retained austenite γ_R in the dual-phase steels is usually desirable because it appears to increase the ductility at a particular strength level by means of a TRIP mechanism.[197,198] In some DP steels (e.g., with composition: Fe-0.12C-5.10Mn-0.009Si-0.004P-0.004S-0.064Al-0.018N), the ductility of the specimen tested is proportional to the amount of γ_R.[200] The retained austenite of moderate stability thus transforms to martensite during straining, which produces increased uniform elongation, strength, and work-hardening rate. This increased work-hardening arises from the gradual strain-induced transformation of $\gamma_R \rightarrow \alpha'$ over a large strain range.[201]

Generally, high dislocation densities due to the volume and shape change associated with martensitic transformation are observed in the ferrite grains contiguous with the martensite or M-A constituent (i.e., retained austenite). That is, the dislocation density is high adjacent to the martensite and low in the interior of the ferrite grains.[202]

15.5.3 Properties

The shape, size, amount, and distribution of ferrite and martensite; the carbon content of martensite; and the volume fraction of retained austenite have distinct influence on the properties of dual-phase steels.[203] The dual-phase steels—comprising essentially fine dispersion of hard, strong martensite (with ~0.3% C) but sometimes also retained austenite or even bainite or pearlite in a soft and fine-grained ferrite matrix—are considered as the composite materials. They are characterized by continuous yielding (i.e., no yield point elongation), low yield strength (the YS/UTS ratio being around 0.5), high ultimate tensile strength (Fig. 15.16), high initial work-hardening rate, and high total elongation. In addition, they possess greater resistance to the onset of necking (i.e., plastic instability) in uniaxial tension test and multiaxial sheet metalforming processes to provide large uniform strain.[155,175,187,204]

Dual-phase steels can also be strengthened significantly by static or dynamic strain aging.[186] Additionally, these steels containing low carbon content (e.g., AISI 1006, 1008, 1010) have been shown to exhibit excellent resistance to fatigue crack propagation—especially at low growth rates approaching the fatigue threshold intensity range (ΔK_{th}), below which "long cracks" remain dormant.[205]

DP steels usually have poor *hole expansion ratio* values, which may, however, be enhanced by using Ti addition to provide precipitation strengthening of the ferrite matrix, thereby decreasing the hardness difference between the two phases present.[206] To enhance surface quality, free of red-scale, recently Si addition is partially replaced by the addition of higher P or even Al contents.[22]

15.5.4 Structure-Property Relationships

The variation in tensile strength of DP steels, in terms of volume fraction of the M-A constituent, has been described according to the rule of mixtures of composite materials, which assumes (1) random dispersion of nondeformable particles (the M-A constituent) within the soft ferrite matrix, (2) occurrence of partitioning of strain between the phases, and (3) control of tensile properties by the strength and volume fractions of ferrite and the M-A constituents only.[26]

For a particular volume fraction of M-A particles f_{MA}, both applied flow stress σ_f and the work-hardening rate $d\sigma/d\varepsilon$ at a given true strain can be related to $\lambda^{-1/2}$ by the Hall-Petch equation, where λ is the average size or diameter of M-A particles. Typical expressions for a value of $f_{MA} = 0.2$ have been given by Balliger and Gladman and Lanzilotto and Pickering[207] as

$$\sigma_{f(\varepsilon=0.2)} = 350 + 18.1\lambda^{-1/2} \tag{15.26}$$

$$\frac{d\sigma}{d\varepsilon_{(\varepsilon=0.2)}} = 40.1\lambda^{-1/2} \tag{15.27}$$

Figure 15.23 shows the effects of $\lambda^{-1/2}$ on both σ_f and $d\sigma/d\varepsilon$. Note that $d\sigma/d\varepsilon$ increases more rapidly than does σ_f, as illustrated by the larger coefficient of $\lambda^{-1/2}$ in Eq. (15.27) when compared to Eq. (15.26). Hence an optimum value of uniform strain e_u can

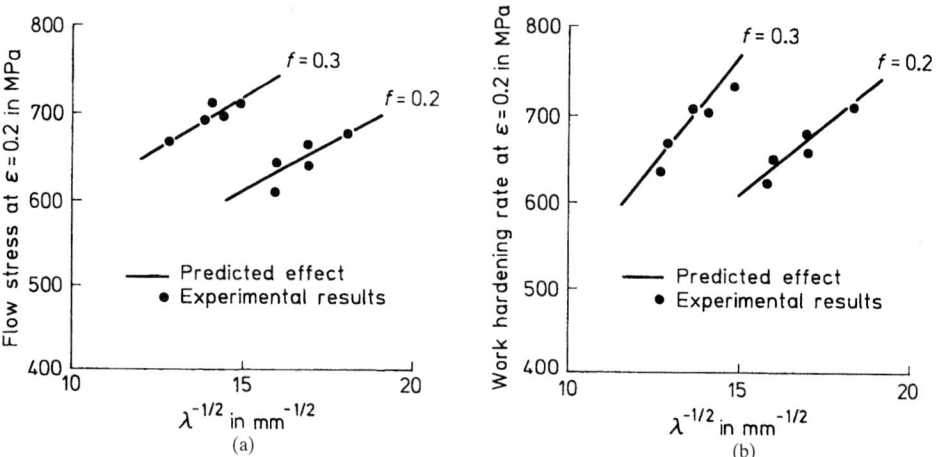

FIGURE 15.23 The effect of size λ and the volume fraction f_{MA} of the M-A particles in a dual-phase steel on (*a*) the flow stress σ_f at a true strain ε of 0.20 and (*b*) the work-hardening rate $d\sigma/d\varepsilon$ at a true strain ε of 0.20.[26] (*Courtesy of F. B. Pickering.*)

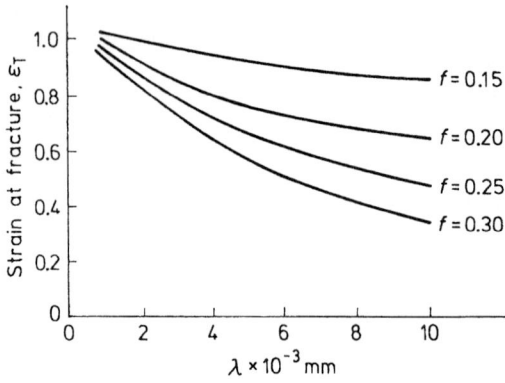

FIGURE 15.24 Effect of size λ and volume fraction f_{MA} of the M-A particles in a dual-phase steel on true strain at fracture ε_T.[26] (*Courtesy of F. B. Pickering.*)

be obtained by refining/decreasing the size of M-A particles. The theoretical and experimental observations also support the fact that $d\sigma/d\varepsilon$ increases with increasing f_{MA} and refinement of the M-A particle size at a given f_{MA} to provide optimum combinations of strength and formability, as measured by uniform strain e_u prior to the onset of necking.[26]

The true strain at fracture ε_T is also an important factor in determining formability, and it should be maximized. Based on the assumptions that: (1) the cracking of the M-A particles is a function of stress[208,209] and (2) nucleation stage for particle cracking, and therefore the nucleation strain plays a significant role for ε_T, a relationship between fracture stress σ_{fr} and ε_T can be given by

$$\sigma_{fr} = 500 + 1420 f_{MA} + \left[40 \left(\frac{f_{MA}}{\lambda} \right)^{1/2} + 75 \right] (\varepsilon_T - 0.2) \qquad (15.28)$$

where

$$\sigma_{fr} = 720 + 24.4 L^{-1/2} \left(\frac{\lambda}{\lambda + L} \right)^{1/2} \qquad (15.29)$$

where L is the mean surface-surface distance between M-A particles, and an interrelationship among L, λ, and f_{MA} is given by

$$L = \lambda \left(0.87 f_{MA}^{-1/2} - 0.98 \right) \qquad (15.30)$$

Equations (15.28) to (15.30) can then be considered to illustrate the effect of both λ and f_{MA} on ε_T of DP steels (Fig. 15.24). Figure 15.24 shows that a combination of smaller λ and smaller f_{MA} contributes to higher ε_T values, and, for a given f_{MA}, ε_T increases with the decrease of λ values.

15.6 ULTRAHIGH-STRENGTH LOW-ALLOY STEELS

Recently, several ultrahigh-strength low-alloy (UHSLA) steels have been commercially produced from the cold-rolled dual-phase low-silicon C-Mn steels as well as from cold-rolled, B-Ti-microalloyed silicon-free C-Mn steels involving appropriate continuous annealing, quenching, and tempering treatments followed by final small temper pass. The nonmicroalloyed UHSS grades are characterized by a high volume percent of tempered martensite microstructure within a ferrite matrix, leading to high YS/UTS ratios in the range of 0.79 to 0.6. They are used in both roll-formed and moderate stamping applications. The B-Ti-microalloyed UHSLA steel grades are characterized by the fully martensitic microstructures obtained by continuous annealing at a temperature just above Ac_3, offering very high YS/UTS ratios in the range of 0.88 to 0.85. The variation in strength levels and low total elongation percent of these UHSLA steel grades are functions of their carbon contents. Table 15.5 lists the typical compositions, mechanical properties, and applications of cold-rolled continuously annealed proprietary ultrahigh-strength steels.[210]

Bag and coworkers[180] have also produced a series of B-V-microalloyed Fe-0.16C-1.32Mn-0.44Si-0.056V-0.0019B-0.4N dual-phase steels (see Table 15.4) containing finely dispersed martensite with different volume fractions of martensite by soaking the hot-rolled plate at 920°C for $1/2$ hr, quenching in 9% iced brine water, inter-critical annealing between 730 and 850°C for 1 hr, and final oil-quenching. An optimum combination of high strength (YS, 680 MPa or 99 ksi; and UTS, 960 MPa or 139 ksi), high ductility (total elongation 22% and uniform elongation 16%), and high impact toughness (Charpy V-notched impact energy, 95 J) was observed at ~55 vol% volume of martensite. These unusual properties were attributed to the finer microstructural constituents and finer precipitation-free ferrite and a possible absence of average internal stress over the composite microstructure volume.[180]

15.7 ULTRAHIGH-CARBON (UHC) STEELS

Ultrahigh carbon steels, which are essentially hypereutectoid steels with 1.0 to 2.1% carbon (15 to 32 vol% cementite) (Al, Cr, and Mn), have been studied extensively within the last two decades[211–214] because of their unique mechanical properties, which include superplasticity at elevated temperatures; high strength, hardness, and wear resistance; and good ductility at room temperature. The reason for this success is attributed to the elimination of proeutectoid carbide network and the development of microstructure consisting of fine spheroidized carbides (or pearlite) and the ultrafine (0.5- to 2-μm) equiaxed ferrite grains.

Four important TMT routes have been developed to break up lamellar pearlite matrix and proeutectoid carbide networks, including hot- and warm-working (HWW), isothermal rolling (ITR), divorced eutectoid transformation (DET), and divorced eutectoid transformation with associated deformation (DETWAD).[211,212] Both DET and DETWAD processing routes are faster and economical because they eliminate the isothermal deformation with intermediate anneals. Although such routes have proved successful in obtaining desirable microstructures and mechanical properties in laboratory experiments, they may not be readily manageable to current mass-production steel processing. For example, mechanical working at intermediate temperature (700 to 850°C), as required in the DETWAD processing, is a drastic change in current production procedure.[215] Among them, three describe the processing for wrought products, and one, the fourth process, describes it for P/M products:[211–215]

1. The first TMT route consists of: (*a*) heating a UHC steel casting to a temperature range of 1100 to 1150°C to dissolve all cementite in the austenite phase; (*b*) continuous deformation by rolling or forging during cooling from 1150°C to a

TABLE 15.5 Typical Compositions, Mechanical Properties, and Applications of Cold-Rolled Continuously Annealed Ultrahigh-Strength (Low-Alloy) Steels[210]

Proprietary grades	Composition, wt%						Mechanical properties			Typical applications
	C	Mn	Si	P	S	Others	YS, MPa (ksi)	UTS, MPa (ksi)	Total (uniform) elong., %	
Nonmicroalloyed dual-phase steel grades										
CAL HI-FORM	0.15	1.45	0.3	0.015	0.007	—	606 (88)	765 (111)	17†	Roll-forming and stamping of such parts as door intrusion beams, bumper reinforcement beams, and seating components (e.g., pillars, tracks, towers, risers, etc.)
DI-FORM	0.15	1.45	0.3	0.015	0.007	—	634 (92)	1033 (150)	13‡	Roll-forming and moderate stamping of such parts as bumper reinforcement beams, door intrusion beams, seating components, and structural cross members
B-Ti-microalloyed fully martensitic steel grades										
MartInsite M-130	0.08	0.45	—	0.01	0.015		923 (134)	1054 (153)	5.4†	Bumper reinforcement beams, door intrusion beams, seating components, and structural cross members
MartInsite M-160	0.12	0.45	—	0.01	0.015		1020 (148)	1178 (171)	5.1†	
MartInsite M-190	0.19	0.45	—	0.01	0.015		1213 (176)	1419 (206)	5.1†	
MartInsite M-220	0.25	0.45	—	0.01	0.015		1350 (196)	1585 (230)	4.7†	

† 2-in. gauge length. ‡ 8-in. gauge length.
Courtesy of Ispat Inland.

15.54

temperature below A_1 (e.g., 650°C). (This HWW process completely eliminates the original massive cementite network and produces microstructure consisting of spheroidized carbide particles and a pearlite matrix. In addition, deformation below the A_1 spheroidizes the pearlite formed during austenite decomposition.) (*c*) The final step is isothermal rolling (ITR) or forging the steel below the A_1 temperature (e.g., 650°C), if required, to break up pearlite lamellae, to spheroidize any remaining pearlite, and to refine the ferrite grains. These processing steps need a height reduction of the castings on the order of 8 to 1 (with a true strain ε of ~2.1) and a final structure comprising very fine cementite particles (0.1 to 1.5 μm diameter) in a very fine-grained ferrite matrix (0.5 to 1.0 μm diameter).

2. The second TMT route is called a *DET process*, in which soaking at ~50°C above the A_1 temperature (e.g., 790°C) for a short time, usually 1 hr, is followed by air-cooling to form divorced eutectoid transformation rather than normal pearlite transformation because undissolved carbides and nonuniform distribution of carbon during heating act as preferred nucleation sites for carbide growth. Hence, the structure consists of a fine and uniform distribution of spheroidized cementite in a fine-grained ferrite matrix. In this second TMT route, the UHC steel is subjected to the same two initial steps (HWW) of the first TMT route. This is then followed by a DET step. The HWW + DET processed material possesses a ferrite grain size of 6 μm and is not superplastic at 700°C. However, its room-temperature properties are excellent with a UTS of 790 MPa and 35% elongation to failure.

3. In the third route, called a *DETWAD process*, the UHC steels are also subjected to the same two initial steps (HWW) as in the first TMT route. This is followed by soaking above the A_1 temperature and subsequent deformation during air-cooling. Here the amount of spheroidized carbide formation increases, and ferrite grain size is further refined. It is reported that HWW or HWW + DETWAD processing leads to superplastic behavior of the UHC-1.5C steel at 700°C due to the presence of the fine ferrite grain size (~1.5 μm).

4. The fourth route involves P/M products which rely on UHC steel powders or white cast iron powders produced by liquid atomization or rapid solidification treatment (RST). These powders are pressed and annealed at 600 to 700°C, and they develop the desired fine-grained structure. Alternatively, hot-pressing, involving simultaneous application of low compact pressure and intermediate temperature for short times, causes densification of material to its near-theoretical value as well as the development of ultrafine structure.

Effect of Alloying Elements. In UHC steels, the addition of (1) Al hinders the formation of a hypereutectoid carbide network and stabilizes the ferrite phase, (2) Mn restricts the harmful effects of S and P, and (3) Cr assists in preventing graphitization and stabilizing the carbides, making it more coarsening-resistant.[216] Extremely fine ferrite grains (<1 μm) have been obtained in the (3%) Al-bearing UHC steels, which is attributed to: (a) a significant reduction of the formation of proeutectoid cementite network and (b) finer pearlite microstructure after the HWW. The fine-grained UHC steels containing Al can be heat-treated to extreme hardness (>Rc 60) and strength (up to 2130 MPa) with a reasonable amount of ductility (up to 5.5%) at room temperature.[217]

Applications. UHC steel wire up to 0.9% C can be used in tires for automobiles and light trucks. The ultrahigh-strength wire is the main component that sets the structural performance of the tire and its weight and rolling resistance. It is possible to attain a very high strength approaching 6000 MPa in UHC-1.8C steel wire.[218]

Structure-Property Relationships. The yield strength in both spheroidized and pearlitic UHC steels has been estimated by combining the strengthening contributions from solid-solution strengthening, grain size strengthening, and particle strengthening, as given, in MPa, by

$$\sigma_y = (\sigma_0)_{ss} + k_1 (= 145)(d_s^*)^{-1/2} + k_2 (= 460)d^{-1/2} \qquad (15.31)$$

where $(\sigma_0)_{ss}$ denotes the solid-solution strengthening (and can be considered to also include the strength of pure iron), and k_1 and k_2 are constant strength coefficients for the interparticle spacing of carbide particles d_s^* (μm) and ferrite grain size d (μm), contributions to strength. Taleff et al.[216] have calculated $(\sigma_0)_{ss}$ values for highly alloyed UHC steels, moderately alloyed UHC steels and eutectoid steels, low-alloy eutectoid steels, and unalloyed eutectoid and hypereutectoid steels which are 330, 170, 60, and 20 MPa, respectively. The relation between hardness and yield strength for spheroidized and pearlitic UHC steels has been given by[216]

$$Rc = 0.0381\sigma_y \qquad \text{MPa} \qquad (15.32)$$

Equations (15.32) and (15.31) can be used together to predict hardness values based on microstructural parameters. This predicted hardness value agrees well with the measured value in the range of 20 to 40 Rc.

15.8   DUCTILE IRON

Syn et al.[219] have reported the thermomechanical working of ductile cast irons (such as Fe-3.6C-2.6Si-0.045Mg and Fe-3.5C-2.5Si-0.4Mn-0.04Mg) by continuous hot- and warm-rolling and by one-step large strain press-forging from a temperature in the γ range (900 to 1100°C) to a temperature below A_1. The press-forged ductile iron exhibits higher strength than the rolled ductile iron for the same strain. Cementite-free ferrite zones around graphite stringers occur in the continuous hot- and warm-worked condition but not in the press-forged condition.

15.9   NONFERROUS TMT

15.9.1   Ni-Base Superalloys

In general, in both Ni-base and Al-base alloys TMT involves warm deformation after solution treatment, followed by aging.[220] Warm-working after aging produces further refinement of the aged substructure. A control of TMT parameters has been used to develop a "necklace" microstructure of fine recrystallized grains in Rene 95 (containing Ni-0.15C-3.5Al-2.5Ti-3.5Nb-14.0Cr-8.0Co-3.5W-0.01B) comprising both MC [that is, (TiNb)C] and overaged γ' particles in the necklace region (Fig. 15.25). This duplex structure exhibits superior stress-rupture, notched low-cycle fatigue, and crack-propagation resistance properties compared to those obtained in conventionally processed fine-grained material. As a result of these properties, this alloy has been adopted as compressor and turbine disks for gas-turbine engines.[220] A typical TMP schedule of this high-strength wrought superalloy consists of: (1) reheating of homogenized-annealed ingot material to a temperature below γ' solvus in the range of 1093 to 1138°C (2000 to 2080°F) for the initial 40 to 50% reduction;

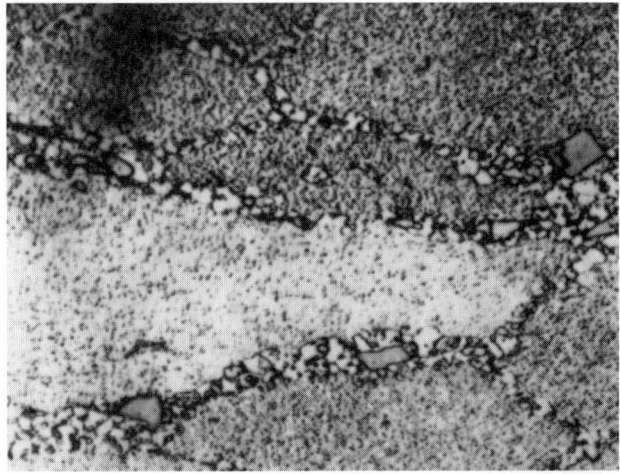

21.7μm
├─────────┤

FIGURE 15.25 Typical necklace microstructure obtained in a thermo-mechanically processed René 95.[220] (*Reprinted by permission of The Metallurgical Society, Warrendale, Pa.*)

(2) recrystallization treatment at a temperature just above the γ' solvus, usually at 1163°C (2125°F) to produce uniform grains; (3) subsequent cooling to a temperature range of 1079 to 1107°C (1975 to 2025°F) for the final 40 to 50% work reduction.[220]

A different TMT route has been found for another nickel-base superalloy, Udimet 700 (containing Ni-17.2Co-15.3Cr-4.95Mo-3.97Al-3.53Ti-0.08C-0.027B-0.17Fe), in achieving large increases in yield and tensile strengths, creep resistance, and fatigue strength (both low-cycle and high-cycle fatigue) between room temperature and 760°C (1400°F) with very little loss of ductility.[221] The applications of this high-strength wrought alloy in this temperature range thus include disks, blades, and shafts in gas-turbine engines. The typical process comprises: (1) initial solutionizing at 1177°C (2150°F) for 4 hr; (2) aging at 1066°C (1950°F) for 4 hr to form γ' [the γ' solvus temperature is 1132°C (2070°F)]; (3) 78% total warm-working reduction at 1066°C (1950°F) (this causes the formation of a stable dislocation substructure aligned in polygonal cell boundaries, the cell size being dependent on the γ' precipitate distribution); and (4) final aging at 843°C (1550°F) for 4 hr and at 760°C (1400°F) for 16 hr.

In a PM Astroloy (containing Ni-17Co-15Cr-5.25Mo-3.5Ti-4Al-0.03C-0.025B) consolidated by hipping or extrusion, the necklace microstructure has been observed to contain fine (4- to 6-μm) grains along the boundaries of larger (30- to 40-μm) warm-worked grains, which results in high strength of fine grains and good creep and crack growth behavior of coarse grains for applications such as gas-turbine disks. For a necklace microstructure, the processing step must aim for partial recrystallization and fulfill the requirements of preferred grain boundary recrystallization and continued recrystallization to avoid excessive growth of the newly formed grains. This recrystallization process is determined by the morphology and distribution of the grain boundary γ' precipitates as well as the strain rate and strain, which, in turn, are functions of the starting microstructure and the temperature of the deformation process.[222]

In IN-718, forging and TMP have produced fine-grained structure with low cycle fatigue and fracture properties for the turbine disk applications.[223,224] The maximum hot tensile ductility tested in the temperature range of 1000 to 1050°C occurs at a strain rate of $2.5\,s^{-1}$.[225] Using processing maps, the hot-working characteristics of IN-718 in the temperature range of 900 to 1200°C and strain rate range of 10^{-3} to $10\,s^{-1}$ are characterized by (1) the presence of two regimes of DRX, one with peak efficiency of 40% occurring at 1200°C and $10^{-1}\,s^{-1}$ and the other with a peak efficiency of 38% occurring at 950°C and $10^{-3}\,s^{-1}$; (2) δ-phase (Ni_3Nb) serving as nucleant for DRX at lower temperatures, whereas interstitial carbon is present due to carbide dissolution causing DRX at higher temperatures; and (3) production of fine grain by initial hot-working stage at 1200°C at $10^{-1}\,s^{-1}$ followed by finishing operations in the lower-temperature DRX regime.[226]

15.9.2　Al-Base Alloys

Unlike hot-rolling of steel sheet, hot-rolling of Al-base alloys is accomplished at much lower temperature without the occurrence of phase transformation. However, a combination of hot deformation and recrystallization leads to the desired microstructural transformations. Dynamic recrystallization that occurs during high-temperature rolling of steel and low-SFE alloys is rarely found in hot-rolling of Al alloys. Here the amount of stored energy is reduced by dynamic recovery, and *postdynamic* recrystallization must be considered as the main processes, e.g., in self-annealing of hot band. They are strongly dependent on alloy composition, temperature, strain rate, strain, and interpass time, particularly in multistand hot-rolling lines with thin gauges, relatively high temperature, and high rolling speed.[227]

In Al-base alloys, TMTs have been aimed at optimizing both mechanical properties, especially tensile and fatigue strengths, fracture toughness, and stress corrosion cracking (SCC) resistance. An improvement in these properties has usually been attributed to (i) the effects of TMT in optimizing grain size and shape and (ii) a more uniform distribution of precipitate particles and dislocation produced by the TMT.[228] There are several types of TMTs for grain refinement of medium- to high-strength Al alloys, based on the variation in the starting structure and thermal and plastic deformation treatments.

Final Thermomechanical Treatment.　One class of TMT, called *final thermomechanical treatment* (FTMT), is applied to high-strength Al-Zn-Mg-Cu (7xxx series, e.g., 7075, 7049), Al-Mg-Si (6xxx series), and Al-Cu-Mg (e.g., 2024) alloys. This treatment involves[229] (1) solution treatment at 465 to 470°C (869 to 878°F); (2) water-quenching at room temperature; (3) natural aging for 2 to 3 days; (4) first artificial aging at 105°C (221°F) for 6 hr; (5) 10 to 50% warm deformation; and (6) final aging (usually overaging) at 105 to 120°C (221 to 248°F) for various times, depending on the extent of prior deformation. An excellent example of improved SCC (and fatigue) resistance in these alloys by ITMT is the study of Paton and Sommer.[230] In the case of 7075 alloy, they found that aging to the peak-hardened (that is, T6) condition (i.e., typical solution treatment at 475°C or 887°F, water-quenching, and aging at 120°C or 248°F for 24 hr), followed by 15% deformation at 193°C (380°F) (above GP zone solvus temperature), and then final overaging at 163°C (326°F) for 3 hr,

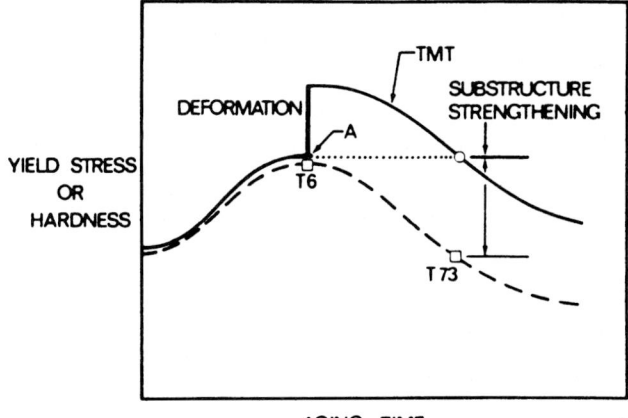

AGING TIME

FIGURE 15.26 Schematic comparison of strength as a function of aging time for conventional heat treatment (dashed line) and for TMT (solid line) in 7075 alloy.[230] (*Courtesy of The Institute of Metals, England.*)

TABLE 15.6 Typical Mechanical Properties of 7075 Al Alloy[230]

Conditions	Heat-treatment cycle[†]	Yield stress, MPa (ksi)	Ultimate tensile stress, MPa (ksi)	Threshold stress,[‡] MPa (ksi)	K_{IC}, MPa $\sqrt{in.}$ (ksi $\sqrt{in.}$)
T6	ST 475°C, WQ age 120°C/24 hr	510 (74)	579 (84)	69 (10)	25.8 (23.5)
T73	T6 + 163°C/24 hr	427 (62)	503 (73)	276 (40)	29.1 (26.5)
TMT	T6 + 15% deformation at 193°C, age, 163°C/3 hr	517 (75)	558 (81)	276 (40)	26.4 (24)

[†] ST, solution treatment; WQ, water quench.
[‡] For 30-day life in stress corrosion test.
Courtesy of the Institute of Metals, England.

gave strength and fracture toughness as good as the conventional T6 temper with SCC resistance equivalent to the observed T73 condition (but better than the T6 condition) (Table 15.6). This process is schematically illustrated in Fig. 15.26. The dotted curve represents the single aging curve; T6 condition denotes the peak hardness without the need of SCC resistance; and T73 condition shows the SCC resistance requirement for the overaged material. The solid curve demonstrates the substructure effect, which provides an increment of strength usually lost by overaging to achieve a required level of SCC resistance.[9,230,231] Note that the initial aging in this treatment cycle produces uniformly distributed nuclei, and 15% deformation at slightly above the normal aging temperature preserves a more uniform and homogeneous dislocation substructure (or slip mode) and thus the reduction of slip planarity[232] (dislocation banding), a characteristic of room-temperature deformation.[194] The slip planarity mode is susceptible to SCC and hence is undesirable, while uniform and homogeneous dislocation tangle structure (or slip mode) is nonsusceptible.[233] However, incorporation of final overaging yields enhanced SCC resistance at the expense of decreased strength.

Retrogression and Reaging (RRA) Treatment.   A post-forming tempering by retrogression and reaging (RRA, involving T6 age + 220°C for 6 min + T6 age) treatment[234] has been reported in high-strength 7475 Al alloy to provide SCC resistance equivalent to that of T73 temper together with T6 strength level[235] so that it can be applied in air frame construction by means of superplasticity.[236] The RRA tempered condition can give a larger size of grain boundary precipitates (GBPs) of η (MgZn$_2$) to reduce SCC susceptibility[235] and a greater volume fraction of coherent matrix precipitates to arrest the strength loss when compared to the T6 tempered condition. Note that the benefit of increased SCC resistance of the superplastically formed components by post-forming T73 or RRA temper would be decreased with the increase of the amount of superplastic deformation.[232]

Intermediate Thermomechanical Treatment.   Another class of TMT is intermediate thermomechanical treatment (ITMT), which has been reported by DiRusso et al.[237,238] to obtain refinement and desired morphology of grains and precipitate particles in high-purity ingot of 7075 alloy (containing Al-5.54Zn-2.5Mg-1.57Cu-0.2Cr-0.01Si) for improved ductility and toughness. This process is categorized into two groups: One is Frankford Arsenal ITMT, or FA-ITMT. The processing steps involve (Table 15.7)[229] (1) an initial complete homogenization treatment of the 7075 Al ingot at a high temperature to precipitate Cr as E-phase particles; (2) slow (furnace) cooling to precipitate Zn, Mg, and Cu as coarse particles; (3) warm-working at a low temperature and subsequent recrystallization homogenization treatment to eliminate original grain boundaries in the cast ingot and produce the fine, equiaxed grains. However, it can be further hot-worked, if needed, to complete a continuous recrystallization process and obtain the elongated grains.[238] Another class of ITMT developed at Institute Sperimantale dei Metalli Leggeri (ISML) under a United States/Italy research program and called ISML-ITMT, relies on preserving most of the Cr in the supersaturated solid solution in the Al-rich matrix during both the partial homogenization and the low-temperature deformation step. This process involves the following steps: (1) partial homogenization; (2) warm-working (70 to 80%) at a relatively low temperature (260 to 330°C, or 500 to 626°F); (3) recrystallization while preventing grain boundary migration by a fine dispersion of E-(Al$_{18}$Mg$_3$Cr$_2$) precipitates from supersaturated solid solution; (4) homogenization; and (5) conventional hot deformation (rolling) to obtain a fine, equiaxed grain structure (Table 15.7).[229] Like these two classes of ITMT, the conventional ITMT (Table 15.7) does not yield a fine-grained product due to (a) the precipitation of Cr during the initial homogenization treatment and (b) the occurrence of dynamic recovery during the working operation; both of these factors are effective in retarding recrystallization of the deformed product into a fine-grained structure.[239] Table 15.8 compares the tensile properties of conventionally and ITMT-processed 7075 plate 1 in. (or 25.4 mm) thick in the longitudinal (L) direction and long transverse (LT) direction.[229]

Four-Step TMT.   A third class of TMT for Al alloys, developed in recent years, is a four-step TMT, utilizing static recrystallization, for commercial-purity 7075 and 7475,[240-243] Li-bearing Al alloys,[244,245] and others. A significant feature of this treatment is that conventionally available plate materials can be processed into very fine-grained sheets, especially to achieve superplastic properties. (See the next section

TABLE 15.7 Ingot Processing Treatment Used on 7075 Alloy[238]

Process	Billet thickness, in.	Initial homogenization treatment, hr/°F	Initial deformation, %/°F	Thickness after initial deformation, in.	Recrystallization homogenization treatment, hr/°F	Final deformation, %/°F
			Sheet			
Experimental (FA-ITMT)	1.5	7/860 + 17/900, FC to 775, 5/775, FC to 500, 4/500	70.5/500, AC to RT	0.443	7/860 + 17/900, FC to 800	64/800, AC to RT
Experimental	1.5	7/860 + 17/900, Q, 1/625	70.5/625, AC to RT	0.443	7/860 + 17/900, FC to 800	64/800, AC to RT
Conventional	1.5	7/860 + 17/900, FC to 800	89.5/800, AC to RT	—	—	—
ISML-ITMT	1.5	1/625	70.5/625, AC to RT	0.443	7/860 + 17/900, FC to 800	64/800, AC to RT

Notes: AC to RT, air-cooled to room temperature; Q, cold water-quenched; FC, furnace-cooled. All rolling carried out without intermediate reheats except for rolling at 800°F.
Reprinted by permission of The Metallurgical Society, Warrendale, Pa.

TABLE 15.8   Comparison of Tensile Properties of Conventionally- and ITMT-Processed 7075 Plate (1 in. or 25.4 mm Thick) in the L (Longitudinal) Direction (Long Transverse, or LT, Direction)[229]

Treatment	YS (0.2%), MPa	UTS, MPa	ε [2 in $(5 \times 10^{-2}$ m)], %	RA, %	K, MPa·m$^{1/2}$
(a) T6 temper					
(1) Conventional	526 (502)	587 (568)	10 (9.5)	14–17 (14–16)	28.1 (22.6)
(2) FA-ITMT	507 (509)	574 (572)	18 (19.0)	29.8 (35.1)	30.9 (27.9)
(3) ISML-ITMT	514 (508)	576 (573)	17.5 (18.2)	29.4 (29.6)	30.4 (33.8)
(b) T73 temper					
(1) Conventional	457 (445)	528 (516)	12 (10.5)	29 (20.0)	34.7 (31.0)
(2) FA-ITMT	468 (458)	528 (518)	16.5 (16.0)	48.5 (45.1)	51.3[†] (44.4)[†]
(3) ISML-ITMT	454 (450)	520 (513)	16.5 (14.5)	50.0 (38.4)	51.6[†] (43.5)[†]
(c) FTMT treatment					
(1) FA-ITMT	574 (561)	607 (603)	13.7 (12.2)	37.4 (34.6)	24.9 (22.9)
(2) ISML-ITMT	608 (573)	630 (614)	13.2 (11.2)	28.2 (25.2)	27.9 (22.6)

[†] K_Q (ASTM E-399).

for more details.) These materials become superplastic in their deformation behavior and have, therefore, emerged as valuable materials with great potential for aerospace structural materials. Among these alloys, Li-bearing alloys are of considerable importance because of their lower density and high elastic modulus (with reasonable strength and ductility) when compared to conventional aerospace Al alloys.[246]

This processing sequence used for grain refinement of 7075 alloy is schematically shown in Fig. 15.27, and a typical processing parameter for 7075 alloys is given in Table 15.9. This consists of (1) solution treatment in a molten salt bath to produce a standard starting material by dissolving soluble precipitates, leaving dispersoid particles undissolved; (2) overaging to precipitate a high density of uniform and large second-phase particles with 0.5 to 1 μm in diameter to act as nucleation sites for recrystallizing grains;* (3) warm-working below the recrystallization temperature to ~85% reduction to introduce defects into the alloy and produce local intense deformation zones around the hard, large precipitate particles serving as nucleation sites during recrystallization; and (4) rapid heating of the alloy above the recrystallization temperature and solution treatment in order to dissolve large precipitates, leaving the fine dispersoid particles in the fine recrystallized grained matrix by effectively pinning grains and thus preventing grain growth.[245]

Since both 7075 and 7475 Al alloys are identical in composition, differing in the latter for Fe and Si as the impurity concentration, their TMT parameters are similar. As a result, the superplastic properties of 7475 are superior to those of 7075, because the four-step TMT produces much finer grain size (10 to 14 μm in diameter); and, under optimum TMT conditions, this material exhibits superplastic properties up to ~516°C and a total elongation of 1200% for a strain rate of 2×10^{-4} s^{-1}.

Double Recrystallization Processing.   Shin and his coworkers[246a] and Sherby and Wadsworth[246b] have reported that the modification of four-step TMT schedule

---

* This also produces an increased density for fine (0.1-μm) Cr-rich dispersoid particles in 7xxx series alloys or Al₃Zr particles in Al-Li-bearing alloys containing Zr.

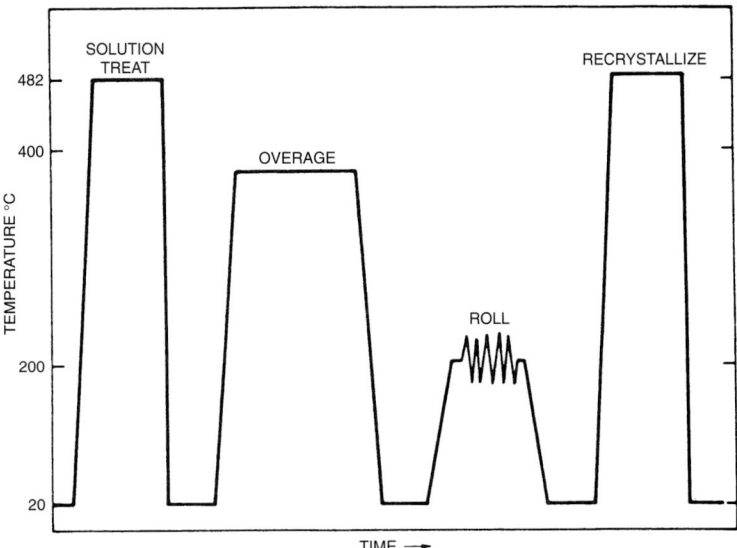

FIGURE 15.27   Schematic diagram showing the four-step thermomechanical treatment developed for grain refinement in 7075 Al alloy.[240] (*Reprinted by permission of The Metallurgical Society, Warrendale, Pa.*)

involving the double recrystallization (DR) stage produces a more stable final fine-grain size and significant improvement in the superplasticity of 7475 Al alloy and P/M processed Al-5Mg-1Cr alloy, respectively.

Five-Step TMT or TMPA.   A new TMT route is developed to produce a stable fine grain structure in the 7075 Al alloy. The method consists of:[247] (1) solution treatment at 490°C for 3 hr, followed by air-cooling; (2) overaging at 350°C for 2 hr, followed by water-quenching; (3) warm-rolling at 200 to 220°C to 85% reduction thickness; (4) recrystallization at 490°C for 30 min; and (5) artificial aging at 100°C for 1 hr, followed by water-quenching. Artificial aging after recrystallization tends to precipitate M phase to improve the distribution of the second-phase particles. During superplastic deformation, both the M-phase particles which have not been dissolved and the dispersion of insoluble particles containing Cr and Mn slow down the growth of grains. The dissolved M phase can produce some element concentration on grain boundaries, which may also be beneficial in reducing the grain boundary migration.

After grain refinement by the TMPA treatment proposed by Jiang et al., grain sizes of 7075 Al alloy are more refined, and the stability of grain size is reinforced so that excellent superplasticity is achieved. The maximum elongation reported by this process is much higher (2100%), and the optimum strain rate is $8.33 \times 10^{-4}\,s^{-1}$ which is higher than the $2 \times 10^{-4}\,s^{-1}$ reported by Paton et al. This treatment may by advantageous for SPF of 7075 Al alloy. Thus the overall time for completion of the entire treatment is shorter than that for the TMP process proposed by Wert et al.[240,247]

Reciprocating Extrusion.   Recently, a new method called *reciprocating extrusion* has been developed which provides refinement of phases, grains, and inclusions,

TABLE 15.9 Typical Thermomechanical Process Parameters for Grain Refinement of Important Al-Alloys[241,243-245]

Step	Parameters		
	Al-Zn-Mg-Cu alloys	Al-Li alloys	
	7075 and 7475 alloys	Al-4Cu-2Li-0.15Zr alloy (containing Al-3.56 Cu-1.7Li-0.15Zr)	Al-3Cu-2Li-1Mg-0.15Zr alloy (containing Al-2.9Cu-1.9 Li-1.0Mg-0.15Zr)
Solution treatment	482°C, 3 hr, WQ	538°C, 0.5 hr, WQ	460°C, 16 hr + 500°C, 16 hr
Overaging	400°C, 8 hr, WQ	400°C, 16 hr, WQ	400°C, 8 hr, WQ
Warm-working	Rolling at 200°C, 90% reduction	Extrusion at 300°C to a true strain of $\varepsilon = -2.1$	Extrusion at 300°C to a true strain of $\varepsilon = -2.3$
Recrystallization	482°C, 0.5 hr, WQ	500°C, 0.3 hr, WQ (discontinuous recrystallization in the SPF temperature range 450–500°C)	500°C, 0.3 hr, WQ (discontinuous recrystallization in the SPF temperature range 400–520°C)

WQ, water quench.

redistributing all the constituents uniformly in the matrix and welding up all prior boundaries. It is noticed that the grain refining stops after 5 passes with limiting size of $3.5\,\mu m$, whereas the inclusion refining continues up to 20 extrusion passes. The inclusion at 5 passes remains $<1\,\mu m$. The grain refinement arises from an overall effect of dynamic recrystallization, fragmentation, and redistribution of grains during the plastic flow in reciprocating extrusion. The inclusion refinement is due to fragmentation and redistribution of inclusion particles. Although strength level decreases by ~10%, the ductility and toughness increase remarkably with increasing extrusion passes. Strength loss may be related to the higher volume fraction of precipitate-free zone (PFZ) possessed by fine-grained alloys, whereas ductility improvement is attributed to the fineness of grains and inclusions. Reciprocating extruded 7075-T6 alloys (involving solution treatment at 470°C for 30 min, water-quenching, and aging in oil bath at 120°C) possess superior strength toughness to the conventional alloys. The available YS to be used in unstable crack growth condition could increase up to 560 MPa.[248]

The fine-grained Al-Li-based alloys containing low (0.15 to 0.20%) and high (0.5%) Zr content can be developed by *ingot metallurgy* (IM) or by *powder metallurgy* (P/M) via *rapid solidification process* (RSP) route. In IM Al-4Cu-2Li-0.15Zr alloys, a proper TMT procedure (Table 15.9) is adopted to obtain a very fine grain size ($2.3\,\mu m$). This fine-grained structure is attributed to the pinning action by the extremely fine Al_3Zr dispersoids (200 Å). This thermomechanically processed material results in superplastic ductilities well over 1000% when deformed at strain rates of 40% min^{-1} ($6.7 \times 10^{-3}\,s^{-1}$) in the temperature range of 450 to 500°C.[244] Likewise, in IM Al-3Cu-2Li-1Mg-0.15Zr alloy, it is possible to achieve a grain size of 3 to $4\,\mu m$ by using an appropriate TMT (see Table 15.9). Figure 15.28 shows a transmission electron photomicrograph of such a fine-grained structure, which appears to be associated with high-angle boundaries. In some cases, subgrain boundaries have been found within fine grains. In this Al-Li-bearing alloy, ductilities on the order of 300 to 400% can be found in the temperature range of 450 to 500°C at strain rates [20 to 50% min^{-1} (3.3×10^{-3} to $8.3 \times 10^{-3}\,s^{-1}$)]. Figure 15.29a shows the optimum superplastic properties found in Al-Li-based alloys containing low and high Zr levels at higher strain rates (20 to 50% min^{-1}) compared to those in the 7475 Al alloy [1 to 2% min^{-1} (1.7×10^{-1} to $3.3 \times 10^{-4}\,s^{-1}$)] and P/M processed Al-5Mg-1.2Cr alloy (up to $10\,s^{-1}$). These increased strain rates are advantageous for commercial SP forming operations.[246b,249] This increased strain rate range for superplasticity is due to ultrafine grain size (3 to $4\,\mu m$) of Al-Li alloys with respect to those observed in 7475 Al alloy (10 to $14\,\mu m$). Note that a small amount of grain growth occurs during superplastic deformation.

Figure 15.29b compares the DR processing data developed by Shin and coworkers[246a] with those obtained on the 7475 Al and Al-Li-based alloys in Fig. 15.29a. Among all the data, the DR 7475 alloy data exhibit the highest ductility, and optimum superplastic properties are observed at much higher strain rates than for other 7475 alloys. These improvements observed in DR 7475 alloy are attributed to the fine grain size of the DR 7475 Al specimens and to the presence of fine Cr dispersoids, which act as effective barriers against grain growth during SP deformation. Figure 15.30 shows the occurrence of equiaxed grains and a small amount of grain growth in the superplastic condition when deformed to 100% at 500°C and a strain rate of 20% min^{-1} ($3.3 \times 10^{-3}\,s^{-1}$).[244]

An Al-6Cu-0.5Zr (supral) alloy has also been thermomechanically processed to develop an ultrafine grain size (2 to $5\,\mu m$) by a continuous (dynamic) recrystallization process. This alloy also exhibits superplastic behavior. The standard TMT route for this alloy is (1) casting of the alloy in chill molds, (2) annealing at 300°C for

5 μm

FIGURE 15.28 TEM micrograph of thermomechanically treated Al-3Cu-2Li-1Mg-0.15Zr alloy.[245] (*Reprinted by permission of The Metallurgical Society, Warrendale, Pa.*)

1 hr, and (3) rolling in 20% reduction steps with interstage annealing of 10 min at the same temperature.[250] This alloy and its modifications were developed by British Aluminum and resulted in their large-scale production.

An Al-4.2Mg-0.14Mn-0.57Cu max. (modified AA5182) has been hot- and cold-rolled after a modified heat treatment. After a salt bath annealing, the cold-rolled material can be cold-deformed without the tendency of Lüder marks. In the annealed condition it has better mechanical properties than the classical AA5182. During a typical paint bake cycle (180°C for 30 min), the yield and tensile strengths increase due to precipitation hardening.[251]

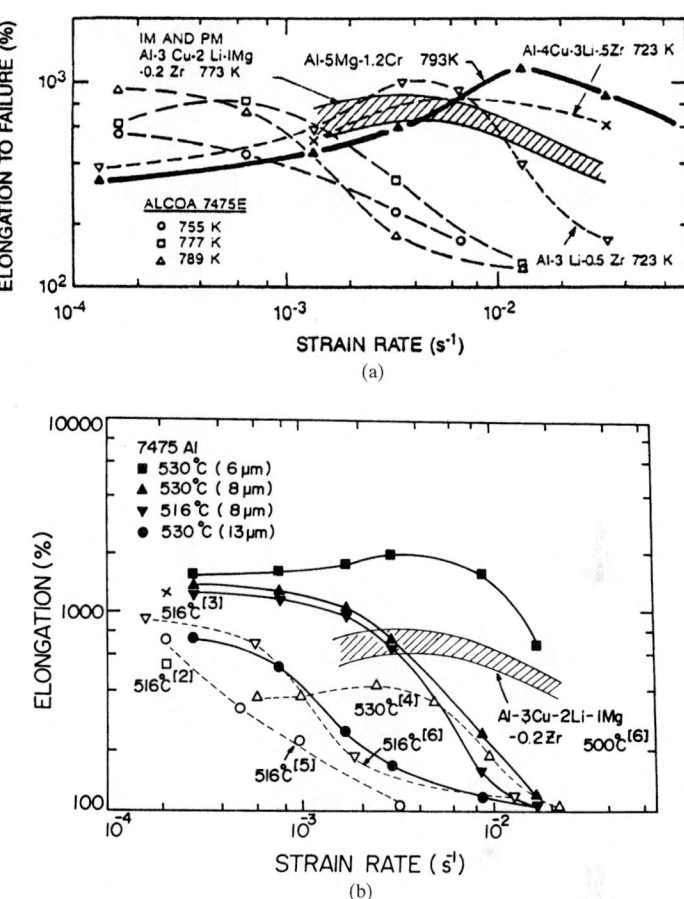

FIGURE 15.29 (a) Elongation to failure versus strain rate for 7475 Al, Al-5Mg-1.2Cr, and Al-Li-based alloys.[246b,249] (*Courtesy of ASM International, Materials Park, Ohio.*) (b) Comparison of data of earlier study[249] shown in (a) with those of Shin and coworkers.[246a] (*Courtesy of The Metallurgical Society, Warrendale, Pa.*)

The modified 5083 Al alloy with 1.6% Mn and lower Fe and Si impurity contents has been found to exhibit fine-grained structure by TMT and moderately high superplasticity with elongation at failure in excess of 500% and 10^{-3} s^{-1}. The second-phase particles help in grain refinement by serving as nucleating sites for recrystallization or by pinning migrating grain boundaries, depending on particle size.[252] Elongation up to 700% has been reported in another modified Al-5083 alloy where 0.6% Cu has been added as a grain refiner.[253]

15.9.3 Zn-Base Alloys

Eutectoid Zn-22Al alloy containing small additions of Cu and/or Mg can readily be cast into thick blocks and then thermomechanically processed to provide them with

5 μm

FIGURE 15.30 TEM microstructure of Al-3Cu-2Li-1Mg-0.15Zr alloy after superplastic deformation to 100% at 500°C and a strain rate of $3.3 \times 10^{-3}\,s^{-1}$.[245] (*Reprinted by permission of The Metallurgical Society, Warrendale, Pa.*)

superplastic properties [i.e., very high ductility (~2000%)]. One important advantage of these alloys is that they can be fabricated with ease into the desired shape and then heat-treated in order to obtain good mechanical properties at room temperature. The optimum forming temperature is 250°C, and the strain rate is low, which restricts its applications to plastic molding. Another limitation is poor creep properties, which limit its use in non-load-bearing applications. However, inexpensive molds can be produced with ease from this alloy at the SPF temperature, in which case die wear is minimal.[254]

15.9.4 Ti-Base Alloys

15.9.4.1 Ti-6Al-4V Alloy. Titanium and its alloys are important structural materials for aerospace applications. Unlike Al and its alloys, titanium and its alloys have higher strength-to-weight ratio, better fracture toughness, better fatigue life, and much higher temperature strength retention, and they do not corrode under typical aircraft usage. Because of these advantageous properties, the use of Ti alloys has tremendously increased in the fabrication of high-performance aircraft structures. However, its high strength, coupled with low elastic modulus, causes many difficulties in the fabrication of components. This fact, in combination with the rising cost of Ti alloys, leads to very high fabrication cost.[255,256]

Among all titanium alloys, Ti-6 Al-4V is the dominant alloy not only for aerospace applications but also for automotive, marine, energy, chemical, prosthetics, and sports industries; and it accounts for more than 50% of all applications. Aerospace applications include critical parts such as bulkheads, wing spars, skin (airframe), compressor disks, engine blades, panels, and gas bottles (plate and sheet components). Note that 36% of the weight of the U.S. Air Force F-22 Raptor fighter aircraft constitutes Ti-6Al-4V.[257]

Ti-6Al-4V alloy has received the greatest attention because it exhibits excellent superplastic properties in the conventionally produced form.[258] Its superplastic behavior and flow stress at elevated temperature are largely dependent upon the morphologies of the α and β phases in the microstructure and the α-grain diameter. For example, when the average α-grain diameter of the processed parts was 3.3 μm, elongation greater than 800% at 850 to 900°C and lower flow stress were observed. These behaviors appeared to inhibit the void formation and localized necking.[256] Similarly, variation of thermomechanical processing parameters such as temperature, strain rate, and strain produces a large variety of microstructures and textures in Ti-6Al-4V alloys, which greatly influences its mechanical properties. The microstructures of these $\alpha - \beta$ alloys (during TMT) are grouped into lamellar, equiaxed, and bimodal structures which largely depend on not only the processing parameters but also the history of the material (chemistry and prior microstructure) and the heat treatment.

Microstructures. Figure 15.31 is a schematic phase diagram of a Ti-6Al-4V alloy.[259] It has a two-phase ($\alpha + \beta$) microstructure. The temperature at which ($\alpha + \beta$) transforms into β (β transus) is a function of the interstitial grade, where oxygen acts as a stabilizer. The transus plays a crucial role for plastic deformation and heat treatment of this alloy. The transus temperature is about 1010 to 1020°C (1850 to 1870°F) for the commercial Ti-6Al-4V grade (containing 0.16 to 0.20 wt% O) and about 970 to 980°C (1780 to 1800°F) for the extra low interstitial (ELI) Ti-6Al-4V grade (containing 0.1 to 0.13 wt% O).[257]

LAMELLAR STRUCTURE. Slow cooling from the β-solution temperature into the two-phase ($\alpha + \beta$) field produces a coarse lamellar structure (with coarse α plates), often called Widmanstätten, acicular, or *β-annealed structure* (Fig. 15.32).[260,261] However, water-quenching from the β-phase field, and subsequent annealing in the $\alpha + \beta$ phase field, results initially in the formation of a much finer lamellar structure over the martensite, and then β-phase precipitates at the martensitic plate boundaries. This structure is often called *β-quenched* or *β-transformed.*[262]

EQUIAXED STRUCTURE. High-temperature deformation within the $\alpha + \beta$ phase field produces an equiaxed structure (Fig. 15.33a),[263] but subsequent annealing at ~700°C results in the formation of a *"mill-annealed"* microstructure (Fig. 15.33b).[260,264] More recently, a reproducible, but coarse, equiaxed structure (with an

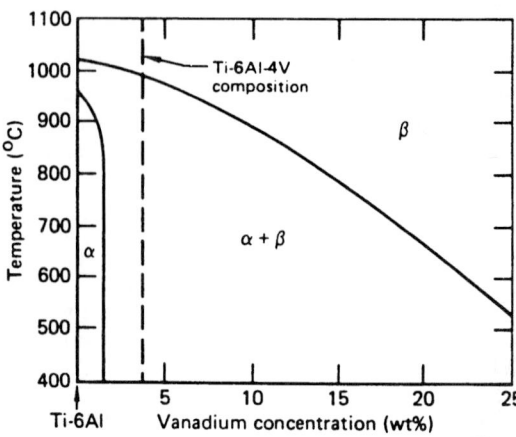

FIGURE 15.31 Schematic phase diagram of Ti-6Al-4V system.[259] (*Reprinted by permission of The Metallurgical Society, Warrendale, Pa.*)

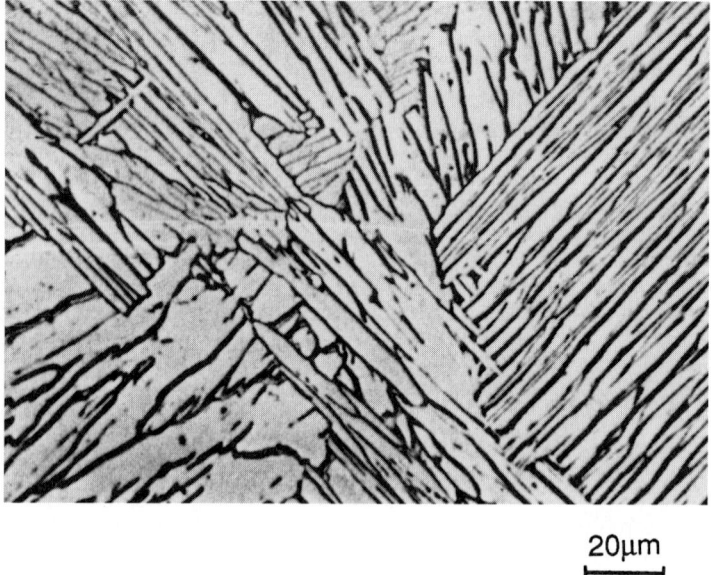

20μm

FIGURE 15.32 Microstructure of Ti-6Al-4V alloy after holding above the β transus followed by furnace-cooling. The white plates are α, and the intergranular dark regions are β. This is a typical Widmanstätten structure.[260,261] (*Reprinted by permission of ASM International, Materials Park, Ohio.*)

α grain size of ~15 to 20 μm), often referred to as a *recrystallization-annealed structure* (Fig. 15.33c),[263] has been developed using an appropriate heat treatment comprising annealing at 925°C for an appropriate time followed by slow cooling.

BIMODAL STRUCTURE.   The third type, called a *bimodal microstructure*, is found when the alloy is water-quenched after recrystallization annealing at 955°C for 1 hr

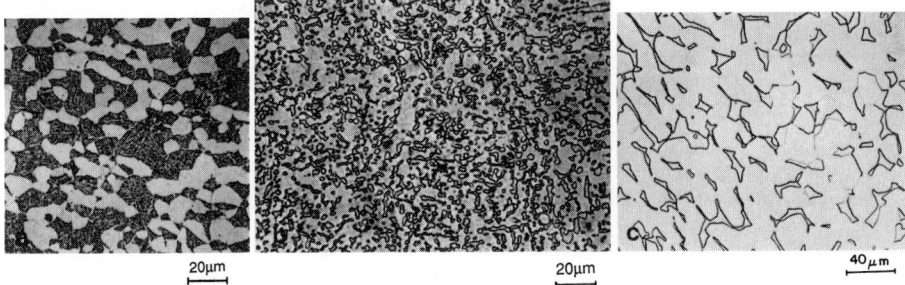

FIGURE 15.33 Optical micrographs of Ti-6Al-4V alloy. (*a*) Forged at 955°C (1950°F) after holding for 1 hr at the same temperature, water-quenched, and annealed for 2 hr at 705°C (1300°F). Equiaxed α (light) in transformed β matrix (dark) containing fine acicular α. Kroll's reagent (ASTM 192).[263] (*b*) Mill-annealed structure (4 hr at 750°C, furnace-cooled). The structure is globular β particles in α matrix.[260,264] (*c*) Recrystallize-annealed plate at 925°C (1700°F) for 1 hr, cooled to 760°C (1400°F) at 50 to 55°C/hr (90 to 100°F/hr), then air-cooled. Equiaxed α with intergranular β. The α-α boundaries are not deformed. Etchant: 50 ml oxalic acid in H_2O, 50 ml 1% HF in H_2O.[263] (*Reprinted by permission of ASM International, Materials Park, Ohio; after J. C. Chestnut.*)

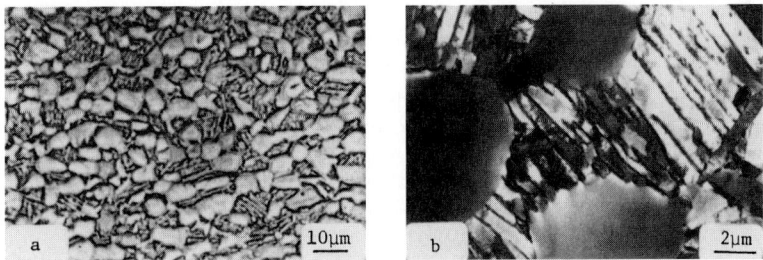

FIGURE 15.34 Bimodal microstructure of Ti-6Al-4V alloy. (*a*) Optical photomicrograph. (*b*) TEM photomicrograph. Solution treatment at 1050°C, 15 min; WQ, deformed at 800°C; recrystallize-annealed at 955°C, 1 hr; WQ, aged at 600°C, 1 hr.[262] (*Reprinted by the permission of The Metallurgical Society, Warrendale, Pa.*)

and then aged (or overaged) at 600°C (Fig. 15.34).[262] The final primary α grain size in this case is again ~15 to 20 μm.

The room-temperature mechanical properties representing commercial and ELI Ti-6Al-4V grades and two extreme microstructures such as lamellar (also called transformed β) and equiaxed ($\alpha + \beta$) are given in Table 15.10. It is clear that the commercial grade is the usual choice where high tensile strength is required, whereas ELI grade is preferred where damage tolerance is a critical concern, such as for bulkheads in fighter aircraft. Likewise, equiaxed microstructure is the usual choice for rotating components such as compressor disks, where resistance to low-cycle fatigue life is important. However, the lamellar structure possesses better fracture toughness and higher-temperature creep strength.[257]

Thermomechanical Treatment. A typical TMT route of Ti-6Al-4V, which is usually carried out in argon atmosphere, consists of the following processing steps:[265]

TABLE 15.10 Comparison of Mechanical Properties of Ti-6Al-4V According to Grade and Microstructure[257]

Grade/ Microstructure	YS, MPa	UTS, MPa	Elong., %	RA, %	K_{IC}, MPa√m	Fatigue life[†]	Creep strength, MPa
Commercial	924	993	10	25	71	—	—
Extralow interstitial (ELI)	827	896	11	27	88	—	—
β	863	932	5.9	—	52.8	42,720	340
$(\alpha + \beta)$	828	897	10	—	40.5	84,370	250

[†] Taken as the number of cycles to initiate crack.

1. Solution treatment at 1050°C for 15 min, followed by water-quenching to produce a fine lamellar $\alpha - \beta$ phase.

2. Subsequent heating for 5 min at 950°C for deformation (rolling) in four steps (passes), followed by water-quenching to produce controlled microstructure (plastically deformed lamellar structure) with controlled texture. The type of texture is a function of mode and temperature of deformation. For example, high-temperature uniaxial rolling produces a transverse texture (i.e., basal planes are parallel to the rolling direction and perpendicular to the rolling plane).

3. Finally, annealing is done at 800°C for 1 hr, followed by water-quenching. This produces the uniform equiaxed microstructure by a recrystallization process, and the α-grain size has been found to be ~2 μm. This structure may be further refined by using the finer starting lamellar structure (β-quenched condition). Aging of this fine equiaxed structure at 500°C for 24 hr, and subsequent air cooling, produces a high yield stress (due to the precipitation of fine Ti$_3$Al particles in the α phase) without the loss of high ductility. It is now realized that both hot deformation parameters and post-deformation annealing can be selected to achieve the desired final microstructure and properties conforming to the shape and workability limits.[265]

Second Variation of TMT. Another variation of TMT for Ti-6Al-4V (containing Ti-6.7Al-4.1V-0.01C-0.013N-0.18Fe-0.164O-0.0044H) utilizes slow cooling in the first and last steps and air-cooling after hot deformation (in this case, extrusion with 10:1 ratio is the reduction stage).[266]

Third Variation of TMT. Alternative TMT route of Ti-6Al-4V consists of the following processing steps:[257]

1. *Primary deformation* (involving forging, side pressing, and cogging operations) of the as-cast (vacuum arc remelted/vacuum induction melted) ingots above the β transus to break up the as-cast microstructure (transformed β) and provide chemical homogenization.

2. *Beta processing* followed by faster cooling rates including air cooling to produce acicular/Widmanstätten/lamellar microstructures with a thin α layer at the prior β boundaries. A prior β grain size of 100 to 200 μm max. and the α layer thickness of ~5 μm max. are desirable.

3. *Secondary processing* consisting of several steps of cogging in the two-phase regime, followed by a homogenization treatment to ensure that the transformed β microstructure is completely converted to a very fine-grained equiaxed $(\alpha + \beta)$ structure via a globularization process. TMT routes for component production involve β forging and heat treatment or $(\alpha + \beta)$ forging, followed by stress-relief annealing.

PROCESSING MAPS. Processing maps are temperature versus strain rate plots which are based on a dynamic materials modeling method, dealing with the design and optimization in hot-working. They rely on the flow stress data over wide ranges of temperature, strain rate, and strain and exhibit regions of safe (or preferred) and detrimental microstructural mechanisms or processes. The safe process includes superplasticity, dynamic recovery and recrystallization, and globularization whereas detrimental process includes void formation, grain boundary cracking, adiabatic shear band formation, and flow localization or instability.[257]

COMMERCIAL GRADE. Figure 15.35 is the processing map for hot-working of commercial Ti-6Al-4V grade with transformed β (lamellar) starting microstructure, showing various microstructural responses. This map indicates that microstructural conversion by cogging is best accomplished in the globularization regime, typically at a temperature of 960°C and strain rate of 10^{-3} to $10^{-2} s^{-1}$ (hydraulic press). Similarly, the higher-temperature limits are bounded by transus, and the lower-temperature limits by the formation of prior β boundary cracking.[257]

Figure 15.36 is the processing map for hot-working of commercial grade with equiaxed $(\alpha + \beta)$ starting microstructure, showing various microstructural processes. Unlike the transformed β preform, globularization, prior β boundary cracking, and lamellar kinking regimes are absent and are replaced by a regime comprising superplasticity at lower strain rates and a broad regime of dynamic recovery at intermediate strain rates.

For optimum superplasticity, a temperature of 825°C (1520°F) and strain rates below $10^{-3} s^{-1}$ under isothermal conditions and a starting material with ~25 vol% of β phase are necessary. Alternatively, Ti-6Al-4V components may be forged at higher strain rates by mechanical presses under nonisothermal situations (hot-die forging) for increased productivity; and flat products may be produced by hot-rolling. These conditions fall into dynamic recovery regime (Fig. 15.36) and are termed *transient processes*. The limit of higher strain rate (for increased productivity) is defined by the condition that avoids the formation of adiabatic shear bands. As seen in Fig. 15.36, the higher strain rate limit may be broadened by raising the deformation temperature up to ~950°C (1740°F). After the deformation process in this regime, the material is mill-annealed in the two-phase field.

ELI GRADE. Figure 15.37 is the processing map for hot-working of ELI grade Ti-6Al-4V with starting transformed β microstructure. The optimum hot-working condition for $(\alpha + \beta)$ cogging is found at 925°C and $10^{-3} s^{-1}$. Unlike for the commercial grade, this map shows extra regimes of void nucleation near the transus and large-grained superplasticity of β. This implies the void formation near the transus representing the upper limit to the cogging temperature and the prior β boundary cracking representing the lower limit to the cogging temperature. Furthermore, if the ELI grade is cogged under the same conditions as the commercial one (i.e., at 960°C, 1760°F), the tendency for the occurrence of void nucleation and growth in the midplane region will greatly enhance due to closer temperature toward its transus.

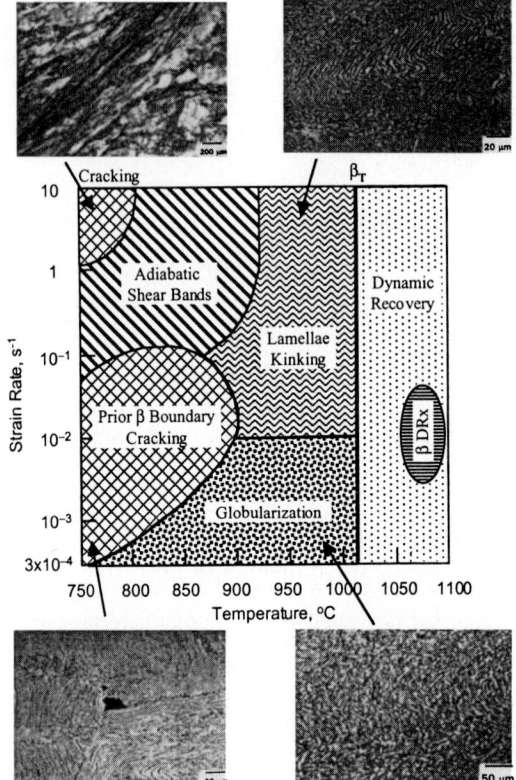

FIGURE 15.35 Processing map for hot-working of commercial-grade Ti-6Al-4V with transformed β starting microstructure showing microstructurally safe and detrimental processes. Microstructures: bottom left—prior β boundary cracking; bottom right—globularization; top left—adiabatic shear band cracking; and top right—lamellae kinking.[257] (*Courtesy of S. C. Madeiros, Y. V. R. K. Prasad, T. Seshacharyulu, W. G. Frazier, J. T. Morgan, and J. C. Malas.*)

Hence, the temperature limits for two void generation processes require the temperature at the surface of the ingot to be higher to prevent prior β boundary cracking and lower at the midplane to avoid void formation. One method to overcome this problem is to design a temperature-differential cogging, so that heating and resoaking cycles are controlled to preserve the differential until the process is completed.[257]

The hot deformation behavior of the ELI grade with equiaxed $(\alpha + \beta)$ starting microstructure is similar to that of the commercial grade, except that the regimes shift to lower temperature.[257]

15.9.4.2 *Ti-Al-Fe-Based Alloys.* The Ti-Al-Fe-based $\alpha - \beta$ alloys such as Ti-5Al-2.5Fe, TIMETAL 62S, and Ti-5.5Al-1Fe can be subjected to solution treatment and aging (STA) following hot-working to increase the strength. Ti-5.5Al-1Fe is a low-cost alternative for Ti-6Al-4V and has nearly the same strength, β transus, and M_s temperature as the Ti-6Al-4V alloy. The best combination of strength and ductility in this alloy can be obtained by solution treatment at 940°C followed by water-quenching and aging at 500°C for 4 to 8 hr. The introduction of less expensive TMP which does not require the formation of brittle hydrides produces excellent

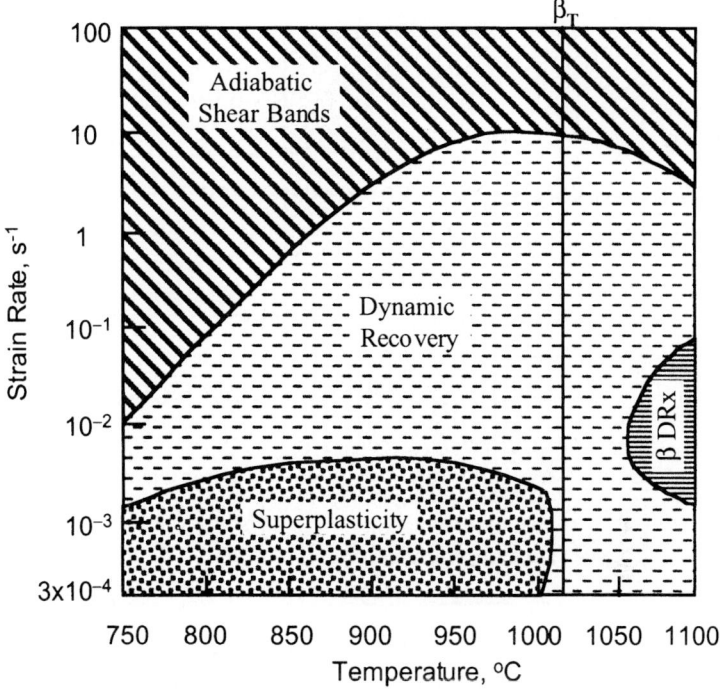

FIGURE 15.36   Processing map for hot-working of commercial grade Ti-6Al-4V with equiaxed ($\alpha + \beta$) starting microstructure.[257] (*Courtesy of S. C. Madeiros, Y. V. R. K. Prasad, T. Seshacharyulu, W. G. Frazier, J. T. Morgan, and J. C. Malas.*)

mechanical properties of the dehydrogenated material compared to the material having so-called β-annealed structures.[267]

15.9.4.3   Ti-Al-Sn-Zr-Based Alloys.   Recently, a high-temperature Ti alloy, called IMI 834 (composition: Ti-5.8Al-4Sn-3.5Zr-0.7Nb-0.5Mo-0.35Si-0.06C), has been developed which has a bimodal microstructure with a good combination of creep and fatigue properties and is used as a replacement for heavy Ni-based alloys for applications as disks and blades in jet engines of modern aircrafts with intended service temperature up to 600°C. To obtain this desired microstructure, the heat treatment consists of (1) solution treatment of hot-forged materials at 1020°C (in α + β phase field) for 2 hr in vacuum, (2) a rapid oil quench to produce ~15% equiaxed primary α grains (~14μm) in a fine-grained matrix of lamellar transformed β (12μm), and (3) aging at 700°C for 2 hr, followed by air-cooling to precipitate fine silicide and ordered Ti₃Al particles.[268]

15.9.4.4   Alpha-Titanium Alloys.   Alpha-titanium alloys are divided into three groups: commercially pure (CP) titanium, alpha alloys, and near-alpha alloys. The unalloyed titanium and alpha alloys containing Al, Sn, Zr, and O as alpha stabilizers have a single-phase, hcp crystal structure at room temperature. They are characterized by relatively low tensile strength and high thermal stability, i.e., good creep

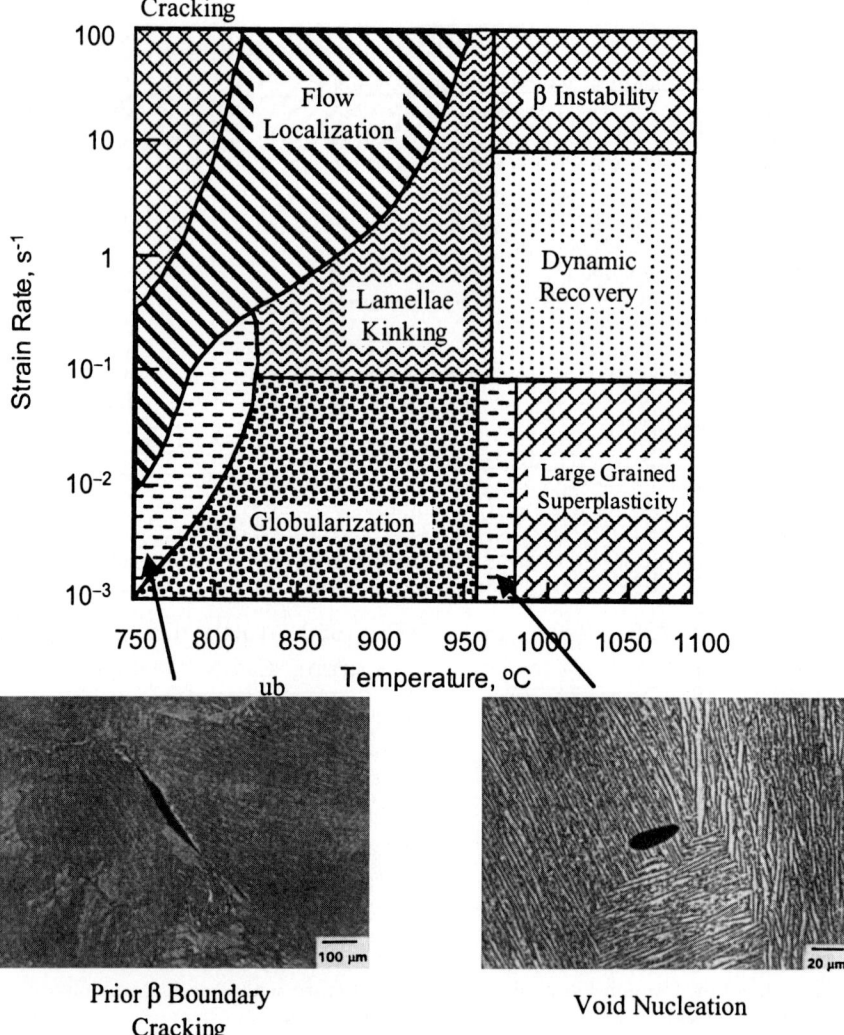

Prior β Boundary Cracking

Void Nucleation

FIGURE 15.37 Processing map for cogging process for ELI Ti-6Al-4V grade with transformed β starting microstructure, showing the preferred window of globularization and its limits of prior β boundary cracking at lower temperatures. The microstructure on the left shows prior β boundary cracking at lower temperatures, and that on the right shows void nucleation and growth at higher temperatures.[257] (*Courtesy of S. C. Madeiros, Y. V. R. K. Prasad, T. Seshacharyulu, W. G. Frazier, J. T. Morgan, and J. C. Malas.*)

resistance and toughness. Near-α alloys possess up to 2% β stabilizers such as Mo and/or V addition to produce a small amount of β phase in the microstructure. They are characterized by higher tensile strength and highest creep resistance of all titanium alloys above 400°C. The TMP of these alloys involves primary working (ingot breakdown) and hot-rolling.[268a]

15.9.5 Cu-Base Alloys

The upsurge of interest in the use of TMT for copper-base alloys has been attributed to the attainment of a combination of the highest possible mechanical strength and conductivity. A TMT schedule was developed for a Cu-2.2Ni-0.6Be alloy to obtain high yield strength (760 MPa) with conductivities of 60% IACS. This TMT consisted of (1) solution treatment at 900°C for 1 hr, followed by water-quenching; (2) cold-rolling to a suitable thickness of the plate; (3) solution treatment at 955°C for 1 hr, followed by water-quenching; (4) cold-rolling to 35% reduction; (5) initial aging at 400°C for 3 hr, and then water-quenching; (6) cold-rolling to 60% reduction; and (7) final aging at 425°C for 6 hr. The microstructures were characterized by the occurrence of two different types of beryllides with (a) similar composition $Cu_{1-x}Ni_xBe$ but different sizes and distributions and (b) random orientation with respect to the matrix.[269]

The TMTs of commercially available low-cost Be-free Cu alloys have also been developed to replace the expensive Cu-Be alloys with improved strength, ductility, formability, stress relaxation properties, and conductivity. Six alloys, for example, Cu-2.4Fe, Cu-0.2Cr, Cu-9Ni-6Sn, Cu-6Ni-7Sn, Cu-4.6Ni-4Al-1Si, and Cu-2Ni-0.4Si alloys were found as leading candidates for this purpose, which, after TMT, produced high yield and tensile strengths and moderate to high conductivity. The two alloys Cu-2.4Fe and Cu-0.2Cr are not high-strength while the two precipitation-hardening ones, Cu-2Ni-0.4Si and Cu-4.6Ni-4Al-1Si, show a combination of high strength and conductivity. The thermomechanically processed Cu-2.4Fe and Cu-0.2Cr alloys, starting from as-cast strips, are characterized by very high recrystallization temperature and high conductivity. Effective distributions of fine stable precipitates that increase the recrystallization temperature, to very high values in Cu-2.4Fe (~725°C) and in Cu-0.2Cr (>625°C), are obtained due to a change from coherent to incoherent precipitates when the material is plastically deformed or further aged. Cu-0.2Cr alloy is used as serpentine fins in brazed Cu/brass heat exchangers. This fin material has, after brazing at ~625°C for 2 min, good mechanical properties and surprisingly high (94% IACS) conductivity.[270]

An example of how a compromise between mechanical properties and conductivity is reached for the precipitation-hardening alloy Cu-2Ni-0.4Si is illustrated in Fig. 15.38.[271] To maximize the strength, a suitable TMT schedule for Cu-4.6Ni-4Al-1Si alloy is solution treatment, cold-working (to 50% reduction), and subsequent aging at 350°C for 2 hr. This produces a yield strength of 135 ± 6 ksi, a UTS of 144 ± 7 ksi, and an elongation of 1.5 to 3%.[272] Stress relaxation is very limited even at 200°C for this alloy. The conductivity is 10 to 20% IACS.[273]

The continuous modulated structures of spinodal Cu-Ni-X alloys (where X can be Sn, Cr, Mn, and Fe)[271] prevail and do not change appreciably by TMT, which provides very high resistance to dislocation movement, i.e., very high strength but lower conductivity and good corrosion resistance. A typical TMT of Cu-9Ni-6Sn alloy comprises:[274] (1) homogenization and annealing at 825°C for varying times, depending upon the section thickness; (2) (prior) cold-working (99% reduction); and (3) aging at 350°C for 5 hr. This treatment yields the optimum mechanical properties, namely, yield strength (0.1% offset), 174 ksi; tensile strength, 202 ksi; and fracture ductility (area reduction), 54%. The drawback with production of this alloy type has been the severe crack sensitivity during casting and hot-working.[271] Powder metallurgy routes have thus been used[275] with much better results. Continuous casting of rod that is further processed through cold-working is another way to avoid the cracking problem.[270] A Cu-9Ni-6Sn alloy reached a YS, 145–167 ksi, UTS, 167–181 ksi; stress relaxation, 90% remaining stress after 1000 hr; and conductivity, 11 to 13%

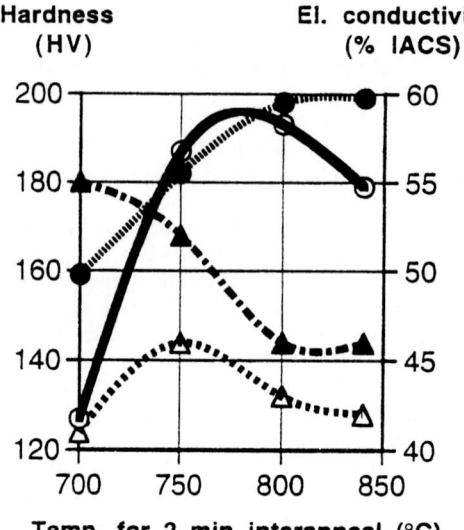

Hardness   **El. conductivity**
(HV)   **(% IACS)**

Temp. for 2 min interanneal (°C)

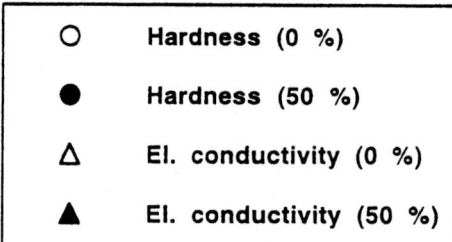

○	**Hardness (0 %)**
●	**Hardness (50 %)**
△	**El. conductivity (0 %)**
▲	**El. conductivity (50 %)**

FIGURE 15.38 Hardness and conductivity versus temperature after annealing at 460°C/4 hr for Cu-2Ni-0.4Si alloy with 0 and 50% reduction which was previously annealed for 2 min at high temperatures.[271] (*Courtesy of R. Sundberg.*)

IACS. The alloy Cu-6Ni-7Sn got the same mechanical properties and somewhat improved conductivity, 13 to 17% IACS.[270]

The main reason for introducing heavy cold-working prior to aging in the above TMT schedules is that the cold-working process creates a fine dislocation substructure and a fine distribution of precipitates, which inhibit the coarsening of the structure, thereby preserving the cold-worked structure even after longer aging times (~100 hr).[7,272]

All the above Cu-base alloys find wide applications in modern electronic devices and heat exchangers such as electric switchgear, connectors, diaphragm members, spring parts, and so forth.

15.10 SUPERPLASTICITY

Superplasticity is a metal deformation phenomenon usually associated with fine grains, grain boundary sliding, and exceptionally high tensile ductility or elongation (of hundreds or even thousands of percent) and is found in specific polycrystalline materials (metallic, ceramic, intermetallic, or composite) at elevated temperature and under certain conditions of deformation. It is, therefore, defined as the ability of polycrystalline material to achieve very high tensile elongation prior to failure (usually greater than 100%) during elevated-temperature deformation ($T > T_m/2$, where T_m is the melting point).

The first description of superplasticity in a metallic material was reported by Bengough in 1912 in $\alpha + \beta$ brass, which exhibited an elongation of 163% at 700°C.[276] However, Pearson[277] was credited with the first observation of this phenomenon in 1934, when he dramatically showed the occurrence of exceptionally large (1950%) tensile ductility in eutectic Sn-Bi alloy. His sensational result and further work on superplasticity were largely ignored until 1962, when Underwood[278] published a review paper describing the experimental work carried out on superplastic materials in the Soviet Union. Later, Backofen et al. studied Zn-Al eutectoid and Pb-Sn eutectic alloys in 1964.[279] Of great significance is that Backofen and his coworkers[280] showed that the superplastic Zn-Al could be formed into a useful shape by a simple air pressure operation like glass blowing. Since then, there was a great upsurge in the field of superplasticity. Subsequently, Lee[281] reported a tensile ductility of 2100% in Mg-Al eutectic alloy, and Ishikawa et al.[282] obtained an elongation of 2900% in Zn-Al eutectoid alloy. Higashi and coworkers[283] found a maximum tensile elongation of 5500% in Al-bronze alloy. Honda et al.[284] reported superplasticity in Cu-38Zn-1.9Sn-1.9Pb alloy in the temperature range of 773 to 873 K and at initial strain rates of $8.3 \times 10^{-4}\,\mathrm{s}^{-1}$ which showed shape memory effect after heat-treating at 1073 K, forming thermoelastic martensite. Ma and Langdon[285] found a tensile elongation up to 7550% in Pb-62Sn eutectic alloy during testing at 413 K. Grant[286] showed a fine grain size of ~0.3 μm in PM (rapidly solidified) Ni-Cr stainless steel (containing 12% B) with an elongation of 300 to 400% at $10^{-2}\,\mathrm{s}^{-1}$ and 600% at $10^{-1}\,\mathrm{s}^{-1}$ at 1000°C.[287] The current world record of superplastic ductility is 8000%, obtained in a thermomechanically processed commercial bronze alloy based on the composition Cu-10Al-4.5Fe-6Ni-2Mn.[288]

Currently, superplasticity is extensively studied to understand the fundamental plastic flow and failure mechanisms and for its technological significance in superplastic forming (SPF) operation at a suitable temperature, called the *SPF temperature*, and within a given strain rate.

There are now more than 100 alloy systems which have been shown to exhibit superplasticity.[289,290] Among them, the main commercial alloys are: (a) the 7xxx series (e.g., 7475 and 7075 Al alloys); (b) 5xxx (e.g., 5083 Al alloy); (c) Al-Cu-Mg;[291] (d) Al-6Cu-0.15Zr (Supral); (e) Al-Li alloys (e.g., 2090, weldalite, and 8090 alloys); (f) Al-5Ca-5Zn alloy; (g) Zn-22Al alloy; (h) Ti and Ti-based alloys (e.g., Ti-6Al-4V);[292] (h) nickel-based superalloy (such as IN100 and IN718) (prepared by conventional or powder metallurgy processing);[276] (i) ultrahigh-carbon steels; (j) stainless steels;[293] and so forth. A number of commercial SP alloys are listed in Table 15.11 together with the SPF temperature, total elongation, and strain rate sensitivity m.[277,292,294] Among the Al alloys, Al-Cu-Mg, Supral, Al-Li, and the 7xxx series are the high-strength heat-treatable alloys. However, the lower operating strain rate in the 7xxx series leads to increased forming times and costs and probably limits its application to the aerospace industry. The eutectic 5083 Al and Al-Ca alloys are non-heat-treatable alloys and may be cast and heavy rolled to produce excellent superplastic ductility.

Al-6Cu-0.15Zr (Supral) is the only alloy that can be superplastically deformed in the severely cold-worked state. In this case, the deformation is associated with the strain-induced continuous (dynamic) recrystallization process, where grain refinement slowly increases with strain. This results in a fine-grained structure without the usual nucleation and growth. Supral 210 and 220 (another modification of

TABLE 15.11 Superplastic Alloys and Their Associated Properties[254,294]

Alloy/grain size (µm)/grade	Superplastic temperature, °C	Elongation, %	Strain rate, s⁻¹	m	Flow stress, MPa	Ref.
Ti-6Al-4V	927	1000–2000	2×10^{-4}	0.8	10	1
Ti-5.5Al-1Fe	827	>400	1×10^{-3}	0.5		2
Ti-6Al-2Mo-4Zr-2Sn (6242)	900	538	2×10^{-4}	0.7		—
Ti-6Al-4Mo-5Zr-1Cu-0.25Si (IMI 700)	800	300	2×10^{-4}	0.7		
IMI 834	990	300	10^{-4}	>0.6		3
Al-20Cu (strip made by casting/rolling)	520	354	4.2×10^{-3}	0.32	10	4
Al-33Cu	380–520	1150		0.9		5
Modified 2004						
Al-6Cu-0.5Zr (Supral)	450	2000		0.5		
Supral 100 (T6)	450	600–1000	10^{-3}	0.38	9	6
Supral 210† (T6)	440	1350	10^{-3}			
Supral 220† (T6)	450	1100	10^{-3}			
Al-Ca	550	1500	10^{-3}			
Alcan 08050 (Al-4.5Zn-4.5Ca)	565	500	10^{-3}	0.3	2.8	
Al-5Ca-5Si	600	900	1.2×10^{-1}	0.63		7
7050 Al [+(0.08–0.12)Zr, (0.17–0.33)Sc]	477	744–1108	1×10^{-2}		18.7–20	
7475 Al (fine-grain)	516	600–1200	2×10^{-4}	0.75	2	8
7475 Al (DR alloy) (6 µm)	530	~2000	2.8×10^{-3}	0.67		
7075 Al (T6)	510	600	10^{-4}	~0.5		9
7091 Al (PM) (3 µm)	300	450	8×10^{-3}	0.38		
Al-4Cu-3Li-0.5Zr	450	900				
Al-4Cu-2Li-0.152Zr (or Al-3Cu-2Li-1Mg-0.1Zr)	450					
Al-1.2Cu-2.4Li-0.6Mg-0.1Zr (8090)	510	1000	2×10^{-4}	0.6		10
Al-10Mg-0.1Zr (1.9 µm)	300	1100	2×10^{-3}			10
Al-10Mg-0.6Zr (0.7 µm)	300	600	8×10^{-3}			11
Al-4.7Mg-1.6Mn (modified 5083)	550	500	4.2×10^{-3}	0.5		12
Al-4.2Mg-0.7Mn (commercial 5083)	550	>400	5×10^{-4}–1×10^{-3}	0.53–0.61		
5083 (PM)	550	465	3×10^{-5}			
Al-4Mg-3Zn (Formall 570, ~7 µm)	500	300–800	2×10^{-4}–2×10^{-3}	0.6		13
Al-16Si-5Fe (PM)	520	~400	1.38×10^{-1}			14
Sn-38Pb	140	4850		~0.61		
Zn-22Al	250	2000	10^{-2}	0.5	10	

Zn-0.3Al	R.T.					15
Cu-10Al-4.5Fe-5.98Ni-1.6Mn (Al-bronze)	800	1400	2×10^{-4}	~0.64		16
Mg-Al eutectic	375	5500		0.8		17
Mg-AZ31 alloy (extruded condition)	325	2100	1×10^{-4}	0.5		
Mg-ZK 60 alloy	270	608		0.52		
Mg-ZK 61 alloy (P/M)	200	1700	1×10^{-3}			18
Fe-1.6C-1.5Cr	650	659	10^{-4}	0.46	45	
Fe-1.51C-3.38Al (DETWAD, 1 to 2 μm)	750	1200	8×10^{-3}			19
Fe-1.4C-1.5Cr-6.7Al (1 to 3 μm)		783				20
Fe-26Cr-6.5Ni (IN744)	900	~800	5×10^{-5}			
Fe-23Cr-5.6Ni-1.3Mo-0.12N (duplex)	900	1000	2×10^{-4}	0.76	21.2	21
Fe-23.5Cr-5.7Ni-1.4Mo (duplex)	900	2240		0.76	21.2	22
Fe-24Cr-7Ni-3Mo (duplex)	850	2240	3×10^{-3}			23
SAF 2304 (2 to 3 μm) (duplex)	970	750	10^{-4}	0.4–0.75	~30	24
IN-100 (P/M)	1010	450		0.5	35	
IN-718	954	350–750	1.3×10^{-3}–3.3×10^{-5}	0.28–0.83	72 – <28	25

† Supral 210 and Supral 220 are two modified 2004 Al alloys produced by Superform Metals.

1. J. Koike et al., Scripta Materialia, vol. 38, no. 8, 1998, pp. 1009–1114.
2. Ref. 295.
3. A. Wisbey and P. G. Partridge, Superplasticity in Advanced Materials, JSRS, Osaka, Japan, 1991, pp. 465–470.
4. N. Tsuji, T. Nakamura, and Y. Saito, Jn. Jap. Inst. Light Metals, vol. 49, no. 8, 1999, pp. 395–400.
5. G. Rai and N. J. Grant, Met. Trans., vol. 6, 1975, p. 385.
6. I. I. Novikov et al., Superplasticity in Advanced Materials, eds. S. Hori et al., JSRS, Osaka, Japan, 1991, p. 145.
7. D. J. Chakrabarti, Superplasticity and Superplastic Forming, 1998, eds. A. K. Ghosh and T. R. Bieler, TMS, Warrendale, Pa., 1998, pp. 155–164.
8. D. H. Shin, K. S. Kim, D. W. Kum, and S. W. Nam, Met. Trans., vol. 21A, 1990, pp. 2729–2737.
9. H. N. Azari, G. S. Murty, and G. S. Upadhyaya, Met. and Mats. Trans., vol. 25A, 1994, pp. 2153–2160.
10. T. R. McNelley and S. J. Hales, Superplasticity and Superplastic Forming Proc., 1995, TMS, Warrendale, Pa., 1995, pp. 57–66.
11. K. Kannan, C. H. Johnson, and C. H. Hamilton, Met. and Mats. Trans.., vol. 29A, 1998, pp. 1211–1220.
12. R. Verma, P. A. Friedman, A. K. Ghosh, S. Kim, and C. Kim, Met. and Mats. Trans., vol. 27A, 1996, pp. 1889–1898.
13. P. Fernandez, Superplasticity in Advanced Materials, JSRS, Osaka, Japan, 1991, pp. 675–680.
14. H. S. Chao, H. G. Jeong, M. S. Kim, and H. Yamagata, Scripta Materialia, vol. 43, no. 3, 2000, pp. 221–225.
15. T. K. Ha, W. B. Lee, C. G. Park, and Y. W. Chang, Met. and Mats. Trans., vol. 28A, 1997, pp. 1711–1713.
16. Z. X. Guo, K. Higashi, and N. Ridley, Met. Trans., vol. 21A, 1990, pp. 2957–2965.
17. H. Watanabe, T. Mukai, K. Ishikawa, Y. Okanda, M. Kohzu, and K. Higashi, Jn. Jap. Inst. Light Metals, vol. 49, no. 8, 1999, pp. 401–404.
18. H. Watanabe et al., in 8th Int. Conf. Creep and Fracture of Engineering Materials and Structures, 1999, pp. 171–174.
19. D. W. Kum, H. Kang, and S. H. Hong, Superplasticity in Advanced Materials, JSRS, Osaka, Japan, 1991, pp. 503–508.
20. D. Hernandez, J. A. Jimenez, and G. Frommeyer, ISIJ Int., vol. 37, no. 1, 1997, pp. 93–95.
21. H. Kitano, K. Nakamura, K. Shirakawa, H. Yoshida, and K. Osada, Ref. 13, pp. 1861–1866.
22. K. Osada and H. Yoshida, Mats. Sc. Forum, vols. 170–172, 1994, pp. 715–724.
23. Y. S. Han and S. H. Hong, Scripta Materialia, vol. 36, no. 5, 1997, pp. 557–563.
24. J. L. Song and P. L. Blackwell, Mats. Sc. and Tech., vol. 15, 1999, pp. 1285–1292.
25. G. D. Smith and F. H. Flower, Conf. Proc.: Superplasticity and Superplastic Forming, 1995, TMS, Warrendale, Pa., 1995, pp. 117–124.

2004 alloys) have been developed by Superform Metals; the former contains 0.35 Mg and 0.14 Si, whereas the latter possesses an additional amount of Ge to further improve strength and SCC resistance properties.[254] These fine-grained (2- to 3-μm diameter) Supral (or Superform) alloys are superplastic material in sheet form, having the lower superplastic deforming temperature and faster forming rates when compared to other commercial superplastic aluminum materials. In these alloys, the second-phase particles are (3 vol%) Al_3Zr with or without (8 vol%) Mg_2Si.[295]

15.10.1 Types of Superplasticity

There are two types of superplasticity based on the deformation characteristics and/or the microstructural mechanisms:[294,296–299]

1. *Fine structural superplasticity* (FSS) involves static or dynamic (continuous) recrystallization process, observed in microcrystalline material.

2. *Internal stress superplasticity* (ISS) or transformation superplasticity is observed during:

(a) thermal (or pressure) cycling of alloys (steel, white cast iron, Fe, Co, Ti, Zr, Ti-6Al-4V, Zr-2.5Nb, and intermetallic Ti_3Al-based super α_2) or compositions (0 to 10% TiC particulate-reinforced Ti, and $NiAl$-ZrO_2 system) through an allotropic phase transformation; (b) thermal cycling of polycrystalline pure metals or single-phase alloys (Zn and α-uranium) that have anisotropic thermal expansion coefficients; and (c) thermal cycling of composite materials (20% SiC whiskers-reinforced 6061 Al alloy) in which the constituent phases have dissimilar, mismatched, or different thermal expansion coefficients.[276,296–298] In this case, a small strain develops during each thermal cycle between 100 and 430°C at a rate of 1 cycle in 100 s which may be accumulated to a large elongation (1400%) after extensive repeated cycling. In ISS materials, m may be high up to 1, that is, may exhibit ideal Newtonian viscous behavior. Such ISS materials deform by slip-controlled mechanism and need not be fine-grained. Although ISS has been widely studied in metallic alloys and their composites, it has been reported in a single intermetallic alloy, super α_2.[299a] Invariably, SP intermetallics are of the FSS group.[290]

Of these types of superplasticity, fine structure superplasticity (FSS) is of great technological importance and is receiving the most attention in research and development work[291] and hence forms the basis of our discussion in this section. Interested readers may like to see the recent reviews on internal stress and/or transformation superplasticity by Sherby and Wadsworth[276,299] and Dunand.[300] Note that FSS cannot be readily achieved in some cases, notably (1) in ceramics and intermetallics where fine grain structure is produced with difficulties, (2) in metal-matrix composites in which cavitation occurs, (3) in some cast or PM near-net shape components in which extensive TMT required to produce fine grains is not feasible, and (4) in metals where a fine grain size is not needed or is unstable at SP temperature due to coarsening. In these cases, ISS is a viable alternative as a forming process.[300]

15.10.2 Fine Structural Superplasticity

The characteristic true flow stress σ-true strain rate $\dot{\varepsilon}$ relationship which describes the superplastic behavior can be given by

$$\sigma = K \dot{\varepsilon}^m \tag{15.33}$$

where K is a constant and m is the strain rate sensitivity factor or the slope of the flow stress-strain rate characteristics; both K and m are functions of material and deformation temperature.

Prerequisites. The prerequisites or conditions for fine structural superplasticity (FSS) are as follows:

1. A fine-grained structure (usually $<\sim 10\,\mu m$ for metals and $<1\,\mu m$ for ceramics) is needed at high temperature and low strain rates.

2. There must exist duplex (or microduplex) microstructures (often eutectics or eutectoids) or quasi- (or pseudo-) single-phase microstructure (such as 7475 Al, 5083 Al, and 8090 and 2090 Al-Li alloys) (where dispersoid second-phase particles maintain and stabilize a fine grain size by pinning migrating grain boundaries against grain growth at the narrow SPF temperature range). The second phase in the duplex structure (e.g., superplastic Ti and duplex stainless steel alloys), which varies between a few percent and 50%,[276] should be similar in strength to, or weaker than, the matrix phase. If the second-phase particles are hard, cavitation may occur at the matrix/hard (or second-phase) particle interfaces or grain boundary, and the materials may become prone to premature failure,[301] particularly at low strain rates, even if no macroscopic necking exists.

3. The grain boundaries should be mobile and high-angled (disordered), and the morphology of the grains should be equiaxed. The high mobility of grain boundaries allows large reduction of stress concentrations normally developed at the triple-point junctions and at the other obstacles along the grain boundary. Equiaxed fine-grains cause extensive grain boundary sliding and superplasticity.[276]

4. There must be high values of m and low flow stress (typically, 10 to 20 MPa) during tensile deformation, which imply strain rates not exceeding $10^{-2}\,s^{-1}$. (In contrast, conventional forming methods are associated with stresses in the material at least one order greater.)

5. Temperature must be above $\sim 0.5 T_m$ of the alloy. The conventional superplasticity in pseudo-single-phase alloys is found at a strain rate of $<10^{-3}\,s^{-1}$ and at a temperature of $\sim 0.8 T_m$.

6. If the second-phase particles are considerably harder than the matrix, they should be distributed uniformly and finely. This inhibits and minimizes cavitation during SP flow of a fine-grained alloy.[276]

7. Grain boundaries in the matrix must resist tensile separation.[276]

The characteristic superplastic behavior depends on the fact that the flow stress of superplastic materials is very sensitive to strain rate at the SPF temperature.

Strain Rate Sensitivity m. The strain rate sensitivity exponent, or index, m may be defined as

$$m = \frac{\partial \ln \sigma}{\partial \ln \dot{\varepsilon}} \tag{15.34}$$

which is a function of both strain rate and temperature.[302,302a] The strain rate sensitivity of metals is the result of the Newtonian viscous nature of the deformation process. This viscosity arises from the resistance offered by internal obstacles within the material. For newtonian viscous flow ($\sigma \propto \dot{\varepsilon}$) with $m = 1$ (such as glass and polymer), the material is said to behave as a viscous liquid.[303,304] Since m is related to the capability of a material to resist plastic instability or necking, a higher value of m for superplastic materials implies greater resistance to catastrophic necking.[305]

Characterization of superplastic properties is often made by measuring m by a step strain rate test method. In this method, a tensile test is initially commenced with a slow strain rate; and each time after the attainment of maximum load, the strain rate is gradually augmented. The flow stress values for each load maxima are plotted against the corresponding strain rates on a log-log scale, and a best-fit curve is drawn from these data. The slope of this curve gives a measure of the strain rate sensitivity index m,[306] as defined by Eq. (15.34). Figure 15.39a shows the $\log \sigma$–$\log \dot{\varepsilon}$ plots for various grain-sized Ti-6Al-4V alloys at 927°C, exhibiting a clear dependence of flow stress on the initial grain size of the material with flow stress increasing with grain size. Figure 15.39b represents the corresponding graph of m as a function of $\log \dot{\varepsilon}$, revealing a maximum value of m with the fine-grain-sized alloy for a given strain rate or strain rate range, thereby producing a maximum superplastic ductility.[294] Figure 15.39c shows the strain rate sensitivity of Al-6Mg-0.3Sc alloy as function of both temperature and strain rate.[302a] Experimental evidence indicates that reducing the grain size increases the strain rate (Fig. 15.40) and/or decreases the temperature for optimum SP flow.[299,307–309] In general, the m values of most commercial superplastic materials are in the range of 0.4 to 0.9 while most of the non-superplastic metals and alloys have usually $m < 0.3$.[301] Note that m does not precisely determine elongation, nor does maximum m always imply maximum elongation. Other factors such as strain, strain rate, grain size, cavitation, initial inhomogeneity, and strain rate path also have important bearing on ductility and the value of m. High m value is associated with grain boundary sliding, which is the dominant SP deformation mechanism in FSS materials.

Usually the $\dot{\varepsilon}$, based on the particular rate-controlling mechanism, is inversely related to the grain size d, raised to the grain size exponent p, according to the creep equation[292,310]

$$\dot{\varepsilon} = A\sigma^n d^{-p} \exp \frac{-Q}{RT} \tag{15.35}$$

where σ is the applied stress, A is a structural factor, $p = 2$ (for lattice diffusion-controlled flow) or 3 (for grain boundary diffusion-controlled flow), $n = 1/m$ is the stress exponent, Q is the creep activation energy, and R and T have the usual meanings. Figure 15.41 shows the variation in superplastic strain rate as a function of inverse of grain size.[311] Hence, grain size refinement appears to be the most effective means of achieving high-strain-rate superplasticity or low-temperature superplasticity (Figs. 15.40 and 15.41).[299,311] Actually, several techniques such as mechanical alloying, inert gas condensation, flame synthesis and sol-gel processing, severe plastic deformation (SPD) [by equal-channel angular (ECA) pressing to a strain of, say, ~12, or torsion straining under high pressure], and crystallization from amorphous solids are available to develop ultrafine, submicron-grained (SMG) (<1 μm), or nanocrystalline materials, which exhibit SP at very high strain rates (10^{-2} to $10^{-3}\,\mathrm{s}^{-1}$) (see Fig. 15.41).[290,312]

15.10.2.1 Mechanism Associated with Superplasticity.

Grain boundary sliding (GBS) has been widely accepted as the primary mode of deformation during the superplastic flow of materials. However, such GBS between two grains cannot occur continuously on all interfaces without being associated with some accommodation

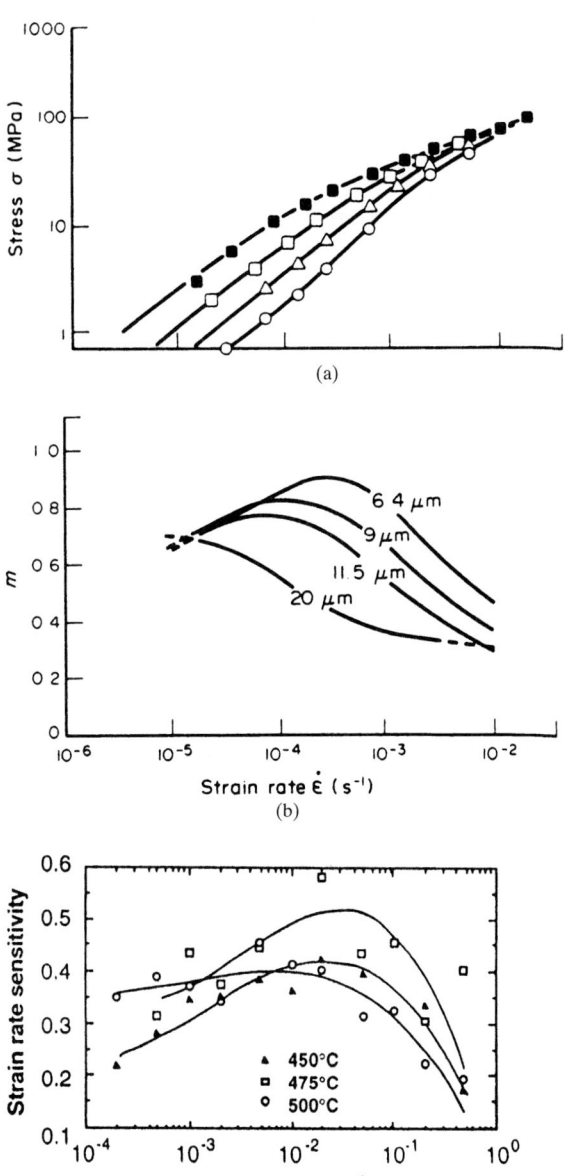

FIGURE 15.39 (*a*) Flow stress versus strain rate for Ti-6Al-4V alloy at 927°C of various grain sizes: ○, 0.64 μm; △, 9 μm; □, 11.5 μm; ■, 20 μm. (*b*) Strain rate sensitivity index *m* as a function of strain rate for various grain-sized Ti-6Al-4V alloy at 927°C.[292,294] (*Reprinted by permission of ASM International, Materials Park, Ohio.*) (*c*) Strain rate sensitivity of Al-6Mg-0.3Sc alloy as a function of deformation strain at different temperatures.[302a]

processes or mechanism such as grain rotation, grain boundary migration (GBM), grain boundary diffusion, solute diffusion, diffusion creep, dynamic recovery, dynamic recrystallization, and/or transverse dislocation motion (or intragranular slip) occurring at intersections with a third grain.[289] The specific accommodation process generally appears to be slower than the grain boundary sliding, and is thus

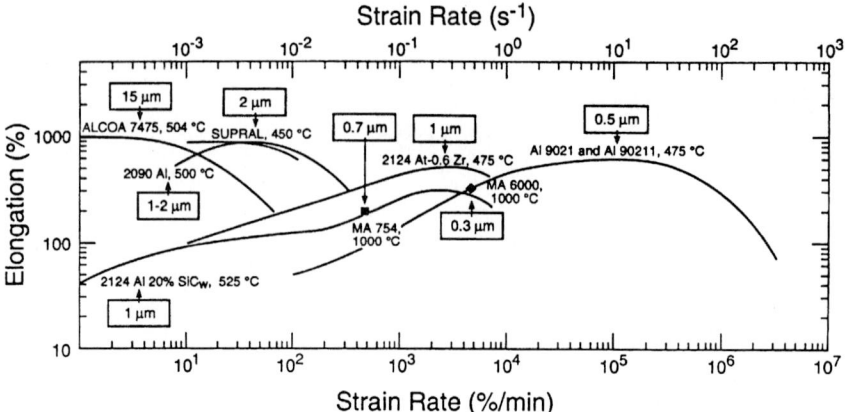

FIGURE 15.40 Elongation-to-failure as a function of strain rate for various superplastic Al alloys. Also included are data points for fine-grained nickel-base superalloys (MA 754 and MA 6000). (*Taken from T. K. Gregory, J. C. Gibeling, and W. D. Nix, Met. Trans., vol. 16A, 1985, p. 777.[299]*)

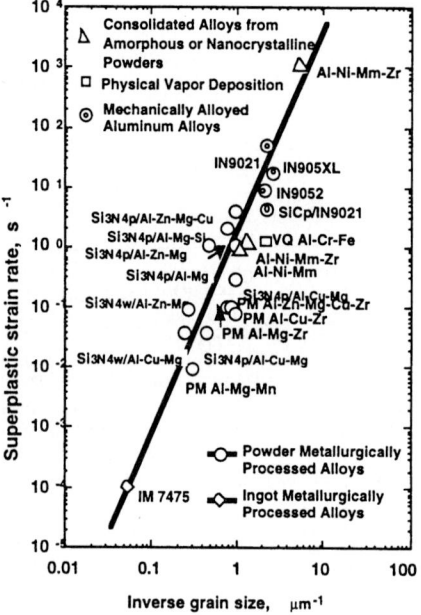

FIGURE 15.41 The variation in superplastic strain rate as a function of inverse of grain size for aluminum alloys.[311] (*Courtesy of TMS, Warrendale, Pa.*)

the rate-controlling step controlling the kinetics of SP deformation. Since these accommodation processes are difficult to determine, it is not an easy task to establish the true nature of the mechanism.

Diffusional flow has now become a well-accepted mechanism.[313] Figure 15.42a illustrates a local diffusional transport for an irregularly shaped grain under tension

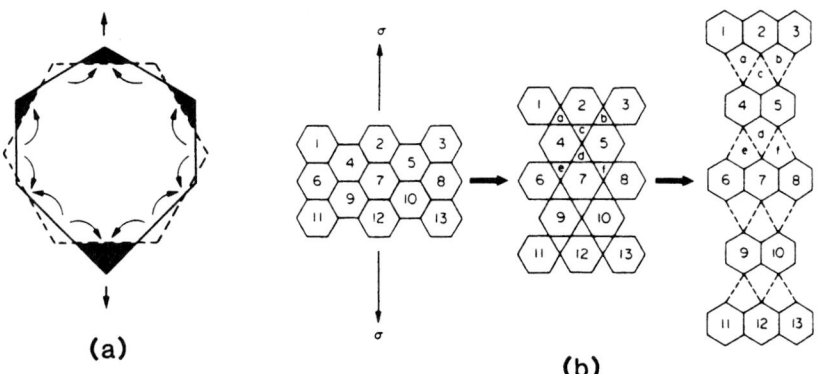

(a)

(b)

FIGURE 15.42 Schematic illustration of (*a*) diffusional flow in a grain under tension;[314] (*b*) grain switching process as proposed by Rachinger, in which grains progressively change their neighbors. Here a–f grains are those which are initially completely below the free surface but progressively become visible at the surface.[313] (*Reprinted by permission of Pergamon Press, Plc.*)

to allow the temporary grain shape changes.[313,314] These proposed processes are Herring-Nabarro creep, involving lattice diffusion occurring at higher temperature, and Coble creep, involving grain boundary as the preferred diffusion path prevailing at lower temperature. In both these cases, the solid is Newtonian-viscous ($m = 1$), and the viscosity is a function of grain size d, being proportional to d^2 or d^3, respectively. This means the strain rate achievable is strongly dependent on the diffusion rates. Thus, a superplasticity theory based on diffusion creep process with grains rotating and rearranging or interchanging (also called *grain switching*) appears to be realistic grain movement where a group of four adjacent grains mutually move at grain boundaries, changing their shape to maintain continuity with each other (and so not elongating appreciably and not developing a texture) (Fig. 15.42*b*). This operation preserves the equiaxed grains without a change in the surface area of the specimen.[315] A considerable grain boundary migration occurs continuously during superplastic deformation.

Some researchers believe that GBS and GBM, accommodated by some dislocation slip (climb or glide) processes at regions adjacent to the grain boundaries, as shown schematically in Fig. 15.43,[290] are the most agreed upon mechanism of deformation during superplastic flow of fine-grained materials.[276] Table 15.12 lists a summary of quantitative slip accommodation theories developed by several researchers. Readers interested in any particular mechanism are referred to the original papers cited by Nieh and Wadsworth.[290] All the theories cited in Table 15.12 have some features that agree with experimental observations in superplastic materials.[290] Others believe, based on the *transitional model concept*, that superplasticity flow, in fine-grained materials, involves both slip and diffusion. This model, however, does not explain the grain switching events during GBS or the retention of equiaxed grain structure during SP flow.[315] The complexity and contradictory nature of the experimental observations on superplastic flow suggest that no single mechanism is adequate to explain all the facts.[316]

15.10.2.2 Internal Cavitation.

Most of the known superplastic alloys, when superplastically deformed, lead to the cavitation phenomenon, in which nucleation,

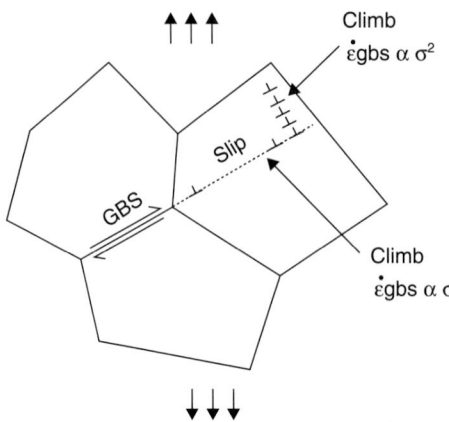

FIGURE 15.43 Schematic diagram of grain boundary sliding (GBS) accommodated by slip process (climb or glide).[290]

growth, coalescence, and interlinkage of cavities or voids occur. Increased level of cavitation and enhancement in the tendency for the transverse interlinkage of cavities may reduce the plastic stability and the superplastic formability (and ductility); terminate the forming process; and may lead to premature failure of extensively deformed regions of a component and degradation of its service properties level (e.g., tensile, creep, fatigue, and stress corrosion behavior).[291,317–320] Superplastic alloys which are prone to internal cavitation are Cu alloys containing Zn or Al, commercial Al-Mg alloys, 7475 Al alloys, nickel-based alloys, microduplex steels, and Zn-22%Al alloys. However, this superplastic cavitation phenomenon is mildly observed in Ti- and Pb-based alloys and UHC steels.[276,304,321]

The cavity nucleation stage differs in different commercial superplastic alloys. Models of cavity nucleation suggest grain boundary particles, grain boundary ledges, preexisting pores (during the extensive TMT required to produce a fine structure and/or P/M processing), and triple-point junctions as preferred sites for the creation of cavity. Recently, intersections of cooperative grain boundary sliding (CGBS) surfaces, in which grain groups slide as blocks, have been proposed as potential sites for cavity formation.[291,322] The observed alignment of cavities at 45° with respect to the sheet midplane, i.e., the same orientation as that predicted for grain group sliding, validates this proposal.[323]

Both theoretical and experimental studies have demonstrated that plane strain and biaxial deformation are much more damaging with respect to the evolution of cavitation than is uniaxial deformation.[324]

Several factors such as alloy cleanliness (or impurity contents), grain size, flow stress, superposition of hydrostatic pressure (during forming process), forming temperature, strain rate, and total strain can influence the cavitation (and ductility) behavior of SP materials (such as 7475 Al alloy and Zn-22Al alloy) during superplastic deformation.[291] Any factor that enhances the flow stress of the alloy enhances the tendency for cavitation.[295,325] Nucleation of cavitation during superplastic deformation of 7475 Al tends to be linked with internal hydrogen outgassing (arising from the casting stage of alloy production) and decohesion of one phase (resulting from an incomplete peritectic reaction during alloy solidification) in a two-phase structure. Hydrogen-enhanced superplastic cavity formation and stabilization may be eliminated or decreased by increasing the high-temperature soaking time in the TMT schedule (prior to superplastic forming). Harmful second-phase particles

TABLE 15.12 Summary of Proposed Models of Grain Boundary Sliding[290]

Ref.	Equation	Year	Remarks
Slip accommodation (rate controlling)			
Ball-Hutchison[1]	$\dot{\varepsilon} = K_1\left(\dfrac{b}{d}\right)^2 D_{gb}\left(\dfrac{\sigma}{E}\right)^2$	1969	Sliding of group of grains
Langdon[2]	$\dot{\varepsilon} = K_2\left(\dfrac{b}{d}\right)^2 D_L\left(\dfrac{\sigma}{E}\right)^2$	1970	Movement of dislocations adjacent to grain boundary
Mukherjee[3]	$\dot{\varepsilon} = K_3\left(\dfrac{b}{d}\right)^2 D_{gb}\left(\dfrac{\sigma}{E}\right)^2$	1971	Grains slide individually
Gifkins[4]	$\dot{\varepsilon} = K_4\left(\dfrac{b}{d}\right)^2 D_{gb}\left(\dfrac{\sigma}{E}\right)^2$	1976	Pile-up at triple points (core/mantle)
Gittus[5]	$\dot{\varepsilon} = K_5\left(\dfrac{b}{d}\right)^2 D_{IPB}\left(\dfrac{\sigma-\sigma_0}{E}\right)^2$	1977	Pile-up at interphase boundary
Sherby-Wadsworth[6]	$\dot{\varepsilon} = 6\times10^8\left(\dfrac{b}{d}\right)^3 \dfrac{D_{gb}}{b^2}\left(\dfrac{\sigma}{E}\right)^2$	1984	Phenomenological $T = 0.4$ to $0.6T_m$
	$\dot{\varepsilon} = 2\times10^9\left(\dfrac{b}{d}\right)^2 \dfrac{D_L}{b^2}\left(\dfrac{\sigma}{E}\right)^2$		Phenomenolgial $T > 0.6T_m$
Kaibyshev et al.[7]	$\dot{\varepsilon} = \dfrac{A}{kT}\left(\dfrac{\sigma-\sigma_0}{G}\right)^2\left(\dfrac{b}{d}\right) D_o\exp(-Q/RT)$	1985	Hardening and recovery of dislocations at grain boundary sliding
Fukuyo et al.[8]	$\dot{\varepsilon} = K_s\left(\dfrac{b}{d}\right)^2\left(\dfrac{D_{chem}}{b^2}\right)\left(\dfrac{\sigma}{E}\right)$	1990	Grain boundary sliding accommodated by dislocation glide
Diffusional accommodation (rate controlling)			
Ashby-Verrall[9]	$\dot{\varepsilon} = K_7\left(\dfrac{b}{d}\right)^2 D_{eff}\left(\dfrac{\sigma-\sigma_0}{E}\right)$	1973	$D_{eff}=D_L 1+\dfrac{3\cdot w}{d}\dfrac{D_{gb}}{D_L}$

D_{gb}, D_p, and D_{IPB} represent diffusivities along grain boundaries, dislocation pipes, and interfacial phase boundaries, respectively; D_L and D_{chain} are lattice and chemical diffusivities, respectively.
$K_1 - K_7$ = materials constants.
w = grain boundary width.
σ_0 = threshold stress.

1. A. Ball and M. M. Hutchinson, *Met. Sci. J.*, vol. 3, 1969, pp. 1–7.
2. T. G. Langdon, *Phil. Mag.*, vol. 22A, no. 178, 1970, pp. 689–700.
3. A. K. Mukherjee, *Mater. Sci., Eng.*, vol. 8, 1971, pp. 83–89.
4. R. C. Gifkins, *Met. Trans.*, vol. 7, 1976, pp. 1225–1232.
5. J. H. Gittus, *J. Eng. Mater. Technol., Trans. ASME*, vol. 99, 1977, pp. 244–251.
6. Ref. 310
7. O. K. Kaibyshev. R. Z. Valiev, and A. K. Emaletinov, *Phys. Status Solidi*, (a), vol. 90, 1985, pp. 197–206.
8. H. Fukuyo, H. C. Tsai, T. Oyama, and O. D. Sherby, *ISIJ Int*, vol. 31, no. 1, 1991, pp. 76–85.
9. Ref. 314.

associated with cavity formation may be modified to less damaging forms by appropriate changes to alloy processing.[321] For example, harder second-phase particles, if distributed uniformly and in fine particles form within the matrix can actually hinder the cavitation during SP deformation by various recovery mechanisms occurring near the particles. However, coarse, hard particles can cause cavitation.[290] Recently, it has been shown that superimposition of hydrostatic pressure varying between 50 and 100% of the uniaxial flow stress during forming can lead to suppression or elimination of nucleation and growth of cavitation in titanium and aluminum and other SP alloys, thereby extending the ductility and increasing the post-forming properties.[276,295,324]

Cavity growth in SP alloys may be controlled by (1) vacancy diffusion into a cavity (2) plastic deformation of the material around a cavity (called power-law growth mechanism), (3) surface diffusion, and (4) other processes including annealing.[319] The latter stages of cavity growth, interlinking of cavities, and subsequent final rupture are the common features in many alloy systems.

Ma and Langdon[325] have determined the important differences in the cavitation behavior in typical SP metals (such as Cu-based CDA 638) and ceramics (such as yttria-stabilized tetragonal zirconia, or Y-TZP). Both materials exhibit extensive cavitation. However, the metal exhibits increased cavitation and more rounded cavities at the lower testing strain rates, while ceramic exhibits decreased cavitation and few rounded cavities at the lower testing strain rates. These results indicate that cavity growth by vacancy diffusion plays an important role in SP metals while diffusion plays a minor role in the cavity growth in SP ceramics.

15.10.2.3 Superplasticity in Titanium Alloys.

Maximum superplasticity in ($\alpha + \beta$ type) titanium and Ti-4.5Al-3V-2Fe-2Mo (called SP-700) alloys[325a] occurs where the phase ratio ranges between 50 and 70% and the corresponding SPF temperature ranges between about 850 and 927°C. Among the titanium alloys, Ti-6Al-4V is the most commercial SP alloy for which extensive data are available. Other $\alpha + \beta$ type alloys in this category can also be quite superplastic, provided they have very fine-grain structure.[292]

It was observed by Wert and Paton that the all-α and all-β alloys exhibit very limited high-temperature ductility. The modified Ti-6Al-4V alloy with 2% Fe, Cr, or Ni can produce SP deformation at the reduced temperature (~815°C), which is quite low compared to the usual Ti-6Al-4V alloy, provided that the β-phase content is about 50% (Table 15.13).[326]

The application of Ti-6Al-4V ranges from aerospace materials to biological and dental materials (such as hip prostheses and dental bases and dental implant super-

TABLE 15.13 Superplastic Properties of Modified Ti-6Al-4V Alloys[326]

Alloy	Temperature, °C	Volume fraction, v_β	Flow stress,[†] Mpa	Elongation,[†] %
Ti-6Al-4V	927	0.57	9.6	600
Ti-6Al-4V	815	0.22	65	225
Ti-6Al-4V-2Fe	815	0.49	19	650
Ti-6Al-4V-2Co	815	0.49	26	670
Ti-6Al-4V-2Ni	815	0.47	13	720

[†] At $\dot{\varepsilon} = 2 \times 10^{-4} \text{s}^{-1}$.

Reprinted by permission of The Metallurgical Society, Warrendale, Pa.

structures). The development of Ti-5.5Al-1Fe alloy as a substitute for Ti-6Al-4V for biological applications was reported to display SP elongation of >400% in the temperature range of 777 to 927°C and strain rate range of 10^{-4} to $10^{-2}\,s^{-1}$ (Table 15.11). The optimum elongation was observed at 827°C and $1 \times 10^{-3}\,s^{-1}$. The m was ~0.5, and the apparent activation energy was 314 kJ/mol.[327]

15.10.2.4 Superplasticity in Aluminum Alloys.
Conventionally, for Al alloys, two quite distinct SP microstructures have been achieved by ingot metallurgy. One is the unrecrystallized structure that undergoes dynamic "continuous" recrystallization during SPF, while the other uses a statically recrystallized structure prior to SPF that remains unchanged except for modest grain growth.[328,329]

Commercial available SP aluminum alloys are standard or slightly modified alloys (e.g., 2004, 2219, 5083, 5251, 6082, 7075, and 8090) and unconventional eutectics (Al-Pt and Al-Ca).[327a]

Fine-grain microstructure in high-strength (e.g., 7075 and 7475) Al alloys can be developed through the application of TMT and associated recrystallization process and applied in aerospace industry by means of superplasticity. The reduced elongation of 7075 Al alloys compared to that of 7475 Al alloys is attributed to the presence of more Fe and Si impurities. The formation of rows of fibers usually observed at the surface of superplastically bulged 7475 Al alloy hemispherical, cylindrical, and conical domes has been suggested to be due to the folding and lamination of the oxide film with sequential stretching of folds.[330]

The TMT schedule for PM 7091 [Al-(5.7–7.1)Zn-(2–3)Mg-(1.1–1.8)Cu-(0.2–0.6)Co-0.12Si-0.15Fe-(0.2–0.5)O] used for improved superplasticity consists of three steps: solutionizing (490°C/3 hr), overaging (330°C/36 hr), and warm-rolling (80 to 82% reduction at 150°C). This SP material has been reported to exhibit 450% elongation with a grain size of 1 to $3\,\mu m$ at 300°C and $8 \times 10^{-5}\,s^{-1}$.[331]

Enhanced superplasticity in Al-10Mg-0.1Zr alloy at 300°C is obtained by selection of appropriate TMT variables to induce the creation of well-defined subgrains so stabilized by β-phase (Al_8Mg_5) particles that the boundary dislocation density promotes (continuous) recrystallization.[332]

Recently, the superplasticity behavior of fine-grained 5083 Al alloys has been extensively studied due to a good combination of ductility, formability, cost, strength, corrosion resistance, and weldability. Following a 5:1 cold-rolling reduction, a commercial grade of 5083 alloy has been reported to provide elongation of 300% at 510°C.[333] Higher elongation (>500%) has been obtained with specific SPF-grade containing Mn, Cr, Cu, and Zr alloying elements and lower Fe and Si impurity contents. The fine-grained structure required for the modified 5083 (Al-4.7Mg-1.6Mn) alloy with 0.2% Zr addition is usually accomplished by heavy cold reductions (~80%) of the alloy. Elongation up to 700% at 550°C has been reported in a modified 5083 Al alloy with 0.6% Cu addition as a grain refiner.[334] Currently, 5083 alloy is superplastically formed for body panels on several low-volume automobiles and applied to motorcycle parts such as frames and fuel tanks by using aluminum hot blow-forming.[335]

Recently the superplastic properties (tensile elongation, 500 to >1000%) in a cold-rolled Al-6Mg-0.3Sc alloy have been observed over a wide temperature range (475 to 525°C) and strain rate ranges (10^{-3} to $10^{-1}\,s^{-1}$). In this case, a uniform distribution of fine coherent Al_3Sc precipitates occurs which effectively pins grain and subgrain boundaries during static and dynamic recrystallization. Consequently, the alloy maintains its fine grain size (~$7\,\mu m$) even after extensive superplastic deformation (>1000%). Note that this alloy is not prone to cavitation during superplastic deformation.[302a]

Supral ranges of Al alloys have been used in the aerospace, automotive, rail, and other industries, whereas Al-Li alloys (such as 8090, 2090, and 2095) are used mostly in the aerospace industry.

15.10.2.5 Superplasticity in Inconel Alloy. There is significant need for complex-shaped airframe and engine components for commercial and military aircraft applications requiring high Ni-base alloys to withstand a combination of high strength, high temperature, and hot gas corrosion. Fine-grained (1 to 10 μm) SP Inconel 718 alloy [with varying proportions of the γ and δ (Ni$_3$Nb) phases] can be used in these applications, which is reported to exhibit, at 954°C (1750°F), (1) 350% elongation at an initial strain rate of $1.3 \times 10^{-3}\,\text{s}^{-1}$ with flow stress of <72 MPa (10.4 ksi) and $m = 0.28$ and (2) 750% elongation at an initial strain of $3.3 \times 10^{-5}\,\text{s}^{-1}$ with flow stress of <28 MPa (4 ksi) and $m = 0.83$.[336] Some processes may be needed after SPF to coarsen grains and use the alloy at high temperature.

15.10.2.6 Superplasticity in Microduplex Stainless Steels. The term *microduplex steels* refers to the high plasticity in the presence of δ/γ phases in fine-grained form, with grain sizes of 2 to ~3 μm, which find commercial applications in the chemical industry and offshore techniques due to their high strength and resistance to both stress corrosion and general corrosion. The first study of superplasticity of a duplex stainless steel was reported by Hayden et al.[337] which exhibited 500% elongation at 1000°C and trapping of N atoms as coarse TiN particles by a small amount of Ti addition. Gibson et al.[338] found a fine-grained duplex microstructure by the recrystallization process during TMT operation, whereas Maehara[339] found a fine-grained duplex microstructure by the precipitation of second σ-phase particles at test temperature after solution treatment above 1250°C, followed by 50% reduction by rolling. In the latter case, very large SP elongation was observed, for example, >2500% in cold-rolled 25Cr-7Ni-3Mo-0.14N steel at 950°C and a strain rate of $4 \times 10^{-3}\,\text{s}^{-1}$ with $m = 0.6$; however, high strain rate superplasticity (HSRSP) at 1000°C was reported, for example, with 500% and 300% elongations at initial strain rates of $5 \times 10^{-1}\,\text{s}^{-1}$ and $1\,\text{s}^{-1}$, respectively. Han and Hong[340] found the superplasticity in Fe-24Cr-7Ni-3Mo duplex stainless steel with a maximum elongation of 750% at 850°C and $3 \times 10^{-3}\,\text{s}^{-1}$ after TMT. However, the superplastic duplex Fe-23Cr-5.6Ni-1.3Mo-0.12N stainless steel exhibited elongation >2200% at 900°C and $2 \times 10^{-4}\,\text{s}^{-1}$ due to its optimum chemical composition and a specific manufacturing route (see Table 15.11).[341] This SP duplex stainless steel is capable of *complementary wrap forming* (CWF) to produce auxiliary power unit (APU) components with properties comparable to those of Ti-6Al-2Sn-4Zr-2Mo SP alloy.[341] The fine-grained (2- to 3-μm) Fe-22Cr-5Ni-3Mo-0.3N duplex stainless steel is found to exhibit maximum elongation of ~800% at 950°C and $10^{-3}\,\text{s}^{-1}$. Its elongation of ~300% at 900 to 1050°C and its high strain rates of about $5 \times 10^{-2}\,\text{s}^{-1}$ allow hot-forming operations such as deep drawing, blow forming, and die forging of complex parts at high deformation rates and low stresses.[342]

15.10.3 Recent Advances in Superplasticity

This section describes new developments in a few specific areas where noteworthy progress has been made. These areas include superplasticity in nanocrystalline materials, intermetallics, and ceramics; high-strain-rate superplasticity (HSRSP); and low-temperature superplasticity (LTSP). The HSRSP and LTSP may be achieved by reducing the grain size below 1 μm and can be applied in the aerospace and automotive industries to increase productivity and reduce production cost.

TABLE 15.14 Superplasticity Observations in Nanocrystalline Materials[344]

Alloy	Grain size,[†] nm	$\dot{\varepsilon}$, s^{-1}	T, °C	Stress, MPa $\varepsilon = 0.1$	Max.	Elongation, %	Ref.
Ti-6Al-3.2Mo	60	2×10^{-4}	575	150	—	1200	1,2
Zn-22Al	80	10^{-2}	120	18	—	250	3
Ni$_3$Al alloy	50	1×10^{-3}	650	400	1530	380	4
Ni$_3$Al alloy			725	270	790	560	
Electrodeposited nickel	20	1×10^{-3}	350	108	360	300	5
Electrodeposited nickel			420	25	67	900	
Al-Mg-Li-Zr alloy (1420)	100	1×10^{-1}	250	106	154	330	6
Al-Mg-Li-Zr alloy (1420)			300	33	146	850	
Al-Cu alloy (2124)	100	1×10^{-3}	350	6	50	405	7
Ti-6Al-4V	70	1×10^{-3}	575	165	—	215	7

[†] Describes the starting average grain size.
Reprinted by permission of the Metallurgical Society, Warrendale, Pa.
1. G. A. Salischev et al., *Materials Science Forum*, vols. 170–172, 1994, p. 121.
2. G. A. Salischev et al., *Materials Science Forum*, vols. 243–245, 1997, p. 585.
3. R. S. Mishra, R. Z. Valiev, and A. K. Mukherjee, *Nanostructural Materials*, vol. 9, 1997, p. 473.
4. R. S. Mishra et al., *Mater. Sc. Eng.*, vol. A252, 1998, p. 174.
5. Ref. 26 in Ref. 344.
6. Ref. 27 in Ref. 344.
7. Ref. 28 in Ref. 344.

15.10.3.1 Superplasticity in Nanocrystalline Materials. The nanocrystalline materials can be grouped into two types: (1) D-type produced from ingot or powder by severe plastic deformation and (2) S-type produced by sintering of nanocrystalline powders. Nanocrystalline materials have been extensively studied, which have increased our understanding of grain-size-dependent phenomenon to a much finer scale. The experimental results exhibit higher flow stresses for superplasticity in nanocrystalline materials. It is believed that slip accommodation plays a dominant role during superplasticity of nanocrystalline materials. The flow stress needed for slip accommodation in nanocrystalline materials is found to be higher than those for conventional SP materials.[343]

SP behavior has been observed in several nanocrystalline materials using miniature tensile specimens (Table 15.14),[344] but the level of SP properties is lower than expected for such a small grain size. Formation of metastable states, disordering, and presence of supersaturated solid solution during severe plastic deformation may be possible factors of reduced SP properties in the nanocrystalline alloys investigated.[345]

15.10.3.2 Superplasticity in Intermetallics. Superplasticity has been observed in several intermetallics with fine microduplex structure (such as aluminides and silicides) or coarse-grained quasi-single-phase structures prior to deformation. (Their m values range, typically, from 0.32 to 1, and grain sizes are $>10\,\mu$m.) In the former case, as in many SP alloys and ceramics, grain boundary sliding plays an important role, whereas in the latter, dynamic recrystallization or subgrain formation is needed for SP deformation.[346] Lin and coworkers[347] have reported superplasticity in Fe$_3$Al and FeAl alloys with a grain size of 100 and 350 μm, respectively. They observed the SP behavior in Fe-28Al-2Ti (in at%) alloy exhibiting 620% elongation with $m = 0.4$ at an 850°C and an initial strain rate of $1.25 \times 10^{-3}\,$s^{-1} and in Fe-36.5Al-2Ti alloy exhibiting 297% elongation with $m = 0.34$ at 1000°C under an initial strain rate of $2.08 \times 10^{-2}\,$s^{-1}. The observed SP behavior is attributed to continuous recovery and recrystallization. Lin and his coworkers[348] also reported superplasticity in Fe_3Al and FeAl alloys at 800 to 900°C and 2×10^{-4} to $4 \times 10^{-3} s^{-1}$ and at 900 to 1000°C and 1.4×10^{-4} to $2.78 \times 10^{-2} s^{-1}$, respectively.

Ti aluminides such as TiAl (γ), near-γ, and Ti_3Al (super-α_2) alloys containing various alloying elements are promising candidates for applications as aerospace (and automotive) structural and engine components due to their attractive elevated-temperature (as high as 650 to 800°C) specific strength, good oxidation resistance, and low density. The fine-grained Ti-Al (43 at%) (γ alloy) is superplastic in the temperature range of 1000 to 1100°C with an elongation of 275%.[349] The optimum SP deformation parameters for Ti-25Al-10Nb-3V-1Mo (α_2 alloy) are ~800% elongation in the transverse rolling direction at 950°C and a strain rate of $8 \times 10^{-5} s^{-1}$.[350]

Two interesting areas have emerged in the development of superplasticity of intermetallics—low-temperature superplasticity in TiAl (γ) and coarse-grained (~100-μm) superplasticity in Fe_3Al- and FeAl-based alloys. For low-temperature superplasticity, a soft phase is intentionally introduced, by microstructural design, to accommodate effectively sliding strains at triple-grain junctions to retard and suppress cavitation and fracture. This approach is possibly applicable to many intermetallic alloy systems. For coarse-grained superplasticity in iron aluminides, although the actual SP deformation mechanisms are not established, scientifically and technologically important progress has been made. In the near future, SPF of intermetallics may possibly be employed only by the aerospace industry, mainly because of the high cost of intermetallics and relatively high SPF temperatures.[290] This is of particular interest because intermetallics have very limited ductility at room temperature and cannot be processed by conventional means.

15.10.3.3 Superplasticity in Ceramics.

SP ceramics cover a wide range of chemical composition, crystal structure, and physical properties, but all have the common feature of a stable, very fine-grained microstructure (<0.3 μm) prior to deformation and deformation temperature of $0.4T_m$ to $0.7T_m$. The success of ceramic superplasticity arises from the ability to obtain a dense form with very fine grains by sintering and to prevent grain growth during high-temperature deformation.[351] These requirements are being fulfilled both by advances in ceramic powder synthesis and by novel breakthroughs in powder consolidation practice.[352]

Superplasticity in fine-grained ceramics often exhibits a large tensile elongation at high temperature. For example, superplastic elongations of 800% in 3Y-TZP (3 mol% yttria, stabilized zirconia polycrystals)[353] and 1038% in 2.5 mol% of yttria (termed 2.5Y-TZP) containing 5 wt% SiO_2 have been reported.[354] Superplasticity in ceramics strongly depends on the grain boundary sliding characteristics; hence grain boundary controlled processing plays a very important role in further development of superplasticity in ceramics. Doping of small amounts of glass phase (impurities or metal cation) into the superplastic ceramics (such as tetragonal ZrO_2 polycrystals) is often a useful method of reducing the high-temperature flow stress, where the stress level is a function of glass phase doped.[355]

SP ceramics are technologically important due to their useful electronic, magnetic, optical, biological, and mechanical properties. Superplasticity of structural ceramics such as ZrO_2, Al_2O_3, mullite, Si_3N_4, and SiC and their composites has attracted much interest with the potential to form near-net shape parts for mechanical applications.[356]

15.10.3.4 High Strain Rate Superplasticity.

High-strain-rate superplasticity (HSRSP) is defined as the superplasticity that occurs with a very high strain rate $\geq 10^{-2} s^{-1}$ (i.e., close to the commercial hot-working rates of 10^{-1} to $10^{-2} s^{-1}$) to make

SPF technique appropriate for commercial applications.[357] It is expected to result in economically viable, high-productivity, and near-net shape forming processes [e.g., for metal-matrix composites and mechanically alloyed (MA) aluminum alloys, by superplastic forging] in the fabrication of parts with complicated shapes for the aerospace, automobile, and even semiconductor industries.[311] High-strain-rate superplasticity has been observed in a number of structural materials and their metal-matrix composites, oxide-dispersion strengthening (ODS) alloys, P/M processed or mechanically alloyed materials and alloys (such as MA IN9021, MA IN905XL, and MA IN9052),[357a] consolidation of nanocrystalline powders, physical vapor deposition, and severe plastic deformation; and is associated with an ultra-fine-grained structure ($\leq\sim 3\,\mu m$) (Fig. 15.41 and Table 15.15).[291,311,357a,357b]

It is clear from Table 15.15 that the majority of HSRSP properties have been observed in Al-based materials. For example, the HSRSP 20 vol% SiC whisker-reinforced 2124 Al is reported to exhibit ~300% elongation at 475 to 550°C and 3.3 $\times\ 10^{-1}\,s^{-1}$ with $m = 0.33$; and as-extruded Si_3N_4 particulate-reinforced 6061 Al composite is reported to exhibit 450% elongation at 545°C and higher $10^{-1}\,s^{-1}$ strain rate with low strain-hardening exponent, while as-extruded whisker-reinforced 6061 Al composition exhibited a 450% elongation at a lower strain rate of $4 \times 10^{-2}\,s^{-1}$ with high strain-hardening exponent.[358] The 2024 composite and pure Al composites exhibited HSRSP at 525°C and $10^{-1}\,s^{-1}$ and at 640°C and $1\,s^{-1}$, with an elongation of ~200 and 280%, respectively.[359]

Recently, HSRSP has been found in Mg-base materials and their composites. Mabuchi et al.[360] have shown the HSRSP in PM AZ91 (Mg-9Al-1Zn-0.2Mn) and ZK61 Mg (Mg-6Zn-0.8Zr) alloys (<1-μm grain size) at 250 to 400°C in the range of 10^{-2} to $10^{-1}\,s^{-1}$, where the origin of HSRSP is not related to the presence of a liquid phase. A doubly extruded SiC particulate-reinforced Mg-based composite, ZK60/SiC/17p, has been shown to exhibit SP behavior at high strain rates with 450% elongation at $10^{-1}\,s^{-1}$ and 350°C.[361]

Commercial SP titanium alloys such as Ti-6Al-4V, Ti-6Al-2Sn-4Zr-4Mo, and Ti-13Cr-11V-3Al, which have great potential applications in the aircraft and space industry, do not exhibit structural superplasticity at high strain rates.[361a]

Recent research and development of HSRSP have led to a new concept of *accommodation helper mechanism* (due to the presence of liquid phase or local melting at grain boundaries, though it is not always observed, as in the case of Mg-based and Al-based composites) and a new opportunity of the *superplastic forging process*. Note that the role of liquid phase to serve as accommodation helper will only be possible for GBS, provided GBS is not properly accommodated by diffusion and dislocation movement. This superplastic phenomenon is of vital significance from the viewpoint of materials science and commercial applications.[311]

15.10.3.5 Low-Temperature Superplasticity. One of the most important observations of low-temperature superplasticity in alloys and finer-grain size or nanocrystalline materials is a large reduction of its SPF temperature compared to microcrystalline materials, for example, ~200°C for titanium alloys, ~325°C for Ni_3Al alloys, 400°C for pure Ni, 200 to 250°C for Al and Mg alloys, 175°C for 8090 Al alloy, and 100 to 150°C for Zn-Al alloys (Tables 15.14 and 15.16).[344,362] Such a reduction in SPF temperature offers many advantages. For example, a Ni_3Al alloy with a 6-μm grain size exhibits superplasticity at 1050°C, which is higher than the current commercial practice of ~900°C as an upper SPF temperature for Ti alloys. The reduction in SPF temperature in nanocrystalline Ni_3Al will enable it to superplastically form within the SPF temperature range of Ti alloys, thereby making use of

TABLE 15.15 Superplastic Properties of High-Strain-Rate Superplastic Materials[311]

Cited references correspond to those mentioned in Ref. 311.

Materials	Temp., K	Strain rate, s^{-1}	Flow stress, MPa	m	Elongation, %	Grain size, μm	References
Dynamic Recrystallization							
Al-Cu-Zr	753	3×10^{-3}	9	0.4–0.5	1270	5	5,6
2124-Zr	748	3×10^{-1}	35	0.5	490	1	7
7475-Zr	793	10^{-1}	11	0.6	600	2.5	8
7475-Zr	793	3×10^{-1}			900	1.3	9
Al-Mg-Zr	773	10^{-1}	21	0.3	570		10
Al-Cu-Zr	743	10^{-1}	25	0.3	480		10
Al-Mg-Mn	823	3×10^{-2}	5.7	0.45	500		10
Thermomechanical Treatment							
SiC_p/1100	903	10^{-1}	7.5	0.4	200	2	11
SiC_w/2124	798	3×10^{-1}	10	0.33	300		3
SiC_w/2024	823	1	4	0.5	150		12
SiC_w/2009	798	3×10^{-2}	15	0.5	180		13
SiC_p/2124	748	3×10^{-1}	48	0.17	139		14
SiC_w/6061	823	2×10^{-1}	6.5	0.32	300		15
SiC_w/6061	873	2×10^{-1}	1.7	0.34	440	2	16
SiC_p/6061	853	3×10^{-1}	5.5	0.33	207		17
SiC_p/6061	853	10^{-1}	4.5	0.5	350		18
SiC_p/6061	853	3×10^{-1}	4.2	0.36	342	2	19
SiC_p/7075	793	5×10^{-1}	9	0.42	200	0.6	20
SiC_p/7075	793	1	12	0.37	200	0.5	20
SiC_p/7075	793	10	31	0.36	200	0.5	20
SiC_p/8090	848	2×10^{-1}	4.7	0.53	300	6	21,22
Si_3N_{4w}/pure Al	903	10^{-1}	10	0.47	200	2	23
Si_3N_{4w}/2124	798	2×10^{-1}	8	0.5	250		24
Si_3N_{4w}/2124	818	4×10^{-1}	2.4	0.35	280	4	25
Si_3N_{4p}/2124	773	3×10^{-1}	7.5	0.3	280	1	25
Si_3N_{4p}/2124	788	4×10^{-1}	2.2	0.4	840	2	25

Material							Ref.
Si$_3$N$_{4p}$/5052	818	1	6	0.3–0.5	700	1	26
Si$_3$N$_{4w}$/6061	798	2×10^{-1}	22	0.5	300		27
Si$_3$N$_{4w}$/6061	818	2×10^{-1}	7	0.5	600	3	28
Si$_3$N$_{4w}$/6061	818	2×10^{-1}	11	0.5	260	3.1	29
Si$_3$N$_{4w}$/6061	818	3×10^{-1}	4.7	0.3	150	3.0	29
Si$_3$N$_{4w}$/6061	833	10^{-1}	11	0.3–0.5	480	3.3	30
Si$_3$N$_{4w}$/6061	818	2×10^{-1}	6.2	0.46	600		31
Si$_3$N$_{4w}$/6061	818	2×10^{-2}	15	0.31	173	3	32
Si$_3$N$_{4p}$/6061	833	2	5.2	0.3–0.5	620	1.3	30
Si$_3$N$_{4p}$/6061	833	1	5.5	0.3–0.5	350	1.9	30
Si$_3$N$_{4p}$/6061	818	10^{-1}	5.3	0.3–0.5	450	3.0	30
Si$_3$N$_{4w}$/7064	798	8×10^{-1}	18	0.4	160	3.5	33
Si$_3$N$_{4w}$/7064	818	2×10^{-1}	10	0.4	230	3.6	33
Si$_3$N$_{4w}$/7064	818	5×10^{-1}	14	0.5	240	3.5	34
Si$_3$N$_{4w}$/7064	833	10^{-1}	4.2	0.3–0.5	380	3.5	35
Si$_3$N$_{4w}$/7075	773	2×10^{-1}	24	0.31	260		36
Si$_3$N$_{4p}$/7064	818	1	10	0.45	330		37
AlN$_p$/1N90	923	3×10^{-1}	13	0.3	200		38
AlN$_p$/6061	873	9×10^{-1}	9	0.5	509		39
AlN$_p$/6061	863	8×10^{-1}	5.7	0.45	683	2	40
IN90	913	2×10^{-2}	10	0.47	400		41
Al-Mg-Sc	672	10^{-2}	33	0.4	1020	2 (subgrain)	42
7091	723	10^{-2}	19	0.2	190	0.5 (subgrain)	43
Al-Li	843	1.4×10^{-1}	5.5	>0.3	250	1.8	44
Al-Si	803	10^{-1}	5.3	0.5	275	<4	45
Al-Si-Cu	793	10^{-1}	0.5	0.48	390	1.4	46
Al-Si-Cu	773	1	2.5	0.5	290		47
SiC$_p$/ZK60	723	1.3	33	0.33	360	0.5	48,49
SiC$_p$/ZK60	623	10^{-1}	15	0.5	400	0.3	50
TiC$_p$/Mg-Zn	743	7×10^{-2}	8	0.43	340	1.4	51
Mg$_2$Si/Mg-Al	788	10^{-1}	7.3	0.5	370	1.4	52,53
Mg$_2$Si/Mg-Zn	713	10^{-1}	9.7	0.5	290	0.9	52
AlN$_p$/Mg-Al	698	5×10^{-1}	25	0.4	200	2	54
AZ91	573	10^{-2}	29	0.5	276	1.4	55

TABLE 15.15 Superplastic Properties of High-Strain-Rate Superplastic Materials[311] (*Continued*)

Cited references correspond to those mentioned in Ref. 311.

Materials	Temp., K	Strain rate, s^{-1}	Flow stress, MPa	m	Elongation, %	Grain size, μm	References
ZK61	623	10^{-1}	11	0.5	445	1.6	55
MA754	1373	10^{-1}	57	0.25	200	0.67	56
MA6000	1273	5×10^{-1}	125	0.47	308	0.26	56
δ/γ duplex stainless steel	1273	7×10^{-2}		0.6	1612	$0.5 \sim 1$	57
UHCS-10Al	1193	10^{-1}	65	0.25	350	0.5	58
Consolidation of Amorphous or Nanocrystalline Powder							
Al-Ni-Mm	885	1	15	0.5	650	1	59,60
Al-Ni-Mm-Zr	873	1	15	0.5	650	0.8	59
Mg-Al-Ga	573	10^{-2}	8	0.5	1080	2	61
Mechanical Alloying							
IN9021	723	7×10^{-1}	5	0.3	300		4
IN90211	748	2.5	40	0.3	505	0.5	62
IN9052	863	10	15	0.6	330	0.5	59,63
IN905XL	848	20	12	0.6	190	0.4	59,63
IN9021	823	50	18	0.5	1250	0.5	59,63
SiC$_p$/IN9021	823	5	5	0.5	600	0.5	59,63
Physical Vapor Deposition							
VQ Al-Cr-Fe	898	1	20	0.5	505	0.5	59
Intense Plastic Straining							
Al-Mg-Li-Zr	623	10^{-2}	85		1180	1.2	64
Al-Cu-Zr	573	10^{-2}			970	0.5	64
Zn-Al	473	3×10^{-2}	27		1970	0.6	65

Courtesy of TMS, Warrendale, Pa.

TABLE 15.16 Conditions and Parameters of Superplastic Deformation of Some Alloys[362]

Alloy	d, μm	T, K	$\dot{\varepsilon}$, s^{-1}	σ, MPa	m	Elongation, %	Ref.
Al-4Cu-0.5Zr	8	773	3×10^{-4}	14	0.5	800	
	0.3	493	3×10^{-4}	23	0.48	>250[†]	
PM 7475 Al-0.7Zr	<1	793	2×10^{-3}–3×10^{-1}		~0.6	100	1
IN 9021 PM (MA)	>0.3–3	798	50		0.3	1000	2
		823	100			900	2
Al-Li (8090) alloy	3.7	623	8×10^{-4}		0.33	~700	3
Mg-1.5Mn-0.3Ce	10	673	5×10^{-4}	25	0.42	320	
	0.3	453	5×10^{-4}	33	0.38	>150[†]	
Mg-6Zn-0.5Zr (ZK60)	3.4	423	10^{-5}		0.3	350	4

  [†] Measurements are restricted by test conditions.
  1. K. Matsuki, H. Matsumoto, M. Tokizawa, and Y. Murakami, *Superplasticity in Advanced Materials*, JSRS, Osaka, Japan, 1991, pp. 551–556.
  2. K. Higashi, T. Okada, T. Mukai, and S. Tanimura, *Superplasticity in Advanced Materials*, JSRS, Osaka, Japan, 1991, pp. 551–556, 569–574.
  3. H.-P. Pu and J. C. Huang, *Superplasticity and Superplastic Forming*, eds. A. K. Ghosh and T. R. Bieler, TMS, Warrendale, P., 1995, pp. 33–40.
  4. Ref. 363.

current die and tooling methods.[344] Lower forming temperature also offers the reduction in energy cost and surface oxidation.[363]

15.10.3.6 Quasi-Superplasticity.   High-strain-rate sensitivity may be achieved in metallic materials by a mechanism involving no grain boundary sliding and fine grains. These materials are termed *class I solid-solution alloys*, in which the glide portion of the glide/climb dislocation creep process is rate-controlling due to solute-drag controlled dislocation motion (or creep). These alloys have modest m value ($m = 0.33$) and elongation of 200 to 400%. The extended elongation inherent in these solid-solution alloys would suggest that they may be categorized as *quasi-superplastic materials* (i.e., *superplastic-like*, or *resembling superplasticity*). Since these quasi-superplastic materials do not possess high strength at low temperatures, they are used mostly as secondary structural components rather than primary structural components (such as mechanically alloyed Al-Mg-Li alloys and UHC steel), which are truly superplastic at elevated temperature (made from ultrafine-grained materials with ultrafine second phases) with improved elastic stiffness, lightness, and very good tensile ductility at low temperature.[299,363a]

15.11 SUPERPLASTIC SHEET FORMING PROCESS

SPF is a metalforming process that utilizes the SP material behavior to produce near-net shape components for the aerospace, transportation, and architectural applications. As SPF is a stretching process, the design engineer must take into account the amount of strain the component and forming method will induce into the sheet. This, in turn, will result in material selection and give an estimation of the final thickness of the component and thus the initial thickness. SPF of Al- and Ti-based alloys has become a significant manufacturing method for aerospace engine and fuselage components. Superplastic Inconel 718 can be superplastically formed at 954°C using flow stresses less than 72 MPa (10.4 ksi).

Since SPF materials have low flow stresses, lower gas or air pressure [usually ≤500 psi (3.4 MPa)] is used to form the sheet in a reasonable time. For most superplastic sheet forming operations, the material flow stress should be <10 ksi (70 MPa).[364] In contrast, very large pressure is employed for the deformation of conventional metals and alloys which is effected by using hydraulically or mechanically driven tools. It thus appears that SPF materials behave as vacuum forming of thermoplastic materials.[303]

Increased initial thickness is needed to fulfill the minimum thickness requirements and the light weight, which is the major attractive advantage of SPF. However, very large elongation is not desirable due to increased cavitation and reduced mechanical properties. It appears that the large elongation of ≥300% is rarely used in structural applications despite the material's potential.[365]

The main advantages of SPF over conventional metalforming processes are as follows:[301]

1. Deep complex-shape parts are produced in a single forming operation, rather than the multistage forming steps in conventional (e.g., cold stamping) methods, thereby minimizing or eliminating expensive machining/fabrication costs and saving weight penalties otherwise associated with overlaps and stress raisers.

2. Cavity tooling is simple, and tooling costs are lower because of the use of only a single major tool, which is quite different from the accurately matched pair of tools or multiple tools in most conventional forming operations.

3. The capital cost of the production equipment is lower because of relatively small deformation pressure, lower weight, lower production cost, less wastage, and more efficient design.

4. The accuracy and repeatability of shape and thickness in the finished components result in reduced assembly/fitting costs and minimal variability in structural performance.

5. Processing time is independent of size and number of parts produced in one SPF press operation, thereby increasing the benefit with size or use of multicavity tooling.

6. Tool wear is negligible owing to minimal slippage between the forming members and the tools, resulting in absence of wear action and thereby offering excellent tool life, particularly when metal tools are used.[298]

7. There are no springback effects and no residual stresses.

There are some shortcomings associated with these SP characteristics: These are (*a*) This behavior is limited to a very small number of alloys in a given metallurgical condition (e.g., a very fine-grained structure). (*b*) This behavior can only be established when the strain rate is confined to values below, say, $10^{-2}\,s^{-1}$ and temperature is around one-half the melting point on the absolute scale. (*c*) The formation of cavitation reduces the mechanical properties.

Factors affecting the choice of forming methods are overall dimensions, aspect ratio, and which surface of the component is dimensionally critical as the tool side will provide the net shape.

The SPF process can be successfully simulated by using finite element modeling to predict the forming pressure to maintain optimum strain rate; predict thinning, forming time, and areas of cavitation; and effect shape optimization, thereby reducing the number of prototypes of forming trials needed to produce an acceptable component.[366]

The conventional shaping methods used for a superplastic alloy are blow forming and movable tool-forming methods. The part may be shaped into a female die cavity (*female forming*) or over a male tool (*male forming*) or a combination of these. The hard tooling in SPF is necessary only to clamp and retain the forming diaphragm at its periphery and to provide a gastight seal and confine the expansion of diaphragm to the required configuration.[367] Recently, a new SPF method, referred to as *gas mass forming* (GMF), has been developed and successfully used in aerospace applications.

15.11.1 Blow Forming

In blow forming, a fixed die is maintained at the superplastic temperature, and the gas pressure is usually applied on one side of the superplastic sheet material, according to a schedule designed to hold the average strain rate within the required SP range. This causes the sheet to form down into the die approximately at a constant strain rate or within the superplastic strain rate range. The extent of the applied pressure depends on the part geometry and the given strain rate.[306] The forming time may vary from 5 min to several hours, depending upon the extent of forming (or deformation) and the superplastic strain rate for the specific superplastic alloy being formed. The gas used may be air provided SP alloy is resistant to oxidation or contamination at the forming temperature. For reactive alloys such as Ti alloys, an inert atmosphere is always used to act both as the forming medium and as a protective atmosphere over both sides of the alloy sheet.[368]

During forming, the sheet surfaces are usually coated with a boron nitride layer (in case of 7475 Al sheet) to prevent sticking with the die and to provide sufficient lubrication over the die entry radii. Various shapes, such as hemispherical, conical, and rectangular configurations, can be formed in this manner.[306]

Figure 15.44 shows a schematic diagram of the thinning process during blow forming of a sheet in a rectangular die. No significant thickness variation occurs during the initial free bulge forming. The differential thinning effect increases with the increase in depth of the bulge. When the forming blank or sheet contacts the die, further thinning is inhibited due to frictional effects. The deformation then proceeds in those areas where the die contact has not yet taken place. As a result, with progressively more contacts of the sheet with the die (or tool), the deformation is restricted to the diminishing "free" areas of the sheet. Since the corners of the die are filled in last, a thickness profile developed in the completely formed part shows the existence crown or thickest portions in the bottom center as well as at the upper sides, and the least thickness at the edge radius (i.e., the area near the last contact).[303,304] According to the evaluations made by Thomsen et al.,[369] the lower the m, the greater the thinning effect. The values of m can, however, be varied by changing the test temperature and/or the superplastic alloy. Thus the thickness variation in the die-formed parts is not so pronounced for titanium alloys because of their higher strain rate sensitivity. Complex-shaped components in these alloys can be produced by combining basic superplastic forming and diffusion bonding (discussed later).

To improve the material thickness variation, inherent in female die forming, a technique called *reverse billowing* may be employed (Fig. 15.45). In this method, the alloy sheet is initially bulged away from the female die cavity to a height greater than its depth. The pressure is then reversed to blow the sheet into the die cavity to produce the required shape. During this and other forming processes, excessive surface area should not be created in the intermediate stage; otherwise, folds and wrinkles are formed in the finish stage.[303]

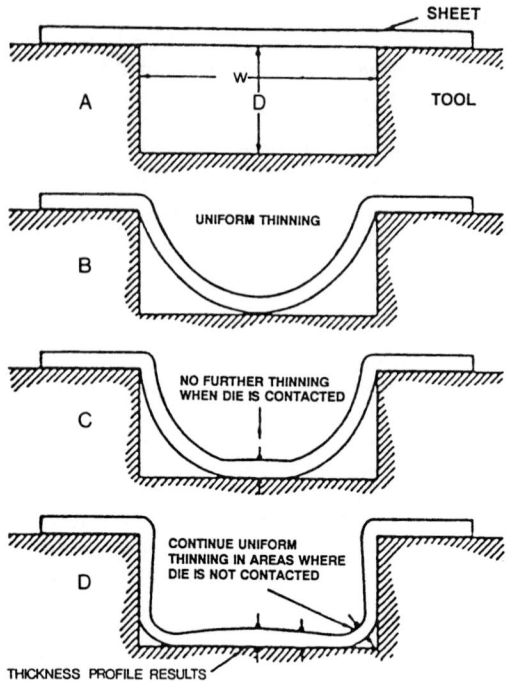

FIGURE 15.44 Schematic diagram of the thinning process during blow forming of a sheet in a rectangular die.[304] (*Reprinted by permission of The Metallurgical Society, Warrendale, Pa.*)

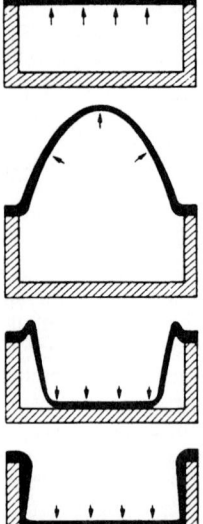

FIGURE 15.45 Reverse billowing sheet forming method.[294,303] (*Reprinted by permission of The Metallurgical Society, Warrendale, Pa.*)

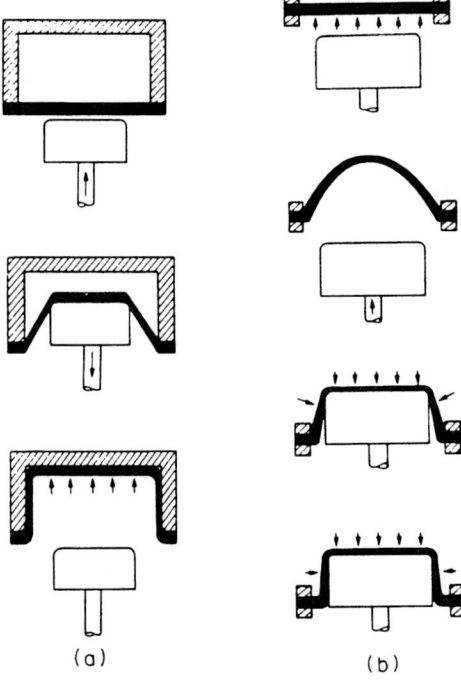

FIGURE 15.46 Movable-tool sheet forming methods: (*a*) Plug-assisted forming method; (*b*) snap-back forming method.[294,303] (*Reprinted by permission of The Metallurgical Society, Warrendale, Pa.*)

(a)   (b)

15.11.2 Movable Tool Forming

For relatively complex and deep shapes, forming techniques involving movable tools, together with gas forming methods, are used which may lead to greater thinning control and larger reduction in forming times when compared to those in the blow forming method. Two such methods used are shown in Fig. 15.46.

In the plug-assisted forming method (Fig. 15.46*a*), a movable tool/die is used initially to push into and stretch the superplastic alloy sheet, which is followed by the withdrawal of the plug. Subsequently, gas pressure is applied on the same side of the sheet as the movable tool to complete forming into the female die. In "snap-back" forming (Fig. 15.46*b*), the sheet is first billowed or stretched by free forming with the application of the gas pressure on the tool side of the sheet. The male tool is then moved into the billowed sheet, and finally the gas pressure is imposed on the opposite side of the sheet in order to conform the superplastic sheet to the shape of the tool.

15.11.3 Gas Mass Forming

In addition to the two primary forming processes, Boeing has developed a proprietary forming process for structural alloy and composite sheet materials to adapt to the needs of the aerospace industry.[370,371] This *gas mass forming* (GMF) process is a new SPF approach that utilizes gas mass flow instead of the forming pressure to control the forming cycle. The advantages of GMF over conventional gas pressure forming are as follows:

1. It is less sensitive to variations in process parameters, such as material property, thickness, and forming temperature irregularities, and has been found to be most appropriate to manufacture complex-shape parts at a higher strain rate (shorter cycle time).
2. Its simplicity, speed, and reliability have led to its use for both single-sheet and multiple-sheet metal forming.
3. It is self-regulating and does not need forming stress or pressure cycle calculations.
4. Its accommodating feature enables anyone to superplastically form parts successfully the first time without any consideration of the SPF parameters, as mentioned in item 1.

15.11.4 Applications of SPF Parts

Superplastically formed parts are used in a wide range of applications for aircraft architectural shapes and machine covers. Examples of aerospace applications of SPF parts include aircraft frames and bulkheads, ribs, beams and compression struts; service and wing access panels, access doors, cowlings, and mounting brackets and supports; ducting, tanks, and vessels; and decorative panels and finishings such as jack cans for the Airbus A310 aircraft.[298] An important aerospace structural application of superplastically formed Ti-6Al-4V is the production hardware. The F-15 aircraft contains more than 70 production hardware parts, which include clips, brackets, and so forth. As for the F-15, there are currently numerous SPF parts in production for the F-18, AV-18, AV-8B, and B-1 aircraft.[372] These SPF parts represent a 40% cost reduction. Examples of aerospace applications of superplastically formed 7475 Al sheets are auxiliary power unit (APU) doors, B-1 fuselage frame bulkheads, B-1 lower deck structure of an avionics compartment,[373] fairing door, B-1 wing flap rib,[374] and so on. Other SPF parts produced from SP Al alloys include small aircraft fuselage fairing female (made from Supral 150); aircraft equipment carrier parts (made from Supral 100 or 2004); electronic closure (formed from 2004 alloy); undercarriage door panel (made from SP 8090 alloy);[327a] weighing machine display covers; railroad luggage racks; Westland helicopter air intake; ejection seats; nose tail; and side fairings of an aviation twin-store carrier. Non-aerospace applications of SPF medium-strength (Supral and 5083) Al parts are small-batch car parts, railway carriage parts, front panels for buildings, and castings for electronic equipment.

15.11.5 Diffusion Bonding

Diffusion bonding[†] (also referred to as *diffusion welding*) depends on diffusion principles for joining two nominally flat and carefully cleaned surfaces. This involves interdiffusion of atoms across the contacting interface when sufficient heat (to forming temperature and for enough time) and pressure (300 to 1000 psi) are applied together in the solid state to obtain a single-piece material (with full closure of cavities at the faying interface, i.e., no abrupt discontinuity in the microstructure

---

[†] *Note*: Diffusion bonding of Ag occurs at 200°C without a need of deformation due to complete dissociation of silver oxide at 190°C (375°F) and its dissolution above this temperature. Likewise Ta, W, Cu, Fe, Zr, and Nb are readily diffusion-bonded due to the high solubility of interstitial contaminants.

and with a minimal deformation); pressure effect causes closure (or collapse) of Kirkendall voids produced during diffusion and/or elimination of remaining voids enclosed within grains. The properties of the joints resemble those of parent materials.[375] Table 15.17 gives the diffusion bonding parameters and bond strengths of metallic and intermetallic joints.[376] It can be applied to join thick and thin sections, similar metals, dissimilar metals, intermetallics, and metal-matrix composites.[376]

15.11.5.1 Diffusion Bonding with Interface Aids. Additional layers of materials in the form of interface foil or coating can be used as a bonding aid to promote diffusivity and solid-state metallic bonding, prevent the creation of brittle intermetallic phases between dissimilar materials, or scavenge impurity elements at the interface, producing in situ clean surfaces. A thin interfacial material (~0.25 mm or 0.0098 in. thick) for such applications includes electroless nickel containing P for high diffusivity in other metallic systems.[376a]

15.11.6   Superplastic Forming/Diffusion Bonding of Ti-6Al-4V Alloy

Currently, a combination of superplastic (gas blow) forming (SPF) and diffusion bonding (DB) (i.e., SPF/DB technology) to Ti-6Al-4V alloy sheets, where diffusion bonding is accomplished before, during, or after the SPF process,[377,378] has been a well-established process in manufacturing a wide variety of complex-shape-three-dimensional parts and sandwich structures for several airframe parts and components of gas-turbine engines, aircraft and airplane structures, including the B-1 bomber, F-15, F-15E, and T-38 fighters,[378] which are cost-effective, approximately 50% cheaper, 25% lighter, and entail a 30% savings of man-hours compared to similar conventionally produced riveted titanium parts. The specific factors contributing to the reduction in cost and weight are the reduced use of tooling; reduced number of detail parts, joints, and fasteners; reduced number of or elimination of holes; reduced machining (or man-hour) requirements due to a high design flexibility, structural efficiency, and increased fly-to-buy ratio of titanium alloys.[379–381] This integral SPF/DB technology is mostly based on the higher deformation temperature (927°C) and bonding pressure of 2 MPa (300 psi) of Ti-6Al-4V alloy.

15.11.7   Basic Shapes of SPF/DB Ti-6Al-4V Structures

There are three established basic forms of SPF/DB structures (Fig. 15.47) that can be fabricated from Ti-6Al-4V alloy sheets.[255,367,382] In each case a titanium alloy sheet, in the initially treated condition, is superplastically formed and diffusion-bonded (for a sufficient period at 927°C under a pressure of ~300 psi)[375] to additional materials during a single processing step. This processing step (i.e., SPF/DB or DB/SPF) is varied, and reinforced sheets, integrally stiffened sheets (where one sheet is extended to form internal cavities), and expanded sandwich structure sheets (where two or more sheets are expanded by gas pressure) are produced.

For the fabrication of reinforced sheet structure (Fig. 15.47), a titanium sheet is superplastically formed into a die cavity containing preplaced titanium pieces where intimate contact and subsequent diffusion bonding between the forming sheets and the titanium details occur. This allows selective thickening of the sheet material for strength, stability, or attachment purposes. This procedure (1) avoids the need to begin with plate, forged, or extended material; (2) eliminates subsequent machining of these materials involving large scrap loss and operating cost; and (3) reduces the assembly costs.

TABLE 15.17 Diffusion Bonding Parameters and Bond Strengths of Metallic and Intermetallic Joints[376]

Material	Technique	Temperature, °C	Pressure, MPa	Time, hr	Joint strength, MPa
Similar metals					
Ti-6Al-4V	Solid phase	930	2–3	1.5	575 (shear)
					990 (tensile)
Ti-6Al-4V	Liquid phase	950	—	1–4	950 (tensile)
(Cu and Ni interlayers)					
IMI 834 (nr α alloy)	Solid phase	990	5	0.5	1040 (tensile)
Al-Zn 7475 T6	Solid phase	516	0.7	4	331 (shear)
Al-Li 8090 T6	Solid phase	560	0.75	4	230 (shear)
Al-Li 8090 T6 (Cu interlayer)	Liquid phase	552	0.75	2	215 (shear)
Intermetallics					
Ti₃Al (Super α₂)	Solid phase	1050	2.1	2	—
Ti-38 mass % Al	Solid phase	1200	15	1.1	225 (tensile)
Metal-matrix composites					
Al-Li 8090/20 wt% SiC particulate T6	Solid phase	560	1.5	4	100 (shear)
Al-Li 8090/20 wt% SiC particulate (8090 interlayer) T6	Solid phase	560	1.5	4	150 (shear)
Al-Li 8090/20 wt% SiC particulate (Cu interlayer) T6	Liquid phase	560	0.75	4	>107 (shear)
Ti-6Al-4V/45 vol% SiC cont. fiber-butt joint	Solid phase	900	10	3	700 (tensile)
Dissimilar materials					
Ti-6Al-4V/304 stainless steel (V and Cu interlayers)	Solid phase	850	10	1	450 (tensile)
Ti-6Al-4V/Stellite HS6 (Ta and Ni interlayers)	Solid phase	900	20	1	406 (shear)
					798 (tensile)
Ti₃Al (Super α₂)/Ni alloy	Solid phase	990	20	1	488 (shear)
HS242 (Ta and Ni interlayer)					

Courtesy of P. G. P. Partridge and A. Wisbey.

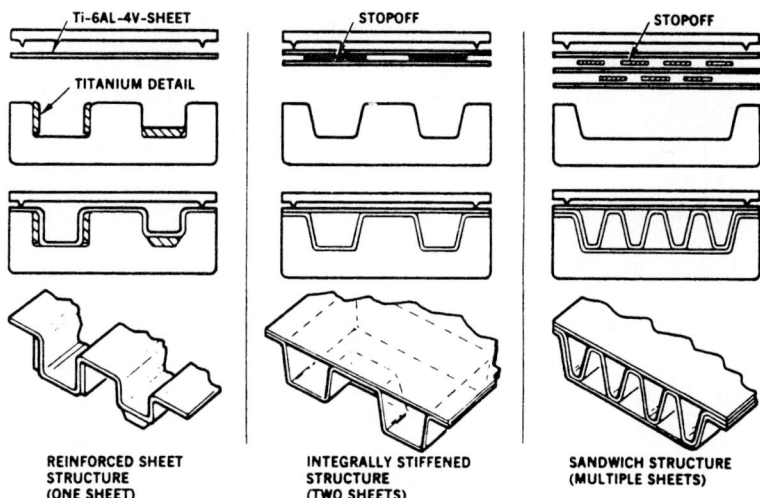

REINFORCED SHEET
STRUCTURE
(ONE SHEET)

INTEGRALLY STIFFENED
STRUCTURE
(TWO SHEETS)

SANDWICH STRUCTURE
(MULTIPLE SHEETS)

FIGURE 15.47 Three basic forms of Ti-6Al-4V structures fabricated by SPF/DB process.[367,380,382] (*Reprinted by permission of The Metallurgical Society, Warrendale, Pa.*)

For the fabrication of integrally stiffened monolithic structure (Fig. 15.47), a pattern for bonding and forming is set up, and a stop-off compound (such as a mixture of yttria and boron nitride in a polymeric binder)[375] is used selectively to one of the sheets of the mating surfaces, to prevent bonding in specific regions corresponding to the die cavity or cavities. Initially, external gas pressure is applied to the sheets at a suitable temperature (927°C) to produce selective diffusion bonding. This is followed by introduction of gas pressure between the sheets to superplastically form the unbonded (unmasked) regions into the tooling cavity or cavities. The boron nitride, used earlier as a stop-off compound, now serves as a lubricant. In some designs, additional details are put in the die cavities for post-form bonding and selective stiffening.[306]

SPF/DB sandwich structure is fabricated by utilizing three or more titanium sheets and selectively bonding and forming with a desired core pattern. Advantages of this structure include[380,382]

1. The outer configuration of the fabricated sandwich part depends on the depth and configuration of the tooling cavity.

2. The core configuration is related with the stop-off pattern, which can be varied and modified to a large extent without a change in tooling.

3. The process inherently renders greater flexibility in edge closure and core designs. Thus a conventional honeycomb sandwich structure or panel can be fabricated with a considerable cost reduction. Moreover, the truss members of the SPF/DB core and the edge member can become an integral unit by using the proper DB pattern. This leads to cost savings (through reduced production steps) and enhanced structural efficiency due to simplified shear ties.

The F-15E Built-up Low Cost Advanced Titanium Structure Program gives a good example of the advantages of SPF and SPF/DB processes. In this program, the

FIGURE 15.48 A variety of structural parts fabricated from Ti-6Al-4V sheet representing three forms of SPF/DB structures.[382] (*Reprinted by permission of The Metallurgical Society, Warrendale, Pa.*)

number of aft fuselage components was reduced from 772 to 46, and 10,000 fasteners were removed. The load factor capability of the aircraft improved, weight reduction was great, and there was an increase of an additional 10 cubic feet of equipment space.[383]

15.11.8 Aerospace Applications of SPF/DB Ti-6Al-4V Structures

Several thousand titanium DB/SPF structural components have been used in service, e.g., as access panels on Airbus aircraft and civil and military applications. Figure 15.48 illustrates a variety of structural parts fabricated from Ti-6Al-4V sheet representing the three forms of SPF/DB structures.[380] Applications of reinforced titanium sheet structures include the nacelle frame from B-1 aircraft (to withstand exposure to high temperature), monolithic helicopter fire-wall, and hemispherical propellant tank. Examples of integrally stiffened SPF/DB sheet structures are hollow structures, B-1 auxiliary power unit (APU) door, F-15 aft nozzle fairing, and toroidal tank.

Important applications of SPF/DB sandwich structures are the B-1 windshield hot-air blast nozzle assembly; the 279 cm (length) × 152 cm (width) × 3.8 cm (height) × 3.8 cm (deep compound curvature) (110 in. × 60 in. × 1.5 in. × 1.5 in.) B-1 engine access door (which is a complex monolithic structure); T-38 main landing gear strut door (which is a flight component of the operating aircraft); F-4 horizontal stabilator inboard trailing edge; F-14A wing glove vane (which is a flight testing aircraft component); 454-kg (1000-lb) wing and fuselage sections of an advanced fighter;

circular sandwich; four-sheet sandwich for Tornado aircraft (for British Aerospace).[255,382] Other applications of the SPF/DB process include hollow fan blades; F-100 aft fan augmenter duct segment; F-100 augmenter seal; and gas-turbine stator vanes.

It is concluded that Ti-alloy SPF/DB structures can replace highly expensive and complex aluminum alloy and stainless steel structures. From a commercial standpoint, there is currently a possibility of more cost-effective designs for an advanced supersonic transport where higher corrosion resistance and strength and longer life are required.[372]

15.11.9 DB/SPF Al Alloys

Previously, aluminum alloy structures have been limited to SPF parts because diffusion bonding of aluminum alloy was not successful[306,375] due to insufficient hot peel strength of Al alloy bonds and the existence of a tenacious Al oxide layer on the surface.

Recently, DB/SPF 7475 Al sheet has been successfully produced with ductile, oxide-free bonds at considerably lower bonding pressure (<1 MPa) and at 510°C and $2 \times 10^{-4} s^{-1}$, in an argon atmosphere.[384] It has been recently reported that diffusion bonding of SP 8090 and 7475 may be achieved by using a copper interlayer, and progress is underway toward concurrent SPF/DB SP Al alloys (Table 15.17). Similarly, the reported optimum parameters for SPF/DB of Al-6Mg (9.7-μm grain size) are as follows: forming temperature 450 to 550°C; forming pressure, 0.5 to 2 MPa; DB temperature, 500°C; DB pressure, 2 MPa; and time for holding temperatures, 0.5 to 2 hr. Here, the oxide film on the welded surface of Al alloys can be readily eliminated by chemical method.[385]

15.11.10 DB/SPF UHC Structures

UHC steels can be diffusion-bonded either to themselves or to other ferrous materials and subsequently superplastically deformed at temperatures below A_1 with the retention of fine-grain structure. These laminated composites have been shown to exhibit superplastic-like behavior.[386] Advantages of these composite structures include the following:

1. Metallurgical bond is developed at the interface between UHC steel and mild-steel laminate, exhibiting tensile ductility above 400% and m values over 0.3.

2. There is the possibility of selective heat treatment; i.e., UHC steels can be transformed to martensite of exceptional hardness without affecting other components.

3. They have much higher impact strengths and much lower DBTT (−140°C) as compared to those obtained for either of the components. The high impact toughness of the composite is attributed to notch blunting of the crack by delamination at the layer interface.[213]

4. Laminations of UHC steel-ferritic stainless steel have been developed (by roll bonding) that exhibit superplasticity at 815°C and $2 \times 10^{-3} s^{-1}$ with $m \sim 0.5$, thereby making coarse-grained nonsuperplastic stainless steels superplastic.[387]

15.11.11 DB/SPF Duplex Stainless Steels

There is a growing interest in the application of DB/SPF in the manufacture of heat exchangers from duplex stainless steels. SP alloys such as 3RE 60 and Zeron 100 have been diffusion-bonded;[388,389] however, the process cycle for diffusion bonding has not offered reliable results.[390] Recently, Avesta 2304 (Fe-23Cr-4Ni-0.1N) steels have been successfully used for DB/SPF manufacturing at temperatures and pressures similar to those employed for Ti-6Al-4V and IMI-550 titanium alloys with maximum elongation of >200% and comparable processing cycle time (<1 hr).[391]

REFERENCES

1. L. M. Kapukite and S. D. Prokoshkin, *Acta Metall.*, vol. 25, 1977, pp. 1471–1483.

2. G. R. Speich et al., in *Proceedings of the International Conference on Phase Transformations in Ferrous Alloys*, eds. A. R. Marder and J. J. Goldstein, TMS-AIME, Warrendale, Pa., 1984, pp. 341–389.

3. L. J. Cuddy, *Met. Trans.*, vol. 15A, 1984, pp. 87–89.

4. M. J. May and D. J. Latham, *Towards Improved Ductility and Toughness*, Climax Molybdenum Development Co., Japan, 1971.

5. R. K. Amin and F. B. Pickering, in *Thermomechanical Processing of Microalloyed Steels*, eds. A. J. DeArdo, G. A. Ratz, and P. J. Wray, *Conference Proceedings*, TMS-AIME, Warrendale, Pa., 1982, pp. 1–31.

6. S. V. Radcliffe and E. B. Kula, in *Fundamentals of Deformation Processing, Sagamore Army Materials Research Conference Proceedings*, eds. W. A. Backofen, J. J. Burke, L. G. Coffin, Jr., N. L. Reed, and V. Weiss, vol. 9, Syracuse University Press, Syracuse, N.Y., 1964, p. 321.

7. V. F. Zackay, *Mater. Sc. Eng.*, vol. 25, 1976, p. 247.

8. G. E. Dieter, in *Encyclopedia of Materials Science and Engineering*, Pergamon Press, Oxford, 1986, pp. 4976–4977.

9. E. B. Kula and M. Azrin, in *Advances in Deformation Processing, Sagamore Army Materials Research Conference Proceedings*, eds. J. J. Burke and V. Weiss, vol. 21, Plenum Press, New York, 1978, p. 245.

10. T. Greyday, in *Strength of Steels and Alloys, Proceedings of the 6th International Conference on Strength of Metals and Alloys*, ed. R. C. Gifkins, Pergamon Press, Oxford, 1983, pp. 1075–1088.

11. *Microalloying '95 Conference Proceedings*, June 11–14, 1995 (Pittsburgh), ISS, Warrendale, Pa., 1995.

12. *International Conference Proceedings on Thermomechanical Processing in Theory, Modelling and Practice [TMP]²*, Sept. 4–6, 1996, Stockholm, The Swedish Society of Materials Technology, 1997.

13. *Thermec '97, International Conference Proceedings on Thermomechanical Processing of Steels and Other Materials*, eds. T. Chandra and T. Sakai, TMS, Warrendale, Pa., 1997.

14. *Proceedings of the International Conference on Microalloying in Steels*, Sept. 7–9, 1998 (Spain), eds. J. M. Rodriguez-Ibabe, I. Gutiérrez, and B. López, *Materials Science Forum*, vols. 284–286, 1998, Trans. Tech. Publications Ltd., Switzerland, 1998.

15. T. Tanaka, T. Emai, K. Kimura, Y. Saito, and T. Hatomura, in *Thermomechanical Processing of Microalloyed Austenite, Conference Proceedings*, TMS-AIME, Warrendale, Pa., 1982, pp. 195–215.

16. L. R. Link and R. W. Armstrong, *Processing Microstructure and Properties of HSLA Steels*, ed. A. J. DeArdo, TMS, Warrendale, Pa., 1988, pp. 203–230.

16a. C. M. Sellars, *Mater. Sc. and Tech.*, vol. 8, 1992, p. 134.

17. *ASM Handbook*, vol. 1, 10th ed., ASM International, Materials Park, Ohio, 1990, pp. 389–423.

18. A. J. DeArdo, *Adv. Mater. Processes*, vol. 133, no. 1, 1988, pp. 71–73.

19. I. Kozasu, in *Constitution and Properties of Steels*, vol. 7 vol. ed. F. B. Pickering, VCH, Weinheim, 1991, pp. 183–217.

20. S. Lee et al., *Met. Trans.*, vol. 26A, 1995, pp. 1095–1100.

21. J. F. Barson, *Adv. Mater. Processes*, March 1996, pp. 27–31.

22. L. Meyer, W. Muschenborn, and U. Schriever, Ref. 12, pp. 93–120.

23. A. M. Sage, *HSLA Steels (Beijing)*, eds. G. Tither and Z. Shouhua, TMS, Warrendale, Pa., 1992, pp. 51–62.

24. A. M. Sage, *Metals and Materials*, vol. 5, no. 10, 1989, pp. 548–588.

25. A. K. Sinha, *Conference Proceedings on Production of Iron and Steel and High Quality Product Mix: Latest Technological Innovations and Processes*, ed. B. R. Nijhawan, ASM International, Materials Park, Ohio, 1992, pp. 195–206.

26. F. B. Pickering, in *Constitution and Properties of Steels*, vol. ed. F. B. Pickering, vol. 7; *Materials Science and Technology*, eds. R. W. Cahn, P. Haasen, and E. J. Kramer, VCH, Weinheim, 1991, pp. 41–94, 335–399.

27. M. Korchynsky, in *International Symposium: Processing Microstructure and Properties of HSLA Steels*, ed. A. J. DeArdo, TMS, Warrendale, Pa., 1988, pp. 169–201.

28. T. T. Hatomura et al., *JISI*, vol. 82, no. 6, June 1996, pp. 532–537.

29. J. K. Solberg, R. Gjengedal, and G. B. Johansen, Ref. 13, pp. 725–731.

30. D. Yu, T. Chandra, and F. J. Barbaro, Ref. 13, pp. 499–505.

31. A. Tamminen, R. Laitinen, L. Myllykoski, D. Porter, and P. Sandvik, Ref. 13, pp. 851–857.

32. A. J. DeArdo, Ref. 14, pp. 15–26.

33. S. Zajac, T. Siwecki, and M. Korchynsky, *International Symposium on Low-Carbon Steels for 90's*, eds. R. Asafahani and G. Tither, TMS, Warrendale, Pa., 1993, pp. 139–149.

34. L. J. Cuddy, in *The Effects of Accelerated Cooling on the Transformation Products That Develop in Deformed and Recrystallized Austenite in Accelerated Cooling of Steels*, ed. P. D. Southwick, TMS-AIME, Warrendale, Pa., 1986.

35. T. Siwecki and G. Engberg, Ref. 12, pp. 121–141.

36. P. D. Hogdson, M. R. Hickson, and R. K. Gibbs, Ref. 14, pp. 63–72.

37. A. Tamminen, Ref. 12, pp. 357–368.

38. K. Nishioka, Y. Hori, S. Ogawa, Y. Mizutani, and A. Kojima, *Nippon Steel Tech. Report, No. 75*, November, 1997, pp. 1–19.

39. E. R. Morgan, T. E. Dancey, and M. Korchynsky, *J. Met.*, vol. 17, 1965, pp. 829–831.

40. K. Tsukada, K. Matsumoto, K. Hirabe, and K. Takeshige, in *Mechanical Working and Steel Processing, Conference XIX*, Iron and Steel Society-AIME, Warrendale, Pa., 1982, pp. 347–371.

41. T. Tanaka, Ref. 11, pp. 165–182.

42. H. Abrams, *Iron and Steel Maker*, vol. 11, no. 12, 1984, pp. 11–17.

43. C. M. Sellars, *HSLA Steels: Metallurgy and Applications Conference Proceedings*, eds. J. H. Gray et al., ASM, Metals Park, Ohio, 1986, pp. 73–81.

44. R. E. Smallman, *Modern Physical Metallurgy*, Butterworths, London, 1985.

45. L. J. Cuddy, in *Encyclopedia of Materials Science and Engineering*, Pergamon Press, Oxford, 1986, pp. 2213–2217.

46. R. W. K. Honeycombe, *HSLA Steels: Metallurgy and Applications Conference Proceedings*, eds. J. H. Gray et al., ASM, Metals Park, Ohio, 1986, pp. 248–260.

47. O. O. Miller, *Trans. ASM*, vol. 43, 1951, pp. 260–289.

48. L. J. Cuddy and J. C. Raley, *Met. Trans.*, vol. 14A, 1983, pp. 1989–1995.

49. T. Gladman, *Proc. R. Soc. London*, vol. 294A, 1966, pp. 298–309.

50. H. Sekine, T. Maruyama, H. Kageyama, and Y. Kawashima, in *Thermomechanical Processing of Microalloyed Austenite*, eds. A. J. DeArdo, G. A. Ratz, and P. J. Wray, *Conference Proceedings*, TMS-AIME, Warrendale, Pa., 1982, pp. 141–161.

51. M. G. Burke, L. J. Cuddy, J. Piller, and M. K. Miller, *Mater. Sc. and Tech.*, vol. 4, no. 2, 1988, p. 113.

52. D. C. Houghton, G. C. Weatherly, and J. D. Embury, in *Thermomechanical Processing of Microalloyed Austenite*, eds. A. J. DeArdo, G. A. Ratz, and P. J. Wray, *Conference Proceedings*, TMS-AIME, Warrendale, Pa., 1982, pp. 267–292.

53. B. Fang, T. Chandra, and D. P. Dunne, *Materials Science Forum*, vol. 13, 1983, pp. 139–146.

54. D. R. DiMicco and A. T. Davenport, in *Thermomechanical Processing of Microalloyed Austenite*, eds. A. J. DeArdo, G. A. Ratz, and P. J. Wray, *Conference Proceedings*, TMS-AIME, Warrendale, Pa., 1982, pp. 59–81.

55. H. Sekine et al., *Proceedings of Thermomechanical Processing of Microalloyed Austenite*, Pittsburgh, TMS-AIME, Warrendale, Pa., 1984, p. 141.

56. S. Yue, C. Roucoules, T. M. Maccagno, and J. J. Jonas, Ref. 11, pp. 355–364.

57. J. J. Jonas, *International Conference Proceedings: High Strength Low Alloy Steels*, eds. D. P. Dunne and T. Chandra, TMS-AIME, Warrendale, Pa., 1985, pp. 80–91.

58. C. M. Sellars, in *Encyclopedia of Materials Science and Engineering*, Pergamon Press, Oxford, 1986, pp. 1270–1274.

59. F. J. Humphrey and M. Hatherly, *Recrystallization and Related Annealing Phenomena*, Pergamon Press, Oxford, 1996.

60. P. W. Roberts and B. Ahlblom, *Acta Metall.*, vol. 26, 1978, pp. 801–813.

60a. *Hot Working Guide*, eds. Y. V. R. K. Prasad and S. Sasidhara, ASM International, Materials Park, Ohio, 1997.

61. E. S. Puchi, J. H. Beynon, and C. M. Sellars, in *International Conference on Thermomechanical Processing of Steels*, Thermec–88, vol. 2, ISI, Tokyo, Japan, p. 572.

62. F. R. Castro-Fernandez et al., *Mater. Sc. and Tech.*, vol. 6, 1990, p. 453.

63. A. Takeuchi and A. S. Argon, *J. Mater. Sc.*, vol. 11, 1976, p. 1547.

64. B. Derby, *Acta Metall.*, vol. 39, 1991, p. 955.

65. D. L. Holt, *J. Appl. Phys.*, vol. 41, 1970, p. 3197.

66. G. H. Edward, M. A. Etheridge, and B. E. Hobbs, *Text. Microstruct.*, vol. 5, 1988, p. 127.

67. H. R. Straker and D. L. Holt, *Acta Metall.* vol. 20, 1972, p. 569.

68. C. R. Barrett, W. D. Nix, and O. D. Sherby, *Trans. ASM*, vol. 59, 1966, p. 3.

69. J. J. Urcola and C. M. Sellars, *Acta Metall.*, vol. 35, 1987, p. 2649.

70. R. W. Cahn, in *Physical Metallurgy*, eds. R. W. Cahn and P. Haasen, 4th ed., Elsevier B. V., 1996, chap. 28, pp. 2399–2501.

71. G. Gottstein and T. Lee, *Mater. Sc. and Eng.*, vol. A114, 1989, p. 21.

72. R. C. Gifkins, *J. Inst. Metals*, vol. 81, 1952, p. 417.

73. J. P. Poirier, *Creep of Crystals*, Cambridge University Press, Cambridge, 1985.

74. D. Ponge and G. Gottstein, *Acta Mater.*, vol. 46, no. 1, 1998, pp. 69–80.

75. T. Sakai, M. G. Akben, and J. J. Jonas, *Acta Metall.*, vol. 31, 1983, pp. 631–642.

76. T. Sakai and J. J. Jonas, *Acta Metall.*, vol. 32, 1984, pp. 189–209.

77. J. J. Jonas, in *Encyclopedia of Materials Science and Engineering*, Pergamon Press, Oxford, 1986, pp. 2218–2223.

78. J. F. Humphreys, *Processing of Metals and Alloys*, vol. ed. R. W. Cahn, vol. 15: *Materials Science and Technology*, eds. R. W. Cahn, P. Haasen, and E. J. Kramer, VCH, Weinheim, 1991, pp. 371–428.

79. T. Maki, K. Akasaka, and I. Timura, in *Thermomechanical Processing of Microalloyed Austenite*, eds. A. J. DeArdo, G. A. Ratz, and P. G. Wray, *Conference Proceedings*, TMS-AIME, Warrendale, Pa., 1982, pp. 217–234.

80. T. Sakai, M. G. Akben, and J. J. Jonas, in *Thermomechanical Processing of Microalloyed Austenite*, eds. A. J. DeArdo, G. A. Ratz, and P. G. Wray, *Conference Proceedings*, TMS-AIME, Warrendale, Pa., 1982, pp. 237–252.

81. M. J. Luton and C. M. Sellars, *Acta Metall.*, vol. 17, 1969, p. 1633.

82. J. P. Sah, G. J. Richardson, and C. M. Sellars, *Met. Sci.*, vol. 8, 1974, p. 325.

83. S. E. Ion, F. J. Humphrey, and S. White, *Acta Metall.*, vol. 30, 1982, p. 1909.

84. K. J. Gardner and R. Grimes, *Metal Sci.*, vol. 13, 1979, p. 216.

85. M. D. Druary and F. J. Humphrey, *Acta Metall.*, vol. 34, 1986, p. 2259.

86. O. Kwon and A. J. DeArdo, *HSLA Steels: Metallurgy and Applications Conference Proceedings*, eds. H. H. Gray et al., ASM, Metals Park, Ohio, 1986, pp. 287–298.

87. T. Tanaka, *Int. Met. Rev.*, vol. 26, 1981, pp. 185–212.

88. R. A. P. Djiac and J. J. Jonas, *Met. Trans.*, vol. 4, 1973, pp. 621–624; *JISI*, April 1972, pp. 255–261.

89. P. D. Hodgson, Ref. 13, pp. 121–131.

90. I. Kozasu, T. Shimizu, and H. Kubota, *Trans. Iron Steel Inst., Jpn.*, vol. 11, 1971, pp. 367–375.

91. T. Gladman, *The Physical Metallurgy of Microalloyed Steels*, The Institute of Materials, London, 1997.

92. C. M. Sellars, in *Hot Working and Forming Processes*, eds. C. M. Sellars and G. J. Davies, The Metals Society, London, 1980, pp. 3–15.

93. W. Roberts, W. Sandberg, T. Siwecki, and T. Werlefors, in *HSLA Steels Technology and Applications*, ed. M. Korchynsky, ASM, Metals Park, Ohio, 1984, pp. 67–84.

93a. R. M. Forbes Jones and L. A. Jackman, *JOM*, January 1999, pp. 27–31.

94. B. Dutta, E. Valdes, and C. M. Sellars, *Acta Metall. Mater.*, vol. 40, no. 4, 1992, pp. 653–662.

95. A. J. DeArdo, Ref. 11, pp. 15–33.

96. A. J. DeArdo, *Microstructural Science*, vol. 24, July 21–24 1996, eds. M. G. Burke, E. A. Clark, and E. J. Palmiere, *Proceedings of the 29th Annual Technical Meeting of the International Metallographic Society*, Pittsburgh, Pa., ASM International, Materials Park, Ohio, 1996, pp. 51–60.

97. J. C. Herman, B. Donnay, and V. Leroy, *ISIJ*, vol. 32, no. 6, 1992, pp. 779–785.

98. C. M. Sellars and J. H. Beynon, *International Conference on High Strength Low Alloy Steels*, Wolongong, Australia, 1984.

99. R. M. Fix, A. J. DeArdo, and Y. Z. Zhang, *HSLA Steels: Metallurgy and Applications Conference Proceedings*, eds. J. H. Gray et al., ASM, Metals Park, Ohio, 1986, pp. 219–227.

100. S. W. Lee, W. Y. Choo, and C. L. Lee, *International Symposium on Low Carbon Steels for 90's*, eds. R. Asafahani and G. Tither, TMS, Warrendale, Pa., 1993, pp. 227–234.

101. M. J. Godden, L. E. Collins, and J. D. Boyd, *International Conference Proceedings: High Strength Low Alloy Steels*, eds. D. P. Dunne and T. Chandra, TMS-AIME, Warrendale, Pa., 1984, pp. 113–118.

102. J. R. Paules, *JOM*, vol. 43, no. 1, 1991, pp. 41–44.

103. S. Zajac, T. Siwecki, B. Hutchinson, and M. Attlegard, *Met. Trans.*, vol. 22A, pp. 2681–2694.

104. M. Korchynsky, Ref. 11, pp. 3–14.

105. J. G. Williams, C. R. Killmore, F. J. Barbaro, A. Meta, and L. Fletcher, Ref. 11, pp. 117–139.

106. Meester, *ISIJ*, vol. 37, 1997, p. 537.

107. R. Taillard, P. Verrier, T. Maurickx, and J. Focht, *Met. and Mats. Trans.*, vol. 26A, 1995, pp. 447–457.

108. J. L. Lee and Y. T. Pan, *Mater. Sc. and Tech.*, vol. 8, 1992, pp. 236–244.

109. P. Manohar and T. Chandra, *ISIJ*, vol. 38, 1998.

110. L. J. Ruiz-Aparicio, M. Hua, C. I. Garcia, and A. J. DeArdo, Ref. 12, pp. 407–414.

111. J. C. Herman, A. De Paepe, and V. Leroy, Ref. 13, pp. 507–514.

112. D. J. Naylor, Ref. 14, pp. 83–94.

113. A. Peters et al., *Steel Research*, vol. 67, no. 10, 1996, pp. 413–418.

114. Y. Jiang, L. Yisa, and I. Tasmura, *HSLA: Processing, Properties and Applications*, eds. G. Tither and Z. Shouhua, TMS, Warrendale, Pa., 1992, pp. 347–352.

115. W. E. Heitman, T. G. Oakwood, and H.-J. Dziemballa, Ref. 11, pp. 395–408.

116. B. Huchenmann, *VDI-Verlag*, Dusseldorf, 1988, pp. 395–409.

117. Y. Ikegami and R. Nemoto, *ISIJ Int.*, vol. 36, no. 7, 1996, pp. 855–861.

118. M. Jahazi, *Metals and Mats.*, vol. 4, no. 4, 1998, pp. 818–822.

119. L. Hyspecka and K. Mazanee, *J. Iron Steel Inst.*, London, vol. 205, 1967, p. 261.

120. M. Tvrdy, L. Hyspecka, and K. Mazanee, *Met. Technol.*, vol. 5, 1978, pp. 73–78.

120a. Y. Tomita, *Met. Trans.*, vol. 22A, 1991, pp. 1093–1102.

121. M. Kh. Shorshorov, *Met. Sci. J.*, vol. 17, 1973, pp. 213–215.

122. G. Kodjaspirov, Ref. 14, pp. 335–342.

123. *Tooling*, August 1978, pp. 42–55.

124. R. W. K. Honeycombe and H. K. D. H. Bhadeshia, *Steels: Microstructure and Properties*, 2d ed., Arnold, London, 1995.

125. A. K. Das and P. K. Gupta, *Met. Mater.*, November/December 1974, p. 429.

126. W. M. Justusson and V. F. Zackay, *Met. Prog.*, vol. 82, no. 6, 1962, pp. 111–114.

127. C. K. Yao, S. Y. Cao, T. Maki, and I. Tamura, *Heat Treatment '83* (Shanghai), The Metals Society, London, pp. 5.80–5.87.

128. O. Johari and G. Thomas, *ASM Trans.*, vol. 58, no. 4, 1965, p. 563.

129. K. J. Pascoe, *An Introduction to Properties of Engineering Materials*, Van Nostrand Reinhold, London, 1972.

130. C. N. Elias and C. S. Da Costa Viana, *Mats. Sc. and Tech.*, vol. 18, 1992, pp. 785–789.

131. S. Iosgawa, H. Yoshida, Y. Hosoi, and T. Tozawa, *Jn. Mats. Process. Tech.*, vol. 74, 1998, pp. 298–306.

132. L. Sanderson, *Tooling*, August 1968, pp. 43–46.

133. M. J. Roberts and W. Jolley, *Met. Trans.*, vol. 1, 1970, pp. 1389–1398.

134. J. J. Irani, *J. Iron Steel Inst., London*, vol. 206, 1968, p. 363.

135. J. J. Irani and D. J. Latham, *Low Alloy Steels*, Special Report no. 114, ISI, London, 1968, p. 55.

136. L. F. Porter and P. E. Repas, *J. Met*, vol. 34, no. 4, 1982, pp. 14–21.

137. J. A. DiCello and D. Aichbhaumik, in *Thermomechanical Processing of Microalloyed Austenite*, eds. A. J. DeArdo, G. A. Ratz, and P. J. Wray, *Conference Proceedings*, TMS-AIME, Warrendale, Pa., 1982, pp. 529–554.

138. S. S. Hansen, J. B. Vander Sande, and M. Cohen, *Met. Trans.*, vol. 11A, 1980, p. 387.

139. G. Tither, *HSLA Steels: Processing, Properties and Applications*, eds. G. Tither and Z. Shouhua, TMS, Warrendale, Pa., 1992, pp. 61–80.

140. R. F. Foley, R. K. Weiss, W. W. Thompson, and G. Krauss, *International Symposium on Low-Carbon Steels for 90's*, eds. R. Asafahani and G. Tither, TMS, Warrendale, Pa., 1993, pp. 243–256.

141. L. F. Porter, in *Encyclopedia of Materials Science and Engineering*, Pergamon Press, Oxford, 1986, pp. 2157–2162.

142. *Carbon Alloy Steels: ASM Specialty Handbook*, ASM International, Materials Park, Ohio, 1996, pp. 3–40; *ASM Handbook*, vol. 1, 10th ed., ASM International, Materials Park, Ohio, 1990, pp. 389–423.

143. A. P. Coldren, V. Biss, and T. G. Oakwood, in *Thermomechanical Processing of Microalloyed Austenite*, eds. A. J. DeArdo, G. A. Ratz, and P. J. Wray, TMS-AIME, Warrendale, Pa., 1982, pp. 591–611.

144. H. Kejian and T. N. Baker, *Mats. Sc. and Engr.*, vol. A169, 1993, pp. 53–65.

145. J. H. Bucher, J. D. Grozier, and J. F. Enrietto, *Fracture*, vol. 6, 1968, p. 244.

146. Z. Nishiyama, in *Martensite Transformation*, eds. M. Fine, M. Meshii, and C. M. Wayman, Academic Press, New York, 1978.

147. C. M. Wayman, in *Solid → Solid Phase Transformations*, eds. H. I. Aaronson, D. E. Laughlin, R. F. Sekerka, and C. M. Wayman, TMS-AIME, Warrendale, Pa., 1982, pp. 1119–1144.

148. I. Tamara, T. Maki, and H. Hoto, *Trans. Iron Steel Inst., Jpn.*, vol. 10, 1970, p. 1163.

149. G. B. Olson and M. Cohen, in *Proceedings of the US/Japan Seminar on Mechanical Behavior of Metals and Alloys Associated with Displacive Phase Transformations*, Troy, N.Y., 1979, p. 7.

150. G. B. Olson, R. Chait, M. A. Azrin, and R. A. Gagne, *Met. Trans.*, vol. 11A, 1980, pp. 1069–1071.

151. B. R. Banerjee, *Application of Fracture Toughness Parameters to Structural Metals*, eds. J. M. Capenos and J. J. Hauser, Gordon and Breach, New York, 1966, pp. 373–406.

152. V. F. Zackay, E. R. Parker, D. Fahr, and R. Bush, *Trans. ASM*, vol. 60, 1967, pp. 252–259.

153. Y. Sakuma, O. Matsumura, and H. Takechi, *Met. Trans.*, vol. 22A, 1991, pp. 489–498.

153a. T. Kobayashi and H. Yamamoto, *Met. Trans.*, vol. 19A, 1978, pp. 319–327.

153b. M. Niinomi, T. Kobayashi, I. Inagaki, and A. W. Thompson, *Met. Trans.*, vol. 21A, 1990, pp. 1733–1744.

153c. T. Kobayashi, M. Niinomi, Y. Koide, and K. Matsunuma, *Trans. Jpn. Inst. Met.*, vol. 27, 1986, pp. 775–783.

154. W. Bleck, K. Hulka, and K. Papamentellos, Ref. 14, pp. 327–334.

155. A. H. Nalkagawa and G. Thomas, *Met. Trans.*, vol. 16A, 1985, pp. 831–840.

156. O. Grassel, G. Frommeyer, C. Derder, and H. Hofmann, *J. Phys. IV France*, vol. 7, 1997, pp. C5-383–388.

157. A. Z. Hanzaki, P. D. Hodgson, and S. Yue, *Met. and Mats. Trans.*, vol. 28A, 1997, pp. 2405–2414.

158. K. Eberle, Ph. Harlet, P. Cantinieaux, and M. V. Populiere, *I&SM*, February 1999, pp. 23–27; *40th MWSP Conference Proceedings*, ISS, Warrendale, Pa., 1998, pp. 251–258.

159. O. Kawano, J.-I. Wakita, K. Esaka, and H. Abe, *Tetsu-To-Haganae*, vol. 82, no. 3, 1996, pp. 232–237.

160. S. Yue, A. DiChiro, and A. Z. Hanzaki, *JOM*, September 1997, pp. 59–61.

161. P. Jacques, X. Cornet, Ph. Harlet, J. Ladriere, and F. Delanny, *Met. and Mats. Trans.*, vol. 29A, 1998, pp. 2383–2393.

162. D. Q. Bai, A. DiChiro, and S. Yue, Ref. 14, pp. 253–260.

163. A. Z. Hanzaki, P. D. Hodgson, and S. Yue, *34th MWSP Conference Proceedings*, ISS-AIME, vol. 30, 1993, pp. 507–514.

164. G. Reisner, E. A. Werner, P. Kerschbaummayr, I. Papst, and F. D. Fischer, *JOM*, September 1997, pp. 62–65, 83.

165. A. Z. Hanzaki and S. Yue, *ISIJ Int.*, vol. 37, no. 7, 1997, pp. 583–589.

166. A. Basuki and E. Aernoudt, *Jn. Mats. Proc. Tech*, vol. 89–90, 1999, pp. 37–43.

167. E. J. Koh, S. K. Lee, S. Lee, S.-H. Park, K. S. Shin, and N. J. Kim, Ref. 13, pp. 795–801.

168. K.-I. Sugimoto, M. Kobayashi, and S.-I. Hashimoto, *Met. Trans.*, vol. 23A, 1992, pp. 3085–3091.

169. O. Matsumura et al., *ISIJ*, vol. 32, no. 12, 1992, pp. 1110–1116.

170. M. Bouet, J. Root, E. Es-Sadiqi, and S. Yue, Ref. 14, pp. 319–326.

171. J. M. Diani, M. Berveiller, and H. Sabar, *Jn. De Physique IV*, vol. 6, January 1996, pp. C1-419–427.

172. F. D. Fischer, E. R. Oberaigner, K. Tanaka, and F. Nishimura, *Int. Jn. of Solids and Structures*, vol. 35, no. 18, June 1998, pp. 2209–2227.

173. T. Gladman, *Met. Technol.*, vol. 10, 1983, pp. 274–281.

174. P. Deb and M. C. Chaturvedi, *Met. Technol.*, vol. 10, 1983, pp. 167–172.

175. R. C. Davies, *Met. Trans.*, vol. 10A, 1979, pp. 113–118.

176. W. Xiaojing, Z. Welong, and H. Baoshan, *HSLA Steels: Processing, Properties, and Applications*, eds. G. Tither and Z. Shouhua, TMS, Warrendale, Pa., 1992, pp. 479–482.

177. S. Xiahong, Z. Xiaoping, C. Xiafei, Y. Weixun, W. Quanshan, Y. Hong, and B. Wei, *HSLA Steels: Processing, Properties, and Applications*, eds. G. Tither and Z. Shouhua, TMS, Warrendale, Pa., 1992, pp. 483–488.

178. Li Chengji, Z. Fugui, C. Fengyu, and Z. Shouhua, *HSLA Steels: Processing, Properties, and Applications*, eds. G. Tither and Z. Shouhua, TMS, Warrendale, Pa., 1992, pp. 469–478.

179. G. R. Speich, *ASM Handbook*, vol. 1, 10th ed., ASM International, Materials Park, Ohio, 1990, pp. 424–429.

180. A. K. Bag, K. K. Ray, and E. S. Dwarkadasa, *Met. and Mats. Trans.*, vol. 30A, 1999, pp. 1193–1202.

181. J. J. Yi, K. J. Yu, I. S. Kim, and S. J. Kim, *Met. Trans.*, vol. 14A, 1983, pp. 1497–1504.

182. J. J. Yi, I. S. Kim, and H. S. Choi, *Met. Trans.*, vol. 16A, 1985, pp. 1237–1245.

183. G. R. Speich, V. A. Demarest, and R. L. Miller, *Met. Trans.*, vol. 12A, 1981, pp. 1419–1428.

184. M. M. Souza, J. R. C. Guimaraes, and K. K. Chawla, *Met. Trans.*, vol. 13A, 1982, p. 575.

185. X.-L. Cai, A. J. G. Reed, and W. S. Owen, *Met. Trans.*, vol. 16A, 1985, pp. 543–557.

186. W. Leslie, *Physical Metallurgy of Steels*, McGraw-Hill, New York, 1981.

187. D. Z. Yang, E. L. Brown, D. K. Matlock, and G. Krauss, *Met. Trans.*, vol. 16A, 1985, pp. 1523–1526.

188. R. Soto, A. Charai, C. Issartel, X. Bano, and G. Rigaut, Ref. 13, pp. 899–906.

189. P. Deb and M. C. Chaturvedi, *Mater. Sc. Eng.*, vol. 78, 1986, pp. L7–L13.

190. F. H. Samuel, *Mater. Sci. Eng.*, vol. 75, 1985, pp. 51–66.

191. R. Pradhan, in *Technology of Continuously Annealed Cold-Rolled Sheet Steel*, ed. R. Pradhan, *Conference Proceedings*, TMS-AIME, Warrendale, Pa., 1985, pp. 297–317.

192. P. H. Chang, *Met. Trans.*, vol. 15A, 1984, pp. 671–678.

193. I. Gupta and P.-H. Chang, in *Technology of Continuously Annealed Cold-Rolled Sheet Steel*, ed. R. Pradhan, *Conference Proceedings*, TMS-AIME, Warrendale, Pa., 1985, pp. 263–276.

194. I. Gupta, S. P. Bhat, and R. S. Cline, *Paint Bake Strengthening in Water Quenched Continuously Annealed Steels*, SAE Paper 840007, 1984.

195. G. Thomas, in *Mechanical Properties and Phase Transformations in Engineering Materials*, eds. S. D. Antolovich, R. O. Ritchie, and W. W. Gerberich, TMS-AIME, Warrendale, Pa., 1986, pp. 147–161.

196. P. H. Chang, *Scr. Met.*, vol. 18, 1984, pp. 1245–1250.

197. N.-R. V. Bangari and A. K. Sachdev, *Met. Trans.*, vol. 13A, 1982, p. 1899.

198. B. V. N. Rao and M. Rashid, *Metallography*, vol. 16, 1983, p. 19.

199. S. Lian and L. Hua, *Mats. Sc. and Tech.*, vol. 11, 1995, pp. 499–507.

200. H. Huang, O. Matsumura, and T. Furukawa, *Mats. Sc. and Tech.*, vol. 10, 1994, pp. 621–626.

201. S. Sengal, N. C. Goel, and K. Tangri, *Met. Trans.*, vol. 16A, 1985, pp. 2023–2029.

202. A. Korzekwa, D. K. Matlock, and G. Krauss, *Met. Trans.*, vol. 15A, 1984, pp. 1221–1228.

203. W. Chuanya, *Heat Treatment '84*, The Metals Society, London, pp. 8.1–8.6.

204. A. R. Marder, *Met. Trans.*, vol. 13A, 1982, pp. 85–92.

205. J. L. Tzou and R. O. Ritchie, *Scr. Met.*, vol. 19, 1985, pp. 751–755.

206. D. T. Llewellyn and R. C. Hudd, *Steels: Metallurgy and Applications*, 3d ed., Butterworth-Heinemann, Oxford, 1998.

207. C. A. N. Lanzilotto and F. B. Pickering, *Met. Science*, vol. 16, 1982, p. 371.

208. E. Smith, *Int. J. Fract. Mech.*, vol. 4, 1968, p. 131.

209. J. T. Barnby, *Acta Metall.*, vol. 15, 1967, p. 903.

210. *Inland Ultra High Strength Steels Selection Guide*, Ispat Inland Inc., East Chicago, Ind.

211. O. D. Sherby, B. Walser, C. M. Young, and E. M. Cady, *Scr. Met.*, vol. 9, 1975, pp. 569–574.

212. E. S. Kayali, H. Sunada, T. Oyama, J. Wadsworth, and O. D. Sherby, *J. Mater. Sci.*, vol. 14, 1979, pp. 2688–2692.

213. D. W. Kum, T. Oyama, O. D. Sherby, O. A. Ruano, and J. Wadsworth, *Superplastic Forming, Conference Proceedings, 1984*, ASM, Metals Park, Ohio, 1985, pp. 32–42.

214. O. D. Sherby and J. Wadsworth, in *Encyclopedia of Materials Science and Engineering*, suppl. vol. 1, ed. R. W. Cahn, Pergamon, Oxford, 1988, pp. 541–545.

215. D. R. Lesuer, C. K. Syn, J. D. Whittenberger, and O. D. Sherby, *Met. and Mats. Trans.*, vol. 30A, 1999, pp. 1559–1568; C. K. Syn, D. R. Lesuer, and O. D. Sherby, in *Thermo-mechanical Processing and Mechanical Properties of Hypereutectoid Steels and Cast Irons*, eds. D. R. Lesuer, C. K. Syn, and O. D. Sherby, TMS, Warrendale, Pa., 1997, pp. 117–125.

216. E. M. Taleef, C. K. Syn, D. R. Lesuer, and O. D. Sherby, in *Thermomechanical Processing and Mechanical Properties of Hypereutectoid Steels and Cast Irons*, eds. D. R. Lesuer, C. K. Syn, and O. D. Sherby, TMS, Warrendale, Pa., 1997, 127–142; *Met. and Mats. Trans.*, vol. 27A, 1996, pp. 111–118.

217. D. E. Kum, in *Thermomechanical Processing and Mechanical Properties of Hypereutectoid Steels and Cast Irons*, eds. D. R. Lesuer, C. K. Syn, and O. D. Sherby, TMS, Warrendale, Pa., 1997, pp. 41–54.

218. D. R. Lesuer, C. K. Syn, and O. D. Sherby, in *Thermomechanical Processing and Mechanical Properties of Hypereutectoid Steels and Cast Irons*, eds. D. R. Lesuer, C. K. Syn, and O. D. Sherby, TMS, Warrendale, Pa., 1997, pp. 175–188.

219. C. K. Syn, D. R. Lesuer, and O. D. Sherby, *Met. and Mats. Trans.*, vol. 28A, 1997, pp. 1213–1218.

220. C. E. Shamblen, R. E. Allen, and F. E. Walker, *Met. Trans.*, vol. 6, 1975, p. 2073.

221. J. M. Oblak and W. Owczarski, *Met. Trans.*, vol. 3A, 1972, pp. 617–626.

222. P. R. Bhowal and N. M. Bhathena, *Met. Trans.*, vol. 22A, 1991, pp. 1999–2008.

223. *Metallurgy and Applications of Superalloy 718*, ed. Edward A. Loria, TMS-AIME, Warrendale, Pa., 1989.

224. A. E. Marsh, *Metallurgia*, vol. 49, 1982, pp. 10–20.

225. R. E. Bailey, *Report No. SP-69-9*, Allegheny Ludlum Steel Research Center, Brackenridge, Pa., 1969.

226. N. Srinivasan and Y. V. R. K. Prasad, *Met. and Mats. Trans.*, vol. 25, 1994, pp. 2275–2284.

227. J. Hirsch, Ref. 12, pp. 78–92.

228. D. W. Chung and M. C. Chaturvedi, *Mater. Sc. Eng.*, vol. 48, 1981, pp. 27–34.

229. Y. Murakami, *Structure and Properties of Nonferrous Alloys*, vol. 8, vol. ed. K. H. Matucha, VCH, Weinheim, 1996, pp. 213–276.

230. N. E. Paton and A. W. Sommer, in *The Microstructure and Design of Alloys, Proceedings, 3d International Conference on Strength of Metals and Alloys*, vol. 1, Institute of Metals, Cambridge, 1973, pp. 101–108.

231. N. E. Paton and A. K. Ghosh, in *Metallurgical Treatise*, eds. J. K. Tien and J. F. Elliott, TMS-AIME, Warrendale, Pa., 1981, pp. 361–384.

232. T. C. Tsai, J. C. Chang, and T. H. Chuang, *Met. and Mats. Trans.*, vol. 28A, 1997, pp. 2113–2121.

233. H. A. Hall, *Corrosion*, vol. 23, 1967, p. 1673.

234. B. M. Cina, U.S. Patent 38567584, Dec. 24, 1974.

235. J. H. Park and A. J. Ardell, *Met. Trans.*, vol. 15A, 1984, pp. 1531–1543.

236. T. C. Tsai and T. H. Chuang, *Met. and Mats. Trans.*, vol. 27A, 1996, pp. 2617–2627.

237. E. DiRusso, M. Conserva, F. Gatto, and H. Markus, *Met. Trans.*, vol. 4, 1973, pp. 1133–1144.

238. J. Waldman, H. Sulinski, and H. Markus, *Met. Trans.*, vol. 5A, 1974, pp. 573–584.

239. H. J. McQueen, in *Thermomechanical Processing of Al Alloys*, ed. J. G. Morris, *Symposium Proceedings*, TMS-AIME, Warrendale, Pa., 1979, pp. 1–24.

240. J. A. Wert, N. Paton, C. H. Hamilton, and M. W. Mahoney, *Met. Trans.*, vol. 12A, 1981, pp. 1267–1276.

241. J. A. Wert, in *Superplastic Forming of Structural Alloys*, eds. N. E. Paton and C. H. Hamilton, *Conference Proceedings*, TMS-AIME, Warrendale, Pa., 1982, pp. 69–83.

242. M. K. Rao and A. K. Mukherjee, *Mater. Sc. Eng.*, vol. 80, 1986, pp. 181–193.

243. C. H. Hamilton, C. C. Bampton, and N. E. Paton, in *Superplastic Forming of Structural Alloys*, eds. N. E. Paton and C. H. Hamilton, *Conference Proceedings*, TMS-AIME, Warrendale, Pa., 1982, pp. 173–189.

244. J. Wadsworth, C. A. Henshall, A. R. Pelton, and B. Ward, *Jn. Mater. Sci. Lett.*, vol. 4, 1985, pp. 674–678.

245. J. Wadsworth, A. R. Pelton, and R. E. Lewis, *Met. Trans.*, vol. 16A, 1985, pp. 2319–2323.

246. A. K. Ghosh and L. K. Gandhi, in *Strength of Metals and Alloys, Proceedings of the 7th International Conference on the Strength of Metals and Alloys, 1985*, vol. 3, eds. H. J.

McQueen, J. P. Bailon, J. I. Dickson, J. J. Jonas, and M. G. Akben, Pergamon Press, Oxford, 1986, pp. 2065–2072.

246a. D. H. Shin, K. S. Kim, D. W. Kum, and S. W. Nam, *Met. Trans.*, vol. 21A, 1990, pp. 2729–2737.

246b. O. D. Sherby and J. Wadsworth, *Mater. Sc. Forum*, vols. 233–234, 1997, pp. 125–138.

247. X. Jiang, Q. Wu, J. Cui, and L. Ma, *Met. Trans.*, vol. 24A, 1993, pp. 2596–2598.

248. J.-W. Yeh and Y.-S. Liao, Ref. 13, pp. 1151–1157.

249. J. Wadsworth, *Superplastic Forming*, ed. S. P. Agrawal, *Conference Proceedings*, ASM, Metals Park, Ohio, 1985, pp. 43–57.

250. B. M. Watts, M. J. Stowell, B. L. Baikie, and D. G. E. Owen, *Met. Sci.*, vol. 10, June 1976, pp. 198–206.

251. B. Verlinden, P. Ratchev, P. De Smet, and P. Van Houtte, Ref. 12, pp. 323–328.

252. K. Kannan, C. H. Johnson, and C. H. Hamilton, *Met. and Mats. Trans.*, vol. 29A, 1998, pp. 1211–1220.

253. M. Furukawa, P. B. Berbon, Z. Horita, M. Nemoto, N. K. Tsenev, R. Z. Valiev, and T. G. Langdon, *Met. and Mats. Trans.*, vol. 29A, 1998, pp. 169–177.

254. R. Sawle, in *Superplastic Forming of Structural Alloys*, eds. N. E. Paton and C. H. Hamilton, *Conference Proceedings*, TMS-AIME, Warrendale, Pa., 1982, pp. 307–317.

255. J. R. Williamson, in *Superplastic Forming of Structural Alloys*, eds. N. E. Paton and C. H. Hamilton, *Conference Proceedings*, TMS-AIME, Warrendale, Pa., 1982, pp. 291–306.

256. Y. Ito and A. Hasegawa, in *Titanium '80, Science and Technology, Proceedings of the 4th International Conference on Titanium, Kyoto, 1980*, vol. 2, eds. H. Kimura and O. Izumi, TMS-AIME, Warrendale, Pa., 1980, pp. 983–992.

257. Y. V. R. K. Prasad, T. Seshacharyulu, S. C. Medeiros, W. G. Frazier, J. T. Morgan, and J. C. Malas, *Adv. Mater. & Processes*, June 2000, pp. 85–89.

258. N. Furushiro, H. Ishibashi, S. Shimoyama, and S. Hori, in *Titanium '80, Science and Technology, Proceedings of the 4th International Conference on Titanium, Kyoto, 1980*, vol. 2, eds. H. Kimura and O. Izumi, TMS-AIME, Warrendale, Pa., 1980, pp. 993–1000.

259. S. M. L. Sastry, P. S. Rao, and K. K. Sankaran, in *Titanium '80, Science and Technology, Proceedings of the 4th International Conference on Titanium, Kyoto, 1980*, vol. 2, eds. H. Kimura and O. Izumi, TMS-AIME, Warrendale, Pa., 1980, pp. 873–886.

260. C. R. Brooks, *Heat Treatment Structure and Properties of Non-ferrous Alloys*, ASM, Metals Park, Ohio, 1982.

261. M. J. Blackburn, W. H. Smyrel, and J. A. Feeney, in *Stress Corrosion Cracking in High Strength Steels and in Ti and Al Alloys*, ed. B. F. Brown, Naval Research Laboratory, Washington, D.C., 1972.

262. M. Peters and G. Luetjering, in *Titanium '80, Science and Technology, Proceedings of the 4th International Conference on Titanium, Kyoto, 1980*, vol. 2, eds. H. Kimura and O. Izumi, TMS-AIME, Warrendale, Pa., 1980, pp. 925–935.

263. *Metals Handbook*, vol. 9, 9th ed., ASM, Metals Park, Ohio, 1985.

264. M. J. Blackburn, J. A. Feeney, and T. R. Beck, in *Advances in Corrosion Science and Technology*, vol. 3, eds. M. G. Fontana and R. W. Staehle, Plenum Press, New York, 1973, p. 100.

265. S. L. Semiatin, J. F. Thomas, Jr., and J. Dadras, *Met. Trans.*, vol. 14A, 1983, pp. 2363–2374.

266. I. Weiss, G. E. Welsch, F. H. Froes, and D. Eylon, in *Strength of Metals and Alloys, Proceedings of the 7th International Conference on the Strength of Metals and Alloys, 1985*, vol. 2, eds. H. J. McQueen, J. P. Bailon, J. I. Dickson, J. J. Jonas, and M. G. Akben, Pergamon Press, Oxford, 1986, pp. 1073–1078.

267. H. Fujii, *Mats. Sc. Eng.*, vol. 243, 1998, pp. 103–108.

268. P. Pototzky, H. J. Maier, and H.-J. Christ, *Met. and Mats. Trans.*, vol. 29A, 1998, pp. 2995–3004.

268a. I. Weiss and S. L. Semiatin, in *Non-Aerospace Applications of Titanium*, eds. F. H. Froes, P. G. Allen, and M. Nitomi, TMS, Warrendale, Pa., 1998, pp. 147–161.

269. W. A. Monteiro, F. Cosandey, and P. Bandaru, Ref. 13, pp. 1769–1775.

270. R. Sundberg and M. Sundberg, Ref. 12, pp. 268–276.

271. R. Sundberg, private communication, 2000.

272. K. E. Amin, P. C. Becker, and R. A. Pisciteli, *Mater. Sci. and Eng.*, vol. 49, 1981, pp. 173–183.

273. *Standard Handbook Wrought Products Alloy Data 2*, 8th ed., Copper Development Association, 1985.

274. J. T. Plews, *Met. Trans.*, vol. 6, 1975, pp. 537–544.

275. C. R. Scorey, S. Chin, M. J. White, and R. J. Livak, *JOM*, vol. 36, November 1984, pp. 52–54.

276. O. D. Sherby and J. Wadsworth, *Mater. Sci. Technol.*, vol. 1, 1985, pp. 925–936; O. D. Sherby and O. A. Ruano, in *Superplastic Forming of Structural Alloys*, eds. N. E. Paton and C. H. Hamilton, *Conference Proceedings*, TMS-AIME, Warrendale, Pa., 1982, pp. 241–254.

277. C. E. Pearson, *J. Inst. Met.*, vol. 54, 1934, p. 111.

278. E. E. Underwood, *J. Met.*, vol. 14, 1962, p. 914.

279. W. A. Backofen, I. R. Turner, and D. H. Avery, *Trans. ASM*, vol. 57, 1964, pp. 980–990.

280. T. H. Thomsen, D. L. Holt, and W. A. Backofen, *Met. Eng. Q.*, vol. 2, 1970, pp. 1–12.

281. D. Lee, *Acta Metall.*, vol. 17, 1969, pp. 1057–1069.

282. H. Ishikawa, F. A. Mohamed, and T. G. Langdon, *Phil. Mag.*, vol. 32, 1975, pp. 1269–1271.

283. K. Higashi, T. Ohnishi, and Y. Nakatani, *Scripta Metall.*, vol. 19, 1985, pp. 821–823.

284. H. Honda, R. Matubara, N. Ashie, K. Nakamura, and S. Miura, *Proceedings of the 1999 International Symposium and Exhibition on Shape Memory Materials (SMM '99)*, May 19–21, 1999, Trans. Tech. Publ., Switzerland, pp. 477–480.

285. Y. Ma and T. G. Langdon, *Met. and Mats. Trans.*, vol. 25A, 1994, pp. 2309–2311.

286. N. J. Grant, *Proc. 5th Int. Conference: Rapidly Quenched Metals*, North-Holland, Amsterdam, 1985.

287. R. W. Cahn and P. M. Hazzledine, *The Encyclopedia of Advanced Materials*, vol. 4, Pergamon Press, Oxford, 1994, pp. 2717–2722.

288. *The Materials World*, Oct. 1999, pp. 67–69.

289. J. W. Edington, K. N. Melton, and C. P. Cutler, *Progr. Mater. Sci.*, vol. 21, no. 2, 1976, pp. 61–265.

290. T. G. Nieh and J. Wadsworth, *Int. Materials Reviews*, vol. 44, no. 2, 1999, pp. 59–75.

291. A. K. Mukherjee, in *Plastic Deformation and Fracture of Materials*, vol. 6, vol. ed. H. Mugharabi, VCH, Weinheim, 1993, pp. 407–460.

292. C. H. Hamilton, in *Superplastic Forming*, ed. S. P. Agrawal, *Conference Proceedings, 1984*, ASM, Metals Park, Ohio, 1985, pp. 13–22.

293. O. D. Sherby and J. Wadsworth, *Encyclopedia of Materials Science and Engineering*, suppl. vol. 1, Pergamon Press, Oxford, 1988, pp. 519–522.

294. C. H. Hamilton, in *Encyclopedia of Materials Science and Engineering*, Pergamon Press, Oxford, 1986, pp. 4783–4787; in *Strength of Metals and Alloys, Proceedings of the 7th International Conference on Strength of Metals and Alloys, 1985*, eds. H. J. McQueen, J. P. Bailon, J. I. Dickson, J. J. Jonas, and M. G. Akben, Pergamon Press, Oxford, 1986, pp. 1831–1857.

295. A. K. Ghosh, *Superplastic Forming*, ed. S. P. Agrawal, *Conference Proceedings, 1984*, ASM, Metals Park, Ohio, 1985, pp. 23–31.

296. C. M. Bedell, P. Zwigl, and D. C. Dunand, *Superplasticity and Superplastic Forming*, eds. A. K. Ghosh and T. R. Bieler, TMS, Warrendale, Pa., 1995, pp. 125–133.

297. M. U. Wu, J. Wadsworth, and O. D. Sherby, *Met. Trans.*, vol. 18A, 1987, pp. 451–462.

298. D. Stephen, in *High Performance Materials in Aerospace*, ed. H. M. Flower, Chapman & Hall, London, 1995, pp. 246–282.

299. O. D. Sherby and J. Wadsworth, *Progress Mats. Sc.*, vol. 33, 1989, pp. 169–221.

299a. C. Schuh and D. C. Dunand, *Acta Mater.*, vol. 46, 1998, pp. 5663–5675.

300. D. C. Dunand, Ref. 13, pp. 1821–1830.

301. O. D. Sherby, R. D. Caligiuri, E. S. Kayali, and R. A. White, *Advances in Metal Processing*, eds. J. J. Burke and R. Mehrabian, Plenum Press, New York, 1981, pp. 133–171.

302. T. F. Montheillet and J. J. Jonas, *Met. and Mats. Trans.*, vol. 27A, 1996, pp. 3346–3348.

302a. T. G. Nieh, L. M. Hsiung, J. Wadsworth, and R. Kaibyshev, *Acta Mater.*, vol. 46, no. 8, 1998, pp. 2789–2800.

303. D. B. Laycock, in *Superplastic Forming of Structural Alloys*, eds. N. E. Paton and C. H. Hamilton, *Conference Proceedings*, TMS-AIME, Warrendale, Pa., 1982, pp. 257–272.

304. A. K. Ghosh and C. H. Hamilton, *Met. Trans.*, vol. 13A, 1982, pp. 733–734.

305. C. H. Hamilton and A. K. Ghosh, *Met. Trans.*, vol. 11A, 1980, pp. 1494–1496.

306. N. W. Mahoney, C. H. Hamilton, and A. K. Ghosh, *Met. Trans.*, vol. 14A, 1983, pp. 1593–1598.

307. A. Uoya, T. Shibata, K. Higashi, A. Inoue, and T. Masumoto, *J. Mater. Sc.*, vol. 11, 1996, p. 2731.

308. T. Imai, S.-W. Lim, D. Jiang, and N. Nishida, *Scripta Metall. Mater.*, vol. 36, 1997, p. 611.

309. T. Mukai, T. G. Nieh, H. Iwasaki, and K. Higashi, *Mater. Sc. Tech.*, vol. 14, 1998, p. 32.

310. O. D. Sherby and J. Wadsworth, in *Deformation, Processing and Structure*, ed. G. Krauss, ASM, Metals Park, Ohio, 1984, pp. 355–389.

311. M. Mabuchi and K. Higashi, *Hot Deformation of Aluminum Alloys II*, eds. T. R. Bieler, L. A. Lalli, and S. R. MacEwen, TMS, Warrendale, Pa., 1998, pp. 87–99.

312. S. Lee et al., *Materials Sc. and Eng.*, vol. 272A, 1999, pp. 63–72.

313. R. W. Cahn and P. M. Hazzledine, in *Encyclopedia of Materials Science and Engineering*, Pergamon Press, Oxford, 1986, pp. 4786–4790.

314. M. F. Ashby and R. A. Verall, *Acta Metall.*, vol. 25, 1973, pp. 149–163.

315. W. D. Nix, in *Superplastic Forming*, ed. S. P. Agrawal, *Conference Proceedings, 1984*, ASM, Metals Park, Ohio, 1985, pp. 3–12.

316. R. C. Gifkins, in *Superplastic Forming of Structural Alloys*, eds. N. E. Paton and C. H. Hamilton, *Conference Proceedings*, TMS-AIME, Warrendale, Pa., 1982, pp. 3–26.

317. M. J. Stowell, in *Superplastic Forming of Structural Alloys*, eds. N. E. Paton and C. H. Hamilton, *Conference Proceedings*, TMS-AIME, Warrendale, Pa., 1982, pp. 321–336.

318. C. C. Bampton, M. W. Mahoney, C. H. Hamilton, A. K. Ghosh, and R. Raj, *Met. Trans.*, vol. 14A, 1983, pp. 1583–1591.

319. A. H. Chokshi, in *Superplasticity in Advanced Materials*, eds. S. Hori, M. Tokizane, and N. Furushiro, JSRS, Osaka, Japan, 1991, pp. 171–180.

320. H. Iwasaki, T. Mori, M. Mabuchi, and K. Higashi, *Materials Sc. and Tech.*, vol. 15, 1999, pp. 180–184.

321. C. C. Bampton and J. W. Edington, *Met. Trans.*, vol. 13A, 1982, pp. 1721–1727.

322. M. G. Zelin, R. S. Yang, R. Z. Valiev, and A. K. Mukherjee, *Met. Trans.*, vol. 24A, 1993, p. 417.

323. M. G. Zelin and S. Guillard, *Materials Sc. and Tech.*, vol. 15, March 1999, pp. 309–315.

324. J. Pilling, in *Superplasticity in Advanced Materials*, eds. S. Hori, M. Tokizane, and N. Furushiro, JSRS, Osaka, Japan, 1991, pp. 181–190.

325. Y. Ma and T. G. Langdon, *Met. and Mats. Trans.*, vol. 27A, 1996, pp. 873–878.

325a. K. Osada and H. Yoshida, *Mats. Sc. Forum*, vols. 170–172, 1994, pp. 715–724.

326. J. A. Wert and N. E. Paton, *Met. Trans.*, vol. 14A, 1983, p. 2535.

327. J. Koike et al., *Scripta Mater.*, vol. 39, no. 8, 1998, pp. 1009–1014.

327a. A. J. Barnes, *Mats. Sc. Forum*, vols. 170–172, 1994, pp. 701–704.

328. N. Ridley, E. Cullen, and F. J. Humphreys, in *Superplasticity and Superplastic Forming 1998*, eds. A. K. Ghosh and T. R. Bieler, TMS, Warrendale, Pa., 1998, pp. 65–74.

329. T. R. McNelley and M. E. McMahon, in *Superplasticity and Superplastic Forming 1998*, eds. A. K. Ghosh and T. R. Bieler, TMS, Warrendale, Pa., 1998, pp. 75–87.

330. M. G. Zelin, *Met. and Mats. Trans.*, vol. 27A, 1996, pp. 1400–1403.

331. H. N. Azari, G. S. Murty, and G. S. Upadhyaya, *Met. and Mats. Trans.*, vol. 24A, 1993, pp. 2153–2160.

332. S. J. Hales, T. C. McNelley, and H. J. McQueen, *Met. Trans.*, vol. 22A, 1991, pp. 1037–1047.

333. J. S. Vetrano, C. A. Lavender, C. H. Hamilton, M. T. Smith, and S. M. Bruemmer, *Scripta Metall.*, vol. 30, 1994, pp. 565–570; C. A. Lavender, J. S. Vetrano, M. T. Smith, S. M. Bruemmer, and C. H. Hamilton, *Mater. Sc. Forum*, vols. 170–172, 1994, pp. 279–286.

334. H. Imamura and N. Ridley, in *Superplasticity in Advanced Materials*, eds. S. Hori, M. Tokizane, and N. Furushiro, JSRS, Osaka, Japan, 1991, pp. 563–568

335. K. Kannan et al., *Met. and Mats. Trans.*, vol. 29A, 1998, pp. 1211–1220.

336. G. D. Smith and H. L. Flower, *Superplasticity and Superplastic Forming*, eds. A. K. Ghosh and T. R. Bieler, TMS, Warrendale, Pa., 1995, pp. 117–124.

337. H. W. Hayden and J. H. Brophy, *Trans. ASM*, vol. 61, 1967, p. 542; H. W. Hayden, R. C. Gibson, H. F. Merrick, and J. Brophy, *Trans. ASM*, vol. 60, 1967, p. 3.

338. R. C. Gibson, H. W. Hayden, and J. H. Brophy, *Trans. ASM*, vol. 61, 1968, p. 85.

339. Y. Maehara, *Met. Trans.*, vol. 22A, 1991, p. 1083; Y. Maehara, in *Superplasticity in Advanced Materials*, eds. S. Hori, M. Tokizane, and N. Furushiro, JSRS, Osaka, Japan, 1991, pp. 563–568; Y. Maehara and Y. Ohmori, *Met. Trans.*, vol. 18A, 1987, pp. 663–672.

340. Y. S. Han and S. H. Hong, *Scripta Mater.*, vol. 36, no. 5, 1997, pp. 557–563.

341. H. Kitano, K. Nakamura, K. Shirakawa, H. Yoshida, and K. Osada, Ref. 13, pp. 1861–1866.

342. D. Hernandez, G. Frommeyer, and H. Hofman, *Steel Research*, vol. 67, no. 10, 1996, pp. 444–449.

343. R. S. Mishra and A. K. Mukherjee, *Superplasticity and Superplastic Forming 1998*, eds. A. K. Ghosh and T. R. Bieler, *Conference Proceedings*, TMS, Warrendale, Pa., 1998, pp. 109–116.

344. R. S. Mishra, S. X. McFadden, R. Z. Valiev, and A. K. Mukherjee, *JOM*, January 1999, pp. 37–40.

345. R. Z. Valiev and R. K. Islamgaliev, *Superplasticity and Superplastic Forming 1998*, eds. A. K. Ghosh and T. R. Bieler, *Conference Proceedings*, TMS, Warrendale, Pa., 1998, pp. 117–126.

346. S. Hanada, W.-Y. Kim, Y. Yoshimi, K. Sato, and T. Sakai, Ref. 13, pp. 1883–1889.

347. D. Lin, D. Li, A. Shan, and Y. Liu, Ref. 13, pp. 1915–1922.

348. T. L. Lin, A. Shan, and D. Li, *Scr. Metall. Mater.*, vol. 31, 1994, pp. 1455–1460.

349. S. C. Cheng, J. Wofenstine, and O. D. Sherby, *Met. Trans.*, vol. 23A, 1992, pp. 1509–1513.

350. A. K. Ghosh and C. H. Cheng, in *Superplasticity in Advanced Materials*, eds. S. Hori, M. Tokizane, and N. Furushiro, JSRS, Osaka, Japan, 1991, pp. 299–310.

351. J. W. Chen, in *The Encyclopedia of Advanced Materials*, vol. 4, Pergamon Press, Oxford, 1994, pp. 2715–2717.

352. J. Wittenauer, in *Superplasticity and Superplastic Forming*, eds. A. K. Ghosh and T. T. Bieler, TMS, Warrendale, Pa., 1995, pp. 85–92.

353. T. G. Nieh and J. Wadsworth, *Acta Metall. Mater.*, vol. 38, 1990, p. 1121.

354. K. Kajihara, Y. Yoshizawa, and T. Sakuma, *Scripta Metall.*, vol. 28, 1993, p. 559.

355. Y. Ikuhara and T. Sakuma, Ref. 13, pp. 1839–1845.

356. F. Wakai, Y. Kodama, N. Murayama, S. Sakaguchi, T. Rouxel, S. Sato, and Tooru Nonami, in *Superplasticity in Advanced Materials*, eds. S. Hori, M. Tokizane, and N. Furushiro, JSRS, Osaka, Japan, 1991, pp. 205–214.

357. K. Higashi, Ref. 13, pp. 1795–1804.

357a. K. Higashi, T. Okada, T. Mukai, S. Tanimura, T.-G. Nieh, and J. Wadsworth, *Mater. Trans., JIM*, vol. 36, no. 2, 1995, pp. 317–322.

357b. O. D. Sherby and J. Wadsworth, *Materials Sc. Forum*, vols. 233–234, 1997, pp. 125–138.

358. M. Mabuchi, K. Higashi, Y. Okada, S. Tanimura, T. Imai, and K. Kubo, in *Superplasticity in Advanced Materials*, eds. S. Hori, M. Tokizane, and N. Furushiro, JSRS, Osaka, Japan, 1991, pp. 385–390.

359. B. L. Lou, J. C. Huang, and H. P. Pu, Ref. 13, pp. 1853–1859.

360. M. Mabuchi, N. Saito, K. Shimojima, M. Nakanishi, Y. Yamada, M. Nakamura, T. Asahina, T. G. Langdon, H. Iwasaki, and K. Higashi, Ref. 13, pp. 1975–1981.

361. T. Mukai, T. G. Nieh, and K. Higashi, in *Superplasticity and Superplastic Forming 1998*, eds. A. K. Ghosh and T. R. Bieler, TMS, Warrendale, Pa., 1998, pp. 179–186.

361a. G. Frommeyer, H. Hofmann, and W. Herzog, *Mater. Sc. Forum*, vols. 170–172, 1994, pp. 483–488.

362. R. B. Valiev, O. A. Kaibyshev, R. I. Kuzhetsov, R. Sh. Musalimov, and N. K. Tsenev, *Doklady Academi Nauk, SSSR*, vol. 301, 1988, p. 186.

363. H. Watanabe, T. Mukai, and K. Higashi, in *Superplasticity and Superplastic Forming 1998*, eds. A. K. Ghosh and T. R. Bieler, TMS, Warrendale, Pa., 1998, pp. 179–186.

363a. O. D. Sherby, T.-G. Nieh, and J. Wadsworth, *Mats. Sc. Forum*, vol. 170–172, 1994, pp. 13–24.

364. R. J. Lederich, S. M. L. Sastry, M. Hayase, and T. L. Mackay, *J. Met.*, vol. 34, no. 8, 1982, pp. 16–20.

365. M. Matsuo, T. Tagata, and N. Matsumoto, Ref. 13, pp. 1953–1959.

366. R. Z. Sadeghi, *MARC Mall*.

367. E. D. Weisert and J. R. Fisher, in *Advanced Processing Methods for Titanium*, eds. D. F. Hasson and C. H. Hamilton, *Conference Proceedings*, TMS-AIME, Warrendale, Pa., 1982, pp. 101–103.

368. C. H. Hamilton, in *The Encyclopedia of Advanced Materials*, Pergamon Press, Oxford, 1994, pp. 2712–2715.

369. T. H. Thomsen, D. L. Holt, and W. A. Backofen, *Met. Eng. Q.*, vol. 10, 1970, p. 1.

370. U.S. Patent 5129248, 1992.

371. U.S. Patent 5309747, 1994.

372. C. J. Pellerin, in *Superplastic Forming*, ed. S. P. Agrawal, *Conference Proceedings 1984*, ASM, Metals Park, Ohio, 1985, pp. 63–69.

373. A. Arieli and R. B. Vastava, in *Superplastic Forming*, ed. S. P. Agrawal, *Conference Proceedings 1984*, ASM, Metals Park, Ohio, 1985, pp. 70–75.

374. C. Bampton, F. McQuilkin, and G. Stacher, in *Superplastic Forming*, ed. S. P. Agrawal, *Conference Proceedings 1984*, ASM, Metals Park, Ohio, 1985, pp. 63–69.

375. D. M. Ward, *Met. Mater.*, September 1986, pp. 560–563.

376. P. G. Partridge and A. Wisbey, in *High Performance Materials in Aerospace*, ed. H. M. Flower, Chapman & Hall, London, 1995, pp. 283–317.

376a. N. W. Mahoney and C. C. Bampton, *ASM Handbook*, vol. 6, 10th ed., ASM International Materials Park, Ohio, 1994, pp. 156–159.

377. D. Subramanayam, M. R. Notis, and J. I. Goldstein, *Met. Trans.*, vol. 16A, 1985, pp. 605–611.

378. H. Morrow, III, and R. F. Lynch, in *Encyclopedia of Materials Science and Engineering*, ed. R. W. Cahn, Pergamon Press, Oxford, 1986, p. 5514.

379. D. S. Darmaid, *Mater. Sc. Eng.*, vol. 70, 1985, pp. 123, 124.

380. D. S. Darmaid, *Mater. Sc. Eng.*, vol. 69, 1984, pp. 105–111.

381. R. V. Safiullin and F. U. Enikeev, in *Superplasticity and Superplastic Forming*, eds. A. K. Ghosh and T. R. Bieler, TMS, Warrendale, Pa., 1995, pp. 213–217.

382. E. D. Weisert and G. W. Stacher, in *Superplastic Forming of Structural Alloys*, eds. N. E. Paton and C. H. Hamilton, *Conference Proceedings*, TMS-AIME, Warrendale, Pa., 1982, pp. 273–289.

383. *Production Bulletin*, McDonnel Douglas Aerospace, St. Louis, Mo.

384. A. Sunwoo, R. Lum, and R. Vandervoort, in *Superplasticity and Superplastic Forming*, eds. A. K. Ghosh and T. R. Bieler, TMS, Warrendale, Pa., 1995, pp. 251–258.

385. J. Niu, J. Zhang, M. Wang, and Z. Wang, Ref. 13, pp. 1923–1928.

386. B. C. Snyder, J. Wadsworth, and O. D. Sherby, *Acta Metall.*, vol. 32, 1984, p. 919.

387. O. D. Sherby and J. Wadsworth, in *Encyclopedia of Materials Science and Engineering*, suppl. vol. 1, ed. R. W. Cahn, Pergamon Press, Oxford, 1988, pp. 541–545, 519–522.

388. H. Kokawa, T. Tsuzuki, and T. Kuwana, *ISIJ Int.*, vol. 35, 1995, p. 1291.

389. Y. Komizo and Y. Maehara, *Trans Jpn. Weld. Soc.*, vol. 19, 1988, p. 83.

390. N. Ridley, M. T. Salehi, and J. Pilling, *Mater. Sc. and Tech.*, vol. 8, 1992, p. 791.

391. J. Pilling, Z. C. Wang, and N. Ridley, *Superplasticity and Superplastic Forming 1998*, eds. A. K. Ghosh and T. R. Bieler, TMS, Warrendale, Pa., 1998, pp. 297–303.

CHAPTER 16
SURFACE HARDENING TREATMENTS

16.1 INTRODUCTION

Many commercial steel parts require hard, wear-resistant surfaces combined with a softer and tougher core; thus the entire part should have considerable impact strength. However, some steel parts require hard, wear-resistant surfaces with a strong and tough core. Hence, specialized techniques, called surface hardening treatments, are necessary in order to achieve the above combination of properties for various applications of steel components. These treatments are divided into two general categories. In the first category, known as *thermochemical surface hardening treatments*, the composition of steel surface is altered and the surface is then hardened with or without quenching. This treatment is divided into two groups: (1) *austenitic thermochemical surface hardening treatments* (e.g., carburizing, carbonitriding, and cyaniding), in which nonmetallic elements, usually carbon, singly or in combination with nitrogen, are diffused into the austenite phase; and (2) *ferritic thermochemical surface hardening treatments* (e.g., nitriding, nitrocarburizing), in which nonmetals, usually nitrogen alone or in combination with carbon, are diffused into the ferrite phase. In the second category, known as *thermal surface hardening techniques*, heat alone is used to bring about changes in metallurgical structures without altering the composition of the steel. The surfaces are heated here either by *conventional methods* of induction and flame hardening processes or by *nonconventional methods* such as laser and electron beam hardening to form austenite to a controlled depth, which is subsequently quenched in water-based, oil-based, or forced-air quenchants for conventional methods or self-quenched for nonconventional methods to produce a hard martensite case. In the former conventional thermal techniques, the components are heated either electrically (induction hardening) or by flame (flame hardening). The nonconventional thermal techniques (lasers and electron beams) have gained popularity in recent years because they involve finely controllable sources of intense energy, promoting extremely fast heating rates.[1]

The surface hardening processes enjoy the following advantages over conventional or through hardening:

1. These methods reduce distortion to a very low level and eliminate cracking, especially in large components.
2. The fatigue life and fatigue strength of the components are increased considerably, owing to the development of compressive stresses in the outer surface layers.

3. A highly wear-resistant surface is produced from a relatively inexpensive steel, which would have been possible by the use of more expensive high-alloy steels.

4. A combination of high surface hardness and wear-resistant surface (case) with ductile or tough core can be produced, which is not possible with conventional treatments.

5. Little, if any, scaling or decarburization arising from surface hardening is desirable in final or finish-machining of parts.

6. Selected areas can be hardened on both large and small components, which would be very difficult with conventional hardening.

The demerits of these treatments are as follows:

1. This is a specialized treatment.

2. Where mass production of heat-treated parts is desired, these methods are not preferred since conventional methods are economical and the most effective.

3. Because very little hardness penetration is achieved, these techniques become undesirable in many applications requiring sufficient strength of the hardened case to withstand the service stresses.

This chapter describes all the categories of surface hardening treatments mentioned above, together with *nonconventional thermochemical treatments* such as *supercarburizing, boriding* (or *boronizing*), and *thermoreactive deposition/diffusion processes*.

16.1.1 Case Depth Measurements

Since the case depth of the surface hardened part is an important engineering requirement and is always specified at some significant location on a part, it is essential to determine this by chemical, mechanical, visual with an acid etch, or nondestructive method. It is also necessary to distinguish between total case depth and effective case depth. Measurements are usually reported as *effective case depth*, *equivalent case depth*, and *total case depth*.

The effective case depth is the distance measured inward and normal to the surface of the hardened part at which a hardness of 50 Rc value is found. The effective case depth is typically in the range of two-thirds to three-fourths of the total case depth. Total case depth is the normal distance from the surface of a hardened or unhardened case to the point where the difference in chemical or physical properties of the case and core is indistinguishable. Sometimes, total case depth is represented as the distance from the surface to the deepest point at which carbon concentration is 0.04% higher than the carbon content of the core.[2]

The conventional mechanical method for inspection and measurement of case depth is expensive and time-consuming since it involves cross-section cutting of the part, polishing, etching, hardness testing, and measurement of the depth to the required hardness (say, 50 Rc) level. Visual methods are macroscopic or microscopic. The macroscopic methods usually employ grinding of the specimens to 600-grit silicon carbide paper, acid etching to develop maximum contrast, water rinsing, and readings of entire darkened area for approximate total case depth, at a magnification up to 20X, whereas the microscopic methods require complete metallographic polishing and suitable etching and reading of case depth at a magnification of 100X, based on the microstructure, representing hardness equivalent to 50 Rc.

Total case depth is read as the distance at which no further change in microstructure occurs.[2]

Microhardness testing is the most precise method of measuring case depth. For this, the Knoop indenter is selected to obtain the highest density of indentations. The square pyramid Vickers indenter is limited to make closely spaced indentations and indentations near the surface. Microhardness traverses are made with loads in the range of 100 to $1000\,g^2$.

Nondestructive techniques of measuring and inspecting the case depth are eddy current testing and resistivity measurements. The former technique has several drawbacks, including high capital cost of equipment, high sophistication demanding trained technicians, complex data requiring experienced interpretation, and difficult applicability to complex-shaped parts.[3] The latter method uses a new device, called a *microhmeter*, developed by AT&T and is based on the relationship between observed resistivity (microhmeter reading) and the depth of the hardened material. It consists of a four-point probe assembly: two outer probes for generating rapidly reversing dc pulses through the hardened part and two inner probes for measuring the resulting voltage and the electronic package which produces the resistivity value.[3]

16.2 THERMAL SURFACE HARDENING

16.2.1 Flame Hardening

This technique was popular in the 1930s and 1940s; but with the advent of induction heating after World War II, most heat treaters switched to induction heating because of its superior control in large production runs. Although flame hardening represents only 5 to 10% of all commercial heat treating in the United States, for many captive and in-house surface flame hardening is still quite popular.[4] Flame hardening consists of rapidly austenitizing the steel surface at $\sim$1550 to 1650°F with the burners using oxyacetylene gas, oxyhydrogen gas, natural gas, propane, or methylacetylene propadiene (MAPP) based fuel gas, followed by immediate quenching in water or polymer quenchants to produce a martensitic structure at the surface, while leaving the core of the part unaffected. Sometimes a forced-air quenchant is used for alloy steels normally suited for oil-quenching. Oxygen is injected into the combustion mixer block under high pressure, usually in the range of 20 to 70 psig, depending on the fuel gas being used and the volume of mixed gases required to maintain the proper ratio and velocity for an economic operation.[4] Some burners are equipped with radiator chambers but mostly spray ports. The quenchant also cools the flame burners without causing overheating or flashback.[5]

The choice of gas depends on (1) the size, shape, and composition of the workpiece; (2) the required case depth; and (3) the relative cost and availability of each gas. Table 16.1 lists the fuel gases used for flame hardening with values of normal burning velocity and the heating values of suitable oxy-fuel gas mixture.[6] This also shows that the flame temperatures of the air-fuel mixtures are comparatively lower than those obtained by oxy-fuel mixtures. Table 16.2 indicates total gas flow per flame port in cubic feet per hour (cfh) at various burning velocities in feet per second, and Table 16.3 provides oxygen/fuel ratios required to meet specific flame conditions which serve as guidelines for the operator to achieve continued control of depth of hardness and repeatability. For flame hardening, first the neutral flame is normally employed, followed by oxidizing (or hotter) flame for less case depth or

TABLE 16.1 Fuel Gas Used for Flame Hardening[6]

Gas	Heating value		Flame temperature				Usual ratio of oxygen to fuel gas	Heating value of oxy-fuel gas mixture		Normal velocity of burning		Combustion intensity[†]		Usual ratio of air to fuel gas
			With oxygen		With air									
	MJ/m³	Btu/ft³	°C	°F	°C	°F		MJ/m³	Btu/ft³	mm/s	in./s	mm/s × MJ/m³	in./s × Btu/ft³	
Acetylene	53.4	1433	3105	5620	2325	4215	1.0	26.7	716	535	21	14,284	15,036	12
City gas	11.2–33.5	300–900	2540	4600	1985	3605	‡	‡	‡	‡	‡	‡	‡	‡
Natural gas (methane)	37.3	1000	2705	4900	1875	3405	1.75	13.6	364	280	11	3,808	4,004	9.0
Propane	93.9	2520	2635	4775	1925	3495	4.0	18.8	504	305	12	5,734	6,048	25.0
MAPP	90	2406	2927	5301	1760	3200	3.5	20.0	535	381	15	7,620	8,025	22

[†] Product of normal velocity of burning multiplied by heating value of oxy-fuel gas mixture.
[‡] Varies with heating value and composition.
Courtesy of ASM International, Materials Park, Ohio.

16.4

TABLE 16.2   Total Gas Flow, per Flame Port at Various Velocities[4]

Flame fort diameter	Burning velocity, ft/s						
	300	350	400	450	500	550	600
0.0292-in. diameter (#69 drill)	5.25[†]	6.0	7.0	8.0	8.75	9.6	10.5
0.035-in. diameter (#64 drill)	7.50	8.75	10.0	11.25	12.5	13.75	15.0
0.040-in. diameter (#60 drill)	9.0	10.5	12.0	13.5	15.0	16.5	18.5
0.0465-in. diameter (#56 drill)	12.75	14.8	17.0	19.5	21.25	23.4	25.5

[†] Total gas flow in cubic feet per hour.
Courtesy of Flame Treating Systems, Inc.

TABLE 16.3   Relationship between Oxygen/Fuel Ratio and Flame Condition[4]

Flame condition	Oxygen/ MAPP	Oxygen/ acetylene	Oxygen/ propane	Oxygen/ natural gas
Very carburizing	2.5/1	1.1/1	3.0/1	1.5/1
Carburizing	3.0/1	1.2/1	3.5/1	1.6/1
Neutral	3.5/1	1.3/1	4.0/1	1.7/1
Oxidizing	4.0/1	1.4/1	4.5/1	1.8/1
Very oxidizing	4.5/1	1.5/1	5.0/1	1.9/1

Courtesy of Flame Treating Systems, Inc.

carburizing flame for deep case depth requirement and for cast iron.[4] To maintain these velocities and ratios, flowmeters for both gases must be used. These coupled with the use of electric solenoid valves controlled from a very small PC controller constitute an economical and precise means for complete quality and production conditions.[4] Figure 16.1 shows the flame heads for use with oxy-fuel gas. Burners are of different designs, based on whether they are fired by an oxy-fuel or an air-fuel gas mixture. Flame heads containing a single orifice or multiple orifices are designed to give a flame pattern to ensure no direct heating of its parts. Flame heads with removable tips have an advantage over fixed-type heads because the mechanically or flame-damaged tips can be corrected without replacing the head. Figure 16.2a and b shows the radiant-type and the high-velocity convection-type air-fuel burners.[6]

16.2.1.1   Techniques of Flame Hardening.   There are different methods of flame hardening, which include the following:

Spot or Stationary Method.   This method uses stationary burners with either a single-orifice or multiorifice design, depending on whether small or large areas are to be hardened. After being flame-heated, the components are usually immersion-quenched or, in some cases, spray-quenched. Irregularly shaped components also can be successfully hardened by this method. In principle, the method does not need elaborate equipment, except probably fixtures and a time device to ensure the uniform processing of each workpiece. Figure 16.2a shows the spot (stationary) method of flame-hardening a rocker arm and the internal lobes of a cam.[6] The method can be automated by indexing the heated parts into either a spray quench

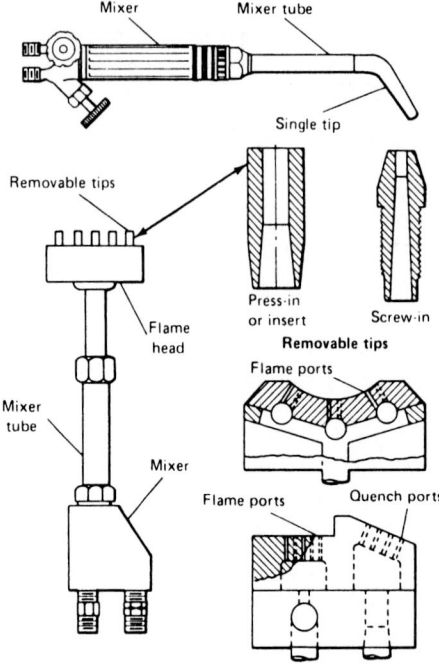

FIGURE 16.1 Flame heads for use with oxy-fuel gas.[6] (*Courtesy of ASM International, Materials Park, Ohio.*)

or a suitable quench bath for large-volume production.[7] This can be applied to scissors, shears, pliers, clippers, and rocker arms.

Progressive Method. This method is normally used to harden large, flat surfaces. The flame head is usually a multiple-orifice type which is closely followed by the (integrated or separate) quench so that only a narrow band is progressively heated to the hardening temperature and subsequently quenched (Fig. 16.2*b*).[6] The tip of the inner core of the flame should be close to the work surface—i.e., only 5 to 6 mm (0.196 to 0.236 in.) away from the workpiece—without touching it. For hardening the entire surface, the flame length should be somewhat shorter and should stop 2 to 3 mm (0.078 to 0.118 in.) from both edges; however, for hardening a narrow band of the entire surface, the flame length should extend, on both sides, 3 to 4 mm (0.118 to 0.157 in.) beyond the width of the band to be hardened due to the lateral dissipation of the heat.[8]

Single passes of up to 1.5-m (60-in.) width can be made; wider areas must be hardened in more than one pass.[6] Simple curved surfaces should be hardened progressively by using contoured flame heads, and some irregular surfaces should be traversed by means of tracer template methods. The rate of travel of the flame head over the surface is a function of the heating capacity of the head, the required case depth, the composition and shape of the workpiece, and the type of the quench employed. Speeds varying between 0.8 and 5 mm/s (2 and 12 in./min) are typical with oxyacetylene heating flame head.[6]

Spinning Method. This method is well suited to round or semiround components such as wheels, cams, gears, short bearing journals or shafts, pins, axles, and

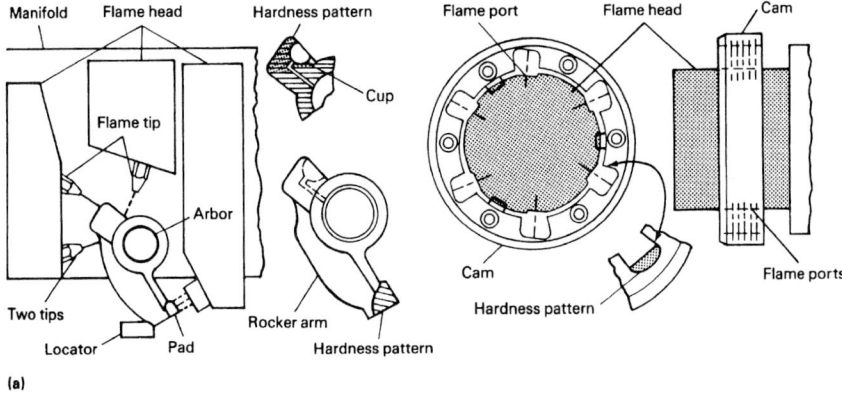

(a)

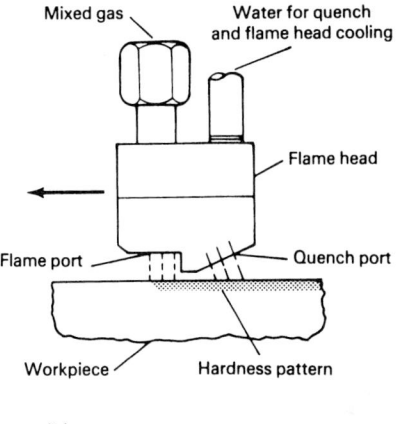

(b)

FIGURE 16.2   Spot (or stationary) and progressive methods of flame-hardening. (*a*) Spot (or stationary) method of flame-hardening a rocker arm and the internal lobes of a cam; quench is not shown. (*b*) Progressive hardening method.[6] [(*a*) *Courtesy of ASM International* (*b*) *Courtesy of Butterworths, London.*]

trunnions. First, heating of the rotating or spinning workpiece surface is carried out with a flame head until the desired hardening temperature is reached; then the flame is withdrawn or extinguished, and the workpiece is quenched by immersion or spray, or a combination of both, in a quench tank (Fig. 16.3).[6] One or more water-cooled heating flame heads, equal in width to the surface to be heated, can be used. Here again the same distance between the tip of the inner core of the flame and the work surface should be maintained, as mentioned above.

This method can be fully automated. This method can also treat parts with irregular sections and mass distribution, e.g., large drive wheels for tracked vehicles, cams and camshafts for marine diesel engines, and crane traveling wheels. Currently, fully automatic integrated flame-heating and quenching equipment is available that is capable of treating round components up to 1.5 m (60 in.) in diameter and up to 2 Mg (2.2 tons) in weight. Mostly it is designed to treat gear wheels.[6]

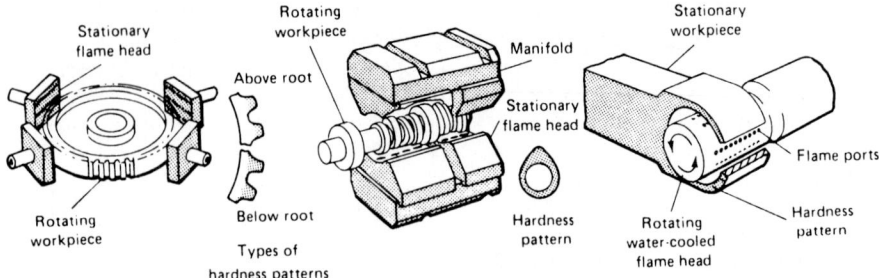

FIGURE 16.3 Spinning methods of flame hardening. In methods shown at left and at center, the workpiece rotates. In method at right, the water-cooled flame head rotates. Quench is not shown.[6] (*Courtesy of ASM International, Materials Park, Ohio.*)

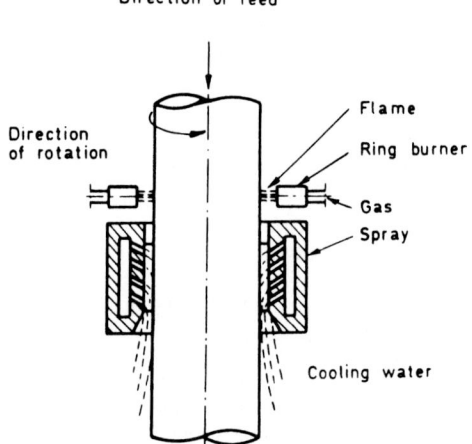

FIGURE 16.4 Principle of progressive spin hardening.[8] (*Courtesy of Butterworths, London.*)

Progressive-Spinning Method. This combines the progressive and spinning methods for hardening long parts, e.g., shafts and rolls. Segments or ring flame heads with attached quenching rings are moved from one end of the work to the other along the rotating workpiece so that hardening is completed after one pass (Fig. 16.4).[8] This method is suited to harden diameters ranging from 20 to 1500 mm (~0.75 to 60 in.).

16.2.1.2 Advantages of Flame Hardening

1. This method is capable of producing case depths ranging from 0.8 to 6.4 mm (0.03 to 0.25 in.), depending on the fuels employed, the design of the flame head, the time of heating, the hardenability of the workpieces, and the quenching method.

2. It is quite rapid and produces a harder and highly wear-resistant surface (deeper case depth) than that provided by thermochemical surface hardening methods. This means that inexpensive (plain carbon) steels can often meet the strength,

hardness, and wear properties of alloy steels at a lesser cost, without scaling or decarburization, thereby eliminating the expensive cleaning operation.[5,6,9]

3. It affords greater control of dimensional stability than is attainable by furnace heating and quenching. A typical example includes a large gear of a complex design where flame hardening of the teeth would not affect the dimensions of the gear.

4. It is versatile and may be used to handle small as well as large parts (e.g., weighing from 1 oz to 25 tons). Typical examples of large parts include large gears, machineways, dies, and rolls where conventional furnace heating and quenching are not possible or uneconomical.[6]

5. Like induction hardening, it can follow any contour; it is very clean without producing any scale or decarburization on hardened surfaces.

6. It can be automated with consistency and repeatability at low capital investment by using automated equipment with an indexing system that reduces dependency on the operator's skills.[9]

7. It has the ability to satisfy stringent engineering requirements with carbon steels.

8. It can be applied to only a small segment, area, or section of a part, e.g., the ends of valve stems and push rods and the wearing surfaces of cams and levers.[6]

16.2.1.3   Disadvantages of Flame Hardening

1. Precise flame hardening requires considerable experience of the operator; however, this dependency can be eliminated by automation of this process.

2. It relies to a great extent on the visual reading, which can be taken care of by the use of infrared pyrometry.

16.2.1.4   Selection of Hardening Methods.   For shafts with deeper splines, the progressive method is preferred because overheating of the topland may be easily avoided. For shafts with shallow splines, where the difference between the outer diameter and root circle may be only a few millimeters, spin hardening is the best choice. Both methods are successfully used for spline shafts, with a small distortion. The same reasoning is used for spline shafts, with a small distortion. The same reasoning is used for spline fittings.[7]

16.2.1.5   Fuel Gases.   The commonly available gases suitable for flame hardening are a mixture of O_2 and a combustible gas, generally coal gas, acetylene, propane, butane, natural gas, methylacetylene propadiene (MAPP) based fuel gas, or hydrogen in a suitable ratio. Producer gas having low heat content is unsuitable for flame hardening.[5,7] The heating time depends on the heating values of the specific fuel mixture as well as on the velocity of fuel burning but is typically 10 to 60 s.[10]

Shallow hardness patterns (<1/8 in. or 3.2 mm deep) can be achieved with oxy-gas mixtures which provide higher-temperature flames and subsequent rapid heat transfer to localize the heat pattern effectively. Deeper hardness patterns may require the use of oxy-gas or air-gas mixtures. Mostly oxy-fuel flame equipment is used to surface-harden gear teeth, rolls, shaft areas, journals, machineways, wear areas of forming dies, internal and external diameters of hubs, and massive sections as well as small parts.[6]

16.2.1.6   Material Selection.   For flame or induction hardening, the areas to be hardened must be free of decarburization; otherwise, they may not harden

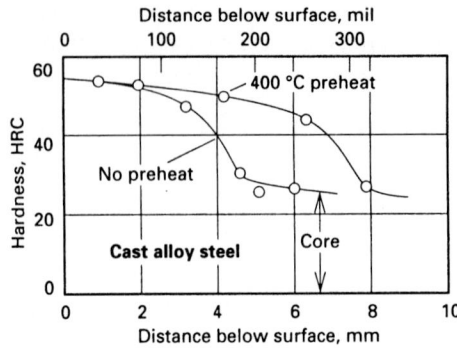

FIGURE 16.5 Effect of preheating on hardness gradient in a ring gear.[6] (*Courtesy of ASM International, Materials Park, Ohio.*)

satisfactorily to the maximum.[11,12] Plain carbon steels containing 0.37 to 0.6% (such as AISI 1045 and 1060), free-machining steels (such as AISI 1145), and alloy steels (typically AISI 4135H, 4140H, 4150, 4340H, 6150H, 8640H, 8642H, and 52100, including martensitic 420C and 440A stainless steels) are widely used for flame hardening. Cast irons [such as gray, ductile, malleable iron with 0.35 to 0.80% combined carbon (as pearlite)] and alloy irons with a fine pearlite matrix are recommended for flame and induction hardening.

16.2.1.7 Pre- and Post-Heat Treatment.

For the best results, steels for induction and flame hardening should be as-rolled, normalized, air-blast-quenched, or quenched and tempered. More intricate parts and steels with >0.40% C should also be stress-relieved at 175 to 240°C (350 to 475°F) before hardening in order to reduce the risk of hardening cracks and of distortion. Prior to hardening, the work surfaces should be slightly preheated by the flame movement to overcome the difficulty in achieving the desired surface hardness and increased hardness penetration (Fig. 16.5).[6] Preheating of prehardened tempered steels as well as cast irons to be flame-hardened is always desirable to minimize cracking. The exact preheat temperature depends on the size and shape of the component and the degree of stress relief or temper desired.

It is usually necessary to temper the components immediately after flame hardening. This is carried out by flame tempering at a temperature between 150 and 200°C (302 and 392°F) for better operating qualities.

Tempering flame heads must possess smaller heat outputs than hardening flame heads; otherwise, cracking of the hardened zone may occur, and critical lower temperature and its control may not be achieved.[6] Large parts hardened to depths of about 6.4 mm (0.25 in.) can be self-tempered by residual heat in the part; this relieves hardening stress and avoids the deployment of a tempering operation.[6]

16.2.1.8 General Procedure Guidelines[6,13]

1. Do not flame-harden more than is required. Keep the flame-hardened zone to the necessary minimum. When a large-shaped flame head is used, a water quench is necessary.

2. Select proper flame head and torch tip for the job. Use the smallest tip as practicable to maintain adequate efficiency.

3. Avoid overheating due to long heating cycle, closer flame head to the work, oversize flame ports, excess O_2 in flame, excessive fuel-gas mixture, or improper pattern of flame tip. The more acute the edge angle, the greater the chance of overheating.[6]

4. Avoid the possibility of distortion due to shape of the part not being well suited to flame hardening, nonuniform heating, nonuniform quenching, very rapid quenching rate, and high material hardenability.

5. Temper the flame-hardened part immediately by flame tempering, self-tempering, or furnace tempering.

6. Surfaces to be flame-hardened should be preheated (discussed earlier).

7. Avoid sharp or small-radius internal angle (or corners), particularly with water-quenching grades.

8. Whenever possible, provide surface breaks to ensure no overlapping of flame-hardening zones.

9. On large parts, particularly when a water quench is necessary, it is better to perform a die penetrant check of the flame-hardened region.

10. For wrought and cast steel parts, the surface defects or conditions likely to be deleterious to successful flame hardening are those that cause localized over-heating, interference with heating or quenching, increase in cracking hazard, soft spots, and weld zone reactions. These are laps, seams, folds, fins (wrought parts), scales, rust, decarburization, pinholes, shrinkage (castings), coarse-grained gate areas (castings), and improper welds.

16.2.1.9 Applications. Carbon and alloy steel castings are extensively used for flame hardening applications. The selection of a specific composition or grade is similar to that for wrought carbon and alloy steels.[6] It finds applications in surface hardening of valve trunnions, gears, stamping dies, large sheaves and rope drums, cast iron automotive cams, and machine cams.[14]

16.2.2 Induction Hardening

16.2.2.1 Fundamental Principles. When an alternating current is passed through an inductor (or work coil)—usually water-cooled copper coil or tubing—a rapidly alternating magnetic field is set up around it. When this magnetic field intersects (or encounters) a conductive body (workpiece), say, a steel part, it induces eddy current I mainly in the surface layers of a workpiece. The flow of eddy current generates heating in the workpiece by the Joule effect, according to the relationship $P = I^2R$, where I is the induced current, R is the electrical resistance of the part, and P is the power generated by the eddy current. In a magnetic workpiece, additional heat is produced as a result of hysteresis losses (arising from the rapid change in orientation of magnetic domain); the magnitude of this latter effect depends upon power density, frequency, and the coercivity of the magnetic material. For most induction heating systems, this is small compared with I^2R power (or eddy current) losses and contributes a little to the temperature rise up to the Curie point of 768°C (1416°F). Beyond this red-hot temperature, bcc ferrite converts to nonmagnetic austenite, and the heating due to hysteresis disappears.[8,15]

The induction heating source distribution depends on a number of factors. In the simplest approach, the extent of induction heating is usually described by the depth of effective penetration of induced current, called the *reference* or *penetration depth* or the *skin depth d*, which is a function of the frequency *f* (in cycle per second, or hertz) of the alternating current in the work coil, electrical resistivity ρ (in ohm-inches or ohm-centimeters), and the relative or effective magnetic permeability μ (dimensionless) of the workpiece material. Reference depth can be calculated with the equation

$$d(\text{in.}) = 3160 \sqrt{\frac{\rho}{\mu f}} \tag{16.1}$$

or

$$d(\text{cm}) = 5030 \sqrt{\frac{\rho}{\mu f}} \tag{16.2}$$

This equation is commonly used in determining the proper frequency for induction heating various conductive materials.[16] For nonmagnetic materials and for magnetic steels heated above the Curie point, $\mu = 1$. For magnetic steels below the Curie point, the value of μ depends on the magnetic field intensity.[17] Table 16.4 shows the value of the reference/penetration depth in AISI 1040 carbon steel at ambient temperature (21°C or 70°F) as a function of frequency and magnetic field intensity **H** at the workpiece surface.[18] The additional depth of heat penetration[†] d_a results from heat flow by conduction toward the core of the component, which can be estimated using the following equation:[19]

$$d_a(\text{in.}) = 0.0387\sqrt{t} \tag{16.3}$$

or

$$d_a(\text{cm}) = 0.0983\sqrt{t} \tag{16.4}$$

where *t* is the time in seconds. Clearly, the total depth of heat penetration becomes $d + d_a$.

It is common practice to use high-frequency current (up to 450 kHz) and shorter heating times (implying a maximum heating effect) when only surface (or shallow) hardening (i.e., small hardness depth ≤1.59 mm, or 0.063 in.) is desired. The higher the frequency (i.e., power density), the more the intensity of the current tends to flow toward the surface, which produces a greater rate of heating near the surface. However, currents at the center are really zero for solid (cylinder) specimens. This type of current (or heat) distribution is sometimes called the *skin effect*.[15] Power rating is selected in such a way that the required energy is supplied to the surface layers within a few seconds. Table 16.5 shows typical power ratings for surface hardening of steel.[20] It is apparent that as the frequency is decreased, the current tends to be more evenly spread from the surface to the interior of the workpiece, thereby providing deeper heating and hardness depths.[15] Frequencies of 3 and 10 kHz are most frequently used for depths in the range of 3.18 to 6.35 mm (0.125 to 0.25 in.) and 1.59 to 3.18 mm (0.063 to 0.125 in.), respectively, while line frequency (50/60 Hz) is used for through hardening of large workpieces[19] where longer heating times are employed. Thus the current frequency, power density, and heating time are the main factors influencing the hardening operation. The depth and area to be hardened are

---

[†] Its value is negligibly small and will be ignored in all discussions throughout the text.

TABLE 16.4 Penetration Depth of AISI 1040 Carbon Steel at Room Temperature (21°C or 70°F)[18]

Magnetic field intensity		Frequency, Hz											
		60		500		3000		10,000		30,000		100,000	
		Penetration depth											
A/mm	A/in.	mm	in.	mm	in.	mm	in.	mm	in.	mm	in.	mm	in.
10	250	2.5	0.100	0.88	0.034	0.36	0.014	0.2	0.008	0.11	0.004	0.06	0.002
40	1000	4.7	0.185	1.63	0.064	0.67	0.026	0.36	0.014	0.21	0.008	0.12	0.005
80	2000	6.3	0.249	2.2	0.086	0.9	0.035	0.49	0.019	0.28	0.011	0.16	0.006
120	3050	7.76	0.306	2.69	0.106	1.1	0.043	0.6	0.024	0.35	0.014	0.19	0.007
160	4050	8.76	0.345	3.03	0.119	1.24	0.049	0.68	0.027	0.39	0.015	0.21	0.008
200	5100	9.63	0.379	3.33	0.131	1.36	0.054	0.75	0.029	0.43	0.017	0.24	0.009
280	7100	11.2	0.442	3.89	0.153	1.59	0.062	0.87	0.034	0.50	0.02	0.27	0.011

Courtesy of Inductoheat, Inc., Madison Heights, Mich.

TABLE 16.5   Power Density Required for Surface Hardening of Steel[20,21]

Frequency, kHz	Depth of hardening[a]		Input[b,c]					
			Low[d]		Optimum[e]		High[f]	
	mm	in.	kW/cm²	kW/in.²	kW/cm²	kW/in.²	kW/cm²	kW/in.²
500	0.381–1.143	0.015–0.045	1.08	7	1.55	10	1.86	12
	1.143–2.286	0.045–0.090	0.46	3	0.78	5	1.24	8
10	1.524–2.286	0.060–0.090	1.24	8	1.55	10	2.48	16
	2.286–3.048	0.090–0.120	0.78	5	1.55	10	2.33	15
	3.048–4.064	0.120–0.160	0.78	5	1.55	10	2.17	14
3	2.286–3.048	0.090–0.120	1.55	10	2.33	15	2.64	17
	3.048–4.064	0.120–0.160	0.78	5	2.17	14	2.48	16
	4.064–5.080	0.160–0.200	0.78	5	1.55	10	2.17	14
1	5.080–7.112	0.200–0.280	0.78	5	1.55	10	1.86	12
	7.112–8.890	0.280–0.350	0.78	5	1.55	10	1.86	12

^{a} For greater depths of hardening, lower kilowatt inputs are used. ^{b} These values are based on use of proper frequency and normal overall operating efficiency of equipment. These values may be used for both static and progressive methods of heating: however, for some applications, higher inputs can be used for progressive hardening. ^{c} Kilowattage is read as maximum during heat cycle. ^{d} Low kilowatt input may be used when generator capacity is limited. These kilowatt values may be used to calculate largest part hardened (single-shot method) with a given generator. ^{e} For best metallurgical results. ^{f} For higher production when generator capacity is available.

Courtesy of ASM International, Materials Park, Ohio.

based upon a combination of power density, frequency, heating time, work coil geometry, and workpiece material. After heating, the specimen is quenched at a time interval depending on time-temperature metallurgical considerations for particular system parameters. These are the primary guidelines for frequency and power selection only. Accurate design of the process and induction coil can be done using computer simulation (see Sec. 16.2.2.11).[22]

16.2.2.2 Induction Hardening Procedure. Induction hardening (i.e., induction heating followed by quenching) is divided into two main groups: single-shot or stationary induction (SI) hardening and scan induction (scanning) or progressive induction (PI) hardening.[23] Either immersion quenching or spray quenching is used. Quenching media may be water, polymer solution, oil, or even forced air, depending on the steel grades and treatment specifications.

Stationary Induction or Single-Shot Hardening. In single-shot or SI hardening, work coils with magnetic flux concentrators are typically used.[22] The multiwinding work coil and integral cooling ring are designed to cover the full length of the area(s) to be hardened so that the entire workpiece is heated at the same time with a one-step process.[24] The workpiece is usually rotated during heating (austenitizing) and, if possible, during quenching to provide increased temperature uniformity.

It is used for components with a rotational symmetry, and machines are frequently built to accommodate a range of similar type of components.[24] It can also be used for different-sized parts with ease. However, the decision to use the same coil or to make the new coil must be proved by technical and economical analyses because the rated power of the induction equipment could not be utilized completely.

Advantages of single-shot hardening include the following:

1. More uniform depth of hardness pattern for parts with diameter changes
2. Increased production rate because of heating of entire part at the same time
3. Reduced cost per part
4. Minimum floor-space requirements
5. Less distortion due to straightening during heating when the material is in the ductile state rather than stressing after surface hardening, as is found in the case of the scanning technique

Disadvantages of single-shot hardening include these:

1. The single-shot work coil is more complex.
2. This is suited to heat-treat only one or similar component(s) (e.g., shaft).[25]
3. More power is required to heat the entire workpiece; hence it is higher in capital cost and lacks flexibility (usually a dedicated system).[26]

Scanning (or Progressive) Hardening. Progressive hardening uses a single-turn or cylindrical work coil,[23] and the long workpiece is fed through the work coil and an integral quenching ring. A portion of the component is heated progressively by moving it at a certain rate, usually automatically, through the coil and is subsequently quenched. Either the component or the work coil may be fixed, depending upon the particular application. Alternatively, a short coil may be moved along the component length once (or several times).

This technique is performed mostly on symmetrical parts (e.g., shafts) and in some cases on nonsymmetrical parts (e.g., gear teeth). This method may continue to serve for more applications, but for mass production the single-shot technique is gradually gaining ground.[25]

16.2.2.3 Gear Hardening. Gears are single-shot induction heated either by encircling the whole gear with the coil or in larger gears by heating them tooth-by-tooth. In the latter case, root and ramp are hardened, leaving the tip ductile and soft, whereas in the former case, through-hardening of teeth occurs. To avoid this through-hardening of gear teeth, *dual-pulse* (or double-frequency) contour hardening was developed, in which the gear is heated in two stages: preheating within an induction coil to a suitable temperature at a medium frequency (3 to 10 kHz) and final heating at a high frequency (30 to 450 kHz) to allow the current to penetrate only an exact repeatable surface layer of gear teeth. Quenching to room temperature is then accomplished, followed by induction tempering at low-to-medium frequency in-line as part of the integral hardening/tempering machine.[18,27,28] The benefits of this method include uniform case depth, superior mechanical properties, and controlled microstructure.

16.2.2.4 Advantages and Disadvantages of Induction Hardening. Induction surface hardening has the following advantages over other conventional surface hardening methods:[29–32]

1. Because only the surface is heated, the possibility of component distortion is minimal—particularly with simple and symmetrical shapes such as shafts. This also offers a substantial energy savings.
2. Selective (or localized) hardening is possible through proper design and manufacturing of the work coil.

3. Extremely rapid heating (on the order of seconds) due to high surface power densities (3 to 5 kW/cm^2) may be achieved.

4. Since induction heating operates at a high power density, the equipment is smaller; i.e., less floor space is needed than if a furnace were employed. Usually no special foundations are needed, or pits as in the case of furnaces.[33]

5. There is minimal surface decarburization and oxidation; i.e., significantly less scale is formed without protective atmosphere. Sometimes on critical parts, vacuum, controlled atmospheres, or shielding gas is required.[30,34] Working conditions are much cooler and cleaner.

6. Increased wear resistance and fatigue strength and improved torsional and bending fatigue life (because of the development of compressive residual stresses, in the surface layers to the order of 6 to 8 ksi, and a refined core) in rear axles and power shafts are produced, and better dimensional characteristics in gears are developed.[30]

7. Control of processing and production is simplified. The process can be adapted to in-line production and can lend itself to full automation, especially for large series of parts, thus eliminating the need to stockpile parts to become ready for heat treatment by a thermochemical process.

8. Setup changes for different parts are quick and simple, and the need for special skilled operators is eliminated.[35]

9. A straightening required can be carried out on the unhardened surface and, to a certain extent, on the hardened surface.

10. Unlike selective carburizing, it eliminates copper masking and extra machining operations.[35]

11. Operating and maintenance costs are lower, with deeper hardening case.[36] Higher efficiency than that provided by some furnace methods can be found.

12. This technique is extremely versatile through proper work coil and quenching process design; a wide range of sizes, shapes, and materials, from small pins to large diameter gears, can be successfully heat-treated in certain locations to a defined hardness depth and pattern.[33]

13. The ease of handling components that are cold immediately before and after hardening is a definite advantage.

14. There is minimal stand-by power.

15. Precision and reproducibility of hardening patterns (for the same heats of steel and other factors), and its high output capability,[31,32] are other advantages.

16. The introduction of induction equipment has also resulted in important "spin-offs," particularly in reduced energy consumption through the use of a high-efficiency power source and improved working conditions.

17. It combines a machine tool type of accuracy with the precise application of heat by means of induced electric power.[32]

18. The amount of energy input may be monitored on a power/time basis, which leads to consistent heat treatment, provided the material is the same.

19. Any deviation from the acceptable quality is soon recognized by quality control and can be quickly corrected. Thus, only a small portion of rejects will occur instead of the entire parts in the case of a batch furnace.

Disadvantages of induction hardening are as follows:

TABLE 16.6 Typical Coil-to-Workpiece Gap with Various Heating Frequencies[18]

	Frequency, kHz		
	1–3	10–25	50–450
Coupling gap			
mm	6–3	3–2	1.5
in.	0.23–0.12	0.12–0.08	0.06

Courtesy of Marcel Dekker, Inc.

1. The process is easy to implement to components having a relatively simple shape.
2. The high capital cost necessitates a high degree of equipment utilization.
3. Only a limited number of grades of steel can be induction hardened.
4. The operation of an effective induction heating plant requires specialized knowledge for running, setup, and maintenance of the induction systems.[22] Full backup facilities such as induction coil and fixture manufacturing must be provided internally or from specialized companies.

16.2.2.5 Induction Coil Design. The term *inductor coil* or *work coil* denotes a current-carrying conductor in close proximity to a workpiece to be heated. Work coils are typically made from oxygen-free high-conductivity (OFHC) copper tubing or solid machined blocks and are almost always water-cooled.[18] An induction coil design must meet the following requirements:[37] (1) provide a specified heating pattern (and metallurgical results) with maximum efficiency, (2) have a sufficient matching to the power supply (with high impedance and power factor[†]), (3) have a geometry to conform to the workpiece and required heat pattern, (4) provide easy loading and unloading of workpieces,[16,22] (5) maintain a proper and uniform coupling (or air) gap between the coil and workpiece, (6) have a satisfactory lifetime, (7) have low sensitivity to changes in the part dimensions and positioning within the specified range, and (8) meet special needs such as quenchant supply, atmosphere, and material handling incorporation into the machine structure, etc.[37] Table 16.6 lists the recommended coupling gaps for three frequency ranges. However, these gaps may vary for different applications.[18]

Low-Frequency Heating Coils. Low-frequency induction heating coils (<3 kHz) have a large number of turns and are usually employed for through-hardening of metals, particularly those with large, simple cross sections. Typical applications include round or round-cornered square (RCS) stock for forging or extrusion and slabs for hot rolling. In these cases, coil design comprises a solenoid coil or its variation to match the basic cross-sectional shape (e.g., square, rectangular, trapezoidal).[21]

Medium to High-Frequency Coils. Medium to high-frequency coil (3 to 25 kHz) designs consist of simple solenoid coils and its variations. These are single turns and multiple turns of copper tube or machined copper blocks with waterways for

---

[†] The *power factor* (cos ϕ) is defined as the cosine of the phase angle between the voltage and current of an electric circuit. It is thus the ratio of the kilowatts to the actual volt-ampere product required from the chosen frequency converter.[18]

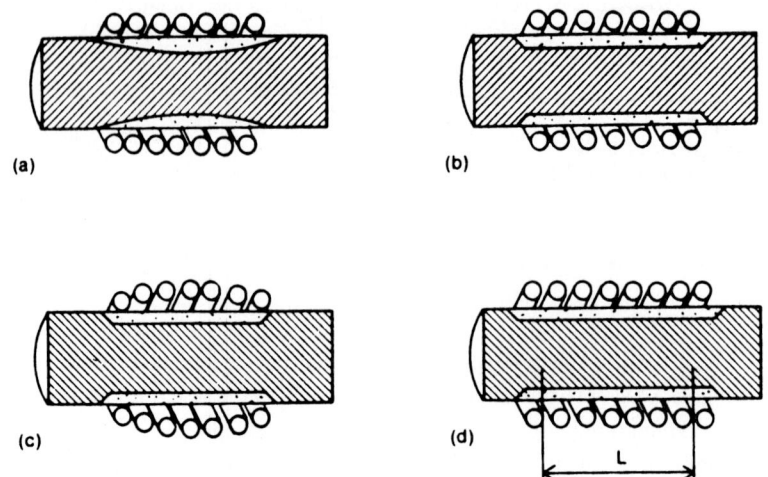

(a)   (b)

(c)   (d)

L

FIGURE 16.6   Typical coil designs used for surface high-frequency induction hardening.[16] (*Courtesy of ASM International, Materials Park, Ohio.*)

cooling. Figure 16.6 shows the typical coil designs, illustrating that control of the heating pattern can be effected by varying the pitch of the coil, contouring the coil, or using a longer coil.[16] Figure 16.7 shows examples of heating patterns in a round bar produced by various designs of induction coils, such as simple solenoid coil (Fig. 16.7a), coil for internal bore heating (Fig. 16.7b), pie-plate type scan inductor (Fig. 16.7c), single-turn scan-inductor with contoured half turn (Fig. 16.7d), and pancake coil (Fig. 16.7e). Bent or contoured induction coils of quite different shapes and designs are used to uniformly concentrate the current into odd-shaped workpieces and can be classified into single-turn inductors, cored inductors, internal coils, split returns, clamshells, and hairpin inductors, split coils, quick-change coil design, folded pancake coil, plate inductor, "Ross" coil, slipper or skid coil, transverse-flux heating (TFH) coils, and longitudinal-flux heating (LFH) coils. They are used for various specialized applications.[16–18,21] On the other hand, when the ends of rectangular coils are bent, they are called conveyor/channel coils and are used to harden the areas beside the center of the coil turn. Some of these coil designs are briefly described below.

Internal (solenoid) coils are used for heating internal bores. They may be of single-turn or multiturn construction and may need addition of core material (or flux concentrator located inside the coil) to intensify heating. Practically, the minimum outside diameter of the coil is limited to ~12 mm. Internal coil coupling should be as tight as possible. An oval or flattened copper conductor should be used in this application for better efficiency.

Clamshell or *clapping-type coils* are usually employed for crankshafts, where part geometry does not permit close coupling, or for unusual-shaped cams (i.e., cams for racing engines). This type of coil possesses a hinge on one side and opens and closes when the part is in position.

Hairpin coils are usually in the form of bent tubing and are used at radio frequency where a single turn would cause problems during loading and unloading of

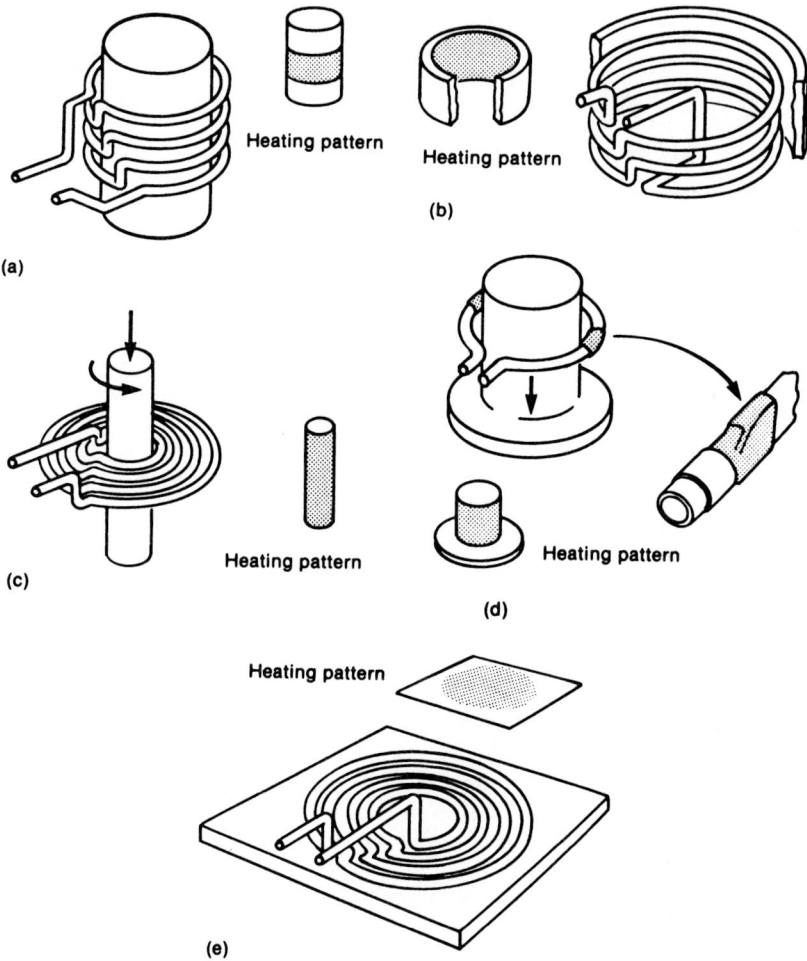

Heating pattern

Heating pattern

(a)

(b)

(c)

Heating pattern

(d)

Heating pattern

Heating pattern

(e)

FIGURE 16.7 Examples of five basic designs of work coils for use with high-frequency (over 200 kHz) units and the heating patterns developed by each: (*a*) A simple solenoid coil for external heating. (*b*) A coil for internal heating of bores. (*c*) A pie-plate type of coil designed to provide high current densities in a narrow band for scanning applications. (*d*) A single-turn coil for scanning a rotating surface, provided with a contoured half-turn that will aid in heating the fillet. (*e*) A pancake coil for spot heating.[16] (*Reprinted by permission of ASM International, Materials Park, Ohio.*)

parts. The part is moved into the coil through sliding from the loading position, and is moved out after the completion of heating.

Split coils are usually employed where a problem exists in providing a high enough power density to the area to be heated without very close coupling and where loading and unloading of the part remain a hurdle. Typical examples of such cases are hardening of journals and shoulders in crankshafts (Fig. 16.8). This design also incorporates the quenching device through the face of the inductor.[21]

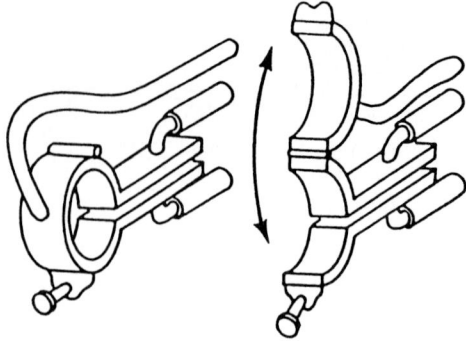

FIGURE 16.8 A split inductor used for heating crankshaft journals.[21] (*Reprinted by permission of ASM International, Materials Park, Ohio.*)

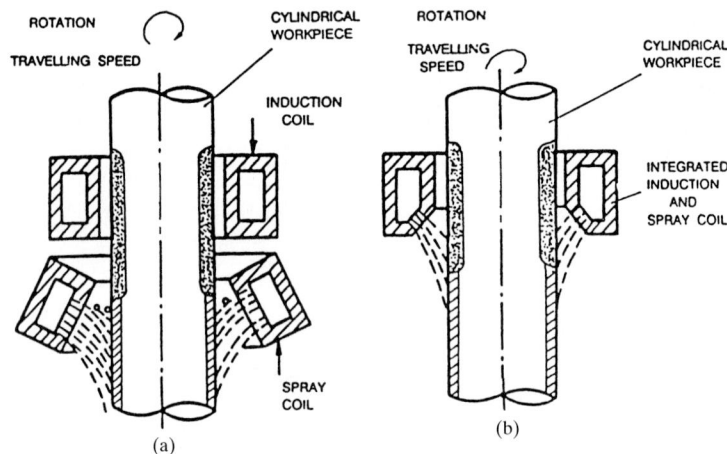

FIGURE 16.9 Inductor/quench designs for progressive (or scanning) hardening of cylindrical workpiece. (*a*) Separate coil and quench design. (*b*) Integrated coil and quench design.[33,33a] (*Courtesy of J. Grum.*)

Quick-change coil design may be recommended if fast or frequent coil change-over is required. This may be in the form of toggle clamp, dovetail, or pneumatic. The toggle clamp and dovetail type designs are generally limited to 300 kW with a frequency range of 3 to 450 kHz. All electrical contact surfaces should be clean, free of nicks or burrs, and be of very good finish with silver plating to ensure good reliable electrical contact.[18]

Scan Inductors. The primary advantages of scanning-type inductor coils are (1) its flexibility in running various lengths of parts, (2) its repeatability, and (3) easy automation process. Scan inductors are usually single-turn with separate or combined inductor/quench design (Fig. 16.9)[33] or multiple-turn with a separate inductor/quench ring design which is mounted on the scanner. They can be machined from a solid copper bar in order to make them very rigid and durable. (See also Sec. 16.2.2.2.)

Transverse-Flux Heating (TFH) Coils. TFH coils are widely used in heating relatively thin metal strips with frequencies from 200 to 1000 Hz (Fig. 16.10*a*)[21] to

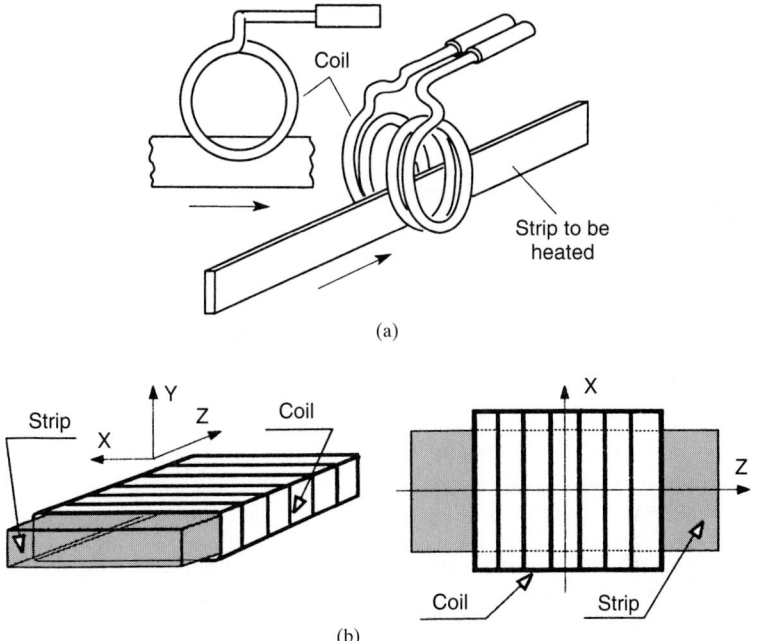

FIGURE 16.10 (*a*) Transverse-flux coil for heating thin sections.[21] (*Reprinted by permission of ASM International, Materials Park, Ohio.*) (*b*) Longitudinal-flux inductor heater.[18] (*Courtesy of Marcel Dekker, Inc, New York.*)

ensure that magnetic flux passes through the flat part at right angles.[38] These TFH coils almost always employ magnetic flux concentrators. This type of work coil does not necessarily require a small coupling gap. A high electrical efficiency can be achieved if the strip thickness is less than 8.5 mm or is 2 to 10 times less than the penetration depth, by using a much lower frequency of the power supply. The reduced operating frequency leads to cost-effective advantages for both capacitor bank and the converter.[39] Sometimes TFH coils are used for local hardening of the edges of knives, hacksaw blades, etc., at high frequency, up to several megahertz.[22]

Longitudinal-Flux Heating (*LFH*) *Coils.*   The LFH coil is a solenoid heater (Fig. 16.10*b*) which surrounds the strip, produces the longitudinally oriented magnetic field, and induces eddy current which circulates within the thickness of the strip, causing heat by the Joule effect (I^2R).[40] Conventionally, LFH inductors have high efficiency and uniform temperature distribution within the strip. High coil efficiency has been found when the ratio of strip thickness d_{st} to penetration depth δ is ~2.7 to 2.9 for magnetic strip and ~3.0 to 3.3 for nonmagnetic strip.[18]

This type of induction heating does not usually require (a) a very tight air gap and (b) large adjustments of the heater for strips with different widths and thicknesses. LFH inductors are effectively employed for low-temperature induction heating of thin magnetic strips when the final temperature of the strip is below the Curie point. However, for non-magnetic thin strip heating (i.e., Al, stainless steels, etc.), or for heating magnetic strip above the Curie point, this type of inductor will need adequate power at high frequencies of up to several hundred kHz or even MHz (M = mega) and sufficient power.[18]

16.2.2.6 Magnetic Flux Concentrators. The application of magnetic flux concentrators (or intensifiers) is the well-recognized and very effective method for work coil improvement. In different applications, they play different roles and have different names, namely, controllers, diverters, cores, impeders, shields, and shunts. In some cases, coils cannot work properly without flux controllers. One purpose of a magnetic flux concentrator is to make the magnetic field more intense in certain workpiece areas (which are difficult to heat) during the induction heating process.

There are several requirements of concentrators in induction heating applications. They must work in a wide range of frequencies, have high permeabilities and saturation flux densities, and have stable mechanical properties and resistance to elevated temperatures caused by magnetic losses in the concentrator and by heat transfer from the workpiece. They must withstand an attack of hot water and quenchants. Machinability is also a very important property for successful application of flux concentrators.[22]

Magnetic materials mostly used for flux concentrators are[22] (1) grain-oriented silicon steel laminations for use at frequencies from below 10 to 50 kHz; (2) ferrites for use in a very wide range of frequencies when properly selected and applied; and (3) magnetodielectric materials (MDM), for use at low frequencies (up to 30 kHz), medium frequencies (between 30 and 200 kHz or even between 10 and 450 kHz), and high frequencies up to 13 MHz. Readers interested in more details should consult other sources.[41–43]

The selection of flux concentrator material depends on the location of the concentrator, its shape, and the applied frequency; the type of service or the particular application; and the extent of exposure to radiated heat. A full understanding of the use of flux concentrators can help to improve coil efficiency; and the use of 2-D and 3-D computer simulators in the initial stages of coil design can greatly reduce the cost associated with coil development and can help to save valuable production time.

Advantages of using properly applied magnetic flux concentrators can be substantial, which can include any or all of the following:[18,22,41–45]

1. Heat pattern development due to better control of distribution of heat sources in the part
2. Better metallurgical properties due to higher hardness achieved for a given coil
3. Better utilization of the power transferred into the workpiece
4. Increased productivity with a given coil due to reduced cycle time
5. Improvement of the coil efficiency and the coil power factor
6. Improvement in the coil matching to power source and in efficiency of the supplying circuitry
7. Ability to treat larger parts than an existing power supply could manage
8. Increased coupling distance between the workpiece and coil part
9. Ability to block, not heat, or selectively heat, specific areas of the part
10. Reduction of number of rejected parts, rework, and scrap
11. Improvement in equipment and coil lives due to reduced power requirements from the generator

12. Minimal geometric distortion of the workpiece

13. Elimination of external magnetic fields in close proximity to the coil

14. Significant energy and capital equipment savings

The technical and economical significance of individual benefits depends on specific conditions of a particular induction heating application. A detailed study with computer simulation and full-scale experiments on the effect of concentrators have shown that the proper application of concentrators is always beneficial in an induction heating process.

Disadvantages of flux concentrators are

1. They are expensive.

2. They require careful maintenance and preventive measures against mechanical damage which increase cost and labor.

Applications include (1) single-shot hardening of long shafts, (2) single-shot hardening of camshafts and drive stem, (3) hardening of valve seat, (4) surface hardening of (nodular iron) rocker arm tip (using split-return type of inductor), and (5) channel coil for continuous annealing (of parts such as shell case and spark plug, moving on a chain conveyor).[46]

16.2.2.7 Coil Impedance Matching. A very important aspect of coil design is the matching (or adjusting) of the inductor coil to the power supply, in order to deliver the required power from the power supply to the workpiece and to obtain sufficient heating in the required time at minimum cost.[16,47,48] For full utilization of the power rating of a generator, full-rated current must be drawn from the power supply, and the power factor of the generator must be close to unity. This is usually termed *impedance matching*.[36] Such impedance matching may be either fixed-frequency source matching or variable-frequency source matching.[36] Variable ratio transformers, capacitors, and sometimes inductors are connected between the output of the power supply and the induction coil.

Typically, an induction heating power supply produces its peak output power when the impedance of the load circuit and the ideal impedance of the power supply are the same. At all other impedance values, the amount of power supply will be different. The impedance of the load can be regarded as the combination of the values and sizes of the workpiece, inductor, and capacitor. Hence as these vary, the power generated by the power supply varies.[48] The transformer serves as an impedance matching device such that the load impedance appears to the power supply as the ideal impedance. The adjustment of these components is usually termed *load matching* or *load tuning*.

16.2.2.8 Power Supplies. In addition to the induction coil and workpiece, the power supplies are perhaps the most significant component of an overall induction heating system. They convert the available utility line frequency power to the useful voltage (or current) and frequency. They are often called converters, inverters, or oscillators, but they are usually a combination of these. The converter portion of the power supply converts the line-frequency (50/60 Hz) ac power supply to dc, and the inverter or oscillator portion changes dc back to single-phase ac at the desired frequency. The new-generation power components developed in the last two decades are power semiconductor devices (such as SCRs, diodes, and MOSFET and IGBT transistors), capacitors, and transformers.

Many different power supply types and models are available to meet the heating requirements of a large variety of induction heating applications. Figure 16.11 shows the graphical representation of various power and frequency combinations that are used for typical heat-treating applications.[18] Figure 16.12 shows the primary design features of the inverter configurations most widely prevalent for induction heating power supplies.[18] The three basic types of inverter circuits are the voltage-fed with a series resonant-load and parallel resonant-load circuits and the current-fed with parallel resonant-load circuits. The chart is further subdivided according to the dc source (fixed or variable), the mode of inverter control, and the load circuit connection (series or parallel). Figure 16.13[18] shows the most commonly used power switching devices, such as thyristor (SCR), IGBT, MOSFET, and vacuum tube, to provide the various power and frequency combinations required for induction heating. The large areas of overlap in this figure provide the opportunity to use more than one type of power supply. The power components are briefly described below.

Thyristor or SCR. High-power SCRs (silicon-controlled rectifiers) having rapid *turn-on/turn-off sets* have largely replaced motor-generator sets for medium-frequency induction heating power supplies.[48] The SCR allows the current to flow in one direction, but only after a positive voltage is applied to its terminal. When the voltage applied is reversed on an SCR, it turns off and blocks the current flow in both the forward and reverse directions. This period is referred to as the *turn-off* time. Its advantages include (1) improved efficiency, (2) low initial and maintenance costs, and (3) availability in a wide variety of sizes and frequencies. The SCR was limited to frequencies up to 10 kHz, but presently with the addition of complex circuitry and techniques it could be operated up to 25 to 50 kHz.[48]

Diode. The diode is the power semiconductor device that allows current to flow only in one direction when voltage is applied in the forward direction. It starts to block the flow of current when the voltage is reversed.

Transistors. These semiconductor devices (such as MOSFETs and IGBTs) allow current to flow from the input terminal to the output terminal by applying a control/signal to the gate or base terminal. For a transistor to be useful in a high-

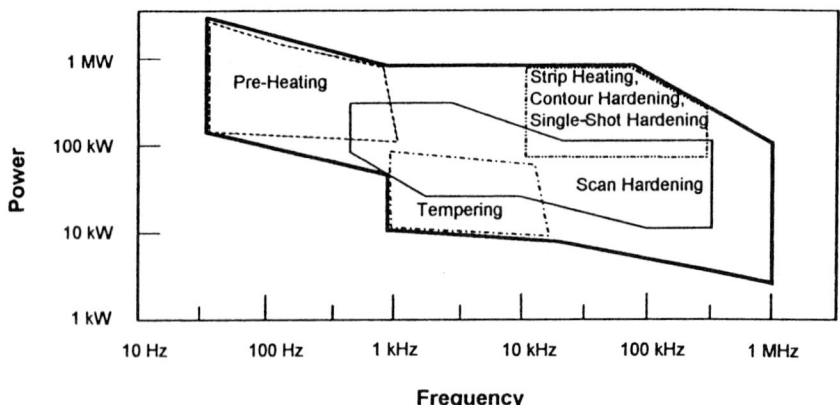

FIGURE 16.11 Typical induction heat treatment applications.[18] (*Courtesy of Marcel Dekker, Inc., New York.*)

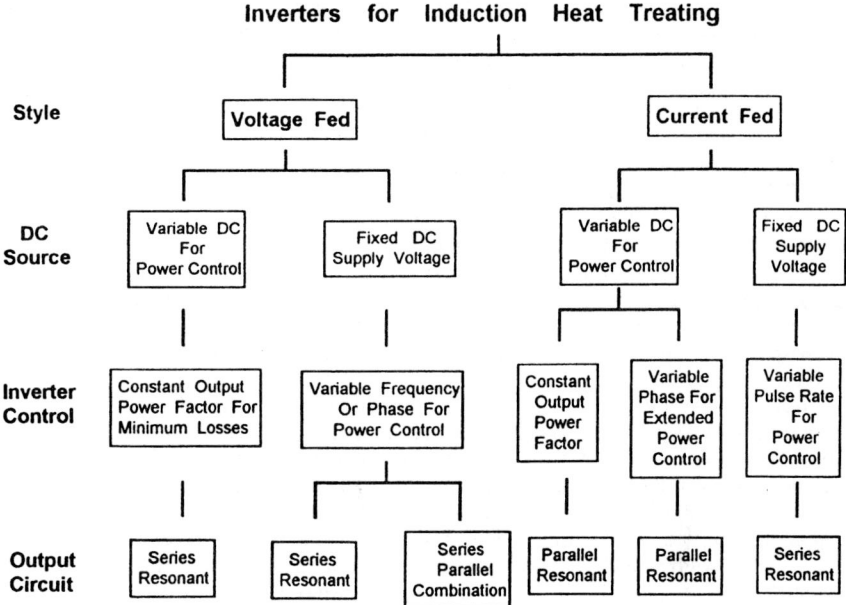

FIGURE 16.12 Induction heat treatment inverters.[18] (*Courtesy of Marcel Dekker, Inc., New York.*)

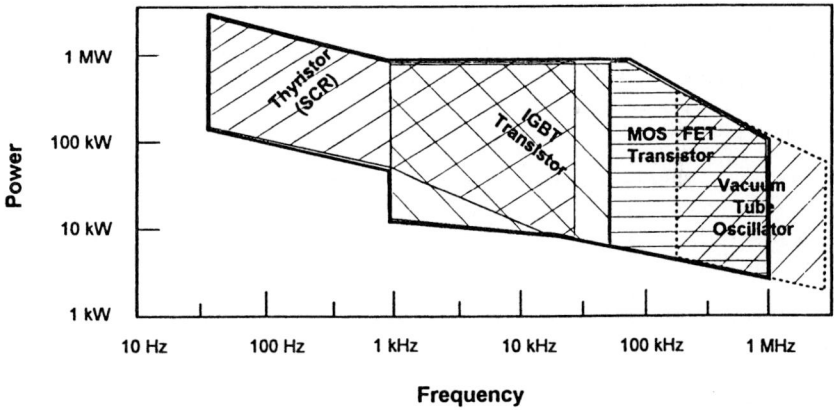

FIGURE 16.13 Modern inverter types for induction heat treatment.[18] (*Courtesy of Marcel Dekker, Inc., New York.*)

frequency induction heating application (10kHz and over), it must (1) block high voltage, (2) carry high current, and (3) switch on and off at any time with an electric signal.

The advantages of transistorized induction heating power supplies are that:

1. They are much smaller and can often be mounted close to the heating coils. This means a reduced shop floor space requirement.

2. There is virtual elimination of interconnections bus or cable losses.

3. Overall system efficiency is higher, i.e., there is a lower electric power and water cooling cost.

4. Their output rating varies from 1 kW to 1.2 MW at heating frequencies from 1 to 800 kHz.[49,50]

MOSFET. The metal oxide silicon field effect transistor technology has offered relatively high voltage, high current, and very fast switching speeds. This is achieved by putting thousands of very small, fast transistors that are all connected in parallel on a single chip of silicon, measuring about $\frac{1}{4}$ in. on each side. Large MOSFET transistor modules combine many of these chips connected in parallel on a common mounting base.

IGBT. The IGBT (insulated gate bipolar transistor) device is a combination of two transistor technologies to provide high voltage, high current, and fast switching speeds compared to an SCR. Bipolar transistors capable of relatively high voltage and high current have been available for about three decades, but are slow switching and need relatively high power control signals. Small, low-power MOSFET transistors with very fast switching speed and low power control requirements have also been available for many years. But the two technologies, i.e., the combination of the MOSFET transister on the control end and the bipolar transistor on the power handling end, have offered the best of both in the IGBT. The IGBT can be operated with ease up to 50 kHz frequencies, but more recently some up to 100-kHz frequency and as low as suited for SCR are also available.[48] Mostly, the IGBT units have been produced for heat-treating applications, where the frequency is higher (>10 kHz), the duty cycle is intermittent, and the power is low when compared to forging/melting applications.[51]

Vacuum Tube Oscillators (Generators). These are widely used in the frequency range of 10 to 27 MHz for different high-frequency applications.

Capacitors are available to meet the needs of new high-frequency transistorized technologies that possess a high kVAR,[†] very low inductance, and low losses; and they are offered in a package that is compact and easy to water-cool.[49]

Transformers, based on ferrite cores, are also available in many unique configurations that have the capability to handle high current at frequencies >3 kHz. These new transformers have low power losses and very low inductance, and enjoy load-matching flexibility.[49]

16.2.2.9 Steel Grades for Induction Hardening.
The induction hardening process is carried out on plain carbon and low-alloy steels containing sufficiently high carbon content, usually in the range of 0.3 to 0.5%, in order to transform the surface layers to martensite after quenching from the austenitizing temperature, which will give rise to surface hardness in the range of 50 to 60 Rc. The upper value is achieved with high carbon content, which may run the risk of quenching or hardening cracks. However, if well-controlled heat treatment is carried out, higher carbon contents are permissible for rolls made of steel containing 0.8% C, 1.8% Cr,

---

[†] This rating term for capacitors identifies the amount of reactive volt-amperes that the capacitor can supply to the circuit when run at a particular voltage and frequency.[18]

and 0.25% Mo.[8] Other materials that can be induction hardened are cast irons and sintered metals.

Plain carbon steels are usually water-quenched, whereas the alloy steels are quenched in air, quenching oils, oil emulsion media, or polymer quenchants (such as PAG-based solution). Before induction hardening, steels should be in either the normalized or the hardened and tempered condition. Steels containing more than 0.5% carbon in the fully spheroidized state should not be chosen for induction hardening due to their poor response to induction hardening; higher temperatures and/or longer heating-up times are needed for dissolution of spheroidal carbides compared to those needed for carbides formed in normalized or hardened and tempered steels. Such conditions may produce coarse austenite grain size at the surface, which will result in coarse martensite structure combined with a high percentage of retained austenite; this may reduce fatigue resistance and promote seizing or galling.[12]

16.2.2.10 Pre- and Post-Heat Treatments. Prior to application of higher-frequency induction heat for rapid austenitization, preheating is carried out using a low-to-medium frequency until the desired temperature ($\sim$400°C) is reached. Preheating makes the attainment of heating and hardening patterns on irregular shapes easier; it increases the hardened depth, and perhaps it provides better size and residual stress control.

Preheating in contour gear hardening ensures a reasonable heated depth at the roots of the gear, the attainment of the required metallurgical results, and a decrease in distortion; reduces the amount of energy needed in the final heat; and permits the high-powered RF current applied during the final stage to follow the contour of the gear.[18]

Tempering of the induction hardened surface may be carried out separately in a conventional furnace or by the use of a second induction coil at lower power densities. The latter process, called *induction tempering*, is a short-time high-temperature treatment rather than a conventional long-time low-temperature treatment. (See Sec. 14.9.1 also for more details.)

Induction tempering, like induction hardening, has become a viable commercial process, replacing the conventional furnace operations in many high-production applications such as oil well pipes, railroad rails, tubes, ball screw, shaft, and bar.[12,16] It may be pointed out that time and temperature are both critical in induction tempering. Hence, it is necessary to establish equivalent time and temperature, using an extension of the Hollomon-Jaffe correlation with appropriate tempering parameter P when induction tempering lines are set up initially or subsequently modified. (See Sec. 14.9.1 also for more details.) This extended relation takes into account rapid heating at fixed temperatures as well as continuous heating and continuous cooling. This is done by measuring an effective tempering time t^* for an isothermal temperature heating cycle which corresponds to the continuous cycle. For continuous heating from room temperature to typical induction tempering temperatures, an increment in t (that is, Δt_i, usually 0.005 to 0.01 times t_{total}, where t_{total} is the total heating time) provides sufficient calculation accuracy.[16]

Induction tempering is a valuable tool for manufacturing cells. The main advantages of induction tempering are[52]

1. Reduced energy cost, because the energy induced into the part is often confined to the hardened region
2. Precise control of power, monitoring of the final temperature of each component, and improved working environment

3. The shortness of the time/temperature relationship to accomplish tempering compared to furnace tempering, thereby minimizing the number of parts in process to just a few.

16.2.2.11   Computer Simulation of Induction Hardening Processes.   Computer simulation is an effective tool for induction heat-treating processes and improved coil design. There is no universal software program available that can accurately simulate all features of the induction heat-treating process. Hence the best approach consists of electromagnetic and thermal field simulation; and if desired, the thermal fields can then be exported to a structural transformation program.[53,54]

There are a number of different software programs available for simulating the induction heating process, but only a few of them, notably Flux 2D (a 2-D coupled electromagnetic plus thermal program) and ELTA (Electro Thermal Analysis) (1-D coupled electromagnetic plus thermal program), are adequate for induction coil optimization. Figure 16.14*a* and *b* shows the temperature distribution in an induction scanning application. Figure 16.14*a* is a map of temperature and isolines at specified temperatures. The white background represents the full austenitization temperature needed for complete martensite formation. In this process, induction was used to harden a 1.8-in.-O.D. shaft for the automotive industry. The frequency for this application was 3 kHz, and the scanning speed was 0.3 in./s. Using ELTA, it was possible to make a good 1-D approximation of this 2-D system.[54]

Some researchers have suggested a hierarchical approach (or a rule of pyramid), whereby the simpler software packages are employed as a foundation for the more complicated simulations, if they are required. This approach has been used in practice and found to significantly reduce design time and process development costs. The main source of error in simulation of induction heat-treating processes lies in the accuracy of the material property and heat-transfer coefficient data. More experimental study is greatly needed in this area to create a database for accurate simulation of induction heat-treating processes.[54]

16.2.2.12   Applications.   This process is now employed as an effective and economical method of hardening a very wide range of components in the general engineering and automotive industries. Components which are usually induction heat-treated range from bolts to crankshafts.[32] Induction heating may also prove to be an ideal technique for annealing, stress relieving, and through-hardening and tempering, and hot forming processes such as forging, extrusion, rolling, swaging, and bending.[17]

Crankshafts, camshafts, splined shafts, transmission shafts, driveshafts, tube-type shafts,[55] universal joints, various gears, valve seats, rocker arms, cylinder heads, wheel spindles, and ball studs are a few of the components chosen for selective surface hardening in the automotive power train industry. Other examples of applications include selective crown hardening of railroad rails and torque hub assembly.[56] The frequencies commonly used range from 10 to 450 kHz for the smaller articles in motor cars and from 1 to 10 kHz in the heavy-vehicle industry.[57] The production of induction-hardened cast pearlitic malleable iron transmission gears has now been widely used in the automotive industry because of various advantages, which include:[36] (1) cheaper casting than an alloy steel forging (material saving); (2) ease of machining cast iron (machinery saving); (3) use of less energy in induction heat treatment; (4) large decrease in noise level; (5) increased wearability due to malleable cast; (6) increased dimensional and distortion control; and (7) cleaner and cooler working conditions.

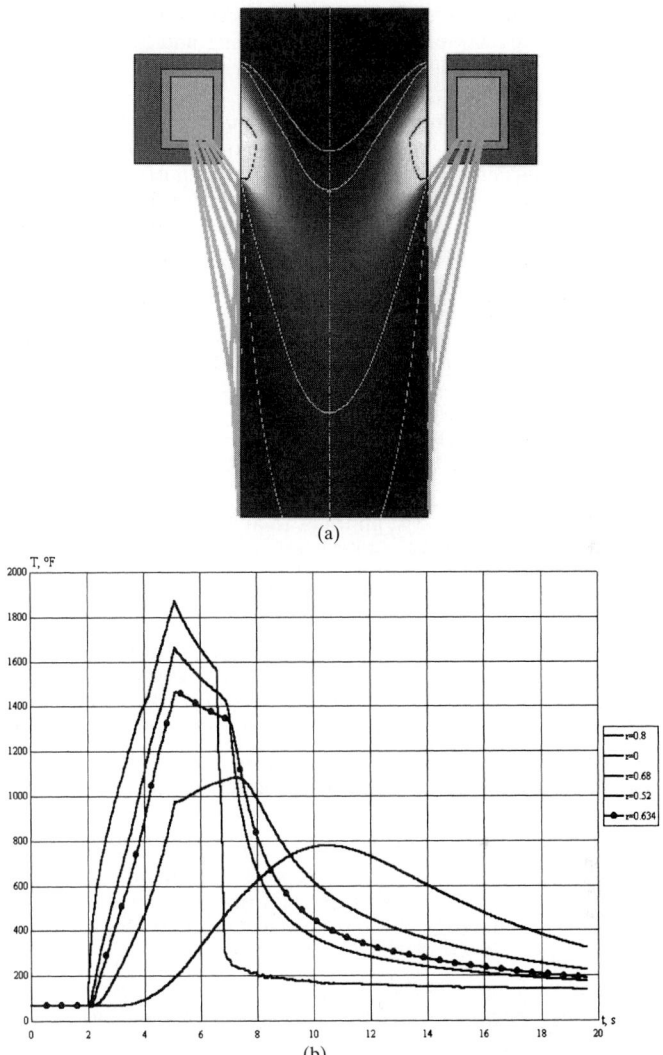

FIGURE 16.14 (*a*) Simulation of induction scanning using ELTA program.[54] (*b*) Temperature graphs at different radii versus time for the induction scanning process using ELTA program.[54] (*Courtesy of Centre for Induction Technology, Inc., Michigan.*)

Surface hardening in the machine tool field includes lathe blade, machine column, and transmission gear and shafts. Other examples include contour gear hardening by the dual-frequency method,[28] tooth-by-tooth hardening of large fraction and other gears up to more than 1422 mm in diameter, and surface hardening of hydraulic press tool holders weighing up to 5000 kg. This clearly demonstrates that size is no problem for the induction hardening process. However, in these cases, specially designed inductors should be used to ensure closer tolerance of localized heating patterns.[57] Applications in the metalworking and hand-tool fields include rolling mill rolls, pliers, hammers, chisels, screw drivers, axe and hatchet products, cutters, and hacksaw blades.

Through-hardening applications include (1) oil country tubular products; (2) structural members; (3) spring steels; (4) chain links; (5) commercial airframe components;[58] (6) head hardening of rails; (7) austempering of lawn mower blades by induction austenitizing and quenching in a salt bath at 650°F to produce the desired bainitic microstructure without distortion; and (8) hardening of sharp edges of axe and hatchet product using channel-type inductor coil, 20- to 50-kW system, and in-line quench product at a rate of 1000 parts/day.

16.2.2.13   Safety Practices of Induction Hardening Equipment.   For better performance of the induction hardening equipment, the following safety practices should be adopted:[17]

1. Regular visual inspection should be made of inverter for evidence of water leakage or cracked hose, arcing or overheating of component and electrical connections.

2. Inverter doors must be closed at all times to avoid the accumulation of contaminants within the enclosure.

3. Operation of all protective circuits—door switches, pressure switches, and temperature switches—should be properly carried out.

4. Air (or coupling) gaps between the coil and workpiece must be constantly maintained once they have been established, because deviations cause under- or overheating. For induction heating of fillet areas, as on axle shaft and for scan inductors, more coupling gap should be provided due to the likelihood of distortion of the workpiece.[59]

5. Where an inductor has been grounded, it is necessary that grounding is not dispensed with or disconnected, for the safety of the operator.

6. All inductors and work coils must have adequate flow,[†] volume, pressure, and temperature of deionized cooling water supplies without undue concentration of solids in the form of salts, carbonates, and atmospheric source pollutants, in order to lessen failure of inductors. A water flow switch should be incorporated if the primary pump fails.

7. Preventive and protective measures must be considered for all dangerous and hazardous factors specific to induction heating systems, such as electric shock at high frequency, magnetic fields, arcing, and melt metal, in order to meet the safety standards.[22]

16.2.3   High-Frequency Resistance Hardening

Recently, a new method of selective surface hardening using high-frequency contact resistance has been developed in which a small area of the component becomes part

---

[†] Adequate water flow for proper cooling of copper coil is given by gpm = $PK_1K_2/K_3\Delta T$, where gpm = gallons per minute; P is total coil power, kW; K_1 is a tubing coefficient (for most high-frequency heat-treating applications), usually 0.5; $K_2 = 3415$ is a conversion constant that is derived from Btu/kWh; K_3 is a conversion constant denoting the heat capacity of water, typically 500; and ΔT is the allowable temperature increase in the cooling water, usually 40°F or less.

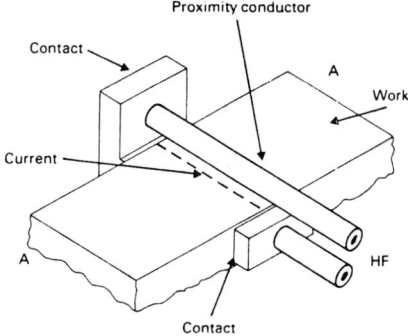

Section A-A

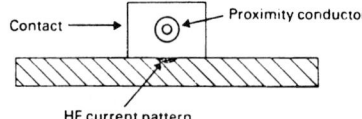

FIGURE 16.15 Basic principles of high-frequency resistance hardening method.[60] (*Courtesy of Wolfson Heat Treating Center, England.*)

of the coil.[60] Figure 16.15 shows the basic principles of this method. High-frequency current in the range of 300 to 500 kHz is supplied directly to the component through two small contacts at the ends of the area to be hardened. The current then flows through a *proximity conductor* placed near the surface to be heated. This produces a high current density and very rapid (usually less than 0.5 s) and highly localized heating of the component surface beneath the conductor. The heated area is self-quenched very rapidly by the adjacent large mass of cold material when the current flow ceases.[60]

In conclusion, when compared to conventional heat treatment, both the energy consumption and distortion are very low. In addition, the process is very fast; i.e., a high production rate of parts in the automotive or appliance industries is achieved. It requires a high-frequency power supply and can be employed with a variety of interchangeable contact systems for hardening a number of different components.[60]

16.2.4 Laser Surface Hardening

Laser surface hardening (LSH), being a promising alternative to conventional thermal surface hardening, has been investigated for more than two decades.[61] In LSH, thermal energy is produced by absorption of the laser radiation at the surface. The increase of temperature in the interior of the workpiece is due to conduction only.

In laser (beam) surface hardening (LSH) treatment, unlike the flame and induction hardening processes, the large mass of unheated iron-base alloys serves as the quenchant. Figure 16.16 shows the principle of laser beam hardening treatment (LBHT) by shaping and integration techniques. An LBHT machine comprises the following separate elements:[62–69]

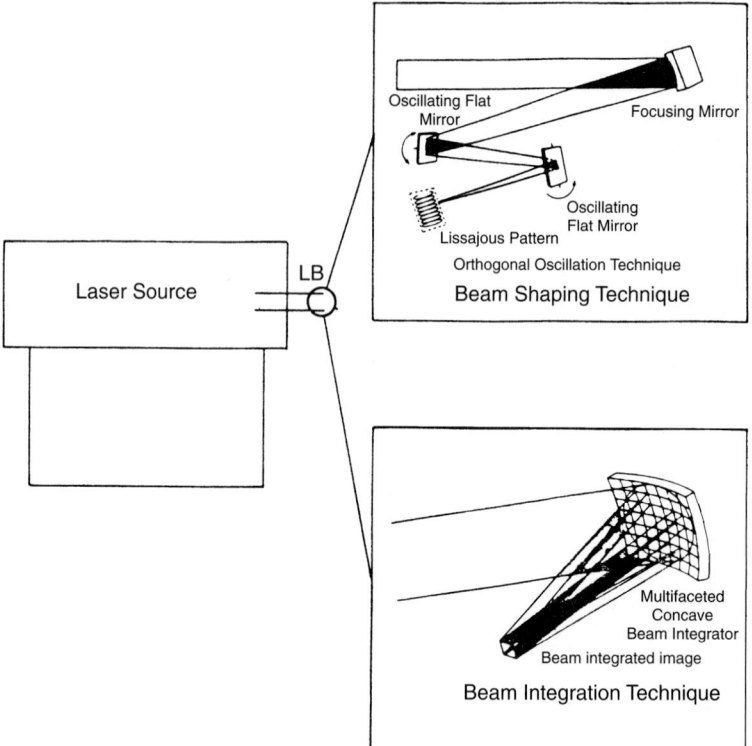

FIGURE 16.16 Principle of laser (beam) hardening treatment by shaping and integrating techniques.[62] (*Courtesy of The Institute of Metals, England.*)

1. The CO_2 laser consists of an optical resonant cavity where a gaseous mixture of CO_2, N_2, and helium flows. An electric discharge excites the gas mixture, emitting photons in the form of a high-power-density coherent beam from the end of the cavity. The Nd:YAG laser consists of a crystal in a gold-plated cavity, excited by flash lamps and emitting high-power-density coherent beams.

2. The laser beam (intense heat source) is brought to the working table by focusing optics, either mirrors or lenses, or by optical fiber (for Nd:YAG only). The focusing optics can shape the beam into a standard pattern such as a 0.25-in. (6.3-mm) square or specialized configurations such as a horseshoe, depending on the process profile.[63] Optical fibers are available for Nd:YAG lasers having shorter wavelength of $1.06\,\mu m$; however, power transmission fibers for the CO_2 laser with longer wavelength of $10.6\,\mu m$ have not yet been developed. Alternatively, hollow waveguide sapphire tubes or shiny tube or square section waveguides may be used.[64]

3. The beam can be focused by manipulation of lenses and mirrors on the surface to be heat-treated. This can be achieved in two ways: (*a*) by the beam oscillating method, where the last two mirrors are oscillated with a certain amplitude and phase shift in two perpendicular directions, so that Lissajous curves can be obtained to cover the surface to be heat-treated; and (*b*) by the beam integration methods (for spots >3 mm). These methods include scanning or rastering the near focused beam at sufficiently high speed to avoid melting; passing the beam through a reflective tube of the required shape, called a kaleidoscope, and imaging the beam at the exit from the kaleidoscope with lenses; and concentrating or focusing each part of the beam on the same region by an integrating mirror, i.e., segmented polished mirror. Another method uses integration by specially shaped lenses or even holographically etched mirrors, called kinoforms.[64]

4. To capture the energy of the beam and to hold the heat in the component to be treated, the component surfaces should be covered with an energy-absorbing black paint, colloidal graphite, spray paint (flat black), india ink, or a coating of zinc-, potassium-, or manganese phosphate $[(Mn,Fe)_5(H_2PO4)_4 \cdot 4H_2O]$, and a few proprietary black oxide coatings. They may also be etched or sand-blasted. Mn phosphate is more popular, but black spray paint and colloidal graphite are usually the most efficient and economical coating. In all cases, the reflectivity of the surface must be reduced to <40% for the stability of the process.[64,65]

Surface coatings should meet the following requirements: (1) proper thickness, (2) good adhesion, (3) chemical inactiveness, (4) high thermal conductivity, (5) high thermal stability, (6) low cost and ease of application and removal,[66] and (7) high degree of absorptivity of the laser beam.[66a]

A high-powered YAG laser (with kaleidoscope) for surface hardening of carbon steel was used without the absorbents. However, the hardened zone produced under similar laser-irradiated conditions was influenced by the type of assist gas and the flow rate. Hence, to obtain a relatively uniform hardened zone, it was necessary to choose the proper type of assist gas and flow rate.[67]

5. Interaction of the laser beam and the part (or the interaction of photons from the laser source and free electrons of the metal) generates a localized heat at the surface by energy transfer (as in thermal conduction), raising the surface temperature of a steel component to the austenitizing temperature (within a fraction of a second); the laser is either switched off or, more usually, translated to diffuse the heat into the bulk component.[68] When the laser beam is moved away, rapid cooling (self- or mass-quenching) of austenite region by conduction of the surrounding cold material produces a hard martensitic surface with increased wear resistance and fatigue properties.[62,69]

Continuous CO_2 lasers with power outputs up to 20 kW, wavelength of 10.6 μm, and typical cross-sectional area of standard 6.3 mm^2 (0.25 in.^{2}) or horseshoe configuration, and power density in the range at 0.5 to 5 kW/cm^2 are available.[70] The surface hardness and the depth of the case achieved are dependent mainly on

1. Size and power of laser beam (output power/optics or energy profile).

2. Relative travel speed between laser beam and the workpiece.

3. Time of impingement (the area to be treated can be scanned by moving the part, the laser, or the optics). This produces a sharp increase in temperature of the surface, without affecting the surrounding area. The sharp temperature gradient between the surface of the sample and surrounding causes self-quenching with a cooling rate much greater than the critical cooling rate.[68]

4. Material properties (prior microstructure, composition, and mass determine the heat sink available for self-quenching).

5. Coating applied prior to the treatment enhances the absorptivity of the surface to infrared radiation produced by CO_2 lasers,[71] without reflecting the laser energy.

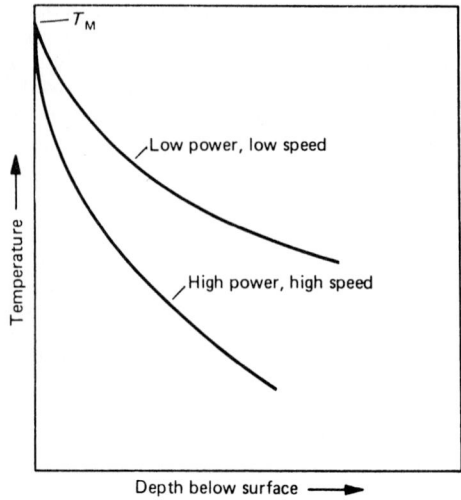

FIGURE 16.17 Effect of processing parameters on heat penetration in laser surface transformation hardening.[72] (*Courtesy of ASM International.*)

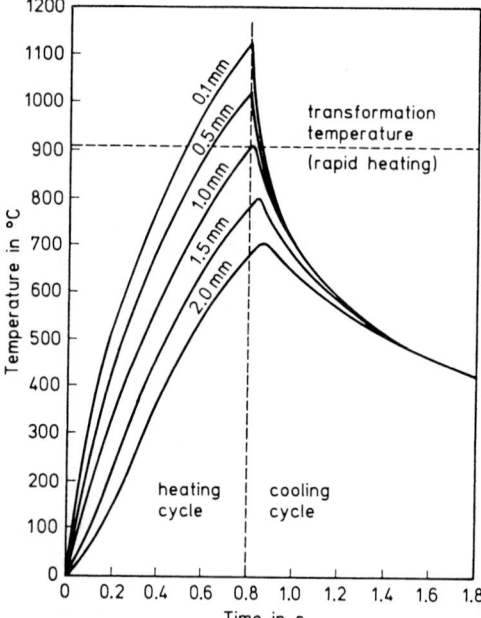

FIGURE 16.18 Heating curves for different depths below the surface.[73] (*Courtesy of VCH, Weinheim.*)

It is thus clear that the use of lower laser power density with correspondingly decreased travel speed will produce a slower rise in surface temperature and less steep temperature gradient, implying a greater austenitization depth (Fig. 16.17).[72] However, the self-quenching cooling rate will be slower, which would limit the material to harden entirely. Likewise, higher power densities are employed together with higher traversing speeds to produce smaller case depth. Figure 16.18 shows the heating curves for different depths below the surface. These curves depend on the power density.[73]

Surface microhardness has been found to increase by 3 times, provided an optimum combination of chemical composition and laser parameters is employed.[61] Such a remarkable improvement in surface hardness is especially significant for increasing the wear or corrosion resistance[74] and critical fatigue life.[75]

The ruby LSH of oil-hardened nonshrinking steel (composition: Fe-0.95C-0.5Cr-0.15V-0.6W) ring has been reported to increase the microhardness 3.3 times when compared to the base material.[76]

Advantages over Conventional Thermal Surface Hardening.   In general, all steels and cast irons, which are capable of hardening by conventional thermal surface hardening techniques, will respond to rapid laser surface hardening treatment. The main benefits claimed for these treatments over the conventional thermal surface hardening methods are as follows:[71,77–80]

1. Minimal or no distortion or warpage of the workpiece and thin heat-affected zones.[80]

2. Reproducibility, reliability, and elimination of post-treatments such as straightening, machining, grinding, sand blasting, and stress-relieving operations.

3. Energy savings.

4. Cleanliness (no pollution of environment, i.e., environmentally friendly).

5. Offers a very rapid, economical hardening method with both shaped and complex and nonsymmetrical figured parts and selective (or confined) areas inaccessible to conventional thermal techniques, e.g., inside of grooves and notches.

6. Flexibility (the writing of a program or its modification can be easily handled; programs, kept on simple tapes, can be used indefinitely).

7. High degree of controllability of the heat source and automation, low heat input, and the ability to self-quench[81] (i.e., no need of external quenchant to form martensite) at higher than the critical cooling rate.

8. Increased process productivity and doubled durability of high-speed steel end cutters compared with standard heat treatment.[82]

9. High hardness and wear and fatigue resistance of treated parts, and low manufacturing cost due to simplified processing sequences.[65]

10. LSH allows conduction of surface heat treatments in cases where conventional hardening methods, e.g., local hardening in notches with high stress concentrations, are deleterious.

11. It is best suited to localized hardening of both (*a*) large components which are not manageable to hardening by other methods and (*b*) small components which experience excessive distortion by conventional methods.[73]

12. Whether the part is small or large, the laser heating process can offer a hardened wear surface at specified locations with depths in the range of 0.020 to 0.14 in. (0.5 to 3.5 mm).[63] The typical case depths for steels and cast irons are 250 to 750 μm and 100 μm, respectively.[72]

13. Induced compressive residual stresses develop at the surface, if surface melting is avoided.[83]

14. Large surface hardening for carbon steel plate has been reported. A large surface hardened zone (60-mm width $\times$ 0.6-mm depth $\times$ 16-mm length) without

tempering has been achieved on a carbon steel plate without tempering, with a high-power (15-kW) CO_2 laser and the truncated pyramid mirror optical system with divided laser beams.[84]

Disadvantages

1. Its low efficiencies of power generation (5 to 10% for CO_2 laser, 1 to 3% for Nd:YAG laser, and 30 to 60% for diode lasers)[85] are the main drawbacks.

2. The intrinsic surface reflectivity of most metals is the main disadvantage. Steels and cast irons, even in a rolled or cast condition, can reflect 50 to 60% of the laser energy. The inexpensive solution is simply to apply flat black paint to absorb nearly all incident beams. For long-run production quantities, the painting operation and positioning of the part for laser treatment can be automated for total control.

3. Laser heat treating is about 15 to 25% more expensive than flame or induction heating. However, to measure the actual value of LSH, the entire routing or process sheet must be taken into account.

4. Capital equipment investment is substantial, due to the higher power levels required. Cost ranges between $200,000 and $500,000, depending on the level of automation and size of the system.[63]

5. LSH is highly sensitive to the process parameters, and it is difficult to set them up properly to produce the desired hardened depth. If the energy input is very high, there is a likelihood of undesirable local melting; or on the contrary, if the heat input is very low, the hardened zone is inadequate.[63]

Applications.   This method is used in aircraft and automotive applications as well as in paper and pulp industries.[63] Components hardened by this method include diesel engine parts made of cast iron,[86] such as pistons, ring grooves, and ring seats; power-steering gear housing and cylinder liners; constructional machinery parts (which have an integral spline and a hydraulic pump motor part[87]); and automotive parts (such as valve guides, seats, and gear teeth) and nonautomotive parts (such as bearings and paperboard die cutting tools).[88] Other areas where LBHT is used are in (1) the annealing of semiconductors to homogenize and improve their properties and in (2) surface alloying of metals.[89]

Other applications include carbon steels (1040, 1050, 1070), alloy steels (4340, 52100), air and oil-hardening types, tool steels, and cast irons (gray, ductile, austem-pered ductile,[90] and malleable); superalloys, depending on carbon contents, mandrel faces, mandrel grip areas, gearboxes, bearing races, camshaft lobes, steel rolling cylinders, crankshaft journals, spline faces, cutting blades, knives and tools, keyways, deflector/idler rolls, tool bed ways, and guide bars.[65] Also included are cutting edges of tools such as milling cutters, drills, and reamers, made of low-alloy steels and high-speed steels, with increased wear resistance by factors of 1.5 to 4,[83] railway car bogie plates,[91] diesel engine piston ring grooves, cylinder liner sealing surfaces, shafts, and racks for rack and pinions.[83]

The advantages of LBSH over electron beam surface hardening (EBSH) are as follows:[62]

1. This method can operate in air (i.e., at atmospheric pressure) without the need for a vacuum. However, if employed, a vacuum (or inert atmosphere) provides an excellent protective environment.

2. The laser beam (LB) is unaffected by stray magnetic fields.

3. LB can be directed with mirrors into areas which are not easily accessible.

4. LB does not produce x-rays.

5. LB can be employed in a time-sharing mode in order to supply many workstations, thereby significantly increasing the machine utilization.

6. LB can be split into many beams that can be employed simultaneously.

7. Capital investment costs of LB facilities with low beam power setting (2 to 3 kW) are comparatively low.

The disadvantages of LBSH over EBSH are as follows:

1. The total energy efficiency of LBSH installation is about 7 to 10 times inferior to that of EBHT.

2. In LBSH, the area to be treated needs to be covered with an energy-absorbing coating, which is an additional operation compared to the EBSH.

3. The flexibility of the energy distribution on the area to be treated is much lower in LBSH than in EBSH. Moreover, mechanical vibration adversely affects the variation of the energy distribution achieved by the electromagnetic alignment of mirrors in the LBHT.

4. The power available in the market for the laser system is limited (<15 kW), whereas it is 100 kW or more in the case of EB systems. This results in lower-volume productivity at beam powers of 10 kW and above.

5. In the industrial environment, EB technology has proved to be more successful than laser technology.

Selection between LB and EB Methods Based on Capital Cost. For applications involving (1) a lower-power beam (<2 kW) (localized and small surfaces), a laser is most economical; (2) between 2- and 3-kW beams, both the LB and EB methods are competitive; and (3) high-power beams (>5 kW), certainly the EB remains the best choice.[62]

Theoretical Models. Theoretical models such as Mazumdar and Steen's finite difference model,[92] Cline and Anthony's Green function approach,[93] Tayal and Mukherjee's approach,[94] and Bokota and Iskierka's approach[95] based on coupling of phase transformations and phase fields have been developed to estimate the temperature distribution during LBSH. But each has one or more drawbacks, such as the complexity of solution, too much simplicity, or long computer hours requirement.

The Tayal and Mukherjee approach is based on the simplification of Green's function in the prediction of temperature distribution on a workpiece during LBSH. They have established a correlation between hardness, hardness profiles, microstructures in the laser-hardened layers of steel, and laser process parameters which can be employed to predict the case depth in laser-hardened steel.[94] Based on numerical simulation data, the effects of laser treatment (the size of particular zone occurring in the surface layer) can be predicted as a function of beam dimensions, power density, and irradiation time affecting the steel (scan speed).[95]

16.2.5 Electron Beam Surface Hardening

The EB method was first recognized in the 1950s, and the production method was available in the 1960s. Further advent of the high-speed scanning (HSS) technique for an areal energy transfer (Fig. 16.19a)[96] has made the methods of EBSH treatment economically more attractive. For example, beam powers of up to 20 kW with an areal throughput of about 20 cm^2/s are now available.[97,98] Like LSH, EBSH is an attractive research field and forms a martensitic structure with a compressive stress

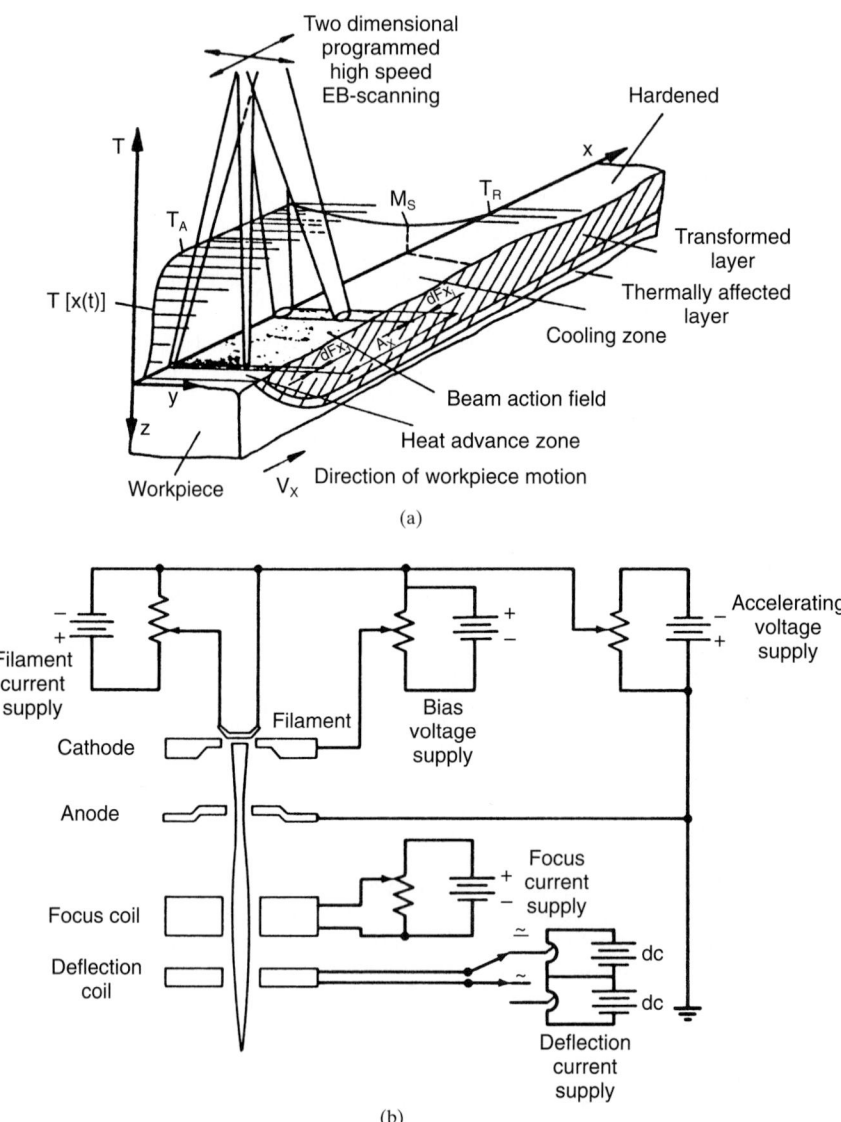

(a)

(b)

FIGURE 16.19   (a) Principle of high-speed scanning technique.[96] (b) A schematic EB generation by an electron gun.[12] (*Reprinted by permission of ASM International, Materials Park, Ohio.*)

in the treated area. The characteristic features that distinguish the EBSH from the LSH are as follows:

1. Unlike LS hardening, EBS hardening units operate in vacuum.
2. Its operating efficiency is significantly higher (about 7 to 10 times better) than that of LSH.
3. Surface coating of the component to absorb energy is not required because there is a direct transmission of kinetic energy of electrons to the atoms of the component surface rather than by absorption.
4. Flexibility of energy distribution on the heated area is much higher in EBSH than in LSH.
5. EBSH is more economical for power higher than 2 kW.

Figure 16.19b shows a schematic EB generation by an electron gun.[12] A similar type of electron gun is also used for EB welding and cutting. When the filament is resistance-heated to an extremely high temperature in a vacuum of 10^{-5} torr, free electrons are emitted which are accelerated and collimated into a dense, high-energy beam by the 60-kV potential maintained between the cathode and anode. After passing through the anode, the electron beam is further collimated by a focus coil. Finally, a deflection coil is used to direct the reconverging beam to a specified location on the workpiece, which is placed in an enclosure at a pressure of $\sim5 \times 10^{-2}$ torr.

When the high-velocity electron beam hits the workpiece, the kinetic energy of the electron beam is changed to heat the workpiece; the charged electron beam is deflected very quickly electromagnetically on the workpiece surfaces, producing both surface and volumetric heating, and with moving the workpiece the area becomes larger (Fig. 16.19a).

In EB hardening, both heating and cooling are very fast. Energy transfer to the interior of the material occurs by conduction, then rapid cooling of the austenite required for martensite formation takes place by self-quenching.

Typical hardening depths achieved by EB are 0.1 to 1.5 mm (0.004–0.06 in.), depending on the materials parameters.[99] Tempering is not usually needed following EB hardening; however, it seems necessary for some high-alloy steel grades.

Figure 16.20 shows the EB heat-treating equipment, where the computer-controlled deflection is used to raster (move) the defocused electron beam over the surface of the workpiece into a shaped microdot pattern. This provides a precise local heating to the austenitizing temperature. When heating ceases, the surface layers rapidly transform into martensite by self-quenching, producing an increase of surface hardness and wear resistance and decrease of grain size when compared to the conventional heat treatment. An increase in EB heat input increases the compressive residual stress in the hardened layer and the fatigue crack growth resistance.[100,101]

In a programmable computer control system, the EB is deflected from one point to another for an adjustable dwell time interval (20×10^{-6} s minimum) at various points of the pattern and power variation of the beam. A cathode-ray tube display shows the pattern as processed by the computer program.[102,103]

The combination of computer and EB renders greater flexibility in the system. Actually, the computer regulates the current, voltage, focus, and total on-time and fault diagnoses in order to obtain the desired pattern.[62]

Advantages of EBSH include the following:[104]

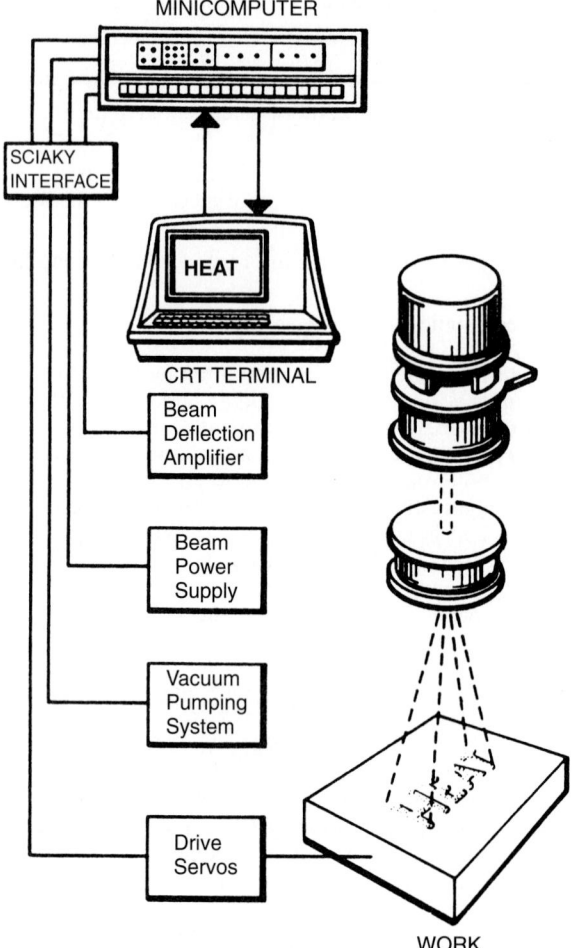

FIGURE 16.20 Schematic of computer-controlled electron beam hardening using beam defocusing method.[102] (*Reprinted by permission of ASM International, Materials Park, Ohio.*)

1. Precise control and reproducibility of the energy input with respect to location and time.
2. Low thermal stress induced in the workpiece with minimum warpage.
3. Consistent hardening depth for both areal and laterally patterned hardening up to a track width of 50 to 100 mm (2 to 4 in.).
4. Clean and bright surface; absence of environmental problems.[96]
5. No surface preparation needed for hardening and for untreated portions.
6. Compatibility and ease of integrating computer numerical control (CNC) or computer-controlled processing such as CAD/CAM (computer-aided

design/computer-aided manufacturing) into mechanical flow lines. Highly suitable for automation.[96]

7. High energy efficiency. About 75% of the power generated is converted into heat when the incident beam falls perpendicularly on a steel surface.

8. High process productivity with available beam power in the range of 20 to 50 kW.

9. No generation of waste products.[104]

10. No post-treatments after hardening.

11. Increased wear resistance of the components due to increase in hardness and decrease in grain size.

12. High flexibility with respect to changing technological tasks.[96]

13. Ability to be easily included into manufacturing systems.[96]

14. EB equipment capability to serve other processes such as deep welding, hardening, and surface melting.

The high capital investment cost of an electron beam facility (between $500,000 and $1,200,000) limits the profitable use of this method to the mass production. However, the multiple use of this equipment is one option, e.g., the modification of existing EB welding equipment in order to carry out the EB surface hardening operations.[104] This enables the user to save investment costs and to ensure a better utilization of the EB facilities.[96]

Recommendations for using EBSH are as follows:

1. The depth of treated layer can be between 0.3 and 2.5 mm (0.01 and 0.1 in.) with best results in the vicinity of 0.75 mm (0.03 in.). For effective self-quenching, thickness of the part must be at least 5 to 10 times the austenitizing depth, depending on the material chosen.

2. EBHT is particularly suitable for complex, nonsymmetrical parts.[103]

3. With regard to the relative position of the beam and part, the incidence of the beam can be precise to ±0.05 mm (0.002 in.).

4. The axis of the beam can hit the surface to be treated at an angle of 20° without an appreciable reduction in beam efficiency.

5. A given result can be obtained by a proper combination of beam power and time of treatment.

Applications. Practically all steels and cast iron components with surfaces that are accessible to the electron beam and with a thermal capacity that is adequate for self-quenching, can be used for EBSH. Significant effects are achieved with carbon steels and low-alloy steels, especially with respect to enhanced abrasive-wear performance. Important applications of surface hardening, some of which are shown in Fig. 16.21,[103] include camshafts, clutch cams, crankshafts, indexing rings, cam rings, cam disk and collars, gears, rotors, rotor spindles, spindles, fasteners, bearing races, machine-tool surfaces (such as milling machine quill), thrust bearings, ball bearings, push-rod ends, tool-joint ends, ball joints, cylinder liners, cutter blades, mower blades, clutch faces, lifters, ring grooves, conveyor links, couplings, output shafts, valve seats, valve lifter caps, fluted rollers, threaded guidance elements (in the textile industry), molds (for the production of glass vessels or plungers for the automotive industry), technical knives, extruding tools, metalworking tools, rolling dies, engrav-

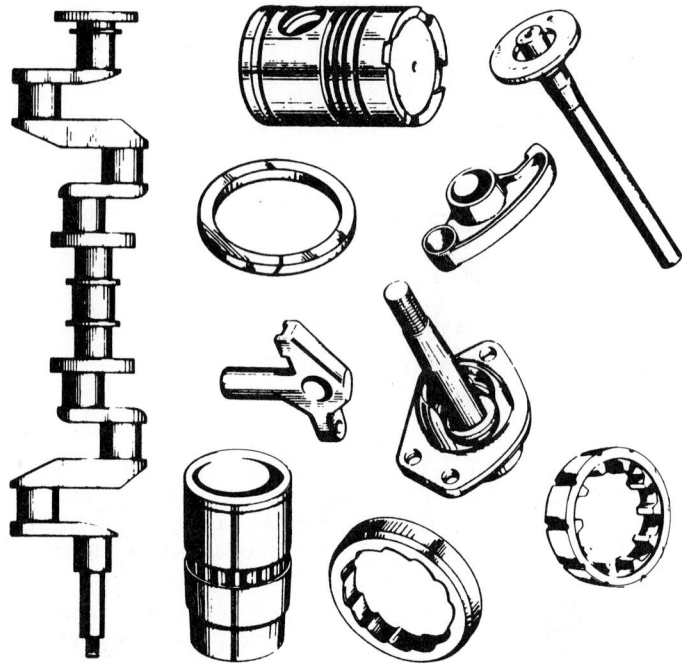

FIGURE 16.21 Typical applications of electron beam surface hardening.[103] (*Courtesy of ASM International, Materials Park, Ohio.*)

ing tools, roller bearing element support, pearlitic cast iron lathe, and (cast iron) valve guides.[96,103]

16.3 AUSTENITIC (OR HIGH-TEMPERATURE) THERMOCHEMICAL SURFACE HARDENING TREATMENT

This section describes carburizing, carbonitriding, cyaniding, and recent developments in the field, notably fluidized beds, alternative atmospheres, vacuum carburizing, and plasma carburizing.

16.3.1 Carburizing

Carburizing is the most widely used surface hardening process in which the production of gears, bushings, bearing races, rock drilling bits, and so forth is desired with improved wear and fatigue resistance while maintaining tough and shock-resistant cores. Typical carburizing (and carbonitriding) grades of plain carbon and low-alloy steels are AISI 1010, 1012, 1018, 1020, 1022, 1025, C1117, 1524, 1527, 2317, 2325, 3115, 3310, 3310H, 4023, 4027, 4118, 4118H, 4320, 4615, 4617, 4620, 4815, 4820, 5015, 8617, 8617H, 8620, 8620H, 8720, 8822H, and 9310.[105] Low-carbon high-alloy carbur-

TABLE 16.7 Chemical Composition of High-Alloy Carburizing Steels[106]

Element	CBS 1000 M	CBS 50 NiL	Pyrowear 53	Vasco X-2M
C	0.13	0.13	0.11	0.14
Mn	0.55	0.25	0.35	0.30
Si	0.50	0.20	0.80	0.90
Cr	1.05	4.20	1.00	5.00
Ni	3.00	3.40	2.25	
Mo	4.50	4.25	3.25	1.40
V	0.40	1.20	0.10	0.45
Cu			2.10	
W				1.35

CBS 1000 M and CBS 50 NiL are produced by Latrobe Steel, Pyrowear by Carpenter Technology, and Vasco X2-M by Teledyne Vasco.
Courtesy of ASM International, Materials Park, Ohio.

izing steels with secondary hardening characteristics such as CBS 100M, CBS 50 NiL, Pyrowear 53, and Vasco X-2M (Table 16.7) have also been developed for gear and bearing and power train applications and are readily carburized in oxygen-free environments (e.g., vacuum carburizing and plasma carburizing). They can also be gas carburized if they are first coated with iron.[106]

Carburizing is a process in which low-carbon steels are heated to a temperature in the austenite phase field, known as the *carburizing temperature*, in contact with a carbonaceous solid (carburizing compound), liquid (salt bath), or gaseous medium, to cause absorption of carbon at the surface and, by diffusion, to create a carbon-concentration gradient between the surface and the interior of the components.[105] Fluidized-bed carburizing, vacuum carburizing, and plasma carburizing are now also being used in the United States to overcome the shortage of natural gas.

Carburizing may be done at a temperature in the range of 800 to 1050°C (1472 to 1922°F), but the widely used carburizing temperatures are around 825 to 925°C (1517 to 1742°F). Carburizing at 1010°C (1850°F) is sometimes used when large reduction in processing times, increased production capability, case depths greater than 0.06 in. (1.5 mm), and more gradual decrease in carbon content of the underlying layers are required.[107–109] However, the lives of the furnace components are shortened somewhat by using high temperatures.[105,109] A carburizing temperature of 982°C (1800°F), therefore, provides the best compromise between productivity and economic operation.[109]

Most commercial carburizing processes are not entirely done under conditions approaching equilibrium but are accomplished in two parts, namely, a carburizing period and a diffusion period. In the first, absorption of free carbon into the surface layers of a piece of low-carbon steel occurs until the solubility limit of carbon in that steel is reached. For example, low-carbon steel can develop a carbon content at the surface layer of about 0.9% at 790°C (1450°F), 1.05% at 845°C (1550°F), 1.2% at 900°C (1650°F), and 1.4% at 955°C (1750°F). In the latter, the carbon atoms diffuse away from the surface toward the core. As carbon content at the surface increases, the driving force for diffusion toward the core increases.

In carburizing carbon case-hardening steels, a case with 0.9% carbon is normally desired, though eventually the surface layers may contain up to 1.2% carbon. Too much carbon will give free cementite and will cause the hard case to be more brittle than usual. Too little carbon will render insufficient hardness in the quenched case.[110]

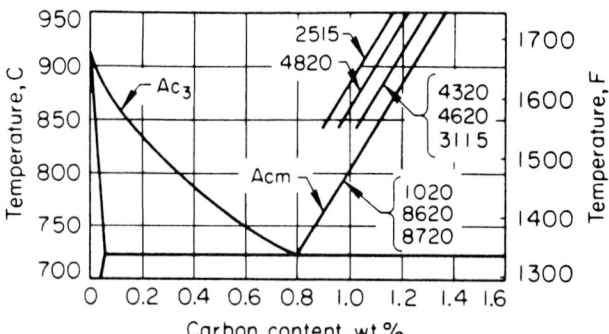

FIGURE 16.22 Fe-Fe$_3$C phase diagram showing shifting of Acm line toward the left of the diagram by alloy addition.[105] (*Reprinted by permission of ASM International, Materials Park, Ohio.*)

Note that the carburizing temperature remains always above the Acm line of the Fe-Fe$_3$C diagram. Moreover, alloying elements shift the Acm line toward the left of the equilibrium diagram, as shown in Fig. 16.22, where approximate limits of carbon content at the surface for AISI 1020 steel and seven low-alloy carburizing grades are superimposed.[105]

Slow cooling from the carburizing temperature will produce a soft and fileable part, whereas the quenched one will form hard martensitic structure in the surface layers. Quenching may be accomplished either from the carburizing temperature or from some lower temperature near, but still above, the solubility limit between γ and γ + Fe$_3$C, Acm. The part is then tempered to increase the toughness of the case.

16.3.1.1 Heat Treatment after Carburizing

Direct Quenching.   Mostly carburized parts are quenched directly from the carburizing temperature of ~925°C (1700°F) or from a temperature of approximately 845°C (1550°F), without being cooled to room temperature. Lowering of the temperature prior to quenching can be accomplished by (1) reducing the temperature in the same furnace, (2) moving the components to another desired temperature zone for quenching, or (3) shifting the components to another furnace.[12] This practice has gained wide acceptance due to the simplicity of the procedure and savings in energy, equipment, labor costs, handling, and operating expenses.

Single Quenching.   The single-quenching technique is carried out for alloy steels such as 3% nickel steel, which are allowed to cool slowly after carburizing to produce a uniform microstructure in the case. These parts are reheated just beyond the A_3 temperature of the case to control the surface carbon content and microstructure[105] and are then quenched in oil (Fig. 16.23a).

Double Quenching.   During carburizing, the grain coarsening is obtained in both the case and core, and so post-carburizing heat treatments are used to produce a fine-grained structure with good impact strength. Such treatments are described as follows:

A double-quenching technique is practiced for mild steel; this technique consists of reheating to just above the A_3 temperature (850 to 900°C, or 1562 to 1652°F) for the core and quenching to refine the core. This constitutes the first-stage quenching sequence. Oil-quenching produces a finer structure of the core. The second stage

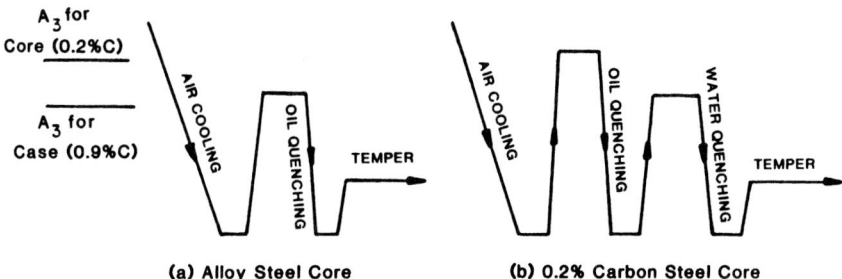

FIGURE 16.23 Schematic diagram for (*a*) single- and (*b*) double-quenching heat treatment after carburizing.

involves heating to just above the A_3 temperature (780 to 820°C, or 1436 to 1508°F) for the case which will result in the formation of fine-grained austenite in the case and a mixture of ferrite and austenite in the core. Subsequent water-quenching will produce hard martensite in the case and ferrite with a small amount of martensite in the core. Finally, the steel is tempered at 150 to 200°C to ensure a tougher core and to relieve quenching stresses present in the case. The operating cycle of this technique is shown schematically in Fig. 16.23*b*. This method may be applied to steels which have not been given prior fine-grain treatment.

16.3.1.2 Measurement of Case Depth. Case depths are specified with a notation of the method of measurement, namely, total case depth, effective case depth to 50 Rc, and case depth to 0.40% carbon. Measurement of case depth can be accomplished by chemical, mechanical, and visual (microscopic and macroscopic) methods.[12] (See Sec. 16.1.1 also for details.)

16.3.1.3 Pack Carburizing. In pack carburizing, also called *solid* or *box carburizing*, steel components with a carbon content not exceeding 0.2% are packed in a steel box in such a way that a uniform, thick, 1-in. (25-mm) layer of commercially available carburizing compound surrounds the parts. To evenly carburize small holes or small teeth, a 6- or 8-mesh material should be employed to ensure good filling. A depth of 13 to 50 mm ($\frac{1}{2}$ to 2 in.) and a minimum depth of 50 mm (2 in.) of carburizing compound for the bottom of the parts and for final layers (on top of the parts), respectively, are required.[111] After packing the box, its lid is sealed with fireclay with the box inverted[†] (Fig. 16.24). It is then heated in the furnace to a proper carburizing temperature in the range of 815 to 1040°C (1500 to 1900°F) for a specified time, depending on the material, desired surface carbon concentration, and case depth. The boxes are then withdrawn from the furnace, and the components removed from the carburizing compound, cooled and subsequently reheated, quenched, and tempered; or the boxes are opened and the parts are quenched directly and tempered. This latter quench method is called *direct quenching*. Figure 16.25 shows the effect of pack carburizing time on carbon gradient and case depth in AISI 3115 steel.

The main advantages of pack carburizing are the following:

1. More or less any type of furnace can be employed, and it does not require any sophisticated knowledge of equipment.

[†] This is not a prerequisite.

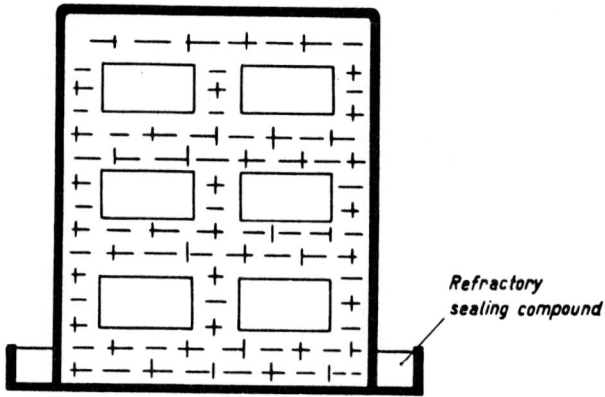

FIGURE 16.24 A pack carburizing box.[8] (*Courtesy of Butterworths, London.*)

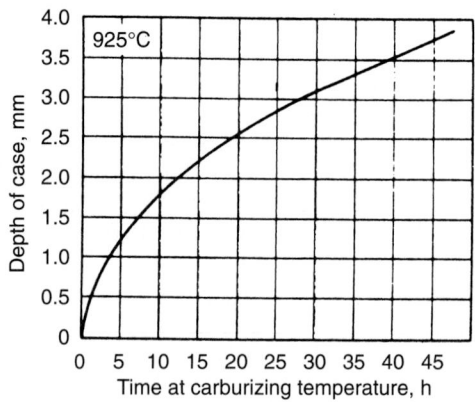

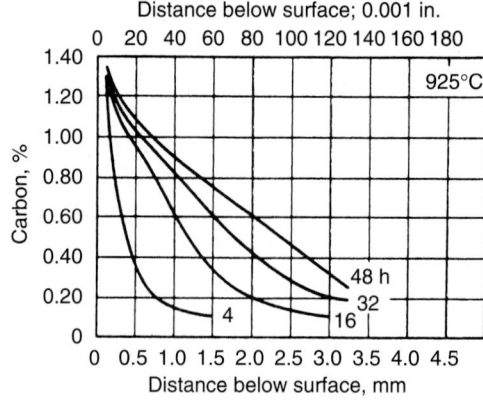

FIGURE 16.25 Effect of pack carburizing time on case depth and carbon gradient in 3115 steel. Carburized compound contains hardwood charcoal, coke, and sodium carbonate.[105] (*Reprinted by permission of ASM International, Materials Park, Ohio.*)

2. Carburizing compound provides good support to the parts, which prevents sagging at high temperatures.

3. The process is especially useful for large, complex-shaped parts and for achieving a deep case.

4. Slow cooling in the pack produces soft workpieces which may be beneficial for machining prior to hardening.

5. It provides a wider choice of stop-off methods for selective carburizing than that of gas carburizing.[111]

Disadvantages of pack carburizing are as follows:

1. The process is labor-intensive.

2. The method is not suitable for large-volume production.

3. The process is inflexible in operation and has poor control over surface carbon concentration and carbon gradient.

4. It is not suitable for the production of shallow case depths with close case depth tolerances.

5. It is not well suited for direct quenching or quenching in die, thus requiring extra handling and processing time.[111]

Carburizing Reactions. If the steel parts are heated with charcoal in the box, the atmospheric oxygen in air enclosed within the box will react with the carbon to form a CO_2-rich mixture, which then continues to react with the hot charcoal and liberates more CO as follows:

$$CO_2 + C \rightleftharpoons 2CO \qquad (16.5)$$

With increasing temperature, the equilibrium is shifted to the right, and thus a CO-rich gas mixture prevails. The CO then reacts with the steel surface, and the cycle of reaction is repeated with CO_2 as follows:

$$2CO \rightarrow CO_2 + C \qquad (16.6)$$

$$3Fe + 2CO \rightarrow Fe_3C + CO_2 \qquad (16.7)$$

or $\qquad 3Fe + 2CO \rightarrow 3Fe + C \text{ (in solution in austenite)} + CO_2 \qquad (16.8)$

The cementite is formed at the very surface of steel. The carbon present in the cementite then dissolves in the austenite and diffuses inward. Thus CO acts as the carrier of carbon from the carbonaceous material to the steel.

When certain carbonates such as $BaCO_3$, Na_2CO_3, K_2CO_3, $CaCO_3$, and so forth are added to the carbonaceous material, they dissociate into metallic oxides and CO_2 at the carburizing temperature. The CO_2 liberated then reacts with the hot charcoal to produce active CO_2 which speeds up the reaction:

$$BaCO_3 \rightarrow BaO + CO_2 \qquad (16.9)$$

$$CO_2 + C \rightarrow 2CO \qquad (16.10)$$

The carbonates thus act as catalytic agents and are called *energizers* because they very rapidly accelerate the CO formation. Because of their beneficial effects, carbonates are included in almost all carburizing compounds. However, the best-known and perhaps the most effective of these energizers is $BaCO_3$.

Carburizing Compounds. Many different carburizing compounds have been employed for carburizing. They include wood charcoal; coke; charred leather; horn; bone; hydrocarbonated bone (finely ground bone and heavy oil); and carbonates such as barium carbonates, calcium carbonate, sodium carbonate, potassium carbonate, oyster shells, and so on, bound by oil, tar, or molasses. There are special proprietary brands of carburizing compound available in the market which can be used with confidence.

A typical chemical composition of a solid commercial carburizing compound is as follows: coke, 25% max.; $BaCO_3$, 12 to 18%; Na_2CO_3, 2% max.; S, 0.5% max.; total inorganic matter exclusive of carbonates, 8% max.; hardwood charcoal, balance. Thus more active carburizing compounds contain both coke and charcoal; if more coke is present, it provides minimal shrinkage, good thermal conductivity, and good hot strength.[111] By adjusting the amount of energizer in the carburizing compound, some degree of control of the case also may be achieved. A good carburizing compound should not shrink appreciably in volume during the operation and should be practically free from sulfur and phosphorus. However, in most cases, carburizing compound diminishes in volume after its use due to the carbon consumption. Hence new material has to be continually added because it will be uneconomical to start up with the fresh carburizing compound for each treatment cycle. The ratio of old to fresh compound varies considerably with the practice, although most companies use from 2 to 5 parts of old to 1 part of fresh compounds. However, old compound is usually screened to remove fines prior to thorough mixing with the fresh compounds.[111]

Applications. This method is still employed in the aircraft and other industries for the carburizing of various gears, wheels, pinions, rolls, rollers, and so forth. The examples are flying-shear timing gear; heavy-duty rolling-mill and industrial gears; calender bull and mine-loader bevel gears; crane cable drums; motor-brake and high-performance crane wheels; continuous-miner drive pinion; kiln-trunnion roller; processor pinch roll; and blooming-mill screws.[111] In general, it is suitable for toolroom applications and for very large parts such as rock-bit drill rods 6 m (10 ft) long and rings 2 m (7 ft) in diameter.

Carburizing Containers. The carburizing container should be made of ordinary mild steel (for occasional use), aluminized mild steel (for high-temperature carburizing), or heat-resistant alloys, for example, 25% Cr-20% Ni steel or 15% Cr-35% Ni steel (for continuous or regular use).[112] They are often equipped with braced lifting hooks or eyes, test pin openings, and special lid-receiving sections. The lid for carburizing containers must be tight enough to avoid air ingress, but not so tight as to hinder easy expulsion of excess gas created within the container. The loose-fitting lids should be partly sealed with clay-based cements.[111]

Prior to placement of alloy carburizing containers in service, it is customary to condition them by "precarburizing" without a workload. This facilitates the carburization of the workload during the first production carburizing cycle.[111]

Since the carburizing compounds have poor heat conductivity, the design of the packing box is a very important consideration. The smaller the container used and the more closely it conforms to the shape of the workpiece, the better and more uniform the carburizing effect.

Carbon Potential and Case Depth. The use of charcoal and $BaCO_3$ readily provides the carbon content as much as 1.2% at the steel surface and the more rapid attainment of required case depth.

The amount of carbon absorption (i.e., surface concentration) and the effective depth of carburized case are dependent on the grade of steel, carbon potential of the carburizing compound, carburizing temperature, and time.

TABLE 16.8 Operating Compositions of Liquid Carburizing Baths[113,114]

	Composition of bath, %	
Constituent	Light case, low temperature 845–900°C (1550–1650°F)	Deep case, high temperature 900–955°C (1650–1750°F)
Sodium cyanide	10–23	6–16
Barium chloride	...	30–55[†]
Salts of other alkaline earth metals[‡]	0–10	0–10
Potassium chloride	0–25	0–20
Sodium chloride	20–40	0–20
Sodium carbonate	30 max.	30 max.
Accelerators other than those involving compounds of alkaline earth metals[§]	0–5	0–2
Sodium cyanate	1.0 max.	0.5 max.
Density of molten salt	1.76 g/cm³ at 900°C (0.0636 lb/in.³ at 1650°F)	2.00 g/cm³ at 925°C (0.0723 lb/in.³ at 1700°F)

[†] Proprietary barium chloride-free deep-case baths are available. [‡] Calcium and strontium chlorides have been employed. Calcium chloride is more effective, but its hygroscopic nature has limited its use. [§] Among these accelerators are manganese dioxide, boron oxide, sodium fluoride, and sodium pyrophosphate.

Courtesy of ASM International, Materials Park, Ohio.

Furnace. Pack carburizing is normally accomplished in batch-type box, car-bottom, or pit furnaces.

16.3.1.4 *Liquid Carburizing.*

This process, also called *salt bath carburizing*, is a method of case hardening steel or iron parts by holding them above their transformation temperature Ac_1 in a molten salt bath. The salt bath decomposes and releases carbon (and sometimes nitrogen) that diffuses into the component surface, which can be hardened by fast quenching from the bath.

Salt bath carburizing processes are mostly based on sodium cyanide salts, which can be grouped into nonactivated and activated types. The former are generally composed of NaCN or KCN, sodium carbonate, and often an alkali metal chloride. The latter contain, in addition, an activated agent such as barium chloride. Table 16.8 lists the operating compositions of liquid carburizing baths.[12] Sometimes silica compounds, manganese dioxide, and phosphates are used as activators. This type of salt bath is used to obtain case depths up to 1.5 mm.[113]

Nonactivated-Type Carburizing Salt Bath, Called Low-Temperature Cyanide-Type Bath (Light-Case Bath). These baths usually operate between 840 and 900°C (1550 and 1650°F) (Table 16.8) and produce shallow cases [0.005 to 0.01 in. (0.13 to 0.25 mm) deep]. Cyanide cases are usually not applied for case depths >0.25 mm (0.010 in.). These baths possess the desirable qualities of easy cleaning, good fluidity, and excellent stability, when compared to the activated carburizing baths.

NaCN or KCN is the active carburizing agent. NaCl improves carburizing activity for a given cyanide content, which is considered to derive from the lower percentage of carbonate; the presence of carbonates inhibits the breakdown of cyanate. These low-temperature (light-case) baths are usually operated [at higher cyanide contents than high-temperature (deep-case baths)] with a protective carbon cover in order to obtain higher carbon content in the carburized case. If the carbon cover becomes thin, the nitrogen content of the carburized case increases and premature depletion of cyanide occurs with reduction in carburization. Carburizing proceeds via the gaseous phase according to the following reactions:

$$2NaCN \rightleftharpoons Na_2CN_2 + C \tag{16.11}$$

and either
$$2NaCN + O_2 \rightleftharpoons 2NaCNO \tag{16.12a}$$

or
$$NaCN + CO_2 \rightleftharpoons NaCNO + CO \tag{16.12b}$$

$$4NaCNO \rightleftharpoons 2NaCN + Na_2CO_3 + CO + 2N \tag{16.13a}$$

and
$$NaCNO + C \rightarrow NaCN + CO \tag{16.13b}$$

or
$$4NaCNO + 2O_2 \rightleftharpoons 2Na_2CO_3 + 2CO + 4N \tag{16.13c}$$

or
$$4NaCNO + 4CO_2 \rightarrow 2Na_2CO_3 + 6CO + 4N \tag{16.13d}$$

$$3Fe + 2CO \rightleftharpoons Fe_3C + CO_2 \tag{16.14a}$$

and
$$3Fe + C = Fe_3C \tag{16.14b}$$

The first reaction is the cyanamide shift. The second reaction forms cyanate and occurs at the interface between the salt bath and the atmosphere; the other two reactions occur at the interface between the salt bath and the steel. A small amount of nascent nitrogen liberated *by* reactions [Eqs. (16.13*a*, *c*, and *d*)] is also absorbed into the steel surface. Equations (16.13*a*, *c*, and *d*) also deplete the activity of the bath and result in an ultimate loss of carburizing effectiveness unless necessary replenishment practice is implemented.[114]

It is thus clear from these reactions that sodium cyanate is an important link in the reaction. When the NaCN content is high, reaction (16.13) occurs more readily to the right side. When the amount of NaCN falls, that of Na_2CO_3 rises and a gradual decrease in carburizing potential or activity is obtained. Thus when a very thin case depth of only a few tenths of a millimeter is desired, a salt bath containing 20% NaCN may be employed. Such a bath is made by melting together equal parts of NaCN and anhydrous Na_2CO_3. After being carburized, the steel parts are quenched directly into water.

Barium Chloride-Activated Carburizing Baths.   These are also termed *high-temperature cyanide-type (deep-case)* baths. Although the reactions involved in low-temperature liquid-carburizing salts apply to some extent, the basic carburizing reaction is given by

$$4NaCN + 2O_2 \rightleftharpoons 4NaCNO \rightleftharpoons Na_2CO_3 + 2NaCN + CO + 2N \tag{16.15}$$

as before we have

$$2NaCN + BaCl_2 \rightleftharpoons 2NaCl + Ba(CN)_2 \tag{16.16}$$

$$Ba(CN)_2 + 3Fe \rightleftharpoons Fe_3C + BaCN_2 \tag{16.17}$$

The carburizing activity of such baths is based upon (1) the cyanide content and (2) the $BaCl_2$ (alkaline earth) content. A typical commercial mixture contains 16 to 20% NaCN and 40% $BaCl_2$. (See also Table 16.8.) An increased amount of one of these substances can compensate for a lower amount of the other.

The activated high-temperature bath is preferred when the parts have to be ground or machined after carburizing because one-half of the case depth will be capable of full hardening in the activated bath compared to that of only about one-third in the nonactivated bath.[113] The deep case consists of carbon dissolved in iron together with superficial nitride-containing skin for improved wear resistance and for improved resistance to softening during tempering and other high-temperature heat-treating operations. These baths are used for producing deeper cases [0.5 to 3.0 mm (0.02 to 0.12 in.)] and operate frequently in the temperature range of 900 to 955°C (1650 to 1750°F) (Table 16.8). It is possible to raise the temperature to as high as 1050°C (1920°F), but it is not recommended because of the very rapid deterioration of the bath and equipment at such high temperatures.

Sometimes a silicate-activated carburizing bath is also useful because it is more highly water-soluble and hence, on quenching, is easily removable from the treated parts. The following reactions are considered to take place:

$$SiO_2 + Na_2CO_3 = Na_2SiO_3 + CO_2 \tag{16.18}$$

$$NaCN + CO_2 = NaCNO + CO \tag{16.19}$$

Silica-based baths produce higher levels of nitrogen than those provided in barium chloride-activated melts; hence, results very similar to those obtained by the gaseous carbonitriding process prevail by using silicon-based salts.[113]

Advantages and Limitations of Liquid Carburizing. The advantages of this process over pack carburizing are that

1. Rapid heat transfer occurs.
2. Simultaneous processing of a variety of shapes and sizes in a single bath can be done.
3. Distortion of parts is very small.
4. Direct quenching with water, oil, or neutral salts is possible.
5. Selective carburizing can be achieved by partial immersion in the bath.
6. Reasonable case depth control and uniformity is possible.
7. Brazing and carburizing can be combined in one operation.

The advantages of salt bath over the gas carburizing treatment are that[113]

1. Heating time is reduced (one-sixth to one-fourth times compared to that in gas).
2. Close temperature control to within ±5°C is possible without difficulty, which is not normally possible with gas.
3. Uniform temperature is maintained in the salt bath.
4. Salt bath protects the workpiece during processing and cooling, whereas gas atmospheres need constant monitoring for this.
5. There is less danger of distortion.
6. It is more flexible (longer and shorter immersion times for deep and shallow case, respectively) than the gas carburizing process.
7. It provides good-quality shallow cases, which are not obtainable by gas carburizing.
8. Regeneration is simple, involving addition of salt every day.

The disadvantages of this process are as follows:

1. Cleaning of complex-shaped components is difficult and troublesome in activated salts. Nonactivated salts are quite water-soluble.

2. It is a labor-intensive operation.

3. Transportation, storing, and handling of poisonous salt are troublesome.

4. There is effluent and disposal of cyanide wastes required.

5. It is not suited for P/M applications due to salt entrapment in the pores.[115]

Furnace. Liquid carburizing is carried out in salt baths which are heated electrically (by a series of resistance heaters surrounding the salt pot) or by gas- or oil-firing. The workpieces are hung on fixtures or on wires and suspended in the salt bath for the appropriate time.

Case Depth and Carbon Concentration. In liquid carburizing, the case depth depends on the carburizing temperature and time. Figure 16.26 shows the influence of three different carburizing temperatures for 1020 steel for 2- to 40-hr time

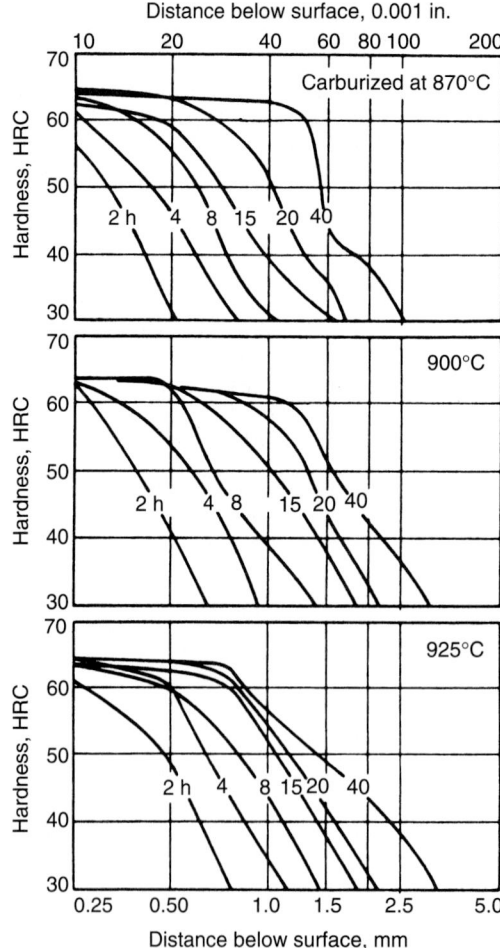

FIGURE 16.26 Influence of three different (liquid) carburizing temperatures on hardness versus depth relationship for 1020 steels for time periods of 2 to 40 hr.[115] (*Reprinted by permission of ASM International, Materials Park, Ohio.*)

cycles.[115] The total case depths (measured to base carbon level) produced in liquid carburizing can be calculated by the simple formula

$$d = k\sqrt{t} \tag{16.20}$$

where d is the case depth, k is a constant which represents the penetration in the early period at carburizing temperature, and t is the time (hours) at a given temperature.

Carbon concentration or the carbon and/or nitrogen content absorbed by steel is controlled mainly by strict adherence to the proper salt bath composition (cyanide content) and to the carburizing temperature and time. The chemical analysis of its bath composition or at least the NaCN content of the bath must be checked daily or at regular intervals in order to maintain the effectiveness of the liquid carburizing bath. Controlled addition of fresh salts must be made every day to maintain the proper carburizing potential. Usually this is practiced by replacing a portion of the used salt bath with an equal amount of fresh salt.

Pre- and Post-Heat Treatment. Prior to the immersion of steel parts into the salt bath, preheating should be carried out between 100 and 400°C (212 and 752°F) to remove any traces of moisture present and to utilize the full capacity of the salt bath with greater efficiency. If, following carburizing, the steel parts are quenched directly from the cyanide salt bath, held at a moderate temperature, into water or otherwise, the thickness of the soft surface zone will increase. Longer treating times would have the same effect on the soft zone thickness as the increased temperature.

Many liquid carburizing facilities case harden the parts (for larger case depths, >0.5 mm) at higher carburizing temperatures (925 to 955°C, or 1700 to 1750°F) followed by direct transferring of the workpieces to a neutral salt bath at 845°C (1550°F) for stabilization, reduced distortion, minimal retained austenite, and finally direct quenching in marquenching oil at 175 to 260°C (347 to 500°F), depending on the alloy content and the hardness desired.[114] When alloy steels are quenched directly, the case hardness is found to be low, and so they are given a double-quenching treatment. The first quench of this process involves reheating followed by quenching in oil or salt bath and a second requenching treatment at a somewhat lower temperature.[8]

Application. Liquid carburizing generally is used for small- and medium-sized parts requiring case depth less than 0.5 mm. Greater economy is obtained with smaller case depth requirement due to very high heating rates attained in this process. This is highly recommended (compared to gas carburizing) for highly stressed components in critical applications such as bearing steels for ball and socket joints, racing machines,[116] transmission shaft with integral gear, and layshaft gear in order to obtain superior fatigue strength. Liquid carburizing is not recommended for parts containing small holes, threads, or recessed areas due to difficulty in cleaning.[114]

Noncyanide Liquid Carburizing. Noncyanide liquid bath contains mainly special grade carbon and carbonates.[12,114] In this bath, mechanical stirring is done by means of one or more simple propeller stirrers to disperse the carbon particles in the molten salt. The chemical reaction between these ingredients may cause the generation of CO and the adsorption of CO on the carbon particles which are presumed to react with steel surfaces in much the same manner as in pack or gas carburizing.[12] The noncyanide liquid carburizing bath is characterized by a higher rate of graphite consumption than a cyanide bath. Normally, replenishment is necessary every hour to maintain a proper bath activity.[114]

This type of bath usually is operated at temperatures higher than those of conventional liquid carburizing (i.e., cyanide-type) baths. Temperatures in the range of 900 to 955°C (1650 to 1750°F) are usually recommended while temperatures below 870°C (1600°F) are avoided, possibly due to the resulting decarburization of the steel.

Advantages of this noncyanide carburizing include the following:

1. Parts that are cooled slowly following noncyanide carburization are easily machinable compared to parts slowly cooled following cyanide carburizing, because of the absence of nitrogen in the former case.

2. As a result of reason 1, parts that are quenched after noncyanide carburizing contain less retained austenite than parts quenched after cyanide carburizing.

Disadvantages of noncyanide carburization are as follows:

1. Control of carbon potential and consistency of carburizing have proved to be difficult.

2. Evenly dispersed solid suspensions are difficult to maintain in large salt baths.[116]

Another variation of a cyanide-free regenerator has been developed recently by Degussa and is suited ideally for automation with molten salt bath quenching. Other advantages of this patented CECONSTANT salt bath carburizing using a low percentage (10%) of cyanide in the salt bath reaction are (1) minimum bailout and waste disposal, (2) safer quenching in molten nitrate salt bath, and (3) attainment of closely controlled carbon potential of 0.5, 0.8, and 1.1% possible with minimum bailout and waste salt disposal.[116]

16.3.1.5 Gas Carburizing. Gas carburizing has now become the most widely used method of case hardening of large volumes of production, its growth increasing every year at the expense of pack or salt bath carburizing. Therefore this process is often referred to as *case carburizing*. This method imparts high fatigue strength, high wear resistance, and retention of a tough and ductile core in the complex parts.[117]

Shallow-case carburizing with case depth <0.5 mm (0.02 in.) is selected for shafts, gears, and races for automatic transmissions, while deep-case carburizing with case depth >1 mm (0.04 in.) is recommended for differential gears, axles, and steering system components. Carburized parts vary in weight from about 45 g (0.1 lb) for planet gear shafts to 3.5 kg (8 lb) for output shafts. Pinion, sun, and ring gears weigh about 180 g (0.4 lb), 450 g (1 lb), and 1100 g (2.5 lb), respectively.[118]

Controlled carburizing atmospheres are produced by blending a carrier gas with an enriching gas, which acts as the source of carbon. The usual carrier, endothermic gas (i.e., endogas), not only is a diluent, but also serves as the accelerator of the carburizing reaction at the component surface. The amount of enriching gas required for gas carburizing depends mainly on the carbon demand, i.e., the rate of absorption of carbon by the workload. Endogas, consisting of CO, H_2, CO_2, and N_2 with smaller amounts of CH_4 and H_2O, is produced by reacting a hydrocarbon gas such as natural gas (mainly methane), propane, or butane with air. For endogas produced in an endogas generator from methane, the air/methane ratio is about 2.5; from pure propane, the air/propane ratio is about 7.5; both of them will produce an O/C ratio of about 1.05 in the endogas. The air/fuel ratio, however, will vary with the composition of the hydrocarbon feed gases and the amount of water vapor in the ambient

TABLE 16.9 Specific Gravity and Composition of Natural Gas in Three Locations of the United States[119]

State	Specific gravity	Composition, vol%			
		CH_4	CH_3CH_3	N_2	CO_2
New York	0.58–0.59	94.1–96.3	1.8–2.0	0.3–1.8	0.83–0.96
Illinois	0.57–0.61	89–97.5	1.6–4.4	0.31–5.7	0.39–0.75
California	0.60–0.63	92–98.8	3.9–5	1.2–1.24	0.76–3.0

CH_4, methane; CH_3CH_3, ethane; N_2, nitrogen; CO_2, carbon dioxide.
Source: American Gas Association.
Reprinted by permission of ASM International, Ohio.

air. Table 16.9 lists typical compositions of natural gas.[119] Sometimes a purified exothermic gas enriched with 5 to 10% hydrocarbon gas (e.g., natural gas or certain propanes) is used as the carrier gas for gas carburizing.

Liquid hydrocarbons are also used as sources of carburizing gas. These liquids are mostly proprietary compounds that range in composition from pure hydrocarbons such as terpenes, dipentene, or benzene to oxygenated hydrocarbons such as alcohols, glycols, or ketones. In general, the liquid is fed in droplet form to a target plate in the furnace, where it volatilizes immediately. The vapors dissociate thermally to provide a carburizing atmosphere comprising carbon monoxide, carbon dioxide, methane, and water vapor. Forced fan circulation causes uniform temperature and even distribution of the atmosphere within the furnace. The control of liquid flow is effected either manually or automatically to ensure the desired carbon potential.

Figure 16.27 is the microstructure of a 1018 steel gas-carburized at 927°C (1700°F) for 12 hr, followed initially by furnace cooling and finally by air cooling to room temperature. This microstructure illustrates a surface carbide network outlining the prior austenite grain boundaries in the pearlite matrix.[120]

Furnaces. Gas carburizing furnaces are classified into two groups: (1) continuous-type furnaces, e.g., mesh belt type, pusher type, shaker hearth, rotary hearth, and integrated pusher/rotary hearth furnace line, and (2) batch-type furnaces, e.g., pit and horizontal sealed quench furnaces. The selection of furnace type depends on the size, shape, quantity, and production run of workpieces and variety of case depth, fixtures, and space requirements.[117] For example, where economical mass production (reduction of energy consumption by 50% or more) with reproducibility and high production rate of similar components of case depths between 0.015 and 0.12 in. (0.4 and 3 mm) (such as in automobile gears and parts) is desired, a continuous-type furnace is the best choice. In these furnaces, the well-separated workpieces, with or without fixtures, enter at one end and pass through preheating, purging, soaking and carburizing, and diffusion zones followed by either cooling or transformation zone for two-stage hardening treatment or oil-quenching zone for single-stage direct hardening treatment. The tempering zone is located last in the continuous line from which workpieces come out in the fully heat-treated condition. If an oil quench is used, parts must be washed prior to tempering to remove oil.

Batch-type furnaces are preferred for small lots with varying case depths, such as large industrial-duty gears, components of machine tools, and material handling equipment. In horizontal batch furnaces, components in small batches are loaded

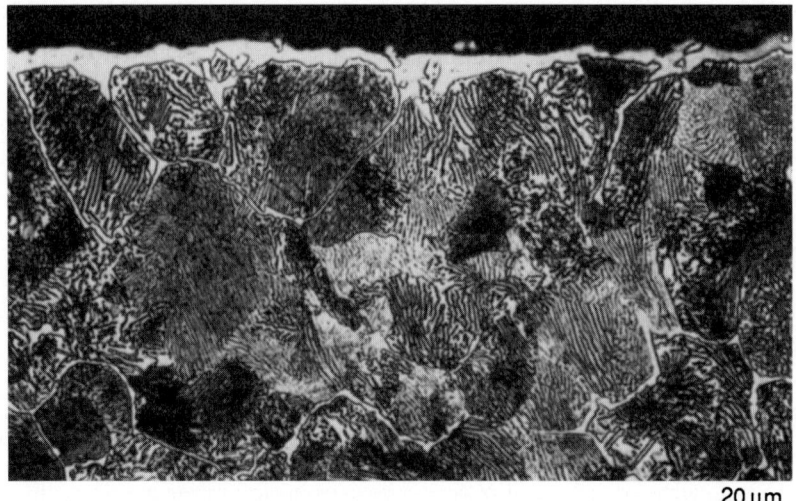

20 μm

FIGURE 16.27 Microstructure of a 1018 steel (gas) carburized and diffused at 927°C (1700°F) for 12 hr, furnace-cooled at 538°C (1000°F) for 2 hr 10 min, and then air-cooled to room temperature. This microstructure shows a high surface carbon content (~1.1%) and a carbide network outlining the austenite grain boundaries in the pearlite matrix.[120] Etched in 1% nital. (*Reprinted by permission of ASM International, Materials Park, Ohio.*)

on the heat-resistant fixtures in the furnaces, preheated, soaked and carburized, and diffused at the required temperature and time. They are then furnace-cooled in the same or different cooling chamber or quenched directly in oil in directly quenched batches. In pit-type furnaces, contact of the hot load with air occurs prior to quenching while in horizontal sealed quench furnaces, air ingress does not occur because quenching is done here under protective atmosphere.[121]

Furnace Atmosphere Parameters. For uniformity of carburizing, the furnace should be equipped with internal fans for good circulation of the atmosphere through the workload. The individual parts must be well separated to allow atmosphere to penetrate the load. Critical parts should be put on fixtures. The furnace should be operated at a positive pressure of 12 to 37 Pa (0.09 to 0.28 torr, or 0.015 to 0.15 in. column of water). It is a good practice to purge air entering the furnace during charging of the parts by using high flow rates of carrier gas. Alternatively, an automatic control system can be used to increase the flow of hydrocarbon enriching gas, compensating for the air entering during door openings.[119]

Gas Carburizing Atmospheres. The carburizing atmospheres are produced by combustion of methane or propane present in natural gas or other hydrocarbon gas in endothermic (or exothermic) gas generators. Endogas is a very complex mixture of CO, N_2, H_2, CO_2, H_2O, and CH_4, and its free and rapid circulation and composition control are very important. In this furnace atmosphere, CO and CH_4 are the sources of carbon; N_2 is inert and acts as a diluent. Many constituents of this atmosphere react with the steel at high temperature in the austenite range, and several reactions take place simultaneously. The most important reversible reaction is

$$2CO \rightleftharpoons CO_2 + C \text{ (in solution in austenite)} \qquad (16.21)$$

After the equilibrium composition of the gas is determined, its carbon potential (or carbon in solution of austenite) at any temperature is obtained by the equilibrium constant at a given pressure K_p, which is written as

$$K_p = \frac{p_{CO_2}}{p_{CO}^2} a_c \qquad (16.22)$$

where p_{CO_2} and p_{CO} are the partial pressures of CO_2 and CO, respectively, and a_c is the activity of carbon in austenite. The equation can be represented in the form

$$K_p = \frac{p_{CO_2}}{p_{CO}^2} f_c(wt\% \, C) \qquad (16.23)$$

where f_c is the activity coefficient of carbon. Rearranging Eq. (16.23), we get

$$wt\% \, C = \frac{K_p}{f_c} \frac{p_{CO}^2}{p_{CO_2}} \qquad (16.24)$$

The sum of partial pressures is equal to the total pressure, so $p_{CO} + p_{CO_2} + p_{inert} = 1$, where p_{inert} is the partial pressure of inert gas, e.g., nitrogen present.

Usually parts are carburized by two-step cycles. During the first step, known as a *carburizing step*, CO content of the atmosphere is greater than the partial pressure required to maintain a required carbon content; the reaction in Eq. (16.21) will proceed to the right direction, and carburizing will take place until a new equilibrium or very high carbon concentration, say 1.2%, is attained. During the second step, $p_{CO_2} \gg p_{CO}$; the reaction will proceed to the left direction, and the steel surface will lose carbon, i.e., decarburization will occur so as to decrease the surface carbon to a lower level, say 0.9%.[122]

Thus in the former step, a high-potential atmosphere results in a rapid carbon penetration and a high surface carbon concentration, say, 1.2% carbon. In the latter step, often called a *diffusion step*, carbon potential is adjusted and maintained in order to produce desired surface carbon concentration in the finished part and diffuse the initially high surface carbon to deeper levels of the case.[123]

Equation (16.24) requires a knowledge of f_c, which varies as a function of temperature and composition of the austenite.[124]

In addition to the carburizing reactions with CO and CO_2, many other reactions may occur as follows:

$$CH_4 \rightleftharpoons 2H_2 + C \text{ (in solution in austenite)} \qquad (16.25)$$

$$CH_4 + CO_2 \rightleftharpoons 2CO + 2H_2 \qquad (16.26)$$

$$CH_4 + H_2O \rightleftharpoons CO + 3H_2 \qquad (16.27)$$

$$CO + H_2O \rightleftharpoons CO_2 + H_2 \qquad (16.28)$$

$$CO + H_2 \rightleftharpoons C \text{ (in solution in austenite)} + H_2O \qquad (16.29)$$

Equation (16.25) represents the principal carburizing reaction for either (1) an atmosphere composed of hydrocarbon gas diluted with nitrogen or (2) a methane atmosphere used in vacuum carburizing. The reaction coefficient for Eq. (16.29) is quite high, suggesting that this reaction controls the carburizing mechanism of the overall process for the gaseous atmosphere containing substantial amounts of hydrogen and carbon monoxide. Note that CO_2 and water vapor present in the reaction product are potent decarburizing agents; therefore, these gases must be removed quickly for the carburizing reaction to proceed. The CO_2 and H_2O content which can be tolerated without causing decarburizing may be calculated from equilibrium data.

In gas carburizing, it is primarily the chemical balance among competing Eq. (16.21) and Eqs. (16.25) through (16.29) that determines the chemical potential of the atmosphere, which, in turn, determines the carbon content at the surface of the component.

The optimum control of carburizing process can be achieved by using computer dynamic control technology, which can accurately control the quality of the case.

Atmosphere Reactions during Gas Carburizing. Carbon transfer from the atmosphere to the workpieces under constant carbon potential can be described by the following scenario:[125]

1. Reactions (16.21), (16.25), and (16.29) are "fast" in both directions. All other reactions are slower.

2. As carbon is transferred to the workpieces, reactions (16.21) and (16.29) consume CO and H_2, respectively, to form CO_2 and H_2.

3. To maintain a constant carbon potential, CH_4 must react with CO_2 and H_2 to restore the gas reactions characteristics of the carbon potential [e.g., Eq. (16.24)].

4. Since reactions (16.26) and (16.27) are "slow," the amount of CH_4 required must be many times the equilibrium amount in order for the "slow" reactions to proceed as quickly as the "fast" reactions.

5. The net result of Eqs. (16.21), (16.26), (16.27), and (16.29) is just Eq. (16.25). Therefore, as carburizing proceeds (at constant carbon potential), CH_4 is consumed and H_2 is generated.

6. An important reason to maintain a flow of carrier gas is to remove the H_2 generated by the carburizing reaction. If the carrier gas flow is too low for the carbon demand, H_2 will build up in the atmosphere and atmosphere CO content will significantly decrease. (A decrease of a few percent is normal for most operations; however, a decrease of 5% is too much.)

7. However, as the carrier gas flow rate increases, the dwell (or residence) time of the atmosphere within the furnace decreases. As a result, the amount of CH_4 needed to drive Eqs. (16.26) and Eq. (16.27) to the right becomes greater.

8. Generation of soot at cold spots within the furnace tends to increase the carbon demand. If sooting becomes so common that brickwork and metal surfaces are covered, all reactions become sluggish because they depend on having catalytically active surfaces available. In a sooted furnace one observes that the atmosphere carbon potential does not respond to changes in enriching gas flows.

In situ Carburizing Atmospheres. Recently, industry's demands for highly flexible heat treatment installations to produce high quality and remain energy-efficient are best met by fully automated batch furnaces running with in situ produced alternative gas carburizing atmospheres. The two atmospheres for which there is the best documentation are as follows:[125]

1. A blend of air and a hydrocarbon gas (such as propane and methane) is introduced directly into the carburizing furnace. If the atmosphere flow rates are relatively low and the furnace temperature is relatively high (e.g., 925°C), the furnace gas composition for a specific carbon potential is the same as when a

traditional endothermic carrier gas is employed. At higher flow rates and/or lower temperatures, the percent of CH_4 in the atmosphere at a given carbon potential will be greater.

2. A blend of nitrogen and methanol, in a $1:2$ molar ratio, is introduced into the furnace to provide the carrier gas.

$$2N_2 + CH_3OH + heat \rightarrow CO + 2H_2 + 2N_2$$

Methane addition serves as an enriching gas to satisfy the carbon demand.

When properly set up, either of these in situ processes can produce results equivalent to conventional gas carburizing with an endothermic carrier gas. The second method is useful primarily when inexpensive sources of hydrocarbon gas are not available.

Advantages claimed in the use of alternative nitrogen-methanol gas atmosphere over the conventional endogas atmosphere are as follows:

1. It has prompt availability.

2. It is economical: (*a*) It takes only a few minutes to start up and introduce this atmosphere in the furnace and to achieve the right chemistry, whereas several hours are wasted in the case of endogas generation; (*b*) It takes less than 10 min to purge and shut down compared to 1 or 2 hr in the case of endogas generation.

3. There is no sooting problem; i.e., it largely contributes cleaner surfaces.

4. There is much lower rate of rework because the results obtained are more reproducible.

5. The system is flexible.

6. There is a substantial decrease in maintenance cost.

7. Given all these savings, it is found that the high cost of the nitrogen-methanol atmosphere is offset to the point that the cost of using this atmosphere matches with that of the endogas atmosphere.[126]

Carburizing Process Variables. Control of three main process variables—temperature, time, and atmosphere composition as carbon potential of the atmosphere—determines the successful operation of the gas carburizing methods. These parameters affect the carbon profile.[127] Other variables that influence the amount of carbon transferred/absorbed to the parts are the alloy content of the parts and the extent of atmosphere circulation.[119]

EFFECT OF TEMPERATURE. Lower temperatures are often used for shallow case depths. The temperature most commonly employed for carburizing is 925°C (1700°F). Sometimes carburizing temperatures are raised up to 980°C (1800°F). Figure 16.28 shows the effect of increasing carburizing temperature from 925 to 1065°C (1700 to 1950°F) for a 3-hr treatment period on case depth and carbon content in AISI 1018 steel. The control of carbon potential for this atmosphere—endothermic gas enriched with natural gas—was done automatically throughout each cycle by the dew point method to produce a surface carbon content of 0.90% and 0.95%.[12]

High-temperature carburizing is defined as carburizing in the temperature range of 982 to 1093°C (1800 to 2000°F), i.e., above the traditional carburizing temperature of 870 to 925°C (1600 to 1700°F). High-temperature carburizing may be recommended where deeper case depths (i.e., over 0.05 in., or 1.3 mm) are required. This increases the production rate by reducing the cycle time, produces more

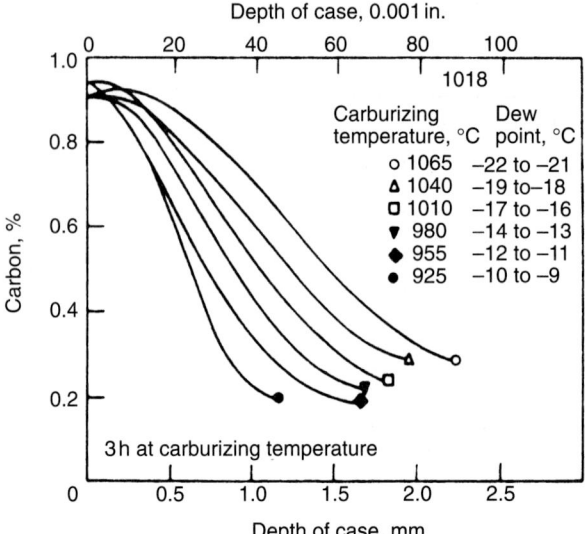

FIGURE 16.28 Effect of increasing carburizing temperature from 925 to 1065°C (1700 to 1950°F) for a 3-hr treatment period on case depth and carbon content in 1018 steel. Natural gas-enriched endothermic gas atmosphere was used, and the carbon potential was automatically controlled by the dew point method to produce a surface carbon of 0.90 to 0.95%.[12,129] (*Reprinted by permission of ASM International, Materials Park, Ohio.*)

gradual carbon gradient between case and core (Fig. 16.29),[127] and cuts down on equipment, energy consumption, and floor space (i.e., furnace size). Thus it offers economic advantages over carburizing at 925°C (1700°F).[107,108] Higher operating temperatures permit higher carbon potential to be employed with less sooting. Moreover, the carbon diffusion reaction is less likely to take place at high carburizing temperatures.[128]

Higher temperatures are hard on furnace fixtures—parts and fixtures are more prone to distortion because they lose strength as the temperature increases.[125]

EFFECT OF TIME. Harris[12] has developed a formula relating the effect of time and temperature on case depth for normal carburizing of plain carbon and alloy steels which can be given as

$$d = 660e^{-8287/T} \sqrt{t} \qquad (16.30)$$

where d is the case depth in millimeters, T is the temperature in kelvins, and t is the time in hours. For a particular temperature the relationship reduces to

$$d(\text{case depth}) = k\sqrt{t} \qquad (16.31)$$

where k is the diffusion coefficient for a given operating temperature and is a function of process temperature, becomes 0.635, 0.533, and 0.457 for 925, 900, and 870°C, respectively, when the case depth is expressed in mm (in.) and the time is expressed in hr. The case depth/time relationship calculated by Harris is shown in Table 16.10

Case Depth, in. (mm)

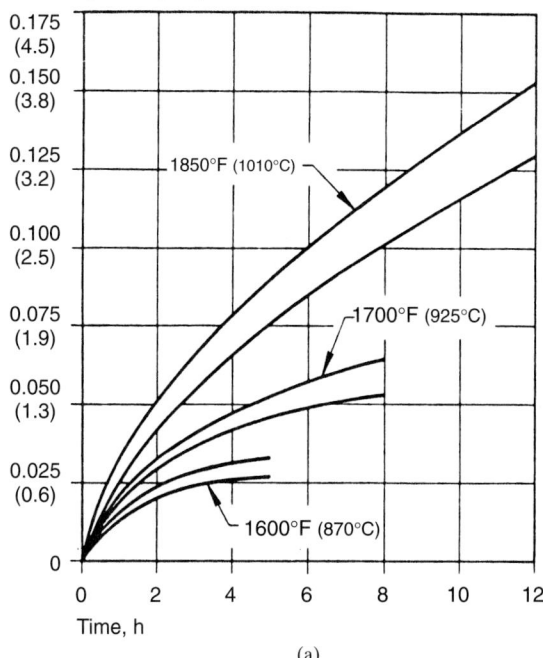

(a)

Carbon Content, %

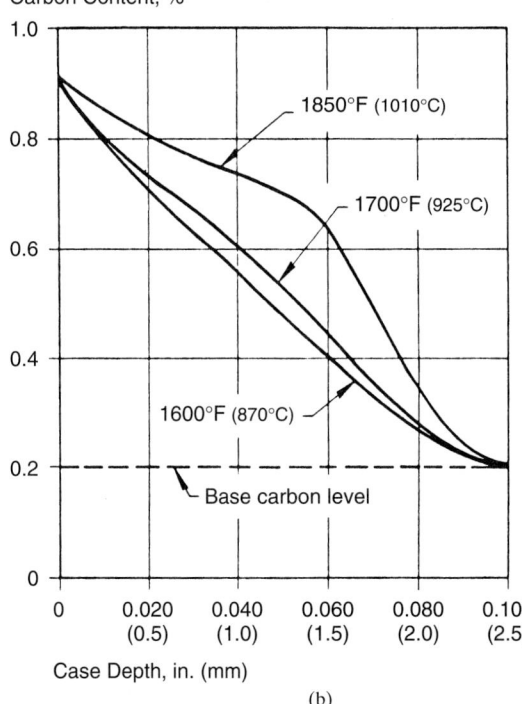

(b)

FIGURE 16.29 Effect of higher carburizing temperature showing (a) reduced cycle time for a given case depth in AISI 8620 steel and (b) a more gradual carbon gradient between case and core (due to heavier cases) than that obtained by carburizing at lower temperatures.[127] (*Reprinted by permission of ASM International, Materials Park, Ohio.*)

TABLE 16.10 Values of Case Depth versus Time Calculated by Harris[12]

	Case depth[†] after carburizing at:					
	870°C (1600°F)		900°C (1650°F)		925°C (1700°F)	
Time t, hr	mm	in.	mm	in.	mm	in.
2	0.64	0.025	0.76	0.030	0.89	0.035
4	0.89	0.035	1.07	0.042	1.27	0.050
8	1.27	0.050	1.52	0.060	1.80	0.071
12	1.55	0.061	1.85	0.073	2.21	0.087
16	1.80	0.071	2.13	0.084	2.54	0.100
20	2.01	0.079	2.39	0.094	2.84	0.112
24	2.18	0.086	2.62	0.103	3.10	0.122
30	2.46	0.097	2.95	0.116	3.48	0.137
36	2.74	0.108	3.20	0.126	3.81	0.150

[†] Case depth, mm $= 0.635\sqrt{t}$ (case depth, in. $= 0.025\sqrt{t}$) for 925°C (1700°F); $0.533\sqrt{t}$ ($0.021\sqrt{t}$) for 900°C (1650°F); $0.457\sqrt{t}$ ($0.018\sqrt{t}$) for 870°C (1600°F). For normal carburizing (saturated austenite at the steel surface while at temperature).
Reprinted by permisssion of ASM International, Metals Park, Ohio.

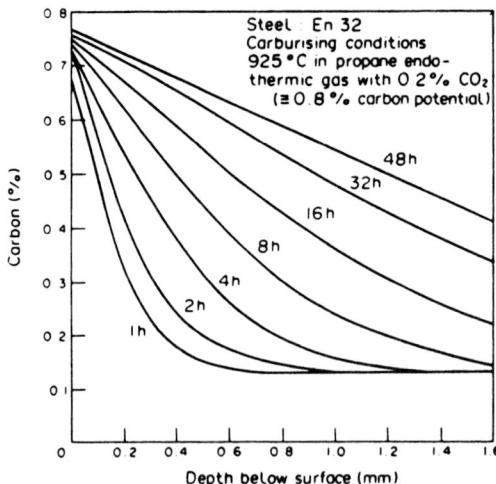

Steel : En 32
Carburising conditions
925 °C in propane endo-
thermic gas with 0.2% CO_2
($\cong 0.8$% carbon potential)

48h
32h
16h
8h
4h
2h
1h

Carbon (%)

Depth below surface (mm)

FIGURE 16.30 Carbon profiles produced in single-stage carburizing at 925°C in times ranging from 1 to 48 hr.[12,129] (*Reprinted by permission of Pergamon Press, Plc; after Still and Child.*)

for three common carburizing temperatures. Figure 16.30 shows the carbon profiles produced in single-stage carburizing at 925°C in times ranging from 1 to 48 hr.[129] This demonstrates that the depth required to obtain a certain carbon level at a certain temperature is directly proportional to the square root of time.

When carburizing is deliberately controlled to produce carbon content of the surface somewhat less than saturated austenite, the calculated case depth will be slightly lower than Eq. (16.31) (or Table 16.10) shows.

In addition to the time at carburizing temperature, several hours may be required to bring large components to operating temperature. When these components are quenched directly from the carburizer, the cycle may be further stretched to permit time for the components to cool from the carburizing temperature to a quenching temperature of about 845°C (1550°F). This period may be treated as a moderate diffusion period; during this period, surface carbon concentration is decreased by maintaining an atmosphere of low carbon potential which is in contact with the component surface.

Harris[12] has also developed a procedure for calculating the carburizing time and diffusion time to produce a given case depth and surface carbon concentration, which is

$$\text{Carburizing time } t_c = \text{total time } t \times \frac{(C - C_0)^2}{C_s - C_0} \qquad (16.32)$$

or $\qquad$ Diffusion time t_d = total time t − carburizing time t_c $\qquad$ (16.33)

where total time t (in hours) is calculated from the equation in Table 16.10, C is the final required surface carbon concentration, C_0 is the original (or core) carbon concentration, and C_s is the surface carbon concentration at the end of the carburizing cycle. This method is most suited for batch-type furnaces.[12]

ALLOY EFFECTS.   The various alloying elements found in carburizing steels affect the activity of carbon dissolved in austenite. Cr seems to decrease the activity coefficient, and Si and Ni seem to increase it. Consequently, the Cr-bearing steel parts equilibrated with a certain furnace atmosphere will take on more carbon than pure iron, while Ni-bearing steels will take on less carbon. In reality, carbides form at lower carbon potentials in Cr-bearing steels than in carbon steels.

CARBON CONCENTRATION GRADIENT AND CARBON POTENTIAL.   The carbon concentration gradient of the carburized parts is a function of the carburizing time, temperature, type of cycle (different combinations of carburizing and diffusion times), carbon potential of the carburizing atmosphere, and composition of steel. The *carbon potential* of a furnace atmosphere at a certain temperature is defined as the percentage of carbon dissolved in iron which is in thermodynamic equilibrium with the furnace atmosphere at that temperature. It is the driving force for the carburizing reaction. The influence of carbon potential of the atmosphere on the carbon concentration gradient, at any given temperature, is shown in Fig. 16.31.[12,129]

Carbon-potential control during carburizing is achieved by changing the flow rate of the hydrocarbon enrichment gas and maintaining a steady flow of endothermic carrier gas. Close control of carbon potential must be obtained because it affects the case depth directly. Too much enriching gas leads to sooting.

To control the carbon potential of hydrocarbon enrichment gas, the concentration of some constituents such as CO_2, water vapor content, and O_2 present in the carburizing atmosphere must be determined; these four basic methods are described below:

1. *Dew point method.* Measurement of average dew point at a given temperature is made by determining the contained water, because the amount of water vapor in the atmosphere is directly related to carbon potential based on the reaction

$$C + H_2O \rightleftharpoons H_2 + CO \qquad (16.34)$$

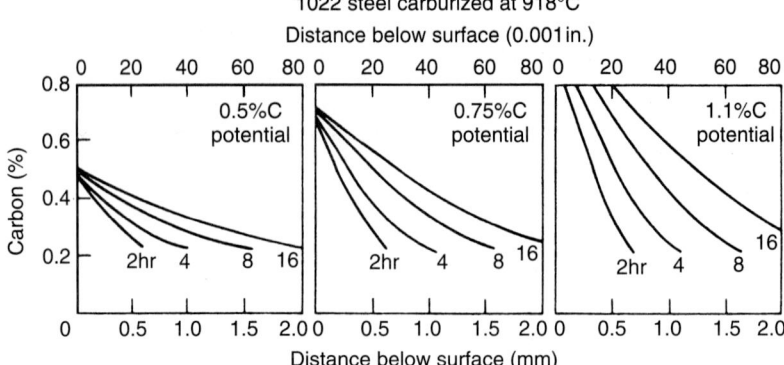

FIGURE 16.31 Carbon concentration gradients in 1022 steel carburized at 920°C (1685°F) with 20% CO, 40% H_2 gas containing sufficient H_2O to produce the carbon potentials shown, namely, 0.50, 0.75, and 1.10% carbon.[12] (*Reprinted by permission of ASM International, Materials Park, Ohio.*)

The main disadvantage of this method is registry of inaccurate results due to either the condensation or presence of hygroscopic materials in the gas sampling system. The two widely used measurement methods are aluminum oxide capacitor and the chilled mirror; both have the drawback of contamination risks from soot and complex vaporized hydrocarbons usually observed in carburizing atmospheres. Consequently, this automatic dew point measurement technique is not reliable for carburizing applications.

2. *Infrared method.* CO_2 present in the atmosphere is measured by an infrared (IR) gas analyzer, based on the reaction $C + CO_2 \rightleftharpoons 2CO$. In this method the absorption of infrared radiation of a gas atmosphere sample is measured by using an infrared gas analyzer. This method can detect changes of CO_2 concentration measuring as low as 0.001%.[130] Since the CO_2 level is very low at high carburizing temperatures, an infrared dew point (dual) system is utilized for greater reliability and in diagnosing conditions within the furnace.[124]

IR analysis provides better uniformity in case depth and helps keep equipment free of soot. IR analysis is especially important when nitrogen-methanol atmospheres are in use,[131] because monitoring of CO indicates the N_2/CH_3OH ratio.

3. *Wire method.* This method involves the measurement of the electric resistance of a wire of iron-nickel alloy (0.003 in. thick and 2 in. long) surrounded by the furnace atmosphere, which is based on the carbon content in the surface.

4. *Oxygen probe method.* Unlike the conventional dew point and infrared analyzers which require a sample of the furnace atmosphere, the oxygen probe (i.e., solid electrolyte oxygen cells) can be located in the furnace chamber in a manner similar to that of a thermocouple[132,133] for measuring the oxygen content of the atmosphere; the CO/CO_2 ratio can be known by using the equilibrium reaction[124] $CO + \frac{1}{2}O_2 = CO_2$.

The most common problems with oxygen probes are as follows: (1) They are dependent on assuming a constant CO, which might not be true. (2) Formation of carbon soot on the probe distorts the bulk reading and causes the automatic control instrument to reduce enriching gas flows, thereby leading to shallow case and low carbon in the parts. (To avoid this situation, an oxygen probe burn-out device is used to periodically oxidize soot away.) (3) There is an unwanted catalytic reaction between probe electrodes and the enriching gas, carrier atmosphere, and hydrocarbon (e.g., leading to the decomposition of CH_4 into CO and H_2 around it). This leads to the same scenario as those for soot contamination. (4) Reference problems are due to (*a*) the contamination of the reference air side of the electrode with the furnace atmosphere gas and (*b*) leakage of the reference air of the probe near the measuring electrode (due to cracks or other flaws in the ceramic materials used in the probe construction—this results in the decrease in millivolt output). (5) Electrode failure occurs over time. (It can sometimes be diagnosed with response time and impedance tests ahead of time, but not with 100% confidence.[134,135])

At present the best method for analysis includes a combination of separate oxygen probe and infrared system to compute carbon potential independently; however, it is not cost-competitive.[135,136]

CONVENTIONAL-CONTROLLED VERSUS COMPUTER-CONTROLLED CARBON PROFILE. The main purpose of a carburizing cycle is to achieve a certain carbon profile, which, in turn, gives rise to a certain hardness profile after quenching. In conventional gas carburizing practice, the carbon control is accomplished by (1) analyzing one of the furnace atmosphere constituents, (2) determining the carbon potential, and (3) maintaining this carbon potential at one or two selected set points for a predefined time period (Fig. 16.32).[137] In contrast, the computer-controlled technique developed by J. Wunning[138] utilizes the microprocessor and a specialized computer program to regulate continuously a desired carbon profile of the workpiece at any carburizing time during different phases of the carburizing cycle by continuously measuring the carbon potential of the atmosphere and temperature. In this case the carbon potential is not regulated at a fixed value for a set time period; instead, it is varied by the microprocessors throughout the processing cycle without the soot or carbide formation (Fig. 16.33). Advantages claimed by computer-controlled processes include (1) optimization of the carburizing process, thereby reducing the furnace idle time[137] and carburizing cycle time (thereby producing cost savings); (2) excellent reproducibility, better quality, and uniformity of carburizing; and (3) greater flexibility based on the choice of different cycle programs.[137,139,140]

16.3.1.6 *Vacuum Carburizing.*

Vacuum carburizing was developed and applied in the United States nearly three decades ago. Vacuum carburizing is a high-temperature, nonequilibrium, boost-diffusion type carburizing process where the steel part being treated is austenitized at 900 to 1040°C (1650 to 1900°F) in a rough vacuum or partial pressure of hydrogen; carburized in a partial pressure of hydrocarbon gas, a mixture of hydrocarbon gases, or hydrocarbon/nitrogen mixtures; diffused in a rough vacuum; and then quenched in either oil or gas. Figure 16.34 describes a typical thermal processing cycle as a function of time.[141] This shows pulsing rather than a constant flow. Note that some systems have constant pressure and gas flow while others use pulse and evacuation cycles to circulate the gas. Table 16.11 provides a comparison of vacuum, gas, and plasma carburizing processes.[142] Both batch and continuous vacuum carburizing equipment have been used throughout the industry.[143]

The furnace atmosphere usually consists of natural gas, pure methane, propane, acetylene, and/or propylene as an enriched gas, and nitrogen as a diluent and inert gas. After loading, the furnace is evacuated to a partial pressure of 13 to 40 Pa (0.1 to 0.3 torr) of hydrogen in a graphite-lined heating chamber and a partial pressure

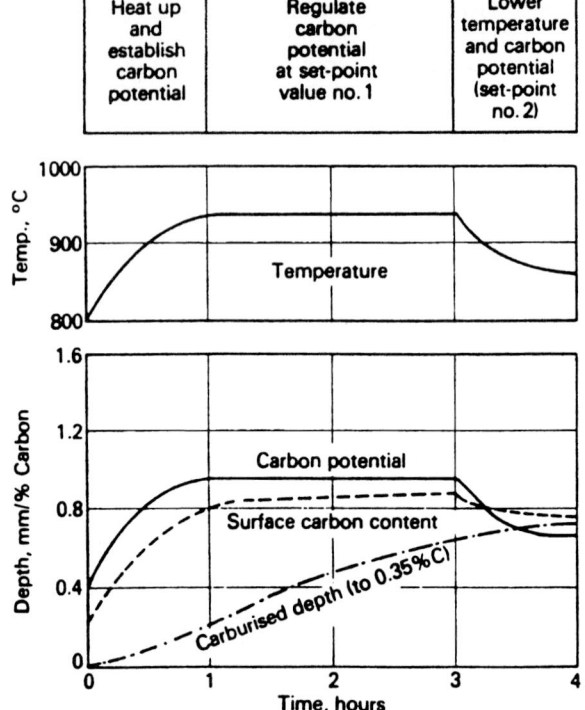

FIGURE 16.32   Carbon profile accomplished in conventional controlled gas carburizing for predefined time periods.[137] (*Courtesy of Wolfson Heat Treatment Center, England.*)

of 40 to 67 Pa (0.3 to 0.5 torr) of hydrogen in a ceramic-lined heating chamber.[141] During the carburizing step, the furnace pressures are kept in the range of 1.3 to 6.6 kPa (10 to 50 torr) in furnaces of graphite construction and 13 to 25 kPa (100 to 200 torr) in furnaces of ceramic construction. Partial pressure exceeding 40 kPa (300 torr) is usually not recommended due to excessive carbon deposition within the furnace associated with higher partial pressure.[141] In this process, only one gas reaction predominates. When methane or propane gas is used, the dissociation reaction at the steel surface is:[144]

$$CH_4 + Fe \rightarrow Fe(C) + 2H_2 \text{ or } C_3H_8 + 3Fe \rightarrow 3Fe(C) + 4H_2 \qquad (16.35)$$

After the appropriate carburizing time t_c, a diffusion time t_d follows during which the carburizing gas is evacuated, and diffusion is allowed to occur in rough vacuum of 67 to 135 Pa (0.5 to 1.0 torr) at the same temperature employed for carburizing prior to cooling to the hardening temperature and subsequent quenching in either oil or gas. The relation between these times at 871°C (1600°F) is given by the equation $t_d = (\frac{2}{3})t_c$.[144]

If carbon potential control was employed during the carburizing (boost) step, the diffusion step might be shortened, or eliminated.[141]

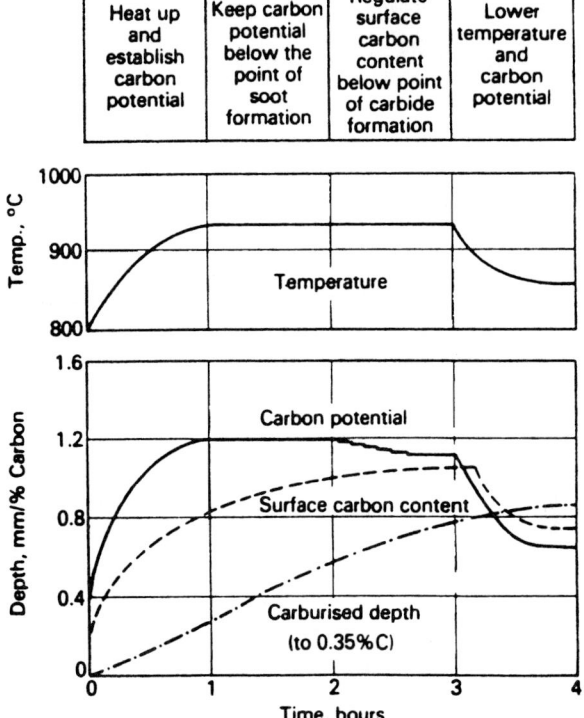

Heat up and establish carbon potential	Keep carbon potential below the point of soot formation	Regulate surface carbon content below point of carbide formation	Lower temperature and carbon potential

FIGURE 16.33 Computer-controlled technique for desired carbon profile of the workpiece at any carburizing time.[137] (*Courtesy of Wolfson Heat Treatment Center, England.*)

Unlike gas carburizing, where carbon potential is a function of the gas atmosphere, the carbon potential in a vacuum furnace is determined by the carbon saturation of steel and its carburizing time at a given temperature (Fig. 16.35a). However, the carbon potential of a propane atmosphere is higher than that of methane due to the availability of more carbon during the (above) cracking reactions. At or below 871°C (1600°F), carburizing or carbonitriding containing CH_4 atmosphere is not practiced due to its poor cracking efficiency compared to propane.[144] Note that large furnace pressure in the carburizing step produces too much carbon soot on the parts, which is not desirable. Hence the recent practice is to pulsate the propane or methane gas in the hot chamber and maintain the vacuum pressure between 20 and 30 torr.[145]

Generally, graphite/carbon fibers are used as the heating materials and fixtures (which are not susceptible to attack by carbon) instead of metals in furnace construction. The use of a higher-temperature carburizing does not (1) create any undue increase in wear or in maintenance costs,[146] (2) cause excessive distortion or warpage, and (3) result in reduction in mechanical properties. The resultant carbon concentration is known by the ratio of carburizing period to diffusion (or boost/

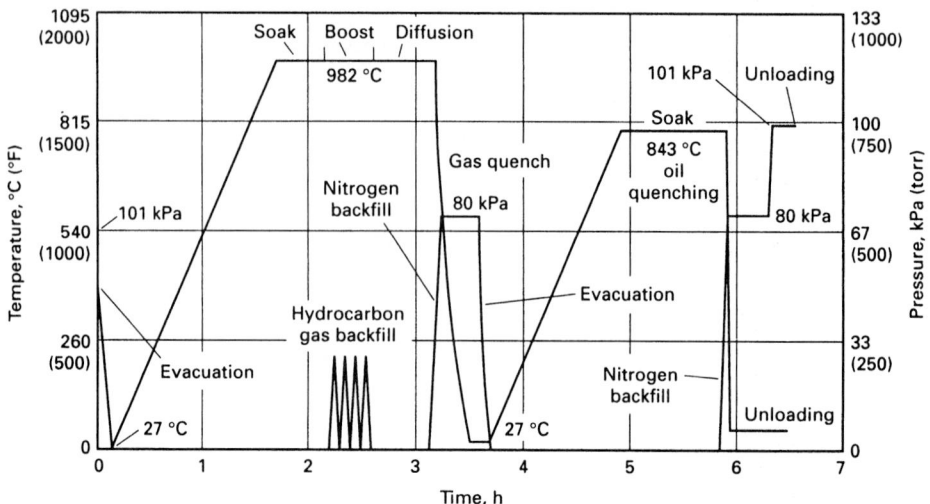

FIGURE 16.34 Plot of temperature and pressure versus time for a typical vacuum carburizing process with a reheat cycle.[141] (*Courtesy of ASM International, Materials Park, Ohio.*)

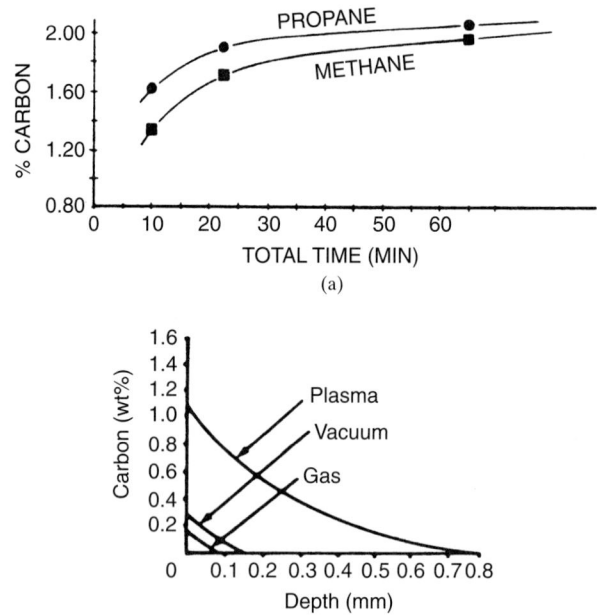

FIGURE 16.35 (*a*) Percent carbon versus total time at 871°C (1600°F) after vacuum carburizing of steel part using propane and methane.[144] (*b*) Carbon profile for plasma, vacuum, and gas after 15 min.[159] [(*a*) *Reprinted by permission of ASM International, Materials Park, Ohio; (b) reprinted by permission of Fairchild Publications, New York.*]

TABLE 16.11 Comparison of Plasma, Vacuum, and Gas Carburizing Processes[142]

	Plasma	Low pressure	Gas
Gas mixtures	CH_4, H_2, Ar	CH_4, C_2H_6	N_2/methanol, endothermic atm., direct gassing
Species for carburizing	Activated by plasma	Thermodynamical dissociation	Thermodynamical dissociation
Pressure range	2–30 mbar	10–600 mbar	1 bar
Gas consumption	<1 m³/hr	~1 m³/hr	5 m³/hr–10 m³/hr
Gas disposal	Pumping	(Burn off)/pumping	Burn oil
Availability of plant	Always	Always	After forming
Thermal emission	None	None	Yes
Control/regulation of the process	LO = plasma parameter	None	O_2-probe/dew point
Carburizing depth (end)	Every, preferential small end's up to about 1.5 mm	Small up to large end's	Middle up to large end's
Partial carburizing	Simple, mechanical masks or shields	Expensive protection with paste	Expensive, protection with paste
Carburizing speed	Small case depth quicker than vacuum or gas carb., also diff. laws	Small case depth quicker than vacuum or gas carb., also diff. laws	$f(C_p, 0)$ diffusion laws
Surface oxidation	None	None	Yes
Geometry of parts	Partly for very complicated shapes (e.g., injection nozzles)	Similar to gas carburizing	Shadow effects, carbon deposition (e.g., blind holes)
Type of load	Ordered load	Ordered load	(Bulk load) ordered load
Quenching media	High-pressure gas > 10 bar	High-pressure gas > 10 bar	Oil

Courtesy of ASM International, Materials Park, Ohio.

16.69

diffusion) period, whereas the required case depth depends on total carburizing time according to the following reactions.

The characteristic difference between endothermic carburizing and vacuum carburizing lies in the fact that in the former the diffusion of carbon occurs intergranularly at the surface while in the latter the diffusion occurs transgranularly at the surface.

The advantages of this method over the conventional gas carburizing include:

1. Preheating and post-carburizing treatment may be accomplished under vacuum, which causes exceptionally clean parts (i.e., bright surface quality).

2. There is elimination of (endothermic) atmosphere generator, which also causes elimination of simultaneous reactions involving CO, CO_2, H_2, and CH_4.

3. There is faster carburizing (i.e., reduced cycle times) and higher effective case depths >0.9 to 1.0 mm (0.035 to 0.040 in.), mainly due to boost-diffusion cycle, greater carbon potential, and use of higher carburizing temperatures, typically 980 to 1040°C (1800 to 1900°F).

4. Carburizing gas consumption is lower. The gas consumption is only about 10% of that in a gas carburizing process.[142]

5. Mechanical properties (such as high fatigue strength, hardenability, etc.) are improved because of no formation of intergranular oxidation products on treated parts.[142]

6. Heating up and shutting down of equipment are done in a few minutes.[145]

7. There is an exceptionally uniform gradation of carbon from the surface in; exact, uniform, repeatable, and predictable surface carbon contents, case depths, and metallurgical properties due to high degree of process control variables, possible with vacuum furnaces.[142,146,147]

8. Use of vacuum media permits the selection of precise levels of carbon availability by varying or constant gas flow, gas pressure, and hydrocarbon concentration in the gaseous mixture.[142,147]

9. There is the ability to avoid external effects of nonequilibrium gas states, surface abnormalities, and temperature differentials.[142,147]

Disadvantages are that

1. Equipment cost is high.

2. A delicate balance exists in vacuum carburizing where the process conditions must be tuned to achieve the best compromise among case uniformity, carburizing rate, and risk of sooting.[148]

Problems of both lack of case depth uniformity and the production of soot in the furnace have been recently reported to be solved by special gas injection techniques and lower gas pressure (usually below 10 mbar).[149]

Advantages of small addition of acetylene in carburizing atmosphere at a level of medium vacuum below 0.13 kPa (1 torr) are (1) considerable reduction in the sooting problem, making it possible to minimize the maintenance cost; (2) high uniformity of case depth even in the complex-shaped parts on loads with high component density and even on bulk loads without increasing the supply amount of the gas; (3) very small thermal losses due to convection;[150] and (4) extremely high productivity due to high load density and the short carburizing times due to high carbon transfer.

TABLE 16.12 Typical Carburizing Constants and Boost/ Diffusion Ratio to Obtain a 0.8 to 0.9% Surface Carbon Content in a Low-Alloy, Low-Carbon Steel[141]

Temperature		Carburizing constant		
°C	°F	$k^{\dagger}$	$k^{\ddagger}$	Boost/diffusion ratio r
840	1550	0.25	0.010	0.75
870	1600	0.33	0.013	0.65
900	1650	0.41	0.016	0.55
925	1700	0.51	0.020	0.50
950	1750	0.64	0.025	0.45
980	1800	0.76	0.030	0.40
1010	1850	0.89	0.035	0.35
1040	1900	1.02	0.040	0.30

[†] To obtain effective case depth (50 HRC hardness) D, in millimeters, when $D = k \sqrt{t}$ and t is in hours. [‡] To obtain effective case depth (50 HRC hardness) D, in inches, when $D = k\sqrt{t}$ and t is in hours.
Courtesy of ASM International, Materials Park, Ohio.

Carbon Gradient and Case Depth Prediction. The carburized case depth d in vacuum (or gas) carburizing can be simply and accurately predicted from the following equations:

$$d = k\sqrt{t} \tag{16.36}$$

$$t_c = rt \tag{16.37}$$

$$t = t_c + t_d \tag{16.38}$$

where k is a temperature-dependent carburizing constant, t is the total carburizing time, r is the boost/diffusion ratio, t_c is the carburizing time, and t_d is the diffusion time. Table 16.12 gives the typical carburizing constants and boost/diffusion ratios to obtain a 0.8 to 0.9% surface carbon content in a low-alloy, low-carbon steel.[141]

Effect of Alloying Elements. Alloying elements affect the rate of carbon absorption; for example, Si and Mn reduce the carbon potential whereas Cr, Mn, and Mo increase the effective carbon absorption and form more stable carbides. Incorrect boost/diffusion ratio may cause the formation of carbide network in these materials. Special higher-temperature cycles must be used on some materials such as stainless steels to depassify the surface before carburizing. The use of high temperature results in dissolution of carbide formers, thereby making carburization effective at high temperatures. Gas composition tends to have a greater effect on case depth and uniformity than does gas partial pressure.[151]

Additional advantages of vacuum carburizing with high-pressure gas quenching are (1) the provision of clean, dry hardened parts; (2) vacuum as protective atmosphere; (3) reduction in servicing, maintenance, and repair costs; (4) ease of high-temperature carburizing; and (5) control of quench intensity, no Leidenfront phenomenon, uniform quenching, and small scatter of distortions.[152]

16.3.1.7 Plasma Carburizing. Plasma carburizing is a nonequilibrium, boost-diffuse method of carburizing where a constant carbon potential is not maintained;

rather, the steel is heated to carbon saturation for the selected temperature. The steel components to be treated are introduced [with a spacing of ~6.4 mm (0.25 in.) in between] into the hot temperature zone, which constitutes the cathode with respect to the furnace wall (anode), where a dc voltage (in the range to 350 V to 1 kV) is applied in an oxygen-free, low-pressure [in the range of 130 Pa to 3.3 kPa (1 to 25 torr) or, preferably, 130 to 670 Pa (1 to 5 torr)] carburizing gas, i.e., a mixture of hydrocarbon gas such as methane (or propane), hydrogen, and argon (or nitrogen). When the flow rate becomes very low at a few cubic feet per minute and pressure is in the range of 10 Pa to 3 kPa, the glow-discharge plasma is produced between the furnace wall (anode) and the workpieces (cathode), which cracks or dissociates methane into carbon and hydrogen, excites the ionized gas atoms (e.g., carbon) to react with the surface of the components,[153] and heats up the component to the processing temperature of 850 to 1040°C (1562 to 1904°F) or usually at about 927°C (1700°F). Thus, all the carbon required is deposited on the surface during this very short carburizing stage. The thickness of the cathode visible glow or luminescence depends on the pressure, gas composition, and temperature. The key factor in determining the necessary operating pressure is the visual observation to ensure that the plasma covers the load and that no hollow cathode effects are obtained. The carburizing gas mixture and the glow discharge supply are shut off, and the workpiece temperature is allowed to fall—typically to 849°C (1560°F); this represents the longer (3 times) carbon diffusion period. (A higher temperature is used to obtain more rapid diffusion rates. This is not favored due to the increased potential for distortion.) Finally, the components are transferred from the hot zone and subsequently oil-quenched to obtain the desired martensitic hard case.[154] Best results are achieved with an integrated oil-quenching facility with component transfer under vacuum.[155] Note that when workloads are heated by a conventional heating process, as used in vacuum furnaces, it reduces the heating-up time during plasma carburizing.[153]

The carbon transfer in propane is very high (Fig. 16.36a);[156] as a result, the carbon saturation at the workload surface is reached within 10 to 15 min. This phenomenon can be exploited to run a multistage cycle with several boost and diffuse periods, as shown in Fig. 16.36b.[157]

Parts in a load can be mechanically masked by not allowing the glow discharge into contact with the area not to be carburized, e.g., by stacking or proper fixturing. Alternatively, copper plating of selected areas is effective for masking.[148]

The increased carburizing rates during plasma carburizing are attributed to the omission of several reaction steps in the dissociation process of hydrocarbon, i.e., (CH_4) gas to produce active soluble carbon owing to the ionizing effect of the plasma. The effective carbon potential is governed by the gas mixture and by its flow rate, gas pressure, plasma power, and temperature. All these parameters may be accurately controlled electrically using experimental calibration from an extensive database.[146]

The advantages of plasma carburizing methods over gas carburizing methods are as follows:[153,155-160]

1. Like plasma nitriding, shorter (door-to-door) processing times are consistently obtained when compared with traditional gas or vacuum carburizing at similar temperatures (e.g., by 20% over gas carburizing and by 5% over vacuum carburizing). This is attributed to the very rapid carbon mass transfer into the component surface (Fig. 16.35b)[159] (Table 16.11).[142]

2. There is predictable surface hardness and case depth, with improved case uniformity because of accurate automatic control of carburizing mechanism. [*Note:*

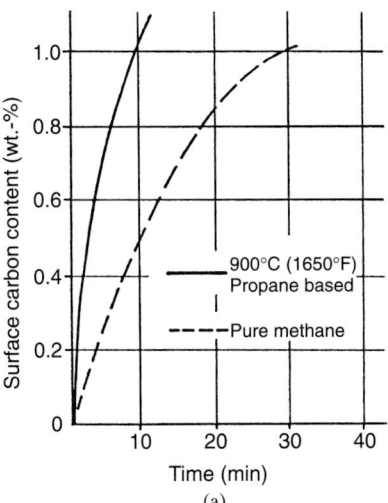

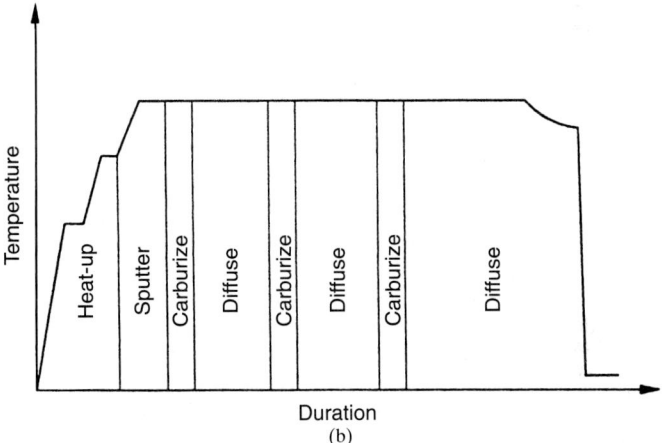

FIGURE 16.36 (*a*) Effect of type of gas on plasma carburizing rate.[156] (*b*) Schematic of a multistage cycle employed for plasma carburizing in propane-base gases.[157] (*Courtesy of Wolfson Heat Treating Center, England.*)

Required uniformity of carburizing can be obtained, provided a minimum carbon mass flow (of 3×10^{-4} g of carbon/min/cm^2 of surface area) into plasma is obtained.[148]]

3. There is improved blind-hole penetration or "downhole" carburizing up to an L/D (length/diameter) ratio of about 12 with respect to L/D ratio up to 9 for atmosphere carburizing.

4. Gas and energy consumption is lower (i.e., very large gas and energy savings) because it is a low-pressure process with very low gas flow. Moreover, there is an absence of furnace idling time and very short shut-down and start-up times.[155]

5. Better fatigue and wear resistance and minimal distortion are observed.

6. Periodic maintenance cost is low due to minimum deterioration of furnace elements and the hearth.

7. Parts are bright and clean (free from soot and internal oxidation)[155] with reproducible and high-quality metallurgical structure (due to uniform penetration of carbon into the surface and no formation of carbides).

8. Plasma in ion carburizing increases the carbon potential at a much reduced absolute pressure.

9. Elimination of intergranular oxides (due to carburizing and oil-quench hardening in a vacuum) improves fatigue properties in the unground state.[148]

10. Like vacuum carburizing, high-temperature plasma carburizing reduces cycle times with higher effective case depths. However, plasma carburizing eliminates two major problems associated with vacuum carburizing, i.e., risk of sooting and carburizing rate.

11. No post-treatment machining is required.

12. There is freedom from pollution.[153]

13. Masking methods are simple for selective carburizing.[158]

14. There is no need for furnace conditioning.[158]

15. Applicability to high-alloy steels such as stainless steels is an advantage.[158]

16. They are integrable into production lines.[158]

The main disadvantages of plasma carburizing methods are (1) they are not applicable to bulk loads (individual loading of components is required), (2) carburizing of entire surface of workpieces is not feasible (due to cathode contact requirement), and (3) capital costs are higher (counterbalanced by low running costs).

Plasma carburizing using integral high-pressure (10- and 20-bar) gas quenching with nitrogen or helium is applied to different core hardening steel parts in the automotive industry. Process parameters have been successfully established for gear parts and for injection nozzles. Other benefits of this integrated system include greatly improved environment-friendliness, no need of cleaning oil residues and disposal of its effluents, and considerable decrease in heat treating costs. The purchase and operation of washing stations to remove oil residues and the cost of disposing of these effluents are no longer necessary. The process cuts heat-treating costs considerably.[161]

Plasma carburizing is used for applications such as constant-velocity joints, gears, hydraulic valve components, diesel engine components, and clutch components.[158]

16.3.1.8 Pulsed Plasma Carburizing. The newly developed pulse plasma carburizing leads to a controllable and reproducible carbon mass flow into the workload, where the pulse length is on the order of microseconds. By varying the plasma parameter, very high mass flow can be achieved, which after just a few minutes causes the saturation of the surface with carbon. A linear increase in the mass flow occurs with increasing pressure. Thus, with a high mass flow at the initiation of the process, a high surface carbon content to just below the saturation point can be achieved by selecting suitable parameter combinations, in order to obtain a steep carbon gradient into the material. Then the mass flow is reduced to such a level that the surface carbon content remains just below the saturation point, but simultaneously avoids the carbide formation in the surface zone. By restricting the mass flow density to the required level, no pauses without plasma for diffusion have to be made during the treatment, a steep carbon gradient is maintained, and process times are as short as possible.[161]

16.3.1.9   Fluidized-Bed Carburizing.   Over the last two decades, design innovations have led to the use of (direct and indirect) fluidized-bed furnaces as a practical tool for carburizing, carbonitriding, nitriding, and nitrocarburizing processes. Mostly direct fluidized beds have been used where the workload is heated by immersion directly in the fluidized bed of small (80-mesh or 180-mesh) dry aluminum oxide particles, which behaves as a liquid and produces a fluidizing effect. This is achieved by feeding a supporting gas up through the bed of particles, which also provides the protective atmosphere for the heat treatment of the workload immersed in the bed. When external gas burners or electric elements are used as the heat source, the bed provides a faster heat-transfer medium. This is integrated with quench and tempering furnaces.

The indirect fluidized-bed furnace comprises a direct fluidized bed which is heated by either internal gas combustion or immersed electrodes where one or a number of radiation walls, typically tubes, are positioned either horizontally or vertically through the bed, separating the workpieces to be treated from the fluidized bed.

The normal atmospheres employed in a conventional fluidized-bed carburizing are (1) endogas atmospheres plus hydrocarbon additions, (2) N/methanol system plus hydrocarbon additions, and (3) direct injection of hydrocarbon and air systems.

It is apparent that the use of fluidized beds has become more widespread with the understanding of the particular advantages of the technique by industry.[162]

The advantages of this process include the following:

1. Uniformity of heating as well as high rates of heating and flow cause the utilization of higher treatment temperatures which, in turn, provide rapid carburizing
2. Close control on temperature uniformity, typically ±5°C, coupled with low capital cost and flexibility are ensured
3. A fluid-bed furnace is very tight (with the upward pressure of the gases, it is difficult for air to leak in)
4. This process produces parts with very uniform finish
5. Elimination of internal oxidation
6. Control of microstructure to improve the wear resistance of the carburized case
7. Development of the uniform carbide dispersion in carburized microstructures/ tempered martensitic structure[163]
8. The rate of atmosphere conditioning and change of atmosphere composition, typically 2 minutes
9. The use of metallic retorts as against ceramic-lined furnaces providing greater atmosphere integrity[163]

16.3.2   Carbonitriding

Carbonitriding is grouped under austenitic thermochemical surface hardening treatments. This is one of the most commonly employed case hardening treatments in which both carbon and nitrogen are absorbed simultaneously into the surface of steel held at an elevated temperature in the austenite phase field by diffusion

TABLE 16.13 Composition and Properties of Sodium Cyanide Mixtures[114]

Mixture grade designation	Composition, wt%			Melting point		Specific gravity	
	NaCN	Na$_2$CO$_3$	NaCl	°C	°F	25°C (75°F)	860°C (1580°F)
96–98[†]	97	2.3	Trace	560	1040	1.50	1.10
75[‡]	75	3.5	21.5	590	1095	1.60	1.25
45[‡]	45.3	37.0	17.7	570	1060	1.80	1.40
30[‡]	30.0	40.0	30.0	625	1155	2.09	1.54

[†] Appearance: white crystalline solid. This grade also contains 0.5% sodium cyanate (NaNCO) and 0.2% sodium hydroxide (NaOH); sodium sulfide (Na$_2$S) content, nil. [‡] Appearance: white granular mixture. Courtesy of ASM International.

process, using a liquid salt bath or gaseous atmosphere, and are then quenched in water, oil, or gas.

16.3.2.1 Cyaniding (or Liquid Carbonitriding). In cyaniding or salt bath carbonitriding, the steel parts are heated above the transformation temperature Ac_1, usually at 843°C (1550°F), in a suitable molten bath containing alkali cyanide to serve as active salt mixed with other salts such as sodium chloride (NaCl) and sodium carbonate (Na$_2$CO$_3$) to provide adequate fluidity and to regulate the melting points of salt mixtures (Table 16.13). The cyanide bath decomposes and releases less carbon and more nitrogen (than those from the nonactivated liquid carburizing bath), which absorb into the part surface. On quenching in water, brine, or oil, a hard case develops.

In these baths, mostly NaCN is used instead of KCN due to its lower cost and higher efficiency. The cyanide concentration is varied, depending on the specific application. However, a normally active cyanide bath contains 30% NaCN, 40% Na$_2$CO$_3$, and 30% NaCl.

The chemical reactions that occur during cyaniding are essentially similar to those of nonactivated salt bath carburizing [Eqs. (16.11) through (16.14)]. The active hardening agents of these baths are CO and N$_2$, which liberate from the decomposition of sodium cyanate. The rate of cyanate decomposition, being a measure of carbonitriding activity of the bath, increases with the higher cyanate concentration and bath temperature, which produce larger case depths with increased carbon concentration.

A fresh cyaniding requires aging for about 12 hr in a molten state to furnish a sufficient amount of cyanate for efficient carbonitriding activity. During aging, any carbon scum formed on the surface must be removed for effective performance. To remove scum, it is necessary to lower the cyanide concentration of the bath to a 25 to 30% level by means of addition of inert salts (sodium chloride and sodium carbonate). The rate of decomposition of the bath at the aging temperature [normally 700°C (1290°F)] is low.[114]

Case depth is thus a function of NaCN concentration but is limited to 0.254 mm (0.01 in.). This process has some drawbacks, as mentioned above for cyanide-type salt bath carburizing.

16.3.2.2 Gaseous Carbonitriding. In this process, a relatively low percentage (2.5 to 8%) of NH$_3$ is introduced in the carburizing, that of atmosphere [e.g.,

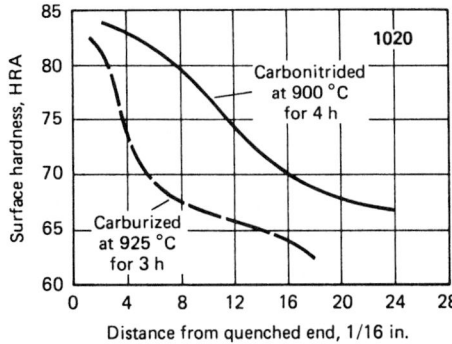

FIGURE 16.37 Comparison of end-quench hardenability curves for 1020 steel carbonitrided at 900°C (1650°F) and carburized at 925°C (1700°F). Hardness was measured along the surface of the as-quenched hardenability specimen. Ammonia and methane contents of the inlet carbonitriding atmosphere were 5%, and the balance was carrier gas.[164] (*After G. W. Powell, M. B. Bever, and C. F. Floe, Trans. ASM, vol. 46, 1954, pp. 1359–1371.*)

carbon-rich gas (or vaporized liquid hydrocarbon) mixture] to diffuse nitrogen into the steel together with carbon. This operation is carried out at a lower temperature and for a shorter time than that of gas carburizing, which results in a relatively shallower case depth, usually from 0.075 to 0.75 mm (0.003 to 0.03 in.). This process may be treated as a modified gas carburizing process, rather than a form of nitriding.

The main advantages of carbonitriding over carburizing are as follows:[113]

1. The carbonitrided case has a greater hardenability than a carburizing case (Fig. 16.37).[164] (Nitrogen increases the hardenability of steel, stabilizes the austenite, and induces retained austenite, especially in alloy steels.) This favors attainment of a hard case and allows the case of the part to be hardened by oil-quenching or even gas-quenching at reduced expense.[113,164]

2. Since absorption of nitrogen by the surface layers of steel decreases the critical cooling rate to form martensite, compared to that in carburizing treatment, less distortion will occur.

3. The resistance of carbonitrided layers to softening during tempering is higher than that of the carburized layer, which is attributed to the precipitation of γ' nitrides. This resistance also increases with an increase in nitrogen content at tempering temperatures of 573 K and above. This resistance is further enhanced by shot-peening, the effect of which is more profound after carbonitriding than after carburizing, as observed in Fe-0.22C-0.12Si-0.78Mn-0.01P-0.018S-1.1Cr steel automatic transmission gear.[165]

4. This gives increased wear resistance as well as a hard and uniform case, when compared to ordinary carburizing.

5. Better fatigue properties than those of equivalent carburized components are obtained, owing to different residual stress pattern in the case.

6. For many applications, carbonitriding of the less expensive steels will furnish properties similar to those obtained in gas carburizing of alloy steels.[164]

7. Sometimes carburizing and carbonitriding are combined to achieve deeper case depths and superior service performance than are feasible by carbonitriding alone.[164]

Disadvantages of carbonitriding are as follows:

1. Compared to carburizing treatment, more time is required to produce a case depth greater than 0.635 mm (0.025 in.) because of lower carbonitriding temperature.

2. Carbonitriding is restricted mostly to case depths up to about 0.75 mm (0.03 in.) or less, whereas no such restriction is applicable to carburizing.

Case Composition. The composition of a carbonitrided case is governed by the time, temperature, atmospheric composition, and type of steel subjected to this treatment. Carbon addition is favored at high temperature. More nitrogen absorption is favored at the lower temperatures, which results in a "compound layer" consisting of Fe-C-N compounds at the surface. This type of case structure is preferred in some wear-resistant applications. For this layer of compounds, a considerable large percentage of NH_3 in the gas mixture is required, and control of furnace atmosphere becomes more critical. Figure 16.38 is the microstructure of carbonitrided and oil-quenched AISI 1020 steel showing an outer white layer of case (left) as cementite followed by retained austenite mixed with martensite and interior martensite matrix (right).[120,164]

Recently, it was reported that the 1 to 2% NH_3 addition at the end of the carburizing cycle is most effective in minimizing the amount of surface nonmartensitic transformation products.[166]

Case Depth. As in the case of carburized parts, carbonitrided parts are usually measured for total case depth or effective case depth.[12] Suitable case depth is governed by service applications and by core hardness. Figure 16.39 shows the effects of time and temperature on effective case depths which are based on a survey of industrial practice.[105,164] Medium-carbon steels with core hardness of 40 to 45 Rc usually require less case depth than steels with core hardness of 20 Rc or below.

20 μm

FIGURE 16.38 Microstructure of a 1020 steel carbonitrided (with high carbon potential) and oil-quenched showing an outer white layer of case (left) as cementite followed by retained austenite mixed with martensite and interior martensite matrix (right). Nital etch.[120] (*Reprinted by permission of ASM International, Materials Park, Ohio.*)

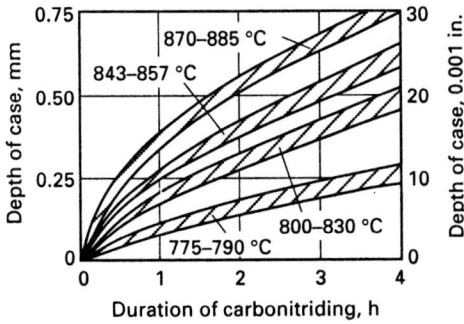

FIGURE 16.39 Effects of time and temperature on case depths based on the results of a survey of industrial practice.[164] (*Reprinted by permission of ASM International, Materials Park, Ohio.*)

Medium-carbon low-alloy steels, i.e., those employed in automotive transmission gears, are often provided the minimum case depth of 0.2 mm (0.008 in.).[164]

Furnaces. Almost any furnace suitable for carburizing treatment can be used for carbonitriding. The furnace must be equipped with a fan to circulate the atmosphere and with protective atmosphere vestibules to the quenching area. The atmospheres generally consist of a mixture of carrier gas, enriching gas, and NH_3. Basically, any atmosphere composition used for gas carburizing can be used for carbonitriding with an additional 2 to 12% anhydrous NH_3 of 99.9+% purity.[12,164]

Temperature Selection. Choice of carbonitriding temperature is based on steel composition, dimensional control, fatigue and wear properties, microstructural constituents, hardness, cost, and equipment. Lower temperature near 704°C (1300°F) causes an explosion hazard as well as superficial, high-nitrogen, brittle cases with low core hardness which are not suitable for most applications. For this reason, most carbonitriding operations are carried out at 790°C (1450°F) or above. Carbonitriding at 900°C (1650°F) appears to be advantageous from the standpoint of favorable combination of hardness, microstructure, wear, and economy. However, other factors, such as distortion and quench cracking, must also be accounted for.[167] The usual compromise is made at about 843°C (1550°F) to achieve both ends.

Void Formation. Case structure may contain subsurface voids or porosity, if the processing conditions are inappropriately adjusted. This problem is often related to excessive NH_3 additions. Table 16.14 summarizes the effect of material and processing variables on the possibility of void formation.[164]

Control of Retained Austenite. Minimum retained γ in the carbonitrided case can be formed by (1) increasing the furnace temperature, (2) reducing the NH_3 flow, (3) maintaining the carbon potential and surface carbon concentration of 0.70 to 0.85%, and (4) restricting the NH_3 content to about 5%. This can also be drastically reduced by cooling the quenched parts to −40 to −100°C (−40 to −150°F). Since the amount of retained γ is usually maximum near the steel surface, it can also be removed from symmetrical parts by grinding. However, these latter methods are expensive.[12]

Quenching Media. Selection of quenching media such as water, oil, or gas is based on steel composition, size, and shape of the part; allowable distortion; desired (case and core) hardness level; and type of furnace equipment employed.

Tempering. Carbonitrided parts developed primarily for wear resistance such as dowel pins, brackets, and washers do not need tempering. Low-carbon steel parts are usually tempered at 135 to 175°C (275 to 350°F) to stabilize austenite and reduce dimensional variations. Most carbonitrided gears are tempered at 190 to 205°C (375

TABLE 16.14   Effect of Material and Variables on the Possibility of Void Formation in Carbonitrided Cases[164]

Material and processing variables[†]	Possibility of void formation
Temperature increase	Increased
Longer cycles	Increased
Higher case nitrogen levels	Increased
Higher case carbon levels	Increased
Aluminum-killed steel	Increased
Severe prior cold working of material	Increased
Increase in alloy content of steel	Decreased
NH_3 addition during heat-up cycle	Increased

[†] All other variables remained constant.
Reprinted by permission of ASM International, Materials Park, Ohio.

to 400°F) to reduce surface brittleness while maintaining a minimum case hardness of 58 Rc. Alloy steel parts requiring surface grinding are tempered to reduce grinding cracks. Tapping screws made of AISI 1020 steel are tempered at 260 to 425°C (500 to 795°F) to minimize breakage in tapping holes in sheet metal. The parts requiring repeated shock loading are frequently tempered at 425°C (795°F) to improve impact (i.e., notch toughness) and fatigue strength.[12]

Applications.   Gas carburizing is widely used on steels such as 1000, 1100, 1200, 1300, 1500, 4000, 4100, 4600, 5100, 6100, 8600, and 8700 series with up to 0.25% carbon content. Also, in many cases, these steel series with a medium carbon (0.30 to 0.50%) content (such as 4140, 4340, 5130, 5140, and 8640) are carbonitrided at 845°C (1550°F) to case depths up to ~0.3 mm (0.010 in.) to obtain a combination of a hard, more wear-resistant surface, and a reasonably tough, through-hardened core (e.g., shafts and transmission gears, heavy-duty gears).[164]

Carbonitriding of P/M Parts.   This process is extensively used in treating ferrous P/M parts with (or without) copper infiltration and with sintered densities of 6.5 g/cm³ (0.23 lb/in.³) minimum. Carbonitriding is accomplished at 790 to 815°C (1450 to 1500°F), which overcomes several problems associated with carburizing of P/M parts made from electrolytic iron powders.[164] The carbonitrided parts are usually tempered at temperatures slightly higher than the temperatures employed for carbonitrided wrought steel parts.[164]

16.3.2.3   Plasma Carbonitriding.   As in plasma carburizing, plasma carbonitriding has been successfully used to produce carbonitrided cases. These cases possess greater hardenability and resistance to tempering and are, therefore, preferred to carburized cases for smaller, less massive parts. In plasma carbonitriding, both carbon and nitrogen are added to the steel surface by using methane/nitrogen or methane/nitrogen/hydrogen glow discharge plasma.[168]

16.3.2.4   Fluidized-Bed Carbonitriding.   The carbonitriding in fluidized beds was studied by Jesinski et al.[169] on AISI 1022 steel with bed material of quartz sand with 0.2- to 0.3-mm particle size and fluidizing gaseous mixture of propane-air-NH_3 or a mixture of N_2-H_2-CO-NH_3 type ammonia-base atmosphere[170] in a suitable proportion to produce an endothermic-type atmosphere in the temperature range of

820 to 940°C (1508 to 1724°F) for 0.5 to 5 hr. Their results were comparable to those for gaseous carbonitrided parts. Their findings clearly demonstrated the attainment of higher-quality hard cases and increased durability of the (fluidized-bed carbonitrided) parts.[169]

16.4 FERRITIC THERMOCHEMICAL SURFACE HARDENING TREATMENTS

The carburizing and carbonitriding treatments discussed in the previous section are austenitic thermochemical treatments, because they involve the addition of alloying elements into the austenite phase and rely on the subsequent transformation of austenite into martensite to produce a high surface hardness. Ferritic thermochemical treatments, on the other hand, involve the diffusional addition of nonmetallic elements into the surface of ferrous parts at temperatures below the eutectoid temperature. Subsequently, the parts are quenched or cooled in the processing medium.[171] In this section, nitriding and nitrocarburizing are considered in detail.

16.4.1 Nitriding

The nitriding process was first used as a commercial heat treatment process during the late 1920s, and since then it has continuously grown to worldwide application. The nitriding process is the result of interactions of the substitutional alloying elements in iron with nitrogen in interstitial solid solution.[172] That is, it involves the diffusion of atomic nitrogen into the surface of steel in the ferritic phase by holding the material at temperatures below 590°C [the eutectoid temperature of Fe-N system (Fig. 16.40)[173]] and usually between 500°C and 590°C, and consequently no phase transformation takes place on cooling to room temperature. The thin nitrided layer (usually 25 μm thick) so developed is usually subdivided into a *compound* or *white layer* near the surface and a *diffusion zone* beneath the compound layer, which are composed of nitrided phases ε-Fe$_{2-3}$N and γ'-Fe$_4$N, respectively. On prolonged nitriding, porosity can be observed in iron nitrides due to their metastability with respect to the molecular nitrogen gas and pure iron at the usually applied temperature (723 to 863 K) and total pressure ($\sim$1 atm).[174]

The extremely hard, wear-resistant iron-alloy nitrogen compounds which are formed on nitriding introduce beneficial compressive stresses at the surface and are resistant to some kinds of corrosion and softening when heated; therefore, quenching or any other treatment is no longer required. Nitriding allows the reduction or even elimination of the need for a subsequent expensive machining operation. The components must be given proper heat treatment prior to nitriding to develop the right kind of structure of the core material and, hence, to obtain better response to nitriding. Since the nitriding process is the final heat treatment and produces growth of material, proper allowance for growth should be taken into account in the final shape. Sharp corners must be avoided; otherwise, growth will occur at the corners, producing projections which are brittle and tend to chip off. The growth is dependent upon temperature and extent of nitriding.

The great hardening produced on nitriding is due to a combination of (1) small (50- to 150-Å), closely spaced particles, (2) a very high dislocation density ($>10^{10}$/cm^2), and (3) a high concentration of N associated with dislocation and with Fe-AlN interfaces.

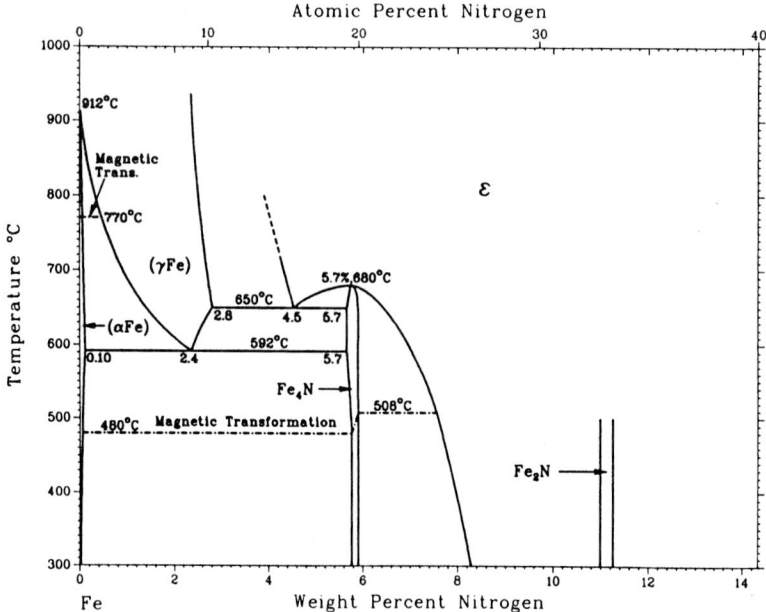

H.A. Wriedt, N.A. Gokcen, and R.H. Nafziger, 1992

Phase	Composition, wt% N	Pearson symbol	Space group
Stable at 0.1 MPa			
(δFe)	0 to ~0.9	cI2	Im$\bar{3}$m
(γFe)	0 to 2.8	cF4	Fm$\bar{3}$m
(αFe)	0 to 0.10	cI2	Im$\bar{3}$m
Fe₄N	5.7 to 5.9	cP5	Pm$\bar{3}$m or P$\bar{4}$3m
ε	~4 to ~11	hP3	P6₃/mmc
Fe₂N	~11.1	o**	...
FeN₆	~61	...	...
FeN₉	~69	...	...
Other phases			
(εFe)(a)	0 to ?	hP2	P6₃/mmc
Martensite	0 to 0.6	cI2	Im$\bar{3}$m
	0.7 to 2.6	(b)	...
Fe₁₆N₂	~3.0	(b)	I4/mmm

(a) Stable at pressures >13 GPa. (b) bct

FIGURE 16.40   Fe-N phase diagram.[173] (*Courtesy of ASM International, Materials Park, Ohio.*)

16.4.1.1   Nitridable Steels.   In principle, all steels can be nitrided. However, since the iron nitrides are unstable, the majority of steels which are nitrided are alloy steels containing a small amount of Al (from 0.85 to 1.5%), Cr, Mo, and V, which will produce stable and hard nitride needles in the outer skin of the steel at the nitriding temperature; these are commonly known in the United States as *Nitralloy* steels. They may also contain other alloying elements such as TI, Nb, W, Mn, or other reasonably strong nitride-forming elements in order to obtain high hardness. Unalloyed carbon steels are not desired for gas nitriding due to the formation of extremely brittle case that spalls easily and the small increment of hardness in the diffusion zone.

Alloy steels used for nitriding can be classified into the following groups:[175]

1. Aluminum-containing low-alloy (Nitralloy) steels (e.g., G, 135M, N, and EZ types), which are generally employed where very high surface hardness and excellent wear resistance are essential (but they provide lower ductility).

2. Medium-carbon, chromium-containing low-alloy steels such as AISI 4100, 4300, 5100, 6100, 8600, 8700, 9300, and 9800 series. These steels produce considerably greater ductility as well as good antigalling and wear resistance properties, but with lower hardness.

3. Hot-work die steels having 5% Cr, for example, H11, H12, and H13. These steels, such as H11 and D2, render substantially high case hardness with exceptionally high core strength.

4. Low-carbon, chromium-containing low-alloy steels such as 3300, 8600, and 9300 series.

5. Air-hardening tool steels such as A2, A6, D2, D3, and S7.

6. High-speed steels such as M2 and M4.

7. Nitronic stainless steels such as 30, 40, 50, and 60.

8. Ferritic and martensitic stainless steels such as AISI 400 series.

9. Austenitic stainless steels such as AISI 200 and 300 series.

10. Precipitation-hardening stainless steels, such as 13-8 PH, 15-5 PH, 17-4 PH, 17-7 PH, A-286, AM 350, and AM 355.

Prior to nitriding, these alloy steel parts are usually austenitized, quenched, and tempered at a high temperature, usually at or above 575°C, to guarantee a structural stability at the nitriding temperature.

The rate of nitriding of the alloy depends on the strength of the interaction of alloying elements with nitrogen, the nitrogen potential of the gas mixture, the composition of the alloying elements, the ease with which precipitates can nucleate and grow, and the nitriding temperature.[172] Figure 16.41 shows a comparison of nitriding characteristics of two conventional nitriding steels and the newly developed nitriding Imanite steel.[176]

The advantages of nitriding over other surface hardening methods are as follows:[177,178]

1. Because low treatment temperature is involved and no quenching is required, distortion can be kept to a minimum, even though some dimensional growth does occur.

2. Increased high surface hardness, higher wear and fatigue resistance, improved corrosion resistance (except for stainless steels) and antigalling properties, good

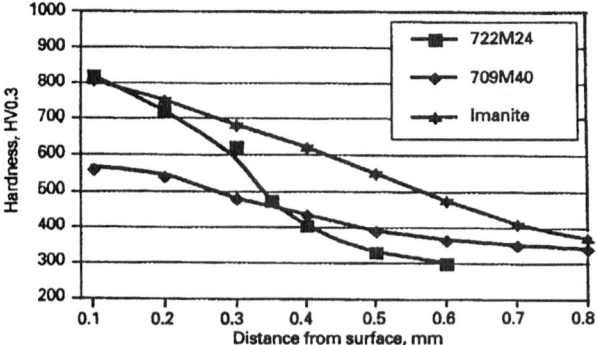

FIGURE 16.41 Comparison of the nitriding characteristics of steels BS 722M24 (3% Cr), BS 709M40 (SAE 4140), and Imanite treated under the same conditions.[176] (*Courtesy of Wolfson Heat Treatment Centre, England.*)

resistance to softening during tempering,[175] and high-temperature hardness of components are obtained.

3. It can be employed safely at a reasonably high temperature, e.g., up to 650°C (1200°F) for short periods and up to 538°C (1000°F) for long periods.

4. Surface contamination encountered in ordinary heat treatment is avoided. Hence components can be machined to final size and hardened by nitriding without any further operation.

5. Selected areas and irregular shapes can be nitrided.

Among the disadvantages of the nitriding process are the following:

1. The process is slow (i.e., requires very long process time, typically 24 to 72 hr).

2. This process requires the use of special steels containing Al and/or Cr.

3. The formation of a brittle and thin white layer (sometimes referred to as a *compound zone*) consists of a mixture of iron nitride phases on treated parts, which can be detrimental on bearing surfaces because it tends to spall (and crack) in service. Removal of this layer by expensive and time-consuming mechanical grinding or shot blasting/peening with fine glass beads at suitable pressure or chemical treatment is necessary[179] before the parts can be put into service. It has been found that complete removal of the white layer is possible by a prolonged hot soaking in a cyanide solution (consisting of 1 lb of NaCN in 1 gal of water). The solution is heated to between 70 and 90°C, and components are immersed in this solution for short periods (which causes the white layer to be friable), followed by blast cleaning (220-mesh grits and 80 psi of pressure).

4. The presence of NH_3 fumes (for gas nitriding) requires the provision of an adequate ventilation system.

5. Allowances must be made for a small amount of growth resulting from an increase in volume at the surface.

6. There is also the danger of inducing temper brittleness; hence steels containing Ni or Mo are considered best, because these elements resist the onset of embrit-

tlement. Parts must be so designed as to avoid sharp corners (by chamfering), which may spall, and dissymmetry, which may cause distortion.

16.4.1.2 Procedure. Nitriding can be carried out on steel by using a gaseous or liquid nitrogenous medium. The former is called *gas nitriding*, and the latter is called *salt bath* (or *liquid*) *nitriding*. Gas nitriding is the more widely used process. Other developments in gas nitriding, such as *Floe process, pressure nitriding, bright nitriding*, and *Nitreg nitriding* as well as the more recent ones such as *plasma nitriding* and *fluidized-bed nitriding*, are also available.

16.4.1.3 Salt Bath or Liquid Nitriding. A liquid nitriding bath is a salt bath such as fused cyanide-cyanate salts which contains mostly alkali cyanide (60 to 70 wt% NaCN and 30 to 40 wt% KCN)—an active ingredient. The process involves melting of salt bath, aging of the molten salt bath, and immersion of the steel parts in the temperature range of 560 to 570°C (1040 to 1060°F) for 1 to 2 hr. During melting of dry salt, the retort is covered or the equipment is completely hooded and vented to guard against explosion or sputtering of the salt. The molten salts should be aged by being held at 565 to 595°C (1050 to 1100°F) for at least 12 hr. Aging produces a decrease of cyanide content of the bath and the formation of a small amount of carbonates (Na_2CO_3) and cyanates (NaCNO). When a level of 5% NaCNO is reached after aging, the bath can be safely used. The NaCN contents for high-speed steels and hot-working tool steels and low-carbon and alloy steels are 15% min. and 20% min., respectively. The alloy steel parts for nitriding must be either in the quenched and tempered or in the stress-relieved condition to develop the required core properties.[12] During nitriding, the NaCN is partially oxidized to form cyanate according to the equation

$$4NaCN + 2O_2 \rightarrow 4NaCNO \qquad (16.39)$$

Sodium cyanate is an unstable compound and is readily decomposed to liberate nascent nitrogen, which is transferred to the surface of the steel part in the following manner:

$$4NaCNO \rightarrow Na_2CO_3 + 2NaCN + CO + 2N \qquad (16.40)$$

The evolution of CO in Eq. (16.40) may produce iron carbide by the following reactions:

$$2CO \rightarrow CO_2 + C \qquad (16.41a)$$

$$3Fe + C \rightarrow Fe_3C \qquad (16.41b)$$

Following nitriding, the parts should be quenched in water, polymer solution, oil, soluble oil solution, or air, depending on the steel compositions.

For satisfactory performance, all parts should be thoroughly cleaned and free of surface oxide. They are preheated prior to immersion in the bath to drive off surface moisture. The bath should be analyzed once or twice per week, and necessary additions should be made to maintain compositions within closer limits in order to achieve consistent nitriding rates. All the contaminants and oxidation products must be removed from the bath. Overheating above 600°C should be avoided. Salts should be completely changed every 3 to 4 months of operation. Usually a titanium-lined or -plated furnace container is recommended for the best result. It is a general practice to cover the bath when not in use. The normal operating temperature for salt bath nitriding is 550 to 750°C (1022 to 1382°F).

The amount of cyanate may be controlled analytically, and its addition can be kept to the desired level in the bath. The optimum limit will vary, depending on the final surface hardness required.

Advantages of salt bath nitriding are as follows:[180,181]

1. Salt baths provide a more uniform nitrogen potential and reduced distortion by supporting the components in the bath.
2. Bath composition varies from the high-speed types containing 90 to 95% cyanide (used in the steel-hardening operation) to those with 20 to 30% cyanide salts for general applications.[180]
3. This treatment might prove cheaper for mass production of components.
4. Post-nitriding treatment improves corrosion resistance and decreases friction coefficient. Steam treatment following nitriding further extends the life of the cutting tools.
5. It involves rapid heating and processing.
6. Good and reproducible nitrided layers can be obtained with ease on low-carbon and low-alloy steels.

Disadvantages of salt bath nitriding include (1) toxicity of salt bath, (2) waste disposal problem, (3) need for thorough washing to remove salt residues to avoid corrosion, (4) lack of in-process control, and (5) limitation to those steels which can be heated to higher temperatures without sacrificing core hardness.[182]

Applications. Salt bath nitriding is applied to a wide variety of carbon and low-alloy steels (e.g., crankshafts, camshafts, cylinder liners, tappet guides, light-duty gears, connecting rods, and clutch plates), tool steels, high-speed tool steels (e.g., end mill drills, reamers, side and face cutters, and form tools), hot-working tool steels (e.g., press forging dies, extrusion dies, and mandrels made from H13 steel), stainless steels, and cast irons.[183] This method offers superior properties to those of other case hardening methods, particularly in the production of automotive parts such as thrust washers (from AISI 1010 steel), shaft, seat bracket (from AISI 1020 steel), and rocker arm shaft (from SAE 1010 steel). However, it is not suitable for many applications requiring deep cases and hardened cores.[181]

Other Salt Bath Compositions. Another typical nitriding salt bath for tool-steel applications has the following composition:[12] 30% (max.) NaCN, 25% (max.) Na_2CO_3 or K_2CO_3, 4% (max.); other active ingredients, 2% (max.) moisture, and balance KCl. A proprietary nitriding salt bath has the following composition: 60 to 61% NaCN, 15 to 15.5% K_2CO_3, and 23 to 24% KCl.[12]

Special Liquid Nitriding Processes. Although there are several commercial proprietary liquid nitriding processes, Table 16.15 represents the basic types.[181] In these special liquid nitriding salt bath processes, proprietary additions, either gaseous or solid, are made to serve several roles such as accelerating the chemical activity of the bath, expanding its applicability to a wide variety of steels that can be processed, and improving the properties obtained after nitriding.[12] Cyanide-free liquid nitriding salt compositions are also available given their wide acceptance, which contain a small amount of cyanides, usually up to 5% in the active bath. The following are the two main processes:

1. *Liquid pressure nitriding.* In this proprietary process, anhydrous NH_3 is introduced through the bottom of the sealed retort into a cyanide-cyanate bath and maintained at a pressure of 7 to 205 kPa (1 to 30 psi) to speed up the nitriding reaction of the bath. The percentage of nascent nitrogen in the bath is controlled by

TABLE 16.15 Liquid Nitriding Processes[181]

Process identification	Operating range composition	Chemical nature	Suggested posttreatment	Operating temperature		U.S. patent number
				°C	°F	
Aerated cyanide-cyanate	Sodium cyanide (NaCN), potassium cyanide (KCN) and potassium cyanate (KCNO), sodium cyanate (NaCNO)	Strongly reducing	Water or oil quench; nitrogen cool	570	1060	3,208,885
Casing salt	Potassium cyanide (KCN) or sodium cyanide (NaCN), sodium cyanate (NaCNO) or potassium cyanate (KCNO), or mixtures	Strongly reducing	Water or oil quench	510–650	950–1200	
Pressure nitriding	Sodium cyanide (NaCN), sodium cyanate (NaCNO)	Strongly reducing	Air cool	525–565	975–1050	
Regenerated cyanate-carbonate	Type A: Potassium cyanate (KCNO), potassium carbonate (K_2CO_3)	Mildly oxidizing	Water, oil, or salt quench	580	1075	4,019,928
	Type B: Potassium cyanate (KCNO), potassium carbonate (K_2CO_3), 1–10ppm, sulfur (S)	Mildly oxidizing	Water, oil quench, or salt quench	540–575	1000–1070	4,006,643

Courtesy of ASM International, Materials Park, Ohio.

maintaining the NH_3 flow rate at 0.6 to $1 m^3/hr$ (20 to $40 ft^3/hr$), which causes the NH_3 dissociation of 15 to 30%.[181] The selection of appropriate pressure is a function of the retort volume, part geometry, surface area to be treated, and process temperature. Maintenance of this bath at an operating temperature of 525 to 565°C (975 to 1050°F) is greatly simplified, because it does not require aging and may be placed into immediate operation using the recommended cyanide/cyanate ratio of 30 to 35% cyanide and 15 to 20% cyanate.[181]

2. *Aerated bath nitriding.* In this process, a measured amount of air is forced through the molten bath to provide agitation and stimulate chemical activity, thereby increasing the rate of nitriding. The bath contains 50 to 60% NaCN, 32 to 38% cyanate, 10 to 30% (usually ~18%) elemental potassium content as cyanide and/or cyanate, and the remainder sodium carbonate. Note that aerated bath nitriding is suited to the plain-carbon (nonalloyed) steels whereas conventional bath nitriding is well suited to only Cr-, Ti-, and Al-alloyed steels.[181]

In aerated low-cyanide nitriding, the base is provided with a cyanide-free mixture of KCNO and a combination of Na_2CO_3 and K_2CO_3, or NaCl and KCl. In the case of nitriding with heavy loading, there is likelihood of the formation of small percentages of cyanide in the bath. This problem can be overcome by quenching in an oxidizing quenching salt bath that destroys the cyanide and cyanate compounds, thereby solving the pollution problem and producing less distortion compared to water quenching. However, in another proprietary cyanide-free salt bath mixture, a very small amount of sulfur (1 to 10 ppm) and lithium carbonate are added to the base salt to keep the cyanide formation below 1.0%.[181]

In aerated cyanide-cyanate nitriding, a high-cyanide, high-cyanate fused salt bath containing 45 to 50% cyanide calculated as KCN and 42 to 50% cyanate calculated as KCNO is used. It is applied to treat carbon and low-alloy steels and stainless steels. This salt composition is also referred to as the Tufftride, according to AMS 2755C-1985.[181]

16.4.1.4 Gas Nitriding. In the conventional gas nitriding process, first the quenched and tempered parts should be thoroughly cleaned, vapor-degreased, and conditioned (by abrasive cleaning with aluminum oxide grit or other abrasives such as garnet or silicon carbide, or by applying a light phosphate coating) immediately prior to nitriding. After loading and sealing the furnace at the start of the nitriding cycle, purging with nitrogen or anhydrous NH_3 is used to expel air from the furnace before the furnace is heated above 150°C (300°F). This prevents oxidation of the parts and the furnace components and, if NH_3 is used as a purging atmosphere, avoids the formation of potentially explosive mixture.[12,175] Subsequently, moisture-free anhydrous NH_3 is allowed to flow into the nitriding furnace over the parts in such a way that all surfaces come in contact with the gas, usually at uniform temperature of 525°C (975°F). However, gas nitriding can be accomplished from 500 to 600°C (930 to 1100°F). The NH_3 gas, when in contact with the hot steel surfaces, dissociates to produce atomic nitrogen, which reacts with the alloying elements in the steel surfaces to produce nitrides according to the equation

$$NH_3 \rightleftharpoons \frac{3}{2}H_2 + N \text{ (dissolved in Fe)} \qquad (16.42)$$

Nitriding is accomplished in an electric furnace with a device for precise temperature control. Most heat treaters employ batch-type furnaces with essential features. They must be equipped to (1) seal the components from air, (2) provide uniform temperature, (3) maintain and circulate the controlled atmosphere throughout the mass of the parts to be nitrided, and (4) brush the nitriding container (i.e., inside of the retort) at intervals in order to clean and remove any deposits such as sulfides which may inhibit the nitriding treatment. Among several types of equipment which can be employed are the vertical retort furnace and movable bell-and-box type furnaces.[180]

Retorts, fixtures, and furnace accessories which are in contact with NH_3 atmosphere are normally constructed of heat-resisting 25Cr-20Ni steels, although nickel, Inconel, Incoloy, and similar alloys are ideal. They frequently require cleaning and care in order to prevent any reaction with the atmosphere. The composition of the exit gas is maintained and measured regularly with the help of a dissociation or absorption pipette; for the first 4 to 10 hr, the dissociation rate of gaseous NH_3 is kept at 15 to 30%, depending on the duration of the total cycle. Usually, a constant degree of dissociation, which is used to control the nitriding process, is maintained in a single-stage technique. However, in two-stage process (Floe process), the temperature and degree of dissociation are varied.[184] On completion of nitriding, the box is taken out of the furnace without interrupting the gas flow. When the charge has cooled to 150°C (300°F), the gas supply is stopped and the gases remaining in the box are expelled with ease with compressed air prior to opening of the box. The nitriding parts now generally exhibit the characteristic matte gray color. Sometimes the color has the shades of blue, yellow, or purple, which are derived from the presence of O_2 in the system, originating from some leak in the box or in the supply tubing or from incomplete drying of the gas. Treatment times are normally between 40 and 60 hr, but for relatively deeper cases up to 90 hr is used.[185]

Note that emergency purging is also required when the supply of NH_3 is cut off or a break occurs in the supply line during the nitriding or cooling cycle.[11]

Advantages of conventional nitriding are (1) simple control methods; (2) low temperature with respect to carburizing; (3) increased surface hardness, wear resistance, fatigue resistance, and corrosion resistance; (4) low distortion; (5) hot hardness; and (6) controlled growth.

Disadvantages include (1) inadequate NH_3 dissociation rate for control of layer properties, (2) possible requirement for removal of brittle white layer, in many instances, (3) a need for copper plating or painting with protective paste for masking, (4) requirement of special activation techniques for stainless steels, (5) requirement of a nitride former in nitridable alloys, and (6) potentially long process time.

Developments in Gas Nitriding. Gas nitriding may be carried out using either a single-stage or double-stage process. In the single stage, which has just been described above, a temperature range of about 500 to 525°C (930 to 975°F) with dissociation rate of 15 to 30% (i.e., an atmosphere containing 70 to 85% NH_3) is used. This process yields a brittle, nitrogen-rich layer, called the white nitride layer, at the surface of the nitrided case. The double-stage process, also called the Floe process, has the benefit of reduced thickness of the white nitrided layer. The first stage of the double-stage process is a duplication of the single-stage process, except for time. The second stage uses a temperature of 550 to 565°C (1025 to 1050°F) and a dissociation rate of 65 to 85% (preferably, 75 to 80%), which has been found to minimize or eliminate the white layer of iron nitride (produced in the single stage) and increase the surface hardness. Figure 16.42 illustrates the microstructures observed when AISI 4140 steel is given single- as well as double-stage nitriding treatment. The microstructure in the single-stage process consists of an outer white surface layer of Fe_2N followed by iron nitride and a matrix of tempered martensite (Fig. 16.42a), whereas the microstructure in the double-stage process consists of only a diffused nitride layer and tempered martensite (Fig. 16.42b).[120]

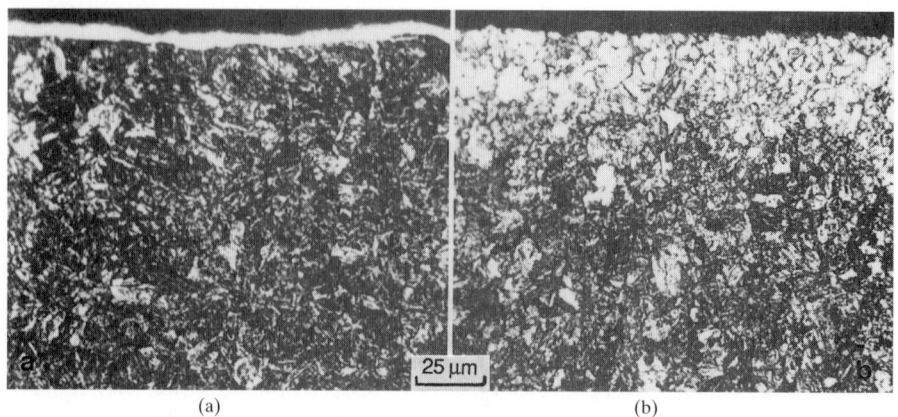

(a)   (b)

FIGURE 16.42 Microstructure of 4140 steel (*a*) single-stage nitrided at 525°C (975°F) for 24 hr with 20 to 30% dissociation showing 0.005- to 0.0075-mm (0.0002- to 0.0003-in.) white surface layer of Fe$_{2-3}$N followed by iron nitride and tempered martensite and (*b*) double-stage nitrided—first-stage nitriding at 525°C (975°F) for 5 hr with 20 to 30% dissociation, followed by second-stage at 565°C (1050°F) for 20 hr with 75 to 80% dissociation. The microstructure shows absence of white layer, diffused nitride layer, and tempered martensite. *Note*: The prenitriding condition involves austenitizing at 845°C (1550°F) followed by oil-quenching, tempering at 620°C (1150°F) for 2 hr, and surface activation with manganese phosphate. 2% nital etch.[120] (*Reprinted by permission of ASM International, Materials Park, Ohio.*)

The total case depth, which depends on the nitriding time and temperature, can be extended to 0.5 mm (0.019 in.) below the surface. The total case depth constitutes several zones. In the light microscope, the outermost *white* or *compound zone*, when etched in 2% nital, appears white. Below this zone, a nitrided zone, called the *diffusion zone*, appears in which nitrogen is assumed to combine with substitutional solutes in the ferrite to form a very fine dispersion of alloy nitrides at the prior grain boundaries perpendicular to the direction of nitriding. Beyond the nitrided (or diffusion) zone, there is a region which is revealed by the black appearance in the microstructure when etched by Oberhoffer's reagent. The extent of hardening within the zone is small.[186]

Pressure Nitriding. This process differs from the conventional gas nitriding in that it demands the use of a sealed retort which will withstand higher pressures than atmospheric. The surfaces to be nitrided are first cleaned and then introduced into the carbon steel retort, which has been evacuated of air and filled with NH$_3$ to a predetermined pressure.

The selection of pressure is a function of the total surface area of the parts to be nitrided and the volume of the retort. The retort is then heated in any furnace where temperature can be controlled for the desired time cycle, after which the retort can be air-cooled, vented, and opened.[187]

The advantages claimed for this nitriding process are (1) the controlled thickness of white layer, (2) production of high surface hardness together with considerable toughness, (3) rapid formation of case during the first few hours, and (4) convenience in nitriding complex-shaped parts that are difficult to handle by other methods.

It also has some drawbacks:

1. It is inconvenient to seal the retort.

2. The dangerous pressure might develop during filling of the retort with gas if enough NH_3 gas is allowed to condense. However, this hazard can be prevented by employing a safety disk or by keeping the retort at a temperature higher than that of the NH_3 supply container.[188]

3. After 45 hr of operation, the NH_3 content expands by about 50%, and further development of the case proceeds at a very slow rate.

4. To limit the depth of white layer to 0.00025 to 0.00050 mm (9.8 to 20 μin.), case depth must not exceed 0.50 to 0.63 mm (0.020 to 0.025 in.).

Application. Gears, pinions, shafts, bushings, seals, cylinder barrels, clutches and piston rings,[180] door sectors (used in car windows), spiral springs, and exhaust valves are hardened by this technique. Gas nitriding is widely employed on steels, e.g., AISI 4130, 4135 modified (with 0.15% V), 4140, 4340, H11, stainless types 302 and 430, and various Nitralloy grades.[177] In gas nitriding of stainless steels (type 300, austenitic), surface oxide film which acts as a barrier to nitrogen penetration must be removed by sand or vapor blasting followed by acid pickling.[189]

However, hard (70 Rc) and superior wear-resistant cases accompanied by a reduced corrosion resistance property are produced after nitriding austenitic stainless steels.[185] The 18% Ni maraging steels in the finish-machined and ground condition are also suitable for nitriding between 420 and 450°C (800 and 850°F) for 20 hr or more, which produces increased wear resistance and surface hardness of the case (67 Rc) due to the combined effect of both nitriding and aging treatment in a single operation.[188]

Bright Nitriding. This is a modified version of gas nitriding using NH_3 and H_2 gases. Atmosphere gas is continually withdrawn from the nitriding furnace and passed through a temperature-controlled scrubber with a water solution of NaOH. Trace amounts of HCN formed in the nitriding furnaces are eliminated in the scrubber, thereby increasing the rate of nitriding. The scrubber also sets up a desired moisture content in the nitriding atmosphere, decreasing the rate of cyanide formation and hindering the cracking of NH_3 to molecular nitrogen and hydrogen. By this process, control over the nitrogen activity of the furnace atmosphere is improved, and nitrided parts are produced containing small or no white layer at the surface. If present, the white layer will consist of only the ductile Fe_4N (γ') phase.[175]

Controlled Gas (or Nitreg) Nitriding. In traditional nitriding, the controlled parameters that include the atmosphere flow rate and the resultant dissociation rate of NH_3 are insufficient because none of these parameters is directly related to the properties of the nitrided layer.[182] Recently, a controlled gas nitriding, called *Nitreg nitriding*, has been developed based on constant monitoring and maintenance of the nitriding potential by a computer-controlled and fully automated gas nitriding system, according to the modified Lehrer phase diagram, which is a plot of nitriding potential versus temperature (Fig. 16.43).[182,190] The nitriding potential is considered as the nitriding capacity of the atmosphere, and its value strictly corresponds to the equilibrium concentration of nitrogen in the steel surface for a given temperature. Mathematically, nitriding power or nitriding potential K_N is given by:

$$K_N = \frac{p_{NH_3}}{(p_{H_2})^{3/2}}$$

(16.43)

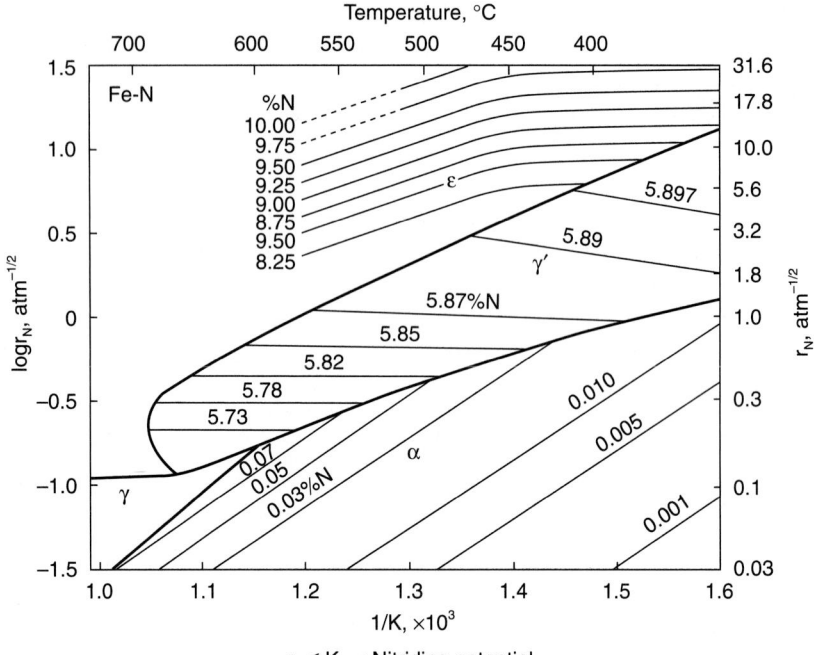

FIGURE 16.43   Lehrer K_N versus T phase diagram, modified by L. Maldzinski.[182,190] (*Courtesy of L. Maldzinski.*)

where p_{NH_3} and p_{H_2} are partial pressures of NH_3 and H_2 in the outgoing atmosphere. To obtain a superficial layer with load-bearing capacity, it is necessary that the combined carbon and nitrogen content be ≤8.5% because higher concentrations are prone to brittle spalling of the layer.[182] Alternatively, the use of very low nitriding potential (usually in the final stage of the nitriding process) provides a case without a compound (white) layer due to the nitrogen concentration below its maximum solubility in ferrite. Thus, control of nitrided layer properties implies control of surface nitrogen and carbon concentration and, consequently, control of phase composition.[191]

Nitriding processing is performed in the extended temperature range of 460 to 600°C (860 to 1112°F).

Advantages of this process include[192] (1) ease of operation, (2) direct relation of controlling parameter (such as nitriding potential) with nitrogen concentration and properties, (3) predictable compound-layer thickness and phase composition, (4) excellent uniformity of layer, irrespective of the part geometry, and (5) no requirement of finish-grinding.[182]

Like other gas nitriding processes, the disadvantages of this process are[182] (1) need for copper plating or protective paste to mask the surface, (2) need of special activation technique for stainless steels, (3) inability to treat satisfactorily sintered P/M components, and (4) frequent occurrence of embrittlement.[176]

Examples of applications of Nitreg process are door sectors (used in car windows), spiral springs, automotive exhaust valve, clutch hubs, rocker arm, cast iron

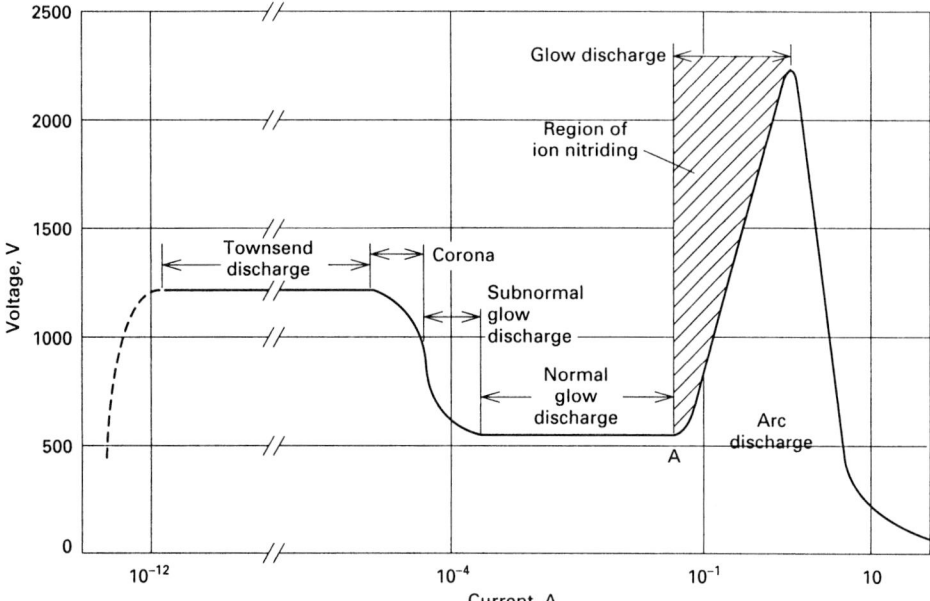

FIGURE 16.44   Paschen curve showing the relationship between voltage and current for nitrogen-hydrogen mixture.

differential housing, hydraulic components of dump truck lifting mechanisms, and crankshafts.[192]

16.4.1.5 Plasma Nitriding.   There are many terminologies used for plasma nitriding, e.g., plasma-ion nitriding, plasma-assisted nitriding, ion nitriding, ionitriding, and glow discharge nitriding. Plasma nitriding is a nitriding process that was developed by Bernhard Berghaus in the 1930s, but the process has gained worldwide popularity in recent years, with about 400 industrial furnaces in operation in Europe and about 150 in North America.

   Plasma Generation.   When a voltage is applied between two electrodes positioned within a nitrogen-hydrogen mixture at some suitable partial pressure (1 to 10 torr), a glow discharge occurs in the abnormal glow discharge range (Fig. 16.44), where total work surfaces conforming to its geometrical form will be completely covered with a uniform purple glow. This glow discharge produces nascent nitrogen in a stable manner by accelerating electrons through an electrical field to give them sufficient kinetic energy to cause the reaction

$$e^- \rightarrow N_2 = N^+ + N + 2e^- \tag{16.44}$$

That is, nitrogen is dissociated, ionized, and accelerated toward the workpiece (cathode), and electrically conductive gas is formed within a short distance of the workpiece. The positively charged nitrogen ions acquire an electron from the cathode (workpiece) and thus emit a photon. This photon emission during the return to the atomic state causes the visible glow discharge that is characteristic of plasma techniques.[193]

Usually the heating of the workpieces occurs by the plasma-driven impact of the nitrogen ions, and no extra heat source is needed. However, in newer system designs, convective heating is employed to shorten the cycle times.

The thickness of the glow envelope can be varied by temperature, time, atmosphere pressure, gas mix, atmosphere composition, dc voltage, and current. Typically, a large or thick glow envelope is created with lower pressure, higher temperature, high hydrogen concentration in the gas mix, and higher dc voltage and current. There is a desirable glow discharge thickness of about 6 mm (0.25 in.), unless parts with holes or slots require a thinner glow envelope.[194]

Equipment. There are two types of widely used plasma nitriding systems: cold-wall (conventional) technology (Fig. 16.45) and hot-wall (pulse-type) technology (Fig. 16.46). Currently dc pulse is becoming widely accepted as a plasma power control system. RF gas discharge has not yet gained any significant acceptance by industry.[194a]

Other types of plasma such as arc plasma, electron cyclotron resonance (ECR) plasma, high-density plasma focusing (PF) with ion energy up to MeV, and laser plasma are used rarely in metallurgical and industrial heat treatment processes due to shortcomings, typical of each type of plasma source.[195]

A typical plasma nitriding installation thus comprises (1) a vacuum chamber, (2) the workload, (3) the gas supply, and (4) a cold-wall continuous dc plasma power supply or hot-wall dc high-frequency pulse plasma power supply (Fig. 16.45) and automatic control of the entire process variables, e.g., power voltage, current, current density, temperature of the workpiece and chamber, time, pressure, and gas mixture.[196–198] The nitriding process consists of five steps: (1) vessel evacuation with a roughing pump or pump-blower combination to a pressure of 0.05 to 0.1 torr to remove air and contaminants, (2) heating to nitriding temperature,[†] (3) introduction of a mixture of nitrogen and hydrogen at a pressure, typically 2.7 to 4 mbar or 2 to 3 torr, and composition after stabilization of load at a uniform temperature, (4) glow discharge processing at nitriding temperature, and (5) cooling.

In this process, the parts to be nitrided are vigorously cleaned (hydrogen cleaned or sputtered, typically, at 1.3 to 4 mbar or 1 to 3 torr for several minutes) after reaching the full-load temperature, or the passivated part surfaces are activated and placed in a vacuum-tight chamber. For the best results (i.e., shorter processing time), auxiliary heating (e.g., radiant heating or convective heating) is used to heat the parts to the ion nitriding temperature, usually in the range of 400 to 565°C (750 to 1050°F). Note that lower ion nitriding temperature below 350°C greatly decreases the diffusion rate of nitrogen, and temperatures above 580°C are not employed because of a structural change that takes place at 592°C, the eutectoid temperature of the binary iron-nitrogen system (Fig. 16.40). Figure 16.45 shows an improved design incorporating convective heating and cooling.[199] During heating and cooling, nitrogen gas at a pressure of 500 torr is used as the convective heat-transfer medium. This unit contains an electrically powered heating element[‡] and a (water-cooled) heat exchanger which is moved into the work chamber only during the cooling stage. An outstanding advantage of this design is that the heating elements are used for radiant heating during the low-pressure plasma nitriding stages. This provides better temperature uniformity, irrespective of the variation of the power density of the

---

[†] At this point usually H_2 is added for both convective heating and sputter cleaning. In a conventional dc system, $N_2 + H_2$ is introduced on the process commencement, and the heating of the part is by continuous dc power. Heating is effected by a combination of both plasma and conductive means.[194a]

[‡] This design has been reported to be problematical due to nitriding of elements and its susceptibility to early failure by mechanical damage.[194a]

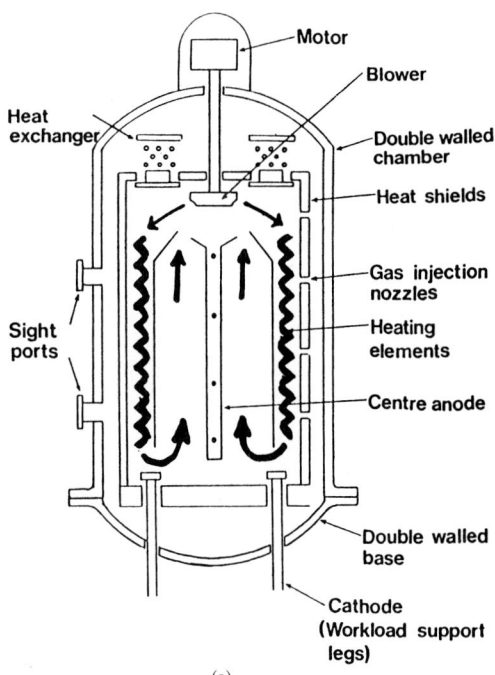

Motor
Blower
Heat
exchanger
Double walled
chamber
Heat shields
Gas injection
nozzles
Heating
elements
Sight
ports
Centre anode
Double walled
base
Cathode
(Workload support
legs)

(a)

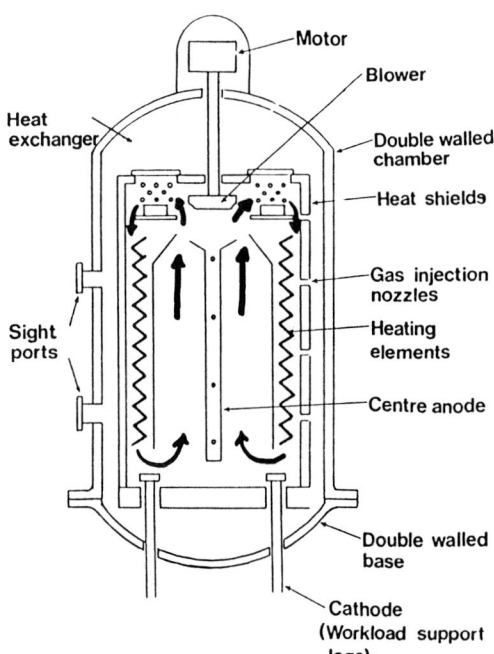

Motor
Blower
Heat
exchanger
Double walled
chamber
Heat shields
Gas injection
nozzles
Heating
elements
Sight
ports
Centre anode
Double walled
base
Cathode
(Workload support
legs)

(b)

FIGURE 16.45 Schematic of a cold-wall plasma nitriding furnace.[199] Gas circulation during (*a*) heating and (*b*) cooling. (*Courtesy of Ipsen Industries International, GmbH.*)

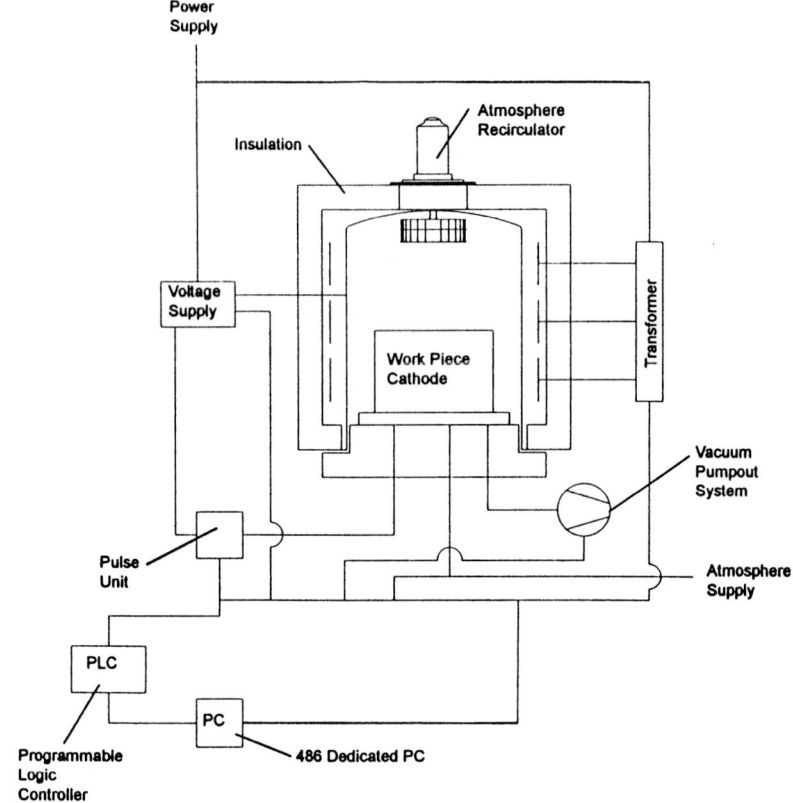

FIGURE 16.46   Schematic of a typical layout of a hot-wall plasma-ion nitriding furnace. (*Courtesy of SECO/WARWICK Corporation.*)

plasma.[197] This allows time-saving (compared to plasma heating), efficient, and simultaneous heating of the large workloads with mixed sizes (Fig. 16.47*b*).

After the load has stabilized at a uniform temperature, the pressure is reduced to 1 to 10 torr (typically, 2.7 to 4 mbar or 2 to 3 torr), and the chamber is backfilled with a mixture of 5 to 50% N_2 and 50 to 95% H_2 gas (and at times a few percent of CH_4).[†] An electric current at a critical negative dc voltage of 700 to 1000 V[‡] is passed between the cathodic workload and the anodic furnace walls, so that a uniform and stable plasma is produced around the workload. That is, nitrogen is dissociated, ionized, and accelerated toward the workpiece (cathode), and electrically conductive gas is formed within a short distance of the workpiece. Upon impact with the

---

[†] CH_4 is used to promote the epsilon phase on low-carbon steels. Medium- to high-carbon steels will promote epsilon phase without the addition of CH_4. It is also used in the ferritic nitrocarburizing process.[194a]

[‡] This voltage range is typical for continuous dc system. In the case of pulsed dc, the range is usually 400 to 700 V because the plasma voltage is only being used to create the plasma and probably contribute a small amount of heating in contrast to continuous dc, in which plasma is the primary heating source. The auxiliary heating element method is an attempt to reduce the process voltage. Of course, pressure and part geometry will influence the necessary voltage.[194a]

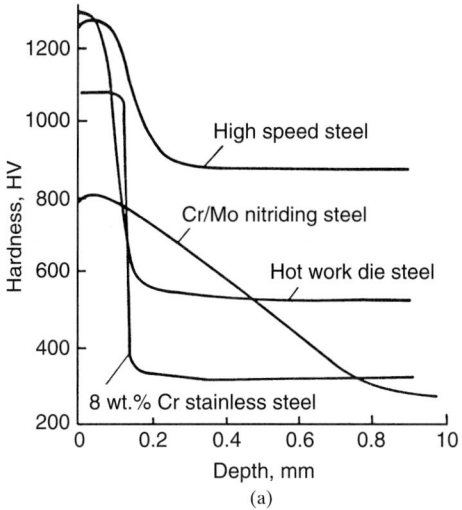

(a)

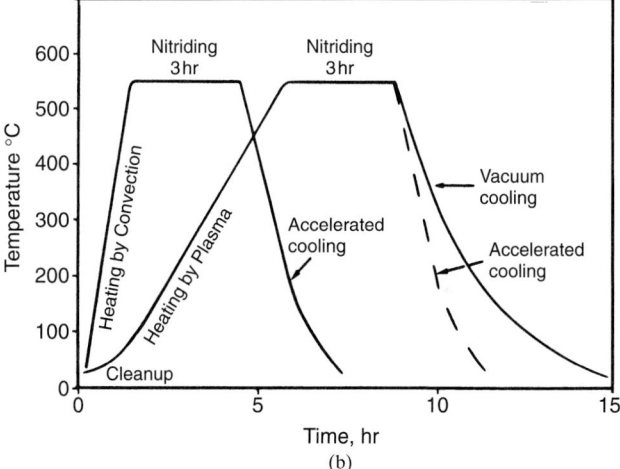

(b)

FIGURE 16.47 (*a*) The hardness versus case depth profile for several steels treated by plasma nitriding. [*Courtesy of K. T. Stevens and A. Davies, in Constitution and Properties of Steels (vol. 7: Materials Science and Technology, F. B. Pickering, vol. ed.), VCH, Weinheim, 1992, p. 786.*] (*b*) Comparison of overall process cycle times for a 3-hr nitriding treatment during resistance heating and fan-forced cooling.[196] (*Reprinted by permission of Fairchild Publications, New York.*)

workpiece, the kinetic energy of nitrogen atoms is also converted into heat, which can totally (or in combination with an auxiliary heating source) bring the load to nitriding temperature. This results in an excellent penetration or absorption of nitrogen atoms into, and a gradual heating up of, the workload.[196,197,199,200]

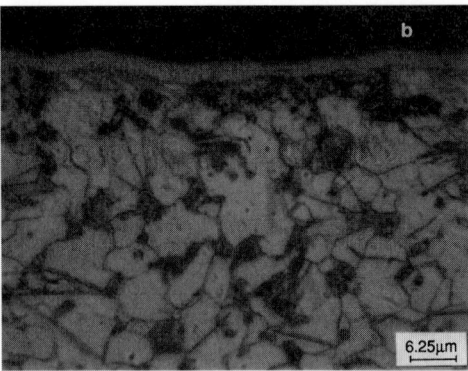

FIGURE 16.48 Optical micrographs of (*a*) ion-nitrided Nitralloy 135 three-point bend test specimen at 510°C for 6 hr in a mixture of 33% N_2 and 67% H_2 at 2-torr pressure and (*b*) microalloyed steel (containing 0.12 C, 0.58 Si, 1.43 Mn, 0.2 Cr, 0.11 Mn, 0.001 V) at 530°C for 5 hr in a mixture of 20% N_2 and 80% H_2 at 3-torr pressure.[196] (*Courtesy of M. H. Jacobs.*)

The plasma current, which ionizes the process gas to produce nascent nitrogen for nitriding, offers a substantial fraction of heat needed to maintain the heated load at temperature, i.e., to compensate for heat losses. However, the auxiliary heating system, i.e., gas burners[†] or resistance heaters, usually supplies the major fraction of the heat needed to bring the load and furnace internals to temperature quickly.

After ion nitriding, power is shut off so that the voltage and process gas flow are terminated. Inert gas circulation at 500 torr may be used to accelerate cooling of the load (by using heat exchanger for heat extraction), or slow vacuum cooling is accomplished (to minimize distortion). The furnace is then ready to be unloaded.

A high alloy content raises the hardness of the nitrided region but reduces the case depth and provides a very abrupt transition in the hardness profile from the case to the core (Fig. 16.47a).

Figure 16.47*b*[196] shows a comparison of two process cycles for a 3-hr nitriding treatment, illustrating time savings during convective heating of workload between room and operating temperatures.[197,200,201] Figure 16.48*a* and *b* shows the ion-nitrided microstructures, respectively, of the Nitralloy 135 and microalloyed steel (containing 0.12 C, 0.58 Si, 1.43 Mn, 0.2 Cr, 0.11 Mo, and 0.001 V) obtained at different process parameters. Note that ion nitriding, like other nitriding processes, produces several distinct structural zones: (1) thin white layers containing mostly ε-iron nitride ($Fe_{2-3}N$) with some γ'-Fe_4N phase;[‡] (2) a diffusion zone of fine iron-alloy nitrides;[§] and (3) a gradient zone of interstitial nitrogen.[190]

The advantages of this process over conventional (gas) nitriding are as follows:

1. Processing time for a given case depth is reduced (as small as one-third to one-half of the time required for traditional gas nitriding). Saturation cycles are short.

[†] There is only one gas-fired unit built which has not been a good commercial venture.[194a]

[‡] The proportion of constituent phases depends on the carbon content of the steel.[194a]

[§] The nitrides formed in the diffusion zone are dependent on the alloy content of the steel. For example, Nitralloy will form nitrides of Cr, Mo, and V. Thus, fine alloy nitrides will form only when low-alloy and plain carbon steels are used.[194a]

2. High-quality product in terms of surface structure and minimal distortion is produced.

3. There is precise control of thickness and composition of the compound zone or white layer.[201] In fact, the white layer produced by ion nitriding is very adherent and coherent, and its microstructure is dense and microcrystalline, unlike the coarse-grained, porous white layers usually formed both by gas and salt bath nitriding.[196] Thus, ion nitriding gives a less brittle case with higher superficial hardness (800 to 1200 HV), low friction coefficient, greater high-cycle fatigue strength, and enhanced wear resistance.

White layer or compound zone development in ion nitriding remains unchanged at 0.0025 to 0.005 mm (0.0001 to 0.0002 in.). At 400°C (750°F) the case depth does not vary appreciably with time.[198]

More precise control of the nitrogen supply at the surface of the workpiece and the ability to choose either an ε- or γ'-monophase layer or to avoid white layer formation completely are benefits.[194]

4. Case depth is very uniform, even on complex-shaped parts; i.e., it achieves a far greater control over product properties than those of the conventional methods. Extreme care is needed when nitriding blind holes.

5. The process can be conducted at comparatively lower temperatures (as low as 375°C, or 700°F) due to plasma activation (which is not present in gas nitriding), enabling high core-strength retention, high surface hardness, low distortion, and improvement in dimensional stability in tools and precision parts.[194,202]

Low-temperature nitriding at 300°C, with a very thin (24-μm) ε-Fe$_{2-3}$N layer and nitrogen solid solution increases the hardness of and doubles the lifetime of the 60MD8 (AFNOR: Fe-0.6C-1.8Si-0.7Mn-0.3Cr-0.5Mo-0.2V) steel cutting knife tools in the beech wood rotary peeling process. Significantly, the friction coefficient of nitrided steel against beech wood is slightly higher than that of non-treated steel. It seems that from the point of view of a wet wood peeling process with a nitrided knife, the knife hardness is more important than the friction coefficient. This is a low-alloy steel commonly used for manufacture of cutting tools for rotary peelers of wet wood (at the speed of about 1 to 2 m/s).[203]

6. Plasma nitriding at 480°C improved the surface life of H13 and D2 steel roll by up to seven-fold by changing substrate material from AISI H13 to D2 steel and a change in surface geometry from a nonconformal (V) to a conformal (U) profile. Shorter nitriding times developed the ε-Fe$_{2-3}$N phase, and longer nitriding times developed γ'-Fe$_4$N with a release of molecular nitrogen, which may be partially responsible for observed increases in porosity.[204] This can be modified by an appropriate ratio of N$_2$/H$_2$.

7. Cleaning and depassivation of surface layers (such as oxide) occur by the sputtering action by high-kinetic-energy ion bombardment.[205,206]

8. Easy masking (or shielding) or selective hardening is an advantage. Steel sheets or sleeves are employed as mechanical masking devices for stopping off, thus eliminating the need for copper or tin plating.[194,202] Sometimes, stop-off paints are used.[194a]

9. There are no environmental pollution, hazard, or disposal problems, and the part usually does not need any other surface finishing operation.[198]

10. There is accurate reproducibility of the operation for batch-type processing.

11. Operating costs are relatively low [due to low energy consumption, high (70 to 90%) energy efficiency of this mechanism, and low gas utilization], floor space and operating supervision are reduced, and there is the ease of operating the equipment.[198,206]

12. There is the ability to automate.[194]

13. There is the ability to treat routinely materials with passive layers such as austenitic stainless steels and titanium alloys.[207]

Limitations of this process include the following:

1. High capital cost (which is offset by lower operating costs and improved metallurgical properties)[198]

2. Need for precision fixtures with electric connections (to avoid localized overheating)[194]

3. Lack of feasibility of liquid quenching for plain carbon steels

4. Long processing times compared to nitrocarburizing processes[11]

5. Problems with temperature measurement and temperature nonuniformity, especially for parts with complex geometry (the temperature uniformity of the load can, however, be promoted by the selection of a suitable pressure and by design of the jigging system and shields, attached to the workpieces[207])

6. Prone to overheating, if pressure, voltage, and current are not accurately controlled (not applicable with modern controls and pulse plasma)

7. Results sensitive to part geometry and arrangement in furnace retort

8. Need for highly skilled and experienced operator[182]

Application. This process can be applied to all steels including a wide variety of alloy steels such as AISI 4140, 4340, Nitralloys; AISI M, A, D, H, and T series of tool steels; 17-4 PH; austenitic and martensitic stainless steels; and cast irons.[211] Examples include gears, crankshafts, cylinder liners, pistons, machinery (plastic extrusion, agricultural, food, etc.), tooling (dies, drills, molds, punches, etc.), and P/M parts (such as transmission gears).

The increased lubricity of white layers, together with higher hardness and fatigue strength, has offered significant growth of the plasma tool and die industry. Hot-work dies, which often fail by thermal fatigue and sticking, have particularly benefited from plasma nitriding following quenching and tempering.

Low-temperature (below 400°C) plasma nitriding of austenitic and duplex stainless steel can be accomplished in the commercial plasma nitriding units to provide improved wear resistance without appreciable drop in the corrosion resistance.[208] However, depassivation of the surface is required prior to nitriding.[198] A wide range of component sizes can be processed, ranging from ball-point pen balls to 10-ton rolls or gears.

Pulse Plasma Nitriding. The hot-wall, pulsed plasma nitriding furnace in conjunction with partial pressure control provides separate power supplies for part heating and for enhancing plasma process control (Fig. 16.46). Comparison of plasma and pulse plasma nitrided 4140 steel shows no difference in steel microstructure after nitriding, and nearly the same surface roughness and surface microhardness values, whereas higher content of nitrogen in nitriding atmosphere produces a higher surface microhardness in the case of pulse plasma nitriding. Pulse plasma nitriding makes it possible to obtain the same nitriding depth in a shorter time compared to conventional plasma nitriding. Compared to normal hardened AISI 4140 steel, the plasma and pulse plasma nitriding in a nitrogen-poor atmosphere improves the tribological properties of the 4140 steel. The results show that surface treatment has practically no effect on the coefficient of friction; on the other hand, plasma and pulse plasma nitriding in a nitrogen-poor atmosphere reduces the wear of AISI 4140 steel compared to the hardened steel.[209]

In general, the advantages of pulse plasma ion nitriding are the following:[193,210]

1. Independent control of process parameters, notably electric current density, ion concentration, surface temperature, and nitrogen diffusion rate.

2. No arc discharge problems linked to the parts not adequately cleaned.

3. Possibility of dense loading of the furnace with parts without any risk of localized overheating or temperature nonuniformity.

4. Desired properties at low operating temperatures and no need for fast cooling.

5. A uniform, nonspalling, reproducible nitrided layer with a specific thickness, surface hardness, and hardness profile can be achieved on complex parts, even comprising narrow slots and deep holes.

6. Desirable metallurgical properties can be consistently produced even on passivated surfaces.

7. Control of the composition and phases of the compound layer. In some cases, the compound layer can be eliminated, leaving only the diffusion zone.

8. It can run automatically without operator supervision or intervention.

9. The process can be stopped at any time during a cycle and cooled to room temperature with no risk or adverse side effect.

10. Electric power and cooling water demands are comparatively low.

11. The total processing time is comparatively shorter without the need for secondary cleaning operations.

16.4.1.6 Fluidized-Bed Nitriding. This process is similar to the standard gaseous nitriding process as far as temperature, precise control of case formation, and surface hardness obtained are concerned. However, advantages claimed include increased deep case (up to 0.07 in. thick) and reduced processing times due to shorter handling and faster heating rates. However, nitriding in a fluidized bed can lead to high NH_3 consumption when used for long cycle times.[211,212]

These costs are offset by the energy savings by using shorter cycles and from higher load densities.[213] During the operation of the furnace, the parts are held in baskets or racks which may be either rested on a load support frame or suspended about 2 in. (50 mm) above the retort bottom. When parts are loaded, nitrogen is the fluidizing gas. After reaching the proper temperature, dry anhydrous NH_3 alone or diluted and mixed with other gases is introduced, through the diffusion plate at the bottom of the bed. The used or effluent atmosphere escapes through a burn-off vent in the furnace lid. At the end of the nitriding cycle, the furnace is purged with nitrogen for about 2 min, and the load is withdrawn from the furnace and put in the fluidized-bed cooler operating on nitrogen to avoid discoloration and maintain the surface characteristics.[213]

Fluidized-bed nitriding is mostly done between 510 and 565°C (950 and 1050°F). It has been shown that the dissociation rate of anhydrous NH_3 in a fluidized bed operating at 534°C (1000°F) is ~50%.[213] After nitriding, parts may be quenched rather than the usual slow-cooled. On some steels, this is advantageous in improving surface hardnesses. Fluidized-bed nitrocarburizing can be made possible by adding methane to NH_3 in the fluidized-bed atmosphere. (However, it is not normally used except to promote the epsilon phase on low-carbon and low-alloy steels.)

The process might be suited to nitriding components with deep small holes or blind holes, which are usually a concern for uniform nitriding by conventional gas nitriding or ion nitriding. The fluid-pulse nitriding (performed in the same fluidized furnace, with an additional electrical pulse signal generator) can produce somewhat the same microhardness profile, case depth, and higher surface hardness, but offers a decrease in the fluid-bed gas consumption by up to 52%.[214]

16.4.2 Nitrocarburizing

Ferritic nitrocarburizing is a thermochemical surface hardening treatment which is rapidly developing, particularly to upgrade the performance of low-carbon steel. This process involves diffusional addition of nitrogen and carbon to the surface of ferrous components (where the enrichment of nitrogen is predominant) at temperatures on the order of 570°C (1060°F), i.e., completely within the ferrite phase field for 1 to 3 hr.[171,215] This produces a thin 0.00055- to 0.0008-in. (0.014- to 0.02-mm) ductile single-phase ε-iron carbonitride ($Fe_{2-3}N,C$), "white layer," or compound layer formed at the component surface.[216] It possesses increased hardness and excellent corrosion resistance and tribological properties (oil retention and high wear and scuffing resistance) after a relatively short treatment time, usually less than 3 hr. In addition, a "diffusion zone" beneath the compound layer provides a considerable increase in tensile strength, stiffness, fatigue resistance, and pitting resistance of treated materials. Fatigue properties can be further improved by quenching in oil or water from the treatment temperature.[217–219]

Nitrocarburizing low-carbon nonalloy steel followed by oil-quenching provides corrosion resistance under conditions of humidity and neutral salt spray, thus diminishing the need for more conventional electroplated nickel finishes.[217]

The part to be nitrocarburized should be quenched and tempered or stress-relieved at 593°C (1100°F) minimum.

16.4.2.1 Gas Nitrocarburizing. Currently, gas nitrocarburizing is, next to gas carburizing, the most widely employed thermochemical surface hardening process.[220] The atmosphere used consists of (1) NH_3 diluted with a carrier gas; (2) 50% NH_3 and 50% endothermic gas (AGA type 302); (3) 35% NH_3 and 65% refined exothermic gas (AGA type 201, nominally 97% nitrogen) enriched with hydrocarbon gas; (4) propane/NH_3/O_2 mixture; or (5) methane/NH_3/O_2 mixture.[171] Any gas leakage in the furnace and around furnace doors must be minimized, and double pilots should be used.

Processing steps are as follows:

1. Cleaning and degreasing the parts thoroughly.
2. Loading the furnace at room temperature, and purging all the air from the furnace with nitrogen.
3. Raising the furnace temperature to 570°C under reduced flow rate.
4. Maintenance of proper nitrocarburizing atmosphere for 3 hr after attaining this temperature.
5. Purging with nitrogen after shutdown.
6. Unloading the furnace and cooling the parts—either oil- or gas-quenching.

Applications. This process finds a wide variety of applications such as on: (1) textile machinery gears, pump cylinder blocks, rocker-arm spacers, and jet nozzles

TABLE 16.16 Atmosphere Systems Used to Produce a Compound Layer Depth of $17 \pm 1 \mu m$ at $570°C^{221}$

System	Ratio of ingoing gases	Residual ammonia, %
Ammonia	—	55
Ammonia/nitrogen	1:0.33	55
Ammonia/endothermic gas	1:1	42
Ammonia/nitrogen/CO_2	1:1.38:0.23	21
Ammonia/nitrogen/CO	1:1.33:0.18	20
Ammonia/nitrogen/CH_4	1:1.43:0.08	22
Ammonia/nitrogen/air	1:1.25:0.25	24
Ammonia/exothermic gas	1:1.5	20

Material: 0.15% C nonalloy steel. Nitriding time: 2 hr, followed by oil quench.
Number of atmosphere volume changes: 7.5 per hour.
Courtesy of Wolfson Heat Treatment Centre, England.

for wear resistance; and (2) crankshafts and driveshafts for improved fatigue properties. This can be successfully applied to wrought and sintered plain carbon and alloy steels, stainless steels, and cast irons. The most significant improvement is, however, found with low-carbon nonalloy steels in both antiscuffing bending and rotational fatigue strength.

Proprietary Methods. *Nitrotec*, a development of the early 1980s in the gas nitrocarburizing process with the addition of a precooling oxidation sequence, is a trademark of Lucas industries. As the name indicates, this process consists of three steps: namely, *nitr*iding, *o*xidizing, and pro*tec*tion (e.g., sealing).[221,222] The salient feature of the process is to use a wide range of atmospheres of NH_3 and a carrier gas but with a minimum residual of 55% NH_3 level for NH_3 or NH_3/N_2 atmosphere, 42% NH_3 level for NH_3/endothermic gas mixture, and 20% NH_3 level for $NH_3/N_2/CO$ atmosphere and 20% NH_3 level for NH_3/exothermic gas atmospheres, shown in Table 16.16.[221] The selection of the atmosphere actually depends on the economics based on local conditions.

The first step of this Nitrotec treatment on steel specimens produces an ε-iron carbonitride nonmetallic compound layer to a depth of 25 to $40 \mu m$ (0.001 to 0.0015 in.) thickness. This compound layer consists of the formation of a substantial void or porosity and ε-carbonitride compound. An increase in porosity level of the compound layer occurs toward the surface in proportion to an increase in the nitrogen content. A dimensional growth is just 2.5 to $8 \mu m$ (0.0001 to 0.0003 in.). A maximum surface nitrogen content of 8% approaches closely to a nonmetallic, hexagonal close-packed ε-nitride phase. A maximum hardness level of 1100 HK (25 g) (or 66 to 67 Rc) is reached just below the extreme surface and represents the limit of the nonporous part of the compound layer.[217] Below this layer lies the nitrogen-rich subsurface/diffusion zone.

The improvements in tensile and fatigue strengths and toughness of the Nitrotec-treated parts are attributed to the greater depth of nitrogen-rich substrate (diffusion zone) beneath the compound layer. Actually, the enhancement of these properties depends on (1) the temperature of the part prior to quenching, (2) treatment time, (3) cooling rate, and (4) thickness of the part.

The combined effect of ε-carbonitride and voids in the nitrocarburized parts gives rise to two important characteristics: (1) better wear resistance than that of either carburized or carbonitrided parts and (2) oil-retention and antiscuffing properties due to the microporous iron carbonitride compound layer similar to the nonferrous sintered porous metal bearings. The second step (i.e., oxidation stage) consists of exposing to an oxidizing atmosphere at a suitable temperature for a short time (flash oxidation) so that a thin (<1-μm, or 40-μin.), corrosion-resistant, black-colored top surface film of Fe_3NO_{3-4} is produced.[222] This oxide structure further imparts a corrosion resistance and wear-resistant characteristics and serves as a carrier for an organic sealant. [These sealants are either water-based mixtures of emulsified microcrystalline and synthetic hydrocarbon waxes or hydrocarbon-solvent-based mixtures of metal soaps (produced from oxidized petroleums and resin acids) with corrosion inhibitors.] As a result, the corrosion resistance properties are greatly enhanced (>500 hr against salt spray exposures). The combined effect of post-nitriding oxidation and quenching together with a specific compound layer and depth produces an attractive aesthetic appearance and wear resistance.

Applications include windshield-wiper linkage assemblies (Fig. 16.49) (for motor vehicles to replace the stainless steel with mild steel),[221] automotive fan motor, viscous slip differential, bumper armatures, seat sliders, steel bars and tubes up to 7.3 m (24 ft) long, piston rods in hydraulic cylinders for such applications as food processing equipment, dry-ice machines, power-shearing systems, and shipboard material-handling equipment. Other potential applications are papermaking and oil and gas drilling industries where both corrosion and wear resistance are in demand.[222]

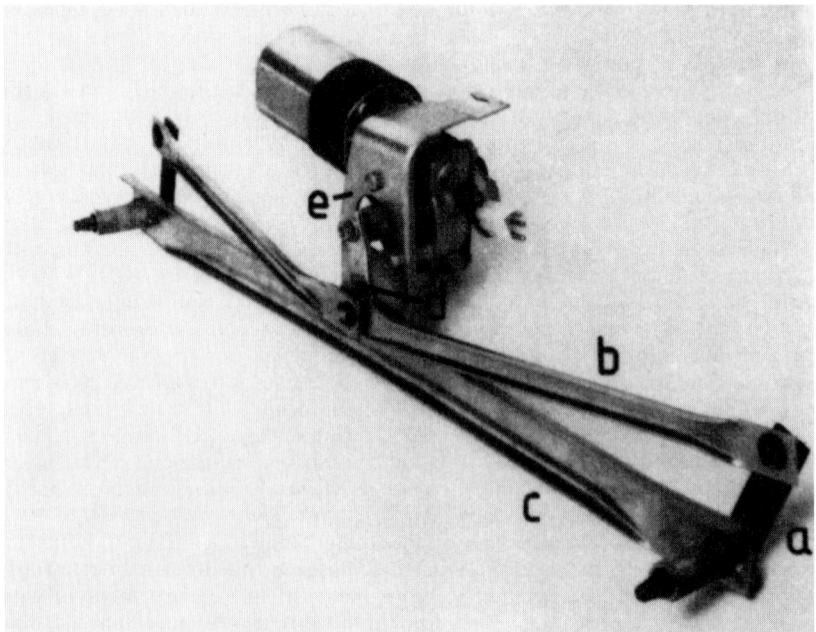

FIGURE 16.49 Windshield wiper linkage system utilizing Nitrotec surface treatment and used in the automotive industry. It contains five component parts: (*a*) spindle link, (*b*) primary link, (*c*) black plate, (*d*) drive link, and (*e*) support plate.[221] (*Courtesy of Wolfson Heat Treatment Centre, England.*)

Oxynitrocarburizing. Oxynitrocarburizing consists of two steps. The first is nitrocarburizing in which an iron carbonitride compound layer, about 10 to 50 μm thick, occurs. In the second oxidation step, <1 to a few μm thick iron oxide forms on top of the compound layer.

The surface layer formed by this process is very hard (800 to 1400 HV) and possesses both the increased resistance to abrasive and adhesive wear and high corrosion resistance. The color of the surface layer lies between anthracite gray and black. It is applied to complex-shaped products out of sheet steel, notably microalloyed interstitial-free steels.[223]

Nitrocarburizing with Integrated and Controlled Post-Oxidation. Figure 16.50 shows a controlled post-oxidation treatment as an integrated part of the nitrocarburizing cycle which provides rapid reduction of the nitrocarburizing temperature to the desired oxidation temperature and rapid changeover of the reducing nitrocarburizing atmosphere to an oxidizing atmosphere. This system uses an oxygen probe within a nitrocarburizing furnace to measure the oxidation potential of the atmosphere and a controller to compare the measured oxidation potential to a set point and to regulate and adjust it at different oxidation temperatures, by intermittently feeding in oxidizing gases (air or water) and reducing gases (hydrogen, endogas, or ammonia). These systems can produce the desired iron oxide phase (Fe_3O_4) with the appropriate thickness of the oxide layer.[224]

16.4.2.2 Salt Bath Nitrocarburizing.
The original salt bath nitrocarburizing process introduced in 1947 involved the use of alkali metal cyanide and cyanates. They depended on an oxidation reaction to transform sodium cyanide to cyanate, followed by catalytic breakdown of the sodium cyanate, releasing carbon and nitrogen to be absorbed at the steel surface in accordance with

$$4NaCNO \rightarrow Na_2CO_3 + 2NaCN + CO + 2N \qquad (16.45)$$

$$2CO \rightarrow [C]_{Fe} + CO_2 \qquad (16.46)$$

The costly detoxification of cyanide-containing effluents and sludges and environmental pollution problems in commercial practice have led to the development of alternative proprietary nontoxic salt bath formulations conforming to AMS 2753. These cyanide-free compositions are basically a mixture of (34 to 38%) alkali cyanates and alkali carbonates, and are marketed under various trade names, for example, TF-1 Tufftride, Tenifer, and Melonite (Houghton Durferrit GmbH); Nu-Tride (Kolene Corporation); and Sursulf (H. E. F.). One main concern of salt bath is the effective and complete removal of residual salt on the processed part, particularly if blind holes and threads are involved.

Proprietary Methods. An adjunct to the salt bath nitrocarburizing treatment, designated as *QPQ* (an acronym for the process sequence of "quench, polish, quench"), was developed in Germany and is carried out in two steps (Fig. 16.51).[225,226] In the first step, the finish-machined parts are preheated in air to 350°C (660°F), followed by nitrocarburizing in accordance with AMS 2753 (in aerated Tufftride TF1 salt bath) at 580°C (1076°F) for about 1 to 2 hr, which produces a ductile ε-nitride ($Fe_{2-3}N$) compound zone (lightly etched) (Fig. 16.52) depth of 0.0003 to 0.0008 in. (10 to 20 μm) thick on the steel surface (with a nitrogen content of 6 to 9% and a carbon content of ~1%, depending on the type and amount of alloying elements or nitride formers). The higher the alloy content, the thinner the layer for the same treatment cycle. The diffusion zone is 0.015 to 0.030 in. thick, which forms beneath the compound layer, depending on the initial composition and

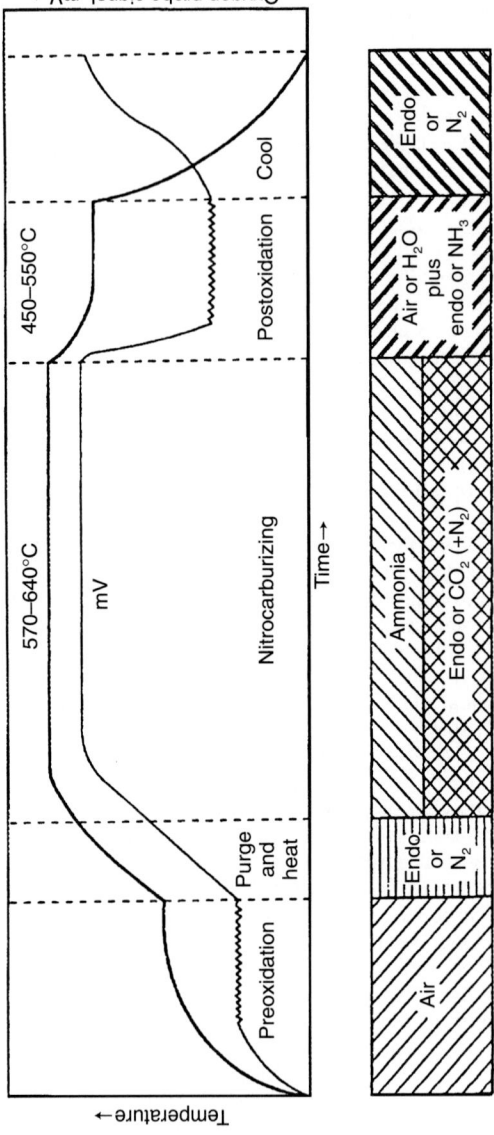

FIGURE 16.50 Typical gaseous nitrocarburizing cycle with integrated and controlled post-oxidation (PRONOX).[224] (*Courtesy of Wolfson Heat Treatment Centre, England.*)

16.106

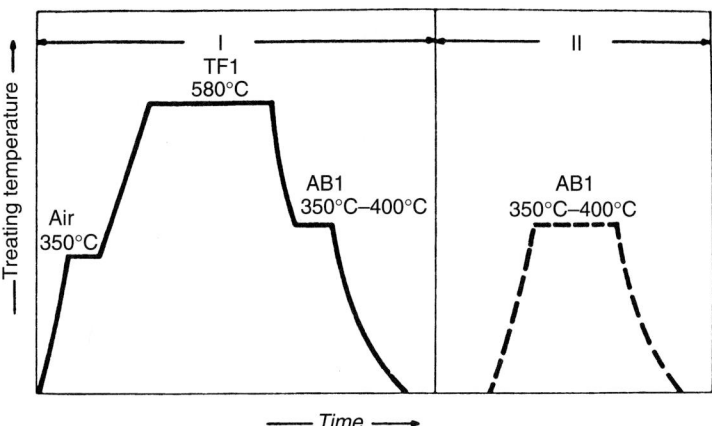

FIGURE 16.51 Schematic diagram illustrating Tufftride TF1/AB1 and QPQ process.[225,226,228,229] (*Courtesy of Houghton Durferrit GmbH.*)

structure of the parent material. The diffusion zone consists of precipitation of γ' Fe_4N nitrides in the outer zone followed by solid solution of nitrogen in ferrite matrix and/or precipitates as very fine needle-shaped nitrides.[227] [A polymeric compound (regenerator REG 1) is employed to regenerate the nitrocarburizing (TF1) bath, either periodically or continuously, to maintain the bath composition at the specified cyanate level.][228]

The cyanate/carbonate salt (TF1) composition changes very slowly and possesses the desired nitrogen and carbon potential. Sometimes the nitrocarburizing temperature is raised from 580 to 630°C (1075 to 1165°F) for the same holding time, which doubles the compound layer thickness and produces an intermediate- or sublayer on the unalloyed and low-alloy steels. This sublayer is a function of the cooling method following tufftriding treatment and consists of carbon-nitrogen bainite or carbon-nitrogen martensite associated with a large amount of retained austenite.[229] Similarly, nitrocarburizing can be produced at a lower temperature of 480°C (895°F) to meet the following: (1) high core hardness requirement in some automotive and machinery parts and tools, (2) possibility of producing very thin compound layer, (3) reduction of volume loss in maraging steels by simultaneously increasing the nitrogen content, (4) elimination of nitride formation at grain boundaries in austenitic stainless steels, and (5) dimensional stability and no distortion of part.[228]

After nitrocarburizing (Tufftride) treatment, the parts are cooled in the oxidizing salt bath composed of alkali hydroxide and alkali nitrate at 350 to 400°C (preferably 370 to 400°C) for 10 to 15 min to allow for temperature equalization and are subsequently water-quenched (<40°C) and washed in hot water for 10 to 20 min. [The oxidative cooling bath may be designated according to company trademarks, that is, AB1 (Houghton Durferrit), KQ-500 (Kolene Corp.), and Oxynit (H.E.F.).] This concludes the process sequence 1, Tufftride Q, shown in Fig. 16.51. Oxidation in the cooling (AB1) bath results in the following: (1) It effectively destroys the cyanides and cyanates produced in the nitrocarburizing bath and adhering to the components, thus making them nontoxic prior to entering the rinsing water. (2) It renders a further enhancement in the properties of the compound layer produced

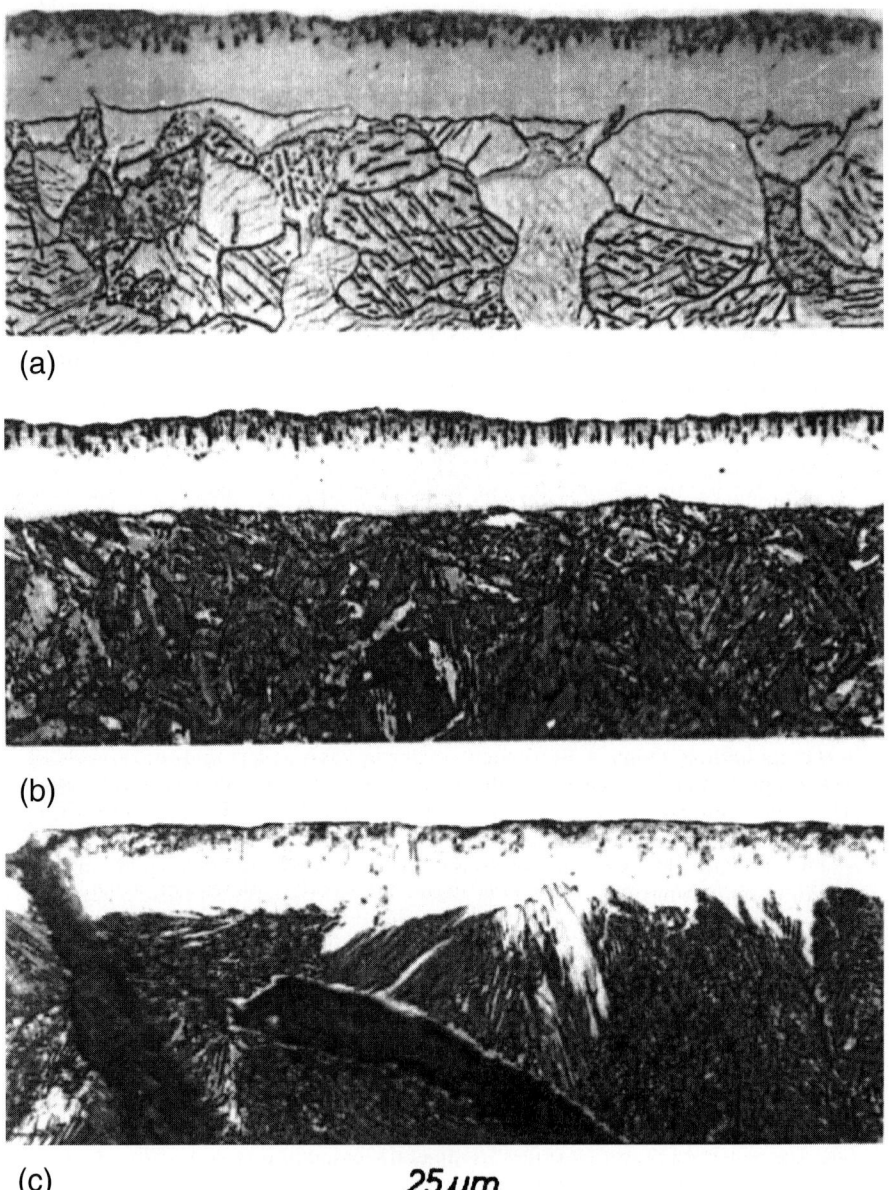

(a)

(b)

(c)   25 μm

FIGURE 16.52 Compound layers formed by salt bath (Melonite) nitrocarburizing on (*a*) SAE 1015, (*b*) SAE 5134, and (*c*) cast iron GG26.[228] (*Courtesy of Houghton Durferrit GmbH.*)

during nitrocarburizing. (3) It retards the cooling rate, thereby minimizing the distortion.[227,228]

In the second step of the QPQ process, the previously (or Q-) treated parts are polished using vibratory polishing, lapping (with emery cloth grade 360 or finer), centerless polishing, blasting with 40- to 70-μm-sized glass beads, or automated blasting with >1-mm-diameter metal shot,[229] to enhance the surface finish. This completes the QP treatment, and at this point wear resistance, gloss, and appearance with respect to non-treated components are improved. The use of solid lubricant following the treatment reduces the friction coefficient while improving seizure and corrosion resistance.

After polishing, the QP components are reimmersed in the oxidative bath, AB1 bath in sequence II (Fig. 16.51), at 350 to 400°C for 20 to 30 min, which constitutes the QPQ treatment. Finally, they are rinsed and oil-dipped. The oxygen present in the form of a layer of Fe_3O_4 is about 3 to 4 μm thick, which produces a lustrous dark finish with an improved corrosion resistance for nonstainless steels.[227] Both the first and second oxidizing treatments in cooling (AB1) bath impart an attractive blue/black surface finish, which provides superior corrosion resistance to that of hard chromium and nickel plating due to enrichment of the compound zone with oxygen.[226] In addition, these treated parts, when obtained manually or automatically with computer control, possess increased wear and antiscuff and increased rotating-bending and rolling fatigue strengths.[225,228,229] Additional benefits of this treatment include: no need of post-machining operation, no need of plating combined with baking treatment, and the ability to use cheaper material.[225,228] Figure 16.52 shows the compound layers produced by salt bath nitrocarburizing on SAE 1015, SAE 5134, and cast iron GG26.[228]

It is suitable for low-carbon steels, low-alloy steels, constructional steels, stainless steels, tool and mold steels, and cast irons.

Applications include windshield-wiper linkage systems, crankshafts, and valves[229] in the automotive industry; domestic appliances; components in offshore technology and plant and machinery;[229] high-speed cutting tools, forming tools, die casting (H13) dies, and extrusion dies; piston rods for gas springs, cylinders, and rods for hydraulic systems, and shock absorber piston rods;[230] and in the photographic and armament fields, undercarriage components on the Tornado jet fighter. The process has been established as a viable alternative for stainless steels with a significant reduction in material and manufacturing costs.[230]

Oxynit Process. Another trade name of salt bath nitrocarburizing, established in the United Kingdom, is the Oxynit process, which is a modified "Sursulf" process.[231] As the name indicates, this treatment consists of two processing steps: *nitr*iding and *oxi*dation. In the first step of the Sursulf process, a sulfur-activated low-toxicity molten salt bath having a high nitriding potential is used for the nitriding/nitrocarburizing of the steel parts. The salt consists of (1) sodium or potassium cyanate with a little amount of Li (to stabilize the chemical composition of the bath) and (2) an organic regenerator salt (to maintain the high nitriding potential of the bath by converting/recycling some amount of carbonate into effective cyanate, thereby liberating NH_3 gas). This treatment produces a compound zone, mainly the stable ε-iron nitride and sulfur compounds with fine porosity at the surface, below which a deep zone of enriched nitrogen diffusion forms. This structure is ideally suited to combat many industrial problems (e.g., wear, abrasion, scuffing, fatigue, fretting, general corrosion, and corrosion-erosion).

The second step is carried out in a strongly oxidizing salt bath containing nitrates, hydroxides, carbonates, and "super" oxidizing agents, capable of decomposing the sulfur compounds present in the surface layers of the compound zone. This step produces a thick magnetic iron oxide, Fe_3O_4, at the surface layers to a depth of 6 to 8 μm, which greatly increases the corrosion resistance properties without sacrificing the excellent tribological properties developed earlier.[231]

16.4.2.3 Plasma Nitrocarburizing. Plasma nitrocarburizing has been performed on an Armco, carbon and alloy steels, and sintered P/M steels in a 30-kW vacuum furnace at a temperature of 570°C and pressure of 4 kPa using a gaseous mixture of 70% N_2, 27% H_2, 3% CH_4.[232] Like other methods, the plasma nitrocarburized surface layers consist of (a) an outer ε-iron nitride compound zone with very high hardness and limited thickness (>5 μm) and a surface hardness of ≥350 HV and (b) an inner diffusion zone composed mainly of a γ'-Fe_4N-type solid solution. This process allows a low-nitrogen and high-carbon addition to occur at the surface layers. As the carbon content in the gas mixture increases, the compound layer thickness increases to a maximum and then decreases. At a very high carbon content, the compound layer does not form at all; instead, deposition of amorphous carbon occurs on the component surface.[233] Following nitrocarburizing, the workpieces are allowed to cool under controlled vacuum conditions.

The advantages over other nitrocarburizing treatments are:[234]

1. Reduced processing time
2. Environmental safety (i.e., less exhaust gas containing N_xO_y compounds and greater energy efficiency)
3. Minimum use of treatment gases
4. Fast cooling that favors a preferable monophased ε-$[Fe_{2-3}(N,C)]$ structure of compound layer at room temperature[235]

Applications include P/M chain gear wheels.[236]

16.4.2.4 Fluid Bed Nitrocarburizing. Although complete replacement of the atmosphere within a fluidized bed can take place within a few minutes, it is a better approach to "condition" the fluidized beds with the process atmosphere for considerably longer periods (5 to 35 min) prior to the introduction of workpieces, for consistent results.[237]

The fluid bed nitrocarburizing method uses a fluid bed furnace in the temperature range of 316 to 649°C (600 to 1200°F) and a 50% NH_3, 50% natural gas atmosphere; 60% NH_3, 23% natural gas, 17% N_2; 50% NH_3, 40% natural gas, 10% N_2; 40% NH_3, 40% natural gas, 20% N_2; 10 to 14% NH_3, N_2, natural gas; or NH_3/CH_4 mixture in varying proportions, depending on the end use of the heat-treated components.[237,238]

Since fluidized-bed nitrocarburizing is characterized by very high nitriding potential, there is the likelihood of severe porosity with subsequent compound layer breakdown and/or occurrence of spalling. In that case, not more than 10 to 14% NH_3 may be needed for fluidized-bed nitrocarburizing, employing NH_3, natural gas, and nitrogen.[237]

The depth of the compound layer can be controlled by varying the atmosphere, time, and temperature. The process imparts a light blue/black lustrous oxide finish which enhances the sliding wear resistance, impact resistance, and in some cases, galling resistance.

The fluidized-bed nitrocarburizing has several advantages over the conventional atmosphere/integral quench nitrocarburizing:

1. The furnace is very tight because of upward pressure of the gases.
2. It results in lower distortion and greater bending strength of the treated parts without fracture.
3. There is a possibility of stuffing the parts together in the basket.
4. Very even finishing is produced due to fluid action.
5. Unlike in salt bath processing, no cyanide process is used and not all the part holes are plugged up after heat treatment.
6. It provides competition for some surface treatments, e.g., titanium nitride coating.

Applications include parts such as hot forged dies and punches; cold-formed sleeves; cutting tools; hobs; milling cutters, typically of D2 or M2 type; paper slitter blades; and finished drills.[238]

16.4.2.5 Austenitic Nitrocarburizing. Austenitic nitrocarburizing can be considered from the viewpoint of atmosphere as the low-temperature carbonitriding process with an option for a higher nitrogen potential and for oxidation. Typically the process is accomplished in the same kind of sealed-quenched furnace as carbonitriding with similar atmosphere except NH_3 content which is increased to 20%. One proprietary austenitic nitrocarburizing atmosphere system is 20%NH_3/40% cracked methanol/40% N.[239]

Standard austenitic nitrocarburizing is accomplished at 700°C for 2 hr followed by an oil quench. However, a 3-hr treatment provides an improved indentation resistance. After the 2-hr treatment, mild steel shows a 25- to 30-μm-thick ε-compound layer and an underlying carbonitrided case to a depth of 125 to 150 μm. The case structure beneath the compound layer, after quenching, comprises two zones: an austenitic zone and a back-up martensitic zone.

Austenitic transformation is achieved either by isothermal treatment to produce lower bainite or by deep freezing and subsequent tempering to form tempered martensite with a microhardness in the range of 700 to 900 HV.[239]

The advantage of this treatment is the increased hardened case below the compound layer which extends the range of its applications.[236]

16.5 SUPERCARBURIZING

The supercarburizing or saturation carburizing process was developed by O. Cullen in 1957. In this process the carbon concentration exceeds its normal solubility limit (e.g., 1.2%) in steel to as much as 4%. However, the usual surface carbon concentration ranges are 1.8 to 2.2%, 2.0 to 2.4%, 2.4 to 2.8%, and over 3%. These carbon levels can be obtained with conventional carburizing methods such as pack, gas (hydrocarbon-enriched gas or methanol plus nitrogen), and so forth, both at atmospheric pressure and in vacuum.[240] The supercarburized surfaces are characterized by high density of globular, granular, or spheroidal carbides with increased wear resistance.[241]

Figure 16.53 shows typical case carbon probes of supercarburized steels. The shapes of these curves can be varied by suitable adjustment in the heat treatment cycle. This process is suited for alloy steels containing substantial quantities of carbide formers, especially Cr, Mo, and W. These include medium-alloyed and more

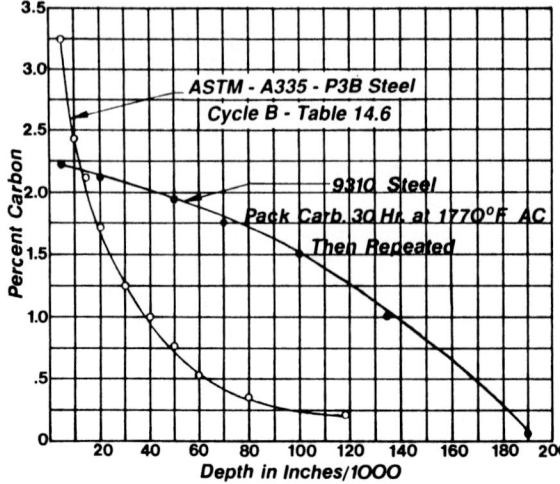

FIGURE   16.53 Typical case carbon probes of super-carburized steels.[240] (*Courtesy of Roy F. Kern.*)

highly alloyed case hardening steels such as AISI 4118, 5120, 8620, 8720, 8822, 9310, and 52100.[240] However, similar treatment applied to 12% Cr steels has produced dispersion of chromium carbides (Cr_7C_3) of 0.5-μm mean diameter constituting 30% of the microstructure.[242]

The process consists of heating to a carburizing atmosphere at 927 to 982°C (1700 to 1800°F) for 2 to 5 hr, to ensure that carbides do not dissolve completely in austenite but remain adjacent to the grain boundaries as leaky carbide peripheries enclosing austenite grains. They dissolve more carbon and form more carbide at temperature. Upon cooling to a subcritical temperature, say, 677°C (1250°F) for ~1 hr, the carbide network thickens. This cycle of reheating to 927 to 982°C and subsequent cooling to 677°C is repeated 2 to 4 times, and finally the parts are oil-quenched after holding at sufficiently high hardening temperature, for example, 816°C (1500°F), for 0.5 to 1 hr to avoid the network carbide, and are subsequently tempered at 177°C (350°F). Table 16.17 lists two typical supercarburizing heat treatment cycles A and B applied to ASTM A335-P3B steel (composition: 0.15% C, 0.48% Mn, 0.26% Si, 0.20% Ni, 2.08% Cr, 0.52% Mo, 0.008% S, 0.02% P, and 0.07% Cu). A typical phase analysis of a low-alloy steel sample after regular carburizing and hardening and after being supercarburized and hardened is as follows:

	Martensite, %	Austenite, %	Carbide, %
Normal carburizing	81	14	5
Supercarburizing	65	10	25

The final hardened structure consists of a large-volume fraction of carbides at the surface with a high surface hardness and a depletion of carbide formers such as Cr and Mo in the matrix. Figure 16.54 shows the surface microstructure of a 9310 steel sample, pack carburized [to a case depth of 0.305 in. (7.75 mm)] and hardened. The case carbon concentration at 0.005 in. (0.127 mm) from the surface was found to be 2.03%.

TABLE 16.17 Two Typical Supercarburizing Cycles A and B for ASTM-A335-P3B Steel[240]

	Cycle A	Cycle B
Carburize at 1700°F for:	2 hr	2 hr
Set control to 1250°F		
Work at 1250°F for:	65 min	75 min
Set control to 1700°F		
Work at 1700°F for:	40 min	75 min
Carburize at 1700°F for:	5 hr	5 hr
Set control to 1250°F		
Work at 1250°F for:	60 min	80 min
Set control to 1700°F		
Work at 1700°F for:	40 min	60 min
Carburize at 1700°F for:	4 hr	5 hr
Set control to 1250°F		
Work at 1250°F for:	—	65 min
Set control to 1700°F		
Work at 1700°F for:	—	60 min
Carburize at 1700°F for:	—	$5^{1}/_{2}$ hr
Set control to 1500°F		
Work at 1500°F for:	15 min	20 min
Hold at 1500°F for:	55 min	30 min
Quench in 120°F oil for:	15 min	20 min
Wash and temper at 350°F for:	3 hr	3 hr
Case hardness:	65 Rc	67 Rc
Surface case carbon content:	2.75%	3.26%

Furnace: Surface combustion all-case
Atmosphere: Furnace purge   400 cfh RX-gen. gas
                               40 cfh natural gas
             Carburize   400 cfh RX gas
                               70 cfh natural gas
             Dew point   9–19°F
             CO_2 0.0%
             CO   17–18%
             CH_4 9.1–10.0%
             Cool and diffuse 400 cfh RX gas
Reprinted by permission of Fairchild Publications, New York.

Advantages of supercarburizing over conventional carburizing, when properly treated, are:[240]

1. Improved abrasive wear resistance in diesel-engine injection pump parts
2. Increased compressive residual stress over −690 MPa (−100 ksi) and greater matrix toughness due to the formation of a large amount of lath martensite[241]
3. Increased long-life bending-fatigue strength by ∼25%
4. Increased contact stress capability in parts such as gears and rolling bearings
5. Increased resistance to scoring, if NH_3 is present in the hardening cycle

Limitations when compared to conventional carburizing are as follows: During grinding, wheel surfaces are abraded away more rapidly. When harder grinding wheels are used, there is always a danger of burning.[240]

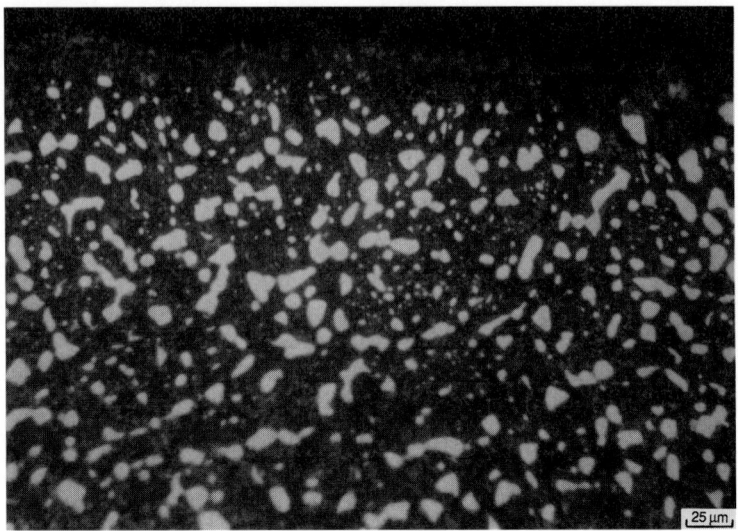

FIGURE 16.54　Surface microstructure of supercarburized and hardened 9310 steel. 4% picral etch.[240] (*Courtesy of Roy F. Kern.*)

16.6 BORIDING (OR BORONIZING)

Boriding, or *boronizing*, is a thermochemical surface hardening process that can be applied to a wide variety of ferrous, nonferrous, and cermet materials. The process involves heating well-cleaned material in the range of 700 to 1000°C (1292 to 1832°F), preferably for 1 to 12 hr, in contact with a boronaceous solid powder (boronizing compound), paste, liquid, or gaseous medium. Other developments in thermochemical boriding include plasma boriding, pulsed-plasma boriding, and fluidized-bed boriding. Currently, multicomponent boriding is also used.

This section describes mainly the various media used for thermochemical boriding, their advantages, limitations, and applications.

16.6.1 Characteristic Features of Boride Layers

During boriding, the diffusion and subsequent absorption of boron atoms into the metallic lattice of the component surface form the interstitial iron-boron compounds (or phases).[243–247] Figure 16.55 is the Fe-B phase diagram showing various phases and their compositions and structures.[248] The resulting layer may consist of either a single-phase boride or a polyphase boride layer. The morphology (Fig. 16.56),[249] growth, and phase composition of the boride layer can be influenced by the alloying elements in the base material. The microhardness of the borided layer also depends strongly on the composition and structure of the boride layer and the composition of the base material (Table 16.18).[245,250,251]

16.6.1.1 Advantages.　Boride layers possess a number of characteristic features with special advantages over conventional case hardened layers.

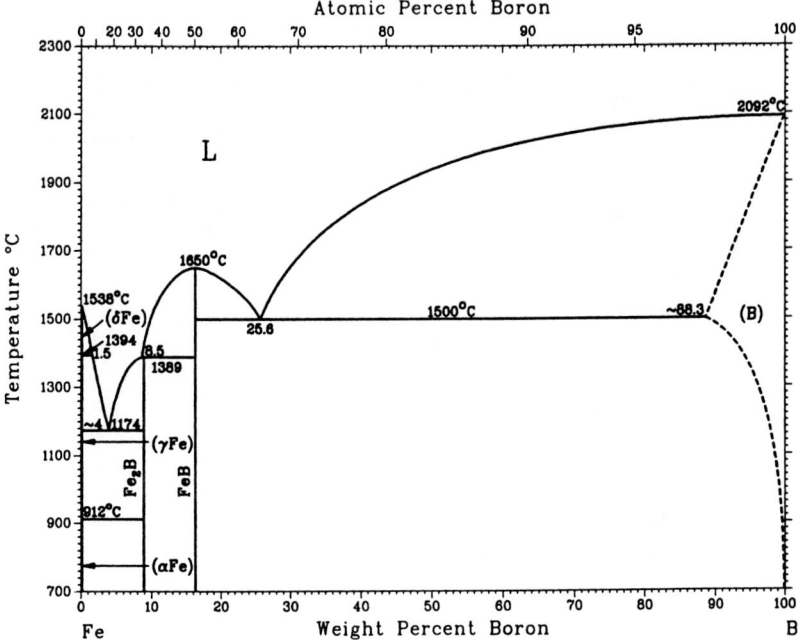

Atomic Percent Boron

P.K. Liao and K.E. Spear, unpublished

Phase	Composition, wt% Fe	Pearson symbol	Space group
(αFe)	0	cI2	Im3̄m
Fe₂B	8.8	tI12	I4/mcm
FeB	16.0 to 16.2	oP8	Pbmn
(βB)	100	hR108	R3̄m
Metastable phases			
Fe₃B	~6	oP16	Pnma
Fe₃B(HT)	~6	(a)	...
Fe₃B(LT)	~6	(b)	...

(a) bct. (b) Tetragonal

FIGURE 16.55 Fe-B phase diagram.[248] (*Courtesy of ASM International.*)

1. Boride layers have extremely high hardness values (between 1450 and 5000 HV) with high melting points of the constituent phases (Table 16.18). The typical surface hardness values of borided steels compared with other treatments and other hard materials are listed in Table 16.19.[245] This clearly illustrates that the hardness of boride layers produced on carbon steels is much greater than that produced by any other conventional surface (hardening) treatments; it exceeds that of the hardened tool steel, hard chrome electroplate, and is equivalent to that of tungsten carbide.

2. The combination of a high surface hardness and a low surface coefficient of friction of the borided layer also makes a significant contribution in combating the

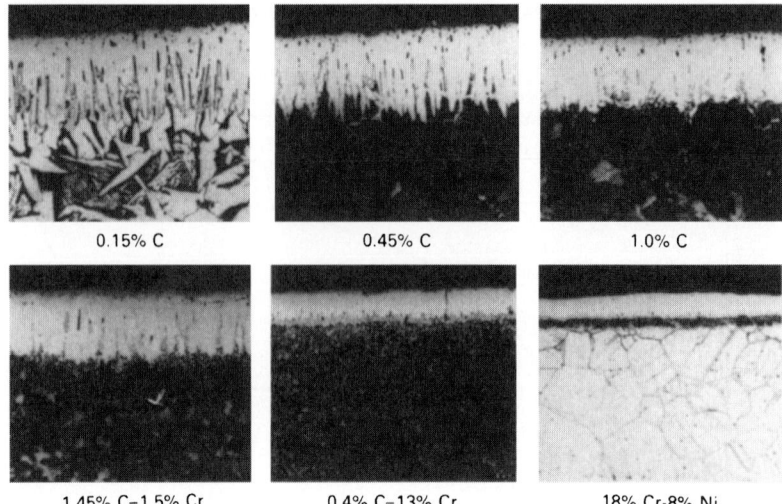

| 0.15% C | 0.45% C | 1.0% C |
| 1.45% C–1.5% Cr | 0.4% C–13% Cr | 18% Cr-8% Ni |

FIGURE 16.56 Effect of steel composition on the morphology and thickness of the boride layer.[249] (*Courtesy of R. Chatterjee-Fischer.*)

TABLE 16.18 Melting Point and Microhardness of Different Boride Phases Formed during Boriding of Different Substrate Materials[245,250,251]

Substrate	Constituent phases in boride layer	Microhardness of layer, HV (or kg/mm^2)	Melting point °C	Melting point °F
Fe	FeB	1900–2100	1390	2535
	Fe$_2$B	1800–2000	—	—
Co	CoB	1850	—	—
	Co$_2$B	1500–1600	—	—
	Co$_3$B	700–800	—	—
Co-27.5Cr	CoB	2200 (100 g)[†]	—	—
	Co$_2$B	~1550 (100 g)[†]	—	—
	Co$_3$B (?)	700–800	—	—
Ni	Ni$_4$B$_3$	1600	—	—
	Ni$_2$B	1500	—	—
	Ni$_3$B	900	—	—
Inco 100	—	1700 (200 g)[††]		
Mo	Mo$_2$B	1660	2000	3630
	MoB$_2$	2330	~2100	~3810
	Mo$_2$B$_5$	2400–2700	2100	3810
W	W$_2$B$_5$	2660	2300	4170
Ti	TiB	2500	~1900	3450
	TiB$_2$	3370	2980	5395
Ti-6Al-4V	TiB		—	—
	TiB$_2$	3000 (100 g)[†]	—	—
Nb	NbB$_2$	2200	3050	5520
	NbB$_4$			
Ta	Ta$_2$B		3200–3500	5790–6330
	TaB$_2$	2500	3200	5790
Hf	HfB$_2$	2900	3250	5880
Zr	ZrB$_2$	2250	3040	5500
Re	ReB	2700–2900	2100	3810

[†] 100-g load. [††] 200-g load.

TABLE 16.19 Typical Surface Hardness of Borided Steels Compared with Other Treatments and Hard Materials[245]

Material	Microhardness, kg/mm^2 or HV
Borided mild steel	1,600
Borided AISI H13 die steel	1,800
Borided AISI A2 steel	1,900
Quenched steel	900
Hardened and tempered H13 die steel	540–600
Hardened and tempered A2 die steel	630–700
High-speed steel BM42	900–910
Nitrided steels	650–1,700
Carburized low-alloy steels	650–950
Hard chromium plating	1,000–1,200
Cemented carbides, WC + Co	1,160–1,820 (30 kg)
Al_2O_3 + ZrO_2 ceramic	1,483 (30 kg)
Al_2O_3 + TiC + ZrO_2 ceramic	1,738 (30 kg)
Sialon ceramic	1,569 (30 kg)
TiN	2,000
TiC	3,500
SiC	4,000
B_4C	5,000
Diamond	>10,000

main wear mechanisms: adhesion, tribooxidation, abrasion, and surface fatigue.[246,252] This fact has enabled the mold makers to substitute easier-to-machine steels for the base metal to still obtain wear resistance and antigalling properties superior to those of the original material.[253] Figure 16.57 shows the effect of boriding on abrasive wear resistance of borided C45 steel, titanium, and tantalum as a function of number of revolutions (or stressing period) based on the Faville test.[254] Figure 16.58 shows the influence of steel composition on abrasive wear resistance.[249,252]

3. Hardness of the boride layer can be retained at higher temperatures than, e.g., that of nitrided cases.

4. A wide variety of steels, including through-hardenable steels, are compatible with the processes.[255]

5. Boriding, which can considerably increase corrosion-erosion resistance of ferrous materials in nonoxidizing dilute acids (Fig. 16.59) and alkali media, is increasingly used to this advantage in many industrial applications.[246]

6. Borided surfaces have moderate oxidation resistance (up to 850°C, or 1550°F) and are quite resistant to attack by molten metals.

7. Borided parts have increased fatigue life and service performance under oxidizing and corrosive environments.

8. The plasma boronized thermally sprayed nickel stellite layer has been reported to cause a significant increase in the surface hardness, the resistance to corrosion and friction wear, and the adhesion of stellite to the substrate, thereby widening the application range of modern surface layers.[257]

9. Boronized steels have increased resistance to molten aluminum and zinc baths degradation at 630°C (1166°F) and 500°C (932°F), respectively, for 6 to 120 hr,

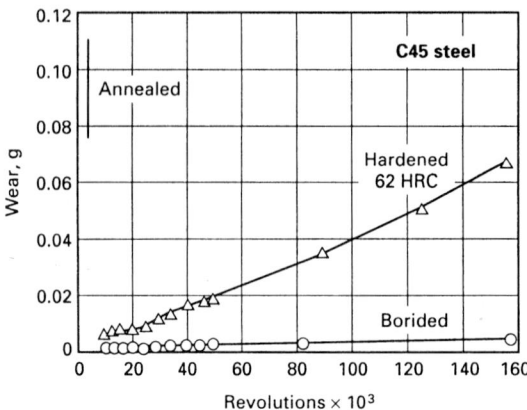

(a)

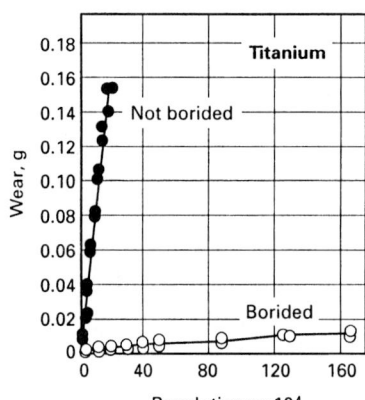

(b)

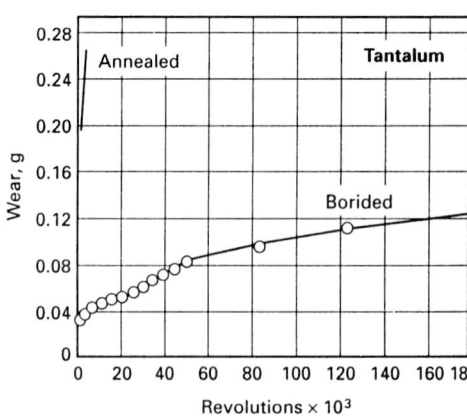

(c)

FIGURE 16.57 Effect of boriding on wear resistance (Faville test). (*a*) 0.45% C steel (C45) borided at 900°C (1650°F) for 3 hr. (*b*) Titanium borided at 1000°C (1830°F) for 24 hr. (*c*) Tantalum borided at 1000°C (1830°F) for 8 hr.[254] (*Courtesy of R. Chatterjee-Fischer and O. Schaaber.*)

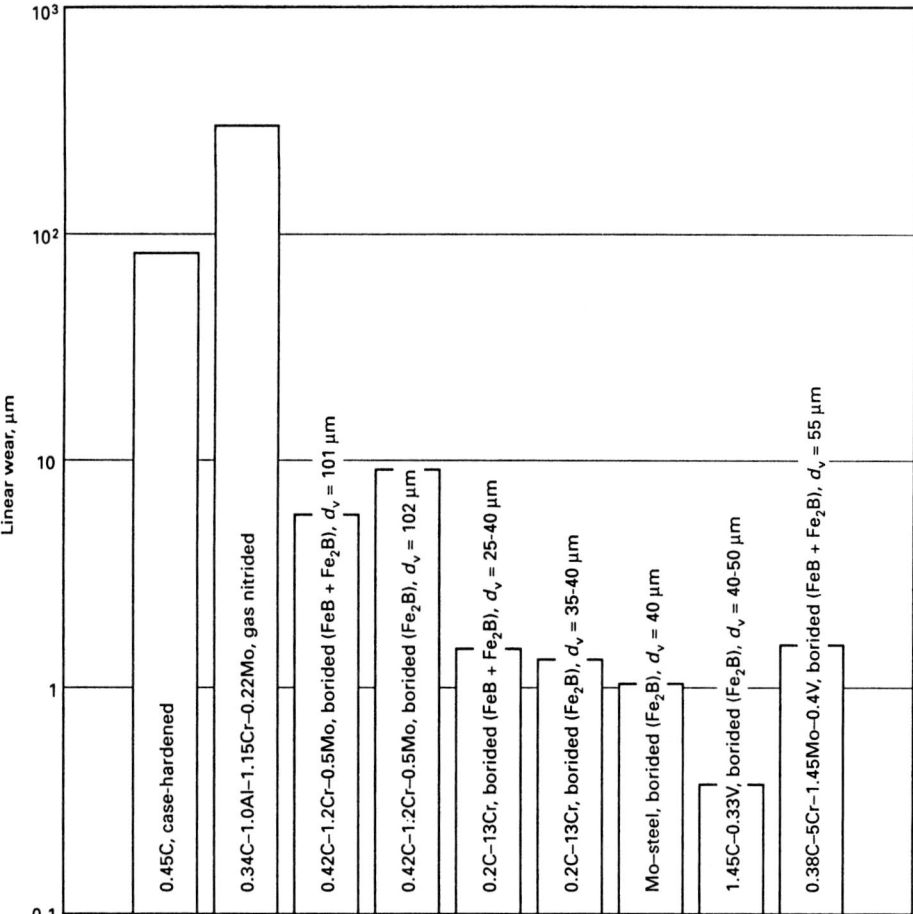

FIGURE 16.58 Effect of steel composition (nominal values in wt%) on wear resistance under abrasive wear (d_v = thickness of the boride layer). Test conditions: DP-U grinding tester, SiC-paper 220, testing time 6 min.[249,252]

which could increase their applicability in various industries handling light molten metals.[258]

16.6.1.2 Disadvantages. Disadvantages of boronizing treatments are as follows:

1. The techniques are inflexible and rather labor-intensive, making the process less cost-effective than other thermochemical surface hardening treatments such as gas carburizing and plasma nitriding. Both gas carburizing and plasma nitriding have the advantage over boronizing because these two processes are flexible systems, offer reduced operating and maintenance costs, require shorter processing times, and are relatively easy to operate. It is, therefore, suited to engineering components that need high hardness and outstanding wear and corrosion resistance of the boride layers, and/or where cheaper labor is available.[245]

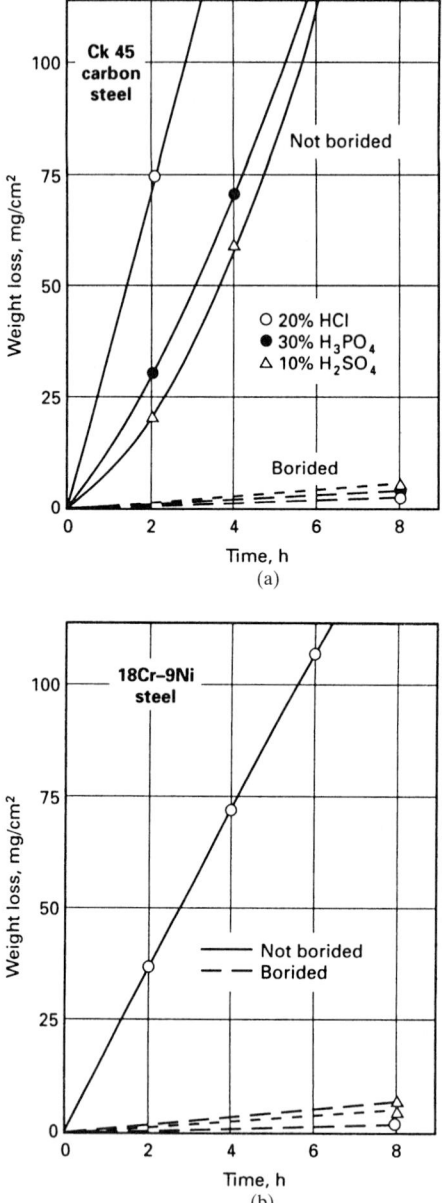

FIGURE 16.59 Corroding effect of mineral acids on boronized and non-boronized (*a*) 0.45% C (Ck 45) steel and (*b*) AISI 321 (X10CrNiTi-18-9) at a temperature of 56°C (130°F).[246,256]

2. The growth (i.e., the increase in volume) resulting from boronizing is 5 to 25% of the layer thickness (or case depth) (e.g., a 25-μm, or 1000-μin., layer would have a growth of 1.25 to 6.25 μm, or 50 to 250 μin.); its magnitude depends on the base material composition but remains consistent for a given combination of material and treatment cycle. However, it can be predicted for a given part geometry and boronizing treatment. For treatment of precision parts, where little stock removal is permitted, an allowance of ~20 to 25% dimensional increase of the final boride layer thickness must be provided.

3. Partial removal of the boride layer for closer tolerance requirements is made possible only by a subsequent diamond lapping because conventional grinding causes fracture of the layer. Thus, precise boronizing is mostly practiced for components with a large cross-sectional area.[245]

4. Boriding of most steels provides a marginal increase, if any, in the bending fatigue endurance limit, although some improvement in the corrosion-fatigue strength has been noticed.

5. In general, the rolling contact fatigue properties of borided alloy steel parts are very poor compared to those of carburized and nitrided steels at high contact loads (2000 N, or 450 lbf). This is why boronizing treatments of gears are limited to those screw designs where transverse loading of gear teeth is minimized.[245]

6. There is frequently a need to harden and temper the tool after boriding,[259] which requires a vacuum or inert atmosphere to preserve the integrity of the boride layer.

16.6.2 Boriding of Ferrous Materials

Unlike the carburizing treatment on ferrous materials, where there is a gradual decrease in composition from the carbon-rich surface to the substrate, the boriding of ferrous materials results in the formation of either a single-phase or double-phase layer of borides with definite compositions. The single-phase boride layer consists of Fe_2B, while the double-phase layer consists of an outer boron-rich, dark-etching phase of FeB and an inner boron-deficient light-etching phase of Fe_2B. The formation of either a single or double phase depends on the availability of boron.[260]

16.6.2.1 Characteristics of FeB and Fe_2B Layers. The formation of a single Fe_2B phase (with a sawtooth morphology due to preferred diffusion direction) is more desirable than a double-phase layer with FeB. The boron-rich FeB phase is considered undesirable, in part, because FeB is more brittle than the iron subboride Fe_2B layer. Also, because FeB and Fe_2B are formed under tensile and compressive stresses, respectively, crack formation is often observed at or in the neighborhood of the FeB/Fe_2B interface of a double-phase layer. These cracks may lead to flaking and spalling when a mechanical strain is applied[247] or even separation (Fig. 16.60) when a component is undergoing a thermal and/or mechanical shock. Therefore, the boron-rich FeB phase should be avoided or minimized in the boride layer.[247] If the formation of FeB is unavoidable, care should be taken to ensure that no closed FeB zones occur.[261]

It has also been reported that the tribological properties depend on the microstructure of the boride layer. The dual-phase, FeB-Fe_2B layers are not inferior to those of monophase Fe_2B layers, provided that the porous surface zone directly beneath the surface is removed.[262] Alternatively, a thinner layer is favored because of less development of brittle and porous surface-zone formation and flaking.

Typical properties of the FeB phase are:

1. Microhardness of ~19 to 21 GPa (2.7×10^6 to 3.0×10^6 psi) (hardness, 1900 to 2100 $HV_{0.1}$)

2. Modulus of elasticity of 590 GPa (85×10^6 psi)

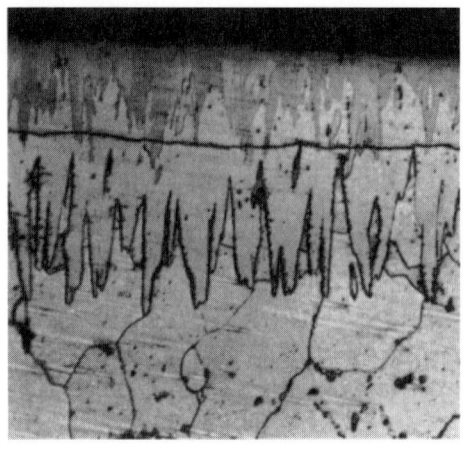

FIGURE 16.60 Separation of two-phase boride layer on a low-carbon (St 37) steel (borided at 900°C, or 1650°F, for 4 hr) caused by grinding with cutting-off disk. 200X.[256]

3. Density of 6.75 g/cm³ (0.244 lb/in.³)

4. Thermal expansion coefficient of 23×10^{-6}/°C (13×10^{-6}/°F) between 200 and 600°C (400 and 1100°F)[249,263]

5. Composition with 16 to 16.2 wt% boron[248]

6. Orthorhombic crystal structure with four iron and four boron atoms per unit cell

7. Lattice parameters: $a = 4.053$ Å, $b = 5.495$ Å, and $c = 2.946$ Å

8. Residual stress on cooling: tensile

Layers of Fe₂B.   The formation of single-phase Fe₂B layers with a sawtooth morphology is desirable in the boriding of ferrous materials.[264] A single Fe₂B phase can be obtained from a double FeB-Fe₂B phase by a subsequent vacuum or salt bath treatment for several hours above 800°C (1470°F), which may be followed by oil-quenching to increase substrate properties.[265]

Typical properties of Fe₂B phase are

1. Microhardness of about 18 to 20 GPa (2.6×10^6 to 2.9×10^6 psi) (hardness 1800 to 2000 $HV_{0.1}$)

2. Modulus of elasticity of 285 to 295 GPa (41×10^6 to 43×10^6 psi)

3. Density of 7.43 g/cm³ (0.268 lb/in.³)

4. Thermal expansion coefficient of 7.65×10^{-6}/°C (4.25×10^{-6}/°F) and 9.2×10^{-6}/°C (5.1×10^{-6}/°F) in the range of 200 to 600°C (400 to 1100°F) and 100 to 800°C (200 to 1500°F), respectively[249,263]

5. Composition with 8.8 wt% boron[248]

6. Body-centered tetragonal structure with 12 atoms per unit cell

7. Lattice parameters: $a = 5.078$ Å and $c = 4.249$ Å

8. Residual stress on cooling: compressive

The solubility of boron in ferrite and austenite is very small (<0.008% at 900°C, or 1650°F), according to Massalski.[266] According to Brown et al.,[267] the phase diagram exhibits an $\alpha/\gamma Fe_2B$ peritectoid reaction at about 912°C (1674°F), based on their findings of higher solubility of boron in ferrite than in austenite at the reaction temperature. However, later work revealed the higher solubility of boron in austenite compared to that in ferrite, thereby suggesting the reaction to be eutectoid in nature.[268]

16.6.2.2 Boriding Reactions. The boriding process consists of two types of reaction.[249] The first reaction takes place between the boron-yielding substance and the component surface. The nucleation rate of the particles at the surface is a function of the boriding time and temperature. This produces a thin, compact boride layer.

The subsequent second reaction is diffusion-controlled, and the total thickness of the boride layer growth at a particular temperature can be calculated by the simple formula

$$d = k\sqrt{t} \qquad (16.47)$$

where d is the boride layer thickness (cm); k is a constant, depending on the temperature; and t is the time (s) at a given temperature. The diffusivity of boron at 950°C (1740°F) is 1.82×10^{-8} cm^2/s for the boride layer and 1.53×10^{-7} cm^2/s for the diffusion zone. As a result, the boron-containing diffusion zone extends more than 7 times the depth of the boride layer thickness into the substrate.[269] It has been proposed that a concentration gradient provides the driving force for diffusion-controlled boride layer growth.[270]

Diffusion case thicknesses range to approximately 0.13 mm (0.005 in.) for ferrous alloys, depending on alloy compositions and configurations. A lower case depth is required for the high-carbon and/or high-alloy steels, whereas higher case depths may be needed for the low- or medium-carbon steels. When case depth is about 320 to 350 μm, subsequent heat treatment is not performed.

16.6.2.3 Ferrous Materials for Boriding. With the exception of aluminum- and silicon-bearing steels, industrial boriding can be carried out on most ferrous materials such as structural steels; case-hardened, tempered, tool, and stainless steels; cast steels; Armco (commercially pure) iron; gray and ductile cast irons; and sintered iron and steel.[271] Because boriding is conducted in the austenitic range, air-hardening steels can be simultaneously hardened and borided. Water hardening grades are not borided because of the susceptibility of the boride layer to thermal shock. Similarly, resulfurized and leaded steels should not be used because they have a tendency toward case spalling and case cracking. Nitrided steels should not be used because of their sensitivity to cracking.[272]

16.6.2.4 Influence of Alloying Elements. The mechanical properties of the borided alloys depend strongly on the composition and structure of the boride layers. The characteristic sawtooth configuration of the boride layer is dominant with pure iron, unalloyed low-carbon steels, and low-alloy steels. As the alloying element and/or carbon content of the substrate steel is increased, the degree of serration of the boride/substrate interface is suppressed, and for high-alloy steels a flat boride layer is formed[273] (Fig. 16.56). Alloying elements mainly retard the boride layer thickness (or growth) caused by restricted diffusion of boron into the steel because of the formation of a diffusion barrier. Figure 16.61 shows the effect of alloying additions in steel on boride layer thickness.[274,275]

Carbon is somewhat insoluble in the boride layer and does not diffuse through the boride layer. During boriding, carbon is driven (or diffused away) from the

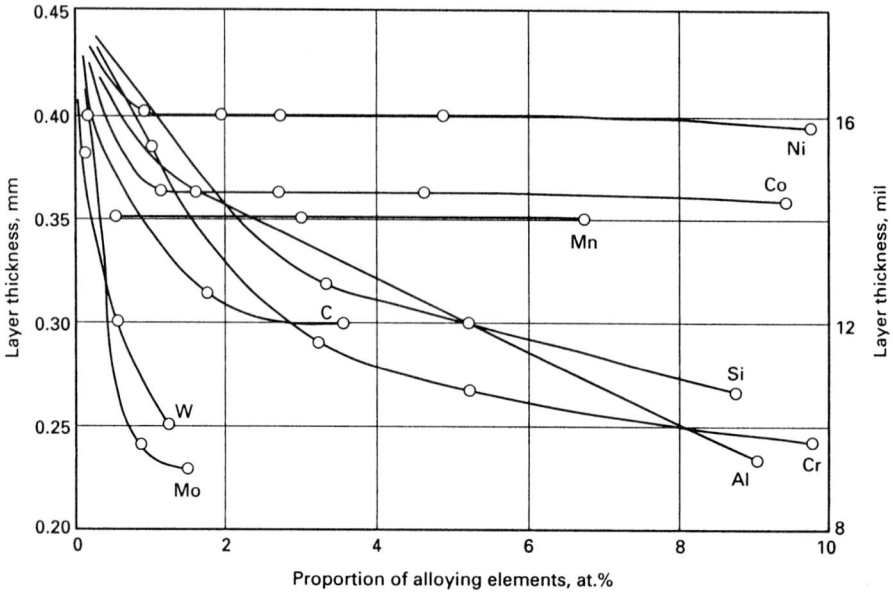

FIGURE 16.61   Effect of alloying elements in steel on boride layer thickness.[274,275]

boride layer to the matrix and forms, together with boron, borocementite $Fe_3(B,C)$ [or more appropriately $Fe_3(B_{0.67}C_{0.33})$ in case of Fe-0.8%C steel] as a separate layer between Fe_2B and the matrix.[276,277]

Silicon, aluminum, and copper, like carbon, are insoluble in the boride layer, and these elements are pushed from the surface by boron and are displaced ahead of the boride layer into the substrate, forming iron silicoborides—$FeSi_{0.4}B_{0.6}$ and Fe_5SiB_2—underneath the Fe_2B layer. Steels containing high contents of these ferrite-forming elements should not be used for boriding because they reduce the wear resistance of the normal boride layer;[243] they produce a substantially softer ferrite zone beneath the boride layer than that of the core.[278] At higher surface pressure, this type of layer buildup results in the so-called *eggshell effect*; i.e., at greater thicknesses, extremely hard and brittle boride layer penetrates into the softer intermediate layer and is consequently destroyed.[279]

Alloying elements such as nickel, chromium, manganese, vanadium, molybdenum, and cobalt are more or less strongly incorporated into the boride layer. A reduction in the degree of both interlocking tooth structure and boride depth can occur at high-nickel-continining steels. Nickel has been found to concentrate below the boride layer; it enters the Fe_2B layer and, in some instances, promotes the precipitation of Ni_3B from the FeB layer.[252,280–282] It also segregates strongly to the surface from the underlying zone corresponding to the Fe_2B layer. This is quite pronounced in both Fe-14Ni and austenitic stainless steels. Consequently, gas boronizing of austenitic stainless steel appears to be a more appropriate treatment for producing a low-porosity, homogeneous, single-phase Fe_2B layer because of the lower boron activity of the gaseous mixture.[282]

Chromium considerably modifies the structure and properties of iron borides. As the chromium content in the base material increases, progressive improvements in the following effects are observed: formation of boron-rich reaction products, decrease in boride depth, and flattening or smoothening of the coating/substrate interface.[283] A reduction of boride thickness has also been noticed in ternary Fe-12Cr-C steels with increasing carbon content.[282] Manganese, tungsten, molybdenum, and vanadium also reduce the boride layer thickness and flatten out the tooth-shaped morphology in carbon steel. The distribution of titanium, cobalt, sulfur, and phosphorus in the boride layer has not been well established.

16.6.2.5   Heat Treatment after Boriding.   Borided parts can be quench-hardened in air, oil, salt bath, or polymer quenchant and subsequently tempered. The heating to hardening temperature is carried out in an oxygen-free protective atmosphere or in a neutral salt bath.[249]

16.6.3   Boriding of Nonferrous Materials

Nonferrous materials such as Ni- , Co-, and Mo-base alloys as well as refractory metals and their alloys and cemented carbides can be borided. Copper cannot; therefore, it provides a good stopping-off material. Of special interest is the boriding of nickel alloys and Ti and its alloys. Ni-base alloys can be boronized using boronizing powder Ekabor Ni to obtain a layer thickness of more than $100\,\mu m$. Permalloy is pack borided with 85% B_4C, 15% Na_2CO_3—or 95% B_4C, 5% $Na_2B_4O_7$ powder mixture at 1000°C (1830°F) for 6 hr in H_2 atmosphere. Figure 16.62 shows the microstructures of some boronized Ni-based alloys with varying layer thickness and hardness values which depend on the composition of the base material. The maximum $HK_{0.1}$ hardness values observed in Inconel 718, Inconel 600, Hastelloy B, Hastelloy C276, Hastelloy C4, Niomonic 80A, and Haynes alloy 242 are 2800, 2200, 2400, 2400, 2000, 1800, and 1800, respectively.[284] Boriding of nickel plate is also done in gaseous BCl_3-H_2-Ar mixture in the temperature range of 500 to 1000°C (930 to 1830°F).[250]

Boriding of pure titanium and its alloys is carried out preferably between 1000 and 1200°C (1830 and 2200°F). Here pack boriding in oxygen-free amorphous boron in combination with high-vacuum (0.0013-Pa, or 10^{-5}-torr) and high-purity argon atmosphere or gas boriding with H_2-BCl_3-Ar gas mixture is preferred. This results in the formation of a very hard compound layer with a negligible diffusion zone. The outer layer is usually TiB_2; however, other borides may also form. The microhardness readings of boride layers formed on Ti and refractory metals are very high compared to those formed on Co and Ni (see Table 16.18). The wear properties of sintered carbides can be increased by boriding because of the acceptance of boron by soft Co and Ni binders.[272]

Boride layers formed on Ta, Nb, W, Mo, and Ni do not exhibit tooth-shaped morphology as with Ti. When cemented carbides (such as WC-Co wire drawing dies) are commercially pack borided with a powder mixture containing 40% B_4C, 45% SiC, and 5% KBF_4, three distinct zones are found to be formed in the boride layer, comprising the exterior (zone I), intermediate (zone II), and interior (zone III) regions (Table 16.20).[245,251,285]

16.6.3.1   Effects of Alloying Elements.   As in iron and steel, suitable alloying additions raise the hardness of the boride layer formed on these metals; this is presumably caused by the formation of solid-solution borides.

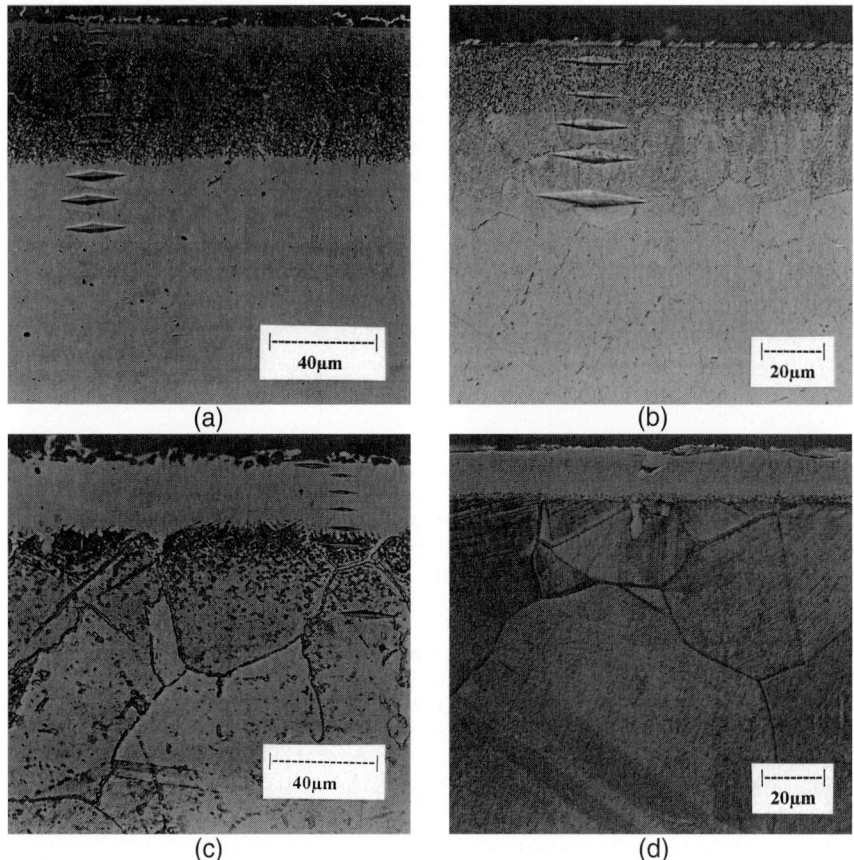

FIGURE 16.62 Photomicrographs of boronized nickel-based alloys such as (a) Hastelloy B (50 μm), (b) Hastelloy C4 (21 μm), (c) Nimonic 80A (30 μm), and (d) Inconel 718 (18 μm) with varying layer thickness in parentheses and hardness values (mentioned above in the text).[284] (*Courtesy of H.-J. Hunger and G. Grute, ESK GmbH.*)

TABLE 16.20 Three Distinct Zones Formed in Borided WC-C Materials[245,251,285]

	Reference 283	Reference 250
Zone I	CoB, W_2B_5, WC	CoB, WC
Zone II	$W_2CoB_2, WCoB, WC$	Co_2B, WC
Zone III	$W_2Co_2B_6, WC, Co$	Co_3B, WC

Additions of alloying elements in Ni, Co, and Ti retard the rate of boride layer growth, and in the case of multiphase layers, the proportion of boride layer containing high boron content (e.g. TiB_2 in Ti) increases. The tooth-shaped morphologies, in the cases of Co and Ti also retarded with alloying addition, and the layers appeared more uniform.[254]

16.6.4 Boriding Techniques

Because extensive investigations have been carried out on boriding of ferrous materials, the boriding described below focuses mainly on ferrous materials.

16.6.4.1 Pack Boriding. Pack boriding[286–288] is the most widely used boriding process because of its relative ease of handling, safety, the possibility of changing the composition of the powder mix, the need for limited equipment, and the resultant economic savings.[243] The process involves packing the annealed, cleaned, smooth parts in a boriding powder mixture contained in a 3- to 5-mm-thick (0.1- to 0.2-in.), heat-resistant steel box so that surfaces to be borided are covered with an approximately 10- to 20-mm-thick (0.4- to 0.8-in.) layer. Many different boriding compounds have been used for pack boriding. They include solid boron-yielding substances, diluents, and activators.

The common boron-yielding substances are boron carbide (B_4C), ferroboron, and amorphous B; the last two have greater boron potential, provide a thicker layer, and are more expensive than B_4C.[288] Silicon carbide (SiC) and alumina (Al_2O_3) serve as diluents, and they do not take part in the reaction. However, SiC controls the amount of B and prevents caking of the boronizing agent. $NaBF_4$, KBF_4, $(NH_4)_3BF_4$, NH_4Cl, Na_2CO_3, BaF_2, and $Na_2B_4O_7$ are the boriding activators. There are special proprietary brands of boriding compounds (powders or granulates) such as Ekabor 1 ($<150\,\mu m$), Ekabor 2 ($<850\,\mu m$), and Ekabor 3 ($<1400\,\mu m$), with various grain sizes (given in parentheses) available on the market that can be used with confidence.[284,289]

Typical compositions of commercial solid boriding powder mixtures are 5% B_4C, 90% SiC, 5% KBF_4; 50% B_4C, 45% SiC, 5% KBF_4; 85% B_4C, 15% Na_2CO_3; 95% B_4C, 5% $Na_2B_4O_7$; 84% B_4C, 16% $Na_2B_4O_7$; amorphous B (containing 95 to 97% B); 95% amorphous B, 5% KBF_4; etc.

The parts conforming to the shape of the container are packed (Fig. 16.63*a*), covered with a lid which rests inside the container, and weighted with an iron slug or stone to ensure an even trickling of the boriding agent during the boriding treatment. It is then heated to the boriding temperature [in the range of 800 to 1000°C (1560 to 1832°F)] for a specified time (1 to 15 hr) in an electrically heated box or pit furnace with open or covered heating coils or a muffle furnace. The container should not exceed 60% of the furnace chamber volume. In principle, boriding should be accomplished in such a way that high internal stresses are relieved, which in turn eliminates cracks or spalling. After boriding treatment, the box containing the workpieces is withdrawn from the furnace and cooled in air. Alternatively, direct quenching of the components in oil or in martempering bath from the boriding temperature, together with tempering, may be carried out.[289] With the packing process, 20 to 50 wt% of fresh boriding powder mixture should be blended with the old after each treatment. In this case, the boriding powder should be discarded after 5 or 6 cycles.

To avoid oxygen-bearing compounds which have a deleterious effect on boronizing, boronizing should be accomplished in a CO-free protective gas atmosphere. This is done either by introducing the packed container into a protective gas retort and heating them in a chamber furnace (Fig. 16.63*b*) or by boronizing directly in a chamber furnace with a protective-gas supply.

The protective gas may be pure argon, pure nitrogen, a mixture of hydrogen and either argon or nitrogen, or, in special instances, pure hydrogen. The component placed in the retort is first flushed with protective gas in order to expel the oxygen. The flow of protective gas must be maintained after boriding until the retort has cooled to about 300°C (570°F).[289]

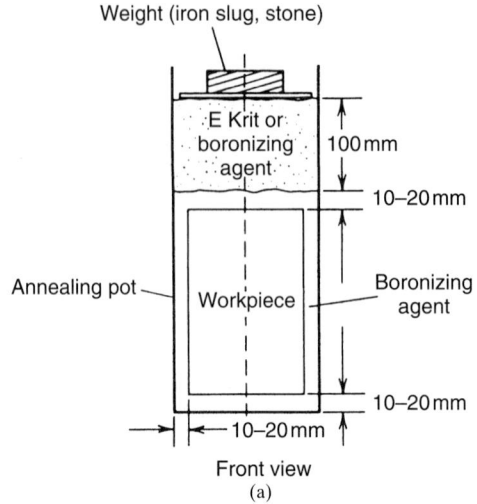

Weight (iron slug, stone)

E Krit or boronizing agent   100 mm

10–20 mm

Annealing pot

Workpiece

Boronizing agent

10–20 mm

10–20 mm

Front view
(a)

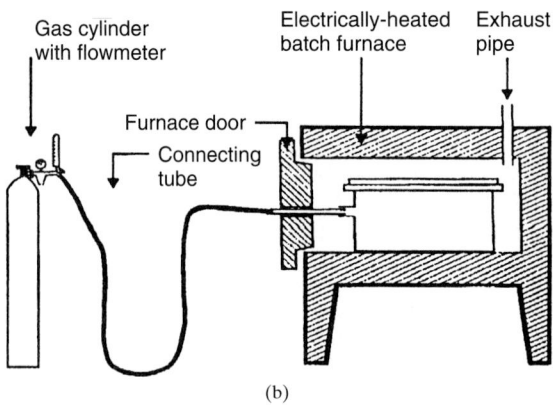

Gas cylinder with flowmeter

Electrically-heated batch furnace

Exhaust pipe

Furnace door

Connecting tube

(b)

FIGURE 16.63 (*a*) Diagram of the packing of a single geometrical part in a pack boriding box.[243] (*Courtesy of A. Graf von Matuschka.*) (*b*) Chamber furnace for pack boriding with protective atmosphere.[289] (*Courtesy of H.-J. Hunger and G. Grute, ESK GmbH.*)

FURNACES.   A large number of furnaces such as retort, pit, chamber, and muffle types can be used for boriding.

CASE DEPTH.   The distance between the surface and the average tooth length has been defined as the case depth.[261] The thickness of the boride layer depends on the substrate material being processed, boron potential of the boronizing compound (Fig. 16.64), boronizing temperature, and time (Fig. 16.65). In ferrous materials, the heating rate, especially between 700°C (1300°F), and the boriding temperature (800 to 1000°C, or 1470 to 1830°F) should be high in order to minimize the formation of FeB.[249]

It is a usual practice to match the case depth with the intended application and base material. As a rule, thin layers (say, 15 to 20 μm) are used for protection against adhesive wear (such as chipless shaping and metal-stamping dies and tools), whereas thick layers are recommended to combat erosive wear (e.g., extrusion tooling for plastics with abrasive fillers and pressing tools for the ceramic industry). The

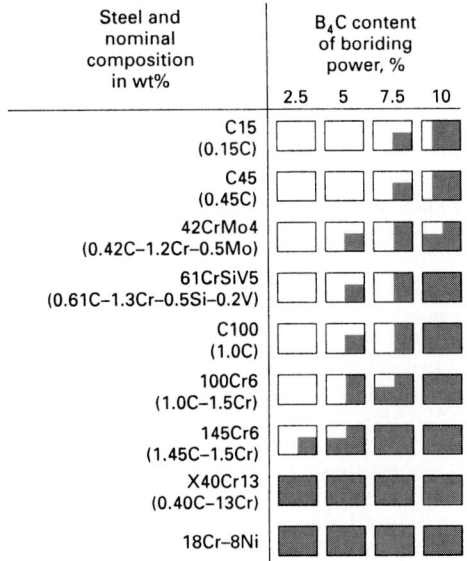

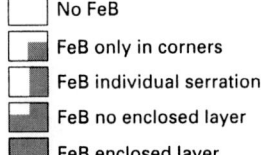

No FeB

FeB only in corners

FeB individual serrations

FeB no enclosed layer

FeB enclosed layer

FIGURE 16.64 Diagram showing the influence of the B_4C content of the boriding powder on the proportion of FeB phase in the boride layer of various steels (borided with pack powder) at 900°C (1650°F) for 5 hr.[254] (*Courtesy of R. Chatterjee-Fischer and O. Schaaber.*)

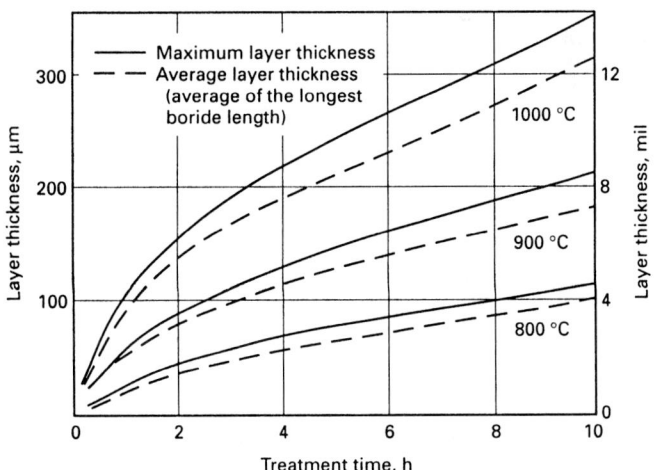

FIGURE 16.65 Effect of pack boriding temperature and time on the boride layer thickness in a low-carbon (Ck 45) steel.[271] (*After W. Fichtl, Oberflaehentechnik mrv Metallpraxis, vol. 11, 1972, p. 434.*)

commonly produced case depths are 0.05 to 0.25 mm (0.002 to 0.01 in.) for low-alloy and low-carbon steels, and 0.025 to 0.076 mm (0.001 to 0.003 in.) for high-alloy steels. However, case depths >0.089 mm (0.0035 in.) are uneconomical for highly alloyed materials such as stainless steel and some tool steels.[272]

BORUDIF PROCESS.   In another modified pack boriding, called the *Borudif process*, steel parts are packed in a 1:4 mixture of B_4C-SiC, and the moderate activator BF_3 plus $(BOF)_3$ gas (formed by passing BF_3 through silica heated to above 450°C, or 840°F) is passed through the pack at 850 to 1100°C (1560 to 2000°F) for 4 hr.[290] The process offers a wide range of boriding potential because of the easy control of $(BOF)_3$ gas concentration, which facilitates the treatment of a wide variety of substrate materials.[290]

16.6.4.2   Paste Boriding.   Paste boriding is used commercially when pack boronizing with powder or granulate is difficult, more expensive, or time-consuming. In this process, the parts are dipped into a paste of 45% B_4C (grain size of 200 to 240 μm) and 55% cryolite (Na_3AlF_6, flux additive),[247] or conventional boronizing powder mixture (B_4C-SiC-KBF_4) in a good binding agent (such as nitrocellulose dissolved in butyl acetate, aqueous solution of methyl cellulose, or hydrolyzed ethyl silicate). Alternatively, the paste is repeatedly applied (i.e., brushed or sprayed) at intervals over the entire or selected portion of parts until, after drying (in a drying oven), a layer about 1 to 3 mm (0.04 to 0.12 in.) thick is obtained. Care should be taken to ensure that no air bubbles form between the part and paste. Subsequently, the ferrous materials are heated (say, at 900°C, or 1650°F, for 4 hr) inductively, resistively, or in a conventional furnace to 800 to 1000°C (1470 to 1830°F) for 5 hr. In this process, a protective atmosphere (for example, 90% N_2/10% H_2, 95% N_2/5% H_2, Ar, cracked NH_3, or N_2) is necessary. A layer in excess of 50-μm thickness may be obtained after inductively or resistively heating to 1000°C (1830°F) for 20 min (Fig. 16.66).[286] This process is of special interest for large components or for those requiring partial (or selective) boriding.[249]

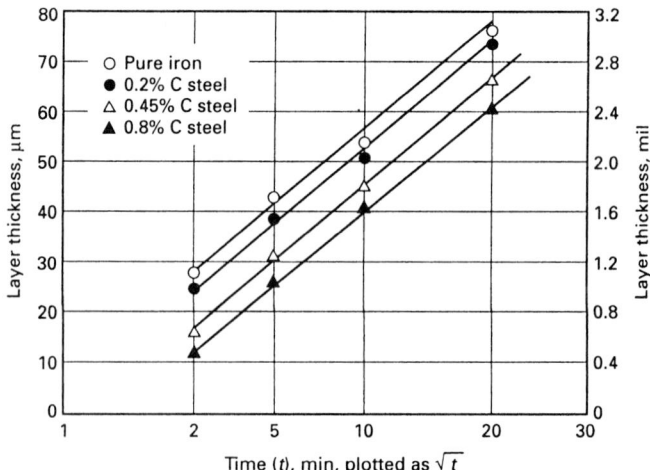

FIGURE 16.66   A linear relationship between boride layer thickness and square root of time for iron and steel boronized with B_4C-$Na_2B_4O_7$-Na_3AlF_6-based paste at 1000°C (1830°F).[286] (*After V. A. Volkov and A. A. Aliev.*)

After the treatment, the removal of paste by brushing, blast cleaning, or washing is necessary. The thickness of the boride layer obtained is a function of the controlled atmosphere. Under identical treating conditions (temperature, time), a much thicker layer is produced in gas than those produced in pack boriding. This means the possibility of reducing treating time up to 50%. On the other hand, pure nitrogen and argon produce a slightly thinner layer than pack boronizing powder or granulate. In this case the boriding time is usually increased by 20%.[261]

16.6.4.3 Liquid Boriding. Liquid boriding is grouped into electroless and electrolytic salt bath processes. These processes have several disadvantages:

1. Removal of excess salt and unreacted boron is essential after the treatment; this step may prove to be expensive and time-consuming.
2. To achieve boronizing reproducibility, bath viscosity is not allowed to increase. This is done by recharging with salt, which involves high maintenance cost.
3. In some cases, protection from corrosive fumes may be required.

ELECTROLESS SALT BATH BORIDING. With ferrous materials this is carried out in a borax-based melt at about 900 to 950°C (1650 to 1750°F), to which about 30 wt% B_4C is added.[291] The boronizing action can be further improved by replacing up to 20 wt% B_4C with ferroaluminum, because it is a more effective reductant. However, superior results have been found by using a salt bath mixture containing 55% borax, 40 to 50% ferroboron, and 4 to 5% ferroaluminum.[292] It has also been shown that a 75:25 KBF_4-KF salt bath can be used at temperature below 670°C (1240°F) for boronizing nickel alloys, and at higher temperatures for ferrous alloys, to develop desired boride layer thickness.

ELECTROLYTIC SALT BATH BORIDING. In this process, the ferrous part acting as the cathode and a graphite anode are immersed in the electrolytic molten borax at 940°C (1720°F) for 4 hr using a current density of about 0.15 A/cm^2.[293] The parts are then air-cooled. In general, the parts are rotated during the treatment to obtain a uniform layer. A high current density produces a thin coating on low-alloy steels in a short time. For high-alloy steels of greater thickness, lower current densities are required for a longer time.[294]

In the fused state, tetraborate decomposes into boric acid and nascent oxygen.

$$B_4O_7 + 2e = 2B_2O_3 + O \qquad (16.48)$$

Simultaneously, sodium ions, after being neutralized in the vicinity of the cathode, react with boric acid to liberate boron.

$$6Na + B_2O_3 = 3Na_2O + 2B \qquad (16.49)$$

In this manner, a high boriding potential is established near the cathode region. Other satisfactory electrolytic salt bath compositions include (*a*) KBF_4-LiF-NaF-KF mixture for parts to be treated at 600 to 900°C (1100 to 1650°F);[294] (*b*) 20KF-30NaF-50LiF-0.7BF_2 mixture (by mole %) at 800 to 900°C (1470 to 1650°F) in 90N_2-10H_2 atmosphere; (*c*) 9:1 (KF-LiF)-KBF_4 mixture under Ar atmosphere;[295] (*d*) KBF_4-NaCl mixture at 650°C (1200°F);[296] (*e*) 90(30 LiF + 70 KF)-10KBF_4 mixture at 700 to 850°C (1300 to 1560°F); and (*f*) 80$Na_2B_4O_7$-20NaCl at 800 to 900°C (1470 to 1650°F).[297]

16.6.4.4 Gas Boriding. Gas boriding may be accomplished with (*a*) diborane (B_2H_6)-H_2 mixture; (*b*) boron halide-H_2 or (75:25 N_2-H_2) gas mixture; or (*c*) organic boron compounds such as $(CH_3)_3B$ and $(C_2H_5)_3B$.

Boronizing with B_2H_6-H_2 mixture is not commercially viable due to the highly toxic and explosive nature of diborane.[243,244] When organic boron compounds are used, carbide and boride layers form simultaneously. Because BBr_3 is expensive and is difficult to handle (with violent reactions with water), and because BF_3 requires high reduction temperature (due to its greater stability) and produces HF fumes, BCl_3 appears to be the only choice for gas boriding.

When parts are gas borided in a dilute $(1:15)$ BCl_3-H_2 gas mixture at a temperature of 700 to 950°C (1300 to 1740°F) and a pressure up to 67 kPa (0.67 bar), a boride layer 120 to 150 μm thick is reported to be produced at 920°C (1690°F) in 2 hr.[298] Recent work has suggested the use of 75:25 N_2-H_2 gas mixture instead of H_2 gas for its better performance because of the production of boride layers with minimum FeB content. The latter phase can be easily eliminated during the subsequent diffusion treatment before hardening.[299] This process can be applied to titanium and its alloys as well.

Disadvantages of gas boronizing with BCl_3 include (1) a tendency of retort material to become borided, (2) an acute problem of corrosion in coating plant due to the formation of chloride, and (3) relatively high consumption of BCl_3 due to conducting the process at atmospheric pressure. In view of these drawbacks, recently gas boronizing with BCl_3 has become an unattractive choice.[300]

16.6.4.5 Plasma Boriding. Both mixtures of B_2H_6-H_2 and BCl_3-H_2-Ar may be used successfully in dc or rf plasma boronizing.[301,302] However, the former gas mixture can be applied to produce boride layer on various steels at relatively low temperatures such as 600°C (1100°F), which is impossible with a pack or liquid boronizing process.[303] It has been claimed that plasma boriding in a mixture of BCl_3-H_2-Ar shows good results, such as better control of BCl_3 concentration, reduction of the discharge voltage, and higher microhardness of the boride films.[304]

The dual-phase layer is characterized by visible porosity, occasionally associated with a black boron deposit. This porosity, however, can be minimized by increasing the BCl_3 concentration. Boride layers up to 200 μm in thickness can be produced in steels after 6 hr of treatment at a temperature of 700 to 850°C (1300 to 1560°F) and a pressure of 270 to 800 Pa (2 to 6 torr).[305]

Advantages of this process are

1. Increased boron potential compared to conventional pack boronizing
2. Control of composition and depth of the borided layer
3. Finer plasma-treated boride layers
4. Reduction in temperature and duration of treatment
5. Elimination of high-temperature furnaces and their accessories
6. Savings in energy and gas consumption

The disadvantages of the process are that (1) the extreme toxicity of the diborane atmosphere presents severe handling problems under experimental conditions and (2) all the disadvantages associated with gas boronizing with BCl_3 (including its highly corrosive effect) are also applicable to plasma boronizing with BCl_3. That is why this process has not gained commercial acceptance.[300] (However, the corrosive action of the substrate surface can be suppressed by stepping up the processing temperature.[306]) To avoid the above shortcomings, boriding from paste (containing a mixture of 60% amorphous boron and 40% liquid borax) in a glow discharge at the impregnating temperature has been developed, which is found to greatly increase the formation of the surface boride layer.[307]

The pulsed-plasma process is based on the principle of generating a high-current glow discharge in equipment capable of operating at temperatures up to 1000°C. The glow discharge occurs at a pressure of several millibars by applying a dc voltage of 400 to 600 V between the walls of the chamber and the workpiece, which acts as the cathode. The pulsed-plasma boronizing produces an optimum spread even for geometrically complex surfaces and the controlled energy input across a wide range via the pulse periods and pulse/pause ratio. The heat required for heating the substrate up to a specified temperature can largely be supplied from an auxiliary heating circuit.[300]

A pulsed-plasma boronizing with BF_3 has been applied on steel and nickel-based alloys with much less severe problem and cost than with using BCl_3. It has been reported to eliminate pores in the steel substrate.[306]

16.6.4.6 Fluidized-Bed Boriding. The boriding in fluidized beds may be accomplished with bed materials of coarse-grained silicon carbide particles, a special boriding powder such as Ekabor WB, and an O_2-free gas such as N_2-H_2 mixture (Fig. 16.67).[246,308] When electricity is used as the heat source, the bed serves as a faster heat-transfer medium. This is usually equipped with quench and tempering furnaces.

This process offers the following advantages:

1. High rates of heating and flow, as well as direct withdrawal of the parts, provide shorter operating cycle times (i.e., rapid boronizing).
2. Temperature uniformity with low capital cost and flexibility is ensured.
3. A fluidized furnace is very tight because of upward pressure of the gas.
4. This process produces reproducibility, close tolerances, and a very uniform finish on mass-produced parts.
5. This process can be adaptable to continuous production and can lend itself to automation as the parts are charged and withdrawn intermittently.
6. Quenching (and subsequent tempering) of the parts directly after this treatment is possible.

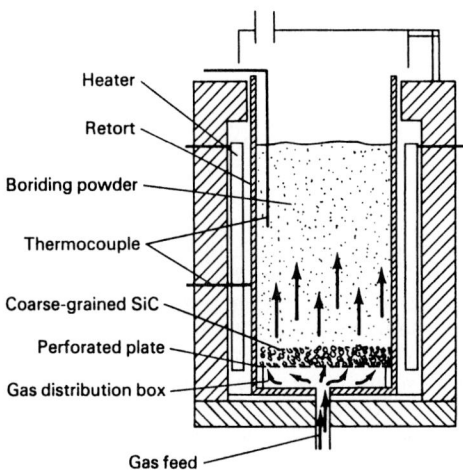

Heater

Retort

Boriding powder

Thermocouple

Coarse-grained SiC

Perforated plate

Gas distribution box

Gas feed

FIGURE 16.67 Diagram showing fluidized-bed boriding facility.[246]

7. Operating cost is low (due to reduced processing time and energy consumption) for mass production of boronized parts.

An important disadvantage lies in the continuous flushing of the boriding agent within the retort by the inert gas. The exhaust gases containing enriched fluorine compounds must be cleaned absolutely, e.g., in an absorber filled with dry $CaCO_3$ chips to avoid environmental problems.[279] Alternatively, pulsed fluidizing process can considerably decrease the amount of exhaust gases.[279]

16.6.4.7 *Multicomponent Boriding.*

Multicomponent boriding is a thermochemical treatment involving consecutive diffusion of boron and one or more metallic elements such as aluminum, silicon, chromium, vanadium, and titanium into the component surface. This process is carried out at 850 to 1050°C (1560 to 1920°F) and involves two steps:

1. Boriding is accomplished by conventional methods, notably pack, paste, and electrolytic salt bath methods.[309] Here the presence of FeB is tolerated and, in some cases, may prove beneficial. Among these methods, much work has been done on the pack method (Table 16.21),[244,309–311] which produces a compact layer at least 30 μm (1 mil) thick.

2. Metallic elements are diffused through the powder mixture or borax-based melt into the borided surface. If the pack method is used, sintering of particles can be avoided by passing argon or H_2 gas into the reaction chamber.

There are six multicomponent boronizing methods:[245,309] boroaluminizing, borosiliconizing, borochromizing, borochromtitanizing, borochromvanadizing, and borovanadizing.

Boroaluminizing layers can be obtained by means of boroaluminizing at 850 to 1000°C (1560 to 1830°F) for 3 to 5 hr where the pasty agent contains mainly boron carbide, aluminum, and potassium borofluoride. The boroaluminized layers have a microhardness of 1500 to 2500 $HV_{0.1}$.[312] When boroaluminizing involves boriding followed by aluminizing, the compact layer formed in steel parts provides good wear and corrosion resistance, especially in humid environments.[249,275,309]

Borosiliconizing results in the formation of FeSi in the surface layer, which enhances corrosion fatigue strength of treated parts.[311]

Borochromizing (involving chromizing after boriding) provides better oxidation resistance than boroaluminizing, the most uniform layer (probably comprising a solid solution boride containing iron and chromium), improved wear resistance compared with traditionally borided steel, and enhanced corrosion-fatigue strength. In this case, a post-heat treatment operation can be safely accomplished without a protective atmosphere.[244,313,314]

Borochromtitanized structural alloy steel provides high resistance to abrasive wear and corrosion as well as extremely high surface hardness of 5000 HV (15-g load).[309] Figure 16.68 shows the microstructure of the case of a borochromtitanized constructional alloy steel part exhibiting titanium boride in the outer layer and iron-chromium boride beneath it.

Borovanadized and *borochromvanadized* layers are quite ductile with their hardnesses exceeding 3000 HV (15-g load). This reduces drastically the danger of spalling under impact loading conditions.[309]

TABLE 16.21 Multicomponent Boriding System[245]

Reference	Multicomponent boriding technique	Medium type	Medium composition(s), wt%	Process steps investigated[†]	Substrate(s) treated	Temperature, °C (°F)
310	Boroaluminizing	Electrolytic salt bath	3–20% Al_2O_3 in borax	S	Plain carbon steels	900 (1650)
311	Boroaluminizing	Pack	84% B_4C + 16% borax 97% ferroaluminum + 3% NH_4Cl	S B-Al Al-B	Plain carbon steels	1050 (1920)
44	Borochromizing	Pack	5% B_4C + 5% KBF_4 + 90% SiC (Ekabor II) 78% ferrochrome + 20% Al_2O_3 + 2% NH_4Cl	S B-Cr Cr-B	Plain carbon steels	Borided at 900 (1650) Chromized at 1000 (1830)
44	Borosiliconizing	Pack	5% B_4C + 5% KBF_4 + 90% SiC (Ekabor II) 100% Si	B-Si Si-B	0.4% C steel	900–1000 (1650–1830)
44	Borovanadizing	Pack	5% B_4C + 5% KBF_4 + 90% SiC (Ekabor II) 60% ferrovanadium + 37% Al_2O_3 + 3% NH_4Cl	B-V	1.0% C steel	Borided at 900 (1650) Vanadized at 1000 (1830)

[†] S, simultaneous boriding and metallizing; B-Si, borided and then siliconized; Al-B, aluminized and then borided; Si-B, siliconized and then borided; B-Cr, borided and then chromized; Cr-B, chromized and then borided; B-V, borided and then varadized.

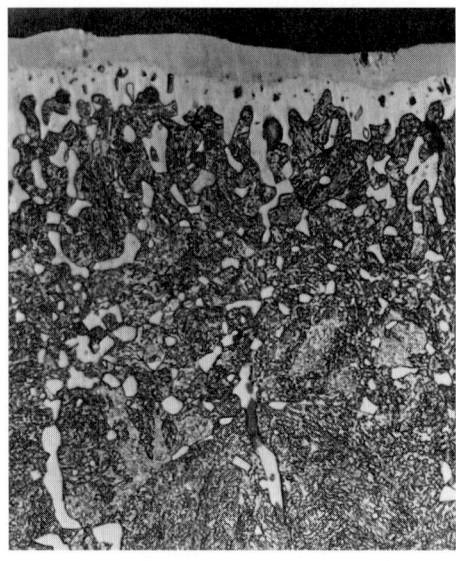

FIGURE 16.68   Microstructure of the case of a borochromtitanized construction alloy steel.[249,309] (*Courtesy of R. Chatterjee-Fischer.*)

16.6.5 Applications of Boriding

Recently borided parts have been used in a wide variety of industrial applications (Table 16.22) because of the numerous advantageous properties of boride layers. In sliding and adhesive wear situations, boriding is applied to

1. Spinning steel rings, steel rope, steel thread guide bushings (made of DIN St 37 steel)
2. Grooved gray cast iron drums (thread guides) for textile machinery
3. Four-holed feedwater regulating valves (made from DIN 1.4571 steel or AISI 316 Ti steel)
4. Burner nozzles, swirl elements, and injector tops (for oil burners) (made from DIN Ck 45 steel) in the chemical industry[246]
5. Drive, worm, and helically toothed gears (made from AISI 4140 steel) in various high-performance vehicle and stationary engines[245]
6. Tools for cold-forming of metals such as aluminum and copper[284]
7. Plungers made from D3 steel for glass manufacture which is less expensive than that of stellite-reinforced plungers[284]

As abrasive wear-resistant materials, borided stainless steels are used for parts such as screw cases and bushings, perforated and slotted hole screens, rollers, valve components, fittings, guides, shafts, and spindles, and borided Ti-6Al-4V for parts such as leading-edge rotor blade cladding for helicopter applications. Other applications in this category include:

1. Nozzles of bag-filling equipment
2. Extruder screws, cylinders, nozzles, and reverse-current blocks in plastic production machinery (extruder and injection-molding machinery)[243]

TABLE 16.22 Proven Applications for Borided Ferrous Materials[246,271]

AISI	BSI	DIN	Application
		St37	Pipe bends, baffle plates, bushes, bolts, nozzles, conveyer tubes, pins for pinned disk mills, base plates, runners, blades, thread guides
1020	. . .	C15 (Ck15)	Gear drives, pump shafts
1043	. . .	C45	Pins, guide rings, grinding disks, bolts
		St50-1	Casting inserts, nozzles, handles
1138	. . .	45S20	Shaft protection sleeves, mandrels
1042	. . .	Ck45	Swirl elements, nozzles (for oil burners), rollers, bolts, gate plates
		C45W3	Gate plates
W1	. . .	C60W3	Clamping chucks, guide bars
D3	. . .	X210Cr12	Bushes, press tools, plates, mandrels, punches, dies
C2	. . .	115CrV3	Drawing dies, ejectors, guides, insert pins
		40CrMnMo7	Gate plates, bending dies
H11	BH11	X38CrMoV51	Plungers, injection cylinders, sprue
H13	. . .	X40CrMoV51	Orifices, ingot molds, upper and lower dies and matrices for hot forming, disks
H10	. . .	X32CrMoV33	Injection-molding dies, fillers, upper and lower dies and matrices for hot forming
D2	. . .	X155CrVMo121	Threaded rollers, shaping and pressing rollers, pressing dies and matrices
		105WCr6	Engraving rollers
D6	. . .	X210CrW12	Straightening rollers
S1	~BS1	60WCrV7	Press and drawing matrices, mandrels, liners, dies, necking rings
D2		X165CrVMo12	Drawing dies, rollers for cold mills
L6	BS224	56NiCrMoV7	Extrusion dies, bolts, casting inserts, forging dies, drop forges
		X45NiCrMo4	Embossing dies, pressure pad and dies
O2	~BO2	90MnCrV8	Molds, bending dies, press tools, engraving rollers, bushes, drawing dies, guide bars, disks, piercing punches
E52100	. . .	100Cr6	Balls, rollers, guide bars, guides
		Ni36	Parts for nonferrous metal casting equipment
		X50CrMnNiV229	Parts for unmagnetizable tools (heat-treatable)
4140	708A42 (En19C)	42CrMo4	Press tools and dies, extruder screws, rollers, extruder barrels and cylinders, plungers, rings, guides non-return valves
4150	~708A42 (CDS-15)	50CrMo4	Non-return valves, dies
4317	. . .	17CrNiMo6	Bevel gears, screw and wheel gears, shafts, chain components
5115	. . .	16MnCr5	Helical-gear wheels, guide bars, guiding columns
6152	. . .	50CrV4	Thrust plates, clamping devices, valve springs, spring contacts
302	302S25 (En58A)	X12CrNi188	Screw cases, bushes
316	~316S16 (En58J)	X5CrNiMo1810	Perforated or slotted hole screens, parts for the textile and rubber industries
		G-X10CrNiMo189	Valve plugs, parts for the textile and chemical industries
410	410S21 (En56A)	X10Cr13	Valve components, fittings
420	~420S45 (En56D)	X40Cr13	Valve components, plunger rods, fittings, guides, parts for chemical plants
		X35CrMo17	Shafts, spindles, valves
Gray and ductile cast iron			Parts for textile machinery, mandrels, molds, sleeves

3. Bends and baffle plates for conveying equipment for mineral-filled plastic granules in the plastics industry

4. Punching dies (for making perforations in accessory parts for cars), press and drawing matrices, and necking rings (made from S1 tool steel)

5. Press dies, cutting templates, punched plate screens (made of DIN St 37 steel)

6. Screw and wheel gears, bevel gears (from AISI 4317 steel)

7. Steel molds (for the manufacture of ceramic bricks and crucibles in the ceramics industry), extruder barrels, plungers and rings (from 4140 steel)

8. Extruder tips, nonreturn valves, and cylinders (for extrusion of abrasive minerals or glass fiber-reinforced thermoplastics, from 4150 steel)

9. Plungers for the glass industry

10. Casting fillers for processing nonferrous metals (from AISI H11 steel)

11. Transport belts for lignite coal briquettes

12. Parts for textile machinery such as thread guides, fins, and drive rollers[260]

Borided parts also find applications in die casting molds, bending blocks, wire-draw blocks, pipe clips, pressing and shaping rollers, straightening rollers, engraving rollers, rollers for cold mills, mandrels, press tools, bushings, guide bars, disks, casting inserts, various types of dies including cold heading, bending, extrusion, stamping, pressing, punching, thread rolling, hot forming, injection molding, hot forging, drawing, embossing, and so on in A2, A6, D2, D6, H10, H11, O2, and other tool steels.[288]

Borided steel parts have also been used as transport pipe for molten nonferrous metals such as aluminum, zinc, and tin alloys (made from DIN St 37), corrosion-resistant transport pipe elbows for vinyl chloride monomer, grinding disks (made from DIN Ck 45), die-casting components, air foil erosion-resistant cladding, data printout components (e.g., magnetic hammers, wire printers), and engine tappets.[255]

Boronized Permalloy is used for magnetic head applications. Boronized cemented carbides are used as drawing dies, guiding parts, and dimensional measurement parts. Some examples of multicomponent boriding include improvement of the wear resistance of austenitic steels (borochromizing), of parts for plastics processing machines (borochromtitanizing), and of dies used in ceramic industry (borochromizing).[309]

16.7 THERMOREACTIVE DEPOSITION/DIFFUSION (TRD) PROCESS

Thermoreactive deposition/diffusion (TRD), also called *carbide coating*, is a method of case hardening/coating of steel parts with a hard, wear-resistant layer of carbides, nitrides, or carbonitrides that is carried out by salt bath, fluidized-bed, or powder pack processing.[315,316]

Salt Bath TRD Process. The salt bath immersion and fluidized-bed processes, developed by Arai of the Toyota Organization, Japan, in 1968,[317] were previously called the Toyota Diffusion (TD) coating processes.[318] In the salt bath TRD process, the steel parts are first preheated at 500 to 700°C (932 to 1292°F) to minimize distortion and reduce the processing time, then heated in an electric salt bath or gas-heated furnace comprising molten (carbon-free) borax ($Na_2B_4O_7$ or $Na_2O \cdot 2B_2O_3$) and appropriate carbide-forming additives and reducing agents, such as boron carbide, aluminum,[†] etc., at 850 to 1050°C (1560 to 1920°F) for 0.5 to 10 hr. The bath temperature is selected to conform to the hardening temperature of the substrate steel. The immersion time is determined according to the desired layer depth and substrate (Fig. 16.69).[315,320] Carbide-forming additives usually take the form of appropriate ferroalloy powders or oxides of carbide-forming elements. Thus, for VC coatings, ferrovanadium or vanadium pentoxide (V_2O_5) can be used as additive.[319] Currently, ferroalloy powder is not being used.[321] The formation of refractory carbide (for example, VC, NbC, TaC, TiC, Mn_5C_2, Cr_7C_3, and $Cr_{23}C_6$)[322] (and/or nitride) of increasing thickness (2 to 20 μm, or 0.08 to 0.8 mil) occurs on the substrate surfaces by a reaction between the carbide-forming (and/or nitride-forming) elements such as Ti, V, Ta, Nb, Mo, W, and Cr dissolved in the molten borax bath and carbon (and nitrogen) atoms diffusing from the substrate into a deposited layer[323] (Fig. 16.70). After TRD processing, the steel parts are quenched in air, salt, or oil, followed by a single- or double-tempering at 150 to 200°C or 500 to 650°C (300 to 390°F or 930 to 1200°F) for better dimensional stability and minimum distortion.[324] High-speed steels and other steels with austenitizing temperature in excess of 1050°C (1920°F) are post-TRD heat-treated in vacuum, gas, or protective salt to achieve the full substrate hardness.[315] A low-temperature TRD coating, which consists of preliminary nitriding followed by immersion into a low-temperature salt bath at nitriding and carbonitriding temperatures, has also been applied.[316,321]

Hard carbide coating is applicable to all types of carbon-containing materials such as high-carbon structural steels, tool steels (say, D2, W1, and M2 types), P/M high-speed steels, stainless steels, cobalt alloys, cemented carbides, carbide-metal cermets, carbide ceramics, and carbon. Prior carburization of carbon-deficient metals such as low-carbon steels and nickel alloys is required for the application of the carbide coating. Nitrided steels can be used for the application of carbonitride coating. The formation of nitride coating can occur on nitride ceramics.[325]

The carbide coating thickness d (cm) is a function of bath temperature T (K), immersion time t (s), composition of substrate steel, carbide type, and section thickness of the components processed. However, the relation between d, T, and t is given by[318,322]

$$\frac{d^2}{t} = K = K_0 \exp\left(-\frac{Q}{RT}\right) \tag{16.50}$$

where K is the layer growth rate constant (cm^2/s), K_0 is the constant term of K (cm^2/s), Q is the activation energy (kJ/mol), and R is the gas constant. Figure 16.69 shows the relation between the thickness of the vanadium carbide and niobium carbide layers formed on W1 and H13 steels, respectively, versus borax bath temperature and immersion time.[315,320]

Advantages.   Numerous advantages are claimed by this process.[324,326,327]

1. Simplicity of equipment, easy operation, easy change of types of carbide coating, and no need for a protective atmosphere[320]

2. Very short immersion time (<1 min) to coat sufficiently small specimens[328]

3. Ability of through-hardening by using suitable quenchant after removal from the bath

---

[†] Aluminum also acts as a deoxidant and prevents excessive oxidation.[319]

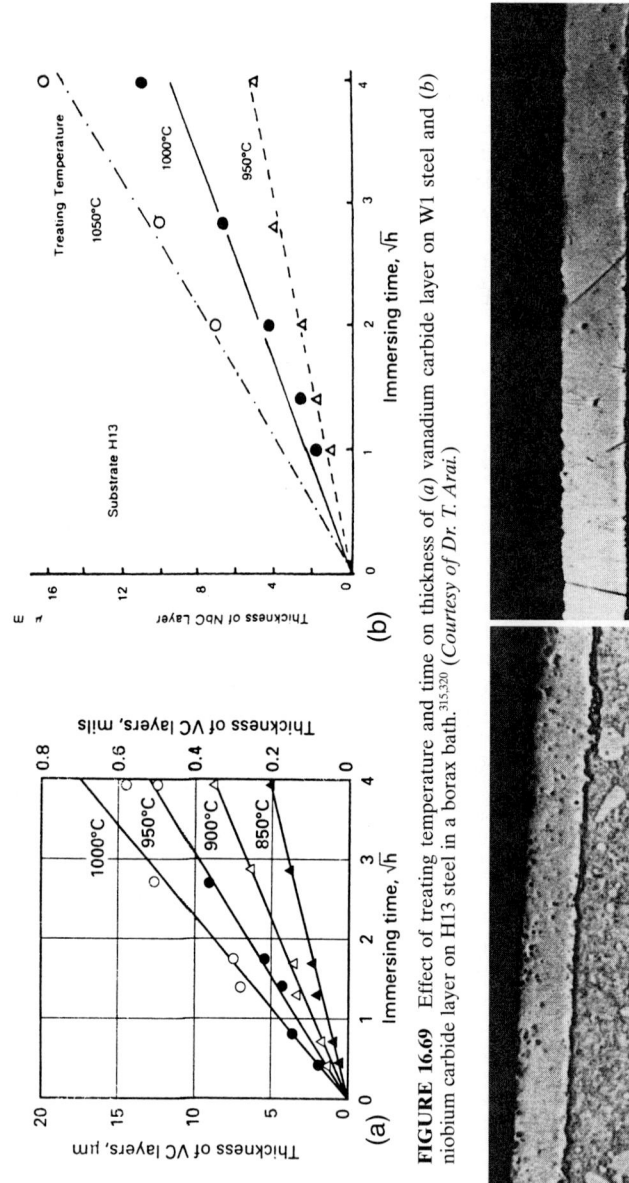

FIGURE 16.69 Effect of treating temperature and time on thickness of (a) vanadium carbide layer on W1 steel and (b) niobium carbide layer on H13 steel in a borax bath.[315,320] (*Courtesy of Dr. T. Arai.*)

FIGURE 16.70 The vanadium carbide layer formed by the TRD process on (a) D2 steel substrate[321] and (b) H13 steel substrate with chromium plating for edge protection.[321] (*Courtesy of T. Arai.*)

16.140

4. Exceptionally high surface hardness (between 1200 and 4000 DPH) without any remarkable change in fatigue strength, fatigue life, or toughness[323,324,329]

5. Excellent resistance to abrasive wear, adhesive wear, fretting wear (equivalent to those produced by the CVD/PVD processes),[315,317] and thermal shock

6. Superior galling resistance to cold forging[326]

7. Much better seizure properties against many running materials, including aluminum[320]

8. Improved corrosion and oxidation resistance, the latter being found at temperatures up to 800°C in the presence of Cr_7C_3 carbides

9. Very good resistance to attack by molten aluminum[320]

10. Absence of cracks, spalls, peels, or flakes, even after oil-quenching, provided steel and processing conditions are optimal together with surface conditions

11. Development of compressive residual stress in the VC and TiC coated D2 steel substrate[330]

12. Selective coating possible by masking with copper, stainless plating, thermal spraying, or wrapping with foils[331]

13. Chemical removal of carbide layer from the area not to be coated[331]

14. Similar applications to those of PVD and CVD coatings[332]

15. Easy repair or retreatment of worn areas possible without removing the sound carbide[315,320]

16. Long bath life

17. Low operating cost

Disadvantage. Being a high-temperature processing treatment, it tends to induce some distortion, especially in large and complex-shaped parts.[320] However, it can be reduced to within $\pm 5\,\mu m$ by judicious control and selection of substrate materials and pretreat-, TRD-, and post-TRD treatments.[333]

Applications. Unlike PVD and CVD coatings which are mostly limited to tool applications, this process has been successfully applied 10 times more on machine components (made of construction steels) in various fields of production machines and various automobile components, pump components, etc., than on tooling applications.[317,332] Table 16.23 lists the applications of TRD processed tooling and machine components.[321]

Fluidized-Bed TRD Process. The fluidized-bed TRD (carbide coating) process consists of introducing clean carbon-containing materials into the mixture of Al_2O_3 powder (~80-mesh) as a major fluidizing powder, Fe-Cr or Fe-V powder (100- to 200-mesh) as carbide-forming element sources, and NH_4Cl pellet as an activator, which were held in a fluidized condition at a temperature of 1073 to 1300 K in an argon gas atmosphere.[327] The substrate is then hardened by various quenching methods.[334]

Advantages. (1) The fluidized carbide coated materials exhibit excellent tribological properties and superior adhesion and high corrosion resistance compared to CVD coated or salt bath TRD process.[334] (2) The method is simple and clean and is much less expensive than the conventional coating process. (3) It has the ability for continuous automation of the whole process, starting from preheating to tempering, leading to increased productivity.[334]

TABLE 16.23 Applications of TRD-Processed Tooling and Machine Components[321]

Application	Tool or machine components	Related products or industries
Sheet metal stamping	Bending dies, drawing dies, dies for various kinds of forming, forming rollers, blanking dies, perforating punches, shearing blades, cam plates, guide blocks, pilot pins, etc.	Automobile components, bicycle components, various machine parts, home appliances, etc.
Cold forging	Piercing punches, hexagonal trimming dies, punches for cross-recessed head screws, rotary swaging dies, extrusion punches and dies, upsetting punches and dies, sizing punches and dies, ironing punches and dies, heading punches, knock-out pins, roll-forming rolls, cropping knives, etc.	Automobile components, bicycle components, bolts and nuts, various machine parts, etc.
Hot forging	Press-forging dies, closed die forging dies, upsetting dies, roll-forming rollers, extrusion dies, upsetting dies, cropping knives, etc.	Automobile components, bicycle components, bolts and nuts, various machine parts, etc.
Warm forging	Piercing punches, rotary swaging dies, extrusion punches and dies, upsetting punches and dies, sizing punches and dies, ironing punches and dies, knock-out pins, cropping knives, etc.	Automobile components, bicycle components, various machine parts, etc.
Wire making	Drawing dies made of cemented carbides, rolling rollers, shearing blades, peeling dies, etc.	Tire steel cord, electric wire, telegram wire, welding wire, various wires, etc.
Wire forming	Bending dies, winding mandrels, shearing blades, etc.	Coil spring, wire brush, stationary, etc.
Powder metallurgy	Compacting dies, center cores, shape forming dies, cutting knife for formed refractory products, mixing rods, plug gauge for size measuring, etc.	Various P/M products, magnets, insulators, etc.
Aluminum extrusion	Extrusion dies, pipe extrusion mandrels, etc.	Door sash, aluminum pipes, etc.
Casting	Cores and core pins for aluminum and zinc die casting, cavity inserts, sprue bushings, sprue spreaders, squeeze pins and bushings, gravity casting pins, deburring tools, wear parts for sand molding, etc.	Automobile components, bicycle components, various machine parts, home appliances, etc.

Application	Examples
Plastic and rubber forming	Extrusion dies, molding dies, core pins, cores, die plates, rollers for films, various cutting tools, breaker plates, pellet knives, various types of cut-off knives, injection sleeves, injection nozzles, small injection screws, extrusion screws, extrusion sleeves, screw tips, check rings, non-return valves, sprue bushings, torpedo, hot runner tips, guide for steel inserts, etc.	Automobile components, bicycle components, various machine parts, home appliances, etc.
Lumber making	Cutting tools, etc.	
Paper processing	Chipper knives, cut-off knives, scissors, rollers for corrugated cardboard, etc.	
Textile/leather processing	Cutoff knives, scissors, etc.	
Agriculture	Cutting blades for cropping rice, etc.	
Pipe forming	Clamping parts, etc.	
Glass molding	Sealing cones for blow molding, conveyor parts for glass bottle, etc.	
Cutting and grinding	Drilling bushes, turning centers, contact plates for checking tool failure, grinding centers, positioning parts, supporting blocks, grinding stone holders, sand blasting nozzles, etc.	
Assembling	Clamping tips, locating pins, wire winding guides, jaws of riveting machine, tips for ultrasonic honing, drive bits, socket wrenches, hallmarking punches, etc.	
Food making	Extrusion plates, parts of homogenizer, knives, etc.	
Automobile	Roller chain components, distributor components, etc.	Automobiles, motorcycles, bicycles
Pumping	Vanes, plungers, cylinders, spray nozzles, side plate of torchoid pump, valves, nozzles	Corrosive liquid, abrasive liquid, high pressure, etc.
Machinery	Various wear parts, etc.	

Courtesy of T. Arai.

Application. The fluidized-bed carbide coating process has been applied to other kinds of tooling and components of some production machines such as metal stamping punches and die-casting core pins.[334]

REFERENCES

1. A. J. Hicks, *Met. Mater. Technol.*, vol. 15, no. 7, 1983, pp. 325–330.
2. C. M. Klaren and J. Nelson, in *ASM Handbook*, vol. 4: *Heat Treating*, 10th ed., ASM International, Materials Park, Ohio, 1991, pp. 454–461.
3. W. R. Hain, *Heat Treating*, vol. 18, no. 8, 1986, pp. 35–37.
4. M. M. Sirrine, private communication, 2000.
5. K. Boiko, *Heat Treating*, vol. 17, no. 8, 1985, pp. 18–21.
6. T. Ruglic, in *ASM Handbook,* vol. 4: *Heat Treating*, 10th ed., ASM International, Materials Park, Ohio, 1991, pp. 268–285.
7. H. W. Gronegress, *Flame Hardening*, Springer-Verlag, Berlin, 1964.
8. K. E. Thelning, *Steel and Its Heat Treatments*, Butterworths, London, 1985.
9. M. M. Sirrine, *Ind. Heat.*, March 1998, pp. 71–74.
10. G. R. Speich, in *Encyclopedia of Materials Science and Engineering*, Pergamon Press, Oxford, 1986, pp. 2285–2288.
11. P. D. Jenkins, *Metallurgia*, vol. 45, no. 4, 1978, pp. 196–199.
12. *Metals Handbook*, 9th ed., ASM, Metals Park, Ohio, 1981.
13. J. Nelson, private communication, 2000.
14. M. M. Sirrine, *Proceedings of 16th ASM Heat Treating Society Conference and Exposition*, 19–21 March 1996, Cincinnati, Ohio, 1996, pp. 497–500.
15. S. Zinn, *Heat Treating*, vol. 16, no. 4, 1984, pp. 20–24.
16. S. L. Semiatin and D. E. Stutz, *Induction Heat Treatment of Steel*, ASM, Metals Park, Ohio, 1986; S. Zinn and S. L. Semiatin, *Elements of Induction Heating*, EPRI, Calif., 1988.
17. *Guide to Induction Heat Treating Equipment*, Inductoheat, Madison Heights, Mich., 1988.
18. V. I. Rudnev, R. L. Cook, D. L. Loveless, and M. R. Black, *Steel Heat Treatment Handbook*, eds. G. E. Totten and M. A. H. Howes, Marcel Dekker, Inc., 1997, chap. 11A, pp. 765–871
19. H. B. Osborn, *Met. Prog.*, vol. 68, no. 6, 1955, pp. 105–109.
20. T. H. Spencer et al., *Induction Hardening and Tempering*, ASM, Metals Park, Ohio, 1964.
21. P. A. Hassell and N. V. Ross, in *ASM Handbook,* vol. 4: *Heat Treating*, 10th ed., ASM International, Materials Park, Ohio, 1991, pp. 164–202.
22. V. S. Nemkov and R. C. Goldstein, private communication, 2000.
23. P. K. Bhargava and K. M. Joseph, *Tool Alloy Steels*, June 1979, pp. 195–198, 200–203.
24. M. Melander, Y. Shanchang, and T. Ericsson, in *Heat Treatment 1983*, The Metals Society, London, pp. 2.75–2.85.
25. R. Priestner, *Met. Mater.*, vol. 8, no. 4, 1974, pp. 225–228.
26. K. D. Spain, *The 1st International Automotive Heat Treating Conference Proceeding*, eds. R. Coals, K. Funatani, and C. A. Stickels, ASM International, Materials Park, Ohio, 1998, pp. 319–335.
27. *Inductoheat Bulletin*, November 1988.
28. J. M. Storm and M. R. Chaplin, *Gear Technol.*, vol. 10, no. 2, 1993, pp. 22–25.

29. A. F. Leatherman and D. A. Stutz, *Met. Treating*, vol. 21, no. 2, 1970, pp. 3–12.

30. C. L. Kirk, *Met. Prog.*, vol. 94, no. 1, 1968, pp. 68–70.

31. A. M. Gutherie and K. C. Archer, *Heat Treat. Met.*, no. 1, 1975, pp. 15–21.

32. B. Harrison, *Metallurgia*, vol. 45, no. 3, 1978, pp. 145–146.

33. J. Grum, *8th International Induction Heating Seminar Proceedings*, Nov. 3–6, 1998, Kissimmee, Fla., Inductoheat, Inc., Madison Heights, Mich.

34. D. Barntkrecht, *17th ASM Heat Treating Society Conference Proceedings*, Sept. 15–18, 1997, eds. D. L. Milan, D. A. Pottet, Jr., G. D. Pfaffmann, V. Rudnev, A. Muehlbauer, and W. B. Albert, ASM International, Materials Park, Ohio, 1998, pp. 525–528.

35. *Induction Heating Equipment & Systems*, TOCCO, Inc., Madison Heights, Mich., 1995.

36. G. D. Pfaffmann, *Heat Treating*, vol. 12, no. 5, 1980, pp. 34–42.

37. R. S. Ruffini, R. T. Ruffini, and V. S. Nemkov, *The 1st International Automotive Heat Treating Conference Proceedings*, eds. R. Coals, K. Funatani, and C. A. Stickels, ASM International, Materials Park, Ohio, 1998, pp. 293–300.

38. W. A. Smutz et al., *Proceedings of International Heat Treating Conference*, eds. G. Totten and R. A. Wallis, ASM International, Materials Park, Ohio, 1994, pp. 219–223.

39. A. Muehlbauer, A. Ruhnke, V. Demidovitch, A. Nikanorov, S. Lupi, and F. Dughiero, *17th ASM Heat Treating Society Conference Proceedings*, Sept. 15–18, 1997, eds. D. L. Milan, D. A. Pottet, Jr., G. D. Pfaffmann, V. Rudnev, A. Muehlbauer, and W. B. Albert, ASM International, Materials Park, Ohio, 1998, pp. 865–875.

40. V. I. Rudnev and D. L. Loveless, *Ind. Heat.*, February 1995, pp. 46–51.

41. R. S. Ruffini, R. T. Ruffini, and V. S. Nemkov, *Ind. Heating*, June 1998, pp. 59–64.

42. R. S. Ruffini, R. T. Ruffini, and V. S. Nemkov, *Ind. Heating*, November 1998, pp. 69–72.

43. R. S. Ruffini, R. T. Ruffini, V. S. Nemkov, and R. C. Goldstein, *Heat Treat. Met.*, vol. 26, no. 4, 1999, pp. 84–89.

44. R. S. Ruffini, *Industrial Heating Clinic*, Sept. 15–26, 1995, Auburn Hills, Ohio, CIT, Mich., 1995.

45. J. Stambaugh and T. Learman, *17th ASM Heat Treating Society Conference Proceedings*, Sept. 15–18, 1997, eds. D. L. Milan, D. A. Pottet, Jr., G. D. Pfaffmann, V. Rudnev, A. Muehlbauer, and W. B. Albert, ASM International, Materials Park, Ohio, 1998, pp. 711–717.

46. R. Cook and V. Rudnev, *9th International Induction Heating Seminar Proceedings*, May 2000, Clearwater Beach, Fla., Inductoheat, Inc., Madison Heights, Mich.

47. J. Kosiniewski, *9th International Induction Heating Seminar Proceedings*, May 2000, Clearwater Beach, Fla., Inductoheat, Inc., Madison Heights, Mich.

48. S. Baskerville, *9th International Induction Heating Seminar Proceedings*, May 2000, Clearwater Beach, Fla., Inductoheat, Inc., Madison Heights, Mich.

49. D. Loveless, *8th International Induction Heating Seminar Proceedings*, Nov. 3–6, 1998, Kissimmee, Fla., Inductoheat, Inc., Madison Heights, Mich.; *Ind. Heating*, March 1998, pp. 53–58; *9th International Induction Heating Seminar Proceedings*, May 2000, Clearwater Beach, Fla., Inductoheat, Inc., Madison Heights, Mich.

50. D. Loveless, *17th ASM Heat Treating Society Conference Proceeding*, Sept. 15–18, 1997, eds. D. L. Milan, D. A. Pottet, Jr., G. D. Pfaffmann, V. Rudnev, A. Muehlbauer, and W. B. Albert, ASM International, Materials Park, Ohio, 1998, pp. 615–620.

51. J. Stambaugh, *9th International Induction Heating Seminar Proceedings*, May 2000, Clearwater Beach, Fla., Inductoheat, Inc., Madison Heights, Mich.

52. G. Welch, *Proceedings 16th Heat Treating Society Conference*, March 19–21, 1996, ASM International, Materials Park, Ohio, pp. 89–93.

53. V. S. Nemkov and R. C. Goldstein, *Ind. Heating*, November 1999, pp. 77–80.

54. V. S. Nemkov, R. C. Goldstein, V. A. Bukasis, A. Zenkov, and D. Koutchnassov, *Proceedings: 3d International Conference on Quenching and Control of Distortion,* March 26–29, 1999, Prague, Czech. Republic, pp. 69–72.

55. P. Nowak and K. H. Miller, *17th ASM Heat Treating Society Conference Proceedings,* Sept. 15–18, 1997, eds. D. L. Milan, D. A. Pottet, Jr., G. D. Pfaffmann, V. Rudnev, A. Muehlbauer, and W. B. Albert, ASM International, Materials Park, Ohio, 1998, pp. 529–537.

56. G. Bobart, *Ind. Heating,* June 1990, pp. 16–19.

57. M. Garaway, *Metallurgia,* vol. 45, no. 3, 1978, pp. 165–167.

58. R. Creal, *Heat Treating,* vol. 16, no. 6, 1984, pp. 28–30.

59. M. R. Black, *9th International Induction Heating Seminar Proceedings,* May 2000, Clearwater Beach, Fla., Inductoheat, Inc., Madison Heights, Mich.

60. H. N. Udall, *Heat Treat. Met.,* no. 4, 1982, pp. 94, 95.

61. C. I. Cerny, I. Furbacher, and V. Linhart, *Jn. Mats. Eng. Perform.,* vol. 7, no. 3, June 1998, pp. 361–366.

62. G. Sayegh, in *Heat Treatment '83,* The Metals Society, London, p. 2.63.

63. J. Wollenweber, *Adv. Mater. and Process.,* 12/96, pp. 25–26.

64. W. M. Steen, in *The Encyclopedia of Advanced Materials,* vol. 4, Pergamon Press, Oxford, 1994, pp. 2755–2761.

65. R. D. Gregory, *Ind. Heating,* January 1998, pp. 43–46.

66. D. S. Gnanamuthu and V. S. Shankar, in *Proceedings: NATO Advanced Studies Institute on Laser Surface Treatment,* Italy, 1985, pp. 413–433; H. G. Woo and H. S. Cho, *Surface and Coatings Technology,* vol. 102, 1998, pp. 205–217.

66a. J. Grum, R. Sturm, and P. Zerovnik, in *Proceedings of the Second International Conference on Quenching and Control of Distortion,* ASM International, Materials Park, Ohio, 1996, pp. 181–191.

67. M. Hino, M. Hiramatsu, K. Akiyama, and H. Kawasaki, *Mats. and Manufact. Processes,* vol. 12, no. 1, 1997, pp. 37–46.

68. R. C. Reed, Z. Shen, J. M. Robinson, and T. Akbay, *Mats. Sc. and Tech.,* vol. 15, January 1999, pp. 109–118.

69. D. S. Gnanamuthu, C. B. Shaw, Jr., W. E. Lawrence, and M. R. Mitchell, in *Conference Proceedings on Laser Solid Interactions and Laser Processing,* American Institute of Physics, New York, 1979.

70. V. Lopez, B. Fernandez, J. M. Bello, J. Ruiz, and F. Zubiri., *ISIJ International,* vol. 35, no. 11, 1995, pp. 1394–1399.

71. W. M. Steen, *Laser Materials Processing,* Springer-Verlag, London, 1996.

72. O. A. Sandven, in *ASM Handbook,* vol. 4, 10th ed., ASM International, Materials Park, Ohio, 1991, pp. 286–296.

73. B. L. Mordike, in *Processing of Metals and Alloys,* vol. 15: *Materials Science and Technology,* vol. ed. R. W. Cahn, VCH, Weinheim, 1991, pp. 111–136.

74. M. L. Escudero and J. M. Bello, *Mater. Sc. Eng.,* vol. 158A, 1992, pp. 227–233.

75. A. Yoshida, K. Fujita, S. Ando, and T. Tani, *Bulletin of JSME,* vol. 28, 1985, pp. 2407–2413.

76. A. Ramalingam, B. J. Kalaiselvi, S. Sridharan, P. K. Palanisamy, and V. Masilamani, *Bull. Mater. Sci.,* vol. 21, no. 3, June 1998, pp. 247–249.

77. C. E. Navara, B. Bengtsson, W.-B. Li, and K. E. Easterling, in *Heat Treatment '83,* The Metals Society, London, p. 2.40.

78. G. Sayegh, *Heat Treat. Met.,* no. 1, 1980, pp. 5–10.

79. T. Fukuda, M. Kikuchi, A. Yamanishi, and S. Kiguchi, *Heat Treatment '83,* The Metals Society, London, p. 2.34.

80. J. Mazumdar, *JOM,* vol. 5, 1983, pp. 18–26.

81. J. C. Ion, T. J. I. Moisio, M. Paju, and J. Johanssson, *Mats. Sc. and Tech.*, vol. 8, 1992, pp. 799–803.

82. A. N. Safonov, N. F. Zelentsova, A. A. Mitrofanov, V. V. Vasil'tsov, and I. N. Il'ichev, *Welding International*, vol. 112, no. 2, 1997, pp. 152–155.

83. H. Bande, G. L'Esperance, M. U. Islam, and A. K. Koul, *Mats. Sc. and Tech.*, vol. 7, May 1991, pp. 452–457.

84. H. Yamamoto, M. Oikawa, K. Minamida, and H. Kawasumi, *Seimitsu Kaogakkai Shi* (=*Jn. of Japan Soc. of Precision*), vol. 63, no. 9, September 1997, pp. 1320–1324; M. Oikawa, H. Yamamoto, K. Minamida, and H. Kawasumi, *Seimitsu Kaogakkai Shi* (=*Jn. of Japan Soc. of Precision*), vol. 63, no. 8, 1997, pp. 1138–1142.

85. E. Beyer, V. Krause, and P. Loosen, *Jn. De Physique IV*, vol. 4, April 1994, pp. C4-13–23.

86. T. Bell, *Metallurgia*, vol. 49, 1982, pp. 103–111.

87. J. H. P. C. Magaw, in *Heat Treatment '84*, The Metals Society, London, p. 2.1.

88. S. Nagasawa, Y. Fukuzawa, Y. Ito, M. Abe, I. Katayama, and A Yoshizawa, *ICALEO '97: Laser Materials Processing*, vol. 83, no. 2, 1997, pp. F91–F97.

89. M. Bamberger, *International Mater. Reviews*, vol. 43, no. 5, 1998, pp. 189–203.

90. S. K. Putatunda, L. Bartosiewicz, R. J. Hull, and M. Lander, in *Surface Modification Technologies, IX*, eds. T. S. Sudarshan, W. Reitz, and J. J. Stiglich, TMS, Warrendale, Pa., 1996, pp. 355–370.

91. Y. Shijian, T. Xinan, and Z. Changchi, *Chinese Jn. of Lasers*, vol. 24, no. 1, January 1997, pp. 87–90.

92. J. Mazumdar and W. Steen, *J. Appl. Phys*, vol. 51, 1980, p. 941.

93. H. E. Cline and T. R. Anthony, *J. Appl. Phys.*, vol. 48, 1977, p. 3895.

94. M. Tayal and K. Mukherjee, *J. Appl. Phys.*, vol. 75, no. 8, 1994, pp. 3855–3861.

95. A. Bokota and S. Iskierka, *ISIJ International*, vol. 36, no. 11, 1996, pp. 1383–1391.

96. R. Zenker and B. Furcheim, *Proceedings: International Heat Treating Conference*, April 18–20, 1994, Illinois, ASM International, Materials Park, Ohio, 1994, pp. 299–301; *16th ASM Heat Treating Society Conference*, March 19–21, 1996, Ohio, 1996, pp. 461–466.

97. S. Panzer, S. Schiller, and B. Furchheim, in *ASM Handbook*, vol. 4, 10th ed., ASM International, Materials Park, Ohio, 1991, pp. 59–75.

98. S. Schiller and S. Panzer, *Conference Proceeding: The Laser, Electron Beam in Welding, Cutting and Surface Treatment: State of the Art, 1993*, ed. R. Bakish, Bakish Materials Corp., Englewood, N.J., 1993, pp. 79–81.

99. T. Jokinen and T. Meuronen, in *4th International. Conference on Advances in Surface Engineering*, University of Northumbria at Newcastle, May 14–17, 1996, Special Publ. no. 206–208, Royal Society of Chemistry, 1997, pp. 186–198.

100. J. R. Hwang and C.-P. Fung, *Surface and Coatings Technology*, vol. 80, 1996, pp. 271–278.

101. R. Zenker, A. Wachowiak, M. Dobielinska, N. Frenkler, W. Frackowiak, and R. Reinhold, in *5th International. Conference: Carbides, Nitrides, and Borides*, The Inst. of Materials, London, October 1990, pp. 57–62.

102. P. H. Warren and R. J. Johnson, in *Heat Treatment '83*, The Metals Society, London, p. 2.58.

103. R. C. Hansson, *J. Heat Treating*, vol. 3, no. 1, 1983, pp. 30–37.

104. S. Schiller, S. Panzer, and B. Furchheim, in *ASM Handbook*, vol. 4, 10th ed., ASM International, Materials Park, Ohio, 1991, pp. 297–311.

105. *Carburizing and Carbonitriding*, ASM, Metals Park, Ohio, 1977.

106. C. A. Stickels, *The 1st International Automotive Heat Treating Conference Proceedings*, July 13–15, 1998, eds. R. Colas, K. Funatani, and C. A. Stickels, ASM International, Materials Park, Ohio, 1999, pp. 32–36.

107. W. E. Jominy, *Metal. Prog.*, vol. 85, no. 5, 1964, pp. 70–73.

108. N. O. Kates, *Met. Prog.*, vol. 97, no. 1, 1970, pp. 90–92.

109. A. M. Aitchison, *Heat Treat. Met.*, no. 3, 1982, pp. 71–72.

110. *Cassel's Manual of Heat Treatment and Case Hardening*, Birmingham, England.

111. R. W. Foreman, *ASM Handbook*, vol. 4: *Heat Treating*, 10th ed., ASM International, Materials Park, Ohio, 1991, pp. 325–328.

112. *Met. Prog.*, vol. 85, no. 5, 1964, pp. 77–79.

113. D. J. Grieve, *Met. Mater. Technol.*, vol. 7, no. 8, 1975, pp. 397–403.

114. A. D. Godding, *ASM Handbook*, vol. 4: *Heat Treating*, 10th ed., ASM International, Materials Park, Ohio, 1991, pp. 329–347.

115. M. Rosso et al., *Advances in P/M and Particulate Materials Proceeding*, June 29–July 2, 1997, APMI, Princeton, N.J., pp. 6-9–6-22.

116. R. Beckett and H. Kunst, in *Heat Treatment '84*, The Metals Society, London, pp. 16.1–16.5.

117. N. P. Milano, *Met. Prog.*, vol. 91, no. 6, 1978, pp. 54–56.

118. R. G. Whitebeck et al., *Adv. Mats. & Process.*, April 1999, pp. H21–H26.

119. C. A. Stickels, *ASM Handbook*, vol. 4: *Heat Treating*, 10th ed., ASM International, Materials Park, Ohio, 1991, pp. 312–324.

120. *Metals Handbook*, vol. 7, 8th ed., ASM, Metals Park, Ohio, 1973.

121. S. M. Tapaswi, *Tool and Alloy Steels*, May 1981, pp. 161–167.

122. W. C. Hiatt and J. P. Crosbie, *Met. Prog.*, vol. 98, no. 4, 1970, p. 143.

123. H. C. Dill, *Met. Prog.*, vol. 102, no. 3, 1972, pp. 84, 85.

124. G. Krauss, *Steels: Heat Treatment and Processing Principles*, ASM International, Materials Park, Ohio, 1990.

125. C. A. Stickels, private communication, 2000.

126. J. Kelso and A. Wilson, *Heat Treating*, vol. 17, no. 3, 1985, pp. 32, 33.

127. G. O. Ratliff and W. H. Samuelson, *Met. Prog.*, September 1975, pp. 75–77.

128. J. A. Lutz, *Adv. Mats. & Process.*, June 1997, pp. 68AA–68CC.

129. G. Parrish and G. S. Harper, *Production Gas Carburizing*, Pergamon Press, Oxford, 1985.

130. E. E. Staples, *Met. Prog.*, vol. 85, no. 5, 1964, pp. 74–76.

131. M. J. Fischer, *Heat Treating: 1994 Conference Proceeding*, eds. G. F. Totten and R. A. Wallis, ASM International, Materials Park, Ohio, 1994, pp. 167–169.

132. N. F. Smith, *Metallurgia*, vol. 50, no. 12, 1983, pp. 502–505.

133. R. G. H. Record, *Metall. Treat. Form.*, December 1972, pp. 413–416.

134. D. W. McCurdy, *Heat Treating: 1994 Conference Proceeding*, eds. G. F. Totten and R. A. Wallis, ASM International, Materials Park, Ohio, 1994, pp. 117–121.

135. M. J. Fischer and D. McCurdy, *Proceedings 16th ASM Heat Treating Society Conference*, eds. J. L. Dorsett and R. E. Luetze, ASM International, Materials Park, Ohio, 1996, pp. 257–258.

136. R. N. Blumenthal and A. T. Melville, *Proceedings 16th ASM Heat Treating Society Conference*, eds. J. L. Dorsett and R. E. Luetze, ASM International, Materials Park, Ohio, 1996, pp. 279–283.

137. B. Edenhoffer, *Heat Treat. Met.*, no. 4, 1985, pp. 87–91.

138. Ref. 1 in Ref. 118.

139. A. Knierem and H. Pfau, in *Heat Treatment '84*, The Metals Society, London, pp. 13.1–13.8.

140. D. Grassl and B. Edenhoffer, in *Heat Treatment '84*, The Metals Society, London, pp. 12.4–12.10.

141. J. S. Pierre, *ASM Handbook*, vol. 4: *Heat Treating*, 10th ed., ASM International, Materials Park, Ohio, 1991, pp. 348–351.

142. F. Preisser, E. Seemann, and W. R. Zenker, *The 1st International Automotive Heat Treating Conference Proceeding*, July 13–15, 1998, eds. R. Colas, K. Funatani, and C. A. Stickels, ASM International, Materials Park, Ohio, 1999, pp. 142–148.

143. D. H. Herring, *Heat Treating: 1994 Conference Proceeding*, eds. G. F. Totten and R. A. Wallis, ASM International, Materials Park, Ohio, 1994, pp. 5–12.

144. R. G. Weber, *Ind. Heat.*, June 1982, pp. 10, 11.

145. C. L. Bronson, *Heat Treat.*, vol. 128, no. 11, 1986, pp. 30, 31.

146. C. H. Luiten, F. Limque, and F. Bless, *Heat Treatment '79*, The Metals Society, London, pp. 69–75.

147. F. Preisser, E. Seemann, and W. R. Zenker, *The 1st International Automotive Heat Treating Conference Proceeding*, July 13–15, 1998, eds. R. Colas, K. Funatani, and C. A. Stickels, ASM International, Materials Park, Ohio, 1999, pp. 135–141.

148. W. L. Grube and S. Verhoff, *ASM Handbook*, vol. 4: *Heat Treating*, 10th ed., ASM International, Materials Park, Ohio, 1991, pp. 352–362.

149. B. Edenhofer, *Heat Treat. Met.*, 1999.1, pp. 1–5.

150. M. Sugiyama, K. Ishikawa, and H. Iwata, *18th Heat Treating Conference Proceeding*, eds. R. A. Wallis and H. W. Walton, ASM International, Materials Park, Ohio, 1999, pp. 49–56.

151. D. H. Herring, *Ind. Heat.*, September 1996, pp. 59, 60, 62, 64, 66.

152. F. Preisser, P. Hellmann, R. Seemann, and W. R. Zenker, *17th ASM Heat Treating Society Conference Proceeding*, Sept. 15–18, 1997, eds. D. L. Milan, D. A. Pottet, Jr., G. D. Pfaffmann, V. Rudnev, A. Muehlbauer, and W. B. Albert, ASM International, Materials Park, Ohio, 1998, pp. 43–51.

153. F. Hombeck and W. Rembges, in *Heat Treatment '84*, The Metals Society, London, pp. 15.1–15.9.

154. H. W. Westeren, *Metall. Met. Form.*, vol. 39, no. 11, 1972, pp. 390–393.

155. M. H. Jacobs and T. J. Law, in *Heat Treatment '84*, The Metals Society, London, pp. 50.1–50.7.

156. J. G. Conybear, *Heat Treat. Met.*, no. 3, 1988, pp. 24–27.

157. B. Edenhofer, J. G. Conybear, and G. T. Legge, *Heat Treating Met.*, no. 1, 1991, pp. 6–12.

158. B. Edenhofer and J. W. Bouwman, *Steel Heat Treatment Handbook*, eds. G. E. Totten and M. A. H. Howes, Marcel Dekker, New York, 1997, pp. 483–525.

159. M. H. Jacobs, *Heat Treat.*, vol. 17, no. 1, 1985, pp. 38–41.

160. M. Booth, T. Farrell, and R. H. Johnson, *Heat Treat. Met.*, vol. 2, 1983, pp. 45–51.

161. F. Preisser, A. Melber, and F. Schnatbaum, *Heat Treating: 1994 Conference Proceeding*, eds. G. F. Totten and R. A. Wallis, ASM International, Materials Park, Ohio, 1994, pp. 263–264; F. Preisser, *Heat Treat. Met.*, 1998.3, pp. 65–71.

162. R. W. Reynoldson, *Heat Treating: 1994 Conference Proceeding*, eds. G. F. Totten and R. A. Wallis, ASM International, Materials Park, Ohio, 1994, pp. 183–189.

163. R. W. Reynoldson, *17th ASM Heat Treating Society Conference Proceeding*, Sept. 15–18, 1997, eds. D. L. Milan, D. A. Pottet, Jr., G. D. Pfaffmann, V. Rudnev, A. Muehlbauer, and W. B. Albert, ASM International, Materials Park, Ohio, 1998, pp. 53–60.

164. J. Dossett, in *ASM Handbook*, vol. 4: *Heat Treating*, 10th ed., ASM International, Materials Park, Ohio, 1991, pp. 376–386.

165. Y. Watanabe, N. Narita, S. Umegaki, and Y. Mishima, *18th Heat Treating Conference Proceeding*, eds. R. A. Wallis and H. W. Walton, ASM International, Materials Park, Ohio, 1999, pp. 401–408.

166. W. E. Dowling, W. T. Donlon, and J. P. Wise, *18th Heat Treating Conference Proceeding*, eds. R. A. Wallis and H. W. Walton, ASM International, Materials Park, Ohio, 1999, pp. 387–397.

167. N. P. Milano, *Met. Prog.*, vol. 88, no. 1, 1965, pp. 79–81.

168. W. L. Grube and D. P. Koistinen, *J. Heat Treating*, vol. 2, no. 3, 1982, pp. 211–216.

169. J. Jesinski, L. Jezvorski, and M. Kubara, *Heat Treat. Met.*, vol. 2, 1985, pp. 41–46.

170. Z. Rogalski and Z. Obuchowicz, *Heat Treat. Met.*, 1994.4, pp. 84–89.

171. T. Bell, in *Heat Treatment Shanghai '83*, The Metals Society, London, 1984, pp. 1.1–1.10.

172. K. H. Jack, *Heat Treatment '73*, The Metals Society, London, pp. 39–50.

173. *Metals Handbook*, vol. 3, 10th ed., ASM International, Materials Park, Ohio, 1992, p. 198.

174. M. A. J. Somers and E. J. Mittemeijer, *17th ASM Heat Treating Society Conference Proceeding*, Sept. 15–18, 1997, eds. D. L. Milan, D. A. Pottet, Jr., G. D. Pfaffmann, V. Rudnev, A. Muehlbauer, and W. B. Albert, ASM International, Materials Park, Ohio, 1998, pp. 321–330.

175. C. H. Knerr, T. C. Rose, and J. H. Filkowski, *Metals Handbook*, vol. 4: *Heat Treating*, ASM International, Materials Park, Ohio, 1991, pp. 387–409.

176. A. M. Staines, *Heat Treat. Met.*, 1996.1, pp. 1–6.

177. R. M. Spencer, *Met. Prog.*, vol. 103, no. 2, 1973, pp. 92–94.

178. C. A. Weymueller, *Met. Prog.*, vol. 102, no. 1, 1972, pp. 38–50.

179. D. A. Dashfield, *Heat Treatment '73*, The Metals Society, London, 1975, pp. 67–70.

180. W. Leeming, *Met. Prog.*, vol. 85, no. 2, 1964, pp. 86–87.

181. Q. D. Mehrkam, J. R. Easterday, B. R. Payne, R. W. Foreman, D. Vukovich, and A. D. Godding, *ASM Handbook*, vol. 4, 10th ed., ASM International, Materials Park, Ohio, 1991, pp. 410–419.

182. W. K. Liliental, C. D. Morawski, and G. J. Tymowski, *The 1st International Automotive Heat Treating Conference Proceeding*, July 13–15, 1998, eds. R. Colas, K. Funatani, and C. A. Stickels, ASM International, Materials Park, Ohio, 1999, pp. 45–52.

183. G. P. Dubal et al., *The 1st International Automotive Heat Treating Conference Proceeding*, July 13–15, 1998, eds. R. Colas, K. Funatani, and C. A. Stickels, ASM International, Materials Park, Ohio, 1999, pp. 90–95.

184. J. Grosch, in *Steel Heat Treatment Handbook*, Marcel Dekker, New York, 1997, pp. 663–719.

185. T. Bell, B. J. Birch, V. Kurotechenko, and S. P. Evans, in *Heat Treatment '73*, The Metals Society, London, pp. 50–57.

186. S. Mirdha and D. H. Jack, *Met. Sci.*, vol. 16, 1982, p. 398.

187. L. Sanderson, *Tooling*, February 1966, pp. 45–48.

188. *Tooling*, July 1978, pp. 13–16.

189. V. J. Cuppola, *Met. Prog.*, vol. 80, no. 1, 1961, pp. 83, 84.

190. L. Maldizinski, G. Tymowski, and J. H. Tacikowski, *Surf. Engineering*, vol. 15, no. 5, 1999, pp. 377–384.

191. A. Nakonieczny, J. Senatorski, T. Tacikowski, G. Tymowski, and W. Liliental, *Heat Treat. Met.*, 1997.4, pp. 81–88.

192. W. K. Liliental, C. D. Morawski, and A. Czelusniak, *Heat Treating: 1994 Conference Proceeding*, eds. G. F. Totten and R. A. Wallis, ASM International, Materials Park, Ohio, 1994, pp. 361–368.

193. U. Huchel and S. Dressler, *Heat Treating: 1994 Conference Proceeding*, eds. G. F. Totten and R. A. Wallis, ASM International, Materials Park, Ohio, 1994, pp. 143–154.

194. J. O'Brien and D. Goodman, *ASM Handbook*, vol. 4, 10th ed., ASM International, Materials Park, Ohio, 1991, pp. 420–424.

194a. D. Pye, private communication, 2000.

195. T. K. Borthakur, A. Sahu, S. R. Mohanty, B. B. Nayak, and B. S. Acharya, *Surf. Eng.*, vol. 15, no. 1, 1999, pp. 55–58.

196. M. H. Jacobs, *Heat Treating*, vol. 18, no. 1, 1986, pp. 26–29.

197. D. M. Hulett and M. A. Taylor, *Met. Prog.*, vol. 128, 1985, pp. 18–21.

198. D. Pye, in *Steel Heat Treatment Handbook*, Marcel Dekker, New York, 1997, pp. 721–764.

199. M. H. Jacobs, Ipsen Industries GmbH, England, private communication, February 1987.

200. M. Weck and K. Schlotermann, *Metallurgia*, vol. 51, no. 8, 1984, pp. 328–332.

201. B. Edenhofer, *Heat Treatment '79,* The Metals Society, London, pp. 52–59.

202. A. J. Hicks, *Metallurgia*, vol. 46, no. 3, 1979, pp. 147–162.

203. R. Rudnicki, P. Beer, A. Sokolowska, and R. Marchal, *Surf. and Coatings Technol.*, vol. 107, 1998, pp. 20–23.

204. M. U. Devi and O. N. Mohanty, *Surf. and Coatings Technol.*, vol. 107, no. 1, 1998, pp. 55–64.

205. T. Bell, *Metallurgia*, vol. 49, no. 3, 1982, pp. 103–111.

206. F. Hombeck, *Heat Treatment '83,* The Metals Society, London, p. 1.41.

207. S. Ruset, *Heat Treat. Met.*, 1994.4, pp. 90–92.

208. B. Larish, H.-J. Spies, U. Brusky, and U. Rensch, *The 1st International Automotive Heat Treating Conference Proceeding*, July 13–15, 1998, eds. R. Colas, K. Funatani, and C. A. Stickels, ASM International, Materials Park, Ohio, 1999, pp. 221–228.

209. B. Podgornik, J. Vizintin, and V. Leskovskel, *Surf. and Coatings Technol.*, vol. 108–109, 1998, pp. 454–460.

210. D. Pye, *Heat Treating: 1994 Conference Proceeding*, eds. G. F. Totten and R. A. Wallis, ASM International, Materials Park, Ohio, 1994, pp. 353–359.

211. *Met. Prog.*, vol. 113, no. 1, 1978, pp. 49–53.

212. *Heat Treat. Met.*, no. 3, 1980, p. 64.

213. A. Dinurozi and R. E. Duffy, *Ind. Heat.*, July 1990, pp. 32–36.

214. W. Ye, A. Yang, and D. Zhang, *Heat Treating: 1994 Conference Proceeding*, eds. G. F. Totten and R. A. Wallis, ASM International, Materials Park, Ohio, 1994, pp. 473–477.

215. T. Bell, *Heat Treat. Met.*, no. 2, 1975, pp. 39–49.

216. C. Dawes and D. F. Tranter, *Met. Prog.*, vol. 124, no. 6, 1983, pp. 17–22.

217. C. Dawes, D. F. Tranter, and C. G. Smith, *Heat Treatment '79,* The Metals Society, London, 1980, pp. 60–68.

218. K. Bennett, Q. Weir, and J. Williamson, *Heat Treat. Met.*, no. 4, 1981, pp. 79–81.

219. C. Dawes, D. F. Tranter, and C. G. Smith, *J. Heat Treating*, vol. 1, no. 2, 1979, pp. 30–42.

220. B. Edenhofer, W. Lerche, and W. Gohring, *Heat Treat. Met.*, 1995.2, pp. 27–33.

221. C. Dawes and D. F. Tranter, *Heat Treatment '84*, The Metals Society, London, pp. 28.1–28.8; *Heat Treat. Met.*, no. 3, 1985, pp. 70–76.

222. W. H. Blossfield, *Adv. Mater. & Process.*, February 1993, pp. 52–53.

223. H. S. Blaauw and J. Post, *Heat Treat. Met.*, 1996.3, pp. 53–56.

224. B. Edenhofer, *Heat Treat. Met.*, 1998.4, pp. 79–85.

225. I. V. Etchells, *Heat Treat. Met.,* no. 4, 1981, pp. 85–88.

226. G. Wahl and I. V. Etchells, in *Heat Treatment '84*, The Metals Society, London, pp. 29.1–29.7.

227. J. R. Easterday, *Salt Bath Ferritic Nitrocarburizing and The Kolene QPQ Process*, Kolene Corporation, Detroit, Michigan, 1996.

228. G. Wahl, *Adv. Mats. & Processes*, April 1996, pp. 37, 38; *Heat Treat. Met.*, 1995.3, pp. 65–73.

229. J. Bosslet and M. Kreutz, *New Developments in Salt Bath Nitrocarburizing and Tufftride-QPQ Process*, Houghton Durferrit GmbH, Germany.

230. S. Yoshida, *The 1st International Automotive Heat Treating Conference Proceeding*, July 13–15, 1998, eds. R. Colas, K. Funatani, and C. A. Stickels, ASM International, Materials Park, Ohio, 1999, pp. 209–212.

231. D. Goldstraw, *Heat Treatment '84*, The Metals Society, London, pp. 30.1–30.9.

232. A. Burdese, D. Firrao, and M. Rosso, *Heat Treatment '84,* The Metals Society, London, pp. 32.4–32.8.

234. G. Husnain, *Adv. Mats. & Processes*, July 1995, pp. 48AA–48CC.

235. C. Ruset, A. Bloyce, and T. Bell, *Heat Treat. Met.*, 1995.4, pp. 95–100.

236. T. Bell, *ASM Handbook*, vol. 4, 10th ed., ASM International, Materials Park, Ohio, 1991, pp. 425–436.

237. K. E. Moore and D. N. Collins, *Heat Treat. Met.*, 1996.3, pp. 61–62.

238. K. Boiko, *Heat Treating*, vol. 18, no. 4, 1986, pp. 65, 66.

239. P. F. Stratton and K. Bennett, *Heat Treat. Met.,* 1996.1, pp. 7–10.

240. R. F. Kern, *Heat Treat.*, vol. 18, no. 10, 1986, pp. 36–38.

241. G. Parrish, *Carburizing: Microstructures and Properties*, ASM International, Materials Park, Ohio, 1999.

242. H. S. Ming, T. Takayama, and T. Nishizawa, *J. Jpn. Inst. Met.*, vol. 45, no. 11, November 1981, pp. 1195–1201.

243. A. Graf von Matuschka, *Boronizing*, Hanser, Munich, 1980.

244. R. Chatterjee-Fischer, *Hart.-Tech. Mitt.*, vol. 36, no. 5, 1981, pp. 248–254.

245. P. Dearnley and T. Bell, *Surf. Engrg.*, vol. 1, no. 3, 1985, pp. 203–217.

246. W. J. G. Fichtl, *Saving Energy and Money by Boronizing*, paper presented at the meeting of the Japan Heat Treating Association, Tokyo, Nov. 25, 1988; *Boronizing and Its Practical Applications*, paper presented at the 33d Harterei-Kolloquium, Wiesbaden, Germany, October 5–7, 1977.

247. A. Galibois, O. Boutenko, and B. Voyzelle, *Acta Met.*, vol. 28, 1980, pp. 1753–1763, 1765–1771.

248. *ASM Handbook*, vol. 3, 10th ed., ASM International, Materials Park, Ohio, 1992, p. 81.

249. R. Chatterjee-Fischer, in *Surface Modification Technologies*, ed. T. S. Sudarshan, Marcel Dekker, New York, 1989, Chap. 8, pp. 567–609.

250. S. Motojima, K. Maeda, and K. Sugiyama, *J. Less-Common Met.*, vol. 81, 1981, pp. 267–272.

251. O. Knotek, E. Lugscheider, and K. Leuschen, *Thin Solid Films*, vol. 45, 1977, pp. 331–339.

252. K. H. Habig and R. Chatterjee-Fischer, *Tribol. International*, vol. 14, no. 4, 1981, pp. 209–215.

253. D. J. Bak, *New Design News*, February 16, 1981, p. 78.

254. R. Chatterjee-Fischer and O. Schaaber, *Heat Treatment' 76*, The Metals Society, London, 1976, pp. 27–30.

255. *Mater. Eng.*, August. 1970, p. 42.

256. W. J. G. Fichtl, *Hart.-Tech. Mitt.*, vol. 29, no. 2, 1974, pp. 113–119.

257. P. Bielinski, K. Sikorski, and T. Wierzhon, *J. Mats. Sc. Letters*, vol. 15, 1996, pp. 1335–1336.

258. D. N. Tsipas, G. K. Triantafyllidis, J. K. Kiplagat, and P. Psillaki, *Mater. Letters*, vol. 37, 1998, pp. 128–131.

259. H. C. Child, *Metall. Mater. Technol.*, vol. 13, no. 6, 1981, pp. 303–309.

260. R. Chatterjee-Fischer, *Powder Metall.*, vol. 20, no. 2, 1977, pp. 96–99.

261. G. Wahl, *Boriding—A Process for Producing Hard Surfaces to Withstand Extreme Wear*, Durferrit GmbH, Germany, February 1997.

262. W. Liluental, J. Tacikowski, and J. Senatorski, *Heat Treatment '81*, The Metals Society, London, 1983, pp. 193–197.

263. H. Kunst and O. Schaaber, *Hart.-Tech. Mitt.*, vol. 22, translation HB 7122-III, 1967, pp. 275–292.

264. D. N. Tsipas, J. Rus, and H. Noguerra, *Heat Treatment '88*, The Metals Society, London, 1988, pp. 203–210.

265. P. A. Dearnley, T. Farrell, and T. Bell, *J. Mater. for Energy Systems*, vol. 8, no. 2, 1986, pp. 128–131.

266. T. B. Massalski, *Binary Alloy Phase Diagrams*, ASM, Materials Park, Ohio, 1986.

267. A. Brown et al., *Metall. Sci.*, vol. 8, 1974, pp. 317–324.

268. T. B. Cameron and J. E. Morral, *Met. Trans.*, vol. 17A, 1986, pp. 1481–1483.

269. H. Kunst and O. Schaaber, *Hart.-Tech. Mitt.*, vol. 22, no. 1, translations HB 7122-I and HB 7122-II, 1967, pp. 1–25.

270. M. J. Lu, *Hart.-Tech. Mitt.*, vol. 38, no. 4, 1983, pp. 156–159.

271. W. Fichtl, N. Trausner, and A. G. Matuschka, *Boronizing with Ekabor*, Elektroschmeltz Kempten, GmbH, Germany.

272. "Boroalloy Process," Process Data Sheet 4, Lindberg Heat Treating Company, Watertown, Wisc.

273. A. J. Ninham and I. M. Hutchings, *Wear of Materials*, vol. 1, 1989, pp. 121–127.

274. M. E. Blanter and N. P. Bosedin, *Metalloved. Term. Obra. Met.*, vol. 6, 1955, pp. 3–9.

275. G. V. Samsonov and A. P. Epik, in *Coatings on High Temperature Materials*, Part 1, ed. H. H. Hausner, Plenum Press, New York, 1966, pp. 7–111.

276. J. J. Smit, Delft University of Technology, Laboratory of Metals, unpublished research, 1984.

277. C. M. Brakman, A. W. J. Gommers, and E. J. Mittemeijer, *Heat Treatment '88*, The Institute of Metals, London, 1988, pp. 211–217.

278. H. C. Fiedler and W. J. Hayes, *Met. Trans.*, vol. 1, 1970, pp. 1070–1073.

279. W. J. G. Fichtl, *Jahr. Oberflachen Tech.*, vol. 45, 1989, Metall-Verlag, Berlin/Heidelberg, pp. 420–427.

280. G. Palombarini, M. Carbucicchio, and L. Cento, *J. Mater. Sc.*, vol. 19, 1984, p. 3732.

281. V. I. Pokmurskii, V. G. Protsik, and A. M. Mokrava, *Sov. Mater. Sc.*, vol. 10, 1980, p. 185.

282. P. Goeurist, R. Fillitt, F. Thevenol, J. H. Driver, and H. Bruyas, *Mater. Sc. Engrg.*, vol. 55, 1982, pp. 9–19.

283. M. Carbucicchio and G. Sambogna, *Thin Solid Films*, vol. 126, 1985, pp. 299–305.

284. H.-J. Hunger and G. Trute, *Wear Protection by Means of Boronizing*, ESK GmbH, March 1996.

285. H. E. Hintermann, *Thin Solid Films*, vol. 84, 1981, pp. 215–243.

286. V. A. Volkov and A. A. Aliev, *Steel USSR*, vol. 5, no. 3, 1975, pp. 180–181.

287. I. N. Kiolin, V. A. Volkov, A. A. Aliev, and A. G. Kuznetsov, *Steel USSR*, vol. 7, no. 1, pp. 53–54.

288. N. Komutsu, M. Oboyashi, and J. Endo, *J. Japan Inst. Met.*, vol. 38, 1974, pp. 481–486.

289. H.-J. Hunger and G. Trute, *Heat Treat. Met.*, 1994.2, pp. 31–39.

290. P. Goeuriot, F. Thevenot, J. H. Driver, and A. Laurent, *Trait. Therm.*, vol. 152, 1981, pp. 21–28.

291. L. S. Lyakhovich, *Improving the Life of Forming Tools by Chemico-Thermal Treatment,* NIITI, Minsk, 1971 (in Russian).

292. K. Hosokawa, Y. Yamashita, M. Veda, and T. Seki, *Kinzoku Hyomen Gitjutsu,* vol. 23, no. 4, translation RTS 7945, 1972, pp. 211–216.

293. H. Orning and O. Schaaber, *Hart.-Tech. Mitt.*, vol. 17, no. 3, translation BISI 3953, 1962, pp. 131–140.

294. H. C. Fiedler and R. J. Sieraski, *Met. Prog.*, vol. 99, no. 2, 1971, pp. 101–107.

295. A. Bonomi, R. Habersaat, and G. Bienvenu, *Surf. Technol.*, vol. 6, 1978, pp. 313–319.

296. V. Danek and K. Matiasovsky, *Surf. Technol.*, vol. 5, 1977, pp. 65–72.

297. K. Matiasovsky, M. C. Paucirova, P. Felner, and M. Makyta, *Surf. Coat. Technol.*, vol. 35, 1988, pp. 133–149.

298. L. P. Skugorawa, V. I. Shylkov, and A. I. Netschaev, *Metalloved. Term. Obra. Met.*, no. 5, 1972, pp. 61–62.

299. F. Hegewaldt, L. Singheaser, and M. Turk, *Hart.-Tech. Mitt.*, vol. 39, no. 1, 1984, pp. 7–15.

300. H.-J. Hunger and G. Lobig, *Thin Solid Films*, vol. 310, 1997, pp. 244–250.

301. E. Filep, Sz. Farkas, and G. Kolozsvary, *Surf. Engrg.*, vol. 4, 1988, pp. 155–158.

302. A. M. Staines, *Met. Mater.*, vol. 1, 1985, pp. 739–745.

303. P. Casadesus, C. Frantz, and M. Gantois, *Met. Trans.*, vol. 10A, 1979, pp.1739–1743.

304. A. Raveh, A. Inspektor, U. Carmi, and R. Avni, *Thin Solid Films*, vol. 108, 1983, pp. 39–45.

305. T. Wierzchon, J. Bogacki, and T. Karpinski, *Heat Treat. Met.*, 1980.3, p. 65.

306. E. R. Cabeo, L. Gunther, S. Biemer, K.-T. Rie, S. Hoppe, and M. Frick, *Vak. Forsch. Prax*, vol. 11, no. 2, 1999, pp. 92–95.

307. S. A. Isakov and S. A. Al'tshuler, *Transl. Metalloved. Term. Obra. Met.*, no. 3, March 1987, pp. 25–27.

308. A. V. Matuschka, N. Trausner, and J. Zeise, *Hart.-Tech. Mitt.*, vol. 43, no. 1, 1988, pp. 21–25.

309. R. Chatterjee-Fischer, *Met. Prog.*, vol. 129, no. 5, 1986, pp. 24, 25, 37.

310. S. Ya Pasechnik et al., in *Protective Coatings on Metals*, vol. 4, ed. G. V. Samsonov, Consultants Bureau, New York, 1972, pp. 37–40.

311. N. G. Kaidash et al., in *Protective Coatings on Metals*, vol. 4, ed. G. V. Samsonov, Consultants Bureau, New York, 1972, pp. 149–155.

312. Z. Zhengxin and L. Fengzhen, *Proceedings 4th International Congress on Heat Treatment of Materials*, June 3–7, 1985, Berlin, pp. 1169–1176.

313. G. V. Zemskov et al., *Izv. V.U.Z. Chernaya Metall.*, vol. 10, translation BISI 15286, 1976, pp. 130–133.

314. R. L. Kogan et al., *Zashch. Pokrytiya Met.*, vol. 10, translation VR/1103/77, 1976, pp. 100–102.

315. T. Arai and S. Harper, *Metals Handbook*, vol. 4, 10th ed., ASM International, Materials Park, Ohio, 1991, pp. 448–453.

316. T. Arai, H. Fujita, Y. Sugimoto, and Y. Ohta, in *Proceedings of the 6th International Congress on Heat Treatment of Materials*, ed. G. Krauss, ASM International, Materials Park, Ohio, 1988, pp. 49–53.

317. T. Arai and N. Kamatsu, an article from research and development in Japan awarded the Okochi Memorial Prize, 1981.

318. T. Arai, *J. Heat Treating*, vol. 1, no. 2, 1979, pp. 15–22.

319. S. B. Fazluddin, K. Koursaris, C. Ringas, and K. Cowie, in *Surface Modification Technologies VI*, eds. T. S. Sudarshan and J. F. Braza, TMS, Warrendale, Pa., 1993, pp. 45–60.

320. T. Arai and T. Iwama, *Proceedings: 11th International Die Casting Congress*, Cleveland, Ohio, June 1981, G–T 81-092, pp. 1–6.

321. T. Arai, private communication, 2000.

322. T. Arai, H. Fujita, Y. Sugimoto, and Y. Ohta, *J. Materials Eng.*, vol. 9, no. 2, 1987, pp. 183–189.

323. T. Arai, *Toughness of Thin Hard Coated Steels*, presented at ASM–TMS Materials Week, Indianapolis, IN, 1997.

324. H. C. Child, *Met. Mater. Technol.*, vol. 13, no. 6, June 1981, pp. 303–309.

325. T. Arai and H. Oikawa, *Proceedings International Institute for Science Symposium*, Nov. 4–7, 1987, pp. 1385–1390.

326. T. Arai, *Wire,* vol. 31, no. 3, pp. 102–104, 208–210.

327. T. Arai, in *Surface Modification Technologies III*, eds. T. S. Sudarshan and D. G. Bhat, TMS, Warrendale, Pa., 1990, pp. 587–598.

328. T. Arai and S. Moriyama, *Thin Solid Films*, vol. 249, 1994, pp. 54–61.

329. S. Hotta, K. Saruki, and T. Arai, *Surf. Coat. Technol.*, vol. 70, 1994, pp. 121–129.

330. S. Hotta, Y. Itou, S. Saruki, and T. Arai, *Surf. Coat. Technol.*, vol. 73, 1995, pp. 5–13.

331. H. Fujita and T. Arai, *4th International Congress on Heat Treatment of Materials*, June 3–7, 1985, Berlin, 1985, pp. 1109–1124.

332. T. Arai, *TRD Method*, presented at International Congress for Surface Finishing, Interfinish 1992, Oct. 5–8, 1992, Sao Paulo.

333. Y. Sugimoto and T. Arai, *Proceedings of the Japan International Tribology Conference*, Nagoya, 1990, pp. 1981–1985.

334. N. Nakanishi, H. Takeda, H. Tachikawa, and T. Arai, in *8th International Conference on Heat Treatment of Materials*, ed. I. Tamura, The Japan Society for Heat Treatment Research Institute for Applied Sciences, 1992, pp. 507–510.

CHAPTER 17
DEFECTS AND DISTORTION IN HEAT-TREATED PARTS

17.1 INTRODUCTION

Quality deficiency of heat-treated parts is attributed to a number of causes, including faulty heat treatment practices, the use of an inappropriate steel grade, defective material, improper machining or finishing, and poor design. Faulty heat treatment or material selection that leads to the attainment of the wrong microstructures and inadequate mechanical properties may develop residual stresses within a part which are responsible for dimensional changes (distortion), and contribute (along with defective material, poor machining, and poor design) to quench cracking. Whereas quench cracking is a fairly rare occurrence to many, shape and size changes and dimensional stability are, in general, very much ongoing problems to most. This chapter deals with residual stresses, their influence on dimensions and on cracking, and their potential to adversely affect service performance.

Heat treaters of steel machine parts do not, as a rule, employ temperatures above 1000°C (1832°F) and, therefore, are not likely to induce either the *overheated* or the *burned* condition. The concern regarding the avoidance of these conditions is more common in the domain of the rolling mills and forgemasters who heat stock to temperatures in excess of 1100°C (2012°F): how much above this temperature depends upon the alloy and the piece size. Nevertheless, steel users should be aware of what *overheating* and *burning* are, their significance, and how to identify either condition. Overheating and burning fall into the defective-material category and, therefore, are included in this chapter.

Some of these conditions result in a characteristic appearance of the treated parts that can be easily recognized by simple inspection. Some of these factors do not produce any distinguishing features in the semifinished or finished part. In particular, some of the visual evidence does not recognize the presence of overheating and burning and the development of residual stresses leading to distortion, quench cracking, and eventual failure of the heat-treated parts; metallurgical laboratory examination is needed to establish those problems that contribute significantly to the service performance of the part. Tool designers must also be aware of the problems and difficulties in manufacture, heat treatment, and use.

17.2 OVERHEATING AND BURNING OF LOW-ALLOY STEELS

When low-alloy steels are heated to high temperature (usually > 1200°C, or 2200°F), prior to hot mechanical working (such as forging) for a long period, a deterioration in the room-temperature mechanical properties (particularly tensile ductility and impact strength or toughness) can be obtained after the steel has been given a final heat treatment (comprising reaustenitizing, quenching, and tempering).[1-3] Linked with the impaired mechanical properties is the appearance of intergranular matte facets on the normal ductile fracture surface of an impact specimen. (These facet surfaces correspond to prior-austenite grain boundaries formed during overheating.) This phenomenon is known as *overheating* and has been a matter of concern, especially in the case of steel forgings. Overheating has also been noticed in steel castings (due to variation in pouring temperature and effectiveness of the proprietary grain inoculants applied to the mold surface for control of cast surface quality or for control of grain size when present in the steel), in heavily ground parts, and in affected zones of welds.[4] The usual practice is to reject the overheated products as being unsuitable for service.

It has been established that overheating is essentially a reversible process. It is caused by the solution of MnS particles in austenite during heating or reheating at high temperatures; the amount of MnS in solution increases with temperature, and its subsequent reprecipitation during cooling occurs at intermediate cooling rates as very fine (~0.5-μm) arrays of α-MnS particles on the austenite grain boundaries. On subsequent heat treatment, the intergranular network of sulfides may provide a preferential, lower-energy fracture path. As a result, when impact-loaded, a ductile intergranular fracture develops due to decohesion of the MnS/matrix interface and progress of microvoid coalescence. Figure 17.1a and b shows the usual appearance of the fracture path surface at different magnifications, revealing small MnS particles within the dimples.[1]

When the low-alloy steel is heated prior to hot working at too high a temperature (normally > 1400°C, or 2550°F), incipient (local) melting occurs at the austenite grain boundaries as a result of the segregation of phosphorus, sulfur, and carbon.[5] During cooling, initially dendritic sulfides (probably type II-MnS) form (Fig. 17.2a)[6] within the phosphorus-rich austenite grain boundary, which then transforms to ferrite. This results in excessively weak boundaries. Subsequent heat treatment provides a very poor impact strength and almost completely intergranular fracture surface after impact failure. This phenomenon is termed *burning*. Burning thus occurs at a higher temperature than overheating and is irreversible. If this occurs during hot working (e.g., forging), it usually leads to tearing or rupture of the steel due to the grain boundary liquation during cooling or subsequent heat treatment.[4,6] Figure 17.2b shows impact energy curves for En 111 alloy (composition: Fe-0.35C-0.7Mn-1.25Ni-0.6Cr) steel specimens heated to different temperatures from the usual soaking range to the burning range (without undergoing hot-working operation).[6]

17.2.1 Detection of Overheating

There are two basic methods for the determination of the occurrence of overheating, namely, fracture testing and metallography (or etch testing). Overheating may also be detected by a decrease in mechanical properties. But such changes are not

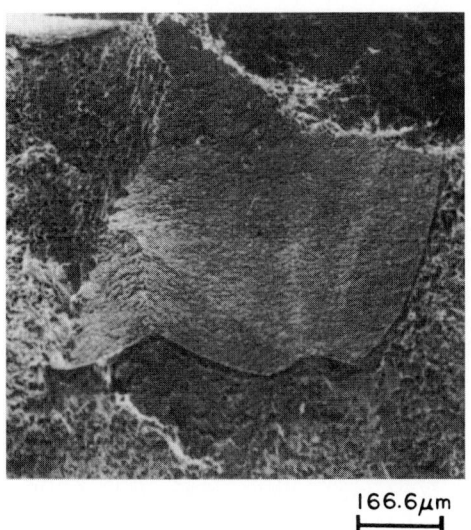

166.6μm

(a)

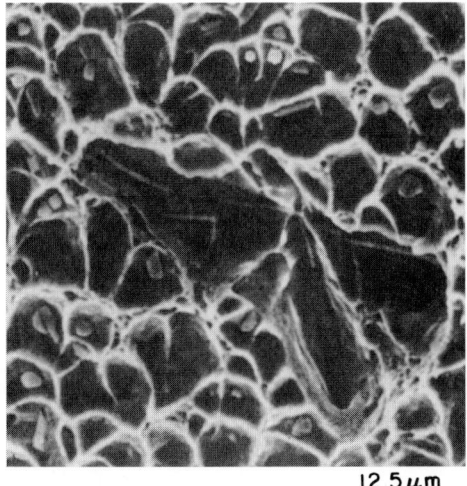

12.5μm

(b)

FIGURE 17.1 Fracture surface of an impact-loaded specimen. (*a*) Appearance of intergranular fracture of 4.25Ni-Cr-Mo steel containing 0.34% Mn and 0.008% S, in fully heat-treated condition but after cooling from 1400°C (2550°F) at 10°C/min (20°F/min). (*b*) Same specimen as in (*a*) but at higher magnification, showing ductile dimples nucleated by MnS particles precipitated at austenite grain boundaries. (*Courtesy of The Institute of Materials, UK.*)

very marked unless the overheating temperature is high or too prolonged; in some instances the mechanical properties do not change, even after the observation of extensive faceting. Usually the two methods mentioned above should be used in conjunction with some measure of toughness by impact or other testing in order to get a clear understanding of the degree and severity of overheating.[2]

17.2.1.1 Fracture Testing. The direction of fracture testing is important in steels manufactured by conventional methods. It has been observed by some workers[7] that

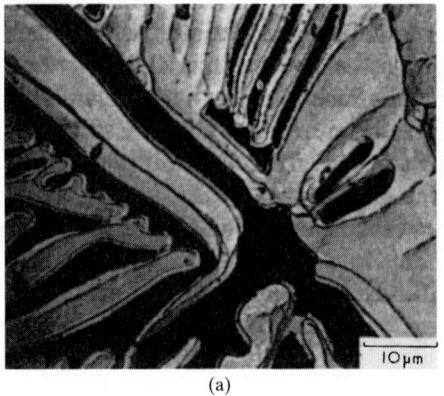

(a)

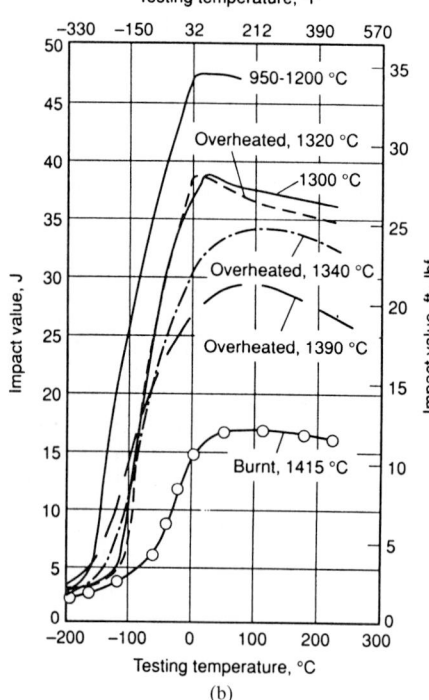

(b)

FIGURE 17.2 (*a*) Fe-1Mn-0.4C-0.02S steel air-cooled from 1445°C (1633°F) and then fractured (*After Brammar*). Extraction replica ×1000. (*Courtesy of R.W.K. Honeycombe.*) (*b*) Impact energy values versus test temperature for En 111 alloy steel specimens heated to the indicated temperature for 1 hr, oil-quenched, and tempered for 1 hr at 675°C (1245°F).[6]

the longitudinal fracture test specimens (parallel to the rolling direction) do not exhibit faceting until the corresponding transverse fractures display extensive faceting. However, the testing direction in high clean steels, including electroslag-refined (ESR) and vacuum arc remelted (VAR) steels, has been found to be insignificant.[8]

The scanning electron microscope is considered to be the best and most convenient tool to detect the facets on the overheated fracture surfaces. These facets are characterized by small, well-defined, ductile dimples; each dimple is usually nucleated, by fine arrays of inclusion particles: α-MnS particles (Fig. 17.1) in Mn-bearing steels[9,10] or chromium sulfides in Mn-free steels.[11,12]

It is now recognized that the fracture test specimen should always be tested in the toughest possible state [e.g., quenched and highly tempered (in the range of 600 to 650°C, or 1100 to 1200°F) steels after high-temperature austenitization] because this condition is most prone to overheating effects. Baker and Johnson[5] have suggested that an increased proportion of facets in the fracture specimens with increasing tempering temperature is attributed to the corresponding increase of the plastic zone size. In this case a slight amount of weakening will be sufficient to impart faceting because the grain boundary strength becomes lower.[2] Note that the existence of facets in the fractured specimens is not always associated with a lowering of impact strength.[13]

17.2.1.2 Metallography (or Etch Testing). The most widely used etchant technique employs Austin's reagent (aqueous solution of 10% nitric acid and 10% sulfuric acid), ammonium persulfate, molten zinc chloride, saturated solution of picric acid at 60°C (140°F), or an electrolytic etch based on saturated aqueous ammonium nitrate. Table 17.1 shows the etching characteristics of overheated and burned steels.[14] The etchant procedure with Austin's etchant is as follows: The sectioned specimen is etched for 30s in the etchant, removed, and washed off; and this sequence is repeated three times. If the steel has been overheated, the original austenite grain boundaries will be preferentially attacked, and a black network of etch pits will be observed under the microscope.[15] According to Preece and Nutting,[14] the best results are obtained when ammonium nitrate etch is applied on the sectioned steel specimen in the fully heat-treated condition where this etchant preferentially attacks the matrix (original austenite grains), leaving the grain boundary unaffected (which appears as a white network). Bodimeade[16] concluded that all these etchants did not cope with mildly overheated low-sulfur steels. Table 17.2 is a summary of the results of potentiostatic etching techniques carried out by McLeod[13] using nitric-sulfuric, saturated aqueous picric acid (at 60°C, or 140°F), and ammonium nitrate etchants. He considered that when the suitable etching conditions were established, the potentiostatic etching method rendered more reliable and reproducible results compared to the conventional techniques. However, the same problem with mildly overheated low-sulfur steels persisted. Hence, the use of etch tests for low-sulfur low-alloy steels is not recommended for the detection of mild overheating.

17.2.2 Detection and Effects of Burning

Burning is not commonly encountered. The two etchants (namely, nitric-sulfuric acid and ammonium nitrate solution) used for overheating can be successfully employed to detect burning. When applied to burned steels, these etchants react in a manner opposite to that of overheated steels. Preece and Nutting[14] found ammonium nitrate solution to be the ideal reagent to detect this phenomenon. Other reagents are Stead's and Oberhoffer's reagents, which may also be used to check the burning effect. However, these etchants are unable to differentiate between overheated and nonoverheated steels.

TABLE 17.1 Etching Characteristics of Overheated and Burned Steels[14]

Reagent	Method	Action on overheated steel	Action on burned steel
2.5% nitric acid in ethyl alcohol	Swab surface for 30 s	May produce grain contrast, but not indicative of overheating	White boundaries outlining preexisting austenite grains
Saturated aqueous solution of ammonium nitrate	Electrolytic, specimen anode, current density 1.0 A cm^{-2} (6.5 A in.$^{-2}$)	White boundaries outlining preexisting grains	Black boundaries outlining preexisting austenite grains
Aqueous 10% nitric acid + 10% sulfuric acid	Etch for 30 s, swab surface; repeat three times, then repolish lightly	Black boundaries outlining preexisting austenite grains	White boundaries outlining preexisting austenite grains
85% orthophosphoric acid (Fine's reagent)	Electrolytic, specimen anode, current density 0.15 A cm^{-2} (1.0 A in.$^{-2}$), etching time 15 min	Does not differentiate between overheated and nonoverheated steel	Attacks inclusions at grain boundaries
Oberhoffer's reagent	Swab surface for 30 s	Does not differentiate between overheated and nonoverheated steel	Shows phosphorus segregation at grain boundaries

Courtesy of The Institute of Materials, UK.

TABLE 17.2 Summary of Potentiostatic Etching Experiments[13]

Solution	Anodic loop voltage, mV	Observed effect	Best etching conditions		Comments
			Voltage, mV	Observed effect	
Saturated aqueous ammonium nitrate	−400	Slight general etching	2200 (for 2 min)	Classic white boundaries on a dark background	Operates best in the transpassive region at >+1500mV; time at any potential is important *Underetching*: random array of black pits *Overetching*: uniform black surface film
Aqueous 10% nitric acid + 10% sulfuric acid	200	Vigorous dissolution of specimen; formation of flaky black film	None	...	Most aggressive etchant of the three examined
	−250	Milder attack; large black pits in mildly etched matrix	About −250 (for 30s)	Discontinuous array of grain boundary pits and some random pits within grains	Polish lightly after etching to eliminate matrix etching effects
Saturated aqueous picric acid at 60°C (140°F)	100	No real, positive indication of overheating	None	...	Anodic loop very weak, necessitating long etching times because current density is very low; Teepol additions gave no improvement

Courtesy of The Institute of Materials, UK.

17.7

17.2.3 Factors Affecting Overheating

The occurrence and severity of overheating depend principally on important factors, notably steel composition (mainly sulfur content), soaking temperature, grain size, cooling rate through the overheating range, degree of hot reduction, and method of manufacture. On the other hand, the amount of faceting found on the test fracture is a function of the heat treatment, particularly the tempering temperature, the test temperature, the test specimen orientation, and the amount of deformation after sulfide reprecipitation.

Composition. Sulfur is the constituent that greatly influences overheating. For steels with less than 0.002 wt% sulfur, overheating does not occur because of the very low volume fraction of sulfides formed. However, the commercial production of such very-low-sulfur steels (for example, ESR steels) is expensive. Above this level of sulfur, the overheating onset temperature rises with the increasing amount of sulfur. It has been explained that steels with low sulfur content (0.01 to 0.02%) are more prone to this defect than those with high sulfur content (>0.3%) because the transgranular strength is high, and therefore a small amount of grain boundary sulfide precipitation is enough to induce intergranular failure.[17] The phosphorus content has been regarded with the greatest concern in connection with burning. At constant phosphorus level, there is an increase in the overheating temperature with the increase of sulfur content, whereas the burning onset temperature decreases. Burning temperature is reduced with the increase in phosphorus content. At low sulfur contents, a wide gap between overheating and burning exists. For example, in the case of vacuum remelted steels, the temperature gap between the onset of overheating and burning is ~300 to 400°C (~570 to 750°F), and there is a remote possibility of burning occurring within the forging range, unless the overheating is severe.[2] However, at high sulfur content, the gap becomes narrow.

Temperature. To avoid overheating, care must be exercised in choosing a correct heating temperature so that uneven heating, flame impingement, and so forth do not occur.[3]

Cooling Rates. The cooling rate through the overheating range affects the size and dispersion of intergranular α-MnS particles. The intermediate cooling rate generally employed, 10 to 200°C/min (20 to 360°F/min), gives rise to maximum faceting as well as to the greatest loss in impact strength. However, slow and rapid cooling rates will suppress overheating. At very slow cooling rates, the sulfide particles become large, fewer in number, and more widely dispersed; and they have no more deleterious effects than the other inclusions already present. At rapid rates, the sulfide inclusions are too fine to produce any damaging effect.[18]

Methods of Manufacture. Electroslag remelted steels are less susceptible than vacuum remelted steels, presumably due to the difference in oxygen level. Similarly, nickel steels are more prone to overheating. Vacuum-remelted steels have a lower overheating temperature than some comparable air-melted steels.

17.2.4 Prevention of Overheating and Burning

To prevent overheating of steels, a properly selected temperature should lie between a temperature low enough for the metal to be safe and one high enough to be sufficiently plastic. The better the temperature control, the better the compromise.

Severe overheating can be reduced to mild overheating by soaking the steel at 1200°C (2200°F); with care, it may be removed completely. Hot-working through the overheating range to a low finish temperature is also reported to remove the effects of overheating.

The alloying additions with a greater sulfide-forming tendency, such as calcium,[†] zirconium, cerium (~0.03% of the melt), or mixed rare earth metal (in the form of misch metal containing 52% Ce, 25% La, and 12% Nd), have been shown to prevent overheating by increasing significantly both the overheating temperature and the mechanical properties of the steel (e.g., ductility and toughness). Provided that a high Ce/S ratio (>2) existed, a complete change in sulfide morphology occurred in low-alloy steels where the elongated MnS inclusion occurring in the untreated steel was totally replaced by small globular type-I rare earth sulfides and oxysulfides of high thermal stability even after austenitizing at 1400°C (2550°F).[2] This treatment does not show intergranular faceting. Burning can also be avoided in the same way by treating with calcium, zirconium, cerium, or mixed rare earth addition to form refractory, less-soluble sulfides.

Control of Cooling Rates. Control of cooling rates is not a practical method for large forgings because extremely slow cooling is prohibitively time-consuming and causes excessive scaling and decarburization, and rapid quenching from high temperatures produces cracking and distortion of the parts.[2]

17.2.5 Reclamation of Overheated Steel

Severely overheated steels can often be completely restored by any of the following heat treatments:

1. Repeated normalizing (as many as six) starting at temperatures up to 50 to 100°C (90 to 180°F) higher than usual, followed by a standard normalizing treatment.[2]
2. Repeated oil-hardening and tempering treatments after prolonged soaking at 950 to 1150°C (1740 to 2100°F) in a noncarburizing (neutral) atmosphere. Rehardening more than three times is not advisable.
3. Soaking at 900 to 1150°C (1650 to 2100°F) for several hours. This causes growth of MnS particles by the Ostwald ripening process and results in an excessive scale formation and a loss of dimensional accuracy of the forgings.

Alternatively, a large extent of hot reduction minimizes or reduces overheating.[19]

17.3 RESIDUAL STRESSES

During heat treatment (while hot), thermal and transformation residual stresses can cause yielding (plastic flow), hence growth and/or distortion of a part. When the part is cold, the residual stresses should not exceed the yield strength of the material, so their effect on dimensions is elastic.[20]

Heat treatment often causes stress- and strain-related problems such as residual stress, quench cracks, and deformation and/or distortion. The *residual stress* may be

---

[†] Calcium-treatment of some machinable steels is for modification of the oxides.

defined as the self-equilibrating internal or locked-in stress remaining within a body with no applied (external) force, external constraint, or temperature gradient.[21,22] There are two types of residual stresses:

1. *Macro-* or *long-range residual stress* is a first-order stress that represents an average of body stresses over all the phases in polyphase materials. Macroresidual stresses act over large regions compared to the grain size of the material. Traditionally, engineers consider only this type of residual stress when designing mechanical parts.

2. *Microresidual stress*, also termed *tesselated stress* or *short-range stress*, is a second-order or texture stress which is associated with lattice defects (such as vacancies, dislocations, and pile-up of dislocations) and fine precipitates (e.g., martensite).[23–25] Microresidual stress is the average stress across one grain or part of the grain of the material. This information is indispensable in studying the essential behavior of material deformation.

These two types of residual stresses may also be classified further as a tensile or compressive stress located near the surface or in the body of a material. This section focuses on the effects, development, control, and measurement of long-range residual stresses.

17.3.1 Effects of Residual Stress

The major effects of residual stress include dimensional changes and either an increased or decreased resistance to crack initiation and propagation, depending on whether the surface residual stresses are compressive or tensile.[20] Dimensional changes occur when the residual stress (or a portion of it) in a body is eliminated. In terms of crack initiation, residual stresses can be either beneficial or detrimental, depending on whether the stress is tensile or compressive.

Compressive Residual Stress.   Because residual stresses are algebraically summed with applied stresses, residual compressive stresses in the surface layers are generally helpful because the built-in compressive stresses can reduce the effects of imposed tensile stresses that may produce cracking or failure. Compressive stresses therefore contribute to the improvement of fatigue strength, fatigue life, and resistance to stress corrosion cracking in a part and an increase in the bending strength of brittle ceramics and glass.[25]

Figure 17.3 shows that the endurance limit fatigue strength of selected steels increases with the surface residual compressive stress developed by specific heat treatment and surface processing. It is also apparent that, in the presence of high compressive stress, a poor microstructure in steel samples has a small influence on good endurance limit fatigue strength.[26–28] These fatigue improvements are of great significance in components, particularly where stress raisers, such as notches, keyways, oil holes, and so forth, are highly desirable in the design of components (e.g., crankshafts, half-shafts, and so on).[29] Many fabrication methods have been developed to exploit this phenomenon. Prestressed parts (including shrink-fits, prestressed concrete, interference fits, bolted parts, coined holes, wire-wound concrete pipe), mechanical surface working processes (such as conventional shot-peening, gravity peening, laser shock peening, roller burnishing, surface rolling, lapping, and so on) of hardened ferrous and nonferrous alloys, and surface harden-

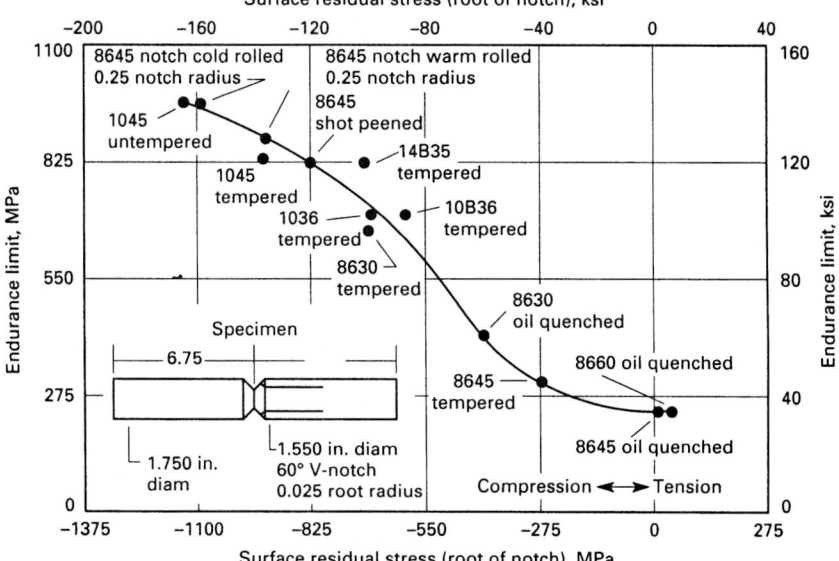

FIGURE 17.3 Effect of surface residual stress on the endurance limit of selected steel. All samples were water-quenched except as shown, and all specimen dimensions are given in inches.[26,27] (*Reprinted by permission of John Wiley & Sons, New York.*)

ing treatments are widely used to produce residual compressive stresses at the component surface.

Shot-peening further improves the surface compressive residual stress (Fig. 17.4*a* and *b*)[30–33] and skin hardness in carburized steel, which leads to substantial increases in bending fatigue performance and inducement of more random finish by eliminating directional machining marks.[34]

Figure 17.5*a* and *b* shows the residual stress distribution produced by air-blast shot-peening, gravity peening, and laser shock peening (LSP) in Ti-6Al-4V and Inconel 718 after final machining. It is clear from these figures that LSP residual stress distribution diminishes linearly with depth, without reducing the surface compression stress.[35]

In roller burnishing, a rolling force is applied to the surface, using either rollers or spherical bearings to locally strengthen the surface of the part (especially at fillets and in grooves) by producing compressive residual stress in a case hardened surface. It may also be applied to non-case-hardened parts (e.g., fillet rolling of some crankshafts).[36]

Tensile Residual Stress. Tensile residual stress at the surface of a part is usually undesirable because it adds to any applied tensile stresses and effectively increases the stress levels; it may cause unpredicted stress-corrosion cracking (due to the combined effect of stress and environment), quench cracking, and grinding checks at low external stresses, and tend to reduce fatigue life and strength of a part. In this case the extent of residual stresses may be closer to or even larger than the strength of the material.

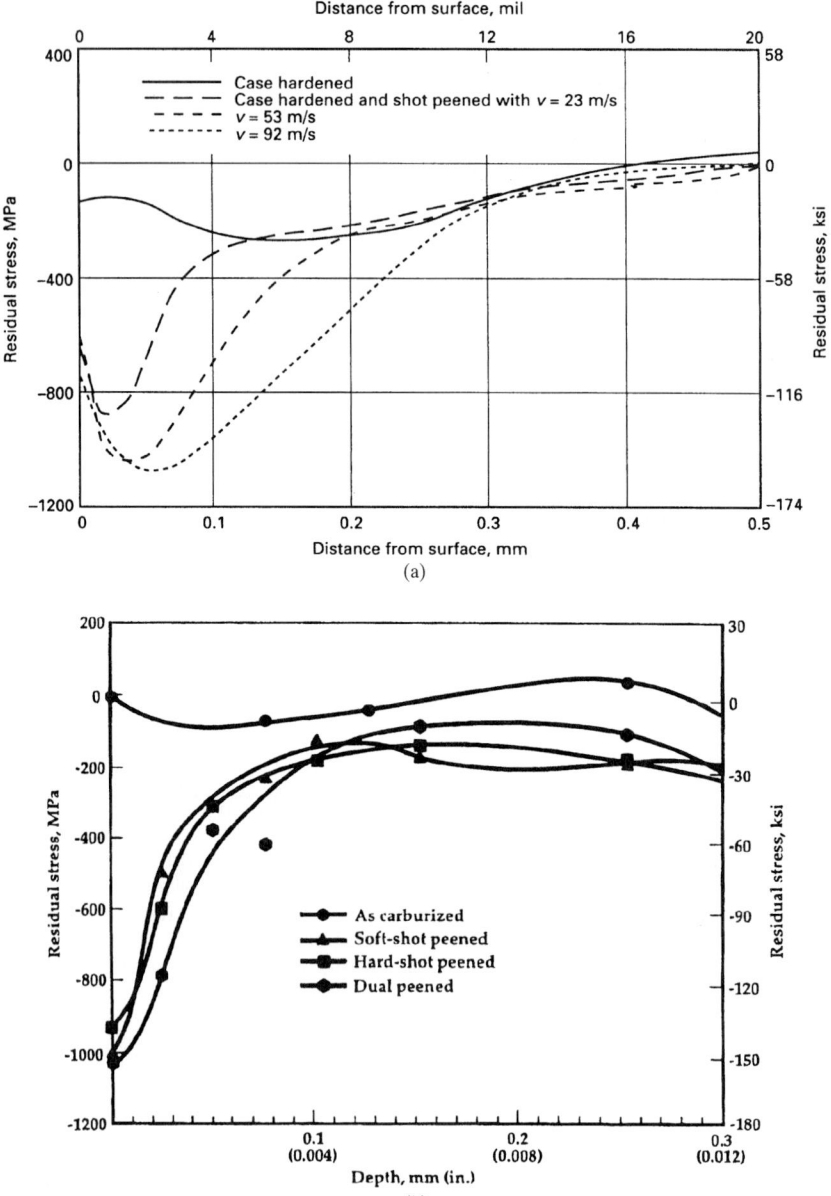

FIGURE 17.4 (*a*) Effect of shot-peening at different shot velocities on compressive residual stress in carburized 16MnCr5 steel (1.23% Mn, 1.08% Cr).[30,31] (*b*) Residual stress as a function of distance from the surface of the carburized SAE 4320 in the as-carburized condition and after various shot-peening treatments.[32,33]

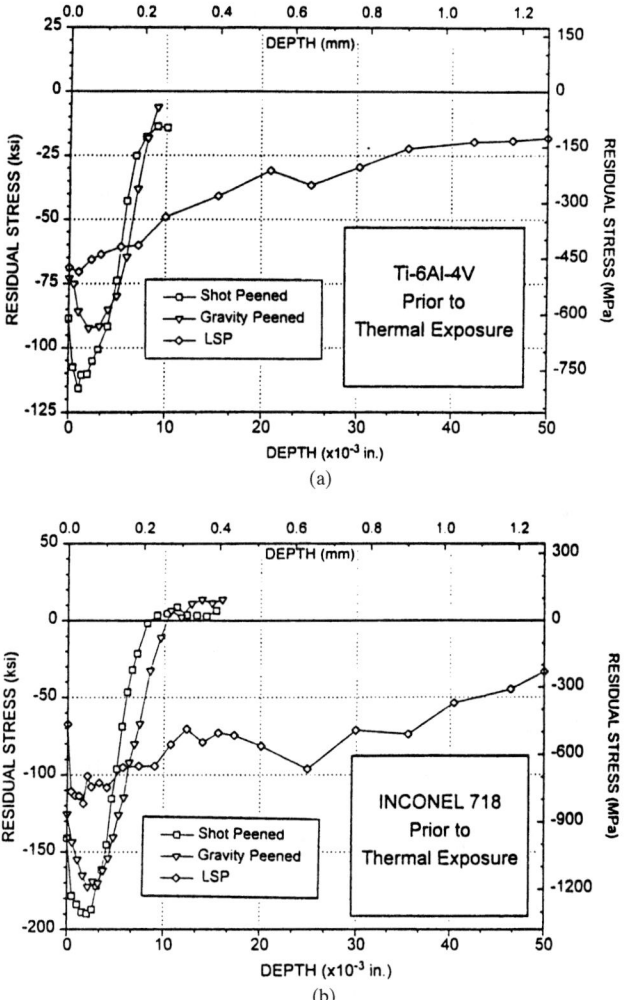

FIGURE 17.5   Residual stress distribution developed in (*a*) Ti-6Al-4V and (*b*) Inconel 718 test specimens.[35] (*Courtesy of P. S. Prevey, D. J. Hornbach, and P. W. Mason.*)

   Residual tensile stresses in the interior of a component also may be damaging because of the existence and consequence of defects that serve as stress raisers in the interior part. The uncommon phenomenon of delayed cracking, in the absence of adverse environments and large applied stresses, has now been attributed to the action of residual stresses on minute defects in the material.[29] For example, a 17.5-cm-diameter (6.9-in.) × 125-cm-long (49.2-in.) steel shaft exploded into several pieces while lying free of any applied loads, on a laboratory floor. Under normal loading, it would have required a tensile strength larger than 150 MPa (22 ksi) to rupture the shaft. Hence, the understanding of residual stress formation is very important, and this must be given due consideration in the manufacture and performance analysis of processed parts.[29]

17.3.2 Development of Residual Stress in Processed Parts

Variations in stresses, temperature, and chemical species within the body during processing cause the production of macroresidual stresses. Various manufacturing processes such as forming, machining and assembling, heat treatment, shot-peening, casting, welding, flame cutting, and plating render their characteristic residual stress pattern to processed parts. Table 17.3 lists a summary of compressive and tensile residual stresses at the surface of parts fabricated by common manufacturing processes.

In heat-treated parts, residual stresses may be classified as those caused by a thermal gradient alone and those caused by a thermal gradient in combination with a structural change (phase transformation). When a steel part is quenched from the austenitizing temperature to room temperature, a residual stress pattern is established due to a combination of thermal gradient and local transformation-induced volume expansion.

Thermal contraction develops nonuniform thermal (or quenching) stress due to the different rates of cooling experienced by the surface and the interior of the steel part. Ferrite-to-austenite transformation involves contraction while the part is thermally expanding, hence the development of residual stresses. Because of the very low yield strength of steel at such temperatures, these residual stresses could con-

TABLE 17.3 Summary of Compressive and Tensile Residual Stresses at the Surface of Parts Created by Common Manufacturing Processes[25]

Compression at the surface	Tension at the surface
Surface working: shot-peening, surface rolling, lapping, and so on	Road or wire drawing with deep penetration
Rod or wire drawing with shallow penetration[†]	Rolling with deep penetration
	Swaging with deep penetration
Rolling with shallow penetration[†]	Tube sinking of the outer surface
Swaging with shallow penetration[†]	Plastic bending of the shortened side
Tube sinking of the inner surface	Grinding: normal practice and abusive conditions
Coining around holes	Direct-hardening steel (through-hardened)[‡]
Plastic bending of the stretched side	Decarburization of steel surface
Grinding under gentle conditions	Weldment (last portion to reach room temperature)
Hammer peening	Machining: turning, milling
Quenching without phase transformation	Built-up surface of shaft
Direct-hardening steel (not through-hardened)	Electrical discharge machining
Case hardening steel	Flame cutting
Induction and flame hardening	
Prestressing	
Ion exchange	

[†] Shallow penetration refers to ≤1% reduction in area or thickness; deep penetration refers to ≥1%.
[‡] Depends on the efficiency of quenching medium. Reprinted by permission of Pergamon Press, Plc.

TABLE 17.4 Changes in Volume during the Transformation of Austenite into Different Phases[4]

Transformation	Change in volume, %, as a function of carbon content, %C
Spheroidized pearlite → austenite	$-4.64 + 2.21 \times (\%C)$
Austenite → martensite	$4.64 - 0.53 \times (\%C)$
Spheroidized pearlite → martensite	$1.68 \times (\%C)$
Austenite → lower bainite	$4.64 - 1.43 \times (\%C)$
Spheroidized pearlite → lower bainite	$0.78 \times (\%C)$
Austenite → upper bainite	$4.64 - 2.21 \times (\%C)$
Spheroidized pearlite → upper bainite	0

Courtesy of the Institute of Materials, U.K.

tribute to distortion.[20] Transformational volume expansion induces transformation stress arising from the transformation of austenite into martensite or other transformation products.[37] Table 17.4 lists the changes in volume during the transformation of austenite into different structural constituents.[38]

17.3.2.1 Thermal (Contraction) Residual Stresses.   The relation between the thermal stress σ_{th} during cooling (quenching) and the corresponding temperature gradient in the component is given by

$$\sigma_{th} = E \, \Delta T \, \alpha \qquad\qquad (17.1)$$

where E is the modulus of elasticity and α is the thermal coefficient of expansion of the material. It is thus apparent that thermal stresses are greatest for materials with high elastic modulus and coefficient of thermal expansion. Temperature gradient is also a function of thermal conductivity. Hence, it is quite unlikely for one to develop high-temperature gradients in good thermal conductors (e.g., copper and aluminum), but it is much more likely in steel and titanium.[39] Another term involving thermal conductivity, called *thermal diffusivity* D_{th}, is sometimes used in context with temperature gradient. It is defined as $D_{th} = k/\rho c$, where k is the thermal conductivity, ρ is the density, and c is the specific heat. It is clear that low D_{th} (or k) promotes large temperature gradient or thermal contraction. It should be emphasized that large size of the part and high heating or cooling rates (severity) of quenching medium also augment temperature gradients, leading to large thermal contraction.

Table 17.5 lists some of the relevant material properties that affect thermal and residual stresses.[39] The variations with temperature are important to heat-treat distortion (and development of residual stress).[36]

17.3.2.2 Residual Stress Pattern due to Thermal Contraction.   Residual stress is developed during heating or cooling of a solid part that involves thermal volume changes without solid-state phase transformation. This situation exists, e.g., when a steel part is heated to or cooled from a tempering temperature below A_1. These residual stresses rely on there being a thermal gradient; once a part has attained a uniform temperature throughout (either once hot or once cooled), there are no thermal gradients and, therefore, no consequential residual stresses.[20]

TABLE 17.5   Relevant Physical Properties in the Development of Thermal Stresses[41]

Metal	Modulus of elasticity		Coefficient of expansion		Thermal conductivity	
	GPa	psi × 10⁶	10⁻⁶/K	10⁻⁶/°F	W m⁻¹ k⁻¹	Btu in./ft²·h·°F
Pure iron (ferrite)	206	30	12	7	80	555
Typical austenitic steel	200	29	18	10	15	100
Aluminum	71	10	23	13	201	1400
Copper	117	17	17	9	385	2670
Titanium	125	18	9	5	23	160

Courtesy of Wolfson Heat Treatment Centre, England.

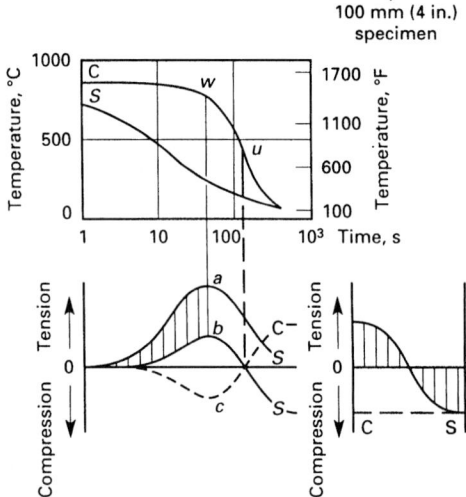

FIGURE 17.6   Development of thermal and residual stresses in the longitudinal direction in a 100-mm-diameter (4-in.) steel bar on water-quenching from the austenitizing temperature of 850°C (1560°F). Transformation stresses are taken into consideration.[42]

When one is considering a part that is already austenitized and then quenched, the stresses developed during the quench are of a high magnitude and in excess of the hot yield strength of cooling austenite. Hence some yielding will occur.[20] Figure 17.6 shows the development of longitudinal thermally induced residual stresses in a 100-mm-diameter (4-in.) steel bar on water-quenching from the austenitizing temperature of 850°C (1560°F).[40] At the start of cooling, the surface temperature S falls drastically compared to the center temperature C (top left sketch of Fig. 17.6). At time w, the temperature difference between the surface and core is at a maximum of about 550°C (1020°F), corresponding to a thermal stress of 1200 MPa (80 tons/in.²) due to linear differential contraction of about 0.6%, if relaxation does not take place. Under these conditions, tensile stresses are developed in the case with a maximum value of a (lower diagram), corresponding to time w in the upper diagram, and the core will contract, producing compressive stresses with a maximum of c (at time w). The combined effect of tensile and compressive stresses on the surface and core, respectively, will result in residual stresses as indicated by curve C, where a complete neutralization of stress will occur at some lower temperature u. Further decrease in temperature, therefore, produces longitudinal, compressive residual stresses at the surface and the tensile stresses at the core, as shown in the lower right-hand diagram of Fig. 17.6. Figure 17.7a is a schematic illustration of the distribution of residual stress over the diameter of a quenched bar due solely to thermal contraction in the longitudinal, tangential, and radial directions.[22]

The maximum residual stress attained on quenching increases as the quenching temperature and quenching power of the coolant are increased. Tempered glass is made by utilizing quenching techniques in which glass is heated uniformly to the annealing temperature and then surface-cooled rapidly by cold air blasts. This produces compressive surface stresses to counteract any tensile bending stress, if developed during loading of the glass, thereby increasing its load-carrying capacity.[43]

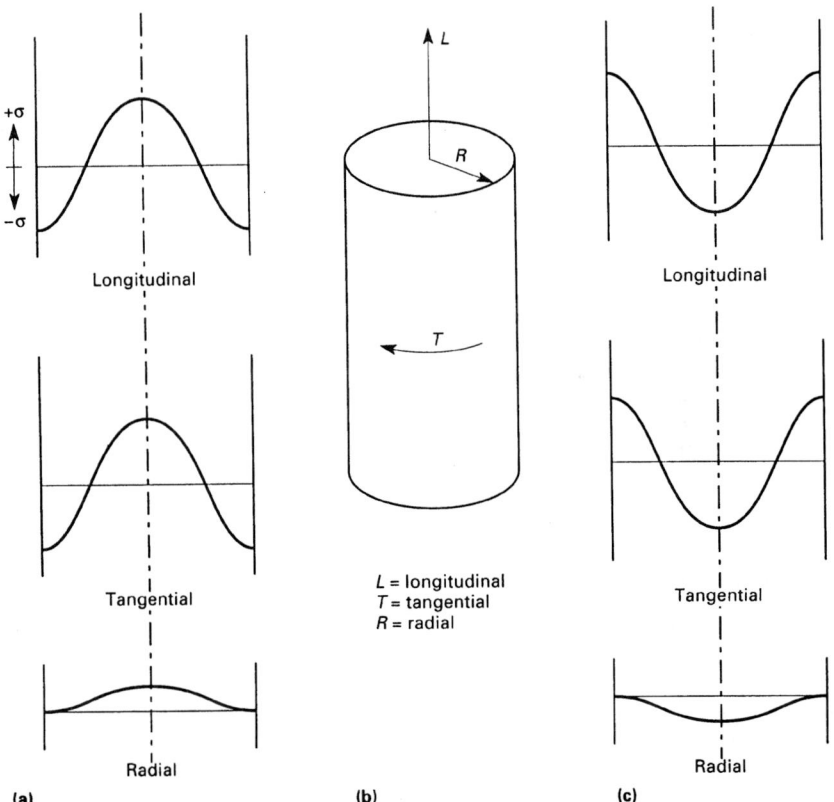

FIGURE 17.7 Schematic illustration of the distribution of residual stress over the diameter of a quenched bar in the longitudinal, tangential, and radial directions due to (*a*) thermal contraction and (*c*) both thermal and transformational volume changes. (*b*) Schematic illustration of orientation of directions.[22] (*Reprinted by permission of The McGraw-Hill Companies, Inc., New York.*)

17.3.2.3   Residual Stress Pattern due to Thermal and Transformational Volume Changes.[42]   During quench hardening of a steel (or other hardenable alloy) part, hard martensite forms at the surface layers, associated with the volume expansion, whereas the remainder of the part is still hot and ductile austenite. Later, the remainder austenite transforms to martensite, but its volumetric expansion is restricted by the hardened surface layer. This restraint causes the central portion to be under compression with the outer surface under tension. Figure 17.7*c* illustrates the residual stress distribution over the diameter of a quenched bar, showing volume expansion associated with phase transformation in the longitudinal, tangential, and radial directions.[22] At the same time during the final cooling of the interior, its thermal contraction is hindered by the hardened surface layers. This restraint in contraction produces tensile stresses in the interior and compressive stresses at the outer surface (Fig. 17.7*a*). However, the situation shown in Fig. 17.7*c* prevails, provided that the net volumetric expansion in the interior, after the surface has hardened, is larger than the remaining thermal contraction. In some particular conditions, these volumetric changes can produce sufficiently large residual stresses that can cause plastic deformation on cooling, leading to warping or distortion of the steel part. While plastic deformation appears to reduce the severity of quenching stresses, in most severe quenching the quenching stresses are so high that they do not get sufficiently released by plastic deformation. Consequently, the large residual stress remaining may reach or even exceed the fracture stress of steel. This localized rupture or fracture is called *quench cracking*.[42,43]

It should be emphasized again that for a given grade of steel, both large size of the part and higher quenching speed contribute to the larger value of thermal contraction, compared to the volumetric expansion of martensite. In contrast, when the parts are thin and the quenching rate is not high, thermal contraction of the part subsequent to the hardening of the surface will be smaller than the volumetric expansion of martensite. Similarly, for a given quenching rate, the temperature gradients decrease with decreasing section thickness, and consequently the thermal component of the residual stress is also decreased.[27]

Figure 17.8*a* shows the continuous cooling transformation diagram of DIN 22CrMo44 low-alloy steel exhibiting austenitic decomposition with the superimposed cooling curves of the surface and center in round bars of varying dimensions. If the large-diameter (100-mm, or 4-in.) bar is water-quenched (i.e., for slack quenching) from 850°C (1562°F), martensite transformation occurs at the surface, and pearlitic + bainitic transformation occurs at the center, resulting in a residual stress pattern (top of Fig. 17.8*b*) similar to that due solely to thermal stress (Fig. 17.7*a*). During the rapid quenching of the medium-size (30-mm, or 1.2-in.) bar diameter, the start of bainite transformation at the center coincides approximately with the transformation of martensite on the surface. This results in compressive stresses at both the surface and center, with tensile stresses in the intermediate region (middle of Fig. 17.8*b*). When the smaller-diameter (10-mm, or 0.4-in.) bar is drastically quenched (e.g., in brine), the entire bar transforms to martensite. This is associated with very little temperature variation between the surface and the center of the part. In this situation, tensile residual stress is developed at the surface and compressive stress at the center of the bar (bottom, Fig. 17.8*b*).[44,45]

Although the shallower hardening steels exhibit higher surface compressive stresses, deep hardening steels may develop moderately high surface compressive stresses with severe water quenching. When these deep hardening steels are through-hardened in a less efficient quenchant, they may exhibit surface tensile stresses.[27,41] Rose has pointed out the importance of transformations of core and surface before and after the stress reversal. According to him, the tensile surface

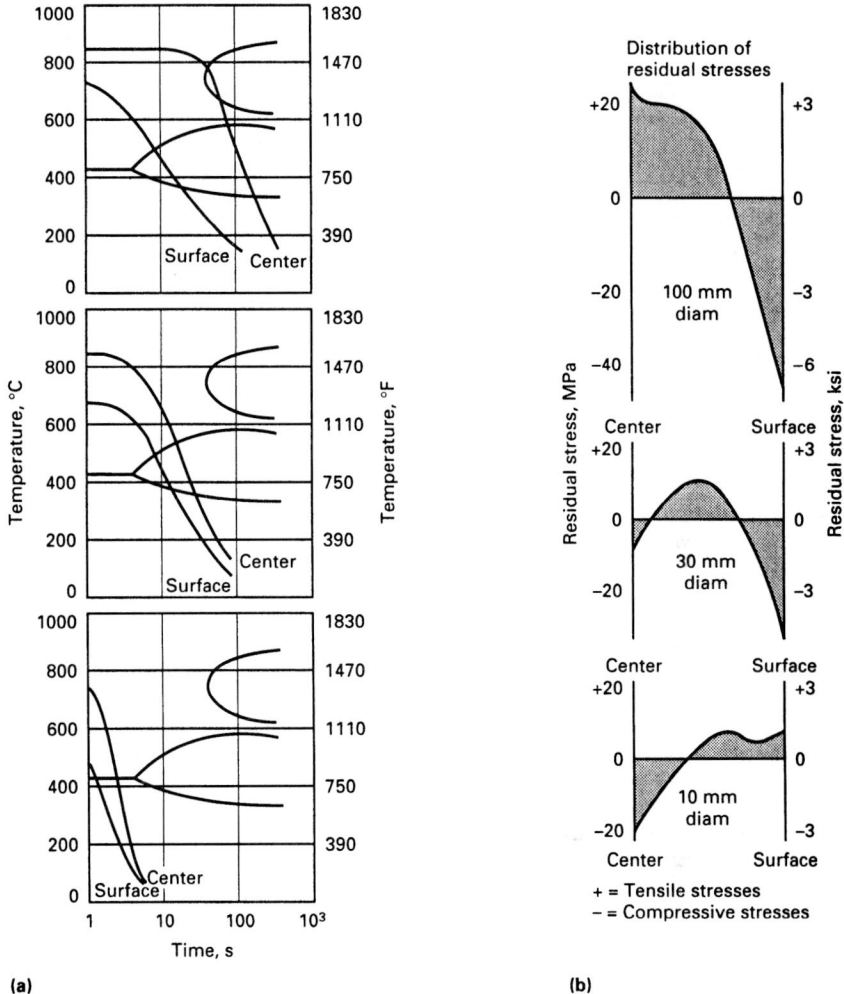

FIGURE 17.8   (*a*) Continuous cooling transformation diagrams of DIN 22CrMo44 steel showing austenitic decomposition with the superimposed cooling curves of the surface and center during water-quenching of round bars of varying dimensions. (*b*) The corresponding residual stress pattern developed because of thermal and transformational volume changes.[44,45] (*Reprinted by permission of Butterworths, London; after A. Rose.*)

residual stress occurs when the core transforms after, and the surface transforms before, the stress reversal (Fig. 17.7*c* and bottom of Fig. 17.8), whereas compressive surface residual stress takes place when the core transforms before, and the surface transforms after, the stress reversal (top of Fig. 17.8*b*). His analysis is capable of explaining complex stress patterns for various combinations of part sizes, quenching rate, and steel hardenability.[24] However, the residual stress pattern in the hardened steels can be modified either with different transformation characteristics or during the tempering and finish-machining (after hardening) operations.

17.3.2.4   Residual Stress Pattern after Surface Hardening.   In general, thermo-chemical and thermal surface-hardening treatments produce beneficial compressive residual stresses at the surface (Fig. 17.9).

   Carburizing and Quenching.   When low-carbon steels are carburized and then quenched, first martensite (as an example) forms at some distance from the surface (Figs. 17.9 and 17.10),[23,46] where the part temperature has dropped below the higher interior M_S temperatures. The volume changes at this stage are quickly accommo-dated by the neighboring austenite due to its low flow stresses and the high tem-peratures. The surface austenite does not transform due to its low M_S (reduced by ~ 470°C, or 878°F for a 1% increase in carbon). When the temperature falls below M_S in the (higher-carbon) surface regions, the expansion of the formed martensite at the surface is constrained by the interior martensite that formed earlier. Conse-quently, the surface microstructure is held in compression. The contributory factors affecting this process and the position of maximum compressive stress include carbon and alloy contents which set M_S temperatures, and steel hardenability; total case depths; quenching severity, which depends on the temperature of start; and the temperature-dependent plastic flow behavior of martensite and austenite. Despite the complexity of the interactions that affect the formation of residual stresses, hardened carburized parts with the martensite-austenite microstructure described earlier usually develop favorable compressive stresses.[47]

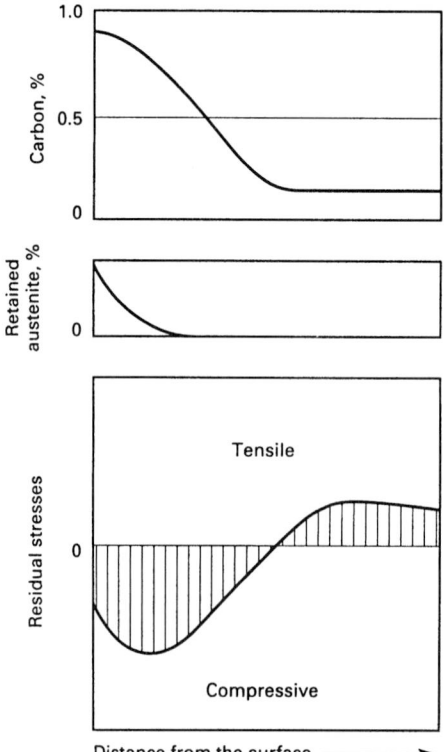

FIGURE   17.9 Relationship between carbon content, retained austenite, and residual stress pattern. It shows the devel-opment of peak compressive stress some distance away from the surface.[23] (*Reprinted by permission of Pergamon Press, Plc.*)

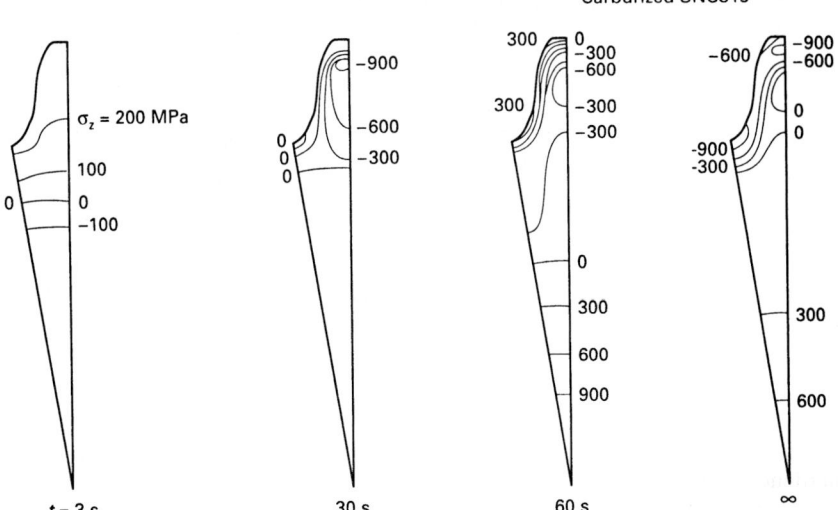

FIGURE 17.10   Axial stress distribution (given in MPa) in carburized gear during quenching process.[46]

According to Koistinen[48] and Salonen,[49] the peak compressive stress takes place at 50 to 60% of the total case depth, corresponding to about 0.5 to 0.6% carbon level, which produces a low retained austenite content and martensite hardness around the maximum. Another factor that might influence this compressive residual stress profile is that the martensite formed in lower-carbon regions of the case is of the lath type, which also affects the retained austenite content.[23] The reversal sign of residual stress takes place at or near the case/core interface. Figure 17.10 shows the details of generation of axial stress distribution of a carburized gear (made from deeper hardening steel) during quenching. In the early stages, the contour lines of equal stress were largely unaffected by the surface profile. Later a zone of high compressive stress distribution occurred in the central portion of the teeth, which remained until the end of the quench.[46]

Nitriding and Nitrocarburizing.   In nitriding, a compressive residual stress is set up in the surface layers due to the volume increase arising from the formation of nitrides in the shallow nitrided layer. High-temperature nitriding produces a little relaxation of stresses, whereas low-temperature nitriding imparts a maximum residual stress. Carburized surface hardening, on the other hand, relies on the martensite transformation for its hardness and its residual stresses.[20]

In nitrocarburizing, improvements in residual surface compressive stress and fatigue strength depend on the hardness and depth of the diffusion zone. These properties, in turn, decrease with increasing carbon and alloy content (i.e., increased hardenability). During quenching, after nitrocarburizing, a (macro-) compressive residual stress is produced in the compound layer and gamma prime phase.[50] When nitrocarburized parts are rapidly quenched, the above properties are further enhanced.[51]

Boriding.   In borided steel processed at 900°C (1650°F), a high compressive residual stress is developed at the surface layers (Fig. 17.11), which consists of FeB and Fe_2B phases.[50] This is attributed to the lower thermal expansion coefficient and the larger specific volume in a borided layer compared to those in a ferrite matrix.[21,52]

Induction Hardening. In an induction-hardened steel part, a compressive surface residual stress is produced when wear-resistant hard martensite (with slightly lower density) is formed on the surface of a section concurrently with volume expansion while the nonhardened core remains essentially unchanged (Fig. 17.12).[53,54] The magnitude of the compressive stress, which is affected by both thermal contraction and martensite formation, may be a considerable fraction of the yield strength, which permits the application of significantly higher stresses than could normally be possible in fatigue loading. As in the carburizing practice, the surface compressive residual stresses are usually found to increase with depth below the surface[54] (Fig. 17.12).[53] A fairly sharp transition to a tensile state takes place near the hardness drop-off between the case and unhardened surrounding

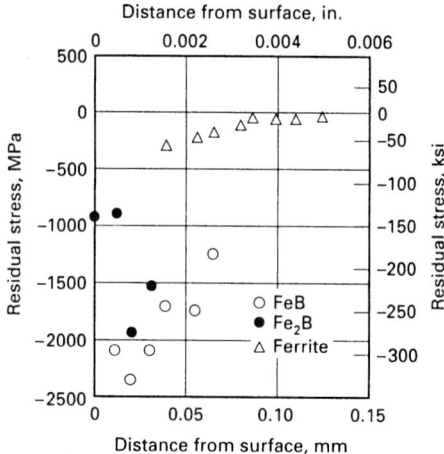

FIGURE 17.11 Residual stress distribution of FeB and Fe_2B layers in borided layers in borided steel processed at 900°C (1650°F).[21,52]

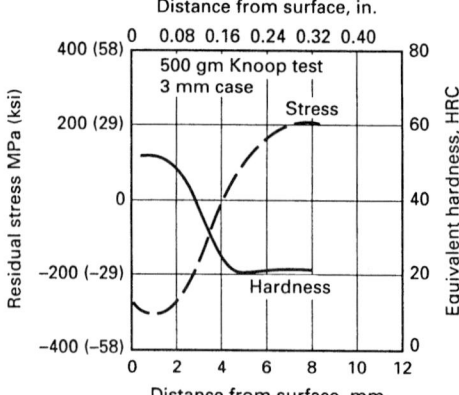

FIGURE 17.12 A typical hardness and residual stress profile in induction hardened (to 3-mm, or 0.12-in. case depth) and tempered (at 260°C, or 500°F) 1045 steel.[53] (*Reprinted by permission of ASM International, Materials Park, Ohio.*)

material. With an increase in distance from the steep transition, the tensile condition gradually fades away toward zero stress.[53] In induction hardening, an increase in hardenability changes the depth at which transition from compressive to tensile stress occurs. The increase in the rate of heating produces an increase in the maximum compressive and tensile residual stresses without affecting the mode of stress distribution.[55]

One of the problem areas with induction hardening of gears is the tendency to develop high-magnitude tensile residual stresses in a narrow zone beneath the hardened layer, thereby providing a potential failure site. This characteristic is taken into consideration when designing gears for induction hardening and selecting a case depth. It is also reflected in the design allowable stress value for bending.[56,57]

Plasma Nitriding and Electron Beam Hardening. In plasma nitrided and plasma nitrided plus electron beam treated steel, compressive residual stresses have been observed. However, the measured values have been reported to be considerably higher in the latter (Table 17.6).[58]

Laser Hardening. In laser hardened medium-carbon steel (AISI 1042), 4xxx series, and 80W Cr V8 tool steel, high surface compressive residual stress has been observed.[59,60] The characteristics of the residual stress field depend on the material, the specimen type, and the laser hardening condition, including the laser processing parameters and the scanning pattern of the laser beam.[60]

17.3.2.5 Residual Stress in Other Processing Steps.

As welding progresses, the temperature distribution in the weldment becomes nonuniform and varies as a result of localized heating of the weldment by the welding heat source. During the welding cycle, comprising heating and cooling, complex strains develop in the weld metal and adjacent areas. As a result, appreciable residual stresses remain after the completion of welding. Since the weld metal and heat-affected zone contract on cooling (Fig. 17.13a),[61] they are restrained by the cool adjacent part. This produces

TABLE 17.6 Compressive Residual Stress in Plasma Nitrided and Plasma Nitrided plus Electron Beam Treated Fe-0.42C-0.96Cr-0.6Mn-0.37Si Steel[58]

Depth, μm	Phase	Retained austenite, wt%	Compressive residual stresses in nitrous ferrite (martensite), MPa
	Plasma nitriding		
50	α solid solution + γ'	—	435
200	α solid solution + γ'	—	638
400	α solid solution	—	520
	Plasma nitriding + electron beam treatment		
50	α solid solution + γ solid solution	14	595
200	α solid solution + γ solid solution	12	785
400	α solid solution	—	682

Note: α solid solution (nitrous martensite); γ'-Fe_4N phase; γ solid solution (nitrous or retained austenite).

tensile residual stress in the weldment region and compressive residual stress in the surrounding base-metal region (Fig. 17.13b).

In general, a steep residual stress gradient is developed because of the steep tendency of the thermal gradient. This may, in turn, lead to hot cracking (between columnar grains) or severe centerline cracking in the weld area.[62] Catastrophic failures of welded bridge and all-welded ships are mostly attributed to the existence of large and dangerous tensile residual stress in them.[63] Poor design features such as square-corner hatches (in ships) are also to be avoided.[20]

The machining and grinding operations in manufacturing are important, since they are always utilized to produce the finished surface.[†] It has been shown that gentle surface grinding, using a soft sharp wheel and slow downfeed, produces compressive residual stress at the surface, whereas conventional (normal practice) and abusive grinding results in surface tensile stresses of very high magnitude (Fig. 17.14).[25,64] However, the gentle grinding method is expensive from the viewpoint of operating time and wear of the wheel.

[†] In gearing, at least, a gearing step is an unwanted groove along the bottom edge of a flank and somewhat in the tooth fillet; it is produced by flank grinding. It can happen when flank grinding is used to correct excessive heat treat distortion.[20]

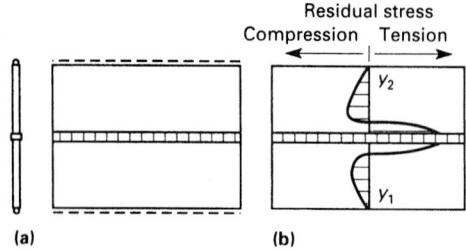

FIGURE 17.13 (a) The transverse shrinkage occurring in butt weldments. (b) Longitudinal residual stress patterns in the weldment and surrounding regions. This also shows longitudinal shrinkage in a butt weld.[61]

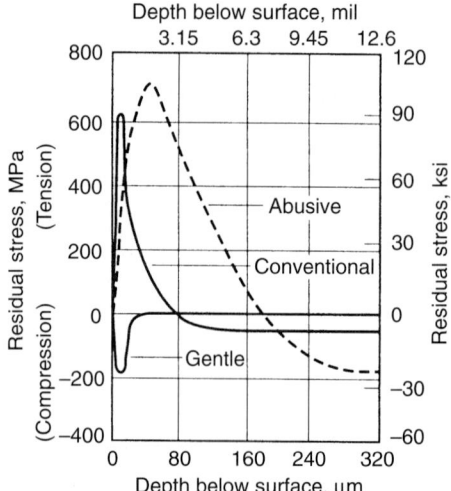

FIGURE 17.14 Residual stress distribution after gentle, conventional, and abusive grinding of hardened 4340 steel.[25] (*Reprinted by permission of Pergamon Press, Plc.*)

As a result of temperature gradients during cooling, castings develop compressive stresses at the surface and tensile stresses in the interior.[25] However, transient temperature gradient and phase transformation occurring during the early stages of solidification and cooling of continuous steel castings in the mold give rise to the development of harmful residual stresses, leading to the formation of cracks.[65]

Chemical processes such as electroplating, scale formation, and corrosion of metals can produce residual stresses due to coherency strains arising from the matching tendency of crystal structures of the outer surface product with the crystal structure of the adjacent layer.[25] Residual stresses are also introduced when heat-treated parts are subjected to successive heating and cooling cycles during service conditions.

17.3.2.6 Residual Stress in the Heat-Treated Nonferrous Alloys. In nonferrous alloys, notably age-hardenable aluminum alloys, copper-beryllium alloys, certain nickel-base superalloys, and so on, a significant amount of thermal stress is generated during quenching prior to precipitation hardening. The quenching process in this condition does not invariably involve a phase change; rather, this is confined to the post-quenching aging treatment. In other nonferrous alloys such as uranium and titanium alloys, the final structural condition is not obtained by a slow cool.

When high-strength titanium alloy is quenched from a solution annealing temperature of 850 to 1000°C (1560 to 1830°F), it develops large residual stress caused by poor thermal conductivity of titanium, leading to high-temperature gradient. This problem can, however, be avoided by stress-relief annealing at 650 to 700°C (1200 to 1290°F), which produces a slight reduction in mechanical properties. When a high-strength aluminum age-hardening alloy is rapidly quenched from the solution temperature, high thermal and residual stresses are induced due to the high coefficient of expansion of aluminum. Uphill quenching from liquid nitrogen temperature (−196°C, or −320°F) in a steam blast alleviates this problem. This induces stresses opposite in sign to those developed on water-quenching from the solutionizing and cancels out their effect. This is followed by aging of the alloy in the conventional manner.[39]

Fast polyalkylene glycol (PAG) quenching of solution-treated aluminum alloys tends to reduce residual stress levels because of its more uniform heat extraction rate (thermal shock is smaller, and thereby machining is less likely to produce further distortion), thereby helping solve major and long-standing distortion problems among aluminum workpieces.[66]

17.3.3 Control of Residual Stresses in Heat-Treated Parts

Table 17.7 lists some typical values of maximum compressive residual stresses developed in the surface hardened steels that have been reported in the literature.[39] It is worth noting that there is a marked influence of tempering on the residual stress level. Tempering must be accomplished at 150 to 180°C (300 to 356°F) to maintain 50 to 60% retention of the residual stress level obtained after quenching because a higher tempering temperature greatly reduces surface compressive stresses. However, a higher thermal stress-relieving (at ~600°C, or 1110°F) of steel parts is used for fabrications, castings, heavy flame-cut sections, and so forth. Alternatively, serious residual tensile stresses may be avoided effectively by gentle grinding of the surface.

TABLE 17.7   Compiled Summary of the Maximum Residual Stresses in Surface Heat-Treated Steels[39]

Steel	Heat treatment	Residual stress (longitudinal)	
		MPa	ksi
832M13 (type)	Carburized at 970°C (1780°F) to 1-mm (0.04-in.) case with 0.8% surface carbon		
	Direct-quenched	280	40.5
	Direct-quenched, –80°C (–110°F) subzero treatment	340	49.0
	Direct-quenched, –90°C (–130°F) subzero treatment, tempered	200	29.0
805A20	Carburized and quenched	240–340[†]	35.0–49.0
805A20	Carburized to 1.1–1.5 mm (0.043–0.06 in.) case at 920°C (1690°F), direct oil quench, no temper	190–230	27.5–33.5
805A17		400	58
805A17	Carburized to 1.1–1.5 mm (0.043–0.06 in.) case at 920°C (1690°F), direct oil quench, tempered 150°C (300°F)	150–200	22–29
897M39	Nitrided to case depth of about 0.5 mm (0.02 in.)	400–600	58.0–87.0
905M39		800–1000	116.0–145.0
Cold-rolled steel	Induction hardened, untempered	1000	145.0
	Induction hardened, tempered 200°C (390°F)	650	94.0
	Induction hardened, tempered 300°C (570°F)	350	51
	Induction hardened, tempered 400°C (750°F)	170	24.5

[†] Immediately subsurface, that is, 0.05 mm (0.002 in.).
Courtesy of Wolfson Heat Treatment Centre, England.

17.3.4   Measurement of Residual Stresses

There are two methods of measuring residual stresses: the destructive method, also called the dissection method, and the nondestructive methods comprising mainly x-ray diffraction, neutron diffraction, and ultrasonic and magnetic methods.

17.3.4.1   Destructive (or Dissection) Method.   The dissection method relies on removing layers of material from a part and measuring the dimensional or strain changes (by strain gauges, traveling microscope, micrometer, and whatever) resulting from each layer removal event.

This method, being old but reasonably accurate, uses well-established methods and can be employed in confined situations at site.[67] However, it is tedious, time-consuming, and expensive.[68] The other drawbacks are the destructive, or at best semidestructive, nature of the method and its ability to measure only the macroresidual stresses.

The hole-drilling method is an example of the dissection technique, and it is used extensively for measuring residual stresses, which are less than nominally one-half the yield strength of the material. It consists of the mounting of strain gauges or a three-element strain-gauge rosette on the surface and measurement of strains. Then a rigidly guided milling cutter is used to drill a small, straight, circular, perpendicular, and flat-bottomed hole not exceeding 3.2 mm (0.125 in.) at the center of the rosette and into the surface of the component being analyzed. Strain redistribution occurring at the surface in the surrounding area of the hole (resulting from the residual stress relief) is then measured with the previously installed strain gauges. The residual stress is calculated at a large number of points in a surface from the strain measurements using the well-established method, involving the magnitude and direction of strain, hole size, and materials properties.[25,38]

Layer removal can be effected by grinding (e.g., the outer layers of a cylinder can be gently removed in increments by grinding) and the strain change at each increment measured by strain gauges attached to the bore of the cylinder. To minimize the introduction of spurious strains by the grinding operation, the rate of metal removal should be less than 3.125×10^{-4} m/s (1.23×10^{-2} in./s), and readings are recorded after 15 min of the end of the grinding process to ensure that any heat generated has been dissipated.[69] The sensitivity of the strain measurement and the accuracy of the stress calculation may be enhanced by employing the reverse-taper hole-drilling method.[70]

17.3.4.2 Nondestructive Methods.

The method is described as nondestructive, but this will depend on what one is trying to achieve. If one wants to determine the surface stresses on a part, it is indeed nondestructive. If one wants to determine the residual stress gradient through a carburized case, it is destructive, i.e., unless it is executed in an area of the part where an incrementally produced groove is of no consequence to the fitness or purpose of the part.[20]

The main difficulty with the nondestructive methods is that measurements of crystallographic lattice parameters, ultrasonic velocities, or magnetization changes are made that are indirectly related to the residual stress. The above quantities are usually dependent on the stress and material parameters (such as metallurgical textures), which are difficult to quantify.[68,71]

The x-ray diffraction (XRD) method is probably the most well-established technique for measuring both macro- and microresidual stresses nondestructively. In most instances, the x-ray diffraction method has been employed to provide quantitative values for residual stress profiles in the surface or fully hardened components.[72] This technique depends on the determination of lattice strains and the stress-induced differences in the lattice spacing. Macroresidual strain is measured from the shift of diffraction lines in the peak position using the so-called nonlinear $Sin^2 C$ method from which residual stress is calculated.[72] For the measurement on microstrain the Voigt single-line method is applied.[73] Precision in lattice strain measurement of the order of 0.2% is possible.

XRD methods are capable of high spatial resolution, on the order of millimeters, and depth resolution, on the order of microns. The macroscopic residual stress and data related to the degree of cold working can be obtained simultaneously by XRD methods. XRD methods are applicable to most polycrystalline materials, metallic, or ceramic and are nondestructive at the sample surface.[74]

Possible sources of x-ray measurement errors are (1) grain size, (2) round surfaces (versus flat), (3) error in peak position, (4) stress relief due to aging, and (5) sample anisotropy.[70]

Portable x-ray diffraction equipment is now commercially available in various forms that allow stress measurements to be made very quickly (ranging from 4 to 30 s). The main drawbacks are that it cannot be applied to noncrystalline materials such as plastics, and it is only capable of measuring residual stresses of materials very close to the surface under examination. That is, the measurement is purely surface-related (a depth of 0.01 mm, or 0.4 mil, is commonly quoted).[75]

Neutron radiography or diffraction, used for polycrystalline materials, has a much deeper penetration than x-rays, but has major safety problems and the disadvantage of being nonportable.

Ultrasonic method for evaluating residual stress involves ultrasonic stress birefringence or sonoelasticity; this depends upon the linear variation of the velocities of sound in a body (i.e., ultrasonic waves) with the stress. This method has the potential for greater capability, versatility, and usefulness in the future.[67,71] However, this has the disadvantage, in common with the magnetic methods, that it requires transducers shaped to match the surface being inspected.[76]

The magnetic method is based on the stress dependence of the Barkhausen noise amplitude. Each time an alternating magnetic field induced in a ferromagnetic material is reversed, it generates a burst of Barkhausen noise. The peak amplitude of the burst, as determined with an inductive coil near the surface of the component material, varies with the surface stress level. Since Barkhausen noise depends on composition, texture, and work hardening, it is necessary in each application to use calibrated standard (reference) samples with the same processing history and composition as the component being analyzed. This method is used to measure residual stresses well below the yield strength of the ferromagnetic material.[†] This method is rapid, and the measurements are made with the commercially available portable equipment. However, this method is limited to only ferromagnetic materials.[71]

Thermal evaluation for residual stress analysis (TERSA) is a new nondestructive method that is in an experimental stage. It has the advantage that it is completely independent, remote, and noncontacting. It consists of merely directing a controlled amount of energy from a laser energy source into the volume of the material being inspected and then making a precise determination of changes in the resulting temperature rise by infrared radiometry. However, the working instrument will also require some form of display to enable visual examination to be made of any high-stressed regions.[76]

17.4   QUENCH CRACKING

Anything that produces excessive tensile quenching stress is the basic cause of quench cracking, which may be considered as severe distortion. Quench cracking is mostly intergranular, and its formation may be related to some of the same factors that cause intergranular fracture in overheated and burned steels. Important contributors to cracking, apart from stress, in heat treatment are (1) part design, (2) steel grades, (3) part defects, (4) heat-treating practice, and (5) tempering practice.[77] (See Sec. 13.11 for more details.)

1.   Part Design.   Features such as sharp corners; the number, location, and size of holes; deep keyways; splines; and abrupt change in section thickness within a part (i.e., badly unbalanced section) enhance the crack formation because while the one (thin) area is cooling quickly in the quenchant, the other (thick) area immediately adjacent to it is cooling very slowly. One solution to this problem is to change the material so that a less drastic quenchant (for example, oil) can be employed. An alternate solution is to prequench, that is, to cool it prior to the rest of the part. This will produce an interior of the hole or keyway that is residually stressed in com-

---

[†] Barkhausen noise property is also being developed to detect grinding burns in ground case hardened steels.[20]

pression, which is always desirable for better fatigue properties.[77] The third solution is a design change, and the fourth is to use a milder quenchant.

Tip cracking is a design-induced problem that begins with the gear designer. If the design cannot be modified, nor the steel changed, then it is up to the manufacturing people to find a way of avoiding tip cracking. Copper plating the blank is a good example.

The Brown method of submerged induction hardening using a Delapena-type inductor, and employing hardened and tempered 4340 steel, and a good-quality quenching oil, very rarely encounters hardening problems.[56,57]

2. Steel Grades. Sometimes this can be checked by means of a spark test, whereas at other times a chemical analysis must be made. In general, the carbon content of plain carbon steels should not exceed the required level; otherwise, the risk of cracking will increase. The suggested average carbon contents for water, brine, and caustic quenching are given below:

Method	Shape	Carbon, %
Induction hardening	Complex	0.33
	Simple	0.50
Furnace hardening	Complex	0.30
	Simple	0.35
	Very simple, such as bar	0.40

A decrease in carbon content from 0.72 to 0.61% has been shown to slightly increase the thermal crack resistance of rim-quenched railroad wheels.[78]

Because of segregation of carbon and alloying elements, some steels are more prone than others to quench cracking. Among these steels, 4140, 4145, 4150, and 1345 appear to be the worst. A good option is to replace the 4100 series with the 8600 series. An additional disadvantage with the use of 1345 steels is the manganese floating effect, which leads to very high manganese content in the steel rolled from the last ingot in the same heat. Similarly, dirty (unclean) steels (that is, steels with more than 0.05% S, for example, AISI 1141 and 1144) are more susceptible to cracking than the low-sulfur grades. The reasons for this are that they are more segregated in alloying elements; the surface of this hot-rolled high-sulfur steel has a greater tendency to form MnS stringers and seams, which act as stress raisers during quenching; and they are usually coarse-grained (for better machinability), which increases brittleness and therefore promotes cracking.[†] If these high-sulfur grades are replaced by calcium-treated steels or cold-finished leaded steels, this problem can be obviated.[77]

3. Part Defects. Surface defect or weakness in the material may also cause cracking, for example, deep surface seams or nonmetallic stringers in both hot-rolled and cold-finished bars. Other defects are inclusions, stamp marks, and so forth. For large-seam depths, it is advisable to use turned bars or even magnetic particle inspection. The forging defects in small forgings, such as seams, laps, flash line, or shearing crack, as well as in heavy forgings, such as hydrogen flakes and internal ruptures, aggra-

---

[†] Cleanliness with respect to these steels is generally taken as oxide cleanliness. In this respect, it is usually (but not always) the case that machinable steels are not aluminum-treated (to ensure no abrasive alumina)—this is why they are generally described as coarse-grained steels.[36]

vate cracking. Similarly, some casting defects, for example, in water-cooled castings, promote cracking.[64]

4. Heat-Treating Practice.   Higher austenitizing temperatures produce a faster cooling rate during quenching and increase the tendency toward quench cracking. Similarly, steels with coarser grain size rate more prone to cracks than fine-grain steels because the latter possess more grain boundary area to stop the movements of cracks, and grain boundaries help to absorb and redistribute residual stresses. In other words, fine-grained steeels are tougher.[36] An outstanding contributor to severe cracking is improper heat-treating practice, for example, nonuniform heating and nonuniform cooling of the component involved in the heat-treatment cycle. It is a good heat-treating practice to anneal alloy steels prior to the hardening treatment (or any other high-temperature treatment, for example, forging, welding, and so forth) because this produces a fine-grained microstructure and relieves stresses.[79]

Water-Hardening Steel.   The water-hardening steels are most susceptible to cracks if they are not handled properly. Soft spots are most likely to occur in the water hardening steels, especially where the tool is grabbed with tongs for quenching. Normally the cleaned surface shows adequate hardening and the scaled surface insufficient hardening, which can be examined with a file. Soft spots may occur from the use of fresh water, or water contaminated with oil or soap. Most large tools emerging from hardening operations contain some soft spots. However, accidental soft spots in the wrong place should be investigated, and steps must be taken to eliminate them.

Figure 17.15 shows the typical appearance of a thumbnail check as a soft spot on chipping chisels, which occurs on the bit near the cutting edge. The cracks enclosing the soft spots should be avoided by switching to brine quench.[80]

Oil-Hardening Steel.   High-carbon low-alloy steels such as AISI 52100 (as well as carburized steels) are susceptible to microcrack during hardening. Lyman[81] has developed a heat treatment cycle to eliminate this defect, which comprises quenching from the austenitizing temperature to a temperature just below the M_S of the steel (sufficient to produce 30 to 40% martensite in the austenite) followed by transfer to a salt bath held at 260°C (500°F) to temper that martensite,[20] and final quenching to complete the formation of martensite.

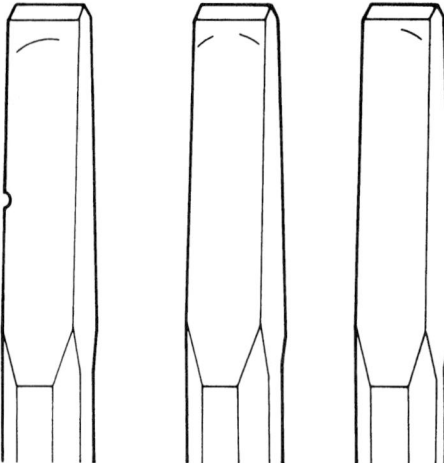

FIGURE 17.15 Typical appearance of thumbnail check as a soft spot on chipping chisel.[80] (*Courtesy of Society of Manufacturing Engineering.*)

Polymer and Salt Bath Quenching. Polymer quenchants have found well-established use in the quenching of solution-treated aluminum alloys, hardening of plain carbon steels with less than 0.6% C, spring steels, boron steels, hardenable stainless steels, all carburizing and alloy steels with section thickness greater than about 50 mm (2 in.), through-hardening steel parts, and induction and flame hardening treatments because of their numerous beneficial effects, including elimination of soft spots, distortion, and cracking problems associated with trace water contamination in quenching oils.[82]

Agitation is an important parameter in polymer quenching applications, both to ensure a uniform polymer film around the quench part and to provide a uniform heat extraction from the hot part to the adjacent area of quenchant, by preventing a buildup of heat in the quench region.

Salt bath cooling of induction-hardened complex-shaped cast iron parts reduces the danger of cracking, which is usually experienced when air cooling followed by hot-water quenching is used.[83]

Air-Hardening Steel. Similarly, when air hardening steels are improperly handled, they are likely to crack. For example, avoidance of tempering treatment or use of oil quenching in air-hardening steel can lead to cracking. However, the common practice in the treatment of air-hardening steels is initially to quench in oil until "black" (about 540°C, or 1000°F), followed by air cooling to 65°C (150°F) prior to tempering. As compared to air cooling right from the quenching temperature, this practice is totally safe and minimizes the formation of scale.

Decarburized Steel. Decarburization can occur at temperatures above about 700°C (1292°F) and usually arises from (1) drop of carbon potential of a furnace atmosphere below that necessary to maintain the carbon content of the surface;[84] (2) insufficient protection as a result of plant failure (for example, leaks, air ingress, defective furnace or container seals, defective valves); (3) poor process control (for example, insufficient atmosphere-monitoring equipment, poor maintenance, and poor supervision); (4) incorrect diffuse stage of boost-diffuse carburization program; or (5) the existence of decarburizing agents such as CO_2, water vapor, H_2, and O_2 in the furnace atmosphere.[34,77,85]

A partially decarburized surface on the part occurring during tool hardening also contributes to cracking because martensite transformation is completed therein well before the formation of martensite in the core. Significant decarburization causes inadequate surface microstructures and reduced residual compressive or even surface tensile residual stress. Decarburized surface on the tools has reduced hardness, which will lead to premature wear and scuffing. If surface carbon is >0.6%, the surface hardness should be acceptable. If surface carbon is ≤0.6%, all the important properties will be adversely affected; for example, bending fatigue limit could be reduced by 50%.[34] Partial decarburization must be avoided, especially on all deep-hardening steels, by providing some type of protective atmosphere during the heating operation, stock removal by grinding, or a carbon restoration process. In addition to protective atmosphere, salt baths, inert packs, or vacuum furnaces may be used to obtain the desired surface chemistry on the tools or dies. The fact that the better and more consistent performance of the tools is observed after grinding reveals the existence of partial decarburization remaining. Other surface features to which this would apply include, for example, severe surface roughness and internal oxidation with associated high-temperature transformation products.[20]

Carburized Alloy Steel. Two types of peculiar cracking phenomena prevail in the carburized and hardened case of the carburized alloy steels: microcracking and tip cracking. Microcracking of quenched steels are small cracks appearing across or

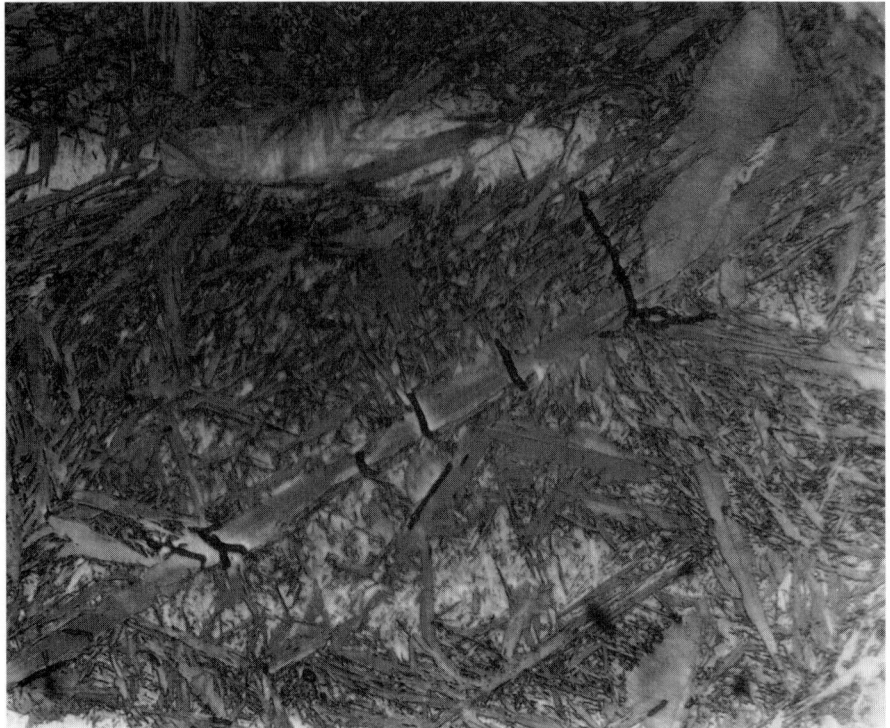

FIGURE 17.16   Microcracking in a Ni-Cr steel.[85] (*Courtesy of G. Parrish.*)

alongside martensite plate (Fig. 17.16)[85] and the prior austenite grain boundaries.[86] They form mostly on those quenched steel parts that contain chromium and/or molybdenum as the major alloying elements with or without nickel content and where the hardening is done by direct quenching.

Microcracks are observed mostly in coarse-grained structures and are associated with large martensite plates. This is presumably because of more impingements of the larger plates of martensite by other large plates. Another cause of microcracking is the increased carbon content of steel (exceeding eutectoid value such as AISI 52100) and that of martensite (i.e., increased hardenability), which is a function of austenitizing temperature and/or time.[85] This finding was established for 8620H steel, which has a higher austenitizing temperature prior to quenching where there is a greater tendency to microcrack.[87] This problem can be avoided by selecting a steel with less hardenability (i.e., with lower austenitizing temperature). Another solution is to change the heat-treating cycle to carburizing, slow cooling to black temperature, reheating to, say, 815 or 845°C (1500 or 1550°F), and quenching.[77] Microcracking in case-hardened surfaces may be aggravated by the existence of hydrogen, which tends to absorb during carburizing and reheating in an end-othermic atmosphere. However, this hydrogen-enhanced microcracking can be eliminated by tempering the carburized parts at 150°C (300°F) immediately after quenching. Tempering exhibits an additional beneficial effect in that it has the ability to heal the smaller microcracks due to the volume changes and associated plastic flow that develop during the first stage of tempering.[88]

Apple and Krauss[87] have observed that the reduction of microcracking leads to better fatigue resistance. On the other hand, Kar et al.[89] have noticed an improvement in fracture resistance after elimination of both undissolved carbides and microcracks from an AISI 52100 steel microstructure.

Tip cracking refers to the cracking that appears in the teeth of carburized and quenched (or nitrided) gears at the case/core interface; these cracks are crescent-shaped and, when subjected to loading, progress in fatigue across the tooth section until breakage occurs. Many heat treaters have solved this problem to a great extent by decreasing the carbon content and case depth to the minimum acceptable design level, or by copper-plating the outer diameter of the gear blank prior to hobbing.[83]

Nitrided Steels. The nitrided cases are very brittle. Consequently, cracking may occur in service prior to realizing any improved wear and galling resistance. This can be avoided by a proper tool design, e.g., incorporating all section changes with a minimum radius of 3 mm (0.125 in.).

Borided Steel. Monophase Fe_2B layers are preferred to avoid brittleness, crack formation, and flaking of boride layer. Rapid heating to 900°C and the incorporation of diffusion anneal of 2 hr at 1000°C in argon during the pack boriding treatment tend to avoid the formation of brittle FeB and convert the FeB/Fe_2B layer to monophase Fe_2B layer, respectively.[90]

5. Tempering Practice. The longer the time the steel is kept at a temperature between room temperature and 100°C (212°F) after the complete transformation of martensite in the core, the more likely the occurrence of quench cracking. This arises from the volumetric expansion of retained austenite into martensite.

There are two tempering practices that lead to cracking problems: tempering too soon after quenching, i.e., before the steel parts have transformed to martensite in hardening, and skin tempering, usually observed in heavy sections (≥50 mm, or 2 in. thick in plates and >75-mm, or 3-in., diameter in round bars).

It is the normal practice to temper immediately after the quenching operations. In this case, some restraint must be exercised, especially for large sections (>75 mm, or 3 in.) in deep-hardening alloy steels. The reason is that the core has not yet completed its transformation to martensite which is accompanied by an expansion, whereas the surface and/or projections, such as flanges, begin to temper, which involves a shrinkage. These simultaneous, opposing volume changes produce radial cracks. This problem can become severe if rapid heating practice (e.g., induction, flame, lead, or molten salt bath) is used for tempering. Therefore, very large and very intricate deep-hardening alloy steel parts should be removed from the quenching medium, and tempering should be started while they are slightly warm to hold comfortably in the bare hands (~60°C, or 140°F).

Skin tempering occurs in heavy section parts when the final hardness is >360 HB (or 39 Rc). This is due to insufficient tempering time and is usually determined when the surface hardness falls by 5 or more Rc points from the core hardness. This cracking often occurs several hours after the component has cooled from the tempering temperature and often runs through the entire cross section. This problem, however, can be removed by retempering for 3 hr at the original tempering temperature (provided no cracking is present), which is associated with a change in hardness of 2 Rc points maximum.[77] Next time, the parts can be tempered for a longer time.

17.5   DISTORTION IN HEAT TREATMENT

Distortion can be defined as an irreversible and usually unpredictable dimensional change in the component during processing from heat treatment and from temperature variations and loading in service. The term *dimensional change* is used to denote changes in both size and shape.[91] *Distortion* is therefore a general term often used by engineers to describe all irreversible dimensional change in a component as a result of heat treatment operations.[92] Although it is recognized as one of the most difficult and troublesome problems confronting the heat treater and heat-treatment industries on a daily basis, it is only in the simplest thermal heat-treatment methods that the mechanism of distortion is understood. Changes in size and shape of ferrous parts may be either reversible or irreversible. Reversible changes, which are produced by applying stress in the elastic range or by temperature variation, neither induce stresses above the elastic limit nor cause changes in the metallurgical structure. In this situation, the initial dimensional values can be restored to their original state of stress or temperature.

Irreversible changes in size and shape of heat-treated parts are those that are caused by stresses in excess of the elastic limit or by changes in the metallurgical structure (e.g., phase changes). These dimensional changes sometimes can be corrected by mechanical processing to remove extra and unwanted material or by heat treatment (annealing, tempering, or cold treatment) to redistribute residual stresses.

When heat-treated parts suffer from distortion beyond the permissible limits, it may lead to scrapping of the article, rendering it useless for the service for which it was intended, or it may require necessary correction. Allowable distortion limits vary to a large extent, depending on service applications; in cases where very little distortion can be tolerated, specially desired tool steels are used. These steels possess metallurgical characteristics that minimize distortion.

17.5.1   Types and Causes of Distortion

Distortion can be classified into two categories: *size distortion*, which is the net change in specific volume between the parent and transformation product produced by phase transformation without a change in geometrical form, and *shape distortion* or *warpage*, which is a change in geometrical form or shape (of the workpiece) and is revealed by changes of curvature (angular relations) or curving, bending, twisting, and/or nonsymmetrical dimensional change without any volume change.[92,93] The former may be considered a natural occurrence with some degree of predictability while the latter, being much less predictable, is a major processing concern. Usually both types of distortion occur during a heat-treatment cycle.

The distortion or the dimensional changes may be attributed to numerous factors introduced into a component before, during, or even after heat treatment which are illustrated in Fig. 17.17.[94] The important factors related to steel, machining, and heat treatment are described here.

Steel-Related Factors.   These include as-cast shape, hardenability, and transformation temperature.

1. *As-cast factor* on distortion is small compared to other factors.
2. *Hardenability.* The effect of hardenability on dimensional change may be explained by the change of transformation temperature. Once the hardenability

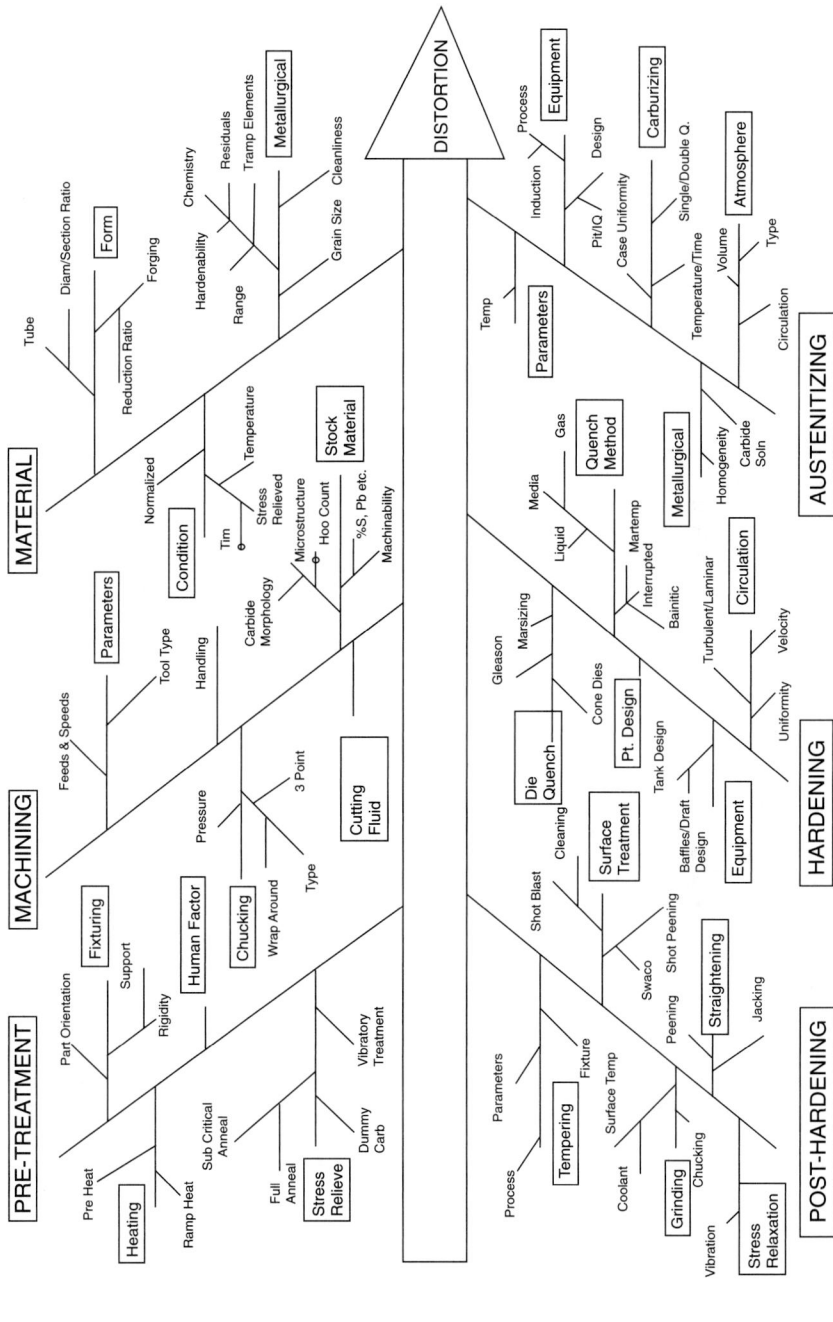

FIGURE 17.17 Ishawaka diagram summarizing some of the many factors that should be considered when faced with a distortion problem.[94]

becomes adequate to produce a fully martensitic structure, the low rate of dimensional change arises from the relatively small effect of alloying elements on the M_s temperature. That is, hardenability control provides an essentially similar transformation in a component of a given size (shape and dimension) under constant heat treatment conditions.[36]

3. *Transformation temperature* plays an important role in dimensional change, thereby establishing the importance of hardenability control in minimizing variability. Since the hardenability is related to the consistent cooling rate, increasing the cooling rate will lead to the occurrence of transformations at lower temperatures. Thus cooling rates have dual roles of changing both thermal strains and transformation temperature.[95]

Machining-Related Factors. During machining, the distortion of a part is a function of geometry, order of material removed, and stress state in the material removed. If the change of shape which takes place is not accommodated, the part may be scrapped during machining. Measurement of the initial residual stress distribution and the use of finite element modeling enable the development of machining procedures which reduce or eliminate distortion.

Distortion occurs due to removal of stressed material from the forging.

Quenching stresses, when machined away, usually cause the largest amount of distortion, not necessarily as a result of the magnitude of the quenching stresses, but due to the large amount of material where quenching residual stresses are present. Turning and shot-peening stresses, although typically higher in magnitude than quenching stresses, are present in a shallow layer of near-surface material, and therefore, they exert less influence on distortion than quenching.[95]

Dimensional Changes Caused by Changes in Metallurgical Structure during Heat Treatment. Various dimensional changes produced by a change in metallurgical structure (or transformation stress) during the heat treatment cycle of steel parts are described below.[96]

1. *Heating (austenitizing).* When annealed steel is heated from room temperature, thermal expansion occurs continuously up to Ac_1, where the steel contracts as it transforms from body-centered cubic (bcc) ferrite to face-centered cubic (fcc) austenite. The extent of decrease in volumetric contraction is related to the increased carbon content in the steel composition (Table 17.4). Further heating expands the newly formed austenite.

2. *Hardening.* When austenite is cooled quickly, martensite forms; at the intermediate cooling rates, bainite forms, and at slow cooling rates, pearlite precipitates. In all these transformation sequences, the magnitude of expansion increases with the decrease in carbon content in the austenite (Table 17.4). However, the net volume change for the ferrite through austenite to martensite condition is an increased expansion as the carbon content increases (assuming that all carbides are dissolved during austenitizing).[20] The volume increase is maximum when austenite transforms to martensite, intermediate with lower bainite, and least with upper bainite and pearlite (Table 17.4). The volume increases associated with the transformation of austenite to martensite in 1 and 1.5% carbon steels are 4.1 and 3.84%, respectively; the volume increases involved in the transformation of austenite to pearlite in the same steels are 2.4 and 1.33%, respectively. Such volume increases are less in alloy steels and least in 2C-12Cr and A10 tool steels. Note that plastic deformation (strain) occurs during such transformations at stresses that are lower than the yield stress for the phases present.[97] The occurrence of this plastic deformation, called the *transformation plasticity effect*, influences the development of stresses during the hardening of steel parts.[98] During quenching from the austenite range, the steel contracts until the M_s temperature is reached, then expands during martensitic transformation; finally, thermal contraction occurs on further cooling to room temperature. As the hardening temperature increases, a greater amount of carbide goes into solution; consequently, both the grain size and the amount of retained austenite are increased. This also increases the hardenability of steel.

More trouble with distortion comes from the quenching or hardening operation than during heating for hardening, in which the faster the cooling rate (i.e., the more severe the quenching), the greater the danger of distortion. When the milder quenchants are used, the extent of distortion is lessened. The severity of quenching thus influences the distortion of components.

The quench distortion remains a complex phenomenon involving numerous factors such as material, thermal, plastic, and elastic properties and phase transformations. The properties of the quenchants and the part/quenchant interface are also very significant and often difficult to characterize or quantify. This process, however, can be successfully predicted by computer simulation provided necessary input data are available and reasonably accurate.[99]

The dependence of volume increase, particularly in steel parts of different dimensions, on grain size (or hardenability) is another important factor. Variations in volume during quenching of a fine-grained shallow-hardening steel in all but small sections are less than for a coarse-grained deep-hardening steel of the same composition.

3. *Tempering.* There is a certain correlation among the tempering temperature, volume change, and its state of stress. The magnitude of the distortion during the tempering process depends on the steel composition and the austenitizing temperature. Tempering causes a continuous reduction in the state of stress and reduces the volume of martensite, but not adequately enough to equalize completely the prior volume increase as a result of martensitic transformation, unless the components are completely softened.

Low-Alloy and Plain (Medium- and High-) Carbon Steels. During the first and third stage of tempering, a decrease in volume occurs that is associated with the decomposition of: high-carbon martensite into low-carbon martensite plus ε-carbide in the former stage, and aggregate of low-carbon martensite and ε-carbide into ferrite plus cementite in the latter stage. In the second stage, however, an increase in volume takes place (due to the decomposition of retained austenite into bainite) that tends to compensate for the early volume reduction. As the tempering temperature is increased further toward A_1, more pronounced volume reduction occurs.

High-Alloy Steels. In some highly alloyed tool-steel compositions, the volume changes during martensite formation are less striking because of the large proportion of retained austenite and the resistance to tempering of alloy-rich martensite. These hardened steels show sharp increases in both hardness and volume between 500 and 600°C (930 and 1100°F) owing to the precipitation of very finely dispersed alloy carbides from the retained austenite. This produces a depleted matrix in alloy content, raising the M_s temperature of retained austenite. During cooling down from the tempering temperature, further transformation of retained austenite into martensite will occur with an additional increase in volume.

TABLE 17.8 Typical Volume Percentages of Microconstituents Existing in Four Different Tool Steels after Their Standard Hardening Treatments[82]

Steel	Hardening treatment	As-quenched hardness, HRC	Martensite, vol%	Retained austenite, vol%	Undissolved carbides, vol%
W1	790°C (1450°F), 30min; WQ	67.0	88.5	9	2.5
L3	845°C (1550°F), 30min; OQ	66.5	90	7	3.0
M2	1225°C (2235°F), 6min; OQ	64	71.5	20	8.5
D2	1040°C (1900°F), 30min; AC	62	45	40	15

Note: WQ, water quenched; OQ, oil quenched; AC, air cooled.
Courtesy of ASM International.

17.5.1.1 Size Distortion. Size distortion is the result of a change in volume generated by a change in microstructural structure during heat treatment. For a given component and a given steel grade, heat-to-heat variations of hardenability affect the relative proportions of microstructural constituents, e.g., martensite, bainite, ferrite, and pearlite, with the consequence that the dimensions of the part will be altered by an amount relating to the predominant microstructure.[20] Table 17.8 shows the typical volume percentages of microconstituents present in four different tool steels after their standard hardening treatments. Typical dimensional changes during hardening and tempering of several tool steels are given in Table 17.9. It is apparent that some steels such as M3 and M41 high-speed steels show appreciable increase in size of about 0.2% after hardening and tempering between 540 and 595°C (1000 and 1100°F) to produce complete secondary hardening. Other types, such as A10, expand very little when hardened and tempered over the entire temperature range up to 595°C (1100°F). Excessive size changes in oil-hardening non-shrinkable tool steel are usually caused by lack of stress relief (when necessary) and hardening and/or tempering at the incorrect temperature. The golden rule is to learn to be suspicious of tools that are seriously off-size in only one dimension. It is further noted that alloying addition in steel brings about a change in the specific volume of many microconstituents, but to a lesser extent than carbon.[100] This table provides comparative data on size distortion in a variety of steels; however, this information cannot be used alone to predict the shape distortion factor.

17.5.1.2 Shape Distortion or Warpage. This is sometimes called *straightness*, or *angularity change*. It is found particularly in nonsymmetrical components during heat treatment. From the practical viewpoints, warpage in water- or oil-hardening steels is normally of greater magnitude than is size distortion and is more of a problem because it is usually not predictable. This is caused by the sum effect of more than one of these factors:

1. Rapid heating (or excessive heating), drastic (or careless) quenching, or nonuniform heating and cooling cause severe shape distortion. Slow heating as well as preheating of the parts prior to heating to the austenitizing temperature yields the most satisfactory result. Rapid quenching produces thermal and

TABLE 17.9 Typical Dimensional Changes during Hardening and Tempering of Several Tool Steels[82]

Tool steel	Hardening treatment Temperature °C	Hardening treatment Temperature °F	Quenching medium	Total change in linear dimensions after quenching, %	Total change in linear dimensions, %, after tempering at 150°C 300°F	205°C 400°F	260°C 500°F	315°C 600°F	370°C 700°F	425°C 800°F	480°C 900°F	510°C 950°F	540°C 1000°F	565°C 1050°F	595°C 1100°F
O1	815	1500	Oil	0.22	0.17	0.16	0.18	...	...	...	...	...	...	...	...
O1	790	1450	Oil	0.18	0.09	0.12	0.13	...	...	...	...	...	...	...	...
O6	790	1450	Oil	0.12	0.07	0.10	0.14	0.10	0.00	-0.05	-0.06	...	-0.07	...	...
A2	955	1750	Air	0.09	0.06	0.06	0.08	0.07	...	0.05	0.04	...	0.06	...	...
A10	790	1450	Air	0.04	0.00	0.00	0.08	0.08	0.01	0.01	0.02	...	0.01	...	0.02
D2	1010	1850	Air	0.06	0.03	0.03	0.02	0.00	...	-0.01	-0.02	...	0.06	...	...
D3	955	1750	Oil	0.07	0.04	0.02	0.01	-0.02	...	...	...	...	...	...	...
D4	1040	1900	Air	0.07	0.03	0.01	-0.01	-0.03	...	-0.4	-0.03	...	0.05	...	...
D5	1010	1850	Air	0.07	0.03	0.02	0.01	0.00	...	0.3	0.03	...	0.05	...	...
H11	1010	1850	Air	0.11	0.06	0.07	0.08	0.08	...	0.3	0.01	...	0.12	...	...
H13	1010	1850	Air	-0.01	...	...	...	...	...	...	0.00	...	0.06	...	...
M2	1210	2210	Oil	-0.02	...	...	...	...	...	...	...	-0.06	0.10	0.14	0.16
M41	1210	2210	Oil	-0.16	...	...	...	...	...	...	...	-0.17	0.08	0.21	0.23

Courtesy of ASM International.

mechanical stresses associated with the martensitic transformation. In the case of low- and high-hardenability steels, respectively, this problem becomes severe or very small.

2. Nonsymmetrical carburizing may also contribute to distortion.[95]

3. Residual stresses are present in the component before heat treatment. These arise from machining, grinding, straightening, welding, casting, spinning, forging, and rolling operations, which will also furnish a marked contribution to the shape change.[101]

4. Residual stresses developed during heat treatment are caused by: (1) thermal gradients within the metal, (2) nonuniform changes in the metallurgical structure, and (3) nonuniformity in the composition of the metal itself, such as that caused by segregation.[102]

5. Applied stress causes plastic deformation. Sagging and creep of the components occur during heat treatment as a result of improper support of components or warped hearth in the hardening furnace. Hence, large, long, and complex-shaped parts must be properly supported at critical positions to avoid sagging or preferably are hung with the long axis on the vertical.

6. Nonuniform agitation/quenching or nonuniform circulation of quenchant around a part results in an assortment of cooling rates that creates shape distortion.[103] Uneven hardening, with the formation of soft spots, increases warpage. Similarly, an increase in case depth, particularly uneven case depth in case hardening steels, increases warpage on quenching.[104]

7. There is tight (or thin and highly adherent) scale and decarburization, at least in certain areas. Tight scale is usually a problem encountered in forgings hardened from direct-fired gas furnaces having high-pressure burners. Quenching in areas with tight scale is extremely retarded compared to that in the areas where the scale comes off. This produces soft spots, and in some cases, severe unpredicted distortion. Some heat treaters coat the components with a scale-loosening chemical prior to their entry into the furnace.[103] Similarly, the areas beneath the decarburized surface do not harden as completely as the areas below the nondecarburized surface. The decarburized layer also varies in depth and produces an inconsistent softer region compared to the region with full carbon. All these factors can cause a condition of unbalanced stresses with resultant distortion.[103]

8. There are long parts with small cross sections ($>L = 5d$ for water quenching, $>L = 8d$ for oil quenching, and $>L = 10d$ for austempering, where L is the length of the part and d is its diameter or thickness).

9. There are thin parts with large areas ($>A = 50t$, where A is the area of the part and t is its thickness).

10. There is unevenness of, or greater variation in, section.

Figure 17.7 summarizes a large number of possible contributory factors.

17.5.2 Examples of Distortion

1. *Ring die.* Quenching of ring die through the bore produces the reduction in bore diameter as a result of formation of martensite, associated with the increased volume. In other words, metal in the bore is upset by shrinkage of the surrounding metal and is short when it cools.[24] However, all-over quenching causes the outside diameter to increase and the bore diameter to increase or decrease, depending upon the precise dimensions of the part. When the outside diameter of the steel part is induction- or flame-hardened (with water quench), it causes the part to shrink in outer diameter.[79] These are the examples of the effect of mode of quenching on distortion.[105]

2. *Thin die.* A thin die, with respect to wall thickness, is likely to increase in bore diameter, decrease in outside diameter, and decrease in thickness when the faces are hardened. If the die has a very small hole, insufficient quenching of the bore may enlarge the hole diameter because the body of the die moves with the outside hardened portion.

3. *Bore of a finished gear.* Similarly, the bore might turn oval or change to such an extent that the shaft cannot be fitted by the allowances that have been provided. Even a simple shape such as a diaphragm or orifice plate may, after heat treatment, lose its flatness in such a way that it may become unusable.

4. *Production of long pins.* In the case of the production of long pins (250 mm long × 6-mm diameter, or 10 × $\frac{1}{4}$ in.) made from medium-alloy steel, it was found, after conventional hardening, that when the pins were mounted between centers, the maximum swing was over 5 mm (0.20 in.). However, the camber could be reduced to within acceptable limits by martempering, intense or press quenching.

5. *Hardening and annealing of long bar.* When a 1% carbon steel bar, 300 mm long (or more) × 25-mm diameter (12 in long or more × 1-in. diameter), is water quenched vertically from 780°C (1435°F), the bar increases in both diameter and volume but decreases in length. When such bars are annealed or austenitized, they will sag badly between the widely spaced supports. Hence, they should be supported along their entire length in order to avoid distortion.

6. *Hardening of half-round files.* Files are usually made from hypereutectoid steel containing 0.5% chromium. Files are heated to 760°C (1400°F) in an electric furnace after being surface coated with powdered wheat, charcoal, and ferrocyanide to prevent decarburization. They are then quenched vertically in a water tank. On their removal from the tank, the files appear like the proverbial dog's tail. The flat side has curved down, the camber becomes excessive, and the files can no longer be used in service. One practical solution is to give the files a reverse camber prior to quenching. The dead flat files could, however, be made possible, and the judgment with regard to the actual camber needed depends upon the length and the slenderness of the recut files.[106]

Similarly, when a long, slender shear knife is heat-treated, it tends to curve as a dog's tail, unless special precautions are taken.

7. *Hardening of chisels.*[79] Chisels about 460 mm (18 in.) long and made from 13-mm (0.5-in.) AISI 6150 bar steel are austenitized at 900°C (1650°F) for 1.5 hr and quenched in oil at 180°C (360°F) by standing in the vertical position with chisel point down in special baskets that allow stacking of two 13-mm (0.5-in.) round chisels per 650 mm^2 (1 in.^{2}) hole. Subsequently, hardened chisels are tempered between 205 and 215°C (400 and 420°F) for 1.5 hr. These heat-treated parts show 55 to 57 Rc hardness but are warped. The reasons for this distortion are that

- The portion of the bar that touches the basket cools slowly, producing uneven contraction and thermal stress.
- The martensite formation is delayed on the inner or abutting side of the bar, causing unequal expansion during transformation. This distortion can be elimi-

nated or minimized by loading the parts in the screen-basket in such a way that the stacking arrangement permits sufficient space between each part and by slightly decreasing the austenitizing temperature.[78] Distortion can also be minimized by austempering the part, provided that the carbon content is on the high side of specification, to produce the lower bainitic structure of 55 to 57 Rc. If higher yield stress is not warranted, only chisel ends need hardening and subsequent tempering.[79]

8. *Hardening of a two-pounder shot.* The hardness of a two-pounder shot was specified at 60 Rc on the nose and 35 Rc at the base. A differential hardening technique was performed on the shot made of a Ni-Cr-Mo steel. This technique consisted of quenching the shot in the ice-cold water by its immersion in a tank up to the shoulder, followed by drawing out the water from the tank at a stipulated rate until the waterline reached the base of the nose. The final step involved withdrawing the shot from the tank when completely cold. The back end was then softened by heating in a lead bath after initial tempering. The first few shots hardened in this way were observed to split vertically across the nose. The failure was, however, avoided by withdrawal of the shot before attaining ice-cold temperature and its subsequent immersion in warm water.[106]

9. *Hardening of a burnishing wheel.* In the manufacture of railway axles, the gearing surface on which the axle rests in the housing has to be given a high burnishing polish, employing a circular pressure tool that is made of 1.2C-01.5Cr steel. For satisfactory results, the hardness of the tool surface should be about 60 Rc. It has been found that the tool usually cracks before its withdrawal from the cold-water quenching bath. This problem may, however, be avoided by quenching the tool in water for 10 s prior to transferring it to an oil bath for finish quenching. Time quenching can be judiciously applied for many heat treatment problems of distortion or cracking. Stress-relieving treatment after the use of the tool for some time may also enhance its performance life. As indicated above, martempering is also one of the solutions for this problem.[105]

10. *Hardening of case-carburized mild steel.* If oil-hardening steels are not available for making a component, mild steel parts are carburized and water quenched to obtain the desired hardness, possibly resulting in excessive distortion.

Press straightening of case hardened shaft is fairly common. Shaft length-to-diameter ratio is important (short, fat shafts are difficult; long slender shafts are easier). Straightening depends very much on the yield strength of the core; the core is caused to yield while the case remains elastic. Therefore, a ferritic core (mild steel) should lend itself to straightening better than a bainitic or martensitic core (alloy steel). Some heat treaters prefer to straighten after quenching and before tempering, because tempering raises the yield strength of the hardened material.[20]

11. *Hardening of carburized low-carbon steel rollers.* The best course of quenching carburized En32 steel rollers (25-mm dia. × ≥600 mm long, or 1-in. dia. × ≥2 ft long), employed in textile printing, is to roll them down skids into water quenching tanks because this produces less warpage than when quenched slowly with the bar either in vertical, horizontal, or inclined positions. These are the procedures adopted for hardening of cylinders with length considerably greater than the diameter.

12. *Hardening of helix gears.* The distortion of the helical gears made of IS 20MnCr1 Grade steel (similar to AISI 5120) used as the third speed gear in the gear box of Tata trucks is an unavoidable natural consequence of the hardening process after carburizing. This type of distortion is linked with increased length and decreased diameter and occasionally increased helical angle.[107] If the extent of distortion can be controlled, a constant correction to the helix angle can be imparted in the soft-stage manufacturing (machining) prior to heat treatment, so that this correction can compensate for the distorted angle and may result in a gear with the desired helix angle. Thus a constant magnitude of distortion without minimization is assured in every job of every batch of production in commercial manufacturing. However, the residual stress system and metallurgical properties such as core strength, case depth, surface hardness, proper microhardness in the surface regions, and so forth are assured.[108] Similarly, when heavy-duty tooth gear is gas carburized and quenched to harden the surface layer, the diameter and tooth span increase and tapering and bending also occur.

13. *Nitriding.* A rolling mill screw, after liquid nitriding, may also show a small decrease in length, which causes pitch errors in the screw.[107]

14. *Induction and flame hardening of spur gears.* Spur gears, after induction and flame hardening, exhibit increased circular pitch, the error being maximum for the tooth groove quenched first. Similarly, in line-heating process, the thin plate undergoes convex bending; and the thick plate, concave bending.[107]

15. *Head hardened rail.* The rail head is hardened for its increased abrasion resistance. This is done by two methods. One is through heating of the rail followed by spray quenching the rail head only. In this case, the quench distortion of the rail tends to be head convex type (Fig. 17.18a), due to faster cooling of rail head compared to the rail base.

The second process involves induction heating of the rail head followed by spray quenching. Here, the quench distortion tends to be a head concave type

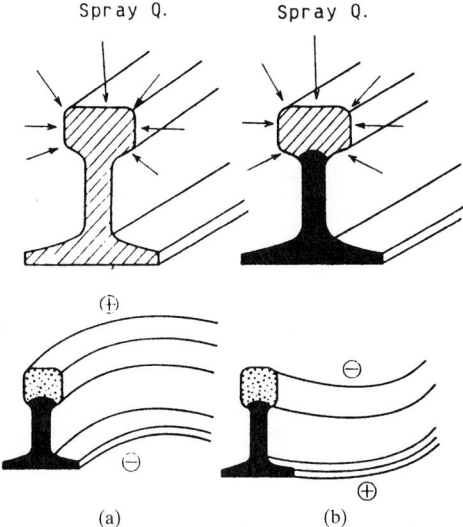

FIGURE 17.18 Quench distortion of head hardened rail.[109] (*Courtesy of ASM International, Materials Park, Ohio.*)

(Fig. 17.18*b*), due to the slower cooling of the rail head compared to the rail base, even though the rail head is water spray quenched.[109]

17.5.3 Precautions

1. Inadequate support during the heat treatment cycle, poorly designed jigs and quenching fixtures, or incorrect loading of the parts may cause distortion.[93] In general, plain carbon and low-alloy steels have such a low yield strength at the hardening temperature that the parts are capable of distorting under their own weight. Every care, therefore, must be taken to ensure that parts are carefully supported or suspended during heating. Long parts are preferably heated in a vertical furnace or with the length in the vertical plane.[110] They should be quenched in the vertical position with vertical agitation of the quenchants. Also, it must be remembered that many tool steels are spoiled by failure to provide enough support when they are taken out of the furnace for quenching. Thus every precaution is taken to ensure that parts are adequately supported during the entire heat treatment by employing well-designed jigs, fixtures, and so on.

2. Large parts must be raised off the hearth plate to ensure adequate heat circulation and more even heating and cooling. Care must be taken in transferring the load. Preferably, the parts should be placed on trays that can be grasped to remove the load. If the individual part must be handled by tongs, avoid holding it at the thinner section, which will lose heat rapidly and might bend more easily.[102]

3. Tool steels should be heated to hardening temperature slowly, or in steps, and uniformly. Hot salt baths are used to render fast, uniform heat input.

4. It is best to heat small sections to the lower region of the recommended hardening temperature range and to heat large sections to the higher temperature range. Overheating by employing too high a temperature or too long a heating time must be avoided.

5. It is a good practice to protect the surface of the component from decarburization by (packing in) cast iron chips, controlled furnace atmosphere, or using a vacuum furnace. If a separate preheating furnace is not available, the part can be put in a cold furnace, after which the temperature is raised to proper preheating temperature and kept at that temperature to attain uniform heating throughout, prior to proceeding to the hardening temperature.[111]

6. With the slower cooling rate, which is consistent with good hardening practice, a lower thermal gradient will be developed, thereby producing less distortion. (*Note*: Slower cooling rate is possible with steels having adequate hardenability to give required microstructure and hardness.[20])

7. Thus rapid heating and cooling rates of irregularly shaped parts must be avoided.

8. Proper selection of quenchant with desirable quenching properties and adequate agitation during hardening must be provided.

9. Recent innovations in quench tank designs are oscillating quench elevators, reversible and variable. Oil circulation, ultrasonic agitation, very high-flow-rate quenching, and flood quenching minimize and ensure uniform breakdown of vapor phase and provide better uniformity and uniform circulation of quench media.[94]

10. The ideal method of cooling is a spray quenching; especially computer-aided quenching (CAQ) is more promising.[109]

17.5.4 Methods of Preventing Distortion[106,112]

Currently, there are various methods of prevention of distortion (or warping) that are employed in the industry: straightening, support and restraint fixtures, quenching fixtures, pressure quenching, press quenching, rolling die quenching, and stress-relieving. These are described below.

1. *Straightening.* Straightening is one method to remove or minimize distortion. Since straightening (after hardening) can largely relieve the desirable residual compressive stresses (in plain-carbon and low-alloy steels) that may cause breakage, it would be better to accomplish this before the steel cools below the M_s temperature, i.e., when the steel is in the metastable austenitic state.[45] This temperature is above 260°C (500°F) for most tool steels and is preferably about 400°C (750°F) for long shear knives, which are usually made of 2C-12Cr steel. Warping on parts such as shafts and spindles can be corrected by straightening during or after hardening, followed by grinding to size.[108] Mostly high-alloy steels are straightened after hardening due to the higher percentage of retained austenite and their comparatively low yield stress. Straightening also can be accomplished during the tempering process.[45] However, straightening of hardened parts with higher strength will cause a loss of fatigue properties and possibly initiation of cracks at the surface. Hence, straightening after the hardening treatment must be very carefully controlled and should be followed by a low-temperature tempering treatment.

   The case-hardened (e.g., nitrided, carburized) parts can be straightened to a very large extent as a result of their lower core hardness. Nitrided parts may be straightened at 400°C (750°F).[45]

   As a rule, the parts subjected to straightening after heat treatment are likely to be distorted at subsequent stages of machining.[113]

2. *Support and restraint fixtures.* Fixtures for holding finished parts or assemblies during heat treatment may be either the support or the restraint type. For alloys that are subjected to very rapid cooling from the solution-treatment temperature, it is common practice to use minimum fixturing during solution treatment and to control dimensional relations by using restraining fixture during aging. Support fixtures are used when restraint type is not needed or when the part itself renders adequate self-restraint. Long, narrow parts are very easily fixtured by hanging vertically. Asymmetrical parts may be supported by placing them on a tray of sand or a ceramic casting formed to the shape of the part.[80] Restraint fixtures may require machined grooves, plugs, or clamps. Some straightening of parts can be accomplished in aging fixtures by forcing and clamping slightly distorted parts into the fixture. The threaded fasteners for clamping should not be used because they are difficult to remove after heat treatment. It is preferable to use a slotted bar held in place by a wedge.[80] The bore of a hub, the most important dimension in the hardening of thin spur gears, can be mechanically plugged to prevent the reduction of the bore and keep the out-of-roundness close to tolerance limits. When hardening large hollows, either restraining bands on the outside during tempering or articulated fillers serve the same purpose.

3. *Quenching fixtures.* When water quenching or oil quenching is essential, distortion can be minimal by employing properly designed quenching fixtures that forcibly prevent the steel from distortion.[114] Figure 17.19 shows a typical impingement-type quenching fixture. The requirements essential for the better design of this type of fixture are as follows:[103]

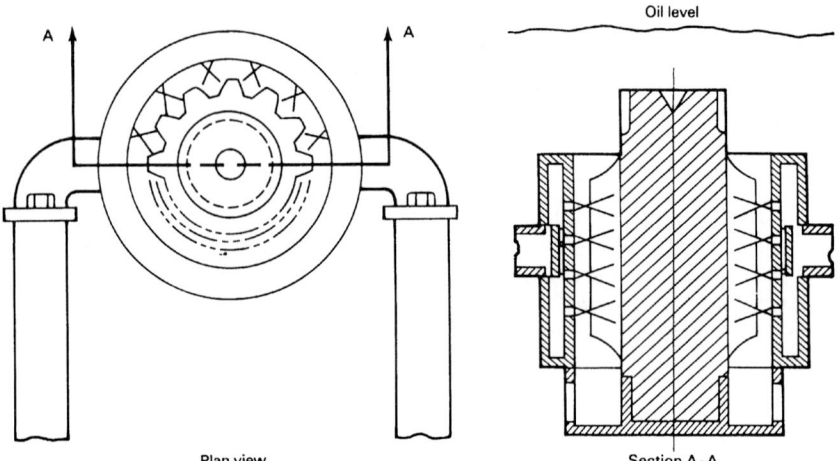

FIGURE 17.19 A typical impingement-type quenching fixture. (*Reprinted by permission of Fairchild Publications, New York.*)

    a. There must be an accurate positioning of the part in the fixture. Whenever possible, round bars should be rotated during quenching to level out variations in jet pressure around the part.

    b. There should be an unhindered flow of quenchant through the sufficiently large holes (3.3 to 6.4 mm, or 0.13 to 0.25 in., in diameter). Jets as large as 12.25 mm (0.50 in.) in diameter may be employed with furnace-heated heavy sections (e.g., plates). A large portion of the excess quenchant with these large jets is for the removal of scale.[115]

    c. Spacing between the holes should be reasonably wide (for example, 4*d*, where *d* is the hole diameter).

    d. For oil-quenching fixtures, the facility to submerge the part is required to reduce fumes and flashing.

    e. There must be the provision for efficient cleaning of the holes.

    f. A facility must be available to drain out the hot quenchant for effective quenching performance with cold quenchant.

4. *High-pressure gas quenching.* High-pressure gas quenching, using helium or hydrogen up to 20 bar, is claimed to approach the efficiency of oil and reduce distortion. However, pressures up to 40 bar are being considered to provide quench rates equivalent to or faster than those of oil.[95] This is economical and fast, provides even cooling, and offers a unique design, minimum distortion, and improved metallurgical qualities. As a result of these beneficial effects, this is suited to quench large-diameter tooling for the aluminum extrusion industry; quench large-diameter carburized gear, large fasteners, and precision gears to be jigged vertically; harden high-speed steel tools (such as saw blades, dies, and other parts with edge configuration) and 718 jet engine compressor blades.[116] This is also employed to quench (vacuum processed) large sections of titanium alloy castings for aircraft applications.[117] Figure 17.20 is a pressure-quench

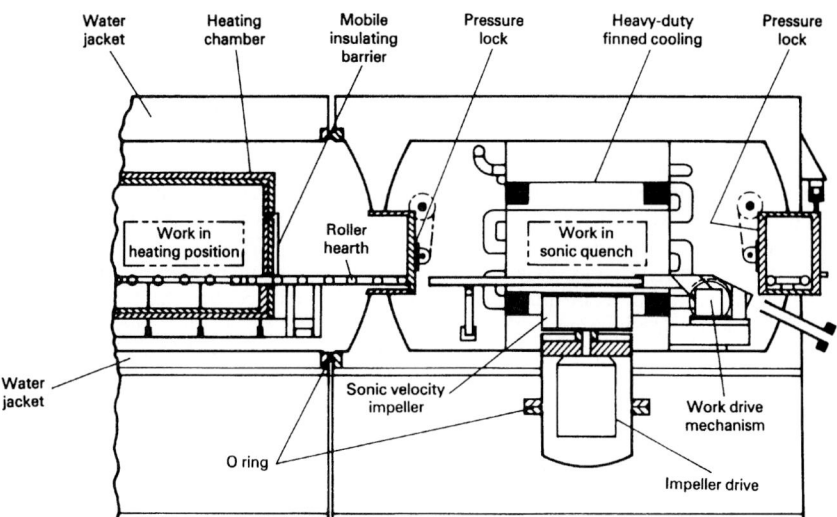

FIGURE 17.20 Pressure-quench module for attachment into standard vacuum-sealed quenched and continuous vacuum furnaces.[116] (*Courtesy of Hayes, Inc.*)

module that may be attached to vacuum-sealed quenched and continuous-vacuum furnace as a replacement for the oil quench section.[116] (See Sec. 13.2.3.7 for more details.)

High-pressure gas quenching coupled with vacuum heat treatments is commercially used. These facilities may be expensive, but they minimize or eliminate post-heat treatment corrective machining.[95]

5. *Press quenching.*   Press quenching is widely employed in preventing and controlling quench distortion in components whose geometry renders them particularly prone to distortion.[118] Although press quenching maintains the geometry (shape, flatness, etc.), "uniform" dimensional changes still occur. Press quenches usually pulsate the restraining tools—to allow the component to move (e.g., contract) during quenching.[36]

   Press quenching is used on bevel and spiral bevel gear wheels, large crown wheels, and annular gears.[20,36] Flat circular diaphragms of spring steel used in the control or measurement of pressure are press quenched between two copper blocks, which cannot be accomplished by direct quenching.[104]

6. *Rolling die quenching.*   A rolling die quench machine can provide uniform water quenching with minimal distortion for large-production runs. When a heated part is placed on the rollers, the die closes and the rolls turn. This removes any distortion incurred during heating. According to manufacturers of rolling die quench machines, symmetrical parts with the following straightness can be achieved in production:

$$\text{TIR} = K\left(\frac{l}{d}\right) \qquad (17.2)$$

where TIR is the *total indicator reading* of straightness, *l* is the length (in.), *d* is the diameter (in.), and *K* is a constant $= 10^{-4}$.

For minimum yield strength requirements of 310 MPa (45 ksi), air hardened or normalized parts with negligible distortion can be produced.[103]

7. *Stress relieving.*   The presence of residual stresses in the parts caused by cold-working, drawing, extrusion, forging, welding, machining, heading operations, or air cooling following normalizing greatly increases the tendency of distortion. However, these residual stresses can be relieved by subcritical annealing or normalizing treatment just before the final machining operation, which decreases the distortion to an appreciable extent. This is of special importance for intricate parts with close dimensional tolerances.[104]

Stress reduction is necessary to avoid distortion during hardening and to avoid cracking resulting from the combination of residual stress to the thermal stress produced during heating to the hardening temperature. In the event that stress relieving is not performed after heat treatment, large distortions of the part can be removed by heavy grinding. However, the drawbacks of this operation are possible elimination of most, if not all, of the hardened case of the carburized and hardened part; and danger of burning and crack formation on the surface layers.

Hence, it is customary to stress-relieve plain carbon or low-alloy steel parts at a temperature of 550 to 650°C (1020 to 1200°F) (for 1 to 2 hr), hot-worked and high-speed steels at 600 to 750°C (1110 to 1380°F), and the heavily machined or large parts at 650°C (1200°F) (for 4 hr) prior to final machining and heat-treatment operations.

Subresonant stress relieving may also be employed to neutralize thermally induced stress without changing the mechanical properties or the shape of the component. These components include: large workpieces, premachined or finish-machined structural or tubular, nonferrous, hardened, nonsymmetrical or varying section thickness, stationary, or assembled. However, this does not work on copper-rich alloys and the edges of burned plates.[119]

Distortion due to residual stress that occurs during nitriding can be minimized or eliminated by previously stress-relieving the workpiece at a sufficiently high temperature.[120]

The stress relief technique has been found to be ineffective for heat-treating asymmetrical parts which require adequate support at critical locations to avoid sagging, or preferably are hung with the long axis in the vertical position.[121] However, fulfilling these requirements cannot stop distortion, because the main reason of distortion, i.e., high plastic deformation of the asymmetrical component after hardening, is not removed. Consequently, straightening is needed. Straightening, however, introduces regions of high stress concentrations, which may initiate development of surface and initial cracks.[113]

17.5.5 Control of Distortion

In order to remove or minimize distortion, the modern trend is to shift from water quenching practice to milder quenching (if the steel in question has adequate hardenability), for example, oil quenching, polymer quenching, martempering, austempering, or even air-hardening practice. One may also need to change the grade of steel. Milder quenchants produce slower and more uniform cooling of the parts, which drastically reduces the potential distortion.

Other strategies of controlling distortion for age-hardening aluminum, beryllium, and other alloys include: alloy and temper selection, fixturing, age hardening temperatures, proper machining, and stamping operations.[121]

The fewer the number of reheats applied to components in case-hardening steels following carburizing, the smaller the distortion on the finished part. When top priority is given to minimum distortion, it is desirable to make the parts from oil-hardening steels with a controlled grain size and to harden them by martempering direct from carburizing.

Currently, polyalkylene glycol-base quenchants, such as UCON quenchants HT and HT-NN, are variously used for direct quenching from the forging treatment, continuous cast quenching, and usual hardening of forged and cast steels and cast iron. In this case, boiling does not take place at the component surface but rather at the external surface of the deposited polymer film. More uniform cooling occurs, and thermal stresses are released. Because of the lower boiling point and high thermal conductivity, UCON quenchants act through the martensite zone more rapidly than oil.[122]

Distortion during ferritic nitrocarburizing is minimal because of low treatment temperature and the absence of subsequent phase transformations.[70]

For many applications the distortion due to induction hardening is small and acceptable, requiring no special measures to control or correct. However, the shape and the ratio of hardened to unhardened thickness are important factors.[20] There are many methods of reducing distortion in induction-hardened components; these methods are usually found by experience with variables such as the hardening temperature and the type and temperature of quenching medium employed. Methods of reducing distortion in induction hardened parts include: the hardening of small spindles held vertically in jigs; the plug-quenching of gears to prevent the bores from closing in; the flattening of cams by clamping them together during tempering; and the selective hardening of complex shapes.[123]

As a replacement of medium- or slow-quenching oils, UCON quenchants E and E-NN can be readily used in induction- and flame-hardening operations, in both spray and immersion types, for high-carbon and most alloy steels and traditional hardening of cast iron and cast or forged steels of complex geometry with better distortion-reduction properties. Agitation of quenchant should be carried out by motor-driven stirrers to move the medium with respect to the part being quenched or by pumps that force the medium through the appropriate orifice. Alternatively, the parts are moved through the medium, and for some applications, spray quenchant is recommended. Water additives are sometimes employed in salt baths to increase heat extraction.[80]

Ultrasonic quenching is also effective in controlling distortion, which involves the introduction of ultrasonic energy (waves with a frequency of 25 kHz) in the quenching bath. This breaks down the vapor film that surrounds the part in the initial stages of water or oil quenching.[111]

17.5.6 Distortion after Heat Treatment

Straightening. When every possible case has been employed to minimize distortion, it may still be essential to straighten after heat treatment, which has already been discussed.

Grinding after Heat Treatment. In the case of carburized or nitrided parts, the metallurgist and designer, together with the production engineer, must collaborate regarding the amount to be removed by grinding after heat treatment. This grinding allowance must be taken into account when determining the initial dimensions and when specification for the case depth is to be applied.

Distortion may also occur after heat treatment, with time, owing to the completion of any unfinished transformation or the effect of increased temperature during grinding. For example, fully hardened components such as blade shears may be damaged by characteristic crazing pattern because of heavy and careless grinding. Local overheating results in the transformation of undecomposed austenite, and the accompanying changes in volume produce sufficient stresses to cause cracking and development of a crazed pattern.

Dimensional Stability. To achieve dimensional stabilization or stability[†] (i.e., retention of exact size and shape) over long periods, which is a vital requirement for gauges and test blocks, the amount of retained austenite in heat-treated parts must be reduced because retained austenite slowly transforms and produces distortion when the material is kept at room temperature, heated, or subjected to stress. Dimensional stabilization also reduces internal (residual) stress, which causes distortion in service.

The excessive retained austenite in steels directly quenched after carburizing arising from high case carbon content, high substitutional alloy content, high carburizing temperature, or geometry may be reduced by several methods, such as tempering, refrigeration or subzero treatment, intercritical temperature reheating, and shot-peening (Fig. 17.21). Process selection is dependent on the required performance level of the carburized part.[32]

Multiple tempering (with prolonged tempering times) is needed to achieve stabilization for, say, high-speed steel tools. However this is not essential for, say, case hardened gears made from conventional steels. The first tempering reduces internal stress and facilitates its transformation to martensite on cooling. The second and third retempering reduce the internal stress produced during the transformation of retained austenite.

It is the usual practice to carry out a single or repeated cold treatment after the initial tempering treatment. In cold treatment, the part is cooled below M_f, which will cause the retained austenite to transform to martensite; the extent of transformation depends on whether the tool part is untempered or first tempered. Cold treatment is normally accomplished in a refrigerator at a temperature of −70 to −120°C (−100 to −184°F). Tools must be retempered immediately after return to room temperature following cold treatment in order to reduce internal stress and increase the toughness of the fresh martensite. Finally, they are ground to size.

Subzero treatment should be employed judiciously, and according to Parrish and Harper, considered as a last resort for case hardened parts, especially commercial gears. Reduced fracture and bending fatigue resistance have been noted in refrigerated-treated carburized steels.[21]

If material selection, carbon gradient, and quenching are right, the retained austenite content and surface hardness will each be acceptable. Hence, there is no need to subzero-treat. The need to use refrigeration on a regular basis could be a sign that something is not quite right; e.g., surface carbon control is suspect, or target surface carbon is high. The heat treater should endeavor to get the processing right rather than resorting to the easier option of refrigeration. On the other hand, there may be a good reason for running with a high carbon for a given application.[20]

A recent innovation in subzero treatments avoids thermal tempering and replaces it with subzero treatment in the presence of a cyclic magnetic field. Rota-

---

[†] For full dimensional stability, the residual stresses in a body should be zero.[20]

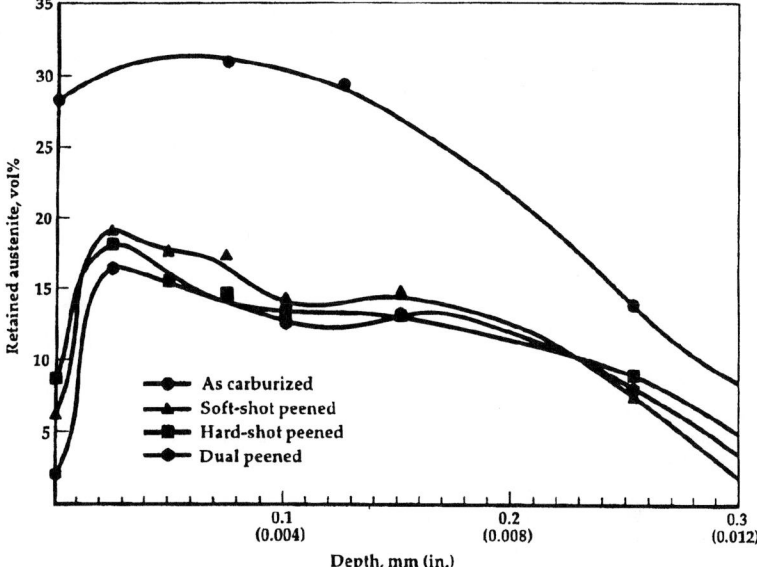

FIGURE 17.21   Retained austenite as a function of distance from the surface of a carburized SAE 4320 in the as-carburized condition and after various shot-peening treatments.[32,33] (*Courtesy of ASM International, Materials Park, Ohio.*)

tion beam test results suggest that with this treatment, fatigue lives comparable to those of conventionally tempered parts could be achieved.[34]

Vibratory stress-relieving process (VSRP) has been claimed to provide the ultimate in precision and long-term dimensional stability and to serve as an effective alternative to thermal treatment for component stabilization in large castings and fabrications. However, it does not offer any metallurgical benefits such as driving off hydrogen and modification of microstructure.[20,124]

17.5.7   Distortion and Its Control in Heat-Treated Aluminum Alloys

The high levels of residual stress and distortion that are produced in the water-quenched aluminum extrusion and forgings (such as 2000, 6000, and 7000 series) and aluminum castings can be reduced 60 to 100% by using proper selection of polyalkylene glycol quenchant or polyvinylpyrrolidone 90 concentration [for example, 25% solutions for wrought alloys, 20 to 30% UCON quenchant A for thicknesses up to 25 mm (1 in.), and 17 to 22% for larger than 25-mm (1-in.), section thickness in casting alloys] with sufficient agitation, lower bath temperature, proper fixture (throughout solutionizing, quenching, and age hardening treatments), and straightening (in the as-quenched state after taking out from the fixture) procedure. The initial cost of these polymer solutions as a replacement to the conventional hot-water quenching method is easily compensated for by other advantages, such as reduced scrap, reduced machining (compared to two machining operations required—one before and another after heat treatment—in the conventional water-quenching method), and increased fatigue life as a result of reduced convective heat transfer or film coefficient between the part and the quenchant, more uniform quench, precise control of quench rates, and improved heat-transfer qualities from the deposition of liquid organic polymer on the surface of the part being quenched.[125–127] This method costs less, therefore saves time, and allows easy shaping, bending, and twisting of the parts without establishing residual stresses. Such parts as leading-edge wing skins, spars, and bulkheads are used in the aerospace industries.[123]

17.6 IMPORTANCE OF DESIGN

The wrong design of the component may result in the establishment of nonuniform heating and cooling of the components, which produces overload and/or internal stresses, leading to distortion and failure during or after hardening. Correct consideration at the design stage plays an important role in lessening the distortion and danger of cracking. The basic principle of successful design is to plan shapes that will minimize the temperature gradient through the part during quenching. Fundamental rules such as maintaining a simple, uniform, regular, and symmetrical section with comparatively few shape changes; ensuring small and smooth cross-sectional size changes; and using large radii are still too frequently overlooked at the design stage. Thus, successful heat treatment demands a rational design that avoids sharp corners as well as sudden and undue changes of section.

It is often possible for tool designers to compensate for size distortion. For example, in preparing precision hobs for gear cutting, dimensional accuracy must be kept within very close tolerances. On linear longitudinal growth, it is the general practice to go out-of-round in the following high-speed steel bars as much as 0.3% in M1 type, 0.2% in M2 type, and 0.15% in T1 type during heat treatment. These data will alter slightly with changes in the design of the hobs, but essentially the growth in tungsten-base high-speed steel is lower than that of the molybdenum-base high-speed steel (M1 and M2). This does not create any difficulty if the growth is compensated for and if the steel is consistent in its growth.[112]

The distortion produced in the surface hardening of long shafts by the scanning method can be a great problem if the equipment is not in very good condition. Due consideration must be given so that locating centers run concentrically, in line, and at the appropriate speed; the coil must be accurately aligned, and the quench must be correctly designed with sufficient number of holes of suitable size and angle. For long shafts with a relatively small diameter (e.g., half shafts, which are likely to distort), the use of hydraulically operated restraining rolls usually overcomes this.[128]

The designer should bear in mind the following rules while designing a die or machine part that is to be heat-treated:

1. Distribution of the material should be as uniform as possible.

2. Provide fillets (larger radii) at the base of keyways, cutter teeth, and gear teeth to avoid stress concentration; semicircular keyways, which permit the use of round-corner keyways, are the right choices. Ideally, drives using involute splines are preferred to keyways.

3. Avoid abrupt changes of section; in other words, provide smooth changes of section.

4. Large holes (such as drawing or cutting openings in die rings or plates) must be centrally located from the outer contour. In some cases, holes are drilled through the heaviest section of the tool in order to help fairly balance the weight of the section rather than to unbalance it.[80] Deep blind holes should always be avoided because they cause nonuniform quenching. If this is not possible, the hole can be ground in after hardening. Drilled hole junctions in a steel part should be avoided because they enhance very high and undesirable cooling conditions. The problem with these cross-holes is to get sufficient quenchant into them. The inside surface of the holes tends to be in a state of high tensile stress, usually leading to cracking, at least with water quenching. As a minimum, the corner at the junction of the holes with outer diameter of the part should be given a generous radius to better distribute the tensile stress.[116] Similarly, grooves and keyways in highly stressed areas should be avoided; or, if possible, they should be located in low-stressed areas of the part. Alternatively, fixtures should be used that make it possible for the hole or the inside of the groove to be quenched in the beginning or more rapidly than the rest of the part.[27]

5. Round off all the holes, corners, and outer edges.

6. If sharp corners are unavoidable, provide relief notches in place of sharp edges.

7. The insertion of identification marks on the hardened component is recommended, preferably after hardening with tools having well-rounded edges and minimum deformation (shallow penetration depth), and at positions far away from the high-stress concentration zones (reentrant angles, bends, and so on).[129]

8. Large intricate dies should be made up in sections, which frequently simplifies heat treatment.[80]

REFERENCES

1. N. P. McLeod and J. Nutting, *Met. Technol.*, vol. 9, 1982, pp. 399–404.
2. G. E. Hale and J. Nutting, *Int. Met. Rev.*, vol. 29, no. 4, 1984, pp. 273–298.
3. R. W. Gardiner, *Met. Technol.*, vol. 4, 1977, pp. 536–547.
4. T. J. Baker and W. D. Harrison, *Met. Technol.*, vol. 2, no. 5, 1975, pp. 201–205.
5. T. J. Baker and R. Johnson, *JISI*, vol. 211, 1973, pp. 783–791.
6. I. S. Brammar, *JISI*, vol. 201, September 1963, pp. 752–761.
7. R. C. Andrews, G. M. Weston, and R. T. Southin, *J. Aust. Inst. Met.*, vol. 21, 1976, pp. 126–131.
8. R. C. Andrews and G. M. Weston, *J. Aust. Inst. Met.*, vol. 22, 1977, pp. 171–176.
9. G. D. Joy and J. Nutting, in *Effects of Second Phase Particles on the Mechanical Properties of Steels*, Iron and Steel Institute, London, 1971, pp. 95–100.
10. R. N. O'Brien, D. H. Jack, and J. Nutting, in *Proceedings of Heat Treatment '76*, Metals Society, 1976, London, pp. 161–168.
11. C. L. Briant and S. K. Banerjee, *Met. Trans.*, vol. 10A, 1979, pp. 1151–1155.
12. B. J. Sultz and C. J. McMahon, Jr., *Met. Trans.*, vol. 4, 1973, pp. 2485–2489.
13. N. P. McLeod, Ph.D. thesis, University of Leeds, 1978.
14. A. Preece and J. Nutting, *JISI*, vol. 164, 1950, pp. 46–50.
15. R. Prestner, *Met. Mater.*, April 1974, p. 229.

16. A. H. Bodimeade, Ph.D. thesis, University of Leeds, 1974.

17. G. D. Joy, Ph.D. thesis, University of Leeds, 1971.

18. D. R. Glué, C. H. Jones, and H. K. M. Lloyd, *Met. Technol.*, vol. 2, 1975, pp. 416–421.

19. *Carbon and Alloy Steels (ASM Specialty Handbook)*, ASM International, Materials Park, Ohio, 1996, pp. 308–328.

20. G. Parrish, private communication, 2000.

21. T. Hanabusa and H. Fujiwara, in *Proceedings 32d Jpn. Congr. Mater. Res.*, 1989, pp. 27–36.

22. G. E. Dieter, *Engineering Design*, McGraw-Hill, New York, 1982.

23. G. Parrish and G. S. Harper, *Production Gas Carburizing*, Pergamon Press, Oxford, 1985.

24. B. Hildenwall and T. Ericsson, in *Proceedings of Hardenability Concepts with Applications to Steel*, eds. D. V. Doane and J. S. Kirkaldy, TMS-AIME, 1978, pp. 579–606.

25. E. B. Evans, in *Encyclopaedia of Materials Science and Engineering*, Pergamon Press, Oxford, 1986, pp. 4183–4188.

26. R. F. Kern and M. E. Suess, *Steel Selection*, Wiley-Interscience, New York, 1979.

27. R. F. Kern, *Selecting Steels and Designing Parts for Heat Treatment*, ASM, Metals Park, Ohio, 1969.

28. R. B. Liss, C. G. Massieon, and A. S. McClosky, *The Development of Heat Treat Stresses and Their Effect on Fatigue Strength of Hardened Steel*, Presented at Society of Automotive Engineers midyear meeting, 1965.

29. R. W. Shin and G. H. Walter, in *Proceedings of Residual Stresses for Engineers and Metallurgists*, ed. J. Vande Walle, ASM, Metals Park, Ohio, 1981, pp. 1–20.

30. *Carbon and Alloy Steels (ASM Specialty Handbook)*, ASM International, Materials Park, Ohio, 1996, pp. 365–389.

31. B. Scholtes and E. Macherauch, in *Case Hardened Steels: Microstructural and Residual Stress Effects*, ed. D. E. Diesburg, TMS-AIME, 1984, pp. 141–151.

32. G. Krauss, *Advanced Mats. & Process.*, July 1995, pp. 48U–48Y.

33. J. A. Sanders, M.S. thesis, Colorado School of Mines, Golden, 1993.

34. G. Parrish, *Carburizing: Microstructures and Properties*, ASM International, Materials Park, Ohio, 1999.

35. P. S. Prevey, D. J. Hornbach, and P. W. Mason, *Proceedings: 17th ASM Heat Treating Conference*, eds. D. Milan et al., 1997, ASM International, 1998, pp. 3–12.

36. W. T. Cook, private communication, 2000.

37. R. W. K. Honeycombe and H. K. D. H. Bhadeshia, *Steels: Microstructure and Properties*, 2d ed., Arnold, London, 1995.

38. B. S. Lement, *Distortion in Tool Steel*, American Society for Metals, Metals Park, Ohio, 1959.

39. H. C. Child, *Heat Treat. Met.*, 1981.4, pp. 89–94.

40. A. Rose and H. P. Hougardy, in *Proceedings of the Transformation and Hardenability in Steels Symposium*, Climax Molybdenum Company, Ann Arbor, Mich., 1967, pp. 155–167.

41. H. P. Kirchner, *Strengthening of Ceramics: Treatment Tests and Design Applications*, Marcel Dekker, New York, 1979.

42. W. Baldvin, Jr., *Residual Stresses*, in *Proceedings of the American Society for Testing and Materials*, vol. 49, 1949, pp. 539–583.

43. R. E. Reed-Hill and R. Abbaschian, *Physical Metallurgy Principles*, 3d ed., PWS-Kent Publishing, Boston, 1992.

44. A. Rose, *Hart.-Tech. Mitt.*, vol. 21, no. 1, 1966, pp. 1–6.

45. K. E. Thelning, *Steel and Its Heat Treatment*, Butterworths, London, 1985.

46. T. Yamaguchi, Z. G. Wang, and T. Inoue, in *Proceedings of the 27th Japan Congress on Materials Research, 1984*, p. 147; *Mater. Sc. Technol.*, vol. 1, 1985, pp. 872–876.

47. D. E. Diesburg, C. Kim, and W. Fairhurst, *Proceedings of Heat Treatment '81*, Metals Society, London, 1983, pp. 178–184.

48. D. P. Koistinen, *Trans. ASM*. vol. 50, 1958, pp. 227–241.

49. L. Salonen, *Acta Polytech. Scand. Ser.*, vol. 109, 1972, pp. 7–26.

50. H. C. F. Rozendaal, P. F. Colijn, and E. J. Mittemeijer, *Surf. Eng.*, vol. 1, 1985, pp. 30–42.

51. *Case Hardening of Steel*, ed. H. E. Boyer, ASM International, Metals Park, Ohio, 1987.

52. T. Endo and M. Kawakami, *J. Soc. Mater. Sci. Jpn.*, vol. 312, 1983, p. 114.

53. E. D. Walker, in *Proceedings of Residual Stresses for Engineers and Metallurgists*, ed. J. Vande Walle, ASM, Metals Park, Ohio, 1981, pp. 41–50.

54. S. L. Semiatin and D. E. Stutz, *Induction Heat Treatment of Steel*, ASM, Metals Park, Ohio, 1985.

55. M. Melander, *Mater. Sci. Eng.*, vol. 1, 1985, pp. 877–882.

56. G. Parrish et al., *Heat Treat. Met.*, vol. 25, 1998.1, pp. 1–8.

57. G. Parrish et al., *Heat Treat. Met.*, vol. 25, 1998.2, pp. 43–50.

58. P. Petrov, D. Dimitrov, M. Aprakova, and S. Valkanov, *Mater. and Manufact. Process.*, vol. 13, no. 4, 1998, pp. 555–564.

59. J. Grum, R. Sturm, and P. Zerovnik, in *Proceedings Second International Conference on Quenching and Control of Distortion*, eds. G. E. Totten et al., ASM International, Materials Park, Ohio, 1996, pp. 181–191.

60. R. L. Peng and T. Ericsson, *Scand. Jn. of Met.*, vol. 27, 1998, pp. 223–232.

61. K. Masabuchi, in *Encyclopaedia of Materials Science and Engineering*, Pergamon Press, Oxford, 1986, pp. 4180–4183.

62. L. Karlsson, in *Thermal Stresses I*, vol. 1, ed. R. Hetnarski, Elsevier, Amsterdam, 1986, pp. 299–389.

63. L. Novikov, *Theory of Heat Treatment of Metals*, Mir Publishers, Moscow, 1978.

64. R. N. Mittal and G. W. Rowe, *Met. Technol.*, vol. 9, 1982, pp. 191–197.

65. J. O. Kristiansson, *J. Therm. Stresses*, vol. 5, 1982, pp. 315–330.

66. *Polymer Quenchant User Report*, Tenaxol, Inc., Milwaukee, Wisc.

67. R. G. Bathgate, *Met. Forum*, vol. 6, 1983, p. 11.

68. L. Mordfin, in *Proceedings of Residual Stresses for Engineers and Metallurgists*, ed. J. Vande Walle, ASM, Metals Park, Ohio, 1981, pp. 189–210.

69. F. Abbasi and A. J. Fletcher, *Mater. Sci. Technol.*, vol. 1, 1985, pp. 770–779.

70. G. E. Totten and M. A. H. Howes, *Steel Heat Treatment Handbook*, eds. G. E. Totten and M. A. H. Howes, Marcel Dekker, New York, 1997, chap. 5, pp. 251–292.

71. L. Mordfin, in *Encyclopaedia of Materials Science and Engineering*, Pergamon Press, Oxford, 1986, pp. 4189–4194.

72. E. J. Mittemeijer, *J. Heat Treat.*, vol. 3, no. 2, 1983, pp. 114–119.

73. T. H. De Keijser, L. J. Langford, E. J. Mittemeijer, and A. B. P. Vogels, *J. Appl. Crystallogr.*, vol. 15, 1982, pp. 308–314.

74. D. J. Horrnbach and P. S. Prevey, *17th ASM Heat Treating Conference*, eds. D. L. Milan et al., ASM International, Materials Park, Ohio, 1997, pp. 13–18.

75. T. R. Finlayson, *Met. Forum*, vol. 6, 1983, pp. 4–10.

76. D. S. Mountain and G. P. Cooper, *Strain*, vol. 25, no. 1, 1989, pp. 15–19.

77. R. F. Kern, *Heat Treat.*, vol. 17, no. 4, 1985, pp. 38–42.

78. D. H. Stone, in *Proceedings of the 1988 ASME/IEEE Joint Railroad Conference*, American Society of Mechanical Engineers, New York, 1988, pp. 43–53.

79. C. E. "Joe" Devis, *Ask Joe*, ASM, Metals Park, Ohio, 1983.

80. Chapter 8, in *Troubleshooting Manufacturing Processes*, 4th ed., ed. L. K. Gillispie, Society of Manufacturing Engineers, Dearborn, Mich., 1988.

81. J. Lyman, *J. Eng. Mats. and Tech.*, vol. 106, January 1984, pp. 253–256.

82. A. K. Sinha, in *ASM Handbook*, vol. 4: *Heat Treating*, ASM International, Materials Park, Ohio, 1992, pp. 601–619.

83. G. Wahl and I. V. Etchells, in *Proceedings of Heat Treatment '81*, Metals Society, 1983, pp. 116–122.

84. M. J. Gilersleeve, *Mats. Sc. and Technol.*, vol. 4, no. 4, 1991, pp. 307–310.

85. G. Parrish, *The Influence of Microstructure on the Properties of Case-Carburized Components*, ASM, Metals Park, Ohio, 1980.

86. R. P. Brobst and G. Krauss, *Met. Trans.*, vol. 5, 1974, pp. 457–462.

87. C. A. Apple and G. Krauss, *Met. Trans.*, vol. 4, 1973, pp. 1195–1200.

88. T. A. Balliett and G. Krauss, *Met. Trans.*, vol. 7, 1976, pp. 81–86.

89. R. J. Kar, R. M. Horn, and V. F. Zackay, *Met. Trans.*, vol. 10A, 1979, pp. 1711–1717.

90. W. Fichtl, *Hart-Tech. Mittelungen*, vol. 33, 1978, pp. 1–8.

91. G. E. Hollox and R. T. Von Bergn, *Heat Treat. Met.*, 1978.2, pp. 27–31.

92. T. Bell, *Survey of Heat Treatment of Engineering Components*, Iron and Steel Institute, 1973, pp. 69–72.

93. K. W. Chambers, *Heat Treatment of Metals*, Iron and Steel Institute, 1966, pp. 94–95.

94. H. W. Walton, in *2d International Conference on Quenching and the Control of Distortion*, eds. G. E. Totten et al., ASM International, Materials Park, Ohio, 1996, pp. 143–148.

95. W. T. Cook, *Heat Treat. Met.*, 1999.2, pp. 27–36; in *Proceedings of the 18th Conference on Heat Treating*, eds. R. A. Wallis and H. W. Walton, ASM International, Materials Park, Ohio, 1999, pp. 12–22.

96. R. W. Wilson, *Metallurgy and Heat Treatment of Tool Steels*, McGraw-Hill, New York, 1975, pp. 93–95.

97. P. G. Greenwood and R. H. Johnson, *Proc. Royal Soc.*, vol. A283, 1965, p. 403.

98. B. L. Josefson, *Mater. Sci. Technol.*, vol. 1, no. 10, 1985, pp. 904–908.

99. D. Huang, K. Arimota, D. Lambert, and M. Narazaki, *Heat Treating Conference and Exposition*, St. Louis, Mo., Oct. 9–12, 2000, pp. 1–5.

100. A. Ferrante, *Met. Progr.*, vol. 87, 1965, pp. 87–90.

101. B. R. Wilding, *Heat Treatment of Engineering Components*, Iron and Steel Institute, 1970, pp. 20–25.

102. B. A. Becherer and L. Ryan, in *ASM Handbook*, vol. 4: *Heat Treating*, ASM International, Materials Park, Ohio, 1992, pp. 761–766.

103. R. F. Kern, *Heat Treat.*, vol. 17, no. 3, 1985, pp. 41–45.

104. D. J. Grieve, *Met. Mater. Technol.*, vol. 7, no. 8, 1975, pp. 397–403.

105. F. D. Waterfall, *Met. Treat Drop Forg.*, April 1985, pp. 139–144.

106. S. Visvanathan, *TISCO J.*, vol. 23, no. 4, 1976, pp. 199–204.

107. Y. Toshioka, *Mater. Sci. Technol.*, vol. 1, no. 10, 1985, pp. 883–892.

108. R. Verma, V. A. Swaroop, and A. K. Roy, *TISCO J.*, October 1977, pp. 157–160.

109. S. Owaku, *2d International Conference on Quenching and the Control of Distortion*, eds. G. E. Totten et al., ASM International, Materials Park, Ohio, 1996, pp. 149–154.

110. Section 8, in *Cassels Handbook*, 9th ed., ICI Ltd., 1964.

111. R. F. Harvey, *Met. Progr.*, vol. 79, no. 6, 1961, pp. 73–75.

112. A. K. Sinha, *Tool Alloy Steels*, August 1980, pp. 219–224.

113. A. Khersonsky, *Met. Heat Treat.*, July/August, 1996, pp. 43–45.

114. G. F. Melloy, *Hardening of Steel*, Lesson 5, in *Heat Treatment of Steels*, Metals Engineering Institute, American Society for Metals, 1979, pp. 1–28.

115. R. F. Kern, *Heat Treat.*, vol. 18, no. 9, 1986, pp. 19–23.

116. Hayes, Inc., private communication, October 2000.

117. J. M. Neiderman and C. H. Luiten, *Proceedings of Heat Treatment '84*, Metals Society, London, 1984, pp. 43.1–43.8.

118. *Met. Mater.*, vol. 9, July/August 1975, pp. 52–53.

119. T. E. Hebel, *Heat Treat.*, vol. 21, no. 9, 1989, pp. 29–31.

120. D. Pye, in *Steel Heat Treatment Handbook*, eds. G. E. Totten and M. A. H. Howes, Marcel Dekker, New York, 1997, Chap. 10, pp. 721–764.

121. F. Dunlevey, *Heat Treat.*, vol. 21, no. 2, 1989, pp. 34–35.

122. *UCON Quenchants for Ferrous and Nonferrous Metals*, Tenaxol, Inc., Milwaukee, Wisc., 1988.

123. R. Creal, *Heat Treat.*, vol. 18, no. 12, 1986, pp. 27–29.

124. R. A. Claxton, *Heat Treat. Met.*, 1991.2, pp. 53–59; 1991.3, pp. 85–89.

125. C. E. Bates, *J. Heat Treat.*, vol. 5, no. 1, 1987, pp. 27–40.

126. *Information on Polymer Quenchants*, Tenaxol, Inc., Milwaukee, Wisc., 1989.

127. C. E. Bates and G. E. Totten, *Heat Treat. Met.*, 1988.4, pp. 89–97.

128. P. D. Jenkins, *Metallurgia*, vol. 45, no. 4, 1978, pp. 196–199.

129. F. Strasser, *Heat Treat. Met.*, 1980.4, pp. 91–96.

CHAPTER 18
SURFACE MODIFICATION AND THIN-FILM DEPOSITION

18.1 INTRODUCTION

Surface modification and thin-film deposition involve the alteration of surface composition or structure through the introduction of high-energy or particle beams, physical vapor deposition (PVD), and/or chemical vapor deposition (CVD) techniques that can serve to improve selected mechanical properties and wear and hardness, reduce friction, and increase fatigue, corrosion, and oxidation resistance of the critical surface area. Ion beam processes are used to produce surface-modified thin-film deposits. PVD is the production of a condensable vapor by physical means and subsequent low-temperature deposition of elements and alloys, as well as compounds using reactive deposition processes at low to ultrahigh vacuum. CVD is the deposition of atoms or molecules by the high-temperature chemical reaction at atmospheric pressure or low vacuum to ultrahigh vacuum. This chapter describes the ion beam, physical vapor deposition, and chemical vapor deposition processes with respect to the principles of operation, the effects on the surface properties, advantages and disadvantages, and applications.

18.2 ION BEAM PROCESSES

Ion beam processes offer various possibilities as industrial surface treatments for improving wear and corrosion resistance of critical components. Ion beam processes can be grouped into four categories: ion implantation, ion beam mixing, ion beam-assisted deposition, and ion plating. All of them allow independent control of process parameters such as particle energy, particle flux, gas pressure, and substrate temperature. Figure 18.1 shows a schematic view of ion implantation, ion beam mixing, and ion beam-assisted deposition after coating.[1]

18.2.1 Ion Implantation

Ion implantation is an important surface modification, precise doping, or surface-alloying process for metals and alloys, semiconductors, ceramics, dielectrics, optical materials, and polymers, which involves the generation of high-energy gaseous or

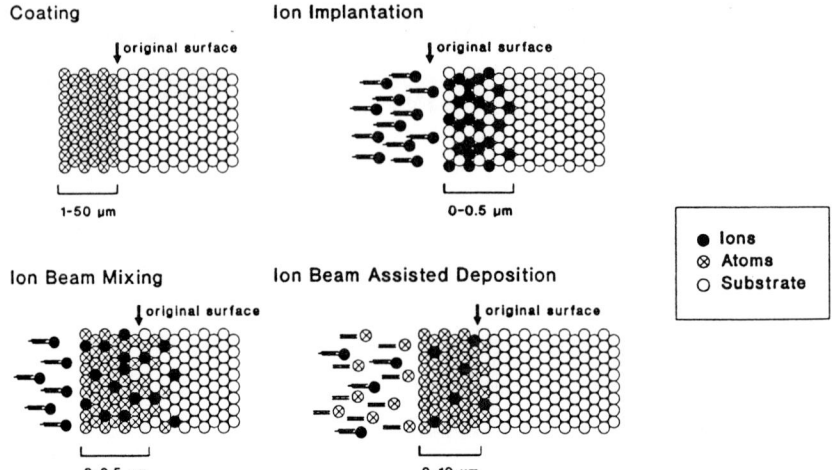

FIGURE 18.1 Schematic illustration of coating, ion implantation, ion beam mixing, and ion beam-assisted deposition.[1] (*Reprinted by permission of The Metallurgical Society, Warrendale, Pennsylvania.*)

metallic ion beams in an ion source or gun and acceleration, under very good vacuum (10^{-6} torr or better) through the scanner plates toward a highly charged substrate, where the subsequent bombardment of the surface of the substrate material occurs (Fig. 18.1), independent of thermodynamic criteria such as solid solubility and diffusivity. The basic equipment requirements include an ion source—an evacuated chamber containing an ionized gas, or plasma, a high-voltage accelerator, a mass separator (usually magnetic deflection), and a beam sweeper. Figure 18.2 is a schematic illustration of an ion beam accelerator.[2] McHargue has described a review dealing with the effect of ion implantation in (a) metals and alloys with respect to microstructure, hardness, wear and friction, fatigue, oxidation, and corrosion and (b) ceramics with respect to structure and surface mechanical properties.[3]

The mean penetration depth of ion-implanted layers is a function of ion energy and atomic mass, as well as the mass and atomic number of the substrate material and the angle of incidence. Typical penetration depths are <1 μm for most implant applications, with ion implantation doses in the range 2 to 6×10^{17} ions/cm^2 at energies of 50 to 100 keV.[4] The ion beam can be rastered over large target areas and dose uniformity is usually excellent ($\pm 1\%$). In some cases focused beams with diameters in the micron range are employed for modification of the selective area of a substrate.[5] A high-energy ion implanter (1 to 10 MeV) is needed for deep penetration ($\sim 5\,\mu$m).

During ion implantation the energetic ions are brought to rest within the first 0.5 μm or less below the surface in less than 10^{-12} s by displacing atoms from their normal sites and by ionizing substrate atoms, which allows the production of a large number of point defects and the formation of many novel/new surface alloys or compounds unattainable by conventional (equilibrium) processing techniques at room temperature.

The total energy loss per unit distance (or stopping power) due to electronic and nuclear collisions can be expressed by[6]

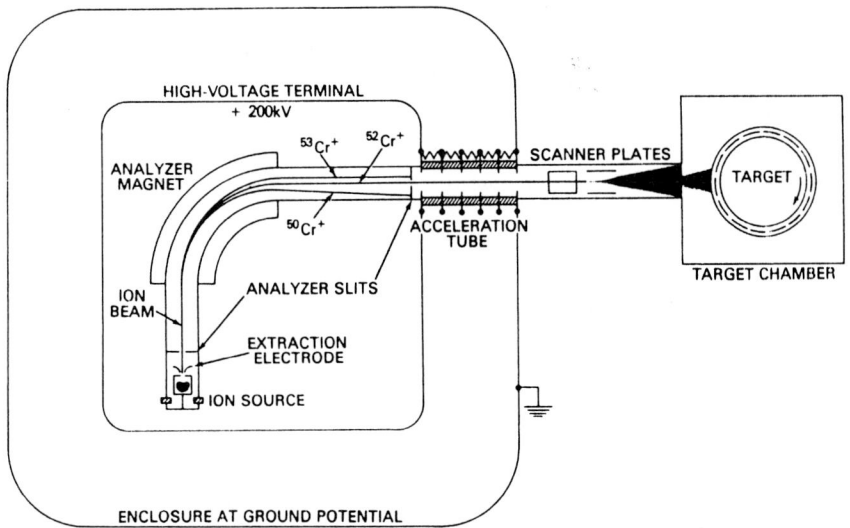

FIGURE 18.2 Schematic diagram of an ion beam accelerator for surface alloying. Mass analyzed $^{52}Cr^+$ ion beam being transported for the implantation of bearing components to increase their corrosion resistance.[2] (*After F. A. Smidt and B. D. Sartwell, Nucl. Instrum. and Methods in Phys. Res., vol. B6, 1985, pp. 70–77.*)

$$\frac{-dE}{dx} = N[S_n(E) + S_e(E)] \tag{18.1}$$

where dE = energy loss, dx = distance traveled by ion, N = target atomic density, $S_n(E)$ = nuclear stopping cross section, and $S_e(E)$ = electronic stopping cross section.

The values of $S_n(E)$ and $S_e(E)$ can be calculated and the total distance traveled by the ion before it comes to rest can be determined. However, the sequence of elastic collisions is a stochastic process. Thus the concentration profiles of both the implanted species and the substrate atom displacements are represented to a first approximation by a Gaussian distribution as a function of depth. Figure 18.3 shows a schematic view of some characteristics of ion implantation by 100-keV N^+ ions in iron (top) with profiles of N^+ ion concentration and damage profiles (lower left) and the formation of a collision cascade of vacancy-interstitial pairs or Frenkel defects by a nitrogen ion (lower right).[5] The implanted ions, the lattice defects, and the associated compressive residual stresses will tend to attain very high strength and hardness of the implanted layer.

The concentration profile of implanted ions, C_x, is given by

$$C_x = \frac{\phi}{N \Delta R_p \sqrt{2\pi}} \exp \frac{-(x - R_p)^2}{2(\Delta R_p)^2} \tag{18.2}$$

where ϕ is the dose or ion fluence (ions/cm^2), x is the distance from the surface (cm), N is the atomic target density, R_p is the projected range onto the incident direction of the ion (cm), and ΔR_p is the mean straggle (cm).

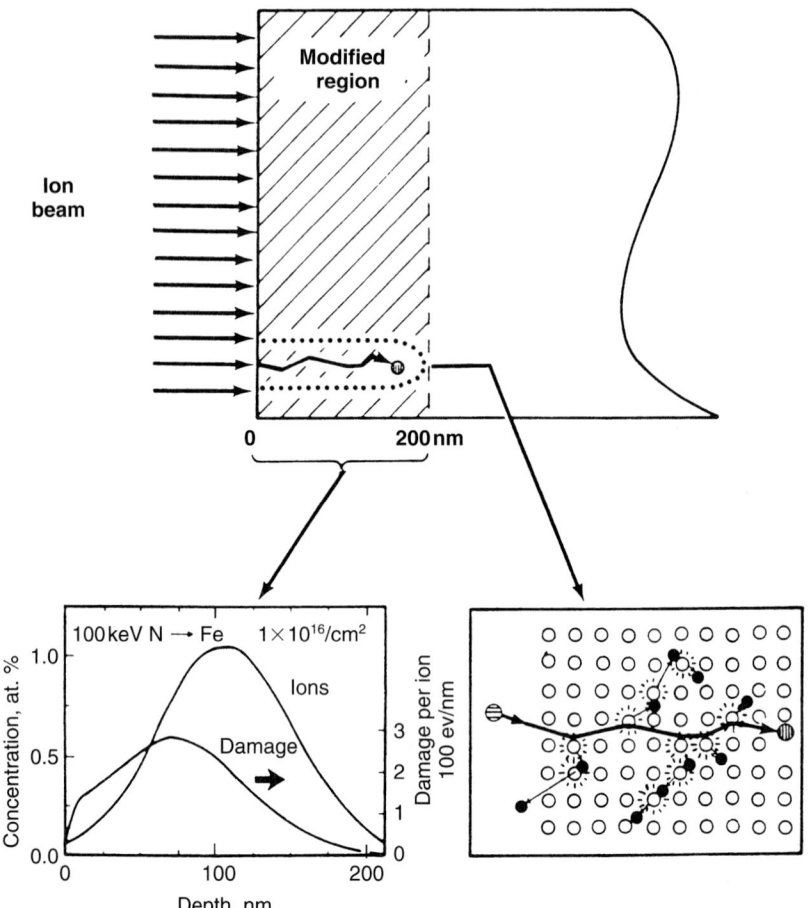

FIGURE 18.3 Schematic view of ion implantation of iron with 100-keV nitrogen ions (top). Nitrogen ion and damage profiles (lower left). Displacement cascade region of high defect density generation (lower right).[5] (*Courtesy of ASM International, Materials Park, Ohio.*)

Practically, the parameters R_p and ΔR_p can be calculated rapidly by employing computer programs developed by Biersack et al.[7] Detailed information about ion and damage profiles or energy disposition can also be found by a Monte-Carlo computer simulation program such as TRIM.[8]

Conventional line-of-sight ion implantation often requires masking of the convex targets in order to minimize sputtering and maximize the retained dose.

The formation of metastable phases (crystalline or amorphous and extended solid solution even in immiscible systems) is the important characteristic of both ion implantation and ion beam mixing. Examples of highly metastable (a) crystalline phases are 8at% Ag in Cu, 7at% Au in Fe, 3at% Pb in Fe, 15at% Sb in Ni, 1 to 3at% W in Cu, and (b) amorphous alloy phases are Ti and C in Fe or steels, transition metals (Fe, Ni, Co) metalloid (B,P) systems, Ta in Cu, Au in Pt, 25at% B in Fe, and

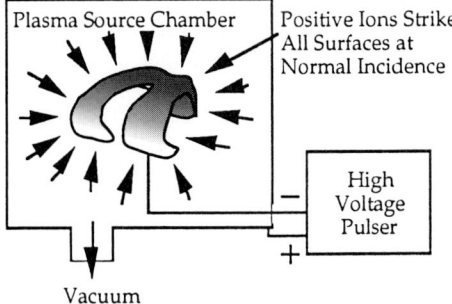

FIGURE 18.4 Schematic of the non-line-of-sight, plasma source ion implantation process.[10]

Bi and Dy in Ni.[7] A critical concentration of 12at% B is required for amorphization in Ni and about 7at% in Nb.[9]

Characterization of ion implanted materials can be performed by *ion channeling* to determine the lattice site of the implanted ion, transmission electron microscopy to determine the microstructures, and x-ray diffraction to analyze the crystal structure.

Plasma immersion ion implantation (PIII or PI[3]) or plasma source ion implementation (PSII) is a non-line-of-sight ion implantation technique with high throughput, which consists of immersing the conducting substrate to be implanted in a plasma and repetitive pulsing to a high negative voltage (Fig. 18.4)[10] (50 to 100 keV), thereby accelerating the ions at high energies from the plasma and bombarding them into the substrate surface from all directions and at normal incidence to curved surfaces. This offers the implantation of entire exposed surfaces of even complex-shaped components without the need for beam rastering, target manipulation, target masking, or more complicated fixturing.[11] This process, however, is limited to readily ionized plasma and conducting substrate, although recently this limitation has been overcome by new developments in metal ion source.[12] This process is substantially less expensive. This has been applied to many low-alloyed and microalloyed steels, stainless steel, tool steel, and so forth.[13] Table 18.1 lists examples of industrial applications of PSII treatment.[14,15,16]

A third variation developed by Braun[17] is to produce heavy ion beams from solids by employing a metal vapor vacuum arc (MEVVA) source in a pulsed mode, providing non-mass-analyzed ion beams accelerated from a broad beam source (up to 50 nm in diameter). These sources have been found to produce pulsed ion contents to 1 A (typically 200 to 300 mA) at pulse repetition frequencies to 50 Hz and extraction voltages from 10 to 100 keV.[18] The commercial development of this source has been expected to offer low new unit costs and high productivity.[19]

Advantages. The advantages of conventional ion implantation over other surface modification techniques are the following:

1. Controllability, uniformity, and reproducibility of impurity dose and penetration depth

2. Ability to accomplish at a low process temperature ($\sim$150°C or 300°F) with the absence of thermal distortion or substrate topography, provided the substrate is cooled when using high beam currents

3. Ability to selectively change the properties of the substrate material by using simple lithographic masking

TABLE 18.1   Examples of Industrial Applications of PSII Treatment[14,15,16]

Part treated	Application	PSII treatment	Improvement factor	Comments
A-2 tool steel score die	Forming aluminum	Species: nitrogen Dose: $3 \times 10^{17} cm^{-2}$ Energy: 50 keV	×1.8	Lowered the propensity for adhesive wear
Cold forming die	Manufacturing steel bolts	Species: nitrogen Dose: $3 \times 10^{17} cm^{-2}$ Energy: 50 keV	×4	Improved retention of surface finish
Thin film media overcoats	—	Species: nitrogen Dose: $5 \times 10^{17} cm^{-2}$ Energy: 50 keV	×100	—
Electromagnetic rail guns	Firing projectiles	TiN, TaN PSII-IBED	Significant improvement in pitting and spark erosion	Ref. 14 characterization by SEM, TEM, and Auger spectroscopy
TiN-coated drills	Drilling alloy steel	Species: carbon Dose: $3 \times 10^{17} cm^{-2}$ Energy: 50 keV	×1.8	Low-friction carbonitride phase may have formed on the surface
TiN-coated cutting tools	Cutting alloy steel	Species: nitrogen Dose: $2.5 \times 10^{17} cm^{-2}$ Energy: 95 keV	×4–8	Ref. 15
17-7 PH stainless steel	Erosion test by high-pressure alumina impingement	Species: nitrogen Dose: $3 \times 10^{17} cm^{-2}$ Energy: 50 keV	11%	Improvement measured at a depth several times the implanted layer
Si-wafers	Fabricating shallow junctions	Species: boron Dose: $7 \times 10^{14} cm^{-2}$ Energy: 1 keV	68-nm deep junction	Ref. 16 implant depth reported after rapid thermal annealing

4. Improvement in wear, hardness, fatigue, and corrosion resistance in steels, chromium and titanium alloys by N⁺ ion or multiple ion implantation[1]

5. Direct formation of supersaturated, intermediate, or metastable structures[20]

6. Flexible nature of resultant alloy impurity distribution achieved by using multiple implantation

7. Substantial increase of high cycle fatigue life in steel and Cu, Ni, and Ti alloys[21]

8. Freedom from introduction of any ionized atomic species into the surface regions of any substrate material irrespective of thermodynamic criteria

9. Ability to perform the treatment without the need for further heat treatment for a variety of applications

10. Negligible changes in dimensions (usually limited to 30 nm) due to subsurface swelling arising from the extra atoms; therefore preferred for the treatment of precision tools

11. Improved surface polish by the preferential sputter erosion of asperities

12. Production of a considerable biaxial compressive stress in the bombarded surface that results in the closure of surface microcracks

13. Excellent adhesion of implanted layer that is resistant to peel off under thermal or mechanical stress[9]; that is, ion implanted layer integral to the substrate with no tendency to detach in service, unlike some coatings

14. No need of ultrahigh-vacuum environment for deposition of high-purity material[22]

15. No elastic modulus mismatch problems between a hard coating and soft substrate

16. Ability to serve as a research tool to develop new alloys[23]

17. Bulk of the component unaffected with respect to composition, microstructure, and thermal softening

Disadvantages. The disadvantages of conventional ion implantation processes are the following:

1. It is a line-of-sight process such that complex geometries and recessed surfaces are difficult to implant uniformly.

2. It produces a large extent of (atomic-scale radiation) damage in the substrate material, requiring a subsequent high-temperature anneal. This (post-implantation) annealing step is essential in semiconductors to recognize the electrical activation of an implanted dopant ion by its movement to a substitutional lattice site and to repair the lattice damage created by the ion stopping process. Redistribution of the implanted ions during heat treatment may also take place.

3. Shallow implantation depth ($\sim0.5\,\mu$m) even at normally practicable energy of ~200 keV. Hence, it is not advisable to treat components subjected to severe wear such as rock cutters, or in cases where the tool surface temperatures are in excess of 500°C.

4. The high cost of equipment necessary for such treatment restricts its application to relatively high-cost parts in which specific surface conditions are very important, for example, in fail-safe systems in which high reliability is a must.[24]

Applications. In spite of the relatively shallow depth of penetration, ion implantation has proved to be a highly versatile and controllable process for modification of the composition, microstructure, and properties of the near surface regions of any material. Implantation has been reported to be technically suitable for the treatment of high-precision, high-cost components such as for extending the service lifetime of aerospace bearings and specific metal cutting tools.[4] In metals and structural ceramics, the concentration of alloying impurities required for ion implantation vary in the range 1 to 20 at %, which corresponds to larger ion doses (10^{15} to 10^{18}/cm^2).

Ion implantation can be successfully applied to numerous materials as described below.

1. *Metals.* High-dose (10^{17} ions/cm^2) implantation of Ni and Cr ions into steels improves corrosion resistance. Likewise the implantation of nitrogen, argon, boron, and carbon into steels increases its hardness. Specifically nitrogen ion implantation of critical parts such as aircraft brake pads and artificial joints for the human body reduces material wear due to friction. Implantation of Au, Pb, or Sn provides solid lubrication in miniature gears in spacecraft.

N$^+$ implantation has been found to significantly increase the hardness and wear resistance of steels and cemented carbide components such as cutting and metal-forming tools and bearings. Successful applications of ion implantation include hard chromium and alloyed steels, tool steels, stainless steels, carbides, chromium coatings, and metals such as aluminum and titanium alloys. Tooling applications include press tools, metal cutting/forming tools such as routers, haggers, end mills, saw blades, slotting saws, and valve stem cutters, mill rolls, piercing tools, forming tools, wire drawing and extrusion dies, molds, screws and nozzles, plastic injection molding tools, punches, broaches, taps and dies, brake knives, rubber slitters, ball bearings, 440C stainless steel bearings, fluid metering pumps, valve components, Co-Cr and Ti-based alloys for orthopedic prostheses (implants) for hips and knee joints,[2,25-27] and crankshafts for racing cars.

Ion implantation is also being used to improve the performance of certain types of coatings. Examples include multiple energy nitrogen ion implantation into (1) physical vapor-deposited TiN coatings, such as on cutting inserts, to increase their lifetimes and (2) electroplated chromium to deposit CrN surface layer to restrain the formation of microcracks, thereby increasing the useful life of the electroplate.[2]

Table 18.2 lists the typical applications of nitrogen implantation to tooling and components.[28]

2. *Semiconductors.* Ion implantation is widely used as the principal method of introducing impurities for doping purposes into elemental and compound semiconductor wafer* during device processing† in the electronics industry.[24,29] The reasons for the popularity of this method are[30] (1) the preciseness of the dopant concentration that can be introduced ($\leq 1\%$ variation over the dose range 10^{11} to 10^{17}/cm^2), (2) the ability to employ simple masking materials (oxides, nitrides, or photoresist) at low implantation temperatures, (3) the clean environment, (4) the ability to tailor the depth distribution of the dopant by varying implant energy, (5) the purity of the ion beam, (6) the ability to far exceed the dopant solid solubility and diffusivity, and (7) the usually small device dimensions that can be fabricated. In addition, for GaAs the lack of a viable shallow junction diffusion technology demands the use of ion implantation for certain critical dopants.[31] Ion implantation of dopants (such as boron, phosphorus, and arsenic) has been performed routinely in laboratories around the world to fabricate junction devices and produce integrated circuits.[30] In VLSI technology, ion implantation is mainly employed to selectively dope surface regions of a wafer.[32] The use of ion implantation in semiconductors requires a postimplantation annealing at elevated temperatures to eliminate implantation-induced defects, activate the dopant, and obtain a single crystal with the desired electrical characteristics.[33]

High-dose ion implantation of semiconductors (silicon wafers) may lead to the formation of a thin amorphous (α) layer at the implanted surface.[34] This effect is called *ion channeling*. Rutherford backscattering (RBS) using the ion channeling approach for a certain crystal direction has been applied extensively to identify the

---

* Wafer is usually a slice of crystalline semiconductor ingot used as a substrate material.
† The energies vary between 50 and 500 keV for most device applications.

TABLE 18.2 Typical Applications of Nitrogen Ion Implantation to Tooling and Components[28]

Tool/part	Material (substrate)	Application	Result
Compacting punches	S7 tool steel	Compacting bronze powder (bearings)	Fatigue life increased 6 times
Compressor shaft	Chrome plate (4140 steel)	Helicopter engines	Wear greatly reduced
Cutting tool inserts	TiN (tungsten carbide)	Finish machining 4140 steel	Life extended 2.5 times
Forming die	D2 tool steel	Flanging 0.080-in. hot-rolled steel (automotive flywheels)	Galling drastically reduced
High-speed stamping die	Tungsten carbide	Stamping 0.016-in. phosphor bronze (electronic connectors)	Life tripled
Mold cavities	Chrome plate (A2 tool steel)	Injection molding 50% glass + mineral-filled phenolic (transformers)	Life extended 10 times
Pistons	440C stainless steel	Hydraulic pump parts for light aircraft	Life extended 22 times
Slitter blades	O1 tool steel	Cutting candy	Life extended 6 times
Tablet compacting punches	S5 tool steel	Compacting pharmaceutical powder (ibuprofen)	Sticking reduced
Valve spring retainers	Ti-6Al-4V	Stock car engines	Life more than doubled

18.9

lattice distortion and its depth of distribution.[35] This is also useful to characterize epitaxial single-crystal multilayers.[36] In addition to postannealing the semiconductor, the difficulty in producing very shallow or very deep junctions sometimes poses a limitation; however, these are relatively minor issues and do not inhibit its wide application. Reactive ion implantation of silicon into natural single-crystal diamond and into polycrystalline CVD (chemical vapor deposition) diamond causes the formation of a SiC_x barrier as a protection against oxidation.[37] In III-V technology, both doping and device isolation are normally achieved by implantation.[38]

3. *Ceramics.* Ion implantation creates a wide variety of microstructures in engineering ceramics (such as Al_2O_3, Si_3N_4, and SiC) from crystalline, metastable solid solutions with large concentrations of point defects and dislocations to amorphous phases which were previously unattainable. The specific defect or phase structure produced (after implantation) is a function of several implantation parameters (ion species, ion fluence, and substrate temperature) and material parameters (chemical bonding type). Generally, high fluences, low temperatures, and covalent bonding favor amorphization. Usually, crystalline implanted surfaces display hardness increases of 10 to 50%, whereas amorphous implanted surfaces display hardness decreases of 40 to 50%. Significant increase in flexural strength, without causing degradation of optical properties, has been shown for infrared windows made of sapphire.[39]

4. *Optical materials.* Ion implantation is extensively employed to change locally the refractive index of light-guiding materials such as glass, quartz, $LiNbO_3$ and III-V semiconductors (GaAs and AlGaAs). In the latter situation, it is possible to integrate the light-guiding regions with lasers, detectors, and couplers on the same chip.[40]

5. *Materials synthesis.* Buried regions of Si_3N_4 or SiO_2 can be produced in single-crystal silicon by high-dose (10^{17} cm^{-2}) implantation of N^+ or O^+ ions, respectively, followed by annealing. Similarly, synthesis of SiC and conducting silicide layers can be performed. Metastable phases with different transition temperatures can be formed in superconducting materials; employing ion implantation and the ability to closely monitor the quantity of the specific impurity makes this an especially useful process.

Magnetic-bubble memories also take advantage of several implantation-induced effects. The stress between the rare-earth garnet layers and their beneath substrates can be decreased by high-dose proton implantation and the conducting paths for the magnetic bubbles can be formed by implantation in selective areas. Several other nonsemiconductor applications have been used for implantation, and specialized machines that rotate even large workpieces with curved surfaces make it possible to achieve good dose uniformity and therefore good control of the particular material modification effect.[40]

6. *Nuclear applications.* It also finds applications as a means of stimulating or accelerating the effects of radiation damage that occurs in nuclear reactor materials in order to study such phenomena as the formation of helium bubbles and irradiation-induced creep.[41,42]

7. *Polymers.* Polymers can also be successfully treated by ion beams. Examples include the modifications of strength, gas permeation, conductivity, and the metallization of "difficult" polymers such as Teflon or Kapton for electronics packaging.[43]

18.2.2 *Ion Beam Mixing.* The ion beam mixing (IBM) process is ion implantation into the interface region of an already coated substrate (Fig. 18.1), which will greatly improve adhesion. However, the attainable coating thickness is limited by the maximum available energy of the ion beam and is therefore usually below 1 μm. IBM is an efficient and flexible process for the production of either crystalline or amorphous and extended surface alloys, where a thin film of chosen material is deposited onto the component and subsequently bombarded with high-energy inert gas ion (Ar$^+$, Kr$^+$, or Xe$^+$) beams to induce implantation by collisional recoils, with atomic displacements as random walk. Heavy ions generate many atomic displacements and mix more readily. More efficient mixing of two or more thin layers of different elements is also obtained if the temperature is adequate to allow radiation-enhanced diffusion to occur.[44] It is thus an alternative process to high-fluence ion implantation due to sputtering restrictions.[9]

In practice, there are two main configurations of deposition. The bilayer structure (Fig. 18.5a) is usually employed to study the extent of intermixing with fluence, while multiple alternating layers (Fig. 18.5b) are employed to form alloys of fixed composition determined by the relative thicknesses of the elemental layers.[45] Ion beam mixing also alloys more atoms per incident ion, for example, fluences of 10^{15} to 10^{16} ions/cm^2 are typical for mixing, whereas >10^{17} ions/cm^2 are mostly required with ion implantation. Thus, high concentrations are obtained with less accelerator time.

The configurations in Fig. 18.5a were employed to alloy Al and Pt by Mayer et al.,[46] as shown in Fig. 18.6. With Xe ion mixing, RBS analysis suggests that the initial sharp interface between the layers (the nearly vertical solid lines at the right edge of the Pt signal and left edge of the Al signal) becomes graded with the formation of a Pt-Al alloy. In this case, a stoichiometric compound (PtAl$_2$) forms, and the mixing proceeds at a constant composition with increasing Xe influence, as shown by the steps in the two signals. The ion beam mixed phases can be quite different from those formed by thermally reacting the layers, and, in this system, PtAl$_4$ forms when the Pt/Al layers are annealed, instead of PtAl$_2$.[47] Many experimental studies have been carried out in many systems to determine the influence of temperature and irradiation parameters. Different mechanisms identified in IBM include the ballistic effects, radiation-enhanced diffusion, thermal spike mixing, and chemical effects.

The ballistic effects denote mass transport. They are independent of temperature and consist of both recoil mixing and cascade mixing. The former occurs by a direct collision between an incident ion on a target atom. However, the number of atoms transported by this mechanism is smaller than that of the cascade mixing. Sigmund et al.[48] have developed a collision model of cascade mixing. If one assumes that the atom relaxation process is basically equivalent to a vacancy-interstitial pair

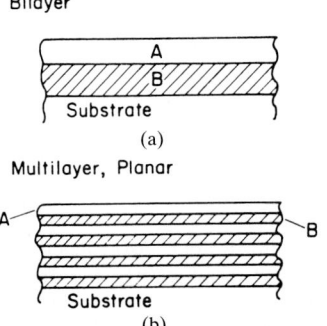

FIGURE 18.5 Schematic configurations used for ion beam mixing: (a) bilayer and (b) multilayer.[45]

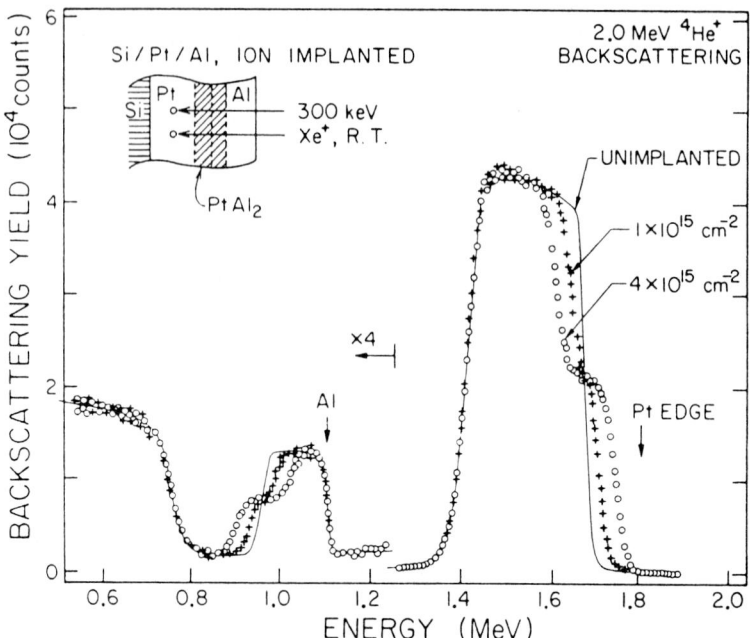

FIGURE 18.6   RBS spectra for Al/Pt layers ion beam-mixed with 300 keV Xe. The steps show the formation of PtAl₂.[46]

formation and that the phenomenon is almost isotropic, an apparent diffusion coefficient D_0 for cascade mixing can be given by the Andersen formula[49]:

$$\frac{D_0 t}{\phi} = \frac{0.42 F_d R_c^2}{6 N E_d} \tag{18.3}$$

where R_c is the distance between a stable vacancy-interstitial pair, E_d is the displacement threshold energy, F_d is the amount of elastic energy lost per unit length, t is the irradiation time, and ϕ is the ion fluence. Since the number of atomic displacements is much greater than the number of implanted ions, the cascade mixing exerts a very efficient effect in homogenizing the multilayers to an atomic scale even at lower temperatures.

When the mixing is carried out at a temperature where radiation effects are mobile, an additional temperature-dependent mechanism takes place due to the extra cascade transport resulting from the long-range migration of radiation-induced defects. The diffusion coefficient, $D^*(T)$, for radiation enhanced diffusion mixing is expressed by

$$D^*(T) = C_v(T) D_v(T) + C_i(T) D_i(T) \tag{18.4}$$

where C_v and C_i are the concentrations of vacancies and interstitials, respectively, and D_v and D_i are the corresponding diffusivities. However, if a large number of low-temperature mixing experiments can be understood on the basis of ballistic mechanism, there appear to be many instances where chemical driving forces have to be considered. Recent calculations suggest that mixing at low temperature must assume a diffusion effect during the thermal spike of the cascade in which local thermal equilibrium is set up and a temperature is defined. Thus the effective diffusion coefficient, D, is described by

$$D = D_0\left[1-\left(\frac{2\Delta H_m}{kT}\right)\right]$$ (18.5)

where D_0 is the diffusion coefficient for the pure ballistic effect and T is an effective temperature defined by $(3/2)\,kT = \theta_D$, where θ_D is the energy density in the cascade and ΔH_m is the heat of mixing. This diffusion coefficient D can be either positive or negative depending on the sign of ΔH_m. The above equation can explain why in binary systems with a sufficiently large positive heat of mixing, the value of D is negative and there are difficulties in mixing, as in the Cu-W system.[9,50,51]

The phases of ion beam-mixed alloys are often the same as those of ion-implanted or -irradiated alloys with the same composition, as well as the immiscible alloy systems.[9]

The advantages of ion beam mixing include (1) the ability to markedly enhance film/substrate adhesion properties due to improvement of the solid solubility and ion-induced compound formation at the film/substrate interfaces and formation of compressive residual stresses within the film, and (2) the independent control of processing parameters such as ion flux and energy.[52]

18.2.3 Ion Beam-Assisted Deposition.

Ion beam-assisted deposition (IBAD) is a process in which ion bombardment and physical vapor deposition are combined. The extra energy imparted to the deposited atoms leads to atomic displacements at the surface and in the bulk, as well as improved atom migration along the surface. These resulting atomic movements result in the improved film properties such as better adhesion and cohesion of the film, higher density, and modified residual stress, when compared to similar films formed by PVD without ion bombardment. When the ion beam or the evaporant is a reactive species, compound semiconductors such as Si_3N_4 can be synthesized at very low temperatures. In addition, adjustment of the ratio of reactive ions to atoms arriving at the substrate surface allows adjustment of the stoichiometry of IBAD films. Detailed reviews of the IBAD process can be found elsewhere.[53–57]

There are two primary ways to accomplish the IBAD process. One way is based on the low-energy (0.5 to 5 keV) ion source such as the broad-beam Kaufman-type ion gun (Fig. 18.7a), and is used without mass separation. The second way is based on high-energy ion implanters. Figure 18.7a and b are schematic IBAD configurations with a simultaneous and alternating evaporation and inert gas ion bombardment arrangement.[1]

IBAD processing can be grouped into three types[4]:

1. *Nonreactive IBAD*, where inert gas ions (Ar^+) are used to influence the nucleation and growth of the deposited elements or compounds.

2. *Reactive IBAD*, i.e., when the vacuum chamber is backfilled with a reactive gas and where the ion beam is used to both influence film growth and provide atoms for the growth of a compound film (e.g., Si or Ti deposition with N^+ ion bombardment to form Si_3N_4 or TiN) in a reactive gas atmosphere.

3. *Modified reactive IBAD*, where atoms in the form of a backfilled molecular gas (e.g., N_2 or O_2) are provided so that these atoms are introduced (i.e., activated) into the growing film by bombardment with energetic ions (inert or reactive).

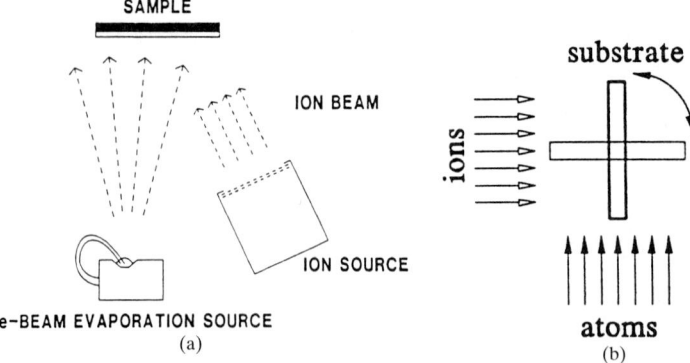

FIGURE 18.7   Classical IBAD geometry with simultaneous evaporation and inert gas ion source (left side) and alternating IBAD arrangement (right side).[1] (*Reprinted by permission of The Metallurgical Society, Warrendale, Pennsylvania.*)

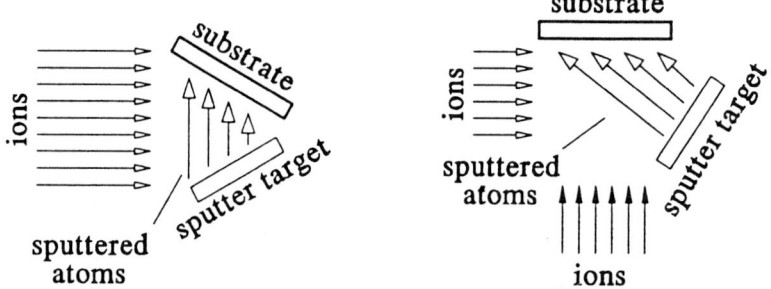

FIGURE 18.8   Single-beam (left side) and dual-beam (right side) arrangement for IBAD with sputter deposition.[1] (*Reprinted by permission of The Metallurgical Society, Warrendale, Pennsylvania.*)

This can be employed to produce a stoichiometric compound when the evaporant is adequately reactive.[58,59] It can also be employed to compensate for the loss of constituent elements (e.g., oxygen) from the evaporating compound (e.g., Al_2O_3 or SiO_2), which tends to decompose at high temperatures yielding a metal-rich coating in the absence of such an O_2 backfill.[4]

The typical ion energy range employed for IBAD/RIBAD is 100 eV to 30 keV, because even in the alternating arrangement the deposited intermediate layer is very thin, thereby requiring less energy. The deposition technique employed for IBAD/RIBAD is evaporation of elements or alloys, but it can also be used for sputter deposition in either dual- or single-beam arrangements (Fig. 18.8).[1] (See Sec 18.3.2 for more details on sputter deposition.)

Recent experimental studies have shown that the ion beam assistance energy and the arriving ions (I) per arriving atoms (A) ratio are important parameters in modifying the properties of the deposited coating.[9]

Advantages and Limitations. The advantages of the IBAD process over competing coating processes are (1) low deposition temperature, (2) high adhesion, (3) coating densification, (4) residual stress reduction, (5) columnar microstructure elimination, and (6) controlled microstructure and crystalline film orientation.[4]

Even polymers with low melting points can be coated, due to lower deposition temperatures (below ~100°C or 212°F). The properties of adhesion, density, and stress are superior to those of PVD films, and there is a greater control over the microstructure. Based on the deposition parameters, films can be deposited as (1) metastable crystalline, (2) amorphous, (3) textured crystalline or epitaxial (for some materials), and (5) nanocrystalline.[53]

Eventually, the composition, or crystalline phases, can be precisely varied as a function of thickness to form functionally gradient materials with properties such as graded hardness, density, tensile strength, stress, coefficient of thermal expansion, and refractive index.[53]

The limitations of the IBAD process are (1) moderately higher cost than PVD, (2) line-of-sight processing, and (3) limited users.[53]

Applications. The widespread industrial growth in ion beam-assisted coatings is justified by their excellent tribological and mechanical properties (hardness and adhesion), corrosion resistance, reproducible and stable optical properties, and decorative value. There is promise of useful dielectric applications. The films are very smooth, with a fine equiaxed microstructure rather than a columnar one, and they are applied at low temperatures.[60]

Applications can be described in the areas of optical films, oxidation and corrosion-protection coatings, and tribological coatings.

1. *Optical films.* The IBAD process has been used in optical thin films for two types of applications.

In the first type of applications, where densification is the primary concern, low-energy argon or oxygen ions are typically employed to bombard the optical thin films during deposition.[61] As a result, the density is enhanced, sometimes approaching the bulk material. The main advantage is certainly not the increase in refractive index to near-bulk values, but rather the stability of the refractive index value under humidity and temperature variations because of the freedom from voids or pores in the film that absorb water vapor. This makes optical-coating design simpler and favors better control and reproducibility in the fabrication process. Another attraction of using low-energy ions is the improved adhesion to the substrate, which also assists in the increased productivity.

In the second type of applications, where graded refractive index profiles is the main requirement, the IBAD process can be extensively used to fabricate graded index antireflection coatings, reflection filters, and mirrors.[62] Some of these devices can be tens of microns thick, which implies that stress control is a major factor. Usually, low energies are preferred for the deposition of optical films to reduce absorption owing to radiation damage.

2. *Diamond-like carbon (DLC) coating.* Currently, IBAD, ion-induced deposition (IID), and filtered cathodic-arc (or vacuum-arc) deposition processes can be used to deposit diamond-like carbon (DLC) and zirconia for the biomedical and corrosion applications due to their high hardness, low coefficient of friction and wear rate, and chemical inertness.[4,63] A low-friction DLC coating is one of the more successful applications of IBAD technology.

In a classical IBAD, using a broad-beam ion source of Kaufman type (Fig. 18.7a), the coating material itself is evaporated with simultaneous bombardment at a low-energy (0.1 to 2 keV) argon ion beam to deposit DLC at rates between 0.1

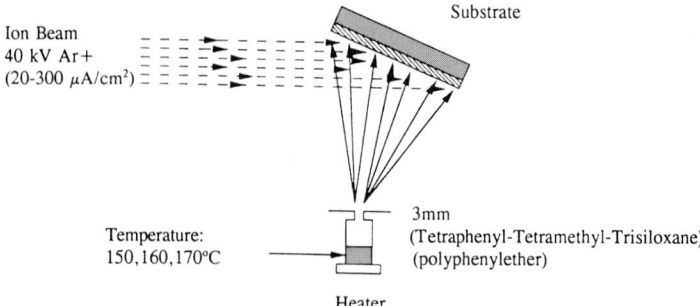

FIGURE 18.9 Setup for depositing DLC by ion-induced deposition using hydrocarbon vapor.[63] (*Reprinted by permission of The Metallurgical Society, Warrendale, Pennsylvania.*)

and 1 nm/s (1 and 10 Å/sec).[63] In ion-induced deposition of DLC, the ion beam is used to both decompose and provide impact energy to an organic evaporant gas which is directed toward the substrate. Figure 18.9 exhibits a setup using a 10- to 40-keV nitrogen or argon ion beam and hydrocarbon vapor to form thin films of DLC that are highly halogenated and may contain some amounts of silicon and oxygen. The films are less hard than the filtered cathodic arc amorphous carbon but are very lubricious with lower residual stress, which permits the deposition of thicker films without peeling off from the substrate.[63]

In filtered cathodic-arc (also called vacuum-arc) DLC, pure-carbon ions with energies of 20 to 30 eV are emitted from graphite cathode and impacted on the negatively biased substrate (see Fig. 18.12). This type of amorphous DLC is characterized by high hardness, high residual stress, and good adhesion with a density of $\geq 3.0\,g/cm^3$ and a specific chemical (sp^3) bonding content of 80 to 90%. (See Sec 18.3.1.3 for more details on vacuum arc deposition.)

The potential applications of DLC coatings include (1) friction- and wear-resistant coatings for engine components, tools and dies, pump and wear components, and bearings and gears; (2) coatings for audio speakers, x-ray windows, and heat sinks for electronic devices; (3) mold release coatings for compression and injection molds; (4) coatings for resistance to chemical attacks; (5) coatings in optical fields for sunglasses, ophthalmic lenses, infrared windows, laser optics, and fiber optics, and magnetic and optical recording disks and heads; and (6) coatings for medical (prosthetic) devices.[28,64]

This process is also used for hard, protective coatings for optics and windscreens on vehicles. Although DLC absorbs strongly in the visible range, thin coatings (20 to 200 nm or 200 to 2000 Å) remain transparent to serve as protective transmission coatings. The advantages of these coatings are high scratch hardness, low porosity (reduced number of pinholes), and superior adhesion to most substrates.[53]

3. *Wear- and corrosion-resistant coatings.* Several researchers have investigated the parameters required to achieve hard wear-resistant IBAD coatings (e.g., TiN coating of 316 stainless steel electric razor screens and on Al_2O_3) by using bombardment with either (1) a nitrogen ion, or (2) an argon ion to accelerate incorporation of ambient nitrogen gas with a reactive species (e.g., TiN).[65,66] They are also attractive for corrosion-protection applications.

Pt, TiC, TiN, B, diamond-like carbon, chromium nitride, boron nitride, chromium oxide, silicon nitride, and silicon coated on metals using the IBAD process offer excellent corrosion resistance.[67–72]

IBAD chromium nitride (Cr_xN_y) is found to be a candidate for electroplated hard chromium (EHC) replacement due to its high hardness, good corrosion resistance, and overall similarity to EHC.[4]

4. *High-temperature oxidation resistance.* Oxidation protection of titanium alloys with IBAD coatings of chromium nitride,[73] TiN,[73] and silicon nitride[74] has been reported.

5. *Ion-induced CVD.* Several researchers have adopted reactive IBAD to produce unique hydrocarbon or ceramic films.[75] This process involves the introduction of a gas into the chamber, cooling the substrate to induce condensation of the gas, and bombarding the surface with an ion beam. During the process, hydrocarbon bonds are collapsed, volatile species are released, and a coating is effected. For silicone oil vapor, the films can vary between very-low-friction solid lubricants and very-hard, corrosion-resistant silicon oxycarbide ($Si_xO_xC_y$) coatings, depending on the arrival ratio of ions to vapor-condensed atoms. This process is akin to CVD, where the high temperature of the substrate provides the energy to begin chemical reactions leading to the film formation. In the ion beam case, the same or similar reactions can be beam-induced at room temperature, opening the likelihood of depositing CVD-like films on polymers and other temperature-sensitive substrates.

6. *Friction and wear resistance.* The IBAD process can be used to deposit solid lubricant coatings (e.g., molybdenum disulfide), which exhibit greater adherence to the substrate and longer lifetime.[76,77]

18.2.4 Ion Plating. Ion plating or ion vapor deposition* is the name given to a class of ion-assisted PVD processes. The bombarding species as well as the depositing species can be from numerous sources. Bombardment can occur in plasma or vacuum conditions. The process is often termed IBAD if bombardment takes place in vacuum.[78] In *plasma-based ion plating*, the substrate remains in contact with a plasma, and the ions are accelerated from the plasma and strike the substrate surface. In *vacuum-based ion plating*, the film deposition occurs in a vacuum and the bombardment is from an ion or plasma "gun." In *reactive ion plating*, the plasma or ion/plasma gun produce ions of a reactive species to both bombard and react with the depositing material to form a compound film material. In some instances, such as when using low-voltage, high-current electron-beam evaporation or arc vaporization, a sufficient amount of the vaporized source material can be ionized to permit bombardment by "film ions." Often the term *ion plating* is associated with modifying terms such as *reactive ion plating*, *sputter ion plating*, *chemical ion plating*, *arc ion plating*, and *alternating ion plating*, which denote the method employed to bombard the film, the source of depositing material, or other specific conditions of the deposition.

Figure 18.10a shows a simple plasma-based ion plating system with a resistively heated vaporization source, in which the negatively biased substrate can be placed in the plasma generation regime or in a remote or downstream location outside the active plasma generation area. Figure 18.10b shows a schematic vacuum-based configuration with an electron-beam evaporation source.[79] Figure 18.10c is a schematic depiction of a cathodic- (multiple) arc (or ion bond) (source) ion plating unit.[78] This

---

* In the case of ion plating, the ionized fraction is usually less than 5%.

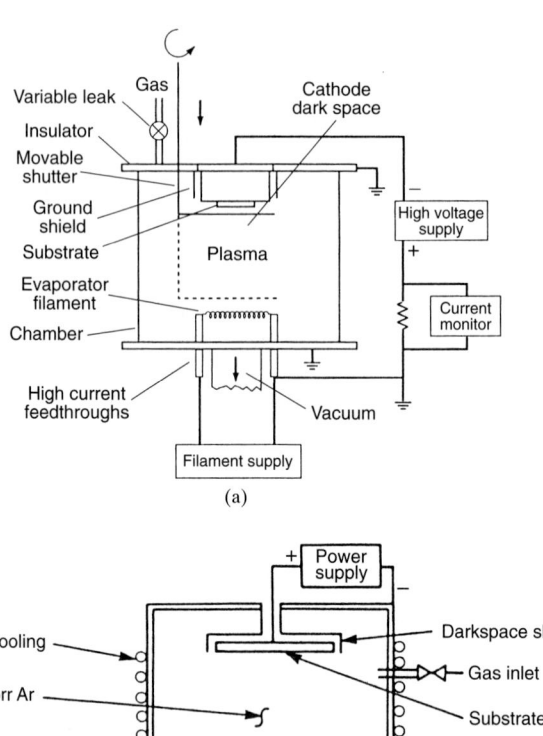

FIGURE 18.10 Schematics showing typical ion plating installations: (*a*) plasma-based configuration with resistively heated vaporization source;[78] (*b*) vacuum-based configuration with electron-beam-heated evaporation source;[79] and (*c*) ion-bond PVD process.[80] [(*a*) *Reprinted by permission of ASM International, Materials Park, Ohio; (b) Courtesy of Institute of Physics Publishing, Bristol; (c) Courtesy of Ion Bond, Inc.*]

is a cost-effective method for hard-coating tools and wear parts, and closely toleranced components made from temperature-sensitive materials.

For the deposition of compound coatings such as TiN, TiC, ZrN, HfN, Si_3N_4, or Al_2O_3, metal vapors are released into a "reactive" plasma produced in argon plus an appropriate "reactive" gas, i.e., N_2 for nitride coating; CH_4, C_2H_6, or C_2H_2 for carbides; O_2 for oxides; and possibly BCl_3 for borides. For the deposition of oxide

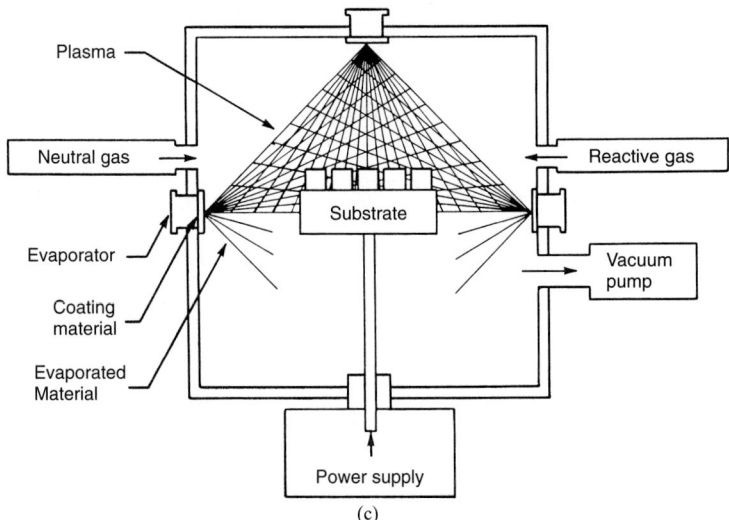

(c)

FIGURE 18.10 *(Continued)* Schematics showing typical ion plating installations: (*a*) plasma-based configuration with resistively heated vaporization source;[78] (*b*) vacuum-based configuration with electron-beam-heated evaporation source;[79] and (*c*) ion-bond PVD process.[80] [*(a) Reprinted by permission of ASM International, Materials Park, Ohio; (b) Courtesy of Institute of Physics Publishing, Bristol; (c) Courtesy of Ion Bond, Inc.*]

coatings or when trying to deposit nonconducting substrates made from glass, ceramic, or plastic, generally the dc power supply is replaced by an rf power supply, usually rated at 13.56 MHz, which minimizes the likelihood of charging or arcing.[81]

The extent of the reaction is a function of the plasma conditions, bombardment conditions, and the availability of the reactive species. By controlling the availability, the film composition can be changed. For example, in the reactive ion plating of TiN, first the availability of the nitrogen is decreased in the plasma at the commencement of deposition, to form an initial layer of titanium; next, the availability of nitrogen in the plasma is increased to form a "graded" interface.[78]

Advantages and Limitations of Plasma-Based Ion Plating. Plasma-based ion plating is the widely used ion plating configuration.

The advantages of plasma-based ion plating are the following[79,82,83]:

1. Superior surface-covering ability ("throwing power") of the evaporant under the appropriate conditions to provide more uniform coating of complex geometry

2. Ability to sputter clean the substrate surface prior to deposition

3. Ability to achieve good adhesion even in many difficult-to-deposit systems

4. Ability to introduce heat and defects into the first few monolayers of the surface to improve nucleation, reaction, and diffusion

5. Improvement of reactive deposition process (activation of reactive gases, bombardment-improved chemical reaction, and adsorption of reactive species)

6. Flexibility in tailoring film properties by adjusting bombardment conditions (such as density, morphology, and residual stress)

7. Equipment requirements similar to those of sputter deposition

8. Deposition source that can be from resistance or induction-heated thermal evaporation, arc vaporization, sputtering, or chemical vapor precursor gases[78].

The limitations of plasma-based ion plating are the following:

1. The intense energy of ion vapor deposition may impart substantial energy to the substrate in excess of that of evaporation or sputtering. This can easily melt plastic or anneal aluminum substrates.

2. The energy of ion deposition rapidly heats the deposition chamber and tooling, driving off the absorbed gas and contaminants. These gases activate and contaminate the plasma and react with the films. Thus ion plating needs the same cleanliness and operating care as sputtering. The heat buildup in the chamber and tooling also necessitates water cooling, means to avoid burns, and extended cooling cycles, slowing production. Additionally, the exceptional throw of evaporant flux drives the film far under deposition masks.[79]

3. To bombard growing films of electrically insulating materials from a plasma, the surfaces must either reach a high self-bias or a bias with pulsed dc power or an rf potential.[78]

4. Uniform bombarding and availability of reactive species may be difficult to achieve with a complex substrate surface.

5. High residual compressive growth stresses may be built into the film layer due to "atomic peening".[78]

Applications. Typical applications of the ion plating process are the following:

1. Good adhesion between a film and substrate (e.g., Ag on steel for mirrors and bearings; Ag on Be for diffusion bonding)

2. Electrically conductive coatings (Al, Ag, and Au) on semiconductors and plastics

3. Hard, wear-resistant, and abrasion-resistant coatings such as TiN, TiN_x, TiC_xN_y, $(TiAl)C_xN_y$, and $Ti_{0.5}Al_{0.5}N$ on metal cutting and working tools, molds, dies, jewelry, and certain aircraft components, and $CrN + Cr_2O_3$ on piston rings[78]

4. Low-shear solid film lubricants (e.g., Ag, Au, and Pb)

5. Corrosion-resistant coatings such as Al on U, mild steel, and Ti, and carbon and tantalum on biological implants

6. Decorative coatings (TiN provides gold-colored deposits, TiC_xN_y provides rose-colored deposits, TiC provides black deposits, ZrN provides brass-colored deposits) applied to hardware, cutlery, jewelry, and guns

7. Deposition of electrically conductive diffusion barriers (such as TiN and HfN on semiconductor devices)

8. Deposition of insulating coatings (such as SiO_2, Al_2O_3, and ZrO_2)

9. Deposition of permeation barriers on webs

10. Deposition of optically clear, electrically conductive films (indium-tin oxide, ITO)

11. Deposition of Al on very large structural parts for corrosion protection (as a substitute for electroplated Cd)

12. Base deposition for further deposition by other means such as electroplating and painting

13. Filling vias and trenches on semiconductor surfaces by sputter deposition[78]

14. Improved strength of the copper/cordierite-type glass ceramic interface relative to evaporation[52]

18.3 PHYSICAL VAPOR DEPOSITION

In physical vapor deposition (PVD) processes, the depositing material is created by physical means (either vacuum evaporation or sputtering) and transported from the source (target) through a vacuum or low-pressure gaseous (plasma) environment to the substrate, where subsequently the vapor phase is condensed onto a substrate material to form a thin film. The basic PVD processes, which can be considered as complementary techniques, fall into three broad categories: thermal evaporation, sputter deposition, and molecular beam epitaxy (MBE). Thermal evaporation is grouped into resistance evaporation, electron beam evaporation, ion vapor evaporation (described in an earlier section), vacuum arc deposition, and laser ablation. Sputter deposition is divided into dc sputtering, rf sputtering, and magnetron sputtering. Molecular beam epitaxy is grouped into conventional MBE, solid-source MBE (SSMBE), metalorganic MBE (MOMBE), and gas source MBE (GSMBE). Table 18.3 is a comparison of typical deposition characteristics of hard coatings produced by electron beam evaporation, sputtering, and cathodic arc evaporation deposition techniques.

Advantages and Disadvantages. The major advantages of PVD processes are the following:

1. The ease with which a wide variety of deposited graded, multicomponent, multilayered, and nonequilibrium coatings and alloys can be produced. For example, nonequilibrium Al alloy PVD coatings show high tensile strength, improved corrosion resistance, and excellent thermal stability.[84]

2. Relatively low temperature on the substrate material surface (100 to 550°C) during deposition[43] allows the use of many substrates including heat-sensitive materials and brazing filler metal to manufacture brazed carbide tools; this also produces crack-free coatings with a uniform fine-grained microstructure.[85,86]

3. PVD coatings exhibit smooth finish and, therefore, produce less friction during machining.[86]

4. PVD coatings produce compressive surface stress, which assists in resisting crack initiation and propagation.[86]

5. PVD coating retains the transverse rupture strength of carbide, while the CVD coating usually reduces the transverse rupture strength. This makes the PVD coatings more desirable for milling inserts, which need higher impact strength due to interrupted cutting.[86]

6. PVD coatings can be applied uniformly over sharp cutting edges, whereas CVD coated inserts need honing.[86]

7. Multilayered PVD coatings of TiN/TiCN and TiN/TiAlN combine the beneficial effects of the individual layers into a multilayered coating.[86]

TABLE 18.3 Characteristics of Hard Coatings Produced by Three PVD Processes

	E-beam	Sputtering	Arc evaporation
State of source	Solid	Solid	Solid
Type of source	Metals	Metals, alloys, compounds	Metals, alloys, compounds
Flux composition	Atoms, ions	Atoms, ions	Ions, atoms, clusters
Heating and conditioning	Argon ions	Argon ions	Metal ions, argon ions
Source voltage (V)	70–100	300–800	10–40
Source current (A)	140	<10	50–400
Mean particle energy (eV)	<50	10–40	50–150
Degree of ionization (I/N %)	10–50	<10	50–80
Substrate voltage (V)	100	0–2000	50–1200
Substrate temperature (°C)	400–550	150–550	200–550
Substrate current density (mA/cm^2)	<5	2–5	<7.5
Deposition rate (μm/min)	<0.05	<0.02	<0.08
Ion to neutral ratio (I/N)	1.0	0.1	2.4
Source location	Bottom	Top, side, bottom	Top, side, bottom
Number of sources	1	Multiple	Multiple to 12
Control of stoichiometry	Narrow range	Narrow range	Wide range
System throughput	Low	Low	High
Operation complexity	High	Moderate	Low
Fixturing complexity	High	High	Low
Ion generation	Indirect solid to vapor	Indirect solid to vapor	Direct solid to vapor
Ion energy levels (I/N)	3 eV	3 eV	50 eV
Process time (3/4 μm TiN)	4 hr	6–8 hr	4 hr
Alloy source evaporation	Not possible	Best control	Good control
Gas phase reaction control	Difficult	Difficult	Good

Courtesy of IonBond, Inc.

8. The parts with complex geometry can be coated uniformly and the chemical composition of the material to be coated is usually not a significant factor on coating. Sometimes, a second coating can be applied on the first coating with Au, Ti, or Cu without affecting the mechanical properties of the first layer.[87]

9. PVD-coated cutting tools allow specific cutting operations like high-speed cutting or cutting without coolants.[88] Improvements in forming tool performance are attributed to increased hardness to resist wear, lubricity to facilitate metal flow, and inertness to resist metal pickup and corrosion.[89]

10. The film characteristics such as hardness, structural constitution, coefficient of friction, or intrinsic stresses can be optimized to meet their application requirements by adapting deposition conditions.[88]

11. It is possible to achieve significant improvements in the performance and properties of cutting materials through the use of multilayer hard coatings or hard coatings with complex composition as compared to simply coated or uncoated grades by using the following mechanisms: (1) mixed crystal strength and stability, (2) precipitation hardening, (3) outer layer formation by partial reaction, (4) essential bonding and grain structure, (5) advantageous lattice structure,

(6) thin-film strength and elasticity, (7) stress-adapted complex property profile, and (8) uniform applicability.[90]

12. PVD is commercially viable for coating three-dimensional engineering components and retail consumer products with high-rate processes that can be easily automated and that offer reliable quality.[91]

13. The thickness of the deposits can vary on the order of angstroms to millimeters; however, a thickness of 10 to 15 μm is used for most hard wear-resistant PVD coatings. Very high deposition rates (25 μm/sec) have been obtained with the advent of electron beam-heated sources.

14. Unlike the CVD process, the PVD processes are clean and pollution-free.

15. PVD film growth usually involves a minimum of parameter development and exploration. Hence it is ideal for exploratory studies.[92]

16. Pulsed vacuum arc deposition offers instantaneous deposition rates of 400 μm/s.[93]

The main disadvantages of PVD processes (with the exception of EB-PVD) are the following:

1. Low deposition rates (1 to 5 μm/hr) and difficulty in applying oxide coating routinely.[93a]

2. It can achieve only line-of-sight depositions.

3. It uses expensive growth apparatus (most use ultrahigh vacuum).

4. It is not definitely adaptable to high-volume film growth on a production scale.

5. PVD coatings are mechanically bonded to the substrate surface and have a weaker bond strength compared to CVD coatings that are chemically bonded.[86]

6. PVD coatings are applied to a thickness of 2 to 4 μm whereas CVD coatings are usually deposited to a thickness of 5 to 12 μm.[86]

7. PVD coatings require a very clean and smooth substrate surface to obtain the highest bond strength.

Applications. Thin films deposited by PVD processes are used for many applications—for example, in anticorrosion coatings for gas turbine blades and vanes, hard coatings (>500 HV) for tool materials (cemented carbides, hot and cold work dies), optical coatings, decorative coatings, microelectronics, high-temperature semiconductors, and the machinery, aerospace, automotive, computer, medical, and appliance industries.[84,85] Table 18.4 summarizes various hard coatings for cutting tools deposited by PVD and CVD techniques.[94]

The representative films deposited by PVD are metals (Cu, Al, Ti, Cr, Ta, Mo, W, Zr, Hf); alloys (Ti-6Al-4V, high-Ni alloys, MCrAlY, nichrome); nitrides (TiN, ZrN, HfN, TaN, CrN, TiAlN, TiZrN, Ti-6Al-4VN); carbides (TiC, TaC, WC, ZrC); oxides (CuO, TiO$_2$, ZrO$_2$, Al$_2$O$_3$); and carbonitrides (TiCN, TaCN, ZrCN).[85]

The most common coatings used presently in the metal cutting and forming tool industry include TiN and ZrN coatings for general purpose cutting and forming steels; TiAlN and AlTiN coatings for cutting and machining hard materials such as Ti, Ni, etc.; TiCN for cutting and forming steels; CrN for forming tool steels only; solid MoS$_2$-Ti lubricant coatings for cutting and forming tool steels; and amorphous a-C type DLC coating (black) for machining nonferrous and abrasive alloys as well as forming tool steels. Of these, TiCN and TiAlN coatings have experienced more growth during this decade.[95] Table 18.5 summarizes the relative cutting performance

TABLE 18.4 Hard Coatings for Cutting Tools Deposited by PVD and CVD Techniques[94]

Coatings	PVD	CVD
Carbides	TiC, B_4C, Cr_xC_y, ZrC, WC	TiC, WC
Nitrides	TiN, ZrN, CrN, NbN, VN, TiAlN	TiN, ZrN, HfN
Carbonitrides	TiCN, ZrCN	TiCN, ZrCN
Oxides	Al_2O_3, ZrO_2	Al_2O_3, ZrO_2
Multilayer coatings	TiCN/TiN, TiAlN/TiN, TiN/NbN (superlattice coating)	TiC/TiN, TiCN/TiN, TiC/Al_2O_3, TiC/Al_2O_3/TiN, TiCN/Al_2O_3, TiCN/Al_2O_3/TiN, [TiCN/Al_2O_3]$_n$
Multiphase coatings	MoS_2/TiN, MoS_2/metal	Al_2O_3 + ZrO_2
Other coatings	MoS_2, WS_2 (soft lubricant coatings), hard DLC (a-C)	Diamond, TiB_2, cBN (potential)

Courtesy of D. G. Bhat.

TABLE 18.5 Relative Cutting Performance of Coatings for High-Speed Tool Steels

Application	Best	Better	Good
High- and low-alloy steels and stainless steels at medium to high cutting speeds	TiAlN	TiCN	TiN
Piercing	MoS_2-Ti lubricant	TiCN	TiN
Blanking	TiCN	—	TiN
Fine blanking	MoS_2-Ti lubricant	TiCN	TiN
Drawing, flanging, forming, extrusion	MoS_2-Ti lubricant	TiCN	CrN
Cold heading/impact extrusion	MoS_2-Ti lubricant	—	TiN
High- and low-alloy steels and stainless steels at low cutting speed and for interrupted cuts	TiCN	TiAlN	TiN
Aluminum and Al-Si alloys, cast aluminum	a-Diamond, DLC	TiAlN	—
Copper, brass, and bronze	CrN	—	—
Titanium-based alloys	TiAlN	—	—
Nickel-based alloys	TiAlN	—	—

Coating (color): TiN and ZrN (gold), TiAlN (bronze), AlTiN (black), TiCN (blue black), CrN (silver), and MoS_2-Ti lubricant (silver).
Source: After O. Knotek et al. and IonBond.

of various coatings for high-speed tool steels. Usually the transition metal nitride PVD coatings are used for components whose dimensional stability is critical for their performance, such as moving parts in a machine.[96]

Although ceramic coatings have admirable properties, they are very thin ($\sim 5\,\mu m$) and, therefore, highly rely on the inherent shear strength of the substrate to provide the necessary mechanical support.[97] Important applications also include the metallization of advanced ceramics for use as chip carriers and parts of hybrid and microwave ceramic circuit packages in thin-film resistors, capacitors, and other devices.[85]

Examples of applications of PVD coatings also include evaporative coatings on webs, sputter coatings on architectural glass, sputter deposition of functional layers in VLSI semiconductors, and various decorative coatings.[93]

Very thin hard coatings ($<0.1\,\mu m$) are used for low-contact force applications such as "flying head" on hard disc drives. Transparent hard coatings such as DLC and SiO_2 are also being employed to enhance the abrasion resistance of transparent plastic surfaces such as those used for aircraft canopies and sunglasses.[78]

18.3.1 Thermal Evaporation

18.3.1.1 Resistance Evaporation. In resistance evaporation, source material is heated to a sufficiently high temperature for significant sublimation or evaporation by applying a low-voltage, high ac current to a heating element fabricated from a refractive conductive material (such as W, Mo, Ta, C, and composite ceramics of BN/TiB_2) in the form of filaments, coils, boats, and special-purpose designs (Fig. 18.11a), to deposit thin film. This method is used for low-vapor-pressure source materials such as Al, Mg, Ba, Cr, silicon monoxide, Sn, and ZnS. The evaporated material condenses on the substrates placed within both a vacuum chamber and a line of sight of the source material, forming thin films.[85]

The advantages of this method are its reliability, its economical cost, and the absence of ionizing radiation damage to the substrate. The disadvantages include possible contamination from the resistive heating element, short filament life, limitations on the film thickness, variable alloy composition of the deposited film, inability to deposit high-melting-point materials, and inability to use reactive materials that would react with the chamber itself.[85]

Figure 18.11a shows some resistively heated source configurations and Fig. 18.11b shows a schematic resistance-heated evaporation system comprising the resistance-heated refractory source. Readers may find a brief review provided by Graper.[98]

18.3.1.2 Electron Beam Evaporation. Electron beam (EB) evaporation is a variant of thermal evaporation using an electron beam to evaporate the target. EB heating of source material can be accomplished by a focused or an unfocused beam. The electron beam-heated source (Fig. 18.11c), comprising a power supply and evaporation source, can be distinguished from a resistance-heated source in two ways: (1) Electron beams generated from electron guns are directed to melt and evaporate the ingots and preheat the substrate within a vacuum chamber, and (2) the evaporant is placed in a water-cooled cavity or hearth. Currently, electron beam-heated sources cannot normally be used for the alloys required for metallization; therefore, other techniques such as magnetron or CVD are more commonly used today.[93]

The following are four widely applicable guidelines for successful deposition of thin films from an electric beam-heated evaporation source[99]:

1. Select a charge form with the largest volume/area ratio. Avoid evaporating powdered or granular materials.

2. Use the largest hearth volume conforming to the evaporant charge and the desired film.

3. Use the largest beam spot area but still realize the required deposition rate.

4. Increase the spot size if increased beam power produces instability or film pinholing.

The advantages of the EB-PVD process are the following[94,99,100]:

1. It is a cost-effective and robust coating technology.

2. It offers many desirable characteristics such as increased deposition rates (up to 100 to 150 μm/mt with an evaporation rate of ~10 to 15 kg/hr), dense coatings,

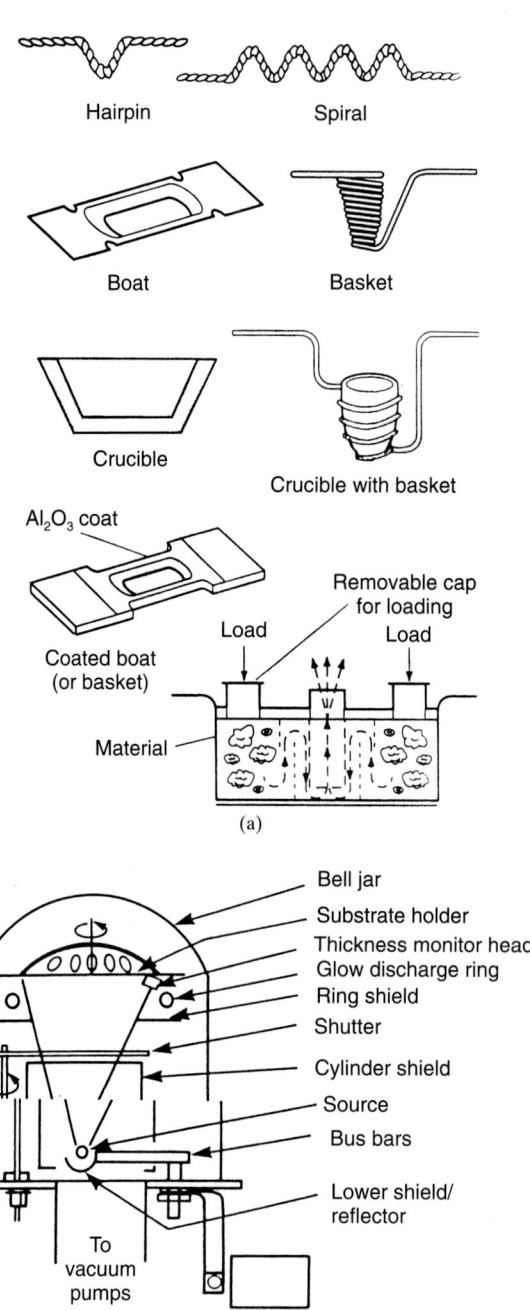

Hairpin   Spiral

Boat   Basket

Crucible

Crucible with basket

Al_2O_3 coat

Coated boat
(or basket)

Load

Removable cap
for loading
Load

Material

(a)

Bell jar
Substrate holder
Thickness monitor head
Glow discharge ring
Ring shield
Shutter
Cylinder shield
Source
Bus bars
Lower shield/
reflector

To
vacuum
pumps

Transformer

(b)

FIGURE 18.11 Schematics: (*a*) resistively heated thermal evaporation source configurations.[78] (*Courtesy of Donald M. Mattox.*) (*b*) Vacuum system for deposition from resistance-heated sources[98] and (*c*) electron beam-heated evaporation source.[99] (*Courtesy of Institute of Physics Publishing, Bristol, U.K.*)

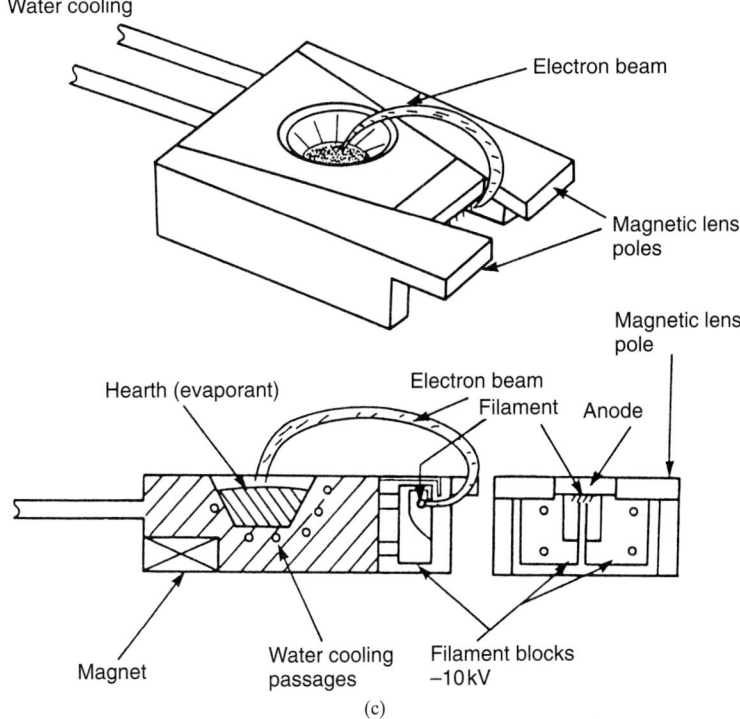

FIGURE 18.11 (*Continued*) Schematics: (*a*) resistively heated thermal evaporation source configurations.[78] (*Courtesy of Donald M. Mattox.*) (*b*) Vacuum system for deposition from resistance-heated sources[98] and (*c*) electron beam-heated evaporation source.[99] (*Courtesy of Institute of Physics Publishing, Bristol, U.K.*)

precise composition control, columnar and polycrystalline microstructure, high-purity coatings, good surface finish and uniform microstructure, and high thermal efficiency.

3. It offers high flexibility in depositing multicomponent and multilayered metallic/ceramic coatings.

4. A variation of the EB-PVD process using ion bombardment of the condensing film, called the IBAD process, offers additional benefits such as dense coatings with improved adhesion. This is discussed in an earlier section.

5. Even elements with low vapor pressure such as C, Mo, and W are readily evaporated by this process.[94]

Applications of the EB-PVD process include metallic and ceramic coatings (oxides, nitrides, and carbides) at relatively low temperatures for numerous applications ranging from microelectronics through machining tools and forging dies, to turbine industries.[94,100]

The use of lift-off metallization technology for gallium arsenide and other high-performance devices has led to a resurgence of EB-deposited semiconductor metallization. The development of the new generation of hard multilayer optical coatings has made EB-PVD the technology of choice in optics.[99]

18.3.1.3   Vacuum Arc Deposition.   Vacuum arc deposition (VAD) is a process for depositing thin films and coatings in which a high-current, low-voltage electrical discharge between a pair of electrodes in vacuum is used to evaporate material from one of the electrodes, which subsequently deposits on a substrate. The discharge is sustained in a plasma atmosphere of the ionized and evaporated electrode material. In the vacuum arc mode most commonly used for deposition, the evaporation is concentrated at minute, naturally occurring hot spots on the cathode surface, referred to as *cathode spots*, and the process of using this arc mode is sometimes called *cathodic arc evaporation* or *deposition* (*CAE* or *CAD*).[93]

The arc current flowing into the cathode is also concentrated at these spots. Each spot is capable of supporting a maximum current in the range 0.1 to 150 A, depending on the cathode material; when the current exceeds this value, multiple spots occur and their number is proportional to the current. The spot diameter is on the order of a μm. The vapor evaporated from the spot is fully ionized (typically forming multiply ionized species) by the concentrated current flowing through the plasma into the spot. The metal vapor plasma thus formed expands away from the cathode surface and is further heated by the electric current. During this expansion, the thermal energy of the plasma is converted into directed motion, much like exhaust gas expanding through a nozzle in a jet engine.[101] The resulting plasma jet is hypersonic, with a typical velocity of 10^4 m/sec. The ions in the jet have typical directed kinetic energies of 30 to 50 eV. The velocity, energy, and average charge multiplicity of cathode spot-produced ions were recently tabulated by Yushkov et al.[102]

The reaction force of the jet on the microscopic liquid pool at the cathode spot ejects the liquid material, forming a spray of liquid metal droplets in a direction nearly parallel to the cathode surface. These droplets, usually known as *macroparticles (MP)*, can contaminate the coatings if they are not removed from the plasma flow.[103]

A coating will form on any solid surface, which intercepts a portion of the metal plasma jet. In high vacuum, the composition of the coating will often be identical to the cathode composition. In a low-pressure reactive gas atmosphere, thin films of nitrides, carbides, and oxides are readily formed.[93]

VAD is characterized by the control of morphology, excellent coating uniformity, coating of line-of-sight surfaces, retention of alloy composition, efficient deposition of compounds (by introducing a reactive gas), and production of a copious amount of energetic ions.[93,104] The quality of the deposited films depends on the properties of the products emitted from the arc source, substrate temperature, and bias.

Advantages and Disadvantages.   VAD offers several important advantages over CVD and other PVD techniques[93]:

1. The intrinsic energy of the depositing particles ($\sim$30 eV) is an order of magnitude greater than in sputtering (typically 3 eV) and two orders of magnitude greater than in thermal evaporation (typically 0.3 eV). The high energy assists in achieving good adhesion and fully dense coatings.

2. The energy of all of the depositing particles can be controlled, by negatively biasing the substrate. This can be employed to obtain films with a preferred crystalline orientation.

3. The metal ions can be used to sputter clean the substrate surface as a first step in the deposition process, by using a high negative bias to the substrate (typically ~1 kV). This insures good coating adhesion.

4. High ionization, high kinetic energy, and high substrate temperature (due to ion bombardment) increase adatom mobility and surface diffusion of the growing film.[105] (See Sec. 18.2.3.)

5. The plasma beam can be collimated, directed, and swept using magnetic fields.

6. The composition of the source electrode can be transposed to the coating.

7. Unlike thermal evaporation, the cathodic arc sources can be mounted in any orientation/configuration.

8. The deposition rates are quite high, typically 10μm/hr in large industrial systems.

9. Because of the high deposition rates, the vacuum requirements are modest.

10. The inexpensive and low-voltage power supplies (e.g., dc welders) used are attractive from a safety viewpoint.[78]

11. This is a fast, efficient, and relatively cost-effective process, offering an alternative for the deposition of functional thin films with improved mechanical properties.

The disadvantages of VAD are the following:

1. MP contamination, if not filtered out, may cause lumps, porosity, and usually poor microstructure in the coating.

2. The high energy per deposited particle implies a higher heat flux to the substrate at a given deposition rate than lower-energy techniques.

Industrial Batch Deposition Systems.   In a typical industrial batch coating system, either one or more cathodes are placed in the wall of a vacuum chamber, or a central rod cathode is used. The anode may be a separate electrode, or the vacuum chamber may serve this function. Means are usually incorporated for rotating the workpieces in order to achieve reasonable uniformity on a complex surface (e. g., drills), for controlling the gas flow, for heating the substrates and monitoring their temperature, and for negatively biasing the substrates.

Filtered Vacuum Arc Deposition.   While good service performance is obtained in arc-deposited coatings, despite the incorporation of MPs, they are not tolerated in many electronic and optical applications. The MPs can be filtered from the plasma jet by magnetically bending the plasma past any obstacle that occludes any straight line path between the cathode and the substrate, thereby blocking the MPs. The process under these conditions is known as *filtered vacuum arc deposition (FVAD).*[106,107] Typically, a toroidal duct is used as the blocking obstacle, as shown schematically in Fig. 18.12. FVAD is currently used commercially to coat coinage dies, and systems are currently being developed for Cu metallization of VLSI semiconductor wafers,[108] DLC coatings on magnetic read/write heads,[109] SnO_2 transparent conducting,[110] and other optical coatings on large glass panels.

Other VAD Variants.   Extremely high instantaneous deposition rates (up to 400μm/s) have been obtained for selective coatings, by using a pulsed arc between

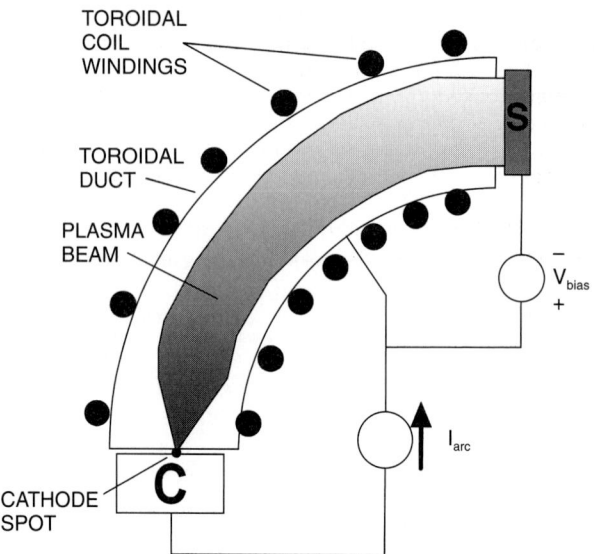

FIGURE 18.12 Apparatus for depositing amorphous DLC films using filtered cathodic arc. A plasma jet of ionized cathode material is emitted from the cathode spot located on the surface of the cathode C. The magnetic field produced by the toroidal coil bends the plasma beam in order to reach the anode substrate surface. A negative bias applied to the substrate accelerates the impinging ions. (*Courtesy of R. L. Boxman.*)

a cathode source electrode and a workpiece acting as the anode and placed in close vicinity to the source. The instantaneous heat flux is also very high, and can be employed for enhancing the coating adhesion, melting the workpiece surfaces, surface alloying, and quenching. All of these have been combined in an example in which 1010 steel was case-hardened in a single 0.1-s pulse applied to a graphite cathode.[111]

Several cathodes of different materials can be mounted in close proximity, and operated either sequentially or simultaneously, to produce multilayer, multicomponent, or graded composition coatings. For example, 150 layer TiN/NbN[112] and superhard (TiNb)N[113] coatings were deposited on carbide substrates.

The vacuum arc can also be used in an approach where one of the electrodes is thermally isolated, allowing it to be grossly heated to a sufficiently high temperature for sublimation, or heating a crucible electrode sufficiently to evaporate a metallic charge placed therein. These hot cathode vacuum arcs or hot anode vacuum arcs have the advantages of being free of MPs in some cases, having a low energy cost per particle evaporated.[114,114a,114b] In another variant, the hot refractory anode vacuum arc, the material emitted from cathode spots, is either reevaporated from a hot refractory anode or hydrodynamically deflected from it.[115] Readers can find the recent reviews elsewhere.[116,117]

Applications. Both cathodic and anodic arc evaporation are widely used to deposit hard and wear-resistant coatings both for decorative and functional applications. Such coatings can also be used as adherent basecoats on which the balance of the coating is formed by sputter deposition or thermal evaporation (cathodic arc). Typically, these coatings are a few μm in thickness.

Reactive arc PVD allows deposition of oxides, nitrides, or compounds for a number of applications such as transparent electric conductors (ITO), high- and low-temperature superconducting coatings, optical coatings (such as thick molybdenum coatings on metal substrate to fabricate very-high-power laser mirrors), AlN coatings on WC-Co substrates for piezoelectric sensors, and amorphous silicon.

The most important commercial applications are wear-resistant hard coatings for corrosion and oxidation protection, solid lubricated bearings, and in electronics.[118]

The applications of deposition of wear-resistant hard coatings include metal-cutting tools (such as TiN, CrN, ZrN, TiCN, and TiAlN); punching and forming tools (such as TiN, CrN, and TiCN); machine parts [such as CrN, TiN, diamond-like carbon coatings (see also Sec. 18.2.3), and metal-carbon (mCr: CH—) coatings]; plastic injection molds (TiN and CrN coatings); and dental tools, surgical tools (such as scissors) and orthopedic implants (such as hip stems). In all these applications the coatings produce spectacular increases in tool (product) life, productivity, and quality.[119,120]

The important automotive applications are engine components, light alloy disk-brake rotors, piston skirt, air conditioning compressor components, and power train components such as gears and plain bearings. Hostile environment tribology applications include MoS_2, Pb, and Ag in spacecraft rolling bearings, stellite and triboloys on less expensive substrate materials, and W-WC multilayer for improved erosion resistance.[121]

Other application fields are electrically active coatings, such as metallization of plastics and ceramics with copper or aluminum; decorative coatings for both the required color and increased abrasion resistance in watches, jewelry, spectacle frames, knives, door knobs, plumbing fixtures, and architectural elements (e.g., church weathercocks), etc.; erosion and hot air corrosion-resistant MCrAlY alloy coatings of turbine elements; and fabrication of composite materials such as deposition of TiAlV on to SiC fibers prior to Hipping.[122]

18.3.1.4   Laser Ablation.   Film deposition by laser ablation has attracted intense interest for the synthesis of high-purity metal, alloy, compound, and semiconducting and insulating materials. Since the late 1980s, these technologies have been applied successfully to deposit the films of high-T_c superconducting ferroelectric and ferromagnetic materials, colossal magnetoresistance, transparent conductors, biomaterials, etc.[123,124] Singly or in combination with magnetron sputtering, it has been used to grow hard coatings, solid lubricants, and nanocomposites as well as unique microstructures and chemistry, solely tailored to meet the challenges experienced by aerospace components in extreme conditions.[125]

The film deposition by laser ablation is performed by irradiation of the target surface by a focused laser beam. The incident laser beam ablates (or removes) a portion of the target (called an *evaporated plume*) material, which is transferred and deposited onto the solid substrate, resulting in a stoichiometric deposition (Fig. 18.13a).[126]

The characteristics of film deposition are dictated by the laser used for the ablation. Table 18.6 summarizes the ablation lasers, showing their wavelengths and approximate pulse widths. Among the lasers, the excimer laser is well suited for laser ablation due to its short wavelength and small pulse width. The wavelength

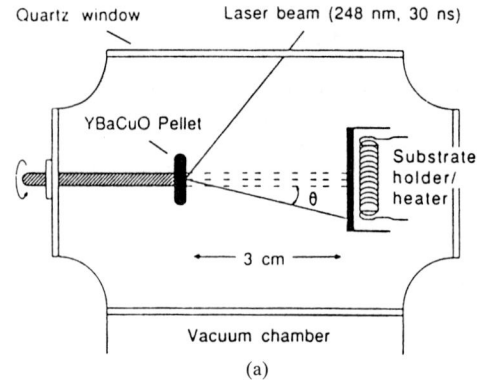

(a)

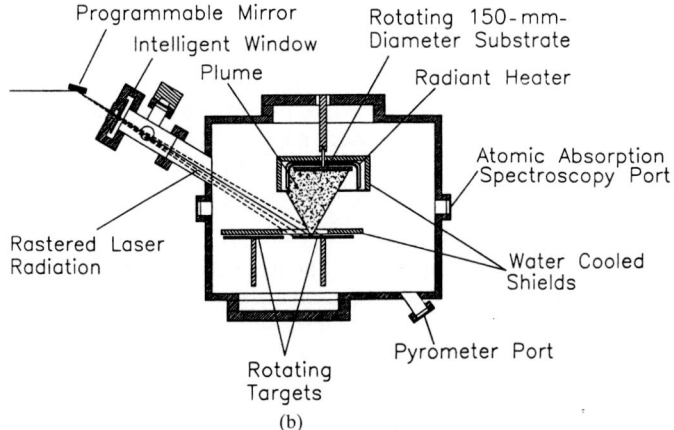

(b)

FIGURE 18.13   Schematic diagram of typical (*a*) laser ablation[126] and (*b*) pulsed laser ablation deposition systems.[29] [*Courtesy of PVD. (a) After T. Venkatesan, X. D. Wu, A. Inam, and J. B. Watchman, Appl. Phys. Lett., vol. 52, 1988, p. 1193.*]

TABLE 18.6   Lasers Used for Laser Ablation[95]

Lasers	Wavelength	Pulse width
Excimer laser	193 nm (ArF)	6–12 ns
	248 nm (KrF)	6–12 ns
	308 nm (XeCl)	6–12 ns
Nd:YAG	266 nm (4th)	5–20 ns
	355 nm (3rd)	5–20 ns
	532 nm (2nd)	5–20 ns
	1064 nm (fundamental)	2–20 ns
Ruby laser	694 nm	1 ms
CO_2 laser	10.6 μm	cw

Courtesy of Institute of Physics Publishing, Bristol.

of the ablation laser determines both the absorption coefficient of the target material and the cross section of ambient gas excitation. The pulse width also plays a key role in the ablation mechanism.[126] Figure 18.13*b* shows pulsed laser ablation, in which the laser is continuously pulsed and moved about the target in order to ensure that no pitting forms, to assist the formation of a uniform film, to produce high deposition rate, and to enable coating on a wide variety of materials.[29] The pulse width, pulse intensity, and repetition rate are modulated for specific applications.[123]

When compared to conventional film deposition processes such as thermal evaporation, sputtering, molecular beam epitaxy (MBE), organometallic vapor deposition (OMCVD), etc., pulsed laser beam ablation possesses the following advantages[126–129]:

1. The ability to melt, evaporate, or vaporize the surface by controlling the input energy density without affecting the interior surface[127]
2. The ability to deposit a film of high melting materials and control the processing region with extreme precision
3. The ability to transfer the pellet stoichiometry to the growing film[128]
4. The ability to form films in an oxidation environment with relative high vapor pressure due to the absence of a heater or filament in the deposition chamber
5. Virtual freedom from contamination,[126] unlike evaporation heater or filament
6. Thermal stability of the target due to low average power
7. Reproducible properties (such as adherence, crystalline structure) of the deposited films at a low temperature using any target material[129]
8. Possibility of fabricating unique coating architectures (e.g., functional gradient, multilayers, and nanostructures) that combine properties such as hardness, toughness, low friction, and wear resistance[125]
9. Extension of the conventional process to include organic materials through a process called matrix-assisted pulsed laser evaporation (MAPLE)[128]

The disadvantages of this method are the following:

1. Use of complex transmitting and focusing systems onto the evaporant placed within the vacuum system, which involves special designs and increases the setup cost.
2. Use of a laser whose wavelength is compatible with the absorption characteristics of the material to be evaporated.
3. Very low-energy conversion efficiency and low production rate.[93]
4. Limitation of large area deposition (≤10- to 20-mm diameter), high cost of the excimer laser, and droplet formation on the deposited film due to high instantaneous evaporation rate.[123]
5. Difficulties in coating nonplanar objects because of a line-of-sight process.[127]
6. Unsuitable for mass production due to the high cost of the laser. However, this technique has also been successfully used to grow high-quality multicomponent thin film (using sophisticated material and device research in laboratories and small scale production of high-cost-performance devices).[125]

18.3.2 Sputter Deposition

The sputter deposition process involves a target and a plasma of neutral working gas such as argon. It is a nonthermal evaporation process in which surface atoms or molecules are physically ejected from the target source by a momentum transfer collision process from the impact of high-energy ions or particles. Here both the target (coating or deposition source) and the substrate are placed in a vacuum chamber and evacuated to a pressure typically in the range 10^{-4} to 10^{-7} torr. The most important parameters controlling the growth and properties of the films by sputter deposition processes include target voltage and current, reactant partial pressure and flow rate, and substrate temperature and substrate bias. Coatings can be modified by reactive sputtering or cosputtering (by using two or more cathodes and introducing a second gas such as nitrogen to react with an evaporated metal) to achieve control of both electrical and optical parameters.[85] Schematic diagrams of the sputter coating processes are shown in Fig. 18.14. The target (cathode) is connected to a negative voltage supply and the substrate (anode) faces the target.[123] Vacuum chamber walls may also serve as the anode. More often, an arrangement in which there is an anode in close vicinity to the cathode is used.

Likewise, in many cases, the substrate is an insulating material.

Advantages and Disadvantages. The advantages of sputter deposition processes over other thin film deposition techniques are the following[130]:

1. Use of very wide range of source and film materials (i.e., metals, alloys, compounds, semiconductors, and insulators)
2. Small sputtering-yield variations from one material to another when compared to the relative variation in the evaporation rates at a particular temperature
3. Ease of low-temperature deposition of refractory materials
4. Ability to deposit in a gas atmosphere with sputtering gas pressure of a few mtorr
5. Unlike thermal evaporation and arc evaporation deposited films, no droplet contamination, provided that source arcing is prevented[131]
6. Ease of formation of multicomponent films
7. Uniformity of film thickness over extended areas
8. Superior film adhesion, but not as good as cathodic arc evaporation
9. Freedom from environmental pollution
10. A stable, reproducible, long-lived, large- and small-area vaporization source provided by the sputtering target[78]
11. Small radiant heating in the system with respect to vacuum evaporation[78]
12. Ease of incorporation of in situ surface preparation into the processing[78]

The disadvantages of sputtering deposition include the following[130]:

1. Requirement that target (source) materials be in sheet or tube form
2. Low target material utilization in some configurations and low deposition rate
3. High substrate heating due to bombardment of high-energy particles
4. High setup costs due to the required vacuum condition
5. A line-of sight process, particularly at low pressures, which may not be suitable for three-dimensional components

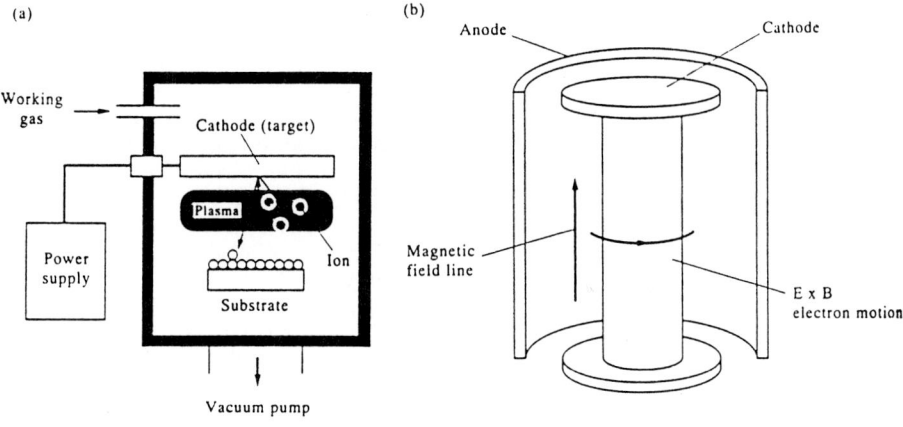

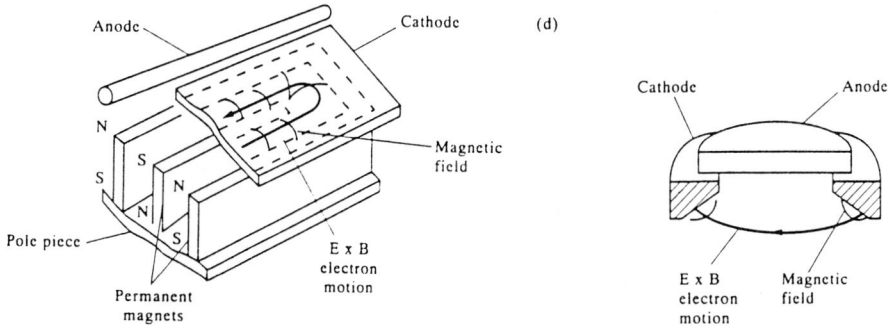

FIGURE 18.14 Schematic diagrams of sputter deposition processes: (*a*) dc glow discharge sputtering system; (*b*) cylindrical magnetron source; (*c*) planar magnetron sputtering source; and (*d*) gun magnetron sputtering source.[123] (*Reprinted by permission of Pergamon Press, Plc.*)

6. Less energy-efficient sputtering[78]

7. Fragility and easily breakable sputter target, especially those of insulators, during handling or by nonuniform heating[78]

Applications. Some applications of sputter deposited films are the following[131a,131b]: (1) single and multilayer conductor films for microelectronics and semiconductor devices (e.g., Al, Mo, Mo/Au, Ta, Ta/Au, Ti, Ti/Au, Ti/Pd/Au, Ti/Pd/Cu/Au, Cr, Cr/Pd/Au, Ni-Cr, W, W-Ti, and W/Au); (2) compound conductor films for semiconductor electrodes (e.g., WSi_2, $TaSi_2$, $MoSi_2$, and $PtSi$); (3) barrier layers for semiconductor metallization (e.g., TiN and WTi); (4) magnetic films for recording (e.g., Fe-Al-Si, Fe-Si, Fe-Ni-Mo, Co-Nb-Zr, Co-Cr, Co-Ni-Cr, and Co-Ni-Si); (5) optical coatings—metallic (reflective, partially reflective) (e.g., Cr, Al, Mg) and dielectric (antireflective and selective reflective) (e.g., MgO, TiO_2, and ZrO_2); (6) transparent electrical conductors [e.g., InO_2, SnO_2, InSnO (ITO)]; (7) electrically conductive compounds (e.g., Cr_2O_3, RuO_2); (8) transparent gas/vapor permeation barrier (e.g., SiO_{2-x}, Al_2O_3); (9) diffraction gratings (e.g., C/W); (10) photomasks (e.g., Cr, Mo, W); (11) wear and erosion resistance (tool coatings) (e.g., TiN, TiAlN, TiCN, CrN, Al_2O_3, and TiB_2); (12) decorative [e.g., Cr, Cr alloys, Cu-based alloys (gold colored)]; (13) decorative and wear resistant coatings (e.g., TiC, TiN, TiCN, TiAlN, Cr, Ni-Cr, CrN, ZrN, and HfN); (14) dry lubricant films—electrically nonconductive (e.g., MoS_2); (15) dry lubricant films—electrically conductive (e.g., WSe_2 and $MoSe_2$); and (16) free-standing structures.

Currently, sputtering techniques are available that can compete with evaporation or other higher-deposition-rate techniques. Deposition rates up to $3\,mg/cm^2/hr$ have been obtained by using dual magnetron sputtering.[132]

18.3.2.1 *Glow Discharge Sputter Deposition.*

The dc glow discharge sputtering system consists of the cathode (the target) and the anode (substrate) facing each other (Fig. 18.14*a*) . The cathode-anode separation is typically a few centimeters. Argon is the most common sputtering gas, at pressures on the order of 1 torr. The target is usually water-cooled, which serves both as the source of coating material and the electrode sustaining the glow discharge. Electrically, the substrates can be grounded, biased, or left floating. The distance between the cathode and anode is usually about 50 to 100 mm.

The voltage-current characteristics of the glow discharge is quite flat for a wide range of currents, and thus, for stable operation, a power supply with controlled current output is generally used to excite the discharge.[93] The sputtering rate increases approximately linearly with the discharge current. (See Fig. 16.44.) The sputtering rate can also be enhanced at a certain voltage if the working gas (or sputtering) pressure increases due to an increase in ion collection by the cathode.[133] However, gas scattering causes the decrease of the sputter deposition rate with the increase of the target-to-substrate separation distance. The incidence of scattering of the sputtered atoms increases with increasing gas pressure because the mean-free path between collisions decreases with the increase of pressure. Typically, for a dc glow discharge a pressure of 20 to 100 mtorr is employed. Voltages in the range of 1 to 5 kV are normally employed during operation, although higher voltages may be required to start the discharge.[132]

The deposition rate is a strong function of the power density at the target surface, the size of the erosion area, the source material, the source-to-substrate distance, and the working gas pressure. Thus the optimum operating condition is achieved by controlling these parameters to obtain the maximum power flux that can be applied to the target without causing cracking, sublimation, or melting. The maximum power limit can be raised if the cooling rate of the target is increased by properly designing the coolant flow channels and enhancing the thermal conductance between the target and the target backing plate.

Even though planar-diode glow discharge sputter deposition techniques are extensively employed to deposit thin conducting films due to their simplicity and the relative ease of fabrication of a large variety of single- and multicomponent target materials, they possess several disadvantages such as[123,130] (1) the requirement of an electrically conductive target, (2) their low deposition rates, (3) the substantial substrate heating owing to bombardment of the electron, (4) their low-energy efficiency due to dissipation of 75 to 95% of the power supplied through target heating, (5) their relatively small deposition surface areas, and (6) their not being suitable for reactive sputtering, particularly when poisoning of the target takes place, resulting in the formation of an insulating layer at the target surface.

18.3.2.2 Triode Sputter Deposition.

The triode sputter deposition involves a common configuration, known as the *hot cathode triode*, in which the electrode system is free of the target. Electrons are released at the cathode by thermionic emission rather than by ion bombardment. These systems can operate at low pressures, 0.5 to 1 mtorr. Some triode systems are magnetically enhanced to limit plasma to a restricted column. This prevents the plasma from contacting the substrate and thereby causing radiation or high-temperature damage. It is suitable for coating temperature-sensitive substrates.[85]

18.3.2.3 RF Sputter Deposition.

The rf sputter deposition technique is used to deposit films from nonconducting or insulating targets (Fig. 18.15).[132] The rf voltage is capacitively coupled from the conducting target holder to the front surface of the insulating target. During each cycle, electrons and ions are attracted to the target surface from the plasma when the target surface is positive or negative, respectively. However, the lighter electrons are far more mobile and, therefore, reach the surface in large numbers, negatively charging it, until a dynamic equilibrium is set up such that the charged surface repels most of the arriving electrons and the net current, averaged over each cycle, becomes zero. This occurs when the average negative surface potential with respect to the adjacent plasma is nearly equal to the amplitude of the rf voltage. In these situations, the target has a negative potential during most of the cycle, which accelerates incoming ions, thereby sputtering material from it.[93]

The amount of the resulting negative bias is thus equivalent to the zero-to-peak voltage of the rf signal. The period for the electrode to serve as an anode is of very short duration and the electrode mostly serves as a cathode during the rf cycle. Hence, one can expect the target to be sputtered as in the dc discharge. The discharge current-voltage characteristics of an rf sputter system are asymmetric and appear as a leaky rectifier or diode (Fig. 18.16). Typical rf frequencies dedicated to industrial, scientific, and medical uses by Federal Communication Commission (FCC) are 13.56 and 27 MHz. These frequencies were designated by international convention.

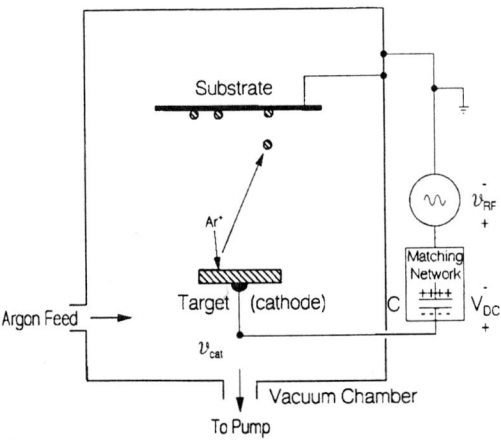

FIGURE 18.15 Schematic diagram of rf sputtering which consists, typically, of a small-area cathode (target) and a large-area anode, in series with a blocking capacitor C. The capacitor is actually part of an impedance-matching network that increases the power transfer from the rf source to the plasma discharge.[134] (*Courtesy of John Wiley, New York.*)

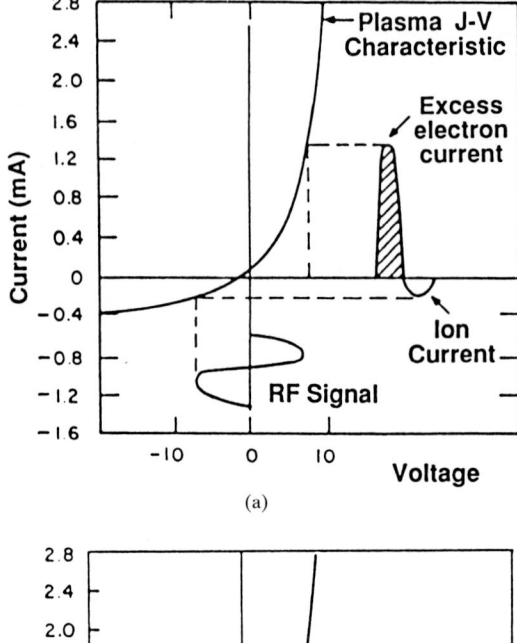

(a)

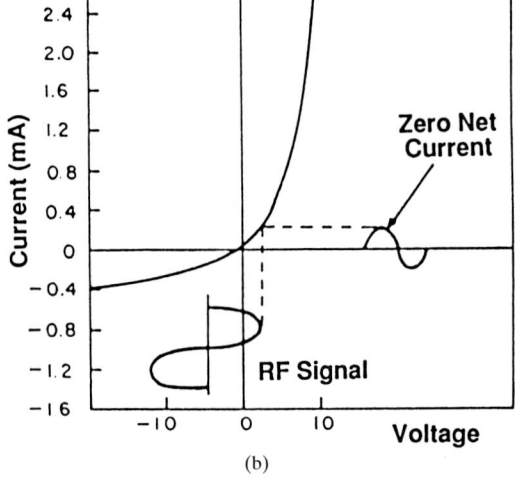

(b)

FIGURE 18.16 Formation of pulsating negative sheath on capacitively coupled cathode in an rf discharge. (*a*) Net current/zero self-bias voltage; (*b*) zero current/nonzero self-bias voltage. (*After H. S. Butler and G. S. Kino, Phys. Fluid, vol. 6, 1963, p. 1346.*)

Typically, rf discharges can be accomplished at very lower pressure (1 to 15 mtorr) as compared to the planar dc glow discharge, because (1) oscillating electrons at high frequencies yield increased ionizing collisions with the sputtering gas, leading to increased ionization and (2) the secondary electrons are not lost. (In contrast, in dc glow discharge many secondary electrons are lost at the anode before contributing most of their energy to the ionization process.)[132]

The main advantages of rf sputtering are (1) the ability to sputter insulators and conducting and nonconducting materials, (2) the accessibility of lower operating pressures, (3) the potential for reduced film contamination due to the use of lower sputtering gas pressure, (4) the lower voltage and sputtering gas pressure, (5) the high deposition rates on large substrate loads, and (6) the more efficient ionization process.[85]

Its disadvantages are (1) low deposition rate (compared to dc sputtering) due to low thermal conductivity of the insulating target materials, (2) formation of "hot spots" on the target material due to its low thermal conductivity, which generates tensile stresses and leads to fracturing of the target material (that is why it is recommended to deposit insulating films reactively from a metal source), (3) the fact that compound materials deposited in an rf discharge may not be representative of the initial target composition, and (4) high cost.

RF sputtering is widely used to deposit various kinds of conducting, semiconducting, and insulating coatings despite the complexity of the rf power source.[123] The application of rf sputtering are quite varied and include deposition of metals, metallic alloys, oxides, nitrides, and carbides. In industrial situations, dc magnetron is normally used and rf (or lately pulsed) sputtering is used if dc will not work (insulating materials or cathode poisoning).[93] RF planar magnetron sputtering may be used at much lower voltages (often under 500-V amplitude) than rf sputtering discharges that do not use magnetic trapping. RF magnetrons are about half as efficient as dc magnetrons, but the rf excitation is necessary for sputtering insulation.[134]

A number of brief reviews and articles on rf discharges can be found elsewhere.[134–137]

18.3.2.4 Magnetron Sputter Deposition.

The term *magnetron sputter** is applied to any device that uses magnetic and electric fields where the so-called $\mathbf{E} \times \mathbf{B}$ drift current (or electron motion), a "Hall effect," forms closed paths.[138] In the magnetron sputter deposition process, a combination of electric and magnetic fields (perpendicular to the cathode-anode path and produced by an assembly of permanent magnets) is applied to the target (cathode) to trap the secondary electrons and intensify the plasma and the primary electron motion in the vicinity of the cathode. This produces longer helical paths of electrons around the magnetic field lines, which provides more opportunities for enhancing ionizing collisions (and ionization efficiency) and, therefore, provides a larger improvement in the sputter deposition rates at much lower pressures ($<10^{-2}$ Pa) and/or higher current densities than the conventional glow discharge sputter deposition.[139]

There are several configurations of magnetron sputter deposition technologies. Figures 18.14b, c, and d show the cylindrical magnetron, the planar magnetron, and the S-gun-type magnetron, respectively. The cylindrical magnetron is particularly useful in providing uniform coatings over large areas, increased target utilization, and higher sputtering rates. The cylindrical-hollow magnetron technique (also called the *inverted magnetron*) is beneficial to uniformly deposit complex-shaped components with nonplanar surfaces. The cylindrical-post magnetron (with a longitudinal magnetic field) is effective in avoiding substrate bombardment by energetic particles, thereby preventing the heating up of the substrate. The planar magnetron source uses a suitably arranged magnetic field to greatly increase the plasma close to the target.[140] It is used to deposit metallic films and dielectric films with high

---

* Magnetron is a magnetically enhanced cathode. Deposition rates of magnetron sputtering are high, and the process does not cause radiation damage.

deposition rates when compared to diode sputtering.* The S-gun-type magnetron is useful to deposit films on thermally sensitive substrates, such as electronic devices, because this technique permits good isolation of the substrate from the glow discharge plasma.[123]

This method is a well-established commercial technique for the deposition of architectural glass (low-emissivity coatings), integrated circuits (metal films), semiconductors [transparent conductive oxide (such as zinc oxide, indium-tin oxide, and tin oxide) films] and hard coatings (TiN), as well as for the deposition of magnetic storage devices, e.g., spin valve giant magnetoresistance (GMR) read heads, due to its high throughput.[141]

The advantages of magnetron sputtering are the following[123,139]:

1. Low substrate temperature (down to room temperature).
2. Reduced substrate heating from electron bombardment during deposition.
3. High deposition rates (up to 12 μm/mt).[104] For precision deposition of multilayers, however, slower rates may often be desired.[121]
4. Good adhesion of films on substrates, but not better than electron beam or arc evaporation techniques.
5. Very good thickness uniformity and high density of the films.
6. Reduced working "gas pressure" requirements.
7. Good controllability and long-term stability of the process.
8. Ease of sputtering alloys and compounds of materials with different vapor pressures.
9. Ability to deposit many compounds from elemental (metallic) targets by reacting sputtering in rare/reactive gas mixtures.
10. Relative cost-effective deposition method.
11. Scalability to large areas (up to $3 \times 6\,\mathrm{m}^2$).

The disadvantages are the following[123]:

1. Utilization of small (25 to 30%) surface area of the target (owing to the choice of target materials and the difficulties in fabrication of the target), which leads to the formation of a "race track" as more material is sputtered. Currently, designs are available where the target and/or magnets are rotated in order to keep the race track moving from place to place on the target to achieve a higher target utilization. Target utilization in magnetron sputtering can also be enhanced by flattening the magnetic field lines parallel to the target surface.[132]
2. The likelihood of the porous film formation from large target-to substrate separation due to reduced electron and ion bombardment at the substrate.
3. An inherent nonuniformity in the plasma due to the magnetic fields of the cathode.

Figure 18.17 shows the balanced and unbalanced magnetron configurations. The former was developed mainly for metallization in microelectronic applications in

---

* One drawback of the planar magnetron configuration is that the plasma is not uniform over the target surface. Hence the deposition pattern is a function of the position of the substrate with respect to the target. This implies that various types of fixturing must be employed to establish position equivalency for the substrate(s). The nonuniform plasma also implies nonuniform target utilization, sometimes only 10 to 30% of the target utilization.[78]

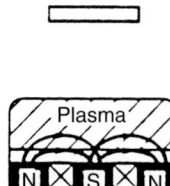

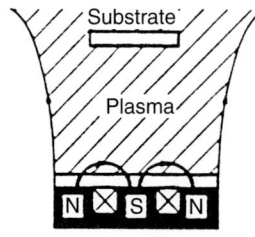

FIGURE 18.17 Schematic diagrams of balanced and unbalanced magnetron configurations.[123] (*Courtesy of Pergamon Press, Plc.*)

which electron-ion bombardment of the growing film is to be avoided. However, in metallurgical applications, ion bombardment of the film is required to produce changes in its structure and properties. Frequently, a separate ion gun is incorporated into the systems to ion-bombard the growing film. In 1986, Window and Saviides[142] developed the unbalanced magnetron, which allows the performance of ion bombardment without the need of an extra ion gun. In the unbalanced magnetron sputtering, the magnetic field is designed in such a way that a portion of the plasma trapped near the cathode leaks out and reaches the substrate. Ions from this plasma can be used to bombard the substrate at high energy, if the substrate is negatively biased.[93, 123]

In the unbalanced magnetron, the magnetic fields are intentionally arranged to allow electrons to escape. Unbalanced magnetron sputter deposition offers high plasma densities and deposition rates and is used to coat curved, rotationally-asymmetric substrates with multilayer coatings, and the number of such applications is rapidly increasing.

Reactive magnetron sputtering has been widely used in a wide range of compound films such as TiN, Al_2O_3, thin-film optical coatings, and high-T_c superconducting oxides, carbides, and fluorides with a variety of microstructures.

18.3.2.5 Closed-Field Unbalanced Magnetron Sputtering. Magnetic sputtering produces little ionization to affect the adhesion or the structure of the coating. The ionization produced by the unbalanced magnetron is about 10 times that of the balanced magnetron that results in a good dense coating structure. Further increases of ionization around 10 times compared to that produced by the unbalanced magnetron can be achieved by using a symmetrical multiple magnetron coating system.[143,144] Figure 18.18 shows a schematic of a four-magnetron closed-field unbalanced sputtering system (CFUMS).[145] The advantages claimed by the increased ionization from CFUMS in combination with the advantage of stability and flexibility of the sputtering system are being exploited to form excellent coatings for cutting tools and complex multilayer and superlattice structures, and coatings with controlled gradations of composition, structure, and properties.[143] This also leads to the great improvement in adhesion to the substrate of a wide range of coatings including DLC coatings and to the successful industrial application of such coatings.

18.3.3 Ionized PVD

Ionized PVD (I-PVD) is defined as the physical sputtering of metal atoms into a dense, inert gas plasma, intense ionization of the sputtered metal atoms (using a

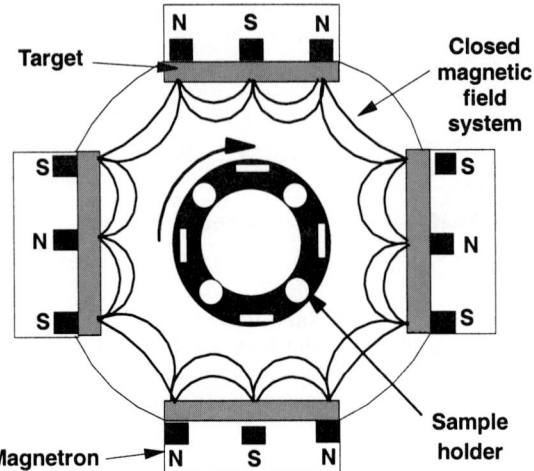

FIGURE 18.18 A schematic diagram of a closed-field unbalanced four-magnetron sputtering system.[145] (*Courtesy of D. G. Teer.*)

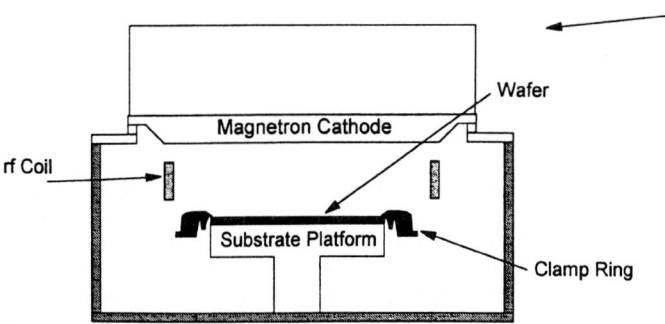

FIGURE 18.19 Schematic diagram of I-PVD tool based on inductively coupled rf plasma ionization of sputtered atoms.[148]

high electron-density discharge, $\sim 10^{12}/cm^3$), and subsequent deposition of the film from these ions.[146,147] The purpose of ionizing the sputtered metal atoms is to control their kinetic energy and change their directional distribution from a "diffuse" to a "collimated" high-density thin plasma sheath, thereby enhancing the degree of ionization of the sputtered material prior to reaching the substrate.[134,147] The flux of the depositing species must consist of >50% ions. Such effects are obtained by biasing the substrate, which accelerates the metal ions toward the substrate with speeds in excess of their random thermal speed.

Figure 18.19 shows a schematic diagram of an I-PVD system based on inductively coupled rf plasma ionization of sputtered atoms. The dc magnetron is placed on top of the substrate/tool in a sputter-down mode.[148] The specific type of I-PVD, whether it depends on the rf inductively coupled plasma (ICP), microwave electron cyclotron resonance (ECR) plasma, dc hollow cathode magnetron ionization, or some other technique, is mostly irrelevant. For more details, readers are referred to a recent publication.[149]

The I-PVD source has been developed from a research tool to a solution for advanced metal deposition in microelectronics manufacturing and for deposition of dielectrics. The successful applications of I-PVD are the deposition of metal and nitride thin films onto high-aspect-ratio features (such as deep, narrow trenches, or vias) found in modern integrated circuits (ICs) and the deposition of conformal coatings on topologically difficult substrates.

The advantages of I-PVD include the following[147,148]:

1. It provides cleaner, more uniform films at a lower cost.
2. It offers increased residence time for sputtered atoms by increasing the number of energetic electrons they encounter along the path.
3. Deposition occurs mainly from ions rather than neutrals.
4. The direction and energy of condensing particles can be controlled electrically.
5. Both the physical and chemical nature of the deposited films can be changed and controlled. Similarly, it allows one to selectively resputter parts of the film and rearrange the deposition topography in a controlled fashion.

18.3.4 Molecular Beam Epitaxy (MBE)

Epitaxy is defined as the growth of thin crystalline layers deposited on a crystalline substrate. In epitaxial growth, the substrate acts as a seed crystal and the epitaxial film duplicates the structure (orientation) of the crystal. Here much greater control of the depth distribution of impurities with limited selected-area control capability is achieved when compared to ion implantation.[150] Epitaxial layer thicknesses usually lie between 0.5 and 20 μm.

Epitaxy permits the production of semiconductor crystals with a high degree of crystalline perfection that is required for the material to be successful in device applications. Semiconductor devices combine two or more layers that have different electrical properties into a structure that can accomplish some useful function. The epitaxy process can be used to combine such different layers. In homoepitaxy, the substrate and epitaxial materials remain the same. In heteroepitaxy, the two are different. Heteroepitaxy is usually more difficult to accomplish because the chemical and structural differences between the substrate and the epitaxial materials can lead to the formation of defects near the interface between the two.[151]

Epitaxial III-V and II-VI semiconductor layers can be deposited or grown by conventional MBE, solid-source MBE (SSMBE), metalorganic MBE (MOMBE), gas-source MBE (GSMBE), chemical beam epitaxy (CBE), and metalorganic vapor phase epitaxy (MOVPE). Solid-source and gas-source MBE methods are used to grow phosphides such as GaInAlAsP alloys. Here, silicon is used for n-type doping and beryllium (sometimes carbon) for p-type doping. The II-VI structures such as ZnMgSSe and ZnCdSe are grown by conventional MBE. In this case, chlorine generated from a $ZnCl_2$ source and nitrogen from an rf plasma source are employed for n-type and p-type doping, respectively. All these materials have been developed and used for device applications, primarily for laser diodes, photodiode detectors, solar cells, and heterojunction bipolar transistors (HBTs).[29,152]

The epitaxial deposition processes can be defined as follows:

Conventional MBE: all sources solid, no P (UHV technique)

SSMBE: all sources solid, including (red) P, cracker sources for P and valved cracker sources for As (UHV technique)

MOMBE: gaseous group III source/solid group V source

GSMBE: solid group III source/gaseous group V source (UHV technique)

CBE: gaseous group III source/gaseous group V source (UHV technique)

MOVPE: gaseous metalorganic III source/gaseous group V sources [low-pressure (typically, ~30 to 80 torr) technique]

18.3.4.1 Conventional Molecular Beam Epitaxy.

Developed in 1970s, *molecular beam epitaxy* is a sophisticated form of evaporation technique to deposit very thin films of perfect single-crystal compound semiconductors. This is accomplished in a stainless steel ultrahigh-vacuum (UHV) chamber with base pressures in the 10^{-10}- to 10^{-11}-torr range (to maintain the highest purity of the growing film), while pressure during growth at the surface is about 10^{-6} torr.[153] The elemental source materials are kept in high-temperature crucibles with small orifices (called *effusion cells, Knudsen cells*, or *k-cells*). This allows the source materials to effuse streams of molecular beams of the constituent elements (such as As_2 or As_4) that are directed at a controlled rate, without scattering, to a hot crystalline substrate, resulting in the formation of epitaxial films.[154] The substrate, held in a heater stage, is usually rotated to enhance flux uniformity onto the growth surface.

Figure 18.20 shows the schematic growth kinetics involved in MBE, CBE, and MOVPE.[155] In MBE, the sticking coefficients of group III elements are unity at <500°C and there is the likelihood of growth down to low temperatures. In CBE, the boundary layer is not present and the growth kinetics are only limited to the pyrolysis reactions of the group III metalorganic on the substrate surface. This results in a sharp decrease in CBE growth rate with decreasing substrate temperature and requires a lower practical limit of ~450°C for GaAs growth using traditional precursors such as triethylgallium (TEG). Hence it is essential to pre-pyrolyze the thermally stable arsine AsH_3 precursor to form As_2 species; otherwise a poor morphology layer may occur. In CBE, unlike MOVPE, "AH_x" species are unlikely to play a key role in eliminating carbon-containing radicals from the growth surface. This can cause excessive carbon uptake if conventional MOVPE precursors are used. In MOVPE, the presence of a hot boundary layer of gas near the substrate causes partial pyrolysis of the group III metalorganic precursor. The presence of a large excess of a group V hydride (e.g., AsH_3) helps group III alkyl decomposition both in the gas phase and at the growth surface. The presence of [AsH_x] species, produced by AsH_3 pyrolysis, favors the clean removal of carbon-containing radicals from the system as stable hydrocarbons.[155]

Figure 18.21 shows schematic diagrams of typical MBE systems used for Si and III-V semiconductor growth.[156,157] The systems comprise UHV sample preparation and growth chamber, sample transfer or buffer chamber, a multi-stage load-lock for loading and unloading samples without breaking a vacuum, and the in situ analysis chamber. In all cases, the k-cells are surrounded by the cryo-panel to minimize heating of the chamber radiation from the glowing heater and sources. The buffer chamber is the next higher stage of vacuum, typically pumped by an ion pump. This contains a higher-temperature bake-out stage (~500°C or 932°F) to remove the remaining absorbed impurities. The deposition chamber may contain numerous diagnostic techniques for in situ analysis of the growing film. These analytical techniques may include reflection high-energy electron diffraction (RHEED), Auger electron spectroscopy (AES), x-ray photoemission spectroscopy (XPS), low-energy electron diffraction (LEED), modulated-beam mass spectrometry (MBMS), secondary ion mass spectroscopy (SIMS), ellipsometry, and pyrometry.[158] RHEED is used during surface preparation as well as for in situ calibration growth rates, obser-

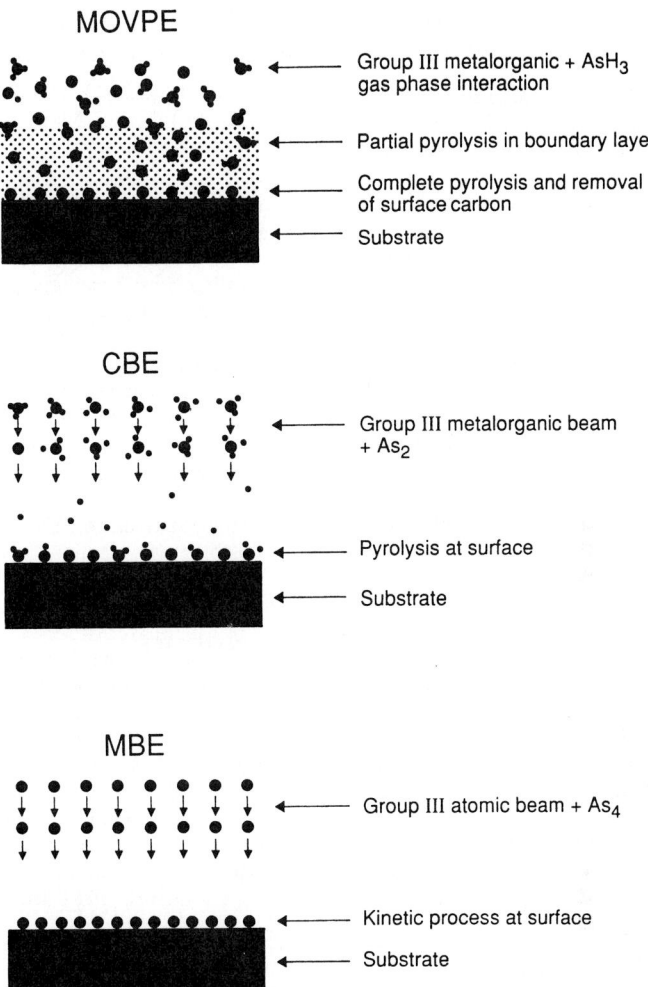

FIGURE 18.20   Schematics of growth kinetics involved in MOVPE, CBE, and MBE, respectively.[155]

vation of removal of oxides from the surface, calibration of the surface temperature, monitoring the arrangement of the surface atoms, and providing information on the crystal imperfection, surface roughness, and surface reconstruction.[159,160] AES is used to record the type of atoms present and LEED to study surface morphology. MBMS allows one to study the chemical species and reaction kinetics.

The MBE process starts with the introduction of high-purity (GaAs) substrates in which the top surface of each substrate is highly polished and chemically prepared for epitaxial growth. Material growth is started with a (GaAs) buffer layer to form an atomically smooth growth front. The remaining device layers are grown on top of the buffer layer to produce the necessary layer structure.[161]

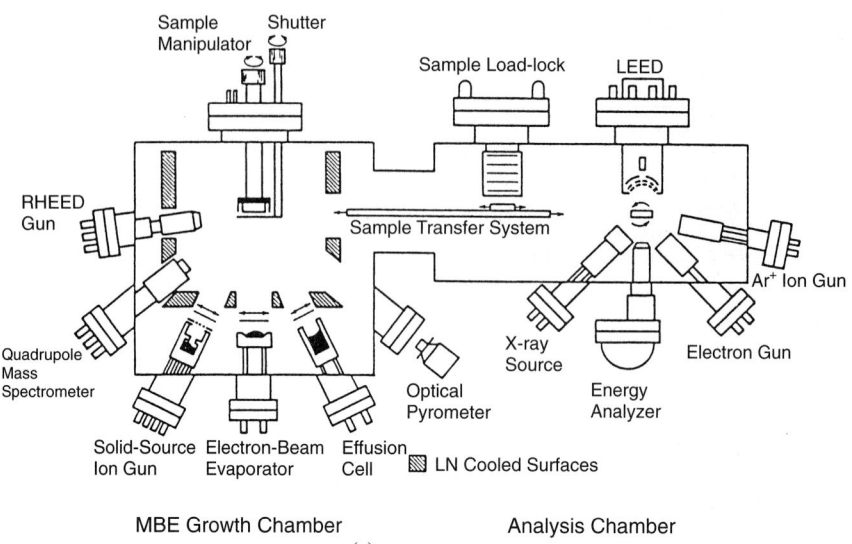

MBE Growth Chamber   Analysis Chamber

(a)

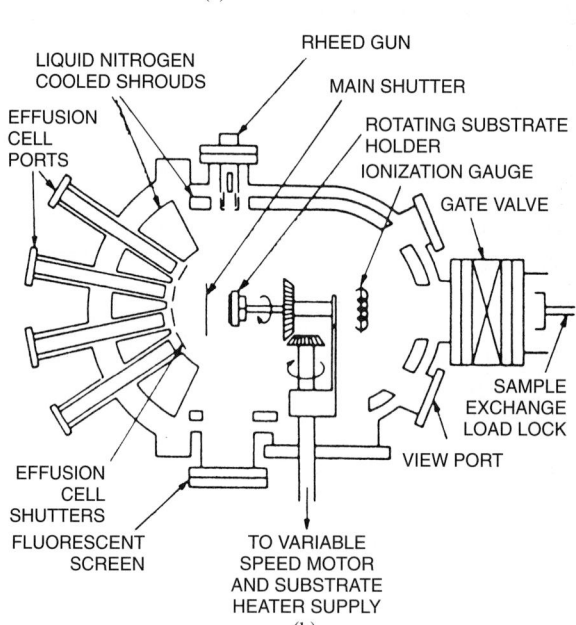

(b)

FIGURE 18.21 Schematic of typical MBE systems used for (*a*) Si and (*b*) III-V growth.[156,157]

The MBE process involves growing the semiconductor alloys one atomic mono-layer at a time, producing highly controlled layer thicknesses and alloy composition and accurate doping concentration. Growth rates are determined by the cell temperatures and are monitored by pregrowth flux measurements using an ion gauge, in situ monitoring using the optical absorption of the atomic beams, or postgrowth characterization. The growth rate can also be controlled by the evaporation rate and, most significantly, switched on and off with shutters in the fraction of a second required to grow 1 ML. Typical growth temperature and growth rates for GaAs are on the order of 600°C (1112°F) and 1 ML/s (a few Å/s) or 1 μm/hr, respectively.[160,162] However, the growth rate can be substantially increased for thicker layers or rapidly reduced for fine subatomic layer control.

Post-growth characterization may be accomplished by various techniques including electric transport, photoluminescence, secondary ion mass spectrometry (SIMS), or other surface-sensitive methods.[162]

Numerous conference proceedings,[163] books,[164–166] and extensive review articles[167–169] have been published on MBE, and the reader should refer them and other references for more details as necessary.

Advantages and Disadvantages. The advantages of MBE process are the following[158]:

1. Its ability to utilize relatively low growth temperatures and slow growth rates
2. The detailed understanding of its growth mechanisms
3. Its near-equilibrium growth process
4. Its lack of gas phase interactions due to the use of the UHV condition, which allows extremely fine control of both the composition and thickness over growth rate of heterostructures down to about 1 ML/s
5. The use of substrate rotation to produce excellent uniformity of ±1.5% across the substrate up to 4 in. in diameter
6. The ability to produce sharp or abrupt interfaces, both for intentional dopants and major constituents, by using shutters over each molecular beam source or to produce interfaces with graded composition and doping profiles[170]
7. Its increased reproducibility and throughput
8. Its flexibility to grow numerous materials
9. Possibility of relatively quick realization of specific device structures
10. Its capability of allowing the use of in situ surface analysis techniques to be used before, during, and after growth
11. Its great effectiveness for research projects and for development and production where a large throughput is not needed[160]

The disadvantages of MBE process, which limit its applicability, include the following[158]:

1. The capital, maintenance, and repair costs associated with operating the system are high, although techniques with similar capabilities such as MOCVD have comparable costs.
2. It has been less successfully used for the fabrication of phosphorus- or zinc-containing structures. However, this problem can be minimized by the advent of valved cracker evaporation sources or As and P.
3. There is also difficulty in the introduction of gaseous group V sources [such as Ga(CH$_3$)$_3$ and PH$_3$] to the MBE growth environment.
4. Its installation and use is complex.
5. Its low growth rate requires prolonged time to grow thick structures such as surface-emitting lasers.

6. Specific surface defects linked with the MBE process, such as oval defects, can adversely affect device processing, in particular, large-area devices such as solar cells.

7. Air ingress is unavoidable for component failure or during the periodic opening of the chamber for addition or replacement of source materials.[158]

8. Its relatively limited ability to grow on a large number of substrates simultaneously. Hence the production of low-cost devices such as LEDs, or large-area devices such as solar cells, are accomplished preferably by other growth techniques such as CVD.[160]

9. Source replacement is a major enterprise in most MBE systems which requires opening the system to the atmosphere, degas baking upon closure to eliminate absorbed impurities which could be outgased during deposition, and subsequent detailed process recalibration.[29]

10. Some elements such as carbon do not get incorporated from thermally generated beams.[170]

Applications. MBE has been extensively studied and used to grow a wide variety of material systems such as III-V semiconductor materials [e.g., (Al,Ga)As/GaAs, GaAs/(In-, Ga)As, and (In,Al)As/ (In,Ga)As], II-VI compound semiconductors, and IV semiconductors. It has also been applied to the single-crystal growth of a wide range of metals and oxides and metal and insular films.[158]

The MBE process has led to significant improvements in the performance of conventional electronic devices such as the (magnetic rare earth) superlattices, III-V quantum-well, quantum-wire, and quantum-dot lasers, and GaAs field effect transistors which, in turn, have also led to their wide use in the fabrication of highly complex, high-performance homo- and heterostructure materials in a wide variety of significant applications from the ubiquitous light-emitting diode (LED) to ultra-high-speed transistors (e.g., pseudomorphic high-electron mobility transistors, or PHEMTs) to high-density, compact semiconductor laser arrays (e.g., vertical cavity surface-emitting layers (VCSELs). These developments have thus led to significant advances in the understanding of semiconductor physics.[141,156,171]

MBE also finds applications for nonsemiconducting materials and for combinations of different materials such as fluoride/semiconductors and metal/semiconductor heterostructures. For example, in situ deposition of metals has been used to produce Schottky barriers that do not suffer from the presence of interface states due to exposure in air.[158]

18.3.4.2 Solid-Source Molecular Beam Epitaxy. Since conventional MBE involves all solid sources but no phosphorus source, it is incapable of phosphide epitaxy. This is a severe disadvantage of conventional MBE, because many important devices such as transmitters for fiber optical telecommunication systems at 1.3 to $1.55\,\mu m$ and visible-light LEDs and laser diodes contain some P.[172]

Because of this drawback, variants of MBE have been developed: namely, SSMBE, CBE, MOMBE, and GSMBE. The SSMBE is more convenient than CBE, GSMBE, or MOCVD, producing state-of-the-art III-V compound (phosphide) semiconductors. Hence, SSMBE is rapidly seizing the market from MOMBE and GSMBE.[172–174]

Finally, research-type single-wafer MBE systems can come with 9- or 10-cell configurations; for example, 10-cell SSMBEs are usually equipped with two Ga, two In, two Al, one P (Cracker), one As (cracker), and two dopant sources. This configuration is very useful in practice when preparing complex devices such as monolithic surface-emitting resonant cavity LEDs (RCLEDs), vertical cavity surface-emitting lasers (VCSELs), modulators, high-efficiency two-junction solar cells, or basically any advanced optoelectronic device that consists of layers of several different compositions.[172–174] Table 18.7 is a compilation of laser diode materials grown by SSMBE.[173]

Despite the importance for optical and electrical devices, there has been relatively few studies of the growth kinetics of InAsP, particularly when grown using SSMBE. Previous SSMBE studies of the growth of InAsP have focused on growth using dimer As and P (As_2 and P_2). Several researchers have compared the use of As_2 and As_4 on the growth of GaAs, AlGaAs, InGaAs, and InGaAsP, but not on the growth of InAsP. Overall, the best quality for the samples grown in the experiment occurred for the InAsP/InP multiquantum wells (MQW) structure grown at 520°C with As_4 flux.[175]

A key component in SSMBE is a phosphorus valved cracker cell (Fig. 18.22) or alternatively a polycrystalline GaP compound source for the production of a stable, molecular beam of P_2, which is preferred over a P_4 beam because of the higher sticking probabilities of P.[173] The main advantage of the valved cracking oven in SSMBE is safety, health, and environmental friendliness; the highly toxic PH_3 is removed. Although high efficiencies have been reported for the valved cracker, the efficiency of the generation of dimer molecules from tetramer molecules in a solid-source cracker may not be as high as for the hydride gas cracker. It has been reported that the As_2/As_4 ratio is 6 to 8 times higher for the hydride gas cracker than for a solid-source cracker measured under similar situations.[176] The controlled growth and reproduction of II-VI quaternary and ternary alloy compositions appear to be difficult using SSMBE.[177]

18.3.4.3 Metalorganic Molecular Beam Epitaxy. Like MBE and MOCVD, MOMBE has been developed mainly for fabrication or deposition of III-V compound semiconductor structures with low defect concentrations. Figure 18.23a is a schematic MOMBE apparatus that is based on the conventional UHV chamber containing liquid N_2-cooled cryopanels surrounding the substrate heater assembly and the source flange.[178] Figure 18.23b is a cross-sectional view of a UHV growth chamber used in MOMBE.[178] The ports used for conventional cells and injector cells are the same. The distance between the entrance of the injector or effusion cells and the substrate surface is typically less than 5 in.

Unlike conventional MBE, where the growth chamber is frequently pumped with an ion pump, the large quantity of hydrogen produced during growth from the group V halides generally used in MOMBE prevents its use. Generally, turbomolecular or diffusion pumps are used as the main pumping units to achieve reliable service performance. In addition to the main continuously operated pumping unit, an additional pump may be required to expedite the removal of gaseous species during growth and reduce background pressure, thereby enabling the growth to reach the level expected with incorporation at a lower flux rate.

Since the group V hydrides do not sufficiently decompose on the substrate during growth, it is customary to crack these gaseous phase compounds prior to approaching at the wafer surface.[179–183] The most popular way of doing this is the catalytic, or low-pressure, method. When heated to temperatures around 800°C (1470°F) or higher, the catalyst (typically, Mo or Ta) permits the effective decomposition of hydrides to group V dimers or monomers,[181,182] despite the fact that pressure within the cracker cell is only $\sim10^{-3}$ torr. These cracker cells are of nearly the same dimensions as the standard MBE effusion ovens, and thus may be employed as direct replacements for other cells in the MBE system.

TABLE 18.7 Compilation of Laser Diode Materials Grown by Solid-Source MBE*[173]

Layer	λ = 680 nm / thickness (nm)/ doping (cm⁻³)	λ = 808 nm / thickness (nm)/ doping (cm⁻³)	λ = 980 nm / thickness (nm)/ doping (cm⁻³)	λ = 1300 nm / thickness (nm)/ doping (cm⁻³)	λ = 1300 nm / thickness (nm)/ doping (cm⁻³)	λ = 1550 nm / thickness (nm)/ doping (cm⁻³)
n-substrate	GaAs (100)—0° off	GaAs (100)—0° off	GaAs (100)—0° off	InP (100)—2° off	InP (100)—2° off	InP (100)—2° off
n-buffer	$Ga_{51}In_{49}P$/50/2E18	None	None	InP/1000/1E18	InP/1000/1E18	InP/1000/1E18
n-cladding	$(Al_{70}Ga_{30})_{51}In_{49}P$/ 900/1E18	$Ga_{51}In_{49}P$/2500/n^{++}	$Ga_{51}In_{49}P$/1700/ 1E18	None	None	None
GRIN, linearly graded bandgap	$(Al_xGa_{1-x})_{51}In_{49}P$, 0.7 > x > 0.3/100	None	None	None	None	None
Waveguide	$(Al_{30}Ga_{70})_{51}In_{49}P$/50	GaInAsP (λ_g = 0.69 µm)/150	GaInAsP (λ_g = 0.72 µm)/100	GaInAsP (λ_g = 1.1 µm)/90	$Ga_{24}In_{76}P$/5, tensile strained, (−1.7%) strained, SCH	GaInAsP (λ_g = 1.25 µm), 70 SCH
Quantum well	$In_{58}Ga_{42}P$/7, compressive strained	GaInAsP/7, tensile strained	$Ga_{81}In_{19}As$/6, compressively strained	GaInAsP/6, compressively (0.7%) strained	$InAs_{43}P_{57}$/8, compressively (1.4%) strained	GaInAsP/6, compressively (0.75%) strained
Number of QWs	1	1	1	5	10	5
Barrier	None	None	None	GaInAsP (λ_g = 1.1 µm)/20	$Ga_{24}In_{76}P$/5, tensile (−1.7%) strained	GaInAsP (λ_g = 1.25 µm)/20, SCH
Interfacial layer	None	None	None	None	InP/2 monolayers between each barrier and quantum well	None
Waveguide	$(Al_{30}Ga_{70})_{51}In_{49}P$/50	GaInAsP (λ_g = 0.69 µm)/150	GaInAsP (λ_g = 0.72 µm)/100	GaInAsP (λ_g = 1.1 µm)/90	$Ga_{24}In_{76}P$/5, tensile (−1.7%) strained, SCH	GaInAsP (λ_g = 1.25 µm)/70, SCH
GRIN, linearly graded bandgap	$(Al_xGa_{1-x})_{51}In_{49}P$, 0.3 > x > 0.7/100	None	None	None	None	None
p-Cladding	$(Al_{70}Ga_{30})_{51}In_{49}P$/ 900/1E18	$Ga_{51}In_{49}P$/ 2500/p^{++}	$Ga_{51}In_{49}P$/2000/ 5E17 → 5E18	InP/1000/1E18	InP/2000/1E18	InP/1000/1E18
p-Barrier reduction	$Ga_{51}In_{49}P$/100/ 2E19	None	None	None	None	None
p-Contact	GaAs/150/5E19	GaAs/150/p^{++}	GaAs/200/low-E19	GaInAs/200/2E19	GaInAs/100/2E19	GaInAs/200/2E19

* λ refers to the nominal wavelength, λ_g is the bandgap wavelength; SCH is the separate confinement heterostructure; and GRIN is a graded index layer. Courtesy of M. Pessa.

18.50

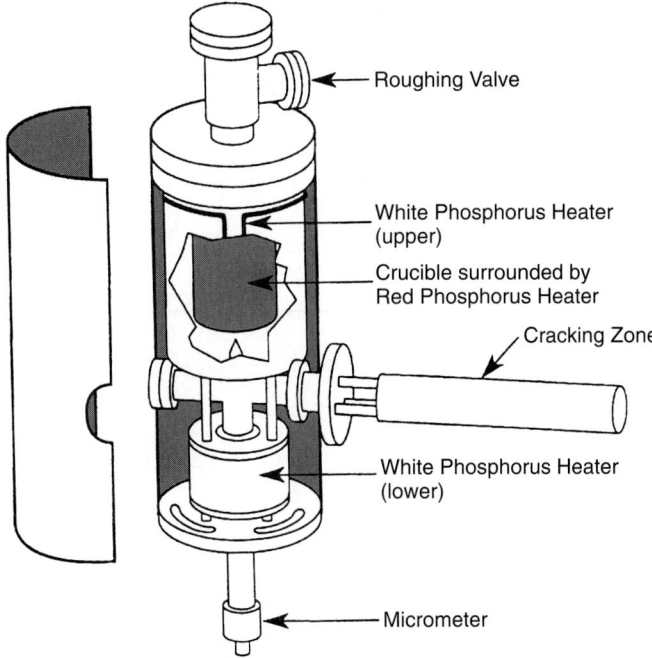

FIGURE 18.22 Schematic cross section of phosporus valved cracker cell comprising three heated zones, a dual-zone crucible with two heaters, and a separate cracking sector.[173] (*After EPI MBE Products Group. Courtesy of M. Pessa.*)

Because of their lower vapor pressures, the gaseous group-III and dopant sources need more complexity in the gas-handling manifold. There are two ways of controlling the flow of compounds into the chamber. The first is by means of mass flow controllers to regulate and direct the flow of carrier gas through the MOS bubbler. The major advantage of this device is that most of the sources can be regulated at or below room temperature.

The alternative approach uses accurate control of the pressure behind an orifice that precludes the use of mass flow controllers or carrier gases. The advantages of this method include reduced pumping loads and simpler flow dynamics. The main disadvantage is the necessity of heating the sources in order to raise the vapor pressure and to give a sufficient flux. For both methods, the MOS are added through a cell heated adequately to avoid condensation on the cell walls. The gases are switched in the same fashion as the group V hydrides described earlier. Usually the switching speed is quite adequate to provide the same nature of adjustment as in MBE.

In situ calibration techniques may be influenced by using gaseous group III sources. Measurements of the group III flux with beam flux gauges are difficult due to the degradation of the ion gauge from exposure to MOS. The other major in situ measurement technique is RHEED. The utility of RHEED has been shown for MOMBE growth of binary compounds. However, its usefulness for the calibration of ternary or quaternary composition is currently doubtful. If RHEED is to be

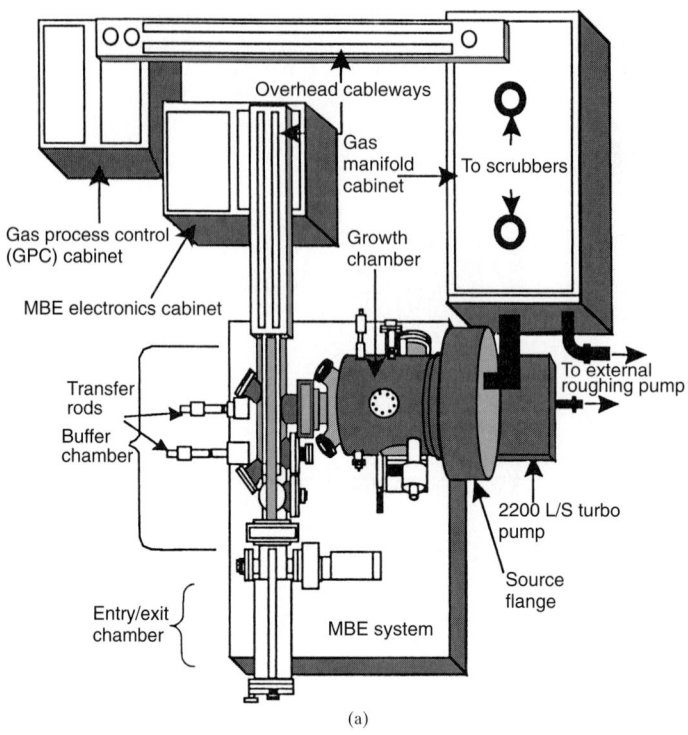

Overhead cableways

Gas
manifold
cabinet

To scrubbers

Gas process control
(GPC) cabinet

Growth
chamber

MBE electronics cabinet

Transfer
rods

To external
roughing pump

Buffer
chamber

2200 L/S turbo
pump

Source
flange

Entry/exit
chamber

MBE system

(a)

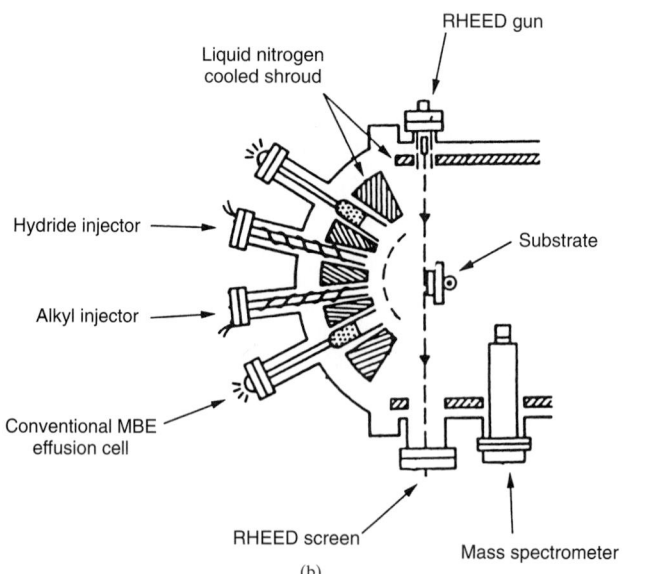

RHEED gun

Liquid nitrogen
cooled shroud

Hydride injector

Substrate

Alkyl injector

Conventional MBE
effusion cell

RHEED screen

Mass spectrometer

(b)

FIGURE 18.23 Schematic diagrams of (*a*) MOMBE growth apparatus and (*b*) cross-sectional view of UHV growth chamber used in MOMBE.[178] [(*a*) *Courtesy of Intevac.* (*b*) *Courtesy of Institute of Physics Publishing, Bristol.*]

employed for the in situ calibration of ternary composition, software algorithms must be developed to balance for the interaction of the MOS on the substrate.[178]

The advantages of MOMBE include the following:

1. Its ability to grow ternary materials and high concentrations of In[184]

2. The possibilities of the growth of wide-gap III-N materials, particularly those containing In

The disadvantages are the following:

1. The process provides much greater complexity with respect to equipment design and growth kinetics.

2. Safety is a major concern in the use of MOMBE. Here, hydrides, AsH_3 and PH_3, are more hazardous than metalorganics. Hence some of the minimum precautions must be exercised while dealing with the MOMBE process.[178]

18.3.4.4 Gas-Source Molecular Beam Epitaxy. GSMBE is a variant of conventional MBE that uses gaseous sources. Figure 18.24 exhibits a GSMBE system with three UHV chambers and a gas-handling system. The UHV system is a stan-

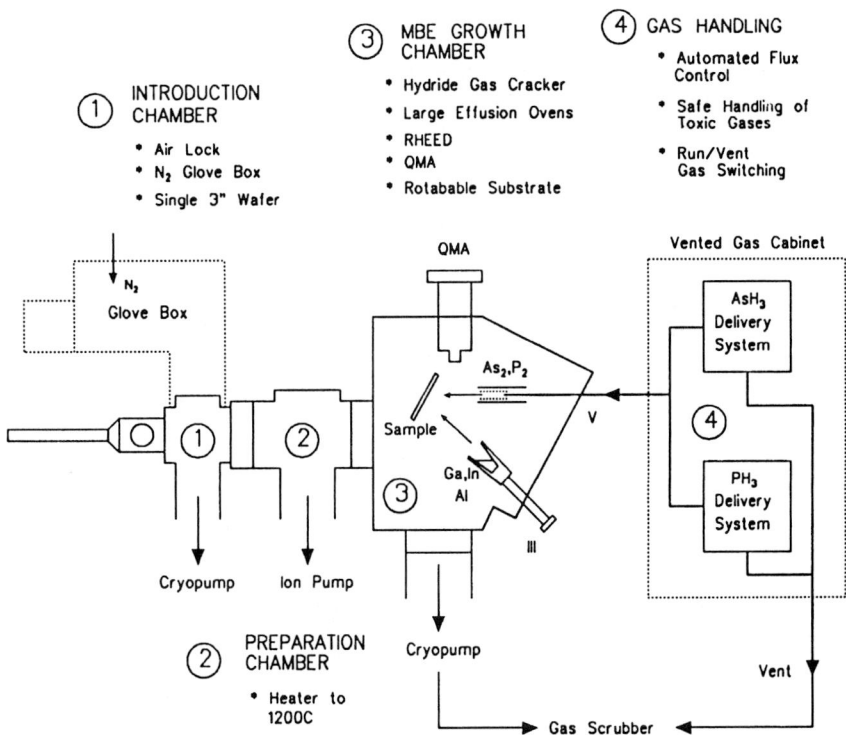

FIGURE 18.24 Schematic GSMBE system with a single growth chamber, two auxiliary UHV chambers, and a gas-handling system.[176] (*Courtesy of Institute of Physics Publishing, Bristol.*)

dard MBE system supplied with the single 3-in. wafer capability.[176] The introduction and preparation chambers are employed for the loading, unloading, and in situ storage of wafers. The N_2-purged glove box reduces the wafer contamination, expedites the pumping down of the introduction chamber by decreasing the noble gases incorporated in the sorption roughing pumps, and offers the additional safety needed for the operator. The growth chamber comprises seven effusion cells,* a single gas-cracking oven for both AsH_3 and PH_3, and a RHEED system for in situ characterization of epitaxial films. A quadrapole mass analyzer (QMA) is used to determine the cracking patterns of AsH_3 and PH_3 and for both leak checking and residual gas analysis. The gas-handling system allows the controlled introduction of AsH_3 and PH_3 at reduced flow rates into the UHV growth chamber from high-pressure gas storage cylinders. A computer controls the entire system including gas flows, effusion cell temperatures, and shutters. All the effluents from the vacuum pump and the gas-handling system are passed through a gas scrubber (for the safe removal of toxic gases).[176]

As in a conventional SSMBE (i.e., old SSMBE) system, a heated substrate is exposed to molecular beams of group III elements, each beam supplied by a separate effusion oven (or Knudsen cell). However, the group V molecular beam(s) is (are) generated by thermal dissociation (i.e., cracking) of the suitable hydride gas such as arsine (AsH_3) and phosphine (PH_3) at the gas entrance into the UHV growth chamber. These gases, after being effectively cracked at temperatures of 800 to 1100°C (1470 to 2010°F), produce the dimer molecules As_2 and P_2, respectively. At the usual growth rate of $1\,\mu m/hr$, a considerable quantity of the byproduct H_2 is generated in the growth chamber, which is instantly removed by operating suitable vacuum pumps.[176]

Both thermal and plasma cracking have been successfully used in a high-temperature "cracking" cell that decomposes or cracks apart the precursor prior to impinging on the substrate. This partial decomposition of the growth sources permits for, at times, lower growth temperatures and higher source utilization. An example of adopting this approach is in the MBE growth of II-VI compounds. II-VI compounds have traditionally been very difficult to dope p-type. Nitrogen is a suitable dopant element in many II-VI semiconductors; however, N_2 is very unreactive. An ECR plasma cracking cell for N_2, which produces atomic nitrogen, has greatly enhanced p-type doping efficiency and directly results in the demonstration of II-VI lasers.[160]

The physical mechanisms controlling film growth in GSMBE are similar to those reported in SSMBE but very different from those found in MOMBE and CBE. For example, in III-III-V ternary alloy growth, with the presence of an excess amount of group V flux, the growth rate and film composition in GSMBE are determined solely by the arrival rate of the group III species and not by the substrate temperature.† Because it is difficult with present technology to control the substrate temperature in a reproducible way, the insensitivity to modest changes in substrate temperature is a principal advantage of GSMBE over the MOMBE and CBE processes. Hence, controlled growth of complex multilayer heteroepitaxial structures becomes easier in GSMBE than in MOMBE and CBE.[176] (Note: This does not hold with new SSMBE; rather it holds with old MBE technology that used red P as

---

* However, a higher-cell reactor (than the one with 7-cell) or as many as any MBE system will be more useful.[172]

† According to Pessa, GSMBE is sensitive to the substrate temperature, and controlling the substrate temperature is necessary for achieving good material quality.[172]

a source. For example, quite complex two-junction solar cells with GaInP/GaAs layers and tunnel junctions exhibit higher conversion efficiencies when grown by SSMBE.)

When compared to conventional SSMBE (i.e., old SSMBE) where solid As_4 and P_4 are employed in heated effusion cells, GSMBE offers some important advantages, as given below:[176]

1. The ability to rapidly produce and accurately control group V beam. For example, just a few minutes are required for changing the V/III ratio to check the transition from an As-stabilized to a Ga-stabilized condition during GaAs growth, whereas several hours are needed for changing the flux from a large-capacity, solid As_4 oven.*

2. The use of gaseous PH_3 instead of solid P improves the MBE growth of phosphide compounds. This might be true in an old SSMBE technology (that used red P as a source), which is no more in use.

3. The dimer molecules generated in the GSMBE have a much better sticking coefficient than the tetramer molecules generated from the original arsenic cells of MBE (which could not produce dimer As fluxes), thereby resulting in more efficient use of source materials.

4. Introduction of the group V source material from outside the UHV chamber can significantly prolong the time between venting of the growth chamber because of the rapid depletion of group V effusion cells at the often needed high V/III ratios in MBE growth.

5. Unlike MOMBE and CBE growth, the carbon incorporation problem (generated from the pyrolysis of metal alkyl) is not produced during GSMBE growth. In reality, the molecular fragments released during cracking of AsH_3 in GSMBE may contribute to preclude low levels of carbon during growth.[177]

6. The retention of commercial effusion cells used in conventional MBE for the group III sources because the current generation of effusion cells offers long life, excellent film uniformity, and adequate group III flux stability for most applications.[176]

7. The capability of growing with relative ease a wide range of III-V materials in one growth chamber while retaining the precise control of layer thicknesses on the atomic scale characteristic of elemental MBE; this aspect serves as a powerful tool for fabricating a wide range of modern III-V heteroepitaxial devices.

8. GSMBE is an alternative technique for the growth of Si and Ge alloys, which tend to reduce problems of segregation of the Ge to the growth surface, due to the existence of hydrogen. For this purpose gaseous precursors such as disilane (Si_2H_6) and germane (GeH_4) are employed.

9. The addition of a gas injector permits the use of a wide variety of gaseous and liquid sources such as AsH_3, PH_3, and so forth.

10. Alternative precursors may allow new growth modes and dopant species, as well as ease in controlling certain growth species. An example of the latter is the growth of mixed III-V arsenide/phosphide materials, which is just a great advantage of SSMBE over GSMBE. (Note: This is particularly true for Al-containing

---

* Rapid and abrupt changing of the V/III flux ratio is an everyday operation in layer growth when valved cracker sources are used.[172]

materials. Al-containing As/P layers cannot be grown by GSMBE, although the reason for this has not yet been well established.) Growth of these III-V materials using solid P is very difficult due to the high pressure of P, leading to high-vacuum capability problems, and the difficulty in controlling the As/P ratio in the solid. Replacing solid sources of As and P with AsH_3 and PH_3, respectively, greatly enhances the ability to grow these materials.

The main drawbacks of GSMBE (and CBE) are the following:

1. The use of gaseous source increases the cost and complexity over that of a solid-source MBE system due to the additional equipment needed to control and store the gases. Much higher gas loads require high-speed pumps such as cryo, turbomolecular, or a diffusion pumping system. However, the incorporation of a fully automated GSMBE system is cost-competitive with respect to a CBE system.

2. Since AsH_3 and PH_3 are very toxic, great care is necessary during gas-handling and storage operations to warrantee a safe environment for the crystal grower and the community. Furthermore, since the P is deposited in the growth chamber and vacuum pumps can create hazardous conditions, special precautions must be taken during system maintenance.

3. It does not usually exhibit selective-area growth. However, recently selective-area growth of InGaAs/InP has been reported by using GSMBE with a high group V flux coupled with atomic hydrogen exposure.[185]

4. The high toxicity of the hydrides leads to a substantial increase in the equipment installation and operation costs and produces a considerable difficulty for the crystal grower. These problems may become less significant in the future as less hazardous ways of producing and storing the hydrides on-site are developed and suitable less toxic, alternative group V source materials become available.[176]

5. The important problems associated with using gaseous hydrides in an MBE system include the choice of vacuum pumps, the operation of the cracking oven, various approaches for hydride delivery, and safety concerns.

Since Panish pioneered the GSMBE technique, very-high-quality layers of $GaAs$[186] and InP have been produced.[187] Currently, GSMBE heterostructures have been employed to fabricate a wide range of high-performance devices such as InGaAsP/InP lasers,[188–190] InGaAs/GaAs/InGaP lasers,[191] InGaAs/InP bipolar transistors,[192] InGaAsP/InP photodiodes,[193] and multiple quantum well optical modulators.[194] The hydride gases AsH_3 and PH_3 have been extensively employed to grow films of GaAs, InP, GaP, AlP, InGaAs, InGaP, InGaAsP, and InGaAlP, whereas NH_3 has been used to produce GaN films[195] and GaPN films.[196] Many laboratories worldwide now produce III-V films by this method.

The commercial recognition of LEDs and the achievement of semiconductor lasers in the III-V nitrides has produced international interest in optoelectronics research and applications in these GSMBE-grown materials. The wide bandgaps of these materials and their strong atomic bonding also make them potential candidates for high-power and high-temperature devices.[197]

The use of GSMBE, using H_2Se and elemental Zn as source materials, has led to the growth of high-quality II-VI compounds such as ZnSe or closely lattice-matched III-V compounds such as GaAs and (In,Ga)P. The undoped ZnSe epilayers are comparable in quality to material grown by MBE.[177]

18.4 CHEMICAL VAPOR DEPOSITION

Chemical vapor deposition (CVD) involves the creation of a vapor, transport, and film growth, all of which take place simultaneously at or near the substrate.[85] CVD is an atomistic and very versatile vapor deposition process based on homogeneous and/or heterogeneous chemical reactions that can be employed in the production of coatings, powders, fibers, and monolithic parts. With CVD, it is possible to deposit a large variety of single-crystalline, polycrystalline, and amorphous thin films of IV, IV-IV, III-V, and II-VI semiconductors; metallic alloys; compounds (such as carbides, nitrides, borides, oxides, and intermetallics); dielectrics; insulators; and superconductors. CVD coating thicknesses are functions of the specific process, but are mostly in the 5- to 10-μm range, that provide adequate wear resistance or electronic properties. In some specific cases, the thickness may go up to 75 μm. Smooth substrate surfaces are usually recommended. With increasing coating thicknesses, the coating surface may become somewhat nodular. Applications range from the fabrication of integrated circuits (microelectronics), optoelectronic devices, sensors, and waveguides, through micromachines and catalysts, to the deposition of hard and protective coatings on ball bearings and cutting tools.[198] This section emphasizes important aspects of CVD related to its principles, its chemical reactions, its advantages and disadvantages, and the processes and equipment involved.

18.4.1 Principles

CVD is the process of chemically reacting one or more gaseous compounds of a material (metalorganic or inorganic precursors) to be deposited, with other vapor or gas phases, to (atomistically) produce a dense solid film or coating onto a heated substrate and a gaseous byproduct.[199,200] Although CVD and PVD are unequivocally separate processes, an important recent trend is that the sharp distinction between these two processes does not hold well. For example, CVD now often incorporates a plasma, a physical phenomenon, whereas PVD often uses a chemical environment (such as reactive evaporation and reactive sputtering). Similarly, CVD and PVD operations are often accomplished in the same integrated equipment (separate modules on a cluster tool) in a successive manner without disturbing the vacuum (thereby minimizing contamination), so that the distinguishing features between the two basic processes becomes equivocal.[201,202]

All CVD systems, irrespective of the classifications, consist of (1) a reactor (reaction chamber), (2) a heating system for the substrate or for the entire chamber, (3) a load-lock, including the wafer/substrate transport, (4) a gas delivery system (gas supply, distribution, and flow-control), (5) a vacuum system, (6) an exhaust system for neutralizing the exhaust gases (scrubber), and (7) an electrical and microprocess control system. The precursor gases are inert gases (N_2 and Ar), reducing gases (H_2), and numerous reactive gases (CH_4, CO_2, H_2O vapor, NH_3, Cl_2, and so forth). Some of the precursors are in the form of a high-vapor-pressure liquid at room temperature ($TiCl_4$, $SiCl_4$, and CH_3SiCl_3). These are heated to a relatively moderate temperature ($< \sim 60°C$), and the vapor is carried into the reactor by bubbling a carrier gas (H_2 or Ar) through the liquid. Some of the precursors are formed by converting a solid metal or compound into a vapor, for example, $AlCl_3$ formed by a reaction of Al metal with Cl_2 or HCl gas. [Very-low-vapor-pressure precursors may be dissolved in a solvent to form a "cocktail," which is subsequently flash-evaporated (e.g., Pb (thd)$_2$ in thf where thd stands for tetramethyl heptadionate and thf stands for heptafluorate).]

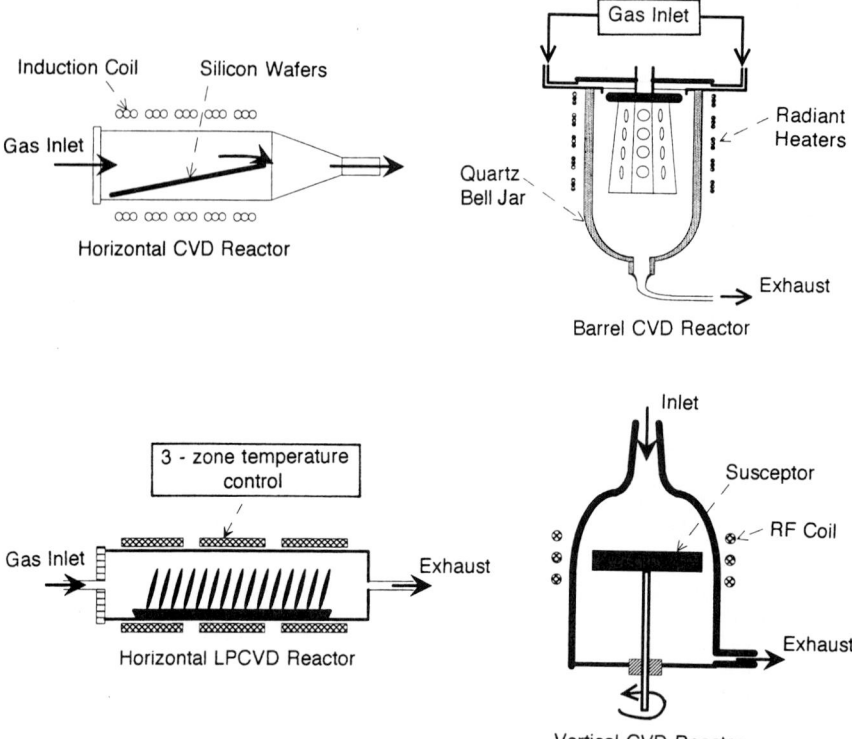

FIGURE 18.25   Schematic of several CVD reactor configurations used in epitaxial growth of semi-conductors.[160] (*Reprinted by permission of VCH, Weinheim.*)

Three types of sealed reactors are available: a horizontal reactor, a vertical reactor, and a barrel reactor, as shown in Fig. 18.25.[160] They may operate at pressures between high vacuum and several atmospheres. Heating the substrate can be carried out in a cold-wall or hot-wall reactor. In the former, only the sample and its fixture are heated by induction. In the latter, the substrate is kept in the furnace, which is heated either resistively or inductively. Close control over the temperature is done by closed-loop feedback systems. Finally, exhaust gases are removed from the reactor and treated in a chemical scrubbing unit.[201]

18.4.2   Criteria for CVD Sources

The CVD reaction sources must fulfill the following requirements for successful deposition:

1. Room temperature stability
2. Adequate volatility and sufficient partial pressure for reasonable growth rates
3. Reaction temperature below the melting point of the substrate

4. Reaction products that possess the required element or compound with readily removable byproducts

5. The availability of electronic-grade purity on a commercial scale

6. Reduced toxicity to facilitate applications on a commercial scale

7. Low cost

8. Convenient growth rate, which can be a high growth rate for coating applications or a low growth rate for semiconducting superlattices or active dielectric films for microelectronics

The choice of source depends entirely on the commercial applications:

1. Hydrides and halides used for low temperature and low growth rate in microelectronics, i.e., for multiple quantum wells

2. Hydrides used for low temperature but high growth rates

3. Halides used for high temperature and high growth rates (hard coatings)

4. Metalorganics

5. Specialty chemicals

Each gas has its advantages and shortcomings. For example, SiH_4 is pyrophoric, costly, and dangerous to handle, but when diluted with N_2 it can be handled readily. It has the advantage of reacting with NH_3 at low temperatures to form Si_3N_4. $SiCl_4$ is less expensive than SiF_4 and can be handled easily, but it forms solid imides with NH_3 at room temperature. NH_4Cl can clog the exhaust, particularly if any moisture is present.

It is noted that metalorganic compound sources are typically less toxic (albeit they are often pyrophoric). Hence, many depositions are now accomplished, particularly for III-V and II-VI semiconductors and high-T_c superconductors, ferroelectrics, or dielectrics by MOCVD.

18.4.3 Chemical Reactions

The chemical processes used in the CVD of thin films can be classified according to the nature of the chemical reaction as follows[198,201]: (1) thermal decomposition (pyrolysis); (2) reduction; (3) oxidation; (4) hydrolysis; (5) nitridation; (6) boride, carbide, and aluminide formation; (7) chemical transport; (8) disproportionation; (9) catalysis; (10) synthesis; (11) photolysis; and (12) combined reactions.

The main classes of chemical compounds used in CVD reactions are inorganic compounds such as halides, hydrides, and halohydrides (of metals and metalloids), and organometallic compounds.

Metalorganic hydride systems such as $Ga(CH_3)_3$-AsH_3-H_2 are advantageous due to the following reasons: (i) Being endothermic reactions, cold-wall reactors with a single temperature zone can be employed. (ii) Lower substrate temperatures than for hydrides or halohydrides are required. Their main disadvantage is that since most metalorganics are volatile liquids, they require accurate pressure control.

Thermal Decomposition or Pyrolysis. Pyrolysis occurs in CVD and MOCVD and involves the thermal decomposition of compounds, namely, hydrides, halohydrides, halides, and organometallic compounds (carbonyls). The compounds used in the high-temperature pyrolysis are halohydrides and halides and in the low-temperature decomposition are hydrides and organometallics (e.g., nickel-, iron-, molybdenum-, tungsten-, and ruthenium-carbonyls) to deposit the respective metallic films.[198,201,203–206]

$$SiH_4(g) \xrightarrow{500–650°C} Si(s) + 2H_2(g) \tag{18.6}$$

$$2ReCl_5(g) \xrightarrow[<20\ torr]{1000–1250°C} 2Re(s) + 5Cl_2(g) \tag{18.7}$$

$$Ni(CO)_4(g) \xrightarrow[up\ to\ 760\ torr]{180–200°C} Ni(s) + 4CO(g) \tag{18.8}$$

$$Fe(CO)_5(g) \xrightarrow{\geq200°C} Fe(s) + 5CO(g) \tag{18.9}$$

$$xFe(CO)_5(g) + yNi(CO)_4(g) \xrightarrow{>200°C} Fe_xNi_y(s) + 5(x+y)CO(g) \tag{18.10}$$

$$Mo(CO)_6(g) \xrightarrow[1–760\ torr]{300–700°C} Mo(s) + 6CO(g) \tag{18.11}$$

$$W(CO)_6(g) \xrightarrow{\sim400°C} W(s) + 6CO(g) \tag{18.12}$$

$$Ru(CO)_{12}(g) \xrightarrow[vacuum]{250–500°C} Ru(s) + 12CO(g) \tag{18.13}$$

where (g) and (s) stand for gaseous and solid states, respectively. This reaction is reversible. Similar pyrolytic reactions are those of other hydrides such as germane (GeH_4), disilane (Si_2H_4), diborane (B_2H_6), ZrI_4, trisobutyl aluminum, and ruthenium acetyl acetonate [$Ru(C_5H_7O_2)_3$]. (See also Table 18.8.)[206–216]

TABLE 18.8 Pyrolytic and Reduction Reactions in CVD with Some Examples of Thin Film Deposition[198]

Reaction	Source	T_{dep} (°C)	Thin film	Reference
Pyrolysis	$Al(C_4H_9)_3$	>250	Al	206
(AB → A + B)	B_2H_6	700–800	B	207
	$B_2H_6 + PH_3$	950	BP	208
	$B_3N_3H_6$	400–700	BN	207
	$GaBr_3 \cdot NH_3$		GaN	209
	SiH_4	600–1200	Si	210
	$SiH_4 + GeH_4$	800–850	$Si_{1-x}Ge_x$	211
	CrI_2	800–1000	Cr	211a
	ZrI_4	1200	Zr	207
Reduction	BCl_3	950–1200	B	207
(AX + H_2 ⇌ A + HX)	$BBr_3 + PCl_3$	1050	BP	212
	$GeCl_4$	850–910	Ge	213
	$FeCl_2$	650	Fe	198
	$MoCl_5$	900–1300	Mo	198
	$NbCl_5 + GeCl_4$	900	Nb_3Ge	206
	$SiCl_4$	850–1250	Si	214
	$SiCl_4 + GeCl_4$	1100–1250	$Si_{1-x}Ge_x$	215
	$TaCl_5 + BBr_3$	1300–1700	TaB	207
	$TaCl_4 + BCl_3$	800–1200	TiB_2	207
AX + M* → A + MX	WF_6	500–800	W	216

* Reducing agent.
Courtesy of Institute of Physics Publishing, Bristol.

Pyrolysis of halohydrides or halides are as follows:

$$SiH_2Cl_2(g) \longrightarrow Si(s) + 2HCl(g) \tag{18.14}$$

$$2AuCl_3(g) \longrightarrow 2Au(s) + 3Cl_2(g) \tag{18.15}$$

$$SnCl_2(g) \longrightarrow Sn(s) + Cl_2(g) \tag{18.16}$$

Generally, pyrolysis reactions with a positive enthalpy of reaction are carried out in cold-wall single-temperature-zone reactors.[206]

The advantages of pyrolysis reactions include (1) reaction efficiency, (2) lower reaction temperature than that of reduction or disproportionation reactions, (3) less chemical attack at the film/substrate interface, (4) the availability of reactant sources, (5) less sensitivity to substrate surface conditions, (6) the possibility of operation in an inert atmosphere, and (7) thickness and uniformity of epitaxial layers.[206] The main disadvantage of these reactions is the gas-phase nucleation at high temperatures, limited reactant purity, nonreversibility of the reactions, high cost of the reactant gas, contamination of deposits (especially for MOCVD), and defective epitaxial layers. However, homogeneous reactions can be avoided by a proper selection of pressure and flow conditions.

Reduction. Chemical reduction reactions usually employ hydrogen gas as the reducing agent to effect the reduction of halides, halohydrides, oxyhalogenides, carbonyl halogenides, or other oxygen-bearing compounds at a lower reaction temperature than that required in the absence of hydrogen.

Like pyrolysis reactions, reduction reactions with a positive enthalpy of reaction are accomplished in cold-wall one-temperature-zone reactors.[206] The most common example is the reduction of silicon tetrachloride ($SiCl_4$) on single-crystal Si wafers to produce homoepitaxial Si films (with growth rates of 0.4 to 1.5 μm/min) according to the reaction:

$$SiCl_4 + 2H_2(g) \xrightarrow{\ 1150-1250°C\ } Si(s) + 4HCl(g) \tag{18.17}$$

Refractory metal films such as W, Mo, and Re can be deposited by the hydrogen reduction of the corresponding hexafluorides, that is[198,201,205]:

$$WF_6(g) + 3H_2(g) \xrightarrow{\ 400-700°C\ } W(s) + 6HF(g) \tag{18.18}$$

$$WCl_6(g) + 3H_2(g) \xrightarrow{\ 600-700°C\ } W(s) + 6HCl(g) \tag{18.19}$$

$$MoF_6(g) + 3H_2(g) \xrightarrow{\ 300°C\ } Mo(s) + 6HF(g) \tag{18.20}$$

$$MoCl_6(g) + 3H_2(g) \xrightarrow{\ 400-1350°C\ } Mo(s) + 6HCl(g) \tag{18.21}$$

$$ReF_6(g) + 3H_2(g) \xrightarrow[\sim 20 \text{ torr}]{\ 500-900°C\ } Re(s) + HF(g) \tag{18.22}$$

$$TaCl_5(g) + \tfrac{5}{2}H_2(g) \xrightarrow{\ 700-1000°C\ } Ta(s) + 5HCl(g) \tag{18.23}$$

Reduction reactions are necessary in many applications such as microelectronics and hard coatings, where the hydrogen reduction of halohydrides or halogenides results in the deposition of thin metal, semiconductor, and dielectric films.[198] (See also Table 18.8.)

Oxidation. Oxidation is the reaction of a vapor-phase substance with oxygen or another oxidant such as CO_2, N_2O, NO, NO_2, or O_3, rendering a solid oxide film. The sources include metal hydrides, halohydrides, halogenides, and metalorganic compounds. For oxidation reactions above 500°C a mixture of CO_2 or nitrogen oxides and oxygen is used as the oxidant whereas below 500°C oxygen is preferred as the oxidant. These reactions may be either endothermic or exothermic; hence they are carried out in cold-wall or hot-wall reactors, respectively. Oxidation reactions are normally used to produce amorphous dielectric films (SiO_2, HfO_2, ZrO_2, BST, Al_2O_3, TiO_2), numerous silicate glasses ($SiO_2P_2O_5$, $SiO_2B_2O_3$), semiconductors (ZnO), conductors (SnO_2, In_2O_3), ferroelectrics (PZT, $SrBi_2Ta_2O_9$, or SBT), superconductors (YBCO), or magnetic films.[206]

Typical examples of oxidation reactions are the following:

$$SiH_4(g) + O_2(g) \xrightarrow{\text{300–450°C}} SiO_2(s) + 2H_2(g) \tag{18.24}$$

$$SiH_4(g) + 2O_2(g) \longrightarrow SiO_2(s) + 2H_2O(g) \tag{18.25}$$

$$2PH_3(g) + 5O_2(g) \xrightarrow{\text{450°C}} 2P_2O_5(s) + 6H_2(g) \tag{18.26}$$

The deposition of SiO_2 is often performed at a stage in the processing of integrated circuits where higher substrate temperatures are not desired. Frequently, about 7% phosphorus is simultaneously introduced into the SiO_2 film by the reaction of Eq. (18.26) in order to provide an easily flowing glass film to produce a planar insulating surface, i.e., planarization.[205] Another important oxidation reaction of producing SiO_2 is the following:

$$SiCl_4(g) + 2H_2(g) + O_2(g) \longrightarrow SiO_2(g) + 4HCl(g) \tag{18.27}$$

The final application here is the production of optical fiber for communication purposes. In this case, the SiO_2 forms as a cotton-candy-like deposit comprising soot particles less than 1000 Å in size.[205] These are subsequently consolidated by high-temperature sintering to produce a fully dense silica rod to be subsequently drawn into fiber. Whether SiO_2 film deposition or soot formation occurs is controlled by process variables promoting heterogeneous or homogeneous nucleation, respectively. For example, homogeneous soot formation occurs due to the high $SiCl_4$ concentration in the gas phase.[205] (See also Table 18.9.)[206,207,217–220]

TABLE 18.9 Oxidation and Hydrolysis Reactions in CVD with Some Examples of Thin-Film Deposition[198]

Reaction	Source	T_{dep} (°C)	Thin film	Reference
Oxidation	$Al(C_2H_5)_3 + O_2$		Al_2O_3	206
$(AX + O_2 \rightleftharpoons AO + XO)$	$Si_2H_4 + O_2$	<400	SiO_2	217
	$SiH_4 + PH_3 + O_2$	<400	$SiO_2P_2O_5$	218
	$SiH_4 + B_2H_6 + O_2$	<400	$SiO_2B_2O_3$	218
	$Si(OC_2H_5)_4 + O_3$	300–400	SiO_2	219
	$Zn(C_2H_5)_2 + O_2$	250–500	ZnO	206
	$TiCl_4 + O_2$		TiO_2	206
Hydrolysis	$AlCl_3 + CO_2 + H_2$	700–1200	Al_2O_3	220
$(AX + H_2O \rightleftharpoons AO + HX)$	$InCl_3 + H_2O$		In_2O_3	206
	$SnCl_4 + H_2O$	450	SnO_2	206
	$SiCl_4 + O_2 + H_2$	800	SiO_2	206
	$TiCl_4 + CO_2 + H_2$		TiO_2	206
	$TaCl_5 + H_2 + O_2$	600	Ta_2O_5	206
	$PbCl_2 + TiCl_4 + H_2O + O_2$	500	$PbTiO_3$	206
	$ZrCl_4 + CO_2 + H_2$	800–1000	ZrO_2	207

Courtesy of Institute of Physics Publishing, Bristol.

Hydrolysis. In this case a reaction occurs between the gaseous compound (external or in situ formed) $(CO_2 + H_2, H_2 + O_2,$ or $NO + H_2)$ and water vapor, resulting in the formation of a solid film, normally an oxide. A typical example is the deposition of alumina by the hydrolysis of aluminum trichloride, according to the following reactions[201]:

$$2AlCl_3(g) + 3CO_2(g) + 3H_2(g) \xrightarrow[\sim 1 \text{ torr}]{800-1300°C} Al_2O_3(s) + 6HCl(g) + 3CO(g) \quad (18.28)$$

$$2AlCl_3(g) + 3H_2O(g) \xrightarrow[H_2]{800-1400°C} Al_2O_3(s) + 6HCl(g) \quad (18.29)$$

In many instances, a reaction between the gaseous reactants and water occurs immediately on contact; in this situation, it is essential to mix them just near the substrate. Hydrolysis is employed to obtain both amorphous and epitaxial films such as insulating or garnet materials, respectively, in cold-wall or hot-wall reactors. (See also Table 18.9.)

Nitridation. Nitridation is the reaction between a gaseous reactant and NH_3, nitrogen, hydrazine (N_2H_4), or another nitrogen-bearing compound, resulting in the formation of a thin nitride film. These reactions are employed to deposit silicon nitride (Si_3N_4), oxynitrides $(Si_xO_yN_z$ and $Al_xO_yN_z)$, semiconducting compounds (AlN, BN, and GaN), metallic nitrides (TiN and TaN), and superconducting nitrides (NbN). (See also Table 18.10.)[206,207,221–227]

Typical examples are the deposition of Si_3N_4 from silicon tetrachloride and NH_3 as well as from dichlorosilane and NH_3; the latter is one of the most important industrial LPCVD processes:

$$SiCl_4(g) + 4NH_3(g) \xrightarrow[\text{up to 760 torr}]{850°C} Si_3N_4(s) + 12HCl(g) \quad (18.30)$$

$$3SiCl_2H_2(g) + 10NH_3(g) \xrightarrow[\text{low pressure}]{} Si_3N_4(s) + 6NH_4Cl(g) + 6H_2(g) \quad (18.31)$$

$$3SiCl_2H_2(g) + 4NH_3(g) \xrightarrow[\text{high } N_2 \text{ dilution}]{755-810°C} Si_3N_4(s) + 6HCl(g) + 6H_2(g) \quad (18.32)$$

TABLE 18.10 Nitridation and Chemical Transport Reactions in CVD with Some Examples of Thin-Film Deposition[198]

Reaction	Source	T_{dep} (°C)	Thin film	Reference
Nitridation	$Al(C_2H_5)_3 + NH_3$	1000–1100	AlN	207
$(AX + NH_3 \rightarrow AN + HX)$	$BCl_3 + NH_3$	900–1900	BN	207
	$GaCl + NH_3$	850	GaN	206
	$NbCl_5 + NH_3 + H_2$	950–1000	NbN	206
	$SiCl_2H_2 + NH_3$	750–900	Si_3N_4	221
	$SiH_4 + N_2H_4$	550–1150	Si_3N_4	206
	$TaCl_5 + NH_3$	900–1300	Ta_3N_5	207
	$TiCl_4 + N_2 + NH_3$	1100	TiN	207
	$ZrCl_4 + N_2$	2500	ZrN	207
Chemical transport	$Ga + AsCl_3 + H_2$	750–850	GaAs	222
	$Ga + AsH_3 + HCl + H_2$		GaAs	223
	$Ga + In + AsH_3 + HCl + H_2$		GaInAs	224
	$Ga + Al + AsH_3 + HCl + H_2$	670–770	$Al_xGa_{1-x}As$	225
	$In + PH_3 + HCl + H_2$	630–700	InP	226
	$Fe + Cl_2 + SiH_4 + H_2$	750–860	$FeSi_2$	227

Courtesy of Institute of Physics Publishing, Bristol.

Deposition of thin films is carried out in cold- or hot-wall reactors at atmospheric or low pressure.[198] The deposition temperature can be lowered to 400 to 600°C (750 to 1100°F) by using a high-frequency (13.56 MHz) plasma. Figure 18.26 exhibits a wide range of surface morphologies of α-Si_3N_4 coatings such as (a) rounded dome-like (or botryoidal), (b) botryoidal crystalline, (c) fine-faceted crystalline, and (d) coarse-faceted crystalline obtained at 1 atmospheric pressure over a range of deposition parameters such as temperature, concentration, and Si/(Si + N) ratio.[228]

Other examples of nitridation include the deposition of BN in a hydrogen atmosphere by the reaction of boron trihalide and NH_3 by the following CVD reactions:[201]

$$BCl_3(g) + NH_3(g) \xrightarrow{\;1300°C\;} BN(s) + 3HCl(g) \qquad (18.33)$$

$$BF_3(g) + NH_3(g) \xrightarrow[760\ torr]{1100-1200°C} BN(s) + 3HF(g) \qquad (18.34)$$

Low-temperature deposition of BN is made by using diborane as shown in the following:

$$B_2H_6(g) + 2NH_3(g) \xrightarrow[<1\ torr]{300-400°C} 2BN(s) + 6H_2(g) \qquad (18.35)$$

Typically, deposition of TiN uses $TiCl_4$ and a mixture of $N_2 + H_2$ or $NH_3 + H_2$, according to the following reactions:

$$TiCl_4(g) + \tfrac{1}{2}N_2(g) + 2H_2(g) \xrightarrow{\;850-1000°C\;} TiN(s) + 4HCl(g) \qquad (18.36)$$

$$TiCl_4(g) + NH_3(g) + \tfrac{1}{2}H_2(g) \longrightarrow TiN(s) + 4HCl(g) \qquad (18.37)$$

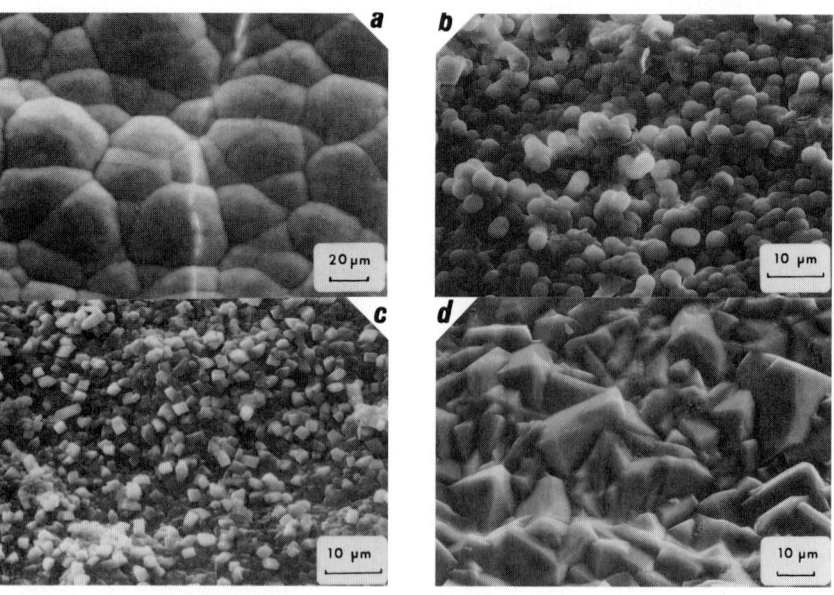

FIGURE 18.26   Surface grain morphology of CVD α-Si_3N_4 coatings: (*a*) botryoidal amorphous, (*b*) botryoidal crystalline, (*c*) fine-faceted crystalline, and (*d*) coarse-faceted crystalline.[228] (*Courtesy of D. G. Bhat and J. E. Roman.*)

Examples of titanium carbonitride deposition are given by the following reactions:

$$2TiCl_4(g) + CH_4(g) + N_2(g) \xrightarrow{1000°C} TiC_xN_y(s) + 4HCl(g) \qquad (18.38)$$

$$2TiCl_4(g) + CH_3CN(g) + 4\tfrac{1}{2}H_2(g) \xrightarrow{700-850°C} TiC_xN_y(s) + CH_3Cl(g) + 8HCl(g) \qquad (18.39)$$

Boride, Carbide, and Aluminide Formation.[201] *Boride* deposition of TiB_2, ZrB_2, rare earth borides, and so forth on steel, refractory metals, and alloys can be made using the following hydrogen reduction of the halides (also called *coreduction*)[229]:

$$MCl_4(g) + 2BCl_3(g) + 5H_2(g) \longrightarrow MB_2(s) + 10HCl(g) \qquad (18.40)$$

Good deposition of some borides is obtained under the conditions listed in Table 18.11. Among these boride coatings, much research work has been directed toward the deposition of TiB_2.

The deposition of TiB_2 or ZrB_2 coatings on various substrate materials is carried out by passing a mixture of $TiCl_4$ (or $ZrCl_4$) and BCl_3-H_2 gas mixture over a heated part placed in a vacuum chamber where the gas decomposition into atomic boron and titanium (or zirconium) and subsequent deposition of TiB_2 (or ZrB_2) occurs at the component surface when the appropriate deposition temperature and gas pressure are maintained (Table 18.11). It is necessary to adjust the gas flow so that the atomic ratios are the following:

$$B/Ti = 1 \text{ to } 2 \text{ and } H/Cl = 6 \text{ to } 10 \text{ for } TiB_2$$

$$B/Zr = 1.0, \text{ and } H/Cl = 20.0 \text{ for } ZrB_2$$

It should be further noted that when $B/(B + Cl) = \sim 0.4$, the TiB_2 deposit becomes dense with a {1010} or {1120} preferred orientation, which is often associated with a columnar appearance[229a] and microhardness values of about 3300 to 4500 HV (50-g load). For a good adherent deposit of TiB_2 on steel and cemented carbides, it is desirable to precoat the substrate with corrosion-resistant layers of cobalt and TiC, respectively.[229a-229c]

This process has several advantages, such as high purity of the deposit; a relatively high rate of deposition; close chemical composition control; high resistance to thermal shock, erosion, and/or corrosion at elevated temperatures; and large economic savings for the mass production of small parts.

Titanium diboride can also be deposited using diborane according to the reaction

$$TiCl_4(g) + B_2H_6(g) \longrightarrow TiB_2(s) + 4HCl(g) + H_2(g) \qquad (18.41)$$

Boron carbide is usually deposited from BCl_3 by the following reaction:

$$4BCl_3(g) + CH_4(g) + 4H_2(g) \xrightarrow[10-20 \text{ torr}]{1200-1400°C} B_4C(s) + 12HCl(g) \qquad (18.42)$$

However, boron carbide can also be deposited from diborane in a plasma, according to the reaction

$$2B_2H_6(g) + CH_4(g) \xrightarrow{400°C} B_4C(s) + 8H_2(g) \qquad (18.43)$$

Silicon carbide coatings can be made by the decomposition of methyl trichlorosilane (MTS):

TABLE 18.11 Chemical Vapor Deposition (CVD) Conditions for Some Borides[229]

Boride	Precursors	Temperature		Pressure		After*
		°C	°F	kPa	torr	
HfB$_2$	HfCl$_4$-BCl$_3$-H$_2$	1400	2550	0.4	3	Gebhardt and Cree (1965)
NbB$_2$	NbBr$_5$-BBr$_3$	850–1750	1560–3180	0.003–0.025	0.025–0.2	Armas et al. (1976)
Ni-B	Ni(CO)$_4$-B$_2$H$_6$-CO	150	300	87	650	Mullendore and Pope (1987)
SiB$_4$	SiH$_4$-BCl$_3$-H$_2$	800–1400	1470–2500	6.5–80	50–600	Dirkx and Spear (1984)
SiB$_x$	SiBr$_4$-BBr$_3$	975–1375	1790–2500	0.007	0.05	Armas and Combescure (1977)
TaB$_2$	TaBr$_5$-BBr$_3$	850–1750	1560–3180	0.003–0.025	0.025–0.2	Armas et al. (1976)
	TaCl$_5$-B$_2$H$_6$	500–1025	930–1875	100	760	Randich (1980)
TiB$_2$	TiCl$_4$-BCl$_3$-H$_2$	1200–1415	2200–2580	0.4–2	3–15	Gebhardt and Cree (1965)
	TiCl$_4$-B$_2$H$_6$	600–900	1100–1650	100	760	Pierson and Mullendore (1980)
	TiCl$_4$-BCl$_3$-H$_2$	750–1050	1380–1920	100	760	Caputo et al. (1985)
	TiCl$_4$-BCl$_3$-H$_2$	1200	2200	6.5	50	Desmaison et al. (1987)
ZrB$_2$	ZrCl$_4$-BCl$_3$-H$_2$	1400	2550	0.4–0.8	3–6	Gebhardt and Cree (1965)

* All references cited in this table can be found in Ref. 229 at the end of this chapter.

$$CH_3SiCl_3(g) \xrightarrow[\text{10--50 torr}]{\text{900--1400°C}} SiC(s) + 3HCl(g) \qquad (18.44)$$

Other precursor combinations include $SiCl_4/CH_4$, $SiCl_4/CCl_4$, SiH_2Cl_2/C_3H_8, and $SiHCl_3/C_3H_8$, usually in an H_2 atmosphere.

Titanium carbide can be deposited by reacting titanium chloride with a hydrocarbon in a hydrogen atmosphere according to the reaction

$$TiCl_4(g) + CH_4(g) \xrightarrow[\text{H}_2,\ 1\ \text{atm}]{\text{900--1100°C}} TiC(s) + 4HCl(g) \qquad (18.45)$$

Other carbon sources include toluene ($CH_3C_6H_5$) and propane (C_3H_8).

Molybdenum carbide can be deposited on diamond particles by a reaction of molybdenum chlorides with diamond.

Chromium carbide can be deposited by reacting chromium chloride with steel in an H_2 atmosphere, according to the reaction

$$(1-x)CrCl_2(g) + 7(1-x)H_2(g) + 3(x\text{Fe-C})(s) \longrightarrow (Cr_{1-x}Fe_x)_7C_3(s) + 14HCl(g) \qquad (18.46)$$

where $x = 0$ to 0.6; Fe-C = carbon dissolved in the austenite.

Nickel aluminide can be deposited by a reaction between $AlCl_3$ in the vapor phase and the nickel or superalloy substrate.

Chemical Transport.   Chemical transport is characterized by a reverse equilibrium chemical reaction in the source and substrate regions maintained at different temperatures within a single reactor (see Table 18.10). Figure 18.27 shows the schematic chemical transport of the epitaxial growth of GaAs and InP. Here Ga or In is transported from the liquid source using HCl according to the equations

$$Ga(l) + HCl(g) \underset{T_3}{\overset{T_1}{\rightleftharpoons}} GaCl(g) + \tfrac{1}{2}H_2(g) \qquad (18.47)$$

$$In(l) + HCl(g) \underset{T_3}{\overset{T_1}{\rightleftharpoons}} InCl(g) + \tfrac{1}{2}H_2(g) \qquad (18.48)$$

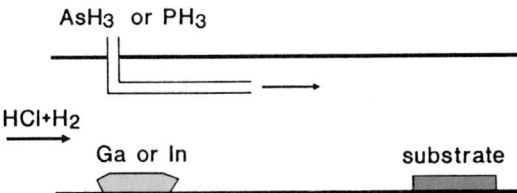

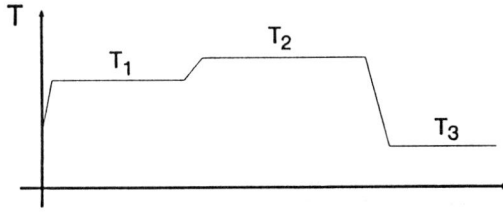

FIGURE   18.27 Schematic diagram of epitaxial growth by chemical transport of GaAs and InP and qualitative temperature profile.[198] (*Courtesy of Institute of Physics Publishing, Bristol.*)

where (l) represents the liquid state. At the source, maintained at higher temperature T_1, the source material transforms into a gaseous metal monochloride and then precipitates on the substrate surface, which is maintained at a lower temperature T_3.[198]

Disproportionation. The disproportionation (or decomposition) reaction is characterized by the dissociation of a reactant species due to its instability at a lower temperature and the formation of another, more stable, higher-valence chemical species, producing simultaneously the elemental form of the reactant to be deposited. Three typical reactions are the disproportionation of GaCl, GeF$_2$, and AlCl:

$$3GaCl(g) \underset{\text{high } T}{\overset{\text{low } T}{\rightleftharpoons}} 2Ga(s) + GaCl_3(g) \tag{18.49}$$

$$2GeI_2(g) \underset{\text{high } T}{\overset{\text{low } T}{\rightleftharpoons}} Ge(s) + GeI_4(g) \tag{18.50}$$

$$3AlCl(g) \underset{\text{high } T}{\overset{\text{low } T}{\rightleftharpoons}} 2Al(s) + AlCl_3(g) \tag{18.51}$$

Reaction (18.49) is used, for example in the deposition of GaAs. GaCl, GeI$_2$, and AlCl are produced at higher temperature by the back reactions. The advantages of these reactions are reaction reversibility, in situ reactant formation, and in situ vapor etching. Disadvantages include the reactor complexity, contamination effects, the use of halides as disproportionation agents, and so forth. Deposition based on disproportionation is usually carried out in a temperature-gradient, hot-wall, multi-zoned system.[198]

Catalysis. The velocity of a chemical reaction can be augmented by a catalyst. PH$_3$ and AsH$_3$ are two important source gases in the deposition of III-V films. It should be noted that the thermal decomposition of these hydrides is catalyzed in the presence of crystalline material with the same anion as the hydride (Fig. 18.28). For example, the complete decomposition of AsH$_3$ in the absence and presence of GaAs occurs above 750°C (1380°F), and as low as 550°C (1020°F), respectively. Another example is hydrogen reduction of WF$_6$:

$$WF_6(g) + 3H_2(g) \longrightarrow W(s) + 6HF(g) \tag{18.52}$$

In the forward reaction, dissociation of H$_2$ into atomic hydrogen occurs at catalytically active surfaces such as freshly deposited W film by the Si reduction. This leads to thicker deposition of W films.[230] An example of a deposition reaction stimulated by a Pt homogeneous catalyst is the Si$_3$N$_4$ film formation according to the following:

$$3SiH_4(g) + 4NH_3(g) \xrightarrow[600°C]{Pt} Si_3N_4(s) + 12H_2(g) \tag{18.53}$$

Synthesis. Synthesis is based on the formation of a solid film arising from the reaction between two or more gaseous compounds. Many III-V and II-VI thin films are deposited by synthesis from metalorganic compounds and hydrides by MOCVD. A typical example is the homoepitaxial (III-V semiconductor) growth/deposition on a GaAs substrate by the interaction between Ga(CH$_3$)$_3$ and AsH$_3$, and the heteroepitaxial (II-VI semiconductor) growth of ZnSe on GaAs by the interaction between Zn(CH$_3$)$_2$ and H$_2$Se, according to the equations[155]

$$Ga(CH_3)_3(g) + AsH_3(g) \longrightarrow GaAs(s) + 3CH_4(g) \tag{18.54}$$

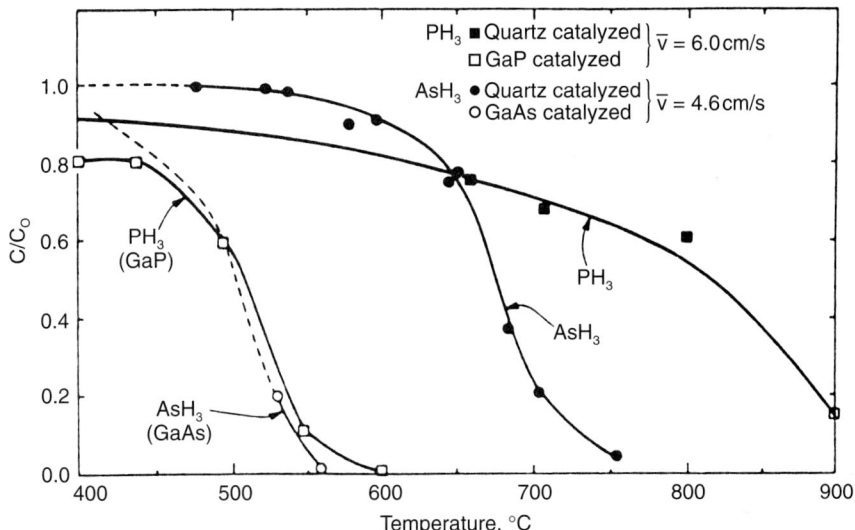

FIGURE 18.28 Decomposition of PH_3 and AsH_3 in hydrogen as a function of temperature in a quartz vessel with and without a GaP or GaAs coating. (*After P. D. Dapkus and J. J. Coleman, in III-V Semiconductor Materials and Devices, R. J. Malik, Ed., North-Holland, Amsterdam, 1989. Reprinted by permission of Elsevier Science.*)

$$Zn(CH_3)_2(g) + H_2Se(g) \longrightarrow ZnSe(s) + 2CH_4(g) \tag{18.55}$$

Photolysis. In photolysis, absorption of ultraviolet light by the precursor molecules occurs, which leads to the dissociation and subsequent deposition. Metal carbonyls and alkyls, as well as hydrides and halogenides, are employed to deposit metals and semiconductors by photolysis. The simplest photolysis reaction of PCVD is the one-photon photoabsorption dissociation reaction. Examples of the photolysis (or photodissociation) of dimethyl cadmium and trimethyl gallium by a one-photon photoabsorption dissociation reaction are given by[231]

$$(CH_3)_2Cd + hv \rightarrow CH_3 + CH_3Cd \rightarrow 2CH_3 + Cd \tag{18.56}$$

$$(CH_3)_3Ga + hv \rightarrow 2CH_3 + CH_3Ga; CH_3Ga + hv \rightarrow CH_3 + Ga \tag{18.57}$$

where hv is the photon energy. The primary photoproduct is [CH_3Cd], which is subjected to a secondary thermal dissociation to Cd and CH_3. The other group II metal dialkyls Me_2Zn and Me_2Hg have the tendency to photodissociate by a similar mechanism.[155]

Combined Reactions.[198] (a) *Pyrolysis + reduction reactions* take place simultaneously, for example, in the formation of (1) SiC films by C_3H_8 pyrolysis and $SiCl_4$ reduction in an H_2 atmosphere; (2) TiC film by CH_4 pyrolysis and $TiCl_4$ reduction [Eq. (18.45)]; and (3) AlN film by NH_3 pyrolysis and $AlCl_3$ reduction:

$$C_3H_8(g) + SiCl_4(g) + 2H_2(g) \longrightarrow 3SiC(s) + 12HCl(g) \tag{18.58}$$

$$AlCl_3(g) + NH_3(g) \longrightarrow AlN(s) + 3HCl(g) \tag{18.59}$$

(b) *Oxidation + nitridation reactions* occur when oxynitrides are deposited. (c) *Oxidation + hydrolysis reactions* occur, for example, in the deposition of epitaxial films of magnesium ferrites[206]:

$$MgCl_2(g) + 2FeCl_2(g) + 3H_2O(g) + \tfrac{1}{2}O_2(g) \longrightarrow MgFe_2O_4(s) + 6HCl(g) \quad (18.60)$$

(d) *Pyrolysis, chemical transport, and reduction reactions* are usually involved when hydrides and halogenides are used. Thus the growth of many III-V and II-VI compounds is based on a combination of group V or group VI hydride decomposition and group III or group II halide reduction reactions. One example is the growth of GaAs[229]:

Pyrolysis: $\qquad\qquad\qquad AsH_3(g) \longrightarrow As(s) + H_2(g)$ $\qquad\qquad$ (18.61)

Chemical transport: $\quad HCl(g) + Ga(l) \longrightarrow GaCl(g) + \tfrac{1}{2}H_2(g)$ $\qquad\qquad$ (18.62)

Reduction: $\qquad\qquad\qquad GaCl(g) + \tfrac{1}{2}H_2(g) \longrightarrow Ga(s)$ $\qquad\qquad$ (18.63)

Hence, in the presence of As vapor, the overall reaction for the GaAs growth can be explained by

$$2H_2(g) + 4GaCl(g) + As_4(g) \longrightarrow 4GaAs(s) + 4HCl(g) \quad (18.64)$$

18.4.4 Advantages

The choice of CVD coating over the more conventional coating techniques is generally based on the following unique characteristics[199–201]:

1. Unlike most PVD and plating processes, it is not a line-of-sight process.
2. High-purity films, typically 99.99 to 99.999% (also semiconductor films of ppm to ppb purity) are achieved because of the availability of reactant gases in high-purity form.
3. Near-theoretical density (typically, >99%) or controlled density of deposit is readily achieved.
4. Ability to grow atomically abrupt interfaces and smooth surfaces.[232]
5. Ability to create metastable phases.[232]
6. Deposition of refractory materials at temperatures far below their melting points or sintering temperatures. For example, W can be deposited at temperatures as low as 280°C by H_2 reduction of WF_6.
7. Economical for the deposition of semiconductor and insulator films in the integrated circuit industry, because many substrates (or wafers) can be coated simultaneously.
8. The required equipment is relatively simple and does not need an ultrahigh vacuum and electrically conductive substrate.[201]
9. Its flexibility permits many composition variations during deposition (to achieve graded deposition or, literally, hundreds of distinct layers), and the codeposition of the compound is easily obtained.[201,229]
10. Deposition is readily achieved for conformal (or uniform) coverage of the substrate with deep recesses, steps, vias, trenches, blind holes, large L/D tubes, high-aspect-ratio holes, porous bodies, and other complex-shaped parts.

11. It is well adapted to the deposition of the doped epitaxial coatings as well as to many process variations.

12. Its high throughput and deposition rate allow one to readily achieve thick coatings (in some cases, centimeters thick).[201]

13. Utilization of relatively high reactant gas (e.g., O_2 or H_2) partial pressure during film growth.[232]

14. Amenability to large-scale and large-area depositions.[232]

15. Good control of film stoichiometry and thickness.[232]

16. The process is usually competitive and, at times, even more economical than other coating processes.

17. Highly preferred grain orientations and fine-grained, equiaxed deposits can be obtained.[229]

18. Deposition at atmospheric pressure is possible in a relatively simple apparatus. (However, reduced pressure renders better results, including reduced risk of hydrogen embrittlement.)

19. Generally good bonding to a substrate and radiation-damage-free deposition are obtained.

20. Coatings on substrates of complex shapes and on particulate materials can be deposited in a gas-fluidized bed.[229]

18.4.5 Disadvantages

1. In the case of hazardous (explosive or pyrophoric) or extremely toxic chemical precursors, a closed system is a prerequisite.

2. Since it is most versatile at high temperatures, $\geq 600°C$ (1100°F), its applications are confined to substrates that are thermally stable at such temperatures. (The use of plasma and laser activation and metalorganic CVD processes partially surmounts this problem, and allows the deposition at very low temperatures.[229])

3. Many reactions either leave solid byproducts or generate solid byproducts with neutralizing solutions. These byproducts can be corrosive and toxic, and require the careful selection of disposal procedures that incur extra costs.

4. It is a labor-intensive process.

5. There is a need of high energy, especially with high deposition temperatures.

6. Sometimes efficiency of the process is low, leading to high costs.

7. It involves higher costs of capital equipment, operations, and (some) reactants, and lower utilization of material.

8. The reactant depletion with the advancing deposition may cause coating inhomogeneities and thickness variations and, in some cases, difficulty in maintaining the substrate at a uniform temperature.

18.4.6 Applications

Wear-, Erosion-, and Corrosion-Resistant Applications.   For the successful application of CVD, one must first consider service conditions such as wear, erosion, friction, the existence of a corrosive environment, temperature, and the nature of any materials to be used with the part. The second important factor to be considered is the compatibility of coating and base materials. These include the relative thermal coefficient of expansion during cooling from the deposition temperature and the possible chemical and metallurgical reactions between the two during the deposition process, causing the formation of soft, ductile, hard, or brittle interlayers.[201]

Table 18.12 lists the most important properties of hard CVD-coating materials to be considered in selecting one of them for a given application. Some materials such as TiB_2, TiC, and SiC offer very low wear rates. MoS_2 and WSe_2 are used for their superior dry lubricant properties. Polycrystalline diamond, the hardest material, may offer the best wear and erosion resistance material once the deposition problems are solved. The CVD diamond and DLC* are used in ultrahard tool and computer hard disk coatings, electronic and semiconductor devices, optics, and so forth.[232a,232b]

Table 18.13 lists specific production applications for the wear, erosion, and corrosion resistance offered by CVD coatings.[201] Table 18.14 is a summary of some results reported for CVD-TiC/TiN-coated tools.[86]

Cutting Tool Industry. CVD-coated cemented carbides constitute over 80% of all coated carbide metal cutting tools in use. CVD coatings such as TiC, TiN, TiC_xN_y, Al_2O_3, and Al-oxynitrides have a large and still increasing importance (thickness 5 to $10\,\mu m$). TiC coatings are variously used on cemented tungsten carbide tools and both TiN and TiC_xN_y are used on high-speed steels and cemented carbide tools.[†] Moderate-temperature TiBCN coatings with low-carbon content on cemented carbide tools can be used for the milling of steel.[232c] Table 18.15 lists the coating materials for cemented carbide tools for different applications.

Figure 18.29 shows the surface and cross-section microstructures of a fine-grained and coarse-grained TiC coating on cemented carbide. It is reported that a CVD-TiC coating with fine, equiaxed grains, a strong (200) preferred orientation, and a minimum of eta-phase is mostly desirable for the cemented carbide cutting tools. This can be accomplished by a proper selection of the important CVD parameters.[233] Figure 18.30*a* shows a typical microstructure of a cemented tungsten carbide tool with multilayer coatings of TiN/TiCN/TiC.[94]

Most coatings on high-speed steel are performed by PVD (Table 18.5), particularly for brazed tools, end mills, reamers, threading inserts, and gear-cutting tools. Both CVD and PVD are used for grooving tools and drills, and milling inserts. For high-speed steels, a combination of CVD-TiC and -TiN coatings gives the best results.

The materials described in Table 18.12 can be employed as multilayer hard coatings to exploit the advantages of the strongest characteristics of each individual layer of material. The multilayer coatings are of nominal total thickness of about $10\,\mu m$ ($400\,\mu in.$) and use TiN for increased lubricity and galling resistance, TiC and TiC_xN_y, for improved abrasion resistance, Al_2O_3 for chemical inertness and thermal insulation, and occasionally HfN and ZrN for electrical insulation and optical transparency. The choice of an optimal combination of materials is a function of the type

---

* The CVD diamond needs the presence of atomic hydrogen (to selectively eliminate graphite and activate and stabilize the diamond structure) and oxygen and a heat energy source (to dissociate the hydrogen). The basic CVD reaction for diamond deposition involves the decomposition of a hydrocarbon such as CH_4; this reaction can be improved by microwave plasma, thermal methods (hot filament), or plasma arc. A common deposition method for DLC is a high-frequency rf gas discharge (13.56 MHz) generated in a mixture of H_2 and a hydrocarbon such as CH_4, n-C_4H_{10}, or C_2H_2.

† TiN is also used as a decorative coating, a thin film resistor, a diffusion barrier in semiconductor metallization schemes, a contact layer for Si, a gate electrode for ICs, and a solar energy absorber and a transparent heat mirror.[232c]

TABLE 18.12 Selected Wear and Corrosion Properties of CVD-Coating Materials[201]

Material	Hardness		Thermal conductivity, W/m·K	Coefficient of thermal expansion at 25°C (77°F), 10^{-6}/K	Remarks
	GPa	10^6 psi			
TiC	31.4	4.5	17	7.6	High wear and abrasion resistance, low friction, high hardness, suitable for reducing mechanical properties and abrasive wear; susceptible to chemical attack; not a good diffusion barrier
TiN	20.6	3.0	33	9.5	High lubricity; stable and inert; excellent diffusion barrier, low coefficient of friction; used for gear components and tube and wire-drawing dies
TiCN	24.5–29.4	3.5–4.3	20–30	8	Stable lubricant; excellent protection against abrasive wear; used to coat tools and dies for the processing of ceramics, graphite, and filled plastics
Cr_7C_3	22.1	3.2	11	10	Resists oxidation to 900°C (1650°F); excellent corrosion and oxidation resistance; used in combination with TiC and TiN base layer
SiC	27.4	4.0	125	3.9	High hardness and conductivity, shock resistant, good oxidation resistance; used in coating of graphite and carbon to impart wear and oxidation resistance
TiB_2	33.0	4.7	25	6.6	High hardness and wear resistance, good protection against abrasion
Al_2O_3	8.8	2.7	34	8.3	Oxidation resistant, very stable
DLC	29–49	4.2–7.1	200	—	Very hard, high thermal conductivity, chemical inertness, variable electrical conductivity
Diamond	98	14.2	180	2.9	Extreme hardness and high thermal conductivity*

* CVD-diamond is cut and processed into cutting tools.
Reprinted by permission of ASM International, Materials Park, Ohio.

TABLE 18.13 Wear-, Erosion-, and Corrosion-Resistance Applications of CVD[201]

Metal forming (noncutting)
  Tube and wire-drawing dies (TiN)
  Stamping, chamfering, and coining tools (TiN)
  Drawing punches and dies (TiN)
  Deep-drawing dies (TiC)
  Sequential drawing dies
  Coating on dressing sticks for grinding wheels (B_4C)
Ceramic and plastic processing
  Molding tools and dies for glass-filled plastics (TiCN)
  Extrusion dies for ceramic molding (TiC)
  Kneading components for plastic mixing (TiC)
Chemical and general processing industries
  Pumps and valve parts for corrosive liquids (SiC) and abrasive liquids (TiB_2)
  Valve liners (SiC)
  Positive-orifice chokes (SiC, TiB_2)
  Packing sleeves, feed-screws (TiC)
  Thermowells (SiC, Al_2O_3)
  Heating elements (SiC)
  Abrasive-slurry transport (WC)
  Sandblasting nozzles (TiC, B_4C, TiB_2)
Solder-handling in printed-circuit processing (TiC, TiN)
Textile-processing rolls and shafts (Al_2O_3, TiC, WC)
Paper-processing rolls and shafts (TiC)
Valves for coal-liquefaction components (TiB_2)
Cathode coating for aluminum production (TiB_2)
Oxidation-resistant coatings for carbon-carbon composites (SiC)

Machine elements
  Gear components (TiN)
  Coating on stainless steel spray-gun nozzles (TiC)
  Components for abrasive processing (TiC)
  Coating on ball bearings (TiC)
  Turbine blades (SiC, TiC)
Nuclear
  Coatings for neutron flux control in nuclear reactors (B_4C)
  Coatings for shielding against neutron radiation (B_4C)
  Coatings for fusion reactor application (SiC)
  Nuclear waste container coatings (SiC)
Instruments
  Radiation sensor
  Thermionic cathodes (W-Th)
  Target coatings for x-ray cathodes (W-Re)
Metal coating
  NbC and TaC for the protection of Nb and Ta metals

Reprinted by permission of ASM International, Materials Park, Ohio.

TABLE 18.14 Application Results for CVD-Coated High-Speed Steel Cutting and Metal Forming Tools[86]

Tool type	Uncoated life	Coated life	Improvement, %
Tap	3,000	9,000	300
Cut-off tool	150	1,000	630
Class "C" hob	1,500	4,500	300
Form tool	4,950	23,000	460
Stamping punch	500,000	3,000,000	600
Drill	1,000	4,500	400

Reprinted by permission of ASM International, Materials Park, Ohio.

TABLE 18.15   Coating Materials for
Cemented-Carbide Tools[201]

Unalloyed steel	TiCN, TiAlN
Stainless steel	TiCN, TiAlN
Cast iron	TiAlN
Al-wrought alloys	TiN
Al-cast alloys	TiCN
Copper	CrN
Brass	TiCN
Bronze	TiCN

Courtesy of D. G. Bhat and P. F. Woerner.

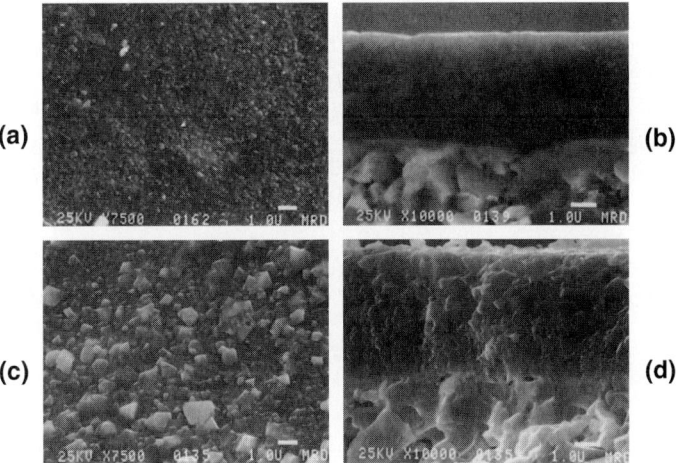

FIGURE 18.29   Scanning electron micrographs of TiC coating on cemented carbide: (*a*) surface and (*b*) cross section of fine-grained (<0.2 μm) TiC; (*c*) surface and (*d*) cross section of coarse-grained (≥1 μm) TiC.[233] (*Courtesy of T. Cho, D. G. Bhat, and P. F. Woerner.*)

of machining operation involved, the material to be machined, and other factors. Table 18.16 lists the criteria for such a selection.

TiN coating is usually combined with a very thin undercoating of TiC or TiC_xN_y to improve adhesion (Fig. 18.30*a*). (These coatings are conducting.) Al_2O_3 coatings are the common choice in high-speed machining applications where oxidation resistance and elevated-temperature stability are critical factors. Like TiN, Al_2O_3 deposition is used on an intermediate TiC layer.[201]

The use of monolithic ceramic cutting tools is increasing because of their good thermal stability and their extreme resistance to deformation at high machining speeds. Silicon nitride is the primary material for high-speed machining of cast iron due to its superior thermal stability, hardness, and wear resistance. When CVD-coated with TiN, it is used for the machining of steels, because TiN gives an additional increase in chemical resistance.

TABLE 18.16 Criteria for Selecting Coating Materials for Cutting Tools[201]

Property	Optimum coating
Oxidation and corrosion resistance; high-temperature stability	Al_2O_3, TiN, TiC
Crater-wear resistance	Al_2O_3, TiN, TiC
Hardness and edge retention	TiC, TiN, Al_2O_3
Abrasion resistance (flank wear)	Al_2O_3, TiC, TiN
Low coefficient of friction and high lubricity	TiN, Al_2O_3, TiC
Fine grain size	TiN, TiC, Al_2O_3

Source: D. G. Bhat and P. F. Woerner, *JOM*, February 1986, p. 68. Courtesy of D. G. Bhat and P. F. Woerner.

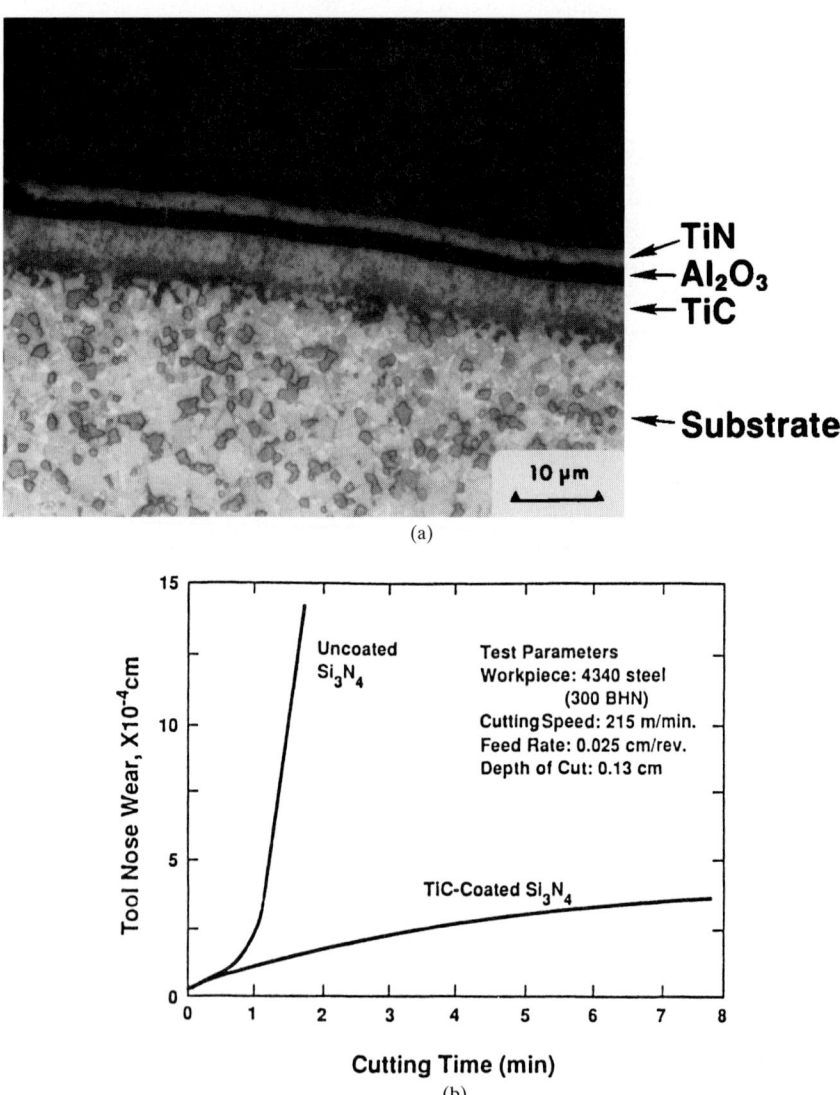

FIGURE 18.30 (*a*) Microstructure of a typical cemented carbide cutting tool.[229] (*b*) Wear of coated and uncoated ceramic tools.[201] [(*a*) Courtesy of D. G. Bhat; (*b*) Reprinted by permission of ASM International, Materials Park, Ohio.]

Other ceramic cutting-tool materials are Al_2O_3, Sialon, alumina-carbide composites, and a composite of Si_3N_4 reinforced with SiC whiskers. The last composite material can be made by chemical-vapor infiltration (CVI) and provides high strength and toughness as shown in Table 18.17.[201] Figure 18.30b shows the appreciable increase in performance of coated ceramic tools over uncoated ones.[201] These coated ceramics find applications in the machining of hard materials such as cast iron and superalloys.

Boride Coatings. CVD-TiB_2 coatings are variously used on cemented carbide cutting inserts,[229c] on graphite electrodes in aluminum reduction cells, and on letdown valves in coal conversion reactors.[233a,233b] ZrB_2 coatings on graphite are sometimes used as a spectrally selective surface at elevated temperatures.[233c]

Monolithic Metallic Structures. Made of refractory materials (such as W, Nb, Re, Ta, Mo, and Ni) and ceramics, these are produced by the CVD process for a variety of applications (see Table 18.18) as rods, tubes, manifolds, nozzles, ordnance

TABLE 18.17 Selected Properties of Commercial Ceramic Cutting Tools[201]

Material	Hardness, GPa (KHN)	Fracture toughness, MPa·m	Rupture modulus, MPa	Thermal conductivity, W·mK	Coefficient of thermal expansion, m/m°C × 10⁻⁶
Al_2O_3*	15.6	2.9	280	32.3	8.22
Al_2O_3/TiC[†]	16.2	3.5	450	32.3	8.7
Al_2O_3/SiC[‡]	17.0	6.0	675	35.2	7.35
Si_3N_4/20% SiC whiskers[§]	13.0	6.0	1050	39.5	3.0

* Kennametal designation K060.
[†] NGK designation NTK-HC2.
[‡] Greenleaf designation WG-300.
[§] GTE Valenite designation Quantum 6.

TABLE 18.18 Free-Standing Components Produced by CVD[201]

High-strength structural parts of nickel alloyed with small amounts of boron
Shapes such as plates, rods, and tubes
Furnace muffles (tungsten)
Hollow spheres with thin walls
Boats and crucibles for liquid-phase and molecular beam epitaxy (pyrolitic graphite), and crucibles for silicon single-crystal processing (silicon nitride)
Electrodes for plasma etching (pyrolytic graphite)
Trays for silicon-wafer handling (pyrolytic graphite)
Heating elements for high-temperature furnaces (Pyrolytic graphite)
Aircraft disk brakes (carbon-carbon)
Re-entry heat shields, rocket nozzles, and other aerospace components (carbon-carbon)
High-temperature turbine blades, and components for internal combustion engines (silicon nitride)
Heart valves and dental implants (pyrolytic carbon*)

Note: Pyrolytic carbon is a generic term commonly used for CVD carbon but usually applied to a low-temperature form (900 to 1400°C, or 1650 to 2550°F) with limited preferred orientation. Pyrolytic graphite is the higher-temperature product (1900 to 2300°C, or 3450 to 4170°F) with a high degree of preferred orientation in the "C" direction.
Reprinted by permission of ASM International, Materials Park, Ohio.

parts, and thrust chambers. They are normally coated on a disposable mandrel of graphite, copper, or molybdenum, which is then removed by machining or chemical etching.[201] Thus CVD Re can be used to fabricate thin-walled, small-diameter, or complex-shaped parts, as well as to coat graphite/carbon, ceramic, and metallic parts.[234]

Ultrafine Powders. These are produced by CVD and are characterized by high purity, small particle size, great uniformity, freedom from aggregation, and lower sintering temperatures, which are of importance in the manufacture of high-quality hot-pressed or sintered ceramic structures with excellent mechanical and electrical properties. Moreover, the sintering temperatures needed for CVD powders are lower than those needed for conventional powders.[201] These powders are now used to produce a wide range of ceramics such as structural components, pigments, catalyst supports, electro-optic and magnetic devices, drug delivery carriers, and so forth.[201]

The important CVD powders made experimentally or on a production-run fall into the following categories:

1. Cubic β-SiC powder is produced mainly by gas-phase processes, e.g., by the reaction of silicon tetrachlorides ($SiCl_4$) with hydrocarbons or pyrolysis of trimethylsilane [$(CH_3)_3SiCH$], trimethylchlorsilane [$(CH_3)_3SiCl$], or tetramethyl-silane[$(CH_3)_4Si$]. The reaction temperatures are between 800 and 1500°C (1470 and 2730°F); either Ar or H_2 is employed as a carrier gas. Particle size and morphology are functions of reaction temperature and the composition of the gas phase. For example, at 1000°C, nanosize particles 10 to 30 nm in size are produced, and below 1100°C amorphous powders are obtained.[235]

2. Submicron β-SiC powder can be produced by reacting silane and acetylene in a 10- to 50-watt continuous-wave CO_2 laser beam.

3. Amorphous Si_3N_4 powder is produced by laser CVD from halogenated silanes and NH_3 with an inert sensitizer such as SF_6.[236] Another production route is the reaction between silicon halides and NH_3 at 1000°C (1832°F) in an rf plasma with a mean particle size of 0.05 to 0.1 μm. CVD Si_3N_4 is of industrial importance due to the improved performance of "bulk" Si_3N_4 such as sintered, reaction-bonded, or hot-pressed material. Its hardness is controllable between 2500 to 3200 kg/mm^2, HV_{500g}.

4. AlN submicron powder is produced from aluminum alkyls and from the reaction of Al powder, Li salt, and nitrogen at 1000°C (1832°F). MgO powder is produced from Mg vapor and O_2 at 800°C (1470°F), and tungsten carbide.[237]

5. Fe, Ni, Co, Mo, and W powders produced by the pyrolysis of the respective metal carbonyls or halides. For example, W is produced from WF_6 to a purity greater than 99.9995%.

6. Fine-grained (1000 Å) refractory (such as W-C and Ta-C alloys) and ceramic (such as TiB_2, and SiC) coatings with superior physical and mechanical properties are made possible by controlled nucleation thermochemical deposition (CNTD), a variant of the CVD process.[238]

CVD ceramic powders such as SiC and Si_3N_4, can be used in applications such as gas turbines, turbochargers, reciprocating engines, bearings, machinery, and process equipment.[201]

CVD-Coated Powders. Both metallic and ceramic powders (with ≥5-μm or 200-μin. particle size) can be uniformly coated with Ni or Fe by CVD using the fluidized bed process. A typical application is the coating of W particles to promote the sintering process in P/M applications. This causes a reduction in sintering temperature, sintering time, and grain growth; minimal contamination; and improvement in properties.[239] Ceramic powders coated with their sintering aids such as zirconia coated with yttria stabilizer, tungsten carbide coated with cobalt or nickel, alumina abrasive powders coated with a relatively brittle second phase such as $MgAl_2O_4$, and plasma spray powders can be produced without the segregation of alloying elements.

Fibers. Boron and SiC fibers are now commercially available. Boron fibers are produced by the reduction of BCl_4 with H_2 at 1300°C (2370°F). Actually, the boron fiber consists of a boron envelope around a tungsten boride core, which typically holds 5% of the fiber cross section. The fiber is characterized by high strength, high modulus, low density, a tendency to grain growth at high temperatures, high reactivity with many metals, and high cost.

CVD SiC fibers are usually produced by the reaction of silane and a hydrocarbon in a tubular glass reactor. The substrate is a carbon (or tungsten) monofilament wire that is precoated with a 1-μm (40-μin.) layer of pyrolytic graphite to ensure a constant resistivity and a smooth deposition surface.

Both CVD boron and SiC fibers have similar properties, except that SiC is more refractory and less reactive than boron, which offers a potential cost advantage. The fibers maintain most of their mechanical properties when exposed to elevated temperatures in air, up to 800°C (1470°F), up to 1 hr. SiC fiber-reinforced SiC is used as a structural material in engineering and space technology.[235] Most applications of CVD SiC fibers are still in the experimental stage, and include the following:

1. Reinforcement for ceramic and polymer composites.

2. Reinforcement for metal-matrix composites with such metals as Ti, titanium aluminide, Al, Mg, and Cu.

   Applications for these fibers have been reported mostly in advanced aerospace programs and include drive shafts, fan blades, and other parts.[201]

CVD-Coated Fibers. Generally, CVD coating is applied to the inorganic fibers and acts as a diffusion barrier and prevents the diffusion reaction at elevated temperatures. Common diffusion barrier materials include pyrolytic carbons, SiC, TiB_2, BN, TiN, and rare-earth sulfides.

Whiskers. CVD whiskers of refractory compounds such as carbides and nitrides of Si, Ti, and Zr and oxides of Al, Si, and Zr are grown by vapor-solid (VS) and/or vapor-liquid-solid (VLS) mechanism(s). For example, the formation of TiC whiskers in the presence of Ni as a catalyst substrate and a gaseous mixture of H_2, $TiCl_4$, and CH_4 in a CVD reactor occurs by the VLS mechanism at temperatures far below the ternary eutectic melting point in the Ti-Ni-C system.[239a,239b]

SiC-whisker particle-reinforced Al_2O_3 ceramic cutting tools find applications in the machining of aerospace alloys such as stainless steels, Inconel, and other exotic alloys with exceptional fracture toughness and high machining speed.[239a,239b]

Electronic Applications. CVD is a primary process in the production of thin films of three types of materials, namely, semiconductors, conductors, and insulators. Table 18.19 lists the electrical characteristics of these materials. The ability to grow thin, epitaxial films and multilayers on numerous semiconductor substrates has led to the development of electronic devices [such as discrete devices and integrated circuits (ICs)]. Table 18.20 defines the functions of these devices. The important semiconductor materials in the production or development are silicon (ultrapure single-crystal ingot or wafer, epitaxial and polysilicon), Ge, III-V and II-VI compounds, SiC, and diamond films.

TABLE 18.19 Electrical Characteristics of Materials[201]

Device	Typical material	Electron mobility	Resistivity, $\Omega \cdot cm$
Conductor	Cu Ag Au W Silicides	Free to move	10^{-5}–10^{-6}

Semiconductor	Si
	Ge
	III-V and II-VI
	  compounds
	Diamond
	SiC
		Partially able to
		  move
			10^{-2}–10^{9}

Insulator (dielectrics)	Al_2O_3 SiO_2 Si_3N_4 Glass	Bound to nucleus	10^{12}–10^{22}
Ferroelectrics*	PZT SBT	Atoms displaced	1.25 (at ~5 V) 1.2 (at ~5 V)

  * After Joe Cuchiaro.

TABLE 18.20 Devices and Their Functions[201]

Devices	Functions
Transistors	Active semiconductor switching (on/off) devices that provide power amplification and have three or more terminals.
Resistors	Passive devices that offer resistance to the flow of electric current in accordance with Ohm's law: $R = V/I$, where R is resistance, V is applied voltage, and I is current. They are usually made of nichrome, Ta, Ti, or W.
Thermistor	They are made of semiconductor material with a resistance that varies rapidly and predictably with temperature.
Capacitor	Devices that can store an electric charge and comprise two conductors usually made of nichrome, Ta, Ti, Pt, or Au, separated by a dielectric, usually made of silica or silicon nitride.
Integrated circuit (IC)	Microcircuit consisting of interconnected parts [such as bipolar transistors, field-effect transistors (FET), resistors, and capacitors] inseparably associated and formed in situ or within a single-crystal substrate (such as a single chip of semiconductor material) to perform an electronic circuit function.
DRAM*	Dynamic random access memory is a monolithic semiconductor memory that operates at extremely fast read-and-write speeds dynamically in an $m \times n$ matrix of memory cells that lose state information when power is removed.
NVRAM*	Random access memory that is read-and-write capable in an $m \times n$ matrix of memory cells that retain state information when the power is removed, resulting in a nonvolatile performance.
Ferroelectric*	Nonlinear dielectric material that polarizes in an electric field and retains polar state after the field is removed, resulting in a remnant charge on capacitor electrodes.

  * After Joe Cuchiaro.
  Reprinted by permission of ASM International, Materials Park, Ohio.

The performance of most ICs (made of Si) is a function, to a large extent, of the integrity and purity of the single-crystal Si chips, fabricated by slicing a cylindrical single-crystal ingot into thin wafers of up to 200 mm in diameter.[240] Epitaxial-Si films are a vital part of every microelectronic system. Important applications include field-effect transistors (FETs), dynamic random access memory devices (DRAMs), and other IC designs.[241,242] To provide the necessary semiconducting properties, epitaxial-Si films are doped with B, As, or P in the silicon structure by the addition of gases such as hydrides (B_2H_6, AsH_3, and PH_3) during the deposition. P as a dopant decreases internal stresses, increases the moisture resistance, and serves as an alkali getter (mostly of Na). Boron as a dopant reduces the etch rate and improves step coverage.[243,244] These elements are deposited from their hydrides. Doping during CVD is gradually replacing doping by diffusion or by ion implantation.

Like CVD-epitaxial Si, CVD-deposited polysilicon is widely used in the fabrication of ICs and is doped in a similar way. Polysilicon films are used as (1) gate electrodes in MOS devices for high-value resistors to guarantee good ohmic contact to crystalline Si, (2) diffusion sources to form shallow junctions, and (3) emitters in bipolar technology.[245]

Germanium is mostly used as an Si-Ge alloy which finds applications in the form of (1) transistors with speeds in excess of 60 GHz and minimum operating voltages (1.5 V), (2) Ge films on Si to tailor the bandgap of heterostructures, and (3) photo-voltaic converters and photodetectors.[201]

Composition variations of III-V and II-VI compound semiconductors allow the tailoring of electronic and optoelectronic properties to match specific applications. For example, the control of stoichiometry of the element (achieved by proper handling of MOCVD reactions) allows one to tailor the bandgap, which provides greater flexibility in the design of transistors and optoelectronic devices.[201] The advantages of III-V compound semiconductors over Si lies mostly in the large bandgap and higher carrier mobilities. Usually, these properties allow operation at higher frequencies and higher temperatures.[150]

Semiconductor diamond may be the most suitable material for many semiconductor applications such as high-power and high-frequency transistors and cold cathodes, or in the harsh environment present in internal combustion and jet engines. In the case of CVD diamond, it is now possible to take advantage of these properties. (Of nearly equal potential is AlGaN, which is much further along in device development.[245a]) CVD diamond (or AlGaN) may find applications in the following kinds of situations:

1. Schottky power diodes capable of operating at 500°C, exhibiting good rectification. Such a design would be appropriate for high-voltage transistors without a need for voltage conversion from line to input. This would substantially reduce the size of the power supply.

2. A series of field-effect transistors (MOSFET and MESFET) that serve better than Si devices and have the features of high power-handling capacity, low saturation resistance, and excellent high-frequency performance.

3. In high-power and high-frequency systems, such as microwave, and in harsh environments such as internal combustion and jet engines.[246]

CVD also plays an increasingly vital role in the design and processing of advanced electronic conductors and insulators and its related structures such as diffusion barriers and high-thermal-conductivity substrates (heat sinks). In these areas, CVD materials such as TiN, Si_3N_4, SiO_2, diamond, and AlN are of special importance.

Optoelectronic Applications. Optoelectronics is a discipline that combines optics and electronics. It deals with optical wavelengths between $0.2\,\mu m$ (UV) and $3\,\mu m$ (near infrared). The properties of optoelectronic materials comprise both electrical and semiconductor properties (electron action) and optical properties such as transmission, reflection, and absorption (phonon action).[201]

Critical properties of optoelectronic materials include bandgap (operating range), carrier lifetime (efficiency), and resistivity (response time). To optimize these properties, the processing parameters such as stoichiometry (bandgap definition), carrier concentration and mobility (resistivity), and defect density (carrier life time) must be controlled. Si finds applications as a photodetector with a response time in the nanosecond range and a spectral response band from 0.4 to $1.1\,\mu m$, which matches the 0.905-μm photoemission line of GaAs. III-V and II-VI semiconductor compounds produced by CVD exhibit superior optical properties and constitute the most vital group of optoelectronic materials. Among the wide range of possible combinations, III-V compounds of Ga such as GaAs, GaAsP, GaAlAs, GaP, GaInP, GaInAs, and GaAlP are popular, followed by other common compounds such as InAs, InP, BP, InAsSb, AlAs, and so forth.

Table 18.21 lists the applications of III-V compound semiconductor materials. The most important LED materials are GaAs, GaP, and ternary alloy $GaAs_{1-x}P_x$. The latter has a direct bandgap for $x < 0.45$. Diodes with $x = 0.4$ emit red radiation. The ternary GaAlAs system can be employed to make highly, effective diodes where the n-type layer comprises $Ga_{0.3}Al_{0.7}As$ and the p-type layer $Ga_{0.6}Al_{0.4}As$.[155]

Table 18.22 provides the properties of some photodiodes.[155] The typical construction of an LED device, a photodiode detector (structure), a solar cell, and a heterojunction bipolar transistor (HBT) device are shown in Fig. 18.31. Of great interest among recent experiments have been the III-nitride materials such as GaN, InGaN, AlGaN, and so on. These materials have been used to produce blue/UV LEDS and lasers (Nichia) and high-temperature electronics. Further, efficient UV

TABLE 18.21 Applications of III-V and III-Nitride Compound Semiconductors[155]

Semiconductor	Applications
GaAs	Solar cells, light-emitting diodes (LEDs)
GaAs/AlGaAs	Heterostructure lasers, solar cells, high-electron-mobility transistors (HEMTs), heterojunction bipolar transistors (HBTs), field-effect transistors (FETs)
GaP	Red LED, photocathodes
GaAsP	Red LED
InGaP	Red LED
AlGaInP	Yellow/green LED
InGaP/AlGaInP	Red laser pointer
GaSb/AlGaSb	Thermal imaging device, environmental sensors
GaN	Blue LED
InGaN	Green LED
InP	Gunn diodes, weather radar devices
InP/InGaAs	Detector in optical fiber technology
InGaAsP	Emitter in optical fiber technology
GaN, InGaN, AlGaN[245a]	UV/blue green emitters, white lighting, high-temperature electronics

Reprinted by permission of VCH, Weinheim.

TABLE 18.22 Properties of Some Photodiodes[155]

Material	Dopant	Peak emission (typical values), nm	Color (commercial diodes)	External quantum efficiencies, %
GaAs	Si	$910 \rightarrow 1020$	Infrared	10
GaP	N	570	Green	0.1
GaP	N, N	590	Yellow	0.1
GaP	Zn, O	700	Red	4
$GaAs_{0.6}P_{0.4}$		650	Red	0.2
$GaAs_{0.35}P_{0.65}$	N	632	Orange	0.2
$GaAs_{0.15}P_{0.85}$	N	589	Yellow	0.05
$Ga_{0.6}Al_{0.4}As$	Zn	650	Red	3
$Ga_xAl_{1-x}As$ $(1 < x < 0.7)$	Si	$870 \rightarrow 890$	Infrared	15

Source: After W. Wold and K. Dwight, *Solid State Chemistry*, Chapman & Hall, New York, 1993, p. 38.

TABLE 18.23 Applications of II-VI Compound Semiconductors[155]

Semiconductor	Applications
ZnS (Ag-doped)	Blue phosphor for TV cathode ray tubes
ZnS (Mn-doped)	Thin film electroluminescent displays (TFELs)
ZnSe, ZnSSe, ZnMgSSe	Blue/green LEDs and lasers (potential application)
ZnSe, ZnS	Interference coatings for optical components
ZnCdS	Solar cells
ZnCdS (Cu-doped)	Green phosphor in TV cathode ray tubes
CdS/CdTe	Solar cells
CdTe	Infrared detector
CdHgTe	Infrared detector, thermal imaging systems

Reprinted by permission of VCH Weinheim.

LEDs coupled with cheap phosphorus make low-wattage, high-brightness, long-lifetime white lights.[245a]

Commonly available II-VI compounds are ZnS, ZnSe, ZnTe, CdS, CdTe, and HgCdTe. Although the wide bandgap and direct-transition nature of II-VI semiconductors make them suitable as optoelectronic and optical materials, their applications, unlike the III-V compounds, have been confined to simple applications such as in optical coatings, gratings, photoconductors, and photovoltaic and electroluminescent displays (see Table 18.23).[155] An important reason is that it is difficult to achieve a p-type doping. HgCdTe is widely used in military night sights, which detect in the 8- to 13-μm spectral band.

Optoelectronic devices are used in a wide variety of consumer products such as televisions, compact disk players, laser communications, laser printers, radar detectors, cellular telephones, direct-broadcast television, and so forth.

Optoelectronic components produced by CVD include semiconductor lasers, light detectors (both photoconductor and photodetector devices), light-emitting diodes (LEDs), photovoltaic cells, imaging tubes, laser diodes, optical waveguides,

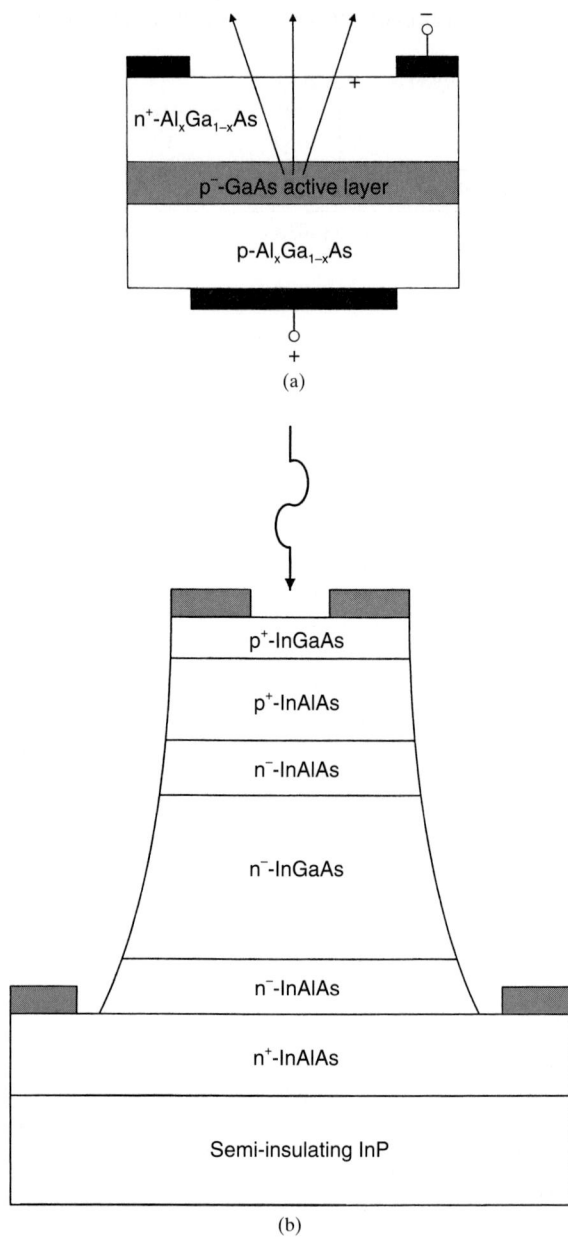

FIGURE 18.31 Schematics: (*a*) A generic LED device structure. It generates light by passing current through a p-n junction formed in a direct-energy bandgap material. (*b*) A photodiode detector (structure). The primary material used for photodetectors is silicon. (*c*) A typical GaAs solar cell. The basic structure is a p-n junction in GaAs with a high-Al-content AlGaAs window. The antireflective coating maximizes the amount of light passing into the GaAs, whereas the front contact grid lines minimize the shadowing effect but hold resistive losses low. (*d*) A heterojunction bipolar transistor (HBT) device. It is used mainly as microwave power transistors, usually in communications.[29] (*Courtesy of Structural Materials Industries, Inc.*)

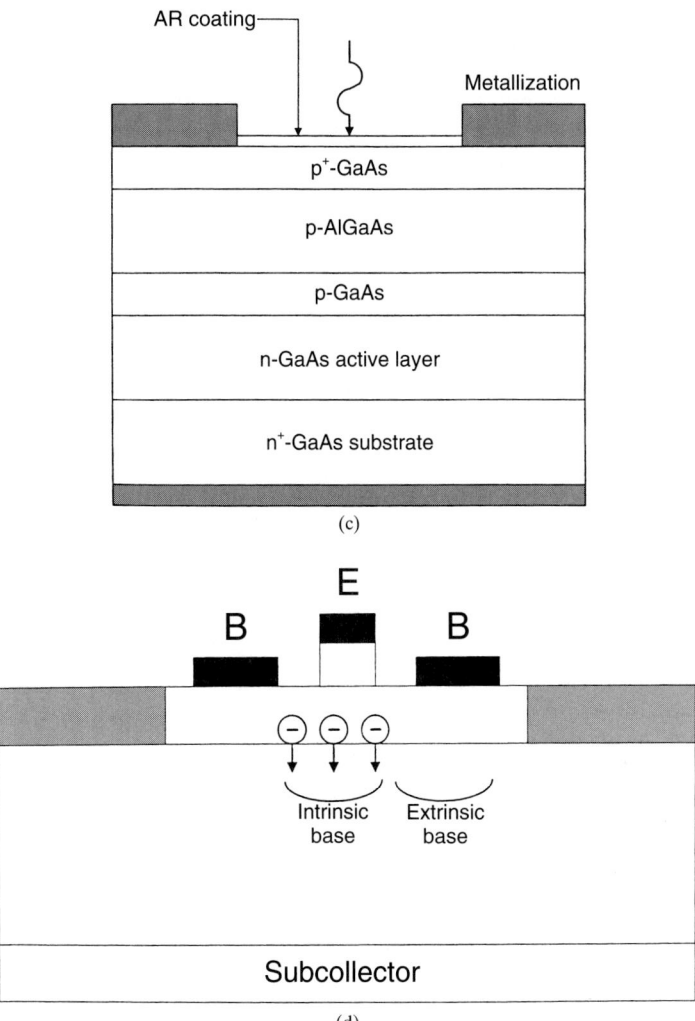

FIGURE 18.31 (*Continued*) Schematics: (*a*) A generic LED device structure. It generates light by passing current through a p-n junction formed in a direct-energy bandgap material. (*b*) A photodiode detector (structure). The primary material used for photodetectors is silicon. (*c*) A typical GaAs solar cell. The basic structure is a p-n junction in GaAs with a high-Al-content AlGaAs window. The antireflective coating maximizes the amount of light passing into the GaAs, whereas the front contact grid lines minimize the shadowing effect but hold resistive losses low. (*d*) A heterojunction bipolar transistor (HBT) device. It is used mainly as microwave power transistors, usually in communications.[29] (*Courtesy of Structural Materials Industries, Inc.*)

TABLE 18.24 Optoelectronic Devices and Applications[247]

Device materials	Applications
Photodiode (LED) (GaAsP, GaAlAs, GaP, GaInP, and GaAlP)	Cameras, strobes, illuminators, remote watches, clocks, scales, calculators, computers, optical transmission devices, controls, IR sensors
Pin photodiode	Fiber-optic communications, fiber links
Laser detector	Digital audio disks, video disks, laser beam position sensors, distance sensors
Photodiode array (LED)	Photodiode chips can be arrayed monolithically or nonmonolothically to any element
Photo transmitters	Optical switches, strobes, toys
IR-emitting diodes (GaAs)	Remote controls, optical switches, optoisolators, choppers, pattern recognition devices
IR-emitting diodes (GaAlAs)	Optical switches, encoders, photo IC sensors
Photo interrupter	Highly precise position sensors, noncontact switches
Reflective sensor	Tape-end sensors, liquid-level sensors
Optic receiver	Optic remote controller
Photocoupler	Isolators, impedance converters, noise suppressors

Courtesy of Lumex, Palatine, Illinois.

impact diodes, Gunn diodes, mixer diodes, varactors, photocathodes, and HEMTs (high electron mobility transistors). Major applications are listed in Table 18.24.[247]

Ferroelectric Applications. The important piezoelectric applications include signal devices, surface acoustic wave devices (resonators, traps, filters, etc.), sensors (pickups, keyboards, microphones, etc.), and electromechanical transducers (actuators, vibrators, etc.). Typical materials are AlN, ZnO, $PbTiO_3$, $Pb(ZrTi)O_3$ (PZT), and $LiTaO_3$. The important pyroelectric applications include those associated with infrared sensing such as fire or heat alarms, cooking controls, door openers, etc. Typical materials include $PbTiO_3$, $LiTaO_3$, and $Pb(ZrTi)O_3$.[201]

Ferroelectric material, in thin film form, can be integrated into capacitor structures that perform nonvolatile random access memory (NVRAM) functions. Ferroelectric nonvolatility arises from a dipole shift in the crystal lattice (most common materials have a perovskite crystalline structure) subjected to an externally applied electric field, where the dipole shift remains after the electric field is removed. The resulting polarization along the axis normal to the electrodes results in a remnant polarization that can be sensed electrically through a resistor, or across a load capacitor. To create a ferroelectric memory cell, a transistor is placed in series and wire connected to the capacitor forming a (1T/1C) cell which raises the potential across the ferroelectric to "write" the memory state. Figure 18.32a shows a common 1T/1C ferroelectric memory capacitor placed over a transistor drain region and wired with a tungsten plug for connection and Fig. 18.32b shows the equivalent circuit diagram. This cell can be simplified by placing the ferroelectric material directly in the gate of the transistor forming a ferroelectric field effect transistor (FeFET) memory cell. However, this simplified configuration may effect the inversion threshold voltage characteristics due to the placement of multiple metals in the channel region.[247a]

Optical Applications. Until recently, optical coatings were produced mostly by PVD processes. CVD techniques such as plasma CVD and MOCVD are now strong

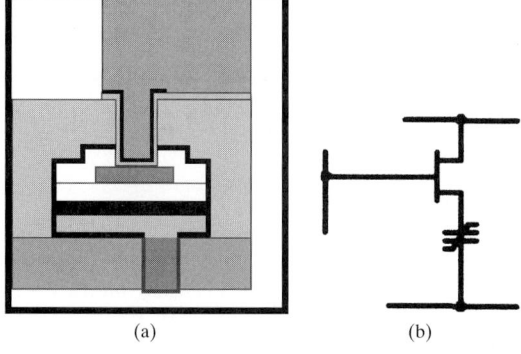

(a)   (b)

FIGURE 18.32 (*a*) A common 1T/1C ferroelectric memory capacitor placed over a transistor drain region and wired with a tungsten plug for connection and (*b*) an equivalent circuit diagram.[247a] (*Courtesy of Joe Cuchiaro.*)

TABLE 18.25   Properties of Transparent Electrically Conductive Coatings[201]

	Noble metals	Tin oxide	ITO
Thickness (nm)	7–35	10–300	10–300
Surface resistivity($\Omega/\square$)	1–30	80–50,000	7.5–1000
Luminous transmission, %	70–85	70–80	75–90
Abrasion resistance	Good	Best	Better
Chemical resistance	Best	Best	Better
Flatness	Conforms to substrate	Distorts substrate	Conforms to substrate
Adhesion	Best	Best	Best
Resistivity stability	Best	Better	Good
Cost	Moderate	Low	Moderate

candidates, especially in high-quality applications where three-dimensional surfaces or deep recesses need to be coated.

Antireflective (AR) coatings. The function of an AR coating is to minimize the surface reflection of optical elements and enhance the amount of light transmitted. A typical AR material is MgF_2. CVD-AR coating is suitable for 3-D surfaces or deep recesses. These coatings find numerous applications such as laser optics, lenses for cameras and binoculars, microscopes, telescopes, range finders, instrument panels, and automotive and architectural glasses.

Hot (infrared)- and cold (visible)-mirror CVD coatings. These are used in projectors to maintain a film gate at low temperature and avoid damaging the film. They are also used extensively in tungsten halogen lamps.

Transparent conductive oxide (TCO) coatings. These are also made by PVD processes (e.g., reactive evaporation, reactive sputtering, ion plating), but are limited by complex processing, high equipment cost, and relatively low throughput. Hence CVD stands a fair chance at a competitive price.[206] The superior properties of these films, particularly high electrical conductivity and optical transmission, absence of contamination, non-stoichiometry, smooth and flawless surface morphology, fine-grained polycrystalline structure, increased environmental stability, and easy etchability, are necessary for sophisticated applications. Table 18.25 lists the properties of TCO coatings.[201] The most important TCO films are fluorine- or Sb-doped SnO_2, In_2O_3, and Sn-doped In_2O_3 (90% In-10% Sn), called *indium tin oxide (ITO) transparent coatings*. They are increasingly used as electric heaters for windshield deicers, windows in heterojunction solar cells, heat-mirror coatings in energy-conserving windows, electrodes in liquid crystal displays, and gates in charge-coupled device imagers.[206]

18.4.7 CVD Processes

CVD processes can be grouped into (1) thermal CVD, (2) plasma-assisted (enhanced) CVD (PACVD, PECVD), (3) laser CVD (LCVD) and photo CVD (PCVD), (4) metalorganic CVD (MOCVD) or organometal vapor phase epitaxy (OMVPE), (5) chemical beam epitaxy (CBE), (6) atomic layer epitaxy (ALE) or almost equally alternating layer deposition (ALD), (7) chemical vapor infiltration (CVI), and (8) fluidized-bed CVD.

18.4.7.1 Thermal CVD. In thermal CVD, the chemical reactions begin only with thermal energy. They require high temperatures, usually between 800 and 2000°C (1472 and 3632°F), which can be produced by high-frequency induction heating, resistance heating, radiant heating, or infrared radiation. Thermal CVD can be grouped into two basic systems, called *cold-wall reactor* and *hot-wall reactor* (with either horizontal or vertical type). They are usually accomplished at atmospheric or low pressure and are used for producing all types of films used in microelectronics and optoelectronics, and for hard and protective coatings. They are of different types such as atmospheric pressure CVD (APCVD), low-pressure CVD (LPCVD), ultrahigh-vacuum CVD (UHVCVD), rapid thermal CVD (RTCVD), and vapor phase epitaxy (VPE).

Atmospheric pressure CVD (APCVD) is a CVD process in which gaseous reactants, diluted by N_2, H_2, Ar, or He, pass over a hot substrate in a reactor at *atmospheric pressure*. Atmospheric pressure reactors can be subgrouped into low-temperature (up to 500°C or 932°F), and high-temperature (>500°C or 932°F) reactors. High-temperature reactors can be further subdivided into horizontal, barrel, continuous, and single wafer or pancake vertical flow, or continuous reactors. They are either cold-wall or hot-wall type and can be rf inductively, resistively, or radiantly heated reactors.

Figure 18.33*a* is a schematic diagram of a cold-wall production reactor using induction heating for the deposition of silicon epitaxy in semiconductor devices. The power is supplied by a solid-state high-frequency (20 kHz) generator. A radiation reflector, shown in detail A, improves uniformity and efficiency of deposition. Pressures can vary between 100 mbar and 1 atm. Figure 18.33*b* is a hot-wall reactor utilizing resistance heating and is used for the coating of cutting tools with TiC, TiN, and Ti(CN). Such reactors are often large and the coating of a hundred parts can be performed under precisely controlled conditions.[201]

Low-pressure CVD (LPCVD) is a CVD process carried out at a pressure lower than 1 atm (usually 0.1 to 10 torr or 10 to 100 torr for GaAs) and at low or high temperatures between 300 and 900°C (572 and 1652°F). They are of horizontal and vertical types. The difference between LPCVD and APCVD is based on the fact that the reduction of pressure greatly influences the limiting step in the deposition. At low pressure, the mass transfer rate of the gaseous reactants and byproducts to the substrate become much greater than the surface reaction rate, whereas at atmospheric pressure these two rates are of the same order of magnitude.[218] However, the

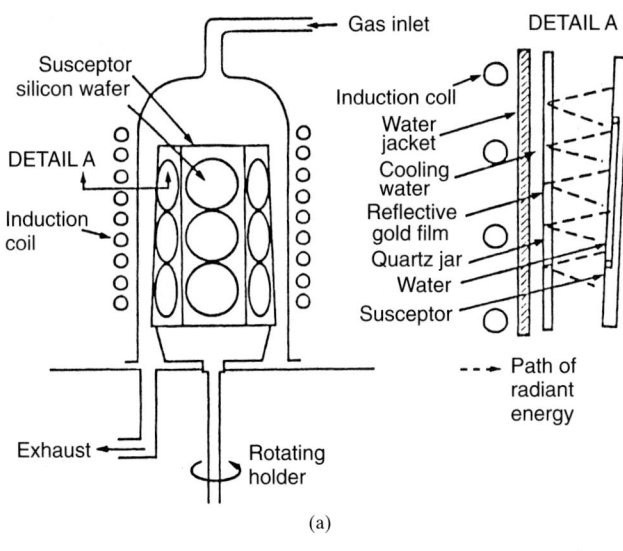

(a)

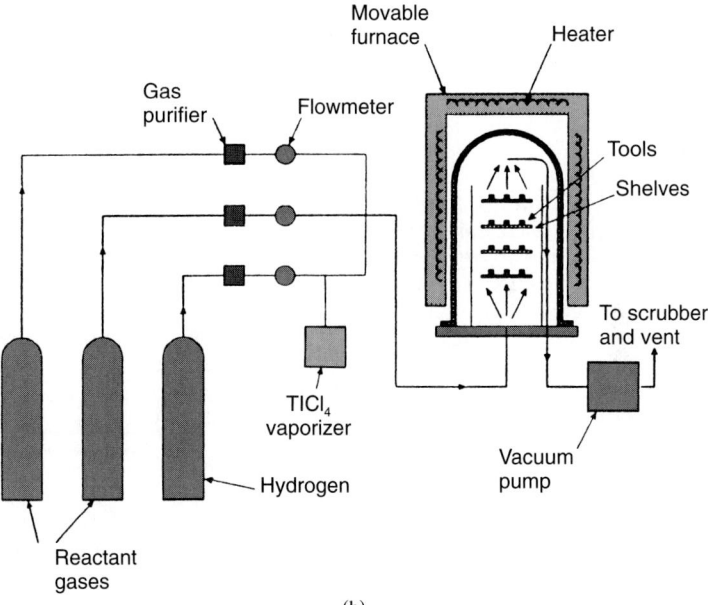

(b)

FIGURE 18.33 (*a*) Schematic of cold-wall production reactor for silicon epitaxy and (*b*) hot-wall production reactor for the coating of cutting tools.[201] (*Courtesy of H. O. Pierson.*)

chemical reactions involved in the deposition process as well as the starting gases used remain practically the same in both APCVD and LPCVD processes. It should be noted that a lower pressure level is linked to a lowering of gas viscosity and an increase of the gas diffusivity, both of which produce a more uniform film. Low-pressure reactions using both inorganic and metalorganic reactants are used to deposit almost any type of CVD films, in the polycrystalline, amorphous, or epitaxial conditions.[198,206]

Figure 18.34 shows the cluster system with a vertical batch-type hot-wall LPCVD reactor. This cluster system permits different process steps in one system such as in situ HF cleaning, oxidation, and LPCVD of poly-Si or Si_3N_4.[248] For poly-emitters, a notable improvement of the interface between substrate and poly-Si was found, with a thickness control of one monolayer.

Ultrahigh-vacuum CVD (UHVCVD) is a CVD process which is accomplished at a pressure below 10^{-3} torr, i.e., very low for conventional CVD processes, but high when compared to MBE processes ($\sim 10^{-5}$ torr). The UHVCVD for epitaxial Si and SiGe employs the normal-flow horizontal hot-wall batch-type configuration[249,250] with a load-locked chamber and a base pressure of 10^{-9} torr. Deposition is accomplished using a turbo-pumped system to prevent hydrocarbon backstreaming. Its main advantages over LPCVD and vacuum evaporation techniques are (a) a comparatively low substrate temperature, (b) removal of halogen vapor back-etching when using halogenide reactions, (c) reduction in film impurities, (d) better control of the deposit structure, and (e) process monitoring by using a residual gas analyzer.[201,206]

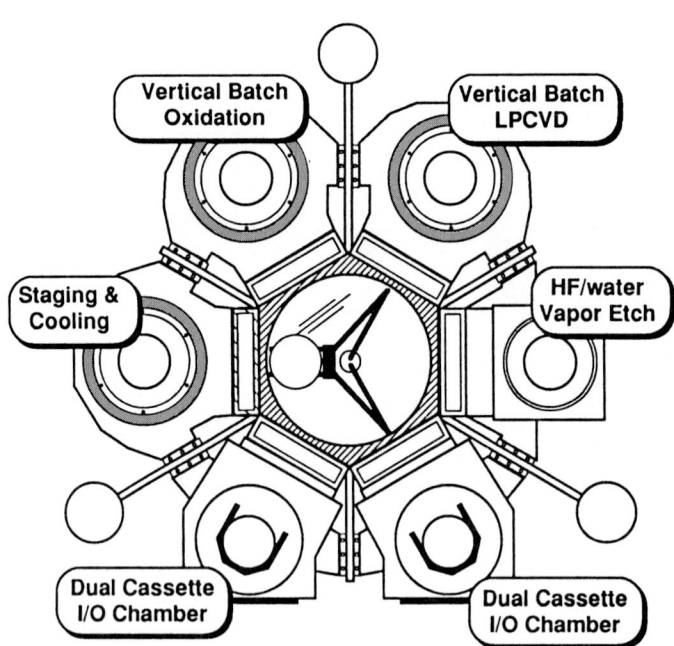

FIGURE 18.34 Plan view of cluster system with LPCVD for poly-Si gate. (*Courtesy of ASM International, Materials Park, Ohio.*)

Rapid thermal CVD (RTCVD) is a low-pressure CVD process where rapid heating and cooling of the substrate is used to start and stop the surface reactions of the CVD process by heat switching. RTCVD is a relatively new silicon wafer heating process that incorporates lamp radiation directly into the wafer. Thus this process is a combination of rapid thermal processing (RTP) (from the outside by lamp radiation) and LPCVD for deposition of Si, poly-Si, SiO_2, SiGe, III-V compound semiconductors, and TiS_2.[251-253] Figure 18.35 shows a typical RTCVD system consisting of a load-locked stainless steel chamber and wafer heating by tungsten halogen lamp banks through water-cooled quartz windows. Pulsed arc lamp sources are also available. Temperature control of 1% between 100°C and 1100°C is obtained with a pyrometer. A microwave plasma is used for in situ cleaning in the load-lock at room temperature.

There are two means of accomplishing RTCVD. The first approach uses control of deposition kinetics by lamp power switching, called *limited reaction processing (LRP)*. The main limitation of LRP is the likelihood of deposition taking place over a wide temperature range (i.e., during ramping or cooling), which may lead to nonideal film microstructures. The second approach uses deposition control by gas flow switching. This necessitates the very careful design of the gas manifold, to limit gas switching volume and thus switching time. Currently, most RTCVD processes are based on both lamp power and gas flow switching to control deposition.[255] Further information on RTCVD reactors and lamp-heated systems can be found elsewhere.[256]

The technique can be used for growing high-quality epitaxial films because of its significant capability of in situ multiprocessing (such as in situ cleaning via excursions to high temperature). RTCVD is also an ideal candidate for the growth of thin layers (<10 nm), because of the short deposition times of only a few seconds arising from the low thermal mass of the RTCVD reactor.[254] Thus, a metal oxide semiconductor (MOS) and SiGe capacitor have been fabricated and semiconductor heterostructural devices,[257] SiGe superlattices, and $Si/P^+(B)$ silicon superlattices[258] have been grown by RTCVD.

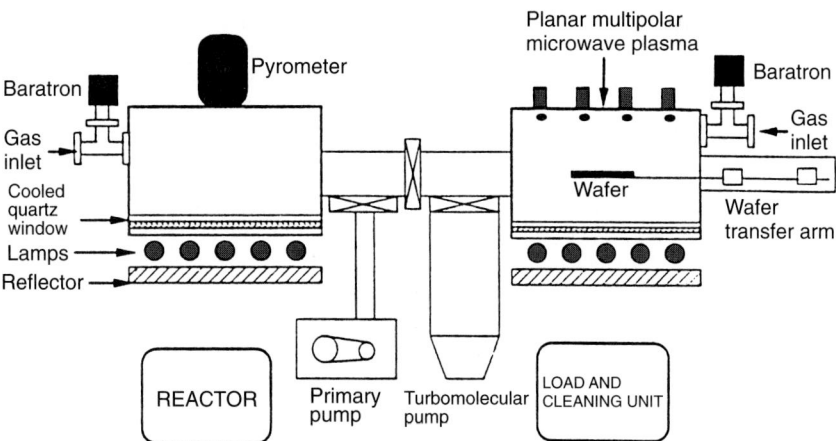

FIGURE 18.35 Schematic of a RTCVD system FUV 4.[198] (*Courtesy of D. Dutarte, P. Warren, I. Berbezier, and P. Perret, Thin Solid Films, vol. 222, 1992, pp. 52–56.*)

This process is competitive with MBE and UHVCVD. Selective deposition of Si[259] and SiGe[260] layers has also been performed. Further, since RTCVD chamber volumes can be very small, gas can be exchanged quickly, permitting the direct growth of multilayer heterostructures. Thus, thin oxidepolysilicon capacitor[261] and epitaxial Si-thin oxide-polysilicon semiconductor structures[262] have also been deposited. RTCVD is not confined to group IV semiconductor processing. Group III-V compounds have also been grown by RTCVD.

RTCVD is fully compatible with the concept of single-wafer processing, which will probably form the foundation of future integrated circuit processing lines. Competing techniques such as MBE and UHVCVD are not likely to eclipse RTCVD since MBE is very complex and, therefore, difficult to commercialize, and UHVCVD is not a single-wafer technology.

However, temperature measurement and uniformity are two challenging problems that RTCVD and RTP technology, in general, must solve before the possibility of their widespread acceptance. Temperature measurement can be done by either pyrometry, thermometry, or thermal expansion methods.[263]

The major advantages of the RTCVD process include (1) the ability for rapid heating (up to 300°C/s) and cooling rates due to the reactor's low thermal mass, (2) absolute temperature and uniformity of wafer temperature, (3) the ability to carry out in situ cleaning steps and the deposition of a variety of multiple layers, and (4) the ability to grow ultrathin epitaxial films with reproducible and extremely sharp dopant transition profiles.[256]

The limitations of the RTCVD process are (1) its commercial application being limited due to less perfection in real-time, accurate temperature measurement; (2) film deposition on the reactor windows due to infrared heating causing inaccurate temperature measurement; and (3) the creation of particulates diminishes wafer yield.[263]

Vapor phase epitaxy (VPE) is an advanced CVD process in which the nearly perfect growth of single-crystal (usually semiconductor) films occurs on the single-crystal substrates of the same or different composition. It is accomplished at atmospheric or low pressure, in cold- or hot-wall reactors by APCVD, LPCVD, or UHVCVD.

Applications. The conventional CVD technique is used to produce (1) metal films (Al, Au, Cu, Mo, Pt, W, TiN, silicides, and so forth) for microelectronics and protective coatings; (2) semiconductors (Si, Ge, $Si_{1-x}Ge_x$, III-V, II-VI) for microelectronics, optoelectronics, and energy conversion devices; (3) dielectrics (SiO_2, Si_3N_4, AlN, and so forth) for microelectronics and hard-coating applications; and (4) ceramic materials (Al_2O_3, BN, MoS_2, TiN, TiB_2, HfN, ZrO_2, and so forth) used for hard coating, protection against corrosion, or diffusion barriers.[198]

18.4.7.2 *Plasma CVD.*

Plasma CVD, also known as plasma-assisted or plasma-enhanced CVD (PACVD or PECVD), is a variant of the conventional CVD process that is activated by a plasma (using either rf, microwave, or photonic excitation) in which the deposition temperature is substantially lower (300 to 700°C or 570 to 1290°F). Plasma CVD thus combines a chemical and a physical process and may tend to bridge the gap between CVD and PVD. In other words, it is akin to PVD processes operating in a chemical environment such as reactive sputtering.[201] The process was developed because the high deposition temperature of thermal CVD does not allow the use of many heat-sensitive substrates such as low-melting-point metals, materials subjected to solid-state phase transformation over the range of deposition temperature, polymers, Al, cermets and steels, deposition of metastable compounds and structures [e.g., (TiAl)N and diamond], and others. Moreover, the thermal expansion mismatch between a substrate and a coating and the result-ing stress causing cracking and delamination or spalling during cooling are re-duced.[264,265] Table 18.26 compares the deposition temperatures for thermal and plasma CVD for some commercially important coatings.

There are two types of plasma used in CVD: glow-discharge (nonisothermal) and arc plasma (isothermal). Table 18.27 summarizes their characteristics.

The main advantages of PECVD are (1) its capability of producing conformal films at relatively low temperature, (2) its relative insensitivity to wafer tempera-ture, and (3) the likelihood of obtaining films with amorphous structure on various heat-sensitive substrates. PECVD processes have been widely used in hard-material deposition applications.[266]

The disadvantages of PECVD are (1) its low deposition rate and low efficiency, (2) the difficulty in controlling film properties, (3) its inability to handle solid or liquid reactants, (4) the exposure of substrate and film to radiation damage, (5) its production of nonstoichiometric and inhomogeneous films, (6) its use of compli-cated and expensive equipment, and (7) its limited use on a production scale.[106,267]

PECVD is used in both semiconductor and nonsemiconductor applications. Most PECVD systems use radio frequency with operating frequencies of 450 kHz or 13.56 MHz. A typical rf plasma reactor consists of parallel electrodes, as illustrated in Fig. 18.36. Microwave glow discharge is also used at a standard frequency of 2.45 GHz.

TABLE 18.26 Typical Deposition Temperatures for Thermal and Plasma CVD[201]

Material	Deposition temperature	
	Thermal CVD	Plasma CVD
Epitaxial silicon	1000–1250°C (1830–2280°F)	750°C (1380°F)
Polysilicon	650°C (1200°F)	200–400°C (390–750°F)
Silicon nitride	900°C (1650°F)	300°C (570°F)
Silicon dioxide	800–1100°C (1470–2010°F)	300°C (570°F)
Titanium carbide	900–1100°C (1650–2010°F)	500°C (930°F)
Titanium nitride	900–1100°C (1650–2010°F)	500°C (930°F)
Tungsten carbide	1000°C (1830°F)	325–525°C (615–975°F)

Reprinted by permission of ASM International, Materials Park, Ohio.

TABLE 18.27 Characteristics of Plasmas[201]

	Glow discharge	Arc
Plasma type	Nonisothermal (nonequilibrium)	Isothermal (equilibrium)
Frequency	3.45 MHz and 2.45 GHz (microwave)	≈ 1 MHz
Power	1–100 kW	1–20 MW
Flow rate	mg/s	None
Electron concentration	10^9–10^{12}/cm^3	10^{14}/cm^3
Pressure	200 Pa–0.15 atm	0.15–1 atm
Electron temperature	10^4 K	10^4 K
Atom temperature	500 K	10^4 K

Source: M. Thorpe, *Chem. Eng. Progress*, July 1989, pp. 43–53.

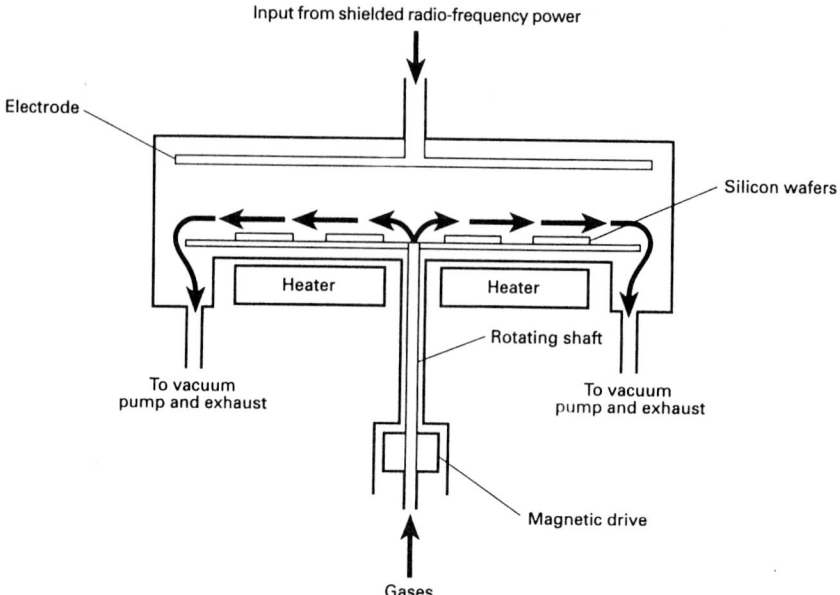

Input from shielded radio-frequency power

Electrode

Silicon wafers

Heater   Heater

Rotating shaft

To vacuum
pump and exhaust

To vacuum
pump and exhaust

Magnetic drive

Gases

FIGURE 18.36   Parallel-electrode RF-plasma deposition apparatus.[201]

Some applications of PECVD films include the manufacture of amorphous Si and other semiconductor films for cheap solar cells, optical antireflective coatings for solar cells and photothermal absorbers, films for optical waveguide fibers as well as insulators, dielectrics, and diffusion mask or photolithographic mask coatings in the production of integrated circuits.[206]

Recently, Ti(NCO) coatings for tool steels have been developed by metalorganic PECVD or PCVD.[268] Nonsemiconductor applications include deposition of (TiAl)N, (TiAl)C, and codeposited TiC/a:C-H layers and TiN-Al_2O_3 multilayers, and crystalline iron borides.[266]

A microwave plasma can also be generated by electron cyclotron resonance (ECR) by the combination of electric and magnetic fields at much lower gas pressures of 10^{-5} to 10^{-3} torr.[269] Cyclotron resonance is reached when the frequency of the alternating electric field matches the natural frequency of the electrons orbiting the lines of force of the magnetic field. This takes place at the standard microwave frequency of 2.45 GHz coupled with a magnetic field of 875 gauss. Figure 18.37 shows a schematic ECR plasma reactor, suitable for the deposition of diamond.[201] ECR and other plasma techniques are used widely in semiconductor production but so far have remained mostly experimental in other areas of applications.[270]

The advantages of ECR plasma are: (1) It reduces the potential substrate damage resulting from high-intensity ion bombardment, usually observed in a standard high-frequency plasma where the ion energy may approach 100 eV, (2) It reduces the risk of damaging heat-sensitive substrates due to its operation at a relatively low temperature.

The disadvantages of ECR plasma are (1) a more difficult process control (due to the addition of variables of the magnetic field) and more costly equipment (due

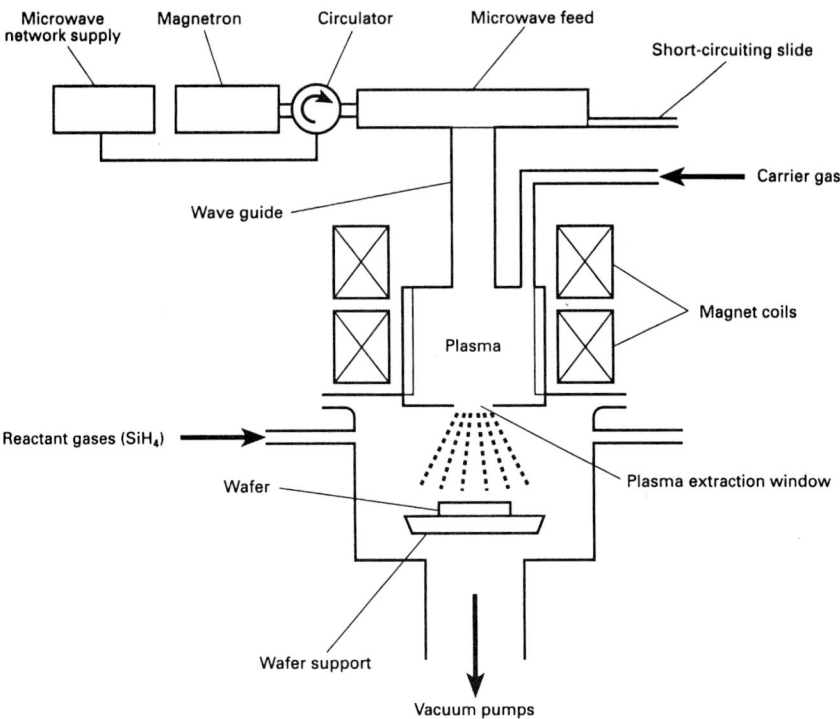

FIGURE 18.37 Schematic of electron cyclotron resonance (ECR) microwave deposition apparatus.[201]

to the addition of a complicated magnetic field), (2) a requirement of a lower pressure (10^{-3} to 10^{-5} torr) than the pressure of 0.1 to 1 torr required for rf plasma deposition, and (3) a need for a high-intensity magnetic field.[201]

18.4.7.3 Laser CVD (LCVD) and Photo CVD (PCVD). Thermal laser CVD occurs when the laser thermal energy contacts and, thereby, heats an absorbing substrate. The wavelength of the laser is such that little or no energy is absorbed by the gas molecules. Since the substrate is locally heated, deposition is restricted to the heated area. Table 18.28 lists some examples of materials deposited by thermal laser CVD process.[201]

High-frequency radiation, used in CVD processes, has been variously called *photo-assisted*, *photo-sensitized*, *photo-chemical*, or, collectively, *photo CVD*.[267] Photo-assisted CVD (PACVD) is a variant of the traditional CVD processes that can operate over a pressure range (typically, 0.01 to 1 atm) for deposition at low temperatures (below the pyrolysis thresholds) or for improved selected area epitaxy (SAE) (deposition) of thin films. Light sources for PCVD include arc lamps, CO_2 lasers, Ar^+ lasers, Nd: YAG lasers, and excimer lasers.[231]

The advantages of PACVD over thermal CVD include (1) the potential for a substantial lowering of deposition temperatures, (2) the low (0.1 to 5 eV) excitation energies needed to prevent film damage, (3) its ability to excite a range of different

TABLE 18.28 Examples of Materials Deposited by Thermal Laser CVD[201]

Materials	Reactants	Pressure	Laser (nm)
Aluminum	$Al_2(CH_3)_6$	10 torr	Kr (476–647)
Carbon	C_2H_2, C_2H_6, CH_4		Ar-Kr (488–647)
Cadmium	$Cd(CH_3)_2$	10 torr	Kr (476–647)
Gallium arsenide	$Ga(CH_3)_3, AsH_3$		Nd: YAG
Gold	Au (ac.ac.)	1 torr	Ar
Indium oxide	$(CH_3)_3In, O_2$		ArF
Nickel	$Ni(CO)_4$	350 torr	Kr (476–647)
Platinum	$Pt(CFCF_3COCHCOCF_3)_2$		Ar
Silicon	SiH_4, Si_2H_6	1 atm	Ar-Kr (488–647)
Silicon oxide	SiH_4, N_2O	1 atm	Kr (531)
Tin	$Sn(CH_3)_4$		Ar
Tin oxide	$(CH_3)_2SnCl_2, O_2$	1 atm	CO_2
Tungsten	WF_6, H_2	1 atm	Kr (476–531)
$YBa_2Cu_3O_x$	Halides		Excimer

Reprinted by permission of ASM International, Materials, Park, ohio.

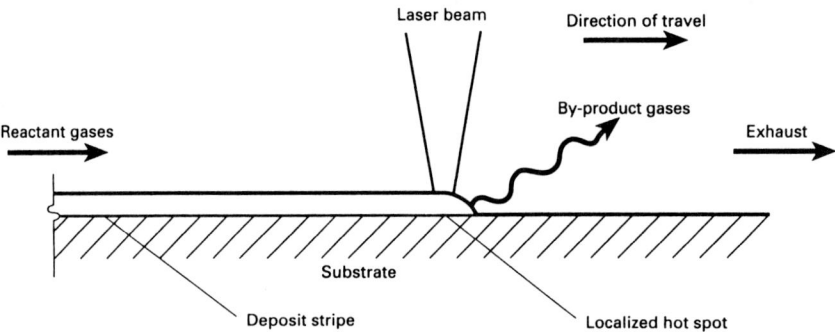

FIGURE 18.38 Schematic of laser-CVD growth mechanism (stripe deposition or laser writing).[201] (*Reprinted by permission of ASM International, Materials Park, Ohio.*)

processes according to the wavelength used and the photoabsorption properties of the chemical precursors and film, (4) the possibility of maintaining a well-defined reaction volume by using either proper optical focusing arrangements or lasers, (5) the possibility of localized deposition—with scanning—to produce direct, maskless, 3-D pattern generation, (6) its ability to provide the possibility of rapid rastering with increased reaction (and deposition) rates by using high-power lasers, and (7) the potential for minimal deleterious side reactions due to the use of monochromatic radiation.[231,267]

The main limitation of PACVD is that it is a slow and potentially inefficient deposition process.

Laser writing (or thin-stripe deposition) of metal and semiconductor lines (such as metal interconnects on an integrated circuit or metal mask repairs) are examples where PCVD finds wide applications.[271,272] This is accomplished by rastering a laser beam across the substrate, as shown in Fig. 18.38.[231] These processes usually depend on the localized heating by the incident laser beam to favor film deposition. Growth of semiconductor films by PCVD has been employed for a wide variety of materials ranging from Si to III-V and II-VI compounds. Both photothermal and photochemical mechanisms can be used to grow high-quality semiconductor films, and these provide advantages over traditional CVD growth mechanisms with the potential for more control over film properties and the localization of deposition.[231]

The photothermal (or thermal laser) mechanism is based on the absorption of laser radiation by the substrate to increase its temperature, leading to the deposition by pyrolytic decomposition of gaseous molecular precursors, to enhance the surface mobility of precursor species to promote epitaxial growth.[155] This mechanism is commonly employed for good selected-area deposition, selective surface reactions, and fast growth rates[155] either by laser scanning or projection imaging using a pulsed laser source. The laser wavelength to be selected must be highly absorbing in the substrate without producing vapor phase absorption. Thus wavelengths may range from uv to the ir, but usually are in the visible range. Infrared such as the 10.6-μm CO_2 line is used for metal substrates but not for semiconductors. Figure 18.39 shows the schematic process of laser photothermal-assisted CVD.[231]

In the photolysis (or photolaser) mechanism, low-temperature, nonlocalized thin-film deposition below the pyrolysis thresholds is obtained by a photolytic laser-induced CVD (reaction). Strong absorption bands in the uv spectrum at wavelengths typically below 250 nm can be employed to nonthermally break (cleave) the chemical bonds in reactant (precursor) molecules to effect deposition. Although uv lamps have been employed, more energy can be achieved from uv lasers, such as excimer (e.g., excited dimer) lasers with photon energies in the range of 3.4 eV (XeF laser) to 6.4 eV (ArF laser). Selective area epitaxy or deposition can also be obtained by projection imaging of a pattern with uv radiation onto the substrate, held at a low temperature.[231]

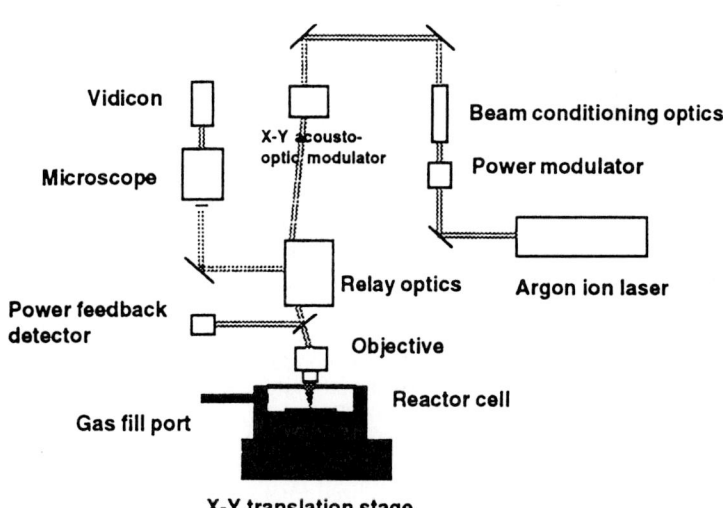

FIGURE 18.39 Schematic of laser photothermal-assisted CVD.[231] (*Courtesy of Institute of Physics Publishing, Bristol.*)

18.4.7.4   Metalorganic CVD.   Metalorganic CVD (MOCVD), also known as *organometallic vapor phase epitaxy (OMVPE)* and *metalorganic vapor phase epitaxy (MOVPE)*, is a version of the CVD process that uses at least one organometallic compound such as metal alkoxides and metal amides (that is, σ or π types) as a deposition precursor, usually in combination with hydrides or other reactants. Although OMVPE is technically the most accurate term, all three essentially imply the same thing and are often used interchangeably, with OMVPE being the most frequently used. This is a nonequilibrium technique in which gaseous precursors of the required material (vapors) are transported to a surface where they react to deposit a solid film on the substrate. Metalorganic compounds are organic combinations of organic ligands with metal atoms and have been developed to give a high-purity, precisely controllable, and relatively safe source of metal atoms for the epitaxy process. Organometallic compounds include metal alkyls and aryls, and oxygen-linked compounds such as metal alkoxides, β-diketonates, 2-ethyl hexanoates, naphthenates, neodecanoates, and stearates. The oxygen-linked organometallics find use in coating various substrates with metal oxide film via MOCVD.

The prerequisite properties in the selection of suitable candidates for OMVPE precursor are the following:[273]

1. The precursor should be a gas, liquid, or solid with adequate vapor pressure at temperatures below 150°C, but preferably a gas or liquid with a vapor pressure greater than a few torr at <50°C.[245a]

2. The precursor should be relatively easy to synthesize at high purity from commercially available chemicals.

3. The precursor should be thermally and chemically stable at its evaporation temperature through extended heating, yet be easily decomposable at process temperature.

Most OMVPE reactions occur in the temperature range of 300 to 800°C (570 to 1470°F). OMVPE growth processes are commonly operated at a low pressure (~30 to 80 torr). The partial pressures of the source gases range from a few mtorr to a few torr, which are much greater than those of background impurity gases such as oxygen, CO_2, or water vapor. These pressures offer the advantage of using relatively inexpensive vacuum equipment and materials and reduce the long downtime needed for such maintenance as opening the growth chamber and pumping the residual gases.[274] A modified version of this process consists of the prior deposition or codeposition of small amounts of catalysts by CVD to facilitate the OMVPE process. This method permits a relatively low-temperature (100 to 200°C) thin-film deposition of various metals to be made on numerous electronic materials such as III-V semiconductors, thin-film dielectrics, organic polymers, and assembled solid-state circuits and microchips.[275]

OMVPE tends to be the dominant growth technique for the high-volume production of high-quality compound semiconductor-based epitaxial wafers and devices (from both III-V and II-VI material systems) and of refractory carbides, nitrides, and oxynitrocarbide Ti(NCO). Recently, the III-V nitrides (such as InAlGaN quaternary systems) have attracted greater attention in the use of visible (blue and green) LEDs, injection lasers, transistors, photodetectors, and photocathodes. These semiconductors are characterized by a direct bandgap over the entire alloy composition range.[276] Currently, the growth capability of the OMVPE technique covers materials with a wide range of energy gaps, solar cells, and In-P devices for fiber communications.

Typical devices produced by OMVPE and their primary (optoelectronic) applications include highly efficient solar cells used in next-generation satellites, high-brightness LEDs in all spectrum colors (or wavelength ranges) as replacements for incandescent lighting, and lasers for a variety of uses in communications conduits, printing, bar-code scanning, higher-capacity CD-ROMs, DVDs, and hard disks in computers.[277,278] The use of OMVPE-based devices based on GaAs compound semiconductors has increased dramatically in a wide range of telecommunications technologies and wireless communications systems, including long-range portable telephones, high-bit-rate optical fiber networks, satellite communications,[274] and cell phones, as they accomplish communication at wavelengths not reachable with silicon-based devices. Table 18.29 exhibits examples of oxide materials produced by LP or RDR (rotating-disk reactor) OMVPE and their applications.[278a]

The major advantages of OMVPE relative to other techniques are (1) its relative simplicity; (2) its great versatility, because of the production of practically all III-V and II-VI semiconductor compounds and alloys; (3) its growth capabilities with uniform layers, excellent conformality of deposited films, tailored doping profiles, low background doping densities, and abrupt interfaces for use in sophisticated electronic devices such as HEMTs and HBTs; (4) its suitability for large-scale production applications; (5) its formation of crystalline and exceptional-purity metallic single and multilayered thin films with enhanced electrical and physical properties; (6) its greater flexibility for VLSI production of integrated microelectronic circuits and new discrete devices;[275] (7) its excellent control over film composition in order to produce heterostructures, multiquantum wells, bandgap variations (for many classes of electronic and optical devices),[279] and superlattices (with very short transitions in compositions); (8) its ability to prepare multilayer structures with layers as thin as a few atomic layers; (9) its ability to grow strain-compensated InGaAsP/InGaP superlattices due to the lattice mismatch of the crystal lattices of the two materials to create a built-in strain in each layer;[280] (10) its elimination of pinhole-type defects; (11) the absence of radiation-process-induced damage; and (12) its low particle counts.[278a]

TABLE 18.29 RDR-OMVPE Tool Demonstrated Material Capabilities and General Application Areas[278a]

Type	Example materials	Applications
Oxides	ZnO:Ga/Al/BF/In; ZnO:N; Zn_2GeO_4:Mn; $ZnIn_2O_4$:Mn/Ga; Zn_2SiO_4:Mn; $ZnGa_2O_4$:Mn/Ga; Al_2O_3; SiO_2; CuAgO/CuAlO/Cu; CuGaO; BST, PZT, and $SrBi_2T_2O_9$; YBa_2Cu_3O; CeO; InO; TCOs; Ta_2O_5; ZrO; MnO; HfO; CeO; MnO; MgO	Displays (TCOs, heaters, phosphors), memory (DRAM, NVRAM, PRAM), varistors, SAW/microwave
Silicon/ carbides	Si, SiGe, SiGeC SiC, diamond	Optoelectronics/waveguides
Nitrides	InGaAlN, Si_3N_4, TiN, BN	Dielectrics
III-Vs	GaAs, InP, InSb, InGaAsSb, AlGaAs, protective coatings	Pyroelectrics, compound semiconductors
II-VIs	HgCdTe, CdTe, ZnTe, ZnSe	
Metals	Cu, W, Al, Pt, Ru, Ir	

The drawbacks of OMVPE include (1) a need for expensive reactants, (2) the precise control of a large number of parameters needed to achieve the necessary uniformity and reproducibility, and (3) the use of large amounts of toxic/hazardous gases such as AsH_3 and PH_3 and the creation of a considerable volume of vapor phase waste stream.[267,281] However, less hazardous precursors such as tertiarybutylphosphine (TBP) and tertiarybutylarsine (TBAs) as well as less reactive species to be used as single sources are being developed to overcome these problems.[155,282]

A schematic LP-OMVPE vertical-geometry RDR is shown in Fig. 18.40a.[29] This reactor has a basic ability to compensate for depletion effects by independently controlling the reactant distribution radially within the downwardly directed carrier gas flow (Fig. 18.40b), thereby achieving high uniformities at high production efficiencies. This degree of reactant distribution uniformity (directly over the entire deposition surface) is difficult to achieve in other reactor geometries, but also attributes an extra degree of process optimization. This reactor technology has been scaled to 400-mm diameters and can also allow multiple wafer production. Materials Research Corporation recently developed its own rotating-disk reactor for TiN, metals, and associated Si films. A recent variant of the rotating disk OMVPE uses a water-cooled closed-space reactor injector. The advantages claimed from placing the cooled shower head injector in the proximity of the substrate are (1) uniform film deposition; (2) efficient reactor utilization, thereby offering material cost reduction and environmental safety hazards reduction; and (3) excellent layer-to-layer abruptness in the deposit (due to the reduced residence time of the reactants over the substrate). Thus a combination of cooled shower head injector and the rotating-disk reactor makes it simpler to scale up to large wafer sizes.[29]

To achieve the growth of III-V compound semiconductors, group III alkyls and group V hydrides can be introduced to the reaction chamber. The substrate, located on a hot susceptor, has a catalytic effect on the decomposition of gaseous products. Growth takes place mainly on this hot surface. TMG, TMA, TMI, and AsH_3 are generally employed to grow GaAs, AlGaAs, and InGaAs thin films, which are widely used for electronic and optical devices in commercial applications. These source materials are preferred for MOVPE because they have controllable and appropriate vapor pressures, and the diffusion-limited epitaxial growth can occur within a wide temperature range (830 to 1000 K) with an adequate uniformity both in compositions and layer thicknesses.[274]

18.4.7.5 *Chemical Beam Epitaxy.*

In chemical beam epitaxy (CBE), both group V sources and metalorganic group III sources are used in the gaseous states under the UHV conditions. CBE can be considered as a modified MBE process where all the solid sources of epitaxial deposition are replaced by gas sources, transported from a MOVPE gas manifold.[283] It may be considered as the bridge between MBE and conventional CVD.[245a] In practice, solid dopant sources are sometimes used for easy availability.[284] Like MOMBE process, CBE aims to combine the advantages of MOVPE (i.e., high throughput, absence of oval defects) with those of MBE (lack of gas-phase reactions, precise interface control).[155] Unlike the MOVPE process, the hydrides are precracked in CBE prior to reaching the hot substrate. The group V incorporation efficiency is greatly increased and the hydride consumption rate is reduced by a factor of >10. When the precracking efficiency of hydrides becomes ~100%, the toxic gases transform into low-vapor-pressure elemental form condensing within the growth chamber. This also diminishes the quantity of toxic gas to be treated in the exhaust gas.

The CBE process is different from GSMBE based on a complex growth mechanism in deriving the group III elements. In the growth of a binary compound such

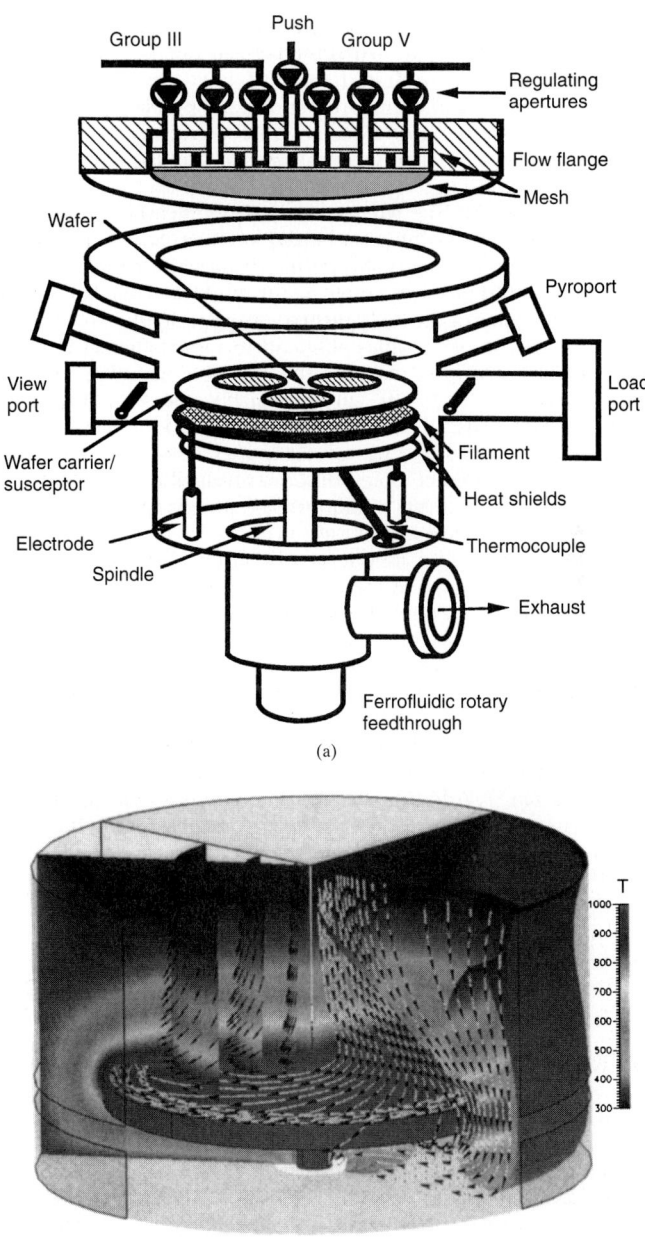

(a)

(b)

FIGURE 18.40 (*a*) Schematic of a LP-OMVPE rotating-disk reactor.[29] (*b*) Flow streamlines and isotherms within such a reactor. (*Courtesy of Structured Materials Industries, Inc.*)

as GaAs, the growth rate[285] and the impurity addition[286] are sensitive to the growth temperature because of its unique surface chemical kinetics, which finds no similarity in GSMBE. In ternary or quaternary $Ga_xIn_{1-x}P_{1-y}$ growth, the stoichiometry is found to be a function of both temperature and the precursor flow rate in a complicated fashion.[287] One important feature of CBE growth is a sharp diminution of notorious oval defects which mainly arises from the splitting of Ga clusters from the elemental Ga effusion cells.[288]

When compared to MOMBE where nominally elemental group V sources are employed, the difference are not pronounced because both CBE and MOMBE share the same group III chemistry. Minor differences are observed in the growth characteristics or material property involving group V sources. For example, in the temperature dependencies of the As/P incorporation ratio, the presence of group V monomers due to the use of hydrides will lead to more efficient incorporation of P.[289] In the case of carbon doping in GaAs, the hydrogen chemistry in CBE reduces the maximum doping level to $1 \times 10^{20} cm^{-3}$,[290] which is about an order of magnitude lower than that in MOMBE.[291] The use of elemental group V also greatly reduces the hazardous situation. For some, however, the two acronyms CBE and MOMBE are interchangeable. It should be noted that a reproducible control of group V flux is very important for the quaternary composition tuning. This is readily achieved in CBE by metering the hydride gas flow rate.

The CBE mode produces a unique combination of some major advantages from both the physical (MBE) and the chemical (OMVPE) deposition techniques. Because of the high vacuum ($<1 \times 10^{-4}$ torr) and the excellent cryopumping speed of the condensable gases, the beam-like characteristics of the injected gases in CBE are insured. In contrast to the viscous flow regime, a complicated gas flow pattern across the substrate surface does not prevail. The film thickness uniformity is thus in direct proportion to the distribution of the beam profile. This diminishes the problem of thickness uniformity to a relatively simple engineering issue of designing a gas injector nozzle that will provide a uniform beam profile over the total substrate surface.[286] Additionally, the injectors for the group III and the dopant source gases are regulated at low temperature ($\sim 50°C$), which permits premixing of the source gases if gas phase reaction is negligibly small within the injector. The homogeneous gas phase mixture guarantees a superior degree of composition and doping uniformity. The premixing concept can also be used for group V hydrides because their cracking temperatures are closer to each other. The beam-like features in CBE allow the use of a mechanical shutter in producing sharp interfaces even in a continuous growth mode,[287] free of any growth interruption, which is a common requisite in MOCVD for purging the residual gases.

A major advantage of the CBE/MOMBE/MBE over the MOVPE process is that UHV conditions permit the use of in situ analytical techniques such as RHEED, AES, modulated-beam mass spectroscopy (MBMS), and residual gas analysis possible in understanding the growth mechanism. However, optical techniques are being routinely utilized in OMVPE tools that yield temperatures (pyrometry), deposition rate (reflectometry), and even composition (spectroscopy and ellipsometry). In regard to system maintenance, the advantage of CBE is the easy replenishment of source materials, which is inherited from MOCVD without breaking the vacuum. In this situation, unlike for the MOCVD reactor, routine system cleaning in not needed,[284] although many modern OMVPE production tools are designed around minimizing maintenance issues.

Dopant gases kept outside the growth system permit easy handling and precise control of the doping level, and provide a fast exchange and a flexible selection of

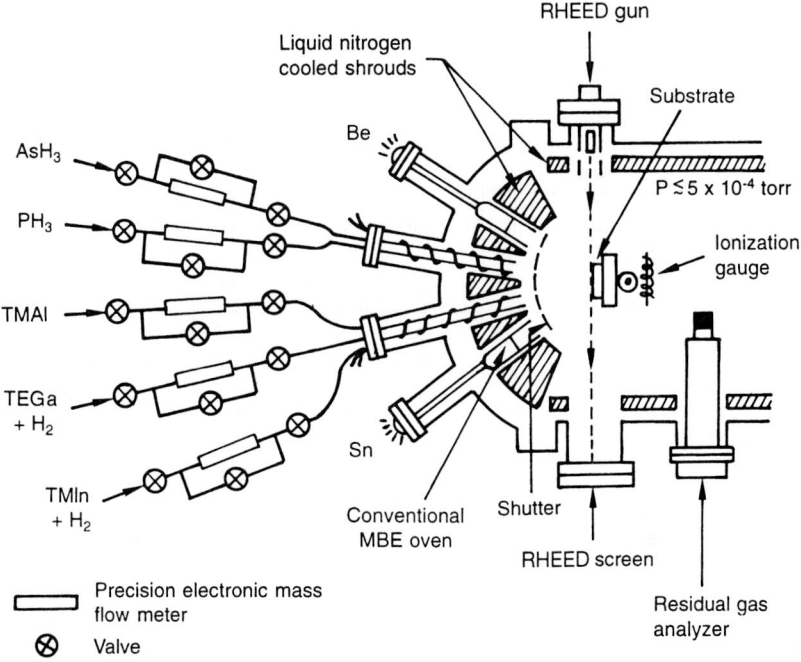

FIGURE 18.41   Schematic of a CBE system showing the growth chamber, the gas-handling manifold, and the in situ surface diagnostic probes. (*Courtesy of Institute of Physics Publishing, Bristol.*)

dopant element, which is especially attractive for system manufacture and device production applications.

A primary challenge in CBE is the development of gaseous sources for both n- and p-type doping instead of evaporated elemental sources.[155]

In CBE, in contrast to MOVPE, AsH_3 species are unlikely to play a significant role in the removal of carbon-containing radicals from the growth surface. This can lead to heavy carbon contamination if conventional MOVPE precursors (e.g., TMG, GMA) are used. There is thus a need for special precursors designed for the CBE process that should pyrolyze efficiently at the relatively low growth temperatures in CBE (i.e., 450 to 550°C) and should eliminate their alkyl radicals cleanly by using unimolecular deposition process. Otherwise, heavy carbon contamination can take place.[155]

Figure 18.41 shows a schematic CBE system with a growth chamber, equipped with a gas-handling manifold, and the in situ surface diagnostic tool.[284] Figure 18.42 shows a CBE system with significant pumping requirements.[29]

18.4.7.6   Atomic Layer Epitaxy.   Atomic layer epitaxy (ALE) [also termed alternating layer deposition (ALD)*][292] is a surface-controlled growth process that can

---

   * Some researchers have used the term "alternating" over "atomic" layer deposition because, in some of the processes, single "atomic" sheets are not really formed, but whole areas are. Furthermore, the formation of single saturation layers or mixed layers occur, through design.[292]

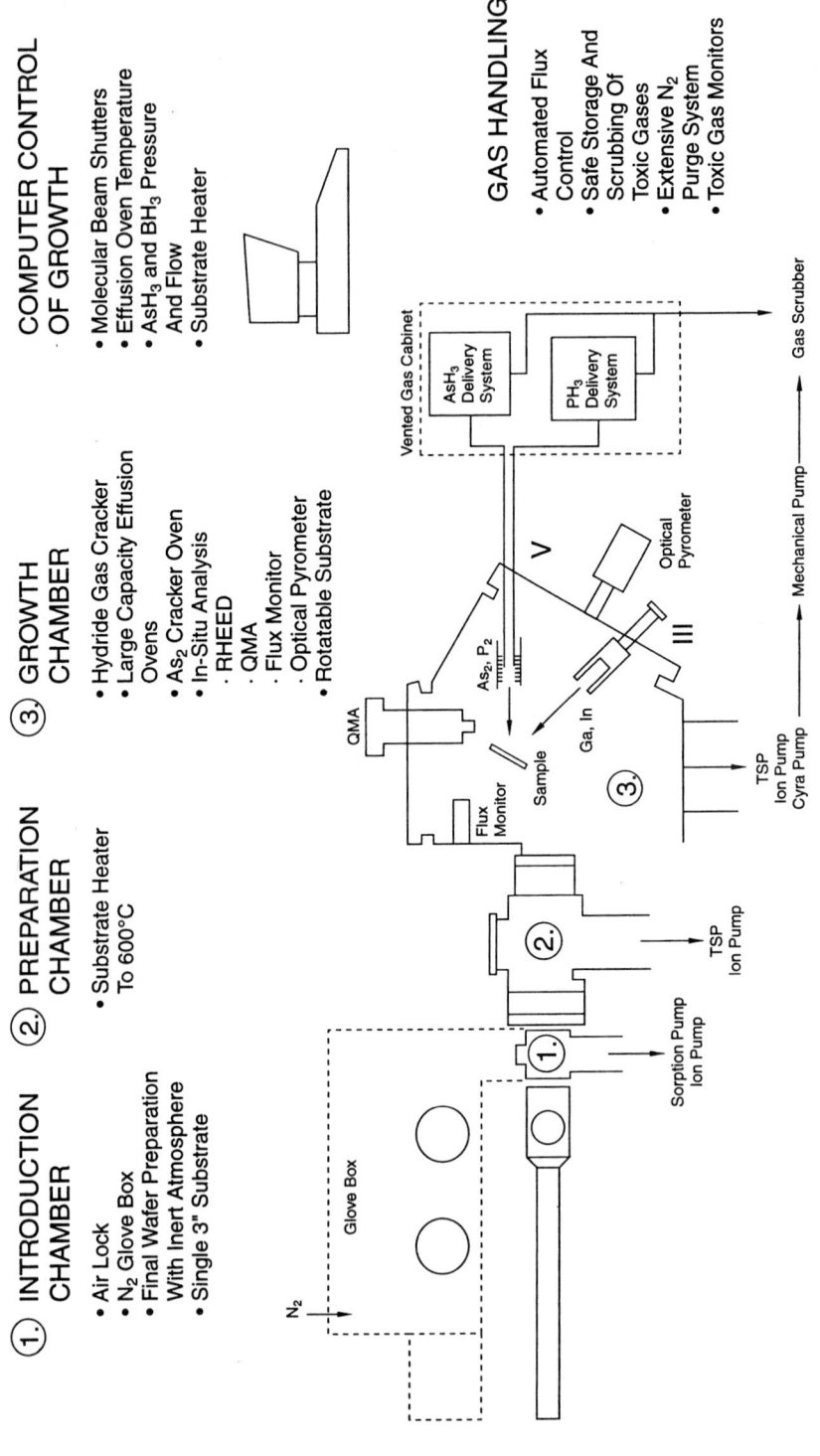

FIGURE 18.42 Schematic of a CBE system showing important pumping requirements.[29] *(Reprinted by permission of VCH, Weinheim.)*

① INTRODUCTION CHAMBER
- Air Lock
- N_2 Glove Box
- Final Wafer Preparation With Inert Atmosphere
- Single 3" Substrate

② PREPARATION CHAMBER
- Substrate Heater To 600°C

③ GROWTH CHAMBER
- Hydride Gas Cracker
- Large Capacity Effusion Ovens
- As_2 Cracker Oven
- In-Situ Analysis
  · RHEED
  · QMA
  · Flux Monitor
  · Optical Pyrometer
- Rotatable Substrate

COMPUTER CONTROL OF GROWTH
- Molecular Beam Shutters
- Effusion Oven Temperature
- AsH_3 and BH_3 Pressure And Flow
- Substrate Heater

GAS HANDLING
- Automated Flux Control
- Safe Storage And Scrubbing Of Toxic Gases
- Extensive N_2 Purge System
- Toxic Gas Monitors

TABLE 18.30   Target Materials for ALE or ALD Deposition[292]

Oxides: dielectrics/pyroelectrics	Al_2O_3, TiO_2, ZrO_2, $Hf(Si)O_2$, Ta_2O_5, MgO, CeO_2, Nb_2O_5, La_2O_3, Y_2O_3, SiO_2, V_2O_5
Transparent conductors/ semiconductors	In_2O_3, In_2O_3:Sn, SnO_2, SnO_2:Sb, ZnO, ZnO:Al, Ga_2O_3
II-VI compounds	ZnS, ZnSe, ZnTe, $ZnS_{1-x}Se_x$, CaS, SrS, BaS, CdS, CdTe, MnTe, HgTe, $Hg_{1-x}Cd_xTe$, $Cd_{1-x}Mn_xTe$
II-VI-based TFEL phosphors	ZnS:(Mn,Tb,Tm,K,Na,O,F), CaS: (Eu,Ce,Tb,Pb), SrS:(Ce,Tb,Pb)
Nitrides/carbides	AlN, GaN, InN, TiN, TaN, MoN, SiC, GeC, diamond
Fluorides	CaF_2, SrF_2, ZnF_2
Perovskites	$LaNiO_3$, $LaCoO_3$
Metals	Cu, Al, W, Pt, Ir, Ru, etc.
Other	LaS_3, PbS, In_2S_3, YBCO

Source: M. Ritala, *Appl. Surf. Sci.*, vol. 112, 1997, p. 223. Courtesy of Structured Materials Industries, Inc.

be conducted under low-pressure OMVPE or high-vacuum CBE conditions. ALE has been developed for the growth of ultrathin-film manufacturing of compound materials (such as III-V and II-VI semiconductors, oxides, and nitrides) and covalent materials; for the epitaxial layer growth of single crystals; and for producing tailored molecular structures on solid surfaces[293] (see Table 18.30).[292]

ALE involves sequential saturated surface reactions that lead to the sequential or stepwise film growth of single monolayers of semiconductor materials.[294] Consequently, the growth rate in ALE is proportional to the repetition rate of the reaction sequences rather than to the flux of reactant. Thus, the required thickness of a thin film can be achieved solely by counting the number of reaction sequences in the process. Figure 18.43 shows a schematic ALE process.[29]

ALE has a self-limiting deposition process in which epitaxial layer growth automatically stops at one or, occasionally, two monolayers. The growth rate is independent of growth parameters such as vapor pressure of the precursors and growth temperature[295] (and the film composition is independent of the number of excess incident molecules),[294] but is dependent on the number of growth cycles and the lattice constant of the material.[295] It provides the user digital deposition control. The digital nature of the process makes ALE promising in the growth of crystalline compound layers, complex layered structures, superlattices, layered alloys, and quantum-effect devices. A good review article on ALE is given by Goodman and Pessa[296] and Suntola.[297] The largest use of the ALE process is in the fabrication of large-area II-VI-based electroluminescent display devices.[29]

The advantages obtainable with ALE are (1) it has a lower epitaxial crystal growth temperature achieved in single-crystal epitaxy ALE; (2) it is well suited for making precise interfaces and material layers needed in superlattice structures and superalloys; (3) in ultrathin-film and large-area uniformity applications, ALE can be used to produce high-quality material with extremely uniform thickness (even to a single atomic layer and over large substrate areas) and it has very good reproducibility of heterointerface abruptness and precisely controlled interface properties[295,297]; (4) it also leads to an effective step coverage and a very low pinhole density in thin films; (5) an extreme in utilizing the conformal coating characteristic is the use of ALE in producing coatings or single molecular structures in micropores of porous materials such as supports for a heterogeneous catalyst; (6) the layer-by-

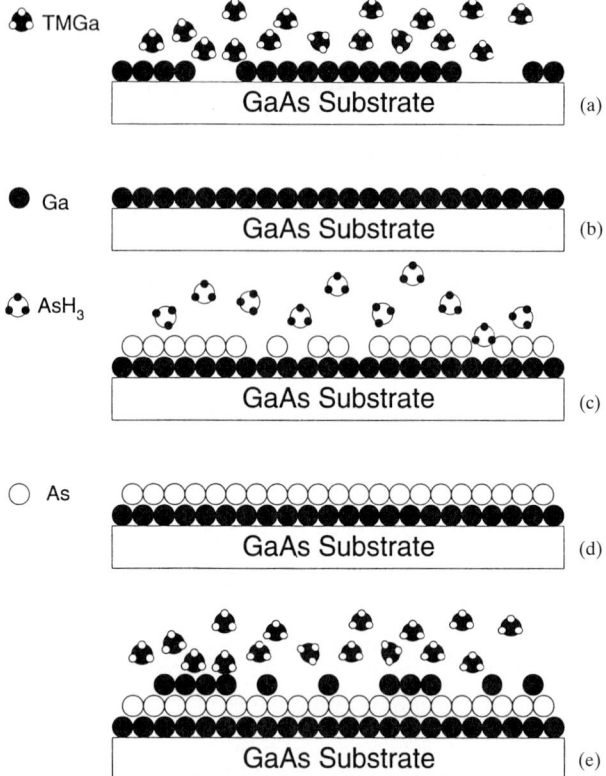

FIGURE 18.43 Representation of the atomic layer epitaxy process.
(*a*) TMGa exposure, (*b*) leaving a Ga layer, (*c*) arsine exposure,
(*d*) leaving an As layer, (*e*) the cycle begins to repeat.[29]

layer growth mechanism permits deposition of alternating elements, providing films
of tailored compositions or complex multilayer structures in a single run; and (7)
process scale-up is relatively simple.[292]

The limitations are: (1) much slower thin-film growth than conventional OMVPE
and (2) heavy carbon contamination in the films. However, combination of standard
SAE and ALE growth modes may lead to some structures not feasible by other
means.

Equipment.   There are two important classifications of equipment for the ALE
process, as given below.[297]

Traveling wave reactor. A "traveling wave" ALE reactor utilizes the saturation
feature of the ALE by incorporating flow dynamics for fast sequencing, high-density
packaging of substrates, and a high material utilization via an effective multihit
condition (Fig. 18.44).

In a traveling wave reactor, the reactants are fed into the reaction zone by a main
inert gas flow through the reactor. Each reactant, in turn, is supplied in the main
gas flow for sequencing. The flow speed, the pressure, and the timing of the inert

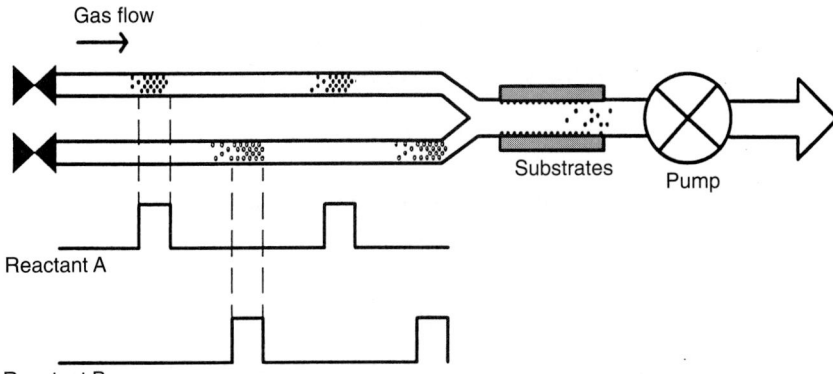

FIGURE 18.44 A traveling wave ALE reactor.[297](*Courtesy of Institute of Physics Publishing, Bristol.*)

gas flow are set to fulfill the requirements of keeping the adjacent sequences separated and to ensure the necessary purging between the reaction sequences.

A significant advantage of the traveling wave reactor is that it is easily scalable to large substrates and to very large batches. At laboratory scale, the traveling wave reactor is an effective tool for studying the reaction kinetics of the ALE process by observing the thickness of the thin films formed in different dosage and reaction conditions.

For in situ observation of the reactions, a mass spectrometer or other detector can be used to monitor the exhaust gases of a traveling wave reactor. In order to detect low-vapor-pressure exhausts, the detector as well as the gas sample line must be heated.

Rotating substrate reactor. A mechanically simple moving substrate reactor employs rotating substrate holders and fixed reactant supplies in different sectors (Fig. 18.45). An advantage of the moving substrate reactors is the simplicity of sequencing: No high-speed valving is required; it is adequate to switch the fluxes on and off when starting and stopping the process, while the rotation acts as a sequencer. The limitations of the rotating substrate reactors are a smaller flexibility in sequence structures and more difficult scaling-up of the reactor to production volumes. Rotating substrate reactors can be either single shot reactors or multishot reactors, depending on the design of the moving part and the feed of the reactants (Fig. 18.45). The flow conditions at the substrates in the reactors in Fig.18.45*b* are similar to those in a traveling wave reactor. The valving function in rotating cassette-type reactors occurs just at the edge of the substrate where the moving part meets the fixed feeding channel. This reduces the area of feeding channel wall between the valving point and the substrates, leading to fast on/off response times in reactant pulses to the substrates.[297]

18.4.7.7 Chemical Vapor Infiltration. Chemical vapor infiltration (CVI) refers to a special CVD process in which the gaseous reactants infiltrate a heated porous fibrous preform, such as SiC fibers (typically Nicalon or Nextel proprietary grades) and react to deposit a solid film, thereby filling the pores to form a matrix of the desired material (e.g., SiC).[201,298] Of six types of CVI processes, the isothermal-

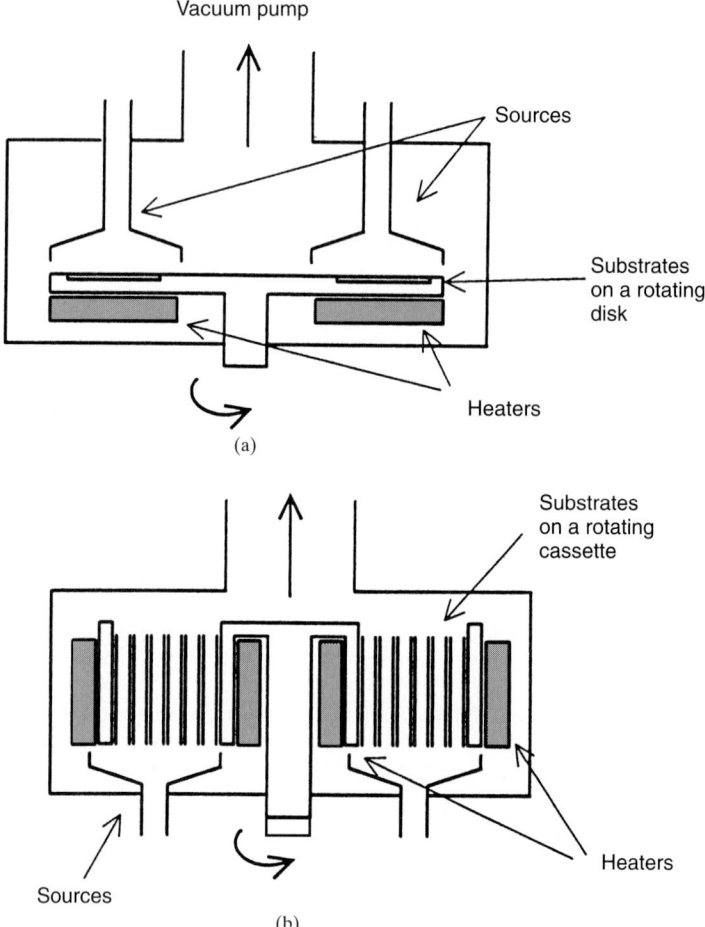

FIGURE 18.45   Rotating cassette reactors. (*a*) Rotating plate. (*b*) Rotating cassette.[297] (*Courtesy of Institute of Physics Publishing, Bristol.*)

diffusion, forced-flow thermal-gradient (FCVI), and isothermal-flow processes are popular.[298] Readers are referred to current review for more details.[298–300] In a typical FCVI system (Fig. 18.46), both the gas inlet and substrate are water cooled, and the top of the substrate is heated by a hot-wall reactor to introduce the temperature gradient. Under pressure, the gaseous precursors enter the cool side of the substrate and flow through it to reach the hot zone, where the deposition reaction occurs.

The process is used to produce high-strength silicon carbide and carbon-carbon composites, as well as other reinforced metal or ceramic composites using oxides (Al_2O_3 and ZrO_2), nitrides, and borides as matrix materials. Infiltration metals for coating enhancers or diffusion barriers may include Ni, Cr, W, Al, or other organometallic precursors.[301] Table 18.31 summarizes the properties of carbon-

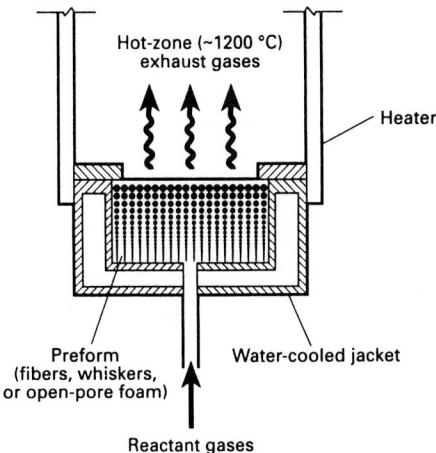

FIGURE 18.46 Chemical vapor infiltration apparatus. (*Source: N. Tai and T. Chou, J. Am. Ceram. Soc., vol. 73, no. 6, 1990, pp. 1489–1498.*)

TABLE 18.31 Properties of Carbon-Carbon Composites Prepared by FCVI[298]

Run number	Initial porosity (v/0)	Infiltration time, hr	Rate of weight gain, g/hr	Final porosity, %	Open porosity, %	Density, g/cm³	Deposition efficiency, %
			Propylene				
PCVI-33	45.64	28.5	0.3683	8.00	4.75	1.677	7.6
PCVI-24	45.65	11.5	0.9581	6.13	3.69	1.713	19.9
PCVI-32	45.86	10.5	0.9776	8.94		1.660	10.1
PCVI-37	44.59	9	1.0902	8.22	4.65	1.672	11.3
PCVI-28	44.76	6	1.5478	10.77	7.86	1.624	16.0
PCVI-27	44.42	21.5	0.4619	7.78	4.69	1.680	4.8
PCVI-25	48.62	7	1.6976	8.04	4.62	1.680	17.6
PCVI-36	46.48	7.75	1.3537	8.96	4.58	1.660	7.0
PCVI-22	44.96	2.75	3.5133	9.81	6.56	1.642	18.2
PCVI-31	44.25	2.75	3.0636	13.15	—	1.578	15.9
PCVI-21	45.67	11.5	0.8913	8.72	5.93	1.664	8.3
PCVI-23	45.95	12.25	0.8171	10.23	6.56	1.635	7.6
PCVI-35	45.63	8.75	1.2481	6.19	3.85	1.712	11.5
PCVI-38	43.90	8.5	1.1791	6.49	4.36	1.704	10.8
			Propane				
PACVI-16	49.7	14	0.88	8.0	3.6	1.68	24.4
			Methane				
MCVI-1	62.5	31.5	≈ 0	—	—	—	≈ 0
MCVI-3	47.8	6.5	≈ 0	—	—	—	≈ 0
MCVI-7	53.7	53	0.16	32.3	—	1.226	10.0
MCVI-11	49.5	38.5	0.25	16.9	5.0	1.514	15.4

Reprinted by permission of VCH, Weinheim.

carbon composites made by the FCVI process. It appears that acceptably low porosities ($\approx 10\%$) were obtained using infiltration times of 3 hr minimum with propylene. Methane tends to need longer infiltration times, or higher temperature to obtain sufficient densification.[298]

The advantages[298] of CVI include the following:

1. In contrast to sintering, CVI utilizes low pressures/stresses and temperatures. Consequently, mechanical and chemical damage to the substrate (i.e., fibers, whiskers, particulates, or other reinforcing materials in the preform) is considerably less.

2. The FCVI process also provides flexibility in the selection of the processing conditions as well as uniform and thorough densification over a wide range of operating conditions.

3. Precoating of the preform constituents (e.g., fibers, for enhanced fiber-matrix bonding) can be accomplished as an initial step of the CVI process using different reagents in the same CVD equipment.

4. It is possible for more densification of materials fabricated by other processes.

5. Superior purity and microstructure of the matrix are obtained when compared to other fabrication methods.

6. Sintering aids are not needed.

The main limitations of this method are the following:

1. There is the need for interdiffusion of reactants and reaction byproducts through relatively long, narrow, or tortuous channels that can take several weeks. It produces a low quality of the fabricated part without achieving full densification due to the formation of closed porosity. Currently, a new method based on volume heating of the preform (by microwave or radio frequency) with pulsed power has been developed to overcome this limitation. This method can offer rapid "inside-out" infiltration of the preform without any residual porosity, thereby achieving high throughput of fabricated parts with uniform density.[302]

2. CVI is a low-productivity process due to slow deposition.

Applications of CVI include carbon fiber-carbon matrix composites in aircraft brake disks, jet engine after-burner nozzles, nose cones, leading edges (such as on the space shuttle), and heat sinks (in electronic devices).[298] SiC fiber-SiC matrix composites produced by the FCVI process can be used for aerospace applications.

18.4.7.8   Fluidized-Bed CVD.   Fluidized-bed CVD, being a combination of conventional fluidized-bed technology and the CVD process, has proven to be a useful method to coat powders of WC for cutting tool applications as well as to coat nuclear-fuel particles such as uranium-thorium carbide particles with pyrolytic carbon and silicon carbide for containing the products of nuclear fission. The fluidized gas consists of a mixture of a nonreactive gas such as CH_4, C_3H_8, C_3H_6, He, or other gas and a reactive gas in order to form in situ the reactive precursors by the reaction with particles or solid components in the bed. Thus methyltrichlorosilane (CH_3SiCl_3) is the desired precursor for silicon carbide. Similarly, a mixed

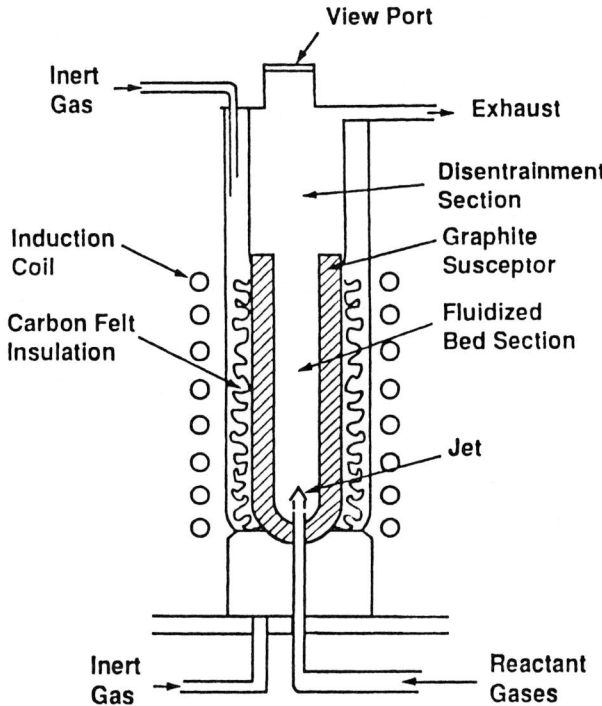

FIGURE 18.47   Fluidized-bed CVD reactor.[201] (*Reprinted by permission of Noyes Publications, Park Ridge, N.J.*)

fluidizing gas of $ZrCl_4$ + hydrocarbon, $HfCl_4$ + CH_4 + C_3H_6, and $TiCl_4$ + C_3H_6 can be used to deposit ZrC, HfC, and TiC, respectively.[201]

Figure 18.47 shows a typical fluidized-bed CVD reactor.[201] A gas mixture containing vapors of reactive precursors is introduced into a fluidized-bed reactor containing inert particles such as Al_2O_3 held at 1000°C. The powders or components to be coated are immersed in or placed above the bed.[303–305]

This process is used to deposit Si, Ti, TiN, TiO_x, Zr, ZrN, Al, and Si_3N_4 coatings on substrates such as steel, copper, Cu-Ni alloy, silica, powdered mica and nickel, and graphite fiber. The technique can also be applied to coating with other materials such as borides, carbides, and sulfides.[303–305]

The advantages of fluidized-bed CVD are the following: (1) Since the parts to be coated are in close contact with the bed, even very unstable, short-lived species can be sufficiently coated with a coherent, adherent, and conformal deposit. (2) Deposition occurs at atmospheric pressure and at lower temperatures than with conventional CVD. (3) Other materials such as carbides, borides, and sulfides can be successfully coated. (4) The fluidized-bed reactor can be used to deposit homogeneous and very uniform, simple, and composite coatings. (5) The combination of short times, low temperature, and fast deposition rates can lead to lower costs and expanded potential for CVD in many applications.[303–306]

REFERENCES

1. R. Emmerich, B. Enders, and W. Ensinger, in *Surface Modification Technologies VI*, ed. T. S. Sudarshan and J. F. Braza, The Metallurgical Society, Warrendale, Pa., 1993, pp. 811–835, 836–858, and 859–880.

2. J. K. Hirvonen and B. D. Sartwell, in *ASM Handbook, vol. 5: Surface Engineering*, 10th ed., ASM International, Materials Park, Ohio, 1994, pp. 605–610.

3. C. J. McHargue, *Int. Met. Reviews*, vol. 31, no. 2, 1986, pp. 49–76.

4. J. K. Hirvonen and J. D. Demaree, in *Advances in Coatings Technologies for Surface Engineering*, eds. C. R. Clayton, J. K. Hirvonen, and A. K. Srivatsa, The Metallurgical Society, Warrendale, Pa., 1997, pp. 53–70.

5. L. E. Rehn, S. T. Picraux, and H. Wiedersich, *Surface Alloying by Ion, Electron and Laser Beams*, ASM International, Metals Park, Ohio, 1987, pp. 1–17.

6. S. K. Ghandi, *VLSI Fabrication Principles*, Wiley, New York, 1983.

7. J. P. Biersack, and L. G. Hoggmark, *Nucl. Instrum. Meth.*, vol. 174, 1980, p. 257.

8. J. F. Ziegler, J. P. Biersack, and W. Littmark, *The Stopping and Range of Ions in Solids, vols. I and II,* Pergamon Press, Oxford, England, 1986.

9. J. P. Riviere, *Second ASM Heat Treatment and Surface Engineering Conference in Europe, Materials Science Forum*, vols. 163–165, 1994, pp. 431–447.

10. J. T. Scheuer, K. C. Walter, W. G. Horne, and R. A. Adler, in *Advances in Coatings Technologies for Surface Engineering*, eds. C. R. Clayton, J. K. Hirvonen, and A. R. Srivatsa, The Metallurgical Society, Warrendale, Pa., 1997, pp. 111–118.

11. X. B. Tian, X. Wang, S. Wang, B. Tang, and P. K. Chu, in *Surface Engineering: Science and Technology I*, eds. A Kumar et al., The Metallurgical Society, Warrendale, Pa., 1999, pp. 177–184.

12. K. Sridharan et al., in *Surface Modification Technologies IX,* eds. T. S. Sudarshan, W. Reitz, and J. J. Stiglich, The Metallurgical Society, Warrendale, Pa., 1996, pp. 401–419.

13. C. W. Ensinger, *Surf. Coat. Technol.*, vol. 100–101, 1998, pp. 341–352.

14. M. A. Otoni, A. Graf, G. Colombo, J. R. Conrad, K. Sridharan, M. M. Shamim, R. P. Fetherston, and A. Cohen, *Materials Research Society Symp.*, vol. 316, 1994, p. 577.

15. J. N. Matossian, *J. Vac. Sci. Technol.*, vol. B12, no. 2, 1994, p. 850.

16. S. F. Felch, T. Sheng, E. Ganin, K. K. Chan, D. L. Chapek, R. J. Matyl, and J. R. Conrad, in *10th International Conference on Ion Implantation Technology Extended Abstract*, Catania, Italy, 1994, p. 6.1.

17. I. Braun, *J. Vac. Sci. Technol.*, vol. 11A, no. 4, 1993, p. 1480.

18. P. J. Evans and F. J. Paoli, *Surface Coat. Technol.*, vol. 65, 1995, pp. 175–178.

19. J. R. Treglio, A. J. Perry, and R. J. Stinner, *Surface Coat. Technol.*, vol. 65, 1994, p. 184.

20. M. Braun, *Vacuum*, vol. 38, no. 11, 1988, pp. 973–977.

21. R. G. Vardiman, in *Ion Plating and Implantation Application to Materials*, ed. R. F. Hochman, American Society for Metals, Metals Park, Ohio, 1986, pp. 1078–1113.

22. D. G. Ingram and R. S. Bhattacharya, in *Ion Plating and Implantation Application to Materials*, ed. R. F. Hochman, American Society for Metals, Metals Park, Ohio, 1986, pp. 55–62.

23. V. A. C. Haanappel, H. J. Schmulzler, and M. F. Stroosnijder, in *Surface Performance of Titanium*, eds. J. K. Gregory, H. J. Rack, and D. Eylon, The Metallurgical Society, Warrendale, Pa., 1997, pp. 129–138.

24. R. F. Hochman, in *Ion Plating and Implantation Application to Materials*, ed. R. F. Hochman, American Society for Metals, Metals Park, Ohio, 1986, pp. 1–6.

25. K. O. Legg and H.-S. Legg, *Trends in Ion Implantation and Ion-Assisted Coatings*, 1988, pp. 1–18.

26. J. I. Onate, F. Alonso, J. L. Viviente, and A. Arizaga, *Surf. Coat. Technol.*, vol. 65, 1994, pp. 165–170.

27. B. Holtkamp, *Adv. Mats. & Processes*, 12/93, pp. 45–47.

28. *Ion Beam Surface Engineering Bulletin,* SRI, San Antonio, Tex.

29. G. S. Tompa, *CVD of Nonmetals*, ed. W. S. Reed, Jr., VCH, Weinheim, Germany, 1996, pp. 193–259.

30. D. K. Sadana, M. Strathman, J. Washburn, C. W. Magee, M. Maenpaa, and G. R. Booker, *Appl. Phys. Lett.*, vol. 37, 1980, p. 615.

31. K. S. Jones, in *Encyclopedia of Advanced Materials*, eds. D. Bloor et al., Pergamon Press, Oxford, England, 1994, pp. 1186–1195.

32. S. Mahajan, in *Processing of Semiconductors, vol. 6*, vol. ed. K. A. Jackson, VCH Weinheim, Germany, 1996, pp. 251–275.

33. T. Bernstein, I. W. Hall, and R. Kalish, *Radiation Effects,* vol. 46, 1980, pp. 31–38.

34. B. Drosd and J. Washburn, *J. Appl. Phys.*, vol. 51, no. 8, 1980, pp. 4106–4110; A. Bloyce, in *Surface Performance of Titanium*, eds. J. K. Gregory, H. J. Rack, and D. Eylon, The Metallurgical Society, Warrendale, Pa., 1997, pp. 155–169.

35. D. K. Sadana, M. Strathman, J. Washburn, and G. R. Booker, *Appl. Phys. Lett.*, vol. 37, no. 2, 1980, pp. 234–236.

36. A. L. Greer and R. E. Somekh, in *Materials Science and Technology, Vol. 15: Processing of Metals and Alloys*, vol. ed. R. W. Cahn, VCH, Weinheim, Germany, 1991, pp. 329–370.

37. A. Kirkpatrick and S. Dallek, in *Surface Modification Technologies IX*, eds. T. S. Sudarshan, W. Reitz, and J. J. Stiglich, The Metallurgical Society, Warrendale, Pa., 1996, pp. 401–419.

38. S. J. Pearton, in *Encyclopedia of Advanced Materials*, eds. D. Bloor et al., Pergamon Press, Oxford, England, 1994, pp. 1180–1186.

39. C. J. McHargue, *JOM*, July 1991, pp. 40–43.

40. S. J. Pearton, in *Encyclopedia of Advanced Materials*, eds. D. Bloor, et al., Pergamon Press, Oxford, England, 1994, pp. 1175–1180.

41. H. T. Clapp and L. E. Rehn, in *Encyclopedia of Materials Science and Engineering*, Pergamon Press, Oxford, England, 1986, pp. 2409–2412.

42. I. W. Hall, *Metallography*, vol. 15, 1982, pp. 105–120.

43. *Plasma Assisted Processes for Surface Engineering*, Brochure, Joint Committee on Plasma & Ion Surface Engineering (PISE), Germany, 2000.

44. G. Dearnaley, *Surf. Eng.*, vol. 2, no. 3, 1986, pp. 213–221.

45. D. A. Lilienfeld, L. S. Hung, and J. W. Mayer, *Nucl. Inst. Meth.*, vol. B19/20, 1987, p. 1.

46. J. W. Mayer, B. Y. Tsaur, S. S. Lau, and L.-S. Hung, *Nucl. Inst. Meth.*, vol. 182/183, 1981, p. 1.

47. D. M. Follstaedt, in *Materials Science and Technology, Vol. 15: Processing of Metals and Alloys*, vol. ed. R. W. Cahn, VCH, Weinheim, Germany, 1991, pp. 247–287.

48. P. Sigmund and A. Gras-Marti, *Nucl. Inst. Meth.*, vol. 182/183, 1981, p. 187.

49. H. H. Andersen, *Appl. Phys.*, vol. 18, 1979, p. 131.

50. Z. L. Wange, J. F. M. Westendorp, and F. W. Saris, *Nucl. Instr. Meth.*, vol. 209/210, 1983, p. 115.

51. Y. T. Cheng, T. Van Rossum, M. A. Nicolet, and J. L. Johnson, *Appl. Phys. Lett.*, vol. 45, 1984, p. 185.

52. P. A. Scott and J. M. Rigsbee, in *Ion Plating and Implantation Application to Materials*, ed. R. F. Hochman, American Society for Metals, Metals Park, Ohio, 1986, pp. 103–105.

53. G. K. Hubler and J. K. Hirvonen, in *ASM Handbook, vol. 5: Surface Engineering*, 10th ed. ASM International, Materials Park, Ohio, 1994, pp. 593–601.

54. F. A. Smidt, *Int. Met. Rev.*, vol. 15, 1990, pp. 61–128.

55. J. K. Hirvonen, *Mater. Sci. Rep.*, vol. 6, 1991, pp. 215–274.

56. G. K. Hubler, *Crit. Rev.*, in *Surf. Chem.*, vol. 2, no. 3, 1993, pp. 169–198.

57. G. K. Wolf, *J. Vac. Sci. Technol. A*, vol. 10, no. 4, 1992, pp. 1757–1764.

58. Y. Baba and T. A. Sasaki, *Mat. Sci. Eng.*, vol. A115, 1989, p. 203.

59. R. A. Kant and B. D. Sartwell, *J. Vac. Sci. Technol.*, vol. A8, 1990, p. 861.

60. G. Dearnaley, *Surf. Coat. Technol.*, vol. 33, 1987, pp. 453–467.

61. P. J. Martin, R. P. Netterfield, and W. G. Sainty, *J. Appl. Phys.*, vol. 55, 1984, p. 235.

62. E. P. Donovan, D. Van Vechten, A. D. F. Kahn, C. A. Carosella, and G. K. Hubler, *Appl. Opt.*, vol. 28, 1989, p. 940.

63. A. J. Armini, S. N. Bunker, and L. A. Stelmack, in *Advances in Coatings Technologies for Surface Engineering*, eds. C. R. Clayton, J. K. Hirvonen, and A. K. Srivatsa, The Metallurgical Society, Warrendale, Pa., 1997, pp. 79–90.

64. R. Jethanandini, *JOM*, Feb. 1997, pp. 63–65.

65. R. A. Kant, S. A. Dillich, B. D. Sartwell, and J. A. Sprague, *Met. Res. Soc. Symp. Proc.*, vol. 128, 1989, p. 165.

66. M. Barth, W. Ensinger, A. Shroer, and G. K. Wolf, in *Surface Modification Technologies II*, eds. T. S. Sudarshan and D. G. Bhat, The Metallurgical Society, Warrendale, Pa., 1990, p. 195.

67. M. Iwaki, *Mater. Sci. Eng.*, vol. A115, 1989, p. 369.

68. W. Ensinger and G. Wolf, *Mater. Sci. Eng.*, vol. A116, 1989, p. 1.

69. G. Wolf, *Nucl. Instrum. Meth. Phys. Res. B*, vol. 46, 1990, p. 369.

70. P. W. Natishan, E. McCafferty, E. P. Donovan, D. W. Brown, and G. K. Hubler, *Surf. Coat. Technol.*, vol. 51, 1992, p. 30.

71. Y. Chen, S. Liu, Z. Shang, C. Xu, Y. Zheng, X. Liu, and S. Zou, *Surf. Coat. Technol.*, vol. 51, 1992, p. 227.

72. E. McCafferty, G. K. Hubler, P. M. Natishan, P. G. Moore, R. A. Kant, and B. D. Sartwell, *Mater. Sci. Eng.*, vol. 86, 1987, p. 1.

73. C. J. Bedell, H. E. Bishop, G. Dearnaley, and J. E. Despout, *Nucl. Instrum. Meth. Phys. Res. B*, vol. 59/60, 1991, p. 245.

74. S. Kiyama, H. Hirano, Y. Domoto, K. Kuramoto, R. Suzuki, and M. Osumi, *Nucl. Instrum. Meth. Phys. Res. B*, vol. 80/81, 1993, p. 1388.

75. Y. Itoh, S. Hibi, T. Hioki, and J. Kawamoto, *J. Mater. Res.*, vol. 6, 1991, p. 871.

76. L. E. Seitzman, I. L. Singer, R. N. Bolster, and C. R. Gossett, *Surf. Coat. Technol.*, vol. 51, 1992, p. 232.

77. H. Kuwano and K. Nagai, *J. Vac. Sci. Technol.*, vol. A4, 1986, p. 2993.

78. D. M. Mattox, in *Handbook of Physical Vapor Deposition (PVD) Processes*, Chap. 8: "Ion Plating and Ion Beam Assisted Deposition," William Andrew, Noyes Publications, Park Ridge, N.J., 1998, pp. 398–443; in *ASM Handbook, vol. 5: Surface Engineering*, ASM International, Materials Park, Ohio, 1994, pp. 582–592.

79. E. B. Graper, in *Handbook of Thin Film Process Technology*, eds. D. A. Glocker and S. I. Shah, Institute of Physics Publishing, Bristol, England 1995, pp. A1.3.1–A1.3.3.

80. *IonBond Physical Vapor Deposition Hard Coating Systems*, IonBond, Inc., Madison Heights, Mich., 1984.

81. P. D. Dearnley, in *Ion Plating and Implantation Application to Materials*, ed. R. F. Hochman, American Society for Metals, Metals Park, Ohio, 1986, pp. 31–38.

82. D. M. Mattox, in *Handbook of Plasma Processing Technology: Fundamentals, Etching, Deposition and Surface Interactions, Chap. 13: "Ion Plating,"* eds. S. M. Rossnagel, J. J. Cuomo, and W. D. Westwood, Noyes Publications, Park Ridge, N.J., 1990.

83. H. K. Pulker, in *Coatings on Glass*, Elesevier, Amsterdam, The Netherlands, 1984, p. 250.

84. B. Shaw, E. Sikora, K. Kennedy, P. Miller, E. Principe, K. Scammon, K. Heidersbach, T. Miller, and J. Singh, in *Advances in Coatings Technologies for Surface Engineering*, eds. C. R. Clayton, J. K. Hirvonen, and A. R. Srivatsa, The Metallurgical Society, Warrendale, Pa., 1997, pp. 287–304.

85. T. Moran, *Physical Vapor Deposition (PVD)*, Business Communications Company, Inc., Norwalk, Conn., 1990.

86. M. Podob, in *Surface Modification Technologies XII*, eds. T. S. Sudarshan, K. A. Khor, and M. Jeandin, ASM International, Materials Park, Ohio, 1998, pp. 15–24.

87. M. Kramis and H. Sert, *Wear*, 1998.

88. E. Lugscheider, C. Barimani, and M. Lake, in *Surface Engineering Science and Technology I*, eds. A. Kumar, Y.-W. Chung, J. J. Moore, and J. E. Smugeresky, The Metallurgical Society, Warrendale, Pa., 1999, pp. 405–413.

89. B. J. Janoss, *Forming and Fabricating*, Oct. 1997.

90. O. Knotek, F. Loffler, and G. Kramer, in *Surface Modification Technologies VI*, eds. T. S. Sudarshan et al., The Metallurgical Society, Warrendale, Pa., 1993, pp. 465–483.

91. M. E. Graham, in *Surface Engineering Science and Technology I*, eds. A. Kumar, Y.-W. Chung, J. J. Moore, and J. E. Smugeresky, The Metallurgical Society, Warrendale, Pa., 1999, pp. 47–58.

92. D. L. Schulz and T. J. Marks, in *CVD of Nonmetals*, ed. W. S. Rees, Jr., VCH, Weinheim, Germany, 1996, pp. 37–150.

93. R. L. Boxman, private communication, 2000.

93a. J. Singh, F. Quli, D. E. Wolfe, J. T. Schriempf, and J. Singh, in *Surface Engineering: Science and Technology I*, The Metallurgical Society, Warrendale, Pa., 1999, pp. 59–74.

94. D. G. Bhat, presented at *II Jornadas de Ingenieria Metallurgica y Ciencia de Los Materieles*, Universidad Central de Venezuela, Caracas, Venezuela, Nov. 1997, pp. 15–25.

95. R. H. Horsfall and M. A. Peliman, *Role of Present and New PVD Coatings to Meet the Needs of the Cutting Tool Industry*, Multi-Arc Inc., Rockaway, N.J., 1996.

96. M. Pakala and R. Y. Lin, in *Surface Modification Technologies IX*, eds. T. S. Sudarshan, W. Reitz, and J. J. Stiglich, The Metallurgical Society, Warrendale, Pa., 1996, pp. 467–475.

97. P. A. Dearnley et al., *Surf. Mod. Tech. VI*, eds. T. S. Sudarshan et al., The Metallurgical Society, Warrendale, Pa., 1993, pp. 143–166.

98. E. B. Graper, in *Handbook of Thin Film Process Technology*, eds. D. A. Glocker and S. I. Shah, Institute of Physics Publishing, Bristol, England 1995, pp. A1.1:1–A1.1:7.

99. E. B. Graper, in *Handbook of Thin Film Process Technology*, eds. D. A. Glocker and S. I. Shah, Institute of Physics Publishing, Bristol, England, 1995, 1.2.:1–A1.2:8.

100. D. E. Wolfe, M. B. Movchan, and U. J. Singh, in *Advances in Coatings Technologies for Surface Engineering*, eds. C. R. Clayton et al., The Metallurgical Society, Warrendale, Pa., 1997, pp. 93–110.

101. B. Juettner, V. F. Pucharev, E. Hantzsche, and I. Beilis, in *Handbook of Vacuum Arc Science and Technology: Fundamentals and Applications*, eds. R. L. Boxman, and D. M. Sanders, Noyes Publications, Park Ridge, N.J., 1995, pp. 73–281.

102. G. Yu. Yushkov, A. Anders, E. M. Oaks, and I. G. Brown, *J. Appl. Phys.*, vol. 88, 2000, pp. 5618–5622.

103. R. L. Boxman and S. Goldsmith, *Surf. Coat. Technol.*, vol. 52, 1992, pp. 39–50.

104. P. A. Lindfors, in *Ion Plating and Implantation Application to Materials*, ed. R. F. Hochman, American Society for Metals, Metals Park, Ohio, 1986, pp. 161–167.

105. A. M. Peters, J. J. Moore, B. Mishra, and R. Weiss, in *Surface Modification Technologies XII*, eds. T. S. Sudarshan, K. A. Khor, and M. Jeandin, ASM International, Materials Park, Ohio, 1998, pp. 37–41.

105a. R. C. Tucker, Jr., *Proceedings: ITSC'95*, Kobe, Japan, May 1995, pp. 253–258.

106. R. L. Boxman, S. Goldsmith, V. N. Zhitomirsky, B. Gidalevich, I. Beilis, and M. Keidar, *Surf. Coat. Technol.*, vol. 86–87, 1996, pp. 243–253.

107. P. J. Martin and A. Bendavid, *Thin Solid Films*, vol. 394, 2001, pp. 1–15.

108. P. Siemroth and T. Schuelke, *Surf. Coat. Technol.*, vol. 133–134, 2000, pp. 106–113.

109. A. Anders, F. W. Ryan, W. Fong, and S. S. Bhatia, in *Proc. XIXth Int. Symp. on Discharges and Electrical Insulation in Vacuum*, Xi'an, China, September 2000, pp. 541–547.

110. L. Kaplan, I. Rusman, R. L. Boxman, S. Goldsmith, M. Nathan, and E. Ben-Jacob, *Thin Solid Films*, vol. 290–291, 1996, pp. 355–361.

111. R. L. Boxman, S. Goldsmith, S. Shalev, H. Yaloz, and N. Brosh, *Thin Solid Films*, vol. 139, 1986, pp. 41–52.

112. V. N. Zhitomirsky, I. Grimberg, L. Rapoport, N. A. Travitzky, R. L. Boxman, S. Goldsmith, and B. Z. Weiss, *Surf. Coat. Technol.*, vol. 120–121, 1999, pp. 2199–2225.

113. R. L. Boxman, V. N. Zhitomirsky, I. Grimberg, L. Rapoport, S. Goldsmith, and B. Z. Weiss, *Surf. Coat. Technol.*, vol. 125, 2000, pp. 257–262.

114. H. Erich, B. Hasse, M. Mausback, and K. G. Mueller, *IEEE Trans. Plasma Phys.*, vol. 18, 1990, pp. 895–903.

114a. A. M. Dorodnov, A. N. Kuznetson, and V. A. Petrosov, *Sov. Phy. Lett.*, vol. 5, 1979, p. 418.

114b. A. M. Dorodnov and V. A. Petrosov, *Sov. Phys.–Tech. Phys.*, vol. 26, 1981, p. 304.

115. I. I. Beilis, R. L. Boxman, and S. Goldsmith, in *Proc. XIXth Int. Symp. on Discharges and Electrical Insulation in Vacuum*, Xi'an, China, September 2000, pp. 226–230.

116. D. M. Sanders, D. B. Boercker, and S. Falabella, *IEEE Trans. Plasma Phys.*, vol. 18, 1990, pp. 883–894.

117. R. L. Boxman, in *Proc. XIXth Int. Symp. on Discharges and Electrical Insulation in Vacuum*, Xi'an, China, September 2000, pp. 226–230.

118. P. J. Martin and D. R. McKenzie, in *Handbook of Vacuum Arc Science and Technology: Fundamentals and Applications*, eds. R. L. Boxman, D. M. Sanders, Noyes Publications, Park Ridge, N.J., 1995, pp. 467–493.

119. P. J. Martin, in *Handbook of Thin Film Process Technology*, eds. D. A. Glocker and S. I. Shah, Institute of Physics Publishing, Bristol, England, 1995, pp. A1.4:1–1.4:16; in *Handbook of Vacuum Arc Science and Technology: Fundamentals and Applications*, eds. R. L. Boxman and D. M. Sanders, Noyes Publications, Park Ridge, N.J., 1995, pp. 367–396.

120. J. Vetter, in *Surface Modification Technologies IX*, eds. T. S. Sudarshan et al., The Metallurgical Society, Warrendale, Pa., 1996, pp. 455–466.

121. S. Ramalingam, in *Handbook of Vacuum Arc Science and Technology: Fundamentals and Applications*, eds. R. L. Boxman and D. M. Sanders, Noyes Publications, Park Ridge, N.J., 1995, pp. 519–559.

122. J. Vetter and A. J. Perry, in *Handbook of Vacuum Arc Science and Technology: Fundamentals and Applications*, eds. R. L. Boxman and D. M. Sanders, Noyes Publications, Park Ridge, N.J., 1995, pp. 493–519.

123. R. F. Bunshah, in *Encyclopedia of Advanced Materials*, eds. D. Bloor et al., Pergamon Press, Oxford, England, 1994, pp. 2008–2017; *IEEE Trans. on Plasma Science*, vol. 18, no. 6, December 1990, pp. 846–854.

124. D. B. Christey and G. K. Hubler, eds., *Pulsed Laser Deposition of Thin Films*, Wiley, New York, 1994.

125. J. S. Jabinski, A. A. Voevodin, J. J. Nainaparaampil, S. V. Prasad, and N. A. Pierce, *Surface Engineering: Science and Technology I*, The Metallurgical Society, Warrendale, Pa., 1999, pp. 127–142.

126. A. Morimoto and T. Shimizu, in *Handbook of Thin Film Process Technology*, eds. D. A. Glocker and S. I. Shah, Institute of Physics Publishing, Bristol, England, 1995, pp. A.1.5:1–A.1.5:10.

127. W. Reitz and J. Rawers, in *Surface Modification Technologies VI*, eds. T. S. Sudarshan and J. F. Braza, The Metallurgical Society, Warrendale, Pa., 1993, pp. 521–530.

128. D. B. Christey, A. Pique, R. C. Y. Auyeung, R. A. McGill, R. Chung, S. Lakeou, P. Wu, J. FitzGerald, H. D. Wu, and M. Dunignan, in *Surface Engineering Science and Technology I*, eds. A. Kumar, Y.-W. Chung, J. J. Moore, and J. E. Smugeresky, The Metallurgical Society, Warrendale, Pa., 1999, pp. 143–154.

129. J. S. Horwitz, in *ASM Handbook, vol. 5: Surface Engineering*, 10th ed., ASM International, Materials Park, Ohio, 1994, pp. 621–626.

130. S. L. Rohde, in *ASM Handbook, vol. 5: Surface Engineering*, 10th ed., ASM International, Materials Park, Ohio, 1994, pp. 573–581.

131. R. L. Boxman, S. Goldsmith, V. N. Zhitornisky, B. Alterkop, E. Gidalevich, I. Bellis, and M. Keider, *Surf. Coat. Technol.*, vol. 86–87, 1996, pp. 243–253.

131a. Section X, in *Handbook of Thin Film Process Technology*, eds. D. A. Glocker and S. I. Shah, Institute of Physics Publishing, Bristol, England, 1995.

131b. E. L. Paradis, *Thin Sold Films*, vol. 72, 1980, p. 327.

132. S. I. Shah, in *Handbook of Thin Film Process Technology*, eds. D. A. Glocker and S. I. Shah, Institute of Physics Publishing, Bristol, England, 1995, pp. A.3.2:1–A3.2:18.

133. S. Schiller, U. Heisig, and K. Goedicke, *Thin Sold Films*, vol. 40, 1977, pp. 327–334.

134. J. E. Mahan, *Physical Vapor Deposition of Thin Films*, John Wiley, New York, 2000.

135. R. J. Hill, ed., *Physical Vapor Deposition*, Temescal, Fairfield, Calif., 1986.

136. J. L. Cecchi, in *Handbook of Plasma Processing Technology*, eds. S. M. Rossnagel, J. J. Cuomo, and W. D. Westwood, Noyes Publications, Park Ridge, N.J., 1990, pp. 14–69.

137. J. S. Logan, in *Handbook of Plasma Processing Technology*, eds. S. M. Rossnagel, J. J. Cuomo, and W. D. Westwood, Noyes Publications, Park Ridge, N.J., 1990, pp. 140–159.

138. A. S. Penfold, in *Handbook of Thin Film Process Technology*, eds. D. A. Glocker and S. I. Shah, Institute of Physics Publishing, Bristol, England, 1995, pp. A.3.2:1–A3.2:27.

139. K. Ellmer, *J. Phys. D, Appl. Phys.*, vol. 33, no. 4, 2000, pp. R17–R32.

140. A. L. Greer and R. E. Somokh, in *Materials Science and Technology, vol. 15: Processing of Metals and Alloys*, vol. ed. R. W. Cahn, VCH, Weinheim, Germany, 1991, pp. 329–370.

141. R. C. Farrow, *IBM J. Res. Devel.*, vol. 42, no. 1, 1998.

142. B. Window and N. Saviides, *J. Vac. Sci. Technol.*, vol. A4, 1986, pp. 196–202.

143. D. G. Teer, private communication, 2000.

144. D. G. Teer, US Patent No. 5 556 519, September 1996.

145. D. P. Monaghan, D. G. Teer, P. A. Logan, I. Efeoglu, and R. D. Arnell, *Surf. Coat. Technol.*, vol. 60, 1993, pp. 525–530.

146. S. M. Rossnagel, *J. Vac. Sci. Technol.*, vol. B16, no. 6, 1998, p. 3008.

147. J. A. Hopwood, in *Thin Films*, ed. J. A. Hopwood, Academic Press, San Diego, 2000, pp. 1–7.

148. S. Rossnagel, in *Thin Films*, ed. J. A. Hopwood, Academic Press, San Diego, 2000, pp. 37–66.

149. J. A. Hopwood, ed., *Thin Films*, Academic Press, San Diego, 2000.

150. J. M. Parsey, Jr., in *Materials Science and Technology, vol. 16: Processing of Semiconductors*, vol. ed. K. A. Jackson, VCH, Weinheim, Germany, 1996, pp. 475–587.

151. D. J. Kapolnek, *Selected Epitaxy of GaN*, PhD. thesis, University of California, Santa Barbara, 1997.

152. M. Pessa, *Epitaxial Layers,* Tampere University of Technology, Tampere Finland, 1996.

153. F. J. Bruni and L. T. Nuyen, *JOM*, Aug. 1998, pp. 34–36.

154. W. E. Stanchina and J. F. Lam, in *Materials Science and Technology, vol. 16: Processing of Semiconductors*, vol. ed. K. A. Jackson, VCH, Weinheim, Germany, 1996, pp. 377–392.

155. A. C. Jones and P. O'Brien, *CVD of Compound Semiconductors: Precursor Synthesis, and Development and Applications*, VCH, Weinheim, Germany, 1997.

156. M. A. Hasan, J. Knall, S. A. Barnett, A. Rockett, J.-E. Sundgren, and J. E. Greene, *J. Vac. Sci. Technol.*, vol. 5B, 1987, p. 1332.

157. A. Y. Cho and K. Y. Cheng, *Appl. Phys. Lett.*, vol. 38, 1981, p. 360.

158. S. A. Barnett and I. T. Ferguson, in *Handbook of Thin Film Process Technology*, eds. D. A. Glocker and S. I. Shah, Institute of Physics Publishing, Bristol, England, 1995, pp. A2.0:1–A2.0:35.

159. A. Anselm, *An Introduction to MBE Growth*, University of Texas, 1997. (http://www.ece.utexas.edu/projects/ece/mrc/groups/streat_mbe/mbechapter.html)

160. T. F. Kuech and M. A. Tischler, in *Materials Science and Technology, vol. 16: Processing of Semiconductors*, vol. ed. K. A. Jackson, VCH, Weinheim, Germany, 1996, pp. 107–172.

161. D. Hartzell, L. K. Leung, and F. J. Towner, *JOM*, Aug. 1998, pp. 37–39.

162. D. R. Leadley, Warwick University, 1997. (http://www.warwick.ac.uk/~phsbm/mbe.htm)

163. *18th North American Conference on Molecular Beam Epitaxy*, Montreal, Canada, October, 10–13, 1999.

164. E. H. Parker, *The Technology and Physics of Molecular Beam Epitaxy*, Plenum, New York, 1985.

165. M. A. Herman and H. Sitter, *Molecular Beam Epitaxy: Fundamentals and Current Status*, Springer, New York, 1989.

166. J. Y. Tsao, *J. Crystal Growth*, vol. 110, 1991, p. 595.

167. C. T. Foxon and B. A. Joyce, *Current Topics Mater. Sci.*, vol. 7, 1981, p. 1.

168. E. Kasper and J. C. Bean, *Silicon Molecular Beam Epitaxy*, vols. I and II, CRC Press, Boca Raton, Fla., 1988.

169. B. A. Joyce, D. D. Vvedensky, and C. T. Foxon, in *Handbook on Semiconductors*, vol. 3, ed. S. Mahajan, North-Holland, Amsterdam, The Netherlands, 1994, p. 275.

170. C. R. Abernathy, J. D. McKenzie, and S. M. Donovan, *J. Cryst. Growth*, vol. 78, 1997, pp. 74–86.

171. J. M. Parsey, Jr. *JOM*, Dec. 1995, p. 24.

172. M. Pessa, private communication, 2000.

173. M. Pessa, M. Toivonen, M. Jalonen, P. Savolainen, and A. Salokatve, *Thin Solid Films*, vol. 306, 1997, pp. 237–243.

174. M. Pessa, M. Toivonen, P. Savolainen, S. Orsila, P. Sipila, M. Saarinen, P. Melanen, V. Vilokkinen, P. Uusimaa, and J. Haapamaa, *Thin Solid Films*, vol. 367, 2000, pp. 260–266.

175. G. Dagnall, J.-J. Shen, T.-H. Kim, R. A. Metzer, A. S. Brown, and S. R. Stock, *J. Electron. Mat.*, vol. 28, no. 8, 1999, pp. 933–938.

176. G. Y. Robinson, in *Handbook of Thin Film Process Technology*, eds. D. A. Glocker and S. I. Shah, Institute of Physics Publishing, Bristol, England, 1995, pp. A2.2:1–A2.2:22.

177. C. A. Coronado, E. Ho, P. A. Fisher, J. L. House, K. Lu, G. S. Petrich, and L. A. Kolodziejski, *J. Electron. Mat.*, vol. 23, no. 3, 1994, pp. 269–273.

178. C. R. Abernathy, in *Handbook of Thin Film Process Technology*, eds. D. A. Glocker and S. I. Shah, Institute of Physics Publishing, Bristol, England, 1995, pp. A2.1:1–A2.0:24.

179. M. B. Panish, *J. Electrochem. Soc.*, vol. 127, 1980, p. 2730.

180. M. B. Panish, H. Temkin, and S. Sumski, *J. Vac. Sci. Technol.*, vol. 3B, 1985, p. 657.

181. A. R. Calawa, *Appl. Phys. Lett.*, vol. 38, 1981, p. 701.

182. M. B. Panish and S. Sumski, *J. Appl. Phys.*, vol. 55, 1984, p. 3571.

183. D. Hyuet, M. Lambert, D. Bonnevie, and D. Defresne, *J. Vac. Sci. Technol.*, vol. 3B, 1985, p. 823.

184. C. R. Abernathy, *J. Cryst. Growth*, vol. 1997, pp. 74–86.

185. N. Kuroda, S. Sugou, T. Sakaki, and M. Kitamura, *Proceedings of the 5th International Conference on InP and Related Materials*, Paris, 1993.

186. J. E. Cunningham, T. H. Chiu, G. Timp, E. Agyekum, and W. T. Tsang, *Appl. Phys. Lett*, vol. 53, 1988, p. 1285.

187. M. Lambert, A. Perales, R. Vergnaund, and C. Stack, *J. Cryst. Growth*, vol. 105, 1990, p. 97.

188. D. Huert and M. Lambert, *J. Electron. Mat.*, vol. 15, 1986, p. 37.

189. M. B. Panish and H. Temkin, *Appl. Phys. Lett.*, vol. 44, 1984, p. 785.

190. H. Temkin, S. N. G. Chu, M. B. Panish, and R. A. Logan, *Appl. Phys. Lett.*, vol. 50, 1987, p. 956.

191. M. C. Wu, Y. K. Chen, M. A. Chin, and A. M. Sergent, *IEEE Photon. Technol.*, vol. 4, 1992, p. 676.

192. R. N. Nottenburg, Y. K. Chen, M. B. Panish, D. A. Humphrey, and R. A. Hamm, *IEEE Electron. Dev. Lett.*, vol. 10, 1989, p. 30.

193. W. G. Wey, K. S. Gibony, J. E. Bowers, M. J. W. Rodwell, P. Silvestre, P. Thiagarajan, and G. Y. Robinson, *IEEE Photon. Technol. Lett.*, vol. 5, 1993, p. 1310.

194. H. Temkin, D. Gershoni, and M. B. Panish, *Appl. Phys. Lett.*, vol. 50, 1987, p. 1776.

195. R. C. Powell, N.-E. Lee, and J. E. Green, *Appl. Phys. Lett.*, vol. 60, 1992, p. 2505.

196. J. N. Baillargeon, K. Y. Cheng, G. E. Hofler, P. J. Pearah, and K. C. Hsieh, *Appl. Phys. Lett.*, vol. 60, 1992, p. 2540.

197. R. F. Davis, M. J. Paisley, Z. Sitar, D. J. Kester, K. S. Ailey, K. Linthicum, L. B. Rowland, S. Tanaka, and R. S. Kern, *J. Cryst. Growth*, vol. 178, 1997, pp. 87–101.

198. L. Vescan, in *Handbook of Thin Film Process Technology*, eds. D. A. Glocker and S. I. Shah, Institute of Physics Publishing, Bristol, England, 1995, pp. B1.0:1–B1.0:12 and pp. B1.4:1–B1.4:41.

199. J. M. Blocher, Jr., in *Encyclopedia of Materials Science and Engineering*, Pergamon Press, Oxford, England, 1986, pp. 644–649.

200. M. L. Green and R. A. Levy, *Met. Progr.*, June 1985, pp. 63–69.

201. H. O. Pierson, in *ASM Handbook, vol. 5: Surface Engineering*, 10th ed., ASM International, Materials Park, Ohio, 1994, pp. 510–516; *Handbook of Chemical Vapor Deposition: Principles, Technology, and Applications*, vol. ed., Noyes Publications, Norwich, N.Y., 1999.

202. M. B. Bader et al., *Integrated Processing Equipment*, May 1990, pp. 149–154.

203. R. A. Levy and M. L. Green, *Mat. Res. Soc. Symp. Proc.*, vol. 71, 1986, pp. 229–247.

204. M. L. Green, M. E. Gross, L. E. Papa, K. J. Schnoes, and D. Brason, *J. Electrochem. Soc.*, vol. 132, no. 11, 1985, pp. 2677–2685.

205. M. Ohring, *The Materials Science of Films*, Academic Press, San Diego, Calif., 1992.

206. C. E. Morosanu, *Thin Films by Chemical Vapor Deposition*, Elsevier, Amsterdam, The Netherlands, 1990.

207. F. S. Glasso, *Chemical Vapor Deposition Materials,* Chemical Rubber Company, Boca Raton, Fla., 1991.

208. T. L. Chu, J. M. Jackson, A. E. Hyslop, and S. S. C. Chu, *J. Appl. Phys.*, vol. 42, 1971, p. 420.

209. T. L. Chu, *J. Electrochem. Soc.*, vol. 118, 1971, p. 1200.

210. B. J. Baliga, Ed., *Epitaxial Silicon Deposition*, Academic Press, Orlando, Fla., 1986.

211. G. M. Oleszek and R. L. Anderson, *J. Electrochem. Soc.*, vol. 120, 1973, p. 554.

211a. J. M. Blocher, Jr., in *Encyclopedia of Advanced Materials*, Pergamon Press, Oxford, England, 1994, pp. 644–664.

212. E. Yamaguchi and M. Minakata, *J. Appl. Phys.*, vol. 55, 1984, p. 3089.

213. E. F. Cave and B. R. Czorny, *RCA Review*, vol. 24, 1963, pp. 523–545.

214. S. Nakamura, *Ext. Abstr. Electrochem. Soc. Spring Meeting*, vol. 14, 1965, pp. 110–112.

215. H. Aharoni, I. A. Bar-Lev, and S. Margalit, *Thin Solid Films*, vol. 11, 1972, p. 313.

216. W. A. Bryant, *J. Electrochem. Soc.*, vol. 125, 1978, p. 1534.

217. R. S. Rosler, *Solid State Technol.*, vol. 20, 1977, p. 63.

218. W. Kern and R. S. Rosler, *J. Vac. Sci. Technol.*, vol. 14, 1977, pp. 1082–1099.

219. K. Maeda and S. M. Fischer, *Solid State Technol.*, June 1993, p. 83.

220. Landolt-Bornstein, *Numerical Data and Functional Relationships in Science and Technology*, Springer, Berlin, 1984, p. 17c.

221. R. C. Rossi, in *Handbook of Thin Film Deposition Processes and Techniques*, ed. K. K. Schuergraf, Noyes Publications, Park Ridge, NJ, 1988, pp. 80–111.

222. E. Effer, *J. Electrochem. Soc.*, vol. 112, 1965, p. 1020.

223. J. Tietjen and J. Amick, *J. Electrochem. Soc.*, vol. 113, 1966, p. 724.

224. P. Kordos, R. Schmbera, M. Heyden, and P. Balk, *GaAs and Related Compounds (Inst. Phys. Conf. Series)*, vol. 63, 1981, p. 131.

225. K. H. Bachem and M. Heyen, *J. Cryst. Growth*, vol. 55, 1981, p. 330.

226. H. Jurgensen, J. Korec, M. Heyden, and P. Balk, *J. Cryst. Growth,* vol. 66, 1984, pp. 73–82.

227. J. L. Regolini, F. Trincat, I. Berbezier, and Y. Shapira, *Appl. Phys. Lett.*, vol. 60, 1992, p. 956.

228. D. G. Bhat and J. E. Roman, in *10th International Conference on Chemical Vapor Deposition*, ed. G. Cullen, Honolulu, Hawaii, October 22, 1987, pp. 579–587A.

229. D. G. Bhat, in *Surface Modification Technologies*, Chap. 2, ed. T. S. Sudarshan, Marcel Dekker, New York, 1989, pp. 141–208.

229a. T. Takahachi and H. Kamiya, *J. Cryst. Growth*, vol. 26, 1974, pp. 203–209.

229b. H. O. Pierson and A. W. Mullendore, *Thin Solid Films*, vol. 95, 1982, pp. 99–104.

229c. K. Voigt and H. Westphal, *Proceedings of the Tenth Plansee Seminar*, vol. 2, Risley Translation 4877, 1981, pp. 611–622.

230. M. L. Green, Y. S. Ali, T. Boone, B. A. Davidson, C. L. Feldman, and S. Nakahara, *J. Electrochem. Soc.*, vol. 134, no. 9, 1987, pp. 2285–2292.

231. S. J. C. Irvine, in *Handbook of Thin Film Process Technology*, eds. D. A. Glocker and S. I. Shah, Institute of Physics Publishing, Bristol, England, 1995, pp. B1.3:1–B1.3:17.

232. D. L. Schultz and T. J. Marks, in *CVD of Nonmetals*, ed. W. S. Rees, Jr., VCH, Weinheim, Germany, 1996, pp. 37–150.

232a. H. O. Pierson, *Handbook of Carbon, Graphite, Diamond and Fullerences*, Noyes Publications, Park Ridge, N.J., 1994.

232b. W. A. Yarborough, *J. Mater. Res.*, vol. 7, no. 2, 1992, pp. 379–383.

232c. H. Holzschuh, *J. Phys. IV, France*, vol. 10, 2000, pp. Pr-49–Pr-54.

233. T. Cho, D. G. Bhat, and P. F. Woerner, in *ASM International Conference on Surface Modifications and Coatings*, 1985, paper No. 8512-016, pp. 1–4.

233a. D. G. Bhat, in *Surface Modification Technologies*, eds. T. S. Sudarshan and D. G. Bhat, The Metallurgical Society, Warrendale, Pa., 1988, pp. 1–21.

233b. H. O. Pierson, in *Chemically Vapor Deposited Coatings*, ed. H. O. Pierson, The American Ceramic Society, Westerville, Ohio, 1981, pp. 27–45.

233c. H. E. Rebenne and D. G. Bhat, *Surf. Coat. Technol.*, vol. 63, 1994, pp. 1–13.

234. A. J. Sherman, R. H. Huffies, and R. B. Kaplan, *JOM*, July 1991, pp. 20–23.

235. F. Thummler and R. Oberacker, *An Introduction to Powder Metallurgy*, The Institute of Materials, London, England, 1993.

236. R. Bauer, R. Smulders, E. Geus, J. Van der Put, and J. Schoomman, *Ceram. Eng. Sci. Proc.*, vol. 9, no. 7–8, 1988, pp. 949–956.

237. A. Kato, *Ceram. Bulletin*, vol. 66, no. 4, 1987, pp. 647–650.

238. D. G. Bhat and R. A. Holzl, *Thin Solid Films*, vol. 95, 1982, pp. 105–112; J. J. Stiglich, Jr, and D. G. Bhat, *Thin Sold Films*, vol. 72, 1980, pp. 503–509.

239. S. Mulligan and R. J. Dowding, *U. S. Army Mater. Tech. Lab Report 90-56*, Watertown, Mass., 1990.

239a. D. G. Bhat and K. Narasimhan, *Materials & Manufacturing Processes*, vol. 7, no. 4, 1992, pp. 613–624.

239b. K. Narasimhan and D. G. Bhat, *Surf. Coatings Technol.*, vol. 61, 1993, pp. 171–176.

240. S. P. Wang and R. C. Bracken, *Proceedings of the 10th International Conference on CVD*, ed. G. Cuillen, Electrochemical Society, Pennington, N.J., 1987, pp. 755–787.

241. J. Borland, *JOM*, Oct. 1991, pp. 23–27.

242. M. Green, *JOM*, Oct. 1991, p. 22.

243. D. Pramanic, *Semiconductor International*, June 1988, pp. 94–99.

244. J. K. Elliott, *Semiconductor International*, April 1988, pp. 150–153.

245. M. Venkatesan and L. Beinglass, *Sold State Technol.*, March 1993, pp. 49–53.

245a. G. S. Tompa, private communication, 2001.

246. M. W. Geis, in *Diamond Related Materials I*, 1992, pp. 684–687.

247. *Optoelectronic Devices*, Catalog 86-1, Issue III, Lumex, Palatine, Ill., 1986.

247a. J. Cuchiaro, private communication, 2001.

248. M. Hendricks, C. J. Werkhoven, F. Huussen, and E. Granneman, *Int. Conf. Electron. Mater.* (EMRS 1992 Meeting), 1992.

249. B. S. Meyersen, *App. Phys. Lett.,* vol. 48, 1986, pp. 797–799.

250. B. S. Meyerson, K. J. Uram, and F. K. LeGoues, *Appl. Phys. Lett.*, vol. 53, 1988, p. 2555.

251. J. F. Gibbons et al., *Mater. Res. Soc.*, vol. 74, 1987, p. 629.

252. J. L. Regolini, D. Dutartre, and D. Bensahel, *Solid State Technol.*, Feb., 1991, p. 47.

253. A. Kermani and F. Wong, *Solid State Technol.*, July 1990, pp. 41–43.

254. M. L. Green et al., *J. Appl. Phys.*, vol. 65, 1989, 2558–2560.

255. R. Singh, *J. Appl. Phys.*, vol. 63, 1988, pp. R59–R114.

256. K. H. Jung, T. Y. H. Sieh, and D. L. Kwong, *JOM*, Oct. 1991, pp. 38–43.

257. C. A. King, J. L. Hoyt, and J. F. Gibbons, *IEEE Trans. Electron Devices*, vol. 36, 1989, pp. 2093–2104.

258. C. M. Gronet et al., *Appl. Phys. Lett.*, vol. 48, 1986, pp. 1012–1014.

259. J. L. Regolini et al., *Appl. Phys. Lett.*, vol. 54, 1989, pp. 658–659.

260. Y. Zhong et al., *Appl. Phys. Lett.*, vol. 57, 1990, pp. 2092–2094.

261. V. Murali et al., *J. Electron. Mater.*, vol. 18, 1989, pp. 731–736.

262. J. C. Sturm et al., *IEEE Electron Device Lett.*, vol. 7, 1986, pp. 577–579.

263. M. L. Green, in *Encyclopedia of Advanced Materials*, Pergamon Press, Oxford, England, 1994, pp. 2200–2203.

264. P. K. Bachman, G. Gartner, and H. Lydtin, *MRS Bulletin*, Dec. 1988, pp. 1–59.

265. D. G. Bhat, in *Surface Modification Technologies—An Engineer's Guide*, ed. T. Sudarshan, Marcel Dekker, New York 1989, pp. 141–208.

266. K. Bartsch and A. Leonhardt, in *Surface Engineering: Science and Technology I*, eds. A. Kumar et al., The Metallurgical Society, Warrendale, Pa., 1999, pp. 87–98.

267. W. S. Rees, Jr., in *CVD of Nonmetals*, ed. W. S. Rees, Jr., VCH, Weinheim, Germany, 1996, pp. 1–35.

268. S. K. Kim, in *Surface Engineering: Science and Technology I*, eds. A. Kumar et al., The Metallurgical Society, Warrendale, Pa., 1999, pp. 105–110.

269. S. Matsuo, in *Handbook of Thin-Film Deposition Processes and Techniques*, ed. K. K. Schuegraf, Noyes Publications, Park Ridge, N.J., 1988, pp. 147–169.

270. R. F. Bunshah, ed., *Handbook of Deposition Technologies for Films and Coatings*, 2nd ed., Noyes Publications, Park Ridge, N.J., 1994.

271. I. P. Herman et al., *Mater. Res. Soc., Symp. Proc.*, vol. 29, 1984, p. 29.

272. P. Burggrapp, *Semiconductor International,* May 1998, p. 116.

273. S. M. Zemskova et al., *J. Phys. IV, France*, vol. 10, 2000, pp. Pr2-35–Pr2-42.

274. M. Hata, *JOM*, Aug. 1998, pp. 40–43.

275. Ref. UCLA Case No. 1992-517; US Patent No. 5,403,620.

276. R. D. Dupuis, *J. Crystal Growth*, vol. 178, 1997, pp. 56–73.

277. Compound Semiconductor.com.

278. P. Grodzinski, S. P. Den Baars, and H. C. Lee, *JOM*, Dec. 1995, pp. 25–31.

278a. G. S. Tompa, L. G. Provost, and J. Cuchiaro, in *Transport Conductive Oxides*, June 19–20, 2000, Denver, Colo.

279. M. Shimizu et al., *J. Cryst. Growth*, vol. 145, 1994, pp. 209–213.

280. R. S. Goldman and R. M. Feenstra, *J. Vac. Sci. Technol.*, vol. B15, 1997, p. 1027.

281. G. B. Stringfellow, *Organometallic Vapor-Phase Epitaxy: Theory and Practice*, Academic Press, Boston, Mass., 1989.

282. M. Razeghi, in *ASM Handbook, vol. 5: Surface Engineering*, 10th edition, ASM International, Materials Park, Ohio, 1994, pp. 517–531.

283. W. T. Tsang, *Appl. Phys. Lett.*, vol. 45, 1984, p. 1234.

284. T. H. Chiu, in *Handbook of Thin Film Process Technology*, eds. D. A. Glocker and S. I. Shah, Institute of Physics Publishing, Bristol, England, 1995, pp. A.2.3:1–A2.3:18.

285. W. T. Tsang, T. H. Chiu, J. E. Cunningham, and A. Robertson, Jr., *Appl. Phys. Lett.*, vol. 50, 1987, p. 1376.

286. T. H. Chiu et al., *J. Elect. Mat.*, vol. 17, 1987, p. 217

287. T. H. Chiu et al., *J. Cryst. Growth*, vol. 124, 1992, p. 165.

288. W. T. Tsang, *Appl. Phys. Lett.*, vol. 46, 1985, p. 1086.

289. J. E. Cunningham et al., *J. Cryst. Growth*, vol. 136, 1994, p. 282.

290. T. H. Chiu et al., *Appl. Phys. Lett.*, vol. 57, 1990, p. 171.

291. M. Kongai et al., *J. Cryst. Growth*, vol. 98, 1989, p. 167.

292. *ALD Overview*, Structured Materials Industries, Inc. Piscataway, N.J., January, 2001.

293. A. L. Green and R. E. Somekh, in *Materials Science and Technology, Vol. 15: Processing of Metals and Alloys*, vol. ed. R. W. Cahn., VCH, Weinheim, Germany, 1991, pp. 329–370.

294. S. M. Bedair, in *Encyclopedia of Advanced Materials*, Pergamon Press, Oxford, England, 1994, pp. 142–153.

295. M. Pessa, R. Mäkela, and T. Suntola, *Appl. Phys. Lett.*, vol. 38, no. 3, 1981, pp. 131–132.

296. C. H. L. Goodman and M. V. Pessa, *J. Appl. Phys.*, vol. 60, no. 3, 1986, pp. R65–R81.

297. S. Suntola, in *Handbook of Thin Film Process Technology*, eds. D. A. Glocker and S. I. Shah, Institute of Physics Publishing, Bristol, England, 1995, pp. B.1.5:1–B1.5:17.

298. W. J. Lackey, in *CVD of Nonmetals*, ed. W. S. Rees, Jr., VCH, Weinheim, Germany, 1996, pp. 321–366.

299. W. J. Lackey and T. L. Starr, in *Fiber Reinforced Ceramics: Fabrication of Fiber-Reinforced Ceramic Composites by Chemical Vapor Infiltration: Processing, Structure and Properties*, ed. K. S. Mazdiyasni, Noyes Publications, Park Ridge, N.J., 1990, pp. 397–450.

300. S. Vaidyaraman, W. J. Lackey, G. B. Freeman, P. K. Agrawal, and M. D. Langman, *J. Mater. Res.*, vol. 10, 1995, p. 1469.

301. P. H. Lee, *Bulletin,* IonBond, Inc., Madison Heights, Mich., 1997.

302. V. Midha and D. J. Economou, *J. Electrochem. Soc.*, vol. 144, 1997, p. 4062.

303. A. Sanjurjo et al., *Surf. Coatings Technol.*, vol. 49, 1991, pp. 110–115.

304. B. J. Wood et al., *Surf. Coatings Technol.*, vol. 49, 1991, pp. 228–232.

305. A. Sanjurjo, M. C. H. McKubre, and G. D. Craig, *Surf. Coatings Technol.*, vol. 39/40, 1989, pp. 691–700.

306. A. Sanjurjo, K. Lau, and B. Wood, *Surf. Coatings Technol.*, vol. 54/55, 1992, pp. 219–223.

CHAPTER 19
THERMAL SPRAY COATINGS

19.1 INTRODUCTION

Thermal spraying is a method in which metallic, ceramic, cermet, and some polymeric materials in the form of powder, wire, or rod are fed to a torch or gun with which they are heated to near, or somewhat above, their melting point. The resulting molten or nearly molten finely divided droplets of materials are accelerated in a gas stream by hot gas or plasma jets, and successively impacted on a (colder) substrate, leading to lateral flattening and rapid solidification and cooling onto a substrate, or onto a previously deposited layer.[1,2] The resulting microstructure of the solidified coating exhibits thin lamellar "splats" adhering to the substrate surface by mechanical bonding (i.e., overlapping and interlocking of many splats). The quality of the interfaces between these splats directly influences the properties of coatings, particularly the coating adhesion, bond strength, wear resistance, and thermal transport properties.[3] The total coating thickness, usually produced by multiple passes of the coating device, is in the range of 50 to 1000 μm, but both thinner and thicker coatings are possible for some applications.[1,2] Example microstructures are shown in Fig. 19.1.[1,2]

Thermal spraying is a very powerful and flexible surface modification technique for surface protection and extension of its service life in many industries. In recent years, it has become very extensively used for performance-critical applications due to the increased reliability and reproducibility of the thermal spray coatings.[4] Numerous thermal sprayed coatings have been produced with added functional properties such as high wear and corrosion resistance and high-temperature properties.

Table 19.1 shows the comparison of coating characteristics among thermal spray, CVD, and PVD coatings.[2] Tables 19.2 and 19.3 list comparisons of coating characteristics among different thermal spray processes.[5]

This chapter describes the advantages and disadvantages, important processes and recent developments, post-spray treatments, coating characteristics, and applications of thermal spraying techniques.

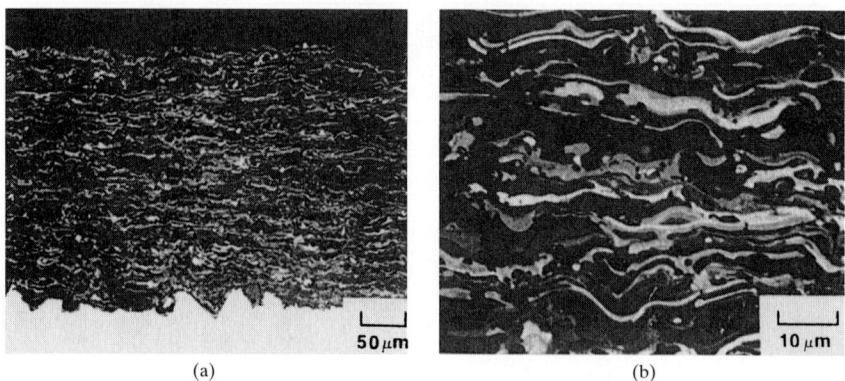

(a)   (b)

FIGURE 19.1   (a, b) Microstructures of as-polished Al_2O_3-TiO_2 denotation gun (D-Gun®) coating, at different magnifications, showing the lamellar nature of a thermal spray coating.[1] (*Reprinted by permission of ASM International.*)

TABLE 19.1   Comparison of Different Surface Coating Characteristics[2]

Characteristics	Thermal spray	CVD	PVD
Compositions	Metals and ceramics	Limited	Metals and ceramics
Coating thickness, typical	50–500 μm	5–75 μm	3–15 μm
Throwing power	Line-of-sight	Omnidirectional	Line-of-sight to moderate with bias
Substrate type	Metals, some cermets, ceramics, and polymerics	Metals, some ceramics, and cermets	Metals, ceramics, cermets, and polymerics
Substrate temperature	Ambient	High (800–1000°C)	Ambient to moderate
Adherence	Poor to excellent mechanical	Metallurgical to very good chemical	Moderate mechanical to excellent chemical
Deposition cost, $/$\mu$m·cm^2	0.002–0.05	0.001–0.5	0.05–1

Courtesy of the High Temperature Society of Japan.

19.2 ADVANTAGES AND DISADVANTAGES

The advantages of thermal spray processes are the following[1]:

1. An extremely wide variety of substrate and coating material selections can be made. Coatings can be applied over large areas. Virtually any material that melts without decomposing can be sprayed.

2. The coatings are inexpensive and relatively simple to produce (compared to chrome plating) and many of them provide excellent wear and corrosion resistance.

TABLE 19.2 Typical Characteristics of Thermal Spray Processes[5]

Process	Gas temperature, °C	Particle velocity, m/s	Adhesion, MPa	Oxide content, %	Porosity, %	Spray rate, kg/hr	Relative cost, low = 1	Typical deposit thickness, mm
Flame	3,000	40	8	10-15	10-15	2-6	1	0.1-15
Arc wire	N/A*	100	12	10-20	10	12	2	0.1->50
Air plasma spray	12,000	200-400	4->70	1-3	1-5	4-9	4	0.1-1
Vacuum plasma spray	12,000	400-600	>70	ppm	<0.5	4-9	5	0.1-1
HVOF (Jet Kote)	3,000	800	>70	1-5	1-2	2-4	3	0.1->2
D-Gun®	4,000	800	>70	1-5	1-2	0.5	N/A*	0.05-0.3

* N/A: not applicable.
Source: S. Grainger, ed., *Engineering Coatings: Design and Application*, Abington Publishing, 1989, p. 77. Reprinted by permission of ASM International.

TABLE 19.3 Comparison of Typical Thermal Spray Processes[1]

Process	Materials	Feed material	Surface preparation	Substrate temperature		Particle velocity	
				°C	°F	m/s	ft/s
Powder flame spray	Metallic, ceramic, and fusible coatings	Powder	Grit blasting or rough threading	150–160	302–325	65–130	200–400
Wire flame spray	Metallic coatings and fusible coatings	Wire	Grit blasting or rough threading	95–135	200–275	230–295	700–900
Ceramic rod spray	Ceramic and cermet coatings	Rod	Grit blasting	95–135	200–275	260–360	800–1100
Two-wire electric arc	Metallic coatings	Wire	Grit blasting or rough threading	50–120	125–250	240	800
Nontransferred arc plasma	Metallic, ceramic, compounds, and plastics	Powder	Grit blasting or rough threading	95–120	200–250	240–560	800–1850
High-velocity oxy-fuel	Metallic, ceramic, and cermet coatings	Powder	Grit blasting	95–150	225–300	100–550	325–1800
D-Gun®	Metallic, ceramic, and cermet coatings	Powder	Grit blasting or as-machined	95–160	225–300	730–790	2400–2600
Super D-Gun®	Metallic, ceramic, and cermet coatings	Powder	Grit blasting or as-machined	95–160	225–300	850–1000	2800–3300
Transferred arc plasma	Metallic fusible coatings	Powder	Light grit blasting or chemical cleaning	Fuses base metal	Fuses base metal	490	1600

Reprinted by permission of ASM International.

3. The coating is applied without significantly heating the substrate (below 150°C) at a fast deposition rate. Thus, materials with very high melting points can be applied to finely machined, fully heat-treated parts without changing the properties and thermal distortion of the parts.

4. Lower maintenance cost, extension of service life, better product quality, and higher production capacity (by reduction in production stop loss) can be achieved.[6]

5. It is possible, in some cases, to strip with high-pressure water jets or other methods and recoat worn or damaged coatings without changing the properties or dimensions of the part.[7]

6. Thickness of the modified layer can be in the order of mm.

7. It is an appropriate process for the surface modification of Al alloys and for improvement in the functionality of mechanical equipment by imparting necessary properties to their surface alone.[8]

8. Unlike chrome plating, it is an environmentally friendly process.

9. In some cases [e.g., plasma, wire arc, or high-velocity oxy-fuel (HVOF) processes], it is possible to form near-net and free-standing shaped parts.

10. It can be applied to welded surfaces to provide additional protection to the component.[9]

The disadvantages of thermal spray processes are the following[1]:

1. It is a line-of-sight process. It can only coat what the torch or gun can "see."

2. It is not possible [with the exception of D-Gun® spray (DGS)] to coat very smooth or mirror-like surfaces without, to some extent, roughening the surface by grit blasting and/or machine threading.

3. The application of thermal sprayed coatings are sometimes limited by their poor coating properties such as inhomogeneities, layered microstructures with defects (porosity, cracks, and oxide inclusions), and low adhesion. (See Table 19.2.)

4. There are size limitations on coating small and deep cavities based on where a torch or gun will fit.

5. There are limitations in regard to many electronic and optical applications and special corrosion and wear applications.[2]

6. The production of pit-free surfaces over large areas can be a challenge, despite very low surface roughness, with the exception of coatings with the highest density and cohesive strength.[2]

7. The capital cost is very high and the payout period is very long. (However, compared to PVD or most CVD, neither is excessive.)

8. Thermal spraying is not successful with sharp angles, sharp ends, and so forth, due to the high concentration of thermal stresses developed in these coated areas.

9. To achieve improved thermal sprayed coatings, it is occasionally essential to use different post-spray treatments such as laser alloying (or remelting), rolling, squeezing, impregnation, and so forth[10] (to be discussed later).

As shown in Fig. 19.2, thermal spray processes are distinguished by the heat source (combustion, electrical, or plasma), feedstock material (powder, wire, or rod), and surrounding environment (e.g., air, low pressure, or vacuum). Thermal spray is

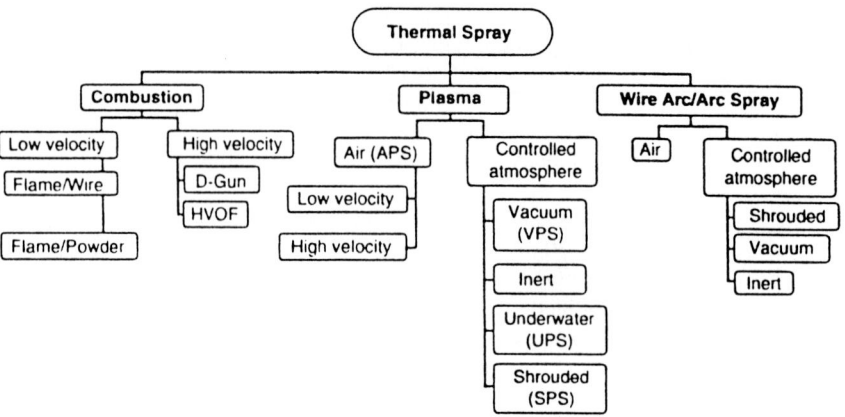

FIGURE 19.2   General types of thermal spray processes.[11,12] (*Reprinted by permission of ASM International.*)

typically classified into three main categories, namely (1) combustion spray [flame-powder/wire/rod, high-velocity oxy-fuel gas, and detonation-gun®], (2) electric/wire-arc or arc spray (air shroud, vacuum, inert), and (3) plasma spray (atmospheric, controlled atmosphere, low pressure, vacuum).[11,12] Major thermal spray processes used to produce a variety of coatings for various industrial applications are described below. Included is also a brief discussion of some recent developments.

19.3   PROCESSES

Thermal spray coating processes differ mostly in the manner by which the droplets of the material to be deposited are heated and accelerated. In most cases, the sprayed surfaces should be cleaned and degreased, masked, and roughened (by rough threading and/or grit blasting) prior to spraying to ensure adequate bond strength between the coating and the substrate. Table 19.4 shows the process parameters of various thermal spraying techniques.[13]

Flame Spray. *Flame spray (FS)*, also known as *combustion flame spray*, involves the combustion of fuel gas [such as acetylene, propane, methyl-acetylene-propadiene (MAPP) gas, and H_2] in compressed O_2 with an O_2:fuel ratio of 1:1 to 1.1:1] to heat and melt the feedstocks/coating material (in powder, wire, or rod form) (Figs. 19.3a and 19.3b) and propel the molten particles. In many flame spray guns, several combinations of gases are used to balance out the operating cost and coating properties. Usually, changing the nozzle and/or air gap is adequate to adapt the gun to different alloys, wire sizes, or gases. Figure 19.3a and 19.3c shows powder and wire/rod flame spray guns, respectively.[11,12]

    An FS process is characterized by an axial gas inlet, a low flame velocity, a transport gas flow rate of 3 to 5 Nl/min, a spray distance of 120 to 250 mm, a spray angle of 90°, and a powder feed rate of 50 to 100 g/min. Powders (typically, spherical, 5 to 100 µm in diameter) are introduced axially or perpendicular to the torch. Wire

TABLE 19.4 Process Parameters of Various Thermal Spraying Techniques[13,35]

Thermal spraying techniques	Working flame	Flame temperature, K	Flame velocity, m/s	Powder particle sizes, μm	Powder injection feed rate, g/min	Spray distance, mm
FS (powder)	Fuel + O_2 (g)	3000–3350	80–100	5–100	50–100	120–250
APS	Ar or mixture or Ar + H_2, Ar + He and Ar + N_2 (g)	Up to 30,000	800	5–100	50–100	60–130
AS	Various electrically conductive wires (e.g., Zn, Al)	Arc temperature of 6100 K by an arc current of 280 A*	Velocity of molten particles formed can reach up to 150	N/A	50–300	50–170
DGS	Detonation wave from a mixture of acetylene + O_2	Up to 4500 with 45% acetylene	2930[†]	5–60	16–40[‡]	100[§]
HVOF	Fuel gases (g) (acetylene, kerosene, propane, propylene, or H_2) with O_2	Up to 3440 K at ratio of O_2:acetylene (1.5:1 by volume)	2000	5–45	20–80	150–300
VPS	Ar mixed with H_2, He, or N_2	Temperature expressed in electron temperature of 10,000 to 15,000	Velocity of plasma in the range 1500–3000	5–20	50–100 (spraying in vacuum; during spraying pressure ~ 655–133 Pa)	300–400
CAPS	Same as APS	Same as APS	Same as APS	Same as APS	Same as APS	100–130 in shrouded plasma spray[¶]

* D. R. Marantz, in *Science and Technology of Surface Coating*, eds. B. N. Chapman and J. C. Anderson. Academic Press, London, 1974, p. 308.
[†] C. W. Smith, in *Science and Technology of Surface Coating*, eds. B. N. Chapman and J. C. Anderson. Academic Press, London, 1974, p. 262.
[‡] Y. S. Borisov, *Detonation Spraying; Equipment, Materials and Applications*, Essen, Germany: Thermische Spritzkonferenz, 1990.
[§] E. Schwartz, in *9th International Thermal Spraying Conference*, Instituut voor Lastechniek, The Hague, The Netherlands, 1980, p. 91.
[¶] M. Okada and H. Maruo, *British Welding J.*, vol. 15, 1968, p. 371.
Source: After M. L. Lau.

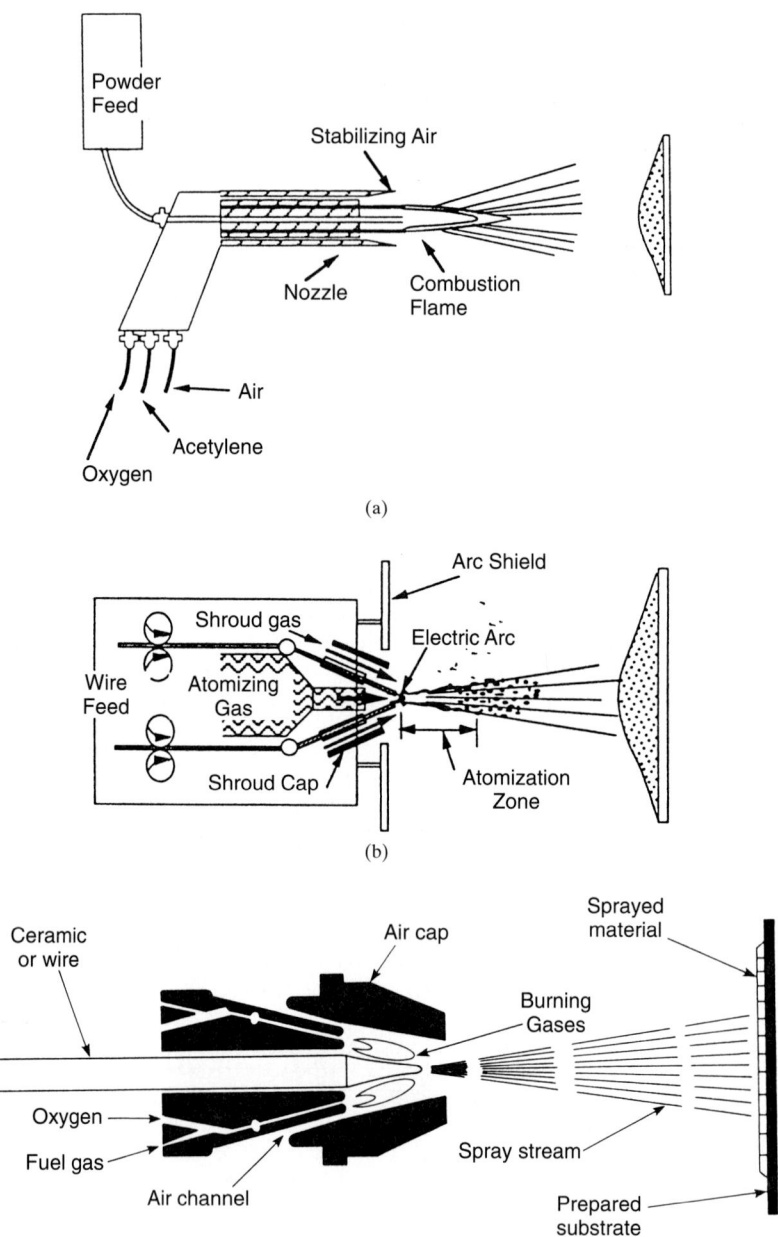

FIGURE 19.3 Schematics of (*a*) powder flame spray,[11,12] (*b*) wire-arc spray, and (*c*) wire/rod flame spray.[1] (*Reprinted by permission of ASM International.*)

TABLE 19.5 Maximum Temperature of Heat Sources[1]

Heat source	Approximate temperature (stoichiometric combustion)
Propane-oxygen	2526°C (4579°F)
Natural gas-oxygen	2538°C (4600°F)
Hydrogen-oxygen	2660°C (4820°F)
Propylene-oxygen	2843°C (5240°F)
Methylacetylene/propadiene- oxygen (MAPP)	2927°C (5301°F)
Acetylene-oxygen	3087°C (5589°F)
Plasma arc	2200–28,000°C (4000–50,000°F)

Source: Adapted from Publication 1G191, National Association of Corrosion Engineers. Reprinted by permission of ASM International.

(made of Mo, Zn, or Al, 3 to 6 mm in diameter, with a feed rate of 80 to 650 g/min) or rod (made of Al_2O_3 or Cr_2O_3) can also be used instead of the powder.[13]

Flame temperatures and characteristics are functions of the O_2:fuel ratio and pressure. Table 19.5 shows the approximate temperatures for stoichiometric combustion at 1 atm for some oxy-fuel combinations. Generally, as-deposited (or cold spray) flame-sprayed coatings are characterized by lower bond strengths (typically, 60 MPa for flame-sprayed NiAl coatings, 70 MPa for self-fluxing alloys, and 15 to 30 MPa for ceramic coatings), higher (10 to 20%) porosity, a narrower working temperature range, coating thicknesses typically in the range of 100 to 2500 μm, and higher heat transmittal to the substrate than most other thermal spray processes.

Advantages. The *advantages* of the flame spray process include its low capital cost, high deposition rates and efficiencies, and relative ease of operation and cost of equipment maintenance. As a result, this process is used worldwide.

Disadvantages. Usually the process yields low-performance coatings and is not used where high-density, high-bond strength coatings are required.

Applications. The flame spray process finds applications for (1) the reclamation of worn or out-of-tolerance parts, frequently using nickel-base alloys; wear-resistant coatings using tungsten carbide/nickel-base alloys; corrosion-resistant Zn coating on bridges and other structures; and some bearings using bronze alloys as sealants.[1]

Recently, flame spray BN-based cermet (5.5 BN/Ni-14Cr-8Fe-3.5Al) coating has been successfully applied to a hearth roll in a stainless strip annealing-pickling line which is exposed to a temperature of 1200°C in an oxidative atmosphere. This abradable coating has surface cleaning (i.e., buildup/pickup resistance) characteristics over the surface layer and has more high-temperature wear resistance than ceramic (44Si-28CaO-17MgO-MnO) coating, which prevents pickup and provides four times more serviceable life of hearth rolls.[14]

Flame and plasma processes can also be used to produce very specialized powders for medical applications.[15]

Flame Spray and Fuse. This is a variation of the flame spray method in which the coating materials are self-fluxing* (i.e., these compounds contain elements that

---

* Self-fluxing or fusible alloys are basically characterized as a group of Ni(Cr)BSi alloys with high-temperature brazing characteristics, thereby giving these alloys the highest cohesive and adhesive strength of thermal sprayed coatings. In some cases, Ni is substituted by Co and in rare cases by Fe. After fusion, they represent some of the toughest-wear alloy systems available.

react with oxygen or oxides to form low-density oxides or borosilicates that float to the surface, thereby improving density to near 100%, bonding, etc.). Usually, the coating materials are spherical nickel- or cobalt-based alloy powders that contain boron, phosphorus, or silicon, either singly or in combination, as melting-point depressants and fluxing agents. In reality, parts are prepared and coated as in other thermal spray processes and then fused. There are two variants: spray and fuse, and spray-fuse. In the former, the fusion is done after spray coating using one of several techniques such as flame or torch, induction, or vacuum, inert, or hydrogen furnaces. In the latter, both the spray coating and fusion are conducted simultaneously.[1]

The alloys used generally fuse at 1010 to 1175°C (1850 to 2150°F), based on the composition. Reducing atmosphere flames are used to ensure a clean, well-bonded coating. In vacuum and hydrogen furnaces, there is a tendency for the liquid to wick, sag, or run onto neighboring areas, which is removed by several commercially available brushable stopoff materials. These coatings are fully dense and exhibit metallurgical bonds. Excessive porosity and nonuniform bonding usually imply insufficient heating.[1]

Spray-and-fuse (hardfacing) coatings exhibit several excellent properties such as high wear and corrosion resistance, high hardness (up to 65 Rc), and high density, but the main drawbacks are that the high temperatures limit the substrate that can be coated[15a] and that high stresses at the coating/substrate (due to high heat input) lead to the distortion of components.[16] Spray and fuse coatings, applied in a thickness of 0.5 to 2 mm, show a high bond strength and a very low porosity level. These coatings find applications in many industries for wear and corrosion protection and/or to minimize friction. Typical applications include plungers, sleeves, tools, extruders, glass molds, tubes, rolls, and guides.[17] Other important applications are sucker rods in the oil industry and plowshares in the agriculture industry. Blended tungsten carbide or chromium carbide powders are commercially available to increase resistance to wear from abrasion, fretting, and erosion. Grinding is normally required for machining a fused coating due to its high hardness. Use of spray-and-fuse coatings is confined to substrate materials that can withstand the 1010 to 1175°C (1850 to 2150°F) fusion temperatures. Slower cooling rates may be required to prevent cracking in very thick sections or where a substantial difference in the thermal expansion coefficients between the coating and the substrate is present.[1]

Electric- (or Wire-) Arc Spray or Arc Spray. The *electric-* (or *wire-*) *arc*, or *arc spray* (*AS*) process, also known as the *cold process*, uses two relatively ductile, electrically conductive, opposed charged wires (such as Zn, Al, 85Zn-15Al, 316 or 440 stainless steels, or cored, 1.5 to 5 mm or 0.06 to 0.2 in. in diameter), which are arc melted. The molten metal on the wire tips is atomized [using compressed air, reactive, or inert atmosphere, with gas flow rates of 1 to 80 m³/hr and gas pressures of 0.2 to 0.7 MPa] and accelerated (with velocity up to 150 m/s) toward the substrate surface[13] (Fig. 19.3*b*). Alloy coating can be produced provided the wires are made of different materials.[13] The structure and properties of the coatings are strong functions of the processing parameters such as arc voltage, particle size, particle velocity, atomizing gas pressure, nozzle design, and so forth.[18] The coating thickness varies between 100 and 1500 μm, and the bond strength lies in the range of 10 to 30 MPa for Zn and Al coatings and up to 70 MPa for NiAl coatings.[13]

Advantages. The advantages of arc spray over flame spray process are the following[13]:

1. Higher coating bond strengths, in excess of 69 MPa (10 ksi) for some materials
2. High deposition/spray rates and efficiency up to 40 kg/hr (88 lb/hr)[19]
3. Ability to deposit less expensive wear-resistant coatings
4. Lower substrate heating than in flame spray processes due, primarily, to the absence of a flame touching the substrate
5. In most instances, one of the most inexpensive and most portable coating techniques
6. Well adapted in many diverse industries and for on-site coating of large equipment
7. Denser coatings
8. Low electrical power requirements (5 to 30 kW), and, with few exceptions, no need of expensive gas such as argon
9. Potential to deposit composite coatings containing carbides or oxides using cored wires

Disadvantages. (1) The use of compressed air during arc spraying results in severe oxidation in the molten metal droplets and a consequent reduction of carbon level and increase of brittle and porous iron oxide inclusions in the coating. However, carbon steel coatings with enhanced mechanical properties have been developed by increasing carbon retention (by 80%) through the use of N_2 as an atomizing gas. In this situation, an increase (by 40%) in the microhardness of the final carbon steel coating has been found compared to the one using compressed air.[20] Moreover, the abrasive wear resistance of the nitrogen-sprayed coating was approximately 25% better than that of the air-sprayed coating.[20] (2) Only electrically conductive wires can be sprayed, and for substrate preheating, a separate heating source is required.[19]

Applications. Electric-arc sprayed coatings are widely employed in the high-volume, low-cost production of zinc corrosion-resistant coatings. In specialized cases, metal-face molds can be made by using a fine spray attachment available from some manufacturers, which can duplicate extremely fine detail, such as the relief lettering on a printed page.

They find applications as carbon steel coatings for automotive engines and worn machine parts such as shafts, journals, and rolls.[20]

It is possible to produce tungsten carbide coatings more economically as an alternative for some applications. However, the main problem with that method is the relatively low carbide (<30%) content, which may be inadequate for severe wear resistance applications.[21]

Arc spray can also be used as a brazing method for assembly of the pump impeller in automotive industry.[22]

Plasma Spray. *Plasma spray (PS)* is one of the most versatile and economical thermal spray coating processes used to produce protective wear-, heat-, and corrosion-resistant coatings of virtually any metallic, cermet, or ceramic materials (including oxides, carbides, borides, and refractory materials) and to deposit *functionally graded materials* such as NiAl-8 wt% Y_2O_3-stabilized ZrO_2 (YSZ), also called *thermal barrier coatings (TBCs)*, for turbine blades. In this process, the temperature and velocity of particles are the two most important parameters. Hearth rolls (in the continuous annealing line of a steel plant) plasma-sprayed with Cr_3C_2-NiCr (clad powder) have shown higher buildup resistance than those sprayed with sintered and crushed Cr_3C_2-NiCr powders.[23] Plasma-sprayed coatings of Cr_2O_3 on hardened steel drilling components are used in petroleum and mining to increase the service lifetime of the components,[13] and coatings of Cr_2O_3/TiO_2 are used on the Al piston rings of automotive engines.[24] Plasma-sprayed copper exhibits exceptional conductivity, thereby making it an ideal tool for producing highly efficient electrical connections.

The advantages of plasma spraying over CVD and some other methods include high coating rates, an ease of coating thickness control, cleanliness, and no limitations in selecting coating materials and substrates.[25] The advantages of plasma spraying for producing free-standing refractory linings as high-temperature ceramic thermal shields for aerospace and nuclear applications are (1) radiative diffuse behavior at high temperature (higher emissivity and reflectivity), (2) large choice of candidate materials, and (3) good thermomechanical behavior due to the specific structure of the coatings.[26]

The drawbacks of plasma-sprayed coatings are their inherent porosity and the presence of oxidized and unmelted particles in the coatings. Some of these problems are, however, being taken care of by the use of HVOF process (to be discussed later).[27]

Plasma spray can be categorized into (a) air or atmospheric plasma spray (APS) and (b) controlled atmosphere plasma spray (CAPS), which also includes vacuum plasma spray (VPS), low-pressure plasma spray (LPPS®), inert gas plasma spray, and shrouded plasma spray (SPS), depending on the environment used. All commercially important plasma spray processes employ a nontransferred dc electric arc (due to arc confinement to the plasma gun) to heat gases to peak temperatures >25,000 K (just outside the nozzle exit), creating jets of ionized plasma with a temperature range of 3000 to 25,000 K. At these temperatures, the plasma gases (Ar, He, H_2, or N_2) are dissociated and ionized into an equilibrium mixture of positive ions and electrons as energy is pumped into them by the confined electric arc discharge.[11]

Atmospheric Plasma Spray. In *air* or *atmospheric plasma spray (APS)*, the plasma-forming gas (Ar, a mixture of Ar + H_2, Ar + He, or Ar + N_2) is allowed to flow between a cylindrical, thoriated tungsten cathode and an annular water-cooled copper anode (Fig. 19.4a).[11] An electric arc is initiated between the two electrodes using a high-frequency discharge and then maintained using dc electric power (30 to 120 kW) and applied voltage (typically 30 to 80 V). The arc ionizes the continuous flow of Ar gas (flow rate of 40 to 80 Nl/min), creating a high-pressure gas plasma. The resulting increase in gas temperature increases the volume of the gas and, hence, its pressure and velocity at the exiting nozzle.[1]

Powder (usually spherical, in the size range of 5 to 100 μm and with a tighter size distribution) is usually introduced into the injection port located inside or outside the nozzle at a feed rate in the range of 50 to 100 g/min, with an injection angle of 90° or sometimes 60° forward or backward from the plasma flame; the carrier gas flow rate is in the range 3 to 10 Nl/min. The powder is both heated and accelerated by the high-temperature, high-velocity plasma gas stream. Usually, the powder velocities are in the range of ~300 and 550 m/s and temperatures are at or slightly above the melting point. The torch design and operating parameters play critical roles in determining the temperature and velocity achieved by the powder particles. The operating parameters include the gas flows, carrier gas flow, powder feed rate, power levels, spraying distance (standoff), and angle of deposition.[1]

APS coatings are characterized by much higher density (than that of flame spray coatings), porosities in the range of 5 to 20%, bond strengths in the range of 15 to 25 MPa (but reaching ≥70 MPa for NiAl, NiCrAl, or Mo coatings), and coating thicknesses in the range of 50 to 500 μm (~0.05 to 0.50 mm or 0.002 to 0.020 in.) [but even higher for some applications (e.g., dimensional restoration or thermal barriers)].[1,13]

APS is capable of producing coatings of almost any meltable or heat-softenable materials. Moreover, the coatings can be deposited onto a very wide variety of substrate materials. Like other thermal spray processes, two or more materials with widely differing properties can be APS-sprayed simultaneously to obtain aggregate coatings with a desirable combination of properties that are not possible to obtain with other coating processes. APS is recommended for processing materials that do not undergo oxidation during spraying or if a certain extent of oxidation is permissible.[28] The process produces a higher-density, uniform and consistent coating.[29]

The *drawbacks* of APS are coatings with lower densities, higher porosities (which also results in reduced corrosion resistance and wear resistance), lower bond strengths, and quite rough surfaces. In addition, materials reactive with O_2 or N_2 cannot be sprayed in some applications. A few of these drawbacks can be overcome, if VPS, LPPS®,* or SPS processes are used.[30] It should be noted that with DG- and HVOF-spray coatings the abrasion wear resistance increases significantly with respect to the plasma-sprayed coatings due to better cohesion of the sprayed particles and denser microstructures of these coatings, resulting from the high particle velocities in these processes.[31]

Controlled Atmosphere Plasma Spray (CAPS). This is a variation of APS that allows spraying enclosed in a controlled atmosphere (inert atmosphere or reactive atmosphere) and pressures in the range of low pressure (30 kPa) to high pressure (300 kPa). VPS and LPPS®, being batch and continuous processes, respectively, may be considered as special categories of CAPS that operate at reduced pressure to produce high-performance coatings with very low porosity and oxide content.[32]

Vacuum Plasma Spraying (VPS). This consists of a plasma jet stream produced by heating an inert gas by an electric arc generator (with power >80 kW) (Fig. 19.4b).[13] The powders (with size distribution in the range −45 + 10 μm)[31a] are introduced via a port into the plasma jet working in a vacuum (pressures prior to and during spraying are about 1.3 Pa and ~655 Pa to 13.3 kPa, respectively),[13] undergo melting, and are accelerated toward the substrate.[13,33] The position of the injection port in the nozzle plays an important role in the VPS system because the pressure of the powder injector must be higher than the pressure in the nozzle in order to propel the powders adequately.[34,35] The speed and the length of the vacuum plasma forming gas are higher (1500 to 3500 m/s) and longer (250 mm), respectively, than those of APS.[13] Spraying distance is in the range of 300 to 400 mm. The bond strength of alloy coatings is >80 MPa. Higher particle velocities lead to denser deposits. The oxide content in the resulting coating is very small and porosities are usually lower than 1 to 2%. However, it is possible to deposit a porosity-free coating by carefully controlling the processing parameters. The coatings have a typical thickness of 150 to 500 μm.

In a specific system, transferred arc preheating and sputter-cleaning are used before loading, and, once loaded, the batch is completely automatic (and the parts are then plasma sprayed in a nontransferred manner. VPS can handle up to 24 parts like turbine vanes per batch. The technique is used to develop oxidation- and corrosion-resistant coatings and thermal barrier coatings for gas turbine engine components. A typical cycle time is ~4 to 5 hr including pumpdown and cooling.

---

* Sometimes VPS and LPPS® are considered the same. They are two trademarks for basically the same system. The difference is that VPS is a batch process while the LPPS is a continuous process.

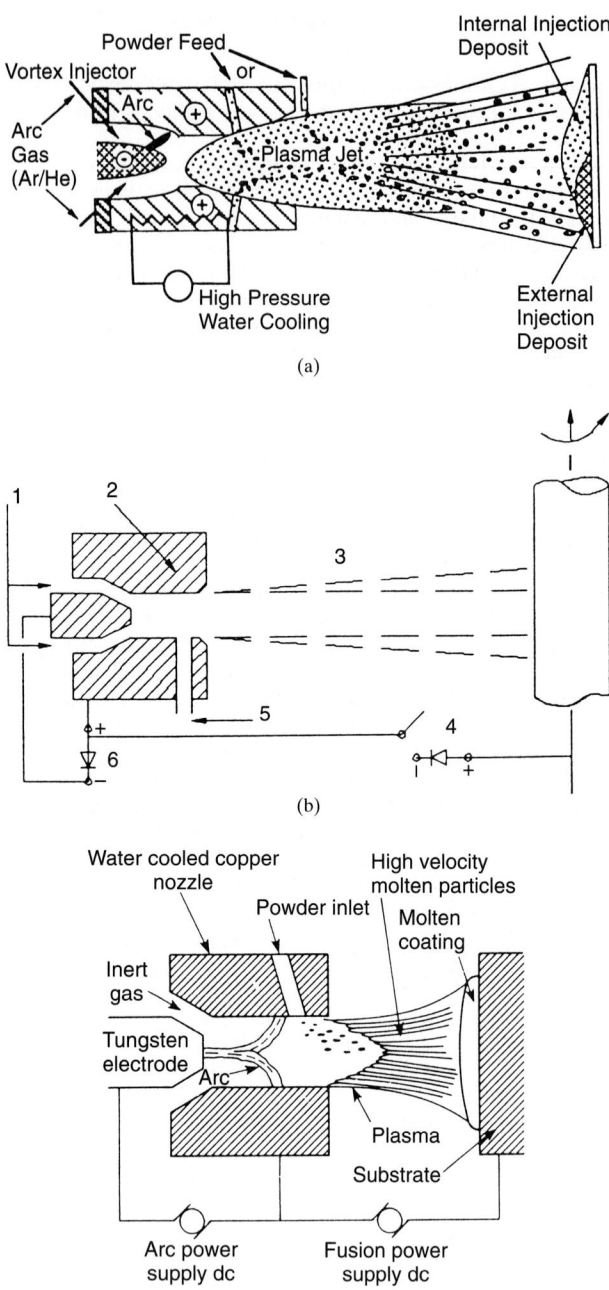

FIGURE 19.4 Schematic of (*a*) an atmospheric plasma spraying process[11]; (*b*) vacuum plasma spraying process[13] [legends: 1. working gases inlet; 2. anode; 3. vacuum environment; 4. transferred arc generator (used to clean up, if negatively polarized, otherwise to heat up the substrate prior to spray); 5. powder injection port; and 6. plasma arc generator]; and (*c*) transferred plasma-arc spraying.[1] [*(a) and (c): Reprinted by permission of ASM International. (b): Reprinted by permission of John Wiley & Sons, England.*]

VPS is limited to line-of-sight. Here the individual parts are fixtured on a manipulator within a load-locked chamber. For reactive metals and intermetallic compounds such as nickel aluminides and MoS_2, VPS is capable of producing dense, free-standing forms with excellent mechanical properties.[36]

Low-Pressure Plasma Spraying (LPPS)®.   This has become a widely accepted practice, particularly in the aircraft engine industry. It has proven to be an effective means for applying complex, hot corrosion-resistant Ni-Co-Cr-Al-Y-type coatings to high-temperature aircraft engine components without oxidation of the highly reactive constituents. Normal operating procedures of inert-atmosphere, LPPS® systems require that the spray chamber be pumped down to low pressures, as noted above, or be repeatedly cycled after pumping to ~55 MPa (0.4 torr) and then be backfilled with inert gas to about 40 kPa (300 torr). Once the system has been sufficiently purged to achieve an acceptable inert atmosphere, the plasma spray operation is resumed and the chamber pressure is adjusted to the appropriate level for spraying. The entire spray operation is accomplished in a soft vacuum of ~6.7 kPa (50 torr). Plasma preheating in the transfer chamber optimizes the use of the plasma jet, which runs on a continuous basis and improves bonding strength, preventing oxidation of the substrate.[32,37] Because of the complexity of low-pressure spraying, the entire process is best controlled by computer to ensure complete reproducibility and uniformity throughout the coating. Improved productivity can be achieved by using load/lock prepumping and venting chambers and robotics.[1]

Inert-atmosphere, LPPS® chamber provides several advantages over conventional inert-atmosphere APS. Because of the lower pressure, the plasma gas stream temperature and velocity profiles are prolonged to larger distances, so the coating properties become less sensitive to standoff. Moreover, the substrate can be preheated without oxidation. This permits better control of residual stress and higher bond strength values. Deposition efficiency can be increased due to increased particle dwell time in the longer heating zone of the plasma and higher substrate temperature. The closed system also reduces environmental problems such as dust and noise.[1]

The basic characteristics of inert-atmosphere LPPS include the following:

1. Its ability to produce denser and higher-bond-strength coatings with little impurity contamination from the atmosphere is attributed to the lengthening, higher velocity, and lower temperature of the plasma jet.

2. Like VPS, the quality of the sprayed coatings is also a function of the fundamental plasma characteristics such as ion and neutral particle temperature, electron temperature and density, and plasma velocity.

3. LPPS jets are characterized by longer heating and accelerating zones compared with the APS jets.[38]

4. LPPS coating exhibits better high-temperature oxidation resistance than APS coating.[39]

5. It can process typically 5 to 10 pieces per hour and is suited to the continuous production of large and heavy parts to be handled by robots.

Inert Plasma Spraying (IPS).   The complex LPPS process is not required for all applications. Plasma spray using an inert-gas shroud around the plasma gas effluent can be just as effective in preventing oxidation during deposition as spraying in an inert-gas, low-pressure chamber. IPS involves the plasma spraying into an inert-gas (He, N_2) chamber. *Shrouded plasma spraying (SPS)* is developed for thermal barrier coatings in which the plasma jet is protected from the atmosphere. The shielding nozzle is connected to the anode of the plasma torch, and the nozzle is in close proximity (100 to 130 mm) to the substrate.

The IPS/SPS process has been used extensively to spray Ni-Co-Cr-Al-Y alloys on turbine blades, vanes, and outer air seals, and on thermal barriers as an undercoat. Compared to chamber spraying, it has a much lower capital cost but greater sensitivity to standoff. It is difficult to preheat the substrate to high temperatures without oxidizing the substrate, a technique used within low-pressure chamber spray to control the residual stress in some high-temperature, oxidation-resistant coatings. However, residual stress in these coatings can be controlled when using inert-gas shrouding through control of deposition rates, auxiliary cooling, and so forth.

A simple IPS/SPS process can also be used to confine hazardous materials. Hazardous materials are grouped into two categories: toxic and pyrophoric. Toxic materials include beryllium and its alloys. Pyrophoric materials include Mg, Ti, Li, Na, and Zr, which tend to burn readily when in a finely divided form or when purified by the plasma process.

Reactive Plasma Spraying. This is mostly concerned with the *in situ* formation of nitrides, carbides, and cermets by the reaction between the spraying powder and atmospheric gas.[40]

Cr-nitride composite coatings can be fabricated by reactive LPPS® (RLPPS) using Cr powder as a spray material. The application of transferred arc is aimed at accelerating the nitriding reaction in the spray processing.[41]

High-Pressure Plasma Spraying (HPPS). This is found to be suitable for high-melting-point materials such as ZrO_2, with improved deposition efficiency and coating hardness.[42]

Plasma Transferred-Arc (PTA) Spraying (Hardfacing). This, using powder metal alloys, has emerged as an ideal process for quality deposition (hardfacing) mainly due to very low dilution in the thin single-layer deposit and the ability to completely automate the process. Figure 19.4c is a schematic representation of the process. A secondary arc current is established through the plasma and substrate that controls surface melting and penetration depth. The method of heating and heat transfer in the PTA process removes many of the problems related to using powders with wide particle size distributions or large particle sizes.[1]

Several advantages resulting from this direct heating include superior wear resistance, exceptional corrosion resistance, ability to produce graded carbide microstructure, metallurgical bonding, oxide-free coating and bond lines, high-density coatings, high quality yield, high thicknesses per pass, smooth as-deposited surface, fast processing time, and ability to preheat parts in an inert atmosphere.[1,43]

Coating thicknesses of 0.50 to 6.35 mm (0.020 to 0.250 in.) and widths up to 32 mm (1.25 in.) can be made in a single pass at a powder feed rate of 9 kg/hr (20 lb/hr). In addition, less electrical power is needed than with nontransferred arc processes. For example, for an 88% tungsten carbide, 12% Co material, plasma spray deposition of 0.30 mm (0.012 in.) thick and 9.50 mm (0.375 in.) in width might require 24 passes at 40 to 60 kW to obtain the maximum coating properties. This same material can be applied, using the TPA process, in one pass at about 2.5 kW.[1]

The *disadvantages* of PTA include (1) the possibility of variation in its microstructure[44] and (2) its limitation to substrates that are electrically conductive and can sustain some melting.

Applications of the PTA process include coating the face areas of exhaust valves in spark- and compression-ignited internal combustion engines with wear-resistant powder materials.[45] The PTA process is also used in hardfacing applications such as valve seats, plowshares, oil field components, and mining machinery.[1]

High-Velocity Oxy-Fuel Spraying. HVOF, being a variation of combustion spraying, has been developed into an extremely versatile coating process. The HVOF process is characterized by (1) spraying with high powder velocity [due to supersonic gas velocities achieved by higher combustion pressure (~6 bar)] toward the substrate and low thermal energy (i.e., relatively low-temperature combustion flame) when compared to plasma spraying, (2) elimination of electric power and a vacuum system, (3) smooth surface finish, (4) tight small spray pattern suited for small parts, and (5) high mechanical bond.

Six types of gas-fueled HVOF systems are currently popular: CDS, JP-5000, Jet-Kote, Diamond Jet (DJ), Top Gun, and HV2000.[46] Figure 19.5 shows the schematic HVOF processes of two different guns: the standard Diamond Jet gun (Fig. 19.5a)

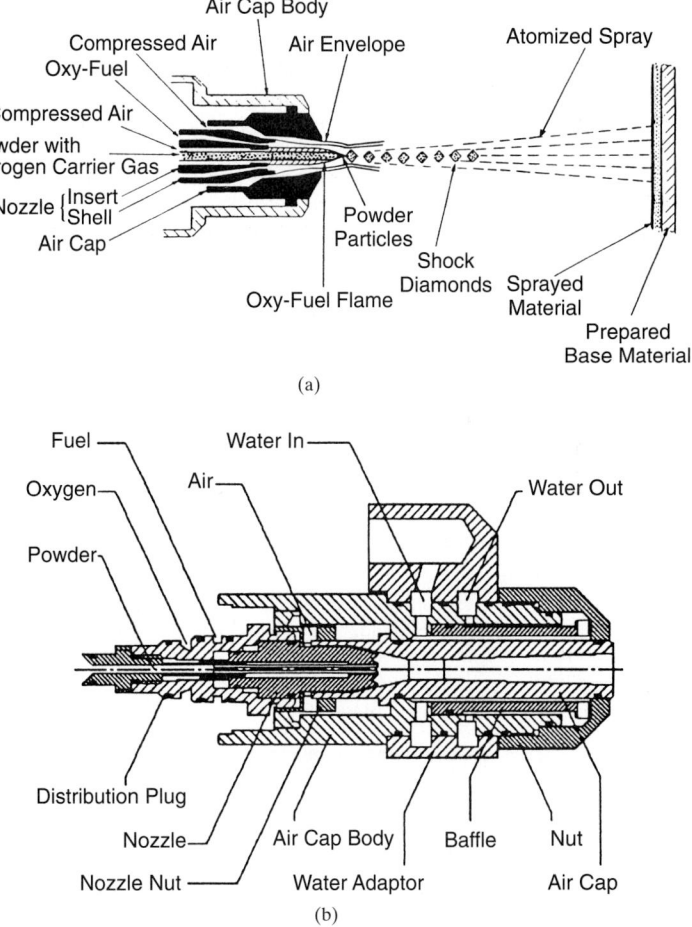

FIGURE 19.5 (a) Standard Diamond Jet gun. (b) Hybrid Nozzle Diamond Jet 2600. (*Courtesy of Sulzer Metco.*)

and the Hybrid Diamond Jet 2600 (Fig. 19.5b).[47] Fuel, usually propane, propylene, MAPP, acetylene, hydrogen, or kerosene, is mixed with oxygen and continuously burned in a combustion chamber at a high pressure, which creates a supersonic velocity of ~1800 m/s, with a characteristic multiple shock diamond pattern that is visible in the flame. Flame speeds of 2000 m/s and particle velocities of 600 to 800 m/s are claimed by some HVOF equipment suppliers.[46] Powder (5 to 45 μm) is introduced (into the combustion chamber in the CDS, HV2000, and Top Gun systems, and into the exhaust barrel in the Jet-Kote and Diamond Jet systems at different distances from the chamber) mostly axially into the nozzle (with a powder feed rate of 20 to 80 g/min) as suspension in the carrier gas (such as nitrogen) and is heated and accelerated by the hot gas flow. The powder is usually fully or partially melted in either an oxidizing or reducing environment. The HVOF spray distance (or standoff) is much greater than that of plasma (e.g., ~150 to 300 mm).[1] This means that a shorter spray distance with HVOF can lead to overheating of the substrate. Table 19.6 shows the characteristics of the two guns.

The temperature of the oxygen-acetylene flame reaches a maximum of 3167°C if mixed at 1.5:1 (by volume) and the oxygen-propylene reaches a maximum of 3170°C at the ratio of 4:1. The fuel gas flow rates are in the range of 40 to 60 Nl/min and the O_2 flow rate is correspondingly greater (e.g., 430 Nl/min for the HV2000 torch). The flame velocity of the exhaust jet in the Jet-Kote torch is ~2000 m/s.

With the appropriate equipment, operating parameters, and choice of powder, coatings with high density and with bond strengths can reach 90 MPa (13 ksi), particularly in carbide coatings. Coating thicknesses are usually in the range of 0.05 to 0.50 mm (0.002 to 0.020 in.), but substantially thicker coatings can occasionally be used when required with some materials.[1] It takes only 20 mt to deposit a 100-μm-thick coating onto a 0.5-m-long, 0.1-m-diameter cylinder.

Advantages. The advantages of HVOF spray over flame spray and APS coatings are the following:

1. It uses a wide range of coating materials such as metal alloys, cermet, and polymers.
2. It makes for an economical deposition of dense coatings made of metals and hard metals for thin coatings (~50 to 100 μm) and small parts.
3. The equipment it uses to deposit coatings of metallic and ceramic materials is very portable, and allows a wide range of melting temperatures from Al at 650°C to Mo with 2620°C in spray enclosures and open-field applications.[48]

TABLE 19.6 Characteristics of the Standard DJ (Air-Cooled) and DJ 2600 (Water-Cooled) Guns

	Standard DJ	DJ 2600 (DJ + Hybrid)
Fuel gases	Hydrogen/propane/propylene	Hydrogen
Combustion	Outside the nozzle	Inside the nozzle
Spray rate, kg/hr	1.4–6.8	1.4–6.8
Flame speed, m/sec	1250–1500	2300
Flame Temperature, °C	2650–2700	2800
Gun cooling	Air	Air/water
Special characteristics		Shroud effect (N_2)

Courtesy of Sulzer Metco.

4. Its capability of forming very dense coatings with reduced oxidation (oxide content <1%) and decomposition, homogeneity, reduced variation in the phase composition, low porosity, high hardness and adhesive strength, and high bond strength (>70 MPa or 10 ksi).[28,49,50]

5. It is a straightforward process with such features as high particle velocities, low flame temperature, and qualities comparable to VPS coatings[47] [i.e., more economical coatings of similar quality (as compared to the VPS process) with a less complicated process (atmosphere)].

6. It offers a potential flexibility by a combination of lower process temperatures (typically, 2800 to 3200°C), lower substrate temperature (<175°C or 350°F), higher flame velocity (2000 m/s), and particle velocities (500 to 800 m/s) in the Top Gun HVOF system.[51]

7. It involves lower particle temperatures when compared to the particle temperatures produced by plasma spray, which leads to carbide coatings with less carbide loss than plasma-sprayed coatings.

The advantages of HVOF process over other coating and thermal spray processes are the following:

1. Uniform particle heating due to high microturbulence.

2. Shorter exposure time for particles to heat in flight due to high particle velocities (but longer standoff).

3. Higher kinetic energy upon particle impact.

4. Improved wear resistance due to homogeneous distribution of carbides.[52]

5. Environmentally safe, cleaner, and less expensive alternative to chromium plating for wear resistance.[53]

Disadvantages. The disadvantages of HVOF are the following:

1. High capital and operating costs (not compared to PVD or VPS).

2. Spray pattern limited to 0.5 in. in diameter.

3. The lower temperature of the process relative to plasma and detonation gun spray has created a concern over its ability to spray ceramic coatings. However, it is reported that Top Gun and HVOF systems with acetylene and propylene fuels can produce Al_2O_3, Al_2O_3-TiO_2, and Cr_2O_3 coatings.[46]

4. It is uneconomical with increasing coating thickness. For thick coatings and complete coating of large parts, VPS spraying is less expensive.[47]

5. The continuous nature of the combustion and high stability of the flame at large distances from the exit nozzle, creating high thermal flux to the substrate.

6. The reaching of the substrate temperature, in some cases to 1200°C, which has an undesirable effect.

7. The ability to alter the chemical composition of the substrate and cause thermal distortion and residual stress of the precision component, thereby making it difficult to spray thin samples or easily melted materials such as plastics and composites.[54]

Applications. The HVOF processes can produce coatings of virtually any metallic or cermet material and, for some HVOF processes, many ceramics. HVOF coatings have primarily been used in a diverse range of engineering applications (such as the repair and new manufacture of certain industrial and aero-gas turbine components, and the refurbishment of equipment used in the oil and gas industries) to prolong component life by increasing wear and corrosion resistance.[55]

HVOF-WC-Co coatings are widely used to reduce wear or modify friction in many sliding, abrasive, and corrosive applications such as compressor piston rods, pump plungers, shaft sleeves on centrifugal pumps and fans, and midspans of compressor blades in gas turbines. They find applications, both for the original manufacture and the repair of components, in industrial and aerospace markets. Optimum properties of this coating have been reported by using a high-pressure/high-velocity oxygen fuel (HP/HVOF) process.[56–58]

HVOF-Cr_3C_2-NiCr cermet coatings are used for high-temperature wear resistance and corrosion resistance applications, in the chemical, aerospace, and other industries [e.g. the use of Cr_3C_2-20 wt% NiCr cermet coating on piston rings (carbon steel) of reciprocating engines]. (This sprayed coating appears to be a promising candidate to replace the conventional hard chromium plating film.[59,60]) Despite lower mechanical and wear-resistance properties than those of WC-Co coatings, the Cr_3C_2-NiCr system exhibits a high stability, even at temperatures in the range of 700 to 900°C, where WC-Co shows high decarburization of WC coupled with the formation of undesirable W_2C.[59] Those few HVOF systems that use acetylene as a fuel (for the highest available flame temperature) are necessary to apply the highest melting-point ceramics such as zirconia or some carbides.

HVOF-mild steel (wire) sprayed coatings on the cylinder walls of aluminum automobile engines provide wear resistance with metal oxidation, which has been reported to exhibit a beneficial effect.[61]

HVOF-ceramic-polymer nanocomposite coatings have been successfully applied either for surface protection, providing low friction and inert corrosion barriers, or where tailored electrical and magnetic properties with enhanced abrasion and wear resistance are necessary.[62]

Detonation Gun (D-Gun®) Spray.   In the detonation gun (D-Gun® and Super D-Gun®) processes, shown schematically in Fig. 19.6, a mixture of oxygen and acetylene, together with a charge of powder (5 to 60 μm in particle size) is injected (at a feed rate of 16 to 40 g/min) into the (long water-cooled) barrel and detonated by using a spark with nitrogen or air as a carrier gas. The high-temperature (up to 4500 K with 45% acetylene), high-pressure detonation wave (up to 2930 m/s) moving down the barrel heats the powder particles to their melting points or above and accelerates them to a supersonic velocity of about 750 m/s in the D-Gun® and about 1000 m/s in the Super D-Gun®.[13] This is a cyclic or intermittent process that involves the purging of the barrel with nitrogen after each detonation and a repetition of the process up to 15 Hz. In this case, the spray distance is held at about 100 mm and the allowable angle of spray is lower than in APS. A spray spot diameter of about 25 mm (1 in.) and a few (3 to 10) μm in thickness is deposited with each detonation cycle. A uniform coating thickness on the part is obtained by properly overlapping the spot diameters of the coating in many layers. Typical coating thicknesses are in the range of 0.05 to 0.50 mm (0.002 to 0.02 in.), but thinner and much thicker coatings can be obtained. The thickness of the D-Gun® coatings are typically 50 to 300 μm.[1]

In the D-Gun® spray, the flame temperature is relatively low (3000°C) and the gas velocity is very high. Hence, the melting behavior of the coating powder plays an important role in achieving poreless and high-bond-strength coatings. The detonation gun sprayings are characterized by the best quality of the coatings, i.e., the

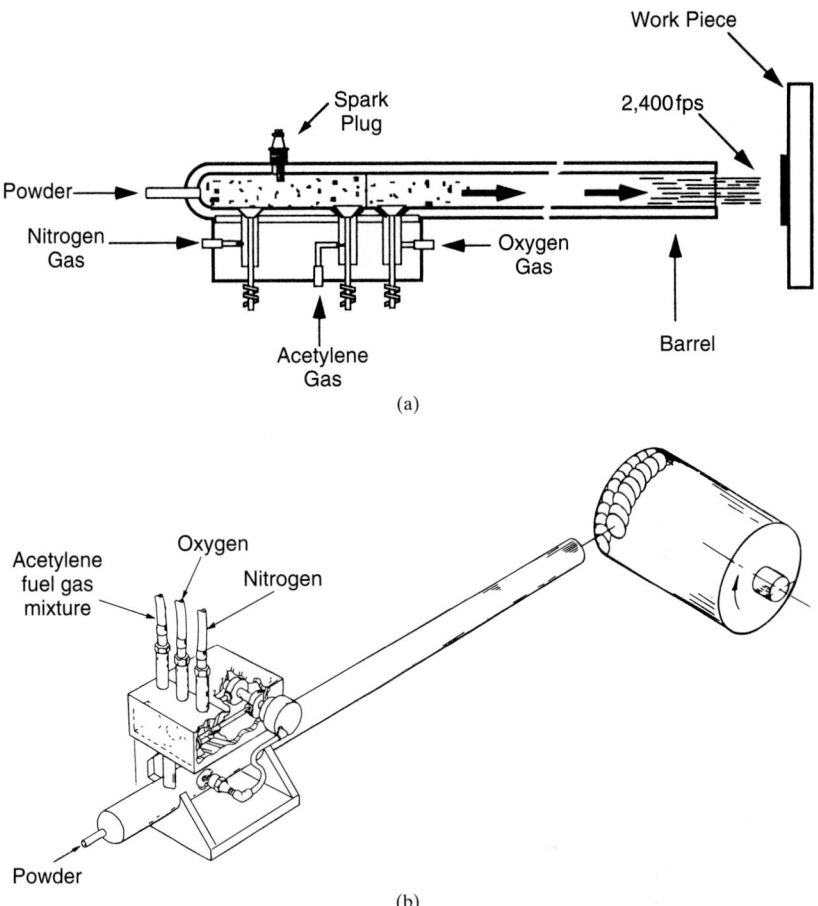

FIGURE 19.6 Schematics of detonation gun processes: (*a*) D-Gun® and (*b*) Super D-Gun®.[13] (*Courtesy of Praxair Surface Technologies, Inc.*)

highest bond strengths (e.g., >120 MPa for Mo coatings, >83 MPa for WC-Co coatings, and 70 MPa for Al_2O_3 coatings) and lowest porosities (usually no porosity for Mo coatings, ~0.5% for WC-Co coatings, and <2% for Al_2O_3 coatings). The quality of the coatings is a function of the powder properties such as its morphology and size distribution.

Careful control of the gases used generally contributes to negligible oxidation of the metallics or carbides. The extremely high velocities and consequent kinetic energy of the particles in the Super D-Gun® process allow most of the coatings to be deposited with residual compressive stress, rather than tensile stress as is typical of most of the other thermal spray coatings. This is of particular significance relative to coating thickness limitations and the influence of the coatings on the fatigue properties of the substrate.[1]

Practically all metallic, ceramic, and cermet materials can be deposited using detonation gun deposition. Detonation gun coatings are used extensively for wear

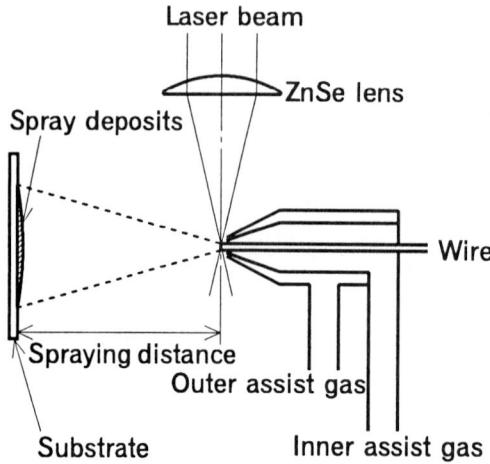

Laser beam

ZnSe lens

Spray deposits

Wire

Spraying distance

Outer assist gas

Substrate   Inner assist gas

FIGURE 19.7 Schematic diagram of the laser spraying process.[65] (*Reprinted by permission of High Temperature Society of Japan.*)

and corrosion resistance, for preventing the erosion of heat exchangers (in coal-fired boilers and fluidized-bed combustion boilers).[63] Detonation-sprayed coatings can extend the service life of components by two to three times or more.[64] They are frequently specified for the most demanding applications, but often can also be the most economical choice because of their long life.

Laser Spray (LS).   Recently, a laser spraying method, using wire with Ar gas (as a chemically inactive gas) or ($N_2 + O_2$) (as a chemically reactive gas) has been developed to produce compositional gradient (or thermal barrier) coatings as well as metals or ceramics such as nitrides and oxides. Figure 19.7 shows a schematic diagram of the laser spraying method with the directions of gas flow/spraying and laser beam crossed. The spraying gun comprises a center outlet for a pure Ti wire (0.9 mm in diameter), two coaxial ring nozzles (inner and outer) for assist gases (such as Ar, N_2, and its mixtures), and a laser beam focused by a ZnSe lens.[65]

Cold Gas Dynamic Spray (CGDS).   This has recently emerged as a unique thermal spray process in which the deposition of deformable solid particles occurs free of any contamination, recrystallization, and grain growth, at temperatures much lower than the powder flame, arc spray, plasma arc, and HVOF coating processes. In this process, a coating is formed by exposing a metallic or dielectric substrate to a high-velocity jet of solid-phase particles, which have been accelerated by a supersonic gas jet at a temperature significantly lower than the melting or softening temperature of the particle material. Figure 19.8 is a schematic diagram of the CGDS process. Compressed gas (He or N_2) is injected into a manifold system containing a gas heater and powder metering vessel. The pressurized gas is heated electrically to a relatively low temperature (200 to 500°C). The feedstock particles are accelerated by a careful control of the gas dynamics linked to a converging/diverging (de Laval) nozzle. The coating is produced due to low-temperature ballistic impingement of the powders on the substrate with or without a prior grit-blasting treatment of the substrate surface. Using this process, Ni-Al bronze coating on a 2618-T61 Al piston has been produced with increased wear resistance and high hardness. In this

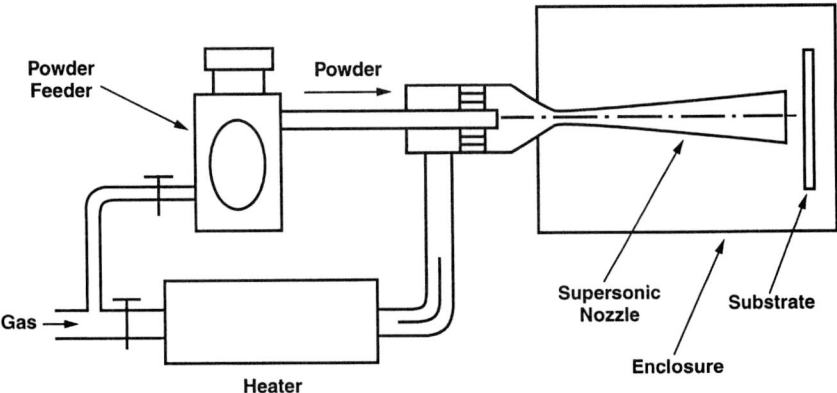

FIGURE 19.8 Schematic diagram of the cold gas-dynamic spray (CGDS) process.[66] (*Reprinted by permission of ASM International.*)

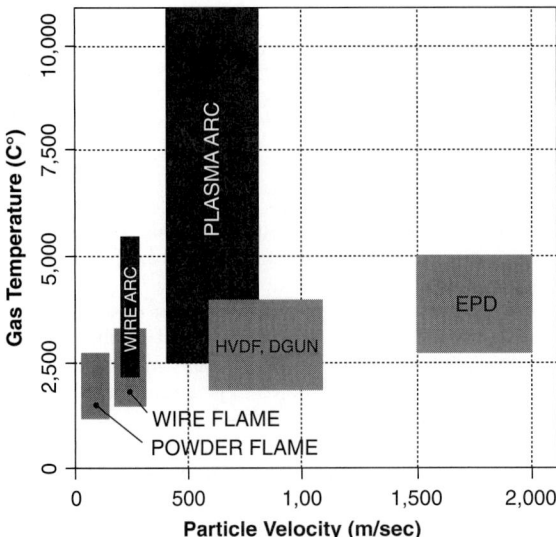

FIGURE 19.9 A comparison of particle velocity and gas temperature for several thermal spray processes.[67] (*Reprinted by permission of ASM International.*)

process, the powder particle size distribution of the feedstock and the impingement velocity are significant factors in order to obtain dense coatings with minimal porosity, high bond strength, hardness, and wear resistance.[66]

Electromagnetic Powder Deposition (EPD). This process is a potential alternative to the current thermal spray technologies (HVOF and D-Gun® spraying) and is based on the well-established "railgun" technology, powered by a capacitor bank, where powder particles are accelerated by a high-velocity argon gas column to a final velocity up to 2 km/s (Fig. 19.9) and impact the substrate in a solid state, result-

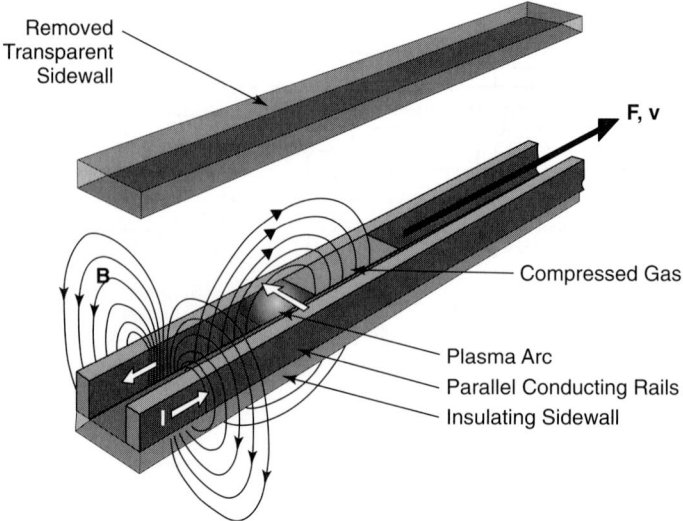

FIGURE 19.10 The plasma armature is accelerated down the length of the railgun by an electromagnetic Lorentz force created by the interaction of the magnetic fields surrounding the rails and the current flowing through the armature.[67] (*Reprinted by permission of ASM International.*)

ing in the melting of both the metal powder particles and an equal amount of substrate mass. This interaction offers the potential for the formation of a fusion bond between the deposited material and the substrate.

Figure 19.10 shows a simple railgun and the electrical current and magnetic fields that interact to generate a Lorentz force (the cross product of the current density vector and magnetic field vector). As the armature moves down the railgun, it pushes a column of gas and its subsequent shock compression in front of it. A system is under development that provides 10 to 100 discharges at a 30-Hz rate.[67]

19.4 POST-SPRAY TREATMENT

Post-spraying techniques consist of heat treatment (such as annealing, laser remelting, or laser engraving), impregnation or sealing, finishing operations (such as grinding, polishing, and lapping), and repair of coatings. Heat treatment is used to alter the coatings' phase composition, to decrease the porosity level, to improve strength and ductility, or to increase other coatings' properties. Impregnation (sealing) is used for added corrosion resistance[15a] and for electrical applications of the coatings (such as *corona rolls*). Post-spray treatments such as laser engraving, grinding, and polishing are an integral part of the production of ceramic *anilox rolls* and others.[13]

19.4.1 Heat Treatment

Among the heat treatment processes, annealing is the most widely used, particularly in research laboratories. The laser treatment or HIPing process is less often used due to its high equipment and operating costs and its limited availability.[13]

19.4.1.1 Annealing Treatment *Metal and Alloy Coatings.* The furnace annealing of thermally sprayed metal and alloy coating is usually employed in vacuum or a hydrogen atmosphere to reduce the electrical resistivity of conductor coatings or to improve the adhesion of the bond coatings (deposited onto the turbine blades) (Table 19.7).

Ceramic Coatings. Annealing treatment is used to sinter the coating, previously separated from the substrate. This treatment may increase its coating density and elastic modulus or restore superconduction in high-T_c semiconductors (Table 19.7). Thermal spray processes associated with annealing treatment are also used to produce free-standing shaped components.[13]

19.4.1.2 Laser Treatment. Laser treatment involves laser remelting (also known as glazing) and laser engraving.

Laser Remelting. Laser surface remelting is used for the surface sealing of porous materials such as thermal-sprayed coatings, homogenization, microstructural refinement (to harden the surface by rapid self-quenching), improved solid solubility, and metallic glass formation. Table 19.8 shows examples of laser remelting in thermal sprayed coatings, which seems to be a preferred method to reduce porosity and increase the coating density. In the case of surface-melting ceramics, the use of sharp pulses (e.g., 20 ns, 20 MW) from an excimer laser (249 nm, in the ultraviolet) has been reported to both melt a 10-μm layer (and, therefore, to seal against corrosion) and smoothen the surface, thereby reducing the heat transfer coefficient.[68]

Laser Engraving. This is made mostly with a pulsed CO_2 laser beam and observed mostly on APS chromia coatings (Fig. 19.11a) for the anilox rolls of a fairly small density of lines. Further work has been reported on spraying ceramics such as Al_2O_3-TiO_2 (Fig. 19.11b) and a much higher density of lines.[13]

Laser engraving serves three functions in the production of thermal barrier coatings (TBCs):

1. It alters the crack growth mechanism that takes place at the TBC/bond interface.

2. The TBC applied on the laser grooved parts produces a better adhesion with the substrate. That is, laser grooving of the bond coat at periodic intervals is preferred to enhance the adhesion between the TBC and the substrate.

3. The size of the laser groove (i.e., depth and width) dictates the columnar microstructures' grain size of the TBC. The grain size of the TBC plays a dominant role in the service performance of the components.[69]

19.4.1.3 Hot Isostatic Pressing (HIP). Hot isostatic pressing (HIP) consists of the simultaneous application of high pressure (up to 300 MPa by compressors) and high temperature (up to 2300 K) on the treated components, held in a closed vessel using Ar as the working gas. HIP treatment results in a decrease in porosity. It is suitable for small sprayed samples.[13] Table 19.9 shows some examples of this treatment.[13] Figures 19.12a and 19.12b show the bond strength of functionally graded

TABLE 19.7 Examples of the Furnace Annealing of Thermal Sprayed Coatings[13]

Sprayed material	Spraying process	Annealing temperature, K	Annealing time, hr	Annealing atmosphere	Property improvement	Reference*
Al_2O_3	APS	1422–1866	0.5–4	Air	Density increase from 3.38 to 3.60 g/cm³	a
ZrO_2	APS	1370–1870	1–36	Air	Young's modulus more than four times	b
$YBa_2Cu_3O_x$	APS	1223 673–773	1–100 24–36	Air O_2	Restoring superconduction	c
Cu	VPS	1070	Not given	H_2	Decreases resistivity from 14 to 7 $\mu\Omega$	d
NiCoCrAlY	VPS	1323	4	Vacuum	Improve adhesion	e

* (a) V. S. Thompson and O. J. Whittemore, Jr., *Ceramic Bull.*, vol. 47, 1968, pp. 637–641; (b) H. E. Eaton and R. C. Novak, *Surf. Coat. Technol.*, vol. 32, 1987, pp. 227–236; (c) L. Pawlowski, A. Gross, and R. McPherson. *J. Mater. Sci.*, vol. 26, 1991, pp. 3803–3808; (d) M. Braguier, J. Bejat, R. Tueta, M. Verna, G. Aubin, and C. Naturel, Improvements of plasma spraying processes for hybrid microelectronics, *Conference on Hybrid Microelectronics*, Canterbury, U.K., 25–27 September, 1973, pp. 15–37; (e) T. Cosack, L. Pawlowski, S. Schneiderbanger, and S. Sturlese, Thermal barrier coatings on turbine blades by plasma spraying with improved cooling, *37th ASME International Gas Turbine & Aeroengine Congress and Exposition*, Cologne, Germany, June 1–4, 1992, paper 92-GT-319.

TABLE 19.8 Some Data about Laser Remelting of Thermally Sprayed Coatings[13]

Sprayed material (composition, wt. %)	Spraying process	Laser maximum power	Power density, W/cm²	Overlap, %	Property improvement	Reference[‡]
Ti	VPS	CO_2, TEA*	2.3×10^6	50	Corrosion resistance	a
Stellite (Co-29Cr-4W-3Ni-3Fe-1C-1Mn-1Si)	APS	CO_2,CW,† 10 kW	1.9×10^5	25	Porosity reduction	b
(Ni-23Co-20Cr-8Al-4Ta-0.6Y)	VPS	YAG, 30 W			Microstructure modification	c
Al_2O_3-13TiO₂	APS	CO_2,CW,† 2 kW	5.6×10^4	30-40	Wear resistance	d
Al_2O_3	APS	CO_2,CW†	1.27×10^4	70	Porosity	e
ZrO_2-20Y₂O₃	APS	CO_2,CW†	0.4-0.7×10^4		Porosity	f

* Transverse excitation at atmospheric pressure.
† Continuous wave.
‡ (a) R. J. Pangborn and D. R. Beaman, *J. Appl. Phys.*, vol. 51, 1980, pp. 5992–5993; (b) M. L. Capp and J. M. Rigsbee, *Mat. Sci. Eng.*, vol. 62, 1984, pp. 49–56; (c) R. Streiff, M. Pons, and P. Mazars, *Surf. Coat. Technol.*, vol. 32, 1987, pp. 85–95; (d) W. Aihua, T. Zengyi, Z. Beidi, F. Jiangmin, M. Xianyao, D. Shijun, and C. Xudong, *Surf. Coat. Technol.*, vol. 52, 1992, pp. 141–144; (e) A. Gorecka-Drzazga et al., *Revue Internationale des Hautes Temperatures et Refractaires*, vol. 21, 1984, pp. 153–165; (f) K. A. Jasim, R. D. Rawlings, and D. R. F. West, *Surf. Coat. Technol.*, vol. 53, 1992, pp. 75–86.
Reprinted by permission of John Wiley & Sons, England.

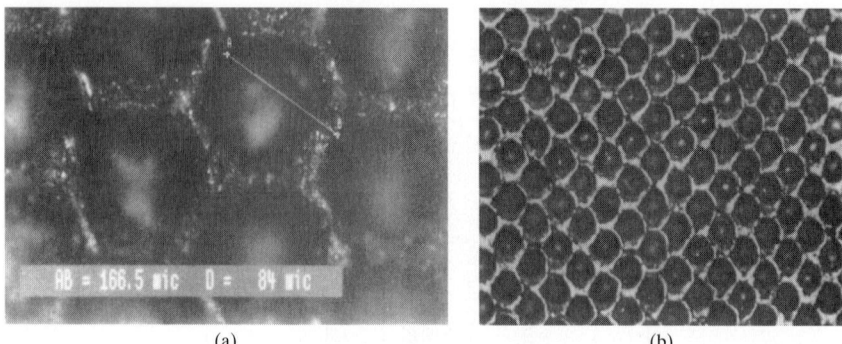

(a)   (b)

FIGURE 19.11   The surface of a laser-engraved APS coating of (*a*) Cr_2O_3 [the density of lines (LD) = 60/cm, engraving angle = 60°] and (*b*) Al_2O_3 + 13 wt% TiO_2 (LD = 285/cm, engraving angle = 45°).[13] (*After L. Pawlowski, R. Zacchino, R. Dal Maschio, V. M. Sglavo, J. Andresen, and F. J. Driller, in Thermische Spritzkonferenz, Aachen, Germany, March 3–5, 1993, pp. 132–138.*) (*Reprinted by permission of John Wiley & Sons, England.*)

TABLE 19.9   Some Data of Hot Isostatic Pressing Treatment of Thermal Sprayed Coatings[13]

Sprayed materials	Spraying process	Pressure, MPa	Temperature, K	Time, hr	Property improvement
Fe-20Cr-9Al-1.5Y*	APS				Hot corrosion resistance
Ni-21.5Co-16.3Cr-0.1Al[†]	VPS	180–350	1173–1453	3	Microhardness less scattered, yield strength higher (60 MPa)
Al_2O_3[‡]	APS	100–130	1370–1570	1–1.7	Microhardness increase from 700 to 1200_{HV3}
ZrO_2-4Y_2O_3[‡]	APS	100–130	1370–1570	1–1.7	Microhardness increase from 400 to 1200_{HV3}; tensile strength increase from a few to 60 MPa
TiC[‡]	APS	100–130	1370–1570	1–1.7	Microhardness increase from 700 to 1200_{HV3}; tensile strength increase from a few to 60 MPa

*   C. Burman et al., *Surf. Coat. Technol.*, vol. 32, 1987, pp. 127–140.
[†] H. D. Steffens, R. Dammer, and U. Fischer, *Surf. Eng.*, vol. 4, 1988, pp. 39–42.
[‡] H. Kuribayashi et al., *Ceramic Bull.*, vol. 65, 1986, pp. 1306–1310.
Reprinted by permission of John Wiley & Sons, England.

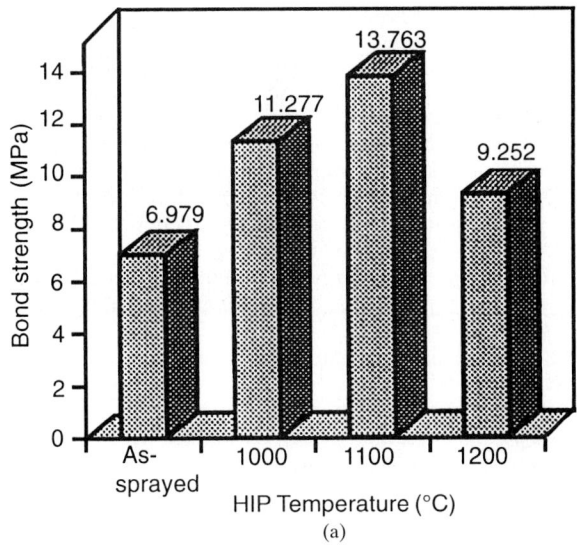

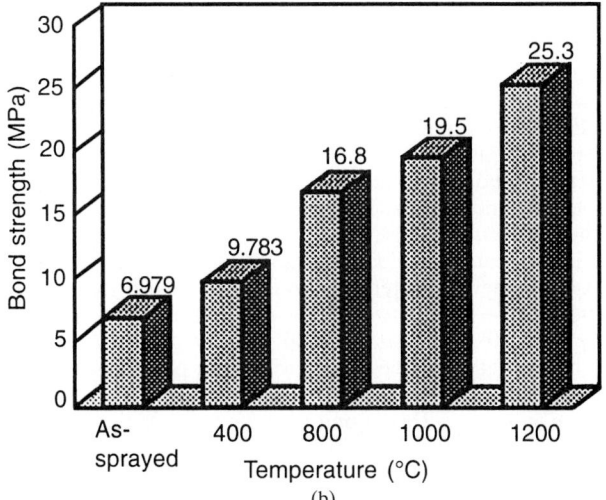

FIGURE 19.12 Bond strength of functionally graded material (FGM) coatings after (*a*) HIP treatment and (*b*) vacuum heat treatment.[70] (*Reprinted by permission of ASM International.*)

$ZrO_2/NiCoCrAlY$ coatings before and after HIP treatment and vacuum heat treatment.[70] The increase in bond strength after each treatment can be attributed to (1) the densification of microstructures and reduction in porosity (especially in HIP treatment), (2) an interdiffusion bond in the coating, and (3) a decrease in thermal residual tensile stress. However, the improvement in properties is more pronounced with the vacuum heat treatment.[70]

19.4.2   Impregnation (or Sealing)

Impregnation (or sealing) is used primarily to seal the interconnected porosity in thermally sprayed coatings prior to finishing, assist in reducing particle pullout from the surface during finishing for coatings with low cohesive strength, and increase the coating's reliability with respect to the wear and corrosion behavior. The degree or effectiveness of the impregnation depends on many coating characteristics such as overall sample density, amount of closed versus interconnected porosity, size and distribution of inherent porosity, amount of absorbed sealants in the coatings, overall coating thickness, depth of penetration of sealants into the coatings, and insulation of pores.[10] To ensure optimum sealing of the coating surface, sealants must be applied just after the deposition of the coating and prior to surface finishing. However, for corrosion protection of thermal sprayed coatings, it is better to seal after the final mechanical treatment.[10] Sealants are of two types: inorganic sealers and organic sealers.[1,13]

   Inorganic Sealers.   The ceramic coatings can be sealed with glass-forming constituents such as SiO_2 or other oxides during the heat treatment at the glass-softening temperature. Cr_2O_3 impregnation has been reported for APS-ZrO_2 coatings for applications in adiabatic engines. The sealing process involved spraying, immersion, or painting of an aqueous solution of Cr^{6+} ions on the coated surface, followed by heat treating at 810 K that converts Cr ions into Cr_2O_3. As a result, the microhardness increased from 300 to 1500 VPN (after undergoing 14 impregnation cycles). Similarly, the wear resistance of the sealed coatings was improved.[71]

   Cu foil sealant has been used by Ito et al.[72] for APS-Ti-Mo coated mild steel in which vacuum heat treatment at 0.1 Pa pressure and 1370 K temperature (just above $Cu_{melting\ point}$) for 10 to 30 min produced sealed coatings with better adhesion and increased corrosion resistance.

   Organic Sealers.   For thermally sprayed coatings, organic sealers are commercially available that include waxes, epoxies, phenolics, or silicones. They can be applied to the sprayed coatings, based on their application. The wax sealants are effective in preventing liquid infiltration at low service temperatures. Resin-based sealants are beneficial at temperatures up to ~260°C (500°F). Some silicone-based sealants may be useful to provide effective protection in salt-spray tests conducted according to MIL standards up to 480°C (900°F). Epoxy and phenolic sealants are more effective for coatings with high porosity contents within their limits of stability (up to 300°C, or 570°F).[1]

   Vacuum Impregnation.   This is one of the most effective (but not extensively used in production) sealing methods for coating porosity which usually fills all interconnected pores open to the exterior surface. This is done by immersing the parts in the sealant, placing them together in a vacuum chamber, and drawing a soft vacuum. During the vacuum release, air pressure forces the sealants into the pores. However, most applications do not need this method. Low-viscosity anaerobic sealants may also be effective.

   The depth of penetration of some sealants may be >1.8 mm (0.070 in.). Irrespective of the method or type of sealant employed, pores or interconnected porosities that are not linked to the exterior surface are incapable of being sealed, and machining or wear in service may open these with a consequent loss of corrosion protection.[1] Table 19.10 lists a comparison of dielectric constant and electrical loss factors of APS-Al_2O_3 coatings impregnated with silicone under vacuum and at atmospheric pressure.[13]

TABLE 19.10 Comparison of Dielectric Constants and Loss Factors for APS-Al_2O_3 Coatings Impregnated with Silicone under Vacuum and at Atmospheric Pressure[13]

Pressure in the chamber while impregnating, hPa	Dielectric constant	Loss factor at 1 kHz
1011	6.24 ± 0.43	0.039 ± 0.008
0.13	4.31 ± 0.17	0.019 ± 0.002

Reprinted by permission of John Wiley & Sons, England.

19.4.3 Finishing

Thermally sprayed coatings must be finished for most service applications. The common finishing methods include grinding, lapping, polishing, machining, abrasive brushing, peening, or vibratory finishing. However, special care is needed in such endeavors to avoid damage of the coatings, resulting in excessive surface porosity because of extraction of coating particles or thermal stress-related cracking. The ultimate surface finish obtained with a thermal spray coating depends on its composition, deposition parameter (which are solely responsible for the size and amount of true porosity in the coating), and the cohesive strength or particle-to-particle bonding within the coating.

19.4.3.1 Grinding. Table 19.11 provides typical machining parameters for some thermal sprayed metallic coatings. Generally, lower in-feed rates are used with wrought materials. Burnishing is frequently employed with soft materials such as Sn, Zn, and babbit metals to form a smooth, dense bearing surface; they are often more cost-effective and provide better finishes for many other coatings.

If grinding does not make a very smooth surface, it may be worthwhile to lap the coating after grinding. Again, it is advisable to consult with the manufacturers of lapping materials for specific recommendations.

Diamond or CBN (cubic boron nitride) wheels are used to grind the ceramic coatings while SiC wheels are used to grind metallic coatings.

19.4.3.2 Polishing and Lapping. Polishing is used with the same machines as used for grinding, but with a finer abrasive wheel. Close control of the polishing parameters leads to surfaces with $R_a < 0.2 \mu m$ (for chromium oxide coatings). Smoother surfaces can be obtained with lapping (using very fine diamond or Al_2O_3 particles in an oil or grease mixture).[13]

19.4.4 Other Methods

Ohmori et al.[73] reported a heat treatment at 1073 K coupled with electric current conduction through APS-8% Y_2O_3-stabilized ZrO_2 coatings that led to increased adhesion with copper and stainless steel substrates. Similarly, Burman et al.[74] have reported the elimination of oxide present in the as-sprayed APS-FeCrAlY coatings by exposing it to electron beam remelting.[13]

TABLE 19.11 Typical Ranges of Speeds and Feeds Used in Machining Thermal Sprayed Metal Coatings[1]

| | High-speed steel tool | | | | Carbide tool (WC-6 Co) | | | |
| | Speed | | Feed | | Speed | | Feed | |
Coating metal	m/s	sfm	mm/rev	in./rev	m/s	sfm	mm/rev	in./rev
Steels								
Low carbon, medium carbon, low alloy	0.25–0.50	50–100	0.075–0.125	0.003–0.005	0.25–0.50	50–100	0.075–0.125	0.003–0.005
High carbon, stainless	—	—	—	—	0.15–0.20	30–40	0.075–0.100	0.003–0.004
Nonferrous metals								
Brass, bronze, nickel, copper, monel	0.50–0.75	100–150	0.075–0.125	0.003–0.005	1.25–1.80	250–350	0.050–0.150	0.002–0.006
Lead, tin, zinc, aluminum, babbit	0.75–1.00	150–200	0.075–0.175	0.003–0.007	1.25–1.80*	250–350*	0.050–0.100	0.002–0.004*

* Aluminum only.
Reprinted by permission of ASM International.

19.4.5 Coating Repair

The repair of thermal spray coatings by coating over service-worn or in-process damaged coatings is usually not recommended, even though the predeposited coating is reference ground, cleaned, and grit blasted. Hence the recommended method is to strip the existing coating and apply a completely new coating. It should be noted that when applying a multilayered coating, it is necessary to apply each new layer over the as-deposited surface of the previous layer; grinding or grit blasting between layers is not recommended.[1]

19.5 COATING CHARACTERISTICS

Coatings exhibit lamellar or flattened grains tending to flow parallel to the substrate. The coating structure is heterogeneous with respect to wrought and cast materials. This is attributed to variations in the condition of the individual particles on impact. The variations in porosity and oxide content as well as chemical composition of the coating significantly influence the coating's properties, and in the case of corrosion, the underlying substrate.[5] Coating characteristics can be classified into the following:

1. Microstructure [complex, composite structures consisting of a metal substrate (usually), a metallic bond coat, and a ceramic overlay, porosities, deposit cohesion and adhesion, unmelted particles, and oxidation levels]
2. Mechanical properties (hardness, modulus, bond strength)

19.5.1 Microstructures

Thermal spray coatings consist of a layered structure composed of splats along with pores, oxides, cracks, deposit cohesion and adhesion, and unmelted particles, which lead to their anisotropic and heterogeneous structure. The service performance of these coatings depends strongly on the microstructure of the coating and the properties of the coating/substrate interfaces.[65] A detailed examination of microstructural details by both optical and scanning electron microscopy is necessary in order to understand the mechanism of coating formation and the coating properties such as bond strength, hardness, and corrosion resistance, and thereby the coating or bulk properties.

Cross-sectional microstructures of several thermal sprayed coatings are shown in Figs. 19.1 and 19.13. Usually, the higher-particle-velocity coating processes provide the densest and superior bonded coatings, both cohesively (splat-to-splat) and adhesively (coating-to-substrate). Metallographically calculated porosities for detonation gun coatings and some HVOF coatings are $\leq 2\%$, whereas for most plasma-sprayed coatings porosities are in the range of 5 to 15%.[1]

*Porosity.** Porosity is a characteristic feature and structural index of thermally sprayed coatings.[75] Porosity strongly influences the ultimate physical and mechani-

---

* There are two types of porosity. (1) Open porosity comprises coarse pores formed between the lamellae and are interconnected through small channels to the surface. (2) Closed porosity, so called microporosity, comprises spherical pores formed by entrapped gas within the lamellae. Porosity (or voids), cracks, and coating decohesion are usually recognized by metallographic techniques.

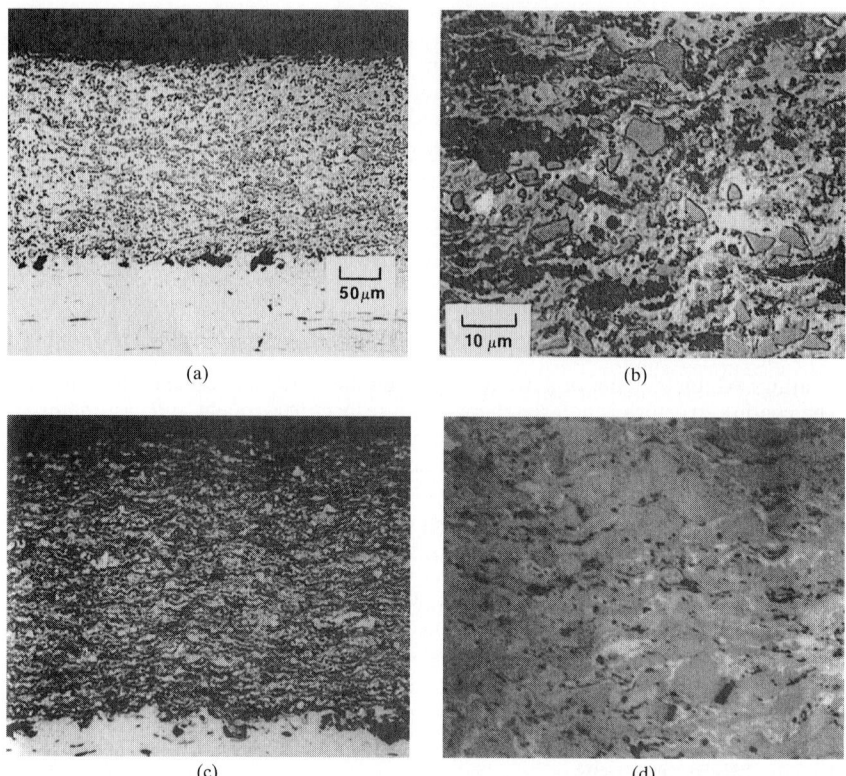

(a)   (b)

(c)   (d)

FIGURE 19.13 Microstructures of a detonation-gun-deposited tungsten carbide/cobalt cermet coating: (*a*) as polished and (*b*) etched. A mechanically mixed chromium carbide/NiCr cermet coating: (*c*) as polished and (*d*) etched.[1] (*Reprinted by permission of ASM International.*)

cal properties of thermally sprayed deposits; it is a function of thermal, fluid flow, and solidification conditions. Most thermally sprayed coatings (except VPS, post-heat-treated, or fused coatings) contain porosity (up to 25%). Porosity can be harmful in coatings with respect to (1) corrosion (sealing of coatings recommended), (2) machined finish, and (3) strength, microhardness, ductility, and wear characteristics. However, porosity can be beneficial with respect to (1) lubrication (porosity appears as a reservoir for lubricants), (2) reduction of stress levels and increase of thickness limitations, (3) increase of thermal barrier properties, (4) abradability in clearance control coatings, (5) increase of shock resistance properties, and (6) applications in prosthetic devices, etc.[19]

Porosity can be reduced to a minimum by optimizing the spray-deposited conditions (such as sprayed material, spraying parameters, and pressure induced on the surface of substrate during droplet impingement) or by thermomechanical processes.[75a] It has been shown that porosity decreases with increasing particle velocity and temperature. High droplet temperature prior to impingement, generated by short spraying distance, reduces the porosity level due to increased filling of the cavities by the molten particles.[76] In other words, the apparent increase in the porosity level is primarily attributed to the reduction in impact energy and temperature prior to impingement,[77] shadowing effect (unmelted particles/spray angle), and shrinkage and stress-relief effects.[19]

Plasma-sprayed ceramic coatings are usually associated with porosity and some microcracks, which may be both beneficial and harmful to the performance of ceramic coatings. For example, pores can serve as crack arresters, and thus improve fracture toughness, whereas cracks give rise to either transformation toughening (e.g., for PSZ) or fracture initiation and fatigue failures in the part.[36]

Unmelted Particles. These can sometimes be observed in the coating microstructure, due to partial melting during the spraying process.[13] Unmelted particles seem to break the chemical homogeneity and the overall microstructure. In many instances, unmelted particles reduce the cohesive strength of the coating.[13] They also create shadow porosity.[31a]

Oxides. Oxidation of thermal sprayed metals can significantly affect the microstructure, phase composition, properties, and performance of sprayed coatings. In many applications, metal oxides (often dominated along splat boundaries) can be detrimental toward corrosion, strength, and machinability.[19] (Some oxides are lubricious.[31a]) The extent of oxidation occurring during spray coating is a function of the material being deposited, the method of deposition, and the particular deposition process. Oxidation may occur due to the oxidizing potential of the fuel-gas mixture in flame spraying, HVOF, or detonation gun spraying or due to air inspirated into the gas stream in plasma spraying or any other methods. It is noted that the latter cause can be corrected by using inert-gas shrouds or low-pressure chambers with plasma spraying. Use of carbon-rich gaseous mixtures in oxy-fuel processes can produce carburization rather than oxidation with some metallic coatings. Metallic coatings are perhaps most prone to oxidation, but carbide coatings may suffer a remarkable loss of carbon that is not especially prominent in metallographic examination. Oxidation during deposition can result in higher porosity, weaker coatings due to brittleness, incompatible thermal coefficients of expansion, and disruption of the chemical uniformity of surfaces exposed to a corrosive environment.[1,78]

Most of the thermal spray processes result in very rapid quenching of the particles on impact. Quench rates have been approximated to be 10^6 to 10^8 °C/s for metallics and 10^4 to 10^6 °C/s for ceramics. Consequently, the materials deposited may be in thermodynamically metastable states, and the grains within the splats may be submicron in size or even amorphous. The metastable phases present may not possess the predicted characteristics, especially corrosion characteristics, of the material, and this factor should be borne in mind during the selection of coating compositions.[1]

Cooling and solidification of most materials are associated with shrinkage or contraction. The tensile strength in the coating increases with coating thickness up to a level exceeding the bond or cohesive strength and leads to coating failure. High-strength materials such as austenitic stainless steels are prone to a high degree of stress buildup and are therefore limited to low coating thicknesses. Usually thin coatings are more durable than thick coatings.[19]

The stress buildup in coatings is a function of the spraying process and coating microstructure. Dense coatings usually exhibit more stress buildup than porous coatings. It should be noted that FS coatings usually have greater thickness limitations than PS coatings. On the other hand, the systems with high kinetic energy and low thermal energy [HVOF and high-energy plasma (HEP)] can produce very dense and relatively stress-free coatings. This is attributed to the compressive stresses developed from mechanical deformation (as in shot peening) during particle impact deposition offsetting the tensile shrinkage stresses due to cooling and solid-ification.[19]

19.5.2 Mechanical Properties

The mechanical properties of thermal spray coatings are not well documented except for their hardness and bond strength. The bond strength between the coating and the substrate depends on the selection of coating parameters, the properties of the materials used for spraying, the thermophysical properties, and the surface conditions of the substrate materials.[13]

Substrate temperature during spraying has a significant effect on the mechanical properties (strength and ductility) of sprayed coatings. Substrate preheating improves the adhesion of ceramic coating and also modifies the splat morphology. The splat morphology and particularly the splat/splat and splat/substrate interface are critical to properties such as wear, erosion, corrosion, and bond strength.[5] Coating hardness seems to be determined by the highest temperature to which the coatings are subjected.

Industrially, rough surfaces for thermal spray coating adhesion are usually made by grit blasting, or high-pressure water-jet. EDM* machining of Al alloys can be used as an industrial surface preparation method prior to thermal spray coating for higher coating/substrate adhesion strength.[79]

Table 19.12 lists the typical mechanical properties data for a wide range of plasma-sprayed coatings.[5] However, the sensitivity of the properties of the coatings to particular deposition parameters makes general cataloging of properties by simple chemical composition and common process (e.g., WC-12Co by plasma spray) practically meaningless. The situation becomes more complex because the proper-ties of coatings on test specimens may vary somewhat from those on parts due to differences in geometry and thermal conditions. However, coatings made by com-petent suppliers using adequate quality control will be quite reproducible, and thus the measurement of various mechanical properties of these standardized coatings may be of great importance in the selection of coatings for specific applications. Properties that may be of importance include the modulus of elasticity, strain-to-fracture, modulus of rupture, and hardness. Examples of some of these are listed in Table 19.13.[1]

The anisotropic structure of coatings results in a difference in mechanical prop-erties in the longitudinal and transverse directions. Strength in the longitudinal direction can be 5 to 10 times that of the transverse direction.

19.6 APPLICATIONS

Thermal spray technologies have expanded their applications to many industries. The thermal sprayed coatings offer various properties such as tribological (wear resistance, abradable or abrasive wear resistance, corrosion resistance, oxidation resistance, and heat resistance), thermal behavior, electrical conductivity or resis-tivity, textured surfaces, dimensional restoration, copying of intricate surfaces,

---

* Electric discharge machine or electric discharge machining.

TABLE 19.12 Typical Mechanical Properties of Plasma-Sprayed Coatings[5]

Material	Bond tensile strength*		Rockwell macro/micro hardness	Density	
	MPa	ksi		g/cm³	lb/ft³
Pure metals					
Aluminum	8.3	1.2	45/58 HRH	2.48	155
Copper	21.4	3.1	65/142 HRB	7.20	449
Molybdenum (fine)	57.2	8.3	70/1450 HR15N	9.90	618
Molybdenum (coarse)	55.2	8.0	65/1448 HRA	8.96	559
Nickel (fine)	23.4	3.4	84/... HR15T	7.95	496
Nickel (coarse)	33.1	4.8	81/... HR15T	7.48	467
Niobium	54.5	7.9	61/1344 HRC	7.06	441
Tantalum	46.9	6.8	65/1585 HRA	14.15	883
Titanium	41.4	6.0	78/... HR15N	4.17	260
Tungsten	40.0	5.8	50/500 HRA	16.90	1055
Alloy metals					
304 stainless	17.6	2.55	88/... HR15T	7.22	451
316 stainless	23.4	3.4	70/... HR30T	6.80	425
431 stainless	31.0	4.5	35/... HRC	6.25	390
80Ni-20Cr (fine)	31.0	4.5	90/... HR15	7.48	467
80Ni-20Cr (coarse)	29.0	4.2	90/... HR15T	7.19	449
40Ni-60Cu	24.1	3.5	72/... HRB	7.89	493
35Ni-5In-60Cu	24.1	3.5	83/... HR15T	7.94	496
10Al-90Cu (fine)	28.3	4.1	88/... HR15T	6.73	418
10Al-90Cu (coarse)	22.1	3.2	81/... HR15T	6.30	393
Hastelloy 31 (fine)	41.4	6.0	79/... HR15T	7.65	478
Hastelloy 31 (coarse)	23.4	3.4	79/... HR15T	7.83	489
5Al-95Ni	68.3	9.9	80/490 HRB	7.51	469
20Al-80Ni	47.6	6.9	80/510 HRB	6.92	432
6Al-19Cr-75Ni	49.6	7.2	90/250 HRB	7.51	469
12Si-88Al	16.5	2.4	78/60 HR15T	2.49	155
5Al-5Mo-90Ni	37.9	5.5	80/200 HRB	7.43	464
Hastelloy X	42.7	6.2	89/... HR15T	7.65	478
Hastelloy C	42.1	6.1	90/... HR15T	8.25	515
420 stainless	22.1	3.2	70/... HR15N	7.10	443
0.9C stainless	33.8	4.9	35/... HRC	7.05	440
Cast iron	35.9	5.2	28/... HRC	7.00	437
Ti-6Al-4V	33.1	4.8	35/... HRC	4.30	268
Monel	44.8	6.5	35/... HR15N	8.50	531
0.2C steel	22.1	3.2	95/... HRB	6.90	431
Metal composites					
95Ni-5Al	33.8	4.9	80/500 HR15T	7.39	461
80Ni-20Al	32.4	4.7	86/500 HR15T	7.02	438
65Ni-35Ti	32.1	4.65	72/660 HR15N	6.62	413
75Ni-19Cr-6Al	42.7	6.2	92/250 HR15T	7.71	481
75Ni-9Cr-7Al-5Mo-5Fe	27.6	4.0	80/250 HRB	6.90	431
90Ni-5Al-5Mo	48.3	7.0	80/200 HRB	7.40	462

TABLE 19.12   Typical Mechanical Properties of Plasma-Sprayed Coatings[5] (*Continued*)

Material	Bond tensile strength*		Rockwell macro/micro hardness	Density	
	MPa	ksi		g/cm^3	lb/ft^3
Carbide powders and blends					
88WC-12Co (cast, fine)	44.8	6.5	88/ . . . HR15N	13.75	858
88WC-12Co (cast, coarse)	44.8	6.5	81/ . . . HR15N	12.41	775
88WC-12Co (sintered)	55.2	8.0	85/ . . . HR15N	14.55	908
83WC-17Co	68.9	10.0	85/950 HR15N	11.10	693
75Cr$_3$C$_2$-25NiCr (fine)	41.4	6.0	84/950 HR15N	6.41	400
75Cr$_3$C$_2$-25NiCr (coarse)	34.5	5.0	80/1850 HR15N	6.23	389
75Cr$_3$C$_2$-25NiCr (composite)	—	—	. . ./1850 HR15N	—	—
85Cr$_3$C$_2$-15NiCr	—	—	80/1850 HR15N	5.80	362
Ceramic oxides					
ZrO$_2$ (calcinated)	44.8	6.5	70/ . . . HR15N	5.30	331
Chromium oxide	44.8	6.5	90/ . . . HR15N	4.80	300
80ZrO$_2$-20yttria	15.2	2.2	80/ . . . HR15N	5.00	312
TiO$_2$	—	—	87/ . . . HR15N	4.10	256
Al$_2$O$_3$ (white)	44.8	6.5		—	—
87Al$_2$O$_3$-13TiO$_2$	15.5	2.25	90/ . . . HR15N	3.50	218
60Al$_2$O$_3$-40TiO$_2$	27.6	4.0	90/850 HR15N	3.50	218
50Al$_2$O$_3$-50TiO$_2$	—	—	85/ . . . HR15N	4.0	250
Al$_2$O$_3$-gray (fine)	6.9	1.0	87/193 HR15N	3.30	187
Al$_2$O$_3$-gray (coarse)	—	—	85/ . . . HR15N	3.30	187
Magnesium zirconate	17.2	2.5	75/ . . . HR15N	4.20	262

* Over a grit-blasted surface roughened to 2.5 to 4.1 μm (100 to 160 μin.) AA (arithmetic average).
Source: Ref. 28 in Ref. 5.
Reprinted by permission of ASM International.

catalytic and prosthetic properties, and so forth, based on metallic, carbide-containing, ceramic, or composite materials.[19,29]

Wear Resistance.   Thermal spray coatings are used to resist practically all types of wear, such as abrasive, erosive, and adhesive, in virtually any type of industry. The materials used vary from soft metals to hard metal alloys to carbide-based cermets to oxides. Usually, the wear resistance of the coatings increases with their density and cohesive strength; for example, the higher-velocity coatings such as HVOF and especially D-gun® coatings offer the best wear resistance for a given composition, in contrast to plasma spray coatings (Table 19.14). Table 19.15 shows examples of erosive wear data, obtained by various laboratory tests for some thermal spray coatings.[1]

     Wear-resistant WC-*M* (*M* = Ni, Co, Cr, Mo, or Co-Cr) coatings, produced by HVOF or APS techniques, are used to reduce wear or modify friction in many sliding, abrasive, and corrosive applications such as compressor piston rods, pump plungers, shaft sleeves on centrifugal pumps and fans, and midspan dampers on jet engine compressor fans and blades.[80] These coatings are primary candidates for replacing chrome plating for use on aircraft landing gear components.

     The plain and alloyed WC-Co and Cr$_3$C$_2$-NiCr cermet coatings using HVOF,

TABLE 19.13 Mechanical Properties of Representative Plasma, D-Gun®, and High-Velocity Oxy-Fuel Coatings[1]

Parameter	Type of coating: Tungsten-carbide-cobalt				Alumina	
	W-7Co-4C	W-9Co-5C	W-11Co-4C	W-14Co-4C	Al_2O_3	Al_2O_3
Nominal composition, wt%	Detonation gun	High-velocity combustion	Plasma	Detonation gun	Detonation gun	Plasma
Thermal spray process						
Rupture modulus, 10^3 lb/in.²*	72	—	30	120	22	17
Elastic modulus, 10^6 lb/in.²*	23	—	11	25	14	7.9
Hardness, kg/mm², HV$_{300}$	1,300	1,125	850	1,075	>1,000	>700
Bond strength, 10^3 lb/in.²†	>10,000‡	>10,000‡	>6500	>10,000	>10,000‡	>6500

* Compression of free-standing rings of coatings.
† ASTM C633-89, "Standard Test Method for Adhesion or Cohesive Strength of Flame-Sprayed Coatings," ASTM, 1989.
‡ Epoxy failure.
Source: Publication 1G191, National Association of Corrosion Engineers. Reprinted by permission of ASM International.

TABLE 19.14 Abrasive Wear Data for Selected Thermal Spray Coatings[1]

Material	Type	Wear rate, mm³/1000 rev
Carballoy 883	Sintered	1.2
WC-Co	Detonation gun	0.8
WC-Co	Plasma spray	16.0
WC-Co	Super D-Gun	0.7
WC-Co	HVOF	0.9

ASTM G65 dry sand/rubber wheel test, 50/70-mesh Ottawa silica, 200 rpm, 30-lb load, 3000-revolution test duration.
Reprinted by permission of ASM International.

TABLE 19.15 Erosive Wear Data for Selected Thermal Spray Coatings[1]

Material	Type	Wear rate, μm/g
Carballoy 883	Sintered	0.04
WC-Co	Detonation gun	1.3
WC-Co	Plasma spray	4.6
AISI 1018 steel	Wrought	21

Silica-based erosion test; particle size, 15 μm; particle velocity, 139 m/s; particle flow, 5.5 g/min; ASTM Recommended Practice G75.
Reprinted by permission of ASM International.

plasma, and D-Gun® processes are more popular (for protection against abrasive fretting and erosive wear environments). D-Gun® and HVOF-Cr_3C_2-NiCr cermet coatings are commonly used due to their high-temperature (530 to 815°C) wear and erosion protection applications in the aircraft engine industry such as for exhaust flaps on turbine engines, turbine compressors, midspan stiffeners, pump seals and liners, and knife edge seals. Such coatings using DG processes are used in numerous industrial sectors such as steel plant machinery, printing rolls, and the petrochemical industry.[36] WC-Co cermet coating is preferably used in service conditions below 530°C due to the degradation of WC by oxidation.[81]

Recently, TiC-Ni coatings using HVOF and D-Gun processes have generated more interest due to their combination of wear and corrosion resistance, high-density, low-friction properties, good sprayability by different thermal spray processes, low production cost, etc., using optimized spray conditions and powder compositions.[82] They appear to be an alternative to plain and alloyed WC-Co and Cr_3C_2-NiCr coatings which, in APS, are subjected to phase changes due to oxidation and decarburization.

Mo-based coatings have been used for several adhesive wear conditions. Flame-sprayed Mo-wire and plasma-sprayed Mo-pseudo-alloys are employed in the automotive, paper and pulp, and aerospace industries. For example, Mo-based coatings find applications in automotive piston rings to provide scuff resistance and to decrease adhesive sliding wear.[36,83]

FeCrMo coating is used for adhesive wear resistance on large-bore cylinder liners and stamping dies, NiCrMo coating for both wear and corrosion resistance, and

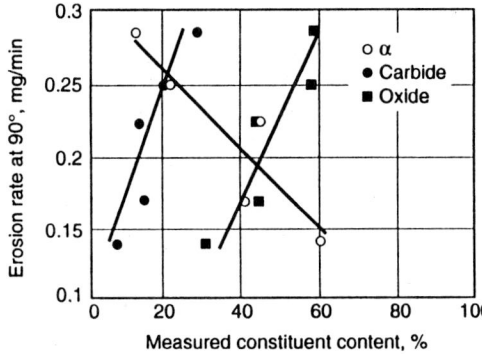

FIGURE 19.14 Steady-state erosion rates versus constituent composition.[86] (*Reprinted by permission of The Metallurgical Society.*)

WC-Co/NiCrMo coatings for a high degree of abrasive and/or adhesive wear resistance on rotors for positive displacement pumps, and on knife edges.[84]

Co-based alloys such as CoCrW and CoCrMo alloys, when deposited by HVOF and PTA processes, are expected to have superior resistance to a combination of harsh environments such as wear and corrosive conditions at elevated temperatures. They provide better erosion and erosion-corrosion properties than the Ni-based metal phase coatings. The metal composition providing the best erosion and corrosion properties is an 8.5Cr-6.5Co matrix.[85]

A comparison between the post-sprayed powder and the actual coating microstructure has revealed that the retention of the FeCrAlY matrix is much better than that of the Cr_3C_2 particles, which can form oxides during HVOF thermal spraying.[86] Erosion tests show that both oxides and carbides increase the erosion rate of the coating (Fig. 19.14) and that the small amounts of hard constituents are desirable for erosion resistance.[5,86]

The use of chromia and Al_2O_3 coatings have some drawbacks: toxicity of certain chromium oxides, the likelihood of metallic Cr in chromia coatings, and the brittleness of pure Al_2O_3 coatings.[87]

D-Gun® sprayed Al_2O_3-ZrO_2-TiO_2 coatings have been reported to exhibit very high abrasive wear resistance; the corresponding APS coatings have also shown superior abrasion wear resistance to plain APS-Al_2O_3 coatings.[88]

Friction Control. Many industries such as the automotive, aircraft, and pulp and paper industries, require sprayed coatings to reduce the frictional coefficient and improve the scuff resistance properties of the substrate material. Mo and Mo-based alloys are usually selected for these applications.[89]

Thermal spray coatings are employed in some applications to give specific frictional characteristics to a surface, ranging from low to high friction (with a coefficient of friction up to 1.42 for the plasma-sprayed WC-Co coatings).[90] The textile industry offers, as an example, a variety of applications involving the complete range of friction characteristics and surface topographies to handle very abrasive synthetic fibers. Oxide coatings such as alumina are mostly used with surfaces that change from very smooth to very rough, depending on the coefficient of friction required.[1] Low-friction Co-based coatings find specific applications in the protection against adhesive and fretting wear, galling and seizure of gas turbine and jet engine parts, or the like from Ti alloys.[91]

The application of APS-Al_2O_3 and/or Cr_2O_3 coatings allows the replacement of steel by Al as the rotating drum material. The APS-Al_2O_3-TiO_2 coating is used for Al friction dampers in buildings to decrease their failures during or after earthquakes.

Corrosion Resistance. Flame-sprayed Zn, Al, Zn-Al, and Zn-Al-Mg alloy coatings (in the 5- to 20-mil thickness range) are often used to provide long-lasting, low-maintenance protection from direct chemical and sacrificial galvanic corrosion (with or without a passive barrier layer)[92] on bridges, ships, tanker trucks, hulls of fishing vessels, the interiors of steel fresh water tanks and conduits, and other large steel structures. Other corrosion-resistant applications for thermal spray coatings are oxidation and sulfidation resistance in power boilers, shielding exhaust manifolds and stacks, industrial flue gas stacks and ducts, automotive valve seats, turbine blades of jet engines, and cylinder heads of marine diesel engines.[93] In a comparison of several different thermal spray processes, it was reported that high-temperature corrosion-resistant coatings must incorporate compositions that favor the formation of protective oxides at splat boundaries, be sufficiently dense to form protective oxides within and to fill voids, and be sufficiently thick to postpone the diffusion of corrosive species to the substrate material along the fast diffusion paths of the coating porosity.[94]

Infrastructure Maintenance. For infrastructure maintenance, thermal spraying is used for underwater tunnels, river dams, concrete and steel gates on a river lock and gate system, river dams, steel piping running underground and aboveground, off-shore oil rigs, water towers, railroad cars, and ships.[36]

Dimensional Restoration. Thermal spray coatings are frequently used to restore the dimensions of worn parts (such as valves and giant turbines used to regulate flow at hydroelectric dams, fuser rolls in copier machines, nonrotating air seals in aircraft turbine engines, carbon steel railway motor turbocharger shafts, and surfaces of printing press cylinders.[93] Sometimes, a coating with low residual stress and/or low cost is employed to restore the worn area, followed by the application of a thin, more wear-resistant coating over it. In any buildup application, it should be noted that the properties of the coatings are perhaps quite different from those of the substrate, and that the coating will not contribute any structural strength to the part. Obviously, if care is not exercised, the coating may reduce the fatigue strength of the part.[1]

Thermal Barrier Coatings. The conventional duplex thermal barrier coating (TBC) comprises a metallic bond coat (typically, MCrAlY usually applied by VPS or EB-PVD) and a thick ($\geq$125 μm) ceramic top coat (CaO-, MgO-, Y_2O_3-, or CeO_2-stabilized ZrO_2 by APS or by EB evaporation). This class of ceramic has a low thermal conductivity to effectively insulate the underlying superalloy substrate from the high-temperature environment. For gas turbine applications, the thickness of the TBCs lies in the range of 125 to 250 μm (0.005 to 0.010 in.). At this thickness range, for hot sections operating at temperatures around 1000°C, the surface of the superalloy component can be lowered by ~100°C, thereby extending the lifetime at the turbine temperature or allowing the turbine to function at a higher, more efficient temperature. The tenacious feature of the TBC is attributed to the porosity, microcracks, and inherent toughness of the specific ceramic (YSZ).[36] However, ceramic coatings sprayed onto metals have certain drawbacks: (1) high residual and thermal stress and (2) relatively low bond strengths. Surface cracking and debond-

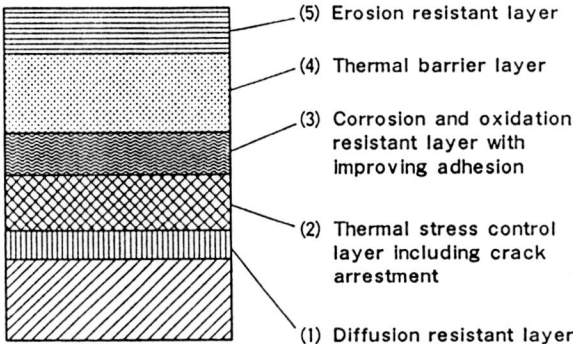

(5) Erosion resistant layer

(4) Thermal barrier layer

(3) Corrosion and oxidation resistant layer with improving adhesion

(2) Thermal stress control layer including crack arrestment

(1) Diffusion resistant layer

FIGURE 19.15 Concept of multi-layered thermal barrier coatings for gas turbine components used at high temperature.[95] (*Reprinted by permission of ASM International.*)

ing have been common experiences of mechanical failure. One way to overcome these problems may be to fabricate functionally gradient materials (FGMs).

Figure 19.15 exhibits the concept of a multi-layered TBC on a superalloy substrate with different unique functions corresponding to the individual layers.[95]

Thermal cycle tests have established that the lifetimes of four-layer TBC are about twice those of two-layer TBC comprising a bond layer and a ceramic layer. The temperatures of substrates with the four-layer TBC are reported to be about 95°C lower than those of uncoated substrates.[96]

Functionally Graded Materials (FGMs).* Functionally graded materials with either continuously or stepwise variations of composition and/or microstructure provide solutions to numerous engineering problems confronting coating systems with large differences in the coefficient of thermal expansion (CTE). A classical example is the TBCs, in which large differences in the CTEs between the substrate and coating can result in failure during the thermal cycle. By continuously grading the composition of the coating from the metallic bond coat at the substrate/coating interface to that of the ceramic TBC at the outer surface, the stresses resulting from mismatch are reduced. One of the major goals in the fabrication of FGMs is that the final structure should vary in a regular and consistent fashion.[97,98]

The growing applications of FGMs include turbine components, rocket nozzles, chemical reactor tubes, burner nozzles, molds, furnace walls,[99] advanced batteries, solid oxide fuel cells (SOFCs) (to produce power cost-effectively),[100] thermoelectric devices (comprising alternate layers of semiconductor materials like $FeSi_2$), and tubular laminate-superalloy composite gun barrels.[101] Examples of thermally sprayed FGMs for burner nozzle applications and for thermoelectric devices are shown in Figs. 19.16 and 19.17.[11,12]

Electrical Applications. Like thermal properties, the electrical conductivity of thermal spray coatings is anisotropic and lower than that of their wrought or

---

* The fabrication of FGMs can also be realized by powder metallurgy which allows dissimilar materials to be integrated while minimizing the stress and allowing normally compatible properties, such as hardness and corrosion resistance, to be incorporated in the same material.

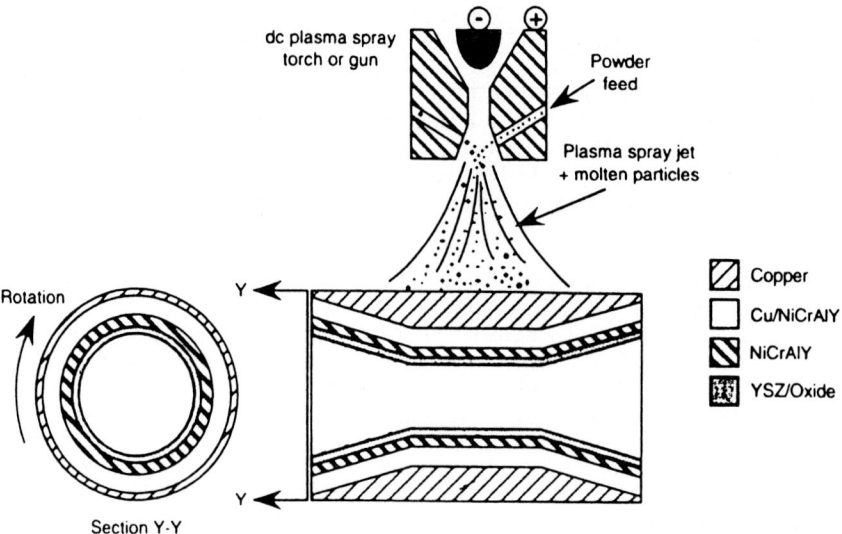

FIGURE 19.16 Schematic of a thermally sprayed FGM for burner nozzle applications.[11] (*Reprinted by permission of ASM International.*)

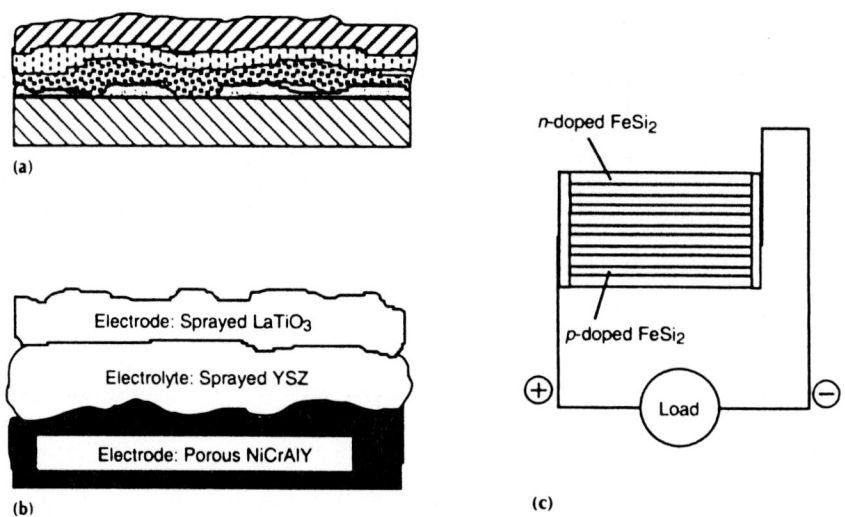

FIGURE 19.17 Other applications of thermally sprayed FGMs for thermoelectric devices: (*a*) strain-control coating, (*b*) sprayed electrodes for SOFCs, and (*c*) plasma-sprayed alternating layers.[11] (*Reprinted by permission of ASM International.*)

sintered counterparts due to their lamellar microstructure and porosity. Metallic or conductive cermet coatings are, however, employed as electrical conductors where both wear resistance and electrical conductivity are required. Thermal spray oxide coatings are used as electrical insulators, but it is essential to seal the coating to prevent moisture, even from the air, from penetrating the coating and reducing its insulating propensity. Thermal spray coatings have also been used to make high-temperature thermocouples and strain gages. Electromagnetic or radio-frequency shielding can also be achieved by flame or electric-arc spray coatings of zinc, tin, or other metals.[1]

Electronics Industry. Examples of applications here include: metal- (Cu, Al, steel, Kovar™) ceramic (Al_2O_3) substrates for hybrid microelectronics; plasma-sprayed sputtering sources of Ta-Hf nitrides and $YBa_2Cu_3O_x$ targets (in the deposition of high-quality PVD films); microwave integrated circuits (MICs) (such as using plasma-sprayed Mg-Mn ferrite inserted into an Al_2O_3 substrate and striplines with sprayed ferrites and dielectrics); capacitor electrodes (such as APS-Al coating in a small-size double-layer capacitor); heater rolls [such as APS-NiAlMo (bond coat)/$MgO.Al_2O_3$ (electrical insulating spinal coat)/(Cu-Zn ferrite, cermet TiO_2 + NiCr, or cermet Al_2O_3 + NiCr) heater coat] for use in energy-saving small office machines (such as laser printers, photocopier machines, faxes); conductor paths for hybrid electronics (such as VPS- and APS-copper coatings used as conductor paths, on sintered Al_2O_3 substrates); and plasma-sprayed Al_2O_3-28% MgO coated integrated circuit (IC) brackets for electrical insulation.[13]

Energy Industry. Boilers in power generation plants (APS- or HVOF-310 stainless steel coatings or APS-NiCr coatings on boiler tubes); MHD generators (APS-TBC deposit comprising NiCrAl, $25\,\mu m$ thick as the base coat and Y_2O_3-stabilized ZrO_2 (YSZ), 100 to $150\,\mu m$ thick, as the top coat for conversion of thermal energy of a plasma (1800 K, 800 to 900 m/s) to electricity; stationary gas turbines [VPS-powder (Ni-38Cr-11Al-0.25Y) coatings on stationary gas turbine (Udimet 520 alloy) blades]; SOFCs (such as VPS-$La_{0.84}Sr_{0.16}MnO_3$ coatings on porous sintered ceramics with APS-perovskite coatings as the cathode, dense ZrO_2-8% Y_2O_3 as a solid electrolyte, and ZrO_2-30 vol% Ni composites as the anode, for conversion of chemical into electrical energy with a high fuel utilization factor of 87.1% and a high power generation efficiency of 38%)[102]; and perovskite as the oxidizer electrode (cathode), rf induction plasma spray coating of a dense Y_2O_3-stabilized-ZrO_2 (YSZ) membrane as the electrolyte, and Ni/YSZ cermet as the fuel electrode (anode).[103]

Petrochemical and Chemical Industries. Tools in petroleum search installations such as super D-Gun® spray-WC-15 Co powder coatings on drill bit cones; flame spray and fuse (with powder self-fluxing alloy of composition, Ni-43.5W-6Cr-1.35B-1.9Fe-6.25Co-1.8Si-3.1C) coatings on polycrystalline diamond cutters and post-spraying fusion in furnace; super D-Gun® spray-powder coatings of composition W-20 Cr-7Ni-6C on rotors; APS-Cr_2O_3 coatings on drilling components (diameter ~45 mm) made of hardened steel (Rc 52 to 59) for substantially prolonged service lifetimes; HVOF-316 stainless steel coatings on the chemical refinery vessels to provide a good corrosion resistance against sulfur and NH_3;[13] HVOF-sprayed-stainless steel 316L powder coatings ($760\,\mu m$ thick) and Hastelloy C-276 coatings on the ends of gas-well tubing for the extension of the corrosion life of vessels in chemical refineries and gas-well tubing in gas research drilling, respectively.[13]

Thermal spray Ti coatings are being used in various applications, especially to produce corrosion-resistant surfaces in the chemical process industries.

Automotive Industry. Examples of applications here are: APS-Al_2O_3 coating on aluminum midplates for diode assembly in automotive alternators to provide resistance against salt corrosion and moisture absorption; pinion shafts; ZrO_2-coated disk brake pads; electric arc spray coating of Fe-C wire on Al valve lifters; turbocompressor housing (of combustion engines); cylinder head gaskets; AlSi-50% Mo plasma spraying of synchronizer ring and hydraulic torque converter pump impeller; plasma-sprayed ceramic (Al_2O_3 + TiO_2) coating at distributor rotors' discharge tips (to reduce electromagnetic interference and suppress noise); and thermal sprayed Fe_3O_4 coating on torque sensors.[104]

Aerospace Industries. Thermal barrier coatings have found widespread applications in the aerospace and other industries: TBCs of 8 wt% Y_2O_3-stabilized-ZrO_2 (top ceramic layer, 1000 μm thick), CoCrAlY as bond coatings (and also as corrosion-resistant layers) and particulate composites containing volume ratios of 85/15 and 40/60 of both components and thickness ~500 μm as intermediate coatings for gas turbine combustors, shrouds, blades, nozzles, and vanes, and on internal combustion cylinders and valves as well as for piston crowns and cylinder heads in adiabatic diesel engines. In other cases, they may be used to diffuse heat as either surface conductors or thermal emitters. Because of their unique lamellar microstructure and porosity, the thermal conductivity of thermal spray coatings is usually anisotropic and significantly lower than that of their wrought or sintered counterparts.[1]

The duplex TBCs, which comprise a MCrAlY bond coating and a Y_2O_3, CaO, or MgO-stabilized-ZrO_2 top coating, are useful for having high durabilities. The concept of multilayered TBCs is shown in Fig. 19.15.[105]

The advantages and improvements by the use of TBCs on various parts of gas turbines (such as blades, nozzles, vanes, and combustion chambers) are (1) an increase of engine power and efficiency due to increased operating temperature and increased thermal cycling (between ambient and operational temperatures over 1100°C) lifetime, (2) an increase in compressor efficiency due to a reduced air flow for turbine cooling, and (3) a longer service life of the metallic substrate material due to a decreased thermal fatigue load.[106]

MCrAlY alloy coatings are used for increased oxidation and hot-gas corrosion resistance of components operating in high temperatures and corrosive environments. For example, LPPS®-sprayed MCrAlY alloy coatings are used in many hot parts in gas turbines and aircraft engines as the undercoatings of thermal barrier coatings.[107]

High-temperature materials such as gas turbine blades and diesel engine components are usually protected by plasma spray or PVD techniques with an intermediate NiCoCrAlY alloy bond coating to improve adhesion strength and to decrease oxidation. The primary TBC material is (6 to 8%) YSZ for thermal insulation in the hot areas of gas turbine components. They are characterized by considerable toughness, low micro-porosity, high melting points, low thermal conductivity, and good thermal shock resistance.[108]

Iron and Steel Industries. Applications here include the following: APS-Al_2O_3 + 25% ZrO_2 coating on cooling rolls of a continuous annealing line (CAL), with the increased roll lifetime from 3 months (for a roll with an electroplated Cr coating) to up to 2 years; APS-(Al_2O_3 + TiO_2) (300-μm thick) coating on stave cooling pipe of blast furnace shells for extended protection against carburization; D-Gun flame-sprayed CoCrAlY-Y_2O_3/CrB_2 coating on hearth rolls for an extended service life to 6 years (compared to 1.5 to 3 years with the conventional ceramic/cermet coating)[109];

thermal spray (>1 mm thick) coating on chute liners, grizzly bars (subjected to abrasive wear by ore fines), sintering fans and cooler fans (subjected to attack by dust-laden gas), both at sinter plants[109]; and HVOF-WC-cermet coating onto the rolling surface of steel rolls used in the steel industry to increase the abrasion and friction resistance of the rolls.[110]

Nonferrous Metals Industries. Applications in nonferrous metals industries include the following: Multicoating with a 50-μm NiCr bond coating, 100-μm Al_2O_3 top coat and between them three 100-μm particulate composite coatings using NiCr and Al_2O_3 powders blended in ratios of 25:75, 50:50, and 75:25 for hot extrusion dies for extended life and remarkable cost savings; thermal spray Cr_2O_3-TiC composite/ NiCrAl bond protective coatings onto mild steel substrate against liquid copper, for an extended life up to 5 hr; multicoating with a 400-μm-thick Mo coating sprayed with IPS under Ar atmosphere as the bond coat, a 200-μm-thick particulate composite of Mo + 50 wt% ZrO_2 (stabilized with calcia), and a 400-μm-thick ZrO_2 stabilized with calcia as the top coat for protection against liquid zirconium[13]; HVOF-WC-Co cermet (particle size ≤45 μm) sprayed coatings (~200 μm in thickness) on mild steel sink rolls for protection against the molten zinc bath and for maintaining a smooth surface longer[111]; plasma spray ceramic [such as triple spinel (MgO-Al_2O_3-ZrO_2) and zircon ($ZrSiO_4$)] coatings on H13 steel and cast iron parts in squeeze casting machines, cast iron heating elements, thermocouple protectors, piston molds, and ladles (as replacements for ceramic and ceramic-slurry coated tools) for the longest life in molten aluminum and for lower aluminum casting operation costs.[112]

Shipbuilding Industry. Applications in the shipbuilding industry include the following: SPS- and VPS-Co-25Cr-10.5Al-2.5Hf-5Pt coating on the blades of the second-stage high-pressure turbines of the LMN2500 marine engine in a U.S. Navy test ship for more corrosion resistance (at high-temperature, low-contamination conditions) than provided by PVD coatings; APS-multicoating with a Ni-9Cr-7Al-5.5Mo-5Fe bond coat, ZrO_2-18TiO_2-10Y_2O_3 top coat, and sealing on 410-type stainless steel valve stems of U.S. Navy aircraft carrier ships for refurbishment with extended performance life; and AS-aluminum coatings on nonskid helicopter flight decks of U.S. Navy ships for increased corrosion resistance and service life.[13]

Machine Building Industry (Textile, Pump Construction, Agroalimentary, etc.). Applications here include the following: APS-Cr_2O_3 coating on rotating aluminum drum, changing the direction of the fiber in the textile machine to achieve increased productivity and fiber velocity; APS-Al_2O_3-TiO_2 coating onto the fiber guides; APS-biocompatible ceramic coating on piston and piston liners of agroalimentary pumps for transporting liquids (e.g., yogurt, chocolate, etc.) to achieve increased three-body abrasion (due to particles in the liquid) wear protection; and APS-multicomponent ceramic coatings on the surfaces of piston and piston liners of vacuum pumps for improved sliding adhesive wear.[13]

Printing Industry. Applications in the printing industry include the following: APS-Al_2O_3 (up to 2 mm thick) coating on *Corona rolls* for sufficient wear-resistant properties; APS-Cr_2O_3 coating on *Anilox rolls (ARs)* for applications in printing machines working with flexographic system. [The PlazJet (Praxair) high-power plasma-sprayed Al_2O_3-40% TiO_2 and high-purity Cr_2O_3 coatings at 12 lb/hr (5.5 kg/hr) and 18 lb/hr (8.2 kg/hr), respectively, yield significant cost and time savings compared to conventional plasma spraying, providing important advantages in terms of paper making and print rolls.[113]]

TABLE 19.16 Tentative Coating Specification for the Production of Anilox Rolls[114]

Condition number	Coating property	Property specification	Remarks
1	Porosity	Low (e.g., <5%)	To prevent ink penetration to the metallic roll
2	Pore size	Small (say < half of the cell diameter, i.e., $15\,\mu m$ for LD 300 lines/cm)	To prevent ink accumulation inside the cells
3	Microhardness	HV greater than 1000	To promote wear resistance
4	Cohesion	High Young's modulus, high toughness	To promote coating integrity
5	Metallic impurities	No impurities	To promote corrosion resistance
6	Light absorption	High light absorption	For a wavelength of $10.6\,\mu m$ (CO_2 laser)

Source: After L. Pawlouski, et al., "Structure-Properties Relationship in Plasma Sprayed Chromium Oxide Coatings, *Thermische Sprizkonfenenz,* Aachen, Germany, March 3–5, 1993, pp. 132–138. Reprinted by permission of John Wiley & Sons, England.

Table 19.16 lists the tentative coating specification for the production of anilox rolls. The conditions 1 to 4 of the coating property specifications can be easily achieved, whereas condition 5 is obtained by an application of agglomerated powder produced by spray drying using electrical power >50 kW, and condition 6 by the use of appropriate grinding and polishing methods.[13,114]

Paper Industry. Examples of applications in the paper industry are as follows: APS multicoatings with Ni-6.5Al-6Mo bond coat (80 to $130\,\mu m$ thick), 8% Y_2O_3- and 1.7% HfO_2-stabilized (17% porous, 380-μm-thick) ZrO_2, and 50-μm-thick, 5 to 7% porous ZrO_2 of the same ZrO_2 on Yankee dryers to eliminate the delamination of the paper (actually, thermally sprayed coatings on experimental rolls offer increased drying efficiency and production of good-quality paper); HVOF-WC-NiCr or WC-Co (wear-resistant) coatings [of 100-μm thickness and mirror finish (R_a = 0.01 to $0.03\,\mu m$)] on *gloss Calendar rolls* for achieving two-years' service lifetime; FS-(kanthal M, i.e., Fe-22Cr-6Al alloy) wire coating on steel tubing in boilers* for superior corrosion-erosion resistance;[13] plasma-sprayed Al_2O_3-2–4% TiO_2 coatings (up to 10 mm thick and diamond-ground) on steel and cast iron rolls to impart high hardness and hydrophilic properties for more uniform release of processed paper with an improved worker safety environment.[115]

Decorative Coatings. Applications in decorative coating industries are the following: FS-copper coatings onto glass artware during the glass-blowing stage for achieving different color deposits; APS-ceramic oxide coatings (Al_2O_3 for white, TiO_2 for gray, Al_2O_3-TiO_2 for blue, Cr_2O_3 for black, and ZrO_3 for yellow coloration).[13]

---

* Boilers are used in the pulp industry to fire black liquor and inorganic compounds. The boiler serves as a chemical reactor, in which the organic portion of the fuel, i.e., black fuel, is burned and the inorganic portion is reduced to sodium sulfide.

Mining Industry. Applications in the mining industry include the following: APS-(400-μm) composite coatings [of 60 vol% self-fluxing NiCrSiB, 35 vol% bronze (Cu-10Sn), and 5 vol% MoS_2] on the internal parts of *hydraulic steel props* used in coal mines for excellent corrosion resistance in aggressive mine water[116] and for increased service life[13]; APS-Cr_2O_3 coatings on hardened-steel drilling components in the petroleum mining industry to increase the service life times of those parts[13].

Medical. Medical applications include the following: plasma-sprayed Zr-reinforced hydroxyapatite [$Ca_2(PO_4)_6(OH)_2$]*[117,118] composite coatings (of a chemical composition similar to that of bone) of thickness 50 to 400 μm with a range of porosity levels and compositions on (1) pure Ti or Ti-6Al-4V for orthopedic implants (for hip, knee, tooth, and other prostheses) and (2) ceramic substrates of ZrO_2 and Al_2O_3 for use in artificial joints and dental roots.[119]

Ceramic Industry. Applications in the ceramic industry are as follows: APS-ceramic rolls and high-energy plasma (HEP) and sprayed ceramic (Y_2O_3) tubes as free-standing bodies; APS-WC-17Co coatings on replaceable mild steel wear plates to be applied on brick clay (cast mild steel) extruders; APS-W (630-μm), Ta-W, or Re-W coatings on graphite crucibles (to melt oxide ceramics such as Al_2O_3, Al_2O_3.ZrO_2, and Al_2O_3.Y_2O_3) and HEP-sprayed membranes.[13] Table 19.17 shows the properties of ceramic rolls prepared by thermal spraying.[13]

Nuclear Industry. The nuclear industry uses plasma-sprayed coatings of both B_4C (moderator) and W in electron beam facilities and in advanced fusion devices for low erosion rates,[120] and VPS-W or IPS-Be coatings onto stainless steels, thereby offering remote repair work on damaged walls of magnetic *fusion energy devices.*

Miscellaneous Applications. A variety of other applications have been developed for thermal spray coatings as given below:

1. Coatings used as nuclear moderators, catalytic surfaces, parting films for hot isostatic presses, and freestanding components such as rocket nozzles, crucibles, and molds

2. Thermal sprayed Al_2O_3-based ceramic coated bicycle rims for improved brake performance and wear resistance

3. Thermal sprayed Al_2O_3 and WC coated and filled with a phenolic-based sealer to penetrate and fill the porosity in yacht sailboat winches for traction control[111]

4. Thermal sprayed Cr_3C_2/NiCr-coated brake poles for lawn mower brake clutches with improved wear resistance[115]

5. Thermal spray on stainless steel fan blades for land-based gas turbines.[93]

6. Thermal sprayed polymers (such as polyethylene, polypropylene, polyester, polyamides, polyvinylidine fluoride, polytetrafluoroethylene, and ethylene methacrylic acid copolymer)[83] in a thickness range of 0.04 to 6.35 mm (0.002 to 0.25 in.), on a wide variety of substrate materials for applications such as plow blades, tank linings, pump impellers, external pipe coatings, structural steel coatings, transfer chutes, light poles, vacuum systems, etc.

---

* Hydroxyapatite is classed as an excellent calcium phosphate bioceramic with unique bioactive properties that promotes rapid chemical bonding to natural bone (i.e., bone joint integrity), enhancing bone growth on its surface, offering long-term joint stability, and exhibiting osteoconductive properties.

TABLE 19.17 Properties of Ceramic Rolls Prepared by Thermal Spraying[13]

	Powder characteristics		Process characteristics		Ceramic characteristics (after all treatments)		Ceramic properties		
No.	Grain size, μm	Chem. composition, wt%	Method	Postspraying	Density, kg/m^3	Phase content	E, GPa	Fracture stress σ_f, MPa	TEC (300–1300), 10^{-6}/K
1*	–180	Y_2O_3	APS		4450				8
2†	–180	Al_2O_3–$0.02SiO_2$	HEP	1823 K/4hr (air)	3580	α-Al_2O_3	165	79.4 ± 9.9	7.4–9.0
3†	–200	Al_2O_3–$22.3ZrO_2$	HEP	As above	3373	α-Al_2O_3, (t + m) ZrO_2	95	76 ± 4.6	7–9
4†		Sillimantin 60™, bal. Al_2O_3, 25–27SiO_2	Sintering				56 ± 12	37.2 ± 2.1	4.6–5.7

* C. E. Holcombe, Jr., *Ceramic Bull.*, vol. 57, 1978, p. 610.
† E. H. Lutz, *Powder Metallurgy International*, vol. 25, 1993, pp. 131–137.
Reprinted by permission of John Wiley & Sons, England.

Process Combinations

1. CO_2-gas laser beam to thermal sprayed WC-12% Co coatings provides dense film, metallurgical bond to the base metal, and a higher hardness.[121]

2. (Laser beam) nitriding after plasma-sprayed Al_2O_3 coatings or Ti coatings show a lower porosity, smoother surface (only one finishing process), increased wear, and corrosion resistance.[122] The advantages of special gas nitriding after plasma-sprayed Al_2O_3 ceramic coatings are no masking during nitriding and no removal of nitrided layer.

3. Thermal sprayed coating used as an undercoat for painting steel structures, anti-abrasive components for machine parts, repairing materials by overlay, and heat-resistant materials, exploiting the advantages of its strong adhesion to substrate and superior applicability.

4. For corrosion protection purposes, however, the conventional thermal spray has been used as a sacrificial protection material, not as an insulating material from the environment due to the porous structure of the coating.[123]

5. Ni electron brush plating on the arc-sprayed coating for improved wear properties of the mold coating surface.[124]

REFERENCES

1. R. C. Tucker, Jr., *ASM Handbook*, vol. 5, 10th ed., ASM International, Materials Park, Ohio, 1994, pp. 497–509.

2. R. C. Tucker, Jr., *Conference Proceedings: Thermal Spraying—Current Status and Future Trends*, ed. A. Ohmori, High Temperature Society of Japan, Osaka, 1995, pp. 253–258.

3. S. Boire-Lavigne, C. Moreau, and R. G. Saint-Jacques, in *Thermal Spray Industrial Applications*, eds. C. C. Berndt and S. Sampath, ASM International, Materials Park, Ohio, 1994, pp. 621–626.

4. L. Pejryd, J. Wigren, and N. Hanner, in *Thermal Spray: A United Forum for Scientific and Technological Advances*, ed. C. C. Berndt, ASM International, Materials Park, Ohio, 1998, pp. 445–450.

5. A. R. Marder, *ASM Handbook, vol. 20: Materials Selection and Design*, ASM International, Materials Park, Ohio, 1997, pp. 470–490.

6. M. Sawa and J. Oohori, in *Conference Proceedings: Thermal Spraying—Current Status and Future Trends*, ed. A. Ohmori, High Temperature Society of Japan, Osaka, 1995, pp. 37–42.

7. H. Kreys, T. Ahlorn, J. Niebergali, and R. Schwetzke, in *Conference Proceedings: Thermal Spraying—Current Status and Future Trends*, ed. A. Ohmori, High Temperature Society of Japan, Osaka, 1995, pp. 127–131.

8. T. Miyamoto et al., in *Conference Proceedings: Thermal Spraying—Current Status and Future Trends*, ed. A. Ohmori, High Temperature Society of Japan, Osaka, 1995, pp. 3–8.

9. A. Kumar, J. Boy, R. Zatorski, and P. March, in *Thermal Spray: A United Forum for Scientific and Technological Advances*, ed. C. C. Berndt, ASM International, Materials Park, Ohio, 1998, pp. 83–90.

10. S. Steinhauser, B. Wielage, U. Hofmann, and G. Zimmermann, in *Thermal Spray: A United Forum for Scientific and Technological Advances*, ed. C. C. Berndt, ASM International, Materials Park, Ohio, 1998, pp. 491–497.

11. R. Knight and R. W. Smith, *ASM Handbook*, vol. 8, 10th ed., ASM International, Materials Park, Ohio, 1998, pp. 408–419.

12. R. W. Smith and R. Knight, *JOM*, vol. 47, no. 8, 1995, pp. 32–39.

13. L. Pawlowski, *The Science and Engineering of Thermal Spray Coatings*, John Wiley & Sons, Chichester, England, 1995.

14. S. Midorikawa et al., in *Conference Proceedings: Thermal Spraying—Current Status and Future Trends*, ed. A. Ohmori, High Temperature Society of Japan, Osaka, 1995, pp. 43–46.

15. H. Yara, K. Miyagi, and A. Ikuta, in *Conference Proceedings: Thermal Spraying—Current Status and Future Trends*, ed. A. Ohmori, High Temperature Society of Japan, Osaka, 1995, pp. 163–167.

15a. R. C. Tucker, Jr., private communication, 2001.

16. J. Youngchang and Y. Feng, in *Conference Proceedings: Thermal Spraying—Current Status and Future Trends*, ed. A. Ohmori, High Temperature Society of Japan, Osaka, 1995, pp. 827– 832.

17. I. Kretschmer, P. Heimgartner, R. Polak, and P. A. Kammer, in *Thermal Spray: A United Forum for Scientific and Technological Advances*, ed. C. C. Berndt, ASM International, Materials Park, Ohio, 1998, pp. 199–202.

18. J. Sheard, J. Heberlein, K. Stelson, and E. Pfender, in *Thermal Spray: A United Forum for Scientific and Technological Advances*, ed. C. C. Berndt, ASM International, Materials Park, Ohio, 1998, pp. 613–618.

19. *Thermal Spray Coatings Page Index*, Gordon England, Surrey, England.

20. Z. Zurecki, D. Garg, and D. Bowe, *J. Thermal Spray Tech.*, vol. 6, 1997, p. 417.

21. K. Wira, C. W. Lim, and N. L. Loh, in *Conference Proceedings: Thermal Spraying—Current Status and Future Trends*, ed. A. Ohmori, High Temperature Society of Japan, Osaka, 1995, pp. 465–470.

22. T. Nakano, R. Uchino, and T. Kusano, in *Conference Proceedings: Thermal Spraying—Current Status and Future Trends*, ed. A. Ohmori, High Temperature Society of Japan, 1995, pp. 15–20.

23. S. Y. Hwang and B. G. Seong, in *Conference Proceedings: Thermal Spraying—Current Status and Future Trends*, ed. A. Ohmori, High Temperature Society of Japan, Osaka, 1995, pp. 59–63.

24. J. M. Park, S. W. Lee, and B. K. Kim, in *Thermal Spray: A United Forum for Scientific and Technological Advances*, ed. C. C. Berndt, ASM International, Materials Park, Ohio, 1998, pp. 121–126.

25. H. Takeda, *Ceramic Coating*, Nikkan-Kogyo, 1987, pp. 179–205; H. Yajima, Y. Kimura, and T. Yoshioka, in *Conference Proceedings: Thermal Spraying—Current Status and Future Trends*, ed. A. Ohmori, High Temperature Society of Japan, Osaka, 1995, pp. 621–626.

26. T. Priem, R. Ranc, E. Rigal, J. Giral, G. Olalde, and J. F. Robert, in *Conference Proceedings: Thermal Spraying—Current Status and Future Trends*, ed. A. Ohmori, High Temperature Society of Japan, Osaka, 1995, pp. 725–729.

27. J. Larsen-Basse and P. Kodali, in *Surface Modification Technologies XII*, eds. T. S. Sudarshan, K. A. Khor, and M. Jeandin, ASM International, Materials Park, Ohio, 1998, pp. 501–506.

28. E. Lugscheider, P. Remer, C. Herbst, K. Yuschenko, Y. Borisov, A. Chernishov, P. Vitiaz, A. Verstak, B. Wielage, and S. Steinhauser, in *Conference Proceedings: Thermal Spraying—Current Status and Future Trends*, ed. A. Ohmori, High Temperature Society of Japan, Osaka, 1995, pp. 235–240.

29. G. Barbezat, S. Keller, and K. H. Wegner, in *Conference Proceedings: Thermal Spraying—Current Status and Future Trends*, ed. A. Ohmori, High Temperature Society of Japan, Osaka, 1995, pp. 9–13.

30. A. S. Khanna, C. Coddet, C. S. Harendranath, and K. Anuja, in *Conference Proceedings: Thermal Spraying—Current Status and Future Trends*, ed. A. Ohmori, High Temperature Society of Japan, Osaka, pp. 577–583.

31. K. Niemi, P. Vuoristo, E. Kumpulainnen, P. Sorsa, and T. Mantyla, in *Conference Proceedings: Thermal Spraying—Current Status and Future Trends*, ed. A. Ohmori, High Temperature Society of Japan, Osaka, 1995, pp. 687–692.

31a. R. Knight, private communication, 2001

32. S. Keller, P. Tommer, R. Clarke, and A. R. Nicoll, in *Conference Proceedings: Thermal Spraying—Current Status and Future Trends*, ed. A. Ohmori, High Temperature Society of Japan, Osaka, 1995, pp. 275–281.

33. T. S. Srivatsan and E. J. Lavernia, *J. Mat. Sci.*, vol. 27, 1992, p. 5965.

34. M. E. Vinayo, in *7th International Symposium on Plasma Chemistry*, Eindhoven, Netherlands, 1985, p. 1161.

35. M. L. Lau, *Thermal Spraying of Nanocrystalline Materials*, Ph.D. thesis, University of California, Irvine, 2000.

36. H. Herman and S. Sampath, *Thermal Spray Coatings,* Web Page, pp. 1–13.

37. K. Honda, I. Chida, M. Saito, Y. Itoh, and K. F. Kobayashi, in *Conference Proceedings: Thermal Spraying—Current Status and Future Trends*, ed. A. Ohmori, High Temperature Society of Japan, Osaka, 1995, pp. 411–416.

38. H. J. Kim, B. L. Choi, and S. H. Hong, in *Conference Proceedings: Thermal Spraying— Current Status and Future Trends*, ed. A. Ohmori, High Temperature Society of Japan, Osaka, 1995, pp. 435–440.

39. H. Yajima, Y. Kimura, and T. Yoshioka, in *Conference Proceedings: Thermal Spraying— Current Status and Future Trends*, ed. A. Ohmori, High Temperature Society of Japan, Osaka, 1995, pp. 621–626.

40. S. Oki, S. Gohda, E. Lugscheider, P. Jokiel, and R. W. Smith, in *Conference Proceedings: Thermal Spraying—Current Status and Future Trends*, ed. A. Ohmori, High Temperature Society of Japan, Osaka, 1995, pp. 561–564.

41. Y. Tsunekawa, M. Okumiya, and T. Kobayashi, in *Conference Proceedings: Thermal Spraying—Current Status and Future Trends*, ed. A. Ohmori, High Temperature Society of Japan, Osaka, 1995, pp. 755–760.

42. S. Sodeoka, M. Suzuki, T. Inoue, X. Ono, and K. Ueno, in *Conference Proceedings: Thermal Spraying—Current Status and Future Trends*, ed. A. Ohmori, High Temperature Society of Japan, Osaka, 1995, pp. 283–288.

43. R. J. DuMola and G. R. Heath, in *Thermal Spray: A United Forum for Scientific and Technological Advances*, ed. C. C. Berndt, ASM International, Materials Park, Ohio, 1998, pp. 427–434.

44. P. J. Meyer, in *Conference Proceedings: Thermal Spraying—Current Status and Future Trends*, ed. A. Ohmori, High Temperature Society of Japan, Osaka, 1995, pp. 217–222.

45. R. Chattopadhyay, in *Conference Proceedings: Thermal Spraying—Current Status and Future Trends*, ed. A. Ohmori, High Temperature Society of Japan, Osaka, 1995, pp. 31–34.

46. A. J. Sturgeon, M. D. F. Harve, F. J. Blunt, and S. B. Dunkerton, in *Conference Proceedings: Thermal Spraying—Current Status and Future Trends*, ed. A. Ohmori, High Temperature Society of Japan, Osaka, 1995, pp. 669–673.

47. M. C. Nestler, H. M. Hohle, W. M. Balbach, and T. Koromzay, in *Conference Proceedings: Thermal Spraying—Current Status and Future Trends*, ed. A. Ohmori, High Temperature Society of Japan, Osaka, 1995, pp. 101–106.

48. H. Kreye, S. Zimmermann, and P. Heinrich, in *Conference Proceedings: Thermal Spraying—Current Status and Future Trends*, ed. A. Ohmori, High Temperature Society of Japan, Osaka, 1995, pp. 393–398.

49. M. Thorpe and H. Richter, *J. Thermal Spray Technol.*, June 1992.

50. A. J. Sturgeon and M. D. F. Harvey, in *Conference Proceedings: Thermal Spraying— Current Status and Future Trends*, ed. A. Ohmori, High Temperature Society of Japan, Osaka, 1995, pp. 933–938.

51. M. Dvorak and J. A. Browning, in *Conference Proceedings: Thermal Spraying—Current Status and Future Trends*, ed. A. Ohmori, High Temperature Society of Japan, Osaka, 1995, pp. 405– 409.

52. M. L. Thorpe and H. J. Richter, in *Thermal Spray: International Advances in Coating Technology*, ed. C. C. Berndt, ASM International, Materials Park, Ohio, 1992, p. 137.

53. J. A. DeBarro and M. R. Dorfman, in *Conference Proceedings: Thermal Spraying— Current Status and Future Trends*, ed. A. Ohmori, High Temperature Society of Japan, Osaka, 1995, pp. 651–656.

54. K. Kadyrov and V. Kadyrov, in *Conference Proceedings: Thermal Spraying—Current Status and Future Trends*, ed. A. Ohmori, High Temperature Society of Japan, Osaka, 1995, pp. 417–424.

55. G. Naisbitt, T. Alderton, and C. Bruce, in *Thermal Spray: A United Forum for Scientific and Technological Advances*, ed. C. C. Berndt, ASM International, Materials Park, Ohio, 1998, pp. 59–63.

56. D. W. Parker and G. L. Kutner, in *Adv. Mater. Process*, vol. 140, 1991, p. 68.

57. K. Sakaki, Y. Shimizu, and N. Saito, in *Conference Proceedings: Thermal Spraying— Current Status and Future Trends*, ed. A. Ohmori, High Temperature Society of Japan, Osaka, 1995, pp. 301–306.

58. S. Y. Hwang, B. G. Seong, and M. C. Kim, in *Proceedings of the 9th National Thermal Spray Conference*, October 7–11, 1996, p. 107.

59. G. M. Guilemany and J. A. Calero, in *Thermal Spray: A United Forum for Scientific and Technological Advances*, ed. C. C. Berndt, ASM International, Materials Park, Ohio, 1998, pp. 717–721.

60. H. Fukutome, H. Shimizu, N. Yamashita, and Y. Shimizu, in *Conference Proceedings: Thermal Spraying—Current Status and Future Trends*, ed. A. Ohmori, High Temperature Society of Japan, Osaka, 1995, pp. 21–26.

61. S. E. Hartfield-Wunsch and S. C. Tung, in *Proceedings 1994, National Thermal Spray Conference*, Boston, MA, June 20–24, 1994, ASM International, Materials Park, Ohio, pp. 19–24.

62. E. Petrovicova, R. Knight, R. W. Smith, and L. S. Schadler, in *Thermal Spray: A United Forum for Scientific and Technological Advances*, ed. C. C. Berndt, ASM International, Materials Park, Ohio, 1998, pp. 877–883.

63. Y. Fukuda and M. Kumon, in *Conference Proceedings: Thermal Spraying—Current Status and Future Trends*, ed. A. Ohmori, High Temperature Society of Japan, Osaka, 1995, pp. 107–110-1.

64. K. Yushehenko, E. Astakhov, Yu Borisov, G. Holmberg, and P. Kaski, in *Conference Proceedings: Thermal Spraying—Current Status and Future Trends*, ed. A. Ohmori, High Temperature Society of Japan, Osaka, 1995, pp. 137–140.

65. A. Utsumi, J. Matsuda, M. Yoneda, M. Katsumura, and T. Araki, in *Conference Proceedings: Thermal Spraying—Current Status and Future Trends*, ed. A. Ohmori, High Temperature Society of Japan, Osaka, 1995, pp. 325–330.

66. R. B. Bhagat, M. F. Amateau, A. Papyrin, J. C. Conway, Jr., B. Stutzman, and B. Jones, in *Thermal Spray: A United Forum for Scientific and Technological Advances*, ed. C. C. Berndt, ASM International, Materials Park, Ohio, 1998, pp. 361–367.

67. J. L. Bacon, D. G. Davis, R. L. Sledge, J. R. Uglum, R. C. Zowarka, and R. J. Polizzi, in *Thermal Spray: A United Forum for Scientific and Technological Advances*, ed. C. C. Berndt, ASM International, Materials Park, Ohio, 1998, pp. 399–406.

68. W. M. Steen, *The Encyclopedia of Advanced Materials*, Pergamon Press, Oxford, England, pp. 2755–2761.

69. D. E. Wolfe, M. B. Movchan, and J. Singh, *Advances in Coating Technologies for Surface Engineering*, eds. C. R. Clayton et al., The Metallurgical Society, Warrendale, PA, 1997, pp. 93–110.

70. K. A. Khor, Y. W. Gu, C. H. Quek, and Z. L. Dong, *Surface Modification Technologies XII*, eds. T. S. Sudarshan, K. A. Khor, and M. Jeandin, ASM International, Materials Park, Ohio, 1998, pp. 495–500.

71. J. Carr and J. Jones, *Post Densified Cr₂O₃ Coatings for Adiabatic Engine*, SAE Technical Paper 840432, Society of Automotive Engineers Warrendale, PA, 1984.

72. H. Ito, R. Nakamura, and M. Shroyama, *Surf. Eng.*, vol. 4, 1988, pp. 35–38.

73. A. Ohmori, K. Aoki, S. Sano, Y. Arata, and N. Iwamoto, *Thin Solid Films*, vol. 207, 1992, pp. 153–157.

74. C. Burman, T. Ericsson, I. Kvernes, and Y. Lindblom, *Surf. Coat. Technol.*, vol. 32, 1987, pp. 127–140.

75. D. Matejka and B. Benko, *Plasma Spraying of Metallic and Ceramic Materials*, John Wiley & Sons Ltd., West Sussex, UK, 1989.

75a. H. Hu, Z. H. Lee, D. R. White, and E. J. Lavernia, *Met. and Mats. Trans.*, vol. 31A, 2000, pp. 723–733.

76. V. V. Sobolev, J. M. Guilemany, and A. J. Martin, in *Proceedings of the 15th International Thermal Spray Conference*, Nice, France, ed. C. Coddet, ASM International, Materials Park, Ohio, 1998, p. 503.

77. L. C. Erickson, T. Troczynski, and H. M. Hawthorne, in *Conference Proceedings: Thermal Spraying—Current Status and Future Trends*, ed. A. Ohmori, High Temperature Society of Japan, Osaka, 1995, pp. 743–748.

78. M. F. Smith, R. C. Dykhuizen, and R. A. Neiser, in *Thermal Spray: A United Forum for Scientific and Technological Advances*, ed. C. C. Berndt, ASM International, Materials Park, Ohio, 1998, pp. 885–893.

79. O. O. Popoola, M. J. Zaluzec, and H. Haack, *Surface Modification Technologies XII*, eds. T. S. Sudarshan, K. A. Khor, and M. Jeandin, ASM International, Materials Park, Ohio, 1998, pp. 75–84.

80. J. Wigren, L. Pejryd, D. J. Greving, J. R. Shadley, and E. F. Rybicki, in *Conference Proceedings: Thermal Spraying—Current Status and Future Trends*, ed. A. Ohmori, High Temperature Society of Japan, Osaka, 1995, pp. 113–118.

81. L. Russo and M. Dorfmann, in *Conference Proceedings: Thermal Spraying—Current Status and Future Trends*, ed. A. Ohmori, High Temperature Society of Japan, Osaka, 1995, pp. 681–686.

82. P. Vuoristo, T. Mantyla, L.-M. Berger, and M. Nebelung, in *Thermal Spray: A United Forum for Scientific and Technological Advances*, ed. C. C. Berndt, ASM International, Materials Park, Ohio, 1998, pp. 909–915.

83. B. J. Taylor and T. S. Eyre, *Tribology*, April 1979, p. 79.

84. J. A. DeBarro and M. R. Dorfman, in *Conference Proceedings: Thermal Spraying—Current Status and Future Trends*, ed. A. Ohmori, High Temperature Society of Japan, Osaka, 1995, pp. 651–656.

85. T. Rogne, T. Solem, and J. Berget, in *Thermal Spray: A United Forum for Scientific and Technological Advances*, ed. C. C. Berndt, ASM International, Materials Park, Ohio, 1998, pp. 113–119.

86. K. J. Stein, B. S. Schorr, and A. R. Marder, *Elevated Temperature Coatings: Science and Technology II*, eds. N. B. Dohrte and J. M. Hampikian, The Metallurgical Society, Warrendale, PA, 1996, p. 99.

87. E. Lugscheider, H. Jungklaus, P. Remer, and J. Knuuttila, in *Conference Proceedings: Thermal Spraying—Current Status and Future Trends*, ed. A. Ohmori, High Temperature Society of Japan, Osaka, 1995, pp. 833–838.

88. K. Niemi, P. Vuoristo, T. Mantyla, E. Lugscheider, J. Knuuttila, and H. Jungklaus, in *Conference Proceedings: Thermal Spraying—Current Status and Future Trends*, ed. A. Ohmori, High Temperature Society of Japan, Osaka, 1995, pp. 675–680.

89. J. DeFalco, L. Russo, and M. R. Dorfmann, in *Thermal Spray: A United Forum for Scientific and Technological Advances*, ed. C. C. Berndt, ASM International, Materials Park, Ohio, 1998, p. 990.

90. Z. H. Tong, C. X. Ding, B. Qiao, and K. W. Huang, in *Conference Proceedings: Thermal Spraying—Current Status and Future Trends*, ed. A. Ohmori, High Temperature Society of Japan, Osaka, 1995, pp. 713–717.

91. K. Hajmrie and A. P. Chilkovich, in *Thermal Spray: A United Forum for Scientific and Technological Advances*, ed. C. C. Berndt, ASM International, Materials Park, Ohio, 1998, pp. 127–130.

92. T. Lester, S. J. Harris, D. Kingerley, and S. Matthews, in *Thermal Spray: A United Forum for Scientific and Technological Advances*, ed. C. C. Berndt, ASM International, Materials Park, Ohio, 1998, pp. 183–189.

93. Sultzer Metco, *Thermal Spray Coatings Bulletin*, Westbury, N.Y.

94. S. T. Bluni and A. R. Marder, *Corrosion*, vol. 52, 1996, p. 213.

95. M. Tamura, M. Takahashi, J. Ishii, K. Suzuki, M. Sato, and K. Shimomura, in *Thermal Spray: A United Forum for Scientific and Technological Advances*, ed. C. C. Berndt, ASM International, Materials Park, Ohio, 1998, pp. 323–328.

96. Y. Kojima, K. Wada, T. Teramae, and Y. Furuse, in *Conference Proceedings: Thermal Spraying—Current Status and Future Trends*, ed. A. Ohmori, High Temperature Society of Japan, Osaka, 1995, pp. 95–99.

97. W. D. Swank, J. R. Fincke, D. C. Haggard, S. Sampath, and W. Smith, in *Thermal Spray: A United Forum for Scientific and Technological Advances*, ed. C. C. Berndt, ASM International, Materials Park, Ohio, 1998, pp. 451–458.

98. J. R. Finke, W. D. Swank, and D. C. Haggard, in *Thermal Spray: A United Forum for Scientific and Technological Advances*, ed. C. C. Berndt, ASM International, Materials Park, Ohio, 1998, pp. 451–458.

99. A. H. Bartlett, R. G. Castro, D. P. Butt, H. Kung, J. J. Petrovic, and Z. Zurecki, *Ind. Heat.*, vol. LXIII, no. 1, 1996, pp. 33–36.

100. F. Fendler, R. Henne, and M. Lang, in *Proceedings of the 8th National Thermal Spray Conference*, ASM International, Materials Park, Ohio, 1995, pp. 533–537.

101. L. J. Westfall, in *Thermal Spray: Advances in Coating Technology, Proceedings of the Second National Thermal Spray Conference*, ASM International, Metals Park, Ohio, 1988, p. 417.

102. A. Notomi and N. Hisatome, in *Conference Proceedings: Thermal Spraying—Current Status and Future Trends*, ed. A. Ohmori, High Temperature Society of Japan, Osaka, 1995, pp. 79–82.

103. K. Mailhot, F. Gitzhofer, and M. I. Boulos, in *Thermal Spray: A United Forum for Scientific and Technological Advances*, ed. C. C. Berndt, ASM International, Materials Park, Ohio, 1998, pp. 21–25.

104. T. Miytamoto and S. Sugimoto, in *Conference Proceedings: Thermal Spraying—Current Status and Future Trends*, ed. A. Ohmori, High Temperature Society of Japan, Osaka, 1995, pp. 3–8.

105. M. Takahashi, Y. Itoh, and M. Miyazaki, in *Conference Proceedings: Thermal Spraying—Current Status and Future Trends*, ed. A. Ohmori, High Temperature Society of Japan, Osaka, 1995, pp. 83–88.

106. H. D. Steffens, J. Wilden, and I. A. Josefiak, in *Thermal Spray: A United Forum for Scientific and Technological Advances*, ed. C. C. Berndt, ASM International, Materials Park, Ohio, 1998, p. 988.

107. R. Yamasaki, J. Takeuchi, A. Nakahira, M. Saitoh, and Y. Itoh, in *Conference Proceedings: Thermal Spraying—Current Status and Future Trends*, ed. A. Ohmori, High Temperature Society of Japan, Osaka, 1995, pp. 863–868.

108. K. S. Ravichandran, K. An, and R. Taylor, in *Thermal Spray: A United Forum for Scientific and Technological Advances*, ed. C. C. Berndt, ASM International, Materials Park, Ohio, 1998, pp. 291–298.

109. M. Sawa and J. Oohori, in *Conference Proceedings: Thermal Spraying—Current Status and Future Trends*, ed. A. Ohmori, High Temperature Society of Japan, Osaka, 1995, pp. 37–42.

110. Y. Matsubara and A. Tomiguchi, in *13th International Thermal Spraying Conference*, Orlando, FL, May 28–June 5, 1992, ed. C. C. Berndt, ASM International, Materials Park, Ohio, pp. 637–641.

111. Y. Kobayashi et al., in *Conference Proceedings: Thermal Spraying—Current Status and Future Trends*, ed. A. Ohmori, High Temperature Society of Japan, Osaka, 1995, pp. 211–216.

112. Y. Wang, in *Surface Modification Technologies XII*, eds. T. S. Sudarshan, K. A. Khor, and M. Jeandin, ASM International, Materials Park, Ohio, 1998, pp. 525–532.

113. G. Irons, D. Poirier, and A. Roy, in *Conference Proceedings: Thermal Spraying—Current Status and Future Trends*, ed. A. Ohmori, High Temperature Society of Japan, Osaka, 1995, pp. 205–209.

114. L. Pawlowski, R. Zacchino, R. Dal Maschio, V. M. Saglavo, J. Andersen, and F. J. Driller, in *Thermische Spritzkonferenz*, Aachen, Germany, March 3–5, 1993, pp. 132–138.

115. W. J. Lenling, P. R. Gilson, and D. L. Ohmann, in *Surface Modification Technologies XII*, eds. T. S. Sudarshan, K. A. Khor, and M. Jeandin, ASM International, Materials Park, Ohio, 1998, pp. 519–524.

116. D. Matejka et al., in *1st Plasma Technik Symposium,* Lucerne, Switzerland, May 18–20, 1988, pp. 247–257.

117. K. A. Khor and P. Cheang, in *Thermal Spray: A United Forum for Scientific and Technological Advances*, ed. C. C. Berndt, ASM International, Materials Park, Ohio, 1998, pp. 769–774.

118. A. J. Sturgeon and M. D. F. Harvey, in *Conference Proceedings: Thermal Spraying—Current Status and Future Trends*, ed. A. Ohmori, High Temperature Society of Japan, Osaka, 1995, pp. 933–938.

119. T. Kameyama, M. Ueda, K. Onuma, A. Motoe, K. Ohsaki, H. Tanizaki, and K. Iwasaki, in *Conference Proceedings: Thermal Spraying–Current Status and Future Trends*, ed. A. Ohmori, High Temperature Society of Japan, Osaka, 1995, pp. 187–192.

120. W. K. W. M. Mallener, H. Gruhn, and H. Hoven, in *Conference Proceedings: Thermal Spraying—Current Status and Future Trends*, ed. A. Ohmori, High Temperature Society of Japan, Osaka, 1995, pp. 229–233.

121. N. Takasaki, M. Kumagawa, K. Yairo, and A. Ohmori, in *Conference Proceedings: Thermal Spraying—Current Status and Future Trends*, ed. A. Ohmori, High Temperature Society of Japan, Osaka, 1995, pp. 987–992.

122. H. D. Steffens, J. Wilden, and C. Buchmann, in *Conference Proceedings: Thermal Spraying—Current Status and Future Trends*, ed. A. Ohmori, High Temperature Society of Japan, Osaka, 1995, pp. 981–986.

123. T. Suzuki, K. Ishikawa, and Y. Kitamura, in *Conference Proceedings: Thermal Spraying—Current Status and Future Trends*, ed. A. Ohmori, High Temperature Society of Japan, Osaka, 1995, pp. 1033–1038.

124. L. Xianjun and M. Dan, in *Thermal Spray: A United Forum for Scientific and Technological Advances*, ed. C. C. Berndt, ASM International, Materials Park, Ohio, 1998, p. 986.

APPENDIX A

CONVERSION TABLE FOR UNITS, CONSTANTS, AND FACTORS IN COMMON USE

Quantity	Symbol	Traditional Unit	SI Unit
1 Atmosphere (pressure)	atm		$101\ 325\ \mathrm{N\,m^{-2}}$
Avogadro's constant	N_A	6.0225×10^{23}	$6.0225 \times 10^{23}\ \mathrm{mol^{-1}}$
1 Angstrom	Å	10^{-8} cm	10^{-10} m
1 bar	bar		$10^5\ \mathrm{N\,m^{-2}}$
Boltzmann's constant	k	$1.380 \times 10^{-16}\ \mathrm{erg\,deg^{-1}}$	$1.380 \times 10^{-23}\ \mathrm{J\,K^{-1}}$
1 calorie	cal	2.61×10^{19} eV	4.184 J
		4.186 joules	
1 dyne	dyn		10^{-5} N
1 dyne cm^{-2}		$1.45 \times 10^{-5}\ \mathrm{lb/in^2}$	$10^{-1}\ \mathrm{N\,m^{-2}}$
1 day		86 400 s	86.4 ks
1 degree (angle)		0.017 rad	17 m rad
1 erg		6.24×10^{11} eV	10^{-7} J
		2.39×10^{-8} cal	
1 erg cm^{-2}		$6.24 \times 10^{11}\ \mathrm{eV\,cm^{-2}}$	$10^{-3}\ \mathrm{J\,m^{-2}}$
gas constant	R	$8.3143 \times 10^7\ \mathrm{erg\,g\text{-}atom^{-1}}$	$8.3143\ \mathrm{J\,mol^{-1}\,K^{-1}}$
		$1.987\ \mathrm{cal\,deg^{-1}\,g\text{-}atom^{-1}}$	
electronic charge	e	4.8×10^{-10} e.s.u.	1.6021×10^{-19} C
1 electron volt	eV	3.83×10^{-20} cal	
		1.6021×10^{-12} erg	1.6021×10^{-19} J
Faraday constant	$F = N_A e$		$9.6487 \times 10^4\ \mathrm{C\,mol^{-1}}$
1 inch	in	2.54 cm	25.4 mm
1 kilocalorie	kcal	4.186×10^{10} erg	
1 kilogram	kg	2.21 lb	1 kg
1 kilogram cm^{-2}	kg cm^{-2}	$14.22\ \mathrm{lb/in^2}$	$10^4\ \mathrm{kg\,m^{-2}}$
1 litre	l		$1\ \mathrm{dm^3}$
mass of electron	m_e	9.1091×10^{-28} g	9.1091×10^{-31} kg
1 micron	μm	10^4 Ångstroms	10^{-6} m
		10^{-4} cm	
1 minute (angle)	min	2.91×10^{-4} radians	$\mathrm{min} = 2.91 \times 10^{-4}$ rad
Planck's constant	h	6.6256×10^{-27} erg s	6.6256×10^{-34} J s
1 pound	lb	453.59 g	0.453 59 kg
1 pound (force)	lbf		4.448 22 N
1 p.s.i.	lbf/in^2		$6894.76\ \mathrm{N\,m^{-2}}$
1 radian	rad	57.296 degrees	1 rad
1 ton (force)	1 tonf		9.964 02 kN
1 t.s.i.	1 tonf/in^2	$1.5749\ \mathrm{kg/mm^2}$	$15.4443\ \mathrm{MN\,m^{-2}}$
			15.443 MPa
			$(\mathrm{Pa} = \mathrm{Pascal} = \mathrm{N\,m^{-2}})$
1 ton	t	1000 kg	10^3 kg
1 torr	torr	1 mm Hg	$133.322\ \mathrm{N\,m^{-2}}$
velocity of light	c	$2.997\ 925 \times 10^{10}$ cm/s	$2.997\ 925 \times 10^8\ \mathrm{m\,s^{-1}}$

APPENDIX B
TEMPERATURE CONVERSIONS

Look up temperature to be converted in middle column. If in degrees Centigrade, read Fahrenheit equivalent in right-hand column; if in Fahrenheit degrees, read Centigrade equivalent in left-hand column. (Source: Republic Steel.)

	−459.4 to 0						0 to 100				
°C	°F		°C		°F	°C		°F	°C		°F
−273.0	−459.4	...	−73.3	−100	−148.0	−17.8	0	32.0	10.0	50	122.0
−267.8	−450	...	−72.2	−98	−144.4	−17.2	1	33.8	10.6	51	123.8
−262.2	−440	...	−71.1	−96	−140.8	−16.7	2	35.6	11.1	52	125.6
−256.7	−430	...	−70.0	−94	−137.2	−16.1	3	37.4	11.7	53	127.4
−251.1	−420	...	−68.9	−92	−133.6	−15.6	4	39.2	12.2	54	129.2
−245.6	−410	...	−67.8	−90	−130.0	−15.0	5	41.0	12.8	55	131.0
−240.0	−400	...	−66.7	−88	−126.4	−14.4	6	42.8	13.3	56	132.8
−234.4	−390	...	−65.6	−86	−122.8	−13.9	7	44.6	13.9	57	134.6
−228.9	−380	...	−64.4	−84	−119.2	−13.3	8	46.4	14.4	58	136.4
−223.3	−370	...	−63.3	−82	−115.6	−12.8	9	48.2	15.0	59	138.2
−217.8	−360	...	−62.2	−80	−112.0	−12.2	10	50.0	15.6	60	140.0
−212.2	−350	...	−61.1	−78	−108.4	−11.7	11	51.8	16.1	61	141.8
−206.7	−340	...	−60.0	−76	−104.8	−11.1	12	53.6	16.7	62	143.6
−201.1	−330	...	−58.9	−74	−101.2	−10.6	13	55.4	17.2	63	145.4
−195.6	−320	...	−57.8	−72	−97.6	−10.0	14	57.2	17.8	64	147.2
−190.0	−310	...	−56.7	−70	−94.0	−9.4	15	59.0	18.3	65	149.0
−184.4	−300	...	−55.6	−68	−90.4	−8.9	16	60.8	18.9	66	150.8
−178.9	−290	...	−54.4	−66	−86.8	−8.3	17	62.6	19.4	67	152.6
−173.3	−280	...	−53.3	−64	−83.2	−7.8	18	64.4	20.0	68	154.4
−169.5	−273	−459.4	−52.2	−62	−79.6	−7.2	19	66.2	20.6	69	156.2
−167.8	−270	−454.0	−51.1	−60	−76.0	−6.7	20	68.0	21.1	70	158.0
−162.2	−260	−436.0	−50.0	−58	−72.4	−6.1	21	69.8	21.7	71	159.8
−156.7	−250	−418.0	−48.9	−56	−68.8	−5.6	22	71.6	22.2	72	161.6
−151.1	−240	−400.0	−47.8	−54	−65.2	−5.0	23	73.4	22.8	73	163.4
−145.6	−230	−382.0	−46.7	−52	−61.6	−4.4	24	75.2	23.3	74	165.2
−142.8	−225	−373.0	−45.6	−50	−58.0	−3.9	25	77.0	23.9	75	167.0
−140.0	−220	−364.0	−44.4	−48	−54.4	−3.3	26	78.8	24.4	76	168.8
−137.2	−215	−355.0	−43.3	−46	−50.8	−2.8	27	80.6	25.0	77	170.6
−134.4	−210	−346.0	−42.2	−44	−47.2	−2.2	28	82.4	25.6	78	172.4
−131.7	−205	−337.0	−41.1	−42	−43.6	−1.7	29	84.2	26.1	79	174.2
−128.9	−200	−328.0	−40.0	−40	−40.0	−1.1	30	86.0	26.7	80	176.0
−126.1	−195	−319.0	−38.9	−38	−36.4	−0.6	31	87.8	27.2	81	177.8
−123.3	−190	−310.0	−37.8	−36	−32.8	0.0	32	89.6	27.8	82	179.6
−121.0	−185	−301.0	−36.7	−34	−29.2	0.6	33	91.4	28.3	83	181.4
−117.8	−180	−292.0	−35.6	−32	−25.6	1.1	34	93.2	28.9	84	183.2
−115.0	−175	−283.0	−34.4	−30	−22.0	1.7	35	95.0	29.4	85	185.0
−112.2	−170	−274.0	−33.3	−28	−18.4	2.2	36	96.8	30.0	86	186.8
−109.5	−165	−265.0	−32.2	−26	−14.8	2.8	37	98.6	30.6	87	188.6
−106.7	−160	−256.0	−31.1	−24	−11.2	3.3	38	100.4	31.1	88	190.4
−103.9	−155	−247.0	−30.0	−22	−7.6	3.9	39	102.2	31.7	89	192.2
−101.1	−150	−238.0	−28.9	−20	−4.0	4.4	40	104.0	32.2	90	194.0
−98.3	−145	−229.0	−27.8	−18	−0.4	5.0	41	105.8	32.8	91	195.8
−95.6	−140	−220.0	−26.7	−16	+3.2	5.6	42	107.6	33.3	92	197.6
−92.8	−135	−211.0	−25.6	−14	+6.8	6.1	43	109.4	33.9	93	199.4
−90.0	−130	−202.0	−24.4	−12	+10.4	6.7	44	111.2	34.4	94	201.2
−87.2	−125	−193.0	−23.3	−10	+14.0	7.2	45	113.0	35.0	95	203.0
−84.4	−120	−184.0	−22.2	−8	+17.6	7.8	46	114.8	35.6	96	204.8
−82.0	−115	−175.0	−21.1	−6	+21.2	8.3	47	116.6	36.1	97	206.6
−78.9	−110	−166.0	−20.0	−4	+24.8	8.9	48	118.4	36.7	98	208.4
−75.8	−105	−157.0	−18.9	−2	+28.4	9.4	49	120.2	37.2	99	210.2
−73.3	−100	−148.0	−17.8	0	+32.0	10.0	50	122.0	37.8	100	212.0

100 to 1000

°C		°F	°C		°F	°C		°F	°C		°F
38	100	212	93	200	392	219	425	797	354	670	1238
39	102	216	94	202	396	221	430	806	357	675	1247
40	104	219	96	204	399	224	435	815	360	680	1256
41	106	223	97	206	403	227	440	824	363	685	1265
42	108	226	98	208	406	229	445	833	366	690	1274
43	110	230	99	210	410	232	450	842	368	695	1283
44	112	234	100	212	414	235	455	851	371	700	1292
46	114	237	102	215	419	238	460	860	374	705	1301
47	116	241	104	220	428	241	465	869	377	710	1310
48	118	244	107	225	437	243	470	878	379	715	1319
49	120	248	110	230	446	246	475	887	382	720	1328
50	122	252	113	235	455	249	480	896	385	725	1337
51	124	255	116	240	464	252	485	905	388	730	1346
52	126	259	119	245	473	254	490	914	391	735	1355
53	128	262	121	250	482	257	495	923	393	740	1364
54	130	266	124	255	491	260	500	932	396	745	1373
56	132	270	127	260	500	263	505	941	399	750	1382
57	134	273	130	265	509	266	510	950	402	755	1391
58	136	277	132	270	518	268	515	959	404	760	1400
59	138	280	135	275	527	271	520	968	407	765	1409
60	140	284	138	280	536	274	525	977	410	770	1418
61	142	288	141	285	545	277	530	986	413	775	1427
62	144	291	143	290	554	279	535	995	416	780	1436
63	146	295	146	295	563	282	540	1004	418	785	1445
64	148	298	149	300	572	285	545	1013	421	790	1454
66	150	302	152	305	581	288	550	1022	424	795	1463
67	152	306	154	310	590	291	555	1031	427	800	1472
68	154	309	157	315	599	293	560	1040	429	805	1481
69	156	313	160	320	608	296	565	1049	432	810	1490
70	158	316	163	325	617	299	570	1058	435	815	1499
71	160	320	166	330	626	302	575	1067	438	820	1508
72	162	324	169	335	635	304	580	1076	441	825	1517
73	164	327	171	340	644	307	585	1085	443	830	1526
74	166	331	174	345	653	310	590	1094	446	835	1535
76	168	334	177	350	662	313	595	1103	449	840	1544
77	170	338	179	355	671	316	600	1112	454	850	1562
78	172	342	182	360	680	319	605	1121	460	860	1580
79	174	345	185	365	689	321	610	1130	466	870	1598
80	176	349	188	370	698	324	615	1139	471	880	1616
81	178	352	191	375	707	327	620	1148	477	890	1624
82	180	356	193	380	716	330	625	1157	482	900	1652
83	182	360	196	385	725	332	630	1166	488	910	1670
84	184	363	199	390	734	335	635	1175	493	920	1688
86	186	367	202	395	743	338	640	1184	499	930	1706
87	188	370	204	400	752	341	645	1193	504	940	1724
88	190	374	207	405	761	343	650	1202	510	950	1742
89	192	378	210	410	770	346	655	1211	516	960	1760
90	194	381	213	415	779	349	660	1220	521	970	1778
91	196	385	216	420	788	352	665	1229	527	980	1796
92	198	388	219	425	797	354	670	1238	532	990	1814
93	200	392							538	1000	1832

1000 to 2000						2000 to 3000					
°C		°F	°C		°F	°C		°F	°C		°F
538	1000	1832	816	1500	2732	1093	2000	3632	1371	2500	4532
543	1010	1850	821	1510	2750	1099	2010	3650	1377	2510	4550
549	1020	1868	827	1520	2768	1104	2020	3668	1382	2520	4568
554	1030	1886	832	1530	2786	1110	2030	3686	1388	2530	4586
560	1040	1904	838	1540	2804	1116	2040	3704	1393	2540	4604
566	1050	1922	843	1550	2822	1121	2050	3722	1399	2550	4622
571	1060	1940	849	1560	2840	1127	2060	3740	1404	2560	4640
577	1070	1958	854	1570	2858	1132	2070	3758	1410	2570	4658
582	1080	1976	860	1580	2876	1138	2080	3776	1416	2580	4676
588	1090	1994	866	1590	2894	1143	2090	3794	1421	2590	4694
593	1100	2012	871	1600	2912	1149	2100	3812	1427	2600	4712
599	1110	2030	877	1610	2930	1154	2110	3830	1432	2610	4730
604	1120	2048	882	1620	2948	1160	2120	3848	1438	2620	4748
610	1130	2066	888	1630	2966	1166	2130	3866	1443	2630	4766
616	1140	2084	893	1640	2984	1171	2140	3884	1449	2640	4784
621	1150	2102	899	1650	3002	1177	2150	3902	1454	2650	4802
627	1160	2120	904	1660	3020	1182	2160	3920	1460	2660	4820
632	1170	2138	910	1670	3038	1188	2170	3938	1466	2670	4838
638	1180	2156	916	1680	3056	1193	2180	3956	1471	2680	4856
643	1190	2174	921	1690	3074	1199	2190	3974	1477	2690	4874
649	1200	2192	927	1700	3092	1204	2200	3992	1482	2700	4892
654	1210	2210	932	1710	3110	1210	2210	4010	1488	2710	4910
660	1220	2228	938	1720	3128	1216	2220	4028	1493	2720	4928
666	1230	2246	943	1730	3146	1221	2230	4046	1499	2730	4946
671	1240	2264	949	1740	3164	1227	2240	4064	1504	2740	4964
677	1250	2282	954	1750	3182	1232	2250	4082	1510	2750	4982
682	1260	2300	960	1760	3200	1238	2260	4100	1516	2760	5000
688	1270	2318	966	1770	3218	1243	2270	4118	1521	2770	5018
693	1280	2336	971	1780	3236	1249	2280	4136	1527	2780	5036
699	1290	2354	977	1790	3254	1254	2290	4154	1532	2790	5054
704	1300	2372	982	1800	3272	1260	2300	4172	1538	2800	5072
710	1310	2390	988	1810	3290	1266	2310	4190	1543	2810	5090
716	1320	2408	993	1820	3308	1271	2320	4208	1549	2820	5108
721	1330	2426	999	1830	3326	1277	2330	4226	1554	2830	5126
727	1340	2444	1004	1840	3344	1282	2340	4244	1560	2840	5144
732	1350	2462	1010	1850	3362	1288	2350	4262	1566	2850	5162
738	1360	2480	1016	1860	3380	1293	2360	4280	1571	2860	5180
743	1370	2498	1021	1870	3398	1299	2370	4298	1577	2870	5198
749	1380	2516	1027	1880	3416	1304	2380	4316	1582	2880	5216
754	1390	2534	1032	1890	3434	1310	2390	4334	1588	2890	5234
760	1400	2552	1038	1900	3452	1316	2400	4352	1593	2900	5252
766	1410	2570	1043	1910	3470	1321	2410	4370	1599	2910	5270
771	1420	2588	1049	1920	3488	1327	2420	4388	1604	2920	5288
777	1430	2606	1054	1930	3506	1332	2430	4406	1610	2930	5306
782	1440	2624	1060	1940	3524	1338	2440	4424	1616	2940	5324
788	1450	2642	1066	1950	3542	1343	2450	4442	1621	2950	5342
793	1460	2660	1071	1960	3560	1349	2460	4460	1627	2960	5360
799	1470	2678	1077	1970	3578	1354	2470	4478	1632	2970	5378
804	1480	2696	1082	1980	3596	1360	2480	4496	1638	2980	5396
810	1490	2714	1088	1990	3614	1366	2490	4514	1643	2990	5414
816	1500	2782	1093	2000	3632	1371	2500	4532	1649	3000	5432

	3000 to 4000				
°C		°F	°C		°F
1649	3000	5432	1927	3500	6332
1654	3010	5450	1932	3510	6350
1660	3020	5468	1938	3520	6368
1666	3030	5486	1943	3530	6386
1671	3040	5504	1949	3540	6404
1677	3050	5522	1954	3550	6422
1682	3060	5540	1960	3560	6440
1688	3070	5558	1965	3570	6458
1693	3080	5576	1971	3580	6476
1699	3090	5594	1977	3590	6494
1704	3100	5612	1982	3600	6512
1710	3110	5630	1988	3610	6530
1715	3120	5648	1993	3620	6548
1721	3130	5666	1999	3630	6566
1727	3140	5684	2004	3640	6584
1732	3150	5702	2010	3650	6602
1738	3160	5720	2015	3660	6620
1743	3170	5738	2021	3670	6638
1749	3180	5756	2027	3680	6656
1754	3190	5774	2032	3690	6674
1760	3200	5792	2038	3700	6692
1765	3210	5810	2043	3710	6710
1771	3220	5828	2049	3720	6728
1777	3230	5846	2054	3730	6746
1782	3240	5864	2060	3740	6764
1788	3250	5882	2065	3750	6782
1793	3260	5900	2071	3760	6800
1799	3270	5918	2077	3770	6818
1804	3280	5936	2082	3780	6836
1810	3290	5954	2088	3790	6854
1815	3300	5972	2093	3800	6872
1821	3310	5990	2099	3810	6890
1827	3320	6008	2104	3820	6908
1832	3330	6026	2110	3830	6926
1838	3340	6044	2115	3840	6944
1843	3350	6062	2121	3850	6962
1849	3360	6080	2127	3860	6980
1854	3370	6098	2132	3870	6998
1860	3380	6116	2138	3880	7016
1865	3390	6134	2143	3890	7034
1871	3400	6152	2149	3900	7052
1877	3410	6170	2154	3910	7070
1882	3420	6188	2160	3920	7088
1888	3430	6206	2165	3930	7106
1893	3440	6224	2171	3940	7124
1899	3450	6242	2177	3950	7142
1904	3460	6260	2182	3960	7160
1910	3470	6278	2188	3970	7178
1915	3480	6296	2193	3980	7196
1921	3490	6314	2199	3990	7214
1927	3500	6332	2204	4000	7232

	4000 to 5000				
°C		°F	°C		°F
2204	4000	7232	2482	4500	8132
2210	4010	7250	2488	4510	8150
2215	4020	7268	2493	4520	8168
2221	4030	7286	2499	4530	8186
2227	4040	7304	2504	4540	8204
2232	4050	7322	2510	4550	8222
2238	4060	7340	2515	4560	8240
2243	4070	7358	2521	4570	8258
2249	4080	7376	2527	4580	8276
2254	4090	7394	2532	4590	8294
2260	4100	7412	2538	4600	8312
2265	4110	7430	2543	4610	8330
2271	4120	7448	2549	4620	8348
2277	4130	7466	2554	4630	8366
2282	4140	7484	2560	4640	8384
2288	4150	7502	2565	4650	8402
2293	4160	7520	2571	4660	8420
2299	4170	7538	2577	4670	8438
2304	4180	7556	2582	4680	8456
2310	4190	7574	2588	4690	8474
2315	4200	7592	2593	4700	8492
2321	4210	7610	2599	4710	8510
2327	4220	7628	2604	4720	8528
2332	4230	7646	2610	4730	8546
2338	4240	7664	2615	4740	8564
2343	4250	7682	2621	4750	8582
2349	4260	7700	2627	4760	8600
2354	4270	7718	2632	4770	8618
2360	4280	7736	2638	4780	8636
2365	4290	7754	2643	4790	8654
2371	4300	7772	2649	4800	8672
2377	4310	7790	2654	4810	8690
2382	4320	7808	2660	4820	8708
2388	4330	7826	2665	4820	8726
2393	4340	7844	2671	4840	8744
2399	4350	7862	2677	4850	8762
2404	4360	7880	2682	4860	8780
2410	4370	7898	2688	4870	8798
2415	4380	7916	2693	4880	8816
2421	4390	7934	2699	4890	8834
2427	4400	7952	2704	4900	8852
2432	4410	7970	2710	4910	8870
2438	4420	7988	2715	4920	8888
2443	4430	8006	2721	4930	8906
2449	4440	8024	2727	4940	8924
2454	4450	8042	2732	4950	8942
2460	4460	8060	2738	4960	8960
2465	4470	8078	2743	4970	8978
2471	4480	8096	2749	4980	8996
2477	4490	8114	2754	4990	9014
2482	4500	8132	2760	5000	9032

INDEX